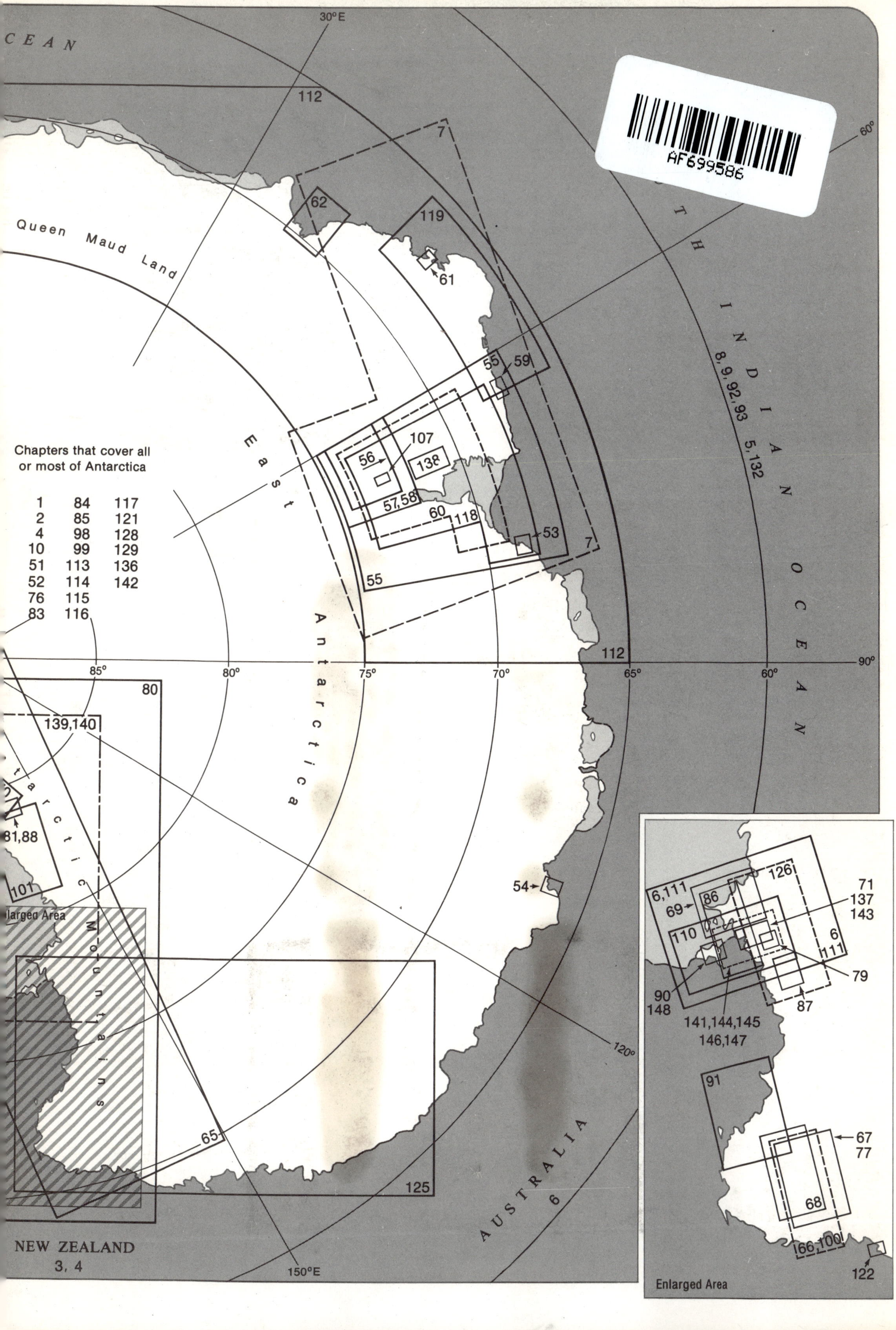
CEAN
Queen Maud Land
East Antarctica
Chapters that cover all or most of Antarctica
1 84 117
2 85 121
4 98 128
10 99 129
51 113 136
52 114 142
76 115
83 116
30°E
60°
90°
120°
150°E
85°
80°
75°
70°
65°
60°
INDIAN OCEAN
8,9,92,93
5,132
Mountains
Enlarged Area
NEW ZEALAND
3, 4
AUSTRALIA
6
112
7
62
119
61
55
59
107
56
138
57,58
60
118
53
139,140
80
81,88
101
54
65
125
6,111
69
86
126
71
137
143
110
6
111
79
90
148
141,144,145
146,147
87
91
67
77
68
66,100
122

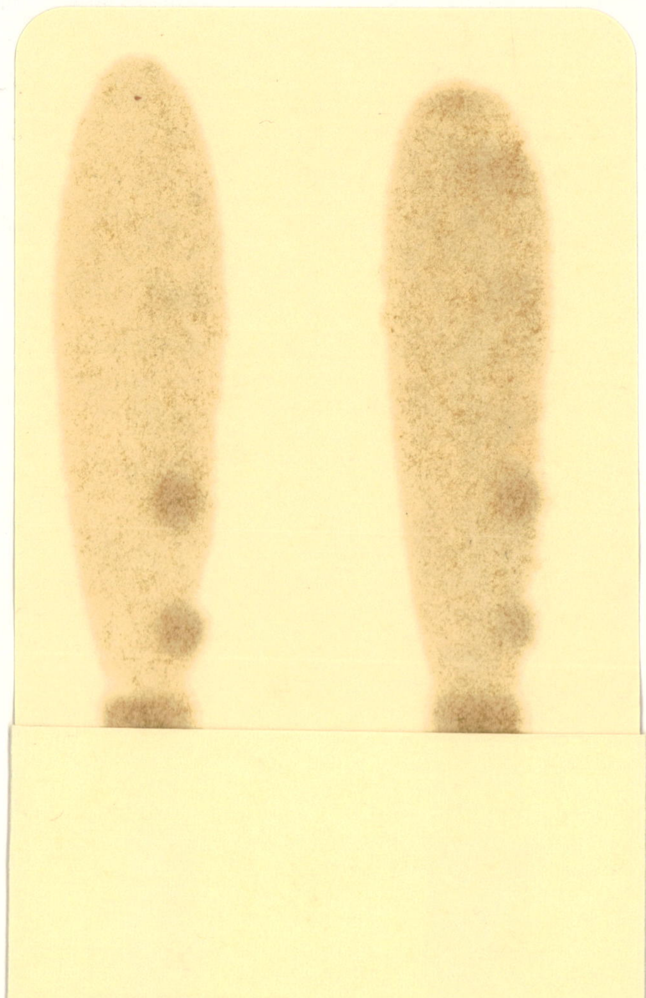

ANTARCTIC GEOSCIENCE

SUPPORTING ORGANIZATIONS

AMAX Foundation
Amoco Foundation
Atlantic Richfield Company
Continental Oil Company
Getty Oil Company
Holmes and Narver, Inc.
International Union of Geological Sciences
Inter-Union Commission on Geodynamics
National Science Foundation
Phillips Petroleum Corporation
Scientific Committee on Antarctic Research
United States Steel Corporation
University of Wisconsin

SYMPOSIUM STEERING COMMITTEE

Campbell Craddock (U.S.A.), chairman
Robin D. Adams (N.Z.)
Charles R. Bentley (U.S.A.)
Ian W. D. Dalziel (U.S.A.)
Tore Gjelsvik (Norway)
Ian R. McLeod (Australia)
Mikhail G. Ravich (U.S.S.R.)

U.S. ORGANIZING COMMITTEE

Campbell Craddock, chairman
Charles R. Bentley
Ian W. D. Dalziel
Robert H. Dott, Jr.
James H. Zumberge
W. Timothy Hushen, secretary

INTERNATIONAL UNION OF GEOLOGICAL SCIENCES
SERIES B — NUMBER 4

ANTARCTIC GEOSCIENCE

Edited by CAMPBELL CRADDOCK
Assistant Editors
Janet K. Loveless
Teri L. Vierima
Kathy A. Crawford

Symposium on Antarctic Geology and Geophysics
Madison, Wisconsin, U.S.A.
August 22–27, 1977
Sponsored by the
SCIENTIFIC COMMITTEE ON ANTARCTIC RESEARCH
INTERNATIONAL UNION OF GEOLOGICAL SCIENCES
INTER-UNION COMMISSION ON GEODYNAMICS

The University of Wisconsin Press
Madison, Wisconsin 1982

Published 1982

The University of Wisconsin Press
114 North Murray Street
Madison, Wisconsin 53715

The University of Wisconsin Press, Ltd.
1 Gower Street
London WC1E 6HA, England

First printing

Printed in the United States of America

For LC CIP information see the colophon

ISBN 0-299-08410-8

Publication of this volume is supported by National Science Foundation Grant No. DPP77-22935; however, any opinions, findings, conclusions, or recommendations expressed herein are those of the authors and do not necessarily reflect the views of the Foundation.

Contents

Part VII. Structural Geology and Tectonics

Part VIII. Mineral Deposits

Part IX. Crustal Structure

In Memoriam

MIKHAIL GRIGOREVICH RAVICH was born 21 October 1912 in Gomel, Belorussia. His mother died when he was three years old, and he moved to Leningrad with his father in 1923. He began his career as a geologist at age 17, carrying out over the years extensive work in the Kola Peninsula, the Taimyr Peninsula, and other parts of the Soviet Arctic. He graduated from the Leningrad Mining Institute in 1936, and he was awarded the C.Sc. degree in 1945 and the D.Sc. degree in 1954. In 1955 he became vice-director of the Research Institute of Arctic Geology in Leningrad. Just two years later he made the first of his four trips to Antarctica. For more than twenty years he was the recognized leader of Soviet Antarctic geologists, and he contributed scores of maps, papers, and books to the Antarctic literature. Professor Ravich was a valued member of the SCAR Working Group on Geology, and also served as vice-president for Antarctica of the Commission for the Geological Map of the World. The bounce in his step and the twinkle in his eye, both reflecting his passion for Antarctic geology, remained with him to his last day.

FRANKLIN ALTON WADE—Al to his friends—was born 5 February 1903 in Akron, Ohio. By 1926 he had received the B.S. and M.A. degrees from Kenyon College, with a major in chemistry and a minor in geology. After a brief term as an industrial chemist, he enrolled at Johns Hopkins University to seek a doctorate in chemistry. Supporting himself by instructing in both chemistry and geology, he eventually shifted his interest and his degree program to geology. In 1933 he made the first of seven trips to Antarctica as a member of the second Byrd expedition. He received his Ph.D. degree from Johns Hopkins in 1937, with a dissertation on the petrology and structure of the Edsel Ford Range. In 1936 he joined the faculty of Miami University and ten years later established their summer field camp in Wyoming. He left Miami in 1954 to head the Department of Geology at Texas Technological University, completing his career there as Horn Professor of Geology and director of the Antarctic Research Center. He led five Antarctic expeditions from Texas Tech to the Transantarctic Mountains, Marie Byrd Land, and Ellsworth Land; he and his students published their results in a series of maps, papers, and reports. Al survived a helicopter crash in 1966 in Marie Byrd Land. During his last years he was chairman of the Antarctic panel of the Circum-Pacific Map Project.

The year 1978 marked the passing of two distinguished and widely known Antarctic scientists, the Soviet geologist M. G. Ravich in April and the American geologist F. A. Wade in October. Both men attended the 1977 symposium, and each contributed to its success. Indeed, a highlight of the week occurred after the symposium banquet when Professor Ravich, acting on behalf of the Soviet Academy of Sciences, awarded Professor Wade the Bellingshausen-Lazarev medal. Their warm embrace in the moment that followed reaffirmed for everyone present the strength of the bonds that unite Antarctic scientists of all nations.

Antarctic geoscientists find inspiration in the life and works of these two outstanding polar geologists. We shall cherish their memory.

C. C.

Preface

The papers in this volume were presented at the third Symposium on Antarctic Geology and Geophysics, held in Madison, Wisconsin, during the week of August 22–26, 1977. Sponsored by the Scientific Committee on Antarctic Research (SCAR), the International Union of Geological Sciences (IUGS), and the Inter-Union Commission on Geodynamics (ICG), this symposium followed in the tradition of earlier meetings in Cape Town (1963) and Oslo (1970). The scientific sessions were held in the Wisconsin Center, overlooking Lake Mendota, on the Madison campus of the University of Wisconsin; official hosts were the University's Department of Geology and Geophysics and the National Academy of Sciences. More than 200 people from 15 countries attended the symposium.

Preparations for the Madison symposium began in 1972 with a proposal submitted to the U.S. Polar Research Board (PRB). Endorsement by PRB was followed in 1973 by an invitation to the SCAR Working Groups on Geology and Solid Earth Geophysics, and in 1974 by SCAR sponsorship for the symposium. Soon thereafter, on invitation, IUGS and ICG joined SCAR as international sponsors. The National Science Foundation (NSF) made a generous grant in support of the symposium to the National Academy of Sciences for use by PRB. Additional support was sought and obtained from U.S. corporations and foundations. An international steering committee and a U.S. organizing committee were established. The steering committee determined the program format, solicited certain review papers, and invited the submission of abstracts. The steering committee met in Sydney, Australia, in August, 1976, to review all preliminary abstracts, and accepted those appropriate for the symposium. Final abstracts were received in the spring of 1977, and most manuscripts were submitted just before the meeting.

The symposium began with a six-day field trip, attended by 35 scientists, to the Precambrian rocks of the Lake Superior district. The field trip returned to Madison Sunday evening, August 21, in time for a welcoming reception at the Wisconsin Center. The scientific sessions began Monday morning, with the papers grouped into twelve categories. Sixteen half-day sessions were required, so simultaneous sessions had to be scheduled on three days of the week. On Monday evening all participants were invited to dinners held in the homes of five University faculty members. Late Tuesday afternoon a tour of Weeks Hall of Geological Sciences, the new home of the Department of Geology and Geophysics, was followed by an informal reception. The symposium banquet Thursday evening was highlighted by an address on Scott's last expedition by Professor Robert L. Nichols, and by presentations of Soviet polar medals to Professor Nichols and Professor F. A. Wade by Professor M. G. Ravich. The symposium closing ceremonies were held late Friday afternoon, and on Saturday many participants enjoyed a field trip to the classical Baraboo district. The SCAR Working Groups on Geology and Solid Earth Geophysics held meetings on Sunday and Monday.

Serious discussions with publishers took place just after the symposium when the magnitude of the task had become clear. A book subsidy proposal was submitted to NSF, and they responded with a grant to the University of Wisconsin, under the direction of the editor. After negotiations with several organizations, the University of Wisconsin Press was selected to publish the book. Discussions between the editor and the director of the University of Wisconsin Press produced a letter of agreement which was acceptable to NSF, PRB, and the international sponsoring organizations. The machinery was finally in place.

Editorial work on this book began in the summer of 1977. Assistant editor Jan Loveless read each paper on its arrival in Madison and identified major problems for the editor to discuss with those authors during symposium week. After the symposium, assistant editor Teri Vierima and the editor read each manuscript for content, clarity, and style, and all figures were examined for problems that necessitated modification or redrafting. The manuscript was then returned to the author with our suggestions and queries for his revision and approval. The returned manuscripts were carefully reviewed by assistant editor Kathy Crawford and the editor to achieve clarity and uniformity of style. The manuscripts were set in type, and the galley proofs were sent to the authors for final approval.

In an early stage of planning it was decided to include both a location map and a geologic map of Antarctica. The location map on the front endsheets of the book shows the approximate area treated by each complete chapter in the book. The American Geographical Society 1:5,000,000 Geologic Map of Antarctica is in the rear pocket for reference by the reader.

The length of time to reach publication is regretted, but we have encountered many obstacles and delays in our quest for a quality book at a reasonable price. It is a large volume, involving many individuals—all of them busy with competing demands on their time. Many negotiations and consultations about business matters were necessary, and organizations can be slow in coping with the unusual. Authors have been generally cooperative, but some correspondence has been slowed by change of address, irregular mail service, or absence in the field. With contributors from many nations writing in but a single language, care and patience have been needed to achieve clarity. Finally, the length of the text, and especially the numerous and complex tables, have provided a major challenge to our typesetter.

We hope that the final product will be worth the wait, and we thank our authors and readers for their patience and understanding along the way. As a restless contributor to earlier symposia slow to appear in print, I used to wonder if the editor had died. I will never wonder again.

Campbell Craddock, editor
Department of Geology and Geophysics
University of Wisconsin-Madison
Madison, Wisconsin 53706
February 1981

Acknowledgments

As convenor of the symposium and editor of the proceedings volume, I express my sincere thanks to the many organizations and individuals whose participation and service helped bring about the meeting and this book. The Scientific Committee on Antarctic Research, the International Union of Geological Sciences, and the Inter-Union Commission on Geodynamics sponsored the symposium and assisted with the travel costs of some key speakers. The National Science Foundation provided generous grants to the National Academy of Sciences (Polar Research Board) in support of the symposium and to the University of Wisconsin in support of the proceedings volume. Certain corporations and foundations, listed elsewhere, made important contributions in support of the symposium. The symposium Steering Committee established the program, reviewed the abstracts, and assisted the editor in handling the manuscripts. The U.S. Organizing Committee provided advice and assistance in obtaining financial support and arranging the symposium and related events. The National Academy of Sciences (Polar Research Board) and the University of Wisconsin were the official hosts for the symposium. Special thanks are offered to the following units on the Madison campus of the University of Wisconsin: the faculty, staff, and students of the Department of Geology and Geophysics, the Wisconsin Center, the University of Wisconsin Press, the Cartography Laboratory, and the Office of the Chancellor.

It is not feasible to cite here each of the scores of people who helped in this task, but the special contributions of some must be noted. L. DeGoes and W. T. Hushen of the Polar Research Board and R. H. Rutford and M. D. Turner of the National Science Foundation provided major assistance in organizing the symposium and securing the necessary support. Innumerable problems, large and small, before and during the symposium were handled cheerfully and efficiently by M. Woods and S. Johannsen. R. H. Dott, Jr., and I. W. D. Dalziel organized and led the Baraboo field trip, and M. G. Mudrey, Jr., W. W. Montgomery, and C. Smith were my co-leaders on the Lake Superior field trip. The following faculty members and their wives graciously hosted dinners at their homes for symposium participants: C. R. Bentley, C. J. Bowser, C. Craddock, R. H. Dott, Jr., L. R. Laudon, T. S. Laudon, and M. G. Mudrey, Jr. Professor R. L. Nichols kindly agreed to deliver the banquet address and to prepare the text of it for this book.

Director T. Webb and Chief Editor E. Steinberg and their many associates at the University of Wisconsin Press have given patient guidance and unfailing support in bringing this large work to publication. Assistant Editors K. Crawford, J. Loveless, and T. Vierima, each holding a graduate degree in geology or physics, made major contributions to the editing of the book. R. Klein, S. Smith, and B. Charly gave important assistance in processing manuscripts and detecting and correcting errors. T. W. Bastien kindly granted permission to use the photograph from the Ellsworth Mountains on the dust jacket.

Finally, I thank my wife and my children for their helpfulness and for their patience with my preoccupation during the nine-year cycle from initial planning of the symposium to publication of this book.

Campbell Craddock

SCAR/IUGS/ICG Third Symposium on Antarctic Geology & Geophysics, Madison, Wisconsin, 22–27 August 1977

ROW 1 (left to right) E. N. Kamenev, W. E. LeMasurier, R. H. Rutford, T. S. Laudon, A. G. Dralkin, I. W. D. Dalziel, C. Craddock, R. Adams, C. R. Bentley, M. G. Ravich, T. Gjelsvik, I. McLeod, C. Bull, T. W. Gevers, P. Grew, E. S. Grew, J. Splettstoesser, A. Klaver, and J. C. Pratsch.

ROW 2 (left to right) F. H. C. Hotchkiss, C. C. Plummer, J. D. Bradshaw, A. B. Ford, D. Skinner, D. A. Coates, D. H. Elliot, N. H. Fourcade, T. VanAutenboer, V. I. Bardin, H. Plume, R. Plume, R. Kyle, and P. Kyle.

ROW 3 (left to right) P. D. Rowley, D. J. Drewry, H. L. Pollak, C. Smith, G. Grindley, L. Harrington, R. Oliver, M. R. A. Thomson, G. F. Webers, E. Perry, C. P. Andrews-Speed, J. LaBrecque, J. F. Lovering, J. F. Sutter, P. F. Ciesielski, and F. Weaver.

ROW 4 (left to right) G. Grikurov, C. W. Edwards, J. P. Bloch, E. F. Roots, O. González-Ferrán, R. D. Winn, Jr., R. H. Dott, Jr., E. H. Colbert, P. Tasch, M. D. Turner, M. Winslow, and M. Weaver.

ROW 5 (left to right) B. A. Hall, K. E. Rose, J. Nougier, T. S. Winsnes, Y. Ohta, T. Torii, R. H. N. Steed, C. K. Scharnberger, K. L. Russell, D. C. Rex, P. W. G. Tanner, M. M. Schneider, B. A. Hall, T. J. Wilson, E. Nelson, R. B. Allen, M. L. Cole, and D. Kurtz.

ROW 6 (left to right) L. DeGoes, G. L. Johnson, A. Hjelle, B. A. Fossum, H. Decleir, C. Vieira, P. N. Webb, M. G. Laird, S. Snelson, H. R. Katz, R. Forsythe, C. Stern, J. B. Anderson, S. W. Wise, and B. L. Ward.

ROW 7 (left to right) F. Tessensohn, J. Lameyre, R. Schlich, H. Miller, T. Prestvik, V. Covacevich, J. E. Gabites, S. L. Cande, M. G. Laub, M. G. Mudrey, P. D. Clarkson, J. G. Hattner, R. D. Powell, M. T. Ledbetter, E. Stump, and G. S. Schatz.

ROW 8 (left to right) W. T. Hushen, J. Goslin, M. C. M. Arnould, W. J. Zinsmeister, A. N. Federman, K. Haugland, K. E. LaPrade, B. G. Lopatin, W. P. Siard, A. D. Saunders, R. K. H. Falconer, P. F. Barker, S. D. Weaver, R. J. Pankhurst, D. W. Cooke, D. J. DeMaster, D. R. Watts, N. Wells, K. Kaminuma, and P. J. Barrett.

ROW 9 (left to right) G. Dreschhoff, E. J. Zeller, L. G. Marshall, R. L. Nichols, J. C. Behrendt, R. J. Adie, R. J. Tingey, A. G. Reid, R. G. Alexander, Jr., F. J. Adler, A. D. Cumming, K. Hinz, G. Renner, J. Burrell, D. H. Griffiths, I. Norton, P. Simpson, D. R. Hunter, L. G. Wolmarans, M. Yoshida, J. H. Wrenn, B. McKelvey, H. Brady, and H. Harris.

Symposium Closing Ceremonies
The Wisconsin Center
Friday, August 26, 1977

Craddock: For our brief closing ceremonies we will have presentations by three of our most distinguished participants. Professor Wade has asked me to convey his apologies to the group for not being able to come to this ceremony, along with his appreciation for the warmth and fellowship in the events of last evening which made him feel very happy indeed. The three speakers who we will have, in order, will be Tore Gjelsvik, Mikhail Ravich, and Tom Gevers. I will introduce first Dr. Tore Gjelsvik, who is director of the Norsk Polarinstitutt in Oslo and also the president of SCAR. He spoke to us Monday to open the symposium, and I will ask him also to address us now at the close.

Gjelsvik: I shall not attempt to give a review of the very interesting papers and discussions which we have been listening to here in the Wisconsin Center during this week. Looking back at the two earlier symposia on Antarctic Geology and Geophysics, I can see a trend away from preliminary descriptive investigations to more detailed, mature, and varied scientific studies. This is the way we should go. I am particularly pleased to note the increasing information from the ice-covered areas of the continent and from the floor of the seas surrounding Antarctica.

Yet, the areas selected for investigations are governed by their accessibility and the logistic possibilities at hand, and we see that there are still great areas about which we know far too little. I am particularly referring to the coast of the Antarctic continent outside the Antarctic Peninsula and the Ross Sea area. We need to know much more about the conditions and processes on the Antarctic continental shelf, as was clearly demonstrated by the review paper given by Len Johnson yesterday. Thus, there is still a challenge in Antarctica for scientists with the pioneering spirit which Bob Nichols so ably described in his banquet speech last night. It is my hope that we will have much more information from this part of Antarctica when we meet at the fourth symposium on Antarctic Geology and Geophysics in seven years' time.

Others are going to say thanks and farewell from the participants of the symposium, but before stepping down, however, I would like on behalf of SCAR to thank the members of the U.S. Organizing Committee and particularly the chairman, Professor Craddock, for their efforts to ensure the success of the symposium. You have all reason to relax and be happy with what you have done. I also want to thank the faculty members and their wives for their kind and splendid hospitality. Thank you for your attention.

Craddock: Thank you very much, Tore, for these generous remarks. I'd like now to introduce a man who has requested a special opportunity to speak to this group. I feel sure that as always his remarks will be heartfelt and very interesting to us. Professor Mikhail Ravich of the Soviet Union.

Ravich: I thank you for giving me the opportunity to speak at the concluding session of the third symposium. We have listened to 140 papers, and it will be quite a while before we can analyze all the material presented. As in all great affairs, there were a few absences; we will analyze these and will discuss them at the Working Group sessions so that at the next symposium there won't be these omissions.

I would now like to talk about the great accomplishment of this symposium as opposed to the two preceding ones. Let us pay attention to the great geophysical work that has been done in Antarctica to understand better what lies under the ice. We have obtained much information by aeromagnetics, by radar sounding, and by gravity. This was reflected in the papers given in the last few days. I wish to say that deep seismic sounding has not received the attention it deserves. I call all geologists to make further efforts to develop this method which can be used with great success in understanding the deep crust and deep structure of the earth.

In our symposium we reached a new stage in the study of the Precambrian. We saw at the symposium the first attempts to divide up and establish a stratigraphy in the Precambrian. A great accomplishment was in the stratigraphy of the Upper Precambrian where we have better controls. In this way we must pursue further the stratigraphy of the Precambrian.

A new stage was also reached at this symposium in the field of structural geology. We not only show the present structures on maps, but we have tried to see also how they've been developed in time from the Archean to the present. I am a great believer in the primeval continent of Gondwanaland, and at this symposium, in contrast to the two preceding ones, there was much more information on the reconstruction of Gondwanaland and showing the similarities between the various Gondwana continents.

It was very pleasant to hear so many more papers on volcanism in Antarctica. Antarctica is the largest volcanic province in the world. I wish to express my regret that there wasn't more on plutonism, on plutonic rocks, and magmatism. We need to learn more about intrusive rocks as they can be used as a sign to determine the geological history of an area. I also wish to mention that with intrusive rocks there may occur deposits of useful economic minerals. At this symposium we are not only seeing a description of the geology but also an understanding of the geological structure of Antarctica and its role in understanding the geology of the world as a whole.

I wish to take this opportunity that I have to speak at the concluding session to thank the National Science Foundation and the U.S. Academy of Sciences and the University of Wisconsin for inviting the Soviet scientists to come to this meeting. With great pleasure, I express my thanks to Professor Campbell Craddock. I wish to mention his organizational talents, his geological erudition, and his ability to get along with everybody and make it a good meeting. I wish to thank all the helpers and other people who helped Professor Craddock to make this a great meeting for all the Soviets and other foreign scientists who came. They made all possible arrangements so that it would be a very pleasant meeting for us and foresaw all our needs and requests. Thank you.

Craddock: Thank you very much, Professor Ravich, for these kind and gracious remarks. As our final speaker, I will call now on one of the most distinguished of our members, Professor Emeritus Tom Gevers from the University of Witwatersrand in South Africa. I first met Tom at the occasion of the first Antarctic symposium in 1963, and one of the most enduring impressions that I carried away from that meeting was the great eloquence which he brought to his remarks. Although he has not labored so much of his life in Antarctica, he has lived and worked in a region very closely involved in Gondwanaland. I think it is fitting that we call on Tom Gevers, as a countryman of Alec du Toit, to make the final remarks for our symposium.

Gevers: It is indeed an honor to have the valedictory finale of the symposium delegated to me. Since I so obviously cannot pose as an important Antarctic geologist, this can only be due to my being the most antique member of this gathering, adorned by a long white beard grown in the Darwin Mountains of the Transantarctic Range at the expense of the National Science Foundation in Washington. Antiquity has its advantages. From the height of nearly 80 years, I have a long vista of history back to the beginning of this century. But do not be alarmed! All I wish to refer to in this connection is that I remember the years when those interested in polar exploration waited with bated breath for news of the trek to the South Pole by the Norwegians, under Amundsen, and the British, under Scott; and grieved over the loss of the latter leader with four companions.

The first symposium on Antarctic Geology held in Cape Town in 1963 was indeed a memorable occasion. It was there that I for the first time met Dr. Laurence Gould, who played such a prominent role in the first American expeditions, being second in command to Admiral Byrd in the first of 1928; and that tough old timer Alton Wade from Texas, with whom I later spent a regrettably only short time at Byrdland Camp West. And it was at Cape Town that I listened with awe to Bob Nichols performing with that dramatic verve which he displayed last night in his description of Scott's ill-fated expedition to the Pole. As a venue, Cape Town has the advantage of its spectacular scenic setting; so everything went fine and everybody seemed happy, except the opponents of Continental Drift. The great Alec du Toit, alas, was no longer among the living. That so many of Antarctic geologists and geophysicists I then met in Cape Town are with us here in Madison today is testimony to the enduring spell which the White Continent casts on all who have once beheld its scenic and scientific splendors.

At the second symposium in Norway, held in 1970 under the able organization of Tore Gjelsvik, the number of participants and papers presented had grown very considerably, eloquent testimony to the effectiveness of the lure of the Deep South on men of stamina and adventurous spirit. A particularly pleasant memory was the invitation of participants by members of the Norsk Polarinstitutt and the University of Oslo to their homes. I myself had the privilege of being invited to the house of that great petrologist Tom Barth.

Now, for the third symposium, we are here in Madison, Wisconsin. The latter state, of course, had not been a closed book to any of us. We knew of the beer that made Milwaukee famous; and of the beer that makes Milwaukee jealous. We knew of the Lake Superior iron and copper ores. All of us had also known of the University of Wisconsin as one of the great centers of learning in the United States. But only now, through this symposium, have Wisconsin, Madison, its great university and especially the able men of its Department of Geology and Geophysics become live entities to most of us.

The large number of participants, over 200, and papers presented, around 140, has, of course, greatly increased the work of organizing this symposium; yet it nevertheless ran like clockwork, thanks to the untiring effort of Professor Craddock and his zealous assistants. But more of this anon! Drs. Gjelsvik and Ravich have dealt with the scientific side of this symposium; so let me speak, with enthusiasm and, if you will allow me, at times levity, of its human aspects. To begin with, let others who have not had the privilege of experiencing it imagine a horde (excuse me! this term does not apply to the ladies!) of 200 geologists transported in four buses invading the homes of six members of staff! And everyone left with the happy feeling that he had been the center of attention, an impression for which the gracious wives and daughters of our hosts were in no mean measure responsible. A later reception in the Department of Geology and Geophysics itself not only added to these happy memories, but also made many of those connected with universities feel envious.

The pre-symposium Lake Superior excursion was an eye-opener in many ways; unfortunately, not of the iron and copper mines themselves, which were all barred by a strike. Having done most of my field work in the arid regions of southern Africa, I was, of course, used to vastly more abundant and extensive outcrops. All the greater is my admiration for those untiring and skillful geologists of olden and more recent times who have managed to unravel the geology in such detail. To Cam Craddock, our leader, moreover, the obscuring forests very obviously were an added attraction. Most of us who went on this excursion will in retrospect think not only of Precambrian geology and iron and copper ores, but just as much of the calm beauty of the landscape, of the wooded hills and secluded lakes, so beloved by Cam and his family.

But, nevertheless, I also have to voice some criticism. Cam's lucid but detailed account of the Keweenawan lavas and tuffs was rather more succinctly put by a young guide in the Arcadia copper mine, which through tourism has been turned into a gold mine. He told us the whole story in simple terms readily understood by ordinary geologists. All these enormously thick lavas had come, more than 600 million years ago, from one huge volcano on the site of present Lake Superior; and, as is the habit of volcanoes, this supergiant one fine day collapsed to form a huge hole subsequently filled with water to form the world's largest and deepest lake. The young guide, however, was not from Texas. With some reluctance I disclose that Cam also landed some of us in jail, Dr. Ravich and I, the most venerable members of the excursion, being locked up in maximum security cells for one night. Discreetly, I am not disclosing details.

Dr. Ravich has referred to the notable increase in geophysical research and the enthusiasm—not as yet very evident at the first symposium in Cape Town—of Antarctic geologists—with one notable exception—for Continental Drift in its modern, refurbished guise of ocean-floor spreading and plate tectonics. In a more lighthearted vein, let me add that a close scrutiny of the human element manifest at this symposium reveals another type of spreading, that of hirsuteness. For those who do not know any Latin, this means the cover of the cranium and neck with long locks and concealment of the face under a thick growth of fungus. The overall impression rather lends support to Darwin's theory of the Descent of Man. However, the delightful aspect and captivating charm of the wives and daughters of our hosts, of our own female consorts, and of the young ladies at the information center lead me to believe all the more firmly in my own theory of dual evolution, which is that only Adam is of arboreal simian origin, but that Eve is of inspired divine creation. One can picture Eve, radiantly beautiful fresh from the Creator's hands, looking on in shocked amazement as Adam came climbing down from the trees. For the ease of mind of those of younger years, let me divulge that being close on 80 years of age by no means implies imperviousness to the everlasting charms of the divine creation. That is why I was so often to be seen at the information center!

Now—at long last—let me close in more serious vein. Even university professors have to stop talking some time! I know that I am speaking on behalf of all participants in this symposium when I call it an outstanding success. Many have worked hard to this end, and we extend our thanks to all. I should like to mention specially Miss Mim Woods, the Chairman's unobtrusive assistant. Furthermore, practically every one of the 140 odd papers presented was illustrated by slides, and the hard work and unfailing patience of the projectionists were greatly appreciated. In particular we wish to thank you personally, Professor Craddock, for your outstanding leadership, based, apart from personal qualities, on wide erudition and a detailed knowledge of Antarctic geology. We leave the University of Wisconsin in Madison with vivid personal experience of the proverbial American friendliness and generosity, such as many of us have already enjoyed in the Antarctic.

We visitors, about to scatter to the four corners of the earth, are also saying goodbye to each other. In addition to scientific links, human bonds have also been formed and strengthened to influence the immediately following SCAR meeting, further work in the Antarctic, and the next symposium. Whether at my age I shall be there, I do not know; but I do hope that I shall be able once more to experience the good fellowship and warm humanity that has been the hallmark of these gatherings of Antarctic geologists and geophysicists.

To our hosts we say goodbye and many thanks! Do svidaniya; bolshoe spasiba! Farvel og mange tak för ålt! Au revoir et merci mille fois pour toute votre bonté! Totsiens en baie dankie vir al die vriendelikheid! Saionera; domo arigato! Tot weerziens en dank U wel! Adios y muchas gracias por todos atenciones! Auf Wiedersehen und vielen Dank!

Craddock: You will agree that Tom Gevers is a very tough act to follow, and I think he has said it all in almost every language. On behalf of all those who have played a part in this, and there are many of them, I would like just to express our thanks for your very generous remarks. If you have enjoyed yourself here and have learned something about Antarctic geology and geophysics you didn't know before, and if you've seen old friends and renewed acquaintances, and if you've made new friends as well, then we have in fact accomplished our task. I might finish with just two observations. For many years I have been privileged as a geologist to visit many lands—I think almost every one that is represented here at our symposium. I have traveled with many geologists, and I've seen many rocks. If in some small way I have repaid this debt for my own continuing education through the years, then I will feel satisfaction. Finally, I have believed for many years that it is both a privilege and a pleasure to spend one's life as a student of the earth. And if one is a student of the earth in Antarctica, then he must indeed be a member of a very special fellowship of people. Thank you for your attendance, and we wish you a safe journey home. Our symposium is concluded.

Figure 1. Captain Robert F. Scott (1868–1912), R.N., C.V.O.

Captain Scott and His Last Expedition

ROBERT L. NICHOLS
15 Pleasant Street W-22, Harwich Port, Massachusetts 02646, U.S.A.

Students of Antarctic exploration agree that the three greatest British Antarctic explorers were Scott, Shackleton, and Mawson. If one had to use one word to characterize Shackleton, the word would be *leadership.* Young men from all over the world flocked around him in the hope that he would take them on one of his Antarctic expeditions. If we had to characterize Sir Douglas Mawson, the eminent Australian geologist, with one word, the word would be *strength.* He started out on a man-hauling journey with two companions, both of whom died on the trip. It was only Mawson's great strength, both physical and mental, that finally enabled him to return safely, though with great difficulty, to his base station. If one had to characterize Robert Falcon Scott, the greatest of these, with one word, the word would be *character.*

Scott (Figure 1) was the leader of the British National Expedition of 1901–1904, often called the Discovery Expedition after the name of its ship. On this expedition he man-hauled about 2000 miles and pushed "Farthest South," reaching a point some 200 miles closer to the Pole than anyone before him. Before the *Terra Nova* Expedition, his last, he was a captain in the British Navy and had commanded a succession of battleships. His objectives on his last expedition were to reach the South Pole and to carry out a complex and ambitious scientific program. His scientific staff was the largest of any Antarctic expedition up to that time and consisted of a meteorologist, a physicist, three geologists, and four biologists. The shelf of tomes published by the expedition is proof of its scientific success. To reach the Pole, Scott used dogs, ponies, motor sledges, and man-hauling. But man-hauling had the greatest appeal for him. Cherry-Garrard (a member of Scott's last expedition) thought that Scott sledged harder than anyone he had ever traveled with, had the strongest combination of strong mind and strong body that he had ever known, and would carry weight in any gathering of human beings. Admiral Markham wrote that "Scott must have been a man after God's own heart for his most noble quality was his unselfishness."

Dr. Edward A. Wilson, B.A., M.B. (Cantab), Edgar Evans, Petty Officer, R.N., Lawrence E. G. Oates, Captain 6th Inniskilling Dragoons, and Henry R. Bowers, Lieutenant, R.I.M. were Scott's companions on the Polar Journey.

Bill Wilson was Scott's right-hand man. He was the chief scientist, and M.D., a productive zoologist, a talented artist, and Scott's dearest friend, and they had been man-hauling companions on the "Farthest South Trip" of the Discovery Expedition. Men loved Bill Wilson.

Edgar Evans was the biggest and strongest man of the Polar Party and a giant of a worker. He and Scott had man-hauled together for more than a thousand miles when on the Discovery Expedition. He was responsible for the sledges, tents, sleeping bags, crampons, skis, and other pieces of trail equipment.

Titus Oates was in charge of the ponies, and it was because of his devoted care that they were able to move heavy loads of food and fuel almost to the foot of the Beardmore Glacier.

Birdie Bowers was the meteorologist, navigator, observer, and photographer. Scott called him a marvel and noted that he was apparently oblivious to the cold. A friend wrote of Bowers that "he was one of those rare spirits who exercise a boundless if unknown influence for good wherever they are."

Scott and his four gallant companions started for the Pole from Cape Evans, on the lower slopes of Mount Erebus, and sledged approximately 400 miles across the Ross Ice Shelf. This put them at the foot of the dangerously and cruelly crevassed Beardmore Glacier, which they climbed, slowly and with great difficulty. This put them on the Polar Plateau, 10,000 feet above sea level. They then marched more than 350 miles across the Polar Plateau, and on January 17, 1912, reached the South Pole (Figure 2), only to find that Amundsen, using Siberian huskies, had beaten them by little more than a month.

Scott wrote, "Jan. 15.—Height 9220. T. −25°. . . . It is wonderful to think that two long marches would land us at the Pole. . . . it ought to be a certain thing now, and the only appalling possibility the sight of the Norwegian flag forestalling ours. . . . Only 27 miles from the Pole. We *ought* to do it now. . . .

"January 16.—The worst has happened, or nearly the worst. . . . About the second hour of the march Bowers' sharp eyes detected what he thought was a cairn; he was uneasy about it, but argued that it must be a sastrugus. Half an hour later he detected a black speck ahead. Soon we knew that this could not be a natural snow feature. We marched on, found that it was a black flag tied to a sledge . . . near by the remains of a camp; sledge tracks and ski tracks going and coming and the clear trace of dogs' paws—many dogs. This told us the whole story. The Norwegians have forestalled us and are first at the Pole. It is a terrible disappointment, and I am very sorry for my loyal companions. Many thoughts come and much discussion have we had. To-morrow we must march on to the Pole and then hasten home with all the speed we can compass. All the day dreams must go; it will be a wearisome return. . . .

"January 17.— . . . *The Pole.* Yes, but under very different circumstances from those expected. We have had a horrible day—add to our disappointment a head wind with a temperature −22°, and companions labouring on with cold feet and hands.

"We started at 7:30, none of us having slept much after the shock of our discovery. . . . the wind is blowing hard, and there is that curious damp, cold feeling in the air which chills one to the bone in no time. . . . Great God! this is an awful place and terrible enough for us to have laboured to

Invited lecture, presented after the official symposium dinner, the evening of Thursday, August 25, 1977.

Figure 2. Scott's party at the South Pole, January 17, 1912 (front: Bowers left, Wilson right; standing: Oates left, Scott center, Evans right).

it without the reward of priority. Well, it is something to have got here. . . . Now for the run home and a desperate struggle. I wonder if we can do it."

By the time they began the long march back, the season was unduly advanced. Appalling sledging conditions were encountered—blizzards, dangerous crevasses, sastrugi, sandy snow, soft snow, and poor visibility which made it very difficult at times to find the tracks and the food and fuel caches made on the outward march to the Pole. Disappointment, low temperatures, starvation, and death marched with them; so, too, did determination, courage, comradeship, and a love of England. All five men died on the return journey.

Edgar Evans died first. He had received a nasty cut in repairing a sledge. He suffered from frozen fingers and toes, and finally from a concussion resulting from a bad fall on the Beardmore Glacier.

Titus Oates was the next to go. He suffered terribly from frozen feet, and finally, with the midday temperature down to $-43°F$, he said on March 15th that he could not go on and proposed that he should be left in his sleeping bag. The others could not agree to this and urged him to struggle on. Scott wrote on March 16 or 17—"In spite of its awful nature for him he struggled on and we made a few miles. At night he was worse and we knew the end had come.

"Should this be found I want these facts recorded. Oates' last thoughts were of his Mother, but immediately before he took pride in thinking that his regiment would be pleased with the bold way in which he met his death. We can testify to his bravery. He has borne intense suffering for weeks without complaint, and to the very last was able and willing to discuss outside subjects. He did not—would not—give up hope to the very end. He was a brave soul. This was the end. He slept through the night before last, hoping not to wake; but he woke in the morning—yesterday. It was blowing a blizzard. He said, 'I am just going outside and may be some time.' He went out into the blizzard and we have not seen him since. . . . We knew that poor Oates was walking to his death, but though we tried to dissuade him, we knew it was the act of a brave man and an English gentleman. We all hope to meet the end with a similar spirit, and assuredly the end is not far."

On March 19th, Scott, Wilson, and Bowers reached a point eleven miles from One Ton Depot (the 60th camp from the Pole) where there was an abundance of food and fuel. From March 20th to the 29th a blizzard kept them from leaving their tent. On March 23rd, Scott wrote, ". . . must be near the end. *Have decided it shall be natural—we shall march for the depot with or without our effects and die in our tracks."* And then on March 29,

> Since the 21st we have had a continuous gale. . . . We had fuel to make two cups of tea apiece and bare food for two days on the 20th. Every day we have been ready to start for our depot *11 miles* away, but outside the door of the tent it remains a scene of whirling drift. I do not think we can hope for any better things now. We shall stick it out to the end, but we are getting weaker, of course, and the end cannot be far.
>
> It seems a pity, but I do not think I can write more.
>
> R. Scott.
>
> Last Entry.
>
> For God's sake look after our people. (Figure 3).

Between March 20th and 29th Scott wrote his magnificent "Message to the Public" and letters the like of which had never been written before.

My dear Mrs. Wilson,

If this letter reaches you Bill and I will have gone out together. We are very near it now and I should like you to know how splendid he was at the end—everlastingly cheerful and ready to sacrifice himself for others, never a word of blame to me for leading him into this mess. He is not suffering, luckily, at least only minor discomforts.

His eyes have a comfortable blue look of hope and his mind is peaceful with the satisfaction of his faith in regarding himself as part of the great scheme of the Almighty. I can do no more to comfort you than to tell you that he died as he lived, a brave, true man—the best of comrades and staunchest of friends.

My whole heart goes out to you in pity.

Yours,
R. Scott.

My dear Mrs. Bowers,

I am afraid this will reach you after one of the heaviest blows of your life.

I write when we are very near the end of our journey, and I am finishing it in company with two gallant, noble gentlemen. One of these is your son. He had come to be one of my closest and soundest friends, and I appreciate his wonderful upright nature, his ability and energy. As the troubles have thickened his dauntless spirit ever shone brighter and he has remained cheerful, hopeful, and indomitable to the end.

The ways of Providence are inscrutable, but there must be some reason why such a young, vigorous and promising life is taken.

My whole heart goes out in pity for you.

Yours,
R. Scott.

To the end he has talked of you and his sisters. One sees what a happy home he must have had and perhaps it is well to look back on nothing but happiness.

He remains unselfish, self-reliant and splendidly hopeful to the end, believing in God's mercy to you.

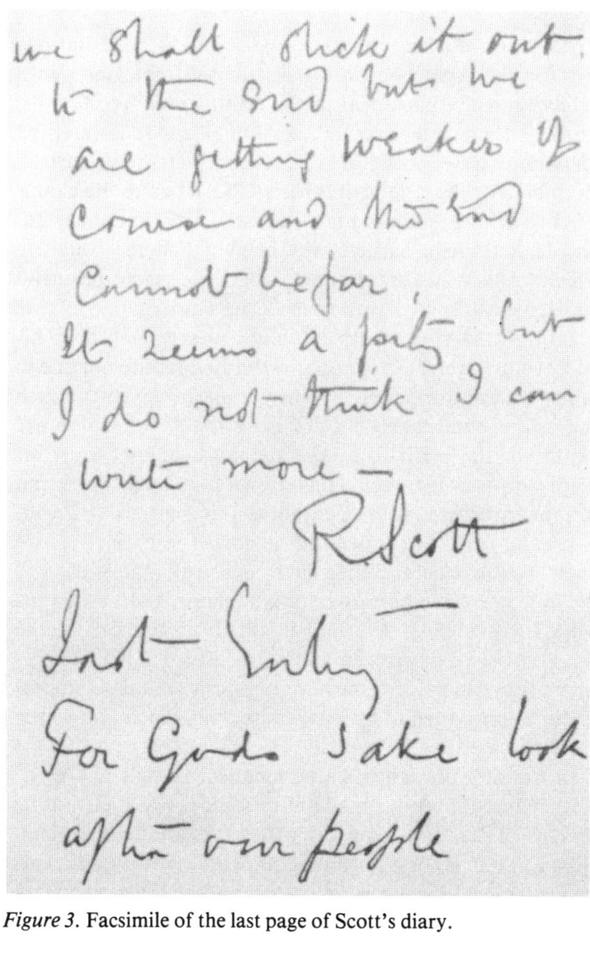

we shall stick it out
to the end but we
are getting weaker of
course and the end
cannot be far.
It seems a pity but
I do not think I can
write more —
R. Scott
Last Entry
For God's sake look
after our people

Figure 3. Facsimile of the last page of Scott's diary.

My dear Barrie, [To Sir J. M. Barrie, the author]

We are pegging out in a very comfortless spot. Hoping this letter may be found and sent to you, I write a word of farewell. . . . More practically I want you to help my widow and my boy—your godson. We are showing that Englishmen can still die with a bold spirit, fighting it out to the end. It will be known that we have accomplished our object in reaching the Pole, and that we have done everything possible, even to sacrificing ourselves in order to save sick companions. I think this makes an example for Englishmen of the future, and that the country ought to help those who are left behind to mourn us. I leave my poor girl and your godson, Wilson leaves a widow, and Edgar Evans also a widow in humble circumstances. Do what you can to get their claims recognised. Goodbye. I am not at all afraid of the end, but sad to miss many a humble pleasure which I had planned for the future on our long marches. I may not have proved a great explorer, but we have done the greatest march ever made and come very near to great success. Goodbye, my dear friend,

Yours ever,
R. Scott.

We are in a desperate state, feet frozen, &c. No fuel and a long way from food, but it would do your heart good to be in our tent, to hear our songs and the cheery conversation as to what we will do when we get to Hut Point.

Later.—We are very near the end, but have not and will not lose our good cheer. . . . We did intend to finish ourselves when things proved like this, but we have decided to die naturally in the track.

As a dying man, my dear friend, be good to my wife and child. . . . I never met a man in my life whom I admired and loved more than you, but I never could show you how much your friendship meant to me, for you had much to give and I nothing . . .

On November 12th the search party found Scott's tent. It was nearly covered with snow. Scott, Wilson, and Bowers had died eight months earlier, on or about the 29th of March. Scott was in the center, and Wilson and Bowers on each side. Everything was tidy and the tent had been well pitched. Wilson and Bowers had apparently died while sleeping in their bags. Scott had evidently died last. He had thrown back the flaps of his bag and his left arm was stretched over Wilson, his closest friend. His diary and letters were on the floor of the tent. When the sledge was dug out, 35 pounds of geologic specimens collected on the moraines of the Beardmore Glacier were found. At Wilson's request these had been carried up to the very end, although it increased his party's labors. Above the tent a snow cairn was built and on top of it was placed a cross, made from two skis (Figure 4). A metal cylinder containing the following record was left on the spot.

November 12, 1912, lat. 79 degrees, 50 mins. South. This cross and cairn are erected over the bodies of Captain Scott, R.N., C.V.O., Doctor E. A. Wilson, M.B., B.C., Cantab., and Lieutenant H. R. Bowers, Royal Indian Marine—a slight token to perpetuate their successful and gallant attempt to reach the Pole. This they did on January 17, 1912, after the Norwegian Expedition had already done so. Inclement weather with lack of fuel was the cause of their death. Also to commemorate their two gallant comrades, Captain L. E. G. Oates of the Inniskilling Dragoons, who walked to his death in a blizzard to save his comrades about eighteen miles south of this position; also of Seaman Edgar Evans, who died at the foot of the Beardmore Glacier. "The Lord gave and the Lord taketh away; blessed be the name of the Lord."

Figure 4. Grave site of Scott, Wilson, and Bowers.

The search party then sledged on south in the hope of finding Oates's body. They found his sleeping bag, but he had been buried by the snow. Near where Oates walked to his death amid an Antarctic blizzard, a cairn surmounted by a cross was built, to which these words were attached:

> Hereabouts died a very gallant gentleman, Captain L. E. G. Oates of the Inniskilling Dragoons. In March 1912, returning from the Pole, he walked willingly to his death in a blizzard, to try and save his comrades, beset by hardships. This note is left by the Relief Expedition of 1912.

The following is a poem about Oates by an unknown author.

"No Surrender Oates"*

It was not in the fury and the foam,
The swift, earth-shaking tumult, and the shout
Of close-knot squadrons riding hard and home,
That he went out.

For him no trumpets called with jubilant blast,
Only the ice-winds everlasting moan;
Alone into the solitude he passed,
Yet not alone.

For joyfully the long line of his peers,
Most joyfully those staunch old bands and true,
Which rode at Balaclava, in far years, and Waterloo, . . .
Watched him go out into the deathly wild,—
Ay, many valiant souls of mighty men
Saw that, and smiled.

Later a great memorial service was held in Saint Paul's cathedral in London. King George attended in admiral's uniform, along with a notable gathering of statesmen, government officials, army and navy officers, and representatives of all that is best in Britain.

* Nickname Captain Oates earned in the Boer War.

Figure 5. Statue of Capt. Scott at Christchurch, New Zealand, sculptured by his wife, Lady Scott. (Photo by R. L. N.)

It was from Christchurch, New Zealand, that Scott had left for the Antarctic on his last expedition. Here, in a lovely little city park next to the meandering Avon River, stands an imposing marble statue of Scott, sculptured by his wife, Lady Scott (Figure 5). It depicts Scott in trail clothing. All the details are there—the ski pole, the gauntlet gloves that kept his hands warm at $-45°$, and the windproofs, the glove harness, the helmet, and the wallet in which he kept his diary and in which he wrote beautiful and enduring English prose even as he was dying.

Scott's "Message to the Public" was written near his end. It starts out, "The causes of the disaster are not due to faulty organisation, but to misfortune in all risks which had to be undertaken." Further on he writes, "We are weak, writing is difficult, but for my own sake I do not regret this journey, which has shown that Englishmen can endure hardships, help one another, and meet death with as great a fortitude as ever in the past. We took risks, we knew we took them; things have come out against us, and therefore we have no cause for complaint, but bow to the will of Providence, determined still to do our best to the last. But if we have been willing to give our lives to this enterprise, which is for the honour of our country, I appeal to our countrymen to see that those who depend on us are properly cared for."

In London there is a statue identical to that in Christchurch except that it is made of bronze. It was a gift to the citizens of London from the officers of the British fleet. Here Captain Robert Falcon Scott, who lost his life in Antarctic exploration, looks diagonally across the street at a man who lost his life in Arctic exploration; Sir John

Franklin returns his gaze. On a tablet attached to Scott's statue are these words from his "Message to the Public."

HAD WE LIVED, I SHOULD HAVE HAD A TALE
TO TELL OF THE HARDIHOOD, ENDURANCE, AND
COURAGE OF MY COMPANIONS WHICH WOULD
HAVE STIRRED THE HEART OF EVERY
ENGLISHMAN. THESE ROUGH NOTES AND OUR
DEAD BODIES MUST TELL THE TALE.

There are a score of memorials to Scott and his companions in various parts of the world. One in Queenstown, New Zealand, reads:

ERECTED
TO COMMEMORATE
THE PATIENT, STUBBORN, INVINCIBLE COURAGE,
THE LOYAL COMRADESHIP,
AND BRILLIANT ACHIEVEMENT
OF
CAPTAIN ROBERT FALCON SCOTT
DR. EDWARD ADRIAN WILSON
CAPTAIN LAWRENCE E. G. OATES
LIEUT. HENRY R. BOWERS
AND
PETTY OFFICER EDGAR EVANS
WHO REACHED THE
SOUTH POLE
ON 17TH JANUARY 1912 AND PERISHED
ON THE RETURN JOURNEY

THEY REST IN THE GREAT WHITE SILENCE OF
ANTARCTICA AMID THE SCENES OF
THEIR TRIUMPHS, WRAPPED IN
THE WINDING SHEETS OF
THE ETERNAL SNOWS.

At the summit of Observation Hill near McMurdo Station, Antarctica, there is a nine-foot wooden cross erected by members of the expedition on January 20 and 21, 1913 in memory of Scott's Polar Party. It is the southernmost cross on planet Earth. It faces the south and looks toward the Ross Ice Shelf and the Beardmore Glacier where Scott and his companions died (Figure 6). It bears these words:

IN MEMORIAM
CAPT. R. F. SCOTT, R. N.
DR. E. A. WILSON
CAPT. L. E. G. OATES, INS. DRGS.
LT. H. R. BOWERS, R. I. M.
PETTY OFFICER E. EVANS, R. N.
WHO DIED ON THEIR
RETURN FROM THE
POLE, MARCH
1912
TO STRIVE, TO SEEK,
TO FIND,
AND NOT TO
YIELD*

Acknowledgments. Figures 1, 2, 3, and 4 are reproduced from *Scott's Last Expedition,* arranged by Leonard Huxley, with a preface by Sir Clements R. Markham, and published by Smith, Elder and Co., 1913. The long quotations from Scott's diary and letters are also from *Scott's Last Expedition.* The diary and letters are now in the British Museum.

* Last line of Tennyson's "Ulysses."

Figure 6. The wooden cross at the summit of Observation Hill, McMurdo Station, Antarctica, in memory of Scott's Polar Party. (Photo by R. L. N.)

Addendum

The Rewards from Antarctic Field Work

I am often asked why I have continued to do geologic field work in cold, barren, and remote areas like Alaska, Northwest Territories, Greenland, Patagonia, and Antarctica. My answer is:

I like the masculinity of the explorer's world.

I like the physical adventure.

I like the intellectual adventure.

I like the close companionship with carefully selected and well-adjusted field men.

The food tastes so good.

The sleeping bag feels so warm.

I like the simplicity of the life. The field man's world consists only of eating, sleeping, and working.

I like the scenery, the warm tent, a little wind in my face, and the warmth of the sun of my wind-proofs.

I like to step on land no man has even trod on before.

I take no little satisfaction in the fact that this business has made it possible for me to give to a very few Tufts boys a never-to-be-forgotten experience.

But above all

I like the challenge, the physical, the mental, and the moral challenge; and my greatest reward has come from knowing that I came very, very close to my greatest possible potential almost every day during a field season, something which unfortunately I did not do at Tufts University too often.

Robert L. Nichols
Professor Emeritus of Geology
Tufts University
Medford, Massachusetts

I Gondwanaland

1

Antarctica and Gondwanaland

Review Paper

CAMPBELL CRADDOCK
Department of Geology and Geophysics, University of Wisconsin, Madison, Wisconsin 53706, U.S.A.

Abstract. Antarctica is a central piece in the Gondwanaland reconstruction, with ties to all the other major fragments. A strong case for the reality of Gondwanaland rests upon such diverse evidence as (1) similarities in rocks of Precambrian–Mesozoic age, (2) distribution of upper Paleozoic glacial deposits, (3) the Paleozoic–Mesozoic fossil record, (4) Antarctic orogens with continuations in other continents, (5) widespread Jurassic basaltic volcanic rocks, (6) magnetic anomaly belts in the oceanic regions, and (7) the morphology of continental margins. The Gondwanaland reconstruction is generally satisfactory although some problems remain, such as the dimensions of the Indian fragment and the origin and age of certain oceanic plateaus. The history of Gondwanaland is inferred from (1) the geology and structure of the now dispersed fragments, (2) paleomagnetic polar wander curves, and (3) ocean floor ages obtained from magnetic anomalies and deep sea drilling.

The early history of Gondwanaland is poorly known, but some fragments were joined even in Archean time. The main landmass had formed by late Precambrian time, and Gondwanaland was largely intact from early Paleozoic until Jurassic time. The Pacific border of Gondwanaland was an active continental margin, and the major orogenic episodes (early Paleozoic, middle Paleozoic, early Mesozoic) formed belts which are traceable along the edges of the South American, African, Antarctic, and Australian shields and which record interactions of the Gondwanaland and Pacific plates.

Although it may have begun earlier locally, breakup was underway by Jurassic time; the northwest coast of Australia underwent rifting and sea-floor spreading, and extensive basaltic eruptions occurred in Brazil, southeastern Africa, and Antarctica. The India-Antarctica separation began in latest Jurassic or Early Cretaceous time. A spreading axis advanced westward across the southern Pacific during Cretaceous time, eventually separating New Zealand, West Antarctica, and East Antarctica–Australia. Australia separated from East Antarctica in early Tertiary time, and the bending of the New Zealand geosyncline and the inception of the Alpine fault followed later in the Tertiary Period. The isolation of Antarctica was completed in middle Tertiary time by the separation of South America from the Antarctic Peninsula. This separation allowed the development of the Antarctic Circumpolar Current, and the modern ice sheet probably dates from about that time.

The geologic history of Antarctica must be studied in the context of its place in Gondwanaland. Although the idea of drifting continents had appeared earlier, it was not until late in the nineteenth century that geologists first suggested the existence of a large ancestral continent in the Southern Hemisphere in the geologic past. This hypothetical protocontinent—now termed Gondwanaland—included present-day South America, Africa, Arabia, Madagascar, Ceylon, India, Antarctica, Australia, and New Zealand (Figure 1.1). Because of its central position in the reassembly, Antarctica clearly must play an important role in determining the reality and history of Gondwanaland. Despite its great importance to the concept, however, the geology of Antarctica was almost unknown when Gondwanaland was first postulated. Only in the past few years has knowledge of Antarctica, the last fragment of the supposed supercontinent to be explored, advanced to a point where meaningful geologic tests of the Gondwanaland hypothesis are possible.

Antarctica was probably first sighted about 1820 from ships sailing near the tip of the Antarctic Peninsula; this discovery confirmed the long-suspected existence of a southern polar continent. Later expeditions collected continental rocks such as granite and sandstone from icebergs, but the first geologic specimens obtained on the continent were collected in the 1890's at Cape Adare, south of New Zealand. The next twenty years saw vigorous exploration by expeditions from several nations, and the broad pattern of Antarctic geology began to emerge. Other private and national expeditions before and after World War II added more geologic information, but large areas of the continent still remained unexplored as late as 1955. Most of these early geologic observations were confined to the coastal fringe of Antarctica; except for the South Pole parties of Ernest Shackleton, Roald Amundsen, and Robert F. Scott—and the Laurence M. Gould party in support of Byrd's polar flight—the vast interior of the continent remained geologically unknown.

Since the International Geophysical Year in 1957–58, ski-equipped aircraft and oversnow tracked vehicles have opened the entire continent to geologic study. Only a few significant rock exposures remain unvisited by geologists today. Our knowledge of the geology of this remote continent has more than doubled during the past twenty years.

In 1937 the great South African geologist Alexander du Toit published *Our Wandering Continents,* which must be rated among the most important books on geology ever written. In this brilliant synthesis he set forth in detail the geologic evidence then available for continental drift and for the existence of Gondwanaland. Regarding Antarctica, he stated: "The role of the Antarctic is a vital one. As will be observed . . . the shield of East Antarctica constitutes the 'key-piece'—shaped surprisingly like Australia, only larger—around which, with wonderful correspondences in outline, the remaining 'puzzle-pieces' of Gondwanaland can with remarkable precision be fitted."

Because so little was then known about the geology of Antarctica, Du Toit's Gondwanaland reassembly in effect predicted the geologic patterns to be expected in the Antarctic interior. It is a tribute to his genius and foresight that subsequent studies in Antarctica have largely confirmed these predictions. When Du Toit's book was published, the Gondwanaland hypothesis had few advocates, particularly among Northern Hemisphere scientists. Unfortunately, Du Toit did not live to see the wide respect and acceptance accorded his work today.

A generally satisfactory reassembly of Gondwanaland can be made today, quite similar to that of Du Toit (1937), but some problems remain. The shape of the Antarctic Peninsula and its position relative to southern Africa and South America are poorly understood. The age and origin of certain submarine plateaus, such as the Kerguelen Plateau, are not clear. The nature of the area between India and Australia has long been debated; Veevers and others (1975) have proposed that Greater India adjoined western Australia (see Figure 1.1).

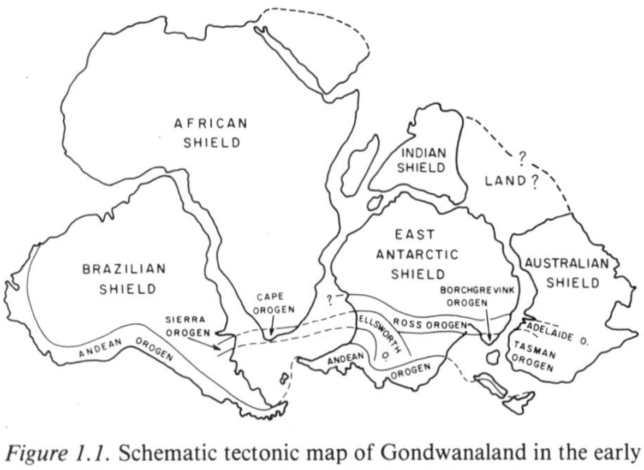

Figure 1.1. Schematic tectonic map of Gondwanaland in the early Mesozoic before fragmentation, with the addition of the younger late Mesozoic–early Cenozoic Andean orogen. The shapes and structures of the Gondwanaland fragments as they exist today have been used in this reconstruction. Areas in the western Pacific which today lie north of Australia–New Zealand and which may have been part of Gondwanaland are not included. The possible landmass between peninsular India and Australia is that postulated by Veevers and others (1975).

Geology of Antarctica

Early in this century, after geologic data became available from the Antarctic Peninsula, the Ross Sea region, and other coastal localities, it was recognized that Antarctica could be divided into two major geologic provinces. The first comprises the larger part of the continent that faces mainly upon the Atlantic and Indian Oceans; since most of this province lies in the area of east longitudes, it is commonly known as East Antarctica. The second province makes up the smaller part of the continent that faces mainly upon the Pacific Ocean; it is commonly known as West Antarctica. East Antarctica is a typical continental shield or stable platform, consisting of a foundation of igneous and metamorphic rocks overlain by a sequence of younger, flat-lying sedimentary and volcanic rocks. By contrast, West Antarctica is composed of generally younger rocks that are widely deformed and metamorphosed; the age and nature of the basement rocks are poorly known. Intrusive and extrusive igneous rocks are abundant, and some volcanic activity continues there today. The geology of Antarctica has been portrayed on a 1:5,000,000 geologic map of the continent (Craddock, 1972*a*).

Rock exposures comprise less than 5% of the area of the continent, and those in East Antarctica occur in an oval belt that includes the coastal region and the Transantarctic Mountains. The nunataks and mountain ranges within this belt reveal a basement complex composed mainly of high-grade metamorphic rocks and intrusive igneous rocks. Gneisses of the granulite facies are the most abundant rocks, but lower grade metamorphic rocks of the amphibolite and greenschist facies occur in some localities. A wide range of igneous rocks has been reported, but felsic varieties such as granite are most common. These crystalline basement rocks record a complex geologic history involving several cycles of deformation, regional metamorphism, and emplacement of igneous intrusives.

In contrast to the other continents, Antarctica has so far revealed few rocks with apparent ages greater than 1.8 billion years. This fact may imply that (1) the oldest rocks in the basement of East Antarctica are truly younger than those of other continental shields; (2) insufficient sampling has so far prevented the discovery of the oldest rocks present; or (3) later metamorphic events, especially in early Paleozoic time, have reset the mineral clocks, causing apparent radiometric ages younger than the true ages of the rocks. The youngest known rocks in the East Antarctic basement complex are fossiliferous upper Precambrian and Cambrian strata, generally folded and metamorphosed, which occur in the Transantarctic Mountains and at a few coastal localities.

Overlying the East Antarctic basement complex is a succession of mainly flat-lying sedimentary and volcanic rocks. These beds comprise the Beacon Supergroup (sedimentary) and the Ferrar Group (igneous), and they represent the Gondwana sequence—as described from peninsular India and the other southern continents—in Antarctica. The Beacon Supergroup contains rocks as old as Devonian and the Ferrar Group rocks as young as Jurassic. Beacon and Ferrar strata are widely exposed in the Transantarctic Mountains, but they have been found in place at only a few localities along the coast of East Antarctica.

The geologic history of West Antarctica is complex and not well established. All rocks whose ages are known appear to have formed during the Phanerozoic; no definitely Precambrian rocks have yet been discovered. In much of West Antarctica the oldest rocks are igneous intrusive and metamorphic varieties that form a basement complex of Paleozoic (?) age. Sedimentary and volcanic sequences of probably Paleozoic and Mesozoic age are widely distributed, and most of these rocks are strongly folded and somewhat metamorphosed. Intrusive igneous rocks were emplaced throughout much of West Antarctica during Mesozoic time and perhaps early Cenozoic time. Upper Cenozoic volcanic and sedimentary layers on the Antarctic Peninsula and to the west are flat-lying and undisturbed. Volcanism which began in middle Tertiary time has continued into recent times in much of coastal West Antarctica.

The geologic composition and history of a continent can be effectively studied form a tectonic map, which shows the age and distribution of the major structural units that define the architecture of the continent (Craddock, 1970, 1972*b*). In particular, such maps portray shields (stable areas with a basement complex of ancient igneous and metamorphic rocks overlain by younger flat-lying strata) and orogens (belts of folded and metamorphosed strata, commonly intruded by large bodies of granite). The major tectonic units of Antarctica are the East Antarctic shield, four orogens in the Transantarctic Mountains and West Antarctica, and a Cenozoic volcanic province. The delineation of these tectonic provinces allows a major test of the Gondwanaland hypothesis—the geologic compatibility of the opposing coasts.

Evidence for Gondwanaland

In fitting Antarctica into his Gondwanaland reassembly, Du Toit (1937) had to depend largely on the shape of the continent because so little was then known about its geology. Since that time, progress in Antarctic geology has yielded a number of discoveries that bear on the Gondwanaland problem. Some lines of evidence that seem especially significant are summarized below.

1. The basement rocks of coastal East Antarctica are similar, both as a terrane and in some details, to those along the matching coasts of the other Gondwanaland fragments in Du Toit's reassembly. Work by geologists of several nations has shown that the structural grain in these ancient Antarctic rocks is compatible with that in like rocks of the suggested matching coast. The basement rocks of all these areas are compositionally similar, consisting of high-grade metamorphic rocks such as granulite gneisses, along with igneous intrusives such as granite. In particular, the unusual rock charnockite is widespread in eastern Africa, Ceylon, eastern India, and coastal East Antarctica.

2. The rocks of the Beacon Supergroup in Antarctica are similar to the Paleozoic and Mesozoic Gondwana sedimentary sequences on the other southern continents and continental islands. The lower Beacon strata consist of detrital sedimentary rocks as old as Devonian in some localities. Beds of ancient tillite were first discovered in the Transantarctic Mountains in 1958, and many other Paleozoic tillite localities are now known in Antarctica. These tillites occur in the Beacon sequence and are considered Carboniferous or Permian in age. Overlying these glacial beds are younger Permian strata which commonly include coal beds and bear the *Glossopteris* flora. This succession of distinctive rock types can be matched, at least in part, in Australia, India, Madagascar, Africa, the Falkland Islands, and South America.

3. With the discovery of the tillites in Antarctica, evidence for late Paleozoic glaciation is now known from all the major Gondwanaland fragments. Similar deposits had been previously found in South America, Africa, India, and Australia, and their existence in Antarctica had been predicted. The character and distribution of these ancient glacial beds imply the presence of Paleozoic ice sheets of continental dimensions. However, along some coasts, directional indicators show that the ice flowed onto the present-day continent from the adjacent ocean basin. If we accept modern geography as that of the Permian, both the anomalous flow directions and the wide latitudinal range of continental glaciation are awkward to explain. Recent studies have shown, however, that the Gondwanaland reconstruction presents an attractive alternative in understanding this early glacial period (Crowell, 1977).

4. The Paleozoic and Mesozoic fossil record in Antarctica, only now emerging in detail, bears a strong resemblance to that found in the other southern continents. Marine faunas are as old as Cambrian and include archaeocyathids, trilobites, gastropods, bryozoans, fish, echinoderms, brachiopods, and pelecypods. These animals flourished in shallow waters close to land under conditions similar to those on modern continental shelves. Although these fossil faunas are similar to those of Australia, South America, and South Africa, it is unlikely that such animals migrated across deep ocean basins. Land animals are less abundant but also show close affinities to the faunas of the other southern continents. Recent discoveries of Triassic reptiles and amphibians in the Transantarctic Mountains are of great importance because these animals, such as *Lystrosaurus,* must have moved over land routes. Antarctic fossil floras, especially the Permian–Carboniferous *Glossopteris* flora, bear a strong resemblance to other southern floras. The pronounced overlap between the *Glossopteris* floras of Antarctica and India, for instance, poses two questions if we suppose that Permian and modern geography are identical. Can a reasonable dispersal mechanism, such as wind or water currents, be found to connect these distant lands? And is it possible that these two widely separated lands, one polar and the other tropical, could support nearly identical floras when their climates would have been so different? The fossil animal and plant record in Antarctica strongly suggests that the present geographic isolation of the continent did not exist during Paleozoic and early Mesozoic time.

5. The deformed and metamorphosed rocks underlying the Beacon strata in the Transantarctic Mountains can be compared to rocks in southeastern Australia. Strata in both areas were folded and intruded by granitic rocks during Paleozoic time. Each continent reveals an early Paleozoic orogen paralleled by a middle Paleozoic orogen lying to the east, or away from the Precambrian shield (Figure 1.1). Structural trends in both areas are anomalous in being nearly perpendicular to the present shoreline, but they roughly parallel the coast in the Gondwanaland reassembly.

6. The Ellsworth orogen formed in early Mesozoic time. It probably represents the continuation of the Cape orogen of South Africa and a part of Du Toit's (1937) Samfrau geosyncline. Between the Transantarctic Mountains of East Antarctica and the coastal belt of West Antarctica lies a large region that has been explored only during the last twenty years. Bedrock exposures in this area are found in the Ellsworth Mountains and in nunatak groups and

small ranges to the south and west. On the basis of similarity of rock types and structural continuity, most of these outcrops have been assigned to a new tectonic province, the Ellsworth orogen. The thick sedimentary sequence is mainly Paleozoic in age (some Precambrian strata may be present) and has undergone strong post-Permian folding. Some of these formations resemble the Beacon strata in the Transantarctic Mountains but differ in being both considerably thicker and strongly folded. In part of the province, these deformed Paleozoic strata are invaded by granitic bodies that were emplaced during late Triassic to early Jurassic time. In its present setting, the Ellsworth Mountains fold belt is a discordant tectonic fragment resting between East Antarctica and coastal West Antarctica. Both in stratigraphy and in structural style, however, it bears strong resemblance to the Cape fold belt of southern Africa and the fold mountains of eastern Argentina. In the Gondwanaland reassembly, it represents the natural continuation of a fold belt that begins in Argentina and continues across southern Africa into Antarctica.

7. Jurassic igneous rocks, mainly basaltic in composition, are widespread in Antarctica. These rocks occur both as volcanic deposits and as shallow intrusive sills and dikes. They are common throughout the length of the Transantarctic Mountains, where they have been termed the Ferrar Group. These Jurassic mafic igneous rocks may be compared to rocks of similar age and composition that occur over large areas of Brazil, southern Africa, and Tasmania. In the Gondwanaland reassembly, it is reasonable to interpret the rocks of all these areas as belonging to a single igneous province—one perhaps related to the initial fragmentation of the protocontinent. Jurassic volcanic rocks of more varied composition are abundant in the Antarctic Peninsula and common along the coast of West Antarctica. These latter rocks, along with counterparts in Argentina and Australia, may be the products of tectonism along an active margin of Gondwanaland.

8. Magnetic anomaly belts parallel to and symmetrical about the mid-ocean ridges strongly suggest that sea-floor spreading and continental displacement have occurred. Within the last twenty years, oceanographic surveys have revealed this unexpected pattern of parallel belts of high and low magnetic intensity, a pattern which initially defied explanation. Concurrent studies of paleomagnetism from both lava flow and marine sediment sequences, however, produced evidence that the earth's magnetic field undergoes periodic reversals of its polarity. If the parallel high- and low-intensity belts are ascribed to these polarity reversals, then the slowly spreading sea floor can be thought of as a magnetic tape which freezes in the effects of the existent magnetic field as new crust is formed at the mid-ocean ridges by the cooling of silicate melts from the earth's interior. If this interpretation is correct, it is possible to determine both the rates of sea-floor spreading and the times when the various fragments of Gondwanaland began to separate.

9. Better knowledge of the bathymetry of the Southern Ocean allows more critical testing of the morphologic fit of the Gondwanaland fragments. The earliest speculations about continental drift were fostered by the similarity in shape of opposing coasts, especially the Atlantic coasts of South America and Africa. The coastline, however, may undergo significant modification with only a small change in sea level, and a more realistic comparison is obtained by using a submarine contour near the edge of the continental shelf or part way down the continental slope. Although subject to some modification by erosional and depositional processes, such a contour is a natural border for the continent and may have changed little in shape since the time of separation from its adjoining landmass. Recent oceanographic surveys have yielded greatly improved bathymetric maps of the Antarctic continental shelf and slope. Studies using these new maps have shown that good morphologic matches can be obtained between Africa and Antarctica, and between Australia and Antarctica, as required by the Gondwanaland hypothesis.

Du Toit's reassembly of Gondwanaland, if correct, predicts that certain geologic features should exist in Antarctica. The basement rocks of coastal East Antarctica must resemble in composition and structure those of the matching coats of the other Gondwanaland fragments. The Beacon strata should be present throughout the length of the Transantarctic Mountains and should resemble the Gondwana sequences of the other southern continents. Upper Paleozoic glacial beds should occur in Antarctica. The Paleozoic and Mesozoic fossil record should consist of forms resembling those elsewhere in Gondwanaland rather than unique forms developed on an isolated continent. The eastward extension of the Samfrau geosyncline and the Cape fold belt from southern Africa should be found in Antarctica. The southward-trending Paleozoic orogens of southeastern Australia should continue in the northern Transantarctic Mountains. Jurassic mafic igneous rocks can be expected in the Transantarctic Mountains along with the strata of the Beacon Supergroup.

The rapidly accumulating geologic data from Antarctica strongly suggest that each of these predictions will be proven correct. The accuracy of Du Toit's predictions, the striking geologic similarities between Antarctica and the other southern lands, and the new evidence from the oceans for sea-floor spreading combine to provide a very strong case for the reality of Gondwanaland.

Recent delineation of the tectonic provinces of Antarctica allows comparison with those other landmasses and permits a more accurate reconstruction (Figure 1.1). Tectonic ties between Antarctica and Australia are provided by the early Paleozoic Ross and Adelaide orogens, and by the middle Paleozoic Borchgrevink and Tasman orogens. The early Mesozoic Ellsworth orogen in Antarctica is the continuation of the Cape orogen in southern Africa.

Origin of Gondwanaland

Although Gondwanaland seems to have existed by late Precambrian time, its early history remains obscure. Uncertainty surrounds the mode and place of formation of the individual Precambrian shields. Embleton and McElhinny (1975) have presented paleomagnetic evidence for positioning Madagascar off Kenya and Tanzania, as shown by Du Toit (1937). McElhinny (1973) has cautioned that separate blocks within a present-day continent may have moved independently, and Embleton and others (1974) postulate a large rotation of southeast Australia since Silurian time. Harrington and others (1973) have shown the likelihood of two phases of large lateral displacement in the Australia–New Zealand–East Antarctica sector. The existence and extent of Precambrian shield in West Antarctica and in southernmost South America remain problematic.

The development of the Gondwanide shields is part of the general problem of the early Precambrian history of the lithosphere. Hargraves (1976) has proposed a model in which shields are created from an early lithospheric segregation by convection during the early Precambrian and by plate tectonics processes during the late Precambrian; some Precambrian orogens are considered as probably intracratonic, but others formed along the edges of the cratonic nuclei. All the main shields within Gondwanaland contain early Precambrian rocks, and some of the oldest rocks yet discovered occur in southern Africa. However, the study of all these Precambrian terranes, and especially the key central fragments of Africa and East Antarctica, is complicated by a widespread late Precambrian–early Paleozoic thermal event which clouds the meaning of radiometric ages and paleomagnetic determinations. Thus, while these shields clearly began to form during early Precambrian time, their relative positions at various times during the Precambrian are poorly known.

The identification of tectonic and/or metamorphic belts within one shield, and their probable continuations in other shields, is an important tool for refining the Gondwanaland reconstruction and establishing at what times now-separated shields were continuous. Such belts, however, are presently poorly defined in the Precambrian rocks of the Gondwanide shields. Rutland (1976) has recognized three main basement provinces in Australia, with the oldest in the west and the youngest in the east. Ravich and Kamenev (1972) have identified three kinds of tectonic elements in the East Antarctic shield, but Elliot (1975) concluded that insufficient data exist yet to define clearly the tectonic-chronologic provinces. Orogenic and metamorphic zones in the Precambrian of Africa have been reviewed by Clifford (1974), who emphasized the complications created by widespread retrograde metamorphism during the Pan-African event 600 m.y. ago. A few attempts have been made to correlate Precambrian belts between Gondwanide shields. M. B. Katz (1972, 1974) postulated ties between granulite belts in India, Madagascar, and Africa, and Crawford (1974) matched granulite belts in India and Ceylon with those in Enderby Land, Antarctica. Possible continuations of African Precambrian structures in South America are shown on a map by Hurley and Rand (1973).

Paleomagnetic polar wander curves from different crustal blocks provide some information about the early history of Gondwanaland, though much remains to be learned. Briden (1973) and Piper and others (1973) have argued that Africa was intact and joined to South America after about 2300 m.y. ago. Facer (1974), however, has emphasized differences between the Australian Precambrian curve and those of other shields and has suggested the likelihood of relative movements during Precambrian time. Curves from 2400 m.y. ago for Africa, India, and Australia are given by Giddings and others (1973); convergence of these curves at about 800 m.y. may indicate the time of origin of Gondwanaland. Veevers and McElhinny (1976) concluded that the Australian platform was intact between 1800 m.y. and 450 m.y. ago, and McElhinny and others (1974) indicated that Gondwanaland had formed by 750 m.y. ago or earlier. Hailwood (1974) has inferred from the curves that Gondwanaland was intact by early Paleozoic time, but some separations may have existed during Cambro-Ordovician time (Embleton, 1972; Klootwijk, 1973). In his comprehensive review, McElhinny (1973) concluded that Gondwanaland had certainly formed by the Ordovician Period and that it remained essentially intact into the Mesozoic Era, although some minor disruptions may have occurred during late Paleozoic or Triassic time.

Comparison of fossil faunas (Runnegar, 1979) and floras can provide tests for the continuity of Gondwanaland at different intervals during Phanerozoic time. Lower Paleozoic fossils, however, are rare through much of Gondwanaland. Firm conclusions are not possible at present about Cambrian paleogeography; additional comparative studies are needed, especially for the Cambrian faunas of South America, Antarctica, and Australia. Palmer (1973) has pointed out that the dissimilarity of Cambrian trilobite faunas from Argentina and Antarctica may be due to simple latitudinal differences or to separation of one or both areas from the rest of Gondwanaland at that time. Studies of Devonian fossils, however, indicate the unity of Gondwanaland by that time. Devonian marine faunas of Antarctica are typical of a province that includes southern Africa, the Falkland Islands, and southern South America; ties between Antarctica and New Zealand are also probable (Boucot and others, 1967). Devonian fish fossils in southeastern Australia are similar to those in the Transantarctic Mountains. The Permian floras and the Triassic and younger vertebrate faunas (Colbert, 1979) indicate that Gondwanaland remained essentially intact well into the Mesozoic.

Tectonic evolution of Gondwanaland

Although the early history of Gondwanaland remains obscure, some facets of its tectonic evolution from late Precambrian time until its breakup in the Mesozoic Era can be described. The Pacific border of Gondwanaland, along the edges of the Brazilian, African, East Antarctic, and Australian shields, was an active continental margin during this time interval (Craddock, 1975). Three main orogenic episodes are defined by deformation, igneous activity, and regional metamorphism, and these orogens can be traced intermittently along this restored continental margin. The importance of these belts in refining the Gondwanaland reconstruction is seen in the accompanying map (Figure 1.1).

Thick sedimentary and volcanic sequences of late Precambrian–early Paleozoic age, which underwent Cambro-Ordovician orogeny, occur along much of the Pacific border. In South Australia, the rocks of the Adelaide geosyncline were deformed during the Delamerian orogeny, and in addition, the late Precambrian Penguin orogeny has also been recognized in Tasmania (Rutland, 1976). Harrington and others (1973) have postulated large strike-slip displacements along latitudinal faults in this region, and thus the Tasmanian segment of this belt may be well east of its original position. The Ross orogen can be traced almost across Antarctica, where it roughly parallels the Transantarctic Mountains, but its continuation in Africa is in doubt. It may be represented by the Cambrian Cape granites, or it may be related to the east-northeast-trending Damaran belt farther north. The tectonic history of southern South America is poorly known, but the early Paleozoic orogen may have reached this part of Gondwanaland as well (Borrello, 1969).

Evidence for a second orogenic peak during the Devonian can be found along the Gondwanaland border from Australia to South America. The main deformation in the western Tasman orogen was the Devonian Tabberabberan orogeny (Rutland, 1976), and Devonian plutonism and regional metamorphism have been identified in western New Zealand (Suggate and Grindley, 1972). Evidence for this orogeny in Antarctica is found in the Borchgrevink orogen of northern Victoria land and at scattered localities along the coastal belt of West Antarctica (Craddock, 1972*a,b*). Some deformations near the southern tip of Africa appear to be pre-Karroo (Haughton, 1969), and these structures may also record Devonian orogeny. The Paleozoic history of southern South America is poorly known, but Harrington (1967) has reported deformation in the Andean geosyncline of Late Devonian–Carboniferous age.

A third orogenic cycle of Permo-Triassic age, the Gondwanide orogeny of Du Toit (1937), affected rocks along the Pacific border of Gondwanaland from Australia to South America. The principal phase of deformation in the eastern Tasman orogen was the Permian Hunter-Bowen orogeny (Rutland, 1976), and some granite intrusives also of Permo-Triassic age occur in New Zealand (Suggate and Grindley, 1972). The Ellsworth orogen in Antarctica was described by Craddock (1964), and its significance was discussed by Craddock (1972*a*) and Ford (1972). The Cape orogen in southernmost Africa developed in strata of the Cape and Karroo systems, probably during Permian and early-middle Triassic time (Bishopp and Van Eeden, 1971; De Beer and Van Zijl, 1974). The westward continuation of this belt in the Sierra orogen of Argentina was identified by Du Toit (1937), and evidence for probably related deformation in the Andean belt is given by Borrello (1969) and Gansser (1973*a*). Dalziel (1972) emphasized the role of early Mesozoic orogeny in southern South America, the Scotia arc, and the Antarctic Peninsula.

Thus, the Pacific border of Gondwanaland was an active margin which underwent successive cycles of deposition, orogeny, and cratonization. In Australia these orogenic belts are progressively younger from the shield toward the Pacific Ocean, but this pattern is not so clear elsewhere. The three orogens discussed here have indistinct boundaries in time and space; they represent deformation peaks during a long interval of tectonic activity. In most cases the ensimatic or ensialic character of the orogen has not been established, and some orogens may have formed in marginal basins with continental crust on the seaward side. Thus, this active margin of Gondwanaland was the site of extensive folding, regional metamorphism, igneous activity, faulting, and probably horizontal displacement of some crustal blocks. Nevertheless, Gondwanaland seems to have remained essentially intact from late Precambrian well into Mesozoic time, although the first indications of the eventual breakup may have appeared during late Paleozoic time (McElhinny, 1973; Klootwijk, 1975; Valencio, 1975*a,b*).

Fragmentation of Gondwanaland

Recent geologic and geophysical studies on land and at sea have clarified the history of the breakup of Gondwanaland. It remains difficult, however, to establish exactly the time of separation of one continental block from another. In some cases, early manifestations of crustal extension, such as faulting, volcanic activity, subsidence, and marine transgression, preceded by millions of years the ultimate parting and formation of new sea floor. Hence, for any given case, separation dates based on land geology may differ considerably from dates based, for example, on marine magnetic anomalies. Twelve events in the Gondwanaland fragmentation are discussed briefly below, roughly in chronologic order.

1. Possibly the first separation to occur was that of West Gondwanaland (South America–Africa) from East Gondwanaland (Antarctica-India-Australia). On the basis of paleomagnetic polar wander curves, Valencio (1975*a,b*) suggested that this separation may have begun in as early as Permian time. Ravich (1973) favors a Triassic age, and several writers argue that the widespread Jurassic volcanicism marks the start of separation. Marine magnetic anomalies in this area are too poorly defined at present to clarify this uncertainty, but see valuable recent work by Bergh and Norton (1976) and Barrett (1977). It should also be noted that Shields (1977) has argued for an India-Arabia separation during Paleocene time, and that Colbert (1979) has inferred interconnection between India, Madagascar, and Africa even in Late Cretaceous time on the basis of the vertebrate fossil record.

2. Northwestern Australia was the site of rifting and faulting in Early Jurassic time, and sea-floor spreading began there in Late Jurassic time (Larson, 1975; Exon and Willcox, 1976). It is not clear, however, whether a continental fragment was detached from this coast.

3. More than a dozen papers since 1970 have offered estimates of the time of separation of Africa and South America, based on paleomagnetic curves, sea-floor magnetic anomalies, or geologic evidence. Larson and LaFountain (1970) inferred from paleomagnetic data that separation began during Triassic time, but most subsequent studies have yielded a Late Jurassic or Early Cretaceous estimated age for this event. Herz (1977) postulated doming, rifting, and eruption beginning 147 m.y. ago during Jurassic time, with the formation of new ocean floor commencing in middle Cretaceous time.

4. The time of separation of India from Antarctica is poorly defined, but it was probably Late Jurassic or Early Cretaceous. Laughton and others (1973) summarized the history of the Indian Ocean and dated this separation as older than 75 m.y. Veevers and McElhinny (1976) placed the time at 130 m.y. ago, and Casshyap (1976) reported evidence for a possible seaway off the modern east coast of India during Early Cretaceous time. Crustal extension normal to the east coast of India may have caused subsidence, block faulting, and marine transgression in Late Jurassic time (Aditya and Sinha, 1976; Sastri and others, 1976).

Johnson and others (1976) have dated this separation at about 130 m.y. ago and Larson (1977) at 135-120 m.y. ago.

5. Broken Ridge separated from the Kerguelen Plateau by early Tertiary time, and it is possible that these features may in part represent continental fragments detached from Gondwanaland. Quilty (1973) reported Cenomanian foraminifera from sedimentary rocks northeast of the Kerguelen Plateau and concluded that the initial separation of Broken Ridge had occurred by 92 m.y. ago. Watkins and others (1974) studied Kerguelen Island and found that it is not older than late Oligocene in outcrop; no evidence for a continental origin of the island was discovered. Luyendyk (1976) postulated that both Broken Ridge and the Kerguelen Plateau began to form more than 100 m.y. ago through the activity of the Kerguelen Islands hot spot.

6. The history of the spreading ridge in the southern Pacific and Indian Oceans, an important factor in the breakup of Gondwanaland, was discussed by Bowin (1974). In his model, a spreading ridge had formed in the southeastern Pacific by Late Cretaceous time and had migrated westward into the Indian Ocean by early Tertiary time and northwestward to the Gulf of Aden by late Tertiary time. This ridge system played an important role in the movements of New Zealand, Australia, and India away from Antarctica. Johnson and others (1976) have postulated that the ridge between Australia and Antarctica developed as a branch from a ridge in the Indian Ocean.

7. On morphologic grounds, it is probable that the New Zealand Plateau formerly lay adjacent to West Antarctica. Some geologic similarities exist, but the geology of coastal West Antarctica is too poorly known to permit a detailed comparison. Magnetic anomalies south of the New Zealand Plateau suggest separation from Antarctica during Late Cretaceous time, and the date has been estimated at 81-85 m.y. ago (Falconer, 1973; Molnar and others, 1975; Wellman, 1976).

8. New Zealand is a microcontinent that was probably attached to Australia before the opening of the Tasman Sea. This separation seems to have begun in Late Cretaceous time and to have ended in early Tertiary time. The period of separation has been placed at 81-60 m.y. ago by Molnar and others (1975) and at 85-55 m.y. ago by Griffiths (1976).

9. The low region of deep troughs and largely unknown geology between the Transantarctic Mountains and coastal West Antarctica, along with the unusual trend of the Ellsworth Mountains, make horizontal displacement of West Antarctica relative to East Antarctica highly probable. If such movement occurred, it was probably during Late Cretaceous and early Tertiary time (Craddock, 1975; Molnar and others, 1975).

10. The separation of Australia from East Antarctica may have begun as long ago as the Jurassic Period (Harrington and others, 1973), but the final break with the formation of new sea floor took place 53-55 m.y. ago (Sclater and Fisher, 1974; Molnar and others, 1975; Veevers and McElhinny, 1976). There have been about 500 km of dextral bending of the New Zealand geosyncline and another 500 km of dextral strike-slip displacement on the Alpine fault, both probably since Eocene time (Carter and Norris, 1976), although earlier movements have been argued (Jones and Roots, 1974; Grindley, 1976). The bending may have begun only about 40 m.y. ago (Molnar and others, 1975) and movement on the Alpine fault about 20 m.y. ago (Griffiths, 1976; Wellman, 1976).

11. The northward-moving Indian shield eventually collided with the mainland of Asia, but it is difficult to establish the exact time of this event. Although Ravich (1973) has advocated connection in latest Cretaceous time, most writers argue for collision during late Eocene or early Oligocene time (Raiverman, 1972; Athavale, 1973; Colchen, 1975; Molnar and Tapponnier, 1975; Blow and Hamilton, 1975; Sclater and others, 1976). If the Siwalik beds were derived from the rising Himalayas, then collision was underway by Miocene time (Gansser, 1973*b*; Verma, 1973; Dutta and Grigorescu, 1975). Most of the northward movement of India may have occurred during early Tertiary time (Klootwijk, 1976).

12. The separation of South America from Antarctica may have been the last major event in the breakup of Gondwanaland. Many writers have stressed geologic similarities between southern South America and the Antarctic Peninsula, but H. R. Katz (1972, 1973) has argued for contrasting histories since Late Jurassic time. Dalziel and Elliot (1973), however, have postulated a narrow land connection between South America and West Antarctica at the beginning of the Tertiary period. Dalziel (1974) inferred crustal extension from the widespread Middle and Upper Jurassic volcanic rocks around the Scotia Sea, perhaps related to the separation of South America (and Antarctica?) from Africa. Suarez (1976) has argued for the continuity of three Late Jurassic-Early Cretaceous tectonic belts from South America into the Antarctic Peninsula. Barker and Griffiths (1972) have postulated detachment of South America from the Antarctic Peninsula in early Tertiary time, and sea-floor magnetic anomalies suggest opening of the Drake Passage during late Oligocene time (Barker and Burrell, 1976). Thus differing interpretations have been offered for the Mesozoic-Cenozoic history of this complex region. New information and insights should emerge from the numerous papers on the Scotia Arc to follow in this symposium volume.

Isolation of Antarctica

Severance of the South America-Antarctic Peninsula connection was probably the last step in the isolation of Antarctica. Leg 35 of the Deep Sea Drilling Project (DSDP) was conducted in the Southeast Pacific Basin during 1974 (Craddock and Hollister, 1976). Some of the objectives of that cruise were to determine the age and character of the oceanic basement, the age and nature of the marine sediments, the history of deep and surface paleocirculation patterns, and the history of continental glaciation of West Antarctica. In particular, in order to clarify the tectonic evolution of the region, it was important to discover whether any pre-Cretaceous sea floor exists in the Southeast Pacific Basin.

Four holes were drilled south of lat. 60° S in the area northwest of the Antarctic Peninsula. Two of these holes were on the Antarctic continental rise; one failed at shallow depth in Pliocene sand, but the other reached lower Miocene or Oligocene strata. The other two holes

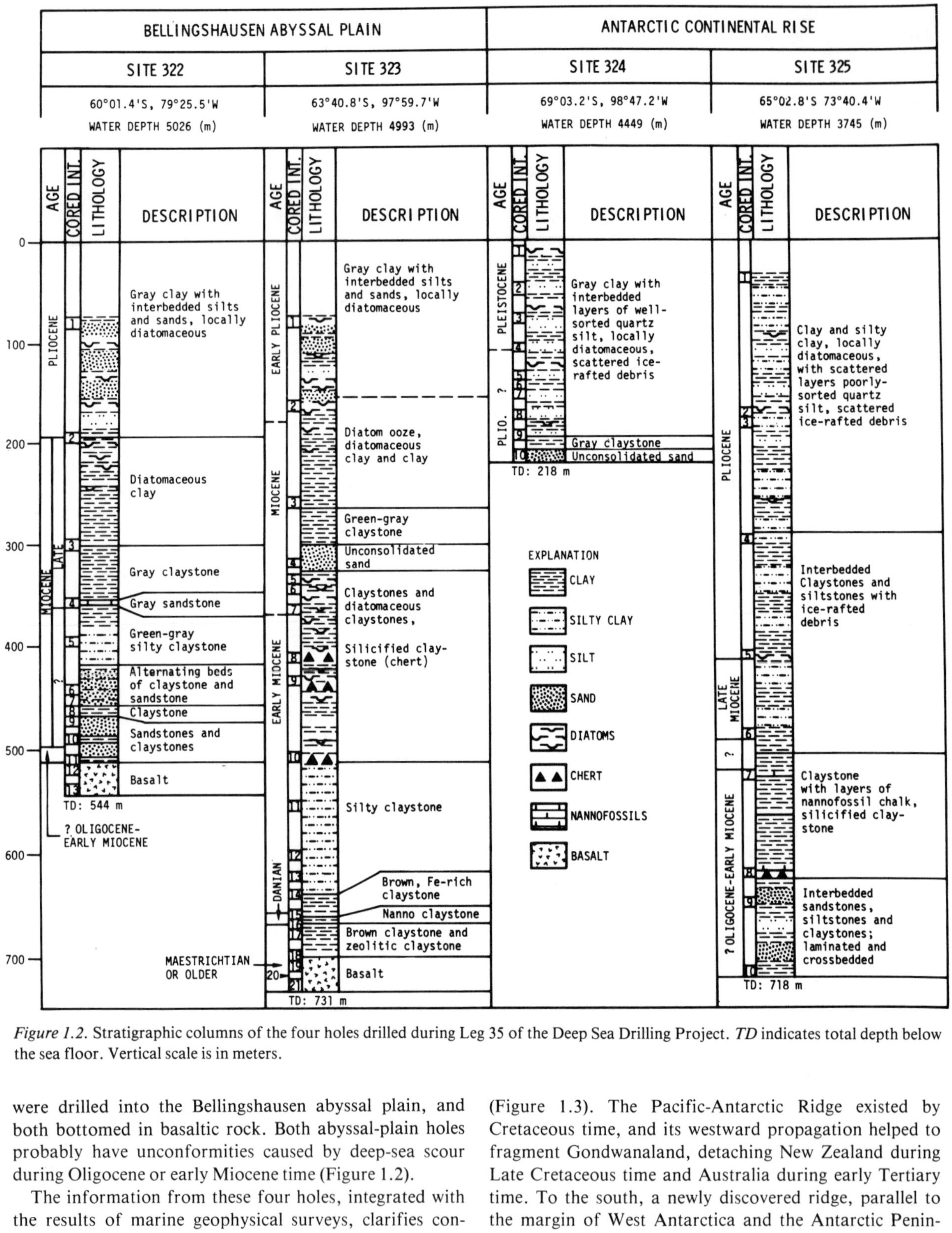

Figure 1.2. Stratigraphic columns of the four holes drilled during Leg 35 of the Deep Sea Drilling Project. *TD* indicates total depth below the sea floor. Vertical scale is in meters.

were drilled into the Bellingshausen abyssal plain, and both bottomed in basaltic rock. Both abyssal-plain holes probably have unconformities caused by deep-sea scour during Oligocene or early Miocene time (Figure 1.2).

The information from these four holes, integrated with the results of marine geophysical surveys, clarifies considerably the geologic history of this interesting region. The Southeast Pacific Basin seems to have been a deep oceanic region since late Paleozoic time or earlier, a part of the vast sea that formerly bordered Gondwanaland. No oceanic crust older than Late Cretaceous has been found, however, and it is unlikely that any pre-Cretaceous oceanic crust exists here.

The tectonic history of the region during the last 100 m.y. has been dominated by three spreading-ridge axes (Figure 1.3). The Pacific-Antarctic Ridge existed by Cretaceous time, and its westward propagation helped to fragment Gondwanaland, detaching New Zealand during Late Cretaceous time and Australia during early Tertiary time. To the south, a newly discovered ridge, parallel to the margin of West Antarctica and the Antarctic Peninsula, may have formed during the Cretaceous Period and seems to have propagated northeastward. The southwestern segment of this ridge was subducted beneath West Antarctica during Eocene time, but the segment to the northeast was active during Eocene time and was subducted beneath the Antarctic Peninsula in early Miocene time. The southern Chile Ridge may also have formed in Eocene or Oligocene time and may have been subducted beneath southern South America in early Miocene time.

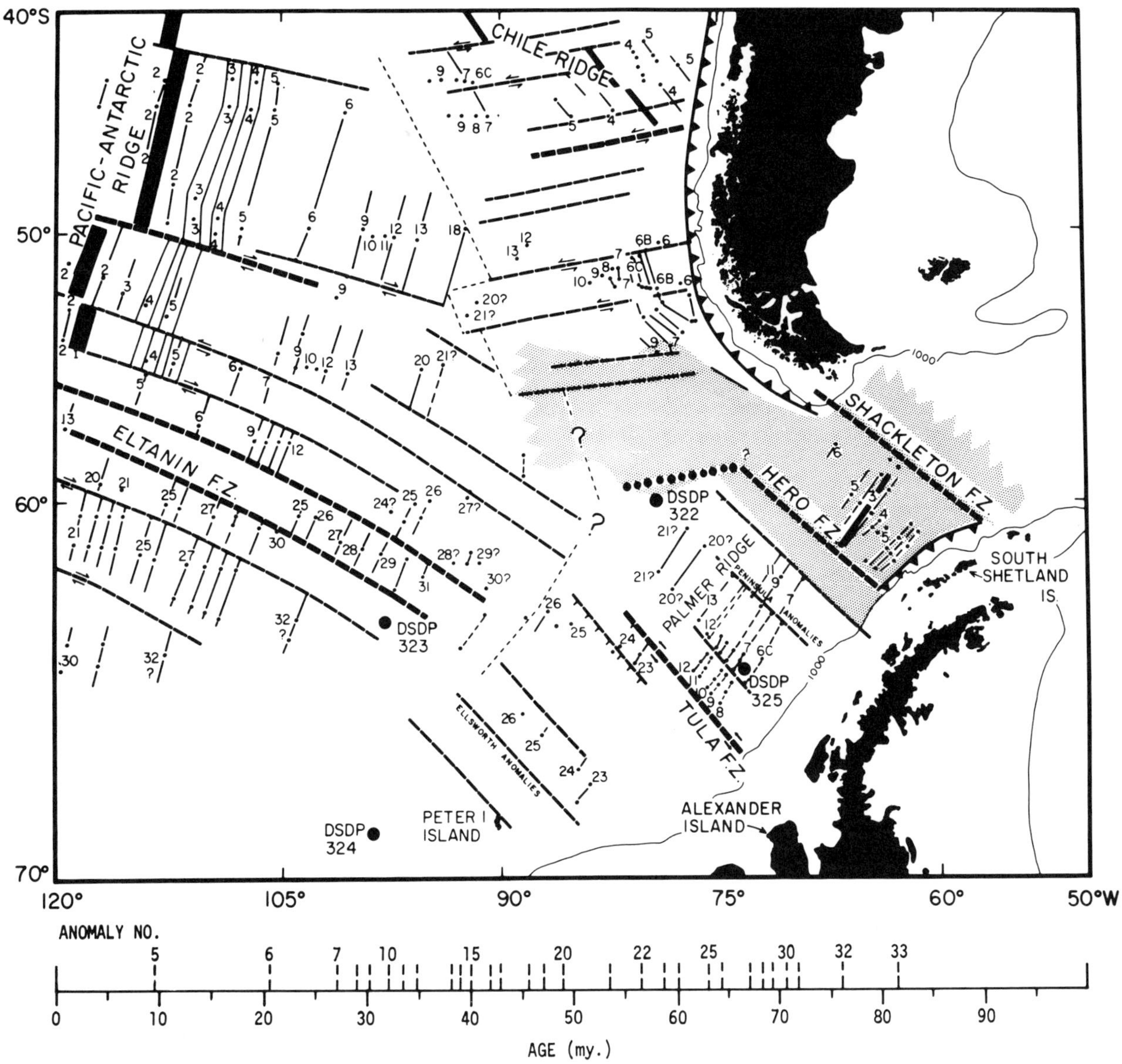

Figure 1.3. DSDP Leg 35 drill sites, magnetic anomalies, and principal tectonic features in the Southeast Pacific Basin (after Herron and Tucholke, 1976).

Deep-sea sediment erosion during Oligocene or early Miocene time, inferred at the abyssal plain sites 322 and 323, is attributed to the initiation of the Antarctic Circumpolar Current. The age of this event is in agreement with the Scotia Sea magnetic anomalies in suggesting that the final opening of the Drake Passage occurred during middle Tertiary time.

Volcanic rocks at least 7 m.y. old overlie a glaciated surface in the Jones Mountains of coastal West Antarctica. Ice-rafted debris occurs in strata as old as early Miocene (about 16–17 m.y.) at site 325 off the Antarctic Peninsula. Interpretation of the cored sediments from DSDP Leg 35 suggests that Antarctic glaciation was weak in the early Miocene but reached full development sometime during late Miocene time.

Hence, although Antarctica may have been in a polar position since early Mesozoic time (Holden, 1976), the present glaciation seems to have developed during the last 25 m.y.—after the continent achieved its total geographic isolation and the modern patterns of oceanic circulation had been established.

References cited

Aditya, S., and Sinha, R. N. 1976. Evolution of the coastal sedimentary basins, south of Latitude 18° N., India. (*In Proceedings of Workshop on Coastal Sedimentaries of India—South of 18° N. Latitude.* Madras, Oil and Natural Gas Commission, 23.)

Athavale, R. N. 1973. Inferences from recent Indian paleomagnetic results from the northern margin of the Indian plate and the tectonic evolution of India. (*In* Tarling, D. H., and Runcorn, S. K., *eds. Implications of Continental Drift to the Earth Sciences. 1.* London, Academic Press, 117–30.)

Barker, P. J., and Burrell, J. 1976. The opening of the Drake Passage. *Abstr. 25th Int. Geol. Cong., Sydney, 1976,* 881.

Barker, P. J., and Griffiths, D. H. 1972. The evolution of the Scotia Ridge and Scotia Sea. *Phil. Trans. R. Soc. Lond.,* Ser. A, *271,* 151–83.

Barrett, D. M. 1977. Aqulhas Plateau off southern Africa: a geophysical study. *Geol. Soc. Am. Bull., 88,* No. 6, 749–63.

Bergh, H. W., and Norton, I. O. 1976. Prince Edward fracture zone and the evolution of the Mozambique Basin. *J. geophys. Res., 81,* No. 29, 5221–39.

Bishopp, D. W., and Van Eeden, O. R. 1971. Explanatory text for the tectonic map of Africa: areas coordinated by the Southern Group. (*In Tectonics of Africa.* Paris, UNESCO, 561–85.)

Blow, R. A., and Hamilton, N. 1975. Paleomagnetic evidence from DSDP cores of northward drift of India. *Nature, Lond., 257,* 570–72.

Borrello, A. 1969. Los geosinclines de la Argentina. *Annls Dir. Nac. Geol. Miner., 14,* 188 pp.

Boucot, A. J., Doumani, G. A., Johnson, J. G., and Webers, G. F. 1967. Devonian of Antarctica. (*In* Oswald, D. H., *ed. International Symposium on the Devonian System. 1.* Calgary, Alberta Society of Petroleum Geologists, 303–7.)

Bowin, C. 1974. Migration of a pattern of plate motion. *Earth planet. Sci. Lett., 21,* No. 4, 400–4.

Briden, J. C. 1973. Applicability of plate tectonics to pre-Mesozoic time. *Nature, Lond., 244,* No. 5416, 400–5.

Carter, R. M., and Norris, R. J. 1976. Cenozoic history of southern New Zealand: an accord between geological observations and plate-tectonic predictions. *Earth planet. Sci. Lett., 31,* 85–94.

Casshyap, S. M. 1976. Paleocurrents and continental assembly: a comparison from Paleozoic coalfields of India, Antarctica, South Africa, and Australia. *Abstr. 25th Int. Geol. Congr., Sydney, 1976,* 238–39.

Clifford, T. N. 1974. Review of African granulites and related rocks. *Spec. Pap. geol. Soc. Am., 156,* 49 pp.

Colbert, E. H. 1979. Gondwana vertebrates. *Proc. 4th Int. Gondwana Symposium, Calcutta, 1977,* Section 3, 135–43.

Colchen, M. 1975. Paleogeographic and structural evolution of the Tibetan area of the Nepal Himalaya (Annapurna region). *Himal. Geol., 5,* 83–103.

Craddock, C. 1964. The structural relation of East and West Antarctica. (*In* Sundaram, R. K., *ed.* Rock Deformation and Tectonics. *22nd Int. Geol. Congr., New Delhi, 1964,* Part 4, 278–92.)

Craddock, C. 1970. Tectonic map of Gondwanaland. (*In* Bushnell, V. C., and Craddock, C., *eds.* Geologic Maps of Antarctica. *Antarct. Map Folio Ser.,* Folio 12, Pl. XXIII.)

Craddock, C. 1972*a*. *Geologic Map of Antarctica. 1:5,000,000.* New York, American Geographical Society.

Craddock, C. 1972*b*. Antarctic tectonics. (*In* Adie, R. J., *ed. Antarctic Geology and Geophysics.* Oslo, Universitetsforlaget, 449–55.)

Craddock, C. 1975. Tectonic evolution of the Pacific margin of Gondwanaland. (*In* Campbell, K. S. W., *ed. Gondwana Geology.* Canberra, Australia National University Press, 609–18.)

Craddock, C., and Hollister, C. D. 1976. Geologic evolution of the Southeast Pacific Basin. (*In* Hollister, C. D., Craddock, C., and others. *Initial Reports of the Deep Sea Drilling Project,* Vol. 35. Washington [U. S. Government Printing Office], 723–43.)

Crawford, A. R. 1974. Indo-Antarctica, Gondwanaland, and the distortion of a granulite belt. *Tectonophysics, 22,* 141–57.

Crowell, J. C. 1977. Problems concerning the late Paleozoic glaciation of Gondwanaland. *Abstr. 4th Int. Gondwana Symposium, Calcutta, 1977,* Section 4, 10 pp.

Dalziel, I. W. D. 1972. Large-scale folding in the Scotia arc. (*In* Adie, R. J., *ed. Antarctic Geology and Geophysics.* Oslo, Universitetsforlaget, 47–55.)

Dalziel, I. W. D. 1974. Margins of the Scotia Sea. (*In* Burk, C. A., and Drake, C. L., *eds. The Geology of Continental Margins.* New York, Springer-Verlag, 567–80.)

Dalziel, I. W. D., and Elliot, D. H. 1973. The Scotia arc and Antarctic margin. (*In* Nairn, A. E. M., and Stehli, F. G., *eds. The Ocean Basins and Margins. 1. The South Atlantic.* New York, Plenum Press, 171–246.)

De Beer, J. H., and Van Zijl, J. S. V. 1974. Plate tectonic origin for the Cape fold belt. *Nature, Lond., 252,* No. 5485, 675–76.

Du Toit, A. L. 1937. *Our Wandering Continents.* Edinburgh, Oliver and Boyd, 366 pp.

Dutta, A. K., and Grigorescu, D. A. 1975. Towards a correlation of the Mammalian fauna of the Siwaliks with their equivalents in southeastern Europe, including Crimea and Caucasus. (*In* Verma, V. K., *ed. Recent Researches in Geology. 2.* Delhi, Hindustan Publishers, 265–74.)

Elliot, D. H. 1975. Tectonics of Antarctica: a review. *Am. J. Sci., 275-A,* 45–106.

Embleton, B. J. J. 1972. On the unity of Gondwanaland during the lower Paleozoic. *Search, 3,* No. 9, 338–39.

Embleton, B. J. J., and McElhinny, M. W. 1975. The Palaeoposition of Madagascar: palaeomagnetic evidence from the Isalo Group. *Earth planet. Sci. Lett., 27,* 329–41.

Embleton, B. J. J., McElhinny, M. W., Crawford, A. R., and Luck, G. R. 1974. Palaeomagnetism and the tectonic evolution of the Tasman orogenic zone. *J. geol. Soc. Aust., 21,* No. 2, 187–93.

Exon, N. F., and Willcox, J. B. 1976. Structure and tectonic evolution of the Exmouth Plateau area off western Australia. *Abstr. 25th Int. Geol. Congr., Sydney, 1976,* 345–46.

Facer, R. A. 1974. Apparent polar wander relative to Australia during the Precambrian. *Earth planet. Sci. Lett., 22,* 44–50.

Falconer, R. K. H. 1973. Numerical studies of Cretaceous magnetic anomalies in the southwest Pacific Ocean. (*In Oceanography of the South Pacific.* Wellington, New Zealand Commission for UNESCO, 257–62.)

Ford, A. B. 1972. Weddell orogeny—latest Permian to early Mesozoic deformation at the Weddell Sea margin of the Transantarctic Mountains. (*In* Adie, R. J., *ed. Antarctic Geology and Geophysics.* Oslo, Universitetsforlaget, 419–25.)

Gansser, A. 1973*a*. Facts and theories on the Andes. *J. geol. Soc. Lond., 129,* 93–131.

Gansser, A. 1973*b*. Ideas and problems on Himalayan geology. *Seminar on Geodynamics of the Himalayan region.* Hyderabad, National Geophysics Research Institute, 97–103.

Giddings, J., Crawford, A., Embleton, J. J., and McElhinny, M. W. 1973. Precambrian paleomagnetism in Australia: evidence for the date of formation of Gondwanaland. *Abstr. with Programs, geol. Soc. Am., 5,* No. 7, 635.

Griffiths, J. R. 1976. A plate tectonic model for the southwest Pacific. *Abstr. 25th Int. Geol. Congr., Sydney, 1976,* 84.

Grindley, G. W. 1976. Palaeomagnetic evidence for New Zealand's Cretaceous–Cenozoic drift and rotation. *Abstr. 25th Int. Geol. Congr., Sydney, 1976,* 84–85.

Hailwood, E. A. 1974. Paleomagnetism of the Msissi Norite (Morocco) and the Paleozoic reconstruction of Gondwanaland. *Earth planet. Sci. Lett., 23,* No. 4, 376–86.

Hargraves, R. B. 1976. Precambrian geologic history. *Science, N.Y., 193,* No. 4251, 363–71.

Harrington, H. J., Burns, K. L., and Thompson, B. R. 1973. Gambier-Beaconsfield and Gambier-Sorrell fracture zones and the movement of plates in the Australia–Antarctica–New Zealand region. *Nature Phys. Sci., 245,* No. 146, 109–12.

Harrington, H. J. 1967. Devonian of South America. (*In* Oswald, D. H., *ed. International Symposium on the Devonian System. 1.* Calgary, Alberta Society of Petroleum Geologists, 651–71.)

Haughton, S. H. 1969. *Geological History of Southern Africa.* Cape Town, Geological Society of South Africa, 535 pp.

Herron, E., and Tucholke, B. E. 1976. Sea-floor magnetic patterns and basement structure in the southeastern Pacific. (*In* Hollister, C. D., Craddock, C., and others. *Initial Reports of the Deep Sea Drilling Project,* Vol. 35. Washington [U.S. Government Printing Office], 263–78.)

Herz, N. 1977. Timing of spreading in the South Atlantic: information from Brazilian alkalic rocks. *Geol. Soc. Am. Bull., 88,* No. 1, 101-12.

Holden, J. C. 1976. Permian-Triassic continental configurations and the origin of the Gulf of Mexico: comment. *Geology, 4,* No. 6, 324-25.

Hurley, P. M., and Rand, J. R. 1973. Outline of Precambrian chronology in lands bordering the South Atlantic, exclusive of Brazil. (*In* Nairn, A. E. M., and Stehli, F. G., *eds. The Ocean Basins and Margins. 1. The South Atlantic.* New York, Plenum Press, 391-410.)

Johnson, B. D., Powell, C. M., and Veevers, J. J. 1976. Spreading history of the eastern Indian Ocean and Greater India's northward flight from Antarctica and Australia. *Geol. Soc. Am. Bull., 87,* No. 11, 1560-66.

Jones, J. G., and Roots, W. D. 1974. Evolution of the Tasman Sea. *Nature, Lond., 252,* No. 5484, 613-14.

Katz, H. R. 1972. Plate tectonics and orogenic belts in the southeast Pacific. *Nature, Lond., 237,* No. 5354, 331-32.

Katz, H. R. 1973. Tectonic setting and evolution of continental margins in the Southeast Pacific. (*In Oceanography of the South Pacific.* Wellington, New Zealand Commission for UNESCO, 340 pp.

Katz, M. B. 1972. Paired metamorphic belts of the Gondwanaland Precambrian and plate tectonics. *Nature, Lond., 239,* No. 5370, 271-73.

Katz, M. B. 1974. Paired metamorphic belts in Precambrian granulite rocks in Gondwanaland. *Geology, 2,* No. 5, 237-41.

Klootwijk, C. T. 1973. Paleomagnetism of upper Bhander sandstones from central India and implications for a tentative Cambrian Gondwanaland reconstruction. *Tectonophysics, 18,* No. 1-2, 123-45.

Klootwijk, C. T. 1975. Paleomagnetism of upper Permian red beds in the Wardha valley, central India. *Tectonophysics, 25,* No. 1-2, 115-37.

Klootwijk, C. T. 1976. The drift of the Indian subcontinent; an interpretation of recent palaeomagnetic data. *Geol. Rdsch., 65,* No. 3, 885-908.

Larson, E. E., and LaFountain, L. 1970. Timing of the breakup of the continents around the Atlantic as determined by paleomagnetism. *Earth planet. Sci. Lett., 8,* No. 5, 341-51.

Larson, R. L. 1975. Late Jurassic sea-floor spreading in the eastern Indian Ocean. *Geology, 3,* No. 2, 69-71.

Larson, R. L. 1977. Early Cretaceous breakup of Gondwanaland off western Australia. *Geology, 5,* No. 1, 57-60.

Laughton, A. S., Sclater, J. G., and McKenzie, D. P. 1973. The structure and evolution of the Indian Ocean. (*In* Tarling, D. H., and Runcorn, S. K., *eds. Implications of Continental Drift to the Earth Sciences. 1.* London, Academic Press, 203-12.)

Luyendyk, B. P. 1976. Deep Sea Drilling on the NinetyEast Ridge: synthesis and a tectonic model. *Abstr. 25th Int. Geol. Congr. Sydney, 1976,* 895-96.

McElhinny, M. W. 1973. *Paleomagnetism and Plate Tectonics.* Cambridge, Cambridge University Press, 357 pp.

McElhinny, M. W., Giddings, J. W., and Embleton, B. J. J. 1974. Paleomagnetic results and late Precambrian glaciations. *Nature, Lond., 248,* No. 5449, 557-61.

Molnar, P., and Tapponnier, P. 1975. Cenozoic tectonics of Asia: effects of a continental collision. *Science, N.Y., 189,* No. 4201, 419-26.

Molnar, P., Atwater, T., Mammerickx, J., and Smith, S. W. 1975. Magnetic anomalies, bathymetry and the tectonic evolution of the South Pacific since the Late Cretaceous. *Geophys. Jnl, 40,* No. 3, 383-420.

Palmer, A. R. 1973. Cambrian trilobites. (*In* Hallam, A., *ed. Atlas of Palaeobiogeography.* Amsterdam, Elsevier, 3-11.)

Piper, J. D. A., Briden, J. C., and Lomas, K. 1973. Precambrian Africa and South America as a single continent. *Nature, Lond., 245,* No. 5423, 244-48.

Quilty, P. C. 1973. Cenomanian-Turonian and Neogene sediments from northeast of Kerguelen Ridge, Indian Ocean. *J. geol. Soc. Aust., 20,* No. 3, 261-71.

Raiverman, V. 1972. Time series and stratigraphic correlation of Cenozoic sediments in foothills of Himachal Pradesh. *Himal. Geol., 2,* 82-101.

Ravich, M. G. 1973. Superkontinent iuzhnogo polushariia. Chelovek i stikhiia; nastol'nyi gidrometeorologicheskii kalendar, 201-2.

Ravich, M. G., and Kamenev, E. N. 1972. *Kristallicheskii fundament Antarkiticheskoi platformy* [Crystalline basement of the Antarctic platform]. Leningrad, Gidrometeoizdat, 658 pp. [English translation: New York, John Wiley, 1975, 582 pp.]

Runnegar, B. 1979. Marine fossil invertebrates of Gondwanaland: palaeogeographic implications. *Proc. 4th Int. Gondwana Symposium, Calcutta, 1977,* Section 3, 144-59.

Rutland, R. W. R. 1976. Orogenic evolution of Australia. *Earth Sci. Rev., 12,* No. 2-3, 161-96.

Sastri, V. V., Raju, A. T. R., Sinha, R. N., Venkatachala, B. S., and Banerji, R. K. 1976. Stratigraphy and evolution of Cauvery basin, India. (*In Proceedings of Workshop on Coastal Sedimentaries of India—South of 18° N. Latitude.* Madras, Oil and Natural Gas Commission, 34-35.)

Sclater, J. G., and Fisher, R. L. 1974. Evolution of the east central Indian Ocean with emphasis on the tectonic setting of the Ninety East Ridge. *Geol. Soc. Am. Bull., 85,* No. 5, 683-702.

Sclater, J. G., Luyendyk, B. P., and Meinke, L. 1976. Magnetic lineations in the southern part of the Central Indian Basin. *Geol. Soc. Am. Bull., 87,* No. 3, 371-78.

Shields, O. 1977. A Gondwanaland reconstruction for the Indian Ocean. *J. Geol., 85,* No. 2, 236-42.

Suarez, M. 1976. Plate-tectonic model for southern Antarctic Peninsula and its relation to southern Andes. *Geology, 4,* No. 4, 211-14.

Suggate, R. P., and Grindley, G. W. 1972. *Geological Map of New Zealand, 1:1,000,000.* Wellington, New Zealand Department of Scientific and Industrial Research.

Valencio, D. A. 1975*a*. The paleomagnetism of South American rocks and its significance for the fragmentation of Gondwanaland. (*In* Campbell, K. S. W., *ed. Gondwana Geology.* Canberra, Australia National University Press, 3-8.)

Valencio, D. A. 1975*b*. The South American paleomagnetic data and the main episodes of the fragmentation of Gondwanaland. *Phys. Earth Planet. Interiors, 9,* No. 3, 221-25.

Van Eysinga, F. W. B. (comp.) 1975. *Geological Time Table.* 3rd ed. Amsterdam, Elsevier.

Veevers, J. J., and McElhinny, M. W. 1976. The separation of Australia from other continents. *Earth Sci. Rev., 12,* No. 2, 3, 139-59.

Veevers, J. J., Powell, C. M., and Johnson, B. D. 1975. Greater India's place in Gondwanaland and in Asia. *Earth planet. Sci. Lett., 27,* 383-87.

Verma, R. K. 1973. Paleomagnetism of Indian rocks and the birth of the Himalaya. (*In Seminar on Geodynamics of the Himalayan Region.* Hyderbad, National Geophysics Research Institute, 49-58.)

Watkins, N. D., Gunn, B. M., Nougier, J., and Baksi, A. K. 1974. Kerguelen: continental fragment or oceanic island? *Geol. Soc. Am. Bull., 85,* 201-2.

Wellman, H. W. 1976. Sea floor reconstruction for the last 85 million years and the bending and faulting of the New Zealand region. *Abstr. 25th Int. Geol. Congr., Sydney, 1976,* 106-7.

2

The Reconstruction of New Zealand, Australia, and Antarctica

Review Paper

G. W. GRINDLEY
New Zealand Geological Survey, Department of Scientific and Industrial Research, Lower Hutt, New Zealand

F. J. DAVEY
Geophysics Division, Department of Scientific and Industrial Research, Wellington, New Zealand

Abstract. Reconstruction of the Southwest Pacific sector of Gondwanaland has been approached from four independent sets of criteria: (1) marine magnetic anomalies dating from 85 m.y. to the present which enable approximate positioning of the three major lithospheric plates; (2) bathymetric matching of continental margins (2-km isobath); (3) paleomagnetic studies of Upper Cretaceous-Cenozoic volcanic rocks which allow construction of polar-wandering paths for the separate plates; and (4) matching of linear pre-Late Cretaceous geologic features such as orogenic belts, granitic batholiths, volcanic rift zones, island arcs, and associated mafic-ultramafic plutonic rocks. An acceptable reconstruction can be drawn, consistent with criteria 1–3, thus permitting a relative positioning of the three plates in Late Cretaceous time. Matching of geologic markers (criterion 4) across some continental margins (Victoria-Tasmania-Antarctica) is possible with minor discrepancies. However, across other margins (South Australia–Antarctica; Campbell Plateau–West Antarctica; New Zealand–Lord Howe Rise–Eastern Australia) the geologic evidence is equivocal.

Some geotectonic conclusions that emerge from this and previous studies are:

1. The Adelaide geosyncline of South Australia apparently does not continue into East Antarctica.
2. Although similar stratigraphic sequences are present in the lower Paleozoic rocks of eastern Australia, East Antarctica, and New Zealand, they cannot be aligned without disruption of the 85 m.y. reconstruction.
3. A belt of magnetic anomalies crossing Campbell Plateau from southeastern New Zealand may provide a link if it continues into West Antarctica.
4. The configuration of New Zealand in Late Cretaceous time is controversial. Some plate tectonic reconstructions show development of the oroclinal bend during late Cenozoic time, whereas geologic and paleomagnetic evidence suggests that bending took place earlier during the Early Cretaceous Rangitata orogeny.
5. The change 90 m.y. ago from acid-intermediate calc-alkaline volcanism to alkaline mafic volcanism coincides with the inception of sea-floor spreading and with a change from taphrogenic sedimentation in rifts and grabens to open-water sedimentation in troughs and basins.

It is now widely accepted that New Zealand, Antarctica, and Australia were once joined as part of the supercontinent of Gondwanaland. Several reconstructions of the Southwest Pacific region have appeared during the last few years, showing possible reconstructions of the three major lithospheric plates (Pacific, Indian, and Antarctic) in Late Cretaceous time prior to the initial separation of the Campbell Plateau–Lord Howe Rise–New Zealand microcontinent from Australia and Antarctica. These reconstructions fall into several categories: (1) those produced by reversing sea-floor spreading using oceanic magnetic anomalies and fracture zones (see e.g., Molnar and others, 1975); (2) those produced by bathymetric matching of the continental margins, usually at the 2-km isobath (see e.g., Griffiths, 1971*b*; Sproll and Dietz, 1969); and (3) those produced by analysis of the sea-floor magnetic anomalies, with adjustments made to the finite rotation poles in the early stages of separation to permit a reasonable bathymetric fit (see e.g., Crook and Belbin, 1975, 1978).

So far, there has been little attention paid to the geology of the once adjacent continental margins, except between northern Victoria Land in Antarctica and southeast Australia where preliminary data indicate that matching of geologic features may provide a means of testing the various reconstructions (Harrington and others, 1973; Cooper, 1975; Laird and others, 1977). During the past three years International Geological Correlation Programme (IGCP) Project 7 has been actively investigating the upper Precambrian–lower Paleozoic geology of the region with the objectives of improving regional correlations and providing a firmer basis for matching of geologic features.

The two maps presented here (Figures 2.1 and 2.2) are modified versions of the Crook-Belbin (1978) 90 m.y. computer-drawn reconstruction. Crustal blocks are defined by

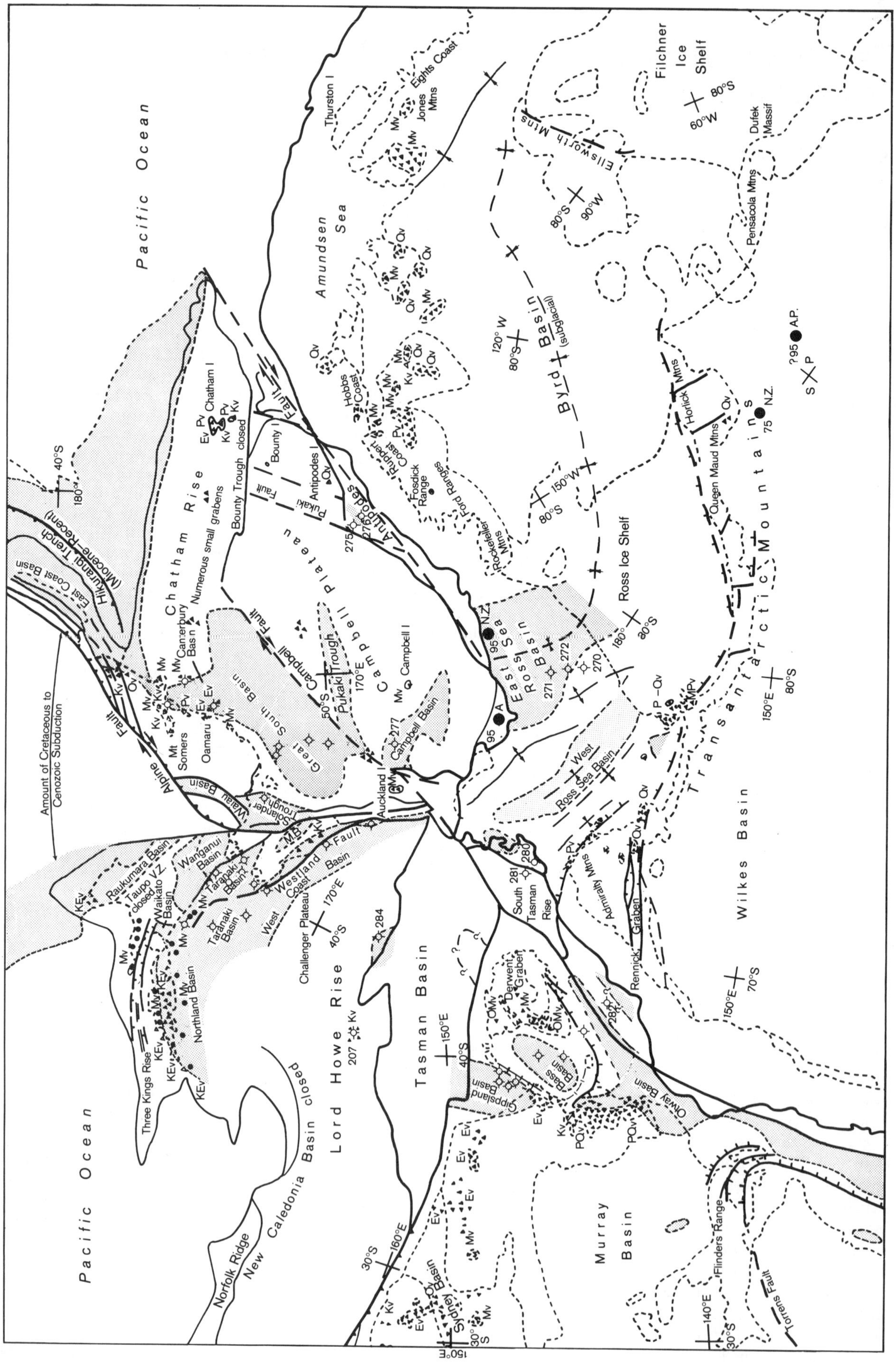

Pacific Ocean
Pacific Ocean
Chatham Rise
Chatham I
Bounty Trough closed
Numerous small grabens
Bounty I
Antipodes I
Pukaki Fault
Campbell Plateau
Campbell Fault
Pukaki Trough
Campbell I
Campbell Basin
Great South Basin
Canterbury Basin
Oamaru
Mt Somers
Alpine Fault
Amount of Cretaceous to Cenozoic Subduction
Hikurangi Trench (Miocene–Recent)
East Coast Basin
Waiau Basin
Solander Trough
Auckland I
Westland Coast Basin
Fault
Wanganui Basin
Raukumara Basin
Taupo V.Z. closed
Waikato Basin
Taranaki Basin
West Taranaki Basin
Challenger Plateau
Northland Basin
Three Kings Rise
Lord Howe Rise
New Caledonia Basin closed
Norfolk Ridge
Tasman Basin
South Tasman Rise
Derwent Graben
Bass Basin
Gippsland Basin
Otway Basin
Sydney Basin
Murray Basin
Flinders Range
Torrens Fault
Wilkes Basin
Rennick Graben
Admiralty Mtns
West Ross Sea Basin
East Ross Sea Basin
Ross Ice Shelf
Rockefeller Mtns
Ford Ranges
Fosdick Range
Ruppert Coast
Hobbs Coast
Amundsen Sea
Byrd Basin (subglacial)
Transantarctic Mountains
Queen Maud Mtns
Horlick Mtns
Pensacola Mtns
Dufek Massif
Filchner Ice Shelf
Ellsworth Mtns
Jones Mtns
Eights Coast
Thurston I

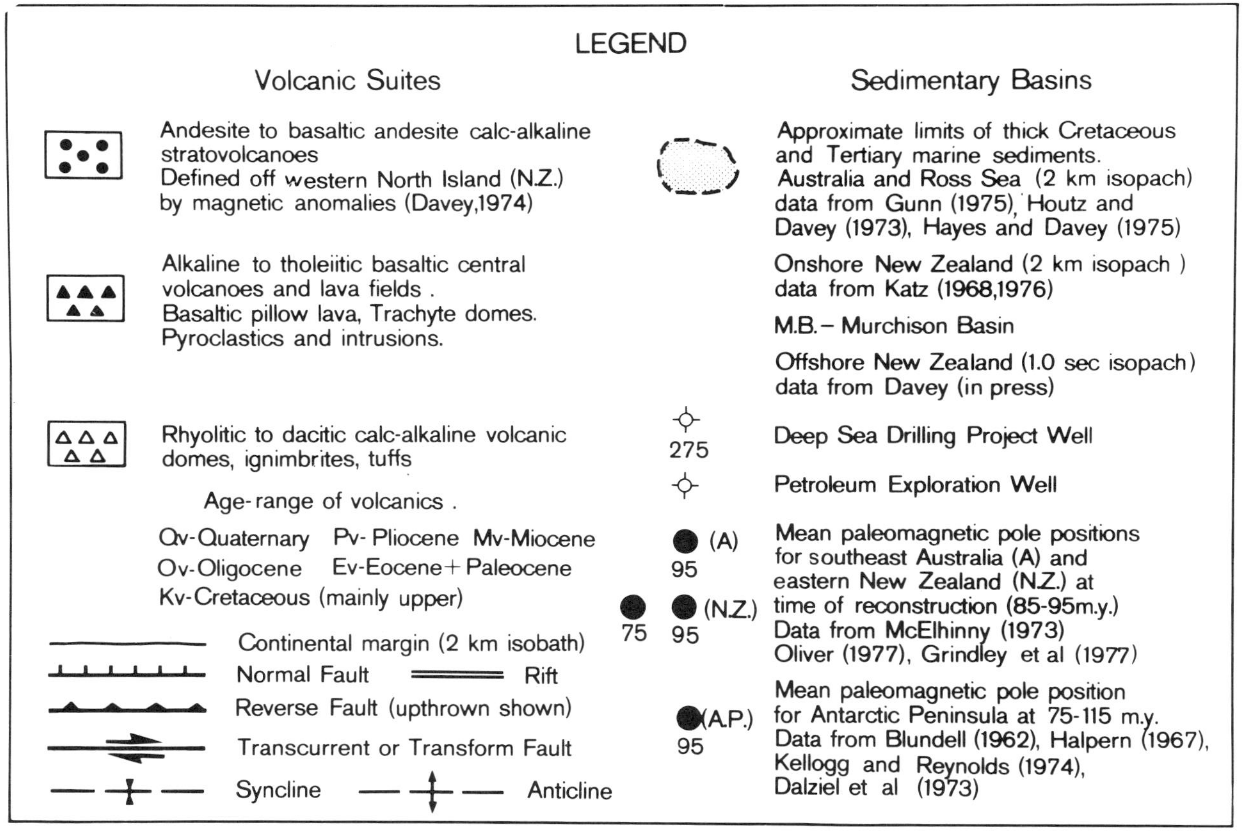

Figure 2.1. Continental reconstruction of southeast Australia, New Zealand, and adjoining regions of Antarctica prior to continental separation in Late Cretaceous time (85 m.y. ago) modified from Crook and Belbin (1978). Major post-separation sedimentary basins and volcanic suites, major tectonic features, DSDP and other significant drillholes, and paleomagnetic pole positions for East Australia (A), New Zealand (N.Z.), and the Antarctic Peninsula (A.P.) are shown. Stereographic projection; scale 1:20,000,000. Data sources are shown in legend; also Craddock (1972*b*), LeMasurier and Wade (1976), Australia, Bureau of Mineral Resources (1965), and New Zealand Geological Survey (1971).

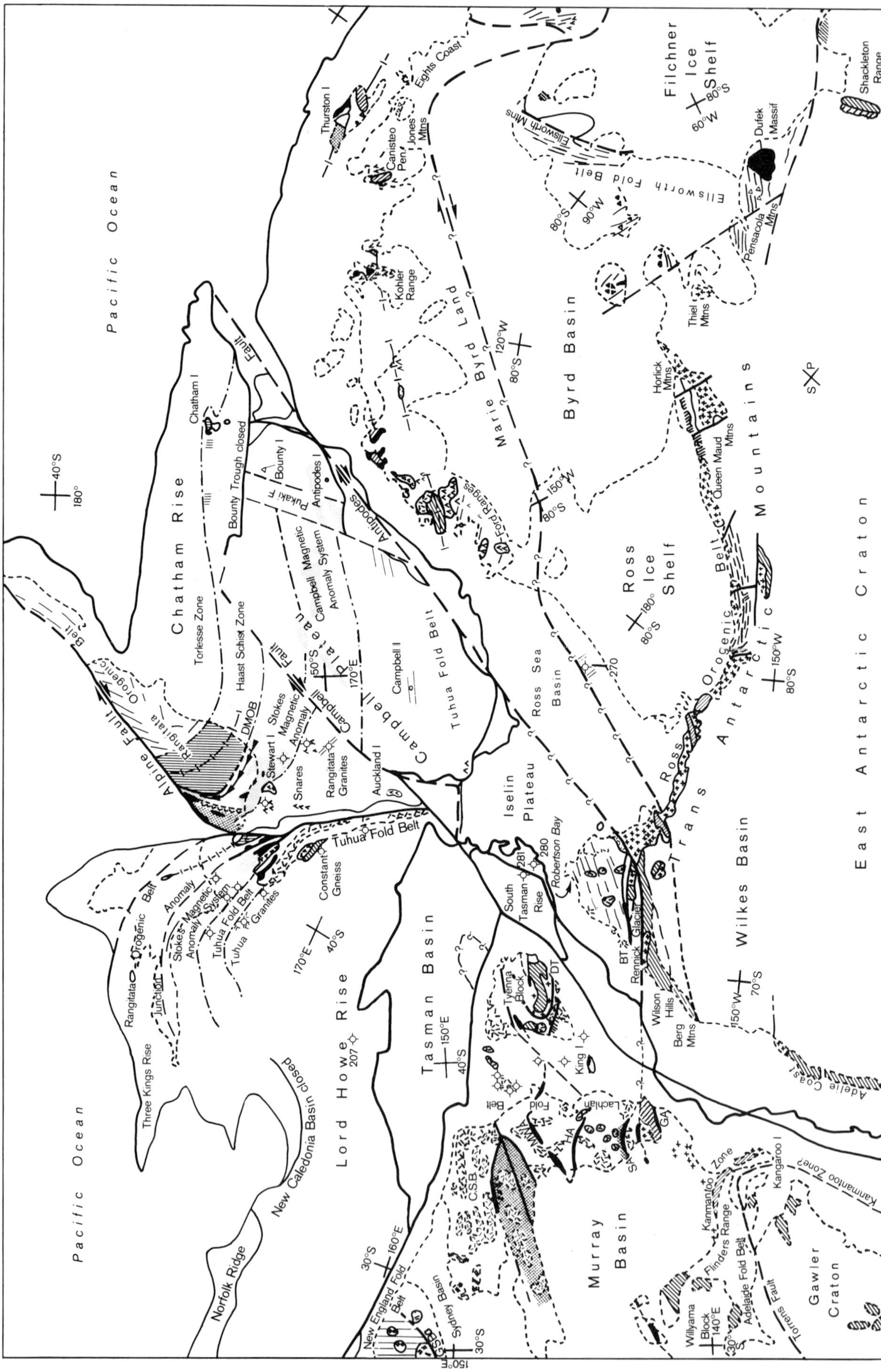

Pacific Ocean
Pacific Ocean
Norfolk Ridge
New Caledonia Basin closed
Three Kings Rise
Lord Howe Rise
207
Rangitata Orogenic Belt
Junction
Stokes Magnetic Anomaly System
Tuhua Fold Belt
Tuhua Granites
Constant Gneiss
170°E
40°S
Tasman Basin
150°E
40°S
Alpine Fault
Rangitata Orogenic Belt
DMOB
Stewart I
Snares
Stokes Magnetic Anomaly
Rangitata Granites
Auckland I
Campbell Plateau
Campbell Fault
50°S
170°E
Haast Schist Zone
Torlesse Zone
Chatham Rise
Campbell Magnetic Anomaly System
Campbell I
Tuhua Fold Belt
Bounty Trough closed
Bounty I
Pukaki F
Antipodes I
Antipodes
Chatham I
Fault
40°S
180°
Iselin Plateau
Ross Sea Basin
South Tasman Rise
281
280
Robertson Bay
Tyenna Block
DT
King I
Lachlan Fold Belt
C.S.B.
HA
SA
GA
New England Fold Belt
Sydney Basin
30°S
160°E
30°S
150°E
Murray Basin
Kanmantoo Zone
Kanmantoo Zone?
Kangaroo I
Flinders Range
Adelaide Fold Belt
Willyama Block
140°E
30°
Torrens Fault
Gawler Craton
Adelie Coast
Berg Mtns
Wilson Hills
Rennick Glacier
BT
150°W
70°S
Wilkes Basin
East Antarctic Craton
Trans Antarctic Mountains
Ross Orogenic Belt
150°W
80°S
Ross Ice Shelf
180°
80°S
270
Ford Ranges
150°W
80°S
Marie Byrd Land
120°W
80°S
Byrd Basin
Queen Maud Mtns
Horlick Mtns
S P
Thiel Mtns
Kohler Range
Pensacola Mtns
Dufek Massif
Ellsworth Fold Belt
Ellsworth Mtns
90°W
80°S
60°W
80°S
Filchner Ice Shelf
Shackleton Range
Thurston I
Canisteo Pen
Jones Mtns
Eights Coast

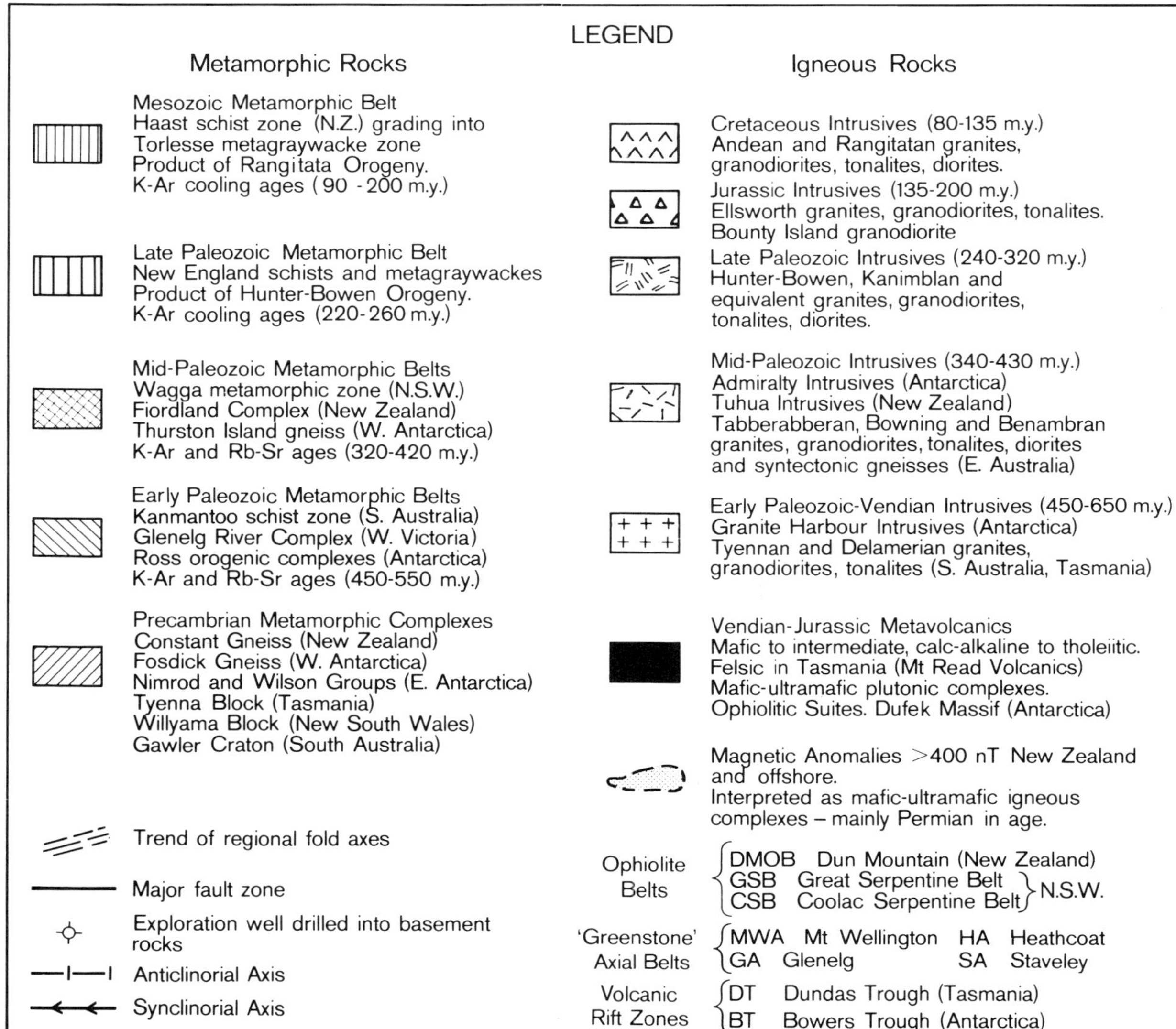

Figure 2.2. Continental reconstruction of southeast Australia, New Zealand, and adjoining regions of Antarctica prior to continental separation in Late Cretaceous time (85 m.y ago). Modified from Crook and Belbin (1978). Generalized pre–Upper Cretaceous basement geology and other important structural features useful for testing the reconstruction are shown. Also shown are aeromagnetic anomaly belts, associated with the southwest margin of the New Zealand geosyncline and their probable offshore extensions. Stereographic projection; scale 1:20,000,000. Data sources are various geologic and tectonic maps, including Craddock, 1972*b,c*; Australia, Bureau of Mineral Resources (1965), and New Zealand Geological Survey (1971).

the 2-km isobaths. The finite rotation poles used are given in Crook and Belbin (Table 2; 1978). The projection used is stereographic with projection centers at lat. 40° S, long. 170° E for Australia–Lord Howe Rise–New Zealand–Campbell Plateau and at the South Pole for Antarctica. The scale of the maps is 1:20,000,000 at lat. 71° S (Antarctica) and at lat. 70° S, long. 170° E (remaining continents). The scales and projections were chosen to facilitate the transfer of geologic data from published maps. The scale mismatch between the two projections is small between lat. 60° and 80° S and between long. 120° E and 140° W.

Oceanic magnetic anomalies

Southeast Indian Ocean

McKenzie and Sclater (1971) and Weissel and Hayes (1971, 1972) have shown that the southeast Indian Ocean formed during the north-south rifting of Australia from East Antarctica. Sea-floor magnetic lineations corresponding to anomaly 21, and perhaps 22, were identified, indicating that continental separation began about 55 m.y. ago. Between anomaly 22 and the base of the continental slope of Australia and Antarctica exists a magnetically quiet zone, 100–250 km wide, which they believe may represent foundered or altered continental blocks resulting from initial rifting. However, Roots (1976) has suggested that magnetically quiet zones could form by mutual interference of normally and reversely magnetized blocks of oceanic crust in the early stages of spreading, that is, oblique to the continental margins. Both explanations are consistent with the observation that the seaward boundaries of the quiet zones appear to cross the magnetic lineations and thus do not represent isochrons. The magnetic lineations suggest a complex spreading history and, south of Tasmania, suggest large southerly offsets of the spreading axis to the east where it meets the Macquarie triple junction. The Weissel-Hayes reconstruction of Antarctica and Australia assumes the morphologic fit of Sproll and Dietz (1969) and Smith and Hallam (1970), which leads to an unacceptable overlap of the continental South Tasman Rise and the Iselin Plateau in the north-central Ross Sea. A westerly movement of Australia relative to East Antractica of about 250 km would eliminate the overlap (Griffiths, 1974) and is consistent with the known geology (Laird and others, 1977), and with the magnetic lineation data (Norton and Molnar, 1977). The present reconstruction accepts a more westerly position for Tasmania than the Crook-Belbin reconstruction, sufficient to accommodate the South Tasman Rise without overlap.

Tasman Sea

The Tasman Sea is underlain by oceanic crust generated during Late Cretaceous and early Tertiary time (Ringis, 1970; Hayes and Ringis, 1973). Anomalies 24 to 36 (60–80 m.y.) have been identified, but older oceanic crust without recognizable magnetic lineations may have formed earlier during the Cretaceous Normal Interval (Helsley and Steiner, 1969). Only younger oceanic crust (60–70 m.y. old) is found in the north, implying later spreading and either a major shear zone between the northern and southern anomalies, or westward subduction of older crust under eastern Australia combined with destruction of the eastern half of the anomaly pattern by the volcanism of Dampier Ridge in the east. Only slight differences in spreading to either side of the Tasman axis have been detected, but spreading rates differed appreciably to either side of some transform fractures. The central anomaly coincides with a basement ridge masked by later sedimentation and aligned northwest, oblique to the Australian margin where subduction for a period of 6–7 m.y. is inferred (Hayes and Ringis, 1973). The southern limit of spreading is fairly well defined by magnetic and seismic reflection data, the latter showing thick sediments overlying the Tasman Sea anomalies but only thin sediments over the more rugged morphology associated with the younger anomalies to the south. This boundary zone is presently aligned with the southwest extremity of the Alpine Fault in New Zealand and extends to the South Tasman Rise, marking a major shear zone that was active 60–80 m.y. ago. Weissel and others (1977) have since revised the poles of rotation for the opening of the Tasman Sea, which makes the hypothesized subduction of Tasman Sea crust under the Australian margin unnecessary.

Southwest Pacific Ocean

Magnetic anomalies demonstrate that the formation of the Southwest Pacific Ocean probably took place in three main episodes (Christoffel and Falconer, 1972, 1973; Molnar and others, 1975). Anomalies 25 to 36 (63–80 m.y.) are well defined along the southern margin of the Campbell Plateau. Anomalies 6 to 25 (21–63 m.y.) are poorly defined, the spreading being slow because of the proximity of the pole of rotation. The direction of spreading at that time was also different, and no anomalies of that age have been identified west of long. 170° W. Spreading rates increased during the formation of anomalies 1 to 6 (0–21 m.y.), which are conspicuous to either side of the Pacific-Antarctic Ridge. At the western end of the ridge, the Pacific-Antarctic Fracture Zone, 1200 km long, links the ridge with the Macquarie triple junction and is probably a series of en echelon fracture zones joined by short spreading axis segments. The Cretaceous anomalies do not extend west of the Campbell Plateau margin into the Emerald Basin, where DSDP drilling has shown that the oceanic crust is overlain by middle Oligocene sediments (Leg 29, 1973). Anomalies older than 5 have not been recognized south of the Pacific-Antarctic Ridge to the west of about long. 140° W. If symmetrical spreading is assumed, reconstructions produced by reversing sea-floor spreading give a good fit along Western Antarctica, with some overlap in the Ross Sea (Christoffel and Falconer, 1973).

Palinspastic problems

Problems arise in the reconstruction of the whole region because the timing and amount of movement on the Macquarie Ridge–Alpine Fault boundary are unknown, as are the mode of formation of the oceanic crust in the Emerald Basin and the existence of a postulated pre-Oligocene plate boundary through Antarctica. For example, based on relative plate motions, the Emerald Basin may have been extensional during Eocene time, becoming progressively more oblique and finally compressive during late Cenozoic time (Weissel and others, 1977). Alternatively, it may have been extensional during Oligocene time, becoming compressional in the north and extensional in the south by late Cenozoic time (Molnar and others, 1975).

To eliminate the problems of overlap of adjoining continental blocks, various authors have postulated relative motions at a plate boundary between East Antarctica and Marie Byrd Land (Wellman, 1976) or between East and West Antarctica as a whole (Hayes and Ringis, 1973; Molnar and others, 1975). Others have also suggested a degree of internal deformation along the Alpine Fault zone in pre-Oligocene time (Grindley, 1974; Crook and Belbin, 1978) or in the Bounty Trough (Wellman, 1976). Weissel and others (1977) consider that a plate boundary through Antarctica is not essential, as all necessary movements can be taken up along the boundary through New Zealand.

The precise relocation of continental fragments based solely on oceanic magnetic data is not possible, partly because of the incomplete record of the sea-floor anomalies (e.g., north of the Ross Sea), partly because of ambiguities in the interpretation of the age of crust formed in the Cretaceous Normal Interval (e.g., oldest part of Tasman), and partly because of distortion of continental crust and unknown movements during early rifting. The detailed positions of the continental fragments prior to breakup must be determined by other means.

Cretaceous-Cenozoic depositional basins

New Zealand region

In New Zealand, the major Late Cretaceous and Cenozoic basins flank either side of the basement geanticline formed mainly of Paleozoic and Mesozoic rocks and raised during the Kaikoura orogeny. These basins and associated rises extend offshore under the continental shelf to the northwest (Lord Howe Rise and New Caledonia Basin) and to the southeast (Campbell Plateau, Bounty Trough, and Chatham Rise) of New Zealand (Figure 2.1).

The Northland Basin lies off the northwest coast and contains more than 4 km of sediments out to the shelf edge. Marine Upper Cretaceous to lower Tertiary sedimentary beds are overlain by thick Oligocene-Miocene olistostromes of Cretaceous-Oligocene sedimentary rocks and Cretaceous-Eocene basaltic volcanic rocks, all probably emplaced from the northeast (Brothers, 1974; Grindley, 1974). Large volcanic centers within the offshore sedimentary sequence are probably related to the onshore andesitic lower Miocene Waitakere Group (Davey, 1974; Hayward, 1976).

In western Cook Strait and western North Island are the Taranaki and Wanganui Basins. The Wanganui Basin on the east contains as much as 4 km of Plio-Pleistocene marine sediments overlying basement and is separated from the Taranaki Basin by a faulted basement ridge. The Taranaki Basin contains more than 5 km of sediments—an almost complete Eocene–Pliocene marine sequence overlying coal measures resting on varied basement rocks (Cope and Reed, 1967), and thinning westward toward the Challenger Plateau.

The West Coast Basin has a stratigraphic sequence similar to that of the Taranaki Basin into which it grades offshore. Onshore it consists of elongate, north-trending, fault-angle depressions separated by basement ranges which themselves were the sites of early Tertiary depressions with 1–3 km of sediments. These were uplifted in early Miocene time, providing sediments for the parallel late Tertiary depressions. Offshore to the west of the Westland Fault, an undeformed Eocene–Pliocene sequence more than 2 km thick indicates a more stable tectonic environment.

The Waiau Basin extends south of New Zealand as a structural feature into the Solander Trough, although they may be separate depositional features (Katz, 1974). The Solander Trough contains more than 2.5 km of Eocene and younger marine sedimentary beds in a rifted depression between the faulted western edge of the Campbell Plateau and the uplifted early Miocene ocean floor of the Macquarie Ridge (Varne and Rubenach, 1972). It passes southward into the Emerald Basin, an area of Oligocene–Quaternary pelagic sedimentation on late Eocene or early Oligocene oceanic crust (Summerhayes, 1969; Houtz, 1975). The Waiau Basin contains Upper Cretaceous–Eocene non-marine sedimentary rocks overlain by more than 2 km of Oligocene–Pliocene marine strata. The Moonlight Tectonic Zone marks a persistent mobile zone along the eastern edge of the basin, with a history of pre-Oligocene uplift, thick Oligocene and Miocene marine sedimentation from rising fault scarps, and intense Plio-Pleistocene folding and uplift (Turnbull and others, 1975).

The East Coast Basin extends northeast from Marlborough along eastern North Island and continues into the Raukumara Basin. Both basins contain as much as 4.5 km of upper Lower Cretaceous–Cenozoic marine sedimentary rocks that extend to the base of the continental slope (Katz, 1974). The East Coast is divided into two parts by eastern Cook Strait where thick, undisturbed, upper Cenozoic flysch in the Hikurangi Trench passes westward into intensely folded and faulted older Cenozoic and Cretaceous sedimentary beds.

In the reconstruction, the East Coast Basin is attached to the Pacific Plate and the Raukumara Basin to the Indian Plate. The reconstruction refers, of course, to Late Cretaceous time and is greatly influenced by the choice of a position for the Indian-Pacific plate boundary through North Island, connecting the Alpine Fault system with the Kermadec Trench. The present plate boundary seems to be a wide zone of distributed deformation extending eastward to the Hikurangi Trench, and this complex situation may well have existed in the past. Because the Raukumara Basin shows close stratigraphic and tectonic similarities to the Northland Basin into Oligocene time (Brothers, 1974; Grindley, 1974), it has been found convenient to attach it to the same plate during that interval. Its wide separation from the East Coast Basin purely reflects the amount of Cretaceous–Cenozoic subduction in the Kermadec-Hikurangi trench system. The true palinspastic pattern must be more complex.

The Canterbury Basin contains as much as 2 km of sedimentary beds, and extends from eastern South Island to the continental shelf edge where it may pass into the Great South Basin. Deposition started during Late Cretaceous time and has been marked by slow transgression westward over a peneplain covered by quartzose coal measures, followed by slow marine sedimentation until Oligocene time. Regression commenced in Miocene time with uplift and tilting away from the Alps and the deposition of thick marine, and finally terrestrial, conglomerates in Plio-Pleistocene time.

On the Campbell Plateau, three other major sedimentary basins have been identified by seismic reflection studies (Katz, 1976; Davey, 1977) and drilling of DSDP (Leg 29, 1973) and petroleum wells. The Great South Basin, nearest South Island, contains mainly lower Tertiary and Upper Cretaceous sedimentary beds, most probably in excess of 3 km thick. The Pukaki Trough, a graben-like offshoot of the Great South Basin, contains at least 2 km of mostly pre-Eocene terrigenous sedimentary rocks overlain by pelagic Eocene-Oligocene strata. Further to the southwest, the Campbell Basin contains primarily pre-Miocene sedimentary rocks—pelagic Eocene-Oligocene beds overlying older terrigenous rocks. These three basins and the Bounty Trough were probably formed during the early rifting between New Zealand and Antarctica, and may contain sedimentary units older than Late Cretaceous, by analogy with the rift basins of South Australia.

Eastern Australian region

The major Cretaceous-Cenozoic sedimentary basins in southeast Australia are considered to be related to the Australian-Antarctic and Tasman rifting. The Otway Basin in the west is separated from the Bass Basin by the Mornington Peninsula-King Island basement ridge. The Bass Basin in turn is separated from the Gippsland Basin in the east by the Bassian Rise. The early phases of rifting in the Otway and Gippsland Basins, and possibly the Bass Basin (Robinson, 1974), were marked by deposition of Lower Cretaceous non-marine terrigenous sedimentary beds (Otway and Strzelecki Groups) in rift valleys within the continental basement (Falvey, 1974; Brown, 1975). Some basalts were extruded, and volcaniclastic sediments are important.

Following middle Cretaceous uplift and development of an unconformity, marine incursion occurred in the west during Cenomanian-Turonian time; shallow seas covered the Otway Basin by Campanian time, and regression followed during Maestrichtian time. In the Gippsland Basin, non-marine sedimentation continued from Late Cretaceous into Paleocene time, accompanied by volcanism; the first marine incursion from the Tasman Sea was delayed unti late Eocene time, partly because of rapid deltaic sedimentation (Hocking, 1972). Subsidence continued in all three basins during the Tertiary Period, controlled mainly by wrench movements and normal faulting associated with the opening of the Southern Ocean.

Ross Sea region

In the region from Victoria Land to West Antarctica, only two sedimentary basins have been delineated, both in the Ross Sea. The East Ross Sea Basin contains as much as 4 km of sediments (Houtz and Davey, 1973), probably all of post-middle Oligocene age (Hayes and Frakes, 1975), and mainly glaciomarine in origin. The West Ross Sea Basin also contains more than 3 km of sediments in a north-south trough (Hayes and Davey, 1975). The sedimentary beds are undated but may correspond in part with the lower Tertiary (Eocene-Oligocene) sedimentary rocks found as erratics in the McMurdo Sound region (Cranwell and others, 1960; Wilson, 1967; McIntyre and Wilson, 1966). The Ross Sea basins are undoubtedly continuous with the subglacial Byrd Basin and Bentley Trench between the Ellsworth Mountains and Marie Byrd Land (Bentley and Clough, 1972).

Origin of basins

Two origins have been suggested for the sedimentary basins at and near rifted continental margins: (1) a failed arm or splay rift of the original main rift, aligned normal or oblique to the continental margin (Burke and Dewey, 1973), or (2) a result of crustal flow during post-rifting sea-floor spreading, subparallel to the continental margin (Bott, 1971).

Basins of the first type could include the Strzelecki (Gippsland) Basin, resulting from the Tasman Sea opening (Gunn, 1975), and perhaps the Otway Basin, resulting from the early opening of the Indian Ocean in Late Jurassic-Early Cretaceous time (Griffiths, 1971*a*). Basins of the second type could include later development of the Otway, Gippsland, and Bass Basins, resulting from the opening of the Southern Ocean, and the basins on the Campbell Plateau. The Ross Sea basins trend normal to the continental margin at the Ross Sea, but if linked with the Byrd Basin, could be considered intracratonic rifts. A linear gravity anomaly trending north-south in the Ross Sea has been correlated with early rifting by Hayes and Davey (1975). Geophysical studies (Woollard, 1962) indicate thinner than average crust (25-30 km) beneath the Byrd Basin compared to adjoining Marie Byrd Land (35-40 km) and the Transantarctic Mountains (40-45 km), indicative of subcrustal flow away from the Byrd Basin toward coastal Marie Byrd Land and the Transantarctic and Ellsworth mountain systems (see Grindley, 1967, Figure 13).

A third possible origin, particularly relevant to basins near major transcurrent faults such as the Alpine and Campbell, is transform basins that are produced along or at a low angle ($<45°$) to the controlling transform fault. Many of the basins in the New Zealand region fall into this category, e.g., the Murchison Basin and the Pukaki Trough (Figure 2.1). If the Campbell Fault joins the rift south of Australia (Figures 2.1 and 2.2) the formation of the Otway, Bass, and Strzelecki Basins may also be linked to early dextral motion along this fault.

Testing the reconstruction

Geologic markers

Unequivocal geologic reference lines are hard to find and most geologic fits are based on generalized comparisons of stratigraphy, volcanism, plutonism, metamorphism, and deformation. In Figure 2.2, basement features used most in such comparisons are shown.

South Australia-Antarctica fit

Traditionally the Adelaide geosyncline of South Australia has been linked with the Ross geosyncline of Antarctica on the basis of late Precambrian-Cambrian sedimentation patterns and terminal orogenies (Ross and Delamerian) in Late Cambrian time followed by Early Ordovician granitic plutonism (Craddock, 1972*a*). Tasmania has been moved laterally to form part of this belt (Harrington and others, 1973; Cooper, 1975; Veevers and McElhinny, 1976) and to bring the Tuhuan Fold Belt of western New Zealand closer to Victoria. This hypothesis utilized a fracture zone (Gambier-Beaconsfield Fracture Zone of Harrington and others, 1973) connecting the South Australian margin through Bass Strait with northeast Tasmania. Crawford and Campbell (1973), the original proponents of the fracture zone, suggested dextral instead of sinistral displacement to account for supposed dextral oroclinal bending of the Adelaide Fold Belt and Dundas Trough (Tasmania). Daily and others (1973), showed that the Adelaide geosyncline and Dundas Trough formed as curved depositional troughs marginal to their respective cratons (Gawler and Tyennan Blocks) and developed into curved fold belts by later cratonward folding and thrusting. The orocline hypothesis in South Australia has also been disproved paleomagnetically (Giddings and Embleton, 1974). Daily and others (1973) proposed that the Adelaidean and parallel Early Cambrian Kanmantoo Troughs probably continued along strike westward from Kangaroo Island, around the southern margin of the Gawler craton, and thence northward into the Officer Basin. Only Precambrian gneiss, schist, and marble are known from the Adelie Coast of Antarctica (Stillwell, 1918) lying opposite Kangaroo Island (Figure 2.2). The belt of Ordovician granite plutonism does, however, continue into East Antarctica from western Victoria and South Australia (Craddock, 1972*b*).

Victoria-Tasmania-Antarctica fit

Northern Victoria Land is a critical piece in the Gondwana jigsaw. Similarities in Cambrian stratigraphy of the Bowers Trough with the Dundas Trough of Tasmania have recently been emphasized by Laird and others (1977), who align these troughs in their reconstruction, placing Tasmania 250 km further northwest than in Figure 2.2. Important similarities cited include Vendian–Lower Cambrian mafic volcanic rocks, Middle-Upper Cambrian fossiliferous shales and sandstones, and Upper Cambrian–Lower Ordovician quartzose conglomerates. Differences are apparent in the age and nature of the Precambrian basement and in the absence in Antarctica of the felsic Mount Read volcanic arc of Cambrian age. The apparent absence east of the Dundas Trough of the widespread upper Proterozoic (Vendian?) Robertson Bay Group of semischistose quartzose flysch, deformed during Ordovician time and intruded by Devonian–Carboniferous granitic plutons, is also surprising. However, mica schist drilled in DSDP well 281 on the South Tasman Rise has been dated (K-Ar) at 305 m.y. (Ovenshine and others, 1974) and correlated with the Robertson Bay Group. Other possible correlatives are the Rocky Cape Group of northwest Tasmania (Solomon and Griffiths, 1974) or undated quartzose flysch in the Stawell Trough of western Victoria. A useful reference line in this sector may be the boundary between the Lower Ordovician (Ross-Delamerian) granites to the west and the Lower Devonian (Admiralty-Bowning) granites to the east. This boundary is aligned on our reconstruction, while the Bowers Trough trends towards the Stavely "greenstone axial belt" in western Victoria.

Eastern Australia-New Zealand fit

Few comparisons have been made of eastern Australian pre-Mesozoic rocks with those of New Zealand, mainly because of the intervening submerged Lord Howe Rise continental block. Harrington and others (1973) and Cooper (1975) have concluded that most similarities are between New Zealand and Victoria in the Cambro-Ordovician rocks, and have moved New Zealand and Tasmania along the Gambier-Beaconsfield Fracture Zone to align New Zealand with Victoria and Tasmania with South Australia. A Silurian–Early Devonian sinistral displacement of 1100 km is inferred along the fracture zone (Harrington and others, 1973; Veevers and McElhinny, 1976).

K-Ar dating of basement rocks from oil wells in Bass Strait and interpretation of aeromagnetic surveys (Haematite Exploration Proprietary Pty. Ltd., 1965) point to a continuation of the Waratah Bay "greenstone belt" of Cambrian age across Bass Strait toward the Cambrian Beaconsfield mafic-ultramafic belt in northern Tasmania (W. D. M. Hall, personal communication). The Beaconsfield Fault in Tasmania does not trend west through Bass Strait but continues northwestward toward Central Victoria. The Gambier-Beaconsfield Fracture Zone, therefore, appears to link two unrelated fault systems, and hypotheses requiring large translations of microcontinents along this zone should be viewed with caution.

The Tuhua Fold Belt (Suggate, 1965) can be traced northward from northern South Island (Grindley, 1971) into western Cook Strait where lower Paleozoic rocks were drilled in the Moa-1 oil well (Wodzicki, 1974). The Tuhua Fold Belt may thus underlie much of the eastern Lord Howe Rise and continue into eastern Queensland where deformed Paleozoic quartzose flysch and ophiolitic rocks occur in nappes thrust from the east (Murray, 1974).

New Zealand-Marie Byrd Land fit

Correlation of basement rock belts eastward from New Zealand into Marie Byrd Land faces similar difficulties because of the intervening submerged Campbell Plateau region. Several parallel pre-Cretaceous rock belts on the South Island of New Zealand (Grindley, 1974) are shown in Figure 2.2 with possible eastward continuations. The Haast Schist Zone, forming the axial zone or collision suture of the Rangitata Orogenic Belt, reappears at Chatham Islands (Hay and others, 1970) and gives K-Ar ages indicative of early to Middle Jurassic deformation and metamorphism (160–190 m.y.) (Adams and Robinson, 1977), 50 m.y. earlier than on the mainland. Even with the Bounty Trough closed, the Haast Schist Zone strikes east, to the north of Marie Byrd Land, and is unlikely to be recognized there.

Other basement rocks on the Campbell Plateau include the Complex Point phyllites of Campbell Island (Oliver and others, 1950), the Paleozoic graywackes dredged offshore from the Bounty Islands (Cullen, 1975), and the Lower Jurassic granodiorite that intrudes them (Wasserburg and others, 1963). The Complex Point Formation has recently been dated (K-Ar) at 440 m.y. (C. J. Adams, personal communication); this age undoubtedly records Silurian deformation as in the Greenland Group of western South Island (Adams and others, 1975). The Complex Point Formation is a phyllite-slate-quartzose sandstone unit (Beggs, 1978), similar in composition to the Ordovician rocks of the Aorere and Greenland Groups in western South Island (Grindley, 1961; Nathan, 1976) and also similar to dredged samples from the Bounty Islands. The cleavage strikes north of east, aligning with magnetic lineations at the continental margin (Davey and Christoffel, 1978), and pointing to a continuation of the Tuhua Fold Belt across the southern Campbell Plateau. Lower Cretaceous granites crop out on Stewart Island and on Snares and Auckland Islands (Denison and Coombs, 1977) and were encountered in two oil wells drilled in the Great South Basin by Hunt Petroleum Ltd. (W. A. Watters, personal communication).*

The age of the Marie Byrd Land basement depends on the interpretation of large numbers of radiometric age determinations (both K-Ar and Rb-Sr). The only macrofossils are Upper Jurassic ammonites, belemnites, and pelecypods from eastern Ellsworth Land (Stevens, 1967), which establish the ages of Middle and Upper Jurassic folded sedimentary beds and calc-alkaline felsic volcanic rocks deformed in the Andean orogeny (Laudon, 1972). Undeformed Upper Cretaceous (80–90 m.y.) calc-alkaline felsic volcanic rocks crop out in localized areas in central Marie Byrd Land (LeMasurier and Wade, 1976). In western Marie Byrd Land, quartzose sandstones and slates of the Edsel Ford Ranges (Wade, 1969; Wade and Wilbanks, 1972) have yielded acritarchs, referred to the late Proterozoic(?) (Riphean?) by Iltchenko (1972), and possible Early Cambrian microphytoliths (Grikurov and Lopatin, 1975). Correlation with the Berg Group (west of Rennick Glacier) and possibly the Robertson Bay Group is suggested and is supported by similar Early Ordovician (Ross orogeny) K-Ar ages.

The migmatized gneiss of the Fosdick Range represents either an older basement (Klimov, 1967) or a lower, more highly metamorphosed infrastructure (Wilbanks, 1972). Rb-Sr radiometric data are inconclusive because of partial resetting during Early Cretaceous time (Halpern, 1972). Possible correlatives are the Constant Gneiss of western South Island (Adams, 1975) and the Wilson Group gneiss-migmatite sequence of northern Victoria Land (Gair and others, 1969).

The metasedimentary rocks in the Edsel Ford Ranges are intruded by numerous Lower Cretaceous granites (Wade, 1969), and a few Carboniferous granites (Rb-Sr, 320–350 m.y., Halpern, 1968; Grikurov and Lopatin, 1975) are also exposed.

On the Ruppert Coast north of the Fosdick anticlinorium, intermediate to mafic metavolcanic rocks (Klimov, 1967) are underlain by pelitic and quartzose metasedimentary rocks (G. W. Grindley, personal communication) that locally contain Devonian plant fossils (Grindley and Mildenhall, 1978) and are intruded by Lower Cretaceous granites.

A basement of mafic to intermediate orthogneiss (Klimov, 1967) underlies the metavolcanic rocks on the Hobbs Coast and has provided early to middle Paleozoic Rb-Sr model ages (Craddock, 1972*b*). In eastern Marie Byrd Land, similar orthogneiss and amphibolite with early Paleozoic radiometric ages are overlain by intermediate metavolcanic rocks, locally as old as Late Devonian (Wade and Wilbanks, 1972; Lopatin and others, 1974), intruded by Permian quartz diorite (Rb-Sr, 265 m.y., Halpern, 1972) and Mesozoic granites. On Thurston Island, basement gneisses with early Paleozoic radiometric ages are intruded by upper Paleozoic gabbros and diorites and lower Mesozoic granites (Lopatin and Orlenko, 1972; Craddock, 1972*b*).

The above synopsis suggests that Marie Byrd Land basement geology resembles that of New Zealand to the west and south of the Median Tectonic Line of Landis and Coombs (1967). Lopatin and others (1974) have summarized the Precambrian–Mesozoic geologic history of Marie Byrd Land in terms of several broadly dated igneous and metamorphic age groups, most of which are potentially correlative with New Zealand rock units, as described in the New Zealand Geological Survey 1:250,000 map series (Grindley, 1961; Bowen, 1964; Wood, 1960, 1962; Watters and others, 1968) and the relevant chapters of *The Geology of New Zealand* (Grindley, 1978; Wood, 1978).

Geophysical markers

Magnetic data over the exposed rocks and their concealed offshore extensions can provide guides to the trend of basement rock belts of different magnetic expression, e.g., the above-mentioned correlation of Cambrian mafic-ultramafic belts across Bass Strait. Magnetic surveys in New Zealand have defined a large linear magnetic anomaly system, the Stokes Magnetic Anomaly (Wellman, 1973) which can be traced throughout New Zealand, except where it is displaced by the Alpine Fault. The anomaly system is associated with two of the geotectonic zones of Grindley (1974): the Marginal Syncline Zone of volcaniclastic Permian–Jurassic sedimentary beds, and the Rotoroa-Fiordland Zone of lower Paleozoic gneiss, schist, and amphibolite, with upper Paleozoic to Mesozoic intrusions. A small continuous magnetic anomaly, the Junction Anomaly of Hatherton (1969), is correlated with the Dun Mountain Ophiolite Belt (Coombs and others, 1976), marking the junction and possible suture between the eutectonic Haast Schist Zone and Permian metagraywacke equivalents and the miotectonic, volcanic arc–derived sedimentary beds of the Marginal Syncline. This anomaly can be traced from Northland Peninsula south to the Alpine Fault (Wellman, 1959; Hatherton, 1967) and from northwest Otago to east of the Otago coast (Hatherton, 1969). Larger anomalies are associated both with mafic-ultramafic subvolcanic plutonic bodies exposed in the gneissic Fiordland and Rotoroa Complexes, below the Permian island-arc volcanic and volcaniclastic sequences, and with their buried extensions within the Marginal Syncline (Hatherton, 1966).

The Stokes Anomaly System has been deformed into a major dextral orocline and displaced 480 km by the Alpine Fault (Wellman, 1973; Woodward and Hatherton, 1975). A similar linear anomaly system across the northern Campbell Plateau (Christoffel and McMullen *in* Fleming, 1970) has been identified with intrabasement magnetic rocks and correlated with the Stokes Anomaly System (Wellman, 1973), although it has been separately named the Campbell Anomaly System (Davey and Christoffel, 1978). Jurassic granodiorite at the Bounty Islands to the north of the anomaly system, however, is not in accord with simple extension of the New Zealand structural zones

* In August 1977, a third Hunt oil well in this area drilled indurated quartzose sandstones closely similar to the Greenland and Aorere Groups (Ordovician) of the Tuhua Fold Belt in western South Island, thus providing a link between southwest Firodland and Campbell Island (Figure 2.1).

across the Campbell Plateau. On the mainland, no granitic intrusive rocks are known north or east of the anomaly belt, the northern edge of which is marked by the Junction Anomaly (= Dun Mountain Ophiolite Belt). This problem would be compounded if the graywackes dredged off Bounty Island prove to be lower Paleozoic in age (Cullen, 1975).

The Campbell Anomaly System intersects the eastern edge of the Campbell Plateau obliquely between the Bounty and Antipodes Islands and might be found along strike in eastern Marie Byrd Land or Thurston Island, where exposed basement rocks include mafic-intermediate orthogneiss, similar to older parts of the Fiordland and Rotoroa Complexes, which are also intruded by granite, diorite, and gabbro plutons. Possibly, future aeromagnetic surveys will locate a continuation of the Campbell Anomaly System in this area.

In northwest New Zealand, the Junction Anomaly of the Stokes Anomaly System has been traced by aeromagnetic surveys into Northland Peninsula (Hatherton and Sibson, 1970). Any basement anomalies further west are mainly obscured by the large magnetic anomalies associated with postulated Miocene andesitic volcanic centers (Davey, 1974) and by mafic bodies intrusive into Cenozoic sedimentary rocks on the Norfolk Ridge (Hochstein, 1967; Davey, 1977). Continuation of the Stokes Anomaly System along the eastern Lord Howe Rise might be expected.

Tectonic implications

The linear magnetic anomaly systems have been used as indicators of the finite post-Jurassic dextral strain undergone by the New Zealand geosynclinal system (Wellman, 1973). Correlation of anomalies across the Campbell Plateau implies a dextral transcurrent fault between the Auckland Islands and the Bounty Trough with a displacement of at least 350 km (Davey and Christoffel, 1978), 250 km of which is probably pre-Late Cretaceous. This fault, already mapped morphologically south of the Auckland Islands by Summerhayes (1969), links with the head of Bounty Trough and does not seem to offset Chatham Rise. The latest movement of ~ 100 km may have opened up the Bounty Trough during Cenomanian-Senonian time (Grindley and others, 1977).

To the southwest, the Campbell Fault may link with the major Gambier-Sorell Fracture Zone (Harrington and others, 1973) along the Australian/Tasmanian-Antarctic margin. Pre-Late Cretaceous dextral displacement of at least 250 km along this fault may have offset the Dundas and Bowers Cambrian troughs (Figure 2.2), and Harrington and others (1973) suggest as much as 900 km of dextral shift to offset the Tasmanian Jurassic dolerites from their correlatives in northern Victoria Land. The parallel Antipodes Fracture Zone along the Campbell Plateau margin may also have had some dextral strike-slip motion in Early Cretaceous time prior to rifting, as suggested by Cullen (1967) and Houtz (1975). A link across the Ross Sea to join the Rennick Fault system in northern Victoria Land is suggested. More speculatively, a major dextral strike-slip fault between Marie Byrd Land and the Ellsworth Mountains might account for the latter's anomalous northerly trend (Ford, 1972; Craddock, 1975). This fault would cross the Ross Sea north of DSDP well 270, where Cambrian(?) calc-silicate gneiss of the Ross orogenic belt was encountered (Ford and Barrett, 1975).

The chief drawback of the Crook-Belbin reconstruction and of the reconstructions in Figures 2.1 and 2.2 is that the Alpine Fault is required to have had some 400 km of sinistral displacement before the earliest sea-floor anomalies were produced 85 m.y. ago. This supposed sinistral offset could be reduced or eliminated by bending the New Zealand geosynclinal system during Cenozoic time as argued by Wellman (1975, 1976). Paleomagnetic data are presently being collected to determine the timing of dextral rotation of Cretaceous volcanic sites within the New Zealand shear belt in western South Island and eastern North Island. First results from the Mount Somers Volcanics (about 90 m.y. old) of Canterbury, between 60 and 90 km from the Alpine Fault, suggest little or no rotation (Oliver, 1977; Oliver and others, 1979) during the past 90 m.y. (Figure 2.1). If this result is confirmed at other sites, especially those west of the Alpine Fault, and pre-Late Cretaceous bending becomes necessary, some other explanation will be required for the sinistral offset.*

Finally, the timing of events in the Rangitata orogeny and early rifting episodes prior to sea-floor spreading is still uncertain. About 100 m.y. ago, subduction may have ceased along the Pacific margin, with minor granitic emplacement and calc-alkaline volcanism continuing in localized areas (e.g., DSDP well 207, Lord Howe Rise, 94 m.y ago) (McDougall and Van der Lingen, 1974). Rifting with tholeiitic volcanism started much earlier—in Middle Jurassic time in East Antarctica and Tasmania—and continued into Early Cretaceous time in the Otway and Strzelecki basins. Alkaline basalt volcanism followed both calc-alkaline volcanism and granite intrusion in the West Coast Basin aproximately 90 m.y. ago (Adams and Nathan, 1978). The profound structural disturbance, with accompanying igneous activity and metamorphism, of the Rangitata orogeny is still little understood in the context of plate movements, although several scenarios are available (Fleming, 1970; Suggate, 1972; Landis and Bishop, 1972; Grindley, 1974; Griffiths, 1975; Waterhouse, 1975; Coombs and others, 1976). Displacements during the fundamental change in tectonic regime from subduction related orogenesis to rift-valley taphrogenesis and drift are largely unknown, yet critical to the palinspastic restoration of continental blocks in the Southwest Pacific.

Acknowledgments. The authors are grateful to the Steering Committee of the Third SCAR-IUGS Symposium on Antarctic Geology and Geophysics for the invitation to write this review paper and to their colleagues in the Department of Scientific and Industrial Research (D.S.I.R.) and elsewhere for support and encouragement. One of us (G. W. G.) is indebted to other members of the International Geological Correlation Programme working group on South-west Pacific Basement Correlation for guidance in the field around many of the critical southeastern Australian and Tasmanian sequences in 1976 and 1977. The illustrations were drafted in the Cartographic Section of D.S.I.R. under the supervision of Robin Phillips, and the manuscript was typed by M. Ruthven. R. I Walcott (Geophysics Division) and R. P. Suggate (Geological Survey) critically reviewed the manuscript.

* A paleomagnetic study by Grindley and Oliver (1979) has shown little or no rotation of 80 m.y. mafic dikes at Buller Gorge, Westland, and confirmed pre-Late Cretaceous oroclinal bending for the region west of the Alpine Fault.

References cited

Adams, C. J. D. 1975. Discovery of Precambrian rocks in New Zealand: age relations of the Greenland Group and Constant Gneiss, West Coast, South Island. *Earth planet. Sci. Lett., 28,* No. 1, 98–104.

Adams, C. J. D., and Nathan, S. 1978. Cretaceous chronology of the lower Buller Valley, South Island, New Zealand. *N.Z. Jl Geol. Geophys., 21,* No. 4, 455–62.

Adams, C. J. D., and Robinson, P. 1977. Age of metamorphism of schists, Chatham Islands, South-west Pacific. *N.Z. Jl Geol. Geophys., 20,* No. 2, 287–302.

Adams, C. J. D., Harper, C. T. , and Laird, M. G. 1975. K-Ar ages of low-grade metasediments of the Greenland and Wainta Groups in Westland and Buller, New Zealand. *N.Z. Jl Geol. Geophys., 18,* No. 1, 39–48.

Andrews, P. B., Gostin, V. A., Hampton, M. A., Margolis, S. V., and Ovenshine, A. T. 1974. Synthesis-sediments of the southwest Pacific Ocean, southeast Indian Ocean and South Tasman Sea. DSDP Leg 29. (*In* Kennett, J. P., Houtz, R. E., and others. *Initial Reports of the Deep Sea Drilling Project,* Vol. 29. Washington [U.S. Government Printing Office], 1147–53.)

Australia. Bureau of Mineral Resources, Geology and Geophysics. 1965. *Geological Map of the World, Australia and Oceania, 1:5,000,000.* 1st ed. Hobart, Aust., Mercury Press for Commonwealth Map of the World, Paris, International Geological Congress, sheet 12.

Beggs, J. M. 1978. Geology of the metamorphic basement and Late Cretaceous to Oligocene sedimentary sequence of Campbell Island, Southwest Pacific Ocean. *J. R. Soc. N.Z., 8,* No. 2, 161–77.

Bentley, C. R., and Clough, J. W. 1972. Antarctic subglacial structure from seismic refraction measurements. (*In* Adie, R. J., *ed. Antarctic Geology and Geophysics.* Oslo, Universitetsforlaget, 683–91.)

Blundell, D. J. 1962. Paleomagnetic investigations in the Falkland Islands Dependencies. *Scient. Rep. Br. antarc. Surv., 39,* 24 pp.

Bott, M. H. P. 1971. Evolution of young continental margins and the formation of shelf basins. *Tectonophysics, 11,* 319–27.

Bowen, F. E. 1964. Sheet 15, Buller. *Geological Map of New Zealand, 1:250,000.* Wellington, New Zealand Department of Scientific and Industrial Research.

Brothers, R. N. 1974. Kaikoura orogeny in Northland, New Zealand. *N.Z. Jl Geol. Geophys., 17,* No. 1, 1–18.

Brown, B. R. 1975. New deep-water data: Otway Basin. *Bull. Aust. Soc. Explor. Geophys., 6,* 42–45.

Burke, K., and Dewey, J. F. 1973. Plume-generated triple junctions: key indicators in applying plate tectonics to old rocks. *J. Geol., 81,* 406–33.

Christoffel, D. A., and Falconer, R. K. H. 1972. Marine magnetic measurements in the southwest Pacific Ocean and the identification of new tectonic features. (In Hayes, D. E., *ed. Antarctic Oceanology 2. The Australian-New Zealand Sector.* Washington, American Geophysical Union, 197–210.) [Antarctic Research Series, Vol. 19.]

Christoffel, D. A., and Falconer, R. K. H. 1973. Changes in the direction of sea floor spreading in the south-west Pacific. (*In* Fraser, R., *comp. Oceanography of the South Pacific.* Wellington, New Zealand National Commission for UNESCO, 241–47.)

Coombs, D. S., Landis, C. A., Norris, R. J., Sinton, J. M., Borns, D. J., and Craw, D. 1976. The Dun Mountain Ophiolite belt, New Zealand, its tectonic setting, constitution, and origin, with special reference to the southern portion. *Am. J. Sci., 276,* 561–603.

Cooper, R. A. 1975. New Zealand and south-east Australia in the early Paleozoic. *N.Z. Jl Geol. Geophys., 18,* No. 1, 1–20.

Cope, R. A., and Reed, J. J. 1967. The Cretaceous palaeogeology of the Taranaki–Cook Strait area. *Proc. Australas. Inst. Min. Metall., 222,* 63–72.

Craddock, C. 1972*a*. Antarctic tectonics. (*In* Adie, R. J., *ed. Antarctic Geology and Geophysics.* Oslo, Universitetsforlaget, 449–55.)

Craddock, C. 1972*b*. *Geological Map of Antarctica, 1:5,000,000.* New York, American Geographical Society.

Craddock, C. 1972*c*. *Tectonic Map of Antarctica, 1:5,000,000.* New York, American Geographical Society.

Craddock, C. 1975. Tectonic evolution of the Pacific margin of Gondwanaland. (*In* Campbell, K. S. W., *ed. Gondwana Geology.* Canberra, Australia National University Press, 609–618.)

Cranwell, L. M., Harrington, H. J., and Speden, I. G. 1960. Lower Tertiary microfossils from McMurdo Sound, Antarctica. *Nature, Lond., 186,* 700–2.

Crawford, A. R., and Campbell, K. S. W. 1973. Large-scale horizontal displacement within Australian-Antarctica in the Ordovician. *Nature, Lond., 241,* 11–14.

Crook, K. A. W., and Belbin, L. 1975. Plate tectonic and geometric reconstruction of the southwest Pacific margin of Gondwanaland. (*In* Veevers, J. J., *ed. Deep Sea Drilling in Australasian Waters.* Sydney, Challenger Symposium, 31.)

Crook, K. A. W., and Belbin, L. 1978. The southwest Pacific area during the last 90 million years. *J. geol. Soc. Aust., 25,* No. 1, 23–40.

Cullen, D. J. 1967. The Antipodes fracture zone, a major structure of the south-west Pacific. *N.Z. Jl Mar. Freshwat. Res., 1,* No. 1, 16–25.

Cullen, D. J. 1970. A tectonic analysis of the south-west Pacific. *N.Z. Jl Geol. Geophys., 13,* No. 1, 7–20.

Cullen, D. J. 1975. Autochthonous rocks of the Bounty Islands region, southwest Pacific Ocean. *N.Z. Jl Geol. Geophys., 18,* No. 6, 767–85.

Daily, B., Jago, J. B., and Milnes, A. R. 1973. Precambrian blocks and arcuate tectonic trends in South Australia and Tasmania. *Nature phys. Sci., 244,* 61–64.

Dalziel, I. W. D., Kligfield, R., Lowrie, W., and Opdyke, N. D. 1973. Paleomagnetic data from the southernmost Andes and Antarctandes. (*In* Tarling, D. H., and Runcorn, S. K., *eds. Implications of Continental Drift to the Earth Sciences.* New York, Academic Press, 87–101.)

Davey, F. J. 1974. Magnetic anomalies off the west coast of Northland, New Zealand. *J. R. Soc. N.Z., 4,* No. 2, 203–16.

Davey, F. J. 1977. Marine seismic measurements in the New Zealand region. *N.Z. Jl Geol. Geophys., 20,* No. 4, 719–78.

Davey, F. J., and Christoffel, D. A. 1978. Magnetic anomalies across Campbell Plateau, New Zealand, *Earth planet. Sci. Lett., 41,* 14–20.

Davey, F. J., and Houtz, R. E. 1977. The Campbell Plateau and its relationship with the Ross Sea, Antarctica. *Mar. Geol., 25,* 61–72.

Denison, R. E., and Coombs, D. S. 1977. Radiometric ages for some rocks from Snares and Auckland Islands, Campbell Plateau. *Earth planet. Sci. Lett., 34,* 23–29.

Elliott, J. L. 1972. Continental drift and basin development in south eastern Australia. *J. Aust. petrol. Expl. Ass., 12,* No. 2, 46–51.

Falvey, D. A. 1974. The development of continental margins in plate tectonic theory. *J. Aust. petrol. Expl. Ass., 14,* No. 1, 95–106.

Fleming, C. A. 1970. The Mesozoic of New Zealand; chapters in the history of the circum-Pacific Mobile Belt. *Q. Jl geol. Soc. Lond., 125,* 125–70.

Ford, A. B. 1972. Fit of Gondwana continents—drift reconstruction from the Antarctic continental viewpoint. *Proc. 24th Int. Geol. Congr., Montreal, 1972,* Sect. 3, 113–21.

Ford, A. B., and Barrett, P. J. 1975. Basement rocks of the south-central Ross Sea, Site 270, DSDP Leg 28. (*In* Hayes, D. E., Frakes, L. A., and others. *Initial Reports of the Deep Sea Drilling Project,* Vol. 28. Washington [U.S. Government Printing Office], 861–68.)

Gair, H. S., Sturm, A., Carryer, S. J., and Grindley, G. W. 1969. The geology of northern Victoria Land. Sheet 13. (*In* Bushnell, V. C., and Craddock, C., *eds.* Geologic Maps of Antarctia. *Antarct. Map Folio Ser.,* Folio 12, Plate XII.)

Giddings, J. W., and Embleton, B. J. 1974. Large-scale horizontal displacement in southern Australia—contrary evidence form palaeomagnetism. *J. geol. Soc. Aust., 21,* No. 4, 431–36.

Griffiths, J. R. 1971*a*. Continental margin tectonics and the evolution of southeast Australia. *J. Aust. petrol. Expl. Ass., 12,* 75–79.

Griffiths, J. R. 1971*b*. Reconstruction of the south-west Pacific margin of Gondwanaland. *Nature, Lond., 234,* 203–7.

Griffiths, J. R. 1974. Revised continental fit of Australia and Antarctica. *Nature, Lond., 249,* 336–38.

Griffiths, J. R. 1975. New Zealand and the southwest Pacific margin of Gondwanaland. (*In* Campbell, K. S. W., *ed. Gondwana Geology.* Canberra, Australia National University Press, 619–38.)

Grikurov, G. E., and Lopatin, B. G. 1975. Structure and evolution of the West Antarctic part of the circum-Pacific mobile belt. (*In* Campbell, K. S. W., *ed. Gondwana Geology.* Canberra, Australia National University Press, 639–50.)

Grindley, G. W. 1961. Sheet 13, Golden Bay. *Geological Map of New Zealand 1:250,000.* Wellington, New Zealand Department of Scientific and Industrial Research.

Grindley, G. W. 1967. The geomorphology of the Miller Range, Transantarctic Mountains, with notes on the glacial history and neotectonics of East Antarctica. *N.Z. Jl Geol. Geophys., 10,* No. 2, 557–98.

Grindley, G. W. 1971. Sheet S8, Takaka. *Geological Map of New Zealand 1:63,360.* Wellington, New Zealand Department of Scientific and Industrial Research.

Grindley, G. W. 1974. New Zealand. (*In* Spencer, A. M., *ed. Mesozoic-Cenozoic Orogenic Belts: Data for Orogenic Studies. Spec. Publs geol. Soc. Lond.,* 387–416.)

Grindley, G. W. 1978. Chapter 2. West Nelson, pp. 80–99; Chapter 3. Tectonism of the early Geosynclinal Cycle, the Tuhua orogeny and the New Zealand geanticline, pp. 117–35. (*In* Suggate, R. P., *ed. The Geology of New Zealand.* Wellington, Government Printer.)

Grindley, G. W., and Mildenhall, D. C. 1978. Discovery of fossils in Marie Byrd Land, Antarctica. *Newsl. geol. Soc. N.Z., 45,* 33–34.

Grindley, G. W., and Oliver, P. J. 1979. Paleomagnetism of Upper Cretaceous dikes, Buller Gorge, North Westland, in relation to the bending of the New Zealand Orocline. *Bull. R. Soc. N.Z., 18,* 131–47.

Grindley, G. W., Adams, C. J. D., Lumb, J. T., and Watters, W. A. 1977. Paleomagnetism, K-Ar dating and tectonic interpretation of Cretaceous and Cenozoic volcanic rocks from Chatham Islands, New Zealand. *N.Z. Jl Geol. Geophys., 20,* No. 3, 425–67.

Gunn, P. J. 1975. Mesozoic-Cenozoic tectonics and igneous activity, south eastern Australia. *J. geol. Soc. Aust., 22,* 215–21.

Haematite Explorations Proprietary Pty. 1965. Bass Strait and Encounter Bay aeromagnetic survey, 1960–61. *Publs Bur. Miner. Resour. Aust., 60,* 38 pp., 6 pl., 4 figs.

Halpern, M. 1967. Rubidium-strontium isotopic age measurements of plutonic igneous rocks in eastern Ellsworth Land and northern Antarctic Peninsula, Antarctia. *J. geophys. Res., 72,* 5133–42.

Halpern, M. 1968. Ages of Antarctic and Argentine rocks bearing on continental drift. *Earth planet. Sci. Lett., 5,* 159–67.

Halpern, M. 1972. Rb-Sr total-rock and mineral ages from the Marguerite Bay area, Kohler Range and Fosdick Mountains. (*In* Adie, R. J., *ed. Antarctic Geology and Geophysics.* Oslo, Universitetsforlaget, 197–204.)

Harrington, H. J., Burns, K. L., and Thompson, B. R. 1973. Gambier-Beaconsfield and Gambier-Sorell fracture zones and the movement of plates in the Australia-Antarctica-New Zealand region. *Nature Phys. Sci., 245,* 109–12.

Hatherton, T. 1966. A geophysical study of the Southland syncline. *Bull. N.Z. Dep. scient. ind. Res., 168,* 48 pp.

Hatherton, T. 1967. A geophysical study of Nelson-Cook Strait region, New Zealand. *N.Z. Jl Geol. Geophys., 10,* No. 6, 1330–47.

Hatherton, T. 1969. Geophysical anomalies over the eu- and miogeosynclinal systems of California and New Zealand. *Geol. Soc. Am. Bull., 80,* No. 2, 213–30.

Hatherton, T., and Sibson, R. H. 1970. Junction magnetic anomaly north of Waikato River. *N.Z. Jl Geol. Geophys., 13,* No. 3, 655–62.

Hay, R. F., Mutch, A. R., and Watters, W. A. 1970. Geology of the Chatham Islands. *Bull. geol. Surv. N.Z., n.s., 83,* 86 pp.

Hayes, D. E., and Davey, F. J. 1975. A geophysical study of the Ross Sea, Antarctica. (*In* Hayes, D. E., Frakes, L. A., and others. *Initial Reports of the Deep Sea Drilling Project,* Vol. 28. Washington [U.S. Government Printing Office], 887–907.)

Hayes, D. E., and Frakes, L. A. 1975. General synthesis, Deep Sea Drilling Project Leg 28. (*In* Hayes, D. E., Frakes, L. A., and others. *Initial Reports of the Deep Sea Drilling Project,* Vol. 28. Washington [U.S. Government Printing Office], 919–42.)

Hayes, D. E., and Ringis, J. 1973. Sea-floor spreading in the Tasman Sea. *Nature, Lond., 243,* 454–58.

Hayward, B. W. 1976. Lower Miocene stratigraphy and structure of the Waitakere Ranges and Waitakere Group (new). *N.Z. Jl Geol. Geophys., 19,* No. 6, 871–96.

Helsley, C. E., and Steiner, M. B. 1969. Evidence for long periods of normal polarity during the Cretaceous period. *Earth planet. Sci. Lett., 5,* 325–32.

Hochstein, M. P. 1967. Interpretation of magnetic anomalies across Norfolk Ridge. *N.Z. Jl Geol. Geophys., 10,* No. 5, 1302–8.

Hocking, J. B. 1972. Geological evolution and hydrocarbon habitat, Gippsland Basin. *J. Aust. petrol. Expl. Ass., 12,* 132–37.

Houtz, R. E. 1975. South Tasman Basin and borderlands: A geophysical summary. DSDP Leg 29. (*In* Kennett, J. P, Houtz, R. E., and others. *Initial Reports of the Deep Sea Drilling Project,* Vol. 29. Washington [U.S. Government Printing Office], 1135–46.)

Houtz, R. E., and Davey, F. J. 1973. Seismic profiler and Sonobuoy measurements in Ross Sea, Antarctica. *J. geophys. Res., 78,* No. 17, 3448–68.

Iltchenko, L. N. 1972. Late Precambrian acritarchs of Antarctica. (*In* Adie, R. J., *ed. Antarctic Geology and Geophysics.* Oslo, Universitetsforlaget, 599–602.)

Katz, H. R. 1968. Potential oil formations in New Zealand and their stratigraphic position as related to basin evolution. *N.Z. Jl Geol. Geophys., 11,* No. 5, 1017–33.

Katz, H. R. 1974. Margins of the Southwest Pacific. (*In* Burk, C. A., and Drake, C. L., *eds. The Geology of Continental Margins.* New York, Springer-Verlag, 549–65.)

Katz, H. R. 1976. Sedimentary basins and petroleum prospects onshore and offshore New Zealand. (*In* Halbouty, M. T., Maher, J. C., and Lian, H. M., *eds.* Circum-Pacific Energy and Mineral Resources. *Mem. Am. Ass. Petrol. Geol.,* 25, 217–28.)

Kellogg, K. S., and Reynolds, R. L. 1974. Paleomagnetic study of igneous rocks of the northern Lassiter Coast, Antarctic Peninsula. *Antarct. Jnl U.S., 9,* No. 2, 38–40.

Klimov, L. V. 1967. Nektorye rezul'taty geologicheskikh issledovarnii na Zemle Meri Berd v 1966–1967 gg. [Some results of geological observations in Marie Byrd Land, 1966–67]. *Inf. Byull. sov. antarkt. Eksped.,* No. 66, 18–25. [English translation: *Inf. Bull. Sov. Antarct. Exped., 6,* No. 6, 555–59.]

Laird, M. G., Cooper, R. A., and Jago, J. B. 1977. New data on the lower Palaeozoic sequence of northern Victoria Land, Antarctica, and its significance for Australia-Antarctic relations in the Palaeozoic. *Nature, Lond., 265,* 107–10.

Landis, C. A., and Bishop, D. G. 1972. Plate tectonics and regional stratigraphic-metamorphic relations in the southern part of the New Zealand geosyncline. *Geol. Soc. Am. Bull., 83,* 2267–84.

Landis, C. A., and Coombs, D. S. 1967. Metamorphic belts and orogenesis in southern New Zealand. *Tectonophysics, 4,* 501–18.

Laudon, T. S. 1972. Stratigraphy of eastern Ellsworth Land. (*In* Adie, R. J., *ed. Antarctic Geology and Geophysics.* Oslo, Universitetsforlaget, 215–24.)

LeMasurier, W. E., and Wade, F. A. 1976. Volcanic history in Marie Byrd Land: implications with regard to Southern Hemisphere tectonic reconstructions. (*In* Gonzalez-Ferran, O., *ed. Andean and Antarctic Volcanology Problems.* Rome, International Association of Volcanology and Chemistry of the Earth's Interior, 398–424.)

Lopatin, B. C., and Orlenko, E. M. 1972. Outline of the geology of Marie Byrd Land and the Eights Coast. (*In* Adie, R. J., *ed. Antarctic Geology and Geophysics.* Oslo, Universitetsforlaget, 245–50.)

Lopatin, B. G., Krylov, A. Ya., and Aliapyshev, O. A. 1974. Osnovnye tektono-magmaticheskie etapy v razvitii zemli Meri Berd i berega Eitsa (Zapadnaia Antarktida) [Main tectonomagmatic stages of development of Marie Byrd Land and Eights Coast (West Antarctica)]. *Antarktika, Doklady Komissii, 13,* 52–60.

McDougall, I., and Van der Lingen, G. J. 1974. Age of the rhyolites of the Lord Howe Rise and the evolution of the southwest Pacific Ocean. *Earth planet. Sci. Lett., 21,* 117–26.

McElhinny, M. W. 1973. *Paleomagnetism and Plate Tectonics.* Cambridge, Cambridge University Press, 358 pp.

McIntyre, D. J., and Wilson, G. J. 1966. Preliminary palynology of some Antarctic Tertiary erratics. *N.Z. Jnl Bot., 4,* No. 3, 315–21.

McKenzie, D. P., and Sclater, J. G. 1971. The evolution of the Indian Ocean since the Late Cretaceous. *Geophys. J. R. astr. Soc., 25,* 437–528.

Molnar, P., Atwater, T., Mammerickx, J., and Smith, S. M. 1975. Magnetic anomalies, bathymetry and the tectonic evolution of the South Pacific since the Late Cretaceous. *Geophys. J. R. astr. Soc., 40,* 383–420.

Murray, C. G. 1974. Alpine-type ultramafics in the northern part of the Tasman geosyncline—possible remnants of Palaeozoic ocean floor. (*In* Denmead, A. K., Tweedale, G. W., and Wilson, A. F., *eds. The Tasman Geosyncline—a Symposium.* Brisbane, Geological Society of Australia, Queensland Division, 161–84.)

Nathan, S. 1976. Geochemistry of the Greenland Group (early Ordovician), New Zealand. *N.Z. Jl Geol. Geophys., 19,* No. 5, 683–706.

New Zealand Geological Survey. 1971. *Geological Map of the World, Australia and Oceania, 1:5,000,000.* Wellington, N.Z., Government Printer, sheet 13.

Norton, I., and Molnar, P. 1977. Implications of a revised fit between Australia and Antarctica for the evolution of the eastern Indian Ocean. *Nature, Lond., 267,* 338–40.

Oliver, P. J. 1977. *The Mesozoic Geology of the Mt. Somers Area, Canterbury including Geochemical and Paleomagnetic Studies of the Cretaceous Calc-Alkaline Mt. Somers Volcanics.* Ph.D. dissertation, University of Canterbury.

Oliver, P. J., Mumé, T. C., Grindley, G. W., and Vella, P. 1979. Paleomagnetism of the Upper Cretaceous Mt. Somers Volcanics, Canterbury, New Zealand. *N.Z. Jl Geol. Geophys., 22,* No. 2, 199–212.

Oliver, R. L., Finlay, H. J., and Fleming, C. A. 1950. The geology of Campbell Island. *Bull. Cape Exped. Ser., 3,* 62 pp.

Ovenshine, A. T., Csejtey, B., Jr., Smith, J. G., and Andrews, P. B. 1974. Petrography and age of quartz-muscovite-chlorite semischist, Site 281, South Tasman Rise. DSDP Leg 29. (*In* Kennett, J. P., Houtz, R. E., and others. *Initial Reports of the Deep Sea Drilling Project,* Vol. 29. Washington [U.S. Government Printing Office], 1117–21.)

Reynolds, R. L., and Kellogg, K. S. 1974. Paleomagnetism of igneous rocks of the central Lassiter Coast, Antarctic Peninsula. *Antarct. Jnl U.S., 9,* No. 5, 227–28.

Ringis, J. 1970. Magnetic lineations in the Tasman. *Geoexploration, 8,* 250.

Robinson, V. A. 1974. Geologic history of the Bass Basin. *J. Aust. petrol. Expl. Ass., 14,* No. 1, 45–49.

Roots, W. D. 1976. Magnetic smooth zones and slope anomalies: a mechanism to explain both. *Earth planet. Sci. Lett., 31,* 113–18.

Smith, A. G., and Hallam, A. 1970. The fit of the southern Continents. *Nature, Lond., 225,* 139–41.

Solomon, M., and Griffiths, J. R. 1974. Aspects of the early history of the southern Tasman Orogenic Zone. (*In* Denmead, A. K., Tweedale, G. W., and Wilson, A. F., *eds. The Tasman Geosyncline—a Symposium.* Brisbane, Geological Society of Australia, Queensland Division, 19–46.)

Sproll, W. P., and Dietz, R. S. 1969. Morphological continental drift fit of Australia and Antarctica. *Nature, Lond., 222,* 345–48.

Stevens, G. R. 1967. Upper Jurassic fossils from Ellsworth Land, West Antarctica, and notes on Upper Jurassic biogeography of the South Pacific region. *N.Z. Jl Geol. Geophys., 10,* No. 2, 345–93.

Stillwell, F. L. 1918. The metamorphic rocks of Adelie Land. *Scient. Rep. Aust. antarct. Exped.,* Ser. A, *3,* No. 1, 1–230.

Suggate, R. P. 1965. The tempo of events in New Zealand geological history. *N.Z. Jl Geol. Geophys., 8,* No. 6, 1139–48.

Suggate, R. P. 1972. Mesozoic-Cenozoic development of the New Zealand region. *Pacif. Geol., 4,* 113–20.

Summerhayes, C. P. 1969. Marine geology of the New Zealand Subantarctic sea floor. *Bull. N.Z. Dep. scient. ind. Res., 190,* 92 pp. [*Mem. N.Z. Oceanog. Inst., 50.*]

Turnbull, I. M., Barry, J. M., Carter, R. M., and Norris, R. J. 1975. The Bobs Cove beds and their relationship to the Moonlight Fault Zone. *J. R. Soc. N.Z., 5,* No. 4, 355–94.

Varne, R., and Rubenach, M. J. 1972. Geology of Macquarie Island and its relationship to oceanic crust. (*In* Hayes, D. E., *ed. Antarctic Oceanology 2. The Australian-New Zealand Sector.* Washington, American Geophysical Union, 251–66.) [Antarctic Research Series, Vol. 19.]

Veevers, J. J., and McElhinny, M. W. 1976. The separation of Australia from other continents. *Earth Sci. Rev. 12,* No. 2–3, 139–59.

Wade, F. A. 1969. Sheet 18, Marie Byrd Land. (*In* Bushnell, V. C., and Craddock, C., *eds.* Geologic Maps of Antarctica. *Antarct. Map Folio Ser.,* Folio 12, Pl. XVIII.)

Wade, F. A., and Wilbanks, J. R. 1972. Geology of Marie Byrd and Ellsworth Lands. (*In* Adie, R. J., *ed. Antarctic Geology and Geophysics.* Oslo, Universitetsforlaget, 207–14.)

Wasserburg, G. J., Craig, H., Menard, H. W., Engel, A. E. J., and Engel, C. 1963. Age and composition of a Bounty Island granite and age of a Seychelles Island granite. *J. Geol., 71,* No. 6, 785–89.

Waterhouse, J. B. 1975. The Rangitata Orogen. *Pacif. Geol., 9,* 35–73.

Watters, W. A., Speden, I. G., and Wood, B. L. 1968. Sheet 26, Stewart Island. *Geological Map of New Zealand 1:250,000.* Wellington, New Zealand Department of Scientific and Industrial Research.

Weissel, J. K., and Hayes, D. E. 1971. Asymmetric sea floor spreading south of Australia. *Nature, Lond., 231,* 518–22.

Weissel, J. K., and Hayes, D. E. 1972. Magnetic anomalies in the Southeast Indian Ocean. (*In* Hayes, D. E., *ed. Antarctic Oceanology 2. The Australian-New Zealand Sector.* Washington, American Geophysical Union, 165–96.) [Antarctic Research Series, Vol. 19.]

Weissel, J. K., and Hayes, D. E. 1977. Evolution of the Tasman Sea reappraised. *Earth planet. Sci. Lett., 36,* No. 1, 77–84.

Weissel, J. K., Hayes, D. E., and Herron, E. M. 1977. Plate tectonics synthesis: the displacements between Australia, New Zealand and Antarctica since the Late Cretaceous. *Mar. Geol., 25,* 231–77.

Wellman, H. W. 1959. Geological interpretation of airborne magnetometer observations from Nelson to Waikato River, New Zealand. *Geol. Mag., 96,* No. 2, 118–24.

Wellman, H. W. 1973. The Stokes Magnetic Anomaly. *Geol. Mag., 110,* No. 5, 419–29.

Wellman, H. W. 1975. New Zealand 60 m.y. ago. *Bull. Aust. Soc. Explor. Geophys., 6,* 55–56.

Wellman, H. W. 1976. Sea floor reconstructions for the last 85 million years and the bending and faulting of the New Zealand region. *Abstr. 25th Int. Geol. Cong., Sydney, 1976,* 106–7.

Wilbanks, J. R. 1972. Geology of the Fosdick Mountains, Marie Byrd Land. (*In* Adie, R. J., *ed. Antarctic Geology and Geophysics.* Oslo, Universitetsforlaget, 277–84.)

Wilson, G. J. 1967. Some new species of Lower Tertiary dinoflagellates from McMurdo Sound, Antarctica. *N.Z. Jnl Botany, 5,* No. 1, 57–83.

Wodzicki, A. 1974. Geology of the pre-Cenozoic basement of the Taranaki–Cook Strait–Westland area, New Zealand, based on recent drillhole data. *N.Z. Jl Geol. Geophys., 17,* No. 4, 747–58.

Wood, B. L. 1960. Sheet 27, Fiord. *Geological Map of New Zealand, 1:250,000.* Wellington, New Zealand Department of Scientific and Industrial Research.

Wood, B. L. 1962. Sheet 22, Wakatipu. *Geological Map of New Zealand, 1:250,000.* (1st ed.) Wellington, New Zealand Department of Scientific and Industrial Research.

Wood, B. L. 1978. Chapter 2: Fiordland. (*In* Suggate, R. P., *ed. The Geology of New Zealand. 1.* Wellington, Government Printer, 100–110.)

Woodward, D. J., and Hatherton, T. 1975. Magnetic anomalies over southern New Zealand. *N.Z. Jl Geol. Geophys., 18,* No. 1, 65–82.

Woollard, G. P. 1962. Crustal structure in Antarctica. (*In* Wexler, H., Rubin, M. J., and Caskey, J. E., *eds. Antarctic Research: The Matthew Fontaine Maury Memorial Symposium.* Washington, American Geophysical Union, 53–73.) [Geophysical Monograph No. 7.]

3

West Antarctica and New Zealand: A Geologic Test of the Model of Continental Split

H. R. KATZ
New Zealand Geological Survey, Department of Scientific and Industrial Research, Lower Hutt, New Zealand

Abstract. Geologic and geophysical evidence indicates that a New Zealand–West Antarctica split, as has been postulated from oceanic magnetic anomalies, cannot be explained by usual models of continental separation with formation of Atlantic-type passive margins. Several objections are presented:

1. Prebreakup orogenic and plutonic histories are very different on the two sides.
2. While typical prerift and/or rift-valley sequences began forming on the New Zealand side about 30 m.y. before breakup, this early rifting occurred widely across the whole New Zealand block and not just near the area of later breakup. No prebreakup rift sequences are known on the Antarctic side, where uplift and erosion seem to have prevailed.
3. No postbreakup sedimentary wedge occurs on the New Zealand margin, but a fully developed one is found along Antarctica.
4. The early subsidence of the large Campbell Plateau area, about synchronous with continental breakup, is anomalous both in its timing and scale. In contrast, the Antarctic side seems to have undergone continuous uplift and erosion.
5. The close association of continental separation, in time and space, with orogenies that include granite intrusion and high *T/P* ratio metamorphism does not fit any known model.

Thus, if continental separation has indeed occurred, as suggested by the magnetic anomalies, a revised model is needed. The different patterns of tectonic evolution may be due to the inequality of the partners: a relatively small area—which became the New Zealand microcontinent—was split off from a much larger and more stable continental mass along a preexisting line of weakness which was the original junction between two different tectonic provinces. This may explain why the New Zealand–Antarctica split cannot readily be compared with continental separation elsewhere.

Recent plate tectonic models juxtapose New Zealand and Antarctica up to about Santonian time (Griffiths, 1971, 1974; Christoffel and Falconer, 1973; Hayes and Ringis, 1973; Molnar and others, 1975). In particular, the Campbell Plateau and Chatham Islands are placed against the eastern Ross Sea and Marie Byrd Land of West Antarctica. A sequence of magnetic anomalies, up to number 36, which is immediately adjacent to the continental margin of the Campbell Plateau (Christoffel and Falconer, 1972), suggests that breakup began just prior to 81 m.y. ago.

If these reconstructions are correct, the trends and patterns of the pre-Senonian geology should be compatible in the two regions whose original positions were close together. Also, Late Cretaceous rifting and breakup should have influenced the tectonic evolution on both sides during and after that time. The following review and discussion examines to what degree the known geology supports these reconstructions, and highlights some problems inherent in a case that is atypical and therfore does not fit the usual model of passive margins. The New Zealand–Antarctica continental split may in fact be unique.

Geology of West Antarctica

Pre-Cenozoic bedrock in the Ross Sea–Marie Byrd Land–Ellsworth Land region

While relatively good exposures exist in the Ford Ranges and on the Edward VII Peninsula adjacent to the Ross Sea, outcrops along the Ruppert and Hobbs Coasts, in the Kohler Range, along the Eights Coast, and on Thurston Island (Figure 3.1) are few and discontinuous. There is little doubt, however, that the oldest formation in the entire region is an extremely thick and intensely folded sequence of metasedimentary rocks, mainly graywacke, quartzite, slate, and phyllite (Wade and Wilbanks, 1972; Lopatin and Orlenko, 1972), which reportedly contains poorly preserved *Acritarcha* assigned a Riphean age by Iltchenko (1972). The rocks have yielded metamorphic dates ranging between 445 and 475 m.y. (Krylov and

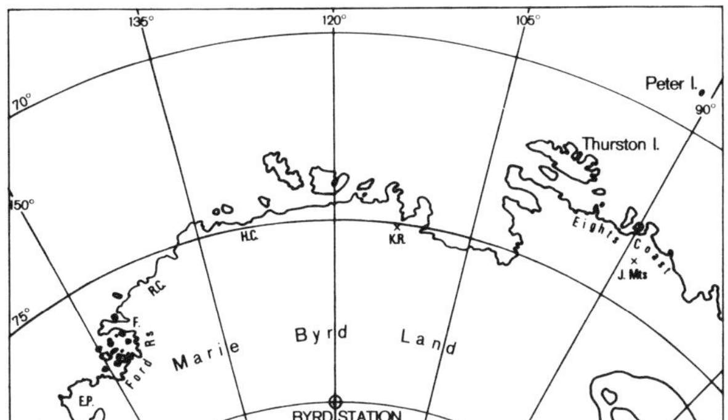

Figure 3.1. Locality map for Marie Byrd Land and Eights Coast. E.P.: Edward VII Peninsula. F: Fosdick Mountains. R.C.: Ruppert Coast. H.C.: Hobbs Coast. K.R.: Kohler Range. J. Mts: Jones Mountains.

others, 1970). Because of the sparsity of data, however, the age of the sequence and its time of deformation are still open to question (see Wade, 1969; and Craddock, 1972, who regard them as possibly upper Paleozoic). It appears more probable, though, that these beds correlate with the upper Precambrian to lower Paleozoic formations in the Transantarctic Mountains. This is supported by the occurrence of marble in the metamorphic basement in the eastern Ross Sea, which is similar to the lower Paleozoic Koettlitz Marble of southern Victoria Land (Ford and Barrett, 1975). Thus, Marie Byrd Land may be part of the Ross orogen (Elliott, 1975) of Ordovician age.

In many places the metasedimentary rocks are overlain by, or interbedded with metavolcanic rocks. Although the correlation of these metavolcanic rocks is critical to understanding the tectonic evolution, their age and relationship remain uncertain. They have been regarded in places as Paleozoic (Craddock, in Elliott, 1975; Kilmov, 1967; Lopatin and Orlenko, 1972; Wade and Wilbanks, 1972) and Mesozoic (Craddock, 1970*b*, 1972; LeMasurier and Wade, 1976), but it is unlikely that they are as young as Late Cretaceous. It appears doubtful that they correlate, as held by LeMasurier and Wade (1976), with a rhyodacite ash flow and a vitrophyric rhyolite at Mount Petras and the USAS Escarpment, respectively, which are not metamorphosed and have given ages of 80.8 ± 4 m.y. and 87.6 ± 4 m.y.

Plutonic rocks are widespread and are of both middle-late Paleozoic and Mesozoic in age. The intermediate to acid Paleozoic plutonism (Craddock, 1970*a*; Halpern, 1968; Lopatin and Orlenko, 1972; Wade and Wilbanks, 1972) may represent the Borchgrevink orogeny established in Northern Victoria Land (Craddock, 1970*c*), while one single granite of Late Triassic age in the Jones Mountains (Figure 3.1) (Rutford and others, 1972) is the only evidence that could hint at an extension into this region of the early Mesozoic Gondwanide orogeny, which is known from the Antarctic Peninsula (Katz, 1973; Elliott, 1975).

Extensive plutonism occurred in late Mesozoic (Jurassic to mid-Cretaceous) time (Boudette and others, 1966; Craddock, 1970*a*; Drake and others, 1964; Halpern, 1968, 1972; Lopatin and Orlenko, 1972; Munizaga, 1972; Rutford and others, 1972; Wade, 1969, 1972). In western Marie Byrd Land the ages of adamellites and granodiorites concentrate around 90–105 m.y. (Albian to Turonian). Deep-seated deformation with syntectonic high-temperature metamorphism, migmatization, and granitization in the Fosdick Mountains (Wilbanks, 1972) is thought to be the same age (Halpern, 1972). Here, and in the Kohler Range and at Mount Petras in eastern Marie Byrd Land, this younger orogenic phase manifests itself by west-northwest structural axes superimposed on, and partly obliterating, older northeast-trending axes in the basement rocks. Such a Cretaceous orogenic event is of particular interest, as it forms part of a regional circum-Pacific belt that also occurs in South America and New Zealand. In West Antarctica, perhaps misleadingly, it has been termed the "Andean orogeny" (Elliott, 1975; see Katz, 1973).

In fact, the so-called orogenies in Marie Byrd Land of middle Paleozoic to Cretaceous age are represented only by thermal events, which includes the production of significant volumes of calc-alkaline magmas. While these may suggest active subduction (Elliott, 1975), no sedimentary beds definitely related to such orogenies have been located so far. Unlike South America and New Zealand, Marie Byrd Land and Ellsworth Land west of about long. 80° W seem to have no sedimentary record of late Paleozoic and Mesozoic time. Hence, the regional implication and tectonic significance of these orogenies are poorly known. Nor is their timing well defined, as the intrusion or cooling of the magmas may not be contemporaneous with tectonic deformation.

Cenozoic rocks of Marie Byrd Land and adjacent areas

A very low relief erosion surface (Craddock and others, 1964; LeMasurier, 1972) separates basement rocks from the products of Cenozoic volcanism. This highly alkaline, bimodal, basalt-trachyte suite of rocks (LeMasurier, 1972) represents a profound change in geochemical character from the older calc-alkaline volcanic rocks in West Antarctica (reference in Katz, 1973). Numerous radiometric dates on these volcanic rocks (Craddock and others, 1964; Doumani 1964; LeMasurier, 1972; Rutford and others, 1972; Halpern and González-Ferrán, 1973; LeMasurier and Wade, 1976), give ages consistently younger than 13–15 m.y., i.e., middle Miocene to Pleistocene. Only three Oligocene to early Miocene ages are known (LeMasurier and Wade, 1976), from the basal volcanic unit of mainly palagonite breccias. Yet LeMasurier and Wade (1976) suggest that Cenozoic volcanism began even earlier, and they interpret the Eocene Dorrel Rock gabbro (lat. 75° 40′ S, long. 111° 30′ W), because of its alkaline character, as a subvolcanic pluton associated with the volcanism; the Dorrel Rock gabbro, however, may represent a late phase of Mesozoic magmatism (Lopatin and Orlenko, 1972).

Widespread alkaline igneous activity is typically associated with an extensional or rift tectonic environment. Faults which result in different elevations of the pre-Cenozoic erosion surface, and in migration of volcanic eruption centers along fault lines, have been reported (Doumani, 1964; LeMasurier and Wade, 1976). Recent vertical adjustments are inferred from elevated, subaquatic Pleistocene deposits (hyaloclastites; LeMasurier, 1972), even allowing for considerable changes in ice sheet level. Seismic investigations show ice-filled grabens parallel to the volcanic ranges (Bentley and others, 1960; Bentley, 1965). The isostatically adjusted surface topography below the ice cover (Bentley, 1972) indicates many areas deeply depressed below sea level and fragmentation of West Antarctica into an island archipelago separated by narrow channels and extensive marine basins (Figure 3.2). The Byrd Basin in the very center of the region, probably connecting the Ross Sea with the Weddell Sea, descends to more than 2500 m below sea level. Crustal attenuation and regional breakup are indicated by seismic wave dispersion studies which show that West Antarctica is underlain by a crust 10 km thinner than East Antarctica and likely represents an assemblage of smaller crustal units (Adams, 1972). This is supported by the probable derivation of the Cenozoic volcanic rocks from the upper mantle, as suggested by their strontium isotope composition (Halpern and González-Ferrán, 1973).

The age of extension and crustal breakup in West Antarctica is mainly late Cenozoic but, as inferred from the volcanic record, possibly began in Oligocene time. This is consistent with the sedimentary record in nearby basins. Evidence from seismic profiles and drilling suggests that

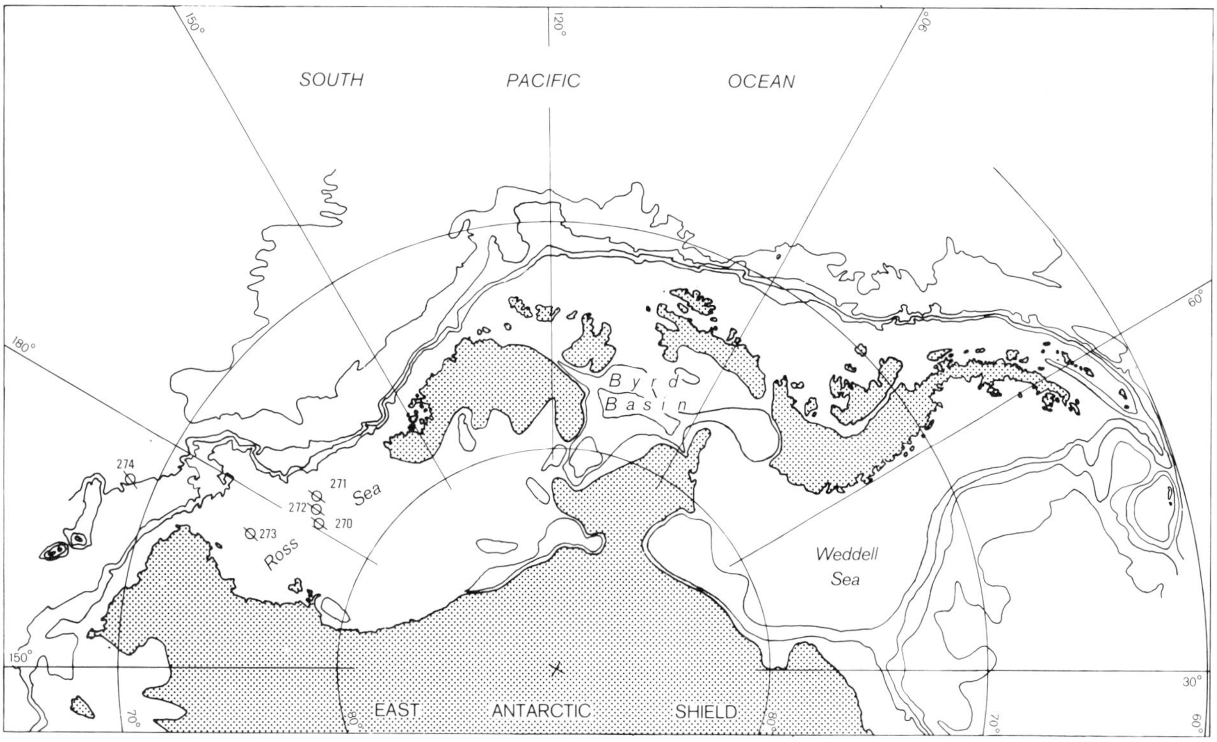

Figure 3.2. Island archipelago of West Antarctica. Land areas (gray) are shown isostatically adjusted after removal of ice (after Bentley, 1972). Bathymetric contours are at 1000 m intervals. Numbers 270–274 are Deep Sea Drilling Project (DSDP) well sites.

the sediments in the eastern basin of the Ross Sea, as much as 4 km thick, are Oligocene and younger (Hayes and others, 1975; Hayes and Davey, 1975); older beds with Paleogene and Late Cretaceous fossils, found as glacial erratics in the McMurdo Sound region (Wilson, 1968; Webb and Neall, 1972), may be related to an older rift structure in the western Ross Sea (Hayes and Davey, 1975). It is separated from the eastern basin by a series of shallow basement highs along the 180th meridian.

In the Byrd Basin (Figure 3.2) layered rocks as much as 3 km thick are inferred from their low seismic velocity to be Cenozoic sedimentary rocks, possibly interbedded with volcanic rocks (Bentley and Clough, 1972). To the northeast of Marie Byrd Land, seismic surveys and drilling in the Bellingshausen Basin have shown very thin Cretaceous and lower Paleocene pelagic claystone overlain by thick, mainly upper Cenozoic, silty claystone, indicating an abrupt increase in terrigenous detritus derived from Antarctica, probably originally transported down the continental slope by turbidity currents (Hollister and others, 1974; Vanney and Johnson, 1976).

Thus widespread uplift and more vigorous erosion in West Antarctica is suggested since Oligocene time but particularly in Miocene through Pleistocene time when voluminous debris was shifted out to sea by a large continental ice sheet. Both uplift and contemporaneous Cenozoic alkaline volcanism are probably related to active block tectonics in the course of continental fragmentation.

Geology of the Campbell Plateau–Chatham Rise

Marine seismic investigations (Davey, 1977) show a consistent, strong reflecting horizon called reflector *X* across most of the region (Figure 3.3). In the southwest it correlates with a middle Paleocene horizon in Deep Sea Drilling Project (DSDP) hole 277, while at DSDP site 275 in the southeast, it corresponds to the top of Upper Cretaceous sedimentary rocks (Kennett and others, 1975). In the Great South Basin, toward Snares Depression, it is older than dredged phosphorite samples of late Eocene to early Miocene age (Summerhayes, 1969). Thus reflector *X*, although possibly slightly time-transgressing on a regional scale, lies near the base of the Tertiary sequence.

In several areas of the Campbell Plateau, profiler records indicate sedimentary beds older than reflector *X*. Basement in the Campbell Basin (Figure 3.3) is as much as 1.2 seconds beneath it, while in the graben-type Pukaki Trough, pre-*X* sedimentary layers are at least 1.2 seconds thick. They onlap the Pukaki Rise north of the trough, across which younger sediments show little or no thinning. Also in the Great South Basin, which is filled with more than 2000 m of Cenozoic sedimentary beds, there appears to be a considerable thickness of pre-*X* sedimentary rocks. Cretaceous strata were drilled into for 750 m on its northern flank (Endeavour-1; Wildig and Sweetman, 1971) (Figure 3.3), reaching down into Santonian–Campanian beds (Harris, 1971); basement was not reached. In fact, onshore geology along the northwestern flank of the Great South Basin suggests that a thick Cretaceous sequence extends into the subsurface offshore (Katz, 1974*a*, 1976). Cretaceous rocks also occur on Campbell Island, which lies on an old structural high; peneplained basement schist is here overlain by Maestrichtian quartz-pebble conglomerate and carbonaceous mudstone (Garden Cove Formation) (Oliver and others, 1950; Wilson, 1972). The mainly fine-grained terrigenous sediments in DSDP hole 275, where basement was not reached, are upper Campanian (Kennett and others, 1975).

The basement on the Campbell Plateau consists of Mesozoic, mainly Cretaceous, granite, as well as graywacke and schist of probably late Paleozoic to early Meso-

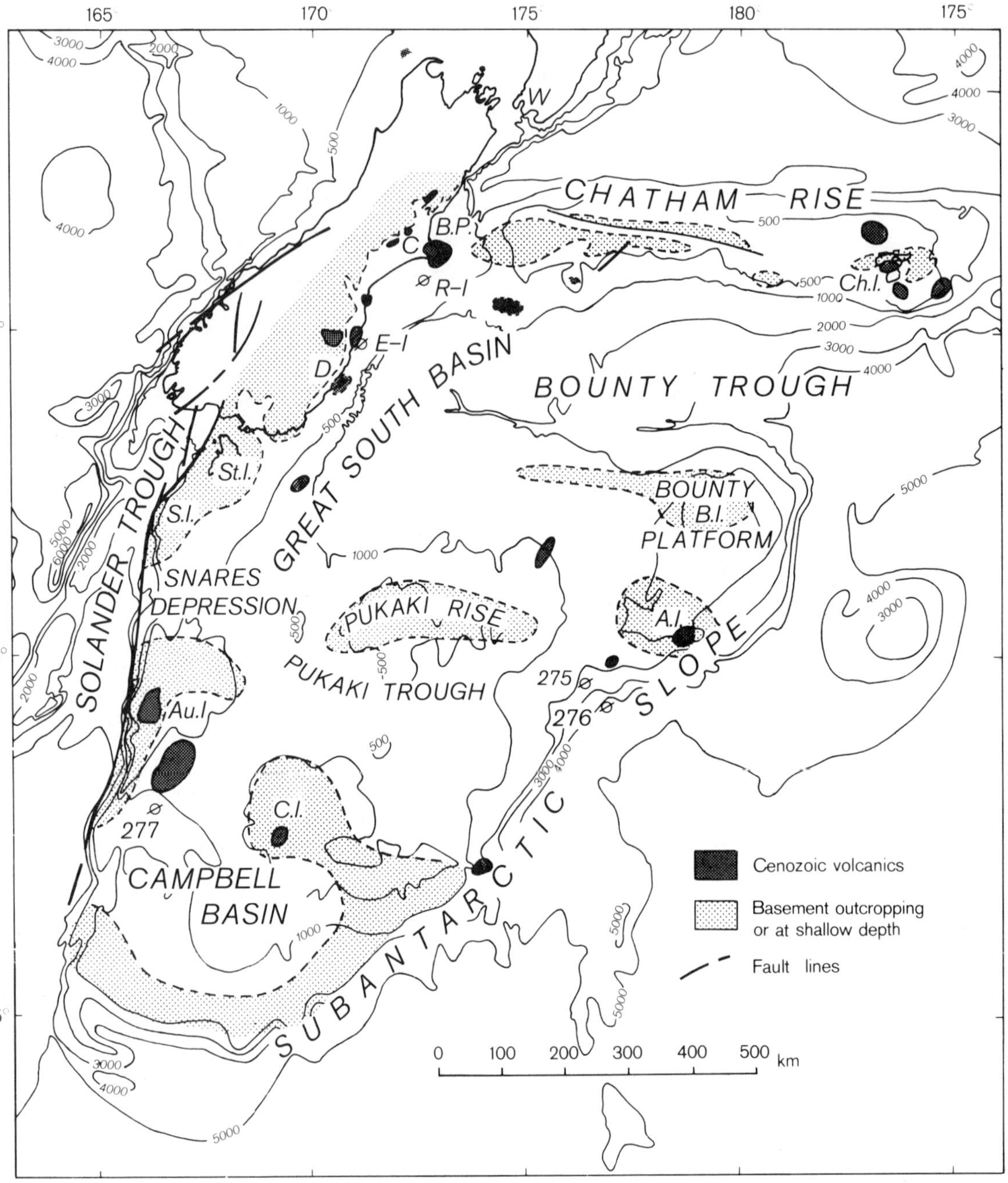

Figure 3.3. Locality map of Campbell Plateau–Chatham Rise region, New Zealand. W, C, and D: Cities of Wellington, Christchurch, and Dunedin. B.P.: Banks Peninsula. C.I.: Campbell Island. St.I: Stewart Island. Ch.I.: Chatham Islands. S.I.: Snares Islands. B.I.: Bounty Islands. Au.I: Auckland Islands. A.I.: Antipodes Islands. E-1, R-1: well sites Endeavour-1 and Resolution-1. Numbers 275–277: DSDP well sites. Bathymetric contours in meters.

zoic age, which are assumed to be part of the New Zealand geosyncline (Fleming, 1970). Vague petrographic similarities between Bounty Platform graywackes and lower Paleozoic Greenland Group rocks in western South Island, New Zealand (Cullen, 1975), do not prove the presence of older rocks. The Bounty Island granites are the only instrusive rocks dated as Late Triassic (189 m.y.) (Wasserburg and others, 1963), and this area may indeed be a special situation.

Uplift, and erosion with extensive peneplanation, occurred during Early to early Late Cretaceous time. In Senonian time the region became gradually depressed and submerged, with sediments accumulating in some basins to very great thicknesses; marine deposition had already begun locally by Aptian-Albian time (see below). Seismic evidence suggests that the main structural configuration of the Campbell Plateau (Figure 3.3) was in existence by Senonian time; also in existence were the Bounty Trough and Chatham Rise, where seismic horizon *A*, which is at the base of the Tertiary sequence (Austin and others, 1973), roughly coincides with reflector *X* of the Campbell Plateau (Davey, 1977). Thus, the conglomerates and sandstones of the Cretaceous Waihere Bay Group on Pitt Island—one of the Chatham Islands (Figure 3.3) (Hay and others, 1970)—cannot be derived from the southwest as suggested by Austin and others (1973) and by Grindley (1976); indeed,

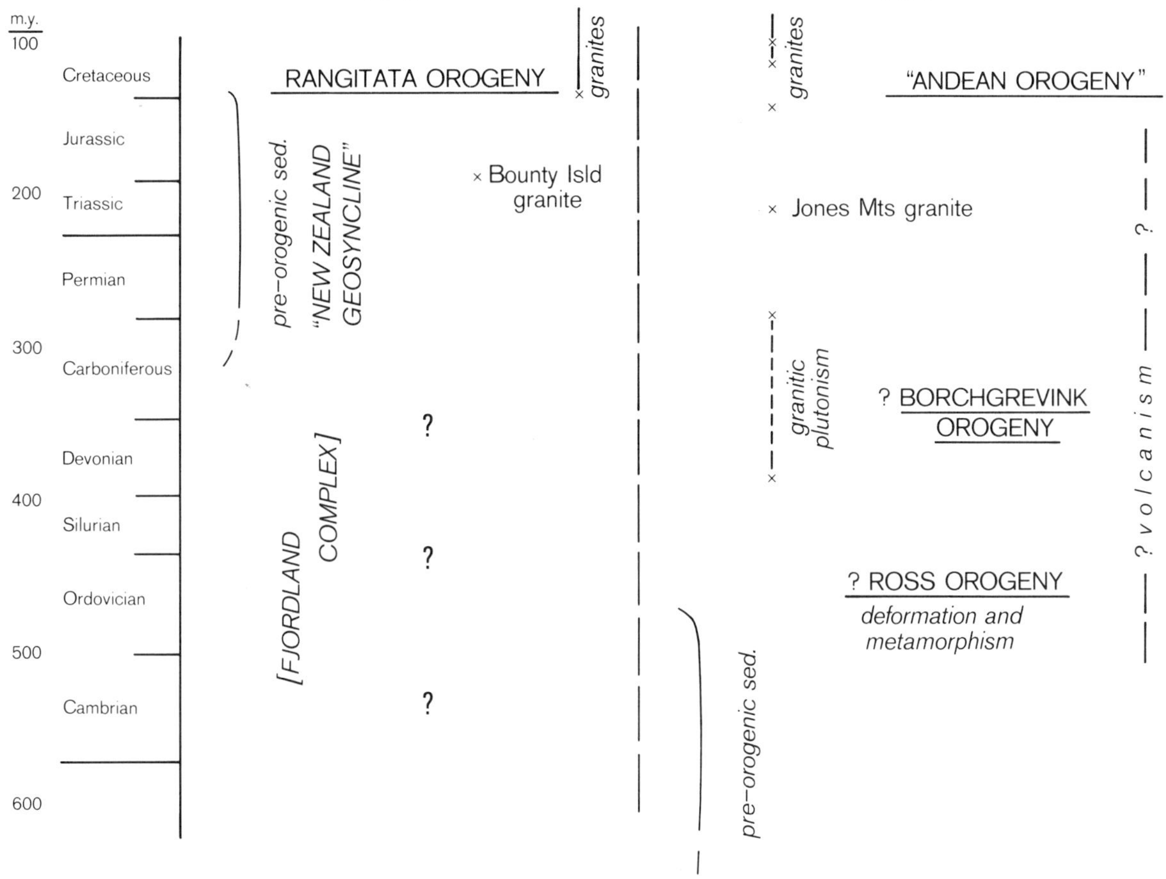

Figure 3.4. Paleozoic to Mesozoic geologic histories in the Campbell Plateau–New Zealand (N.Z.) and Marie Byrd Land (M.B.) regions.

recent sedimentologic studies (M. G. Laird, personal communication) indicate a land area and sediment source in the north to northeast. The Waihere Bay Group as a whole represents a continuous, mainly shallow-marine sequence from Aptian–Albian through Senonian time (Mildenhall, 1977; Speden, 1976; Wilson, 1976; H. R. Katz, M. G. Laird, and others, unpublished data), indicating a very quiet tectonic environment during this period.

Tectonism in general is very restricted in the whole region. In the Chatham Islands, late Senonian volcanism produced alkaline olivine basalt (Hay and others, 1970) and was followed by block faulting and erosion; the unconformably overlying lower Tertiary sequence is but slightly deformed. Elsewhere on the Campbell Plateau, subsidence generally continued from Senonian through early Tertiary time, a notable last episode of deformation occurring at the end of Oligocene time (Katz, 1974*b*). A regional unconformity exists near the Oligocene–Miocene boundary, with a varying thickness of beds missing on either side, as seen in the offshore wells Endeavour-1 (Wildig and Sweetman, 1971) and Resolution-1 (Milne and others, 1975) and in DSDP hole 277 (Kennett and others, 1975). This unconformity may be partly due to a change in oceanic current regime as postulated by Kennett and others (1972), but in this region it is also definitely controlled tectonically; evidence for deformation is seen on Campbell Island (Oliver and others, 1950) and in the Chatham Islands and Chatham Rise where Oligocene beds are mostly absent (Hay and others, 1970; Katz 1974*b*, 1976).

Apart from the Late Cretaceous volcanism in the Chatham Islands, there was widespread Cenozoic volcanic activity, mainly of alkali basalt, throughout the region. Along the northwestern margin between Dunedin and Banks Peninsula, there is evidence of basaltic volcanism in Paleocene time (Wildig and Sweetman, 1971), Eocene and Oligocene time, (Gage, 1957), middle to late Miocene time (McDougall and Coombs, 1973; Stipp and McDougall, 1968) and early Pleistocene time (Gair and Rickwood, 1965). Neogene volcanics are known from southwest of the Campbell Plateau (Figure 3.3) (Summerhayes, 1969; Denison and Coombs, 1977), while Antipodes Island on its eastern margin consists of alkali basalt of probably Quaternary age (Cullen, 1969). Cenozoic volcanism on the Chatham Islands occurred in two separate phases, both limburgitic in character, with radiometric ages of 35–50 m.y. and 2–6 m.y. (Grindley, 1976). These periods of volcanism correlate with north-south crustal tension and accompanying faulting along the Chatham Rise (Katz, 1976), which resulted in its northward down-bending and eventual burial beneath flat-lying Plio-Pleistocene sediments (Katz, 1974*b*).

Comparison of tectonic evolutions

Before the time of supposed rifting and breakup

In the two regions, the ages and nature of basement rocks are to a great extent very different, as are the timing and nature of tectonic events (Figure 3.4). The only rocks comparable in age are the Triassic granites in the Jones

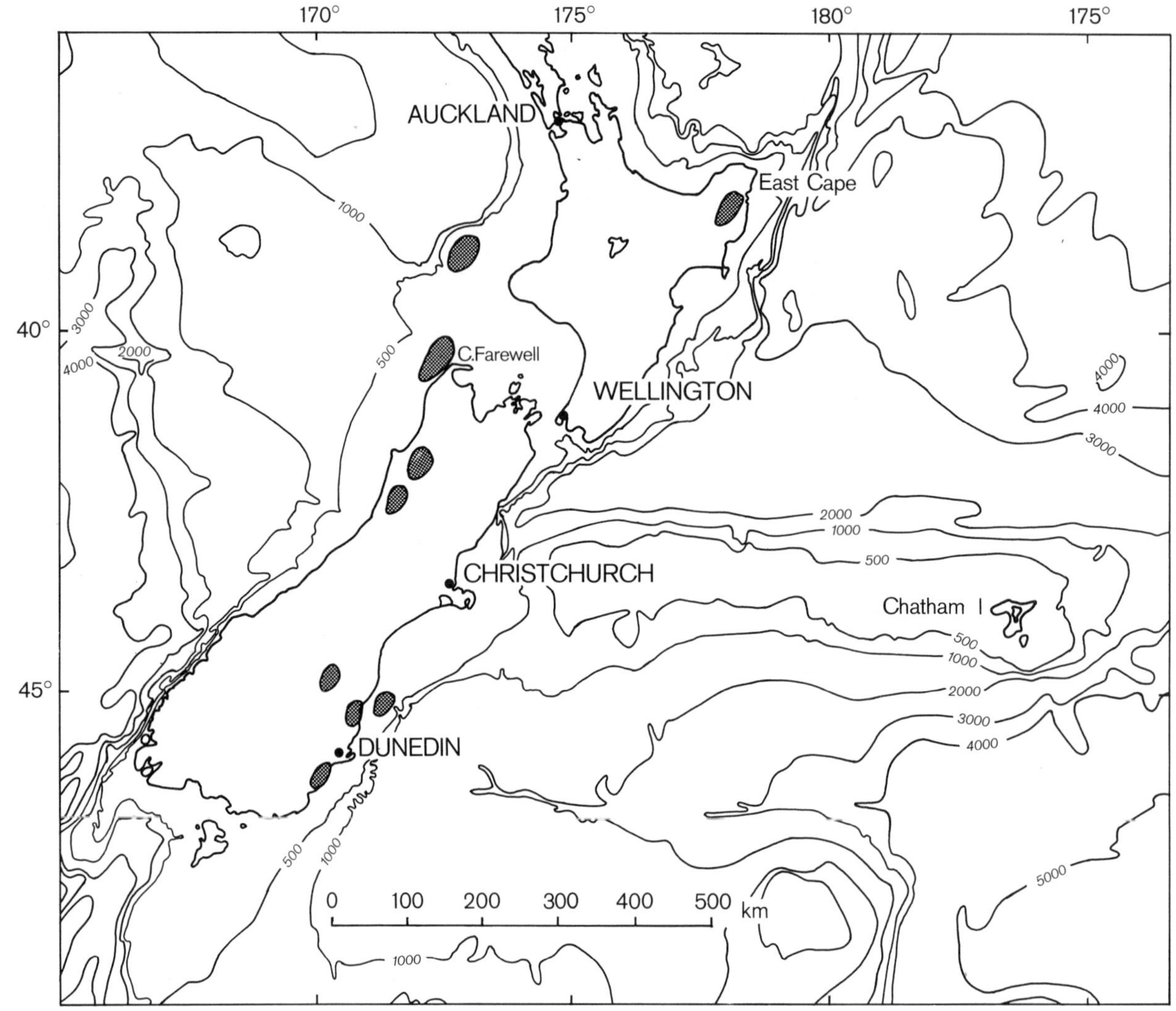

Figure 3.5. Location of middle to Late Cretaceous rift deposits in New Zealand (cross-hatched).

Mountains and Bounty Islands and, in particular, the Cretaceous granodiorite plutons which are widespread both in Marie Byrd Land and along the western margin of the Campbell Plateau (Denison and Coombs, 1977). This latter plutonism, however, is a typical circum-Pacific feature occurring independently in widely separated areas, and it is therefore not considered of much significance for possible reconstructions.

On the other hand, it may be that geologic elements and features in two regions separated in the course of continental split do not match simply because the *locus* of their separation may have followed an original line of weakness that formed the junction between different tectonic provinces; but in this case, the basis for a geologic test of proposed reconstructions is very tenuous. The classical models of passive margins, such as on both sides of the Atlantic, have been built on the premise that the prebreakup basements do in fact match.

In principle, then, it is not so much the matching of individual geologic features but rather the basic trend and patterns of tectonic evolution, which here may serve as a guide to reconstruction. In this respect, a recent solution has been proposed by way of an "across-strike facies reconstruction" (LeMasurier and Wade, 1976), where the interrelationship of elements that supposedly occurred parallel to the line of later breakup is investigated. Thus the New Zealand geosyncline is placed along the present coast of Marie Byrd Land, which is interpreted as the associated magmatic arc that was the source of the volcanogenic arc-trench gap assemblage of the New Zealand geosyncline (Landis and Bishop, 1972); this presupposes their being contemporaneous. However, there is no known volcanomagmatic activity in Marie Byrd Land concurrent with Hokonui rock deposition (Permian to Jurassic). LeMasurier and Wade (1976) furthermore postulate a converging plate boundary and associated subduction zone along coastal West Antarctica, thereby reconstructing a paired metamorphic belt, with the New Zealand geosynclinal assemblage forming a low *T/P* ratio outer belt with respect to the high *T/P* ratio "Andean intrusive suite" of Marie Byrd Land. In their model this subduction would have persisted until after 80 m.y. ago; obviously there is a conflict here with the continental separation model, which puts the West Antarctica–New Zealand split at precisely the same time.

Furthermore, the traditional interpretation and paleotectonic significance of the two facies belts of the New Zealand geosyncline—the Hokonui (west) and Torlesse (east) assemblages (Fleming, 1970; Landis and Bishop, 1972)—are debatable and have recently come under strong

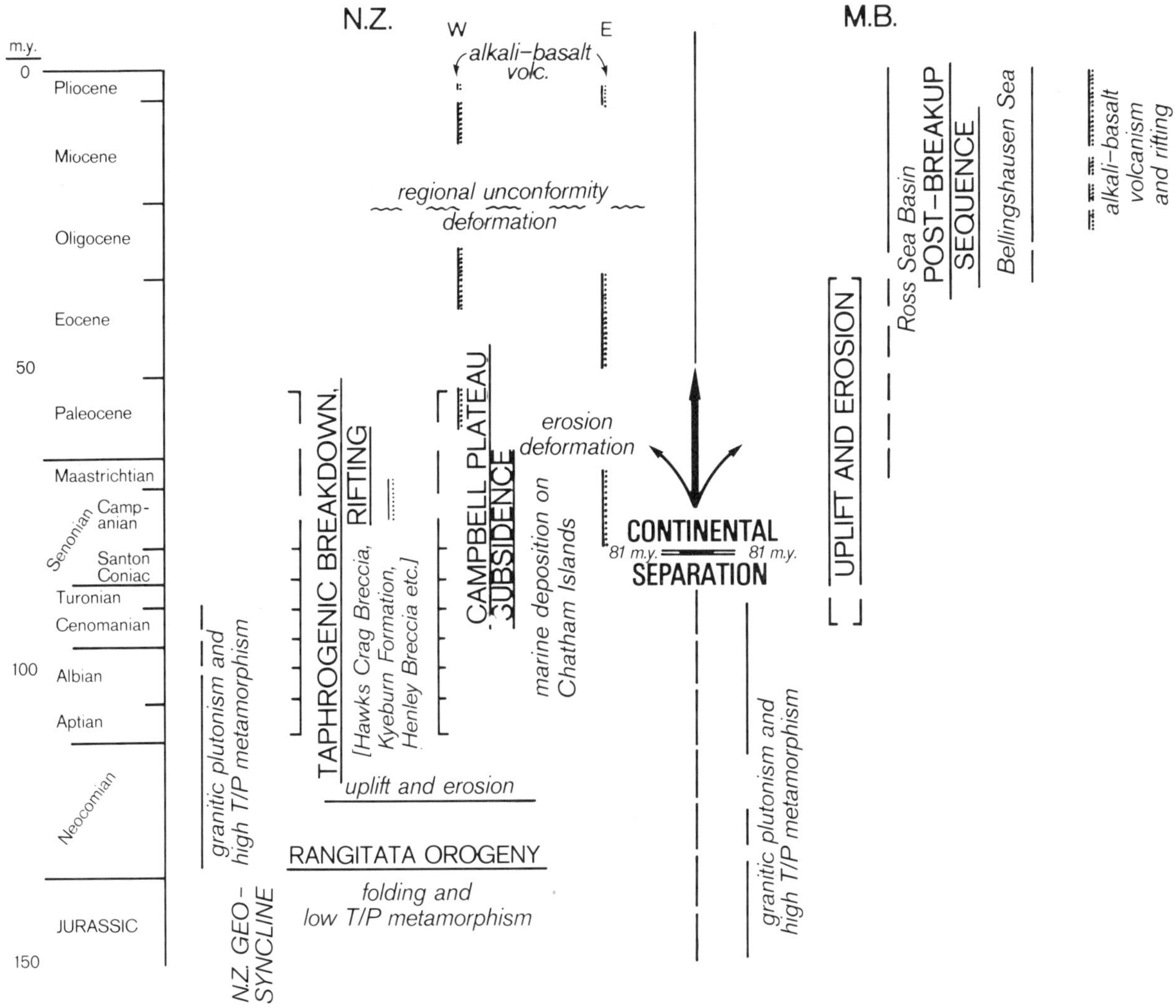

Figure 3.6. Mesozoic to Cenozoic geologic histories in the Campbell Plateau-New Zealand (N.Z.) and Marie Byrd Land (M.B.) regions.

attack (Bradshaw and Andrews, 1973; Austin, 1975). The two belts may actually have formed along two different and opposing continental margins which collided during the Early Cretaceous Rangitata orogeny and were welded together (Blake and others, 1974; Coombs and others, 1976). This model implies that not the Hokonui rocks, but rather the nonvolcanogenic Torlesse rocks, the source of which clearly was a crystalline-plutonic terrane, were derived from Marie Byrd Land. (It also implies, of course, that Marie Byrd Land was far distant and separated by an ocean from New Zealand in the Triassic and Jurassic—thus defying the model of Gondwanaland. In quick succession, then, the two blocks or parts of them were welded together in Early Cretaceous time and separated again in Late Cretaceous time.)

Thus on various counts, the reconstruction proposed by LeMasurier and Wade (1976), which is based on their view that the sedimentary, metamorphic, and volcanic facies in New Zealand are complementary to those in Marie Byrd Land, is considered improbable.

The Rangitata orogeny in New Zealand, however, has wider implications with regard to the tectonic evolution before continental breakup. Here the Cretaceous granites, which are considered to be a late-stage manifestation of this orogeny (Landis and Coombs, 1967), have a K-Ar age of about 93 m.y. (Denison and Coombs, 1977), which indicates that final cooling and uplift occurred in Turonian time. Seismic sections suggest that extensive peneplanation and deroofing of these granites were accomplished by Senonian time when regional foundering of the Campbell Plateau and formation of extensive sedimentary troughs and basins took place. Likewise in many other areas across the New Zealand region, rapid subsidence of mainly graben-type sedimentary basins occurred during Late Cretaceous time, a process which in some places had already begun in Aptian time (about 110 m.y. ago) (Figures 3.5 and 3.6). In South Island the Henley Breccia, Kyeburn Formation, Hawks Crag Breccia, and Horse Range Breccia are all early post-Rangitata sedimentary deposits formed in active graben structures (Beck and Nathan, 1978; McKellar, 1978). A similar paleotectonic environment is indicated by locally thick quartzose coal measures, as at Kaitangata (Katz, 1976), at Shag Point, and in the bottom of Endeavour-1 well (Wildig and Sweetman, 1971). In northwestern South Island the Upper Cretaceous Paparoa coal measures are locally controlled by fault-bounded grabens associated with extrusive alkaline olivine basalts (e.g., Morgan Formation; Gage, 1952) that are mineralogically similar to those in the Upper Cretaceous of the Marlborough region in northeastern South Island

(Challis, 1964). Local, abrupt subsidence is also implied by the Pakawau coal measures which, particularly in some offshore wells (e.g., Cook-1; Van Oyen, 1970), attain very great thicknesses. In northeastern North Island, the breccias and conglomerates of the Aptian to Albian Koranga and TeWera Sandstones at the base of the Cretaceous sequence (Speden, 1975) may also represent deposits shed into local grabens that were developing in the Torlesse terrane.

Thus widespread rifting and taphrogenic breakdown followed soon after the Rangitata orogeny (Figure 3.6). In fact, in some areas rifting and subsidence began while orogenic uplift was still continuing nearby. (Fleming, 1970, calls these sediments "late synorogenic"; they are, however, clearly postorogenic at any one place.) In terms of a continental breakup model (Falvey, 1974), these formations can also be interpreted as prerift or rift-valley sequences, which formed immediately before final breakup. But while their time span, which here covers the last 30 m.y. prior to breakup, corresponds well with the model of an Atlantic-type continental separation, their geographic occurrence is not, in the New Zealand region, confined to the zone of later breakup as in the normal model. On the contrary (and contrary to the statement by Grindley, 1976), the middle to Late Cretaceous Waihere Bay Group on the Chatham Islands (Hay and others, 1970), which is the closest to the zone of supposed rifting and pull-apart from Antarctica, does, upon recent detailed investigations, in no way indicate tectonic control with rifting and graben formation (H. R. Katz, M. G. Laird, and others, unpublished data). Such is suggested only by the occurrence of alkali basalt volcanics at the very top of the sedimentary sequence which were erupted at the same time as and immediately prior to actual breakup (70–80 m.y. according to Grindley, 1976; their upper age limit is not defined and may be still younger). Similar basalts of comparable age, however, also occur in the New Zealand region in areas far removed from the site of continental split.

After supposed rifting and breakup

Passive continental margins of the classical Atlantic-type include a postbreakup, prograding sedimentary sequence which forms the new shelf and continental slope. No such sequence can be recognized along the Campbell Plateau–Chatham margin. Strong lateral shearing during the early stages of separation (Houtz and Markl, 1972) could have initially, but not completely, inhibited the formation of a postbreakup sequence. Alternatively, it may have been displaced by later faulting along the "Antipodes Fracture Zone" (Cullen, 1967); the latter's existence, however, has been postulated solely on morphologic grounds. Katz (1974*b*) has argued that the lack of a continental margin clastic sedimentary wedge may be due mainly to the great distance from any source area and the action of a strong northeast-flowing western boundary current.

In West Antarctica, where the Ross Basin subsided during Oligocene time (Hayes and Davey, 1975), the continental slope and rise are of a very gentle gradient and are formed by a wedge of sediments which are the seaward extension of the sedimentary basin underlying the Ross Sea shelf. Although seismic data are sparse (Houtz and Davey, 1973), it appears that this wedge can be considered a postbreakup sequence related to a Late Cretaceous continental rift margin. Davey and Houtz (1977) examined the possibility that the Ross Basin was originally continuous with the Campbell Basin, the two having been rifted apart by continental separation, but they correctly concluded that there was no evidence to support this. Indeed, the Campbell Basin formed during Senonian time and terminated its basin history by Oligocene time, whereas the Ross Basin formed during Oligocene time and lasted throughout the Neogene, with only indirect evidence of any older history. Thus, whereas the Campbell Basin formed during the time of supposed rifting and breakup, the Ross Basin formed later and is a postbreakup basin (Falvey, 1974) with a prograding sequence extending across the shelf edge and continental slope. A similar sequence, also Oligocene–Miocene to Plio-Pleistocene in age, underlies the continental slope and rise to the east in the Bellingshausen Sea (Hollister and others, 1974; Vanney and Johnson, 1976) and probably also in the Amundsen Sea. No analogous sequence is known on the Campbell margin, where near DSDP hole 276 (Kennett and others, 1975) oceanic basement with its thin sediment cover is flat right to the base of the continental rise and slope, which itself is practically devoid of sediment (Houtz and Markl, 1972).

On the Campbell Plateau and Chatham Rise, intermittent Neogene sedimentation produced thin and incomplete sequences (Katz, 1974*b*, 1976). Mainly on the basis of structure and alkali-basalt volcanism, which are similar in both West Antarctica and Campbell Plateau, widespread extensional tectonics is inferred for both regions. Crustal structure is clearly continental, but crustal thickness is less than in adjacent "fully continental" areas in both cases.

Conclusions

The geologic histories of Marie Byrd Land and the Campbell Plateau–Chatham Rise region in many ways do not follow the patterns of better known models of continental separation.

The following observations are made:

1. *The age and nature of prebreakup basement rocks and the prebreakup orogenic and plutonic histories are basically different in the two regions.* This may be due to the existence, in a marginal area or appendix of the original supercontinent, of a separate tectonic province which had undergone an independent evolution. The ancient suture along which the two parts were joined may have become reactivated later to serve as the *locus* of subsequent breakup.

2. *Typical prerift and/or rift-valley sequences developed widely across the entire New Zealand block but are not in evidence near the zone of later breakup. No prebreakup rift sequences are known on the Antarctic side, where uplift and erosion seem to have prevailed.* This different behavior in the two regions may be the result of an imbalance between the continental masses involved. The process of splitting off a relatively small area—which became the New Zealand microcontinent—from a much larger continental mass such as Antarctica (at that time still combined with Australia, from which the New Zealand–Lord Howe Rise region separated at about the same time as the Campbell Plateau from Marie Byrd Land; Hayes and Ringis, 1973) may produce a tectonic situation

different from the classical continental separation model, where partners of nearly equal size are involved. Thus, while on the more stable Antarctic side the prebreakup rift-valley stage is hardly developed, if at all, it occurs widely across the New Zealand block which, about 30 m.y. before final breakup, began cracking and forming rifted grabens from one end to the other (Figures 3.5 and 3.6).

3. *The early subsidence of the Campbell Plateau, which submerged below the Late Cretaceous sea at the same time as continental breakup occurred and a deep ocean basin between Antarctica and New Zealand was formed, appears anomalous both in its timing and scale. In contrast, no subsidence occurred at this time in West Antarctica, which apparently underwent continuous uplift and erosion.* If rifting and continental breakup is caused by a thermal anomaly leading to an uplift-erosion-subsidence cycle (Dewey and Byrd, 1970; Falvey, 1974), the present contrast is not explained, nor is the occurrence of general subsidence synchronous with breakup over an area reaching 800 km from the *locus* of the breakup. Normally in such situations, an immediate consequence of rifting and breakup is differential uplift and tilting (e.g., Rhine Graben, Red Sea–Afar Rift, northwestern Australia, and Gabon and Brazil basins on the Atlantic margin; see Falvey, 1974).

As the small New Zealand block drifted away from an apparently more stable Antarctica, perhaps such subsidence at an early stage was due to a stronger regime of extension caused by higher heat flow and greater thermal expansion in this area. Density changes due to metamorphism in the deeper crust (Falvey, 1974), and possibly also an increase in lithospheric mantle density (Hsu, 1965), would have resulted in early collapse and subsidence. Indeed, greater mean density had developed in the process of crustal thinning during the uplift-erosion stage of the preceding Rangitata orogeny and had led to some subsidence at least locally. This may have been accentuated by regional stretching and extension (Bott, 1971) in the course of continental separation. Whatever the reason, the reduction in crustal thickness of the Campbell Plateau (17–23 km) (Adams, 1962) clearly did not occur in the later part of Cenozoic history, but must have developed mainly in the period between the Rangitata orogeny and continental breakup.

4. *There is no post-breakup sequence on the New Zealand margin, but a fully developed one is found along Antarctica.* The stratigraphic sequence in DSDP holes 275 and 277 near the margin of the Campbell Plateau (Kennett and others, 1975) shows a marked upward change from terrigenous to hemipelagic material. Obviously, the terrigenous sediment supply was sharply diminished at an early stage, which was probably due to the great distance from source areas. Therefore, because of the early subsidence of the large Campbell Plateau block, no continental margin sedimentary wedge was formed on the New Zealand side.

5. *Granite intrusion and/or high temperature metamorphism* are manifest along the western side of Campbell Plateau and in Marie Byrd Land and have been related to the collision-type Rangitata and "Andean" orogenies. Their last cooling and uplift may be as young as 88–90 m.y. ago, i.e., immediately before Senonian time. Since breakup and the generation of new oceanic crust were, according to magnetic anomalies, accomplished in early Senonian time (81 m.y. ago), there seems to be a *close time and space relationship between orogenic magmatism and continental separation.* This differs from usual separation models and places constraints on the possible mechanisms.

While in New Zealand, this calc-alkaline, granodiorite magmatism is confined to the orogenic foreland (Landis and Coombs, 1967), the notion that the Torlesse rocks were derived from Marie Byrd Land and "rafted into their present position by sea-floor spreading" (Blake and others, 1974) during the Early Cretaceous Rangitata orogeny, implies that similar magmatism also occurred on the opposite side of the colliding orogenic belt.

Furthermore, it is unexplained why an early sea-floor spreading episode caused collision-type orogeny with late-stage granodiorite magmatism, soon to be followed in the same general region by another episode of sea-floor spreading of entirely passive Atlantic type.

6. *In later Cenozoic time, extensional tectonics, faulting, and rifting, as well as alkali basalt volcanism, continued intermittently across Campbell Plateau and West Antarctica.* Both areas by then may have reached a similar state of evolution and comparable tectonic conditions. It may be significant that, compared to New Zealand, the postbreakup evolution of West Antarctica is protracted and delayed (Figure 3.6). Volcanism, rifting, fragmentation, and probably also the relative (though much less marked) thinning of the crust with local subsidence may only have occurred in Marie Byrd Land from Oligocene–Miocene time onward.

In summary, continental separation between New Zealand and West Antarctica, if it did occur, was based on peculiar geodynamic conditions which differ from those in other examples. The main difference may have been the situation created by the split of two unequal partners. The tentative explanations and interpretation offered here should be critically evaluated in the light of future studies.

Acknowledgments. This paper was much improved by comments on an early draft from M. G. Laird (Christchurch), B. Sporli (Auckland) and, in particular, R. W. Murphy (Singapore) who suggested a basic reorientation of its organization and scope. The final version was critically read by B. W. Collins and P. J. Oliver. Margaret Phelan (N.Z. Geological Survey) patiently typed a number of draft alterations as well as the final manuscript. Illustrative drawings were prepared at the Department of Scientific and Industrial Research Cartographic Section with advice and supervision from R. Phillips.

References cited

Adams, R. D. 1962. Thickness of the earth's crust beneath the Campbell Plateau. *N.Z. Jl Geol. Geophys., 5,* No. 1, 74–85.

Adams, R. D. 1972. Dispersion-wave studies in Antarctica. (*In* Adie, R. J., *ed. Antarctic Geology and Geophysics.* Oslo, Universitetsforlaget, 473–80.)

Austin, P. M. 1975. Paleogeographic and paleotectonic models for the New Zealand geosyncline in eastern Gondwanaland. *Geol. Soc. Am. Bull., 86,* No. 9, 1230–34.

Austin, P. M., Sprigg, R. C., and Braithwaite, J. C. 1973. Structure and petroleum potential of eastern Chatham Rise, New Zealand. *Bull. Am. Ass. Petrol. Geol., 57,* No. 3, 477–97.

Beck, A. C., and Nathan, S. 1978. Nelson and Westland. (*In* Suggate, R. P., Stevens, G. R., and Te Punga, M. T., *eds. The Geology of New Zealand. 2.* Wellington, Government Printer, 375–79.)

Bentley, C. R. 1965. The land beneath the ice. (*In* Hatherton, T., *ed. Antarctica.* Wellington-Auckland, Reed, 259–77.)

Bentley, C. R., 1972. Subglacial rock surface topography. (*In* Heezen, B. C., Tharp, M., and Bentley, C. R., *eds.* Morphology of the Earth in the Antarctic and Subantarctic. *Antarct. Map Folio Ser.,* Folio 16, Pl. VII.)

Bentley, C. R., and Clough, J. W. 1972. Antarctic subglacial structure from seismic refraction measurements. (*In* Adie, R. J., *ed. Antarctic Geology and Geophysics.* Oslo, Universitetsforlaget, 683–91.)

Bentley, C. R., Crary, A. P., Ostenso, N. A., and Thiel, E. C. 1960. Structure of West Antarctica. *Science, N.Y., 131,* 131–36.

Blake, M. C., Jr., Jones, D. L., and Landis, C. A. 1974. Active continental margins: contrasts between California and New Zealand. (*In* Burk, C. A., and Drake, C. L., *eds. Geology of Continental Margins.* New York, Springer-Verlag, 853–72.)

Bott, M. H. P. 1971. Evolution of young continental margins and formation of shelf basins. *Tectonophysics, 11,* No. 5, 319–27.

Boudette, E. L., Marvin, R. F., and Hedge, C. E. 1966. Biotite, potassium-feldspar, and whole-rock ages of adamellite, Clark Mountains, West Antarctica. *Prof. Pap. U.S. geol. Surv.,* 550-D, 190–94.

Bradshaw, J. D., and Andrews, P. B. 1973. Geotectonics and the New Zealand geosyncline. *Nature Phys. Sci., 241,* 14–16.

Challis, A. 1964. Basalt, Core sample from 5,693 feet, Arahura. (*In* Oortman Gerlings, H. *Taramakau-1 and Arahura-1.* Shell BP and Todd Oil Services. *Open-File Petrol. Rep., N.Z. geol. Surv.,* No. 451, Appendix 2, Unpublished.)

Christoffel, D. A., and Falconer, R. K. H. 1972. Marine magnetic measurements in the southwest Pacific Ocean and the identification of new tectonic features. (*In* Hayes, D. E., *ed. Antarctic Oceanology 2. The Australian–New Zealand Sector.* Washington, American Geophysical Union, 197–209.) [Antarctic Research Series, Vol. 19.]

Christoffel, D. A., and Falconer, R. K. H. 1973. Changes in the direction of sea floor spreading in the southwest Pacific. (*In* Fraser, R., *comp. Oceanography of the South Pacific 1972.* Wellington, New Zealand National Commission for UNESCO, 241–48.)

Coombs, D. S., Landis, C. A., Norris, R. J., Sinton, J. M., Borns, D. J., and Craw, D. 1976. The Dun Mountain ophiolite belt, New Zealand, its tectonic setting, constitution, and origin, with special reference to the southern portion. *Am. J. Sci., 276,* 561–603.

Craddock, C. 1970*a*. Radiometric age map of Antarctica (*In* Bushnell, V. C., and Craddock, C., *eds.* Geologic Maps of Antarctica. *Antarct. Map Folio Ser.,* Folio 12, Pl. XIX.)

Craddock, C. 1970*b*. Geologic map of Antarctica (*In* Bushnell, V. C., and Craddock, C., *eds.* Geologic Maps of Antarctica. *Antarct. Map Folio Ser.,* Folio 12, Pl. XX.)

Craddock, C. 1970*c*. Tectonic map of Antarctica (*In* Bushnell, V. C., and Craddock, C., *eds.* Geologic Maps of Antarctica. *Antarct. Map Folio Ser.,* Folio 12, Pl. XXI.)

Craddock, C. 1972. Antarctic tectonics. (*In* Adie, R. J., *ed. Antarctic Geology and Geophysics.* Oslo, Universitetsforlaget, 449–55.)

Craddock, C., Bastien, T. W., and Rutford, R. H. 1964. Geology of the Jones Mountains area. (*In* Adie, R. J., *ed. Antarctic Geology.* Amsterdam, North-Holland Publishing Company, 171–87.)

Cullen, D. J. 1967. The Antipodes fracture zone, a major structure in the south-west Pacific. *N.Z. Jl Mar. Freshwat. Res., 1,* No. 1, 16–25.

Cullen, D. J. 1969. Quaternary volcanism at the Antipodes Islands: its bearing on structural interpretation of the south-west Pacific. *J. geophys. Res., 74,* No. 17, 4213–20.

Cullen, D. J. 1975. Autochthonous rocks of the Bounty Islands region, south-west Pacific Ocean. *N.Z. Jl Geol. Geophys., 18,* No. 6, 767–85.

Davey, F. J. 1977. Marine seismic measurements in the New Zealand region. *N.Z. Jl Geol. Geophys., 20,* No. 4, 719–77.

Davey, F. J., and Houtz, R. E. 1977. Campbell Plateau and its relationship with the Ross Sea, Antarctica. *Mar. Geol. 25,* 61–72.

Denison, R. E., and Coombs, D. S. 1977. Radiometric ages for some rocks from Snares and Auckland Islands, Campbell Plateau. *Earth planet. Sci. Lett., 34,* 23–29.

Dewey, J. F., and Bird, J. M. 1970. Mountain belts and the new global tectonics. *J. geophys. Res., 75,* 2625–47.

Doumani, G. A. 1964. Volcanoes of the Executive Committee Range, Byrd Land. (*In* Adie, R. J., *ed. Antarctic Geology.* Amsterdam, North-Holland Publishing Company, 666–75.)

Drake, A. A., Stern, T. W., and Thomas, H. H. 1964. Radiometric ages of zircon and biotite in quartz diorite, Eights Coast, Antarctica. *Prof. Pap. U.S. geol. Surv.,* 501-D, 50–53.

Elliot, D. H. 1975. Tectonics of Antarctica: a review. *Am. J. Sci., 275-A,* 45–106.

Falvey, D. A. 1974. The development of continental margins in plate tectonic theory. *J. Aust. petrol. Expl. Ass., 14,* No. 1, 95–106.

Fleming, C. A. 1970. Mesozoic of New Zealand: chapters in the history of the circum-Pacific Mobile Belt. *Q. Jl geol. Soc. Lond., 125,* 125–70.

Ford, A. B., and Barrett, P. J. 1975. Basement rocks of the south-central Ross Sea, site 270, DSDP leg 28. (*In* Hayes, D. E., Frakes, L. A., and others. *Initial Reports of the Deep Sea Drilling Project,* Vol. 28. Washington [U.S. Government Printing Office], 861–68.)

Gage, M. 1952. The Greymouth Coalfield. *Bull. geol. Surv. N.Z., n.s., 45,* 240 pp.

Gage, M. 1957. Geology of Waitaki Subdivision. *Bull. geol. Surv. N.Z., n.s., 55,* 135 pp.

Gair, H. S., and Rickwood, P. C. 1965. Timaru and Geraldine Tholeiitic Basalt Sheets. (*In* Thompson, B. N., and Kermode, L. O., *eds.* New Zealand Volcanology South Island. *Information Ser., N.Z. Dep. sci. ind. Res., 51,* 34–38.)

Griffiths, J. R. 1971. Reconstruction of the south-west pacific margin of Gondwanaland. *Nature, Lond., 234,* 203–7.

Griffiths, J. R. 1974. Revised continental fit of Australia and Antarctica. *Nature, Lond., 249,* 336–38.

Grindley, G. W. 1976. Palaeomagnetic evidence for New Zealand's Cretaceous–Cenozoic drift and rotation. *Abstr. 25th Int. Geol. Cong., Sydney, 1976, 1,* 84–85.

Halpern, M. 1968. Ages of Antarctic and Argentine rocks bearing on continental drift. *Earth planet. Sci. Lett., 5,* 159–67.

Halpern, M. 1972. Rb-Sr Total-rock and mineral ages from the Marguerite Bay Area, Kohler Range and Fosdick Mountains. (*In* Adie, R. H., *ed. Antarctic Geology and Geophysics.* Oslo, Universitetsforlaget, 197–206.)

Halpern, M., and González-Ferrán, O. 1973. Edad del Volcán Andrus y relacion inicial Sr^{87}/Sr^{86} de rocas volcanicas de Tierra Maria Byrd e Islas Shetland del Sur, Antarctica. *Revta bras. Geoc. 3,* 141–48.

Harris, W. F. 1971. Palynological investigation of samples from Endeavour—1. (*In* Wildig, A., and Sweetman, I. A. D. *Well Résumé Endeavour—1.* BP Shell Todd (Canterbury) Services. *Open-File Petrol. Rep., N.Z. geol. Surv.,* No. 303, Appendix 3, Unpublished.)

Hay, R. F., Mutch, A. R., and Watters, W. A. 1970. Geology of the Chatham Islands. *Bull. geol. Surv. N.Z., 83,* 86 pp.

Hayes, D. E., and Davey, F. J. 1975. Geophysical study of the Ross Sea, Antarctica. (*In* Hayes, D. E., Frakes L. A., and others. *Initial Reports of the Deep Sea Drilling Project,* Vol. 28. Washington [U.S. Government Printing Office], 887–907.)

Hayes, D. E., and Ringis, J. 1973. Seafloor spreading in the Tasman Sea. *Nature, Lond., 243,* 454–58.

Hayes, D. E., Frakes, L. A., and others. 1975. *Initial Reports of the Deep Sea Drilling Project,* Vol. 28. Washington (U.S. Government Printing Office), 1017 pp.

Hollister, C. D., Craddock, C., and others. 1974. Deep drilling in the Southeast Pacific Basin. *Geotimes, 20,* No. 8, 16–19.

Houtz, R. E., and Davey, F. J. 1973. Seismic profiler and sonobuoy measurements in Ross Sea, Antarctica. *J. geophys. Res., 78,* No. 17, 3448–68.

Houtz, R. E., and Markl, R. G. 1972. Seismic profiler data between Antarctica and Australia. (*In* Hayes, D. E., *ed. Antarctic Oceanology 2. The Australian-New Zealand Sector.* Washington, American Geophysical Union, 147–164.) [Antarctic Research Series, Vol. 19.]

Hsu, K. J. 1965. Isostasy, crustal thinning, mantle changes, and the disappearance of ancient land masses. *Am. J. Sci., 263,* 97–109.

Iltchenko, L. N. 1972. Late Precambrian acritarchs of Antarctica. (*In* Adie, R. J., *ed. Antarctic Geology and Geophysics.* Oslo, Universitetsforlaget, 599–602.)

Katz, H. R. 1973. Contrasts in tectonic evolution of orogenic belts in south-east Pacific. *J. R. Soc. N.Z., 3,* No. 3, 333–62.

Katz, H. R. 1974*a*. Offshore petroleum potential in New Zealand. *J. Aust. petrol. Expl. Ass., 14,* No. 1, 3–13.

Katz, H. R. 1974*b*. Margins of the southwest Pacific. (*In* Burk, C. A., and Drake, C. L., *eds. Geology of Continental Margins.* New York, Springer-Verlag, 549–65.)

Katz, H. R. 1976. Sedimentary basins and petroleum prospects onshore and offshore New Zealand. (*In* Halbouty, M. T., Maher, J. C., and Lian, H. M., *eds.* Circum-Pacific Energy and Mineral Resources. *Mem. Am. Ass. Petrol. Geol., 25,* 217–28.)

Kennett, J. P., Burns, R. E., Andrews, J. E., Churkin, M., Davies, T. A., Domitrica, P., and others. 1972. Australian-Antarctic continental drift, paleocirculation changes and Oligocene deep-sea erosion. *Nature Phys. Sci., 239,* No. 91, 51–55.

Kennett, J. P., Houtz, R. E., and others. 1975. *Initial Reports of the Deep Sea Drilling Project,* Vol. 29. Washington (U.S. Government Printing Office), 1197 pp.

Klimov, L. V. 1967. Nektorye rezul'taty geologicheskikh issledovanii na Zemle Meri Berd v 1966–1967 gg. [Some results of geological observations in Marie Byrd Land, 1966–1967]. *Inf. Byull. sov. antarkt. Eksped.,* No. 66, 18–25. [English translation: *Inf. Bull. Sov. Antarct. Exped., 6,* No. 6, 555–59.]

Krylov, A. Ya., Lopatin, B. G., and Mazina, T. I. 1970. Vozrast porod gor Forda; Berega Rupperta (Zapadnaya Chast' Zemli Meri Berd) [The age of the rocks from the Ford Mountains and Ruppert Coast (the western part of Marie Byrd Land)]. *Inf. Byull. sov. antarkt. Eksped.,* No. 80, 34–37 [English translation *Inf. Bull. Sov. Antarct. Exped., 8,* No. 2, 64–66.]

Landis, C. A., and Bishop, D. G. 1972. Plate tectonics and regional stratigraphic-metamorphic relations in the southern part of the New Zealand geosyncline. *Geol. Soc. Am. Bull., 83,* 2267–84.

Landis, C. A., and Coombs, D. S. 1967. Metamorphic belts and orogenesis in southern New Zealand. *Tectonophysics, 4,* 501–18.

LeMasurier, W. E. 1972. Volcanic record of Cenozoic glacial history of Marie Byrd Land. (*In* Adie, R. J., *ed. Antarctic Geology and Geophysics.* Oslo, Universitetsforlaget, 251–59.)

LeMasurier, W. E., and Wade, F. A. 1976. Volcanic history in Marie Byrd Land: implications with regard to Southern Hemisphere tectonic reconstructions. (*In* González-Ferrán, O., *ed. Andean and Antarctic Volcanology Problems.* Rome, International Association of Volcanology and Chemistry of the Earth's Interior, 398–424.)

Lopatin, B. G., and Orlenko, E. M. 1972. Outline of the geology of Marie Byrd Land and the Eights Coast. (*In* Adie, R. J., *ed. Antarctic Geology and Geophysics.* Oslo, Universitetsforlaget, 245–49.)

McDougall, I., and Coombs, D. S. 1973. Potassium-argon ages for the Dunedin volcano and outlying volcanics. *N.Z. Jl Geol. Geophys., 16,* No. 2, 179–88.

McKellar, I. C. 1978. Otago. (*In* Suggate, R. P., Stevens, G. R., and Te Punga, M. T., *eds. The Geology of New Zealand. 2.* Wellington, Government Printer, 393–96.)

Mildenhall, D. C. 1977. Upper Cretaceous palynomorphs from the Waihere Bay Group and Kahuitara Tuff, Chatham Islands, New Zealand. *N.Z. Jl Geol. Geophys., 20,* No. 4, 655–72.

Milne, A. D., Simpson, C., and Threadgold, P. 1975. Well Completion Report Resolution-1, Offshore Canterbury, New Zealand. (*In* BP Shell Todd (Canterbury) Services. *Open-File Petrol. Rep., N.Z. geol. Surv.,* No. 648, Unpublished.)

Molnar, P., Atwater, T., Mammerickx, J., and Smith, S. M. 1975. Magnetic anomalies, bathymetry and the tectonic evolution of the South Pacific since the Late Cretaceous. *Geophys. J. R. astr. Soc., 40,* 383–420.

Munizaga, F. 1972. Rb-Sr Isotopic ages of intrusive rocks from Thurston Island and adjacent islands. (*In* Adie, R. J., *ed. Antarctic Geology and Geophysics.* Oslo, Universitetsforlaget, 205–6.)

Oliver, R. L., Finlay, H. J., and Fleming, C. A. 1950. Geology of Campbell Island. *Bull. Cape Exped. Ser., 3,* 62 pp.

Rutford, R. H., Craddock, C., White, C. M., and Armstrong, R. L. 1972. Tertiary glaciation in the Jones Mountains. (*In* Adie, R. J., *ed. Antarctic Geology and Geophysics.* Oslo, Universitetsforlaget, 239–43.)

Speden, I. G. 1975. Cretaceous stratigraphy of Raukumara Peninsula. *Bull. geol. Surv. N.Z., 91,* 70 pp.

Speden, I. G. 1976. *Inoceramus opetius* in the Kahuitara Tuff, Chatham Islands, New Zealand. *N.Z. Jl Geol. Geophys., 19,* No. 3, 385–87.

Stipp, J. J., and McDougall, I. 1968. Geochronology of the Banks Peninsula Volcanoes, New Zealand. *N.Z. Jl Geol. Geophys., 11,* No. 5, 1239–60.

Summerhayes, C. P. 1969. Marine geology of the New Zealand Subantarctic sea floor. *Bull. N.Z. Dep. scient. ind. Res., 190,* 92 pp.

Vanney, J. R., and Johnson, G. L. 1976. The Bellingshausen-Amundsen Basins (southeastern Pacific): major sea-floor units and problems. *Mar. Geol., 22,* 71–101.

Van Oyen, F. H. 1970. Cook no. 1 well completion report. (*In* New Zealand Aquitaine Petroleum. *Open-File Petrol. Rep., N.Z. geol. Surv.,* No. 513, Unpublished.)

Wade, F. A. 1969. Geology of Marie Byrd Land. (*In* Bushnell, V. C., and Craddock, C., *eds.* Geologic Maps of Antarctica. *Antarct. Map Folio Ser.,* Folio 12, Pl. XVII.)

Wade, F. A. 1972. Geologic Survey of Marie Byrd Land. *Antarct. Jnl U.S., 7,* No. 5, 144–45.

Wade, F. A., and Wilbanks, J. R. 1972. Geology of Marie Byrd and Ellsworth Lands. (*In* Adie, R. J., *ed. Antarctic Geology and Geophysics.* Oslo, Universitetsforlaget, 207–14.)

Wasserburg, G. J., Craig, H., Menard, H. W., Engel, A. E., and Engel, C. G. 1963. Age and composition of a Bounty Island granite, and age of a Seychelles Island granite. *J. Geol., 71,* 785–89.

Webb, P. N., and Neall, V. E. 1972. Cretaceous Foraminifera in Quaternary deposits from Taylor Valley, Victoria Land. (*In* Adie, R. J., *ed. Antarctic Geology and Geophysics.* Oslo, Universitetsforlaget, 653–57.)

Wilbanks, J. R. 1972. Geology of the Fosdick Mountains, Marie Byrd Land. (*In* Adie, R. J., *ed. Antarctic Geology and Geophysics.* Oslo, Universitetsforlaget, 277–84.)

Wildig, A., and Sweetman, I. A. D. 1971. Well Résumé Endeavour—1. BP Shell Todd (Canterbury) Services. *Open-File Petrol. Rep., N.Z. geol. Surv.,* No. 303 (Unpublished).

Wilson, G. J. 1968. On the occurrence of fossil microspores, pollen grains, and microplankton in bottom sediments of the Ross Sea, Antarctica. *N.Z. Jl Mar. Freshwat. Res., 2,* No. 3, 381–89.

Wilson, G. J. 1972. Age of the Garden Cove Formation, Campbell Island. *N.Z. Jl Geol. Geophys., 15,* No. 1, 184–85.

Wilson, G. J. 1976. Late Cretaceous (Senonian) dinoflagellate cysts from the Kahuitara Tuff, Chatham Islands. *N.Z. Jl Geol. Geophys., 19,* No. 1, 127–30.

4

Geologic History and Regional Patterns in New Zealand and West Antarctica—Their Paleotectonic and Paleogeographic Significance

R. A. COOPER
New Zealand Geological Survey, Department of Scientific and Industrial Research, Lower Hutt, New Zealand

C. A. LANDIS
Department of Geology, University of Otago, Dunedin, New Zealand

W. E. LEMASURIER
Department of Natural and Physical Sciences—Geology, University of Colorado, Denver, Colorado 80202, U.S.A.

I. G. SPEDEN
New Zealand Geological Survey, Department of Scientific and Industrial Research, Lower Hutt, New Zealand

Abstract. The geologic history and distribution of major rock units in New Zealand and West Antarctica are generally consistent with a sea-floor spreading reconstruction that places the southern part of the New Zealand region adjacent to the Ross Sea—Marie Byrd Land margin of Antarctica. Marie Byrd Land has a history comparable with that of the western province of New Zealand. A middle Paleozoic orogenic episode, with accompanying metamorphism and granitic emplacement, deformed a sequence of upper Precambrian to lower Paleozoic sedimentary rocks in each area. Mesozoic calc-alkaline igneous activity and subsequent widespread planation during late Mesozoic to early Cenozoic time were strikingly similar in the two areas. Marie Byrd Land and the western province of New Zealand together are likely to have formed part of a Paleozoic-Mesozoic continental block, a portion of Gondwanaland. However, this block may not have been a major source for sediments of the New Zealand geosyncline.

New Zealand's eastern (or geosynclinal) province has some similarities with eastern Ellsworth Land and the Antarctic Peninsula. Triassic, Jurassic, and Early Cretaceous macrofaunas of Ellsworth Land and the Antarctic Peninsula have close affinities with New Zealand faunas, and somewhat similar volcaniclastic sedimentary facies are developed in each area. The quartzofeldspathic graywacke suite of New Zealand's Alpine (Torlesse) Assemblage may be matched by thick, upper paleozoic-lower Mesozoic graywacke-shale sequences on the Antarctic Peninsula (Trinity Peninsula Series). However, there are some significant differences in the nature and timing of Mesozoic sedimentation in the two regions; distinct belts equivalent to the Alpine and Hokonui Assemblages of New Zealand are not apparent in the Antarctic Peninsula region.

During the Mesozoic, a broad belt from the Andean chain through West Antarctica to New Zealand underwent orogenesis, but deformation and magmatism were not everywhere uniform and coeval. In New Zealand the main late Jurassic-early Cretaceous phase of the Rangitata orogeny is expressed as intense tectonism in the eastern province and magmatism in the western province. In Antactica, an equivalent event is represented by tectonism and magmatism at the base of the Antarctic Peninsula, and by magmatism only, throughout Marie Byrd Land.

Most geologists now accept the idea that New Zealand once formed part of Gondwanaland (Figure 4.1). Geologic similarity between the New Zealand and eastern Australian lower Paleozoic sequences has long been recognized (Keble and Benson, 1929; Brown and others, 1968; Harrington and others, 1974; Cooper, 1975). Comparison between New Zealand and West Antarctica, on the other hand, is less persuasive. This is in part because of lack of rock outcrop and consequent paucity of geologic data throughout most of West Antarctica, but also because of the apparent lack of clearly correlative units (Halpern, 1968; LeMasurier and Wade, 1976; Davey and Houtz, 1977). In this paper we briefly examine the available geophysical, geologic, and faunal data of New Zealand and West Antarctica (particularly Marie Byrd Land and Ellsworth Land) to see if they are consistent with, and provide constraints on, the reassembly of the two regions prior to late Cretaceous time.

The geologic history of New Zealand from Permian time to the present is summarized in table form (Figure 4.2); a simple nomenclature which identifies the major geologic units and events is adopted. The geologic history of West Antarctica is summarized in Table 4.1.

Sea-floor spreading and New Zealand's position in Gondwanaland

Sea-floor magnetic and bathymetric data in the Southwest Pacific Ocean provide reasonable constraints on reconstructions involving the New Zealand region (New

Table 4.1. Stratigraphy and geologic history of regions in West Antarctica. Data from Dalziel and Elliot (1973), Katz (1973),

	Northern Antarctic Peninsula	SW Antarctic peninsula	Bowman and Wilkins Coast
Quaternary	Alkali basalts (4.6–1.4 m.y.)		
	Pecten Conglomerate	Olivine basalts, tuffs, (submarine)	
	Dikes and volcanics (age ?)	Block faulting, tilting, planation	
Tertiary	James Ross Is Volcanic Gp (Miocene-Pliocene ?)	Dolerite-camptonite dikes (15 m.y.)	
	Uplift, erosion, block faulting	Block faulting, open folding	Block faulting, open folding, uplift
	Deposition of mostly locally derived, early Tertiary (Paleocene–Oligocene ?) and Campanian sediments		Granite-diorite intrusives
		Intrusive and extrusive tuffaceous sequence (70 m.y.); plant fragments	Feldspathic and calcareous sandst and siltst of volcanic origin (Crabeater point)
Cretaceous	Open folding, block faulting, erosion	------?------?------	
	(Derived Aptian fossils) ------?------?------	Qtzose sdst, siltstone, congl, pebbly mudstone, and andesitic tuff derived from granodioritic and volcanic terrane to E; low-grade regional metamorphism to prehnite-laumontite facies; penecontemporaneous deformation (faulting, folding)	Thrusting (to NE) and regional folding Diorite and metagabbro intrusives Green and purple volcanic arkose, acidic tuff
	Andesitic-rhyolite lava and tuff, non-marine plant-bearing sediments (derived Jurassic fossils) hornfelses, gabbros		
Jurassic	------?------?------	------?------?------	------?------?------
		Dioritic intrusives	Synkinematic orthogneiss
	Low-grade regional metamorphism, polyphase deformation, uplift, and erosion (age uncertain)		(185 m.y.)
	------?------?------	------?------?------	------?------?------
Triassic	Feldspar or qtz-rich sdst, siltstone, congl, lmst, greenschist, and minor penecontemporaneous volcanic sediment (Trinity Peninsula Series, Legoupil Fm); derived from NW ?, from acid-intermediate and plutonic terrane	Arkosic and lithic (local) sdst, siltstone, congl, pillow lava; derived from plutonic and metamorphic qtzose terrane, with minor penecontemporaneous volcanism	Purple and black acidic volcanic rocks (minor hiatus ?) Shallow-water qtzite, congl, and shale derived from eastern granitic and sedimentary terrane; minor volcanic detritus
Late Paleozoic	------?------?------	------?------?------	------?------?------
	Granite, volcanics		
	Regional metamorphism, polyphase deformation, uplift and erosion (age uncertain)		
Pre-late Paleozoic (age uncertain)	Orthogneiss, etc.	Granodiorite, tonalite gneiss	Schist, orthogneiss

Elliot (1975), Mehnert and others (1975), and references given in text.

Lassiter Coast	Eastern Ellsworth Land	Western Ellsworth Land	Marie Byrd Land
		Alkaline basalts	
	Olivine basalts		Alkaline basalts (30–0.2 m.y.)
	Arching and block faulting		
	Peneplanation	Peneplanation	Peneplanation
Lamprophyre and dolerite dikes (age?)			
Diorite-granodiorite intrusives and dikes (119–199 m.y.), minor folding	Gabbro-quartz monzonite intrusives (102-109 m.y.) Mafic dikes	Granite, leucogranite, adamellite, felsic dikes (104 m.y.)	Granite, adamellite, monzonite (98 m.y.), felsic flows
Tight folding (to SE)	Complex folding		
------?------?------	------?------?------		
Shallow marine siltst, lithic sdst, qtzite, congl, and lmst derived from granitic and acidic volcanic terrane to W; intruded by sills and dikes; overlain by andesitic-dacitic lava and tuff	Graywacke, arkose, siltst, congl,and tuff derived from acid- intermediate volcanic terrane and crystalline complex; older, in part, contemporaneous dacitic flows		
------?------?------	------?------?------		
		Granite, granodiorite (230–180 m.y.)	
		Gabbro-diorite-granodiorite intrusion (265–300 m.y.)	Granodiorite emplacement and deformation (352 m.y.)
			Quartz-rich metagraywacke, slate, phyllite, amphibolite gneiss

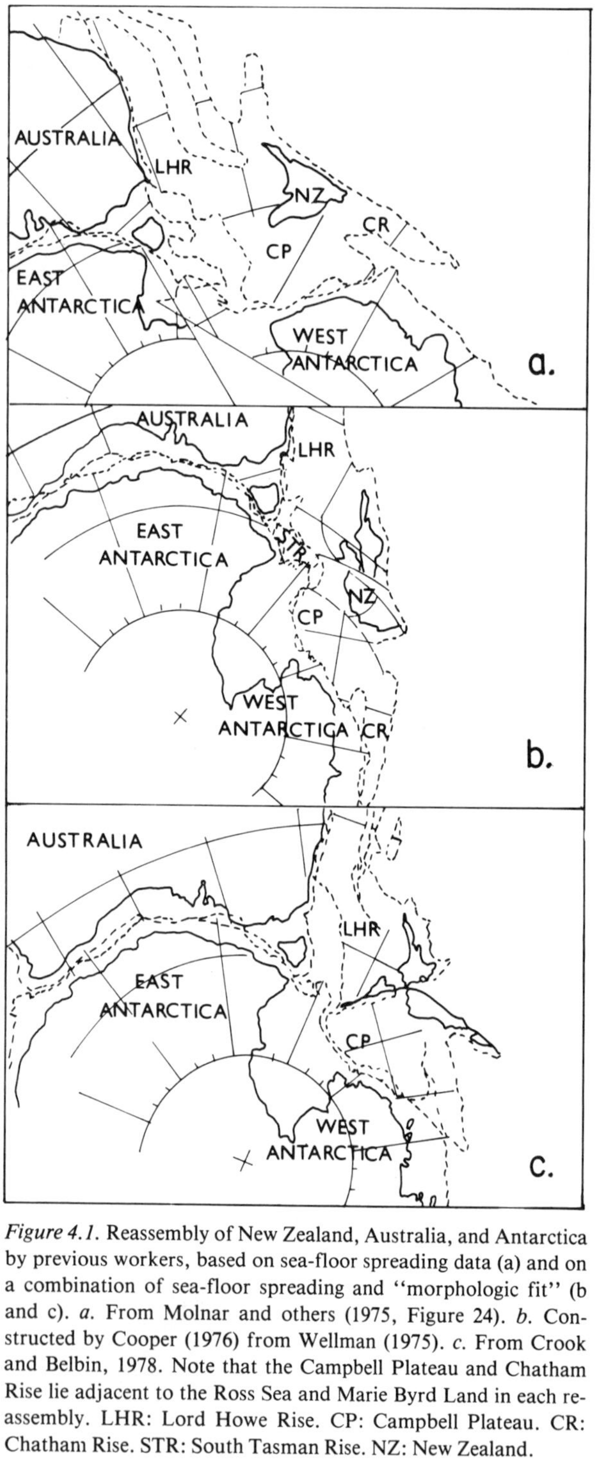

Figure 4.1. Reassembly of New Zealand, Australia, and Antarctica by previous workers, based on sea-floor spreading data (a) and on a combination of sea-floor spreading and "morphologic fit" (b and c). *a.* From Molnar and others (1975, Figure 24). *b.* Constructed by Cooper (1976) from Wellman (1975). *c.* From Crook and Belbin, 1978. Note that the Campbell Plateau and Chatham Rise lie adjacent to the Ross Sea and Marie Byrd Land in each reassembly. LHR: Lord Howe Rise. CP: Campbell Plateau. CR: Chatham Rise. STR: South Tasman Rise. NZ: New Zealand.

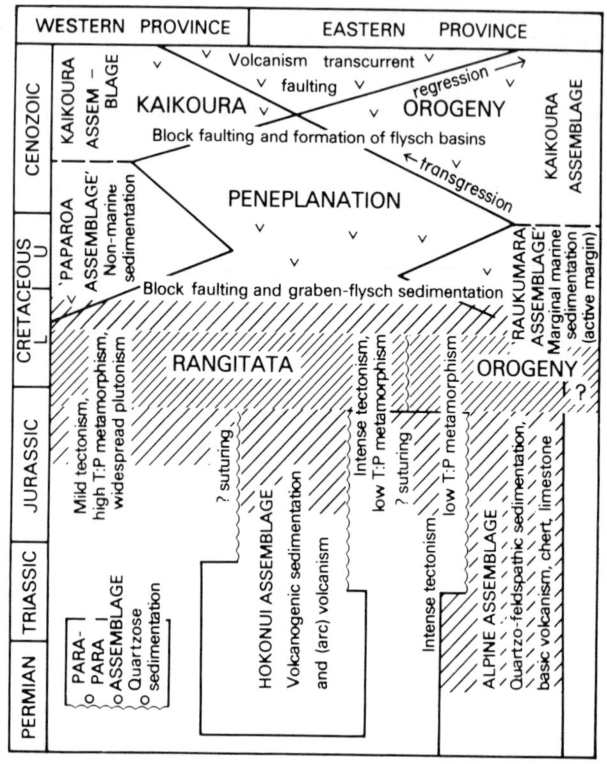

Figure 4.2. Outline of the late Paleozoic to recent geologic history of the New Zealand region showing major units, features, and events. "Paparoa assemblage" and "Raukumara assemblage" are used here as informal names for distinctive sedimentary sequences. T:P stands for temperature to pressure ratio. Note that the Alpine Assemblage extends down into the Carboniferous. The diagram is intended as a general outline only; relationships of the "Paparoa" and "Raukumara" assemblages to adjacent units are not discussed. Based on Aronson (1968), Fleming (1970), Bishop (1974), Carter and others (1974), Bishop and Laird (1976), Coombs and others (1976), Andrews and others (1977), Stevens and Speden (1977), and Suggate and others (1978). V-shapes: volcanic rocks; oblique hatching: deformation; circles: quartzose conglomerate.

Zealand, Lord Howe Rise, Norfolk Ridge, Campbell Plateau, and Chatham Rise), Australia, and Antarctica prior to separation in late Mesozoic time (Christoffel and Falconer, 1972, 1973; Hayes and Ringis, 1973; Weissel and Hayes, 1972, 1977; Molnar and others, 1975; Weissel and others, 1977) (Figure 4.1).

A model of the history of plate interactions in the South Pacific during the last 75 m.y. has been derived from a reassessment and synthesis of the magnetic and bathymetric data by Weissel and others (1977). According to their model, a three-plate system has been active in the New Zealand region, with a plate boundary in New Zealand itself, for most of the time interval from late Cretaceous to present. Net relative motion on the plate boundary in New Zealand is inferred to be about 1000 km in a right-lateral sense. No significant amount of relative motion between East and West Antarctica is proposed. A late Mesozoic (~80 m.y.) reconstruction of Australia, Antarctica, and New Zealand is not presented by Weissel and others but can be inferred from the finite rotation poles given; it is similar to that derived by Crook and Belbin (1978) based on morphologic fit and sea-floor spreading data (see Figure 4.1*c*).

An improbably large right-lateral shift (~900–1000 km) on the Alpine Fault is postulated in both models, together with an implied earlier (pre-80 m.y.) left-lateral shift of about 450 km which is required to reunite the late Paleozoic to early Mesozoic reference lines across the fault (Wellman, 1956; Suggate, 1963). A possible solution to the problem, outlined by Wellman (1975), accommodates the 1000 km right-lateral displacement partly by shear throughout a broad zone (producing oroclinal bending of the "New Zealand Geosyncline") and partly by slip (of 450 km) along the Alpine Fault. Wellman's model for New Zealand is incorporated into a reconstruction of the region in Figure 4.1*b* (after Cooper, 1976).

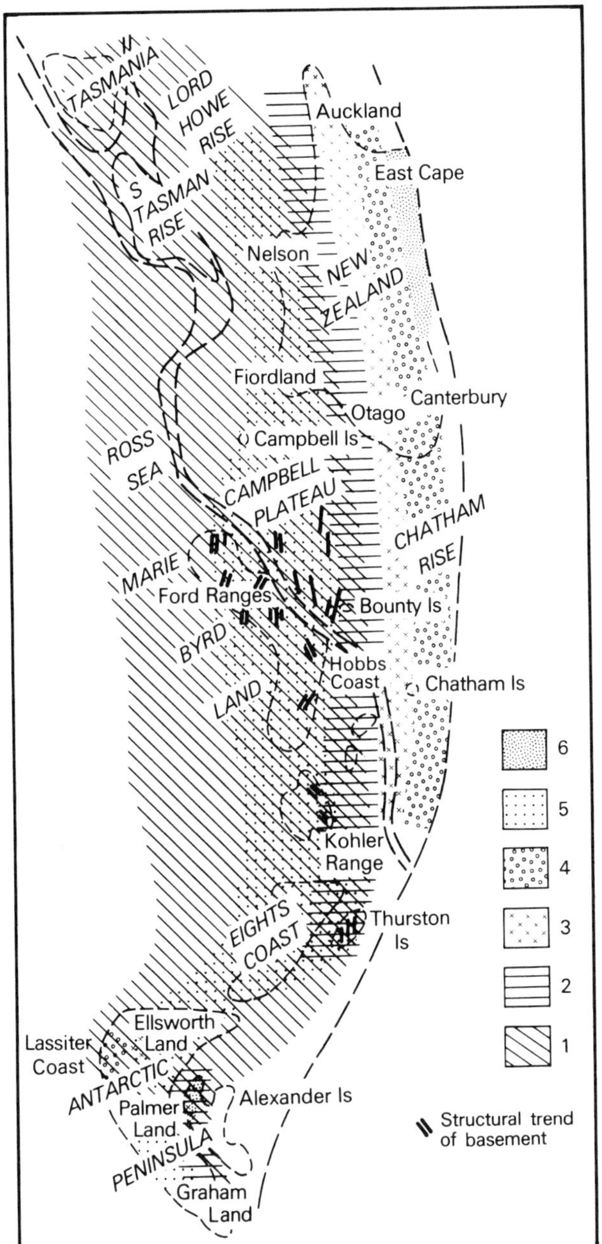

Figure 4.3. Proposed Cretaceous reassembly of New Zealand and West Antarctica, based mainly on the reassembly shown in Figure 4.1*b* and accommodating an inferred subsequent crustal extension of 10–15% in West Antarctica (Figure 4.1). Zones are numbered as follows: 1. Late Precambrian to middle Paleozoic "geosynclinal" sedimentation and volcanism; middle Paleozoic metamorphism and acid plutonism; 2. Late Paleozoic–Jurassic plutonism and volcanism (mainly intermediate to basic) and related metamorphism; associated with strong magnetic anomalies; 3. Late Paleozoic–Jurassic volcanogenic (arc-derived) sedimentation, deformation, and variable metamorphism in middle Mesozoic time; 4. Late Paleozoic–Jurassic graywacke sedimentation and associated submarine volcanism; sediment dominantly quartzofeldspathic; 5. Early to middle Cretaceous plutonism (mainly granitic), local acid volcanism, and metamorphism; 6. Middle to Late Cretaceous "Raukumara assemblage"; minor volcanism.

In an attempt to integrate the available bathymetric, magnetic, and on-land geologic data, we present a Cretaceous reconstruction of New Zealand and West Antarctica (Figure 4.3) based largely on the Wellman-Cooper model (Figure 4.1*b*). This map is the basis of the following discussion of the geologic framework. Three main regions are distinguished. An inner or cratonward region (Zone 1) comprises mainly upper Precambrian and lower Paleozoic sedimentary and igneous rocks (Western Province of New Zealand [Landis and Coombs, 1967] and Marie Byrd Land). A beltlike middle region (Zone 2) comprises mainly upper Paleozoic and lower Mesozoic basic to acidic intrusive, volcanic, and metasedimentary rocks (Brook Street terrane of Coombs and others [1976], Rotoroa Igneous Complex and correlatives of Suggate and others [1978], and western Ellsworth Land) and is characterized in New Zealand by a distinctive magnetic anomaly pattern (see e.g., Wellman, 1973). An outer or oceanward region (Zones 3 and 4) comprises upper Paleozoic and Mesozoic "geosynclinal" sedimentary sequences ("New Zealand Geosyncline" or Eastern Province of Landis and Coombs, 1967). Comparable rocks, but without clear-cut zonation and polarity, are recognized in eastern Ellsworth Land and along the Lassiter Coast. In addition, a large area (Zone 5) characterized by Cretaceous plutonism coincides mainly with the region of older crystalline basement (Zone 1) but extends locally into the upper Paleozoic–lower Mesozoic volcano-plutonic terrane (Zone 2). A small peripheral strip (Zone 6) is represented by the "Raukumara assemblage."

Geophysical patterns

Crustal thickness

Crustal thickness models in the New Zealand region, inferred from surface-wave dispersion studies and bathymetry (summarized by Cullen, 1970), show the Lord Howe Rise and Campbell Plateau to be underlain by thin ("subcontinental") continental crust, 17–25 km thick, and New Zealand and the Chatham Rise to be underlain by continental crust of "normal" continental thickness (greater than 30 km). Crustal-thickness studies in Antarctica (Bentley and others, 1960; Woollard, 1962) show that whereas coastal Marie Byrd Land is underlain by crust of "normal" continental thickness (30–38 km), the Ross Sea region as well as the Byrd subglacial basin and Filchner Ice Shelf (ice-filled regions lying poleward of Marie Byrd Land, Eights Coast, and Ellsworth Land in Figure 4.3) are underlain by somewhat thinner crust (28–30 km). These regions of thin crust form a continuous, large area in our reassembly (Figure 4.3). Crustal thinning is likely to have taken place during the early stages of continental separation (Hayes and Davey, 1975).

Magnetic surveys

A magnetic survey of the Campbell Plateau by Davey and Christoffel (1978) has revealed a distinctive east-northeast-trending belt of large-amplitude anomalies. They suggest that the anomaly belt represents the displaced offshore continuation of the Stokes Magnetic Anomaly (Wellman, 1973), a major zone of magnetic anomalies that extends through New Zealand. In Antarctica, the continuation of this anomaly belt would be expected to run parallel to the

Marie Byrd Land coast along the continental shelf (Davey and Houtz, 1977). Clearly, magnetic surveys of this part of the Antarctic coast are needed.

STRUCTURAL TREND

Apart from the linear anomaly belt, Davey and Christoffel (1978) show a general east-west trend in magnetic anomalies over central and eastern Campbell Plateau, reflecting the trend of basement structures (see also Fleming, 1970, Figure 23). In our reassembly (Figure 4.3), the Campbell Plateau magnetic trend agrees well with the dominant structural trend of metamorphic rocks in Marie Byrd Land (Wade and Wilbanks, 1972).

Zone 1—upper Precambrian to lower Paleozoic rock region

In Marie Byrd Land the oldest sedimentary rocks are metagraywackes, slates, and phyllites; they are inferred to represent a sequence at least 4300 m thick (Warner, 1945; Wade and Wilbanks, 1972; Wade, 1975) and are probably late Precambrian to early Paleozoic in age (Iltchenko, 1972; Lopatin and Orlenko, 1972; Lopatin and others, 1974). Rare tuffs, spilites, and carbonate rocks are known (Wade, 1975). The graywackes are highly silicic and are composed of quartz, feldspar, and lithic clasts in a muscovite-chlorite matrix. In the range of rock types and particularly in the silicic nature of the graywackes, these sedimentary rocks resemble the upper Precambrian to lower Paleozoic clastic sequence of West Nelson and Westland in New Zealand (Laird, 1972; Cooper, 1979) as well as basement rocks at Campbell Island (Beggs, 1978) and Bounty Islands (Cullen, 1975). Migmatites and amphibolite-facies gneisses of unknown age in the northern Edsel Ford Ranges (Fosdick Mountains) may represent the more metamorphosed equivalents of these rocks (Wilbanks, 1972) and are correlated with similar metamorphic complexes in central and eastern Marie Byrd Land (Wade and Wilbanks, 1972). Similar rocks underlie much of Fiordland in New Zealand (Wood, 1960, 1962).

K-Ar dates on metashales of the Edsel Ford Ranges suggest a deformational event during early or middle Ordovician time (Lopatin and others, 1974). Synkinematic and postkinematic granodiorite plutons were emplaced in western Marie Byrd Land about 350 m.y. ago (Halpern, 1968; Lopatin and Orlenko, 1972; Lopatin and others, 1974) and are inferred to indicate a late Devonian to Carboniferous orogeny (Wade and Wilbanks, 1972; Lopatin and Orlenko, 1972). The event matches the late Devonian to Carboniferous Tuhua orogeny in New Zealand and its accompanying granitic intrusions (Landis and Coombs, 1967; Grindley, 1974; Suggate and others, 1978).

With the exception of some rocks in the Kohler Range (discussed below), no rocks of definitely late Paleozoic to early Mesozoic age are known in Marie Byrd Land (LeMasurier and Wade, 1976), and the next youngest rocks are Mesozoic acid volcanic and intrusive rocks (discussed below). In New Zealand's Western Province, this time interval is represented only by quartzose shallow-marine and nonmarine sedimentary rocks of Permian and Triassic age (Parapara assemblage; Figure 4.2) (I. Raine, personal communication).

In summary, the geologic history of Marie Byrd Land, in so far as it is known, compares well with that of the Western Province of New Zealand and is compatible with their juxtaposition during Paleozoic and early Mesozoic time.

Zone 2—upper Paleozoic to lower Mesozoic igneous rock region

In eastern Marie Byrd Land and western Ellsworth Land, the late Paleozoic-Mesozoic intrusive and metamorphic history is complex. A gabbro-diorite-granodiorite suite in the Eights Coast–Thurston Island region gives radiometric ages of 265–300 m.y. (Lopatin and others, 1974). A whole-rock-mineral isochron age of 265 m.y. has been reported by Halpern (1972) on quartz diorite in the Kohler Range.

In the Eights Coast–Thurston Island region, granite and granodiorite give emplacement ages of 230–180 m.y. (Craddock, Bastien, and Rutford, 1964; Craddock, 1970*a,b*; Lopatin and others, 1974). In the Jones Mountains, Craddock and others (1964) have described the rocks as quartz-perthitic orthoclase-albite-muscovite granite intruded by widespread basalt and felsite dikes. The granite gives a K-Ar (muscovite) age of 199 m.y.

In New Zealand, comparable intrusive and calc-alkaline volcanic rocks of Permian to Jurassic age are widespread in the Brook Street terrane (Devereux and others, 1968; Aronson, 1968; Coombs and others, 1976; Williams and Harper, 1978), an area also characterized by strong magnetic anomalies (Wellman, 1973). Our reconstruction places the Bounty Islands along strike from Thurston Island. The Bounties consist of granodiorite (189 m.y.), and microdiorite, felsic intrusive rocks, and quartzose graywacke have been dredged from the seafloor nearby (Cullen, 1975; Wasserburg and others, 1963).

Zone 2 thus forms a belt of upper Paleozoic to lower Mesozoic intrusive and volcanic rocks extending through New Zealand and the Campbell Plateau to the Eights Coast–Amundsen Sea region of West Antarctica.

Zones 3 and 4—upper Paleozoic–Mesozoic sedimentary rock region

SEDIMENTARY ASSEMBLAGES

In the eastern province of New Zealand, two distinct major Permian–Jurassic sedimentary assemblages or belts are recognized. A western andesitic-volcanogenic Hokonui Assemblage (Zone 3) ranges from Permian to Jurassic in age, and an eastern quartzofeldspathic Alpine Assemblage (Zone 4; also referred to as Torlesse) ranges from Carboniferous to early Cretaceous in age but is quite incomplete (Fleming, 1970; Coombs and others, 1976; Stevens and Speden, 1977; Andrews and others, 1977). Coombs and others (1976) have suggested that the two assemblages were brought into juxtaposition during the late Jurassic to early Cretaceous Rangitata orogeny (main phase) by suturing, the suture zone incorporating the intervening Haast Schist terrane.

As in New Zealand, there are in West Antarctica, within upper Paleozoic to Lower Cretaceous sequences, two distinct major suites based on detrital mineral composition.

First, there is the older Carboniferous(?) to Middle Triassic (Grikurov and Dibner, 1968; Schopf, 1973; Thomson, 1975) quartzofeldspathic Trinity suite (Trinity Peninsula Series and correlatives, including Miers Bluff Formation, Legoupil Formation, and Greywacke-Shale Series) derived largely from acid to intermediate plutonic and/or volcanic terranes (Elliot, 1965; Aitkenhead, 1965), locally with a component from metamorphosed sedimentary rocks (Fleet, 1968) or penecontemporaneous volcanic rocks (Bell, 1973). Second, there is a Middle Jurassic to Albian acid-intermediate volcanogenic marine and nonmarine suite which is well developed at Alexander and Livingston Islands (Horne, 1968; Tavera, 1970; Elliot, 1974; Covacevich, 1976), eastern Ellsworth Land (Laudon and others, 1969; Laudon, 1972), and the Lassiter Coast (Williams and others, 1972). These suites are grossly similar in composition and in metamorphic rank to the New Zealand Alpine and Hokonui Assemblages, respectively.

However, whereas the two suites are generally contemporaneous in New Zealand, those of West Antarctica differ in age and are separated by a major late Triassic–early Jurassic orogeny (Gondwanian; Elliot, 1975). Deformation of the Antarctic suites is comparable to that of their New Zealand equivalents, except for the Middle–Late Jurassic volcanogenic suite in eastern Ellsworth Land and along the Lassiter Coast, where deformation is more intense. There is in New Zealand no equivalent to the widespread Upper Jurassic Volcanic Suite of West Antarctica (Adie, 1955), although altered acidic tuffs have been reported in marine sedimentary sequences (Fleming and Kear, 1960, Figure 16).

In the Holonui and Alpine Assemblages, detrital mineral composition varies both geographically and stratigraphically (Reed, 1966; Mayer, 1969; Landis and Bishop, 1972; Skinner, 1972; Boles, 1974; Force, 1974; Andrews and others, 1977). Similar marked geographic and possibly stratigraphic variation in composition is also reported for the Trinity Peninsula Series (Aitkenhead, 1965; Fleet, 1968; Fraser and Grimley, 1972; West, 1974) and for the Jurassic–Lower Cretaceous volcanogenic suites (Bibby, 1966; Rowley, 1973; Elliot, 1974; Bell, 1975) in West Antarctica, with the latter being in part derived from the former in places (Bibby, 1966; Elliot, 1967).

In both New Zealand and West Antarctica, the provenance and source directions of the sediments, particularly for the grossly similar quartzofeldspathic suites, were more variable in space and time than generally has been accepted. The source of quartzofeldspathic detritus comprising New Zealand's Alpine graywackes remains a first-order problem. The crystalline basement of West Antarctica has been proposed as a possible source (e.g., Bradshaw and Andrews, 1973; Blake and others, 1974). Our reconstruction (Figure 4.3) throws little light on the problem.

Suarez (1976) has related the Mesozoic assemblages in the Alexander Island–Palmer Land region to an island arc–marginal basin system, which extended along the Pacific margin of the Antarctic Peninsula and the southern Andes in late Jurassic–early Cretaceous time. This depositional environment probably continued westward along the Mesozoic coast of West Antarctica but changed its character somewhat within the New Zealand sector.

In summary, although the two sedimentary suites form mainly contemporaneous but discrete belts in New Zealand, they are of two different ages in Antarctica and apparently are not distributed in distinct belts.

Faunas

The Legoupil Formation contains a small, cosmopolitan fauna (Thomson, 1975). *Bakevelloides* aff. *hekiensis* (Kobayashi and Ichikawa) is very similar to specimens identified as *Bakevelloides* n. sp. (Campbell and Force, 1972) from the Kaihikuan Stage of New Zealand (Ladinian; Grant-Mackie and others, 1976) and provides a good age correlation for the Antarctic formation.

Middle and Late Jurassic faunas and floras of West Antarctica (Halle, 1913; Schopf in Elliot, 1975, p. 77) have strong affinities with those of New Zealand, as evidenced by the large number of West Antarctic species (including dominant taxa and those most useful for correlation—*Buchia, Inoceramus,* ammonites, and belemnites) identified with or compared to New Zealand species (Stevens, 1965, 1967; Imlay in Williams and Rowley, 1971; Thomson and Willey, 1972; Willey, 1975; Quilty, 1977). The similarity will be strengthened once New Zealand faunas are described more fully. Both regions were sited within a marginal Tethyan province characterized by a middle latitude, warm, temperate climate (Stevens, 1971).

In addition, Aptian and Albian faunas of the two regions have many described (e.g., *Dimitobelus macgregori* Glaessner (Willey, 1972) and *Aucellina* cf. *radiatostriata* Bonarelli) and undescribed species in common, but faunal affinities are less marked. Continuing differentiation of New Zealand faunas from West Antarctic faunas during Late Cretaceous and early Cenozoic time is indicated by the expansion of the endemic element in faunas and the development of subprovinces (Fleming, 1967; Kauffman, 1973). Yet, faunas in both regions continued to share identical species, and there are other species with close affinities, of both Campanian age (Webb, 1968; Henderson, 1970; Stevens and Speden, 1977) and early Tertiary age (Elliot and others, 1975; Zinsmeister, 1976).

The faunas, then, suggest that geographic links were strongest during Middle and Late Jurassic time; they began to weaken in Early Cretaceous time and continued to weaken through Late Cretaceous and early Tertiary time. The faunas indicate that both New Zealand and West Antarctica were located within the same faunal province, especially prior to Late Cretaceous time.

Early Cretaceous orogenesis, and subsequent erosion and sedimentation

The main phase of the Rangitata orogeny (late Jurassic–early Cretaceous) (Figure 4.2) in New Zealand was expressed as intense tectonism and high *P-T* ratio metamorphism in the Eastern Province, and by widespread granitic plutonism and localized low *P-T* ratio metamorphism in the Western Province (Landis and Coombs, 1967).

In West Antarctica, comparable late Jurassic to early Cretaceous deformation is widespread at the southern end of the Antarctic Peninsula in eastern Ellsworth Land and along the Lassiter Coast (Halpern, 1972; Laudon, 1972; Williams and others, 1972) (Table 4.1), and possibly northward along the east flank of Palmer Land to the Bowman Coast (Fraser and Grimley, 1972). Calc-alkaline plutonism and volcanism were widespread in the same region and also throughout Marie Byrd Land, where associated deformation is much less apparent (LeMasurier and Wade, 1976). Thus, the distinct separation of magmatism and tectonism in New Zealand seems to have persisted across the Marie Byrd Land sector of reconstructed Mesozoic West Antarctica (Figure 4.3), but no farther. In eastern Ellsworth Land and on the Antarctic Peninsula, magmatism and tectonism occurred together. The magmatic, tectonic, and sedimentary history of the northern and southwestern Antarctic Peninsula differs in many other important respects from that of New Zealand (see Table 4.1) and is not discussed in detail here.

Mid-Cretaceous to early Tertiary sedimentation in New Zealand is represented by mainly nonmarine fault-controlled basin sequences in the Western Province ("Paparoa assemblage") (Bishop, 1971; Nathan, 1974), and by mainly shelf-marine to marginal flysch-basin sequences in the Eastern Province ("Raukumara assemblage") (Carter and others, 1974; Suggate and others, 1977; Stevens and Speden, 1977). In the Antarctic Peninsula, locally derived, fault-controlled basinal sequences similar to those in the New Zealand Western Province characterize the upper Cretaceous to lower Tertiary succession (Bibby, 1966; Fraser and Grimley, 1972; Elliot, 1975). However, the northern Antarctic Peninsula differs from New Zealand in its incomplete stratigraphic record, the lack of well-developed flysch basins, and the continuation of Mesozoic plutonism into middle Cenozoic time (Rex, 1972; Elliot, 1975).

Latest Cretaceous–early Cenozoic planed surfaces are well developed in Marie Byrd Land, in Ellsworth Land, on the Lassiter Coast, and elsewhere in the Antarctic Peninsula (Table 4.1). Similar extensive planed surfaces are present in both the Western and Eastern Provinces of New Zealand, where they commonly underlie uppermost Cretaceous or lower Cenozoic sedimentary rocks (Wilson, 1963; Bishop, 1971). These surfaces represent (at least in New Zealand) subaerial erosion which began prior to, and finished after, separation of the two regions. The regional extent and stratigraphic position of "peneplanation" is compatible with original geographic continuity.

Late Mesozoic–Cenozoic development

Development of the East Pacific Rise–Pacific-Antarctic Ridge and separation of the New Zealand block from West Antarctica are generally believed to have begun roughly 80 m.y. ago. Thereafter, the geologic characteristics of the two regions began to evolve along independent lines. Fluvial (or marine) planation came to an end in both regions some time after the separation, and was succeeded by a varied history of volcanism, sedimentation, and block faulting. While much of Cenozoic history has no significance for predrift reconstructions, the history of tectonic displacements bears directly on the validity of any reconstruction. In the case of New Zealand, each sea-floor spreading reconstruction has attempted to incorporate the complex Cenozoic strike-slip displacements that have affected this block, as discussed earlier. The Antarctic block, however, is commonly represented by its present-day outline.

A considerable body of evidence indicates that there has been significant Cenozoic tectonic activity in West Antarctica, but the nature and magnitude of displacements have been largely speculative because they are difficult to measure directly. On the basis of sea-floor spreading data, Molnar and others (1975) have postulated the existence of a spreading plate boundary between East and West Antarctica from 81 to 63 m.y. ago. They also postulate 500 km of convergence across the same boundary from 63 to 38 m.y. ago, followed by cessation of plate-boundary activity in Antarctica. Volcanic history, displacement of the prevolcanic erosion surface, and a variety of geophysical characteristics suggest that the Ross Sea–Byrd subglacial basin region is the axis of a major intracontinental rift system, and it is possible that this rift originated as an arm of the Pacific-Antarctic Ridge which subsequently failed (LeMasurier, in press). However, there is no geologic evidence as yet that supports a period of Cenozoic convergence between East and West Antarctica, and it seems likely that much of the vertical and lateral displacement related to the rift system has taken place since the development of the continental ice sheet (LeMasurier and Rex, Chapter 89, this volume). Acknowledging the persistent difficulties in determining the tectonic history of the Ross Sea–Byrd Basin region, we think it most likely that this region has evolved primarily, if not entirely, by tectonic extension, that most of the extension has taken place in the past 80 m.y., and that this process is responsible for the zones of thin crust beneath the Byrd Basin and the Ross Sea, and probably beneath the Lord Howe Rise and the Campbell Plateau. Thus, to restore the Marie Byrd Land–Ellsworth Land region to its predrift configuration, some compensation for late Mesozoic–Cenozoic extension seems to be required. In the absence of more specific constraints, we have arbitrarily compressed the Antarctic portion of the region shown in Figure 4.3 by 10–15%, which approximately "restores" the Marie Byrd Land continental crust to East Antarctic thicknesses.

Conclusions

1. Geologic history and structural trends in Marie Byrd Land compare well with those in the Western Province of New Zealand and are compatible with juxtaposition of the two regions in our reconstruction (Zone 1, Figure 4.3).

2. Upper Paleozoic to lower Mesozoic basic to acid igneous rocks of the Eights Coast generally match those of the Brook Street terrane of New Zealand and may represent a continuation of the same belt (Zone 2, Figure 4.3). It should be possible to test this conclusion by an aeromagnetic survey of the Marie Byrd Land–western Ellsworth Land coastal regions.

3. Rocks similar in composition and age to the major assemblages (Hokonui and Alpine) of the "New Zealand Geosyncline" are present in the Antarctic Peninsula, but contemporaneous paired sedimentary belts cannot be recognized. Thus, extrapolation of the Hokonui and Alpine belts into the Antarctic Peninsula cannot be demonstrated at this time. No remnant of these belts has been found, as yet, in the eastern Marie Byrd Land–western Ellsworth Land region, which must have lain inland of this province in Mesozoic time.

4. Mesozoic faunas suggest close links between New Zealand and the Antarctic Peninsula during Jurassic time, but faunal links weakened during Cretaceous and early Tertiary time.

5. An orogenic event equivalent to the Rangitata orogeny of New Zealand can be recognized widely throughout Marie Byrd Land, Ellsworth Land, and the southern Antarctic Peninsula. The distinct separation of late Mesozoic tectonism and plutonism in paired belts in New Zealand probably persisted across Marie Byrd Land and Ellsworth Land, but not into the southern Antarctic Peninsula, where tectonism and magmatism occurred together.

6. Cenozoic tectonic activity in West Antarctica has apparently been dominated by vertical displacements and lateral extension centered over a Ross Sea–Byrd Basin axis. Postdrift tectonic displacements in Marie Byrd Land and Ellsworth Land have been accommodated in our reconstruction (Figure 4.3) by correcting for an assumed 10–15% extension, roughly parallel to longitude lines.

Acknowledgments. The preparation of this paper was greatly facilitated by the provision of advance copies of manuscripts of K. A. W. Crook and L. Belbin (Australian National University, Canberra); F. J. Davey (Geophysics Division, Department of Scientific and Industrial Research) and D. A. Christoffel (Department of Physics, Victoria University of Wellington, New Zealand); F. J. Davey and R. E. Houtz (Lamont-Doherty Geological Observatory, Columbia University, U.S.A.); and J. R. Weissel, D. E. Hayes, and E. M. Herron (Lamont-Doherty Geological Observatory). We thank B. Scott, Elma MacGregor, and Carolyn Landis for assistance in the preparation of the manuscript.

References cited

Adie, R. J. 1955. The petrology of Graham Land: II. The Andean granite-gabbro intrusive suite. *Scient. Rep. Falkld Isl. Depend. Surv.*, No. 12, 39 pp.

Aitkenhead, N. 1965. The geology of the Duse Bay–Larsen Inlet area, northeast Graham Land (with particular reference to the Trinity Peninsula Series). *Scient. Rep. Br. antarc. Surv.*, No. 51, 62 pp.

Andrews, P. B., Speden, I. G., and Bradshaw, J. D. 1977. Lithological and paleontological content of the Carboniferous–Jurassic Canterbury Suite, South Island, New Zealand. *N.Z. Jl Geol. Geophys.*, *19*, No. 6, 791–819.

Aronson, J. L. 1968. Regional geochronology of New Zealand. *Geochim. cosmochim. Acta*, *32*, 669–97.

Beggs, J. M. 1978. Geology of the metamorphic basement and Late Cretaceous to Oligocene sedimentary sequence of Campbell Island, southwest Pacific Ocean. *J. R. Soc. N.Z.*, *8*, 161–77.

Bell, C. M. 1973. The geology of southern Alexander Island. *Bull. Br. antarc. Surv.*, No. 33–34, 1–16.

Bell, C. M. 1975. Structural geology of parts of Alexander Island. *Bull. Br. antarc. Surv.*, No. 41–42, 43–58.

Bentley, C. R., Crary, A. P., Ostenso, N. A., and Thiel, E. C. 1960. Structure of West Antarctica. *Science, N.Y.*, *131*, 131–36.

Bibby, J. S. 1966. The stratigraphy of part of north-east Graham Land and the James Ross Island Group. *Scient. Rep. Br. antarc. Surv.*, No. 53, 37 pp.

Bishop, D. G. 1971. Sheet S1, S3, and Pt. S4, Farewell-Collingwood. *Geological Map of New Zealand 1:63360.* Wellington, New Zealand Department of Scientific and Industrial Research.

Bishop, D. G. 1974. Stratigraphic, structural and metamorphic relationships in the Dansey Pass Area, Otago, New Zealand. *N.Z. Jl Geol. Geophys.*, *17*, No. 2, 301–55.

Bishop, D. G., and Laird, M. G. 1976. The stratigraphy and depositional environment of the Kyeburn Formation, central Otago (With an appendix by D. C. Mildenhall). *J. R. Soc. N.Z.*, *6*, 55–71.

Blake, M. C., Jr., Jones, D. L., and Landis, C. A. 1974. Active continental margins: contrasts between California and New Zealand. (*In* Burk, C. A., and Drake, C. L., *eds. The Geology of Continental Margins.* New York, Springer-Verlag, 853–72.)

Boles, J. R. 1974. Structure, stratigraphy, and petrology of mainly Triassic rocks, Hokonui Hills, Southland, New Zeland. *N.Z. Jl Geol. Geophys.*, *17*, No. 2, 337–74.

Bradshaw, J. D., and Andrews, P. B. 1973. Geotectonics and the New Zealand geosyncline. *Nature Phys. Sci.*, *241*, No. 105, 14–16.

Brown, D. A., Campbell, K. S. W., and Crook, K. A. W. 1968. *The Geological Evolution of Australia and New Zealand.* Oxford, Pergamon Press, 409 pp.

Campbell, J. D., and Force, E. R. 1972. Stratigraphy of the Mount Potts Group at Rocky Gully, Rangitata Valley, Canterbury. *N.Z. Jl Geol. Geophys.*, *15*, No. 1, 157–67.

Carter, R. M., Landis, C. A., Norris, R. J., and Bishop, D. G. 1974. Suggestions towards a high-level nomenclature for New Zealand rocks. *J. R. Soc. N.Z.*, *4*, No. 1, 5–18.

Christoffel, D. A., and Falconer, R. K. H. 1972. Marine magnetic measurements in the southwest Pacific Ocean and the identification of new tectonic features. (*In* Hayes, D. E., *ed. Antarctic Oceanology 2. The Australian–New Zealand Sector.* American Geophysical Union, 197–204.) [Antarctic Research Series, Vol. 19.]

Christoffel, D. A., and Falconer, R. K. H. 1973. Changes in the direction of sea floor spreading in the southwest Pacific. (*In* Fraser, R., *comp. Oceanography of the South Pacific, 1972.* Wellington, New Zealand National Commission for UNESCO, 241–47.)

Coombs, D. S., Landis, C. A., Norris, R. J., Sinton, J. M., Borns, D. J., and Craw, D. 1976. The Dun Mountain ophiolite belt, New Zealand, its tectonic setting, constitution, and origin, with special reference to the southern portion. *Am. J. Sci.*, *276*, 561–603.

Cooper, R. A. 1975. New Zealand and south-east Australia in the early Paleozoic. *N.Z. Jl Geol. Geophys.*, *18*, No. 1, 1–20.

Cooper, R. A. 1976. The fit of New Zealand against Australia-Antarctica in the late Mesozoic. *Newsl. geol. Soc. N.Z.*, *42*, 31–34.

Cooper, R. A. 1979. Lower Paleozoic rocks of New Zealand. *J. R. Soc. N.Z.*, *9*, 29–84.

Covacevich, V. 1976. Fauna Valanginiana de Peninsula Byers, Isla Livingston, Antarctica. *Revta Geol. Chile*, No. 3, 25–56.

Craddock, C. 1970*a*. Radiometric age map of Antarctica. (*In* Bushnell, V. C., and Craddock, C., *eds.* Geologic Maps of Antarctica. *Antarct. Map Folio Ser.*, Folio 12, Pl. XVIX.)

Craddock, C. 1970*b*. Geologic map of Antarctica. (*In* Bushnell, V. C., and Craddock, C., *eds.* Geologic Maps of Antarctica. *Antarct. Map Folio Ser.*, Folio 12, Pl. XX.)

Craddock, C., Anderson, J. J., and Webers, G. F. 1964. Geologic outline of the Ellsworth Mountains. (*In* Adie, R. J., *ed. Antarctic Geology.* Amsterdam, North-Holland Publishing Company, 155–70.)

Craddock, C., Bastien, T. W., and Rutford, R. H. 1964. Geology of the Jones Mountains area. (*In* Adie, R. J., *ed. Antarctic Geology.* Amsterdam, North-Holland Publishing Company, 171–87.)

Crook, K. A. W., and Belbin, L. 1978. The Southwest Pacific Area during the last 90 m.y. *J. geol. Soc. Aust.*, *25*, 23–40.

Cullen, D. J. 1970. Tectonic map of the southwest Pacific, 1:10,000,000 at lat. 46° S. *Misc. Ser. N.Z. oceanogr. Inst.*, *20.*

Cullen, D. J. 1975. Autochthonous rocks of the Bounty Islands region, south-west Pacific Ocean. *N.Z. Jl Geol. Geophys.*, *18*, No. 6, 767–85.

Dalziel, I. W. D., and Elliot, D. H. 1973. The Scotia arc and Antarctic margin. (*In* Nairn, A. E. M., and Stehli, F. G., *eds. The Ocean Basins and Margins. 1. The South Atlantic.* New York, Plenum Press, 171–246.)

Davey, F. J., and Christoffel, D. A. 1978. Magnetic anomalies across Campbell Plateau, New Zealand. *Earth planet. Sci. Lett., 41,* 14–20.

Davey, F. J., and Houtz, R. E. 1977. The Campbell Plateau and its relationship with the Ross Sea, Antarctica. *Mar. Geol., 25,* 61–72.

Devereux, I., McDougall, I., and Watters, W. A. 1968. Potassium-argon mineral dates on intrusive rocks from the Foveaux Strait area. *N.Z. Jl Geol. Geophys., 11,* No. 5, 1230–35.

Elliot, D. H. 1965. Geology of north-west Trinity Peninsula, Graham Land. *Bull. Br. antarc. Surv., 7,* 1–24.

Elliot, D. H. 1967. The geology of Joinsville Island. *Bull. Br. antarc. Surv., 12,* 23–40.

Elliot, D. H. 1975. Tectonics of Antarctica: a review. *Am. J. Sci., 275-A,* 45–106.

Elliot, D. H., Rinaldi, C., Zinsmeister, W. J., Trautman, T. A., Bryant, W. A., and del Valle, R. 1975. Geological investigations on Seymour Island, Antarctic Peninsula. *Antarct. Jnl U.S., 10,* No. 4, 182–86.

Elliot, M. H. 1974. Stratigraphy and sedimentary petrology of the Ablation Point Area, Alexander Island. *Bull. Br. antarc. Surv.,* No. 39, 87–113.

Fleet, M. 1968. The geology of Oscar II Coast, Graham Land. *Scient. Rep. Br. antarc. Surv.,* No. 59, 46 pp.

Fleming, C. A. 1967. Biogeographic change related to Mesozoic orogenic history in the Southwest Pacific. *Tectonophysics, 4,* 419–27.

Fleming, C. A. 1970. The Mesozoic of New Zealand: chapters in the history of the circum-Pacific mobile belt. *Q. Jl geol. Soc. Lond., 125,* 125–70.

Fleming, C. A., and Kear, D. 1960. The Jurassic sequence at Kawhia Harbour, New Zealand (Kawhia Sheet N73). *Bull. geol. Surv. N.Z., 67,* 1–50.

Force, E. R. 1974. A comparison of some Triassic rocks in the Hokonui and Alpine belts of South Island, New Zealand. *J. Geol., 82,* No. 1, 37–49.

Fraser, A. G., and Grimley, P. H. 1972. The geology of parts of the Bowman and Wilkins coasts, Antarctic Peninsula. *Scient. Rep. Br. antarc. Surv.,* No. 67, 59 pp.

Grant-Mackie, J. A., Paris, J. P., Freneix, S., Campbell, J. D., Speden, I. G., and Stevens, G. R. 1976. Advances in the correlation of Mesozoic sequences of New Zealand and New Caledonia. *Abstr. 25th Int. Geol. Congr., Sydney, 1976,* 1, 268–70.

Grikurov, G. E., and Dibner, A. F. 1968. Novye dannye o serii Triniti (C_{1-3}) v Zapadnoi Antarktide [More information on the Trinity Series (C_{1-3}) of western Antarctica]. *Dokl. Akad. Nauk SSSR,* Ser. geol., *179,* No. 2, 410–12.

Grindley, G. W. 1974. New Zealand. (*In* Spencer, A. M., *ed. Mesozoic-Cenozoic Orogenic Belts: Data for Orogenic Studies.* London, Geological Society of London, 387–416.)

Halle, T. G. 1913. The Mesozoic flora of Graham Land. *Wiss. Ergebn. Schwed. Sudpolarexped., 1901–1903, 3,* No. 14, 123 pp.

Halpern, M. 1968. Ages of Antarctic and Argentine rocks bearing on continental drift. *Earth planet. Sci. Lett., 5,* 159–67.

Halpern, M. 1972. Rb-Sr total-rock and mineral ages from the Marguerite Bay area Kohler Range and Fosdick Mountains. (*In* Adie, R. J., *ed. Antarctic Geology and Geophysics.* Oslo, Universitetsforlaget, 197–204.)

Harrington, H. J., Burns, K. L., Thompson, B. R., and Ozolins, A. P. 1974. Regional geology of Victoria in relation to satellite imagery; a preparatory study. *Rep. CSIRO Min. Res. Lab. Invest., 106,* 37 pp.

Hayes, D. E., and Davey, F. J. 1975. A geophysical study of the Ross Sea, Antarctica. (*In* Hayes, D. E., Frakes, L. A., and others. *Initial Reports of the Deep Sea Drilling Project,* Vol. 28. Washington [U.S. Government Printing Office], 887–907.)

Hayes, D. E., and Ringis, J. 1973. Seafloor spreading in the Tasman Sea. *Nature, Lond., 243,* 454–58.

Henderson, R. A. 1970. Ammonoidea from the Mata Series (Santonian-Maastrichtian) of New Zealand. *Spec. Pap. paleont. Ass.,* No. 6, 1–82.

Horne, R. R. 1968. Petrology and provenance of the Cretaceous sediments of southeastern Alexander Island. *Bull. Br. antarc. Surv.,* No. 17, 73–82.

Iltchenko, L. N. 1972. Late Precambrian acritarchs of Antarctica. (*In* Adie, R. J., *ed. Antarctic Geology and Geophysics.* Oslo, Universitetsforlaget, 599–602.)

Katz, H. R. 1973. Contrasts in tectonic evolution of orogenic belts in the southeast Pacific. *J. R. Soc. N.Z., 3,* No. 3, 333–62.

Kauffman, E. G. 1973. Cretaceous Bivalvia. (*In* Hallam, A., *ed. Atlas of Paleobiogeography.* Amsterdam, Elsevier, 353–83.)

Keble, R. A., and Benson, W. N. 1929. Ordovician graptolites of north-west Nelson. *Trans. N.Z. Inst., 59,* 840–63.

Laird, M. G. 1972. Sedimentology of the Greenland group in the Paparoa Range, West Coast, South Island. *N.Z. Jl Geol. Geophys., 15,* No. 3, 372–93.

Landis, C. A., and Bishop, D. G. 1972. Plate tectonics and regional stratigraphic-metamorphic relations in the southern part of the New Zealand geosyncline. *Geol. Soc. Am. Bull., 83,* 2267–84.

Landis, C. A., and Coombs, D. S. 1967. Metamorphic belts and orogenesis in southern New Zealand. *Tectonophysics, 4,* 501–18.

Laudon, T. S. 1972. Stratigraphy of eastern Ellsworth Land. (*In* Adie, R. J., *ed. Antarctic Geology and Geophysics.* Oslo, Universitetsforlaget, 215–23.)

Laudon, T. S., Lackey, L. L., Quilty, P. G., and Otway, P. M. 1969. Geology of Eastern Ellsworth Land. (*In* Bushnell, V. C., and Craddock, C., *eds.* Geologic Maps of Antarctica. *Antarct. Map Folio Ser.,* Folio 12, Pl. III.)

LeMasurier, W. E. In press. Rift valley volcanism in Marie Byrd Land, West Antarctica. (*In* Green, J., *ed. Volcanoes and Volcanology.* Dowden, Hutchinson and Ross.)

LeMasurier, W. E., and Wade, F. A. 1976. Volcanic history in Marie Byrd Land. Implications with regard to Southern Hemisphere tectonic reconstructions. (*In* Gonzalez-Ferran, O., *ed. Andean and Antarctic Volcanology Problems.* Rome, International Association of Volcanology and Chemistry of the Earth's Interior, 398–422).

Lopatin, B. G., and Orlenko, E. M. 1972. Outline of the geology of Marie Byrd Land and the Eights Coast. (*In* Adie, R. J., *ed. Antarctic Geology and Geophysics.* Oslo, Universitetsforlaget, 245–50.)

Lopatin, B. G., Krylov, A. Ya., and Aliapyshev, O. A. 1974. Osnovnye tektono-magmaticheskie etapy v razvitii Zemli Meri Berd i Berega Eitsa, Zapadnaia Antarktida [The principal tectonic-magmatic stages in the development of the Marie Byrd Land and Eights Coast (western Antarctica) according to radioactive data]. *Antarktika, Doklady Komissii, 13,* 52–60.

Mayer, W. 1969. Petrology of the Waipapa Group, near Auckland, New Zealand. *N.Z. Jl Geol. Geophys., 12,* No. 2–3, 412–35.

Mehnert, H. H., Rowley, R. D., and Schmidt, D. L. 1975. K-Ar ages of plutonic rocks in the Lassiter Coast area, Antarctica. *J. Res. U. Geol. Surv. 3,* No. 2, 233–36.

Molnar, P., Atwater, T., Mammerickx, J., and Smith, S. M. 1975. Magnetic anomalies, bathymetry, and tectonic evolution of the South Pacific since the Late Cretaceous. *Geophys. J. R. astr. Soc., 49,* 383–420.

Nathan, S. 1974. Stratigraphic nomenclature for the Cretaceous–lower Quaternary rocks of Buller and north Westland, West Coast, South Island, New Zealand. *N.Z. Jl Geol. Geophys., 17,* No. 2, 423–45.

Quilty, P. G. 1977. Late Jurassic bivalves from Ellsworth Land, Antarctica: their systematics and paleogeographic implications. *N.Z. Jl Geol. Geophys., 20,* No. 6, 1033–80.

Reed, J. J. 1966. Mineralogy and petrology in the New Zealand Geological Survey 1865–1965. *N.Z. Jl Geol. Geophys., 8,* No. 6, 999–1087.

Rex, D. C. 1972. K-Ar age determinations on volcanic and associated rocks from the Antarctic Peninsula and Dronning Maud Land. (*In* Adie, R. J., *ed. Antarctic Geology and Geophysics.* Oslo, Universitetsforlaget, 133–36.)

Rowley, P. D. 1973. Geologic observations on the northern Lassiter Coast and southern Black Coast. *Antarct. Jnl U.S., 8,* No. 4, 154–55.

Schopf, J. M. 1973. Plant material from the Miers Bluff Formation of the South Shetland Islands. *Rep. Inst. polar Stud. Ohio State Univ.,* No. 45, 45 pp.

Skinner, D. N. B. 1972. Subdivision and petrology of the Mesozoic rocks of Coromandel (Manaia Hill Group). *N.Z. Jl Geol. Geophys., 15,* No. 2, 203–27.

Stevens, G. R. 1965. The Jurassic and Cretaceous belemnites of New Zealand and a review of the Jurassic and Cretaceous belemnites of the Indo-Pacific region. *Pal. Bull. geol. Surv. N.Z., 36,* 283 pp.

Stevens, G. R. 1967. Upper Jurassic fossils from Ellsworth Land, West Antarctica, and notes on Upper Jurassic biogeography of the South Pacific region. *N.Z. Jl Geol. Geophys., 10,* No. 2, 345–93.

Stevens, G. R. 1971. Relationship of isotopic temperatures and faunal realms to Jurassic-Cretaceous paleogeography, particularly of the southwest pacific. *J. R. Soc. N.Z., 1,* No. 2, 145–58.

Stevens, G. R., and Speden, I. G. 1977. Mesozoic of New Zealand. (*In* Moullade, M., and Nairn, A. E. M., *eds. Mesozoic of the World.* Amsterdam, Elsevier, 251–328.)

Suarez, M. 1976. Plate-tectonic model for southern Antarctic Peninsula and its relation to southern Andes. *Geology, 4,* 211–14.

Suggate, R. P. 1963. The Alpine Fault. *Trans. R. Soc. N.Z.,* Geol., *2,* No. 7, 105–29.

Suggate, R. P., Stevens, G. R., and TePunga, M. T., *eds.* 1978. *The Geology of New Zealand* (2 vols.). Wellington, Government Printer.

Tavera, J. 1970. Fauna Titonica-Neocomiana de Isla Livingston, Islas Shetland del Sur, Antarctica. *Ser. cient. Inst. antart. chil., 1,* No. 2, 175–86.

Thomson, M. R. A. 1975. New palaeontological and lithological observations on the Legoupil Formation, northwest Antarctic Peninsula. *Bull. Br. antarc. Surv.,* No. 41–42, 169–85.

Thomson, M. R. A., and Willey, L. E. 1972. Upper Jurassic and Lower Cretaceous *Inoceramus* (Bivalvia) from South-east Alexander Island. *Bull. Br. antarc. Surv.,* No. 29, 1–19.

Wade, A. 1975. Swanson Group, Ford Ranges, Marie Byrd Land. *Antarct. Jnl U.S., 10,* No. 5, 244–45.

Wade, F. A., and Wilbanks, T. R. 1972. Geology of Marie Byrd and Ellsworth Lands. (*In* Adie, R. J., *ed. Antarctic Geology and Geophysics.* Oslo, Universitetsforlaget, 207–14.)

Warner, L. A. 1945. Structure and petrography of the southern Edsel Ford Ranges, Antarctica. *Proc. Am. Phil. Soc., 89,* 78–122.

Wasserburg, G. J., Craig, H., Menard, H. W., Engel, A. E., and Engel, C. G. 1963. Age and composition of a Bounty Island granite, and age of a Seychelles Island granite. *J. Geol., 71,* 785–89.

Webb, P. N. 1968. Comments on Late Cretaceous marine climates in New Zealand and adjacent areas. *Tuatara, 16,* No. 1, 8–10.

Weissel, J. K., and Hayes, D. E. 1972. Magnetic anomalies in the southeast Indian Ocean. (*In* Hayes, D. E., *ed. Antarctic Oceanology 2. The Australian-New Zealand Sector.* Washington, American Geophysical Union, 165–96.) [Antarctic Research Series, Vol. 19.]

Weissel, J. K., and Hayes, D. E. 1977. Evolution of the Tasman Sea reappraised. *Earth planet. Sci. Lett., 36,* 77–84.

Weissel, J. K., Hayes, D. E., and Herron, E. M. 1977. Plate tectonics synthesis: the displacements between Australia, New Zealand, and Antarctica since the Late Cretaceous. *Mar. Geol., 25,* 231–77.

Wellman, H. W. 1956. Structural outline of New Zealand. *Bull. N.Z. Dep. scient. ind. Res., 121,* 36 pp.

Wellman, H. W. 1973. The Stokes Magnetic Anomaly. *Geol. Mag., 110,* No. 5, 419–29.

Wellman, H. W. 1975. New Zealand 60 m.y. ago. *Bull. Aust. Soc. Explor. Geophys., 6,* No. 2–3, 55–56.

West, S. M. 1974. The geology of the Danco Coast, Graham Land. *Scient. Rep. Br. antarc. Surv., 84,* 58 pp.

Wilbanks, J. R. 1972. Geology of the Fosdick Mountains, Marie Byrd Land. (*In* Adie, R. J., *ed. Antarctic Geology and Geophysics.* Oslo, Universitetsforlaget, 277–84.)

Willey, L. E. 1972. Belemnites from southeast Alexander Island: I. the occurrence of the family Dimitobelidae in the Lower Cretaceous. *Bull. Br. antarc. Surv.,* No. 28, 29–42.

Willey, L. E. 1975. Upper Jurassic and Lower Cretaceous Grammato-dontinae (Bivalvia) from southern Alexander Island. *Bull. Br. antarc. Surv.,* No. 41–42, 1–22.

Williams, J. Y., and Harper, C. T. 1978. Age and status of the Mackay Intrusives in the Eglington-upper Hollyford area. *N.Z. Jl Geol. Geophys., 21,* 733–42.

Williams, P. L., and Rowley, P. D. 1971. Geological studies of the Lassiter Coast. *Antarct. Jnl U.S., 6,* No. 4, 120.

Williams, P. L., Schmidt, D. L., Plummer, C. C., and Brown, L. E. 1972. Geology of the Lassiter Coast area, Palmer Land, Antarctica: preliminary report. (*In* Adie, R. J., *ed. Antarctic Geology and Geophysics.* Oslo, Universitetsforlaget, 143–48.)

Wilson, D. D. 1963. Geology of Waipara Subdivision (Amberley and Motunau Sheets S68 and S69). *Bull. geol. Surv. N.Z., 64,* 122 pp.

Wood, B. L. 1960. Sheet 27 Fiord. *Geological Map of New Zealand 1:250,000.* Wellington, New Zealand Department of Scientific and Industrial Research.

Wood, B. L. 1962. Sheet 22 Wakatipu. *Geological Map of New Zealand 1:250,000.* Wellington, New Zealand Department of Scientific and Industrial Research.

Woollard, G. P. 1962. Crustal structure in Antarctica. (*In* Wexler, H., Rubin, M. J., and Caskey, J. E., *eds. Antarctic Research: The Matthew Fontaine Maury Memorial Symposium.* Washington, American Geophysical Union, 53–73.) [Geophysics Monography No. 7.]

Zinsmeister, W. J. 1976. A new genus and species of the gastropod family Struthiolariidae, *Antarctodarwinella ellioti,* from Seymour Island, Antarctica. *Ohio J. Sci., 76,* No. 3, 111–14.

5

Seismicity, Fracture Zones, and Poles of Rotation of the Pacific-Antarctic and Indian-Antarctic Plate Boundaries

R. K. H. FALCONER
Atlantic Geoscience Centre, Geological Survey of Canada, Bedford Institute of Oceanography, Dartmouth, Nova Scotia B2Y 4A2 Canada

R. FERIEL FALCONER
35 Slayter Street, Dartmouth, Nova Scotia B3A 2A2, Canada

Abstract. The seismicity of the entire Pacific-Antarctic plate boundary and of the Indian-Antarctic plate boundary east of the Australian-Antarctic discordance is discussed. Virtually all the activity is on ridge-ridge transform faults, and 30% of it occurs as pairs or triplets of events of similar magnitude; the remainder is random in time. The earthquake magnitude-frequency relationship for the fracture zones on the Pacific-Antarctic boundary gives $b = 1.75$, and for those on the Indian-Antarctic boundary, $b = 1.99$, compared with $b = 0.99$ for Mid-Atlantic Ridge fracture zones. The higher Antarctic boundary *b*-values probably reflect lower stresses and smaller source areas related to higher temperatures. The Antarctic temperatures are higher because age offsets across the Antarctic fracture zones are less than 10 m.y. as compared with 20 m.y. in the Atlantic. The epicenters of each fracture zone have been fitted to a small circle, enabling the pole of relative plate rotation to be determined. The standard deviation of epicenter scatter about these small circles ranges from 6 to 28 km, averaging 17 km; all the scatter can be attributed to epicenter location uncertainties. The active parts of the fracture zones, which average 250 km in length, are thus very narrow. The rotation poles determined from the epicenters are: Indian-Antarctic, at lat. 30° S, long. 166° W; and Pacific-Antarctic, either at (a) lat. 71° S, long. 122° E or at (b) lat. 76° S, long. 144° E, the two solutions depending on whether the 1000-km-long Pacific-Antarctic fracture zone at the western end of the boundary is considered to be several sections or one section. The Indian-Antarctic pole and the Pacific-Antarctic pole (b) fit what is known of the boundary topography, magnetic anomalies, and focal mechanisms. They do not fit the spreading rates averaged over the last 1.8 m.y. The epicentral poles, however, apply to movements of only the last 13 years. They are significantly different from poles determined from data reflecting several million years' plate movement. Thus, the rotation poles appear to have moved during the last few million years.

In this paper we present studies of the seismicity of the boundary of the Antarctic plate with both the Pacific plate and the eastern part of the Indian plate (Figure 5.1). The positions of the plate boundaries in the area are reasonably well delineated by bathymetric, magnetic, and epicentral data; and in most places the nature of the boundaries (ridge or transform) is clear. However, the western end of the Pacific-Antarctic boundary from long. 180° westward to the Indian-Pacific-Antarctic triple junction is complex and may be either one large fracture zone or a series of ridge and fracture zone segments (Falconer, 1972; Christoffel and Falconer, 1972). We hoped that a closer study of seismicity would help elucidate the nature of that boundary. Our initial aim has not been accomplished, but the study has yielded results of general significance in two fields: the structure and tectonics of active fracture zones and the determination of poles of relative plate rotation.

We examined the statistics of earthquake occurrence, particularly earthquake magnitude-frequency relationship *b*-values, as Francis (1968*a,b*) has shown that for the Mid-Atlantic Ridge, the *b*-value of fracture zones differs considerably from that of the ridge axis. The *b*-value results did not prove to be useful for defining plate boundary type, but comparison of the Atlantic and the Antarctic boundary data provided information on the stresses at active fracture zones.

Critical examination of the epicenter distribution showed that wherever the nature of the boundary is clear, epicenters are not only associated with ridge-ridge transform faults, but they also accurately define them. This led to the development of a numerical technique for determining a pole of relative plate rotation from the epicenters. Interpretation of the complex western section of the Pacific-Antarctic boundary proved to be crucial in determining the Pacific-Antarctic pole, and two alternative poles were derived. One of these, and the Indian-Antarctic pole, differ significantly from poles previously determined from other data, indicating changes in plate motions.

Most of the work presented here is discussed more fully in Falconer (1974), although the seismicity data base for *b*-value calculations is now up to June 1976. More detailed work on both the *b*-values and the epicentral method, including analysis of the Indian-Pacific boundary, is being prepared for publication elsewhere.

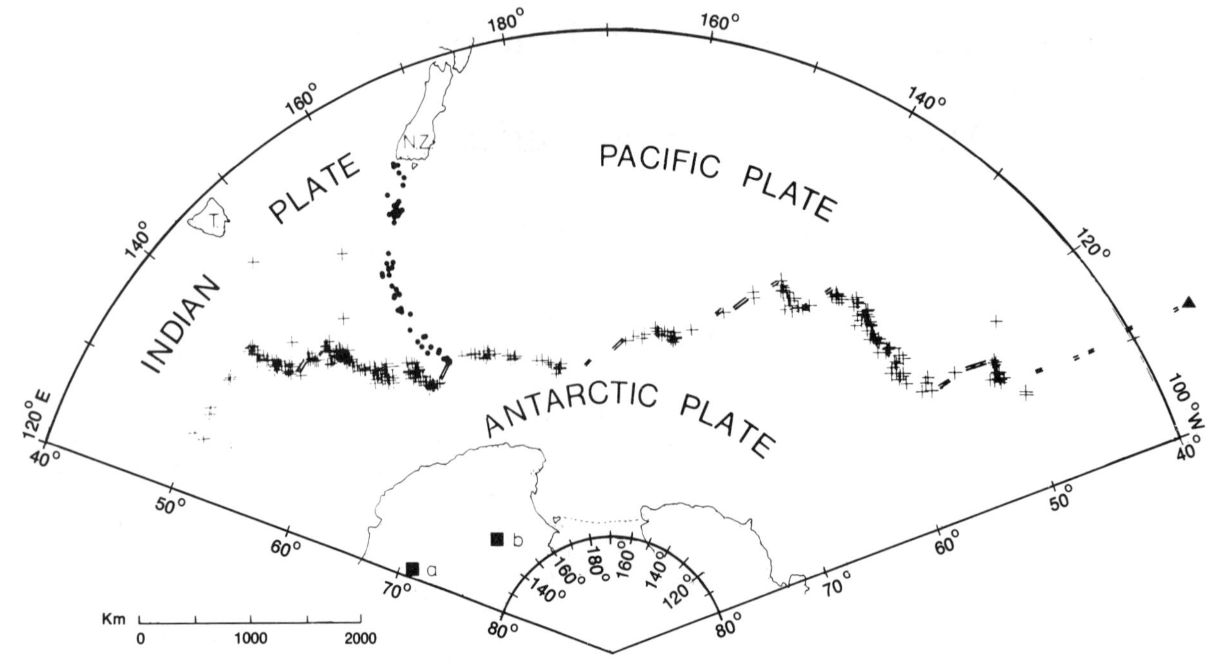

Figure 5.1. The Antarctic plate boundary, long. 120° E–100° W. USCGS epicenters for 1964 through June 1976 shown by crosses on Indian-Antarctic and Pacific-Antarctic boundaries and by dots on the Macquarie Ridge complex portion of the Indian-Pacific boundary. Double lines indicate ridge axes where definitely identified. Triangle is Pacific-Nazca-Antarctic triple junction. Squares (a) and (b) are alternative Pacific-Antarctic epicentral poles. T: Tasmania. NZ: New Zealand.

Seismicity data

We discuss a 120° sector of the Antarctic plate boundary, from long. 130° E to 110° W (Figure 5.1). This covers all of the Pacific-Antarctic plate boundary and the Indian-Antarctic plate boundary east of the Australian-Antarctic discordance (Hayes and Conolly, 1972). Figure 5.1 also shows the epicenters along the Macquarie Ridge complex, which is the Indian-Pacific plate boundary from the Indian-Pacific-Antarctic triple junction northward to New Zealand (Hayes and Talwani, 1972; Christoffel and R. F. Falconer, 1973).

The data source is earthquakes located by the U.S. Coast and Geodetic Survey (USCGS) (and its successors) for the period January 1964 through March 1972 for the epicentral method, and through June 1976 for the *b*-values. Earlier data were not used as the detection level and epicenter accuracy were not as good. It is probable that most earthquakes of magnitude greater than 4.5, and almost certainly all greater than 5.5, were detected (Evernden, 1969). The magnitudes used are body wave magnitudes, m_b. Reported depths of all the earthquakes are shallower than 49 km.

The accuracy of the epicenter locations is of concern in this study. Evernden's (1969) theoretical analysis of the Worldwide Standard Network indicated errors of at least 30 km in the area. However, he considered a configuration of only four stations, whereas the minimum number used by the USCGS for any event was six, and for only 13% of the events were fewer than ten stations used. Sykes (1970*a*) relocated earthquakes west of long. 180° and noted that his locations are less scattered than those of the USCGS. Differences between the Sykes and USCGS positions for 66 common events show no directional bias; seven differences are more than 30 km, and the standard deviation is 24 km. Given that Sykes considered only the better observed events, it is reasonable to assume that epicenter locations may be uncertain by at least 25 km.

The earthquakes are not uniformly distributed in time. Thirty percent of them occur in clusters consisting of two or three adjacent events occurring within hours or days of each other. Spatially the events in each cluster could originate at the same place, since they lie within 40 km of each other. The events in each pair or triplet vary in magnitude by less than 0.4, so they are not typical foreshock or aftershock sequences associated with one main event (Sykes, 1970*b*).

Throughout the area, wherever the ridge axis has been identified from magnetic anomalies and/or bathymetry, it is aseismic. This applies for the USCGS seismicity and for the more sensitive, but less positionally accurate, T-phase data (Northrop, 1973). Virtually all the activity is on fracture zones between the offset ridge segments, i.e., on the ridge-ridge transform faults. There are, however, some events away from the active transforms. These epicenters are mostly near morphologic features and magnetic anomaly offset lines that are extensions of the active fracture zones. The activity could be due to differential thermal subsidence across the offsets (DeLong and others, 1977). We do not consider these mid-plate events in the analysis discussed below.

Epicentral pole method

Study of the seismicity on the fracture zones led to the development of a new method for finding the pole of relative motion of two plates. The method is based on the fact that ridge-ridge transform faults should be along small circles about the rotation pole for the plate boundary. Earthquakes associated with a transform fault should therefore be along the arc of a small circle, centered on the

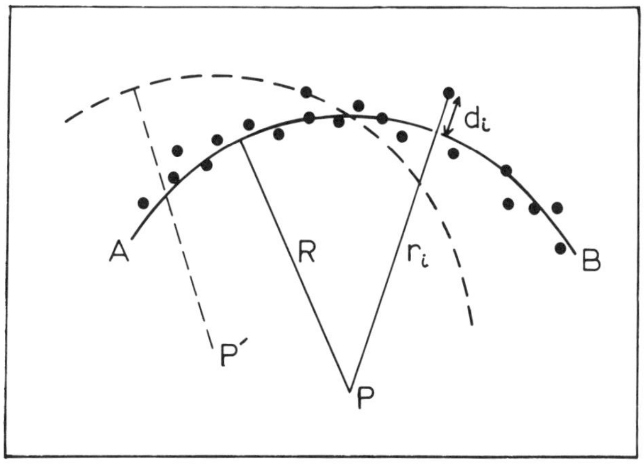

Figure 5.2. The epicentral method. Earthquakes associated with a transform fault scatter about *AB,* the arc of a small circle centered on rotation pole *P.* Arc *AB* is defined by the radius *R,* the mean of distances r_i of epicenters from *P.* Standard deviation of distances from circle, d_i, is less about true arc (center *P*) than about any other (point *P′*).

pole (Figure 5.2). However, due to location uncertainties, and possibly to a finite width of the zone of tectonic activity, the epicenters should scatter randomly about the arc. In practice, the transform fault, i.e., the small circle arc, is not precisely defined. If the mean distance, *R,* of the epicenters from the pole *P* is taken to define the arc, the standard deviation, *S,* of the epicenter scatter about the arc can be calculated. Now *S* can be calculated about any point *P′*, but the minimum value of *S* will be obtained for calculation about *P,* the true center of the arc. This minimum property is used to find the rotation pole position. In general, there may be several transform faults along a boundary, each one with seismicity scattered about a separate small circle arc about the same pole. The *S* that is used then is a weighted mean of the standard deviation of each transform fault. The value of *S* is calculated over a wide grid of points and the position giving minimum *S* is considered the best estimate of the pole, which we call the epicentral pole.

Application of the epicentral method to the Indian-Antarctic and Pacific-Antarctic plate boundaries is described below. The results are from Falconer (1974).

Indian-Antarctic plate boundary

The Indian-Antarctic plate boundary extends from the Indian-Pacific-Antarctic triple junction at lat. 61.5° S, long. 161° E (Falconer, 1972) to the African-Indian-Antarctic triple junction in the central Indian Ocean at lat. 25° S, long. 65° E (McKenzie and Sclater, 1971). We consider here primarily the section east of long. 135° E (Figure 5.3), which contains the majority of the seismicity on the boundary (Sykes, 1970*a*). The boundary is basically a series of northward-trending fracture zones that offset aseismic westward-trending spreading ridge segments. The bathymetry and magnetics of the area are known on at least the reconnaissance level (Hayes and Conolly, 1972; Weissel and Hayes, 1972), but some of the ridge axis segments and active fracture zone sections are identified only indirectly from data obtained near them.

In applying the epicentral method to the boundary, a division into three sections of arc, ignoring the small fracture zone at long. 144° E, would be sensible (Figure 5.3). However, at long. 140° E there is a spreading ridge segment at lat. 52.3° S (Weissel and Hayes, 1972), so the 140° E section is considered to be two parts, giving a total of four sections. These four arcs yield an epicentral pole at lat. 30° S, long. 166° E, with a mean standard deviation of 21.3 km. The standard deviations of the individual arcs range from 18 to 25 km.

Note that the standard deviations, that is the scatter of the epicenters about single arcs, is comparable to the probable uncertainty in epicenter locations. The transform activity could thus be considered to lie along a zero-width line. Similar "line activity" within broader fracture zones, where the broader zone is defined from topography and magnetic anomaly offsets, has been revealed by detailed microseismicity studies in the FAMOUS area of the Mid-Atlantic Ridge at lat. 37° N (MacDonald and Luyendyk, 1977).

On the Indian-Antarctic boundary west of the area shown in Figure 5.3, epicenter lineations are not striking. There are, however, seven areas of activity probably associated with fracture zones (McKenzie and Sclater, 1971; Weissel and Hayes, 1972). Application of the epicentral method to just those seven zones gave a pole at lat. 28° S, long. 162° W, very close to the pole determined from only the eastern part of the plate boundary. Using data from all of the boundary (total 11 sections) gives a pole at lat. 30° S, long. 166° W.

The arcs in Figure 5.3 are the small circles about lat. 30° S, long. 166° W as defined by the epicenters in each zone. The epicenter location uncertainties and the lack of detailed bathymetric and magnetic coverage, particularly of the active transform faults themselves, preclude detailed comparisons of the "activity lines" with the fracture zones, other than to say the general agreement is good. The long zone from about lat. 55° to 60° S is complex, and Hayes and Conolly (1972), from topography, and Weissel and Hayes (1972), from magnetics, consider it to be several fractures, although they differ in where they place the fractures. Epicenter data sets based on several different subdivisions of this zone yield epicentral poles less than 4° from lat. 30° S, long. 166° W. Statistically, however, the fit within the 50°- 60° S section is better for one long zone than for several short ones. Two focal mechanism solutions (Banghar and Sykes, 1969) from the northern end of that section give predominantly strike-slip motion in directions agreeing to within 10° with the direction define by the epicentral pole (Figure 5.3).

Pacific-Antarctic plate boundary

The Pacific-Antarctic plate boundary extends from the complex Pacific-Nazca-Antarctic triple junction near lat. 35° S, long. 110° W (Herron, 1972) to the Indian-Pacific-Antarctic triple jucntion (Figure 5.1). Bathymetric and magnetic data (Molnar and others, 1975; Christoffel and Falconer, 1972, 1973; Herron, 1971) reveal topographically "smooth" aseismic ridge segments offset by fracture zones; most but not all of the latter are seismically active. East of long. 150° W, six significant arcs are evident (Figure 5.4). From long. 150° W to 180° (Figure 5.1), the seismicity is somewhat scattered, and only near long. 160° W is there any linear zone. This sector is not well surveyed, and there are probably more fracture zones than have been

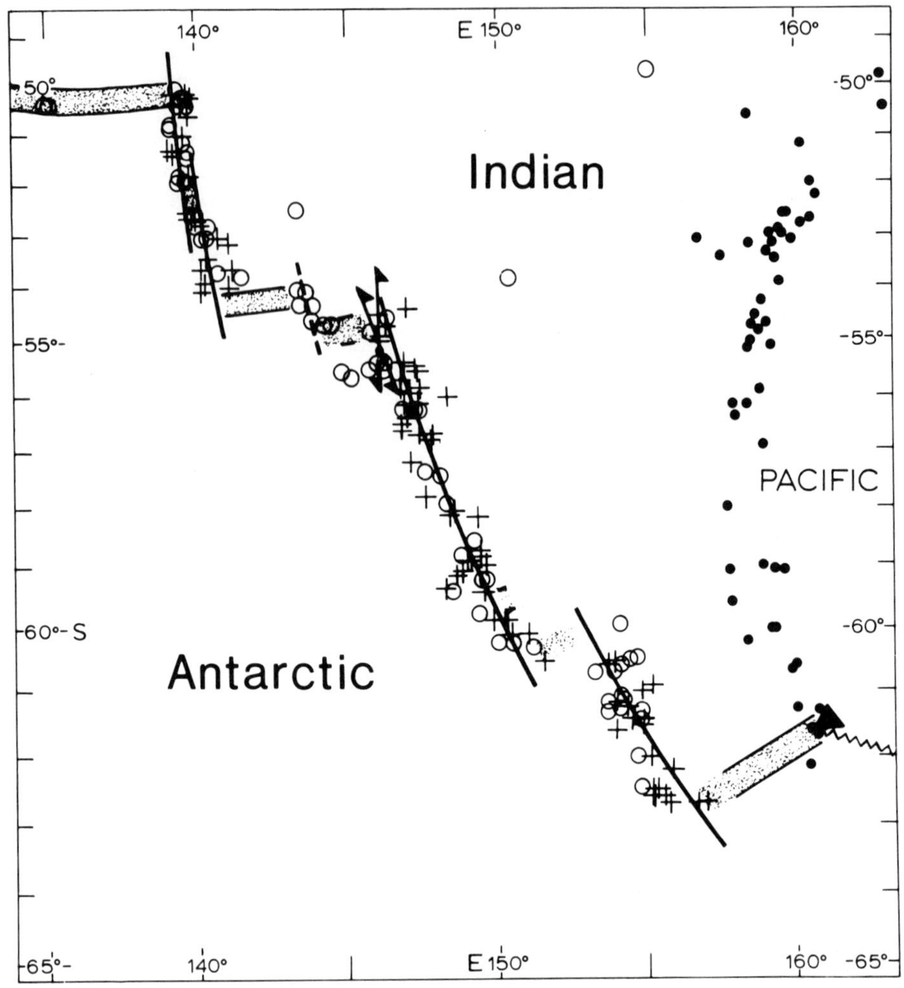

Figure 5.3. The eastern part of the Indian-Antarctic plate boundary. Circles and crosses are Indian-Antarctic boundary epicenters and dots are Indian-Pacific boundary epicenters. Heavy arcs represent the small circles about the Indian-Antarctic epicentral pole at lat. 30°S, long. 166° W, defined by the epicenters shown by crosses. Ridge axes are shown by the stippled pattern and lines where definite and by only the stippled pattern where inferred. The triangle is the Indian-Pacific-Antarctic triple junction (Falconer, 1972). Arrows show direction of strike-slip motion from focal mechanism solutions (Banghar and Sykes, 1969).

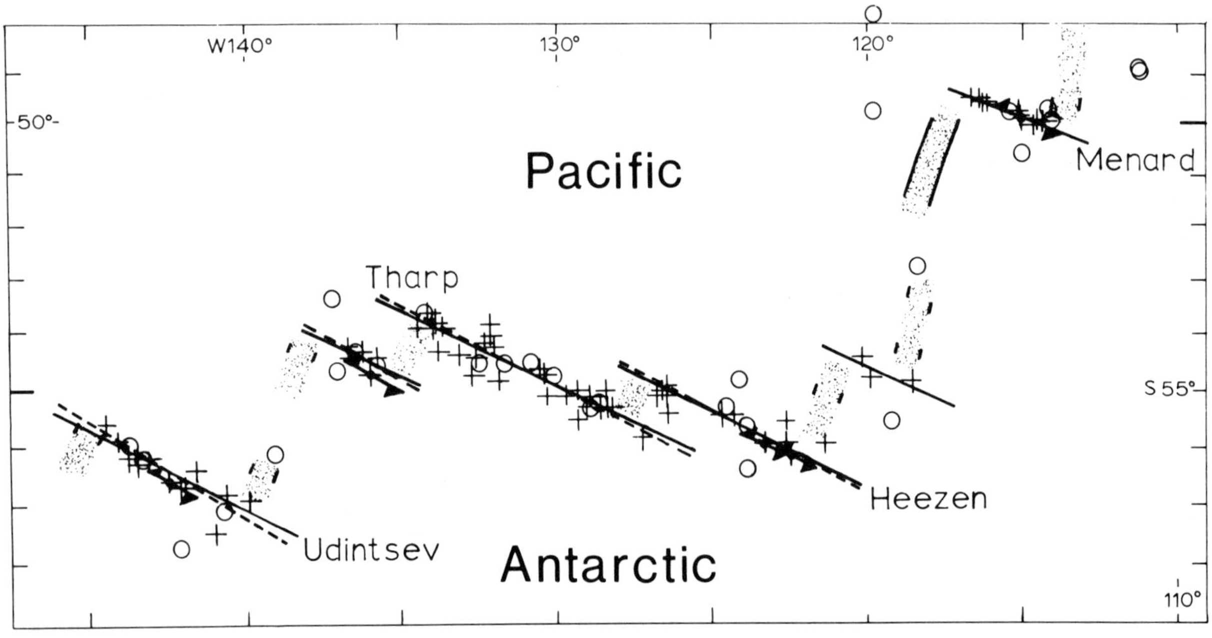

Figure 5.4. The eastern part of the Pacific-Antarctic plate boundary. Solid arcs about the Pacific-Antarctic epicentral pole (b) at lat. 76° S, long. 144° E; dashed arcs are about the pole (a) at lat. 71° S, long. 122° E. Other symbols same as in Figure 5.3. Focal mechanisms are from Forsyth (1972) and Molnar and others (1975).

mapped. From long. 180° westward to the Indian-Pacific-Antarctic triple junction (Figure 5.5), there is a 1000 km zone of seismicity and disturbed bathymetry and magnetics named the Pacific-Antarctic fracture zone by Christoffel and Falconer (1972). Whether it is one long fracture or several shorter ones with ridge segments between them, was one of the major questions we sought to answer.

To apply the epicentral method to the boundary we take the seven sections east of long. 180° plus the Pacific-Antarctic fracture zone as either (a) three sections based on apparent seismic gaps, or (b) one section (Figure 5.5). With (a), a total of ten sections, the epicentral pole is at lat. 71° S, long. 122° E, with S = 14.9 km. With (b), eight sections, we get lat. 76° S, long. 144° E and S = 15.5 km. Standard deviations of individual zones range from 5.8 to 28 km. Here also the scatter about small circles is comparable to epicenter location uncertainties, suggesting "zero-width" transform faults.

On the eastern half of the boundary, the two epicentral pole solutions yield transform faults which differ very little (Figure 5.4). Five earthquake focal mechanism solutions (Forsyth, 1972; Molnar and others, 1975) yield strike-slip motion in directions which differ from the transform directions by less than 5°. Bathymetric profiles across the active parts of the fracture zones (Molnar and others, 1975) mostly show deep troughs close to the epicenter-defined transform faults. The coincidence is particularly good for the Udintsev fracture zone and less so for the Tharp zone, but the troughs are not very marked there. The Menard fracture zone is remarkably linear. It is 170 km long, the epicenter standard deviation is 5.8 km, and only two epicenters are more than 8 km from the epicenter transform fault. However, a deep trough on the one bathymetric profile through the active section is 25 km north of the transform. This "displacement" may reflect a systematic error in epicenter locations.

In the Pacific-Antarctic fracture zone, the two epicentral solutions differ considerably (Figure 5.5). The standard deviation for one long zone is 20.3 km; for the three zones it is, from east to west: 8.2 km, 15.7 km, and 25 km. In the whole zone no magnetic anomalies characteristic of active ridge segments are identifiable (Christoffel and Falconer 1972, 1973; Falconer, 1974; Molnar and others, 1975). The bathymetric relief of as much as 3 km is typical of major fracture zones, but the disturbed area is 200 km wide compared with less than 50 km at most other fracture zones (Molnar and others, 1975), suggesting that several fracture zones are present. Preliminary bathymetric contours of Hayes and Conolly (1972) and Molnar and others (1975), and the more detailed compilation of Vanney and others (in press), favor the shorter northwestward-trending fractures, although there are long troughs with a more westerly trend. None of the inferred epicenter transforms can be closely linked to a particular bathymetric feature for much of its extent. The focal mechanism solutions (Molnar and others, 1975) fit the single long zone interpretation better. Considering also the extra seismicity that was not used in the epicentral method, and the fact that the long zone intersects the Indian-Pacific-Antarctic triple junction, we favor the single long zone interpretation.

Earthquake *b*-values

In an attempt to investigate further the Pacific-Antarctic fracture zone, we studied the earthquake magnitude-frequency relationship

$$\log N = a - bM,$$

where N is the number of events of magnitude M or greater, and a and b are constants. The constant a is determined by the level of activity and is not considered here; it is b which is of interest. Its value depends on the proportion of large to small events: a high b-value indicates a greater proportion of smaller events, and a low b indicates a greater proportion of larger events.

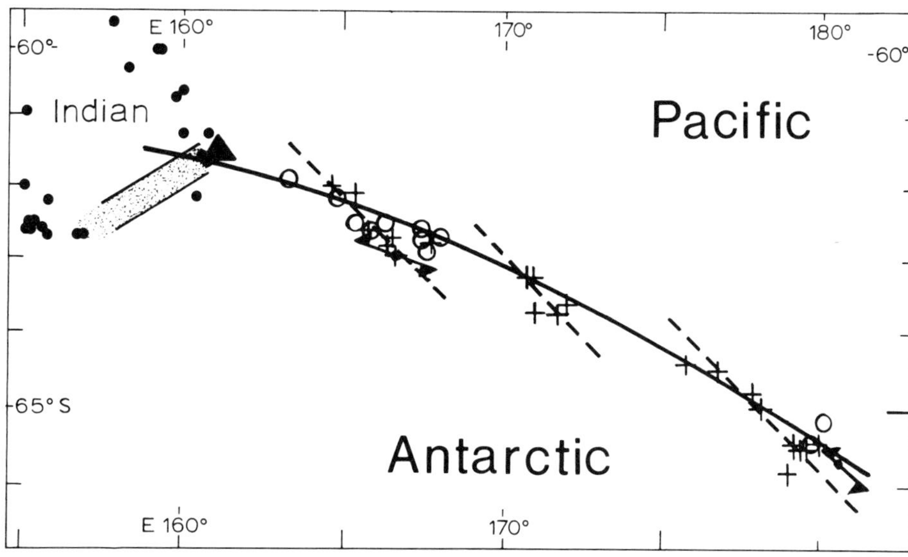

Figure 5.5. The Pacific-Antarctic fracture zone. Solid arcs are about the Pacific-Antarctic epicentral pole (b) at lat. 76° S, long. 144° E; dashed arcs are about the pole (a) at lat. 71° S, long. 122° E. Dots are epicenters on Indian-Pacific and Indian-Antarctic boundaries. The triangle is the Indian-Pacific-Antarctic triple junction. Other symbols same as in Figure 5.3. Focal mechanisms are from Molnar and others (1975), but the solution at lat. 62.8° S, long. 166.3° E was incorrectly given by them as being at lat. 62.8° S, long. 166.3° W.

The impetus for our *b*-value study was the work of Francis 1968*a,b*) who showed that for the Mid-Atlantic Ridge, *b*-values are significantly higher for the ridge axis (1.72) than for the fracture zones (0.99) (Figure 5.6). If this difference is general, and if some of the seismicity of the Pacific-Antarctic fracture zone is associated with spreading ridge segments, it might be reflected in a high *b*-value. Ridge segments might be seismically active there as the expected half-spreading rate is 24–28 mm/yr (Minister and others, 1974), which is below the 30 mm/yr rate above which ridges appear to become aseismic (Menard, 1967; Cann, 1968).

To calculate *b* we use Utsu's (1965, 1966) maximum likelihood estimator. For the calculated *b*-value to be valid, the relationship between log *N* and *M* must be reasonably linear. Figure 5.6 shows that this is not so for all magnitudes, mainly due to reduced sensitivity of the detection network for low magnitudes, but also partially due to magnitudes not being given for 23% of the earthquakes, probably the smaller ones. We thus consider only magnitudes 5.3 and greater for calculation of *b*. For statistical tests we use the 90% confidence limits on *b*, using the fact that *b* is chi-square distributed with 2*N* degrees of freedom.

To check the general applicability of Francis's (1968*a,b*) results we first calculated *b* for all Antarctic boundary seismicity and obtained for *b,* with 90% confidence limits, a value of 1.86 ± 0.33 (Figure 5.6). Since the seismicity is probably all on fracture zones, the *b*-value difference from the Atlantic fracture zones ($b = 0.99$) and its similarity to the Atlantic ridge ($b = 1.72$) is initially surprising. For the Indian-Antarctic boundary east of long. 135° E (Figure 5.3), the seismicity is all on fracture zones. The *b* value is

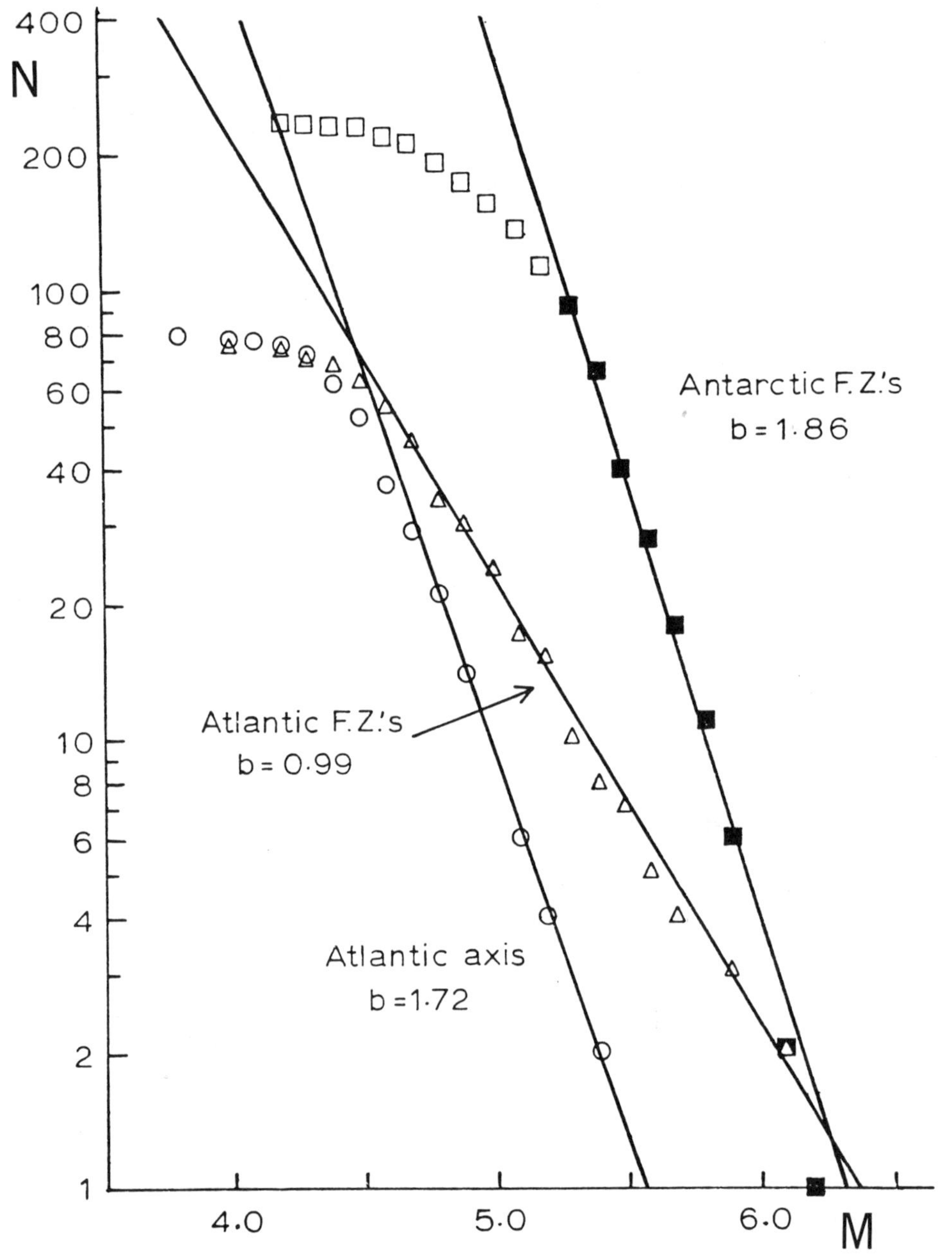

Figure 5.6. Earthquake magnitude-frequency data. Atlantic data are from Francis (1968*a*). Antarctic data are from this study: line *b* = 1.86 from magnitudes 5.3 and greater (solid squares). *N* is number of events of magnitude *M* and greater.

1.99 ± 0.52 (Figure 5.7), which is greater than the Atlantic fracture zone value at the 98% level, but it is similar to the Atlantic ridge axis value. Although there is no Indian-Antarctic ridge axis seismicity to provide an axis *b*-value, it is clear that a high *b*-value is not just an axis characteristic.

For the Pacific-Antarctic boundary, the *b*-value of the whole boundary is 1.75 ± 0.45. East of long. 180° the seismicity is all on fracture zones and that area gives $b = 1.60 \pm 0.48$. Again the *b*-value is significantly higher than the Atlantic fracture zone value, but similar to the Atlantic ridge axis value. The *b*-value of the Pacific-Antarctic fracture zone is 2.06 ± 0.92, which is not significantly different from the *b*-values of the known fracture zone areas of the Antarctic plate boundary. However, it also is not significantly different from the Atlantic axis value (1.72). Thus the *b*-value is not an unequivocal indicator of the nature of the boundary.

Comparison of the Atlantic and Antarctic *b*-values may provide information on the tectonic stress in both areas. A relationship between *b*-value and average stress, stress drop, and earthquake source area has been shown theoretically (Wyss, 1973), experimentally (Scholz, 1968), and observationally (Gibowicz, 1973; Francis, 1974); high *b* reflects low average stress, low stress drop, and small source area, and these in turn reflect high *b*-values. Basically the Atlantic fracture zone *b*-value is low, while the Atlantic axis and Antarctic fracture zones have high *b*-values (Figure 5.6). The Antarctic fracture zones therefore seem to be under lower stresses than the Atlantic fracture zones and to have smaller source areas. This could be due to different thermal conditions. For each of the Antarctic boundary fracture zones the age of the ridge axis offset in relation to the spreading rate is such that the age of the fractures is less than 10 m.y. In the Atlantic the age of

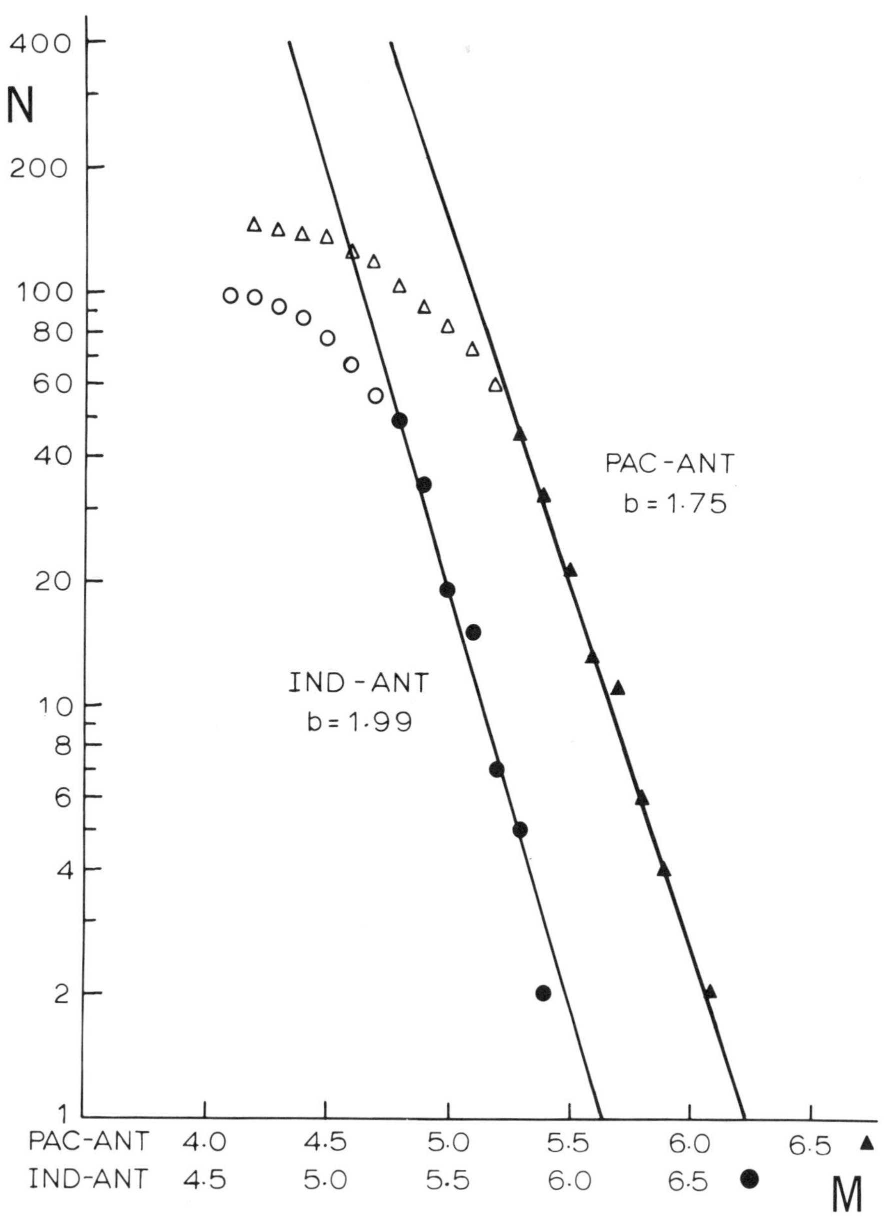

Figure 5.7. Cumulative earthquake magnitude-frequency data for Pacific-Antarctic and Indian-Antarctic plate boundaries. Lines from magnitudes 5.3 and greater (solid symbols). *N* is number of events of magnitude *M* and greater.

the oldest crust along most of the active transform faults is greater than 20 m.y. Since crustal temperature is inversely related to age (McKenzie and Sclater, 1969), the Antarctic fracture zones are at higher temperatures. Stresses are likely to be lower and the thickness of brittle crust less than in the Atlantic. The *b*-values seem to reflect this.

Poles of rotation

The epicentral method provides the pole of relative rotation for the boundary. To evaluate the calculated poles, we consider briefly what the poles predict on the boundaries and compare the poles with those determined by others. It is important to note that the epicentral method uses only data for 1964 to 1972, and so yields an effectively instantaneous pole, and defines instantaneous transform faults. However, the topographic structure of the transform faults probably reflects the tectonic patterns over a much longer period; the published poles discussed below pertain to periods of several million years, as they are based on magnetic anomaly offsets, spreading rates, and/or fracture zone trends.

The Indian-Antarctic epicentral pole is at lat. 30° S, long. 166° W. We have shown that the transform faults predicted by the pole adequately fit what is known of the boundary topography, magnetic anomalies, and focal mechanisms. The pole is significantly removed from other well-determined poles (e.g., McKenzie and Sclater, 1971; Weissel and Hayes, 1972; Minister and others, 1974), which group near lat. 10° S, long. 150° W. The transform fault directions predicted by those poles are within 10° of those of the epicentral pole; all "fit" available structure data. The epicentral method does not provide a spreading rate, but observed spreading rates should be proportional to the sine of the angular distance from the pole. For the 30° S, 166° W pole, however, they are not (Figure 5.8).

The Pacific-Antarctic epicentral pole is either at (a) lat. 71° S, long. 122° E, if the Pacific-Antarctic fracture zone is three sections, or at (b) lat. 76° S, long. 144° E, if it is one long section. As discussed earlier, either pole fits the known boundary structure east of long. 180°, and the Pacific-Antarctic fracture zone is the critical area. The data favor, although not strongly, solution (b), the single zone. The majority of published Pacific-Antarctic poles (e.g., Morgan, 1968; Chase, 1972; Minister and others, 1974) are near pole (a) or slightly west of it; they predict the model of several more northward-trending fracture zones for the geometry of the Pacific-Antarctic fracture zone. However, these pole determinations were not based on any data from that area. They used data only from east of long. 180°, and there all the poles predict transform direc-

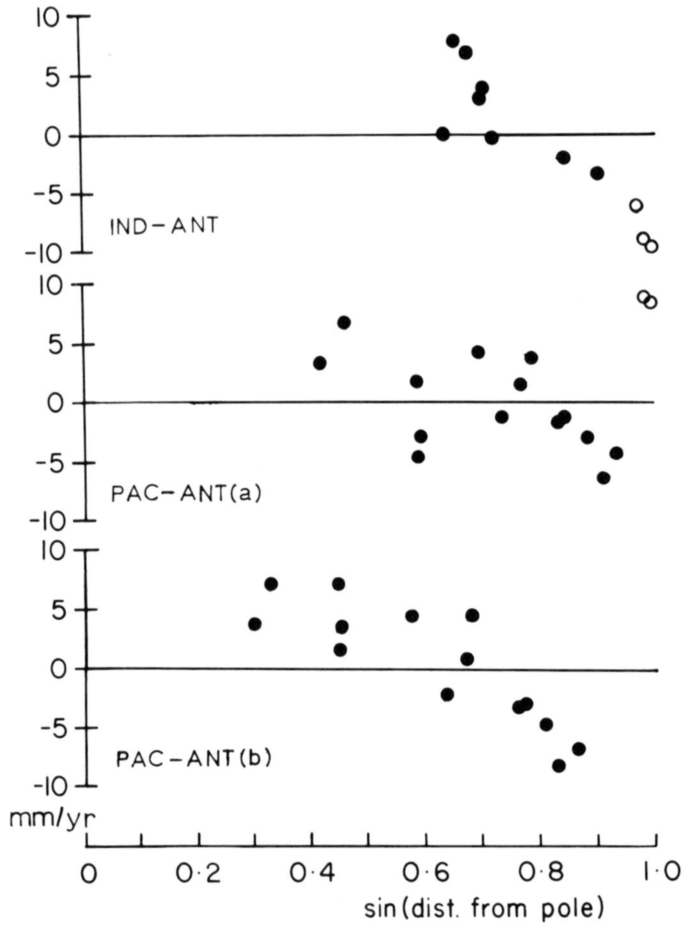

Figure 5.8. Deviation of ridge half-spreading rates from linear relationship v = vmax sin (distance) as a function of angular distance from pole. Indian-Antarctic vmax = 4.5 cm/y, Pacific-Antarctic (a) vmax = 5.8 cm/y, and Pacific-Antarctic (b) vmax = 6.5 cm/y. Rates are based on anomaly 2 (1.79 m.y. ago), except Ind-Ant open circles which are based on anomaly 5 (9.3 m.y. ago) from McKenzie and Sclater (1971). Data "should" be randomly scattered about the zero line.

tions within 10° of each other. With respect to spreading rates, epicentral pole (a) fits, while the favored solution, pole (b), does not (Figure 5.8).

In summary, an Indian-Antarctic pole at lat. 30° S, long 166° W and a Pacific-Antarctic pole at lat. 76° S, long. 144° E fit virtually everything that is known of the relevant plate boundaries except the spreading rates. The spreading rates, however, are averaged over the last 1.8 m.y. or more, whereas these epicentral poles are based on the last 13 years. We cannot say what the current spreading rates are. To fit the epicentral poles, spreading rates on parts of the boundaries would have to have changed from the 1.8 m.y. average rates by up to 30%. Changes of that magnitude during a short time have occurred elsewhere (e.g., Weissel and Hayes, 1972).

The shifts in pole position from poles reflecting several million years' plate motion to the instantaneous epicentral poles are such that on the Indian-Antarctic boundary and the Pacific-Antarctic boundary east of long. 180°, the directions of relative plate motion change little. Only on the Pacific-Antarctic fracture zone are direction changes substantial, and that is the most disrupted area of the two boundaries. Minimum energy considerations suggest that when there are changes in plate motions, those which minimize changes of direction are preferred (Lachenbruch and Thompson, 1972). This seems to be the case for the Indian-Antarctic and Pacific-Antarctic plate-movement changes during the last few million years.

Acknowledgments. This work was begun while we were with the Physics Department, Victoria University of Wellington, New Zealand, where the advice and encouragement of D. A. Christoffel were very much appreciated. F. F. Evison and D. vere Jones of the university, and R. D. Adams and S. J. Gibowicz of the Geophysics Division, Department of Scientific and Industrial Research, also provided much useful help. We thank R. A. Folinsbee for critical reading and helpful comments.

References cited

Banghar, A. R., and Sykes, L. R. 1969. Focal mechanisms of earthquakes in the Indian ocean and adjacent regions. *J. geophys. Res., 74,* 632-49.

Cann, J. R. 1968. Geological processes at mid-ocean ridge crests. *Geophys. J.R. astr. Soc., 15,* 331-41.

Chase, C. G. 1972. The *N* plate problem of plate tectonics. *Geophys. J.R. astr. Soc., 29,* 117-22.

Christoffel, D. A., and Falconer, R. F. 1973. Magnetic measurements in the Macquarie Ridge region. (*In* Fraser, R., *comp. Oceanography of the South Pacific 1972.* Wellington, New Zealand National Commission for UNESCO, 233-40.)

Christoffel, D. A., and Falconer, R. K. H. 1972. Marine magnetic measurements in the south-west Pacific Ocean and the identification of new tectonic features. (*In* Hayes, D. E., *ed. Antarctic Oceanology 2. The Australian-New Zealand Sector.* Washington, American Geophysical Union, 197-209.) [Antarctic Research Series, Vol. 19.]

Christoffel, D. A., and Falconer, R. K. H. 1973. Changes in the direction of sea floor spreading in the south-west Pacific. (*In* Fraser, R., *comp. Oceanography of the South Pacific 1972.* Wellington, New Zealand National Commission for UNESCO, 241-47.)

DeLong, S. E., Dewey, J. F., and Fox, P. J. 1977. Displacement history of oceanic fracture zones. *Geology, 5,* 199-202.

Evernden, J. F. 1969. Precision of epicentres obtained by small numbers of world-wide stations. *Bull. seism. Soc. Am., 59,* 1365-98.

Falconer, R. K. H. 1972. The Indian-Antarctic-Pacific triple junction. *Earth planet. Sci. Lett., 17,* 151-58.

Falconer, R. K. H. 1974. *Geophysical Studies of the South-west Pacific.* Ph.D. dissertation. Victoria University of Wellington, New Zealand, 2 vols.

Forsyth, D. W. 1972. Mechanisms of earthquakes and plate motions in the east Pacific. *Earth planet. Sci. Lett., 17,* 189-93.

Francis, T. J. G. 1968*a*. The detailed seismicity of mid-oceanic ridges. *Earth planet. Sci. Lett., 4,* 39-46.

Francis, T. J. G. 1968*b*. Seismicity of mid-oceanic ridges and its relation to the properties of the upper mantle and crust. *Nature, Lond., 220,* 899-901.

Francis, T. J. G. 1974. A new interpretation of the 1968 Fernandina Caldera collapse and its implications for the mid-oceanic ridges. *Geophys. J.R. astr. Soc., 39,* 301-18.

Gibowicz, S. J. 1973. Stress drop and aftershocks. *Bull. seism. Soc. Am., 63,* 1433-46.

Hayes, D. E., and Conolly, J. R. 1972. Morphology of the southeast Indian Ocean. (*In* Hayes, D. E., *ed. Antarctic Oceanology 2. The Australian-New Zealand Sector.* Washington, American Geophysical Union, 125-45.) [Antarctic Research Series, Vol. 19.]

Hayes, D. E., and Talwani, M. 1972. Geophysical investigation of the Macquarie Ridge Complex. (*In* Hayes, D. E., *ed. Antarctic Oceanology 2. The Australian-New Zealand Sector.* Washington, American Geophysical Union, 211-34.) [Antarctic Research Series, Vol. 19.]

Herron, E. M. 1971. Crustal plates and sea floor spreading in the southeastern Pacific. (*In* Reid, R. L., *ed. Antarctic Oceanology 1. Biology of the Antarctic Seas.* Washington, American Geophysical Union, 229-37.) [Antarctic Research Series, Vol. 15.]

Herron, E. M. 1972. Two small crustal plates in the South Pacific near Easter Island. *Nature Phys. Sci., 240,* 35-37.

Lachenbruch, A. H., and Thompson, G. A. 1972. Oceanic ridges and transform faults: their intersection angles and resistance to plate motion. *Earth planet. Sci. Lett., 15,* 116-22.

MacDonald, K. C., and Luyendyk, B. P. 1977. Deep-tow studies of the structure of the Mid-Atlantic Ridge crest near lat. 37° N. *Geol. Soc. Am. Bull., 88,* 621-36.

McKenzie, D. P., and Sclater, J. G. 1969. Heat flow in the eastern Pacific and sea floor spreading. *Bull. volcan., 33,* 205-26.

McKenzie, D. P., and Sclater, J. G. 1971. The evolution of the Indian Ocean since the Late Cretaceous. *Geophys. J.R. astr. Soc., 25,* 437-528.

Menard, H. W. 1967. Sea floor spreading, topography and the second layer. *Science, N.Y., 157,* 923-24.

Minister, J. B., Jordan, T. H., Molnar, P., and Haines, E. 1974. Numerical modeling of instantaneous plate tectonics. *Geophys. J.R. astr. Soc. 36,* 541-76.

Molnar, P., Atwater, T., Mammerickx, J., and Smith, S. M. 1975. Magnetic anomalies, bathymetry and the tectonic evolution of the South Pacific since the late Cretaceous. *Geophys. J.R. astr. Soc., 40,* 383-420.

Morgan, W. J. 1968. Rises, trenches, great faults, and crustal blocks. *J. geophys. Res., 73,* 1959-82.

Northrop, J. 1973. Seismicity gaps in the Pacific-Antarctic Ridge and East Pacific Rise. (*In* Fraser, R., *comp. Oceanography of the South Pacific 1972.* Wellington, New Zealand National Commission for UNESCO, 285-90.)

Scholz, C. H. 1968. The frequency-magnitude relationship of microfracturing in rock and its relation to earthquakes. *Bull. seism. Soc. Am., 58,* 399-415.

Sykes, L. R. 1970*a*. The seismicity of the Indian Ocean and a possible nascent island arc between Ceylon and Australia. *J. geophys. Res., 75,* 5041-55.

Sykes, L. R. 1970*b*. Earthquake swarms and sea floor spreading. *J. geophys. Res., 75,* 6598–6611.

Utsu, T. 1965. A method for determining the value of *b* in a formula log $n = a - bM$ showing the magnitude frequency relation for earthquakes. (In Japanese with English abstract.) *Geophys. Bull. Hokkaido Univ., 13,* 99–103.

Utsu, T. 1966. A statistical significance test of the difference in *b*-values between two earthquake groups. *J. Phys. Earth, 14,* 37–40.

Vanney, J. R., Falconer, R. K. H., and Johnson, G. L. In press. Geomorphology of the Ross Sea and adjacent oceanic provinces. *Mar. Geol.*

Weissel, J. K., and Hayes, D. E. 1972. Magnetic anomalies in the southeast Indian Ocean. (*In* Hayes, D. E., *ed. Antarctic Oceanology 2. The Australian-New Zealand Sector.* Washington, American Geophysical Union, 165–96.) [Antarctic Research Series, Vol. 19.].

Wyss, M. 1973. Towards a physical understanding of the earthquake frequency distribution. *Geophys. J.R. astr. Soc., 31,* 341–59.

6

Comparison of the Folded Complexes of the Adelaide (Australia) and Ross (Antarctica) Geosyncline Areas

M. G. RAVICH†
Department of Antarctic Geology, SEVMORGEO, Moyka 120, Leningrad 190121, U.S.S.R.

Abstract. A comparison of the folded complexes of the Adelaide and Ross geosynclines shows their similarity in chemical and mineralogic composition, and mode and time of origin.

The oldest formations consist of rocks regionally metamorphosed in the amphibolite facies environment. Such rocks on the Eyre Peninsula have been derived by recrystallization from granulite facies assemblages. The age of the ancient metamorphic basement of both geosynclines ranges from 2000 to 1700 m.y.

The younger folded sequence of the geosynclines is 6–7 km thick; it was formed 550–500 m.y. ago. It contains zonal metamorphic complexes, formed under greenschist and amphibolite facies conditions and pierced by syngenetic granitoid intrusions of the same age, as well as by younger (450–400 m.y.) post-tectonic intrusions.

The striking similarity of the folded sequences of the Ross and Adelaide geosynclines suggests the presence in the geologic past of a single Ross-Adelaide geosyncline which extended for a distance of more than 8000 km through both continents, on the sites of the modern Transantarctic Mountains and ranges of southeastern Australia. This speculation is valid only under the assumption that Antarctica and Australia were adjacent parts of the Gondwanaland supercontinent during late Precambrian and Paleozoic time.

In early September 1973, the author, accompanied by R. L. Oliver, Adelaide University, and G. E. Grikurov, Research Institute of Arctic Geology, visited southern Australia to study the basement of the Adelaide geosyncline area. We studied bedrock outcrops in the Adelaide and Port Lincoln areas situated in the Mount Lofty Range in a region underlain by folded strata with zonal metamorphism. Outcrops of the older crystalline basement complex on the Eyre Peninsula were also visited. More than 70 specimens were collected and then studied at the Research Institute of Arctic Geology, Leningrad. Rocks of the Adelaide geosyncline basement were found to be similar to rocks of the Ross geosyncline basement. In this paper the author has attempted to make a comparative study of these rocks.

Comparison of the Adelaide folded complex and Ross complex

Adelaide folded complex

A section through the folded sequence of the Adelaide geosyncline area, starting in the vicinity of Adelaide, was traced eastward for approximately 40 km (Figure 6.1) across the strike of metamorphic zones. The section crosses upper Proterozoic–Cambrian rocks with a total stratigraphic thickness of about 5–7 km.

From west to east these zones are the following: (1) chlorite-sericite; (2) biotite-sericite; (3) andalusite-staurolite; (4) sillimanite-muscovite; and (5) sillimanite-orthoclase (migmatite). These metamorphic zones are not persistent either in thickness or along the strike. The zone of migmatites makes up the highest grade metamorphic rocks. The other zones commonly wedge out and replace each other both across and along the strike. The zones vary in width from 4–5 km to 25–30 km.

The chlorite-sericite zone consists of phyllites and mica schists of metasiltstone and metaclaystone type. They are intercalated with fine-grained dolomites and quartz and arkosic sandstones. The interlayers of sandstones and dolomites vary in thickness from 1 m to 3 m; they account for about 15–20% of the volume of the zone. Eighty percent of the typical mica schist (specimens 1 and 2)* consists of almost equal proportions of angular granules of quartz and sodic plagioclase; thin flakes of muscovite, forming 15–20% of the volume, are randomly scattered throughout the rock. Fine-grained sandstones (specimen 3*b*) consist of 40–50% angular quartz fragments, with a considerable admixture of sodic plagioclase and microcline (as much as 30%). Poorly crystallized dolomitic cement accounts for 20% and neogenic sericite and chlorite flakes, about 10% of this rock. All these terrigenous and carbonate rocks have been metamorphosed under low-grade greenschist facies conditions.

The biotite-sericite zone consists of two rock units: (a) distinctive arkosic metasandstones with intercalations of dolomites and phyllites (the main section); and (b) metagritstones, grading into arkosic and quartzitic sandstones with intercalations of micaceous schists, and schists containing ore minerals and chlorites (20 km south of the main

† Deceased.

* Here and elsewhere in this chapter, specimen numbers coincide with those of outcrops on the maps of Figures 6.1 and 6.2, and these numbers are also given in Tables 6.1 and 6.2.

section). Boundaries with the chlorite-sericite zone are obscured, and the zone is tentatively recognized by the first appearance of biotite aggregates. The major rock of the zone (specimens 3*c* and 3*d*) consists of angular fragments of microcline and microcline-microperthite (as much as 40–60% of the rock) and finer quartz grains (20%). All the fragments are cemented or, more precisely, embedded in a sericite aggregate, which constitutes 20–40% of the rock. The author has the impression that the sandstones are strongly disintegrated granites (similar to alaskite, and perhaps, normal granites), if it is assumed that sericite has replaced plagioclase due to metamorphism. About 5 km east of these quasi-sandstones is a rocky cliff 50 m high consisting of strongly disintegrated (as small as sand-size grains), cataclastic, granitoid rock, whose fragments have been scattered and cemented with flaky biotite-sericite aggregates (specimen 5). Fragments of plagioclase, microcline, and quartz of different sizes make up about 60% of the rock. Feldspar fragments retain the shape of primary laths, while quartz grains have angular and splintery forms. The sericite-pelitic matrix accounts for 30% and aggregates of fine dark green biotite flakes make up 10% of the rock. Such granitoid cataclasites underwent superimposed metamorphism of the greenschist facies, accounting for the composition of their peculiar matrix. They have obviously resulted from older granites, crystallized prior to the zonal metamorphism. Fine-grained dolomites do not differ from those described for the chlorite-sericite zone, but their layers may be 40–50 m thick. This is also true for micaceous schists, which mainly consist of sericite aggregates, but everywhere contain clastic grains of plagioclase, microcline, and quartz (20–40% of the rock).

The section of this same metamorphic zone to the south is somewhat different in nature. There it is represented by gritstones (grading into arkosic sandstones), intercalated with sericite slates and slates containing ore minerals and chlorite (outcrop 13, Figure 6.1). In gritstones subangular clastic grains (0.5–2.0 cm) make up 60–80% of the whole rock, and 20–30% of the rock is sericite matrix with abundant grains of ore minerals, mainly pyrite. Quartz dominates the detrital fraction of the gritstone—it comprises almost half the rock; sodic plagioclase and microcline in equal amounts make up the remaining half. It is obvious that granitoids have also been the source rocks for gritstone; however, they have not only been strongly disintegrated, but also well rounded, the fragments having been transported for a considerable distance. Chlorite slate with abundant ore minerals is typical of this section (specimen 13*e*). The slate is composed of 70% almost colorless chlorite flakes, forming fine radial aggregates, and 30% disseminated fine grains of ore minerals. The slate seems to have formed from metamorphism of basic effusive rocks.

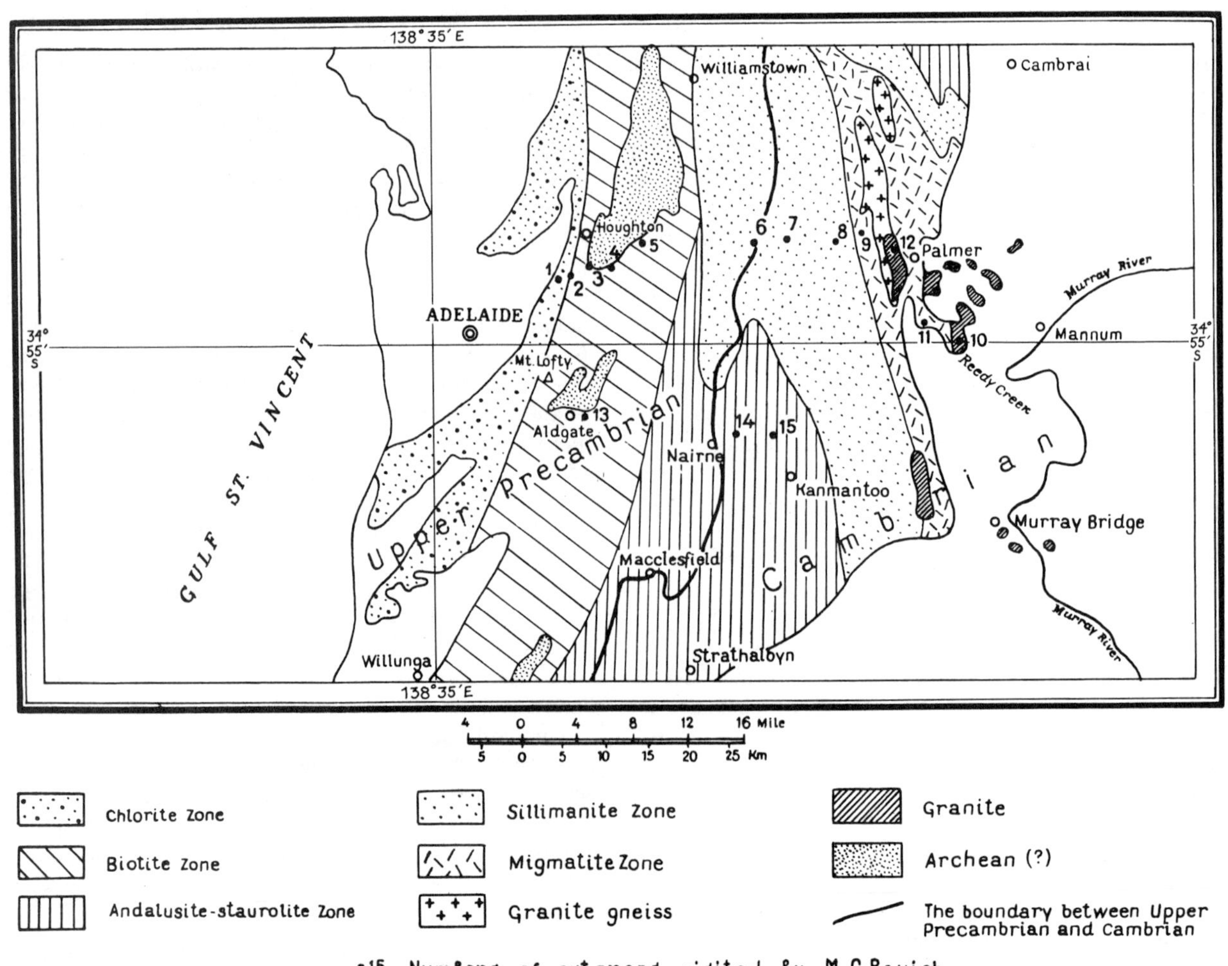

Figure 6.1. Geologic sketch map of the metamorphic zones of the Mount Lofty Ranges. Numbers on map correspond to specimen numbers given in text.

A small outcrop of diorite-like bedrock, which is considered to be basement beneath the folded strata of the Adelaide geosyncline (R. L. Oliver, oral communication), stands apart. Virtually all the rock of this outcrop is typical lamprophyre (specimen 13), similar to lamprophyric dikes of the so-called "Palmer Granite" described below.

The andalusite-staurolite zone has been studied 20 km south of the main section. It consists of biotite-quartz schists with staurolite and garnet and rare andalusite crystals. Typical rocks of the zone are microschists (specimen 14), with silty granules of quartz embedded in an aggregate of thin flakes of brown biotite. Usually these minerals occur in equal amounts. Staurolite crystals are characterized by a lattice texture, as they are filled with inclusions of fine quartz granules. Where garnet is present, the schist is better crystallized and consists of quartz (38%), biotite (25%), staurolite (20%), garnet (10%), and muscovite (7%) (specimen 15*a*).

The sillimanite-muscovite zone is composed of crystalline schists, quartzites, and marbles. Schists contain quartz, oligoclase, and muscovite; biotite with a 3-10% admixture of sillimanite is the prevalent dark mineral. Quartzites (specimen 6) contain 90% clastic quartz grains, impregnated with sillimanite crystals (3%) and aggregates of muscovite flakes (7%). Marbles (specimen 8) are 90% strongly cataclased, small and large crystals of dolomite with coiled polysynthetic twins. In marbles, sillimanite needles with muscovite flakes occur. This zone is marked by the appearance of granite veins, more common in the next zone, where they will be described as "Palmer Granite."

The sillimanite-orthoclase zone is actually an area of migmatites (outcrops 9 and 11, Figure 6.1) granitic in composition, locally containing admixtures of sillimanite. Fine-grained granite gneisses (specimen 9) are typical rocks of the zone. In addition to gneissic structure, these rocks are characterized by granoblastic texture. The composition (specimen 9) is as follows: 35% oligoclase (An_{28}), 23% microcline, 28% quartz, 12% biotite, 1% ore minerals, and less than 1% sillimanite. All minerals are fresh; only microcline grains with distinct twinning lattice are slightly altered to clay. Thin (as much as 0.5 m) veins of leucocratic granite with muscovite flakes are abundant in these granite gneisses. Dendritic, nebulitic, and ptygmatic migmatites occur along with the granite gneisses in this zone.

The paleosome of migmatites is generally represented by biotite plagiogneisses, and the neosome by alaskitic granites. The plagiogneisses consist of oligoclase (An_{12-15})—50%, quartz—28%, biotite—20%, apatite—1%, and sillimanite—1%. Chemical analysis of the paleosome (specimen 11) displays its affinity to granite: SiO_2—71.33%, TiO_2—0.56% Al_2O_3—12.93%, Fe_2O_3—1.26%, FeO—2.65%, MgO—1.63%, CaO—0.80%, K_2O—4.24%, Na_2O—3.40%, P_2O_5—0.20%, and losses—0.7%, totaling 99.71%.

Granitoids of the Adelaide Geosyncline (Palmer Granite)

Granite bodies, variable in size, penetrate the migmatite zone (White and others, 1967). Their length varies from 2-3 km to 10-11 km and more. One of these bodies, cropping out along the Reedy Creek beaches, has been studied in detail. It consists essentially of granodiorite, grading into skialithic migmatites of a granite gneiss type. The granodiorites are partly broken and slightly gneissified, but have retained their hypidiomorphic grandular texture. Their composition in several specimens is rather uniform (outcrop 10): oligoclase (An_{24-29})—45-51%, orthoclase—3-5%, quartz—15-20%, hornblende—8-10%, biotite—10-12%, sphene—3-4%, opaque minerals—1-2%, and apatite—as much as 1%.

Lamprophyre dikes, several meters wide, cut the granodiorites. The lamprophyres are similar to kersantites in composition. In different dikes (outcrop 10) they vary only slightly in composition: labradorite—50-60%, quartz—7-10%, biotite—18-20%, hornblende—13-19%, augite—0-2%, and opaque minerals—2-3%. Lamprophyres have prismatic texture, grading into diabasic texture. Plagioclase laths (3-7 mm) occur as phenocrysts; they commonly show zonal structure with andesine or labradorite (An_{48-53}) in the center and andesine (An_{35-39}) in the margin.

The surrounding rocks are of particular interest. Usually they consist of migmatites with uniform 5-7 mm thick interlayers of different composition. Melanocratic layers, forming a paleosome, consist of oligoclase (An_{20}) (50%), quartz (25%), and biotite (25%). The leucocratic layers, representing the neosome, consist of oligoclase (An_{15}) (45%), quartz (50%), and biotite (5%). The migmatites are cut by numerous veins of alaskitic granite. Cordierite is common in the migmatite paleosome.

Another larger granite body, approximately 8 km northwest of that just described, is made up of biotite granites (outcrop 12). Granites consist of oligoclase (An_{17-19}) (40-45%), microcline (20-25%), quartz (25-30%), biotite (4-5%), and sphene (1%).

To complete the description of intrusive rocks it is necessary to mention again the diorite-like rock occurring in outcrop 13, situated in arkosic metasandstones, approximately 40 km west-southwest of the swarm of lamprophyre bodies in the Palmer Granite. On the geologic map of this area (Barker sheet, scale 1:250,000), this outcrop (no more than 10 m in size) lies near the northern edge of the sheet; it is defined as lower Precambrian (Archean?) and interpreted as a tectonic block of the basement complex beneath the Adelaide folded sequence. This rock, however, proved to be lamprophyre, compositionally similar in chemistry to salite odinite, but close to cuselite. Fresh rock, with hypidiomorphic granular texture, consists of oligoclase (An_{12-15}) (48-50%), microcline (13-15%), quartz (7-8%), clinopyroxene (28-30%), sphene (1%), and small, rare zircon crystals.

Chemical analyses of the Palmer granitoids (Table 6.1) show that they are represented not only by normal granites (White and others, 1967), but also by granodiorites, close in composition to quartz diorite.

According to their magnesium-calcium content, the granodiorites occupy an intermediate position between granodiorites and quartz diorites (Zavaritskii, 1941), but in all other aspects they are similar to quartz diorites with a rather high Na_2O content; the plagioclase in them is closer to oligoclase than to andesine. The dominant rock in the batholith-like massif from the McMurdo Sound region in the Transantarctic Mountains is also granodiorite, grading to quartz diorite. Comparison of chemical analyses of both these rocks (Table 6.1) shows their nearly complete likeness. The higher content of iron and calcium oxide in the biotite-amphibole quartz diorite from the McMurdo

Sound area as compared to similar rock from the Adelaide area is reflected in the higher content of hornblende in the former.

The Palmer Granite is rather uniform in composition and similar to an intermediate normal granite. In this connection it is similar to post-tectonic granite from the McMurdo Sound region in Antarctica (Table 6.1).

It should be emphasized that enclosing migmatites, especially thin-layered and skialithic ones, are also similar to granites in composition; therefore their contacts with granitoid bodies are obscure. Migmatites seem to be gradually replaced by granites, but this condition reflects the contact influence of granites on migmatites.

As for the dikes and veins associated with the Palmer granitoids, especially the lamprophyres, the situation is complex. The lamprophyres were regarded in the field as orthoamphibolites because of their dark color and prismatic texture. Kersantite varieties prevail among lamprophyres, as indicated by the similarity in chemical composition of the rocks under discussion and typical kersantites from Brittany (Rosenbusch, 1934). Unlike the latter, however, the Palmer kersantites are characterized by considerably higher calcium content due to the composition of the plagioclases, which are closer to labradorite than to andesine. The other slight difference is the higher iron and low magnesium content in the Palmer kersantites, which is explained by general higher ferruginosity of the dark minerals in them.

Lamprophyres occurring in phyllites (outcrop 13) apparently have no genetic connection with the Palmer granitoids. Chemically (Table 6.1, specimen 13) they are more similar to cuselites, than to odinites (Rosenbusch, 1934).

The Rb-Sr age of the Palmer Granite is 490 ± 15 m.y. (White and others, 1967), and a new K-Ar age on the same granite (specimen 12) is 400 m.y.

A whole-rock K-Ar age on lamprophyre of the odinite series (specimen 13) is 565 m.y. However, the Rb-Sr age of the same rock is 800 m.y. (R. L. Oliver, oral communication); this age is thought to be more correct. Thus, according to the age determinations, this lamprophyre should not be grouped with the Palmer granitoids; it should be considered older, namely, late Proterozoic.

Ross folded complex

The geologic sequences of the Ross geosynclinal area have been carefully studied in southern Victoria Land in the coastal part of the Transantarctic Mountains near McMurdo Sound. Metamorphic and intrusive formations outcrop throughout the entire region near McMurdo Sound; only on the summits of mountains more than 2000 m high are they overlain by almost flat-lying terrigenous rocks of Paleozoic–Mesozoic age (the Beacon Supergroup) (Lopatin, 1972).

The metamorphic complex in the McMurdo Sound region apparently has a zonal structure, but not as distinct as that in the Adelaide area. The chlorite-sericite and biotite-sericite zones are not recognized, and a complex zone that consists of phyllites, and biotite-actinolite and graphite schists with abundant marble beds is assigned to the greenschist facies. Here, unlike the Adelaide area, there are widespread carbonate formations and a generally higher degree of metamorphism. The transition from greenschist to amphibolite facies zones seems to be gradual. The low-temperature subfacies of the amphibolite facies is represented by a sillimanite-muscovite zone. It consists of two-mica schists (with sillimanite, muscovite,

Table 6.1. Chemical compositions (in wt. %) of the Palmer granitoids and similar rocks

	1	2	3	4	5	6	7	8.	9	10	11	12	13	14
SiO_2	59.95	59.28	66.0	59.5	73.23	72.88	72.0	77.22	73.27	51.21	52.30	60.78	61.70	56.78
TiO_2	1.26	0.98	0.5	0.6	0.31	0.14	0.3	0.29	0.63	1.50	—	0.68	—	1.04
Al_2O_3	16.18	17.51	15.3	16.5	13.42	13.84	13.1	11.61	10.97	18.45	17.45	15.54	16.36	17.85
Fe_2O_3	2.98	1.33	2.0	2.6	0.99	0.49	1.5	1.05	0.96	3.75	1.80	1.68	2.29	0.44
FeO	4.08	5.18	2.9	4.1	1.17	1.62	1.8	0.83	3.85	5.90	4.80	2.62	2.75	5.89
CaO	4.97	5.57	3.8	6.2	1.61	1.50	1.5	0.77	0.55	6.88	4.78	6.61	2.65	5.13
MgO	2.91	2.69	1.8	3.8	0.64	0.64	0.6	0.69	2.16	4.35	8.50	2.47	1.94	3.46
MnO	0.14	0.09	0.1	0.1	0.02	0.03	0.1	0.04	0.06	0.18	—	0.13	—	—
K_2O	2.10	2.63	2.8	1.9	4.34	5.00	4.8	3.64	4.51	2.00	3.23	2.01	3.05	3.30
Na_2O	4.14	3.45	3.5	3.0	3.64	3.30	3.5	3.03	2.00	3.96	2.12	6.30	5.05	3.33
P_2O_5	0.38	0.26	0.3	0.3	0.07	0.06	0.1	0.3	0.18	0.49	—	0.19	—	—
Loss on ignition	1.05	1.00			0.21	0.56		0.46	0.93	1.15	2.90	0.39		
Total	100.14	99.97			99.65	100.06		99.71	100.01	99.82		100.00		

1. Specimen 10: Granodiorite, Adelaide area.
2. Specimen 134*c*: Biotite-hornblende quartz diorite, McMurdo Sound area; analyses from Lopatin, 1972.
3. Average composition of granodiorite; analyses from Zavaritskii, 1941.
4. Average composition of quartz diorite; analyses from Zavaritskii, 1941.
5. Specimen 12: Biotite granite, Adelaide area.
6. Specimen 119: Biotite granite, McMurdo Sound area; analyses from Lopatin, 1972.
7. Average granite; analyses from Zavaritskii, 1941.
8. Specimen 7: Leucogranite, Adelaide area.
9. Specimen 10*d*: Skialithic migmatite, Adelaide area.
10. Specimen 10*b*: Lamprophyre kersantite, Adelaide area.
11. Specimen 15: Kersantite; Brittany analysis, from Rosenbusch, 1934.
12. Specimen 13: Lamprophyre odinite, Adelaide area.
13. Specimen 2: Cuselite; Pfaltz analysis, from Rosenbusch, 1934.
14. Specimen 15: Odinite; Lausitz analysis, from Rosenbusch, 1934.

See reference to specimen numbers in text and their locations plotted on maps in Figures 6.1 and 6.2

and garnet) and marbles (with amphibole and diopside). There are no migmatites and skialithic granites in this zone. The high-temperature subfacies of the amphibolite facies is represented by the sillimanite-orthoclase zone. It consists of various plagiogneisses and schists (with sillimanite and almandine, or with amphibole and clinopyroxene), marbles (with diopside and forsterite), and even more varied migmatites, including skialithic granites. This zone is marked by intensive migmatization and granitization.

GRANITES OF THE ROSS GEOSYNCLINE

Syntectonic batholith-like massifs, consisting mainly of granodiorites and quartz diorites, prevail among the intrusive rocks. Biotite and biotite-hornblende granites are also widespread, but gabbro-diorites are not common. Relatively small post-tectonic granitoid stocks occur as well, with normal granites locally grading into quartz diorites and quartz monzonites. Mineral compositions and chemical analyses of the main representative intrusive rocks from the McMurdo Sound region are similar to tne intrusive rocks from the Adelaide geosyncline area (Table 6.1).

Isotopic ages of the intrusive rocks are 550–500 m.y. and 490–430 m.y. for syntectonic and post-tectonic granitoids, respectively (Lopatin, 1972). These ages are close to those of the granitoids from the Adelaide geosyncline area, namely, the Palmer Granite, which are probably post-tectonic.

Zonal and contact metamorphism, migmatization, and granite magmatism in the rocks of both geosynclinal systems under discussion are believed to have resulted from a complex process of tectono-magmatic activization, superimposed mainly on the upper Proterozoic and lower Cambrian rocks within the time period from 550 to 500 m.y. ago. The activity is associated with the culmination in development of the Ross and Adelaide orogenies which gave rise to the folded belts in place of the former geosynclines.

Comparison of the ancient basements of the Adelaide and Ross geosynclines

THE ANCIENT BASEMENT OF THE ADELAIDE GEOSYNCLINE

The author visited some large outcrops situated on the Eyre Peninsula coast about 30 km south-southwest of the city of Port Lincoln (Figure 6.2). The localities studied are (1) a massif of old cataclastic alaskite granites at the contact with a thick dike of orthoamphibolites; (2) part of a massif of hypersthene granites (charnockites); and (3) biotite-garnet plagiogneisses and biotite-hypersthene gneisses with a dike of pyroxenite.

Fine- or medium-grained leucocratic alaskite granites cut a dike of orthoamphibolite, with the latter partially assimilated. The granites are markedly cataclased. They consist of orthoclase (50–55%), oligoclase (An_{18-20}) (15–20%), quartz (25–28%), biotite (2%), hornblende (0–2%), and fine-grained opaque minerals (1%) (specimens 16*d* and 16*e*).

The orthoamphibolite is a nearly black, fine-grained, crystalline rock of prismoid blastic texture. It consists of andesine (An_{31-35}) (40–50%), amphibole (35–40%), quartz (8–15%), biotite (2–4%), and opaque minerals (1%) (specimens 16 and 16*a*). Contact alteration of the orthoamphibolite is expressed by quartz enrichment (up to 15%) and by development of orthoclase and sodic plagioclase.

Within 2–3 m of the contact with the granite, the orthoamphibolite is changed into peculiar biotite-amphibole gneiss (specimen 16*c*), which consists of plagioclase (30%), orthoclase (25%), quartz (20%), hornblende (15%), biotite (10%), and some very small sphene crystals.

Orthopyroxene granites of intrusive charnockite type are exposed in the southern Eyre Peninsula (specimen 17). They are massive or slightly gneissic, dark-gray and bluish, medium-grained rocks. Their texture is porphyrohypidiomorphic grading into porphyrogranoblastic. The rocks consist of microcline (35–40%), oligoclase (An_{24-27}) (20–25%), quartz (25–30%), orthopyroxene (3–5%), opaque minerals (1–2%), and sparse clinopyroxene (1%), garnet (1–2%), and hornblende (as much as 6–7%) (specimens 17 and 17*a*); rare grains of fayalite occur in one specimen of granite (specimen 17).

Inclusions of pyroxene orthoamphibolites (several meters in size) occur in this granite-charnockite. They consist of amphibole (60%), clinopyroxene (15%), andesine (An_{33-35}) (20%), biotite (3–4%), and opaque minerals (1–2%) (specimen 17*b*).

Metamorphic rocks are exposed on the coast of the Eyre Peninsula, about 7 km south of outcrop 17. These rocks include garnet-biotite plagiogneisses and hypersthene-biotite plagiogneisses (with sillimanite), enclosing dikes of pyroxenites and veins of pegmatoid alaskite granite.

Dikes of pyroxenite type, varying from 2 to 5 m in width, consist of black, fine-grained rock with prismatic blastic texture and are obviously recrystallized. The rock consists of clinopyroxene (45%), orthopyroxene (12%), bytownite (An_{77-78}) (38%), opaque minerals (3%), and garnet (2%) (specimen 18*a*).

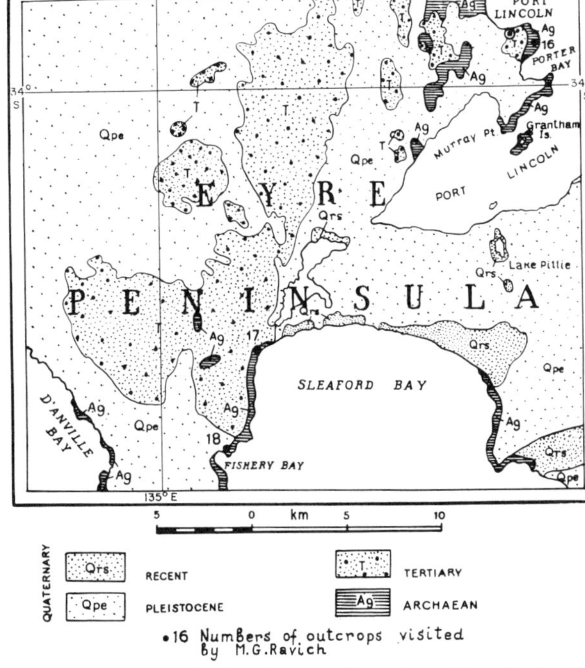

Figure 6.2. Geologic sketch map of southern Eyre Peninsula. Numbers on map correspond to specimen numbers given in text.

Granite veins, 1–2 m wide, penetrate gneisses in every part of the outcrop. The granite is medium-grained and leucocratic with hypidiomorphic texture. It consists of orthoclase (25%), oligoclase (45%), quartz (30%), biotite (4%), and opaque minerals (1%) (specimen 18*b*).

According to chemical analyses (Table 6.2), alaskite granite from the southern Eyre Peninsula (specimen 16*a*) is more leucocratic and alkaline than ordinary alkaline granite, as the former has a higher (by 2%) K_2O content and a lower iron content. Charnockite granite (specimen 17) differs from average granite only in a higher (by 1%) content of ferruginous components and a lower (by 1%) content of alumina. In other elements the compositions of both rocks are similar, which indicates that this charnockite is close to a true granite. However, the origin of the charnockite is completely different, but discussion of this problem is beyond the scope of this paper. Nonetheless, it should be emphasized that the appearance of charnockites in the Archean formations of the Eyre Peninsula suggests a profound difference of these rocks from the zonal metamorphic and granitoid formations in the area of Adelaide. Comparison of both regions in Australia makes the author conclude that the Precambrian formations of the Eyre Peninsula are much more deep seated and much older than the zonal metamorphic and granitoid formations constituting the folded sequence of the Adelaide geosyncline area.

Rocks of the gneiss formation of the Eyre Peninsula, on the whole, are similar in composition to granodiorites, except for the high K_2O and low CaO and Na_2O contents. This does not mean, however, that they necessarily formed from granodiorites; various origins are possible, including a sedimentary-terrigenous origin. However, it may be assumed that in the process of metamorphism, the source rocks simultaneously underwent granitization, which made their composition close to that of granodiorites, enriched (even compared to granites) in potassium, the most mobile component. A comparison of compositions of both types of gneisses reveals that the garnet-biotite gneiss has undergone a higher degree of granitization than the hypersthene-biotite gneiss. This is reflected in a higher content in the garnet-biotite gneiss of K_2O, SiO_2, and FeO, which enter into the composition of dark minerals, while the hypersthene-biotite gneiss is characterized by higher Fe_2O_3 and TiO_2 contents held in titanomagnetite. In addition, hypersthene-biotite gneisses contain two and one half times more opaque minerals than garnet-biotite gneisses. Garnet-biotite gneisses seem to have formed due to further granitization of hypersthene-biotite gneisses, especially since indistinct relics of orthopyroxene have been found in the garnet-biotite gneisses.

Rb-Sr isochrons show that the age of charnockites, similar to specimen 17, is 1853 ± 22 m.y. (Cooper and others, 1976), and the Rb-Sr age of enclosing gneisses is 2643 ± 134 m.y. (Cooper and others, 1976).

Whole-rock K-Ar age data on Precambrian specimens of the Eyre Peninsula are questionable, since they may represent rejuvenated ages. However, the following ages have been obtained: alaskite granite—1285 m.y., and charnockite granite—1300 m.y.

The ancient basement of the Ross geosyncline

The oldest formation in the Transantarctic Mountains is the Nimrod Group, found in the Miller and Geologist Ranges along the upper Nimrod Glacier. This group is represented by various schists, marbles, and gneisses, regionally metamorphosed in the amphibolite facies. It is intruded by granite stocks of different sizes (Gunner and Faure, 1972).

The Nimrod Group differs from the Eyre Peninsula Archean gneiss formation not only in wide development of carbonate rocks, but in the absence of hypersthene-bearing varieties among the gneisses. The latter fact may reflect the more intense granitization of gneisses of the Nimrod Group, resulting in complete replacement of hypersthene-bearing gneisses by garnet-biotite varieties.

Table 6.2. Chemical compositions (in wt. %) of Precambrian rocks of the Eyre Peninsula and similar rocks

	1	2	3	4	5	6	7	8	9	10
SiO_2	75.31	73.2	73.64	72.0	46.85	48.2	51.11	66.93	62.47	66.0
TiO_2	0.16	0.2	0.35	0.3	2.64	1.0	0.81	0.63	1.31	0.5
Al_2O_3	12.67	12.5	11.96	13.1	13.18	17.9	14.80	15.04	14.66	15.3
Fe_2O_3	0.57	1.9	1.34	1.5	3.50	3.2	2.24	1.26	4.61	2.0
FeO	0.52	1.3	2.47	1.8	13.02	6.0	8.98	4.08	2.25	2.9
CaO	0.79	0.5	1.53	1.5	9.15	11.0	11.65	1.57	3.48	3.8
MgO	0.48	0.2	0.51	0.6	5.24	7.5	7.83	2.34	1.80	1.8
MnO	0.02	0.1	0.05	0.1	0.27	0.1	0.23	0.01	0.08	0.1
K_2O	7.00	4.9	4.96	4.8	1.49	0.9	0.54	4.76	4.00	2.8
Na_2O	2.01	4.6	2.82	3.5	2.58	2.5	0.80	2.58	3.06	3.5
P_2O_5	0.03	0.0	0.07	0.1	0.37	0.3	0.12	0.04	0.73	0.3
Loss on ignition	0.30		0.02		1.59		1.15	0.49	1.28	
Total	99.86		99.72		99.98		100.26	99.73	99.73	

1. Specimen 16*a*: Alaskite granite, Eyre Peninsula.
2. Average alkaline granite; analyses from Zavaritskii, 1941.
3. Specimen 17: Orthopyroxene granite, Eyre Peninsula.
4. Average granite; analyses from Zavaritskii, 1941.
5. Specimen 16: Orthoamphibolite, Eyre Peninsula.
6. Average gabbro; analyses from Zavaritskii, 1941.
7. Specimen 18*a*: Pyroxenolite, Eyre Peninsula.
8. Specimen 18: Garnet-biotite plagiogneiss, Eyre Peninsula.
9. Specimen 18*c*: Hypersthene-biotite plagiogneiss, Eyre Peninsula.
10. Average granodiorite; analyses from Zavaritskii, 1941.

See reference to specimen numbers in text and their locations plotted on maps in Figures 6.1 and 6.2.

Rb-Sr age determinations on gneisses, veined granites, and schists of the Nimrod Group in more than 20 specimens yield three isochrons with the following ages: 2027 ± 38 m.y., 1950 ± 153 m.y., and 1828 ± 141 m.y. (Gunner and Faure, 1972). These data may reflect rejuvenation and yield the age of the main phase of superimposed metamorphism of the Nimrod Group; the oldest isochron age here differs from isochron ages of the Eyre Peninsula gneiss formation by 500 m.y. Thus the Eyre Peninsula gneisses are upper Archean and represent a real crystalline basement beneath the folded sequence of the Adelaide geosyncline, while the Nimrod Group rocks are considerably younger.

Conclusions

A comparison of characteristic rocks of the folded sequences of the Adelaide and Ross geosynclines shows that they are similar in chemical and mineralogic composition, and in the conditions and time of their origin.

The basement complex beneath the folded sequence consists of regionally metamorphosed rocks of the amphibolite facies. It has been discovered that the amphibolite facies of the Eyre Peninsula formed due to metamorphism superimposed on primary rocks of the granulite facies. The age of the formation of the basement that underlies these geosynclinal areas ranges from 2600 to 2000 m.y.

The folded geosynclinal sequence, 6–7 km in thickness, was deformed and metamorphosed in the time interval from 550 to 500 m.y. ago. It consists of contact-metamorphic rocks and regionally metamorphosed strata, locally with zonal structure, belonging to the greenschist and amphibolite facies.

Some similarity of the folded sequences of the Ross and Adelaide geosynclines, as well as some similarity of their crystalline basements, suggests the presence in the geologic past of a single Ross-Adelaide geosyncline system, extending for a distance of more than 8000 km through the modern Transantarctic Mountains and across southeastern Australia. This speculation has validity only if we assume that Antarctica and Australia were adjacent parts of Gondwanaland during Precambrian and early Paleozoic time (Ravich, 1972).

References cited

Cooper, J. A., Fanning, C. M., Flook, M. M., and Oliver, R. L. 1976. Archaean and Proterozoic metamorphic rocks on southern Eyre Peninsula, South Australia. *J. geol. Soc. Aust., 23,* Pt. 3, 287–92.

Gunner, J., and Faure, G. 1972. Rb-Sr geochronology of the Nimrod Group, central Transantarctic Mountains. (*In* Adie, R. J., *ed. Antarctic Geology and Geophysics.* Oslo, Universitetsforlaget, 305–11.)

Lopatin, B. G. 1972. Basement complex of the McMurdo "Oasis," south Victoria Land. (*In* Adie, R. J., *ed. Antarctic Geology and Geophysics.* Oslo, Universitetsforlaget, 287–92.)

Ravich, M. G. 1972. *Zagadki Gondvany* [Enigma of Gondwanaland]. Moscow, Znanie, 64 pp.

Rosenbusch, H. 1934. *Opisatel'naia petrologiia* [Descriptive petrology]. Leningrad, Gosgeolnefteizdat, 720 pp.

White, A. J. R., Compston, W., and Kleeman, A. W. 1967. The Palmer Granite—a study of a granite within a regional metamorphic environment. *J. Petrology, 8,* No. 1, 29–50.

Zavaritskii, A. N. 1941. *Pereschiet khimicheskikh analizov izverzhennykh gornykh porod* [Conversion of chemical analyses of igneous rocks]. Moscow, Izdatel'stvo Akad. Nauk SSSR, 108 pp.

7

Geologic Comparison of Southeastern Peninsular India and Sri Lanka with a Part of East Antarctica (Enderby Land, MacRobertson Land, and Princess Elizabeth Land)

L. V. FEDOROV and M. G. RAVICH†
Department of Antarctic Geology, SEVMORGEO, Moyka 120, Leningrad 190121, U.S.S.R.

J. HOFMANN
Section of Geosciences, Mining Academy, 92 Freiberg, German Democratic Republic

Abstract. New geologic evidence from East Antarctica obtained recently by Soviet and Australian geologists gives strong support for correlation between central East Antarctica and southeastern peninsular India and Sri Lanka. Substantial similarity in composition of rocks (ranging in age from Archean to Permian) and their mode of occurrence, identical nature of major structural elements, and compatibility of geologic history all suggest that evolution of the geologic structure of both regions took place within a single province of the hypothetical Gondwana super-continent.

The Mahanadi Graben (India), filled by the Permian coal measures of Gondwanian type, is thought to have continued into the rift zones of the Amery Ice Shelf and Lambert Glacier (East Antarctica), where similar coal measures with a *Glossopteris* flora are wedged among the crystalline basement blocks.

Intracratonic depressions of an aulacogen type form a regular pattern on both continents and contain similar middle and upper Proterozoic sequences of slightly metamorphosed terrigenous rocks with thick jaspilite horizons at the base and a similar pattern of zonal metamorphism. The simultaneity of formation of these sequences is reliably established by absolute age determinations.

The similarity of the crystalline basement of the ancient Indian and Antarctic platforms may indicate existence of a single Indo-Antarctic platform before the fragmentation of Gondwanaland. The similarity is observed not only in the composition of rocks and the degree of their granulite facies metamorphism, but also in the identical type of fold structures and deep-seated faults.

In Du Toit's reconstruction of early Mesozoic Gondwanaland (Figure 7.1), peninsular India was shown adjacent to central East Antarctica, between long. 35° and 90° E (Du Toit, 1937). Such a reconstruction is supported by matching of a 1000-fathom isobath (Smith and Hallam, 1970), and paleomagnetic studies suggest that the two regions were a single part of Gondwanaland as far back as Paleozoic or even late Precambrian time (McElhinny and Embleton, 1974). Crawford (1974) describes the similarity of Archean granulite facies metamorphic rocks of southern and southeastern peninsular India, Sri Lanka, and East Antarctica (in the sector between long. 35° and 90° E), and infers the existence of a stable region of Archean age, which split into two separate "granulite coastal belts" as a result of fragmentation of Gondwanaland in late Paleozoic–Mesozoic time. Paleogeographic maps from the Permian Period show a great similarity between the Upper Permian deposits of the Mahanadi Valley (India) and the Amery Group of Antarctica (Dietz and others, 1972; Elliot, 1975). The microflora from the Amery Group is much closer to that of India, than to the Australian microflora (Balme and Playford, 1967).

New geologic evidence from East Antarctica recently obtained by Soviet and Australian geologists has considerably advanced understanding of the geology of the sector under consideration (Tingey, 1972; Tingey and England, 1973; England and Langworthy, 1975; Fedorov and Tarutin, 1977; Ravich, Jr., and others, 1977). That evidence is analyzed here in terms of the stratigraphy, metamorphism, and tectonics of southeastern peninsular India and Sri Lanka; from this comparison emerges a striking and convincing similarity between the geologic evolution of the Indian and Antarctic fragments of Gondwanaland.

Figure 7.1. Du Toit's (1937) reconstruction of Gondwanaland.

† Deceased.

Mahanadi and Godavari Grabens (Southeastern peninsular India) and similar structures in MacRobertson Land (East Antarctica)

The recent geophysical and geologic investigations carried out by the Soviet and Australian Antarctic expeditions in MacRobertson and Princess Elizabeth Lands have positively demonstrated that the graben occupied by Lambert Glacier and the Amery Ice Shelf is associated with a major rift zone (Figure 7.2). The Beaver Lake graben, where the Permian sedimentary rocks (Amery Group) are concentrated, is a subsidiary structure of the rift zone delineating the western boundary of the axial rift. Thus the most apparent similarity between the Permian deposits of southeastern peninsular India and of MacRobertson Land in Antarctica is their structural position within the rift grabens superposed on the Precambrian terranes.

Cross-bedding measurements indicate that the sediments of the Amery Group were derived from the southeast (Mond, 1972; Ravich, Jr., and others, 1977). The possibility of subglacial continuation of the Amery Group southward is also suggested by discoveries of Permian sandstones with a *Glossopteris* flora in Mount Rymill moraines 300 km south of the Beaver Lake locality (Trail and McLeod, 1969), as well as by geophysical evidence for the presence of a thick cover of low-density deposits in the subglacial depressions.

Precise correlation of the Amery Group with the upper Paleozoic deposits of peninsular India is not established (Elliot, 1975). Several authors (Balme and Playford, 1967; Mond, 1972; Ravich, Jr., and others, 1977) stress a Late Permian age for the microflora from the Amery Group, but some (Ravich, Jr. and others, 1977) prefer to assign its lower part (the Radok Conglomerate) to the Early Per-

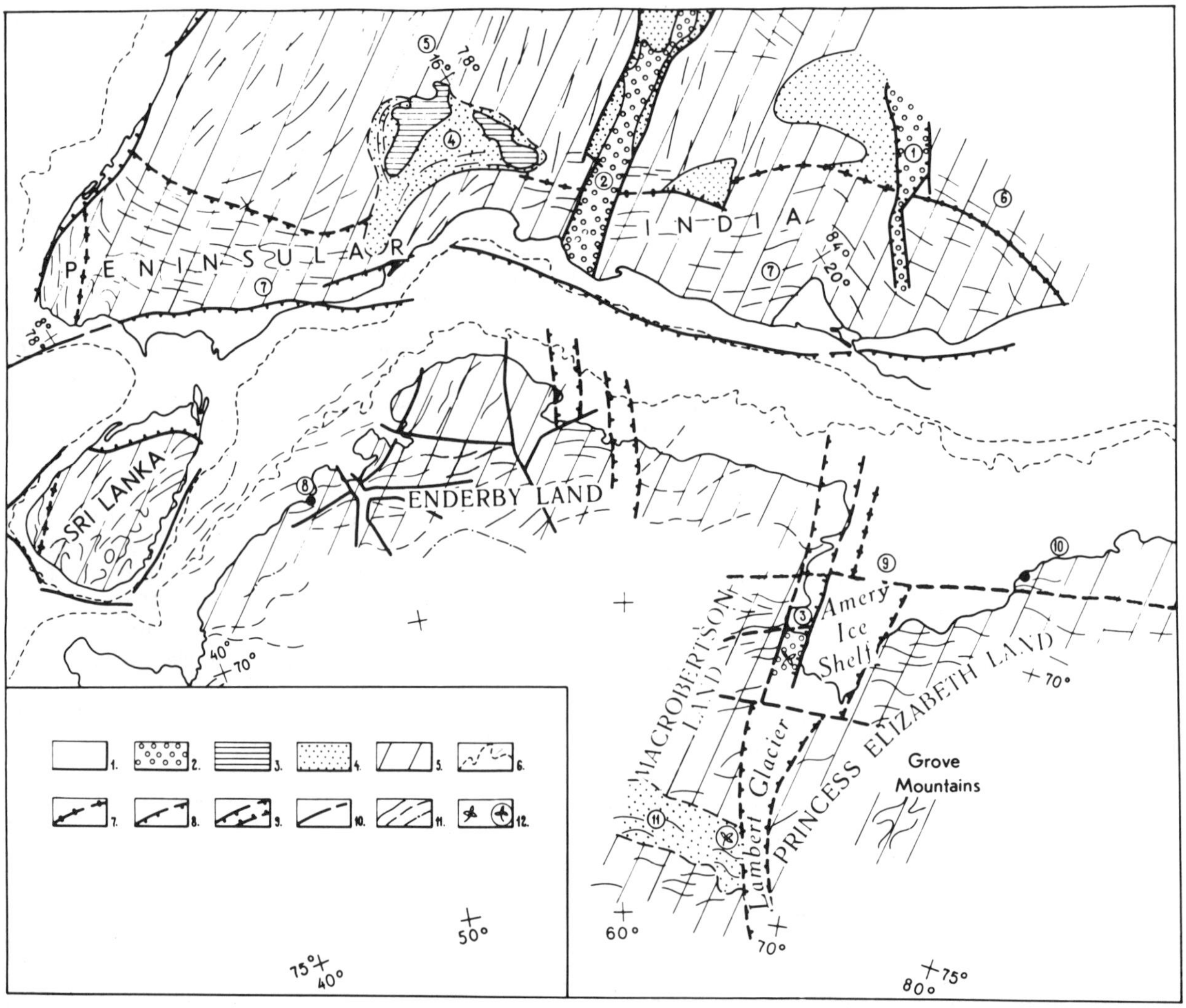

Figure 7.2. Partial Gondwanaland reconstruction: southeastern India and Sri Lanka against Enderby Land, MacRobertson Land, and Princess Elizabeth Land, East Antarctica. 1. Neogene–Quaternary deposits in India and ice sheet in Antarctica; 2. Permian and/or Permian–Carboniferous coal measures; 3. Proterozoic Karnool and Sullavai Series; 4. Proterozoic Pakhals Series, Singareni Series, and Cuddapah System in India, and the Sodruzhestvo Series, Ruker Series, and Menzies Series in Antarctica; 5. The lower Precambrian crystalline rocks; 6. The 1000 fathom isobath; 7. Deep crustal ruptures; 8. Faults; 9. Grabens; 10. Unclassified disjunctive dislocations; 11. Structural lines: continuous for fold axes, dashed for positive magnetic anomalies, and dashed-dotted for negative magnetic anomalies; 12. Sites of discoveries of Permian flora in Antarctica: in bedrock outcrops and in moraines (circled).

Circled numbers: 1. Mahanadi Valley (graben); 2. Godavari Valley (graben); 3. Beaver Lake graben; 4. Cuddapah Basin; 5. Dharwar Belt; 6. Satrupa Belt; 7. Eastern Ghat Belt; 8. Molodezhnaya Base; 9. Prydz Bay; 10. Davis Base; 11. Ruker Complex.

mian on the following grounds: (1) the presence of both *Glossopteris* and *Gangamopteris* genera, which is typical of the lower part of the "Lower Gondwana" (Upper Carboniferous to Lower Permian); (2) the presence of large leaves, suggesting the optimal conditions of growth which prevailed in Early Permian time (Plumstead, 1964); and (3) the presence of cordaites, which are typical of the entire "Lower Gondwana" section. On the other hand, the composition of the Bainmedart Coal Measures (which overlie the Radok Conglomerate in a continuous section) reveals more similarity to the Upper Permian Raniganj Series than to the Lower Permian Barakar Series of peninsular India (Table 7.1).

Table 7.1. Comparison of coals from the Barakar Series (Bennett and Taylor, 1972), from the Raniganj Series (Krishnan, 1968), India, and from the Bainmedart Coal Measures (Ravich, Jr., and others, 1977), East Antarctica

Stratigraphic unit and age	Water content (%)	Volatiles (%)	Fixed carbon (%)	Ash content (%)
Barakar Series, Lower Permian	1-3	20-30	56-65	18[a]
Raniganj Series, Upper Permian	3-8	30-36	50-60	16[a]
Bainmedart Coal Measures, Permian	3.0-6.5	28.7-36.9	36-52.5	15.0-30.0[b]

[a] Coals from mines.
[b] Strongly oxidized coals from outcrops.

The above evidence lends strong support to the possibility that the Mahanadi Graben of peninsular India continued into East Antarctica in the Lambert Glacier area.

There are no direct geologic data to confirm a similar continuation of the Godavari Graben (Figure 7.2). However, aeromagnetic anomalies over eastern Enderby Land form a pattern rather like that over the Lambert Glacier rift zone. They indicate the existence of a sublongitudinal fault zone with graben- and horst-like structures which range from 130 to 150 km in width and extend inland for at least 300 km between long. 57°30′ and 60° E.

Juxtaposition of the major disjunctive features of East Antarctica and southeastern peninsular India, as shown in Figure 7.2, is further supported by a facies analysis of the Lower Permian coal measures of the Mahanadi and Godavari Valley in India (Mitra and others, 1975). These rocks seem to have received sediments derived from the southeast (present geography), i.e., probably from the MacRobertson Land area then adjacent to peninsular India.

Proterozoic complexes in the southern Prince Charles Mountains (East Antarctica) and in southeastern India

Proterozoic jaspilite-bearing sequences, metamorphosed in the greenschist and low-grade amphibolite facies, are exposed in the southern Prince Charles Mountains and have been extensively studied (Grikurov and Soloviev, 1974; England and Langworthy, 1975; Fedorov and Tarutin, 1977). Rocks of similar age, composition, and structural setting are also known in southeastern peninsular India. Comparison of these complexes is undertaken here in order to test the hypothesis that they were formed within intracratonic depressions that developed on a common Indo-Antarctic crystalline basement after its consolidation at about the end of Archean time. The high degree of geologic similarity between the upper Precambrian rocks and structures of southeastern India and those of the southern Prince Charles Mountains is illustrated in Table 7.2 and discussed below.

In both regions the ancient crystalline basement is overlain with distinct structural unconformity by lithologically similar, zonally metamorphosed complexes. In the Prince Charles Mountains, the zoned complex is represented by the Ruker and Menzies Series.* The Ruker Series occupies the central part of the intracratonic fold belt and is characterized by a volcanic-sedimentary sequence that has undergone high-grade greenschist facies metamorphism; jaspilites are typical of this unit. The Menzies Series is believed to represent a lateral time-equivalent of the Ruker Series (at the margin of an intracratonic basin); it differs locally from the Ruker Series in its higher metamorphic grade, which reaches the stress accented staurolite-almandine-kyanite subfacies of the amphibolite facies.

In southeastern peninsular India, the lower supracrustal complex is represented by the Singareni Series (the lower "Cuddapah System"). Within this unit, greenschist sequences with jaspilites grade laterally into staurolite-, kyanite- and almandine-bearing schists of the almandine-amphibolite facies; the latter are associated with a tectonized thrust zone at the contact of the supracrustal mica schists and the basement gneisses.

Both in Antarctica and India, the Ruker, Menzies, and Singareni Series are intruded by numerous dikes and sheet-like bodies of amphibolites and various metabasites which are probably the products of a tectono-thermal activization of the crystalline basement in middle late Proterozoic time.

The upper parts of the epicratonic fold complexes are represented by the Sodruzhestvo Series in the Prince Charles Mountains and by the Pakhals Series (the upper "Cuddapah System") in southeastern peninsular India. These units display good correlation in their original rock type and in their metamorphic rank, which does not exceed the low-grade greenschist facies; an important feature is the presence in both units of conglomerates with jaspilite pebbles.

The Pakhals Series is overlain by the upper Proterozoic–lower Paleozoic(?) Karnool Series, which is not folded and metamorphosed, and which represents the lower unit of the platform cover. This series is synchronous with the Sodruzhestvo Series in age of accumulation, but its structural equivalents are as yet unknown in the Prince Charles Mountains area.

The data on the internal tectonic structure of the Proterozoic fold complexes are not adequate enough in both areas under consideration to allow for detailed correlation. Nevertheless, there are some common features in the general structural pattern: (1) folds in the supracrustal rocks strike parallel to the major structural elements in the crystalline basement; (2) intensity of deformation (folding) and metamorphism increases towards the thrust contacts of the

* Following Russian usage, *series* here means a lithostratigraphic unit formed during a period or longer.

Table 7.2. Correlation of the Proterozoic of southeastern India and the Prince Charles Mountains

Southeastern India			Prince Charles Mountains		
System and/or Series	Composition and thickness	Structure	Series	Composition and thickness	Structure
Sullavai (Karnool)	Conglomerate, sandstone, and quartzite 360 to 480 m	Undeformed			
Angular unconformity					
Pakhals (upper Cuddapah)	Metasandstone; quartzite; chlorite-sericite schist; phyllite; metamorphosed carbonate rock; basal conglomerate with pebbles of ferruginous quartzite ~2500 m	Gentle folding	Sodruzhestvo	Metasandstone; phyllite; micaceous quartzite; mica-chlorite-quartz and mica-carbonate-quartz schist; meta-conglomerate with jaspilite pebbles ~3500 m	Tight folding
Angular unconformity				Angular unconformity	
Singareni (lower Cuddapah)	Quartzite; phyllite; talc, chlorite, and biotite schist; grunerite gneiss; ferruginous quartzite (jaspilite) Quartzite; phyllite; meta-morphosed carbonate rock; andalusite-mica and garnet-staurolite-kyanite-mica schist; ferruginous quartzite (jaspilite)	Tight linear folding parallel to sub-longitudinal structural trends in the crystalline basement. Locally the crystalline basement is overthrust on the Singareni Series	Menzies Ruker	Plagioclase, mica, garnet, sillimanite, kyanite, staurolite, and am-phibole schist; am-phibolite; metacon-glomerate Micaceous chlorite and carbonate-quartz schist; metasand-stone and quartzite; jaspilite; metacon-glomerate 3500 to 4000 m	Tight folding along sub-latitudinal and NNW–SSE axes; schistosity parallel to lamination; strong inter-granular de-formation Tight folding along lati-tudinal axes, in-tense secondary schistosity and cleavage
Angular unconformity				Angular unconformity	
Crystalline basement					

Proterozoic supracrustal complexes with the basement gneisses.

Additional similarities can be inferred from geochronologic evidence which confirms the presence of synchronous geologic events in southeastern peninsular India (Crawford, 1973) and in the Prince Charles Mountains (Halpern and Grikurov, 1975). Crawford proposes the following stages of development for the Cuddapah Basin: 2590 ± 40 m.y.—metamorphism of the basement underlying the Cuddapah System; 1700–1600 m.y.—initiation of sedimentation of the Cuddapah System; 1580 m.y.—volcanic rocks in the basal part of the Cuddapah section; 1370 m.y.—orogeny followed by slight metamorphism and folding of the lower Cuddapah System; 1225 m.y.—termination of accumulation of the upper Cuddapah System by folding and metamorphism; 1150 m.y.—emplacement of kimberlites and early dolerites; 1140–900 m.y.—emplacement of dolerites; 900 m.y.—initiation of accumulation of the Karnool Series; 600–450 m.y.—tectonic activization, emplacement of granites and pegmatites into pre-Karnool complexes, and "rejuvenation" of the Cuddapah metamorphic rocks.

Reconnaissance geochronology in the Prince Charles Mountains area showed that the minimum age of granites intruding the crystalline basement is 2500 m.y. The next group of ages, close to 900–1000 m.y., was obtained on synkinematic metabasites in the Ruker and Menzies Series; these ages probably indicate the time of refolding and superposed metamorphism responsible for development of metamorphic zonation in the epicratonic fold complex. The time of accumulation of the Sodruzhestvo Series can be roughly deduced from the age of 800 m.y. obtained on granitic pebbles from the intraformational conglomerates in the middle part of the series, as well as by the presence in this sequence of late Precambrian acritarchs and unidentifiable fossil remains. The time of folding and metamorphism in the Sodruzhestvo Series is apparently indicated by numerous ages of about 500 m.y. obtained on both the Sodruzhestvo Series phyllites and the muscovite granites which intrude both the crystalline basement and the Ruker-Menzies Series.

Thus, the Proterozoic histories of southeastern peninsular India and the Prince Charles Mountains area in East Antarctica have much in common, and there is a distinct coincidence in major orogenic, magmatic, and metamorphic episodes affecting the supracrustal middle and upper Proterozoic complexes in both areas. Such a coincidence is much easier understood under the assumption that these regions possessed a common lower Proterozoic and Archean crystalline basement.

Structural position of the Archean Eastern Ghat Belt and the Archean complexes in Enderby Land and the Lambert Glacier area

It has been observed (Crawford, 1974) that the tectonic setting of the Archean granulite facies complexes is suggestive of the contiguous position of the Indo-Antarctic regions during early Precambrian time. Additional data (Yanshin, 1965; Ravich and Kamenev, 1972) and recent evidence from the Lambert Glacier area of East Antarctica permit the introduction of some structural details essential for the suggested reconstruction (Figure 7.2). It can be seen that the major structural elements of both plicative (folds) and disjunctive (faults) nature form a common pattern in both subcontinents and on Sri Lanka; the same is true for geophysical anomalies which apparently reflect deep-seated structural features. Therefore, the proposed Indo-Antarctic unity appears feasible even as long ago as early Precambrian time.

Geotectonic evolution of the Indian and Antarctic subcontinents during early Precambrian time requires further investigation and can at present be comparatively analyzed only in general terms. The granulite facies terranes in Antarctica are believed to represent the stable lower Archean shields surrounded by mobile belts of lower metamorphic grade (polymetamorphic and amphibolite facies). The Napier Complex in central Enderby Land consists of the oldest granulite facies rocks and yielded isotopic ages in the range of 3800–4100 m.y. (Sobotovich and others, 1974). The adjacent Rayner Complex, also of granulite facies, is believed to be somewhat younger. The surrounding Humboldt and Insel Complexes yield isotopic ages that fall near the end of Archean time, and they are composed of highly migmatized and granitized polymetamorphic and high-grade amphibolite facies rocks.

In the Eastern Ghat Belt of peninsular India the oldest isotopic dates range from 2900 to 3300 m.y. (Sarkar and others, 1967). The younger ages are thought to be caused by granitization, migmatization, and pegmatite veining which took place 2300–2600 m.y. ago at deep levels of the Eastern Ghat complex and reset the original isotopic equilibrium (Moralev and Perfil'ev, 1972). These processes were apparently associated with orogeny and metamorphism in the adjacent Satrupa and Dharwar Belts which seem very similar to the Antarctic lower Precambrian mobile belts both in composition of the original rocks and in their metamorphic histories.

Discussion

New data from East Antarctica give strong support for geologic correlation between the central part of the East Antarctic coastal region and southeastern peninsular India. Substantial similarity in the composition of rocks and their mode of occurrence, in the strike and nature of major structural elements, and in the main features of geologic history suggests that both regions formed as a single unit within the hypothetical supercontinent of Gondwanaland. Hence Du Toit's reassembly of Gondwanaland with Antarctica at the center gains new confirmation by the geologic matching of these once contiguous Indo-Antarctic regions.

The presence of a relatively small outcrop of Permian coal measures with a *Glossopteris* flora in the Beaver Lake graben among vast crystalline basement terranes becomes comprehensible in the light of this proposed reassembly which brings the Lambert Glacier rift zone of East Antarctica into juxtaposition with a feature of close geologic similarity—the Mahanadi Graben of southeastern peninsular India.

Proterozoic intracratonic depressions of an aulacogen type also formed a regular pattern on both continents; they contain similar sequences of slightly metamorphozed terrigenous rocks with thick jaspilite units at the base. The similar nature of the zonal metamorphism and the simultaneity of structural evolution of those sequences in Antarctica and in southeastern peninsular India are reliably shown by the radiometric age determinations.

A common parentage of the crystalline basements of the ancient Indian and Antarctic platforms is displayed not only in the composition of rocks and in the nature of their granulite facies metamorphism, but also in the identical style of fold deformations and in their common patterns of faults. A more detailed study of the lower Precambrian rocks in both subcontinents in order to establish more precise correlation is certainly required, but it may be anticipated that such work will reveal a closer similarity between the crystalline basements of southeastern India and East Antarctica than between those of India and adjacent Northern Hemisphere terranes.

This paper does not attempt to review all the evidence for the existence of a single Indo-Antarctic craton. It is intended to show only the essential facets of such a unity and to encourage a further study of geologic aspects relevant to the Gondwanaland problem. A careful geologic exploration of the continents surrounding Antarctica, and especially of their marginal parts, will advance the understanding of the subglacial geology of Antarctica and will allow the prediction of the future discoveries of certain geologic phenomena.

References cited

Balme, B. E., and Playford, G. 1967. Later Permian plant microfossils from the Prince Charles Mountains, Antarctica. *Revue Micropaléont., 10,* No. 3, 179–92.

Bennett, A. J. R., and Taylor, G. H. 1972. Coals from the vicinity of the Prince Charles Mountains. (*In* Adie, R. J., *ed. Antarctic Geology and Geophysics.* Oslo, Universitetsforlaget, 591–98.)

Crawford, A. R. 1973. The age of the Cuddapah and Karnool Systems, southern India. *J. geol. Soc. Aust., 19,* 453–64.

Crawford, A. R. 1974. Indo-Antarctica, Gondwanaland, and the distortion of a granulite belt. *Tectonophysics, 22,* 141–57.

Dietz, R. S., Holden, J. C., and Sproll, W. D. 1972. Antarctica and continental drift. (*In* Adie, R. J., *ed. Antarctic Geology and Geophysics.* Oslo, Universitetsforlaget, 837–42.)

Du Toit, A. L. 1937. *Our Wandering Continents.* Edinburgh, Oliver and Boyd, 366 pp.

Elliot, D. H. 1975. Gondwana basins of Antarctica. (*In* Campbell, K. S. W., *ed. Gondwana Geology.* Canberra, Australia National University Press, 493–536.)

England, R. N., and Langworthy, A. P. 1975. Geological work in Antarctica—1974. *Rec. Bur. Miner. Resour. Geol. Geophys. Aust.,* 1975/30, 28 pp. (Unpublished.)

Fedorov, L. V., and Tarutin, O. A. 1977. Geologicheskoe stroenie gory Ruker (iuzhnaia chast' gor Prins-Charl'z, Vostochnaia Antarktida) [Geological structure of Mount Ruker, southern Prince Charles Mountains, East Antarctica]. *Antarktika, Doklady Komissii, 16,* 93–99.

Grikurov, G. E., and Soloviev, D. S. 1974. Geologicheskoe stroenie gornogo obramlenia lednika Lamberta [Geological structure of the mountain fringe of Lambert Glacier]. *Inf. Byull. sov. antarkt. Eksped.,* No. 88, 21–29.

Halpern, M., and Grikurov, G. E. 1975. Rubidium-strontium data from the southern Prince Charles Mountains. *Antarct. Jnl U.S., 10,* No. 1, 9–15.

Krishnan, M. S. 1968. *Geology of India and Burma.* Madras, Higginbothams, 536 pp.

McElhinny, M. W., and Embleton, B. J. J. 1974. Australian palaeomagnetism and the Phanerozoic plate tectonics of eastern Gondwanaland. *Tectonophysics, 22,* 1–29.

Mitra, N. D., Laskar, B., and Basu, U. K. 1975. Pattern of clastic dispersal in the lower Gondwana coalfields of peninsular India. (*In* Campbell, K. S. W., *ed. Gondwana Geology.* Canberra, Australia National University Press, 53–59.)

Mond, A. 1972. The Permian sediments of the Beaver Lake area, Prince Charles Mountains. (*In* Adie, R. J., *ed. Antarctic Geology and Geophysics.* Oslo, Universitetsforlaget, 585–89.)

Moralev, V. M., and Perfil'ev, Yu. S. 1972. K geologii dokembriya iuzhnoi Indii [Contribution to the Precambrian geology of southern India]. *Sov. Geol.,* No. 6, 98–107.

Plumstead, E. P. 1964. Paleobotany of Antarctica. (*In* Adie, R. J., *ed. Antarctic Geology.* Amsterdam, North-Holland Publishing Company, 637–54.)

Ravich, G. M., Gor, Yu. G., Dibner, A. F., and Lobanova, O. V. 1977. Stratigrafiia verkhnepaleozoiskikh uglenosnykh otlozhenii Vostochnoi Antarktidy (raion ozera Biver) [The stratigraphy of upper Paleozoic coal measures (Beaver Lake area, East Antarctica)]. *Antarktika, Doklady Komissii, 16,* 62–75.

Ravich, M. G. 1972. *Zagadki Gondvany* [Enigma of Gondwanaland]. Moscow, Znanie, 64 pp.

Ravich, M. G., and Kamenev, E. N. 1972. *Kristallicheskii fundament Antarkticheskoi platformy* [Crystalline basement of the Antarctic platform]. Leningrad, Gidrometeoizdat, 658 pp. [English translation: New York, John Wiley, 1975, 582 pp.]

Sarkar, S. N., Saha, A. K., and Miller, J. A. 1967. Potassium-argon ages from the oldest metamorphic belt in India. *Nature, Lond., 215,* 946–48.

Smith, A. G., and Hallam, A. 1970. The fit of the southern continents. *Nature, Lond., 225,* 139–44.

Sobotovich, E. V., Kamenev, E. N., Komaristy, A. A., and Rudnik, V. A. 1974. Drevneishie porody Antarktidy (Zemlia Enderbi) [The oldest rocks of Antarctica (Enderby Land)]. *Izv. Akad. Nauk SSSR,* Ser. geol., No. 11, 30–50.

Tingey, R. J. 1972. Geological work in Antarctica—1971. *Rec. Bur. Miner. Resour. Geol. Geophys. Aust.,* 1972/132, 49 pp. (Unpublished.)

Tingey, R. J., and England, R. N. 1973. Geological work in Antarctica—1972. *Rec. Bur. Miner. Resourc. Geol. Geophys. Aust.,* 1973/161. 17 pp. (Unpublished.)

Trail, D. S., and McLeod, I. R. 1969. Geology of the Lambert Glacier Regions (Sheet 12 Lambert Glacier Region). (*In* Bushnell, V. C., and Craddock, C., *eds.* Geologic Maps of Antarctica. *Antarct. Map Folio Ser.,* Folio 12, Pl. XI.)

Yanshin, A. L. 1965. Tektonicheskoe stroenie Evrazii [Tectonic structure of Eurasia]. *Geotektonika,* No. 5, 7–35.

8

Structural Limits of the Southern Crozet Basin and its Relation to the Enderby Basin and the Kerguelen-Heard Plateau

J. GOSLIN* and R. SCHLICH**
Institut de Physique du Globe, Laboratoire de Géophysique Marine, 94100 Saint Maur des Fossés, France

Abstract. Three oceanographic cruises *Marion Dufresne* 1 (1973), *Marion Dufresne* 5 (1975), and *Marion Dufresne* 11 (1976) have provided the opportunity for a reconnaissance survey southwest of the Kerguelen-Heard Plateau. A major topographic feature, the Marion Dufresne Seamount, has been discovered near lat. 54° S and long. 50° E, roughly in line with the Ob and Lena Seamounts. These aligned seamounts separate two deep ocean basins: to the north, the Crozet Basin, where magnetic anomalies and fracture zones are quite well known—the sea floor of the Crozet Basin is dated as Late Cretaceous (anomaly 34) in the vicinity of the seamounts; and to the south, the Enderby Basin, where neither magnetic anomalies nor fracture zones have yet been mapped. Near the seamounts, the floor of both basins is blanketed by a uniform sediment cover. These seamounts were probably formed in Late Cretaceous time during a major episode of anomalous volcanism in the vicinity of the Southeast Indian Ridge. This tentative mechanism and date of formation are based on study of seismic reflection profiles and visual examination of free-air gravity anomaly data. A broad basement rise lies southwest of the Kerguelen Islands. It is separated, however, from the Ob–Lena–Marion Dufresne Seamount chain by a major structural feature and therefore seems unrelated to the chain itself. This structural feature follows a north-northeast trend and therefore could be related to the fracture zone system of the Crozet Basin.

Several aseismic ridges are known in the world ocean. The great variety in their structures has caused them to be termed ridges, plateaus, rises, banks, or seamount chains. Although several papers have dealt with certain aspects of their tectonic history (e.g., Detrick and others, 1977), no agreement among different workers has been reached on the origin and formation mechanism of such structures. The lack of seismic activity may be their only common feature, and they seem to be the group of structures least readily explained by straightforward plate tectonic theory.

Many such aseismic ridges are found in the Indian Ocean. Some of them have been extensively studied, for example, the Ninety East Ridge (Udintsev, 1965; Francis and Raitt, 1967; Laughton and others, 1970; McKenzie and Sclater, 1971; Bowin, 1973; Luyendyk and others, 1974; Sclater and Fisher, 1974; Luyendyk and Rennick, 1977). Others have received less attention, such as the Chagos-Laccadive Ridge (Fisher and others, 1971; McKenzie and Sclater, 1971; Ben Avraham and Bunce, 1977), the Seychelles Bank (Francis and Shor, 1966; Matthews and Davies, 1966; Francis and others, 1966), the Agulhas Plateau, the Madagascar Plateau and the Mozambique Ridge, the Kerguelen-Heard Plateau and the Broken Ridge, and the Crozet Plateau. (See Figure 9.7 in Schlich, Chapter 9, this volume, for location of regional features.)

Despite the differences in structure and tectonic setting of these ridges, they might provide a critical test of the validity of proposed reconstructions of the Indian Ocean, particularly since some of the aseismic ridges are thought to have been formed during major changes in spreading history (McKenzie and Sclater, 1971) or during episodes of anomalous volcanism along mid-ocean ridges (Norton and Molnar, 1977; Ben Avraham and Bunce, 1977; Detrick and others, 1977).

In this paper we shall present a newly discovered structure, the *Marion Dufresne Seamount* (Schlich, 1975), and discuss its possible relationships with other neighboring structures (the Ob and Lena Seamounts) and with a recently discovered broad basement rise southwest of the Kerguelen-Heard Plateau.

Structural framework

The Ob, Lena, and Marion Dufresne Seamounts are lined up along a roughly west-northwest trend. The Ob and Lena Seamounts culminate at depths of 247 and 254 m, respectively (Udintsev and others, 1975). The seamount chain lies between two deep ocean basins: the Crozet Basin to the north, in which several bathymetric and magnetic profiles have been obtained (Schlich, 1975; and others), and the less well-known Enderby Basin to the south. The Crozet Basin reaches depths of 4000 to 4500 m in the vicinity of the seamounts. The Enderby Basin deepens to more than 5000 m just south of the Marion Dufresne Seamount (Figure 8.1).

Magnetic lineations as old as anomaly 34 have been mapped in the southern Crozet Basin. These anomalies are oriented N 120° E and are offset right-laterally by several fracture zones trending N 28° E in this area (Schlich, 1975). An attempt to follow the anomaly sequence farther

* Present address: Laboratoire de Géophysique Interne, Université de Bretagne Occidentale, Avenue le Gorgeu, 29283 Brest Cedex, France.

** Present address: Laboratoire de Géophysique Marine, Institut de Physique du Globe de Strasbourg, 5, rue René Descartes, 67084 Strasbourg Cedex, France.

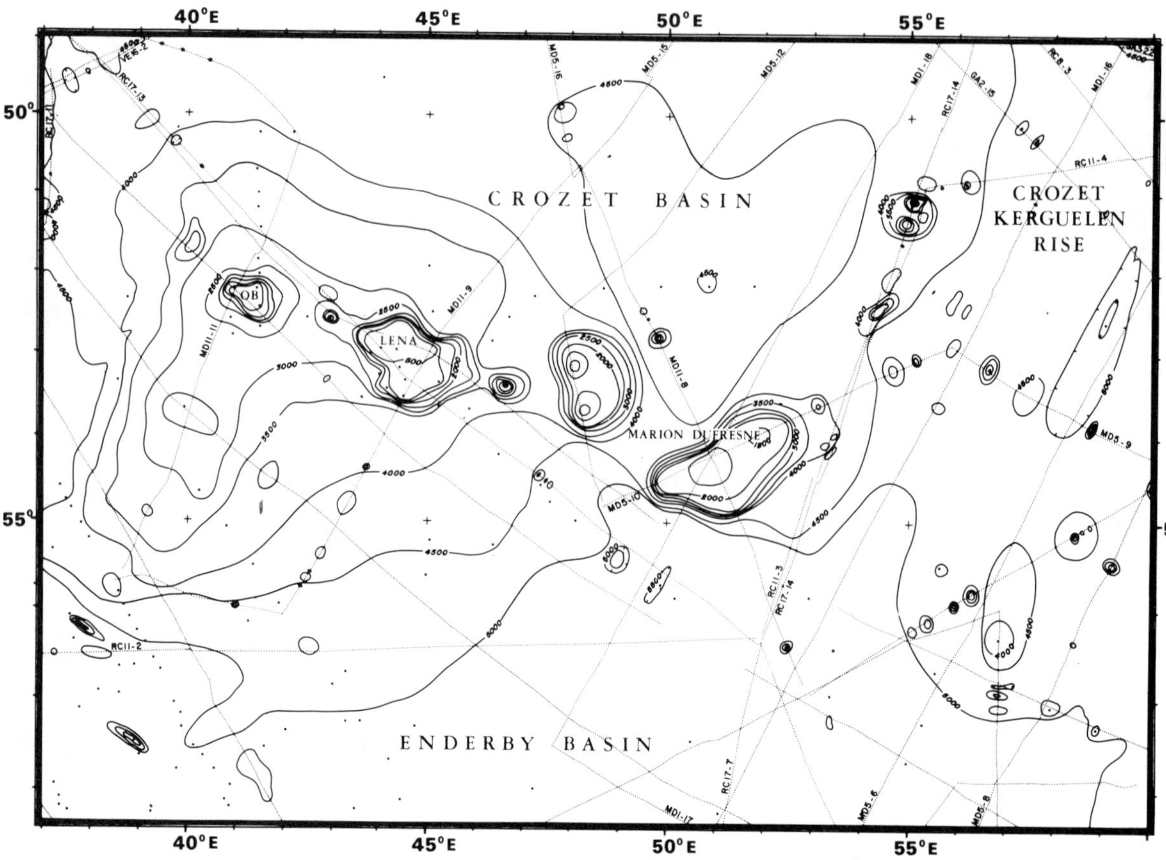

Figure 8.1. Bathymetric chart of Marion Dufresne Seamount and its environs. Soundings are in uncorrected meters. U.S.S.R. ships' tracks are unlabeled. The initials RC stand for R/V *Robert Conrad* and VE for R/V *Vema,* both of the Lamont-Doherty Geological Observatory. Initials GA stand for M/V *Gallieni* and MD for M/V *Marion Dufresne,* both of the Terres Australes et Antarctiques Françaises. Contour interval is 500 m. (For scale, 1° of latitude ≅ 111 km.)

south in the northern Enderby Basin explains the orientation chosen for the profiles MD1-16, MD1-18, MD5-6, MD5-8, MD11-9, MD11-11 (Figure 8.1). Examination of these profiles did not allow us to identify any magnetic anomalies in the northern Enderby Basin, and even less, any fracture zone orientation in this area. Farther east, in the vicinity of the Kerguelen-Heard Plateau, the anomaly sequence can only be followed back to anomaly 11 (early Oligocene) north of the Kerguelen-Heard Plateau (Schlich and others, 1971) and to anomaly 19 (early Eocene) east and west of the Kerguelen-Heard Plateau (Schlich, 1975).

Topography of the Marion Dufresne Seamount and adjacent region

The ships' tracks along which bathymetric data are available are shown on the bathymetric map of Figure 8.1. The map was compiled from all the data available along these tracks, implementing older soundings by oceanographic ships from the U.S.S.R. (shown as dots on Figure 8.1). The vast dimensions of the area studied (9° × 23°) should be noted along with the relative paucity of available bathymetric data.

The Marion Dufresne Seamount appears to consist of two separate bathymetric highs. The eastern high, roughly 260 km × 130 km, culminates at about 1300 m below sea level and follows a southwest-northeast trend. The western high, 140 km × 120 km, reaches a shallower depth of 970 m and appears to be roughly north-south in orientation, but other lines are needed to confirm this trend. The Marion Dufresne Seamount rises from the adjoining basins with relatively very steep slopes (10 to 15%), except toward the northeast where it is tied through a broad rise to the Crozet-Kerguelen saddle.

The bathymetric map (Figure 8.1) suggests an *en echelon* structure. The steps of the *en echelon* structure do not seem to be controlled by the Crozet Basin fracture zones; the fracture zones offset the identified magnetic anomalies to the right, while the bathymetric highs appear to be offset to the left. An alternative explanation is that the steep southern flanks of the seamounts are tectonically controlled by northwest-southeast trending fracture zones in the northern Enderby Basin, but, as mentioned above, no such fractures could be identified from the profiles available in this area.

The Crozet-Kerguelen Rise has many scattered seamounts, probably related to the fracture zones of the Crozet Basin (Figure 8.1). An elongated deep also follows the north-northeast trend of these fracture zones. South of the rise, toward the Enderby Basin, the sea floor deepens in several steps; these steps are especially clear along profile MD1-16.

Shallow seismic structure of the Marion Dufresne Seamount and of the adjoining basins

Few continuous seismic profiles have been successfully taken because of the prevailing bad sea conditions in the area of study (Figure 8.2). Only three expendable-sono-

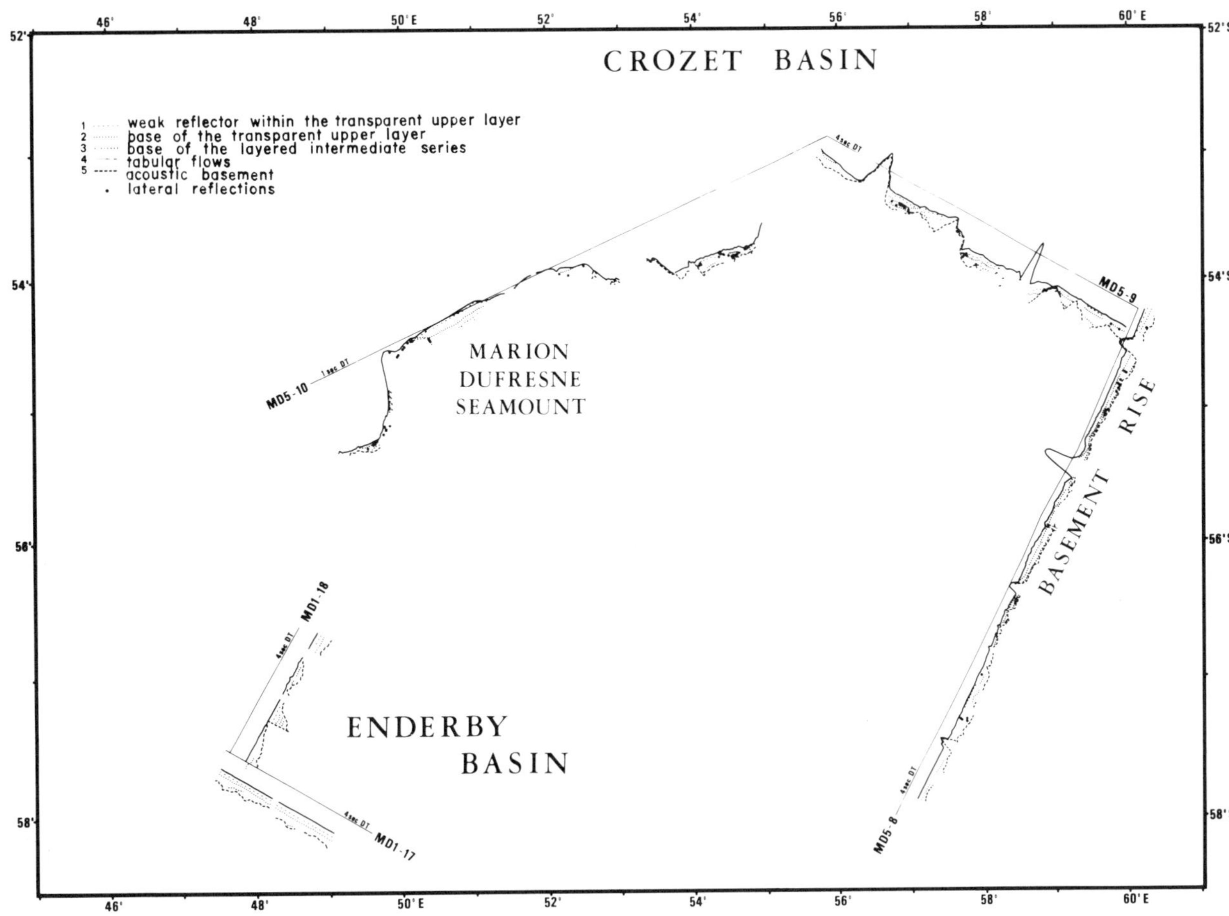

Figure 8.2. Interpreted seismic reflection profiles. Thin lines show the "averaged" ship's tracks. The meaning of the labeled sedimentary layers is explained more fully in the text.

buoy refraction profiles have been taken over the Marion Dufresne Seamount; moreover, the quality of the refraction data is poor.

Marion Dufresne Seamount (profile MD5-10)

The acoustic basement was not reached on the central part of the seamount itself (profile MD5-10, Figure 8.2). This basement is overlain by more than 1.3 seconds of sediment DT (double-way time). The sediments are regularly layered, but none of the sedimentary units that were identified in the adjacent basins, as discussed below, seem to be present over the top of the seamount. A sedimentary rise forms the topographic culmination of the seamount. The acoustic basement crops out to form the southwestern flank, against which the sediment appears to have been dammed in a rather complex manner. The sedimentary sequence down to the acoustic basement looks undeformed at the foot of the seamount's flanks. This suggests that the formation of the Marion Dufresne Seamount did not occur after deposition of the sediment and, therefore, the seamount is probably a relatively old structure.

Northern Enderby Basin (profiles MD1-17 and MD1-18)

The acoustic basement in the northern Enderby Basin (labeled 5 in Figure 8.2) shows the numerous reflection hyperbolas typical of a rough oceanic basement. It is at least 6 km deep, which indicates an age for this area of greater than 100 m.y. (Sclater and others, 1971). This limit is the only estimate we can make for the age of the northern Enderby Basin in the absence of any identifiable magnetic anomaly, as noted above. The sediment overlying the basement is commonly more than 1 second DT thick. A few tabular (basaltic?) flows or sills interrupt the layering of the sediment cover; they are labeled 4 in Figure 8.2. The sediment cover can be divided into three distinct units. The uppermost layer, in some places more than 0.5 second DT thick, is very transparent to the seismic waves, and only weak subhorizontal reflectors can be seen (labeled 1 in Figure 8.2). The intermediate unit, on the other hand, is densely layered with strong reflectors. This unit in most places shows no more tectonic deformation than the uppermost layer. The lowest sedimentary unit closely follows the basement relief and thus presents a rough topography. Its thickness rarely exceeds 0.2 second DT, and it is topped by a very strong seismic reflector (labeled 3 in Figure 8.2). A major basement offset is visible on profile MD1-18, near lat. 57° 20′ S and long. 48° E (Figure 8.2). This offset probably indicates the presence of a major fracture zone in the Enderby Basin whose direction is not determined at present.

Crozet-Kerguelen Rise (profile MD5-9)

The basement topography is very rough along this profile. Numerous basement highs disturb the sedimentary units in their vicinity, suggesting that some tectonic activity occurred after sediment deposition. Two tabular (basaltic?) flows or sills of some importance are present. The three sedimentary units described in the Enderby Basin also seem to form the sedimentary cover in this area. The lower unit follows the rough basement topography while the two upper units seem nearly horizontal. A important jump in the basement depth occurs near lat. 53° 30′ S and long. 57° 40′ E, coincident with a topographic step, the lower side of which is the elongated deep mentioned above. This feature indicates the location of one of the major fracture zones of the South Crozet Basin.

Southwest Kerguelen Province (profile MD5-8)

A most striking basement feature is to be noticed along this profile: a broad (several hundreds of kilometers) basement high is situated south of the culmination of a long-wavelength topographic high (the former culminating near lat. 56° 40′ S, the latter near lat. 55° 30′ S, along profile MD5-8). This southern basement high could be part of an elongated structure whose trend would be close to the azimuth of profile MD5-9. If such a structure does exist, it would bear no relationship to the Ob, Lena, and Marion Dufresne Seamount chain: no topographic expression is apparent along profile MD1-16 (Figure 8.1) between this structure and the seamount chain. The major fracture zone mentioned in the preceding paragraph probably separates two distinct domains at this latitude: to the west, the seamount chain, constituting the southern boundary of the Crozet Basin; and to the east, a region of more subdued topography lying southwest of the Kerguelen Islands. In spite of this change in tectonic setting, the sediment cover can still be divided here into three units, their characteristics resembling those described in the other areas. The lower layer thickens toward the north as it passes from the Enderby Basin on to the Crozet-Kerguelen Rise, where it reaches in places a thickness of 0.3 second DT, filling the basement troughs (Figure 8.2).

Gravity data

Three gravity profiles, one over each seamount, are shown in Figure 8.3. These profiles represent "total gravity" data, and neither the Eötvös correction (due to the ship's motion over the rotating earth) nor the regional gravity field were subtracted. As the ship's speed and heading were nearly constant along each profile, the Eötvös correction can be considered as constant for these short profiles (the change in the Eötvös correction at constant speed and heading when going from lat. 57° S to lat. 50° S would be at most 5 mgal). The regional gravity field was computed for each even degree along the profiles, and straight lines were fitted through the resulting points for each of the three profiles. These lines are shown as dashes in Figure 8.3. Thus, the gravity profiles represent roughly the free-air gravity anomaly when taken with respect to the "zero" lines just mentioned. The irregularities of very short wavelength which affect the MD11-8 profile are due to very rough sea conditions. A sedimentary basin, similar to the one apparent on the seismic profile MD5-10 taken over the Marion Dufresne Seamount, seems also to be present under the central part of the Lena Seamount; the gravity profile MD11-9 shows a distinct low over this area, which cannot be accounted for solely by the slight topographic low.

A free-air anomaly profile over the aseismic Walvis Ridge (Goslin and Sibuet, 1975) is shown in Figure 8.3 for comparison. Though the *Marion Dufresne* profiles are of rather poor quality because of bad sea conditions, it seems that no long-wavelength gravity anomaly is present on the sides of the Ob–Lena–Marion Dufresne Seamount chain. This suggests that the seamounts do not represent a younger load superimposed on old lithosphere, and it therefore favors creation of this chain contemporaneously with the adjoining sea floor (about Late Cretaceous? time). Computations of 2-dimensional gravity models are in progress to test this interpretation, which is as yet a hypothesis based only on the visual examination of the profiles. The gravity edge effect is similar to the one over the Walvis Ridge for which a shallow compensating mechanism was postulated (Goslin and Sibuet, 1975), but this must also be checked by computation.

Magnetic profiles

As mentioned above, no linear magnetic anomalies have been identified and mapped in the northern Enderby Basin. The Ob and Lena Seamounts magnetic signatures are characterized by high-frequency anomalies indicating the presence of shallow magnetic material (Figure 8.4, profiles MD11-11 and MD11-9).

The Marion Dufresne Seamount itself causes quite a different magnetic anomaly pattern. The absence of short-wavelength anomalies may be partially explained by the greater depth of the magnetic sources. Nevertheless, the overall amplitude of the anomalies makes a continental origin for the Marion Dufresne Seamount improbable, but the amplitude is not characteristic of an uplifted, standardly magnetized, oceanic crust. The basaltic flanks of the seamount can be tentatively correlated with negative magnetic anomalies (along profiles MD1-18 and MD5-10 in particular). At the latitude of the area studied, the magnetic field inclination is large enough so that such a correlation is probably not an artifact.

Concluding remarks

The chain formed by the Ob, Lena, and Marion Dufresne Seamounts separates two deep ocean basins. On the north, the Crozet Basin grows older toward the south, reaching Late Cretaceous age (magnetic anomaly 34). The sea floor is offset by well-defined fractures zones trending north-northeast. The Enderby Basin seems older. No magnetic anomaly or fracture-zone trends could be mapped in this area, and moreover, the Crozet Basin fracture zones are interrupted toward the south by the seamount chain. A thick sediment cover blankets both basins in the vicinity of the chain. This sedimentary sequence can be divided into three distinct layers which are present throughout all the deep areas.

The seamounts seem to have been formed in Late Cretaceous time together with the adjacent oceanic crust. They could be the expression of anomalous volcanic processes occurring at the ridge crest (Detrick and others, 1977) while the plate system underwent major changes in configuration and direction of spreading. A major tectonic episode possibly occurred in the world ocean during Late Cretaceous time and bore consequences for the structure of the aseismic ridges of the Indian Ocean (Luyendyk and

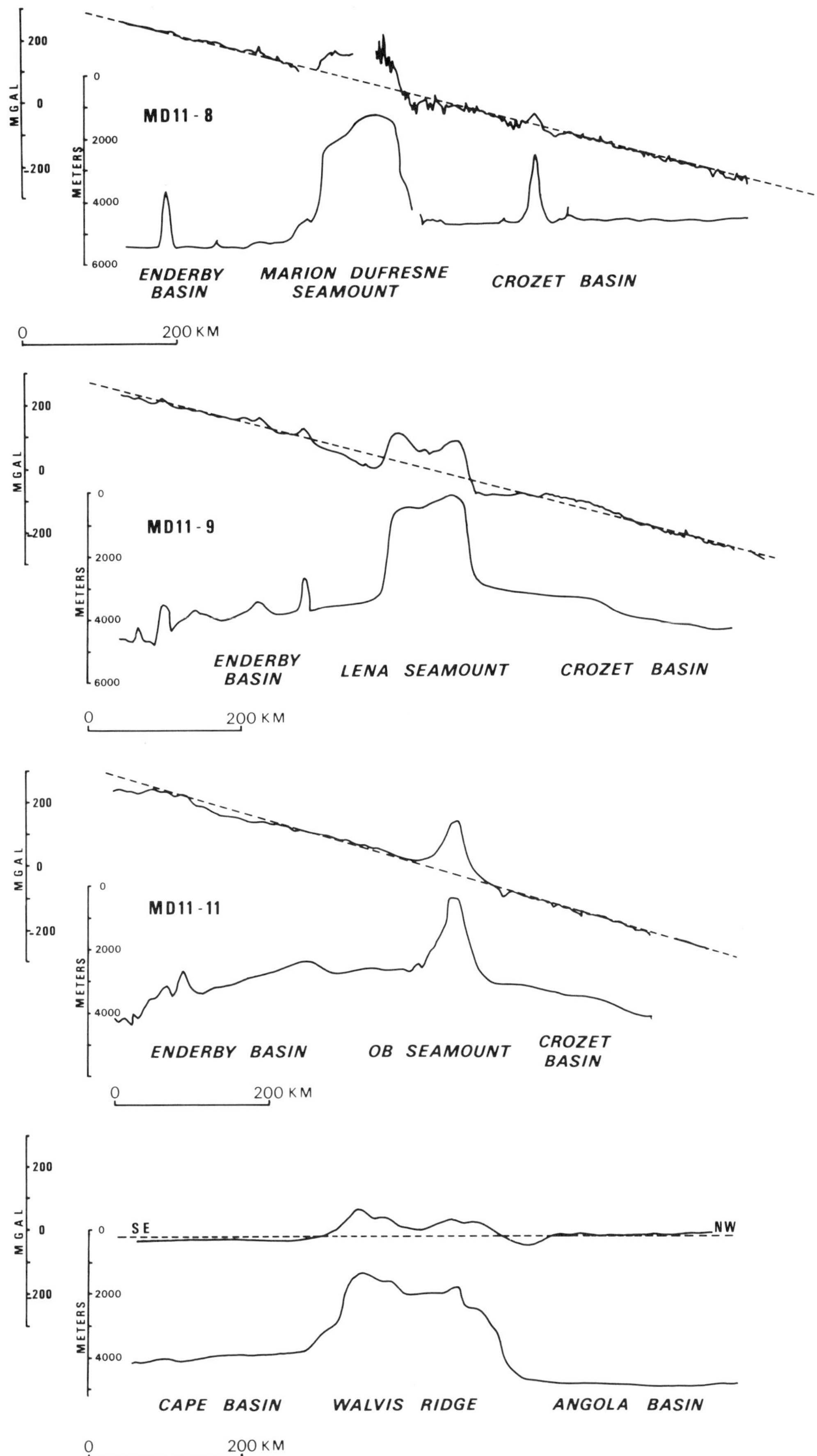

Figure 8.3. Bathymetry and total gravity profiles. Position of the profiles are shown in Figure 8.1. A profile over an old aseismic ridge (Walvis Ridge, South Atlantic) is also shown for comparison. The dashed lines represent the regional gravity field computed in even degrees along the profiles.

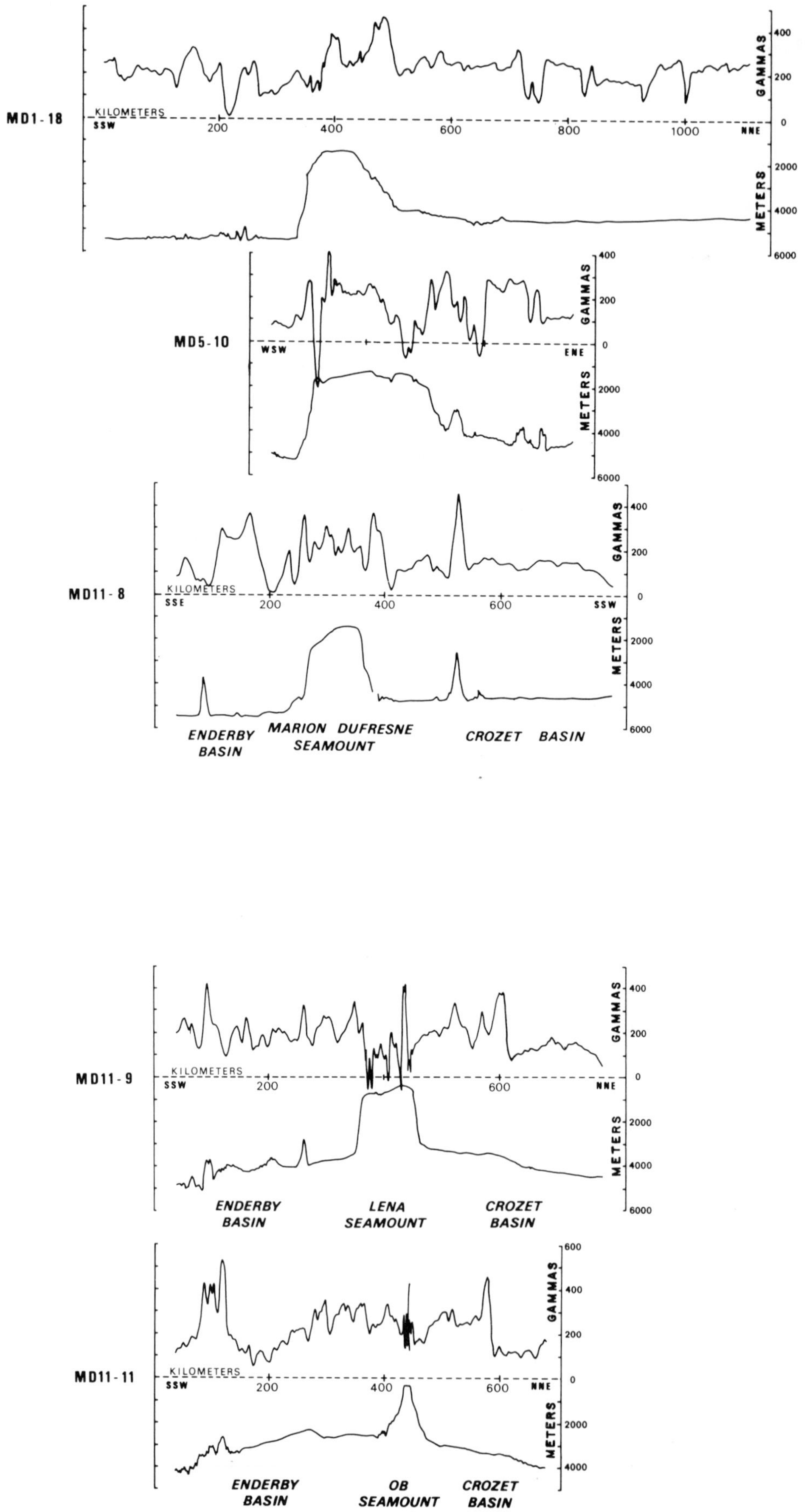

Figure 8.4. Bathymetry and magnetic anomaly profiles. The regional magnetic field was subtracted using the model of Cain and others (1964).

Rennick, 1977). The Nikitin Seamount, a much smaller structure mapped in the northern Indian Basin, has been roughly dated at the same age from the magnetic anomaly pattern (McKenzie and Sclater, 1971). The apparent *en echelon* pattern of the Ob–Lena–Marion Dufresne Seamount chain does not seem to be related to the fracture zones of the Crozet Basin.

A major structural feature terminates the chain at its eastern end. This feature follows the Crozet Basin fracture-zone trend in the northern part of the study area, but its trace is lost toward the south under the sediment cover of the Enderby Basin. To the east of this feature, the Kerguelen domain is complicated by the occurrence of a long-wavelength, low-amplitude basement high, unrelated to the Ob–Lena–Marion Dufresne Seamount chain.

Gravity model computations now in progress may yield some insight into the deep structure of the seamount chain. They represent one possible means of testing our hypothesis of an ancient date of formation for the chain, about in Late Cretaceous times.

Acknowledgments. We are grateful to D. Hayes for the soundings from the *Vema* 16, *Conrad* 8, *Conrad* 11, and *Conrad* 17 cruises, which contributed to the bathymetric chart presented in this paper.

References cited

Ben Avraham, Z., and Bunce, E. T. 1977. Geophysical study of the Chagos Laccadive Ridge, Indian Ocean. *J. geophys. Res., 82,* 1295–1305.

Bowin, C. 1973. Origin of the NinetyEast-Ridge from studies near the equator. *J. geophys. Res., 78,* 6029–43,

Cain, J. C., Hendricks, S., Daniels, W. E., and Jensen, D. C. 1964. Computation of the main geomagnetic field from spherical expansion. *Nasa Rep.,* No. x-611-64-316.

Detrick, R. S., Sclater, J. G., and Thiede, J. 1977. The subsidence of aseismic ridges. *Earth planet. Sci. Lett., 34,* 185–96.

Fisher, R. L., Sclater, J. G., and McKenzie, D. P. 1971. Evolution of the central Indian Ridge, western Indian Ocean. *Geol. Soc. Am. Bull., 82,* 553–62.

Francis, T. J. G., and Raitt, R. W. 1967. Seismic refraction measurements in the southern Indian Ocean. *J. geophys. Res., 72,* 3015–42.

Francis, T. J. G., and Shor, G. G. 1966. Seismic refraction measurements in the northwestern Indian Ocean. *J. geophys. Res., 71,* 427–49.

Francis, T. J. G., Davies, D., and Hill, M. N. 1966. Crustal structure between Kenya and the Seychelles. *Phil. Trans. R. Soc. Lond.,* Ser. A, *259,* 240–61.

Goslin, J., and Sibuet, J. C. 1975. Geophysical study of the easternmost Walvis Ridge, South Atlantic: deep structure. *Geol. Soc. Am. Bull., 86,* 1713–24.

Laughton, A. S., Matthews, D. H., and Fisher, R. L. 1970. The structure of the Indian Ocean. (*In* Maxwell, A., *ed. The Sea.* New York, Wiley Interscience, *4,* 543–86.)

Luyendyk, B. P., and Rennick, W. 1977. Tectonic history of the aseismic ridges in the eastern Indian Ocean. *Geol. Soc. Am. Bull., 88,* 1347–56.

Luyendyk, B. P., Davies, T. A., and others. 1974. Results of DSDP leg 26 and the geologic history of the southern Indian Ocean. (*In* Davies, T. A., Luyendyk, B. P., and others. *Initial Reports of the Deep Sea Drilling Project,* Vol. 26. Washington [U.S. Government Printing Office], 909–43.)

Matthews, D. H., and Davies, D. 1966. Geophysical study of the Seychelles Bank. *Phil. Trans., R. Soc. Lond.,* Ser. A, *259,* 227–39.

McKenzie, D. P., and Sclater, J. G. 1971. The evolution of the Indian Ocean since the Late Cretaceous. *Geophys. J. R. astr. Soc., 25,* 437–528.

Norton, I., and Molnar, P. 1977. Implications of a revised fit between Australia and Antarctica for the evolution of the eastern Indian Ocean. *Nature, Lond., 267,* 338–40.

Schlich, R. 1975. Structure et âge de l'océan Indien occidental. *Mem. Soc. Geol. Fr.,* No. 6, 103 pp.

Schlich, R., Delteil, J. R., Moulin, J., Patriat, P., and Guillaume, R. 1971. Mise en évidence d'une sédimentation de marge continentale sur le plateau de Kerguelen-Heard. *C.r. Acad. Sci., Paris, 272,* 2060–63.

Sclater, J. G., and Fisher, R. L. 1974. Evolution of the east central Indian Ocean, with emphasis on the tectonic setting of the NinetyEast Ridge. *Geol. Soc. Am. Bull., 85,* 683–702.

Sclater, J. G., Anderson, R. N., and Bell, M. L. 1971. Elevation of ridges and evolution of the central eastern Pacific. *J. geophys. Res., 76,* 7888–915.

Udintsev, G. B. 1965. New data on the bottom topography of the Indian Ocean. *Oceanologiya, 5,* 933–98.

Udintsev, G. B., and others. *eds.* 1975. *Geologo-geophysitcheskie atlas Indianskovo Okeana* [Geological-geophysical atlas of the Indian Ocean]. *Moscow, Akademia Nauk SSSR,* 151 pp.

9

Structure and Age of the Western Indian Ocean—Results of Deep Sea Drilling

R. SCHLICH*
Institut de Physique du Globe, Laboratoire de Géophysique Marine, 94100 Saint Maur des Fossés, France

Abstract. The Indian Ocean is characterized by a complex system of mid-ocean ridges which extends from the Gulf of Aden southward to near the Mascarene Islands, where it divides into two distinct branches: the Southwest Indian Ridge and the Southeast Indian Ridge. These ridges separate Antarctica from Africa and Australia. Numerous submarine plateaus or aseismic ridges, together with the active ridges and the adjacent continents, outline several deep ocean basins.

Approximately 450,000 km of bathymetric and magnetic profiles from the Indian Ocean are now available. The structure of these plateaus, active ridges, and associated ocean basins has been investigated—in the north mainly by English and American workers and in the south by the *Galliéni* (1967–72), *Marion Dufresne* (1973–77), and *Suroít* (1977) cruises. The major structural features have been defined, and linear magnetic anomalies back to anomaly 34 (Santonian) have been identified in several ocean basins. Although large areas are still unexplored, particularly the entire region south of lat. 56° S, use of fracture zones and magnetic lineations has made it possible to distinguish the major events which have contributed to the formation of the Indian Ocean, at least since Late Cretaceous time. Changes in spreading rates and/or directions are recorded for the following: between anomalies 31 and 30 (Campanian to Maestrichtian), just before anomaly 27 (Maestrichtian to early Paleocene), near anomalies 22/21 (late Paleocene to early Eocene), during the time interval separating anomalies 18 and 15 (middle to late Eocene), and, finally, at the time of anomaly 5 (late Miocene). These different spreading phases observed since Late Cretaceous time are synchronous in all the large oceanic basins.

The deep sea drilling by the *Glomar Challenger* has made important contributions to knowledge of the complex geologic history of this ocean. Paleontologic dating of sediments cored at the basalt contact in several large ocean basins has confirmed the proposed geophysical interpretation and has suggested some revisions in the time scale for magnetic reversals of Late Cretaceous and Paleogene age.

The Indian Ocean is bounded by the African, Indian, Australian, and Antarctic continents, which were formed after the breakup of Gondwanaland. The search for the fragmentation pattern of this ancient continent, which is linked to the history of the formation of the Indian Ocean, is very difficult in comparison to the problems of the other major oceans of the world. Unlike the Atlantic Ocean, where two distinct blocks, America and Africa, are separated by the Mid-Atlantic Ridge, the Indian Ocean includes a large number of continental fragments separated by a complex system of mid-ocean ridges. Any attempt at reconstruction is made even more difficult by the existence of many submarine plateaus, which are in some cases capped by islands and are commonly quite extensive. These plateaus have been interpreted as uplifted blocks of oceanic crust, as volcanic structures, and as continental fragments. In fact, the limited extent of active trenches leads to the assumption that nearly all of the oceanic crust formed at the spreading ridges is still present. Within the concepts of sea floor spreading, continental drift, and plate tectonics, it should be possible, from studies of the fracture zones and the magnetic lineations, to trace the movements of the continents or continental fragments and thus establish the evolution of the Indian Ocean and the structure and age of its main ocean basins. The long cores obtained by the *Glomar Challenger* (Deep Sea Drilling Project [DSDP] Legs 22 to 27) in the Indian Ocean have greatly contributed to knowledge of the geologic history of this ocean. They have made it possible to check the validity of the proposed reconstructions and to verify the chronology of the major events which have contributed to the formation of the Indian Ocean, at least since Late Cretaceous time.

This paper reviews the structure of the major ocean basins in the western Indian Ocean, including some new results mainly from the southern and northern Crozet basins, and compares the proposed geophysical interpretations to the geologic data collected by the *Glomar Challenger* in 1972.

The data

The results presented for the northern part of the western Indian Ocean, north of the Mascarene Islands, are mainly from English and American workers. They have been derived to a great extent from the observations made during the International Indian Ocean Expedition (1959 to 1966) and from the data collected after this period, mainly by Scripps Institution of Oceanography and Woods Hole Oceanographic Institution. The data covering the area south of lat. 20° S, gathered mainly after 1966, have been obtained by the following institutions: the Institut de

* Present address: Laboratoire de Géophysique Marine, Institut de Physique du Globe de Strasbourg, 5, rue René Descartes, 67084 Strasbourg Cedex, France.

Physique du Globe de Paris, first with the *Galliéni* cruises and, after 1972, with the *Marion Dufresne* and the *Suroit* cruises; the Universities of Witwatersrand and Cape Town with the *R.S.A.* and the *T. B. Davie* cruises; the U.S. National Science Foundation with the *Eltanin* cruises; the Lamont-Doherty Geological Observatory with the *R. Conrad* and the *Vema* cruises; the Scripps Institution of Oceanography with the *Melville* cruises; and the Woods Hole Oceanographic Institution with the *Chain* and *Atlantis II* cruises. The oceanographic cruises carried out between 1967 and 1977 in the western Indian Ocean by the Laboratoire de Géophysique Marine of the Institut de Physique du Globe de Paris provided about 250,000 km of geophysical profiles (Figure 9.1).

The western Indian Ocean

The western part of the Indian Ocean is bounded on the north and northwest by India, Arabia, and Africa. The Ninety East Ridge, which approximately follows the 90° E meridian, marks its eastern boundary. On the south and

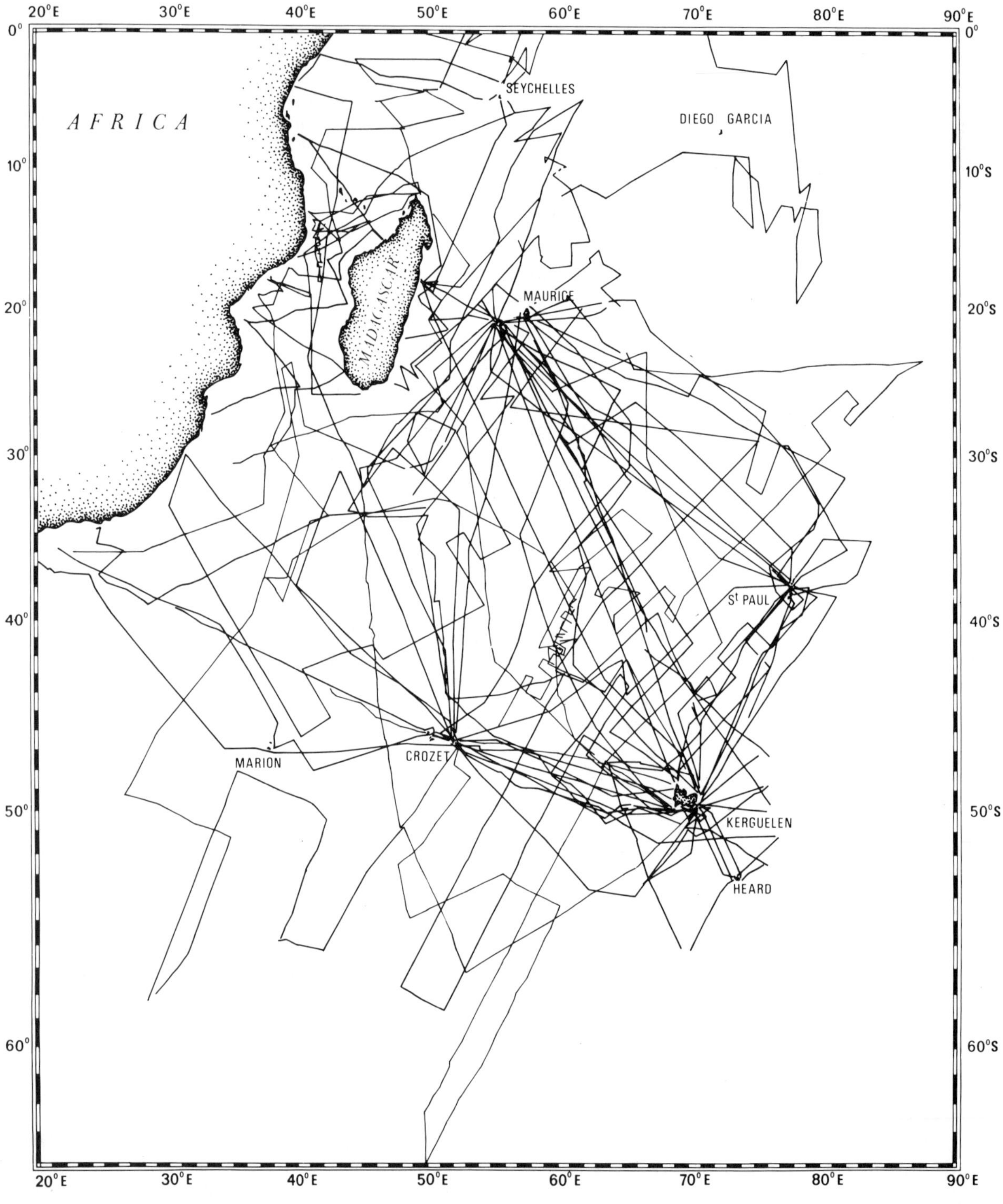

Figure 9.1. Ship tracks of *Galliéni* (1967–72), *Marion Dufresne* (1973–77), and *Suroit* (1977) oceanographic cruises in the western Indian Ocean.

southeast, beyond the active ridges, its boundary is defined by the Kerguelen-Heard Plateau and the Antarctic continent. The mid-ocean ridge system of the western Indian Ocean, which includes the Sheba Ridge, the Carlsberg Ridge, and the Central Indian Ridge, extends from the Gulf of Aden to near the Mascarene Islands. Here it subdivides into two distinct branches, the Southwest Indian Ridge and the Southeast Indian Ridge, which separate Antarctica from Africa and from Australia. In the western Indian Ocean there are numerous submarine plateaus or aseismic ridges which are mainly oriented in a north-south direction, including the Ninety East Ridge, the Chagos-Laccadive Ridge, the Seychelles and Mascarene Plateaus, the Kerguelen-Heard and Crozet Plateaus, the Madagascar and Mozambique Ridges, the Agulhas Plateau, and the Ob, Lena, and Marion Dufresne Seamounts. These features, along with the active ridges and the surrounding continents, demarcate a number of deep ocean basins. The Arabian Sea is located in the north. The Somali Basin lies to the southwest of the Carlsberg Ridge, and the Central Indian Basin extends south of India between the Chagos-Laccadive and Ninety East Ridges. To the east of Madagascar and Madascar Ridge are the Mascarene and Madagascar Basins and, further southwest, the Mozambique and Agulhas Basins. Between the Southwest and Southeast Indian Ridges lies the Crozet Basin, which extends south of the Ob, Lena, and Marion Dufresne Seamounts as the Enderby or African-Antarctic Basin. These basins are very low relief areas characterized by relatively thin, homogeneous sedimentary deposits, acoustically transparent with few internal reflectors. Sediment thicknesses generally reach 200 to 1000 m (Francis and Shor, 1966; Francis and Raitt, 1967; Ewing and others, 1969; Goslin and Schlich, Chapter 8, this volume). Exceptions are in the northern part of the Central Indian Basin (Neprochnov, 1961; Naini and Leyden, 1973), the western Somali Basin (Francis and others, 1966; Bunce and others, 1967; Schlich and others, 1972), the Arabian Sea (Neprochnov, 1961; Bungenstock and others, 1966; Closs and others, 1969), and the Mozambique Channel (Lort and others, 1979), where the sediment cover reaches several kilometers in thickness.

The Arabian Sea and Somali Basin

North of the Mascarene Islands, the Central Indian Ridge, bounded on the east by the Chagos-Laccadive Ridge and on the west by the Mascarene Plateau, extends northward almost to the equator, where it meets the southern end of the Carlsberg Ridge, which in turn extends northward into the Gulf of Aden as the Sheba Ridge. These structures, as well as the adjacent basins (Arabian Sea and Somali Basin), have been studied in some detail by English and American researchers. Langseth and Taylor (1967) and Fisher and others (1967, 1968, 1971) determined the major structural trends, the principal fracture zones, the ridge axis, and a spreading (half) rate of 1.8 to 2.4 cm/yr for the Central Indian Ridge; they also defined the relative motion of the Indian plate with respect to the African plate. The oldest magnetic anomaly which has been identified is anomaly 5 (~ 10 m.y. old). This leads to the conclusion that the Central Indian Ridge was formed in the Miocene, subsequent to the breakup of Gondwanaland, at the site of an old transform fault in connection with a change in spreading direction. Rifting along the ridge has separated the Mascarene Plateau and the Chagos-Laccadive Ridge symmetrically (Fisher and others, 1971; McKenzie and Sclater, 1971). The Carlsberg Ridge has been studied chiefly by Matthews (1963, 1966), Matthews and others (1965), Vine (1966), Cann and Vine (1966), Le Pichon and Heirtzler (1968), and Fisher and others (1968). These workers have determined the direction of the fracture zones and of the magnetic lineations (~ 130°) and calculated a spreading (half) rate of 1.2 to 1.3 cm/yr, based essentially on the axial anomaly. Beyond anomaly 5 (~ 10 m.y. old), the magnetic anomaly sequence seems to be interrupted. As for the Central Indian Ridge, most authors accept a slowdown or even a complete halt of spreading in Miocene time.

In the Arabian Sea north of the Carlsberg Ridge and in the Somali Basin between the Carlsberg Ridge and the Seychelles Bank, Matthews (1966), Bowin and Vogt (1966), and Taylor (1968) observed magnetic anomalies with very large amplitudes. In the Arabian Sea, these anomalies trend in an almost east-west direction, and in the Somali Basin the trend is about 120°; these directions differ significantly from that of the active Carlsberg Ridge axis which trends 130°-140°. McKenzie and Sclater (1971) identified these anomalies and recognized a complete sequence of Late Cretaceous and Paleocene magnetic anomalies (29 to 23) in the two basins. On the basis of these observations, and using the Heirtzler and others (1968) time scale of magnetic field reversals, they calculated a spreading (half) rate of 6.5 cm/yr. Considering the distance which separates these anomalies from the Carlsberg Ridge anomalies, they concluded that a slowdown in spreading, and perhaps even a complete halt, must have occurred between Eocene time (~58 m.y. ago) and Miocene time (~ 10-30 m.y. ago).

The Madagascar and Mascarene Basins

The Madagascar Basin is bounded on the south and southeast by the extremely rough topography of the Southwest Indian Ridge, and to the west by the gentle upward slope of the Madagascar Ridge. To the northeast, the southern part of the Central Indian Ridge and its triple junction with the Southwest and Southeast Indian Ridges, marks another boundary for this basin. North of the Madagascar Basin, the Mascarene Basin lies between Madagascar and the Mascarene Plateau and extends toward the Farquhar Group and the Amirante and Seychelles Banks. The structure of these two basins has been defined mainly by data obtained on the *Galliéni* and the 1973 *Marion Dufresne* cruises (Schlich, 1974, 1975). Several major fracture zones have been mapped south of lat. 15° S. Their directions are about 048° to 050° in the southern parts of the Madagascar and Mascarene Basins, but they change progressively to 038° in the central parts of these basins. In the northern part of the Madagascar Basin, east of Mauritius and La Reunion Islands, the directions turn back to about 047°.

A complete sequence of Late Cretaceous, Paleocene, and early Eocene magnetic anomalies (30 to 19) has been recognized in the Madagascar Basin (Schlich, 1974, 1975). The trend of the magnetic lineations is 130° for anomalies 30 to 28 and 127° for anomalies 27 to 23. For the more recent anomalies (22 to 19), the directions are poorly defined and are assumed to be about 135°. In the southern part of the Mascarene Basin east of Madagascar, Late Cretaceous magnetic lineations (34 to 30), oriented 138°, have been identified (Schlich, 1974, 1975).

In conjunction with these different changes in direction, several distinct phases of spreading have been determined in the Madagascar and Mascarene Basins. According to the Heirtzler and others (1968) magnetic time scale, the first phase corresponds to the Late Cretaceous magnetic anomalies 34 to 31, with a spreading (half) rate of 4.0 cm/yr observed in the Mascarene Basin. In the Madagascar Basin, however, this phase of spreading is quite poorly defined; in the southern part of the basin, close to the Madagascar Ridge, anomaly 30 is the oldest anomaly that can be recognized. The second phase corresponds to anomalies 30, 29, and 28 in the Madagascar Basin, with an unusually high spreading (half) rate of 12.2 cm/yr. The following phase, which is observed only in the Madagascar Basin, corresponds to the Paleocene anomalies 27 to 23, with a spreading (half) rate of 7.1 cm/yr. Beyond anomaly 23, the spreading (half) rate in the Madagascar Basin, determined from anomalies 22, 21 and 19, may be estimated at 3 or 4 cm/yr, if the identifications are correct. Farther north, Fisher and others (1971) have observed recent magnetic anomalies (5 to 1) associated with the Central Indian Ridge and have calculated a spreading (half) rate of about 2.0 cm/yr. If one assumes the continuity of these two systems, which has not been established since no Eocene or Oligocene anomalies have been recognized, a mean spreading (half) rate of 1.6 cm/yr is obtained between anomalies 22 and 5. This result is comparable to the values found for the Central Indian and Crozet Basins (see next section).

The Central Indian and Crozet Basins

The Central Indian Basin and the Crozet Basin, separated by the presently active Southeast Indian Ridge, are the largest ocean basins in the western Indian Ocean. They have been created by symmetric sea floor spreading since Late Cretaceous time. The Central Indian Basin, which is bounded to the east by the Ninety East Ridge, to the west by the Chagos-Laccadive Ridge, and to the north by India, was mainly studied by McKenzie and Sclater (1971) and later by Sclater and Fisher (1974). The Crozet Basin is bounded on the northeast and northwest by the Southeast and Southwest Indian Ridges, respectively, and on the south, beyond the Kerguelen and Crozet Islands, by a chain of marked topographic features—the Ob, Lena, and Marion Dufresne Seamounts. The Crozet Basin has been surveyed in great detail by the *Galliéni* and *Marion Dufresne* cruises.

McKenzie and Sclater (1971) have identified a complete sequence of Late Cretaceous, Paleocene, and early Eocene magnetic anomalies (33 to 22) for the Central Indian Basin. They have located the principal fracture zones and have determined a direction of about 096° for the magnetic lineations. For these Late Cretaceous and Paleozene anomalies, they calculated (using the Heirtzler and others time scale) a spreading (half) rate of 5.6 cm/yr for the older anomalies (33 to 30) and a rate of about 8.0 cm/yr for anomalies 30 to 23. Thus, they have explained the rapid movement of India northward and its collision with Asia 40–50 m.y. ago, which gave rise to the Himalayas. McKenzie and Sclater (1971) did not observe anomalies younger than 22 and, as for the Carlsberg and Central Indian Ridges, inferred a considerable slowdown of spreading during Eocene and Oligocene time. These conclusions appeared rather surprising in view of the observations made by Schlich and Patriat (1971) in the northern part of the Crozet Basin. Later, Sclater and Fisher (1974) and Sclater and others (1976) recognized younger anomalies in the southern part of the Central Indian Basin and estimated a spreading (half) rate of 4.0 cm/yr for anomalies 22 to 17, a rate of 2.5 cm/yr for anomalies 16 to 5, and a rate of 3.0 cm/yr for anomalies 5 to 1. The Eocene magnetic anomalies (22 to 18) trend approximately 100°; the younger ones, anomalies 15 to 7 and anomalies 5 to 1, strike respectively 135° and 138°. The change in anomaly trend between the times of anomalies 18 and 15 (45 to 40 m.y. ago according to the Heirtzler and others time scale) can be explained by a re-orientation of the direction of the Southeast Indian Ridge from east-west to northwest-southeast when India collided with Eurasia (Sclater and others, 1976). For the Late Cretaceous and Paleocene magnetic anomalies, Sclater and Fisher (1974) finally proposed a spreading (half) rate of 5.7 cm/yr for anomalies 33 to 30, a rate of 12.0 cm/yr for anomalies 30 to 27, and a rate of 8.1 cm/yr for anomalies 27 to 23.

The Crozet Basin, along with the Southeast Indian Ridge, has been studied in much detail since 1967. Preliminary results have been published by Schlich and Patriat (1967, 1971), and a first synthesis has been proposed by Schlich (1975). New data, collected between 1975 and 1977, have been added to these results and are presented in this paper. The fracture zones have been located precisely throughout the whole basin. In the southern part of the Crozet Basin, north of the Ob, Lena, and Marion Dufresne Seamounts, the fracture zones are oriented 036°. They change progressively to a 028° orientation in the central part of the basin between Crozet and Kerguelen Islands, and abruptly to a 045° orientation in the northern part of the basin. The complete sequence, from anomaly 34 (Late Cretaceous) to anomaly 1 (present), has been recognized in the Crozet Basin. The trend of the magnetic lineations is 126° for anomalies 34 to 31, 118° for anomalies 30 to 23, and 135° for anomalies 19 to 1. Anomaly 34 (Santonian) has been observed just north of the Ob, Lena, and Marion Dufresne Seamounts, and anomaly 31 (Campanian or Maestrichtian) has been mapped north of the Crozet Islands. Anomalies 10 and 11 (early Oligocene) were recognized at the northern edge of the Kerguelen-Heard Plateau, and older anomalies 17 and 18 (middle to late Eocene) occur on either side of this structure (Schlich, 1975). The axial magnetic anomalies associated with the Southeast Indian Ridge were recognized and located by the earliest expeditions of the *Galliéni*. It was possible to demonstrate that the ridge axis is offset several times by fracture zones oriented 045° (Schlich and Patriat, 1971). From the Heirtzler and others (1968) time scale, supplemented beyond anomaly 30 by McKenzie and Sclater (1971), several distinct spreading phases can be identified. They are generally associated with changes in the lineation orientations. To determine these different phases, the observed magnetic profiles, projected along straight lines perpendicular to the magnetic anomaly strikes, were compared to synthetic magnetic anomaly profiles computed for different spreading rates. The first phase of spreading cor-

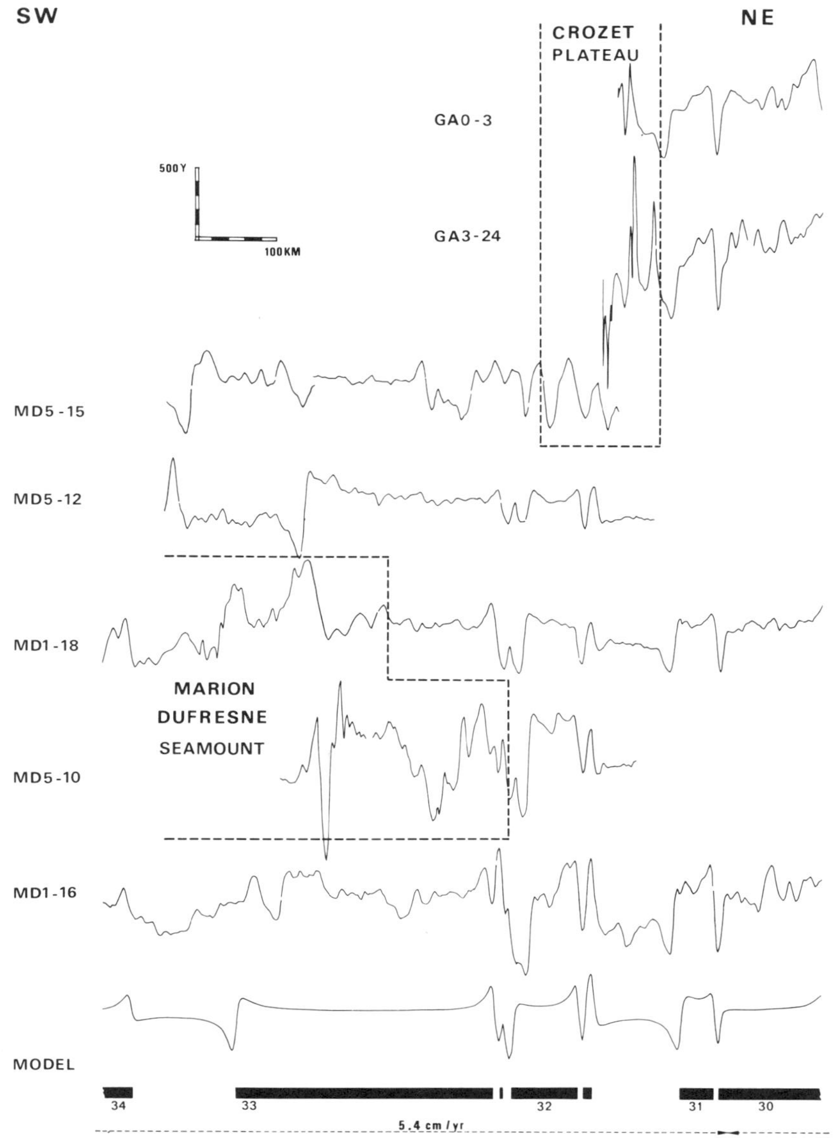

Figure 9.2. Magnetic anomaly profiles in the southern Crozet Basin, projected at 036°, compared with a synthetic profile computed for a ridge striking 120° which was created at lat. 50° S, with a spreading (half) rate of 5.4 cm/yr, and observed at lat. 51° S and 54° S.

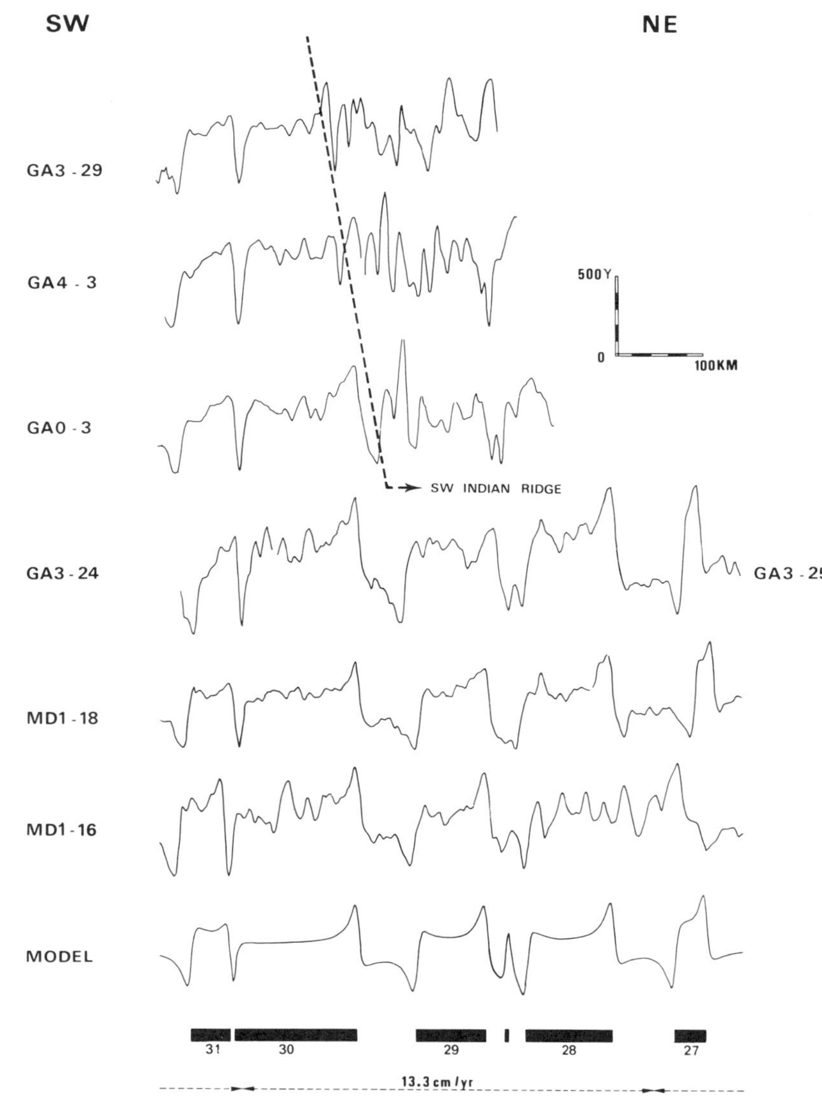

Figure 9.3. Magnetic anomaly profiles in the southern Crozet Basin, projected at 028°, compared with a synthetic profile computed for a ridge striking 120° which was created at lat. 50° S, with a spreading (half) rate of 13.3 cm/yr and observed at lat. 47° S and 56° E.

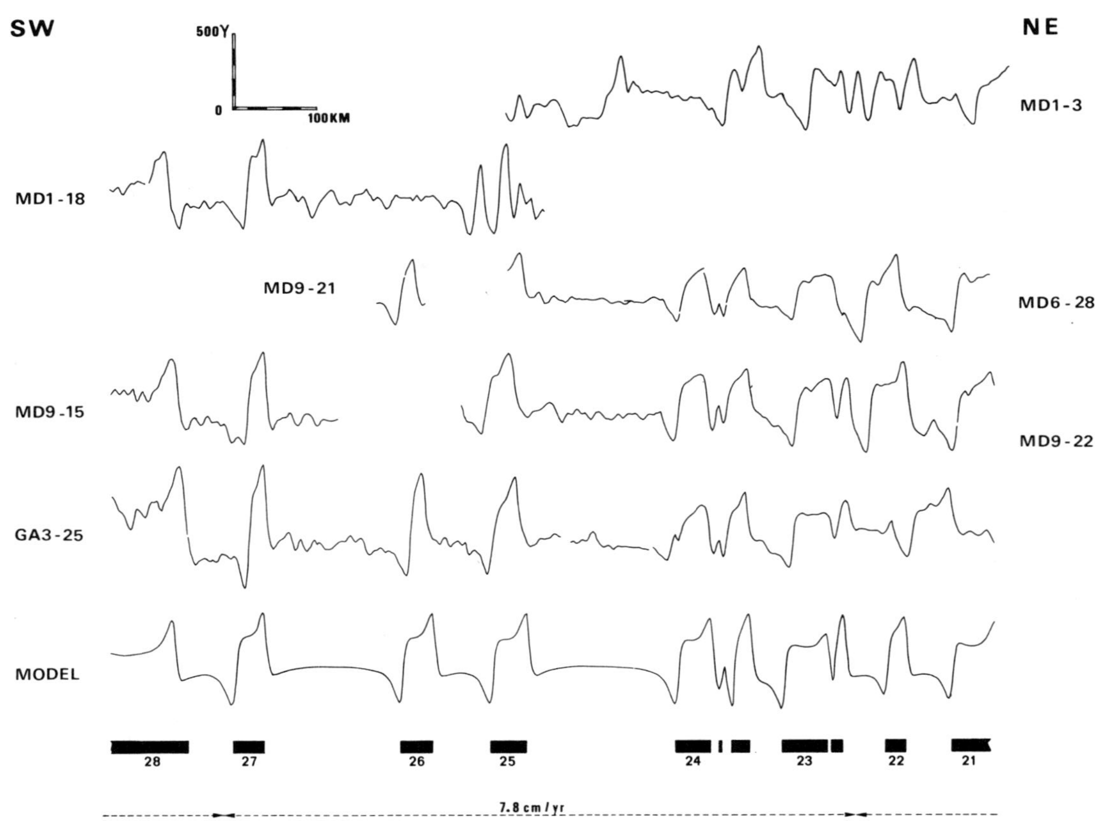

Figure 9.4. Magnetic anomaly profiles in the central Crozet Basin, projected at 028°, compared with a synthetic profile computed for a ridge striking 120° which was created at lat. 50° S, with a spreading (half) rate of 7.8 cm/yr, and observed at lat. 40° S and 60° E.

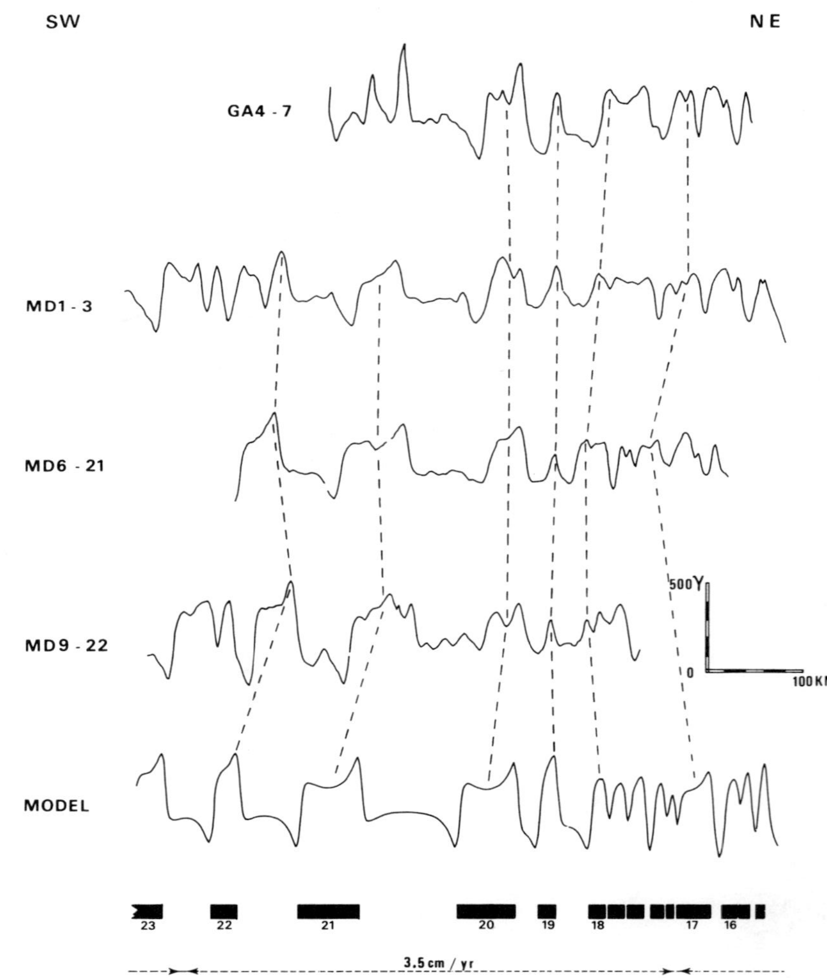

Figure 9.5. Magnetic anomaly profiles in the central Crozet Basin, projected at 045°, compared with a synthetic profile computed for a ridge striking 135° E which was created at lat. 40° S, with a spreading (half) rate of 3.5 cm/yr, observed at lat. 40° S and 65° E.

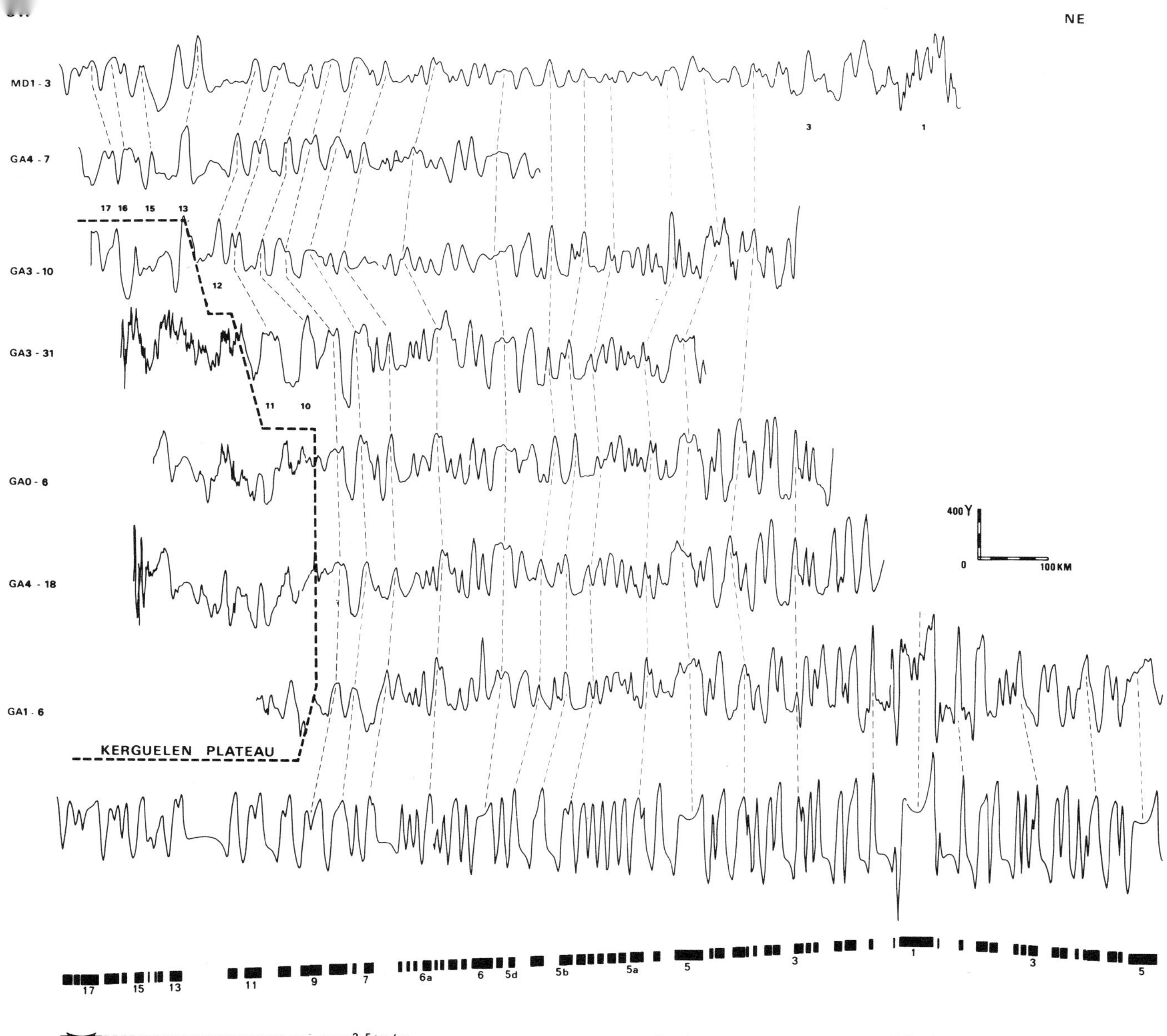

Figure 9.6. Magnetic anomaly profiles in the northern Crozet Basin, projected at 045°, compared with a synthetic profile generated from a ridge striking 135° at lat. 40° S, and observed between lat. 40° S and 35° S. The spreading (half) rates are 2.5 cm/yr for anomalies 17 to 5 and 3.4 cm/yr for anomalies 5 to 1.

responds to anomalies 34 to 31, with a spreading (half) rate of 5.4 cm/yr (Figure 9.2). The magnetic profiles MD1-18 and MD5-10 cross the Marion Dufresne Seamount, while profiles GA0-3 and GA3-24 coming from the north and MD5-15 coming from the south, terminate at the Crozet Plateau. This explains the lack of correlation with the model for part of these profiles. The second phase of spreading corresponds to anomalies 30 to 28, with a spreading (half) rate of 13.3 cm/yr (Figure 9.3). The magnetic profiles GA0-3, GA3-29, and GA4-3 continue to the north into the Southwest Indian Ridge, and the correlations with the model are interrupted. The third phase of spreading corresponds to the Paleocene anomalies 27 to 23/22, with a spreading (half) rate of 7.8 cm/yr (Figure 9.4). The following phase includes anomalies 22 to 18, with a spreading (half) rate of approximately 3.5 cm/yr (Figure 9.5). The correlations are somewhat difficult to establish. The time interval between anomalies 22/21 and 19 (~54 to 47 m.y. ago, according to the Heirtzler and others time scale) corresponds to the abrupt change of spreading direction which affected the Southeast Indian Ridge in early middle Eocene time. In the Central Indian Basin, Sclater and others (1976) located this abrupt change of direction between anomalies 18 and 15 (~46 to 40 m.y. ago). In fact, the lineations shown for anomalies 20, 19, and 18 in figures 1 and 3 published by Sclater and others (1976, p. 372–74) are very poorly defined, and it appears that complementary data should be collected in this area. The last two spreading phases, clearly oriented southwest-northeast, correspond to anomalies 17 to 5 and to anomalies 5 to 1, and are associated with the Southeast Indian Ridge axis. The spreading (half) rates are respectively 2.5 cm/yr and 3.4 cm/yr, with some minor changes in these two intervals (Figure 9.6).

Summary of major results

The different spreading phases observed in the western Indian Ocean since Late Cretaceous time are synchronous in all the large ocean basins. The only difference cor-

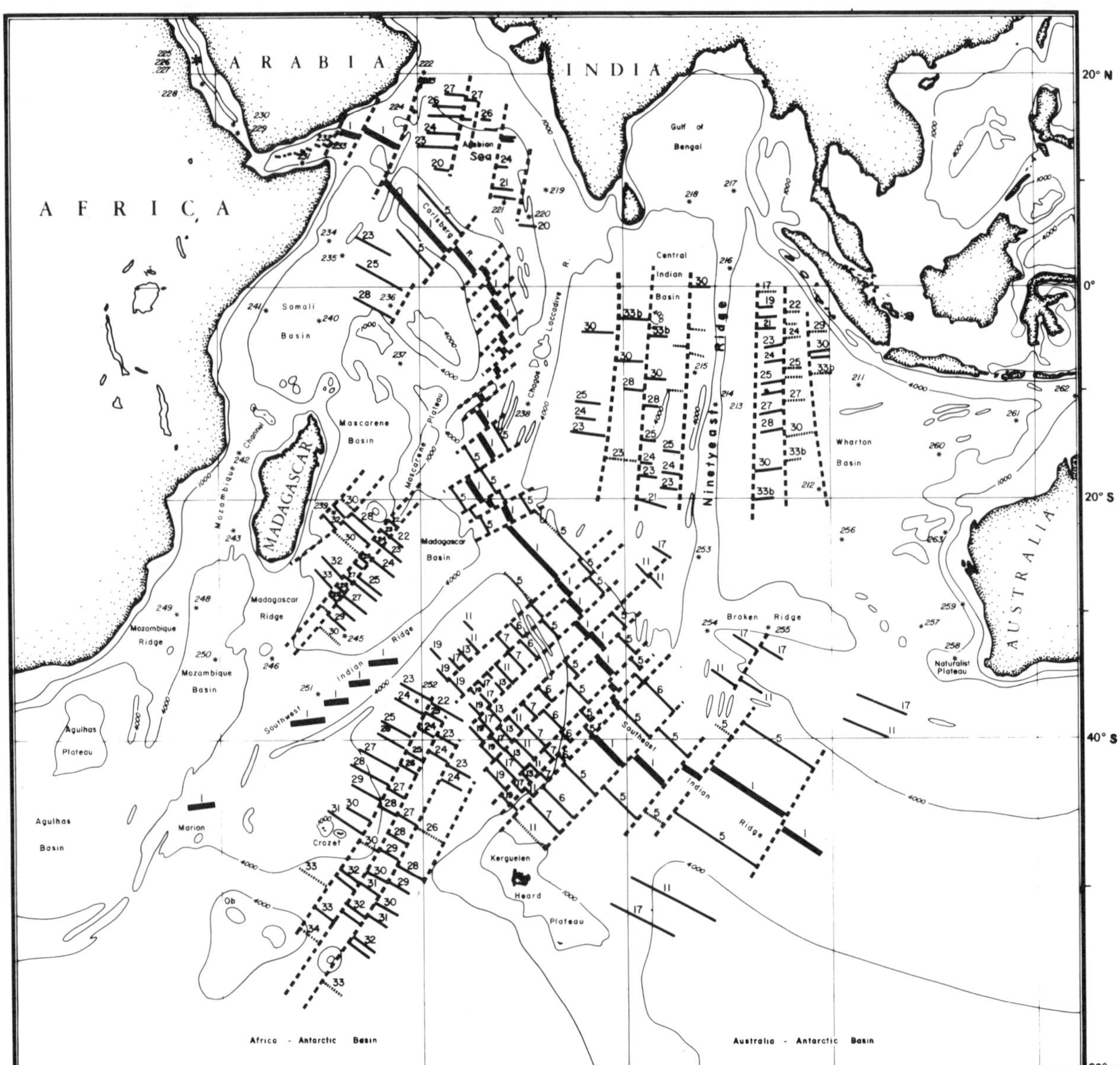

Figure 9.7. Fracture zones and magnetic lineations in the Indian Ocean. The numbering of the magnetic lineations follows the scheme proposed by Pitman and Heirtzler (1966) and Heirtzler and others (1968). DSDP sites are shown by numbers next to stars.

responds to the time of the abrupt change in lineation orientation which is observed in the Central Indian Basin between anomalies 18 and 15 and in the Crozet Basin between anomalies 22/21 and 19. As already discussed, the change in direction in the Central Indian Basin needs further investigation.

The main results of this synthesis are summarized in Figure 9.7. The numbering of the magnetic anomalies follows the scheme proposed by Pitman and Heirtzler (1966) and Heirtzler and others (1968), and that of McKenzie and Sclater (1971) for the Late Cretaceous anomalies older than anomaly 30. The corresponding numerical data for the western Indian Ocean are shown in Tables 9.1 and 9.2.

Glomar Challenger deep sea drilling in the Indian Ocean

The deep sea drilling carried out by the *Glomar Challenger* in the Indian Ocean in 1972 has confirmed, to a first approximation, the proposed structural scheme, whether we are referring to the interpretations of McKenzie and Sclater (1971) and Sclater and Fisher (1974) for the northern part of the Indian Ocean, or to our own conclusions for the more southerly portion of that ocean. A more refined analysis enables us to note, however, that in the case of certain drill holes, there are some systematic differences between the biostratigraphic ages determined from sediments cored directly at the contact of the oceanic crust, and the magnetic ages deduced from the interpretation of the magnetic anomalies and from the time scale of geomagnetic field reversals (Heirtzler and others, 1968; McKenzie and Sclater, 1971). In the Indian Ocean, and including the Wharton Basin in the east, there are essentially three, and perhaps five, sites which can be used to evaluate these age differences: site 213 in the Wharton Basin, just east of the Ninety East Ridge; site 221 in the Arabian Sea; site 236 in the Somali Basin, north-northeast of the Seychelles Bank; site 239 in the southern Mascarene Basin; and site 245 in the southern Madagascar Basin.

Site 213, situated in the Wharton Basin, lies close to, but just north of, anomaly 26 (Sclater and Fisher, 1974). The magnetic age, according to the Heirtzler and others (1968) time scale, is 64 m.y. The sediments sampled directly at the basalt contact were dated by nannofossils and foraminifera as late Paleocene and correspond to an age between 56 and 58 m.y. (Berggren, 1972; Hardenbol and Berggren, in press). This determination is particularly well established, since the biostratigraphic age was confirmed by calcareous inclusions found in the cored basalt (Sclater and others, 1974).

Table 9.1. Magnetic anomalies, magnetic lineations, and spreading (half) rates in the northwestern Indian Ocean

Location	Anomaly	Lineation	Spreading (half) rate (cm/yr)	References
Carlsberg Ridge	1-5	~130°	~1.2	Matthews (1963, 1966) Matthews and others (1965) Cann and Vine (1966) Le Pichon and Heirtzler (1968) Fisher and others (1968)
Arabian Sea	23-29	90°	~ 6.5	Matthews (1966)
Somali Basin	23-29	120°	~6.5	Bowin and Vogt (1966) Taylor (1968) McKenzie and Sclater (1971)
Central Indian Ridge	1-5	~135°	~2.0	Fisher and others (1967, 1968, 1971) McKenzie and Sclater (1971)
Southeast Indian Ridge (northeast flank)	1-5	138°	3.0	Le Pichon and Heirtzler (1968) Sclater and others (1976)
Central Indian Basin	5-16	135°	2.5	McKenzie and Sclater (1971) Schlater and Fisher (1974) Sclater and others (1976)
Central Indian Basin	17-22	~100°	~4.0	
Central Indian Basin	23-27	~90°	8.1	
Central Indian Basin	27-30	~90°	12.0	
Central Indian Basin	30-33	~90°	5.7	

Table 9.2. Magnetic anomalies, magnetic lineations, and spreading (half) rates in the southwestern Indian Ocean

Location	Anomaly	Lineation	Spreading (half) rate (cm/yr)	References
Madagascar Basin	19-22	~135°	~3.5	Schlich (1974, 1975)
Madagascar Basin	23-27	127°	7.1	
Madagascar Basin	28-30	130°	12.2	
Mascarene Basin	31-34	138°	4.0	Schlich (1974, 1975)
Southeast Indian Ridge (southwest flank)	1-5	135°	3.4	Le Pichon and Heirtzler (1968) Schlich and Patriat (1971) Schlich (1975)
Crozet Basin	5-17	135°	2.5	Schlich and Patriat (1971) Schlich (1974, 1975)
Crozet Basin	18-22	~135°	~3.5	
Crozet Basin	23-27	118°	7.8	
Crozet Basin	28-30	118°	13.3	
Crozet Basin	31-34	126°	5.4	

Site 221 in the Arabian Sea is located, according to Whitmarsh (1974), close to, but just south of, anomaly 20. The proposed identification is questionable (Schlich, 1975), and it is suggested that this site should not be considered (Schlich and others, 1979).

Site 236 has been dated as late Paleocene (middle Thanetian), or 56–58 m.y. old (Berggren, 1972; Hardenbol and Berggren in press), from the sediments cored at the contact with the basalt (Vincent, 1974). The magnetic age is somewhat uncertain because site 236 lies between anomalies 25 and 28, which have been identified by McKenzie and Sclater (1971). An interpolation, using the spreading rate proposed by these authors, places site 236 close to anomaly 27, which would give it an age of 65 to 67 m.y. according to the Heirtzler and others (1968) time scale. New data collected in 1977 with the *Suroit* did not permit improvement of this magnetic age determination.

Drill hole site 239 in the Mascarene Basin is situated south of and immediately next to anomaly 31 (Schlich, 1974, 1975). The interpretation leaves little doubt that, based on the magnetic reversal time scale (Heirtzler and others, 1968; Mckenzie and Sclater, 1971), the magnetic age is 72–73 m.y. The sediments cored just above the basalt contact were dated by nannofossils and foraminifera as late Campanian, first on board the *Glomar Challenger.* Later on shore, the paleontologists were able to make more precise determinations and proposed a late Maestrichtian age for the oldest recovered sediments. This corresponds to an age between 66 and 68 m.y., according to the Obradovich and Cobban (1975) biostratigraphic scale.

Site 245 in the Madagascar Basin is located next to and south of anomaly 29. Based on the Heirtzler and others (1968) time scale, the magnetic age of the oceanic crust here is 69–70 m.y. (Schlich, 1974, 1975). The sediment cored at the bottom of the drill hole was dated as early Paleocene (middle Danian) and corresponds to an age of 61–63 m.y., according to the Berggren (1972) or Hardenbol and Berggren (in press) biostratigraphic scale. The last few meters of sediment at the basalt contact were probably not recovered, and if the average sedimentation rate for the Paleocene series is considered, it becomes necessary to extend the probable age range by about 1 m.y. Under these conditions, the minimum possible age for the oceanic crust would fall between 61 and 64 m.y.

Considering these results, it is clear that the time scale proposed in 1968 by Heirtzler and others should be adjusted and made younger by a factor of about 8–10% for Late Cretaceous and Paleocene time.

A new chronologic time scale

The deep drill holes made in all of the oceans which have reached the oceanic crust enable us to calibrate the time scale of magnetic field reversals and, consequently, to date more precisely the oceanic basins by the magnetic lineations. There are relatively few drill holes in the immediate vicinity of clearly identified magnetic anomalies for which the age of the corresponding oceanic crust has been defined confidently. Thus, while the biostratigraphic ages are commonly well determined at the DSDP drill sites, the same is not true for the magnetic ages. The published magnetic profiles only rarely permit a critical examination of the proposed interpretations. For the younger magnetic anomalies, especially those for Miocene and Oligocene time, the

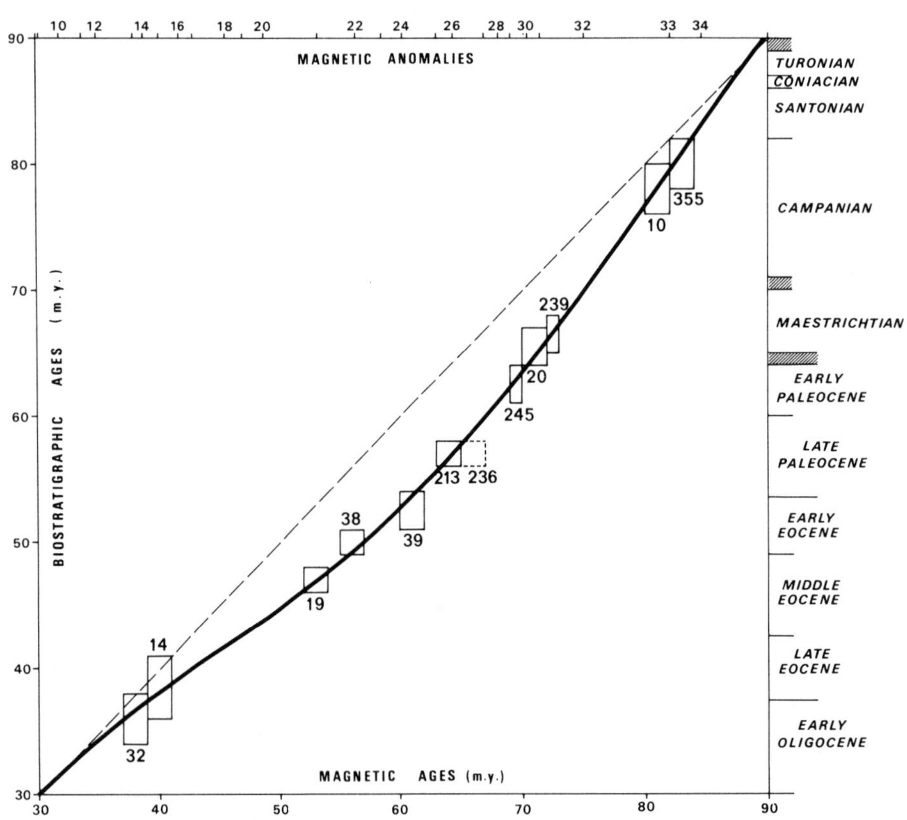

Figure 9.8. Comparison of magnetic ages and biostratigraphic ages for Eocene, Paleocene, and Late Cretaceous time. The square in dashed lines corresponds to a questionable data point. The numbers next to the squares refer to DSDP sites. The vertical and horizontal extent of each square corresponds to the accuracy of each data point.

correlation deduced from the Heirtzler and others (1968) time scale between the biostratigraphic ages and magnetic ages is good. For the older magnetic anomalies of Eocene, Paleocene, and Late Cretaceous age, there are, as already shown, some appreciable differences which can now be specified by data from 10 drill holes.

In the North Atlantic, the drill hole at site 10 lies on the older side of anomaly 33, which corresponds to a magnetic age of 80–82 m.y. (Cande and Kristoffersen, 1977). The sediment cored at the basalt contact was dated as middle to early Campanian (76–80 m.y. old). In the South Atlantic, a number of drill holes provide some precise data for the dating of Cretaceous and Tertiary anomalies. Site 19, about 1000 km west of the ridge axis, is situated in the immediate vicinity of anomaly 21 (52–54 m.y.). The sediment cored at the basalt contact there is dated as middle Eocene (46–48 m.y.). Site 20, about 1300 km from the ridge axis, is centered on anomaly 30 (70–72 m.y.), and the underlying basalt was dated from calcareous inclusions as late Maestrichtian in age (64–67 m.y.). Site 355, drilled by the *Glomar Challenger* in the Brazil Basin, is located next to anomaly 33 in the reversed polarity interval between anomalies 33 and 34 (Supko and Perch-Nielsen, 1977), and it corresponds to a magnetic age of 82–84 m.y. The sediment cored directly at the contact with the basalt has been dated as early Campanian (78–82 m.y.). In the northeastern Pacific, two drill holes, the first at site 38 centered on anomaly 22 (55–57 m.y.), and the second at site 39 in the immediate vicinity of anomaly 24 (60–62 m.y.), yield early Eocene biostratigraphic ages (49–51 m.y. and 51–54 m.y., respectively).

Figure 9.8 gives the results of this comparison between magnetic ages derived from the Heirtzler and others (1968) time scale, supplemented by McKenzie and Sclater (1971) beyond anomaly 30 and by Larson and Pitman (1972) for anomalies 33 and 34, and the micropaleontologic ages obtained from nannofossils and foraminifera found in the sediment cored at the basalt contact, as deduced from the biostratigraphic time scales proposed by Berggren (1972) and Hardenbol and Berggren (in press) for the Cenozoic, and by Obradovich and Cobban (1975) for the Cretaceous. On the basis of these results, Schlich and others (1979) have proposed a new chronologic time scale of geomagnetic field reversals for Eocene, Paleocene, and Late Cretaceous time (Figure 9.9).

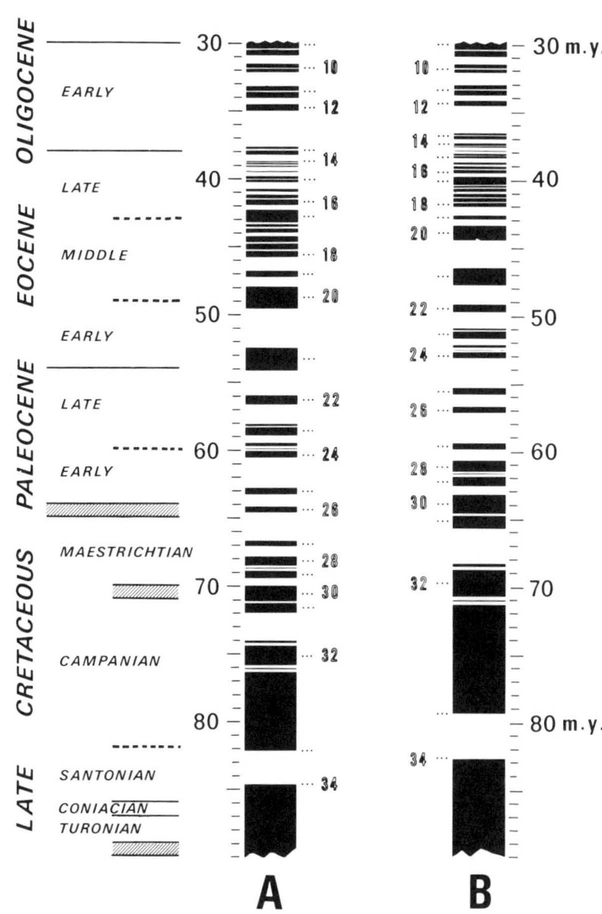

Figure 9.9. Chronology of geomagnetic field reversals for Eocene, Paleocene, and Late Cretaceous time. A. According to Heirtzler and others (1968), McKenzie and Sclater (1971), and Larson and Pitman (1972); B. According to interpretation of deep sea drilling sites of the *Glomar Challenger.* Shown from left to right are the geologic periods and epochs, the ages in m.y., the geomagnetic field reversals with the periods of normal magnetization in black, and the magnetic anomaly numbers.

Concluding remarks

The new time scale enables us to define more precisely the geologic events which have affected the evolution of the world oceans. In the western Indian Ocean it is interesting to note that the very high spreading (half) rate of 12–13 cm/yr calculated for anomalies 30 to 28 is considerably reduced if the new chronologic time scale is used. In fact, this spreading phase becomes continuous with the following phase associated with anomalies 27 to 23. This observation is favorable to the new proposed time scale, since no marked change in direction affects the anomaly pattern between spreading phases two and three, that is, between anomalies 30–28 and anomalies 27–23. In the Crozet Basin, the first spreading phase still corresonds to anomalies 34 to 31, with a slightly reduced spreading (half) rate of about 3.9 cm/yr. The second phase of spreading now corresponds to anomalies 30 to 23, and the spreading (half) rate for this whole sequence becomes 8.6 cm/yr. The following phase includes anomalies 22 to 18 as before, but now corresponds to a spreading (half) rate of 4.0 cm/yr.

In conclusion, the geologic history of the western Indian Ocean, at least since Late Cretaceous time, is becoming clear, and the first paleogeographic reconstructions can be attempted. There are still some important limitations and problems in defining the ancient continent of Gondwanaland. In particular, it is necessary to determine the relative position of India with respect to Africa and to study further the separation of Africa and Antarctica, that is, to determine the Mesozoic evolution of the Somali, Mozambique, and Enderby Basins.

Acknowledgments. This research was supported by Université Pierre et Marie Curie Paris, Centre National de la Recherche Scientifique, Institut National d'Astronomie et de Géophysique, Terres Australes et Antarctiques Françaises, and Centre National pour l'Exploitation des Océans. Ship time involved in the French Indian Ocean research program was provided by Terres Australes et Antarctiques Françaises (*Galliéni* and *Marion Dufresne*) and Centre National pour l'Exploitation des Océans (*Suroit*).

References cited

Berggren, W. H. 1972. A Cenozoic time scale. Some implications for regional geology and paleobiogeography. *Lethaia, 5,* 195–215.

Bowin, C. O., and Vogt, P. R. 1966. Magnetic lineations between Carlsberg Ridge and Seychelles Bank, Indian Ocean. *J. Geophys. Res., 71,* 2625–30.

Bunce, E. T., Langseth, M. G., Chase, R. L., and Ewing, M. 1967. Structure of the western Somali Basin. *J. Geophys. Res., 72,* 2547–55.

Bungenstock, H., Closs, H., and Hinz, K. 1966. Seismische Untersuchungen im nördlichen Teil des Arabischen Meeres (Gulf von Oman). *Erdöl Kohle Erdgas Petrochem., 19,* No. 4, 237–43.

Cande, S. C., and Kristoffersen, Y. 1977. Late Cretaceous magnetic anomalies in the North Atlantic. *Earth planet. Sci. Lett., 35,* 215–24.

Cann, J. R., and Vine, F. J. 1966. An area on the crest of the Carlsberg Ridge: petrology and magnetic survey. *Phil. Trans. R. Soc. Lond.,* Ser. A, *259,* 198–217.

Closs, H., Bungenstock, H., and Hinz, K. 1969. Ergebnisse seismischer Untersuchungen im nördlichen arabischen Meer, ein Beitrag zur internationalen indischen Ozean Expedition. *Meteor Forsch. Ergebn.,* Ser. C, No. 2, 1–28.

Ewing, M., Eittreim, S., Truchan, M., and Ewing, J. I. 1969. Sediment distribution in the Indian Ocean. *Deep Sea Res., 16,* 231–48.

Fisher, R. L., Engel, C. G., and Hilde, T. W. C. 1968. Basalts dredge from the Amirante Ridge, western Indian Ocean. *Deep Sea Res., 15,* 521–34.

Fisher, R. L., Johnson, G. L, and Heezen, B. C. 1967. Mascarene Plateau, western Indian Ocean. *Geol. Soc. Am. Bull., 78,* 1247–66.

Fisher, R. L., Sclater, J. G., and McKenzie, D. P. 1971. Evolution of the central Indian Ridge, western Indian Ocean. *Geol. Soc. Am. Bull., 82,* 553–62.

Francis, T. J. G., and Raitt, R. W. 1967. Seismic refraction measurements in the southern Indian Ocean. *J. Geophys. Res., 72,* 3015–42.

Francis, T. J. G., and Shor, G. G. 1966. Seismic refraction measurements in the northwestern Indian Ocean. *J. Geophys. Res., 71,* 427–49.

Francis, T. J. G., Davies, D., and Hill, M. N. 1966. Crustal structure between Kenya and the Seychelles. *Phil. Trans. R. Soc. Lond.,* Ser. A, *259,* 240–61.

Hardenbol, J., and Berggren, W. A. In press. A new Paleogene numerical time scale. *Am. Ass. Petrol. Geol.*

Heirtzler, J. R., Dickson, G., Herron, E. M., Pitman, W. C., III, and Le Pichon, X. 1968. Marine magnetic anomalies, geomagnetic field reversals and motions of the ocean floor and continents. *J. Geophys. Res., 73,* No. 6, 2119–36.

Langseth, M. G., and Taylor, P. T. 1967. Recent heat flow measurements in the Indian Ocean. *J. Geophys. Res., 72,* 6249–60.

Larson, R. L., and Pitman, W. C., III 1972. World-wide correlation of Mesozoic magnetic anomalies, and its applications. *Geol. Soc. Am. Bull., 83,* 3645–62.

Le Pichon, X., and Heirtzler, J. R. 1968. Magnetic anomalies in the Indian Ocean and sea-floor spreading. *J. Geophys. Res., 73,* 2101–17.

Lort, J. M., Limond, W. Q., Segoufin, J., Patriat, P., Delteil, J. R., and Damotte, B. 1979. New seismic data in the Mozambique Channel. *Mar. geophys. Res., 4,* 71–89.

McKenzie, D. P., and Sclater, J. G. 1971. The evolution of the Indian Ocean since the Late Cretaceous. *Geophys. J. R. astr. Soc., 25,* 437–528.

Matthews, D. H. 1963. A major faultscarp under the Arabian sea displacing the Carlsberg Ridge near Socotra. *Nature, Lond., 198,* 950–52.

Matthews, D. H. 1966. The Owen fracture zone and the northern end of the Carlsberg Ridge. *Phil. Trans. R. Soc. Lond.,* Ser. A, *259,* 172–86.

Matthews, D. H., Vine, F. J., and Cann, J. R. 1965. Geology of an area of the Carlsberg Ridge, Indian Ocean. *Geol. Soc. Am. Bull., 76,* 675–82.

Naini, B. R., and Leyden, R. 1973. Ganges Cone: a wide angle seismic reflection and refraction study. *J. Geophys. Res., 78,* 8711–20.

Neprochnov, Y. P. 1961. Moshchnost osadkov kotloviny Aravyskogo Morya [Sediment thickness of the Arabian Sea Basin]. *Dokl. Akad. Nauk SSSR, 139,* No. 1, 177–79.

Obradovich, J. D., and Cobban, W. A. 1975. A time scale for the Late Cretaceous of the western interior of North America. *Spec. Pap. Geol. Ass. Can., 13,* 31–54.

Pitman, W. C., III, and Heirtzler, J. R. 1966. Magnetic anomalies over the Pacific-Antarctic Ridge. *Science, N.Y., 154,* 1164–71.

Schlich, R. 1974. Sea-floor spreading history and deep sea drilling results in the Madagascar and Mascarene Basins, Western Indian Ocean. (*In* Simpson, E. S. W., Schlich, R., and others. *Initial Reports of the Deep Sea Drilling Project,* Vol. 25. Washington [U.S. Government Printing Office], 663–78.)

Schlich, R. 1975. Structure et âge de l'océan Indien occidental. *Mém. Soc. Géol. Fr., 6,* 1–103.

Schlich, R., and Patriat, P. 1967. Profils magnétiques sur la dorsale médio-indienne "Indo-Pacifique." *Annls Géophys., 23,* 629–33.

Schlich, R., and Patriat, P. 1971. Anomalies magnétiques de la branche est de la dorsale médio-indienne entre les îles Amsterdam et Kerguelen. *C. r. Acad. Sci., Paris, 272,* 773–76.

Schlich, R., Muller, C., and Sigal, J. 1979. Le Maestrichtien du site 239 (D.S.D.P. Océan Indien) dans son cadre régional et son incidence sur l'échelle chronologique des inversions du champ magnétique terrestre. *Bull. Soc. Géol. Fr., 7,* No. 21, 57–63.

Schlich, R., Aubertin, F., Delteil, J. R., Leclaire, L., Magnier, P., Montadert, L., Patriat, P., and Valery, P. 1972. Données nouvelles sur le substratum du bassin de Somalie à partir d'un profil de sismique-réflexion. *C. r. Acad. Sci., Paris, 275,* 1331–34

Sclater, J. G., and Fisher, R. L. 1974. Evolution of the east central Indian Ocean with emphasis on the tectonic setting of the NinetyEast Ridge. *Geol. Soc. Am. Bull., 85,* 683–702.

Sclater, J. G., Luyendyk, B. P., and Meinke, L. 1976. Magnetic lineations in the southern part of Central Indian Basin. *Geol. Soc. Am. Bull., 87,* 371–78.

Sclater, J. G., Jarrard, R. O., McGowran, B., and Gartner, S. 1974. Comparison of the magnetic and biostratigraphic time scales since the Late Cretaceous (*In* Von Der Borch, C. C., Sclater, J. G., and others. *Initial Reports of the Deep Sea Drilling Project,* Vol. 22. Washington [U.S. Government Printing Office], 381–86.)

Supko, P. R., and Perch-Nielsen, K. 1977. General synthesis of Central and South Atlantic drilling results, Leg 39, Deep Sea Drilling Project. (*In* Supko, P. R., Perch-Nielsen, K., and others. *Initial Reports of the Deep Sea Drilling Project,* Vol. 39. Washington [U.S. Government Printing Office], 1099–1131.)

Taylor, P. T. 1968. Interpretation of the north Arabian Sea, aeromagnetic survey. *Earth planet. Sci. Lett., 4,* 232–36.

Vincent, E. 1974. Cenozoic planktonic biostratigraphy and paleoceanography of the tropical western Indian Ocean. (*In* Fisher, R. L., Bunce, E. T., and others. *Initial Reports of the Deep Sea Drilling Project,* Vol. 24. Washington [U.S. Government Printing Office], 1111–50.)

Vine, F. J. 1966. Spreading of the ocean floor: new evidence. *Science, N.Y., 154,* 1405–15.

Whitmarsh, R. B. 1974. Some aspects of plate tectonics in the Arabian Sea. (*In* Whitmarsh, R. B., Weser, O. E., and others. *Initial Reports of the Deep Sea Drilling Project,* Vol. 23. Washington [U.S. Government Printing Office], 527–35.)

10

Paleomotion between Africa, South America, and Antarctica, and Implications for the Antarctic Peninsula

I. O. NORTON*
Department of Earth and Planetary Sciences, Massachusetts Institute of Technology, Cambridge, Massachusetts 02139, U.S.A.

Abstract. One of the major problems in reconstructions of Gondwanaland has been what to do with the Antarctic Peninsula. Various solutions have been proposed, usually involving some motion between West and East Antarctica, but the number of possible reconstructions is a reflection of the lack of suitable constraints that can be placed on paleomotion between South America, Africa, Antarctica, and the Antarctic Peninsula. In this study a proposed history of motion between Africa and Antarctica is described, which when combined with the relatively better known Africa-South America motion, yields a history of South America-Antarctica motion. The Africa-Antarctica motion is constrained by fracture zones and magnetic lineations mapped on the flanks of the Southwest Indian Ridge and is consistent with an evolutionary model for the rest of the Indian Ocean. It is shown that the Drake Passage has opened since Oligocene time and that it is not necessary for any major movement between West and East Antarctica to have occurred since Early Cretaceous time. Before this time, however, the marine geophysical data require an overlap of the Antarctic Peninsula onto South America. The Gondwanaland reconstruction supported by these data is that of Du Toit, and Smith and Hallam. This means that the northern part of the Antarctic Peninsula could have been emplaced in its present position relative to the rest of Antarctica during the early stages of the breakup of Gondwanaland.

Relative motion between Africa, South America, and Antarctica has continued ever since the breakup of Gondwanaland some time during the Mesozoic Era. The present-day boundaries between these plates are fairly well understood—the Mid-Atlantic Ridge between Africa and South America (Dickson and others, 1968; Ladd, 1974), the Southwest Indian Ridge between Africa and Antarctica (Bergh and Norton, 1976; Norton, 1976), and the more complicated America-Antarctica Ridge between South America and Antarctica (Barker and Griffiths, 1972; Forsyth, 1975). Mapping of sea-floor magnetic anomalies has shown that the Mid-Atlantic and Southwest Indian Ridges have had relatively simple spreading histories (Ladd, 1974; Rabinowitz and LaBrecque, 1979; Bergh and Norton, 1976; Bergh, 1977), but that of the Scotia Sea area has been much more complicated (Barker and Griffiths, 1972; Barker, 1972*a*; Dalziel and Elliot, 1973; de Wit, 1977). Figure 10.1 shows the present positions of the continents and major plate boundaries.

Africa and Antarctica have been separating in a north-northeast south-southwest direction since Jurassic time. Evidence for this comes from mapping of magnetic anomalies generated at the Southwest Indian Ridge which date back to Early Cretaceous time (Bergh and Norton, 1976; Bergh, 1977). All mapped lineations strike approximately east-west, and although fracture zones are poorly known, those that have been mapped trend slightly east of north. Thus, although spreading rates have varied considerably (the present half-rate on the Southwest Indian Ridge is about 0.8 cm/yr [Norton, 1976], while in Late Cretaceous time it was about 4.5 cm/yr [Bergh and Norton, 1976]), all evidence indicates a fairly simple paleomotion between Africa and Antarctica. This information, together with fracture-zone and magnetic-anomaly data from the rest of the Indian Ocean, has been used by Norton and Sclater (1979) to develop a history of the evolution of the Indian Ocean and the breakup of Gondwanaland. These reconstructions will be briefly described, and from reconstructions of the relative positions of South America and Africa (Le Pichon and Hayes, 1971; Ladd, 1974), some inferences can be drawn about South America–Antarctica motion.

A theorem of geometry due to Euler states that any movement of a rigid body on a sphere can be described by a rotation about an axis passing through the center of the sphere. As applied to the earth, which is assumed to be spherical, the points where the axis intersects the surface are termed rotation poles; only one is required to define the rotation (Bullard and others, 1965; Morgan, 1968). Movement of a plate (a rigid body on the earth's surface) is therefore completely described by a rotation pole and an angle of rotation. These parameters are derived from fracture zones and magnetic anomalies, or in some cases, by fitting isochrons on plate pairs (a continental edge can be an isochron) using standard plate tectonic techniques, such as is described in Le Pichon and others (1973).

In this study a series of maps showing the relative positions of the Gondwanaland continents at selected times is presented (Figures 10.1–10.7). With six plates moving relative to one another, it is convenient to hold one fixed for reference; Africa is fixed in its present coordinates for this

* Present address: University of Texas Science Institute, 700 The Strand, Galveston, Texas 77550.

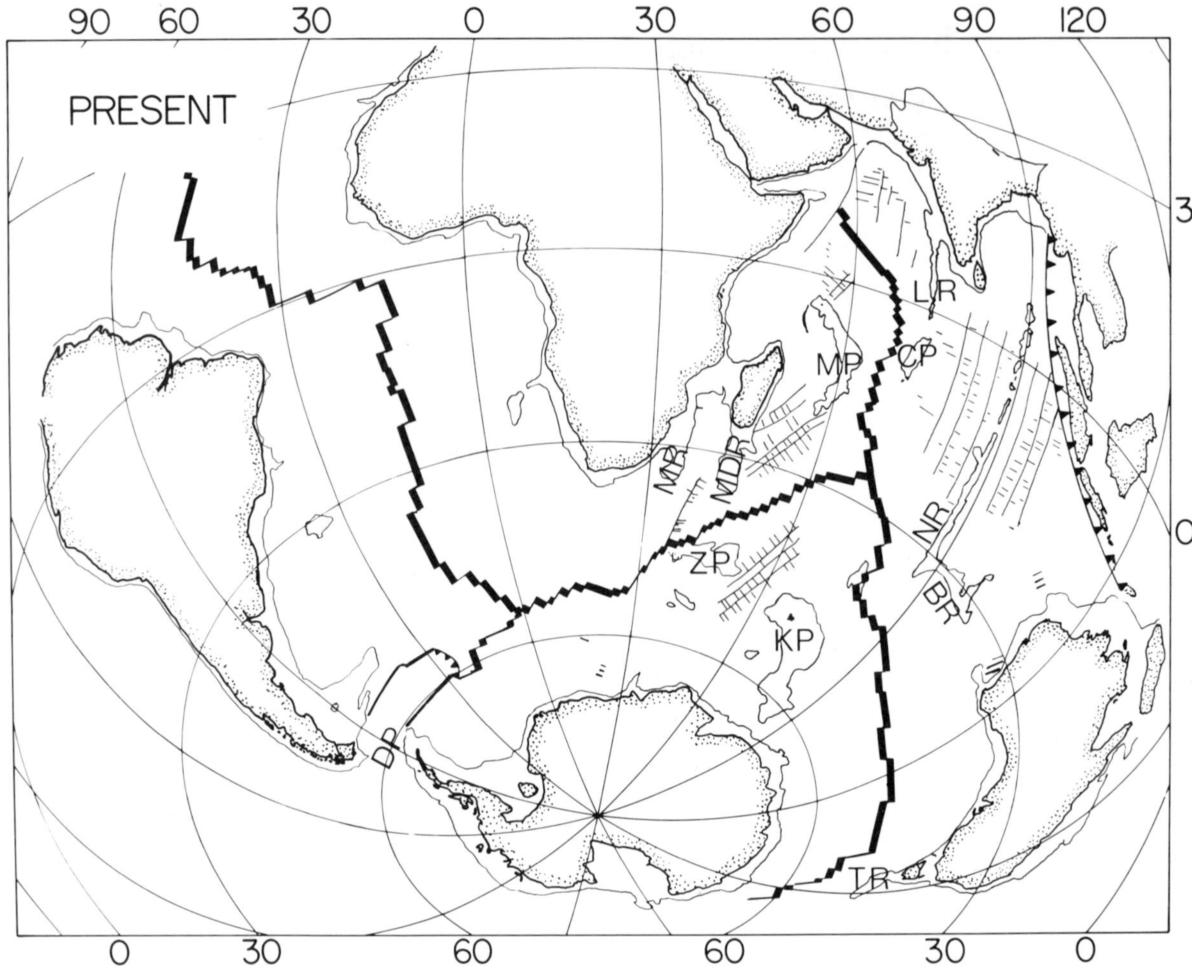

Figure 10.1. The present configuration of the Gondwanaland continents with their 2000 m isobaths. Ridge axes are shown by heavy lines, transform faults by lighter lines. Magnetic lineations and fracture zones identified in the ocean basins and used in the reconstructions are shown by the grid patterns. The hachured line is a trench. (Same format is used for Figures 10.2 through 10.7.) Sources are Bergh and Norton (1976), Schlich (1975), McKenzie and Sclater (1971), Sclater and Fisher (1974), Ladd (1974), Markl (1974), and Bergh (1977).
MR: Mozambique Ridge. MDR: Madagascar Ridge. MP: Mascarene Plateau. LR: Laccadive Ridge. CP: Chagos Plateau. NR: Ninety East Ridge. BR: Broken Ridge. ZP: Crozet Plateau. KP: Kerguelen Plateau. TR: Tasman Rise. DP: Drake Passage. Lambert equal-area projection.

study. The magnetic reversal time scale used is that of LaBrecque and others (1977) for Late Cretaceous time to the present, and that of Larson and Hilde (1975) for Jurassic through Early Cretaceous time.

Early Oligocene (anomaly 15) reconstruction

In the first reconstruction shown (Figure 10.2), Antarctica has been moved toward Africa along the Prince Edward fracture zone (Bergh and Norton, 1976) in the Mozambique Basin. India has been rotated with respect to Africa to restore Chagos Plateau against Mascarene Plateau, and Antarctica has been rotated with respect to India using rotation parameters derived by Weissel and others (1977) from magnetic anomalies in the Southeast Indian Ocean. The India–Africa rotation is dated by DSDP site 238 (Fisher and others, 1974) at the foot of Chagos Plateau, as no magnetic lineations as old as Oligocene in age have been identified on the Central Indian Ridge (Fisher and others, 1971).

As can be seen in Figure 10.2, the Drake Passage has been closed. This is a result of the Antarctica–Africa and South America–Africa rotations; no constraints from the Scotia Sea have been put into the calculations. Note that this reconstruction does not indicate the actual time of opening of Drake Passage—only that it has opened since early Oligocene time. This is because the finite rotations used in the reconstruction are an average of many small and possibly different rotations. Thus, although the Cenozoic history of the South Atlantic, for example, can be adequately described by one rotation pole and angle (Ladd, 1974), at least four poles and angles are needed for the same time period in the Southwest Indian Ocean (Bergh and Norton, 1976). There is extensive evidence from the Scotia Sea region that Drake Passage has opened since Oligocene time, in accord with the reconstruction in Figure 10.2. Barker (1972*a*) and Barker and Griffiths (1972) have described magnetic lineations in the western Scotia Sea which trend northeast-southwest. The anomalies reportedly date the opening as approximately 29 m.y. ago (Barker and Burrell, 1977), but the possibility of subduction beneath the South Shetland Islands may complicate the picture (Barker and Griffiths, 1972). The trend of the anomalies is in accord with simple extension having produced the Drake Passage.

Further evidence for the time of opening comes from

paleocurrent analysis made possible by DSDP cores in the South Pacific (Craddock and Hollister, 1976) and Scotia Sea area (Barker and others, 1976). These studies show that the Antarctic Circumpolar Current has probably been in existence since late Oligocene time. Note that a true circumpolar current also needs a clear passage between Australia and Antarctica, but Barker and Burrell (1977) show that this passage was probably always larger than Drake Passage.

In this and subsequent reconstructions, the plate boundary between South America and the Antarctic Peninsula has been drawn as a patterned line. This is because the inferred motion between these blocks is so small that the nature of the boundary cannot be assessed with any certainty.

Early Eocene (anomaly 22) reconstruction

Australia separated from Antarctica at approximately anomaly 22 time (Weissel and Hayes, 1972) so the next reconstruction is for this time. When fitting lithospheric plates together in plate tectonic reconstructions, it is preferable to derive rotation parameters for each pair of plates independently. If, however, there is inadequate data on one plate pair, rotations for this pair can be derived. In the case of the anomaly 22 reconstruction, there is not sufficient data from the Southwest Indian Ridge to derive an Africa-Antarctica rotation (Bergh and Norton, 1976). We therefore derived the rotation by first fitting Antarctica to Australia (this is the predrift fit), then rotating Australia with respect to India, and then rotating India with respect to Africa. Antarctica is thus rotated with respect to Africa around the "circuit" Antarctica to Australia to India to Africa. It turns out that there has been no significant movement between Australia and India for the last 50 m.y. (Norton and Molnar, 1977), so the continental fit between Antarctica and Australia also yields the position of Antarctica relative to India 50 m.y. ago.

The derived Antarctica-Africa rotation agrees with the data of Bergh and Norton (1976) in the Mozambique Basin, which indicate very little motion between anomalies 15 and 22. Motion between South America and the Antarctic Peninsula has also been minimal through this time span, a result in accord with geologic information from the Scotia Sea region (Dalziel and Elliot, 1973; Elliot, 1975).

Early Paleocene (anomaly 28) reconstruction

At the beginning of Cenozoic time, spreading patterns in the Indian Ocean underwent a small change when the Seychelles platform separated from India. The reconstruction shown in Figure 10.4 is for this time. It was derived from the anomaly 22 fit by rotating India with respect to Antarctica and Africa along fracture zones in the Crozet and Madagascar Basins, respectively. These fracture zones were mapped by Schlich (1975). Comparison between Figures 10.3 and 10.4 shows that there was very little motion between South America and Antarctica during early Cenozoic time.

Late Cretaceous (anomaly 34) reconstruction

There is a 30 m.y. gap in the magnetic-reversal time scale corresponding to the middle Cretaceous normal polarity interval. Anomaly 34 (Campanian–Santonian) is the first reversal after this interval, so the next reconstruction (Figure 10.5) is for this time. Fracture zones and magnetic anomalies in the Madagascar and Crozet Basins (Schlich, 1975) and Mozambique Basin (Bergh and Norton, 1976) have been used to derive the Antarctica-Africa rotation.

About 200 km of overlap between the Antarctic Peninsula and South American is indicated in the reconstruction. It is probably beyond the occuracy of the reconstruction to say whether this overlap is real or a result of errors in the Antarctica-Africa rotation. Since the reconstructions are being described backwards through time, if the overlap is real, it implies some separation between South America and Antarctica during latest Cretaceous time.

Early Cretaceous (Barremian, anomaly M-1) reconstruction

Anomaly M-1 has been mapped in the South Atlantic Ocean (Larson and Ladd, 1973), in the eastern Weddell Sea (Bergh, 1977), and off western Australia (Markl, 1974; Larson, 1977). Some assumptions about the directions of relative motion between the Gondwana continents through the Cretaceous normal interval allow a reconstruction for this time to be derived from the anomaly-34 reconstruction. We assume that Mesozoic anomalies mapped by Bergh (1977) in the eastern Weddell Sea were formed by Africa-Antarctica motion, and that their N 60° W strike, approximately perpendicular to the trend of the Prince Edward fracture zone, shows that the direction of relative motion between Africa and Antarctica did not change much through the Cretaceous normal interval.

Following Markl (1974) and Larson (1977) we assume that Mesozoic lineations off western Australia date the separation between India and Australia-Antarctica. These anomalies strike about N 30° E, while Late Cretaceous anomalies in the Wharton Basin strike east-west (Sclater and Fisher, 1974); this implies a major change in spreading direction between India and Australia-Antarctica during the Cretaceous normal interval (Larson and others, 1978). In the South Atlantic Ocean, we use poles given by Rabinowitz and LaBrecque (1979).

Figure 10.6 shows the reconstruction. To accommodate the Mesozoic anomalies between Africa and Antarctica and those between India and Australia, Madagascar has been pushed to the north along the Davie Ridge, a feature in the Mozambique Channel which Heirtzler and Burroughs (1971) suggested may have been a transform fault between Africa and Madagascar. A northerly position for Madagascar in Gondwanaland is supported by fracture zone patterns in the Somali Basin (Bunce and Molnar, 1977) and by paleomagnetic evidence (Embleton and McElhinny, 1975; McElhinny and others, 1976).

Overlap of the Antarctic Peninsula onto South America is slightly greater compared to the anomaly 34 overlap (Figure 10.5), but the geometry of the reconstruction suggests that this may be real. It is not possible to move Antarctica around by an amount sufficient to eliminate the overlap and still keep the Caird Coast against the Mozambique Ridge and India at the right orientation with respect to Antarctica.

Unless there has been movement between West and East Antarctica since 115 m.y. ago, the explanation for the overlap must be sought in the Scotia Sea region. Creta-

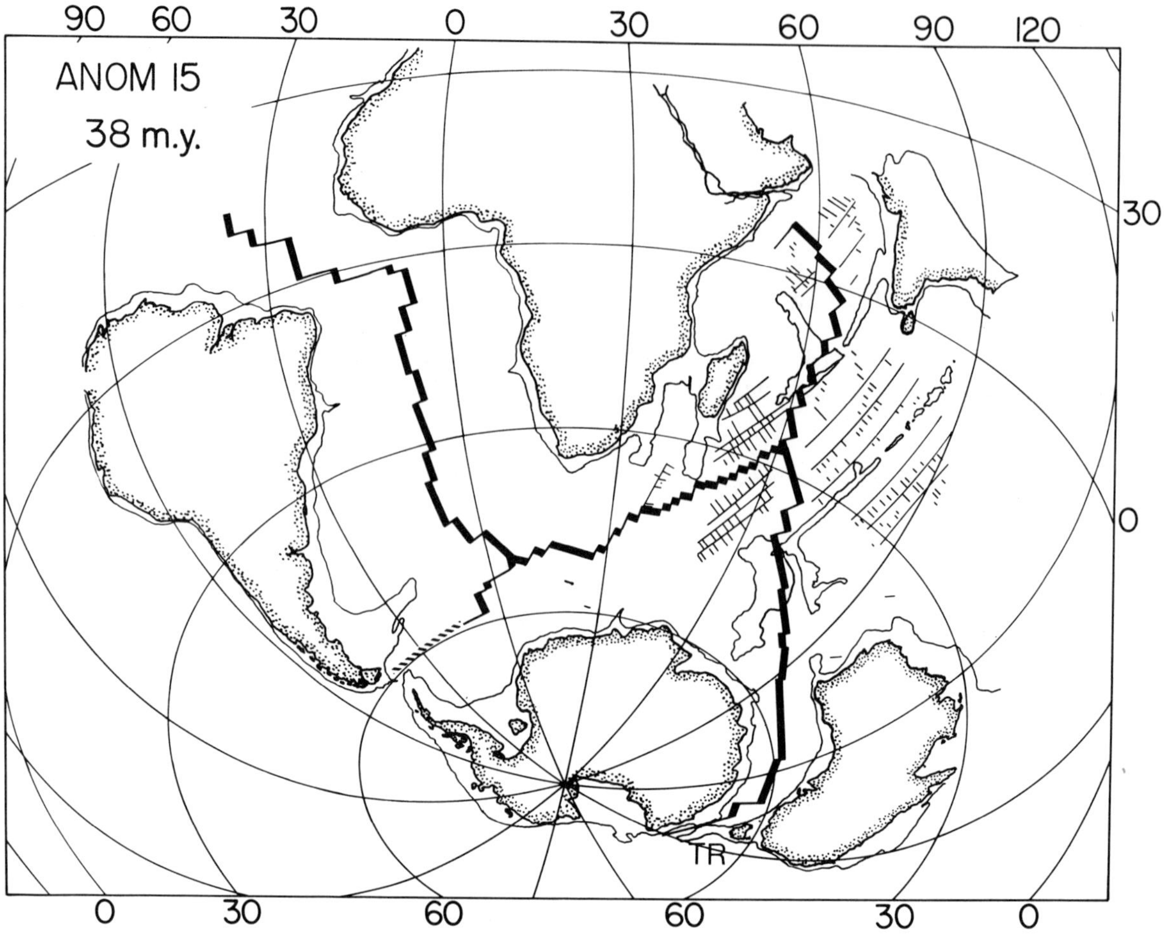

Figure 10.2. Early Oligocene reconstruction. TR: Tasman Rise.

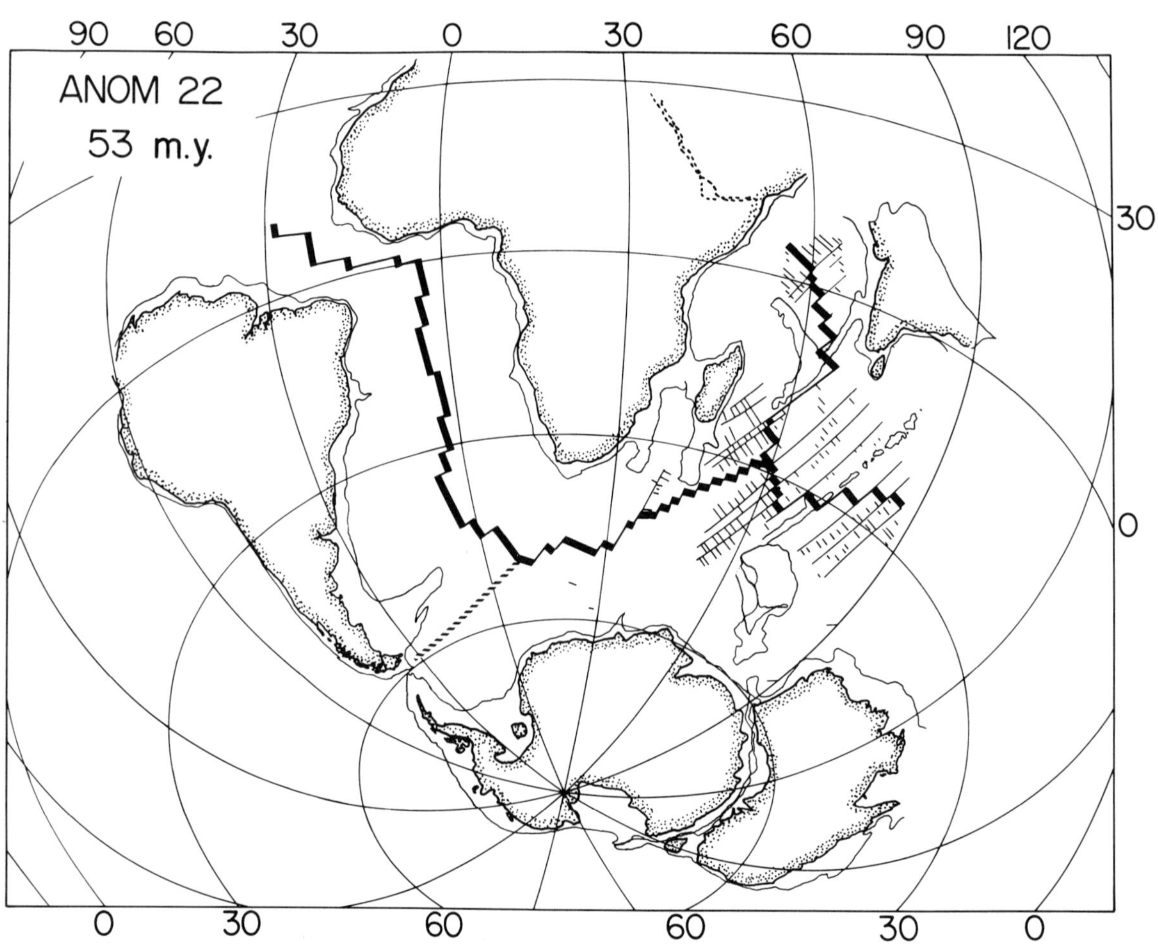

Figure 10.3. Early Eocene reconstruction.

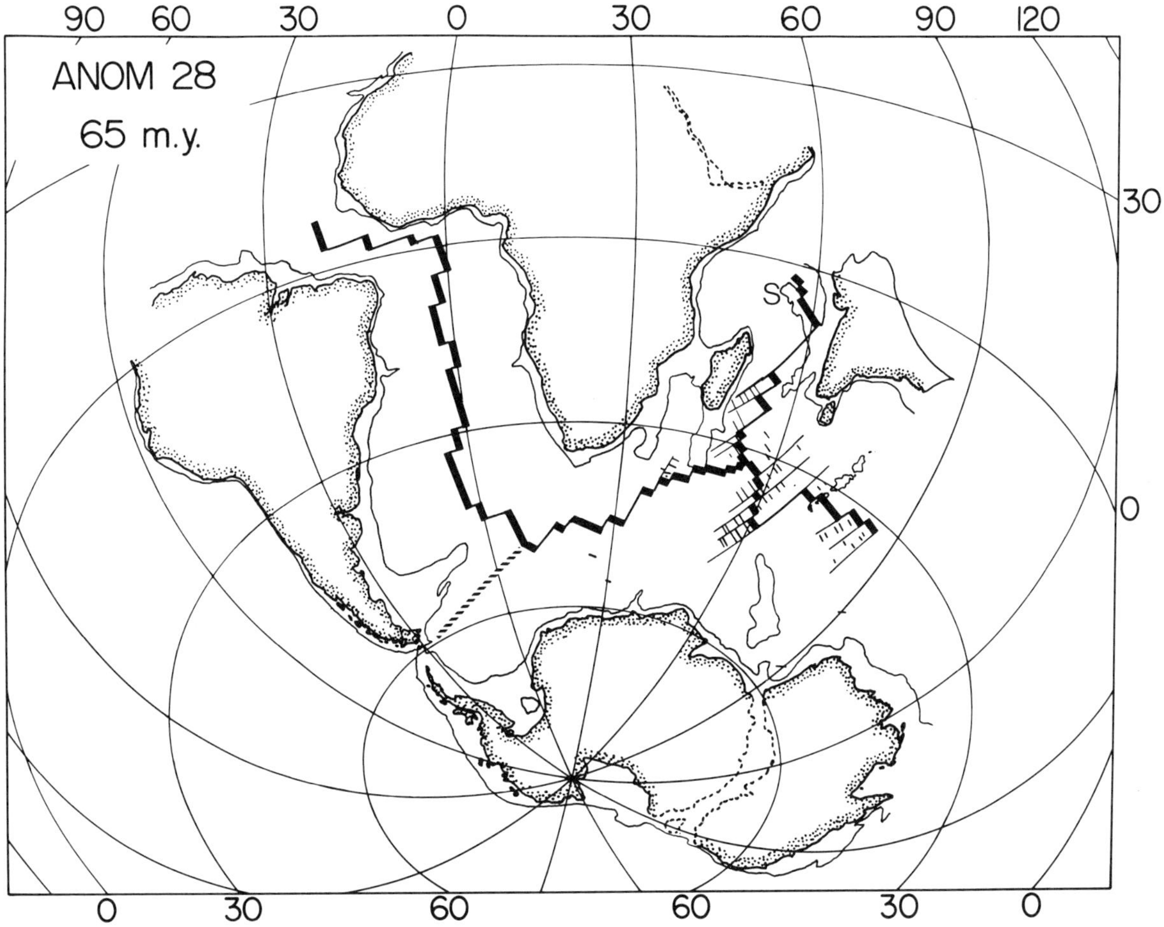

Figure 10.4. Early Paleocene reconstruction. S: Seychelles platform.

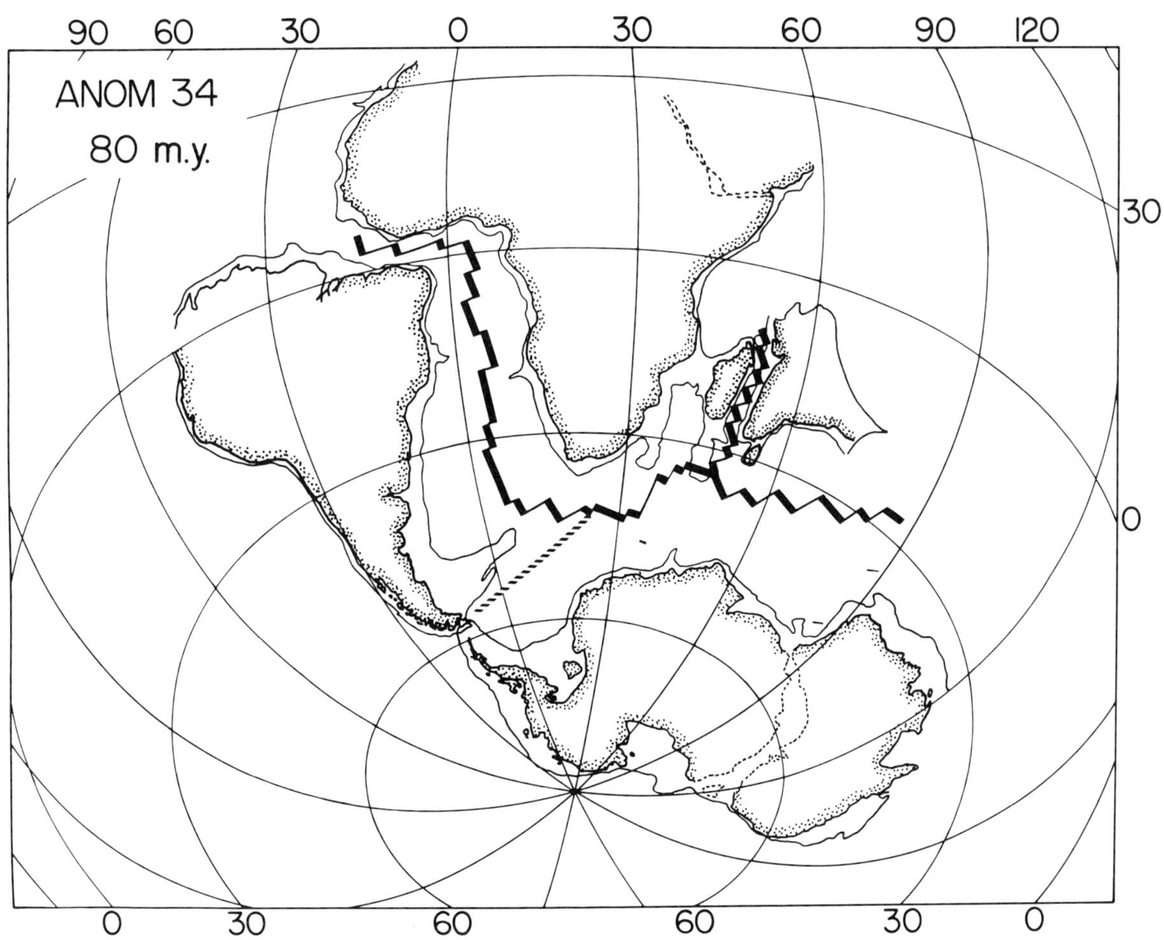

Figure 10.5. Late Cretaceous reconstruction.

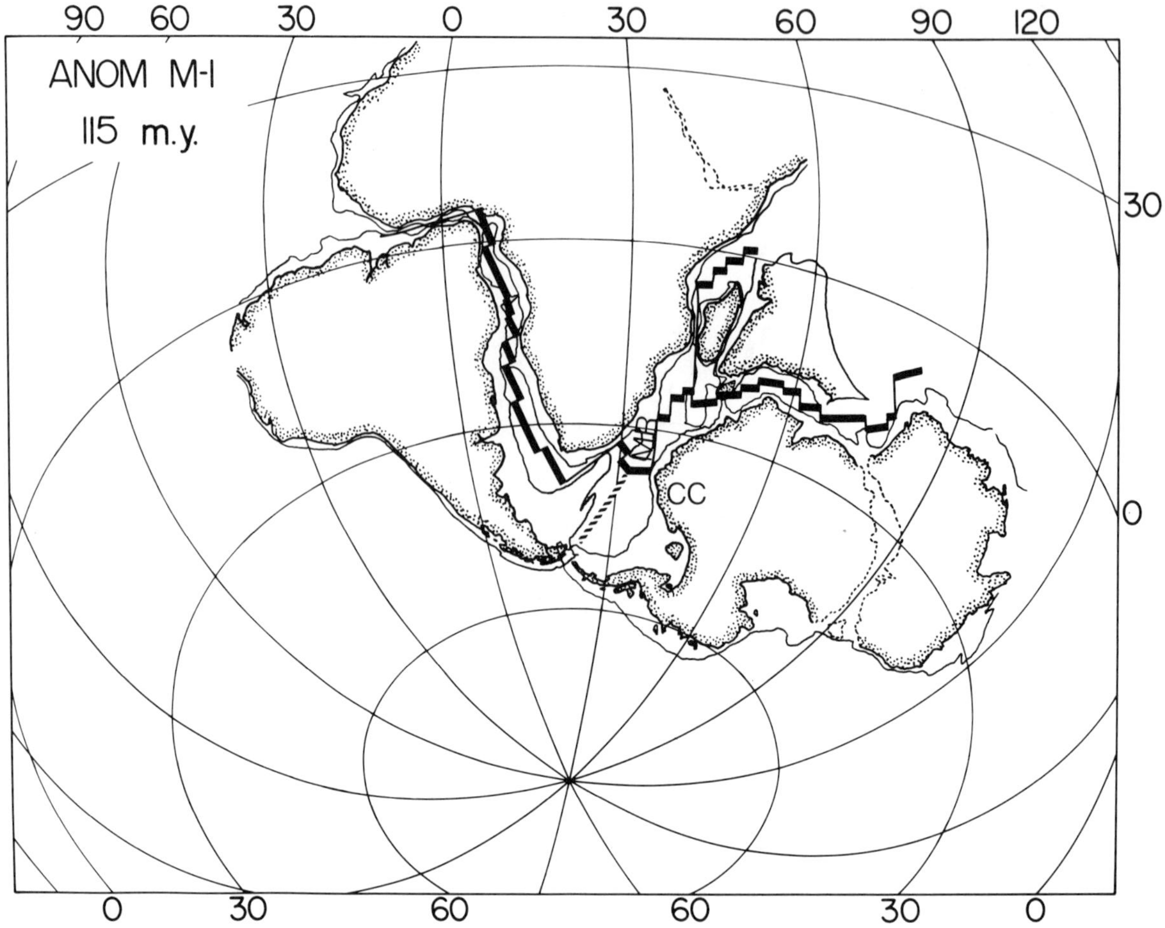

Figure 10.6. Early Cretaceous reconstruction. CC: Caird Coast. MR: Mozambique Ridge.

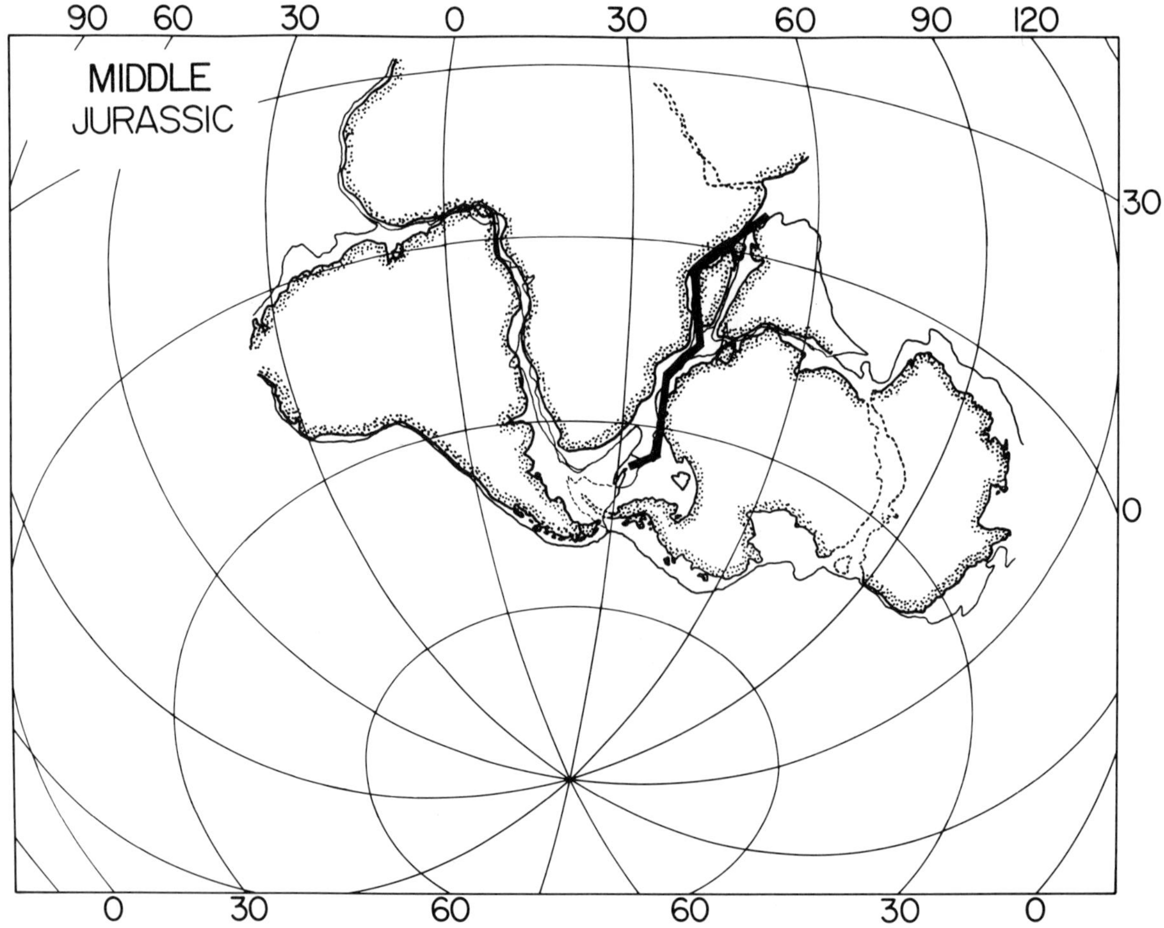

Figure 10.7. Gondwanaland in Middle Jurassic time, when the first break was separating East and West Gondwanaland. The heavy line denotes the break.

ceous deformation in the northern Antarctic Peninsula seems to have been minimal (Dalziel and Elliot, 1973; Elliot, 1975), but in the southern Andes a marginal sea existed during Early Cretaceous time (Dalziel and others, 1974). It formed during a time when Pacific sea floor was being subducted beneath southern South America, but it closed later in Cretaceous time. It is therefore probable that at the time represented by Figure 10.6, southern South America had a very different shape to that shown; some of the overlap could be eliminated if the correct shape was known.

Gondwanaland reconstruction

Shown in Figure 10.7 is a reconstruction of Gondwanaland based on the sequence of reconstructions previously described. It is basically that of Du Toit (1937) and Smith and Hallam (1970), but it differs from the latter in that India is farther south relative to Madagascar. This pushes Antarctica farther south relative to Africa (the India-Antarctica fit is that of Smith and Hallam) and leaves room for the Mozambique Ridge to be continental in character. This is possible, as it lies adjacent to the Falkland Plateau, which is known to be continental (Barker and others, 1976). As in the Smith and Hallam reconstruction, however, there is a large overlap of the Antarctic Peninsula onto South America. No solution to this problem is proposed here.

In Figure 10.7, the thick line separating East and West Gondwanaland represents the line along which Gondwanaland is inferred to have first broken. This is suggested by stratigraphy in eastern Africa. Kent (1974) and Kent and others (1971) have gathered a great deal of information from exploration boreholes and outcrops in coastal Kenya, Tanzania, and Mozambique which indicates that open marine conditions existed there by Middle Jurassic time; these marine rocks become younger to the south. India's separation from Australia appears to have occurred in Early Cretaceous time (Markl, 1974; Larson, 1977), so the whole of East Gondwanaland must have separated from West Gondwanaland in Jurassic time or earlier. A southward migration of marine conditions is consistent with a model for the breakup in which East Gondwanaland moved south relative to Africa in the early stages of breakup, with subsequent sundering of India from Antarctica and South America from Africa.

Discussion

A set of reconstructions tracing a proposed history of evolution of the South Atlantic and Indian Oceans has been described. It is easier to follow the sequence of events by going forward in time, so the following is a summary of the stages.

1. East Gondwanaland (Madagascar, India, Antarctica, and Australia) separated from Africa and South America in Early Jurassic time and moved south relative to Africa along a system of long transform faults and short ridge segments.
2. At approximately anomaly M-11 time (127 m.y. ago; Valanginian), South America separated from Africa (Larson and Ladd, 1973; Rabinowitz, 1976), and possibly at the same time India separated from Antarctica-Australia (Markl, 1974; Larson, 1977). Motion continued throughout the Cretaceous Period, but by Barremian time it is possible that the bulk of the Antarctic Peninsula was in its present position relative to East Antarctica.
3. In Cretaceous time, Madagascar became fixed to Africa, and in Late Cretaceous time, India left Madagascar, resulting in a reorganization of spreading directions. The Andean orogeny, related to subduction of Pacific sea floor, continued along the southern Andes and the northern Antarctic Peninsula (Dalziel and Elliot, 1973).
4. The spreading regime established in Late Cretaceous time continued into the Tertiary Period, with a reorganization in the northwest Indian Ocean in early Paleocene time when the Seychelles platform separated from India.
5. Australia separated from Antarctica in early Eocene time (Weissel and Hayes, 1972) and became attached to the Indian plate (Norton and Molnar, 1977).
6. India collided with Asia during late Eocene–early Oligocene time (Molnar and Tapponnier, 1975), and the resultant reorganization of plate boundaries caused spreading to commence on the Central Indian Ridge. Accelerated motion between Africa and Antarctica resulted in the opening of Drake Passage, but whether this was a direct result of the India-Asia collision is not clear; there is, however, a coincidence in timing.
7. Since Oligocene time, plate motions have been fairly regular, except between South America and the Antarctic Peninsula where the South Sandwich Trench commenced subduction about 10 m.y. ago (Barker and Griffiths, 1972). A further complication in this area was the eastward movement of South Georgia relative to South America (Dalziel and Elliot, 1973), but this could have occurred any time in the Tertiary Period.

Acknowledgments. I am grateful to H. W. Bergh, P. Rabinowitz, J. K. Weissel, and E. T. Bunce for preprints of their papers. I thank John Sclater for the opportunity to work on this project. Research was supported by grants DES 75-21196 and DPP 76-19053 from the National Science Foundation.

References cited

Barker, P. F. 1972*a*. Magnetic lineations in the Scotia Sea. (*In* Adie, R. J., *ed. Antarctic Geology and Geophysics.* Oslo, Universitetsforlaget, 17–26.)

Barker, P. F. 1972*b*. A spreading centre in the East Scotia Sea. *Earth planet. Sci. Lett., 15,* 123–32.

Barker, P. F., and Burrell, J. 1977. The opening of Drake Passage. *Mar. Geol., 25,* 15–34.

Barker, P. F., and Griffiths, D. H. 1972. The evolution of the Scotia Ridge and Scotia Sea. *Phil. Trans. R. Soc. Lond.,* Ser. A, *271,* 151–83.

Barker, P. F., Dalziel, I. W. D., and others. 1976. Evolution of the southwestern Atlantic Ocean basin: Results of leg 36. (*In* Barker, P. F., Dalziel, I. W. D., and others. *Initial Reports of the Deep Sea Drilling Project,* Vol. 36. Washington [U.S. Government Printing Office], 993–1014.)

Bergh, H. W. 1977. Mesozoic sea floor off Dronning Maud Land, Antarctica. *Nature, Lond., 269,* 686–87.

Bergh, H. W., and Norton, I. O. 1976. Prince Edward fracture zone and the evolution of the Mozambique Basin. *J. geophys. Res., 81,* 5221–39.

Bullard, E. C., Everett, J. E., and Smith, A. G. 1965. The fit of the continents around the Atlantic Ocean. (*In* Blackett, P. M. S., Bullard, E., and Runcorn, S. K., *eds.* Symposium on Continental Drift. *Phil. Trans. R. Soc. Lond.,* Ser. A, *258,* 41–51.)

Bunce, E. T., and Molnar, P. 1977. Seismic reflection profiling and basement topography in the Somali Basin: possible fracture zones between Madagascar and Africa. *J. geophys. Res., 82,* 5305–11.

Craddock, C., and Hollister, C. D. 1976. Geologic evolution of the Southeast Pacific Basin. (*In* Hollister, C. D., and Craddock, C., *eds. Initial Reports of the Deep Sea Drilling Project,* Vol. 35. Washington [U.S. Government Printing Office], 723–43.)

Dalziel, I. W. D., and Elliot, D. H. 1973. The Scotia arc and Antarctic Margin. (*In* Nairn, A. E. M., and Stehli, E. G., *eds. The Ocean Basins and Margins. 1. The South Atlantic.* New York, Plenum Press, 171–245.)

Dalziel, I. W. D., de Wit, M. J., and Palmer, K. F. 1974. Fossil marginal basin in the southern Andes. *Nature, Lond., 250,* 291–94.

de Wit, M. J. 1977. The evolution of the Scotia arc as a key to the reconstruction of southwestern Gondwanaland. *Tectonophysics, 37,* 53–81.

Dickson, G. O., Pitman, W. C., III, and Heirtzler, J. R. 1968. Magnetic anomalies in the South Atlantic and ocean floor spreading. *J. geophys. Res., 73,* 2087–2100.

Du Toit, A. L. 1937. *Our Wandering Continents.* Edinburgh, Oliver and Boyd, pp. 366.

Elliot, D. H. 1975. Tectonics of Antarctica: A review. *Am. J. Sci., 275A,* 45–106.

Embleton, B. J. J., and McElhinny, M. W. 1975. The palaeoposition of Madagascar: palaeomagnetic evidence from the Isalo Group. *Earth planet. Sci. Lett., 27,* 329–41.

Fisher, R. L., Sclater, J. G., and McKenzie, D. P. 1971. Evolution of the Central Indian Ridge, western Indian Ocean. *Geol. Soc. Am. Bull., 82,* 553–62.

Fisher, R. L., Bunce, E. T., and others. 1974. Site 238. (*In* Fisher, R. L., Bunce, E. T., and others, *eds. Initial Reports of the Deep Sea Drilling Project,* Vol. 24. Washington [U.S. Government Printing Office], 469–587.)

Forsyth, D. W. 1975. Fault plane solutions and tectonics of the South Atlantic and Scotia Sea. *J. geophys. Res., 80,* 1429–43.

Heirtzler, J. R., and Burroughs, R. H. 1971. Madagascar's paleoposition: New data from the Mozambique Channel. *Science, N. Y., 174,* 488–90.

Kent, P. E. 1974. Continental margin of Africa—a region of vertical movements. (*In* Burk, C. A., and Drake, C. L., *eds. The Geology of Continental Margins.* New York, Springer-Verlag, 313–20.)

Kent, P. E., Hunt, J. A., and Johnstone, D. W. 1971. The geology and geophysics of coastal Tanzania. *Geophys. Pap., Inst. Geol. Sci.,* No. 6, 101 pp.

LaBrecque, J. L., Kent, D. V., and Cande, S. C. 1977. Revised magnetic reversals for the Early Cretaceous and Late Jurassic. time. *Geology, 5,* 330–35.

Ladd, J. W. 1974. *South Atlantic Sea-floor Spreading and Caribbean Tectonics.* Ph.D. dissertation, Columbia University, 251 pp.

Larson, R. L. 1977. Early Cretaceous breakup of Gondwanaland off western Australia. *Geology, 5,* 57–60.

Larson, R. L., and Hilde, T.W.C. 1975. A revised time scale of magnetic reversals for the Early Cretaceous and Late Jurassic. *J. geophys. Res., 80,* 2586–94.

Larson, R. L., and Ladd, J. W. 1973. Evidence for the opening of the South Atlantic in the Early Cretaceous. *Nature, Lond., 246,* 209–12.

Larson, R. L., Carpenter, G. B., and Diebold, J. B. 1978. Geophysical and geological studies of the Wharton Basin near the Investigator fracture zone. *J. geophys. Res., 83,* 773–82.

Le Pichon, X., and Hayes, D. E. 1971. Marginal offsets, fracture zones and the early opening of the South Atlantic. *J. geophys. Res., 76,* 6283–93.

Le Pichon, X., Francheteau, J., and Bonnin, J. 1973. *Plate Tectonics.* Amsterdam, Elsevier, 300 pp. [Developments in Geotectonics, Vol. 6]

McElhinny, M. W., Embleton, B. J. J., Daly, L., and Pozzi, J. P. 1976. Paleomagnetic evidence for the location of Madagascar in Gondwanaland. *Geology, 4,* 455–57.

McKenzie, D. P., and Sclater, J. G. 1971. The evolution of the Indian Ocean since the Late Cretaceous. *Geophys. J. R. astr. Soc., 25,* 437–528.

Markl, R. G. 1974. Evidence for the breakup of eastern Gondwanaland by the Early Cretaceous. *Nature, Lond., 251,* 196–200.

Molnar, P., and Tapponnier, P. 1975. Cenozoic tectonics of Asia: effects of a continental collision. *Science, N.Y., 189,* 419–26.

Morgan, W. J. 1968. Rises, trenches, great faults, and crustal blocks. *J. geophys. Res., 73,* 1959–82.

Norton, I. O. 1976. The present relative motion between Africa and Antarctica. *Earth planet. Sci. Lett., 33,* 219–30.

Norton, I. O., and Molnar, P. 1977. Implications of a revised fit between Australia and Antarctica for the evolution of the eastern Indian Ocean. *Nature, Lond., 267,* 338–39.

Norton, I. O., and Sclater, J. G. 1979. A model for the evolution of the Indian Ocean and the breakup of Gondwanaland. *J. geophys. Res., 84,* 6803–30.

Rabinowitz, P. D. 1976. Geophysical study of the continental margin of southern Africa. *Geol. Soc. Am. Bull., 87,* 1643–53.

Rabinowitz, P. D., and LeBrecque, J. 1979. The Mesozoic South Atlantic Ocean and evolution of its continental margins. *J. geophys. Res., 84,* 5973–6002.

Schlich, R. 1975. Structure et age de l'Océan Indien Occidental. *Mém. Soc. Géol. Fr., 6,* 103 pp.

Sclater, J. G., and Fisher, R. L. 1974. Evolution of the east Central Indian Ocean, with emphasis on the tectonic setting of the NinetyEast Ridge. *Geol. Soc. Am. Bull., 85,* 683–702.

Smith, A. G., and Hallam, A. 1970. The fit of the southern continents. *Nature, Lond., 225,* 139–44.

Weissel, J. K., and Hayes, D. E. 1972. Magnetic anomalies in the southeast Indian Ocean. (*In* Hayes, D. E., *ed. Antarctic Oceanology 2. The Australian-New Zealand Sector.* Washington, American Geophysical Union, 165–96. [Antarctic Research Series, Vol. 19.]

Weissel, J. K., Hayes, D. E., and Herron, E. M. 1977. Plate tectonics synthesis: displacement between Australia, New Zealand and Antarctica since the Late Cretaceous. *Mar. Geol., 25,* 231–77.

11

Precambrian Age for Gneisses at Cape Meredith in the Falkland Islands

D. C. REX
Department of Earth Sciences, University of Leeds, Leeds LS2 9JT, U.K.

P. W. G. TANNER
British Antarctic Survey, Natural Environment Research Council, Madingley Road, Cambridge CB3 0ET, U.K.

Abstract. Concordant K-Ar ages of 953 and 977 m.y. on biotite and hornblende from gneisses of the Cape Meredith Complex confirm their Precambrian age. These ages are consistent with those of 1100 ± 250 m.y. given by areas of basement gneisses in southern Africa and Antarctica which lay close to the Falkland Islands prior to the breakup of Gondwanaland.

The gneissic rocks at Cape Meredith (Figure 11.1) in the Falkland Islands occupy a narrow coastal strip about 5 km long. They were first described by Andersson (1907) who recognized their probable pre-Devonian age and assigned them to the Archean. Baker (1924) studied them in more detail and identified schists, gneisses, granites, and pegmatites, cut by lamprophyre dikes and uncomformably overlain by Lower Devonian sandstones. He was the first to suggest stratigraphic correlations between the Falkland Islands and southern Africa: these were later amplified by du Toit (1937) and Adie (1952).

This paper is based upon radiometric dating of material collected by C. G. Smith of the British Antarctic Survey during a brief visit to Cape Meredith in 1969. Most of the specimens are of feldspathic gneiss, granite, and pegmatite with a few specimens of amphibolite, but the relative ages of the different rock types are not known.

Radiometric dating

Hornblende separated from an amphibolite with the mineral assemblage hornblende-plagioclase-quartz gives a K-Ar apparent age of 977 ± 40 m.y.; biotite from a foliated quartz-plagioclase-biotite gneiss gives 953 ± 30

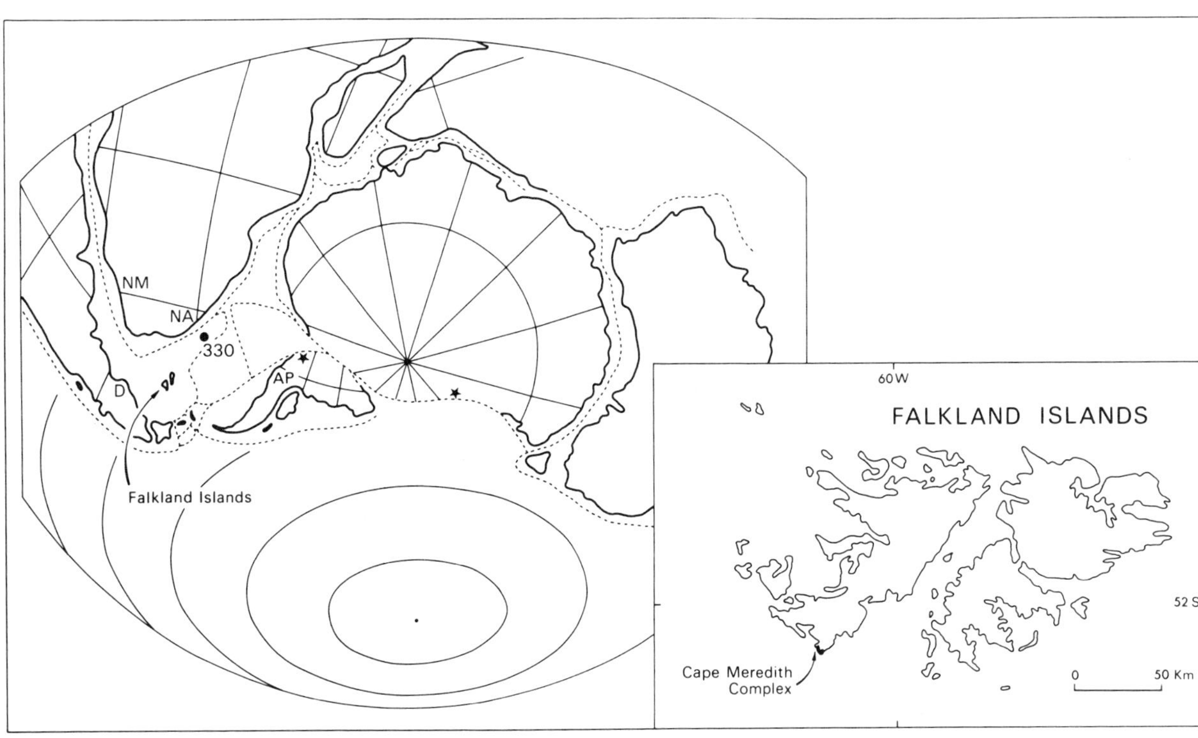

Figure 11.1. Location of the Falkland Islands on the Gondwanaland (B) reconstruction of Barker and Griffiths (1977). AP: Antarctic Peninsula. D: Deseado Massif. NA: Natal. NM: Namaqualand. 330: DSDP site 330. Locations of 1000 m.y. ages from Antarctica mentioned in the text are shown by asterisks.

Table 11.1. K-Ar age data on minerals from gneisses at Cape Meredith, Falkland Islands

Specimen number	K (wt %)	^{40}Ar (radiogenic) scc^a/g × 10^{-4}	^{40}Ar (radiogenic) % total	Age (m.y.)
Biotite FI106	7.13	3.5393 3.5241	99.3 98.6	953 ± 30
Hornblende FI112	0.585	0.2991	94.5	977 ± 40

Constants: λ_e = 0.584 × $10^{-10}yr^{-1}$; λ_β = 4.72 × $10^{-10}yr^{-1}$; and $^{40}K/K$ = 0.0119 atomic %.

[a] Standard cubic centimeter (cm^3 at S.T.P.).

m.y. (Table 11.1). These roughly concordant ages from different components of the complex confirm a minimum age of late Precambrian for the gneisses.

Discussion

The Gondwanaland reconstruction (B) of Barker and Griffiths (1977) is used as a basis for a brief review of possible age comparisons between the Cape Meredith Complex and gneissic terranes in neighboring parts of Africa, South America, and East Antarctica.

Africa

A close comparison may be drawn between the Cape Meredith 950–980 m.y. ages and those from southern Africa. Both in Namaqualand and on the Natal Coast there are extensive areas of gneisses belonging to an older craton (>2500 m.y.) which have been intruded by granite plutons and affected by the Kibaran (1100 ± 200 m.y.) orogenic event. They give K-Ar ages of 900–1250 m.y. (Namaqualand) and 950–1150 m.y. (Natal) (Nicolaysen and Burger, 1965).

South America

The K-Ar ages from the Cape Meredith Complex lie between those from the Trans-Amazonian nucleus (~2000 m.y.) (Cobbing and others, 1977) and the Brazilian orogenic cycle (450–650 m.y.) (Cordani and others, 1973). Ages of 1100–1400 m.y. are reported by Cordani and others from part of the São Francisco craton in eastern Brazil, but their significance is not understood at present. No radiometric data are available from the gneisses of the Deseado Massif, the nearest outcrop of such rocks to the Falkland Islands (Figure 11.1).

East Antarctica

Ages of 991–1031 m.y. on biotite and hornblende have been reported from gneisses considered to be once part of East Antarctica (Clarkson and Brook, 1977) and close to its point of previous nearest proximity to the Falkland Islands (Figure 11.1). Similar ages of 1000 m.y. on hornblende from the Nimrod Group (Grindley and McDougall, 1969) farther south are considered to be possibly reset dates in considerably older rocks (Gunner and Faure, 1972).

Comparison with the Falkland Plateau

Deep sea drilling at site 330 on the Falkland Plateau (Figure 11.1) by *Glomar Challenger* has shown that gneissic and granitic rocks occur beneath a thin (556 m) cover of Jurassic and younger sediments. Rb-Sr whole-rock dating of four specimens of gneiss (Beckinsale and others, 1977) gave an age of 535 ± 66 m.y., and K-Ar whole-rock dating of gneiss and K-feldspar pegmatite gave ages of 399 and 287 m.y. The Rb-Sr data compare well with geochronologic evidence from the Malmesbury Formation and the Cape Granites (550 m.y.) in southern Africa for an early Paleozoic orogeny, and with ages of 500 ± 50 m.y. from Argentina and from the Ross orogen in East Antarctica (Beckinsale and others, 1977).

Thus the gneissic and granitic rocks forming the basement on the Falkland Plateau record two major isotopic events, at roughly 960 and 550 m.y. ago, both of which can be identified in adjoining parts of a reconstructed Gondwanaland.

References cited

Adie, R. J. 1952. Representatives of the Gondwana system of the Falkland Islands. *Symposium sur le Séries de Gondwana, 19th Int. Geol. Congr., Algiers, 1952,* 385–92.

Andersson, J. G. 1907. Contributions to the geology of the Falkland Islands. *Wiss. Ergebn. Schwed. Südpolarexped., 1901–3, 3,* No. 2, 1–38.

Baker, H. A. 1924. *Final Report on Geological Investigations in the Falkland Islands, 1920–22.* Stanley (Government Printer), 1–38.

Barker, P. F., and Griffiths, D. H. 1977. Towards a more certain reconstruction of Gondwanaland. *Phil. Trans. R. Soc. Lond.,* Ser. B., *279,* 143–59.

Beckinsale, R. D., Tarney, J., Darbyshire, D. P. F., and Humm, M. J. 1977. Rb-Sr and K-Ar age determinations on samples of the Falkland Plateau basement at Site 330, DSDP. (*In* Barker, P. F., Dalziel, I. W. D., and others. *Initial Reports of the Deep Sea Drilling Project,* Vol. 36. Washington [U.S. Government Printing Office], 923–27.)

Clarkson, P. D., and Brook, M. 1977. Age and position of the Ellsworth Mountains crustal fragment, Antarctica. *Nature, Lond., 265,* No. 5595, 615–16.

Cobbing, E. J., Ozard, J. M., and Snelling, N. J. 1977. Reconnaissance geochronology of the crystalline basement rocks of the Coastal Cordillera of southern Peru. *Geol. Soc. Am. Bull., 88,* 241–46.

Cordani, U. G., Amaral, G., and Kawashita, K. 1973. The Precambrian evolution of South America. *Geol. Rdsch., 62,* 309–17.

Du Toit, A. L. 1937. *Our Wandering Continents.* Edinburgh, Oliver and Boyd, 366 pp.

Grindley, G. W., and McDougall, I. 1969. Age and correlation of the Nimrod Group and other Precambrian rock units in the central Transantarctic Mountains, Antarctica. *N.Z. Jl Geol. Geophys., 12,* No. 2–3, 391–411.

Gunner, J., and Faure, G. 1972. Rb-Sr geochronology of the Nimrod Group, Central Transantarctic Mountains. (*In* Adie, R. J., *ed. Antarctic Geology and Geophysics.* Oslo, Universitetsforlaget, 305–11.)

Nicolaysen, L. O., and Burger, A. J. 1965. Note on an extensive zone of 1000 million-year old metamorphic and igneous rocks in southern Africa. *Sciences Terre, 10,* No. 3–4, 447–518.

II *Scotia Arc Region*

12

The Early (Pre–Middle Jurassic) History of the Scotia Arc Region: A Review and Progress Report*

Review Paper

IAN W. D. DALZIEL
Lamont-Doherty Geological Observatory of Columbia University, Palisades, New York 10964, U.S.A.

Abstract. The building of the present Pacific Ocean margin cordillera in southernmost South America, the islands of the North and South Scotia Ridges, and the Antarctic Peninsula was initiated in Middle to Late Jurassic time with the widespread extrusion of calc-alkaline, dominantly silicic, volcanic rocks throughout the Scotia arc region. These volcanic rocks everywhere rest with profound unconformity on a previously deformed and metamorphosed "basement" complex. This paper is a review of our state of knowledge of this basement, based on the literature and on largely unpublished field observations by the author and his associates. The pre–Middle Jurassic history of the Scotia arc region has an important bearing on the reconstruction of Gondwanaland and on the early history of the southeastern Pacific Ocean basin, as well as on the evolution of the Scotia arc itself.

The petrotectonic assemblage of the basement, considered in relation to the geology of the South Atlantic–Weddell Sea area as a whole, leaves little doubt that the complex represents portions of the fore-arc and main magmatic arc of a pre–Middle Jurassic subduction zone along the Pacific margin of Gondwanaland. Taken with the distribution of undoubtedly Precambrian crystalline rocks, the analysis also suggests that the Scotia arc may be located in an old reentrant in the margin of the supercontinent and thus calls to question the long-assumed original physical continuity of the Andes and the West Antarctic cordillera.

It is hard to escape the conclusion that the important uplift and erosion of the basement complex during Late Triassic to Early Jurassic time, following deformation and metamorphism, was causally related to the initial breakup of Gondwanaland.

Along the shores of Seno Agostini in the northern part of the Andes of Tierra del Fuego and on Mount Flora at the tip of the Antarctic Peninsula, similar gently dipping sequences of interlayered silicic volcanic and sedimentary strata of Jurassic age rest unconformably upon previously deformed and metamorphosed rocks (Figure 12.1) (Dalziel, 1972*a*, figures 6 and 7). This unconformity beneath the Jurassic sequences represents the most significant known break in the geologic history of the Scotia arc region. In classical terms, the break represents the division between the early Mesozoic "Gondwanide" or "Gondwanian" deformation of middle Paleozoic to lower Mesozoic rocks in the "Samfrau geosyncline" (Du Toit, 1937) and late Mesozoic and Cenozoic "Andean" volcano-tectonic events related to the building of the present continental margin cordillera in South America and West Antarctica. (For reviews see Dalziel and Elliot, 1973, and Elliot, 1975.) It will be demonstrated here, however, that the pre–Middle Jurassic geology seems to indicate that Du Toit's Gondwanide deformation resulted from the same basic cause as the subsequent Andean events, namely subduction of Pacific Ocean floor lithosphere beneath Gondwanaland (Barker, Dalziel, and others, 1976; de Wit, 1977).

The pre–Middle Jurassic rocks of the Scotia arc region are widely referred to as the "basement" of the region. South of the Drake Passage this basement is commonly subdivided into (1) a metamorphic complex with greenschist-blueschist to amphibolite facies assemblages exposed mainly in the South Orkney and South Shetland Islands, and (2) the Trinity Peninsula Formation (formerly the "Trinity Peninsula Series"; see Table 12.1) of the Antarctic Peninsula and its supposed equivalents in the South Orkneys (Greywacke-Shale Formation), South Shetlands (Miers Bluff Formation), the Antarctic Peninsula (Legoupil Formation), and Alexander Island (LeMay Formation). In the islands of the South Scotia Ridge and in the northern areas of the peninsula, the Trinity Peninsula Formation and its correlatives have undergone only zeolite or prehnite-pumpellyite facies metamorphism; hence, the distinction from the "metamorphic" complex. Southward along the peninsula, however, British Antarctic Survey geologists have reported that the metamorphic grade increases (see, for example, Aitkenhead, 1965; Elliot, 1966; Stubbs, 1968; Skinner, 1973). Radiometric data from Marguerite Bay reported in this volume by Gledhill and others (Chapter 35) indicate that the gneiss there could be the metamorphic equivalent of the Trinity Peninsula Formation. Blueschist metamorphism of the LeMay Formation on Alexander Island is reported by Edwards (Chapter 36, this volume). Recent observations in the South Orkney Islands make it clear that, there at least, rocks lithically

* Lamont-Doherty Geological Observatory contribution no. 3050.

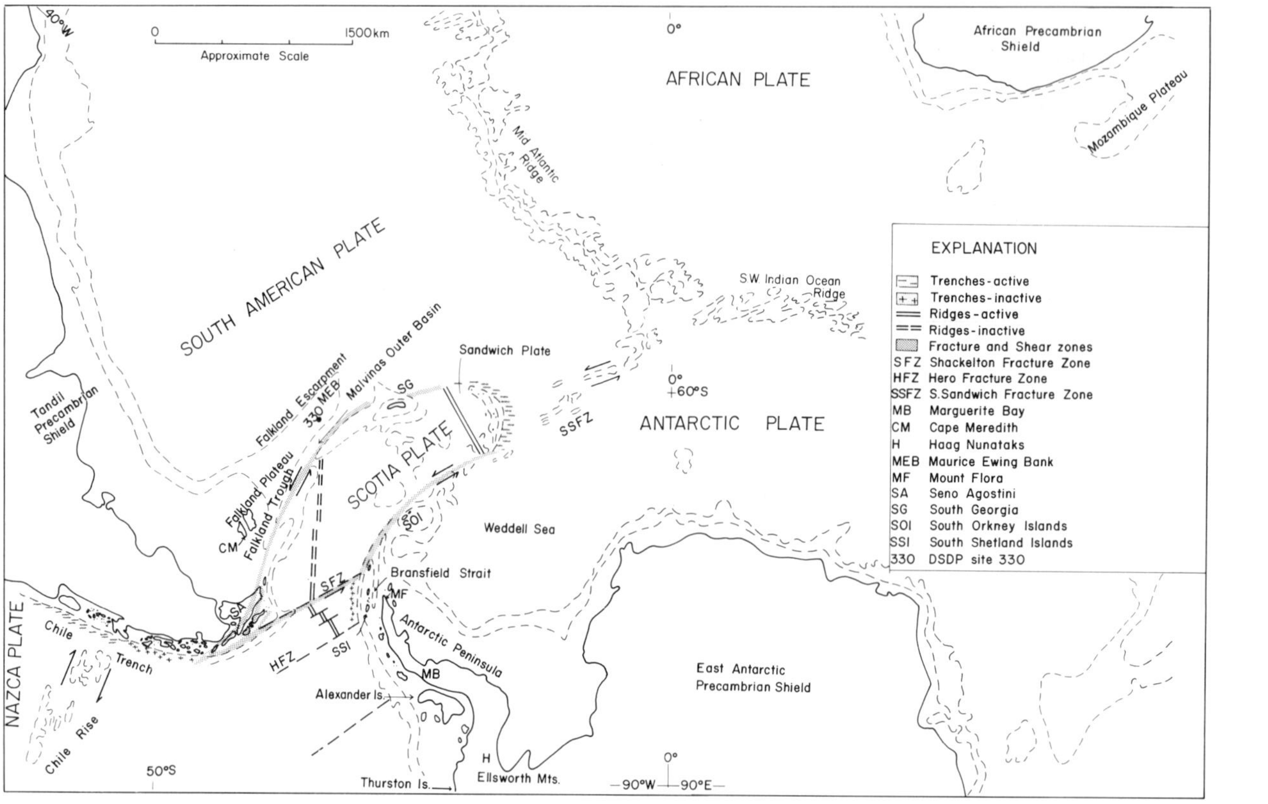

Figure 12.1. Present tectonic setting of the Scotia arc region.

comparable to the Trinity Peninsula Formation of the Antarctic Peninsula underwent the same deformation as rocks of the so-called metamorphic complex (Dalziel and others, 1977). Thus, the only evidence for two, separate, pre-Middle Jurassic "basement complexes" now lies in limited radiometric data from the South Orkney Islands metamorphic complex and from a few rocks in the Antarctic Peninsula that suggest the presence of a pre-Gondwanide complex (Gledhill and others, Chapter 35, this volume). This will be discussed below.

The writer has been studying the pre-Middle Jurassic rocks of the region since 1968. With students and associates, I have mapped the geologic structure of all the occurrences of basement rock around the coasts of the South Orkney and South Shetland Islands, seen much of the basement in southernmost South America, visited critical basement localities on the Antarctic Peninsula, and participated in drilling into the pre-Jurassic metamorphic and igneous basement of the eastern Falkland (Malvinas) Plateau. These rocks are crucial to the understanding of the original configuration of Gondwanaland, the early history of the Pacific Ocean margin, and the evolution of the Scotia arc. Hence, it seems timely to present a review of the complex, an outline of our progress to date in unraveling its history, and my ideas on its significance at the time of the symposium.

Geotectonic setting

As in crystalline "basement" complexes in other parts of the world, the pre-Middle Jurassic metamorphic rocks of the Scotia arc region have been extensively referred to as "Precambrian" or "Precambrian(?)" in the literature. In fact, the only evidence of a Precambrian age for any of the rocks south of the Falkland (Malvinas) Plateau and outside East Antarctica and the Ellsworth Mountains region comes from the Elephant Island group of the South Shetlands (Figure 12.1). An assemblage of acritarchs comparable to one found in upper Precambrian strata of the Soviet Union has been reported from phyllite on Clarence Island (Iltchenko, 1972). Despite reservations mainly based on the lack of any other evidence of pre-Carboniferous rocks in the region, I am not personally qualified to dispute this report. In fact, since I believe the Clarence Island metamorphic rocks represent part of a pre-late Mesozoic fore-arc complex (see below), any age older than Middle Jurassic seems possible. The Clarence Island phyllites are locally among the least metamorphosed parts of the Scotia arc basement rocks, so the preservation of microfossils is not surprising. However, if the Precambrian age is correct, it need not imply that the rocks are, or ever were, part of the crystalline shield of Gondwanaland. In my interpretation, they would be "exotic," having been rafted into the region on a downgoing lithospheric plate.

In considering the geotectonic setting of the basement rocks in the Scotia arc, it is important to identify the undoubtedly Precambrian crystalline rocks in the South Atlantic–Weddell Sea area as a whole. These are located in the Tandil Shield of northeastern Argentina (>2000 m.y. old), the Cape Meredith complex of the southwestern Falkland (Malvinas) Islands (approximately 1000 m.y. old), beneath the Maurice Ewing Bank at the eastern end of the Falkland Plateau (>535 ± 65 m.y. old), in southernmost Africa (>1000 m.y. old), in East Antarctica (reportedly as old as 4000 m.y.), and in the Haag Nunataks near Ellsworth Mountains (>1000 m.y. old) (Sobotovich and

Table 12.1 Basement and probable basement rocks of the Scotia arc region

Rocks and location	Stratigraphic evidence	Paleontologic evidence	Radiometric evidence
South America			
Metamorphic rocks along present Pacific margin (Cecioni, 1955; Dalziel and Elliot, 1973; Dalziel and others, 1975; Douglass and Nestell, 1976; Forsythe and Mpodozis, 1979)	Deformed and metamorphosed prior to emplacement of upper Mesozoic Patagonian batholith	Include fusulinid limestone containing Upper Pennsylvanian to Lower Permian fauna (Cecioni, 1955; Douglass and Nestell, 1976)	
Metamorphic rocks a Andean Cordillera (Kranck, 1932; Katz, 1973; Dalziel and Cortés, 1972)	Unconformably overlain by Upper Jurassic Tobifera Formation		
Metamorphic rocks east of Andean Cordillera (Dalziel and Elliot, 1973 Halpern, 1973; Natland and others, 1974)	Unconformably overlain by Upper Jurassic Tobifera Formation		Halpern (1973): *306 ± 156 m.y.* (based on Rb-Sr whole rock isochron from schist); *254 ± 10* and *251 ± 12 m.y.* (based on Rb-Sr and K-Ar data, respectively, on biotite from gneiss) Halpern and others (1972): *315 ± 25 m.y.* and *250 m.y.* (based on Rb-Sr whole-rock isochron and Rb-Sr mineral data, respectively)
South Georgia			
Metasedimentary rocks in southeast part of the island (Storey and others, 1977)[a]			
South Orkney Islands			
Metamorphic rocks of Coronation, Powell, Signy and Inaccessible Islands (Matthews and Maling, 1967; Thomson, 1968, 1971, 1973 and 1974; Dalziel, 1971; Dalziel and others, 1977)	Coronation and Powell Islands: unconformably overlain by Upper Jurassic–Lower Cretaceous conglomerate, sandstone, and shale (Thomson, 1975*a*; Thomson and Willey, 3975). Signy and Inaccessible Islands[b]		Miller (1960): *180–200 m.y.* (based on K-Ar biotite data from Signy Island schist) Grikurov and others (1967): cite a *200 m.y.* age (based on whole-rock and plagioclase K-Ar data) Rex and Baker (1973): *300 ± 50 m.y.* (based on Rb-Sr whole rock isochron from Signy Island schist) Rex (1976): *286 ± 75 m.y.* (based on eight-point whole rock isochron from Signy Island schist): *177–189 m.y.* (based on K-Ar data from schist)
Greywacke-Shale Formation of Laurie Island and immediately adjacent islets (Pirie, 1905 & 1906; Dalziel, 1971)	Structural style and metamorphic state similar to pre-Upper Jurassic–Lower Cretaceous graywacke and shale of Powell Island[b]	"Graptolites" reported by Scottish National Antarctic Expedition (Pirie, 1905, 1906) and by Cordini (1955) widely believed to be plant fragments	
Graywacke and shale of Fredriksen Island (Thomson, 1973)	Structural style and metamorphic state similar to pre-Upper Jurassic–Lower Cretaceous graywacke and shale of Powell Island[b]		
Graywacke and shale of Powell Island (Dalziel, 1971; Thomson, 1973)	Unconformably overlain by Upper Jurassic–Lower Cretaceous (?) conglomerate, sandstone, and shale (Thomson, 1973).[c] Structurally overlies metamorphic complex along sheared contact; deformed during main phase of deformation of metamorphic rocks (Dalziel and others, 1977)		

(continued on following pages)

Table 12.1. (continued)

Rocks and location	Stratigraphic evidence	Paleontologic evidence	Radiometric evidence
South Shetland Islands			
Metamorphic rocks of southern Elephant Island (Araya and Hervé, 1966; Dalziel, 1972*b*; Dalziel, 1976)	*b*		Dalziel (1972*b*): *78–88 m.y.* (based on K-Ar data from schist collected in southwest) Rex and Baker (1973): *100 m.y.* (based on K-Ar data from mica)
Metamorphic rocks of Clarence and northern Elephant Islands (Wordie, 1921; Tyrrell, 1945; Grikurov and others 1970; Dalziel, 1976)	*b*	Acritarchs reported from Clarence Island phyllite correlated with uppermost Riphean flora from Soviet Union (Iltchenko, 1972)	Grikurov and others (1970): *30–99 m.y.* (based on K-Ar data from Clarence Island phyllite) Dalziel and others (unpublished data): *104–105 m.y.* (based K-Ar data on muscovite concentrate from glaucophane schist)
Metamorphic rocks and ultramafic body on Gibbs Island (Tyrrell, 1945; de Wit and others, 1977)	Structural style and metamorphic state comparable to metamorphic rocks of southern Elephant Island[b]		
Metamorphic rocks of Aspland, Eadie, and O'Brien Islands (Dalziel, 1976)	Structural style and metamorphic state comparable to metamorphic rocks of southern Elephant and Gibbs Islands[b]		
Metamorphic rocks of Smith Island (Smellie and Clarkson, 1975; Rivano and Cortés, 1975; Dalziel, 1976)	Structural style and metamorphic state comparable to metamorphic rocks of northern Elephant Island[b]		Dalziel and others (unpublished data): *58 m.y.* (based on K-Ar data from glaucophane concentrate)
Miers Bluff Formation (graywacke and shale) of Livingston Island (Hobbs, 1968; Dalziel, 1972*a*)[a]	*b*	Plant fragments of post-Carboniferous and possibly Mesozoic age (Schopf, 1973)	Dalziel (1972*a*): *197 ± 23 m.y.* (based on Rb-Sr five-point whole-rock isochron from shale)
Antarctic Peninsula (including Alexander Island)			
Trinity Peninsula Formation[d] (Adie, 1957; Dalziel and Elliot, 1973)	Unconformably overlain by Jurassic volcanic and sedimentary strata (Adie, 1957; Bibby, 1966)[c] Apparent equivalents in southern Graham Land and northern Palmer Land more highly metamorphosed (Aitkenhead, 1965; Elliot, 1966; Skinner, 1973)	Plant fragments believed to be Carboniferous (Croft, in Adie, 1957, p. 22) Spores comparable to Lower to Middle Carboniferous flora of Soviet Union (Grikurov and Dibner, 1968)	Grikurov and Dibner (1968) and Grikurov (1972): *105–110 m.y.* and *165 m.y.* (based on K-Ar data) Dalziel (1972*a*): *242 ± 50 m.y.* (based on two-point Rb-Sr whole-rock isochron from shale)
Legoupil Formation ≡ (Trinity Peninsula Formation?) (Halpern, 1964 and 1965; Thomson, 1975*b*)	Comparable in rock types and structural style to Trinity Peninsula Formation (Elliott, 1965 and 1966; Halpern, 1965) and Miers Bluff Formation (Dalziel, 1972*a*); unconformably (?) overlain by andesite dated radiometrically as Coniacian (Halpern, 1964)[b]	Permo-Triassic (probably Lower–Middle Triassic) Bivalvia (Thomson, 1975*b*)	Halpern (1964): *Between 116 m.y. and 74 m.y.* (based on K-Ar data on hornblende from a diorite pebble from the formation and K-Ar data on pyroxene from a crosscutting dike), but more recent paleontologic evidence indicates older than *116 m.y.*
LeMay Formation of Alexander Island ≡ (Trinity Peninsula Formation?) (Grikurov and Dibner, 1968; Grikurov, 1971; Edwards, Chapter 36, this volume)	Broadly comparable in rock types and structural style to Trinity Peninsula Formation, but contains ophiolitic fragments and is locally metamorphosed to to glaucophane schist facies (Edwards, Chapter 36, this volume)	Palynomorphs with Lower to Middle Carboniferous affinities (Grikurov and Dibner, 1968). Benthonic invertebrates of probable Middle–Late Triassic age (Edwards, 1977 and Chapter 36, this volume). Fauna has circumPacific affinities unconformably overlain by Upper Jurassic–Lower Cretaceus Fossil Bluff Formation (Edwards, 1977 and Chapter 36, this volume)	Grikurov and others (1967): *165 m.y.* (based on K-Ar analysis of a biotite concentrate)

Table 12.1. (continued)

Rocks and location	Stratigraphic evidence	Paleontologic evidence	Radiometric evidence
Metamamorphic complex at Marguerite Bay (Adie, 1954; Hoskins, 1963; Fraser, 1965)	Unconformably overlain by Jurassic volcanic and sedimentary strata		Grikurov and others (1966): *90–133 m.y.* (based on seven whole-rock K-Ar analyses of schist and gneiss) Halpern (1971): *108 ± 5 m.y.* (based on Rb-Sr data from biotite in a diorite gneiss) Halpern (1972): *200 m.y.* (based on a two-point whole-rock Rb-Sr isochron) Rex (1976): *240 m.y.* and *160 m.y.* (based on K-Ar data from coexisting hornblende and biotite, respectively) Gledhill and others (Chapter 35, this volume) *175 ± 7 m.y.* (based on six-point whole-rock Rb-Sr isochron for granite-gneiss); *111 m.y.* (based on K-Ar analysis of biotite separate from leucocratic gneiss)
Lizard Hill (Tabarin Peninsula) intrusive rocks			*Rex (1976): 384 ± 15* and *351 ± 14 m.y.* (based on biotite) and *374 ± 15 m.y.* (based on muscovite; all K-Ar analyses)
Target Hill (Jason Peninsula) gneiss			Rex (1976): *237 and 243 m.y.* (based on K-Ar data from hornblendes)

[a] The Sandebugten sedimentary sequence of South Georgia and the False Bay schists of Livingston Island are not considered by the author to be part of the basement complex (see text).

[b] Despite lack of contact with Jurassic or Cretaceous sedimentary and volcanic strata and therefore a definite unconformable relationship, these rocks are much more highly deformed and metamorphosed than the upper Mesozoic sequences of the South Scotia Ridge and Antarctic Peninsula.

[c] Actual contact unexposed or faulted but unconformable relationship clear from clasts in basal breccia or conglomerate and overall field relationships.

[d] Classically the Trinity Peninsula "Series," but in fact a mappable lithic unit with no time-stratigraphic significance (see Edwards, Chapter 36; Gledhill and others, Chapter 35; and Thomson, Chapter 37, all this volume).

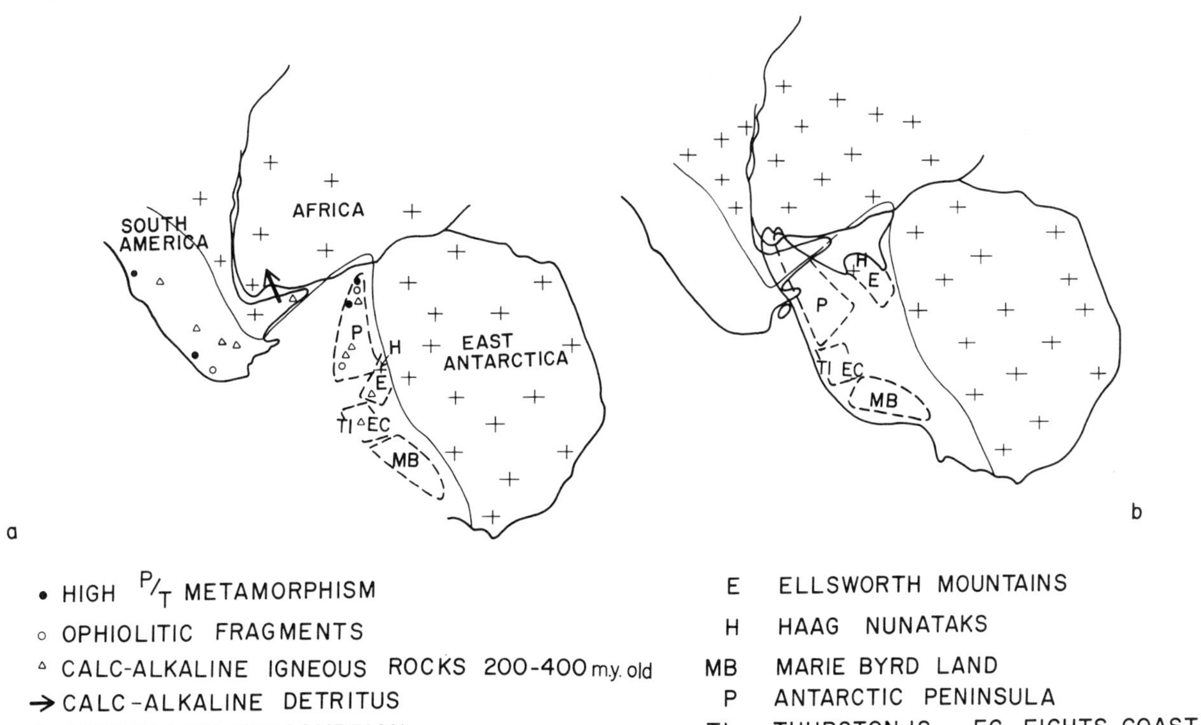

Figure 12.2. Reconstructions of the Scotia arc region with respect to Gondwanaland prior to Middle Jurassic time. *a.* The author's suggested model, based on the reconstruction of Smith and Hallam (1970). *b.* Reconstruction of Smith and Hallam (1970).

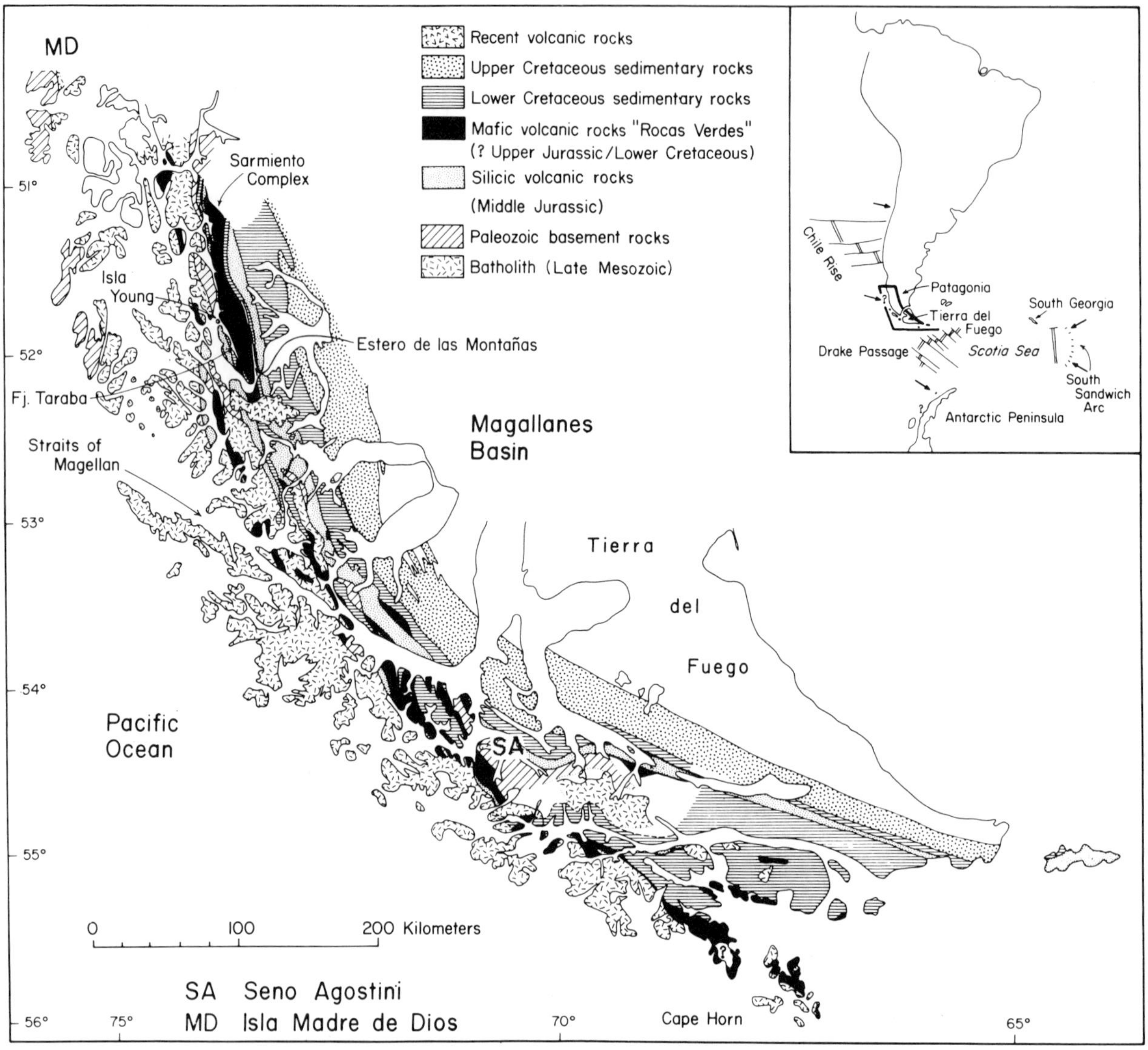

Figure 12.3. Geologic map of southern South America with location (inset).

others, 1974; Cingolani and Varela, 1976; Barker, Dalziel, and others, 1976; Clarkson and Brook, 1977) (see Figure 12.1). The geology of the Ellsworth Mountains region indicates that the rocks there are the exposed part of a small lithospheric plate that probably rotated counterclockwise away from a position along the Transantarctic margin of the East Antarctic craton during late Mesozoic or Cenozoic time (Schopf, 1969; Barker, Dalziel, and others, 1976; Clarkson and Brook, 1977; de Wit, 1977). This means that the Pacific limit of known Precambrian shield rocks in Gondwanaland defines a pronounced reentrant into the supercontinent on the reconstruction of Smith and Hallam (1970) (see Figure 12.2). This reconstruction is now widely accepted as essentially correct apart from their (acknowledged) neglect of late Mesozoic and Cenozoic oroclinal bending and microplate movement, in West Antarctica particularly (Barker, Dalziel, and others, 1976; de Wit, 1977). Thus the basement of the Scotia arc region lies entirely oceanward, with respect to Gondwanaland, of Precambrian crystalline massifs, regardless of the precise geometry of the reconstructed Scotia arc region (Figure 12.2). Apart from the acritarchs mentioned above, fossils in the Scotia arc basement rocks range in age only back to Carboniferous, while radiometric data indicate no ages older than middle to late Paleozoic and few older than Triassic (Table 12.1).

Stratigraphy

The field, paleontologic, and radiometric evidence for dating the rocks assigned to the basement complex of the Scotia arc region is outlined, with critical references, in Table 12.1.

South America

The pre–Middle Jurassic basement rocks in South America to the south of the Chile Trench–Chile Rise triple junction (approximately lat. 46° S; here taken as the arbitrary northern limit of the Scotia arc region) are known in several areas (Figure 12.3).

Along the present Pacific margin there are major outcrops of a metamorphic complex that includes massive Upper Pennsylvanian–Lower Permian fusulinid limestone beds (Cecioni, 1955; Douglass and Nestell, 1976). Field relationships of the various lithic units have only recently been established. In addition to the limestone these units

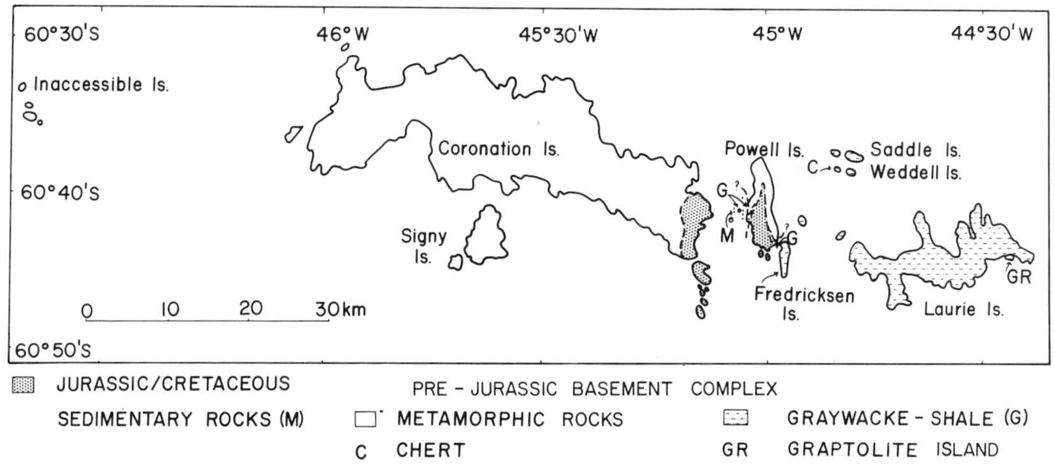

Figure 12.4. Geologic map of South Orkney Islands.

include: mafic pillow lava; red, green, and white chert (locally overlying pillow lava); a graywacke-shale turbidite sequence or sequences; thick-bedded mature sandstone; and pebbly mudstone (Cecioni, 1955; Dalziel and Elliot, 1973; Dalziel and others, 1975). Slices of ocean floor rocks (the pelagic sedimentary sequences, mafic pillow lavas, and massive fusulinid limestones) form the blocks of a "mega-mélange" within a matrix of graywacke and shale (Forsythe and Mpodozis, 1979).

In the southernmost part of the Andean Cordillera there are numerous exposures of pre-Middle Jurassic basement rocks (Kranck, 1932; Dalziel and Cortés, 1972). Some of these rocks, however, were tectonically reactivated during the mid-Cretaceous climax of the Andean deformation in the cordillera and are difficult to distinguish from a deformed Upper Jurassic and Lower Cretaceous sedimentary and igneous rocks (Bruhn and Dalziel, 1977). Definitely pre-Middle Jurassic sequences consist mostly of phyllite or recrystallized white or red chert and interlayered argillite. Volcanic rocks are present, however, in the basement of Cordillera Darwin in Tierra del Fuego, possibly indicating proximity to a pre-Middle Jurassic continental margin magmatic arc (Nelson, Dalziel, and Milnes, 1979; Nelson, Dalziel, and Ridley, 1979).

Boreholes in the pre-Middle Jurassic basement at the eastern end of the Straits of Magellan have penetrated a foliated metamorphic terrane which has yielded Paleozoic radiometric ages (Halpern, 1973). Exposures of basement rock farther north in eastern Argentina consist of calc-alkaline igneous rocks of Paleozoic to earliest Mesozoic age (Halpern and others, 1972).

South Georgia

Correlation of the quartzose graywacke and shale of the Sandebugten sedimentary sequence on South Georgia with the pre-Jurassic Trinity Peninsula Formation of the Antarctic Peninsula has been tentatively made on lithologic grounds (Adie, 1964). While no fossil or radiometric dating of the Sandebugten rocks is yet available, structural and sedimentologic evidence indicate that they are late Mesozoic in age (Dalziel and others, 1975; Thomson and others, Chapter 19, this volume). They will not, therefore, be further considered here. Recently, however, metasedimentary rocks considered correlative with the pre-Middle Jurassic basement in Tierra del Fuego have been reported from southeastern South Georgia (Storey and others, 1977).

South Orkney Islands

Metamorphic rocks exposed in the Inaccessible, Coronation, Signy, and Powell Islands are unconformably overlain by Upper Jurassic to Lower Cretaceous(?) sedimentary rocks on Coronation and Powell Islands (Figure 12.4. Paleozoic to early Mesozoic radiometric ages from the schists on Signy Island probably reflect the age of deformation and metamorphism or uplift and cooling of the complex (for details and references, see Table 12.1). The rocks consist mainly of quartzo-feldspathic, argillaceous, and mafic phyllite and schist, with marble, calc-silicate beds, and chert (Matthews and Maling, 1967; Thomson, 1968, 1973, and 1974). The deformed sedimentary sequences of Laurie, Fredriksen, and Powell Islands and those of the smaller Saddle and Weddell Islands consist of graywacke rich in quartz, feldspar, and silicic igneous detritus, and of shale (Pirie, 1905, 1906; Dalziel, 1971; Thomson, 1973). Thick-bedded chert occurs on a small islet of the Weddell group for which the name "Scapa Rock" has been proposed (unpublished observation made by scientific party in course of R/V *Hero* Cruise 77-1, see Dalziel and others, 1977). Recently, a Late Triassic radiolarian fauna has been recovered from the chert (D. L. Jones, personal communication). Deformation of the graywacke and shale on Powell Island along with the underlying metamorphic rocks (Dalziel and others, 1977) and clasts of graywacke within the Jurassic and/or Cretaceous Powell Island conglomerate (Thomson, 1973) provide the only other reliable clues to the age of the graywacke and shale sequences or sequences (Table 12.1). A lithic, metamorphic, and structural transition between the metamorphic complex and the graywacke and shale on Powell Island seems to rule out the possibility that the metamorphic rocks once represented the crystalline basement on which the graywacke and shale were deposited, as implied in the interpretation of Rb-Sr data by Gledhill and others (Chapter 35, this volume).

South Shetland Islands

The basement rocks of the South Shetland Islands (Figure 12.5) have not been seen in contact with any other tectono-stratigraph unit of known age. Sailing directions

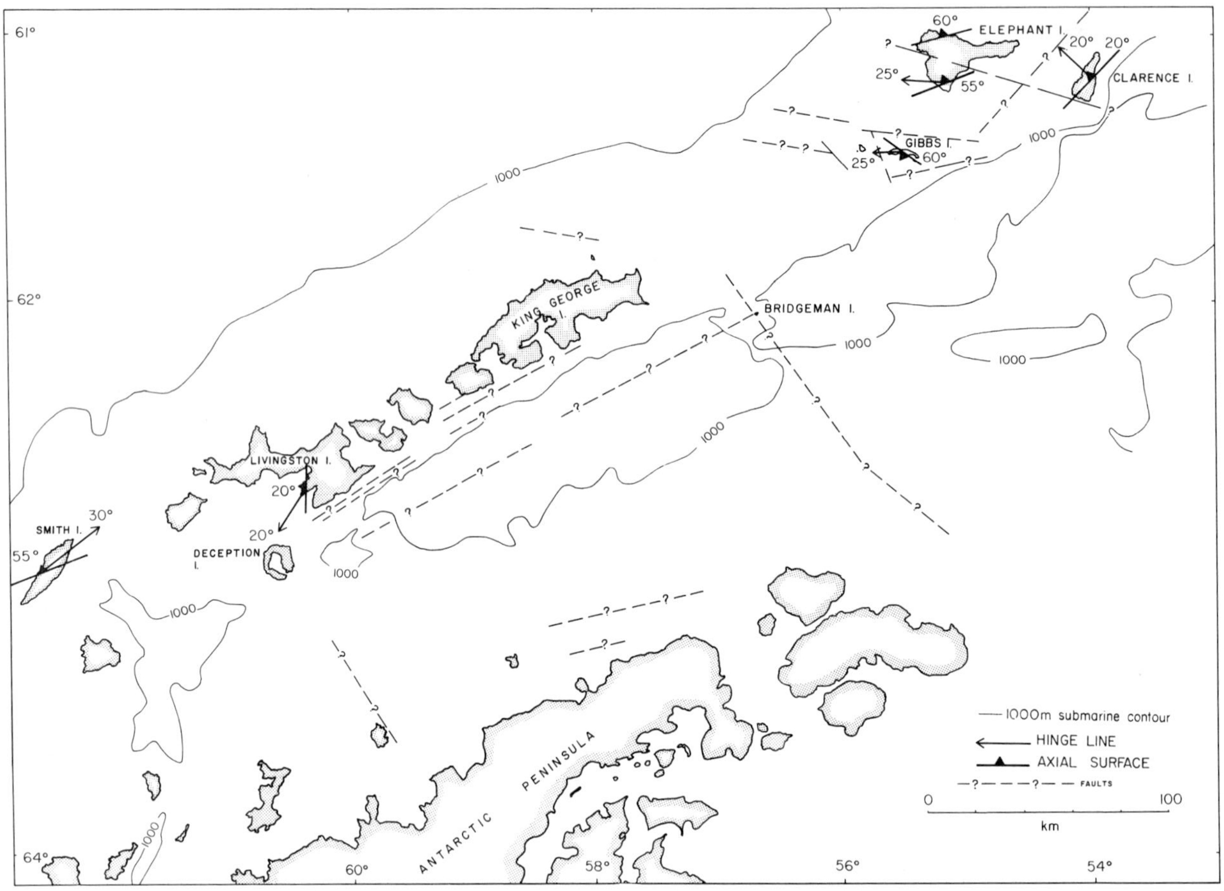

Figure 12.5. a. South Shetland Islands, showing faults and dominant structural trends in probably pre-Jurassic rocks.

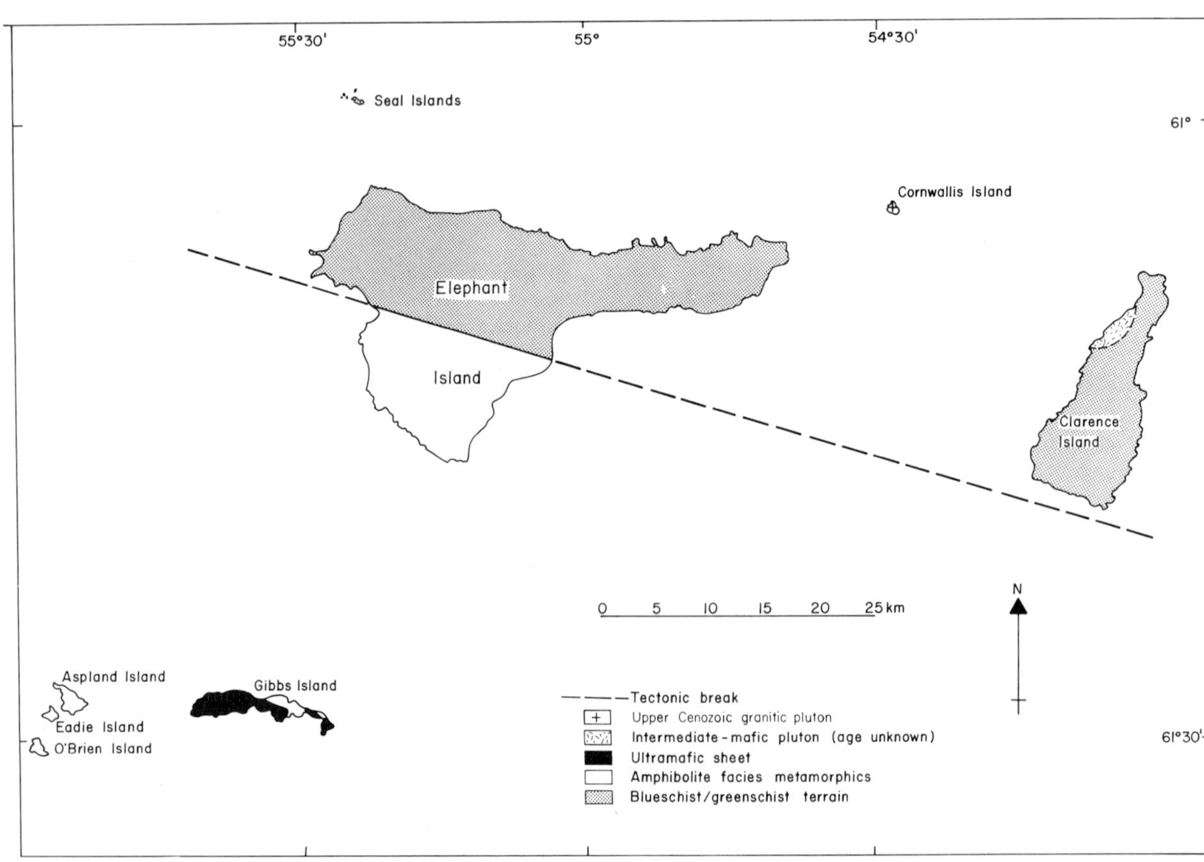

Figure 12.5. b. Elephant Island group, South Shetland Islands, showing main rock types.

for Antarctica mention "good coal" on Smith Island (presumably from 19th century sealers' observations). This has yet to be confirmed.

Elephant Island group

The rocks here consist of deformed gray and green phyllite (including metatuff); gray, green, and blue schist; amphibolite; chert; calc-silicate rocks; and a dunite-serpentinite complex. As mentioned above, Iltchenko (1972) has reported Precambrian acritarchs from phyllite on Clarence Island. Radiometric ages range only as old as 105 m.y. (Table 12.1), yet the structural style and metamorphic grade suggest correlation with the Paleozoic-lower Mesozoic complex of the South Orkneys. Hence the K-Ar dates are generally ascribed to reheating effects during late Mesozoic or Cenozoic time (Dalziel, 1972*b*; Rex, 1976). A granite that yielded a 9 m.y. K-Ar age forms Cornwallis Island just east of Elephant Island (Rex and Baker, 1973).

Smith Island

Brief landings at Cape Smith by British and Chilean parties during the 1974–75 season (Smellie and Clarkson, 1975; Rivano and Cortés, 1975) and two days of landings around the island by four of us from Columbia University during the subsequent season (Dalziel, 1976) revealed that the island is almost entirely composed of highly deformed blue and green schist with interlayers of chert and calc-silicate rock as much as several tens of meters thick. Although limited K-Ar radiometric data indicate an age of only 58 m.y., the lithic and structural similarity to the northern part of Elephant Island makes these rocks seem likely to be part of the pre-Middle Jurassic basement complex.

Livingston Island

A narrow strip of medium- to high-grade schist within the Barnard Point intrusive rocks on Livingston Island (False Bay schist of Hobbs, 1968) appears to be some sort of autolith rather than a basement xenolith. An Rb-Sr age of 45 m.y. was obtained from the pluton (Dalziel and others, 1973).

The graywacke-shale sequence on Hurd Peninsula, known as the Miers Bluff Formation, has yielded late Paleozoic or Mesozoic plant fragments and a Late Triassic Rb-Sr isochron (Table 12.1) (Hobbs, 1968; Dalziel, 1972*a*; Schopf, 1973). The latter is interpreted as the age of diagenesis or deformation and metamorphism (Dalziel, 1972*a*). From field appearance and published descriptions, the rocks are so far petrographically indistinguishable from the Greywacke-Shale Formation of the South Orkneys and from the Trinity Peninsula and Legoupil Formations of the Antarctic Peninsula (see Edwards, Chapter 36, this volume; Thomson, Chapter 37, this volume).

Antarctic Peninsula

The principal basement localities of the Antarctic Peninsula are shown in Figure 12.6.

Legoupil Formation

The Legoupil Formation on the west coast of Graham Land is, as mentioned above, lithologically identical to the Miers Bluff Formation of Livingston Island (Halpern, 1965; Dalziel, 1972*a*). Previous suggestions of an Early Cretaceous age for the strata at Cape Legoupil have now been superseded by the recovery of Permo-Triassic Bivalvia (Thomson, 1975*b*). As Thomson notes, it is probably significant that the same outcrops have been mapped as both Trinity Peninsula Formation (Elliot, 1965) and Legoupil Formation (Halpern, 1965).

Trinity Peninsula Formation

This unit of dominant graywacke and shale, with chert, limestone, various other sedimentary rock types, and possibly mafic metavolcanic rocks, is now generally regarded as late Paleozoic-early Mesozoic(?) in age. (See Dalziel and Elliot, 1973, and Thomson, 1975*b* and Chapter 37, this volume, for reviews.) As in the case of the Greywacke-Shale Formation of the South Orkneys, the detritus in the graywacke beds of the Trinity Peninsula Formation seems to have been derived from a dominantly silicic igneous terrane (Aitkenhead, 1965; Elliot, 1965, 1966; Thomson, Chapter 37, this volume).

Mapping by the British Antarctic Survey seems to have established a transition into higher temperature metamorphic equivalents of the Trinity Peninsula Formation farther south along the Antarctic Peninsula (Aitkenhead, 1965; Elliot, 1966; Stubbs, 1968; Skinner, 1973). This includes the LeMay Formation of central Alexander Island (Edwards, Chapter 36, this volume).

Marguerite Bay

Apart from a banded biotite gneiss interpreted by Hoskins (1963) to be of sedimentary origin, the dioritic and granitic gneiss and the amphibolite and hornblende schist exposed at Marguerite Bay are all believed to represent originally igneous rocks of pre-Jurassic age (Table 12.1) (see Gledhill and others, Chapter 35, this volume).

Tabarin Peninsula

Limited radiometric (K-Ar) dating suggests that granitic plutons of the Tabarin Peninsula (on the Weddell Sea side of the tip of the Antarctic Peninsula) are Carboniferous or older (Rex, 1976).

Jason Peninsula

Limited K-Ar data indicate a maximum age of 237–243 m.y. for high-grade gneiss at this locality where the Trinity Peninsula Formation is of distinctly lower grade (Rex, 1976; Gledhill and others, Chapter 35, this volume).

Deformation and metamorphism

South America

The problem in South America is to distinguish between pre-Middle Jurassic metamorphism and deformation and the effects of Andean reactivation in the same basement rocks. Where the unconformity is exposed, the difference can be clearly seen. Elsewhere, the distinction is based on lithology, a somewhat dubious criterion in view of the similarity of "basement" and "cover." Along the Pacific margin the effects of the Andean events were less pronounced than in the High Cordillera inland (Dalziel and others, 1974; Suárez, 1976; Bruhn and Dalziel, 1977; Nelson, Dalziel, and Milnes, 1979; Nelson, Dalziel, and Rid-

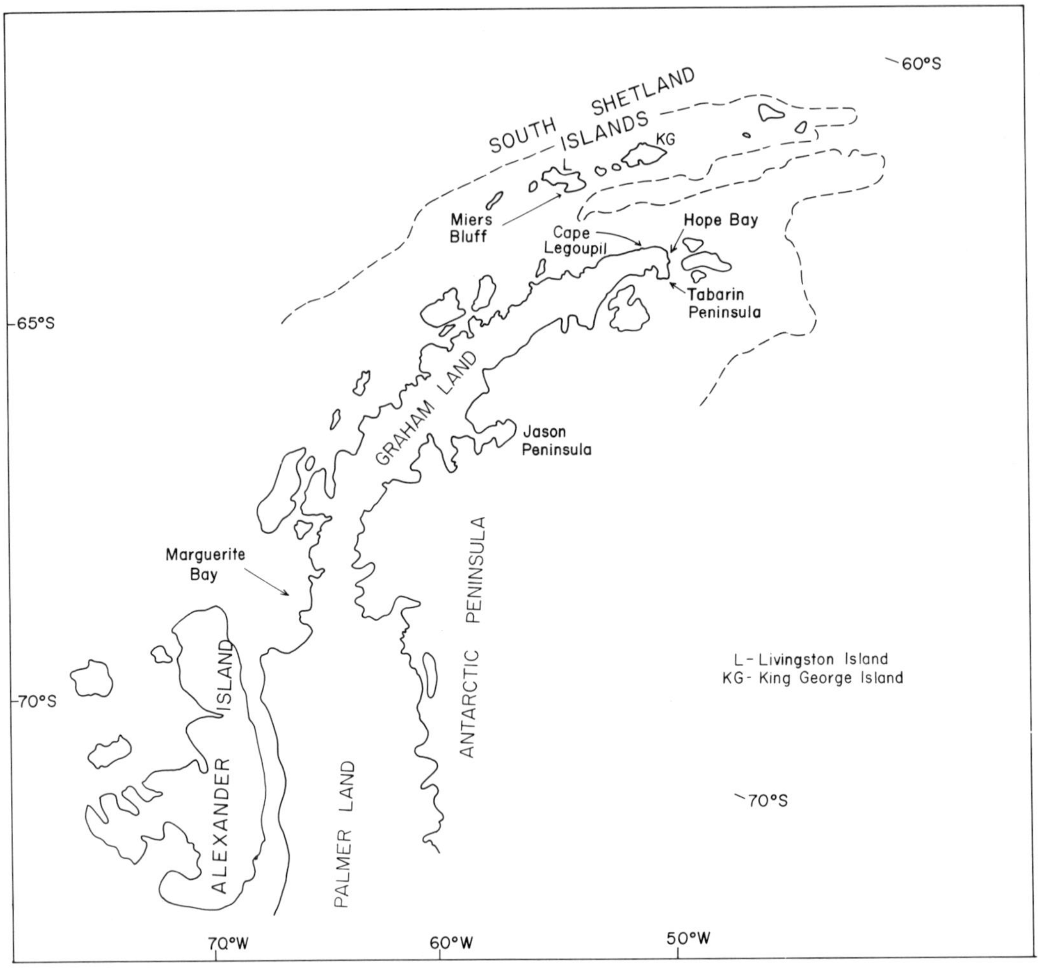

Figure 12.6. Antarctic Peninsula, showing localities mentioned in text.

ley, 1979). In the basement near the west end of the Straits of Magellan, there are major asymmetric folds with southeast-trending hinge lines and axial surfaces oblique to the more northerly late Mesozoic-Cenozoic Andean trend (e.g., on Isla Desolacion; observation made in the course of R/V *Hero* Cruise 75-6; see Dalziel and others, 1975). Otherwise it can only be said that pre-Middle Jurassic thrusts and tight to subisoclinal and isoclinal folds dominate the structure (Forsythe and Mpodozis, 1979).

To an even greater extent, it is presently difficult to distinguish between the metamorphic effects of Andean and pre-Andean events. It is not, for example, possible to delineate a spatial pattern of pre-Andean metamorphic facies. However, we do know that we are dealing with comparatively low-grade metamorphism, probably no higher than greenschist facies in the basement of the cordillera. Pre-Jurassic gneissic rocks do occur beneath the Magallanes basin on the Atlantic side of the cordillera (Halpern, 1973).

South Orkney Islands

The Greywacke-Shale Formation of the South Orkney Islands has suffered zeolite or prehnite-pumpellyite facies metamorphism at most. More laboratory work needs to be undertaken in order to establish the precise grade, but apart from a northwest-trending shear zone through Laurie Island, where greenschists occur, the metamorphism was clearly below greenschist facies conditions. The structure is complex and will be described in detail elsewhere. Major asymmetric folds verge northward in central Laurie Island and southwestward in the western part of the island. The two vergences are separated by a tectonic shear zone. Two entirely different structural regimes are present in the graywacke and shale of eastern Laurie and Powell Islands. Eastern Laurie is dominated by tight, upright, north-trending folds, while the thin graywacke and shale beds on southern Powell are dominated by a gently to moderately south-dipping foliation parallel to that in the structurally underlying metamorphic complex (Dalziel and others, 1977).

The exact relationships between these various structural domains is not yet clear. It should be emphasized, however, that there is no evidence that one of these sets of structures overprints another. They merely occur in different domains and are the earliest and dominant tectonic structure within their own domain.

The metamorphic complex on Coronation and Signy Islands reached albite-epidote amphibolite facies, and locally amphibolite facies, prior to the deposition of the unconformably overlying Jurassic to Cretaceous sedimentary sequence (Thomson, 1968, 1974). The grade is slightly lower (greenschist facies) in eastern Coronation and Powell Islands, and it clearly decreases upward on Powell Island into the structurally overlying graywacke and shale beds in the southern half of the island.

The structure of the metamorphic complex is dominated by "main phase," tight, asymmetric folds with south-southeast-trending subhorizontal hinge lines and variably dipping axial surfaces. Pre-main phase isoclinal folds are locally present, and the main phase structures are deformed by later, almost coaxial, and more open, asymmetric folds. As the upper Mesozoic sedimentary strata are virtually undeformed, all these structures (in addition to various late-stage open folds and kink bands) are pre-Andean.

South Shetland Islands

Elephant Island group

Blueschist was dredged from the sea floor south of Clarence Island by the *Discovery* expedition in 1927 (Tyrrell, 1945). It was not until the landings on Smith Island in 1974–75, however, that the real significance of the dredge haul was appreciated (Smellie and Clarkson, 1975; Rivano and Cortés, 1975). Subsequently, blueschist has been found on northwestern Elephant Island (Dalziel, 1976). Moreover, we have found a distinct break in metamorphic facies that trends south of east through the main drainage glaciers of Elephant Island and to the south of Clarence Island (Figure 12.5*b*). This line, presumably a major fault such as is common in other relatively high-pressure metamorphic belts around the Pacific (Miyashiro, 1973), separates the blueschist-greenschist terrane of northern Elephant and Clarence Islands from the higher temperature epidote-amphibolite and amphibolite facies terrane of southern Elephant, Gibbs, Aspland, Eadie, and O'Brien Islands (Figure 12.5*b*). The dunite-serpentinite complex of Gibbs Island occurs in the latter terrane (de Wit and others, 1977).

We seem to be able to distinguish the same set of "main phase" structures north and south of the tectonic break. These structures consist mainly of tight, asymmetric folds with a strong mineral alignment lineation that is usually parallel to the hinge lines, which have a generally east-west trend although it varies from locality to locality (Figure 12.5*b*. Earlier (isoclinal) folds are distinguishable in southern Elephant Island, and various post-main phase folds are important on Elephant, Aspland, Eadie, and O'Brien Islands. Tectonic emplacement of the Gibbs Island ultramafic sheet accompanied the main phase of deformation (de Wit and others, 1977).

Smith Island

The blueschist mapped around the coast of Smith Island seems to identify the metamorphic terrane there with that of northern Elephant Island and Clarence Island (Dalziel, 1976). Despite its remoteness from the Elephant Island group (it is approximately 300 km to the southwest; see Figure 12.5*a*), the structure is dominated by what seems to be the same main phase deformation with tight, asymmetric folds accompanied by a mineral alignment lineation, predated by isoclinal folds and postdated by various conjugate folds and kink bands. The dominant trend of the axial surfaces of the main phase folds is 70° and the gently plunging hinge lines trend approximately 55°.

Livingston Island

The Miers Bluff Formation lies on the inverted limb of a near-recumbent, asymmetric anticline with its hinge line essentially parallel to the trend of the South Shetlands ridge; vergence is toward the Antarctic Peninsula (Dalziel, 1972*a*). As in the case of the Greywacke-Shale Formation of the South Orkneys, the metamorphic grade is below greenschist facies.

Antarctic Peninsula

Trinity Peninsula Formation

From my very limited observations (at Hope Bay only) and from a study of the British Antarctic Survey Reports, it seems clear that the structure of the Trinity Peninsula Formation is dominated by one set of large asymmetric folds with hinge lines subparallel to the peninsula and verging toward the Weddell Sea (Dalziel, 1972*a*) (see Figure 12.6 for locations). While the probably equivalent Legoupil Formation seems to have a compatible structural geometry (Halpern, 1965), low-grade metamorphic rocks near the Chilean Baso O'Higgins, of uncertain relationship to the Legoupil Formation, have a more complex polyphase structural history (Miller, 1966 and Chapter 13, this volume).

Like the graywacke-shale sequences of the South Orkney and South Shetland Islands, the metamorphic grade of the Trinity Peninsula Formation is sub-greenschist facies in the northern Antarctic Peninsula. Higher grade assemblages appear southward where the structure is less well known (Aitkenhead, 1965; Elliot, 1965 and 1966; Stubbs, 1968; Skinner, 1973; Edwards, Chapter 36, this volume).

Marguerite Bay

The orthogneiss of the basement at Marguerite Bay has undergone albite-epidote amphibolite and amphibolite facies metamorphism (Hoskins, 1963).

Alexander Island

Edwards (Chapter 36, this volume) reports blueschist facies metamorphism of the LeMay Formation on Alexander Island (Figure 12.6).

Discussion

Geotectonic interpretation of the Scotia arc basement rocks

It has been established that the entire Scotia arc region from the Chile Trench–Chile Rise triple junction to the base of the Antarctic Peninsula lies on the Pacific side of any undoubtedly Precambrian crystalline rocks. This is true, independent of the precise geometry of Gondwanaland reconstruction in the Antarctic Peninsula–southern South American area (Barker, Dalziel, and others, 1976; de Wit, 1977) (see Figure 12.2). The stratigraphy of the pre-Middle Jurassic basement of the region is poorly known. However, the assemblage of rock types and their field relationships is meaningful. The main elements of this petrotectonic assemblage are as follows:

1. Extensive graywacke and shale sequences derived from a volcano-plutonic terrane of granitic composition and affected mainly by sub-greenschist facies metamorphism; tightly folded with slaty cleavage; metamorphic grade higher in extreme south, including blueschist facies in rocks with chert and fragmented ophiolitic rocks.
2. Extensive chert-argillite sequences, mostly affected by greenschist facies or lower grade metamorphism and subisoclinal to isoclinal folds.

3. Extensive metasedimentary–metavolcanic–chert–marble–calc-silicate sequences affected by pre-Middle Jurassic polyphase deformation; blueschist-greenschist facies metamorphism on the Pacific side and epidote-amphibolte to amphibolite facies on the continental side.
4. Massive fusulinid limestone.
5. Chert overlying extensive mafic pillow lava.
6. Tectonically emplaced dunite-serpentinite.
7. Granitic rocks, apprently limited to continental side of other elements.

Given their location Pacificward of the Precambrian crystalline rocks of Gondwanaland and pre–Middle Jurassic calc-alkaline igneous rocks elsewhere in the South Atlantic–Weddell Sea area (Figure 12.2), the basement rocks of the Scotia arc region seem to represent portions of the fore-arc and the main magmatic arc of a Paleozoic–early Mesozoic subduction zone along the Pacific margin of Gondwanaland (Dalziel and Forsythe, 1978; Forsythe and Mpodozis, 1979). In this interpretation, the various units in the petrotectonic assemblage are taken to represent the following:

1. Graywacke-shale sequences: possible trench and/or midslope basin turbidite deposits, now part of an "accretionary prism," as suggested by Suárez (1976).
2. Chert-argillite sequences: pelagic sedimentary deposits accreted to the continent at the inner wall of a trench.
3. Metasedimentary–metavolcanic–chert–marble–calc-silicate sequences: pelagic and hemipelagic sedimentary deposits with interlayered volcanic (commonly pyroclastic) deposits from midplate or arc volcanism—also part of a trench inner wall accretionary prism.
4. Fusulinid limestone: perhaps originally a shallow-water outer arc along the continental margin or else a more exotic Bahama Bank–type feature "scraped off" the downgoing slab.
5. Mafic pillow lavas with overlying bedded chert and other fragmented ophiolitic rocks: a slice of the top of oceanic basement with a veneer of pelagic sediments sheared off a downgoing slab.
6. Dunite-serpentinite: a slice of oceanic lower crust or upper mantle (see de Wit and others, 1977).
7. Granitic rocks of Marguerite Bay, Jason Peninsula, Tabarin Peninsula (if their pre-Jurassic apparent age is confirmed), and the South American continent east of the Andes: parts of the unroofed volcanic arc that supplied some of the detritus for the turbidites ([1] above).

Geotectonic significance

As pointed out elsewhere (Smellie and Clarkson, 1975; Rivano and Cortés, 1975; Barker, Dalziel, and others, 1976; de Wit, 1977), the existence of a Paleozoic–early Mesozoic subduction complex along the Antarctic Peninsula margin of Gondwanaland would be in keeping with evidence of Paleozoic subduction of Pacific floor lithosphere beneath central Chile (Gonzalez-Bonarino, 1971). It would also complete a highly complex but essentially continuous zone of Paleozoic–early Mesozoic subduction along the entire Pacific margin of Gondwanaland from northern South America through Antarctica to New Zealand and indeed strengthen what seems to be growing evidence that subduction has been going on intermittently around the entire margin of the Pacific Ocean basin throughout Phanerozoic time.

Thus, I believe that Du Toit's "Samfrau geosyncline" (1937, Figure 2) represents a complex assemblage of sedimentary and igneous rocks in arc-trench gap, volcanic-arc, and back-arc settings along the Pacific margin of Gondwanaland during middle to late Paleozoic and early Mesozoic time. "Gondwanide" folding of the "geosynclinal" rocks within the "Samfrau orogenic zone" (Du Toit, 1937, Figure 10) represents deformation in these same environments. Deformation was comparatively short-lived and demonstrably episodic in the back-arc environment of the Sierra de la Ventana (Argentina), the Falkland (Malvinas) Islands, and the Cape Fold Belt that constituted the "foredeep" of the Samfrau geosyncline (Du Toit, 1937). However, by analogy with modern trench environments (see Moore and Karig, 1976), deformation seems likely to have been drawn out over much longer periods of time in the arc-trench gap accretionary prism now comprising the basement of most of the Scotia arc region. Hence, care should be taken before distinguishing "Gondwanian" and "pre-Gondwanian" deformational events.

Uplift and erosion of the entire arc-trench gap, arc, and back-arc complex throughout the South Atlantic–Weddell Sea area occurred in Late Triassic to Early Jurassic time, with erosion of all these tectonic elements prior to the Middle to Late Jurassic deposition of sedimentary strata and widespread extrusion of silicic volcanic rocks. Inasmuch as this uplift and erosion immediately preceded the initiation of sea-floor spreading in the South Atlantic and southwest Indian Oceans and the start of the breakup of Gondwanaland, it is likely that these events are in some way causally related. The simultaneous uplift and erosion of the entire Samfrau orogenic zone may reflect a distinct and important hiatus in subduction along the Pacific margin of Gondwanaland that allowed the isostatic "rebound" of the subduction complex and thus temporarily halted deformation in the arc-trench gap. In the North American Cordillera, however, uplift of this type is being tentatively related to changes in the dip of the subduction zone, the rate of subduction, and migrating triple junctions. In either case, plutonism may have been continuous.

Reconstruction of the Samfrau orogen

The relationships of the Sierra de la Ventana of northeastern Argentina on the South American plate to the Cape Fold Belt on the African plate are now well established, and the general continuity of the Cape Fold Belt into the Transantarctic Mountains somewhere north of the Pensacola Mountains and into Australia and New Zealand seems clear. It is the various parts of West Antarctica whose relationships are uncertain. Because of the lack of definitive sea-floor spreading data from the Weddell Sea, the complex history of the Scotia Sea and Drake Passage, and the ice cover of the Weddell Sea–Ross Sea embayment, the basement rocks provide the best available clues to the detailed reconstruction of the Samfrau orogenic belt and hence of the margin of Gondwanaland, in the South Atlantic–Weddell Sea area.

South America

A northwest trend is distinguishable in the South American Paleozoic basement from the Brazilian shield south-

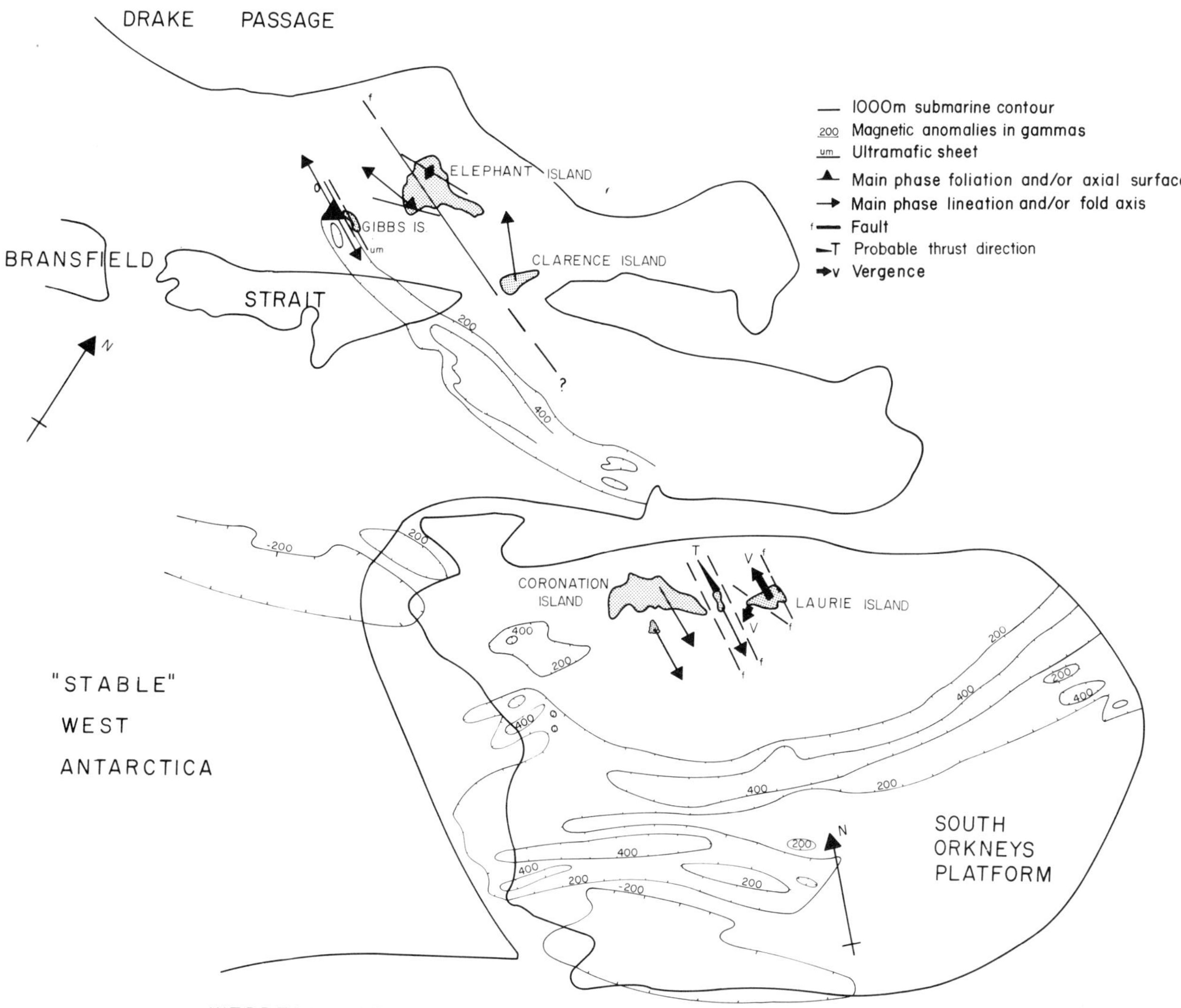

Figure 12.7. Suggested original (pre-Cenozoic?) relationship of South Orkney Islands platform and northern Antarctic Peninsula.

ward. It can be traced across the Andes near lat. 50° S, and major folds with northwest- or west-northwest-trending hinge lines occur in the basement south of the west end of the Straits of Magellan (Dalziel and others, 1975). Recently initiated structural studies in the Andean foothills around the Patagonian "orocline" have not revealed any of the compressional structures to be expected on the concave side of a cordillera bent or buckled about a vertical axis (M. Winslow, personal communication, 1979). Hence, given also the fact that the orocline parallels a major bend (promontory) in the margin of the Precambrian crystalline shield (Figures 12.1 and 12.2), I am now inclined to believe that, with only minor Mesozoic–Cenozoic tectonic modifications, the bend in the southernmost Andes is an original one.

South Scotia Ridge and Antarctic Peninsula

The close similarity of the physiography and geology on either side of Drake Passage has, of course, long been recognized (Barrow, 1831; Arctowski, 1895). This led many geologists, including myself, to propose various reconstructions of the initial relationship of these two parts of Gondwanaland, all involving a continuous cordilleran connection (see Dalziel and Elliot, 1973, for review). The comparatively recent recognition of an uplifted Cretaceous marginal basin floor in the southernmost Andes (Katz, 1973; Dalziel and others, 1974) led to the appreciation that the Weddell Sea might be the equivalent of this basin in the Antarctic continent (Dalziel, 1974; Suárez, 1976). Hence, the equivalent in South America of the Antarctic Peninsula would not be the whole Andean Cordillera but would instead be only the upper Mesozoic, Pacific margin, volcanic arc and arc-trench gap rocks with their underlying pre-Jurassic basement (Dalziel, 1974; Suárez, 1976; Barker, Dalziel, and others, 1976; de Wit, 1977). Indeed, as plate tectonic processes have become better understood, I have come to wonder if the geologic similarities between the southern part of the Andean Cordillera and the Antarctic Peninsula might perhaps reflect similarities between contemporaneous processes rather than the disruption of a once-continuous cordillera. The "connection" may well have been comparable to that between the Aleutian island arc and the Kamchatka Peninsula today.

The trend of main phase structures in the basement of the South Shetland Islands is rather variable but distinctly oblique to the orientation of the South Shetland Ridge (Figures 12.5 and 12.7). The metamorphic facies of the South Orkney Islands basement rocks, together with the structural trends and the bathymetry of the platform on which they are situated, leads me to propose the reconstruction of the South Scotia Ridge shown in Figure 12.7. The resulting compatibility of the magnetic signatures of the South Orkney Islands platform and the southern part of the ridge on which the South Shetlands are situated (Harrington and others, 1972; Watters, 1972) may, however, be in part at least coincidental. The pronounced magnetic high south of Gibbs Island seems to be associated with the south-southwest-dipping dunite-serpentinite complex on the island, but the anomalies south of the South Orkneys (present coordinates) have a different trend in the reconstruction. This trend indicates that the anomalies on the South Orkneys platform are related either to the late Mesozoic arc along the Antarctic Peninsula, as suggested by Harrington and others (1972), or else to the opening of the Weddell Sea. They could well be related to both if the Weddell Sea is in fact a back-arc marginal basin spreading center behind the late Mesozoic–Cenozoic magmatic arc of the Antarctic Peninsula.

The exact geometric configuration of the South America–Africa–Antarctica continental mass prior to the initiation of sea-floor spreading in Early Cretaceous time remains uncertain. For the reaons mentioned above, I no longer believe in oroclinal bending of the entire southernmost Andes through 90° and, hence, in an originally near-rectilinear Andean–West Antarctic cordillera (Dalziel and Elliot, 1973). On the other hand, the east-closing cordilleran cusp reconstructed by Barker and Griffiths (1972) on the basis of the late Cenozoic spreading in Drake Passage need not necessarily be the original (i.e., pre-Cretaceous) configuration of the continental margin of Gondwanaland. Indeed it seems unlikely to be because it leaves South Georgia, part of the sedimentary infilling and floor of the Early Cretaceous marginal basin in the southern Andes, isolated from the Pacific margin. The suggestion of de Wit (1977) that the Andean-Antarctandean cordillera trended southeastward (relative to present South American coordinates) from Tierra del Fuego into the Antarctic Peninsula was based on the belief that basement structures south of Drake Passage trend parallel to the South Scotia Ridge and the Antarctic Peninsula. Our recent work has shown this to be incorrect (Figures 12.5 and 12.7). My current belief is that prior to the opening of the South Atlantic Ocean and Weddell Sea, the South Scotia Ridge and Antarctic Peninsula were immediately adjacent to the East Antarctic craton. This satisfies the limited paleomagnetic data that indicate a pre-Cenozoic bend in the Antarctic Peninsula (Dalziel and others, 1973). It also satisfies my prejudice that the Western Weddell Sea is a marginal basin comparable to the one which opened in the southernmost Andes in Early Cretaceous time (Dalziel, 1974; Suárez, 1976). This is in keeping with the emerging symmetry of pre-Middle Jurassic basement structures across the Scotia Sea and Drake Passage: on either side the basement structures are more east-west trending (present coordinates) than were the late Mesozoic and Cenozoic cordilleras.

During opening of the Weddell Sea, the South Scotia Ridge–Antarctic Peninsula arc on the Pacific side of the Weddell Sea marginal basin would have rotated counterclockwise away from a position on the margin of the East Antarctic craton, just as the Ellsworth Mountains crustal fragment seems to have done (Schopf, 1969; Barker, Dalziel, and others, 1977; Clarkson and Brook, 1977; de Wit, 1977). The Thurston Island–Eights Coast region of West Antarctica seems from its rock types and radiometric ages (Craddock, 1970) to represent an additional fragment of pre-Jurassic Pacific margin volcanic arc.

Hence, at the time of writing I am inclined to the view that the reentrant in the Precambrian crystalline shield of Gondwanaland in the vicinity of what is now the Scotia arc was of primary importance in the development of the arc, having been perpetuated in the configuration of the Paleozoic–early Mesozoic Gondwanide orogen as well as in the late Mesozoic–Cenozoic Andean-Antarctandean orogens and the Scotia arc.

Acknowledgments. I am particularly grateful to the Division of Polar Programs of the National Science Foundation and to Mort D. Turner, Program Manger for Earth Sciences, for the funds and logistic opportunities to undertake this work over several Antarctic field seasons. Albert Betzel, Robert L. Dale, Robert Elder, Price Lewis, Kendall Moulton, Philip Smith, and many others have helped in the logistic planning. Without the enthusiasm and seamanship of Captain Pieter J. Lenie, Master of R/V *Hero* since 1972, the work could never have been accomplished. I am grateful also to various other masters and the crews of R/V *Hero,* USCGC *Edisto* (1968–69), and USCGC *Glacier* (1970–71). Several of my colleagues, research associates, and graduate students have participated in the field work and discussions, notably David Elliot, Richard Schweickert, Maarten de Wit, Gerry Stirewalt, David Price, Richardson Allen, Stephan and Alice Brocoum, Randall Forsythe, and Roy Kligfield. Finally I wish to thank Robert H. Dott, Jr. for encouraging me to become involved in the Scotia arc region and for his continuing advice, discussion, and constructive criticism.

References cited

Adie, R. J. 1954. The petrology of Graham Land: I. The basement complex; early Palaeozoic plutonic and volcanic rocks. *Scient. Rep. Falkld Isl. Depend. Surv.,* No. 11, 22 pp.

Adie, R. J. 1957. The petrology of Graham Land: III. Metamorphic rocks of the Trinity Peninsula Series. *Scient. Rep. Falkld Isl. Depend. Surv.,* No. 20, 26 pp.

Adie, R. J. 1964. Stratigraphic correlation in western Antarctica. (*In* Adie, R. J., *ed. Antarctic Geology.* Amsterdam, North-Holland Publishing Company, 307–13.)

Aitkenhead, N. 1965. The geology of the Duse Bay-Larsen Inlet area, northeast Graham Land (with particular reference to the Trinity Peninsula Series). *Scient. Rep. Br. antarc. Surv.,* No. 51, 62 pp.

Araya, R., and Hervé, F. 1966. Estudio geomorphico y geologico de las Islas Shetlands del sur, Antarctica. *Publnes Inst. antárt. chil.,* No. 8, 76.

Arctowski, H. 1895. Observations sur l'interêt que présente l'exploration géologique des terres australes. *Bull. Soc. Géol. Fr.,* Ser. 3, *23,* 589–91.

Barker, P. F., and Griffiths, D. H. 1972. The evolution of the Scotia Ridge and Scotia Sea. *Phil. Trans. R. Soc. Lond.,* Ser. A, *271,* 151–83.

Barker, P. F., Dalziel, I. W. D., and others. 1976. The evolution of the southwestern Atlantic Ocean basin: Leg 36 data. (*In* Wise, S. W., Jr., *ed. Initial Reports of Deep Sea Drilling Project,* Vol. 36. Washington [U.S. Government Printing Office], 993–1014.)

Barrow, J. 1831. Introductory note (to an article on the islands of Deception). *R. Geogr. Soc., 1,* 62.

Bibby, J. S. 1966. The stratigraphy of part of north-east Graham Land and the James Ross Island Group. *Scient. Rep. Br. antarc. Sur.,* No. 53, 37 pp.

Bruhn, R. L., and Dalziel, I. W. D. 1977. Destruction of the early Cretaceous marginal basin in the Andes of Tierra del Fuego. (*In* Talwani, M., and Pitman, W. J., III, *eds. Maurice Ewing Series, 1: Island Arcs, Deep Sea Trenches and Back-arc Basins.* Washington, American Geophysical Union, 395–405.)

Cecioni, G. 1955. Prime notizie sepra l'esistanza del Paleozoico Superiore nell'Arcipelago Patagonico tra: paralleli 50° y 52° S. *Mem. Atti Soc. tosc. Sci. nat.,* Sec. A, *62,* 201–4.

Cingolani, C. A., and Varela, R. 1976. Investigaciones Geológicas y Geochronologicas en el extremo sur de la Isla Gran Malvina, sector de Cabo Belgrano (Cape Meredith), Islas Malvinas. *Act. 6 Cong. Geol. Argent., Buenos Aires,* 18 pp.

Clarkson, P. D., and Brook, M. 1977. Age and position of the Ellsworth Mountains crustal fragment, Antarctica. *Nature, Lond., 265,* 615–16.

Cordini, I. R. 1955. Contribución al conocimiento del Sector Antarctico Argentino. *Publnes Inst. antárt. argent.,* No. 1, 277 pp.

Craddock, C. 1970. Radiometric age map of Antarctica. (*In* Bushnell, V. C., and Craddock, C., *eds.* Geologic Maps of Antarctica. *Antarct. Map Folio Ser.,* Folio 12, Pl. XIX.)

Dalziel, I. W. D. 1971. Structural studies in the Scotia arc: the South Orkney Islands. *Antarct. Jnl U.S., 6,* 124–26.

Dalziel, I. W. D. 1972*a*. Large-scale folding in the Scotia arc. (*In* Adie, R. J., *ed. Antarctic Geology and Geophysics.* Oslo, Universitetsforlaget, 47–55.)

Dalziel, I. W. D. 1972*b*. K-Ar dating of rocks from Elephant Island, South Scotia Ridge. *Geol. Soc. Am. Bull., 83,* 1887–94.

Dalziel, I. W. D. 1976. Margins of the Scotia Sea. (*In* Burke, C. A., and Drake, C. L., *eds. The Geology of Continental Margins.* New York, Springer-Verlag, 567–80.)

Dalziel, I. W. D. 1976. Structural studies in the Scotia arc: "basement" rocks of the South Shetland Islands. *Antarct. Jnl U.S., 11,* No. 2, 75–77.

Dalziel, I. W. D., and Cortés, R. 1972. The tectonic style of the southernmost Andes and the Antarctandes. *Proc. 24th Int. Geol. Cong., Montreal, 1972,* 316–27.

Dalziel, I. W. D., and Elliot, D. H. 1973. The Scotia arc and Antarctic margin. (*In* Nairn, A. E. M., and Stehli, F. G., *eds. The Ocean Basins and Margins: 1. The South Atlantic.* New York, Plenum Press, 171–246.)

Dalziel, I. W. D., and Forsythe, R. D. 1978. Du Toit's "Samfrau geosyncline" and plate tectonics. *Abstr. with Program, geol. Soc. Am.,* annual meeting, Toronto, 385.

Dalziel, I. W. D., de Wit, M. J., and Palmer, K. F. 1974. Fossil marginal basin in the southern Andes. *Nature, Lond., 250,* 291–94.

Dalziel, I. W. D., de Wit, M. J., and Ridley, W. I. 1975. Structural and petrologic studies in the Scotia arc: the Patagonian Andes. *Antarct. Jnl U.S., 10,* No. 6, 307–10.

Dalziel, I. W. D., Kligfield, R., Lowrie, W., and Opdyke, N. D. 1973. Paleomagnetic data from the southernmost Andes and the Antarctandes. (*In* Tarling, D. H., and Runcorn, S. K., *eds. Implications of Continental Drift to the Earth Sciences.* New York, Academic Press, 1, 87–101.)

Dalziel, I. W. D., Elliot, D. H., Thomson, J. W., Thomson, M. R. A., Wells, N. A., and Zinsmeister, W. J. 1977. Geologic studies in the South Orkney Islands: R/V *Hero* Cruise 77-1, January 1977. *Antarct. Jnl U.S., 12,* No. 4, 98–101.

de Wit, M. J. 1977. The evolution of the Scotia arc as a key to the reconstruction of southwestern Gondwanaland. *Tectonophysics, 37,* 53–81.

de Wit, M. J., Dutch, S., Kligfield, R., Allen, R., and Stern, C. 1977. Deformation, serpentinization and emplacement of a dunite complex, Gibbs Island, South Shetland Islands: possible fracture zone tectonics. *J. Geol.,* 85, 745–62.

Douglass, R. C., and Nestell, M. K. 1976. Late Paleozoic Foraminifera from southern Chile. *Prof. Pap. U.S. geol. Surv.,* No. 858, 49 pp.

Du Toit, A. L. 1937. *Our Wandering Continents.* Edinburgh, Oliver and Boyd, 366 pp.

Edwards, C. W. 1977. Early Mesozoic marine fossils from central Alexander Island. *Bull. Br. antarc. Surv.,* No. 49.

Elliot, D. H. 1965. Geology of northwest Trinity Peninsula, Graham Land. *Bull. Br. antarc. Surv.,* No. 7, 1–24.

Elliot, D. H. 1966. Geology of the Nordenskjöld Coast and a comparison with north-west Trinity Peninsula, Graham Land. *Bull. Br. antarc. Surv.,* No. 10, 1–43.

Elliot, D. H. 1975. Tectonics of Antarctica: A review. *Am. J. Sci., 275A,* 45–106.

Forsythe, R. D., and Mpodozis, C. 1979. Proto-Pacific crust in southern Chile. *Abstr. with program, geol. Soc. Am.,* Cordilleran section meeting, San Jose, California, 78.

Fraser, A. G. 1965. The petrology of Stonington and Trepassey Islands, Marguerite Bay. *Scient. Rep. Br. antarc. Surv.,* No. 52, 51.

Gonzalez-Bonarino, F. 1971. Metamorphism of the crystalline basement of central Chile. *J. Petrology, 12,* 149–75.

Grikurov, G. E. 1971. Geologischeskoe stroenie tsentral'noi chasti Zemli Aleksandra I [The geological structure of the central part of Alexander I Land]. *Antarktika, Doklady Komissii,* 13–42.

Grikurov, G. E. 1972. Tectonics of the Antarctandes. (*In* Adie, R. J., *ed. Antarctic Geology and Geophysics.* Oslo, Universitetsforlaget, 163–67.)

Grikurov, G. E., and Dibner, A. G. 1968. Novye dannye o Serii Triniti (C_{1-6}) v zapadnoy Antarktide [More information on the Trinity Series (C_{1-2}) in western Antarctica]. *Dokl. Akad. Nauk SSSR, 179,* No. 2, 410–12. [English translation: Dokl. (Proc.) Acad. Sci. U.S.S.R., Geological sciences sect. 179, 39–41.]

Grikurov, G. E., Krylov, A. Ya., and Silin, Yu. I. 1966. Absoliutnyi vozrast nekotorykh porod iz raiona zaliva Margerit, Antarkticheskii poluostrov [Absolute age of certain rocks in the Marguerite Bay region of the Antarctic Peninsula]. *Dokl. Akad. Nauk SSSR,* Geology, *171,* No. 6, 1399–401. [English translation: Dokl. (Proc.) Acad. Sci. U.S.S.R., Geological sciences sect., 171, 127–30.]

Grikurov, G. E., Krylov, A. Ya., and Silin, Yu. I. 1967. Absoliutnyi vozrast nekotorykh porod dugi skotiya i Zemli Aleksandra I (Zapadnaia Antarktika) [Absolute age of some rocks from the Scotia arc and Alexander I Land (western Antarctica)]. *Dokl. Akad. Nauk SSSR,* Geology, *172,* No. 1, 168–71. [English translation: Dokl. (Proc.) Acad. Sci. U.S.S.R., Geological sciences sect., 172, 19–22.]

Grikurov, G. E., Krylov, A. Ya., Polyakov, M. M., and Tsovbun, Ya. N. 1970. Vozrast porod v severnoi chasti Antarkticheskogo poluostrova i na Yudznykh Shetlandskikh ostrovakh (po dannym kaliyargonovogo metoda) [Age of rocks from the northern part of the Antarctic Peninsula and the South Shetland Islands (from data of the potassium-argon method)]. *Inf. Byull. sov. antarkt. Eksped.,* 80, 30–34. [English translation: 8, 61–63.]

Halpern, M. 1964. Cretaceous sedimentation in the "General Bernardo O'Higgins" area of northwest Antarctic Peninsula. (*In* Adie, R. J., *ed. Antarctic Geology.* Amsterdam, North-Holland Publishing Company, 334–47.)

Halpern, M. 1965. The geology of the General Bernardo O'Higgins area, northwest Antarctic Peninsula. (*In* Hadley, J. B., *ed. Geology and Paleontology of the Antarctic.* Washington, American Geophysical Union, 177–209.) [Antarctic Research Series, Vol. 6.]

Halpern, M. 1971. Evidence for Gondwanaland from a review of West Antarctic radiometric ages. (*In* Quam, L. O., *ed.* Research in the Antarctic. *Am. Ass. Adv. Sci.,* Publn No. 93, 717–30.)

Halpern, M. 1972. Rb-Sr total rock and mineral ages from the Marguerite Bay area, Kohler Range and Fosdick Mountains. (*In* Adie, R. J., *ed. Antarctic Geology and Geophysics.* Oslo, Universitetsforlaget, 197–204.)

Halpern, M. 1973. Regional geochronology of Chile south of 50° latitude. *Geol. Soc. Am. Bull., 84,* 2407–22.

Halpern, M., Urguart, M. U., and Linares, E. 1972. Radiometric ages of crystalline rocks from southern South America as related to Gondwana and Andean geologic provinces. *Upper Mantle Symposium, Buenos Aires, 1972, 2,* 345–56.

Harrington, P. K., Barker, P. F., and Griffiths, D. H. 1972. Crustal structure of the South Orkney Islands area from seismic refraction and magnetic measurements. (*In* Adie, R. J., *ed. Antarctic Geology and Geophysics.* Oslo, Universitetsforlaget, 27–32.)

Hobbs, G. J. 1968. The geology of the South Shetland Islands: IV. Livingston Island. *Scient. Rep. Br. antarc. Surv.,* No. 47, 34 pp.

Hoskins, A. K. 1963. The basement complex of Neny Fjord, Graham Land. *Scient. Rep. Br. antarc. Surv.* No. 43, 49 pp.

Iltchenko, L. N. 1972. Late Precambrian acritarchs of Antarctica. (*In* Adie, R. J., *ed. Antarctic Geology and Geophysics.* Oslo, Universitetsforlaget, 599–602.)

Katz, H. R. 1973. Contrasts in tectonic evolution of orogenic belts in the south-east Pacific. *J. R. Soc. N. Z., 3,* 333–62.

Kranck, E. H. 1932. Geological investigations in the Cordillera of Tierra del Fuego. *Acta geogr. Helsingf., 4,* No. 2, 231 pp.

Matthews, D. H., and Maling, D. H. 1967. The geology of the South Orkney Islands, I. Signy Island. *Scient. Rep. Br. antarc. Surv.,* No. 25, 32 pp.

Miller, H. 1966. Kleintektonische Untersuchungen in der Umgebung der Station "General Bernardo O'Higgins," Nordwest Spitz der Antarktischen Halbinsel. *Geol. Rdsch., 55,* No. 3, 809–19.

Miller, J. R. 1960. Potassium-argon ages of some rocks from the South Atlantic. *Nature, Lond., 187,* 1019–20.

Miyashiro, A. 1973. *Metamorphism and Metamorphic Belts.* New York, Wiley, 492 pp.

Moore, J. C., and Karig, D. E. 1976. Sedimentology, structural geology and tectonics of the Shikoku subduction zone, southwestern Japan. *Geol. Soc. Am. Bull., 87,* 1259–68.

Natland, M. L., Gonzalez, E., Cañon, A., and Ernst, M. 1974. A system of stages for correlation of the Magallanes Basin sediments. *Mem. Geol. Soc. Am. 139,* 1–126.

Nelson, E. P., Dalziel, I. W. D., and Milnes, A. G. 1979. Deformation in Cordillera Darwin, southern Chile. *Abstr. with Program, geol. Soc. Am.,* Cordillera section meeting, San Jose, California, 119.

Nelson, E. P., Dalziel, I. W. D., and Ridley, W. I. 1979. Geological studies in Cordillera Darwin, Tierra del Fuego, southern Chile: R/V *Hero* Cruises 77/4 and 78/2. *Antarct. Jnl U.S., 14,* 14–15.

Pirie, J. H. H. 1905. On the graptolite-bearing rocks of the South Orkneys. *Proc. R. Soc. Edinb., 25,* 463–70.

Pirie, J. H. H. 1906. Sledge and boat journeys. (*In* Brown, R. N. R., Mossman, R. C., and Pirie, J. H. H., *eds. The Voyage of the Scotia. Being the Record of a Voyage of Exploration in the Antarctic Seas.* Edinburgh, Blackwood, 145–79.)

Rex, D. C. 1976. Geochronology in relation to the stratigraphy of the Antarctic Peninsula. *Bull. Br. antarc. Surv.,* No. 43, 49–58.

Rex, D. C., and Baker, P. E. 1973. Age and petrology of the Cornwallis Island granodiorite. *Bull. Br. antarc. Surv.,* No. 32, 55–61.

Rivano, S., and Cortés, R. 1975. Nota preliminar sobre el Hallazgo de rocas metamorphicas en la Isla Smith (Shetland del Sur Antarctica Chilena). *Ser. cient. Inst. antárt. chil., 111,* No. 1, 9–14.

Schopf, J. M. 1969. Ellsworth Mountains: position in West Antarctica due to sea-floor spreading. *Science, N. Y., 164, 63–66.*

Schopf, J. M. 1973. Plant material from the Miers Bluff Formation in the South Shetland Islands. *Rep. Inst. polar Stud. Ohio State Univ.,* No. 45, 45 pp.

Skinner, A. C. 1973. Geology of north-western Palmer Land between Eureka and Meiklejohn Glaciers. *Bull. Br. antarc. Surv.,* No. 35, 1–22.

Smellie, J. L., and Clarkson, P. D. 1975. Evidence for pre-Jurassic subduction in western Antarctica. *Nature, Lond., 258,* 701–2.

Smith, A. G., and Hallam, A. 1970. The fit of the southern continents. *Nature, Lond., 225,* 139–44.

Sobotovich, E. V., Kamenev, E. N., Komaristy, A. A., and Rudnik, V. A. 1974. Drevneishie porody Antarktidy (Zemlia Enderbi) [The oldest rocks of Antarctica (Enderby Land)]. *Izv. Akad. Nauk SSSR,* Ser. geol., No. 11, 30–50.

Storey, B. C., Mair, B. F., and Bell, C. M. 1977. The occurrence of Mesozoic oceanic floor and ancient continental crust on South Georgia. *Geol. Mag., 114,* 203–8.

Stubbs, G. M. 1968. *Geology of Parts of the Foyn and Bowman Coasts, Graham Land.* Ph.D. dissertation, University of Birmingham, England, 244 pp.

Suárez, M. 1976. Plate-tectonic model for southern Antarctic Peninsula and its relation to southern Andes. *Geology, 4,* 211–14.

Thomson, J. W. 1968. The geology of the South Orkney Islands: II. The Petrology of Signy Island. *Scient. Rep. Br. antarc. Surv.,* No. 62, 30 pp.

Thomson, J. W. 1971. The geology of Matthews Island, South Orkney Islands. *Bull. Br. antarc. Surv.,* No. 26, 51–57.

Thomson, J. W. 1973. The geology of Powell, Christoffersen and Michelsen Islands, South Orkney Islands. *Bull. Br. antarc. Surv.,* No. 33–34, 137–67.

Thomson, J. W. 1974. The geology of the South Orkney Islands: III. Coronation Island. *Scient. Rep. Br. Antarc. Surv.,* No. 86, 39 pp.

Thomson, M. R. A. 1975*a*. Fossils from the South Orkney Islands: II. Matthews Island. *Bull. Br. antarc. Surv.,* No. 40, 75–79.

Thomson, M. R. A. 1975*b*. New paleontological and lithological observations in the Legoupil Formation, north-west Antarctic Peninsula. *Bull. Br. antarc. Surv.,* No. 41–42, 169–85.

Thomson, M. R. A., and Willey, L. 1975. Fossils from the South Orkney Islands: I. Coronation Island. *Bull. Br. antarc. Surv.,* No. 40, 15–21.

Tyrrell, G. W. 1945. Report on rocks from West Antarctica and the Scotia arc. *"Discovery" Rep., 23,* 37–102.

Watters, D. G. 1972. Geophysical investigation of a section of the South Scotia Ridge. (*In* Adie, R. J., *ed. Antarctic Geology and Geophysics.* Oslo, Universitetsforlaget, 33–38.)

Wordie, J. M. 1921. Shackleton Antarctic Expedition, 1914–1917: geological observations in the Weddell Sea area. *Trans. R. Soc. Edinb., 53,* Pt. 1, 17–27.

13

Geologic Comparison between the Antarctic Peninsula and Southern South America

HUBERT MILLER
Geologisch-Paläontologisches Institut der Universität Münster, 4400 Münster, Federal Republic of Germany

Abstract. The Carboniferous(?) to Triassic "Trinity Peninsula Series" includes the most common basement rocks of the northern Antarctic Peninsula. However, the basement rocks of the southernmost Andes are predominantly of early Paleozoic age. The author's investigations in Patagonia demonstrate polyphase tectonic superimposition within this slightly to moderately metamorphosed basement, the sedimentation age of which is partly documented to be Early to Middle Devonian. Major orogeny took place during late Paleozoic time in the Patagonian Andes, whereas in the Antarctic Peninsula, Late Triassic to Early Jurassic folding and metamorphism is known. This is proof of the different tectonic histories of Patagonia and the Antarctic Peninsula.

Above all, the Cretaceous-Cenozoic strata of the Antarctic Peninsula differ from those of the Magellan region (southernmost Andes) by lack of a flysch facies preceding a typical molasse facies. Andean deformation is significantly less in the Antarctic Peninsula than in the Magellan region. Comparison of the Andean orogeny between West Antarctica and the central Andes would seem to be of more importance.

There is no necessity to follow mountain belts around the whole earth. The continental fracture pattern opposite the Chile Ridge (Chonos triple junction) is related only to the fracture pattern of the Hercynian and Andean orogenies of this region. The oceanic Chile Ridge is not structurally manifested toward the east in the Andean chain. A transverse segmentation exists in the Patagonian Andes along distinct lines at lat. 40°, 46°, and 50° S and in the region of the Scotia Sea. This segmentation cannot be entirely explained by plate-tectonic models.

Three major problems that have been extensively investigated by the author during recent years are discussed in this paper:

1. The Paleozoic development of the Antarctic Peninsula and southernmost South America (Figure 13.1), based on new data from the Patagonian Andes;
2. The Mesozoic development of the Patagonian–Fuegian Andes in comparison to the Antarctic Peninsula and the South Shetland Islands; and
3. The boundary between the Antarctic plate, the American plate, and the Nazca plate (the Chonos triple junction).

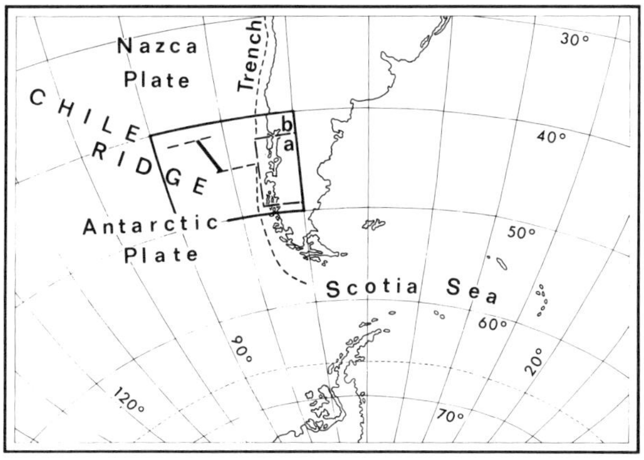

Figure 13.1. Index map of the Scotia arc with locations of investigated areas. For *a* see Figure 13.2 and for *b* see Figure 13.3.

Precambrian and Paleozoic geology

Table 13.1 shows a comparison of the southern Andean pre-Mesozoic rocks (lat. 44° S) with those of the Antarctic Peninsula (including the South Shetland Islands). Finds of fusulinids (Cecioni, 1956) on the islands between lat. 50° and 52° S are the basis for upper Paleozoic rock classification in South America. In addition, a Lower to Middle Devonian fauna was discovered recently on Potranca Island from the Chonos Archipelago (approximately lat. 45° S) by the author (Miller and Sprechmann, 1978).

South America

The question of the age of the slightly to moderately metamorphosed rocks of the southernmost Andes has been recently resolved by the fauna mentioned above. Two series are distinguished in the Chonos Archipelago: the Canal King Formation (presumably Cambrian–Ordovician) and the Potranca Formation (Devonian). The age of the clastic, very weakly metamorphosed sedimentary rocks of the Potranca Formation is established by the brachiopod *Derbyina whitiorum* Clarke. From structural considerations, the Canal King Formation is older than the fossiliferous beds of the Potranca Formation; it shows a higher degree of metamorphism and additional folding. Furthermore, typical poikiloblastic albites of the Canal King Formation are found as clasts in the graywackes of the Devonian Potranca Formation. Permo-Carboniferous sedimentary rocks, similar to the fusulinid-bearing beds around Madre de Dios Island, are preserved in a small band at the eastern margin of the Chonos Archipelago.

In the area of Lago General Carrera and Lago Cochrane (lat. 46°–48° S, on the eastern flank of the Patagonian cordillera), Lagally (1975) and Miller (1976) named the Paleozoic Lago Cochrane and Lago General Carrera Formations. The Devonian age of the Lago Cochrane Formation is known from previous literature (Lesta and Ferello, 1972, p. 606).

The Permo-Carboniferous Hercynian orogeny affected all these formations. The rocks show very complicated deformation (Figure 13.2). The oldest folding is known only from the Canal King and Lago General Carrera Formations. A further folding, presumably at the Devonian–Carboniferous boundary, affected the entire region. A younger folding developed vigorously in only some places; this is the only folding within the sedimentary rocks of the eastern margin of the Chonos Archipelago that is regarded as equivalent to the Madre de Dios series. Metamorphism reaches the greenschist-amphibolite facies boundary (beginning of development of garnet, barroisite). Stilpnomelane is abundant in places.

The pre-Mesozoic metamorphic rocks of Tierra del Fuego may well be compared with the Canal King Formation (Miller, 1970, p. 52).

Antarctic Peninsula and South Shetland Islands

Previous studies of the Antarctic Peninsula have considered the upper Precambrian metamorphic rocks and Carboniferous Trinity Peninsula Series to be basement units (Grikurov, 1972; Adie, 1972). Recent work (Thomson, 1975*a,b;* Chapter 37, this volume; Edwards, Chapter 36, this volume) has shown that "Trinity Peninsula Series–like" rocks at two important localities (LeMay Formation and Legoupil Formation) are of Triassic age. Rocks of high-pressure paragenesis have been described from different places in the South Shetland Islands (Smellie and Clarkson, 1975; Rivano and Cortés, 1976; Dalziel, 1976*a*) and have so far been only vaguely classified stratigraphically. Gledhill and others (Chapter 35, this volume) found very few indications of pre-Mesozoic orogenies in the Antarctic Peninsula.

As shown in Table 13.1, at present no agreement is evident in the inferred Paleozoic history of the basement rocks of the southernmost Andes and those of the Antarctic Peninsula. In particular, the thick, monotonous Devonian series of Patagonia (Potranca Formation and Lago Cochrane Formation) have not been found in the Antarctic Peninsula. Likewise, an exact equivalent to the Trinity Peninsula Series is absent in Patagonia. Here, the rocks of Madre de Dios Island are lithologically very similar to the Trinity Peninsula Series (Dalziel and Elliot, 1973, p. 190 and following), but are older than the fossiliferous Trinity Peninsula Series of Alexander Island and Cape Legoupil. Thus, the manifold Patagonian history of Paleozoic time (Miller, 1976) cannot be adapted easily into the scheme of a uniform "Samfrau geosyncline" with exclusively early Mesozoic folding.

The age of folding of the pre-Andean rocks on both sides of the Scotia Sea

In the literature, the age of deformation of the "Trinity Peninsula Series" and comparable sedimentary rocks ("Gondwanian orogeny") is almost universally recorded as being Late Triassic to Early Jurassic (de Wit, 1977, p. 68 and following). This was, in principle, based on radio-

Table 13.1. Basement rocks of the southernmost Andes and Antarctic Peninsula (including the southern branch of the Scotia arc)

	Southern South America (Andean region only)		Antarctic Peninsula	
Triassic			Sandstone Graywacke Argillite Volcanic rocks Chert (weakly metamorphosed) ? ?	"Trinity Peninsula Series"
Permian	Shale Quartzite Metavolcanic rocks Limestone Chert	Madre de Dios series		
Carboniferous				
Devonian	Shale Graywacke Sandstone (weakly metamorphosed)	Potranca and Lago Cochrane Formations	? Gneiss	Radiometric ages, after Gledhill and others, Chapter 35, this volume
Silurian				
Ordovician	Phyllite Micaschist Greenschist Marble	Canal King and Lago General Carrera Formations		
Cambrian	?			
	?			
Precambrian	?			

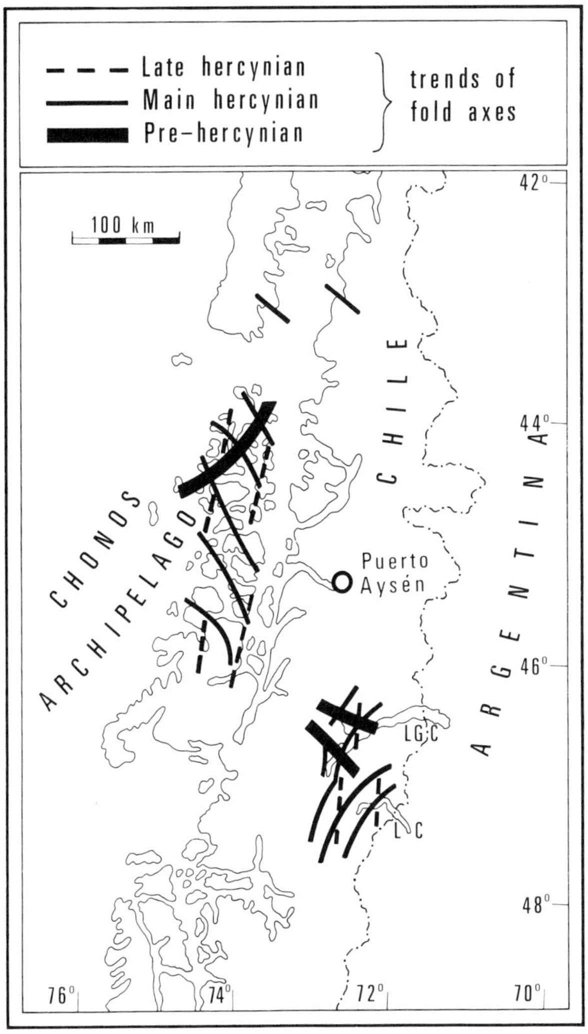

Figure 13.2. Trends of fold axes in the Paleozoic basement of the Aisén region, Chile. LGC: Lago General Carrera (= Lago Buenos Aires). LC = Lago Cochrane (= Lago Pueyrredón). Continental area after Lagally (1975).

metric ages (e.g., Adie, 1972; Dalziel, 1972*a*; Grikurov, 1972; Miller, 1960). However, Dalziel (1972*a*) and Rex and Baker (1973) have demonstrated that isolated age determinations are of little reliability in stratigraphically poorly known regions. New geochronologic work (Gledhill and others, Chapter 35, this volume) has finally shown the importance of Late Triassic–Early Jurassic metamorphism and plutonism in the Antaractic Peninsula. Together with the Triassic age of at least part of the "Trinity Peninsula Series," evidence now exists for a strong orogeny of Late Triassic–Early Jurassic age in the Antarctic Peninsula and Alexander Island.

The Permo-Carboniferous age of the major folding in Patagonia is proved by the marked flysch and molasse sedimentation of this age in the Tepuel-Languiñeo area. Further north, the main orogeny of the basement is also commonly of Permo-Carboniferous age (Hervé, 1976; Hervé and others, 1974; Hervé and others, 1976; Miller 1973, 1976; Munizaga and others, 1973). The "Gondwanian" folding of the Sierras Australes de Buenos Aires in Argentina, often quoted as being of Early Jurassic age, seems to have been completed mainly during Paleozoic time, with only slight aftereffects during Triassic time (Kilmurray, 1975). In contrast, the Late Triassic–Early Jurassic orogeny of the southern branch of the Scotia arc emphasizes another fundamental difference in development between South America and West Antarctica.

In previous work (Miller, 1966), the author had interpreted the structures in the Cape Legoupil area as being Alpidic, starting from the assumption that the deformed beds were Cretaceous in age (Halpern, 1964, 1965). Recent finds of Triassic fossils near Cape Legoupil, in rocks considered so far as Cretaceous or late Paleozoic (Thomson, 1975*a,b;* Halpern, 1964, 1965; Elliot, 1965), show that the complicated deformation in the surroundings of Cape Legoupil is probably of Late Triassic–Early Jurassic ("Gondwanian") age.

The Triassic plants from Livingston Island (Orlando, 1968; Hobbs, 1968, p. 25; Thomson, Chapter 37, this volume) did not take part in the metamorphism and strong folding of the Miers Bluff Formation (Dalziel, 1972*a*). If their Triassic age is confirmed, then the deformation of the Miers Bluff Formation would be pre-Triassic or Early Triassic.

Mesozoic and Cenozoic geology of the Scotia arc

All geologists working at the southern end of South America, on the Antarctic Peninsula, or on the islands of the Scotia arc, have used the overall geologic relations in discussing the Mesozoic-Cenozoic development of these regions. Dalziel and Elliot (1971, Figure 3; 1973, Figure 9), for instance, presumed a previous linear connection between South America and West Antarctic and, therefore, arranged their structural units according to this hypothesis. The author, however (Miller, 1967, p. 649 and following) has emphasized the differences between the South American Magellan geosyncline (including South Georgia) and the West Antarctic geosyncline (including the South Shetland Islands) (Table 13.2). Katz (1972, 1973) has also

Table 13.2. Andean development of the Andes and the Antarctic Peninsula. The similarity between the Antarctic Peninsula (including the South Shetland Islands) and the Central Andes is evident, just as is the contrast between the Magellan region and the other areas

	Magellan region	Antarctic Peninsula	Central Andes	Aisén region
Volcanism	First acid, then basic	Manifold	Manifold	Manifold
Folding	Strong	Mostly weak	Weak to moderate	Weak
Regional metamorphism	Weak to moderate	Lacking to weak	Lacking	Lacking
Typical flysch basin	Well developed	Lacking	Lacking	Lacking

pointed to general contrasts between the Mesozoic–Cenozoic history of the southern point of South America and the Antarctic Peninsula. Katz (1973, p. 356) and Dalziel (1976*b*) have attributed these differences to the lack of collisional events in West Antarctica.

Suárez (1976) and Suárez and Pettigrew (1976)—insisting on the close relationship between the continents—tried to explain the different geologic character of the southern end of South America and the Antarctic Peninsula by connecting geotectonic zones in an unusual way. They declared the Magellan geosyncline, in its classic sense (together with the Yaghan Formation of Tierra del Fuego and South Georgia Island), to be a back-arc basin, and they extended this basin along the eastern margin of the Antarctic Peninsula into the Lassiter Coast region. They regard the sedimentary sequences of Alexander Island as deposits of a fore-arc and intra-arc basin formerly situated west of the South American coast but not present there today.

The differences in development between the continents are not eliminated by this interpretation. The Antarctic Peninsula lacks the flysch-molasse sequence of the Upper Cretaceous and Tertiary which is typical for the Magellan region and which would be expected on the east side of the peninsula. It would be very unusual if these zones, together with a previous continental foreland, were completely consumed by subduction without causing strong deformation in the remaining regions.

Why, in particular, does it seem necessary to search for a synonymous development between the Mesozoic–Cenozoic of South Patagonia and the Antarctic Peninsua? In the opinion of the author, we should recognize that geosynclines, island arcs, and mountain belts do not surround the earth in a uniform, continuous belt.

The northern margin of the Antarctic plate in South America

Mesozoic and Cenozoic continental geology between lat. 40° and 50° S.

The Andean geosyncline terminates at about lat. 40° S. Farther south, typical geosynclinal sequences of Late Jurassic and Cretaceous age followed by intensive folding occur from lat. 50° S to the south. Between lat. 40° and 50° S in the Andean chain, however, there is a region of reduced activity coinciding with the termination of the Chile Ridge. The sequence there starts with acidic volcanic rocks, typical in the Upper Jurassic of the whole of Patagonia. It consists mainly of different types of volcanic rocks and a few clastic sedimentary rocks and, from the Upper Jurassic to the Miocene, it is hardly more than 2000 to 3000 m thick. Folding is minimal and fracture tectonics predominate; only at the eastern margin of the mountains (easily observed south of the Lago Pueyrredón) are the beds locally raised in gentle flexures. In the area of the islands and fjords, the Mesozoic is represented only by the Cretaceous Traiguén Formation (Espinosa and Fuenzalida, 1971, p. 17), consisting mainly of volcanic rocks. Southward to lat. 50° S no flysch whatsoever has developed; only farther south, below lat. 50° S, does the typical flysch-molasse sequence of the Magellan geosyncline begin at the eastern flank of the mountains. On the Pacific side of the Magellan region, intensive basic volcanism is found in the Lower Cretaceous, regarded by Dalziel and others (1974) as having the characteristics of a fossil marginal basin resulting from a mantle diapir (de Wit and others, 1976). The author believes that this short-lived basin, only about 20 km wide, is a little too small for any comparison with the recent, rather late orogenetic marginal basins of the western Pacific. The Traiguén Formation seems to correspond lithologically to the "oceanic" rocks of this "marginal basin." The rocks of the Traiguén Formation are very gently folded and overlie Paleozoic basement rocks.

No typical Mesozoic geosyncline whatsoever exists between lat. 40° and 50° S. This zone should not be termed a "transition zone," as is done by Aubouin (1975, Plate 1) and Cecioni and Charrier (1974, Figure 1), because it does not form a transition from the Andean to the rather different Magellan geosyncline. It constitutes a separate, independent element between two other geotectonic zones. Only the numerous granitic rocks continue as a common element from central Chile through the Aisén region into Tierra del Fuego.

Fracture pattern in Patagonia opposite the Chile Ridge

The Chile Ridge has been investigated repeatedly during recent years (Morgan and others, 1969; Herron and Hayes, 1969; Klitgord and others, 1973). The ridge is divided into short crestal segments by numerous transform faults. It represents the northern margin of the Antarctic plate and meets the South American continent at lat. 45° S. At this point, the Chilean earthquake zone ends abruptly (Figure 13.3). An evaluation of all recorded earthquakes with $M > 2.0$ (National Geophysical and Solar-Terrestrial Center, Boulder, Colorado) by Schmidt (1976) revealed a nearly east-west trend for the transform faults next to the continent, and a value of 335° for the approximate trend of the easternmost spreading axis.

The earthquakes nearly end at the continental margin south of lat. 46° S, although the Chile Trench can be followed gravimetrically without interruption to the south (Hayes, 1966, Figure 5). The slight seismicity south of lat. 46° S is a result of the very low subduction rate between the Antarctic and American plates in this region (Chase, 1972, p. 121). The differences in subduction activity (north of lat. 46° S, several cm/yr and south of lat. 46° S, nearly zero) suggest that pronounced lateral movements occurring between the Nazca plate in the north and the Antarctic plate in the south, at the southernmost transform fault of the Chile Ridge, would also be manifested on the continent.

Some authors (Stiefel, 1970, p. 978; 1972, p. 1123; Marangunić, 1971), therefore, have suggested that fracture patterns on the continent opposite the Chile Ridge are directly related to oceanic tectonics in the region of the Chile Ridge. To check these relationships, the author, in collaboration with Lagally (1975) and Schmidt (1976), investigated such fracture systems in two different areas (the Chonos Archipelago in the coastal area and the Lago General Carrera–Lago Cochrane in the interior). The results are as follows:

1. Joint fabrics in Paleozoic slates are directly connected with their Hercynian deformation. Their arrangement changes according to the direction of the corresponding fold axes (Figure 13.4*a*, *b*, and *c*).
2. Joint fabrics in Mesozoic granites directly depend on a stress system with east-west-trending principal horizontal stress. Within this system of deformation, north-south- and east-west-striking fractures are outstanding, as well as shear fractures striking diagonal to these (Figures 13.4*d* and *e*).
3. A simple system of shear fractures has developed in volcanic rocks of the Cretaceous(?) Traiguén Formation. It also fits well with the Andean system of deformation (Figure 13.4*f*).
4. Furthermore, evaluation of joint systems, without regard to rock formation, reveals a predominance of directions which, regarded as a whole, clearly correspond to the Andean orogeny, neither exhibiting relationships to the north-northwest-south-southeast strike of the spreading axes of the Chile Ridge, nor any prominence of the east-west direction of the transform faults (Figures 13.4*g*, *h*, *i*, and *j*).

The widespread occurrence of the east-west direction in granites, larger photolineaments, and Mesozoic andesitic dikes cannot be related to transform faults as this direction is almost absent in Paleozoic slates of the continental margin; however, in the sense of *ac*-joints, this direction may simply be attributed to the Andean deformation.

Therefore, it seems that the oceanic structure of the Chile Ridge ceases at the western margin of the Chile Trench. The present Chile Rise has no effect on continental fracture tectonics.

Tectonic segmentation of the Andean chain

It has been pointed out repeatedly that the South American Andes are divided into segments separated by transverse boundaries (Gerth, 1955, p. 224, p. 230; Stauder, 1973, p. 5954; Sillitoe, 1974; Barazangi and Isacks, 1976). One of these segment boundaries appears at about lat. 40° (at the northern margin of the north Patagonian massif), the next occurs at about lat. 46° S (at the southern margin of the seismic zone and the termination of the Chile Ridge), another appears at about lat. 50° S (beginning of the Magellan geosyncline with ophiolitic volcanism, metamorphism, and folding, a typical flysch trough, and a molasse trough), and the next boundary separates the northern and southern branches of the Scotia arc.

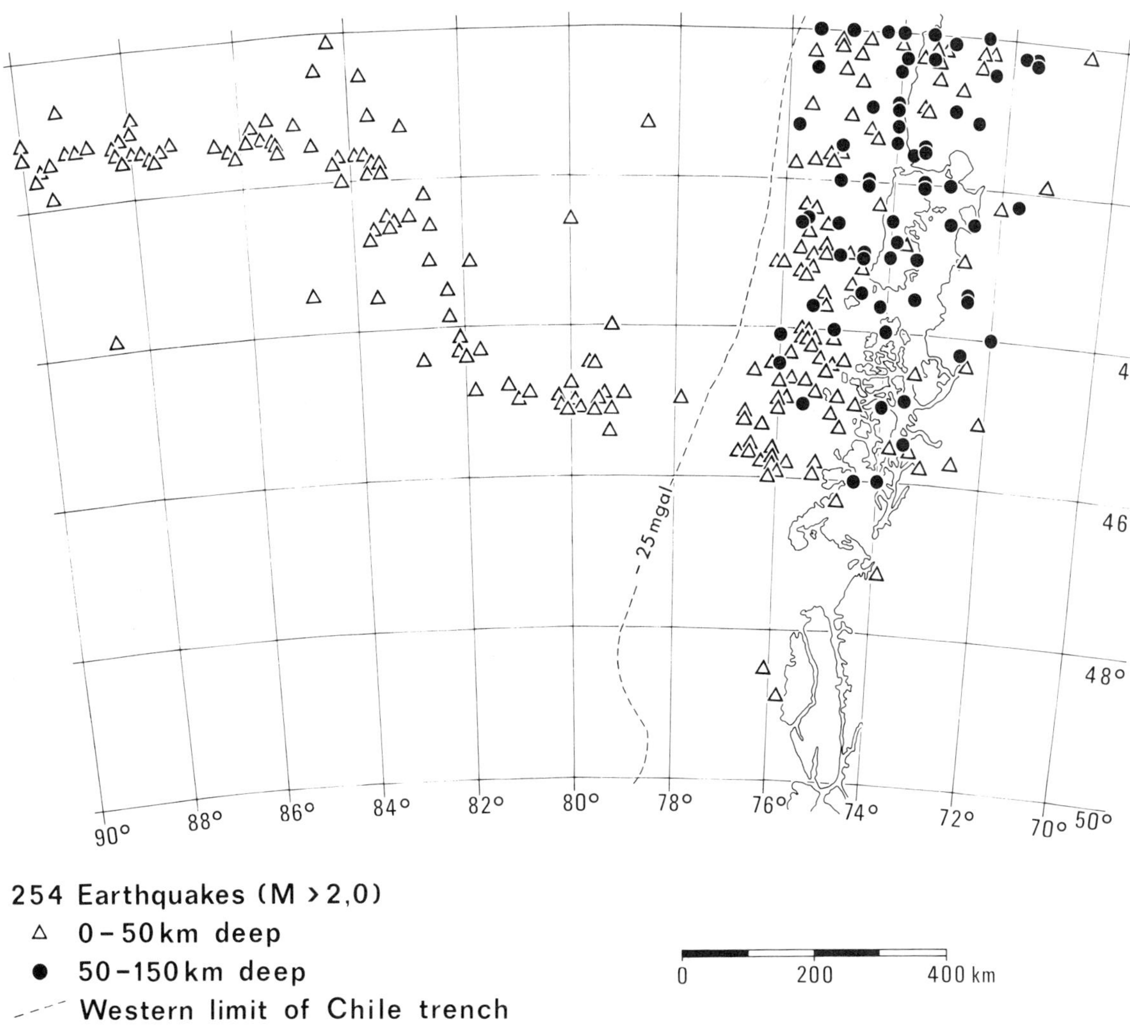

Figure 13.3. Earthquake distribution at the Chonos triple junction (Aisén region, Chile), after Schmidt (1976). The −25 mgal line after Hayes (1966).

This differentiated development is partly evident by Triassic time (there is a lack of Triassic rocks in the Andes south of lat. 41° S); it becomes more pronounced during Jurassic and Cretaceous time. As indicated by the seismicity, this segmentation is still evident today. Only the wide range of magmatic processes occurred in all parts. Some differences are still present, however, between magmatic provinces (Katsui, 1972; González-Ferrán, 1972, Figure 1*a*; Stern and others, 1976).

Data are known from the central Andean region that oppose the exclusive application of plate-tectonic models for Andean events (Scholl and others, 1970; Scholl and Marlow, 1974*a,b*). Also, a geologic history exclusively controlled by Pacific oceanic plates does not fit the geologic facts in the immediate area of the Scotia Sea. Subduction zones, constantly developing at the western coast of southernmost South America since Cretaceous time (Barker and Griffiths, 1972, Figure 12; Larson and Pitman, 1972, Figure 7) cannot explain the difference in geologic development of the Andean and West Antarctic region, even if temporary changes in spreading velocity are taken into account (Larson and Pitman, 1972).

Attempts to explain the complicated nature of this region by dividing the Scotia arc into numerous single plates are, therefore, appreciated (Barker and Griffiths, 1972, Figure 14). This paper will not present another microplate puzzle, but the author feels that first it is necessary to investigate further the still poorly known stratigraphic and tectonic facts of the Antarctic Peninsula and southernmost South America before emphasizing if and how these investigation results may fit plate tectonic models.

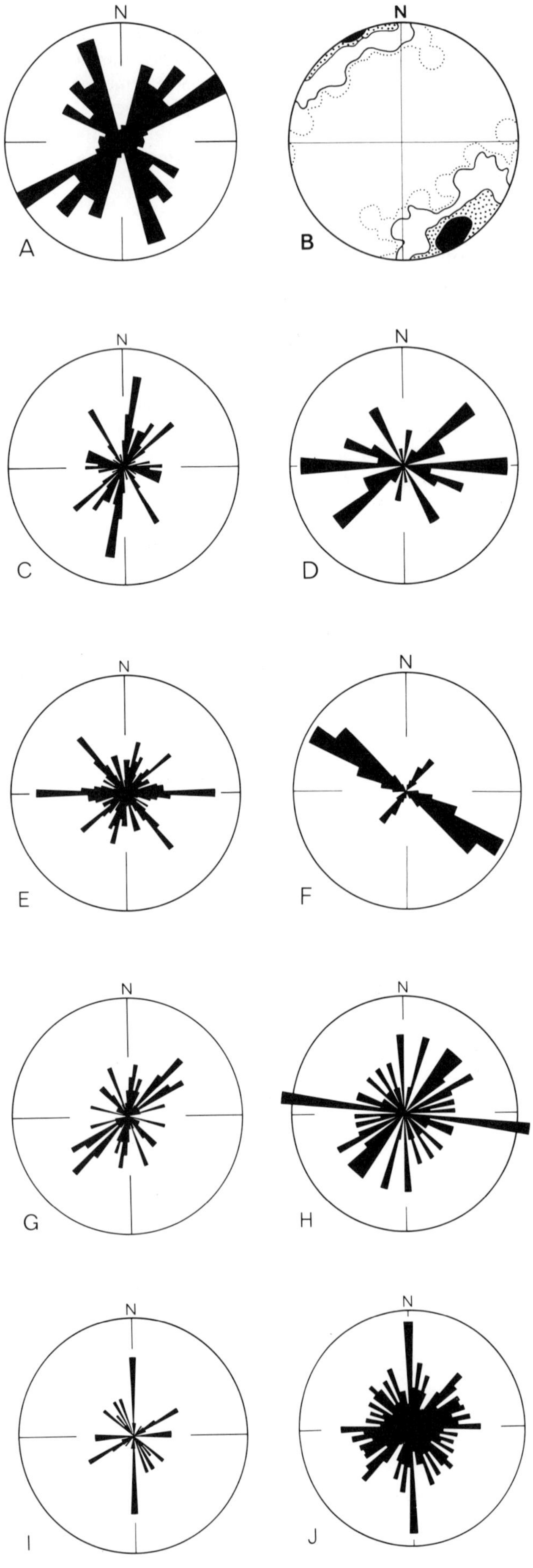

Figure 13.4. Structural pattern of some continental areas opposite the Chile Ridge. A. Trends of 1179 photolineaments from 11 areas of Paleozoic rocks on Chonos Archipelago, after Schmidt (1976). The 140° to 160° fractures are parallel to the main Hercynian fold axes (Figure 13.4, B). The 60° fractures are cross fractures (*ac*) of the main Hercynian folding. B. Maxima of main Hercynian fold axes from 146 locations on Chonos Archipelago. Contours are at 0.6, 1.4, 5.5, and 11%. C. Fractures (69) in Paleozoic schists at Bahía Murta, Lago General Carrera, after Lagally (1975). The main peak (10°) coincides with the maximum of *b*-axes of folds on the site of measurement. The 100° maximum is attributed to cross fractures (*ac*). D. Trends of 470 photolineaments from four areas of Mesozoic-Cenozoic granites on Chonos Archipelago, after Schmidt (1976). E. Photolineaments (170) of granites from Lago General Carrera–Lago Cochrane area, after Lagally (1975). F. Photolineaments (90) from the Cretaceous(?) Traiguén Formation east of Chonos Archipelago, after Schmidt (1976). G. Faults (64) from Lago General Carrera–Lago Cochrane area, after Lagally (1975). H. Andesitic dikes (48) from Lago General Carrera–Lago Cochrane area, after Lagally (1975). I. Directions of 39 rivers and lakes from Lago General Carrera–Lago Cochrane area, after Lagally (1975). J. Photolineaments (2637) from Lago General Carrera–Lago Cochrane area, after Lagally (1975). No clear relationship to transform faults or spreading axes of the Chile Ridge is visible, but the symmetry of the Andean orogeny is well expressed.

Acknowledgments. I first met with the problems concerning West Antarctica during my 1964 stay at the Base General Bernardo O'Higgins at Cape Legoupil, and I wish to thank the workers at the Instituto Antártico Chileno and all their associates for their efforts to make this stay possible. I am grateful to the Universidad Austral de Chile, Valdivia, the Instituto des Investigaciones Geólogicas, Santiago, and the Deutsche Forschungsgemeinschaft, Bonn (Grant Mi 120/4-5), for providing assistance to carry out research work in Patagonia. Furthermore, I wish to thank M. Hervé, M. Suárez, and U. Lagally, as well as the crew of the small boat *Guamblín* for their assistance in the field. I am grateful to Pedro Sprechmann, Tübingen, for the determination of the poorly preserved fauna of Potranca Island.

References cited

Adie, R. J. 1972. Recent advances in the geology of the Antarctic Peninsula. (*In* Adie, R. J., *ed. Antarctic Geology and Geophysics.* Oslo, Universitetsforlaget, 121-24.)

Aubouin, M. J. 1975. Réflexion sur les bordures Pacifiques: l'exemple des Cordillères sud-américaines. *C. r. Acad. Sci. Paris, 280,* D, 2633-36.

Barazangi, M., and Isacks, B. L. 1976. Spatial distribution of earthquakes and subduction of the Nazca plate beneath South America. *Geology, 4,* 686-92.

Barker, P. F., and Griffiths, D. H. 1972. Evolution of the Scotia Ridge and Scotia Sea. *Phil. Trans. R. Soc., Lond.,* Ser. A, *271,* 151-83.

Cecioni, G. 1956. Primeras noticias sobre la existencia del Paleozoico Superior en el Archipiélago Patagónico entre los paralelos 50° y 52° S. *An. Fac. Cienc. Fís. mat. Univ. Chile, 13,* 183-202.

Cecioni, G., and Charrier, R. 1974. Relaciones entre la Cuenca Patagónica, la Cuenca Andina y el Canal de Mozambique. *Ameghiniana, 11,* 1-38.

Chase, C. G. 1972. *N* plate problem of plate tectonics. *Geophys. J., 29,* No. 2, 117-22.

Dalziel, I. W. D. 1972*a*. Large-scale folding in the Scotia arc. (*In* Adie, R. J., *ed. Antarctic Geology and Geophysics.* Oslo, Universitetsforlaget, 47-55.)

Dalziel, I. W. D. 1972*b*. K-Ar dating of rocks from Elephant Island, South Scotia Ridge. *Geol. Soc. Am. Bull., 83,* No. 6, 1887-94.

Dalziel, I. W. D. 1976*a*. Structural studies in the Scotia arc: "basement" rocks of the South Shetland Islands. *Antarct. Jnl U.S., 11,* No 2, 75.

Dalziel, I. W. D. 1976*b*. Structural evolution of the southern limb of the Scotia arc. *Abstr. 25th Int. Geol. Congr., Sydney, 1976, 3,* 680.

Dalziel, I. W. D., and Elliot, D. H. 1971. Evolution of the Scotia arc. *Nature, Lond., 244,* 246-52.

Dalziel, I. W. D., and Elliot, D. H. 1973. Scotia arc and Antarctic margin. (*In* Nairn, A. E. M., and Stehli, F. G., *eds. The Ocean Basins and Margins. 1. The South Atlantic.* New York, Plenum Press, 171-246.)

Dalziel, I. W. D., de Wit, M. J., and Palmer, K. R. 1974. Fossil marginal basin in the southern Andes. *Nature, Lond., 250,* No. 5464, 291-94.

de Wit, M. J. 1977. The evolution of the Scotia arc as a key to the reconstruction of southwestern Gondwanaland. *Tectonophysics, 37,* No. 1-3, 53-81.

de Wit, M. J., Dalziel, I. W. D., Stern, C. R., and Bruhn, R. 1976. The development and destruction of a Mesozoic ensialic marginal basin. *Abstr. 25th Int. Geol. Congr., Sydney, 1976, 1,* 107-8.

Elliot, D. H. 1965. Geology of north-west Trinity Peninsula, Graham Land. *Bull. Br. antarc. Surv., 7,* 1-24.

Espinosa, W., and Fuenzalida, R. 1971. Geología de las hojas Riveros, Puerto Aisén y Balmaceda entre los paralelos 45° y 46° de latitud Sur. *Unpubl. Rep., Inst. Invest. Geol., Santiago, Chile,* 36 pp.

Gerth, H. 1955. *Der Geologische Bau der Südamerikanischen Kordillere.* Berlin, Borntraeger, 264 pp.

González-Ferrán, O. 1972. Distribution, migration and tectonic control of Upper Cenozoic volcanism in West Antarctica and South America. (*In* Adie, R. J., *ed. Antarctic Geology and Geophysics.* Oslo, Universitetsforlaget, 173-78.)

Grikurov, G. E. 1972. Tectonics of the Antarctandes. (*In* Adie, R. J., *ed. Antarctic Geology and Geophysics.* Oslo, Universitetsforlaget, 163-67.)

Halpern, M. 1964. Cretaceous sedimentation in the "General Bernardo O'Higgins" area of north-west Antarctic Peninsula. (*In* Adie, R. J., *ed. Antarctic Geology.* Amsterdam, North-Holland Publishing Company, 334-47.)

Halpern, M. 1965. The geology of the General Bernardo O'Higgins area, northwest Antarctic Peninsula. (*In* Hadley, J. B., *ed. Geology and Paleontology of the Antarctic.* Washington, American Geophysical Union, 177-209.) [Antarctic Research Series, Vol. 6.]

Hayes, D. E. 1966. A geophysical investigation of the Peru-Chile Trench. *Mar. Geol., 4,* 309-51.

Herron, E. M., and Hayes, D. E. 1969. A geophysical study of the Chile Ridge. *Earth planet. Sci. Lett., 6,* 77-83.

Hervé, F. 1976. Superimposed folding and metamorphism in the Laraquete-Colcura area (metamorphic basement of central Chile). *Münster. Forsch. Geol. Paläont., 38-39,* 99-110.

Hervé, F., Munizaga, F., Godoy, E., and Aguirre, L. 1974. Late Paleozoic ages of blueschists from Pichilemu, central Chile. *Earth planet. Sci. Lett., 23,* 261-64.

Hervé, F., Munizaga, F., Mantovani, M., and Hervé, M. 1976. Edades Rb/Sr neopaleozoicas del basamento cristalino de la Cordillera de Nahuelbuta. *Act. 1 Congr. Geol. Chil., 2,* 19-26.

Hobbs, G. J. 1968. The geology of the South Shetland Islands. IV. The geology of Livingston Island. *Scient. Rep. Brit. antarc. Surv., 47,* 34 pp.

Katsui, Y. 1972. Late Cenozoic volcanism and petrographic provinces in the Andes and Antarctica. *J. Fac. Sci. Hokkaido Univ., 15,* No. 1-2, 27-39.

Katz, H. R. 1972. Plate tectonics and orogenic belts in the south-east Pacific. *Nature, Lond., 237,* No. 5354, 331-32.

Katz, H. R. 1973. Contrasts in tectonic evolution of orogenic belts in the south-east Pacific. *J. R. Soc. N. Z., 3,* No. 3, 333-62.

Kilmurray, J. O. 1975. Las Sierras Australes de la provincia de Buenos Aires. Las fases de deformación y nueva interpretación estratigráfica. *Revta. Asoc. geol. argent., 30,* 331-48.

Klitgord, K. D., Mudie, J. D., Larson, P. A., and Grow, S. A. 1973. Fast sea-floor spreading on the Chile Ridge. *Earth planet. Sci. Lett., 20,* 93-99.

Lagally, U. 1975. *Geologische Untersuchungen im Gebiet Lago General Carrera—Lago Cochrane, Prov. Aisén/Chile, unter besonderer Berücksichtigung des Grundgebirges und seiner Tektonik.* Ph.D. dissertation, University of Munich, 131 pp.

Larson, R. L., and Pitman, W. C., III 1972. World-wide correlation of Mesozoic magnetic anomalies, and its implications. *Geol. Soc. Am. Bull., 83,* 3645-63.

Lesta, P. J., and Ferello, R. 1972. Región extraandina de Chubut y Norte de Santa Cruz. (*In* Leanza, A. F., *ed. Geología Regional Argentina.* Córdoba, Academia Nacional de Ciencias, 601-53.)

Marangunić, Č. 1971. Sistemas de fracturas en los Canales Patagónicos y posible relación con la mesodorsal de Chile Occidental. *Publnes Dep. Geof. Geol. Univ. Chile, 104,* 11 pp.

Miller, H. 1966. Kleintektonische Untersuchungen in der Umgebung der Station "General Bernardo O'Higgins," Nordwestspitze der Antarktischen Halbinsel. *Geol. Rdsch., 55,* No. 3, 809-19.

Miller, H. 1967. Kritische Betrachtungen zum geologischen Zusammenhang zwischen den chilenischen Anden und der Westantarktis. *Geol. Rdsch.. 56,* No. 2, 643-56.

Miller, H. 1970. Vergleichende Studien an prämesozoischen Gesteinen Chiles unter besonderer Berücksichtigung ihrer Kleintektonik. *Geotekt. Forsch.,* 36, 64 pp.

Miller, H. 1973. Características estructurales del basamento geológico chileno. *Act. 4 Congr. Geol. Argent., 4,* 101–15.

Miller, H. 1976. El basamento de la provincia de Aysén (Chile) y sus correlaciones con las rocas premesozoicas de la Patagonia Argentina. *Act. 6 Congr. Geol. Argent., 1,* 125–41.

Miller, H., and Sprechmann, P. 1978. Eine devonische Faunula aus dem Chonos-Archipel, Region Aisén, Chile, und ihre stratigraphische Bedeutung. *Geol. Jb., 28,* 37–45.

Miller, J. A. 1960. Potassium-Argon ages of some rocks from the South Atlantic. *Nature, Lond., 187,* No. 4742, 1019–20.

Morgan, W. J., Vogt, P. R., and Falls, D. F. 1969. Magnetic anomalies and sea-floor spreading on the Chile rise. *Nature, Lond., 222,* 137–42.

Munizaga, F., Aguirre, L., and Hervé, F. 1973. Rb/Sr ages of rocks from the Chilean metamorphic basement. *Earth planet. Sci. Lett., 18,* 87–92.

Orlando, H. A. 1968. A new Triassic flora from Livingston Island, South Shetland Islands. *Bull. Br. antarc. Surv., 16,* 1–13.

Rex, D. C., and Baker, P. E. 1973. Age and petrology of the Cornwallis Island Granodiorite. *Bull. Br. antarc. Surv.,* No. 32, 55–61.

Rivano, S., and Cortés, R. 1976. Note on the presence of the lawsonite-sodic amphibole association on Smith Island, South Shetland Islands, Antarctica. *Earth planet. Sci. Lett., 29,* 34–36.

Schmidt, W. 1976. *Lineamenttektonik und Seismizität am Ostende des Chile-Rückens (Südamerika).* Diplom-thesis, Münster University, 99 pp.

Scholl, D. W., and Marlow, M. S. 1974*a*. Global tectonics and the sediments of modern and ancient trenches: some different interpretations. (*In* Kahle, C. F., *ed.* Plate Tectonics—Assessments and Reassessments. *Mem. Am. Ass. Petrol. Geol., 23,* 255–72.)

Scholl, D. W., and Marlow, M. S. 1974*b*. Sedimentary sequence in modern Pacific trenches and the deformed circum-Pacific eugeosyncline. (*In* Dott, R. H., Jr., and Shaver, R. H., *eds.* Modern and Ancient Geosynclinal Sedimentation. *Spec. Publs Soc. econ. Paleont. Miner., Tulsa, 19,* 193–211.)

Scholl, D. W., Christensen, M. N., v. Huene, R., and Marlow, M. S. 1970. Peru-Chile trench sediments and sea-floor spreading. *Geol. Soc. Am. Bull., 81,* 1339–60.

Sillitoe, R. H. 1974. Tectonic segmentation of the Andes: implications for magmatism and metallogeny. *Nature, Lond., 250,* No. 5467, 542–45.

Smellie, J. L., and Clarkson, P. D. 1975. Evidence for pre-Jurassic subduction in western Antarctica. *Nature, Lond., 258,* No. 5537, 701–2.

Stauder, W. 1973. Mechanism and spatial distribution of Chilean earthquakes with relation to subduction of oceanic plate. *J. geophys. Res., 78,* No. 23, 5033–61.

Stern, C., Skewes, M. A., and Durán, M. 1976. Volcanismo orogénico en Chile Austral. *Act. 1 Congr. Geol. Chil., 2,* F 195–212.

Stiefel, J. 1970. Das Andenprofil im Bereich des 45. südlichen Breitengrades. *Geol. Rdsch., 59,* No. 3, 961–79.

Stiefel, J. 1972. Zur tektonischen Entwicklung Chiles im Känozoikum. *Geol. Rdsch., 61,* No. 3, 1109–25.

Suárez, M. 1976. Plate-tectonic model for southern Antarctic Peninsula and its relation to southern Andes. *Geology, 4,* No. 4, 211–14.

Suárez, M., and Pettigrew, T. H. 1976. Upper mesozoic island-arc-back-arc system in the southern Andes and South Georgia. *Geol. Mag., 113,* No. 4, 305–28.

Thomson, M. R. A. 1975*a*. First marine Triassic fauna from the Antarctic Peninsula. *Nature, Lond., 257,* No. 5527, 557–78.

Thomson, M. R. A. 1975*b*. New palaeontological and lithological observations on the Legoupil Formation, north-west Antarctic Peninsula. *Bull. Br. antarc. Surv.,* No. 41–42, 169–85.

14

The Petrochemistry of the Patagonian Batholith, Ultima Esperanza, Chile

CHARLES R. STERN*
Lamont-Doherty Geological Observatory of Columbia University, Palisades, New York 10964, U.S.A.

JANET B. STROUP
Department of Geological Sciences, State University of New York at Albany, Albany, New York 12222, U.S.A.

Abstract. Petrochemical study of specimens collected from a traverse of the Patagonian batholith, Ultima Esperanza, Chile, indicates that the plutons of the batholith belong to the calc-alkaline rock series, the major rock type is tonalite, and the average composition of the batholith is comparable to the average circum-Pacific andesite. The batholith displays transverse petrochemical variations similar to those of other circum-Pacific batholiths. For plutons of similar SiO_2 content, both the K_2O content and the Ce/Yb ratio increase to the east. These chemical data support the suggestion that the batholith originated from subduction-related igneous processes. Trace-element data suggest that within plutons lying progressively farther east, amphibole fractionation has been increasingly more important than clinopyroxene fractionation in producing the observed chemical variations.

The Patagonian batholith crops out along the Pacific margin of southernmost South America as an essentially linear feature on a regional scale, rarely exceeding 100 km in width while extending from Cape Horn northward for more than 2000 km. Reconnaissance field studies (Krank, 1932) indicate that the batholith is a complex composite of individual plutons; tonalite and granodiorite are the dominant plutonic rock types, with lesser amounts of diorite, adamellite, and rare gabbro. Individual plutons have yielded radiometric ages spanning the interval from 166 to 12 m.y. ago (Halpern, 1973; Skarmeta, 1974). Detailed studies of individual plutons within the batholith are lacking, but its general similarity in extent and rock type to other circum-Pacific batholiths has been the basis for the suggestion that it represents the root of a subduction-related volcanic arc. Although uplift and erosion have exposed the batholith to a deep level, the possible volcanic equivalents of the plutonic rocks of the batholith have been preserved as the basalt-andesite-dacite volcanic rocks that form part of the Hardy Formation in the vicinity of Cape Horn (Suarez and Pettigrew, 1976).

In this paper we present the results of a geochemical study of specimens collected in a west-to-east traverse of the batholith, to the west of and within the Ultima Esperanza region of southern Chile, between lat. 51° and 52° S (Figures 14.1 and 14.2). The data indicate that the plutons of the batholith belong to the calc-alkaline rock series. For plutons of similar SiO_2 content, the batholith displays an eastward increase as both K_2O content and Ce/Yb ratio, which is similar to the asymmetric petrochemical variations observed within other circum-Pacific batholiths. These data are consistent with the hypothesis that the batholith originated as a result of subduction-related igneous processes involving a complex sequence of partial melting and crystal-liquid interaction of subducted oceanic lithosphere and of both the overlying mantle and the lower continental crust, as well as crystal-liquid fractionation within both subcrustal and crustal magma chambers.

Field relationships of the batholith

Figure 14.1 shows the regional outcrop pattern of the Patagonian batholith and other major rock units south of lat. 50° S, while Figure 14.2 depicts the geology between lat. 51° and 52° S in more detail, along with published radiometric ages for plutons in this region and specimen localities of some of the rocks collected for this study. The geology of the southernmost Andes has been discussed in a number of recent publications (Dalziel and others, 1974; Suarez and Pettigrew, 1976; de Wit, 1977; Bruhn and others, 1978). The building of the present Andean Cordillera south of lat. 50° S was initiated in Middle Jurassic time with the extrusion over the whole of southern South America of a vast field of dominantly silicic volcanic rocks (Bruhn and others, 1978). These volcanic rocks everywhere rest unconformably on a previously deformed and metamorphosed basement complex. Between lat. 51° and 52° S, this basement complex flanks the batholith on its Pacific margin (Figure 14.2). A radiometric age of 166 m.y. for a pluton located near lat. 47° S (Skarmeta, 1974) suggests that in some areas plutonism along the Pacific continental margin was contemporaneous with silicic volcanism during Middle Jurassic time.

A number of plutons dated radiometrically fall in the age bracket of 155–120 m.y., which led Halpern (1973) to suggest that this time interval represents an episode of plutonism. South of lat. 49° S, these plutons are located on the Pacific side of the outcrop belt of ophiolite complexes, the "rocas verdes," which have been interpreted as the mafic floor of a marginal basin that existed during Early Cretaceous time (Figures 14.1 and 14.2) (Dalziel and others, 1974). These plutons are intruded into and contain many roof pendants of pre-Jurassic metasedimentary rocks and Jurassic silicic volcanic rocks, both of which

* Present address: Department of Geological Sciences, University of Colorado, Boulder, Colorado 80309, U.S.A.

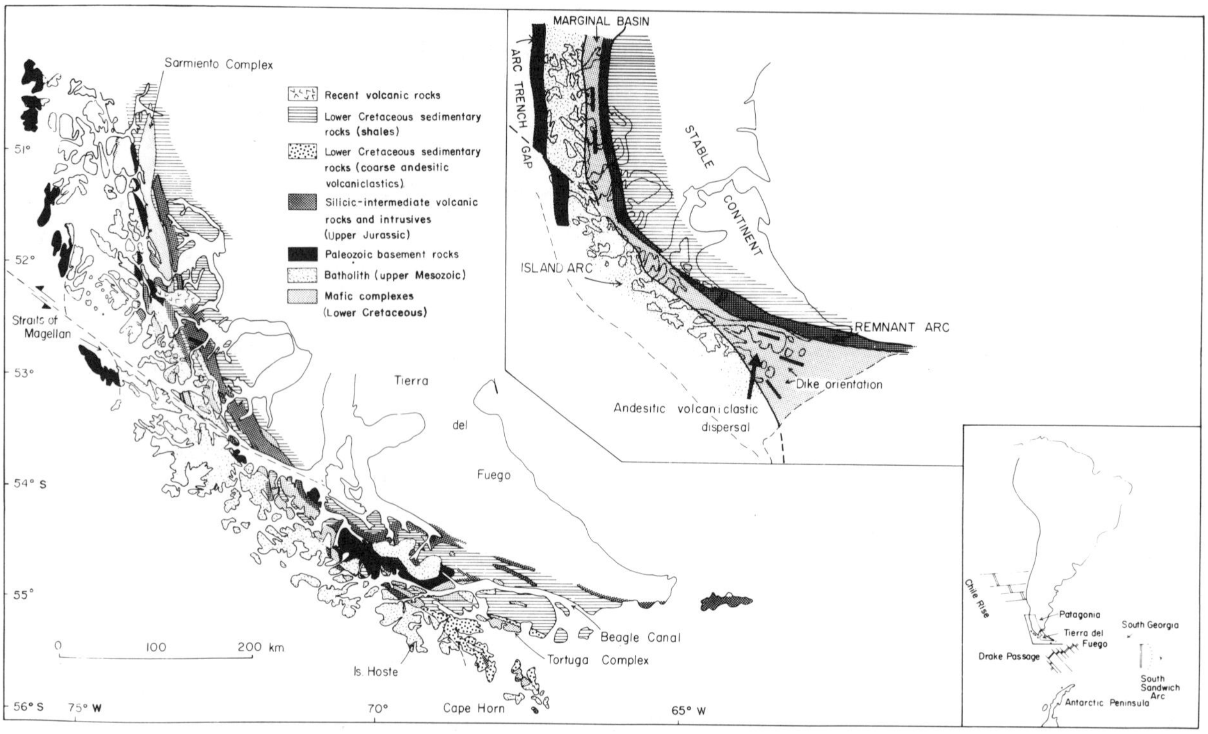

Figure 14.1. Geologic map of southern Chile. Inset shows major stratigraphic-tectonic units during Early Cretaceous time.

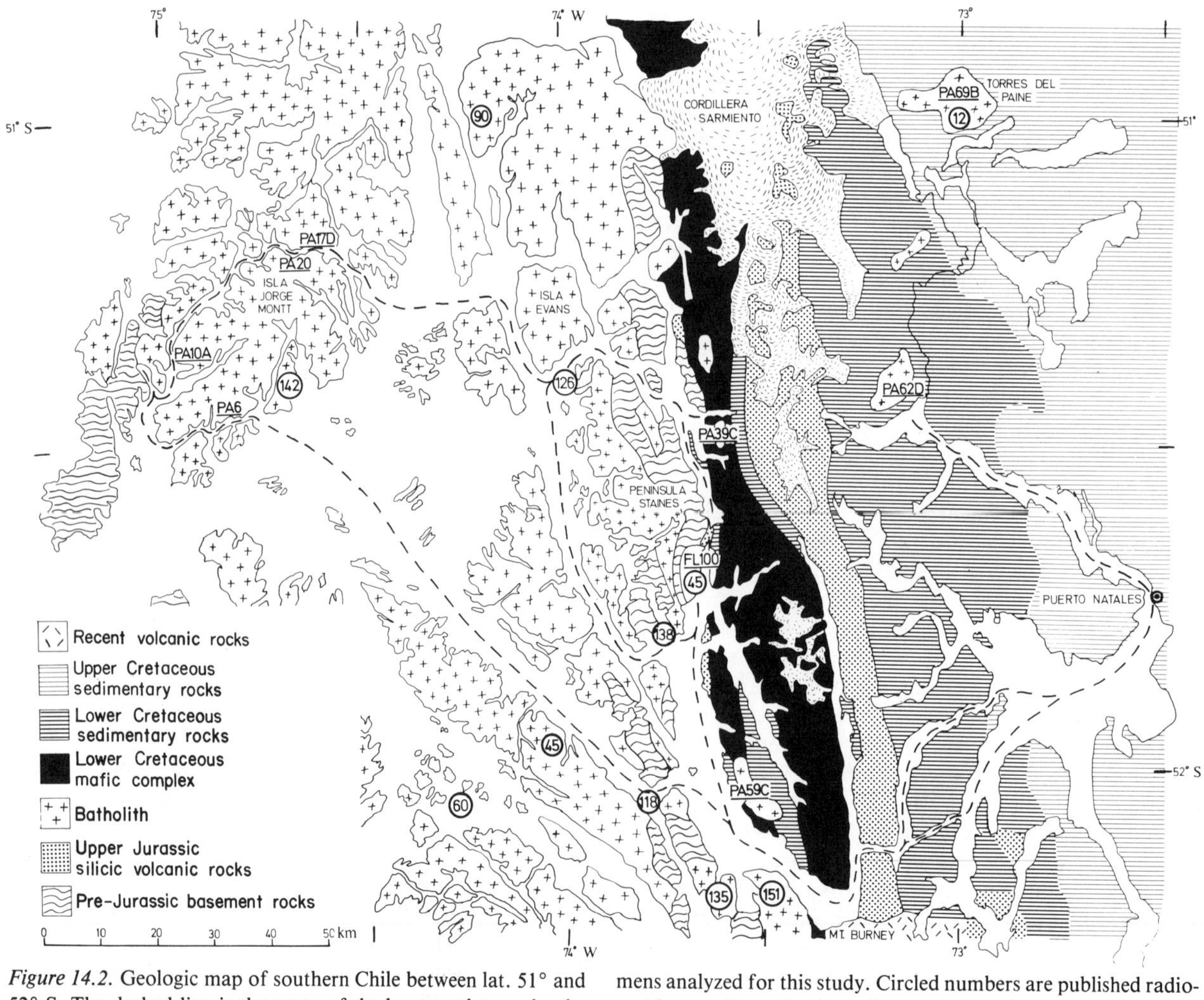

Figure 14.2. Geologic map of southern Chile between lat. 51° and 52° S. The dashed line is the route of the boat used to make the sampling traverse. Underlined numbers show locations of specimens analyzed for this study. Circled numbers are published radiometric age determinations (in m.y.) for plutons in this region (Halpern, 1973).

have counterparts in the pre-Cretaceous South American continental crust lying to the east of the ophiolites. This indicates that the plutons south of lat. 49° S were emplaced into a section of continental crust which was separated from stable South America to the east by the formation of the marginal basin (Dalziel and others, 1974; Suarez and Pettigrew, 1976). During Early Cretaceous time, this marginal basin was the site of deposition of volcaniclastic flysch derived dominantly from the Pacific side of the basin. Basalt, andesite, and dacite lava and breccia occur within the Hardy Formation, a facies of the basin sedimentary rocks found on the Pacific side of the ophiolite belt in the vicinity of Cape Horn (Figure 14.1) (Suarez and Pettigrew, 1976). This suggests that the Lower Cretaceous plutons were emplaced beneath an active volcanic island arc.

Deformation of the volcaniclastic infilling of the marginal basin occurred in mid-Cretaceous time, between Aptian and Coniacian time. An eastward migration of the easternmost extent of plutonism is indicated by the many plutons which are found cutting both the deformed basin sedimentary rocks and, in places, the ophiolites (Figures 14.2). These post-deformation plutons have radiometric ages of less than 100 m.y.

The deformation and uplift of the marginal basin in mid-Cretaceous time resulted in the formation of a tectonic basin farther to the east which became the site of deposition of Upper Cretaceous and Cenozoic sedimentary beds (Bruhn and Dalziel, 1977; Dott and others, Chapter 21, this volume). These foreland basin strata now crop out to the east of the spine of the southern Andean Cordillera, which is composed of an uplifted belt of deformed Jurassic silicic volcanic rocks (Figures 14.1 and 14.2). Continued eastward migration of the easternmost extent of plutonism is suggested by the occurrence of plutons within the precordillera, cutting these foreland basin sedimentary rocks. The famous Torre del Paine pluton (Figure 14.2), dated as 12 m.y. in age, and the Fitz Roy Massif are located within the precordillera.

Uplift and deformation initiated in mid-Cretaceous time and continuing, perhaps episodically, through the Cenozoic Era up to the present (Bruhn and Dalziel, 1977; Winslow, Chapter 15, this volume) have dissected the western edge of the Patagonian batholith to a deep level. Bruhn and Dalziel (1977) have suggested that 5 km or more of uplift of the western margin of the batholith occurred during mid-Cretaceous time alone, with additional, uplift continuing to the present. Bruhn and Dalziel (1977) noted that deformation within the Lower Cretaceous volcanic arc rocks is limited and that pretectonic plutons commonly exhibit no deformation textures. The margins of many plutons exhibit planar fabrics, but these may be due to processes of intrusion rather than to regional deformation; they cannot be used as criteria for assigning a pretectonic or posttectonic age to a pluton.

Along the western margin of the batholith, contacts between plutons and country rock or other plutons are subvertical. Here the country rock rarely exhibits contact-metamorphic effects, which is consistent with the present deep level of exposure and possibly with the relatively dry nature of the magmas (Dalziel and others, 1975). Plutons which intrude the Lower and Upper Cretaceous sedimentary rocks along the eastern margin of the batholith commonly have subhorizontal contacts with the country rock. Contact metamorphism is more prevalent in the vicinity of these plutons.

Many of the plutons of the batholith are cut by thin, steeply inclined diabase dikes. Synplutonic diabase dikes are common, displaying various stages of disruption and assimilation. Katz (1962) has discussed the tectonic significance of such dikes cutting the Miocene Torre del Paine pluton.

In this paper, we divide that part of the batholith cropping out between lat. 51° and 52° S into three separate zones: the western belt lying west of the outcrop belt of ophiolites, the central belt where plutons cut ophiolites and Lower Cretaceous sedimentary rocks, and the eastern belt within the precordillera where plutons cut Cretaceous and Cenozoic sedimentary rocks.

As noted above, the easternmost limit of plutonism seems to have migrated eastward with time. The central and eastern belts do not contain Lower Cretaceous plutons, but the western belt contains Upper Cretaceous and Cenozoic as well as Lower Cretaceous plutons (Figure 14.2). This indicates that the eastward migration of the easternmost limit of plutonism resulted from a widening rather than an eastward shift in the zone of plutonism. In this regard, it is significant that Mount Burney, an active calc-alkaline volcano at lat. 52° S (Stern and others, 1976), is located within the central zone of plutons and not within the eastern zone (Figure 14.2).

Petrography

Figure 14.3 shows a classification based on modal mineralogy of the rock types found in the Patagonian batholith between lat. 51° and 52° S. Western, central, and eastern plutons are shown by different symbols. Table 14.1 lists the modal mineralogy of selected specimens identified in Figure 14.3 by solid symbols and located in Figure 14.2. Besides the mineralogic variations shown, textures are variable, and both equigranular and plagioclase-phyric varieties are present.

Figure 14.3 indicates that, within the western part of the batholith, the greatest proportion of rocks are tonalites

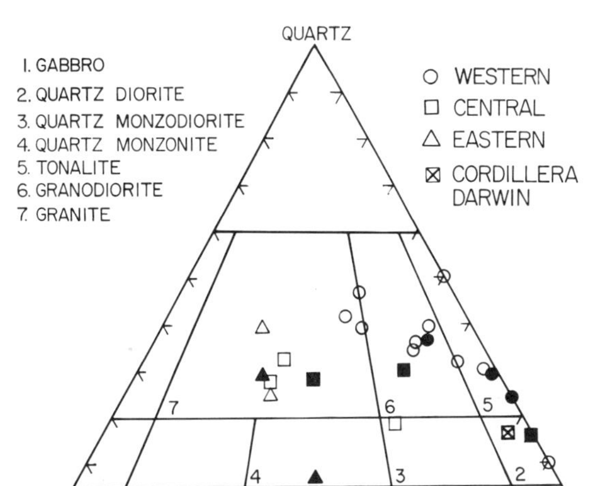

Figure 14.3. Classification based on modal mineralogy (Streckeisen, 1976) of plutons of the Patagonian batholith. Solid symbols are speciments listed in Table 14.1 and located on Figure 14.2. Note that western plutons are dominantly composed of tonalite and granodiorite.

and granodiorites. In the central and eastern sections of the batholith, granites become more common. Nevertheless, the western part of the batholith is volumetrically much more significant because of its greater areal extent (Figure 14.2). Tonalite and granodiorite are thus volumetrically the most important rock types of the Patagonian batholith, as they are in other circum-Pacific batholiths.

Table 14.1 lists the volumetric proportion of the selected rocks as estimated from our traverse, the route of which is illustrated in Figure 14.2. In the region studied, plutons lying west of the outcrop belt of ophiolites comprise 80% of the exposed area of the Patagonian batholith. Farther south, in the vicinity of the Cordillera Darwin at lat. 54° S (Figure 14.1), plutons within the central zone are more significant, comprising at least 40% of the exposed area of batholith.

Tonalite specimen PA20 makes up 75% of the western batholith or 60% of the total batholith between lat. 51° and 52° S. Almost all of Isle Jorge Montt, Isle Evans, and the western part of Peninsula Staines (Figure 14.2) are composed of this rock type. More mafic rocks in most places occur as border phases of large tonalite or granodiorite plutons. Where contacts were observed, the mafic phases are usually intruded by the more intermediate phases. This is similar to the sequence observed in other circum-Pacific batholiths (Cobbing and Pitcher, 1972).

Major-element chemistry

Table 14.2 lists the major-element chemistry of the same rocks whose modal mineralogy is tabulated in Table 14.1. Major-element abundances were determined by microprobe analysis of unfluxed glasses fused from splits of powdered rock samples, using minerals as primary standards. Fusion was done rapidly (<30 sec), at approximately 1700° C on a molybdenum strip under 40 psi argon pressure in order to minimize alkali loss. The accuracy and precision of the method as determined by repeated analysis of U.S.G.S standard rock specimens BCR-1, AGV-1, and W-1 are comparable to that reported by Mazzullo and Bence (1976). Table 14.2 also lists the average major-element composition of the Patagonian batholith between lat. 51° and 52° S calculated by taking the average of the selected specimens weighted according to their volumetric proportion, as listed in Table 14.1.

Figure 14.4 is an alkalis silica versus diagram comparing plutons within the Patagonian batholith to the tholeiitic,

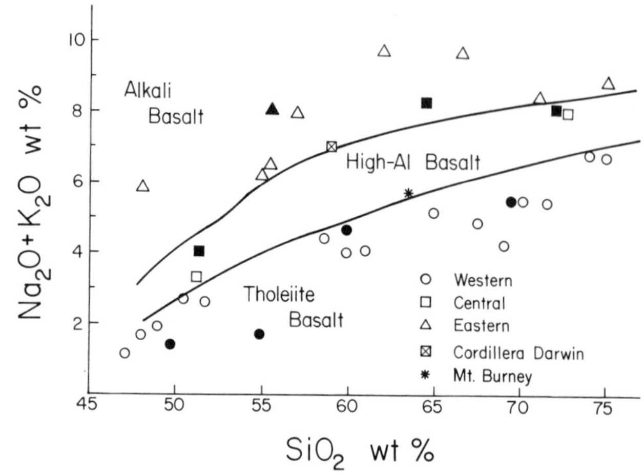

Figure 14.4. Total alkalis versus silica content for plutons of the Patagonian batholith. Solid symbols are specimens listed in Table 14.2. Lines separate tholeiitic, high-Al, and alkali basalt fields of Japan (Kuno, 1966). Note that western plutons are in the tholeiite basalt series field, central plutons are in the high-Al basalt series field, and eastern plutons are in the alkali basalt series field.

Table 14.1. Modal mineralogy and volumetric proportion of selected specimens from the Patagonian batholith

	Western Zone				Central Zone			Eastern Zone	
	Gabbro PA17D	Quartz diorite PA6	Tonalite PA20	Granodiorite PA10A	Quartz diorite PA59C	Granodiorite PA39C	Granite FL100	Quartz Monzonite PA62D	Granite PA69B
Quartz	0	14.5	25.9	32.9	9.3	27.4	24.3	4.0	27.5
Plagioclase	43.4	47.2	61.5	49.8	51.8	50.3	33.9	44.3	25.5
K-spar	0	0	0.4	9.0	0	15.4	36.7	30.6	44.3
Biotite	3.6	0.4	6.7	11.9	9.4	0	4.7	1.5	2.6
Amphibole	15.2	37.5	4.5	2.0	26.2	5.8	0.3	12.6	0
Pyroxene	31.3	0.2	0	0.2	0	0	0	3.5	0
Opaque minerals	6.5	0.1	0	0	0	0	0	0.1	0
Accessory minerals	0	0.1	1.0	0.2	3.3	1.1	0.1	3.4	0.1
Volumetric proportion	4%	8%	60%	8%	2.5%	5%	2.5%	5%	5%

Table 14.2. Major-element compositions (in wt %) of selected specimens from the Patagonian batholith and the average composition of the batholith between lat. 51° and 52° S

	Western Zone				Central Zone			Eastern Zone		Average batholith
	PA17D	PA6	PA20	PA10A	PA59C	PA39C	FL100	PA62D	PA69B	
SiO_2	49.77	54.93	59.42	69.42	51.31	64.46	71.95	55.52	66.45	59.98
TiO_2	1.05	0.43	0.59	0.34	1.26	0.35	0.34	1.60	0.33	0.61
Al_2O_3	13.98	14.43	18.49	15.45	18.73	17.78	14.19	15.13	18.16	17.42
FeO (total Fe)	12.36	9.03	5.92	3.15	9.16	2.81	2.33	11.37	2.87	6.20
MgO	11.12	8.71	2.87	0.93	4.48	1.36	0.36	1.52	0.51	3.23
MnO	0.24	0.20	0.18	0.07	0.16	0.07	0.02	0.24	0.04	0.18
CaO	10.67	10.93	5.91	3.52	8.68	3.14	1.03	4.84	1.85	5.94
Na_2O	1.25	1.26	3.38	3.35	3.23	5.54	4.59	5.46	5.08	3.45
K_2O	0.13	0.39	1.39	2.14	0.78	2.90	3.58	2.68	4.61	1.66
P_2O_5	0.14	0.16	0.16	0.07	0.33	0.11	0.10	0.52	0.05	0.16
Total	100.71	100.47	98.31	98.44	98.12	98.52	98.49	98.88	99.95	

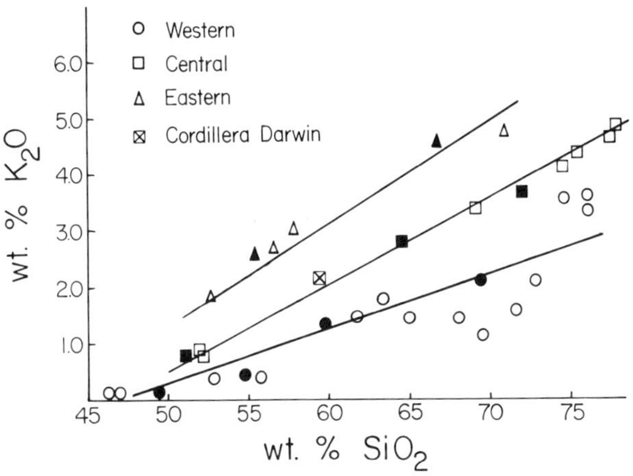

Figure 14.5. K_2O versus SiO_2 diagram showing that plutons to the east have higher K_2O at a given SiO_2 content. Solid symbols are specimens listed in Table 14.2.

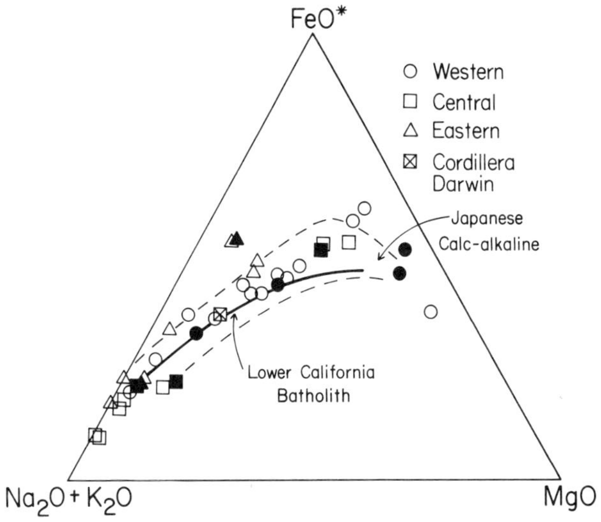

Figure 14.6. A-M-F diagram comparing the trend of the Patagonian batholith with the Japanese calc-alkaline suite and the trend exhibited by the plutons of the Southern California batholith (Carmichael and others, 1974). Solid symbols are specimens listed in Table 14.2. (FeO* is total Fe.)

high-Al, and alkali basalt fields of Japan (Kuno, 1966). Plutons in the western section of the batholith fall within the tholeiite basalt series field, central plutons fall dominantly within the high-Al basalt series field, and eastern plutons fall within the alkali basalt series field. Figure 14.5 is a K_2O versus SiO_2 diagram which indicates that sections of the batholith to the east have progressively higher K_2O content for a given SiO_2 content. Note also in Figures 14.4 and 14.5, as well as Table 14.2 and Figure 14.3, that the western part of the batholith contains more basic rocks and the eastern sections contain more silicic and alkali-rich rocks, although the range of silica and alkali content overlaps. Within each zone of the batholith, iron, magnesium, and calcium decrease and potassium and to some extent sodium increase with increasing silica content. In Figure 14.5 it can be seen that the rate of increase of potassium with increasing silica content is greater in plutons to the east.

Figure 14.6, an A-M-F diagram, compares plutons of the Patagonian batholith with the Southern California batholith and the calc-alkaline (hypersthenic) rock series of Japan (Kuno, 1966). The lack of iron enrichment in intermediate plutonic rocks, illustrated for plutons from each zone of the Patagonian batholith in Figure 14.6, is characteristic of the calc-alkaline rock series.

Trace elements

The Sr, Rb, Zr, Y, and rare-earth element contents of the same speciments listed in Tables 14.1 and 14.2 are given in Table 14.3 along with the K/Rb, Rb/Sr, and Ce/Yb ratios. Table 14.3 also lists the average trace-element abundance of the Patagonian batholith, calculated by taking a weighted average of the selected specimens. Sr, Rb, Zr, and Y abundances were determined by X-ray fluorescence techniques (Norrish and Chappell, 1967). Rare-earth element (REE) abundances were determined by isotope-dilution mass spectrometry methods outlined by Kay (1970).

Rb and Sr generally increase with increasing SiO_2, and the Rb/Sr ratio decreases slightly. The comparatively low Rb content and Rb/Sr ratio for gabbro specimen PA17D suggest plagioclase accumulation. The low Sr and high Rb contents for granite specimen FL100 suggest extensive plagioclase fractionation, K/Rb ratios fall predominantly within the 100–300 range but do not correlate with SiO_2 or K_2O content.

Table 14.3. Trace-element abundances (in ppm) of selected specimens from the Patagonian batholith and the average trace-element content of the batholith between lat. 51° and 52° S

	Western Zone				Central Zone			Eastern Zone		Average batholith
	PA17D	PA6	PA20	PA10A	PA59C	PA39C	FL100	PA62D	PA69B	
Rb	5	21	40	92	71	104	171	55	190	57
Sr	175	93	283	740	408	849	39	322	260	326
Zr	19	57	64	89	79	95	178	352	244	92
Y	12	11	16	20	28	16	17	49	23	18
Ce	—	13.8	21.9	31.2	39.0	51.7	—	93.7	102.5	31.9
Nd	—	7.03	9.81	11.7	17.7	23.6	—	51.0	33.6	—
Sm	—	1.66	2.21	2.32	4.49	3.29	—	12.1	6.49	—
Eu	—	.55	.79	.66	1.39	.98	—	3.21	1.26	—
Gd	—	—	2.27	2.11	4.67	2.55	—	—	5.16	—
Dy	—	1.85	2.46	2.22	5.27	2.38	—	9.32	3.55	—
Er	—	1.05	1.43	1.30	2.98	1.17	—	4.74	1.89	—
Yb	—	.87	1.05	.92	2.24	.90	—	2.74	1.44	1.15
K/Rb	216	154	290	194	92	232	174	406	202	242
Rb/Sr	0.029	0.226	0.141	0.124	0.174	0.122	4.38	0.171	0.730	0.175
Ce/Yb	—	15.9	20.9	33.9	17.4	57.4	—	34.2	71.2	27.7

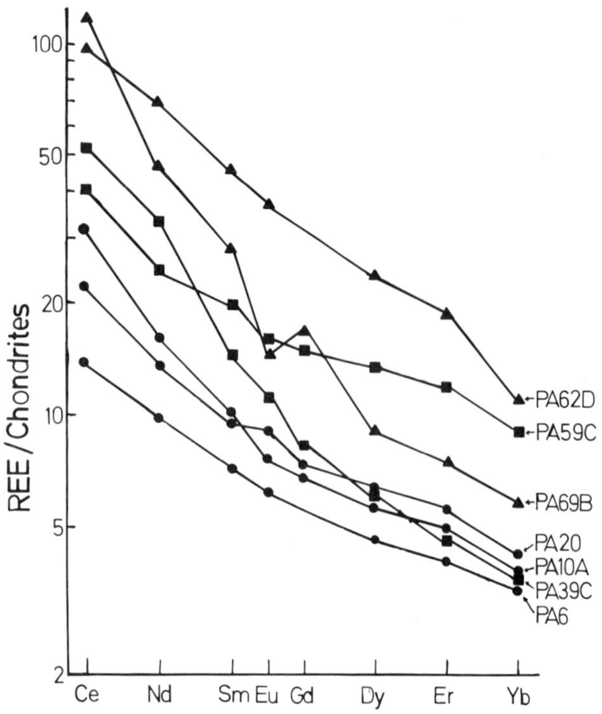

Figure 14.7. The REE abundances, normalized to the Leedy chondrite, of selected specimens from the Patagonian batholith. Circle: specimens from the western zone of the batholith. Squares: specimens from central zone. Triangles: specimens from eastern zone. The data are listed in Table 14.3.

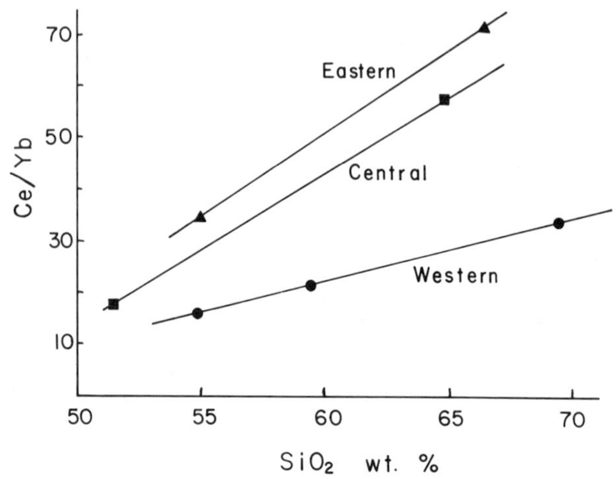

Figure 14.8. A diagram of the Ce/Yb ratio versus SiO_2 content showing that for a given SiO_2 content, Ce/Yb is greater in plutons lying to the east.

Rare-earth element abundances normalized relative to the rare-earth element content of the Leedy chondrite are plotted in Figure 14.7. The Ce/Yb ratios are plotted against SiO_2 content in Figure 14.8. These figures indicate a number of important characteristics for plutons of the Patagonian batholith. First, Figure 14.8 shows that for a given SiO_2 content, Ce/Yb is higher for plutons farther east. Secondly, for plutons within each section of the batholith, Ce/Yb increases with increasing SiO_2. Figure 14.7 indicates that for plutons in the western section of the batholith, Ce/Yb increases with increasing SiO_2 because Ce increases markedly, although Yb remains nearly constant. Within the central and eastern sections of the batholith, the increase in the Ce/Yb ratio with increasing SiO_2 is due to both the decrease of Yb as well as to the increase in Ce. Thirdly, Figure 14.7 shows that the Yb and total REE abundance in plutons of similar SiO_2 content increases from west to east. Finally, negative Eu anomalies occur in some rocks with high SiO_2, suggesting plagioclase fractionation.

Y behaves like a heavy rare-earth element (Lambert and Holland, 1974), as can be seen in Figure 14.9, a plot of CaO versus Y content. For plutons within the western section of the batholith, Y remains nearly constant as CaO decreases, behavior equivalent to the nearly constant content of Yb with increasing SiO_2 in western plutons. In central and eastern plutons, Y decreases as CaO decreases, equivalent to the decrease in Yb with increasing SiO_2 observed in these plutons.

Discussion

The Patagonian batholith resembles other circum-Pacific batholiths in its extent, regionally linear form, and composite nature. It is chemically comparable to other circum-Pacific batholiths in its calc-alkaline character, asymmetric chemical variations, and average composition. The chemical similarity between circum-Pacific batholiths, including the Patagonian batholith, and active volcanic arcs at convergent plate margins supports the suggestion that these batholiths originated as a result of subduction-related processes.

For the Lower Cretaceous plutons of the Patagonian batholith, direct evidence for a contemporaneous overlying comagmatic volcanic arc is preserved as basalt-andesite-dacite lava and breccia within the Upper Jurassic to Lower Cretaceous Hardy Formation cropping out in the vicinity of Cape Horn (Figure 14.1). Indirect evidence of the existence of contemporaneous volcanism is contained in the volcaniclastic detritus within the Lower Cretaceous sedimentary rocks which were deposited in a basin flanking the locus of intrusion of Lower Cretaceous plutons.

The similarity of the Patagonian batholith to other

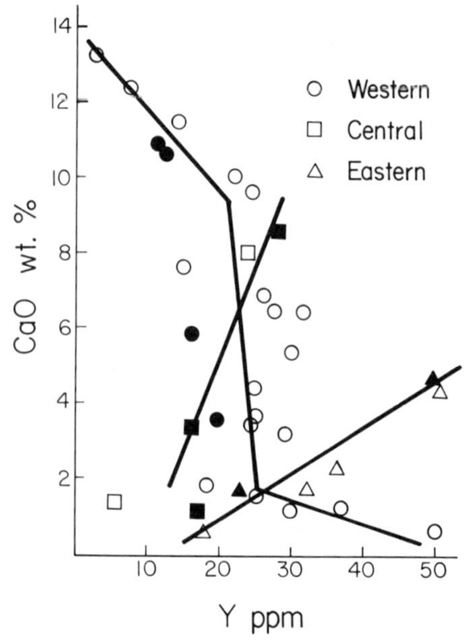

Figure 14.9. A CaO versus Y diagram showing that for plutons to the east, Y decreases as CaO decreases. In the western section of the batholith, Y remains constant as CaO decreases.

circum-Pacific batholiths, as well as to active volcanic arcs located along convergent plate boundaries, plus the evidence for contemporaneous overlying volcanism, supports the suggestion that the Patagonian batholith represents the root of a subduction-related volcanic arc. Plutonism began along the present South American continental margin during Middle Jurassic time, following a hiatus in subduction along the margin of the southeast Pacific basin (Dalziel, Chapter 12, this volume). Subduction under the present South American continental margin has continued, perhaps episodically, up to the present, as evidenced by the range of pluton ages as well as the presence of recent calc-alkaline volcanism as far south as lat. 52° S (Stern and others, 1976). Plutonism during Late Jurassic time occurred along the continental margin of western South America. During Early Cretaceous time, plutonism south of lat. 49° S occurred within a sliver of South American continental crust which had been separated from the stable continent by the formation of a marginal basin partly floored by oceanic-type crust. The basin was uplifted during mid-Cretaceous time, and during Late Cretaceous and Cenozoic time, the easternmost extent of plutonism migrated eastward. Plutons intruded both the Lower Cretaceous sedimentary rocks deposited in the marginal basin and subsequently, the Upper Cretaceous foreland basin sedimentary beds deposited as a result of the mid-Cretaceous orogeny.

Recent experimental and empirical observations suggest that multistage, rather than single-stage, processes are involved in the petrogenesis of the calc-alkaline rock series (Ringwood, 1974; Stern, 1974; Perfit, 1977; Lopez-Escobar, 1974). Small-scale differences in trace-element abundances and initial Sr-isotope ratios between plutonic and volcanic calc-alkaline rocks require further elucidation, but the general chemical equivalence of plutonic and volcanic rocks suggests that sources and processes of magma generation are similar (Perfit, 1977). Proposed multistage models for the generation of the calc-alkaline magma series involve the following: varying degrees of partial melting of subducted oceanic lithosphere, followed by reaction of this partial melt with the overlying peridotite mantle; partial melting of this modified mantle; high-pressure fractional crystallization of these primary melts; mixing of these liquids with partial melts of the lower continental crust; and finally, low-pressure differentiation and contamination with small amounts of continental crust within crustal-level magma chambers. Within a composite batholith such as the Patagonian batholith, each individual pluton must be studied in detail in order to determine the extent to which these different sources and processes have played a part. Lacking such detailed data, we restrict our discussion here to a general consideration of the possible variations in sources and processes that might have resulted in the transverse petrochemical variations we have observed across the Patagonian batholith.

The cause of the increasing K_2O content and Ce/Yb ratio at a given SiO_2 content for plutons located progressively eastward in the Patagonian batholith may be a time-dependent process, since plutons in the eastern zone of the batholith are on the average younger than those to the west. Alternatively, the origin of these transverse variations may be related only to the position of the plutons. The fact that young (<100 m.y.) plutons occur in the western part of the batholith and chemically resemble other western plutons and not plutons of similar age to the east suggests that the cause of the chemical variations is related to position and not age. Andesitic lava from Mount Burney, a recently active calc-alkaline volcano located within the central zone of plutons at lat. 52° S, is chemically consistent with its position (see Figure 14.4, where the average composition of Mount Burney lava is indicated with a star).

The origin of the observed west-to-east variations in K_2O content and Ce/Yb ratio within the Patagonian batholith may be in part dependent on variations in the relative contributions of the different possible sources: subducted oceanic lithosphere, the overlying mantle wedge, and the continental crust. Halpern (1973) determined the initial $^{87}Sr/^{86}Sr$ ratio to be 0.705 for the 151-m.y.-old pluton located at lat. 52° S (Figure 14.2) within the western zone of the batholith. This low initial Sr-isotope ratio precludes anatexis of "ancient" continental crust as a source for this pluton. However, the pre-Jurassic basement rocks of southern South America have been interpreted as Paleozoic arc-trench gap deposits (de Wit, 1977; Dalziel, Chapter 12, this volume) which would not have developed significant radiogenic ^{87}Sr by Mesozoic time and cannot be eliminated as a possible source for the plutons of the Patagonian batholith. For the 45-m.y.-old pluton (FL100; Figure 14.2) within the central zone of the batholith, Halpern (1973) determined an initial Sr-isotope ratio of 0.707. This suggests an increased contribution of the continental crust for plutons to the east. However, data on the regional variations in initial Sr-isotope ratios and other isotopic data are insufficient to constraint the variations in the contributions of the different possible sources.

Eastwardly increasing K_2O content at constant SiO_2 content (Figure 14.5) may be due to smaller degrees of partial melting of a single source, increasing depth of the source region of the primary magmas accompanied by decreasing stability of K_2O-bearing phases, or a more complicated zone-refining mechanism involving a potassium-rich mineral such as phlogopite (Best, 1975). Eastwardly increasing Ce and Yb contents and Ce/Yb ratio for rocks of similar silica content is consistent with lower degrees of partial melting. The observed increase of Yb (Figure 14.7) for rocks of similar SiO_2 content suggests that garnet, a phase that retains Yb, was not more important within the source region of the plutons lying progressively eastward. Since garnet becomes more stable at increasing pressure, this suggests that any possible change in the depth to the source region for the plutons to the east did not reach into the garnet stability field.

Within each zone of the batholith, K_2O and Ce content and Ce/Yb ratio increase with increasing SiO_2 conent (Figures 14.4 and 14.8). This may be related to crystal-liquid fractionation, which has been shown to be important in the generation of the chemical diversity observed within calc-alkaline plutons (Perfit, 1977). In the western section of the batholith, Yb content remains nearly constant as SiO_2 content increases, suggesting that differentiation here may involve olivine, clinopyroxene, or plagioclase. The removal of these phases increases Ce/Yb in residual liquids without decreasing the Yb content. In the central and eastern section of the batholith, as SiO_2 content increases, Yb decreases (Figure 14.7), suggesting that differentiation here involves phases such as amphibole or garnet, which decrease the Yb content of residual liquids.

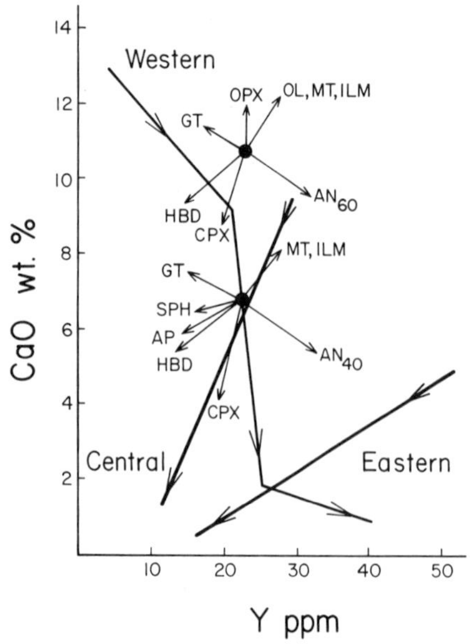

Figure 14.10. A CaO versus Y diagram showing the trends for each section of the batholith (taken from Figure 14.9) and the direction of the trends that would be produced by fractional crystallization involving different mineral phases separating from a basaltic and an andesitic composition (large dots) (Lambert and Holland, 1974). For western plutons, plagioclase and clinopyroxene control the fractionation trend; for eastern plutons, amphibole replaces clinopyroxene as a dominant phase. OPX: orthopyroxene. CPX: clinopyroxene. OL: olivine. MT: magnetite. ILM: ilmenite. GT: garnet. HBD: hornblende. SPH: sphene. AP: apatite.

Similar conclusions are reached with respect to the behavior of yttrium. Figure 14.10 illustrates the trends of CaO versus Y for the different sections of the batholith, taken from Figure 14.9. Also shown are the trends, indicated by the direction of the arrows, that would be produced by removal of different mineral phases (Lambert and Holland, 1974). For western plutons, Y content remains nearly constant as CaO decreases, which is consistent with fractionation involving clinopyroxene and plagioclase. For central and eastern plutons, however, decreasing Y content with decreasing CaO is consistent with fractionation involving amphibole along with plagioclase. Y content decreases more rapidly with decreasing CaO content for plutons lying progressively eastward, indicating eastward increase in the important of amphibole fractionation.

In summary, the variations in trace elements across the Patagonian batholith suggest that chemical variations within the western section of the batholith are in part controlled by clinopyroxene fractionation. Amphibole fractionation becomes increasingly more important for plutons farther east. Studies of the regional variation in initial Sr and other isotopic ratios are needed to evaluate variations in the contribution made by the different possible source regions. Only more detailed studies of individual plutons can fully evaluate all the possible variables that may be involved in the generation of this composite calc-alkaline batholith.

Acknowledgments. We are grateful to M. A. Skewes for assistance in the field. I. W. D. Dalziel, S. E. DeLong, D. Elthon, and W. I. Ridley critically reviewed the manuscript. The research was supported by National Science Foundation grants DES 74-04076, EAR 76-82456, and DDP 74-21415.

References cited

Best, M. G., 1975. Migration of hydrous fluids in the upper mantle and potassium variation in calc-alkalic rocks. *Geology, 3,* 429–32.

Bruhn, R. L., and Dalziel, I. W. D. 1977. Destruction of the Early Cretaceous marginal basin in the Andes of Tierra del Fuego. (*In* Talwani, M., and Pitman, W. C., III, *eds. Maurice Ewing Series, I: Island Arcs, Deep Sea Trenches, and Back-arc Basins.* Washington, American Geophysical Union, 395–406.)

Bruhn, R. L., Stern, C. R., and de Wit, M. J. 1978. The bearing of field and geochemical data on the origin and development of a volcano-tectonic rift zone and back-arc basin in southernmost South America. *Earth planet. Sci. Lett., 41,* 32–46.

Carmichael, I. S. E., Turner, F. J., and Verhoogen, J. 1974. *Igneous Petrology.* New York, McGraw-Hill, 739 pp.

Cobbing, E. J., and Pitcher, W. S. 1972. The coastline batholith of central Peru. *J. Geol. Soc. Lond. 128,* 421–60.

Dalziel, I. W. D., de Wit, M. J., and Palmer, K. F. 1974. Fossil marginal basin in the southern Andes. *Nature, Lond., 250,* 291–94.

Dalziel, I. W. D., de Wit, M. J., and Ridley, W. I. 1975. Structure and petrology of the Scotia arc and the Patagonian Andes. *Antarct. Jnl U.S., 10,* 307–10.

de Wit, M. J. 1977. The evolution of the Scotia arc as a key to the reconstruction of southwestern Gondwanaland. *Tectonophysics, 37,* 53–81.

Halpern, M. 1973. Regional geochronology of Chile south of 50° latitude. *Geol. Soc. Am. Bull., 84,* 2407–22.

Katz, H. R. 1962. Fracture patterns and structural history in the sub-Andean belt of southernmost Chile. *J. Geol., 70,* 595–603.

Kay, R. W. 1970. *The Rare Earth Geochemistry of Alkaline Basaltic Volcanics.* Ph.D. dissertation, Columbia University, 160 pp.

Krank, E. H. 1932. Geological investigations in the Cordillera of Tierra del Fuego. *Acta Geogr. Helsingf., 4,* No. 2, 231 pp.

Kuno, H. 1966. Lateral variations of basalt magma type across continental margins and island arcs. *Bull. volcan., 29,* 195–222.

Lambert, R. St. J., and Holland, J. G. 1974. Yttrium geochemistry applied to petrogenesis utilizing calcium-yttrium relationships in minerals and rocks. *Geochim. cosmochim. Acta, 38,* 1393–1414.

Lopez-Escobar, L. 1974. *Plutonic and Volcanic Rocks from Central Chile (33°–42° S): Geochemical Evidence Regarding their Petrogenesis.* Ph.D. dissertation, Massachusetts Institute of Technology, 269 pp.

Mazzullo, L. J., and Bence, A. E. 1976. Abyssal tholeiites from DSDP Leg 34: the Nazca plate. *J. geophys. Res., 81,* 4327–52.

Norrish, K., and Chappell, B. W. 1967. X-ray fluorescence spectography. (*In* Zussman, J., *ed. Physical Methods in Determinative Mineralogy.* New York, McGraw-Hill, 161–214.)

Perfit, M. R. 1977. *The Petrochemistry of Igneous Rocks from the Cayman Trench and the Captains Bay Pluton, Unalaska Island: their Relation to Tectonic Processes at Plate Margins.* Ph.D. dissertation, Columbia University, 307 pp.

Ringwood, A. E. 1974. The petrological evolution of island arc systems. *J. Geol. Soc. Lond., 130,* 193–204.

Skarmeta, J. M. 1974. *Geologia de la region continental de Aisin entre los 45°–46° latitude sur.* Dissertation, Universidad de Chile, 324 pp.

Stern, C. R. 1974. Melting products of olivine tholeiite basalt in subduction zones. *Geology, 2,* 227–31.

Stern, C., Skewes, M. A., and Duran, M. 1976. Volcanismo orogeníco en Chile austral. *Act. 1 Congr. Geol. Chil., Santiago, Chile, 2,* 195–212.

Streckeisen, A. L. 1976. To each plutonic rock its proper name. *Earth Sci. Rev., 12,* 1–33.

Suarez, M., and Pettigrew, T. H. 1976. An upper Mesozoic island-arc–back-arc system in the southern Andes and South Georgia. *Geol. Mag., 113,* 305–28.

15

The Structural Evolution of the Magallanes Basin and Neotectonics in the Southernmost Andes

MARGARET A. WINSLOW
Lamont-Doherty Geological Observatory and Department of Geological Sciences of Columbia University, Palisades, New York 10964, U.S.A.

Abstract. The Andean foothills and Atlantic lowlands in southernmost South America are underlain by deformed Lower Cretaceous to middle Tertiary sedimentary rocks of the Magallanes Basin. Early Cretaceous sedimentation there commenced along a subsiding shelf edge coeval with turbidite deposition in an adjacent marginal basin, immediately behind the Pacific margin volcanic arc. The marginal basin and volcanic terrane were deformed and uplifted during middle Cretaceous time, upwarping the western and southern boundaries of the shelf edge to form a boundary to a foredeep or foreland basin. During Late Cretaceous time, the now well-defined Magallanes Basin subsided rapidly, receiving thick deposits of turbidites whose clasts were derived from the rising marginal basin and remnant arc terranes.

Lower Cretaceous to Paleogene sedimentary rocks from the deepest parts of the Magallanes Basin to the shelf-foreland basin hinge were folded during middle-Tertiary time. The folds are generally asymmetric and are cut by high-angle faults on the overturned limb. Fold vergence is toward the craton. There is an overall tightening and increase in degree of overturn proceeding along the Magallanes Basin axis from northwest to southeast. There also seems to be a diachronous involvement of sedimentary rocks along strike such that the deformation appears to have begun earlier in the southeast than in the northwest.

Two strike-slip fault systems dissecting the tip of the South American continent and shelf show abundant evidence of present activity. The faults clearly displace and rotate Tertiary fold axes. Although there is evidence of ductile shearing along vertical planes of Cretaceous age, the displacement histories of the presently active faults are unclear because of near parallelism to both Cretaceous and Tertiary structural trends. These fault trends can be projected off the Atlantic and Pacific margins into major transform plate boundaries. The pattern of magnetic anomalies from the now-consumed sections of the Chile Rise suggests a possible correlation in space and time between the cessation of middle Tertiary compressional tectonics and the initiation of late Tertiary to Holocene transcurrent faulting throughout the region.

Four major physiographic provinces lie parallel to the arcuate Pacific margin of the South American continent from lat. 47° S to Cape Horn at lat. 56° S, changing gradually from a southerly trend to due eastward across Tierra del Fuego (Figure 15.1). From the Pacific margin to the Atlantic, the following physiographic units are recognized in the southern Andes, which roughly correspond to geologic provinces in the Andean Cordillera and lowlands (cf. Kranck, 1932; Harrington, 1965): (1) the Pacific archipelago, consisting mainly of calc-alkaline intrusive rocks of the Patagonian batholith, which comprises the root zone of a Late Jurassic–Cretaceous magmatic arc, intruded into older continental basement rocks; (2) deformed Upper Jurassic silicic volcanic rocks and Lower Cretaceous ophiolitic and volcaniclastic assemblages interpreted as the floor and infilling of an Early Cretaceous marginal basin (Dalziel, Dott, and others, 1975; Bruhn and Dalziel, 1977); (3) the Andean foothill belt of deformed Lower Cretaceous through Tertiary sedimentary rocks; these highlands, which lie along the Atlantic side of the present Andean Cordillera, are also referred to as the Precordillera or sub-Andean fold belt by Chilean geologists (Katz, 1962); and (4) the Atlantic lowlands which are underlain by the undeformed rocks of the stable platform and the Magallanes Basin, whose deformed western and southwestern margins are part of the Andean foothill belt. The Magallanes Basin, a well-defined foreland trough or foredeep, received sediments shed from the rising Andean Cordillera from middle Cretaceous to middle Tertiary time.

Sedimentation in the Magallanes Basin

The Magallanes Basin is bounded on its Atlantic margin by the Springhill Platform, a stable shelf which has been slowly subsiding since Late Jurassic time. The western and southern boundaries of the Magallanes Basin correspond to the outcrop of Tobifera volcanic rocks, which are conformably overlain by uppermost Jurassic through Tertiary sedimentary rocks. The Tobifera volcanic rocks now form highly strained, fault-bounded slivers between the Magallanes Basin and the deformed marginal basin rocks, obscuring the nature of the boundary between the two basins. As discussed below, there was no major structural or sedimentologic boundary between the marginal basin and the continental shelf edge until the marginal basin was destroyed and uplifted.

The Magallanes Basin therefore encompasses all of the sedimentary rocks between the stable shelf and the marginal basin. Thus it includes both the gently warped sedimentary rocks that underlie the Atlantic lowlands as well as the more intensely deformed sedimentary rocks of the Andean foothills (cf. Natland and others, 1974).

More than 9000 m of Cretaceous and Tertiary sediments were deposited into the deepest parts of the Magallanes Basin on top of Upper Jurassic Tobifera Series volcanic rocks (Figure 15.2). Nearly continuous sedimentation commenced with Upper Jurassic black mudstones and continued with deeper water Cretaceous turbidites, lower Tertiary shallow-marine sediments, and finally Miocene–Pliocene tuffs and littoral sandstones. The source rocks for these dominantly deep-water foreland sediments were: (1) the magmatic arc, with debris reaching the subsiding foreland basin as a distal facies; (2) the uplifted and deformed Lower Cretaceous rocks from the marginal basin; (3) the eroded horsts and inliers of the Jurassic volcanic terrane; and finally, (4) the uplifted and eroded western margin of the foreland basin itself. The Pacific boundary of the basin shed second-generation Tertiary conglomerates into local successor basins which rest with pronounced angular unconformity upon deformed older sedimentary strata within the Magallanes Basin.

The earliest sediment deposited along the western margin of the stable Springhill Platform was the Upper Jurassic Springhill sand which represents locally reworked Tobifera volcanic detritus (see Dott and others, chapter 21, this volume). Following this sand was black mudstone deposited under euxinic conditions. Thicknesses (Figure 15.2) of Upper Jurassic sedimentary layers do not suggest that there was a definable Magallanes Basin between the magmatic arc and the stable platform at this time. A marginal basin was initiated between the magmatic arc and the stable platform sometime between latest Jurassic and earliest Cretaceous time. Deep-marine flysch-type sedimentation in the marginal basin was coeval with deposition of black mudstone toward the shelf edge. Limey shale, limestone, and algal reef debris are intercalated with the black mudstone along the shelf-basin hinge on the Atlantic side (Figure 15.2).

Uplift and deformation of the marginal basin began by late Early Cretaceous time (Dalziel and others, 1974), forming a steep western boundary to the developing and now well-defined Magallanes Basin. At the same time, rapid subsidence from less than 500 m to more than 1500 m water depth occurred in the central part of the Magallanes Basin (Natland and others, 1974). This subsidence was followed by a sudden influx during Cenomanian–Campanian time of detritus which formed immature breccia, conglomerate, and volcaniclastic sandstone, followed in turn by the first appearances of slate clasts. The clasts in the conglomerate are arc-derived pebbles consisting of mafic dike material, cleaved slate, schistose rocks, and foliated Tobifera volcanic rocks. Dalziel and others (1974) have suggested 5–10 km of uplift for parts of the deformed marginal basin, resulting in a continuous source terrane for the remainder of Cretaceous time. Rhythmically bedded sandstone and shale continued to be deposited until Maestrichtian time. From late Eocene to early Miocene time, regressive facies became widespread as the present Andean Cordillera rose.

Deposition of open-marine mudstone and sandstone continued in the east and southeast until middle Miocene time (Natland and others, 1974) when most of the basin rose above sea level. Except for an upper Miocene transgressive facies along the Atlantic coast, the basinal sedimentary strata were folded and uplifted in all parts of the region. Late Tertiary sedimentation was more isolated, forming troughs and basins on the Atlantic side of thrusts, reflecting local reworking of older foreland basin sources.

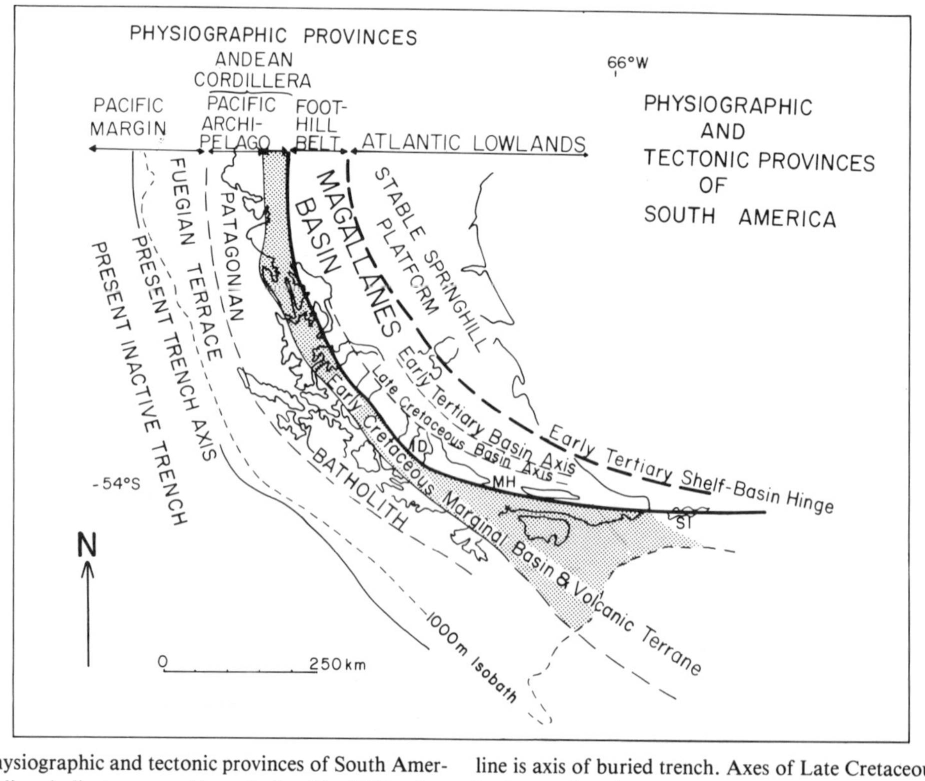

Figure 15.1. Physiographic and tectonic provinces of South America. Light solid lines indicate mapped boundaries. The 1000 m isobath is shown in the west and south by the dotted line. Heavy solid line is axis of buried trench. Axes of Late Cretaceous and Tertiary basins are based on maximum thicknesses of sedimentary strata. ID: Isla Dawson. MH: Monte Hope. SI: Isla de los Estados.

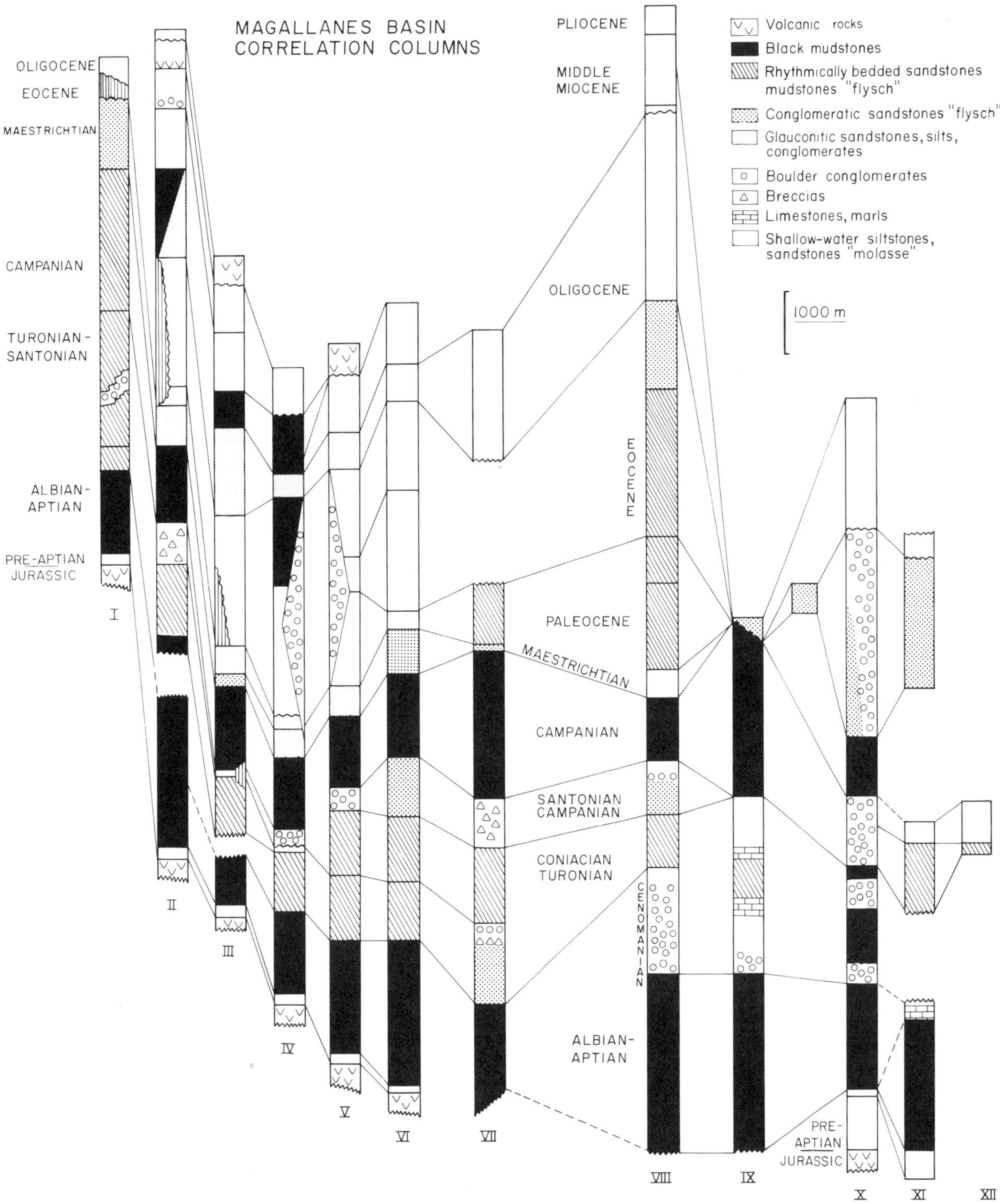

Figure 15.2. Correlation of series and stages in the deformed strata of the Magallanes Basin. Locations of columns I-XII are shown on Figure 15.3. Sources of data include Empresa Nacional del Petroleo (ENAP) (unpublished data), Katz (1961), Cortés and Cañon (1963), Scott (1966), Cespedes (1963), and Natland and others (1974).

However, fresh andesitic detritus, indicative of local calc-alkaline magmatic activity, continued to accumulate in the northern part of the basin until at least Miocene time.

There is a great contrast in sedimentation rates between Early and Late Cretaceous time in the Magallanes Basin. While thick sediments (3000–4000 m) accumulated in the marginal basin during Early Cretaceous time at rates exceeding 150 m/m.y., the region toward the stable platform subsided slowly, receiving only 500–1000 m of sediments on the shelf, with as much as 3000 m of black mudstone nearer the marginal basin. Although the nature of the boundary between the marginal basin and Magallanes Basin is unknown, the contrasts in sediment types and thicknesses indicate a subsiding shelf-edge which dropped off into a deep marginal basin. In contrast, the deposition of thick sequences of deformed marginal-basin debris at bathyal depths (Natland and others, 1974) suggests that the Magallanes Basin was a deep foreland trough by at least

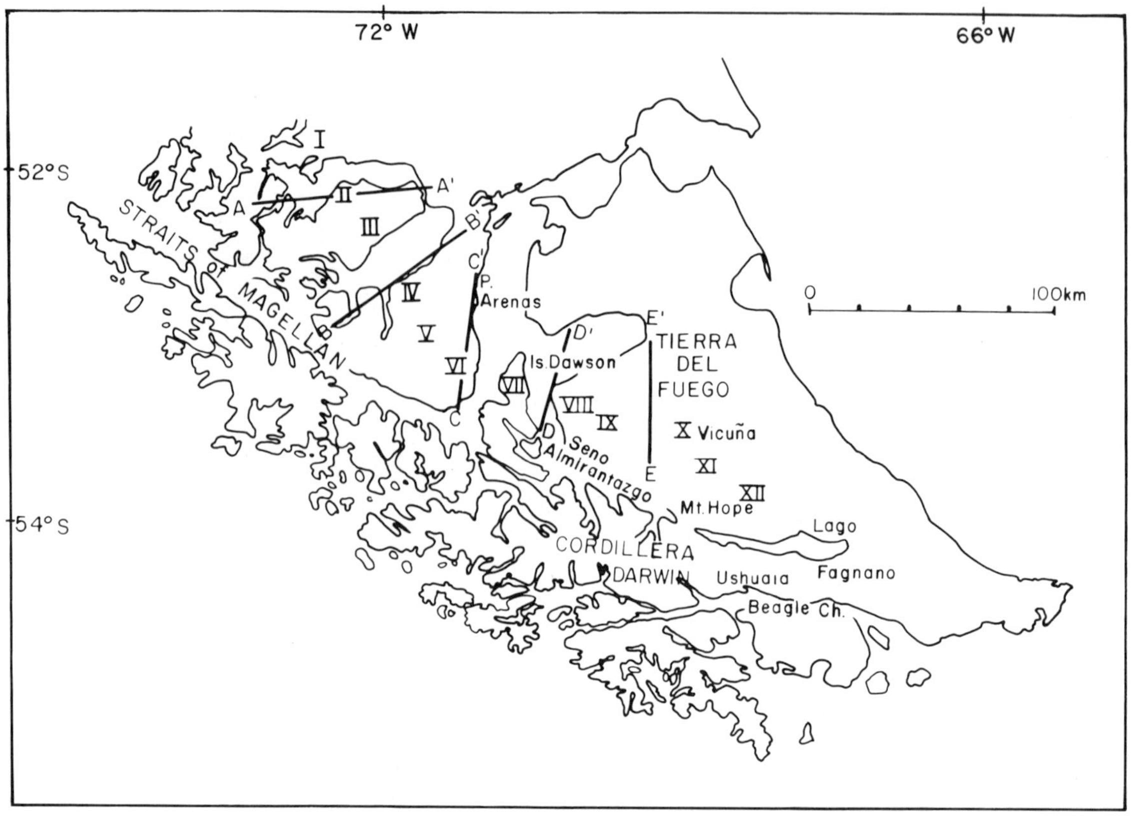

Figure 15.3. Geographic map of southernmost South America. I-XII indicate localities of stratigraphic columns from Figure 15.2. Lines of cross sections (A–A′, etc.) refer to Figure 15.4.

Cenomanian time. During Late Cretaceous time, 4000–6000 m of deep-water sediments were deposited in the deeper part of the basin and were succeeded by more than 3800 m of lower Tertiary mudstone and sandstone in the southern part. Over the same period, the Springhill Platform (Figure 15.1) remained stable, receiving a maximum of 500 m of sediments after Albian time.

Deformation of the Magallanes Basin

The rocks of the Magallanes Basin record progressive deformation from middle Cretaceous through middle Tertiary time. After the cessation of folding and thrusting in the foothills and lowlands, strike-slip faulting continued and is still present throughout the southern tip of the continent.

Andean deformation

Deformation of volcaniclastic rocks of the Early Cretaceous marginal basin occurred before 80–90 m.y. ago (Halpern and Rex, 1972; Halpern, 1973) and involved sedimentary rocks as young as Aptian in age (Dalziel and others, 1974). The boundary between the deformed marginal basin and the Magallanes Basin is now marked by a long outcrop of Jurassic Tobifera volcanic rocks. In certain areas these rocks occur as fault-bounded horsts. Northeast of this inferred deformed remnant-arc terrane (Bruhn and Dalziel, 1977), Lower Cretaceous sedimentary rocks of the Magallanes Basin lie conformably on the Tobifera volcanic rocks. The boundary between the two sedimentary basins may have been defined in part by the eroded volcanic terrane, but the original contacts have been strongly deformed, as the Andean phase was concentrated near the marginal basin–remnant arc contact. Along the Pacific boundary of the Magallanes Basin, Lower Cretaceous sedimentary rocks as young as Albian–Aptian in age are involved in tight folds with Tobifera volcanic rocks. These deformed rocks now crop out as the highest peaks in the Andean foothill belt. Here, the structures consist of tight, overturned to recumbent folds with a penetrative flattening cleavage. Phyllosilicates are concentrated and aligned along cleavage planes.

In contrast, in the central part of the Magallanes Basin (Figure 15.3) near the middle Cretaceous shelf-basin hinge, there is no evidence of Andean-age deformation. Drilling and field mapping have revealed pre-Jurassic gneiss with a radiometric age of 240 m.y. overlain by undeformed Tobifera volcanic rocks and Lower Cretaceous sandstone (Cortés, 1960; Halpern, 1973). A subhorizontal plane of décollement separates the sandstone form overlying folded Lower Cretaceous to middle Eocene sedimentary rocks which are cut by imbricate thrusts (Cortés, 1960). Therefore, there is no evidence for a pre-Tertiary deformation in the central part of the basin. The main Andean deformation which affected the marginal basin and the Lower Cretaceous sedimentary strata along the Pacific border of the Magallanes Basin seems to die out toward the middle Cretaceous shelf-foreland basin hinge, on the Atlantic side.

Tertiary deformation

Upper Cretaceous through middle Tertiary sedimentary rocks of the central and Atlantic regions of the Magallanes Basin have been folded and thrust northeastward toward the stable platform. The deformed rocks crop out in the

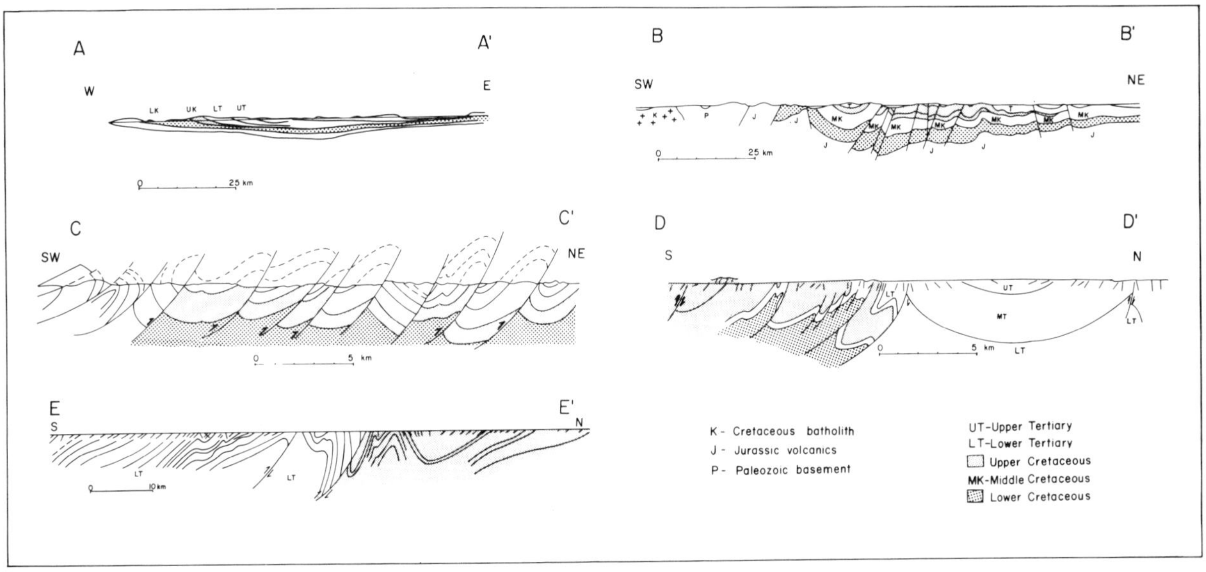

Figure 15.4. Structural sections across foreland fold and thrust belt. Locations of cross sections are shown on Figure 15.3. Note the systematic increase in asymmetry in the southeastern sections. A–A′: northernmost cross section from Thomas (1949). B–B′: schematic section from ENAP and Muñoz Christie (1956). C–C′: section modified after Cespedes (1963). Sections D–D′ and E–E′ are from author's data in Tierra del Fuego. Unshaded areas are Tertiary rocks.

eastern foothills and underlie the Quaternary alluvium of the Atlantic lowlands (Figure 15.1). It is not known if the Tertiary deformation affected the previously deformed Pacific edge of the Magallanes Basin, although there is some evidence there of late-stage open folding of Lower Cretaceous rocks comparable to that affecting the Tertiary sedimentary layers to the east. In contrast to the Andean deformation in the cordillera, the Tertiary phase seems to have been "thin-skinned." Basement gneiss from drill holes in the basin reveals no post-Jurassic radiometric ages (Halpern, 1973). Also, both fold amplitudes and the inclination of thrust planes seem to decrease with depth down to a décollement plane above the Tobifera rocks in the southeastern part of the basin.

The effects of the Tertiary deformation in the cordillera are unknown, presumably because the "strain hardening" that resulted from the middle Cretaceous deformation there prevents the previously tectonized rocks from responding in the same ductile manner as the semi-indurated Upper Cretaceous and Tertiary strata in the Magallanes Basin. This results in an apparent progression of deformation in time toward the stable craton, as is observed in other mountain belts. There is a clear diachroneity of Tertiary deformation. Across the strike of the axis of the basin from the Pacific side toward the Atlantic, increasingly younger rocks are involved in folding, and the amplitude of folds decreases. However, there also seems to be a diachronous involvement along the axis from southeast to northwest in that the youngest beds involved in folding are older toward the southeast along strike.

Briefly, to the southeast in Tierra del Fuego, middle Eocene conglomerates are the youngest strata involved in folds. Nearby, undeformed and poorly indurated Miocene siltstones overlie, with pronounced angular unconformity, tightly deformed and cleaved Maestrichtian mudstones (Figure 15.2). To the northwest (Figure 15.3) the youngest beds involved in the thrusts are post-middle Miocene coal-bearing beds which are unconformably overlain by undeformed Pliocene tuffs. Pliocene–Pleistocene strata do not appear to be folded in this zone. However, farther north, Pliocene beds are tilted and gently warped, and Quaternary glacial deposits lie upon them with slight unconformity. Miocene hypabyssal intrusions form a chain in the foothills parallel to the cordillera; one (Cerro Paine) has been radiometrically dated at 12 ± 2 m.y. (Halpern, 1973). The intrusions are cut by mafic dikes, all of which have been warped and offset by strike-slip faults (Katz, 1962). Also in the north, strongly folded Pliocene tuffs and pyroclastic rocks are overlain with sharp angular unconformity by basaltic flows (Figure 15.2), which are covered by and intercalated with glacial moraines. Post-intrusive Pliocene–Pleistocene mafic dikes and olivine basalt flows are gently folded and faulted to the north of the study area (Katz, 1962).

Tertiary structures

Large-scale folds with wavelengths as great as 2 km extend across the eastern foothills and Atlantic lowlands in the Upper Cretaceous and Tertiary strata. Nearer the cordillera, they are relatively tight and asymmetric, with a regionally pervasive, steeply dipping fracture or spaced cleavage. The cleavage, which is commonly parallel to steeply dipping reverse faults, shows pressure solution effects along discrete planes and rotation of fossils without distortion. The overturned limbs of folds are commonly cut by high-angle reverse faults that dip steeply (50°–80°) toward the cordillera. The folds decrease in amplitude and become more open toward the Atlantic, although high-angle reverse faults offset middle Tertiary strata under the Atlantic lowlands.

Along the shelf-basin hinge, however, the structural styles of Tertiary folds vary between the northwestern and southeastern parts of the Magallanes Basin (Figures 15.3 and 15.4). There is an increase both in shortening and in degree of overturn or vergence from northwest to southeast. From lat. 51° to 52° S, the folds change from very open folds with wavelengths of 20 km and dips rarely exceeding 20°, involving sedimentary rocks as young as Mio-

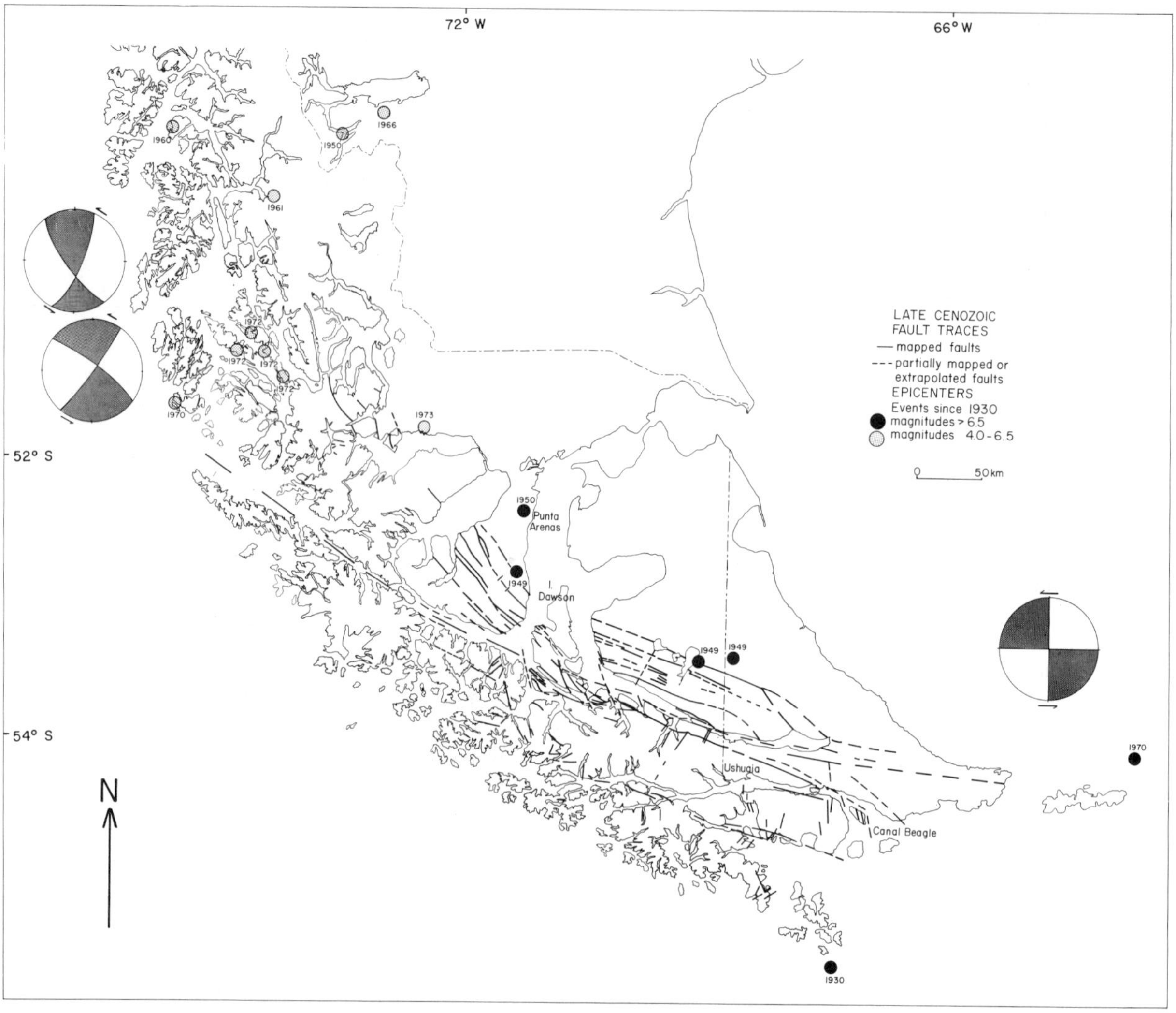

Figure 15.5. Late Cenozoic fault systems and seismicity. Solid lines indicate fault traces with documented recent motion (Bruhn and others, 1976; Winslow, 1976). Focal mechanisms are from Forsyth (1975). Epicenters are relocated from Gutenberg and Richter (1954).

cene, to symmetric, tight anticlines and synclines lacking a penetrative cleavage. From lat. 52° to 53° S, the post-Oligocene structures consist of asymmetric, relatively tight anticlines with overturned east limbs separating gentle, open synclines (east limbs less than 30°). The anticlines have half-wavelengths of hundreds of meters while the synclines are extremely open (>5 km across), suggesting high-angle fault control. A spaced cleavage is especially well developed on the overturned limb and lies subparallel to high-angle reverse fault planes that dip steeply toward the cordillera. Farther southeast in Tierra del Fuego, overturned, tightly folded anticlines and synclines of more uniform wavelength show pronounced overturning toward the Atlantic, with axial surfaces dipping from 22° to 55° southward. In this latter area, imbricate thrusts control the tight folds which decrease in amplitude downward to a gently southward-dipping plane of décollement. A spaced cleavage fans around the folds and is locally truncated by a more penetrative and shallower dipping spaced cleavage that closely parallels the thrust planes. Whether or not the increasing tightness and asymmetry suggest an actual increase of regional shortening (Figure 15.3) is still untested.

It is evident that uplift of the cordilleran border of the Magallanes Basin was accompanied by deep subsidence of the basin floor. Lower Cretaceous strata of the basin were involved in folding at the same time as the marginal basin strata. Upper Cretaceous and Paleogene sedimentary rocks were deformed sometime after the deposition of lower Eocene beds. Late Cenozoic brittle shearing truncates Tertiary structures in the foothills, with clear evidence of Pleistocene to modern activity. However, the time of initiation of strike-slip faulting is less clear because of the paucity of definite ages for Tertiary and earlier structures and because of the near parallelism between the faults, and earlier structural trends. Prevalence of subhorizontal slickensides on Tertiary high-angle faults and Cretaceous mylonitic foliations in narrow, subvertical shear zones (de Wit, 1977; I. W. D. Dalziel, personal communication) suggests a possible strike-slip component to the tectonism throughout late Mesozoic and Cenozic time. The apparent southeast to northwest diachroneity in the Tertiary deformation may correlate with offshore geophysical evidence along the Pacific margin, as discussed below.

Late Cenozoic tectonics

Superimposed upon the deformed Magallanes Basin and Andean Cordillera are fracture systems whose topographic expressions include linear fjords, river valleys, and lakes traceable for 50–100 km along strike from lat. 51° S to the tip of South America (Figure 15.5). Preliminary studies of aerial photos have revealed that many segments of the lineaments have undergone recent strike-slip displacements. The evidence includes offset stream valleys, talus slopes, and moraines, truncated stream meanders (Figure 15.6), as well as sag ponds, fresh scarps, and linear truncations of vegetation. Field mapping since 1975 (Bruhn and others, 1976; Winslow, 1976) has revealed that the fractures are high-angle fault traces that partly exploit early Tertiary and Cretaceous structural trends, reactivating high-angle reverse faults and shear zones.

The faults are less prominent north of lat. 51° S. Furthermore, there seem to be two distinct fault systems: one which trends south to southeast (140°–168°), the other trending more easterly (90°–125°). The southerly trending faults (Figure 15.5) are limited to the Pacific margin and do not occur east of long. 70° W, while the easterly faults run across Tierra del Fuego and do not appear at the surface north of lat. 53° S (Figure 15.7).

The Magellan shear zone and related faults

Included in the easterly trending fault system is the spectacular topographic lineament formed by the western end of the Straits of Magellan and the narrow passages, fjords, and lakes to the east (Figure 15.5). This steep-walled passage, defined in part by the Magellan fault (Katz, 1964; Fuenzalida, 1972), can be traced 600 km along strike, apparently offsetting the western margin of the Patagonian batholith by 80 km in a left-lateral direction. Although the magnitudes of strike-slip displacements along east-west faults range from several hundreds of meters to perhaps as much as 80 km, there are significant dip-slip components on reactivated thrust faults, with recent northward-directed reverse motion of several tens of meters on some faults north of the Magellan fault.

Southeasterly trending faults

The more southerly trending lineaments correspond to a wide zone of faults with post–middle Miocene to Holocene displacements. Individual faults can be followed for as far as 50 km. Both left- and right-lateral offsets of several tens of meters have been mapped (Katz, 1964), with possibly greater displacements obscured by the near-parallelism with earlier structural trends. In addition to strike-slip components on the order of several hundreds of meters to more than 5 km along individual faults, the recent southerly trending fault scarps also show normal components with the Pacific side downthrown as much as 500 m.

Near the Straits of Magellan, the two fault systems converge. Here the two fracture sets that parallel each of the major fault systems intersect, with the easterly trending fracture sets consistently offset by the more southeasterly ones. This evidence suggests an earlier age for the former. In the center of the straits, a major easterly trending fault defines Isla Dawson's southern coast. Southeasterly trending faults which control major valleys rotate early Tertiary fold axes in a direction consistent with left-lateral shear (Figure 15.8). Brecciated zones separate rotated blocks that result from as much as 2 km of Tertiary differential uplift and tilting. Complex minor structures, including kink bands, late-stage open folds, and fracture patterns, possibly result from the interaction of the two transcurrent trends.

Seismicity

Seismic data from lat. 46° to 56° S since 1930 indicate low activity with infrequent major ($M \geq 7.0$) earthquakes (Figure 15.5). The epicenters are located along major faults of both shear systems, particularly along a 350 km segment of the Magellan fault (Forsyth, 1975; Bruhn and others, 1976; Winslow, 1976). Ground displacements associated with the events include landslides, tilting, differential subsidence, and formation of scarps. Focal mechanisms for these events (Figure 15.5) yield left-lateral strike-slip solutions with a nodal plane parallel to the mapped fault traces. Two focal mechanisms from a cluster near lat. 52° S on the Pacific margin also show predominantly left-lateral motion aligning with the southeasterly faults (Forsyth, 1975).

Offshore faults

Southeasterly trending faults continue off the Pacific margin offsetting perched basins of the Fuegian Terrace as well as the axis of the inactive, sediment-filled trench (Herron and others, 1977). The Shackleton Fracture Zone, which also has a southeasterly trend (137°), intersects the trench and continental margin near lat. 57° S. Here recent *en echelon* faults along the inner wall of the trench parallel the faults that offset the perched basins on the terrace. In contrast, north of lat. 51° S, deformation offshore is confined to shortening and block faulting along the inner wall of the trench in response to slow convergence between the Antarctic and South American plates. Onshore data indicate block faulting and uplift from lat. 51° S northward, which is consistent with a simple conpressional model (Katz, 1973). Extensions of the North Scotia Ridge westward and the Shackleton Fracture Zone northwestward onshore along continental faults intersect near the Pacific margin. The Shackleton Fracture Zone and North Scotia Ridge are assumed to form the western and northern transform boundaries of the Scotia plate (Forsyth, 1975). Teleseismic studies by Forsyth (1975) and marine geophysical data (Barker, 1970, 1972; Ewing and others, 1971; Barker and Griffiths, 1972; Herron and Tucholke, 1976; Herron and others, 1977) suggest a trench-fault-fault triple junction near lat. 52° S with left-lateral motion along both transform boundaries. The seismicity pattern indicates infrequent but large earthquakes located along mapped faults. There is a cluster of shallow strike-slip epicenters near lat. 52° S, long. 74° W, which may represent the location of the triple junction, if it can be so narrowly defined on a continent. Therefore, the tip of the continent lies within the zone in which the relative motion among the South America, Antarctic, and Scotia plates is taken up (Dott, 1976; Bruhn and others, 1976), causing the dissection of the continental crust along subvertical shear zones. The age of the youngest magnetic anomaly from the now-consumed section of the South Chile Rise along the Pacific coast is Miocene (Figure 15.9). Since 30 m.y. ago, the Chile Rise has progressively collided with the South American continental margin, from lat. 56° S northward to its present intersection at about lat. 46° S. The obliteration of the

Figure 15.6. Truncated stream meanders along active fault. Left-lateral motion indicated by field evidence. Rio San Juan, Magallanes, Chile.

Figure 15.7. Lago Fagnano–Monte Hope–Seno Almirantazgo. Large graben defined by Magellan fault system. Monte Hope is a tilted sliver, bounded on both sides by active faults.

Figure 15.8. Fold axes showing left-lateral drag on Isla Dawson in Straits of Magellan. Long ridge between two linear valleys is a tilted horst.

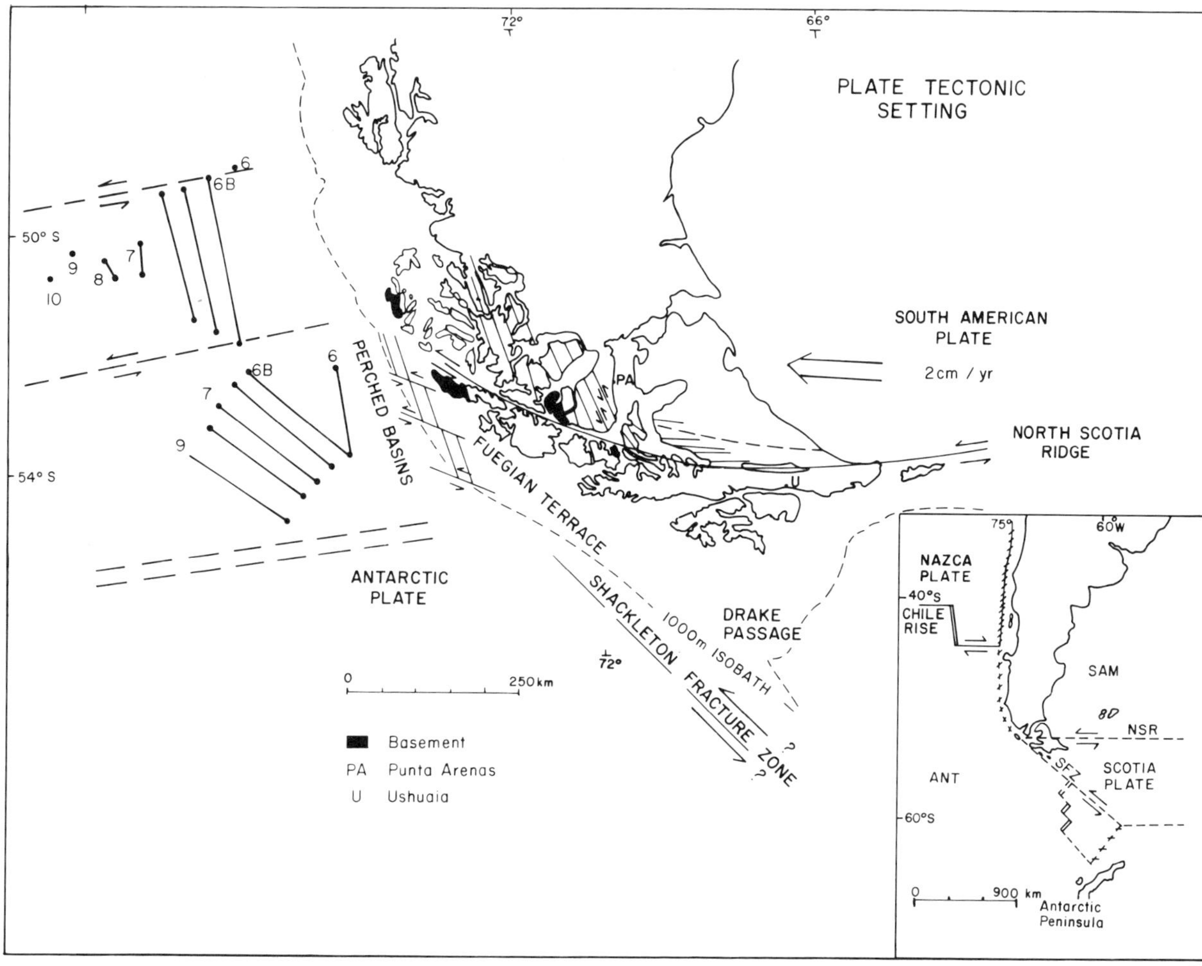

Figure 15.9. Plate tectonic setting of southernmost South America. Inset is modified after Forsyth (1975). Sources for Pacific margin data are from Herron and Tucholke (1976). Numbered bars are magnetic anomalies. Note that the anomalies become younger *toward* the continent, indicating subduction of a ridge at about anomaly-6 time.

southern end of the Chile Trench as a well-defined subduction zone was followed by the initiation of strike-slip motion as South America—Antarctic motion was transferred to South America-Scotia plate decoupling. The cessation of major compression in the foothill belt, with its apparent northward diachroneity, and the Neogene to Holocene strike-slip faulting throughout the region (Dott, 1976) may correlate with this northwest-migrating ridge collision. However, the initiation of the strike-slip faulting may have accompanied or even predated the Tertiary folding.

Summary and conclusions

Sedimentation commenced on the western shelf margin of the southern tip of the South American continent at the same time that deep-water flysch was deposited in the adjacent marginal basin. The Magallanes Basin did not become a well-defined foredeep or foreland basin until the Andean orogenesis that resulted in the initial uplift and deformation of the marginal basin and remnant arc, as well as the Pacific edge of the foredeep. The axis of deposition migrated toward the stable platform as thick sediments were shed from the rising cordillera. A Tertiary phase of deformation is recorded in the Upper Cretaceous and Paleogene sedimentary rocks that were folded and thrust toward the stable shelf. There is a diachroneity of fold initiation across the axis to the northeast and possibly along the axis of the Magallanes Basin from southeast to northwest. Active strike-slip faults which dissect the entire tip of the continent have offset and rotated older structures. The timing of their initiation is problematic, as evidence for middle Cretaceous shearing along subvertical planes has been reported. The prevalence of Neogene to Holocene strike-slip motion in the region, combined with Mesozoic shearing, suggests a major component of transcurrent motion throughout the Mesozoic and Cenozoic deformational history, rather than simple compression.

In time and space, the progressive collision of the Chile Rise from south to north with the South American continental margin near lat. 56° S after anomaly-6 time might be correlated with the diachronous Tertiary folding. The cessation of folding was followed (or possibly accompanied) by continued or renewed strike-slip motion. Since the collision of the ridge, the tip of the continent has been dominated by strike-slip faults, with some minor normal and reverse components.

The transform faults that define the Scotia plate's northern and western boundaries converge near the continental margin. Therefore, the part of the continent involved in shearing is within a wide zone of decoupling between the two plates and may be responding as a semi-independent wedge among the South America, Antarctic, and Scotia plates. Furthermore, the change in strike of all structural trends in southern South America (the "oroclinal bend" of Carey, 1955), regardless of its age or origin, has been accentuated by drag along the recent transcurrent fault systems, which have rotated and displaced regional structural trends eastward. There is no doubt that early Tertiary fold geometries have been affected by strike-slip faults. This suggests that the distinctive eastward bend of the tip of South America has been exaggerated by late Cenozoic decoupling over a broad zone between the South America and Scotia plates along the North Scotia Ridge. The relationship between strike-slip faults on land and the two transform boundaries of the Scotia plate must be further examined in order to understand the dissection of a continental margin occurring near the converging arms of a trench-fault-fault triple junction.

Acknowledgments. I wish to thank Ian W. D. Dalziel for his support and criticisms as advisor on my project in Chile as part of National Science Foundation Grant DPP-74-21415. R. H. Dott, Jr. was an especially supportive teacher as well as critic during my initial field work. I am extremely grateful for the logistic support of field assistants, jeeps, and boats, and the supplying of base camps by the Departamento de Exploraciones of the Empresa Nacional del Petroleo (ENAP), Punta Arenas, Chile, which allowed me to do detailed field mapping in this region of difficult access over the last four field seasons.

I would also like to thank Raúl Cortés and Carlos Castro of ENAP for their direct support and friendship in the field. Their discussions and advice were invaluable to all phases of my work.

This manuscript has been critically reviewed by Walter Alvarez, Walter Pitman, William Ludwig, Ian W. D. Dalziel, R. H. Dott, Jr., C. P. Andrews-Speed, and R. L. Bruhn.

References cited

Barker, P. F. 1970. Plate tectonics of the Scotia Sea region. *Nature, Lond., 228,* 1293–97.

Barker, P. F. 1972. A spreading centre in the east Scotia Sea. *Earth planet. Sci. Lett., 15,* 123–32.

Barker, P. F., and Griffiths, D. H. 1972. The evolution of the Scotia Ridge and Scotia Sea. *Phil. Trans. R. Soc. Lond.,* Ser. A, *271,* 151–83.

Bruhn, R. L., and Dalziel, I. W. D. 1977. Destruction of the early Cretaceous marginal basin in the Andes of Tierra del Fuego. (*In* Talwani, M., and Pitman, W. J., III, *eds. Maurice Ewing Series, 1: Island Arcs, Deep Sea Trenches and Back-arc Basins.* Washington, American Geophysical Union, 395–406.)

Bruhn, R. L., Winslow, M. A., and Dalziel, I. W. D. 1976. Late Tertiary to Recent structural evolution of southernmost South America. *EOS, 57,* No. 4, 334.

Carey, S. W. 1955. The oroclinal concept in geotectonics, part I. *Pap. Proc. R. Soc. Tasmania, 89,* 255–88.

Cespedes, S. 1963. Reconocimiento geologico del area compendida entre el Seno Skyring y el Estrecho de Magallanes. Santiago, Empresa Nacional del Petroleo, (Unpublished.)

Cortés, R. 1960. Revisión de algunos problemas geológicos en parte de Ultima Esperanza y las expectativas petrolíferas de la región. Santiago, Empresa Nacional del Petroleo, (Unpublished.)

Cortés, R., and Cañon, A. 1964. Rango Estratigraphico y Frecuencia de los Principales Fosiles del Cretaceo de Ultima Esperanza. Santiago, Empresa Nacional del Petroleo, (Unpublished.)

Dalziel, I. W. D., de Wit, M. J., and Palmer, K. F. 1974. Fossil marginal basin in the southern Andes. *Nature, Lond., 250,* 291–94.

Dalziel, I. W. D., Dott, R. H., Jr., Winn, R. D., Jr., and Bruhn, R. L. 1975. Tectonic relations of South Georgia Island to the southernmost Andes. *Geol. Soc. Am. Bull., 86,* 1034–40.

de Wit, M. J. 1977. The evolution of the Scotia arc as a key to the reconstruction of southwestern Gondwanaland. *Tectonophysics, 37,* 53–81.

Dott, R. H., Jr. 1976. Contrasts in tectonic history along the eastern Pacific rim. (*In* Sutton, G. H., Manghnani, M. H., and Moberly, R., *eds. Geophysics of the Pacific Ocean Basin and its Margin.* Washington, American Geophysical Union, 299–308.) [Geophysics Monograph 19.]

Ewing, J. I., Ludwig, W. J., Ewing, M., and Bittreim, S. L. 1971. Structure of the Scotia Sea and Falkland Plateau. *J. geophys. Res., 76,* 7118–37.

Forsyth, D. W. 1975. Fault plane solutions and tectonics of the South Atlantic and Scotia Sea. *J. geophys. Res., 80,* No. 11, 1429–43.

Fuenzalida, R. H. 1972. *Geological Correlation between the Patagonian Andes and Antarctic Peninsula and some Tectonic Implications.* Master's thesis, Stanford University, 75 pp.

Gutenberg, B., and Richter, C. F. 1954. *Seismicity of the Earth.* 2nd ed. Princeton, N.J., Princeton University Press, 310 pp.

Halpern, M. 1973. Regional geochronology of Chile south of 50° latitude. *Geol. Soc. Am. Bull., 84,* 2407–22.

Halpern, M., and Rex, D. C. 1972. Time of folding of the Yahgan Formation and age of the Tekenika beds, southern Chile, South America. *Geol. Soc. Am. Bull., 83,* 1881–86.

Harrington, H. J. 1965. Personal communication cited in Ludwig, W., Ewing, M., and Ewing, J. Seismic refraction measurements in the Magallanes Straits. *J. geophys. Res., 70,* No. 8, 1855–76.

Herron, E. M., and Tucholke, B. E. 1976. Sea-floor magnetic patterns and basement structure in the southeastern Pacific. (*In* Hollister, C. D., Craddock, C., and others. *Initial Reports of the Deep Sea Drilling Project,* Vol. 35. Washington [U.S. Government Printing Office], 263–78.)

Herron, E. M., Bruhn, R., Winslow, M., and Chuaqui, L. 1977. Post Miocene tectonics of the margin of southern Chile. (*In* Talwani, M., and Pitman, W. C., III, *eds. Maurice Ewing Series, 1: Island Arcs, Deep Sea Trenches and Back-arc Basins.* Washington, American Geophysical Union, 273–84.)

Katz, H. R. 1961. Algunas notas acerca de la intrusion granitica en la Cordillera Paine—Provincia de Magallanes. *Revta Minerales, 74,* 1–15.

Katz, H. R. 1962. Fracture patterns and structural history in the sub-Andean belt of southernmost Chile. *J. Geol., 70,* No. 5, 595–603.

Katz, H. R. 1964. Some new concepts on geosynclinal development and the mountain building at the southern end of South America. *Proc. 22nd Int. Geol. Cong., New Delhi, 1964, 4,* 241–55.

Katz, H. R. 1973. Contrasts in tectonic evolution of orogenic belts in the south-east Pacific. *J. R. Soc. N. Z., 34,* 333–62.

Kranck, E. H. 1932. Geological investigations in the Cordillera of Tierra del Fuego. *Acta geogr. Helsingf., 4,* No. 2, 1–231.

Muñoz Christie, J. 1956. Chile. (*In* Jenks, W. F., *ed.* Handbook of South America Geology. *Mem. geol. Soc. Am., 65,* 189–214.)

Natland, M. L., Gonzalez, E., Cañon, A., and Ernst, M. 1974. A system of stages for correlation of Magallanes Basin sediments. *Mem. geol. Soc. Am., 139,* 126 pp.

Scott, K. M. 1966. Sedimentology and dispersal pattern of a Cretaceous flysch sequence, Patagonian Andes, southern Chile. *Bull. Am. Ass. Petrol. Geol., 50,* 72–107.

Thomas, C. R. 1949. Geology and petroleum exploration in Magallanes Province, Chile. *Bull. Am. Ass. Petrol. Geol., 33,* 1553–68.

Winslow, M. A. 1976. Active transcurrent shear zones in southern Chile as landward expressions of transform plate boundaries. *Abstr. with Programs, geol. Soc. Am.,* 1173–74.

16

Paleontologic and Paleoenvironmental Synthesis for the Southwest Atlantic Ocean Basin Based on Jurassic to Holocene Faunas and Floras from the Falkland Plateau

SHERWOOD W. WISE, P. F. CIESIELSKI*, D. T. MacKENZIE, and FRANK H. WIND
Antarctic Marine Geology Research Facility, Department of Geology, Florida State University, Tallahassee, Florida 32306, U.S.A.

KAREN E. BUSEN
Department of Geological Sciences, University of Southern California, Los Angeles, California 90007, U.S.A.

A. M. GOMBOS
Chevron Oil Company, Lafayette, Louisiana 70501, U.S.A.

BILAL U. HAQ, G. P. LOHMANN, and R. C. TJALSMA
Woods Hole Oceanographic Institution, Woods Hole, Massachusetts 02543, U.S.A.

W. K. HARRIS
Department of Mines, Eastwood, S.A. 5063, Australia

RICHARD W. HEDLUND and D. N. BEJU
Amoco Production Company, Tulsa, Oklahoma 74102, U.S.A.

DAVID L. JONES, GEORGE PLAFKER, and WILLIAM V. SLITER
U.S. Geological Survey, Menlo Park, California 94025, U.S.A.

Abstract. The following paleontologic and paleoenvironmental summary is based on three Deep Sea Drilling Project drill core sequences and a reconnaissance study of more than 75 piston cores taken on the Falkland (Malvinas) Plateau. A Middle(?) to Late Jurassic inland sea transgressed the southwestern part of Gondwanaland and became progressively more restricted, so that by Oxfordian time, predominantly pelagic and nektonic forms were preserved, particularly coccoliths, belemnite rostra and arm hooks (onychites), and rare decapod remains. Stagnant conditions continued into Early Cretaceous time, and well-preserved marine palynomorphs and phytoplankton contributed to the high organic content of Aptian black sapropelic claystones deposited in a shallow, quiet-water environment. Abrupt loss of all palynomorphs near the Aptian–Albian time boundary coincides with the ventilation of this segment of the incipient South Atlantic Basin. Sharp changes in the benthonic foraminiferal populations in late Albian time are attributed to down-flank subsidence of the Falkland Plateau as sea-floor spreading widened the South Atlantic Basin. A prominent stratigraphic hiatus encompassing most of Cenomanian–Santonian time suggests erosion and dissolution of calcareous microfossils by cold bottom currents. A return to normal pelagic sedimentation and a sharp lowering of the carbonate compensation depth (CCD) is evidenced by a thick upper Campanian–Maestrichtian chalk sequence deposited at paleodepths close to those of the present day. The Tertiary sequence is characterized by sharp fluctuations of the CCD (low stands during late Paleocene–early Eocene, Oligocene, Miocene, and late Quaternary time) and strong erosional events (Cretaceous–Tertiary boundary, Miocene, and Miocene–Pliocene boundary). Conspicuous reworking of microfossils throughout the Miocene sequence occurred as a result of an increase in current velocity associated with the opening of Drake Passage and the establishment of the Antarctic Circumpolar Current. A sharp change from calcareous ooze to diatom ooze and glacial marine sedimentation near the Miocene–Pliocene time boundary, along with erosional loss of considerable section, records intensification of the circumpolar current during the severe late Miocene Antarctic glaciation (terminal Miocene event).

Presently situated at intermediate depth in the southwest Atlantic sector of the Southern Ocean, the marine sediments of the Falkland (Malvinas) Plateau span at least 150 m.y. of Cenozoic–Mesozoic time (Scientific Staff, DSDP Leg 36, 1974), predating in part the separation of the plateau from the southern margin of Africa during the breakup of southwestern Gondwanaland. Subsequently,

* Present address: Department of Geology, University of Georgia, Athens, Georgia 30602, U.S.A.

the plateau has been ideally located to record major oceanographic events in this part of the world, such as the mid-Cretaceous anoxic event, the Turonian erosional crises, the Cretaceous–Tertiary boundary event, and more recently, the opening of Drake Passage and the severe late Miocene glaciation of Antarctica.

This paper summarizes paleontologic and paleoenvironmental data from recent exploratory activities on the eastern extremity of the plateau (the Maurice Ewing Bank; see Figure 16.1) by the Deep Sea Drilling Project (DSDP) and cruises 07-75 and 11-76 of the *Islas Orcadas*. Data are drawn largely from the DSDP Leg 36 *Initial Reports* volume (Barker, Dalziel, and others, 1977) and from a recent summary of piston core studies by Ciesielski and Wise (1977).

A geologic map of older (pre-Pliocene) sediments of the Maurice Ewing Bank is shown in Figure 16.1. These beds are overlain unconformably by Pliocene to Holocene glacial marine sediments and pelagic oozes sufficiently thin (generally less than 1 m) to allow penetration by piston cores. Figure 16.2 is a geologic cross section of the older sediments drawn along the seismic profile lines in Figure 16.1. Included are the sedimentary sequences drilled by DSDP Leg 36. These sections, along with basic paleontologic and sedimentologic information from the cores, are reproduced in greater detail in Figures 16.3 and 16.4. The following narrative describes and interprets the information displayed in chronologic order, beginning with the deepest drill hole, DSDP 330 (Figure 16.3).

Jurassic Period

Core 15 of hole 330 contains a clean, olive-gray sandstone at the top of the basal sandy unit which Thompson (1977) cited as evidence for a marine transgression over the calcreted Precambrian basement (Barker and others, 1977, p. 996) of what was in Jurassic times the southwestern sector of Gondwanaland. Marine palynomorphs in the beach sand are rare (less than 1% of the flora), but Harris (1977) found enough similarity between these and overlying floras to date the entire Jurassic sequence (cores 15 to 5) as Oxfordian to Kimmeridgian in age. Hedlund and Beju (1977) disagreed in part, placing cores 15 and 14 in the Middle Jurassic. Although terrestrial palynomorphs are overwhelmingly the dominant fossils throughout this sequence, other marine elements gradually appear, beginning with bivalves in core 14, followed by a few belemnites, radiolarians, calcareous benthic foraminifers, and calcareous nannofossils in cores 13 and 12. As the basin deepened, grain size, terrigenous fraction, and bioturbation all decreased. This effect can be seen upward through the first 120 m of section to core 11 where the sediment grades into an extremely fine-grained Oxfordian sapropelic claystone deposited in a highly restricted environment far removed

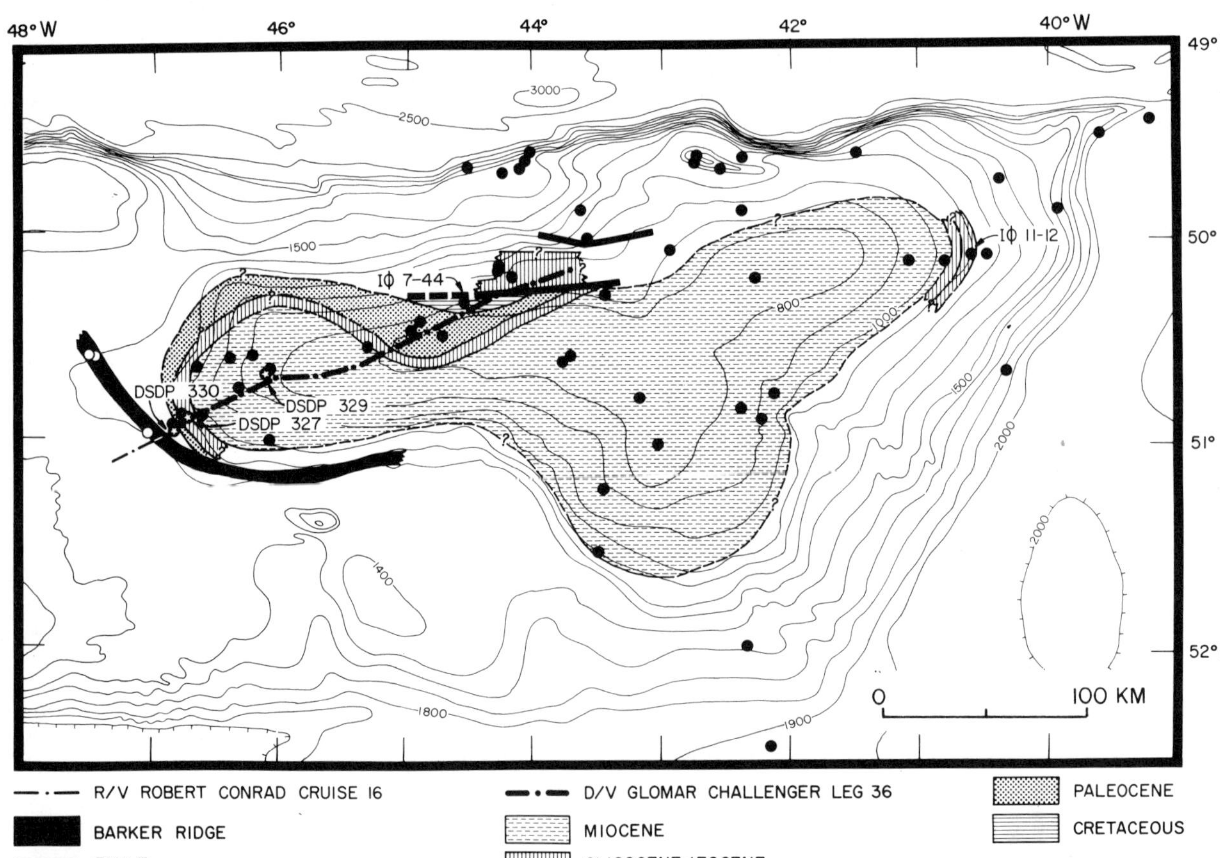

Figure 16.1. Geologic map of older (pre-Pliocene) strata sampled on the Maurice Ewing Bank directly beneath a thin (1–2 m) widespread cover of Plio-Pleistocene siliceous ooze and glacial marine clastic sediments (after Ciesielski and Wise, 1977). Starred dots mark the three DSDP sites shown in Figure 16.2. Solid dots mark piston core localities; the two labeled *Islas Orcadas* cores (IO 7-44 and IO 11-12) are represented in Figure 16.3. Superimposed on the map are *Glomar Challenger* Leg 36 and *Robert Conrad* 16-06 seismic reflection profile tracks used in constructing the geologic cross section in Figure 16.2. Bathymetry (in fathoms) from Lonardi and Ewing (1971).

from an active source area. There, the sedimentation rate drops from more than 100 m/m.y. to a quarter of that value, and the organic content of the sediments rises appreciably (average 3‰).

Increasingly restricted, perhaps by active volcanic chains along the Pacific margin of southwestern Gondwanaland (Barker and others, 1977), the shallow interior Oxfordian sea became sufficiently stagnant to inhibit bioturbation by infauna, but not enough to prevent epifaunal bivalves from reaching local abundance (although benthic foraminifers are absent in cores 5 to 7). Bivalves present include *Inoceramus* and a new Buchiidae described by Jones and Plafker (1977) as *Jeletzkiella falklandensis*. These authors suggested, for all molluscan assemblages recovered from the drill holes, a probable water depth no greater than 200 m and possibly considerably less. Some nektonic molluscs flourished in the better oxygenated surface waters. Juvenile (but not adult) belemnites are common to abundant in cores 330-8 to 330-5; an ammonite segment was noted in core 8.

A further record of the cephalopod population is provided by some 150 specimens of microscopic scolecodont-like objects identified by Wind and others (1977) as onychites. Also known as cephalopod arm hooks, these objects fit around sucker disks to increase grasping efficiency. They were particularly well developed in belemnites (A. Seilacher, personal communication). No evidence of true scolecodont-bearing polychaetes was found (probably due to the presence of deoxygenated bottom waters), but a problematic clawlike microfossil from core 330-6 has been identified as the clasper from a walking appendage of a decapod of the family Glypheidae (Wind and others, 1976). This group flourished during the Jurassic Period, but was generally believed to have been extinct since the Eocene Epoch. Recently, however, a modern specimen was described from the South China Sea as a "living fossil" (Forest and others, 1976).

Although sparse, calcareous phytoplankton are extremely well preserved toward the top of the Jurassic section (cores 8 to 5) and easy to extract for scanning electron microscope (SEM) study from the fine, dark clayey matrix. Assemblage diversity compares favorably with that encountered in the higher latitudes of the Northern Hemisphere (England and northern France). Marine palynomorphs (dinoflagellates, acritarchs, and Prasinophyceae) increase steadily toward the top of the Jurassic section with the peculiar exception of core 8 where their numbers jump sharply. Harris (personal communication), however, notes that 95% of that flora consists of *Tasmanites* which dominate the samples to form a tasmanite or oil shale. The environmental preferences of these forms are not well understood.

In summary, the Jurassic section records a major marine transgression into the southwestern sector of Gondwanaland (correlative with transgressions into the Magallanes Basin of South America and the Oteniqua and Algoa Basins of South Africa; see Thompson, 1977). The epeiric sea became progressively more restricted until by Oxfordian time, the marine community consisted primarily of plankton, nekton, and some byssate attached molluscs that lived

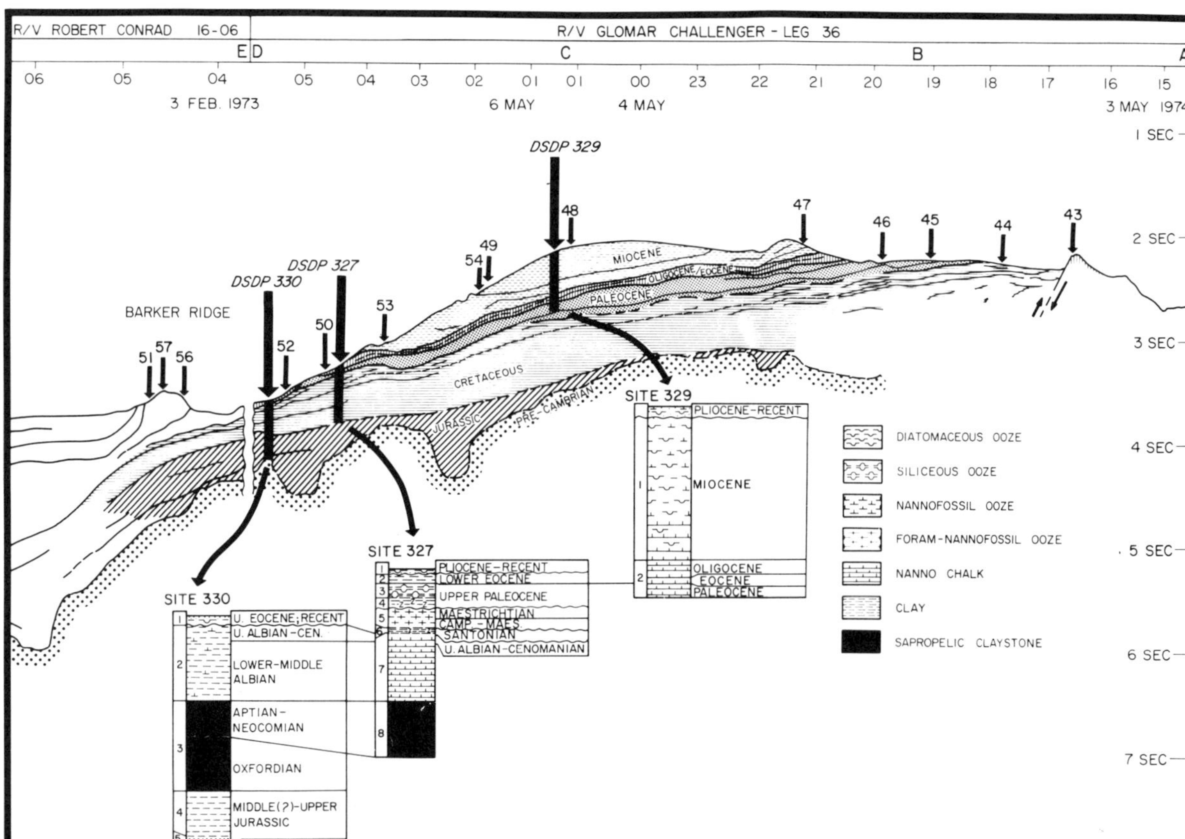

Figure 16.2. Geologic cross section across the Maurice Ewing Bank, as interpreted from piston- and drill-core data and from seismic profiles (from Ciesielski and Wise, 1977). This ENE–WSW section follows the seismic tracks in Figure 16.1. Vertical scale: two way reflection time in seconds. Horizontal scale: ship's travel time in hours.

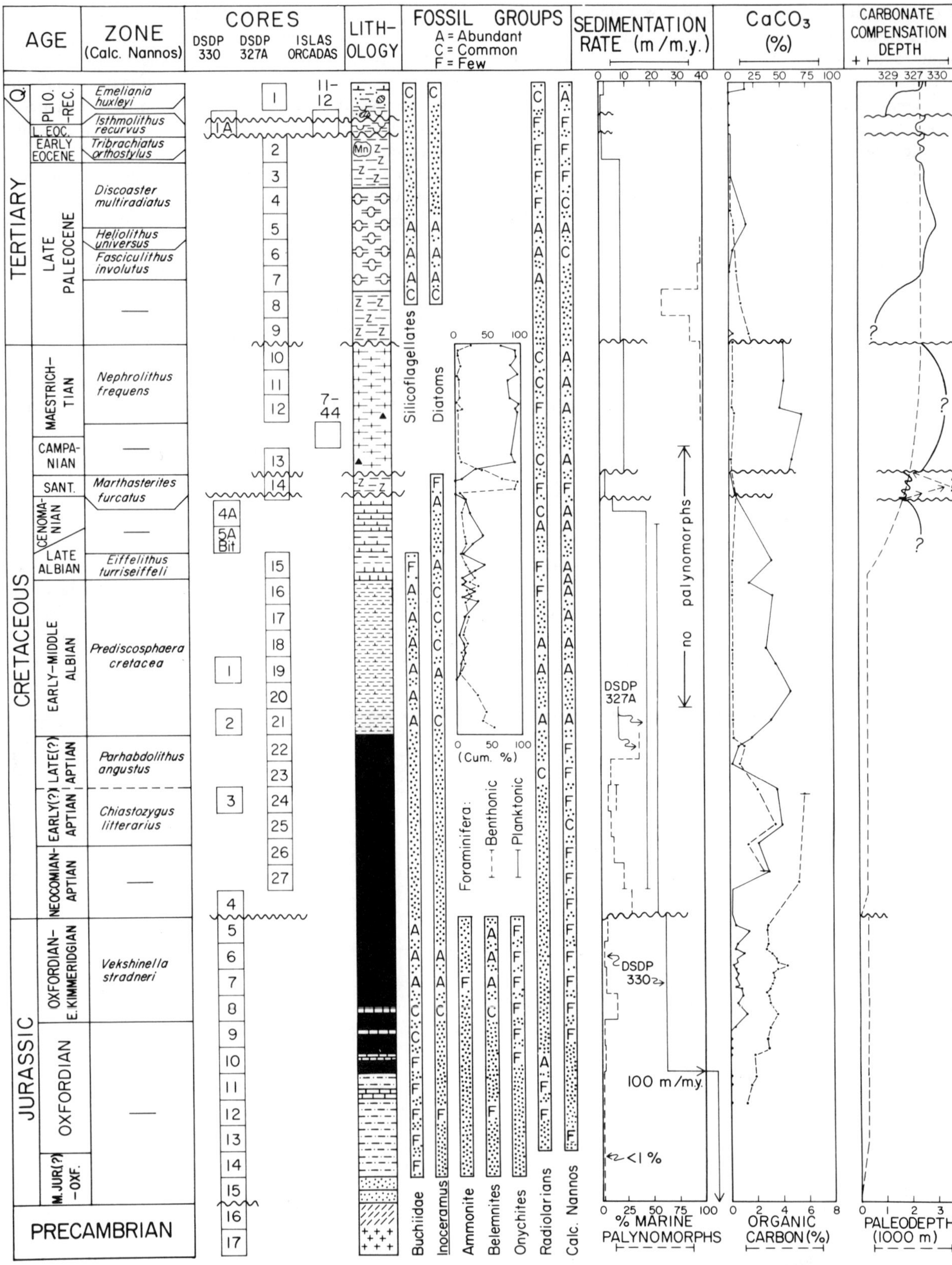

GRANITE GNEISS SANDSTONE ZEOLITIC CLAY SILTY CLAY LIMESTONE

▲CHERT ⊘ROCK CLAST

Figure 16.3. Paleontologic and lithologic properties of cores from Deep Sea Drilling Project holes 327A and 330, and from *Islas Orcadas* stations 07-75-44 and 11-76-12. See Figure 16.2 for lithologic symbols not given in key (note: nannos means nannofossils).

exclusively in neritic waters. Although far from an active source area, at no time did the environment become open marine. Basin depths remained quite shallow. Several new molluscs, onychites, and coccolith taxa are described from this part of the section in the Leg 36 *Initial Reports* volume.

Cretaceous Period

Neocomian–Aptian anoxic event

The depositional environment represented by the Lower Cretaceous cores is remarkably similar to that of the euxinic Oxfordian beds. The organic content rises to nearly 6% in DSDP core 330-4. The microflora contains Early Cretaceous palynomorphs (Harris, 1977) along with a heavy mixture of reworked Oxfordian pollens, spores, and coccoliths (Wise and Wind, 1977, Table 6C). Despite lack of evidence of any structural or lithologic discontinuity, cores 330-4 and 330-5 are separated by a biostratigraphic hiatus representing as much as 50 m.y. The basin floor must have been uplifted to near sea level or slightly above, perhaps as the Falkland Plateau was being rifted from the southern margin of Africa (at about 127 m.y. ago, according to Larson and Ladd, 1973).

Stagnation continued through Aptian time, with conditions highly unfavorable for the existence of most forms of marine life. There is no evidence of benthonic epifauna or infauna. The planktonic microfloras are generally impoverished except for sporadic blooms of stress-tolerant coccoliths which, in one instance, produced a thin chalk lamina (core section 330-3-2). The dominant constituent of this chalk, *Braarudosphaera,* is an encystment form, tolerant of hyposaline and fluctuating continental margin environments. It has been labeled a "disaster form" by Fischer and Arthur (1977) who note its proliferation during times of environmental crisis (e.g., the Cretaceous-Tertiary boundary event).

A notable feature of the braarudosphaerid chalk is the exceptional preservation of the other coccolith species. An unusually large number of coccospheres are preserved, including those with fragile nonarticulated coccoliths found clustered, essentially intact. This requires deposition in extremely quiet, shallow waters with no subsequent disturbance by bioturbation, a further indication of the highly stagnant bottom conditions.

The development of euxinic depositional environments throughout most of the world's oceans during Aptian time has been carefully documented by Fischer and Arthur (1977). They postulated the expansion and intensification of the oxygen-minimum layer (an oxygen-poor zone a few hundred meters below the ocean's surface) during what they labeled a "polytaxic" phase in the earth's climatic cycle. This condition develops during times of warm, globally uniform climates, resulting in reduced oceanic convection rates. They postulated similar conditions for Oxfordian-Kimmeridgian time.

Following much the same reasoning, Barker and others (1977) and Schlanger and Jenkyns (1976) presented similar models which may be applied to the development of euxinic conditions on the Falkland Plateau and to the north in the adjacent basins of the incipient South Atlantic. Barker and others (1977) postulated the restriction of deep-water circulation into the basins to the north due to the presence of the Falkland Plateau which was still in contact with the southern tip of South Africa during Aptian time. The plateau therefore served as a sill, preventing the introduction of deep, oxygenated water into the South Atlantic from the Indo-Pacific. A connection between the Mozambique Plateau and Antarctica may have served the same purpose (Thompson, 1977). The submerged crests of these plateaus would have also intersected the greatly thickened oxygen-minimum layer where highly euxinic conditions prevailed. In the restricted South Atlantic Basin, the rain of organic detritus from surface plankton and the influx of organic matter from terrestrial runoff along the narrow basin margins eventually depleted the available oxygen supply to the point that the entire basin may have become anoxic.

Albian ventilation

On the plateau, anoxic conditions ended rather abruptly in early Albian time with the deposition of light-colored, well-oxygenated nannofossil claystones. The carbonate content of the cores rises sharply from that in the Aptian beds as organic carbon content plummets (Figure 16.3). Palynomorphs disappear entirely. Bivalves (*Inoceramus* and Buchiidae) reappear and become abundant (their fragments represent as much as 70% of the coarse fraction of the cores) as do coccoliths and, in selected samples, radiolarians. Coccolith diversity more than doubles. *Aucilina,* an important bivalve guide fossil in austral and boreal regions, appears in DSDP 327A, core 18, and persists through the upper Albian beds.

An austral assemblage comprised of numerous small planktonic foraminifera dominated by hedbergellids occurs throughout the Albian sequence (Sliter, 1977). These are accompanied by a few ostracods and a relatively diverse benthonic foraminiferal assemblage of large agglutinated and calcareous species which indicate a paleodepth of deposition between 100 and 400 m. A gradual deepening of the depositional site, however, is indicated by the slight reduction in benthonic specimens from hole 327A, cores 21 to 15 (Figure 16.3).

The dramatic change in depositional environment and in the attendant faunas and floras at the Albian–Aptian time boundary is attributed to the passage of the eastern tip of the Falkland Plateau past the tip of Africa, allowing deep exchange of oxygenated waters through the developing seaway between the Indian and Atlantic Oceans (Barker and others, 1977, Figure 5*b*).

Late Albian subsidence and a Cenomanian–Santonian hiatus

With continued sea-floor spreading, the plateau began subsidence down the ridge flank, and the effects of this deepening are strongly evident in the cores beginning in the upper Albian–Cenomanian beds. In contrast to the thick (140 m) lower–middle Albian nannoclaystone section present in hole 327A, the upper Albian–Cenomanian section spans only 40 m. The Cenomanian sequence (core section 327A-14-6) consists of a greenish-gray clay with a carbonate content of only 5%. Foraminifers and calcareous nannofossils are moderately etched and poorly preserved, indicating deposition in deeper water near or below the lysocline. *Inoceramus* prisms dominate the samples, but other bivalve fragments are rare.

A sharp erosional hiatus separates the Cenomanian beds from the overlying 12 m of Santonian variegated brown and mottled zeolitic clay, interbedded with diffuse zones of light greenish-gray micrite. These contain a few strongly corroded calcareous microfossils deposited near the CCD, which fluctuated above and below the site of deposition. At this level the foraminiferal planktonic/benthonic percentages reverse sharply (Figure 16.3), carbonate content approaches zero, and the sedimentation rate is minimal. *Inoceramus* prisms reach peak abundance in the lower samples of this unit, then fall off sharply in the upper samples where they are heavily corroded. Other organisms include rare ostracods and fish teeth, and abundant small (70 μm), problematic calcareous objects that closely resemble marine algae (Sliter, 1977, Plate 14). The limited benthonic foraminiferal assemblages consist primarily of rare agglutinated and resistant calcareous species which are equivalent to modern low-latitude assemblages from abyssal environments (2500 to 4000 m). Another sharp hiatus separates the Santonian beds from the overlying Campanian carbonate ooze which contains bathyal rather than abyssal benthonic foraminiferal assemblages.

Despite the deep-water habitat suggested by the benthonic foraminifers, no tectonic evidence exists to suggest that the Santonian dissolution facies was produced by pronounced structural subsidence (Barker and others, 1977). Instead, it is assumed that the Falkland Plateau followed a normal subsidence curve as it spread laterally away from the ridge crest (Sclater and others, 1977). Distant tectonic and global paleoclimatic events during middle Cretaceous time, therefore, are believed to have combined to change the symmetry of the world's oceans, intensify cooler bottom-water flow, alter basic circulation patterns, increase dissolution, and raise the CCD in this study area. The abundant *Inoceramus* debris in the Cenomanian and lower Santonian samples probably represents winnowing and downslope transport of these objects from higher elevations on the plateau due to increased current velocities or change in flow directions.

The same series of events is believed to account for the missing Cenomanian, Turonian, and Coniacian sediments represented by the hiatus below the Santonian. This hiatus is now widely recognized in DSDP drill sites throughout the Southern Hemisphere (Sliter, 1977, Figure 10) as well as in other parts of the world. Accordingly, Fischer and Arthur (1977) recognize Cenomanian–Turonian time as a crisis or an "oligotaxic" episode in the world's oceans. The various models postulated to account for these mid-Cretaceous events are too numerous and complex to outline here. Instead, the reader is referred to Fischer and Arthur (1977) and Sliter (1977) who explore these matters in some detail.

Campanian–Maestrichtian carbonate sedimentation

Normal pelagic sedimentation resumed during Campanian–Maestrichtian time when an appreciably lower CCD combined with a moderately high rate of clay and biogenic silica deposition allowed exceptionally good preservation of calcareous microfossils. Nineteen new coccolith taxa are described from this interval by Wise and Wind (1977). A number of these seem provincial to the cooler high latitudes. Foraminiferal assemblages in these beds at Site 327 are also strongly austral in character. Palynomorph assemblages reappear in the Maestrichtian sequence, and are 89% marine, probably a reflection of the worldwide reduction of terrestrial source area during this stage.

Part of a coring gap spanning the Campanian–Maestrichtian boundary has been filled by *Islas Orcadas* core 11-76-12 taken 52 km north of the DSDP drill sites. A comparison of the calcareous faunas and floras with those of hole 327A led Ciesielski and others (1977) to conclude that an exceptionally strong latitudinal surface-water thermal gradient existed across the plateau during Late Cretaceous time. With circulation from the Pacific impeded or prohibited by the then existing South American–Antarctic isthmus, the plateau apparently served as a major barrier in the Atlantic between cold (possibly upwelling) waters to the south and warmer surface waters flowing from the north, a pattern that may have persisted well into Tertiary time. This interpretation may in part resolve some paradoxical questions concerning the existence of austral foraminiferal faunas at the drill sites in close latitudinal juxtaposition to transitional and Tethyan faunas in southern South America and South Africa (see Sliter, 1977). Similar problems of apparent provincial mismatch among molluscan faunas (Jones and Plafker, 1977) can now be resolved by new age data provided by Stapleton and Beer (1976) for sediments of the Cape Province of South Africa. They have determined that sediments in the Knysna Basin previously dated as "Late Jurassic" are actually Early Cretaceous (Hauterivian) in age. Thus present paleontologic data are in agreement with the tectonic reconstruction model presented by Barker and others (1977), which fits the eastern Falkland Plateau prior to rifting snuggly against the Mozambique Plateau. Alternative tectonic models by Sliter (1977) and by Jones and Plafker (1977) which position the plateau initially in the region of the Antarctic Peninsula are, therefore, not necessary.

Tertiary Period

The thick Campanian–Maestrichtian chalk sequence is capped by a hardground and an erosional disconformity which marks an abrupt end of carbonate sedimentation. Current erosion must have been intense, since palynomorphs indicate the absence of Danian sediments. The carbonate compensation depth stood well above the site of deposition since calcareous nannofossils are absent in the next 30 m of section at hole 327A. Paradoxically, carbonates were deposited intermittently during Danian time at DSDP site 323 in the Bellingshausen Sea on the Pacific side of the South American–Antarctic isthmus (Weaver and others, 1976). This is in contrast with the worldwide rise in the CCD boundary which Worsley (1974) believes occurred prior to Danian time.

Paleocene deposition began with 20 m of dark, zeolitic clay containing a significant component (10–46%) of terrestrially derived organic matter (spores, pollen, and tracheids). Upward in the section, these clays give way to diatom oozes in core 327A-8, and coccoliths finally appear in core 6 as the CCD gradually descended to approximately 2500 m (assuming that the plateau then was near its present-day water depth). The Paleocene diatoms, silicoflagellates, and radiolarians are exceptionally well preserved due to the thin overburden (less than 100 m) and moderately high clay content of the sediment. Calcareous microfossils, poorly preserved at site 327 due to their close proximity to the CCD, are better represented higher on the plateau at site 329 and in contemporaneous piston-core sediments (IO

775-45 and RC15-84), where carbonate values reach 80% (Figure 16.4). Since calcareous planktonic assemblages are somewhat restricted in diversity because of the high-latitude site of deposition, somewhat provincial zonations were utilized for the coccoliths and diatoms by DSDP shipboard scientists. As in the Cretaceous beds, some coccoliths endemic to these latitudes are present in the Tertiary section (e.g., *Hornibrookina australis* and *Dictyococcites antarcticus*).

Carbonate values in hole 327A peak in the upper Paleocene and lower Eocene beds, then decline upward in section to near zero. Because of drilling problems in cherty limestones at site 329, most of our knowledge of the remaining condensed Eocene section (less than 20 m) comes from piston cores which recovered middle Eocene zeolitic clays and upper Eocene diatomaceous ooze. All these sediments are deficient in calcareous microfossils, an indication that the CCD fluctuated at shallow depths (~1500–2500 m) throughout this epoch.

During Oligocene time the CCD must have dropped appreciably as relatively thick Oligocene chalks were deposited. These chalks contain reasonably well preserved siliceous microfossils in their upper part. Above an inferred Oligocene–early Miocene hiatus, an extraordinarily

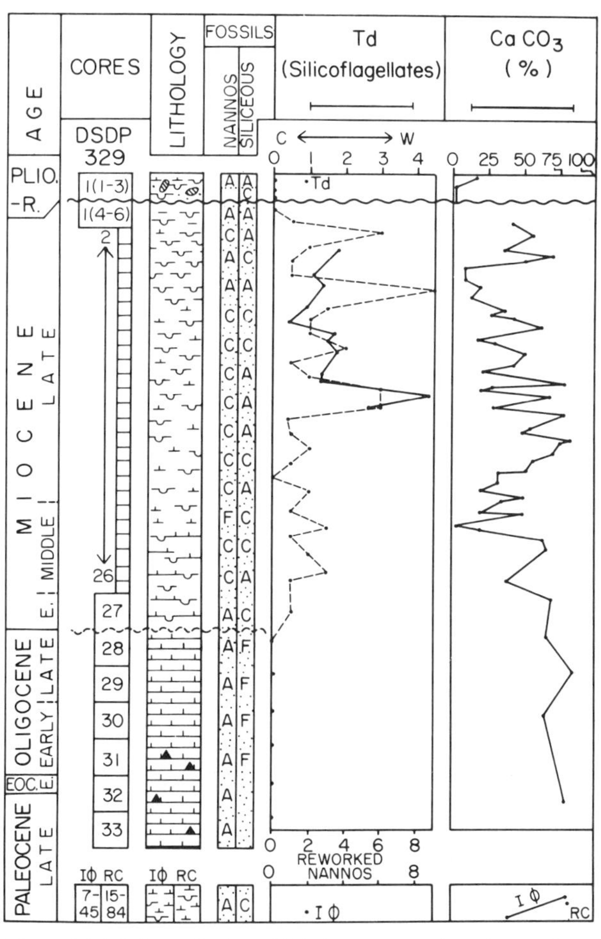

Figure 16.4. Paleontologic and lithologic properties of cores from Deep Sea Drilling Project hole 329, *Islas Orcadas* (IO) station 07-75-45, and *Robert Conrad* (RC) station 15-84. Dashed line in Td column: number of reworked calcareous nannofossils (nannos) per microscope field of view at 500× (bottom scale). Td: relative paleotemperature. C: Cold. W: Warm. Explanations for all other symbols are given in Figures 16.2 and 16.3.

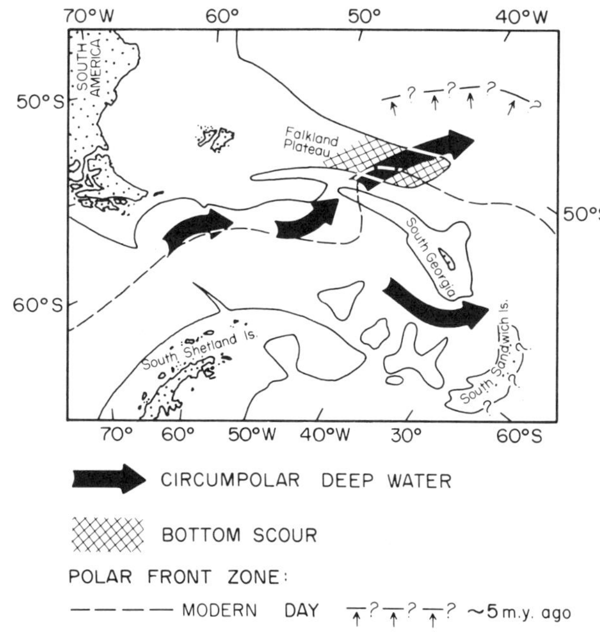

Figure 16.5. Scouring effects of circumpolar deep water during the "terminal Miocene event" as visualized by Ciesielski and Wise (1977). Land positions, bathymetry, and the present-day position of the Polar Front are from Barker and others (1977, Figure 10).

thick sequence of middle to upper Miocene diatomaceous coccolith oozes accumulated on the higher elevations of the Maurice Ewing Bank. The calcarous microfloras of these deposits are extremely restricted in diversity, dominated by two or three cold-water species (Haq and others, 1977). Another important characteristic of the Miocene sequence is the large number of older coccoliths, diatoms, and radiolarians reworked throughout the sequence. The incidence of reworked coccoliths (Cretaceous, Paleocene, and primarily Oligocene) is indicated in Figure 16.4 where the number of specimens per 10 fields of view of the microscope (at 500×) is recorded. One peak in this curve coincides with a warm peak in the silicoflagellate relative-paleotemperature curve of Busen and Wise (1977), an indication of a possible fluctuation in the dominant current direction. The abrupt change from Miocene carbonate sediments to noncalcareous Plio-Pleistocene diatomaceous glacial marine clastic sediments at site 329 marks a major erosional disconformity seen throughout this region. Calcareous nannofossils are not present higher in the section until the top few centimeters where an upper Pleistocene–Holocene coccolith ooze is presently accumulating.

The Tertiary sequence of the Maurice Ewing Bank records not only increasingly severe climatic deterioration from late Eocene time onward, but the continuation of a complex erosional-depositional history which sculptured much of the present topography of the bank. This erosional history is recounted by Ciesielski and Wise (1977) who recognized three primary episodes, two climatic and one tectonic in origin: (1) the Cretaceous–Tertiary boundary event, which caused deep scouring and exposure of older Mesozoic units; (2) the opening of Drake Passage, which Barker and Burrell (1977; and Chapter 43, this volume) believe began to form a deep water conduit about 29 m.y. ago for the Antarctic Circumpolar Current which was established by early Miocene time (this event is responsible for the inferred Oligocene–Miocene hiatus and the extensive reworking of Oligocene material into the Miocene deposits as the circumpolar current established a flow direction that impinged on the Maurice Ewing Bank); and (3) the particularly severe late Miocene glaciation of Antarctica (Hayes and Frakes, 1975; Meyewski, 1973) which increased the intensity of circumpolar flow to the point that circumpolar deep water impinging on the bank from the southwest stripped away the upper sediment cover, exposing units at least as old as Maestrichtian (Figure 16.5). The concomitant northward expansion of the Antarctic water mass forced the Polar Front (Antarctic Convergence) well to the north of the bank, shutting off appreciable carbonate deposition until the southern limit of the Polar Front zone migrated to its present position along the southern margin of the region in late Pleistocene–Holocene times (Figure 16.5).

The late Miocene hiatus, so well developed on the Maurice Ewing Bank (Ciesielski and others, 1977; Ciesielski and Wise, 1977), has now been documented in many sectors of the Southern Ocean and its environs (Figure 16.6). The agent of erosion, however, was not everywhere the same. In the Ross Sea the unconformity was formed by the grounded ice sheet (Hayes and Frakes, 1975); along the adjacent abyssal margin (DSDP sites 266 and 274), by bottom-current winnowing (Frakes, 1975); in New Zealand, by glacial eustatic regression (Kennett and Watkins [1974] as reinterpreted by Van Couvering and others [1976] and Weaver [1976]); and along the Agulhas Plateau, by climatically induced intensification of local currents (Tucholke and Carpenter, 1977). At DSDP site 328 (Malvinas Outer Basin, adjacent to the Falkland Plateau; Gombos, 1977), the agent was probably Antarctic bottom water rather than circumpolar deep water. Correlations of this "terminal Miocene event" elsewhere in the

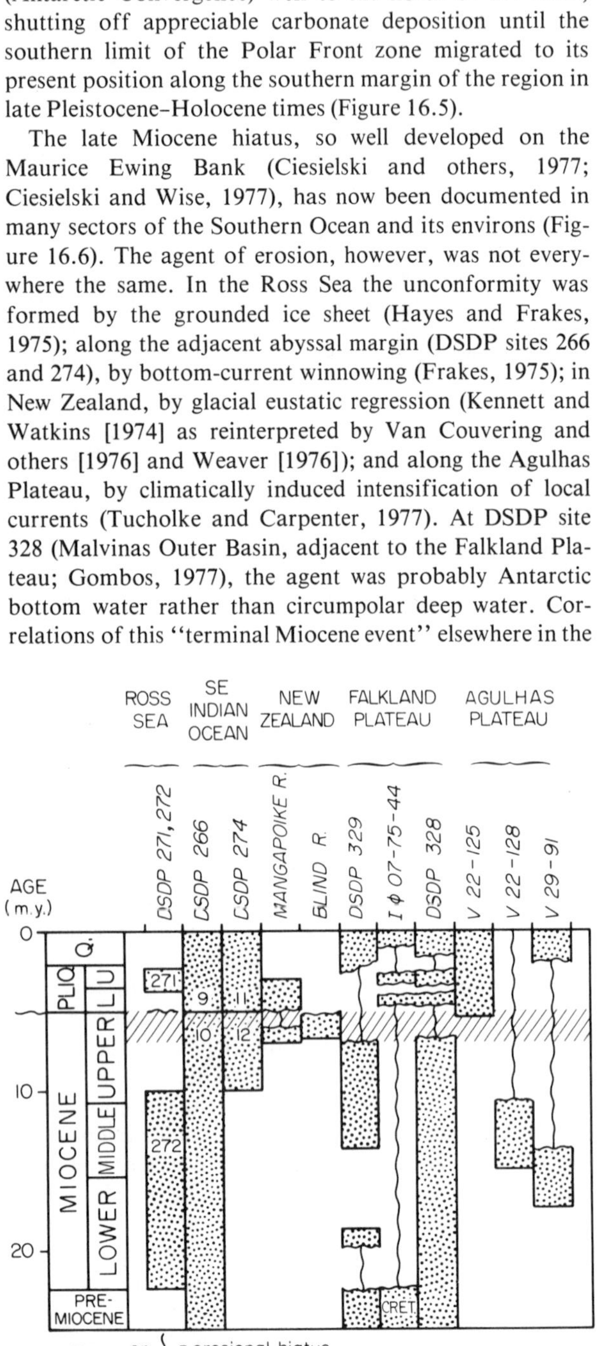

Figure 16.6. Late Miocene erosional and biostratigraphic hiatuses, a result of the "miocene Terminal event" as recorded in cores from the Southern Ocean and its environs. Shaded interval represents time of these hiatuses.

world are given by Van Couvering and others (1976) and Peck and others (1976).

The Pliocene–Holocene history of the study area is complex and poorly understood from the DSDP drill records. It is, however, being worked out from the piston cores by Ciesielski (Chapter 17, this volume).

Acknowledgments. The senior author thanks all who contributed data to this paper. The assistance and support of the Argentine Naval Hydrographic Office, the Deep Sea Drilling Project, and the piston-core curatorial facilities at the Lamont-Doherty Geological Observatory and at the Florida State University Antarctic Marine Geology Research Facility are gratefully acknowledged. S. W. W. was supported by National Science Foundation grants DPP 77-19360 and DPP 78-07183.

References cited

Barker, P. F., and Burrell, J. 1977. Note added in proof. (*In* Barker, P. F., Dalziel, I. W. D., and others. *Initial Reports of the Deep Sea Drilling Project,* Vol. 36. Washington [U. S. Government Printing Office], 1009.)

Barker, P. F., Dalziel, I. W. D., and others. 1977. *Initial Reports of the Deep Sea Drilling Project,* Vol. 36. Washington (U. S. Government Printing Office), 1080 pp.

Barker, P. F., Dalziel, I. W. D., Elliott, D. H., von der Broch, C. C., Thompson, R. W., Plafker, G., and others. 1977. Evolution of the southwestern Atlantic Ocean Basin: results of Leg 36, Deep Sea Drilling Project. (*In* Barker, P. F., Dalziel, I. W. D., and others. *Initial Reports of the Deep Sea Drilling Project,* Vol. 36. Washington [U. S. Government Printing Office], 993–1014.)

Busen, Karen L., and Wise, S. W. 1977. Silicoflagellate biostratigraphy, Deep Sea Drilling Project Leg 36. (*In* Barker, P. F., Dalziel, I. W. D., and others. *Initial Reports of the Deep Sea Drilling Project,* Vol. 36. Washington [U. S. Government Printing Office], 697–744.)

Ciesielski, P. F., and Wise, S. W. 1977. Geologic history of the Maurice Ewing Bank of the Falkland Plateau (southwest Atlantic sector of the Southern Ocean) based on piston and drill cores. *Mar. Geol., 25,* 175–207.

Ciesielski, P. F., Sliter, W. V., Wind, F. H., and Wise, S. W. 1977. Paleoenvironmental analysis and correlation of a Cretaceous *Islas Orcadas* core from the Falkland Plateau, southwest Atlantic. *Mar. Micropaleontol., 2,* No. 1, 27–34.

Fischer, A. G., and Arthur, M. A. 1977. Secular variations in the pelagic realm. (*In* Cook, H. S., and Enos, P., *eds.* Basinal Carbonate Sediments. *Spec. Publs Soc. econ. Palaeont. Miner., Tulsa, 25,* 19–50.)

Forest, J., de Saint Laurent, M., and Chace, F. A. 1976. *Neoglyphea inopinata:* a crustacean "living fossil" from the Philippines. *Science, N.Y., 192,* No. 4242, 884.

Frakes, L. A. 1975. Paleoclimatic significance of some sedimentary components at Site 274. (*In* Hayes, D. E., Frakes, L. A., and others. *Initial Reports of the Deep Sea Drilling Project,* Vol. 28. Washington [U. S. Government Printing Office], 785–89.)

Gombos, A. M. 1977. Paleogene and Neogene diatoms from the Falkland Plateau and Malvinas Outer Basin: Leg 36, Deep Sea Drilling Project. (*In* Barker, P. F., Dalziel, I. W. D., and others. *Initial Reports of the Deep Sea Drilling Project,* Vol. 36. Washington [U. S. Government Printing Office], 575–688.)

Haq, B. U., Lohmann, G. P., and Wise, S. W. 1977. Calcareous nannoplankton biogeography and its paleoclimatic implications: Cenozoic of the Falkland Plateau (DSDP Leg 36) and Miocene of the Atlantic Ocean. (*In* Barker, P. F., Dalziel, I. W. D., and others. *Initial Reports of the Deep Sea Drilling Project,* Vol. 36. Washington [U. S. Government Printing Office], 745–60.)

Harris, W. K. 1977. Palynology of cores from Deep Sea Drilling Sites 327, 328, and 330, South Atlantic Ocean. (*In* Barker, P. F., Dalziel, I. W. D., and others. *Initial Reports of the Deep Sea Drilling Project,* Vol. 36. Washington [U.S. Government Printing Office], 761–816.)

Hayes, D. E., and Frakes, L. A. 1975. General synthesis, Deep Sea Drilling Project Leg 28. (*In* Hayes, D. E., Frakes, L. A., and others. *Initial Reports of the Deep Sea Drilling Project,* Vol. 28. Washington [U. S. Government Printing Office], 919–42.)

Hedlund, R. W., and Beju, D. 1977. Stratigraphic palynology of selected Mesozoic samples DSDP Hole 327A and Site 330. (*In* Barker, P. F., Dalziel, I. W. D., and others. *Initial Reports of the Deep Sea Drilling Project,* Vol. 36. Washington [U. S. Government Printing Office], 817–28.)

Jones, D. L., and Plafker, G. 1977. Mesozoic megafossils from DSDP Hole 327A and Site 330 on the eastern Falkland Plateau. (*In* Barker, P. F., Dalziel, I. W. D., and others. *Initial Reports of the Deep Sea Drilling Project,* Vol. 31. Washington [U. S. Government Printing Office], 845–56.)

Kennett, J. P., and Watkins, N. D. 1974. Late Miocene–early Pliocene paleomagnetic stratigraphy, paleoclimatology, and biostratigraphy in New Zealand. *Geol. Soc. Am. Bull., 85,* No. 10, 1385–98.

Larson, R. L., and Ladd, J. W. 1973. Evidence for the opening of the South Atlantic in the Early Cretaceous. *Nature, Lond., 246,* No. 5430, 209–12.

Lonardi, A., and Ewing, M. 1971. Bathymetry chart of the Argentine Basin (Map, no scale). (*In* Ahrens, L. H., Press, F., Runcorn, S. K., and Urey, H. C., *eds. Physics and Chemistry of the Earth, 8,* London, Pergamon Press.)

Mayewski, P. A. 1973. *Glacial Geology and Late Cenozoic History of the Transantarctic Mountains, Antarctica.* Ph.D. dissertation, Ohio State University, 215 pp.

Peck, D., Missimer, T. M., and Wise, S. W. 1976. Glacial-eustatic control of an upper Miocene transgressive-regressive sequence in a low latitude (southeastern U. S.) coastal plain. *Abstr. with Programs, geol. Soc. Am., 8,* 1044.

Schlanger, S. O., and Jenkyns, H. C. 1976. Cretaceous oceanic anoxic events: causes and consequences. *Geologie Mijnb., 55,* No. 3-4, 179–84.

Scientific Staff. 1974. Southwestern Atlantic Leg 36. *Geotimes, 19,* No. 11, 16–18.

Sclater, J. G., Hellinger, S., and Tapscott, C. 1977. The paleobathymetry of the Atlantic Ocean from the Jurassic to the present. *J. Geol., 85,* 509–52.

Sliter, W. V. 1977. Cretaceous foraminifers from the southwestern Atlantic Ocean, Leg 36, Deep Sea Drilling Project. (*In* Barker, P. F., Dalziel, I. W. D., and others. *Initial Reports of the Deep Sea Drilling Project,* Vol. 36. Washington [U. S. Government Printing Office], 519–74.)

Stapleton, R. P., and Beer, E. M. 1976. "Upper Jurassic" sediments of South Africa. *Nature, Lond., 264,* No. 5581, 49.

Thompson, R. W. 1977. Mesozoic sedimentation on the eastern Falkland Plateau. (*In* Barker, P. F., Dalziel, I. W. D., and others. *Initial Reports of the Deep Sea Drilling Project,* Vol. 36. Washington [U. S. Government Printing Office], 877–92.)

Tucholke, B. E., and Carpenter, G. B. 1977. Sediment distribution and Cenozoic sedimentation patterns on the Agulhas Plateau. *Geol. Soc. Am. Bull., 88,* No. 9, 1337–46.

Van Couvering, J. A., Berggren, W. A., Drake, R. E., Aguirre, E., and Curtis, G. H. 1976. The terminal Miocene event. *Mar. Micropaleont., 1,* No. 1, 263–86.

Weaver, F. M. 1976. *Late Miocene and Pliocene Radiolarian Paleobiogeography and Biostratigraphy of the Southern Ocean.* Ph.D. dissertation, Florida State University, 175 pp.

Weaver, F., Rogl, F., Haq, B. U., and Schrader, H. J. 1976. Paleontological summary of Deep Sea Drilling results from Leg 35, Southeast Pacific Basin. (*In* Hollister, C. D., Craddock, C., and others. *Initial Reports of the Deep Sea Drilling Project,* Vol. 35. Washington [U. S. Government Printing Office], 531–38.)

Wind, F. H., Dinkelman, M. G., and Wise, S. W. 1976. Oxfordian onychites and a possible decapod microappendage from the Falkland (Malvinas) Plateau, southwest Atlantic. *Antarct. Jnl U. S., 11,* 167–69.

Wind, F. H., Dinkelman, M. G., and Wise, S. W. 1977. Jurassic scolecodont-like microfossils from the Falkland Plateau. (*In* Barker, P. F., Dalziel, I. W. D., and others. *Initial Reports of the Deep Sea Drilling Project,* Vol. 36. Washington [U. S. Government Printing Office], 829–44.)

Wise, S. W., and Wind, F. H. 1977. Mesozoic and Cenozoic calcareous nannofossils recovered by DSDP Leg 36 drilling on the Falkland Plateau, Atlantic sector of the Southern Ocean. (*In* Barker, P. F., Dalziel, I. W. D., and others. *Initial Reports of the Deep Sea Drilling Project,* Vol. 36. Washington [U. S. Government Printing Office], 269–492.)

Worsley, T. 1974. The Cretaceous-Tertiary boundary event in the ocean. (*In* Hay, W. W., *ed.* Studies in Paleooceanography. *Spec. Publs Soc. econ. Palaeont. Miner., 20,* 94–125.)

17

The Maurice Ewing Bank of the Falkland (Malvinas) Plateau: Late Miocene–Holocene Depositional and Erosional History and Its Paleoenvironmental Implications

Abstract

PAUL F. CIESIELSKI*
Antarctic Marine Geology Research Facility, Department of Geology, Florida State University, Tallahassee, Florida 32306, U.S.A.

A micropaleontologic and sedimentologic analysis of 56 piston cores provides the basis of a geologic study of the late Miocene to Holocene depositional and erosional history of the intermediate-depth Maurice Ewing Bank (MEB) located at the eastern extremity of the Falkland (Malvinas) Plateau, southwest Atlantic Ocean.

The position of the Polar Front and strong bottom currents beneath the axis of the Antarctic Circumpolar Current (ACC) are presently the factors most affecting sedimentation on the MEB. Fluctuations in the position of the Polar Front and in the intensity of the ACC are inferred to have largely controlled deposition on the MEB since the Miocene initiation of ACC flow over the bank. The sedimentary record of the MEB is used to infer the late Miocene to Holocene oceanographic and climatic conditions in the South Atlantic sector of the Southern Ocean.

The MEB is shown to have suffered major erosion by intense ACC flow during late Miocene time. Late Miocene erosion exposed most Cretaceous–Miocene sediment now near the surface and shaped the present configuration of the bank. This erosional event is inferred to have occurred during late Miocene to early Pliocene time (~7.4–4.65 m.y. ago), with the major phase of erosion taking place between middle paleomagnetic epoch 7 and late epoch 6 (~7.4–6.2 m.y. ago). A decrease in ACC velocity during an early Pliocene amelioration of climate led to resumption of deposition on the bank from 4.6–4.4 to 3.9 m.y. ago. Intensification of the ACC during the late Gilbert and early Gauss paleomagnetic epochs again resulted in limited deposition and widespread erosion and/or nondeposition over most of the MEB from about 3.9 to 3.0 m.y. ago. ACC velocity again apparently began to wane during the middle Gauss epoch and allowed widespread deposition on the bank during late Gauss time (2.8–2.43 m.y. ago) and over more limited areas during earliest Matuyama time (2.43 to about 2.13 m.y. ago). During middle to late Matuyama time, bottom-water intensity reached its greatest post-Miocene level, forming a regional disconformity between sediments of about 1.8 and 1.0 m.y. in age. This increased circumpolar circulation and associated erosion is inferred to have peaked during the late Matuyama epoch (1.2–1.0 m.y. ago), and is roughly time equivalent with the Great Patagonian Glaciation in nearby southern Argentina. Between about 1.0 and 0.7 m.y. ago, the bank was blanketed with a coarse, erosion-resistant layer of ice-rafted detritus which "armored" the older sediment thereby protecting it from major subsequent erosion. Sedimentation on the MEB during the Brunhes epoch (690,000 yr ago to present) was intermittent and finally culminated with the deposition of a veneer of carbonate ooze during or since latest Pleistocene time.

* Present address: Department of Geology, University of Georgia, Athens, Georgia 30602, U.S.A.

18

Geologic Evolution of South Georgia

P. W. G. TANNER*
British Antarctic Survey, Natural Environment Research Council, Madingley Road, Cambridge CB3 0ET, U.K.

Abstract. A late Jurassic–Early Cretaceous island arc–marginal basin system correlated with that in the southern Andes has recently been recognized on South Georgia. The component parts of this system are: (1) the mafic complex and continental crust of the probable basin floor, (2) turbidite facies successions occupying the basin, and (3) the contemporary island-arc assemblage. A major reverse fault, the Cooper Bay dislocation, separates the turbidite sequences (Cumberland Bay and Sandebugten Formations) from units that probably make up the floor of the basin. The Cumberland Bay Formation consists primarily of volcanic andesitic detritus; the Sandebugten Formation is more siliceous and contains abundant quartz and silicic volcanic fragments. Regional metamorphism and deformation of the Cumberland Bay Formation increase from southwest to northeast accompanied by the onset of tight chevron folds and development of a north-northeast-trending elongation lineation on the slaty cleavage. It is confirmed that this formation everywhere overlies, and is in thrust contact with, the Sandebugten Formation. Two minerals new to South Georgia are reported: pumpellyite, from various rocks including those of the Cumberland Bay Formation, and stilpnomelane, developed widely in the Sandebugten Formation. Preliminary results of paleocurrent analysis of part of the Cumberland Bay Formation in northeast South Georgia affected by chevron folding show that sediment dispersal directions were mainly to the west or northwest and that apparently anomalous directions are due to the effects of a second phase of folding. Evidence is presented that some of the tool marks in these turbidites are caused by shale or siltstone fragments. A new synthesis of directional data, both structural and sedimentary, is used to fit the South Georgia continental block back into its approximate predrift position south of Burdwood Bank. This reconstruction requires a $<10°$ counterclockwise rotation of the block and suggests an original maximum width of the marginal basin here (after Andean deformation) of not more than 200 km. It argues against a recent reconstruction in which it is proposed that the marginal basin widens markedly to the southeast in this area and that the Scotia Sea is floored by Lower Cretaceous oceanic crust.

Since 1974, field work by geologists of the British Antarctic Survey and by Dalziel and others (1975) has led to the recognition that the turbidite facies rocks making up much of the island of South Georgia (Trendall, 1959) were deposited in a marginal basin of Late Jurassic–Early Cretaceous age. This basin had a composite floor of oceanic and continental segments (Storey and others, 1977), and island-arc assemblages from the ancient shelf area have been described from Annenkov Island (Suárez and Pettigrew, 1976) (Figure 18.1). These stratigraphic-tectonic units form a continuation of the island arc–marginal basin system of the same age in the southern Andes described by Katz (1972, 1973) and Dalziel and others (1974, 1975).

Five major stratigraphic-tectonic units have been recognized on South Georgia (Table 18.1). Where exposed, contacts between these units are tectonic, and it is important to note that the two turbidite sequences, the Cumberland Bay and Sandebugten Formations, are everywhere separated from the other units lying to the south by a steeply dipping fault zone, the Cooper Bay dislocation (Figure 18.1). This zone is occupied by mylonitic rocks in the southeast (Trendall, 1959), but the amount and direction of movement along it are not known (Stone, in press, *a*).

Floor of the Mesozoic marginal basin

Drygalski Fjord Complex

The geology of this complex (formerly the southeastern igneous complex of Trendall, 1959) has recently been described in some detail (Bell and others, 1977; Storey and others, 1977). It consists of multiply deformed siliceous metasedimentary rocks and gneisses cut by a wide variety of igneous rocks, largely gabbroic, but including some granite. Radiometric dating (Tanner and Rex, 1979) suggests that many, possibly most, of the igneous plutonic rocks are older than the marginal basin and that only the dike swarms and some of the gabbros may be of Late Jurassic–Early Cretaceous age and comparable in age to the adjacent mafic complex.

Two main sets of mafic dikes cut the complex (Trendall, 1959): the earlier set trends northwestward and has a steep to vertical dip; it is crossed and locally displaced by a set of less common dikes trending between northeastward and eastward. Adjacent to the inferred contact with the Larsen Harbour Formation, dikes make up 60% of the rock and

* Present address: Department of Geology, University of Glasgow, Glasgow G12 8QQ, U.K.

Table 18.1. Major stratigraphic-tectonic divisions of South Georgia. (Unit numbers are keyed to Figure 18.2.)

Age	Rocks units and contacts	Divisions
	5. Annenkov Island Formation and associated plutonic and hypabyssal rocks	Island-arc assemblage
	——— Faulted contact? ———	
	4. Cumberland Bay Formation	
Late Jurassic–middle Cretaceous	——— Thrust fault ———	Marginal basin turbidite sequences
	3. Sandebugten Formation, Cooper Bay Formation, and part of the Ducloz Head Formation	
— — — — — — — —	— Contact not exposed — — —	— — —
Late Jurassic	2. Larsen Harbour Formation	Mafic complex and continental crust of basin floor
— — — — — — — —	— Contact not exposed — — —	
Pre-Jurassic to Early Cretaceous	1. Drygalski Fjord Complex	

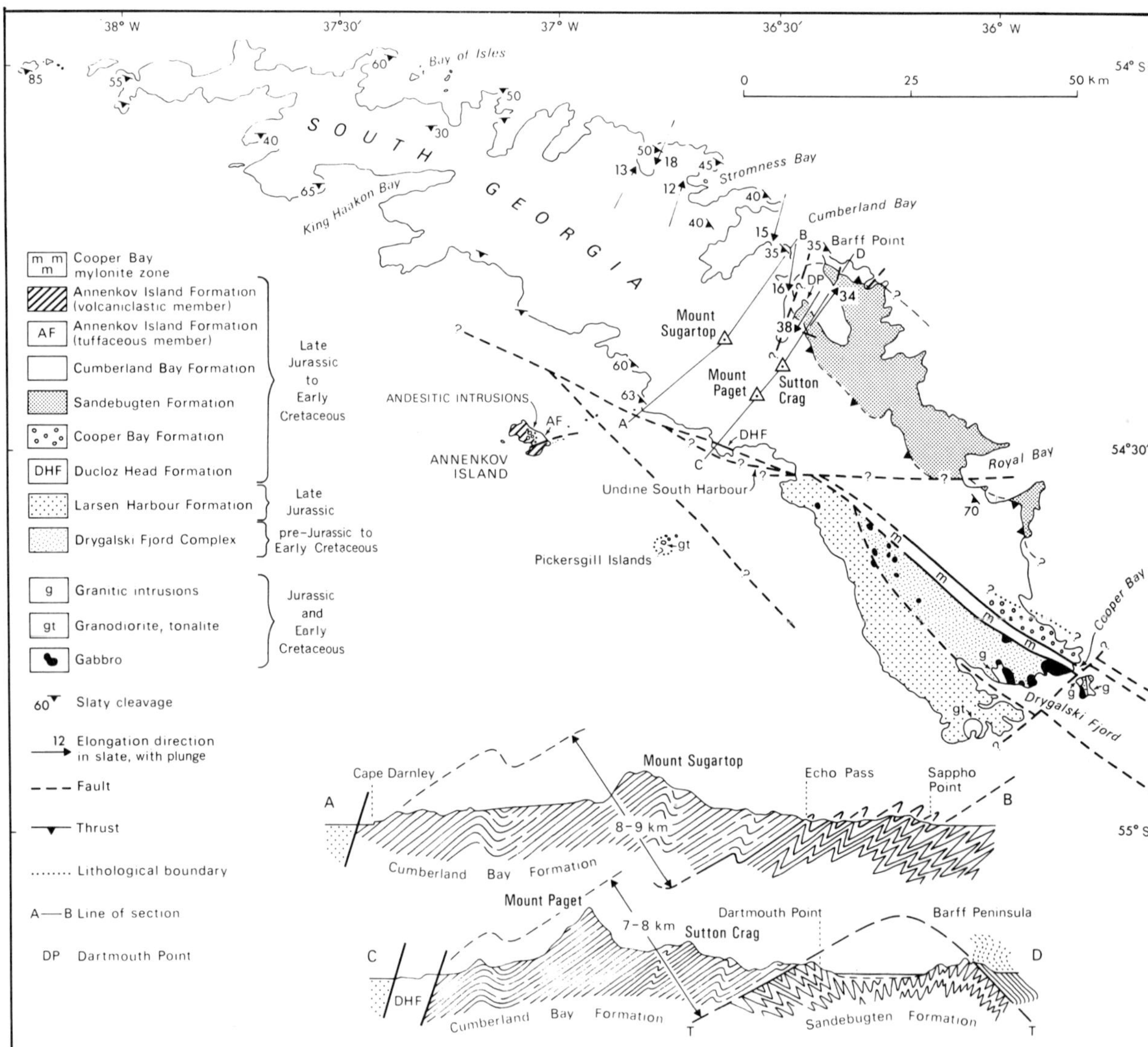

Figure 18.1. Geologic map of South Georgia. Profiles AB and CD show folds of bedding in each case and have the same vertical and horizontal scale (2.6 × map scale). Folds in the Sandebugten Formation are schematic only, but correctly depict the change in dip of the axial planes from southwest to northeast.

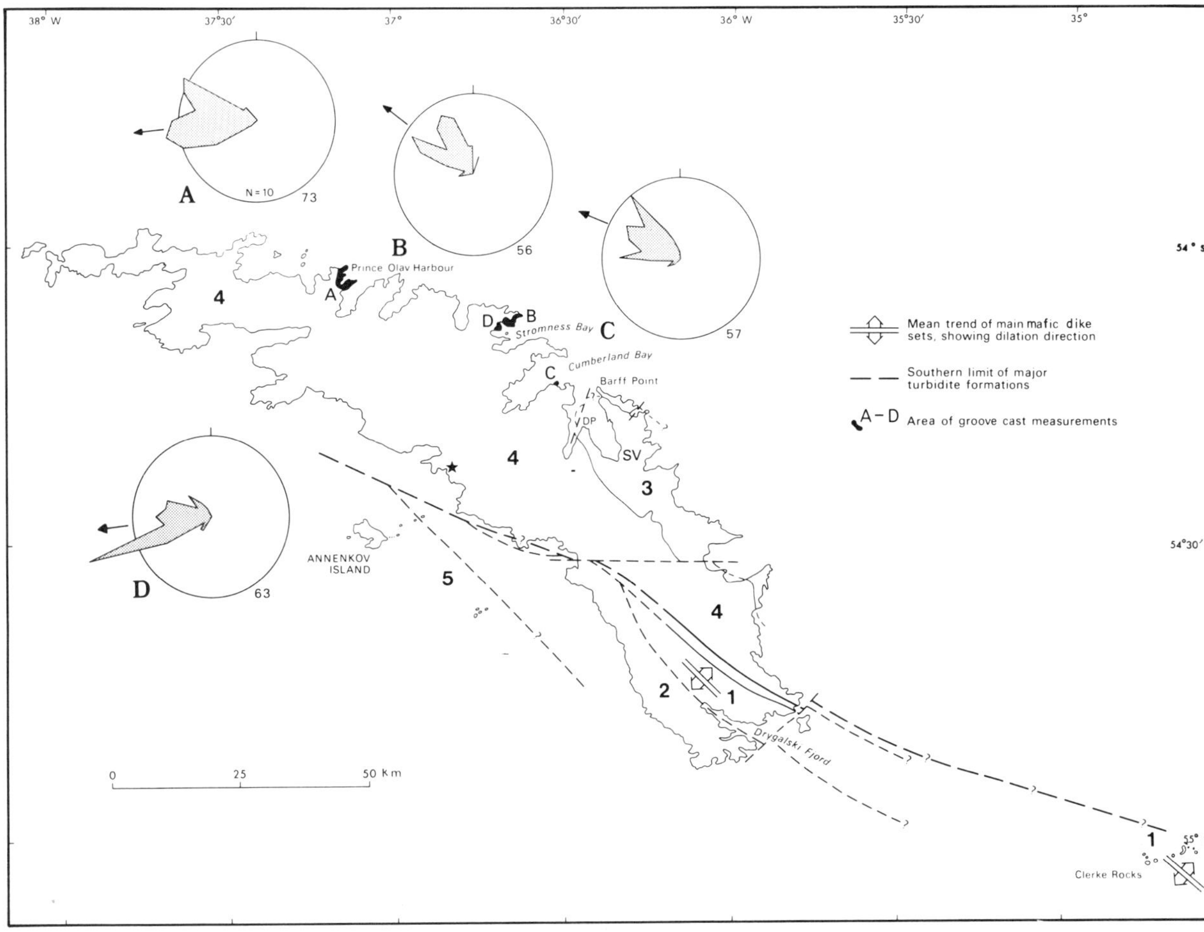

Figure 18.2. Rose diagrams of paleocurrent directions in Cumberland Bay and Sandebugten Formations of northeast South Georgia. Sediment dispersal directions in diagrams A-D from the Cumberland Bay Formation are based upon groove mark orientations (for explanation, see text); resultant vector directions are shown by arrows. Total number of observations are given for each diagram; radius of circle $N = 10$ observations. Units 1-5 refer to numbers on Table 18.1. Star shows location of pumpellyite-bearing assemblage in Jacobsen Bight. SV: Sörling Valley.

include sheeted dikes (Trendall, 1959); they decrease in number away from this contact (Storey and others, 1977).

On Clerke Rocks, some 65 km to the east-southeast (Figure 18.2), a large granite pluton and other rocks correlated with the Drygalski Fjord Complex are cut by closely spaced swarms of mafic dikes; the dikes trend northwestward and all dip steeply to the southwest. Only a few percent of the dikes are oblique to the main trend.

The dike orientations from these two widely separated parts of the complex suggest that the maximum extensional principal stress was oriented northeast-southwest during emplacement of the adjacent mafic complex (Larsen Harbour Formation) (Figure 18.2).

Larsen Harbour Formation

This formation consists mainly of pillow lavas, stratiform breccias, rhyolites, and thin volcaniclastic beds, cut by numerous dikes (Mair, in press). The sequence dips at about 25°, generally to the southwest, but is internally undeformed. Thickness was initially estimated at 5 km (Bell and others, 1977), but subsequent mapping and allowance for dilation by dikes has reduced this to 2-3 km (B. F. Mair, personal communication). Preliminary work showed that the main dike swarms trend between northwest and north.

Greenschist facies assemblages are common, with the widespread development of actinolitic hornblende, epidote, chlorite, and sodic plagioclase (Mair, in press), and the rocks are better termed metabasalts rather than true spilites. They are similar in petrography to those described by Stern and others (1976) from layer 2B of the Chilean ophiolites; mineral assemblages there are thought to result from hydrothermal circulation of heated seawater.

Recognition of this formation as the upper part of an ophiolite complex relies, in the absence of associated ultramafic rocks, upon: (1) the presence, in specific localities at the lowest tectonic level, of rocks with 90% dikes by volume (Storey and others, 1977), which possibly represent the top of a sheeted dike complex; and (2) comparison with more fully developed ophiolite complexes, particularly in the southern Andes (Dalziel and others, 1974; Stern and others, 1976). Geochemical work is being carried out at present on rocks from the Drygalski Fjord Complex and the Larsen Harbour and Annenkov Island Formations, in particular to test this hypothesis of an oceanic origin for the South Georgia mafic complex.

A further problem is that turbidite facies rocks belonging to the Cumberland Bay Formation are nowhere seen to actually overlie the Larsen Harbour Formation. As there is no evidence to suggest that the mafic complex was emplaced *after* deposition of the Cumberland Bay Formation, it is suggested that the Drygalski Fjord Complex and the Larsen Harbour Formation were uplifted as a single block by reverse fault movement on the Cooper Bay dislocation and were subsequently stripped of any sediment cover.

COOPER BAY DISLOCATION

This major tectonic break, which possibly crosses the entire South Georgia block, is marked in the south by a zone of cataclastic rocks (Cooper Bay mylonite zone, Figure 18.1). This zone is probably bounded by later faults, and it separates the turbidite sequences to the northeast from other rock units to the southwest (Figures 18.1 and 18.2). At Cooper Bay, where it was first recognized by Holtedahl (1929, p. 56), Trendall (1959) described the rocks within the zone as well-foliated gneiss and noted that they had undergone "intense cataclastic shearing." The cataclastic layering dips steeply to the southwest or is nearly vertical.

Sedimentary rocks of the Cooper Bay Formation to the north (Stone, in press, *a*) (Figure 18.1), gneisses and igneous rocks of the complex (Storey and others, 1977), and mafic dikes are all strongly deformed within the zone. No dikes cross into the Cooper Bay Formation or cut the mylonitic layering. Development of mylonitic fabrics occurred early in the tectonic history, and Stone (in press, *a*) has recognized that cataclasis took place early in the structural history of the Cooper Bay sedimentary beds beneath the tectonic break. The mylonitic banding in the fault zone is folded by late folds, and it appears likely that a large, northeasterly directed, reverse fault movement on the Cooper Bay dislocation occurred sometime during the main deformation of the turbidite sequences.

Members of the Ducloz Head Formation are separated from the Cumberland Bay Formation by a major fault (B. C. Storey, personal communication), which may possibly be an extension of the Cooper Bay dislocation. Both pillowed and massive lavas with sedimentary rocks were reported from Ducloz Head by Trendall (1959, p. 23); possible correlations between these rocks, the Cooper Bay Formation, and other major units is under study.

On a large scale, the Cooper Bay dislocation appears to cut across the junction between the Larsen Harbour Formation and Drygalski Fjord Complex (Figure 18.1). When extrapolated to Clerke Rocks along the northern edge of a submarine magnetic anomaly (Barker and Griffiths, 1972), this break also appears to cut across the trend of the dikes in the Clerke Rocks granite (Figure 18.2). It does not, therefore, appear to be parallel to primary junctions between continental and oceanic crust but is a tectonic break which developed after the composite floor of the basin had formed.

Marginal basin flysch sequences

CUMBERLAND BAY FORMATION

This formation is a volcaniclastic sequence of sandstone and slate with individual sandstone and grit beds reaching 1-2 m in thickness. It is of turbidite facies, with a total vertical thickness probably on the order of 8 km (Figure 18.1 cross sections), a thickness close to the earlier estimate of 10 km by Trendall (1959). The arenaceous rocks consist mainly of andesitic rock fragments, plagioclase clasts, and rare quartz and silicic igneous grains in a partially recrystallized matrix. Other detrital grains include pale green clinopyroxene and epidote; brown hornblende and detrital mica flakes (Thomson and others, Chapter 19, this volume) are reported for the first time.

Amygdaloidal lava fragments are common and in some rocks are accompanied by abundant pyroclastic debris, notably devitrified glass shards, altered pumice fragments, and volcanic lapilli (Clayton, in press, *a*; Stone, in press, *b*), all testifying to explosive volcanicity of island-arc type in the source region. Geochemical study of 160 specimens from different levels in the sediment pile confirms their overall calc-alkaline parentage (Clayton, in press, *b*), as first suggested by Tyrrell (1930) from petrographic studies.

Structure

The progressive development of cleavage in these rocks from mudstone with hackly cleavage on the southwest coast to well-crystallized slate along the central northeast coast has been described by the author (Thomson and others, Chapter 19, this volume). Cleavage varies in dip from steep to vertical in the southwest to gently dipping ($<30°$) in the northeast (Figure 18.1 cross sections); it has an arcuate trend along the island, parallel to both major fold traces and to the trend of gently plunging fold hinges. The angle between bedding and slaty cleavage is greatest in the southwest and decreases progressively toward the northeast where the two elements are locally parallel.

A true elongation lineation is rarely developed on the slaty cleavage and is difficult to see in the field, but it has been found in a sufficient number of localities to show that it has a constant trend (Figure 18.1). Mineral and clast elongation lineations reported from the Cumberland Bay area by Dalziel and others (1975, Figure 4) are more variable in orientation, and some appear to be parallel, or at a low angle, to the bedding/cleavage intersection.

On the northeast side of the island the Cumberland Bay Formation is strongly affected by early (F_1) chevron folds associated with the development of the slaty cleavage. The transition from unfolded rocks to rocks with true chevron folds, having planar limbs and narrow hinge zones (Ramsay, 1974), is marked by upright fold pairs of short wavelength; they have steeply dipping middle limbs and show strongly developed divergent cleavage fans in slaty rocks throughout the structure. With northeasterly decrease in interlimb angle to 60° or less, the axial planes of the folds become more gently dipping, and fanning of the cleavage is more difficult to detect except in the hinge zones. Wavelengths of these folds are on the order of hundreds of meters.

At a number of localities, the early chevron folds are affected by later folding (F_2). For example, four major folds of this age trend west-northwest through Stromness Bay. They are large open structures, associated with a well-developed crenulation cleavage and a lineation plunging gently to about 290°, which is at a small but variable angle to the hinges of the early folds. These large F_2 folds cause marked changes in the trend of the F_1 cleavage; these either have not been described previously (Skidmore, 1972) or have been recognized only on a small scale (Dalziel and others, 1975, Figure 3, inset). Farther south, Stone (in press, *b*) has noted two deformation episodes, and Clayton (in press, *c*) has recognized three phases of folding from the area northeast of King Haakon Bay (Figure 18.1).

Where the F_1 elongation lineation occurs in rocks unaffected by later folding it plunges to the south-southwest (Figure 18.1); however, where the slaty cleavage is folded by F_2 folds, such as to the west of Stromness Bay, the lineation plunges gently to the north-northeast. The elongation lineation always lies in the plane of the slaty cleavage but the dip of the cleavage varies considerably, especially where the rocks are affected by the second folding. This explains the apparent discrepancy between the plunges of the elongation lineations and the dip values for slaty cleavage as shown for adjacent areas on Figure 18.2.

Sedimentology

The Cumberland Bay Formation is a classic turbidite sequence with the graywacke showing well-developed Bouma divisions. Interlayering of thinly bedded, more distal sequences and thickly bedded, more proximal sequences is common (Mortimore, 1979). The general features of the sequence have been well described by Trendall (1959), Frakes (1966), Skidmore (1972) and Dalziel and others (1975); emphasis is given here to new observations made in the type area of Cumberland Bay.

Beds with chaotic, slump-folded, and disrupted internal lamination, large rafts of sediment, and angular fragments of mudstone and shale set in a much finer grained matrix are common. They reach at least 1–1.5 m in thickness, are ungraded or only graded within the basal centimeter or less, and appear to have been deposited by flowing, liquified debris (Lowe, 1976); tool marks or "slide" marks are locally prominent on the bases of these beds. Folded or disrupted calcareous lenses occur within these beds, supporting an early diagenetic origin for the lenses (Trendall, 1959). Beds affected locally by small slump folds are also found throughout the sequence.

Many of the graded beds have planar bases; classifiable bottom structures, apart from trace fossils, are absent. Elsewhere in the sequence, bottom structures consist largely of groove casts (tool marks) and load structures; current marks such as flute casts, longitudinal furrows and ridges (Figure 18.3), and fleur-de-lys structure are uncommon. Other structures, not described previously, include chevron marks, water-escape structures, washout channels, load-casted ripple marks, and frondescent marks.

Most tool marks result from pieces of shale or siltstone being ripped from the surface of underlying beds and dragged along at the base of the flow. Shale fragments have been found at the end of tool marks at three localities (Figure 18.4) and were clearly responsible for the marks (cf. Dżuliński and Walton, 1965, Figure 66A). Some tool marks, particularly in fine parallel sets, were caused by small pebbles and rock fragments carried by the turbidity current, and one pebble was noted at the end of a tool

Figure 18.3. Longitudinal furrow-and-ridge structure showing current directed from right to left. Cumberland Bay Formation, Tønsberg Point, Stromness Bay. Scale in cm.

mark. Shale and mudstone fragments are commonly found in the lower parts of graded beds which have prominent tool marks on the base. Prod and bounce marks were sought but are rare, and this agrees with the conclusion that shale fragments rather than wood or deeply ribbed ammonites are the tools responsible.

Two, and locally three, sets of groove marks occur on the soles of individual beds in each of the subareas (A–D, Figure 18.2) described below. The maximum divergence of the trend was 35° between extremes, but it was generally 10°–30°.

For statistical reasons care was taken to count only one groove direction per bed (a mean if there was some variation in direction), or two directions if two distinct crossing sets were observed. Direct measurement of the bedding/cleavage intersection for unfolding purposes was made for each bed or set of beds, if possible from an area not more than 10 m across, for this parameter was found to vary considerably and to affect the precision of the unfolding procedure.

Paleocurrent analysis

Preliminary results of work carried out by the author between Dartmouth Point and Prince Olaf Harbour (Figure 18.2) during 1973–74 and 1975–76 are discussed; it follows previous work, largely by Trendall (1959) and Dott (1974), which has been reviewed by Dalziel and others (1975).

The diagrams in Figure 18.2 (A–D) show only currents directed to the quadrant between west and north. Although groove marks do not give the direction of current flow, 34 other unambiguous indicators found on the soles of beds from these areas show (with the exception of one set of flute marks from the east side of Fortuna Bay) that the currents flowed in this direction. The indicators are: 19 sets of flute casts, 4 tools lying at the end of tool marks, 2 chevron groove casts, and 9 longitudinal ridge-and-furrow marks. Flute casts and longitudinal ridge-and-furrow marks are generally parallel to groove marks where these occur on nearby bedding surfaces. When restored to their original orientation, the current direction indicators from a particular area occur randomly throughout the whole range given by groove marks from the same area.

The possible effect of F_2 on the unfolding procedure was recognized early in the present work, and areas are divided into those affected by only F_1 and those affected by both F_1 and F_2. Prince Olav Harbour is typical of an area affected only by major F_1 folding; standard tectonic unfolding procedures result in a bimodal distribution of current directions (Figure 18.2, diagram A). A similar pattern is seen farther south in other areas not affected by second folds (B and C), but the trend shows a pronounced clockwise rotation in resultant vector direction from 265° (circular standard deviation, or CSD, 30°) at Prince Olaf Harbour to 307° (CSD, 32°) at Leith Harbour and 294° (CSD, 25°)

Figure 18.4. Siltstone fragments at the end of tool marks. Cumberland Bay Formation, 1.5 km southeast of Cape Saunders, Stromness Bay. Scale in cm.

at Maiviken. This change in orientation occurs in conjunction with the change in trend of the F_1 cleavage (Figure 18.1; and Thomson and others, Chapter 19, this volume, Figure 19.2) and suggests either that some feature of the morphology of the sedimentary basin exerted a control over the orientation of the tectonic structures, or that both the F_1 folds and the current directions were originally colinear and have been affected by later deformation. Structural and sedimentologic data from the whole of South Georgia west of long. 36° 20′ W are at present being analyzed by the author and by D. I. M. Macdonald. Rigorous statistical tests are being carried out on both structural and sediment dispersal data to test the above possibilities.

Southeast of Leith Harbour (Figure 18.2, diagram D) the rocks lie in the core of an F_2 synform which has an axis oblique to the hinges of local major F_1 folds. The result of unfolding the groove directions solely about F_1 is shown in Figure 18.2, diagram D; the pattern is closely similar to that obtained by Dott (1974, Figure 9) from the same locality. However, when the effects of the F_2 deformation are removed before unfolding about F_1, a "normal" northwest-directed distribution results. Thus the anomalous result shown in Figure 18.2, D (with a resultant vector direction of 264° (CSD, 27°) compared with 315° (CSD, 32°) for a directly adjacent area unaffected by F_2 folding) is the result of an incomplete unfolding procedure.

There is broad agreement between these data and the total sole-mark data (mean 300° from 225 observations) of Dalziel and others (1975). However, current directions given by small-scale cross-bedding are from both north and south (Dott, 1974; Dalziel and others, 1975) and have a mean vector azimuth at 32° to that for the sole-mark data. Although the main current dispersal directions given by sole marks in rocks on the northeast side of the island is to the northwest, recent work on rocks from the southwest coast between King Haakon Bay and Undine South Harbour suggests that a significant amount of sediment in this area was derived from the southwest and that the sediment dispersal directions were complex (D. I. M. Macdonald, personal communcation).

Sandebugten Formation

This formation is composed of turbidite facies rocks (Trendall, 1959; Dalziel and others, 1975) which are more siliceous than those of the Cumberland Bay Formation and appear on the average to be less thickly bedded and more strongly deformed. In thin section the most noticeable feature is the abundance of clastic quartz grains (Tyrrell, 1930), which constitute up to 50% of the detrital content (Stone, in press, *b*), a few trachytic and dacitic fragments, and the presence of felsitic and granitic clasts.

Structure

These rocks are strongly deformed into flattened chevron folds which commonly have an interlimb angle of only 19°–25° and wavelengths of tens of meters, an order of magnitude less than those of folds in the Cumberland Bay Formation. Early structures on the Barff Point Peninsula are affected by a crenulation cleavage (F_2), and the change in dip of the axial planes of the F_1 folds between Barff Point and Dartmouth Point (shown diagrammatically on Figure 18.1 cross section) probably results from the effects of regional F_2 deformation. Farther south the structure is dominated by major second folds with gently dipping axial planes (Stone, in press, *b*).

Sedimentology

The entire coastal section between Sörling Valley and Barff Point has been examined. Although the bases of many inverted beds are exposed, bottom structures are rare, and only five measurable groove casts were found. Restoration of these data by unfolding gives original nondirectional trends of 263°, 266°, 274°, 290°, and 325°, which lie within the same range as those from the Cumberland Bay Formation. Paleocurrent analysis, using small-scale cross-bedding orientations, gives either north- or south-directed patterns (Dott, 1974; Dalziel and others, 1975, Figure 7) which are not dissimilar from patterns reported by Dott from separate localities in the Cumberland Bay Formation.

Relationship between the two formations

The separate identity and approximate outcrop areas of the formations were first suggested by Tyrrell (1930). Trendall (1959) recognized that the critical contact zone lay on the Dartmouth Point peninsula (Figure 18.1) and correctly analyzed the tectonic structures on either side of it (1959, Figure 12), but he failed to find the actual contact, probably because of snow cover. A thrust contact (modified unconformity?) between the Sandebugten Formation and the overlying Cumberland Bay Formation was first reported by Aitkenhead and Nelson (1962, in Skidmore, 1972) some 10 km to the north-northeast near Barff Point (Figure 18.1), and it was subsequently mapped in detail by Stone (in press, *b*). A thrust contact between the two formations was described by Dalziel and others (1975) from the Dartmouth Point peninsula, but these workers considered that all of the rocks on the Barff Peninsula belonged to the Sandebugten Formation.

Both thrust contacts, on Barff Point and Dartmouth Point, were located by the author and structural mapping was carried out for several kilometers on either side of them. This work has shown that:

1. An outcrop of the Cumberland Bay Formation occurs on the Barff Peninsula (Figure 18.1); it is folded by a few, large open folds with marked cleavage fans and is lithologically and structurally similar to Cumberland Bay Formation rocks on the Grytviken Peninsula to the west and to those in the southern part of the Dartmouth Point peninsula.
2. The Sandebugten Formation on the Barff Peninsula has been intensely folded by numerous tight folds with steeply dipping axial planes, and is lithologically identical to rocks found beneath the thrust plane on Dartmouth Point (cf. Trendall, 1959).
3. The contact between the two formations near Barff Point is sharply defined and dips northward at 47°. Bedding in the Cumberland Bay Formation above the thrust plane is approximately parallel to the contact which, as reported by Stone (in press, *b*), cuts across the hinge zones of folds in the underlying Sandebugten Formation.
4. In most respects the thrust contact on the Barff Peninsula is the mirror image of that on Dartmouth Point (Figure 18.1, cross section CD), as described by Dalziel and others (1975).

Island-arc assemblage

The Annenkov Island Formation has recently been described in detail by Suárez and Pettigrew (1976) and Pettigrew (in press). Unpublished work by the author, D. I. M. Macdonald, and B. C. Storey on the Hauge Reef Islands (the line of islands between Annenkov Island and South Georgia) and the Pickersgill Islands (Figure 18.1) has confirmed earlier suggestions by Pettigrew that rocks belonging to the Annenkov Island Formation (mainly finely banded tuffs) and associated plutonic and hypabyssal rocks of the island-arc assemblage occur in these areas. A large monzodiorite pluton exposed in the Pickersgill Islands has yielded a Rb-Sr isochron age of 81 ± 10 m.y. (Tanner and Rex, 1979).

Regional metamorphism

A reassessment of metamorphic assemblages in rocks from South Georgia is still at a preliminary stage, but two new minerals—pumpellyite and stilpnomelane—are reported here; their significance is briefly discussed.

Pumpellyite

This mineral is present as minute, colorless to blue-green crystals in volcanic rock fragments as large as 1 cm across in a rock from Jacobsen Bight (starred on Figure 18.2); the identification has been confirmed by electron microprobe analyses. It occurs with prehnite and, together with other possible pumpellyite in isolated grains from prehnite-bearing rocks elsewhere in the Cumberland Bay Formation, indicates that the formation has been metamorphosed to the prehnite-pumpellyite facies. Pumpellyite has also been found:

1. associated with biotite in narrow veinlets cutting the Clerke Rocks granite; an adjacent gabbro is cut by prehnite veins; and
2. in an epidosite, a rock type previously described by Tyrrell (1916, 1918), collected from the Larsen Harbour Formation (Figure 18.1) by the author when carrying out reconnaissance work in 1973–74.

Stilpnomelane

Stilpnomelane has been found by the author to occur widely in rocks of the Sandebugten Formation, where it forms radiating groups and small clusters of brown or green-brown laths. Identification has been checked by microprobe, the mineral having previously been misidentified as biotite (Stone, in press, *b*). It also occurs in banded clinozoisite-actinolite-albite rocks and albite-rich veins from Shag Rocks (Tanner, in press, Figure 3), a small group of islets 250 km west-northwest of South Georgia.

These new mineral occurrences on South Georgia support the suggestion by Suárez and Pettigrew (1976) that metamorphic grade increases northeastward from rocks of the island-arc assemblage to the turbidite sequences. Laumontite-bearing zeolite facies rocks of the Annenkov Island Formation (Suárez and Pettigrew, 1976) are succeeded northeastward by the prehnite-pumpellyite assemblages (Cumberland Bay Formation), and these in turn overlie rocks containing stilpnomelane-epidote-(prehnite) (Sandebugten Formation).

Discussion

Recent work on South Georgia supports earlier geologic correlations with the southern Andes (Dalziel and Elliot, 1971, 1973; Dalziel and others, 1975; Suárez and Pettigrew, 1976), and the new data enable the South Georgia continental block to be fitted back with increased precision into its predrift position. These correlations are:

1. Igneous plutons and stocks of the island-arc assemblage in southwest South Georgia are equated with those forming the northeastern margin of the root zone of the Early Cretaceous island arc in South America (Patagonian batholith).
2. The Annenkov Island Formation is equated with the Hardy Formation of the southern Andes; both are volcaniclastic deposits formed on the flank of an active island arc (Suárez and Pettigrew, 1976).
3. A mafic complex on South Georgia is equivalent to mafic rocks which occur geographically between the Patagonian batholith and the Yahgan Formation in many places in the southern Andes (cf. Stern and others, 1976, Figure 1). These rocks are probably oceanic crust and part of the floor of the marginal basin (Dalziel and others, 1975).
4. The sliver of continental crust on South Georgia (Drygalski Fjord Complex) is similar in age to rocks forming the roots of a Jurassic magmatic arc in South America (de Wit, 1977); it was probably split off from the main outcrop to the northeast by opening of the Early Cretaceous marginal basin.
5. The Cumberland Bay Formation is equated with the Yahgan Formation (for discussion and full bibliography, see Thomson and others, Chapter 19, this volume).

Directional features used to guide the reconstruction are the trends of (1) boundaries between stratigraphic-tectonic units, (2) axial traces of major early folds in the flysch sequences, and (3) dikes in the Drygalski Fjord Complex. Linear features which can be used when more data are available from the Yahgan Formation include the elongation lineation on the slaty cleavage and sediment dispersal patterns in the turbidite facies rocks.

The proposed fit is shown in Figure 18.5; it requires only a $<10°$ counterclockwise rotation of the South Georgia block in its present location. This reconstruction supports that figured by Suárez and Pettigrew (1976) and Dalziel and others (1975) but differs from the speculative reconstruction of de Wit (1977) in which he places the South Georgia block entirely within the marginal basin and proposes that the Scotia Sea itself is floored by Lower Cretaceous oceanic crust.

Shag Rocks consist of metamorphic rocks with a steeply dipping east-west foliation and are correlated with the pre-Jurassic metamorphic complex (Tanner, in press). The Shag Rocks continental block on Figure 18.5 is positioned such that (1) the metamorphic rocks making up the block do not intervene between the outcrops of the Yahgan and Cumberland Bay Formations and (2) the block has not been rotated by an amount greater than that which has affected the South Georgia block (the present coordinates of both blocks are shown around the margins of dotted units on Figure 18.5). A closer fit, with the South Georgia block butted against the 1000 m submarine contour east of Cape

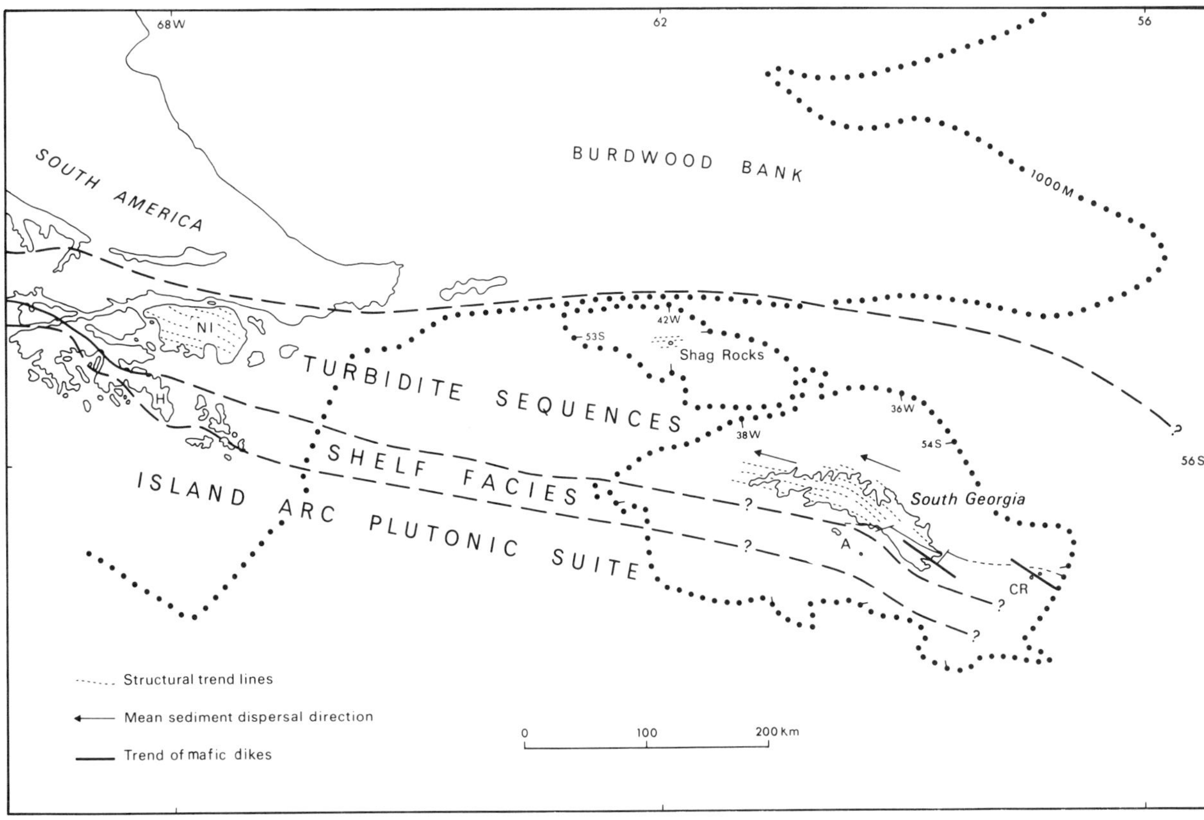

Figure 18.5. Inferred predrift position of South Georgia adjacent to the Burdwood Bank. The island-arc plutonic suite is equivalent to the Patagonian batholith; the shelf facies comprises the Annenkov Island Formation (A) and Hardy Formation (H) intruded by numerous discrete plutons; the outcrop of the turbidite sequences includes belts of mafic complex and older continental crust. Bathymetry from Barker and Griffiths (1972); geology of southern Andes from de Wit (1977) and Stern and others (1976); tectonic trends on Navarino Island from Katz and Watters (1966). CR: Clerke Rocks. NI: Navarino Island.

Horn, is possible if the Shag Rocks block is rotated counter-clockwise through 25° and would eliminate the problem of locating the "missing blocks" between South Georgia and Tierra del Fuego (Figure 18.5).

The present width of the Mesozoic marginal basin after Andean deformation of the sedimentary infill probably does not exceed 200 km, and sediment dispersal directions in the turbidites on South Georgia may indicate that the basin is becoming narrower, or closing, to the east. South of Drake Passage and the Scotia Sea the only rocks which are of directly equivalent age, and possibly occur in the same tectonic setting, are the sedimentary beds at Crabeater Point (Thomson and others, Chapter 19, this volume). The marginal basin is replaced far to the north in Peru by a pair of ensialic Mesozoic basins founded on Precambrian and Paleozoic continental crust (Cobbing, 1976). Upper Jurassic rocks (Latady Formation) in the southern Antarctic Peninsula (for review, see Thomson and others, Chapter 19, this volume) possibly occupy a similar ensialic basin to the south.

Acknowledgments. I wish to thank my colleagues D. I. M. Macdonald, B. F. Mair, and B. C. Storey for much helpful discussion; N. H. Woodcock and D. I. M. Macdonald for assistance with the statistical analysis; R. J. Boud for drafting the diagrams; and G. M. Tew for typing the manuscript. Particular thanks are due to Captains Cole, Elliott, Phelps, and Woodfield and the officers and crew of the RRS *John Biscoe* and RRS *Bransfield* for their invaluable logistic support of geologic teams on South Georgia.

References cited

Aitkenhead, N., and Nelson, P. H. H. 1962. The geology of the area between Cumberland West Bay and Cape George, South Georgia. *Prelim. Geol. Rep. Br. antarc. Surv.,* No. 15, 13 pp. (Unpublished).

Barker, P. F., and Griffiths, D. H. 1972. The evolution of the Scotia Ridge and Scotia Sea. *Phil. Trans. R. Soc. Lond.,* Ser. A, 271, No. 1213, 151–83.

Bell, C. M., Mair, B. F., and Storey, B. C. 1977. The geology of part of an island arc-marginal basin system in southern South Georgia. *Bull. Br. antarc. Surv., 46,* 109–27.

Clayton, R. A. S. In press, *a.* The geology of north-western South Georgia, III. Petrology of the Cumberland Bay Formation. *Bull. Br. antarc. Surv.*

Clayton, R. A. S. In press, *b.* The geochemistry of greywackes from South Georgia. *Bull. Br. antarc. Surv.*

Clayton, R. A. S. In press, *c.* The geology of north-western South Georgia, IV. Structural geology. *Bull. Br. antarc. Surv.*

Cobbing, E. J. 1976. The geosynclinal pair at the continental margin of Peru. *Tectonophysics, 36,* No. 1, 157–65.

Dalziel, I. W. D., and Elliot, D. H. 1971. The evolution of the Scotia arc. *Nature, Lond., 233,* No. 5317, 246–52.

Dalziel, I. W. D., and Elliot, D. H. 1973. The Scotia arc and Antarctic margin. (*In* Nairn, A. E. M., and Stehli, F. G., *eds. The Ocean Basins and Margins. 1. The South Altantic,* New York, Plenum, 171–245.)

Dalziel, I. W. D., de Wit, M. J., and Palmer, K. F. 1974. Fossil marginal basin in the southern Andes. *Nature, Lond., 250,* No. 5464, 291–94.

Dalziel, I. W. D., Dott, R. H., Jr., Winn, R. D., Jr., and Bruhn, R. L. 1975. Tectonic relations of South Georgia Island to the southernmost Andes. *Geol. Soc. Am. Bull., 86,* 1034–40.

de Wit, M. J. 1977. The evolution of the Scotia arc as a key to the reconstruction of southwestern Gondwanaland. *Tectonophysics, 37,* No. 1-3, 53-81.

Dott, R. H., Jr. 1974. Paleocurrent analysis of severely deformed flysch-type strata—a case study from South Georgia island. *J. sedim. Petrol., 44,* No. 4, 1166-73.

Dżuliński, S., and Walton, E. K. 1965. *Sedimentary Features of Flysch and Greywackes.* Amsterdam, Elsevier, 300 pp.

Frakes, L. A. 1966. Geologic setting of South Georgia Island. *Geol. Soc. Am. Bull., 77,* 1463-68.

Holtedahl, O. 1929. On the geology and physiography of some Antarctic and Sub Antarctic islands. *Scient. Res. Norw. antarct. Exped.,* No. 3, 172 pp.

Katz, H. R. 1972. Plate tectonics and orogenic belts in the southeast Pacific. *Nature, Lond., 237,* No. 5354, 331-32.

Katz, H. R. 1973. Contrasts in tectonic evolution of orogenic belts in the South-east Pacific. *J. R. Soc. N.Z., 3,* No. 3, 333-62.

Katz, H. R., and Watters, W. A. 1966. Geological investigations of the Yahgan Formation (upper Mesozoic) and associated igneous rocks of Navarino Island, southern Chile. *N.Z. Jl Geol. Geophys., 9,* No. 3, 323-59.

Lowe, D. R. 1976. Subaqueous liquified and fluidized sediment flows and their deposits. *Sedimentology, 23,* 285-308.

Mair, B. F. In press. The Larsen Harbour Formation and associated intrusive rocks of southern South Georgia. *Bull. Br. antarc. Surv.*

Mortimore, R. N. 1979. Distal and proximal turbidites at Nilse Hullet, western South Georgia. *Bull. Br. antarc. Surv., 47,* 117-28.

Pettigrew, T. H. In press. The geology of Annenkov Island. *Bull. Br. antarc. Surv.*

Ramsay, J. G. 1974. Development of chevron folds. *Geol. Soc. Am. Bull., 85,* 1741-54.

Skidmore, M. J. 1972. The geology of South Georgia. III. Prince Olav Harbour and Stromness Bay areas. *Scient. Rep. Br. antarc. Surv.,* No. 73, 50 pp.

Stern, C., de Wit, M. J., and Lawrence, J. R. 1976. Igneous and metamorphic processes associated with the formation of Chilean ophiolites and their implication for ocean floor metamorphism, seismic layering, and magnetism. *J. geophys. Res., 81,* No. 23, 4370-80.

Stone, P. In press, *a.* Geological observations in the Cooper Bay-Wirik Bay area, South Georgia. *Bull. Br. antarc. Surv.*

Stone, P. In press, *b.* The geology of South Georgia: IV. Barff Peninsula and Royal Bay areas. *Scient. Rep. Br. antarc. Surv.*

Storey, B. C., Mair, B. F., and Bell, C. M. 1977. The occurrence of Mesozoic oceanic floor and ancient continental crust on South Georgia. *Geol. Mag., 114,* No. 3, 203-8.

Suárez, M., and Pettigrew, T. H. 1976. An upper Mesozoic island-arc-back-arc system in the southern Andes and South Georgia. *Geol. Mag., 113,* No. 4, 305-28.

Tanner, P. W. G. In press. Geology of Shag Rocks, part of a continental block on the north Scotia Ridge, and possible regional correlations. *Bull. Br. antarc. Surv.*

Tanner, P. W. G., and Rex, D. C. 1979. Timing of events in an Early Cretaceous island arc-marginal basin system on South Georgia. *Geol. Mag., 116,* No. 3, 167-79.

Trendall, A. F. 1959. The geology of South Georgia—II. *Scient. Rep. Falkld Isl. Depend. Surv.,* No. 19, 48 pp.

Tyrrell, G. W. 1916. Further notes on the petrography of South Georgia. *Geol. Mag.,* Decade VI, *3,* 435-41.

Tyrrell, G. W. 1918. Additional notes on the petrography of South Georgia. *Geol. Mag.,* Decade VI, *5,* 483-89.

Tyrrell, G. W. 1930. The petrography and geology of South Georgia. (*In Report on the Geological Collections Made during the Voyage of the "Quest" on the Shackleton-Rowett Expedition to the South Atlantic and Weddell Sea in 1921-22.* London, Trustees of the British Museum, 28-54.)

19

Fossil and Radiometric Evidence for Ages of Deposition and Metamorphism of Sedimentary Sequences on South Georgia

M. R. A. THOMSON and P. W. G. TANNER
British Antarctic Survey, Natural Environment Research Council, Madingley Road, Cambridge CB3 0ET, U.K.

D. C. REX
Department of Earth Sciences, University of Leeds, Leeds LS2 9JT, U.K.

Abstract. Fossil evidence and K-Ar whole-rock age determinations on slates and mudstones are used to assess the ages of the major sedimentary sequences of South Georgia. Marine fossils in the Cumberland Bay and Annenkov Island Formations indicate a general Early Cretaceous age, although a belemnite found in an erratic block of Cumberland Bay rock type suggests that this sequence may be as old as Late Jurassic in age. K-Ar ages from the Cumberland Bay and Sandebugten Formations range between 51–135 m.y. and reflect times of uplift and cooling following regional metamorphism. Anomalously high ages of 102–135 m.y. in the Cumberland Bay Formation are thought to result from the presence of detrital white mica, reported here for the first time from these rocks, which has retained premetamorphic argon. The best estimate for the minimum age of the deformation and metamorphism of the Cumberland Bay Formation is 82–91 m.y., with postorogenic uplift having continued at least until 50–60 m.y. ago. The radiometric data agree well with the fossil evidence, and together they reinforce previous arguments for a close correlation between the Cumberland Bay Formation and the Yahgan Formation of southern Patagonia. However, they indicate that the Cumberland Bay Formation was deposited after the Upper Jurassic Latady Formation of southeastern Palmer Land and that the two thus cannot be regarded as equivalent. A closer correlative may be found in the Lower(?) Cretaceous Crabeater Point sedimentary rocks of northeastern Palmer Land.

The Cumberland Bay Formation on South Georgia, an island in the Scotia Sea, is a sequence of volcaniclastic turbidites which overlies and is in thrust contact with a similar but more siliceous sequence, the Sandebugten Formation (Figure 19.1). Both formations are separated from other rock units on South Georgia by a major fault zone (see Tanner, Chapter 18, this volume); neither is cut by igneous plutons which could be dated, nor is the base or top of either sequence exposed.

In early attempts to assess the age of these rocks some investigators were possibly influenced by the slaty nature of many of the finer grained sedimentary rocks and by general similarities to Paleozoic or older rocks in Europe with which they were familiar (Thurach, 1890; Ferguson and others, 1914). However, they would have done well to heed Darwin's (1846) cautionary remarks on the clay-slate formation of Patagonia (part of which is now known as the Yahgan Formation), when he pointed out that, although the rocks themselves had a Paleozoic aspect, they contained Cretaceous fossils. Identifications of possible Paleozoic fucoids and sponges (Gregory, 1915) were dismissed by Wilckens (1932) as misidentifications of trace fossils, and Gregory's *Omphyma*-like coral is probably a pseudofossil, as it has features which are very similar to the internal structures seen in prehnite veins and nodules.

Over the past few years, much of South Georgia has been geologically mapped in detail by the British Antarctic Survey, and this is an appropriate time to reassess the age of the sedimentary sequences on the basis of evidence from new fossil collections and from the whole-rock K-Ar dating of slates. We conclude that both sets of data are consistent with a latest Jurassic–Early Cretaceous age for the Cumberland Bay Formation. Only trace fossils (mainly *Chondrites*), which are not diagnostic of age, have so far been obtained from the Sandebugten Formation, and the strong deformation and recrystallization of this sequence makes the possibility of finding recognizable invertebrate fossils unlikely. Radiometric dating likewise only yields minimum ages and does not enable us to confirm or refute the possibility that the two sequences are roughly equivalent in age (Trendall, 1959; Stone, in press; Dalziel and others, 1975).

Paleontology

South Georgia

The turbidite origin of the Cumberland Bay Formation (Trendall, 1959) is reflected in the sporadic distribution of its invertebrate macrofossils. With the exception of trace fossils, the formation is essentially unfossiliferous, and occurrences of hard-shelled species are rare; at only four localities (A, B, C, and O, Figure 19.1) have relatively large numbers of fossils been found.

The most common fossil is the bivalve *Aucellina* which occurs as single valves (localities E and M) or accumulations of dissociated valves, as at Rosita Harbour (A), Prince Olav Harbour (C), and Queen Maud Bay (O). Although not all of these are specifically identifiable, forms similar to *A. andina* Feruglio and the more elongate *A. radiatostriata* Bonarelli are recognizable among the better material (Figure 19.2*a*). Both species were regarded by Cox (1953) as being Aptian in age, but mapping in Alexander Island suggests that they range up into the Albian and may even occur below the Aptian.

Three fragments of large ammonites have been collected from Stromness Bay (D), Cumberland Bay (E), and Grytviken (G), and a fourth was photographed in Jacobsen Bight (M). This last fossil is unidentifiable, but the collected fragments all represent large heteromorphs with coarse trituberculate ribbing (Figure 19.2*c*). Although it is not possible to identify such specimens with certainty, they are reminiscent of some species of *Tropaeum* Sowerby or *Australiceras* Whitehouse, two essentially Aptian genera which had an almost worldwide distribution. An ammonite, collected at Prince Olav Harbour by Filchner's German expedition of 1910–12, was cautiously identified by Pompeckj as a possible acanthoceratid of Cretaceous age (Salomon in Heim, 1912, footnote on p. 453). Many of the Acanthoceratidae have multituberculate ribbing, and it is more than likely that the specimen was a heteromorph like those discussed above or like those from Annenkov Island (see below).

Less diagnostic material includes eight small procerithiid gastropods (*Rhabdocolpus(?)* sp.) from Prince Olav Harbour (B) (Skidmore, 1972), a fragment of the serpulid worm *Rotularia* from locality C, an indeterminate gastropod from locality N, and a crushed *Inoceramus* from locality E. All of these are consistent with a Mesozoic age, Jurassic or Cretaceous being the most probable.

Belemnite fragments, in cobbles of Cumberland Bay rock type from morainal deposits at Little Moltke Harbour (H), are the only reasonably firm paleontologic indicators of the occurrence of pre-Aptian sedimentary beds on South Georgia. The presence of a shallow median groove on one specimen suggests affinity to *Belemnopsis* Bayle, which has a known range of Upper Jurassic to lowest Cretaceous in West Antarctica (Stone and Willey, 1973).

The only fossils which may be regarded as common are trace fossils, amongst which the ichnogenera *Chondrites, Planolites, Arenicolites,* and *Zoophycos* have been recognized (e.g., Skidmore, 1972; Clayton, in press). Radiolaria occur commonly in thin sections, but they are all poorly preserved and are of little stratigraphic use. A few echinoderm fragments have also been observed in thin section. Impressions of fossil wood occur sporadically at all levels in the sequence.

Annenkov Island and Hauge Reef

The most varied fossil marine faunas in the South Georgia area are those of Annenkov Island (Table 19.1). Early collections were studied in detail by Wilckens (1932, 1947), but the preservation of all the material is disappointing, and it is not easy to identify with certainty. However, the faunas generally have a Lower Cretaceous aspect, and Wilckens assigned a fauna from the southeastern corner of

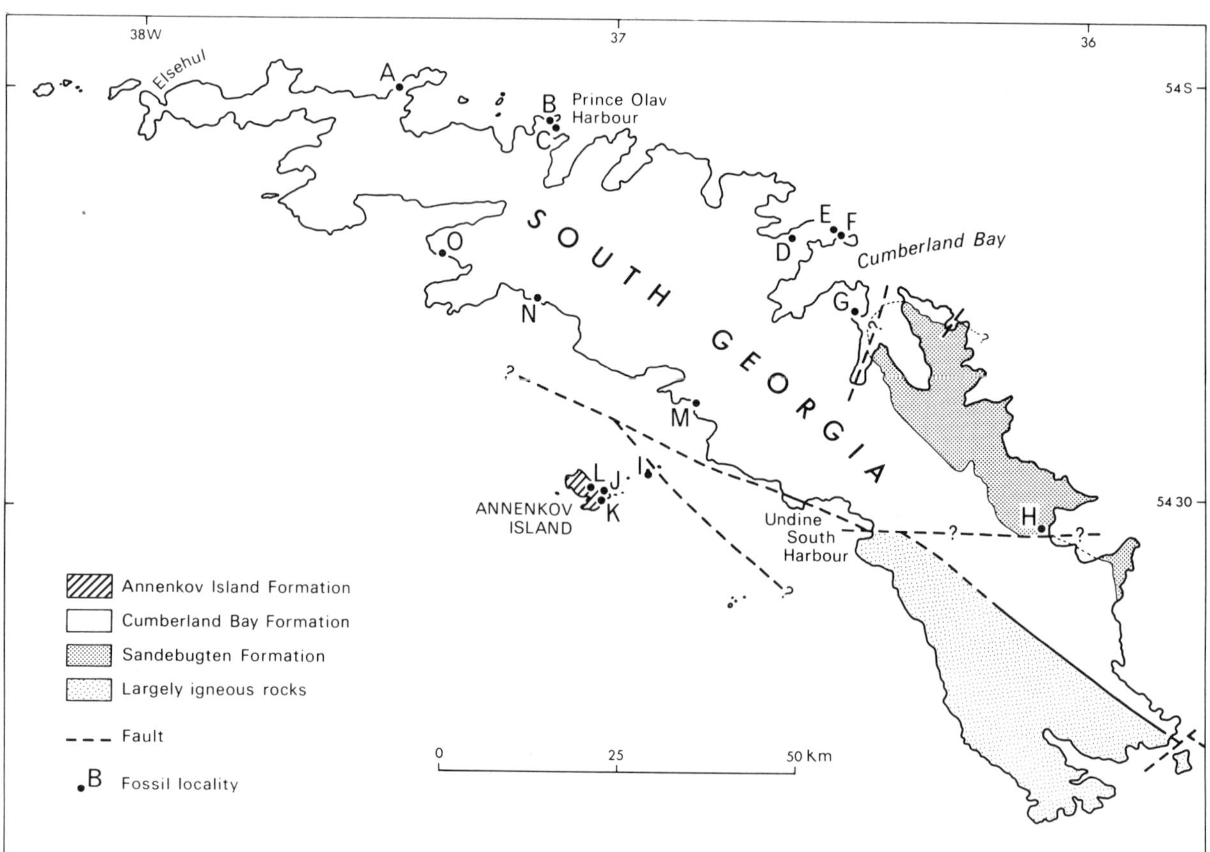

Figure 19.1. Geologic sketch map of South Georgia, showing the distribution of the sedimentary Cumberland Bay, Annenkov Island, and Sandebugten Formations.

the island to the Upper Aptian, largely on the basis of ammonites. His identifications were queried by Casey (1961, footnote on p. 56) who suggested that Wilckens' *Georgioceras* and "*Puzosia matheroni* (d'Orbigny)?" resembled Neocomian species; it is also significant that Wilckens' identification of *Sanmartinoceras* cf. *patagonicum* was based on extremely poor material.

New collections have been made from the lower tuff member of the Annenkov Island Formation at the locality first described by Wilckens (J), from a slightly higher level in the same member on Lawther Knoll (L), and from a bed of unknown stratigraphic position in the upper breccia member on the southern coast (K) (Figure 19.1). Additions to the fauna at locality J include a smooth species of *Aconeceras* with a long, delicate lappet (Figure 19.2*e*) and an evolute ammonite with irregular, blunt ribbing reminiscent of some species of *Kilianella* Uhlig (Figure 19.2*d*) or even an evolute hoplitid. Ammonite fragments found loose at locality L are mostly of a large, open-coiled heteromorph with tuberculate ribbing, which is spinose in the later stages (Figure 19.2*h*). The loose coiling and tuberculate ribbing recall some of the Neocomian Crioceratitidae (e.g., *Paracrioceras* Spath), but the presence of an intermediate nontuberculate stage also suggests affinity to the Aptian genus *Australiceras* Whitehouse. The same heteromorphs are accompanied by the bivalve *Inoceramus* (Figure 19.2*g*), and an involute ammonite (Figure 19.2*b*) with close, narrowly bifurcate, flexuous ribs which are paired at a row of umbilical tubercles (cf. *Neocomites* Uhlig).

Notwithstanding the poor preservation of these fossils, it is difficult to find comparable faunas elsewhere and, with few exceptions (e.g., *Aucellina*), it would seem that many of the constituent species are endemic to Annenkov Island.

Figure 19.2. Representative invertebrate fossils from South Georgia (*a* and *c*) and Annenkov Island. *a. Aucellina* sp., latex cast; *b. "Neocomites"* sp., latex cast; *c.* Latex cast of a whorl fragment of a large heteromorph ammonite; *d. "Kilianella"* sp.; *e. Aconeceras* sp.; *f. Bochianites(?)* sp., latex cast; *g. Inoceramus* sp., latex cast; *h.* Spinose heteromorph ammonite, latex cast. All figures are natural size.

Table 19.1. Principal fossil Mollusca from Annenkov Island

Species and occurrence	Age significance and/or remarks
Upper breccia member, locality K	
Dimitobelus sp.[a]	Late Aptian–Albian or younger
Lower tuff member, locality L	
Inoceramus sp.[a]	Early Cretaceous
Crioceratitid heteromorph[a]	Early Cretaceous
"Neocomites" sp.[a]	—
Lower tuff member, locality J	
Inoceramus spp.	Reported by Wilckens (1947)
Aucellina, andina/radiatostriata group	Aptian (probably has greater age range in Alexander Island
Lytoceras sp.[a]	Giant form
Georgioceras kohllarseni	Genus only known from Annenkov Island
Tropaeum(?) *antarcticum*	Aptian (if generic assignment is correct)
"Puzosia" spp.	—
Aconeceras sp.[a]	Late Neocomian–Aptian
Sanmartinoceras cf. *patagonicum*	Based on very poor material
"Kilianella" sp.[a]	—

[a] Species not previously reported from Annenkov Island.

In view of the correlations suggested by Suárez and Pettigrew (1976), comparable faunas should be expected in the Hardy Formation of southern Patagonia, but no descriptions are available for the few fossils known from these rocks. The incomplete preservation of the Annenkov Island ammonites produces a confused stratigraphic picture because it is possible to identify the various species with genera of different, if not exclusive, age ranges. Several of the ammonites would not appear out of place in European Neocomian faunas, but it has already been shown that the Neocomian aspect of some ammonites from Alexander Island (Thomson, 1972*a*) can be misleading (Taylor and others, 1979; Thomson, 1974). It would thus be unwise to suggest anything other than a broad Early Cretaceous age for the lower tuff member at present.

Belemnite fragments, from the overlying breccia member, probably represent a species of *Dimitobelus* Whitehouse (Pettigrew and Willey, 1975), and they suggest a late Aptian or more probably Albian (or younger) age for that member.

A variety of other fossils from the lower tuff member include Foraminifera, cirripedes, fish scales and bones, fragments of a badly crushed and zeolitized spinal column(?) about 3 cm across, and the fronds of fossil cycads.

A fragment of a small heteromorph ammonite (Figure 19.2*f*) was obtained from a large sedimentary rock xenolith in a microgranite intrusion on Hauge Reef (I). It is a small, straight piece of shaft with relatively coarse, straight prorsiradiate ribbing. The specimen resembles a great many other poorly preserved heteromorph fragments, but the oblique ribbing suggests possible affinity with the genus *Bochianites* Lory, rather than some of the small *"Ancyloceras"* fragments from Alexander Island (Thomson, 1974, plate III) or the Hamitidae. An Early Cretaceous age seems most probable for this fragment.

Summary

Although the great thickness of the Cumberland Bay Formation suggests that it was deposited over a period of tens of millions of years (see below), all fossils obtained *in situ* suggest an approximately Aptian age. The presence of older (Late Jurassic to earliest Cretaceous) beds is suggested by the occurrence of the belemnite *Belemnopsis* in erratics, but their source has not been located. Ammonites from the lower tuff member of the Annenkov Island Formation are poorly preserved, and only a general Early Cretaceous age can be suggested for them at present. The upper breccia member is probably no older than late Aptian in age. The occurrence of the bivalve *Aucellina* in both the Cumberland Bay Formation and the lower tuff member of the Annenkov Island Formation indicates that the two formations are at least in part laterally equivalent, but a precise stratigraphic correlation has yet to be established.

Radiometric dating

Geologic setting

As no material suitable for radiometric dating was obtained from the Annenkov Island Formation, this section is concerned solely with the interpretation of data from the Cumberland Bay and Sandebugten Formations.

Rocks of the Cumberland Bay Formation become increasingly deformed and metamorphosed from southwest to northeast across the island; slaty cleavage (F_1) dips to the south or southwest (Figure 19.3) at 30°–65° and has an arcuate trend. The least deformed and metamorphosed turbidites are found between Jossac Bight and Undine South Harbour on the south side of the island, where the slaty cleavage is hackly and poorly developed and where few large folds exist. Toward the northeast, intense chevron folds are developed, and cleavage becomes the dominant fabric in the argillaceous rocks, accompanied by an elongation lineation. One or two episodes of later folding, associated with crenulation of the F_1 cleavage, occurred locally in areas along the northeast coast (Tanner, Chapter 18, this volume).

The Sandebugten Formation is more intensely deformed than the Cumberland Bay Formation with an early slaty or phyllitic fabric in the argillaceous rocks and strongly developed elongation lineation. These rocks contain stilpnomelane and have been thoroughly recrystallized.

Petrography

Previous descriptions of the petrography of the Cumberland Bay Formation (i.e., Tyrrell, 1916, 1918; Trendall, 1953, 1959; Skidmore, 1972; Dalziel and others, 1975; Stone, in press; Clayton, in press) have all emphasized the volcaniclastic nature of the rocks, the abundance of volcanic (mainly andesitic) fragments, and the scarcity or absence of clasts of silicic igneous or metamorphic rocks. We therefore decided to attempt K-Ar whole-rock dating of the slates, guided by the premise that if rock fragments containing micaceous minerals were absent from the coarse-grained and silty Bouma divisions of turbidite units, then such minerals were unlikely to be present in the interlayered muddy beds. This would eliminate the problem of retention of premetamorphic argon during metamorphism of these muddy beds to form slates; all ages could be regarded as minimum ages for the regional metamorphism.

The analyzed slate specimens are all from Bouma E divisions of turbidite units; none are truly pelagic, as they all contain abundant minute clastic grains. Thin section examination shows that these slates from the Cumberland Bay Formation may be divided into two main types:

1. those with poorly developed cleavage in hand specimen, weak sericite fabric, no elongation of detrital and authigenic opaque granules (mainly pyrite), and little or no evidence of pressure solution effects around grains (designated group A); and
2. those with good cleavage and fabric, deformed and elongate opaque grains and clasts, and marked effects of pressure solution (designated group B).

In the group A slates (numbers 1, 2, 3, 4, 6, and 8 in Table 19.2), one of us (P.W.G.T.) has found clear evidence, contrary to our expectations, of the presence of detrital flakes of white mica. These specimens are nearly all from the west end or southwest side of South Georgia (Figure 19.3), an area which to date has not been studied in as much detail as that around Cumberland Bay and in which the rocks may prove to be petrographically distinct. The clastic micas may be distinguished from those defining the slaty cleavage by their much larger size and by the fact that they are bent around other clastic grains, are commonly crenulated and show strained extinction, and in some cases actually form part of detrital clasts. Examination of thin sections of silty and laminated beds from these localities confirms that they too contain clasts bearing white mica around which subsequent development of the slaty cleavage has given rise to pressure-solution effects.

The group B slates have a well-defined fabric of minute sericite laths, which gives good parallel extinction under

Table 19.2. K-Ar whole-rock age data on slates from South Georgia. (Numbers in left-hand column are keyed to Figure 19.3.)

Specimen number	K (wt %)	^{40}Ar (radiogenic) scc[a]/g × 10^{-4}	^{40}Ar (radiogenic) % total	Age (m.y.)
Cumberland Bay Formation				
1. M.1300	2.47	0.10356	82.4	102 ± 4
2. M.1404	1.45	0.06347	72.3	108 ± 4
3. M.1410	1.77	0.09885	59.1	135 ± 5
4. M.1412	2.03	0.10263	78.5	123 ± 5
5. M.1488	2.50	0.08487	89.1	83 ± 3
6. M.1522	2.38	0.09985	86.4	102 ± 4
7. M.3016	1.968	0.06573	73.0	82 ± 3
8. M.3085	2.696	0.13295	90.5	120 ± 5
9. M.3279	2.619	0.09730	74.7	91 ± 4
10. M.3326	2.786	0.09533	68.0	84 ± 3
11. M.3442	3.087	0.11244	70.4	89 ± 4
12. M.3712	1.917	0.03934	66.8	51 ± 2
Sandebugten Formation				
13. M.3399	2.459	0.06628	29.9	66 ± 3
14. M.3494	0.710	0.02426	31.1	84 ± 3
15. M.3522	2.966	0.08052	69.5	67 ± 3
16. M.3595	4.407	0.12262	65.3	68 ± 3
17. M.3615 A	0.792	0.02074	40.7	64 ± 3
18. M.3646	1.776	0.04959	82.6	67 ± 3

Constants: $\lambda_e = 0.584 \times 10^{-10}$/yr; $\lambda_\beta = 4.72 \times 10^{-10}$/yr; and $^{40}K/K = 0.0119$ atomic %.

For analytical techniques, see Gledhill and others, Chapter 35, this volume.

[a] Standard cubic centimeter (cm^3 at S.T.P.).

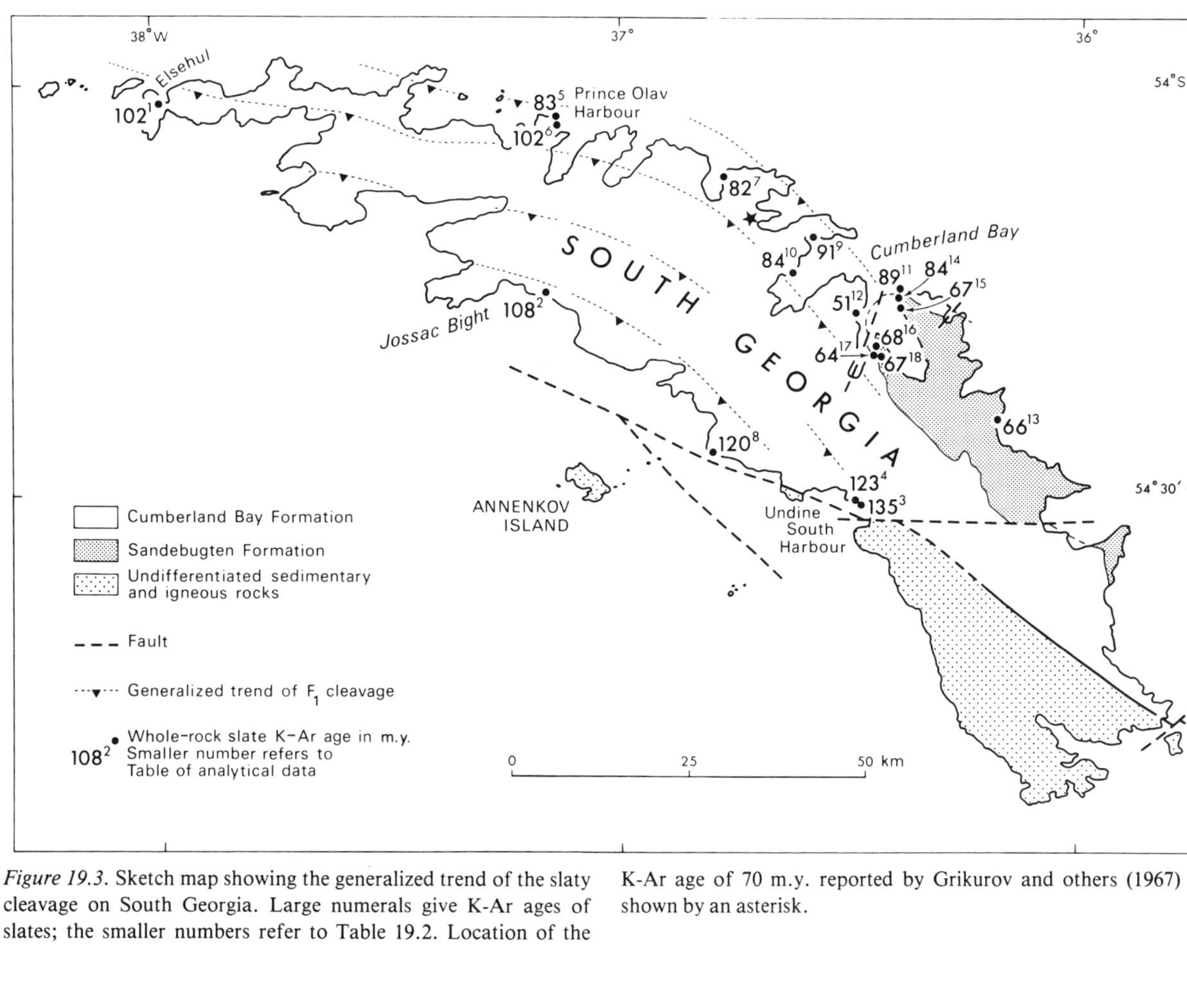

Figure 19.3. Sketch map showing the generalized trend of the slaty cleavage on South Georgia. Large numerals give K-Ar ages of slates; the smaller numbers refer to Table 19.2. Location of the K-Ar age of 70 m.y. reported by Grikurov and others (1967) is shown by an asterisk.

crossed nicols. A few possibly detrital micas can still be recognized.

Specimens selected for dating from the Sandebugten Formation consist of well-crystallized slates. Sericite gives parallel extinction under crossed polars in specimens 16 and 18 (Table 19.2); specimen 15 shows a strong crenulation cleavage superimposed upon the slaty cleavage, specimen 13 is a phyllite, and specimens 14 and 17 are carbonate rich.

RADIOMETRIC DATA*

The only previously published radiometric age from South Georgia is one of 70 m.y. given for a "mylonitized" siltstone-shale (K_2O = 2.73%) from Husvik in Cumberland Bay (for location see Figure 19.3) (Grikurov and others, 1967). In the present work, an age range of 51–135 m.y. has been obtained from slates of the Cumberland Bay Formation and 64–84 m.y. from slates and phyllites of the Sandebugten Formation (Table 19.2). When plotted on a map (Figure 19.3), these data form a distinct pattern: high ages (102–135 m.y.) are restricted to the area of least deformed and metamorphosed Cumberland Bay Formation rocks, and low ages (51–91 m.y.) to the more strongly recrystallized slates of both formations.

This age pattern could reasonably be interpreted in terms of times of cooling following uplift and erosion (Harper, 1964, 1967), with rocks at the highest tectonic levels and lowest metamorphic grade cooling first. However, this interpretation would imply a minimum age of 135 m.y. for the regional metamorphism and deformation of the Cumberland Bay Formation and would conflict with the paleontologic evidence presented earlier, which strongly supports an essentially Early Cretaceous age (120–135 m.y.) for the *sedimentation* even of the lower parts of the sequence that crop out on the northeast side of the island. The formation has a minimum thickness of 7–8 km (Tanner, Chapter 18, this volume); even allowing for the fact that deposition of turbidite aprons in back-arc basins can be unusually rapid (>100 m/m.y.), this process would require some tens of millions of years. Adding the time required for orogenesis and for subsequent cooling of the rocks to the temperature at which radiogenic argon is retained in sericite only increases the discrepancy between paleontologic and radiometric data.

Both sets of data may, however, be reconciled by our discovery of clastic grains with white mica and of detrital mica flakes in the mudstones and poorly crystallized slaty rocks from the southwest side of the island. We infer that these white micas, which are present at the higher stratigraphic levels if not throughout the whole sequence, have contributed premetamorphic argon to the slates. By comparing petrographic data and radiometric ages, we see that the group A mudstones and slates give ages of 102–135 m.y., whereas group B slates give ages of 51–91 m.y. It would therefore seem that the older group B K-Ar ages of 83–91 m.y., given by thoroughly recrystallized slates, represent a true minimum age for the metamorphism and deformation of the Cumberland Bay Formation and that ages of 120–135 m.y. result from incomplete outgassing of argon from the detrital micas. Ages of 66–84 m.y. for the Sandebugten Formation seem to be ages of uplift and cooling; these rocks commonly contain fragments of silicic igneous and metamorphic rocks, yet they have been sufficiently metamorphosed to remove all premetamorphic argon.

The significance of the 102 and 108 m.y. ages is difficult to determine; either one may be a true minimum age for the metamorphic event or both may be mixed ages; we have no criteria to help us decide between the two alternatives. As one of the thoroughly recrystallized slates from the Sandebugten Formation gives an age of 84 m.y. (in Table 19.2) and five ages of 82–91 m.y. have been obtained from the Cumberland Bay Formation where it is most strongly deformed and metamorphosed, it seems reasonable to accept provisionally 82–91 m.y. as a sound minimum age for the metamorphism—with the proviso that it may be as old as 108 m.y.

The source of the detrital mica has not yet been identified. It was largely derived either from igneous rocks belonging to the island arc, which was active during the period when the Cumberland Bay Formation was being deposited, or from uplift and erosion of older continental crust, which formed the foundation to this arc.

In the first case, for material from contemporaneous (120–140? m.y.) volcaniclastic or extrusive igneous rocks to significantly affect the apparent K-Ar age of the slates, it would have to comprise much of the micaceous mineral in these rocks. These grains would also have to retain a very high proportion of their premetamorphic argon; it is possible that argon lost from the detrital micas in the early stages of metamorphism would be taken up by sericite growing in the slaty cleavage, so that no argon would actually leave the local system at that time. However, the amount of white mica available from igneous rocks, even from unroofed early silicic rocks, would be small; furthermore, it is possible that some of the white mica, particularly in the larger clasts, may be of authigenic origin.

In the second case, the detrital micas could be derived from pre-Cretaceous continental crust found on the Pacific side of the Cretaceous (Andean) marginal basin. Muscovite from these rocks could give K-Ar ages ranging from 140 m.y. to more than 250 m.y. If only 30% of the mica in the slates with anomalously high ages came from this source, retention of most of the premetamorphic argon would raise the apparent age of the slate from the inferred true minimum age of 90 m.y. to the observed 130–140 m.y.

Preliminary petrographic work leads us to favor the first interpretation, in which the bulk of the detrital white mica is considered to be of about the same age as the sediments.

The younger radiometric ages from both formations reflect times of final uplift and cooling of this part of the orogenic belt. They suggest that postorogenic uplift continued until 50–60 m.y. ago or later.

Discussion

RELATIONSHIP WITH THE YAHGAN FORMATION

On the basis of rock type and stratigraphic similarities (Wilckens, 1932; Matthews, 1959; Katz and Watters, 1966; Dalziel and others, 1975), the Cumberland Bay Formation of South Georgia has been long compared to the Yahgan Formation of Patagonia. Our investigations have enabled

* For details of our analytical techniques, refer to Gledhill and others, Chapter 35, this volume.

us to test further this correlation from a stratigraphic viewpoint.

The Yahgan Formation is sparsely fossiliferous and its age is imprecisely known; it is separated from the underlying Middle to Upper Jurassic Tobífera Formation by an ophiolite complex of uncertain age (Suárez and Pettigrew, 1976). Direct evidence for its age comes from rare fossil finds, notably the bivalve *Inoceramus* from Navarino Island (Katz and Watters, 1966) and the ammonite *Favrella* from Gardiner Island (Halpern and Rex, 1972). The *Inoceramus* is probably conspecific with *I. pseudosteinmanni* from Alexander Island and the Lago Argentino area (Thomson and Willey, 1972) and suggests an early Neocomian age. The age range of *Favrella* is uncertain; a careful stratigraphic review by Riccardi (1970) suggested that it is probably late Neocomian in age, but on the basis of associated Foraminifera, Natland and others (1974) concluded that the "Estratos con *Favrella*" of the Magellanes Basin are Portlandian to Hauterivian in age.

M. Halpern kindly enabled one of us (M.R.A.T.) to examine the *"Favrella"* from Gardiner Island. It is an external mold of a badly distorted whorl fragment in a cleaved mudstone. It has stout, distant, simple ribs which appear to be somewhat projected on the venter; however, this projection could have been largely induced by distortion. There are hints of ventral tubercles on some ribs, and there is no impression of the venter of the preceding whorl on its dorsum, suggesting a loose mode of coiling. Thus an identification as *Favrella* (Favre, 1908; Douvillé, 1909) seems unlikely, and it is probable that the specimen represents some kind of heteromorph.

Although there is yet no conclusive evidence for the age of the Yahgan Formation, an Early Cretaceous age is consistent with the few fossils so far found, and a stratigraphic correlation with the Cumberland Bay Formation is likely.

Rb-Sr and K-Ar dates of 77–81 m.y. on hornblende and biotite from two dioritic plutons which cut folded rocks of the Yahgan Formation (Halpern and Rex, 1972) provide a minimum age for the deformation and regional metamorphism. Regional considerations indicate that the folding occurred during early Senonian time (about 80–90 m.y. ago) (Katz, 1973). These data agree well with the minimum age of 83–91 m.y. proposed here for the deformation and metamorphism of the Cumberland Bay Formation.

Correlation with the Antarctic Peninsula

Although sedimentary rocks occupying the former back-arc basin on South Georgia can be compared closely with the Yahgan Formation of southern Patagonia, their southern continuation is less certain. Suárez (1976, Figure 2) implied that the Cumberland Bay Formation was equivalent to the Latady Formation of southeastern Palmer Land (Williams and others, 1972). However, the Cumberland Bay Formation is largely Early Cretaceous in age, whereas the Latady Formation is Late Jurassic (middle and late Kimmeridgian (Williams and Rowley, 1971). Thus, although the two formations may be related to the same Late Jurassic–Early Cretaceous magmatic arc, they developed at significantly different times.

The only close correlative may be found in the deformed sedimentary rocks at Crabeater Point, northeastern Palmer Land (Thomson, 1967). Although the fossil fauna from this locality is poorly preserved, a loosely coiled heteromorph ammonite suggests that an Early Cretaceous age for the sequence is most likely (Taylor and others, 1979).

Acknowledgments. In this paper M.R.A.T. was responsible for the paleontology, P. W. G. T. for the petrography and geologic interpretation, and D. C. R. for the K-Ar work. We are grateful to M. H. Dodson and R. J. Pankhurst for constructive criticism and advice concerning the K-Ar interpretation. Much of the fossil material reported from South Georgia was collected by P.W.G.T., but we would like to acknowledge with thanks D. I. M. Macdonald for *Aucellina* collections from localities A and O, D. Orchard and J. Tallowyn for ammonite finds (D and G), and, in particular, T. H. Pettigrew and A. F. Trendall for the fossils collected from Annenkov Island. We thank G. M. Tew for typing the manuscript and R. J. Boud for drafting the diagrams.

References cited

Casey, R. 1961. A monograph of the Ammonoidea of the Lower Greensand. Part 2. *Monogr. Palaeontogr. Soc.*, 45–118.

Clayton, R. A. S. In press. The geology of north-western South Georgia. Part III. Petrology of the Cumberland Bay Formation. *Bull. Br. antarc. Surv.*

Cox, L. R. 1953. Lower Cretaceous Gastropoda, Lamellibranchia and Annelida from Alexander Island. (Falkland Islands Dependencies). *Scient. Rep. Falkld Isl. Depend. Surv.*, No. 4, 14 pp.

Dalziel, I. W. D., Dott, R. H., Jr., Winn, R. D., Jr., and Bruhn, R. L. 1975. Tectonic relations of South Georgia Island to the southernmost Andes. *Geol. Soc. Am. Bull.*, *86*, No. 7, 1034–40.

Darwin, C. 1846. *Geological Observations on South America.* London, Smith Elder, 279 pp.

Douvillé, R. 1909. Sur des ammonites provenant des environs de Lima et sur une nouvelle coupure de la famille des Hoplitidés: *Favrella* n.gen. *C.r. Soc. géol. Fr.*, Ser. 4, *9*, 164–66.

Favre, F. 1908. Die Ammoniten der unteren Kreide Patagoniens. *Neues Jb. Miner. Geol. Paläont. BeilBd.*, *25*, 601–47.

Ferguson, D., Tyrrell, G. W., and Gregory, J. W. 1914. The geology of South Georgia. *Geol. Mag.*, Decade 6, *1*, 53–64.

Gregory, J. W. 1915. The geological relations and some fossils of South Georgia. *Trans. R. Soc. Edinb.*, *50*, Pt. 4, 817–22.

Grikurov, G. E., Krylov, A. Ya., and Silin, Yu. I. 1967. Absoliutnyi vozrast nekotorykh porod dugi Skotiia i Zemli Aleksandra I (Zapadnaya Antarktika) [Absolute age of some rocks from the Scotia arc and Alexander I Land (western Antarctica)]. *Dokl. Akad. Nauk SSSR*, Geology, *172*, No. 1, 168–71. [English translation: *Dokl. (Proc.) Acad. Sci. U.S.S.R.*, Geological sciences sect., *172*, 19–22.]

Halpern, M., and Rex, D. C. 1972. Time of folding of the Yahgan Formation and age of the Tekenika Beds, southern Chile, South America, *Geol. Soc. Am. Bull.*, *83*, No. 6, 1881–86.

Harper, C. T. 1964. Potassium-argon ages of slates and their geological significance. *Nature, Lond.*, *203*, No. 4944, 468–70.

Harper, C. T. 1967. The geological interpretation of potassium-argon ages of metamorphic rocks from the Scottish Caledodonides. *Scott. J. Geol.*, *3*, Pt. 1, 46–66.

Heim, F. 1912. Geologische Beobachtungen über Süd-Georgien. *Z. Ges. Erdk. Berl.*, No. 6, 451–56.

Katz, H. R. 1973. Contrasts in tectonic evolution of orogenic belts in the south-east Pacific. *J. R. Soc. N.Z.*, *3*, No. 3, 333–62.

Katz, H. R., and Watters, W. A. 1966. Geological investigation of the Yahgan Formation (upper Mesozoic and associated igneous rocks of Navarino Island, southern Chile. *N.Z. Jl Geol. Geophys.*, *9*, No. 3, 323–59.

Matthews, D. H. 1959. Aspects of the geology of the Scotia arc. *Geol. Mag.*, *96*, No. 6, 425–41.

Natland, M. L., González, E., Canón, A., and Ernst, M. 1974. A system of stages for correlation of Magallanes Basin sediments. *Mem. geol. Soc. Am.*, No. 139, 126 pp.

Pettigrew, T. H., and Willey, L. E. 1975. Belemnite fragments from Annenkov Island. *Bull. Br. antarc. Surv.*, No. 40, 33–36.

Riccardi, A. C. 1970. *Favrella* R. Douvillé, 1909 (Ammonitina, Cretácico inferior): edad y distribución. *Ameghiniana, 7,* No. 2, 119–38.

Rex, D. C., and Dodson, M. H. 1970. Improved resolution and precision of argon analysis using an MS10 Mass Spectrometer. *Eclog. geol. Helv., 63,* No. 1, 275–80.

Skidmore, M. J. 1972. The geology of South Georgia: III. Prince Olav Harbour and Stromness Bay areas. *Scient. Rep. Br. antarc. Surv.*, No. 73, 50 pp.

Stone, P. In press. The geology of South Georgia: IV. Barff Peninsula and Royal Bay areas. *Scient. Rep. Br. antarc. Surv.*

Stone, P., and Willey, L. E. 1973. Belemnite fragments from the Cumberland Bay type sediments of South Georgia. *Bull. Br. antarc. Surv.*, No. 36, 129–32.

Suárez, M. 1976. Plate-tectonic model for southern Antarctic Peninsula and its relation to southern Andes. *Geology, 4,* 211–14.

Suárez, M., and Pettigrew, T. H. 1976. An upper Mesozoic island-arc–back-arc system in the southern Andes and South Georgia. *Geol. Mag., 113,* No. 4, 305–28.

Taylor, B. J., Thomson, M. R. A., and Willey, L. E. 1979. The geology of the Ablation Point-Keystone Cliffs area, Alexander Island. *Scient. Rep. Br. antarc. Surv.*, No. 82, 65 pp.

Thomson, M. R. A. 1967. A probable Cretaceous invertebrate fauna from Crabeater Point, Bowman Coast, Graham Land. *Bull. Br. antarc. Surv.*, No. 14, 1–14.

Thomson, M. R. A. 1972. Ammonite faunas of south-eastern Alexander Island and their stratigraphical significance. (*In* Adie, R. J., *ed. Antarctic Geology and Geophysics.* Oslo, Universitetsforlaget, 155–60.)

Thomson, M. R. A. 1974. Ammonite faunas of the Lower Cretaceous of south-eastern Alexander Island. *Scient. Rep. Br. antarc. Surv.*, No. 80, 44 pp.

Thomson, M. R. A., and Willey, L. E. 1972. Upper Jurassic and Lower Cretaceous *Inoceramus* (Bivalvia) from south-east Alexander Island. *Bull. Br. antarc. Surv.*, No. 29, 1–19.

Thurach, H. 1890. Geognostische Beschreibung der Insel Süd-Georgien. (*In* Neumayer, G., *Die internationale Polarforschung 1882–1883. Die deutschen Expeditionen und ihre Ergebnisse. 2.* Berlin, A. Asher, 109–66.)

Trendall, A. F. 1953. The geology of South Georgia—I. *Scient. Rep. Falkld Isl. Depend. Surv.*, No. 7, 26 pp.

Trendall, A. F. 1959. The geology of South Georgia—II. *Scient. Rep. Falkd Isl. Depend. Surv.*, No. 19, 48 pp.

Tyrrell, G. W. 1916. Further notes on the petrography of South Georgia. *Geol. Mag.*, Decade 6, *3,* 435–41.

Tyrrell, G. W. 1918. Additional notes on the petrography of South Georgia. *Geol. Mag.*, Decade 6, *5,* 483–89.

Wilckens, O. 1932. Fossilien und Gesteine von Süd-Georgien. *Scient. Res. Norw. Antarct. Exped.*, No. 8, 28 pp.

Wilckens, O. 1947. Paläontologische und geologische Ergebnisse der Reise von Kohl-Larsen (1928–29) nach Süd-Georgien. *Abh. senckenb. naturforsch. Ges.*, No. 474, 66 pp.

Williams, P. L., and Rowley, P. D. 1971. Geologic studies of the Lassiter Coast. *Antarct. Jnl U.S., 6,* No. 4, 120–21.

Williams, P. L., Schmidt, D. L., Plummer, C. C., and Brown, L. E. 1972. Geology of the Lassiter Coast area, Antarctic Peninsula: preliminary report. (*In* Adie, R. J., *ed. Antarctic Geology and Geophysics.* Oslo, Universitetsforlaget, 143–48.)

20

The Structure of the South Georgia Continental Block

P. SIMPSON and D. H. GRIFFITHS
Department of Geological Sciences, University of Birmingham, Birmingham B15 2TT, U.K.

Abstract. The results of a marine geophysical survey of the South Georgia continental shelf are presented. The area can be roughly divided by an east-west line into two distinct regimes. The northern half is characterized by low values of the Bouguer gravity anomaly (<100 mgal) of long wavelength and by little magnetic expression. Both geology and gravity data indicate that the Cumberland Bay-Sandebugten sedimentary rocks extend under the northern half of the area, thickening to a maximum at or near the shelf edge. The southern half of the continental block is characterized by high values of the Bouguer gravity anomaly (>100 mgal) of shorter wavelength. In general, the rocks here are highly magnetic, the anomalies reaching their maximum amplitudes (~1000 nT) in the southeastern corner of the shelf. The gravity data, the known geology of South Georgia, and a comparison of magnetic profiles from the area with Project Magnet flights across southern South America strongly support the views already advanced on the geologic similarities of the two areas. It is suggested that the marked magnetic high lying along the southeastern edge of the shelf may be evidence of an underlying intrusive belt comparable with the Patagonian batholith. The ophiolite suite, mapped as the Larsen Harbour Formation, is not strongly magnetized; it has no consistent and recognizable magnetic signature but seems to be associated with a gravity high.

Over the past few years, geologic work in southern Chile and South Georgia has pointed increasingly to the marked similarities between the two areas (Katz and Watters, 1966; Dalziel and others, 1975; Suárez and Pettigrew, 1976; Storey and others, 1977). Better understanding of the structural history of the region (Dalziel and others, 1975; Barker and Griffiths, 1972) has strengthened the view that as late as Cretaceous time the separation of the two areas was much less than at present, South Georgia possibly lying south of Burdwood Bank. The history of sea-floor spreading in Drake Passage (Barker and Burrell, 1977) is not inconsistent with such a location for this continental block, though the mechanism of its translation eastward is difficult to conceive. The marine geophysical surveys described here, and carried out in the course of a larger study of the Scotia Sea, were aimed at extending geologic and structural information over the South Georgia shelf and so providing an improved basis for comparisons with South America.

Tanner (Chapter 18, this volume) reviews the geologic evolution of the island of South Georgia. Much of the island is occupied by the Cumberland Bay Formation, a series of flyschlike interlayerings of quartz-poor volcaniclastic turbidites and slatey mudstones. Dalziel and others (1975) have interpreted this formation as derived from an active volcanic arc which lay to the south of the island, the dominant direction of depositional transport being northwestward. They also advance strong evidence for the lithologic and age equivalence of this formation to at least part of the Yahgan Formation of southern Chile. The Sandebugten Formation, seen on the northeast coast of the island, is somewhat similar, but is predominantly quartzitic sandstone and shale. The pattern of dispersal here is dominantly southward. Trendall (1959) proposed, and subsequent authors have supported the view, that the two types represent a single Mesozoic succession and are merely lateral facies variants. Together with the Yahgan Formation of southern Chile, which lay to the west along the strike, they seem to have been deposited in an Early Cretaceous marginal basin of small size, lying between a continental calc-alkaline volcanic arc and the main continental mass of South American (Dalziel and others, 1974), the sediments extending beyond the oceanic crust and overlapping onto the basin margins. Subsequently, the Cumberland Bay Formation has been thrust to the northeast over the Sandebugten Formation. On the South Georgia shelf, evidence of island-arc volcanism is seen in the tuffs and lavas, volcanic breccias, and minor intrusions of Annenkov Island (Suárez and Pettigrew, 1976). These volcanic rocks have their equivalent in South America in the Hardy Formation of Peninsula Hardy (Suárez and Pettigrew, 1976). Storey and others (1977) have correlated the Larsen Harbour Formation, cropping out along the southeast side of South Georgia, with the narrow, discontinuous, ophiolite belt extending along the spine of the Andean Cordillera, which has been proposed by Dalziel and others (1974) to be the essentially allochthonous floor of the marginal basin, subsequently deformed and uplifted during middle Cretaceous time. Storey and others (1977) have interpreted the Drygalski Fjord Complex to be older continental crust containing fragments of the remanent arc.

The geophysical data presented here have been drawn from several sources. The vast majority was collected during three cruises of the R.R.S. *Shackleton* in the period 1971–76. This has been supplemented by data collected during cruises by the H.M.S. *Endurance* and the R. R. S. *Bransfield.* The data are presented as contour maps in Figures 20.1, 20.2, and 20.3.

The bathymetry, shown in Figure 20.1, has been contoured at 250 m intervals. Contouring of the major part of the continental shelf was originally carried out on a scale of 1:250,000. In the northeast corner of the area, east of long. 36° 30′ W and north of the 250 m contour, it has

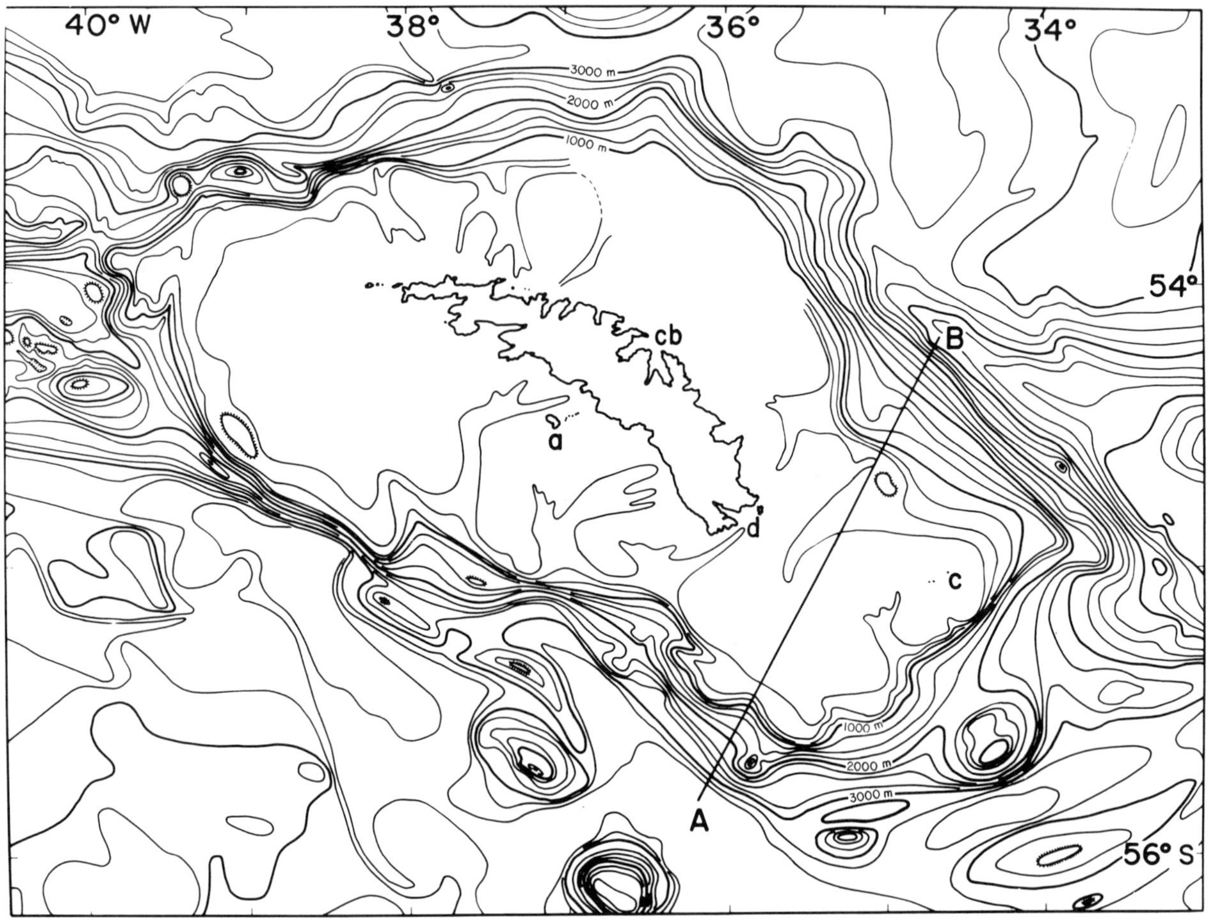

Figure 20.1. Bathymetry of South Georgia. Contour interval is 250 m. c: Clerke Rocks. a: Annenkov Island. d: Drygalski Fjord. cb: Cumberland Bay. For line AB, see text and Figures 20.4*b* and 20.5.

been redrawn from Lonardi and Ewing (1971). The shelf is relatively smooth at both the western and southeastern ends but more irregular in the central section, both to the north and south of the island. It is cut by several channels of about 100 m relief deepening toward the shelf edge. These valleys are apparently of glacial origin as their shoreward ends show a marked correlation with existing glaciers, especially along the north coast. The channels surrounding Clerke Rocks point to the once greater extent of the landmass in that area. To the south of the island the relief of the channels is even greater, being about 150 m. Elsewhere on the central shelf, relief is about 50 m. The continental slope is steepest at the western end of the block, both to the north and south; gradients of at least 18° occur between the 1000 m and 2000 m contours. At the eastern end of the block the slope is more gradual and averages about 5°.

Magnetic and gravity data

To produce the magnetic contours in Figure 20.2, the International Geomagnetic Reference Field (I.G.R.F.) (IAGA, 1976) was first removed and a further arbitrary adjustment of the baseline was made. The magnetic profiles were then smoothed by eye to remove the high-frequency component that would have made contouring impossible with the wide track spacing (15–20 km) of the original data. No attempt was made to eliminate the diurnal variation. On the northern shelf the field is generally smooth and mainly negative, though it rises to positive values as the shelf edge is approached. South of the island the field is mainly positive and is much broken up by anomalies of considerable size and amplitude. This region is also one in which the field possesses a high-frequency component: the northern limit of these small, sharp anomalies is shown by the heavy broken line in Figure 20.2. On the southern shelf there is some indication of a west-northwest to east-southeast trend to the major anomalies. The most noticeable feature is the large 600 nT positive anomaly lying off the southeast corner of the island, near and more or less parallel to the shelf edge. To the west it is interrupted by a bight in the shelf edge. Less pronounced anomalies occur in two places near the shelf edge farther to the west. Elsewhere on the southern shelf, the field, though still mainly positive, is relatively smooth. The general trend appears to be interrupted by two large anomalies to the west of Annenkov Island, though the absence of data in that area makes the contours tentative.

The gravity data were all measured with a LaCoste and Romberg gravimeter and have been reduced using the 1930 International Gravity Formula. A two-dimensional Bouguer correction has been applied using a density contrast of 1.81 g/cm^3. The reduced values were contoured at 10 mgal intervals and are shown in Figure 20.3. The general level of the Bouguer anomaly is high for a continental area lying near sea level (there are no land gravity traverses), since the average value is about +100 mgal. The general trend is similar to that of the magnetic anomalies. The gravity field

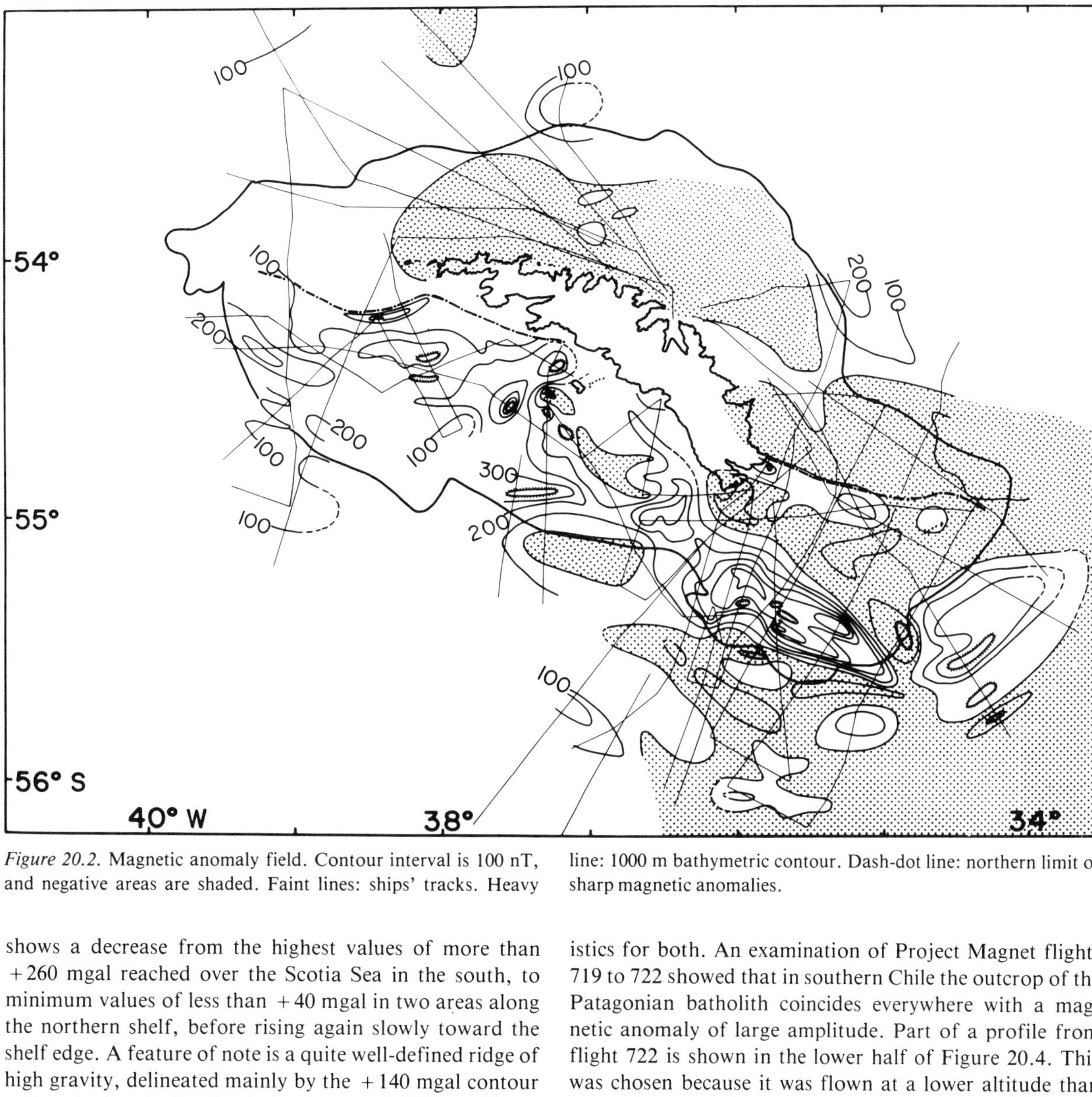

Figure 20.2. Magnetic anomaly field. Contour interval is 100 nT, and negative areas are shaded. Faint lines: ships' tracks. Heavy line: 1000 m bathymetric contour. Dash-dot line: northern limit of sharp magnetic anomalies.

shows a decrease from the highest values of more than +260 mgal reached over the Scotia Sea in the south, to minimum values of less than +40 mgal in two areas along the northern shelf, before rising again slowly toward the shelf edge. A feature of note is a quite well-defined ridge of high gravity, delineated mainly by the +140 mgal contour and enclosing the local culmination marked H on Figure 20.3, which extends across the shelf from west to east and intersects the island on its southeast corner. Continuity is uncertain as the data are sparse in places, but only the low east of the island interrupts the feature to any marked extent. The high generally has an amplitude of 30 mgal and reaches its maximum at Clerke Rocks, though the field here is distorted by the edge effect of the continental block. There is some indication that the ridge is offset in places due to crossfaulting. This appears to be true to the west of Annenkov Island. Just off the southeast coast of South Georgia, the form of the gravity contours is strongly suggestive of faulting and marked displacement of the structure northward. To the south of this high lies a series of unconnected troughs showing minima of about +100 mgal, the trend of these being roughly subparallel to the line of the high.

Discussion and interpretation

The exposed geology of South Georgia and southern Chile leads one to expect similar geophysical characteristics for both. An examination of Project Magnet flights 719 to 722 showed that in southern Chile the outcrop of the Patagonian batholith coincides everywhere with a magnetic anomaly of large amplitude. Part of a profile from flight 722 is shown in the lower half of Figure 20.4. This was chosen because it was flown at a lower altitude than most of the other flight lines and because it directly crosses the strike of the structure. The corresponding geology is shown beneath (I. W. D. Dalziel, personal communication). Throughout southern Chile the contiguity of the batholith and the ophiolites causes a merging of their anomalies and thus obscures those due to the ophiolites, which in general seem to be magnetized relatively weakly. The undoubted correlation between the large magnetic anomaly on the aeromagnetic profile and the position of the Patagonian batholith is quite clear. The upper half of Figure 20.4 shows a traverse across the large magnetic high on the southeast corner of the South Georgia shelf; the position of the traverse is shown by line AB in Figure 20.1. The similarities in amplitude, form, and width of the magnetic anomalies in both traverses strongly suggest the presence of a plutonic intrusion similar to the Patagonian batholith beneath the southeastern part of the South Georgia shelf. If this explanation is correct, the somewhat smaller anomalies to the west may be due to parts of such an intrusive body bordering and truncated by the shelf edge. By comparison with southern Chile, the positive anomaly identified here as batholithic occupies the correct

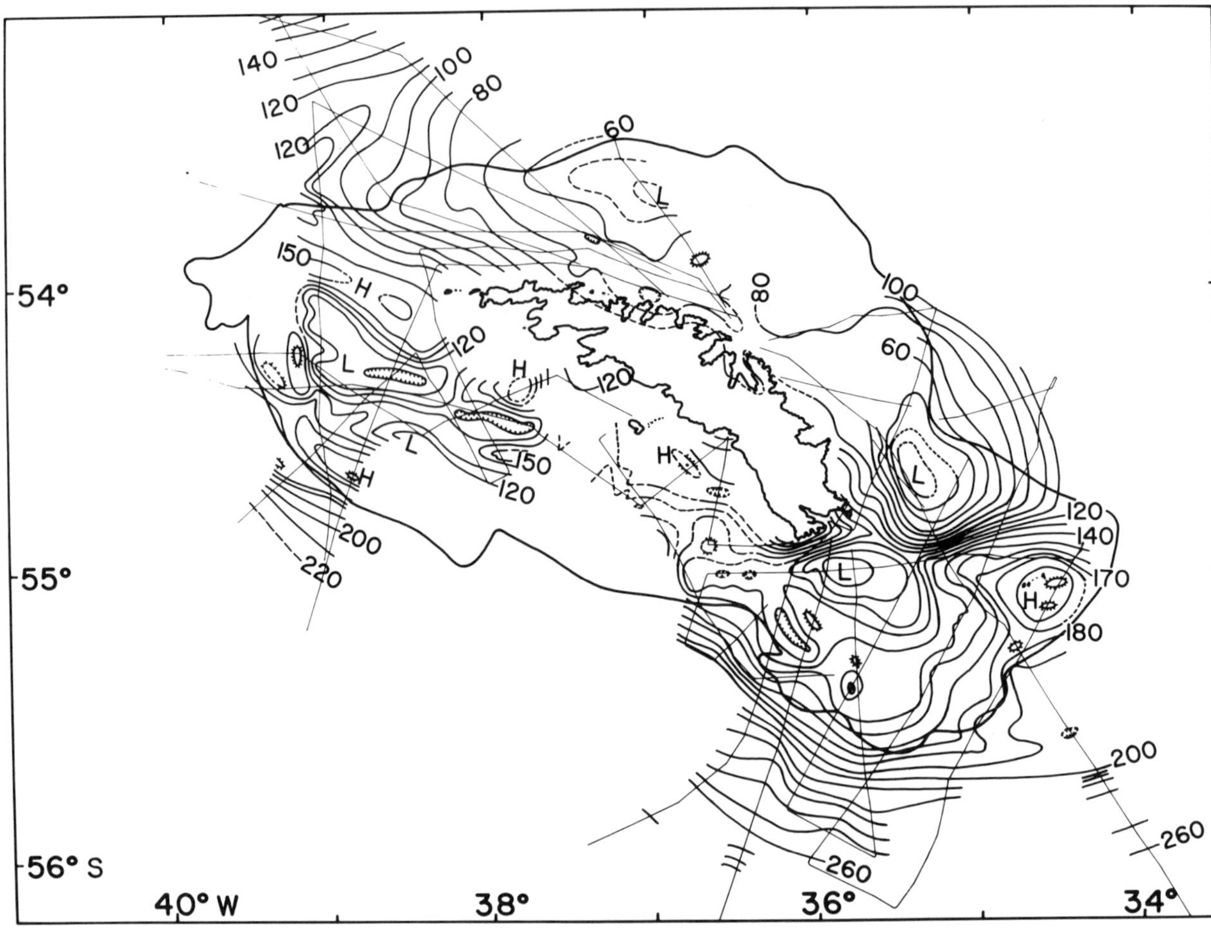

Figure 20.3. Bouguer gravity anomalies (positive), with density contrast 1.81 g/cm³. Contour interval is 10 mgal. Faint lines: ships' tracks. Heavy line: 1000 m bathymetric contour.

structural position to the south of the ophiolites of the Larsen Harbour Formation. Our magnetic measurements on rock specimens from the Larsen Harbour Formation show a low magnetic remanence. Average intensities of magnetization for specimens of dikes and lavas showed valued of 0.23 A/m, with the highest value of 1.08 A/m being obtained for the chilled margin of a dike. These values are consistent with the low values of the magnetic anomaly directly off the eastern end of the island where the ophiolites crop out. The position of the narrow belt of high gravity with respect to the outcrop of the Larsen Harbour Formation and its relationship to the assumed batholith are some justification for considering the high as due to this zone of dense oceanic-type rocks, the equivalent of the ophiolite complex of southern Chile. The high continues eastward to Clerke Rocks, though as previously mentioned, it is apparently displaced northward as a result of transcurrent faulting. The eastward extension of the ophiolite complex that this implies seems to be consistent with what is known of the geology. If Clerke Rocks are, as believed, an outcrop along the strike of the Drygalski Fjord Complex (Tanner, Chapter 18, this volume), then the ophiolites should lie just to the south, paralleling the general trend of the gravity contours. In fact, the geologic evidence points to a position for the ophiolite belt slightly offset to the south of the axis of the gravity high (Tanner, Chapter 18, this volume).

The Drygalski Fjord Complex seems to be characterized by a small positive magnetic anomaly extending eastward

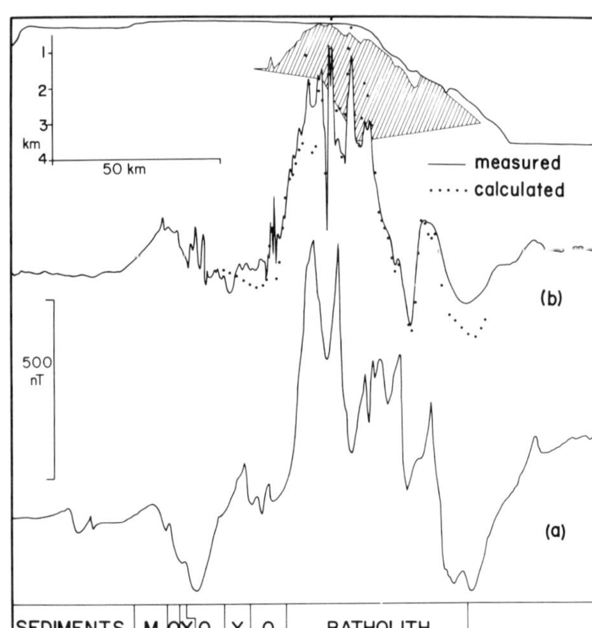

Figure 20.4. Comparison of magnetic profiles for southern Chile and South Georgia. *a.* Project Magnet flight 722 southward from Punta Arenas. The general geology of southern Chile (from I. W. D. Dalziel) is shown at the bottom of the figure: O: ophiolite. Y: Yahgan Formation. M: Magellan Straits. *b.* Line AB of Figure 20.1. The fitted model is shown dotted; the body, shown in the structural profile at the top, is shaded.

from the island to Clerk Rocks, but the contours of the field show no indication of cross faulting.

The large gravity low off the southeast end of the island cannot be satisfactorily explained. Profiler data give no hint of downfaulted sedimentary rocks, and it seems necessary to postulate a block of granitic material of relatively lower density lying between the ophiolite belt and the denser batholithic rocks to the south.

The gravity high is well developed in the west, suggesting that the ophiolite sequence extends out to the shelf edge. Here, however, it seems to occur at depth, beneath a sedimentary cover. In the east, the broken line of Figure 20.2, which marks the northern limit of the short-wavelength magnetic anomalies, roughly defines the boundary between Cumberland Bay nonmagnetic sedimentary rocks and the more magnetic Drygalski Fjord Complex "cropping out" to the south. In the west, the broken line is also believed to define the approximate northern limit of magnetic rocks at outcrop, and so here the sedimentary rocks apparently extend farther south and overlap the ophiolites. This is also indicated by the southward bend of the magnetic anomalies away from the trend of the gravity high. Seismic reflection profiles also suggest that the magnetic rocks of this area, which show east-west–oriented magnetic anomalies, are caused by lavas interbedded with gently folded sedimentary rocks probably belonging to the Annenkov Island Formation, perhaps overlying a less magnetic phase of the batholith. This sedimentary rock distribution also gives rise, at least in part, to the shallow gravity troughs on the western shelf, which like the magnetic anomalies here have an east-west trend.

To the north, the more uniform area of low gravity and smooth magnetic field correlates with the presumed seaward extension of the Cumberland Bay and Sandebugten Formations. An unreversed seismic refraction line (see Figure 20.6 for position) indicated an apparent sea-bed velocity of 5.7 km/sec to the north of Royal Bay. The assumption of underlying rocks with a velocity of 6.6 km/sec gives a minimum thickness of 8.5 km for the upper layer. A velocity of 5.7 km/sec is appropriate for the rocks of the Cumberland Bay Formation, which have densities of 2.65 to 2.70 g/cm^3 (R. G. B. Renner, personal communication).

The gravity low may therefore indicate a deep basin in these sedimentary rocks, though their high density would suggest only a small density contrast with the underlying basement. Lateral density changes to account for the gravity field in this area cannot therefore be discounted.

By two-dimensional modeling, a magnetic body has been fitted to the magnetic profile along line AB of Figure 20.1. The body is shown shaded at the top of Figure 20.4. To obtain a satisfactory fit to the profile, the constraints of a high intensity of magnetization (6.0 A/m), a near-surface top, and a shallow base have to be assumed. Lack of definite knowledge of the position and orientation of South Georgia in Early Cretaceous time made the choice of magnetization direction somewhat uncertain, but geometric constraints indicate the need to use an inclination as steep and possibly steeper than that of an axially dipolar field appropriate to South Georgia in its present position. An apparent inclination of $-75°$ gave satisfactory results. Cobbing and Pitcher (1972) have shown the structure of the South American coastal batholith to be complex; the presence in South Georgia of a near-surface magnetic body

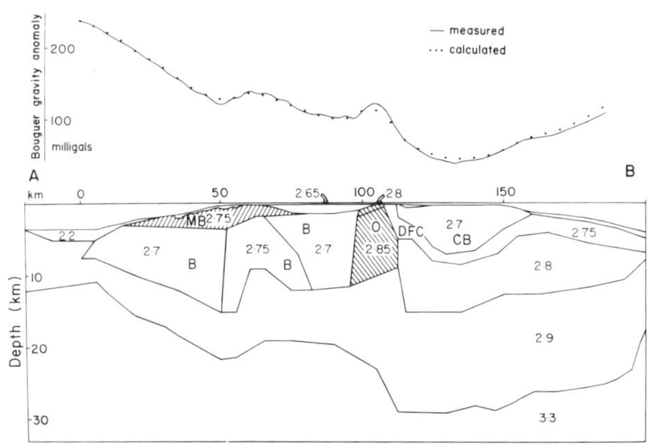

Figure 20.5. Schematic model to fit Bouguer gravity anomalies along line AB of Figure 20.1. Densities in g/cm^3. Average crustal density is 2.99/cm^3, and depth of compensation is 33 km. MB: magnetic phase of batholith. B: non-magnetic batholith. DFC: Drygalski Complex. O: ophiolite complex. CB: Cumberland Bay Formation.

with a shallow base rather than one of extended depth could result from the late intrusion of a more silicic and weakly magnetic phase into a preexisting mafic and more magnetic pluton. The shallow magnetic part of such a batholith has been incorporated into a crustal cross section based on the known geology of South Georgia, its presumed similarity to South America, and the ideas on the nature of the gravity field as outlined above (Figure 20.5). The cross section has been adjusted to fit the gravity field and incorporates many assumptions; it must, therefore, be regarded as schematic or qualitative rather than rigorous. It has been necessary to assume density variations within the batholith to obtain a fit to the measured gravity values. The data also suggest that the batholith may be flanked on its northern edge by a strip of continental basement equivalent to the Drygalski Fjord Complex. A relatively shallow Moho is required under the southern shelf to account for the high observed values of the gravity field. The high-density block of crust at 100 km represents the wedge of oceanic rocks seen at the surface in the Larsen Harbour Formation; it gives rise, in part, to the ridge of high gravity values. To the north the outcrop of the Drygalski Fjord Complex is assumed to be continuous with the continental basement rocks that probably lie beneath the Cumberland Bay Formation. Down the continental slope to the north, the basement is shown to rise, as suggested by the positive values of the magnetic field found toward the shelf edge. The small unmarked bodies in Figure 20.5 are small pockets of recent sediments.

Conclusions

Available geophysical data confirm the geologic similarity between the South Georgia block and southern Chile. This interpretation is based on recognition of the geologic significance of a very limited number of geophysical features, but it is reasonably convincing. That the magnetic anomalies on the southern shelf edge are due to parts of a major intrusion similar to the coastal batholith of Chile is difficult not to accept. The interpretation of the subparallel ridge of high gravity values as resulting from a wedge of dense mafic rocks, seen in outcrop as the ophiolite suite, also seems sound. The identification of the mag-

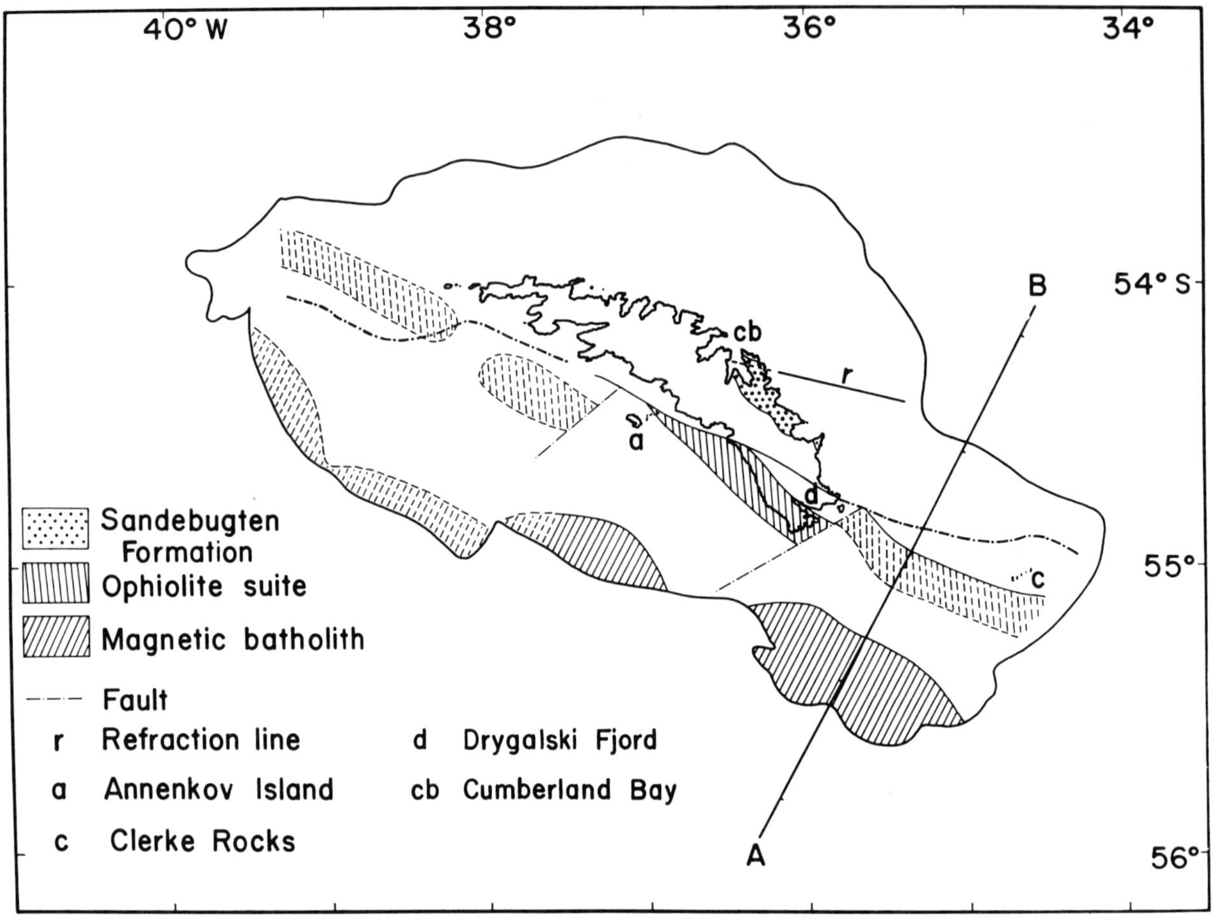

Figure 20.6. Geologic interpretation. The geology of South Georgia is from P. W. G. Tanner (Chapter 18, this volume). The heavy dot-dash line is the northern limit of sharp magnetic anomalies.

netically quiet zone and the area of low gravity on the northern shelf as a deep basin of sedimentary rocks of the Cumberland Bay and Sandebugten Formations is reasonable, but lack of data north of the entrance to Cumberland Bay leaves doubt about its east to west continuity. After the identification of these key features, arguments involving the known geology and comparisons with southern Chile were used to infer the presence of other elements of the structure. No difficulty was found in developing a realistic model to fit profile AB of Figure 20.1, though because of the lack of constraints, the computed cross section has to be regarded as very generalized.

In detail, much remains unresolved. Trends of the gravity contours suggest cross faulting in a few places, but the evidence is very weak, except perhaps at the eastern end of the island. There are also some minor inconsistencies between the gravity and the magnetic data, and as is to be expected, some anomaly features cannot be convincingly explained. The geologic interpretation is summarized in Figure 20.6, in which are shown the areas underlain by the different formations (not everywhere at outcrop); it is necessarily a gross oversimplification. The interpretation advanced, however, lends further credence to the argument for a Cretaceous position of South Georgia much closer to South America than at present. A point of some interest is that in southern Chile there is an apparent widening of the marginal basin toward the south and east. In width, the geologic features on South Georgia seem to fit more closely to a cross section of South America between long. 68° and 70° W rather than to one at the eastern extremity of Tierra del Fuego. Since the marginal basin appears to be widening as the tip of the continent is reached, this suggests that it narrowed again eastward, with an as yet unidentified continental fragment fitting the gap between South America and South Georgia prior to separation.

Acknowledgments. We wish to thank the Captain and ship's company of the R.R.S. *Shackleton* for their help in collecting the data, D. W. G. Tanner for the loan of rock specimens for magnetic measurements, and P. F. Barker for his constructive criticism. The work was financed by a grant from the Natural Environment Research Council.

References cited

Barker, P. F., and Burrell, J. 1977. The opening of Drake Passage. *Mar. Geol., 25,* 15–34.

Barker, P. F., and Griffiths, D. H. 1972. The evolution of the Scotia Ridge and Scotia Sea. *Phil. Trans. R. Soc. Lond.,* Ser. A., *271,* 151–83.

Cobbing, E. J., and Pitcher, W. S. 1972. The coastal batholith of central Peru. *J. Geol. Soc. Lond., 128,* 421–60.

Dalziel, I. W. D., de Wit, M. J., and Palmer, K. R. 1974. Fossil marginal basin in the southern Andes. *Nature, Lond., 250,* 290–94.

Dalziel, I. W. D., Dott, R. H., Jr., Winn, R. D., Jr., and Bruhn, R. L. 1975. Tectonic relations of South Georgia Island to the southernmost Andes. *Geol. Soc. Am. Bull., 86,* 1034–40.

International Association of Geomagnetism and Aeronomy. 1976. International Geomagnetic Reference Field 1975. *EOS, 57,* 120–21.

Katz, H. R., and Watters, W. A. 1966. Geological investigations of the Yahgan Formation (upper Mesozoic) and associated igneous rocks of Navarino Island, southern Chile. *N.Z. Jl Geol. Geophys., 9,* 323–59.

Lonardi, A. G., and Ewing, M. 1971. Sediment transport and distribution in the Argentine Basin, part 4. Bathymetry of the continental margin, Argentine Basin and other related provinces. Canyons and sources of sediments. *Phys. Chem. Earth, 8,* 79–122.

Storey, B. C., Mair, B. F., and Bell, C. M. 1977. The occurrence of Mesozoic oceanic floor and ancient continental crust on South Georgia. *Geol. Mag., 114,* 203–8.

Suárez, M., and Pettigrew, T. H. 1976. An upper Mesozoic island-arc-back-arc system in the southern Andes and South Georgia. *Geol. Mag., 113,* 305–28.

Trendall, A. F. 1959. The geology of South Georgia: II. *Scient. Rep. Falkld Isl. Depend. Surv., 19,* 48 pp.

21

Relationship of Late Mesozoic and Early Cenozoic Sedimentation to the Tectonic Evolution of the Southernmost Andes and Scotia Arc

R. H. DOTT, JR., R. D. WINN, JR.*, and C. H. L. SMITH**
Department of Geology and Geophysics, University of Wisconsin, Madison, Wisconsin 53706, U.S.A.

Abstract. Thick late Mesozoic and early Cenozoic sedimentation occurred on the foreland side of the southernmost Andes, which then included South Georgia island. As in many other orogenic belts, the record indicates profound subsidence of a craton margin as the juvenile cordillera rose; but here, most sedimentation was in unusually deep water and consisted largely of deep-sea fan turbidites. Moreover, Lower Cretaceous flysch was deposited in a back-arc rift basin upon quasi-oceanic crust, while Upper Cretaceous and Cenozoic deposition was largely within the craton margin (Magallanes Basin or exogeosyncline). Deltaic and shallow-marine (molasse) deposition did not become widespread until mid-Cenozoic time. Contemporaneous fore-arc basin rocks apparently underlie the present continental shelf. In the Antarctic Peninsula, a Cretaceous back-arc basin is poorly (if at all) preserved, but fore-arc or intra-arc deposits seem better represented (Alexander and South Shetland Islands). In both regions, calc-alkaline volcanic rocks and associated dioritic batholiths represent a late Mesozoic–early Cenozoic arc, and some andesitic detritus persists through adjacent Lower Cretaceous to Miocene strata. Some Pliocene-Pleistocene volcanic cones exist in southwestern Chile. Facies zones generally match between the two continents through lower Cenozoic strata, but peninsula rocks tend to be little deformed. Dismemberment of a formerly continuous Andean belt and formation of the Scotia Sea apparently occurred during late Cenozoic time (last 30 m.y.). Collision of the future Scotia Sea region with a former triple junction of the East Pacific Ridge system apparently caused the initial breakup and subsequent slowing of subduction beneath the southern Andes.

The thick upper Mesozoic and lower Cenozoic sequence of the Andean foothills of southernmost Chile and Argentina superficially resembles the clastic wedges of most foreland basins (exogeosynclines). In all such cases there is recorded profound subsidence of the edge of a craton apparently causally related to the rise of an adjacent cordillera. The southern Andean example is unusual, however, in having mostly deep-water deposits, whereas shallow-marine and even non-marine deposits are more typical of foreland basins. Moreover, in Tierra del Fuego the lowest deposits of the clastic wedge—thick Upper Jurassic and Lower Cretaceous flysch—were created largely in a back-arc or marginal ocean basin floored in part by quasi-oceanic (ophiolitic) rather than entirely continental crust. Tectonic deformation of the rocks was also of a different sort, with very tight compression and cleaving against the cratonward margin of that basin. Deformation of later Cretaceous and Paleogene strata of the southern Andes produced folds and thrust faults more like those of the foreland margins of most other orogenic belts, although strike-slip deformation is more prominent here than in most foreland belts (see Winslow, Chapter 15, this volume).

This paper discusses the late Mesozoic and early Cenozoic sedimentation and paleogeography of southernmost Chile, Argentina, and South Georgia island, and presents brief comparisons with the Antarctic Peninsula. In late Mesozoic time, South Georgia island lay near present Cape Horn. Apparently the South Orkney Islands and the Antarctic Peninsula were also more closely grouped and tied in some manner to South America (Dalziel and Elliot, 1973; Dalziel and others, 1975; Suárez and Pettigrew, 1976; Winn and Dott, 1978). Departures of our interpretations from prior ones for the environments and processes of deposition of some of the strata therefore have implications for the interpretation of the tectonic evolution of the entire Scotia Sea region.

Jurassic volcanic rocks

Southern South America and the Antarctic Peninsula once composed a land underlain by varied metamorphic rocks that were deformed and exposed during early Mesozoic time as a result of Gondwanian orogenesis. Locally some of those rocks contain upper Paleozoic (Hollingsworth, 1954) and Triassic fossils (Thomson, 1975; Edwards, Chapter 36 this volume). Scattered isotopic dates indicate Permian–Jurassic metamorphism (Grikurov and others, 1966; Grikurov, 1972; Halpern, 1972, 1973; Thomson, Chapter 37, this volume; Gledhill and others, Chapter 35 this volume). The basement rocks are unconformably overlain by very widespread Middle to Upper Jurassic silicic and intermediate volcanic rocks. In southern South America the Serie Tobifera (Figure 21.1) is composed of sodic rhyolite, dacite, and trachyandesite (Stepanicic and Reig, 1955; Dalziel and others, 1974). These volcanic rocks extend beneath the Atlantic shelf and

* Present address: Denver Research Center, Marathon Oil Company, Littleton, Colorado 80160, U.S.A.

** Present address: Atlantic Richfield Company, Denver, Colorado 80201, U.S.A.

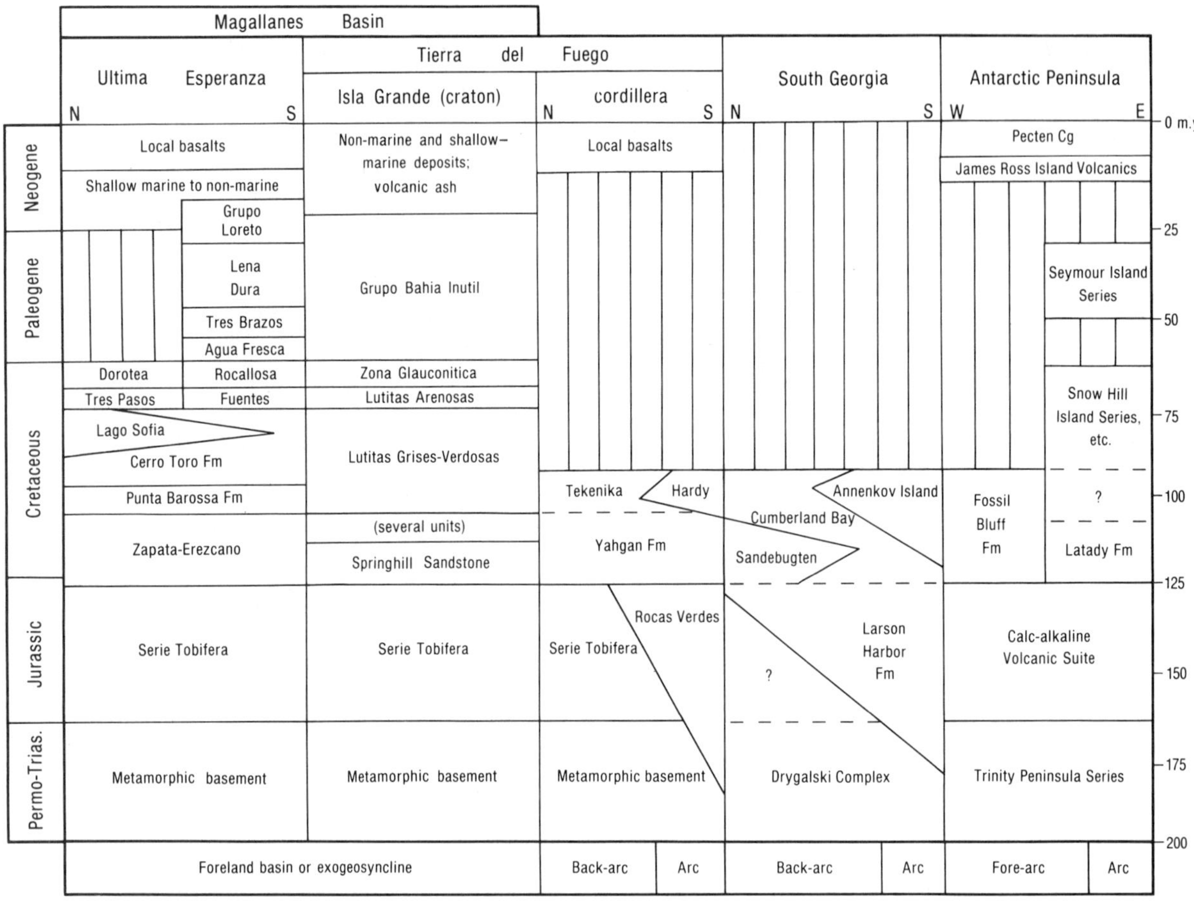

Figure 21.1. Mesozoic-Cenozoic correlation chart for the Scotia Sea region.

Burdwood Bank (Figure 21.2*a*). Locally interstratified with Tobifera volcanic rocks are thin black mudstones and minor sandstones with Upper Jurassic marine fossils (Natland and others, 1974; Dalziel and others, 1974). The presence of ignimbrites and accretionary lapilli, however, indicate that much of the volcanic field lay above sea level.

Composition of the Tobifera suggests derivation from partial melting of continental crust, but otherwise its genesis is unclear. It is related in time to Jurassic calc-alkaline volcanic rocks of the Antarctic Peninsula, which have been interpreted as ensialic volcanic-arc products (Adie, 1972; Suárez, 1976). Because of its age, great extent, and association with block faulting, the Tobifera may have had some relation to the initial breakup of Gondwanaland (Dalziel, 1974). More recently, however, Bruhn and Dalziel (1977) have suggested that it represents a remnant arc formed near the southwestern edge of the South American craton.

Late Mesozoic sedimentology

The Patagonian craton

At the end of the Jurassic Period, most of the region was inundated, for marine Kimmeridgian-Tithonian sedimentary rocks overlie the volcanic rocks; foraminifera suggest maximum water depths of about 200 m (Natland and others, 1974). The basal Springhill Sandstone found on the eastern shelf margin of the Magallanes Basin in the subsurface (Figures 21.2*b* and 21.3) is the principal petroleum reservoir rock of Isla Grande. It formed by weathering and erosion of the felsic volcanic rocks on a land surface with moderate relief. Because of its quartz-rich nature, it is of special interest as a possible first-cycle, mature quartz sandstone. Thin section and X-ray analysis, however, show it to be more heterogeneous, angular, and poorly sorted than typical quartz arenites (Figure 21.4). In addition to monocrystalline quartz (much of it with embayments), there is a wide variety of polycrystalline quartz grains, considerable muscovite, and conspicuous kaolinite occurring both as coarse detritus and as cement.

Back-arc basin

In Late Jurassic time, a sliver of continental metamorphic and Tobifera rocks was rifted away from South America to form the base of a new volcanic arc. The Patagonian batholith, the oldest dated portion of which is about 150–155 m.y. (Halpern, 1973), is presumed to be the roots of that arc. Rifting created a back-arc basin floored by zones of gabbro, sheeted dikes, and pillow basalts (Rocas Verdes), lying between and partially covering horst blocks of Tobifera (Figure 21.3), This quasi-oceanic basin had a V-shape, widening toward the southeast with a width in Tierra del Fuego probably on the order of 200–250 km. (Figure 21.2*a*; Winn, 1975, 1978). Mafic rocks (Larson Harbor) and a continental-type metamorphic complex (Drygalski) seem to represent the marginal basin floor on South Georgia (Storey and others, 1977; Tanner, Chapter

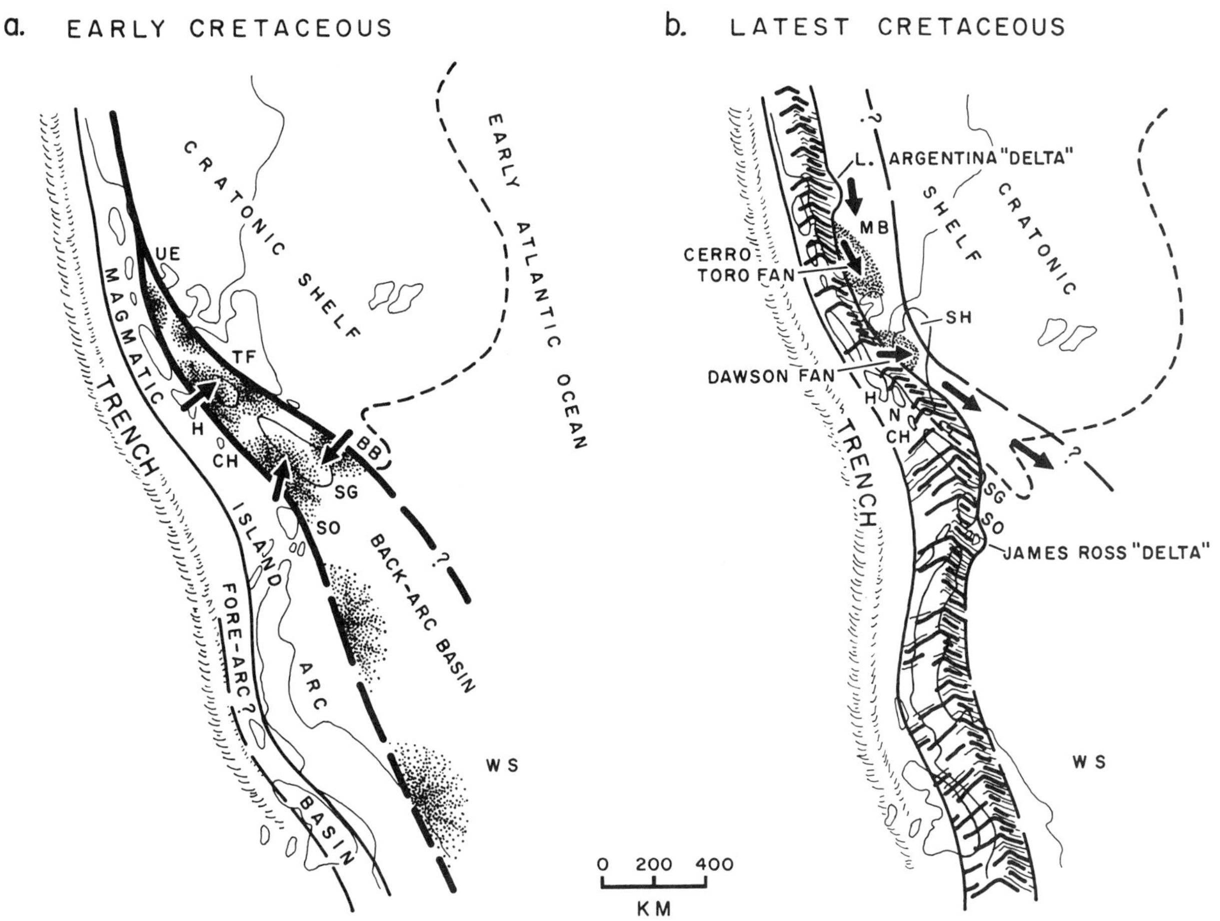

Figure 21.2. Paleogeographic restoration maps of the Scotia Sea region: *a.* Early Cretaceous, showing inferred deep-sea fans in the back-arc basin, and *b.* Latest Cretaceous, with submarine fans in the Magallanes Basin. UE: Ultima Esperanza region. TF: Tierra del Fuego (Isla Grande). BB: Burdwood Bank. H: Hoste Island. CH: Cape Horn. SG: South Georgia. SO: South Orkney Islands. WS: Weddell Sea. MB: Magallanes Basin. SH: Springhill Platform. N: Navarino Island.

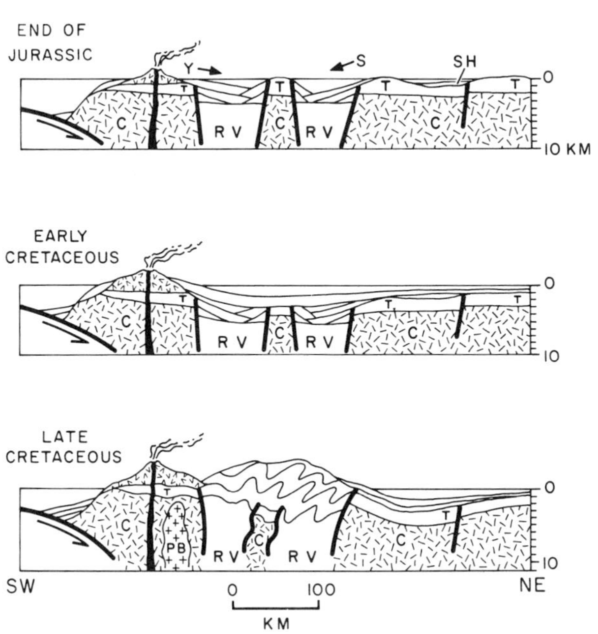

Figure 21.3. Restored cross sections showing our conception of the late Mesozoic evolution of Tierra del Fuego. C: continental crust. RV: Rocas Verdes ophiolites. T: Serie Tobifera. PB: Patagonian batholith. Y: Yahgan sedimentary rocks. S: Sandebugten-type sedimentary rocks. SH: Springhill Sandstone.

18, this volume). Dalziel (1974), Suárez (1976), and deWit (1977) have suggested that the Weddell Sea originated as an extension of this late Mesozoic back-arc (marginal) basin of Tierra del Fuego (Figure 21.2). The striking contrast between severely deformed Jurassic–Cretaceous strata in southern South America and their almost undeformed equivalents in most of the Antarctic Peninsula might be due to non-closure of most of the Weddell Sea portion of such a basin; but moderate deformation at the southern end of the peninsula remains a puzzle (Laudon, 1972; Rowley and Williams, Chapter 38, this volume). Also puzzling is the apparent narrowness of the ophiolite zone of South Georgia (Simpson and Griffiths, Chapter 20, this volume).

Mafic rocks accumulated rapidly, and then 5–6 km of Upper Jurassic–Lower Cretaceous flysch were deposited in the back-arc basin; however, only 1 km of equivalent-aged strata accumulated on the craton, while thicknesses of 2–3 km characterize the intermediate zone seen today in the cordilleran foothills (See Winslow, Chapter 15, this volume). Superficially the flysch all looks similar. In reality, however, there are two petrographically distinct sequences with different patterns: a southerly, arc-derived andesitic sequence (most of Yahgan-Tekenika and Cumberland Bay strata; see Figure 21.1), and a northerly, largely craton-derived siliceous one (basal Yahgan and

Sandebugten strata) derived chiefly from the Tobifera volcanic rocks (Winn, 1975, 1978; Winn and Dott, 1978). Each is well represented both in Tierra del Fuego and on South Georgia Island (Figures 21.1, 21.2*a,* and 21.5). They intermingled in the central zone of the basin.

The flysch is made up of interbedded graywackes and mudstones with diamictites, small slumps, conglomerates, and sedimentary breccias. Grading is common in the sandstones, as are sole marks, flame structures, convolute lamination, and rip-up mudstone intraclasts. The sandstones are regarded as turbidites. Yahgan-Cumberland Bay strata also contain pure quartz-prehnite beds that represent altered tuffs. The andesitic flysch tends to be coarser and more variable in texture and bedding characteristics. Its upper part, both in Tierra del Fuego and on South Georgia, contains conspicuous sedimentary breccias and conglomerates (Suárez and Pettigrew, 1976; Winn and Dott, 1978). They are either massive or show crude grading, are locally tens of meters thick, and contain mostly cobble to small boulder sizes. Most clasts are of intermediate volcanic rocks. Parallel lamination and medium-scale cross-stratification characterize intercalated coarse, green-weathering volcanic sandstones, which also show some graded bedding.

Trace fossils were clearly made by systematic deposit-feeders (Winn, 1975). The assemblage generally corresponds to the *Chondrites* facies of Chamberlain (1971*a,b*), which he believed typified bathyal depths. No definitive shallow-water forms or sedimentary structures were found, but resedimented land plants (including logs as much as 4 m long) and shallow-marine shell debris are present, especially in the Tekenika graded conglomerate sequence of southern Tierra del Fuego (Dott and others, 1977).

The greatest volume of modern turbidites occurs in fans built at the bases of submarine slopes, and the setting of Lower Cretaceous flysch in Tierra del Fuego and South Georgia is most compatible with deposition on such fans. We have postulated that the basin was filled by fans growing inward from both sides (Figures 21.2*a* and 21.3), with the southern, arc-fed fans being larger (Winn and Dott, 1978).

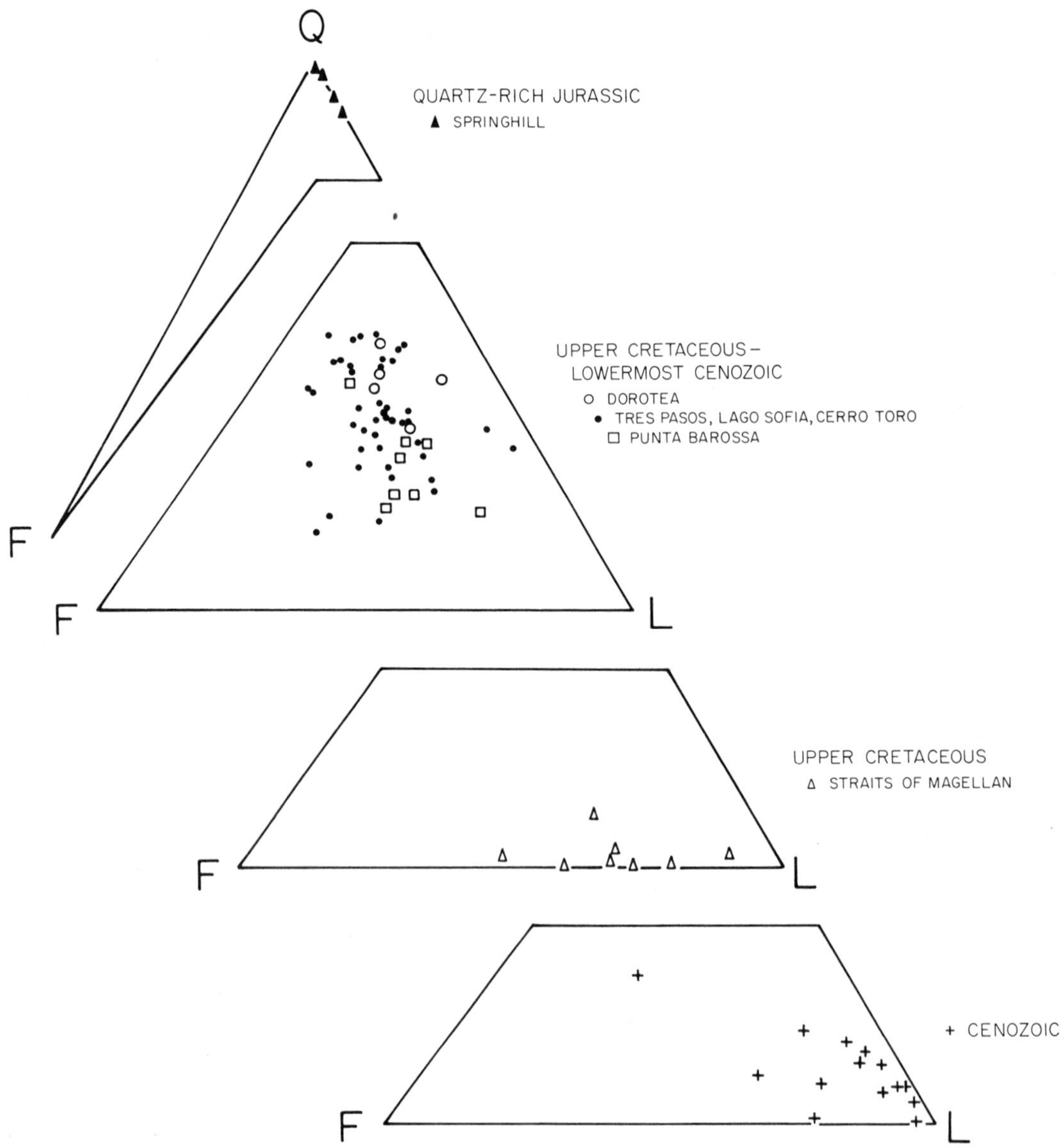

Figure 21.4. Comparative triangular diagrams of quartz, feldspar, and lithic (chiefly volcanic) sand grain proportions for Jurassic, Upper Cretaceous, and Cenozoic formations of the Magallanes Basin (mostly after Smith, 1977).

THE ARC

On southern Hoste, Wollaston, and Cabo de Hornos Islands, batholith rocks are intimately associated with silicic to mafic extrusives and volcaniclastic sedimentary rocks, which contain some Lower Cretaceous fossils. The sedimentary rocks, which were recently named the Hardy Formation (Suárez, 1976; Suárez and Pettigrew, 1976), consist of heterogenous mixtures of graded flysch, coarse green cross-bedded volcanic sandstone, and coarse volcanic conglomerate. Interstratified are pillow basalts, vesicular basalt, andesitic breccias and conglomerates, and silicic ignimbrites, accretionary lapilli tuffs, and glassy flow rocks and breccias. A similar assemblage of volcanic rocks also occurs on southern South Georgia (Annenkov Island Formation of Suárez and Pettigrew, 1976), and geophysical data suggest that batholith rocks occur farther south beneath the shelf edge (Simpson and Griffiths, Chapter 20 this volume). These rocks are taken to represent the volcanic-arc edifice that was the southern source of the andesitic flysch. While the relationships between the non-marine and shallow-marine arc-flank rocks to the deeper marine fan deposits has not been established in detail, we infer an intertonguing relationship on the basis of fossiliferous flysch interstratified with coarse volcanic sandstone, conglomerate (including diamictites), and both submarine and subaerial extrusive igneous rocks (Suárez and Pettigrew, 1976; Dott and others, 1977). Transitional between the typical Yahgan mid-fan flysch and the Hardy arc-flank sedimentary rocks in Tierra del Fuego are the conglomeratic Tekenika beds, which we regard as having been deposited by resedimentation high on a submarine fan rather than on a delta or shoreline (Dott and others, 1977).

FORE-ARC(?) BASIN OF THE ANTARCTIC PENINSULA

Simultaneously, mudstone and sandstone of Late Jurassic to at least Albian ages were deposited in fore-arc and intra-arc basins on the Pacific side of the Antarctic Peninsula (Alexander and South Shetland Islands; Valenzuela and Hervé, 1972; Suárez, 1976; Thomson, Chapter 37, this volume). Their exact counterparts are unknown in South America because the Patagonian batholith lies at the Pacific coast; fore-arc strata may lie beneath the southern Chilean shelf, however.

ANDEAN OROGENY

Structural and stratigraphic evidence indicate that the back-arc basin was closed after existing for about 50 m.y., for its Upper Jurassic–Lower Cretaceous contents were deformed and metamorphosed to prehnite-pumpellyite grade in mid-Cretaceous time. The craton acted as a but-

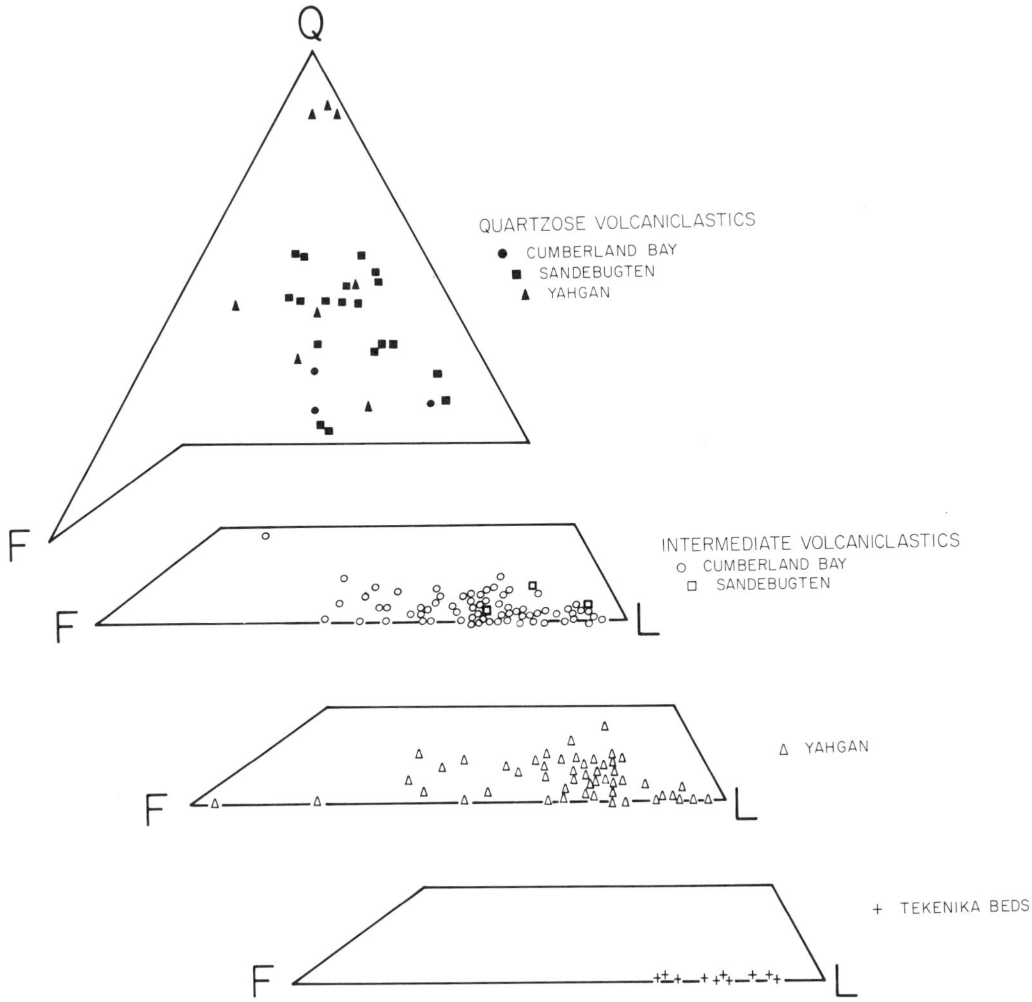

Figure 21.5. Comparative triangular diagrams (as in Figure 21.4) for Lower Cretaceous formations of the Tierra del Fuego-South Georgia back-arc basin (after Winn, 1978; Winn and Dott, 1978).

tress, and the rocks along its margin were folded, faulted, and cleaved most severely, as seen today in Tierra del Fuego (Figure 21.3); deformation greatly diminished southward toward the arc (Dalziel and others, 1975; Bruhn and Dalziel, 1977). The youngest known rocks affected within the main cordillera are the Aptian-Albian(?) Tekenika beds at the top of the Yahgan Formation, while deformed Yahgan strata themselves are cut by 80 m.y. old posttectonic plutons (Halpern and Rex, 1972). On South Georgia deformation of the Cumberland Bay Formation also was post-Albian (Thomson and others, Chapter 19, this volume).

As the Early Cretaceous back-arc basin and arc were deformed and uplifted against the southern edge of the continent, the Magallanes Basin was rapidly downwarped within the edge of the Patagonian craton (Figure 21.3). It became the main locus of subsidence and sedimentation until Miocene time. Foraminifera suggest a sudden deepening of water from a few hundred to perhaps as much as 2000 meters in Albian–Cenomanian time (Natland and others, 1974). Stratigraphic thicknesses also reflect this change. Lower Cretaceous strata of the central Magallanes Basin have a maximum thickness of 1 km, but Upper Cretaceous strata, which represent a similar time span, are 6–7 km thick. If some causal relationship exists between this Late Cretaceous subsidence and the compression and uplift to the southwest, closure of the back-arc basin in Albian time (about 100 m.y. ago) is implied. There was also an increased influx of sand and even minor gravel to the Magallanes Basin from the southwest in Cenomanian time (Punta Barrosa; Katz, 1963), and younger Cretaceous conglomerates contain clasts of diverse, deformed marginal-basin rocks. Dalziel (1974) has suggested that the main Andean orogeny probably occurred at this time with 5–10 km of uplift of the arc and back-arc basin rocks.

As viewed within the cordillera, penetrative deformation and metamorphism certainly were concentrated in mid-Cretaceous time (pre-80 m.y. old plutons). Plutonism, however, spanned the entire interval from about 155 to 12 m.y. ago (Halpern, 1973), and rocks as young as Campanian in the western Magallanes Basin show local penetrative cleavage, while rocks as young as Paleogene have been involved in folding and thrusting all along the pre-cordillera region (see Winslow, Chapter 15, this volume). Furthermore, there is sedimentologic evidence of contributions of coarse clastic material to the Magallanes Basin from the cordillera throughout Late Cretaceous and Paleogene time, and cordillera-derived andesitic detritus appears in rocks at least as young as Miocene. Therefore, we believe that deformation and (especially) elevation of the southern Andes was nearly continuous from mid-Cretaceous at least to mid-Cenozoic time. Through this entire interval, the Magallanes Basin subsided rapidly and received deep-water sediments over all but its eastern margin (Springhill platform), which persisted as a more stable rim.

The Magallanes Basin

Within the Magallanes Basin proper, latest Jurassic and Early Cretaceous (Tithonian–Albian) sedimentation consisted almost exclusively of black mudstone (Zapata Formation). It is less than 1000 m thick over most of the basin, but about twice that thick on the southern side on Isla Grande, Tierra del Fuego. This is the lateral equivalent of the marginal-basin flysch. Very fine-grained, thin sandstone and siltstone laminae, calcareous concretions, and very siliceous (probably tuffaceous) mudstone break the monotony of black argillites. These sediments predate the clear delineation of the Magallanes Basin. Significant sand deposition within the basin began approximately in Cenomanian time, but the sandstone (Punta Barrosa) is clearly diachronous (Katz, 1963). This was the harbinger of thick, rapidly deposited Upper Cretaceous (Santonian–Campanian) flysch of the Cerro Toro Formation (2000 m in the western Magallanes Basin), a succession of turbidite sandstones punctuated by lenses of spectacular deep-water conglomerates and massive sandstones (the Lago Sofia Member). Foraminifera suggest deposition at depths of 1000–2000 m, which seems consistent with a sparse *Chondrites* trace fossil assemblage.

Scott (1966) established that both the normal Cerro Toro flysch and Lago Sofia conglomerates were deposited by currents flowing from north to south. Contemporaneous flysch in Argentina 100–150 km north of Ultima Esperanza shows the same paleocurrent pattern and lithologic characteristics (Furque, 1973). From the geometries of slump folds in zones as much as 30 m thick within the lower Cerro Toro, Scott also suggested a simple sea-bottom slope from west to east. While a regional eastward slope (averaging about 1°) from tectonic islands of the juvenile cordillera into the Magallanes Basin, with its depths of perhaps 1–2 km, seems inescapable, Scott's restoration produced some sedimentologic dilemmas. Conglomerate pebbles include foliated metamorphic, granitic, mafic, and silicic (Tobifera) rock types, which demand a western or cordilleran source for the detritus, yet the transport was north to south. This required lateral supply down the paleoslope but reworking *even of coarse gravel,* with clasts up to 30 cm long, by across-slope currents of uncertain cause. The deposits suggested gravity-driven debris flow and turbidity current processes, but an eastward slope would preclude a simple gravity drive for north-to-south *across-slope* currents!

Graded sands of the Cerro Toro Formation are clearly gravity-flow turbidite deposits. The Lago Sofia deposits, while much coarser, also show many features of gravity-flow mechanisms (diamictite, graded conglomerates, flute marks as much as 1.5 m wide and 1.5 m deep; Scott, 1966) as well as of tractive mechanisms (large-scale cross-bedding and dune forms as well as low-angle inclined bedding; Winn and Dott, 1977). Since Scott's excellent detailed work, much has been learned about subaqueous fans and deep-water gravel deposition. We have recently reexamined the Cerro Toro–Lago Sofia rocks in light of this work and have concluded that they were deposited upon a large deep-sea fan system that prograded from northwest to southeast in the western, probably deepest part of the Magallanes Basin and parallel to its axis. The fan must have been fed by a canyon system with its head near the tectonic islands of the embryonic cordillera, no more than 50 km north-northwest of Ultima Experanza. Such a locus is constrained by the fact that Lago Sofia conglomerate, where represented at all, occurs only as thin zones in Campanian flysch in the Lago Argentino region 100 km north-northeast of Ultima Esperanza. Also the northernmost obvious source of mafic (ophiolitic) con-

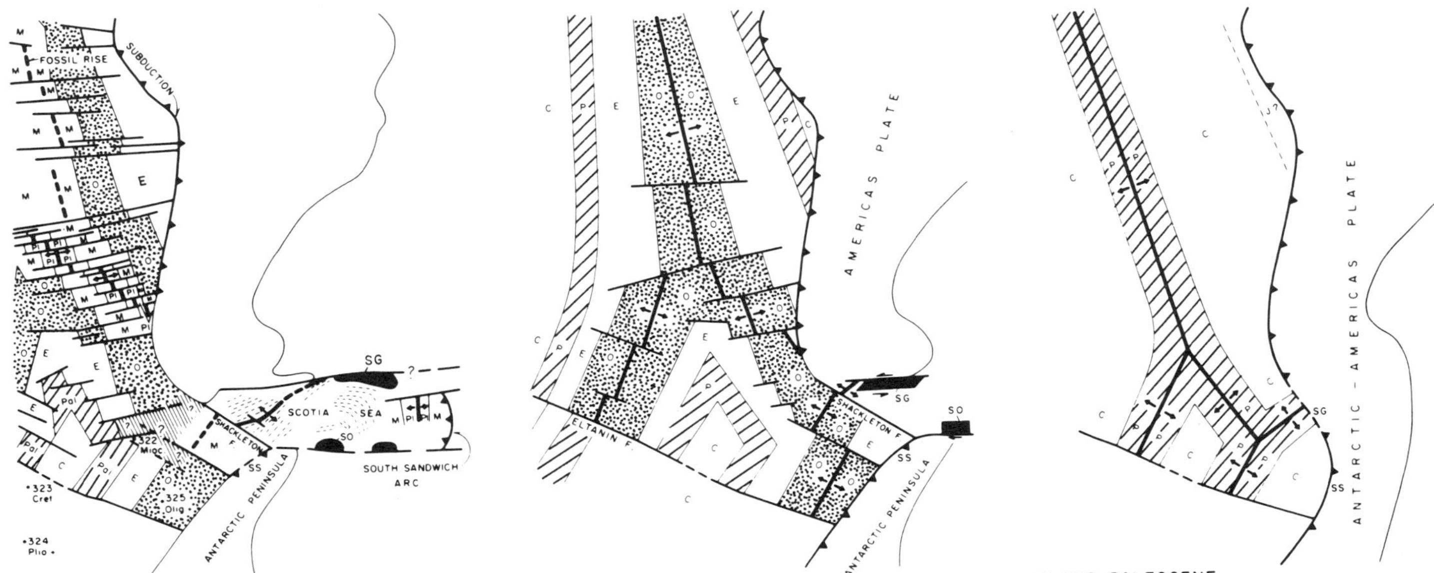

Figure 21.6. Restorations of the southeastern Pacific showing a hypothesis for the opening of the Scotia Sea based chiefly upon sea-floor magnetic anomaly patterns (after Dott, 1976). SS: South Shetland Islands. SG: South Georgia Island. Numbers in A are DSDP holes with age of rocks at bottom indicated.

glomerate clasts is the northern end of the Early Cretaceous marginal basin west of Ultima Esperanza (Figure 21.2*a*). Therefore, we infer that the current patterns in all rock types record the down-fan transport of material of all sizes by gravity-driven mechanisms. The course Lago Sofia is interpreted by us as delinating fan-axis conduits, the largest of which was 6–10 km wide and at least 120 km long; it was characterized by braided channels extending down the fan. The typical Cerro Toro rock types represent fan-surface (overbank) turbidites (Winn and Dott, 1977). The modern Astoria deep-sea fan of offshore Oregon, U.S.A., although somewhat larger, seems a close analogue (see Nelson and Nilsen, 1974).

Conglomerates of the same general age as the Lago Sofia also occur in the Straits of Magellan area (especially Dawson Island) and on western Isla Grande (Figure 21.2*b*). Winn and Smith studied some of these and found them to have a very different clast composition characterized by a preponderance of intermediate to mafic volcanic rocks and a different paleocurrent pattern (Figures 21.2*b* and 21.4). Our conclusion is that those conglomerates represent a separate fan system with different source rocks.

In the western Magallanes Basin, Cerro Toro strata grade upward into the Tres Pasos Formation (2000 m thick) of Campanian and Maestrichtian age. The Tres Pasos has long been referred to as a molasse sequence, presumably because of prominent sandstones in the middle member (Katz, 1963). Scott's and our data demonstrate that the paleocurrent pattern for the lower shaly and middle sandy members is the same as that for the underlying flysch. We interpret these members as sub-wave-base deposits that, like the Lago Sofia, were related to a fan conduit system in which gravel deposition had become minor in this area; the sandstones of these two units are virtually identical. Bedding characteristics (e.g., abundant parallel stratification, scours, some graded beds, paucity of cross-bedding, and paucity of fauna) are more suggestive of rapid sediment gravity flows than shallow-marine tractive processes. These sandstones are very similar in overall aspect to deep-sea fan channel sands described elsewhere (e.g., Piper and Normark, 1971; Nelson and Nilsen, 1974).

Molasse deposits

The upper Tres Pasos Formation shows marked increases in the abundances of body fossils and of tractional structures (Smith, 1977). It is a transition to the truly shallow-marine, glauconitic sedimentary rocks of the overlying Dorotea Formation of Maestrichtian age. At this time, shoaling seems to have occurred only in Ultima Esperanza and to the north in Argentina, where dinosaur bone fragments and some red beds occur (Furque, 1973). Depths on the order of 1000 m still prevailed over much of the remainder of the basin according to Natland and others (1974). Those authors believe that the axis of thickest sedimentation began to migrate eastward at this time. While shelf, littoral, and deltaic deposits appeared earliest (Maestrichtian) in Ultima Esperanza, they did not appear until about Oligocene time farther south near the Straits of Magellan when extensive coaly deposits formed in deltaic facies (Charrier and Lahsen, 1969; Natland and others, 1974). In Ultima Esperanza, Paleogene strata are missing, apparently because of a subtle disconformity beneath coarse volcaniclastic Miocene sandstones and conglomerates containing resedimented molluscan shells apparently deposited on a fan delta. Clearly the molasse deposits are very diachronous over the Magallanes Basin.

At the northeast end of the Antarctic Peninsula, Upper Cretaceous and lower Cenozoic deposits include conglomerate, sandstone, and shale believed to be shallow marine, deltaic, and probably alluvial fan in origin (Howarth, 1966; Grikurov, 1972; Elliot and Trautman, Chapter 32, and Elliot and Wells, Chapter 26, both this volume). These seem to be closely comparable to the upper Tres Pasos, Dorotea, and early Cenozoic formations of Chile and equivalents in Argentina.

Late Cenozoic dismemberment

Magnetic anomaly data suggest that the western and central Scotia Sea are less than 30 m.y. old (Barker, 1972; Pitman and others, 1974; Barker and Burrell, 1977; LaBrecque and Herron, Chapter 48, this volume). Regional stratigraphic continuity suggests that the cordillera remained more or less intact, with subduction occurring from the Pacific side, at least until mid-Cenozoic time. Stratigraphic evidence also indicates a widespread mid-Cenozoic discontinuity in Patagonia (e.g., Hatcher, 1897; Winslow, Chapter 15, this volume) and in the northern Antarctic Peninsula (e.g., Grikurov, 1972). In addition to new patterns of sedimentation thereafter, widely scattered late Cenozoic basaltic volcanism, with little relationship to earlier tectonic patterns, also occurred in Patagonia and in the Antarctic Peninsula. Lastly, deep-sea drilling data suggest inception of the Antarctic Circumpolar Current in Oligocene or Miocene time (Craddock and Hollister, 1976), and marine magnetic anomalies are interpreted to indicate a sufficient opening of the Drake Passage by 23.5 m.y. ago for initiation of the Antarctic Circumpolar Current (Barker and Burrell, 1977).

Herron (1974) and Dott (1976) have argued that spreading on an eastern arm of a former triple junction that lay west of the present site of the Scotia Sea caused the initial dismemberment of the Andean Cordillera about 30 m.y. ago, which also was closely related in time to a collision of southern South America with the East Pacific Rise. Major transform faults related to that spreading arm resulted in the eastward displacement of South Georgia, the South Orkneys, and other portions of the cordillera (Figure 21.6) (Herron, 1974; Dott, 1976; Forsyth, 1975; Herron and others, 1977). Bruhn and Dalziel (1977) and Winslow (1976, and Chapter 15, this volume) have documented continuing left-lateral movements along prominent fault zones in Tierra del Fuego that can be related to major earthquakes associated with the northern Scotia Sea transform fault.

It has been widely assumed that both seismicity and subduction-related volcanism ceased in southern Chile after collision of the continent with the Chile Rise, but apparently the situation is not so simple. At least five active or dormant andesitic volcanoes exist in the cordillera south of the Chile Rise intersection with the continental margin at lat. 47° S (Stern and others, 1976), and we find andesitic detritus persisting at least into Miocene strata in southern Chile. All of this tends to underscore Forsyth's (1975) inference that plate boundaries and relative interactions are today very poorly defined in the Scotia Sea region.

Acknowledgments. The research for this paper has been made possible by grants from the National Science Foundation to R. H. Dott, Jr., dating back to 1961 (most recently grant DDP 72-05799). M. Turner and other foundation staff members always have been extremely helpful. We thank the British Antarctic Survey for making it possible for us to go to South Georgia in 1973. We also thank the crew of the National Science Foundation's R/V *Hero* for supporting a cruise in Chilean Tierra del Fuego in 1974. We greatly appreciate the permission of the governments of Argentina and Chile to work in their countries. Our gratitude to S. Harambour, R. Cortes, and C. Castro of Empresa Nacional del Petroleo, Punta Arenas, Chile, is expressed for their generous support of field work in Magallanes Province in 1976. The Empresa Nacional del Petroleo also provided us with several specimens used in our petrographic studies. Colleagues at Columbia University—I. W. D. Dalziel, R. L. Bruhn, M. J. DeWit, and M. A. Winslow—have provided valuable criticism, encouragement, and patience at all times.

References cited

Adie, R. J. 1972. Evolution of volcanism in the Antarctic Peninsula. (*In* Adie, R. J., *ed. Antarctic Geology and Geophysics.* Oslo, Universitetsforlaget, 137–42.)

Barker, P. F. 1972. Magnetic lineations in the Scotia Sea. (*In* Adie, R. J., *ed. Antarctic Geology and Geophysics.* Oslo, Universitetsforlaget, 17–26.)

Barker, P. F., and Burrell, J. 1977. The opening of the Drake Passage. *Mar. Geol., 25,* 15–34.

Bruhn, R. L., and Dalziel, I. W. D. 1977. Destruction of the Early Cretaceous marginal basin in the Andes of Tierra del Fuego (*In* Talwani, M., and Pitman, W. C., III, *eds. Maurice Ewing Series, 1: Island Arcs, Deep Sea Trenches, and Back-arc Basins.* Washington, American Geophysical Union, 395–405.)

Chamberlain, C. K. 1971*a*. Bathymetry and paleoecology of Ouachita geosyncline of southeastern Oklahoma as determined from trace fossils. *Bull. Am. Ass. Petrol. Geol., 55,* 34–50.

Chamberlain, C. K. 1971*b*. Morphology and ethology of trace fossils from the Ouachita Mountains, southeast Oklahoma. *J. Paleont. 45,* 212–46.

Charrier, R., and Lahsen, A. 1969. Stratigraphy of Late Cretaceous-Early Eocene, Seno Skyring-Strait of Magellan area. *Bull. Am. Ass. Petrol. Geol., 53,* 568–90.

Craddock, C., and Hollister, C. D. 1976. Geologic evolution of the Southeast Pacific Basin. (*In* Hollister, C. D., Craddock, C., and others. *Initial Reports of the Deep Sea Drilling Project,* Vol. 35. Washington [U. S. Government Printing Office], 723–43.)

Dalziel, I. W. D. 1974. Margins of the Scotia Sea. (*In* Burk, C. A., and Drake, C. L., *eds. The Geology of Continental Margins.* New York, Springer-Verlag, 567–80.)

Dalziel, I. W. D., and Elliot, D. H. 1973. The Scotia arc and Antarctic margin. (*In* Nairn, A. E. M., and Stehli, F. G., *eds. The Ocean Basins and Margins. 1. The South Atlantic.* New York, Plenum Press, 171–246.)

Dalziel, I. W. D., Caminos, R., Palmer, K. F., Nullo, F., and Casanova, R. 1974. South extremity of Andes: geology of Isla de los Estados, Argentine Tierra de Fuego. *Bull. Am. Ass. Petrol. Geol., 58,* 2502–12.

Dalziel, I. W. D., Dott, R. H., Jr., Winn, R. D., Jr., and Bruhn, R. L. 1975. Tectonic relations of South Georgia Island to the southernmost Andes. *Geol. Soc. Am. Bull., 86,* 1034–40.

de Wit, M. J. 1977. The evolution of the Scotia arc as a key to the reconstruction of southwestern Gondwanaland. *Tectonophysics, 37,* 53–81.

Dott, R. H., Jr. 1976. Contrasts in tectonic history along the eastern Pacific rim. (*In* Sutton, G. H., Manghnani, M. H., and Moberly, R., *eds. The Geophysics of the Pacific Ocean Basin and its Margin.* Washington, American Geophysics Union, 299–308.) [Geophysics Monograph No. 19.]

Dott, R. H., Jr., Winn, R. D., Jr., de Wit, M. J., and Bruhn, R. L. 1977. Tectonic and sedimentary significance of Cretaceous Tekenika beds of Tierra del Fuego. *Nature, Lond., 266,* No. 5603, 620–22.

Forsyth, D. W. 1975. Fault plane solutions and tectonics of the South Atlantic and Scotia Sea. *J. geophys. Res., 80,* 1429–43.

Furque, G. 1973. Descripcion geologica de la Hoja 58b, Lago Argentino. *Boln minist. Econ. Buenos Aires, 140,* 1–50.

Grikurov, G. E. 1972. Tectonics of the Antarctandes. (*In* Adie, R. J., *ed. Antarctic Geology and Geophysics.* Oslo, Universitetsforlaget, 163–67.)

Grikurov, G. E., Krylov, A. Ya, and Silin, Yu. I. 1966. Absoliutuni vozrast nekotorykh porod iz raiona zaliva Margerit, Antarkticheskii poluostrov [Absolute age of certain rocks in the Marguerite Bay region of the Antarctic Peninsula]. *Dokl. Akad. Nauk SSSR,* Geol., *171,* No. 6, 1399–1401. [English translation: *Dokl. (Proc.) Acad. Sci. U.S.S.R.,* Geol., *171,* 127–30.]

Halpern, M. 1972. Rb-Sr total-rock and mineral ages from the Marguerite Bay area, Kohler Range and Fosdick Mountains. (*In* Adie, R. J., *ed. Antarctic Geology and Geophysics.* Oslo, Universitetsforlaget, 197–204.)

Halpern, M. 1973. Regional geochronology of Chile south of 50° latitude. *Geol. Soc. Am. Bull., 84,* 2407–22.

Halpern, M., and Rex, D. C. 1972. Time of folding of the Yahgan Formation and age of the Tekenika Beds, southern Chile, South America. *Geol. Soc. Am. Bull., 83,* 1881–86.

Hatcher, J. B. 1897. On the geology of southern Patagonia. *Am J. Sci., 4* (Ser. 4), 327–54.

Herron, E. M. 1974. Magnetic lineations in the extreme southeast Pacific and their bearing upon the evolution of the Scotia arc and Patagonia. *Abstr. with Program, geol. Soc. Am.,* 791.

Herron, E. M., Bruhn, R., Winslow, M., and Chuaqui, L. 1977. Post Miocene tectonics of the margin of southern Chile. (*In* Talwani, M., and Pitman, W. C., III, *eds. Maurice Ewing Series, 1: Island Arcs, Deep Sea Trenches, and Back-arc Basins.* Washington, American Geophysical Union, 273–84.)

Hollingsworth, R. V. 1954. *Preliminary Report on Fusulinids from Patagonia.* Midland, Tex., Paleontological Laboratory.

Howarth, M. K. 1966. Ammonites from the Upper Cretaceous of the James Ross Island Group. *Bull. Br. antarc. Surv., 10,* 55–69.

Katz, H. R. 1963. Revision of Cretaceous stratigraphy in Patagonian Cordillera of Ultima Esperanza, Magallanes Province, Chile. *Bull. Am. Ass. Petrol. Geol., 47,* 506–24.

Laudon, T. S. 1972. Stratigraphy of eastern Ellsworth Land. (*In* Adie, R. J., *ed. Antarctic Geology and Geophysics.* Oslo, Universitetsforlaget, 215–24.)

Natland, M. L., Gonzalez, E., Canon, A., and Ernst, M. 1974. A system of stages for correlation of Magallanes basin sediments. *Mem. geol. Soc. Am., 139,* 1–126.

Nelson, C. H., and Nilsen, T. H. 1974. Depositional trends of modern and ancient deep-sea fans. (*In* Dott, R. H., Jr., and Shaver, R. H., *eds.* Modern and Ancient Geosynclinal Sedimentation. *Spec. Publs Soc. econ. Paleont. Miner., Tulsa,* No. 19, 69–91.)

Piper, D. J. W., and Normark, W. R. 1971. Re-examination of a Miocene deep-sea fan and fan-valley, southern California. *Geol. Soc. Am. Bull., 82,* 1823–30.

Pitman, W. C., III, Larson, R. L., and Herron, E. M. 1974. Age of the ocean basins determined from magnetic anomaly lineations. *Charts geol. Soc. Am.*

Scott, K. M. 1966. Sedimentology and dispersal pattern of a Cretaceous flysch sequence, Patagonian Andes, southern Chile. *Bull. Am. Ass. Petrol. Geol., 50,* 72–107.

Smith, C. H. L. 1977. *Sedimentology of the Late Cretaceous (Santonian-Maestrichtian) Tres Pasos Formation, Ultima Esperanza District, Southern Chile.* Master's thesis, University of Wisconsin-Madison, 129 pp.

Stepanicic, P. N., and Reig, O. 1955. Breve noticia sobre el hallazgo de anuros en el denominado Complejo Porfirica de la Patagonia extra-Andina con consideraciones acerca de la composicion geologica del mismo. *Revta Asoc. geol. argent.*, *10*, 215–33.

Stern, C., Skewes, M. A., and Duran, A. M. 1976. Volcanismo orogenico en Chile Austral. *Act. 1 Congr. Geol. Chil., Santiago*, F195–212.

Storey, C., Mair, B. F., and Bell, C. M. 1977. The occurrence of Mesozoic oceanic floor and ancient continental crust on South Georgia. *Geol. Mag.*, *114*, 203–8.

Suárez, M. 1976. Plate-tectonic model for southern Antarctic Peninsula and its relation to southern Andes. *Geology*, *4*, 211–14.

Suárez, M., and Pettigrew, T. H. 1976. An upper Mesozoic island-arc–back-arc system in the southern Andes and South Georgia. *Geol. Mag.*, *113*, 305–28.

Thomson, M. R. A. 1975. New palaeontological and lithological observations on the Legoupil Formation, north-west Antarctic Peninsula. *Bull. Br. antarc. Surv.*, No. 41–42, 169–85.

Valenzuela, E., and Hervé, F. 1972. Geology of Byers Peninsula, Livingston Island, South Shetland Islands. (*In* Adie, R. J., *ed. Antarctic Geology and Geophysics*. Oslo, Universitetsforlaget, 83–90.)

Winn, R. D., Jr. 1975. *Late Mesozoic Flysch of Tierra del Fuego and South Georgia Island: a Sedimentological Approach to Lithosphere Plate Restoration.* Ph.D. dissertation, University of Wisconsin-Madison, 160 pp.

Winn, R. D., Jr. 1978. Upper Mesozoic flysch of Tierra del Fuego and South Georgia Island: a sedimentological approach to lithosphere plate restoration. *Geol. Soc. Am. Bull.*, *89*, 533–47.

Winn, R. D., Jr., and Dott, R. H., Jr. 1977. Large-scale traction-produced structures in deep-water fan-channel conglomerates in southern Chile. *Geology*, *5*, 41–44.

Winn, R. D., Jr., and Dott, R. H., Jr. 1978. Submarine-fan turbidites and resedimented conglomerates in a Mesozoic arc-rear marginal basin in southern South America. (*In* Stanley, D. J., and Kelling, G., *eds. Symposium on Submarine Canyon, Fan and Trench Sedimentation.* Stroudsburg, Dowden, Hutchison, and Ross, 362–73.)

Winslow, M. A. 1976. Active transcurrent shear zones in southern Chile as landward expressions of transform plate boundaries. *Abstr. with Program, geol. Soc. Am.*, 1173–74.

22

The Origin of the Eastern South Scotia Ridge as an Intraoceanic Island Arc

Review Paper

P. F. BARKER, I. A. HILL*, and S. D. WEAVER**
Department of Geological Sciences, University of Birmingham, Birmingham, B15 2TT, U.K.

R. J. PANKHURST
British Antarctic Survey, Natural Environment Research Council, c/o Institute of Geological Sciences, London WE1X 8NG, U.K.

Abstract. A reconnaissance survey of the little-known eastern South Scotia Ridge was carried out aboard the RRS *Shackleton* in March 1976. The region shows considerable topographic relief, being occupied by several discrete crustal blocks with tops at 350 to 2200 m depth and steep sides reaching 5000 m in places. The blocks are strongly magnetized at shallow depth, and recent sediment cover is thin and discontinuous. Three dredge hauls from steep scarps on two of the blocks yielded a homogeneous collection of lavas and agglomerates. Chemical analyses show the lavas to be low-K tholeiites virtually indistinguishable from rocks of the active South Sandwich island arc, and K-Ar dates of 12–20 m.y. have been obtained from representative samples. We conclude that the blocks sampled, as well as other similar features in the region, were part of an intraoceanic island arc, the Discovery arc, similar to the South Sandwich arc but active before the present episode of back-arc extension, which began 7–8 m.y. ago. By analogy with present-day behavior of the Sandwich plate, earlier subduction was probably coupled to back-arc extension; such a spreading feature located in the central Scotia Sea would explain the anomalous crustal structure and complex magnetic character of that region. Subduction beneath this southern part of the Discovery arc may have stopped because of its failure to consume oceanic lithosphere of the Antarctic plate. Such a situation would arise after collision of a section of the Southwest Atlantic spreading ridge with the trench.

The Scotia Ridge is a series of islands and shallow submarine ridges linking South America with the Antarctic Peninsula in a long east-facing loop around the Scotia Sea. Except for the South Sandwich Islands, an active intraoceanic volcanic island arc, the geology of all the land areas of the Scotia Ridge is consistent with their being continental fragments, originally parts of a continuous, compact link between South America and the Antarctic Peninsula at the Pacific margin of Gondwanaland. It is generally accepted that the present length of the Scotia Ridge greatly exceeds the original length of the connection and that the crust beneath the Scotia Sea is not older Pacific ocean floor trapped within the loop, but was formed in the wake of the separating fragments. The exact shape of the original connection, the locations of known fragments within it, and the history of its disruption and dispersal, however, are still in dispute (Hawkes, 1962; Dalziel and Elliot, 1971, 1973; Barker and Griffiths, 1972, 1977; Dalziel, 1974; de Wit, 1977).

To improve reconstruction attempts, marine geophysics has been used to extend the known land geology to the shelf edge and to infer the structure of those fragments which are entirely submarine (see, e.g., Harrington and others, 1972; Simpson and Griffiths, Chapter 20, this volume). One part of the Scotia Ridge, however, because of its remoteness and almost permanent ice cover, has remained very poorly mapped. The region between the South Orkney Islands and the southernmost South Sandwich Islands was first crossed by James Weddell in 1823 in search of land (Weddell, 1825), but the existence of shallow submarine ridges there was unknown until the RRS *Discovery* made two sounding traverses in 1932 (Herdman, 1948). Very few soundings have been added since, and there is only one published geophysical profile through the region (*Eltanin 22*—Hayes and others, 1969).

In March 1976, the RRS *Shackleton* undertook at short notice a two-week reconnaissance of the area between lat. 59° and 62° S, and long. 34° and 39 ° W. Eight days of satellite-fixed survey (3900 km of magnetics, gravity, and bathymetry, including 330 km of reflection profile) preceded a dredging program which was curtailed after two days by bad weather.

The bathymetric and magnetic character, and the results of chemical analysis and K-Ar dating of dredged rocks described here, show that a great part of the eastern South Scotia Ridge originated as an intraoceanic island arc in early or middle Miocene time. This unexpected result has considerable regional significance.

Bathymetry and magnetics

From the distribution of satellite-fixed sounding tracks in Figure 22.1, it can be seen that, although there is little doubt about the reality of the major bathymetric features, contours in many places are speculative. Very few addi-

* Present address: Department of Geology, University of Leicester, Leicester, LE1 7RH, U.K.
** Present address: Department of Geology, University of Canterbury, Christchurch, New Zealand.

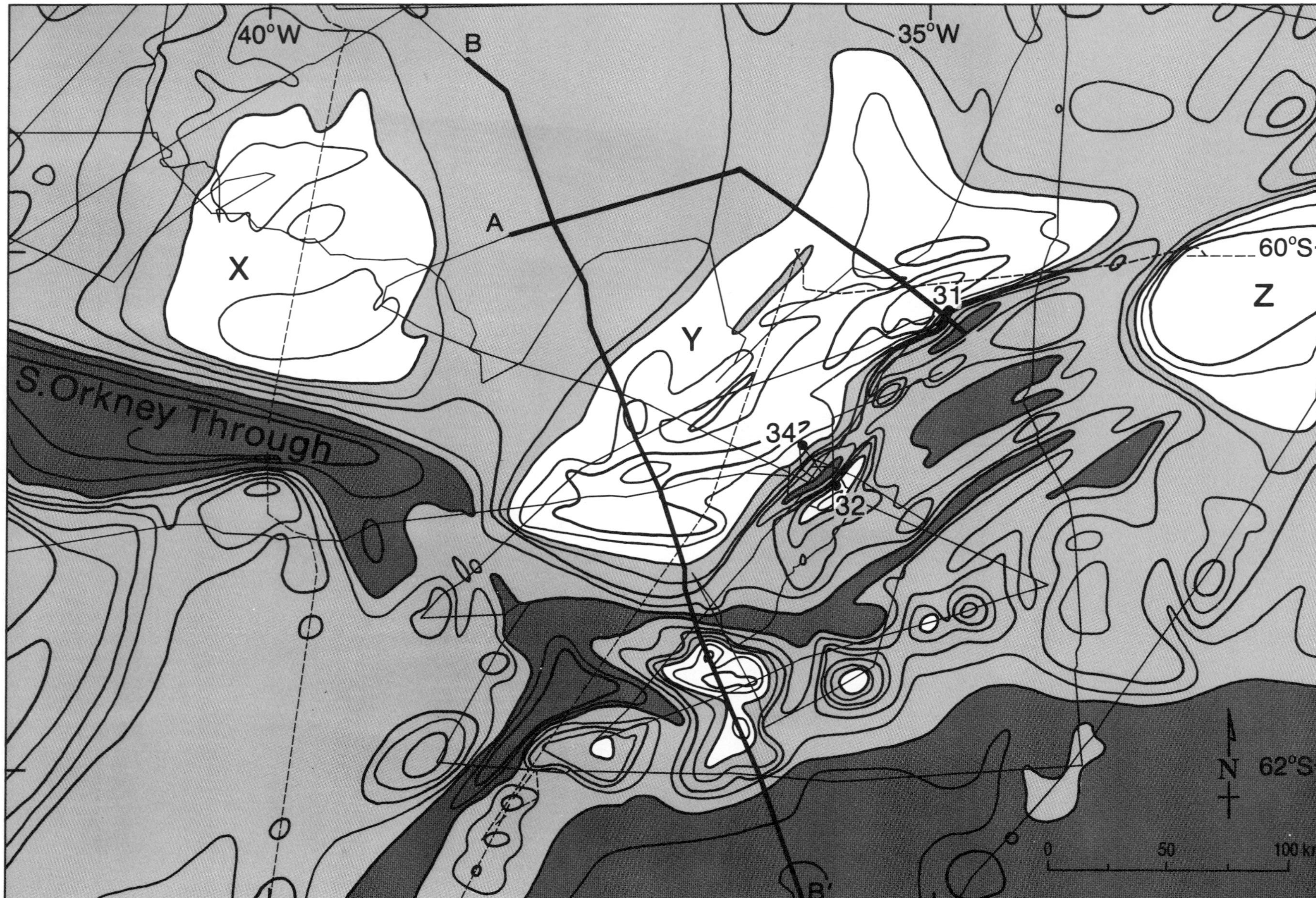

Figure 22.1. Bathymetric chart of eastern South Scotia Ridge. Contour interval is 500 m. Shaded areas: below 2000 m in depth; darker areas: below 4000 in depth. Thin solid lines: satellite-fixed ship tracks with soundings (and magnetics). Dashed line: *Eltanin 22* TRACK. AA′ and BB′: tracks of profiles shown in Figure 22.2. All soundings corrected using Matthews' (1939) tables. Dredge stations shown with numbers. X, Y, and Z: major topographic features described in text. For scale, 1° of latitude = 108 km.

tional, non-satellite-fixed data (from British Admiralty 1:1,000,000 master plotting sheets) were used, and in places large errors in their siting were apparent.

Essentially, the Scotia Ridge forms the boundary between the Scotia Sea to the north and the Weddell Sea to the south, and the active Antarctic-Scotia plate boundary runs approximately east-west through the region (Barker, 1970; Forsyth, 1975). The central Scotia Sea is shallow for an oceanic area and in its southern part contains many elevated crustal blocks, such as the three labeled X, Y, and Z in the upper (northern) part of Figure 22.1. These blocks have rough tops, at 350 to 2200 m depth, suggesting faulting (AA′ and BB′, Figure 22.2). Their margins with the deeper, smoother areas that lie between them and to the north are gentle slopes, although the northern margin of block Z is not well known and may be steeper.

The southern boundary of block X is the deep, steep-sided South Orkney trough, extending from long. 39° to 48° W below 4500 m in depth. Farther southward lies the South Orkney Islands continental fragment, the eastern end of which appears in Figure 22.1. To the east, an irregular area of extreme topographic relief intervenes between block Y to the north and the deep, sediment-covered Weddell Sea floor to the south. Within this area, 3000 m scarps

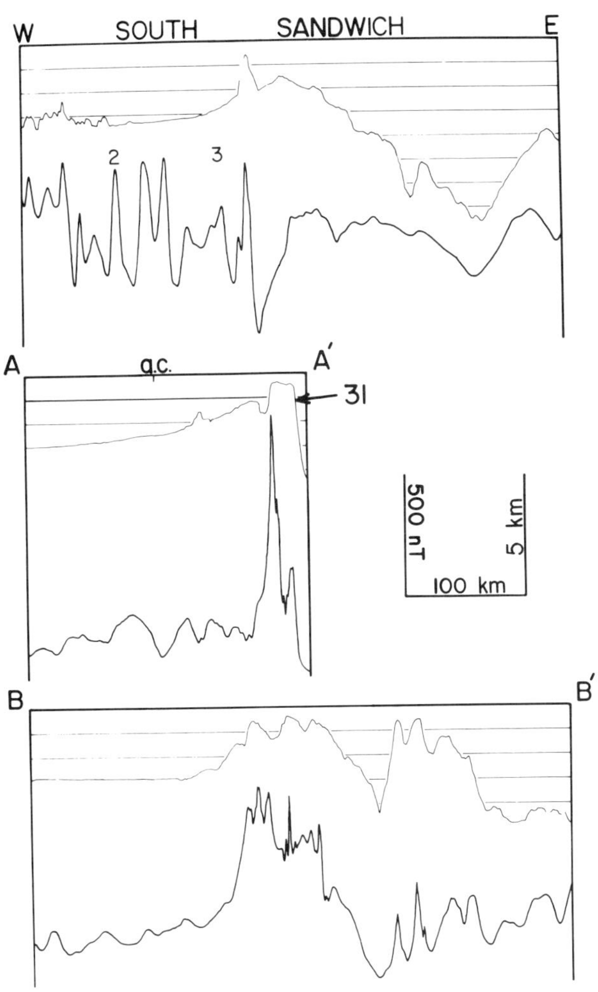

Figure 22.2. Bathymetric and magnetic profiles AA′ and BB′ across the eastern South Scotia Ridge (for location, see Figure 22.1). Comparative profile crosses Sandwich plate (Profile DD′ of Figure 22.4) (Barker 1972). Lamont magnetic anomalies are numbered. a.c.: altered course; see Figure 22.1.

are common, and steep-sided bodies of small extent occur on every track, making it likely that others remain undetected where tracks are sparse. Within this irregular area we can distinguish two provinces: (1) a series of shallow-topped, steep-sided bodies bordering the Weddell Sea, which trend at about 070° and become more subdued eastward; and (2) an area of near-oceanic depths containing fewer and smaller elevated features, which separates the first province from the spectacularly steep southern margin (AA′, Figure 22.2) of block Y. This margin trends 050° overall, but we recognize a major trend of 070° offset repeatedly over small distances to produce the overall strike. With rather less justification, the "oceanic" province to the south has been contoured to show the same 070° trend.

Profiles AA′ and BB′ of Figure 22.2 cross these major provinces (described above) nearly perpendicularly to the strike. As they show, the deeper parts of the Scotia Sea and the northern flanks of the elevated blocks possess only subdued magnetic anomalies, less than 200 nT in amplitude. The rougher, more southerly parts of the blocks, however, and the smaller steep-sided features farther south are strongly magnetized, at least in their upper parts.

Little seismic reflection profiling was attempted during the reconnaissance and no profiles are displayed here. As much as 1 second (~1 km thickness) of sediment occurs in the deeper parts of the Scotia Sea, and thick sediments are presumed to underlie the smooth Weddell Sea floor. Elsewhere, however, sediment cover is apparently thin or absent.

Dredged rocks

Dredge stations

The reconnaissance established that with abundant steep scarps and with sediment thin or absent, the region was ideally suited to further investigation by dredging. Fewer of the elevated blocks than intended were sampled because of bad weather; many specimens of hard rock were recovered at three out of six dredge stations in two days. Two of these (specimens D31 and D34, Figure 22.1) lay on the steep southern margin of block Y about 85 km apart, and the third (D32) lay 20 km southeast of D34 on the equally steep northwestern margin of a smaller block (minimum depth, 1283 m) separated from block Y by depths of at least 3000 m.

Dredge D31

This dredge (lat. 60° 16′ S, long. 34° 52′ W) sampled the bottom between 2200 and 1800 m depth over a 1.5 km horizontal distance; seventy-one discrete specimens of cobble size and greater, weighing 110 kg, were recovered. Forty-one specimens are coarse agglomerate containing large (150 mm) clasts of altered, fine-grained basalt with plagioclase, pyroxene, and olivine phenocrysts and basaltic glass, in a sand-sized matrix of the same components. Eighteen specimens are similar to the agglomerate but much finer grained, and eleven specimens are of the porphyritic basalt found in the agglomerate. No nonvolcanogenic detritus is present, and only one specimen—a fresh, rounded, aphyric basalt—is considered probably ice rafted.

Table 22.1. Sixteen analyses representative of the rock types of the southeast Scotia Ridge dredge hauls, with six comparative analyses of island-arc tholeiites (three each from the South Sandwich and Tonga Islands)

	D31.19	D31.20	D31.22	D32.01	D32.21	D32.22	D32.52	D32.78	D34.13	D34.21	D34.34	D34.36	D34.46	D34.47	D34.49	D34.52	SSS10.1	SSR1.3	SSM4.1	E7	E15	L20
SiO_2	47.60	47.80	50.30	48.30	49.30	48.80	50.30	48.50	50.70	49.30	49.40	49.90	49.60	51.50	50.20	51.00	49.94	50.22	52.53	49.19	51.40	53.45
TiO_2	0.62	0.59	0.44	0.71	0.83	0.91	0.50	0.73	0.56	0.84	0.59	0.58	0.50	0.59	0.59	0.54	0.80	0.56	0.70	0.42	0.96	0.51
Al_2O_3	14.82	14.89	16.34	15.83	15.91	14.46	21.64	16.48	16.87	15.67	19.64	19.71	20.80	18.52	20.31	20.04	17.53	19.59	19.29	20.61	16.76	15.86
tFe_2O_3 (total Fe)	9.58	9.59	9.27	10.03	9.96	11.50	8.19	10.27	10.93	10.22	9.81	9.67	9.29	10.12	9.98	9.46	10.45	10.60	9.92	8.85	11.22	10.19
MnO	0.21	0.24	0.10	0.22	0.16	0.24	0.14	0.20	0.19	0.16	0.16	0.16	0.15	0.18	0.17	0.17	0.17	0.17	0.18	0.18	0.34	0.19
MgO	9.52	8.95	9.52	9.07	8.50	12.01	3.79	9.32	6.17	8.39	4.70	4.57	4.85	5.00	3.77	4.39	6.93	5.60	3.88	5.93	5.46	6.06
CaO	12.93	13.86	11.37	4.00	11.50	2.68	12.75	5.81	11.51	11.46	13.09	12.49	12.52	11.39	12.24	11.85	11.52	12.53	10.88	11.78	9.32	11.38
Na_2O	2.39	2.23	2.23	5.05	2.49	1.85	2.08	4.73	2.26	2.50	2.30	2.16	1.84	2.40	2.12	2.18	2.12	1.76	2.02	1.21	2.61	1.65
K_2O	0.23	0.22	0.18	1.50	0.18	2.53	0.23	0.43	0.33	0.28	0.20	0.26	0.29	0.31	0.27	0.31	0.41	0.15	0.24	0.24	0.21	0.38
P_2O_5	0.16	0.17	0.09	0.08	0.12	0.13	0.05	0.07	0.07	0.12	0.10	0.06	0.06	0.08	0.05	0.11	0.07	0.05	0.00	0.04	0.11	0.08
Total	98.06	98.54	99.84	94.79	98.95	95.11	99.67	96.54	99.59	98.94	99.99	99.56	99.90	100.09	99.70	100.05	99.04	101.23	99.64	98.45	98.46	99.75
									Trace elements (ppm)													
Ba	34	30	44	73	34	99	24	23	46	29	26	34	32	46	27	30	59	46	44	14	44	96
Ce	8	10	9	5	15	15	9	12	9	8	7	6	11	10	10	10	0	0	0	2	0	0
Cr	79	74	67	34	80	50	36	78	24	10	13	8	18	13	10	18	220	65	30	75	76	78
Cu	88	190	100	140	194	157	104	110	82	191	98	145	158	135	136	127	140	75	210	51	12	150
Ga	15	14	15	16	17	19	21	20	17	17	21	19	19	17	18	19	12	13	13	13	20	15
La	6	4	5	3	5	7	8	4	1	5	1	3	6	1	2	2	0	0	0	1	0	0
Nb	2	1	1	1	1	1	3	2	2	2	1	1	1	3	1	1	2	2	1	1	0	0
Ni	11	11	11	9	13	12	11	14	10	3	1	3	3	4	1	4	80	11	5	25	41	26
Rb	2	2	1	16	1	20	3	5	4	4	2	3	5	5	4	4	10	1	5	1	1	6
Sr	178	174	162	76	215	100	214	197	253	216	239	233	221	229	229	205	132	143	138	115	120	215
Y	11	12	9	16	16	17	11	17	12	17	12	12	12	13	10	15	19	11	17	12	27	15
Zr	33	29	24	40	59	55	45	59	30	29	25	29	29	31	27	31	47	23	34	20	47	25
Rock type	2	2	2	2	2	2	1	2	1-2	2	1	1	1	1	1	1	SS	SS	SS	T	T	T
% plagioclase phenocrysts	20	0	25	35	30	0	0	20	25	25	40	40	45	35	40	55	40	45	25	40	11	25

References: South Sandwich—Pearce (1973) and Baker (1978); Tonga—Ewart and Bryan (1972) and Ewart and others (1973).

Dredge D32

This dredge (lat. 60° 56′ S, long. 35° 46′ W) sampled between 2600 and 2000 m depth over 2.9 km horizontally and recovered 119 specimens weighing 330 kg. Sixty-two specimens are basaltic lavas similar to those of D31 but with a greater variability, and thirty-five are agglomeratic, also like those from D31 and including eight finer grained specimens. All these rocks show alteration, and zeolite-filled vesicles are common. Eleven specimens are pale siliceous biogenic ooze or chert, and eleven others are exotic lithologies (including quartzite, pegmatite, diorite, and aphyric basalt), all rounded and striated and probably ice rafted.

Dredge D34

This dredge (lat. 60° 56′ S, long. 35° 46′ W) sampled between 1700 and 1500 m depth over only 0.8 km horizontally and recovered 180 kg of rock, including 84 specimens of basalt and 46 of agglomerate. Both lithologies are similar in hand specimen to those from D31 and D32, but less altered.

No basalt from any dredge haul shows any features suggesting submarine extrusion, in either hand specimen or thin section.

PETROLOGY AND CHEMISTRY OF DREDGED ROCKS

Sixteen of the least altered of the obviously *in situ* lavas from dredges D31, D32, and D34 were selected for chemical analysis. Major and trace elements were determined on powder pellets by standard analytical procedures using a Philips PW 1450 x-ray fluorescence spectrometer. These analyses, together with six others for comparison, are presented in Table 22.1.

On the basis of petrography and chemistry, two rock types have been recognized among the analyzed samples. Basalts in D31 are all of type 2, and those in D32 and D34 are of both types 1 and 2.

Type 1 basalts

These rocks are distinctive in containing 35–55% by volume of calcic plagioclase phenocrysts. These are strongly zoned and occur as discrete phenocrysts as much as 6 mm long and in glomerophyric aggregates as much as 20 mm in diameter. Phenocrysts of olivine make up 1–10% and clinopyroxene 1–4%. Olivines are invariably pseudomorphed by serpentine or red-brown "iddingsite." Olivine-rich varieties (specimen D34.49) contain no orthopyroxene; olivine-poor rocks contain as much as 5% c.En_{70} orthopyroxene as small phenocrysts and as cores in clinopyroxene crystals. Specimen D34.52 contains apparently cognate plagioclase-orthopyroxene-magnetite glomerophyric aggregates.

Groundmasses are fine grained with intergranular or intersertal textures. Excepting the olivine phenocrysts, these rocks show little alteration.

Type 1 basalts have distinctively high CaO and Al_2O_3 contents (Table 22.1), consistent with a high proportion of modal plagioclase. MgO content ranges from 5.0 to 3.8 wt %, K_2O content is low (0.33%), and Na_2O/K_20 ratios are high (6.3–11.5). Abundances of incompatible trace elements (Zr, Nb, Rb, Ba, and REE) are very low.

Type 2 basalts

These rocks differ from type 1 in having a greater proportion of mafic minerals and less modal plagioclase. Feldspar phenocrysts make up 15–30% by volume, olivine 2–15%, and clinopyroxene 1–15%; no orthopyroxene was seen.

Type 2 basalts are generally more altered than type 1. Olivine is completely pseudomorphed; plagioclase may be fresh or replaced by sericite and carbonate along cracks. Ground masses are oxidized and most specimens contain irregular amygdules of serpentine, chlorite, and carbonate minerals.

Chemically, type 2 basalt is lower in Al_2O_3 and higher in MgO (8.4–9.5 wt %). However, the similarities in Na_2O/K_2O ratio, and in contents of K_2O and incompatible trace elements, suggest that types 1 and 2 are petrogenetically related. Moreover, specimen D34.13 appears petrographically and chemically intermediate between types 1 and 2.

Three specimens (D32.01, D32.22, and D32.78) in which plagioclase is extensively zeolitized are severely altered varieties of type 2. The low analytical totals (Table 22.1) suggest that these rocks contain 4–6% H_2O and CO_2; moreover, they are richer in alkalies (Na_2O, K_2O, and Rb) and have less Ca and Sr than type 2, features consistent with sea-water interaction at moderately high temperatures.

K-AR DATING

Six of the freshest lava specimens were selected for K-Ar whole-rock dating: one each of types 1 and 2 from D32, and three type 1 specimens and one type 2 from D34. Most were prepared by separating a 30–80 mesh (A.S.T.M.) size fraction, which tends to contain a lower proportion of altered material than either coarse fragments or fine dust. Two samples were analyzed as genuine "whole-rocks" by taking 1–4 mm chips for argon extraction and grinding a subsample for potassium determination. Potassium was determined in duplicate or triplicate by flame photometry using a Li internal standard with a reproducibility of about 2%. Argon was extracted by vacuum fusion of 5–10 g of material and analyzed by on-line mass-spectrometric isotope dilution with a ^{38}Ar-enriched spike (reproducibility about 1%).

Ages for untreated samples (F or C only in Table 22.2) ranged from 12 to 34 m.y. High atmospheric argon in all

Table 22.2. K-Ar analytical results

D32.52	C	0.468	0.2355	2.6	12.9 ± 6.7
	F/A	0.470	—	0.0	—
D32.78	F	0.110	0.0522	8.9	12.1 ± 1.8
	F/A	0.081	—	0.0	—
D34.21	F	0.139	0.1875	12.1	34.4 ± 3.7
	F/A	0.075	0.0579	11.2	19.7 ± 4.5
D34.34	F	0.205	0.1210	14.9	15.1 ± 1.6
D34.36	C	0.233	0.1132	14.9	12.4 ± 1.1
	F/B	0.209	0.0786	16.6	9.7 ± 2.2
	F/A	0.174	0.1305	13.1	19.2 ± 2.2
D34.52	F	0.248	0.1665	24.8	17.2 ± 1.4
	F/B	0.248	0.1497	36.3	15.5 ± 1.2
	F/A	0.213	0.1364	20.1	16.0 ± 1.3

C: coarse fraction for Ar (1–4 mm chips), reground for K. F: fine fraction for both Ar and K (30–80 A.S.T.M.). A: acid washed. B: prebaked to 300°C.

Constants: $\lambda_\beta = 4.962 \times 10 yr^{-10}$; $\lambda_e = 0.581 \times 10^{-10} yr^{-1}$.

cases produced large analytical errors, but the observed range of ages nevertheless exceeded calculated two-standard-deviation error limits. Two procedures were adopted to test whether this could represent a real range of ages of eruption. First, two samples were prebaked overnight at 300°C (as opposed to the normal 180°C). Atmospheric argon was only slightly reduced, indicating that much of it lies within primary igneous minerals rather than in zeolites or on grain surfaces. Moreover, radiogenic argon was lost, particularly by D34.36, resulting in erroneously young ages. Secondly, the fine fractions of five specimens were washed thoroughly in hot, dilute HCl to remove carbonates, then reanalyzed for K and Ar. The two most altered samples, from D32, lost all radiogenic Ar. The sample from D34.52 lost only small amounts of K and radiogenic Ar, and the age was essentially unchanged. The computed ages of the other two samples from D34, which lost greater amounts, changed considerably but converged from either direction upon 19 m.y. This indicates that the age of 34 m.y. previously obtained from specimen D34.21 is not meaningful, possibly reflecting excess ^{40}Ar in carbonate or zeolite, and casts doubt upon the age of 12 m.y. obtained initially from D32.52, D32.78, and D34.36, which may reflect excess K in secondary minerals. The consistent age of 16 m.y. from D34.52 seems the most reliable, and the age of 19 m.y. from D34.21 and D34.36 could be close to eruption ages, although some K may have been leached from ^{40}Ar-retentive phases during treatment. Given the field disposition of the samples, a spread of true ages over a few million years seems likely. We conclude that eruption ages lie between 12 and 20 m.y., but may all be closer to 16 m.y.

Rare earth element and Sr-isotope data

In view of the possible effects of alteration on trace-element chemistry, care was taken to select the freshest material for mass-spectrometric analysis. Fresh clinopyroxene separated from specimen D34.13B was analyzed for 10 rare earth elements (Table 22.3). Sr-isotopic composition was determined for this clinopyroxene sample and for two whole-rock samples (D34.52, acid washed, and D34.36). $^{87}Sr/^{86}Sr$ ratios (relative to 0.70800 for Eimer and Amend $SrCO_3$) are as follows:

D34.52 WR	0.70358 ± 7
D34.36 WR	0.70362 ± 8
D34.13B cpx	0.70368 ± 7

Interpretation and discussion

Geochemical significance of rock types

The fresh, unzeolitized rocks of types 1 and 2 have low contents of K_2O, TiO_2, Zr, Nb, REE, Y, Ba, Cr, and Ni, high Na_2O/K_2O ratios and low Ce/Y ratios (implying unfractionated rare-earth patterns). Their mineralogy and chemistry are distinctly tholeiitic, and whereas the low K_2O contents are like those of ocean floor basalts, the TiO_2, Zr, Cr, and Ni contents are significantly lower in the dredged rocks. These characteristics are those of the dominantly basaltic rocks known as low-K tholeiites or the "island-arc tholeiite series" (Jakeš and Gill, 1970; Jakeš and White, 1972), which are thought to represent the earliest stages in

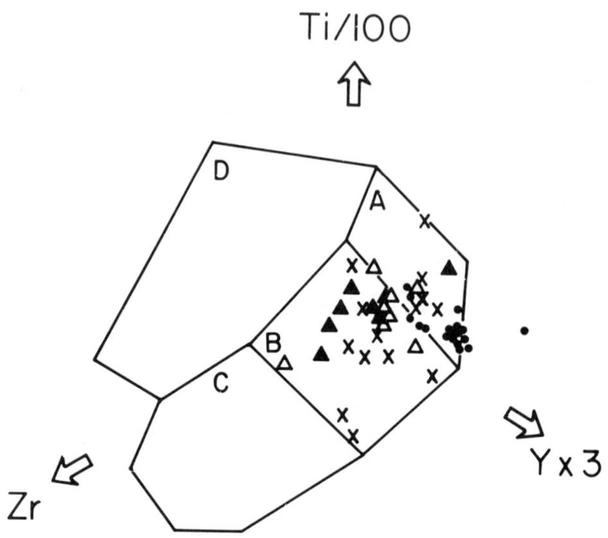

Figure 22.3. Ti-Zr-Y triangular discrimination diagram (after Pearce and Cann, 1973). Open triangles: southeast Scotia Ridge dredged rocks, type 1. Solid triangles: southeast Scotia Ridge dredged rocks, type 2. Dots: basaltic lavas of the island-arc tholeiite series from the Tonga Islands. Crosses: basaltic lavas of the island-arc tholeiite series from the South Sandwich Islands. Low-K tholeiites plot in fields A and B, ocean-floor basalts in field B, calc-alkaline basalts in fields B and C, and intraplate basalts in field D. Triangular boundary omitted from diagram.

island-arc evolution (Baker 1968; Ringwood, 1975). Localities where these rocks are known to occur include the Mariana, Tonga, Izu, and South Sandwich island arcs.

The Ti-Zr-Y discrimination diagram (Figure 22.3) of Pearce and Cann (1973) displays conveniently the chemical affinities of the dredged rocks from the South Scotia Ridge. Types 1 and 2 plot within fields A and B, together with basalts from the Tonga and South Sandwich arcs. Island-arc tholeiites plot throughout both fields A and B, but ocean floor basalts are restricted to field B.

The six analyses of basalts from Tonga and the South Sandwich Islands (Table 22.1) show a striking similarity to analyses of the dredged rocks. Only by their slightly higher Sr contents (a difference outside the probable range of interlaboratory variation) can the dredged rocks be distinguished from South Sandwich lavas, but even this difference is not thought to reflect a significantly different tectonic environment.

On the basis of published rare-earth element distribution coefficients (Grutzeck and others, 1974), the clinopyroxene data given in Table 22.3 can be used to reconstruct concentrations in the parent magma with which it would have

Table 22.3. Rare earth element concentrations (in ppm) in specimen D34.13 clinopyroxene

Rare earth element	Concentration (ppm)
La	0.12
Ce	0.72
Nd	1.47
Sm	0.80
Eu	0.35
Gd	1.44
Dy	1.88
Er	1.20
Yb	1.04
Lu	0.13

been in equilibrium. Expressed as chondrite-normalized abundances, these data exhibit a flat heavy-REE pattern with significant light-REE depletion (Ce/Yb ≈ 0.5), characteristic of the most primitive ocean-ridge basalts and low-K island-arc tholeiites. In the eastern Scotia Sea region, light-REE depletion is found in rocks from the South Sandwich Island arc, but not in abyssal tholeiites from the South Sandwich spreading center (see Pankhurst, Chapter 25, this volume, Figure 25.2).

The $^{87}Sr/^{86}Sr$ ratios of the present samples are even more diagnostic of an island-arc environment. The values reported above are identical within analytical error, implying no significant sea-water alteration. Their average, 0.70363, is well above the range for normal mid-ocean ridge basalts (<0.7030) and the more extended range for dredged basalts from the South Sandwich spreading center (Saunders and others, Chapter 23, this volume). It is only slightly below the minimum value reported from the South Sandwich Islands and is well within the range for other island-arc tholeiites (see Pankhurst, Chapter 25, this volume, Figure 25.1).

In summary, the geochemistry of the dredged rocks indicates that they represent an intraoceanic island arc, in a younger stage and perhaps chemically more primitive than the present South Sandwich arc.

Tectonic setting of dredge sites

The chemical homogeneity of lavas from all three sites suggests that they are typical of the features sampled and perhaps of other steep-sided elevated blocks within the survey area (Figure 22.1). Furthermore, the magnetic character of all the blocks is consistent with the suggested composition (which, however, is not a truly diagnostic criterion, since magnetic material of other compositions and ages occurs elsewhere along the Scotia Ridge).

Figure 22.2 compares profiles AA′ and BB′ with a west-east crossing of the Sandwich plate (Barker, 1972). By analogy we suggest that the deeper area of the Scotia Sea between blocks X and Y was an extensional back-arc basin northwest of the sampled island arc of block Y. The basin's subdued magnetic character is in keeping with this hypothesis; it is really the South Sandwich spreading center which is unusual, displaying the best-formed magnetic lineations yet identified in a back-arc basin. It follows that the island arc we have sampled faced either eastward like the South Sandwich arc or perhaps southeastward. Magnetic anomaly orientations in the inferred back-arc basin are unknown, but would not necessarily indicate exactly the direction of subduction. As shown by section AA′ and by Figure 22.1, the arc-trench gap of our proposed island-arc system is no longer in place. Later dissection, approximately along the line of the islands, created the steep southern scarp of block Y, exposing the deeper part of the lava pile. Other fragments produced during this dissection now lie scattered to the south and east of block Y (including that sampled by D32) and presumably represent the arc-trench gap. There is no trace of the trench itself and no deep earthquake activity which is not obviously associated with South Sandwich trench subduction (Forsyth, 1975; Brett, 1977). To aid further discussion, we have given this newly defined island arc the name "Discovery arc," after the source of the earliest soundings.

Implications for the central Scotia Sea

Although hitherto there has been no direct evidence of South Sandwich-like subduction before 8 m.y. ago, Barker and Griffiths (1972) have speculated that some such mechanism is required to explain central Scotia Sea magnetic anomaly orientations, which were different from those in Drake Passage and in the southwest Atlantic, and thus do not fit the simplest model of South American-Antarctic plate separation. Also de Wit (1977) noted the similarity of central Scotia Sea crustal structure (Allen, 1966; Ewing and others, 1971) to that of back-arc basins elsewhere. The opening of Drake Passage is now understood (Barker and Burrell, 1977), and the possibility remains that, for some time after opening started (29 m.y. ago), it was very simply related to the southwest Atlantic opening. Central Scotia Sea anomalies remain undated, however, and suggested ages range from within the last 40 m.y. (Barker, 1970) to earliest Cretaceous (de Wit, 1977). Detection of early to middle Miocene (20 to 12 m.y. ago) intraoceanic island-arc volcanism in the eastern South Scotia Ridge therefore argues strongly in favor of a similar age for the central Scotia Sea, although subduction (and therefore back-arc extension) could have started before 20 m.y. ago and continued after 12 m.y. ago.

Implications for reconstructions

Several elevated blocks similar to block Y occur in the southern part of the central Scotia Sea, as far west as long. 51° W. Until now, an origin as foundered continental fragments similar to the South Georgia or Shag Rocks blocks seemed likely, and as such, they were included in some reconstructions of the original South America-Antarctic Peninsula continental connection (see, e.g., Barker and Griffiths, 1972; de Wit, 1977). However, if these blocks are fragments of a young intraoceanic island arc isolated by successive back-arc extensional episodes, a negligible fraction of their present volume will be of continental origin, and they can be neglected in a reconstruction. More extensive dredging might resolve this problem.

Relationship to the South Sandwich arc

The Discovery arc was apparently very similar to the present South Sandwich arc in orientation, chemistry, and probably therefore in function but, at least where sampled, is no longer active. Before the current 7–8 m.y. period of rapid back-arc extension began, the South Sandwich arc would have lain directly north of block Y. However, the present South Sandwich Islands appear to lie on oceanic crust created *during* the last 7–8 m.y. (Barker, 1972), so the relationship between the arcs is not simple.

One intriguing aspect of Sandwich plate behavior is its apparent independence of major plate motions. In the southwest Atlantic Ocean, the slow (approximately 17 mm/year along 280°—Forsyth, 1975) separation of the South American and Antarctic plates occurs at a spreading center with long transform offsets, the most southerly of which meets the southern end of the South Sandwich trench (Figure 22.4*a*). This interplate motion then divides in unknown proportions between the North and South Scotia Ridge, thereby defining the Scotia plate (Forsyth, 1975). The Sandwich plate is bounded to the east and west by the South Sandwich Trench and the South Sandwich

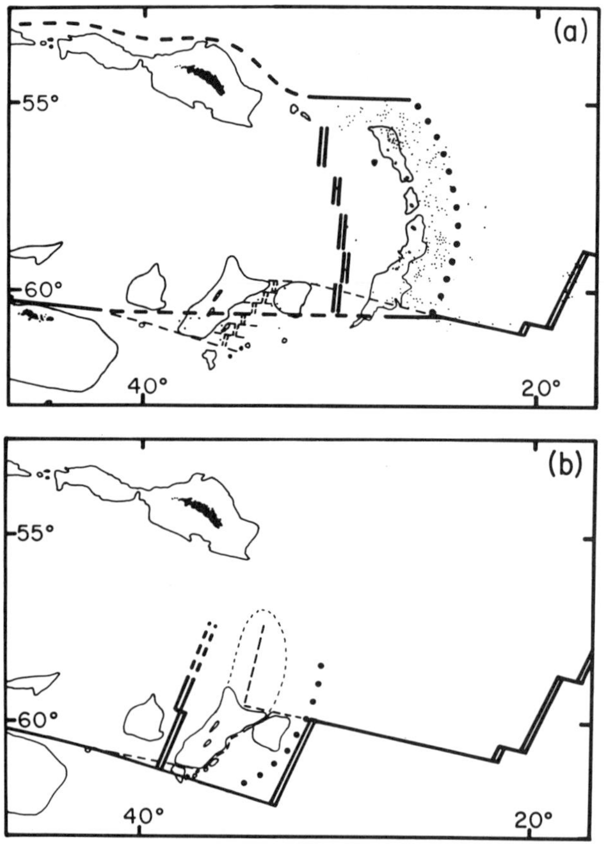

Figure 22.4. Plate motions and boundaries in the Scotia Ridge area. Heavy lines: plate boundaries (dashed where uncertain). Solid contours: present 2000-m-deep contours. Large dots: trench axes. *a.* Present-day, with small dots showing earthquake epicenters (after Brett, 1977). Thin dashed line: hypothetical early boundary (6 m.y. ago?). *b.* Postulated, immediately before 8 m.y. ago, showing Discovery arc and trench (northward extent uncertain) and more southerly section of the southwest Atlantic spreading center. Thin dashed line: suggested new boundary after 7 m.y. ago, with dissection of the Discovery island arc and formation of South Sandwich back-arc spreading.

back-arc spreading center, respectively, and its northern and southern margins are transform faults linking these two features. Its motion is much more vigorous, with fast back-arc extension (70–90 mm/year—Barker, 1972) almost exactly complementing fast consumption at the trench. Possibly the only influence being exerted presently by the major plates is in defining the southern edge of the Sandwich plate. All available oceanic crust of the South American plate is being consumed at the trench, but none from the Antarctic plate.

It is extremely unlikely that subduction and back-arc extension could have started together 7–8 m.y. ago under these conditions of almost complete independence from major plate motion. A more reasonable hypothesis is that subduction of the South American plate has been continuous from perhaps 20 m.y. ago to the present. Since this process embodies its own major driving force, in the excess weight of the cold, sinking slab, it probably persisted through whatever event transformed the originally active Discovery arc into the South Sandwich arc. If subduction continued, we may expect that part of the Discovery trench and arc-trench gap also survived, to be incorporated into the South Sandwich system, but we cannot say at present how much may have survived, nor where it now lies.

We outline here one possibility for the transforming event. As suggested by magnetic lineations off Baja California (Chase and others, 1970; Atwater and Menard, 1970) and off the Antarctic Peninsula and southern Chile (Herron and Tucholke, 1976), subduction can stop when a spreading center arrives at a trench if the newly opposed plate is not already being consumed elsewhere along the trench. Figure 22.4*a* makes clear that if the present rapid eastward motion of the Sandwich plate continues, in about 5 m.y. it will overtake the most southerly ridge section of the southwest Atlantic spreading center, the eastern side of which is on the Antarctic plate. We suggest (Figure 22.4*a*) that the southern end of the Discovery arc and trench lay originally along a more southerly transform of the southwest Atlantic spreading center, its trace joining the eastern end of the South Orkney trough. Back-arc extension within the central Scotia Sea allowed the trench to overtake the more slowly migrating ridge crest, whereupon subduction stopped for the southern part of the Discovery system. Back-arc extension would have stopped also, but then would probably have been renewed at a more easterly site behind the remainder of the arc under the influence of continued subduction of the South American plate.

We suggest that this event took place shortly before the onset 7–8 m.y. ago of the present back-arc extensional episode. At the same time a new path would have been required for the boundary between the Antarctic and Scotia plates to accommodate the northward jump of the southern end of the active trench. Interestingly, this path did not follow the trench or the older back-arc extensional axis. Instead it chose a line (dashed in Figure 22.4*b*) close to the islands of the by then defunct Discovery arc, which is the line taken in other circumstances by *new* back-arc extension (Karig, 1974). The relative motion of the Scotia and Antarctic plates, if similar to the present left-lateral motion, would have generated extension at such an offset, causing dissection of the island arc and the dispersal of the fragments over short distances.

Figure 22.4*a* shows schematically (thin dashed line) the likely extent of the zone of dissection. The earthquake distribution indicates that some tectonic dissection still takes place, but not over the entire zone, suggesting that the plate boundary has migrated over the last 7–8 m.y., with perhaps a tendency toward straightening and simplification (heavy dashed line).

This model predicts that the sea-floor ages southeast of the survey area should increase from about 8 m.y., in an eastward direction. By analogy with the Antarctic Peninsula, however (Herron and Tucholke, 1976), neither residual gravity anomalies nor subsediment topography are likely to survive to mark the dead trench.

Although Figure 22.4 contains a plausible model for conditions immediately before subduction stopped at the Discovery trench, it is not certain for how long before 9 m.y. ago those conditions had existed. In particular, an even *more* southerly section of the southwest Atlantic ridge crest could, even earlier, have choked a section of trench along the eastern side of the South Orkney block. This suggestion is highly speculative but would imply that the Powell Basin, between the block and the Graham Land shelf, formed by back-arc extension and that the enigmatic South Orkney magnetic zone (Harrington and others, 1972) could be the eroded remnant of a magmatic arc.

Acknowledgments. The authors are grateful to the ship's company of the RRS *Shackleton* for invaluable help with data collection, often in trying conditions. Geophysical data reduction and geochemical analysis were carried out in the Geology Department at Birmingham; radiometric, rare earth, and Sr-isotope measurements were made using facilities of the Institute of Geological Sciences Geochemical Division. J. Burrell and P. Simpson assisted with data reduction and took part in useful discussion. The work was financed by a Natural Environment Research Council research contract.

References cited

Allen, A. 1966. Seismic refraction investigations in the Scotia Sea. *Scient. Rep. Br. antarc. Serv.*, No. 55, 44 pp.

Atwater, T., and Menard, H. W. 1970. Magnetic lineations in the northeast Pacific. *Earth planet. Sci. Lett.*, *7*, 445–50.

Baker, P. E. 1968. Comparative volcanology and petrology of the Atlantic island arcs. *Bull. volcan.*, Ser. 2, *32*, No. 1, 189–206.

Baker, P. E. 1978. The South Sandwich Islands: II. Petrology and geochemistry. *Scient. Rep. Br. antarc. Surv.*, *93*, 1–34.

Barker, P. F. 1970. Plate tectonics of the Scotia Sea region. *Nature, Lond.*, *228*, 1293–97.

Barker, P. F. 1972. A spreading centre in the east Scotia Sea. *Earth planet. Sci. Lett.*, *15*, No. 4, 123–32.

Barker, P. F., and Burrell, J. 1977. The opening of Drake Passage. *Mar. Geol.*, *25*, 15–34.

Barker, P. F., and Griffiths, D. H. 1972. The evolution of the Scotia Ridge and Scotia Sea. *Phil. Trans. R. Soc.*, Ser. A, *271*, 151–83.

Barker, P. F., and Griffiths, D. H. 1977. Towards a more certain reconstruction of Gondwanaland. *Phil. Trans. R. Soc.*, Ser. B, *279*, 143–59.

Brett, C. P. 1977. Seismicity of the South Sandwich Islands region. *Geophys. J.R. astr. Soc.*, *51*, 453–64.

Chase, C. G., Menard, H. W., Larson, R. L., Sharman, G. F., and Smith, S. M. 1970. History of sea-floor spreading west of Baja California. *Geol. Soc. Am. Bull.*, *81*, 491–98.

Dalziel, I. W. D. 1974. Margins of the Scotia Sea. (*In* Burke, C. A., and Drake, C. L., *eds. The Geology of Continental Margins.* New York, Springer, 567–80.)

Dalziel, I. W. D., and Elliot, D. H. 1971. Evolution of the Scotia arc. *Nature, Lond.*, *233*, 246–52.

Dalziel, I. W. D., and Elliot, D. H. 1973. The Scotia arc and Antarctic margin. (*In* Nairn, A. E. M., and Stehli, F. G., *eds. Ocean Basins and Margins, 1. The South Atlantic.* New York, Plenum Press, 171–246.)

de Wit, M. J. 1977. The evolution of the Scotia arc as a key to the reconstruction of southwestern Gondwanaland. *Tectonophysics*, *37*, 51–81.

Ewart, A., and Bryan, W. B. 1972. The petrology and geochemistry of the igneous rocks from Eua, Tongan Islands. *Geol. Soc. Am. Bull.*, *83*, 3281–98.

Ewart, A., Bryan, W. B., and Gill, J. B. 1973. Mineralogy and geochemistry of the younger volcanic islands of Tonga, S.W. Pacific. *J. Petrology*, *14*, 429–65.

Ewing, J. I., Ludwig, W. J., Ewing, M., and Eittreim, S. L. 1971. Structure of the Scotia Sea and Falkland Plateau. *J. geophys. Res.*, *76*, No. 29, 7118–37.

Forsyth, D. W. 1975. Fault plane solutions and tectonics of the South Atlantic and Scotia Sea. *J. geophys. Res.*, *80*, 1429–43.

Grutzeck, M., Kridelbaugh, S., and Weill, D. F. 1974. The distribution of Sr and REE between diopside and silicate liquid. *Geophys. Res. Lett.* *1*, 273–75.

Harrington, P. K., Barker, P. F., and Griffiths, D. H. 1972. Crustal structure of the South Orkney Islands area from seismic refraction and magnetic measurements. (*In* Adie, R. J., *ed. Antarctic Geology and Geophysics.* Oslo, Universitetsforlaget, 27–32.)

Hawkes, D. D. 1962. The structure of the Scotia arc. *Geol. Mag.*, *99*, 85–91.

Hayes, D. E., Heirtzler, J. R., Herron, E. M., and Pitman, W. C., III. 1969. Preliminary Report of Volume 21, U.S.N.S. Eltanin Cruises 22-27. *Tech. Rep. Lamont-Doherty Geol. Observ.*, No. 2-CU-2-69.

Herdman, H. F. P. 1948. Soundings taken during the Discovery Investigations, 1932–39. *'Discovery' Rep.*, *25*, 39–106.

Herron, E. M., and Tucholke, B. E. 1976. Sea-floor magnetic patterns and basement structure in the southwestern Pacific. (*In* Hollister, C. D., Craddock, C., and others. 1976. *Initial Reports of the Deep Sea Drilling Project,* Vol. 35, Washington, [U.S. Government Printing Office], 263–78.)

Jakes, P., and Gill, J. B. 1970. Rare earth elements and the island arc tholeiite series. *Earth planet. Sci. Lett.*, *9*, 17–28.

Jakeš, P., and White, A. J. R. 1972. Major and trace element abundances in volcanic rocks of orogenic areas. *Geol. Soc. Am. Bull.*, *83*, 29–40.

Karig, D. E. 1974. Evolution of arc systems in the western Pacific. *Ann. Rev. Earth planet. Sci.*, *2*, 51–75.

Matthews, D. J. 1939. *Tables of the Velocity of Sound in Pure Water and Sea Water.* London, Ministry of Defence, Hydrographic Department, 52 pp.

Pearce, J. A. 1973. *Some Relationships between the Geochemistry and Tectonic Setting of Basic Volcanic Rocks.* Ph.D. dissertation. University of East Anglia.

Pearce, J. A., and Cann, J. R. 1973. Tectonic setting of basic volcanic rocks determined using trace element analyses. *Earth planet. Sci. Lett.*, *19*, 290–300.

Ringwood, A. E. 1975. *Composition and Petrology of the Earth's Mantle.* New York, McGraw-Hill, 618 pp.

Weddell, J. 1825. *A Voyage Towards the South Pole Performed in the Years 1822-24 Containing an Examination of the Antarctic Sea.* London, Longman, Rees, Orme, Brown, and Green, 324 pp.

23

Scotia Sea Floor: Geochemistry of Basalts from the Drake Passage and South Sandwich Spreading Centers

ANDREW D. SAUNDERS*, JOHN TARNEY**, STEPHEN D. WEAVER***, and P. F. BARKER
Department of Geological Sciences, University of Birmingham, Birmingham B15 2TT, U.K.

Abstract. Marine geophysical studies have demonstrated the complex nature of the floor of the Scotia Sea. Magnetic anomaly patterns indicate that spreading has taken place within Neogene time both to the east and west of the Shackleton Fracture Zone in Drake Passage and also that active spreading presently occurs behind the South Sandwich island arc. Specimens dredged from these three spreading axes by the R.R.S. *Shackleton* enable the geochemical characteristics of basalts from the minor slow-spreading centers in Drake Passage to be compared with basalts from the fast back-arc spreading center in the eastern Scotia Sea and with basalts from the major mid-ocean ridge systems.

Basalts from the western Drake Passage (dredge 3) are fine- to medium-grained olivine basalts with rare plagioclase and rare olivine phenocrysts, whereas those east of the Shackleton Fracture Zone (dredge 7) have abundant plagioclase phenocrysts. Chemically, both these groups are very similar to mid-ocean ridge basalts (MORB), though the basalts of dredge 3 are part of a fractionated sequence exhibiting trends of iron, titanium, and incompatible trace-element enrichment.

Basalts dredged from four different sites along the South Sandwich back-arc spreading center are fresh and include both aphyric basalts and basalts with variable amounts of plagioclase, olivine, and rare clinopyroxene phenocrysts. They range from olivine-normative to mildly quartz-normative tholeiites. They are chemically similar to some basaltic suites from the Mid-Atlantic Ridge in having higher K, Rb, Ba, and Sr contents than MORB, rare-earth patterns which are slightly enriched in light REE, and $^{87}Sr/^{86}Sr$ ratios in the range 0.7028 to 0.7033. Compositional variations within and between the four dredge suites are largely attributable to differing degrees of partial melting of a nondepleted mantle source which has suffered variable degrees of K, Rb, Ba, Sr and possibly H_2O metasomatism. These metasomatic fluids may have been derived from the adjacent subducting slab.

There is general agreement that the isolated continental blocks of the Scotia arc once formed a continuous connection between South America and the Antarctic Peninsula (Matthews, 1959; Hawkes, 1962; Barker and Griffiths, 1972, 1977; Dalziel and Elliot, 1973; de Wit, 1977). Fragmentation and dispersal of this connection was a Cenozoic event (Barker and Burrell, 1977) resulting in the formation of the Scotia Sea basin. Marine geophysical studies have shown that the Scotia Sea floor has a complex structure, but that, with the exception of the shallow or emergent North and South Scotia Ridges, it is predominantly oceanic in character (Barker, 1970; Ewing and others, 1971). The history of sea-floor spreading in two areas of the Scotia Sea—namely, Drake Passage and the eastern Scotia Sea—is now reasonably well understood (Barker, 1972; Barker and Griffiths, 1972; Barker and Burrell, 1977).

The southeastward separation of the South Scotia Ridge from Burdwood Bank and Tierra del Fuego was accomplished by sea-floor spreading in Drake Passage east of the Shackleton Fracture Zone (Figure 23.1). Barker and Burrell (1977) have demonstrated that coherent spreading started in the eastern Drake Passage about 29 m.y. ago, although a period of chaotic or ill-defined spreading preceded this. Spreading in this area stopped about 6 m.y. ago (Barker and Burrell, 1977). West of the Shackleton Fracture Zone, the ages of three segments of spreading sea floor have been identified by Barker (1970); they give a full spreading rate of 54 mm/yr, significantly faster than the rate (20–46 mm/yr) deduced for the eastern Drake Passage. In western Drake Passage, only about 20 km of new ocean crust seem to have been generated during the last 4 m.y. (Figure 23.1). The occurrence of a partly filled trench northwest of the South Shetland Islands, and the discordant relationship of the magnetic anomalies with the continental margin, indicate that western Drake Passage ocean crust has been subducted beneath the islands. Spreading virtually stopped about 4 m.y. ago, and subduction no longer seems to be taking place (Barker, 1976).

In the eastern Scotia Sea, 440 km west of the South Sandwich Islands, a series of well-defined magnetic lineations (Figure 23.2) indicates that sea-floor spreading has been in progress for about 8 m.y. at the South Sandwich spreading center. The spreading rate has increased from 50–60 mm/yr about 4 m.y. ago to 70–90 mm/yr at present (Barker, 1972). The South Sandwich Islands lie directly east of ocean crust which is only 5–6 m.y. old, and it there-

* Present address: Department of Geology, Bedford College, Regents Park, London NW1 4NS, U.K.

** Present address: Department of Geology, University of Leicester, Leicester, LE1 7RH, U.K.

*** Present address: Department of Geology, University of Canterbury, Christchurch, New Zealand.

fore seems likely that the islands have been built-up on ocean crust generated at the South Sandwich spreading center. The islands themselves are erupting lavas of the island-arc tholeiite series. These lavas are chemically similar to those from Tonga (Ewart and others, 1973) and have low concentrations of Ti, Zr, P, K, Rb, and Ba; variable light rare-earth element (REE) depleted patterns, with both positive and negative europium anomalies; and rather uniform $^{87}Sr/^{86}Sr$ ratios of about 0.704 (Baker, 1968, 1978; Hawkesworth and others, 1977; Pankhurst, Chapter 25, this volume).

The tectonic setting of the South Sandwich arc and the associated back-arc spreading center is similar to that of many western Pacific arcs. The eastern Scotia Sea is composed of an active marginal basin lying behind an intra-oceanic volcanic arc. Similar basins are found behind the Tonga, Kermadec, Mariana, Bonin, and New Hebrides arcs. The formation of marginal basins seems to be a consequence of subduction (Sleep and Toksoz, 1971; Kanimori, 1971; Packham and Falvey, 1971; Moberly, 1972; Karig, 1974), although there is no consensus as to the exact relationships involved.

In this paper, geochemical data for seven suites of basalts dredged from Drake Passage and the eastern Scotia Sea are presented. The compositions described not only confirm that the sea floor in these two areas is truly oceanic, but they also permit detailed geochemical comparisons between oceanic ridge and back-arc basin magmas. A full account of the mineralogy and chemistry of basalts from the South Sandwich spreading center has been presented elsewhere (Saunders and Tarney, 1979).

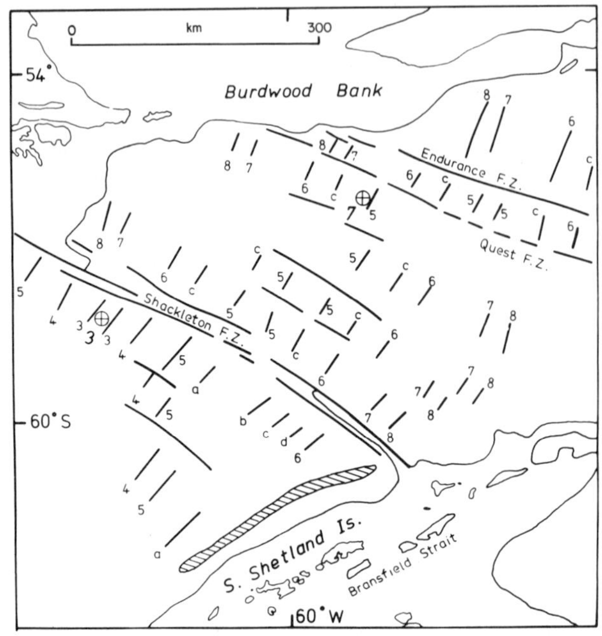

Figure 23.1. Map of Drake Passage in the western Scotia Sea, lying between the southern tip of South America and the northern end of the Antarctic Peninsula. Dredge sites 3 and 7 are shown by circled crosses. The identified magnetic lineations, with Lamont anomaly numbers, are from Barker and Burrell (1977). Anomalies 3 and 5 represent crust approximately 4 and 10 m.y. old, respectively. The locations of the South Shetland trench (shaded area) and of the 2000 m isobath (faint lines) are shown.

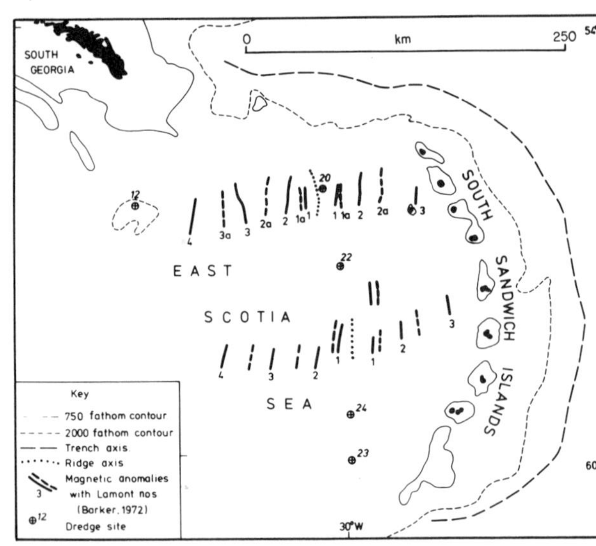

Figure 23.2. Map of the eastern Scotia Sea, showing the locations of dredge sites 12, 20, 22, 23, and 24, and the relative positions of the South Sandwich Islands and the back-arc spreading center (after Barker, 1972).

Dredge locations and specimen descriptions

Dredging was carried out aboard the R.R.S. *Shackleton* between 1972 and 1974. Latitudes and longitudes for all dredge sites are given in the notes to Tables 23.2 and 23.3, and the sites are shown in Figures 23.1 and 23.2. All rock specimens have been coded as to their dredge site, and an arbitrary number assigned to each fragment recovered at that site (e.g., D24.14).

Dredges 3 and 7 were sited on crust generated at spreading centers in Drake Passage (Barker, 1976; Barker and Burrell, 1977). Dredge 3, immediately west of the Shackleton Fracture Zone (Figure 23.1), sampled crust generated by very slow spreading (less than 5 mm/yr) during the last 4 m.y. Dredge 7 was located northeast of the Shackleton Fracture Zone (Figure 23.1) on anomaly 5 (age about 9–10 m.y.), where spreading is taking place at 20 mm/yr or less. Dredge hauls from along the South Sandwich spreading center in the eastern Scotia Sea (dredges 20, 22, 23, and 24; Figure 23.2) yielded abundant fragments of fresh pillow basalt, but off-axis dredges (dredges 12 to 19) yielded almost entirely ice-rafted debris. Of the off-axis dredges, only dredge 12, sited immediately to the west of the oldest magnetic lineation, contained any pillow basalt (one specimen).

Dredge 3 yielded three rock types—conglomerate, hyaloclastite, and basalt. The conglomerate consists of fragments of gray and brown-gray aphyric basalt and olive-colored mudstone embedded in light-gray consolidated ooze, whereas the hyaloclastite is composed of numerous "droplets" of black sideromelane embedded in poorly consolidated buff-colored mudstone. The presence of an aureole of baked mudstone around each droplet suggests injection of fluid basalt into the ooze. The majority of specimens from dredge 3, however, are sparsely porphyritic basalts; both pillows and rounded cobble-forms occur. Alteration is usually restricted to an outer zone 1 to 2 cm wide where development of calcite and clay minerals is clearly visible; the interiors of the fragments are usually fresh. Dredge 7 consists of several fragments of very fresh, plagioclase-phyric pillow basalt with unweathered glassy selvages.

The basalt specimens from the five dredges in the eastern Scotia Sea show fresh fracture surfaces and many have a subpyramidal or pillow form with convex outer surfaces, usually sideromelane-coated. Evidence of alteration is generally absent, but where present, it it usually restricted to the outer zone. Specimens from dredge 20 show the most evidence of alteration, but it is not severe. Dredges 22 and 24 contain strongly vesicular basalts, some resembling basaltic clinker, but vesicles are scarce in the other dredges (including dredges 3 and 7) and are restricted to a zone 1 to 2 cm beneath the sideromelane rind.

Petrography and mineralogy

Dredge 3 contains fine-grained basalts and lavas of microdoleritic texture. All are composed of plagioclase, olivine, pyroxene, titanomagnetite, and ilmenite, but chrome spinel is not found. Selected samples have been analyzed using the Cambridge Instruments microprobe at Leicester University. The dominant phenocryst phase is plagioclase (An_{85} in D03.04), although its modal abundance is less than 5% (Table 23.1). Groundmass plagioclase (An_{60-63}) and phenocryst rims (An_{57-61}) have more sodic compositions but contain less than 0.3% Or component. The other phenocryst phase is olivine which in specimen D03.04 has a composition of Fo_{82} (a lower Fo content than in the olivines from dredges 20, 22, and 23). Pyroxene phenocrysts are not present, although clinopyroxene is a major compoent of the groundmass.

Dredge-7 basalts are strongly plagioclase-phyric, with labradorite-bytownite phenocrysts embedded in a fine-grained, variolitic groundmass. The mineral assemblage is otherwise similar to that of dredge-3 basalts.

The textures of the Scotia Sea basalts are similar to those reported elsewhere for quenched basalts (Bryan, 1972). Three distinct generations of olivine and plagioclase can usually be seen (phenocrysts, both partially resorbed and unresorbed; microphenocrysts; and groundmass). The groundmass phases exhibit a wide variety of textures, depending upon the inferred rate of cooling. Thus, there occurs a complete gradation from straw-colored sideromelane-quenched margins to microdoleritic (subophitic or granular) pillow interiors. Intermediate zones may be spherulitic or variolitic.

Basalts from dredges 20 and 23 are fine grained and are composed of olivine, plagioclase, clinopyroxene, chrome spinel, and titanomagnetite. They are sparsely porphyritic with large, commonly resorbed phenocrysts of plagioclase (as calcic as An_{90}) and olivine (Fo_{86-88}). Basalts from dredge 22 have a similar mineralogy but are largely aphyric. The basalt from dredge 12, specimen D12.19, contains numerous microphenocrysts of plagioclase and olivine, and groundmass plagioclase is abundant. In dredges 3, 7, 12, 20, 22, and 23, pyroxene is restricted to the groundmass. However, dredge 24 has three distinct phenocryst assemblages. One assemblage (e.g., D24.14) has no plagioclase phenocrysts and only rare olivine phenocrysts. The second (e.g., D24.19) contains olivine (Fo_{84}), plagioclase, and clinopyroxene ($Wo_{41}En_{50}Fs_9$) phenocrysts and microphenocrysts. The third assemblage (e.g., D24.13) has abundant plagioclase (An_{85}), rare olivine (Fo_{80}), and no pyroxene phenocrysts.

As in dredges 3 and 7, no zoning can be discerned in the olivine (or pyroxene) phenocrysts, but plagioclase commonly shows both continuous and oscillatory zoning, and the zoning is commonly reversed in the groundmass feldspars. Where found, chrome spinel occurs as minute (10–20 micron) reddish-brown octahedra and cubes, and it is usually embedded within the olivine phenocrysts. It would therefore seem to have formed early in the paragenetic sequence. Chrome spinel also occurs freely in the groundmass, and in specimen D24.19 it occurs within clinopyroxene phenocrysts. Titanomagnetite is a groundmass phase and is found as discrete cubes or octahedra in the variolitic zones or in pillow interiors and as disseminated crystallites in the spherulitic zones. Rare skeletal and dendritic ilmenite occurs locally in the subophitic centers of pillows of dredge 3.

Alteration effects are slight and are indicated only by a yellowish limonite staining or by smectite growth in basalts from dredges 12 and 20. Vesicles within glassy margins may contain fine encrustations of green smectite and palagonite, but mineral grains are completely fresh. Dredge-3 basalts tend to be more altered, with development of calcite as well as smectite (Table 23.1, e.g., D03.07). The basalts from dredges 7, 22, 23, and 24 are unusually fresh, with only slight limonite staining of glassy selvages.

Geochemistry

Seventy-five basalt samples have been analyzed for major and trace elements using a Philips 1450 automatic X-ray fluorescence spectrometer; techniques used are out-

Table 23.1. Representative modal analyses of basalt specimens from the Drake Passage and the eastern Scotia Sea (in vol %)

	Drake Passage					Eastern Scotia Sea							
	D03.04	D03.07	D03.20	D03.48	D07.03	D12.19	D20.11	D20.17	D22.2	D23.1	D24.13	D24.14	D24.19
Phenocrysts and microphenocrysts													
Plagioclase	0.3	3.3	0.1	0.1	15.5	6.3	2.0	0.4	1.6	0.2	8.5	ns	1.4
Olivine	0.4	1.2	0.4	0.1	2.5	7.2	3.3	0.3	3.0	0.6	1.4	1.0	3.5
Pyroxene	ns	ns	ns	ns	ns	ns	ns	ns	ns	ns	ns	ns	0.9
Chromite	ns	ns	ns	ns	ns	0.01	0.01	ns	0.2	0.01	0.01	0.4	0.5
Ground mass													
Plagioclase	31.0	26.6	30.0	46.7		43.8	33.4	52.3	21.6	20.8	11.1	8.7	2.50
Olivine	6.2	3.3	9.4	2.1		0.7	1.0	5.6		3.4			
Pyroxene	54.4	49.0	52.4	37.9	82.0	42.8	60.5	28.9	73.6	75.1	79.0	90.0	68.7
Opaque minerals	6.2	7.8	7.0	8.1				5.9					
Smectites	1.5	8.4	0.7	5.1	ns	ns	ns	7.0	ns	ns	ns	ns	ns
Calcite	ns	1.0	ns	0.2	ns	ns	ns	ns	ns	ns	ns	ns	ns

ns: not seen.

lined in Saunders and others (Chapter 34, this volume). The rare-earth analyses and strontium-isotope measurements were carried out by isotope-dilution mass spectrometry. The methods were similar to those described by Hooker and others (1975) and Pankhurst and O'Nions (1973). The rare-earth element (REE) and Sr-isotope data have been reported elsewhere (Hawkesworth and others, 1977; Saunders and Tarney, 1979).

MAJOR ELEMENTS

Representative analyses are given in Table 23.2 and 23.3.

The basalts of dredge 3 are subalkaline olivine tholeiites (Chayes, 1966), which plot in the Di-Ol-Hy field of the normative tetrahedron, using an assumed Fe_2O_3/FeO ratio of 0.15 (Figure 23.3). D07.03 is mildly quartz normative, and the high Al_2O_3 and CaO contents of this sample are consistent with the high modal plagioclase content. With respect to major elements, the basalts of dredge 3 are indistinguishable from mid-ocean ridge basalts (MORB) (Table 23.2). Although D07.03 has a higher Al_2O_3 content than most MORB, aluminous plagioclase-phyric basalts have been recorded in mid-ocean ridge environments (see, for example, Muir and others, 1964; ARCYANA, 1977).

The basalts of dredge 3 seem to comprise a fractionated sequence. This may be illustrated by plotting oxides and elements against a trace element which has a low bulk distribution coefficient in basaltic systems. Such a trace element is zirconium (Weaver and others, 1972; Tarney and others, 1977). The consanguineous association of the dredge-3 basalts can thus be clearly seen (Figures 23.4 and 23.5): FeO, TiO_2, K_2O, and P_2O_5 increase sympathetically

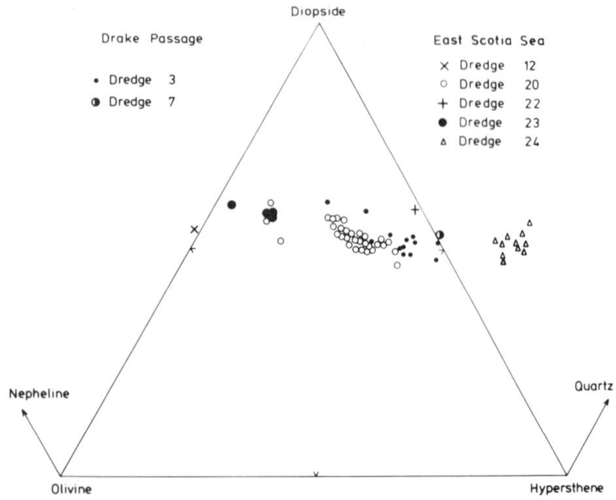

Figure 23.3. Normative plot of major element abundances in dredged Scotia Sea basalt specimens. For the normative calculation, an Fe_2O_3/FeO ratio of 0.15 was assumed.

Table 23.2. Analyses of basalt specimens dredged from the Drake Passage and average analyses of ocean-ridge basalts (oxides in wt %)

	1	2	3	4	5	6	7	8	9	10	11
SiO_2	49.58	49.89	49.23	49.35	49.69	49.80	50.69	49.91	49.92	49.21	49.61
TiO_2	1.36	1.42	1.37	1.83	2.06	2.08	2.13	1.38	1.53	1.39	1.43
Al_2O_3	15.24	15.83	15.55	15.54	15.33	15.47	15.80	17.26	15.63	15.81	16.01
tFe_2O_3 (total Fe)	9.14	9.58	9.47	11.49	11.36	11.53	10.68	9.45	10.75	10.19	12.76
MnO	0.16	0.15	0.18	0.19	0.18	0.19	0.17	0.16	0.17	0.16	0.18
MgO	7.76	8.09	7.11	7.04	6.60	6.64	6.16	6.14	7.65	8.53	7.84
CaO	11.74	11.30	11.58	11.02	10.76	10.69	10.51	11.88	11.17	11.14	11.32
Na_2O	2.75	2.84	2.94	2.60	2.91	2.85	3.15	2.58	2.75	2.71	2.76
K_2O	0.15	0.05	0.23	0.26	0.38	0.41	0.42	0.22	0.16	0.26	0.22
P_2O_5	0.15	0.17	0.18	0.20	0.23	0.24	0.30	0.14	0.13	0.15	0.15
Total	98.03	99.32	97.84	99.52	99.50	99.90	99.81	99.12	—	—	—
					Trace elements (ppm)						
Cr	331	330	319	281	226	227	220	288	—	296	—
Ni	78	62	68	71	82	75	86	59	110	123	—
Rb	2	0.4	4	5	5	6	6	3	1.1	—	—
Ba	35	30	30	48	66	59	60	42	10.5	12	—
Sr	171	140	165	116	158	143	150	128	135	123	—
Zr	98	100	98	136	159	161	165	94	—	100	—
Nb	3	2	2	5	6	5	5	2	—	—	—
Y	30	31	30	43	47	49	48	33	—	43	—
La	5	5	7	3	8	11	9	4	—	—	—
Ce	8	11	11	12	8	22	18	10	—	—	—
Pb	4	1	6	3	3	5	1	2	—	—	—
Th	1	1	1	3	1	1	1	1	—	—	—
Zn	71	74	70	87	88	92	93	81	—	—	—
Ga	17	17	17	18	19	19	22	18	—	—	—
tFe/Mg	1.37	1.37	1.55	1.89	2.00	2.02	2.01	1.79	1.63	1.39	1.89
K/Rb	623	1038	477	432	631	567	581	609	1207	—	—
Ce_N/Y_N	0.55	0.73	0.75	0.57	0.35	0.92	0.77	0.62	—	—	—
Zr/Nb	33	50	49	27	27	32	33	47	—	—	—

1. D03.07: dredged basalt, from lat. 58° 11.56′ S, long. 65° 57.8′ W and 3268 m depth.
2. D03.25A: dredged basalt (fresh interior), from same locality and depth as D03.07.
3. D03.25B: dredged basalt (weathered rim), from same locality and depth as D03.07.

4–7. D03.44, D03.04, D03.02, and D03.48: dredged basalts, from same locality and depth as D03.07.

8. D07.03: dredged plagioclase-phyric basalt, from lat. 56° 18.6′ S, long. 57° 24.2′ W and 3162 m depth.
9. AV1: average mid-ocean ridge basalt (Hart and others, 1972).
10. AV2: Mean of 33 basalts from the Mid-Atlantic Ridge (Melson and Thompson, 1971).
11. AV3: Mean of 94 selected ocean-floor basalts (Cann, 1971).

with Zr. The levels of TiO_2, P_2O_5, and K_2O in the least altered and least fractionated samples (e.g., D03.25A) closely resemble those found in MORB. Note especially the low K_2O and Rb, features characteristic of MORB (Engel and others, 1965; Cann, 1971; Hart and others, 1972).

The major- and trace-element data reveal the essentially homogeneous nature of the basalts from each dredge site in the eastern Scotia Sea, although it is apparent from Table 23.3 and Figures 23.4 and 23.5 that those from dredge 24 comprise a consanguineous fractionated sequence. Normative compositions demonstrate significant differences in the petrologic characteristics of the basalts at different sites (Figure 23.3). Rocks from dredges 20 and 23, with high Al_2O_3 and Na_2O contents, are subalkaline olivine tholeiites which plot in an elongate field across the basalt tetrahedron. Dredge-24 basalts are mildly quartz-normative, whereas the basalt from dredge 12 is mildly nepheline-normative. As pointed out by Chayes (1972), quartz-normative and nepheline-normative basalts are uncommon among typical MORB.

Dredge-24 basalts are high in normative plagioclase, consistent with the occurrence of abundant modal plagioclase in the more fractionated members of this sequence. Several other aspects of their chemistry—notably a high SiO_2 content and Fe/Mg ratio, and low TiO_2, P_2O_5, Zr, Nb, Y, and Ni contents—distinguish them from the other dredged basalts. These characteristics are, however, typical of tholeiites from primitive intraoceanic island arcs (Gill, 1970; Jakeš and White, 1972; Ewart and others, 1973; Baker, 1978). Basalts with such island-arc tholeiite characteristics have also recently been described from the Lau Basin (Gill, 1976).

Trace elements

The basalts of dredge 3 show progressive, sympathetic enrichment of the incompatible elements (i.e., those elements with bulk distribution coefficients, D, less than 0.1)—Rb, Ba, Zr, Nb, Ce, La, and Y as well as TiO_2, K_2O, and P_2O_5 (Figures 23.4 and 23.5). Absolute trace-element abundances, especially those in the least fractionated samples, closely resemble those found in MORB in having low K and Rb contents and light-REE-depleted patterns ($Ce_N/Y_N < 1$) (Engel and others, 1965; Frey and others, 1974). Similarly, the Zr/Nb ratios of dredges 3 and 7 are as high as those found in MORB (Erlank and Kable, 1976). However, the amount of Ba is higher than in most MORB (see, for example, Hart, 1971). Dredges 3 and 7 both contain high levels of Cr. Ni abundances are generally lower than in MORB ("unfractionated" MORB have approximately 200 ppm Ni; Kay and others, 1970), and there is no systematic decrease of Ni within the group. Similarly, Sr content does not behave in a systematic fashion (Figure

Table 23.3. Representative analyses of unaltered basalts from the eastern Scotia Sea (oxides in wt %)

	1	2	3	4	5	6	7	8	9	10
SiO_2	48.90	50.45	50.58	50.70	50.62	51.00	50.36	53.84	53.01	53.08
TiO_2	1.25	1.25	1.27	1.29	1.28	1.02	1.46	0.61	0.74	0.91
Al_2O_3	17.96	16.67	16.62	16.58	16.49	17.90	16.36	14.51	15.77	18.24
tFe_2O_3 (total Fe)	9.43	8.29	8.30	8.43	8.30	7.54	9.07	9.24	9.43	8.99
MnO	0.20	0.15	0.15	0.16	0.15	0.14	0.16	0.17	0.17	0.15
MgO	6.75	7.43	7.69	7.67	7.64	7.37	7.36	7.71	7.13	4.81
CaO	10.86	10.94	11.02	11.12	11.05	10.70	10.84	10.80	11.00	10.48
Na_2O	3.50	2.99	3.03	3.15	3.18	2.72	3.39	1.79	2.32	2.78
K_2O	0.19	0.46	0.42	0.34	0.30	0.57	0.43	0.24	0.24	0.38
P_2O_5	0.17	0.17	0.18	0.19	0.18	0.15	0.20	0.08	0.11	0.12
Total	99.21	98.79	99.26	99.63	99.19	99.11	99.63	98.99	99.92	99.94
					Trace elements (ppm)					
Cr	184	263	274	270	267	196	270	295	214	102
Ni	77	73	76	63	73	66	64	42	27	26
Rb	1	5	5	5	4	8	6	4	3	5
Ba	67	50	56	49	51	83	77	55	40	70
Sr	340	188	189	193	193	195	212	123	149	164
Zr	114	108	111	107	111	84	130	40	55	68
Nb	1	5	5	3	4	3	8	0.2	2	1
Y	24	28	28	29	26	24	30	14	18	23
La	5	8	7	8	7	7	8	3	4	5
Ce	10	15	16	17	16	14	18	7	8	9
Pb	5	2	1	3	4	4	3	1	4	2
Th	1	1	1	4	4	1	1	1	1	1
Zn	69	69	64	68	67	69	68	71	70	74
Ga	16	16	15	17	18	13	15	12	13	14
tFe/Mg	1.62	1.29	1.25	1.27	1.26	1.19	1.43	1.39	1.53	2.17
K/Rb	1577	764	697	564	623	591	595	498	664	631
Ce_N/Y_N	0.85	1.09	1.17	1.20	1.26	1.19	1.23	1.02	0.91	0.80
Zr/Nb	114	22	22	36	28	28	16	200	28	68
Sr^{87}/Sr^{86}	—	0.70281	0.70281	0.70282	—	0.70317	0.70297	0.70325	0.70310	0.70316

1. D12.19: pillow basalt from lat. 56° 42.0′ S, long. 35° 42.0′ W and 2560 m depth; from immediately to the west of the oldest dated magnetic lineation (see Figure 23.2).

2–5. D20.1C, D20.34, D20.43B, and D20.35: pillow basalts from lat. 56° 23.6′ S, long. 30° 39.5′ W and 3290 m depth; from the South Sandwich spreading center.

6. D22.2: pillow basalt from lat. 57° 31.4′ S, long. 30° 16.8′ W and 3072 m depth; from the South Sandwich spreading center.

7. D23.5: pillow basalt from lat. 60° 3.6′ S, long. 29° 57.0′ W and 2516 m depth; from the South Sandwich spreading center.

8–10. D24.14, D24.19, and D24.13B: pillow basalts from lat. 59° 28.5′ S, long. 30° 0.4′ W and 2250 m depth.

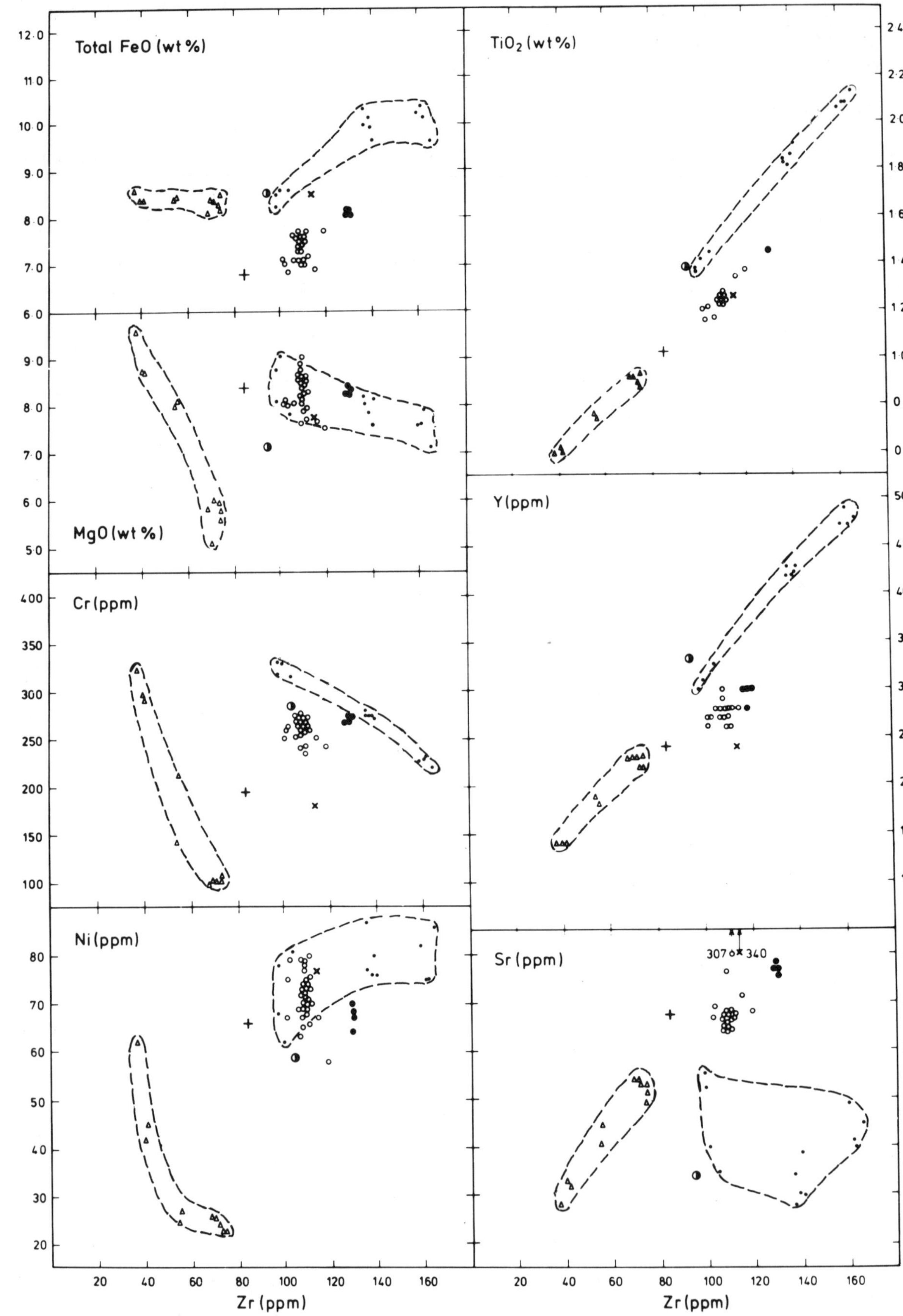

Figure 23.4. Selected oxides and elements versus Zr for Scotia Sea basalts. Symbols as in Figures 23.3 and 23.5.

23.4), although its distribution may be complicated by alteration effects.

It is apparent from Figure 23.4 that there is a fractionation sequence among the eastern Scotia Sea dredge basalts, with increasing Zr from dredge-24 through dredges-22 and -20 to dredge-23 basalts. Incompatible elements such as Ti, P, and the light REE vary in a colinear fashion with Zr. However, other elements behave in a less incompatible manner—Na, Sr, Y, and the heavy rare earths begin to decrease relative to Zr at higher concentrations. Ni and Cr do not show the expected negative correlation, nor is there a systematic relationship between Zr and K, Rb, and Ba when considering dredges 20, 22, 23, and 24 as a group. K, Rb, and Ba are in fact highest in D22.2 which has a low Zr content and Fe/Mg ratio, but there is no indication from Sr-isotope data that this is in any way related to sea water alteration. The basalts from dredges 22 and 24 have higher K/Zr, Ba/Zr, Rb/Zr, and Sr/Zr ratios than those from dredges 20 and 23.

Although dredge-24 basalts seem to be more highly fractionated with respect to major elements, their low trace-element abundances suggest that this may not be the case. The values of Ni and Cr are low in dredge-24 basalts, and the contents of both elements fall sharply in the samples of more evolved rocks. It is important to note, however, that the dredge-24 basalts lie on the same incompatible-element trends (except for K, Rb, Sr, and Ba) as the other dredge suites from the spreading center. They may, therefore, be petrogenetically related.

Absolute abundances of certain trace elements in the eastern Scotia Sea basalts are similar to those found in the Drake Passage basalts and in MORB. However, Na, K, Rb, Ba, and Sr contents are slightly but significantly higher than in the Drake Passage basalts at equivalent Zr levels, and the TiO_2/Zr ratios are lower (Figures 23.4 and 23.5).

Rare-earth data have proved particularly valuable in studying the petrogenesis of mid-ocean ridge basalts (Gast, 1968; Kay and others, 1970; Schilling 1971, 1975; Frey and others, 1974). Chondrite-normalized REE patterns for the eastern Scotia Sea basalts are shown in Figure 23.6. They may be regarded as "undepleted," with Ce_N/Yb_N ranging from 0.92 to 1.43. There are no Eu anomalies. The absolute REE abundances vary systematically with the abundances of other incompatible elements (e.g., Zr), such that

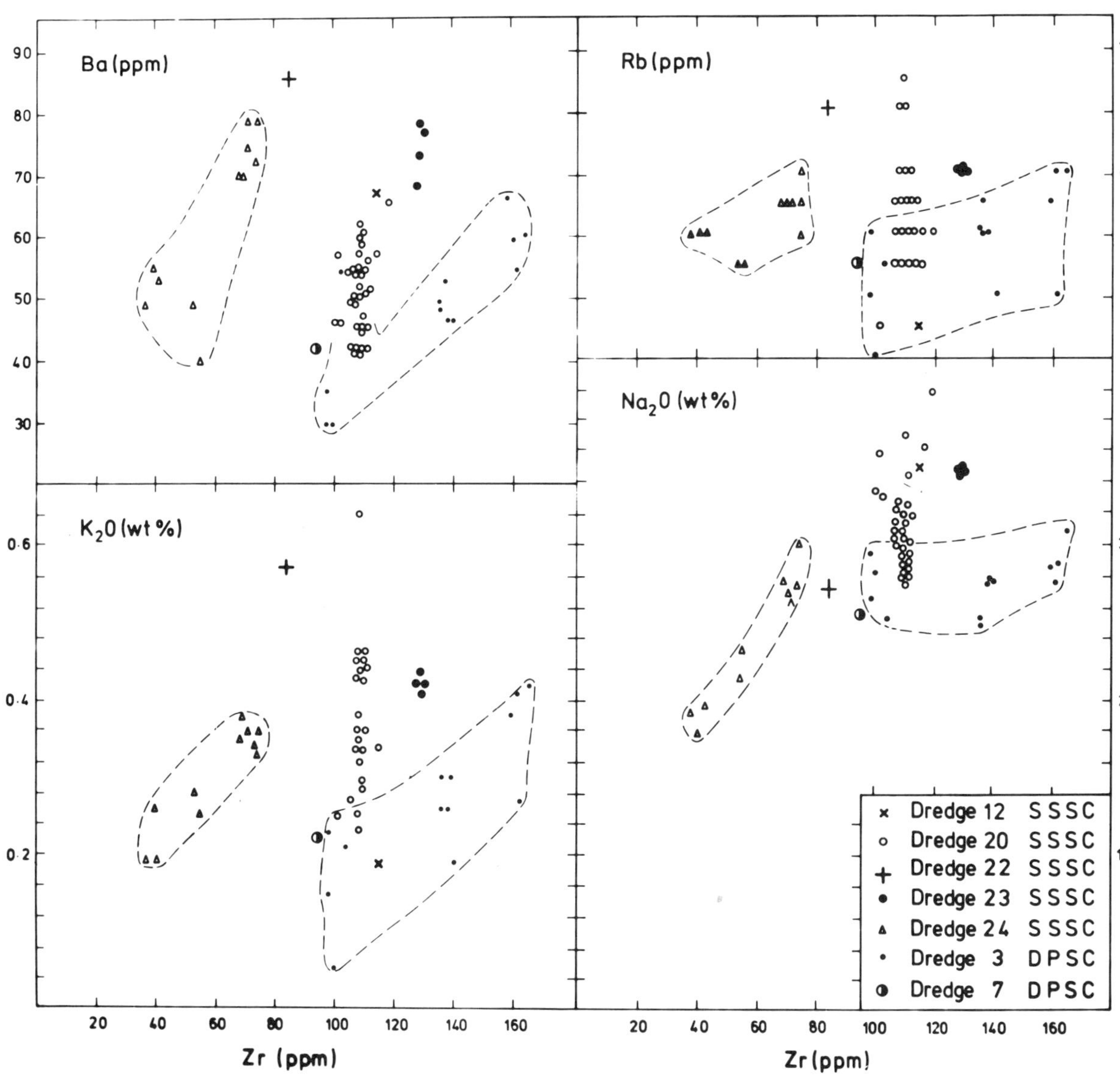

Figure 23.5. Alkalis and alkali-earth oxides and elements versus Zr for Scotia Sea basalts. DPSC: Drake Passage spreading center. SSSC: South Sandwich spreading center.

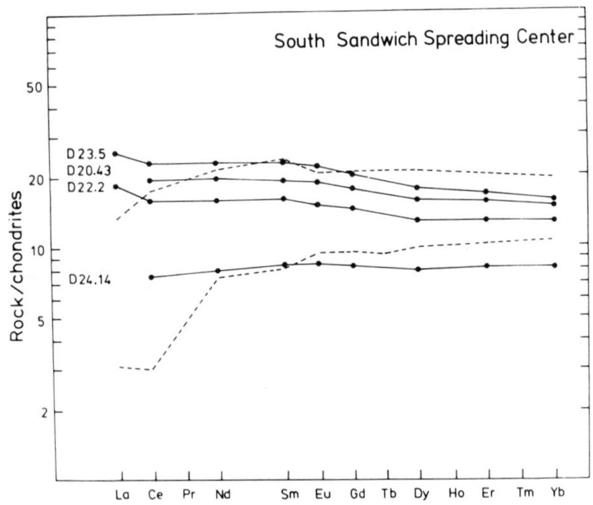

Figure 23.6. Chondrite-normalized rare-earth patterns for South Sandwich spreading center basalt specimens. The dashed lines indicate the range (upper and lower limits) of normal mid-ocean ridge basalts (MORB).

D24.14 has low concentrations of REE and a flat chondrite-normalized pattern, whereas D23.5 has the highest REE levels and highest Ce_N/Yb_N ratio. Depletion of the heavy REE and Y relative to the light REE and Zr has occurred in the more evolved samples (Saunders and Tarney, 1979). The Scotia basalt REE patterns are unlike those of MORB but are within the range of transitional mid-ocean ridge basalts (Frey and others, 1974), Icelandic basalts (O'Nions and Gronvold, 1973), and some back-arc basin basalts (Hart and others, 1972).

Similarly, the data given in Table 23.3 and Saunders and Tarney (1979) show that the South Sandwich spreading-center basalts have $^{87}Sr/^{86}Sr$ ratios slightly higher (0.70281 to 0.70336) than those in fresh MORB (0.7020 to 0.7030; Hedge and Peterman, 1970; O'Nions and Pankhurst, 1974) and that they more closely resemble basalts from Iceland (0.7029 to 0.7032; O'Nions and others, 1976) and "transitional" basalts from the Mid-Atlantic Ridge (0.7033; Erlank and Kable, 1976). It is clear, however, that the basalts from the South Sandwich spreading center have Sr-isotope ratios significantly lower than those in basalts from the adjacent South Sandwich island arc (0.704; Gledhill and Baker, 1973; Hawkesworth and others, 1977).

Seawater alterations

Because the levels of K_2O, Rb, Ba, and Sr are higher in the eastern Scotia Sea basalts than in MORB, it is important to establish that this is not merely an effect of seawater alteration. Full details of the effects of alteration on these basalts are given in Saunders and Tarney (1979), but it is relevant to note here the effects of alteration in D03.25 (Table 23.2). In the altered sample (D03.25B) MgO decreases, but K_2O, Rb, and Sr increase. Ba is not affected. These results are similar to those reported by Hart and others (1974) for ocean ridge basalts and may explain the erratic behavior of Sr in some of the dredge-3 basalts. In the far less altered eastern Scotia Sea basalts, there is little evidence of movement of Ba or Sr; K and Rb tend to vary sympathetically with Ba. Moreover, where K may seem anomalously high (as in D22.2), the groundmass plagioclases have high orthoclase contents. These factors, together with the consistent Sr-isotope values and the fresh appearance of the pillowed basalts, lead us to believe that the observed abundances of K, Rb, and Ba are primary.

Discussion

There are many different and interrelating factors that may have been involved in the generation of the Drake Passage and South Sandwich spreading-center basalts; these factors include (1) the composition of the source, (2) the degrees of partial melting and the conditions under which partial melting occurred (P, P_{H_2O}, etc.), and (3) any possible mineral fractionation in subridge magma chambers. Space does not permit a rigorous examination of these processes, but it is important to assess some of these effects before making comparisons between the different basalt suites.

The basalts of dredge 24 are unusual in that such compositions have not been previously recorded in a spreading-center environment. It has been mentioned that these rocks chemically resemble island-arc tholeiites (e.g., Jakeš and White, 1972), but the Sr-isotope ratios are far lower than those recorded for rocks from island-arc settings. Saunders and Tarney (1979) suggest that dredge-24 basalts represent the fusion products of mantle material in which olivine was refractory. Green (1973) showed that, under water-saturated conditions, partial melting of pyrolitic mantle could yield quartz-normative basalts at pressures as great as 20 kb. Water has the effect of extending the depth range over which orthopyroxene will melt incongruently (Kushiro, 1972, 1975) and also suppresses plagioclase crystallization (Yoder and Tilley, 1962). The rarity of quartz-normative basalts in the ocean-ridge environment may reflect the low P_{H_2O} conditions under which melting normally occurs (Kay and others, 1970). Dredge-24 basalts are unusually vesicular and thus may have been generated under higher P_{H_2O} conditions than normal. Although there is no direct evidence that the subducting slab presently extends beneath the spreading center at depth (Forsyth, 1975; Brett, 1977), the possibility that dehydration of the lithosphere may have contributed to higher P_{H_2O} conditions in the mantle must be borne in mind. Water significantly reduces the solidus of mantle peridotite (Mysen and Boettcher, 1975); thus at constant temperature, hydrous conditions will allow a greater degree of partial melting. The fact that incompatible-element contents are lower in dredge-24 basalts than in the basalts from the other dredges supports this explanation.

If removal of olivine has played an important role in the generation of the dredge-24 basalts, then because of the very low distribution coefficients for most elements between olivine and coexisting magmatic liquid, the rare-earth pattern of D24.14 must be similar to that of the parent mantle. The mantle source seems, therefore, to be undepleted, unlike that predicted for most MORB. The basalts of dredges 20, 22, and 23 could be derived from the same source as those of dredge 24, but under differing degrees and conditions of partial melting. Thus, dredge-23 basalts, which are the most fractionated, probably represent only about 9% partial melting of the mantle source, whereas D24.14 would represent about 30% melting (leaving olivine as the only refractory mineral phase). The increasing Ce_N/Yb_N ratios in dredges 22, 20, and 23 could be a result of the increasing refractory behavior of clinopyroxene in the source as progressively smaller melt fractions are produced, this mineral having higher distribution coefficients for the heavy REE than for the light REE (e.g., Schnetzler and Philpotts, 1970). Hart and others (1972) suggested that the Mariana back-arc basin basalts were derived by small degrees of partial melting at greater depths than normal MORB, and this may be the case for the South Sandwich spreading-center basalts.

Although the levels of some incompatible elements—Ti, P, Zr, Nb, the REE, and Y—are low in the dredge-24 basalts, the levels of K, Rb, and Ba are not much lower than in dredges 20 and 23. Similarly, the basalts of dredge 22 have higher K/Zr, Rb/Zr, Ba/Zr, and Sr/Zr ratios than do the basalts of dredges 20 and 23. It is possible that these variations may be related to the activity of a mineral phase which contained significant quantities of K, Rb, and Ba, namely phlogopite, and that the phlogopite remained in the source during the generation of the basalts sampled by dredges 20 and 23. Thus, at inferred higher degrees of melting, phlogopite would have broken down and contributed K, Rb, and Ba to the liquids that gave rise to the basalts of dredges 22 and 24. It is unlikely, however, that significant quantities of phlogopite could remain refractory within the mantle source during the degrees of melting inferred for the generation of the basalts of dredges 20 and 23 (approximately 10%). Very little phlogopite (probably less than 2 modal % of the mantle source) would be necessary to produce the observed K, Rb, and Ba contents of dredges 22 and 24, and at 10% melting it is quite probable that all of this phlogopite would have broken down and entered the melt.

Saunders and Tarney (1979) have suggested that either the sources of the dredge-22 and -24 basalts have suffered K- and Rb-metasomatism (Lloyd and Bailey, 1975) at some earlier period, or more likely, dewatering of the slab subducting beneath the South Sandwich Islands has locally added K, Rb, Ba, and more radiogenic Sr to the source region of the eastern Scotia Sea basalts. The transitional island arc tholeiite characteristics of the dredge-24 basalts and the high vesicularity of the dredge-20 and -24 basalts lend support to the latter suggestion. The hydrous fluids arising from the subducting lithosphere would also allow a higher degree of partial melting than is normally observed in normal mid-ocean ridge basalts.

Finally, it is noted that although the seismic evidence (Brett, 1977) indicates that the subducting South American lithosphere does not extend much farther westward than beneath the South Sandwich Islands, it is apparent from plate tectonic reconstructions that consumption of lithosphere at the South Sandwich trench must have complemented the generation of 440 km of new lithosphere at the South Sandwich spreading center. The deepest seismic event in the South Sandwich arc system is at 265 km depth, so it is probable that the slab extends to greater depths and farther west than the available data indicate. In addition, the subducting slab must have underlain the South Sandwich spreading center in the past, thus allowing slab-derived fluids to contaminate the mantle source feeding the spreading center to a greater degree than is possible from the present-day position of the subduction zone.

Conclusions

Drake Passage is floored by basalts which closely resemble those occurring at mid-ocean ridges. The basalts of dredge 3 are part of a fractionated sequence exhibiting trends of iron, titanium, and incompatible trace-element enrichment. However, K and Rb levels and Ce_N/Y_N ratios are as low as those in most MORB.

Basalts from the South Sandwich spreading center are enriched in K, Rb, Ba, and Sr relative to MORB, although the Sr-isotope ratios are similar to those found in MORB. However, the mantle source is not depleted in light rare-earth elements, unlike the mantle source for MORB. The basalts of dredges 22 and 24 could be products of fusion of mantle periodotite under more hydrous conditions than that which existed at the sites of either dredges 20 and 23 or most ocean ridge basalts. The high K, Rb, Ba, and Sr contents relative to Zr in the dredge-22 and -24 basalts suggest the involvement of hydrous, alkali, and alkaline-earth element enriched fluids during the formation of these basalts. It is possible that these fluids arose from the dehydrating oceanic lithosphere subducting beneath the South Sandwich Islands, the fluids ascending and thus locally penetrating the source of the back-arc basin basalts.

Acknowledgments. We would like to thank the captain, officers, and crew of the R.R.S. *Shackleton* who provided dredging facilities in the Scotia Sea during the 1971–72 and 1973–74 field seasons. Microprobe facilities were provided by Leicester University, and rare earths and strontium isotopes were determined at the Isotope Dilution Laboratory, Oxford University and were made available by R. J. Pankhurst. Studies in the Scotia arc region are financed by the Natural Environment Research Council, U.K.

References cited

ARCYANA. 1977. Rocks collected by bathyscaph and diving saucer in the FAMOUS area of the Mid-Atlantic Rift Valley: petrological diversity and structural setting. *Deep Sea Res., 24,* 565–89.

Baker, P. E. 1968. Comparative volcanology and petrology of the Atlantic-island arcs. *Bull. volcan., 32,* 189–206.

Baker, P. E. 1978. The South Sandwich Islands: II. Petrology and geochemistry. *Scient. Rep. Br. antarc. Surv., 93,* 1–34.

Barker, P. F. 1970. Plate tectonics of the Scotia Sea region. *Nature, Lond., 228,* 1293–97.

Barker, P. F. 1972. A spreading centre in the east Scotia Sea. *Earth planet. Sci. Lett., 15,* 123–32.

Barker, P. F. 1976. The tectonic framework of Cenozoic volcanism in the Scotia Sea region—A review. (*In* González-Ferrán, O., *ed. Andean and Antarctic Volcanology Problems.* Rome, International Association of Volcanology and Chemistry of the Earth's Interior, 330–46.)

Barker, P. F., and Burrell, J. 1977. The opening of Drake Passage. *Mar. Geol., 25,* 15–34.

Barker, P. F., and Griffiths, D. H. 1972. The evolution of the Scotia Ridge and Scotia Sea. *Phil. Trans. R. Soc. Lond.,* Ser. A, *271,* 151–83.

Barker, P. F., and Griffiths, D. H. 1977. Towards a more certain reconstruction of Gondwanaland. *Phil. Trans. R. Soc. Lond.,* Ser. B, *279,* 143–59.

Brett, C. P. 1977. Seismicity of the South Sandwich Islands region. *Geophys. J. R. Astr. Soc., 51,* 453–64.

Bryan, W. B. 1972. Morphology of quench crystals in submarine basalts. *J. geophys. Res., 77,* 5812–19.

Cann, J. R. 1971. Major element variation in ocean floor basalts. *Phil. Trans. R. Soc. Lond.,* Ser. A, *268,* 495–505.

Chayes, F. 1966. Alkaline and subalkaline basalts. *Am. J. Sci., 265,* 128–45.

Chayes, F. 1972. Silica saturation in Cenozoic basalt. *Phil. Trans. R. Soc. Lond.,* Ser. A, *271,* 285–96.

Dalziel, I. W. D., and Elliot, D. H. 1973. The Scotia arc and Antarctic margin. (*In* Nairn, A. E. M., and Stehli, F. G., *eds. The Ocean Basins and Margins. 1. The South Atlantic.* New York, Plenum Press, 171–246.)

de Wit, M. J. 1977. The evolution of the Scotia arc as a key to the reconstruction of southwestern Gondwanaland. *Tectonophysics, 37,* 53–81.

Engel, A. E. J., Engel, C. G., and Havens, R. G. 1965. Chemical characteristics of oceanic basalts and the upper mantle. *Geol. Soc. Am. Bull., 76,* No. 7, 719–34.

Erlank, A. J., and Kable, E. J. D. 1976. The significance of incompatible elements in Mid-Atlantic Ridge basalts from 45° N with particular reference to Zr/Nb. *Contr. Miner. Petrol., 54,* 281–91.

Ewart, A., Bryan, W. B., and Gill, J. B. 1973. Mineralogy and geochemistry of the younger volcanic islands of Tonga, S.W. Pacific. *J. Petrology, 14,* 429–65.

Ewing, J. I., Ludwig, W. J., Ewing, M., and Eittreim, S. L. 1971. Structure of the Scotia Sea and Falkland Plateau. *J. geophys. Res., 76,* 7118–37.

Forsyth, D. W. 1975. Fault plane solutions and tectonics of the South Atlantic and Scotia Sea. *J. geophys. Res., 80,* 1429–43.

Frey, F. A., Bryan, W. B., and Thompson, G. 1974. Atlantic ocean floor: geochemistry and petrology of basalts from Legs 2 and 3 of the Deep Sea Drilling Project. *J. geophys. Res., 79,* 5507–27.

Gast, P. W. 1968. Trace element fractionation and the origin of tholeiitic and alkaline magma types. *Geochim. cosmochim. Acta, 32,* 1057–86.

Gill, J. B. 1970. Geochemistry of Viti Levu Fiji and its evolution as an island arc. *Contr. Miner. Petrol., 27,* 179–203.

Gill, J. B. 1976. From island arc to oceanic islands: Fiji, southwestern Pacific. *Geology, 4,* 123–26.

Gledhill, A., and Baker, P. E. 1973. Strontium isotope ratios in volcanic rocks from the South Sandwich Islands. *Earth planet. Sci. Lett., 19,* 369–72.

Green, D. H. 1973. Contrasted melted relations in a pyrolite upper mantle under mid-oceanic ridge, stable crust and island arc environments. *Tectonophysics, 17,* 285–97.

Hart, S. R. 1971. K, Rb, Cs, Sr and Ba contents and Sr isotope ratios of ocean floor basalts. *Phil. Trans. R. Soc. Lond.,* Ser. A, *268,* 573–88

Hart, S. R., Erlank, A. J., and Kable, E. J. D. 1974. Sea floor basalt alteration: some chemical and Sr isotopic effects. *Contr. Miner. Petrol., 44,* 219–30.

Hart, S. R., Glassley, W. E., and Karig, D. E. 1972. Basalts and sea-floor spreading behind the Mariana Island arc. *Earth planet. Sci. Lett., 15,* 12–18.

Hawkes, D. D. 1962. The structure of the Scotia arc. *Geol. Mag., 99,* 85–91.

Hawkesworth, C. J., O'Nions, R. K., Pankhurst, R. J., Hamilton, P. J., and Evensen, N. M. 1977. A geochemical study of island-arc and back-arc tholeiites from the Scotia Sea. *Earth planet. Sci. Lett., 36,* 253–62.

Hedge, C. E., and Peterman, Z. E. 1970. The Sr-isotope composition of basalts from the Gorda and Juan de Fuca Rises, northeast Pacific Ocean. *Contr. Miner. Petrol., 27,* 114–20.

Hooker, P. J., O'Nions, R. K., and Pankhurst, R. J. 1975. Determinations of rare-earth elements of U.S.G.S. standard rocks by mixed solvent ion exchange and mass spectrometric isotope dilution. *Chem. Geol. 16,* 189–96.

Jakeš, P., and White, A. J. R. 1972. Major and trace element abundances in volcanic rocks from orogenic areas. *Geol. Soc. Am. Bull., 83,* 29–40.

Kanimori, H. 1971. Great earthquakes at island arcs and the lithosphere. *Tectonophysics, 12,* 187–98.

Karig, D. E. 1974. Evolution of arc systems in the western Pacific. *A. Rev. Earth planet. Sci., 2,* 51–75.

Kay, R. W., Hubbard, J. J., and Gast, P. W. 1970. Chemical characteristics and origin of oceanic ridge volcanic rocks. *J. geophys. Res., 75,* 1585–613.

Kushiro, I. 1972. Effect of water on the composition of magmas formed at high pressures. *J. Petrology, 13,* 311–34.

Kushiro, I. 1975. On the nature of silicate melts and the significance of magma genesis: regularities in the shift of the liquidus boundaries involving olivine, pyroxene, and silicate minerals. *Am. J. Sci., 275,* 411–31.

Lloyd, F. E., and Bailey, P. K. 1975. Light element metasomatism of the continental mantle: the evidence and consequences. (*In* Ahrens, L. H., Press, F., Runcorn, S. K., and Urey, H. C., *eds. Physics and Chemistry of the Earth.* New York, Pergamon Press, 389–416.)

Matthews, D. H. 1959. Aspects of the geology of the Scotia arc. *Geol. Mag., 96,* 425–41.

Melson, W. G., and Thompson, G. 1971. Petrology of a transform fault zone and adjacent ridge segments. *Phil. Trans. R. Soc. Lond.,* Ser. A, *268,* 423–41.

Moberly, R. 1972. Origin of lithosphere behind island arcs, with reference to the western Pacific. *Mem. geol. Soc. Am., 132,* 35–56.

Muir, I. D., Tilley, C. E., and Scoon, J. H. 1964. Basalts from the northern part of the rift zone of the Mid-Atlantic Ridge. *J. Petrology, 5,* 409–34.

Mysen, B. O., and Boettcher, A. L. 1975. Melting of hydrous mantle. 1. Phase relations of natural peridotite at high pressures and temperatures with controlled activities of water, carbon dioxide and hydrogen. *J. Petrology, 16,* 520–48.

O'Nions, R. K., and Gronvold, K. 1973. Petrogenetic relationships of acid and basic rocks in Iceland: Sr-isotopes and rare-earth elements in late and post glacial volcanics. *Earth planet. Sci. Lett., 19,* 397–409.

O'Nions, R. K., and Pankhurst, R. J. 1974. Petrogenetic significance of isotope and trace element variations in volcanic rocks from the mid-Atlantic. *J. Petrology, 15,* 603–34.

O'Nions, R. K., Pankhurst, R. J., and Gronvold, K. 1976. Nature and development of basalt magma sources beneath Iceland and the Reykjanes Ridge. *J. Petrology, 17,* 315–38.

Packham, G. H., and Falvey, D. A. 1971. An hypothesis for the formation of marginal seas in the western Pacific. *Tectonophysics. 11,* 79–110.

Pankhurst, R. J., and O'Nions, R. K. 1973. Determination of Rb/Sr and ^{87}Sr/^{86}Sr ratios of some standard rocks and evaluation of X-ray fluorescence spectrometry in Rb-Sr geochemistry. *Chem. Geol., 12,* 127–36.

Saunders, A. D., and Tarney, J. 1979. The geochemistry of basalts from a back-arc spreading centre in the east Scotia Sea. *Geochim. cosmochim. Acta, 43,* 555–72.

Schilling, J-G. 1971. Sea-floor evolution: rare earth evidence. *Phil. Trans. R. Soc. Lond.,* Ser. A, *268,* 663–706.

Schilling, J-G. 1975. Rare-earth variations across 'normal segments' of the Reykjanes ridge, 60–53° N, the Mid-Atlantic ridge, 29° S, the East Pacific rise, 2–19° S, and evidence on the composition of the underlying low-velocity layer. *J. geophys. Res., 80,* 1459–73.

Schnetzler, C. C., and Philpotts, J. A. 1970. Partition coefficients of rare-earth elements between igneous matrix material and rock-forming mineral phenocrysts—II. *Geochim. cosmochim. Acta, 34,* 331–40.

Sleep, N. H., and Toksoz, M. N. 1971. Evolution of marginal basins. *Nature, Lond., 233,* 548–50.

Tarney, J., Saunders, A. D., and Weaver, S. D. 1977. Geochemistry of volcanic rocks from the island arcs and marginal basins of the Scotia arc region. (*In* Talwani, M., and Pitman, W. C., III, *eds. Maurice Ewing Series, 1: Island Arcs, Deep Sea Trenches, and Back-Arc Basins.* Washington, American Geophysical Union, 367–77.)

Weaver, S. D., Sceal, J. S. C., and Gibson, I. L. 1972. Trace-element data relevant to the origin of trachytic and pantelleritic lavas in the East African rift system. *Contr. Miner. Petrol., 36,* 181–94.

Yoder, H. S., and Tilley, C. E. 1962. Origin of basalt magmas: an experimental study of natural and synthetic rock systems. *J. Petrology, 3,* 342–532.

24

Scotia Arc Volcanism Recorded in Abyssal Piston Cores Downwind from the Islands

A.N. FEDERMAN*, N. D. WATKINS†, and H. SIGURDSSON
Graduate School of Oceanography, University of Rhode Island, Kingston, Rhode Island 02881, U.S.A.

Abstract. Piston cores from cruises of the R/V *Vema,* the USNS *Eltanin,* and the ARA *ISLAS Orcadas* have been used to study the chemistry and other properties of atmospherically transported volcanic ash originating from the South Sandwich Islands during the last 150,000 years.

Electron microprobe analysis has been performed on 78 megascopically visible ash layers from 13 piston cores east of the islands. The chemical composition of glass and pumice shards is similar to whole-rock compositions reported from several of the islands. More than 90% of the glasses examined can be classified as low-K andesites, with SiO_2 content from 53% to 63% and $K_2O < 0.7\%$. Visible tephra layers are not only due to atmospheric transport of explosion products, but also result from marine dynamic processes such as ice and pumice rafting and turbidity currents. Initial results indicate that volcanism in the arc has been mildly explosive for about the last 150,000 years, with at least one major eruption within the last 30,000 years which produced a visible tephra layer 700 km downwind from the source.

Ice cover, short field seasons, inaccessibility, and other logistic difficulties inhibit the study of the South Sandwich Islands (Figure 24.1). In particular, little is known about the recent volcanic history of the Scotia arc (Baker, 1972). Useful insights can be obtained from abyssal piston cores in the region of the Scotia arc; they contain tephra layers which can be recovered for study. Tephra layers in deep-sea cores may be used to reconstruct the explosive volcanic history of a region, if consideration is given to all the factors that govern the emplacement of such layers in the marine environment. In principle, the South Sandwich Islands are favorably located for examining the downwind dispersal of volcanic ash by study of abyssal piston cores. Steady winds flow normal to the trend of the Scotia arc and trench. Any major slumps of ash that would generate ash turbidites would tend to be trapped in the trench, and any such deposits should be absent in the area east of the trench. In practice, several factors, such as bottom water flow and tephra rafting, severely complicate the sedimentary record and any inferred volcanic history of the region.

The Weddell Sea is located directly south of the Scotia arc (Figure 24.1) and is the source of northward-flowing Antarctic Bottom Water (AABW). The flux of AABW and its effect on sediments varies through time; periodically it has reached sufficient velocity to erode and modify abyssal sediments around the Antarctic continent (Kennett and Watkins, 1975). The effect of such modification in the sediments near the South Sandwich Islands may be to concentrate stratigraphically dispersed coarse debris into lag deposits by the removal of finer, lighter silt and clay particles. These lag deposits might well appear to be visible tephra layers, where the material being concentrated is mostly volcanic glass.

Fluctuations in the relative input of ice-rafted debris (IRD) may also complicate the sedimentary record. Tephra present on the surface of pack ice has been observed near Saunders and Thule Islands (Baker, 1968). Increased fluxes of IRD, believed to originate from the South Sandwich Islands, have been correlated with increases in winter sea ice cover as recorded by the near-ice diatom *Eucampi balaustium* (Cooke and others, 1977). The flux of IRD to the sea floor is likely to vary systematically with climate.

A pumice raft was observed in the region of the South Sandwich Islands in 1962 and was attributed to a submarine eruption near Zavodovski Island (Gass and others, 1963). This pumice eventually reached the shores of Australia (Sutherland, 1965). While pumice clasts are found as dispersed fragments throughout the cores, large fragments

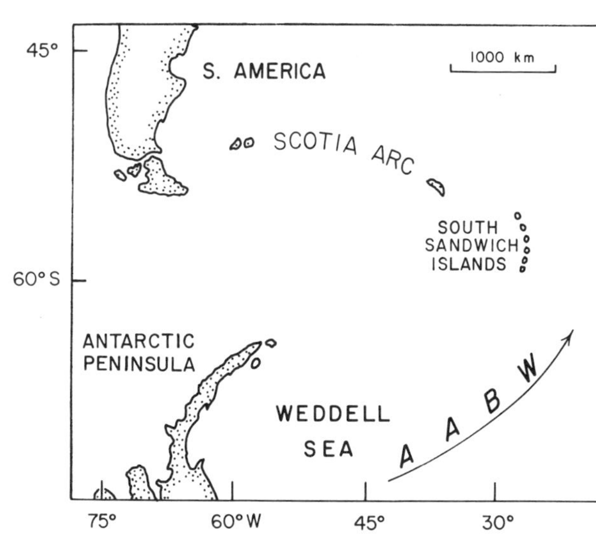

Figure 24.1. The southern Atlantic Ocean, showing the position of the South Sandwich Islands and the direction of flow of the Antarctic Bottom Water. The islands are not to scale.

* Present address: School of Oceanography, Oregon State University, Corvallis, Oregon 97331, U.S.A.
† Deceased.

(>1 cm) are found only on the submarine slopes of the islands themselves. Small fragments of pumice (<1 cm) occur dispersed throughout the cores but not in great abundance. Pumice rafting does not appear to form visible layers, but it may add to layers formed by other processes.

Based on theoretical modeling of volcanic clouds, the greatest accumulation of any particular size class of tephra would not necessarily be at the source but, depending on the wind field and cloud height, at some distance downwind of that source (Shaw and others, 1974). When all the above-mentioned complications are considered, distinguishing tephra deposits produced by explosive eruptions from those produced by other means can be a difficult matter.

Prior to this work, three megascopically visible tephra layers in cores taken on *Vema* cruise 14 had been discovered, and they were correlated on the basis of the index of refraction of glass shards (Ninkovich and others, 1964). One core with a visible layer was 700 km away from the inferred source.

A single tephra layer in sea-floor sediments may show a high degree of chemical heterogeneity for many reasons. Inhomogeneity in a single layer may be caused by a mixed-magma eruption (Sparks and others, 1977), and such layers have been found in abyssal cores (Carey and Sigurdsson, 1977). The chemical compositions of the products of a single eruption can vary significantly over a short period of time, resulting in a heterogeneous layer on the sea floor (Jezek, 1976). Two different sources erupting simultaneously can also produce a mixed layer (Huang and Watkins, 1976). Sedimentary processes can also introduce heterogeneity (Federman, 1977). This study extends previous work by examination of cores taken on R/V *Vema* cruise 14, USNS *Eltanin* cruise 8, and ARA *Islas Orcadas* cruise 07-75. The chemical composition of the glass shards is determined by electron microprobe analysis, primarily to determine the volcanic history of the region.

Table 24.1. Location and pertinent data on cores used in this study

Cruise and core number	Latitude south	Longitude west	Depth (m)	Core length (cm)
Vema 14				
53	56°43′	27°31′	2283	485
54	56°51′	25°08′	1805	435
55	57°36′	23°51′	2212	490
56	57°36′	23°58′	1432	438
57	57°33′	17°08′	1312	597
58	57°30′	13°40′	968	490
Eltanin 8				
14	55°08′	26°05′	2322	285
17	58°15′	25°51′	602	221
18	58°31′	22°24′	1301	974
19	60°03′	22°34′	1359	1084
27	58°36′	23°47′	1870	223
Islas Orcadas 7-75				
27	57°03′	23°34′	1463	1113

Methods

The initial step in this study was to examine the visible tephra layers in selected cores. Glass from a single explosive eruption is generally believed to be homogeneous and a good indicator of the final liquid composition for the erupted magma (Smith and Westgate, 1968). Electron microprobe analysis has been shown to characterize accurately such tephras if care is used to avoid microlites occurring in glass and pumice, since these would produce mixed crystal-glass analyses (Smith and Westgate, 1968; Jezek, 1976).

Megascopically visible ash layers were sampled in 13 cores (Figure 24.2, Table 24.1). About 100 to 300 mg of

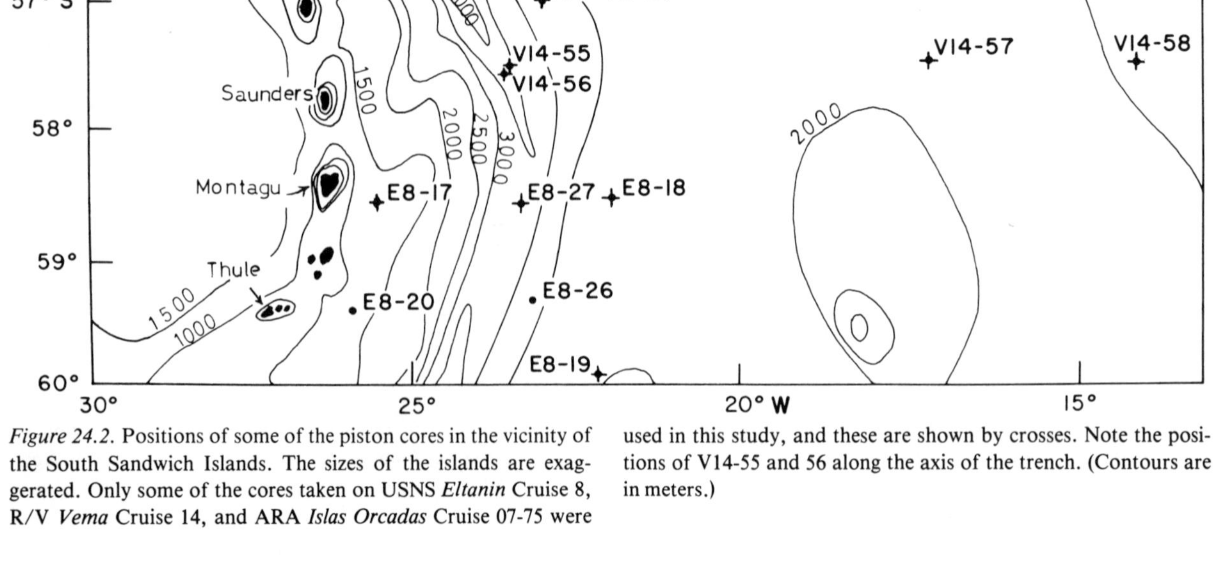

Figure 24.2. Positions of some of the piston cores in the vicinity of the South Sandwich Islands. The sizes of the islands are exaggerated. Only some of the cores taken on USNS *Eltanin* Cruise 8, R/V *Vema* Cruise 14, and ARA *Islas Orcadas* Cruise 07-75 were used in this study, and these are shown by crosses. Note the positions of V14-55 and 56 along the axis of the trench. (Contours are in meters.)

material were removed from each of the 78 layers. Carbonate was dissolved with 10% buffered glacial acetic acid. Samples were disaggregated in distilled water and calgon solution using ultrasonic cleaning. Each sample was wet-sieved, and the >43 μ size fraction was retained.

Glass-crystal-lithic proportions were determined by examination of sieved samples with a petrographic microscope. To avoid microlites in the microprobe analysis of glass, polished thin sections were made of all samples.

The concentration of nine major elements in glass and pumice shards was determined with a Jeol JXA-50A automated electron microprobe analyzer, using the Bence-Albee correction method for major element oxides. Operating conditions were: a beam current of 0.0140 microamps, accelerating voltage of 15 kV, and a beam diameter of 1 to 2 μ. Tests showed little loss of Na or other volatiles in the first 30 seconds of analysis. During each analysis, counts were first collected for Na in order to reduce error due to Na loss.

Variations within each sample were determined by coefficients of variation (% CV) for eight major elements, with Mn excluded because of low concentrations bordering on the precision of the method (Borchardt and others, 1971). The % CV was used to test the homogeneity of single layers within a core. If such a layer proved to be homogeneous, the average of the analyses for that layer was compared for correlation purposes to the average value for each layer using the same CV method.

Results

Chemical compositions of tephra glass are presented in Tables 24.2 and 24.3. A silica frequency histogram is given in Figure 24.3, a plot of alkalis versus silica is shown in Figure 24.4, and the AFM diagram is presented in Figure 24.5.

Of the 78 layers that were examined in 13 cores, only 14 had a mean coefficient of variation less than 12. The averages for these analyses are shown in Table 24.2. We assigned a % CV of less than 12 to indicate a single volcanic event. This value of 12 was chosen after a study of four ash layers in the Mediterranean Sea (Federman and Carey, 1980) revealed that layers which were well correlated to each other by independent stratigraphic methods had a mean % CV of 12 or less when using the CV method. Layers that were inhomogeneous, or not correlated, however, had coefficients of 15 or greater.

Ninkovich and others (1964) used the optical index of refraction of glass and pumice shards to correlate ash layers (Figure 24.6*b*). Using glass chemical compositions to correlate layers, however, produces different results. Layers previously correlated by index of refraction can be shown to have distinctly different chemical compositions (Figure 24.6*a*, Table 24.3). The index of refraction is a property of the total glass and may be sensitive to parameters other than chemical composition. For example, small changes in the oxidation state of Fe may have a potentially large effect on the index of refraction. The amount of hydration is also known to affect the index (Bonatti, 1965). We conclude that the index of refraction is inadequate to accurately characterize or correlate tephra layers which have a small compositional range.

Discussion

More than 90% of the glasses examined are classified as low-K andesites with SiO_2 contents of 54–63% and K_2O < 0.7%. This is consistent with the petrochemistry of rock specimens from the arc. The islands consist mostly of effusive basalts of the low-K oceanic island-arc tholeiite type, and they are believed to represent an early stage in the evolution of an island-arc group (Baker, 1968). Three layers of dacitic glass were found in the cores, and a few shards of quartz-normative tholeiitic composition occur in mixed layers. The sole occurrence of rhyolite known in the arc is the pumice raft sighted in 1962 (Gass and others, 1963). No glass of similar composition has been found in any of the samples examined, either as layers or as dispersed material. The possibility of individual pumice clasts of rhyolitic composition existing dispersed throughout the cores still

Table 24.3. Comparison of correlated and mixed uncorrelated tephra layers (in wt. %)

	Correlated pair of layers		Uncorrelated pair of mixed layers	
Cruise and core number	V14 57	V14 58	V14 57	V14 58
Depth to layer (cm)	64	130	570	315
SiO_2	54.21	54.63	60.75	55.99
Al_2O_3	14.10	14.29	14.29	15.55
FeO (total Fe)	12.26	13.49	11.25	12.60
MgO	4.68	3.94	2.10	2.21
CaO	8.16	8.97	6.16	8.08
Na_2O	2.40	2.48	3.57	3.34
K_2O	0.43	0.44	0.21	0.50
TiO_2	1.23	1.26	1.43	1.36
MnO_2	0.18	0.24	0.21	0.28
Mean % CV	11.2	12.3	20.1	32.5
% CV between layers	11.8		17.6	

Table 24.2. Mean chemistry (in wt. %) and % CV for selected tephra layers

Cruise and core number	E8 18	E8 18	E8 19	E8 18	E8 18	E8 18	V14 54	V14 54	V14 54	V14 55	V14 55	V14 55	V14 56
Depth to layer (cm)	730	790	908	460	518	658	85	160	169	110	130	350	20
SiO_2	68.57	54.81	54.43	55.25	58.15	59.99	57.15	59.06	55.11	53.72	55.44	57.13	58.99
Al_2O_3	13.06	15.58	15.92	14.47	15.07	15.33	14.12	14.23	14.18	15.08	15.05	14.76	15.03
FeO (total Fe)	5.66	11.12	11.61	12.80	11.29	9.58	10.92	9.74	12.20	11.07	11.47	11.08	9.65
MgO	0.62	3.89	4.56	3.81	3.06	2.73	4.19	2.47	4.60	5.58	3.68	3.97	2.69
CaO	3.71	8.13	8.93	8.37	7.39	7.12	8.01	6.89	8.16	9.53	8.57	8.13	7.30
Na_2O	2.51	2.98	2.82	2.86	3.25	2.78	2.86	2.78	2.55	1.88	3.00	2.72	3.23
K_2O	0.62	0.69	0.48	0.27	0.46	0.54	0.33	0.48	0.42	0.37	0.49	0.35	0.36
TiO_2	0.55	1.23	0.91	1.18	1.07	0.86	0.94	1.03	1.20	1.00	1.20	1.03	0.96
MnO_2	0.10	0.20	0.16	0.24	0.25	0.21	0.18	0.16	0.21	0.23	0.24	0.21	0.20
Mean % CV	24.3	9.9	11.1	8.8	11.4	8.2	3.4	10.6	11.2	7.5	10.7	12.1	9.7

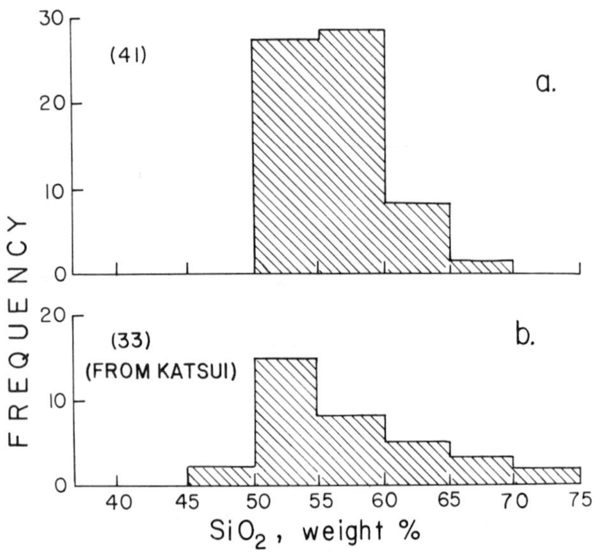

Figure 24.3. Silica frequency histograms for (*a*) South Sandwicn abyssal tephra in 5 wt. % intervals and (*b*) South Sandwich Islands whole-rock data in 5 wt. % intervals (from Katsui, 1972).

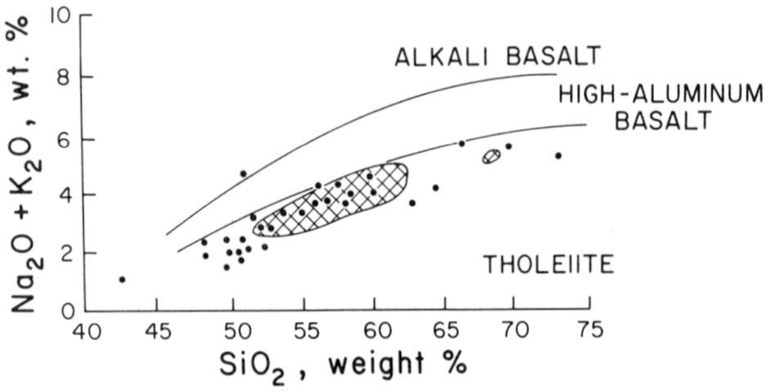

Figure 24.4. South Sandwich abyssal tephra (crosshatch) plotted on the total alkalis versus silica diagram of Baker (1968) for South Sandwich Islands whole-rock data (black circles).

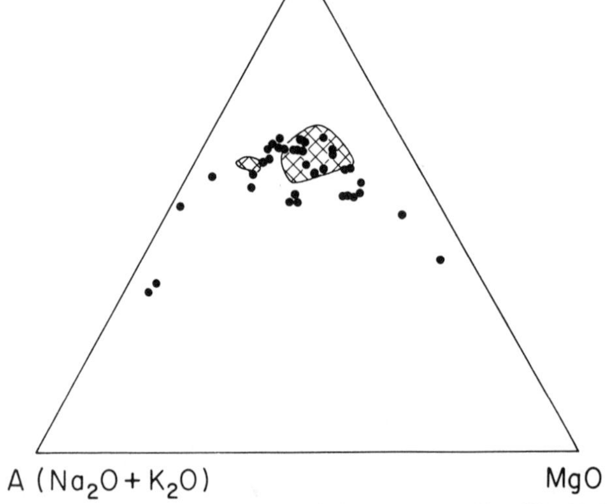

Figure 24.5. AFM diagram for South Sandwich Islands rocks, from Baker (1968), with abyssal tephra superimposed as the cross-hatched field.

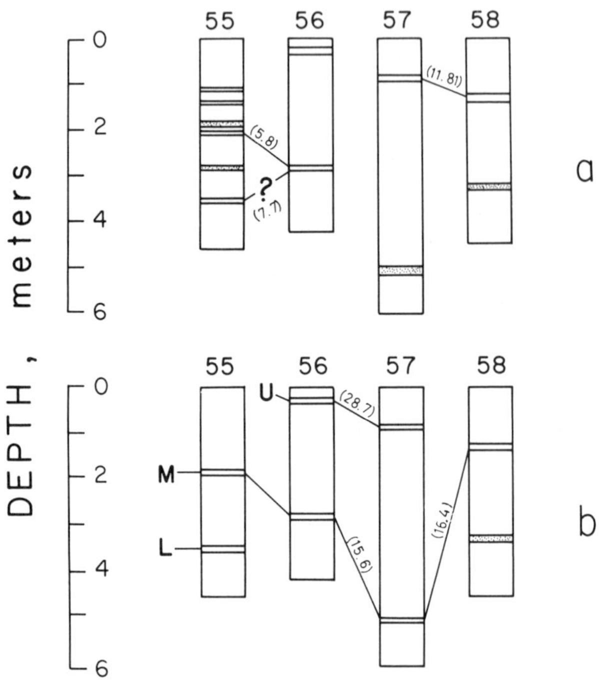

Figure 24.6. a. Cores *Vema* 14–55, 56, 57, and 58; layers with % CV > 20 are stippled. Solid lines show tephras which are chemically correlated. The numbers in parentheses above these lines are the coefficient of variations (% CV) for the means of the layers. The question mark shows a possible, but less probable, correlation. *b.* Index of refraction correlation after Ninkovich and others (1964). U, M, and L represent upper, middle, and lower tephras, respectively. Compare to Figure 24.6*a*. Thicknesses of the layers are exaggerated.

persists, but rhyolitic eruptions are probably extremely rare throughout the South Sandwich Islands. Tephra-producing eruptions of basaltic magma are also apparently rare in this arc.

Due to the vigorous production of planktonic flora and fauna resulting from the proximity of the Antarctic Convergence, sedimentation rates in the region are high, in some cases >8 cm/1000 yr (D. W. Cooke, personal communication, 1977). With such high rates, the longest core in the collection can only record at most 150,000 years of volcanic history. Since the South Sandwich Islands are believed to be 4 m.y. old (Baker, 1968), only a very small fraction of the total explosive volcanic history can be recovered from a study of piston cores. Nevertheless, the Scotia arc produces material of a restricted range in composition, although the explosive products of several islands over a period of 150,000 years were sampled. Highly fractionated silicic volcanism is rare in this young arc. This can be contrasted with the much greater range of compositions produced by older arcs (Baker, 1968).

The total alkalis versus silica diagram (Figure 24.4) shows an overlap between tephra glass and whole-rock compositions. The AFM diagram (Figure 24.5) shows that extreme compositions are not represented in the tephra glass and that the main trends of glass and rock coincide. The narrow range of glass compositions and the restricted range of the field in the AFM diagram may be due to the elimination of highly fractionated compositions from tephra-producing eruptions, or they may be a reflection of the lack of these chemical extremes in the arc as a whole.

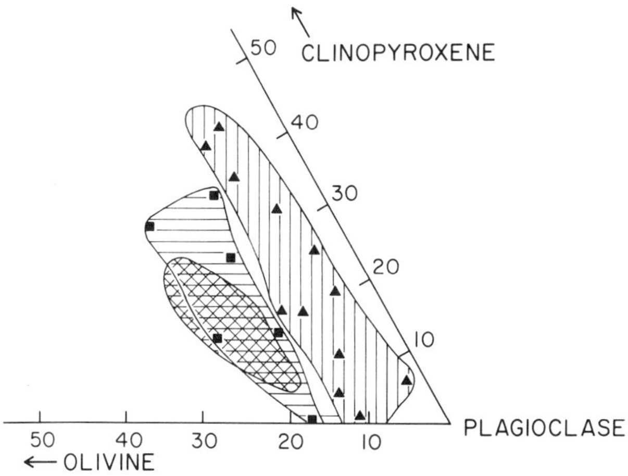

Figure 24.7. The quartz projection to the plane of clinopyroxene, plagioclase, and olivine in the basalt tetrahedron of Yoder and Tilley (1962). Vertical stripes and triangles: hydrous mantle compositions produced by experiments of partial melting. Horizontal stripes and squares: compositions of liquids produced by partial melts of hydrous basalts. Crosshatch: South Sandwich tephra and whole-rock data after Baker (1972).

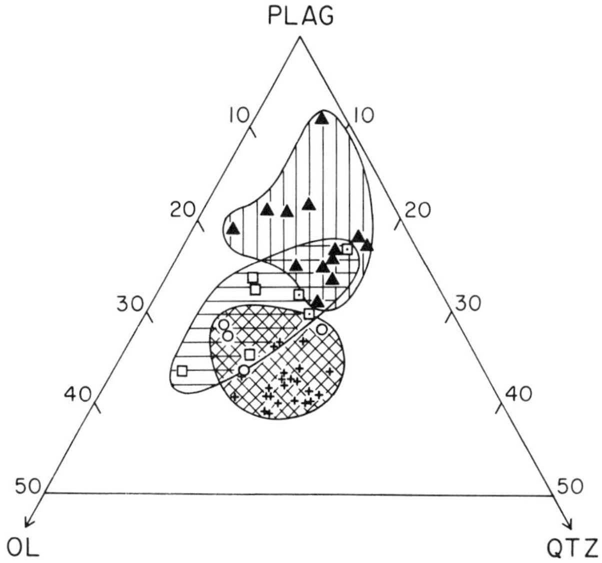

Figure 24.8. The clinopyroxene projection to the plane of plagioclase, olivine, and quartz. Crosses: South Sandwich tephra data. Circles: whole-rock data. (See Figure 24.7 for explanation of other symbols.)

Evidence for a major explosive eruption appears only in cores V14-57 and 58, at about 30,000 years ago (Figure 24.6*a*). The ash layer correlated in cores 57 and 58 becomes finer in grain size, and the glass becomes more abundant, with increasing distance from the source.

Two possible simplified models of island-arc genesis are: (1) hydrous melting of mantle material brought about by degassing of a subducted slab of oceanic crust or (2) hydrous melting of the oceanic crust itself. Liquids have been produced and analyzed in partial melting experiments of hydrous mantle material (Mysen and Boetcher, 1975) and hydrous basalts (Holoway and Burnham, 1972; Helz, 1976). The chemical compositions of the experimental liquids, together with the analyses of the South Sandwich tephra and whole-rock data (Tyrrell, 1945; Baker, 1968), were plotted as projections in the system CMAS (O'Hara, 1968) and the basalt tetrahedron (Yoder and Tilley, 1962).

Analyses of projected planes in the system CMAS are ambiguous, but in the basalt tetrahedron the quartz projection (Figure 24.7) and the clinopyroxene projection (Figure 24.8) show that if hydrous mantle material were melting, a substantial amount of plagioclase fractionation would have to take place to produce the South Sandwich compositions. Liquids produced from partial melts of basaltic crust, however, would not need to fractionate any phases to approach the composition of the South Sandwich tephra or volcanic rocks. Isotopic evidence from the Scotia arc also indicates that at least some degree of partial melting of oceanic crust is taking place (O'Nions and others, 1977).

Conclusions

The study of tephras in a series of abyssal sediment cores east of the Scotia arc leads to the following conclusions:

1. Chemical evidence confirms the existence of tephra from the South Sandwich Islands, as both discrete fragments and as widely dispersed layers, at distances as great as 700 km from their source.
2. In any region with a dynamic sedimentary environment, visible-ash layer stratigraphy must be supplemented by studies of other independent parameters (e.g., nonvolcanogenic ice-rafted debris, manganese micronodule accumulation rates, and particle sizes). These parameters, together with study of the dispersed tephra component and a reliable biostratigraphic or absolute time control framework, are needed to reconstruct completely the history of volcanic activity from the study of abyssal piston cores.
3. The explosive products of the Scotia arc have been for the most part low-K andesites of narrow compositional range, with no significant chemical variation during the last 150,000 years.
4. There has been at least one explosive eruption of major significance in the South Sandwich Islands within the last 30,000 years.
5. It seems likely that the andesites of the South Sandwich Islands originate from partial melting of oceanic crust, rather than from hydrous melting of mantle material.
6. The short time span covered by the piston cores limits this study, but much additional information could be obtained from the study of Deep Sea Drilling Project cores from the region. However, no such cores have yet been taken.

Acknowledgments. The *Vema* core collection was sampled with the kind assistance of F. McCoy, D. Ninkovich, and D. Cooke. The *Eltanin* core collection was sampled with the help of D. Cassidy and his staff. The URI microprobe is supported by NSF grant no. OCE-75-13296 to J-G. Schilling and H. Sigurdsson. The cooperation of the Argentine Naval Hydrographic Office is acknowledged. This research is supported by the National Science Foundation Office of Polar Programs through grant no. OPP-75-19222.

References cited

Baker, P. E. 1968. Comparative volcanology and petrology of the Atlantic island-arcs. *Bull. Volcan., 32,* 189–206.

Baker, P. E. 1972. Recent volcanism and magmatic variation in the Scotia arc. (*In* Adie, R. J., *ed. Antarctic Geology and Geophysics.* Oslo, Universitetsforlaget, 57–60.)

Bonatti, E. 1965. Palagonite, hyaloclastites, and alteration of volcanic glass in the ocean. *Bull. Volcan., 28,* 257–69.

Borchardt, G. A., Harward, M. E., and Scmitt, R. A. 1971. Correlation of volcanic ash deposits by activation analysis of glass separates. *Quat. Res., 1,* 247–60.

Carey, S., and Sigurdsson, H. 1977. The eruptions of the Sofriere and Mt. Pelee in 1902 and a mixed surface tephra layer in the Tobago Trough. *Abstr. Trans. Am. Geophys. Un., 58,* 528.

Cooke, D. W., Burckle, L. H., and Hayes, J. D. 1977. Winter sea ice cover during the late Pleistocene in the Antarctic Ocean. *Abstr. Trans. Am. Geophys. Un., 58,* 416.

Federman, A. 1977. The chemical characterization of abyssal tephras from the South Sandwich Islands. *Abstr. Trans. Am. Geophys. Un., 58,* 528.

Federman, A., and Carey, S. N. 1980. Electron microprobe correlation of tephra layers from eastern Mediterranean abyssal sediments and the island of Santorini. *Quat. Res., 13,* 160–71.

Gass, I. G., Harris, P. G., and Holdgate, M. W. 1963. Pumice eruption in the area of the South Sandwich Islands. *Geol. Mag., 100,* 321–30.

Helz, R. T. 1976. Phase relation of basalts in their melting ranges at PH_2O = 5kb. Part II Melt Compositions. *J. Petrology, 17,* 140–93.

Holoway, J. R., and Burnham, C. W. 1972. Melting relations of basalt with equilibrium water pressure less than total pressure. *J. Petrology, 13,* 1–29.

Huang, T-C., and Watkins, N. D. 1976. Contribution of volcanic ash to abyssal sediments downwind from the Azores. *Abstr. Trans Am. Geophys. Un., 57,* 935.

Jezek, P. 1976. Compositional variation within and among volcanic ash layers in the Fiji plateau area. *J. Geol., 84,* 595–616.

Katsui, Y. 1972. Late Cenozoic petrographic provinces of the volcanic rocks from the Andes to Antarctica. (*In* Adie, R. J., *ed. Antarctic Geology and Geophysics.* Oslo, Universitetsforlaget, 181–85.)

Kennett, J. P., and Watkins, N. D. 1975. Deep-sea erosion and manganese nodule development in the Southeast Indian Ocean. *Science, N.Y., 188,* 1011–13.

Mysen, B. O., and Boetcher, A. L. 1975. Melting of a hydrous mantle: II. Geochemistry of crystals and liquids formed by anatexis of mantle peridotites at high pressures and temperatures as a function of controlled activities of water, hydrogen and carbon dioxide. *J. Petrology, 16,* 549–93.

Ninkovich, D., Heezen, B. C., Conolly, J. R., and Burckle, L. H. 1964. South Sandwich tephra in deep-sea sediments. *Deep Sea Res., 11,* 605–19.

O'Hara, M. J. 1968. The bearing of phase equilibria studies in synthetic and natural systems on the origin and evolution of basic and ultrabasic rocks. *Earth Sci. Rev., 4,* 69–133.

O'Nions, R. K., Evanson, N. M., Hamilton, P. J., and Carter, S. R. 1977. Nd- and Sr-isotope compositions of altered oceanic crust and island arc tholeiites. *Abstr. Trans. Am. Geophys. Un., 58,* 533.

Shaw, D. M., Watkins, N. D., and Huang, T-C. 1974. Atmospherically transported volcanic glass in deep-sea sediments: theoretical considerations. *J. geophys. Res., 79,* 3087–91.

Smith, D. G. W., and Westgate, J. A. 1968. Electron-probe technique for characterizing pyroclastic deposits. *Earth planet. Sci. Lett., 5,* 313–19.

Sparks, R. S. J., Sigurdsson, H., and Wilson, L. 1977. Magma mixing: a tool for triggering acid explosive eruptions. *Nature, Lond., 267,* 315–18.

Sutherland, F. L. 1965. Dispersal of pumice supposedly from the South Sandwich Islands eruption, on southern Australian shores. *Nature, Lond., 297,* 1332–35.

Tyrrell, G. W. 1945. Report on rocks from West Antarctica and the Scotia arc. *"Discovery" Rep., 23,* 37–102.

Yoder, H. S., and Tilley, C. E. 1962. Origin of basalt magmas: an experimental study of natural and synthetic rock systems. *J. Petrology, 3,* 342–532.

25

Sr-Isotope and Trace Element Geochemistry of Cenozoic Volcanic Rocks from the Scotia Arc and the Northern Antarctic Peninsula

R. J. PANKHURST
British Antarctic Survey, Natural Environment Research Council, c/o Institute of Geological Sciences, London WC1X 8NG, U.K.

Abstract. Sr-isotopes and rare-earth element (REE) data for the Pliocene–Holocene volcanic rocks of the eastern Scotia Sea and the northern Antarctic Peninsula are reviewed. Clear distinctions are noted between the major rock types of the region, indicating tectonic control of magma geochemistry. The island-arc tholeiites of the South Sandwich Islands are characterized by a wide range of REE concentrations with variable Eu-anomalies (both positive and negative), ascribed to the effects of low-pressure crystal fractionation. The parent magmas have REE patterns with pronounced depletion in light REE. $^{87}Sr/^{86}Sr$ ratios are rather high, in the range 0.7038–0.7041. These features suggest a major contribution to the magmas from the ocean floor subducted along the South Sandwich Trench. In contrast, the mildly alkaline volcanic rocks of the James Ross Island area and Seal Nunataks have REE patterns strongly enriched in light REE but relatively low in heavy REE, with fairly constant Ce/Yb ratios and no negative Eu anomalies. This suggests small degrees of partial melting of a deep garnet periodite source region followed by rapid eruption without crystal fractionation, typical of volcanism away from convergent plate margins. In these rocks $^{87}Sr/^{86}Sr$ ratios of 0.7029–0.7035 correlate positively with Rb/Sr ratios and are interpreted as being inherited from long-lived heterogeneities in the thick lithospheric mantle. The Quaternary volcanic rocks of Bransfield Strait, which are geographically and tectonically intermediate between the above two groups, show geochemical and isotopic features common to both groups. The data presented and reviewed here show that the geochemical distinction of Cenozoic magma types in this region reflects spatial and temporal changes in both source composition and the physical mechanisms of magma generation, especially the depth and degree of partial melting.

Trace element and isotope data are widely used to characterize basalt magma types. They provide complementary constraints on the petrogenesis of magmas in terms of source region composition, degree of partial melting, and subsequent magmatic evolution through crystal fractionation. Sr-isotope ratios are usually considered to be unaffected in melting or crystallization processes and are thus a property of the source material. Variations in the $^{87}Sr/^{86}Sr$ ratio of the mantle are ascribed to long-term differences in Rb/Sr ratios, and this may be used to infer the relative variation in other lithophile trace elements. Trace elements, however, are often strongly fractionated in magmatic processes, and variation within a given volcanic suite can provide information on the type of crystal-liquid equilibria involved. Rare-earth elements (REE) are particularly useful in such studies since on the whole their distribution patterns in different minerals vary smoothly and predictably, with a major discontinuity caused by the preferential substitution of divalent Eu for Sr in plagioclase. Thus petrogenetic conclusions may be based on the relative abundance levels of different rare-earth elements, especially for the light REE (La to Nd) compared to the heavy REE (Dy to Lu), as much as on their absolute concentrations. Application of this approach is gradually providing a better understanding of magma generation and, hence, tectonic controls.

A considerable amount of Sr-isotope and REE data is now becoming available for the Holocene volcanic rocks of the Scotia Sea (Tarney and others, 1977; Hawkesworth and others, 1977) and Bransfield Strait (Weaver and others, 1979). These are here supplemented by new data for the volcanic rocks of the James Ross Island area (Table 25.1). $^{87}Sr/^{86}Sr$ ratios are summarized as histograms in Figure 25.1, and chondrite-normalized REE patterns are shown in Figure 25.2.

Scotia Sea

Basalts dredged from the behind-arc spreading center of the eastern Scotia Sea are olivine tholeiites whose geochemistry is discussed by Hawkesworth and others (1977) and Saunders and others (Chapter 23, this volume). They show some features quite distinct from typical mid-ocean ridge basalts (MORB), notably relatively high incompatible element contents and a significant lack of light rare-earth element depletion (see Table 25.2, column 1). Their measured $^{87}Sr/^{86}Sr$ ratios range from 0.7028 to 0.7034, whereas most fresh abyssal tholeiites have values less than 0.7030. Sea-water alteration could easily account for the higher $^{87}Sr/^{86}Sr$ ratios, since this is a commonly observed phenomenon in submarine basalts (e.g., Hart and others, 1974; O'Nions and Pankhurst, 1976), and the primary magmas may well have had values within the range for other abyssal tholeiites.

It seems unlikely that alteration is responsible for light REE concentrations, however, since these are resistant to

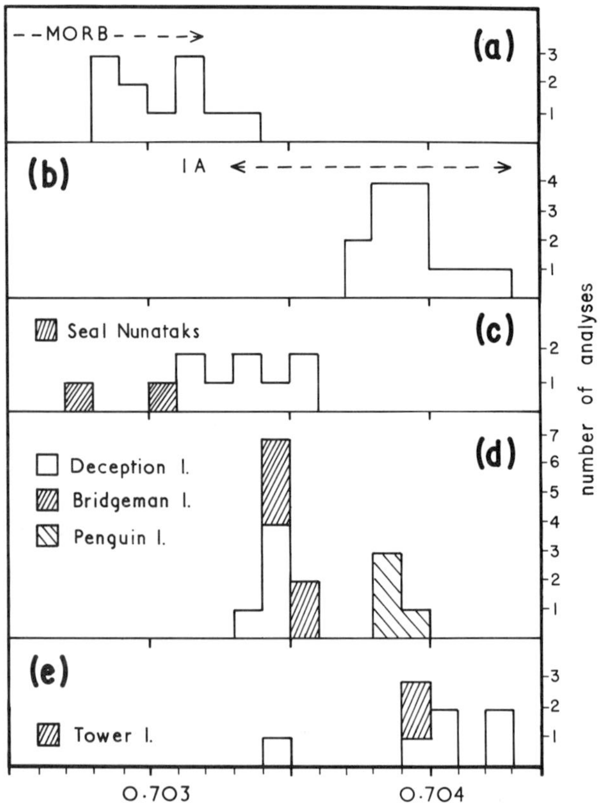

Figure 25.1. $^{87}Sr/^{86}Sr$ ratios for (a) 11 basalts dredged from the Scotia Sea, (b) 12 basalts and 1 andesite from the South Sandwich Islands, (c) 10 alkali basalts and hawaiites from the James Ross Island group and Seal Nunataks, (d) 14 basalts and andesites from the volcanic islands of Bransfield Strait, and (e) 8 basalts from Tower Island and Byers Peninsula, Livingston Island (South Shetland Islands). All are based on high-precision analyses so that a spread of more than two columns represents real variation; in case (a) this may simply be due to sea-water alteration. MORB: mid-ocean ridge basalts. IA: averages for island arcs (Hart and Brooks, 1977). Data sources: (a) Saunders and others (Chapter 23, this volume), (b) Hawkesworth and others (1977), (c) this paper, (d) Weaver and others (1979), and (e) unpublished data (see acknowledgments).

alteration in marine environment (e.g., Kay and Senechal, 1976). Moreover, in the present case, $^{143}Nd/^{144}Nd$ ratios are clearly unaffected, being indistinguishable from those of fresh MORB with $^{87}Sr/^{86}Sr$ less than 0.7030 (Hawkesworth and others, 1977). Crystal fractionation alone also seems a rather unlikely explanation since the absence of Eu-anomalies rules out significant plagioclase removal. The overall increase in light REE concentrations in Figure 25.2*a* would require crystallization of about 70% of the primary magma as olivine and pyroxenes and would still be unable to account for the increase in the ratio of light to heavy REE. Hart and others (1972) showed that the back-arc tholeiites of the Mariana Basin were slightly enriched in lithophile elements compared with the otherwise similar typical MORB. They ascribed this to somewhat smaller degrees of melting at greater depths than beneath major oceanic ridges and suggested that this was associated with slower spreading rates. Clearly this idea cannot apply to the Scotia Sea back-arc center, where the spreading rate is considerably faster than on most mid-ocean ridges (Barker, 1972). Nevertheless, variable degrees of melting of a common source could account for most of the observed variation in Figure 25.2*a*, with the most enriched rock representing about one-third the degree of melting of the least enriched. For a quantitative estimation it is necessary to know the absolute REE abundance level in the source; thus, 7–20% melting would be appropriate for source concentrations of 1–2 times that of chondrite. Even so, the Ce/Yb ratios observed could not reasonably be obtained from a source in which light REE were significantly depleted relative to heavy REE. Hence, the upper mantle beneath the eastern Scotia Sea must differ at least in this one respect from that beneath the Atlantic and Pacific mid-ocean ridges, although some other abyssal basalts from the major oceans have trace element and isotope characteristics much closer to those described here, e.g., basalts from Deep Sea Drilling Project Leg 37 drilled in the Atlantic Ocean crust just south of the Azores (e.g., O'Nions and Pankhurst, 1976) and those from the Bouvet triple junction (Dickey and others, 1977).

Table 25.1. New analytical data for Cenozoic volcanic rocks from the northern Antarctic Peninsula

Sr isotopes					
Sample number	Locality	Rock type	Rb (ppm)	Sr (ppm)	$^{87}Sr/^{86}Sr$
27788	Paulet Island	Hawaiite	16	1000	0.70326
27803	Paulet Island	Hawaiite	17	995	0.70319
27812	Red Island	Hawaiite	19	590	0.70334
27826	Rosamel Island	Hawaiite	17	830	0.70319
27827	Dundee Island	Hawaiite	22	700	0.70353
27831	Tabarin Peninsula	Hawaiite	19	640	0.70346
27834	Tabarin Peninsula	Hawaiite	17	535	0.70353
27839	Egg Island	Hawaiite	19	715	0.70337
29138	Larson Nunatak	Hawaiite	11	375	0.70307
29139	Åkerlundh Nunatak	Hawaiite	8	585	0.70279
23957	Tower Island	Tholeiite	3	470	0.70394
23959	Tower Island	Olivine tholeiite	8	490	0.70390

Rare-earth elements (ppm)										
Sample number	La	Ce	Nd	Sm	Eu	Gd	Dy	Er	Yb	Lu
27788	53.8	108.9	49.6	9.10	3.39	9.51	7.90	4.40	3.68	0.51
27826	37.2	73.3	41.4	8.23	3.12	9.13	7.74	4.18	3.45	0.44
29139	24.9	49.3	30.2	6.70	2.76	7.66	6.04	2.73	1.90	0.20

Rb and Sr determined by X.R.F.
$^{87}Sr/^{86}Sr$ is relative to 0.70800 for E & A standard, ± 0.00005.

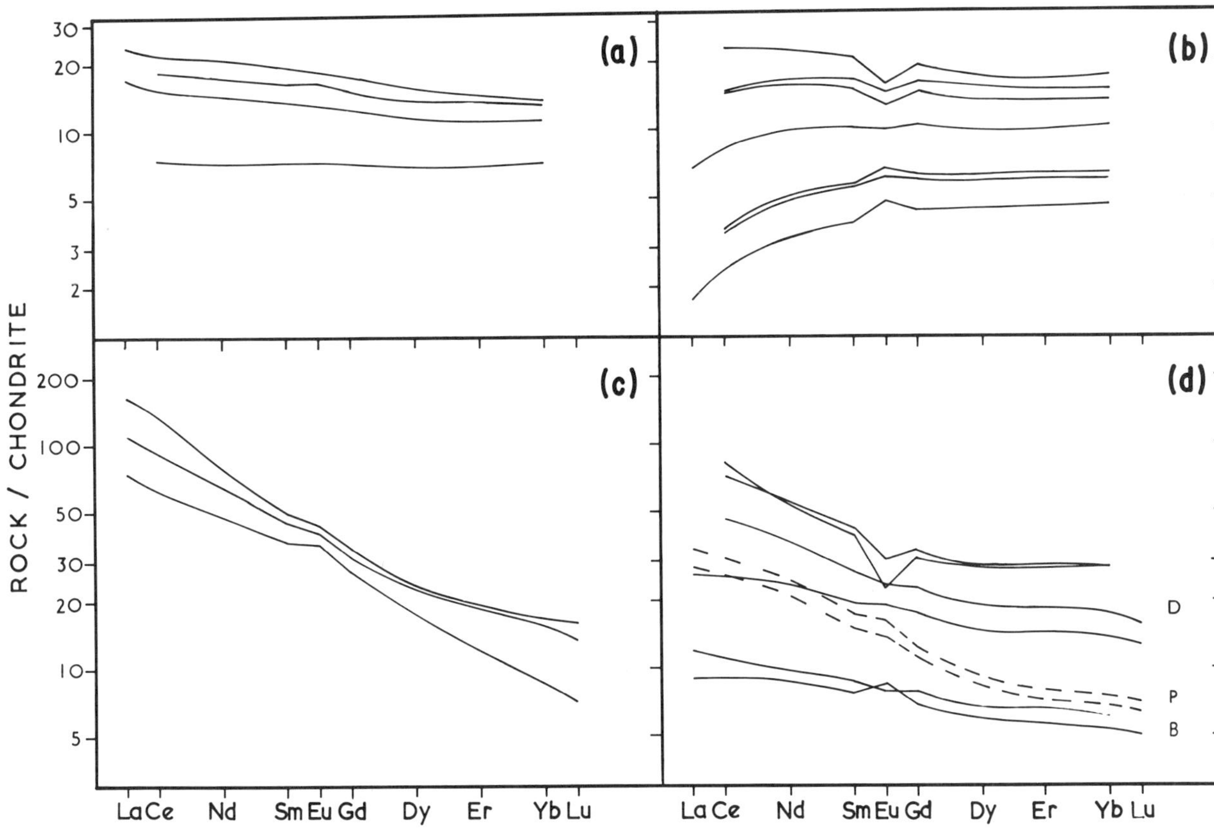

Figure 25.2. Rare-earth element abundances divided by those for average chondrite meteorites, for rock groups (a)-(d) as in Figure 25.1 D: Deception Island. B: Bridgeman Island. P: Penguin Island. References as in Figure 25.1, except for (a) Hawkesworth and others (1977).

South Sandwich Islands

The South Sandwich Islands constitute an entirely intra-oceanic island arc in the early stages of development (Baker, 1968). The volcanic rocks are predominantly low-alkali basalts belonging to the island-arc tholeiite series of Jakeš and Gill (1970), although an iron-enrichment differentiation trend has led to andesitic rocks, particularly in the southern part of the arc (Baker and others, 1977). Uniform $^{87}Sr/^{86}Sr$ ratios close to 0.704 were first reported by Gledhill and Baker (1973), and this has been confirmed by recent high-precision analyses (Hawkesworth and others, 1977) (see Figure 25.1).

The REE patterns are also highly distinctive (Figure 25.2*b*) in that they show both the light REE depletion characteristic of the most magnesian abyssal tholeiites and island-arc tholeiites and also the development of significant negative Eu-anomalies indicating extensive plagioclase fractionation. Hawkesworth and others (1977) have suggested that the sample with no Eu-anomaly, an aphyric basalt from Saunders Island, could be viewed as parental

Table 25.2. Comparative trace element data (in ppm) for least evolved volcanic rocks

	1	2	3	4	5	6	7
Ce_N[a]	7.5	2.4	<10	26	62	137	≥100
Ce_N/Sm_N	1.03	0.51	<1	1.7	1.7	2.7	2.8
Ce_N/Yb_N	1.04	0.62	<1	3.9	7.2	8.2	12
K_2O	0.30	0.12	0.14	0.48	1.35	1.34	1.5
Rb	4	2.4	1.1	5.2	8	16	31
Sr	148	112	136	534	585	1000	900
Ba	63	53	10.5	148	114	205	450
K/Rb	619	420	1050	770	1400	685	400
K/Sr	17	9.0	8.5	7.7	19	11	14
Rb/Sr	0.027	0.021	0.008	0.010	0.014	0.016	0.04
Sr/Ba	2.4	2.1	13	3.3	5.1	4.9	2
$^{87}Sr/^{86}Sr$	0.7032	0.7039	≤0.7030	0.7039	0.7028	0.7033	0.7031-0.7034
$^{143}Nd/^{144}Nd$	0.51305	0.51314	≈0.5131	—	—	—	<0.5129

1. Scotia Sea tholeiite (dredge 24); Hawkesworth and others (1977) and Tarney and others (1977).
2. South Sandwich Islands "oceanite" (SSM1.3); Hawkesworth and others (1977).
3. Average MORB; Hart (1971) and O'Nions and others (1977).
4. Penguin Island basalt; Weaver and others (1979).
5. Seal Nunataks hawaiite; this paper, except K_2O from Baker and others (1977).
6. Paulet Island hawaiite; this paper, except K_2O from Baker and others (1977).
7. Average Ross Island alkali basalt; Sun and Hanson (1975*b*). Nd-isotope composition for alkali basalts analyzed by O'Nions and others (1977).

[a] Subscript N signifies chondrite-normalized concentrations.

to the remainder, with the more evolved patterns being derived by crystal fractionation of olivine, plagioclase, and clinopyroxene (the latter being necessary to account for the increase in Ce/Yb ratios), and the more primitive patterns being partly due to crystal accumulation. However, as pointed out by Baker and others (1977), the geochemistry of the South Sandwich Islands rocks is too diverse to permit but a single parental magma type, and there is no mineralogic evidence for cumulus plagioclase in the rocks which show positive Eu-anomalies. The latter feature could equally be due to residual clinopyroxene in the source, since this rejects divalent Eu relative to other REE.

Thus, even the rock with the most depleted pattern, an "oceanite" dike from Montagu Island (the lowest line in Figure 25.2*b*), might be regarded as representing a liquid composition. On the assumption that it was formed by not more than 40% partial melting, its source region would have had *maximum* REE concentrations of less than one times that of chondrite for Ce, and two times for Yb. The geochemical characteristics of such a magma differ from those of mid-ocean ridge basalts only in having rather higher Rb and Ba contents (Table 25.2, columns 2 and 3) and a higher $^{87}Sr/^{86}Sr$ ratio. The latter feature could be due to their source region having had a higher Rb/Sr ratio over a considerable period of time, but this is not consistent with the strong light-REE depletion. The cause of this depletion is commonly thought to be the removal of incompatible elements (i.e., those preferentially concentrated into silicate liquids) during a previous episode of partial melt extraction (Gast, 1968), and this process would apply equally to Rb. Alternatively, Rb, Ba, and ^{87}Sr could have been introduced into the source region from the subducted lithosphere before or during magma generation, either by volatile transfer or by infiltration of small partial melts. Altered ocean floor basalts are enriched in these components as a result of sea-water interaction (Hart, 1971), but their REE patterns and Nd-isotope ratios are essentially unaffected (O'Nions and others, 1977).

South Shetland Islands

The volcanic rocks of the South Shetland Islands are basalt, andesite, and rhyolite of low-K calc-alkaline character (Weaver and others, Chapter 29, this volume). Their evolution appears to be related to extended Mesozoic–Cenozoic subduction of the Pacific Ocean floor beneath the Antarctic Peninsula. Current collaboraive work between the British Antarctic Survey and Birmingham University is aimed at establishing a detailed radiometric time scale on which to base interpretation of chemical changes. As yet, no REE data are available, but Figure 25.1*e* shows initial $^{87}Sr/^{86}Sr$ ratios for a few preliminary samples: Cretaceous volcanic rocks from Byers Peninsula, Livingston Island (Pankurst and others, 1980) and two lower Tertiary tholeiitic basalts from Tower Island (Rex, 1976; Baker and others, 1977). These are comparable with the values obtained from the South Sandwich Islands, with only a hint of a somewhat wider range—as usual in more complex, evolved island arcs.

Bransfield Strait

Recent volcanicity along the axis of Bransfield Strait is interpreted by both Tarney and others (1977) and Baker and others (1977) as being associated with incipient marginal basin development behind the South Shetland island arc. Full discussion of the geochemistry and petrogenesis of rocks from the volcanic islands of Deception, Bridgeman, and Penguin has been presented elsewhere (Weaver and others, 1979). The overall character of the rocks from the first two is intermediate between abyssal and calc-alkaline basalts, whereas rocks from Penguin Island are distinctly more alkaline. This is confirmed by the REE data (Table 25.2, column 4, and Figure 25.2*d*). Basalts from Deception and Bridgeman Islands have rather flat REE patterns and, in the latter case, low overall abundances consistent with moderate degrees of melting of a source comparable in some respects with that supplying the Scotia Sea spreading center. Only in the more evolved andesites of Deception Island are light-REE enriched patterns developed, eventually with negative Eu-anomalies, which indicate differentiation by crystal fractionation similar to that observed in the South Sandwich Islands. Penguin Island basalts on the other hand have rather steeper REE patterns, though not as steep as those for the James Ross Island group. $^{87}Sr/^{86}Sr$ ratios are also distinctive for each island (Figure 25.1*d*), those for Penguin Island being similar to those of the South Sandwich and South Shetland samples so far analyzed, but the others being somewhat lower.

James Ross Island group

Baker and others (1977) describe the Pliocene–Holocene volcanic rocks of James Ross Island and the smaller islands of Prince Gustav Channel as hawaiites with relatively high Na_2O, which become progressively more undersaturated through time, with the youngest lavas, from Paulet Island, being quite strongly Ne-normative. In this respect they show affinities with the even more strongly undersaturated basanitoids and alkali basalts of the Ross Sea and Marie Byrd Land provinces. Recent hawaiites also crop out in the Seal Nunataks and have been previously correlated with the James Ross Island group.

REE contents have been determined for three samples (Table 25.1), using the same mass-spectrometric isotope dilution method as for the results shown in Figures 25*a* and *b*, but at the Institute of Geological Sciences in London. The three samples were chosen to represent the range of undersaturated compositions from the Seal Nunataks to Paulet Island. Chondrite-normalized REE patterns (Figure 25.2*c*) show the marked linear light-REE enrichment characteristic of alkali basalts, with very slight positive Eu-anomalies. Significant low-pressure crystal fractionation is thus ruled out, and indeed, most alkali basalts have rather high Ni contents, suggesting not more than about 10% separation of olivine. In comparison with the basanitoids of Ross Island, considered by Sun and Hanson (1975*b*) to be essentially unmodified primary magmas, these samples have marginally lower light-REE and higher heavy-REE contents, resulting in significantly lower Ce/Yb ratios (Table 25.2, columns 5–7). In both cases, however, rather small degrees of melting, with garnet as a prominent phase in the residuum, are indicated, since this is the only mantle mineral with a strong preference for heavy REE.

From published data for the distribution of REE between basaltic melts and common minerals (e.g., Hanson, 1977), it is possible to model the theoretical enrichment of each element in the melt relative to the source. To a first approximation, the enrichment of Ce is inversely proportional to the degree of melting, whereas that of Yb is essentially buffered at a constant level which decreases with the proportion of garnet. A simple model relating these compositions would be that the James Ross Island samples were formed by about 1–2% melting of garnet peridotite with 5% garnet and 4–5 times chondritic abundances of REE, and that the Ross Island basanitoids were formed by 1% or less melting of a chemically similar source but with about 10% garnet. The hawaiite from the Seal Nunataks would then represent about 3% melting with 10% garnet in the initial peridotite. However, this specific model has extreme consequences if oceanic nephelinites with high Ni contents are also considered as primary magmas derived from the same type of source. These show levels of incompatible element enrichment more than twice that shown by the most evolved basanitoid and would therefore require 0.5% melting or less (Kay and Gast, 1973). To avoid such low degrees of melting it is necessary either to invoke a secondary enrichment process such as zone refining (Harris, 1974) or else to propose that the source region is already rich in incompatible elements. Sun and Hanson (1975*b*) have suggested that the Ross Island basanitoids were produced by about 10% melting of garnet peridotite with incompatible element concentrations about 4 times higher than usually ascribed to the suboceanic mantle, Ce about 10.5 times chondrite, and Yb about 3 times chondrite. Correspondingly, the mildly alkaline volcanic rocks of the James Ross Island group could be thought of as resulting from either (1) greater degrees of melting (as much as 15%) of a similar incompatible-element enriched source, or (2) comparable degrees of melting of a rather less enriched source. A small proportion of garnet in the residual mantle is necessary in every case.

The $^{87}Sr/^{86}Sr$ ratios of rocks from this area show a considerable range (Table 25.1 and Figure 25.1*c*) which overlaps with that of MORB but not with the volcanic arcs of the region. Oceanic island alkali basalts also show wide variation, from values comparable with MORB up to about 0.706 (Peterman and Hedge, 1971), but with relatively little local variation, at least for the smaller islands (O'Nions and Pankhurst, 1974). Sun and Hanson (1975*a*) have shown that for oceanic alkali basalts overall there is a correlation between the $^{87}Sr/^{86}Sr$ ratio and the Rb/Sr ratio, and they have interpreted this as due to the long-lived geochemical heterogeneity of the suboceanic mantle, persisting for 1000 and 3000 m.y. Thus, mantle source regions which yield volcanic rocks with high Rb/Sr ratios also have more radiogenic Sr. A similar covariation exists in the present case, but *two* linear trends exist in an Rb-Sr isochron plot, both corresponding to an age of about 500 m.y. but with a $^{87}Sr/^{86}Sr$ intercept of about 0.7029 for the Prince Gustav Channel volcanic rocks and a lower one of 0.7025 for those of the Seal Nunataks (Pankhurst, 1977, Figure 4). This could be explained either by long-lived heterogeneity or by variable contamination with a chemically uniform component with high Rb/Sr and $^{87}Sr/^{86}Sr$ ratios. The former alternative is preferred here; although the local crustal structure is not well studied, there is no clear evidence for pre-Carboniferous crust in the Antarctic Peninsula (Gledhill and others, Chapter 35, this volume). The significance of the age of 500 m.y. depends upon the degree of melting involved in magma production. If this is about 10%, then the Rb/Sr ratios of the basalts will be essentially the same as those of their individual sources, and the age would roughly date a differentiation event in the lithospheric mantle. If the degree of melting is as low as 1 or 2%, however, the Rb/Sr ratios of the melts would be increased over that of the source, perhaps by a factor of 2, so that the apparent age would be too young.

Compared to the oceanic tholeittes of the region, K, Sr, and light REE are most enriched in these rocks (Table 25.2). Their K/Rb and Sr/Ba ratios are actually considerably higher than those of the Ross Island basanitoids. This could be due to partial depletion of Rb and Ba in the source region during the proposed mantle differentiation, or it could reflect retention of these elements in a residual phase such as phlogopite during melting. Alternatively, the sequence of falling K/Rb ratios and increasing Rb/Sr ratios shown in Table 25.2 (columns 4–7) might suggest that there is also an increasing amount of Rb caused perhaps by contamination with continental crust.

Conclusions

Many previous authors (e.g., González-Ferrán and Vergara, 1972; Katsui, 1972; Baker and others, 1977) have commented on the marked differences in volcanic types between the oceanic island-arc and calc-alkaline basalts of the South Sandwich and South Shetland Islands and the continental alkali-olivine basalts of Marie Byrd Land and Victoria Land. Baker and others (1977) have shown that in the northern Antarctic Peninsula region the character of volcanicity has tended to change from the former to the latter through time, although both are represented in Holocene erupted rocks.

It is clear that these transitions in magma type, both spatial and temporal, are tectonically controlled. In the entirely oceanic environments of the Scotia Sea and the South Sandwich Islands, moderate degrees of partial melting and considerable fractional crystallization of the resultant magmas are associated with mantle sources relatively depleted in lithophile trace elements (though less so than some other mid-ocean ridge sources). Secondary enrichment, especially in ^{87}Sr, is most noticeable where altered ocean floor is being subducted and contributes to the magma source either by volatile transfer upon dehydration or by direct addition of small partial melts. Although little geochemical data has so far been published for the Cenozoic volcanic rocks of the South Shetland Islands, their higher $^{87}Sr/^{86}Sr$ ratios probably reflect a somewhat more lithophile-enriched source where subduction has occurred over a long period beneath the continental margin. Volcanic rocks from the incipient marginal basin of Bransfield Strait show trace-element affinities with the Scotia Sea basalts (especially on Deception and Bridgeman Islands) but also show features due to smaller degrees of melting at greater depth (on Penguin Island). These latter characteristics become even more marked for the intracontinental subalkaline magmas of the James Ross Island area and Seal Nunataks, and this trend is apparently continued into Marie Byrd Land and into the Ross Sea petrologic province. For these rocks there is trace element and isotope evidence of derivation at greater depths (residual garnet) and mantle sources with variable lithophile-element enrichment resulting from long-lived heterogeneities. If a single unifying tectonic factor is sought to explain Cenozoic magma generation throughout West Antarctica in a general way, it would appear to be the increasing thickness (in both space and time) of stabilized subcontinental lithosphere, together with a consequent reduction of geothermal gradients.

Acknowledgments. P. E. Baker kindly provided samples from the James Ross Island area. Geochemical studies of rocks from the South Shetland Islands and Bransfield Strait are in collaboration with S. D. Weaver, A. D. Saunders, and J. Tarney of the University of Birmingham, to whom I am also grateful for comments on the first draft of this manuscript.

References cited

Baker, P. E. 1968. Comparative volcanology and petrology of the Atlantic island-arcs. *Bull. volcan., 32,* 189–206.

Baker, P. E., Buckley, F., and Rex, D. C. 1977. Cenozoic volcanism in the Antarctic. *Phil. Trans. R. Soc. Lond.,* Ser. B, *279,* 131–42.

Barker, P. F. 1972. A spreading centre in the east Scotia Sea. *Earth planet. Sci. Lett., 15,* 123–32.

Dickey, J. S., Jr., Frey, F. A., Hart, S. R., Watson, E. B., and Thompson, G. 1977. Geochemistry and petrology of dredged basalts from the Bouvet triple junction, South Atlantic. *Geochim. cosmochim. Acta, 41,* 1105–18.

Gast, P. W. 1968. Trace element fractionation and the origin of tholeiitic and alkaline magma types. *Geochim. cosmochim. Acta, 32,* 1057–86.

Gledhill, A., and Baker, P. E. 1973. Strontium isotope ratios in volcanic rocks from the South Sandwich Islands. *Earth planet. Sci. Lett., 19,* 369–72.

González-Ferrán, O., and Vergara, M. 1972. Post-Miocene volcanic petrographic provinces of West Antarctica and their relation to the Southern Andes of South America. (*In* Adie, R. J., *ed. Antarctic Geology and Geophysics.* Oslo, Universitetsforlaget, 189–95.)

Hanson, G. N. 1977. Geochemical evolution of the sub-oceanic mantle. *J. Geol. Soc. Lond., 134,* 235–53.

Harris, P. G. 1974. Anatexis and other processes within the mantle. (*In* Sorensen, H., *ed. The Alkaline Rocks.* New York, Wiley, 427–36.)

Hart, S. R. 1971. K, Rb, Cs, Sr and Ba contents and Sr isotope composition of ocean floor basalts. *Phil. Trans. R. Soc. Lond.,* Ser. A, *268,* 573–88.

Hart, S. R., and Brooks, C. 1977. The geochemistry and evolution of early Precambrian mantle. *Contrib. Mineral. Petrol., 61,* 109–28.

Hart, S. R., Erlank, A. J., and Kable, E. J. D. 1974. Sea floor basalt alteration: some chemical and Sr isotopic effects. *Contrib. Mineral. Petrol., 44,* 219–30.

Hart, S. R., Glassley, W. E., and Karig, D. E. 1972. Basalts and sea-floor spreading behind the Mariana Island arc. *Earth planet. Sci. Lett., 15,* 12–18.

Hawkesworth, C. J., O'Nions, R. K., Pankhurst, R. J., Hamilton, P. J., and Evensen, N. M. 1977. A geochemical study of island-arc and back-arc tholeiites from the Scotia Sea. *Earth planet. Sci. Lett., 36,* 253–62.

Jakeš, P., and Gill, J. 1970. Rare earth elements and the island arc tholeiitic series. *Earth planet. Sci. Lett., 9,* 17–28.

Kay, R. W., and Gast, P. W. 1973. The rare earth content and origin of alkali-rich basalts. *J. Geol., 81,* 653–82.

Kay, R. W., and Senechal, R. G. 1976. The rare earth geochemistry of the Troodos ophiolite complex. *J. geophys. Res., 81,* 964–70.

Katsui, Y. 1972. Late Cenozoic petrographic provinces of the volcanic rocks from the Andes of Antarctica. (*In* Adie, R. J. *ed., Antarctic Geology and Geophysics.* Oslo, Universitetsforlaget, 181–85.)

O'Nions, R. K., and Pankhurst, R. J. 1974. Petrogenetic significance of isotope and trace element variations in volcanic rocks from the Mid-Atlantic. *J. Petrology, 15,* 603–34.

O'Nions, R. K., and Pankhurst, R. J. 1976. Sr isotope and rare earth element geochemistry of DSDP Leg 37 basalts. *Earth planet. Sci. Lett., 31,* 255–61.

O'Nions, R. K., Hamilton, P. J., and Evensen, N. M. 1977. Variations in $^{143}Nd/^{144}Nd$ and $^{87}Sr/^{86}Sr$ ratios in oceanic basalts. *Earth planet. Sci. Lett., 34,* 13–22.

Pankhurst, R. J. 1977. Strontium isotope evidence for mantle events in the continental lithosphere. *J. Geol. Soc. Lond., 134,* 255–68.

Pankhurst, R. J., Weaver, S. D., Brook, M., and Saunders, A. D. 1980. K-Ar chronology of Byers Peninsula, Livingston Island, South Shetland Islands. *Bull. Br. antarc. Surv., 49,* 277–82.

Peterman, Z. E., and Hedge, C. E. 1971. Related strontium isotope and chemical variations in oceanic basalts. *Geol. Soc. Am. Bull., 82,* 493–99.

Rex, D. C. 1976. Geochronology in relation to stratigraphy of the Antarctic Peninsula. *Bull. Br. antarc. Surv., 43,* 49–58.

Sun, S. S., and Hanson, G. N. 1975*a*. Evolution of the mantle: geochemical evidence from alkali basalts. *Geology, 3,* 297–302.

Sun, S. S., and Hanson, G. N. 1975*b*. Origin of Ross Island basanitoids and limitations upon the heterogeneity of mantle sources for alkali basalts and nephelinites. *Contrib. Mineral. Petrol., 52,* 77–106.

Tarney, J., Saunders, A. D., and Weaver, S. D. 1977. Geochemistry of volcanic rocks from the island arcs and marginal basins of the Scotia Arc region. (*In* Talwani, M., and Pitman, W. C., III, *eds. Maurice Ewing Series. 1: Island Arcs, Deep Sea Trenches and Back-Arc Basins.* Washington, American Geophysical Union, 367–77.)

Weaver, S. D., Saunders, A. D., Pankhurst, R. J., and Tarney, J. 1979. A geochemical study of magmatism associated with the initial stages of back-arc spreading: The Quaternary volcanics of Bransfield Strait, South Shetland Islands, *Contrib. Mineral. Petrol., 68,* 151–69.

26

Mesozoic Alluvial Fans of the South Orkney Islands*

DAVID H. ELLIOT and NEIL A. WELLS**
Institute of Polar Studies and Department of Geology and Mineralogy,
The Ohio State University, Columbus, Ohio 43210, U.S.A.

Abstract. Mesozoic conglomeratic rocks crop out on both sides of the Lewthwaite Strait in the South Orkney Islands. They were deposited by a variety of processes, including landslide, mudflow, sheet debris flow, sheetflood, braided stream, and shallow marine transport. The conglomerate beds on Powell Island and adjacent islands consist mainly of sheetflood and sheet debris flow sediments, together with fluvial sediments and other minor facies. Clasts indicate derivation from the Greywacke-Shale Formation which crops out to the east and north. Bedding is poorly developed. The pattern of attitudes suggests alluvial fan deposits that have been tilted westward. These conglomerates are interpreted as coalescing alluvial fan deposits banked against an uplifted fault block to the east. The conglomerates on eastern Coronation Island and on the Robertson Islands are largely debris flow deposits, with other facies subordinate. Clasts indicate derivation from the metamorphic complex to the west. Bedding is poorly developed. The deposits represent a bajada, but individual alluvial fans cannot be identified. Attitudes of beds have been modified by tilting to the east.

The conglomeratic rocks of the Lewthwaite Strait area represent alluvial fans; sediment was shed from adjacent highlands into block-faulted basins that apparently trended north-south. The block faulting may be related to uplift following the early Mesozoic Gondwanian orogeny.

The conglomeratic rocks cropping out adjacent to Lewthwaite Strait (Figure 26.1) were discovered in 1903 by members of the Scottish National Antarctic Expedition who landed at Spence Harbour. No further collecting was done until 1933 when landings were made on Powell and Michelsen Islands during investigations by RRS *Discovery* II. The petrography of the rocks collected was described by Tilley (1935). In more recent years, geologists with the Falkland Islands Dependencies Survey, and subsequently the British Antarctic Survey, have visited most outcrops of these conglomerates. The geology has been summarized by J. W. Thomson (1971, 1973, 1974). Preliminary results of fieldwork conducted in the Lewthwaite Strait area during January 1977 are presented here.

The conglomeratic rocks to the west of Lewthwaite Strait have been designated the Spence Harbour Conglomerate and those to the east of the strait on Powell Island and the islands to the south have been named the Powell Island Conglomerate (Figure 26.2). Neither conglomerate has been formally described, partly because proper type sections cannot be designated.

The Spence Harbour Conglomerate (Thomson, 1971, 1974) consists almost entirely of conglomeratic rocks, except on the southern arm of Matthews Island where a thick sandstone sequence crops out. The clasts in the conglomerate are almost entirely of metamorphic rocks derived from the Coronation Island metamorphic complex. The Spence Harbour Conglomerate rests on the metamorphic complex at the Robertson Islands, Divide Peaks, The Menhir, and East Cape, whereas at the southwest corner of Gibbon Bay, it overlies the Gibbon Bay Shale. The shale has yielded a sparse and rather poorly preserved assemblage of invertebrates which is tentatively assigned an Early Cretaceous age (Thomson and Willey, 1975; M. R. A. Thomson, personal communication). The matrix of the conglomerate has not yet yielded any identifiable fossils; however, sandstone erratics from a moraine on the southern side of Matthews Island contain a poorly preserved ammonite fauna, which although was previously assigned a Middle to Late Cretaceous age (Thomson, 1975), is now thought to be Early Cretaceous in age (M. R. A. Thomson, personal communication). The only other fossils occur in boulders of calcareous grit within the conglomerate at Rayner Point; these are believed to be Late Jurassic–earliest Cretaceous in age (Thomson and Willey, 1975). Further collections were made by M. R. A. Thomson and W. J. Zinsmeister during January 1977, and these may allow the present age assignments to be refined.

The Powell Island Conglomerate (Thomson, 1973) consists of conglomeratic rocks with sparse interbedded sandstone and granulestone lenses; however, plant-bearing

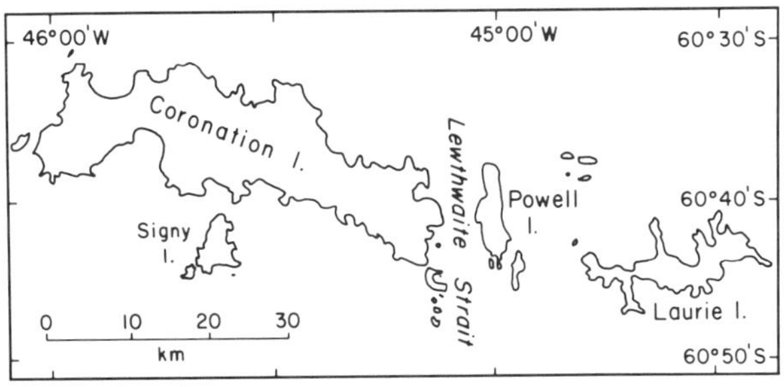

Figure 26.1. Location map for the Lewthwaite Strait, South Orkney Islands.

* Contribution no. 328 of the Institute of Polar Studies, The Ohio State University.

** Present address: Department of Geology and Mineralogy, University of Michigan, Ann Arbor, Michigan 48109, U.S.A.

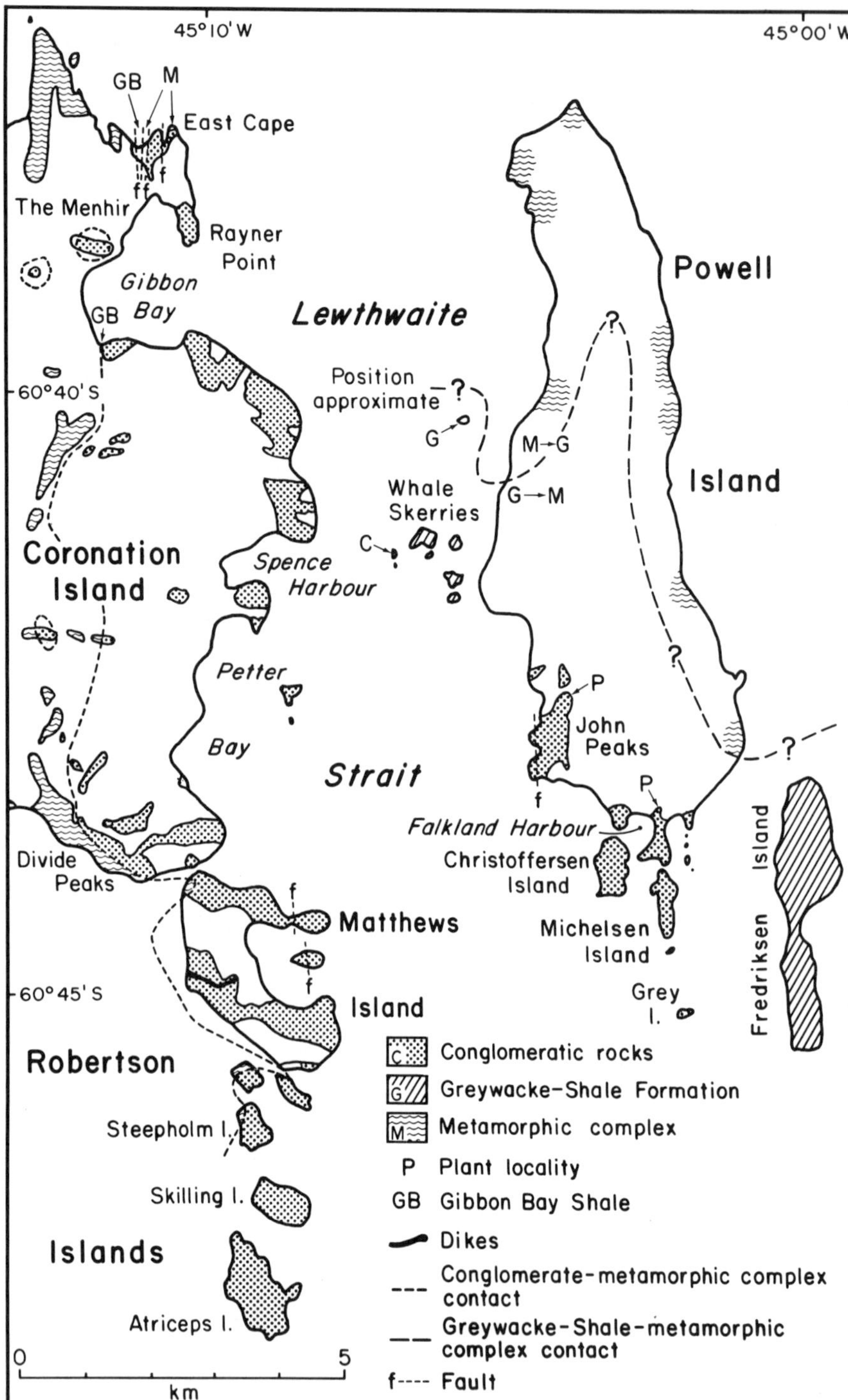

Figure 26.2. Geologic sketch map of the Lewthwaite Strait area. Modified from Dalziel and others (1977).

sandstone crops out at the north end of John Peaks and northeast of Falkland Harbour. The conglomerate is not seen in stratigraphic contact with the underlying Greywacke-Shale Formation from which it was derived. The age of the conglomerate is unclear; a Jurassic age has been inferred (see Thomson, 1973), but material collected in 1977 was examined by J. M. Schopf (personal communication, 1978) who believed a more precise age than Jurassic-Cretaceous could not be supported.

A detailed description of the Mesozoic conglomerate is in press (Wells, in press).

Sedimentary facies

The inferred depositional processes, the facies patterns, and the attitudes of beds lead to the interpretation of the conglomerates as alluvial fan deposits. Products of a variety of depositional processes—including rock falls, mudflows, sheet debris flows, sheetfloods, and braided streams—were recognized.

Landslide blocks

Rock masses as large as 15 × 20 × 30 m and in nonplanar contact with the conglomerate are interpreted as

Figure 26.3. Landslide block of Greywacke-Shale Formation rocks in Powell Island Conglomerate; contact indicated by dotted line. Southeast point of Powell Island (see Figure 26.9 for location). Note person for scale.

landslide blocks (Figure 26.3). The contacts can commonly be traced around and beneath the blocks. Soft-sediment deformation was observed beneath two blocks.

The boulder-sized clasts (1–4 m across) present elsewhere could have been deposited by rockfalls (Rapp, 1960) or, more likely, by some other process (e.g., carried in or on a mudflow).

Mudflow deposits

Those deposits formed by mudflows (Bull, 1972; Sharp and Nobles, 1953) are characterized by clay-sized matrix, matrix-supported clasts, very poor sorting, a disorganized fabric, vertical orientation of some clasts, irregular upper surfaces, and lack of internal stratification (Figure 26.4). The thickness of the individual mudflow deposits ranges between 0.5 and 4 m. Mudflows are uncommon.

Most mudflows are not confined to channels, but are instead spread out over the fan surface. Scouring of the underlying beds is usually slight, if present; however, several flows have scoured underlying channelform sandstones, indicating that they flowed down preexisting stream beds.

Sheet debris flow deposits

Debris flow deposits (Bull, 1972) are regularly bedded (usually planar), unchanneled sheets of disorganized sediment (Figure 26.5). Internally they are, at best, poorly stratified and little sorted, although the larger clasts may be piled together, and individual beds may differ somewhat in modal size. Clast orientation varies from random to subparallel to bedding; a few have their long axes perpendicular to bedding. These deposits are clast-supported and apparently contain very little or no clay-sized matrix; the lack of fine-grained matrix is interpreted to mean that they were originally debris flows and not mudflows or that the fines were washed out by percolating meteoric water (as suggested by Hooke, 1967). Clasts show a wide range of roundness, but the majority fall in the subrounded to subangular class.

Figure 26.4. Mudflow deposit exhibiting dispersed, randomly oriented clasts in a muddy matrix. Spence Harbour Conglomerate, near East Cape. Pocket knife is 8 cm long.

Figure 26.5. Crudely stratified conglomerate deposited by sheet debris flow; bedding indicated by sandstone partings. Powell Island Conglomerate, east coast of Christoffersen Island.

The thickness of individual deposits ranges from 10 cm to several meters. Locally the base shows a slight scour, indicating minor early channelization, and in a few places the debris flow is draped over and around clasts protruding from the underlying deposit.

Deposits similar to those just described, but generally much thinner and more lenticular, occur in a marine sandstone sequence on Matthews Island, and in that case are considered to have been deposited subaqueously.

SHEETFLOOD DEPOSITS

Deposits that are regarded as having sheetflood origin are, like the sheet debris flow deposits, laterally and longitudinally extensive sheets of sediment 0.5–2 m thick and are characterized by being clast-supported and having a sand- to granule-sized matrix (this is the sheetflood of Bull, 1972, rather than of Blissenbach, 1954, and Bluck, 1967). The range of clast sizes is similar to that of the debris flows. Scour at the base is more common than in debris flows and rarely exceeds 15 cm. These rocks are moderately well organized in that a crude vertical sorting of the clasts gives a rough stratification, and most clasts are oriented subparallel to bedding. In a few places, normal grading is poorly developed (reverse grading is very rare). The clasts are slightly more rounded than those in debris flow depositis. Cross-beds are present in some places (Figure 26.6), and fine-grained material is entirely absent. Both of these factors indicate deposition by traction.

The difference between sheetflood and sheet debris flow is the difference between traction and mass movement. As Bull (1972) has noted, transition to debris flow occurs when the bed load is entrained by a sufficient increase in the volume of water and the velocity of the flood. The reverse occurs on dewatering of the debris flow. In a lower flow regime, sheetfloods grade into braided streams (Bluck, 1967).

SANDSTONE PARTINGS

On almost any vertical cliff face, sandstone and granulestone partings, at intervals of 0.5 to 3 m, can be seen. These are regularly present between sheetflood and sheet debris flow conglomerate. They are in some places gradational with the top of the underlying unit, but in a few others they scour into it. They are not persistent laterally because they are scoured into and eroded by the succeeding unit or pass laterally into indeterminate fractures which represent contacts between successive deposits. The partings are as much as 0.75 m thick and are in many places the major (and in some, the only) clue to stratification. Features such as broad troughs, trough and tabular cross-beds, good sorting, and grading indicate that the sandstone was deposited by traction currents. They usually contain many unidentifiable wood fragments.

As Bluck (1967) has suggested, the almost uniform interspersal of partings with conglomerate units could indicate that many are part of a flood or flow event. Possibly they are formed in the waning stages of either a debris flow

Figure 26.6. Tabular cross-bedding in a sheetflood conglomerate unit. Powell Island Conglomerate, southwest Michelsen Island. See penguins (0.5 m tall) for scale.

Figure 26.7. Conglomeratic braided stream deposits. Cliff face in photograph is 3 m high. Powell Island Conglomerate, southwest coast of Christoffersen Island.

Figure 26.8. Marine sandstone with ripples overturned to the east. Spence Harbour Conglomerate, southeast Matthews Island.

after dewatering or a sheetflood (which may itself have been derived from a debris flow) after the bed load has been dropped. In either case, this produces a thin sheet of water or a rapidly shifting braided stream capable of moving and reworking only granules, sand, and plant debris. Some sandstone beds may have been deposited by separate sheetfloods.

Braided stream deposits

These deposits consist of well-stratified, well-sorted, fine-grained conglomerate and pebbly granulestone and sandstone (Bluck, 1967). They are channelform; the many broad channels, abundant crosscutting relationships, planar to low angle cross-beds, tabular cosets, trough cross-beds, and pebble lenses show them to be braided stream deposits (Figure 26.7). Cross-bedding is uncommon, however, and few beds show imbrication.

Marine sandstone

These sandstone beds are coarse grained, micaceous, and lack sedimentary structures except for rare ripple marks. Their micaceous character imparts a flaggy aspect to the outcrops (Figure 26.8). Bed thickness ranges from 1 to 5 cm, and as many as 40 beds occur in a meter of section. Beds appear to be couplets or gradational units with the medium- to coarse-grained sandstone below and fine-grained sandstone above, but the weathered character precludes obtaining unequivocal evidence. A few fragmentary ammonites were found in the talus below the sandstone outcrops, and therefore the sandstone is regarded as marine in origin.

Facies patterns and bedding attitudes

Other lines of evidence for an alluvial fan interpretation include the attitudes of beds and the facies patterns. The attitudes of the beds appear to be highly variable, but collection of detailed data shows that they could be sorted into groups with radiating dips; this pattern would be expected for surfaces and bedding of radiating fans. The measured dips are mostly between 10° and 25°, steeper than most reported in the literature (e.g., as high as 14°, Hooke, 1967). Some postdepositional tilting of the fans seems probable and is demonstrable at southern Matthews Island and Gibbon Bay. The amount of tilting is limited because all dips but one are less than the angle of repose of sand, and in the Powell Island Conglomerate the trough axes and dip directions are approximately concordant. This pattern seems consistent with uplift along a mountain front, tilting the fans outward. Although earlier workers have mapped a fold axis on Michelsen Island (Thomson, 1973), our observations lead us to conclude that the rocks are not folded and that the fold axis mapped is a reflection of the primary dips within an alluvial fan that has been tilted.

Facies patterns showed that downdip, distal deposits tend to be fluvial complexes, deposited by braided streams. In addition, rocks are finer grained and better sorted down-dip. These attributes, however, have caused them to be largely eroded. The landslide blocks are proximal only; those on southeastern Powell Island are described below. Some interfan deposits are also present; these are flatter and comprise stream deposits and channeled mud and debris flows. They contain plant debris, including small, nearly complete root systems in muddy sandstone.

Powell Island fans

The Powell Island Conglomerate crops out on southern Powell, Michelsen, Christoffersen, and Grey Islands and on surrounding islets. It consists of rounded to subrounded clasts of graywacke, shale, and metasedimentary rocks (slate and quartz-veined metamorphosed sandstone and shale) which were derived from the Greywacke-Shale Formation. Sheetflood and sheet debris flow deposits are by far the most widespread facies and consist of laterally extensive bodies of rock that change thickness only gradually. The individual units can be distinguished readily where sandstone partings occur, but otherwise, contacts between units are obscure. In some places these deposits are stacked one on top of another with only a little variation in clast size and bedding thickness. Interfan deposits (e.g., channeled mudflows) occur east and northeast of Falkland Harbour. Traction (sheetflood and braided stream) deposits predominate on southwestern Christoffersen Island and on John Peaks and are present on southern and western Michelsen Island. Landslide blocks occur on southeastern Powell Island. However, the overall dominance of sheetflood and debris flow deposits clearly indicates that the preserved rocks are primarily midfan deposits (see Hooke, 1967).

At John Peaks, however, the presence of a plant-bearing sequence and a high proportion of braided stream and river channel conglomerate (Bluck, 1967) points to a slightly different environment, farther from the source and with gentler slopes. Since the John Peaks beds are the youngest rocks in the area, these changes may indicate retreat of the mountain front.

The above-mentioned landslide blocks, which consist of outcrops of Greywacke-Shale Formation rocks, had previously been mapped as occurring in fault contact with the conglomerate (see Thomson, 1973, Figure 9). This mapped "fault" contact on the east side of the headland dips steeply to the west; however, at low tide the contact on the north side can be seen to flatten out, and the graywacke-shale rocks can be seen to rest on top of the conglomerate. Southward the "fault" contact can be traced over the top of the headland and back to the northwest, then westward

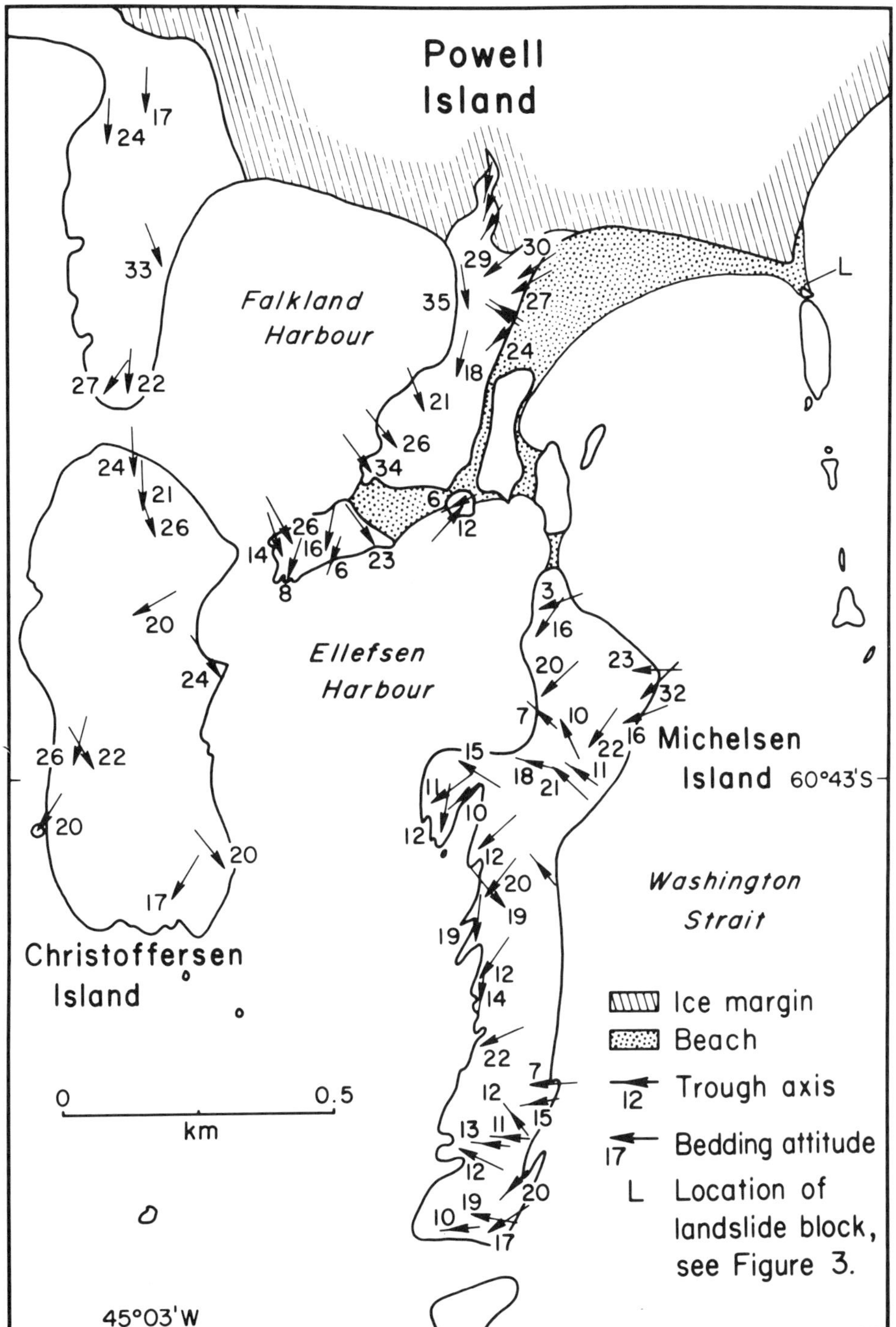

Figure 26.9. Attitudes of strata in the Powell Island Conglomerate. Arrows show direction and amount of dip of bedding. This sketch map has been compiled from oblique photographs taken from Fredriksen Island and northern John Peaks; it differs somewhat from previously published maps, particularly with respect to Michelsen Island.

beneath beach deposits. Just to the north of this previously mapped contact, there is another outcrop of graywacke similarly enclosed in the conglomerate (Figure 26.3). Here about 80% of the contact can be traced around the boulder. Other outcrops of the Greywacke-Shale Formation on the string of islets to the south are probably also landslide blocks and perhaps indicate the trend of a former fault scarp.

Fluvial paleocurrent structures are almost entirely lacking, but paleocurrent directions can be inferred assuming that attitudes reflect convex fan surfaces. This suggests (Figure 26.9) a southward-sloping fan on Christoffersen Island and Powell Island adjacent to Falkland Harbour and a westward- to southwestward-oriented fan, of which the landslide blocks are a part, on southeastern Powell Island and northern Michelsen Island; interfan deposits are located between the two. A fan with westward orientation occurs on southern Michelsen Island, and another fan may be present in the north-central part of the island. Variable directions in the west quadrant occur on John Peaks, and westward orientations were recorded on the Whale Skerries (see Figure 26.2 for locations). These directions

Figure 26.10. Large angular clasts of metamorphic rocks forming a sedimentary breccia. Spence Harbour Conglomerate, northern arm of Matthews Island.

are taken to infer the former existence of coalescing fans with a general southwestward to westward slope.

The relationships of the facies on Christoffersen Island, John Peaks, southeastern Powell Island, and Michelsen Island suggest upward and southward or westward changes in environments dominated first by debris flows, then by sheetfloods, and finally by braided streams. This is in accordance with Bluck's (1967) model for the evolution of fans.

Coronation Island fans

The Spence Harbour Conglomerate crops out in a belt on the east side of Coronation Island and on the Robertson Islands. It is substantially different from the Powell Island Conglomerate. It is composed almost entirely of angular fragments of metamorphic rocks (Figure 26.10) derived from the Coronation Island metamorphic complex (Thomson, 1971, 1974). Debris flow deposits dominate the conglomerate. Many outcrops consist entirely of massive, poorly stratified debris flows with indeterminate partings that in some cases are demonstrable contacts between units. Sandstone partings are few, which makes bedding difficult to determine; wood fragments have not been found in these partings. Marine sandstone and subordinate subaqueous debris flow deposits form an important part of the section on Matthews Island (Figure 26.11); other facies (mudflows, fluvial conglomerate, sheetwash conglomerate, and possible talus deposits) are scattered sparsely throughout the outcrops.

The angularity of clasts in some stratigraphically low deposits suggests a possible talus origin (see Nilsen, 1969). Such deposits crop out at East Cape and south of Matthews Island; they overlie metamorphic basement that appears to be *in situ* and are themselves overlain by debris flow deposits. They are largely unbedded though crudely stratified by vertical sorting, weather quite differently from the overlying rocks, and contain the most angular clasts observed anywhere. The bedding in the overlying debris flow units is nearly parallel to the apparent stratification in the "talus deposits" (cf. Rapp, 1960).

Figure 26.11. View from the north of the southern arm of Matthews Island. The marine sandstone shown in Figure 26.8 crops out on the east side of the saddle; the cliffs to the east consist of massive sheet debris flow deposits (elevation of summit is about 275 m).

Mudflow deposits were observed only at the southeastern corner of Matthews Island and at East Cape. A large landslide block was also noted at East Cape.

Braided stream deposits crop out on the north shore of Spence Harbour and form a 10-m-thick sequence of sandstone and granulestone. The conglomerate is stratified, and beds are 0.5 to 1.5 m thick. Dispersed large clasts as much as 1.6 m across occur in the sequence, and layers of pebbles and cobbles are interbedded. The only good example of imbrication was found in this section, and it indicates flow to the east. This fluvial sequence overlies and is overlain by debris flow units.

A 200-m-thick sequence of sandstone interbedded with thin debris flow units crops out on the southern arm of Matthews Island. The sandstone is flaggy, micaceous, and coarse grained. Sedimentary structures are notably absent, although ill-defined ripple marks occur at a number of levels, and slumped or overturned ripples were observed at one locality (Figure 26.8). Broad, ill-defined scours were also noted. The interbedded debris flow units range in thickness from 50 to 200 cm; their characteristics are very similar to those already described—clasts are polymodal, randomly oriented (but including some individual clasts standing on end), and lack size-sorting and stratification. The sandstone is regarded as marine in origin because of the discovery of marine invertebrates in the talus below the outcrops.

The Spence Harbour Conglomerate of the southern arm of Matthews Island is unique because of its great variability in sedimentary rock types; elsewhere the rocks are rather uniform. The field data favor a nonmarine environment for all sedimentary facies except the flaggy sandstone sequence.

Because of ice cover and steep cliffs, data points are limited, and individual fans cannot be determined. However, dispersal was generally eastward; this is suggested by the attitudes of the conglomerate beds (Figure 26.12) and of a few sandstone partings, by the direction of overturning of ripple marks in the marine sequence, and by one large slump preserved at East Cape. The sparse evidence suggests that the fans coalesced, forming a bajada.

Deposition must have been close to a coastline to allow a marine incursion with accompanying sandstone deposition. The debris flow units interbedded with the sandstone imply steep slopes adjacent to the coast. The local depositional environment for the sandstone must have been near shore, as evidenced by the broken ammonites and dispersed large pebbles. Evidence for wave or current action is limited to sparse ripple marks (see Figure 26.8). It is possible that the sandstone units resulted from differential settling of sediment swept off the fans into a quiet lagoon by storm floods.

The conglomeratic rocks rest on older rocks at a number of localities (see Thomson, 1974), but the contact was examined only at East Cape, at the southwest corner of Gibbon Bay, and on the island immediately south of Matthews Island. At Gibbon Bay the Spence Harbour Conglomerate, overlying the Gibbon Bay Shale, consists entirely of debris flow material. The base of the conglomerate is highly irregular and appears to consist of load cast structures impressed in the shale. The conglomerate there is dark, as noted by J. W. Thomson (1974); however, on the basis of field observation, no special sedimentologic significance is attached to the color variation, which is probably the result of the incorporation of dark shale at the base. At East Cape the conglomerate overlies what appears to be a weathered schist surface that might have been a bare rock pediment surface (cf. Williams, 1969).

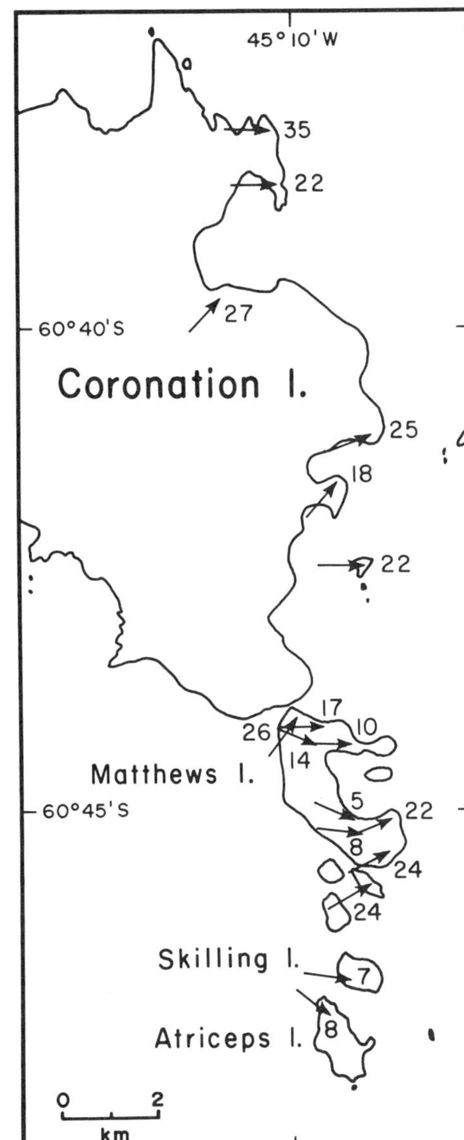

Figure 26.12. Attitudes of strata in the Spence Harbour Conglomerate. Arrows show direction and amount of dip of bedding.

The Spence Harbour Conglomerate is believed to be a set of alluvial fans or a bajada that formed near sea level and adjacent to an east-facing fault scarp. There is some evidence of prograding sequences: the "talus deposits" are overlain by debris flow units that would form at a later stage in the evolution of a fan, and debris flow and sheetflood deposits overlie the Matthews Island marine sequence.

Conclusions

The sedimentary facies recognized in the Powell Island and Spence Harbour Conglomerates lead to the conclusion that both sequences represent alluvial fans. The nature of the record suggests that both sequences must be related to fault scarps and that the source of the clasts must have been local. The Powell Island Conglomerate is interpreted as having originated from coalescing alluvial fans banked against west-facing fault scarps which placed Greywacke-Shale Formation rocks at relatively high elevations. The Spence Harbour Conglomerate is believed to be a bajada deposit (a series of coalescing alluvial fans) in which sediment was dispersed in a generally eastward direction. In both cases, the outcrop distribution and sparse dispersal data suggest that the faults trended north-south.

Further laboratory studies of the fossils will probably refine the ages of the two conglomerate sequences, but at present the Powell Island Conglomerate is dated as Jurassic-Cretaceous, whereas the Spence Harbour Conglomerate may be Early Cretaceous in age. It is not possible to say whether more than one period of fault block tectonism occurred. The conglomerate units represent erosion of different structural and metamorphic levels in the pre-Jurassic basement, but the available data do not allow an unambiguous sequence of events to be reconstructed.

The inferred north-south faults are approximately parallel to the dominant structural trend in the adjacent metamorphic complex, but these faults lie approximately perpendicular to the apparent distribution of the major tectonostratigraphic units of the South Orkney Islands (see Dalziel and Elliot, 1971, 1973). Block faulting and conglomerate deposition are possibly related to uplift following the early Mesozoic Gondwanian orogeny; alternatively, or in addition, the faulting may be related to the younger Andean events of late Mesozoic age.

Acknowledgements. We would like to express our appreciation for the assistance given by the master and crew of the R/V *Hero* during cruise 77-1; it would have been difficult to accomplish the objectives of the project without their enthusiastic support. The field work was supported by Grant No. DPP74-21509 from the Division of Polar Programs, National Science Foundation.

References cited

Blissenbach, E. 1954. Geology of alluvial fans in semiarid regions. *Bull. geol. Soc. Am., 65,* No. 2, 175–90.

Bluck, B. J. 1967. Deposition of Old Red Sandstone conglomerates in the Clyde area: a study in the significance of bedding. *Scot. J. Geol., 3,* No. 2, 139–67.

Bull, W. B. 1972. Recognition of alluvial-fan deposits in the stratigraphic record. (*In* Rigby, J. K., and Hamblin, W. K., *eds.* Recognition of Ancient Sedimentary Environments. *Spec. Publs Soc. econ. Paleont. Miner., Tulsa,* No. 16, 63–83.)

Dalziel, I. W. D., and Elliot, D. H. 1971. Evolution of the Scotia arc. *Nature, Lond., 233,* No. 5317, 246–52.

Dalziel, I. W. D., and Elliot, D. H. 1973. The Scotia arc and Antarctic Margins. (*In* Nairn, A. E. M., and Stehli, F. G., *eds. The Ocean Basins and Margins. 1. The South Atlantic.* New York, Plenum Press, 171–246.)

Dalziel, I. W. D., Elliot, D. H., Thomson, J. W., Thomson, M. R. A., Wells, N. A., and Zinsmeister, W. J. 1977. Geologic studies in the South Orkney Islands: R/V *Hero* Cruise 77-1, January 1977. *Antarct. Jnl U.S., 12,* 98–101.

Hooke, R. L. B. 1967. Processes on arid-region alluvial fans. *J. Geol., 75,* No. 3, 438–60.

Nilsen, T. H. 1969. Old Red sedimentation in the Buelandet-Vaerlandet Devonian District, western Norway. *Sedim. Geol., 3,* No. 1, 35–57.

Rapp, A. 1960. Recent development of mountain slopes in Karkevagge and surroundings, northern Scandinavia. *Geogr. Annlr, 42,* No. 2–3, 1–197.

Sharp, R. P., and Nobles, L. H. 1953. Mudflow of 1941 at Wrightwood, southern California. *Bull. geol. Soc. Am., 64,* No. 5, 547–60.

Thomson, J. W. 1971. The geology of Matthews Island, South Orkney Islands. *Bull. Br. antarc. Surv.,* No. 26, 51–57.

Thomson, J. W. 1973. The geology of Powell, Christoffersen and Michelsen Islands, South Orkney Islands. *Bull. Br. antarc. Surv.,* No. 33–34, 136–67.

Thomson, J. W. 1974. The geology of the South Orkney Islands: III. Coronation Island. *Scient. Rep. Br. Antarc. Surv.,* No. 86, 1–39.

Thomson, M. R. A. 1975. Fossils from the South Orkney Islands: II. Matthews Island. *Bull. Br. antarc. Surv.,* No. 40, 75–79.

Thomson, M. R. A., and Willey, L. E. 1975. Fossils from the South Orkney Islands: I. Coronation Island. *Bull. Br. antarc. Surv.,* No. 40, 15–21.

Tilley, C. E. 1935. Report on rocks from the South Orkney Islands. *'Discovery' Rep.,* No. 10, 383–90.

Wells, N. A. In press. Mesozoic alluvial fan conglomerates and other sedimentary rocks of the South Orkney Islands. *Rep. Inst. polar Stud. Ohio State Univ.,* No. 71.

Williams, G. E. 1969. Characteristics and origin of a Precambrian pediment. *J. Geol., 77,* No. 2, 183–207.

27

New Bird Ichnites from Fildes Peninsula, King George Island, West Antarctica

VLADIMIR COVACEVICH
Instituto de Investigaciones Geológicas, Casilla 10465, Santiago, Chile

PAT VICKERS RICH
Monash University, Earth Sciences Department, Clayton, Victoria 3168, Australia

Abstract. Isolated footprints and trackways, probably made by birds, are known from continental lacustrine sedimentary rocks on King George Island in West Antarctica. Because the fossil flora in rocks associated with the track-bearing strata is indicative of a Tertiary age (probably Oligocene or Miocene), a dinosaurian origin for the tracks is ruled out. Three new types of tracks are present, one indicative of a large ground bird with broad toes (maximum print length (mpl) = 180 mm), probably either a ratite or a gruiform (cf. phororhacoid). A second morphotype is characterized by marked webbing of three relatively broad, moderately elongate digits (mpl = 86 mm), probably representing a large duck or small gooselike form. The third morphotype has four elongate, slender, unwebbed digits (mpl = 102.6 mm) and could belong to any of a number of avian groups.

In this collection, the tracks of the large ground bird are of greatest significance in that they indicate dispersal of such a nonvolant type between South America or east Antarctica and King George Island in West Antarctica. Although such paleoornithologic data cannot establish whether such a dispersal route between East Antarctica–Australia and South America was an archipelagic or a land-bridge route, it does indicate that obligate land dwellers were able to utilize this route for dispersal. This route appears to have been important in the movement of organisms between the separating pieces of the Gondwana jigsaw.

Fossil bird footprints and trackways, ichnites, of mid-Tertiary age (Covacevich and Lamperein, 1972) have been known from the Fildes Peninsula of King George Island, West Antarctica (Figure 27.1*a*) since at least 1969, but until recently only one type, *Antarctichnus fuenzalidae,* thought to be related to rails, had been reported (Covacevich and Lamperein, 1969, 1970, 1972). Now at least three more morphotypes have been found (some preliminarily reported by Covacevich and Hernández, 1971), one representing a large ground bird, perhaps a ratite or a phororhacoid, a second representing a bird with marked webbing and relatively short, stout digits suggestive of ducks and geese, and a third that has four well-developed digits and that could have been produced by a number of different avian families. These latter three morphotypes are described and illustrated in this paper, and their taxonomic and biogeographic significance is considered.

Bird ichnites descriptions

Morphotype I

Material

Two isolated, natural cast footprints of a left and a right foot in a good state of preservation; one natural cast of an incomplete digit. See Figures 27.2*a* and *d* and 27.3*e*).

Measurements

See Figures 27.2*b* and *c* and below.

Description

The two main footprints consist of three digits with a maximum midtoe length of 180 mm and a maximum approximate width across the digits of 150 mm. Each track has a distinct heel impression behind and continuous with the midtoe (III); the heel has a maximum width of 32.0 mm (track 2; Figure 27.2*a* and *b*); lateral digits (II and IV) do not connect directly with the heel impression. All digits exhibit several narrow and wide segments, where some of the narrow segments may represent the midsections of individual phalanges, which tend to be waisted, especially the more proximal, larger elements. The angle between the internal(?) toe (II) and the midtoe (III) is larger in both instances (41°) than the angle between the midtoe (III) and the lateral(?) digit (IV) (34°–35°). On the slab with track 2, a smooth surface with worm trails extends posteriorly from the heel for 35 mm with no indication of an imprint of a fourth (I) digit on the avian footprint; unfortunately, without a larger area of rock slab preserved behind the heel, the presence or absence of a fourth digit cannot be absolutely ascertained. The only evidence for claws is present on digit III of track 1 (Figures 27.2*c* and 27.3*e*) and on the isolated digit (track 3, Figure 27.2*d*) and perhaps on digit II(?) of both tracks 1 and 2.

Comment

Footprints of morphotype I represent a large ground bird, whose possible dinosaur affinities have been ruled out because of the age of the enclosing rocks and the whole assemblage of ichnites, and not on the basis of a broad survey of comparative morphologies within dinosaurs and birds. Of the large ground birds, ostriches (Struthionidae) can be dismissed immediately because they have only two digits, whereas the large Fildes tracks have at least three. Cassowaries (Casuariidae) possess an extremely elongate claw on digit II that does not appear on the Fildes prints.

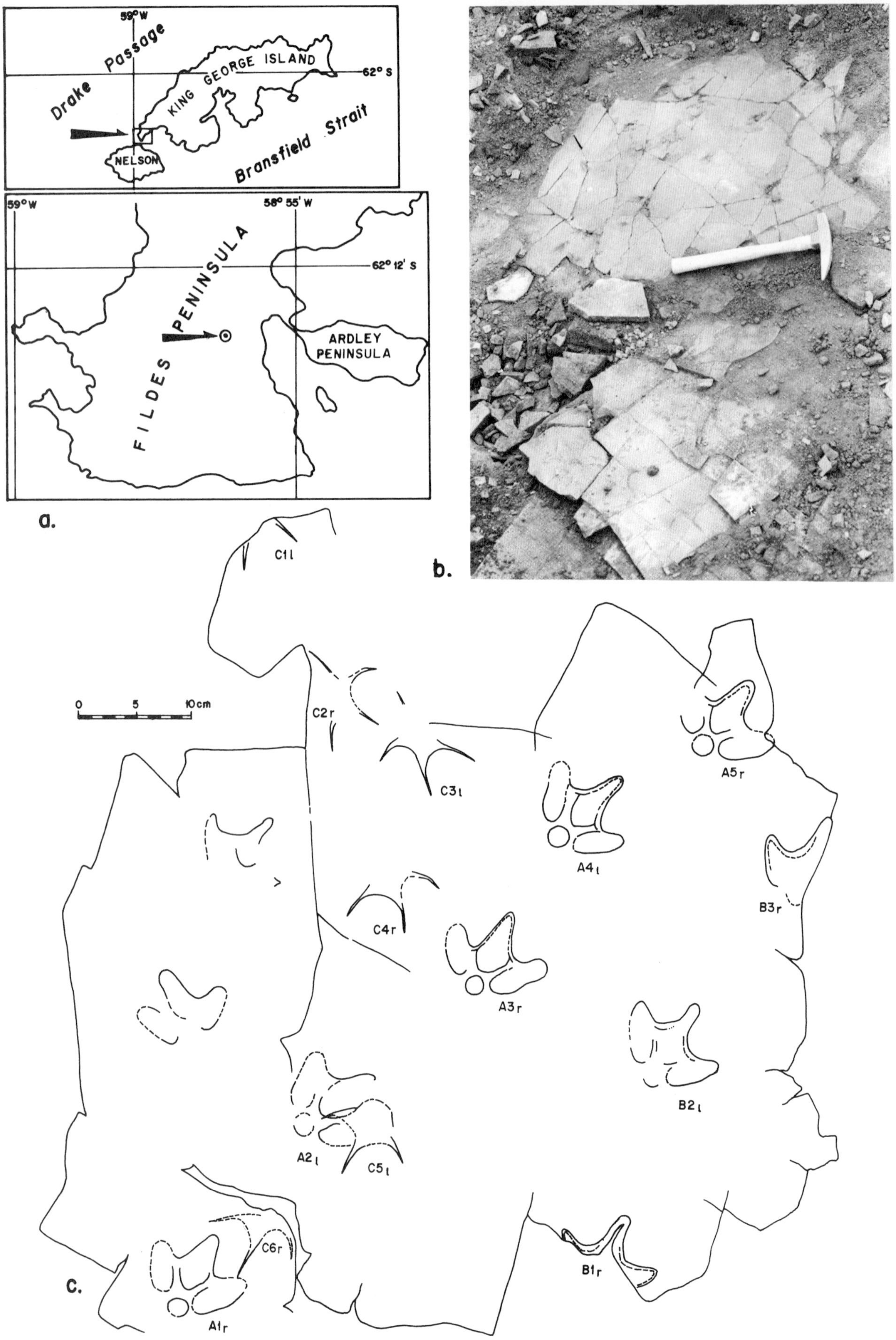

Figure 27.1. a. Sketch map showing the location of the investigated area and the place where ichnites were found. *b.* Rock surface with prints of morphotypes II and III. Compare with Figure 27.1*c. c.* Schematic drawing of bird ichnites morphotypes II (A1R to A5R and B1R to B3R) and III (C1L to C6R). Morphotype III corresponds to natural cast prints from the adverse side of the sample. Dashed lines indicate less accuracy.

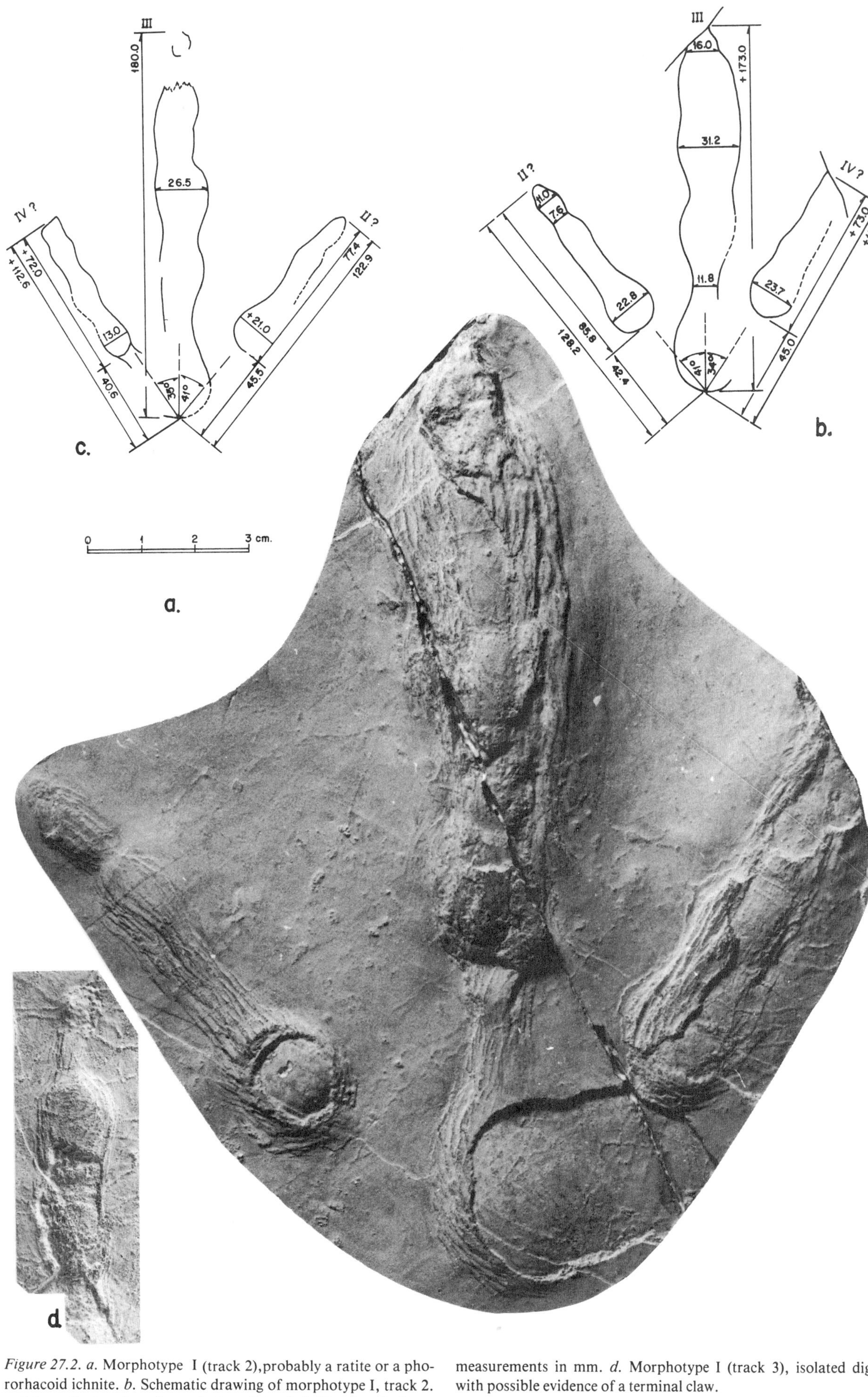

Figure 27.2. a. Morphotype I (track 2), probably a ratite or a phororhacoid ichnite. *b.* Schematic drawing of morphotype I, track 2. *c.* Schematic drawing of morphotype I, track 1. Numbers indicate measurements in mm. *d.* Morphotype I (track 3), isolated digit with possible evidence of a terminal claw.

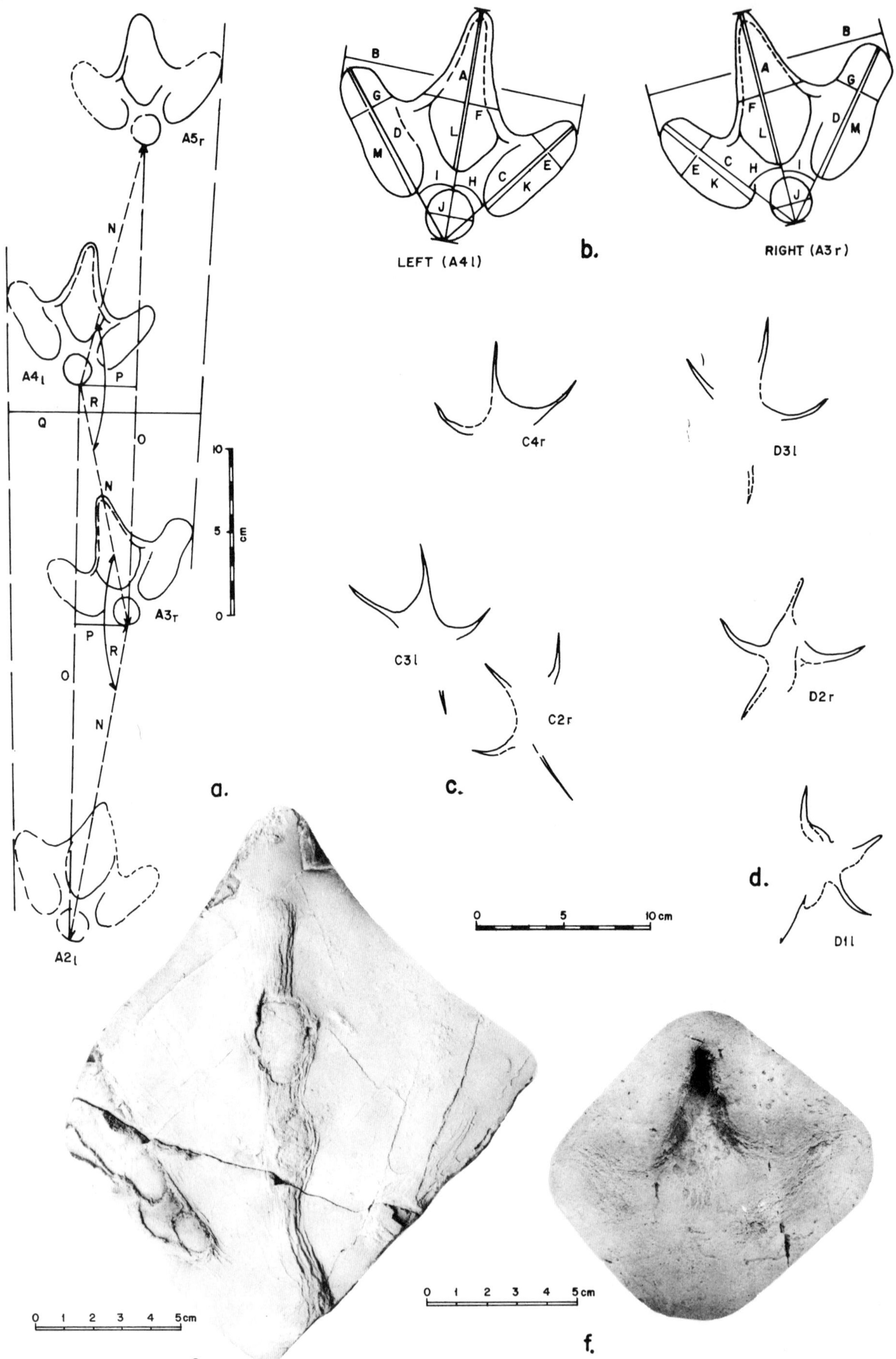

Figure 27.3. a and *b.* Schematic drawings of part of a trackway and isolated right and left footprints of morphotype II in their natural position, showing the measurements taken for these ichnites (for further explanation see Table 27.1). *c* and *d.* Schematic drawings of morphotype III trackways D1L to D3L and C2R to C4R. In C2R and C3L the bird may have stopped and stood. *e.* Morphotype I (track 1), probably a ratite or a phororhacoid. *f.* Morphotype II (track A3R), probably an anatid.

Mihirungs (Dromornithidae), extinct forms from Australia, in genera where the digits are known, tend to have shorter, broader terminal phalanges than those indicated in the large Fildes tracks (see Rich and Green, 1974; Rich and Gill, 1976). Members of the South American genus *Rhea* have an elongate midtoe (III) and two short side toes with a marked disparity between the mid- and side toes; the Fildes tracks have relatively longer side toes. It is difficult, however, to exclude any of the remaining ratites on the basis of anything but size. Adult emus (Casuariidae; see Rich and Green, 1974) produce prints that are somewhat larger than the Fildes prints, but certainly juveniles would render this point of difference useless. Kiwis (Apterygidae) and tinamous (Tinamidae) produce smaller prints that may or may not show digit I, which is present in both forms. Because there is very little rock preserved behind the heel impression in the better of the two most complete morphotype I prints (track 2), it is impossible to determine if a first digit was present, although we strongly suspect that it was not. Prints of the moas (Dinornithidae and Emeidae), likewise, may or may not show the presence of the first digit; certainly there would be size overlap between the Fildes tracks and the smaller members of the Dinornithiformes. Known adult members of the Aepyornithiformes (including both *Aepyornis* and *Mullerornis*) would produce prints larger than the Fildes prints, but overlap could occur if juveniles were involved, although more robust toes might be predicted for the Aepyornithiformes.

Another group besides the ratites that possibly could have made the Fildes tracks is in the order Gruiformes. The Phororhachidae, a carnivorous, presently extinct, New World family (primarily South American) had a first digit (and thus four toes), but because the hallux was elevated, it may not have left an impression. The cariamas (Cariamidae), also a family of gruiforms, can probably be excluded because of the stubby-toed appearance of their foot and the fact that they are semipalmate (van Tyne and Berger, 1961). Bustards (Otididae) as well as gruiforms, both with only three toes, might be excluded because the toes are short and broad.

Tracks of each of the following groups that include larger forms can be excluded from consideration because of the reasons enumerated.

1. Megapodes (Megapodidae), curassows (Cracidae), turkeys (Meleagridae), cranes (Gruidae), trumpeters (Psophidae), herons (Ardeidae), whale-billed storks (Balaenicipitidae), hammerheads (Scopidae), storks (Ciconiidae), ibises (Threskiornithidae), and screamers (Anhimiidae) all have a first digit and thus a total of four, not three, toes. In most cases, the toes are more slender than those indicated in the Fildes tracks, and thus the tracks would not be as robust. Some groups, for instance the cranes, have much broader angles between the digits.
2. Flamingoes (Phoenicopteridae) and limpkins (Aramidae) have partial webbing between digits and slender, not broad, toes.
3. Secretarybirds (Sagittariidae) have digits that are short, not elongate, and which are semipalmate and sharply clawed distally.
4. Penguins (Spheniscidae) have three digits that are nearly the same length, and webbing to the end of the digits is usually preserved.

Other avian families can be dismissed either because the arrangement of toes, number of toes, or less importantly, size differs markedly from the Fildes prints.

In conclusion, then, because the prints were made by a large, broad- and elongate-toed bird, they seem quite likely, when compared with tracks of modern and fossil forms, to belong either to the ratites or to phororhacoids, thus signaling the presence of an obligate ground-dwelling bird in mid-Tertiary beds of the South Shetland Islands. This occurrence has significant biogeographic implications. Evidently ground birds of mid-Tertiary time were able to disperse between either South America or Antarctica and King George Island, thus making the Antarctic Peninsula area an important dispersal route for obligate terrestial forms.

Morphotype II

Material

Two trackways of natural foot impressions on one rock slab are numbered as follows: A1R-A2L-A3R-A4L-A5R (five footprints) and B1R-B2L-B3R (three footprints). These eight tracks are well enough preserved to give detailed information on the shape of the bird's foot and mode of walking. (See Figures 27.1*b* and *c* and 27.3*f*.)

Measurements

See Table 27.1, Figures 27.3*a* and *b*, and below.

Pace (Oblique pace, N): straight-line measurement between two consecutive heels (A1R to A5R) or between the end of consecutive midtoes (digit III; B1R to B3R).

A1R to A2L, 195 mm — A4L to A5R, 151 mm
A2L to A3R, 196 mm — B1R to B2L, 195 mm
A3R to A4L, 149 mm — B2L to B3R, 200 mm

Stride (O): straight-line measurement between two consecutive heels of the same foot.

A1R to A3R, 386 mm — A3R to A5R, 290 mm
A2L to A4L, 338 mm — B1R to B3R, 390 mm

Width of pace (P): see Figure 27.3*a* for explanation of measurement.

A1R to A4L, 30.0 to 35.5 mm — B1R to B3R, 37 mm
A2L to A5R, 35.5 to 40.0 mm

Width of tracks (Q): maximum width of a trackway.

A1R to A3R—A2L to A4L, 105 mm
A2L to A4L—A3R to A5R, 119 mm

Pace angulation (R): angle made by three consecutive tracks.

A1R-A2L-A3R, 162° — A3R-A4L-A5R, 150°
A2L-A3R-A4L, 156° — B1R-B2L-B3R, 162°

Description

Morphotype II footprints are about half the size of morphotype I prints, with the maximum length ranging from 79.0 to 86.0 mm and maximum width from 88.0 to 94.5 mm. Morphotype II prints are characterized by having three digits with the midtoe longest (mean, 82.3 mm; range, 79.0-86.0 mm), the internal digit (II) the shortest (mean, 64.2 mm; range, 62.0-66.8 mm), and the lateral digit (IV) intermediate (mean, 71.2 mm; range, 68.0-74.8 mm). This relationship is also valid for each finger internal length impression (K,L,M; Figure 27.3*b*). The digits are joined by an interdigital web that does not extend all the way to the end of the digits. The heel imprint is preserved as a circular unit, separate from the digital and web im-

Table 27.1. Measurements of cf. Anseriform footprints (morphotype II) from the Fildes Peninsula, King George Island, Antarctica. All measurements in mm unless marked as degrees

	A	*B*	*C*	*D*	*E*	*F*	*G*	*H*	*I*	*J*	*K*	*L*	*M*	*N*	*S*
A1R	80.0	88.0[a]	63.8[a]	68.8[a]	20.4[a]	27.5	21.0	40°	36°	16.0	43.0[a]	55.0	51.5[a]	195.0	8.0
A2L	86.0	—	—	73.0[a]	—	28.0[a]	21.0[a]	39°[a]	40°	19.0[a]	—	59.0	52.0[a]	196.0	—
A3R	80.0	94.5	62.0	74.8	18.0	25.6	18.8	41°	40°	16.0	45.0	57.5	55.0	149.0	12.3
A4L	85.0	91.2	66.8	69.8	19.0	24.0	18.0[a]	39°	41°	18.0	46.0	61.0	51.0[a]	151.0	11.3
A5R	79.0	92.0[a]	—	68.0[a]	—	28.0	22.0	45°[a]	44°[a]	17.0	—	56.0	51.0[a]	—	7.0
B1R	—	91.0	—	—	—	25.0	20.5	—	—	—	—	—	—	195.0	14.5
B2L	84.0	91.5	66.6	73.0	22.0	26.0	21.0	42°	38°	—	48.0[a]	59.5[a]	52.5	200.0	16.0
B3R	82.0[a]	—	62.0[a]	—	18.0	—	—	—	—	—	45.0	59.0	—	—	12.5
n	7	6	5	6	5	7	7	6	6	5	5	7	6	6	7
Range	79.0–86.0	88.0–94.5	62.0–66.8	68.0–74.8	18.0–22.0	24.0–28.0	18.0–22.0	39°–45°	36°–44°	16.0–19.0	43.0–48.0	55.0–61.0	51.0–55.0	149.0–200.0	7.0–16.0
Mean	82.3	91.4	64.2	71.2	19.5	26.3	20.3	41°	40°	17.2	45.4	58.1	52.2	181.0	11.7
Standard deviation	2.5	1.9	2.1	2.5	1.5	1.5	1.3	2.1	2.5	1.2	1.6	1.9	1.4	—	3.0
Standard error	6.5	3.6	4.5	6.2	2.4	2.1	1.7	4.3	6.1	1.4	2.6	3.8	1.9	—	9.0
Coeff. of variation	3.1	2.1	3.3	3.5	7.9	5.5	6.4	5.1	6.2	6.8	3.6	3.3	2.6	—	25.7

A: Maximum length of the track (= length of midtoe—III).
B: Maximum width of the track.
C: Length of internal toe (II).
D: Length of external toe (IV).
E: Width of internal toe (II).
F: Width of midtoe (III).
G: Width of external toe (IV).
H: Angle between internal toe (II) and midtoe (III).
I: Angle between midtoe (III) and external toe (IV).
J: Width of heel.
K: Internal length of internal toe (II).
L: Internal length of midtoe (III).
M: Internal length of external toe (IV).
N: Pace (oblique pace).
O: Stride (see Morphotype II section, page 249).
P: Width of pace (see Morphotype II section, page 249).
Q: Width of trackway (see Morphotype II section, page 249).
R: Pace angulation (see Morphotype II section, page 249).
S: Maximum depth of midtoe (III) impression.

[a] Approximate measurements.

pressions. The impression of the terminal phalanx of the midtoe (III) is decidedly deeper than those of the other digits.

Comment

Of the modern birds, the following families are known to have well-developed interdigital webs:

with three digits,
- Spheniscidae (penguins)
- Diomedeidae (albatrosses)
- Procellariidae (shearwaters)
- Hydrobatidae (storm petrels)
- Pelecanoididae (diving petrels)
- Alcidae (auks; hallux absent or vestigial)

with four digits,
- Gaviidae (loons)
- Phaethontidae (tropic birds)
- Pelecanidae (pelicans)
- Sulidae (boobies)
- Phalacrocoracidae (cormorants)
- Anhingidae (darters)
- Phoenicopteridae (flamingoes)
- Anatidae (ducks, geese, and kin)
- Stercorariidae (skuas; hallux small)
- Laridae (gulls; hallux small or vestigial)
- Rynchopidae (skimmers; hallux very small)

A few additional families have small webs at the base of the digits:

with three digits,
- Haematopodidae (oystercatchers)
- Recurvirostridae (avocets)
- Burhinidae (thick-knees)

with four digits,
- Fregatidae (frigatebirds)
- Ciconiidae (storks)
- Threskiornithidae (ibises)
- Anhimidae (screamers)
- Sagittariidae (secretarybirds)
- Chionididae (sheathbills)
- Aramidae (limpkins)
- Cariamidae (cariamas)
- Phalaropodidae (phalaropes)
- Glareolidae (pratincoles; webbing between III and IV only)

Webbing in the Fildes prints of morphotype II appears to be more extensive than just a basal web and yet does not extend the full length of the digits. Only three digits appear to be present, and all three are involved in the webbing. Thus, a number of the cited families can be eliminated, and only the forms in the first and second categories above need be seriously considered. The Hydrobatidae and Pelecanoididae, both fossil and recent, are small, delicate birds that seem unlikely to have produced the Fildes prints. Although of large size, albatrosses (Diomedeidae) have elongate, slender, delicate digits, and like the Hydrobatidae and Pelecanoididae, the webbing is continuous to the end of the digits. Penguins (Spheniscidae) possess digits that are all about the same length, and webbing extends to the distal end of these digits. Most of the remaining groups in the first and second categories can be eliminated because the digits leave relatively slender impressions, except for the Anatidae (ducks and kin) and perhaps flamingoes (Phoenicopteridae). Flamingoes also have somewhat more slender digits, whereas the morphotype II prints from the Fildes Peninsula consist of broad digits suggestive of a rather heavy bird with respect to foot size. A large duck or small goose seems to be the most likely possibility for the bird that made these prints.

MORPHOTYPE III

Material

Two trackways, one consisting of six footprints (C1L to C6R) preserved as natural cast impressions and a second consisting of three well-preserved natural track impressions (D1L to D3L) (see Figures 27.1*c* and 27.3*c* and *d*).

Measurements

See below.

Description

Ichnites consisting of four slender digits, which range in maximum length from 84.8–102.6 mm and maximum width from 76.0–82.8 mm. In both trackways, the distance between homologous points on each of the tracks (i.e., pace, *N*) varies greatly, from 124 to 216 mm, with the exception of the distance between C2R and C3L (N = 77 mm) where the bird may have stopped and stood. Angles between the digits are large, approaching 90° in most cases. The digits are elongate with the midtoe (III) being the longest; in some prints the side toes are of nearly equal length, whereas in others, one toe is slightly longer than the other. The first digit is well developed, elongate, and variable in length with respect to the other digits, but in all cases, shorter than digit III. In most prints, the anterior margins of digits II and IV are distinctly concave anteriorly. No separate, distinct ungual prints on any of the digits are indicated.

Comment

It is impossible to determine very specifically which group of birds made the morphotype III prints based on the fossil evidence available. Not only is the fossil evidence limiting, but so too is the lack of knowledge on modern avian trackways.

Morphotype III prints were made by a bird probably without much webbing between the slender digits. (With the several prints represented, if webbing had been present, it probably would have appeared in at least one print; it should be kept in mind, however, that if the substrate had had a certain consistency, webbing might not have left an impression.) The bird had four digits, all of which were well developed, including digit I. Birds with this foot arrangement occur in several families: Ardeidae (herons), Cochlearidae (boat-billed herons), Balaenicipitidae (whale-billed storks), Scopidae (hammerheads), Cathartidae (New World vultures), Cracidae (curassows), Opisthocomidae (hoatzins), Mesoenatidae (mesites), Rallidae (rails), Eurypygidae (sunbitterns), Columbidae (pigeons), Musophagidae (touracos), and perhaps even some of the Passeriformes such as the Menuridae (lyrebird). Even those forms with an elevated hallux of moderate length could have produced such tracks, as most of the fossil tracks do not have a continuous connection between digit I and the remainder of the sole of the foot. Thus, the following groups need also be considered: Tinamidae (tinamous), Megapodidae (megapodes), Tetraonidae (grouse), Numididae (guinea fowls), Phasianidae (pheasants, peafowls), Meleagridae (turkeys), Psophiidae (trumpeters), Rhynochetidae (kagus), perhaps Cariamidae (cariamas, but unlikely because of large claws on digits of this group and the very small hallux), Rostratulidae (painted snipes), Scolopacidae (sandpipers), Dromadidae (crab plovers), possibly Glareolidae (pratincoles; but because of elongate claw on digit III, probably not), Thinocoridae (seed snipes), and Chionididae (sheathbills).

Birds with four elongate toes that have only a basal webbing which might not appear in the fossil tracks, include: Ciconiidae (storks), Threskiornithidae (ibises), Anhimidae (screamers), and Aramidae (limpkins). These, too, need to be seriously considered.

On the other hand, the Falconiformes, except for the Cathartidae, can be eliminated because the Fildes prints lack any evidence of the terminal claws that would be present in tracks made by these birds. Prints produced by members of the Sagittariidae (secretarybirds), Accipitridae (hawks and kin), Pandionidae (ospreys), Falconidae (falcons), and the Strigidae (regular owls) and Tytonidae (barn owls) would tend to have the terminal claw marks preserved. Members of the Jacanidae have extremely elongate digits, with a particularly elongate ungual phalanx on digit I, and can thus be eliminated.

If only the modern species in the above families were to be considered, size and present distribution could be used to eliminate a number of forms. Neither of these criteria, however, should enter into a morphologic analysis, and thus no single family can be identified from the Fildes morphotype III prints. They might have been made by individuals in one of several avian families. Additionally, there are a number of exclusively fossil groups that need be considered but are not here, simply because the possibilities are already so varied that adding more is of little value at this stage.

Trackways and isolated footprints assigned to the family Rallidae (Covacevich and Lamperein, 1970), *Antarctichnus fuenzalidae,* although distinctly smaller than morphotype III prints, may also belong to a similar range of families along with this new morphotype.

Some cautionary remarks based on interpretation of modern avian footprints

During our studies involving the Fildes Peninsula avian ichnites, one of us (P. V. R.) was able to photograph and measure footprints of several modern species of birds in the families Spheniscidae, Rheidae, Casuariidae, Phasianidae, Anatidae, Gruidae, Stercorariidae, and Laridae. In this brief survey, it became evident that the type of footprints made by an individual bird depended not only on the conformation of the bird's foot but also on the substrate and the speed with which the bird was moving. Pseudowebs can result from birds lacking them (e.g., the peafowl, *Pavo cristatus*) moving quickly through rather liquid mud. On the other hand, if the substrate is relatively compacted, webbing, although present, may not be apparent, even though digits may show up well (e.g., Cape Barren Goose, *Cereopsis novaehollandiae,* in our sample). Consistency of the substrate often determines whether the digits in the print connect directly to the heel or are separated from it by a space. A short hallux may give the impression of being longer if the substrate allows the foot to sink in deeply, rather than if the foot leaves only a slight impression, where the hallux may leave no imprint on a harder substrate.

Features that tend to remain consistent independent of substrate are angles between the toes (e.g., 90° in *Grus rubicanda,* but small acute angles in *Rhea americana* between digits II, III, and IV); relative lengths of digits II and IV to III; and widths of the digits with respect to length (either slender or robust).

Thus, based on this very limited survey, we believe that a much more detailed, careful survey of modern avian footprints is necessary before several of the Fildes Peninsula tracks can be confidently assigned to groups any more precisely than they are here. Even then, definite assignment to one family may still be impossible. Although some keys to identification of avian footprints already exist (Lambrecht, 1933), they are far from complete or satisfactory.

Biogeographic significance of the Fildes avian ichnites

During the past 15 years, biogeographers have been seriously reviewing animal and plant distributions in light of theories involving mobile continents. Today, it is no longer generally accepted that continents have always been where they are now. Although East Antarctica, for instance, is thought to have remained relatively close to its present position through much of Mesozoic and Cenozoic time, its neighbors, South America, Africa, and Australia, are thought to have slowly moved away from it during the past 200 m.y. During Mesozoic time, when these continental fragments were probably joined to form much of Gondwanaland, Antarctica (both East and West) could have served as an important connecting route between many parts of this supercontinent (see Figures 27.4, 27.5, and 27.6; also see Rich 1975*a,b* for a summary of evidence supporting such reconstructions). To date, terrestrial vertebrate evidence cited in support of such a dispersal route has consisted of comparisons between faunal elements, mostly modern, on each of the Gondwanic fragments (South America, Australia, Africa, India, Madagascar, and New Zealand) and calculations of faunal similarities. The Fildes Peninsula footprints, then, are some of the first strongly suggestive evidence (e.g., morphotype I tracks) that terrestrial vertebrates were actually present in some places along this dispersal route during the mid-Tertiary and that Antarctica was serving as a biogeographic link for at least some forms.

Additional information on radiometric ages for Fildes Peninsula Tertiary volcanic rocks and clearer models for western Gondwanaland paleogeography and the spreading history of Drake Passage and the Scotia arc will firmly support the importance of this new finding.

Conclusions

At least four types of bird tracks occur in lacustrine sedimentary rocks of mid-Tertiary age on King George Island of West Antarctica. One morphotype apparently repre-

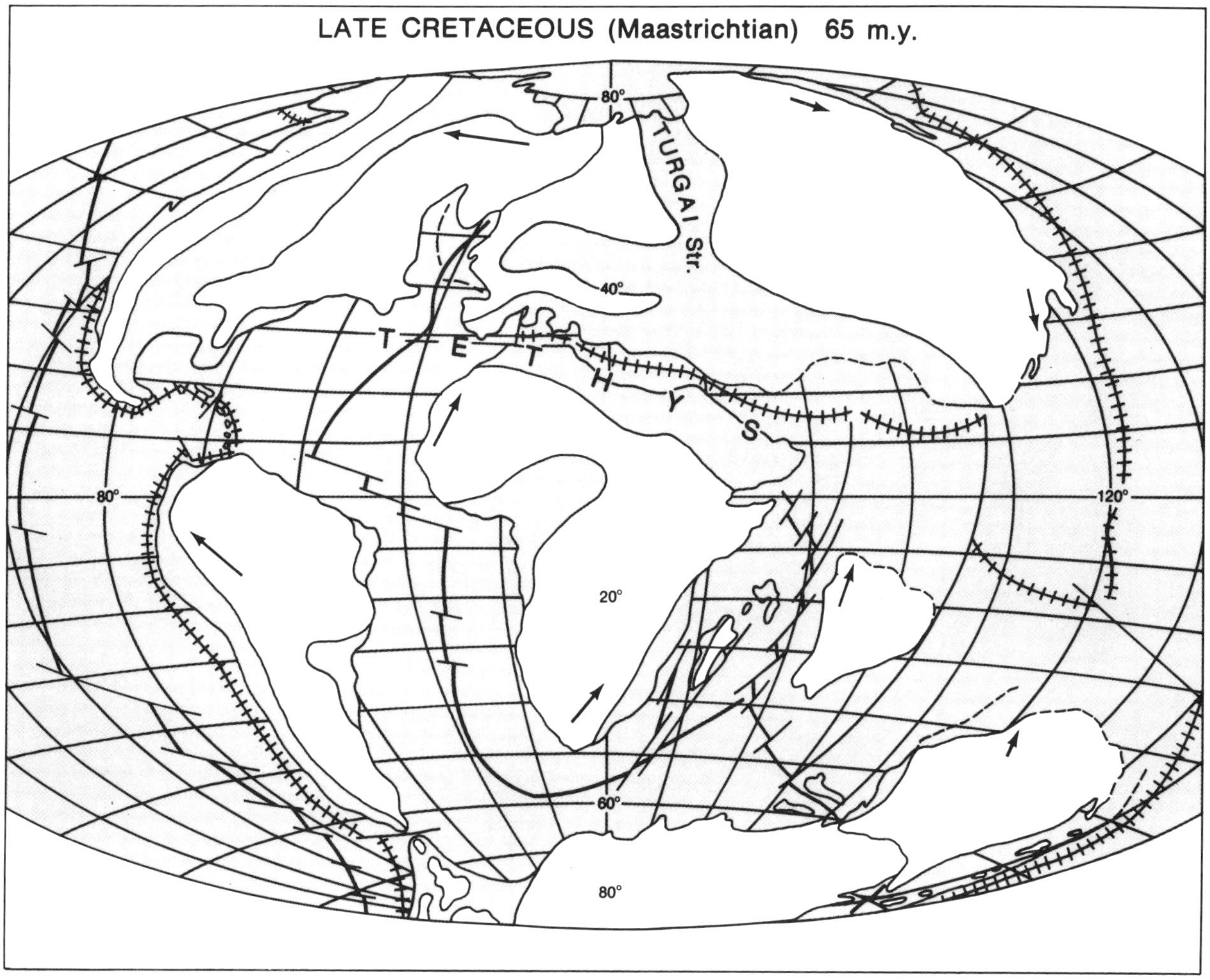

Figure 27.4. Reconstruction of continental arrangement during Late Cretaceous time (from Rich, 1975*a*, modified from maps prepared by R. H. Tedford). Hatched lines: trenches. Heavy solid lines: spreading axes. Light solid lines: transform faults.

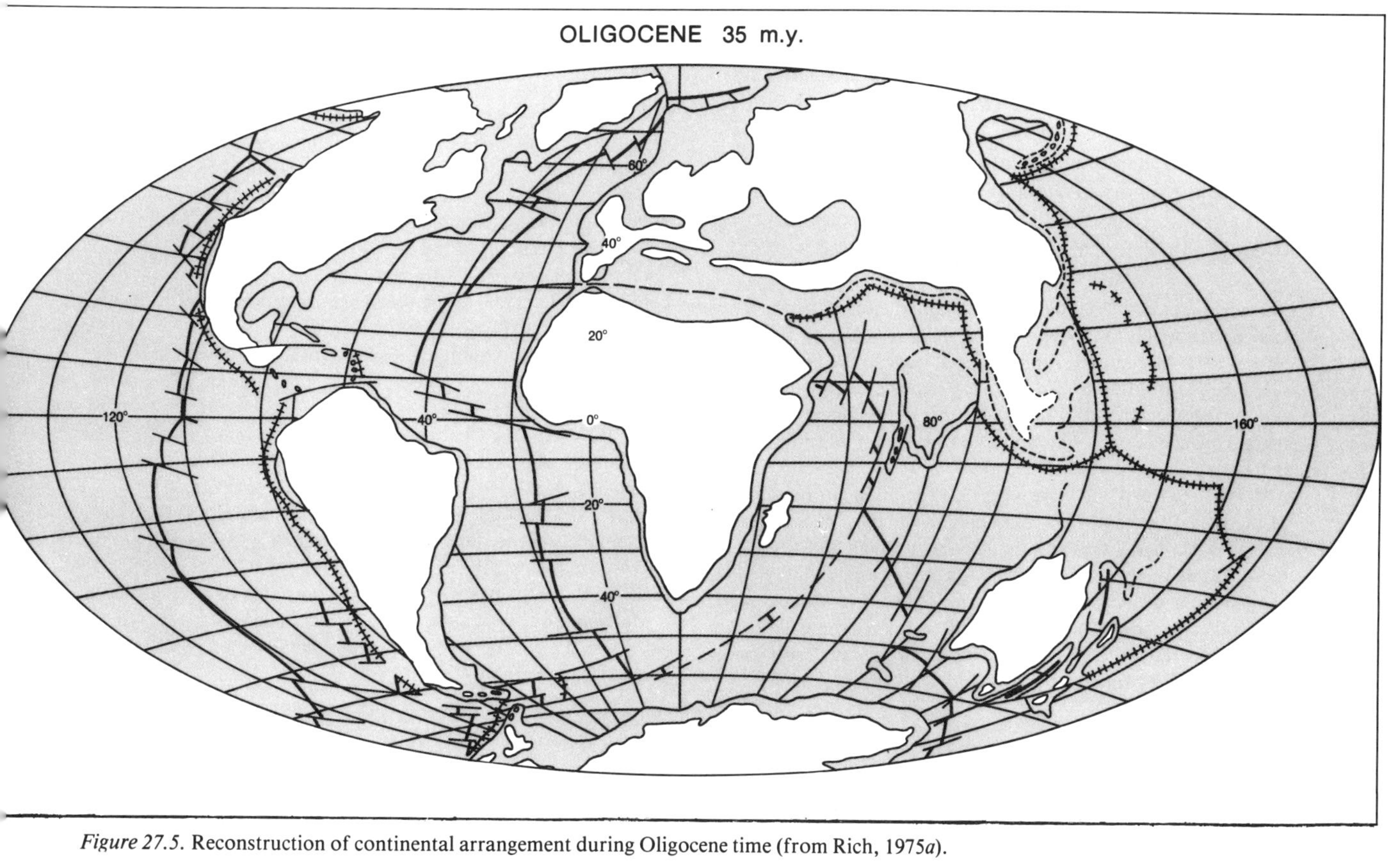

Figure 27.5. Reconstruction of continental arrangement during Oligocene time (from Rich, 1975*a*).

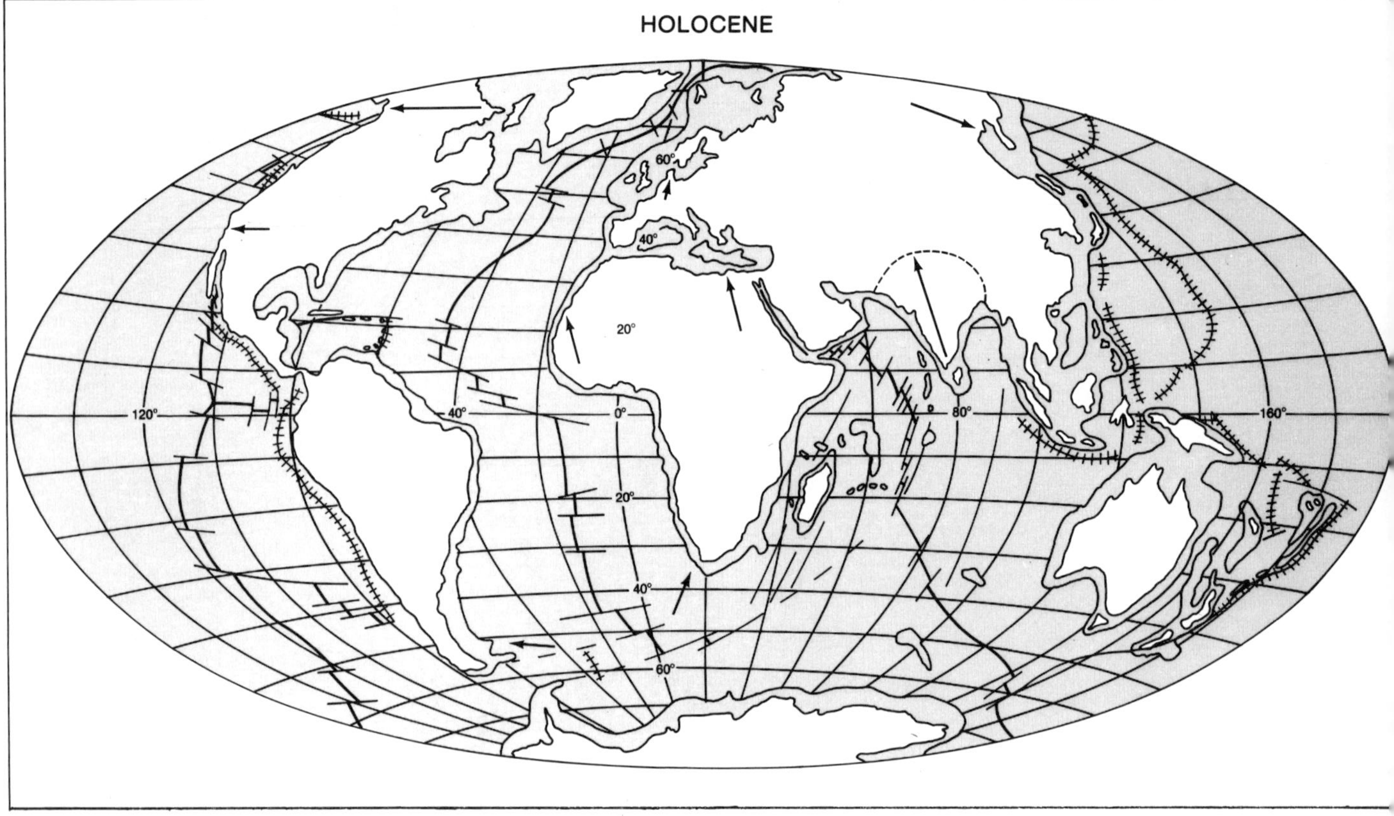

Figure 27.6. Continental arrangement at present (Holocene) (from Rich, 1975*a*).

sents a nonvolant ground bird, probably either a ratite or a gruiform bird. A second morphotype probably represents an anatid. The two other morphotypes (including tracks previously assigned to the rails, *Antarctichnus fuenzalidae*) could belong to a number of avian families.

The main importance of the King George Island avian ichnites is that morphotype I strongly indicates that at least part of the Antarctic dispersal route (West Antarctica) was utilized by obligate terrestrial forms. This discovery gives additional support to the hypothesis that this route could have been an important one in dispersal of organisms between South America and Australia.

Acknowledgments. We would like to thank the Instituto Antártico Chileno (INACH) for sponsoring the field work during the 1968–69 and 1970–71 Antarctic summer expeditions to the Fildes Peninsula; the Instituto de Investigaciones Geológicas (IIG) for partly supporting the scientific work; and the Departamento de Geología, Universidad de Chile, for loan of the ichnites, now housed in their type paleontologic collections.

We wish to thank the Melbourne Zoo for aid in our brief survey of modern avian footprints, especially the director J. Sullivan, as well as B. Scribano, H. Peltz, and G. Slater. Allan McEvey, Curator of Birds at the National Museum of Victoria, kindly made available his avian osteology collection. Mary Lee Vickers typed the final manuscript. Teresa Engelhardt translated some important literature pertinent to this study. The M. A. Ingram Trust kindly provided funding for a part of this study.

References cited

Covacevich, V., and Hernández, P. 1971. Investigaciones paleontológicas en las Islas Shetland del Sur, Antártica. *Boln Inst. antárt. chil., 6,* 3–6.

Covacevich, V., and Lamperein, C. 1969. Nota sobre el hallazgo de icnitas fósiles de aves en Península Fildes, Isla Rey Jorge, Shetland del Sur, Antártica. *Boln Inst. antárt. chil., 4,* 26–28.

Covacevich, V., and Lamperein, C. 1970. Hallazgo de icnitas en Península Fildes, Isla Rey Jorge, Archipiélago Shetland del Sur, Antártica. *Ser. cient. Inst. antárt. chil., 1,* No. 1, 55–74.

Covacevich, V., and Lamperein, C. 1972. Ichnites from Fildes Peninsula, King George Island, South Shetland Islands. (*In* Adie, R. J., *ed. Antarctic Geology and Geophysics.* Oslo, Universitetsforlaget, 71–74.)

Lambrecht, K. 1933. *Handbuch der Palaeornithologie.* Berlin, Borntraeger, 1024 pp.

Rich, P. V. 1975*a*. Antarctic dispersal routes, plate tectonics and the origin of Australia's non-passeriform avifauna. *Mem. Natn. Mus. Vict., 36,* 63–126.

Rich, P. V. 1975*b*. Changing continental arrangements and the origin of Australia's non-passeriform continental avifauna. *Emu, 75,* No. 3, 97–112.

Rich, P. V., and Gill, E. 1976. Possible dromornithid footprints from Pleistocene dune sands of southern Victoria, Australia. *Emu, 76,* No. 4, 221–23.

Rich, P. V., and Green, R. H. 1974. Footprints of birds at South Mt. Cameron, Tasmania. *Emu, 74,* No. 4, 245–48.

van Tyne, J., and Berger, A. J. 1961. *Fundamentals of Ornithology.* London, John Wiley, 624 pp.

28

Potassium-Argon Ages and Paleomagnetic Results from King George Island, South Shetland Islands*

D. R. WATTS**
Institute of Polar Studies, The Ohio State University, Columbus, Ohio 43210, U.S.A.

Abstract. Twenty-nine oriented drill cores and seventeen oriented hand specimens were collected from eight sites in a reconnaissance paleomagnetic survey of King George Island. Rock units samples in this survey include Tertiary andesite flows of the Ezcurra Inlet Group and the Fildes Peninsula Group, a Tertiary granodiorite intrusion, and a basaltic dike and lava flow from rocks assigned to the Upper Jurassic Volcanic Group. A total of ten potassium-argon age determinations were made on this rock collection, with at least one age determination per site. Age determinations from the Ezcurra Inlet Group, the Fildes Peninsula Group, and the granodiorite intrusion range between 58 m.y. and 46 m.y. The meaning of the apparent ages from the Upper Jurassic Volcanic Group is not clear because of the metasomatic alteration that has affected these rocks.

Acceptable paleomagnetic results were obtained from five of the six sites in the lower Tertiary units, permitting the calculation of a mean pole position. The early Tertiary pole calculated for these sites lies at lat. 81° S, long. 73° W, which agrees well with a previous determination from the Fildes Peninsula Group. The two virtual geomagnetic poles determined from the Upper Jurassic Volcanic Group agree rather well with the Jurassic pole from East Antarctica; however, in view of the uncertainty about the age of these rocks, the significance of this agreement is unknown. Although the mean early Tertiary pole calculated from all available South Shetland data (lat. 77° S, long. 78° W) shows excellent agreement with an early Tertiary pole predicted for East Antarctica by closing Australia and Antarctica and using poles from Australia, this apparent agreement may not be real because of the rather large scatter in the results from the South Shetland Islands. If the mean pole determined in this study is accurate, however, two interpretations are possible. Either Antarctica has moved as a single plate since early Tertiary time, or any rotation of the South Shetland Islands since early Tertiary time has been about a pole located approximately at the present South Pole.

During the 1973–74 austral field season, a reconnaissance collection of oriented drill cores and hand specimens was taken from eight sites in various igneous rock units exposed on King George Island in the South Shetland Islands. The objective in making this collection was to obtain paleomagnetic data and potassium-argon ages that could be used to establish paleopole positions of Cenozoic and possibly Mesozoic age for the northern Antarctic Peninsula. Such data are potentially useful in reconstructing the tectonic history of this area.

The geology of King George Island has been summarized by Barton (1965), and the petrology of the volcanic rocks has been described by Hawkes (1961). The following geologic summary is based mainly on these two studies.

King George Island consists almost entirely of plutons and lava flows, with sedimentary rocks occurring only as discontinuous beds between lava flows. Thick sequences (as much as 915 m) of pyroxene andesite lavas, which have been correlated with the Upper Jurassic Volcanic Group of the Antarctic Peninsula, are regarded as the oldest rocks exposed on King George Island. These rocks crop out mainly on Fildes Peninsula, Barton Peninsula, and in the area of Admiralty Bay (Figure 28.1). The correlation of these rocks is based principally on original composition, state of alteration, and style of structural deformation. Particularly in the area of Admiralty Bay, the Upper Jurassic Volcanic Group has been extensively mineralized and metasomatized during events associated with the emplacement of plutons of the Andean Intrusive Suite and during volcanic activity in Tertiary time. The alteration is manifest in the replacement of plagioclase phenocrysts by sericite, clay minerals, and calcite and in the development of secondary epidote and pyrite. A basaltic dike and a basaltic lava flow from rocks exposed on the southern end of Fildes Peninsula, mapped by Barton (1965) as the Upper Jurassic Volcanic Group, were sampled in this study.

Plutons assigned to the Andean Intrusive Suite crop out sparsely on King George Island and range in composition from quartz gabbro to granodiorite. Potassium-argon results, to be discussed later, support an early Tertiary age of emplacement for some of these plutons in the South Shetland Islands. Oriented specimens were collected from the pluton exposed at Noel Hill. Hawkes (1961) has reported that both quartz diorite and granodiorite are represented in this body and, on that basis, has suggested that it may be a composite intrusion.

Barton (1965) has described six sequences of Tertiary volcanic and sedimentary rocks on King George Island, two of which, the Ezcurra Inlet Group and the Fildes Peninsula Group, were sampled in this study. The Ezcurra Inlet Group is a 593 m succession of agglomerate, andesite,

* Contribution no. 321 of the Institute of Polar Studies, The Ohio State University.

** Present address: Department of Geology and Mineralogy, University of Michigan, Ann Arbor, Michigan 48109, U.S.A.

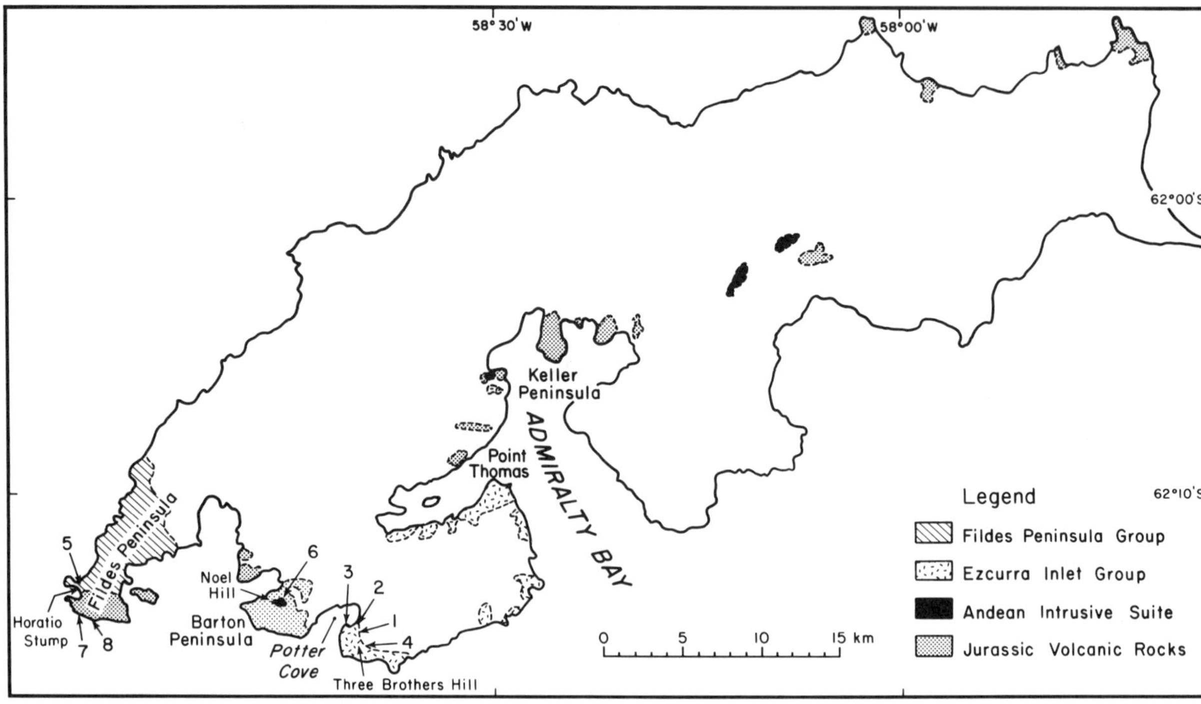

Figure 28.1. Location and geologic map of King George Island (modified from Barton, 1965). The numbered sites correspond to the following units: 1. Upper lava, lower part of the Ezcurra Inlet Group, Potter Cove; 2. Middle lava, lower part of the Ezcurra Inlet Group, Potter Cove, 3. Lower lava, lower part of the Ezcurra Inlet Group, Potter Cove; 4. Volcanic plug, Three Brothers Hill, Potter Cove; 5. Volcanic plug, Horatio Stump, Fildes Peninsula; 6. Andean intrusion, Noel Hill; 7. Dike, Fildes Peninsula; 8. Lava, Fildes Peninsula. White area on island is ice covered.

and tuff with minor sedimentary beds between the volcanic units. Plant remains from the sedimentary units are not diagnostic and indicate only a post-middle Cretaceous age for this sequence. Although Barton (1964, 1965) has suggested a middle Tertiary age for these rocks, potassium-argon dating indicates that they are early Tertiary in age (58–49 m.y.). The most abundant lavas are hypersthene-augite andesites. The plugs at Three Brothers Hill and Point Thomas were probably volcanic centers from which some of these lavas were erupted. The rocks assigned to the Ezcurra Inlet Group are unaltered and were apparently unaffected by the emplacement of the Andean plutons.

The Fildes Peninsula Group is exposed mainly on Fildes Peninsula, though a few outcrops have been mapped by Barton (1965) on northern Nelson Island. Rock types include andesite, agglomerate, and tuff with subordinate interbedded sedimentary rocks. The principal rock type is a basaltic andesite, though a few flows of hypersthene-augite andesite resembling those of the Ezcurra Inlet Group are present. Barton (1965) suggested that the lavas of the Fildes Peninsula Group were erupted from volcanic centers near Horatio Stump. Plant fragments found in the sedimentary beds have been interpreted by Orlando (1964) as Miocene in age, and Barton (1964) has suggested an Oligocene–Miocene age. Potassium-argon evidence, discussed below, indicates an early Tertiary age (54–45 m.y.) for at least part of this group.

Potassium-argon age determinations

Ten potassium-argon age determinations of whole-rock samples, with at least one age determination per sampling site, were carried out in the Potassium-Argon Geochronology Laboratory at Ohio State University. The rock units examined in this study include three lava flows and the Three Brothers Hill plug from the Ezcurra Inlet Group, the Noel Hill intrusion, the Horatio Stump plug from the Fildes Peninsula Group, and a basaltic dike and lava flow from the so-called Upper Jurassic Volcanic Group at the southern end of Fildes Peninsula. The collecting sites are indicated in Figure 28.1, and the results of the potassium-argon determinations are summarized in Table 28.1.

Lavas from the Ezcurra Inlet Group were sampled at a well-exposed section, and the dated units are here referred to as the "upper lava," the "middle lava," and the "lower lava" to indicate relative stratigraphic position. All three flows, however, are from the lower exposed part of the Ezcurra Inlet Group along the north-facing shore of Potter Cove. These lava flows are visible from the sea as three cliff-forming outcrops above the beach, immediately east of the Argentine hut. The three flows are very similar macroscopically. Thin section analysis reveals numerous subhedral to anhedral plagioclase phenocrysts (labradorite) in a crystalline matrix with a trachytic texture. The composition of the plagioclase ranges from An_{57} to An_{51}. The groundmass consists of numerous needles of plagioclase surrounding irregular grains of pyroxene and abundant opaque mineral grains. Trace amounts of volcanic glass are present in the groundmass of the middle lava. Minor secondary alteration is present in all three of the units, and it includes sericite and clay minerals, which are largely confined to the plagioclase phenocrysts, and chlorite in the interstices of the matrix.

Although the unaltered appearance of the dated samples from these outcrops suggests that they are suitable for

Table 28.1. K-Ar whole-rock age data from King George Island

Rock unit	% K[a]	Moles ^{40}Ar/gm (x 10^{-11})	$\%_{40}$ radiogenic ^{40}Ar	Age (m.y.)
Ezcurra Inlet Group				
Upper andesite lava, Potter Cove	0.3324	2.945	52.4	49.1 ± 0.9
Middle andesite lava, Potter Cove	0.4640	4.859	41.4	57.9 ± 0.8
Lower andesite lava, Potter Cove	0.4701	4.218	29.9	49.7 ± 1.7
Basaltic plug, Three Brothers Hill	1.796	16.42	62.3	50.6 ± 0.7
Andean Intrusive Suite				
Granodiorite, Noel Hill, specimen 1	2.320	19.24	55.6	46.0 ± 0.7
Granodiorite, Noel Hill, specimen 2	2.173	19.24	40.8	50.2 ± 0.6
Fildes Peninsula Group				
Basaltic plug, Horatio Stump	0.3598	3.3531	48.9	54.3 ± 0.6
Upper Jurassic Volcanic Group				
Basaltic dike[b]	0.6423	9.247	55.6	79.2 ± 2.6
Basaltic lava, specimen 1[b]	0.2487	2.902	28.6	64.4 ± 0.8
Basaltic lava, specimen 2[b]	0.2403	4.663	36.3	106.0 ± 1.2

Constants used: $1/\lambda = 1.885 \times 10^9$ yr; $^{40}K/K = 0.0119$ atom %; $(^{40}Ar/^{39}Ar)_{atm} = 1.581 \times 10^{-3}$; $(^{36}Ar/^{38}Ar)_{atm} = 5.35$. Error limits represent an estimate of one standard deviation for each individual analysis calculated in a manner similar to that described by Cox and Dalrymple (1967).
Argon isotope analyses were performed by J. F. Sutter.
[a] Value represents mean of duplicate fractions determined by flame photometer method.
[b] Treated in HCl.

potassium-argon age determination, examination of Table 28.1 shows that the apparent ages do not correlate with the observed stratigraphic succession. The middle lava shows the greatest apparent age of 57.9 m.y., older than the lower lava by 8 m.y.

A single age was obtained from a basaltic andesite specimen collected from the Three Brothers Hill plug at Potter Cove. Thin-section analysis reveals abundant phenocrysts of plagioclase, augite, and hypersthene embedded in a fine-grained crystalline matrix. Many of the plagioclase phenocrysts exhibit normal zoning from labradorite to andesine. Alkali feldspar is present between the plagioclase laths. The plagioclase phenocrysts all exhibit some degree of alteration to sericite and chlorite. The crystalline groundmass is composed predominantly of needles and laths of plagioclase, along with abundant irregular opaque minerals and minor pyroxene. Chloritic flakes are dispersed throughout the groundmass.

The 50.6 m.y. age of the Three Brothers Hill plug is in excellent agreement with the age determinations from the upper lava (49.1 m.y.) and the lower lava (49.7 m.y.). However, the apparent age of 57.9 m.y. for the middle lava is difficult to interpret. It could be the result of excess argon, but this cannot be demonstrated. It is concluded that 50 m.y. is a minimum age for the Ezcurra Inlet Group, and the possibility that these rocks are older cannot be precluded.

Two specimens from the Noel Hill intrusion, collected 50 m apart in the field, were dated. Some alteration was detected in the dated granodiorite samples: orthoclase has a cloudy appearance, due to the presence of clay minerals and sericite; plagioclase contains minor amounts of sericite, epidote, and carbonate; and biotite is chloritized around the rims of grains. The calculated ages of 46.0 m.y. and 50.2 m.y. are discordant, but they are interpreted as yielding a mean age of 48 m.y., the assumed minimum age for the time of cooling of this unit. The possibility of argon loss must be considered, however, as Grikurov and others (1970) have reported an age of 55 m.y. for a granodiorite at Marion Cove (probably the same unit). The reported ages of the Noel Hill intrusion are somewhat older than the 40 m.y. age reported by Dalziel and others (1973) for a tonalite intrusion on Livingston Island. The dates suggest that at least part of the Andean Intrusive Suite of the South Shetland Islands was emplaced during Eocene time.

The dated sample from the Horatio Stump plug on Fildes Peninsula is composed of unaltered irregular laths of plagioclase (An_{61}) in a finely crystalline matrix of plagioclase, pyroxene, and opaque minerals. The 54.3 m.y. age given in Table 28.1 is older than the age of 45 m.y. reported by Grikurov and others (1970). As the Horatio Stump plug is probably one of the eruptive centers for the Fildes Peninsula Group, the age determinations imply that at least part of the Fildes Peninsula Group is early Tertiary (54–45 m.y.) in age.

The clustering of the potassium-argon dates from the Ezcurra Inlet Group, the Fildes Peninsula Group, and the Andean intrusion at Noel Hill (58–45 m.y.) suggests that these rock units are genetically related and may be the result of the same magmatic event. Further work is needed to substantiate this interpretation.

The basaltic dike and the lava flow from the Upper Jurassic Volcanic Group have been affected by metasomatism. The vertical dike, 1 m in width and cutting steeply dipping agglomerates, is traceable for about 50 m. Thin-section study shows phenocrysts of unaltered plagioclase (An_{61}) in a groundmass of needle-shaped plagioclase, with a trachytic texture. Irregular pyroxene grains, opaque minerals, alkali feldspar, and interstitial calcite are also present in the groundmass.

The basaltic lava is flat-lying and forms the upper part of a cliff overlooking Fildes Strait. Thin sections show

Table 28.2. Paleomagnetic directions and statistics on oriented rock cores from King George Island

							Pole		
Rock unit	n	N	D	I	k	α_{95}	Latitude	Longitude	Treatment[a] (oersted)
Ezcurra Inlet Group									
Upper lava, Potter Cove	11	11	11.2°	−69.1°	55	6.2°	79°S	157°E	300
Middle lava, Potter Cove	5	5	103.4°	−83.3°	168	5.9°	57°S	84°W	200
Lower lava, Potter Cove	7	7	53.3°	−81.6°	253	3.8°	68°S	97°W	100
Three Brothers Hill	6	6	301.8°	−72.2°	52	9.4°	61°S	19°E	200
Andean Intrusive Suite									
Noel Hill	5	10	228.6°	80.6°	333	2.6°	70°S	102°W	300
Fildes Peninsula Group									
Horatio Stump[b]	5	10	—	—	—	—	—	—	300
Upper Jurassic Volcanic Group									
Basaltic dike	3	6	29.1°	−59.0°	372	3.5°	56°S	172°E	200
Basaltic lava	4	8	55.7°	−65.4°	100	5.6°	56°S	148°W	300

n: number of independently oriented specimens collected from a unit.
N: number of cylindrical specimens that were demagnetized and measured.
D: mean site declination.
I: mean site inclination.
k: estimate of Fisher's precision parameter.
α_{95}: radius of the circle of 95% confidence.
[a]The peak demagnetization field which gave minimum within-site dispersions.
[b]The results from the Horatio Stump plug were random at all levels of treatment and are not listed.

irregular phenocrysts of plagioclase (An_{54}) and pseudomorphs of a ferromagnesium mineral in a matrix of plagioclase, pyroxene, and secondary calcite. Few of the plagioclase phenocrysts exhibit zoning, and alteration is confined to sericite and clay minerals in the cores of the phenocrysts. Spherulites of quartz are present and are invariably surrounded by rims of secondary calcite. Chlorite is abundant in the groundmass.

Removal of calcite from these rocks was attempted by grinding whole-rock samples to 200 mesh size, and treating this fraction in concentrated hydrochloric acid for 2 hours. The potassium analysis and the argon isotope determination were performed on the fraction of the washed sample that remained after this treatment. One difficulty with this technique is the possibility of atmospheric argon contamination during the drying of the powder.

The results of these determinations for a single specimen from the dike and two specimens from the lava flow are given in Table 28.1. Because of the lack of agreement between the ages obtained from the two specimens from the same unit, reflecting either argon loss or failure of the above technique, it is difficult to apply a geologic interpretation to the results from the Upper Jurassic Volcanic Group. It is tentatively proposed that 79.2 m.y. is a minimum age for cooling of the dike and that 106.0 m.y. is a minimum age for cooling of the basaltic lava. These ages may be compared with one of 85 m.y. reported by Grikurov and others (1970) from a specimen that was probably collected on southern Fildes Peninsula.

Paleomagnetic results

Twenty-nine oriented 2.54 cm diameter drill cores and seventeen oriented hand specimens were collected from eight sites on King George Island (Figure 28.1) in the units discussed in the previous section. A single cylindrical specimen was cut from each drill core, and two cylindrical specimens were cut from each hand specimen. Each specimen was treated by alternating-magnetic field cleaning in a model GSD-1 Schonstedt A.C. Geophysical Specimen Demagnetizing Unit in steps of 50 oersteds from 50 to 300 oersteds. The magnetic moments, including natural remanent magnetization, were measured using a model SSM-1A Schonstedt Spinner Magnetometer in the paleomagnetic laboratory at Ohio State University. Mean site directions were computed for each level of treatment. The set of directions from the level of treatment which yielded the highest value of the estimate, *k*, for Fisher's precision parameter (McElhinny, 1973, pp. 80–83) was used to compute the mean direction of magnetism for each site.

The paleomagnetic results, along with the computed virtual geomagnetic pole positions and associated statistics, are given in Table 28.2. The results from the Horatio Stump plug are random ($k < 5$) and are not considered further. The data from the other units are regarded as reliable because the values of *k* range between 52 and 372.

The minimum dispersion directions from the Ezcurra Inlet Group lower lava and from the Noel Hill intrusion are plotted on the equal-area projection in Figure 28.2 to illustrate the excellent agreement of the results from these two units. As the magnetization of the Noel Hill intrusion is reversed, and the magnetization of the Ezcurra Inlet Group lower lava is normal, the excellent agreement indicates that the minimum dispersion directions from these two units are unaffected by secondary magnetization.

The minimum dispersion directions from the Ezcurra Inlet Group upper lava and middle lava and from the Three Brothers Hill plug are plotted in Figure 28.3. The directions from these units are all of normal polarity, and the possibility that they are affected by present field magnetization cannot be precluded. Both Figure 28.2 and 28.3 illustrate the steep inclinations of the magnetic directions obtained from the lower Tertiary units.

The minimum dispersion directions from the Upper Jurassic Volcanic Group are plotted in Figure 28.4. Both the dike and the lava flow are of normal polarity. A com-

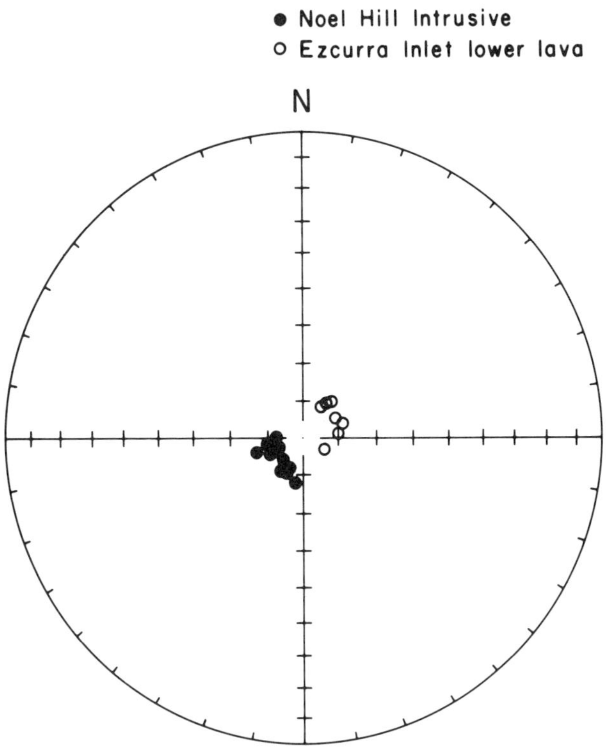

Figure 28.2. Minimum dispersion directions from the Noel Hill intrusion and the lower lava from the Ezcurra Inlet Group. The projection is equal-area. The Noel Hill directions are projected on the lower hemisphere and the Ezcurra Inlet Group directions are projected on the upper hemisphere.

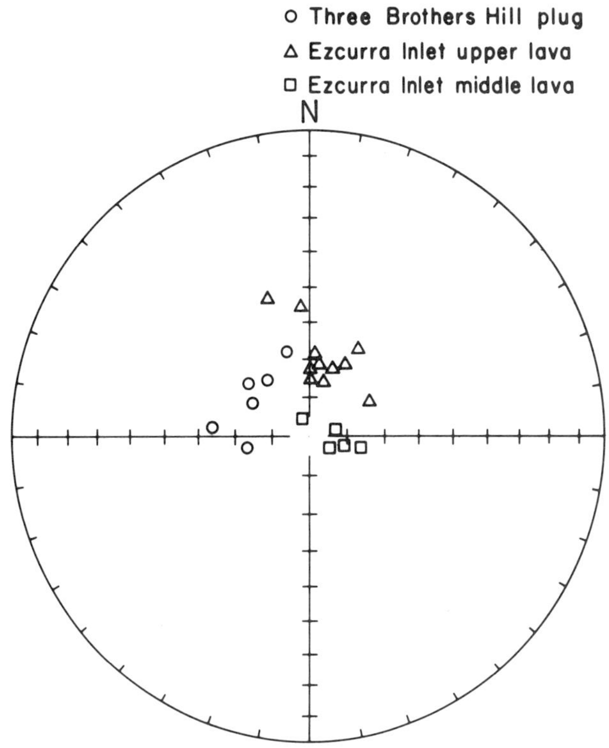

Figure 28.3. Minimum dispersion directions from the upper lava and the middle lava from the Ezcurra Inlet Group and from Three Brothers Hill volcanic plug. The projection is equal-area, upper hemisphere.

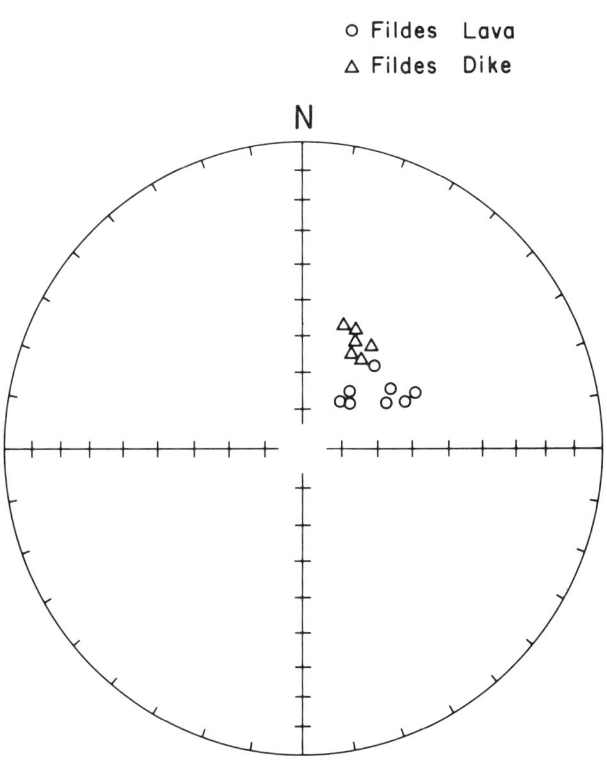

Figure 28.4. Minimum dispersion directions from the so-called Upper Jurassic Volcanic Group. The projection is equal-area, upper hemisphere.

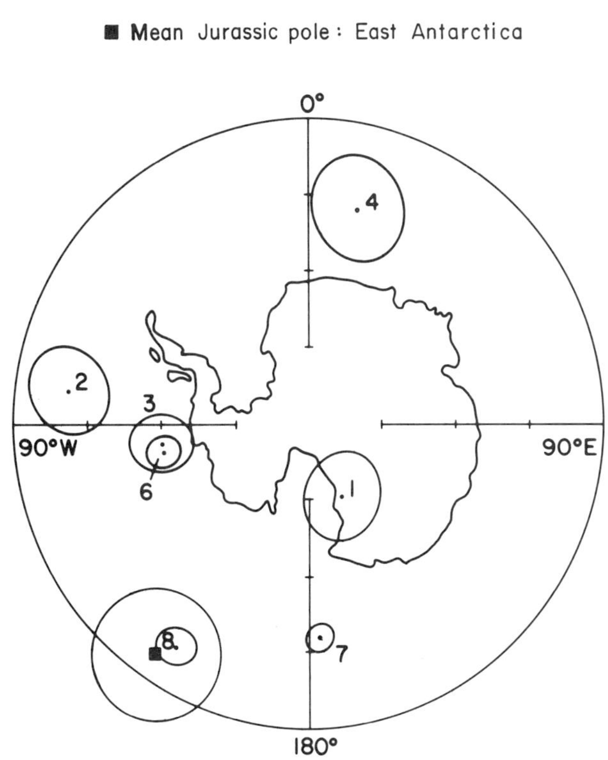

Figure 28.5. Virtual geomagnetic pole positions with associated ovals of 95% confidence for units sampled in this study, and the mean Jurassic pole from East Antarctica with associated circle of 95% confidence (Beck, 1972). The numbers indicate the geologic units from which the poles were obtained, as listed in the caption of Figure 28.1.

Table 28.3. Early Tertiary paleomagnetic pole results from the South Shetland Islands

Rock unit	Location	Age (m.y.)	Sites	Specimens	Pole Latitude	Pole Longitude	Source
Andean intrusion	Livingston Island	40	2	11	83°S	28°E	Dalziel and others (1973)
Fildes Peninsula Group	King George Island	45[a] 54[b]	20	174	83°S	63°W	Scharon and others (1970)
Fildes Peninsula Group	King George Island	45[a] 54[b]	1	22	86°S	126°W	Valencio and Fourcade (1969)
Igneous rocks	King George Island	58–49[a]	5	34	77°S	78°W	This study

[a] Age determination by Grikurov and others (1970).
[b] Age determination, this study.

parison of Figure 28.4 with Figure 28.2 and 28.3 (see also Table 28.1) shows that the supposedly Upper Jurassic Volcanic Group specimens yield shallower inclinations than the lower Tertiary rocks.

The pole positions calculated from the minimum dispersion directions of the units sampled in this survey are plotted in Figure 28.5, along with the mean Jurassic pole position for East Antarctica (Beck, 1972). From Figure 28.5 it is evident that the paleopole positions obtained from the rocks of the Upper Jurassic Volcanic Group agree rather closely with the well-documented Jurassic pole from East Antarctica. Because of the uncertainty about the ages of the rocks collected from the so-called Upper Jurassic Volcanic Group and the unknown effects of metasomatic alteration on the paleomagnetic directions, the significance of the coincidence of the pole positions with the mean Jurassic pole from East Antarctica is not clear.

The poles from the lower Tertiary units (sites 1, 2, 3, 4, and 6 in Figure 28.5) form a rather scattered group (k = 12.4, α_{95} = 22.5°) with a mean pole position of lat. 81° S, long. 73° W.

The mean early Tertiary pole for the South Shetland Islands

The dating of the Fildes Peninsula Group has important implications for paleomagnetic studies in the South Shetland Islands. The Fildes Peninsula Group has been extensively sampled, and two independently determined pole positions have been reported (Scharon and others, 1970; Valencio and Fourcade, 1969). The results from this unit, however, have previously been interpreted to represent Miocene pole positions. The potassium-argon dating of the Horatio Stump plug indicates that this group is probably early Tertiary rather than Miocene in age, and therefore the paleomagnetic results are more likely to represent late Paleocene to Eocene (54–45 m.y.) pole positions.

Dalziel and others (1973) have described limited paleomagnetic results from the tonalite intrusion on Livingston Island dated at 40 m.y. Other paleomagnetic results have been reported from units of possibly early Tertiary age (Scharnberger, 1974; Blundell, 1962), but they are not considered here because of the uncertainty about the ages of the units sampled.

Table 28.3 lists the well-dated early Tertiary paleomagnetic pole positions available from the South Shetland Islands. The mean pole position calculated from the Tertiary data in Tables 28.2 and 28.3 lies at lat. 77° S, long. 78° W. The scatter is rather large (k = 19.3, α_{95} = 12.9°), and the cone of 95% confidence overlaps the present pole position.

Unfortunately, no early Tertiary pole positions are available for East Antarctica. Pole positions for East Antarctica may be estimated, however, by rotating the well-documented Australian poles, using the parameters required to restore Australia to one of the proposed predrift configurations (Sproll and Dietz, 1969; Smith and Hallam, 1970; McKenzie and Sclater, 1971). Table 28.4 shows the result of applying these rotations to the late Paleocene pole and the Eocene (52 m.y.) pole from Australia, as listed in McElhinny (1973); these calculated poles are not significantly different in terms of this study. The Eocene poles calculated for East Antarctica from the various reconstructions all lie within 2° of the mean early Tertiary pole for the South Shetland Islands determined in this study. In Figure 28.6, the pole estimated from the Australia-Antarctica reconstruction, arbitrarily that of Smith and Hallam (1970), can be compared with the calculated South Shetland pole. Although Figure 28.6 indicates excellent agreement between the early Tertiary pole position for the South Shetland Islands and the rotated Australian Eocene pole, the significance is not yet clear because of the rather large cone of 95% confidence that surrounds the South Shetland pole position.

Table 28.4. Estimated paleomagnetic pole positions for East Antarctica using early Tertiary poles from Australia

Australian pole position: latitude	Australian pole position: longitude	Age	Plate rotation parameters[a]: latitude	Plate rotation parameters[a]: longitude	Plate rotation parameters[a]: θ	Estimated pole position: latitude	Estimated pole position: longitude
63°S	140°E	Paleocene	3.4°S	38.8°E	31°	83°S	101°W
			3.6°S	40.0°E	31°	84°S	99°W
			6.0°S	40.5°E	31.6°	83°S	94°W
70°S	126°E	Eocene (52 m.y.)	3.4°S	38.8°E	31°	79°S	47°W
			3.6°S	40.0°E	31°	79°S	44°W
			6.0°S	41.5°E	31.6°	79°S	43°W

[a] The latitude and longitude of the pole of spreading are given; θ is the angle of rotation. Sources for data are Sproll and Dietz (1969), Smith and Hallam (1970), and McKenzie and Sclater (1971).

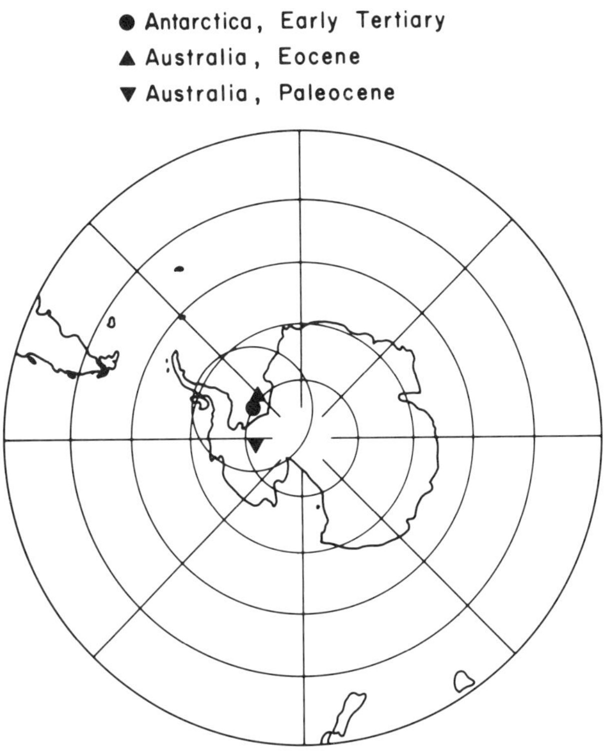

Figure 28.6. The mean early Tertiary paleomagnetic pole from the South Shetland Islands, with associated circle of 95% confidence, compared with early Tertiary pole positions for East Antarctica as calculated from Australian pole positions.

Conclusions

Examination of the available paleomagnetic data from the South Shetland Islands in the light of potassium-argon age data reported in this study gives a mean early Tertiary pole position for the South Shetland Islands at lat. 77° S, long. 78° W. This mean pole position must be regarded as tentative and as having a considerable degree of uncertainty because of the comparatively large scatter of the results. However, if it is approximately correct, the South Shetland pole agrees very well with the Eocene pole predicted for the East Antarctic craton by the rotation of the Australian results. In view of the steep inclinations of the magnetic directions in the South Shetland lower Tertiary rocks, the interpretation of this coincidence must also be tentative; without exhaustive data, only gross tectonic displacements could be resolved. The agreement of the poles, however, suggests at least two interpretations: (1) that East Antarctica and the South Shetland Islands have moved as a single plate since early Tertiary time, and (2) that any movement of the South Shetland Islands since early Tertiary time has been about a pole of spreading located approximately at the present South Pole.

Acknowledgments. The author wishes to thank D. H. Elliot for his critical comments and assistance in examining the thin sections and preparing the manuscript. J. F. Sutter performed the argon-isotope analysis, and H. C. Noltimier graciously made the paleomagnetic facilities available. Ronald Coffman drafted the figures on short notice. This work was supported by National Science Foundation Grant No. DPP 74-21509-A02.

References cited

Barton, C. M. 1964. Significance of the Tertiary fossil floras of King George Island, South Shetland Islands. (*In* Adie, R. J., *ed. Antarctic Geology.* Amsterdam, North-Holland Publishing Company, 603–8.)

Barton, C. M. 1965. The geology of the South Shetland Islands: III. The stratigraphy of King George Island. *Scient. Rep. Br. antarc. Surv.,* No. 44, 33 pp.

Beck, M. E. 1972. Paleomagnetism and magnetic polarity zones in the Jurassic Dufek Intrusion, Pensacola Mountains, Antarctica. *Geophys. J.R. astr. Soc., 28,* 49–63.

Blundell, D. J. 1962. Paleomagnetic investigations in the Falkland Islands Dependencies. *Scient. Rep. Br. antarc. Surv.,* No. 39, 24 pp.

Cox, A., and Dalrymple, G. B. 1967. Statistical analysis of geomagnetic reversal data and the precision of potassium-argon dating. *J. geophys. Res., 72,* No. 10, 2603–14.

Dalziel, I. W. D., Kligfield, R., Lowrie, W., and Opdyke, N. D. 1973. Paleomagnetic data from the southernmost Andes and the Antarctandes. (*In* Tarling, D. H., and Runcorn, S. K., *eds. Implications of Continental Drift to the Earth Sciences. 1.* New York, Academic Press, 87–101.)

Grikurov, G. E., Krylov, A. Ya., Polyakov, M. M., and Tvsobun, Ya. N. 1970. Vozrast porod v severnoi chasti Antarkticheskogo poluostrova i na Yudznykh Shetlandskikh ostrovakh (po dannym kaliy-argonovogo metoda) [Age of rocks from the northern part of the Antarctic Peninsula and the South Shetland Islands (from data of the potassium-argon method)]. *Inf. Byull. sov. antarkt. Eksped.,* No. 80, 30–34. [English translation: *Inf. Bull. Sov. Antarct. Exped., 8,* 61–63.]

Hawkes, D. D. 1961. The geology of the South Shetland Islands: I. The petrology of King George Island: Falkland Islands Dependencies Survey. *Scient. Rep. Br. antarc. Surv.,* No. 27, 42 pp.

McElhinny, M. W. 1973. *Paleomagnetism and Plate Tectonics.* Cambridge, Cambridge University Press, 358 pp.

McKenzie, D. P., and Sclater, J. G. 1971. The evolution of the Indian Ocean since the Late Cretaceous. *Geophys. J.R. astr. Soc., 25,* 437–528.

Orlando, H. A. 1964. The fossil flora of the surroundings of Ardley Peninsula [Ardley Island], 25 De Mayo Island [King George Island], South Shetland Islands. (*In* Adie, R. J., *ed. Antarctic Geology.* Amsterdam, North-Holland Publishing Company, 629–36.)

Scharnberger, C. K. 1974. Paleomagnetism of rocks from Graham Land, Antarctica. *EOS, 55,* 225.

Scharon, L., Scharnberger, C., Early, T., and Shimoyama, A. 1970. International co-operation for paleomagnetic insight into Antarctic tectonic history. *Antarct. Jnl U.S., 5,* 219–24.

Smith, A. G., and Hallam, A. 1970. The fit of the southern continents. *Nature, Lond., 225,* 139–44.

Sproll, W. P., and Dietz, R. S. 1969. Morphological continental drift fit of Australia and Antarctica. *Nature, Lond., 222,* 345–48.

Valencio, D. A., and Fourcade, N. H. 1969. Estudio paleomagnetico y petrographico de algumas formacions Cenozoicas de las islas Shetland del sur. *Contr. Inst. antárt. argent.,* No. 125, 10 pp.

29

Mesozoic–Cenozoic Volcanism in the South Shetland Islands and the Antarctic Peninsula: Geochemical Nature and Plate Tectonic Significance

STEPHEN D. WEAVER*, ANDREW D. SAUNDERS**, and JOHN TARNEY***
Department of Geological Sciences, University of Birmingham, Birmingham, B15 2TT, U.K.

Abstract. The South Shetland Islands form an ensialic island arc with a volcanic history extending back at least 140 m.y. The island block separated from the Antarctic Peninsula during the last 2 m.y. or so with the opening of Bransfield Strait, an incipient marginal basin now 65 km wide. Mesozoic volcanism in the South Shetland Islands was dominated by the eruption of low-K calc-alkaline basalt and basaltic andesite. In the Antarctic Peninsula, calc-alkaline igneous rocks yield ages as old as about 180 m.y. and are related to subduction of Pacific Ocean lithosphere along the west side of the peninsula. Mesozoic volcanic rocks vary from dominantly basaltic on the west side of the peninsula to dominantly rhyolitic on the east. The tectonic setting and the chemistry of the east coast silicic volcanic rocks are both similar to that of the extensive Tobifera silicic volcanic rocks of southernmost South America.

During Tertiary time, subduction and related magmatism probably ceased progressively from south to north along the Antarctic Peninsula as sections of a spreading ridge arrived at the continental margin and as the early Pacific and Antarctic plates became coupled. Spreading in western Drake Passage and subduction opposite the South Shetland Islands continued during Tertiary time. Lower Tertiary volcanic rocks are intermediate in character between island-arc tholeiite and calc-alkaline types. However, Pliocene–Pleistocene volcanism on the South Shetland Islands had an incompatible-element enriched and a more distinctly calc-alkaline nature and may have heralded the opening of Bransfield Strait, perhaps as spreading in western Drake Passage and subduction at the South Shetland trench ceased. The Quaternary volcanoes of Deception and Bridgeman Islands are situated on the axis of back-arc spreading, and their lavas have geochemical characteristics intermediate between calc-alkaline and ocean-floor types, consistent with ensialic marginal basin development.

The Antarctic Peninsula is a long, narrow sliver of continental crust bordered by oceanic crust of the Pacific to the west and the Weddell Sea to the east. There are many geologic similarities between the Antarctic Peninsula and the Andean Cordillera of southernmost Chile. Indeed, the abrupt termination of structural and geologic features at Cape Horn and at the northern end of the Peninsula suggests that these two areas were in some way continuous during early Mesozoic time (Barker and Griffiths, 1972; Dalziel and Elliot, 1973; Barker and Griffiths, 1977). The details of this relationship are still not well known. In particular, evidence of early plate motions is largely dependent upon interpretation of the sedimentary, igneous, and tectonic records.

The aim of this paper is to reconstruct the geochemical pattern of volcanism in the peninsula, both in space and time, in relation to the progressively changing plate tectonic regime following the breakup of Gondwanaland. Comparisons will also be made with southernmost South America. There was widespread calc-alkaline igneous activity in both areas during Mesozoic time; this continued into late Tertiary time in the northern part of the Antarctic Peninsula, with the last vestiges occurring on the South Shetland Islands.

The magmatic record of subduction along the west coast of the Antarctic Peninsula extends at least as far back as Early Jurassic time (Rex, 1976). Magmatism persisted until middle Tertiary time when subduction ceased along most of the Peninsula as the early Pacific and Antarctic plates became coupled (Hayes and Ewing, 1970; Herron and Tucholke, 1976).

The identification of magnetic anomalies in the Southeast pacific, to the southwest of the Hero Fracture Zone (Figure 29.1), indicates that ocean crust has been generated in this area since early Tertiary time (Herron and Tucholke, 1976). Since late Paleocene time there seems to have been a progressive northeastward consumption of segments of the spreading center (Aluk Ridge) and concomitant cessation of subduction along the western margin of the Antarctic Peninsula.

The formation of the Scotia Sea is considered to be an essentially Tertiary event (Barker and Griffiths, 1972: Barker and Burrell, 1977), accomplished by spreading in eastern Drake Passage from about 30 m.y. ago. Late Tertiary lineations west of the Shackleton Fracture Zone are subparallel to the line of the South Shetland Islands. Only extremely slow spreading has been taking place since about 4 m.y. ago (Barker, 1976), and the South Shetland trench (Figure 29.1) has become partially buried.

* Present address: Department of Geology, University of Canterbury, Christchurch, New Zealand.

** Present address: Department of Geology, Bedford College, London NW1 4NS, U.K.

*** Present address: Department of Geology, University of Leicester, Leicester LE1 7RH, U.K.

The islands of the South Shetland group (Figure 29.1) constitute an ensialic crustal block separated from the tip of the Antarctic Peninsula by Bransfield Strait and lying at the western extremity of the South Scotia Ridge. Bransfield Strait occupies an asymmetric grabenlike structure of extensional origin (Ashcroft, 1972; Barker, 1976) and began opening approximately 2 m.y. ago (P. Roach and D. H. Griffiths, personal communication). The formation of the basin seems to be causally related to the geologically recent subduction at the South Shetlands trench and may still be in progress. The Quaternary volcanoes of Deception and Bridgeman Islands lie astride the axial magnetic anomaly along the line of Bransfield trough (Barker, 1976). Penguin Island lies at the northwestern faulted margin of Bransfield Strait. Quaternary volcanic activity has also occurred in the vicinity of James Ross Island and Seal Nunataks, off the east coast of the peninsula (Figure 29.1).

Volcanic rocks from four main areas have been studied:

1. a Mesozoic succession of the Danco Coast on the west coast of the Antarctic Peninsula;
2. a Mesozoic succession of the Oscar II, Foyn, and Bowman Coasts on the east coast of the Antarctic Peninsula;
3. Mesozoic volcanic rocks of the Tobifera Series of the Islas de los Estados, Tierra del Fuego; and
4. Mesozoic–Cenozoic successions of the South Shetland Islands, Bransfield Strait, and the James Ross Island Volcanic Group.

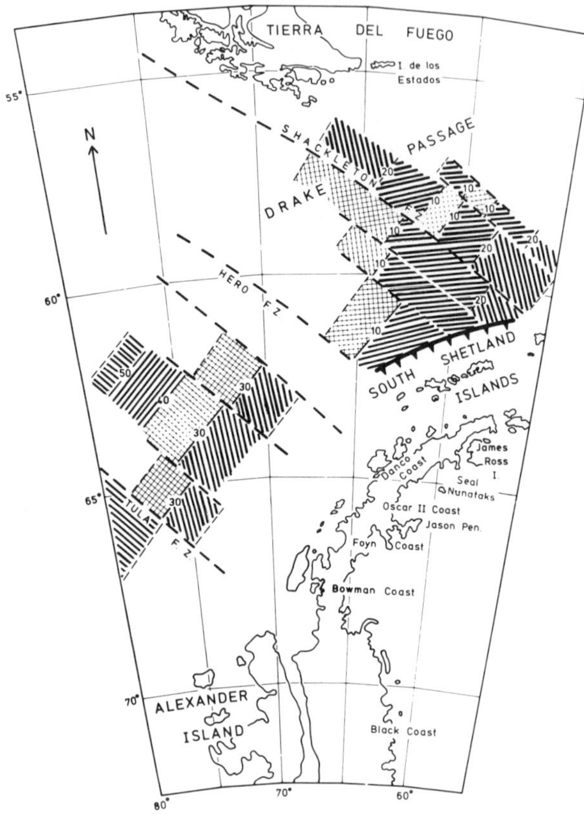

Figure 29.1. Locality map of the Antarctic Peninsula and Tierra del Fuego. Ages of sea floor are given in m.y.: southwest of the Hero Fracture Zone, after Herron and Tucholke (1976), and in Drake Passage, after Barker and Burrell (1977). Dashed lines: transform faults. Toothed line: South Shetlands trench.

Mesozoic volcanic rocks of the Antarctic Peninsula

Calc-alkaline volcanic rocks, referred by British Antarctic Survey geologists to the "Upper Jurassic Volcanic Group," have widespread occurrence on both the west and east coasts of the Antarctic Peninsula (Adie, 1972). These volcanic rocks comprise a basalt-andesite-rhyolite suite and have usually near-horizontal dispositions. The maximum recorded thickness of about 3000 m is on the east coast of the Antarctic Peninsula.

At Hope Bay, on Trinity Peninsula, a sequence of rhyolites lies conformably above and is interbedded with sedimentary rocks of Middle Jurassic age (Adie, 1964; Bibby, 1966). Andesitic pyroclastic rocks of Aptian age occur on Alexander Island (Horne, 1968). Rex (1976) reports K-Ar ages on rocks from the Upper Jurassic Volcanic Group in the range 86–186 m.y., but of course young apparent ages may represent thermal resetting by "Andean" intrusions. Ages in the range 90–140 m.y. may be confidently inferred for the volcanic succession of Byers Peninsula, Livingston Island. The term *Upper Jurassic Volcanic Group* seems therefore to be no longer justified, and we prefer the informal description *Mesozoic volcanic rocks.*

Rock specimens used in this study from the Danco Coast (Figure 29.1) have been described by West (1974); those from the Oscar II, Foyn, and Bowman Coasts (Figure 29.1) have been described by Fleet (1968), Marsh (1968), and Stubbs (1968).

Tobifera Series of Islas de los Estados

During Middle–Late Jurassic time, vast extrusions of dacite-rhyolite lavas and pyroclastic rocks occurred over much of southern South America, giving rise to a lithostratigraphic unit of widespread continuity (Ludwig and others, 1968) which extends from the Patagonian massif to Tierra del Fuego and from the western cordillera to the eastern continental shelf. These volcanic rocks are known as the "Serie Tobifera" and include silicic tuffs, ignimbrites, lavas, and associated quartz-feldspar porphyry flows and intrusive rocks.

At the southern margin of South America, the silicic volcanic rocks are abruptly terminated by the North Scotia Ridge. On Islas de los Estados, east of Tierra del Fuego (Figure 29.1), a thick sequence of silicic volcanic rocks is interbedded with shallow-marine sedimentary rocks (Dalziel and others, 1974). The structure of Islas de los Estados is representative of a considerable length of the High Cordillera of Tierra del Fuego (Dalziel and others, 1974), but the Pacific side of the cordillera is absent south of the island, presumably having been translated along the North Scotia Ridge.

Although the range of rock types within the Serie Tobifera is not well known, the volcanic rocks appear to become progressively less silicic across the strike of the Andean Cordillera (I. W. D. Dalziel, personal communication), a situation analogous to that in the Antarctic Peninsula.

South Shetland Islands

The islands are composed largely of Mesozoic–Cenozoic igneous rocks but have had an earlier geologic history similar to that of the Antarctic Peninsula (Adie, 1964). The

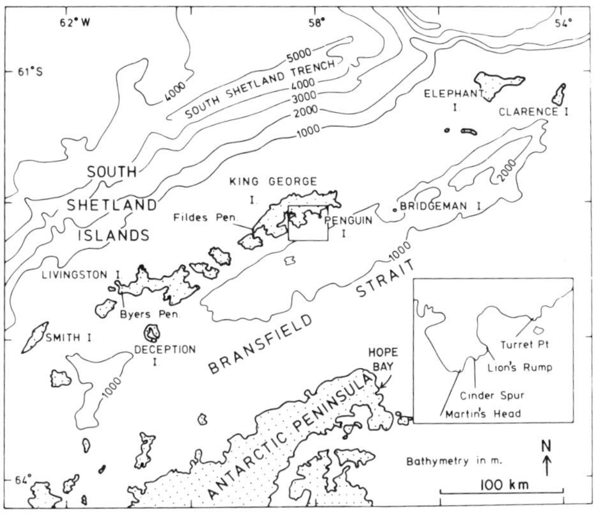

Figure 29.2. Locality and bathymetric map of the South Shetland Islands and Bransfield Strait.

detailed stratigraphy of the volcanic successions of the South Shetland Islands can be found in papers by Hawkes (1961 *a,b*), Barton (1965), Hobbs (1968), and Baker and others (1975).

Four groups of volcanic rocks on the South Shetland Islands have been studied (Figure 29.2):

1. a Mesozoic basalt-rhyolite suite of Byers Peninsula, Livingston island;
2. lower Tertiary basalts of Fildes Peninsula, King George Island;
3. Pliocene–Pleistocene basalts and andesites of the south coast of King George Island; and
4. Quaternary volcanic rocks of Deception, Bridgeman, and Penguin Islands.

In addition, five specimens have been obtained from Seal Nunataks as representatives of the James Ross Island Volcanic Group.

Byers Peninsula, Livingston Island

The geology of Byers Peninsula has been described by Valenzuela and Hervé (1972) and by Pankhurst and others (1980). Marine shale and sandstone containing a Tithonian–Valanginian fauna (Covacevich, 1976) are conformably overlain by a nonmarine sequence of dominantly arenaceous deposits. Basalts, andesites, and rhyolites are interbedded with the sedimentary rocks, which are cut by numerous plugs, sills, and dikes. K-Ar age determinations indicate that the intrusive and extrusive rocks from the nonmarine sequence have ages in the range of 90–125 m.y. (Pankhurst and others, 1980).

Zeolitized flows interbedded with the underlying fossiliferous marine sedimentary rocks are unsuitable for dating, but an age range of about 90–140 m.y. (Late Jurassic–Late Cretaceous) may be inferred for the total succession.

Fildes Peninsula, King George Island

This area is composed of basalt and andesite lavas with subordinate tuffaceous sedimentary rocks. The volcanic pile is cut by numerous dikes and several large dolerite plugs. Recently obtained K-Ar ages suggest that the whole of the peninsula is Tertiary in age; most lavas have yielded dates in the range 56–60 m.y., and the intrusive rocks, 46–50 m.y.

Pliocene—Pleistocene lavas of southern King George Island

Basaltic and andesitic lavas from four small areas on the southern margin of King George Island are considered here. Those at Lion's Rump (Figure 29.2) are probably the oldest of the group, immediately underlying the Pecten Conglomerate of Pliocene age (M. Thomson, personal communication). Lavas and pyroclastic debris are banked against older, presumably Tertiary, volcanic rocks at Cinder Spur, Martin's Head, and Turret Point (Figure 29.2). The presence of readily recognizable volcanic vents and the unconsolidated nature of the associated tephra suggest that these deposits can be no more than a few million years old. In each case, the Pliocene–Pleistocene volcanic rocks appear to represent remnants of the flanks of volcanoes which lay to the south of the present outcrop and have been downfaulted into what is now Bransfield Strait.

Deception Island

The island is an active, composite stratovolcano about 14 km in diameter which possesses a caldera 10 km in diameter. Its geology has been described by Hawkes (1961*b*), González-Ferrán and Katsui (1970), and Baker and others (1975). The deposits may be ascribed to pre- and post-caldera events. Pyroclastic rocks predominate, lavas accounting for only 10–20% of the exposed part of the volcano. Pre-caldera lavas are predominantly basaltic, and there has been a general trend toward more evolved dacitic compositions with time, although basaltic magma seems to have been available throughout the history of the volcano.

Bridgeman Island

Occupying an area of less than 1 km², Bridgeman Island is a remnant of a much larger volcanic edifice. The morphology and state of degradation suggest an age similar to that of the older, post-caldera deposits of Deception Island. González-Ferrán and Katsui (1970) have described an upper series of basaltic andesites mantling older basalts, basaltic andesites, and pyroclastic rocks.

Penguin Island

Penguin Island is situated about 1 km off the south coast of King George Island (Figure 29.2). It is dominated by the regular, uneroded slopes of a basalt scoria cone built up on a platform of olivine-basalt lavas (González-Ferrán and Katsui, 1970).

James Ross Island Volcanic Group

Alkalic olivine basalts and hawaiites occur on and around James Ross Island (Nelson, 1975; Pankhurst, Chapter 25, this volume) and at Seal Nunataks on the Larsen Ice Shelf. K-Ar dates reported by Rex (1976) indicate ages for lavas in the vicinity of James Ross Island in the range of 1–6 m.y., and those from Seal Nunataks are <1 m.y. old. Specimens from Seal Nunataks used in this study have been described by Fleet (1968).

Table 29.1. Representative chemical analyses of Mesozoic volcanic rocks from the South Shetland Islands, the west coast of the Antarctic Peninsula, the east coast of the Antarctic Peninsula, and the Tobifera Series of Islas de los Estados (oxides in wt %)

	South Shetland Islands					West coast of the Antarctic Peninsula					East coast of the Antarctic Peninsula						Tobifera Series of Islas de los Estados			
	1	2	3	4	5	6	7	8	9	10	11	12	13	14	15	16	17	18	19	20
SiO_2	48.27	50.60	55.00	61.41	78.32	49.63	53.51	57.42	65.50	71.16	48.97	50.05	54.11	66.19	72.87	75.76	66.63	71.80	75.90	79.60
TiO_2	1.07	1.59	1.59	0.56	0.13	0.68	1.51	0.85	1.01	0.38	0.81	1.54	0.67	0.56	0.18	0.12	0.48	0.24	0.19	0.09
Al_2O_3	16.54	15.70	15.60	16.39	12.33	15.46	16.64	16.91	15.86	13.80	17.14	16.59	16.54	15.49	14.47	13.18	16.65	14.64	12.96	10.96
tFe_2O_3 (total Fe)	10.99	14.30	11.55	5.97	1.12	8.81	10.96	8.04	4.53	3.17	10.39	10.37	8.49	4.93	2.81	1.48	4.69	3.22	2.26	1.60
MnO	0.20	0.25	0.26	0.25	0.02	0.15	0.17	0.13	0.18	0.11	0.16	0.16	0.13	0.10	0.09	0.03	0.03	0.02	0.01	0.01
MgO	9.58	5.34	4.65	3.55	0.33	10.87	4.13	3.35	1.68	0.53	7.41	8.38	6.37	1.64	0.17	0.25	0.92	0.42	0.19	0.19
CaO	10.49	9.57	7.00	5.71	0.49	11.91	8.22	7.17	3.87	1.19	11.26	6.73	8.00	4.40	1.27	1.56	1.45	0.31	0.21	0.18
Na_2O	2.50	2.50	3.25	4.21	3.19	2.00	3.32	3.28	5.30	4.92	2.36	3.56	2.35	3.36	2.95	1.92	3.31	3.15	3.02	2.00
K_2O	0.24	0.60	1.50	1.03	4.41	0.34	1.15	1.71	1.98	3.71	0.60	1.87	2.11	2.26	5.12	4.80	5.26	5.50	5.44	4.82
P_2O_5	0.22	0.18	0.37	0.23	0.01	0.10	0.28	0.20	0.31	0.07	0.21	0.54	0.24	0.18	0.04	0.03	0.13	0.02	0.02	0.02
Total	100.10	100.38	100.51	99.31	100.35	99.95	99.89	99.06	100.22	99.04	99.31	99.79	99.01	99.11	99.97	99.13	99.45	99.24	100.20	99.47
Trace elements (ppm)																				
Cr	256	44	13	10	6	211	10	56	<1	3	127	98	98	14	3	6	30	4	8	6
Ni	84	13	3	4	<1	60	<1	10	<1	<1	29	28	23	<1	<1	<1	16	<1	3	4
Rb	2	10	28	27	119	7	36	49	52	127	14	102	76	78	208	171	189	246	233	182
Ba	71	105	367	237	670	88	460	386	579	834	208	644	515	595	1107	818	1176	2595	942	850
Sr	350	469	329	478	60	446	523	347	426	145	610	570	636	460	200	175	128	94	65	65
Zr	69	77	149	147	108	55	165	226	309	422	77	198	172	226	284	110	324	660	334	198
Nb	3	2	6	6	8	<1	8	9	11	18	5	12	9	13	16	9	16	19	13	10
Y	18	25	32	18	32	12	28	32	48	46	10	28	23	22	36	24	32	51	29	24
La	4	8	21	13	42	4	23	16	25	31	13	35	22	23	39	18	47	72	30	29
Ce	16	20	36	38	64	14	43	42	60	77	19	57	46	48	78	44	79	130	52	59
Pb	2	6	10	2	9	3	10	11	19	18	5	8	18	11	37	78	14	11	31	14
Th	1	<1	3	4	17	<1	2	6	7	17	<1	<1	6	11	20	16	19	26	25	20
Zn	—	86	93	66	41	66	96	62	93	64	82	91	102	54	156	33	82	124	85	34
Ga	—	21	21	16	5	20	22	20	24	20	21	22	23	21	20	12	29	30	16	8
Na_2O/K_2O	10.42	4.17	2.17	4.09	0.72	5.88	2.89	1.92	2.68	1.33	3.93	1.90	1.11	1.49	0.58	0.40	0.63	0.57	0.56	0.41
K_2O/SiO_2	0.005	0.012	0.027	0.017	0.056	0.007	0.021	0.030	0.030	0.052	0.012	0.037	0.039	0.034	0.070	0.063	0.079	0.077	0.072	0.061
tFe/Mg	1.33	3.11	2.88	1.95	3.94	0.94	3.08	2.78	3.13	6.94	1.63	1.44	1.33	3.49	19.17	6.87	5.91	8.89	13.80	9.77
K/Rb	996	498	445	317	308	403	265	290	316	242	356	152	230	241	204	233	231	186	194	220
Rb/Sr	0.01	0.02	0.08	0.06	1.98	0.02	0.07	0.14	0.12	0.88	0.02	0.18	0.12	0.17	1.04	0.98	1.48	2.62	3.58	2.80
Ba/Sr	0.20	0.22	1.12	0.50	11.17	0.20	0.88	1.11	1.36	5.75	0.34	1.13	0.81	1.29	5.53	4.67	9.19	27.61	14.49	13.08
Ce_N/Y_N	1.82	1.64	2.30	4.32	4.09	2.39	3.14	2.68	2.56	3.42	3.86	4.16	4.09	4.46	4.43	3.75	5.05	5.21	3.67	5.03
Zr/Nb	23.0	38.5	24.8	24.5	13.50	>55	20.6	25.1	28.1	23.4	15.4	16.5	19.1	17.4	17.8	12.2	20.2	34.7	25.7	19.8

1. P726.5: basalt, Byers Peninsula.
2. P864.1: basalt, Byers Peninsula.
3. P726.3: basaltic andesite, Byers Peninsula.
4. P848.1: andesite, Byers Peninsula.
5. P848.16: rhyolite, Byers Peninsula.
6. O527.2: basalt, Danco Coast.
7. O990.2: basaltic andesite, Danco Coast.
8. O934.4: andesite, Danco Coast.
9. O543.4: dacite, Danco Coast.
10. O993.1: rhyodacite, Danco Coast.
11. JP6: basalt, Jason Peninsula.
12. TL554.4: basalt, Foyn Coast.
13. TL45.1: basaltic andesite, Oscar II Coast.
14. TL734.1: dacite, Oscar II Coast.
15. TL61.1: rhyolitic tuff, Oscar II Coast.
16. TL611.1: rhyolite, Oscar II Coast.
17. ST65.2: dacitic tuff, Islas de los Estados.
18. ST304: rhyodacite, Islas de los Estados.
19. ST258.1: rhyolite, Islas de los Estados.
20. ST218.3: rhyolitic tuff, Islas de los Estados.

Geochemistry

This review of the geochemical characteristics of Antarctic Peninsula volcanism is based on a sample population of about 500 specimens analyzed between 1974 and 1977, using standard X-ray fluorescence procedures (summarized in Saunders and others, Chapter 34, this volume). Representative analyses of Mesozoic and Cenozoic volcanic rocks are given in Tables 29.1 and 29.2. The following rock classifiction, based on SiO_2 content, has been used: <53%, basalt; 53–56%, basaltic andesite; 56–62%, andesite; 62–68%, dacite; 68–72%, rhyodacite; and >72%, rhyolite.

Major-element geochemistry

Mesozoic volcanic rocks

The observation of Adie (1972) that andesites are volumetrically dominant on the west coast of the peninsula and rhyolites on the east coast is confirmed and amplified. Figure 29.3 shows the relative abundance of rock types, based on silica content, from three different areas arranged in order from west to east. The Mesozoic volcanic rocks of Byers Peninsula on the South Shetland Islands, lying closest to the trench, are dominated by basalts, although more silicic volcanic rocks are also present. The Danco Coast volcanic rocks on the western side of the main peninsula have a much higher proportion of andesites and dacites, whereas those from the east coast of the peninsula are dominated by rhyolites and dacites. Also included in Figure 29.3 are samples of Tobifera volcanic rocks from Islas de los Estados; these appear to be a continuation of this trend and are made up almost entirely of dacites and rhyolites.

Analyses of Mesozoic volcanic rocks from the four areas are presented in Table 29.1. They have definite calc-alkaline characteristics. The more mafic lavas are plagioclase-phyric high-alumina tholeiites, and the suite shows only moderate iron-enrichment on an AFM diagram (West, 1974).

Significant differences between east coast and west coast volcanic suites are also apparent from the Or-Ab-An diagram (Figure 29.4). To simplify relationships, the samples have been divided into east coast and west coast groups, irrespective of age, with the South Shetland lavas included in the latter group. It is clear that compositionally equivalent rock types are distinctly more sodic on the west coast. Several high-K_2O volcanic rocks occur among the east coast samples but seem to have no recorded counterparts in

Table 29.2. Representative chemical analyses of Cenozoic volcanic rocks from the South Shetland Islands (1–13) and Seal Nunataks (James Ross Island Volcanic Group) (14) (oxides in wt %)

	1	2	3	4	5	6	7	8	9	10	11	12	13	14
SiO_2	49.44	52.54	62.07	56.18	51.44	51.96	57.60	51.64	52.71	62.57	67.73	54.61	49.05	49.56
TiO_2	0.83	0.87	0.89	0.93	0.87	0.86	0.75	1.57	1.68	1.18	0.59	0.60	1.11	2.01
Al_2O_3	15.81	17.00	16.29	16.31	19.83	18.84	15.86	15.62	16.12	15.59	14.59	18.68	15.64	14.43
tFe_2O_3 (total Fe)	10.49	11.03	6.11	7.68	7.97	8.41	6.18	9.63	9.95	7.04	5.16	6.91	9.98	11.48
MnO	0.20	0.22	0.19	0.15	0.14	0.18	0.14	0.18	0.18	0.20	0.18	0.13	0.18	0.14
MgO	7.69	4.58	2.36	5.17	4.05	5.70	7.01	6.43	4.92	1.10	0.40	5.03	9.04	9.19
CaO	10.97	8.59	5.01	6.84	9.90	9.24	7.21	9.83	9.09	3.58	1.81	9.71	10.06	8.44
Na_2O	2.69	3.84	5.49	3.66	3.74	3.71	3.67	4.16	4.90	7.02	7.28	3.55	3.88	4.00
K_2O	0.33	0.58	0.99	1.81	0.81	0.83	1.20	0.31	0.45	1.22	1.59	0.58	0.53	1.02
P_2O_5	0.14	0.15	0.34	0.31	0.20	0.18	0.20	0.21	0.29	0.42	0.10	0.07	0.30	0.21
Total	98.59	99.40	99.74	99.04	98.95	99.91	99.82	99.58	100.29	99.92	99.43	99.87	99.77	100.48
Trace elements (ppm)														
Cr	43	36	9	37	23	28	90	139	52	6	7	67	488	315
Ni	10	6	2	15	14	16	34	35	15	4	2	21	165	144
Rb	1	8	46	44	12	12	14	3	5	21	30	12	5	14
Ba	151	176	516	437	223	211	252	114	132	270	311	108	189	136
Sr	554	532	446	511	486	476	613	342	383	262	132	318	541	416
Zr	54	68	185	237	115	105	145	156	190	492	651	76	80	159
Nb	2	<1	6	3	<1	<1	3	2	7	12	14	2	1	24
Y	13	14	29	23	23	18	15	28	33	65	71	9	9	20
La	5	6	26	25	11	9	12	8	10	25	27	5	10	20
Ce	12	19	39	51	26	20	26	23	29	56	61	9	22	28
Pb	6	8	11	10	12	6	10	6	2	7	11	2	6	4
Th	3	5	3	8	2	2	4	<1	<1	2	2	3	<1	3
Zn	—	—	—	82	88	72	74	76	80	116	132	62	72	104
Ga	—	—	—	23	28	25	25	21	22	24	27	18	22	24
Na_2O/K_2O	8.15	6.62	5.55	2.02	4.62	4.47	3.06	13.42	10.89	5.75	4.58	6.12	7.32	3.92
K_2O/SiO_2	0.007	0.011	0.016	0.032	0.016	0.016	0.021	0.006	0.008	0.019	0.023	0.011	0.011	0.021
tFe/Mg	1.58	2.79	3.00	1.72	2.28	1.71	1.02	1.74	2.35	7.42	14.96	1.59	1.28	1.45
K/Rb	2740	602	179	341	560	574	712	858	747	482	440	401	880	605
Rb/Sr	0.002	0.015	0.103	0.086	0.025	0.025	0.023	0.009	0.013	0.080	0.227	0.038	0.009	0.034
Ba/Sr	0.27	0.33	1.16	0.86	0.46	0.44	0.41	0.33	0.34	1.03	2.36	0.34	0.35	0.33
Ce_N/Y_N	1.89	2.78	2.75	4.53	2.31	2.27	3.54	1.68	1.80	1.76	1.76	2.04	5.00	2.86
Zr/Nb	27.0	>68	30.8	79.0	>115	>105	48.3	78.0	27.1	41.0	46.50	38.0	80.0	6.6

1. P600.3: basalt, Fildes Peninsula.
2. P604.1: basalt, Fildes Peninsula.
3. P611.1: dacite, Fildes Peninsula.
4. P438.1: andesite, Lion's Rump.
5. P821.1: basalt, Cinder Spur.
6. P820.3: basalt, Martin's Head.
7. P815.1: andesite, Turret Point.
8. B138.1: basalt, Deception Island.
9. B213.2: basalt, Deception Island.
10. B184.3: dacite, Deception Island.
11. P870.2: dacite, Deception Island.
12. P642.2: basaltic andesite, Bridgeman Island.
13. P720.1: basalt, Penguin Island.
14. D4689.2: basalt, Seal Nunataks.

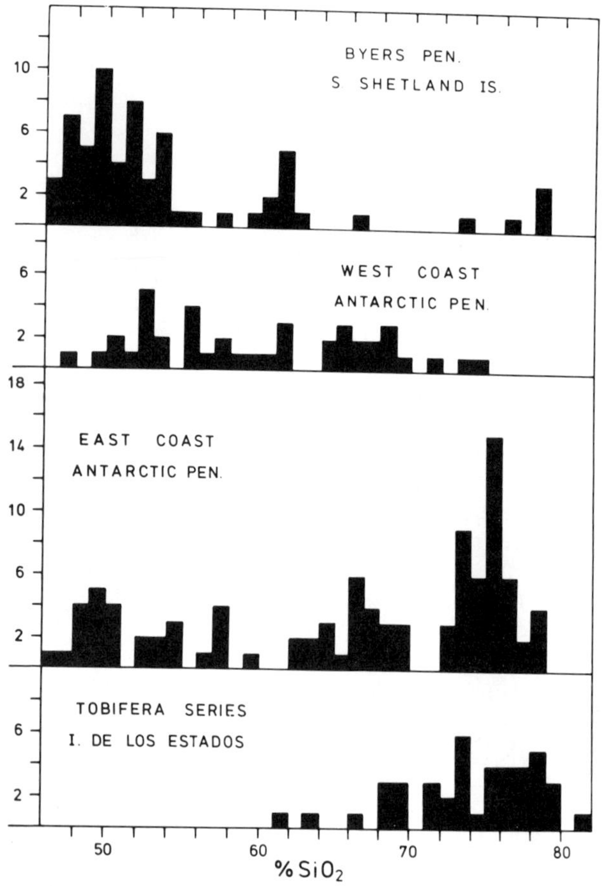

Figure 29.3. Histograms of SiO_2 content of all analyzed Mesozoic volcanic rocks from the South Shetland Islands, the Antarctic Peninsula, and Islas de los Estados.

the associated plutonic suites (Saunders and others, Chapter 34, this volume).

In many volcanic suites there is a well-marked positive correlation between K_2O and SiO_2. There is also a correlation between the K_2O contents of volcanic rocks of equivalent composition and the depth to the Benioff zone (Dickinson and Hatherton, 1967), although this varies in different island-arc and continental margin volcanic suites (Nielson and Stoiber, 1973; Dickinson, 1975). As yet there is no accepted explanation for the K_2O-depth relationship (Cawthorn, 1977), although the zone-refining/scavenging hypothesis (Best, 1975) seems to us most plausible. Rather scattered trends emerge on the K_2O-SiO_2 diagram (Figure 29.5) for the Antarctic Peninsula Mesozoic volcanic rocks. Nevertheless, mean K_2O values at 57.5% SiO_2 vary across the peninsula as follows: Byers Peninsula, 0.8%; Danco Coast, 1.5%; and east coast, 1.8%. The equivalent value for the Tobifera volcanic rocks of Islas de los Estados is 2.4% K_2O. These values are consistent with increasing depths to the Benioff zone going from west to east across the peninsula, and they confirm that subduction was from the Pacific side of the peninsula during Mesozoic time.

Cenozoic volcanic rocks

Cenozoic volcanism is restricted mainly to the northern end of the peninsula, particularly the South Shetland Islands. The Tertiary lavas of Fildes Peninsula, mainly plagioclase-phyric olivine tholeiites, have low K_2O contents and relatively high Na_2O/K_2O ratios (Table 29.2); they may be regarded as members of a low-K high alumina calc-alkaline suite (Tarney and others, 1977). The Pliocene–Pleistocene basalt-andesite lavas of King George Island, however, have higher K_2O levels and lower Na_2O/K_2O ratios (Table 29.2 and Figure 29.6) and have more normal calc-alkaline characteristics.

The Quaternary volcanic rocks in the area show a return to low-K characteristics. Deception Island olivine and quartz tholeiites and their more siliceous differentiates have low K_2O contents (Figure 29.5) and display a marked, linear trend of points near the Ab apex in the Or-Ab-An diagram (Figure 29.4). The very high Na_2O/K_2O ratios of Deception Island basalts (Table 29.2) are matched only by basalts from oceanic spreading centers, and as can be

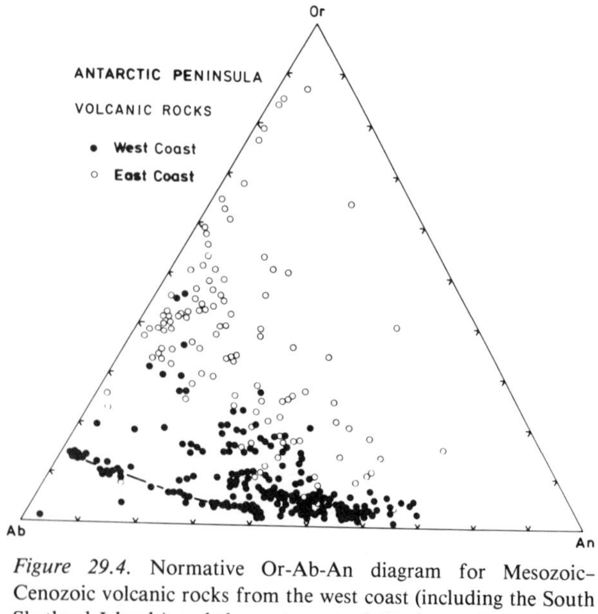

Figure 29.4. Normative Or-Ab-An diagram for Mesozoic–Cenozoic volcanic rocks from the west coast (including the South Shetland Islands) and the east coast of the Antarctic Peninsula. The dashed line indicates the trend of Deception Island lavas.

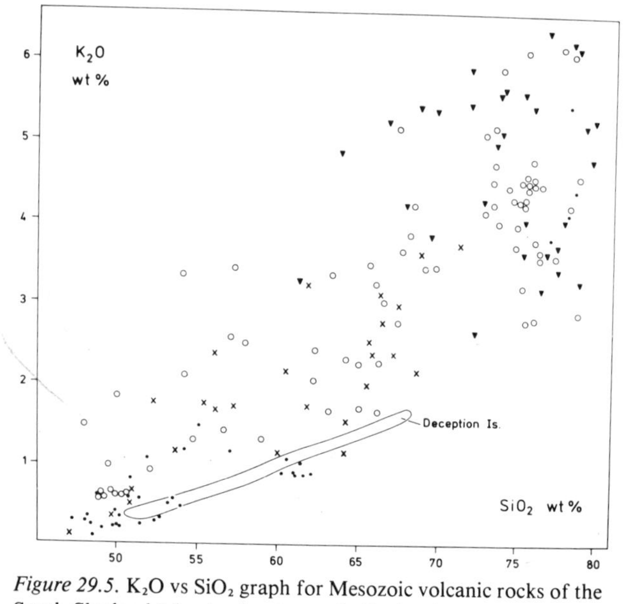

Figure 29.5. K_2O vs SiO_2 graph for Mesozoic volcanic rocks of the South Shetland Islands, the Antarctic Peninsula, and Islas de los Estados. Dots: Byers Peninsula. Crosses: Danco Coast. Open circles: Oscar II and Foyn Coasts. Solid triangles: Islas de los Estados. The field for Deception Island is indicated.

seen from Figure 29.6, a smooth curve could be drawn from ocean-floor basalts through the whole sequence of Deception Island differentiates. Although the Deception suite is distinctive in showing marked soda enrichment (Na_2O reaching 7.5%), peralkalinity is not achieved, and in fact, the Na_2O/K_2O ratio decreases with increasing SiO_2 content, a feature consistent with plagioclase fractionation.

Bridgeman Island lavas are basaltic andesites having high Ca and Al contents, reflecting the abundance (and accumulation) of plagioclase phenocrysts. The olivine-rich magnesian basalts of Penguin Island are slightly silica undersaturated (as much as 5% *ne*) and are distinctly less potassic than the mildly alkaline basalts and hawaiites of the James Ross Island Volcanic Group (Table 29.2 and Figure 29.6) situated on the east coast of the peninsula.

High Na_2O/K_2O ratios characterize the volcanic rocks of the South Shetland Islands from Mesozoic time to the present day, a feature which seems to correlate with their inferred proximity to the trench and to the Benioff zone.

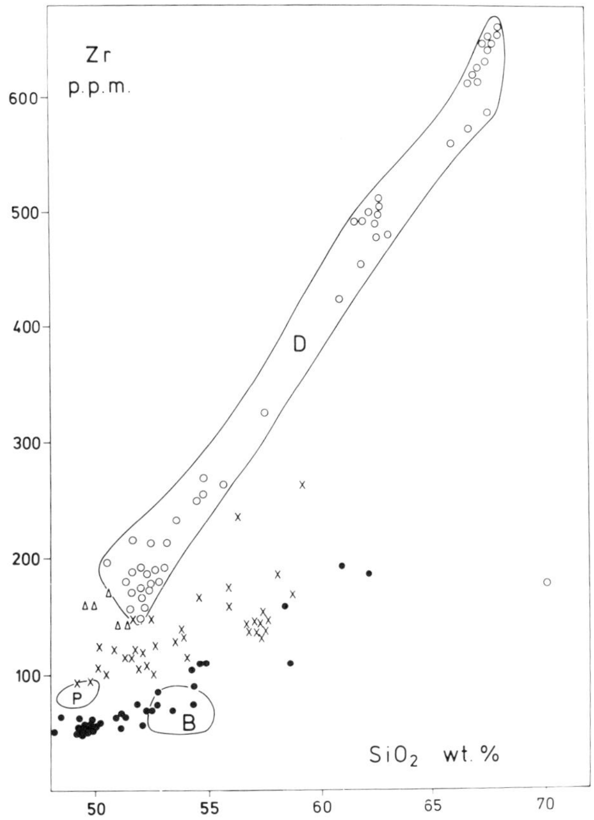

Figure 29.7. Zr vs SiO_2 graph for Cenozoic lavas of the South Shetland Islands and Seal Nunataks (symbols as for Figure 29.6). Fields for Deception (D), Penguin (P), and Bridgeman (B) Islands are indicated.

TRACE ELEMENT VARIATIONS

The progressive west to east increase in the K_2O content of the Mesozoic volcanic rocks is also mirrored by Rb, Th, and to a lesser extent, Ba. K/Rb ratios of most of the Mesozoic volcanic rocks are fairly low (Table 29.1) except for the mafic lavas of the South Shetland Islands, which together with most of the Tertiary lavas have high K/Rb ratios and low Rb/Sr and Ba/Sr ratios. These characteristics are typical of island-arc tholeiites (Jakeš and Gill, 1970). However, the abundances of Sr, Ba, Zr, Cr, and Ni are much higher than in arc tholeiites. Moreover, Ce/Y ratios in these rocks (Tables 29.1 and 29.2) suggest that their rare-earth patterns are moderately light-REE–enriched, in contrast to the flat or light-REE–depleted patterns reported for most island-arc tholeiites (Jakeš and Gill, 1970).

In many volcanic suites the abundance of an incompatible element (i.e., a trace element having a crystal/liquid bulk distribution coefficient approaching zero) such as zirconium serves as a useful index of fractionation (Weaver and others, 1972; Tarney and others, 1977). Smooth trends are obtained when chemical parameters are plotted against Zr (as in the Deception Island volcanic rocks; Figure 29.7), suggesting control by a simple petrogenetic process such as fractional crystallization. An equivalent Zr-SiO_2 graph is presented in Figure 29.8 for the Mesozoic lavas. It is clear that, despite the scatter of data points, the nature of the variation is quite different from that in the Deception Island suite. Zr increases in abundance from the basic rocks up to about 65% SiO_2, beyond which the Zr/SiO_2 ratio falls sharply with increasing silica such that the rhyolites are significantly poorer in Zr than the dacites. One reason for this could be separation of zircon from the magma, but an equally adequate explanation would be that zircon was residual in the source during partial melting or that the siliceous melts are filter-press products from a zircon-bearing crystalline residuum. Of interest

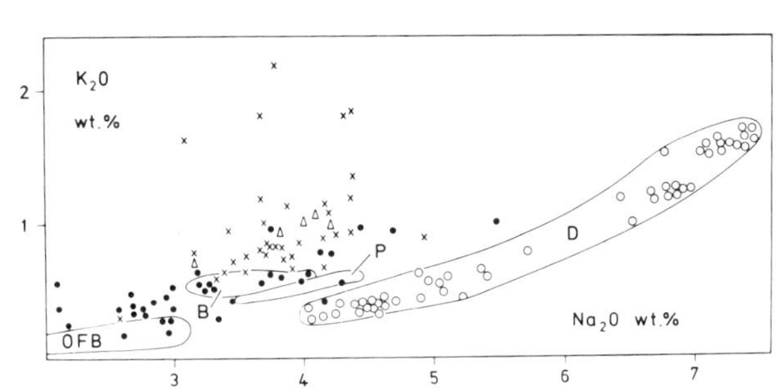

Figure 29.6. K_2O vs Na_2O graph for Cenozoic lavas of the South Shetland Islands and Seal Nunataks. Solid circles: Fildes Peninsula. Crosses: Pliocene-Pleistocene group. Open circles: Deception Island. Open triangles: Seal Nunataks. Fields for Deception (D), Penguin (P), and Bridgeman (B) Islands and for 150 ocean-floor basalts (OFB) are indicated.

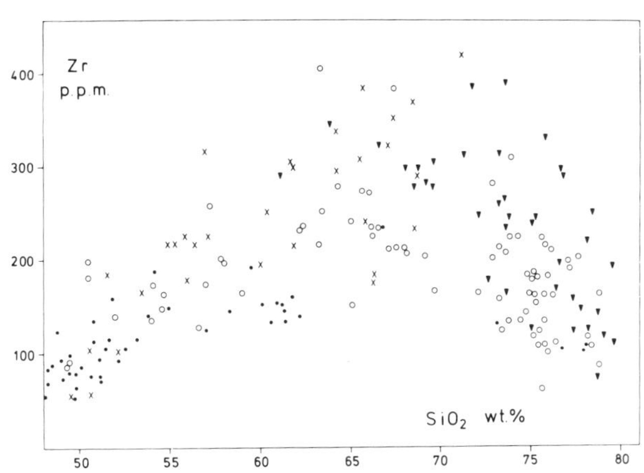

Figure 29.8. Zr vs SiO_2 graph for Mesozoic volcanic rocks from the South Shetland Islands, the Antarctic Peninsula, and Islas de los Estados (symbols as for Figure 29.5).

here is the most silicic Deception Island specimen, a rhyodacite pumice from the 1967 eruption (Baker and others, 1975), which is impoverished in Zr (Figure 29.7). With only a 2% increase in SiO_2, Zr has fallen from 650 to 180 ppm. However, other incompatible elements (Rb, Ba, Y, Th, and REE) are correspondingly low, suggesting that zircon fractionation is not responsible for these discrepancies. It seems unlikely, therefore, that this rhyodacite is consanguineous with the other Deception Island lavas, but it may instead represent partial melting of subjacent crustal material.

The ratio Zr/Nb is a useful discriminant between alkaline and tholeiitic magma types. Zr/Nb is less than 8 in alkali basalts and associated rocks (Weaver and others, 1972; S. D. Weaver, unpublished data) but reaches values in excess of 40 in "depleted" mid-ocean ridge basalts (Erlank and Kable, 1976) and some calc-alkaline suites (see, for example, Table 29.1). Basalts from Seal Nunataks have low Zr/Nb ratios (Table 29.2) and are truly alkaline on this basis. However, the other volcanic rocks, particularly those from the South Shetland Islands, have high Zr/Nb ratios.

The rare-earth elements have proven to be valuable petrogenetic indicators. So far, complete REE data are available only for the Quaternary volcanic rocks of Bransfield Strait (Weaver and others, 1979). However, a very good indication of rare-earth characteristics in all the volcanic suites can be obtained from X-ray fluorescence data on La, Ce, and Y. Because of its similarity in ionic radius and charge to Ho, the behavior of Y closely parallels that of the heavy-REE; hence, plots of Ce against Y on a chondrite-normalized basis are broadly equivalent to using Ce versus Yb. Figure 29.9 summarizes Ce-Y relationships of Antarctic Peninsula and Tobifera volcanic suites.

Chondrite-normalized Ce/Y ratios for the Mesozoic volcanic rocks of Byers Peninsula are rather low; ratios for the basalts lie between 1.5 and 3, with slightly higher ratios for the more silicic rocks. Ratios for the Danco Coast volcanic rocks are within the range of 2–4, and the east coast lavas, in the range of 3–7, whereas the Tobifera volcanic rocks show rather more fractionation with ratios rising from 3 to more than 8 (Figure 29.2*a*). Thus there seems to be a significant, though still moderate, transverse variation in Ce/Y ratios across the peninsula, which correlates with the other geochemical variations noted above. The Ce/Y variations may be depth related and could indicate the increasing influence of garnet in the source; however, much higher Ce/Y ratios would be expected if garnet-rich eclogite were the source because of the strong affinity that residual garnet would have for Y. If subducted oceanic crust were the source for the volcanic rocks, then the magmas were either generated before conversion to eclogite, or they interacted with the overlying mantle wedge to such an extent that their rare-earth patterns assumed the characteristics of that mantle.

The Tertiary basaltic lavas (Fildes Peninsula) from the South Shetland Islands also have low chondrite-normalized Ce/Y ratios (<2). It would seem that these and the Mesozoic basaltic lavas from Byers Peninsula were generated at relatively shallow depths from a source which contained little or no garnet.

The Quaternary lavas (Deception Island, Bridgeman Island, Penguin Island, and Seal Nunataks) display in-

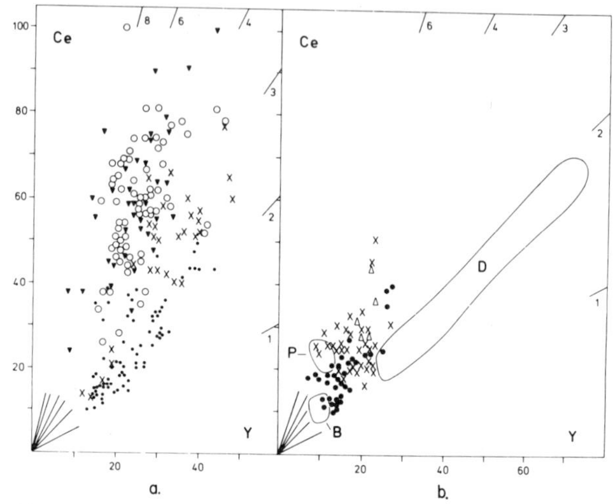

Figure 29.9. a. Ce vs Y graph for Mesozoic volcanic rocks from the South Shetland Islands, the Antarctic Peninsula, and Islas de los Estados (symbols as for Figure 29.5). Indicated ratios are chondrite-normalized values. *b.* Ce vs Y graph for Cenozoic lavas from the South Shetland Islands and Seal Nunataks (symbols as for Figure 29.6). Indicated ratios are chondrite-normalized values.

teresting Ce/Y relationships. Deception lavas produce a remarkably linear trend of rare-earth enrichment, with the Ce/Y ratio remaining constant within the range 1.5 to 2.0 (Figure 29.9*b*). This linear trend is characteristic of tholeiitic differentiation at relatively shallow depths (Tarney and others, 1977; Saunders and others, 1979) where separation of minerals such as olivine, pyroxene, and plagioclase is involved. Bridgeman Island lavas have the same Ce/Y ratios as Deception Island lavas, but all have lower levels of Ce and Y than the most primitive Deception basalts. This implies a much higher degree of partial melting of a similar source to produce the Bridgeman Island lavas, an explanation that would be consistent with the lower abundances of other incompatible elements. Penguin Island basalts, however, have higher Ce/Y ratios than either Deception and Bridgeman Island lavas, although their high Cr and Ni contents and high Mg/Fe ratios imply relatively high degrees of partial melting. It is probable that Penguin Island basalts were generated at greater depths than those of Deception or Bridgeman Island, in a mantle source where garnet was stable. The alkalic basalts of Seal Nunataks have higher chondrite-normalized Ce/Y ratios (2.5–4) and also seem to have been generated from a garnet-bearing mantle source.

Rare-earth data (Figure 29.10) tend to confirm these relationships. The patterns of Bridgeman Island lavas and Deception Island lavas are essentially parallel, with normalized Ce/Yb ratios betwen 1.7 and 2.5, and with Bridgeman lavas having lower REE levels than those of Deception lavas. The more evolved Deception Island lavas have higher REE levels and show the development of distinct negative Eu anomalies consistent with plagioclase fractionation. Penguin Island basalts have steeper REE patterns with normalized Ce/Yb ratios of about 4.0.

Strontium isotope data also indicate time-integrated geochemical differences between the Penguin and Deception-Bridgeman magma sources (Weaver and others, 1979). $^{87}Sr/^{86}Sr$ ratios for three Bridgeman Island and four

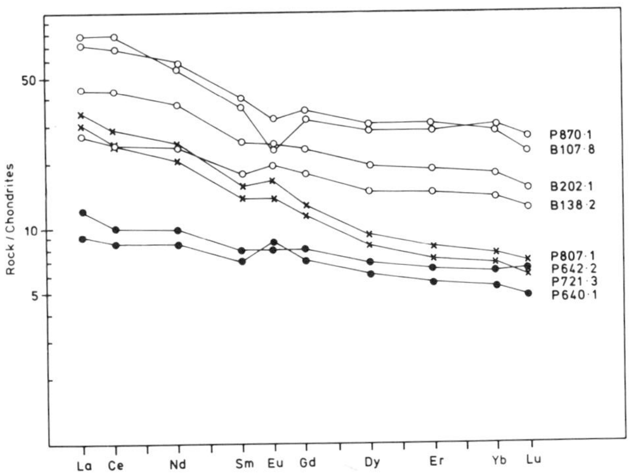

Figure 29.10. Rare-earth element patterns for the Quaternary lavas of Bransfield Strait. Open circles: Deception Island (B138.2 is basalt, B202.1 is basaltic andesite, B107.8 is andesite, and P870.1 is dacite). Crosses: Penguin Island (P807.1 and P721.3 are basalts). Solid circles: Bridgeman Island (P642.2 and P640.1 are basaltic andesites). From Weaver and others (1979).

Deception Island specimens fall within the range of 0.7034–0.7035, whereas the ratios for Penguin Island basalts are distinctly higher at about 0.7039. The ratios for Penguin basalts are only slightly lower than the initial Sr-isotope ratios of the Byers Peninsula Mesozoic volcanic rocks, which average about 0.7040. These isotopic relationships indicate a source for the Deception-Bridgeman lavas which is different from that of the Penguin Island lavas and the Mesozoic volcanic rocks. Although Bridgeman and Deception Island lavas have similar $^{87}Sr/^{86}Sr$ ratios, they do, surprisingly, have very different Rb/Sr ratios. Bridgeman lavas have higher K and especially higher Rb contents than do most primitive Deception Island basalts, but Ba and Sr levels are similar. As a result, Bridgeman Island lavas have equivalent Ba/Sr ratios, but K/Rb and Ba/Rb ratios are lower and Rb/Sr ratios are higher (Table 29.2) (Weaver and others, 1979). This could indicate a relatively recent addition of Rb and K to the Bridgeman magma source (from the subducting plate?). Alternatively, it is necessary to postulate that a K- and Rb-rich mineral such as phlogopite was residual in the source during the generation of Deception Island lavas but was consumed during the generation of Bridgeman Island lavas. Penguin Island volcanic rocks have K/Rb and Ba/Rb ratios higher than, and Rb/Sr ratios as low as, the Deception basalts, but this is mainly a result of higher K, Ba, and Sr levels rather than lower Rb. Nevertheless, this gives rise to a paradox in that there is an apparent inverse relationship between Rb/Sr and $^{87}Sr/^{86}Sr$ ratios in the three recent volcanic suites. It would seem that either phlogopite was residual in the source during generation of Penguin Island basalts, which detailed modeling has shown to be unlikely (Weaver and others, 1979), or that any original phlogopite was removed from the mantle source during earlier episodes of calc-alkaline activity in the South Shetland area.

It is evident that, compared to the relative uniformity of Mesozoic and Tertiary volcanic processes in the South Shetland area, the recent volcanism associated with the extensional opening of Bransfield Strait requires more complex petrogenetic schemes. Tarney and others (1977) have argued that the sodic Deception and Bridgeman Island magmas, with low K, Rb, Rb/Sr, $^{87}Sr/^{86}Sr$, and Ce/Yb and with high K/Rb, Ba/Rb, and Zr/Nb, have close geochemical affinities with mid-ocean ridge basalts. Thus, they may be more directly connected with mantle diapirism (back-arc spreading), which has separated the South Shetland Island block from the Antarctic Peninsula. The Mesozoic–Tertiary calc-alkaline lavas, on the other hand, are related to subduction of Pacific Ocean crust under the South Shetlands, although the lavas themselves may have their origin (in whole or in part) in the mantle wedge above the Benioff zone.

Conclusions

The volcanic record of subduction under the Antarctic Peninsula is perhaps not as complete or continuous as the plutonic record (Saunders and others, Chapter 34, this volume). Certainly there were vast outpourings of lava throughout the peninsula during Mesozoic time ("Upper Jurassic Volcanic Group") just as there were in southern South America (Serie Tobifera). Geochemical variations in the lavas across the peninsula are consistent with subduction from the Pacific side. Similarities in the pattern of Mesozoic volcanism and, in particular, the geochemical similarities between the dominantly silicic volcanic rocks from the east coast of the peninsula and the Tobifera silicic volcanic rocks agree with continental reconstructions which place the Antarctic Peninsula in continuity with southern South America during Mesozoic time. There is no reason to regard the volcanic rocks in the peninsula as being other than subduction related, although in South America the Tobifera volcanic rocks extend well back from the active margin. However, the fact that so much of the magmatic activity was expressed as volcanism may demand special tectonic conditions; it may be significant that this period was characterized by extensive rifting connected with the breakup of Gondwanaland (Barker and Griffiths, 1977).

The strong asymmetry in the present age pattern of the ocean floor in the Pacific suggests that vast amounts of Mesozoic ocean floor (complementary to that in the northwest Pacific) may have been consumed below South America, the Antarctic Peninsula, and West Antarctica during Mesozoic and early Tertiary time. From about 65 m.y. ago (anomaly 26), ocean crust generated at the Aluk Ridge (Herron and Tucholke, 1976) must also have disappeared beneath the peninsula until this ridge itself was progressively consumed and subduction ceased (except at the northern end of the peninsula). Apart from that in the South Shetland Islands, very little Tertiary volcanism has been recognized in the Antarctic Peninsula. To what extent this is a function of erosion or lack of exposure, or only the fact that magmatism does not always lead to extrusive activity, is not known.

The most complete record of volcanism in the peninsula occurs on the South Shetland Islands, and in fact, subduction under the South Shetland Islands has ceased only recently. Byers Peninsula lavas (90–140 m.y. old) are near-trench equivalents of the main Mesozoic volcanic group in the Antarctic Peninsula and may be regarded as members of a low-K calc-alkaline suite derived by relatively shallow-level melting of the mantle wedge above the subduction zone. They are chemically distinct from the arc-tholeiite suites of intraoceanic island arcs (Jakeš and Gill, 1970), but of course, at the time they were generated, the South Shetland area was situated at a major continental margin.

The Tertiary Fildes Peninsula volcanic rocks (40–60 m.y. old) are compositionally very similar to the older Byers Peninsula lavas and indicate steady-state, subduction-related magmatism in the area. The former, however, represent a pulse of magmatic activity at about 60 m.y. ago. This activity may be connected with the initiation of spreading at the Aluk Ridge at about 65 m.y. ago (Herron and Tucholke, 1976) which may well have produced an increase in the rate of subduction beneath the peninsula.

The Pliocene–Pleistocene group of lavas of King George Island is generally richer in incompatible elements (K, Rb, Ba, and others) than the Fildes and Byers suites and also more typically calc-alkaline. However, the age and position of these rocks, close to the fault-bounded southern margin of King George Island, suggest that they may be connected with the initial stages of the rifting open of Bransfield Strait. The opening of the strait heralds a major change in the 140-m.y.-old pattern of volcanism in the area. The cause of this incipient back-arc spreading is not fully known, but it may be that with the apparent cessation of spreading in Drake Passage (Barker, 1976), new spreading occurred within the volcanic arc to compensate for continuing but waning subduction under the South Shetland Islands. Quaternary volcanicity on Deception, Bridgeman, and Penguin Islands is more directly related to this mantle diapirism. Deception and Bridgeman Islands lie along the axial magnetic anomaly in Bransfield Strait. Their volcanic products have geochemical and isotopic characteristics which are transitional between ocean-ridge and calc-alkaline magmas (Weaver and others, 1979), as may be expected where back-arc spreading is at an early stage of evolution. Penguin Island represents an off-axis volcano, erupting mildly alkaline basalts which have been derived from greater depth and from a different mantle source than the Bridgeman-Deception volcanic group.

The James Ross Island–Seal Nunataks alkali basalts could represent subduction-related magmatism occurring well behind the arc, as in Japan. However, alkalic magmas are more commonly associated with ensialic rifting and faulting. Such rifting may well be a secondary extensional feature connected with active mantle diapirism beneath Bransfield Strait.

Acknowledgments. We gratefully acknowledge the logistic support provided by the British Antarctic Survey and the Royal Navy during field work in the South Shetland Islands in 1975. Financial support for this work was received from the Natural Environment Research Council and the University of Birmingham, U.K.

R. J. Adie and C. W. M. Swithinbank kindly allowed us to select Antarctic Peninsula rock specimens from the collections of the British Antarctic Survey. P. E. Baker (Leeds) and I. W. D. Dalziel (Lamont-Doherty) kindly provided specimens from Deception Island and Islas de los Estados, respectively.

Isotopic work on the South Shetland Islands is in collaboration with R. J. Pankhurst (B.A.S.), who supplied the REE and Sr-isotope data on the Bransfield Strait lavas.

References cited

Adie, R. J. 1964. Geological history. (*In* Priestley, R. E., Adie, R. J., and Robin, G. de Q., *eds. Antarctic Research.* London, Butterworths and Company, 118–62.)

Adie, R. J. 1972. Evolution of volcanism in the Antarctic Peninsula. (*In* Adie, R. J., *ed. Antarctic Geology and Geophysics.* Oslo, Universitetsforlaget, 137–41.)

Ashcroft, W. A. 1972. Crustal structure of the South Shetland Islands and Bransfield Strait. *Scient. Rep. Br. antarc. Surv.,* No. 66, 43 pp.

Baker, P. E., McReath, I., Harvey, M. R., Roobol, M. J., and Davies, T. G. 1975. The geology of the South Shetland Islands: V. Volcanic evolution of Deception Island. *Scient. Rep. Br. antarc. Surv.,* No. 78, 81 pp.

Barker, P. F. 1976. The tectonic framework of Cenozoic volcanism in the Scotia Sea region—A review. (*In* González-Ferrán, O., *ed. Proceedings of the Symposium on Andean and Antarctic Volcanology Problems.* Rome, International Association of Volcanology and Chemistry of the Earth's Interior, 330–46.)

Barker, P. F., and Burrell, J. 1977. The opening of Drake Passage. *Mar. Geol., 25,* 15–34.

Barker, P. F., and Griffiths, D. H. 1972. The evolution of the Scotia Ridge and Scotia Sea. *Phil. Trans. R. Soc. Lond.,* Ser. A, *271,* 151–83.

Barker, P. F., and Griffiths, D. H. 1977. Towards a more certain reconstruction of Gondwanaland. *Phil. Trans. R. Soc. Lond.* Ser. B, *279,* 143–59.

Barton, C. M. 1965. The geology of the South Shetland Islands: III. The stratigraphy of King George Island. *Scient. Rep. Br. antarc. Surv.,* No. 44, 33 pp.

Best, M. G. 1975. Migration of hydrous fluids in the upper mantle and potassium variation in calc-alkalic rocks. *Geology, 3,* 429–32.

Bibby, J. S. 1966. The stratigraphy of part of north-east Graham Land and the James Ross Island Group. *Scient. Rep. Br. antarc. Surv.,* No. 53, 37 pp.

Cawthorn, R. G. 1977. Petrological aspects of the correlation between potash content of orogenic magmas and earthquake depth. *Mineralog. Mag., 41,* 173–82.

Covacevich, V. 1976. Fauna Valangíniana de Peninsula Byers, Isla Livingston, Antarctica. *Revta geol. Chile,* No. 3, 25–56.

Dalziel, I. W. D., and Elliot, D. H. 1973. The Scotia arc and Antarctic margin. (*In* Nairn, A. E. M., and Stehli, F. G., *eds. The Ocean Basins and Margins. 1. The South Atlantic.* New York, Plenum Press, 171–246.)

Dalziel, I. W. D., Caminos, R., Palmer, K. F., Nullo, F., and Casanova, R. 1974. South extremity of Andes: Geology of Isla de los Estados, Argentine Tierra del Fuego. *Bull. Am. Ass. Petrol. Geol., 58,* 2502–12.

Dickinson, W. R. 1975. Potash-depth (K-*h*) relations in continental margin and intra-oceanic magmatic arcs. *Geology, 3,* 53–56.

Dickinson, W. R., and Hatherton, T. 1967. Andesitic volcanism and seismicity around the Pacific. *Science, N.Y., 157,* 801–3.

Erlank, A. J., and Kable, E. J. D. 1976. The significance of incompatible elements in Mid-Atlantic Ridge basalts from 45° N with particular reference to Zr/Nb. *Contr. Miner. Petrol., 54,* 281–91.

Fleet, M. 1968. The geology of the Oscar II Coast, Graham Land. *Scient. Rep. Br. antarc. Surv.,* No. 59, 46 pp.

González-Ferrán, O., and Katsui, Y. 1970. Estudio integral del volcanismo Cenozoico Superior de las Islas Shetland del Sur, Antarctica. *Ser. cient. Inst. antart. chil., 1,* 123–74.

Hawkes, D. D. 1961*a*. The geology of the South Shetland Islands: I. The petrology of King George Island. *Scient. Rep. Falkld Isl. Depend. Surv.,* No. 26, 28 pp.

Hawkes, D. D. 1961*b*. The geology of the South Shetland Islands: II. The geology and petrology of Deception Island. *Scient. Rep. Falkld Isl. Depend. Surv.,* No. 27, 43 pp.

Hayes, D. E., and Ewing, M. 1970. Pacific boundary structure. (*In* Maxwell, A. E., *ed. The Sea.* New York, John Wiley, 29–72.)

Herron, E. M., and Tucholke, B. E. 1976. Sea-floor magnetic patterns and basement structure in the southeastern Pacific. (*In* Hollister, C. D., Craddock, C., and others. *Initial Reports of the Deep Sea Drilling Project,* Vol. 35. Washington [U.S. Government Printing Office], 263–78.)

Hobbs, G. J. 1968. The geology of the South Shetland Islands: IV. The geology of Livingston Island. *Scient. Rep. Br. antarc. Surv.,* No. 47, 34 pp.

Horne, R. R. 1968. Petrology and provenance of the Cretaceous sediments of south-eastern Alexander Island. *Bull. Br. antarc. Surv.,* No. 17, 73–82.

Jakeš, P., and Gill, J. 1970. Rare earth elements and the island arc tholeiite series. *Earth planet. Sci. Lett., 9,* 17–28.

Ludwig, W. J., Ewing, J. I., and Ewing, M. 1968. Structure of Argentine continental margin. *Bull. Am. Ass. Petrol. Geol., 52,* 2337–68.

Marsh, A. F. 1968. *Geology of Parts of the Oscar II and Foyn Coasts, Graham Land.* Ph.D. dissertation, University of Birmingham, England, 291 pp.

Nelson, P. H. H. 1975. The James Ross Island Volcanic Group of north-east Graham Land. *Scient. Rep. Br. antarc. Surv.,* No. 54, 62 pp.

Nielson, D. R., and Stoiber, R. E. 1973. Relationship of potassium content in andesitic lavas and depth to the seismic zone. *J. geophys. Res., 78,* 6887–92.

Pankhurst, R. J., Weaver, S. D., Brook, M., and Saunders, A. D. 1980. K-Ar chronology of Byers Peninsula, Livingston Island, South Shetland Islands. *Bull. Br. antarc. Surv.,* No. 49, 277–82.

Rex, D. C. 1976. Geochronology in relation to the stratigraphy of the Antarctic Peninsula. *Bull. Br. antarc. Surv.,* No. 43, 49–58.

Saunders, A. D., Tarney, J., Stern, C. R., and Dalziel, I. W. D. 1979. Geochemistry of Mesozoic marginal basin floor igneous rocks from southern Chile. *Geol. Soc. Am. Bull., 90,* 237–58.

Stubbs, G. M. 1968. *Geology of Parts of the Foyn and Bowman Coasts, Graham Land.* Ph.D. dissertation, University of Birmingham, England, 245 pp.

Tarney, J., Saunders, A. D., and Weaver, S. D. 1977. Geochemistry of volcanic rocks from the island arcs and marginal basins of the Scotia arc region. (*In* Talwani, M., and Pitman, W. C., III, *eds. Maurice Ewing Series, 1: Island Arcs, Deep Sea Trenches, and Back-arc Basins.* Washington, American Geophysical Union, 367–77.)

Valenzuela, E., and Hervé, F. 1972. Geology of Byers Peninsula, Livingston Island, South Shetland Islands. (*In* Adie, R. J., *ed. Antarctic Geology and Geophysics.* Oslo, Universitetsforlaget, 83–89.)

Weaver, S. D., Sceal, J. S. C., and Gibson, I. L. 1972. Trace-element data relevant to the origin of trachytic and pantelleritic lavas in the East African rift system. *Contr. Miner. Petrol., 36,* 181–94.

Weaver, S. D., Saunders, A. D., Pankhurst, R. J., and Tarney, J. 1979. A geochemical study of magmatism associated with the initial stages of back-arc spreading: the Quaternary volcanics of Bransfield Strait, from South Shetland Islands. *Contr. Miner. Petrol., 68,* 151–69.

West, S. M. 1974. The geology of the Danco Coast, Graham Land. *Scient. Rep. Br. antarc. Surv.,* No. 84, 58 pp.

30

The Stratigraphy of Cape Lamb and The Naze, Vega and James Ross Islands, Antarctica

R. A. DEL VALLE, N. H. FOURCADE, and F. A. MEDINA
Instituto Antárctico Argentino, Buenos Aires, República Argentina

Abstract. The Mesozoic sedimentary rocks that crop out on James Ross Island and on Vega island belong to the filling of a broad sedimentary basin developed on the eastern border of Trinity Peninsula. The outcrops of this sedimentary basin are widespread over an area of more than 10,000 km², and this basin continues southeastward under the Weddell Sea. The west edge of the basin is on the northwest border of James Ross Island, and under the outcrops of Seymour (Marambio) Island the sediments have accumulated up to 4500 m in thickness. The exposed sequence is formed usually by fine-grained clastic sediments that were deposited during Late Cretaceous time, showing good development of the biologic distribution in shallow water of a marine environment near the Antarctic continent that was characterized by numerous deltas and estuaries.

Two stratigraphic sections were studied, both of which are made up of Cretaceous sandstone belonging to the lower part of the Marambio Group (Rinaldi, Chapter 31, this volume), with an approximate composite thickness of 265 m and covered in a discordant way by tuffs and effusive rocks of the James Ross Volcanic Group of late Tertiary age (Figure 30.1). The regional stratigraphic sequence of the James Ross insular group is summarized in Table 30.1.

Cretaceous: Marambio Group, López de Bertodano Formation

Rinaldi (Chapter 31, this volume) proposes the name of Marambio Group for a stratigraphic unit defined on Seymour Island (lat. 64° 15′ S, long. 56° 45′ W), grouping together all sedimentary units there which are assigned to the Campanian and to the Maestrichtian. By extension, the Marambio Group would also include all the sedimentary formations of the James Ross insular group assigned to the Upper Cretaceous.

Table 30.1. Regional stratigraphic sequence of the James Ross insular group

Age	Groups	Formations
Miocene–Recent (?)	James Ross Volcanic Group	
Oligocene (?) Eocene	Seymour Group	La Meseta Formation
Paleocene		Cross Valley Formation
Maestrichtian (?)	Marambio Group	Sobral Formation
Middle and late Campanian		Lopez de Bertodano Formation
Early Campanian		Hidden Lake Formation
		Stoneley Point Formation
		Upper Kotick Point Formation
		Lower Kotick Point Formation
		Lagrelius Point Formation
Pre-Jurassic, Late Jurassic, and Early Cretaceous Basement Rocks		

The López De Bertodano Formation at Cape Lamb, Vega Island

A sedimentary sequence approximately 205 m in thickness crops out in this area (Figure 30.2). It consists of green, poorly consolidated, medium- to fine-grained sandstone, locally iron bearing and glauconitic. The sandstone contains abundant calcareous concretions of considerable hardness, which commonly contain invertebrate fossils.

Interbedded with this friable sandstone are some lenticular beds of well-indurated, very fine grained, highly calcitic sandstone, less than half a meter in thickness, which stand out morphologically because of their resistance to erosion. Friable sandstone clearly predominates over consolidated beds, which represent only 1% of the sedimentary rocks of the López de Bertodano Formation.

Between 155 m and 175 m in the stratigraphic column (Figure 30.2), greenish-yellow, fine-grained sandstone is intercalated with approximately 5% of boulder and pebble conglomerate in lenses 10 m to 15 m in thickness.

The whole sedimentary sequence occupies a slightly homoclinal position. Its layers, which have been little disturbed, dip between 5° and 10° to the south.

At Cape Lamb the base of the Cretaceous sedimentary sequence does not crop out, and upwards these sedimentary beds are covered discordantly by the rocks of the James Ross Volcanic Group.

The López De Bertodano Formation at The Naze, James Ross Island

At The Naze, a homogeneous sedimentary sequence approximately 60 m in thickness (Figure 30.2) crops out, and the strata generally lie in a subhorizontal position. This section consists of medium- to fine-grained, greenish, poorly consolidated sandstone. Calcareous concretions, richly fossiliferous, occur throughout the sequence. In the mainly friable beds are intercalated a few thin beds of hard, calcareous, medium-grained sandstone, as well as some units of glauconitic mudstone that contain up to 10% iron and calcareous sandstone.

As at Cape Lamb, the base of the Cretaceous sedimentary sequence does not crop out here. However, through the distribution of the outcrops and their regionally homo-

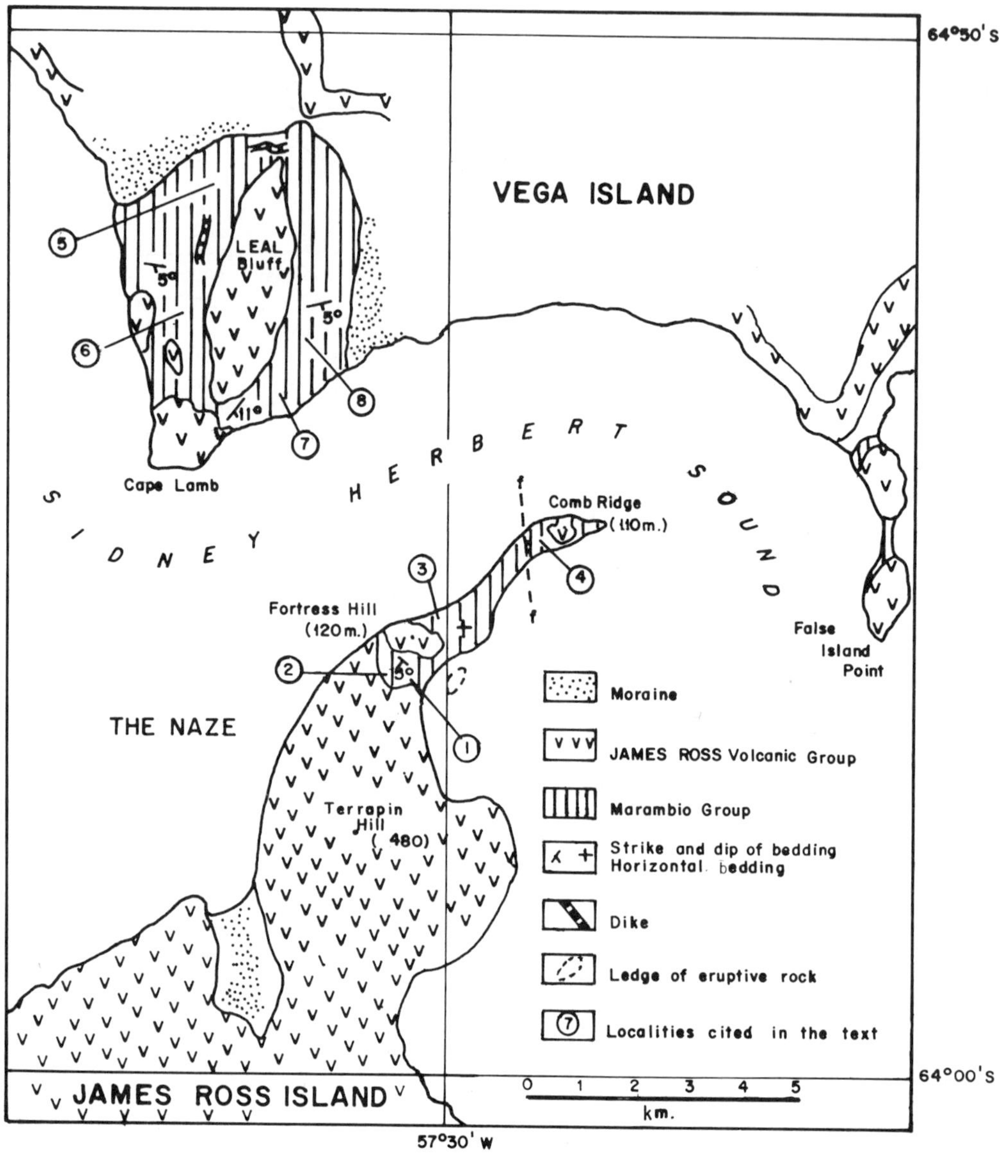

Figure 30.1. Geologic map of Cape Lamb and The Naze, Vega and James Ross Islands.

clinal attitude, it can be established that stratigraphic position is at a slightly higher level than the sequence of Vega Island.

The James Ross Volcanic Group rests discordantly on the Cretaceous sedimentary sequence at The Naze. A good contact exists at the base of Terrapin Hill, where locally it dips 17° to the southeast.

Sedimentary environment and source area

The lithologic and paleontologic similarity that exists between Mesozoic sedimentary rocks of both islands demonstrates, in agreement with other sedimentologic indications, that the sediments were deposited as parts of the same geologic event and that the environmental features remained stable during the whole sedimentary period with few local variations.

These Mesozoic sedimentary rocks are part of the filling of an extensive sedimentary basin that developed eastward from the eastern border of Trinity Peninsula. The west limit of the sedimentation is approximately on the east coast of Prince Gustav Channel that separates Trinity Peninsula from James Ross Island. The outcrops of this basin are spread out over an area of approximately 10,000 km^2. This area represents only a minimal part of the basin inasmuch as it extends east–southeastward of Seymour Island under the Weddell Sea.

The western limit of the basin is located on the western border of James Ross Island, and the basal members of the sedimentary sequence crop out there where strata have their minimal thickness. In contrast, under the Seymour Island exposures, which are the easternmost and stratigraphically highest outcrops of the whole basin, the sedimentary beds attain close to 4500 m in thickness, according to recent geophysical investigations not yet published.

This Mesozoic sequence consists of fine-grained clastic sediments, usually poorly consolidated sandstones and mudstones deposited in shallow-water marine environments within the euphotic zone and near the eastern border of the Trinity Peninsula. Apparently many deltas and estuaries existed in the same time period in which the sedimentation recorded in the outcrops occurred.

Sedimentation took place in temperate waters, generally under low to moderately high energy conditions with continuous continental contributions. On account of the great

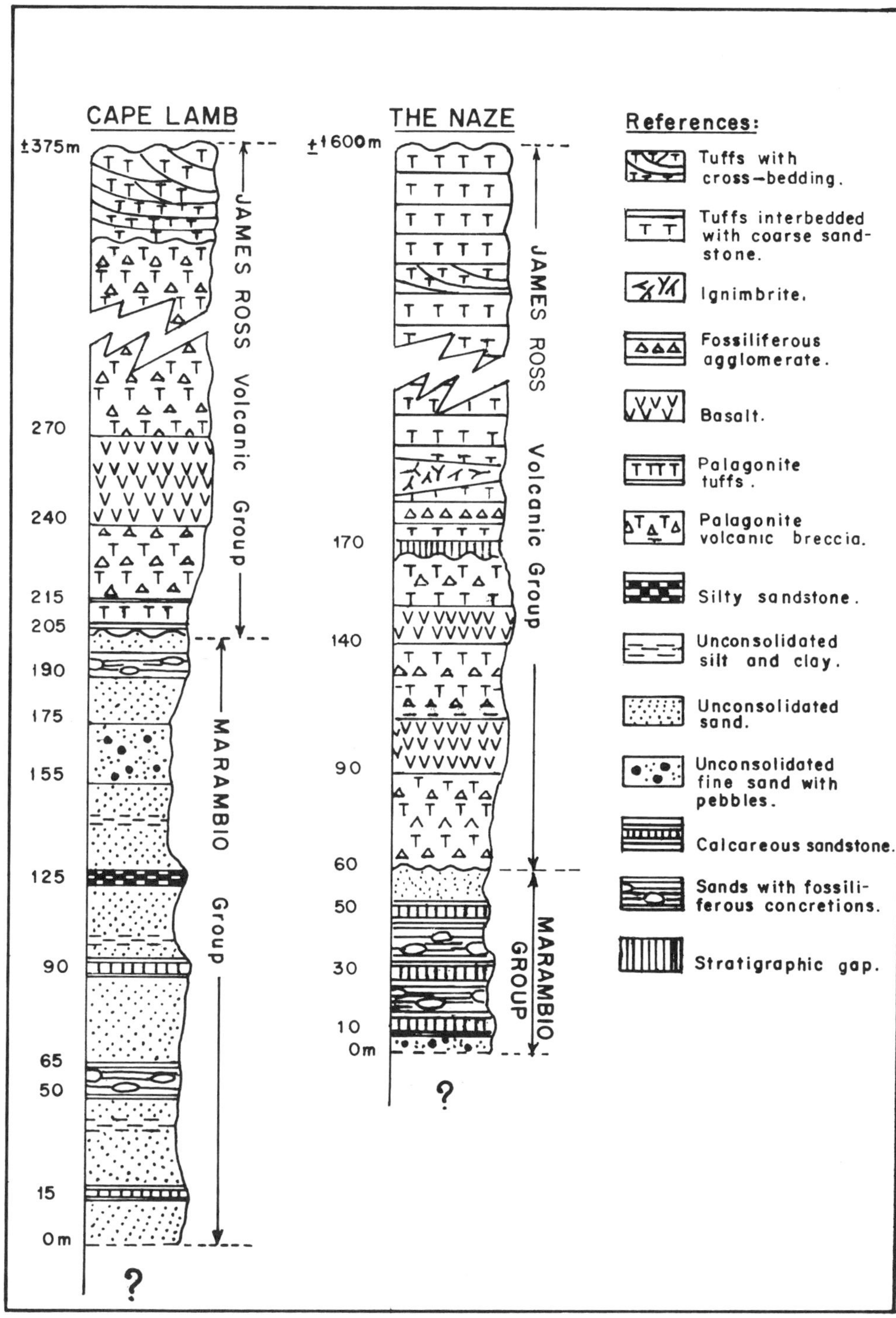

Figure 30.2. Generalized geologic columns for Cape Lamb and The Naze.

development of life on the floor of the basin where the sedimentation took place, deposition probably occurred under reducing conditions.

The composition of the sandstone grains, and especially that of the boulders and pebbles, indicates that their sources were rocks assignable to petrographic units located in the Antarctic Peninsula; these include the Trinity Peninsula Complex, the Upper Jurassic Volcanic Group, and probably certain Middle Jurassic and lower Cretaceous conglomerates. Other clasts are from plutonic sources, however, and do not correspond to the Andean Intrusive Complex.

Age of the López De Bertodano Formation at Cape Lamb and The Naze

To ascribe a paleontologic age to the sedimentary layers, *Ammonoidea,* spread homogeneously in both stratigraphic columns, were used. These cephalopods constitute rather reliable correlation elements because they have been previously studied in depth by several specialists.

The biostratigraphic and paleontologic investigations on the class *Ammonoidea,* carried out on collections from the northeastern region of the Antarctic Peninsula since the beginning of the century, make it possible to establish: a) that genus *Gunnarites* indicates a mid-Campanian age; b)

Table 30.2. Geographic distribution of the *Ammonoidea* species in rocks from The Naze and Cape Lamb (numbers indicate localities shown in Figure 30.1)

	The Naze				Cape Lamb			
Species	1	2	3	4	5	6	7	8
Gunnarites antarcticus	x	x	x	x	x			x
G. kalika	x	x	x	x	x			x
Maorites densicostatus		x				x	x	x
M. seymourianus		x				x		
Maorites sp.				x				
Eupachydiscus grossouvrei		x	x	x		x	x	
Jacobites crofti		x		x	x			
Diplomoceras lambi			x	x	x			x
Pseudophyllites peregrinus				x		x	x	
Phyllopachyceras forbesianus				x			x	
Kitchinites darwini						x	x	
Eoscaphites sp.	x							
Grossouvrites gemmatus					x			
Gaudryceras varagurense						x		

that the biochron of the species *Eupachydiscus grossouvrei* comprises from the upper lower Campanian to the uppermost mid-Campanian; c) that genera *Grossouvrites, Gaudryceras, Pseudophyllites,* and *Maorites* existed from early Campanian to late Campanian time; and d) that genera *Phyllopachyceras, Diplomoceras* and *Kitchinites* commenced in mid-Campanian time and continued through the entire late Campanian.

The *Ammonoidea* assemblages indicate a mid-Campanian age for the strata of the localities treated in this paper. Their geographic distribution is shown in Table 30.2.

Because of the widespread geographic distribution of the diverse genera of *Ammonoidea*, it is not possible to establish any chronologic difference between the two localities. The same situation is manifest in regard to the collections of other phyla. On the strength of the preceding statement, and in order to be able to arrange chronologically the insular outcrops, recourse had to be taken to their geographic position, together with the general homoclinal attitude of the layers and the lack of great structures which could cause stratigraphic repetitions at the regional level. Thus, a simple analysis leads to the conclusion that the sedimentary section which crops out in The Naze occupies a stratigraphic position slightly higher than the outcrops of Cape Lamb.

The systematic study of the paleontologic specimens collected on both islands permitted the identification of seven new species of *Pelecypoda*, one new genus of *Gastropoda*, and one specimen of *Ammonoidea* of probably Early Cretaceous age found not *in situ* but in a boulder within the Campanian strata of The Naze. The following new species of *Pelecypoda* were established: *Modiolus (Modiolus) pontocensis* del Valle–Medina; *M. (M.) fourcadei* del Valle–Medina; *Thracia (Thracia) cretaceous* del Valle–Medina; *Pholadomya gracilis* del Valle–Medina; *Homomya rinaldii* del Valle–Medina; *Aphodina (Tikia) spikermanni* del Valle–Medina; and *Eriphyla camachoi* del Valle–Medina (Table 30.3).

From the Vega Island collection, a new genus of gastropod was established named *Biala* gen. nov. with the genotype *Biala smiti* del Valle–Medina. The new species and the new genus were described by del Valle and Medina (in press).

Table 30.3. Fossiliferous content of strata from Cape Lamb and The Naze

Invertebrates	Cape Lamb	The Naze
Annelidea		
Rotularia callosa (Stoliczka)	x	x
R.shackletoni (Wilckens)	x	x
Serpula sp.	x	x
Mollusca		
Bivalvia		
Nucula suboblonga Wilck.	x	x
Nordenskjoldia Wilck.	x	x
Cucullaea grahamensis Wilck.	x	x
Linotrigonia (Oistotrigonia) antarctica (Wilck.)	x	x
Nototrigonia (Nototrigonia) regina (Wilck.)	x	x
Cytherea antarctica Wilck.	none	x
Lahillia (Lahillia) luisa Wilck	x	x
Panopea (Panopea) clausa Wilck.	x	x
Pinna (Pinna) anderssoni Wilck.	x	x
Lima snowhillensis Wilck.	x	none
L. antarctica Wilck.	none	x
Pycnodonte vesicularis (Lambert)	x	x
Phelopteria (?) sp.	x	none
Modiolus (Modiolus) fourcadei del Valle-Medina	none	x
Modiolus (M.) pontocensis del Valle-Medina	x	x
Thracia (Thracia) cretaceous del Valle-Medina	x	x
Pholadomya gracilis del Valle-Medina	x	x
Homomya rinaldii del Valle-Medina	none	x
Aphrodina (Tikia) spikermanni del Valle-Medina	x	none
Eriphyla camachoi del Valle-Medina	x	none
Gastropoda		
Amberleya spinigera Wilck.	x	none
Fusus charcotianus Wilck.	x	none
Perissoptera nordenskjoldi Wilck.	x	x
Eunaticina arctowskiana Wilck.	x	x
Biala smiti del Valle-Medina	x	none
Cephalopoda		
Gunnarites antarcticus (Weller)	x	x
G. Kalika (Stol.)	x	x
Maorites densicostatus (Kilian and Reboul)	x	x
M. seymourianus (K. and R.)	x	x
Maorites sp.	x	none
Grossouvrites gemmatus (Huppe)	x	none
Kitchinites darwini (Steinmann)	x	none
Phyllopachyceras forbesianum (d'Orbigny)	x	x
Pseudophyllites peregrinus Spath	x	x
Gaudryceras varagurense (Kossmat)	x	none
Eupachydiscus grossouvrei (Koss.)	x	x
Diplomaceras lambi Spath	x	x
Jacobites crofti Spath	none	x
Eoscaphites sp.	none	x
Nautiloidea		
Eutrephoceras simile Spath	x	x
Arthropoda		
Hoploparia stokesi Well.	x	x
Cnidaria		
Cycloceris deltoidophora Felix	x	x
Echinodermata		
Sea urchin spines	none	x
Vertebrates		
Fish		
Ptychodus invertebral disks	x	x
Reptiles		
Plesiosauria remains[a]	x	x

[a] See del Valle and others, 1977.

Among all the specimens of *Ammonoidea* collected, one was found belonging to the genus *Eoscaphites*. This is particularly interesting because that genus is ascribed to the European and North African upper Albian. This finding of an *Ammonoidea* specimen of Albian age, although not *in situ*, indicates that the Lower Cretaceous rocks must have been eroded early enough to furnish clasts to the Cretaceous basin.

Tertiary: James Ross Volcanic Group

To the south of The Naze, at Terrapin Hill (Figure 30.1), the James Ross Volcanic Group is represented by palagonized, very homogeneous, fine- to medium-grained tuffs with colors from brown to yellow, and also by a variable amount of dark olivine-hyalobasalt fragments. The monotony of the pyroclastic accumulations is broken toward the base of the tuff sequence by lenses of fine, sharp-edged vitreous material of yellowish-gray color. In addition, within the same lower sector, the tuffs are interstratified with thin beds of conglomerate. The clastic material of the conglomerate is made up of pebbles and angular blocks of friable greenish sandstone similar to those of the Marambio Group and, in a lesser proportion, by fragments of olivine hyalobasalts similar to those that crop out at Fortress Hill.

Inside the sedimentary blocks occur macrofossils (e.g., *Pinna anderssoni, Gunnarites antarcticus, Jacobites* sp., etc.) which indicate a mid-Campanian age, in agreement with that established for the sedimentary rocks immediately underlying these tuffs (the López de Bertodano Formation). The presence of this agglomerate interstratified at the base of the tuffs, indicates accumulation in a moderately high-energy environment, with relatively close source areas made up of sedimentary rocks of the Marambio Group and volcanic rocks of the James Ross Volcanic Group of earlier phases.

The thin volcanic sequence which crops out at Fortress Hill (Figure 30.1) includes effusive units lithologically more heterogeneous than those found at Terrapin Hill; they vary from fine tuffs and palagonitized breccias to the olivine hyalobasalts which are intercalated in several layers. These rocks represent effusive phases slightly older than the Terrapin Hill tuffs.

Rocks of the volcanic group are well exposed at Leal Bluff (Figure 30.1) where thick accumulations of palagonitized volcanic rocks are present; agglomerates together with massive brown to yellowish tuffs contain pillow-shaped blocks of up to 100 m^3 of olivine basalts. In some parts there are thick, blanket-like intercalations of olivinic lavas.

The volcanic sequence ends with palagonitized tuffs which have a thick diagonal stratification and which lie in angular unconformity upon the tuffs and massive breccias.

At Cape Lamb the rocks of the James Ross Volcanic Group reach a thickness close to 170 m (see the stratigraphic column in Figure 30.2), while at The Naze, the added thicknesses of rocks at Terrapin Hill and Fortress Hill yield a value close to 1540 m (Figure 30.2).

Some intrusive rocks of the volcanic group

The rocks which form a beach ledge and which appear at low tide at the east side of The Naze (Figure 30.1) deserve a separate analysis. This ledge is made up of massive intrusive rock, with a macroscopic medium grained texture and a light-gray color. This rock is a diabase which microscopically has an intersertal texture formed by a tight network of tabular plagioclase crystals (andesine-labradorite). The angular intertices among the feldspar crystals are occupied by subhedral grains of augite, magnetite, titanite, and, in a lesser proportion, by phillipsite and chabazite, which form tight concentrations.

The location of this diabase within the regional stratigraphic scheme is somewhat uncertain because of the isolated position of the outcrop, which does not permit the establishment of any stratigraphic relationships. Because its texture and its particular mineralogic composition devoid of olivine, so common in all eruptive rocks of the James Ross Volcanic Group, this diabase could be considered as a separate subvolcanic body within the Tertiary eruptive sequence.

Tectonic features of the area

The studied region did not undergo great tectonic disturbances during the sedimentation of the Marambio Group in the last stages of Late Cretaceous time. All evidence indicates that tectonic activity was confined to processes of an epirogenic type, free of major folds and faults. Although Bibby (1966) suggested the presence of some folds in the Cretaceous sedimentary sequence, the axes of which would be arranged in an approximately N 20° E orientation, the presence of such folds would not alter the general features of a continental border of the Atlantic type.

The evidence of tectonic activity preserved in the Mesozoic sedimentary rocks is limted to: a) the gently inclined homoclinal attitude of the layers, b) thick basal conglomerate (Malagnino and others, 1979) with a *Perisphintidae* fauna, and c) sandstone with intercalated boulders in the sections of Cape Lachman and Cape Lamb. On the other hand, in the rocks of the James Ross Volcanic Group evidence of tectonic movements are given by the polymictic conglomerates intercalated in the tuffs, as described in this paper, and by the unconformities which occur within the volcanic group.

The basal conglomerates of the sedimentary basin indicate that their basement was made up of pre-Jurassic metamorphic, sedimentary, and volcanic rocks of Jurassic age, and probably also of sedimentary rocks of Early Cretaceous age.

The enormous subsequent accumulation of sedimentary rocks (about 4500 m in less than 10 m.y.) was dominated by the steady and progressive rise of the Antarctic Peninsula, which in this period underwent the Andean orogeny and eruptivity, while at the same time the bottom of the basin was sinking slowly. The sandstone containing boulders found on Vega Island indicates a gentle rise of the source area (the Antarctic Peninsula of that time), which underwent moderate erosion without affecting the Andean plutonic bodies that were then being intruded.

The psephites intercalated in the James Ross Volcanic Group contain phenoclasts of Tertiary volcanic rocks, igneous rocks of Andean type, and Mesozoic sedimentary rocks of the Marambio Group. The presence of Marambio Group clasts means that during late Tertiary time there were already exposed above sea level sedimentary rocks of

Late Cretaceous age, in locations close to the area where tuffs were being deposited.

The evidence of a strong angular unconformity between the Marambio Group of Campanian–Maestrichtian age and the James Ross Volcanic Group of middle to late Micocene age, suggests crustal movements that could have been the first Andean movement to have occurred in the region 70 m.y. ago (equivalent to the Laramide orogeny). On the other hand, the disconformities which appear in the volcanic group and in the conglomerate beds found in it, must reflect more recent orogenic episodes, possibly occurring less than 10 m.y. ago when the second Andean movement took place.

References cited

Bibby, J. S. 1966. The stratigraphy of part of north-east Graham Land and the James Ross Island Group. *Scient. Rep. Br. antarc. Surv.*, No. 53, 37 pp.

del Valle, R. A., and Medina, F. A. In press. Nuevos fósiles del Cretácico Superior de Cabo Lamb (Isla Vega) y Cabo Morro (Isla James Ross). *Contr. Inst. antárt. argent.*, No. 268.

del Valle, R. A., Medina, F. A., and de Gasparini, Z. B. 1977. Nota preliminar sobre el hallazgo de Reptiles Marinos del Subórden *Plesiosauria* en las Islas James Ross y Vega, Antártida. *Contr. Inst. antárt. argent.*, No. 212, 1–36.

Malagnino, E., Olivero, E., Rinaldi, C., and Spikermann, J. 1979. Geología del Borde Occidental de la Isla James Ross, Antártida. *Act. 7 Cong. Geol. Argent., 1978,* 489–503.

31

The Upper Cretaceous in the James Ross Island Group

C. A. RINALDI*
Universidad de Buenos Aires, Facultad de Ciencias Exactas y Naturales, Ciudad Universitaria—Núñez, Buenos Aires, República Argentina

Abstract. Several problems which have arisen in the stratigraphic nomenclature of the Upper Cretaceous sedimentary sequence from the James Ross Island Group are analyzed. This discussion is based on a lithostratigraphic viewpoint, taking into account the work of Andersson (1906), Kilian and Reboul (1916), Bibby (1966), and Rinaldi and others (1978). According to new evidence from Vicecomodoro Marambio (Seymour) Island, Cerro Nevado (Snow Hill) Island, Hamilton Point, Rabot Point, Ekelof Point, Ula Point, Cape Lamb, The Naze, Lachman Grags, Brandy and Villar Fabre Bays, and Cape Obelisco localities, a rearrangement of Bibby's (1966) original outline of the Upper Cretaceous sequence is proposed. The Sobral Formation (Maestrichtian) and the López de Bertodano Formation (middle-upper Campanian), the last one including part of the Older Seymour Island beds and Snow Hill beds (Andersson, 1906; Kilian and Reboul, 1916), are proposed for the first time. All of these formations, plus a part of Bibby's sequence (Hidden Lake Formation, Stoneley Point Formation, Upper Kotick Formation, and Lower Kotick Formation), are included here in the Marambio Group (middle-upper Campanian and lower Maestrichtian). In connection with research carried out by the Geoantar Group* of the Instituto Antártico Argentino in the area of James Ross Island, and due to the problems which arose when attempting to correlate the outcrops along the coasts of the different islands and those brought about by the differences of nomenclature in the works of Andersson (1906), Kilian and Reboul (1916), and Bibby (1966), it became necessary to reexamine the stratigraphy of the Upper Cretaceous of this area.

Bibby (1966) carried out careful work based on lithostratigraphic concepts. On the western part of James Ross Island, he recognized the different outcrops of the Snow Hill Island beds which are well exposed on the eastern part of the island; he included within this unit the upper strata cropping out on Vicecomodoro Marambio (Seymour) Island. These strata, however, show a remarkable lithologic divergence from the underlying layers which are in continuity on Cerro Nevado (Snow Hill) Island, Hamilton Point, Rabot Point, Ekelof Point, Ula Point, Hump Island, Cape Lamb, Dogger Peak, The Naze, Lachman Crags, Brandy and Villar Fabre Bays, and Cape Obelisco (Figure 31.1).

Taking into account the information supplied by Olivero (1975), Medina (1975), Malagnino and others (1979), and Rinaldi and others (1978), we were compelled to subdivide the Snow Hill Island Series into two formations:

1. Sobral Formation (Maestrichtian?), and
2. López de Bertodano Formation (middle-upper Campanian)

These two formations, together with those established by Bibby (1966), form a higher rank unit, which we have named the Marambio Group (Figure 31.1).

Former investigations

Studies on the stratigraphy of the sedimentary rocks of the James Ross Island Group began with Andersson's work (1906). On the basis of faunal elements, he established three chronostratigraphic units named (1) Snow Hill beds (Cretaceous) for the lower part, (2) Older Seymour beds (Cretaceous) for the middle part, and above them (3) a third unit named Younger Seymour Island beds (Tertiary). Later on, based on the study of paleontologic material collected by the Swedish expedition, Kilian and Reboul (1916) redefined the units proposed by Andersson (1906), ascribing to them ages varying from upper Albian to Maestrichtian.

Bibby (1966) carried out a careful study of the Cretaceous sedimentary sequence of James Ross Island, in general employing lithstratigraphic criteria over chronostratigraphic ones. He grouped the outcrops located in the eastern part of the island with the Snow Hill Island Series, which includes the strata named the Snow Hill beds by Andersson, as found in Cerro Nevado (Snow Hill) Island. The same strata are found in Brandy Bay, Cape Obelisco, Lachman Crags, The Naze, Ula Point, Rabot and Hamilton Points of James Ross Island, Marambio (Seymour) Island, Cockburn Island, Humps Island, Falsa Island, and Cape Lamb of Vega Island. Howarth (1966) indicated an upper Campanian age for the layers of the Snow Hill Island Series, on the basis of ammonite collections from the Antarctic campaigns of 1958 through 1960.

Olivero (1975) and Medina (1975) studied Cerro Nevado (Snow Hill) Island in great detail and produced a profile made on the northwest part of it. They pointed out the similarity between the sedimentary rocks of this island and those of the southern part of Vicecomodoro Marambio (Seymour) Island.

Rinaldi and others (1978) described the geology of Vice-

* Present address: CIRGEO, J. Ramírez de Velasco 847, 1414 Buenos Aires, República Argentina.

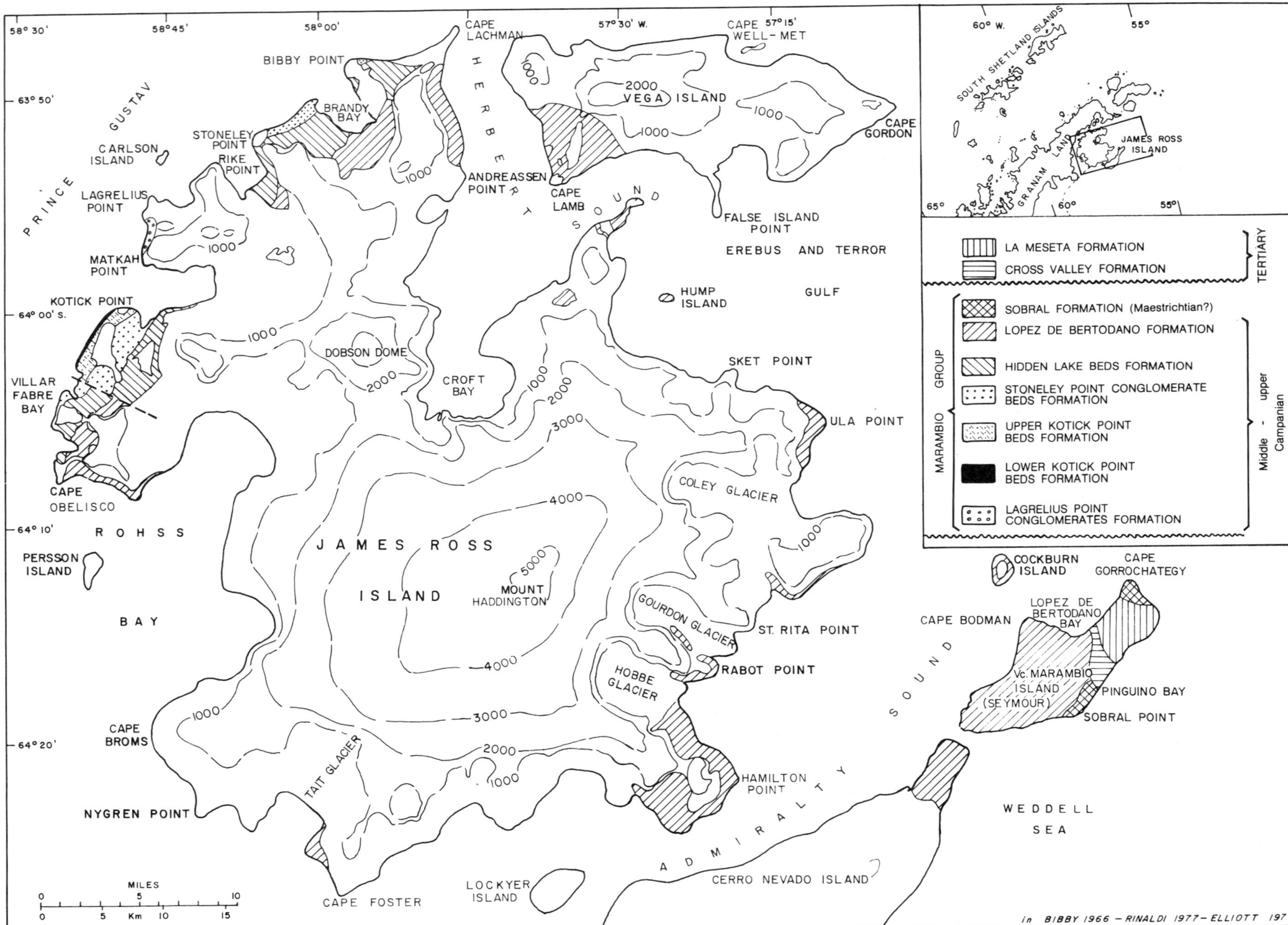

Figure 31.1. A map of James Ross Island and the surrounding area, showing exposures of the Cretaceous Marambio Group and some Tertiary units.

comodoro Marambio (Seymour) Island, establishing a type profile for the sedimentary sequence of the Upper Cretaceous in which the Older Seymour Island beds of Andersson (1906) and Kilian and Reboul (1916) are subdivided into two formations which also include part of the sedimentary beds included in the Snow Hill Island Series of Bibby (1966).

Malagnino and others (1979) studied the outcrops of James Ross Island, establishing the similarity between the Snow Hill Island Series there and the López de Bertodano Formation, which crops out on Cerro Nevado (Snow Hill) Island and on Vicecomodoro Marambio (Seymour) Island. Del Valle and others (Chapter 30, this volume) state likewise the similarity between these sedimentary beds and those of Vega Island and The Naze.

Basis for stratigraphic nomenclature

Andersson (1906) makes a preliminary attempt to divide and name the sedimentary sequence which crops out on Cerro Nevado (Snow Hill) and Vicecomodoro Marambio (Seymour) Islands, mainly on the grounds of chronostratigraphic concepts, even though he points out the lithologic variations between the outcrops of the two islands. He states, "Also the Cretaceous beds on Seymour Island consist of fine-grained soft sandstone and sand." Nodules of the characteristic liver-colored type, "so common on Snow Hill, are never (or only very rarely) met with on Seymour Island."

"The chief constituent of the Cretaceous fauna (considered as a whole) is ammonites, abounding both in specimens and species. Next to them in importance come the bivalves and gastropoda; fish-remains and corals are also tolerably common while echinoids and decapoda are but rarely found. Some scarce pieces of wood were found in the Snow Hill Beds."

Adding to the above, he continues: "Already during the field work we noticed a near relationship between the Cretaceous strata of the two islands, but at the same time such marked differences that a division into the two local groups, Snow Hill beds and Older Seymour Island beds, may prove to be necessary." He reported a list of fossils from the two localities, which was supplied for Kilian, including two respective faunal associations (Andersson, 1906, p. 35–36).

From the summary of Andersson's observations, a preliminary subdivision of the Cretaceous sedimentary sequence can be drawn up, based on chronostratigraphic concepts and mainly supported by paleontologic diversity.

When Kilian and Reboul (1916) published the stratigraphic results of the Swedish Polar Expedition (1902–03), and when they identified ammonites of the Cretaceous formations of the Vicecomodoro Marambio (Seymour) and Cerro Nevado (Snow Hill) Islands, they established the faunal homogeneity in the area, pointing out the great development of the *Kossmaticeras* group, and distinguishing three horizons (Kilian and Reboul, 1916):

1. Black marly calcareous beds on Snow Hill (Cerro Nevado) Island with *Kossmaticeras (Grahamites) skidegatense* Whiteaves sp., *K. (Seymourites) loganianum* Whiteaves, *Desmoseras loryi* Kilian and Reboul, and bryozoa. Red calcareous beds with *Latidorsella (Desmoseras) loryi* and bryozoa, which is considered to be the oldest unit of the sequence and is defined as middle Cretaceous;
2. Layers of kidney-shaped sandstone nodules on Cerro Nevado (Snow Hill) Island with abundant *Kossmaticeras antarcticum* St. Weller; and
3. Gray marly layers on Vicecomodoro Marambio (Seymour) and Cerro Nevado (Snow Hill) Islands with *Pseudophyllites Indra* Forbes sp. and a predominance of *Kossmaticeras Bhavani* Stol. sp. and *K. gemmatus* Huppé sp.

Because of new paleontologic discoveries, the present authors have altered Andersson's scheme, and as it follows from the subdivisions established by them, the change has been made on chronostratigraphic grounds. Formational variations on lithostratigraphic concepts are not defined.

Bibby (1966) established, for the first time in the area, subdivisions on the grounds of lithostratigraphic concepts for the outcrops located to the west of James Ross Island, but beds cropping out in the eastern islands (Cape Lachman, Cape Lamb, Dagger Peak and The Naze, Hump Island, Snow Hill Island, and Seymour Island) were grouped with the Snow Hill Island Series, without specifying if the Older Seymour Island beds of Andersson (1906) and Kilian and Reboul (1916) were included in this group. He also recognized the difficulties which arise when trying to subdivide lithologically this sedimentary unit, establishing an arbitrary sequence in accordance with the abundance of the existing fauna (Bibby, 1966, p. 22).

From the analysis of published literature, it was considered essential to redefine and rearrange the stratigraphic nomenclature of the Upper Cretaceous sandstones of this area. This is why Rinaldi and others (1978) studied the geology of Vicecomodoro Marambio (Seymour) Island, and Olivero (1975) and Medina (1975) studied that of Cerro Nevado (Snow Hill) Island. Later Malagnino and others (1979) reviewed the outcrops located to the west of James Ross Island, while del Valle and others (Chapter 30, this volume) studied the easternmost part of this same island. Analysis of the conclusions resulting from the work carried out by the present author and his collaborators made it possible to confirm the similarity among the sedimentary beds cropping out in the following locations: Cape Obelisco, the central part of the Brandy Bay, Lachman Crags, Herbert Coast, Cape Lamb, The Naze, Ula Point, Cape Cage, Hamilton Point, Cerro Nevado (Snow Hill) Island, and Vicecomodoro Marambio (Seymour) Island. The sedimentary beds at all these localities are included in the López de Bertodano Formation.

Proposed nomenclature

Upper Cretaceous

Marambio Group (after Rinaldi and others, 1978)

The Marambio Group is defined as a unit of upper rank, which includes, from oldest to youngest, the following formations: Lagrelius Point, Lower Kotick Point, Upper Kotick Point, Stoneley Point, Hidden Lake, López de Bertodano, and Sobral.

The first five formational units are made up of concordant marine sedimentary rocks deposited in a neritic environment. They are generally rich in fossils and consist of conglomerates and calcareous sandstones with pelitic intercalations.

The base of the group was not observed, but it is assumed that it corresponds to the sedimentary unit made up of thick beds of conglomerates in the Lagrelius Point Formation, which is located on the coast of the Prince Gustavo Channel. Nearby in Brandy Bay, Malagnino and others (1979) found a fauna of *Perisphinctidae* and pelecypods, which has shed light on the fossil fragments studied by Spath (1953), which were collected not *in situ* in the area of Hidden Lake, to the north of James Ross Island. In the type area (Vicecomodoro Marambio Island or Seymour Island), the same group is overlain in angular discordance by an Eocene–Oligocene sedimentary sequence.

The outcrops of the Marambio Group are located on the coast of James Ross Island at Cape Lachman, Bibby Point, Stoneley Point, Lagrelius Point, Refugio San Carlos, Villar Fabre Bay, Cape Obelisco, Point Hamilton, and Ula Point and at Vega Island (Cape Lamb), Cockburn Island, Marambio (Seymour) Island, and Cerro Nevado (Snow Hill) Island. The formations in the Marambio Group at these places include:

1. *Lagrelius Point Formation* (Bibby 1966)
   Cropping out on Lagrelius Point and consisting of thick conglomerate and interbedded sandstone.
2. *Lower Kotick Point Formation* (Bibby 1966)
   Cropping out on Kotick Point and composed of sandstone containing shale and interbedded limestone.
3. *Upper Kotick Point Formation* (Bibby 1966)
   Found in the same area as the previous formation and composed of conglomerate and thick sandstone in cyclical repetition with the intercalation of shale containing plant remains. It contains ammonite fragments which have been classified by Howarth (1966) as puzosiinids, unidentifiable at the generic level but of Cenomanian–Campanian age.
4. *Stoneley Point Formation* (Bibby 1966)
   Found in the western part of James Ross Island in Brandy Bay, beginning to the north of Tumbledown Ravine. It is a unit of conglomerate, sandstone, and tuff, and its base is not exposed.
5. *Hidden Lake Formation* (Bibby 1966)
   Cropping out in the area between Brandy Bay and Stoneley Point and in the region of Cape Obelisco, it is transitional upward into the López de Bertodano Formation. It is composed of pelite and conglomerate, which contain *Desmophyllites* sp. ind., *Tetragonites (Saghalinites)* cf. *cala* (Forbes), *Gaudryceras varagurense (Kossmaticeras),* and *Submortoniceras chicoense* (Trask). The latter form permits the establishment of its age as lower Campanian (Howarth, 1966).
6. *López de Bertodano Formation* (Rinaldi and others, 1978)
   A thick sedimentary unit made up of consolidated grayish brown sandstone beds which alternate with a few consolidated silty sandstone and thick sandstone beds, some containing calcareous cement. Downwards it becomes silty and gradually passes into fine-grained sandstone and siltstone alternating with calcareous beds. The lower part is represented by medium-grained sandstone, slightly consolidated, which alternate with beds of fine-grained sandstone.

   Regarding the age, Nordenskjöld (1913, p. 8) pointed out that according to the flora and fauna found there, this formation is of Mesozoic age. Kilian and Reboul (1916), on classifying the ammonites, place these rocks in the upper Senonian to upper Cenomanian and point out the presence of older elements such as *Kossmaticeras loganianum* Whiteaves and *K. skidegatense* Whiteaves, which are associated with others of Senonian form. The same authors indicate that the presence of *Kossmaticeras gemmatum* Huppé (= *Grossouvrites gemmatus* [Huppé]), *Lytoceras Kayei* Forbes, *Phylloceras ramosum* (Meek), as well as the existence of a great *Pachydiscus* (*Pachydiscus* aff. *gollevillensis* [d'Orbigny]), very close to *P. Quiriquinae* (Steinmann) and to *P. Colligatus,* suggests a very special relationship with the Quiriquina strata of Chile (Steinmann, 1895). On Vicecomodoro Marambio (Seymour) Island, the forms of the Group Aryaloor (and of the Quiriquina level, Chile) seem to predominate, such as *Kossmaticeras gemmatum* (= *Grossouvrites gemmatus*), *Pachydiscus* aff. *gollevillensis, Gaudryceras, Kossmaticeras bhavani* (*densicostata* and *seymourina* varieties) (= *Maorites densicostatus* and *M. seymourianus*), *Kossmaticeras aemilianum,* and *K. Madrasinum.* Kilian and Reboul (1916) list for the same region, especially Cape Bodman, the existence of *Kossmaticeras antarcticum,* which has never been corroborated. They also state that the presence of *Pseudophyllites indra* Forbes (= *P. peregrinus* Spath) and *Pachydiscus* indicate an upper Cenomanian age for these beds.

Spath (1953, p. 23) points out that *Amm. longanianus* Kilian and Reboul (1916, Plate 18, Figure 2) could be an immature *Gunnarites* or a *Neograhamites,* both of Campanian age. So there are no reports in the present literature of any definitive pre-Campanian ammonoids in the Marambio Group. In addition, Spath states that there is no evidence to support Haug's (1909) hypothesis of a transgression in the Early Cretaceous (Barremian). He also points out that there is little relationship between the ammonites of Antarctica and those of India, finding only two species in common with the Valudayoor Upper Cretaceous. Finally Spath (1953) establishes that the Cretaceous cephalopods of Graham Land are a typical Antarctic fauna in some way related to the Cretaceous provinces of South America, South Africa, and Australia, and that they are probably mixed with the so-called Atlantic and Indo-Pacific elements.

Howarth (1958) agrees with Spath (1953) on the fact that the cephalopods are all of the same age, but adds that the surveys of Collignon (1955) in Madagascar show a more precise correlation, at least for part of the fauna, with the *Diplomaceras bidorsatum* associations of the lower Campanian (Collignon, 1954). For this last author, *Maorites* is, among all the *Kossmaticeratidae,* the most precisely dated genus, and it is confined to the area of Bidorsatum (Madagascar). With respect to the other genera, they are either unknown outside Antarctica, or they are not located in the stratigraphic sequence. Howarth (1958) concluded that all the faunistic associations of the insular group of James Ross Island are of approximately similar age, ranging between lower and upper Campanian.

Howarth (1966), upon studying the fauna of Vicecomodoro Marambio (Seymour) Island, stated that the species found there differ from those existing in the other Antarctic localities (Cape Lachman and Dagger Peak on James Ross Island, Cape Morro, Cape Lamb on Vega Island, Giboso Hump Islet, and Cerro Nevado [Snow Hill] Island). With reference to Kilian and Reboul (1916), he mentions that those authors have established the presence of *Gunnarites antarcticus, Maorites densicostatus, M. seymourianus,* and *Grossouvrites gemmatus,* among other species, and points out that the presence of the first one has not been confirmed by any other author.

Olivero (1975) has found *Gunnarites antarcticus* on Cerro Nevado (Snow Hill) Island. Likewise, del Valle and Rinaldi (1975) point out the abundance of *Grossouvrites gemmatus* and *Maorites densicostatus* in the surroundings of Cape Bodman on Marambio (Seymour) Island. In addition, the presence in the latter place of *Pachydiscus* aff. *gollevilensis,* probably indicates an upper Campanian age, rather than a middle Campanian age.

Howarth (1966) concluded that the fossils found on Cape Lachman, Cape Lamb, Cape Morro, Giboso Islets (Humps), and Cerro Nevado (Snow Hill) Island, can be considered together as contemporaneous fauna with an age between uppermost lower Campanian and middle Campanian. Ammonites of the same age also exist on Vicecomodoro Marambio (Seymour) Island, but the presence of *Pachydiscus* and *Maorites tuberculatus* lead the present author to assign an upper Campanian age for the López de Bertodano Formation.

The ages of the cephalopod fauna of Cape Lamb and Cape Bodman, as assigned by Kilian and Reboul (1916), Spath (1953), Howarth (1958; 1966), and Bibby (1966), fall between lower and middle Campanian, and upper Campanian and lower Maestrichtian, respectively.

On Marambio (Seymour) Island del Valle and Rinaldi (1975) have found *Pachydiscus* aff. *gollevillensis* associated with a cephalopod fauna typically of middle and upper Campanian age. Thus, the fauna of the López de Bertodano Formation is abundant in *Grossouvrites gemmatus, Maorites Seymourianus, Kitchinites darwini, Maorites densicostatus, Pachydiscus* aff. *gollevillensis, Phyllopachyceras forbesianum,* and *Diplomaceras lambi,* this last one in the upper layers of this formation, among which the common occurrence of *Grossouvrites gemmatus* is noteworthy. This fauna makes it possible to compare by parallelism the sedimentary sequence of the López de Bertodano Formation of Vicecomodoro Marambio (Seymour) Island with the middle layers of the Cazador Hill section in the province of Santa Cruz, Argentine Republic.

7. *Sobral Formation* (Rinaldi and others, 1978)
The upper part of the Marambio Group is made up of a monotonous unit of grayish brown, friable, silty sandstone alternating with medium-grained sandstone with calcareous cement, which becomes gray sandstone where abundant glauconitic concretions are present. Towards the base there is well-consolidated, fine-grained gray sandstone intercalated with gray siltstone which pass into thin conglomerate beds of variable thickness. The latter are interbedded with dark gray siltstone, where erratic concretions of different sizes are found which contain remnants of trees trunks and a few bivalves. Many beds have well-developed cross-bedding.

In the basal unit, a well-consolidated bed of conglomeratic sandstone, with a large content of glauconite, stands out. It is characterized by the presence of *Lahilla luisa* and the absence of ammonoids. This tends to justify placing this unit within the lower Maestrichtian, comparable with the sequence of Cazador Hill, Santa Cruz, Argentine Republic.

Bibby (1966) indicates for these sedimentary beds a total thickness of 3580 m for the outcrops of the López de Bertodano Formation on James Ross Island. He also reports observing a folded structure on Humps Island.

The authors of this paper have found no evidence for that type of structure along the layers examined. Rather the structure seems to be monotonously monoclinal along the more than 60 km of outcrops which extend from the east margin of Prince Gustavo Channel to Vicecomodoro Marambio (Seymour) Island. Therefore, reference is made here to sedimentary thickness only in measured sections, perfectly traceable in a homoclinal tectonic model, leaving for another occasion the structural discussion in the light of new observations.

The measured thickness on Vicecomodoro Marambio (Seymour) Island of the sedimentary beds of the López de Bertodano Formation was 1867 m, and 210 m for the Sobral Formation.

In view of all the above-mentioned reasons, we propose the adoption of new stratigraphic nomenclature for the Upper Cretaceous of the James Ross Island Group, according to Table 31.1.

Table 31.1. New stratigraphic nomenclature for the Upper Cretaceous of the James Ross Island Group

Age		Rock unit
Tertiary		
Maestrichtian(?)	Marambio Group	Sobral Formation
Middle–upper Campanian		Lopez de Bertodano Formation
Lower–middle Campanian		Hidden Lake Formation Stoneley Point Formation Upper Kotick Point Formation Lower Kotick Point Formation Lagrelius Point Formation

Acknowledgments. The author wishes to acknowledge the collaboration of his colleagues Juan P. Spikermann, J. Morelli, and A. Massabie for their suggestions and discussions concerning the preparation of this work, and his former students R. del Valle, E. Olivero, E. Malagnino, and F. Medina for their valuable assistance in the field and for their suggestions and contributions to the interpretation of the geology of the area; the Instituto Antártico Argentino, Universidad de Buenos Aires, Comisión Nacional de Energía Atómica, and Centro de Investigación en Recursos Geológicos; and the Argentine Armed Forces for their cooperation, without which this task would have been impossible to achieve.

References cited

Andersson, J. G. 1906. On the geology of Graham Land. *Bull. geol. Instn Univ. Upsala, 7,* 19–71.

Bibby, J. S. 1966. The stratigraphy of part of north-east Graham Land and the James Ross Island Group. *Scient. Rep. Br. antarc. Surv.,* No. 53, 37 pp.

Collignon, M. 1954. Essai de nomenclature stratigraphique des terrains sedimentaires du Madagascar. *Travel Bur. geol. Madagascar, 63,* 1–67.

Collignon, M. 1955. Ammonites Néocretacés du Menabe (Madagascar), II, les Pachydiscidae. *Annls geol. serv. Mines Madagascar, 21,* 1–98.

del Valle, R., and Rinaldi, C. 1975. Sobre la presencia de Diplomoceras Lambi Spath en la Isla Marambio. *Contr. Inst. antárt. argent.,* No. 19, 1–39.

Haug, E. N. 1909. Traité de géologie, Vol. 2. *Les périodes géologiques.* 3rd ed. Paris, Librairie Armand Colin, 1153–1396.

Howarth, M. K. 1958. Upper Jurassic and Cretaceous ammonite faunas from Alexander Land and Graham Land. *Scient. Rep. Falkld Isl. Depend. Surv.,* No. 21, 1–16.

Howarth, M. K. 1966. Ammonites from the Upper Cretaceous of the James Ross Island group. *Bull. Br. antarc. Surv.,* No. 10, 55–69.

Kilian, W., and Reboul, R. 1916. Les Céphalopodes Néocrétacés des Iles Seymour et Snow Hill. *Schwed. Südpolarexped., 3,* No. 6, 1–5.

Malagnino, E., Olivero, E., Rinaldi, C., and Spikermann, J. 1979. Geología del Borde Occidental de la Isla James Ross. Antártida. *Act. 7 Cong. Geol. Argent., 1978,* 489–503.

Medina, F. 1975. Contribución al conocimiento geológico del Extremo N.E. de la Isla Cerro Nevado, Grupo de Islas Ross, Sec. Antártico Argentino. Trabajo final de Licenciatura. *Fac. Cs. Exts. y. Nat. U.B.A.,* Argentina, 1–69. [Unpublished.]

Nordenskjöld, O. 1913. Antarktis. (*In Handbuch der Regionalen Geologie,* Vol. 8. Heidelberg, Carl Winter's Universitätsbuchhandlung, 1–28.)

Olivero, E. 1975. Perfil Geológico, Descripción de la Fauna de Ammonites y Geomorfología del Extremo N.E. de la Isla Cerro Nevado, Grupo de Islas James Ross, Sector Antártico Argentino. Trabajo final de Licenciatura. *Fac. Cs. Exts. y Nat. U.B.A.,* 1–48. [Unpublished.]

Rinaldi, C., Massabie, A., Morelli, J., Roseman, L., and del Valle, R. 1978. Geología de la Isla Vicecomodoro Marambio. *Contr. Inst. antárt. argent.,* No. 217, 1–37.

Spath, L. 1953. The Upper Cretaceous Cephalopod. *Scient. Rep. Falkld Isl. Depend. Surv.,* No. 3, 1–60.

Steinmann, G. 1895. Die Cephalopoden der Quiriquina-Schichten. (*In* Steinmann, G., Decke, W., and Möricke, W., Das Alter und die Fauna der Quiriquinaschichten. *Neues Jb. Min. Geol. Paläont., 3,* 64–94.)

32

Lower Tertiary Strata on Seymour Island, Antarctic Peninsula*

DAVID H. ELLIOT and TIMOTHY A. TRAUTMAN**
Institute of Polar Studies and Department of Geology and Mineralogy,
The Ohio State University, Columbus, Ohio 43210, U.S.A.

Abstract. Lower Tertiary strata on Seymour Island comprise two formations that are in fault contact. The older (Paleocene) Cross Valley Formation is a 105-m-thick sequence of sandstone and pebbly sandstone, representing distributary channel fills, and interbedded silt and silty sand with carbonaceous remains, which was deposited in interdistributary areas. The younger (late Eocene to possibly early Oligocene) La Meseta Formation has a composite thickness of about 450 m and is divisible into three units. Unit I consists of laminated and thin-bedded fine-grained sand and silty sand deposited in a prodelta or delta-slope environment. Unit II represents deposition in a tide-dominated environment; fossiliferous pebbly sandstone and well-sorted fine-grained sand are interbedded with ripple-laminated and bioturbated fine-grained sand and silty clay. Unit III consists of fine-grained sand with intervals of laminated fine-grained sand and clay and interbedded fossiliferous sandstone and pebble horizons; this is interpreted as being deposited in a lagoonal environment. The major megastructure in this formation is large-scale "slump-and-fill" formed by mass movement of sediment and subsequent infilling of the submarine valley that resulted. Deltaic environments are suggested for the two formations. Glass shards and pumice fragments in the older formation indicate contemporaneous volcanism. Declining ratios of plagioclase to total feldspar and changing lithic clast composition and heavy mineral suites document a provenance change from dominantly volcanic in the older formation to dominantly plutonic and metamorphic, and comparable with the present cordillera, in the younger formation.

These Tertiary rocks are the youngest exposed strata of the late Mesozoic–early Cenozoic depositional basin on the southeast flank of the northern Antarctic Peninsula.

Seymour Island (Isla Vice Comodoro Marambio on Argentinean maps), which lies to the east of the northern end of the Antarctic Peninsula, was visited by members of the Swedish South Polar Expedition in 1901–03. Rocks and fossils were collected from both the Cretaceous and Tertiary sequences (Andersson, 1906). Two distinct rock associations were recognized in the Tertiary sequence, which crops out only in the northeastern part of the island. The rocks in Cross Valley were described as calcareous sandstone with tuff fragments, whereas the rest were described as soft sandstone with conglomerate intercalations. The calcareous sandstone was found to contain plants, and the other beds, an abundant invertebrate fauna. The whole Tertiary sequence was designated the Younger Seymour Island beds. Adie (1958) reported that an unconformable relationship between the Tertiary and Cretaceous beds had been established during subsequent work by the Falkland Islands Dependencies Survey. Additional fossil collections were also made (e.g., Marples, 1953). Matthews (1959) seems to have been the first to use the term *Seymour Island Series,* although it was not defined formally. This report is based on fieldwork conducted during the 1974–75 field season (Elliot and others, 1975) and on subsequent laboratory studies (Trautman, 1976).

* Contribution no. 329 of the Institute of Polar Studies, The Ohio State University.

** Present address: Amoco Production Company, P.O. Box 3092, Houston, Texas 77001, U.S.A.

Stratigraphy

Revision of the stratigraphic nomenclature is proposed here as a result of the recent fieldwork. The Tertiary beds can be divided into two formations, corresponding broadly to the two rock associations recognized by the Swedish South Polar Expedition. It is recommended, therefore, that the use of "Seymour Island Series" be discontinued and that the two formations described below be designated the *Seymour Island Group.* The two formations are the Cross Valley Formation, which consists of the older Tertiary strata exposed in Cross Valley, and the La Meseta Formation, which consists of the younger strata exposed around the flanks of the mesa in the northern part of the island.

Cross Valley Formation

The name Cross Valley Formation is proposed for the sequence of nonmarine sandstone, pebbly sandstone, plant-bearing fine-grained sandstone, and unconsolidated sand and clayey sand* that crops out in Cross Valley (Figure 32.1). Rocks that are possibly correlative crop out near Cape Wiman and are discussed below. The Cross Valley Formation is regarded as Paleocene in age (see Zinsmeister and Camacho, Chapter 33, this symposium).

* The use of the words *sand* and *sandstone* throughout this chapter is intentional; sandstone is lithified and forms more resistant ledges than sand, which is loose and extremely friable and can be collected only as loose aggregate.

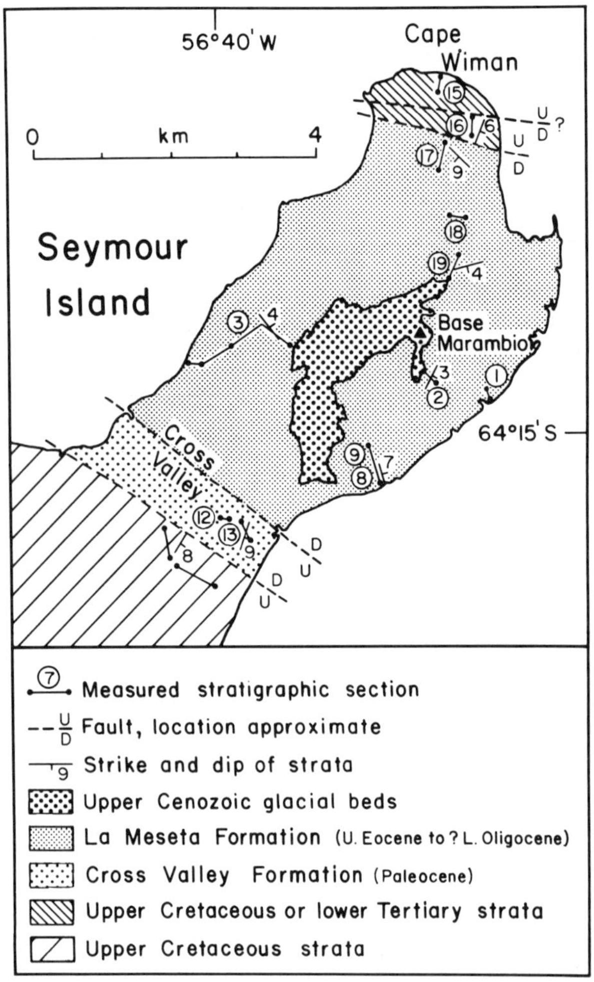

Figure 32.1. Geologic sketch map of the northeastern part of Seymour Island. The outcrop belt of the Cross Valley Formation includes fault slivers of Cretaceous rocks; the fault contacts here are simplified and generalized.

Contacts

A possible lower contact with the Upper Cretaceous Snow Hill Island Series was observed at only one place in Cross Valley; however, poor exposure precluded definite identification as a fault or an unconformity. Elsewhere the Cross Valley Formation was observed only in fault contact with the Cretaceous beds, as well as with younger rocks.

Lithologic summary

The type section (sections S-12 and S-13), which is the only well-exposed section, starts at the base of the slope below a prominent bluff in Cross Valley (Figure 32.2). The type section (Figure 32.3) does not begin at the lowest exposed strata in the valley because of poor exposure, and an estimated minimum of 30 m of section lies below the base of the measured section. The measured section below the bluff (see Figure 32.2) consists of 36 m of friable coarse-grained sandstone. Bedding thickness is difficult to ascertain, although a range is suggested by slightly resistant units 10 to 20 cm thick. The coarse-grained sandstone is poorly sorted, has a large silt-plus-clay fraction, and carries scattered clasts as much as 10 cm across. Thin, limonite-stained, sandy clay intervals and scattered fossil wood fragments occur in the lower part of this 36 m interval.

The friable coarse-grained sandstone passes up into 34 m of bluff-forming, poorly sorted, pebbly coarse-grained sandstone (Figure 32.2), which grades into 12 m of bedded, pebbly, medium-grained sandstone. The bluff-forming pebbly sandstone contains abundant clasts as much as 15 cm across and, like the coarse sandstone below, has a high silt-plus-clay fraction. Volcanic rocks, mainly andesitic with subordinate silicic fragments, predominate in the clasts. Bed thickness in the bluff-forming interval is 1 to 5 m, and individual beds display poorly developed graded bedding. In the 12 m of bedded sandstone, bed thickness decreases to as little as 2 cm; internal structures were not observed. Carbonized wood fragments are sparsely distrib-

Figure 32.2. Bluff-forming pebbly sandstone in the lower part of the Cross Valley Formation (36–70 m interval in section S-12; see Figure 32.3). Bluff is 34 m high; view to the east.

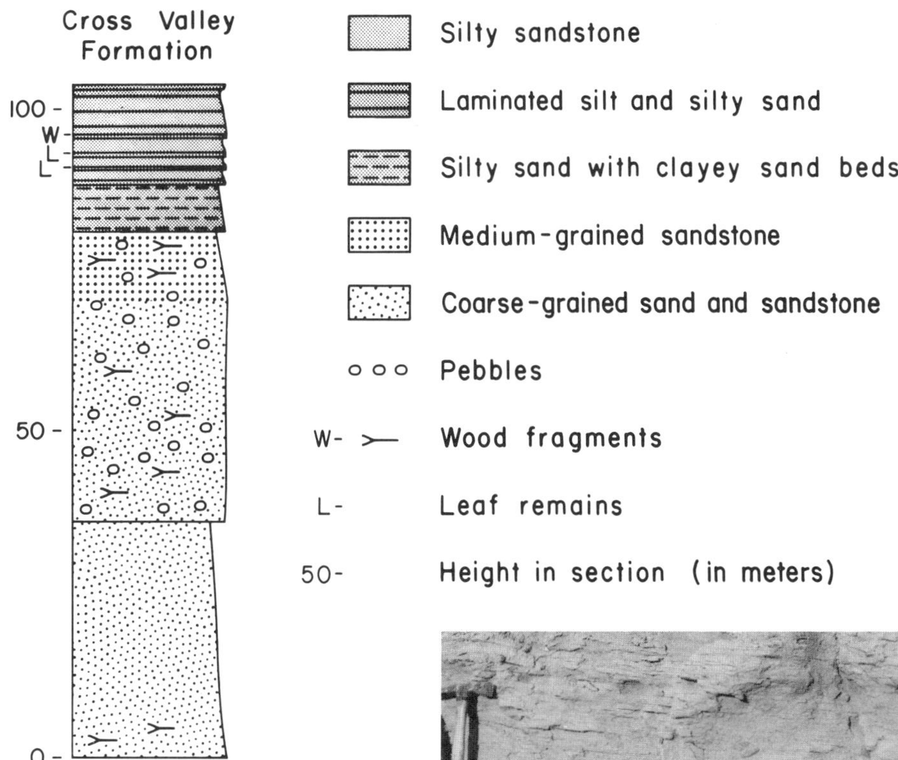

Figure 32.3. Stratigraphic column of the type section of the Paleocene(?) Cross Valley Formation (sections S-12 and S-13).

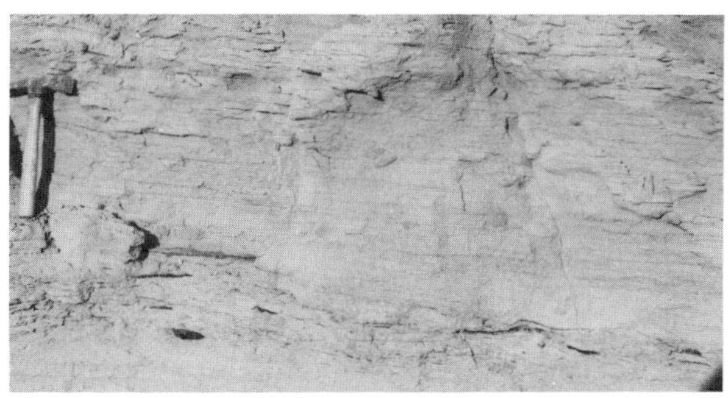

Figure 32.4. Laminated silty sand with coalified plant remains in the upper part of the Cross Valley Formation (98.5 m level in the type section.).

Figure 32.5. Ripple-laminated beds of Unit II, La Meseta Formation; the person is standing on the shell bank at the 165 m level in section S-3. View to the west.

uted except in the upper 12 m where they occur locally concentrated on bedding planes.

The upper 23.5 m of section consists of 4 m of moderately poorly sorted, fine- to medium-grained, silty sandstone with thin, limonitic, clayey sand beds, overlain by 19.5 m of unlithified, laminated (3–10 mm), light gray silt and silty sand with sparse interbeds of resistant, medium-bedded (10–30 cm), calcareous silty sandstone. The unlithified strata contain sparsely distributed concretions and limonitic horizons and locally show traces of cross-bedding. Fossil wood and coalified plant remains occur sparingly in the friable beds, and locally the plant debris is concentrated in laminae (Figure 32.4). More abundant and better preserved fossil leaves occur on bedding surfaces in the resistant silty sandstone beds. The formation of siderite concretions is attributed to the presence of the plant debris. Molluscs were found in the uppermost beds.

La Meseta Formation

The name *La Meseta Formation* (already used informally by Argentinean geologists) is proposed for the sequence of unconsolidated marine sand and mud exposed on the flanks and to the north of the mesa in the northeastern part of the island (Figure 32.5). The sand and mud contain very thin clay beds, pebble horizons, and resistant calcareous sandstone beds and conglomeratic sandstone shell beds. The formation can be divided into three informal members, here referred to as Units I, II, and III. The composite thickness of the formation is about 450 m. Fossils suggest a late Eocene to possibly early Oligocene age (Zinsmeister, 1977; Zinsmeister and Camacho, Chapter 33, this symposium). The faunal differences between the uppermost beds (Unit III) and those below (Units I and II) sug-

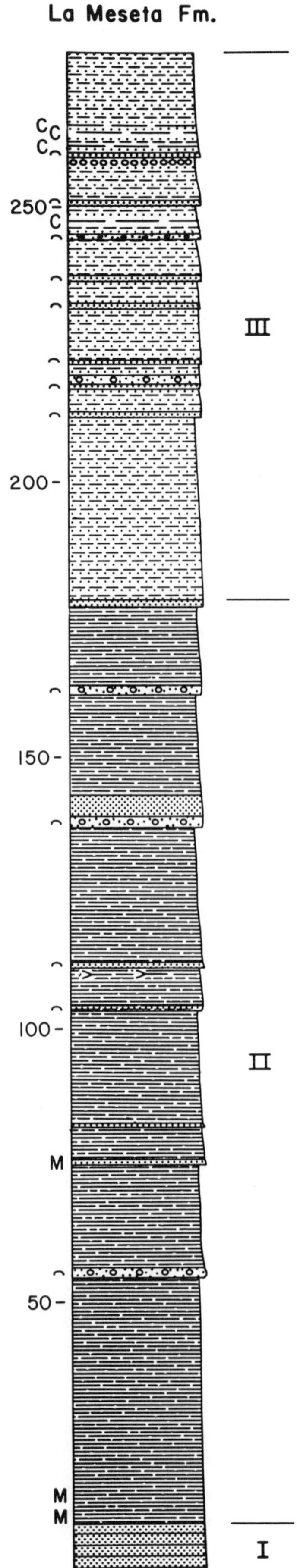

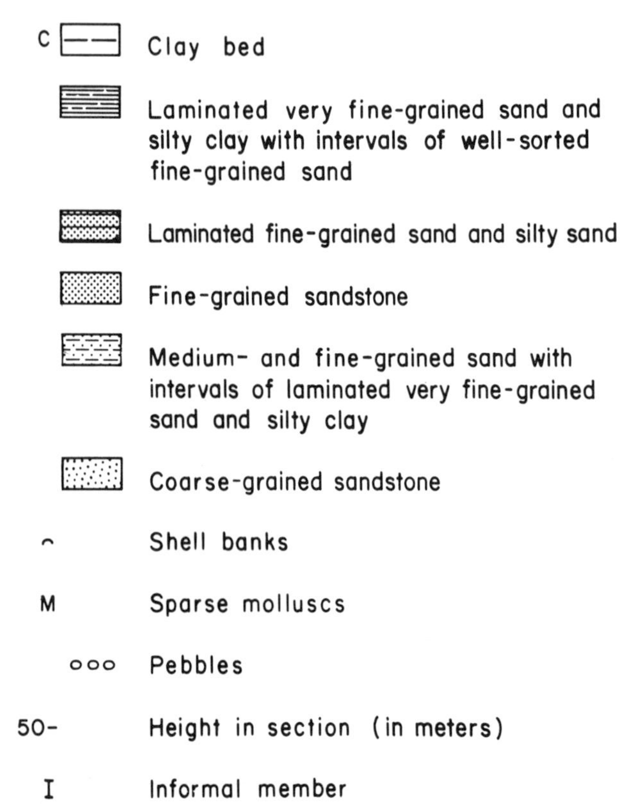

Figure 32.6. Stratigraphic column of the type section of the upper Eocene–lower Oligocene(?) La Meseta Formation (section S-3).

Figure 32.7. Fossiliferous, pebbly, coarse-grained sandstone at the 56 m level in section S-3, La Meseta Formation.

Figure 32.8. Pebble horizon in Unit II of La Meseta Formation (88 m level in section S-8).

gest a more significant time break between Units II and III than is apparent in the stratigraphy. A section (S-3) exposed on the southwestern flank of the mesa has been designated the type section (Figure 32.6).

Contacts

The lower contact of La Meseta Formation is not exposed. It is in fault contact with the Cross Valley Formation and the strata near Cape Wiman. La Meseta Formation beds are overlain by upper Cenozoic glacial deposits that cover the mesa top.

Lithologic summary

Unit I is exposed at the northern end of the island where it is in fault contact with older strata, along the coast at the western end of Cross Valley, and on the southeast coast north of Cross Valley. Unit I consists of at least 160 m of unconsolidated, fine-grained sand and silty sand with sparse silty clay beds. Parallel lamination on a scale of 0.5 to 10 cm occurs throughout the section. Ripple lamination and mottling due to bioturbation were noted in the upper 50 m. Only 10 m of Unit I were measured at the type section, although much more is exposed in the faulted and slightly deformed cliff section along the beach to the southwest.

Unit II consists of laminated, very fine-grained sand and silty clay with interbedded conglomeratic sandstone and well-sorted fine-grained sand (Figure 32.5). At the type section this unit is 170 m thick. Wavy bedding with continuous layers of interbedded mud and sand predominates, but locally the sand is reduced to discontinuous thin lenses isolated in the silty clay. Vertical and oblique burrows are common, and with increasing degrees of bioturbation, the laminated beds grade, both laterally and vertically, into mottled, brownish-gray, silty sand. Flaser bedding and ripple lamination are common. Concretionary horizons are widespread. Scattered silicified and coalified plant remains occur in these beds.

Thick beds (1–5 m) of well-sorted, fine-grained, gray sand are intercalated with the laminated beds. Ripple-drift cross-lamination, small-scale channels as large as 3 m across and 1 m in depth, and convolute bedding as much as 2 m in height are also present in these beds.

The conglomeratic beds (Figure 32.7) are calcium carbonate–cemented, pebbly, medium- to coarse-grained, well-sorted sandstone containing an abundant invertebrate fauna dominated by bivalves and gastropods (Zinsmeister and Camacho, Chapter 33, this symposium). These sandstone shell banks are 0.5 to 1.5 m thick, but although locally sheetlike, they become thinner and pinch out laterally. In a few places the beds have erosional bases and contain low-angle cross-beds. One pebble horizon similar to those in Unit III was noted in the field (Figure 32.8).

Unit III consists of medium- to fine-grained sand with intervals of laminated very fine-grained sand and silty clay and sparsely distributed beds of carbonate-cemented fossiliferous sandstone. The medium- to fine-grained sand is homogeneous, moderately to well-sorted, and contains sparse invertebrate remains. Small-scale cross-beds were noted in a few places. The laminated beds pass vertically and laterally into beds showing increasing degrees of bioturbation until all trace of bedding is lost. The intercalated resistant layers are thin (0.2–1.0 m thick), moderately sorted, fine- to medium-grained sandstone beds which contain turritella gastropods, lingulas, and articulated molluscs (Figure 32.9). Some of these beds also contain dispersed clasts as long as 13 cm. In addition, there are a few 2–6 cm thick clay beds in the uppermost 100 m of the type section and a few widely distributed pebble horizons. These horizons are normally one pebble thick and have clasts which are as large as 30 cm in diameter; they are similar to the example illustrated in Figure 32.8. Fossil wood and plant remains occur sparsely. Unit III is 80 m

Figure 32.9. Fossiliferous sandstone with articulated molluscs at the 126 m level in section S-19, Unit III of La Meseta Formation. Pocket knife is 8 cm long.

thick at the type section but attains a measured maximum of 125 m (sections S-1 and S-2). This unit is exposed only around the mesa.

Beds at and near Cape Wiman

The lack of paleontologic controls prevents assignment of the rocks exposed at and near Cape Wiman to either of the Tertiary formations or to the Cretaceous sequence exposed in the southwestern part of the island. Two sections have been measured. The lower one (section S-15) consists of 120 m of unconsolidated, homogeneous silt and silty sand with sparse interbeds of resistant silty sandstone. The silt and silty sand beds contain concretionary horizons and are locally seen to be finely laminated; the sand shows mottling, and limonitic stains are common. The resistant beds are calcareous, fine grained, and moderately well sorted. Planar cross-bedding in sets 1 m thick occurs in the uppermost sandstone. Carbonized wood fragments are also common.

The overlying 130 m of section (section S-16) consists, in the lower part, of unconsolidated silt and fine-grained sand with interbedded resistant sandstone beds; the grain size increases upward, with coarse-grained sand predominating in the upper 50 m. The textural maturity decreases markedly in the same direction so that the coarse-grained sand is pebbly and has a high clay-plus-silt fraction. The unconsolidated sand in the lowest 25 m contains concretionary horizons and shows extensive burrow mottling. Higher in the section, the sand exhibits much multistory, planar cross-bedding which contains carbonized wood on foreset planes. Pebble horizons and dispersed pebbles as much as 1 cm across occur in the upper part of the section, particularly in the interval from 60 to 90 m, which is the part of the section where limonitic staining is widespread and woody remains are locally abundant. The more resistant beds have calcareous cement but otherwise differ little from the loose sand layers. Low-angle cross-bedding, dispersed plant debris, and burrowing have been observed in a few of these beds.

Depositional environments

Cross Valley Formation

The coarse-grained sandstone at the base of the formation is poorly sorted, having a large silt-plus-clay fraction and containing scattered pebbles as large as 10 cm across and scattered fossil wood fragments. The overlying bluff-forming, pebbly, coarse sandstone (Figure 32.2) lacks well-defined bedding but exhibits poorly developed normal grading. The sandstone and pebbly sandstone are regarded as distributary channel fill deposits, dumped as floods receded. The overlying bedded sandstone, which is slightly better sorted, is believed to represent better washed and sorted sediment deposited adjacent to the main channel. The upper section of unlithified, laminated silt and silty sand with intercalated resistant silty sandstone beds represents deposition in low-energy interdistributary areas; the unlithified beds contain carbonized wood fragments (Figure 32.4), and the resistant silty sandstone has fossil leaves. The molluscs present in the uppermost part of the sequence indicate that marine conditions prevailed locally during deposition.

Thus the Cross Valley Formation is thought to represent distributary channels and interdistributary areas of a delta.

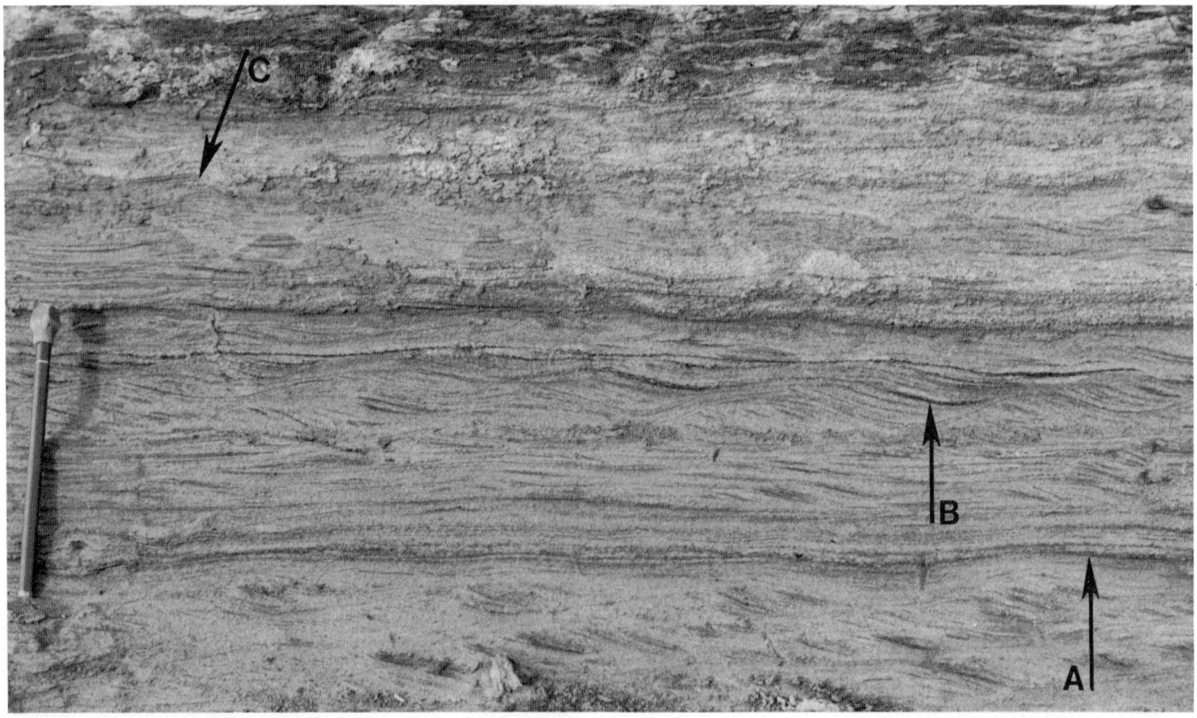

Figure 32.10. Ripple-laminated beds overlain by continuous laminae of mud (arrow A) indicating low-current velocities. Laminae pass upward into climbing ripples (ripple-in-drift type 2) that steepen upward (note ripples climbing to the right). Mud flasers occur in ripple troughs (arrow B) and mud laminae are draped over some ripple crests. The overlying beds show ripple-in-phase lamination (arrow C) that climbs to the left indicating reversal of current flow. Unit II of La Meseta Formation (18 m level in section S-10, which is adjacent to section S-2).

La Meseta Formation

Unit I of the formation consists, where observed, of a monotonous sequence of laminated fine-grained sand and silty sand with sparse silty clay beds. Ripple lamination and burrow mottling occur in the upper part. By association with Unit II, for which the depositional environments are better documented, this member is regarded as a prodelta deposit.

The middle member of this formation, Unit II, consists of laminated, very fine-grained sand and silty clay with interbedded conglomeratic sandstone and well-sorted fine-grained sand. The laminated sediment consists of even and wavy, parallel, very thin to thin, interlayered beds of sand and silty clay. Within this general description, the sediment exhibits flaser and wavy bedding of the type illustrated by Reineck and Singh (1975). The sand (Figure 32.10) also exhibits ripple-in-phase and ripple-in-drift lamination (Reineck and Singh, 1975). The climbing ripples indicate high bed load; the flaser and wavy bedding, fluctuating current strength; and the bioturbation (Figure 32.11), an active bottom fauna. These features are consistent with a tide-dominated environment (Reineck, 1972).

The pebbly coarse-grained sandstone shell banks carry disarticulated, heavy *Cucullaea* shells, many of which lie convex upward. Robust gastropod shells and shark teeth are also abundant. Pebble sizes are as great as 15 cm. The shell banks were deposited in a high-energy environment in which there was little transport of the shell material. The one pebble horizon is a lag deposit.

The third lithofacies of Unit II is the well-sorted, fine-grained sand which forms thick beds that have ripple-drift lamination, small-scale channels, and sparse convolute lamination. The small-scale channels which occur at the top of thick, homogeneous beds point to reworking. The convolute lamination is probably the result of liquefaction (Reineck and Singh, 1975). These sand beds are regarded as sandbar or channel sand deposits in a tidal environment.

Figure 32.11. Burrowed sediment in Unit II of La Meseta Formation (150 m level in section S-3). Pocket knife is 8 cm long.

On the southeast coast, the beds of Unit II exhibit large slump structures. The observed width of the largest slump is about 500 m, and the depth is at least 50 m (Figure 32.12). The slump is filled by laminated sediment of the same composition as that adjacent; the laminated sediment is draped in the slump, and bed thicknesses diminish toward the slump edge. Pebbly sandstone shell banks, as described above, may occur within these slump fills, and the shell banks appear to be aligned longitudinally in the slump fill. These structures are similar to those of the Mississippi River delta described by Shepard (1955, 1956).

The principal lithofacies of Unit II indicates a tide-dominated environment. The large slump structures suggest a delta-front environment, and the thick beds of homogeneous fine-grained sand suggest channel deposits. The depositional environment of the sandstone shell banks remains a problem. High energies and a proximal source of pebble- and cobble-sized clasts are required; the clasts have

Figure 32.12. Large-scale "slump-and-fill" structure formed by slumping of sediment on the delta front; the dotted line indicates the base of the slump structure. La Meseta Formation, east coast of Seymour Island between sections S-1 and S-8. Note person for scale.

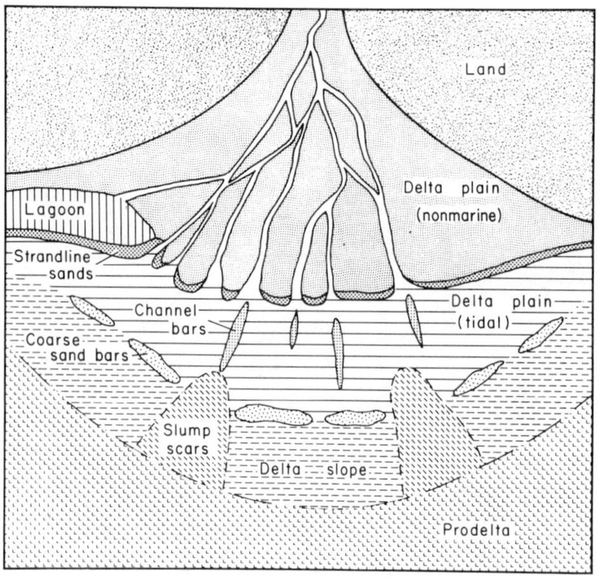

Figure 32.13. Possible depositional environments for La Meseta Formation strata.

not been concentrated by winnowing of underlying sediment. The depositional environment tentatively proposed for Unit II is a tide-dominated delta, in part comparable with that described by Fisher and others (1969). However, the lack of major sand bodies suggests either that the area was marginal to the delta or that rivers discharging into the delta seldom carried sediment coarse enough to form such sand.

Unit III is characterized by medium- to fine-grained sand with intervals of laminated fine-grained sand and silty clay, intercalated clay beds, pebble horizons, and thin, resistant, fossiliferous sandstone beds. The resistant sandstone contains articulated molluscs, turritella gastropods, and lingulas, all indicative of a low-energy environment. The laminated intervals and the few clay beds also suggest low energies. Bioturbation of the laminated beds indicates an active bottom fauna. The homogeneous medium- to fine-grained, moderately to well-sorted sand points to shallow-marine or strandline conditions; these beds contain vertebrate remains. A higher energy environment is illustrated by the pebble horizons, which are regarded as storm lag gravels spread as a result of breaching of sand and gravel bars that protected the environment in which articulated, thin-shelled molluscs could be preserved.

The lithofacies of Unit III could be the result of deposition in lagoonal areas into which streams discharged fine-grained sediment, in which tidal influence was not strong, and which were protected from wave action by sand and gravel bars that were breached only under storm conditions.

Thus La Meseta Formation sedimentary deposits are tentatively regarded as having formed in deltaic environments (Figure 32.13). Unit I represents the prodelta or delta-slope environment; Unit II is the tide-dominated delta top. The large-scale slump structures along the east coast of Seymour Island represent massive slumping of water-saturated sediment at the front of the delta top. The lithofacies of Unit III are regarded as possibly lagoonal deposits adjacent to the main area of deltaic deposition.

Beds at and near Cape Wiman

The lower part (section S-15) consists of unconsolidated homogeneous silt and silty sand with burrow mottling, limonitic staining, and fine lamination in places; carbonate-cemented, resistant silty sandstone beds also occur. The fine lamination and bioturbation suggest quiet-water deposition and an active bottom fauna. This could represent deposition in an interdistributary bay; marine conditions are further suggested by the occurrence, although not *in situ,* of the gastropod *Pleurotemaria* sp.

The overlying strata (section S-16) show a gradation upward from unconsolidated, mottled, silt and silty fine-grained sand with interbedded resistant sandstone beds to immature coarse-grained sand with pebble horizons. The silt and silty sand beds may represent low-energy interdistributary areas of a delta top. The coarse-grained sand exhibits multistory, planar cross-bedding; foreset thicknesses range from 1 to 3 m, and foreset inclinations range between 21° and 32°, with an average of 27°. The planar cross-bedding is indicative of prograding foresets of sand dunes. The immaturity of the sediment, the planar cross-bedding, and the abundant fossil wood suggest high-energy delta channel deposition.

Petrology

Cross Valley Formation

The sand and sandstone are litharenites (Figure 32.14) consisting of volcanic rock fragments and quartz with subordinate feldspar grains and minor to trace amounts of other rock fragments and minerals. Calcite cement is present in lithified specimens. Quartz is largely monocrystalline, and a few grains contain inclusions of either apatite or acicular rutile. Embayed volcanic quartz and polycrystalline quartz are also present. Plagioclase is more abundant than orthoclase and includes both a calcic oligoclase

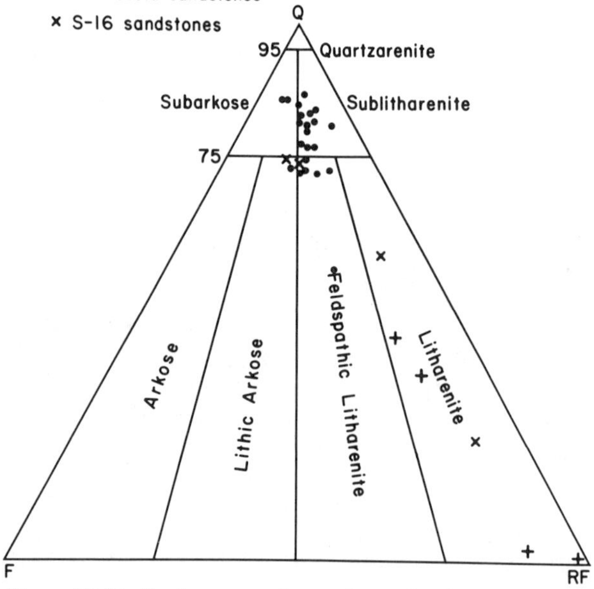

Figure 32.14. Sandstone specimens from the Seymour Island Group and from section S-16 (Cape Wiman beds) plotted on a quartz-feldspar-rock fragments ternary classification diagram.

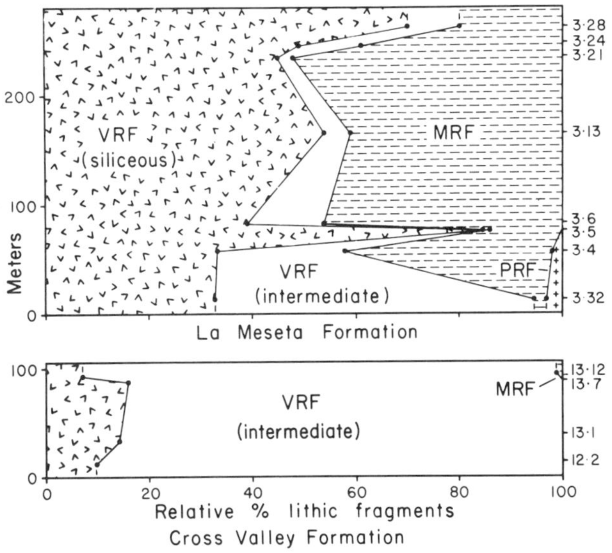

Figure 32.15. Plot of the proportions of different lithic fragments in sandstone from the Cross Valley and La Meseta Formations in sections S-12, S-13, and S-3 (see Figure 32.1 for locations). Specimen numbers on the right-hand side. VRF: volcanic rock fragments. MRF: metamorphic rock fragments. PRF: plutonic rock fragments. Note that the relative percentage of total rock fragments to mineral grains drops from an average of 70% in Cross Valley sandstone to 11% in La Meseta sandstone. La Meseta sandstone is fine to very fine grained, whereas Cross Valley sandstone ranges up to very coarse grained, so the relative proportions of constituents are not strictly comparable.

(An_{20-25}) and a calcic andesine (An_{45-50}). The andesine is more angular than the oligoclase, has Carlsbad-albite twinning and/or oscillatory zoning, and is regarded as volcanic in origin, whereas the albite-twinned oligoclase possibly has a plutonic source. Heavy minerals include opaque grains, brown hornblende, augite and subordinate green hornblende, biotite, muscovite, garnet, and rutile.

Andesitic clasts dominate the lithic fragments. They have a marked reddish brown color caused by oxidation of iron and consist of feldspar laths, mainly plagioclase, set in an aphanitic groundmass. Feldspar laths exhibit a trachytic texture in many clasts; however, some clasts contain microlites lacking flow texture, and some are entirely aphanitic. The larger feldspar crystals are andesine in composition.

Siliceous volcanic fragments with cryptocrystalline or microcrystalline textures are subordinate. Trace amounts of air-fall debris, both tricuspate glass shards and pumice fragments, also occur. Metamorphic rock fragments are present in trace amounts, but no plutonic fragments have been recognized.

La Meseta Formation

The sand and sandstone are mainly sublitharenites (Figure 32.14) but range into adjacent fields. Quartz dominates the clastic grains; lithic grains and feldspars are subordinate, and heavy minerals occur in accessory amounts. Monocrystalline quartz predominates over polycrystalline varieties. Plagioclase is largely oligoclase and occurs in about equal proportion with potassium feldspar. Orthoclase, microcline, and perthite occur in about equal amounts. Lithic fragments include andesitic and silicic volcanic clasts similar to those in the Cross Valley Formation, an equal proportion of two different metamorphic rock fragments—mica schist and metasiltstone—and subordinate plutonic clasts of granitic composition. Heavy minerals include green hornblende, opaque grains, and epidote, with subordinate muscovite, garnet, biotite, zircon, rutile, and apatite.

Gravels

Pebble- and cobble-sized clasts occur in La Meseta Formation beds in sandstone shell banks and as sheetlike concentrations.

Five lithologic subdivisions have been recognized. Among the volcanic clasts, those of silicic composition predominate over intermediate types. Granite and quartz diorite occur in about equal proportions, and gabbro clasts are subordinate. Metasedimentary rock clasts include metasiltstone and subordinate quartzite. Friable sandstone along with some siltstone and shale also occur, and vein quartz is a significant constituent in all gravel counts.

Petrologic trends

The composition of the sandstone in the Seymour Island Group shows significant variation. Volcanic sources dominated during deposition of the Cross Valley Formation, whereas nonvolcanic sources dominated during La Meseta time. The decrease in the ratios of plagioclase to total feldspar and of volcanic rock fragments to total detrital grains indicates decreasing importance of volcanic sources. Sandstone with the highest percentage of volcanic clasts also contains air-fall debris; furthermore, the proportion of plagioclase with oscillatory zoning is highest in the Cross Valley sandstone. In addition to the absolute decrease in abundance of volcanic clasts, the proportion of silicic to andesitic volcanic fragments increases upward in La Meseta Formation sand and sandstone (Figure 32.15). The nonopaque heavy mineral suite also illustrates these trends; augite and brown hornblende predominate in the Cross Valley Formation, but in La Meseta beds the principal minerals are green hornblende and epidote, with zircon, apatite, and tourmaline in accessory to trace amounts.

The contrasting mineralogy and petrology of the two formations points to a dominantly volcanic source being replaced by a plutonic-metamorphic source for the younger strata.

Source rocks

The occurrence in the Cross Valley Formation of andesitic and dactic fragments together with pumice and glass shards suggests a volcanic source, including some active volcanoes, of calc-alkaline composition. Little evidence exists of volcanic rocks associated with a lower Tertiary (Paleocene) volcanic arc situated on the main cordillera of the Antarctic Peninsula; however, on the west flank of the peninsula and on the South Shetland Islands, Tertiary volcanic rocks, some of which may be Paleocene in age, are abundant. The possibility of erosion of preexisting volcanic rocks cannot be excluded.

The petrology of La Meseta sand and sandstone, in the changing ratios of feldspars, proportions of mineral and lithic fragments, and the heavy mineral suites, together with the clast composition of the contained gravels, points to a metamorphic-plutonic complex as the main sediment source. Silicic volcanic rocks, however, were still important.

These trends are interpreted as evidence for the unroofing of a metamorphic-plutonic complex that was subjacent to a volcanic arc which was at least in part active at the time of deposition of the Cross Valley sandstone. The rocks of the subjacent complex, as represented in the sandstone and conglomerate, are comparable with rocks exposed in the Antarctic Peninsula. Metasedimentary rock clasts are similar to the upper Paleozoic–lower Mesozoic Trinity Peninsula Series (Aitkenhead, 1975; Elliot, 1965, 1966; Thomson, 1975) and its thermally metamorphosed equivalents. The silicic clasts have counterparts in the so-called Upper Jurassic Volcanic Group. Plutonic rocks, which are not known in the James Ross Island area in sedimentary rocks older than La Meseta Formation strata, are similar to the Andean intrusions which range in age from Middle Jurassic to early Tertiary (Rex, 1976). Friable sandstone and shale are possibly derived from sedimentary rocks that range from Late Jurassic to Late Cretaceous in age.

Correlation of strata

The lack of paleontologic control on the age of the beds near Cape Wiman makes correlation or identification with other strata difficult. The section (S-16) containing abundant wood remains has some similarity to the Cross Valley Formation, although there is nothing diagnostic. The sandstone petrology suggests a partly volcanic provenance (see Figure 32.14), and the depositional environments are comparable. Therefore, these rocks are tentatively assigned to the same time interval as the Cross Valley beds (late Eocene to possibly early Oligocene).

The other section (S-15) cropping out at Cape Wiman contains no distinctive beds, and it is only on the tenuous basis of the *Pleurotemaria* that they are regarded as Cretaceous(?) and therefore possibly part of the Snow Hill Island Series. All the beds at Cape Wiman have been mapped by the Argentine geologists as Cretaceous; we prefer to leave open the question of their age and correlation.

Structure

The Cretaceous strata of the southwestern part of the island form a homoclinal sequence dipping gently to the southeast and possibly broken by high-angle faults parallel to the strike of the beds.

Although there is clearly an angular discordance between the Cross Valley Formation and the adjacent Cretaceous beds, nowhere has a definitely unconformable relationship been observed. At the type section of the Cross Valley Formation, the beds dip gently (8°) eastward, and in the vicinity of the base of that section, the strata form a shallow syncline striking approximately east-west and flanked to the north and south by Cretaceous rocks. In Cross Valley it seems that there are several fault blocks or slivers which contain both Cretaceous and Tertiary strata, and these blocks are in fault contact with La Meseta Formation rocks.

At Cape Wiman, the beds in section S-15, which is possibly correlative with the upper part of the Cretaceous sequence southwest of Cross Valley, dip gently to the south. If the upper section (S-16) is correlative with the Cross Valley Formation, then a fault probably separates it from section S-15. The upper section dips gently to the east and is in fault contact with La Meseta Formation beds.

La Meseta Formation rocks are disposed in a broad syncline which plunges gently to the southeast. Measured dips on the limbs do not exceed 9°. The contrast in faunas suggests that a hiatus may separate Units II and III (Zinsmeister and Camacho, Chapter 33, this volume), but no break was recognized in the field.

The James Ross Island Basin

The lower Tertiary strata are the youngest exposed beds of the late Mesozoic–early Cenozoic depositional basin that existed on the southeast flank of the northern Antarctic Peninsula. This basin was initiated possibly in Late Jurassic time, probably by Early Cretaceous time, and certainly by Campanian time when clastic sediments were deposited in the James Ross Island area. Relationships between periods of magmatic activity, uplift, and deformation(?) in the cordillera and periods of basin subsidence and filling on the flanks of the cordillera are unclear. Uplift preceded deposition of conglomerate beds that occur at the base of the Campanian sequence, and tilting and erosion within the basin preceded deposition of the lower Tertiary beds.

In early Tertiary time, sediment deposition was renewed with the development of river systems draining highlands to the west and discharging into open seas to the east through distributary systems, one of which was located in the Seymour Island area. The channel sandstone and pebbly sandstone and the interdistributary fine-grained sand and silt with plant remains of the Cross Valley Formation represent deposits of part of such a distributary system. Continued subsidence was accompanied by a marine transgression to the west and northwest, and the shoreline migrated inland until prodelta or delta slope sediments, the fine-grained sand and silty sand of Unit I of La Meseta Formation, were laid down in the Seymour Island area. On progradation of the delta, the prodelta or delta-slope environment was replaced by a tide-dominated delta front with distributary channels and the shell banks of Unit II. From time to time, the delta front collapsed to give large submarine slumps. With time the active parts of the deltas shifted, and the delta-top environments were replaced by lagoonal environments represented by the sand, fossiliferous sandstone, and pebble horizons of Unit III of La Meseta Formation. At the same time, there may have been a sharp deterioration in climate, as shown by the faunal contrasts between Unit III and the older beds (see Zinsmeister and Camacho, Chapter 33, this volume).

Marked provenance changes occurred during early Tertiary time. Volcanic sources, intermediate to silicic in composition and including active volcanoes, dominated during deposition of the Cross Valley Formation. Continued erosion uncovered a plutonic-metamorphic complex that formed the source for La Meseta Formation strata. The provenance was similar to rocks now exposed in the Antarctic Peninsula; however, the relationship of the unroofed plutonic-metamorphic complex to the earlier volcanic terrane is uncertain, and much of the volcanic terrane could be older.

In later Tertiary time, the Seymour Island area was faulted, gently folded, and intruded by dikes assignable to the James Ross Island Volcanic Group of Pliocene–Pleistocene age. The relative ages of these events are not known.

Acknowledgments. The fieldwork on which this paper is based was conducted as part of a joint Argentine–United States investigation of the geology of Seymour Island. We are particularly grateful for the opportunity provided to us by the Instituto Antártico Argentino and the Direccion Nacional del Antártico, Buenos Aires, Argentina, to visit Seymour Island. We would also like to thank Kenneth O. Stanley for much helpful discussion. This research was supported by National Science Foundation grant No. DPP74-21509 to the Ohio State University.

References cited

Adie, R. J. 1958. Geological investigations in the Falkland Islands Dependencies since 1940. *Polar Rec., 9,* No. 58, 3–17.

Aitkenhead, N. 1975. The geology of the Duse Bay–Larsen Inlet area, northeast Graham Land (with particular reference to the Trinity Peninsula Series). *Scient. Rep. Br. antarc. Surv.,* No. 51, 62 pp.

Andersson, J. G. 1906. On the geology of Graham Land. *Bull. geol. Instn Univ. Upsala, 7,* 19–71.

Elliot, D. H. 1965. Geology of Northwest Trinity Peninsula, Graham Land. *Bull. Br. antarc. Surv.,* No. 7, 1–24.

Elliot, D. H. 1966. Geology of the Nordenskjöld Coast and a comparison with north-west Trinity Peninsula. *Bull. Br. antarc. Surv.,* No. 10, 1–43.

Elliot, D. H., Rinaldi, C., Zinsmeister, W. J., Trautman, T. A., Bryant, W. A., and del Valle, R. 1975. Geological investigations on Seymour Island, Antarctic Peninsula. *Antarct. Jnl U.S., 10,* No. 4, 182–86.

Fisher, W. L., Brown, L. F., Scott, A. J., and McGowan, J. H. 1969. *Delta Systems in the Exploration for Oil and Gas, a Research Colloquim.* University of Texas at Austin, Bureau of Economic Geology, 78 pp.

Marples, B. J. 1953. Fossil penguins from the mid-Tertiary of Seymour Island. *Scient. Rep. Falkld Isl. Depend. Surv.,* No. 5, 15 pp.

Matthews, D. H. 1959. Aspects of the geology of the Scotia arc. *Geol. Mag., 96,* No. 6, 425–41.

Reineck, H. E. 1972. Tidal flats. (*In* Rigby, J. K., and Hamblin, W. K., *eds.* Recognition of Ancient Sedimentary Environments. *Spec. Publs Soc. econ. Paleont. Miner., Tulsa,* No. 16, 146–59.)

Reineck, H. E., and Singh, I. B. 1975. *Depositional Sedimentary Environments.* New York, Springer-Verlag, 439 pp.

Rex, D. C. 1976. Geochronology in relation to the stratigraphy of the Antarctic Peninsula. *Bull. Br. antarc. Surv.,* No. 43, 49–58.

Shepard, F. P. 1955. Delta-front valleys bordering the Mississippi distributaries. *Bull. geol. Soc. Am., 66,* No. 12, 1489–98.

Shepard, F. P. 1956. Marginal sediments of the Mississippi delta. *Bull. Am. Soc. Petrol. Geol., 40,* No. 11, 2537–623.

Thomson, M. R. A. 1975. First marine Triassic fauna from the Antarctic Peninsula. *Nature, Lond., 257,* No. 5527, 577–78.

Trautman, T. A. 1976. *Stratigraphy and Petrology of Tertiary Clastic Sediments, Seymour Island, Antarctica.* Masters thesis, The Ohio State University, 170 pp.

Zinsmeister, W. J. 1977. Note on a new occurrence of the Southern Hemisphere aporrhaid gastropod *Struthioptera* Finlay & Marwick on Seymour Island, Antarctica. *J. Paleont., 51,* No. 2, 399–404.

33

Late Eocene (to Possibly Earliest Oligocene) Molluscan Fauna of the La Meseta Formation of Seymour Island,* Antarctic Peninsula

WILLIAM J. ZINSMEISTER
Institute of Polar Studies, The Ohio State University, Columbus, Ohio 43210, U.S.A.

H. H. CAMACHO
Universidad de Buenos Aires, Facultad de Ciencias Exactas y Naturales, Buenos Aires, República Argentina

Abstract. The late Eocene to possibly earliest Oligocene molluscan fauna from the La Meseta Formation of Seymour Island is considerably more diverse than previously reported. Two distinct assemblages are recognized and for convenience are referred to as assemblages A and B. Assemblage A consists of approximately 61 species and is restricted to the lower 250 m of the La Meseta Formation. The molluscs of this assemblage have stronger affinities with the lower Tertiary of New Zealand than of southern South America. Assemblage B is restricted to the upper 100 m of the formation and is considerably less diverse (25 species) and quite distinct from assemblage A. Only 2 species seem to be common to the two assemblages; however, closer study may show that there are no species common to both assemblages. In contrast to assemblage A, assemblage B has closer faunal affinities with southern South America.

A provisional biostratigraphic zonation is proposed here for the molluscan fauna of the La Meseta Formation, based on the first occurrences of members of the gastropod family Struthiolariidae. Assemblage A is divisible into a lower *Antarctodarwinella ellioti* zone and an upper *Antarctodarwinella nordenskjoldi* zone. Only a single zone (*Perissodonta laevis* zone) is presently recognized in assemblage B. These biostratigraphic zones are considered to be provisional and of only local significance at present. Should the same zonal sequence be recognized elsewhere, then there zones would be considered to have temporal significance and would be raised to formal chronostratigraphic zones.

The presence of *Eutrephoceras argentinae* del Valle and Fourcade and an undescribed species of *Aturia* from assemblage A, similar to the Eocene species *A. bruggeni* Ihering from Tierra del Fuego, indicates an Eocene age for the La Meseta Formation. A late Eocene to possible earliest Oligocene age had previously been assigned to the fauna based on fossil penguins. All the fossil penguin material comes from the upper 100 m of the sequence and is associated with molluscs of assemblage B; hence, the uppermost part of the La Meseta Formation may be early Oligocene in age. Palynomorphs suggest that the La Meseta Formation is correlative with the Rio Turbio Formation of southwestern Patagonia and the Leña Dura Formation of the Brunswick Peninsula of Magallanes.

During the austral summer of 1974–75, a joint field party from the Institute of Polar Studies and the Instituto Antarctico Argentino visited Seymour Island, located on the northeast side of the Antarctic Peninsula (Figure 33.1). The original objective of the expedition was to survey the Upper Cretaceous sequence on James Ross Island, but logistic problems prevented any work there. As a result, a geologic survey of the Tertiary sequence at the northern end of Seymour Island was undertaken, and a large collection of fossil invertebrates was made. This collection represents the only fossil material from Seymour Island except for several small collections made by the Larsen Expeditions (1892–93), the Swedish South Polar Expedition (1901–03), and the British Antarctic Survey party in 1946. Preliminary examination of this material provides significant new data on the biogeography of the southern circum-Pacific during Late Cretaceous and early Tertiary time.

* Isla Vice Comodoro Marambio on Argentinean maps.

Stratigraphy

The Tertiary sequence on Seymour Island was first described by Andersson (1906). He proposed the name *Seymour Island Series* for the loosely consolidated sandstone and silty standstone of Tertiary age at the north end of the island. Elliot and Trautman (Chapter 32, this volume) have recommended that the use of Seymour Island Series be dropped in favor of *Seymour Island Group,* which includes the Cross Valley Formation and the La Meseta Formation. Since this paper is primarily concerned with the La Meseta Formation, only a brief description of the Cross Valley Formation will be given.

The Cross Valley Formation crops out in a series of fault slices in Cross Valley, south of the meseta, and at the north end of the island near Cape Wiman. It consists of approximately 105 m of immature sandstone and pebbly sandstone unconformably overlying the Upper Cretaceous Snow Hill Island Series. The sandstone contains a high percentage of volcanic glass and pumice. Locally abundant

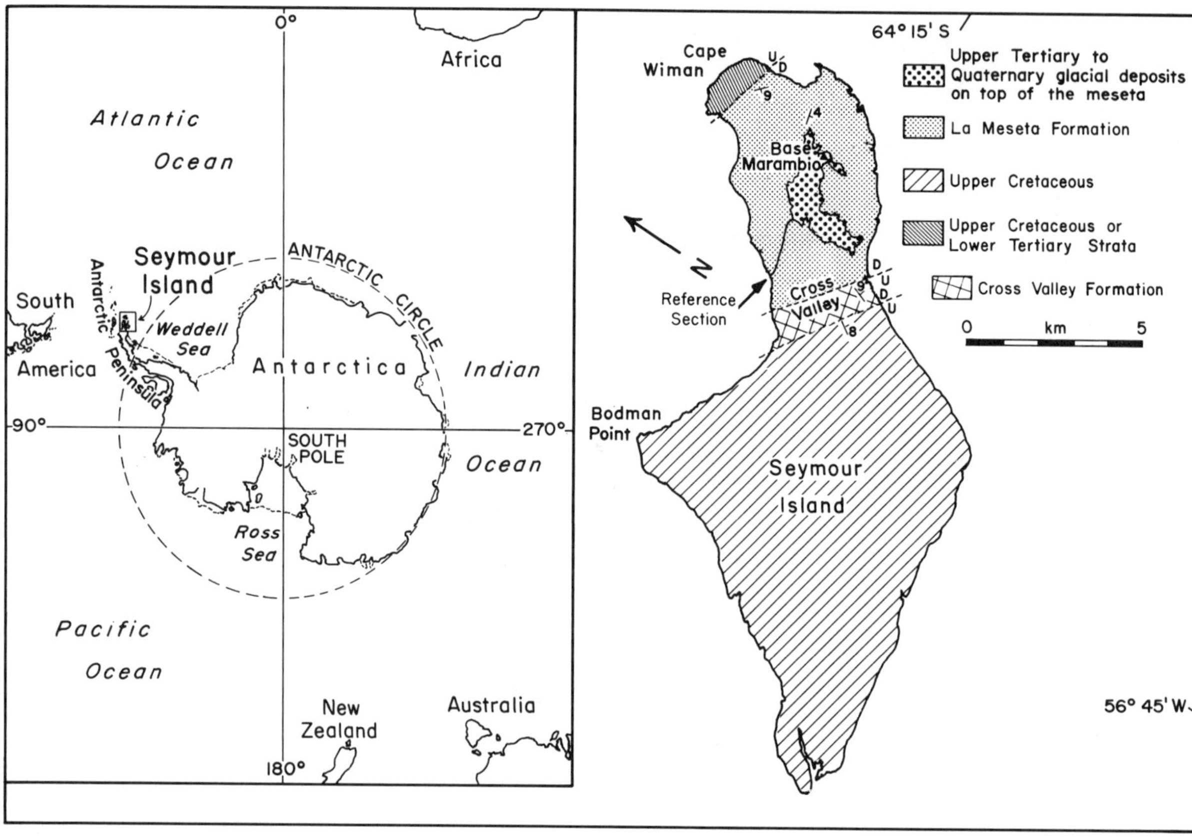

Figure 33.1. Geology and location of Seymour Island.

concentrations of fossil wood and coalified plant debris occur in the lower part of the unit. The Cross Valley Formation is interpreted as representing a nonmarine deltaic deposit (Elliot and Trautman, Chapter 32, this volume). A small collection of marine molluscs and microfossils from a locality near the top of the Cross Valley Formation indicates that the uppermost part is marine in origin.

The La Meseta Formation consists of at least 450 m of loosely consolidated sandstone and sandy siltstone with interbedded conglomerate. Elliot and Trautman have informally divided the formation into three lithologic units. Unit I consists of at least 160 m of unconsolidated fine-grained sand and silty sand with fossils occurring sporadically throughout. Unit II consists of approximately 200 m of fine-grained laminated sand with prominent discontinuous fossiliferous conglomerate beds which may extend for more than 1 km along strike. These shell banks are interpreted as having formed in a high-energy nearshore environment. Local cross-bedding, oscillation ripple marks, and small to large cut-and-fill channels are present in the finer grained facies of this unit, with small conglomeratic lenses occurring locally in the bottom of the cut-and-fill structures. These small concentrations of coarse fossiliferous material are believed to be current deposits in the bottoms of these channels.

The lithic composition of the upper 80 to 125 m of the La Meseta Formation, which corresponds to Unit III of Elliot and Trautman, consists of fine-grained sandstone with intervals of fine clay and sandy, pebbly beds. The fauna of Unit III differs significantly from the fauna of Units I and II; these faunal differences are discussed in detail below. The faunal differences are of such a large magnitude that the authors feel that a hiatus may be present between Units II and III.

Age and correlation of the La Meseta Formation

There has been considerable discussion of the age of the Tertiary sequence on Seymour Island. Dusén (1908) and Wilckens (1911), based on the material collected by the Swedish South Polar Expedition, regarded the Seymour Island Series as Oligocene to Miocene in age. The consensus of later investigators working in the Antarctic Peninsula (Adie, 1957, 1958; Harrington, 1965) was also that the Seymour Island Series was Oligocene to Miocene and more probably the latter. Cranwell (1969), however, has considered the age of the sequence to be Paleocene based on the palynomorphs. Simpson (1971), in a review of fossil penguins, has stated that ages as old as Paleocene or as young as late Miocene were improbable. He has suggested that the penguins were probably late Eocene in age but could be as old as middle Eocene or as young as early Oligocene.

Zinsmeister (1977), in a discussion of the aporrhaid gastropods of Seymour Island and New Zealand, has pointed out that the differences in opinion of the age of the Seymour Island Series stemmed from the lack of stratigraphic data on the collection localities of the Swedish South Polar Expedition. Stratigraphic and faunal data collected during the 1974–75 season indicate that the age of the Seymour Island Group ranges from Paleocene to possible earliest Oligocene.

Cranwell (1959) in her initial paper on the palynomorphs of Seymour Island has described the composition of the material she studied as "tuffaceous . . . (with a) large percentage of curved glass shards." The only place on the island where large amounts of volcanic glass are present is in the Cross Valley Formation (Elliot and Trautman, Chapter 32, this volume). Hall (1977) has examined two specimens from the Cross Valley Formation and stated that the dinoflagellate flora indicated a Paleocene to possibly Maestrichtian age. Since the Cross Valley Formation unconformably overlies the youngest known Cretaceous in the James Ross Island area, it seems likely that the Cross Valley Formation is Paleocene and that the Cretaceous floral elements mentioned by Hall and Cranwell probably represent reworked Cretaceous material. A small molluscan fauna from the uppermost part of the Cross Valley Formation is distinct from the molluscs of the underlying Upper Cretaceous rocks and of the overlying Eocene sedimentary rocks. The small size of the collection and poor state of preservation of the fauna from the Cross Valley Formation prohibit definitive data on the age of the formation.

The molluscan fauna of the La Meseta Formation is exceptionally diverse for high-latitude faunas and provides new insight into the age of the Tertiary rocks on Seymour Island. The presence of two nautiloids (*Aturia* n.sp. and *Eutrephoceras argentinae* del Valle and Fourcade) supports Simpson's (1971) contention of an Eocene age for the La Meseta Formation (Zinsmeister, 1978). *Aturia* n. sp. from Unit II is very similar to *A. bruggeni* Ihering from Eocene rocks of Tierra del Fuego. The occurrence of the genus *Eutrephoceras* in the La Meseta Formation is very important for the determination of the age of the fauna. *E. argentinae* del Valle and Fourcade occurs in Units I and II but in most cases is poorly preserved. One exceptionally well preserved specimen from Unit I was found in the sea cliffs near Cape Wiman. This species is very similar to *E. allani* Flemming from the Eocene of New Zealand. It is, however, believed to be distinct because of differences in suture and shell outline. *E. victorianus* Teichert from the Paleocene of southeast Australia also seems to be closely related, but distinct. The genus *Eutrephoceras,* except for a single species in Oligocene beds from the west coast of North America, is confined to the Paleocene and Eocene of the western hemisphere. Several nautiloids from Oligocene and Miocene strata of Australia, originally described as belonging to the genus *Nautilus,* have recently been referred to the genus *Eutrephoceras* in a catalogue of Australian Tertiary molluscs (Darragh, 1970). A detailed study of these Australian nautiloids is needed to determine if these middle Tertiary species actually belong to the genus *Eutrephoceras.* At the present time, their placement with *Eutrephoceras* is considered questionable. The occurrence of *Aturia* n. sp. and *E. argentinae* del Valle and Fourcade indicates that the age of the La Meseta Formation is Eocene.

Hall (1977) has examined a number of specimens from the La Meseta Formation for palynomorphs. The dinoflagellate floras from Units I and II are characterized by the presence of *Aerosphaeridium kiklyoplokus, Deflandrea fuegiensis, D. macmurodensis, Hystrichasphaeridium tubiferum, Spinidinum aperturum,* and *Tubiosphaera filosa.* Hall has stated that these floras are indicative of a late Eocene–early Oligocene age. The only penguin material collected during the 1974–75 season came from the uppermost parts of Unit II and Unit III. In addition, *Zeuglodon,* a whale known only from the Eocene, was found in Unit III associated with the penguin fossils. Based on all the available fossil evidence, the La Meseta Formation seems to be late Eocene and possibly earliest Oligocene in age.

It should be noted that molluscan material available to Wilckens (1911) came only from Unit III. For some reason, no material was collected from the lower part of the La Meseta Formation by the Swedish South Polar Expedition. Wilckens's incorrect Oligocene–Miocene age assignment for the Seymour Island Series was due to the lack of precise chronostratigraphic knowledge of the Tertiary molluscan faunas of southern South America. He based his comparisons of the Seymour Island fauna on the Patagonian Formation of southern Argentina, which at that time was considered to be Oligocene–Miocene in age. Feruglio (1949–50) accepted Wilckens' correlation of the Tertiary of Seymour Island faunas with the Patagonian Formation but noted that the molluscan resemblance was on a generic rather than a specific level.

Biostratigraphic investigations have since demonstrated that the Patagonian Formation consists of a number of distinct, intricately interbedded, marine and nonmarine units, and that the age is older than previously thought. Camacho and Fernandez (1956) have recorded the presence of several species of bivalves belonging to the cosmopolitan Eocene *Verericardia planicosta* group from the lower part of the Patagonian Formation in central Patagonia near Comodora Rivadavia. The presence of this group of bivalves indicates that the lower part of the Patagonian Formation is Eocene in age. In a later synthesis of the biostratigraphy, Camacho (1974) has considered the lower two-thirds (San Julian Formation and the overlying *Estrata con Monophoraster y Venericor*) to be Eocene in age and the uppermost part (Monte Leon Formation) to be late Oligocene in age. Radiometric data from the Tertiary of Patagonia support an Eocene age for the lower part of the Patagonian Formation (Marshall and others, 1977).

Although information on molluscan faunas is insufficient at present to allow for precise correlation of the Seymour Island fauna with those of southern South America, microfloral evidence suggests that the Seymour Island Group is correlative with the Rio Turbio Formation of southwest Patagonia and the Leña Dura Formation of the Brunswick Peninsula of Magellanes (Hall, 1977; Archangelsky, 1969).

Molluscan fauna of the La Meseta Formation

The molluscan fauna of Seymour Island is considerably more diverse than previously thought. Wilckens (1911) described 26 species of gastropods and bivalves that were collected by the Swedish South Polar Expedition. Preliminary study of the material collected during the 1974–75 season indicates that the total fauna of the La Meseta Formation will eventually approach 100 taxa.

The fauna can be divided into two distinct assemblages. For convenience, these two assemblages will be informally referred to as A and B. Assemblage A, which is restricted to Units I and II, consists of 33 gastropods, 25 bivalves, and 2 nautiloids (Table 33.1). The most characteristic taxa of this assemblage are: *Cucullaea* n. sp., *Antarctodarwinella ellioti* Zinsmeister, *A. nordenskjoldi* (Wilckens), *Struthioptera camachoi* Zinsmeister, and a number of undescribed buccinid-like gastropods. Assemblage B, which is restricted to Unit III, is considerably less diverse and quite distinct from assemblage A (Table 33.2). Only 2 species appear to be common to both assemblages (Table 33.3). None of the characteristic species so abundant in assemblage A occur in B. *Struthiolarella variabilis* Wilckens, *Perissodonta laevis* (Wilckens), and *Cucullaea donaldi* Sharman and Newton are numerically the most abundant in assemblage B, but they are not present in assemblage A.

The abrupt faunal change between Units II and III suggests the presence of a hiatus or a marked change in facies; this is supported by a change in rock type as well. Until additional field work is done to determine the stratigraphic relationships between Units II and III, it is not possible to confirm the existence of an unconformity. Shackleton and Kennett (1975) noted an abrupt drop of 5°C in the temperature of deep-sea bottom water at the end of Eocene time. The change in faunas between Units II and III may reflect this sudden decrease in sea temperatures at the end of Eocene time.

In review of the gastropod family Struthiolariidae from Seymour Island, Zinsmeister and Camacho (1980) have divided the La Meseta Formation into three biostratigraphic units. These units are based on evolutionary and migratory first occurrences within the struthiolariid fauna. They are considered to be informal, local-range zones, valid only on Seymour Island. If a similar succession of units is found elsewhere, these units would then assume temporal significance and represent valid chronostratigraphic units.

Table 33.1. Faunal list — assemblage A (late Eocene)

Bivalvia
Cucullaea n. sp.
Cyclorisimina n. sp.
Eucrassatella n. sp.
Eurhomalia antarctica (Sharman and Newton)
Eurhomalia newtoni (Wilckens)
Eurhomalia n. sp. A
Gaimardia n. sp.
Gomphina n. sp.
Lahillia n. sp.
Mactra irizari Wilckens
Mactra (Longimactra) n. sp.
Modiolus n. sp. A
Modiolus n. sp. B
Mytilus n. sp.
Nielo n. sp.
Nucula nova Wilckens
Nucula n. sp.
Nuculana sp.
Ostrea n. sp. A
Panopea n. sp.
Periploma n. sp.
Portlandia n. sp.
Saxolucina sharmani (Wilckens)
Solemya n. sp.
Tawera lunulifera (Wilckens)
Gastropoda
Antarctodarwinella ellioti Zinsmeister
Antarctodarwinella nordenskjoldi (Wilckens)
Antepepta n. sp.
Austrotoma n. sp. A
Austrotoma n. sp. B
Bathytoma n. sp.
n. gen. n. sp. aff. *Buccinium*
Bulla n. sp.
Calyptraea (Trochita) n. sp.
Cominella aff. *C. obesa* Steinman and Wilckens
Comitas n. sp.
Cancellaria n. sp.
n. gen. n. sp. aff. *Cosmasyrinx*
n. gen. n. sp. aff. *Falsicolus*
n. gen. n. sp. A aff. *Ficus*
n. gen. n. sp. B aff. *Ficus*
Globisium n. sp.
Marshallaria n. sp.
Melampus n. sp.
Melonoides n. sp.
n. gen. n. sp. aff. *Mitra*
Monilopsis n. sp.
Pristimerica n. sp.
Proscaphella n. sp. A
Polinices n. sp.
Pseudofax n. sp.
Pyramidella n. sp.
Steptochetus n. sp.
Struthioptera camachoi Zinsmeister
Struthiolarella n. sp.
n. gen. n. sp. aff. *Speightia*
Turbinella (Ptycheulimella) n. sp.
Turbinella n. sp.
Cephalopoda
Aturia n. sp.
Eutrephoceras argentinae del Valle and Fourcade
Scaphopoda
Cadulus n. sp.

Table 33.2. Faunal list — assemblage B (latest Eocene to possibly earliest Oligocene)

Bivalvia
Cucullaea donaldi Sharman and Newton
Euburneopecten? sp.
Entodesma tenuis (Wilckens)
Eurhomalia n. sp. B
Eurhomalia n. sp. C
Lahillia sp.
Ostrea n. sp. B.
Lucina sp.
Panopea nucloides
Pinna n. sp.
Gastropoda
Bulla glacialis Wilckens
Calyptraea n. sp.
Chrysodomus divericosta Wilckens
Cominella sp.
Cylichnina n. sp. A
n. gen. n. sp. aff. *Ficus*
Fusinus n. sp.
n. gen. aff. *Nassa nordenskjoldi* Wilckens
Perissodonta laevis (Wilckens)
Polinices cf. *P. subtenuis* Ihering
Proscaphella n. sp. B
Struthiolarella variabilis Wilckens
Turricula n. sp.
Turritella n. sp.
Scaphopoda
Dentalium pulchrum Wilckens

Table 33.3. Species common to assemblages A and B

Gastropoda
Turritella (Colposigma) exigua Ortmann
Cylichnina n. sp. B

The type section for this biostratigraphic zonation is located on the northwestern side of Seymour Island, extending from the mouth of Cross Valley northeastward to the top of the meseta near the southeast end of the aircraft runway at the Argentine station of Vicecomodoro Marambio (Figure 33.1). This section corresponds in part to Elliot and Trautman's section 3 (Chapter 32, this volume). The lowest unit is defined by the local range of *Antarctodarwinella ellioti* Zinsmeister. The *Antarctodarwinella ellioti* zone is characterized by the occurrence of *Nucula nova* Wilckens, *Mytilus* n. sp., *Periploma* n. sp., *Austrotoma* n. sp., and *Cucullaea* n. sp. Specimens of *Struthioptera camachoi* Zinsmeister are occasionally encountered, but they become very abundant in the *Antarctodarwinella nordenskjoldi* zone. The upper boundary of the *A. ellioti* zone is defined by the evolutionary first occurrence of *A. nordenskjoldi* (Wilckens). The fauna of the *A. nordensjoldi* zoni is characterized by the occurrence of *Lahillia* n. sp., *Struthiolarella* n. sp., *Aturia* n. sp., *Calyptraea* n. sp., and *Nielo* n. sp. The upper boundary of the *A. nordenskjoldi* zone is recognized by a migratory first occurrence of *Perissodonta laevis* (Wilckens). The third uppermost zone encompasses Unit III. The upper boundary of this *Perissodonta laevis* zone cannot be defined because of the unconformable relationship between the La Meseta Formation and the overlying Holocene glacial deposits.

The extreme individuality of the La Meseta fauna not only specifically, but even generically, makes faunal comparisons with other Southern Hemisphere faunas extremely difficult. In two short papers, Zinsmeister (1976, 1977) did note some striking generic similarities between the Seymour Island fauna and the Paleocene Wangaloan faunas of New Zealand. These notes, along with comments of earlier workers about close similarities between Late Cretaceous and early Tertiary faunas of the Southern Hemisphere, led Zinsmeister (1979) to propose a broad, shallow-water faunal province along the southern circum-Pacific belt, extending from southern South America along West Antarctica to southeast Australia and New Zealand. The name Weddellian Province was proposed for this Late Cretaceous–early Tertiary marine province.

Preliminary examination of the La Meseta fauna reveals that although Seymour Island and southern South America were geographically close during early Tertiary time, the faunal affinities are surprisingly weak, while affinities with New Zealand are very strong. Many of the genera (*Perissodonta, Struthioptera, Antarctodarwinella,* and a number of undescribed genera) which are abundant on Seymour Island are absent in equivalent faunas from Tierra del Fuego and Patagonia. These apparent faunal differences may be due either to incomplete knowledge of the fossil record or to ecologic factors related to differences in latitude between the regions. Although the faunas of southern South America have not been studied in detail since the early part of the century, the absence of so many Seymour Island elements in South America when shallow-water conditions presumably existed suggests that environmental factors due to differences in latitude were primary causes for the observed faunal dissimilarities. This condition seems to have changed abruptly during latest Eocene and earliest Oligocene time. Assemblage B, however, has a number of elements closely related to forms in South America. This change in faunal affinities may be related to sudden cooling in the southern oceans near the end of Eocene time. The sudden decrease in temperature may have reduced the environmental differences between the Antarctic Peninsula and South America during early Tertiary time.

Conclusions

The late Eocene molluscan fauna of the La Meseta Formation is considerably more diverse than previously thought. Preliminary examination of this fauna provides significant new data concerning the Late Cretaceous–early Tertiary biogeography of the southern circum-Pacific region.

The fauna of the La Meseta Formation can be divided into two distinct assemblages. Assemblage A, restricted to Units I and II of Elliot and Trautman, is quite diverse (61 species) and has strong affinities with lower Tertiary faunas of New Zealand. The affinities of assemblage A with the faunas of southern South America are surprisingly weak. Assemblage B, restricted to Unit III, is considerably less diverse (25 species) and is distinct from assemblage A. Assemblage B has much closer faunal affinities with southern South American faunas than does A. Two possible causes for the sudden change in faunas between Units II and III were considered. The faunal differences between assemblages A and B may be due to an unconformity, or they may reflect a sharp decrease in ocean temperature which occurred at the end of Eocene time.

Three provisional biostratigraphic zones (*Antarctodarwinella ellioti, Antarctodarwinella nordenskjoldi,* and *Perissodonta laevis* zones) are recognized in the La Meseta Formation based on evolutionary and migrational first occurrences in the gastropod family Struthiolariidae. Limited knowledge of the molluscan faunas in southern South America prevents precise correlation between rocks there and on Seymour Island. Palynomorphs from Seymour Island, however, suggest that the Rio Turbio Formation of southwestern Patagonia and the Leña Dura Formation of Tierra del Fuego are correlative with the La Meseta Formation.

Acknowledgments. We thank David Elliot, Institute of Polar Studies, Ohio State University, for his help and advice during the course of this study. We also express gratitude to Argentina's Direccion Nacional del Antartico and Instituto Antartico Argentino for field support during the 1974–75 field season. This work was supported by National Science Foundation Grant OPP74-21509 to the Ohio State University, Institute of Polar Studies.

References cited

Adie, R. J. 1957. Geological investigations in the Falkland Island Dependencies before 1940. *Polar Rec., 8,* No. 57, 502–13.

Adie, R. J. 1958. Geological investigations in the Falkland Island Dependencies since 1940. *Polar Rec., 9,* No. 58, 3–17.

Andersson, J. G. 1906. On the geology of Graham Land. *Bull. geol. Instn Univ. Upsala, 7,* 19–71.

Archangelsky, S. 1969. Estudio del paleomicroplancton de la formacion Rio Turbio (Eocene), Provincia de Santa Cruz. *Ameghiniana, 4,* No. 3, 181–218.

Camacho, H. H. 1974. Biostratigrafía de la formaciones marinas del Eoceno y Oligoceno de La Patagonia. *Annls Acad. Nac. cienc. Exact. Fis. Nat., Buenos Aires, 26,* 39–57.

Camacho, H. H., and Fernandez, J. A. 1956. La transgresíon en la costa Atlántica entre Comodoro Rivadavia y el curso inferior del rio Chubut. *Revta Asoc. geol. argent., 11,* 23–45.

Cranwell, L. M. 1959. Fossil pollen from Seymour Island, Antarctica. *Nature, Lond., 184,* No. 4701, 1782–85.

Cranwell, L. M. 1969. Antarctic and circum-Antarctic palynological contribution. *Antarct. Jnl U.S., 4,* 197–98.

Darragh, T. 1970. Catalogue of Australian Tertiary Mollusca (except Chitons). *Mem. Natn Mus. Vict.,* No. 31, 125–212.

Dusén, P. 1908. Uber die Tertiare Flora der Seymour Insel. *Wiss. Ergebn. Schwed. Sudpolarexped. 1901–1903, 3,* No. 3, 1–27.

Feruglio, E. 1949–1950. *Descripcíon geologica de la Patagonia.* 3 vols. Buenos Aires, Direccion General de Yacimentos Petroliferos Fiscales.

Hall, S. A. 1977. Cretaceous and Tertiary dinoflagellates from Seymour Island, Antarctica. *Nature, Lond., 267,* No. 5608, 239–41.

Harrington, H. J. 1965. Geology and morphology of Antarctica. (*In* Oye, P. Van, and Mieghem, J. Van, *eds.* Biogeography and Ecology in Antarctica. *Monographia biol., 15,* 1–71.)

Marshall, L. G., Pascual, R., Curtis, G. H., and Drake, R. E. 1977. South American geochronology: radiometric time scale for Middle to Late Tertiary mammal-bearing horizons in Patagonia. *Science, N.Y., 195,* 1325–28.

Shackleton, N. J., and Kennett, J. P. 1975. Paleotemperature history of the Cenozoic and the initiation of Antarctic glaciation: oxygen and carbon isotope analysis in DSDP Sites 277, 279, and 281. (*In* Kennett, J. P., and Houtz, R. E. *Initial Report Deep Sea Drilling Project,* Vol. 27. Washington [U.S. Government Printing Office], 743–56.)

Simpson G. G. 1971. Review of fossil penguins from Seymour Island. *Trans. R. Soc. Lond.,* Ser. B, *178,* 357–87.

Wilckens, O. 1911. Die mollusken der antarktischen Tertiar formation. *Wiss. Ergebn. Schwed. Sudpolarexped., 1901–1903, 3,* 1–62.

Zinsmeister, W. J. 1976. A new genus and species of the gastropod family Struthiolariidae *Antarctodarwinella ellioti,* from Seymour Island, Antarctica. *Ohio J. Sci., 76,* 111–14.

Zinsmeister, W. J. 1977. Note on a new occurrence of the Southern Hemisphere aporrhaid gastropod *Struthioptera* Finlay and Marwick on Seymour Island, Antarctica. *J. Paleont., 51,* No. 2, 399–404.

Zinsmeister, W. J. 1978. Eocene nautiloid fauna from the La Meseta Formation of Seymour Island, Antarctic Peninsula. *Antarct. Jnl U.S., 13,* 24–25.

Zinsmeister, W. J. 1979. Biogeographic significance of the late Mesozoic and early Paleogene molluscan faunas of Seymour Island (Antarctic Peninsula) to the final breakup of Gondwanaland. (*In* Gray, J., and Boucot, A. J., *eds.* Historical biogeography, plate tectonics, and the changing environment. *37th Biological Colloquim of the Oregon State University.* Corvallis, University of Oregon Press, 347–55.)

Zinsmeister, W. J., and Camacho, H. H. 1980. Late Eocene Struthiolariidae (molluscan: Gastropoda) from Seymour Island, Antarctic Peninsula and their significance to the biogeography of early Tertiary shallow-water faunas of the Southern Hemisphere. *J. Paleont., 54,* 1–14.

34

The Pattern of Antarctic Peninsula Plutonism

ANDREW D. SAUNDERS*, STEPHEN D. WEAVER**, and JOHN TARNEY***
Department of Geological Sciences, University of Birmingham, Birmingham B15 2TT, U.K.

Abstract. Calc-alkaline cordilleran plutons dominate the geology of the northern part of the Antarctic Peninsula (Graham Land). The plutons have previously been divided into two groups: one is older than the extensive but poorly dated "Upper Jurassic Volcanic Group"; the other is younger and generally termed the "Andean Intrusive Suite." Radiometric dates indicate that plutonism was active at least from early Jurassic until late Tertiary time, a range similar to that for plutons of southern South America. A concentration of dates between 90 and 110 m.y. ago is thought to represent a peak of plutonic activity. Spatially and temporally distinct belts of intrusive rocks can be identified: an older group (155–185 m.y. old), restricted to the east coast, and a younger group (45–60 m.y. old), restricted to the northern part of the west coast of the peninsula. Plutons of intermediate ages (70–145 m.y.) occur throughout the peninsula. Existing data are insufficient to establish whether the 45–185 m.y. plutonic episode was essentially continuous or occurred as a number of separate magmatic pulses. However, on a broad scale the foci of plutonic activity seem to have migrated both across and along the peninsula through time, from east to west and from south to north.

Plutonic rocks range in composition from gabbro to granite, with west coast acid intrusive rocks tending to be more trondhjemitic in composition than those from the east coast. East coast plutons have higher K_2O, Rb, Th, and Ba contents than west coast plutons of equivalent composition. Contents of the light rare-earth elements are higher in the east coast rocks, resulting in higher Ce/Y ratios and more fractionated rare-earth patterns. Such variation is consistent with subduction along the west (Pacific) coast of the peninsula and derivation of calc-alkaline magmas from progressively greater depths away from the trench. The petrogenesis of the plutonic suite as a whole is complex, and it is difficult to relate geochemical variations within single age or regional groups to simple igneous fractionation processes. Multistage petrogenetic models are proposed for the generation of cordilleran batholiths.

Plutonic igneous rocks are an important feature of the Antarctic Peninsula, although individual plutons do not assume the dimensions of the batholiths of southern South America, even allowing for the cover of snow and ice. This may of course be simply a function of erosion level. Previous regional studies have been hampered by lack of quantitative data, particularly radiometric dates, and consequently there has been strong reliance on field evidence concerning the relationships between the plutons and the adjacent volcanic and sedimentary rocks. Ages of the adjacent strata in many cases are poorly known. Comparisons with southern Chile are useful in that the geologic histories of the two areas are broadly similar and have led to suggestions that the two areas were in some way continuous during Mesozoic and early Tertiary time (see, for example, Dalziel and Elliot, 1973).

The recently published compendium of new radiometric ages of Antarctic Peninsula igneous rocks by Rex (1976), taken in conjunction with earlier studies, now provides a general chronologic framework on which to base a geochemical study of Antarctic Peninsula plutonism. Accordingly, we have recently analyzed several hundred samples of plutonic rock specimens, collected by British Antarctic Survey geologists over the last decade, in order to provide an overall picture of the geochemical characteristics of Antarctic Peninsula intrusive rocks. This paper is not, therefore, concerned with the geochemical features of individual plutons, but rather with regional differences both across and along the peninsula, and with any time-dependent variations.

The tectonic setting of the peninsula is unusual in that it represents a relatively narrow strip of sialic crust bordered on both sides by oceanic crust. The equivalent plutonism in southern Chile is, on the other hand, set against a major continental margin. Nowhere along the margin of southern South America or along the peninsula is subduction presently taking place, although the active volcanism in Bransfield Strait at the northwest end of the peninsula may be indirectly connected with the last stages of subduction of Drake Passage ocean crust beneath the South Shetland Islands (Barker, 1976; Weaver and others, Chapter 29, this volume). The widespread plutonism may be linked to the subduction of Pacific Ocean crust under the peninsula during Mesozoic and Cenozoic time. By studying the magnetic lineations in the Southeast Pacific Ocean, Herron and Tucholke (1976) have shown that subduction must have occurred during early Cenozoic time and probably earlier, but that subduction stopped during late Paleocene to early Miocene time when the West and East Antarctic plates became coupled.

* Present address: Department of Geology, Bedford College, London NW1 4NS, U.K.

** Present address: Department of Geology, University of Canterbury, Christchurch, New Zealand.

*** Present address: Department of Geology, University of Leicester, Leicester, LE1 7RH, U.K.

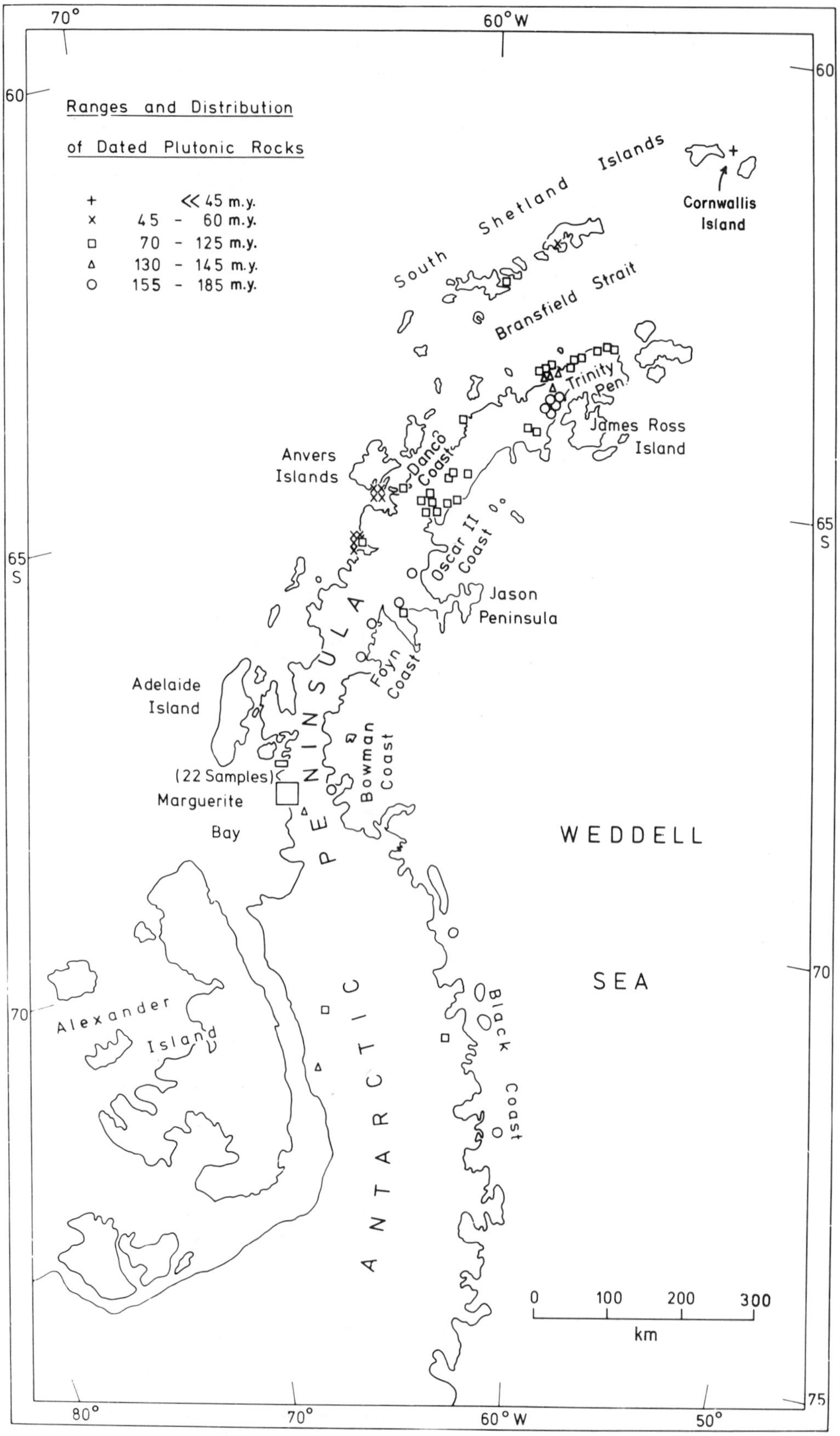

Figure 34.1. Map of the Antarctic Peninsula showing the locations of radiometrically dated plutons. The four age groups between 45 and 185 m.y. old were taken from data shown in Figure 34.2. Data sources are given in the text.

Age relationships

Adie (1954, 1955) presented the first extensive accounts of the plutonic rocks of the Antarctic Peninsula. On the basis of field criteria, especially the relationship of the plutonic rocks to the widespread "Upper Jurassic Volcanic Group," he divided the intrusions into two groups: an earlier "late Paleozoic suite" and a later "Andean Intrusive Suite." Comparisons with Chile led Adie (1955, p. 4) to consider that the second group might be late Cretaceous or early Tertiary in age.

There is as yet little positive evidence for Paleozoic magmatism in the Antarctic Peninsula. Similarly, there is little evidence for Paleozoic plutonic activity in southernmost Chile, although Paleozoic plutons are common in central and northern Chile (Farrar and others, 1970). Field relationships in the Orford Cliff area on the Loubet Coast (Goldring, 1962) and in the Marguerite Bay area (Adie, 1954) suggest a Paleozoic age for some intrusive rocks, but in fact the "Paleozoic" granites from Marguerite Bay have given Rb-Sr whole-rock ages between 98 and 117 m.y. (Halpern, 1972). Rex (1976) has reported three K-Ar ages ranging from 351 to 384 m.y. on mica from a granite on Tabarin Peninsula at the northern end of the Antarctic Peninsula, but these dates await confirmation by Rb-Sr methods.

Later workers have substantiated Adie's (1955) claim of widespread occurrence of plutonic rocks on the peninsula (see Goldring, 1962; Hooper, 1962; Elliot, 1964; Curtis, 1966; Fleet, 1968; Marsh, 1968; Stubbs, 1968; Dewar, 1970; West, 1974; Aitkenhead, 1975), although most rocks belong to the Andean Intrusive Suite on the basis of field and radiometric criteria. However, Fraser and Grimley (1972) have suggested a late Jurassic to Tertiary time span for the Andean Intrusive Suite on the Bowman and Wilkens Coasts, and the occurrence of earlier Lower to Middle Jurassic plutons on the Oscar II, Foyn, and Bowman Coasts has been reported by Marsh (1968) and Stubbs (1968).

A total of 85 radiometric dates on Antarctic Peninsula plutons is now available (Halpern, 1965, 1967; Scott, 1965; Grikurov and others, 1966, 1970; Rex, 1976; and R. J. Pankhurst, personal communication), and these ages are summarized in Figures 34.1 and 34.2. The distribution of radiometric dates throughout the peninsula (Figure 34.1) shows two important features. First, the older intrusive rocks (155–185 m.y. old) are restricted to the east coast (Foyn, Oscar II, Bowman, and Wilkens Coasts), and second, the younger intrusive rocks (45–60 m.y.) are restricted to the west coast, particularly the northern sector. The youngest dated plutonic rock in the peninsula region is the grandiorite of Cornwallis Island near Elephant and Clarence Islands (9.5 m.y.; Rex and Baker, 1973).

In Figure 34.1 the dated intrusive rocks are split into four age groups. These divisions are based on the frequency histogram (Figure 34.2), but whether the groupings are valid or only a function of sampling bias will become apparent with further age determinations. Nevertheless, the maximum at 90 to 110 m.y. on the frequency histogram seems to be a result of the occurrence of many plutons in this broad age range throughout the peninsula. This peak of plutonic activity in mid-Cretaceous time also roughly coincides with a concentration of ages within the range 75–100 m.y. in the Patagonian batholith of southern Chile (Halpern, 1973).

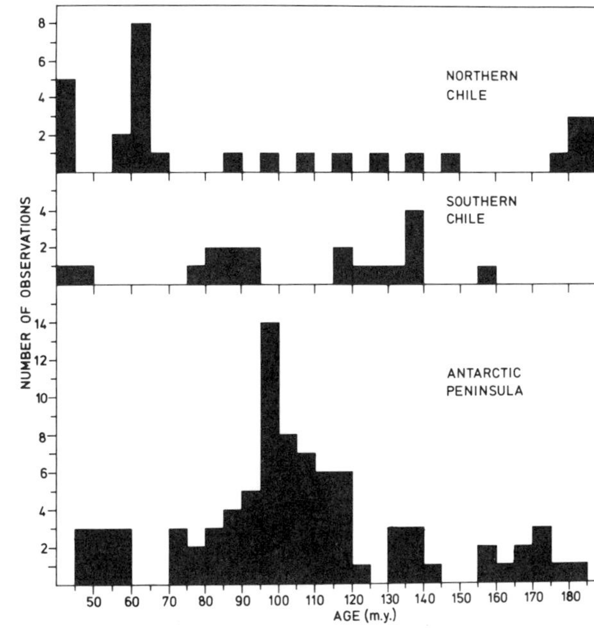

Figure 34.2. Frequency histograms of available age determinations from northern Chile (Farrar and others, 1970; McNutt and others, 1975), southern Chile (Halpern, 1973), and the Antarctic Peninsula (Halpern, 1965, 1967; Scott, 1965; Grikurov and others, 1966, 1970; Rex, 1976; and R. J. Pankhurst, personal communication). All data have been standardized to the following decay constants: $\lambda_\beta = 4.72 \times 10^{-10}$/yr and $\lambda_e = 0.584 \times 10^{-10}$/yr.

A brief comparison with western South America is pertinent at this point since there is a consensus that the Antarctic Peninsula was continuous with the southern Andes before being separated by Cenozoic plate movements (Barker and Griffiths, 1972; Dalziel and Elliot, 1973; Barker, 1976; de Wit, 1977). In northern Chile the magmatic centers in the Copiapo region are arranged into five different age belts parallel to the coast; they decrease in age eastward from Lower Jurassic plutons near the coast to currently active inland volcanoes (Farrar and others, 1970; McNutt and others, 1975). Thus, the focus of plutonic activity has migrated over 120 km eastward since early Jurassic time. In southern Chile the batholith is more than 1000 km long and 100 km wide in places. Radiometric dating (Halpern, 1973) indicates three distinct climaxes of magmatic activity: in late Jurassic to early Cretaceous time (160–115 m.y.), in late Cretaceous time (95–75 m.y.), and in Tertiary time (50–10 m.y.) (see Figure 34.2). The Tertiary intrusions are located mainly in the *eastern* (continental) side of the High Cordillera (Dalziel, 1974). The youngest recorded date (12 m.y.) is from the Cerro del Paine quartz diorite. Halpern (1973) considers the intrusive history in southern Chile to be complex, and that, although the batholith clearly represents several phases of activity, it is difficult to recognize any significant movement of foci through time.

In summary, it seems from Figures 34.1 and 34.2 that, whereas there is broad correspondence in the age patterns of plutonism along the Pacific margin of South America and the Antarctic Peninsula, there is a difference in the spatial arrangements. In northern and central Chile, the plutonic centers have migrated eastward with time, whereas in the Antarctic Peninsula the spatial arrangement is reversed, and the youngest plutonic (and volcanic) rocks

Table 34.1. Representative analyses of Mesozoic-Cenozoic plutonic rocks from the the Antarctic Peninsula (oxides in wt %)

	1	2	3	4	5	6	7	8	9	10	11	12	13	14	15	16	17	18	19	20
SiO_2	69.52	75.39	49.23	56.39	67.87	72.66	76.17	75.68	52.75	59.64	65.64	69.22	73.45	70.24	76.01	50.60	54.00	68.65	70.40	66.52
TiO_2	0.32	0.08	1.43	1.81	0.76	0.32	0.13	0.11	0.57	0.82	0.60	0.39	0.23	0.30	0.18	0.98	1.64	0.46	0.30	0.41
Al_2O_3	15.65	12.67	16.36	16.32	15.25	13.85	12.48	12.13	15.30	17.07	14.15	14.94	14.11	15.45	12.91	18.52	14.91	15.57	15.32	16.29
tFe_2O_3 (total Fe)	3.06	0.95	10.71	8.90	3.71	2.32	1.82	1.48	9.40	6.10	5.39	3.41	1.99	3.02	1.46	7.74	10.04	3.67	2.26	3.42
MnO	0.07	0.05	0.23	0.23	0.13	0.03	0.02	0.04	0.15	0.14	0.10	0.11	0.05	0.06	0.03	0.14	0.19	0.07	0.04	0.11
MgO	0.84	0.53	8.47	4.23	1.13	0.75	0.05	0.03	6.25	1.89	3.43	0.87	0.38	0.33	0.15	6.06	4.66	1.16	0.32	0.21
CaO	2.52	0.59	9.77	7.08	2.44	1.79	0.24	0.75	9.95	5.07	3.32	2.06	1.74	1.81	0.87	10.43	7.41	2.49	0.90	0.59
Na_2O	4.29	4.36	3.80	4.64	6.22	4.32	4.95	4.17	3.07	3.21	2.75	3.79	3.54	4.62	3.13	3.69	4.65	3.84	5.40	6.69
K_2O	3.17	4.34	0.54	0.84	2.43	3.36	4.07	4.20	0.65	2.80	3.20	3.72	4.40	3.72	5.34	0.91	1.73	3.72	4.99	5.24
P_2O_5	0.11	0.03	0.39	0.78	0.20	0.08	0.03	0.03	0.14	0.31	0.12	0.14	0.07	0.07	0.05	0.24	0.46	0.12	0.06	0.07
Total	99.55	98.99	100.93	101.22	100.14	99.48	99.96	98.62	98.23	97.05	98.70	98.65	99.96	99.62	100.13	99.31	99.69	99.75	99.99	99.55
Trace elements (ppm)																				
Cr	<1	2	77	5	4	8	3	4	173	11	220	5	7	3	5	103	39	37	4	3
Ni	6	<1	46	<1	<1	<	<1	<1	26	<1	12	<1	<1	<1	<1	30	3	3	<1	<1
Rb	118	144	20	23	49	86	150	170	21	103	136	130	151	132	328	55	63	137	151	99
Ba	560	584	155	271	752	863	331	705	251	801	661	700	737	786	195	158	280	780	432	37
Sr	261	40	550	587	295	263	22	85	285	682	263	345	231	339	48	586	440	244	59	5
Zr	165	90	86	159	340	192	365	147	92	334	154	198	135	413	128	143	182	218	505	1287
Nb	13	13	6	11	15	7	19	13	2	13	12	13	13	14	12	7	12	15	30	41
Y	22	24	18	43	42	12	45	21	21	29	25	28	23	28	43	23	34	30	36	47
La	13	22	13	22	29	10	56	27	14	26	27	33	22	58	34	11	24	30	36	25
Ce	32	54	28	48	61	31	103	64	32	55	46	67	50	106	74	28	50	62	65	62
Pb	14	21	7	9	10	12	16	17	10	15	21	16	16	21	25	9	19	14	10	10
Th	11	20	<1	2	8	14	23	20	2	9	15	19	20	23	41	2	6	19	22	11
Zn	22	20	87	84	59	20	34	28	81	95	56	50	23	41	17	96	107	27	18	65
Ga	18	15	22	24	23	14	22	18	16	26	18	21	15	22	17	21	23	21	22	33
Na_2O/K_2O	1.35	1.00	7.04	5.52	2.56	1.29	1.22	0.99	4.72	1.15	0.86	1.02	0.80	1.24	0.59	4.05	2.69	1.03	1.08	1.28
K_2O/SiO_2	0.046	0.058	0.011	0.015	0.036	0.046	0.053	0.055	0.012	0.047	0.049	0.054	0.060	0.053	0.070	0.018	0.032	0.054	0.071	0.079
tFe/Mg	4.22	2.08	1.47	2.44	3.81	3.59	42.20	57.19	1.74	3.74	1.82	4.54	6.07	10.61	11.28	1.48	2.50	3.67	8.19	18.88
K/Rb	233	250	224	303	412	324	225	205	257	226	195	238	242	234	135	137	228	225	274	439
Rb/Sr	0.45	3.60	0.04	0.04	0.17	0.33	6.82	2.00	0.07	0.15	0.52	0.38	0.65	0.39	6.83	0.09	0.14	0.56	2.56	19.80
Ba/Sr	2.15	14.60	0.28	0.46	2.55	3.28	15.05	8.29	0.88	1.17	2.51	2.03	3.19	2.32	4.06	0.27	0.64	3.20	7.32	7.40
Ce_N/Y_N	3.55	5.50	3.80	2.73	3.55	6.31	5.59	7.45	3.72	4.63	4.50	5.85	5.31	9.25	4.21	2.97	3.59	5.05	4.41	3.22
Zr/Nb	12.7	6.9	14.3	14.5	22.7	27.4	19.2	11.3	46.0	25.7	12.8	15.2	10.4	29.5	10.7	20.4	15.2	14.5	16.8	31.4

1. O576.2: Adamellite, prevolcanic acid pluton; western Charlotte Bay.
2. O971.1: Granite, prevolcanic acid pluton; eastern Charlotte Bay.
3. O537.1: Gabbro, postvolcanic pluton; 1.0 km to the south of Porro Bluff.
4. O540.1: Tonalite, postvolcanic pluton; 2.5 km to the south of Porro Bluff.
5. O531.3: Granophyre, postvolcanic pluton; northeast of Beneden Head.
6. O904.1: Granophyre, postvolcanic pluton; southern Ronge Island.
7. O940.4: Microgranite, postvolcanic pluton; southwest of Meithe Glacier.
8. TL553.1: Granite (Red Granite); south of Cape Casey.
9. TL593.1: Hypersthene Gabbro, Jurassic pluton; Seligman Inlet.
10. TL644.4: Hornblende-biotite tonalite, Jurassic pluton; south of Flask Glacier.
11. TL514.1: Heterogeneous granodiorite, Jurassic pluton; Cape Casey.
12. TL804.1: Biotite granodiorite, Jurassic pluton; east of "C-Glacier."
13. TL653.1: White adamellite, Jurassic pluton; near Bildad Peak.
14. D4687.1: "White granite," early pluton; Melville and Mapple Glaciers.
15. TL34.2: "White granite," early pluton; Starbuck and Stubb Glaciers.
16. TL567.9: Contact gabbro, "Andean" pluton; Adie Inlet.
17. TL567.7: Contact microgabbro, "Andean" pluton; Adie Inlet.
18. TL584.1: Pink granodiorites, "Andean" pluton; 3.2 km east of Cape Choyce.
19. TL567.2: Pink microgranite, "Andean" pluton; Adie Inlet.
20. TL569.1: Granite, "Andean" pluton; Adie Inlet.

Analyses 1-7 are from the Danco Coast (West, 1974); analyses 8, 9, 11, and 18 are from the Foyn Coast (Marsh, 1968); analyses 10, 12, 13, 16, 17, 19, and 20 are from the Oscar II Coast (Marsh, 1968); and analyses 14 and 15 are from the Oscar II Coast (Fleet, 1968).

lie along the western side of the peninsula. The situation in southern Chile is less well known but seems to be transitional to that in the areas to the north and south, thus possibly lacking a clear pattern.

Geologic considerations

Plutonic rocks analyzed in this study were chosen from British Antarctic Survey collections, mainly from the Danco Coast on the western side of the Antarctic Peninsula, described by West (1974), and from Oscar II, Foyn, and Bowman Coasts on the eastern side, described by Fleet (1968), Marsh (1968), and Stubbs (1968). On the Danco Coast, a suite of early intrusive rocks consisting of granodiorite, adamellite, and granite predates a thick sequence of subaerially erupted tuffs and lavas (West, 1974). These tuffs and lavas have been referred to the "Upper Jurassic Volcanic Group" (Adie, 1972), and hence these early plutons are considered to be pre-late Jurassic in age (see, however, Weaver and others, Chapter 29, this volume). Plutons intruding the Mesozoic volcanic rocks on the Danco Coast range in composition from gabbro to granite, but intermediate rocks seem to be volumetrically unimportant.

Field evidence on the Oscar II and Foyn Coasts suggests that a group of cataclastically deformed red granites predate the main series of plutonic rocks, which ranges from tonalite to adamellite in composition, but with minor gabbros, and which has yielded ages in the range 163–182 m.y. (Rex, 1976). Silicic pyroclastic rocks postdate these plutonic rocks and are themselves cut by a gabbro-microgranite laccolithic body at Adie Inlet and a composite granodiorite-adamellite pluton at Seligman Inlet. The gabbro-microgranite laccolith has yielded an age of 83 m.y. (Rex, 1976). The gabbro is thought to have differentiated, yielding alkali-rich microgranite and granite fractions (Marsh, 1968). To the north along the Oscar II Coast, Fleet (1968) has distinguished a suite of "white granites" which intrude the Mesozoic volcanic rocks. Younger plutons, ranging from gabbro to granite in composition, are referred by Fleet (1968) to the "Andean Intrusive Suite" and have yielded ages in the range 92–105 m.y. (Rex, 1976).

Geochemistry

For the present study, a total of 224 Antarctic Peninsula plutonic rocks has been analyzed using a Philips PW1450 automatic X-ray fluorescence spectrometer. Major elements were determined on 46-mm, pressed-powder disks, calibrations being effected by preparing lithium tetraborate fusion beads on a representative selection of samples and comparing these against fusion beads of international standard and other analyzed rocks. Trace elements were determined on powder disks to give maximum sensitivity and were calibrated against international standard rocks; corrections for mass absorption effects were made using WL_β and MoK_α Compton scatter lines. Rare-earth elements were determined using standard nonseparative neutron activation techniques.

Table 34.1 contains 20 representative analyses covering the range of compositions found, but all of the data have been used in Figures 34.3 to 34.8. Initially, the data were divided into age groups as well as spatial groups in order to determine whether or not there were any significant time-dependent variations. West (1974), for instance, noted that the prevolcanic silicic plutons of the Danco Coast were chemically different from the later (postvolcanic) granophyres, which were found to have higher Ba, Zr, Cu, Zn, Pb, Ce, La, and Y contents. While this may be the case for the Danco Coast area, we have on the other hand noted few significant time-dependent variations among the east coast plutons, where much more data are available. For the purposes of the present paper, therefore, the prevolcanic and postvolcanic plutons are not distinguished separately on the element plots, and we will restrict our comparisons to the compositional differences between east coast and west coast plutons.

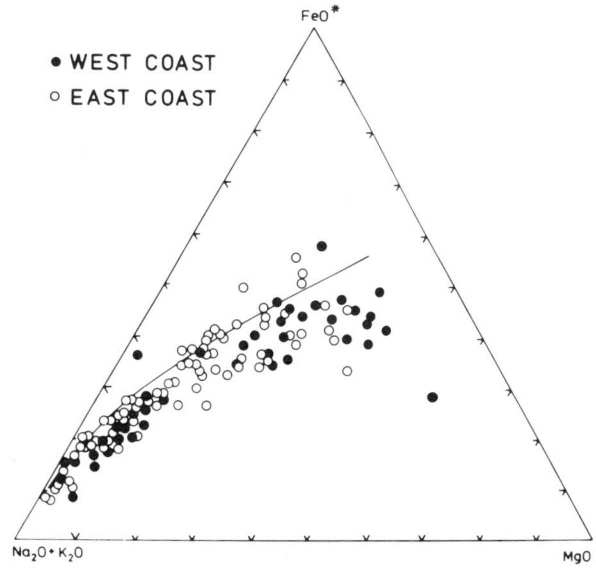

Figure 34.3. AFM diagram showing the calc-alkaline trend of Antarctic Peninsula plutonic rocks. The trend for the Mount Lassen Suite, Cascade Range, U.S.A. (Nockolds and Allen, 1953) is given by the curved line. $FeO^* = FeO + 0.9Fe_2O_3$.

The major element compositions of these plutons are typical of calc-alkaline provinces (Nockolds and Allen, 1953), as shown by the AFM diagram (Figure 34.3). This chemical trend has been noted previously (Adie, 1955; West, 1974) and has been cited by many workers as evidence for consanguinity of rock types (e.g., Adie, 1955). Figure 34.4 shows a normative Or-Ab-An plot for silicic plutons with more than 80% salic constituents and illustrates the relative proportions of rock types, according to the classification of O'Connor (1965). There is a distinct tendency for the west coast plutons to plot nearer the Ab apex of the diagram. West coast plutons are thus more trondhjemitic in composition, whereas east coast intrusive rocks correspond more closely to tonalite, granodiorite, adamellite, and granite compositions.

Plots of selected elements against silica (Figures 34.5 to 34.7) show systematic variations, but there is considerable scatter of data points. The use of other differentiation-index parameters does not reduce this scatter appreciably. Much of the scatter probably arises from the fact that we are comparing plutons over a very large area, but some of the trace-element differences between rocks of similar major-element composition are very large and must indicate varying processes controlling the generation of the plutons.

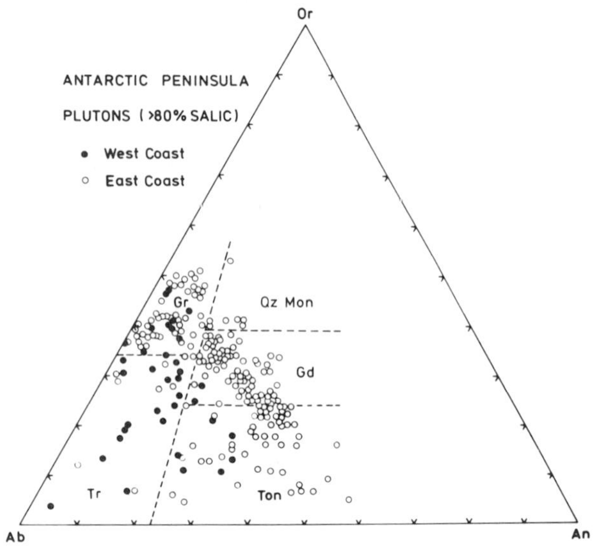

Figure 34.4. Antarctic Peninsula plutons (with more than 80% salic components) plotted on a normative orthoclase-albite-anorthite diagram. Fields modified after O'Connor (1965). Gr: granite. Tr: trondhjemite. Qz Mon: quartz monzonite. Gd: granodiorite. Ton: tonalite.

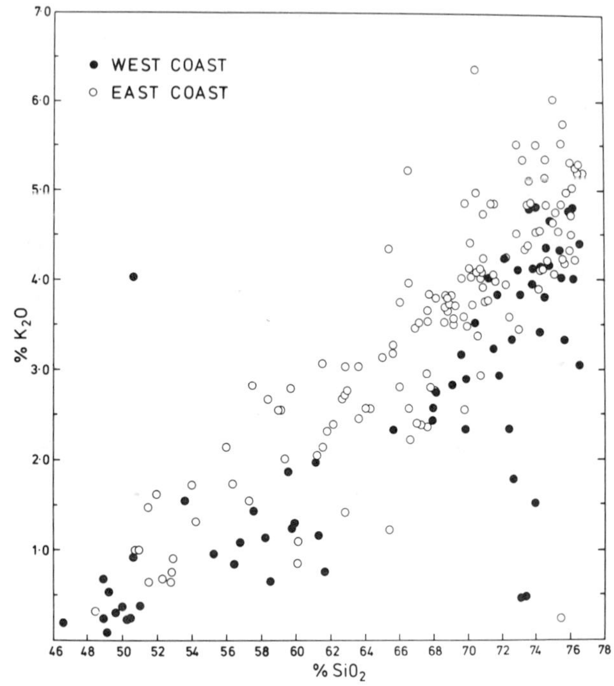

Figure 34.5. Wt % K_2O versus wt % SiO_2 for Antarctic Peninsula plutonic rocks.

Na_2O/K_2O ratios are generally lower in the east coast plutons (Table 34.1). This is also obvious from the SiO_2-K_2O plot (Figure 34.5) where K_2O increases systematically with SiO_2. The range of K_2O at 57.5% SiO_2 (the "K" value of Dickinson, 1975) is approximately 0.9–1.7% for the west coast rocks, but 1.5–2.6% for the east coast plutons. Rubidium varies in a similar fashion. In the west coast plutons, Rb increases from 5–25 ppm at 50% SiO_2 to a maximum of 250 ppm at 76% SiO_2, whereas in the east coast plutons, Rb lies between 20 and 60 ppm at 50% SiO_2 but reaches 433 ppm in the "white granites" of Fleet (1968). The distribution of thorium is similar to that of K and Rb, and Th levels are also slightly but significantly higher in the east coast rocks.

Barium has a different distribution. In the west coast plutons, it increases systematically from about 60 ppm at 50% SiO_2 to a maximum of 920 ppm at 70% SiO_2. However, there is clear evidence of a decrease in Ba concentration in many silicic plutons, beginning at 68% SiO_2. The distribution of Ba in the east coast plutons is very similar, although there is more scatter. The progressive decrease in Ba/SiO_2 ratio at high SiO_2 levels is especially well marked and is a feature which could indicate extensive fractionation of intermediate plagioclase.

Plagioclase fractionation is also suggested by the behavior of strontium (Figure 34.6), which shows a negative (inverse) correlation with SiO_2. The Danco Coast trend is

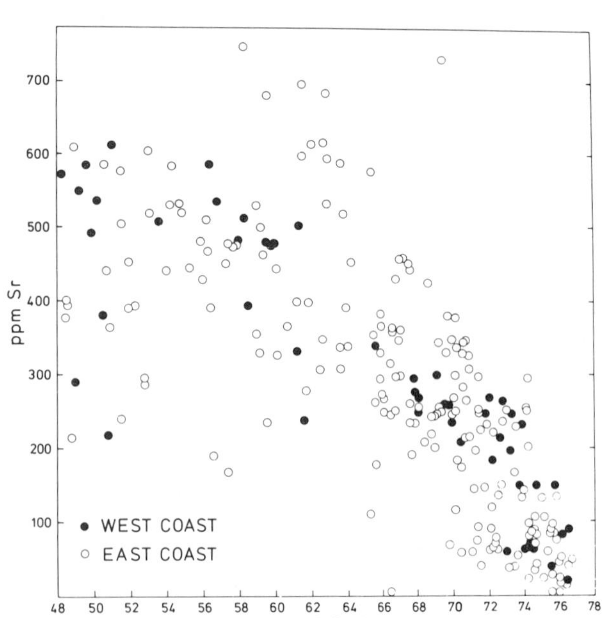

Figure 34.6. Sr versus wt % SiO_2 for Antarctic Peninsula plutonic rocks.

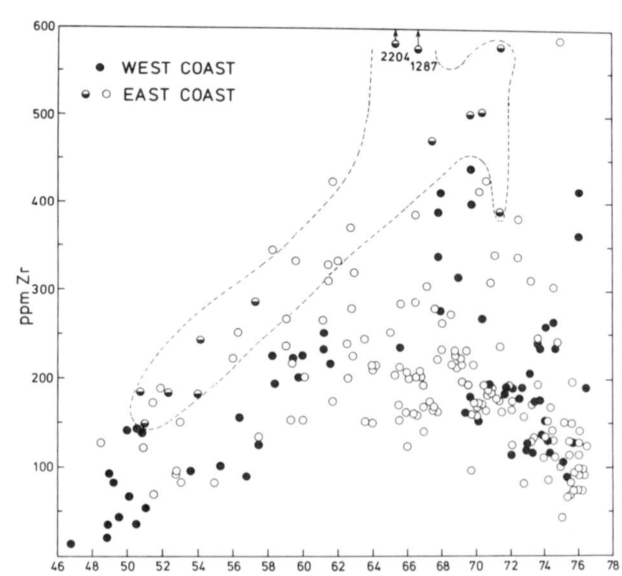

Figure 34.7. Zr versus wt % SiO_2 for Antarctic Peninsula plutonic rocks. The dashed line outlines the range of data (represented by half-filled circles) from the layered gabbro-microgranite complex at Adie Inlet, Oscar II Coast (Marsh, 1968).

particularly clear, with Sr ranging from about 600 ppm at 51% SiO_2 to about 20 ppm at high silica levels. There is no appreciable difference between east and west coast plutonic rocks with regard to Sr levels, although the east coast data show more scatter.

Zirconium distribution (Figure 34.7) is of a type intermediate between that of K_2O and Rb and that of Sr. Zr first increases with increasing SiO_2 but then shows a sharp decrease. In the west coast plutons, Zr lies between 40 and 100 ppm at 50% SiO_2 and increases to more than 400 ppm at the 68–70% SiO_2 level. Thereafter, although some silicic plutons retain high Zr/SiO_2 ratios, the majority show a sharp fall in zirconium concentration to less than 100 ppm at the 75% SiO_2 level. It is notable that the plutons predating the Mesozoic volcanic group in the Danco Coast area have lower Zr/SiO_2 ratios than most of the post-volcanic granophyres. Overall, however, there seem to be little difference between east coast and west coast suites with regard to Zr levels.

In basaltic suites, Zr behaves as an almost ideally incompatible element (i.e., showing a strong preference for entering the liquid phase) (Weaver and others, 1972; Tarney and others, 1977) and is retained in residual liquids so that there is commonly a well-marked positive correlation between Zr and SiO_2. This trend is seen in many tholeiitic rock suites, such as those from Deception Island (Weaver and others, Chapter 29, this volume) and from the Sarmiento ophiolite complex (Saunders and others, 1979), both of which produce Zr-rich siliceous differentiates. Such a relationship is also observed in the fractionated gabbro-microgranite laccolith of Adie Inlet (outlined in Figure 34.7), where Zr levels exceed 2000 ppm at 66% SiO_2. This pattern of Zr behavior is predictable in that most igneous minerals have low distribution coefficients for Zr, and thus it is only when zircon precipitates from a magma that Zr levels will fall sharply in residual liquids. Apparently, however, zircon separation rarely occurs in tholeiitic or alkaline differentiates, even though Zr levels may reach several times those seen in the present calc-alkaline plutons. Contemporaneous calc-alkaline volcanic suites in the Antarctic Peninsula show a similar Zr distribution to that in the plutons (Weaver and others, Chapter 29, this volume) indicating that the pattern is not simply a characteristic of plutonic crystallization.

If zircon separation is the cause of Zr depletion in high-silica rocks, then there must be some complementary high-Zr cumulates. We have no knowledge of any such cumulates ever having been described (on the contrary, most cumulates have unusually low Zr levels). The alternative explanation would suggest that the high-silica, low-zirconium plutons are partial-melting or filter-press products where Zr has been retained (presumably as zircon) in the source.

One hypothesis considered by many workers in the peninsula is that the intermediate rocks are hybrids between mafic and salic magmas. The Zr data show, at least in the general case, that this cannot be so: intermediate rocks have consistently higher Zr (and Ba) contents than associated granitic and more mafic rocks.

The light rare-earth elements Ce and La increase slowly with increasing SiO_2, and their abundance levels are similar to those in Andean plutons from northern and southern

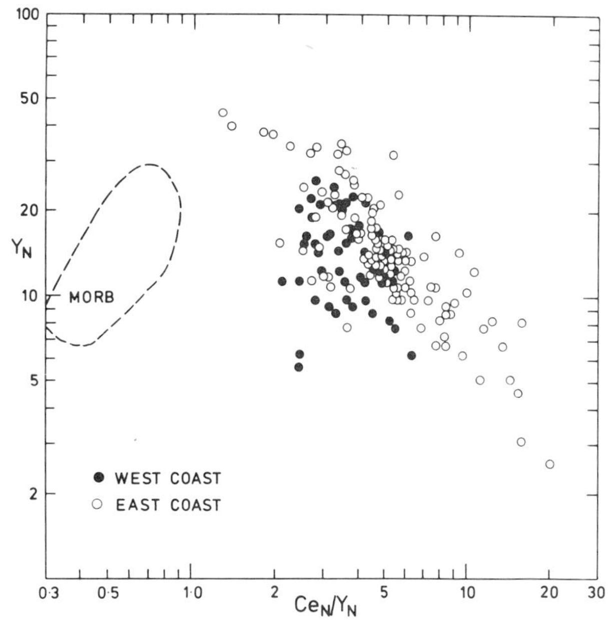

Figure 34.8. Chondrite-normalized Y versus chondrite-normalized Ce/Y for Antarctic Peninsula plutonic rocks. The field for mid-ocean ridge basalts (MORB) is taken from Saunders and others (1979).

Chile (J. Tarney, unpublished data). Their concentration is approximately 45 times chondrite levels in the basic rocks, rising to 100 times chondrite levels in the granitic plutons. Both Ce and La tend to have higher levels in rocks from the east coast than in rocks with comparable SiO_2 contents from the west coast. Yttrium, which closely parallels the heavy rare-earth elements (REE) in its behavior, is significantly lower in some of the east coast plutons, particularly at high silica levels. A plot of Y against Ce/Y (on a chondrite-normalized basis) summarizes the probable light-REE/heavy-REE distributions (Figure 34.8). The REE patterns are clearly light-REE enriched, but not excessively so (Ce_N/Y_N = 2–5 for most plutons). The east coast plutons display a distinct negative correlation between Y and Ce/Y. Normal igneous fractionation by precipitation of minerals such as pyroxene and plagioclase is unlikely to produce such a trend (Saunders and others, 1979). Instead it implies the involvement of a mineral such as garnet or hornblende which selectively takes the heavy REE's and Y into its structure (Arth and Barker, 1976). The trend toward Y depletion is more commonly displayed, and to a greater degree, in Archean tonalites, trondhjemites, and dacites (Arth and Hanson, 1975; Tarney and Windley, 1977).

Rare-earth patterns for a selection of gabbros, diorites, and granitic plutons from the Antarctic Peninsula are shown in Figure 34.9. The patterns show moderate light-REE enrichment and are essentially parallel. The gabbros show only minor europium anomalies, and the diorites and tonalites, small negative or positive anomalies, but the silicic plutons all have appreciable negative anomalies. It is apparent from Figure 34.8 that these patterns must be fairly representative of the range of Antarctic Peninsula plutons. The negative Eu anomalies in the intermediate and silicic plutons may indicate that plagioclase fractionation has taken place.

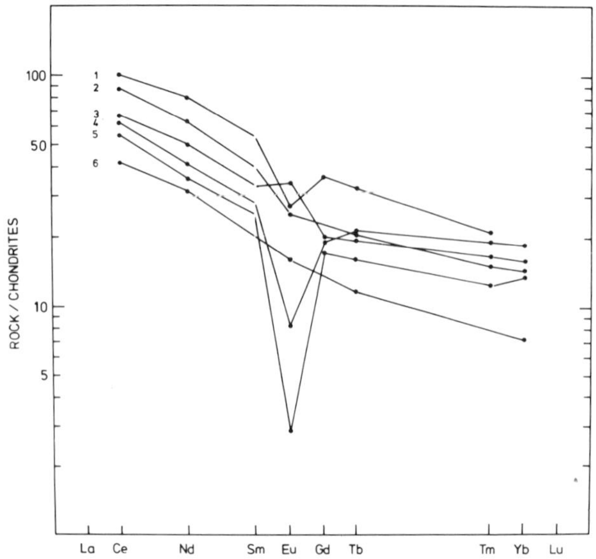

Figure 34.9. Chondrite-normalized rare-earth patterns for Antarctic Peninsula plutons. 1. West coast "Andean" granophyre (O559.2); 2. East coast prevolcanic tonalite (TL645.1); 3. East coast prevolcanic tonalite (TL800.1); 4. West coast prevolcanic granite (O938.9); 5. East coast "Andean" granite (D4613.1); 6. East coast prevolcanic gabbro (TL571.1). Data points for Sm are interpolated.

Discussion

The Antarctic Peninsula, in common with large sectors of the circum-Pacific belt, has been the site of extensive volcanism and plutonism since at least early Jurassic time. This activity is related to consumption of Pacific Ocean floor beneath the peninsula. Plutonic activity seems to have reached its peak at about 90–110 m.y. ago, but following this the foci of intrusive activity seem to have migrated from east to west across the peninsula and also from south to north along the peninsula. No active subduction appears to be taking place at present along the peninsula, and the only unequivocal evidence of a trench is found bordering the northwest coast of the South Shetland Islands (Barker and Griffiths, 1972).

Herron and Tucholke (1976) have presented magnetic data which indicate that Pacific Ocean crust (in the form of the West Antarctic and Aluk plates) was actively subducting beneath the western edge of the continent at least as far back as early Cenozoic time. Subduction seems to have ceased as various sections of the Aluk Ridge spreading center reached the margin of West Antarctica, and this may have led to the present coupling of the West Antarctic and East Antarctic plates. Magnetic anomaly data suggest, in fact, that there has been progressive northeastward consumption of the Aluk spreading center along the Pacific margin of West Antarctica, beginning in late Paleocene time in the south and occurring along the Antarctic Peninsula during early Miocene time. It is interesting to note that such a sequence of events correlates very well with the age pattern of dated plutons in the peninsula, as outlined above. Thus, as spreading and associated subduction ceased progressively from south to north, so igneous activity died out progressively in the same direction. It may be no coincidence that the youngest dated pluton (Cornwallis Island granodiorite; Rex and Baker, 1973) occurs at the northern end of the peninsula, well to the northeast of the Hero Fracture Zone (Herron and Tucholke, 1976) where evidence of more recent subduction exists (Barker, 1976; Barker and Burrell, 1977). One could also speculate that the approach of a ridge about to be subducted might cause the observed trenchward migration of intrusive foci with time, as a result of the higher thermal input to the subducting slab.

From the data presented here and in the companion paper on Antarctic Peninsula volcanic rocks (Weaver and others, Chapter 29, this volume), it would seem that contemporaneous and adjacent plutonic and volcanic rocks of the Antarctic Peninsula have broadly comparable chemistries. In general, element trends are very similar, although metasomatism of the dominantly tuffaceous peninsula volcanic rocks has led to more scatter in the volcanic data. It seems reasonable to assume that the extrusive and intrusive igneous rocks have similar origins, as has been suggested by Myers (1975) and Brown (1977) for equivalent rocks in the Peruvian Andes.

Many authors have discussed the relationship between the potash content of calc-alkaline volcanic rocks and the depth to the Benioff zone: the so-called K-h variation (Dickinson and Hatherton, 1967; Nielson and Stoiber, 1973; Dickinson, 1975; Best, 1975; Cawthorn, 1977). This is usually expressed as an increase in K_2O content at a chosen SiO_2 value, with increasing distance from the trench, but the exact relationship varies in different tectonic environments (Nielson and Stoiber, 1973; Dickinson, 1975). There has been little previous discussion as to whether other incompatible elements (e.g., Rb, Ce, La, Th, and Ba) show comparable variations or whether plutons show the same features as the associated volcanic rocks.

A comparison between the west coast and east coast plutonic suites (Figure 34.5) clearly shows the higher K_2O content of the east coast plutons at all silica levels. This corresponds to K_2O differences between east and west coast volcanic rocks (Weaver and others, Chapter 29, this volume) and is consistent with subduction beneath the peninsula from the western (Pacific) side rather than from the Weddell Sea. Because east and west coast differences seem to be independent of age, it would seem that there was subduction from the west throughout Mesozoic and Cenozoic time. The potash-depth relationship is essentially a spatial function, but it has also been demonstrated that K_2O increases with time in some areas (Jakes and White, 1969). This is not unexpected if, as in northern Chile, magmatic foci move away from the trench with time. In the Antarctic Peninsula, however, there is a trenchward migration of magmatic foci with time and a corresponding decrease in K_2O. This is emphasized even more clearly by the very low potash contents of the youngest volcanic suites on the western side of the peninsula, those in the South Shetland Islands (Weaver and others, Chapter 29, this volume). It would appear that the K-h relationship is essentially spatial and not time dependent.

With respect to elements other than potassium, significant east coast–west coast differences are shown by Rb and Th, and there is only slight evidence for spatial differences in other incompatible elements, such as Ce and La, within the general scatter of data points. Chondrite-normalized

Ce/Y ratios, indicative of rare-earth element behavior, show a considerable overlap between east and west coast rocks, although there is a distinct tendency for Ce/Y ratios to be higher on the east coast. These higher Ce/Y ratios are more a consequence of higher Ce levels than lower Y contents. We conclude that the K-h relationship applies only to elements such as Rb and Th, which are normally close geochemical associates of potassium, and possibly to Ce and La. If a smaller degree of partial melting was the cause of the higher K_2O levels, then other incompatible elements should show comparable variations. The fact that they do not indicates that a zone-refining mechanism, as suggested by Best (1975), may be the explanation.

The origin of the large volumes of orogenic or cordilleran silicic magmas continues to be a subject for much debate. Part of the problem is that there are so many source possibilities in the subduction environment (i.e., the downgoing slab, the overlying mantle wedge, and the lower and upper sialic crust). Add to this, the facts that the subducting basaltic crust is chemically heterogeneous (Kay and Senechal, 1976; Saunders and others, 1979), that ocean sediments may be involved, that there may be differing degrees of melting under varying T, P, and P_{H_2O} conditions, and that there may be extensive crystal fractionation, zone-refining, contamination, and assimilation; then it is not surprising that regional geochemical data such as that obtained for the Antarctic Peninsula conform to no simple patterns of variation.

The present study provides a broad geochemical framework upon which subsequent investigations may be based. A large number of questions remain open. For example, to what extent is sialic contamination or remelting important in the genesis of calc-alkaline rocks? Although it seems likely that at least some diorites, tonalites, and granodiorites in orogenic belts are derived directly from the mantle (Brown, 1977), the origin of granitic rocks is more equivocal. Several approaches to these problems can be taken. Detailed petrologic and geochemical studies of individual plutons may greatly increase our understanding of the effects of high-level processes such as crystal fractionation, metasomatism, contamination, and the mechanisms of intrusion. Rare-earth data, combined with other high-precision trace-element data, may be used to construct models of magma genesis, and important constraints may be added by lead, strontium, oxygen, and neodymium isotopic determinations.

Conclusions

There is a regional pattern to Antarctic Peninsula plutonism. Geochemical differences between east coast and west coast plutons can be related to the depth of the Benioff Zone for Pacific Ocean floor subducted beneath the peninsula since early Mesozoic time. The age distribution of plutons across the peninsula and the progressive cessation of plutonism northeastward along the peninsula during Cenozoic time can be linked, at least in broad terms, to the consumption of sections of the Aluk Ridge spreading center beneath the peninsula. Much more detailed work is needed to confirm and refine these broad relationships and to derive a comprehensive model of calc-alkaline petrogenesis.

Acknowledgments. Studies in the Scotia arc and the Antarctic Peninsula are supported by the Natural Environment Research Council, U.K. We are grateful to R. M. Laws, R. J. Adie, and C. W. M. Swithinbank for allowing us to make use of British Antarctic Survey collections of Antarctic Peninsula material. Neutron activation facilities at Bedford College (London University) were provided by I. L. Gibson; we are grateful to Giselle Marriner for assistance with the analyses.

References cited

Adie, R. J. 1954. The petrology of Graham Land: 1. The basement complex; early Palaeozoic plutonic and volcanic rocks. *Scient. Rep. Falkld Isl. Depend. Surv.*, No. 11, 22 pp.

Adie, R. J. 1955. The petrology of Graham Land: II. The Andean Granite-Gabbro Intrusive Suite. *Scient. Rep. Falkld Isl. Depend. Surv.*, No. 12, 39 pp.

Adie, R. J. 1972. Evolution of volcanism in the Antarctic Peninsula. (*In* Adie, R. J., *ed. Antarctic Geology and Geophysics.* Oslo, Universitetsforlaget, 137–41.)

Aitkenhead, N. 1975. The geology of the Duse Bay–Larsen Inlet area, northeast Graham Land (with particular reference to the Trinity Peninsula Series). *Scient. Rep. Br. antarc. Surv.*, No. 51, 62 pp.

Arth, J. G., and Barker, F. 1976. Rare-earth partitioning between hornblende and dacitic liquid and implications for the genesis of trondhjemitic-tonalitic magmas. *Geology, 4,* 534–36.

Arth, J. G., and Hanson, G. N. 1975. Geochemistry and origin of the early Precambrian crust of north-eastern Minnesota. *Geochim. cosmochim. Acta, 39,* 325–62.

Barker, P. F. 1976. The tectonic framework of Cenozoic volcanism in the Scotia Sea region—A review. (*In* Gonzalez-Ferran, O., *ed. Andean and Antarctic Volcanology Problems.* Rome, International Association of Volcanology and Chemistry of the Earth's Interior, 330–46.)

Barker, P. F., and Burrell, J. 1977. The opening of Drake Passage. *Mar. Geol., 25,* 15–34.

Barker, P. F., and Griffiths, D. H. 1972. The evolution of the Scotia Ridge and Scotia Sea. *Phil. Trans. R. Soc. Lond.*, Ser. A, *271,* 151–83.

Best, M. G. 1975. Migration of hydrous fluids in the upper mantle and potassium variation in calc-alkalic rocks. *Geology, 3,* 429–32.

Brown, G. C. 1977. Mantle origin of Cordilleran granites. *Nature, Lond., 265,* 21–24.

Cawthorn, R. G. 1977. Petrological aspects of the correlation between potash content of orogenic magmas and earthquake depth. *Mineralog. Mag., 41,* 173–82.

Curtis, R. 1966. The petrology of the Graham Coast, Graham Land. *Scient. Rep. Br. antarc. Surv.*, No. 50, 51 pp.

Dalziel, I. W. D. 1974. Margins of the Scotia Sea. (*In* Burk, C. A., and Drake, C. L. *eds. The Geology of Continental Margins.* New York, Springer-Verlag, 567–80.)

Dalziel, I. W. D., and Elliot, D. H. 1973. The Scotia arc and Antarctic margin. (*In* Nairn, A. E. M., and Stehli, F. G., *eds. The Ocean Basins and Margins. 1. The South Atlantic.* New York, Plenum Press, 171–246.)

Dewar, G. J. 1970. The geology of Adelaide Island. *Scient. Rep. Br. antarc. Surv.*, No. 57, 66 pp.

de Wit, M. J. 1977. The evolution of the Scotia arc as a key to the reconstruction of southwestern Gondwanaland. *Tectonophysics, 37,* 53–81.

Dickinson, W. R. 1975. Potash-depth (K-*h*) relations in continental margin and intra-oceanic magmatic arcs. *Geology, 3,* 53–56.

Dickinson, W. R., and Hatherton, T. 1967. Andesitic volcanism and seismicity around the Pacific. *Science, N.Y., 157,* 801–3.

Elliot, D. H. 1964. The petrology of the Argentine Islands. *Scient. Rep. Br. antarc. Surv.*, No. 41, 31 pp.

Farrar, E., Clark, A. H., Haynes, S. J., Quirt, G. S., Conn, H., and Zentilli, M. 1970. K-Ar evidence for the post-Paleozoic migration of granitic intrusion foci in the Andes of northern Chile. *Earth planet. Sci. Lett., 10,* No. 1, 60–66.

Fleet, M. 1968. The geology of the Oscar II Coast, Graham Land. *Scient. Rep. Br. antarc. Surv.,* No. 59, 46 pp.

Fraser, A. G., and Grimley, P. H. 1972. The geology of parts of the Bowman and Wilkins coasts, Antarctic Peninsula. *Scient. Rep. Br. antarc. Surv.,* No. 67, 59 pp.

Goldring, D. C. 1962. The geology of the Loubet Coast, Graham Land. *Scient. Rep. Br. antarc. Surv.,* No. 36, 50 pp.

Grikurov, G. E., Krylov, A. Ya., and Silin, Yu. I. 1966. Absoliutnyi vozrast nekotorykh porod iz raiona zaliva Margerit, Antarkticheskii poluostrov [Absolute age of certain rocks in the Marguerite Bay region of the Antarctic Peninsula]. *Dokl. Akad. Nauk SSSR,* Geology, *171,* No. 6, 1399–401. [English translation: *Dokl. (Proc.) Acad. Sci. U.S.S.R.,* Geological sciences sect., *171,* 127–30.]

Grikurov, G. E., Krylov, A. Ya., Polyakov, M. M., and Tvsobun, Ya. N. 1970. Vozrast porod v severnoi chasti Antarkticheskogo poluostrova i na Yudznykh Shetlandskikh ostrovakh (po dannym kaliy-argonovogo metoda) [Age of rocks from the northern part of the Antarctic Peninsula and the South Shetland Islands (from data of the potassium-argon method)]. *Inf. Byull. sov. antarkt. Eksped.,* No. 80, 30–34. [English translation: *Inf. Bul. Sov. Antarct. Exped. 8,* 61–68.]

Halpern, M. 1965. The geology of the General Bernado O'Higgins area, northwest Antarctic Peninsula. (*In* Hadley, J. B., *ed. Geology and Paleontology of the Antarctic.* Washington, American Geophysical Union, 177–209.) [Antarctic Research Series, Vol. 6.]

Halpern, M. 1967. Rubidium-strontium isotopic age measurements of plutonic rocks in eastern Ellsworth Land and northern Antarctic Peninsula, Antarctica. *J. geophys. Res., 72,* No. 20, 5133–42.

Halpern, M. 1972. Rb-Sr total-rock and mineral ages from the Marguerite Bay area, Kohler Range and Fosdick Mountains. (*In* Adie, R. J., *ed. Antarctic Geology and Geophysics.* Oslo, Universitetsforlaget, 197–204.)

Halpern, M. 1973. Regional geochronology of Chile, south of 50° latitude. *Geol. Soc. Am. Bull., 84,* No. 7, 2407–22.

Herron, E. M., and Tucholke, B. E. 1976. Sea-floor magnetic patterns and basement structure in the southeastern Pacific. (*In* Hollister, C. D., Craddock, C., and others. *Initial Reports of the Deep Sea Drilling Project,* Vol. 35. Washington [U.S. Government Printing Office], 263–78.)

Hooper, P. R. 1962. The petrology of Anvers Island and adjacent islands. *Scient. Rep. Falkld Isl. Depend. Surv.,* No. 34, 69 pp.

Jakeš, P., and White, A. J. R. 1969. Structure of the Melanesian arcs and correlation with distribution of magma types. *Tectonophysics, 8,* 223–6.

Kay, R. W., and Senechal, R. G. 1976. The rare earth geochemistry of the Troodos ophiolite complex. *J. geophys. Res., 81,* 964–70.

Marsh, A. F. 1968. *Geology of Parts of the Oscar II and Foyn Coasts, Graham Land.* Ph.D. dissertation, University of Birmingham, England, 291 pp.

McNutt, R. H., Crocket, J. H., Clark, A. H., Caelles, J. C., Farrar, E., Haynes, S. J., and Zentilli, M. 1975. Initial $^{87}Sr/^{86}Sr$ ratios of plutonic and volcanic rocks of the central Andes between latitudes 26° and 29° south. *Earth planet. Sci. Lett., 27,* 305–13.

Myers, J. S. 1975. Cauldron subsidence and fluidization: mechanisms of intrusion of the Coastal Batholith of Peru into its own volcanic ejecta. *Geol. Soc. Am. Bull., 86,* 1209–20.

Nielson, D. R., and Stoiber, R. E. 1973. Relationship of potassium content in andesitic lavas and depth to the seismic zone. *J. geophys. Res., 78,* 6887–92.

Nockolds, S. R., and Allen, R. 1953. The geochemistry of some igneous rock series. *Geochim. cosmochim. Acta, 4,* 105–42.

O'Connor, J. T. 1965. A classification for quartz-rich igneous rocks based on feldspar ratios. *Prof. Pap. U.S. geol. Surv.,* 525-B, B79–84.

Rex, D. C. 1976. Geochronology in relation to the stratigraphy of the Antarctic Peninsula. *Bull. Br. antarc. Surv.,* No. 43, 49–58.

Rex, D. C., and Baker, P. E. 1973. Age and petrology of the Cornwallis Island granodiorite. *Bull. Br. antarc. Surv.,* No. 32, 55–61.

Saunders, A. D., Tarney, J., Stern, C. R., and Dalziel, I. W. D. 1979. Geochemistry of Mesozoic marginal basin floor igneous rocks from southern Chile. *Geol. Soc. Am. Bull., 90,* 237–58.

Scott, K. M. 1965. Geology of the southern Gerlache Strait region, Antarctica. *J. Geol., 73,* No. 3, 518–27.

Stubbs, G. M. 1968. *Geology of Parts of the Foyn and Bowman Coasts, Graham Land.* Ph.D. dissertation, University of Birmingham, England, 245 pp.

Tarney, J., and Windley, B. F. 1977. Chemistry, thermal gradients and evolution of the lower continental crust. *J. Geol. Soc. Lond., 134,* 153–72.

Tarney, J., Saunders, A. D., and Weaver, S. D. 1977. Geochemistry of volcanic rocks from the island arcs and marginal basins of the Scotia arc region. (*In* Talwani, M., and Pitman, W. C., III, *eds. Maurice Ewing Series,* 1: *Island Arcs, Deep Sea Trenches, and Back-Arc Basins.* Washington, American Geophysical Union, 367–77.)

Weaver, S. D., Sceal, J. S. C., and Gibson, I. L. 1972. Trace-element data relevant to the origin of trachytic and pantelleritic lavas in the East African rift system. *Contr. Miner. Petrol., 36,* 181–94.

West, S. M. 1974. The geology of the Danco Coast, Graham Land. *Scient. Rep. Br. antarc. Surv.,* No. 84, 58 pp.

35

Rb-Sr and K-Ar Geochronology of Rocks from the Antarctic Peninsula between Anvers Island and Marguerite Bay

A. GLEDHILL* and D. C. REX
Department of Earth Sciences, University of Leeds, LS2 9JT, U.K.

P. W. G. TANNER
British Antarctic Survey, Natural Environment Research Council, Madingley Road, Cambridge CB3 OET, U.K.

Abstract. New radiometric data on rocks from Marguerite Bay show (1) that the pink and white granite suite and related hypabyssal rocks are partly, and perhaps wholly, of Cretaceous age and (2) that the older (so-called Precambrian) gneissic rocks were affected by a major metamorphic event during Early Jurassic time and probably belong to the Trinity Peninsula Formation (Carboniferous–Triassic?). The Debenham Islands coarse, pink granite gives a whole-rock Rb-Sr isochron age of 109 m.y. and seems to belong to the same magmatic cycle as the pink and gray granite sheets on Stonington Island. Reheating of the older gneiss during this cycle has reset both hornblende and biotite K-Ar ages to give Cretaceous dates. A Rb-Sr whole-rock isochron age of 175 m.y. has been obtained from the granitic gneiss in the Debenham Islands, and it provides a minimum age for regional metamorphism associated with the Gondwanian orogeny. A hybrid pluton from Arthur Harbour, Anvers Island, gives a whole-rock Rb-Sr age of 35 m.y., and it is inferred that the adjoining Cape Monaco granite is part of an extensive belt of probably Miocene to Holocene intrusive and volcanic rocks in this area. Dating of a granodiorite from the Argentine Islands by the same method yields an age of 72 m.y. In both cases K-Ar ages of minerals from the plutonic rocks are younger by 15 m.y., and we discuss the possible significance of this result. Review of radiometic ages from the Antarctic Peninsula leads to the conclusion that active subduction has occurred beneath this region since at least 183 m.y. ago and that rocks metamorphosed during the Paleozoic Borchgrevink orogeny probably formed a sialic basement beneath the Trinity Peninsula Formation.

The results of previous radiometric dating (mainly K-Ar) on rocks from the Antarctic Peninsula have recently been summarized by Rex (1976). These ages show an extended period of igneous plutonic activity from 183 to 45 m.y. ago, with the youngest dated intrusion, 9.5 m.y. old, from Cornwallis Island (Rex and Baker, 1973). However, the suggested presence of lower Paleozoic plutons and Precambrian basement complex (Adie, 1954) remained unconfirmed by radiometric dates on rocks from the supposed type area for these older rocks in Marguerite Bay (Grikurov and others, 1966; Halpern, 1972). Likewise, no reliable minimum age had been obtained for the early Mesozoic Gondwanian orogeny in the Antarctic Peninsula.

Previous work on the Antarctic Peninsula was mainly confined to K-Ar dating, and we present the results of the first Rb-Sr analyses based upon extensive field sampling. This material was collected by one of us (P.W.G.T.) during a brief visit to the Antarctic Peninsula in 1973–74. Sampling was necessarily restricted to areas adjacent to British Antarctic Survey and U.S. scientific stations, but despite this limitation, a variety of rock types of supposed Precambrian, Paleozoic, and Andean ages were collected. Geochemical analysis of these rocks is nearly complete and will be published elsewhere.

* Department of Earth Sciences, The Open University, Milton Keynes, U.K.

In this paper we introduce the following informal stratigraphic terms in order to avoid repetition of terminology that does not conform with current knowledge or usage:

Antarctic Peninsula Volcanic Group replaces "Upper Jurassic Volcanic Group."

Trinity Peninsula Formation replaces "Trinity Peninsula Series."

Antarctic Peninsula metamorphic complex replaces "Basement Complex" and "pre-Jurassic metamorphic complex" (Dalziel and Elliot, 1973).

Analytical techniques

The potassium contents were measured by flame photometry (results to two significant figures—EEL Model 100; results to three significant figures—Corning-EEL Model 450 incorporating Li internal standard); the quoted result is the average of three dissolutions. The argon isotopes were measured on a modified AEI MS10 mass spectrometer (Rex and Dodson, 1970) fitted with digital output. The uncertainties on the ages are quoted at the 95% confidence limit and calculated from replicate analyses.

The Rb-Sr ratios were measured by XRF (technique described by Pankhurst and O'Nions (1973), modified by M. Bickle, University of Leeds) with estimated errors of 2%

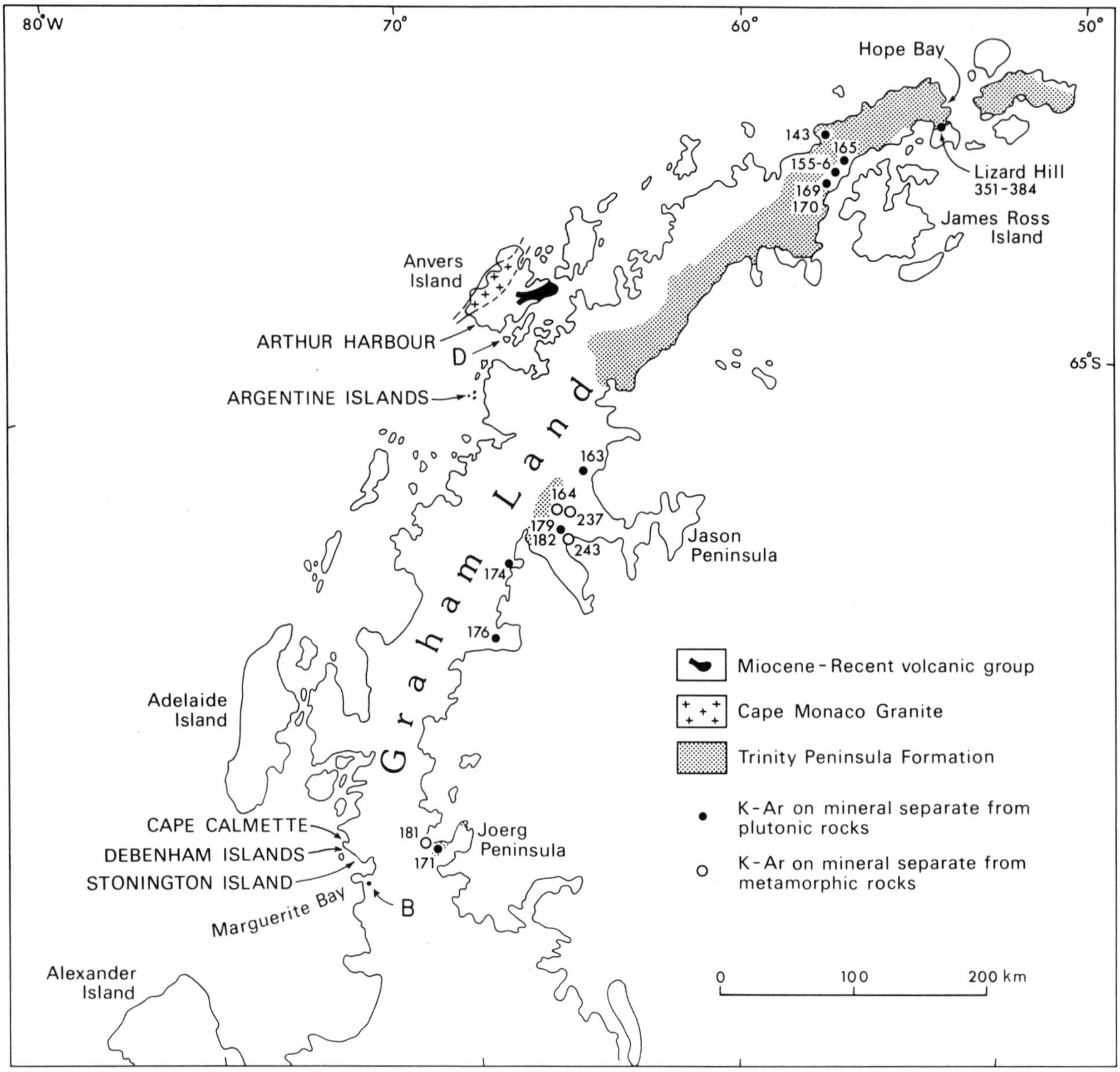

Figure 35.1. Outline map of the Antarctic Peninsula, showing the locations of analyzed rocks. K-Ar data, in m.y., from Rex (1976). B: Blackwall Mountains. D: Doumer Island.

(2S). Rb-Sr values obtained on international standards are as follows: GSP-1, 1.08; AGV-1, 0.102; BCR-1, 0.143; and G-2, 0.350. The Sr isotopes were measured on an AEI MS5 mass spectrometer (results quoted to four significant figures) and on a MM30 mass spectrometer (results quoted to five significant figures). Rb and Sr values on biotite BS 104.2 were measured by isotope dilution. Average values for Eimer and Amend $SrCO_3$ of 0.7080 ± 3(MS5) and 0.70807 ± 1 (MM30) were obtained during the period of study. The data have been regressed according to York (1969) using York Model II. The quoted uncertainties are two sigma; the decay constant of ^{87}Rb used is 1.39×10^{-11}/yr.

Argentine Islands

The geology of the Argentine Islands (Figure 35.1) has been described by Elliot (1964): a granodiorite pluton exposed on the Forge Islands and at The Barchans cuts andesitic lavas and dacite pyroclastic rocks of the Antarctic Peninsula Volcanic Group, resulting in thermal and metasomatic alteration. Specimens were collected from The Barchans between stations 19 and 20 of Elliot (1964, Figure 2).

Petrography

The granodiorite is a medium-grained leucocratic rock with numerous small mafic xenoliths (Elliot, 1964). Analyzed specimens, all collected within a radius of 10 m, fall into two groups: (1) BS 103.7 and 103.12 are from the normal facies, and (2) BS 103.10 and BS 103.11 are from a 1-m-wide pink, potash feldspar–rich dike which has a nonchilled graditional contact with the normal facies.

Granodiorite specimens of the first group consist of perthite-plagioclase-quartz rock with scattered clusters of pyroxene-hornblende-biotite crystals. Particular features are:

1. marked zonation, in some cases oscillatory, of the plagioclase crystals from sodic rims to partially sericitized, cracked, high-anorthite cores;

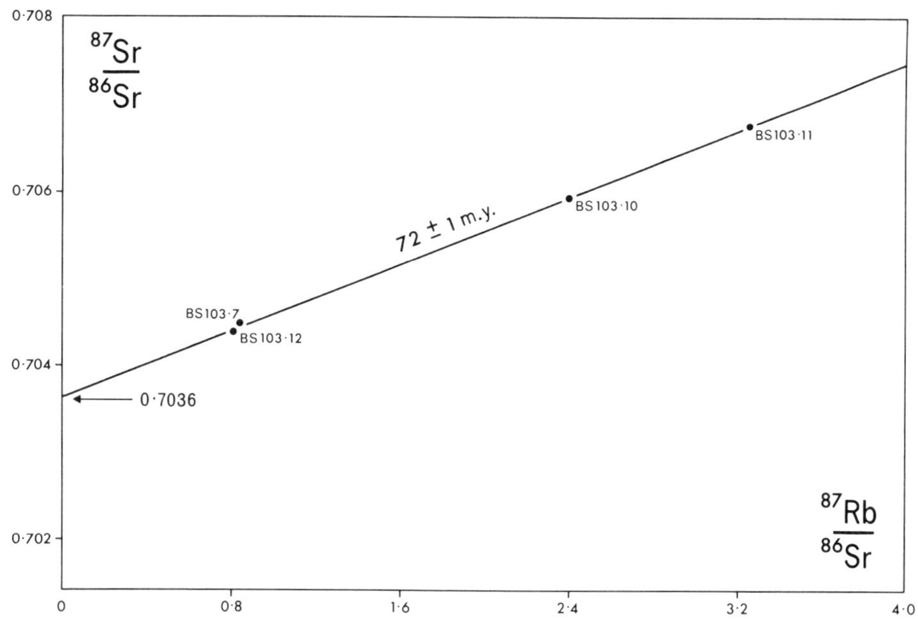

Figure 35.2. Whole-rock Rb-Sr isochron diagram for the granodiorite pluton in the Argentine Islands.

2. sharply defined boundaries between irregularly shaped cores of high-anorthite plagioclase and their mantles of low-anorthite plagioclase, and the presence in the matrix of some crystals of labradorite with very narrow mantles or no mantle;
3. irregular, partially absorbed cores of clinopyroxene within green hornblende, the pyroxene twinning being inherited by the amphibole; the green hornblende is intergrown with and almost certainly replaced by biotite; it is also patchily altered to blue-green amphibole, and biotite is locally chloritized; and
4. perthite in the matrix appears to have partially replaced plagioclase.

These relationships suggest marked chemical disequilibrium during crystallization, with early clinopyroxene and calcic plagioclase occurring as xenocrysts.

The pink dike rock (second group) consists of perthite with quartz-perthite graphic intergrowth, together with large, generally unzoned, plagioclase crystals. Large, thin, randomly oriented plates of biotite are common.

Radiometric data

Biotite from three specimens of granodiorite previously dated from South Island, <0.5 km from the present sampling site, gave K-Ar ages of 56 ± 2, 56 ± 2, and 57 ± 2 m.y. (Rex, 1976, Table III). Two new K-Ar determinations on biotite from specimens of the pink dike rock described above give ages of 57 ± 2 m.y. (Table 35.1). It was not possible to separate hornblende from the granodiorite to check for age concordance, but we conclude that these spatially widely separated, but identical, K-Ar ages from different facies of the pluton are ages of final cooling.

A four-point Rb-Sr isochron gives an age of 72 m.y. (Figure 35.2) compared with 57 m.y. for the K-Ar age. One possible explanation for this age difference is that the Sr isotopes within the magma ceased to reequilibrate on a whole-rock scale at 72 m.y. ago (Roddick and Compston, 1977). The magma then remained above the K-Ar closure temperature of the biotite for 15 m.y. Marked chemical disequilibrium inferred from these rocks, as noted above, supports the isotopic evidence that the body cooled slowly. Another explanation for the age discrepancy is that the magma was intruded at 72 m.y. ago and then suffered a thermal event at 57 m.y. which reset the K-Ar clock; lack of geologic evidence for such an event leads us to favor the slow-cooling explanation.

Yet a third alternative is that the granodiorite magma was emplaced 57 m.y. ago but that isotopic homogenization within the magma was incomplete, resulting in partial inheritance of older Rb-Sr systematics from the source region (Pankhurst and Pidgeon, 1976; Roddick and Compston, 1977). In this case the 72 m.y. age could be geologically meaningless, and the evidence of mineralogic disequilibrium could be related to incomplete equilibrium in the magma.

The K-Ar data concur with Elliot's suggestion that a separate group of fresh-looking lavas and dike rocks in the Argentine Islands are of Tertiary age (Elliot, 1964, p. 25).

Arthur Harbour, Anvers Island

Hooper (1962) has described the geology of Anvers Island. He recognized the presence of the Antarctic Peninsula Volcanic Group, intruded and contact metamorphosed by a suite of "Andean" gabbro and tonalite plutons and their associated hybrid rocks. Specimens were collected by us from near Palmer Station (U.S.A.) sited on the headland about 1 km northeast of Bonaparte Point (Hooper, 1962, Figure 17). Hooper shows this locality to consist of "primary tonalite" but gives no petrographic or field description of rocks from the area.

Petrography

The apparent lack of primary igneous banding and the abundance of mafic xenoliths do not conform with Hooper's (1962) description of "primary tonalite." In thin section, the analyzed rocks, which were collected from an area a few tens of meters across, have many of the features which he describes from the hybrid suite, in particular:

1. plagioclase has well-defined cores of fresh, but cracked

and locally sericitized, high-anorthite plagioclase; the crystals exhibit marked compositional zoning, either oscillatory or diffuse, and microprobe analysis shows that zonation in some crystals is from An_{55} to An_{32} but that those with prominent cores of high relief have centers as calcic as An_{93};

2. perthite is common in some of the specimens, has a replacement texture toward plagioclase, and is accompanied by myrmekite; and
3. clots of pyroxene-hornblende-biotite are scattered throughout the rock.

The biotite is molded upon and replaces hornblende, and the latter encloses and has partially replaced clinopyroxene. Twinning is inherited by hornblende where it replaces pyroxene. Hornblende is altered to secondary blue-green amphibole, and chlorite (after biotite) is present.

Features (1) and (3) are considered by Hooper to be diagnostic of the hybrid suite described from elsewhere on Anvers Island.

RADIOMETRIC DATA

K-Ar dating of biotite (BS 104.2) and hornblende (BS 104.3A) from specimens a few meters apart gives concordant ages of 20 m.y. (Table 35.1). In view of the significantly older Rb-Sr isochron age reported below, the hornblende determination was repeated by making a new separation (BS 104.3B, Table 35.1). Biotite from BS 104.2 was analyzed for Rb-Sr, and a mineral age (21 m.y.) was calculated using the whole-rock Rb-Sr analysis. Both confirmed the original result (Tables 35.1 and 35.2).

A whole-rock Rb-Sr isochron has been constructed from eight data points and gives an apparent age of 35 ± 6 m.y. (Figure 35.3). The initial ratio of 0.7037 is almost identical to that for the Argentine Islands pluton, and the age discrepancy (+15 m.y.) is the same.

In this locality there is some petrographic evidence that the magma may have been emplaced in a semisolid state, namely the broken plagioclase crystals with oscillatory zoning; we therefore suggest that either the magma crystallized 35 m.y. ago and was not emplaced and finally cooled until 20 m.y. ago, or the age discrepancy results from a later thermal event. Younger suites of igneous rocks, including the "Altered Assemblage" and the Cape Monaco granite (Hooper, 1962), which has an elongate outcrop some 8 km wide and 66 km long, occur to the west of Arthur Harbour and could be responsible for this 20 m.y. thermal event.

Table 35.2. Rb-Sr whole-rock age data on rocks from the Antarctic Peninsula

Location and sample number	$^{87}Sr/^{86}Sr$[a]	$^{87}Rb/^{86}Sr$	Sr (ppm)	Rb (ppm)
Argentine Islands				
BS 103.7	0.7044 ±1	0.800	296.5	82.7
BS 103.10	0.7060 ±1	2.386	147.7	122.8
BS 103.11	0.7069 ±1	3.230	122.6	138.1
BS 103.12	0.7045 ±2	0.824	281.6	80.9
Arthur Harbour, Anvers Island				
BS 104.1	0.70412±6	0.7844	269.0	99.9
BS 104.2	0.70400±6	0.6108	417.7	88.2
BS 104.3	0.70408±6	0.7225	209.6	52.3
BS 104.4	0.70413±6	1.0738	286.1	106.2
BS 104.5	0.70393±5	0.3802	439.9	57.8
BS 104.6	0.70394±6	0.5882	439.3	89.3
BS 104.7	0.70439±5	1.3545	222.7	104.3
BS 104.8	0.70391±5	0.3698	434.1	55.5
BS 104.2[b]	0.7646 ±10	186.50	10.3	666.0
Barbara Island, Debenham Islands, Marguerite Bay				
BS 102.23	0.7114 ±1	3.692	132.4	170.3
BS 102.25A	0.7362 ±1	19.841	35.1	242.9
BS 102.25B	0.7419 ±3	23.689	30.2	249.0
BS 102.27	0.7125 ±3	4.180	194.5	133.5
BS 102.28A	0.7130 ±1	4.661	119.1	193.4
Stonington Island, Marguerite Bay				
BS 101.1	0.7083 ±1	1.134	485.7	191.4
BS 101.2	0.7120 ±1	1.956	291.9	198.6
BS 101.3	0.7115 ±1	3.063	226.5	245.8
BS 101.4	0.7084 ±1	1.256	465.6	203.9
BS 101.5	0.7170 ±1	6.441	126.2	283.4
BS 101.6	0.7080 ±1	1.080	514.3	193.6
BS 101.7	0.7159 ±1	4.424	165.1	254.5
BS 101.8	0.7168 ±1	5.206	135.2	245.4
BS 101.9	0.7163 ±1	5.756	126.8	254.4
BS 101.10	0.7188 ±1	5.986	123.9	258.6
BS 101.11	0.7225 ±1	9.681	100.6	339.3
BS 101.12	0.7152 ±1	4.767	140.8	234.0
Barry Island, Debenham Islands, Marguerite Bay				
BS 102.35	0.7145 ±1	2.827	143.9	141.8
BS 102.36	0.7179 ±2	4.202	102.3	149.8
BS 102.38	0.7158 ±1	3.438	133.6	160.1
BS 102.39	0.7153 ±1	3.150	134.1	147.2
BS 102.42	0.7140 ±1	2.578	157.4	141.5
BS 102.43	0.7138 ±1	2.551	166.3	149.7

[a] Numbers following plus-or-minus signs indicate uncertainty in the last decimal place.
[b] Analysis of biotite separate, using $\lambda_\beta = 1.39 \times 10^{-11}$/yr.

Table 35.1. K-Ar age data on minerals from Antarctic Peninsula specimens

Location and specimen number	Mineral	K (wt %)	^{40}Ar (radiogenic) scc[a]/g × 10^{-4}	^{40}Ar (radiogenic) % total	Age (m.y.)
Argentine Islands					
BS 103.11	Biotite	5.32	0.1251	82.8	57±2
BS 103.12	Biotite	5.77	0.1322	86.1	57±2
Anvers Island					
BS 104.2	Biotite	6.949	0.05487	60.9	19.8±0.8
BS 104.3A	Hornblende	0.614	0.004869 0.005042	38.0 33.2	20.1±0.8
BS 104.3B	Hornblende	0.630	0.005239	26.7	20.8±1.0
Marguerite Bay					
BS 101.19	Biotite	6.01	0.2751	86.8	111±4
E 2725.2	Biotite	6.08	0.2167	89.2	87±3

Constants: $\lambda_e = 0.584 \times 10^{-10}$/yr; $\lambda_\beta = 4.72 \times 10^{-10}$/yr; and $^{40}K/K = 0.0119$ atomic %.
[a] Standard cubic centimeter (cm^3 at S.T.P.).

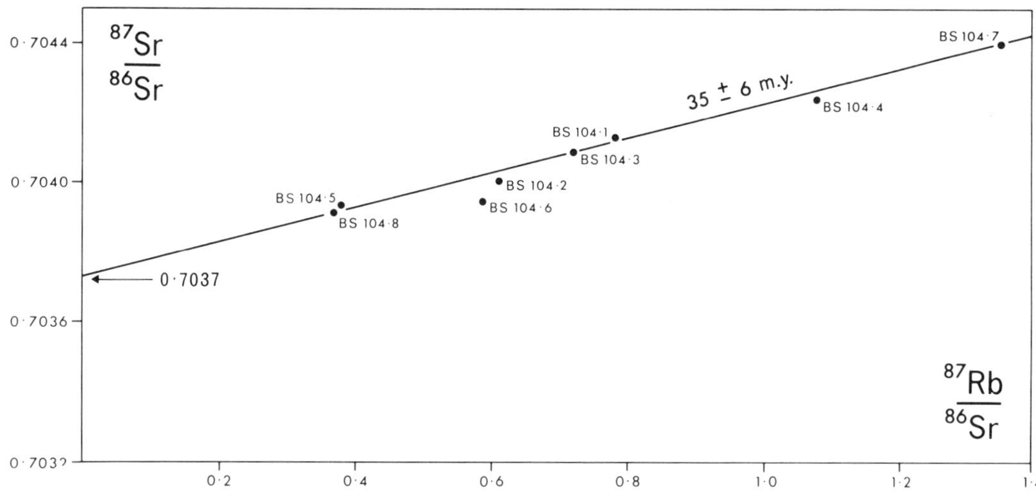

Figure 35.3. Whole-rock Rb-Sr isochron diagram for the hybrid pluton in Arthur Harbour, Anvers Island.

Discussion

The 35 m.y. or younger age for the plutonic rocks in Arthur Harbour, and their close petrographic similarity to Hooper's (1962) widespread hybrid suite, raises the strong possibility that the "Andean" plutonic suite on Anvers Island is, in part at least, of Tertiary age. On Doumer Island (D on Figure 35.1) a quartz diorite pluton, correlated by Hooper with that on Anvers Island, has given a lead-alpha age of 45 ± 5 m.y. and a K-Ar age on biotite of 53 m.y. (Scott, 1965). An Eocene, or more likely Oligocene–Holocene, age for the >1200 m thickness of lavas and pyroclastic rocks which overlie the "Andean" tonalite at the northeast end of Anvers Island (Hooper, 1962) would seem to be confirmed.

A second intriguing possibility is that the Cape Monaco granite (Figure 35.1) is of Miocene (i.e., <22.5 m.y.) or younger age, for Hooper presents detailed field evidence that it was intruded after the "Andean" hybrid and tonalite suites. Sampling of this pluton, together with others in the vicinity, is planned by R. J. Pankhurst in the near future. The conclusions here suggest that emplacement of a varied suite of plutonic rocks as a result of subduction of the Pacific floor in this sector of the Antarctic Peninsula continued well into Miocene time.

Marguerite Bay

Suites of "coarse pink granites" and "white granites" have been described from Graham Land (Adie, 1954) where they intrude a "Basement Complex" for which a Precambrian age was postulated. Analogy between the latter and the basement complexes of South Africa and South America was implied, but no specific comparisons have been made.

Boulders of the coarse pink granite were found in a group of andesites and agglomerates that was (1) thought to rest unconformably upon the white granite in the Blackwall Mountains (Adie, 1954) and (2) inferred to be of Late Jurassic age by comparison with rocks at Hope Bay, at the tip of the Antarctic Peninsula (Figure 35.1). The age of the plutonic suite, lying somewhere between that of the basement and that of the volcanic group, was first suggested to be early Paleozoic (Adie, 1954) and, subsequently, Cambrian (Adie, 1962) or Devonian (Adie, 1972). Although reporting that it was not possible to establish the relationship between the coarse pink granite suite and the Trinity Peninsula Formation, Adie (1954, p. 17) stated that "it is certain that it is earlier than the sediments." He further suggested that mafic xenoliths contained in the pink granites at several localities were derived from *supposedly* lower Paleozoic andesites. Although these early andesites have been referred to subsequently (Adie, 1964, 1972), no evidence of their field relationships, petrology, or age determination has, to our knowledge, been published.

It is on the basis of the above relationships that Precambrian and lower Paleozoic rocks have previously been reported from Marguerite Bay and adjoining areas, although there is no direct evidence bearing upon the ages of the rock units involved.

Debenham Islands coarse pink granite

This pluton is exposed on Barbara Island in the Debenham Islands (Hoskins, 1963, Figure 2) where it cuts with marked discordance across banded gneiss. It was first described by Adie (1954, p. 17) as being a member of the coarse pink granite suite.

Petrography

The granite is coarse grained and consists of perthite-oligoclase-quartz with scattered flakes of greatly chloritized biotite. It is cut by aplitic dikes (BS 102.25A and B) containing a granular assemblage of quartz, perthite, and sodic plagioclase.

Perthite in both granite and aplite is clouded with fine brown dust, giving it a pink color. Plagioclase is slightly sericitized and in some crystals has cores of slightly more calcic composition. Clots of epidote–chlorite–opaque minerals occur as alteration products of biotite. Quartz shows slightly strained extinction, and the rock is cut by fracture zones which locally cross the phenocrysts but generally follow their margins; the fractures are marked by narrow trails of recrystallized plagioclase grains and in places contain epidote.

Radiometric data

A Rb-Sr isochron constructed from five data points—three granite (BS 102.23, 102.27, and 102.28A) and two aplite specimens—gives an age of 109 ± 1 m.y. and an ini-

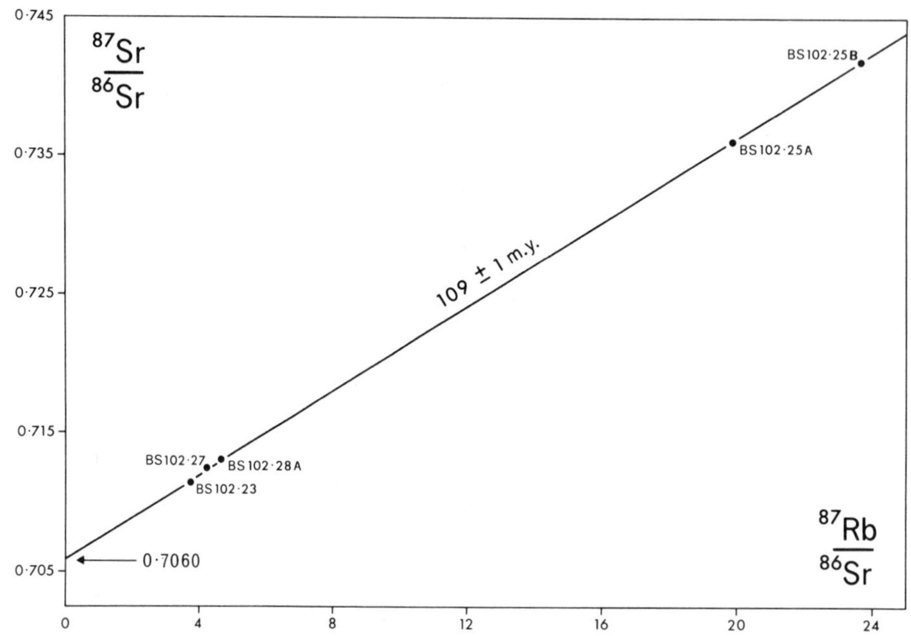

Figure 35.4. Whole-rock Rb-Sr isochron diagram for the coarse pink granite on the Debenham Islands in Marguerite Bay.

tial ratio (I.R.) of 0.7060 (Figure 35.4). The three granite points alone gave an age of 120 ± 25 m.y. with IR = 0.7054.

These dates confirm Halpern's (1972) preliminary Cretaceous age for the pluton. Using an assumed IR = 0.706 (identical, as it happens, to that given by Figure 35.4), he obtained Rb-Sr ages of 98 ± 15 and 115 ± 10 m.y. from two specimens.

GRANITE SHEETS FROM STONINGTON ISLAND

A dike intrusion sequence has been described in detail from Stonington Island by Fraser (1965) who, from lack of evidence to the contrary, followed Adie (1954) in assigning a Precambrian age to the gneiss and to all of the dike suites, except for obviously fresh basalt dikes (Tertiary?). Specimens were collected by us from the pink and gray granite sheets which, according to Fraser (1965), were intruded early in the dike sequence. The specimens were collected from Anemometer Hill on Stonington Island (Fraser, 1965, Figure 4): BS 101.1–101.6 from the gray granitic dike 40 m south of the summit and BS 101.7–101.12 from the thick, pink granitic dike 50 m northeast of the summit.

Petrography

The gray dike is medium grained and contains the assemblage plagioclase–quartz–microcline perthite–potash feldspar–biotite–chlorite. Plagioclase is strongly zoned andesine-oligoclase with sericitized cores. Potash feldspar and perthite vary in amount from common to interstitial and are accompanied by myrmekite. Quartz occurs as small crystals, some with strained extinction. Biotite is largely altered to aggregates of chlorite-epidote, and opaque grains are common.

The pink dike is considerably more coarse grained but has a similar mineral assemblage; in some specimens it

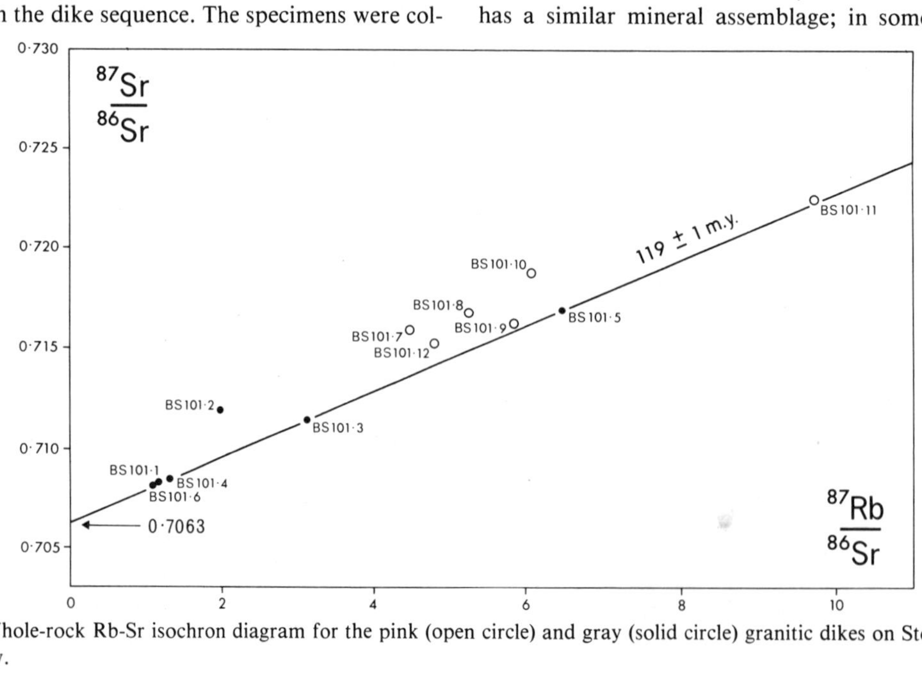

Figure 35.5. Whole-rock Rb-Sr isochron diagram for the pink (open circle) and gray (solid circle) granitic dikes on Stonington Island, Marguerite Bay.

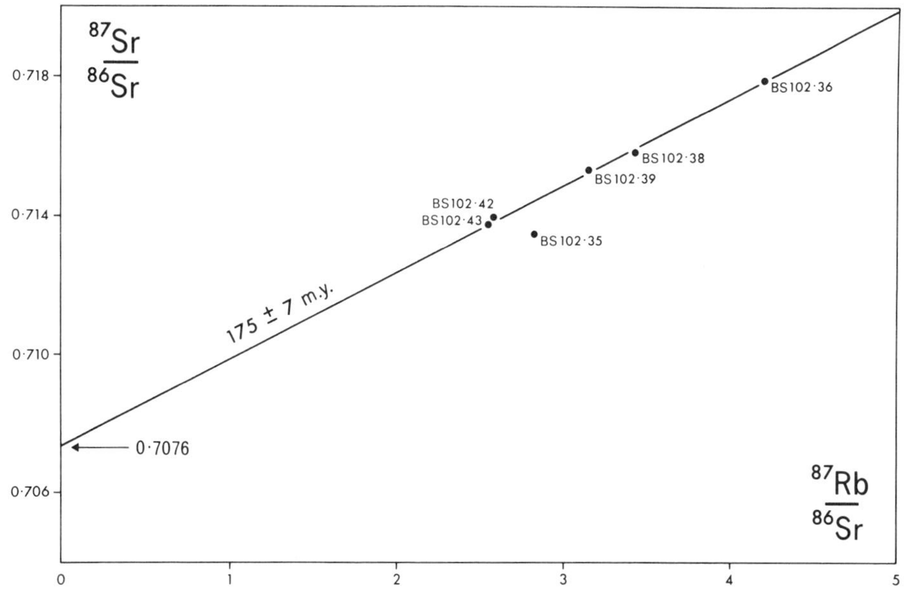

Figure 35.6. Whole-rock Rb-Sr isochron diagram for the pink granitic gneiss in the Debenham Islands, Marguerite Bay.

contains more K-feldspar and perthite than plagioclase. Chlorite with epidote granules occurs after biotite, and muscovite is present in one rock. Quartz occurs in large plates and as interstitial grains, all with strained extinction.

Radiometric data

Five specimens from the gray dike define an isochron with apparent age 119 ± 1 m.y. and IR = 0.7063 (Figure 35.5). BS 101.1 is from the center of the dike, BS 101.2 is from the margin, and the remainder are from the main body of the dike; the only point not falling on the isochron is BS 101.2 from the marginal facies. On the Rb-Sr diagram (Figure 35.5) the six samples from the pink dike give a poorly defined trend roughly parallel to the gray dike isochron.

Our interpretation of these data is that the thin, fine-grained gray dike cooled more quickly than the pink dike, and only specimen BS 101.2 from the contact was contaminated by interchange of Sr isotopes with the country rock. The pink dike is coarse grained and has no chilled margin; it may have cooled slowly or may have been intruded in a partially crystalline state and was possibly more contaminated by Sr isotopes from the wall rocks. This interpretation is supported by the observation that marginal chilling is absent from the early granitic and mafic dikes and sheets also found in this area, which together with other field evidence, suggests that the dikes were intruded into hot country rocks (Fraser, 1965, p. 26).

Grikurov and others (1966) report a K-Ar age* of 110 m.y. on biotite from a xenolith in granite veins from Stonington Island. A whole-rock Rb-Sr analysis on single specimens, using an assumed IR = 0.706, has given ages of 115 ± 10 and 105 ± 15 m.y. on a granodiorite and a quartz-diorite dike (Halpern, 1972).

We conclude that the granitic dikes on Stonington Island, which are petrographically similar to the Debenham Islands granite and have a nearly identical initial $^{87}Sr/^{86}Sr$ ratio, belong to the same Upper Cretaceous plutonic suite. It also follows from the geologic relationships that *all* of the mafic, granitic, and pegmatitic dikes on Stonington Island are of Late Cretaceous or younger age.

GNEISSIC METAMORPHIC COMPLEX IN MARGUERITE BAY

The metamorphic rocks in this area have been described in detail by Hoskins (1963) and Fraser (1965). Hoskins recognized two main groups: banded biotite gneiss of sedimentary origin, and granitic and dioritic gneiss of probably igneous origin.

A suite of specimens was collected from the leucocratic gneiss near Mast Hill on Stonington Island (Fraser, 1965, Figure 4), but the samples showed too little variation in Rb-Sr ratio for the construction of an isochron. However, a second suite collected from the granitic gneiss on Barry Island in the Debenham Islands (Hoskins, 1963, Figure 2, station Y.620) proved suitable.

Petrography

The granite gneiss is a pink, well-foliated rock with a poorly developed fabric in thin section. It consists of a quartz-perthite-plagioclase mosaic with ragged biotite flakes completely or partially altered to chlorite and opaque minerals. The microcline perthite is fresh, and the oligoclase feldspar is only slightly sericitized; the rock also contains antiperthite.

Radiometric data

Six specimens of granite gneiss define an Rb-Sr isochron giving an apparent age of 175 ± 7 m.y. (Figure 35.6). The low initial $^{87}Sr/^{86}Sr$ ratio of 0.7076 suggests that this is a minimum age for the regional metamorphism of a rock with no earlier crustal history. This determination agrees with Halpern's (1972) preliminary Rb-Sr work in which two samples of gneiss from different areas give an approximate age of 200 m.y.

Biotite, separated from the leucocratic gneiss from Stonington Island, gives a K-Ar age of 111 m.y. (BS 101.19, Table 35.1). This result agrees with a number of K-Ar determinations on both hornblende and biotite, separated from various gneisses and amphibolites in Marguerite Bay, which have given ages of 105, 110, 110, 119, and 133 m.y. (Grikurov and others, 1966). A biotite from the diorite

* All K-Ar ages have been recalculated using $\lambda_e = 0.584 \times 10^{-10}$/yr and $\lambda_\beta = 4.72 \times 10^{-10}$/yr.

gneiss on Stonington Island has also given a Rb-Sr age of 108 ± 5 m.y. (Halpern, 1971).

SUMMARY

From the results of our own and previous work in the Marguerite Bay area, we draw the following conclusions:

1. The pink and white granite suites and associated hypabyssal rocks are partly, and perhaps entirely, of Cretaceous age.
2. This area lay within the Andean magmatic arc at this time, and thorough heating of the older gneiss and metamorphic rocks reset both biotite and hornblende K-Ar ages to give Cretaceous dates.
3. The Rb-Sr whole-rock age of 175 m.y. (Early Jurassic) obtained from the orthogneiss is a minimum age for the metamorphism and deformation associated with the Gondwanian orogeny.
4. If it can be demonstrated that the cobbles of pink granite found locally in rocks assigned to the Antarctic Peninsula Volcanic Group are indeed petrographically identical to the Debenham Islands coarse pink granite (Adie, 1954), then those particular volcanic rocks are Late Cretaceous or younger in age, not Late Jurassic.
5. There is no evidence, to data, for the presence of either Precambrian or lower Paleozoic rocks in this area.

Discussion

The term Gondwanian orogeny was proposed by DuToit (1937) for an early Mesozoic (late Triassic–early Jurassic) event affecting thick sedimentary sequences in the southern continents. It has recently been applied in this sense to the Antarctic Peninsula (Dalziel and Elliot, 1973; Elliot, 1975) where the Trinity Peninsula Formation, a thick sequence of turbidite facies sedimentary rocks, was deformed during the Gondwanian event and was subject to uplift and erosion before deposition of the Antarctic Peninsula Volcanic Group. This formation was previously thought to be of late Paleozoic age (Adie, 1957; Grikurov, 1972), but recent work has shown that the Cape Legoupil Formation, which is now correlated with the Trinity Peninsula Formation, contains fossils of probably Early–Middle Triassic age (Thomson, 1975). In addition, rocks on Alexander Island, correlated on lithologic and petrologic similarity with the Trinity Peninsula Formation, contain fossils yielding a Middle–Late Triassic age (Edwards, Chapter 36, this volume).

In the southern part of Graham Land, the Trinity Peninsula Formation has not been recognized with certainty (Figure 35.1), but it has been suggested (Skinner, 1973) that most or all of the metamorphic and migmatitic rocks found in these areas are strongly metamorphosed parts of one formation. This idea finds support from evidence that the degree of regional metamorphism of the Trinity Peninsula Formation increases to the south (Aitkenhead, 1975; Elliot, 1966) and from the fact that the base of the formation has nowhere been observed.

Our radiometric data suggest a minimum age of 175 m.y. for the main metamorphic event in the granitic paragneiss of Marguerite Bay. This is a minimum age for the Gondwanian orogney and, if the stratigraphic range of the Trinity Peninsula Formation does not extend much higher than the lowermost Jurassic (about 200 m.y. in age), then it is possible to correlate the banded biotite gneiss and other orthogneiss in Marguerite Bay with this formation. However, the possibility that the biotite gneiss is older is not excluded; this can only be resolved by further radiometric dating.

An extensive K-Ar dating program on rocks from the Antarctic Peninsula (Rex, 1976) has shown that plutonic rocks of 135–183 m.y. old are widely distributed (Figure 35.1). These ages, from plutons ranging from gabbro to granite, suggest that subduction beneath the western margin of Gondwanaland took place continuously during the period between the Gondwanian event and the opening of the Andean marginal basin in Late Jurassic–Early Cretaceous time (Thomson and others, Chapter 19, this volume). Together with data from younger plutons (Rex, 1976), these dates indicate that the Antarctic Peninsula has been the site of an active magmatic arc for more than 180 m.y. In the absence of stratigraphic markers, subdivision of these igneous rocks into "Andean" and "Gondwanian" suites will depend upon future radiometric dating coupled with geochemical analysis.

Before about 175 m.y. ago the Marguerite Bay area was located within the active Gondwanian magmatic arc, but K-Ar data summarized elsewhere (Rex, 1976) suggest that uplift and cooling of peripheral areas had already occurred by 177–200 m.y. ago. Absence of older, pre-183 m.y. old plutons may be due to the erosion level; it is possible that in Graham Land few outcrops reveal rocks beneath the Trinity Peninsula Formation and its metamorphic equivalents.

From the evidence already presented, it would be reasonable to assume that all of the Antarctic Peninsula metamorphic complex in Graham Land and the Scotia arc is of Gondwanian age. However, two pieces of evidence (Rex, 1976) suggest that this is not so:

1. Rb-Sr whole-rock data from the metamorphic complex on the South Orkney Islands have yielded an "errorchron" of 286 ± 57 m.y., and
2. K-Ar ages of 237 and 243 m.y. have been obtained on hornblende from gneiss at Target Hill west of the Jason Peninsula in Graham Land (Figure 35.1).

The Rb-Sr data from the South Orkney Islands suggest that K-Ar ages there of 176–205 m.y. (for review, see Rex, 1976) are reset ages caused by the Gondwanian event and that the main metamorphism occurred during the Paleozoic Borchgrevink orogeny (Craddock, 1972). The Target Hill K-Ar data are significant in that the hornblende is from strongly metamorphosed gneiss in an area where regional metamorphism of the Trinity Peninsula Formation is low grade. To interpret them as anything but minimum ages for regional metamorphism of the pre-Gondwanian basement is therefore difficult. It should be noted that K-Ar ages on biotite (both coexisting with hornblende and occurring separately) in similar metamorphic rocks seem to have been thermally reset by adjacent plutons to give ages of 164 and 181 m.y. (Figure 35.1). K-Ar ages of 351–384 m.y. from a biotite-muscovite granite on Lizard Hill (Figure 35.1), if substantiated by further work, would be the first evidence of pre-Gondwanian magmatic activity in the Antarctic Peninsula.

In Graham Land and the Scotia arc there is only limited evidence of an older metamorphic basement lying beneath the Trinity Peninsula Formation; it may, however, be more extensively exposed at the deeper structural levels seen farther south in Palmer Land, Ellsworth Land, and Marie Byrd Land.

Acknowledgments. The Rb-Sr and K-Ar analyses were carried out by authors A. G. and D. C. R. in the geochronologic laboratories of the University of Leeds under the general direction of M. H. Dodson. P. W. G. T. was responsible for the geologic interpretation and preparation of the manuscript. We thank J. C. Roddick and R. J. Pankhurst for their cosntructive criticism of the paper, W. Wilkinson for assisting with sample preparation, G. M. Tew for typing the manuscript, and R. J. Boud for drafting the diagrams.

References cited

Adie, R. J. 1954. The petrology of Graham Land. I. The basement complex; early Paleozoic plutonic and volcanic rocks. *Scient. Rep. Falkld Isl. Depend. Surv.,* No. 11, 22 pp.

Adie, R. J. 1957. The petrology of Graham Land. III. Metamorphic rocks of the Trinity Peninsula Series. *Scient. Rep. Falkld Isl. Depend. Surv.,* No. 20, 26 pp.

Adie, R. J. 1962. The geology of Antarctica. (*In* Wexler, H., Rubin, M. J., and Caskey, J. E. *eds. Antarctic Research: The Matthew Fontaine Maury Memorial Symposium.* Washington, American Geophysical Union, 26-39.) [Geophysics Monograph No. 7.]

Adie, R. J. 1964. Geological history. (*In* Priestley, R. E., Adie, R. J., and Robin, G. de Q., *eds. Antarctic Research.* London, Butterworth and Company, 118-62.)

Adie, R. J. 1972. Evolution of volcanism in the Antarctic Peninsula. (*In* Adie, R. J., *ed. Antarctic Geology and Geophysics.* Oslo, Universitetsforlaget, 137-41.)

Aitkenhead, N. 1965. The geology of the Duse Bay-Larsen Inlet area, northeast Graham Land (with particular reference to the Trinity Peninsula Series). *Scient. Rep. Br. antarc. Surv.,* No. 51, 62 pp.

Craddock, C. 1972. Antarctic tectonics. (*In* Adie, R. J., *ed. Antarctic Geology and Geophysics.* Oslo, Universitetsforlaget, 449-55.)

Dalziel, I. W. D., and Elliot, D. H. 1973. The Scotia arc and Antarctic margin. (*In* Nairn, A. E. M., and Stehli, F. G., *eds. The Ocean Basins and Margins. 1. The South Atlantic.* New York, Plenum Press. 171-246.)

DuToit, A. L. 1937. *Our Wandering Continents.* Edinburgh, Oliver and Boyd, 366 pp.

Elliot, D. H. 1964. The petrology of the Argentine Islands. *Scient. Rep. Br. antarc. Surv.,* No. 41, 31 pp.

Elliot, D. H. 1965. Geology of north-west Trinity Peninsula, Graham Land. *Bull. Br. antarc. Surv.,* No. 7, 1-24.

Elliot, D. H. 1966. Geology of the Nordenskjöld Coast and a comparison with north-west Trinity Peninsula, Graham Land. *Bull. Br. antarc. Surv.,* No. 10, 1-43.

Elliot, D. H. 1975. Tectonics of Antarctica: a review. *Am. J. Sci., 275-A,* 45-106.

Fraser, A. G. 1965. The petrology of Stonington and Trepassey Islands, Marguerite Bay. *Scient. Rep. Br. antarc. Surv.,* No. 52, 51 pp.

Grikurov, G. E. 1972. Tectonics of the Antarctandes. (*In* Adie, R. J., *ed. Antarctic Geology and Geophysics.* Oslo, Universitetsforlaget, 163-67.)

Grikurov, G. E., Krylov, A. Ya., and Silin, Yu. 1966. Absoliutnyi vozrast nekotorykh porod iz raiona zaliva Margerit, Antarkticheskii poluostrov [Absolute age of certain rocks in the Marguerite Bay region of the Antarctic Peninsula]. *Dokl. Akad. Nauk SSSR,* Geology, *171,* No. 6, 1399-401. [English translation: *Dokl. (Proc.) Acad. Sci. U.S.S.R.,* Geological sciences sect., *171,* 127-30.]

Halpern, M. 1971. Evidence for Gondwanaland from a review of West Antarctic radiometric ages. (*In* Quam, L. O., and Porter, H. O., *eds.* Research in the Antarctic. *Am. Assoc. Adv. Sci.,* Publn No. 93, 717-30.)

Halpern, M. 1972. Rb-Sr total rock and mineral ages from the Marguerite Bay area, Kohler Range and Fosdick Mountains. (*In* Adie, R. J., *ed. Antarctic Geology and Geophysics.* Oslo, Universitetsforlaget, 197-206.)

Hooper, P. R. 1962. The petrology of Anvers Island and adjacent islands. *Scient. Rep. Falkld Isl. Depend. Surv.,* No. 34, 69 pp.

Hoskins, A. K. 1963. The basement complex of Neny Fjord, Graham Land. *Scient. Rep. Br. antarc. Surv.,* No. 43, 49 pp.

Pankhurst, R. J., and O'Nions, R. K. 1973. Determination of Rb-Sr and $^{87}Sr/^{86}Sr$ ratios of some standard rocks and evaluation of X-ray fluorescence spectrometry in Rb-Sr geochemistry. *Chem. Geol., 12,* 127-36.

Pankhurst, R. J., and Pidgeon, R. T. 1976. Inherited isotope systems and the source region pre-history of early Caledonian granites in the Dalradian Series of Scotland. *Earth planet. Sci. Lett., 31,* 55-68.

Rex, D. C. 1976. Geochronology in relation to the stratigraphy of the Antarctic Peninsula. *Bull. Br. antarc. Surv.,* No. 43, 49-58.

Rex, D. C., and Baker, P. E. 1973. Age and petrology of the Cornwallis Island granodiorite. *Bull. Br. antarc. Surv.,* No. 32, 55-61.

Rex, D. C., and Dodson, M. H. 1970. Improved resolution and precision of argon analysis using an MS10 mass spectrometer. *Eclog. geol. Helv., 63,* No. 1, 257-80.

Roddick, J. E., and Compston, W. 1977. Strontium isotopic equilibration: a resolution to a paradox. *Earth planet. Sci. Lett., 34,* 238-46.

Scott, K. M. 1965. Geology of the southern Gerlache Strait region, Antarctica. *J. Geol., 73,* No. 3, 518-27.

Skinner, A. C. 1973. Geology of north-western Palmer Land between Eureka and Meiklejohn glaciers. *Bull. Br. antarc. Surv.,* No. 35, 1-22.

Thomson, M. R. A. 1975. First marine Triassic fauna from the Antarctic Peninsula. *Nature, Lond., 257,* No. 5527, 577-78.

York, D. 1969. Least squares fitting of a straight line with correlated errors. *Earth planet. Sci. Lett., 5,* 320-24.

36

New Paleontologic Evidence of Triassic Sedimentation in West Antarctica

C. W. EDWARDS*
British Antarctic Survey, Natural Environment Research Council, Madingley Road, Cambridge CB3 0ET, U.K.

Abstract. A new benthonic invertebrate assemblage, collected from a vitric tuff within a metasedimentary sequence in the Lully Foothills of central Alexander Island, provides the first macropaleontologic material for dating this sequence. The fauna consists of coelenterates, gastropods, bivalves, and a number of echinoderm fragments. Although poorly preserved, the invertebrates suggest an early Mesozoic (Middle–Late Triassic) age for at least this part of the metasedimentary sequence, which has been compared with the Trinity Peninsula Formation of northern Graham Land, but which is now formally named the LeMay Formation. The Triassic age of this formation places new constraints on the timing of the tectonic and orogenic events of the Gondwanian orogeny in the Antarctic Peninsula; the sedimentary beds containing the fossils were folded and metamorphosed during this orogeny, and the glaucophane schist facies was reached locally. Comparison of the Lully Foothills fauna with the marine Triassic faunas of the northern Antarctic Peninsula, South America, Australasia, and Japan has revealed similarities in several faunal elements which suggest that they had a circum-Pacific distribution in early Mesozoic time. The age of the Trinity Peninsula Formation as originally defined is critically reexamined in the light of recent investigations; these new studies indicate that large parts of the sequence may be early Mesozoic in age.

Alexander Island is situated on the western side of the Antarctic Peninsula with its eastern margin adjacent to Palmer Land (Figure 36.1). In 1975 a mixed neritic invertebrate fossil assemblage was collected from volcaniclastic sedimentary beds exposed in the central part of the island. This paper describes the occurrence and geologic setting of the fossils and discusses their significance in terms of the stratigraphy of West Antarctica.

Geologic setting

The fossils were collected from a nunatak at lat. 70° 38′ S, long. 69° 44′ W in the Lully Foothills of central Alexander Island (Figures 36.1 and 36.2). They occur within a 2-m-thick vitric tuff, which forms part of a mafic extrusive and volcaniclastic member of a largely metasedimentary formation, exposed over large areas of the island (Bell, 1973*a, b,* 1974, 1976).

These rocks in central Alexander Island were first discussed by Grikurov and others (1967) who obtained a K-Ar age of 165 m.y. for a biotite concentrate from an arkosic sandstone, collected in the LeMay Range east of the Lully Foothills. However, this age was at variance with a palynologic determination by Grikurov and Dibner (1968). They extracted miospores from the metasedimentary rocks at Hope Bay, northern Antarctic Peninsula, and also from the deformed strata of central Alexander Island, and concluded that the palynomorphs at both localities were similar and had Lower and Middle Carboniferous affinities. Previously, W. N. Croft (*in* Adie, 1957) had reported plant fragments within the beds at Hope Bay which he considered to be no older than Carboniferous. On these criteria the rocks at both localities were considered by Grikurov and Dibner (1968) to be part of the same sequence (the Trinity Peninsula Formation of Thomson, Chapter 37, this volume; the Trinity Peninsula Series of Adie, 1957; or the Trinity Series of Grikurov and Dibner, 1968), and they argued that the low grade of metamorphism that the Alexander Island sequence had undergone was sufficient to readjust the radiometric date to give the age of the metamorphism. They preferred the palynologic determination as the indicator of the true age of the sequence. However, the number of specimens of each palynomorph genus obtained by Grikurov and Dibner (1968) from the Antarctic Peninsula was regarded by Schopf (1973) as sparse in relation to the normally accepted palynologic defintion of common or abundant, and he maintained that the possibility of the introduction of derived material from extraneous sources was more likely when, as is the case on Alexander Island, there is a poorly preserved and scant assemblage.

Recent work on sedimentary sequences, supposedly laterally equivalent to the Trinity Peninsula Formation, at Cape Legoupil (Thomson, 1975) and on Alexander Island (Edwards, 1980*b*) has proved the presence of Triassic invertebrate fossils, indicating that the original assignment of a Carboniferous(?) age to the Trinity Peninsula Formation (Adie, 1957, 1964) now requires modification.

In view of the ambiguity of the age of the Trinity Peninsula Formation, and the absence of substantial evidence for the correlation of the sequence at the type locality with the sequence on Alexander Island, it seems appropriate here to name formally the *LeMay Formation,* to include the metamorphosed layered rocks of central Alexander Island and the equivalent sequences in the northern and southern parts of the island. The type area is the northwestern LeMay Range and Lully Foothills. The known and inferred extent of the LeMay Formation on Alexander Is-

* Present address: BP Petroleum Development Brazil Ltd., Avenida Presidente Antonio Carlos, 51–18°, Rio de Janeiro, 20020, R.J. Brazil.

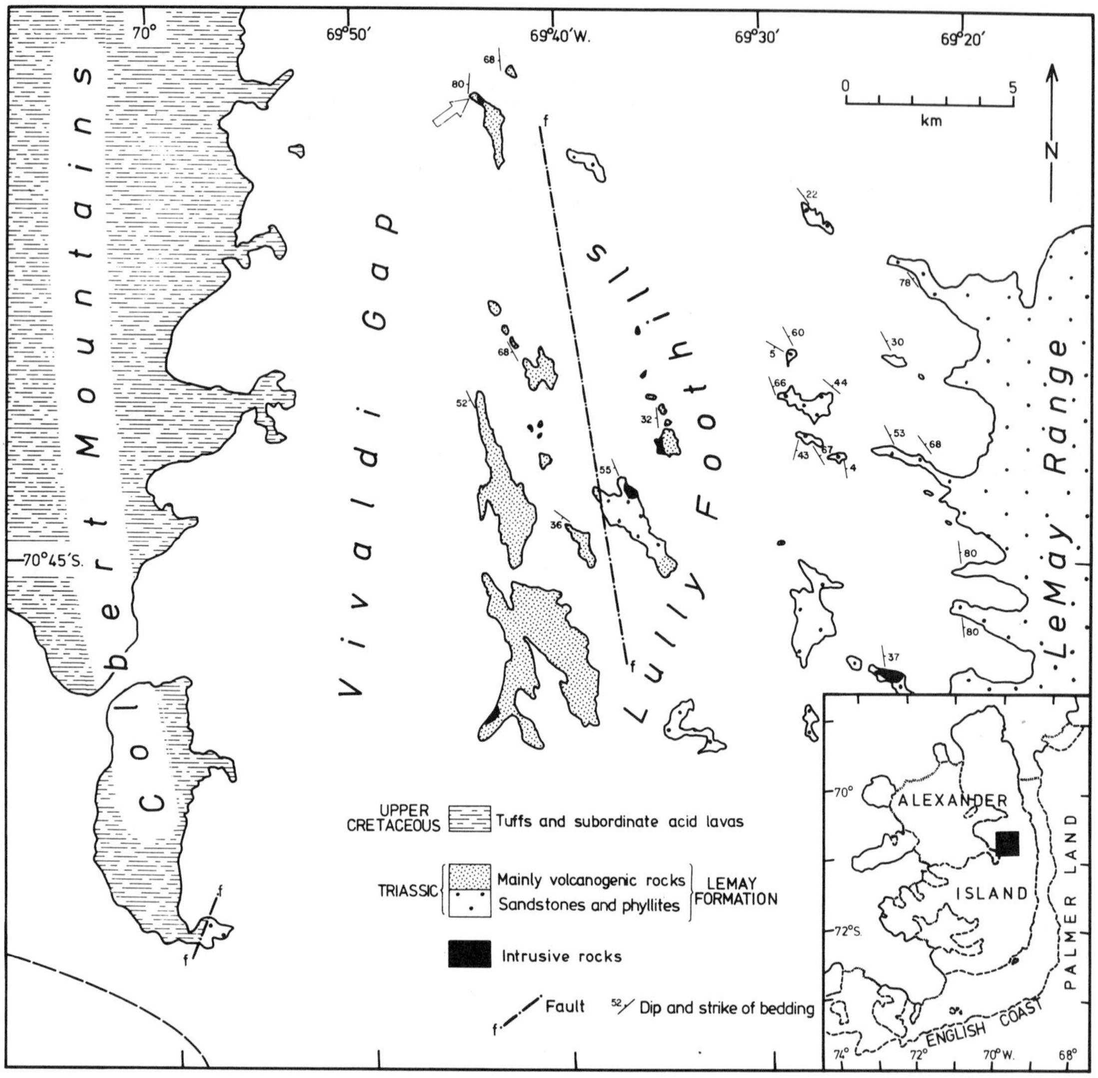

Figure 36.1. Geologic map of the Lully Foothills area of central Alexander Island. The fossil locality is indicated by an arrow. Unpublished geologic data for the Colbert Mountains provided by R. W. Burn and B. W. Care.

Figure 36.2. View eastward from the Colbert Mountains toward the peneplain of the Douglas Range and LeMay Range. Triassic fossils occur in tuffs at the northern end of the nunatak in the middle distance.

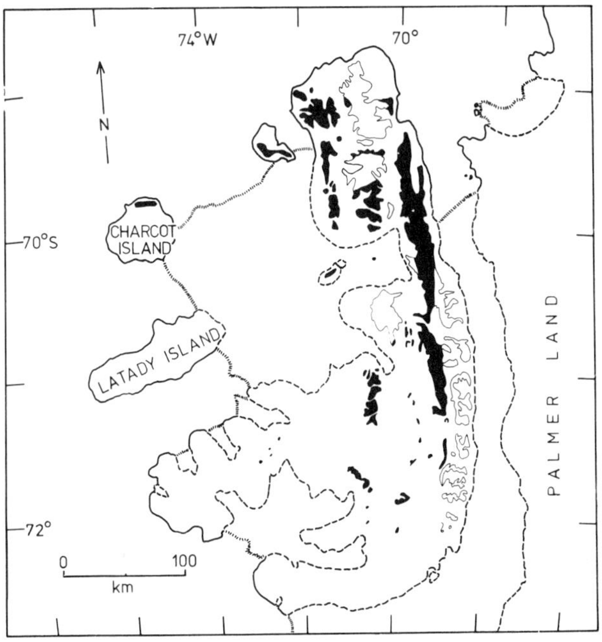

Figure 36.3. Sketch map of Alexander Island showing the approximate extent of the LeMay Formation in black. Outcrops of younger formations and igneous intrusions are shown as unshaded outlines. The remaining areas are ice covered.

land is shown in Figure 36.3. Previous workers have referred to this sequence on Alexander Island as the Trinity Peninsula Series or Trinity Series.

The LeMay Formation is largely composed of flysch-type deposits of sandstone and argillite, with a few outcrops of red chert, serpentinized ultramafic rock, and pillow lavas of uncertain stratigraphic position. Much of the formation is subisoclinally folded with eastward-dipping axial planes which strike approximately north–south, subparallel to the Antarctic Peninsula. Fold relationships indicate at least two major phases of deformation, although open asymmetric folds and kink bands could represent later phases. These "non-sequence type" ophiolites (Miyashiro, 1973, p. 95) and the presence of localized metamorphism of glaucophane schist facies within the LeMay Formation suggest that subduction processes were active in the region, but the ages of the metamorphism and the emplacement of the ophiolites are unknown. Grikurov (1971) divided the sequence in central Alexander Island into lower, middle, and upper terrigenous "series," and a fourth "series" of metasedimentary and metavolcanic rocks, with an aggregate thickness of about 3500 m. In the eastern part of the island the LeMay Formation metasedimentary rocks are unconformably overlain by relatively undisturbed sedimentary beds of the Fossil Bluff Formation (Upper Jurassic–Lower Cretaceous) (Edwards, 1980*a*), and elsewhere on the island by a subhorizontal Upper Cretaceous lava and tuff sequence (Grikurov and others, 1967; Bell, 1974) and by Cenozoic lavas and tuffs (Bell, 1973*a*).

Description of the fauna

The invertebrate fossils obtained from the vitric tuff (Edwards, 1980*b*) all occur as external molds, but despite this unpromising mode of preservation, they show a surprising amount of surface detail (Figure 36.4). The collection includes ten gastropod species belonging to eight genera; five bivalve genera; and echinoid, crinoid, and coelenterate fragments. Additional specimens are undoubtedly present within the indurated sediment, but only those forms on naturally weathered surfaces are immediately obvious. The rock is massive and does not split along any preferred orientation, but at the outcrop it is considerably frost shattered and irregularly broken.

The fossils identified are:

Thamnasteria(?) sp.
Protofusus sp.
Rhabdocolpus sp.
Procerithium (Apicaria) sp.
Procerithiid(?) gastropods
Zygopleura (Katosira) sp.
Omphaloptycha(?) sp.
Amberleya (Eucyclus)(?) sp.
Amphitrochus sp.
Turritellid gastropods
Scurriopsis(?) aranetexta sp. nov. Edwards
Balantioselena aff. *gairi* Speden
Waagenoperna aff. *ozawai* (Kobayashi)
Antiquilima sp.
Indeterminate carditoid bivalve
Indeterminate parallelodontid bivalve
Isocrinus cf. *californicus* Clark
Triadocidaris(?) sp.

The various specimens present in the Alexander Island thanatocoenosis can be compared favorably with similar species from Triassic rocks in New Zealand, Japan, and South America. Of the bivalves in the Lully Foothills assemblage, *Balantioselena* aff. *gairi* has an almost identical counterpart in the Middle Triassic (Ladinian) succession in New Zealand. Likewise, *Waagenoperna* aff. *ozawai* is morphologically similar to *W. ozawai* from the Middle–Upper Triassic (Ladinian–Carnian) Mine Series in Japan. Most of the gastropods are considered to have closely related forms in the Middle and Upper Triassic rocks of Peru, Chile, and northern Argentina (Edwards, 1980*b*).

Comparison with other faunas

The only other Triassic invertebrate fauna from Antarctica is a collection from the Legoupil Formation of northwestern Graham Land (Thomson, 1975). This fauna consists mainly of bivalves but also includes poorly preserved brachiopods (?) and gastropods. The recognition of *Bakevellia (Bakevelloides)* aff. *hekiensis* (Kobayashi and Ichikawa) and a new species of the myophoriid genus *Neoschizodus* suggested a general Triassic age for the formation (Thomson, 1975). The latter form (*Neoschizodus halperni* sp. nov. Thomson) was compared with Lower and Middle Triassic species on the basis of *N. laevigatus* (Ziethen), whereas Thomson (1975) compared the bakevelliid with Upper Triassic species of *Maizuria (Bakevellia)* and *Bakevelloides* from Japan and New Guinea. However, since his paper was written, attention has been drawn to the occurrence of *B. (Bakevelloides)* cf. *hekiensis* at several localities in rocks of the Kaihikuan stage (broadly equated with the Ladinian [Middle Triassic]) at Corbies Creek, North Otago, New Zealand (personal communication from I. G. Speden to M. R. A. Thomson; see Force, 1974),

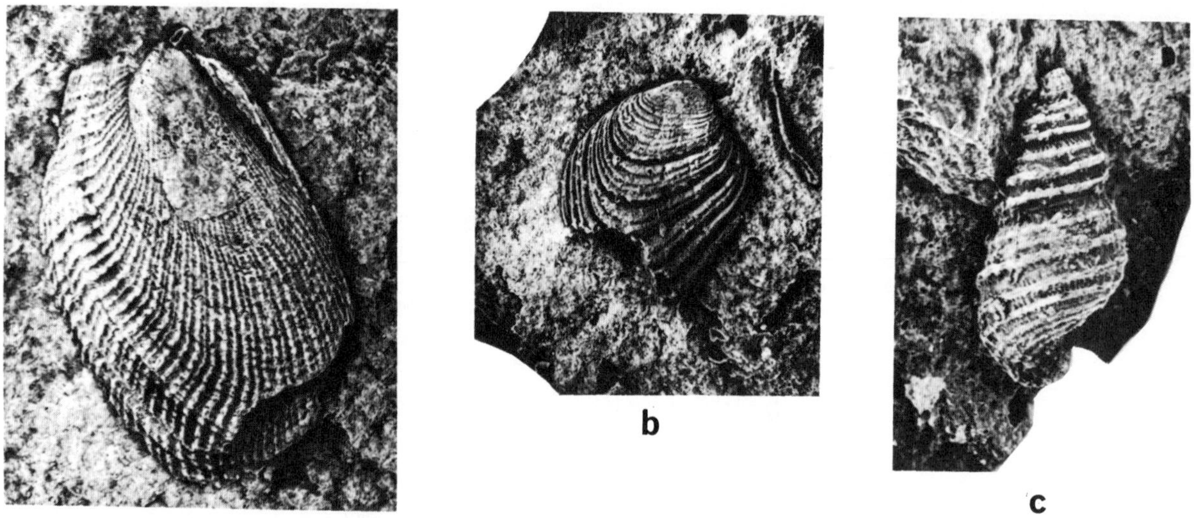

a b c

Figure 36.4. Triassic fossils from the LeMay Formation, northern Lully Foothills, Alexander Island: a. *Antiquilima* sp. ×2.5; b. *Balantioselena* aff. *gairi* Speden ×2.5; c. *Amberleya (Eucyclus)(?)* sp. ×2.5.

which narrows the indicated stratigraphic range of the Legoupil Formation from Lower–Upper Triassic to Lower–Middle Triassic. The beds at Corbies Creek have also yielded *Balantioselena gairi* (Speden, *in* Gair and others, 1962), similar to the Alexander Island species. Thus, from the two West Antarctic Triassic invertebrate localities, two unrelated bivales *(Bakevellia (Bakevelloides)* aff. *hekiensis* and *Balantioselena* aff. *gairi)* are recognized as having similar species counterparts within the same formation in New Zealand. Although the faunas in the LeMay Formation and the Legoupil Formation have no common elements, on the present evidence it is conceivable that the two formations may be approximately coeval.

Delicate leaf impressions were also collected in the Lully Foothills from a siltstone which immediately overlies the fossiliferous tuffaceous unit. Orlando (1968) described a Lower to Middle Triassic flora collected from Williams Point, Livingston Island, South Shetland Islands, and Schopf (1973) examined plant material from the Miers Bluff Formation at an adjacent locality on the same island. Although this material is poorly preserved, Schopf considered it to be probably younger than Carboniferous, and he did not exclude the possibility of a Mezozoic age. However, further comparisons are necessary to determine whether there are any similarities between the plant remains from the Lully Foothills and those from the South Shetland Islands, or indeed from any other Southern Hemisphere locality.

Geochronology of the LeMay Formation

Several elements of the fauna give a Middle–Late Triassic age, but others, primarily because of incomplete preservation, can only be assigned a Mesozoic age. The significance of the age of the LeMay Formation is considerable as it places chronologic constraints on the timing of tectonic and orogenic events of the early Mesozoic (Gondwanian) orogeny (Du Toit, 1937; Elliot, 1975) in this part of West Antarctica. The Triassic age of the invertebrates in the LeMay Formation is significantly different from the age of the formation suggested by both radiometric (Grikurov and others, 1967) and palynologic (Grikurov and Dibner, 1968) means.

Other than the 165 m.y. age of the biotites from the LeMay Range, which Grikurov and others (1967) interpreted as the age of a contact metamorphic event, there are no radiometric age data available from the LeMay Formation. To the northeast in the Marguerite Bay area recent Rb-Sr whole-rock radiometric age determinations give evidence of a major Early Jurassic metamorphic event (Gledhill and others, Chapter 35, this volume), and in the South Orkney Islands schistose rocks underwent a metamorphic event during Late Triassic to Early Jurassic time (199–176 m.y.) (Miller, 1960). A Late Permian age (242 m.y.) (Dalziel, 1972) for the metasedimentary rocks of the Trinity Peninsula Formation at Hope Bay is based on a two-sample Rb-Sr isochron. However, in the South Shetland Islands the supposed lateral equivalent of the Trinity Peninsula Formation, the Miers Bluff Formation (Hobbs, 1968; Thomson, 1975), with a more reliable Rb-Sr five-sample isochron, gives a Late Triassic (197 m.y.) "apparent age" (Dalziel, 1972; personal communication from I. W. D. Dalziel *in* Thomson, 1975, p. 183). These early Mesozoic ages may record a regional metamorphic event recognizable in the LeMay Formation on Alexander Island, which from geologic evidence is post-Ladinian–Carnian and pre-Oxfordian (?) to Kimmeridgian in age.

Paleogeographic considerations

Many of the sedimentary beds within the LeMay Formation were interpreted by Grikurov (1971) and Bell (1975) as being of deep-water origin, and they envisaged a geosynclinal type of environment. Suárez (1976) suggested that the sequence may represent a mélange of sediments scraped from the surface of a subducting oceanic plate, and certainly the presence of sheared pillow lavas, ultramafic rock bodies, and metamorphism of glaucophane schist facies within this sequence endorses this hypothesis. However, clearly some of the LeMay Formation must be of shallow-water origin; the paleontologic material from the Lully Foothills consists entirely of neritic forms which have undergone little transportation. The delicate leaf fragments, coelenterate remains, and echinoid fragments, along with the evidence of rapid changes in the sedimentary environment, suggest nearshore shallow-water deposition for this part of the sequence. The fossiliferous beds occur within a largely volcanic succession of basaltic lavas and tuffs considered to be a facies of the LeMay Formation (Edwards, 1980*b*). Thus it is possible that the lavas and interbedded fossiliferous volcaniclastic sedimentary rocks represent the final stages of the infilling of a forearc basin and are the precursors of the oceanward migration of a volcanic island arc, as suggested in the model proposed by Suárez (1976).

Paleogeographic reconstructions of Gondwanaland during Triassic time (e.g., Smith and others, 1973, Figure 17A; Smith and Briden, 1976) suggest that the distribution of the continental masses may then have provided a continuous shelf environment linking South America, West Antarctica, and Australasia. Less clear, but equally significant in the present discussion, is the relationship between West Antarctica and Japan. As chronoclinal speciation is rarely determinable in fossil species, in part because of gaps in the fossil record, it is proposed that the comparable faunal elements in Japan and in the Lully Foothills of West Antarctica belong to the same topoclinal group. Thus, from the paleogeographic maps presented by Smith and others (1973) and Smith and Briden (1976) and the present evidence of the fossils from Alexander Island and Cape Legoupil, it can be deduced that there probably was a shallow-marine shelf environment marginal to the Gondwana continent and linking South America, Antarctica, and Australasia during Middle to Late Triassic time, similar to that portrayed by Schopf (1974), using information from Kummel (1973), for Early Triassic time. As the gene flow between these widely separated populations may have been greatly reduced solely as a result of the distance separating them, topoclinal speciation may have resulted in the slight variation present in the preserved characteristics of the various elements in the Lully Foothills and their assumed contemporary topoclinal species in Australasia, Japan, and South America.

The proposed shelf region linking the continental landmasses would have ranged in latitude from a subequatorial position to a subpolar (Smith and others, 1973) or midlatitude (McElhinny, 1970; Cox, 1973) position, with an expected latitudinal thermal gradient along the shelves. However, Valentine (1967) indicated that the cosmopolitan nature of species in Early Jurassic time, and inferentially extrapolated back into Triassic time, was made possible by a simple climatic pattern with a relatively uniform worldwide climate. Certainly after the breakup of the Gondwana supercontinent, the establishment of a circumpolar current would have dramatically increased the latitudinal thermal gradient and thus have given rise to faunal realms more distinct than those of Triassic time. Barnard (1973) suggested that the Triassic high latitudes to which elements of the Gondwana *Dicroidium* flora extended may be an indication of the uniformity of the Triassic climate, especially when one considers that at high latitudes there is a period of semi- or total darkness for part of the year. However, this also suggests that perhaps the original configuration of the southern continents is still not fully understood.

Conclusions

Initially the comparison between the Alexander Island metasedimentary succession and the Trinity Peninsula Formation in northern Graham Land was based on rather scant palynologic evidence. However, the discovery of Triassic fossils in the Lully Foothills on Alexander Island and also at Cape Legoupil in northwestern Graham Land, both of comparable age, has thrown doubt on the normally accepted Carboniferous(?) (Adie, 1957, 1964) or Early Carboniferous–Early Permian (Dalziel and Elliot, 1973) age of the Trinity Peninsula Formation.

Both the Miers Bluff Formation of the South Shetland Islands and the Legoupil Formation have been compared structurally (Miller, 1966; Dalziel, 1972) and lithologically (Halpern, 1964) with the Trinity Peninsula Formation of Graham Land. On Alexander Island three phases of deformation were recognized by Bell (1974), but a minimum of two phases has been recognized by this author in the LeMay Range. The Middle–Upper Triassic invertebrate fauna from the Lully Foothills occurs in sedimentary beds which have undergone these deformations, and yet the Upper Jurassic–Lower Cretaceous Fossil Bluff Formation, which is unconformable upon the LeMay Formation, has not. Therefore the intervening period of tectonic activity (Gondwanian orogeny) on Alexander Island must have been restricted to a post-Middle–Late Triassic and pre-Late Jurassic time interval. This is consistent with the age of the deformation or diagenesis of the Miers Bluff Formation (197 m.y. ago). Skinner (1973) tentatively suggested that parts of the metamorphic complex exposed in Palmer Land could represent a high-grade metamorphosed equivalent of the LeMay Formation on Alexander Island. The early Jurassic metamorphic event (Gledhill and others, Chapter 35, this volume), and perhaps also the 200 ± 10 m.y. date (Halpern, 1972), both from Marguerite Bay, Graham Land, may substantiate this interpretation of the Palmer Land complex.

Certainly it seems likely that the LeMay Formation is similar paleontologically to the Legoupil Formation, palynologically to the Trinity Peninsula Formation (irrespective of the absolute identifications of the palynomorphs), and structurally to the Legoupil Formation (Miller, 1966), the Miers Bluff Formation (Dalziel, 1972), and the Trinity Peninsula Formation (Aitkenhead, 1975). The age of the poorly preserved fragments of plants from the Trinity Peninsula Formation at Hope Bay identified by Croft as "no older than Carboniferous" (Adie, 1957) is becoming anachronistic as the evidence points more to an early Mesozoic age for the deposition and subsequent deformation of this apparently widespread sequence.

Acknowledgments. I should like to thank my colleagues at the British Antarctic Survey base on Stonington Island during 1973–75 for encouragement and field support. Discussion of ideas with members of the British Antarctic Survey geology section has proved helpful.

References cited

Adie, R. J. 1957. The petrology of Graham Land: III. Metamorphic rocks of the Trinity Peninsula Series. *Scient. Rep. Falkld Isl. Depend. Surv.*, No. 20, 26 pp.

Adie, R. J. 1964. Geological history. (*In* Priestley, R. E., Adie, R. J., and Robin, G. de Q., *eds. Antarctic Research.* London, Butterworth, 118–62.)

Aitkenhead, N. 1975. The geology of the Duse Bay–Larsen Inlet area, northeast Graham Land (with particular reference to the Trinity Peninsula Series). *Scient. Rep. Br. antarct. Surv.*, No. 51, 62 pp.

Barnard, P. D. W. 1973. Mesozoic floras. (*In* Hughes, N. F., *ed.* Organisms and continents through time. *Spec. Pap. Palaeont.*, *12,* 175–87.)

Bell, C. M. 1973*a*. The geology of Beethoven Peninsula, south-western Alexander Island. *Bull. Br. antarct. Surv.*, No. 32, 75–83.

Bell, C. M. 1973*b*. The geology of southern Alexander Island. *Bull. Br. antarct. Surv.*, No. 33–34, 1–16.

Bell, C. M. 1974. Geological observations in northern Alexander Island. *Bull. Br. antarct. Surv.*, No. 39, 35–44.

Bell, C. M. 1975. Structural geology of parts of Alexander Island. *Bull. Br. antarct. Surv.*, No. 41–42, 43–58.

Bell, C. M. 1976. Sedimentary rock from Charcot Island. *Bull. Br. antarct. Surv.*, No. 44, 103–4.

Cox, C. B. 1973. Triassic tetrapods. (*In* Hallam, A., *ed. Atlas of Palaeobiogeography.* Amsterdam, Elsevier, 213–23.)

Dalziel, I. W. D. 1972. Large-scale folding in the Scotia arc. (*In* Adie, R. J., *ed. Antarctic Geology and Geophysics.* Oslo, Universitetsforlaget, 47–55.)

Dalziel, I. W. D., and Elliot, D. H. 1973. The Scotia arc and Antarctic margin. (*In* Nairn, A. E. M., and Stehli, F. G., *eds. The Ocean Basins and Margins. 1. The South Atlantic.* New York, Plenum Press, 171–246.)

Du Toit, A. L. 1937. *Our Wandering Continents.* Edinburgh, Oliver and Boyd, 366 pp.

Edwards, C. W. 1980*a*. New evidence of major faulting on Alexander Island. *Bull. Br. antarct. Surv.*, No. 49, 15–20.

Edwards, C. W. 1980*b*. Early Mesozoic marine fossils from central Alexander Island. *Bull. Br. antarct. Surv.*, No. 49, 33–58.

Elliot, D. H. 1975. Tectonics of Antarctica: a review. *Am. J. Sci.*, *275-A,* 45–106.

Force, E. R. 1974. A comparison of some Triassic rocks in the Hokonui and Alpine belts of South Island, New Zealand. *J. Geol.*, *82,* 37–49.

Gair, H. S., Gregg, D. R., and Speden, I. G. 1962. Triassic fossils from Corbies Creek, North Otago. *N.Z. Jl Geol. Geophys.*, *5,* No. 1, 92–113.

Grikurov, G. E. 1971. Geologischeskoe stroenie tsentral'noi chasti Zemli Aleksandra I [The geological structure of the central part of Alexander I Land]. *Antarktika, Doklady Komissii,* 1968, 13–42.

Grikurov, G. E., and Dibner, A. F. 1968. Novye dannye o serii Triniti (C_{1-3}) v zapadnoi Antarktide [New data on the Trinity Series (C_{1-3}) in West Antarctica]. *Dokl. Akad. Nauk SSSR,* Geol., *179,* No. 2, 410–12. [English translation: *Dokl. (Proc.) Acad. Sci. U.S.S.R.,* Geol., *179,* 39–41.]

Grikurov, G. E., Krylov, A. Ya., and Silin, Yu. I. 1967. Absoliutnyi vozrast nekotorykh porod dugi Skotiia i Zemli Aleksandra I (Zapadnaia Antarktika) [Absolute age of some rocks from the Scotia arc and Alexander I Land (western Antarctica)]. *Dokl. Akad. Nauk SSSR,* Geol., *172,* No. 1, 168–71. [English translation: *Dokl. (Proc.) Acad. Sci. U.S.S.R.,* Geol., *172,* 19–22.]

Halpern, M. 1964. Cretaceous sedimentation in the "General Bernardo O'Higgins" area of north-west Antarctic Peninsula. (*In* Adie, R. J., *ed. Antarctic Geology.* Amsterdam, North-Holland Publishing Company, 334–47.)

Halpern, M. 1972. Rb-Sr total-rock and mineral ages from the Marguerite Bay area, Kohler Range and Fosdick Mountains. (*In* Adie, R. J., *ed. Antarctic Geology and Geophysics.* Oslo, Universitetsforlaget, 197–204.)

Hobbs, G. J. 1968. The geology of the South Shetland Islands: IV. The geology of Livingston Island. *Scient. Rep. Br. antarct. Surv.*, No. 47, 34 pp.

Kummel, B. 1973. Lower Triassic (Scythian) molluscs. (*In* Hallam, A., *ed. Atlas of Palaeobiogeography.* Amsterdam, Elsevier, 225–33.)

McElhinny, M. W. 1970. Formation of the Indian Ocean. *Nature, Lond.*, *228,* No. 5275, 977–79.

Miller, J. A. 1960. Potassium-argon ages of some rocks from the South Atlantic. *Nature, Lond.*, *187,* No. 4742, 1019–20.

Miller, H. 1966. Kleintektonische Untersuchungen in der Umgebung der Station "General Bernardo O'Higgins," Nordwestspitze der antarktischen Halbinsel. *Geol. Rdsch.*, *55,* No. 3, 809–19.

Miyashiro, A. 1973. *Metamorphism and Metamorphic Belts.* London, George Allen and Unwin, 492 pp.

Orlando, H. A. 1968. A new Triassic flora from Livingston Island, South Shetland Islands. *Bull. Br. antarct. Surv.*, No. 16, 1–13.

Schopf, J. M. 1973. Plant material from the Miers Bluff Formation of the South Shetland Islands. *Rep. Inst. polar Stud. Ohio State Univ.*, No. 45, 45 pp.

Schopf, T. J. M. 1974. Permo-Triassic extinctions; relation to sea-floor spreading. *J. Geol.*, *82,* 129–43.

Skinner, A. C. 1973. Geology of north-western Palmer Land between Eureka and Meiklejohn Glaciers. *Bull. Br. antarct. Surv.*, No. 35, 1–22.

Smith, A. G., and Briden, J. C. 1976. *Mesozoic and Cenozoic Paleocontinental Maps.* Cambridge, Cambridge University Press, 63 pp.

Smith, A. G., Briden, J. C., and Drewry, G. E. 1973. Phanerozoic world maps. (*In* Hughes, N. F., *ed.* Organisms and continents through time. *Spec. Pap. Palaeont.*, *12,* 1–39.)

Suárez, M. 1976. Plate-tectonic model for southern Antarctic Peninsula and its relation to southern Andes. *Geology*, *4,* 211–14.

Thomson, M. R. A. 1975. New palaeontological and lithological observations on the Legoupil Formation, north-west Antarctic Peninsula. *Bull. Br. antarct. Surv.*, No. 41–42, 169–85.

Valentine, J. W. 1967. The influence of climatic fluctuations on species diversity within the Tethyan provincial system. (*In* Adams, C. G., and Ager, D. V., *eds.* Aspects of Tethyan biogeography. *Publ. System. Assoc.*, No. 7, 153–66.)

37

Mesozoic Paleogeography of West Antarctica

M. R. A. THOMSON
British Antarctic Survey, Natural Environment Research Council, Madingley Road, Cambridge CB3 0ET, U.K.

Abstract. With the exception of a few volcanic sequences, bedded rocks of Mesozoic age in West Antarctica are largely confined to the Scotia arc and the Antarctic Peninsula. An interrupted stratigraphic record ranging from Triassic to Upper Cretaceous is preserved. The recognition of two Triassic marine faunas in lateral equivalents of the Trinity Peninsula Formation (originally considered late Paleozoic in age) indicates that at least parts of the area were covered by sea during Triassic time. During Middle Jurassic to Early Cretaceous time, the whole region was the site of a volcanic arc, with depositional troughs located both in front of it and behind it in the Antarctic Peninsula area. Gondwanaland remained essentially coherent throughout much of early Mesozoic time with the first break, the rifting of the South Atlantic, occurring during Neocomian time, and the splits between Africa, India, and Antarctica occurring later in the Cretaceous Period. Probable uplift at the end of Early Cretaceous time seems to have confined Late Cretaceous sedimentation to the eastern coast of Graham Land. Limited Triassic faunas show affinities with other marine faunas marginal to Gondwanaland, and similar relationshps are clearly demonstrated by the Late Jurassic faunas. However, by Early Cretaceous (Aptian) time, clearly marked affinities with the faunas of the Indo-Pacific margin of Gondwanaland were gone. Not until Late Cretaceous (Campanian) time is there again evidence of a wide-ranging interchange of faunas with other parts of the world.

At a time when the reassembly of the greater part of the Gondwanaland jigsaw seems to be largely a matter of fine adjustment, and when it is possible to attempt a stage-by-stage account of the supercontinent's wanderings and final breakup (Smith and Briden, 1976), the geologic setting of West or Lesser Antarctica is far from understood. Deep subglacial troughs (Heezen and Bentley, 1972) divide the region into a number of separate continental blocks whose precise relations with one another are still unknown. Elliot (1972) recognized that these individual fragments have probably moved from original positions closer to the mass of East or Greater Antarctica, but until these can be properly relocated, it will not be possible to attempt a precise paleogeographic reconstruction for the region. The present discussion is therefore primarily concerned with the local situation in the Antarctic Peninsula.

Faunal affinities of Antarctic marine fossils with extra-Antarctic occurrences indicate those regions in the world with which there was easy marine access to or from Antarctica and where migration was uninhibited by land barriers, climatic belts, adverse ocean currents, or other causes. It is possible to explain the pre-Late Cretaceous faunal similarities on the basis of a distribution which was marginal to Gondwanaland, but the development of a trans-Gondwanaland seaway along the eastern margin of Africa as long ago as Early Jurassic time (Hallam, 1973) seems to be more speculative.

Although the Andean orogen is generally shown as being continuous into Marie Byrd Land (e.g., Craddock, 1970), there are no known sedimentary or metasedimentary rocks there which can be assigned to the Mesozoic with any certainty. Jurassic and Cretaceous plutons probably represent the roots of a Mesozoic volcanic arc, and Mesozoic volcanic rocks have been recognized in Ellsworth and Marie Byrd Lands (Craddock, 1972). However, at this stage, little more can be deduced of the paleogeography of Marie Byrd Land, and the information summarized here refers mainly to the Antarctic Peninsula.

The paleogeographic situation of South Georgia is outlined by Suárez and Pettigrew (1976). It previously occupied a position close to Tierra del Fuego, and it is not discussed further in this paper.

Stratigraphic terminology

Many of the old stratigraphic names, proposed during the exploratory phases of Antarctic geology, do not conform to current codes of stratigraphic nomenclature and need to be revised before they become even more deeply entrenched in the literature. Two such terms are "Trinity Peninsula Series" and "Upper Jurassic Volcanic Group"; since they would be in constant use in the present discussion, the opportunity of amending them first is taken.

The term Trinity Peninsula *Series* has no time connotation and has been applied to a variety of deformed sedimentary rocks throughout the Antarctic Peninsula area. It is proposed that it be amended to *Trinity Peninsula Formation* and applied strictly to rocks in the type area of northern Graham Land (Adie, 1957*a*; Aitkenhead, 1975). Other related and possibly equivalent formations have been proposed for specific areas: Miers Bluff Formation (Dalziel, 1972), Legoupil Formation (Halpern, 1964), and LeMay Formation (Edwards, Chapter 36, this volume).

It is becoming increasingly apparent that the Upper Jurassic Volcanic Group encompasses rocks other than Jurassic in age (Taylor and others, 1980) and refers to the rocks of a Mesozoic volcanic arc active over a period extending from at least Middle Jurassic to Early Cretaceous time (see below). This group has never been properly defined, although most Antarctic Peninsula geologists have their own understanding of it and are aware of the need for a suitable name. Although few constituent formations have been recognized, the unit is clearly of group rank and should remain so. The term *Antarctic Peninsula Volcanic Group* is proposed, in deference to its wide distribution throughout the area.

Triassic time

PALEOGEOGRAPHY

Until recent years, Triassic sedimentary rocks were unknown in West Antarctica. The "derived" Triassic flora, reported by Orlando (1967) from Williams Point, Livingston Island, probably did not come from clasts in a nearby conglomerate as was believed but from debris transported to the surface in squeeze-up structures from shales *in situ* beneath a capping sill. Triassic marine faunas have now been identified from the Legoupil Formation of northern Antarctic Peninsula (Thomson, 1975*b*), in rocks similar to those of the nearby Trinity Peninsula Formation, and from the LeMay Formation of central Alexander Island (Edwards, Chapter 36, this volume), which has also been compared with the Trinity Peninsula Formation (Grikurov, 1971). Although it would be rash to suggest that the whole of the Trinity Peninsula Formation and its equivalents (reported throughout the length and breadth of the Antarctic Peninsula area) are Triassic, it is clear that at least parts of it are, and that the area was undergoing sedimentation during Triassic time.

An orogenic episode which folded the Trinity Peninsula Formation has been dated from field relationships and radiometric ages as late Triassic to early or Middle Jurassic (e.g., Adie, 1964; Horne, 1969), or more specifically as Late(?) Triassic (Aitkenhead, 1975). However, evidence of Triassic sedimentation suggests that any Triassic activity from this orogeny was minimal and that the greater part of the deformation occurred in Early(?) Jurassic time. This interpretation agrees well with most of the radiometric data (cf. Horne, 1969, p. 74), especially if the base of the Jurassic is considerably older than the 181 m.y. age deduced by Kulp (1961) and is nearer to 200 m.y. in age (Lambert, 1971).

The flysch-like turbidites and graywackes of the Trinity Peninsula Formation are generally envisaged as having accumulated in a geosynclinal environmental (Aitkenhead 1975; Bell, 1975), but the invertebrate faunas of the Legoupil and LeMay Formations suggest that at least parts of the succession had a shallow-water origin. Perhaps these rocks represent the closing stages of a long period of sedimentation in the area. Little information is available on the location of the source area. However, on the basis of scanty evidence, Aitkenhead (1975, p. 39) tentatively suggested that a northwestern source was the most likely for this sequence in the northern Antarctic Peninsula, and C. W. Edwards (personal communication) has evidence for a western source for some related rocks in Alexander Island. These conclusions are problematic, as they indicate a source area in what is now the Pacific Ocean. The identification of the source area for the Trinity Peninsula Formation must rate as one of the most important unsolved paleogeographic problems in the Antarctic Peninsula.

It is not yet possible to assess the true extent of Triassic sedimentation in West Antarctica, because it is not known how much of the Trinity Peninsula Formation may be Triassic, and there is doubt about whether all or part of the metasedimentary sequence exposed in northern and eastern Palmer Land (southern Antarctic Peninsula) belongs to the Trinity Peninsula "Series" (Adie, 1957*a*) or whether it represents metamorphosed equivalents of the Upper Jurassic Latady Formation (Taylor and others, 1979).

PALEOBIOGEOGRAPHY

Paleontologic data for the Triassic in the Antarctic Peninsula are still somewhat scant. However, as pointed out by Edwards (Chapter 36, this volume), the known species seem to constitute part of a circum-Pacific fauna. Important bivales are *Balantioselena, Waagenoperna, Bakevelloides,* and *Neoschizodus,* which show close similarities to forms from Japan, New Guinea, and New Zealand. A varied assemblage of small gastropods from central Alexander Island has counterparts in Peru, Chile, and northern Argentina.

Jurassic and Early Cretaceous time

PALEOGEOGRAPHY

Sufficient data are available to provide a clearer picture of the distribution of land and sea during Jurassic and Early Cretaceous time (Figure 37.1), although the timing of some of the events remains under discussion. In this summary, the Jurassic and Lower Cretaceous rocks are considered together because they form an uninterrupted sequence in the area. The widespread, mainly andesitic to rhyolitic lavas and pyroclastic rocks of the Antarctic Peninsula Volcanic Group probably represent the remnants of a Mesozoic volcanic arc or elongate land-mass built up on an eroded basement of folded Trinity Peninsula Formation (and equivalent formations) and older rocks.

The discontinuity of exposures makes it difficult to assess the geographic form of this arc. With the exception of a marine embayment at Carse Point, Palmer Land, there is no evidence for any seaways across the structure. It therefore seems to have been principally formed of one, or possibly two, large and elongate islands. In Figure 37.1, the Bransfield Strait (a late Cenozoic feature) has been closed to move the South Shetland Islands approximately 50 km closer to the Antarctic Peninsula. However, these islands still seem to lie outside the main arc, and the complex distribution of subaerial volcanic and marine sequences in the area suggests that a number of small islands may have been present there.

In the northern part of the Antarctic Peninsula, the volcanic rocks are locally conformably underlain by plant-bearing sequences of supposedly Middle Jurassic age (Bibby, 1966; Fleet, 1968; Aitkenhead, 1975). However, published opinions as to the age of the enclosed floras are sufficiently diverse (Taylor and others, 1979) to indicate that their age is still poorly known. These deposits probably accumulated in lacustrine depressions on the Jurassic land surface. There is usually a monogenetic basal conglomerate with Trinity Peninsula Formation clasts, and some sequences contain interbedded lavas and tuffs (Aitkenhead, 1975; Bibby, 1966) suggesting that they may be penecontemporaneous with some of the volcanic rocks which nearby rest directly on the Trinity Peninsula Formation.

In the central and eastern Antarctic Peninsula, the volcanic sequences seem to be entirely terrestrial, and the only known fossils are carbonized plant remains (Adie, 1957*b*). However, on the western coast, much of the sequence is water-lain, and Upper Jurassic marine fossils occur in pyroclastic rocks on Adelaide Island (Thomson, 1972*b*). On Low Island (South Shetland Islands) and at Carse Point (Palmer Land) fossiliferous Upper Jurassic marine sedimentary rocks are respectively interbedded with and underlie the volcanic formations (Smellie, 1980; Thomson, 1975*c*). On Livingston Island (South Shetland Islands), Tithonian-Valanginian shale and sandstone pass conformably upward into marine and then terrestrial volcanic rocks (Valenzuela and Hervé, 1972; Covacevich, 1976). Smellie (1980) deduced that the volcaniclastic rocks of Low Island accumulated within an intra-arc basin or on the flanks of an arc, whereas the lavas were erupted subaerially. Much of the marine sequence on western Livingston Island, typified by barely lithified shale and gritty tuffaceous standstone, represents a shallow-water facies, but flysch-like shales rich in trace fossils, near Start Point, are probably deep-water deposits. It seems likely that the western Livingston Island area underwent considerable uplift during the deposition of the exposed sequence. The eruptive centers of these volcanic rocks have proved difficult to locate. Investigators have usually been limited to the speculation that some of the dikes, which now traverse the volcanic successions, were feeders for some of the lavas (e.g., Dewar, 1970) or that the roots of some vents may now be occupied by plutonic intrusions (Fleet, 1968). However, a number of small vents have been identified in the Marguerite Bay area (Figure 37.1). These all lie on a line parallel to the trend of the Antarctic Peninsula and have the linear distribution expected in a volcanic arc.

In eastern Alexander Island, along the western margin of this volcanic chain, there is a thick (more than 5000 m) sedimentary sequence of dominantly shallow-water origin

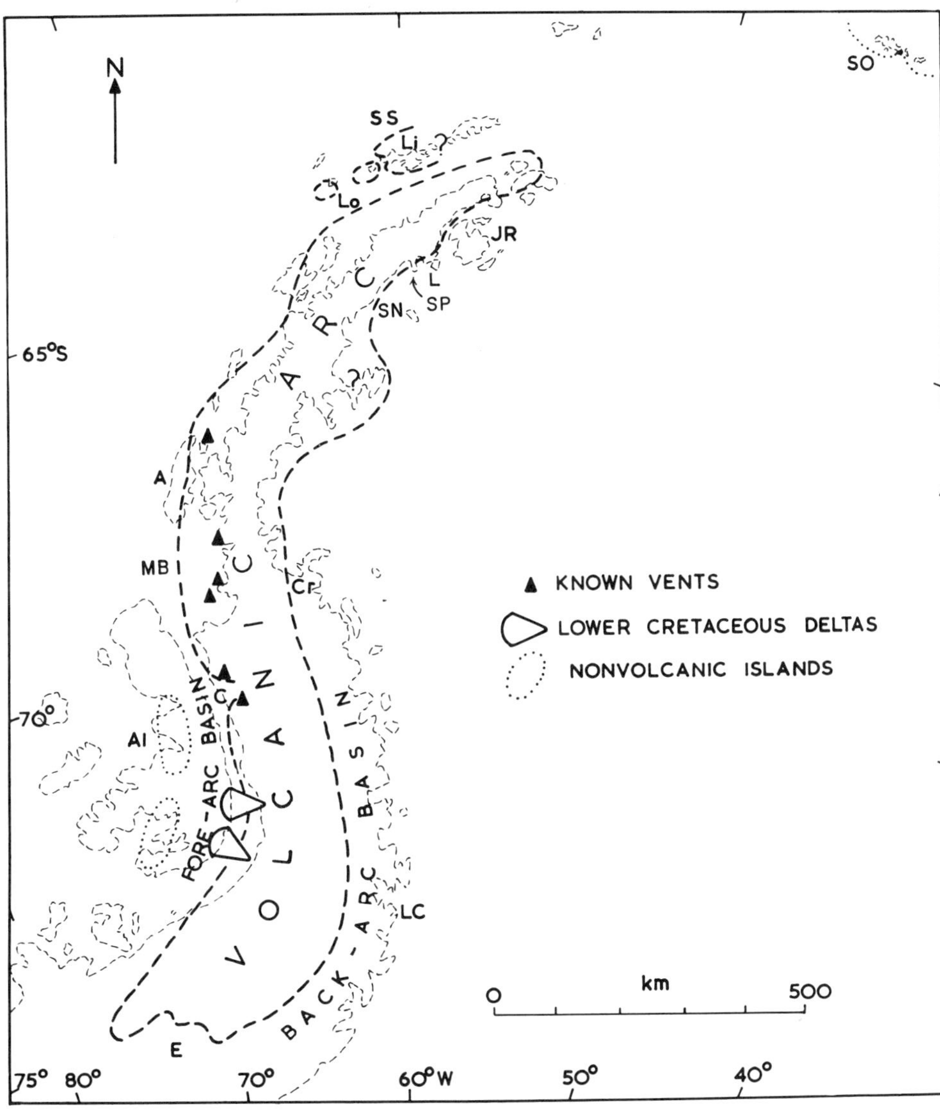

Figure 37.1. Tentative and generalized paleogeographic reconstruction of the Antarctic Peninsula area in Late Jurassic–Early Cretaceous time. A late Cenozoic extensional trough in Bransfield Strait (between the South Shetland Islands and the Antarctic Peninsula) has been closed by approximately 50 km. No allowance has been made for any bending that the peninsula may have undergone. Geographic coordinates relate to the area's present position. A: Adelaide Island. Al: Alexander Island. C: Carse Point. Cr: Crabeater Point. E: eastern Ellsworth Land. JR: James Ross Island. L: Longing Gap. LC: Lassiter Coast. Li: Livingston Island. Lo: Low Island. MB: Marguerite Bay. SN: Seal Nunataks. SO: South Orkney Islands. SP: Sobral Peninsula. SS: South Shetland Islands. Partly after Suárez (1976).

—the Fossil Bluff Formation (Taylor and others, 1979). It rests unconformably on the folded metasedimentary rocks of the LeMay Formation (Edwards, 1980) and spans a time period of at least Kimmeridgian to Albian. Lavas are interbedded in the lower part of the sequence as high as the Tithonian stage (Elliott, 1974), and there is abundant evidence of volcanic debris, introduced by direct aerial fallout, throughout the Cretaceous part of the sequence (e.g., Horne and Thomson, 1972). Rocks include a wide range of mudstones, sandstones, and conglomerates, which were deposited rapidly and which display evidence of instability (notably large-scale slump zones; Bell, 1975) and late-stage remobilization (e.g., clastic dikes). There is no clear division between rocks with Jurassic and Cretaceous faunas. but there are approximately 1000 m of poorly fossiliferous transition beds which include a high proportion of conglomerate (cf. Elliott, 1974). At the southeastern corner of the island, the upper part (Albian or possibly younger) of the sequence is almost entirely composed of plant-bearing sandstone which is partly terrestrial, as indicated by the occurrence of tree trunks and stumps in growth position. There are local cyclothemic coal-measure–type sequences, but these lack any significant development of coal.

It has generally been considered that the bulk of the Fossil Bluff Formation was derived from a source to the east, i.e., the Mesozoic volcanic arc of Palmer Land (Horne, 1969; Taylor and others, 1979), and this still seems most logical in terms of the paleogeographic interpretation by Suárez (1976). However, it is difficult to cite conclusive evidence for an eastern source, and some features even suggest the contrary. Conglomerate has been mapped in the western (supposedly distal) parts of the formation outcrop, and there is evidence that the quartz-rich conglomerate of Stephenson Nunatak (southern Alexander Island) becomes finer grained and better sorted toward the northeast or east, implying a western derivation. C. W. Edwards (personal communication) has noticed that those slump structures from which a paleoslope can be determined seem to indicate a northeast-dipping slope. Thus, the possibility of there being an offshore landmass that also contributed sediment to the Fossil Bluff Formation must be considered.

The author is indebted to C. M. Bell for suggesting that the paleogeographic situation during Fossil Bluff Formation time may have been comparable to that presently found in the Sunda Arc (Le Pichon and others, 1973). Between the volcanic arc (Sumatra, Java, and others) and the trench, there is a chain of islands (Mentawei, Timor, and others) consisting of emergent folded deep-water sedimentary rocks of Eocene to Miocene age, separated from the arc by a slowly subsiding basin. Such a situation would explain the complete lack of Upper Jurassic to Lower Cretaceous rocks in central and western Alexander Island and could account for a westerly source for at least some of the material filling the Fossil Bluff Formation basin.

It now seems that rocks once believed to represent a deep-water axial facies of the Fossil Bluff Formation (Horne, 1969) are really part of the older LeMay Formation (Taylor and others, 1979) and that there is little evidence for Horne's postulated offshore arc. All the rocks known to be part of the Fossil Bluff Formation probably accumulated in a deltaic environment, and include part of Horne's (1969) deltaic, interdeltaic, and shelf facies. In such a regime, local paleogeographic and sedimentation patterns would likely show rapid changes, and this would help explain the observed vertical and lateral variations in rock types.

To the east and south of the ancient volcanic arc, in southeastern Palmer Land and eastern Ellsworth Land, there are thick, highly deformed sequences, thought to represent back-arc accumulations (Suárez, 1976). The Upper Jurassic Latady Formation of the Lassiter Coast consists of a mudstone-shale-sandstone succession with a high proportion of incorporated volcanic debris (Williams and Rowley, 1972). Lavas have not been reported from within the formation, but its upper part is marked by 1000 m of dacitic and andesitic(?) volcanic rocks folded along with the sedimentary rocks (Williams and others, 1972). Mudcracks and ripple marks suggest accumulation in shallow water.

Another thick Jurassic sedimentary sequence in eastern Ellsworth Land to the south has been described as geosynclinal (Laudon, 1972). It consists of shale, sandstone, and conglomerate, which contain fossil faunas ranging from Middle to Late Jurassic in age. Dacitic volcanic rocks, believed to be partly older than a partly contemporaneous with the sedimentary rocks, are also present in the area. Suárez (1976) suggested that the Ellsworth Land sequence represented a continuation of the Latady Formation, but the area between the two has yet to be mapped. Some petrologic differences are apparent in that the Latady Formation sandstone is dominantly lithic (Williams and Rowley, 1972), whereas that of the Ellsworth land sequence is more variable but tends to be arkosic (Laudon, 1972).

Although marine sedimentation on the eastern side of the ancient volcanic arc seems generally to have terminated about the end of Jurassic time, at Crabeater Point, northeastern Palmer Land, there is folded mudstone and conglomerate with a poorly preserved marine fauna of possibly early Cretaceous age (Thomson, 1967; Taylor and others, 1979). The deposits seem to have accumulated in some kind of restricted environment, perhaps an enclosed bay marginal to the retreating Latady Formation sea. Along the coast and eastern hinterland of Palmer Land, between the latitude of Crabeater Point and the Lassiter Coast, there are widespread outcrops of metasedimentary rock of unknown age; some of these rocks may be the equivalent of the Latady Formation (Taylor and others, 1979).

The significance of an isolated outcrop of gently dipping, fossiliferous, upper Jurassic sandstone and shale at Longing Gap is not understood. However, the well-bedded and fissile character of these marine rocks suggests deposition in a quiet-water environment.

The paleogeography of the South Orkney Islands area during Mesozoic time is problematic. Alluvial fan deposits in the central part of the island group (Dalziel and others, 1979) date from late Jurassic or early Cretaceous time (Thomson, 1975*a*; Thomson and Willey, 1975) and indicate that the Coronation Island and Laurie Island areas were parts of a relatively high, mountainous terrain at the time. There is no evidence of Mesozoic volcanicity, and it seems unlikely that the area was ever part of the volcanic arc proper.

The Jurassic and Early Cretaceous paleogeographic history of the Antarctic Peninsula area can be summarized as follows. From Middle Jurassic until at least Early Cretaceous time the Antarctic Peninsula was the site of a volcanic arc. Marine sedimentation occurred both in front of and behind this arc, beginning in eastern Ellsworth Land. Back-arc sedimentary rocks in eastern Palmer Land (Latady Formation) seem to be mainly Upper Jurassic, but Lower(?) Cretaceous deposits at Crabeater Point may represent a continuation of this same episode. Fore-arc sedimentation is represented on Alexander Island (Fossil Bluff Formation) by an apparently continuous sequence of Upper Jurassic to Lower Cretaceous rocks. Sedimentation in this area was strongly influenced by large deltas, which at times became emergent and thus include partly terrestrial sequences. There is some evidence for a chain of offshore, nonvolcanic islands in the area, and intra-arc sequences have been recognized locally.

Paleobiogeography

Although floras, typified by that at Hope Bay (Halle, 1913), constitute an important part of the biological record of the Mesozoic in the Antarctic Peninsula, their identification and true affinities seem to be in doubt. A thorough revision of the Hope Bay flora, together with an investigation of others since discovered in the Antarctic Peninsula area, is required before meaningful biogeographic conclusions can be drawn from them. However, their abundance in both terrestrial and marine deposits indicates that the Mesozoic volcanic arc was forested and presumably received appreciable rainfall. Growth rings in coniferous wood from Upper Jurassic to Lower Cretaceous deposits point to a seasonal climate throughout the period.

Middle Jurassic ammonites from eastern Ellsworth Land (Quilty, 1970) include a variety of stephanoceratids. The specimens are fragmentary and poorly preserved, but Central American and North American affinities are apparent.

Upper Jurassic ammonite faunas, typified by those from Alexander Island, show wide-ranging affinities with corresponding faunas in the Himalayas, Indonesia, and South America (Thomson, 1979). Most of the faunas are dominated by the perisphinctids *Virgatosphinctes* and *Aulacosphinctoides* (Elliott, 1974; Thomson, 1979), but *Blanfordiceras* becomes important in the uppermost Tithonian, and *Spiticeras* and *Bochianites* in the basal Cretaceous. *Virgatosphinctes* of the *andesensis* group occurs in Tithonian beds of the South Shetland Islands, where the uppermost Tithonian–Berriasian sequence is typified by abundant *Spiticeras (Negreliceras?)* with smaller numbers of berriasellids including *Himalayites, Corongoceras(?),* and *Blanfordiceras. Kossmatia* occurs in western Palmer Land (Thomson, 1975*b*) but apparently not in equivalent beds in Alexander Island. Belemnites are represented by the genera *Conodicoelites, Belemnopsis,* and *Hibolithes* (Stevens, 1967; Willey, 1973), which show even wider ranging affinities than the ammonites. This may partially be a reflection of the difficulties of classifying the group at a specific level, but nevertheless, they seem to have been remarkably successful at achieving a rapid radiation and a wide geographic distribution. Close ties with the Indo-Pacific region are also reflected in the bivalves *Inoceramus* (*haasti/subhasti* group) and *Buchia*.

The breakup of Gondwanaland began with the initial rifting of the South Atlantic Ocean in early Neocomian time (Larson and Ladd, 1973; Herz, 1977), although a continuous seaway was probably not formed until Albian time. Despite the initiation of new seaways, and in contrast with the Late Jurassic–earliest Cretaceous faunas, the late(?) Neocomian–Albian faunas seem to show less marked affinities with faunas of presumed comparable age from the Indo-Pacific margin of Gondwanaland. There are some ties with South America and Australia in such genera as *Sanmartinoceras* and *Aucellina* and in the belemnite family Dimitobelidae, but most of the faunas from the three areas apparently show little correspondence. This is well shown by studied of the Ammonoidea (Thomson, 1974) and by an investigation of the genus *Inoceramus,* currently being undertaken by J. A. Crame. Ammonites, resembling such late Neocomian Tethyan genera as *Pseudothurmannia, Emericiceras,* and *Silesites* (Thomson, 1972*a*, 1974) are believed to be late Aptian or younger in age. Assuming that their stratigraphic age has been correctly determined on Alexander Island, they either represent very late arrivals or belong to distinct but homeomorphic forms. Covacevich (1976) has described a small Valanginian fauna from the South Shetland Islands which shows affinities with South African and Madagascan faunas and suggests that there were shelf seas providing migration routes to the east African province. However, their effect on the Alexander Island area has yet to be determined, as faunas of this age have not been detected there (Taylor and others, 1979), even though there seems to be a continuous sedimentary sequence extending in age from late Jurassic to Albian or younger.

An interesting feature of the faunas of the Fossil Bluff Formation of Alexander Island is the occurrence of relatively large numbers of the supposedly deep-water pelagic ammonites *Phylloceras* and *Lytoceras,* whereas these forms are rare or absent in the South Shetland Islands, eastern Plamer Land, and elsewhere in the region. It is likely that this reflects Alexander Island's position on the outside of a volcanic arc and on the border of a large ocean. Although possibly shielded by a chain of intermittent offshore islands, its position would have been more exposed than that of the other localities within or behind the arc.

Late Cretaceous time

Paleogeography

After Early Cretaceous time, the Antarctic Peninsula area seems to have undergone general uplift, and exposures of Upper Cretaceous marine deposits are confined to the eastern side of Graham Land (northern Antarctic Peninsula). The state of the volcanic arc is uncertain, but there are no descriptions of volcaniclastic material introduced by direc aerial fallout into the marine sediments; the inference is that its eruptive activity was much reduced. There are a few reports of Late Cretaceous volcanism (Adie, 1972; Alarcón and others, 1976), but these events are not properly understood and their extent is unknown. Subaerial volcanic rocks from the Colbert Mountains of Alexander Island (Grikurov and others, 1967) may be of Late Cretaceous age, but poorly preserved plant fossils from a similar sequence in the Elgar Uplands (northern Alexander Island) suggest a closer correlation with Tertiary beds on the South Shetland Islands.

Thick sequences of conglomerate, sandstone, and shale in the James Ross Island area were deposited in a basin analogous to the Magellanes Basin of southeastern Patagonia (Dalziel and Elliot, 1973). Although ammonites indicate a Campanian age for the sequence, studies of the bivalve *Inoceramus* by J. A. Crame suggest that beds as old as Turonian are present on James Ross Island and that beds as old as Albian are present on Dundee Island (Crame, 1980). A total thickness of more than 5000 m has been estimated by Bibby (1966) for the exposed sequence in the James Ross Island area. however, there are difficulties in correlating the measured sections, and the true thickness could be substantially different. The sediments were derived from the west, and conglomeratic beds are believed to wedge out eastward and pass into finer grained deposits. Sedimentary clasts from beds of a slightly earlier depositional cycle form a common constituent in the conglomerates of western James Ross Island, together with tuffs and lavas from the Antarctic Peninsula Volcanic Group and lesser amounts of metamorphic and igneous rocks. Farther south, at Sobral Peninsula and Pedersson Nunatak, equivalent conglomerates have an almost entirely volcanic provenance (Elliot, 1966). The beds at Seal Nunataks represent the southernmost known deposits in this Late Cretaceous sedimentary basin.

PALEOBIOGEOGRAPHY

The Upper Cretaceous faunas of the James Ross Island area are varied, but it is the superbly preserved ammonites which have attracted most attention (Spath, 1953; Howarth, 1958, 1966). Characterized by kossmaticeratids, the ammonite faunas also include phylloceratids, lytoceratids, heteromorphs, desmoceratids, and pachydiscids, all of which seem to show more marked cosmopolitan affinities than were admitted by Spath (1953). At a specific level, many constitutents have identical forms in North America, Chile, New Zealand, Japan, Madagascar, and southern India, and these affinities are further enhanced by the number of closely related ammonites species in these same areas. Wilckens' (1910) study of the bivalves and gastropods suggested that these are almost entirely endemic Antarctic forms, but a revision of the faunas, in the light of our increased knowledge of Southern Hemisphere paleontology, would probably reveal more wide-ranging affinities.

Acknowledgments. A paper which is essentially a collation of information froma wide range of sources owes a great debt to all those who have done the ground work. My thanks are also due to numerous past and present members of the British Antarctic Survey who have discussed many of the relevant problems with me over the years.

References cited

Adie, R. J. 1957*a*. The geology of Graham Land: III. Metamorphic rocks of the Trinity Peninsula Series. *Scient. Rep. Falkld Isl. Depend. Surv.*, No. 20, 62 pp.

Adie, R. J. 1957*b*. Geological research in Graham Land. *Advanc. Sci. Lond.*, *13*, No. 53, 545–60.

Adie, R. J. 1964. Geological history. (*In* Priestley, R. E., Adie, R. J., and Robin, G. de Q., *eds. Antarctic Research.* London, Butterworths, 118–62.)

Adie, R. J. 1972. Evolution of volcanism in the Antarctic Peninsula. (*In* Adie, R. J., *ed. Antarctic Geology and Geophysics.* Oslo, Universitetsforlaget, 137–41.)

Aitkenhead, N. 1975. The geology of the Duse Bay-Larsen Inlet area, northeast Graham Land (with particular reference to the Trinity Peninsula Series). *Scient. Rep. Br. antarc. Surv.*, No. 51, 62 pp.

Alarcón, B., Ambrus, J., Olcay, L., and Vieira, C. 1976. Geologia del Estrecho de Gerlache entre los paralelos 64° y 65° lat. sur, Antártica chilena. *Ser. cient. Inst. antárt. chil.*, *4*, No. 1, 7–51.

Bell, C. M. 1975. Structural geology of part of Alexander Island. *Bull. Br. antarc. Surv.*, No. 41–42, 43–58.

Bibby, J. S. 1966. The stratigraphy of part of north-east Graham Land and the James Ross Island Group. *Scient. Rep. Br. antarc. Surv.*, No. 53, 37 pp.

Covacevich, V. 1976. Fauna Valanginiana de Peninsula Byers, Isla Livingston, Antártica. *Revta geol. Chile*, No. 3, 25–56.

Craddock, C. 1970. Tectonic map of Antarctica. (*In* Bushnell, V. C., and Craddock, C., *eds.* Geologic Maps of Antarctica. *Antarct. Map Folio Ser.*, Folio 12, Pl. XXI.)

Craddock, C. 1972. *Geologic Map of Antarctica. 1:5,000,000.* New York, American Geographical Society.

Crame, J. A. 1980. The occurrence of the bivalve *Inoceramus concentricus* on Dundee Island, Joinville Island group. *Bull. Br. antarc. Surv.*, No. 49 (for 1979), 283–86.

Dalziel, I. W. D. 1972. Large-scale folding in the Scotia arc. (*In* Adie, R. J., *ed. Antarctic Geology and Geophysics.* Oslo, Universitetsforlaget, 47–55.)

Dalziel, I. W. D., and Elliot, D. H. 1973. The Scotia arc and Antarctic margin. (*In* Nairn, A. E. M., and Stehli, F. G., *eds. The Ocean Basins and their Margins. 1. The South Atlantic.* New York, Plenum Press. 171–246.)

Dalziel, I. W. D., Elliot, D. H., Thomson, J. W., Thomson, M. R. A., Wells, N. A., and Zinsmeister, W. J. 1979. Geologic studies in the South Orkney Islands R/V *Hero* Cruise 77–1, January 1977. *Antarct. Jnl U.S.*, *12*, 98–101.

Dewar, G. J. 1970. The geology of Adelaide Island. *Scient. Rep. Br. antarc. Surv.*, No. 57, 66 pp.

Edwards, C. W. 1980. New evidence of major faulting on Alexander Island. *Bull. Br. antarc. Surv.*, No. 49 (for 1979), 15–20.

Elliot, D. H. 1966. Geology of the Nordenskjöld Coast and a comparison with north-west Trinity Peninsula, Graham Land. *Bull. Br. antarc. Surv.*, No. 10, 1–43.

Elliot, D. H. 1972. Aspects of Antarctic geology and drift reconstructions. (*In* Adie, R. J., *ed. Antarctic Geology and Geophysics.* Oslo, Universitetsforlaget, 849–58.)

Elliott, M. H. 1974. Stratigraphy and sedimentary petrology of the Ablation Point area, Alexander Island. *Bull. Br. antarc. Surv.*, No. 39, 87–113.

Fleet, M. 1968. The geology of the Oscar II Coast, Graham Land. *Scient. Rep. Br. antarc. Surv.*, No. 59, 46 pp.

Grikurov, G. E. 1971. Geologicheskoe stroenie tsentral'noi chasti Zemli Aleksandra I [The geological structure of the central part of Alexander I Land]. *Antarktika, Doklady Komissii*, 13–42.

Grikurov, G. E., Krylov, A. Ya., and Silin, Yu. I. 1967. Absoliutnyi vozrast nekotorykh porod dugi Skotiia i Zemli Aleksandra I (Zapadnaya Antarktika) [Absolute age of some rocks from the Scotia arc and Alexander I Land (western Antarctica)]. *Dokl. Akad. Nauk SSSR*, Geol., *172*, No. 1, 168–71. [English translation: *Dokl. (Proc.) Acad. Sci. U.S.S.R.*, Geological sciences sect., *172*, 19–22.]

Hallam, A. 1973. Distributional patterns in contemporary terrestrial and marine animals. (*In* Hughes, N. F., *ed.* Organisms and Continents through Time. *Spec. Pap. Paleont.*, No. 12, 93–105.)

Halle, T. G. 1913. The Mesozoic flora of Graham Land. *Wiss. Ergebn. schwed. Südpolarexped.*, *3*, No. 14, 123 pp.

Halpern, M. H. 1964. Cretaceous sedimentation in the "General Bernardo O'Higgins" area of north-west Antarctic Peninsula. (*In* Adie, R. J., *ed. Antarctic Geology.* Amsterdam, North-Holland Publishing Company, 334–47.)

Heezen, B. C., and Bentley, C. R. 1972. Submarine and subglacial topography. (*In* Heezen, B. C., Tharp, M., and Bentley, C. R., *eds.* Morphology of the Earth in the Antarctic and Subantarctic. *Antarct. Map Folio Ser.,* Folio 16, Pl. I.)

Herz, N. 1977. Timing of spreading in the South Atlantic: information from Brazilian alkalic rocks. *Geol. Soc. Am. Bull., 88,* No. 1, 101–12.

Horne, R. R. 1969. Sedimentology and palaeogeography of the Lower Cretaceous depositional trough of south-eastern Alexander Island. *Bull. Br. antarc. Surv.,* No. 22, 61–76.

Horne, R. R., and Thomson, M. R. A. 1972. Airborne and detrital volcanic material in the lower Cretaceous sediments of south-eastern Alexander Island. *Bull. Br. antarc. Surv.,* No. 29, 103–11.

Howarth, M. K. 1958. Upper Jurassic and Cretaceous ammonite faunas of Alexander Island and Graham Land. *Scient. Rep. Falkld Isl. Depend. Surv.,* No. 21, 16 pp.

Howarth, M. K. 1966. Ammonites from the Upper Cretaceous of the James Ross Island Group. *Bull. Br. antarc. Surv.,* No. 10, 55–69.

Kulp, J. L. 1961. Geologic time scale. *Science, N.Y., 133,* No. 3459, 1105–14.

Lambert, R. St. J. 1971. The pre-Pleistocene Phanerozoic time scale—further data. *Spec. Publs geol. Soc. Lond.,* No. 5, 33–34.

Larson, R. L., and Ladd, J. W. 1973. Evidence for the opening of the South Atlantic in the early Cretaceous. *Nature, Lond., 246,* No. 5430, 209–12.

Laudon, T. S. 1972. Stratigraphy of eastern Ellsworth Land. (*In* Adie, R. J., *ed. Antarctic Geology and Geophysics.* Oslo, Universitetsforlaget, 215–22.)

Le Pichon, X., Francheteau, J., and Bonnin, J. 1973. *Plate Tectonics.* Amsterdam, Elsevier, 300 pp. [*Developments in Geotectonics,* Vol. 6.]

Orlando, H. A. 1967. A new Triassic flora from Livingston Island, South Shetland Islands. *Bull. Br. antarc. Surv.,* No. 16, 1–13.

Quilty, P. G. 1970. Jurassic ammonites from Ellsworth Land, Antarctica. *J. Paleont., 44,* No. 1, 110–16.

Smellie, J. L. 1980. The geology of Low Island, South Shetland Islands, and Austin Rocks. *Bull. Br. antarc. Surv.,* No. 49 (for 1979), 239–57.

Smith, A. G., and Briden, J. C. 1976. *Mesozoic and Cenozoic Paleocontinental Maps.* Cambridge, Cambridge University Press, 63 pp.

Spath, L. F. 1953. The Upper Cretaceous cephalopod fauna of Graham Land. *Scient. Rep. Falkld Isl. Depend. Surv.,* No. 3, 60 pp.

Stevens, G. R. 1967. Upper Jurassic fossils from Ellsworth Land, West Antarctica, and notes on Upper Jurassic biogeography of the South Pacific region. *N.Z. Jl Geol. Geophys., 10,* No. 2, 345–93.

Suárez, M. 1976. Plate-tectonic model for southern Antarctic Peninsula and its relation to southern Andes. *Geology, 4,* 211–14.

Suárez, M., and Pettigrew, T. H. 1976. An upper Mesozoic island-arc–back-arc system in the southern Andes and South Georgia. *Geol. Mag., 113,* No. 4, 305–28.

Taylor, B. J., Thomson, M. R. A., and Willey, L. E. 1979. The geology of the Upper Jurassic–Lower Cretaceous succession between Ablation Point and Keystone Cliffs, eastern Alexander Island. *Scient. Rep. Br. antarc. Surv.,* No. 82, 62 pp.

Thomson, M. R. A. 1967. A probable Cretaceous invertebrate fauna from Crabeater Point, Bowman Coast, Graham Land. *Bull. Br. antarc. Surv.,* No. 14, 1–14.

Thomson, M. R. A. 1972*a*. Ammonite faunas of south-eastern Alexander Island and their stratigraphical significance. (In Adie, R. J., *ed. Antarctic Geology and Geophysics.* Oslo, Universitetsforlaget, 155–60.)

Thomson, M. R. A. 1972*b*. New discoveries of fossils in the Upper Jurassic Volcanic Group of Adelaide Island. *Bull. Br. antarc. Surv.,* No. 30, 95–101.

Thomson, M. R. A. 1974. Ammonite faunas of the Lower Cretaceous of south-eastern Alexander Island. *Scient. Rep. Br. antarc. Surv.,* No. 80, 44 pp.

Thomson, M. R. A. 1975*a*. Fossils from the South Orkney Islands: II. Matthews Island. *Bull. Br. antarc. Surv.,* No. 40, 75–79.

Thomson, M. R. A. 1975*b*. New palaeontological and lithological observations on the Legoupil Formation, north-west Antarctic Peninsula. *Bull. Br. antarc. Surv.,* No. 41–42, 169–85.

Thomson, M. R. A. 1975*c*. Upper Jurassic Mollusca from Carse Point, Palmer Land. *Bull. Br. antarc. Surv.,* No. 41–42, 31–42.

Thomson, M. R. A. 1979. Upper Jurassic and Lower Cretaceous ammonite faunas of the Ablation Point area, Alexander Island. *Scient. Rep. Br. antarc. Surv.,* No. 97, 37 pp.

Thomson, M. R. A., and Willey, L. E. 1975. Fossils from the South Orkney Islands: I. Coronation Island. *Bull. Br. antarc. Surv.,* No. 40, 15–21.

Valenzuela, E., and Hervé, F. 1972. Geology of Byers Peninsula, Livingston Island, South Shetland Islands. (*In* Adie, R. J., *ed. Antarctic Geology and Geophysics.* Oslo, Universitetsforlaget, 83–89.)

Wilckens, O. 1910. Die Anneliden, Bivalven und Gastropoden der antarktischen Kreideformation. *Wiss. Ergebn. schwed. Südpolarexped., 3,* No. 12, 132 pp.

Willey, L. E. 1973. Belemnites from south-eastern Alexander Island: II. The occurrence of the family Belemnopseidae in the Upper Jurassic and Lower Cretaceous. *Bull. Br. antarc. Surv.,* No. 36, 33–59.

Williams, P. L., and Rowley, P. D. 1972. Composition of Jurassic sandstones, Lassiter Coast. *Antarct. Jnl U.S., 7,* No. 5, 145–46.

Williams, P. L., Schmidt, D. L., Plummer, C. C., and Brown, L. E. 1972. Geology of the Lassiter Coast area, Antarctic Peninsula: preliminary report. (*In* Adie, R. J., *ed. Antarctic Geology and Geophysics.* Oslo, Universitetsforlaget, 143–48.)

38

Geology of the Northern Lassiter Coast and Southern Black Coast, Antarctic Peninsula

PETER D. ROWLEY and PAUL L. WILLIAMS
U.S. Geological Survey, Denver, Colorado 80225, U.S.A.

Abstract. The northern Lassiter Coast and southern Black Coast consist of Upper Jurassic fine-grained volcaniclastic rocks (Latady Formation) and intertonguing and overlying Upper Jurassic volcanic rocks, all of which were folded tightly along north-northeast-trending axes and intruded by Upper Cretaceous batholiths, stocks, and dikes. The Upper Jurassic volcanic rocks, which are exposed west-northwest of and intertongued with the Latady Formation, were probably deposited in a magmatic arc setting, whereas the Latady Formation was deposited contemporaneously in the adjacent basin to the east-southeast. The Upper Jurassic volcanic rocks consist of rhyodacitic to dacitic ash-flow tuff, dacitic to andesitic lava flows, and minor airfall tuff. The Latady Formation consists of black shale, siltstone, and subordinate sandstone; most clastic material was derived from erosion of the magmatic arc and deposited in a temperate near-shore environment. The exposed folded strata are progressively older to the west-northwest, suggesting that this part of the Antarctic Peninsula is the eastern flank of an anticlinorium. The plutons, 120–95 m.y. old, belong to the widespread calc-alkaline magma series of the Andean Intrusive Suite. Contact metamorphic aureoles, which surround all plutons, have metamorphic assemblages as high as hornblende-hornfels grade. The percentage of K_2O in igneous rocks of the Antarctic Peninsula increases away from the Pacific basin, as is true in other circum-Pacific igneous terranes.

Plutonism and folding recorded in the mapped area represent a phase of subduction of the southern Pacific ocean basin beneath West Antarctica; this subduction started early in Mesozoic time. Subduction is still active under the central Andes, but it ceased under Marie Byrd Land and western Ellsworth Land by early Tertiary time. Starting in early Tertiary time in the Antarctic Peninsula and the southern Andes, transform faulting, sea-floor spreading, and rifting increased in significance, whereas subduction changed in direction and magnitude, decreased in significance, then perhaps ceased locally in late Tertiary time.

The geology of the northern Lassiter coast and southern Black Coast is similar to that of the southern Lassiter Coast, mapped in 1969–70 (Williams, 1970; Williams and others, 1972), and to that of eastern Ellsworth Land, mapped in 1961–62 and 1965–66 (Laudon and others, 1964, 1969; Laudon, 1972). The rocks throughout this region consist of Middle and Upper Jurassic clastic rocks and silicic- to intermediate-composition Upper Jurassic volcanic rocks that were folded tightly and then intruded by Upper Cretaceous plutons and dikes (Figure 38.1). This whole region, forming the southern end of the Antarctic Peninsula, is part of the Andean orogenic belt of Mesozoic and Cenozoic age. This belt extends northward through the Andes of South America, southward and westward through western Ellsworth Land and Marie Byrd Land, and from there into the western Pacific area (Craddock, 1969 and Chapter 1, this volume; Molnar and others, 1975; Craddock and Hollister, 1976; LeMasurier and Wade, 1976).

The mountain ranges of the northern Lassiter Coast and southern Black Coast, at about lat. 72°–74°30′ S, long. 61°–65° W, were first explored by U.S. Geological Survey field parties led by P. L. Williams (1970–71) and P. D. Rowley (1972–73). Jamesway hut base camps on the Wetmore Glacier (1970–71) and in the Ferguson Nunataks (1972–73) were established by air. During most of each 3 month field season, the parties lived in Scott tents and traveled by motor toboggans and Nansen sledges (Williams and Rowley, 1971; Rowley, 1973; Boyer, 1975; Rowley and others, 1976; Boyer and others, 1976).

The region studied (Figure 38.1) is along the Weddell Sea coast and east of the peninsula ice plateau, which has an altitude of about 1800 m. Rock exposures, averaging 5–10% of the region studied, occur as nunataks and mountains with local relief of 800 m or less.

Upper Jurassic Latady Formation

The oldest and most voluminous rock unit exposed in the southern Antarctic Peninsula consists of black and gray, fine-grained, fossiliferous, marine volcaniclastic sedimentary rocks that have been named the Latady Formation (Upper Jurassic) by Williams and others (1972). The type section is in the northern Latady Mountains of the southern Lassiter Coast, where an incomplete section about 600 m thick was measured (Williams and others, 1972). Tight folding, metamorphism, poor exposures, and absence of continuous mappable key units within the formation preclude an accurate estimate of its total thickness. On the basis of a presumed average northward plunge of a few degrees, it seems likely that the Latady Formation is many thousands of meters thick; the base is nowhere exposed. The formation was deposited contemporaneously with volcanic rocks near the shore of a magmatic

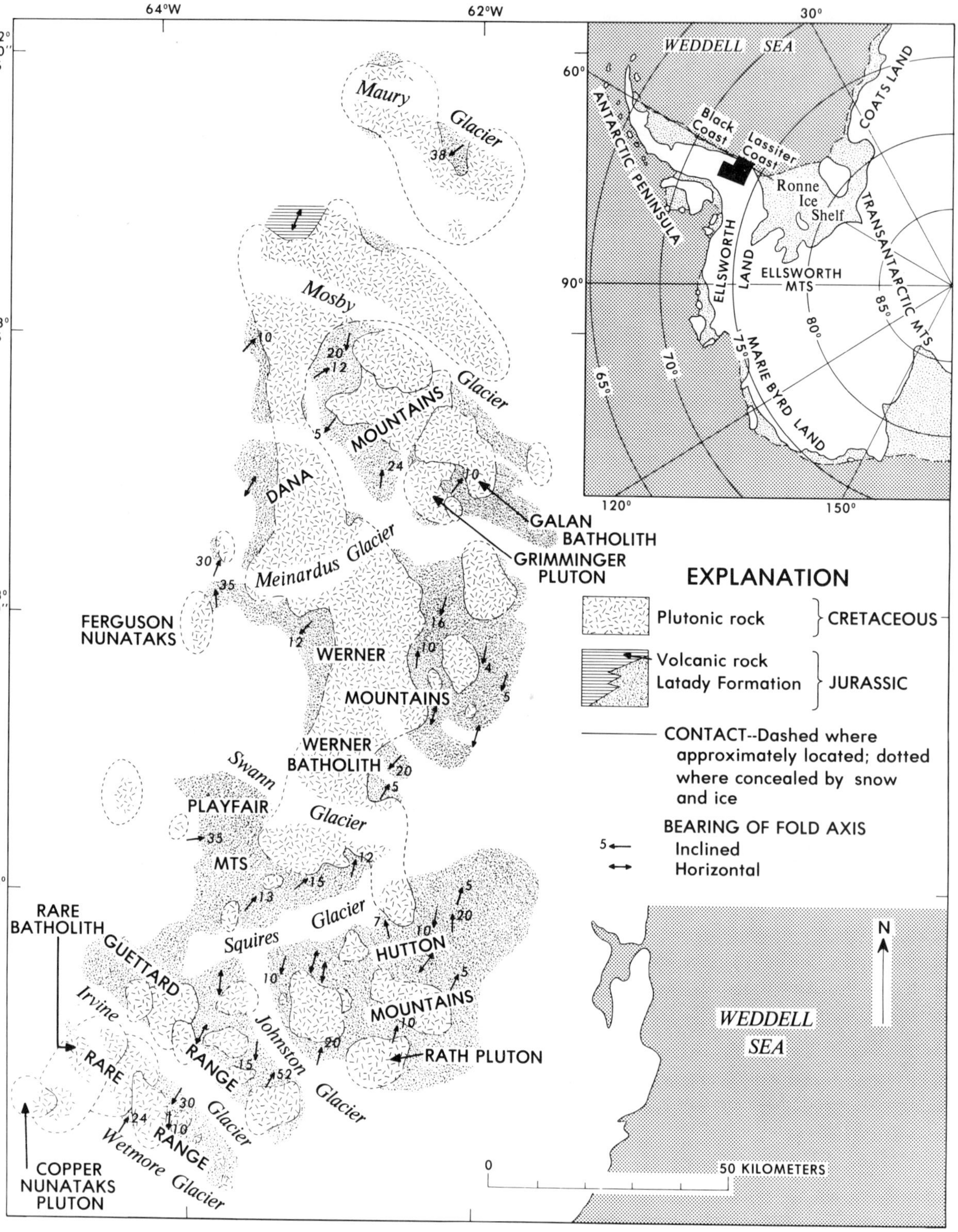

Figure 38.1. Geologic map of the northern Lassiter Coast and southern Black Coast.

arc to the west; the formation appears to intertongue with dacitic to andesitic lava flows and silicic to dacitic ash-flow and airfall tuffs that are now exposed in the western parts of the Lassiter Coast and southern Black Coast.

The Latady Formation is locally rich in invertebrate fossils and plants. Ammonites (*Subdichotomoceras?, Aulacosphinctoides, Virgatosphinctes,* and *Kossmatia*) and several other genera of invertebrates, mostly pelecypods (*Retroceramus, Mytiloceramus, Mytiloides?, Inoceramus, Buchia, Malayomaorca,* and *Trigoniidae*), give an age of Late Jurassic (late Kimmeridgian to late Tithonian) for the Latady Formation on the Lassiter Coast (R. W. Imlay and E. G. Kauffman, written communication, 1975). On the basis of the fossils, Imlay and Kauffman have defined north-northeast-trending units of fixed age within the Latady Formation that are generally younger toward the east. Correlative clastic rocks in eastern Ellsworth Land, however, may be partly older than this; Stevens

(1967) and Quilty (1970, 1972, and Chapter 82, this volume) have assigned them Middle and Late Jurassic ages. Plant fossils of the Lassiter Coast and eastern Ellsworth Land include wood, leaves, shoots, pinnules, leaf sheaths, and fronds of cycadophytes, conifers, gymnosperms, and ginkophytes (J. M. Schopf, written communication, 1973). The plant fossils cannot be used to assign the restricted ages that invertebrate fossils can, but Schopf has concluded that the specimens which have been collected resemble plant fossils described from Middle(?) Jurassic clastic rocks of the northern Trinity Peninsula (Halle, 1913).

The principal rock types in the Latady Formation of the northern Lassiter Coast and southern Black Coast are black and gray siltstone, mudstone, and fissile shale in either massive units as much as 100 m thick or interbedded thin units. Tan and gray volcaniclastic and locally tuffaceous sandstone, which is in most places fine to medium grained, occurs locally as massive beds as much as 10 m thick, although in most places it is thin to medium bedded. Sandstone probably makes up no more than 10% of the section in the western part of the mapped area, but in some places to the east it is more abundant. Thin beds of coarse-grained sandstone and sandy small-pebble conglomerate were observed at several places. Intraformational conglomerate, consisting of fine- to medium-grained sandstone containing black subrounded clay galls as long as 60 cm, occur locally. Coal beds and coal lenses, generally no more than 2 cm thick, were observed at several dozen outcrops. Thin limestone beds occur locally. Volcanic rocks and sills, probably correlative with the Upper Jurassic volcanic rocks, were mapped within the Latady Formation in some places.

Most of the sedimentary structures in the Latady Formation are indicative of deposition in shallow water. Sandstone and siltstone are commonly cross-bedded in gently sloping sets, and in places, sandstone and siltstone contain cut-and-fill structures. Mud cracks and ripple marks were also observed. Waterworn fossil wood fragments, as long as 10 cm but mostly shorter than 1 cm, occur in many sandstone beds; in most places the long axes of the fragments are aligned. Beds of siltstone, mudstone, and fine-grained sandstone locally contain concentrations of delicate plant fragments, including traces of rootlets; some of these fossils have been interpreted by J. M. Schopf (written communication, 1973) as recording a shallow, freshwater pond environment. The invertebrate fossils in the Latady Formation accumulated in shallow marine waters of mid-temperate to cool-temperate climatic zones (R. W. Imlay and E. G. Kauffman, written communication, 1975). Plant fossils indicate a relatively mild, probably frost-free climate on land (Schopf, written communication, 1973).

Most standstone of the Latady Formation is lithic sandstone (Figure 38.2), according to the classification of Folk (1968). Arkose is more abundant in the Jurassic rocks of eastern Ellsworth Land (Laudon, 1972, Figure 38.5). Jurassic sandstone of the southern Antarctic Peninsula is different from sandstone of the Trinity Peninsula Series (upper Paleozoic? and Triassic) of the northern Antarctic Peninsula, most of which is arkose (Elliot, 1965) or quartz-rich lithic sandstone (Aitkenhead, 1975), according to the same classification (Williams and Rowley, 1972). Abundant volcanic rock fragments appear to consist mostly of

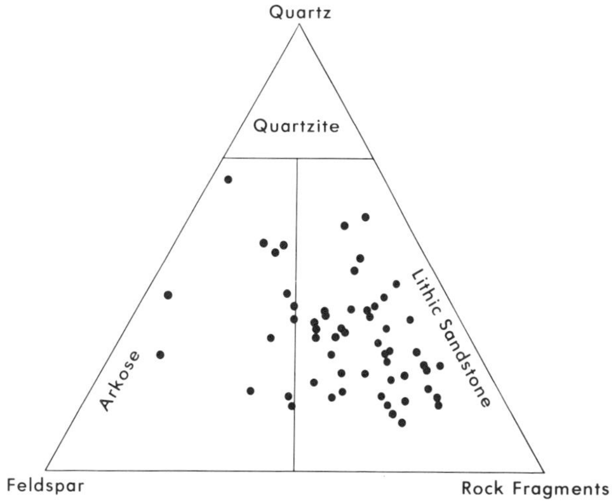

Figure 38.2. Ternary diagram showing composition of clastic particles of sandstone in the Latady Formation, Lassiter Coast and southern Black Coast. Classification generalized from Folk (1968).

dacitic to andesitic lava flows and ash-flow tuff, but basaltic lava flows are also represented, as are silicic pumice and ash-flow tuff. Sedimentary rock fragments, especially chert, are less abundant, and a few examples of metamorphic rocks are also present. Quartz clasts are more abundant than feldspar clasts; plagioclase, mostly oligoclase and andesine, is more abundant than K-feldspar, nearly all of which is orthoclase and sanidine. Most sandstone is well sorted. Sand grains are generally subangular or, less commonly, subrounded. Quartz cement predominates over sericite, calcite, and chlorite cement.

No diagnostic invertebrate fossils were found on the northwestern Lassiter Coast (the western Dana, Werner, and Playfair Mountains), and thus a Late Jurassic age for the rocks here cannot be proven. This has led Kamenev (1975) to suggest that sedimentary rocks of the northwestern Lassiter Coast may be correlative with the Trinity Peninsula Series, following similar earlier hypotheses by Adie (1957, 1969*a,b*) and Grikurov (1973). Kamenev has detected a slightly smaller grain size and slightly greater amount of metamorphic rock in the sedimentary sequence on the northwestern Lassiter Coast as compared to that on the rest of the Lassiter Coast.

We contend, on the contrary, that all these rocks belong to the Latady Formation. The sedimentary rocks on the northwestern Lassiter Coast are identical in sandstone petrographic characteristics, in general lithology, in proportions of rock types, and in structural style to the Latady Formation elsewhere on the Lassiter Coast. The Werner batholith, by far the largest pluton in the southern Antarctic Peninsula, and other plutons in this area (Figure 38.1) have combined to produce more contact metamorphic rock here than elsewhere in the southern Antarctic Peninsula; unfortunately, the few fossils found in these metamorphic rocks are too deformed to be identified. The highest metamorphic grade (hornblende hornfels) is no higher here than it is adjacent to plutons elsewhere in the Lassiter Coast (C. C. Plummer, written communication, 1974), and thus there is no evidence that the rocks of the northwestern Lassiter Coast have been subject to more than one major metamorphic event or to grades of higher temperature and pressure. By far the best collections of fossils were made on the southern Lassiter Coast, on northern Orville Coast, and in eastern Ellsworth Land, where plutons are sparse and small. Yet even the diagnostic fossils found farthest north, in the Watson and Rivera peaks south of the eastern Werner Mountains, indicate a Late Jurassic age. The rocks of the northwestern Lassister Coast pass southward along strike into less metamorphosed Latady rocks containing abundant Upper Jurassic fossils. The rocks of the northwestern Lassiter Coast also pass northward along strike into intertongued volcanic and sedimentary rocks in the northwestern Dana Mountains and volcanic rocks north of the Mosby Glacier.

Upper Jurassic volcanic sequence

Folded volcanic rocks were mapped north of the Mosby Glacier on the southwestern Black Coast (Figure 38.1). These rocks are lithologically and petrographically similar to folded rocks at Mount Poster on the southwestern Lassiter Coast (Williams and others, 1972) and to rocks in northern parts of eastern Ellsworth Land (Laudon and others, 1969; Laudon, 1972). The volcanic rocks occur in the interior part of the southern Antarctic Peninsula and seem to interfinger seaward, toward the Weddell Sea, with the Latady Formation. In the southwestern Black Coast, the volcanic rocks are interbedded with lesser amounts (locally about 20% of the volume) of tuffaceous sandstone and siltstone of the Latady Formation. The volcanic rocks of the southwestern Lassiter Coast largely overlie sedimentary rocks of the Latady Formation (Williams and others, 1972). The volcanic rocks of eastern Ellsworth Land are probably older than or contemporaneous with sedimentary rocks correlative with the Latady Formation (Laudon and others, 1969; Laudon, 1972). Within the Latady Formation, eastward of the main exposures of volcanic rocks, minor volumes of tuff, lava flows, ash-flow tuff, and probable sills were mapped; these exposures occur in the eastern RARE (acronym for Ronne Antarctic Research Expedition) Range, central Guettard Range, northeastern Werner Mountains, and northwestern Dana Mountains, as well as in several places on the southern Lassiter Coast (Williams and others, 1972). The volcanic rocks at these scattered exposures appear to be similar to the main exposures of volcanic rocks; they apparently either represent the eastern feather edges of the main volcanic pile or the local sites of volcanic and hypabyssal rocks presumably related in time and origin to the more abundant volcanic rocks to the west. The apparent intertonguing of volcanic rocks with the Latady Formation suggests that some or all of the volcanic rocks are Late Jurassic in age.

The volcanic rocks of the southwestern Black Coast are about 1200–1500 m thick (Kamenev, 1975), similar to that for volcanic rocks on the southwestern Lassiter Coast (Williams and others, 1972). The rocks consist of dark-green, grayish-green, gray, and black beds of rhyodactic to dacitic ash-flow tuff and amygdaloidal, dacitic to andesitic lava flows and minor beds of gray airfall tuff. These units occur as resistant beds 5–20 m thick. Ash-flow tuff seems to predominate over lava flows; distinction between the two commonly requires microscopic examination of thin sections. The ash-flow tuff contains conspicuous lenticules of collapsed pumice. The lava flows consist in large part of a flow-foliated, amygdaloidal, aphanitic groundmass of hornblende and subordinate plagioclase (oligoclase, andesine, and labradorite) and minor opaque minerals and sphene; some flows contain phenocrysts of pyroxene, lesser hornblende and plagioclase (oligoclase, andesine, and labradorite), and minor biotite and opaque minerals. Most ash-flow tuff is moderately to densely welded and contains small amounts (3–14%) of broken crystals, primarily plagioclase (mostly calcic oligoclase), but with subordinate or minor biotite, sanidine, "beta" quartz, and opaque minerals. Dark-green andesitic dikes were intruded into the sequence, then folded with it.

Volcanic rocks that are similar in composition and origin to those in the southern Antarctic Peninsula are scattered through other parts of the peninsula, where they locally overlie clastic rocks that have been dated palentologically as mainly Middle or Late Jurassic in age (Adie, 1964*a,b,* 1972*a*). Thomson (Chapter 37, this volume) has proposed that they be named the Antarctic Peninsula Volcanic Group.

Depositional environment of Jurassic rocks

The volcanic material within the Latady Formation was probably derived from ash falls and from erosion of a contemporaneous subaerial volcanic highland. On the basis of facies relationships and coarsening of clasts westward toward eastern Ellsworth Land, Williams and others (1972) suggested that the volcanic source for the Latady Formation was west of eastern Ellsworth Land. Suárez (1976), using an analogy with southern South America and partly following the ideas of Dalziel and Elliot (1973), has suggested that the Latady Formation was deposited in a "back-arc" marine basin (a term apparently equivalent to *foreland basin* in the classical sense and to *marginal basin* of de Wit, 1977) located between an ensialic island arc and the continent; this island arc is now represented by the interior part of the Antarctic Peninsula. The present distribution of volcanic rocks on the western parts of the Lassiter Coast and southern Black Coast supports the idea that some kind of a magmatic arc, over a Benioff zone, formerly produced volcanic rocks that intertongued with the Latady Formation and provided the volcaniclastic material within the Latady. This magmatic arc may have persisted here and elsewhere throughout Middle Jurassic to Cretaceous time. However, although the Latady Formation was deposited in shallow water, most of the volcanic rocks in the southern Antarctic Peninsula were deposited subaerially. Thus, here at least, an analogy to an island arc may be misleading. In accordance with an analogy suggested by Stump (1973), Dalziel (1975), and Suárez (1976), the physical setting of the depositional environment of the Jurassic rocks of the Lassiter Coast and southern Black Coast more closely resembles that seen today on the northwestern side of the Alaska Peninsula, which is a near-continent magmatic arc that passes westward into an island arc (Aleutian Islands). Whereas a craton or continental shelf (now Argentina) certainly existed east of the arc off the western side of southern South America during Late Jurassic time (Suárez, 1976), there is no evidence that an emergent craton existed at this same time anywhere near the eastern Lassiter Coast. Based on Gondwana reconstructions (e.g., Craddock, 1969), it seems possible that

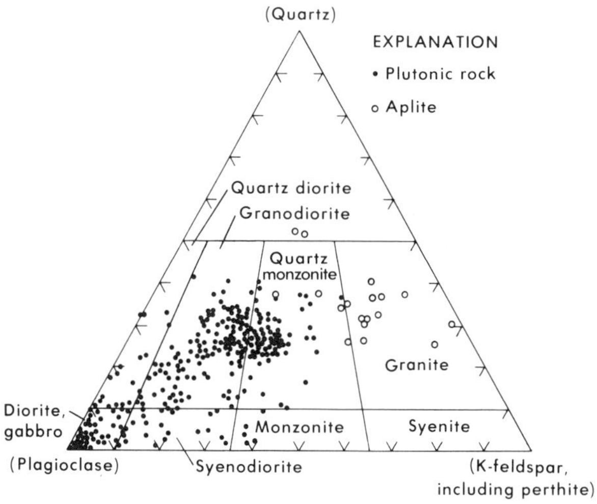

Figure 38.3. Ternary diagram showing modal composition of plutonic rocks of the Lassiter Coast and southern Black Coast. Classification from Bateman (1961).

southernmost Africa, or else the Ellsworth Mountains, eastern Transantarctic Mountains, or Coats Land of Antarctica, may have been the nearest cratonic area.

Cretaceous Andean plutonic rocks

Plutons on the northern Lassiter Coast and southern Black Coast, as in the rest of the southern Antarctic Peninsula (Laudon and others, 1969; Laudon, 1972 and Chapter 95, this volume), intruded folded rocks of the Latady Formation and volcanic sequence about 120–95 m.y. ago (Farrar and others, Chapter 39, this volume). Intrusive contacts are sharp and contain abundant apophyses extending into wall rocks. Emplacement was by both forcible and passive means. Stoped blocks as large as 100 m occur near the contacts of some plutons (Rowley and others, 1977; Vennum, 1978), and mafic inclusions are abundant along most contacts. Most plutons are zoned concentrically by composition; they contain mafic margins that grade into more silicic interiors, and most have fine-grained chilled margins. Some plutons are composite, and separate plutons intrude each other in some places; almost without exception, younger plutonic material is more silicic than older material. Emplacement of each pluton was followed by intrusion of aplite and pegmatite dikes and this event, in turn, was followed by intrusion of intermediate to mafic dikes derived the the same magma chamber that produced the plutons. These latter dikes contain many different textures but are composed of the same minerals as the pluton. Hydrothermal alteration and copper mineralization occur locally as the terminal step in the magmatic process; altered and mineralized rocks are widespread but uncommon except in the Copper Nunataks west of the RARE Range, where a low-grade, noneconomic copper deposit, with affinities to the porphyry-type deposit, is exposed (Rowley and others, 1975, 1977; Rowley and Pride, Chapter 108, this volume).

Plutonic rocks increase in abundance northward through the southern Antarctic Peninsula. Nearly 40 plutons occur in the mapped area (Figure 38.1). Field relationships for many of these plutons have been described by Mehnert and others (1975), Rowley and others (1977),

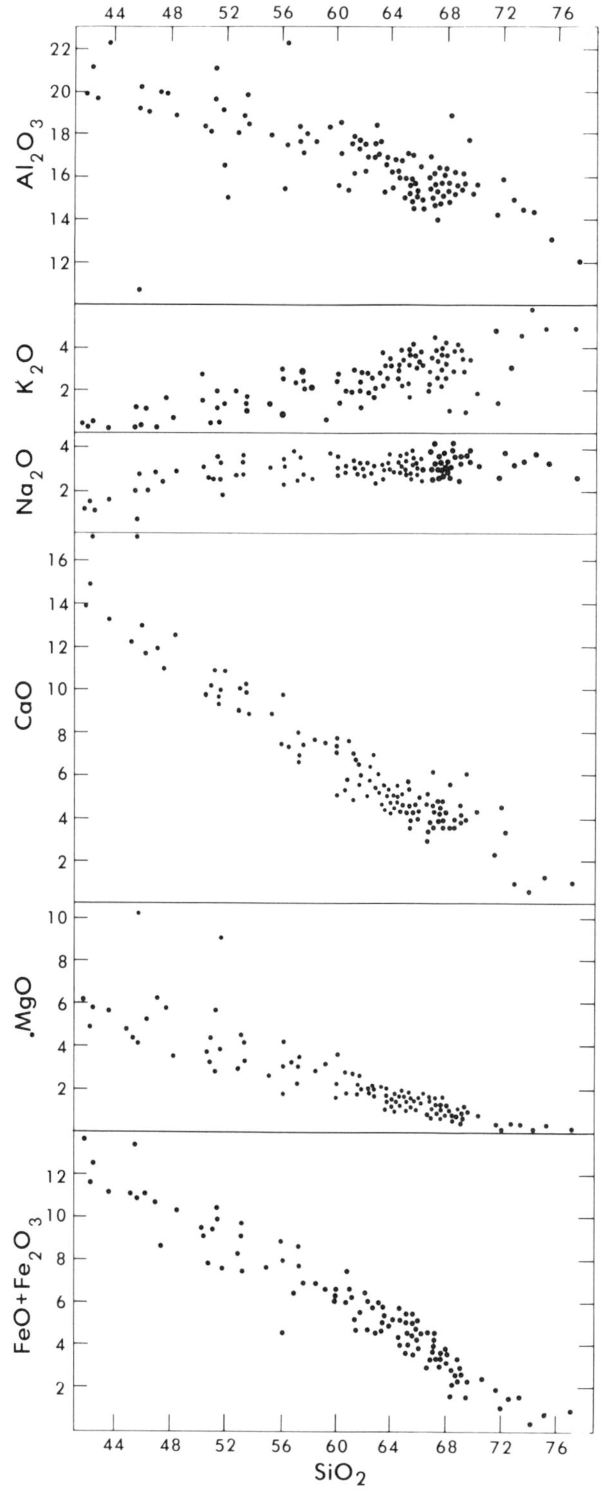

Figure 38.4. Variation diagram of major oxides and silica, in wt. %, for plutonic rocks of the Lassiter Coast and southern Black Coast.

Vennum (1978), and Farrar and others (Chapter 39, this volume). Most plutons are stocks but several are batholiths. By far the largest of these is the Werner batholith, which averages 20 km in width and extends nearly 150 km southward from the southern Black Coast to the central Lassiter Coast.

Modal and chemical analyses (Figures 38.3 and 38.4) of plutonic rocks from the Lassiter Coast—including the area discussed by Williams and others (1972)—and the southern Black Coast indicate that the plutons belong to a calc-alkaline magma series. The alkali-lime index of these rocks is about 61. These rocks are correlated with the Andean Intrusive Suite of Adie (1955) on the basis of geologic setting and composition (Rowley and Williams, 1974).

Rocks of the Lassiter Coast and southern Black Coast are predominantly granodiorite, quartz monzonite, quartz diorite, and diorite (Figure 38.3). Most plutons have medium- to coarse-grained, hypidiomorphic-granular texture. K-feldspar occurs in many plutons as large (as much as several centimeters) phenocrysts or poikilitic masses; K-feldspar is rarely grid-twinned or perthitic except in some northern plutons. Plagioclase ranges from oligoclase to bytownite; most plagioclase is normally zoned. Biotite and green hornblende are the common ferromagnesian minerals, but orthopyroxene and clinopyroxene are abundant in mafic plutons. Opaque minerals consist of magnetite and subordinate ilmenite and pyrite. Sphene is generally an abundant accessory mineral, and apatite is present in most thin sections.

Contact metamorphic rocks

Contact metamorphic aureoles of resistant black hornfels surround all plutons of the Lassiter Coast and southern Black Coast where they intrude Jurassic rocks. In some places phyllite occurs along pluton contacts. As on the southern Lassiter Coast (Plummer, 1974), pelitic rocks in the aureoles contain quartz, plagioclase (oligoclase and subordinate albite), K-feldspar, biotite, muscovite, and graphite. Porphyroblasts of andalusite and subordinate cordierite, garnet, and sillimanite (C. C. Plummer, written communication, 1974) occur in the pelitic rocks adjacent to and generally within 2 km (outcrop width) of contacts. Limestone beds within 2 km of the contacts are converted to marble and contain wollastonite, calcic plagioclase, diopside, garnet (grossularite-andradite?), tremolite, hornblende, actinolite, sphene, and zoisite. Sandstone beds within several kilometers of pluton contacts are recrystallized and bleached and exhibit growth of biotite, muscovite, quartz, and plagioclase (albite and oligoclase). Metavolcanic rocks exhibit growth of hornblende, biotite, chlorite, muscovite, epidote minerals, K-feldspar, quartz, and calcic to sodic plagioclase. Farther outward from contacts, muscovite, biotite, chlorite, epidote, zoisite, and clinozoisite are abundant metamorphic minerals in hornfels. Local retrograde effects include sericitization of andalusite, replacement of cordierite by muscovite and biotite, and replacement of biotite by chlorite (C. C. Plummer, written communication, 1974). More information about the metamorphic assemblages is given by Kamenev and Orlenko (Chapter 40, this chapter).

Most metamorphic minerals are randomly oriented, indicating growth under static conditions. In many rocks, however, crystals of all or some minerals are parallel; in conjunction with streaking and shearing of mineral aggregates and rotation or deformation of porphyroblasts, these textures demonstrate that penetrative deformation occurred locally in association with metamorphism and that plutonism was in part forcible (Plummer, 1974).

Plummer (1974) inferred that on parts of the southern Lassiter Coast the most intense metamorphism reached hornblende-hornfels grade and, in some places, a grade transitional between hornblende hornfels and amphibolite. Less metamorphosed rock is of albite-epidote hornfels grade and, locally, a grade transitional between albite-epidote hornfels and greenschist. Metamorphic grades in rocks of the northern Lassiter Coast and southern Black Coast are the same as those of the southern Lassiter Coast (C. C. Plummer, written communication, 1974).

Kamenev (1975) noted that most country rock west of the Werner batholith is metamorphosed, and he speculated that this resulted from "zonal regional metamorphism," not from contact metamorphism. On the other hand, he interpreted the Latady Formation in the eastern Dana and Werner Mountains to be little metamorphosed and andalusite hornfels—so abundant elsewhere—to be missing from this area. Our collections of metamorphic rocks reveal that hornfels containing andalusite and other typical high- to low-grade metamorphic minerals surround all plutons in these areas. We consider contact metamorphism to be present in both areas, but of higher grade on the western side of the Werner batholith because here the exposed rocks are adjacent to this large igneous body (Figure 38.1).

Where plutons are sparse on the Lassiter Coast and southern Black Coast, there is evidence for a regional zeolite-grade or chlorite-grade metamorphism. Even on the southern Lassiter Coast, coal has reached anthracite grade, as it has elsewhere in the southern Antarctic Peninsula (J. M. Schopf, written communication, 1973). Whether this low grade of metamorphism occurred prior to plutonism—i.e., when tight folding and slaty cleavage occurred (Kamenev and Orlenko, Chapter 40, this volume)—or during burial has not been determined.

Structural geology

Prior to plutonism, the Latady Formation and volcanic rocks on the northern Lassiter Coast and southern Black Coast were folded tightly along north-northeast axes, and a well-developed axial-plane cleavage formed. Folds range from open to isoclinal, but most are asymmetric or overturned. Fold wavelengths range from several meters to about 1 km. Most axial planes dip west-northwestward. The general eastward progression of older to younger units within the folded Latady Formation on the southern Lassiter Coast (R. W. Imlay and E. G. Kauffman, written communication, 1975) suggests that the area lies on the eastern limb of an anticlinorium. Fold axes are horizonal or plunge gently north-northeastward, except in the vicinity of plutons; here forcible intrusion has refolded and sharply displaced regional folds and given strongly divergent directions and steep plunges to fold axes (Kellogg and Rowley, 1974; Kellogg, written communication, 1974).

Thrust faults were not observed but may be present. Laudon and others (1969) suggested that thrust faults occur in the southern Behrendt Mountains of eastern Ellsworth Land; they also occur on the Orville Coast (Rowley, 1978). Northwest-striking, steeply dipping faults having small dip-slip or strike-slip movement have been mapped in the Copper Nunataks (Rowley and others, 1977). This is the only area on the Lassiter Coast where the presence of steeply dipping faults has been proven. Laudon and others (1969), Laudon (1972), and Rowley (1978), however, suggested that Tertiary block faulting or strike-slip faulting may have occurred farther south in the Antarctic Peninsula.

The direction of maximum principal compressive stress in rocks of the northern Lassiter Coast and southern Black Coast was subhorizontal and west-northwestward, perpendicular to the trend of the Antarctic Peninsula, for a long time. Folds trend north-northeastward, parallel to the trend of the peninsula, and the master joint set in folded rocks strikes west-northwestward. This direction of maximum principal compressive stress persisted through and after plutonism during Late Cretaceous time. Subvertical joints in the plutonic rocks consist of a main set that strikes north-northward and a subsidiary extension set that strikes west-northwestward. The west-northwest extension joints generally contain mineral fillings, predominantly epidote; the north-northeast joints are barren of mineral fillings. Most dikes that accompanied Andean plutonism were emplaced into plutonic rocks and strike west-northwestward, perpendicular to the inferred extension direction.

Paleomagnetic studies (Kellogg and Reynolds, 1974, 1978; Reynolds and Kellogg, 1974; Scharnberger and Scharon, Chapter 42, this volume) give paleopole positions for Cretaceous plutons and dikes in the Lassiter Coast and other parts of the Antarctic Peninsula and Ellsworth Land that are very similar to the position of the present South Pole. This similarity indicates that little latitudinal movement or rotation of the southern Antarctic Peninsula has occurred in the last 100 m.y. or more. In contast, paleopole positions from Lower Cretaceous rocks of Marie Byrd Land (Scharnberger and Scharon, 1972) are highly divergent from those of the Antarctic Peninsula and Ellsworth Land. West Antarctica clearly consists of several crustal blocks that had different histories in Mesozoic time (Schopf, 1969; Scharnberger and Scharon, 1972 and Chapter 42, this volume; Stump, 1973; Molnar and others, 1975; Herron and Tucholke, 1976). The Ellsworth Mountains and vicinity (Craddock and others, 1964), located south of the mapped area and showing a structural trend that, perhaps in Late Cretaceous (Elliot, 1975) or Cenozoic time (Schopf, 1969; Clarkson and Brook, 1977), was rotated normal to that of the adjacent Antarctic Peninsula, may represent one of these crustal blocks (Schopf, 1969; Adie, 1970; de Wit, 1977). Recent discovery of high-grade metamorphic rocks at Haag Nunataks (Swithinbank and others, 1976; Clarkson and Brook, 1977) indicates the presence of a crustal block just south of the southern Antarctic Peninsula that may represent a basement complex that underlies the Ellsworth Mountains.

Regional considerations

Folding and plutonism in the southern Antarctic Peninsula are attributed to subduction of the southern Pacific ocean floor under West Antarctica. Such subduction is one part of the Andean orogeny, which in South America spanned Late Triassic through Cenozoic time (Ericksen, 1976). In the central Andes, subduction seems to have been continuous throughout the period of the Andean orogeny. South of where the Chile Ridge intersects the Andes of southern South America (about lat. 45° S), however, transform faulting, spreading, and rifting became more dominant through Tertiary time; subduction occurred through Jurassic and Cretaceous time but apparently became slower and changed direction in Tertiary time and eventually ceased at different times under many parts of West Antarctica (Craddock and Hollister, 1976; Herron and Tucholke, 1976).

Igneous rocks of West Antarctica reveal much about the timing and geometry of the Andean orogeny. The earliest expression of subduction of West Antarctica was emplacement of calc-alkaline plutons, which span an interval from about 180 m.y. to about 80 m.y. (Early Jurassic to Late Cretaceous) throughout West Antarctica (Craddock, 1969; Halpern, 1971; Wade, 1972; Wade and Wilbanks, 1972; Adie, 1972*b*; Krylov, 1972; Lopatin and others, 1974; Grikurov, 1973; Rex, 1976; LeMasurier and Wade, 1976; Farrar and others, Chapter 39, this volume; Saunders and others, Chapter 34, this volume; LeMasurier and Rex, Chapter 89, this volume). These plutons are distinguished, on the basis of lower alkalies and several other traits, from sparsely distributed calc-alkaline Paleozoic(?) plutonic rocks (Adie, 1964*a*; Drake and others, 1971) of West Antarctica. The 180- to 80-m.y.-old plutons range from gabbro to granite but are generally silicic. They are petrologically and chemically related to each other, and together represent a broad calc-alkaline magma series. Isotopic age data and chemical data from these plutonic rocks do not allow us to separate the Gondwana orogeny from the Andean orogeny, as the two orogenies are defined in other parts of Antarctica. Adie (1955) recognized the similar chemical and petrographic characteristics of all the plutonic rocks and called them the Andean Intrusive Suite. Of these Mesozoic Andean plutons, those with ages of 120–80 m.y. are most abundant and most widespread.

Mesozoic silicic to mafic calc-alkaline volcanic rocks are widespread in the Antarctic Peninsula, but sparse in Marie Byrd Land and western Ellsworth Land. All these rocks are similar in chemistry to the Mesozoic intrusive rocks and thus were considered to be the surface expression of the first Mesozoic plutons (Adie, 1964*a*). Like the Mesozoic plutons, the volcanic rocks represent the effects of subduction under West Antarctica. They have traditionally been considered to be Middle or Late Jurassic in age (Adie, 1964*a,b*, 1972*a*), but are probably also Early Jurassic and Cretaceous in age and thus coincide more generally with chemically similar Andean plutons (Grikurov, 1973). Partial confirmation of this suggestion comes from widespread evidence of volcanic material in sedimentary rocks dated paleontologically as Early Cretaceous in age (Horne, 1968; Thomson, Chapter 37, this volume) and from the following K-Ar ages: 186 m.y. (Early Jurassic) on basalt from Jason Peninsula (Rex, 1972), 87–81 m.y. (Late Cretaceous) on sparse silicic calc-alkaline volcanic rocks from Marie Byrd Land (LeMasurier and Wade, 1976), 92–86 m.y. on andesite and a mafic dike from northwestern Palmer Land (Rex, 1972), and 70 m.y. on a calc-alkaline basaltic andesite from the northern Antarctic Peninsula (D. C. Rex, *in* Baker, 1976).

Igneous rocks of West Antarctica that are younger than 80 m.y. seem to reflect different tectonic conditions. This is especially true in Marie Byrd Land and western Ellsworth Land, where volcanic rocks and sparse mafic plutons are distinctly alkalic and were formed under conditions of rifting that started no less than 45 m.y. ago and perhaps sooner (LeMasurier and Wade, 1976; LeMasurier and Rex, Chapter 89, this volume). Subduction under these areas ceased prior to alkaline igneous activity.

More enigmatic is the origin of uppermost Cretaceous and Cenozoic igneous rocks of the Antarctic Peninsula. Plutons of this age (mostly 75–70 and 60–45 m.y. old) have long been assumed to form a substantial proportion of the Andean suite (Adie, 1955, 1972*a*), but recent isotopic dating indicates that they are less common than formerly thought. Rex and Baker (1973) noted that the 10-m.y.-old granodiorite of Cornwallis Island, the youngest pluton known in West Antarctica, is locally of trondhjemitic composition (as this term is used by Ross, 1973) and in petrographic and geochemical respects is different from typical Andean plutons. Sparse modal and geochemical data (Hooper, 1962; Fraser, 1964; Scott, 1965; Elliot, 1964, 1966; Hobbs, 1968; Dewar, 1970; West, 1974; Saunders and others, Chapter 34, this volume) suggest a trondhjemitic affinity for other Tertiary plutons of the peninsula area. Perhaps trondhjemitic character reflects emplacement of Andean plutons above shallow, western (low K_2O) parts of the Benioff zone. But these Tertiary plutons also differ from typical Mesozoic Andean plutons in being generally more mafic (mostly gabbro, diorite, or quartz diorite) and in containing more alumina than Mesozoic Andean plutons. Most of the Tertiary plutons occur in offshore islands along rift zones (Hooper, 1962; Griffiths and others, 1964; Ashcroft, 1972; Davey, 1972; Barker and Griffiths, 1972; Dalziel and Elliot, 1973; Weaver and others, Chapter 29, this volume) west of the northern Antarctic Peninsula, as noted by Grikurov and Lopatin (1975).

Most Tertiary volcanic rocks in the northern Antarctic Peninsula and Scotia arc are less than 8 m.y. old and are either mildly alkalic or tholeiitic (Baker, 1972, 1976; de Wit, 1977). Those from the South Sandwich Islands in the eastern Scotia arc are comparable to arc tholeiites and clearly formed above an active subduction zone under the Scotia arc (Elliot, 1975; Baker, 1976). Insofar as can be determined, these upper Tertiary and Quaternary volcanic rocks are related to subduction, rifting, or transform faulting in and near oceanic settings (Baker, 1972, 1976; Elliot, 1975). Mafic volcanic rocks of the South Shetland Islands span Cenozoic time (Rex, 1972; González-Ferrán, Chapter 84, this volume; Watts, Chapter 28, this volume; Weaver and others, Chapter 29, this volume) and have affinities with both calc-alkaline rocks and arc tholeiites. They reflect a complex history that includes subduction under the Antarctic Peninsula, which ceased progressively from south to north, as well as back-arc spreading of adjacent Bransfield Strait (Baker, 1976; de Wit, 1977; González-Ferrán, Chapter 84, this volume; Weaver and others, Chapter 29, this volume). Subduction under the northern Antarctic Peninsula ceased by late Tertiary time (Weaver and others, Chapter 29, this volume).

Williams and others (1972), Saunders and others (Chapter 34, this volume), and Baker (1976) have found evidence from published petrographic descriptions and geochemical analyses that plutons in the Antarctic Peninsula generally become more silicic eastward. Adie (1964*b*), de Wit (1977), and Weaver and others (Chapter 29, this volume) have reached the same conclusion about the Mesozoic volcanic rocks. Williams and others (1972) have proposed that a quartz diorite line (Moore, 1959) could be drawn approximately axial to the Antarctic Peninsula. We tested this concept further by determining K_2O variation for areas of plutonic and volcanic rock in the Antarctic Peninsula where chemical analyses are sufficiently abundant to determine, for each area, the percentage of K_2O at 60% SiO_2, using the method of Hatherton and Dickinson (1969). Published chemical analyses are sparse, and our conclusions are preliminary. Nonetheless, we note values for K_2O (at 60% SiO_2) of 3.0–3.2% for the eastern coast of the Antarctic Peninsula (Tabarin Peninsula, Nordenskjold Coast, and Oscar II Coast), and values for K_2O of about 1.7–2.5% for the western coast of the peninsula and nearby offshore islands (including Danco Coast, Anvers Island, and Adelaide Island). Chemical analyses for the South Shetland Islands, farther to the west, indicate low values (1.5% or less) for K_2O at 60% SiO_2. Values for the Lassiter Coast and eastern Ellsworth Land generally range from 2.0 to 3.0% and are generally higher eastward, but they show considerable ranges, presumably because the bedrock strip here is too narrow to allow identification of significant regional K_2O variation. It is interesting to note that Mesozoic volcanic rocks and Tertiary plutonic rocks give values comparable to those of Mesozoic plutonic rocks. Even Quaternary volcanic rocks seem to have similar values: volcanic rocks of Deception Island in the South Shetland Islands have a value of 1.3%, while volcanic rocks of James Ross Island, east of the northern end of the peninsula, have a value that may be as high as 5%. Values in the west of 1.5% K_2O and in the east of 3.0% K_2O can be interpreted to suggest depths to an eastward-dipping Benioff zone of about 100 km in the west and 250 km in the east (Hatherton and Dickinson, 1969, Figure 6).

Acknowledgments. Mapping and data reduction were financed by National Science Foundation Grant AG-187, and logistic support was provided by the U.S. Antarctic Research Program of the National Science Foundation and by U.S. Navy Operation Deep Freeze. We thank M. D. Turner and other members of the Division of Polar Programs of the National Science Foundation for support on fiscal and logistics matters. We gratefully acknowledge the field mapping and the later follow-up studies by the other members of the 1970–71 field party—A. B. Ford, W. H. Nelson, and R. L. Reynolds—and of the 1972–73 field party—S. J. Boyer, E. N. Kamenev, K. S. Kellogg, W. R. Vennum, and R. B. Waitt, Jr. We thank C. C. Plummer, K. S. Kellogg, J. M. Schopf, R. W. Imlay, and E. G. Kauffman for contributing unpublished data from their studies in progress. Technical reviews by Warren Hamilton and D. A. Coates greatly improved the manuscript.

References cited

Adie, R. J. 1955. The petrology of Graham Land—II. The Andean granite-gabbro intrusive suite. *Scient. Rep. Falkld Isl. Depend. Surv.*, No. 12, 39 pp.

Adie, R. J. 1957. The petrology of Graham Land—III. Metamorphic rocks of the Trinity Peninsula Series. *Scient. Rep. Falkld Isl. Depend. Surv.*, No. 20, 26 pp.

Adie, R. J. 1964*a*. The geochemistry of Graham Land. (*In* Adie, R. J., *ed. Antarctic Geology*. Amsterdam, North-Holland Publishing Company, 541–47.)

Adie, R. J. 1964*b*. Geological history. (*In* Priestley, R. E., Adie, R. J., and Robin, G. de Q., *eds. Antarctic Research*. London, Butterworths, 118–62.)

Adie, R. J. 1969*a*. Northern Antarctic Peninsula (sheet 2, northern Antarctic Peninsula). (*In* Bushnell, V. C., and Craddock, C., *eds*. Geologic Maps of Antarctica. *Antarct. Map Folio Ser.,* Folio 12, Pl. I.)

Adie, R. J. 1969*b*. Southern Antarctic Peninsula (sheet 2, southern Antarctic Peninsula). (*In* Bushnell, V. C., and Craddock, C., *eds*. Geologic Maps of Antarctica. *Antarct. Map Folio Ser.,* Folio 12, Pl. II.)

Adie, R. J. 1970. Review of Antarctic geology. (*In Proceedings and Papers, International Union of Geological Sciences Second Gondwana Symposium*. Pretoria, Scientific and Industrial Research, Scientia, 15–22.)

Adie, R. J. 1972*a*. Evolution of volcanism in the Antarctic Peninsula. (*In* Adie, R. J., *ed. Antarctic Geology and Geophysics*. Oslo, Universitetsforlaget, 137–41.)

Adie, R. J. 1972*b*. Recent advances in the geology of the Antarctic Peninsula. (*In* Adie, R. J., *ed. Antarctic Geology and Geophysics*. Oslo, Universitetsforlaget, 111–24.)

Aitkenhead, N. 1975. The geology of the Duse Bay-Larsen Inlet area, northeast Graham Land (with particular reference to the Trinity Peninsula Series). *Scient. Rep. Br. antarc. Surv.,* No. 51, 62 pp.

Ashcroft, W. A. 1972. Crustal structure of the South Shetland Islands and Bransfield Strait. *Scient. Rep. Br. antarc. Surv.,* No. 66, 43 pp.

Baker, P. E. 1972. Recent volcanism and magmatic variation in the Scotia arc. (*In* Adie, R. J., *ed. Antarctic Geology and Geophysics*. Oslo, Universitetsforlaget, 57–60.)

Baker, P. E. 1976. Volcanism and plate tectonics in the Antarctic Peninsula and Scotia arc. (*In* González-Ferrán, O., *ed. Andean and Antarctic Volcanology Problems*. Rome, International Association of Volcanology and Chemistry of the Earth's Interior, 347–56.)

Barker, P. F., and Griffiths, D. H. 1972. The evolution of the Scotia Ridge and Scotia Sea. *Phil. Trans. R. Soc. Lond.,* Ser. A, *271,* 151–83.

Bateman, P. C. 1961. Granitic formations in the east-central Sierra Nevada near Bishop, California. *Geol. Soc. Am. Bull., 72,* No. 10, 1521–38.

Boyer, S. J. 1975. Chemical weathering of rocks on the Lassiter Coast, Antarctic Peninsula, Antarctica. *N.Z. Jnl Geol. Geophys., 18,* No. 4, 623–28.

Boyer, S. J., Rowley, P. D., and Kellogg, K. S. 1976. Camping and trail methods used on the Lassiter Coast, Antarctica. *Summit, 22,* No. 4, 8–11, 36–39.

Clarkson, P. D., and Brook, M. 1977. Age and position of the Ellsworth Mountains crustal fragment, Antarctica. *Nature, Lond., 265,* No. 5595, 615–16.

Craddock, C. 1969. Radiometric age map of Antarctica. (*In* Bushnell, V. C., and Craddock, C., *eds*. Geologic Maps of Antarctica. *Antarct. Map Folio Ser.,* Folio 12, Pl. XIX.)

Craddock, C., and Hollister, C. D. 1976. Geologic evolution of the southeast Pacific basin. (*In* Hollister, C. D., Craddock, C., and others. *Initial Reports of the Deep Sea Drilling Project,* Vol. 35. Washington [U.S. Government Printing Office], 723–43.)

Craddock, C., Anderson, J. J., and Webers, G. F. 1964. Geologic outline of the Ellsworth Mountains. (*In* Adie, R. J., *ed. Antarctic Geology*. Amsterdam, North-Holland Publishing Company, 155–70.)

Dalziel, I. W. D. 1975. The Scotia arc tectonics project, 1969–1975. *Antarct. Jnl U.S., 10,* No. 3, 70–81.

Dalziel, I. W. D., and Elliot, D. H. 1973. The Scotia arc and Antarctic margin. (*In* Nairn, A. E. M., and Stehli, F. G., *eds. The Ocean Basins and Margins. 1. The South Atlantic*. New York, Plenum Press, 171–246.)

Davey, F. J. 1972. Marine gravity measurements in Bransfield Strait and adjacent areas. (*In* Adie, R. J., *ed. Antarctic Geology and Geophysics*. Oslo, Universitetsforlaget, 39–46.)

Dewar, G. J. 1970. The geology of Adelaide Island. *Scient. Rep. Br. antarc. Surv.,* No. 57, 66 pp.

de Wit, M. J. 1977. The evolution of the Scotia arc as a key to the reconstruction of southwestern Gondwanaland. *Tectonophysics, 37,* No. 1–3, 53–81.

Drake, A. A., Jr., Marvin, R. F., Stern, T. W., and Hubbard, H. A. 1971. The Landfall Peak adamellite and regional comparison of petrochemical and age data from the Thurston Island–Eights Coast area, West Antarctica. *Prof. Pap. U.S. geol. Surv.,* No. 750-B, B62–B73.

Elliot, D. H. 1964. The petrology of the Argentine Islands. *Scient. Rep. Br. antarc. Surv.,* No. 41, 31 pp.

Elliot, D. H. 1965. Geology of north-west Trinity Peninsula, Graham Land. *Bull. Br. antarc. Surv.,* No. 7, 1–24.

Elliot, D. H. 1966. Geology of the Nordenskjöld Coast and a comparison with north-west Trinity Peninsula, Graham Land. *Bull. Br. antarc. Surv.,* No. 10, 1–43.

Elliot, D. H. 1975. Tectonics of Antarctica—a review. *Am. J. Sci., 275-A,* 45–106.

Ericksen, G. E. 1976. Metallogenic provinces of southeastern Pacific region. (*In* Halbouty, M. T., Maher, J. C., and Lian, H. M., *eds*. Circum-Pacific Energy and Mineral Resources. *Mem. Am. Ass. Petrol. Geol., 25,* 527–38.)

Folk, R. L. 1968. *Petrology of Sedimentary Rocks*. Austin, Texas, Hemphill's, 170 pp.

Fraser, A. G. 1964. Banded gabbros of the Anagram Islands, Graham Land. *Bull. Br. antarc. Surv.,* No. 4, 23–38.

Griffiths, D. H., Riddihough, R. P., Cameron, H. A. D., and Kennett, P. 1964. Geophysical investigation of the Scotia arc. *Scient. Rep. Br. antarc. Surv.,* No. 46, 43 pp.

Grikurov, G. E. 1973. *Geologiia Antarkticheskogo Poluostrova* [Geology of the Antarctic Peninsula]. Moscow, Izdatel'stvo Nauka, 119 pp.

Grikurov, G. E., and Lopatin, B. F. 1975. Structure and evolution of the West Antarctic part of the circum-Pacific mobile belt. (*In* Campbell, K. S. W., *ed. Gondwana Geology*. Canberra, Australia National University Press, 639–50.)

Halle, T. G. 1913. The Mesozoic flora of Graham Land. *Wiss. Ergebn. schwed. Südpolarexped., 1901–1903, 3,* No. 14, 1–123.

Halpern, M. 1971. Evidence for Gondwanaland from a review of West Antarctic radiometric ages. (*In* Quam, L. O., *ed.* Research in the Antarctic. Washington, *Am. Assoc. Adv. Sci.,* Publn No. 93, 717–30.)

Hatherton, T., and Dickinson, W. R. 1969. The relationship between andesitic volcanism and seismicity in Indonesia, the Lesser Antilles, and other island arcs. *J. geophys. Res., 74,* No. 22, 5301–10.

Herron, E. M., and Tucholke, B. E. 1976. Sea-floor magnetic patterns and basement structure in the southeastern Pacific. (*In* Hollister, C. D., Craddock, C., and others. *Initial Reports of the Deep Sea Drilling Project,* Vol. 35. Washington [U.S. Government Printing Office], 263–78.)

Hobbs, G. J. 1968. The geology of the South Shetland Islands—IV. The geology of Livingston Island. *Scient. Rep. Br. antarc. Surv.,* No. 47, 34 pp.

Hooper, P. R. 1962. The petrology of Anvers Island and adjacent islands. *Scient. Rep. Falkld Isl. Depend. Surv.,* No. 34, 69 pp.

Horne, R. R. 1968. Petrology and provenance of the Cretaceous sediments of south-eastern Alexander Island. *Bull. Br. antarc. Surv.,* No. 17, 73–82.

Kamenev, E. N. 1975. Skladchatye osadochnye i intruzivnye obrazovaniia Berega Lassitera i Berega Bleka (Antarkticheskii poluostrova) [Folded sedimentary and intrusive formations of the Lassiter Coast and the Black Coast (Antarctic Peninsula)]. *17th Sov. Antarct. Exped., Winter Res. 1971–1973, 65,* 215–22.

Kellogg, K. S., and Reynolds, R. L. 1974. Paleomagnetic study of igneous rocks of the northern Lassiter Coast, Antarctic Peninsula. *Antarct. Jnl U.S., 9,* No. 38–40.

Kellogg, K. S., and Reynolds, R. L. 1978. Paleomagnetic results from the Lassiter Coast, Antarctica, and a test for oroclinal bending of the Antarctic Peninsula. *J. geophys. Res., 83,* No. B5, 2293–99.

Kellogg, K. S., and Rowley, P. D. 1974. Structural geology of the Lassiter Coast. *Antarct. Jnl U.S., 9,* No. 5, 224–25.

Krylov, A. Ya. 1972. Antarctic geochronology. (*In* Adie, R. J., *ed. Antarctic Geology and Geophysics.* Oslo, Universitetsforlaget, 491–94.)

Laudon, T. S. 1972. Stratigraphy of eastern Ellsworth Land. (*In* Adie, R. J., *ed. Antarctic Geology and Geophysics.* Oslo, Universitetsforlaget, 215–23.)

Laudon, T. S., Behrendt, J. C., and Christensen, N. J. 1964. Petrology of rocks collected on the Antarctic Peninsula traverse. *J. sedim. Petrol., 34,* No. 2, 360–64.

Laudon, T. S., Lackey, L. L., Quilty, P. G., and Otway, P. M. 1969. Geology of eastern Ellsworth Land (sheet 3, eastern Ellsworth land). (*In* Bushnell, V. C., and Craddock, C., *eds.* Geologic Maps of Antarctica. *Antarct. Map Folio Ser.,* Folio 12, Pl. III.)

LeMasurier, W. E., and Wade, F. A. 1976. Volcanic history in Marie Byrd Land: implications with regard to southern hemisphere tectonic reconstructions. (*In* González-Ferrán, O., *ed. Andean and Antarctic Volcanology Problems.* Rome, International Association of Volcanology and Chemistry of the Earth's Interior, 398–424.)

Lopatin, B. G., Krylov, A. Y., and Aliapyshev, O. A. 1974. Osnovnye tektonomagmatecheskiye etapi v rasvetee zemle Marie Byrd e Berega Eightsa (Zapadnaya Antarctida) po radeogennim dannim [Major tectonomagmatic stages in the development of Marie Byrd Land and Eights Coast (West Antarctica) determined radiometrically]. *Antarktika, Doklady Komissii,* No. 13, 52–60.

Mehnert, H. H., Rowley, P. D., and Schmidt, D. L. 1975. K-Ar ages of plutonic rocks in the Lassiter Coast area, Antarctica. *J. Res. U.S. geol. Surv., 3,* No. 2, 233–36.

Molnar, P., Atwater, T., Mammerickx, J., and Smith, S. M. 1975. Magnetic anomalies, bathymetry and the tectonic evolution of the South Pacific since the Late Cretaceous. *Geophys. J. R. astr. Soc., 40,* 383–420.

Moore, J. G. 1959. The quartz diorite boundary line in the western United States. *J. Geol., 67,* No. 2, 198–210.

Plummer, C. C. 1974. Contact metamorphism of the Latady Formation, southern Lassiter Coast, Antarctic Peninsula. *Antarct. Jnl U.S., 9,* No. 3, 82–83.

Quilty, P. G. 1970. Jurassic ammonites from Ellsworth Land, Antarctica. *J. Paleont. 44,* No. 1, 110–16.

Quilty, P. G. 1972. Middle Jurassic brachiopods from Ellsworth Land, Antarctica. *N.Z. Jl Geol. Geophys., 15,* No. 1, 140–47.

Rex, D. C. 1972. K-Ar age determinations on volcanic and associated rocks from the Antarctic Peninsula and Dronning Maud Land. (*In* Adie, R. J., *ed. Antarctic Geology and Geophysics.* Oslo, Universitetsforlaget, 133–36.)

Rex, D. C. 1976. Geochronology in relation to the stratigraphy of the Antarctic Peninsula. *Bull. Br. antarc. Surv.,* No. 43, 49–58.

Rex, D. C., and Baker, P. E. 1973. Age and chronology of the Cornwallis Island granodiorite. *Bull. Br. antarc. Surv.,* No. 32, 55–61.

Reynolds, R. L., and Kellogg, K. S. 1974. Paleomagnetism of igneous rocks of the central Lassiter Coast, Antarctic Peninsula. *Antarct. Jnl U.S., 9,* No. 5, 227–28.

Ross, D. C. 1973. Are the granitic rocks of the Salinian block trondhjemitic? *J. Res. U.S. geol. Surv., 1,* No. 3, 251–54.

Rowley, P. D. 1973. Geologic observations on the northern Lassiter Coast and southern Black Coast. *Antarct. Jnl U.S., 8,* No. 4, 154–55.

Rowley, P. D. 1978. Geologic studies in Orville Coast and eastern Ellsworth Land, Antarctic Peninsula. *Antarct. Jnl U.S., 13,* No. 4, 7–9.

Rowley, P. D., and Williams, P. L. 1974. Plutonic rocks of the Lassiter Coast. *Antarct. Jnl U.S., 9,* No. 5, 225–26.

Rowley, P. D., Williams, P. L., and Schmidt, D. L. 1977. Geology of an Upper Cretaceous copper deposit in the Andean Province, Lassiter Coast, Antarctic Peninsula. *Prof. Pap. U.S. geol. Surv.,* No. 984, 36 pp.

Rowley, P. D., Waitt, R. B., Jr., Boyer, S. J., Kellogg, K. S., and Vennum, W. R. 1976. Expedition to the Lassiter Coast, Antarctica. *Summit, 22,* No. 1, 18–27.

Rowley, P. D., Williams, P. L., Schmidt, D. L., Reynolds, R. L., Ford, A. B., and others. 1975. Copper mineralization along the Lassiter Coast of the Antarctic Peninsula. *Econ. Geol., 70,* No. 5, 982–92.

Scharnberger, C. K., and Scharon, L. 1972. Paleomagnetism and plate tectonics of Antarctica. (*In* Adie, R. J., *ed. Antarctic Geology and Geophysics.* Oslo, Universitetsforlaget, 843–47.)

Schopf, J. M. 1969. Ellsworth Mountains: position in West Antarctica due to sea-floor spreading. *Science, N.Y., 164,* No. 3875, 63–66.

Scott, K. M. 1965. Geology of the southern Gerlache Strait region, Antarctica. *J. Geol., 73,* No. 3, 518–27.

Stevens, G. R. 1967. Upper Jurassic fossils from Ellsworth Land, West Antarctica, and notes on Upper Jurassic biogeography of the South Pacific region. *N.Z. Jl Geol. Geophys., 10,* No. 2, 345–93.

Stump, E. 1973. Earth evolution in the Transantarctic Mountains and West Antarctica. (*In* Tarling, D. H., and Runcorn, S. K., *eds. Implications of Continental Drift to the Earth Sciences. 2.* New York, Academic Press, 909–24.)

Suárez, M. 1976. Plate-tectonic model for southern Antarctic Peninsula and its relation to southern Andes. *Geology, 4,* No. 4, 211–14.

Swithinbank, C., Doalze, C., Wager, A., and Crabtree, R. 1976. Major change in the map of Antarctica. *Polar Rec., 18,* No. 114, 295–308.

Vennum, W. R. 1978. Igneous and metamorphic petrology of the southwestern Dana Mountains, Lassiter Coast, Antarctic Peninsula. *J. Res. U.S. geol. Surv., 6,* No. 1, 95–106.

Wade, F. A. 1972. Geologic survey of Marie Byrd Land. *Antarct. Jnl U.S., 7,* No. 5, 144–45.

Wade, F. A., and Wilbanks, J. R. 1972. Geology of Marie Byrd and Ellsworth Lands. (*In* Adie, R. J., *ed. Antarctic Geology and Geophysics.* Oslo, Universitetsforlaget, 207–14.)

West, S. M. 1974. The geology of the Danco Coast, Graham Land. *Scient. Rep. Br. antarc. Surv.,* No. 84, 58 pp.

Williams, P. L. 1970. Geology of the Lassiter Coast. *Antarct. Jnl U.S., 5,* No. 4, 98–99.

Williams, P. L., and Rowley, P. D. 1971. Geologic studies of the Lassiter Coast. *Antarct. Jnl U.S. 6,* No. 4, 120–21.

Williams, P. L., and Rowley, P. D. 1972. Composition of Jurassic sandstones, Lassiter Coast. *Antarct. Jnl U.S., 7,* No. 5, 145–46.

Williams, P. L., Schmidt, D. L., Plummer, C. C., and Brown, L. E. 1972. Geology of the Lassiter Coast area, Antarctic Peninsula—Preliminary report. (*In* Adie, R. J., *ed. Antarctic Geology and Geophysics.* Oslo, Universitetsforlaget, 143–48.)

39

Ages and Tectonic Implications of Andean Plutonism in the Southern Antarctic Peninsula

EDWARD FARRAR and SANDRA L. McBRIDE
Department of Geological Sciences, Queen's University, Kingston, Ontario, Canada K7L 3N6

PETER D. ROWLEY
U.S. Geological Survey, Denver, Colorado 80225, U.S.A.

Abstract. New and previously published K-Ar (biotite and hornblende) and Rb-Sr (whole-rock and biotite) analyses on 13 plutons and 2 dikes from the southern Antarctic Peninsula yield Late Cretaceous ages ranging between 120 and 95 m.y. The dated rocks are representative of several score of calc-alkaline gabbroic to granitic stocks and batholiths and hundreds of dikes exposed in the region. These rocks are correlated with the Andean Intrusive Suite based on similarities in composition and geologic setting. Within this suite, the younger the pluton, the more silicic it tends to be. Some plutons are locally associated with copper and molybdenum mineralization.

Along most of the Andean orogen in South America, calc-alkaline plutons and volcanic rocks were emplaced during Mesozoic and Cenozoic time in response to subduction that is still active today. In southernmost South America (south of its intersection with the Chile Ridge) and in the Antarctic Peninsula, calc-alkaline Andean plutons and volcanic rocks likewise were emplaced intermittently through most of Mesozoic time. However, in the Antarctic Peninsula in particular, Tertiary calc-alkaline igneous activity was far less common than it was in Mesozoic time, and it had a mafic and trondhjemitic character; this activity ceased by late Tertiary time. The subduction zone above which these rocks were generated was most active during Mesozoic time, but subduction continued at a slower rate and in a different direction in Tertiary time. Transform motion and extensional faulting became significant starting in early to middle Tertiary time and were accompanied by a shift to tholeiitic and mildly alkalic, mafic, igneous activity by middle to late Cenozoic time. To the southwest, Marie Byrd Land and western Ellsworth Land have a calc-alkaline igneous history up to late Cretaceous or early Tertiary time, after which emplacement of alkaline igneous rocks began; this change in rock type occurred simultaneously with a change from a subduction regime to an extension regime.

The southeastern Antarctic Peninsula is divided, from north to south, into the southern Black Coast, the Lassiter Coast, the Orville Coast, and eastern Ellsworth Land (Figure 39.1). Reconnaissance geologic mapping of eastern Ellsworth Land began with the University of Wisconsin Antarctic Peninsula Geophysical Traverse of 1961–62, but most mapping was done during the University of Wisconsin field party of 1965–66 (Laudon and others, 1969; Laudon, 1972). The southern Black Coast, Lassiter Coast, and northern Orville Coast were mapped in reconnaissance by members of the U.S. Geological Survey during the field seasons of 1969–70, 1970–71, and 1972–73 (Williams and others, 1972; Rowley and Williams, Chapter 38, this volume).

Isotopic dating of igneous rocks was undertaken as part of all these geologic mapping programs. Halpern (1967) determined five Rb-Sr ages of whole-rock and biotite samples for three(?) plutons and one dike from eastern Ellsworth Land. Mehnert and others (1975) reported K-Ar ages of biotite and hornblende concentrates for five plutons from the southern and central Lassiter Coast. Rowley and others (1975), Rowley, Farrar, and McBride (1976), and Rowley, Williams, and Schmidt (1977) gave preliminary analytical data on K-Ar ages of biotite and hornblende concentrates for six plutons—including five samples from the large Werner batholith—and one dike, from the central and northern Lassiter Coast and southern Black Coast. The present report gives final analytical data for the ages previously reported in preliminary form, and it summarizes regional tectonic and plutonic implications of all isotopic ages from the southern Antarctic Peninsula.

Geologic setting

The most widespread rock unit in the southern Antarctic Peninsula is a thick sequence of black and gray slate, siltstone, and subordinate sandstone that is characterized by abundant volcaniclastic material and by locally abundant invertebrate and plant fossils. This sequence, named the Latady Formation (Williams and others, 1972) for exposures in the southern Lassiter Coast, is of Late Jurassic age in the southern Black Coast, Lassiter Coast, and northern Orville Coast (R. W. Imlay and E. G. Kauffman, written communication, 1975) and of Middle and Late Jurassic age in eastern Ellsworth Land (Laudon and others, 1969; Laudon, 1972). Silicic, calc-alkaline, ash-flow tuff and andesitic to dacitic lava flows of the southern Antarctic Peninsula (Williams and others, 1972; Rowley and Williams, Chapter 38, this volume; Laudon, 1972) seem to postdate or laterally equate to the Latady Formation; thus most of these volcanic rocks are probably Late Jurassic in age. The Latady Formation and the volcanic rocks were

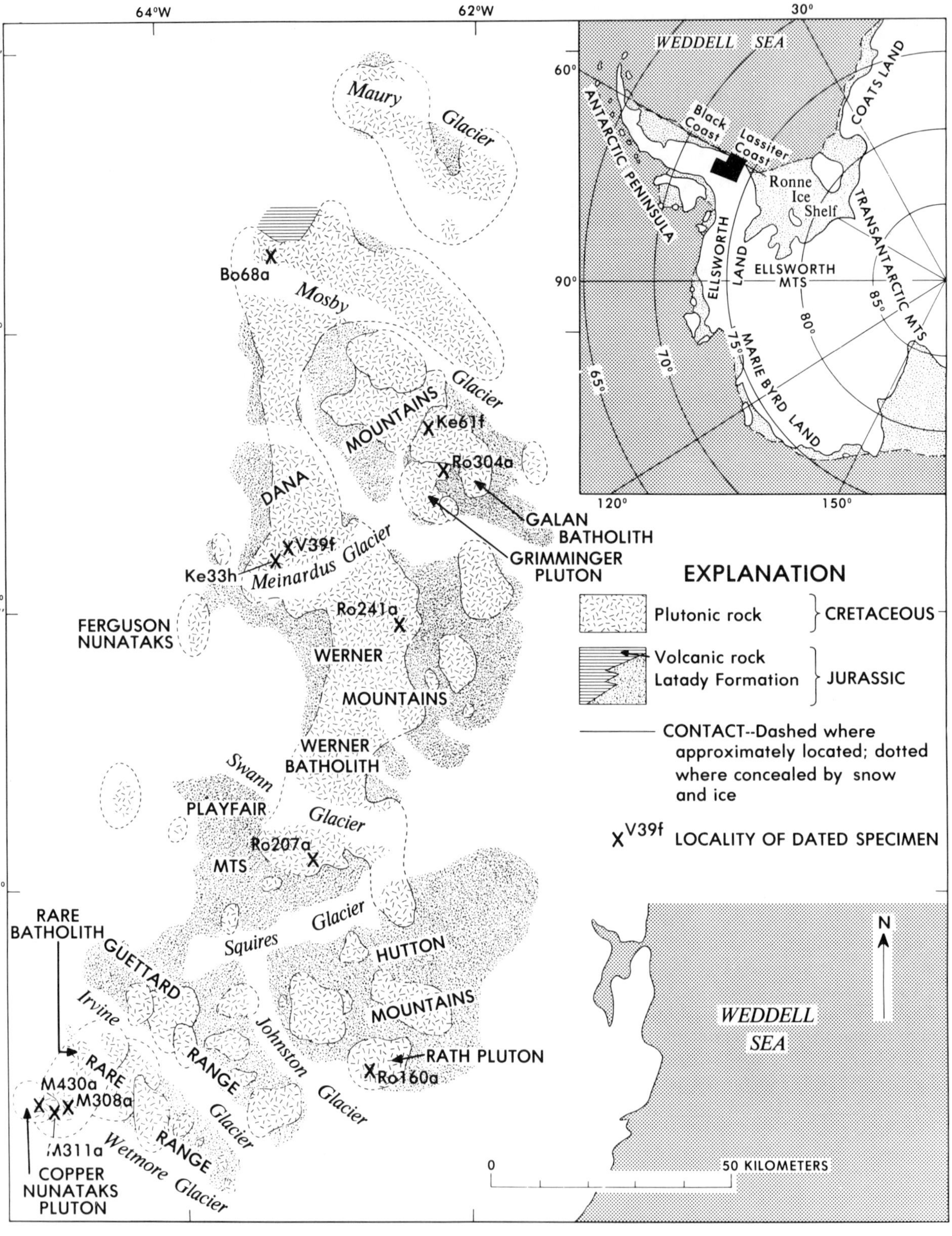

Figure 39.1. Geologic sketch map of central and northern Lassiter Coast and southern Black Coast, showing localities for specimens analyzed in this study. After Rowley and Williams (Chapter 38, this volume).

deformed sometime between Late Jurassic and Late Cretaceous time into open to isoclinal folds, the axial planes of which strike almost northward in the southern Black Coast, northeastward in the southern Lassiter Coast and northern Orville Coast, and west-northwestward in eastern Ellsworth Land.

Calc-alkaline plutonic rocks and related dikes cut the folded rocks throughout the southern Antarctic Peninsula. More than 60 stocks and batholiths, which range in composition from gabbro to granite but most commonly are quartz diorite, granodiorite, and quartz monzonite, have been mapped (Laudon and others, 1969; Laudon, 1972, and Chapter 95, this volume; Williams and others, 1972; Rowley and Williams, Chapter 38, this volume). Igneous rocks increase in volume northward. Intrusive contacts are sharp and commonly chilled, and abundant apophyses extend into the country rock. The plutons formed by magmatic processes, with intrusion in large part passive but in some places forcible. Dikes of mostly intermediate composition were emplaced near the closing stages of intrusion of each pluton. Isotopic ages discussed below indicate that all igenous rocks in the southern Antarctic Peninsula are of Late Cretaceous age.

Detailed modal and chemical data (Laudon and others, 1969; Laudon, 1972, and Chapter 95, this volume; Rowley and Williams, 1974, and Chapter 38, this volume; Rowley, unpublished data, 1977) indicate that igneous rocks in the southern Antarctic Peninsula belong to a single calc-alkaline magma series that is identical to rocks of the "Andean Intrusive Suite" exposed throughout the rest of the Antarctic Peninsula. The Andean Intrusive Suite was defined in the northern Antarctic Peninsula on the basis of a detailed petrologic and geochemical study by Adie (1955), prior to extensive isotopic dating of these rocks. The Andean Intrusive Suite was originally assigned an age of Late Cretaceous and Tertiary on the basis of stratigraphy; as isotopic ages became available, the name was restricted to plutons emplaced during the intervals 75–70 m.y. and 60–45 m.y. (Adie, 1972*a*; Rex, 1976). In keeping with the usage of "Andean orogeny" in the Andes of South America and with the petrologic and geochemical similarity of the rocks, we prefer to use "Andean Intrusive Suite" for rocks emplaced throughout Mesozoic time and, in some places, through much of Tertiary time.

Contact metamorphism of the Latady Formation and overlying volcanic rocks has created aureoles as much as 2 km wide around the plutons. Metamorphism attained grades as high as andalusite hornfels facies (Plummer, 1974). Hydrothermal alteration and copper and molybdenum mineralization, with affinities to porphyry deposits, occurred in scattered places throughout the southern Antarctic Peninsula. Study of these occurrences, plus a literature review of information on other types of minerals and hydrothermally altered rocks elsewhere in the Antarctic Peninsula, demonstrate that the Antarctic Peninsula is part of the circum-Pacific copper province (Rowley and others, 1975, 1977; Rowley and Pride, Chapter 108, this volume).

New age determinations

Analytical K-Ar data for eleven newly dated specimens are summarized in Table 39.1. Three specimens were collected during the 1970–71 season from two plutons and a dike from the Copper Nuntataks of the central Lassiter Coast. The rest of the specimens were collected during the 1972–73 season from four plutons in the northern Lassiter Coast and southern Black Coast. Petrographic data on all analyzed specimens are given in Table 39.2. Volume proportions of normative and modal quartz and feldspars of the eleven specimens are shown on a triangular diagram (Figure 39.2). Rock names are based on modes according to the classification (Figure 39.2) of Bateman (1961, Figure 2). The geology of the collection sites and the ages are discussed below in order of their location from north to south.

Five ages were determined for the largest igneous body in the Lassiter Coast, the Werner batholith (Figure 39.1). This batholith, named by Vennum (1978) for widespread exposures in the Werner Mountains, extends from unnamed mountains in the southern Black Coast at least 140 km south to the Hutton Mountains on the central Lassiter Coast. Field relationships, petrography, and geochemistry support the interpretation that the batholith had a single magma source. However, correlation of some parts of the batholith across major glaciers and ice fields is difficult, and it is possible that the batholith as shown in Figure 39.1 consists of more than one pluton. The batholith generally is concentrically zoned with a more silicic interior. The margins, in most places less than 1 km wide, largely consist of quartz diorite or, locally, diorite. Granodiorite is the most common rock in the interior of the batholith.

The Werner batholith is considered to be composite. The silicic rock of the interior part of the batholith is seen in two places to intrude the mafic margin phases. Thus, there were at least two magmatic pulses in these places. One of these places is near the eastern edge of the batholith in the central Werner Mountains. The other is in an area of extensive roof pendants in the southwestern Dana Mountains (Vennum, 1978). Significant assimilation of wall rocks of the Latady Formation and of the mafic margin phase during emplacement of the silicic interior phase is

Table 39.1. Analytical data for K-Ar ages of plutons and a dike in the northern Lassiter Coast and southern Black Coast, Antarctica

Name of pluton	Specimen number	Mineral analyzed	K (wt %)	^{40}Ar (radiogenic) ccs. NTP/g × 10^{-5}	^{40}Ar (radiogenic) % total	Age ± error[a] (m.y.)
Werner	Bo68a	Biotite	6.092	2.516	3.6	100.6±4.3
Werner	V39f	Biotite	6.825	2.839	13.8	101.3±3.6
		Hornblende	0.406	0.192	15.2	114.4±6.8
Werner	Ke33h	Biotite	7.232	3.119	6.2	104.9±4.6
Werner	Ro241a	Biotite	7.138	3.168	2.8	107.9±5.1
Werner	Ro207a	Biotite	7.597	3.063	6.5	98.2±3.0
		Hornblende	1.006	0.417	7.7	100.9±3.0
Grimminger	Ro304a	Biotite	7.667	3.410	3.7	108.1±2.7
Galan	Ke61f	Biotite	7.446	3.191	8.2	104.3±3.8
Rath	Ro160a	Biotite	6.742	3.013	4.0	108.6±6.8
West RARE	M308a	Biotite	6.547	2.823	6.4	104.9±3.2
Copper Nunataks	M430a	Biotite	7.218	2.817	6.9	95.2±3.0
Copper Nunataks (dike)	M311a	Biotite	7.414	2.909	8.6	95.6±3.0

Constants used: $\lambda_e = 0.585 \times 10^{-10} yr^{-1}$, $\lambda_\beta = 4.72 \times 10^{-10} yr^{-10}$, $^{40}K/K = 0.0119$ atomic percent.
Locations and rock types given in Table 39.2; generally listed by pluton, from north to south (Figure 39.1).
Radiometric analysis done at Queen's University, Kingston, Ontario, by E. F.
[a] Error represents the analytical uncertainty at the 2σ level.

demonstrated by petrographic studies of rocks from the southwestern Dana Mountains by Vennum (1978).

Both the mafic margin phase and the silicic interior phase of the Werner batholith were sampled for geochronologic study (Table 39.1) at the locations shown on Figure 39.1 Both specimens of the mafic margin phase were taken where it was intruded by the interior phase. The age of biotite from diorite of the older mafic phase (specimen Ro241a) in the central Werner Mountains is 107.9 m.y., which gives a minimum age for the older intrusive phase here. The age of biotite from diorite of the older mafic phase in the southwestern Dana Mountains (specimen V39f) is 101.3 m.y., a minimum age of the older intrusive phase as it is exposed here; however, the age on hornblende from the same rock is 114.4 m.y., and this is considered to be closer to the true age of the older intrusive phase. The biotite ages for the older mafic phase in these two places were probably partially reset during emplacement of the main part of the batholith.

Three ages were determined for rocks of the interior of the Werner batholith. At the northern end (specimen Bo68a), an age of 100.6 m.y. (Table 39.1) on biotite from granodiorite was obtained. In the southern part of the batholith (specimen Ro207a), similar ages of 98.2 m.y. on biotite and 100.9 m.y. on hornblende from quartz monzonite were determined. Specimen Ke33h, granodiorite

Table 39.2. Modal data (in vol %) and collection localities of dated specimens; 500 points counted on thin sections and stained rock slabs. Locations from 1:500,000 scale U.S. Geological Survey Antarctic Sketch Map of Ellsworth Land (East Part)–Palmer Land (South Part). Anorthite content of plagioclase determined by *a*-normal method.

	1	2	3	4	5	6	7	8	9	10	11
Plagioclase	55.8	64.5	59.8	62.2	45.6	59.1	49.0	51.5	39.2	29.8	32.3
Quartz	24.7	5.4	23.6	0.6	22.7	5.2	22.9	10.0	30.2	26.6	0.5
K-feldspar	13.9	1.5	7.2	1.4	19.1	0.5	12.1	15.1	11.6	34.4	0
Groundmass	0	0	0	0	0	0	0	0	0	0	57.4
Biotite	2.8	11.6	8.2	13.6	6.2	7.8	12.4	4.6	5.8	4.8	3.4
Hornblende	2.0	15.0	0	7.8	5.0	23.2	2.4	16.2	11.4	3.0	4.8
Pyroxene	0	0	0	11.2	0	1.4	0	trace	0	0	0
Chlorite	trace	0.8	0.8	2.2	0.2	0.4	0.2	0.6	1.2	0.2	0
Sericite	0	0.2	0.4	trace	0	0	0	0	0	0	0
Opaque minerals	0.8	0.4	trace	0.6	0.8	1.8	0.4	1.4	0.2	1.2	1.4
Sphene	trace	0	trace	0	0.4	trace	0	0.6	trace	trace	0.2
Apatite	trace	0	0	0.4	trace	trace	trace	0	0	trace	0
Epidote	0	0.6	0	0	trace	0.6	0.6	0	0.4	0	0

1. Bo68a: Granodiorite of northern Werner batholith. Coarse-grained rock; texture is hypidiomorphic-granular, except for scattered, large (as much as 1.5 cm), light-gray, poikilitic K-feldspar crystals that lack grid twinning. Plagioclase is commonly zoned; rims average An_{25} and cores average An_{35}. Specimen locality: lat. 72° 50′ S, long. 63° 12′ W (approximate, because no small-scale map exists of this area), unnamed mountains of the southern Black Coast; about 30 km N 29° W from Mount Axworthy.
2. V39f: Diorite of older mafic phase from near roof of central Werner batholith. Fine- to medium-grained rock; texture is hypidiomorphic-seriate. Some plagioclase is zoned; rims average An_{38} and cores average An_{75}. Some plagioclase crystals show patchy extinction. Specimen locality: lat. 73° 23′ S, long. 63° 15′ W, southwestern Dana Mountains, Lassiter Coast; about 11 km S 68° W from Mount Barkow.
3. Ke33h: Granodiorite of younger silicic phase from near roof of central Werner batholith. Medium- to coarse-grained rock; texture is hypidiomorphic-seriate. K-feldspar crystal lack grid twinning. Plagioclase is commonly zoned; rims average An_{27} and cores average An_{38}. Specimen locality: lat. 73° 24′ S, long. 63° 19′ W, southwestern Dana Mountains, Lassiter Coast; about 12 km S 63° W from Mount Barkow.
4. Ro241a: Diorite of older mafic phase from near eastern intrusive contact of Werner batholith. Medium-grained rock; texture is hypidiomorphic-granular. K-feldspar crystals lack grid twinning. Plagioclase is commonly zoned; rims average An_{30} and cores average An_{51}. Specimen locality: lat. 73° 32′ S, long. 62° 29′ W, Werner Mountains, Lassiter Coast; about 12 km S 35° W from Mount Fell.
5. Ro207a: Quartz monzonite of southern Werner batholith. Medium- to coarse-grained rock; texture is hypidiomorphic-granular, except for scattered, large (as much as 1.5 cm), light-gray, poikilitic K-feldspar crystals that lack grid twinning. Plagioclase is commonly zoned; rims average An_{24} and cores average An_{36}. Specimen locality: lat. 73° 57′ S, long. 63° 04′ W, Playfair Mountains, Lassiter Coast; about 3 km N 65° W from Mount Kane.
6. Ro304a: Diorite of Grimminger pluton. Medium-grained rock; texture is hypidiomorphic-granular. K-feldspar crystals lack grid twinning. Plagioclase is commonly zoned; rims average An_{30} and cores average An_{55}. Specimen locality: lat. 73° 15′ S, long. 62° 10′ W, eastern Dana Mountains, Lassiter Coast; about 6 km N 22° E from Mount Grimminger.
7. Ke61f: Granodiorite of Galan batholith. Medium- to coarse-grained rock; texture is hypidiomorphic-granular, except for scattered, large (as much as 1 cm), light-gray, poikilitic K-feldspar crystals that generally lack grid twinning. Plagioclase is commonly zoned; rims average An_{23} and cores average An_{38}. Specimen locality: lat. 73° 11′ S, long. 62° 19′ W, eastern Dana Mountains, Lassiter Coast; about 13 km N 09° W from Mount Grimminger.
8. Ro160a: Mafic granodiorite of Rath pluton. Medium-grained rock; texture is hypidiomorphic-granular, except for scattered, large (as much as 1.5 cm), light-gray, poikilitic K-feldspar crystals that lack grid twinning. Plagioclase is commonly zoned; rims average An_{28} and cores average An_{47}. Specimen locality: lat. 74° 19′ S, long. 62° 41′ W, southern Hutton Mountains, Lassiter Coast; about 6 km N 73° W from Mount Rath.
9. M308a: Granodiorite of West RARE batholith. Fine- to medium-grained rock; texture is hypidiomorphic-granular, except for scattered, large (as much as 1 cm), light-gray, poikilitic K-feldspar crystals that lack grid twinning. Plagioclase is commonly zoned; rims average An_{22} and cores average An_{46}. Specimen locality: lat. 74° 23′ S, long. 64° 50′ W, Copper Nunataks, Lassiter Coast; about 16 km N 01° E from Mount Aaron.
10. M430a: Quartz monzonite of Copper Nunataks pluton. Medium- to coarse-grained rock; texture is hypidiomorphic-granular, except for scattered, large (as much as 1 cm), light-pink, K-feldspar crystals that lack grid twinning. Plagioclase is commonly zoned; rims average An_{21} and cores average An_{40}. Specimen locality: lat. 74° 23′ S, long. 65° 02′ W, Copper Nunataks, Lassiter Coast; about 20 km N 17° W from Mount Aaron.
11. M311a: Granodiorite porphyry dike. Medium-grained phenocrysts in an aphanitic groundmass of mostly quartz and K-feldspar. Plagioclase is zoned; rims average An_{33} and cores average An_{40}. Specimen locality: lat. 74° 23′ S, long. 64° 53′ W, Copper Nunataks, Lassiter Coast; about 17 km N 02° W from Mount Aaron.

Petrographer: P. D. Rowley.

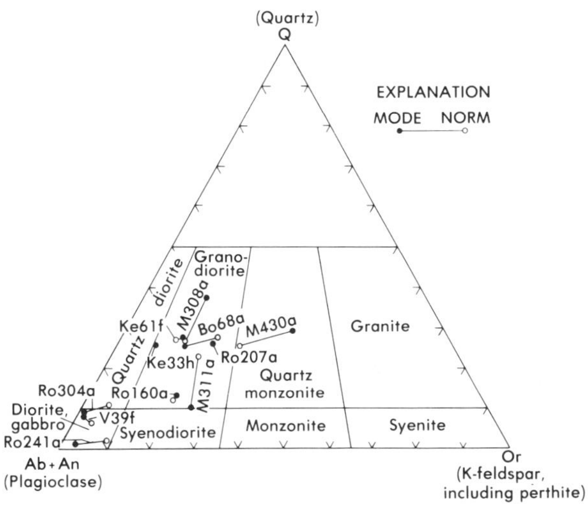

Figure 39.2. Ternary plots of modal and normative quartz-orthoclase-plagioclase (Ab + An) proportions for dated rocks. Modal plutonic rock classification is shown in diagram.

from the central part of the batholith, however, gave a biotite age of 104.9 m.y.. which may be anomalously old.

Assuming that all these rocks belong to the same pluton, we interpret the data to indicate that the batholith cooled from the outside inward; in some or most places, the molten, more silicic interior intruded the already crystallized margin. At least some parts of the thin mafic margin of the composite Werner batholith are 114 m.y. old or perhaps slightly older, while the main interior part of the batholith is as young as about 100 m.y. old. The emplacement of the main part of the batholith locally reset older ages, especially the biotite ages, of the mafic margin. The anomalously(?) old age of specimen Ke33h, collected from granodiorite that intrudes diorite (specimen V39f) in the same area, may be due to incorporation by partial magmatic assimilation of some of the partially reset biotite from the older phase.

The Grimminger pluton is a large stock made up of diorite and quartz diorite. The crosscutting relationships and the general field appearance of rocks of the Grimminger pluton suggest that it is one of the oldest plutons in the Lassiter Coast. Biotite from diorite of the Grimminger pluton has an age of 108.1 m.y. (Table 39.1, specimen Ro304a). As in other intruded plutons in the Lassiter Coast (Mehnert and others, 1975), the true age of the Grimminger pluton may be somewhat older, and the biotite age may reflect partial resetting.

The Galan batholith intrudes the Grimminger pluton (Figure 39.1) and at the intrusive contact, as well as in many other places, it shears and alters the Grimminger pluton. The Galan batholith consists mostly of granodiorite, although in some places along the margins the rock is as mafic as quartz diorite. A biotite age of 104.3 m.y. (Table 39.1, specimen Ke61f) was obtained on granodiorite from the Galan batholith.

The Rath pluton is a concentrically zoned stock that ranges in composition from diorite at the intrustive contacts to granodiorite in the interior. It has an average composition of quartz diorite. A specimen of mafic granodiorite yielded a biotite age of 108.6 m.y. (Table 39.1, specimen Ro160a).

Dates of two plutons and a dike in the Copper Nunataks have been discussed by Rowley and others (1975, 1977). The oldest pluton in the Copper Nunataks is a granodiorite body that is correlated with the West RARE batholith of the nearby RARE (acronym of Ronne Antarctic Research Expedition) Range. This body consists of homogeneous, fine- to medium-grained granodiorite. In the RARE Range it intrudes the North RARE pluton, which also consists of fine- to medium-grained granodiorite but is somewhat more mafic than the West RARE batholith. The North RARE pluton was dated at 113.4 m.y. (biotite) and 119.4 m.y. (hornblende) by Mehnert and others (1975). Biotite from the West RARE batholith in the Copper Nunataks yielded an age of 104.9 m.y. (Table 39.1, specimen M308a). Closely spaced faults and associated slightly hydrothermally altered and pyrite-mineralized rocks formed soon after emplacement of the West RARE batholith in the Copper Nunataks. This in turn was followed by emplacement of the other dated pluton in the Copper Nunataks, the Copper Nunataks pluton, which is a concentrically zoned stock containing quartz diorite and granodiorite margins and a granite interior. Average composition of this pluton is quartz monzonite, and a specimen of this composition gave a biotite age of 95.2 m.y. (Table 39.1, specimen M430a). Plutonism was followed by emplacement of a series of granodiorite porphyry dikes; biotite from one of these dikes was dated at 95.6 m.y. (Table 39.1, specimen M311a). Emplacement of the granodiorite porphyry dikes was followed by shearing, intense hydrothermal alteration, and copper and molybdenum mineralization. This activity formed the Lassiter Coast copper deposit, which has affinities with porphyry deposits. The dike intrusion, alteration, and mineralization are considered to reflect the terminal stages of emplacement of the Copper Nunataks pluton. Mineralized rock is not known to approach ore grade.

Igneous activity in the southern Antarctic Peninsula

K-Ar ages of igneous rocks in the southern Black Coast and northern and central Lassiter Coast (this paper) range from 114 to 95 m.y. The range in K-Ar ages from the central and southern Lassiter Coast is from 119 to 99 m.y. (Mehnert and other, 1975). The range in Rb-Sr ages from eastern Ellsworth Land is from 109 to 102 m.y. (Halpern, 1967).

Mehnert and others (1975) has noted, as we have, that some biotite ages reflect resetting of the true age of a pluton, while most hornblende ages are probably closer to the true age of emplacement. The discrepancy between biotite and hornblende ages is greater for older plutons; many of these older plutons are known by field relationships to be cut by younger plutons. Biotite and hornblende ages are in close agreement only for young plutons that exhibit no evidence of intrusion by later plutons. Plutons selected for isotopic study were chosen as being representative of other plutons in the area. Thus, it is unlikely that any of the undated plutons in the southern Antarctic Peninsula are considerably older or considerably younger than the dated plutons. Based on known field relationships, all plutons in the southern Antarctic Peninsula occupy the same stratigraphic position and postdate the folding. Similarities in field relations, petrology, and geochemistry of all plutonic rocks examined from the southern Antarctic Peninsula

(Laudon and others, 1969: Williams and others, 1972; Rowley and Williams, Chapter 38, this volume) further indicate that all were formed during several phases of the same magma series.

Kamenev (1975), however, suggested that quartz diorite of the Coman pluton, west of the Playfair Mountains, might be preorogenic, that is, predate the folding. His main evidence is the presence within the Coman pluton of some altered rock with chlorite and epidote, somewhat more abundant joints, and local veins of more silicic composition. The intrusive contact of this pluton is not exposed, unfortunately, nor are any rocks that predate or postdate the pluton exposed in the vicinity of the pluton. However, in field appearance and petrographic characteristics, the Coman pluton resembles several other mafic plutons in the Lassiter Coast as well as the intruded mafic margins of the Werner batholith. All of these similar mafic rocks postdate the regional folds, and two of them (North Latady pluton in Mehnert and others, 1975; Grimminger pluton in this report), as well as the intruded mafic margins of the Werner batholith, have K-Ar ages of 119–101 m.y. Thus, we conclude that plutonism in the southern Antarctic Peninsula, including the Coman pluton, and emplacement of all or most dikes occupied a relatively restricted part of Late Cretaceous time (about 120 m.y. or slightly older to 95 m.y.). Partial support for such a narrow time range of plutonism comes from paleomagnetic studies of numerous plutons and dikes in the Lassiter Coast and southern Black Coast; all plutons and dikes studied are normally polarized, suggesting that all were emplaced during the Late Cretaceous epoch of predominantly normal polarity (Kellogg and Reynolds, 1978).

Younger plutons in the southern Antarctic Peninsula are progressively more silicic. The oldest plutons consist of diorite and quartz diorite (Mount Berger pluton, Halpern, 1967; North RARE and North Latady plutons, Mehnert and others, 1975; Rath and Grimminger plutons, mafic margins of the Werner pluton). Except for the reset 101 m.y. K-Ar biotite age of the amfic margin of the Werner batholith, they have a Rb-Sr age of 109 m.y., K-Ar biotite ages of 113–108 m.y., and K-Ar hornblende ages of 119–114 m.y. The Mount Glowa pluton of eastern Ellsworth Land has a Rb-Sr age of 102 ± 6 m.y. (Halpern, 1967) and is considered a quartz diorite by Halpern and by Laudon and others (1969). Modal or chemical evidence for this rock name is not given, however, nor do we know the type of rock classification used. The next youngest plutons consist primarily of granodiorite (McLaughlin pluton, Mehnert and others, 1975; West RARE and Galan plutons). They have K-Ar biotite and hornblende ages of 107–104 m.y. The next youngest plutons consist of granodiorite to quartz monzonite (Luck-Caywood pluton and related dike at Mount Byrd, Halpern, 1967; Terwileger and Crowell plutons, Mehnert and others, 1975; Werner pluton). They have Rb-Sr ages of 103–102 m.y. and K-Ar biotite and hornblende ages of 103–98 m.y. except for the anomalous(?) age (specimen Ke33h) from the central Werner batholith. The youngest pluton consists mostly of quartz monzonite (Copper Nunataks pluton); it is the most silicic pluton recognized in the southern Antarctic Peninsula and contains the most altered and mineralized rocks known in this region. It and an associated granodiorite porphyry dike have K-Ar biotite ages of about 95 m.y.

Regional considerations

Post-Paleozoic igneous rocks are volumetrically the most abundant rock type in West Antarctica. They were emplaced intermittently through Mesozoic and Cenozoic time (Craddock, 1969; Halpern, 1971; Wade and Wilbanks, 1972; Wade, 1972; Adie, 1972*a*; Krylov, 1972; Grikurov, 1973; Lopatin and others, 1974; Rex, 1976; LeMasurier and Wade, 1976; LeMasurier and Rex, Chapter 89, this volume; Saunders and others, Chapter 34, this volume). Compositions and distributions of rock types in the different age ranges tell us much about the plate-tectonic history of West Antarctica (Rowley and Williams, Chapter 38, this volume).

Most igneous rocks in West Antarctica are comprised of plutons and dikes of Mesozoic age. Some formed during plutonic events about 180–160 m.y. ago (Early–Middle Jurassic) and perhaps about 140–130 m.y. ago (Late Jurassic–Early Cretaceous) in the northern and central Antarctic Peninsula (Grikurov, 1973), and about 150–120 m.y. ago (Late Jurassic–Early Cretaceous) in Marie Byrd Land and western Ellsworth Land (Lopatin and others, 1974). Yet by far the most pronounced period of plutonism in West Antarctica occurred 120–80 m.y. ago (Late Cretaceous), coincident with all dated plutons of the southern Antarctic Peninsula. Plutons of Late Cretaceous age are widespread throughout the rest of the Antarctic Peninsula as well as Marie Byrd Land, but—perhaps because of lack of exposure—are uncommon in western Ellsworth Land. All Mesozoic plutons are calc-alkaline and range in composition from gabbro to granite; most are silicic (granite, quartz monzonite, and granodiorite), however. They are similar to each other in petrology and geochemistry and on these similarities were defined as the Andean Intrusive Suite (Adie, 1955).

Folded andesitic to rhyolitic Mesozoic lava flows and ash-flow tuffs, like those in the southern Antarctic Peninsula, are widely scattered throughout the Antarctic Peninsula (Adie, 1969*a,b*; Craddock, 1972). They overlie Middle and Upper Jurassic sedimentary rocks and are commonly contact metamorphosed by Mesozoic plutons. Thus, they have traditionally been considered Middle and Late Jurassic in age (Adie, 1964, 1972*b*). They are calc-alkaline and similar in chemistry to the Mesozoic plutonic rocks; thus, Adie (1964) suggested that they are the surface expression of the first Mesozoic plutons. Probably, however, the Mesozoic volcanic activity that produced these rocks extended from Early Jurassic to late Cretaceous time, and thus, the volcanic rocks are broadly coincident in age with the Mesozoic plutons (Grikurov, 1973; Rowley and Williams, Chapter 38, this volume; Thomson, Chapter 37, this volume).

Igneous rocks of Marie Byrd Land and western Ellsworth Land that are younger than 80 m.y. are considerably different in rock type and geochemistry from those that are older than 80 m.y. Plutons, for example, are scarce; only an alkalic gabbro of 44–42 m.y. (Eocene) (LeMasurier and Rex, Chapter 89, this volume) from Marie Byrd Land and a "granite" pluton of 60 m.y. (Paleocene) (Lopatin and others, 1974) have been reported. Volcanic rocks so far dated range from about 30 m.y. old (Oligocene) to Quarternary, but most are younger than 10 m.y. (Miocene) (Rutford and others, 1972); LeMasurier and Rex, Chapter 89, this volume). Almost without exception, these igneous rocks are alkalic and formed under conditions of extension of continental crust (LeMasurier and Wade, 1976).

Igneous rocks of the Antarctic Peninsula that are younger than 80 m.y. are enigmatic. Plutons are relatively uncommon; most were emplaced during minor pulses at about 75–70 m.y. (Late Cretaceous) and 60–45 m.y. (Paleocene and Eocene) (Grikurov, 1973; Rex, 1976). The youngest pluton dated so far is 10 m.y. old (Rex and Baker, 1973) and occurs in the northwestern part of the South Shetland Islands. Tertiary plutons in the Antarctic Peninsula have traditionally been considered petrologically and geochemically similar to Mesozoic plutons, and thus to be Andean. Many, however, are mafic (gabbro, diorite, and quartz diorite) and trondhjemitic (Rowley and Williams, Chapter 38, this volume; Saunders and others, Chapter 34, this volume). As noted by these authors and by Grikurov (1973) and Grikurov and Lopatin (1975), most are confined to offshore islands of the northern Antarctic Peninsula, near areas of major Cenozoic faulting (Dalziel and Elliot, 1973). Many have been the locus of copper mineralization and hydrothermal alteration (Rowley and Pride, Chapter 108, this volume).

The distribution and origin of volcanic rocks of the Antarctic Peninsula that are younger than 80 m.y. are also poorly known. Most Cenozoic volcanic rocks are less than 5 m.y. old, but basalts of 58–54 and 35 m.y. (Paleocene and Oligocene) on Trinity Peninsula, as well as mafic volcanic rocks that span Cenozoic time in the South Shetland Islands (Hobbs, 1968; Rex, 1972; Baker 1976; González-Ferrán, Chapter 84, this volume; Watts, Chapter 28, this volume; Weaver and others, Chapter 29, this volume), have been mapped. Most Cenozoic volcanic rocks of the peninsula are generally calc-alkaline, mildly alkaline, or tholeiitic; they formed during a complex history of Cenozoic subduction, transform faulting, and extension of both continental and oceanic crust (Baker, 1972, 1976; González-Ferrán, Chapter 84, this volume; Weaver and others, Chapter 29, this volume).

Mesozoic and Cenozoic igneous rocks of West Antarctica can be compared to those of the Andes of South America. In the southern Andes, field relationships and geochemistry of sparsely dated (mostly by Halpern and Carlin, 1971; Halpern, 1973) igneous rocks are not well known, but as might be expected, these rocks have some similarities in ages with igneous rocks of the Antarctic Peninsula. The data (Halpern, 1973) indicate a strong plutonic episode at 100–75 m.y. (Late Cretaceous) and less abundant plutons during episodes between 155–120 m.y. (Middle Jurassic to Early Cretaceous) and 70–10 m.y. (Late Cretaceous to Miocene). This picture contrasts with the ages of better-known calc-alkaline Andean plutons in the central Andes, north of the intersection of the Chile Ridge with South America, where six plutonic episodes (156–137, 128–117, 107–87, 67–59, 44–34, and 23–22 m.y.) span Mesozoic and Cenozoic time (McNutt and others, 1975). Here, there is a local eastward progression of younger ages of plutons (Farrar and others, 1970; James, 1971; McNutt and others, 1975), but such a progression has not been demonstrated in southern South America or in West Antarctica. Saunders and others (Chapter 34, this volume), however, have suggested that northward and westward progressions of younger ages may characterize the Antarctic Peninsula.

We conclude that Andean igneous rocks were emplaced throughout Mesozoic time above an active subduction zone that extended along South America and West Antarctica. Plutonism above the subduction zone varied in time and space; the most abundant plutonism in West Antarctica occurred 120–80 m.y. ago (Rowley and Williams, Chapter 38, this volume). Tertiary subduction continued along the western margin of most of South America, but ceased in Marie Byrd Land and western Ellsworth Land by early Tertiary time and was followed by extensional faulting (LeMasurier and Wade, 1976). South of the intersection of South America with the Chile Ridge, as well as in the Antarctic Peninsula, transform faulting and extensional faulting became dominant by middle to late Tertiary time (Herron and Tucholke, 1976). Subduction in this region slowed and took on a direction different from that north of the Chile Ridge, and by late Tertiary time, it had ceased (Herron and Tucholke, 1976; Craddock and Hollister, 1976).

Acknowledgments. Field work was done by the U.S. Geological Survey and was financed by National Science Foundation Grant AG-187. The U.S. Antarctic Research Program of the National Science Foundation and U.S. Navy Operation Deep Freeze provided logistic support. We are grateful to M. D. Turner and other members of the Division of Polar Programs of the National Science Foundation for assistance with logistics matters. We thank the other geologists of the 1970–71 field party (P. L. Williams, leader, A. B. Ford, W. H. Nelson, and R. L. Reynolds) and the 1972–73 field party (W. R. Vennum, K. S. Kellogg, S. J. Boyer, R. B. Waitt, Jr., and E. N. Kamenev); all belong to the U.S. Geological Survey except Kamenev, who is a member of the Research Institute of the Geology of the Arctic, Leningrad, U.S.S.R. We acknowledge the assistance of D. C. Rex of the University of Leeds, England, for a preprint of his paper. The analytical facilities used in the determination of the K-Ar ages reported herein were supported by a grant from the National Research Council of Canada. We thank W. E. LeMasurier of the University of Colorado at Denver and D. L. Schmidt, E. R. Verbeek, and H. H. Mehnert of the U.S. Geological Survey for technical reviews.

References cited

Adie, R. J. 1955. The petrology of Graham Land—II. The Andean granite-gabbro intrusive suite. *Scient. Rep. Falkld Isl. Depend. Surv.*, No. 12, 39 pp.

Adie, R. J. 1964. The geochemistry of Graham Land. (*In* Adie, R. J., *ed. Antarctic Geology*. Amsterdam, North-Holland Publishing Company, 541–47.)

Adie, R. J. 1969*a*. Northern Antarctic Peninsula. (*In* Bushnell, V. C., and Craddock, C., *eds*. Geologic Maps of Antarctica. *Antarct. Map Folio Ser.*, Folio 12, Pl. I.)

Adie, R. J. 1969*b*. Southern Antarctic Peninsula. (*In* Bushnell, V. C., and Craddock, C., *eds*. Geologic Maps of Antarctica. *Antarct. Map Folio Ser.*, Folio 12, Pl. II.)

Adie, R. J. 1972*a*. Recent advances in the geology of the Antarctic Peninsula. (*In* Adie, R. J., *ed. Antarctic Geology and Geophysics*. Oslo, Universitetsforlaget, 121–24.)

Adie, R. J. 1972*b*. Evolution of volcanism in the Antarctic Peninsula. (*In* Adie, R. J., *ed. Antarctic Geology and Geophysics*. Oslo, Universitetsforlaget, 137–41.)

Baker, P. E. 1972. Recent volcanism and magmatic variation in the Scotia arc. (*In* Adie, R. J., *ed. Antarctic Geology and Geophysics*. Oslo, Universitetsforlaget, 57–60.)

Baker, P. E. 1976. Volcanism and plate tectonics in the Antarctic Peninsula and Scotia arc. (*In* González-Ferrán, O., *ed. Andean and Antarctic Volcanology Problems*. Rome, International

Association of Volcanology and Chemistry of the Earth's Interior, 347–56.)

Bateman, P. C. 1961. Granitic formations in the east-central Sierra Nevada near Bishop, California. *Geol. Soc. Am. Bull., 72,* No. 10, 1521–38.

Craddock, C. 1969. Radiometric age map of Antarctica. (*In* Bushnell, V. C., and Craddock, C., *eds.* Geologic Maps of Antarctica. *Antarct. Map Folio Ser.,* Folio 12, Pl. XIX.)

Craddock, C. 1972. Geologic map of Antarctica: 1:5,000,000. New York, American Geographical Society.

Craddock, C., and Hollister, C. D. 1976. Geologic evolution of the southeast Pacific basin. (*In* Hollister, C. D., Craddock, C., and others. *Initial Reports of the Deep Sea Drilling Projects,* Vol. 35. Washington [U.S. Government Printing Office], 723–43.)

Dalziel, I. W. D., and Elliot, D. H. 1973. The Scotia arc and Antarctic margin. (*In* Nairn, A. E. M., and Stehli, F. G. *eds. The Ocean Basins and Margins. 1. The South Atlantic.* New York, Plenum Press, 171–246.)

Farrar, E., Clark, A. H., Haynes, S. J., Quirt, G. S., Conn, H., and Zentilli, M. 1970. K-Ar evidence for the post-Paleozoic migration of granitic intrusion foci in the Andes of northern Chile. *Earth planet. Sci. Lett., 10,* No. 1, 60–66.

Grikurov, G. E. 1973. *Geologiia Antarkticheskogo polvostrova* [Geology of the Antarctic Peninsula]. Moscow, Izdatel'stvo Nauka, 119 pp.

Grikurov, G. E., and Lopatin, B. G. 1975. Structure and evolution of the West Antarctic part of the circum-Pacific mobile belt. (*In* Campbell, K. S. W., *ed. Gondwana Geology.* Canberra, Australia National University Press, 639–50.)

Halpern, M. 1967. Rubidium-strontium isotopic age measurements of plutonic igneous rocks in eastern Ellsworth Land and northern Antarctic Peninsula, Antarctica. *J. geophys. Res., 72,* No. 20, 5133–42.

Halpern, M. 1971. Evidence for Gondwanaland from a review of West Antarctic radiometric ages. (*In* Quam, L. O., *ed.* Research in the Antarctic. *Am. Ass. Adv. Sci.,* Publn No. 93, 717–30.)

Halpern, M. 1973. Regional geochronology of Chile south of 50° latitude. *Geol. Soc. Am. Bull., 84,* No. 7, 2407–22.

Halpern, M., and Carlin, G. M. 1971. Radiometric chronology of crystalline rocks from southern Chile. *Antarct. Jnl U.S., 6,* No. 5, 191–93.

Herron, E. M., and Tucholke, B. E. 1976. Sea-floor magnetic patterns and basement structure in the southeastern Pacific. (*In* Hollister, C. D., Craddock, C., and others. *Initial Reports of the Deep Sea Drilling Project,* Vol. 35. Washington [U.S. Government Printing Office], 263–78.)

Hobbs, G. J. 1968. The geology of the South Shetland Islands—IV. The geology of Livingston Island. *Scient. Rep. Br. antarc. Surv.,* No. 47, 34 pp.

James, D. E. 1971. Plate tectonic model for the evolution of the central Andes. *Geol. Soc. Am. Bull., 82,* No. 12, 3325–46.

Kamenev, E. N. 1975. Skladchatiye osadochnye i intruzivnye obrazovaniia Berega Lassitera i Berega Bleka (Antarkticheskii polouostrov) [Folded sedimentary and intrusive formations of the Lassiter Coast and the Black Coast (Antarctic Peninsula)]. *17th Sov. Antarct. Exped., Winter Res. 1971–1973, 65,* 215–22.

Kellogg, K. S., and Reynolds, R. L. 1978. Paleomagnetic results from the Lassiter Coast, Antarctica, and a test for oroclinal bending of the Antarctic Peninsula. *J. geophys. Res., 83,* No. B5, 2293–99.

Krylov, A. Ya. 1972. Antarctic geochronology. (*In* Adie R. J., *ed. Antarctic Geology and Geophysics.* Oslo, Universitetsforlaget, 491–94.)

Laudon, T. S. 1972. Stratigraphy of eastern Ellsworth Land. (*In* Adie, R. J., *ed. Antarctic Geology and Geophysics.* Oslo, Universitetsforlaget, 215–23.)

Laudon, T. S., Lackey, L. L., Quilty, P. G., and Otway, P. M. 1969. Geology of eastern Ellsworth Land (*In* Bushnell, V. C., and Craddock, C., *eds.* Geologic Maps of Antarctica. *Antarct. Map Folio Ser.,* Folio 12, Pl. III.)

LeMasurier, W. E., and Wade, F. A. 1976. Volcanic history in Marie Byrd Land: implications with regard to southern hemisphere tectonic reconstructions. (*In* González-Ferrán, O., *ed. Andean and Antarctic Volcanology Problems.* Rome, International Association of Volcanology and Chemistry of the Earth's Interior, 398–424.)

Lopatin, B. G., Krylov, A. Ya., and Aliapyshev, O. A. 1974. Osnovnye tektonomagmatecheskie etapi v rasvitii zemle Meri Berdi Berega Eitsa (Zapadnaia Antarktida) po radeogennim dannim [Major tectonomagmatic stages in the development of Marie Byrd Land and Eights Coast (West Antarctica) determined radiometrically]. *Antartika, Doklady Komissii, 13,* 52–60.

McNutt, R. H., Crocket, J. H., Clark, A. H., Caelles, J. C., Farrar, E. Haynes, S. J., and Zentilli, M. 1975. Initial $^{87}Sr/^{86}Sr$ ratios of plutonic and volcanic rocks of the central Andes between latitudes 26° and 29° south. *Earth planet. Sci. Lett., 27,* No. 2, 305–13.

Mehnert, H. H., Rowley, P. D., and Schmidt, D. L. 1975. K-Ar ages of plutonic rocks in the Lassiter Coast area, Antarctica. *J. Res. U.S. geol. Surv., 3,* No. 2, 233–36.

Plummer, C. C. 1974. Contact metamorphism of the Latady Formation, southern Lassiter Coast, Antarctic Peninsula. *Antarct. Jnl U.S., 9,* No. 3, 82–83.

Rex, D. C. 1972. K-Ar age determinations on volcanic and associated rocks from the Antarctic Peninsula and Dronning Maud Land. (*In* Adie, R. J., *ed. Antarctic Geology and Geophysics.* Oslo, Universitetsforlaget, 133–36.)

Rex, D. C. 1976. Geochronology in relation to the stratigraphy of the Antarctic Peninsula. *Bull. Br. antarc. Surv.,* No. 43, 49–58.

Rex, D. C., and Baker, P. E. 1973. Age and petrology of the Cornwallis Island granodiorite. *Bull. Br. antarc. Surv.,* No. 32, 55–61.

Rowley, P. D., and Williams, P. L. 1974. Plutonic rocks of the Lassiter Coast. *Antarc. Jnl U.S., 9,* No. 5, 225–26.

Rowley, P. D., Farrar, E., and McBride, S. L. 1976. Preliminary interpretation of potassium-argon ages of plutons in northern Lassiter Coast and southern Black Coast. *Antarct. Jnl U.S., 11,* No. 4, 257–58.

Rowley, P. D., Williams, P. L., and Schmidt, D. L. 1977. Geology of an Upper Cretaceous copper deposit in the Andean Province, Lassiter Coast, Antarctic Peninsula. *Prof. Pap. U.S. geol. Surv., 984,* 36 pp.

Rowley, P. D., Williams, P. L., Schmidt, D. L., Reynolds, R. L., Ford, A. B., Clark, A. H., and others. 1975. Copper mineralization along the Lassiter coast of the Antarctic Peninsula. *Econ. Geol., 70,* No. 5, 982–92.

Rutford, R. H., Craddock, C., White, C. M., and Armstrong, R. L. 1972. Tertiary glaciation in the Jones Mountains. (*In* Adie, R. J., *ed. Antarctic Geology and Geophysics.* Oslo, Universitetsforlaget, 239–43.)

Vennum, W. R. 1978. Igneous and metamorphic petrology of the southwestern Dana Mountains, Lassiter coast, Antarctic Peninsula. *J. Res. U.S. geol. Surv., 6,* No. 1, 95–105.

Wade, F. A. 1972. Geologic survey of Marie Byrd Land. *Antarct. Jnl U.S., 7,* No. 5, 144–45.

Wade, F. A., and Wilbanks, J. R. 1972. Geology of Marie Byrd and Ellsworth Lands. (*In* Adie, R. J., *ed. Antarctic Geology and Geophysics.* Oslo, Universitetsforlaget, 207–14.)

Williams, P. L., Schmidt, D. L., Plummer, C. C., and Brown, L. E. 1972. Geology of the Lassiter Coast area, Antarctic Peninsula—Preliminary report. (*In* Adie, R. J., *ed. Antarctic Geology and Geophysics.* Oslo, Universitetsforlaget, 143–48.)

40

Metamorphism of Sedimentary Formations on the Lassiter Coast

E. N. KAMENEV and E. M. ORLENKO
Department of Antarctic Geology, SEVMORGEO, Moyka 120, Leningrad 190121, U.S.S.R.

Abstract. A sequence of metamorphosed terrigenous sedimentary and volcanic rocks, intruded by gabbroids and granitoids, crops out on the northern Lassiter Coast. These rocks were studied during the field season of 1972-73. Rocks resulting from regional, contact, and polymetamorphism have been recognized in the area. Widespread regional metamorphism is represented by phyllites and slates of low-grade greenschist facies. Pl_{10} + Ca + Chl, Pl + Ser + Chl + Q, Ser + Bi + Ep + Q, and Pl + Hm + Bi + Chl + Q + Ca are the dominant mineral associations of the greenschist facies.

Contact metamorphism has formed aureoles around intrusive bodies throughout the area. The width of these contact aureoles ranges from several hundred meters to 1 or 2 km. These metamorphic rocks are assigned to the muscovite and hornblende hornfels facies. The following mineral associations are characteristic: Pl + Bi + Mu + Q, Q + Mu + And + Bi + Chl + Pl, Pl + Cpx + Am + Bi + And, Pl + Am + Bi + Gr (+ Chl), Pl + Cord + Ser + Q, and Cord + And + Sill + Am + Ms + Ksp + Q.

Contact metamorphism of rocks already in the greenschist facies formed polymetamorphic rocks. The wide development of contact-metamorphic and polymetamorphic rocks, and the nature of the mineral assemblages, suggests a similarity between the metamorphic rocks studied here and the products of regional zonal metamorphism of andalusite type; the latter is typical according to current thinking, of the Pacific belt of Mesozoic structure. This fact suggests that regional zonal metamorphism in this belt is associated with gabbro-granite batholiths.

The present paper is based on the rock specimens and data obtained by E. N. Kamenev when he worked with the U.S. Geological Survey party headed by P. D. Rowley. This party studied and mapped all the accessible outcrops in the previously unknown northern Lassiter Coast and southern Black Coast (lat. 72° 30′ to 74° 00′ S, long. 61° 30′ to 64° 30′ W) over an area of about 10,000 km^2. A unit of folded Mesozoic (mainly Jurassic–Cretaceous) sedimentary and volcanic rocks, assigned by American geologists to the Latady Formation (Rowley, 1973), occurs in the area studied. These folded strata, intruded by numerous Upper Cretaceous (Rowley and others, 1976) gabbroids and granitoids, compose about 75% of the mountains in the mapped area. A north-northeast-striking belt of these mainly batholith-like intrusions divides the region into two zones: western and eastern.

Geology of the western region

The western zone is composed of fine-grained slates of phyllitic affinity and phyllites. In most bedrock exposures of slates within this "slate" zone, a barely recognizable bedding is obscured by a well-developed cleavage which roughly parallels the bedding. Less than 2% of the slate unit is composed of fine-grained metagraywacke and polymictic metasandstone forming layers not more than 20 cm thick.

This entire greenschist facies unit is intensely folded; fold limbs dip at steep angles, 40°-80°. The strike changes from one outcrop to another, but the major trend is north-northeast. The folds seem to be isoclinal, and the thickness of the slate unit is approximately 5000 m.

It seems reasonable to designate this unit as a separate Ferguson Member as a part of the Latady Formation. (The most typical section of the member is in the Ferguson Nunataks and adjacent eastern Werner Mountains.) Isolated outcrops of intermediate and felsic metavolcanic rocks in the western Black Coast belong to the western zone. Rowley (1973) compares these metavolcanic rocks with the Poster Volcanics overlying the Latady Formation, according to the data of American workers (Williams and Rowley, 1972). However, unlike the Poster Volcanics, these volcanic rocks are highly sheared and rather uniformly metamorphosed to the greenschist facies. They also contain metasandstone, metatuff, and chert layers as much as 1.5 m thick. The total visible thickness of the metavolcanic unit is about 1200 to 1500 m. Its relationship to the Ferguson Member is unknown.

Geology of the eastern region

The eastern zone consists of coarser grained rocks than the western zone. In addition to shale and siltstone with small jasper lenses, amounting to 50-55% of its volume, it contains 45-50% of fine-grained graywacke and polymictic sandstone with carbonate-clay matrix forming units several tens of meters thick. In places there is a rhythmic interlayering of shale, siltstone, and sandstone, the thickness of each layer being not more than 20-30 cm. Lenses of fine-grained clayey limestone, measuring 0.5 by 1 m make up about 1% of the sandstone units. Sandstones are characterized by cross-bedding. In outcrops nearer to the coast, off the Weddell Sea, some sandstone beds contain isolated, well-rounded quartz pebbles. Both the mean grain size of sandstones and their volume increase eastward across the zone.

Table 40.1. Chemical Compositions (in wt %) of metamorphosed rocks from the northern Lassiter Coast

Zones	Rock type	Specimen number	SiO_2	TiO_2	Al_2O_3	Fe_2O_3	FeO	MnO	MgO	CaO	Na_2O	K_2O	P_2O_5	SO_3	Loss on ignition	Total	H_2O^-
	Metapelites	296	62.94	0.64	18.35	1.40	5.15	0.20	0.49	0.96	1.53	4.38	0.24	0.09	3.16	99.53	0.40
		296k	65.65	0.57	16.50	1.69	2.71	trace	0.67	4.30	4.02	1.10	0.23	0.35	2.09	99.88	0.44
	Metapsammites	297	78.95	0.51	9.96	0.74	2.79	0.03	0.20	0.73	1.34	2.66	0.12	0.15	1.40	99.58	0.18
		297a	76.77	0.35	11.37	0.79	2.12	trace	0.90	1.00	1.95	3.17	0.14	0.12	1.30	99.98	0.20
	Rocks rich in CaO (metabasites?)	297b	54.51	0.66	18.99	1.09	8.32	0.15	3.60	8.01	0.74	1.84	0.29	0.16	1.57	99.93	0.16
	Metapelites	258g	62.26	0.86	16.85	2.04	3.74	0.09	2.34	1.05	3.00	4.00	0.42	—	3.29	99.94	0.22
		257b	65.37	0.75	15.86	1.82	3.31	0.05	2.08	1.95	2.57	3.00	0.45	—	2.60	99.81	0.72
		257a	66.00	0.79	18.16	1.08	3.16	0.04	1.41	1.61	2.90	2.73	0.36	—	1.95	100.19	0.58
		257	65.04	0.86	16.88	1.08	3.88	0.03	1.41	0.72	1.57	3.47	0.41	—	4.59	99.94	0.40
	Metapsammites	267k	72.96	0.43	9.68	0.71	4.17	0.12	2.17	5.43	0.62	0.78	0.24	0.17	2.26	99.74	0.10
		267l	71.01	0.39	9.48	0.88	6.26	0.22	3.71	5.11	0.31	0.88	0.24	—	1.48	99.97	0.18
		277b	73.70	0.60	9.20	2.40	—	0.05	1.00	1.70	1.80	1.80	—	—	—	—	—
	Rocks rich in CaO (metabasites?)	204v	54.77	0.52	8.18	1.00	6.79	0.24	2.70	17.95	0.67	0.20	4.76	0.55	1.24	99.57	0.51
		204e	59.95	0.45	15.70	1.66	4.09	0.13	3.77	6.82	3.01	0.14	0.17	—	4.42	100.31	0.13
	Metapelites	254	65.03	0.64	17.28	1.26	3.91	0.08	0.97	2.09	2.77	3.35	0.17	—	2.90	100.39	0.18
		254zh	61.52	0.76	18.64	2.25	3.62	0.07	1.28	2.08	1.89	3.25	0.33	—	4.60	100.29	0.57
		255b	67.02	0.79	16.37	1.72	3.31	0.08	1.17	1.78	2.10	2.82	0.43	—	2.80	100.39	0.58
		256	63.14	0.94	18.57	1.86	3.45	0.03	1.36	1.22	1.33	3.44	0.62	—	3.82	99.78	0.46
		256a	68.16	0.75	15.15	1.38	3.16	trace	1.99	0.83	2.10	3.52	0.30	—	2.97	100.31	0.28
		256b	63.28	0.86	18.05	1.46	3.45	0.04	2.15	1.05	1.73	3.75	0.29	—	3.68	99.79	0.56
		254b	63.83	0.67	17.80	2.31	1.79	0.03	0.87	1.15	1.79	3.95	0.17	—	5.79	100.15	0.31
		254z	66.22	0.69	17.01	1.14	2.94	0.07	0.97	1.99	1.93	3.53	0.18	—	3.03	99.70	0.37
		255a	62.96	0.77	19.24	1.87	2.37	0.06	0.81	1.54	1.97	4.30	0.24	—	3.92	100.05	0.43
		276o	61.00	0.70	17.00	3.80	—	0.17	1.70	0.90	1.60	4.50	—	—	—	—	—
		276p	61.50	0.60	20.20	4.10	—	0.13	2.30	0.60	1.40	2.90	—	—	—	—	—
	Metapsammites	208d	70.28	0.75	12.97	1.34	3.48	0.10	0.88	4.20	3.45	1.16	0.25	—	0.84	99.70	0.12
		208e	73.96	0.48	11.15	1.36	3.37	0.08	0.90	4.17	2.82	1.15	0.19	—	0.81	100.44	0.19
		254v	69.79	0.33	15.35	0.89	2.98	0.04	0.65	3.26	2.81	2.00	0.13	—	1.35	99.58	0.16
		277b	73.70	0.60	9.20	2.40	—	0.05	1.00	1.70	1.80	1.80	—	—	—	—	—
		277g	68.97	0.73	14.79	0.78	3.18	0.02	0.30	0.65	1.61	6.23	0.22	0.21	2.05	99.74	0.56
	Rocks rich in CaO (metabasites?)	254k	64.58	0.58	13.36	0.78	4.88	0.18	2.23	11.83	0.42	0.20	0.16	—	0.94	100.14	
	Metapelites	225l	62.89	0.52	16.16	0.51	1.79	0.07	0.73	0.49	0.61	4.79	0.09	—	10.88	99.53	0.50
		276i	58.80	0.70	16.60	4.50	—	0.05	1.60	0.50	2.60	3.90	—	—	—	—	—
		276a	67.00	0.60	11.60	3.50	—	0.06	1.60	1.80	1.40	2.10	—	—	—	—	—
		276b	66.70	0.70	12.50	3.20	—	0.05	1.10	0.60	1.70	2.80	—	—	—	—	—
		276l	65.00	0.60	16.30	6.40	—	0.10	2.30	0.50	1.80	4.20	—	—	—	—	—
		276g	68.20	0.70	13.70	5.10	—	0.06	1.40	0.60	2.00	3.30	—	—	—	—	—
	Metapsammites	276v	71.40	0.70	10.40	3.80	—	0.05	1.00	0.70	2.60	2.50	—	—	—	—	—
		276	72.50	0.40	9.30	2.70	—	0.05	1.20	1.00	2.00	2.10	—	—	—	—	
	Rocks rich in CaO (Metabasites?)	268l	45.19	1.00	13.65	3.09	6.43	0.15	4.23	7.48	2.15	2.98	0.43	0.10	13.18	100.06	0.52

Contents of components in specimens 277b, 276o, 276p, 277b, 276i, 276a, 276b, 276l, 276g, 276v, and 276 were obtained by spectrographic analysis. The other specimens were chemically analyzed.

P. D. Rowley found large, poorly preserved pelecypods in sandstone of the Rivera Peaks. According to V. Efremova (personal communication, 1974) these fossils represent casts of several species of *Inocerams* (*Inoceramus* ex. gr. *lamarcki* Park., *I.* cf. *cuvieri* Sow., and *I.* ex. gr. *inconstans* Woods) which allow the assignment of the enclosing silty sandstones to the Upper Cretaceous. The above species occur in the upper Turonian–lower Senonian in western Europe, the Russian Platform, the Caucasus, northern Siberia, Canada, and the U.S.A. In a shale-mudstone unit, E. N. Kamenev also found abundant but poorly preserved remains of fern leaves and stems; N. D. Vasilevskaya identified *Otozamites* (?) sp., suggesting a Jurassic age for the enclosing rocks. However, in specimens from the Werner Mountains, V. D. Korotkevich recognized a spore and pollen spectrum containing spores of *Coniopteris, Osmunda, Leiotreiletes, Lophotriletes, L. triassicus* K.-M., *Acanthotriletes,* and *Periplecotriletes,* and pollen grains of *Piceites, Pinites, Podozamites, Podocarpus, Florinites, Bennettitales, Pseudopodocarpus,* and Cycadaceae, suggesting a Jurassic–Early Cretaceous age for the rocks.

The sandstone-shale unit is folded into open linear folds with widths from tens to hundreds of meters to 1–2 km; the folds are of a parallel type. The fold axes trend northeast, and are commonly curved; some folds are overturned southeastward. The visible thickness of the sandstone-shale unit is about 4500–5000 m. We suggest the name Werner Member for this unit and consider it to be part of the Latady Formation.

The intrusions of the Andean gabbro-granite complex consist of five rock types of the normal calc-alkaline series, i.e., gabbro, diorite, quartz diorite, quartz monzonite, and granite. Each type of rock occurs both as separate intrusive bodies, and as components of complex plutons. Formation of the intrusive massifs can be divided into some stages: the initial stages are represented by basic varieties and the final ones by felsic rocks. Granitoid intrusions are accompanied by numerous dikes and veins. Aplite, granite pegmatite, lamprophyre, and porphyrite occur abundantly; quartz porphyry, felsite, and syenite porphyry are less common. Adjacent folded volcanic and sedimentary strata have undergone both contact and regional metamorphism nearly everywhere.

Regional metamorphism

The rocks formed by regional metamorphism belong to the greenschist facies, and they are best developed in the Rivera Peaks in the Werner Mountains. Greenschist terranes are separated by cordierite and andalusite isograds from the aureoles of contact metamorphism. Mineral assemblages of greenschist rocks are multiphase and diverse. There are not less than 12 assemblage varieties: Q + Pl_{10} + Am + Bi + Chl + Ca + Sph; Q + Pl_{12-22} + Bi ± KFsp ± Mu; Q + Pl_{10} + Am + Ep + Ca; Q + Pl_7 + Ca + Bi + Chl; Q + Pl + Ser + Chl; Q + Ser + Bi + Ep; Q + Pl_{10} Ep; Q + Pl + Chl; Q + Pl + Ca; Pl + Ca + Chl; Ca + Chl + Ser; and Ca + Am.

On the basis of these mineral assemblages, the greenschist rocks may be assigned to the epidote-muscovite or biotite-chlorite subfacies (Dobretsov and others, 1972). Irregular combinations of rocks belonging to these two subfacies, found in certain sections, suggest that rocks of the greenschist facies on the Lassiter Coast have formed under conditions near the boundary between these subfacies. Such conditions exist at 370° to 470°C and pressures of 3 to 7 kb; these conditions were probably produced in the course of folding. Apart from composition, the grade of metamorphism here is related to the pattern of rock folding: the tightly folded slate units are more highly metamorphosed than the massive sandstone units. In isoclinally folded strata of the Ferguson Member, the metamorphism of the greenschist facies is more evident than in the Werner Member, the rocks of which occur in open folds. Hence, the sedimentary rocks of the Lassiter Coast had already been metamorphosed when the Upper Cretaceous intrusions were emplaced.

Contact metamorphism

Contact metamorphic rocks are extensive within aureoles around plutons and within xenoliths in these intrusive rocks. The width of aureoles varies greatly from 10–20 m to 3–4 km. The maximum size of aureoles is known only approximately due to poor exposure throughout the region. The degree of contact metamorphism and the mapped width of contact aureoles depend on many factors, the major ones being: distance from the contact surface and position of the surface in space, orientation of bedding planes with respect to the contact surface, number of veins and dikes in the area, composition of country rocks, and composition of intrusions. The widest aureoles were found near steep contacts or where veins and dikes are abundant. Contact changes in rocks are the greatest near gabbro and diorites. Schist, as a rule, is metamor-

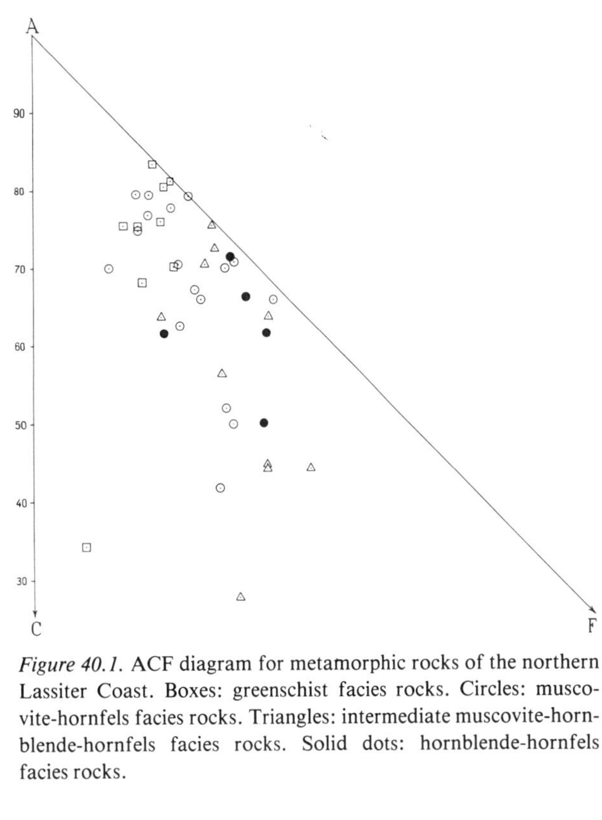

Figure 40.1. ACF diagram for metamorphic rocks of the northern Lassiter Coast. Boxes: greenschist facies rocks. Circles: muscovite-hornfels facies rocks. Triangles: intermediate muscovite-hornblende-hornfels facies rocks. Solid dots: hornblende-hornfels facies rocks.

phosed up to a greater distance away from the contact than sandstone.

Aureoles of contact-metamorphosed rocks have a zonal structure. Zones of muscovite (outer) and hornblende (inner) hornfelses are recognized (after Reverdatto's classification, 1970). In the rocks of the Lassiter Coast, the muscovite hornfels facies is represented by the following mineral assemblages: Q + Pl_{10} + Chl + Bi + Gr; Q + Pl + Bi ± Mu; Q + Pl + Mu + Am_{20} ± Cord ± Bi ± KFsp; Q + Pl + Mu + Bi + Cord + And + Am_{22}; Q + Mu + And ± Bi ± Chl ± Pl; Q + Mu + Bi + Gr ± Pl_{30}; and Q + Mu + Bi_{46} + Cord + Pl_{28} + Chl ± Ep. In addition, such associations as Q + Mu + Bi + And + Cord ± KFsp; Q + Chl + Ep + Cpx ± Pl; Q + Chl + Am_{57} + Bi_{50} ± Mu ± Ep ± Pl; and Q + Mu + Bi_{45-47} + Pl_{56} ± Ca ± Ep ± Chl are transitional to the hornblende hornfels facies.

In the outermost parts of aureoles, contact metamorphism is manifested by the formation of pseudomorphs of biotite after chlorite, andalusite after muscovite, and cordierite after muscovite and chlorite. Metapelites of the outer parts of aureoles are altered to fine-grained porphyroblastic hornfelses, also called spotted slates. Fabric between porphyroblasts is not abundant and is represented by coal-rich clays and coal-rich siliceous material. In the metasandstone, quartz is heavily recrystallized and forms a mosaic aggregate. In inner parts of outer zones, chlorite almost disappears but muscovite and quartz are still present. Porphyroblastic rocks containing the Cord + Bi + Pl assemblage are formed by metamorphism of calcareous phyllites in the outer parts of aureoles. The layers of carbonaceous and carbonaceous-clayey rocks are converted into hornfelses with Q, Am, Pl, Ep and Cpx. Hornfelses of the outer aureole zones usually contain relics of mineral associations of greenschist facies.

In contrast, the inner zones lack any relict products of regional metamorphism. In hornfelses of inner zones there are rather abundant potash feldspar and sillimanite grains formed by progressive dehydration of rocks at the expense of muscovite and biotite. Muscovite was the first to disappear and does not occur anywhere in hornfelses of the innermost parts of aureoles except as the secondary product of postmagmatic activity. Hypersthene and almandine were found in several specimens of hornfels. In the inner aureole zone the following assemblages of the hornblende hornfels facies are the most widespread: Q + Pl_{56} + Cpx + Am_{52} + Bi; Ca + Am_{20} + Cpx (+ Ep); Q + Bi + Cpx + And (+ Ep); Q + Sill + Cord + And + Am + KFsp (+ Mu); Pl + KFsp + Sill + Am + Bi_{53} + Q (+ Mu); Cord + And + Am_{47} + Bi; Q + Pl + Am_{45} + Bi + Gr (+ Mu ± Chl); and Q + Pl_{29-64} + Sill + Cord + Bi ± Gr ± KFsp (+ Mu).

The mineralogic changes during contact metamorphism of pelites on the Lassiter Coast may be represented by the following succession (in space) of mineral assemblages: (1) Mu + Chl + Pl + Q; (2) Mu + Bi + Pl + Q; (3) Mu + Bi + And + Pl + Q; (4) Mu + Bi + Cord + And + Pl + Q; (5) KFsp + Cord + And + Sill + Pl + Q. The first of these is the assemblage in regionally metamorphosed rocks (greenschist facies) beyond the contact aureoles. The subsequent assemblages reflect changes in mineral composition approaching the igneous contacts. The last mineral assemblage is persistent (stable) in the innermost parts of aureoles. This and other mineral assemblages with sillimanite and diopside are similar to those of the amphibolite facies, namely the sillimanite-cordierite-orthoclase-almandine subfacies from the facies series of Abukuma type (Winkler, 1967). These associations form at 670°–680°C and 3.0–3.5 kb.

North of Grimmindger Mount, rocks with high-temperature mineral assemblages comprise a separate block, a relic of the roof in a complex pluton of gabbro, diorite, and granodiorite composition. The rocks of gneissic appearance are characterized by migmatization phenomena which makes them similar to those formed by regional metamorphism. It was noted that in many lower temperature hornfelses an oriented growth of porphyroblasts was prominent and they were commonly deformed. The same phenomena were reported earlier by Plummer (1974) from the southern Lassiter Coast. Marginal parts of some plutons commonly show cataclastic textures. Hence, intrusions have produced strong mechanical effects on the enclosing rocks. Apparently, at least some intrusions and the contact metamorphism which accompanied them were synkinematic.

A comparison of chemical compositions (Table 40.1 and Figure 40.1) of metapelites and metapsammites from zones of contact and regional metamorphism under greenschist facies conditions suggests that during the progressive stage, metamorphism within contact aureoles was mainly isochemical. It was manifested only as dehydration and interaction of mineral phases. Only during the postmagmatic regressive stage did intrusive bodies supply some new components, which resulted in muscovitization, chloritization, epidotization, and carbonization of earlier formed hornfelses.

Discussion

The mineral assemblages described in this report correspond to the regional zonal metamorphism of andalusite type, which is considered typical of the Mesozoic structures of the Pacific belt (Marakushev and others, 1971; Velikoslavinsky, 1972; Glebovitsky, 1973). In particular, they are reminiscent of the metamorphic assemblages observed in the folded sequences in Nevada and California (Compton, 1960; Best and Weiss, 1964). Two stages of metamorphism can be recognized in the studied assemblages. The first stage was responsible for regional greenschist facies metamorphism of the Jurassic sequence. The later contact metamorphism of muscovite and hornblende hornfels facies was superimposed on the greenschist facies assemblages in contact aureoles of the Cretaceous gabbro-granite plutons, thus producing the general zonal pattern of metamorphic facies distribution.

The evidence presented suggests that regional zonal metamorphism of the Mesozoic fold sequences in the Pacific belt is likely to be associated with evolution and emplacement of extensive gabbro-granite plutons.

Acknowledgments. American colleagues in the Lassiter Coast project helped us in collecting material for this study. We are most grateful to them, as well as to the people of the U.S. Antarctic Research Program who facilitated the work of one of us (E. N. K.) during the 1971–73 field season in Antarctica.

References cited

Best, M. G., and Weiss, L. E. 1964. Mineralogical relations in some pelitic hornfelses from southern Sierra Nevada, California. *Amer. Miner., 49,* No. 9-10, 1240-66.

Compton, R. R. 1960. Contact metamorphism in Santa Rosa Range, Nevada. *Geol. Soc. Am. Bull., 71,* No. 9, 1383-416.

Dobretsov, N. L., Sobolev, B. S., and Khlestov, V. V. 1972. *Fatsii regional'nogo metamorfizma umerennykh davlenii* [Regional metamorphism facies of intermediate pressures]. Moscow, Izdatel'stvo Nedra, 286 pp.

Glebovitsky, V. A. 1973. *Problemy evolutsii metamorficheskikh protsessov v podvizhnykh oblastiakh* [Problems of evolution of metamorphic processes in mobile regions]. Leningrad, Izdatel'stvo Nauka, 127 pp.

Marakushev, A. A., Mishkin, M. A., and Tararin, I. A. 1971. *Metamorfizm Tikhookeanskogo poiasa* [Metamorphism of the Pacific belt]. Moscow, Izdatel'stvo Nauka, 134 pp.

Plummer, C. C. 1974. Contact metamorphism of the Latady Formation, southern Lassiter Coast, Antarctic Peninsula. *Antarct. Jnl U.S., 9,* No. 3, 82-83.

Reverdatto, V. V. 1970. Fatsii kontaktovogo metamorfizma [Facies of contact metamorphism]. Moscow, Izdatel'stvo Nedra, 271 pp.

Rowley, P. D. 1973. Geologic observations on the northern Lassiter Coast and southern Black Coast. *Antarct. Jnl U.S., 8,* No. 4, 154-55.

Rowley, P. D., Farrar, E., and McBride, S. L. 1976. Preliminary interpretation of potassium-argon ages of plutons in northern Lassiter Coast and southern Black Coast. *Antarct. Jnl U.S., 11,* No. 4, 257-58.

Velikoslavinsky, D. A. 1972. Sravnitel'naia kharakteristika regional'nogo metamorfizma umerennykh i nizkikh davlenii [Comparative characteristic of regional metamorphism under intermediate and low pressures]. Leningrad, Izdatel'stvo Nauka, 189 pp.

Williams, P. L., and Rowley, P. D. 1972. Composition of Jurassic sandstones, Lassiter Coast. *Antarct. Jnl U.S., 7,* No. 5, 145-46.

Winkler, H. G. F. 1967. Die Genese der metamorphen Gesteine. 4th ed. New York, Springer-Verlag, 334 pp.

41

Aeromagnetic Surveys over the Antarctic Peninsula

R. G. B. RENNER, B. J. DIKSTRA, and J. L. MARTIN
British Antarctic Survey, Natural Environment Research Council, Madingley Road, Cambridge CB3 0ET, U.K.

Abstract. More than 20,000 line-km of regional total-field aeromagnetic profiles have been recorded from the Antarctic Peninsula. Flights were operated at a constant barometric altitude of 2500 m and at right angles to the structural axis. Doppler radar provided navigation control. All data were reduced to the International Geomagnetic Reference Field and corrected for diurnal variations. Selected profiles indicate that the region is divisible into distinct magnetic provinces. Western Palmer Land is characterized by a wide north-south-trending belt of long wavelength anomalies of 500–1000 gamma amplitude; they terminate abruptly on the west at George VI Sound. A similar linear anomaly band 80–120 km in width can also be traced in the offshore area along the west coast of Graham Land and may represent a continuous batholithic feature. Alexander Island exhibits relatively subdued magnetic conditions, indicative of its extensive sedimentary cover. A small anomaly on Beethoven Peninsula in the vicinity of exposed Cenozoic volcanic rocks is tentatively correlated with others of similar magnitude on profiles as distant as 250 km to the north. There is no significant magnetic anomaly associated with the western edge of George VI Sound. Profiles flown over the continental shelf areas of the Bellingshausen Sea, Weddell Sea, and Larsen Ice Shelf show little magnetic relief.

Three geographic units—Palmer Land, Graham Land, and Alexander Island—dominate the arcuate, 1200-km-long Antarctic Peninsula. The physiographic change apparent at about lat. 69° 00′ S forms a natural demarcation between the more massive Palmer Land to the south and the peninsula of Graham Land which tapers gradually to the north. The ice-covered plateau of the main peninsula is bordered by an almost unbroken mountainous escarpment which on the west coast of Graham Land forms an impenetrable barrier to land-operated surveys. The axial plateau itself exhibits a gently undulating topography through which isolated rock outcrops and mountains protrude, the peaks reaching a maximum height of 3500 m in Palmer Land. Throughout the region, glaciers fed by the plateau ice flow outward toward the Bellingshausen Sea and Weddell Sea. The west coast of the peninsula is fringed by a multitude of islands paralleling the axis of the mainland. Alexander Island is the largest and most southerly of the offshore islands and is characterized by north-south-trending mountain ranges, the linearity of which is probably controlled by major north-south block faults (Bell, 1975). Separating Alexander Island from Palmer Land is George VI Sound, a formidable feature extending north-south for 350 km and varying in width from 20 to 50 km. The origin of George VI Sound is still debated, but its tidy coastline and contrasting geology on opposite sides are suggestive of major faulting. There are several areas of ice shelf in the peninsula region; by far the most extensive is the Larsen Ice Shelf.

The difficulties encountered by ground geophysical parties operating in the area since 1959 were partially overcome in 1973 when the British Antarctic Survey acquired an airborne magnetometer. The equipment included a Geometrics model G-803 proton-precession magnetometer, a Cipher model 70 magnetic tape recorder controlled by a Geometrics model G-704 data-acquisition system, and a Geometrics model MARS-6 chart recorder producing in-flight magnetic profiles to which the operator could react if necessary. The recording console and wing-tip sensor were installed in the field in one of the British Antarctic Survey's two de Havilland Twin Otter aircraft. Field operations were based at Adelaide Island (lat. 67° 46 ′ S, long. 68° 55′ W) and Fossil Bluff (lat. 71° 20′ S, long. 68° 17′ W), from which the authors completed two successful seasons of aeromagnetic surveys (Renner, 1976; Dikstra, in press). A third season planned for 1976–77 was cancelled due to the loss of one of the aircraft. More than 20,000 line-km of regional magnetic profiles have been flown over the Antarctic Peninsula (Figure 41.1) while a further 3000 line-km have been recorded in detailed grid networks investigating selected geologic problems.

The present paper is concerned with preliminary findings from the regional surveys.

The survey

The normal total operating range for the aircraft was about 1000 km, although the installation of long-range fuel tanks for part of the survey increased this to 1700 km. Use of fuel supplies at both Adelaide Island and Fossil Bluff made it possible to cover much of southern Graham Land, northern Palmer Land, and Alexander Island (Figure 41.1); this also allowed exploratory flights over the Bellingshausen Sea and Weddell Sea. The flight pattern was controlled by the north-south structural grain of the Antarctic Peninsula, with profiles flown perpendicular to this trend. A constant barometric altitude of 2500 m was maintained; operating altitude was governed by the variable topography and by the need to have adequate terrain clearance to remove the unwanted effect of near-surface

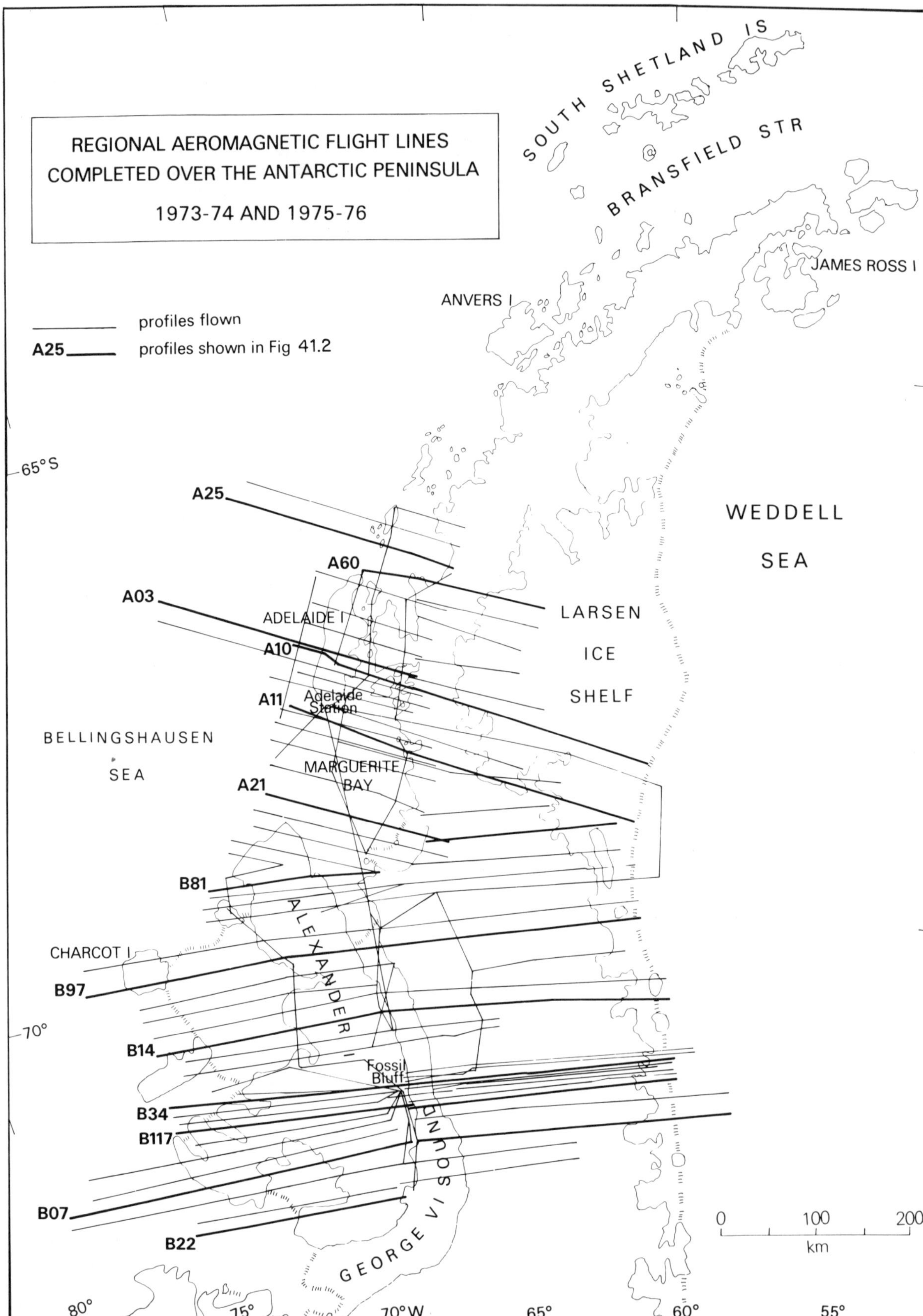

Figure 41.1. Regional aeromagnetic flight lines completed over the Antarctic Peninsula. Bolder, numbered flight lines are those profiles discussed in text and shown in Figure 41.2.

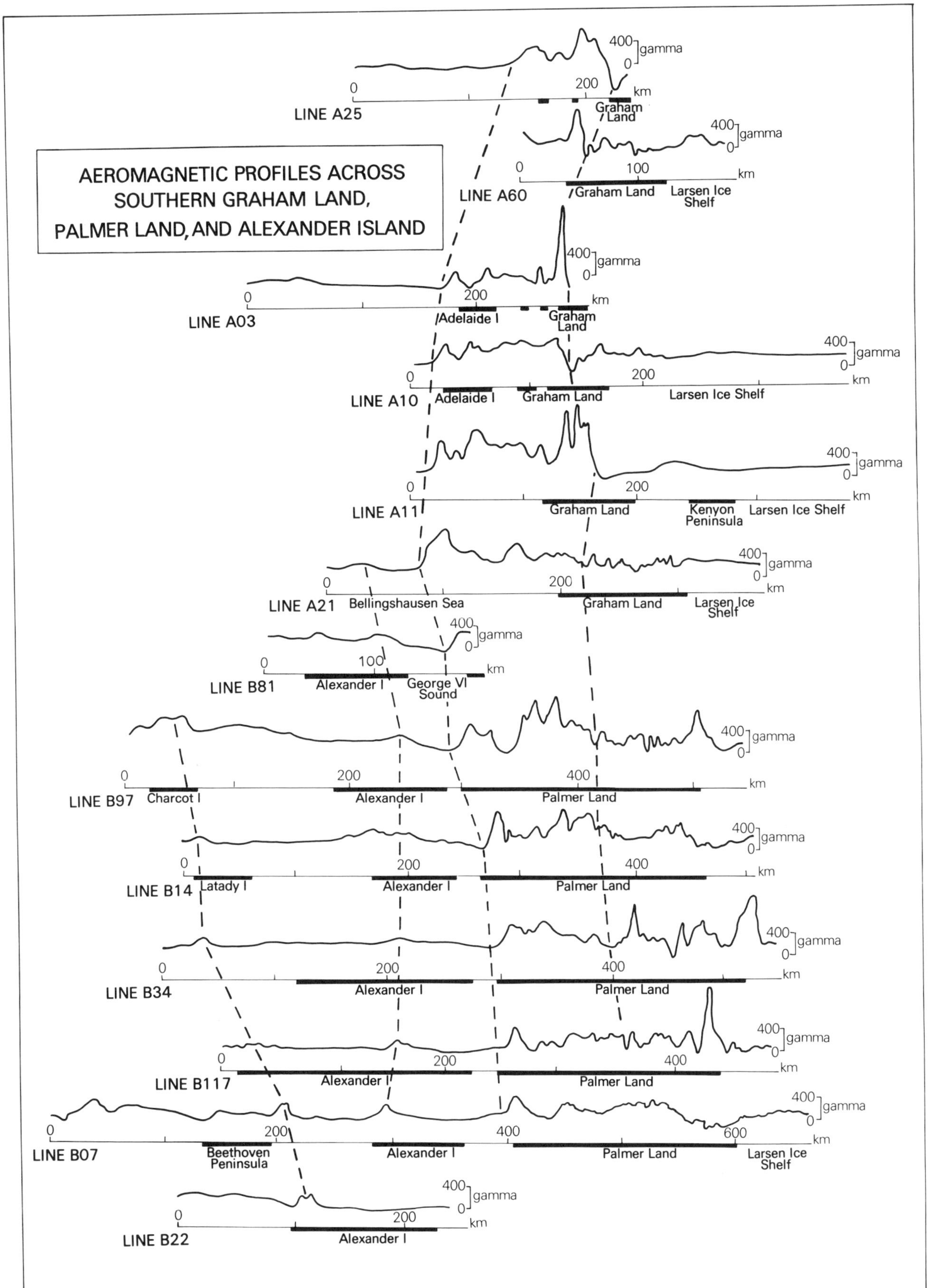

Figure 41.2. Selected aeromagnetic profiles from the Antarctic Peninsula region. Heavy horizontal lines represent land above sea level. Possible correlation between profiles are indicated by dashed lines. Locations of profiles are shown in Figure 41.1.

anomalies. For the reconnaissance survey the line spacing averaged 15–20 km, though near lat. 71° S a more detailed network at 4-km-line spacing was initiated to try to establish a representative crustal cross section based on topographic, geologic, radio-echo sounding, and geophysical information. With a ground speed of 120–140 kt and a 1 second data sampling interval, a total magnetic field reading was recorded for every 62–72 m of ground traversed. Field measurements were recorded in units of 1 gamma (1 gamma = 10^{-9} tesla).

Initially, visual techniques alone were employed to register vertical fixes, running fixes, and position lines relative to identifiable points, while over featureless terrain the navigator relied solely on dead-reckoning. In several areas the available maps were products of reconnaissance surveys, resulting in flight tracks being equated to "floating" landmarks; these tracks had to be adjusted when more accurate maps became available. However, by 1975–76, satellite imagery and trimetrogon photography had provided the basis for more accurate base maps for the areas in question. The aircraft had also been equipped with a Bendix DRA-12 doppler-navigation system, which in conjunction with a Sperry C-12 gyromagnetic compass, calculated the position of the aircraft relative to a starting point. Whenever possible, flight lines were begun over an identifiable feature in order to locate the profile accurately. Constant headings were flown, and en route navigation events were logged. A Shackman 35-mm auto-camera, front mounted in the cockpit, was an additional navigation aid giving, when required, a permanent record of on-line terrain. This facility was of particular advantage where the topography was not immediately identifiable.

Errors in navigation were initially governed by the quality of the maps, some areas being better mapped than others; the doppler system was not installed until 1975–76. It is estimated that the positions of most lines over the mainland are accurate to within 1 km. The largest errors occurred where flight lines extended over the Bellingshausen Sea or the Larsen Ice Shelf. Here, with visual navigation alone, a radius of error of 15 km could be expected toward the ends of the longest lines. The accuracy was greatly improved by the doppler navigation system which reduced the estimated error to less than 2.5 km at the end of a 250 km line.

Continuous diurnal control was available throughout the program from the British Antarctic Survey's geomagnetic observatory at the Argentine Islands (lat. 65° 15′ S, long. 64° 16′ W). During operation from Fossil Bluff, continuous measurements of the earth's magnetic field were also monitored at a temporary observatory established at Adelaide Island. By frequent reference to this record, flying operations could be suspended during magnetically disturbed periods. Regional corrections were applied by the removal of the International Geomagnetic Reference Field (IAGA Commission 2, Working Group 4, Analysis of the Geomagnetic Field, 1969), as computed by Fabiano and Peddie (1969).

Aeromagnetic data from the survey are currently being collated for publication as a regional anomaly map of the Antarctic Peninsula. This will incorporate all current data (Figure 41.1) and that from the field season scheduled for 1977–78. Representative profiles have been selected for discussion from the areas surveyed thus far (Figure 41.2).

Results and discussion

Examination of the selected profiles (Figures 41.1 and 41.2 reveals that several significant magnetic anomaly features are recognizable over parts of the area.

Alexander Island

The greater part of Alexander Island is relatively quiet magnetically, a reflection of the two dominant sedimentary successions: a tightly folded upper Paleozoic–lower Mesozoic succession (the LeMay Formation) with outcrops in northern, western, and central parts of the island, and a relatively undisturbed thick upper Mesozoic succession (the Fossil Bluff Formation) which crops out in the southeast along the west coast of George VI Sound. Secondary magnetic features include:

1. At longitude 78° W on profile B07, 90 km west of the Beethoven Peninsula, an anomaly of 400 gamma amplitude and 40 km wavelength was observed. The anomaly trends northwest to southeast and can be identified on the adjacent tracks (Figure 41.1). Further flight lines are required to delimit its north-south extent.
2. Land traverses have suggested that strong magnetic anomalies are associated with isolated nunataks on Beethoven Peninsula. A more detailed aeromagnetic survey has been partially completed for this area, but the regional flight profiles indicate that the Beethoven Peninsula anomalies may be only part of a more extensive anomaly zone. Along the western margin of Alexander Island, trending approximately north-northwest to south-southeast, is an anomaly which may be correlated through profiles B22, B07, B117, B34, B14, and B97. It has a maximum amplitude of 400 gamma and a wavelength varying between 20 and 40 km; it is traceable for more than 300 km from the Monteverdi Peninsula to Charcot Island. Only on flight B117 over the Beethoven Peninsula is the anomaly associated with exposed rock, the remainder being over sea, ice shelf, or snow-covered features. The geology of the Beethoven Peninsula contrasts sharply with the remainder of Alexander Island; its known outcrops consist of Cenozoic volcanic rocks (Bell, 1973) unconformably overlying arkosic sedimentary rocks. The volcanic rocks consist of palagonite tuffs overlain by subaerially erupted olivine basalts. Similar rocks were collected by R. G. B. Renner in 1975–76 from the Rydberg Peninsula (lat. 73° 19 ′ S, long. 78° 53′ W) on the English Coast. Rydberg Peninsula lies 260 km to the southwest of the Beethoven Peninsula, and both localities probably belong to a broad belt of Cenozoic volcanic rocks which extends across West Antarctica.
3. At long. 70° W on profiles B07 and B117 there is a positive anomaly of 200 gamma amplitude and 15 km wavelength. It may be traced to the north where it broadens, probably influenced by topography. On the more southerly profiles it appears to be associated with local outcrops of dioritic intrusive rocks within the upper Paleozoic–lower Mesozoic sedimentary rocks at Staccato Peaks and Walton Mountains.
4. There is no magnetic anomaly associated with the western edge of George VI Sound; therefore major faulting, if present, may either be confined to the sedimentary sequence, or alternatively, if the basement rocks are

similarly faulted, then the "magnetic basement" may lie at a considerable depth.

LARSEN ICE SHELF AND WEDDELL SEA

Several exploratory flights have traversed the Larsen Ice Shelf and have continued as far as 50 km out over the Weddell Sea. The total magnetic field there exhibits only minor fluctuations of small amplitude and long wavelength occurring near the east coast of the peninsula. Further east the fluctuations give way to an almost constant magnetic field, suggesting that a thick sedimentary cover overlies basement rocks.

GRAHAM LAND AND PALMER LAND

More than 40 profiles have been flown between lat. 66° 30′ S and 72° 20′ S across the west coasts of Graham Land and Palmer Land, and 30 of these continue to the Larsen Ice Shelf. Immediately apparent (Figure 41.2) is the contrast in magnetic signature shown by those profiles over the island archipelago and mainland compared to the relatively quiet magnetic conditions of Alexander Island and the Larsen Ice Shelf. Anomalies both onshore and offshore have amplitudes as much as 1500 gamma and wavelengths as long as 40 km. The complexities of the individual anomalies are superimposed on a broader magnetic pattern which on the west coast is present throughout much of the surveyed area.

The most northerly of the traverses, profile A25 (Figure 41.2), was recorded over Watkins Island and Crystal Sound before terminating on the Loubet Coast short of the plateau escarpment. The composite magnetic anomaly has a total width of 100 km and an amplitude of 1200 gamma. Topography beneath the flight track consisted of low, snow-covered Watkins Island and in Crystal Sound of small jagged islands, the latter being the only surface expression of a rugged submarine morphology known from bathymetric soundings to exist along much of the west coast within the boundaries of the island archipelago. The authors consider profile A25 as typical of this west coast anomaly, and its signature provides a reference for correlations with parallel flight tracks to the north and south. The anomalous band can be expected to vary in detail; however, its basic parameters are of a zone of high magnetic intensity, 80–120 km total width, with individual long-wavelength (>20 km) anomalies averaging 500–600 gamma in amplitude. There is an obvious western onset and commonly an equally distinct eastern cutoff, the location of which at more southerly latitudes gradually transgresses from the coast toward the central axis of the Graham Land plateau. It is evident from profile A21 that in Marguerite Bay the western limb of the anomaly is more pronounced and almost coincident with the eastern edge of a bathymetric trough extending north from George VI Sound. Reference to bathymetric data shows that depths of 1000 m have been measured in the trough, compared to adjacent shelf depths of 200–500 m. To the south the anomaly band widens, with the western limb continuing to be dominant and associated with the east coast of George VI Sound. The amplitudes of the associated high intensity anomalies show a gradual southerly decrease, and identification of the eastern margin becomes increasingly difficult.

Five *Project Magnet* aeromagnetic traverses initiated by the U.S. Naval Oceanographic Office were flown over northern Graham Land and the Weddell Sea in 1965. Three flight tracks—720S, 721N, and 721S—traversed the mainland at altitudes of 2560–2850 m. Reference to the geomagnetic records from the Argentine Islands observatory shows that diurnal variations at the time were less than five gamma and removal of the International Geomagnetic Reference Field produced an additional 1000 km of comparable aeromagnetic profiles over the Antarctic Peninsula. Bregman and Frakes (1970) discussed the significance of the Weddell Sea anomalies and also commented upon a series of larger magnetic anomalies off the north and west coasts of Graham Land. Navigational uncertainties made precise location difficult, but it was suggested that the anomalies reflected "a continuous belt of extreme values or simply isolated centers." Repositioning of the *Project Magnet* flight tracks by correlation with known land magnetic features has provided a more definite alignment of anomalies. Although one of the positive anomalies cited by Bregman and Frakes can now be attributed to the South Shetland Islands and northern Bransfield Strait (Barker and Griffiths, 1972), there are other significant anomalies over the southern slope of Bransfield Strait. Sections of *Project Magnet* flight tracks 720S and 721N are illustrated in Figure 41.3, together with selected profiles of the west coast anomaly south of lat. 66° 30′ S. Flight track 721S (not illustrated) obliquely cuts the area represented by profile 3 and exhibits a similar magnetic signature. The west coast anomaly can be correlated through a distance of 900 km (profiles 1, 2, 3, and 4, Figure 41.3), while south of profile 4, a zone of high magnetic intensity is still observed for at least 300 km farther down the west coast of Palmer Land.

Detailed interpretation of this linear west coast magnetic zone must await additional geophysical evidence. However, off the northwest coast of Graham Land, Davey (1971) correlated marine gravity data with seismic refraction investigations (Ashcroft, 1972) and concluded that major faulting is present within the continental shelf, with the faults downthrown to the northwest and trending parallel to the coast. Since in this area the aeromagnetic data is minimal, the relationship of the magnetic anomalies to gravity anomalies is not known. However, quantitative interpretation on the easternmost of the magnetic anomalies on flight track 720S (profile 1, Figure 41.3) shows that the computed profile from a body 16 km wide, with an upper surface at 2 km depth and a susceptibility of 0.006 emu/cm^3, provides an acceptable fit to the observed values. This postulated body has faulted or steep-sided margins, with that on the east approaching vertical and extending to depths greater than 15 km. To the west the contact is less abrupt, but if faulted, it suggests a downthrow of 2–3 km toward Bransfield Strait. Maximum depth estimates on the aeromagnetic profiles to the south also support the hypothesis that the causative body or bodies has an upper surface within a few kilometers of sea level. Near-shore marine magnetic profiles (unpublished) completed by the Department of Geological Sciences, University of Birmingham, off northwest Graham Land reveal many shorter wavelength anomalies within the main anomaly zone. These reflect near-surface sources probably related to a larger underlying body. Geologic collections made by the British Antarctic Survey on the offshore

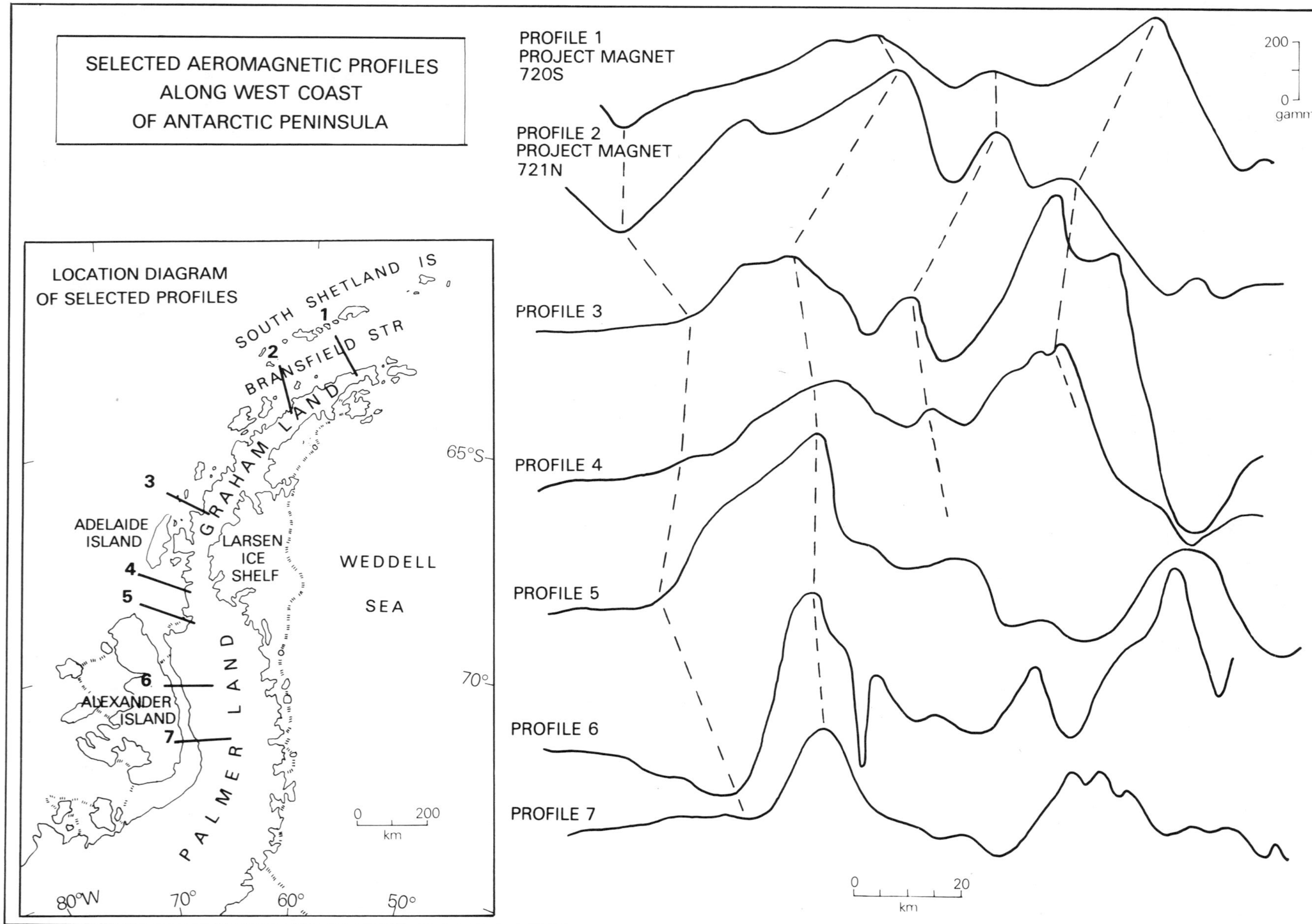

Figure 41.3. Representative aeromagnetic profiles across the west coast anomaly of the Antarctic Peninsula.

islands include basalts, dolerites, gabbros, and mafic extrusive rocks. Over northwestern Palmer Land there is no obvious correlation of the magnetic anomalies with the surface geology although isolated exposures of mafic igneous rocks have been recorded (Skinner, 1973; Rowe, 1973). A gravity survey (Butler and McArthur, in press) has revealed the presence of a positive Bouguer anomaly along part of the west coast of Palmer Land; this may be related to the magnetic anomaly at that latitude.

From the available geophysical, geologic, and bathymetric data it appears that the origin of the linear west coast anomaly lies with or within continental basement rocks. Provided the basement had a sufficiently high magnetic susceptibility or remanent magnetization, then a modified ridge feature could explain the observed anomaly. However, there is no geologic evidence to support the existence of basement rocks of the required magnetic intensity. Furthermore, interpretation of the anomaly off Graham Land does indicate that a strong vertical component is a requisite to fit computed and observed profiles. Intrusive rocks of variable composition and depth intruded into the basement could produce the observed anomaly. Smaller exposed or near-surface intrusive and extrusive rock bodies, possibly related to the underlying parent body, would explain shorter wavelenth anomalies. The total width of the main intrusive body must approach 50–100 km off the west coast of Graham Land and may be even greater in Palmer Land.

Dalziel and others (1974), in discussing island-arc assemblages in the southern Andes, refer to the Patagonian batholith which extends for at least 600 km along the Pacific margin of southern Chile. It has an average width of more than 60 km and a composition varying from tonalites to adamellites. Mesozoic in age, it is intruded into the South American continental "basement," consisting of Paleozoic metasedimentary rocks, and into Upper Jurassic volcanic rocks and is considered to represent the root of an andesitic volcanic chain. Similar tectonic-stratigraphic units have been recognized in both the southern Andes and the Antarctic Peninsula (Suárez, 1976), although not all of the island-arc assemblages are represented on the peninsula. Isolated plutonic bodies have been observed on the peninsula, but there is no record of any body comparable to the Patagonian batholith. From the aeromagnetic evidence, however, a continuous body of batholithic proportions is deduced off the west coast of Graham Land, its seaward boundary being roughly coincident with the outer edge of the island archipelago. South of lat. 69° S the body, though less well defined, continues beneath western Palmer Land trending parallel with, and lying adjacent to, the eastern side of George VI Sound.

Another set of anomalies are observed toward the east of profiles B97, B34, and B117. Although not recorded on all intermediate flight lines, the anomalies have amplitudes of as much as 1500 gamma and wavelengths averaging 30 km. They are situated close to the junction of the Larsen Ice Shelf and the mainland. Comprehensive geologic collections have not been made from the area, although Adie (1955) describes gabbros from Cape Bryant (lat. 71° 14′ S, long. 60° 55′ W) and Cape Christmas (lat. 72° 19′ S, long. 60° 42′ W). Whatever the source rocks may be, the bodies are highly magnetic and of limited areal extent with an upper surface at shallow depth.

Conclusions

The aeromagnetic surveys over Graham Land, Palmer Land, and Alexander Island have shown the area to be divisible into several magnetic provinces.

1. *Alexander Island.* This generally quiet magnetic region reflects the thick sedimentary succession of the upper Paleozoic–lower Mesozoic LeMay Formation and the upper Mesozoic Fossil Bluff Formation. Beethoven Peninsula is associated with a magnetic anomaly which may be correlated with others over a distance of 300 km between Monteverdi Peninsula and Charcot Island. There is an anomaly of unknown origin with a 400 gamma amplitude and 40 km wavelength trending northwest to southeast 90 km west of Beethoven Peninsula.

2. *Larsen Ice Shelf and Weddell Sea.* Minor undulations occur in the magnetic field close to the mainland, but to the east no significant anomalies were observed.

3. *Graham Land and Palmer Land.* A linear anomaly can be traced for 900 km, lying parallel to and off the west coast of Graham Land. It is a composite anomaly 80–120 km in width and an average of 500–600 gamma in amplitude. It occurs within the island archipelago, although to the south of Marguerite Bay it appears as a zone of high magnetic intensity traceable for at least 300 km farther down the west coast of Palmer Land. Using available seismic, gravity, bathymetric, and aeromagnetic data, we conclude that the west coast anomaly is a continuous batholithic feature of variable composition and depth intruded into continental basement. It can be compared to the Patagonian batholith of the southern Andes which represents the root of an andesitic volcanic chain.

Acknowledgments. The authors wish to express their gratitude for the assistance given by all members at the British Antarctic Survey's station at Adelaide Island during 1973–74 and 1975–76, in particular to the aircrew and pilots, Bert Conchie and George Harvey, who so patiently and safely flew the aircraft.

References cited

Adie, R. J. 1955. The petrology of Graham Land: II. The Andean Granite-Gabbro Intrusive Suite. *Scient. Rep. Falkld Isl. Depend. Surv.*, No. 12, 39 pp.

Ashcroft, W. A. 1972. Crustal structure of the South Shetland Islands and Bransfield Strait. *Scient. Rep. Br. antarc. Surv.*, No. 66, 43 pp.

Barker, P. F., and Griffiths, D. H. 1972. The evolution of the Scotia Ridge and Scotia Sea. *Phil. Trans. R. Soc. Lond.*, Ser. A, *271*, No. 1213, 151–83.

Bell, C. M. 1973. The geology of Beethoven Peninsula, southwestern Alexander Island. *Bull. Br. antarc. Surv.*, No. 32, 75–83.

Bell, C. M. 1975. Structural geology of parts of Alexander Island. *Bull. Br. antarc. Surv.*, No. 41–42, 43–58.

Bregman, M. L., and Frakes, L. A. 1970. Magnetic anomalies in the Weddell Sea and vicinity, Antarctica. *Earth planet. Sci. Lett.*, *9*, No. 4, 322–26.

Butler, P. F., and McArthur, M. In press. A regional Bouguer anomaly map of north-western Palmer Land. *Bull. Br. antarc. Surv.*

Dalziel, I. W. D., de Wit, M. J., and Palmer, K. F. 1974. Fossil marginal basin in the southern Andes. *Nature, Lond.*, *250*, 291–94.

Davey, F. J. 1971. Marine gravity measurements in Bransfield Strait and adjacent areas. (*In* Adie, R. J., *ed. Antarctic Geology and Geophysics.* Oslo, Universitetsforlaget, 39–45.)

Dikstra, B. J. In press. An aeromagnetic survey of Alexander Island and parts of Palmer Land: a preliminary report. *Bull. Br. antarc. Surv.*

Fabiano, E. B., and Peddie, N. W. 1969. Grid values of total magnetic intensity IGRF—1965. *ESSA Tech. Rep.*, No. 38, 55 pp.

Renner, R. G. B. 1976. An aeromagnetic survey over Graham Land: a preliminary report. *Bull. Br. antarc. Surv.*, No. 43, 59–61.

Rowe, P. J. 1973. The geology of the area between Riley and Bertram Glaciers, Palmer Land. *Bull. Br. antarc. Surv.*, No. 35, 51–72.

Skinner, A. C. 1973. The geology of north-western Palmer Land between Eureka and Meiklejohn Glaciers. *Bull. Br. antarc. Surv.*, No. 35, 1–22.

Suárez, M. 1976. Plate-tectonic model for southern Antarctic Peninsula and its relation to southern Andes. *Geology, 4,* No. 4, 211–14.

42

Paleomagnetism of Rocks from Graham Land and Western Ellsworth Land, Antarctica

CHARLES K. SCHARNBERGER
Department of Earth Sciences, Millersville State College, Millersville, Pennsylvania 17551, U.S.A.

LeROY SCHARON
NL Industries, Incorporated, Golden, Colorado 80401, U.S.A.

Abstract. Igneous rocks have been sampled for paleomagnetic study at 47 sites in Graham Land and 22 sites in Ellsworth Land. The primary purpose of this study was to elucidate past tectonic relationships between various parts of Antarctica. Cretaceous and Tertiary rocks from both Graham Land and Ellsworth Land yield pole positions close to the present pole, indicating little latitude change for these regions since Cretaceous time. These results contrast with results from Marie Byrd Land. We interpret this to mean that Ellsworth Land is part of an Antarctic Peninsula tectonic unit which was separate from Marie Byrd Land during Mesozoic time. These two regions were probably part of two different lithospheric plates, and plate convergence has produced their present geographic relationship.

The division of Antarctica into two tectonic units—East Antarctica and West Antarctica—has long been recognized (Craddock, 1972). The Antarctic rocks subjected to the most extensive paleomagnetic study are Upper Triassic to Jurassic igneous rocks from the Transantarctic Mountains, a part of East Antarctica (McElhinny, 1973). Scharnberger and Scharon (1972) found that Lower Cretaceous rocks from Marie Byrd Land in West Antarctica produced paleomagnetic pole positions quite different from those of the somewhat older rocks of the Transantarctic Mountains—a difference greater than can easily be explained by the difference in age of the two groups of rocks. This suggested that during the Mesozoic Era, Marie Byrd Land was widely separated from East Antarctica, with a zone of plate convergence between them.

In order to continue the study of the past relationships between East and West Antarctica, rocks in Graham Land (Antarctic Peninsula) and western Ellsworth Land were sampled for paleomagnetic study.

Graham Land

Sample sites and methods

Rocks were sampled at 47 sites, using a portable drill. The sites are in igneous rocks on Deception Island (designated PD), Anvers Island near Palmer Station (PP), Argentine Islands (PG), Adelaide Island (PT), Cono Island (PC), Pourquoi Pas Island (PQ), Horseshoe Island (PH), and Stonington Island (PS) (Table 42.1). From 2 to 9 cores were collected at each site for a total of 361 cores. The cores were oriented with a level and a magnetic compass. From these cores, 860 cylindrical specimens (2.5 cm in diameter by 2.2 cm in length) were prepared. Remanent magnetization (RM) was measured with a PAR-SM1 spinner magnetometer, and alternating field (AF) demagnetization was carried out on apparatus that is described by Doell and Cox (1967).

One pilot specimen from each site was demagnetized in steps at 25, 50, 100, 200, 400, 600, and 800 oersteds. If a specimen showed instability in the direction of the RM vector, then a second specimen was tested. If this second specimen also was unstable, data from that site were not used. Specimens from 15 sites in Graham Land and 19 sites in Ellsworth Land passed this stability test. The minimum level of magnetic cleaning which produced stability in the direction of the RM vector was used to clean three additional specimens from each site. The cleaned specimens were used to determine a mean RM direction for each site. The only exception was for sites on Deception Island, where no magnetic cleaning was required and where natural remanent magnetization (NRM) of all specimens was used. The paleomagnetic data are summarized in Table 42.2 and Figure 42.1.

Results

Deception Island

The sites on Deception Island are in volcanic rocks of the Holocene Neptune Bellows Group and the overlying Pendulum Cove Group (Hawkes, 1961). All cores have normal polarity and stable NRM directions. The reason for the 45° difference between the poles for the two groups is not obvious. It is likely that the rocks sampled do not span a sufficiently long time for the virtual geomagnetic poles to coincide with either an axial or an inclined dipole.

Anvers Island

Cores were collected in tonalite of the "Andean Intrusive Suite" and in diorite and trondhjemite. The geology of Anvers Island is described by Hooper (1962), who concluded that the diorite is metasomatized volcanic rock and the trondhjemite is metasomatized Andean tonalite. Gledhill and others (Chapter 35, this volume) report a Rb-Sr age of 35 ± 6 m.y. for the tonalite and a K-Ar age of 20 ± 1 m.y. on the same rock. The remanent magnetization was probably acquired between these two times. Specimens from only three sites, all in tonalite, passed the stability test. The NRM polarity of these sites was normal, but polarity reversed at 50 oersted cleaning. The pole for the tonalite is not significantly different from the pole calculated for Andean intrusive rocks by Blundell (1962), or from the results of Dalziel and others (1973), also on Andean intrusive rocks.

Argentine Islands

Cores were collected at eight sites in volcanic rocks mapped by Elliot (1964) as part of the "Upper Jurassic Volcanic Group." Only two sites passed the stability test; the RM vectors for these sites are almost exactly reversed from each other. These sites are unusual for the "Upper Jurassic Volcanic Group" of the Antarctic Peninsula in that they produce good paleomagnetic results. Specimens from other sites in these rocks were highly unstable, a fact also noted by Blundell (1962). The pole position for these two sites is very close to that of the Andean intrusives. It is possible that these two sites are in rocks which are not actually part of the "Upper Jurassic Volcanic Group," or that the RM was acquired as a result of reheating during the "Andean" event.

Table 42.1. Geology of sites sampled

Locality and site number	Description	Type of samples	Results used?
Deception Island (PD)			
2–4	Lithic tuffs, Neptune Bellows Group, Holocene	Oriented cores	No
5, 6	Lava flows Pendulum Cove Group, Holocene (younger than Neptune Bellows Group)	Oriented cores	Yes
7	Lava flow, Neptune Bellows Group, Holocene	Oriented cores	Yes
Anvers Island (PP)			
1, 4–6	Tonalite, Tertiary	Oriented cores	Yes, except site 5
7–9, 11	Diorite, probably metasomatic	Oriented cores	No
2, 3, 10	Trondjemite, probably metasomatic	Oriented cores	No
Argentine Islands (PG)			
1–8	Andesite, Upper Jurassic(?)	Oriented cores	Only sites 2 and 8
Adelaide Island (PT)			
1–7	Mafic dikes, Tertiary	Oriented cores	Yes, except site 2
Cono Island (PC)	Mafic dikes, Cretaceous to Tertiary	Oriented cores	Yes
Horseshoe Island (PH)			
1–3	Mafic dikes, Cretaceous to Tertiary	Oriented cores	Yes
Stonington Island (PS)			
1–4	Mafic dikes, Cretaceous(?)	Oriented cores	Yes
Jones Mountains (EJ)			
1–4, 8	Mafic dikes, Cretaceous(?)	Oriented blocks	Yes
5–7, 9	Basalt flows, Tertiary	Oriented blocks	Yes
10	Pink granite, Triassic	Oriented blocks	Yes
Thurston Island (ET)			
1, 3, 4	Mafic dikes, Cretaceous to Tertiary(?)	Oriented blocks	Only site 4
2	Amphibole schist, Paleozoic(?)	Oriented blocks	No
5	Diorite stock, Cretaceous(?)	Oriented blocks	Yes
Dustin Island (ED)			
1	Diorite stock, Cretaceous(?)	Oriented blocks	Yes
Hudson Mountains (EH)			
1–6	Basalt flows, Tertiary(?)	Oriented blocks	Yes

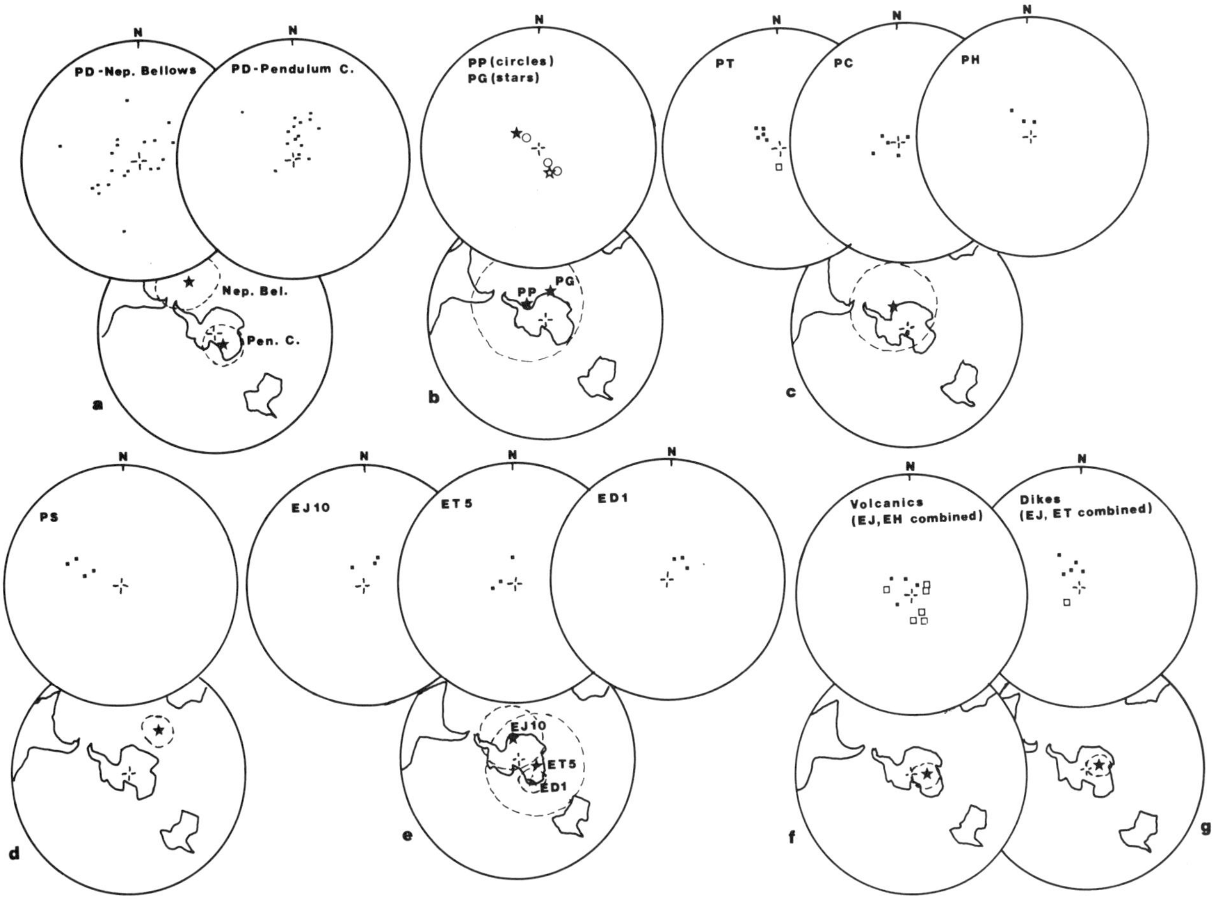

Figure 42.1. Paleomagnetic results: *a.* PD (Deception Island), Holocene volcanic rocks; *b.* PP (Anvers Island), Tertiary tonalite and PG (Argentine Islands), Jurassic (?) andesite; *c.* PT (Adelaide Island), PC (Cono Island), and PH (Horseshoe Island), Cretaceous to Tertiary mafic dikes; *d.* PS (Stonington Island), Cretaceous(?) mafic dikes; *e.* EJ10 (Jones Mountains), Triassic granite; ET5 (Thurston Island), Cretaceous(?) diorite; ED1 (Dustin Island), Cretaceous(?) diorite; *f.* Tertiary volcanic rocks of Jones Mountains and Hudson Mountains combined; *g.* Cretaceous(?) mafic dikes of Jones Mountains and Thurston Island combined. RM vectors are shown as projections on equal-area nets, upper hemisphere, in the upper circle(s) for each group. Solid symbols indicate normal polarity; open symbols indicate reversed polarity. Pole positions are shown with ovals of confidence at 0.95 probability in the lower circle for each group.

Table 42.2. Paleomagnetic data

Site or group	Location							Paleomagnetic Pole					
of sites	Latitude	Longitude	Number	Decl.	Incl.	k	α_{95}	Lat.	Long.	dp	dm	Treatment	Polarity
Deception Island (PD)	62° 57′ S	60° 38′ W											
Neptune Bellows Group			20[a]	247	−82	8	12	55° S	25° W	22	23	NRM	Normal
Pendulum Cove Group			15[a]	8	−69	19	9	80° S	148° E	14	15	NRM	Normal
Anvers Island (PP)	64° 45′ S	64° 10′ W	3[b]	146	81	40	19	75° S	47° W	35	37	AF200	Reversed
Argentine Islands (PG)	65° 15′ S	64° 16′ W	2[b]	324	−73	—	—	72° S	11° E	—	—	AF100	Mixed
Marguerite Bay													
area (PT, PC, PH)	67° 45′ S	68° 00′ W	15[b]	320	−81	70	15	76° S	30° W	28	29	AF100	Mixed
Stonington Island, (PS)	68° 11′ S	67° 00′ W	4[b]	300	−55	75	11	51° S	26° E	11	15	AF200	Normal
Jones Mountains (EJ10)	73° 30′ S	94° 20′ W	3[c]	6	−71	105	12	72° S	96° E	18	21	NRM	Normal
Thurston Island (ET5)	72° 05′ S	99° 20′ W	3[c]	320	−80	35	21	77° S	21° W	29	30	AF100	Normal
Dustin Island (ED1)	72° 34′ S	95° 05′ W	3[c]	39	−74	238	8	70° S	155° E	13	14	AF50	Normal
Mafic dikes—													
Jones Mountains and								78° S	83° E	13	14	NRM, AF50	Mixed
Thurston Island	73° 00′ S	96° 00′ W	6[b]	1	−75	63	8					AF100	
(EJ1, 2, 3, 4, 8, ET4)	73° 00′ S	96° 00′ W	6[b]	1	−75	63	8	78° S	83° E	13	14	NRM, AF50	Mixed
												AF100	
Volcanic rocks—													
Jones Mountains and													
Hudson Mountains													
(EJ5, 6, 7, 9, and all EH)	74° 00′ S	97° 00′ W	10[b]	337	−87	62	6	80° S	89° E	12	12	NRM	Mixed

k: precision parameter (N − 1/N − R), where R is resultant of N unit vectors.
α_{95}: radius in degrees of circle of confidence for the mean vector at 0.95 probability.
dp, dm: semi-minor and semi-major axes of the error ellipse at the 0.95 level of confidence.

[a] Number of cores.
[b] Number of oriented samples.
[c] Number of sites.

Marguerite Bay

Cores were collected at seven sites near Adelaide Station (Base T) on Adelaide Island, where the exposed rocks are lithic tuffs cut by numerous mafic dikes. One site has reversed polarity after magnetic cleaning. Five sites on Cono Island, three sites on Pourquoi Pas Island, and three sites on Horseshoe Island are located in mafic dikes which cut volcanic flows and tuffs. These dikes are of Cretaceous or Tertiary age (Gledhill and others, Chapter 35, this volume), and all passed the stability test. The RM vectors for the three sites on Pourquoi Pas Island are widely dispersed relative to each other and relative to other sites. The reason for this high between-site dispersion is not known, and these sites were not used for paleomagnetic pole calculations. The combined pole position for Adelaide, Cono, and Horseshoe Islands is quite close to that for the Andean intrusive rocks of Anvers Island.

Stonington Island

The four sites on Stonington Island are located in mafic dikes which intrude granitic gneiss. This gneiss is part of the basement complex of the Antarctic Peninsula, which has been regarded as Paleozoic or Precambrian in age (Fraser, 1965; Craddock, 1972; Grikurov and others, 1972). The age of these metamorphic rocks as determined by the Rb-Sr method, however, is only 175 ± 7 m.y., and one of the dikes of Stonington Island has a radiometric age of 119 ± 1 m.y. (Gledhill and others, Chapter 35, this volume). It is not certain, however, whether these are true ages or reset metamorphic ages. The difference between the paleomagnetic results from Stonington Island and those from other islands of Marguerite Bay (Table 42.2 and Figure 42.1) could be due to a significant difference in age between the two groups of rocks. On the other hand, the explanation may lie in tectonic movement of Stonington Island relative to other areas of Marguerite Bay.

Western Ellsworth Land

Sample sites and methods

Igneous rocks of western Ellsworth Land were sampled in the Jones Mountains (EJ), on Thurston Island (ET), on Dustin Island (ED), and in the Hudson Mountains (EH). Three block samples, oriented by level and magnetic compass, were collected at each of 22 sites. Six specimens were prepared from each of the 66 samples and, after appropriate magnetic cleaning, were used to calculate a sample-mean direction of remanent magnetization. Unit weight was given to the sample-means to calculate a site-mean vector. Some sites were combined in groups and five paleomagnetic pole positions were calculated (Table 42.2 and Figure 42.1).

Results

Jones Mountains

In the Jones Mountains a granitic basement complex, dated at 199 ± 6 m.y. (K-Ar, on muscovite), is intruded by mafic and felsic dikes (Craddock, Bastien, and Rutford, 1964). The mafic dikes are cut by the felsic ones, and one of the latter yielded a K-Ar whole-rock age of 104 ± 4 m.y. (Craddock, Bastien, and Rutford, 1964). These rocks are overlain by basaltic volcanic rocks. Site EJ10 is in the granite, sites EJ1, EJ2, EJ3, EJ4, and EJ8 are in the mafic dikes and sites EJ5, EJ6, EJ7, and EJ9 are in the volcanic rocks. Specimens of granite showed stability of the RM direction when tested by AF demagnetization up to a 200 oersted peak field, but pilot specimens became unstable at higher demagnetization levels. NRM was used for paleomagnetic calculations. The paleomagnetic pole is close to the present pole, but in view of the instability which sets in upon demagnetization, it seems likely that this granite has been remagnetized, perhaps during the more recent volcanic eruptions.

Dikes sampled at sites EJ3 and EJ4 had stable NRM directions, but those sampled at sites EJ1, EJ2, and EJ8 required cleaning in 50- or 100-oersted-peak fields. All volcanic rocks sampled showed great stability of NRM directions (see Table 42.2).

Thurston Island and Dustin Island

Thurston Island and nearby Dustin Island consist of a metamorphic complex intruded by granitic to dioritic plutons and younger mafic dikes (Craddock and others, 1969). The metamorphic rocks have been dated at 280 ± 10 m.y. by the Rb-Sr method on biotite separates (Craddock, Gast, and others, 1964). Rubidium-strontium dates on the plutonic igneous rocks (Munizaga, 1972) are 112 ± 5 m.y. (whole-rock) and 132 m.y. (single biotite separate). Sites ET5 and ED1 are in undated diorite plutons, but in view of the radiometric ages of nearby plutons, a Cretaceous age is probable.

Sites ET1, ET3, and ET4 are in mafic dikes of probably Cretaceous to Tertiary age. Pilot specimens from sites ET1 and ET3 showed highly unstable behavior when tested by AF demagnetization. Specimens from site ET4 were used after 50 oersted AF cleaning. Site ET4 was grouped with sites in mafic dikes of the Jones Mountains to calculate a paleomagnetic pole. This pole is located close to both the present pole and the pole positions for the plutonic rocks.

Site ET2 is in metamorphic rock (amphibole schist). The RM vectors of pilot specimens are highly unstable, and specimens from this site were not used.

Hudson Mountains

The Hudson Mountains are basaltic volcanoes, probably late Tertiary to Quaternary in age (Wade and LaPrade, 1969). NRM directions of pilot specimens from all six sites were highly stable. These sites were grouped with sites in volcanic rocks of the Jones Mountains to calculate a paleomagnetic pole for which the oval of confidence just touches the present pole (Figure 42.1*f*). Although both normal and reversed polarity is represented in these sites, lack of absolute ages makes it impossible to infer a time sequence of field reversal from these rocks.

Interpretations

The paleomagnetic data derived from Anvers Island, the islands of Marguerite Bay (except Stonington Island), and western Ellsworth Land suggest a high paleolatitude for the Antarctic Peninsula in Late Cretaceous to early Tertiary time. Similar results have been obtained by Kellogg and Reynolds (1974*a* and *b*) and Reynolds and Kellogg (1974) from Cretaceous igneous rocks of the Lassiter Coast, approximately 600 km south of Marguerite Bay.

The paleomagnetic poles for all rocks sampled in Ellsworth Land, and for all except those on Stonington Island in Graham Land, are within 20° of the present pole. This is in marked contrast to an Early Cretaceous pole position for Marie Byrd Land at lat. 36° S, long. 116° E (Scharnberger and Scharon, 1972). This suggests that in the Cretaceous Period, West Antarctica was separated into two tectonic units: (1) Ellsworth Land–Antarctic Peninsula, and (2) Marie Byrd Land. These units belonged to different lithospheric plates with separate histories of continental drift.

Marie Byrd Land, probably joined to the New Zealand block through much of Cretaceous time, may have occupied a position in Gondwanaland quite different from its present position relative to the Antarctic Peninsula and East Antarctica. Paleomagnetic investigation of Jurassic rocks from the Antarctic Peninsula should help in the reconstruction of the past relationship between the Antarctic Peninsula and East Antarctica.

Acknowledgments. Work in Ellsworth Land was supported by National Science Foundation Grant GA-1138 and in Graham Land by NSF Grant GA-1176. Samples from Ellsworth Land were collected by A. Shimoyama in 1968–69. Work in Graham Land was carried out by L. Scharon while he was an exchange scientist with the 1969–70 British Antarctic Survey expedition. Peter Rowley and Karl Kellogg made several suggestions for improvement of the paper.

References cited

Blundell, D. J. 1962. Paleomagnetic investigations in the Falkland Islands Dependencies. *Scient. Rep. Br. antarc. Surv.*, No. 39, 24 pp.

Craddock, C. 1972. Antarctic tectonics. (*In* Adie, R. J., *ed. Antarctic Geology and Geophysics.* Oslo, Universitetsforlaget, 449–56.)

Craddock, C., Bastien, T. W., and Rutford, R. H. 1964. Geology of the Jones Mountains area. (*In* Adie, R. J., *ed. Antarctic Geology.* Amsterdam, North-Holland Publishing Company, 171–87.)

Craddock, C., White, C. M., and Rutford, R. H. 1969. The geology of the Eights Coast. *Antarct. Jnl U.S., 4,* No. 4, 93–94.

Craddock, C., Gast, P. W., Hanson, G. N., and Linder, H. 1964. Rubidium-strontium ages from Antarctica. *Geol. Soc. Am. Bull., 75,* No. 3, 237–40.

Dalziel, I. W. D., Klingfield, R., Lowrie, W., and Opdyke, N. D. 1973. Paleomagnetic data from the southernmost Andes and the Antarctandes. (*In* Tarling, D. H., and Runcorn, S. K., *eds. Implications of Continental Drift to the Earth Sciences. 1.* London, Academic Press, 87–101.)

Doell, R. R., and Cox, A. 1967. Analysis of alternating field demagnetization equipment. (*In* Collinson, D. W., Creer, K. M., and Runcorn, S. K., *eds. Methods in Paleomagnetism.* Amsterdam, Elsevier, 244–50.)

Elliot, D. H. 1964. The petrology of the Argentine Islands. *Scient. Rep. Br. antarc. Surv.,* No. 41, 31 pp.

Fraser, A. G. 1965. The petrology of Stonington and Trepassey Islands, Marguerite Bay. *Scient. Rep. Br. antarc. Surv.,* No. 52, 51 pp.

Grikurov, G. E., Ravich, M. G., and Soloviev, D. J. 1972. Tectonics of Antarctica. (*In* Adie, R. J., *ed. Antarctic Geology and Geophysics.* Oslo, Universitetsforlaget, 457–68.)

Hawkes, D. D. 1961. The geology of the South Shetland Islands: II. The geology and petrology of Deception Island. *Scient. Rep. Falkld Isl. Depend. Surv.,* No. 27, 43 pp.

Hooper, P. R. 1962. The petrology of Anvers Island and adjacent islands. *Scient. Rep. Br. antarc. Surv.,* No. 34, 69 pp.

Kellogg, K. S. and Reynolds, R. L. 1974*a*. Paleomagnetic study of igneous rocks of the northern Lassiter Coast, Antarctic Peninsula. *Antarct. Jnl U.S., 9,* No. 2, 38–40.

Kellogg, K. S., and Reynolds, R. L. 1974*b*. Paleomagnetic evidence for an undrifted southern Antarctic Peninsula since the middle Cretaceous. *Trans. Am. geophys. Un., 55,* No. 12, 1110.

McElhinny, M. W. 1973. *Paleomagnetism and Plate Tectonics.* Cambridge, Cambridge University Press, 357 pp.

Munizaga, F. 1972. Rb-Sr isotopic ages of intrusive rocks from Thurston Island and adjacent islands. (*In* Adie, R. J., *ed. Antarctic Geology and Geophysics.* Oslo, Universitetsforlaget, 205–6.)

Reynolds, R. L., and Kellogg, K. S. 1974. Paleomagnetism of igneous rocks of the central Lassiter Coast, Antarctic Peninsula. *Antarct. Jnl U.S., 9,* No. 5, 227–28.

Scharnberger, C. K., and Scharon, L. 1972. Paleomagnetism and plate tectonics of Antarctica. (*In* Adie, R. J., *ed. Antarctic Geology and Geophysics.* Oslo, Universitetsforlaget, 843–48.)

Wade, F. A., and LaPrade, K. E. 1969. Geology of the King Peninsula, Canisteo Peninsula, and Hudson Mountains area, Ellsworth Land, Antarctica. *Antarct. Jnl U.S., 4,* No. 4, 92–93.

43

The Influence upon Southern Ocean Circulation, Sedimentation, and Climate of the Opening of Drake Passage

P. F. BARKER and J. BURRELL*
Department of Geological Sciences, University of Birmingham, Birmingham B15 2TT, U.K.

Abstract. A recent analysis of marine geophysical data from Drake Passage concluded that the Antarctic Circumpolar Current (ACC) started 23.5 ± 2.5 m.y. ago (about at the Oligocene-Miocene boundary). In this paper we test this conclusion against the sedimentary record at all suitable DSDP sites in the Southern Ocean. There is close temporal agreement between the inferred onset of the ACC and the emergence in early Miocene time of that siliceous biofacies which today dominates sedimentation between the Antarctic Convergence and Antarctic Divergence. The sedimentary evidence also implies an essential interdependence of the Convergence, the Divergence, and the ACC, as suggested by some numerical models of circulation.

There is considerable speculation about a possible causal relationship between the progressive isolation of Antarctica at the South Pole and the coeval global climatic deterioration, and the part played in this change by ocean currents. We suggest the following crude conceptual model of mid-Cenozoic paleoclimate: an efficient meridional circulation, driven by Antarctic Bottom Water (AABW), arrests a more general climatic deterioration until disrupted by the zonal ACC, so that renewed cooling should follow ACC onset. Paleotemperature data are very sparse, so that their global—as opposed to local—significance is not yet clear; at present, however, there is no striking coincidence between ACC onset and renewed cooling, which apparently started at about 16 m.y. ago rather than 23 m.y. ago. There is a clear need for additional sediment sampling and paleotemperature measurement, guided where possible by realistic numerical models of circulation.

It has been suggested (Cromwell and Frakes, 1970; Hayes, Frakes and others, 1975; Kennett, Houtz and others, 1975; Hammond, 1976) that growth of the southern oceans following the fragmentation of Gondwanaland, with progressive isolation of the Antarctic continent at the south geographic pole, has caused the slow stepwise deterioration in global climate since Late Cretaceous time (Savin and others, 1975). In partial contrast, Donn and Shaw (1977) have argued that in the Northern Hemisphere the climatic deterioration can be explained almost entirely in terms of the changing distribution of land and ocean; in their model the contribution of oceanic circulation is negligible.

We recently presented (Barker and Burrell, 1977) an interpretation of ocean floor magnetic lineations in Drake Passage which discussed the history of separation of the northern Antarctic Peninsula from southern South America, and showed that this opening removed the *final* barrier to complete circumpolar flow of deep water. Consequently, considerably more is now known about the initiation and growth of the Antarctic Circumpolar Current (ACC), which today dominates Southern Ocean circulation. ACC development was the latest and most influential stage in the progressive geographic isolation of Antarctica within oceanic surroundings. For these reasons, and because this event was not accompanied by a large change in the distribution of land and water, it is more amenable than earlier, more subtle changes to both observation and modeling aimed at assessing the susceptibility of climate to changes in oceanic circulation.

In this paper we reevaluate the somewhat fragmentary sedimentary and paleoclimatic record in the Southern Ocean in the light of the new model for the opening of Drake Passage. A reasonably close correspondence emerges between the model and observed sedimentation, which suggests some interesting relationships between various elements of the present circulation. The comparison of ACC growth with the climatic record can only be preliminary at present because of the scarcity of data, but the correspondence is not obvious. We stress the need for further sampling, guided where possible by numerical models.

Drake Passage opening

Coherent sea-floor spreading in Drake Passage began at about anomaly-8 time (~29 m.y. ago). Spreading (in the 120°–300° direction) was rapid at first, at a half-rate of 23 mm/yr until 27 m.y. ago, but later slowed, first to 18 mm/yr and then, at 17 m.y. ago, to 10 mm/yr, before stopping at about 6 m.y. ago (Barker and Burrell, 1977). Figure 43.1 shows the embryonic Drake Passage at anomaly-8 time (with ridge axis located), at anomaly-6 time (~20 m.y. ago), and at present day. Some ocean floor had already formed by 29 m.y. ago, probably during initial disruption of the continuous continental connection between South America and West Antarctica. Fragments of that original connection, such as the South Georgia, Shag Rocks, and South Orkney blocks, now make up the north

* Present address: British Petroleum, Britannic House, Moor Lane, London EC27 9BU U.K.

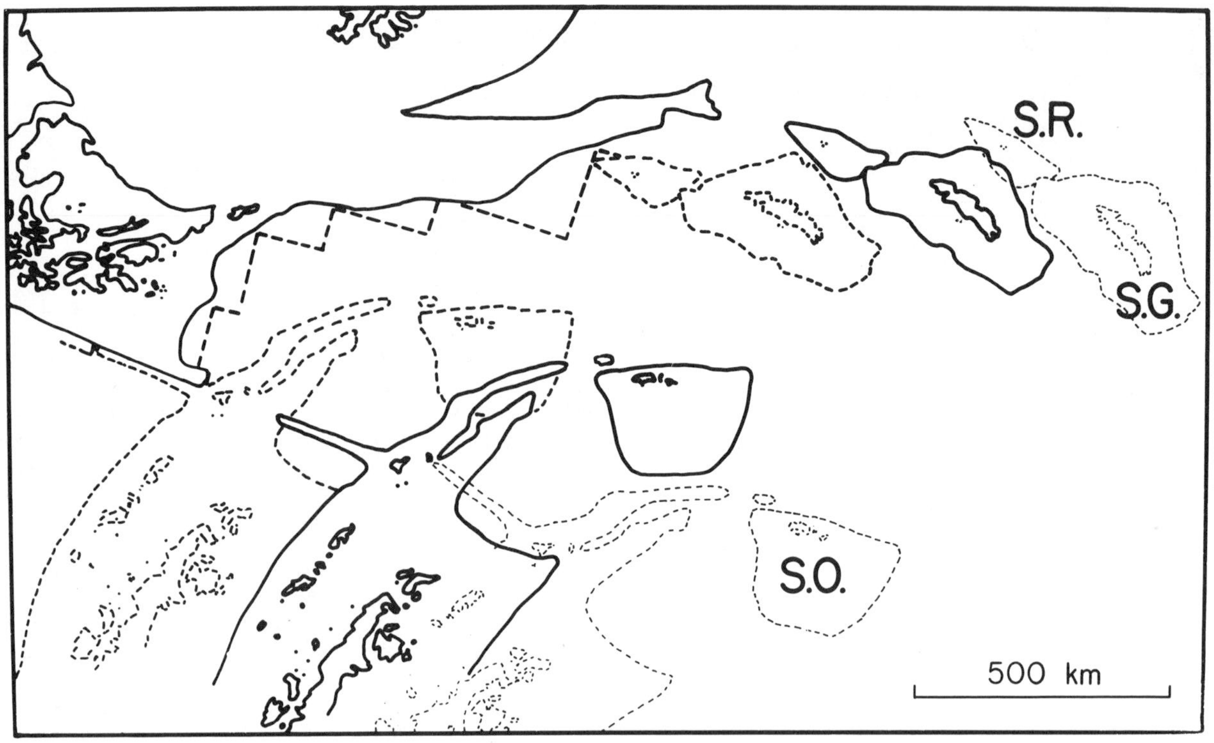

Figure 43.1. Drake Passage opening, showing positions of Antarctic Peninsula relative to South America at times of anomaly 8 (~29 m.y. ago—heavy dashed line), anomaly 6 (~20 m.y. ago—solid line), and present day (light dashed line). Ridge crest and fracture zones at anomaly-8 time are shown by heavy dashed orthogonal set of straight lines close to the South American margin. After Barker and Burrell (1977). Positions of South Georgia (S.G.), Shag Rocks (S.R.), and South Orkney (S.O.) continental fragments are schematic only. Polar stereographic projection.

and south Scotia Ridge. They may have been partial barriers to circumpolar flow at anomaly-8 time, being not so widely dispersed as at present, but a complete barrier was still intact at the western end of Drake Passage. The two submarine ridges extend 60 km southeast of the South American margin and about 200 km northwest of the vicinity of Elephant Island on the South Scotia Ridge, are shown initially overlapping in Figure 43.1. Our contention that these ridges were originally shallow continental fragments and that they subsided only slowly, along with the adjacent ocean crust which formed as Drake Passage opened, is strongly supported by evidence from reflection profiles located farther east in Drake Passage (Barker and Burrell, 1977). Thus, although water at continental-shelf depths could have passed between the Pacific and Atlantic Oceans across the Shackleton Fracture Zone as early as 28 or 29 m.y. ago, a deep gap almost certainly was not in existence until the ends of the two ridges cleared. We estimate that this took place at 23.5 ± 2.5 m.y. ago (Barker and Burrell, 1977).

We showed that other constrictions in the circumpolar path (south of Australia, the Macquarie Ridge Complex, and Kerguelen-Gaussberg) were in existence earlier and have always been wider than the gap at the Shackleton Fracture Zone. Until other regions of the Scotia Sea have been dated, however, the possibility will remain that after 23.5 m.y. ago a fortuitous arrangement of continental fragments and an ancestral South Sandwich island arc (Barker and Griffiths, 1972; Barker and others, Chapter 22, this volume) temporarily produced a constriction in the central Scotia Sea tighter than that at the Shackleton Fracture Zone. For this reason, and others based on oceanographic considerations, we concluded that the time of onset of the ACC was very probably 23.5 ± 2.5 m.y. ago but that growth of the deep-water gap at the Shackleton Fracture Zone since then does not necessarily indicate precisely the growth in water volume transport of the ACC.

Present-day circulation and sedimentation

The major influences on sedimentation in the Southern Ocean today are biogenic productivity and test dissolution, bottom current strength, and the availability of terrigenous debris, all aspects of the general oceanic circulation which is dominated by the ACC. Thus, present-day circulation and sedimentation patterns provide a model for use in checking our inferred history of ACC growth against the sedimentary record.

The ACC is wind driven, although the low attenuation with depth (Maksimov and Vorob'ev, 1964) suggests some reinforcement. Its zonal transport, at about 200 × 10^6 m^3/sec (Reid and Nowlin, 1971; Callahan, 1971), is two to three times the total meridional transport in the Southern Ocean (Gordon, 1971). Between the broad zone of westerly winds which drive the ACC and the Antarctic continental high-pressure area lies a low-pressure zone centered at about lat. 65° S (Deacon, 1963); winds within this zone drive a thin, discontinuous, westward surface current around the Antarctic coast and a more vigorous, more stable cyclonic gyre in the Ross and Weddell Sea embayments. Interaction between this gyre, or current, and the ACC produces upwelling of cold, nutrient-rich Circumpolar Deep Water (CDW) at the Antarctic Divergence (Figure 43.2), and the Antarctic Surface Water (ASW) so

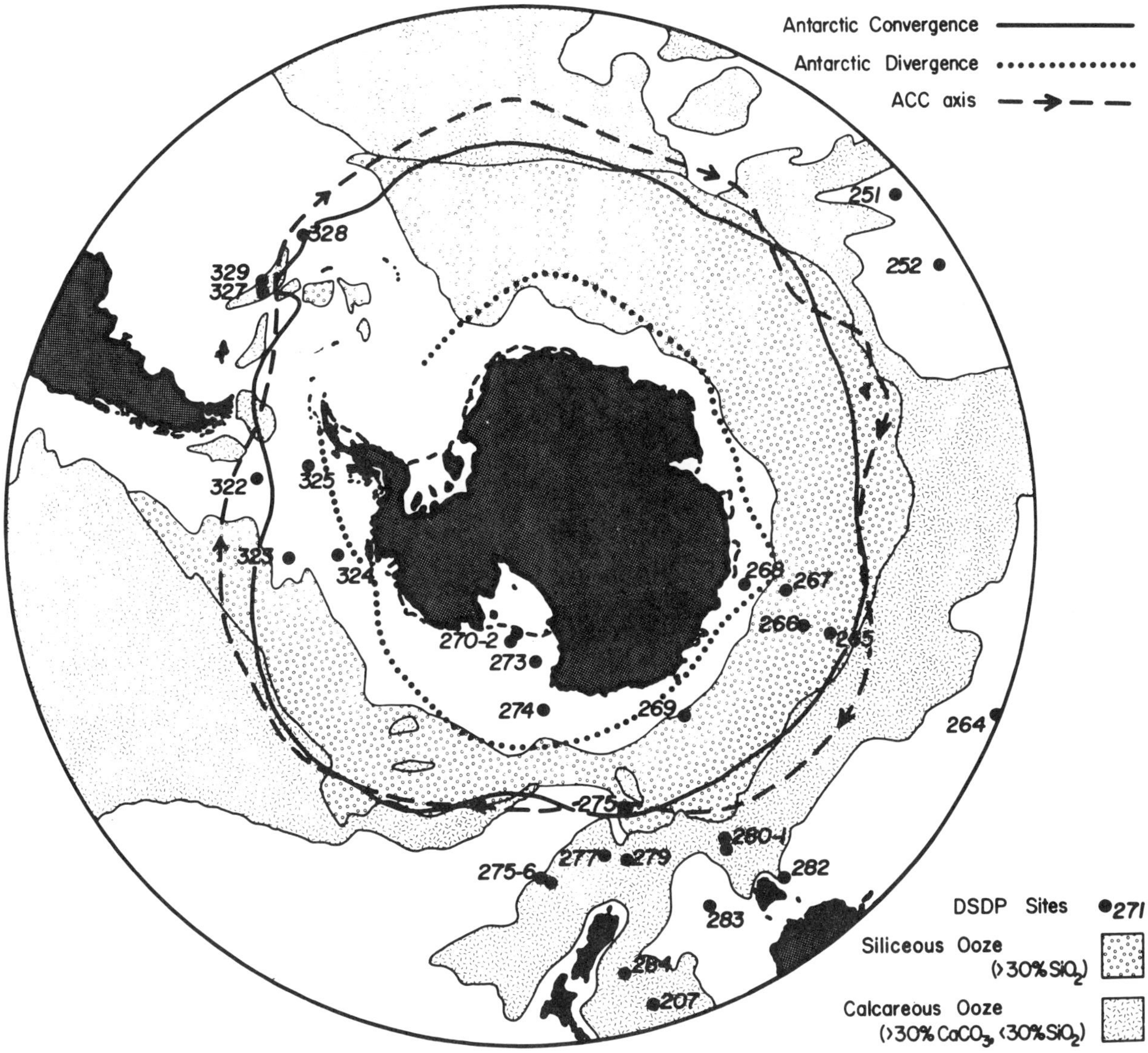

Figure 43.2. Antarctic Convergence, Antarctic Divergence, and simplified biogenic sediment distribution in the Southern Ocean, from Goodell (1973). Approximate ACC axis derived from Tolstikov (1966, plate 103).

formed extends northward to the Antarctic Convergence (or Antarctic Polar Front). Here it mixes with and dives northward beneath Subantarctic Surface Water to form Antarctic Intermediate Water (AIW). This water mass then moves northward beneath the thermocline and above the CDW, which incorporates some North Atlantic Deep Water and rises southward towards the Antarctic Divergence, as noted above. Figure 43.3, an adaptation of a meridional temperature section along long. 115° E (Gordon, 1972*a*), shows these meridional movements which, as the volume transport figures show, are superimposed on the much larger zonal flow of the ACC. Poorly illustrated by Figure 43.3 are the generation and movement of Antarctic Bottom Water (AABW), formed probably by an admixture of CDW and cold salty water from beneath sea ice in the Ross Sea and more especially the Weddell Sea (Gordon, 1971; Carmack and Foster, 1975), neither of which lie on the line of section. AABW flows westward in a narrow, deep contour current around the Antarctic margin (Hollister and Heezen, 1967; Gordon 1972b; Tucholke and Houtz, 1976) and northward where it is diverted by bottom topography, as into the Argentine Basin (Le Pichon and others, 1971). AABW formation and the Ekman Divergence equally share the responsibility for driving the meridional circulation (Gordon, 1971).

A high siliceous and lesser calcareous productivity in the upwelled Antarctic Surface Water gives rise to the band of diatomaceous ooze around Antarctica, between the Antarctic Convergence and Divergence (Figures 43.2 and 43.3). Calcareous skeletons are mostly dissolved in deeper waters, so that foraminiferal ooze survives only on the upper continental slope of Antarctica and on other isolated elevations; the carbonate compensation depth (CCD) is much shallower south of the Antarctic Convergence (Kennett, 1966; Echols, 1971; Echols and Kennett, 1973) than to the north. Beneath and north of the Convergence the siliceous ooze gives way to a less continuous band of calcareous foraminiferal ooze, coinciding largely with crestal areas of the roughly zonal mid-oceanic ridge system around Antarctica. In the path of AABW the bands of ooze are disrupted, both because AABW is particularly corrosive and because its greater bottom current velocities commonly permit deposition of only the coarser, usually terrigenous sediments (Eittreim and others, 1972). Nonde-

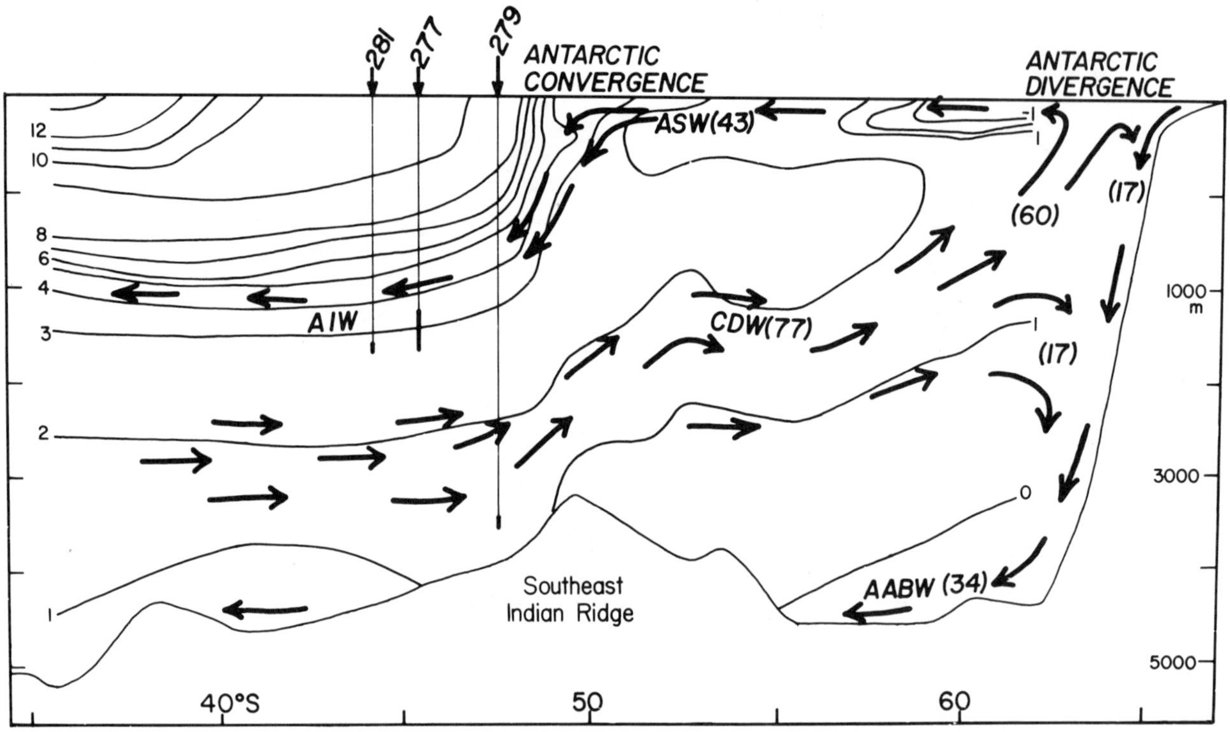

Figure 43.3. Temperature (°C) section for October 1970 along long. 115° E from Gordon (1972*a*), with water masses interpreted by reference to Gordon (1967) and Gordon and Goldberg (1970). Annual average meridional transport values ($\times 10^6$ m^3/sec) from Gordon (1971). DSDP sites 277, 279, and 281 (Kennett, Houtz and others, 1975) projected onto section relative to Antarctic Convergence (Gordon, 1972*a,c*). Sediment types (schematic) from Goodell (1973), as in Figure 43.2. (The top 1000 m of the right-hand scale is expanded, $\times 2$.)

position or erosion is also found in places beneath the ACC axis (Goodell and Watkins, 1968; Goodell and others, 1971; Watkins and Kennett, 1972; Kennett and Watkins, 1976), commonly with manganese nodule formation.

The main mode of dispersal of terrigenous sediment from Antarctica is by ice-rafting (Goodell, 1973). The disposition of ice-rafted debris depends mainly upon iceberg sources, and thus upon the state of continental glaciation, and only secondarily upon circulation. Ice-rafted deposits on the Antarctic continental slope can be disturbed by turbidity currents and the fine fraction entrained as a nepheloid layer in AABW.

Neither the ACC axis, the Antarctic Convergence, nor the siliceous-calcareous ooze boundary form a smooth curve in Figure 43.2, and no two consistently coincide. None are well mapped (the ACC axis being particularly poorly defined) and all are affected in different ways by bottom topography. Their sharpness varies considerably with longitude. The Convergence is thought to migrate seasonally through 4° of latitude (Gordon, 1967), and in places (e.g., Drake Passage—Gordon, 1975; Legeckis, 1977) it is much more complex than shown. It may have moved even greater distances under the influence of short-period climatic changes (Hays, Lozano, and others, 1976) within the period represented by surface sediment samples, particularly when possible erosion or nondeposition beneath the ACC axis is considered. Such temporal variations would tend to obscure any close relationships between the parameters being discussed. It is generally accepted, for example, that the Antarctic Divergence (with the pack-ice edge) controls the southern limit of the zone of high siliceous productivity and that the Antarctic Convergence controls its northern limit. Clearly on the evidence in Figure 43.2, an equally close relationship between the Convergence and ACC axis, for whatever reason, cannot be ruled out. If genuine, such close associations should persist in the sedimentary record throughout the life of the present circulation system.

Siliceous productivity between the Convergence and the Divergence, and strong bottom currents beneath the ACC axis are those elements of the present circulation pattern most likely to have affected the sedimentary record. We need to know, however, whether they could have existed with Drake Passage closed and with the ACC absent. The sedimentary record itself may suggest an answer, but the only independent assessment comes from numerical model studies of circulation.

Numerical models

Gill and Bryan (1971) have modeled the Southern Ocean circulation with and without a barrier-free zone (i.e., Drake Passage). Their ocean basin is of simple shape (70° lat. by 70° long.) and constant depth. One wind-stress and two surface-temperature distributions akin to those of today are imposed, but no specific provision is made for AABW formation at the southern margin or for entry of North Atlantic Deep Water. In different cases, the barrier-free zone (5° lat. in width) is as deep or half as deep as the ocean basin. Despite its simplicity, dictated by computer time limitations, the model predicts many features of present-day circulation. A narrow circumpolar current separates a subtropical and a polar gyre. There is down-

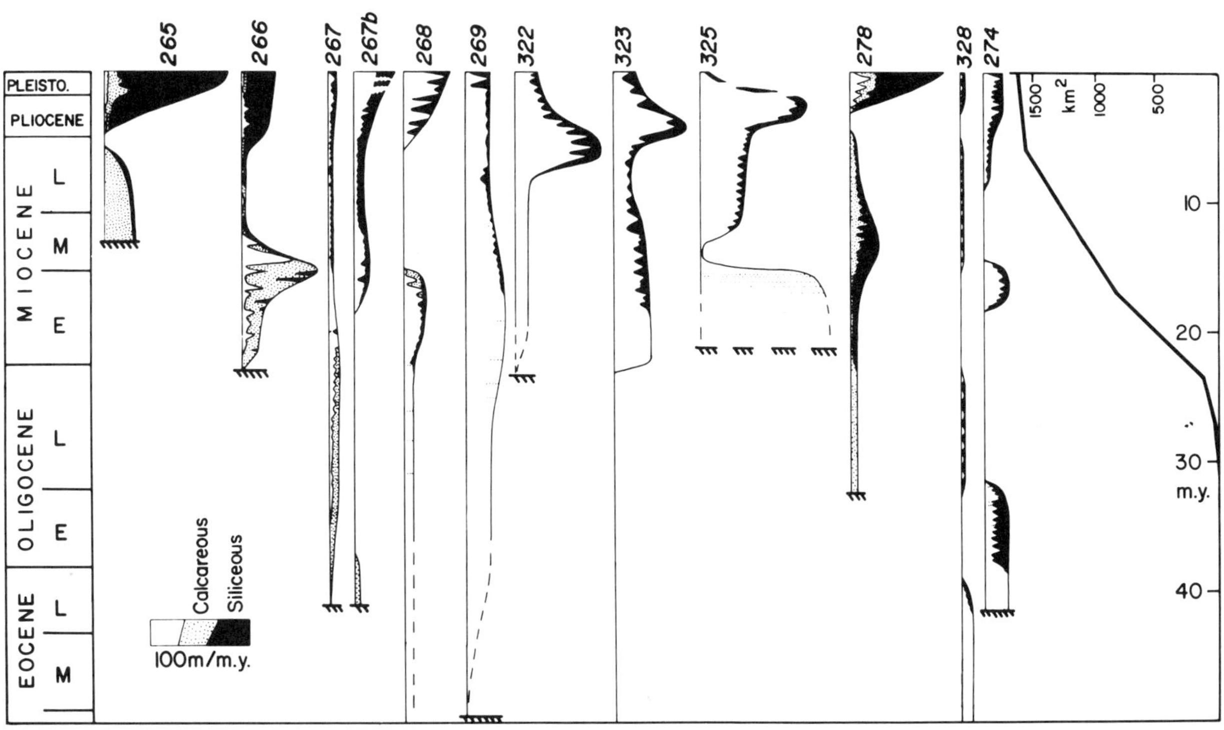

Figure 43.4. Sedimentation at selected DSDP sites in the Southern Ocean (Hayes, Frakes and others, 1975; Kennett, Houtz and others, 1975; Hollister, Craddock and others, 1976; Barker, Dalziel and others, 1977). Black areas: siliceous sediments. Stippled areas: calcareous sediments. Clear areas: clay, turbidites, ice-rafted sediments, and Mn nodules. Estimated growth of cross-sectional area of Drake Passage (Shackleton Fracture Zone) shown on right-hand side is from Barker and Burrell (1977).

welling at the southern margin, and the circumpolar current and subtropical gyre have deep countercurrents. There is a strong convergence slightly north of the gap and a divergence to the south, neither related obviously to the wind shear, which we can equate with the Antarctic Convergence and Divergence. Except in the shallow-gap case, however, these features extend to the ocean floor and meridional transport across the gap is essentially prevented; in that sense there is no Circumpolar Deep Water. In view of the shallow (2400 m below sea level) sill depth at the Shackleton Fracture Zone, the shallow-gap case may be the more appropriate, although the complexity of the ACC path through the disturbed (and evolved) topography of the Scotia Sea region warns against applying the model too closely. The great importance of bottom topography is borne out by other numerical studies (Bryan and Cox, 1972; Johnson and Hill, 1975).

In the case with Drake Passage closed, the Antarctic Convergence and Divergence disappear along with the ACC. A weak convergence occurs farther south, where the imposed wind stress changes sign, but meridional transport is essentially greater and extends to the southern margin. Of course the model, having an imposed surface-temperature distribution and wind stress appropriate to the present day, is vulnerable to exactly the kind of climatic changes which are known to have occurred since Drake Passage opened and which, in part, we are investigating. It gives some justification, however, for believing that the sharply defined band of diatomaceous ooze marking the annulus of cold, nutrient-rich surface water around Antarctica would not have existed before Drake Passage opened; upwelling would have been confined to the vicinities of topographic obstructions, such as the Falkland and Kerguelen Plateaus and, depending upon the strength of the circulation, to the western sides of the ocean basins (Gill and Bryan, 1971).

The diatomaceous ooze therefore seems to be the most useful sedimentary indicator of ACC development. Neither ice-rafting, which depends also on the glacial state of Antarctica, nor hiatuses, which can extend in time beyond their cause and may result also from the (probably preexistent) AABW, are of more than secondary value.

Southern Ocean sediments

Surprisingly few of the Southern Ocean DSDP sites (Figure 43.2) can be used to examine the history of siliceous productivity. The general absence of thick diatomaceous deposits at sites to the north of the present position of the Convergence, for example, supports conclusions drawn from the first set of Southern Ocean drill sites (Kemp and others, 1975) that the Convergence migrated northward to its present position. Also, the sedimentary record is suspect at sites on or close to large topographic irregularities (274, 278, 327, 329, and 330, at or south of the Convergence), which generate their own upwelling or bottom current systems, leading to anomalous productivity or erosion. At other sites (270–73 on the Ross Sea shelf) sedimentation has been dominated by local glacial events.

In Figure 43.4, an envelope representing average sedimentation rates at each of the remaining sites contains the varying proportions of siliceous and calcareous ooze and other sediments (clays or turbidites) found in each core. Many aspects of sedimentation, such as short-term sedimentation rates or sedimentary processes, are thus inadequately, even misleadingly illustrated, but an impression can be gained of overall siliceous productivity. Not all of the sites are ideal; at some only a partial record is available because of inadequate sampling or premature abandonment. The overwhelming dominance of the terrigenous fraction at some sites near Antarctica hinders comparison with pelagic sites, and AABW circulation has apparently controlled sedimentation at sites (328 and 274?) in its present path. Nevertheless, it can be seen that:

1. At several sites (266, 267, 268, 269, and 323) a siliceous biofacies developed in early Miocene time; alternative explanations are available (above) for earlier occurrences. This fixes the initiation of the Antarctic Divergence, which is a component of the ACC.
2. At other sites (265, 322, and 325), generally farther north relative to the present position of the Convergence, the same siliceous biofacies is first seen later, in middle or late Miocene time. This suggests a northward migration through these sites of the Antarctic Convergence (Kemp and others, 1975).
3. At many sites (265, 266, 267, 268, 278, and 325) a hiatus or near-hiatus preceded the initial or a greatly enhanced siliceous sedimentation. This suggests the passage through the site of the Convergence and the ACC axis *together* and supports the notion that they are closely related.
4. The distinct drop in calcareous sedimentation rates accompanying or directly preceding the increase in siliceous deposition (265, 266, 267, 268, 325, and 278?) again suggests northward migration of the Convergence, but also marks in many instances a tectonic subsidence of the site. This again raises the question of the importance of the zonal system of ridge crests in controlling the locations of both the Antarctic Convergence *and* the ACC axis.
5. At many sites (265, 266, 267, 268, 269, 278, and 325) the highest siliceous sedimentation rates were during Plio-Pleistocene time. This probably reflects a real increase in productivity, circulation-controlled, despite our neglect of the effects of compaction and diagenesis in the older sediments and of any tendency in the site selection process to favor (for example) localities with thin, evenly bedded sedimentary sections overlying basement elevations.

Thus, our predicted onset of the ACC 23.5 ± 2.5 m.y. ago (a time indistinguishable from the Oligocene–Miocene boundary) is broadly confirmed by the widespread introduction of a siliceous biofacies in Southern Ocean sediments of early Miocene age. Furthermore, features indistinguishable from the present Antarctic Convergence and Divergence appear to have been part of the ACC from its inception, which provides encouraging support for the numerical model of Gill and Bryan (1971). With some confidence, therefore, we may examine the paleoclimatic record for signs of the influence of the ACC.

Climate and the Antarctic Circumpolar Current

A Southern Hemisphere climatic model

Our crude conceptual model for the effect on climate of the ACC supposes a steady climatic deterioration through early Cenozoic time, as observed, from a cause as yet unspecified. Ultimately, (perhaps in earliest Oligocene time; Shackleton and Kennett, 1975) this would result in the seasonal formation of sea ice around Antarctica, particularly in the Ross Sea and Weddell Sea embayments, and hence in the formation of AABW. The subsequent motion of this bottom water would greatly invigorate the existing sluggish meridional circulation, and the resulting increase in meridional heat transport would slow or even halt the deterioration in climate.

The effect upon this situation of the onset and growth of the ACC would be to prevent the movement of warm surface water southward from the newly created Antarctic Convergence, isolating Antarctica within a ring of cold water and reducing the efficiency of meridional heat transfer to higher latitudes (meridional flow at present is largely ineffectual in this respect, being restricted to the cold, deep CDW and AABW). The original climatic deterioration would therefore be renewed, a dominant factor at this point, and perhaps originally, being the effect on the summer heat balance of high-albedo Antarctic continental ice and snow (Budyko, 1974). The growth of a thick, stable Antarctic ice sheet seems likely to have followed ACC inception, if one were not already in existence.

This model neglects atmospheric circulation, which at present effects a greater global meridional heat transport than the oceans (Manabe and Bryan, 1969), but which in the Southern Ocean region is in many respects coupled to the oceanic circulation (Zillman, 1972). Note also that the short-period climatic variations (10^4–10^5 years) expected from orbitally induced "Milankovitch" insolation changes (Hayes, Imbrie, and Shackleton, 1976) would be superimposed upon the long-period or aperiodic changes of this model.

Paleotemperature data

The most detailed and reliable Cenozoic paleotemperature estimates are those derived from oxygen isotope ($\delta^{18}O$) analyses of planktonic and benthonic foraminifera. At present only two suitable compilations are known, one from nine DSDP and three *Eltanin* sites in the subtropical Pacific and South Atlantic (Savin and others, 1975) and another from DSDP sites 277, 279, and 281 (Figure 43.2) south of Australia and New Zealand (Shackleton and Kennett, 1975). Both studies recognized the apparent temperature drop caused by disequilibrium growth of polar ice sheets (the change in mean oceanic isotopic composition brought about by the present ice cover is equivalent to a temperature drop of about 4°C), but concluded that no significant ice-sheet growth had taken place before middle Miocene time. Both data sets are plotted using the same ice-free temperature scale (Shackleton and Kennett, 1975), which is shown in Figure 43.5. Shackleton and Kennett's data are replotted against time, using available biostratigraphic evidence and sedimentation rates (Kennett, Houtz and others, 1975), to ease comparison and to illustrate better the same distribution. The planktonic-benthonic $\delta^{18}O$ difference (Shackleton and Kennett, 1975) was also plotted, but only if both assemblages were from the same sample. Essentially independent of ice-volume changes and a near-linear function of temperature, this difference shows the vertical temperature difference at a site and is therefore a potentially valuable indicator of current regime.

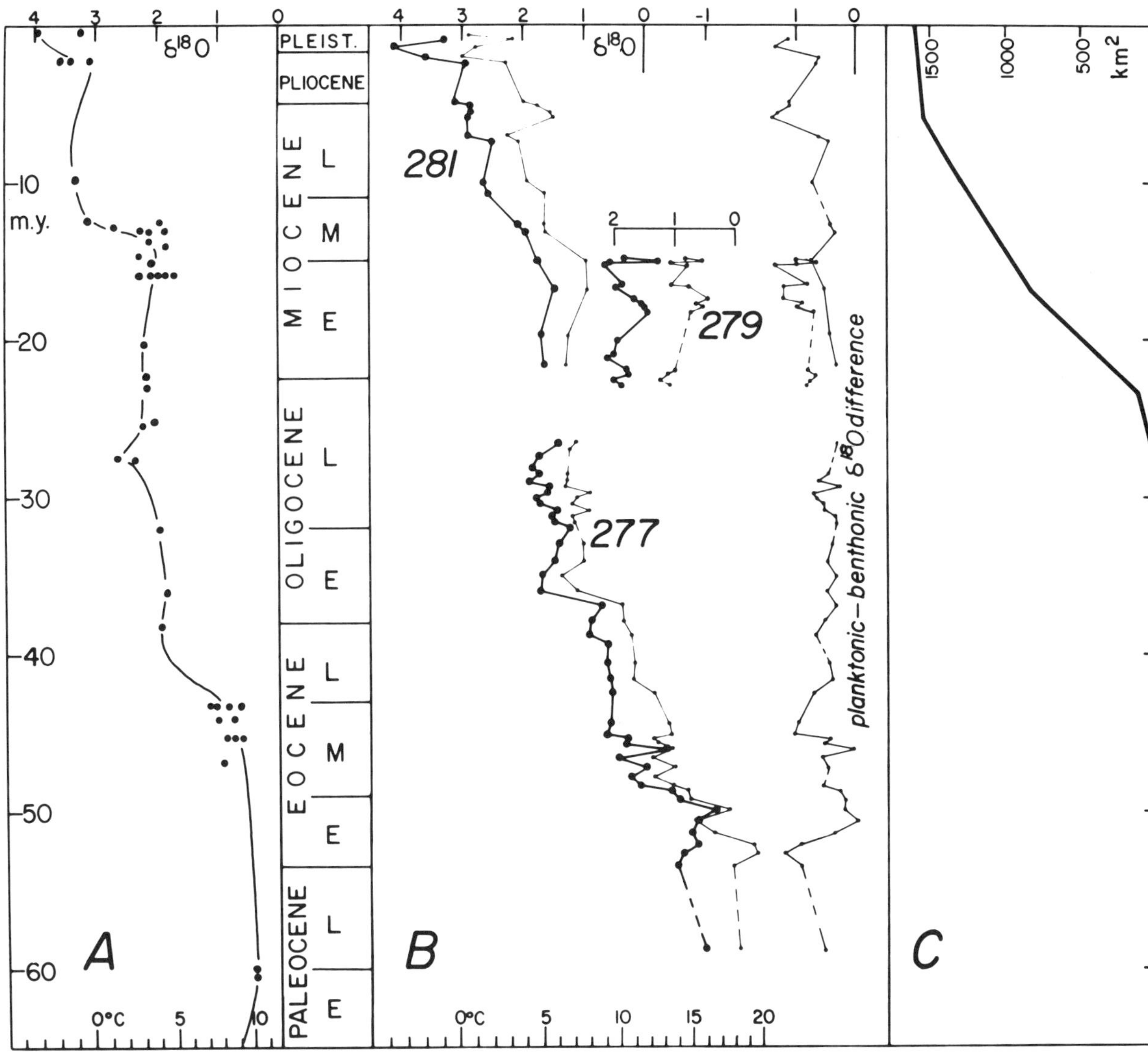

Figure 43.5. Cenozoic oxygen isotope measurements and inferred temperatures (A) on benthonic foraminifera by Savin and others (1975) and (B) on both benthonic (heavy line) and planktonic (light line) foraminifera by Shackleton and Kennett (1975), compared with (C) estimated growth of Drake Passage cross section (Barker and Burrell, 1977). Ice-free temperature scale from Shackleton and Kennett (1975).

Although there are discrepancies of detail, the two data sets are in general agreement, showing a stepwise decline in temperature through the Cenozoic, which permits us to concentrate on the more detailed, serial data set from sites 277, 279, and 281.

In the most general sense, there is reasonable agreement with our crude model. The initial fall in temperature stops at 36 m.y. ago after a sharp 5° drop which has been interpreted as the start of AABW formation (Shackleton and Kennett, 1975; Kennett and Shackleton, 1976). The temperature remains steady (across a time gap of uncertain extent at the change in sites) until about 16 m.y. ago. As much as half of the $\delta^{18}O$ increase between 16 m.y. ago and the present has been attributed to growth of first the Antarctic and then the Arctic ice cover, but whatever the mode, climatic deterioration clearly was renewed. The major discrepancy is a 7.5 m.y. difference between our estimate of ACC onset (23.5 m.y. ago) and the time of renewed cooling (16 m.y. ago).

The significance of this result is uncertain. The null hypothesis is one possibility—that climatic effects of AABW and ACC are negligible and that the (imperfect) correspondence between data and model is coincidental. More probably, however, the 7.5 m.y. delay relates to a minimum size of ACC being required for effective disruption of meridional heat transfer. Sediments suggest that the ACC existed during most of early Miocene time, but constrictions within the central Scotia Sea may have maintained a small effective gap size (Figures 43.4 and 43.5) for a while after 23 m.y. ago. Additional numerical modeling may resolve these uncertainties.

It is also possible that the few existing paleotemperature data are unrepresentative or that their significance is not fully understood. For example, $\delta^{18}O$ measurements on lower latitude benthonic foraminifera are conventionally assumed to yield high-latitude surface temperatures (Emiliani, 1954; Savin and others, 1975; Shackleton and Kennett, 1975), but the best approximation to these is probably provided by AABW, which is highly corrosive and thus unrepresented in the sediments. Figure 43.3 shows sites

277, 279, and 281 projected (parallel with the Convergence—Gordon, 1972*a,c*) onto the 115° E meridian. Possible upwelling aside, the sea bed at the sites lies at present within AIW or CDW, both significantly different from the coldest high-latitude surface water. Through much of Cenozoic time, the benthonic-planktonic $\delta^{18}O$ difference at the shallow sites was less than at present, only 2°–3°C, probably because the sea bed there lay more securely above a deeper thermocline, with AIW and CDW weaker and deeper or absent. The sea bed at site 279 lies deeper, but on one flank of the tectonically active Macquarie Ridge Complex, which greatly perturbs the ACC at present (Gordon, 1972*c*).

Thus, it is not obvious that existing $\delta^{18}O$ data on benthonic foraminifera reflect high latitude surface water temperatures. The planktonic $\delta^{18}O$ data are free of many of these ambiguities, but are nevertheless so few in number that uncertainties remain. For example, the northward passage through a site of a sharp discontinuity such as the Subtropical Convergence is a likely consequence of a steady climatic deterioration but would generate an unrepresentative step in the $\delta^{18}O$ curve. Until more data are obtained, and better (perhaps numerical) models of paleocirculation are available for their interpretation, the degree to which they represent global trends will remain open to question.

Conclusions

A time of initiation of the Antarctic Circumpolar Current (ACC) of 23.5 ± 2.5 m.y. ago, derived from marine geophysical studies in Drake Passage (Barker and Burrell, 1977), is supported by DSDP core evidence of a marked increase in diatomaceous ooze deposition in the southern oceans in early Miocene time. Other features of the sedimentary record at DSDP sites imply a close interdependence through Neogene time of the ACC and the Antarctic Convergence and Divergence. This provides encouraging support for the numerical model of southern ocean circulation investigated by Gill and Bryan (1971).

Our crude conceptual model suggests that renewed high-latitude climatic deterioration should follow ACC onset because of its disruption of oceanic meridional heat transfer. Existing paleotemperature data are sparse and possibly not fully representative, but they do not show renewed cooling until after 16 m.y. ago. Provisionally, we attribute this 7.5 m.y. delay to a threshold size of ACC transport being needed to disrupt meridional flow, and we suggest that a fortuitous arrangement of continental fragments and an island arc in the central Scotia Sea may have restricted ACC transport for some time after 23.5 m.y. ago.

There is a clear need for additional oxygen isotope data from the southern oceans and for further development of numerical models of paleocirculation within which these data can be interpreted.

Acknowledgments. This work was supported by a Natural Environment Research Council research contract.

References cited

Barker, P. F., and Burrell, J. 1977. The opening of Drake Passage. *Mar. Geol., 25,* 15–34.

Barker, P. F., and Griffiths, D. H. 1972. The evolution of the Scotia Ridge and Scotia Sea. *Phil. Trans. R. Soc. Lond.,* Ser. A., *271,* 151–83.

Barker, P. F., Dalziel, I. W. D., and others. 1977. *Initial Reports of the Deep Sea Drilling Project,* Vol. 36. Washington (U.S. Government Printing Office), 1080 pp.

Blakely, R. J. 1974. Geomagnetic reversals and crustal spreading rates during the Miocene. *J. geophys. Res., 79,* 2979–86.

Bryan, K., and Cox, M. D. 1972. The circulation of the world ocean: a numerical study. Part 1, a homogeneous model. *J. Phys. Ocean., 2,* 319–35.

Budyko, M. I. 1974. *Climate and Life.* New York, Academic Press, 508 pp.

Callahan, J. E. 1971. Velocity structure and flux of the Antarctic Circumpolar Current south of Australia. *J. geophys. Res., 76,* 5859–70.

Carmack, E. C., and Foster, T. D. 1975. On the flow of water out of the Weddell Sea. *Deep Sea Res., 22,* 711–24.

Crowell, J. C., and Frakes, L. A. 1970. Phanerozoic glaciation and the causes of ice ages. *Am. J. Sci., 268,* 193–224.

Deacon, G. E. R. 1963. The Southern Ocean. (*In* Hill, M. N., *ed. The Sea; Ideas and Observations on Progress in the Study of the Seas. 2.* New York, Interscience, 281–96.)

Donn, W. L., and Shaw, D. M. 1977. Model of climate evolution based on continental drift and polar wandering. *Geol. Soc. Am. Bull., 88,* 390–96.

Echols, R. J. 1971. Distribution of foraminifera in sediments of the Scotia Sea area, Antarctic waters. (*In* Reid, J. L., *ed. Antarctic Oceanology. 1.* Washington, American Geophysical Union, 93–168. [Antarctic Research Series, Vol. 15.]

Echols, R. J., and Kennett, J. P. 1973. Distribution of foraminifera in the surface sediments. (*In* Goodell, H. G., Houtz, R., Ewing, M., and others, *eds.* Marine Sediments of the Southern Oceans. *Antarct. Map Folio Ser.,* Folio 17, Pl. XIII–XVII.)

Eittreim, S., Gordon, A. L., Ewing, M., Thorndike, E. M., and Bruckhausen, P. 1972. The nepheloid layer and observed bottom currents in the Indian-Pacific Antarctic Sea. (*In* Gordon, A. L., *ed. Studies in Physical Oceanography. 2.* New York, Gordon and Breach, 19–36.)

Emiliani, C. 1954. Temperature of Pacific bottom waters and polar superficial waters during the Tertiary. *Science, N.Y., 119,* 853–55.

Gill, A. E., and Bryan, K. 1971. Effects of geometry on the circulation of a three-dimensional southern-hemisphere ocean model. *Deep Sea Res., 18,* 685–721.

Goodell, H. G. 1973. The sediments. (*In* Goodell, H. G., Houtz, R., Ewing, M., and others, *eds.* Marine Sediments of the Southern Oceans. *Antarct. Map Folio Ser.,* Folio 17, Pl. I–IX.)

Goodell, H. G., and Watkins, N. D. 1968. The palaeomagnetic stratigraphy of the Southern Ocean: 20° west to 160° east longitude. *Deep Sea Res., 15,* No. 1, 89–112.

Goodell, H. G., Meylan, M. A., and Grant, B. 1971. Ferromanganese deposits of the South Pacific Ocean, Drake Passage and Scotia Sea (*In* Reid, J. L., *ed. Antarctic Oceanology. 1.* Washington, American Geophysical Union, 27–92.) [Antarctic Research Series, Vol. 15.]

Gordon, A. L. 1967. Structure of Antarctic Waters between 20° W and 170° W. *Antarct. Map Folio Ser.,* Folio 6, Pl. I–X.

Gordon, A. L. 1971. Oceanography of Antarctic waters (*In* Reid, J. L., *ed. Antarctic Oceanology. 1.* Washington, American Geophysical Union, 169–204.) [Antarctic Research Series, Vol. 15.]

Gordon, A. L. 1972*a*. Introduction: physical oceanography of the southeast Indian Ocean. (*In* Hayes, D. E., *ed. Antarctic Oceanology. 2. The Australian-New Zealand Sector.* Washington,

American Geophysical Union, 2-10.) [Antarctic Research Series, Vol. 19.]

Gordon, A. L. 1972*b*. Spreading of Antarctic bottom waters II. (*In* Gordon, A. L., *ed. Studies in Physical Oceanography. 2.* New York, Gordon and Breach, 1-18.)

Gordon, A. L. 1972*c*. On the interaction of the Antarctic Circumpolar Current and the Macquarie Ridge. (*In* Hayes, D. E., *ed. Antarctic Oceanology. 2. The Australian-New Zealand Sector.* Washington, American Geophysical Union, 71-78.) [Antarctic Research Series, Vol. 19.]

Gordon, A. L. 1975. Contributions of R/V *Conrad* to F DRAKE, 1975. *Antarct. Jnl U.S., 10,* 142-43.

Gordon, A. L., and Goldberg, R. D. 1970. Circumpolar characteristics of Antarctic waters. *Antarct. Map Folio Ser.,* Folio 13, Pl. I-V.

Hammond, A. L. 1976. Paleoceanography: sea-floor clues to earlier environments. *Science, N.Y., 191,* 168-78.

Hayes, D. E., Frakes, L. A., and others. 1975. *Initial Reports of the Deep Sea Drilling Project,* Vol. 28. Washington (U.S. Government Printing Office), 1017 pp.

Hays, J. D., Imbrie, J., and Shackleton, N. J. 1976. Variations in the earth's orbit: pacemaker of the ice ages. *Science, N.Y., 194,* 1121-32.

Hays, J. D., Lozano, J. A., Shackleton, N. J., and Irving, G. 1976. Reconstruction of the Atlantic and western Indian Ocean sectors of the 18,000 B.P. Antarctic Ocean. (*In* Cline, R. M., and Hays, J. D., *eds. Investigation of Late Quaternary Paleoceanography and Paleoclimatology. Mem. geol. Soc. Amer., 145,* 337-72.)

Heirtzler, J. R., Dickson, G. O., Herron, E. M., Pitman, W. C., III, and Le Pichon, X. 1968. Marine magnetic anomalies, geomagnetic field reversals and motions of the ocean floor and continents. *J. geophys. Res., 73,* No. 6, 2119-36.

Hollister, C. D., and Heezen, B. C. 1967. The floor of the Bellingshausen Sea. (*In* Hersey, J. B., *ed. Deep Sea Photography.* Baltimore, Johns Hopkins Press, 177-89.) [Johns Hopkins Oceanographic Studies, 3.]

Hollister, C. D., Craddock, C., and others. 1976. *Initial Reports of the Deep Sea Drilling Project,* Vol. 35. Washington (U.S. Government Printing Office), 929 pp.

Johnson, J. A., and Hill, R. B. 1975. A three-dimensional model of the Southern Ocean with bottom topography. *Deep Sea Res., 22,* 745-51.

Kemp, E. M., Frakes, L. A., and Hayes, D. E. 1975. Paleoclimatic significance of diachronous biogenic facies, Leg 28, Deep Sea Drilling Project. (*In* Hayes, D. E., Frakes, L. A., and others. *Initial Reports of the Deep Sea Drilling Project,* Vol. 28. Washington [U.S. Government Printing Office], 909-18.)

Kennett, J. P. 1966. Foraminiferal evidence of a shallow calcium carbonate solution boundary, Ross Sea, Antarctica. *Science, N.Y., 153,* 191-93.

Kennett, J. P., and Shackleton, N. J. 1976. Oxygen isotopic evidence for the development of the psychrosphere 38 Myr ago. *Nature, Lond., 260,* 513-5.

Kennett, J. P., and Watkins, N. D. 1976. Regional deep-sea dynamic processes recorded by late Cenozoic sediments of the southeastern Indian Ocean. *Geol. Soc. Am. Bull., 87,* 321-39.

Kennett, J. P., Houtz, R. E., and others. 1975. *Initial Reports of the Deep Sea Drilling Project,* Vol. 29. Washington (U.S. Government Printing Office), 1197 pp.

Legeckis, R. 1977. Oceanic polar front in the Drake Passage—Satellite observations during 1976. *Deep Sea Res., 24,* 701-4.

Le Pichon, X., Eittreim, S. L., and Ludwig, W. L. 1971. Sediment transport and distribution in the Argentine Basin, I. Antarctic bottom current passage through the Falkland fracture zone. *Phys. Chem. Earth, 8,* 1-28.

Maksimov, I. V., and Vorob'ev, V. N. 1964. Contribution to the study of deep currents in the Antarctic Ocean. *Inf. Bull. Sov. Antarct. Exped., 4,* 17-19.

Manabe, S., and Bryan, K. 1969. Climate and ocean circulation. *Mon. Weath. Rev., 97,* 740-827.

Reid, J. L., and Nowlin, W. D., Jr. 1971. Transport of water through the Drake Passage. *Deep Sea Res., 18,* 51-64.

Savin, S. M., Douglas, R. G., and Stehli, F. G. 1975. Tertiary marine paleotemperatures. *Geol. Soc. Am. Bull., 86,* 1499-1510.

Shackleton, N. J., and Kennett, J. P. 1975. Paleotemperature history of the Cenozoic and the initiation of Antarctic glaciation: oxygen and carbon isotope analyses in DSDP Sites 277, 279 and 281. (*In* Kennett, J. P., Houtz, R. E., and others. *Initial Reports of the Deep Sea Drilling Project,* Vol. 29. Washington [U.S. Government Printing Office], 743-56.)

Tolstikov, E. I. 1966. *Atlas Antarktiki 1.* Moscow, Glavnoe Upravlenie Geodezii i Kartografii, 225 pls.

Tucholke, B. E., and Houtz, R. E. 1976. Sedimentary framework of the Bellingshausen Basin from seismic profiler data. (*In* Hollister, C. D., Craddock, C., and others. *Initial Reports of the Deep Sea Drilling Project,* Vol. 35. Washington [U.S. Government Printing Office], 197-228.)

Watkins, N. D., and Kennett, J. P. 1972. Regional sedimentary disconformities and upper Cenozoic changes in bottom water velocities between Australia and Antarctica. (*In* Hayes, D. E., *ed. Antarctic Oceanology. 2. The Australian-New Zealand Sector.* Washington, American Geophysical Union, 273-94.) [Antarctic Research Series, Vol. 19.]

Zillman, J. W. 1972. Solar radiation and sea-air interaction south of Australia. (*In* Hayes, D. E., *ed. Antarctic Oceanology. 2. The Australian-New Zealand Sector.* Washington, American Geophysical Union, 11-40.) [Antarctic Research Series, Vol. 19.]

44

A Geophysical Study of the Indo-Atlantic Basin*

JOHN L. LABRECQUE
Lamont-Doherty Geological Observatory of Columbia University, Palisades, New York 10964, U.S.A.

MARCELO KELLER
Dirección Nacional del Antarctico, Serrito 1248, Buenos Aires, Argentina

Abstract. Geophysical data gathered during cruise 1277 of the *Islas Orcadas* provide information for dating the plate tectonic evolution of the Indo-Atlantic Basin. The magnetic anomaly profile observed across the southwest Indian Ocean Ridge is symmetric and correlates with the Campanian to present magnetic reversal time scale. Compensated basement depths date the southern Weddell Basin as Late Cretaceous in age or older. A discontinuous zone of basement relief is observed near lat. 64°–65° S within the Weddell Basin. The area is likely a zone of change in sea-floor spreading direction.

The continental rise and slope of the Dronning Maud Land margin is a zone of anomalously smooth and shallow basement with sediment eroded by canyon incisement and current scour. Sediment cover in the Indo-Atlantic Basin apparently becomes more terrigenous to the west, as deduced from seismic reflection profiles.

The Indo-Atlantic Basin occupies the region flanked to the north by the South Atlantic Ocean and to the east by the Indian Ocean. The sea-floor spreading history of this basin is possibly the last piece of information required to reconstruct ancient Gondwanaland. The breakup of Gondwanaland and the resultant opening of the Indo-Atlantic Basin through the relative motions of the major tectonic plates (Africa, South America, and Antarctica) most likely produced major changes in the world's deep ocean circulation pattern (van Andel and others, 1977). Norton (Chapter 10, this volume) and Norton and Sclater (1979) present a study of the development of the Indo-Atlantic Basin by examining the constraints imposed by the relative motions of the neighboring tectonic plates. Bergh (1977) and Sclater and others (1976, 1978) have recently added, respectively, to our knowledge of the Jurassic and late Cenozoic history of the plate tectonic development of the Indo-Atlantic Basin. A part of this basin, the Weddell Basin, remains largely unexplored. It is likely that the tectonic fabric (i.e., the magnetic anomaly pattern and the basement structure) of the Weddell Sea will be required to unite Gondwanaland with confidence. This paper presents information gathered aboard *Islas Orcadas* cruise 1277, which the authors hope will contribute to the development of the plate tectonic history of the Indo-Atlantic Basin.

During the 1977 field season, *Islas Orcadas* cruise 1277 conducted a reconnaissance survey of the southwest Indian Ocean Ridge, the Dronning Maud Land margin, and the Weddell Sea. Figure 44.1 illustrates the cruise track. The cruise gathered single-channel seismic reflection data (using a 40-cubic-inch air gun), as well as bathymetric (12 and 3.5 kHz), magnetic, and gravimetric data. Satellite navigation and dead reckoning were used to establish the ship's position.

Southwest Indian Ocean Ridge—Magnetic anomaly identifications

Profile AB (Figure 44.2) followed as closely as possible the Africa-Antarctica sea-floor spreading direction predicted by the magnetic anomaly lineation pattern observed in the Mozambique Basin (Bergh and Norton, 1976). The objective was to obtain a continuous sea-floor spreading-type magnetic anomaly pattern across the southwest Indian Ocean Ridge (SWIOR) in order to determine the rates of relative motion between Africa and East Antarctica. The track traversed the southwest Indian Ocean Ridge from the southwest spur of the Agulhas Plateau to the Dronning Maud Land margin. The seismic epicenters for the segment of the SWIOR traversed are not offset in a significant *en echelon* pattern (Figure 44.1), suggesting that only minor offsets across fracture zones might be expected. Profile AB is shown in Figure 44.2 with other data gathered in the region. The magnetic data in those profiles are correlated with the Late Cretaceous and Cenozoic magnetic reversal pattern (LaBrecque and others, 1977; Heirtzler and others, 1968) in Figure 44.3. The comparison of the original profile to that of its phase-shifted (Schouten and McCamy, 1972) and reversed equivalent in Figure 44.3 shows for the first time that the magnetic anomaly pattern is symmetric with respect to the SWIOR rise crest. Though one profile is not sufficient to provide detailed information

* Lamont-Doherty Geological Observatory contribution No. 2901.

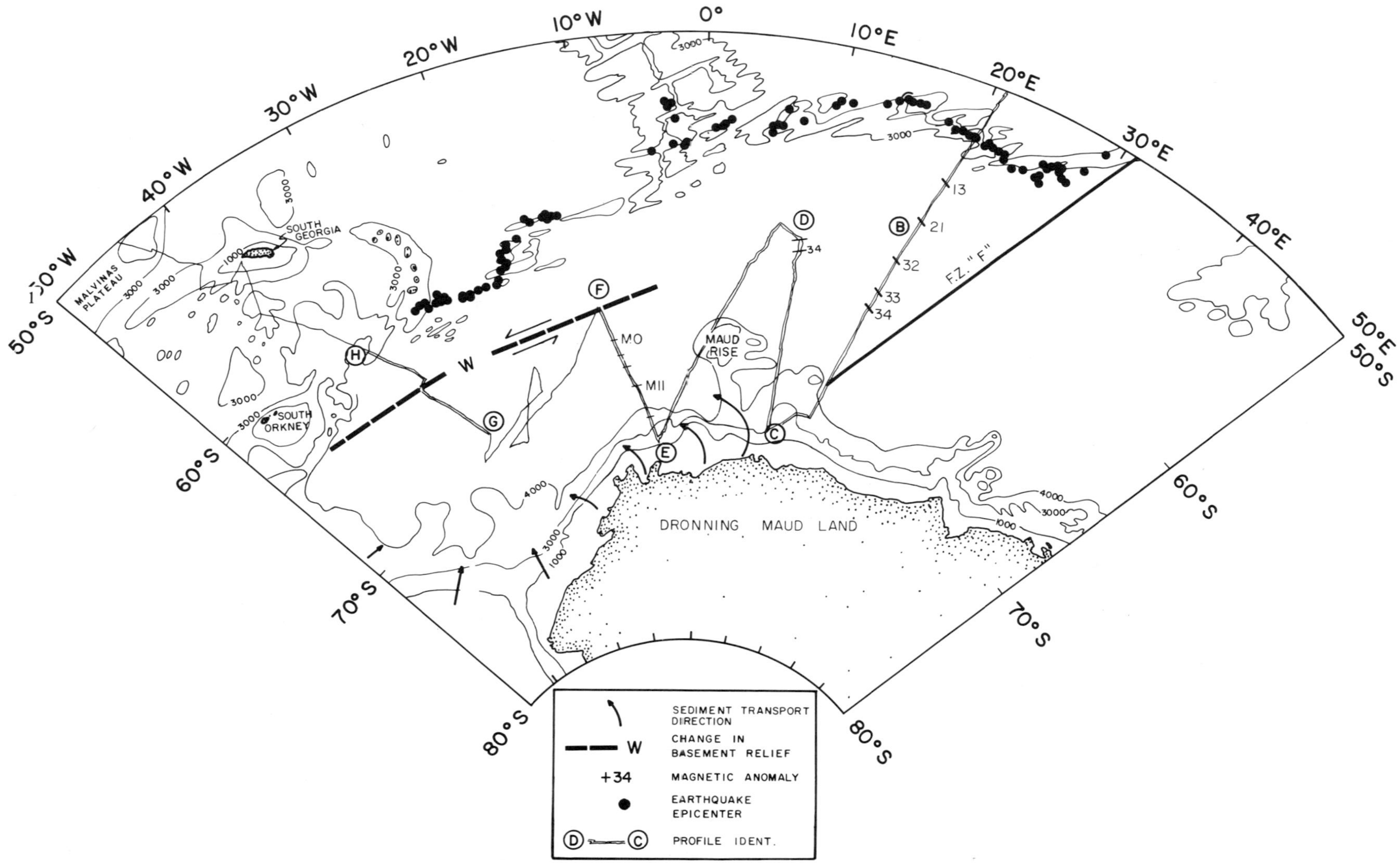

Figure 44.1. Base map showing profiles and locations of marine magnetic anomalies identified in the text. Bathymetry is from Heezen and Bentley (1972). F.Z. "F" represents fracture zone "F" as described by Bergh (1977).

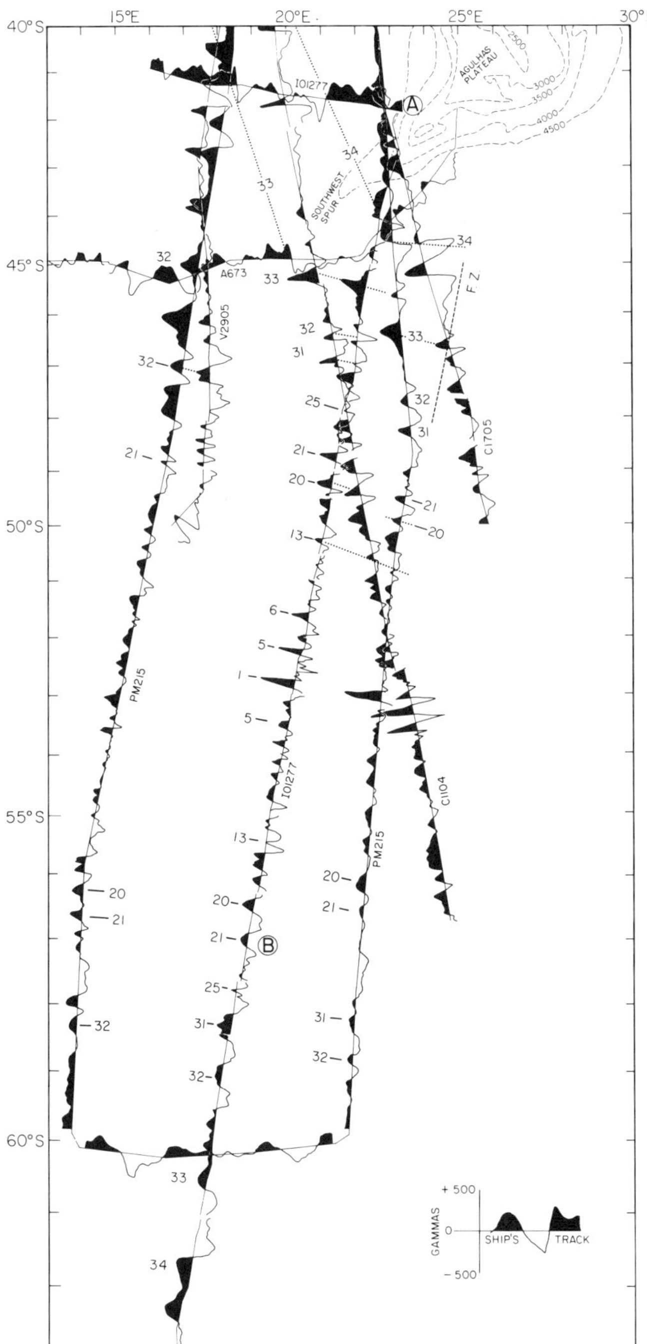

Figure 44.2. Marine magnetic anomaly data are plotted perpendicular to the ship's track at the scale noted on chart. Data are from profile AB of *Islas Orcadas* (IO 1277), *Project Magnet* (PM 215), R/V *Vema* (V 2905), *Conrad* (C 1705 and C 1104), and *Atlantis II* (A 673) cruises. Agulhas Plateau contour intervals are from Tucholke and Carpenter (1977). Possible fracture zone labelled "F.Z." is deduced from magnetic anomaly offset.

on the sea-floor spreading direction across the SWIOR, a comparison to the other available data on the African flank of the SWIOR indicates that the sea-floor spreading direction between Africa and Antarctica has been rather constant since the time of anomaly 34 (Campanian). A similar conclusion was reached by Bergh (1977) based on an analysis of early Mesozoic magnetic anomalies near the Dronning Maud Land margin. Bergh linked bathymetric features observed on both flanks of the SWIOR as conjugate points which have been separated by the relative motion of Africa and Antarctica about a pole of rotation very near to the pole of rotation which describes present-day Africa-Antarctica motion. Bergh hypothesized that those features were the manifestation of a fracture zone "F," which is plotted in Figure 44.1.

Anomaly 34 on the African plate (Figure 44.3) is approximately 288 km south of and subparallel to the southern boundary of the Agulhas Plateau. Anomaly 34 is also observed in profiles BC and CD (Figure 44.4) on the Antarctic plate where it is located approximately 300 km north of the zone of smooth, shoaling basement which parallels the Antarctic continental margin. The smooth acoustic basement observed along the Antarctic margin is comparable to that observed in the region of the Agulhas Plateau (Barrett, 1977; Tucholke and Carpenter, 1977). The type of material which is responsible for this smooth acoustic reflector has yet to be determined. Possible explanations include extensive basaltic flows and sills, high-acoustic velocity sedimentary beds such as limestone and chert, and eroded continental crust. Similar smooth basement is also observed in the Georgia Basin and the northern Agulhas Basin, also of Mesozoic age (Barker and Dalziel, 1976; LaBrecque and Hayes, 1979). Reconstructions of Africa, South America, and Antarctica predict that those regions of observed smooth basement were in close proximity during Mesozoic time (Barron and others, 1978; Norton and Sclater, 1979). These reconstructions imply that the formation of the smooth acoustic basement occurred in a zone common to all the abovementioned regions and should reflect their common environment during Mesozoic time.

North Dronning Maud Land margin

Profiles BC, CD, DE and EF (Figures 44.4, 44.5, and 44.6) traverse the north Dronning Maud Land margin (Princess Astrid Coast). Dense pack-ice prevented studies of the margin landward of the 2000 m isobath on all traverses. The profiles reveal that the continental-rise structure of the north Dronning Maud margin has an unusually shallow and smooth acoustic basement (Figures 44.4 and 44.6, profiles BC–DE) with some large, high-relief structures apparently of fault origin. Large (1000 gamma) anomalies are observed over the smooth acoustic basement of the continental rise. A sonobuoy record (R4, Figure 44.7) in profile CD taken over this smooth reflector did not yield reliable refraction arrivals. However, a subsequent cruise of the R/V *Explorer* in the approximate location of R4 did measure basement velocities of 4.7 to 5.7 km/sec (Hinz, 1978). Reflection profiles from the *Islas Orcadas* 1277 data reveal subbottom reflectors which could be interpreted as basement relief underlying a highly reflec-

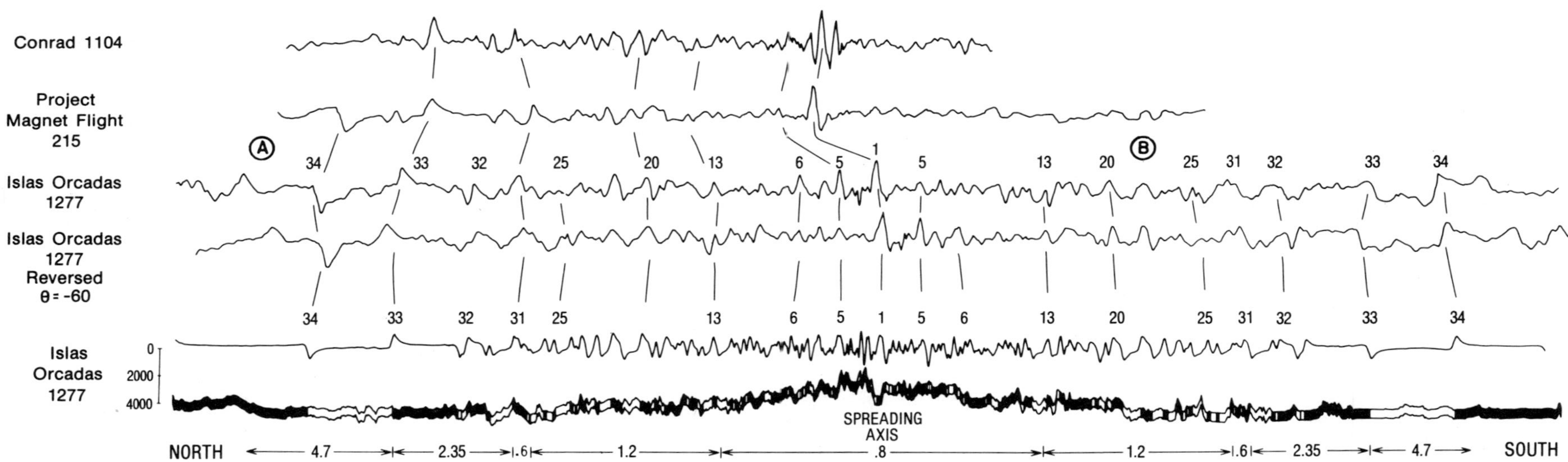

Figure 44.3. Correlations of southwest Indian Ocean Ridge magnetic anomalies with the time scale of LaBrecque and others (1977). Remanent magnetization parameters: Inclination: 64° S. Declination: 0.0. Magnetic intensity: 0.007 emu/cc. Layer thickness: 500 m. Profile azimith: 200°. Half spreading rates: 0.0–34.0 m.y. ago: 0.8 cm/yr; 34.0–62.3 m.y. ago: 1.2 cm/yr; 62.3–67.6 m.y. ago: 0.6 cm/yr; 67.6–76.5 m.y. ago: 2.35 cm/yr; and 76.9–80.0 m.y. ago: 4.7 cm/yr.

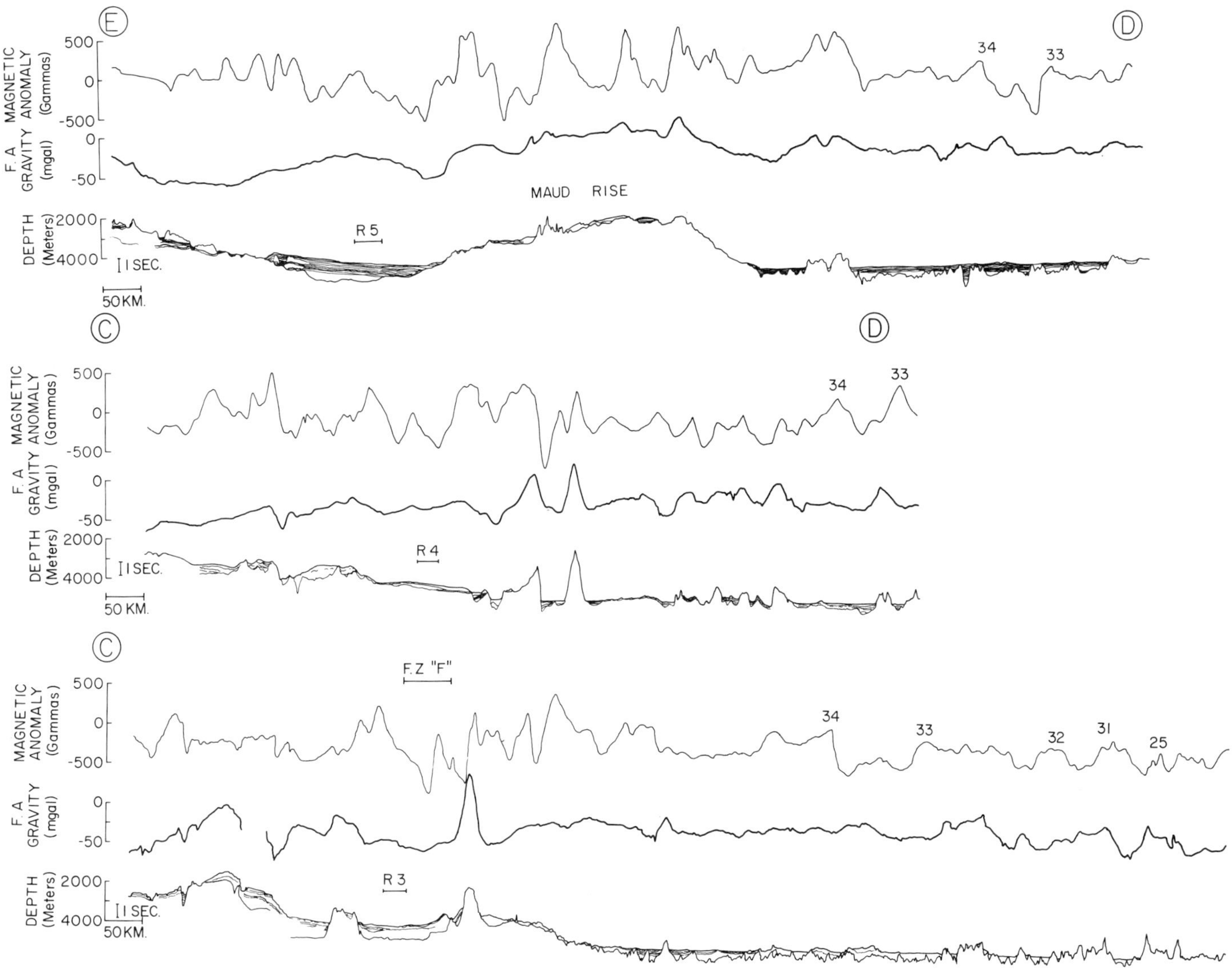

Figure 44.4. Magnetic anomaly, free-air gravity anomaly, and seismic reflection line drawings of profiles BC, CD, and DE projected along an azimuth of N 20° E. Note that subbottom reflections are scaled in seconds of two-way travel time. Sonobuoy profile solutions are given in Figure 44.7.

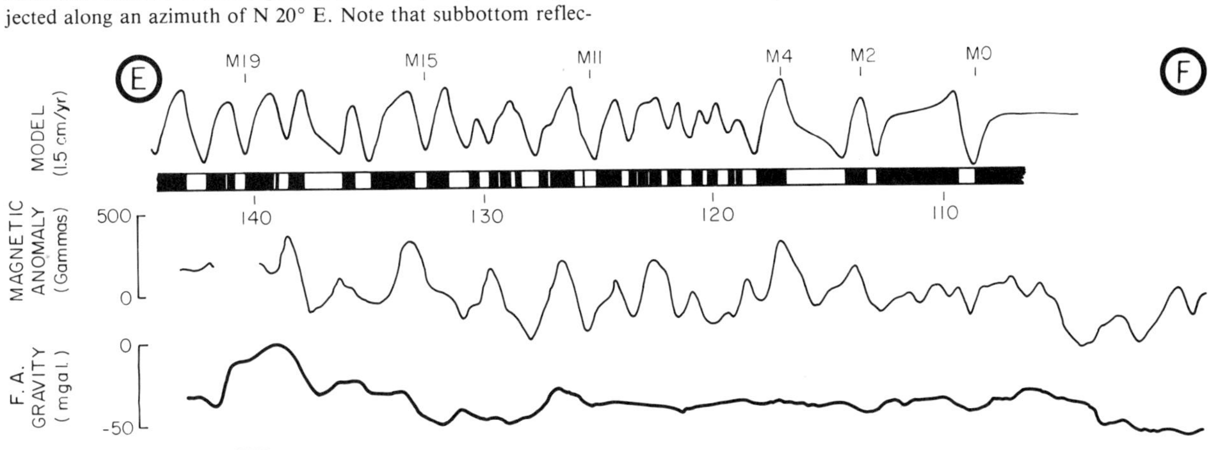

Figure 44.5. Magnetic anomaly, free-air gravity anomaly, and seismic reflection line drawings of profile EF projected along an azimuth of 0° N. The magnetic anomaly sequence has been tentatively correlated to the time scale of Larson and Hilde (1975). Note that the lineation pattern of the sequence has not been established. Remanent magnetization parameters: Inclination: 70° S. Declination: 0.0. Magnetic intensity: 0.01 emu/cc. Azimuth of profile: 0.0. Spreading rate: 1.5 cm/yr. Layer thickness: 500 m.

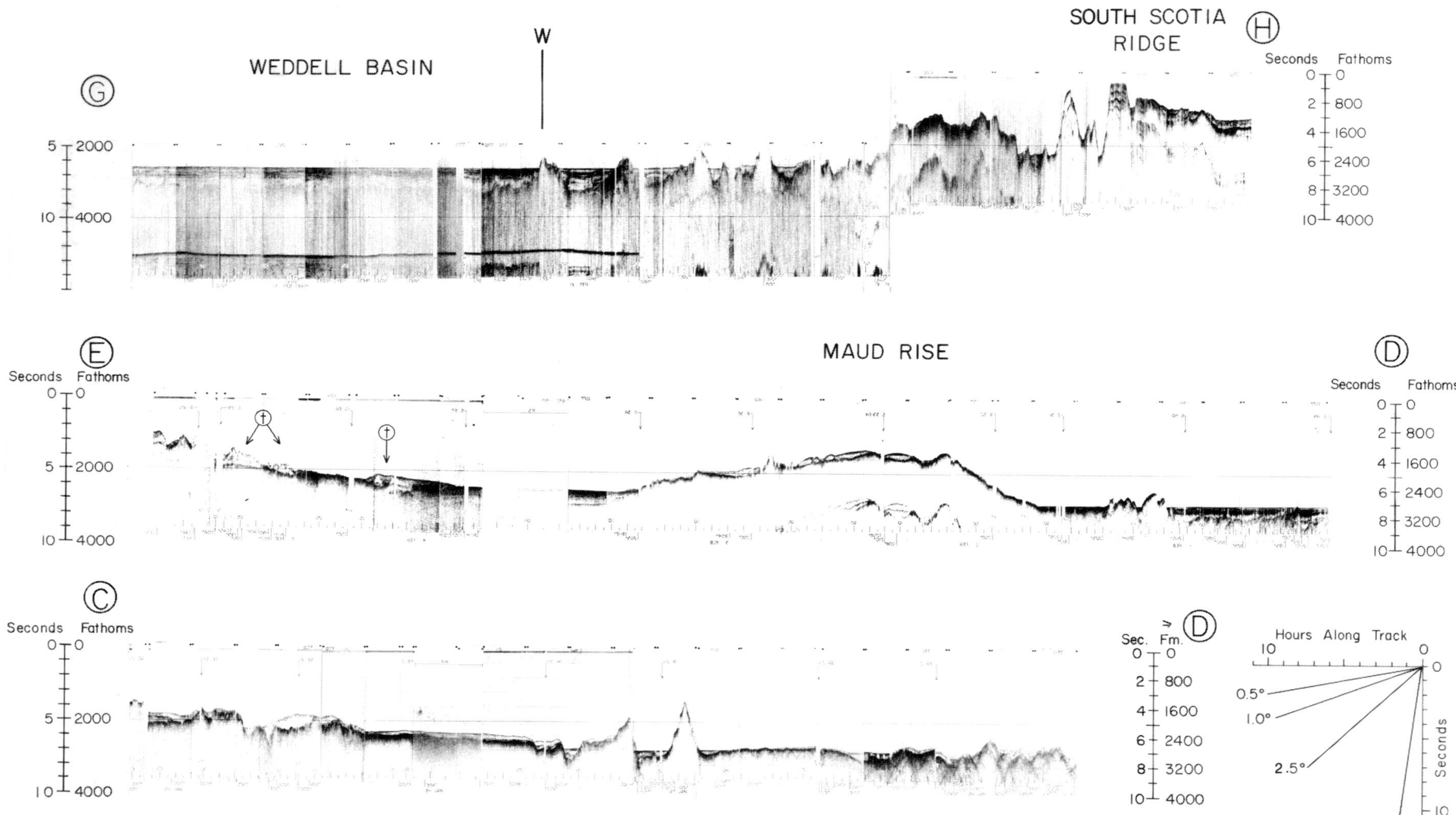

Figure 44.6. Seismic reflection profiles corresponding to the line drawings in profiles BC, DE, and GH.

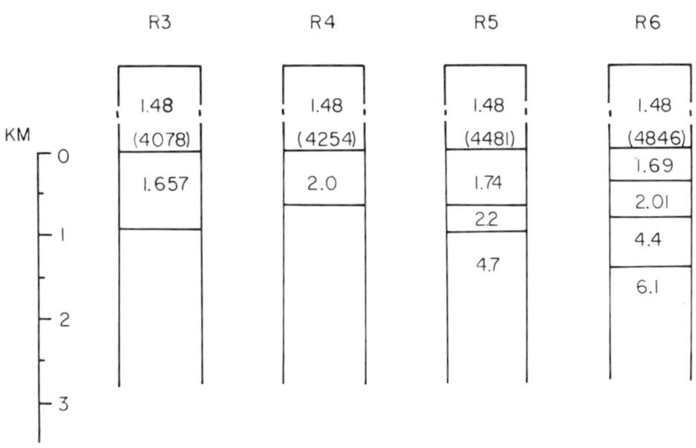

Figure 44.7. Solutions for sonobuoy profiles shown in profiles BC, CD, DE, and EF. Depth of the sediment-water interface (in meters) is indicated in parentheses.

tive, smooth layer. Therefore, it may be that this smooth layer is not oceanic crust but a layer of high-impedance contrast—such as chert, limestone, or basaltic sills—overlying true basement.

The easternmost profile (BC) crosses what appears to be a significant basement structure (marked "F" in profile BC, Figure 44.4). The smooth acoustic basement shoals southward until a very abrupt change in basement depth is observed at "F." The basement to the south of this abrupt change is flat with nearly 1.0 km (see R3, Figure 44.7) of relatively transparent sediment overlying acoustic basement at 5 km depth. The observed zone of abrupt change in basement depth lies along the southward projection of the fracture zone labelled "F" by Bergh (1977). As noted by Bergh this fracture zone can be aligned slightly east of the dramatic 1100 km offset in earthquake epicenters along the SWIOR.

The profile over the Maud Rise (DE, Figures 44.4 and 44.6) reveals a broad, gently sloping bathymetric high with little gravimetric expression. Large magnetic anomalies (1000 gammas) are observed over the Maud Rise, indicating the presence of mafic crust or intrusions. The 300-km-wide basin to the south of the Maud Rise is depicted on most bathymetric maps as a trough-like depression which opens to the Weddell Basin and which parallels the Dronning Maud margin. Sediment reflectors within the basin may be characterized as dense, high-frequency reflectors with acoustically transparent zones of irregular surface morphology. These zones are labelled "t" in seismic reflection profile DE (Figure 44.6).

North Dronning Maud Land margin sediment distribution

The profiles BC to EF indicate a thin (<0.5 sec), transparent sediment cover over the Antarctic plate north of the Dronning Maud margin. Seismic reflection profiles show that the intensity and abundance of reflectors increase toward the west from profiles BC to EF, presumably indicating an increase in the terrigenous component of the sedimentary column.

Westward-flowing surface currents on the order of 10–30 cm/sec have been measured along the margin (Gill, 1973; Tchernia, 1974). These currents are presumably the westward flow of the Weddell gyre. The time of initiation of the Weddell gyre flow is unresolved though its onset is likely to have been contemporaneous with the opening of the Drake Passage in late Oligocene to middle Miocene time (Kennett and others, 1972; Craddock and Hollister, 1976; Dalziel and Elliot, 1973; Barker and Burrell, 1977; LaBrecque and Rabinowitz, 1977). Cormack and Foster (1975) have shown that this present-day westward geostrophic flow should decrease quite abruptly to zero at depths of 2000 m in the region of Cape Norvegia. This rapid westward flow could be responsible for significant scouring of the upper continental margin, as well as for the westward deflection of sediment flow from the continental margin, thereby producing a sediment-starved and eroded margin. Profile BC indicates zones of apparent current-controlled deposition along the bathymetric features near fracture zone "F." The thin veneer of sediment observed on profile CD and in the adjacent basin was most likely caused by the combined effects of current scour, westward deflection of sediment flow, and the relatively shallow basement structure of the continental rise. Profiles DE and EF reveal signs of significant margin denudation. The transparent zones (marked "t" in profile DE, Figure 44.6) are apparently the result of erosional unconformities with younger acoustically opaque sediments overlying eroded sedimentary horizons. These erosional unconformities seem to be due to the incision of submarine canyons. This interpretation is supported by reflectors within the transparent zones which can be traced between these zones. Basement structure apparently protrudes through the margin sediments in at least one zone of profile EF. As with the previous profiles, this sediment cover is unusually thin for a margin which is apparently Late Jurassic or Early Cretaceous in age, as discussed below. The truncation of some reflectors near the top of the continental slope suggests that a more extensive sediment cover may have been eroded away. Hinz (1978) observed similar evidence of erosion during the subsequent cruise of the R/V *Explora.*

Weddell Sea Basin

The southwestern part of the Weddell Sea remains ice covered year round. Study of the basin was therefore limited to the central and northeastern parts of the basin. The seismic reflection records from the Weddell Basin reveal a consistently thick accumulation of sediment. Profile EF and GH (Figures 44.5, 44.6, and 44.8) are representative of the data gathered throughout the Weddell Basin survey. Sediment thicknesses (two-way travel times) of 1.2 and 2.0 seconds were observed, with the thicker deposits in the southwest. Reflection character can be described as dense, high-frequency, and continuous. Sediments with similar reflection characteristics have been observed in the Bering Sea (Ludwig and others, 1971) and were drilled by DSDP hole 190 (Creager and others, 1973). The site yielded silty diatomaceous layers with evidence of graded bedding, indicating the possibility of turbidite transport. Indeed, most cores taken within the Weddell Basin returned sand and silt, commonly with poor penetration of the core pipe.

Insufficient data were gathered during the cruise to provide magnetic anomaly lineations within the Weddell Basin useful for tracing the evolution of the basin. Figure 44.9 attempts to provide an estimate of the age of the basin

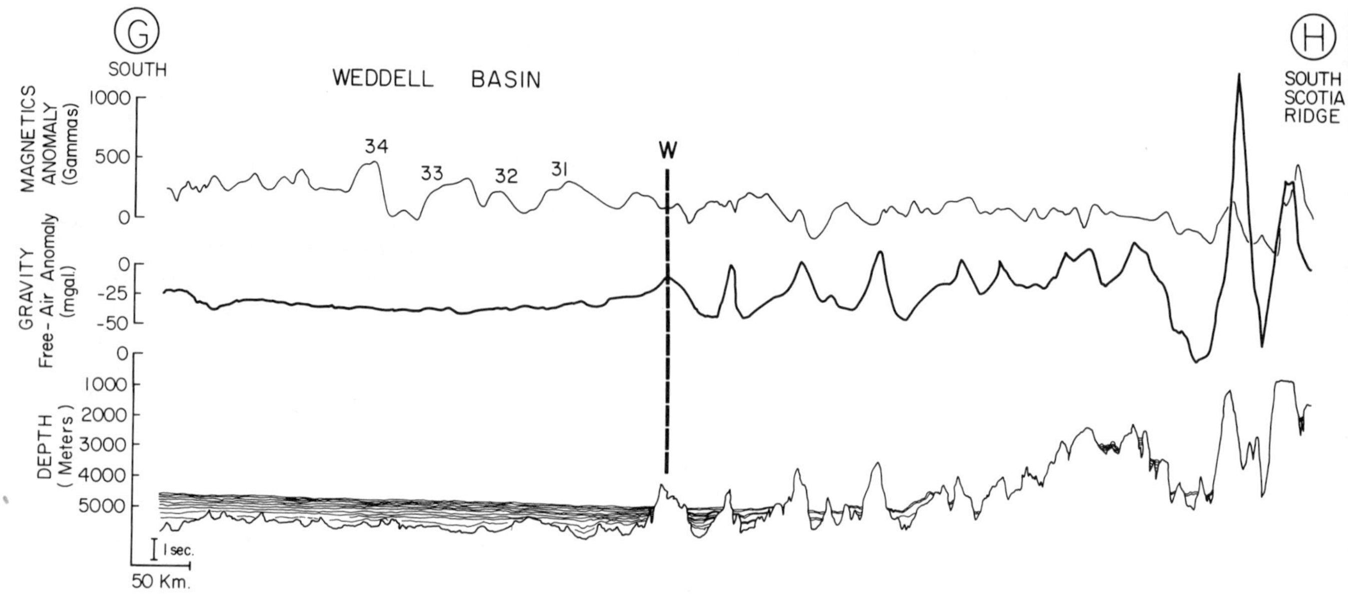

Figure 44.8. Magnetic anomaly, free-air gravity anomaly, and seismic reflection line drawings of profile GH projected along an azimuth of N 20° E. Note that subbottom reflections are scaled in seconds of two-way travel time. Sonobuoy profiles are given in Figure 44.7.

through a comparison of observed basement depth to the Parsons and Sclater (1977) age versus depth curve. Depth and sediment thickness measurements were compiled at two-hour intervals along the ship track in the region west of profile EF and south of zone "W" in Figure 44.1. A total of 87 basement depth estimates were used. An average velocity of 1.9 km/sec and a density of 1.7 gm/cc were assigned to the sediment. The resulting average compensated depth suggests a middle Cretaceous age for the Weddell Basin (Figure 44.9), based on a mean depth of 5725 ± 186 m.

Profile EF recorded a series of large-amplitude magnetic anomalies (Figure 44.5) which have the appearance of sea-floor spreading type magnetic anomalies. The basement beneath these anomalies is deep and apparently undisturbed by faults. It was found that the Mesozoic reversal pattern of Larson and Hilde (1975) could be fitted to the data with a strong correlation by assigning a spreading rate of 1.5 cm/yr. It should be emphasized that the important criterion of anomaly lineation has not been satisfied in this single profile, and therefore we must treat the correlation

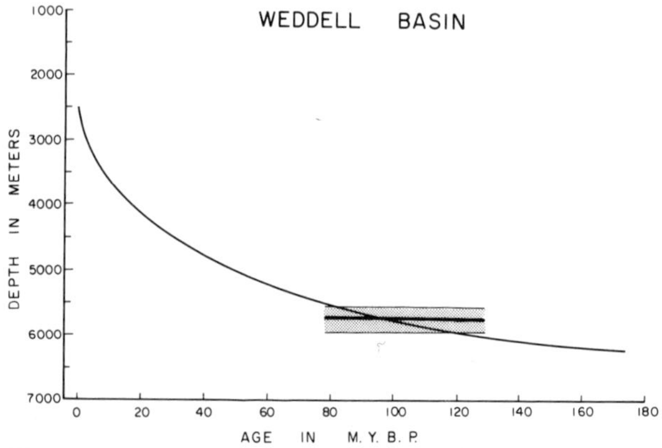

Figure 44.9. Comparison of observed basement depth in Weddell Basin with Parsons and Sclater's (1977) age versus depth curve. Sediment velocity = 1.9 km and sediment density = 1.7 gm/cc.

with skepticism. The data in profile EF have been projected along a north-striking azimuth. The spreading rate and the Late Jurassic age assigned to the sea floor near the margin in profile EF is approximately consistent with that observed for the Dronning Maud margin to the east of fracture zone "F" (Bergh, 1977), although the projected age of the margin is younger and is possibly contemporaneous with the opening of the South Atlantic Ocean (Larson and Ladd, 1973).

Profile GH (Figure 44.8) reveals an abrupt change in the average relief of the Weddell basement in the region marked "W" near lat. 64° S. The sediment reflection character also changes quite abruptly across this boundary as the dense, high-frequency reflections of the Weddell Basin yield to the more acoustically transparent sediment of the Scotia region. A similar variation in sediment reflection character and basement relief is observed near point "F" west of profile EF (Figure 44.5). This transparency is probably due to a more homogeneous sediment cover. This homogeneity may be a result of a greater homogeneity in the composition of distal turbidites or a greater contribution of pelagic sediment.

Zone "W" is interpreted as a zone of major change in the tectonic fabric of the Weddell Basin. A study of tectonic plate motions (Norton, Chapter 10, this volume; Norton and Sclater, 1979) estimated that the relative direction of motion between the South American and Antarctic plates should have changed during Eocene time from a more northerly direction (present coordinates) in the pre-Eocene to an east-west direction in post-Eocene time. Barker and others (Chapter 22, this volume) present evidence for a southern extension of the South Sandwich Trench during middle Cenozoic time. This zone of subduction may have extended as far south as the southern margin of the South Orkney Islands, largely because the South Orkney platform appears to have been displaced by a counterclockwise rotation from the Antarctic Peninsula. No dating or other information is currently available to support this hypothesis. It is thought that back-arc extension might have been the mechanism which produced this rotation.

The rather abrupt change in average relief noted at zone "W" is attributed to a change in the azimuth of fracture zones with respect to the profile direction. Figure 44.1 shows the extension of zone "W" from point "F" to the South Orkney platform. This is done simply to indicate the possible correlation between the zones of abrupt change in basement relief noted in profiles EF and GH (Figures 44.5 and 44.8) and the displaced southern margin of the Orkney Platform. These zones of tectonic disturbance are separated by several hundred kilometers, and the lineation should not be accepted as proof that the zone is well defined or continuous. This lineation likely indicates a zone of change in the regional sea-floor spreading direction rather than a continuous fracture zone. Data gathered during a subsequent cruise of the *Islas Orcadas* (cruise 1578) in the region support the location of this zone of change in basement relief.

Conclusions

A symmetric sequence of sea-floor spreading type marine magnetic anomalies has been observed across the southwest Indian Ocean Ridge. The sequence can be traced from a zone directly south of the Agulhas Plateau to the continental rise of the north Dronning Maud Land margin. The observations of Bergh (1977) and those presented in this paper strongly suggest that the north Dronning Maud Land margin and the Agulhas Plateau–Mozambique Ridge region should be in close proximity in Gondwanaland reconstructions. LaBrecque and Hayes (1979) demonstrate that anomaly 34 forms a magnetic bight around the Agulhas Plateau. It is therefore likely, from the magnetic and structural observations presented in this paper, that the Maud Rise and Agulhas Plateau regions were joined near the site of a plate triple junction in middle Cretaceous time. Cenomanian sediments collected on the Agulhas Plateau (Tucholke and Carpenter, 1977) indicate that the Maud Rise and Agulhas Plateau may have formed in Early Cretaceous time.

The north Dronning Maud Land margin is an eroded and sediment-starved margin. The strong westward current flow in the region, as well as the shallow outlying structure of the Maud Rise, may serve to deflect sediment into the Weddell Basin. Seismic reflection data within the Weddell Basin suggest that the basin is of Mesozoic age and is filled with terrigenous sediment possibly transported by turbidite-type flows. It is likely that much of the sediment eroded from the Dronning Maud Land margin now lies within the Weddell Basin.

Acknowledgments. The authors are indebted for useful discussions with Hugh Bergh and Steven Cande, as well as to the reviewers who suggested significant improvements in this paper. The Argentine Hydrographic Office with the cooperation of Commander Alfredo Yung and the captain and crew of *Islas Orcadas* cruise 1277 were largely responsible for the acquisition of the data. The Division of Polar Programs of the National Science Foundation provided support for this work under grants DPP 74-12862 and DPP 77-15586.

References cited

Barker, P. F., and Burrell, J. 1977. The opening of Drake Passage. *Mar. Geol., 25,* 15–34.

Barker, P. F., Dalziel, I. W. D., and others. 1976. Site 32B. (*In* Barker, P. F., Dalziel, I. W. D., and others. *Initial Reports of the Deep Sea Drilling Project,* Vol. 36. Washington [U.S. Government Printing Office], 1080 pp.

Barrett, D. M. 1977. Agulhas Plateau off Southern Africa: a geophysical study. *Geol. Soc. Am. Bull., 88,* 749–63.

Barron, E. J., Harrison, C. G. A., and Hay, W. W. 1978. A revised reconstruction of the Southern Continents. *Trans. Am. geophys. Un., 59,* 436–49.

Bergh, H. W. 1977. Mesozoic seafloor off Dronning Maud Land, Antarctica. *Nature, Lond., 269,* 686.

Bergh, H. W., and Norton, I. O. 1976. Prince Edward Fracture Zone and the evolution of the Mozambique Basin. *J. geophys. Res., 81,* 5221.

Cormack, E. C., and Foster, T. D. 1975. On the flow of water out of the Weddell Sea. *Deep Sea Res., 22,* 711–24.

Craddock, C., and Hollister, C. D. 1976. Geologic evolution of the southeast Pacific Basin. (*In* Hollister, C. D., Craddock, C., and others. *Initial Reports of the Deep Sea Drilling Project,* Vol. 35. Washington [U.S. Government Printing Office], 711–24.)

Creager, J. S., Scholl, D. W., and others. 1973. Site 190 (*In* Creager, J. S., Scholl, D. W., and others. *Initial Reports of the Deep Sea Drilling Project,* Vol. 19. Washington [U.S. Government Printing Office], 371–412.)

Dalziel, I. W. D., and Elliot, D. H. 1973. The Scotia arc and Antarctic Margin. (*In* Nairn, A.M., and Stehli, G., *eds. The Ocean Basins and Margins. 2. The South Atlantic.* New York, Plenum Press, 171–246.)

Gill, A. E. 1973. Circulation and bottom water production in the Weddell Sea. *Deep Sea Res., 20,* 111–40.

Heirtzler, J. R., Dickson, G. O., Herron, E. M., Pitman, W. C., III, and LePichon, X. 1968. Marine magnetic anomalies, geomagnetic field reversals and motions of the ocean floor and continents. *J. geophys. Res., 73,* 2119–36.

Heezen, B. C., and Bentley, C. R. 1972. Submarine and subglacial topography. (*In* Heezen, B. C., Tharp, M., and Bentley, C. R., *eds.* Morphology of the Earth in the Antarctic and Subantarctic. *Antarct. Map Folio Ser.,* Folio 16, Pl. I.)

Hinz, K. 1978. Bericht über geophysikalische Untersuchungen im Weddell Meer und am ostantarktischen Kontinentalrand mit M/S Explora. Hannover, Bundessanstalt für Geowissenschaften und Rohstoffe (BGR). Archiv no: 79617.

Kennett, J. P., Burns, R. E., Andrews, J. E., Churkin, M. Jr., Davies, T. A., Dumitrica, P., and others. 1972. Australian-Antarctic continental drift, paleocirculation changes and Oligocene deep sea erosion. *Nature Phys. Sci., 239,* 51–55.

LaBrecque, J. L., and Hayes, D. E. 1979. Sea-floor spreading in the Agulhas Basin. *Earth planet. Sci. Lett., 45,* 411–28.

LaBrecque, J. L., and Rabinowitz, P. D. 1977. Magnetic anomalies bordering the continental margin of Argentina. *Am. Ass. Petrol. Geol. Map Series, Cat. No. 826.*

LaBrecque, J. L., Kent, D. V., and Cande, S. C. 1977. Revised magnetic polarity time scale for Late Cretaceous and Cenozoic time. *Geology, 5,* 330–35.

Larson, R. L., and Hilde, T. W. C. 1975. A revised time scale of magnetic reversal for the Early Cretaceous and Late Jurassic. *J. geophys. Res., 80,* 2586–94.

Larson, R. L., and Ladd, J. W. 1973. Evidence for the opening of the South Atlantic in the Early Cretaceous. *Nature, Lond., 246,* 209–12.

Ludwig, W. J., Houtz, R. E., and Ewing, M. 1971. Sediment distribution in the Bering Sea: Bowers Ridge, Shirshov Ridge, and enclosed basins. *J. geophys. Res., 76,* 6367.

Norton, I. O., and Sclater, J. G. 1979. A model for the evolution of the Indian Ocean and the breakup of Gondwanaland. *J. geophys. Res., 84,* 6803–30.

Parsons, B., and Sclater, J. 1977. An analysis of the variation of ocean floor bathymetry and heat flow with age. *J. geophys. Res., 82,* 803–37.

Schouten, H., and McCamy, K. 1972. Filtering marine magnetic anomalies. *J. geophys. Res., 77,* 7089–99.

Sclater, J. C., Bowin, C., Hey, R., Haskins, H., Pierce, J., Phillips, J., and Tapscott, C. 1976. The Bouvet triple junction. *J. geophys. Res., 81,* 1857.

Sclater, J. G., Dick, H., Norton, I. O., and Woodroffe, D. 1978. Tectonic structure and petrology of the Antarctic plate boundary near the Bouvet Triple Junction. *Earth planet. Sci. Lett., 37,* 393–400.

Tchernia, M. P. 1974. Etude de la derive Antarctique est-ouest au moyen d'icebergs suivis par le satellite EOLE. *C. r. Acad. Sci. Paris, 278, B,* 667–70.

Tucholke, B. E., and Carpenter, G. B. 1977. Sediment distribution and Cenozoic sedimentation patterns on the Agulhas Plateau. *Geol. Soc. Am. Bull., 88,* 1337–46.

Van Andel, T. H., Thiede, J., Sclater, J. G., and Hay, W. 1977. Depositional history of the South Atlantic Ocean during the last 125 million years. *J. Geol., 85,* 651–97.

45

Marine Geophysical Research in the Weddell Sea during the Norwegian Antarctic Research Expedition of 1976–77*

B. A. FOSSUM, G. H. MAISEY, and H. O. TORSEN
Continental Shelf Institute, P.O. Box 1883, 7001 Trondheim, Norway

Abstract. This paper reports some results of the marine geophysical research carried out by the Continental Shelf Institute during the Norwegian Antarctic Research Expedition to the Weddell Sea in January and February of 1977. The aim of this program was to gain a better knowledge of (1) the geology beneath the sea bed to a depth of a few hundred meters, (2) the detailed bottom topography, and (3) the magnetic anomalies in the area. The methods chosen to accomplish this program were based on the instrumentation normally used on the Norwegian continental shelf; these included an acoustic sparker system, a sidescan sonar, several bottom samplers, underwater photographic equipment, and a marine magnetometer. To improve and simplify the data collection, the Continental Shelf Institute supplied its high-precision satellite navigation system. All equipment was preinstalled inside rugged containers and showed reliable performance under the rough environmental conditions. Shallow seismic profiling for 1200 km was obtained, as well as 250 km of sidescan sonar recording down to 400 m depths, 28 geologic sampling stations, 7 photo stations, and 4000 km of magnetometer tracks. Examples of typical seismic profiles of glacial deposits and sedimentary and igneous rocks, as well as sidescan sonar profiles showing iceberg ploughing and the underwater face of the barrier, are included.

The Norwegian Antarctic Research Expedition (NARE) of 1976–77, working in the Weddell Sea, consisted of a multidiscipline research group of 22 scientists, including one Argentine visiting scientist, plus a two-man film team. During January and February of 1977 the scientists carried out research both on land and at sea. The ship track near Antarctica is shown in Figure 45.1. The working transport platform was the relatively small icebreaker-sealer M/S *Polarsirkel,* capable of penetrating 3–5-m-thick ice; with a crew of only 12 men, it offered a favorable scientist/crew ratio.

The policy of NARE 1976–77 was to carry out research in accordance with SCAR objectives and recommendations relating to basic knowledge about the Antarctic environment. Instead of concentrating the activity on a few special topics, a broad scientific and instrumental program was undertaken. For the sea party, the research activities included some methods and instrumentation new to Weddell Sea research. The expedition was organized to prevent standby situations during adverse weather and ice conditions or instrument breakdowns. A flexible schedule was defined for the various programs to allow the selection of activities according to conditions.

The NARE was organized under the leadership of the Norsk Polarinstitutt (NPI), but seven individual work groups were organized and equipped by different research institutions in Norway. For a complete description of the expedition, see Orheim (1977).

This paper is a brief summary of the activities of the four-man research group from the Continental Shelf Institute (CSI). The goal of this group was to increase knowledge about (1) the geology beneath the sea bed to a depth of a few hundred meters, (2) the detailed bottom topography, and (3) the magnetic anomalies in the area.

The CSI geophysical and geologic instruments used were the same as those used routinely on the Norwegian continental shelf during studies of glacial geologic phenomena. These instruments included a sparker for shallow seismic profiles, a sidescan sonar for bottom topography, a variety of sampling equipment, underwater photographic equipment, a proton magnetometer, and a high-precision navigation positioning system.

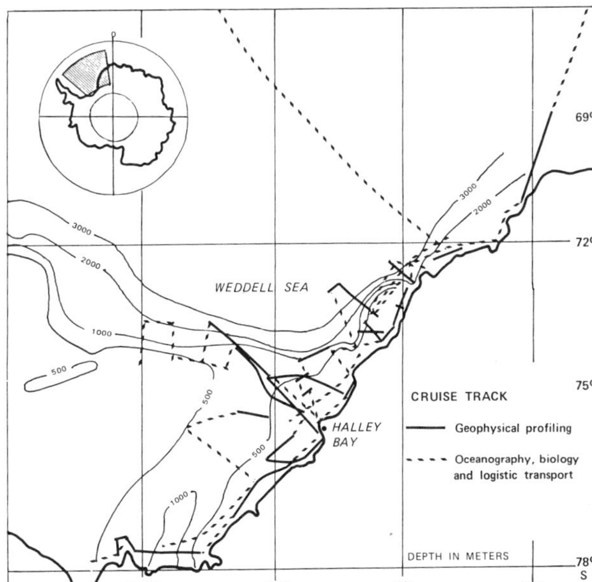

Figure 45.1. Cruise track of the NARE 1976–77.

* Publication no. 7 of the Norwegian Antarctic Research Expedition (1976–77).

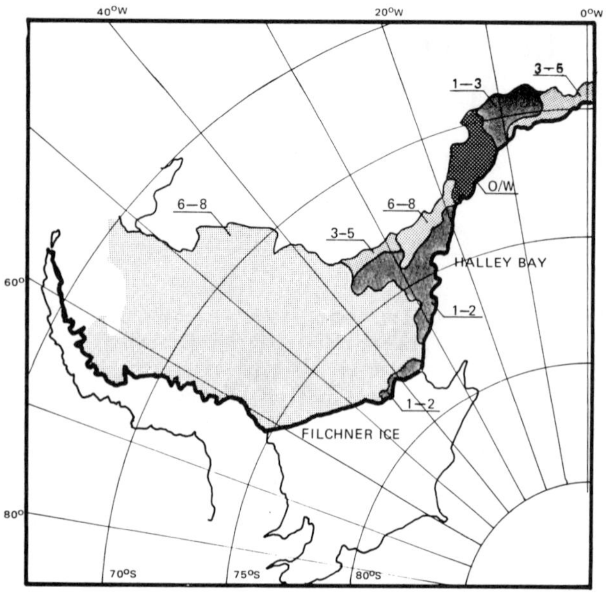

Figure 45.2. Ice conditions in the Weddell Sea, 13 January 1977. Numbers indicate relative amount of the area covered with ice, in eighths.

One important observation on this cruise was that this equipment, which is generally standard on marine surveys, produced good results in an environment as hostile as the Weddell Sea, whenever ice conditions permitted operation. The delicate electronic instrumentation was preinstalled inside rugged 12′ × 8′ × 8′ containers. This technique, used by the CSI for several years, insures optimal working conditions for the instruments and gives maximum reliability.

Characteristics of the operation area for the sea party

The operation area was the continental shelf in the southeastern part of the Weddell Sea, ranging from about lat. 72° S, long. 15° W to lat. 77.5° S, long. 42° W (Gould Bay). The great circle distance is 2000 km. The continental shelf in this area extends from 50 to 400 km in width.

The Weddell Sea has some of the most hostile and dangerous ice conditions in the Antarctic, and to perform marine operations in these areas can be difficult and hazardous, especially in the western and far southern parts. Towing sensors, such as the 1.5-km-long seismic cable, was an additional complication.

The NPI has derived ice-condition statistics based on satellite photos (NIMBUS-NOAA data) from 1967–75 (T. Vinje, personal communication). Ice conditions (Figure 45.2) in the three critical areas where research activity was planned were examined, and a plot of the percentage availability of open water for these areas for the seven years is shown in Figure 45.3. Ice conditions, however, are unpredictable because large annual variations occur. The only quantitative description possible is the statistical behavior of these variations. This information, including data on the worst case situation. gave a good planning background and indicated that the execution of a marine geophysical program using towed sensors was feasible.

Throughout the expedition, ice-condition information was received from U.S. Fleet Weather Facility, Suitland, Maryland, by radio via Palmer Station. The program could thus be adjusted as necessary, and a favorable course maintained at all times.

Since ice conditions were difficult to predict, normal planning of an optimal seismic profile grid for coverage and interpretation was prevented. It was necessary instead to continuously modify the profile grid, based on the current ice conditions in the operations area. The ice conditions demonstrated a typical regional feature: open-water areas expand from east to west along a channel close to the ice shelf during the austral summer season. Reaching as far as long. 40°–50° W, this open-water area nearrows down to a width of only 0.5 to 5 nautical miles in its southwestern parts off the Filchner Ice Shelf and finally terminates (see Figure 45.2 for locations).

The marine geophysical-geologic program—methods and equipment

One of the main programs of the sea party was the marine geophysical-geologic program. Two research groups totaling six men were occupied with nine activities, of which one or more were in operation at all times during the cruise. These activities included the following:

1. deep seismic reflection-refraction measurements;
2. regional gravimetric measurements;
3. shallow seismic reflection;
4. bottom-topography mapping with sidescan sonar;
5. bottom studies with underwater photography;
6. marine magnetic measurements;
7. high-precision navigation and positioning; and
8. bottom sampling.

The Seismological Observatory at the University of Bergen was responsible for the first two activities, while the CSI carried out the remaining activities—the bathymetric program in cooperation with the NPI.

CSI program

Aims

The goals of the CSI program were to define the composition and extent of glacial sedimentary units, to obtain information relevant to the regional glacial history of the Weddell Sea area, and to describe the sea-bed topography as seen on sidescan sonar, echosounder, sparker, and underwater photos.

Methods

The sparker was used for light seismic profiling. The energy source was an electrical discharge over 3 spark elec-

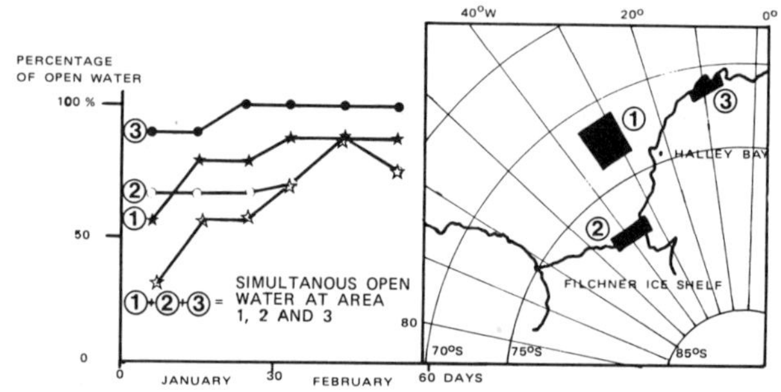

Figure 45.3. Planned research areas and percentage availability of open water for these areas, 1967–75.

trodes, giving a medium-energy acoustical pulse (4500 joules) of 10–20 msec duration. Sea-bed penetration was obtainable down to 400–500 msec. The sparker system was used in three different modes:

1. Normal shallow-water profiling, in water as deep as 750–1000 m, one shot in transit at a time.
2. Deep-water profiling, multishot, from 750 to 2000 m water depth. Because of the long propagation time, the average energy return was low. The multishot mode overcame this problem in that the shooting rate was quicker than the propagation time. Thus, more than one shot was in transit at any time. This asynchronous sequence was used with excellent results, both with two and three shots in transit, for deep-water profiles.
3. Timesharing with deep seismic. This was an experiment in which both deep and shallow seismic measurements were taken simultaneously. To avoid signal interference the sparker was slaved to the DFS-5 deep seismic equipment. A separate energy source and hydrophones were necessary. The experiment aimed for a more efficient use of profiling time and proved successful as a method.

In all modes a 330 m hydrophone cable was used. The sparker results were recorded as analog data on magnetic tape for off-line processing. The records were inspected onboard to determine bottom sample sites.

The sidescan sonar (SSS) gave an acoustical picture of the bottom topography in a 600-m-wide channel. It was used in areas where water depth allowed the towing sensor to operate close to the bottom (within 10 to 60). Having a high lateral resolution of approximately 30 cm, the SSS is a powerful tool for studying iceberg plough marks and other sea-floor features.

Bottom sampling was carried out by gravity coring, grabbing, and dredging. The gravity corer had an immersed weight of 500 kg and a core length of 3 m with an inner liner diameter of 100 mm. The lowering rate was 2 m/sec on soft bottom, but this was increased to 10 m/sec where hard bottom conditions were experienced. Core recovery varied between 270 and 30 cm.

Two types of grabs were used. A large spherical loading grab with a volume of 0.5 m^3 and an empty immersed weight of 500 kg was used to obtain samples of all size fractions up to boulders from the sea bed. In addition, a Smith Macintyre grab taking up to 5 liter samples from an area of 0.1 m^2 of the sea floor was kept in reserve in case well-sorted, fine-grained sand was found. This grab is able to return samples of hour-glass sands.

The dredge had a triangular steel frame with sides of 50 cm length and a tightly knotted trawl net of fiber rope. This type of dredge collects disturbed samples of cohesive soils and particles coarser than granules, in addition to benthos samples. The trawl wire length was twice the water depth, and trawl speeds approximated 1 knot.

The expedition also carried a nonmagnetic bronze piston corer 6 m long and fitted with a self-locking digital compass for obtaining oriented samples. Unfortunately, material failure in the bronze corer top prevented this system from being employed.

Magnetometer data were collected continuously during the expedition, over approximately 4000 km of ship track. The marine proton magnetometer had a 1 gamma resolution, and the analog data were recorded on a paper chart. The program purpose was to take regional measurements, but a denser profile grid in the Halley Bay area together with the seismic data suggest that a local interpretation there may also be of interest.

The ship M/S *Polarsirkel* was an ideal platform from which to operate the marine sensors systems. Well protected under the top deck, a 200 m^2 area was available for the five containerized modules, for housing transmitter-receiver systems, and for laboratories. The deep seismic hydrophone streamer was installed in the rear. On the top deck were located the three winches for the SSS, magnetometer, and shallow-seismic hydrophone cable, along with the geologic sampling and underwater photographic equipment.

The three CSI container modules included: (1) the seismic profiling container with the sparker recording system, the magnetometer transceiver, the SSS transmitter-recorder, plotting facilities for in-field interpretation, and a complete electronic repair and overhaul laboratory; (2) the navigation container with computer communication displays, logging devices, and navigation receivers (this unit had its own stable power supply); and (3) the geologic photo laboratory with facilities for sample description, geotechnical instrumentation, sample storage, and sampler repair, as well as a darkroom for processing underwater photography.

Navigation

Most of the scientific programs of the shipborne party were dependent upon accurate and continuous positioning of the vessel. The navigation and positioning tasks were as follows: to guide the ship during transport (position requirement $\leq$ 1 nautical mile, to guide and control the marine geophysical profiling, and to display off-track parameters from the predetermined profile to the helmsman online (position requirement $\leq$ 20–100 m). All positions referred to the geographic reference system APL (Applied Physics Laboratories reference ellipsoid).

The positioning was obtained by using the Integrated Positioning System (IPS) developed by CSI as a flexible multisensor navigation system for marine geophysical operations. The IPS configuration on this expedition consisted of the following navigation instruments:

1. two-channel satellite receiver, type MX702A TRANSIT;
2. two-channel satellite received, type GEO-II TRANSIT (spare);
3. Omega radio navigation;
4. ship's gyrocompass and speedlog;
5. NORD 10-48 K computer; and
6. remote display and logging devices.

The IPS is installed in a 12′ × 8′ × 8′ container, with temperature regulation and self-contained emergency power source for 1 hour operation.

Advanced processing is necessary to exploit the full potential of the navigation system, allowing use of navigational parameters not usually obtainable during fieldwork. For example, an important factor for deep seismic measurements is the estimated speed over ground; this information is continuously displayed by the IPS.

The IPS software facilities include: an on-line processor for all navigation observations; a position estimator with Kalman filtering; the SAIL system, which displays off-track parameters relative to a defined profile line; and navigational parameter display and logging facilities.

The accuracy during normal profiling is approximately ± 300 m, referred to geographic coordinates. During operation we had the opportunity to check the accuracy when the ship returned to a point previously positioned by the expedition. The accuracy at those revisits was better than ± 100 m.

The satellite system had complete backup equipment based on a two-channel GEO II satellite receiver. This unit functions as a mobile geodetic positioning instrument, and at the shore base, Camp Norway 3, satellite observations over 12–14 hours were read onto paper tape for post-processing. The anticipated accuracy is ± 3–10 m. We hope to return to this site during 1978–79 and reposition it, thereby determining the distance and direction of the ice shelf movement in this area.

In addition, a new position of Bouvetøya was determined with the IPS when NARE landed on the island. The new position lies approximately 3 km east of the old position.

The Omega radio navigation system was planned to be a part of the IPS. Because of the large variations between day and night, the positioning data lacked the necessary precision and stability, giving a typical observed accuracy of ± 3 nautical miles.

Results

As much as conditions permitted, the sparker lines were run perpendicular to the coast. A total profile length of 1200 km was undertaken.

The SSS requires near-sea floor towing (20–60 m) to obtain high-quality recordings; this implies a long towing cable and a slow towing speed. The SSS was used at water depths from 250 m down to 450 m along the ice barrier in the northeastern part of the area. A total of 220 km of profiling was carried out.

During the cruise, 28 stations were occupied for geologic sampling and underwater photography. At these stations, 40 instrument lowerings were carried out and results were obtained from 33 of these. Samples were obtained from sites with water depths as shallow as 240 m and as deep as 1160 m.

At this early stage we are only able to present some preliminary information, mainly based on the sparker and sidescan sonar data.

Sparker data

Sparker recordings indicate three major bedrock provinces on the eastern Weddell Sea continental shelf. In the southern part and to the east of the axis of the Crary trough, we found crystalline rocks associated with large, local magnetic anomalies. The location and extent of this rock complex shows a good fit with the rift zone indicated by Grikurov (Chapter 98, this volume) for this part of the Weddell Sea.

To the west of the axis of the southern part of the Crary trough, the bedrock consists of sedimentary rocks with slightly folded and faulted bedding (Figure 45.4). This region is not associated with steep magnetic anomalies. North of Halley Bay the crystalline rock complex dips steeply below a thick sedimentary sequence with undisturbed bedding and well-developed deltaic sequences (Figure 45.5). The bedding dips away from the coast. On most of the sparker lines these bedrock units are uncomformably overlain by a draped formation with well-developed stratification (Figure 45.4). The draped formation appears to be subdivided into three or more units separated by strong reflectors with great continuity. The top of the draped formation usually shows an erosional contact, either with the sea floor or with overlying till-like deposits. The erosional contact seems to disappear below a water depth greater than 750 m.

A till-like deposit with numerous point reflectors forms an arcuate belt on the shelf north of Halley Bay. It has not been found below a water depth of 450 m. Where the draped formation is present, the till-like deposit overlies it with an erosional contact.

One striking feature on the sparker profiles was the absence of superficial cover over bedrock in the deepest part of the southern Crary trough. The sparker used has a resolution better than 10 m, but no superficial cover can be seen over the bedrock in the deepest part of the trough. Likewise the sea floor between a belt of till-like deposits and the ice barrier in the vicinity of Halley Bay had no visual sedimentary cover over the crystalline rocks that form the sea floor in this area. At this preliminary stage, before interpretive work has been carried out on the data, it would be premature to discuss the seismic lines in relation to depositional models like that of Anderson and others (1977).

SSS data

The sidescan sonar profiles show intensive iceberg ploughing on rises. The plough marks observed are as deep as 5 m or more, more than 200 m wide, and more than 1 km long (Figure 45.6). In many cases the terminal rampart, indicating the arrest of the iceberg, can be seen. The sidescan sonar profiles also show features like boulder clusters, bedrock, and one crater (Figure 45.7) about 150 m in diameter, with crater rims and a depth of about 10 m, which resembles North Sea pockmarks.

On a few profiles the sidescan sonar shows the underwater face of the ice barrier (Figure 45.8). This can be far more irregular than the above water barrier face. A few protruding midwater "noses" or "rams" in the ice-shelf front have been found (Figure 45.9).

Sampling data

The geologic sampling program included coring for micropaleontologic, sedimentologic, and geophysical studies. The core samples have as yet not been studied.

In addition to the coring program, dredging and grabbing were carried out to obtain samples of the gravel and stone fractions in an attempt to identify rocks from local sources. This program was unexpectedly hampered by the dense biological cover on the sea floor, which prevented our equipment in many cases from penetrating into the sea floor. The collected rock specimens in dredge samples show, however, some tendencies toward local variation, e.g., the occurrence of volcanic glass in the vicinity of Maudheim.

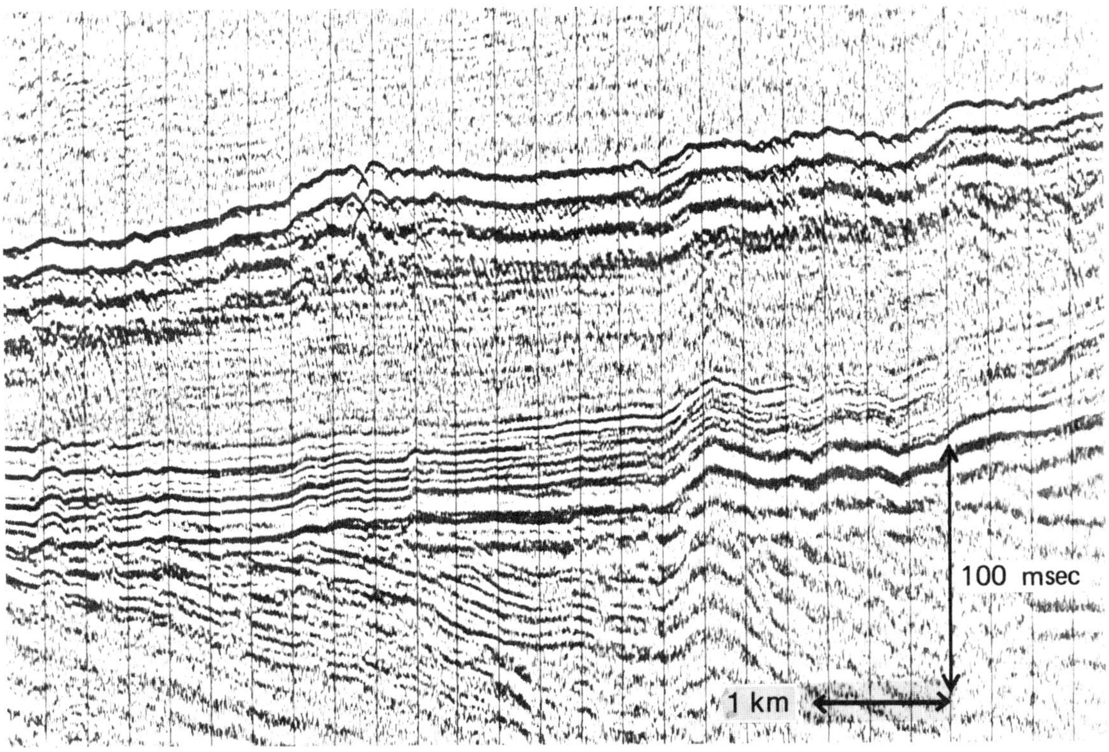

Figure 45.4. Sparker profile showing thin-bedded glacial deposits overlying folded and faulted sedimentary rocks. Location: 77°33.51′ S., 39°58.24′ W. Heading: 137° (left), 317° (right).

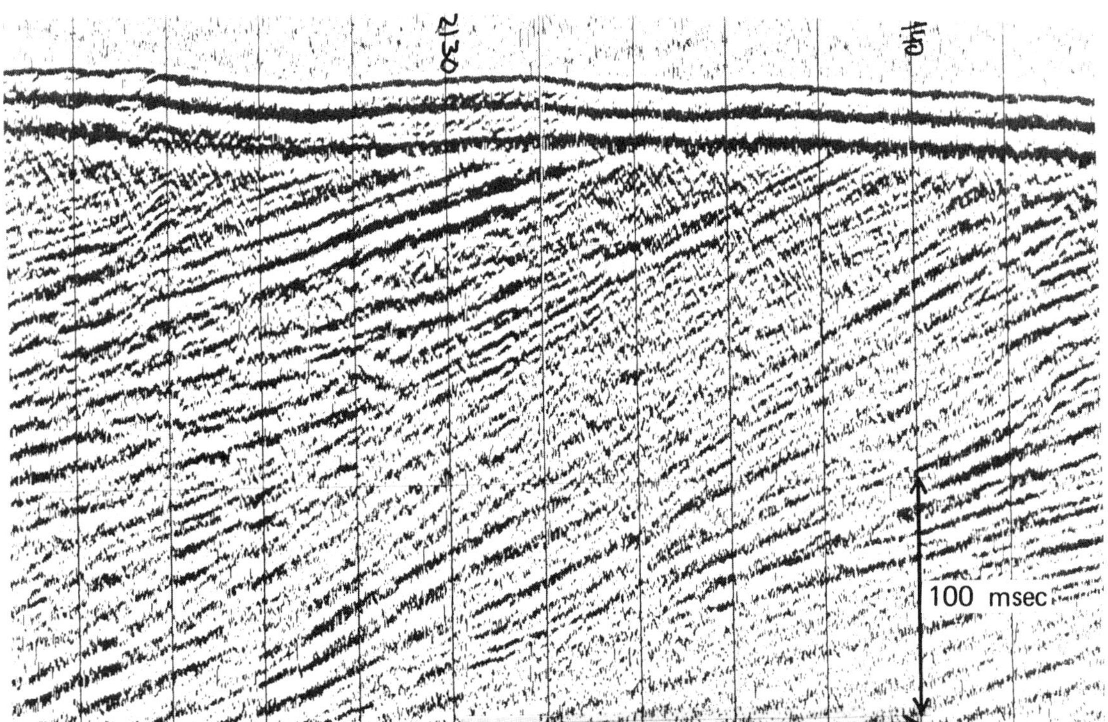

Figure 45.5. Well-bedded sedimentary rocks with a thin cover of glacial deposits. Location: 74°49.87′ S., 24°49.02′ W. Heading: 340° (left), 160° (right).

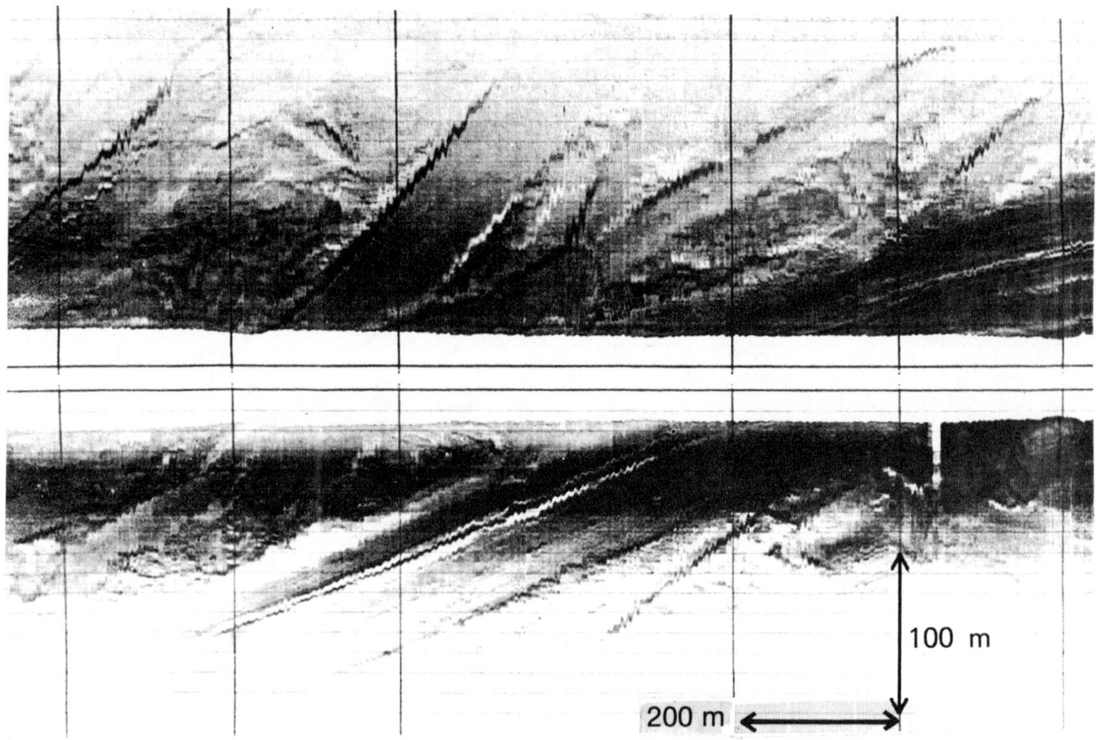

Figure 45.6. Sidescan sonar profile showing iceberg plough marks. Location: 72°32.46′ S., 17°21.29′ W. Heading: 072° (left), 252° (right).

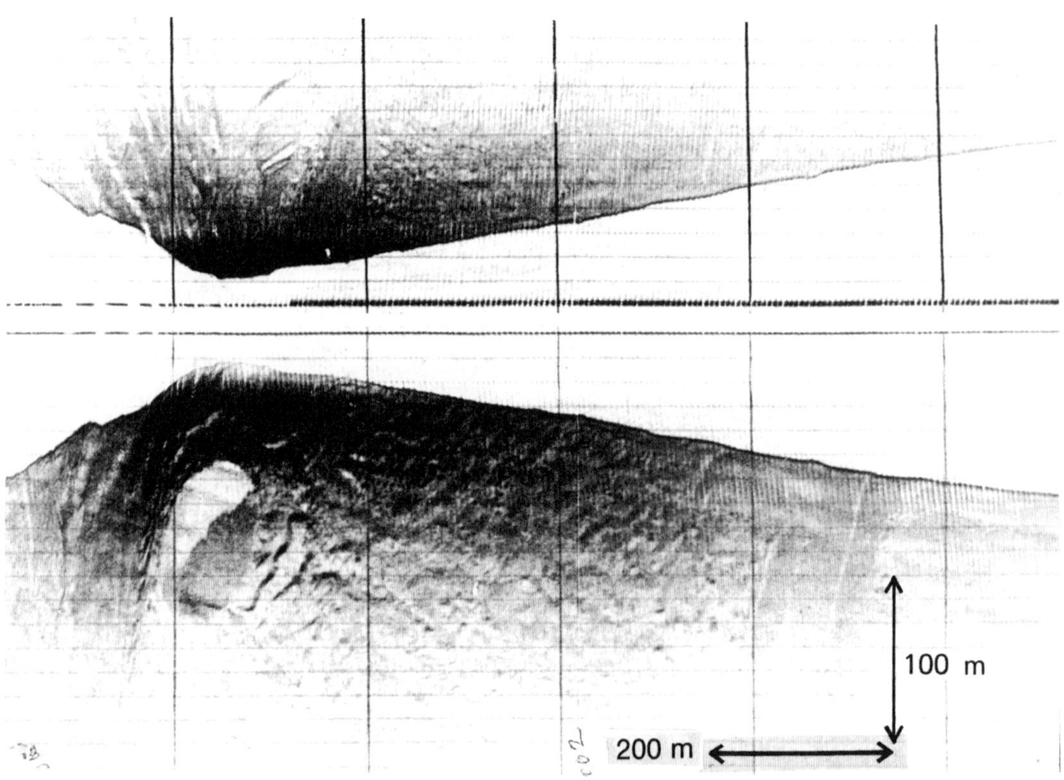

Figure 45.7. Sidescan sonar profile showing a crater in the sea bed. Location: 72°14.76′ S., 16°05.81′ W. Heading: 035° (left), 215° (right).

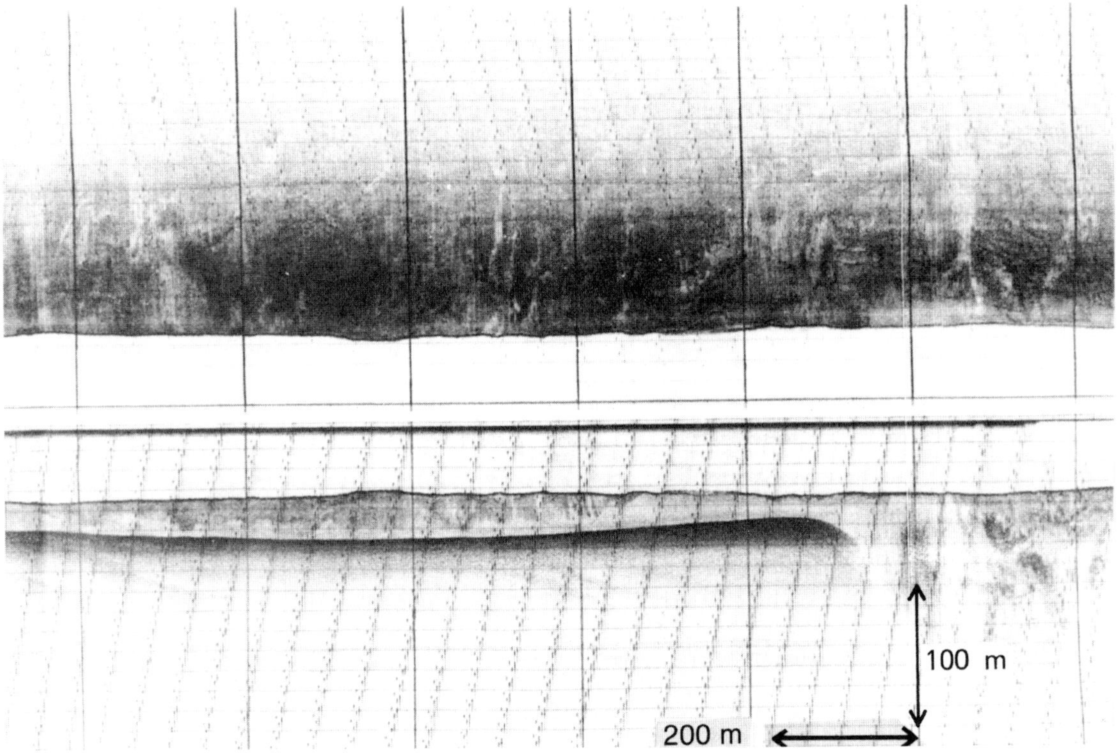

Figure 45.8. Sidescan sonar profile showing the sea floor and an even underwater face of the ice barrier. Location: 74°01.47′ S., 21°40.43′ W. Heading: 264° (left), 084° (right).

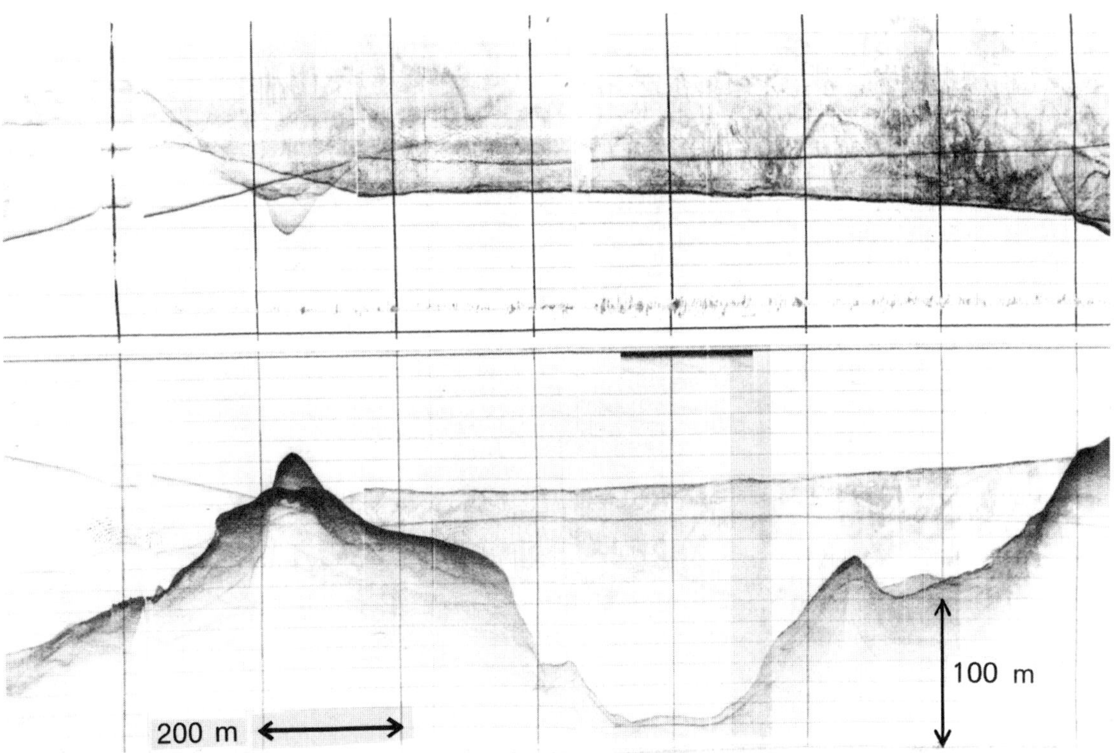

Figure 45.9. Sidescan sonar profile showing an irregular face of the ice barrier, a midwater "nose," and the sea floor. Location: 73°44.50′ S., 20°54.90′ W. Heading: 207° (left), 027° (right).

Acknowledgments. We wish to thank the Norsk Polarinstitutt for financial support and for their effective organization of the expedition. Thanks are due to Kongsberg Vapenfabrikk and Forsvarets Forsknings-Institutt avd. U for lending us vital equipment. We also thank O. Orheim for helpful advice after reviewing the manuscript.

References cited

Anderson, J. B., Clark, H. C., and Weaver, F. M. 1977. Sediments and sediment processes on high latitude continental shelves. Offshore Technology Conference, Houston, May 2–5, 1977. OTC 2739, 91–94.

Orheim, O. 1977. Preliminary Report of the 1976/77 Norwegian Antarctic Research Expedition. *Arbok, Norsk Polarinst. 1976,* 327–32.

46

Seismic Reconnaissance Survey in the Weddell Sea

KRISTEN HAUGLAND
Seismological Observatory, University of Bergen, 5014 Bergen, Norway

Abstract. The first multichannel seismic reconnaissance survey in the Weddell Sea took place as part of the Norwegian Antarctic Expedition in 1976–77. Approximately 1000 km of reflection profiles along with refraction data from 15 sonobuoy stations were recorded from both the continental shelf and the continental slope.

The data show that thick sedimentary layers cover the continental shelf in the southern Weddell Sea and form a prograding continental shelf with no sign of major tectonic activity. There is no evidence of tectonic movement or faults in connection with the Crary trough in the southeastern part of the Weddell Sea.

The position of the continent-ocean boundary in the eastern Weddell Sea is difficult to establish by seismic methods. North of lat. 72° S the data support the existence of the proposed Explora Escarpment. South of this latitude there is no evidence of the escarpment.

The Norwegian Antarctic Expedition to the Weddell Sea and Dronning Maud Land in 1976–77 included a marine seismic reconnaissance program. Approximately 1000 km of multichannel reflection seismics along with refraction data from 15 sonobuoy stations were recorded.

The data were collected from the eastern and southeastern parts of the Weddell Sea, as shown in Figure 46.1. Distribution of the seismic profiles was mainly determined by the ice conditions, but heavy wind was also a problem, especially in the northern part of the study area.

The purpose of the survey was to obtain information on how the continental margin in the eastern Weddell Sea was built up, and also to attempt to determine the position of the continent-ocean boundary in the area. With the time available, it was impossible to do systematic mapping in the Weddell Sea, so an attempt was made to obtain data both from the broad continental shelf in the southern part of the area and from the margin in the northern part of the area where the shelf is ice covered.

Equipment

The acoustic energy pulse was generated by a small air gun array with a total chamber volume of 200 in^3, which was triggered every 20 seconds with a velocity of 5 knots, giving a shotpoint approximately every 50 m. A 16-channel Texas Instruments streamer, each channel 50 m long with 30 hydrophones, served as a detector. The polyurethane skin was flexible, and there were no problems with the streamer due to the cold water. The data were fed from the streamer into a digital field system and were recorded on tape using the SEG "C" format and a sample interval of 4 ms. The near trace was monitored on an EPC recorder. The data from the sonobuoy stations were also digitally recorded on tape.

The Weddell Sea

The Weddell Sea between the Antarctic Peninsula and East Antarctica consists of a central basin with a water depth of 4000–5000 m and a large continental shelf area with a water depth of less than 550 m. In the eastern part of the sea, the shelf edge is close to the front of the ice shelf, and the water depth may reach more than 1000 m on the ice shelf. Here the entire continental shelf is covered by ice, and this makes it difficult or impossible to correlate the marine geophysical data to the geology on land.

To the west, south of lat. 74° S, the continental shelf becomes wider, and the water depth varies between 300 and 500 m, with the shelf edge at approximately 550 m depth. An interesting physiographic feature in the southeastern Weddell Sea is the Crary trough. This is a northeast-southwest-trending trench that parallels the coast between lat. 75° S and 78° S. The trench reaches a water depth of more than 1100 m close to the front of the Filchner Ice Shelf at about long. 38° W. The trough becomes shallower and less distinct toward the north where the threshold depth is approximately 500 m. Its extension beneath the Filchner Ice Shelf is unknown.

Discovery of the continental nature of the Falkland Plateau made it reasonable to introduce the idea of relative movement between East and West Antarctica to avoid the problem of unacceptable overlap of the plates before the breakup of Gondwanaland (Barker and Griffiths, 1977). Consequently, a boundary between the two Antarctic plates must exist or have existed in the past in the Weddell Sea area. If this boundary lies in the eastern Weddell Sea, it might be indicated in a seismic profile crossing the Crary trough. Elliot (1972) has also discussed the possibility that the Crary trough is an expression of a fault system separating East and West Antarctica and that the trench might reach as far south as the Thiel Mountains.

It is usually difficult to detect the continent-ocean boundary by seismic methods. Barker and Griffiths (1977) have postulated that downfaulted continental crust may exist in the eastern and northeastern Weddell Sea. This makes it even more difficult to find the boundary between continental and oceanic crust by normal seismic methods.

Data

A preliminary interpretation of the sonobuoy records has been made by the slope-intercept method based on discrete horizontal layers with constant velocities. The layer thicknesses and velocities are shown in Table 46.1. The velocity at the sea floor was calculated by picking the

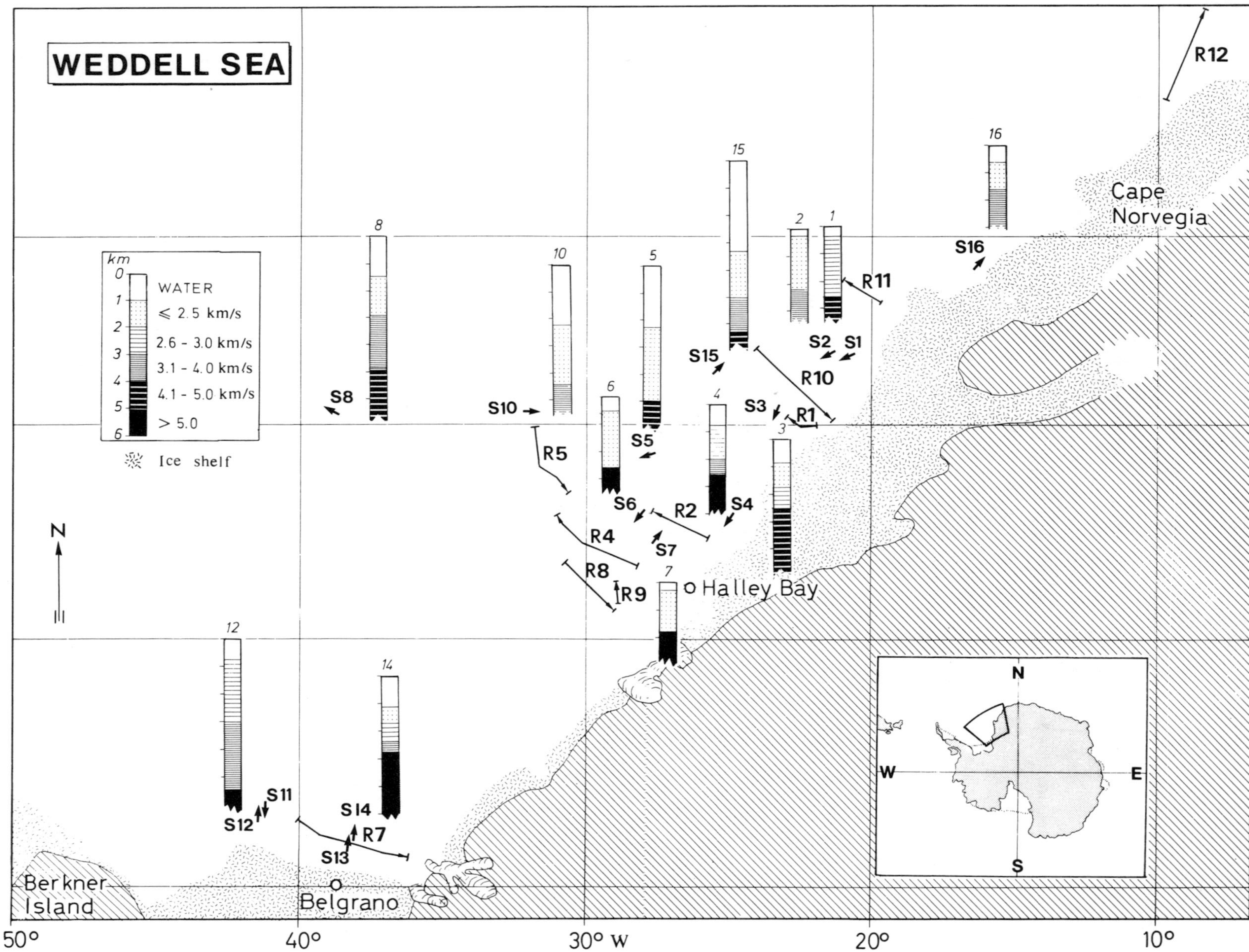

Figure 46.1. The southeastern part of the Weddell Sea, with reflection profiles shown by R and sonobuoy stations shown by S. Velocity-depth results from the sonobuoys are also indicated.

Table 46.1. Preliminary results of the sonobuoys

Sonobuoy	Date	Course	Latitude S	Longitude W	Water depth (km)	Sea-floor velocity (km/s)	Layer 1		Layer 2		Layer 3		Layer 4	
							Velocity (km/s)	Thickness (km)	Velocity (km/s)	Thickness (km)	Velocity (km/s)	Thickness (km)	Velocity (km/s)	Thickness (km)
1	12/01/77	220	73°16.3′	20°35.8′	0.360	1.9	3.0	2.2	4.6	—	—	—	—	—
2	13/01/77	235	73°16.4′	21°20.0′	0.250	2.0	2.4	1.9	4.0	—	—	—	—	—
3	14/01/77	190	73°49.0′	23°20.1′	0.840	1.6	2.1	0.9	3.0	0.8	—	0.8	5.0	—
4	17/01/77	205	74°53.5′	24°54.2′	0.740	1.9	2.7	1.2	3.2	0.6	6.0	—	—	—
5	22/01/77	250	74°17.2′	27°38.2′	2.330	1.7	2.2	2.6	4.8	—	—	—	—	—
6	22/01/77	225	74°49.9′	28°01.0′	0.485	1.8	2.5	2.0	5.6	—	—	—	—	—
7	22/01/77	45	75°07.2′	27°39.3′	0.420	1.7	2.5	1.4	5.8	—	—	—	—	—
8	25/01/77	285	73°53.3′	38°45.1′	1.520	1.7	1.7	0.7	2.4	0.7	3.4	2.0	4.6	—
10	28/01/77	90	73°47.0′	32°25.0′	2.320	1.7	2.4	2.0	3.8	—	—	—	—	—
11	03/02/77	195	77°19.9′	41°12.5′	0.670	2.0	2.4	1.4	2.7	0.7	3.2	—	—	—
12	03/02/77	15	77°30.2′	41°25.8′	0.680	2.0	2.9	2.4	3.4	0.7	4.0	1.8	5.0	—
13	05/02/77	10	77°42.6′	38°15.8′	1.160	1.8	2.4	0.5	2.8	0.6	3.0	—	—	—
14	05/02/77	10	77°38.3′	38°11.3′	1.145	1.8	2.3	0.6	2.7	0.3	3.0	0.3	5.0	—
15	09/02/77	35	73°28.2′	25°38.3′	3.335	1.9	1.9	1.7	3.9	1.2	4.6	—	—	—
16	15/02/77	40	72°22.2′	16°26.9′	0.580	1.7	2.0	0.8	2.2	0.2	3.5	—	—	—

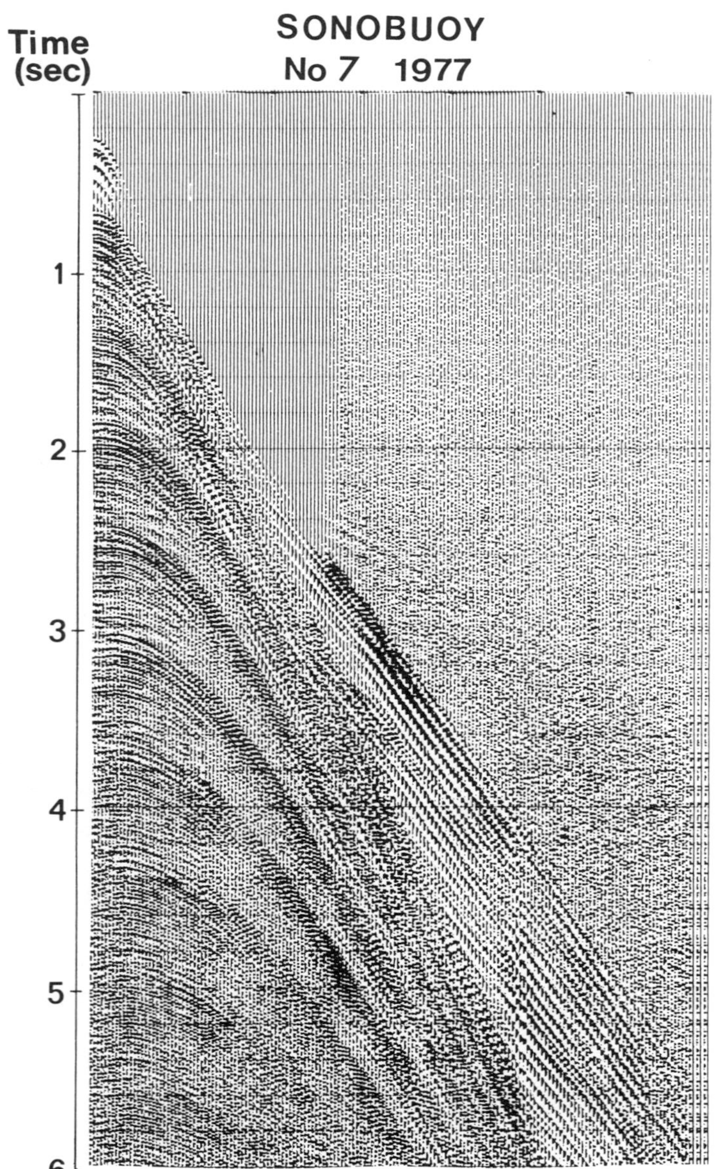

Figure 46.2. The sonobuoy record at station S7. Notice the high-velocity layer between 2 and 3 seconds.

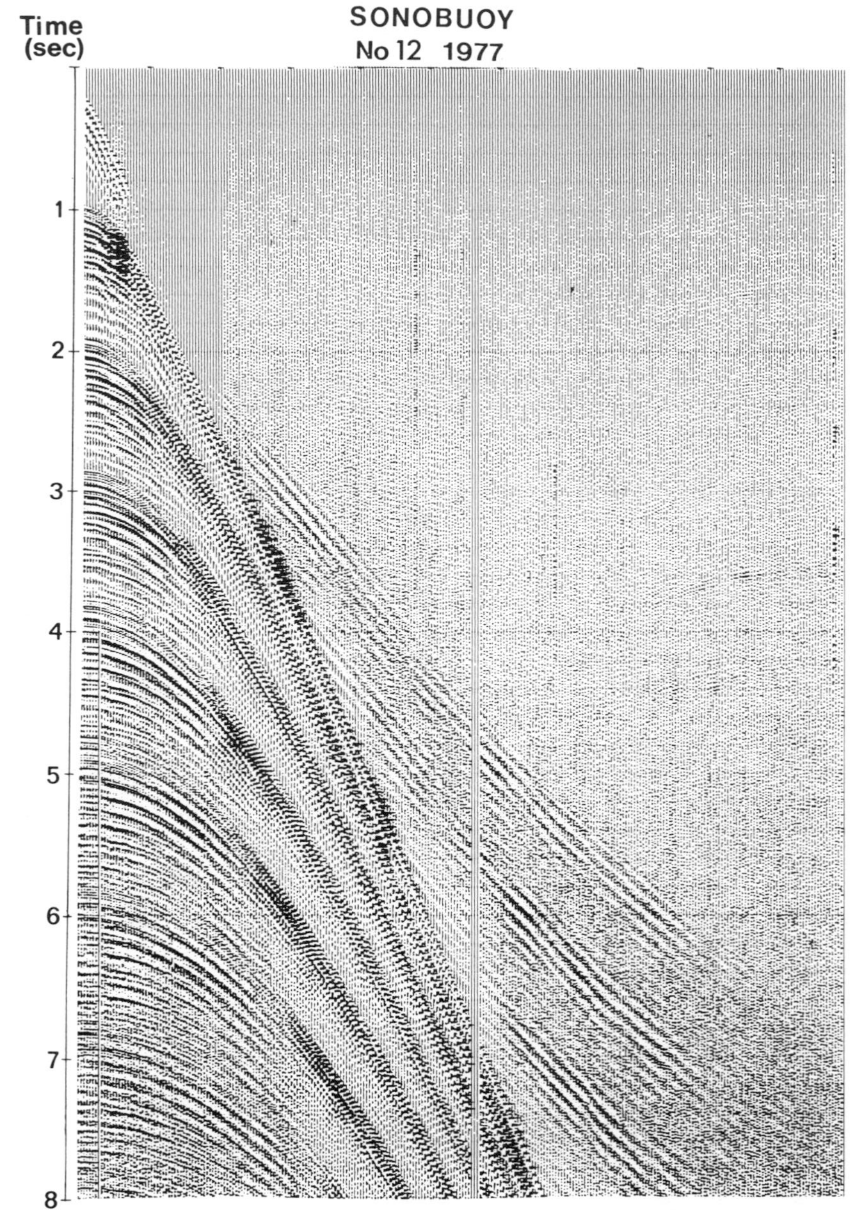

Figure 46.3. The sonobuoy record at station S12 on the western flank of the Crary trough.

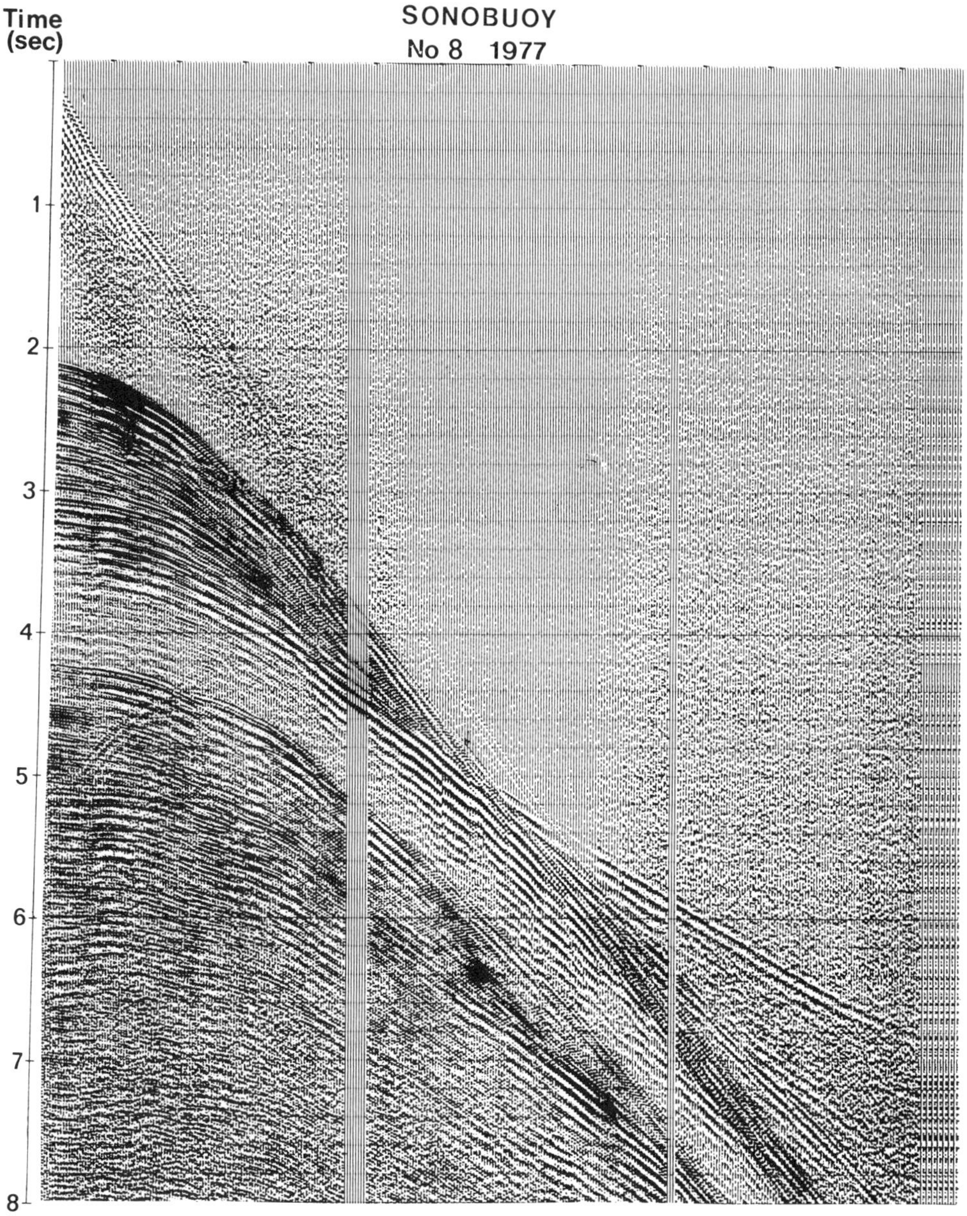

Figure 46.4. The sonobuoy record at station S8 on the continental slope at about lat. 74° S and long. 40° W.

critical distance and using the water depth and velocity. The velocity at the sea floor varies between 1.6 km/s and 2.0 km/s.

Figures 46.2, 46.3, and 46.4 show examples of the sonobuoy records, and Figures 46.5, 46.7, 46.8, and 46.9 show four of the reflection profiles taken from the near-trace monitors without any data processing. Reflection profile R5 (Figure 46.5) covers the outer part of the continental shelf, the shelf edge, and the continental slope down to approximately 1800 m water depth in the area northwest of Halley Bay.

The sedimentary layers are dipping gently, and the profile shows a typical prograding continental shelf with no major tectonic activity. Sonobuoy S10 at the seaward end of R5 shows 2 km of low-velocity sedimentary beds overlying a 3.8 km/s layer. Figure 46.6 shows the combined R5 and R4 profiles without the geographic offset between them. At the inner side of the shelf there is no sign of basement, despite the fact that a shallow reflector with a velocity of 5.6 to 5.8 km/s is clearly seen in sonobuoys S6 and S7.

There is no change in the character of the sea floor between lat. 76° S and 78° S. Most of the seismic energy is trapped in the water layer indicating hard rock at the sea floor. It seems likely that the basementlike reflector in R8 correlates with the reflector wedging out toward the coast in profile R7 (Figure 46.8). Sonobuoy S14 gives a velocity of 5.0 km/s for this reflector. There is a nunatak located only a few kilometers east of profile R7, where upper Precambrian rocks are exposed. It seems reasonable to correlate the 5.0 km/s layer with this nunatak.

The 5.6–5.8 km/s layer recorded in S6 and S7 is a unit with a typical bedrock velocity, and the magnetic data from this area given a shallow depth to the magnetic basement. The 5.0 km/s layer is not recorded in sonobuoys S6 or S7. This means that the 5.0 km/s layer is relatively thin or absent in this area. Reflection profile R7 crosses the Crary trough close to the Filchner Ice Shelf and shows that

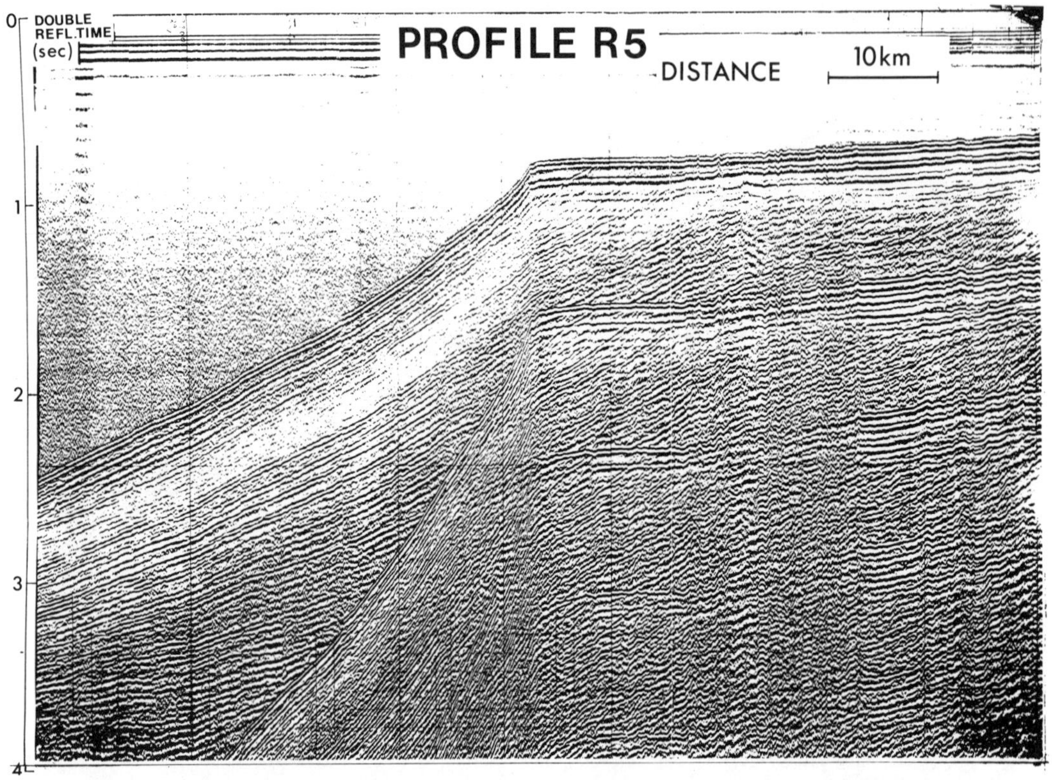

Figure 46.5. The unprocessed near trace of reflection profile R5. The profile includes the outer continental shelf and the upper part of the slope. (See Figure 46.6.)

the trough is mainly eroded. The layers are dipping westward, and there are very few if any recent deposits in the deepest part of the trench. There is no sign even on the processed record of a fault on the eastern flank of the trough, in which case diffraction patterns would be expected.

Sonobuoy S12 (Figure 46.3) on the western flank of the Crary trough shows an approximately 5 km thickness of sediments with velocities ranging from 2 to 4 km/s. This is the thickest cover of sediments recorded in the Weddell Sea on the Norwegian 1976–77 expedition. It has usually been assumed, because of the shallow depth, that the southern part of the Weddell Sea is continental; the thickness of the sediments and the velocity distribution in S12 favor this. The main transport of sediment is most likely from the east, or the east and south, in this part of the sea. We have not been able to detect the continent-ocean boundary in the southern Weddell Sea. Sonobuoy S8 (Figure 46.4) gives velocities that indicate that the boundary is farther north.

The profiles R10, R11, and R12 (Figure 46.9) cross the continental slope and reach 3000 m depth of water or more. Profiles R10 and R12 show the same kind of build up of the continental slope. Profile R12 (Figure 46.9) shows valleys nearly 1000 m deep and on the order of 15 km wide. The truncation of the top layers by the valley sides indicate that the valleys are erosional features, probably caused by strong bottom currents. However, when we consider the lower layers of the reflection beneath the base of the valley, there is indication of weak folding. The continental slope recorded in R12 is rather steep, even though the direction of the profile is only about 40° to the coast. Neethling (1972) has reported slopes of 7.5°–14° in the area north of the SANAE Station, and he has also found the step on the slope that is clearly seen on profile R12.

In early 1978, a German expedition to the eastern Weddell Sea collected geophysical data with multichannel digital equipment (Hinz, 1978). Most of their work was carried out between long. 15° W and 15° E. Hinz (1978) found

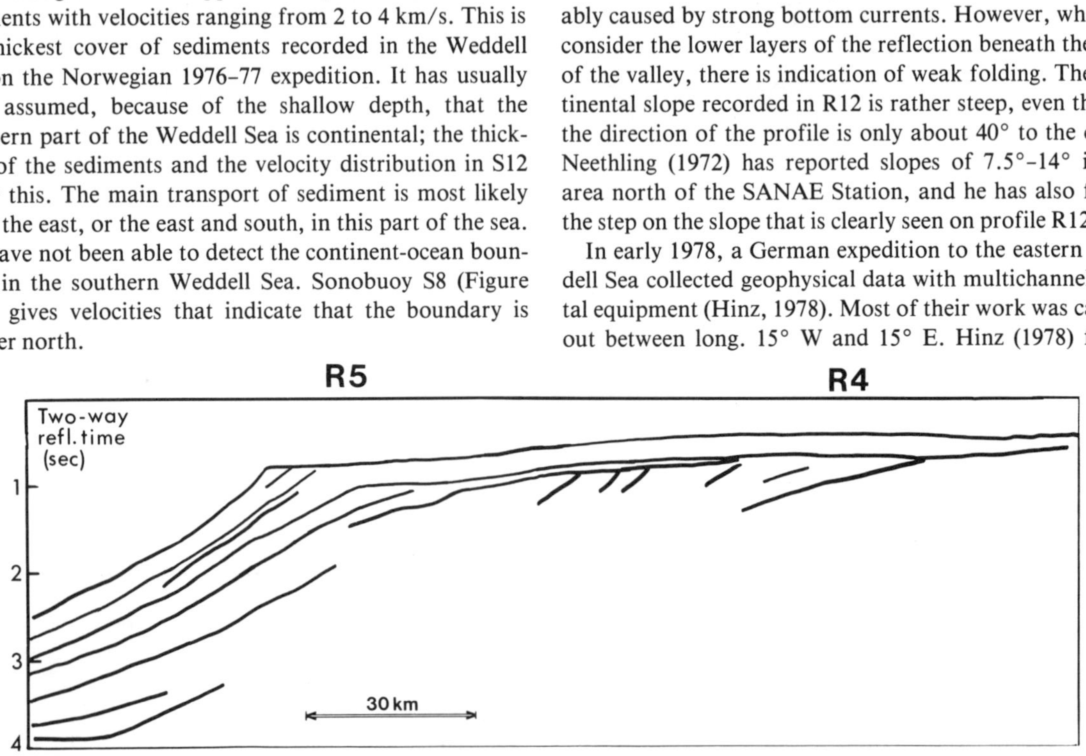

Figure 46.6. Sketch of the main reflectors of profiles R5 and R4 showing the prograding continental shelf northwest of Halley Bay.

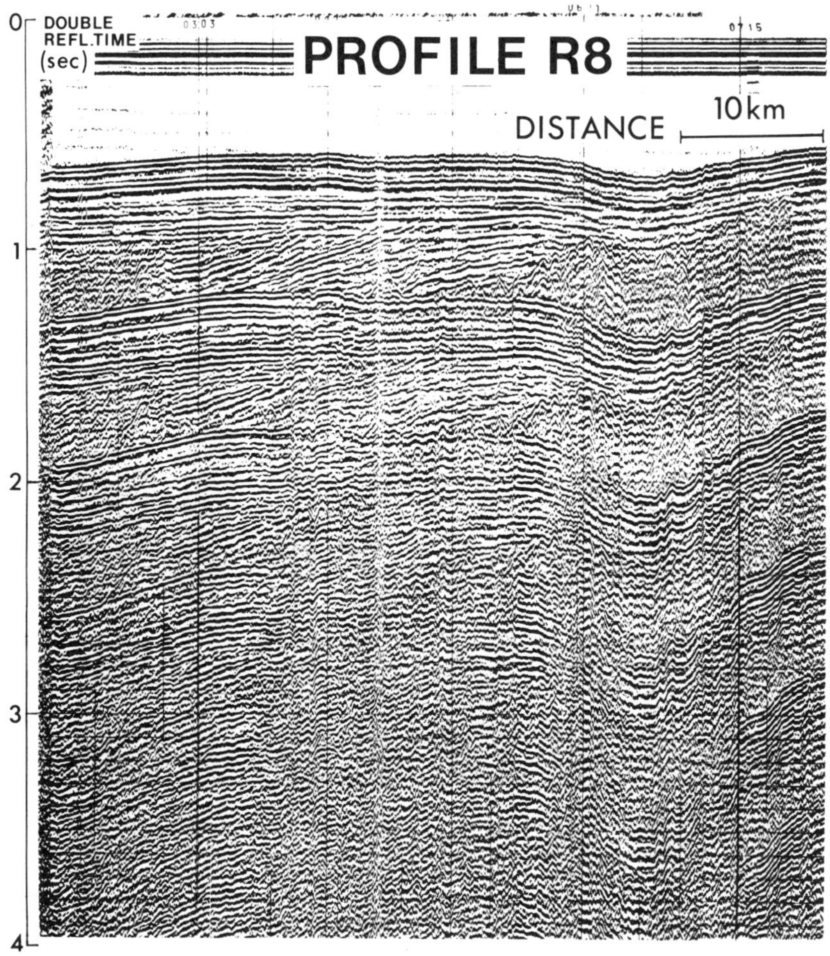

Figure 46.7. Reflection profile R8. Notice the basementlike reflector on the eastern or right side of the profile.

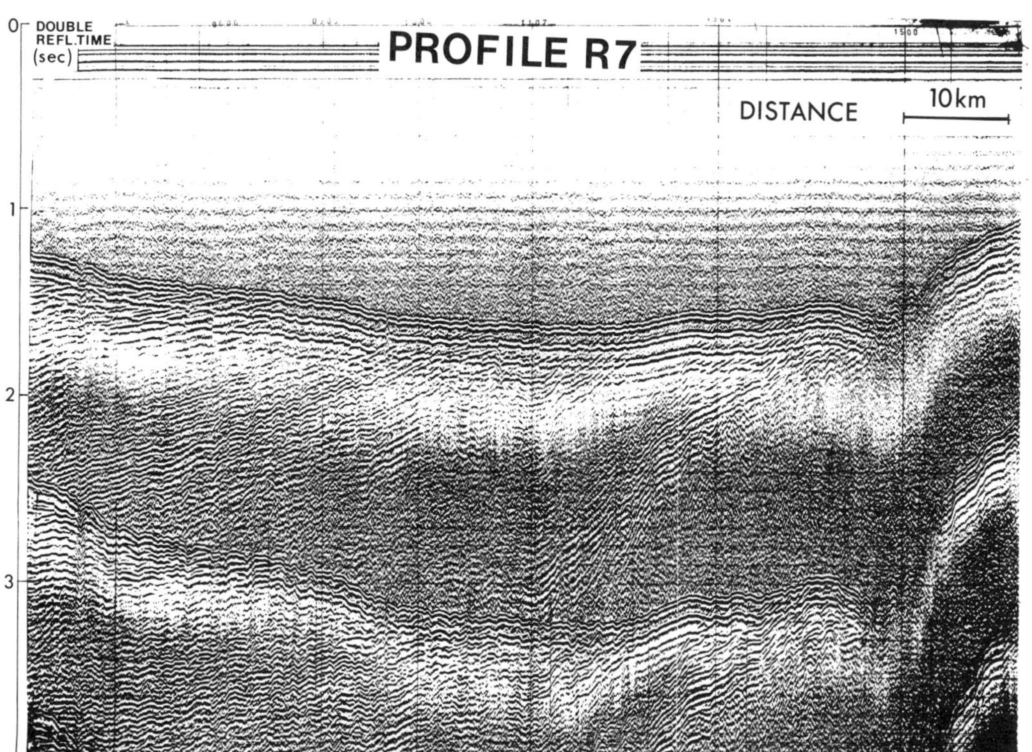

Figure 46.8. A profile (R7) across the Crary trough close to the Filchner Ice Shelf. Notice the layers dipping down to the west (left) and the lack of recent deposits in the deepest part of the trench.

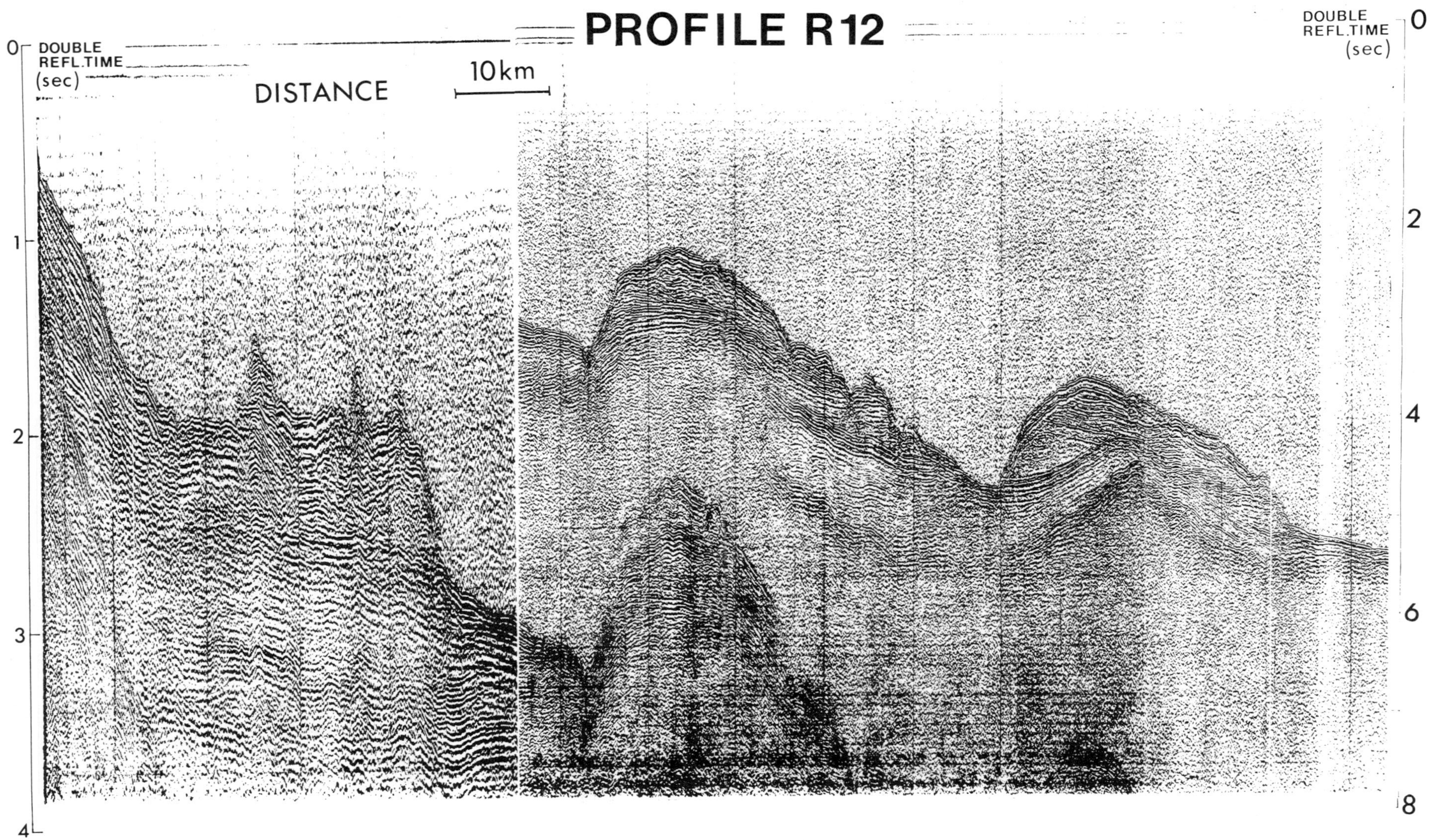

Figure 46.9. Profile R12 at lat. 70°–71° S and long. 10° W. The steep inner slope and the eroded valleys are clearly seen. Note the change in vertical scale.

that the continental margin consists of a narrow shelf and a steep slope down to a plateau at 2000–3500 m depth. The Explora Escarpment, parallel to the coast, is the seaward boundary of this plateau, and it probably represents the boundary between the oceanic and continental crust. Hinz did not rule out that the escarpment may also be a fracture zone.

New data

On the Norwegian Antarctic expedition in early 1979, approximately 1500 km of multichannel seismic reflection data along with refraction data from 30 sonobuoy stations were recorded. This, together with the German results from 1978, considerably improve the data coverage in the eastern and southern Weddell Sea within the last two years. The Norwegian data from 1979 have not yet been processed, but the minor records do not seem to support the existence of the Explora Escarpment west of long. 15° W and south of lat. 72° S. The escarpment is clearly seen on profiles in the Cape Norvegia region.

References cited

Barker, P. F., and Griffiths, D. H. 1977. Towards a more certain reconstruction of Gondwanaland. *Phil. Trans. R. Soc. Lond.*, Ser. B, 279.

Elliot, D. H. 1972. Aspects of Antarctic geology and drift reconstructions. (*In* Adie, R. J., *ed. Antarctic Geology and Geophysics.* Oslo, Universitetsforlaget, 849–57.)

Hinz, K. 1978. Bericht über geophysikalische Untersuchungen im Weddell Meer und am ostantarktischen Kontinentalrand mit M/S Explora. Hannover, Bundesanstalt für Geowissenschaften und Rohstoffe (BGR). Archiv no: 79617.

Neethling, D. C. 1972. Submarine and subglacial morphology, Kronprinsesse Märtha Kyst, Dronning Maud Land. (*In* Adie, R. J., *ed. Antarctic Geology and Geophysics.* Oslo, Universitetsforlaget, 705–11.)

47

Destruction of the Early Cretaceous Back-Arc Basin in the Southern Andes

Abstract

RONALD L. BRUHN*, IAN W. D. DALZIEL, MAARTEN J. DEWIT, and KEITH F. PALMER
Lamont-Doherty Geological Observatory of Columbia University, Palisades, New York 10964, U.S.A.

The upper part of an ophiolite suite preserved in the Andean Cordillera from lat. 51° S to Cape Horn represents the floor of a back-arc basin which existed along the Pacific margin of the South American plate during Early Cretaceous time. Starting in mid-Cretaceous time the basin was destroyed when the frontal arc and basin floor were uplifted and translated toward the plate's interior. These processes initiated the present southern Andean Cordillera. As much as 10 km of vertical displacement of the frontal arc and basin floor occurred relative to the continental shelf. Crustal shortening occurred perpendicular to the longitudinal axis of the basin. Several tens of kilometers of crustal shortening is estimated in the remnant arc in part of Tierra del Fuego where the structural reactivation of continental basement was extensive. Left-lateral horizontal shear was concentrated along narrow belts such as the N.W. Strait of Magellan and the northwest arm of Canal Beagle. The mid-Cretaceous deformation resulted in the development of heterogeneous structures partly due to the nature of the displacements and partly to the originally heterogeneous rheology of the back-arc basin complex. In Tierra del Fuego penetrative deformation was concentrated in the remnant arc while penetrative deformation in the frontal arc and central part of the basin was confined to narrow zones. The deformation in the remnant arc was dominated by the component of progressive simple shear. However, the rock fabrics vary both laterally and vertically, reflecting a heterogeneous distribution of strain. Polyphase deformation occurred in restricted zones throughout the Cordillera indicating that the deformation was neither homogeneous nor synchronous with respect to space or time. For example, the rock fabrics in the remnant arc on Isla de los Estados differ from those in the remnant arc along strike in Tierra del Fuego, although the general nature of the crustal displacements was apparently similar in both areas.

* Present address: Department of Geology and Geophysics, University of Utah, Salt Lake City, Utah 84112, U.S.A.

48

Plate Tectonic Setting and Evolution of the Scotia Sea

Abstract

JOHN L. LaBRECQUE and ELLEN M. HERRON
Lamont-Doherty Geological Observatory of Columbia University, Palisades, New York 10964, U.S.A.

The region of the Scotia Sea is divided into five tectonic plates: the Antarctic, South American, Aluk, Scotia, and Sandwich plates. Magnetic anomalies and fracture zones associated with these various plates may be interpreted to trace the evolutionary history of the Scotia Sea.

The Scotia Sea as we know it has evolved since Oligocene time as a consequence of the relative motion between the South American and Antarctic tectonic plates. Sea-floor spreading magnetic anomalies indicate that the western Scotia Sea was formed during a period of sea-floor spreading which lasted from Oligocene to mid-Miocene time. The cessation of sea-floor spreading in the western Scotia Sea in mid-Miocene time may be related to a shifting of the main zone of relative motion between the South American and Antarctic plates from the North Scotia Ridge to the South Scotia Ridge.

49

A Review of the Geology of the Antarctic Peninsula

Abstract

P. D. CLARKSON, P. W. G. TANNER, and M. R. A. THOMSON
British Antarctic Survey, Natural Environment Research Council, Madingley Road, Cambridge CB3 0ET, U.K.

With the completion of the reconnaissance geological mapping of the Antarctic Peninsula, it is opportune to review progress and to make a critical assessment of the established stratigraphic scheme. To this end, geologists of the British Antarctic Survey are preparing a series of geologic maps at a scale of 1:500,000 to show the known geology. Each of the six map sheets will be reproduced in color to illustrate the outcrop geology with additional structural data and radiometric ages. Plant and animal fossil sites and records of mineralization will also be marked. Data from all available sources, both published annd unpublished, will be incorporated and a detailed geologic summary for each of the rock groups will be printed on each map together with a bibliography.

The aim of this paper is to consider the research presented on the maps with particular reference to the criteria used for the recognition, correlation, age and nomenclature of the major sedimentary, metamorphic and igneous divisions. The age ranges of the major rock units are assessed in the light of present knowledge, and in some cases suggestions are made for their renaming in closer conformity with modern stratigraphic codes. Progress towards an interpretation of the geologic evolution of the Antarctic Peninsula is discussed.

50

Implications Regarding the Evolution of the Scotia Arc from a Plate Tectonic Model of the Dispersion of the Gondwanic Continents

Abstract

B. D. JOHNSON, C. McA. POWELL, and J. J. VEEVERS
School of Earth Sciences, Macquarie University, North Ryde, N.S.W. 2113, Australia

A plate tectonic model has been derived for the dispersion of the Gondwanic continents using available marine magnetic anomalies, deep sea drilling data, and paleomagnetic evidence.

At the present time there are four major lithospheric plates (the South American, African, Indo-Australian, and Antarctic plates) with spreading systems meeting at two triple junctions (the Indian Ocean and Bouvet triple junctions). There is now sufficient sea-floor data to define the relative motions of the Indo-Australian and Antarctic plates, the Indo-Australian and African plates (including Madagascar), and the African and South American plates. By resolving the relative motions of these plates it is possible to obtain the relative motions of the African and Antarctic plates, and the South American and Antarctic plates.

The Scotia arc can thus be examined in terms of a stage-by-stage model of the relative motions of the southern end of South America and the northern end of the Antarctic Peninsula. The model indicates a long history of relatively close juxtaposition of the two continents with a period of lateral motion giving rise to the east-west part of the arc system. The north-south part of the arc appears to be related to a relatively recent dispersion of South America and Antarctica. There appears to be no necessity to invoke large-scale deformations of southern South America (except for the extreme southern tip) or of the Antarctic Peninsula.

III East Antarctica Shield

51

The Lower Precambrian of Antarctica

Review Paper

M. G. RAVICH†
Department of Antarctic Geology, SEVMORGEO, Moyka 120, Leningrad 190121, U.S.S.R.

Abstract. The lower Precambrian of Antarctica, at the present stage of geologic knowledge, can be subdivided into seven structural units (stages) in order of age: (1) lower Archean I, (2) lower Archean II, (3) upper Archean I, (4) upper Archean II, (5) undifferentiated Archean, 6) upper Archean to lower Proterozoic undifferentiated, and (7) lower-middle Proterozoic. The first four units constitute the crystalline basement of the East Antarctic craton; where this basement is poorly exposed or little studied, the rocks are assigned to the undifferentiated Archean unit. The upper Archean to lower Proterozoic stage is recognized only in the Transantarctic Mountains and in West Antarctica where it is believed to represent a reworked basement of the late Precambrian and younger geosynclines. The lower-middle Proterozoic unit is an epicratonic supracrustal sequence.

Recognition of the Archean is based on regional extent of the granulite facies schists. The granulite facies assemblages are well preserved in the lower Archean; the oldest enderbites (about 4000 m.y.) are of the highest temperature grade (pyroxene-granulite subfacies), but the lower Archean II rocks have undergone retrograde metamorphism to a moderate grade of the granulite facies. The upper Archean I is represented by polymetamorphic rocks, with high-grade amphibolite facies rocks regressively derived from those of the granulite facies. The upper Archean II is composed almost entirely of high-grade amphibolite facies rocks with a vast amount of ultrametamorphic migmatites and shadow granites. The undifferentiated upper Archean to lower Proterozoic unit is characterized by amphibolite facies assemblages, and the lower-middle Proterozoic supracrustal rocks attain a low-grade amphibolite facies grade which is zonally superposed on a regionally developed greenschist facies grade.

The most characteristic intrusive formations in the lower Precambrian terrane are the upper Archean to lower Proterozoic anorthosites, metagabbroids, and charnockitoids, the latter including the primary rheomorphic intrusive and secondary apogabbroid varieties.

The Archean–Proterozoic time boundary is marked in the geologic history of the continent by completion of the cratonization of the crystalline basement, which since has become a foundation for diverse suprastructural formations, such as epicratonic platform cover and miogeosynclinal intracratonic and proper geosynclinal pericratonic fold systems.

The early Precambrian as a time unit has an upper age limit at approximately 1700 m.y. ago. It includes the Archean, during which the crystalline basement of Antarctica was formed and cratonized, and the early to middle part of the Proterozoic, when this basement was partly modified by various processes (mainly orogenic and intrusive) and partially covered by supracrustal volcanic and sedimentary formations.

The lower Precambrian (mostly Archean) rocks are particularly widespread within the East Antarctic craton where they comprise almost all the mountain ranges. They crop out along the coast of East Antarctica from Adelie Land to western Queen Maud Land for a distance of 8000 km; the most extensive outcrops are in Enderby, MacRobertson, and Queen Maud Lands.

The main features of the lower Precambrian of Antarctica (Ravich, 1972) are:

1. predominance of granitization products of widespread regional metamorphic and ultrametamorphic processes;
2. presence of typical formations such as (a) the Archean enderbite-charnockite, granite-gneiss and/or migmatite-gneiss, marble-calciphyre, amphibolite-plagiogneiss, and magnetite-quartzite high-grade formations; (b) the Archean to early Proterozoic anorthositic, metagabbroic, and charnockitoid intrusive formations; and (c) the Proterozoic quartzite, dolomite, and jaspilite low-grade metamorphic formations;
3. development of specific tectonic forms, such as large granitized domes and synclinally infolded zones;
4. segmentation of the crust into both large stable and mobile cratonized blocks, the latter continuing to experience multiple tectono-thermal activizations throughout Precambrian geologic history,

These multiple activations in particular, along with a generally insufficient degree of geologic study of the continent, considerably frustrate subdivision of the lower Precambrian into separate rock-stratigraphic units. The modern ice-free terranes of the crystalline basement are mostly neotectonically regenerated orogens which have undergone a number of preceding activizations. Of these,

† Deceased.

the most vigorous and widespread occurred about 500 m.y. ago (the Ross orogeny). It is overimprinted on almost all K-Ar and Rb-Sr age determinations carried out on granites and mica separates, i.e., on material particularly sensitive to superposed events. Therefore, only the few available Pb-determinations on rocks collected from stable cratonic areas can be used for the purpose of true age definitions.

Subdivision of the lower Archean

Available geologic and geochronologic evidence permits recognition of the following structural units (stages) in order of age, within the lower Precambrian (Table 51.1): (1) lower Archean I; (2) lower Archean II; (3) upper Archean I; (4) upper Archean II; 5) undifferentiated Archean; (6) upper Archean to lower Proterozoic, undifferentiated; and (7) lower-middle Proterozoic. The first four units constitute a crystalline basement of the East Antarctic craton; where this basement is poorly exposed or little studied, the rocks are assigned to the undifferentiated Archean unit. The upper Archean to lower Proterozoic stage is recognized only in the upper Precambrian and Phanerozoic geosynclinal areas where it is believed to represent the reworked pre-geosynclinal basement. The last unit represents an epi-cratonic supracrustal sequence.

In general, recognition of the Archean is based on the regional extent of the granulite facies schists which are very typical of the Archean elsewhere in the world. Within the Archean, there is a definite correlation between the age of the rocks and their mineral assemblages (hence the metamorphic facies), as well as their structural position. On the whole, the lower Archean is characterized almost exclusively by granulite facies rocks, the oldest of which preserve the highest grade mineral assemblages of the pyroxene-granulite subfacies. The younger ones, however, are retrograded to a moderate grade of the granulite facies, with consequent development of the granulite facies ultrametamorphic and monomineral rocks. Further decrease of temperature during the late Archean was responsible, first, for the appearance and wide development of polymetamorphic rocks with high-grade amphibolite facies assemblages, regressively superposed onto the older granulite facies rocks, and, later, for generation of large volumes of shadow granites and migmatites formed as typical high-grade amphibolite-facies ultrametamorphic products. Structurally, the lower Archean terranes comprise the nuclei of large stable shields, while the surrounding upper Archean areas occur as smaller, linear forms or as dome-like forms related to higher crustal mobility and granitization phenomena.

The lower Archean I is represented by the Napier Complex, as yet known only in central Enderby Land. This complex is recognized by a unique association of enderbites and mesoperthitic charnockites (enderbite-charnockite formation), as well as by the highest temperature grade of metamorphism. An extremely ancient age for the Napier Complex is indicated by Pb-isochron determinations of close to 4000 m.y. (Ravich and others, 1974). Lithologic variations and structural observations suggest subdivision of the Napier Complex into two "series"*—Raggatt and Tula; their compositions are given in Table 51.1.

The lower Archean II is defined as the Rayner Complex, including the Condon and the Nye Series. Isotopic age determinations on the Rayner Complex rocks usually show the time of tectono-thermal activizations at 2400–2100 m.y. ago (Atrashonok and others, 1967) and at 620–460 m.y. ago (Ravich and Krylov, 1964).

The upper Archean I is represented by characteristic polymetamorphic assemblages which are considerably more widespread than the older granulite facies formations. These polymetamorphic rocks constitute the Napier Complex and are subdivided into two series. The most typical is the Humboldt Series in central Queen Maud Land, while in MacRobertson Land, the equivalent rocks are assigned to the Reinbolt and Larsemann Series, which are probably lithologic facies of the same stratigraphic unit; the Larsemann Series includes rocks of more silicic and aluminous composition.

The upper Archean II is recognized by an abundance of intensely migmatized and granitized amphibolite facies rocks represented by the typical Insel Series in Queen Maud Land and by the synchronous Lambert Series in MacRobertson Land. These units constitute the Insel Complex, the late Archean of which is confirmed by Rb-Sr determinations on the Lambert Series rocks, which range from 2500 m.y. (gneissic granites) to about 3000 m.y. in age (migmatitic granite-gneisses) (Halpern and Grikurov, 1975; Australia. Bureau Mineral Resources, 1975).

Undifferentiated Archean rocks are recognized by the general presence of thick units of ultrametamorphic charnockites and enderbites mixed with migmatized pyroxene-plagioclase schists, migmatites, and granite-gneisses. Isotopic dating of these rocks indicates several tectono-thermal activization epochs during the period from 2000 m.y. to 440 m.y. ago (Ravich and Kamenev, 1972); these epochs took place at 2000–1800 m.y., 1600–1560 m.y., 1200–1100 m.y., 900–800 m.y., 650–600 m.y., and 570–440 m.y. ago. The older age values, obtained by Pb and Rb-Sr methods, are very few, while the great majority of K-Ar dates (more than 200 determinations) fall into the youngest epochs.

An undifferentiated late Archean to early Proterozoic age is tentatively assigned to the reworked basement projections of geosynclinal terranes outside the East Antarctic craton. These projections are composed of amphibolite facies rocks. In the Transantarctic Mountains they are recognized as the Nimrod Series (Group), in northern Victoria Land, as the Wilson Series (Group), in Marie Byrd Land, as the Fosdick Series (Group), and in the Antarctic Peninsula, as the "Basement Complex" (Ravich and Grikurov, 1976*a*; Ravich and Grikurov, 1976*b*). The oldest isotopic age as yet available for these metamorphic rocks is about 2000 m.y. (for the Nimrod Group in the Transantarctic Mountains) (Gunner and Faure, 1972). A few age determinations are close to 1000 m.y., but much more abundant are the reset values related to major orogenic events in surrounding geosynclinal terranes (from 500 to 400 m.y.). Therefore, recognition of all high-grade metamorphic inliers within the Antarctic geosynclinal belts as projections of their older Precambrian crystalline basement remains to a large degree provisional and is thus far based mainly on general structural speculations.

* Following Russian usage, *series* here means a lithostratigraphic unit formed during a period or longer.

Table 51.1. Subdivision of the lower Precambrian in Antarctica

Age	Complex	Series[a]	Composition	Estimated thickness (km)	Type locality
Lower and middle Proterozoic	Ruker	Menzies	Quartzites, micaceous and garnet-micaceous schists; subordinate units to two-mica staurolite-kyanite-fibrolite schists	3-4	Prince Charles Mountains
		Ruker	Quartz-carbonate, quartz-sericite, quartz-chlorite schists and phyllites; quartzites and jaspilites		
Upper Archean-lower Proterozoic	—	Nimrod, Wilson, Fosdick, etc.	Biotite-amphibole and biotite-garnet plagiogneisses and migmatites; garnet-sillimanite-cordierite schists, amphibolites, and marbles	?	Nimrod Glacier in the Transantarctic Mountains, northern Victoria Land, Fosdick Mountains in Marie Byrd Land, etc.
Archean, undifferentiated	—	—	Garnet-biotite plagiogneisses, two-pyroxene schists, migmatites, and shadow charnockites and granites; subordinate units of granite-sillimanite-cordierite schists, pyroxene calciphyres, and quartzites	?	East Antarctic coast
Upper Archean II	Insel	Insel	Biotite, biotite-garnet, biotite-amphibole, biotite-sillimanite plagiogneisses and schists, migmatites and shadow granites; amphibolites, amphibole and humite calciphyres, and marbles	5-6	Queen Maud Land
		Lambert	Biotite, garnet, and amphibole plagiogneisses, shadow granites and migmatites; subordinate units of quartzites, tremolite and diopside calciphyres, and sillimanite-cordierite schists		Southern Prince Charles Mountains
Upper Archean I	Humboldt	Reinbolt and Larsemann	Biotitized and amphibolized shadow charnockites and enderbites, biotite-garnet and sillimanite plagiogneisses, migmatites and shadow granites, biotite-garnet and sillimanite-cordierite-garnet schists; subordinate units of two-pyroxene schists, phlogopite-periclase forsterite calciphyres and quartzites	8-9	MacRobertson Land, Princess Elizabeth Land
		Humboldt	Amphibolized and garnetiferous shadow charnockites and enderbites, amphibole-pyroxene and biotite-garnet-sillimanite-cordierite schists, migmatites and shadow granites, subordinate units of scapolite-diopside and wollastonite-pyroxenite calciphyres and granulites		Humboldt Mountains in Queen Maud Land
Lower Archean II	Rayner	Nye	Pyroxene-plagioclase schists, shadow enderbites and charnockites, biotite-garnet and garnet-pyroxene shadow granites and migmatites intercalated with biotite-garnet schists, quartzites and diopside, forsterite and spinel calciphyres	8-10	Nye Mountains and Condon Hills in southern Enderby Land
		Condon	Biotite-garnet shadow granites and migmatites with subordinate units of enderbites and charnockites, intercalated with pyroxene-plagioclase and garnet-sillimanite cordierite schists and garnet quartzites		
Lower Archean I	Napier	Tula	Garnet mesoperthitic shadow granites with subordinate units of hercynite-hypersthene, hypersthene-cordierite and sillimanite quartzites, mesoperthitic charnockites and enderbites, two-pyroxene schists	>8	Tula and Raggatt Mountains in Enderby Land
		Raggatt	Shadow enderbites and mesoperthitic charnockites, pyroxene-plagioclase schists with rare bands of garnetiferous charnockites and enderbites		

[a] Following Russian usage, "series" here means a litho-stratigraphic unit formed during a period or longer.

The uppermost part of the lower Precambrian (lower-middle Proterozoic) is fundamentally different from the Archean crystalline basement in structural setting and, consequently, in initial composition, as well as in the nature and grade of metamorphism. The lower-middle Proterozoic sequence consists of typical supracrustal formations which either form intracratonic fold systems (graben-synclinoriums) developed from aulacogen-like structures, or constitute extensive underformed ancient platform covers (not included in Table 51.1). In both cases an epicratonic nature of this sequence is distinctly displayed. Characteristic features of the lower-middle Proterozoic formations are: (1) the predominance of original sedimentary rocks, which are not metamorphosed in the platform cover, but are metamorphosed in a greenschist facies environment (with superposition of a low-grade amphibolite facies) in the intracratonic fold systems; (2) the presence of thick jaspilite units (in the intracratonic fold systems); and (3) the abundance of trap magmatic products and absence (or very negligible development) of granitoids.

The lower-middle Proterozoic intracratonic fold system is distinguished in the southern Prince Charles Mountains where two series (probably of the same stratigraphic range) are recognized. The Ruker Series consists of greenschist facies rocks with thick (as much as 300–400 m) jaspilites at the base of the section. The Menzies Series includes higher grade rocks which may represent the Ruker Series greenschists progressively remetamorphosed at a later time.

The oldest platform cover in the Ritscher Upland consists of diverse sandstones interbedded with argillites and siltstones; clastic red beds constitute the top of the section. The lower-middle Proterozoic age of these sedimentary rocks is bracketed by isotopic determinations on mafic sills intruded into them, which yield 1700 m.y. ages (Neethling, 1972), and on granitic basement rocks underlying them, which give 3060–2800 m.y. ages (Soloviev and Halpern, 1975).

Lower Precambrian intrusive formations

Characteristic intrusive formations in the lower Precambrian of Antarctica are anorthosites, metagabbroids, and charnockitoids. Diverse granitoids are more abundant, but they usually represent shadow granites with little or no intrusive mobilization, and therefore properly belong to metamorphic rather than intrusive formations.

Anorthosite

This rock type is represented by a large massif (900 km^2) in central Queen Maud Land and by smaller bodies in some other East Antarctic regions (Humboldt Mountains and MacRobertson Land) (Ravich and Soloviev, 1966). Margins of the large massif consist of metamorphosed melanocratic rocks with banded and gneissic structures of anorthosite-norite or gabbro-norite composition. As a result of recrystallization, the marginal facies changed in mineral composition: there is rather abundant garnet, hornblende, and orthoclase; plagioclase became more sodic; and hypersthene became more ferric. The central part of the massif is composed of normal anorthosites, containing labradorite (93–96%), hypersthene (1–3%), enstatite-augite (1%), quartz and orthoclase (1% of each), titano-magnetite (0.5–1%), and secondary biotite and garnet. Anorthosites have a layered structure due to a higher content of pyroxene in some layers. A few lamprophyre dikes occur within the massif. Its contacts with enclosing rocks are not exposed, but detailed textural observations suggest a lopolithic configuration of the intrusion, with rather steep funnel-like contacts. Distortion of initial inner stratification is probably related to later tectonic movements, which were also responsible for local cataclasis and mylonitization of anorthosites. The age of these anorthosites is presumably late Archean to early Proterozoic, because marginal parts of the massif were apparently metamorphosed in an amphibolite facies environment typical of the final stages of cratonization of the Archean crystalline basement.

Metagabbroids

This rock type is rather widespread in the crystalline basement (Ravich and others, 1965). It occurs mostly in the polymetamorphic and high-grade amphibolite facies terranes among migmatite-gneiss and amphibolite-plagiogneiss metamorphic formations, from which it can usually be distinguished, but only with great difficulty. The reason for this problem is that metagabbroids are also intensely metamorphosed in the amphibolite facies, locally granitized and commonly turned into agmatite-like rocks by abundant veins related to the emplacement of later granites. That is why in the course of geologic mapping they were often mistaken for plagiogneissic and migmatitic formations, and related veins and dikes were described as amphibolites. The most reliably recognized intrusions are in the Wohlthat and Sør-Rondane Mountains (Queen Maud Land) and in the Prince Charles Mountains (MacRobertson Land), but isolated outcrops of metagabbroids and associated dikes of pyroxene orthoamphibolites and highly altered diabases are known in many other regions.

Massifs of metagabbroids are usually dome-shaped or laccolith-like. Their size reaches a few hundred square kilometers; the most common are outcrops a few kilometers across. Intrusive relationships with enclosing rocks have not been observed directly, either due to lack of exposure or because of high alteration of marginal parts of the massifs simultaneously with enclosing migmatite-gneiss terranes. Composition of metagabbroids depends strongly on the intensity of superposed processes and therefore is quite variable. The initial composition of gabbroids was also heterogeneous, as a crystallization process must have caused differentiation and consequent development of not only predominant gabbro and gabbro-norites, but of pyroxenite and even peridotite varieties too; there probably also were varieties close to anorthosites in composition. All compositional varieties studied fall into three major groups: amphibolized gabbroids, granitized metagabbroids, and gneissified amphibole-plagioclase rocks. In these groups, and especially in the first one, the rocks retain relics of primary gabbroic, gabbro-diabasic, and panidiomorphic textures, but the latter on the whole become heteroblastic, granoblastic, and catablastic with irregular development of schistose and gneissose structures.

Amphibolized gabbroids are represented by massive, considerably recrystallized rocks. Plagioclase content ranges from 30% to 60%, and composition from andesine (An_{30-35}) to labradorite (An_{50-55}). There are abundant relics of clinopyroxene (10–40%), partially replaced by normal hornblende (20–40%). There also are relics of orthopyroxene in gabbro-norites and norites, and of olivine in peridotites. Sphene (4–8%) prevails among accessory minerals and has replaced titanomagnetite. The content of quartz and orthoclase is usually low (2–3% of each) but sometimes reaches 15%. Granitized microgabbroids contain an increased amount of orthoclase (10–20%) and biotite (5–15%); plagioclase is more sodic—andesine (An_{30-40})—while all dark minerals (including secondary hornblende and biotite) are more ferric. These rocks chemically approach gabbro-syenites or even monzonites, though their present composition is, in fact, the result of this granitization process. Amphibole-plagioclase rocks are highly schistose or gneissose; their primary gabbroid appearance and composition are almost completely lost. They consist of plagioclase varying in amount from 20% to 70% (usually 35–50%) and hornblende, from 15% to 60% (commonly 30–50%). Plagioclase composition ranges from An_{30} to An_{65}; in some cases the anorthite content varies by 20% in one specimen. Relics of pyroxenes are invariably present (1–2% to 6–7%, and rarely as much as 10–12%). Biotite (1–10%) is also ubiquitous, but orthoclase (1–10%) and quartz (2–3%) are recorded only sporadically. Accessory sphene varies from 2–3% to 6–7%. Chemical analyses of these rocks show little difference from gabbroids; there is some increase in the alkali and silica contents, compensated by a decrease in iron and especially in magnesium.

Extensive zones of diaphthorites reaching many hundreds of meters in width have formed within gabbroid massifs where the latter have been involved in ancient fault dislocations. Diaphthorized gabbroids are banded cataclasites with relics of gabbroic and panidimorphic textures. Their composition is controlled mainly by plagioclase, varying from oligoclase to labradorite, and by amphibole, in which actinolite prevails over the hornblende it replaces. The chemical composition of these diaphthorites differs from that of normal gabbro only in an increased content of constitutional water (as much as 8%) and in a decreased content of alkalis (especially potassium), magnesium, and calcium.

Charnockitoids

In Antarctica this rock type occurs only in the lower Precambrian crystalline basement of the East Antarctic craton (Ravich and Kamenev, 1972). Massifs of charnockitoids are associated with mobile blocks of the crystalline basement and typically occur among polymetamorphic rocks of the Humboldt Complex. Charnockitoid massifs are variable in age, origin, and nature of superposed alterations. Primary intrusive and secondary apogabbroid charnockitoids, for instance, can positively be distinguished. In addition, the multiple activizations of mobile crystalline basement blocks appear to have been responsible for drastic changes in both types of charnockitoids, which partially lost their specific features and were locally turned into granitoids. Crystallization of primary gabbroid intrusions from which the secondary charnockitoids were derived was most likely completed in early Proterozoic time, but their gradual transformation into charnockitoids and granitoids probably extended throughout Proterozoic time. The last event of this polyphase process occurred as late as early Paleozoic time in connection with the Ross orogeny, as indicated by isotopic age determinations ranging from 500 to 550 m.y. (Ravich, 1972).

Intrusive charnockitoids have been found mainly in Queen Maud Land and in the Prince Charles Mountains. Rocks of similar composition (porphyroblastic charnockites) make up extensive bodies in Enderby Land and in MacRobertson Land, but their intrusive origin is dubious. Hence it is preferable to assign these rocks to ultrametamorphic formations. The largest known charnockitoid intrusion is the Lodochnikov Massif in central Queen Maud Land. It is exposed over 4000 km^2 as a vast irregular lensoid plate, extending from west to east for 200 km among the enclosing schists and migmatites. The massif consists mainly of coarse-grained porphyroblastic granosyenites in the central part, and of diorite-syenites, syenites, and even gabbro-norite-syenites in the marginal facies. The latter, however, may represent individual massifs of younger alkaline-mafic rocks intrusive into the major body of granosyenites. Near large faults marked by glacial valleys, coarse-grained two-pyroxene and eulite-fayalite granosyenites are extensively replaced by biotite-hornblende granosyenites and granites, void of specific charnockitoid mineral assemblages. Such secondary alterations of charnockitoids may also be associated with intense activization of the crystalline basement during early Paleozoic time.

The Lodochnikov Massif is characterized by a relatively small proportions of xenoliths and skialiths of enclosing rocks. No distinct contact alteration of enclosing polymetamorphic migmatites and ultrametamorphic charnockites was observed. The dike suite within the intrusive charnockitoids is poorly developed and is represented by rare veins of granites and aplites. The latter may not be related to the formation of charnockitoids but merely superposed upon them. The most widespread variety of massive charnockitoids is represented by porphyroblastic granosyenites of specific composition. They contain about 50% large orthoclase crystals, ranging from 2 to 5 cm, and set in a heterogeneous groundmass consisting of fine (2–4 mm) laths of myrmekitic oligoclase-andesine (An_{25-30}), aggregates of irregular small grains (1–2 mm) of quartz and microcline, and glomeroblastic accumulations of dark minerals, in which the prevailing biotite and amphibole contain relict grains (as long as 1 mm) of pyroxenes and fayalite. The mean mineral composition of the charnockitoids is: orthoclase and microcline—50%, plagioclase—25%, quartz—15%, hornblende—4%, biotite—2%, eulite—2%, ferrosalite—1%, and accessories—1% (sphene, apatite, zircon, and orthite). A high ferruginosity* of dark minerals, as much as 75%–85%, is very characteristic. The chemical composition of the granosyenites is intermediate between granodiorites and syenites.

The formation of charnockitoid intrusions is probably related to late Archean–early Proterozoic tectono-ultrametamorphic processes in the crystalline basement which were responsible for selective melting, transformation of some crustal blocks into a plastic state, and mobilization

* Ferruginosity is estimated as $Fe^{+2} + Fe^{+3}/Fe^{+2} + Fe^{+3} + Mg \times 100$, or determined by optical means when chemical data are not available.

of mixed brittle and plastic masses capable of slow movement and intrusion. Consequently, formation of such rheomorphic intrusions includes: recrystallization within a solid state of refractory components, e.g., pyroxenes, which are turned into secondary amphiboles and biotite; crystallization of melts, with abundant quartz-impregnated orthoclase porphyroblasts as the latest crystallization products; and quartz-feldspar aggregates in the groundmass as the earliest ones. The interaction of brittle, plastic, and melted components produces the anomalous (as compared to normal magmatic rocks) charnockitoid series.

Massifs of apogabbroid charnockitoids are characterized by an absence or negligible amount of granosyenitic rocks and a predominance of various gabbroids feldspathized to diorite-syenite composition. These massifs are relatively small, have initial gabbroic, gabbro-noritic, and even anorthositic composition, and retain the dome-shaped configuration of parental gabbroic intrusions. Massifs of apogabbroid charnockitoids are known in Queen Maud Land (the Humboldt, Sør-Rondane, and Yamato Mountains), in the Vestfold Hills, in the vicinity of Mirny Station, in the Denman and Shackleton Glaciers area, in the Bunger Hills, on the Windmill Islands, in Grearson Oasis, on the Sabrina Coast, and in a number of other locations (Ravich and Kamenev, 1972).

These massifs are rather uniform in composition. The least granitized gabbroids are recorded in the Sør-Rondane and Yamato Mountains, where they are only feldspathized and amphibolized, but not yet turned into true charnockitoids. Elsewhere, there is a continuous series from feldspathized norites and gabbro-norites to quartz diorite-syenites (in some cases to granosyenites and very rarely to granites). There series are characterized by successive change of proportions and chemical compositions of minerals as follows: the amount of plagioclase decreases from 50–60% to 25–40%, and it becomes more sodic in composition (silicic andesine replaces labradorite); the amounts of potassium feldspar, quartz, and fayalite change from 5% to 25%, from 5% to 20%, and from 0% to 5%, respectively; orthopyroxene decreases in amount from 20–25% to 3–10%, but its ferruginosity rises from 40–50% to 80–90%; clinopyroxene decreases in percentage from 10–15% to 0–5% and its ferruginosity increases from 35–40% to 70–80%; and the secondary minerals (amphibole and biotite) also increase in total content from 1–4% to 10–15%. In the chemical composition of the rocks there is a marked change in silica content from 45–50% to 60–65%, but a decrease in calcium and magnesium from 8–10% to 2–3% of each; potassium content increases from 1% to 4%. Aluminum and sodium contents are almost constant, which is typical of granitization of basic rocks. A high content of iron and titanium in charnockitoids distinguishes them from similar magmatic rocks. On the whole, apogabbroic charnockitoids form a continuous granitization series of rocks, which differs from a normal magmatic gabbro-granite series.

Apogabbroid charnockitoids develop through metasomatic recrystallization of originally gabbroic intrusions which were intensely tectonized and brought into interaction with selective melts and high-alkaline solutions. The relics of gabbroids are preserved among intensely altered rocks enriched in orthoclase and albitized. Their dark minerals were enriched in iron, and pyroxene-fayalite assemblages were replaced by biotite and amphibole. Thus, an unstable mineral assemblage was formed, which is very typical of apogabbroic charnockitoids.

Metamorphism

All facies of regional metamorphism are encountered in the lower Precambrian of Antarctica. The most widespread is the granulite facies, onto which the amphibolite facies is common retrogressively superposed; the greenschist facies is subordinate (Ravich and Kamenev, 1972). Metamorphic facies of intermediate pressure considerably dominate over high-pressure facies. Migmatization and granitization processes are of great importance; they begin in the granulite facies environment but culminate in the high-grade amphibolite facies which results in wide development of polymetamorphic rocks with an unstable association of the progressive granulite and retrogressive amphibolite facies. Zonal metamorphism of the epidote-amphibolite and amphibolite facies is also present.

Granulite facies

The granulite facies is represented by three subfacies: (1) medium-pressure and highest temperature pyroxene-granulite subfacies; (2) higher pressure hornblende-granulite subfacies; and (3) highest pressure hornblende-garnet-clinopyroxene subfacies. The first subfacies is responsible for the formation of enderbites (T = 1100–1200°C) and mesoperthitic charnockites (T = 900–1000°C), as yet recorded only within the oldest Napier Complex in the Enderby Land shield and so far not discovered elsewhere in Antarctica. The most stable assemblages of this subfacies are: Pl_{40-80} + Hyp_{25-35} + Cp_{30-45}; Pl_{35-50} + Hyp_{40-55} + Cp_{30-40} + Q; Pl_{25-35} + perthite + Q + Hyp_{45-70}; Pl_{20-30} + perthite + Q + Gr_{60-70} + (Sill).* The second subfacies produced predominantly two-feldspar charnockites (T = 700–800°C) and diverse shadow granites, as well as two-pyroxene schists. The most stable assemblages of the second subfacies are: Pl_{40-60} + Hyp_{30-55} + Cp_{30-40} + Bi_{30-60} + Ho_{40-50}; Pl_{25-45} + Hyp_{40-60} + Bi_{40-60} + Q + (KFsp); Pl_{25-45} + KFsp + Q + Hyp_{55-75} + Bi_{50-70}; Pl_{20-30} + KFsp + Q + Gr_{60-70} + Bi_{60-80} + (Sill + Cor). The third subfacies formed mainly garnet shadow granites (T = 600–800°C and P = 10–13 kbar). Their stable assemblages are: Pl_{35-50} + Cp_{50-65} + Gr_{75-80} (+ Q + KFsp); Pl_{20-30} + Cp_{50-75} + Gr_{75-85} + Ho_{60-80} + Hyp_{50-50} + Bi_{60-80} + (Q + KFsp).

Amphibolite facies

The amphibolite facies is represented by two subfacies: (1) high-temperature and medium-pressure sillimanite-almandine subfacies; and (2) lower temperature and high-pressure staurolite-kyanite subfacies. The first subfacies is represented primarily by migmatized biotite-amphibole and biotite-garnet plagiogneisses, orthoamphibolites, calciphyres, and quartzites. The typical assemblages are: Ho_{45-70} + Bi_{50-65} + Pl_{25-30} + Q (migmatite melanosome); Ho_{70-80} + Bi_{55-80} + Pl_{20-30} + KFsp + Q + Gr_{80-90} (shadow granites); Ho_{25-35} + Bi_{30-35} + Pl_{60-80} + Cp_{20-25} (orthoamphibolites). The staurolite-kyanite subfacies is charac-

* Numbers indicate anorthite content for plagioclase and ferruginosity for mafic minerals.

teristically non-migmatized and consists mainly of plagiogneisses, cummingtonite amphibolites, and quartzites. In plagiogneisses an early assemblage of biotite with sillimanite, cordierite, and garnet coexists with a later assemblage of fibrolite, kyanite, staurolite, and muscovite.

GREENSCHIST FACIES

The greenschist facies assemblages are the youngest within the lower Precambrian. They are developed either in the lower-middle Proterozoic supracrustal sequences or in diaphthorite zones associated with still younger faults. Greenschists of this facies display assemblages of two subfacies. The biotite-chlorite subfacies is characterized by Bi + Chl + Gr (spessartite) + Mt + Q ± Ep + Alb; the sericite-chlorite subfacies is represented by Chl + Ser + Q + Ca + Ep + Hem assemblages.

Problems for further exploration

The lower Precambrian metamorphic and intrusive formations comprise more than half of the exposed areas in Antarctica. Hence, investigation of the lower Precambrian is of extreme importance for understanding the geologic history of the continent. Such investigation can be greatly facilitated by geophysical methods that can penetrate the ice cover and thus can contribute to the understanding of the subglacial geologic structure, primarily the crystalline basement. Therefore, a detailed study of the lower Precambrian metamorphic and intrusive formations, which at present is complete for only 5% of the crystalline basement outcrops, remains a principal task not only for academic purposes, but also for interpretation of future geophysical data.

A reliable stratigraphic correlation between different structural provinces of the crystalline basement must be worked out on the basis of geologic observations and isochron age determinations. The wide range of metamorphic processes that have affected the crystalline basement makes it necessary to continue a detailed study of mineral assemblages in order to work out a genetic classification of Antarctic metamorphic rocks.

The problem of the Archean-Proterozoic boundary is particularly important. It seems appropriate at present to assign to the Archean all variety of formations making up the crystalline basement of the Antarctic craton, and to consider all supracrustal formations (i.e., the epicratonic platform cover, the miogeosynclinal intracratonic fold system, and the proper geosynclinal pericratonic fold system) as Proterozoic. In this definition the Archean-Proterozoic boundary will have clear geologic meaning and will be recognizable in the course of field mapping. The improvement of geochronologic techniques will allow the testing of this suggested classification.

Of special interest is the further study of the enderbite-charnockite formations which are so widespread in Antarctica. These ancient rocks, formed at the boundary between plutonic and ultrametamorphic conditions, are critical for understanding interrelationships of metamorphic and magmatic processes, as well as the early history of the earth.

Comparative studies of structural and petrologic features of the crystalline basements of the Antarctic and other Gondwana platforms are of special value. Such studies can reliably confirm the existence of Gondwanaland (Ravich and Grikurov, 1974) and clarify the differences in the crystalline basements of ancient platforms in the Northern and Southern Hemispheres.

References cited

Atrashonok, L. Ya., Avdzeiko, G. V., Klimov, L. V., Krylov, A. Ya., and Silin, Yu. I. 1967. Sravnitel'nye dannye po absolutnomu vozrastu porod Antarktidy (svintsovyi i argonovyi metody) [Comparative data on the absolute age of the rocks of Antarctica (by lead and argon methods)]. (*In* Afanasiev, G. D., *ed. Voprosy datirovki drevnejshikh geologicheskikh obrazovaniy i osnovnykh porod* [Questions of dating of the oldest geological formations and fundamental rocks]. Moscow, Izdatel'stvo Nauka, 227–29).

Australia. Bureau of Mineral Resources, Geology and Geophysics. 1975. *Preliminary edition, 1:250,000 Geological Series.* Mawson Escarpment South (Sheet 40-42/8).

Gunner, J. D., and Faure, G. 1972. Rb-Sr geochronology of the Nimrod Group, central Transantarctic Mountains. (*In* Adie, R. J., *ed. Antarctic Geology and Geophysics.* Oslo, Universitetsforlaget, 305–11.)

Halpern, M., and Grikurov, G. E. 1975. Rubidium-strontium data from the southern Prince Charles Mountains. *Antarct. Jnl U.S., 10,* No. 1, 9–15.

Neethling, D. C. 1972. Age and correlation of the Ritscher Supergroup and other Precambrian rock units in Queen Maud Land, Antarctica. (*In* Adie, R. J., *ed. Antarctic Geology and Geophysics.* Oslo, Universitetsforlaget, 547–56.)

Ravich, M. G. 1972. Regional metamorphism of the Antarctic platform crystalline basement. (*In* Adie, R. J., *ed. Antarctic Geology and Geophysics.* Oslo, Universitetsforlaget, 505–15.)

Ravich, M. G., and Grikurov, G. E. 1974. Sravnitel'naia tektonika gondvanskikh materikov [Comparative tectonics of Gondwanaland continents]. (*In* Lazurkin, V. M., *ed. Problemy geologii polyarnykh oblastei Zemli* [Problems of geology of polar regions of the Earth]. Leningrad, Research Institute of Arctic Geology, 32–42.)

Ravich, M. G., and Grikurov, G. E., *eds.* 1976*a*. *Explanatory Notes to the Geologic Map of Antarctica, Scale 1:5,000,000.* Leningrad, Research Institute of Arctic Geology, 83 pp.

Ravich, M. G., and Grikurov, G. E., *eds.* 1976*b*. *Geologic Map of Antarctica, Scale 1:5,000,000.* Leningrad, Kartfabrika ob'edineniia, Aerogeologiia.

Ravich, M. G., and Kamenev, E. N. 1972. *Kristallicheskii fundament Antarktichcheskoi platformy* [Crystalline basement of the Antarctic platform]. Leningrad, Gidrometeoizdat, 685 pp. [English translation: New York, John Wiley, 1975, 582 pp.]

Ravich, M. G., and Krylov, A. Ya. 1964. Absolute ages of rocks from East Antarctica. (*In* Adie, R. J., *ed. Antarctic Geology.* Amsterdam, North-Holland Publishing Company, 579–89.)

Ravich, M. G., and Soloviev, D. S. 1966. Geologiia i petrologiia tsentral'noi chasti gor Zemli Korolevy Mod [Geology and petrology of the mountains of central Dronning Maud Land (eastern Antarctica)]. Moscow, Izdatel'stvo Nedra, 290 pp.

Ravich, M. G., Klimov, L. V., and Soloviev, D. S. 1965. Dokembrii Vostochnoi Antarktidy [The Precambrian of East Antarctica]. Moscow, Izdatel'stvo Nedra, 468 pp.

Ravich, M. G., Sobotovich, E. V., Kamenev, E. N., and Rudnik, V. A. 1974. O pozdneazoiskom vozraste drevneyshikh porod Antarktidy [Late Azoic age of the oldest rocks of Antarctica]. *Dokl. Akad. Nauk SSSR, 216,* No. 6, 1368–74.

Soloviev, D. S., and Halpern, M. 1975. Pervye arkheiskie izotopnye vozrasty, poluchennye v Antarktide po porodam kristallicheskogo fundamenta [First Archean isotopic ages, obtained in Antarctica on rocks of the crystalline basement]. *Inf. Byull. sov. antarkt. Eksped.,* No. 90, 23–25.

52

Regional Metamorphism in Antarctica

E. N. KAMENEV
Department of Antarctic Geology, SEVMORGEO, Moyka 120, Leningrad 190121, U.S.S.R.

Abstract. The metamorphic grade of rocks in each major crustal province of Antarctica is a function of their age; the oldest rocks in each province are the most metamorphosed. The largest crustal structures ("geostructures") can probably be correlated with the largest metamorphic zones ("geozones"). It is also possible to distinguish zones of different orders of magnitude of metamorphism. Where the temperature is constant (within the limits of a given facies), a trend from low-pressure to high-pressure metamorphism is observed. Increase in pressure is also characteristic of successive change from one type of metamorphism to another, i.e., within the polymetamorphic complexes. Zoned metamorphic rocks are characterized by an increase in content of bound water from the high-temperature facies zones to the low-temperature ones. The polymetamorphic state builds up simultaneously with the granitization state in the granulite and amphibolite facies zones. Rocks of these zones can be considered as a single polymetamorphic complex, with a regressive trend of evolution. The largest metamorphic zones are lateral and outwardly regressive in the high-temperature field, and both lateral and vertical in the low-temperature field. Lateral regressive zoning reflects prolonged "thermal" tension of a crustal block. These relationships are typical of the process of tectonic cratonization.

Fundamental characteristics of metamorphism of the lower Precambrian crystalline basement of the Antarctic shield and the younger fold belts are: (1) an absence of metamorphic complexes with two-pyroxene gneisses (granulite) facies in the fold regions; (2) the linear nature of metamorphic zones in fold regions, and their non-linear nature in the crystalline basement; and (3) only restricted development of the greenschist (mainly high-grade) facies in the basement within the superimposed depressions and/or fault zones, but extensive development of the greenschist (mainly low-grade) facies in fold belts.

Two-pyroxene (granulite), amphibolite, greenschist, and laumontite-prehnite-pumpellyite facies groups are widely developed in Antarctica. Epidote amphibolite, kyanite schist, and kyanite gneiss facies are uncommon and strictly localized. The jadeite-lawsonite-glaucophane facies is also present in at least one place. Hence, only the eclogite facies is so far unknown in all of Antarctica, but it is unusual and poorly represented on other continents as well.

Metamorphic facies

Two-pyroxene facies

Two-pyroxene (granulite) facies rocks are dominant in most ancient lower Archean terranes of the East Antarctic craton (Ravich and Kamenev, 1972). The low-pressure two-pyroxene facies ($t = 1100°C$, $P_{total} \leq 3$ to 4 kbar, $P_{H_2O} \approx 0.2\ P_{total}$) is known only in the Napier Complex in Enderby Land. Enderbites of this complex are the oldest Antarctic rocks, with an age of 4100 ± 100 m.y. The medium-pressure two-pyroxene facies ($t = 700°$ to $900°C$, $P_{total} \leq 8$ to 9 kbar, $P_{H_2O} \approx 0.3$ to $0.4\ P_{total}$) is developed in the mountain fringe of the Shackleton Ice Shelf and Denman Glacier, in the Prince Charles Mountains, in the Rayner Complex in Enderby Land, and in Queen Maud Land. The high-pressure two-pyroxene facies ($t = 700°$ to $900°C$, $P_{total} \leq 12$ to 13 kbar, $P_{H_2O} \approx 0.3$ to $0.4\ P_{total}$) occurs in the faulted zone (about 300 km wide) on the southern fringe of Wilma and Robert Glaciers in Enderby Land, and in the polymetamorphic complex of the Humboldt Mountains in Queen Maud Land. Investigations on polymetamorphic phenomena in the Archean complexes of the two-pyroxene facies showed a decrease in $\delta T/\delta P$ gradient values in the course of geologic time. Rocks of the medium-pressure two-pyroxene facies are most widely distributed, while those of the high-pressure facies occur rarely.

Amphibolite facies

Amphibolite facies rocks are widely represented in the upper Archean and, to some extent, lower Proterozoic crystalline complexes of the cratonic basement, in the Ross fold system of the Transantarctic Mountains, and within inliers of the Ross system in the West Antarctic orogenic fold system of the Andean type (Gunn and Warren, 1962; Gunner, 1969; Nathan, 1971; Grikurov, 1973; Ravich and Kamenev, 1972; Lopatin and Polyakov, 1976; Kamenev, 1977; Lopatin and others, 1977). The low-pressure amphibolite facies ($t = 550°$ to $700°C$, $P_{total} \leq 4$ to 5 kbar, $P_{H_2O} \approx 0.8\ P_{total}$) is known only in parts of the Transantarctic Mountains basement at the boundary with the craton. The medium-pressure facies ($t = 650°$ to $800°C$, $P_{total} = 7$ to 9 kbar, $P_{H_2O} \approx 0.8\ P_{total}$) is most widely developed. Within the cratonic basement this facies is known in the Insel Complex of Queen Maud Land, in the Sør-Rondane Mountains, in western Enderby Land, and in the central Prince Charles Mountains. Some obvious products of diaphthoresis of rocks of the two-pyroxene facies, partly associated with their ultrametamorphism as well, belong to the medium-pressure amphibolite facies. Rocks of this facies are common in the basement inliers of the Ross and Paleozoic fold systems in Marie Byrd Land, as well as in the Paleozoic and Mesozoic fold belts of West Antarctica as the highest temperature parts of zonal complexes. The high-pressure amphibolite facies (t = 500° to 650°C, P_{total} = 12 to 14 kbar, $P_{H_2O} \approx 0.8\ P_{total}$) is found only in intracratonic fold systems or in deep fault zones, where it is closely associated with the greenschist facies. Rocks of the high-pressure amphibolite facies occur in three regions: in the narrow faulted zone at the boundary of the Weddell plate and the migmatite fields of Queen Maud Land, in the Shackleton Range, and in the southern Prince Charles Mountains. Diaphthorites with kyanite were recently found in the zone of faults separating the Rayner and Napier Complexes of the two-pyroxene facies in Enderby Land. In all the regions mentioned above, the high-pressure amphibolite facies is part of complicated polymetamorphic complexes with metamorphic zoning.

GREENSCHIST FACIES

Greenschist facies rocks are developed in vast Antarctic areas, but so far their paragenesis is little studied (Gunn and Warren, 1962; Ravich and others, 1965; Clarkson, 1972; Grikurov, 1973; Rudiachonok, 1974; Ravich and Kamenev, 1972; Lopatin and Polyakov, 1976; Kamenev, 1977). The high-pressure facies is distinguished in Adelie Land by the presence of lawsonite and sodic amphibole in the metamorphic rocks, whose age is unknown (Stillwell, 1918). These rocks are probably localized in the deep fault zone that separates the crystalline basement region from the fold belt of the Ross system. The low- and medium-pressure greenschist facies occur in the central Transantarctic Mountains and along the Hillary Coast. In all the remaining cases the greenschist facies cannot yet be subdivided by pressure. In the cratonic basement, rocks of the greenschist facies, most commonly associated with rocks of the epidote amphibolite facies, are distributed in relatively narrow zones of diaphthorites, replacing rocks of the higher temperature facies near faults. As a rule, they are the last manifestations of polymetamorphism in the crystalline basement terranes. Rock units deposited in intra-and epi-cratonic troughs, after formation of high-temperature rocks of the medium-pressure two-pyroxene and amphibolites facies, have also been metamorphosed to the greenschist facies. These include the lower and upper greenschist complexes in the southern Prince Charles Mountains, and single greenschist complexes in the upper reaches of Denman Glacier, in the southern Sør-Rondane Mountains, and in western Queen Maud Land. In each case, the formation of these complexes is probably associated with a late stage in development of the underlying polymetamorphic high-temperature crystalline rocks. The greenschist facies temperature in the cratonic basement varies from 300° to 550°C, but high-temperature mineral assemblages prevail. Total pressure values reached 9 to 10 kbar, with $P_{H_2O} > 0.9\ P_{total}$.

In the West Antarctic fold zones, greenschist sequences make up wide and elongate belts within the most deeply exposed parts of the Ross geosynclines; the belts are intruded with granitoid plutons, in some places associated with zonal metamorphic complexes. In the Transantarctic Mountains and in the Ellsworth Mountains the greenschist complexes occur largely in the internal zone of the Ross geosyncline, but in Marie Byrd Land and in the Antarctic Peninsula they occur in the external parts of the Paleozoic and Mesozoic geosyncline zones. Greenschist complexes of fold regions are, as a rule, low-temperature ones (300° to 400 °C), but during their generation the pressure variations may have been rather substantial, probably ranging from 3 to 12 kbar with P_{H_2O} ranging from 0.5 to 0.9 P_{total}. Zonal metamorphic complexes, connected with granitoid plutons, are commonly associated with the greenschist complexes of fold regions. Such near-contact zonal complexes are represented by those surrounding the Admiralty intrusives in northern Victoria Land or plutons of the Andean series on the Lassiter Coast. These zonal complexes all belong to the andalusite facies series.

LAUMONTITE-PREHNITE-PUMPELLYITE FACIES

Laumontite-prehnite-pumpellyite facies rocks (t = 100° to 300°C, $P_{total} \leq 3$ to 5 kbar) are absent in the cratonic basement. This facies occurs in the Precambrian of the Weddell plate cover and in the lower and middle Paleozoic units of the craton cover; these beds have suffered pre-Permian folding. In the Ross system, metamorphism of this facies is inherent in orogenic complexes. Belts of this facies are mapped as extending between belts of the amphibolite and the greenschist facies. In the West Antarctic Andean system, on the other hand, major geosynclinal complexes apparently have been affected by metamorphism of the prehnite-pumpellyite facies (Craddock and others, 1964; Dow and Neall, 1972; Neethling, 1972; Grikurov, 1973; Rudiachonok, 1974; Lopatin and Polyakov, 1976).

Evolution of metamorphism

Fundamental characteristics of metamorphism in the cratonic basement and in fold systems of the Antarctic section of the Pacific belt are: (1) an absence of metamorphic complexes of the two-pyroxene (granulite) facies in the fold systems; (2) the linear nature of metamorphic zones in fold systems, and their non-linear nature in the crystalline basement; and (3) only restricted development of the greenschist (mainly high-grade) facies in the basement within the superimposed depressions and/or fault zones, but extensive development of the greenschist (mainly low-grade) facies in fold regions.

In each separate tectonic province of Antarctica, ancient complexes have formed at high temperatures and younger units at low temperatures. Zonal metamorphism does not change this pattern, because even in provinces that have the same order of magnitude as the metamorphic zones, ancient rocks are more metamorphosed than younger ones. It is possible to distinguish metamorphic zones of different orders of magnitude that correspond to tectonic provinces of the same order of magnitude. For example, the largest structures (first order of magnitude), or "geostructures," could probably be correlated with the largest metamorphic zones, or "geozones," clearly evident on small-scale maps of metamorphic facies. The combined lateral zoning of the Enderby shield and its surrounding epicratonic orogens and intracratonic fold systems furnishes an excellent example of "geozoning" (Figure 52.1). The pyroxene subfacies of the oldest (Azoic) Napier Complex, occupying a coastal part of Enderby Land, is replaced inland successively by hornblende-pyroxene subfacies of the lower Archean Rayner Complex, by the amphibolite facies of the upper Archean rocks of the orogen of the northern Prince Charles Mountains, and finally by the greenschist facies of the lower Proterozoic fold system of the southern Prince Charles Mountains. Zoning in the Transantarctic Mountains furnishes another example of "geozoning" (Figure 52.2). In this Ross fold system the basement projections are represented by amphibolite facies rocks, the geosynclinal complexes proper by the greenschist (mainly low-grade) facies, and the orogen sequences by rocks metamorphosed in the prehnite-pumpellyite facies. Such geozoning is both lateral and vertical. Outer parts of the system seem to have been most metamorphosed, and the inner parts more weakly metamorphosed.

Where metamorphism developed under nearly constant temperature conditions (e.g., within the range of the given facies), a successive transition from low-pressure to high-pressure metamorphism seems to have occurred in Antarctica. Inferred variations in pressure in metamorphic complexes of the two-pyroxene facies in Enderby Land and of the amphibolite facies in the Transantarctic Mountains and in Marie Byrd Land provide excellent examples. High-pressure complexes there are not only the youngest, but most are localized in deep fault zones. Increase in pressure is also characteristic of successive replacement of one type of metamorphism by another in polymetamorphic complexes, as in the Prince Charles Mountains, where the regionally distributed medium-pressure intracratonic greenschist complex is locally affected by repeated metamorphism in the high-pressure amphibolite facies. Obviously, these features are typical of the tectonic cratonization process. Loss of crustal mobility probably leads, on one hand, to closure of paths of fluid transport, thus promoting conditions favorable to the "autoclave" effect (Dobretsov and others, 1970). On the other hand, it also leads to change in the mechanism of stress relief: flowage is replaced by recrystallization and ultimately by dynamic metamorphism.

Metamorphic zoning of all orders of magnitude is characterized by an increase in the amount of bound water in rocks from high-temperature facies zones to low-temperature zones. Charnockitization in the two-pyroxene facies is also accompanied by the appearance in mineral assemblages of biotite, hornblende, and cordierite. Consequently, charnockitization may be attributed to retrograde metamorphism, and charnockitized areas should be considered as polymetamorphic. The amount of water-bearing minerals increases with a gain in the degree of granitization of rocks in the two-pyroxene facies. However, granitiza-

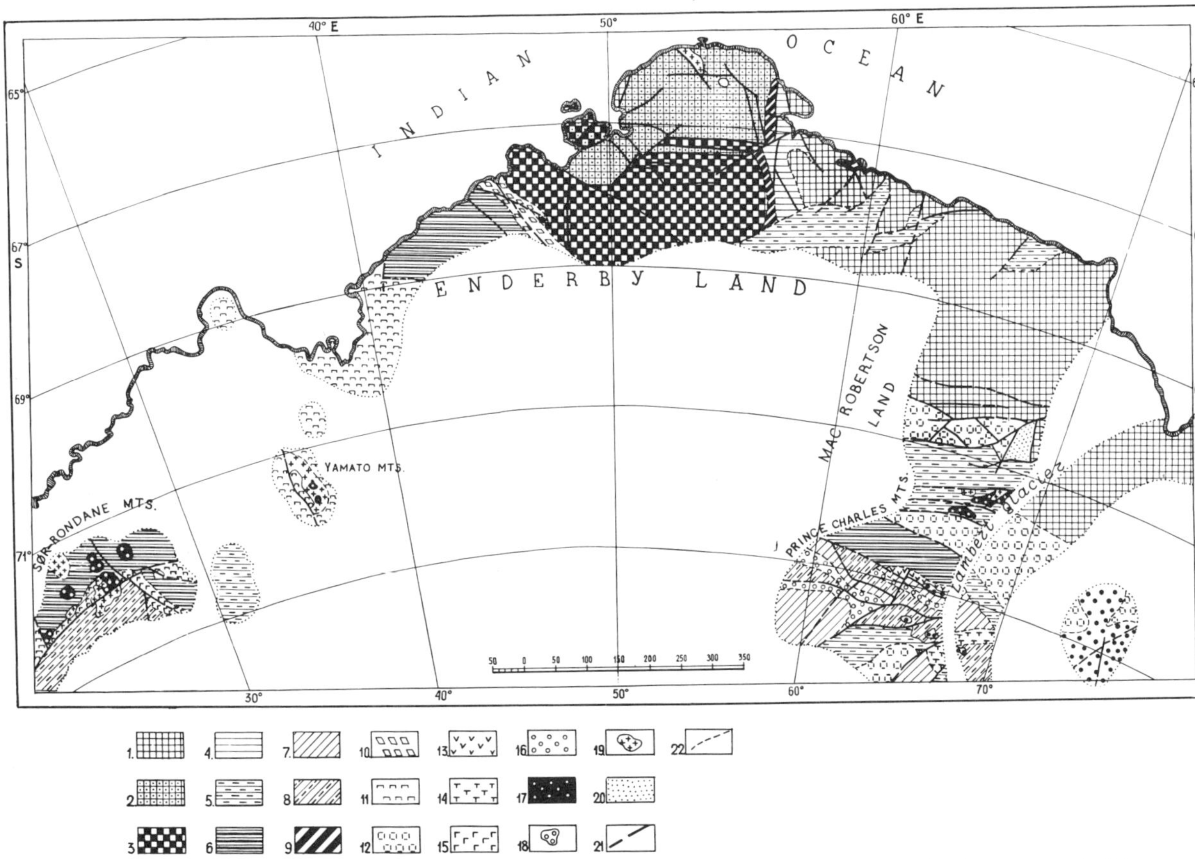

Figure 52.1. Distribution of metamorphic facies in Enderby Land and adjacent terranes. Scale in km. Two-pyroxene facies group: 1. Undifferentiated by pressure; 2. Low pressure; 3. Intermediate pressure; 4. High pressure. Amphibolite (+ epidote amphibolite) facies group: 5. Undifferentiated by pressure; 6. Intermediate pressure; 7. High pressure. Greenschist facies group: 8. Undifferentiated by pressure. Polymetamorphic complexes: 9. Two-pyroxene facies of intermediate and high pressure; 10. Intermediate-pressure two-pyroxene facies and intermediate-pressure amphibolite facies; 11. Undifferentiated two-pyroxene facies and intermediate-pressure amphibolite facies; 12. Undifferentiated two-pyroxene and amphibolite facies; 13. Intermediate- and high-pressure amphibolite facies; 14. Undifferentiated and high-pressure amphibolite facies; 15. Intermediate-pressure amphibolite facies and undifferentiated greenschist facies; 16. Undifferentiated greenschist facies and high-pressure amphibolite facies. Intrusives: 17. Gabbroids; 18. Archean and lower Proterozoic granitoids; 19. Upper Proterozoic and Paleozoic granitoids. Miscellaneous: 20. Unmetamorphosed sedimentary deposits; 21. Faults; 22. Boundaries between facies and complexes.

tion reaches its maximum degree in complexes of the high-temperature amphibolite facies, generally containing relics of mineral associations of the two-pyroxene facies. The following groups of processes are believed to have been simultaneous: (1) charnockitization in all the two-pyroxene facies zones; (2) amphibolization and biotitization in the two-pyroxene facies zones and creation of main mineral assemblages in the high-temperature amphibolite facies zone; and (3) diaphthoresis under the greenschist facies conditions in all the high-temperature facies zones. All this points toward the consideration of the geozones of the two-pyroxene facies and of the amphibolite facies of Antarctic shields as polymetamorphic megacomplexes with a retrogressive development trend. Consequently, the high-temperature parts of the metamorphic geozones (first order of magnitude) are lateral and outwardly retrogressive. However, the low-temperature parts of geozones may be both lateral and vertical: greenschist and prehnite-pumpellyite complexes of the intracratonic fold systems and of the plate cover lie with structural unconformity on the crystalline basement of the craton (vertical zoning), which itself includes local zones of diaphthorites of the greenschist facies (lateral zoning).

Different types of granitoids are closely associated with different types of metamorphic terranes. Ultrametamorphic enderbites and charnockites occur exclusively in the two-pyroxene facies regions, typical only of the cratonic basement. Migmatites and other associated ultrametamorphic granitoids are mainly related to terranes of amphibolite facies rocks in the lower Precambrian complexes of the epicratonic orogens. Granitoid intrusions of a batholithic nature are largely associated with the upper Precambrian and Phanerozoic fold regions, where they occur mostly in areas of the highest temperature metamorphic complexes, predominantly of the amphibolite facies as well. The contemporaneity of granite generation and metamorphism is beyond any doubt. This relationship is not observed in the Antarctic Peninsula, where metamorphic rock complexes of the amphibolite facies contemporaneous with the time of granite generation are not yet reliably known.

The crystalline basement of epicratonic orogens is characterized by great massifs of porphyroblastic granitoids, assigned by M. G. Ravich to so-called "intrusive and apointrusive charnockitoids" (Ravich and Kamenev, 1972). The geologic setting, composition, and structure of the massifs show that "charnockitoids" are somewhat higher temperature analogues of "rapakivi" rocks distributed in the Northern Hemisphere. They are quite unrelated to migmatites, differing in age, in morphology of individual bodies, and in their subordinate fault-block tectonic setting. However, the mineral assemblages of these granitoids, just as those of "rapakivi" rocks, are closely related to the mineral assemblages of enclosing metamorphic rocks, but differ in their higher alkalinity.

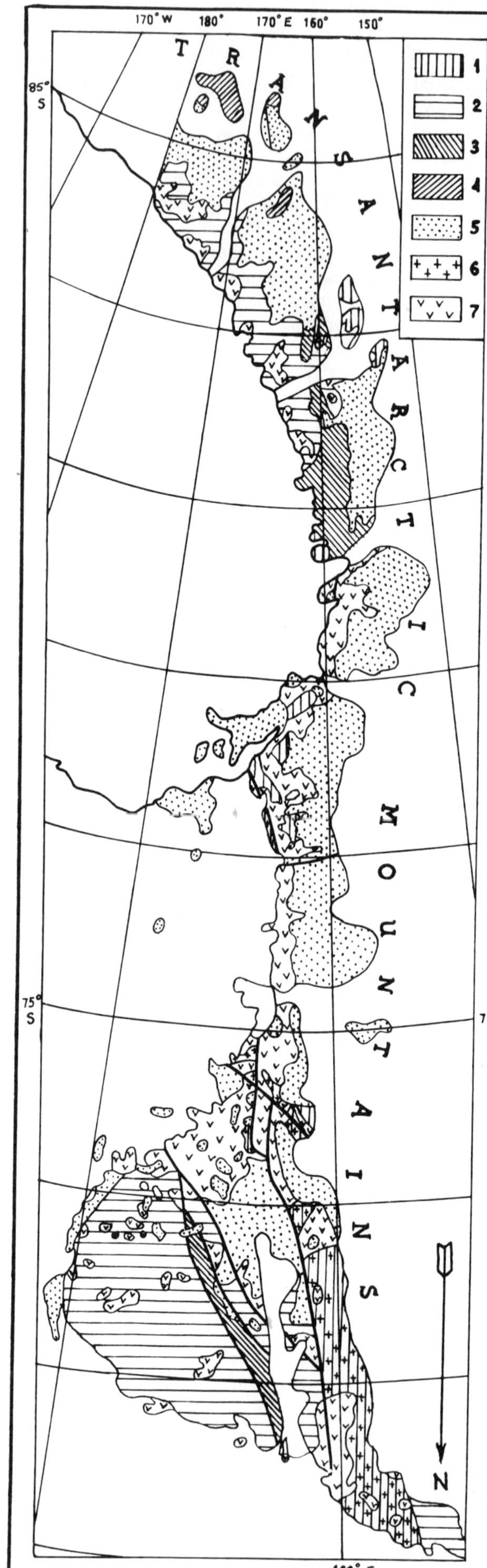

Figure 52.2. Distribution of metamorphic facies in the northern Transantarctic Mountains. (For scale, 1° of latitude equals 108 km.) 1. Low-pressure amphibolite + epidote amphibolite facies; 2. Undifferentiated greenschist facies; 3. Laumontite + prehnite-pumpellyite facies; 4. Laumontite + prehnite-pumpellyite facies without association with higher grade rocks; 5. Unmetamorphosed rocks; 6. Ultrametamorphic granites; 7. Upper Proterozoic and Paleozoic intrusive granitoids.

This suggests a prolonged state of "thermal" tension of the crystalline basement of not only the epicratonic orogens but of the whole Antarctic platform. Long life of the high-pressure metamorphic zones and deep faults, that cut (or restrict) the zones and control magma emplacement, is confirmed by associated anorthosites and alkaline and alkaline-ultrabasic rocks in Queen Maud Land and in the Lambert rift area, whose ages range from Precambrian to Jurassic. Anorthosites and ultrabasic and alkaline igneous rocks are likely to be found in high-pressure zones in Enderby Land, in the Shackleton Range, and in Adelie Land.

Conclusions

The spatial and temporal distribution of metamorphic facies in Antarctica as described above, suggests a definition of the main development of trends of metamorphism through time:

1. smaller oreal extent of metamorphic effects;
2. less high-temperature metamorphism, and more high-pressure metamorphism;
3. retrogressive trend of metamorphism in the crystalline basements of platforms, caused by cooling of the upper crustal layers;
4. progressive trend of metamorphism in fold and orogenic belts, resulting from heating of these belts by plutonic masses, including magmatic, rheomorphic, and tectonic ones;
5. localization of high-pressure metamorphism in deep fault zones at the final stages of consolidation of a mobile region;
6. dependence of regional greenschist metamorphism upon intensity of folding and associated cleavage, probably because of the tectogenetic nature of the heat causing mineral transformation;
7. "submergence" of the thermal source, stimulating regional magmatic and metamorphic transformation and, consequently, deepening of zones of total metamorphism and anatexis.

References cited

Clarkson, P. D. 1972. Geology of the Shackleton Range: a preliminary report. *Bull. Br. antarc. Surv.,* No. 31, 1–15.

Craddock, C., Anderson, J. J., and Webers, G. F. 1964. Geologic outline of the Ellsworth Mountains. (*In* Adie, R. J., *ed. Antarctic Geology.* Amsterdam, North-Holland Publishing Company, 155–70.)

Dobretsov, N. L., Khlestov, V. V., Reverdatto, V. V., Sobolev, N. V., and Sobolev, V. S. 1970. *Fatsii metamorfisma* [The facies of metamorphism]. Moscow, Izdatel'stvo Nedra, 432 pp.

Dow, J. A. S., and Neall, V. E. 1972. Summary of the geology of lower Rennick Glacier, northern Victoria Land. (*In* Adie, R. J., *ed. Antarctic Geology and Geophysics.* Oslo, Universitetsforlaget, 339–44.)

Grikurov, G. E. 1973. *Geologiia Antarkticheskogo poluostrova* [Geology of the Antarctic Peninsula]. Moscow, Izdatel'stvo Nauka, 119 pp.

Gunn, B. M., and Warren, G. 1962. Geology of Victoria Land between the Mawson and Mulock Glaciers, Antarctica. *Bull. geol. Surv. N.Z., n.s.,* No. 71, 157 pp.

Gunner, J. D. 1969. Petrography of metamorphic rocks from the Miller Range, Antarctica. *Rep. Inst. polar Stud. Ohio State Univ.,* No. 32, 44 pp.

Kamenev, E. N. 1977. Nekotorye zakonomernosti raspredeleniia fatsii regional'nogo metamorfizma v Antarktide [Some regularities of facies distribution of regional metamorphism in Antarctica]. *Antarktika, Doklady Komissii, 16,* 100–20.

Lopatin, B. G., and Polyakov, M. M. 1976. *Geologiia i petrologiia Zemli Meri Berd i Berega Eitsa* [Geology and petrology of Marie Byrd Land and Eights Coast]. Moscow, Izdatel'stvo Nauka, 175 pp.

Lopatin, B. G., Fedorov, L. V., and Polyakov, M. M. 1977. Geologiia i metamorfizm proterozoiskikh otlozhenii gor Mak-Koli i Sherger [Geology and metamorphism of the Proterozoic deposits of Mount McCauley and Mount Scherger]. *Antarktika, Doklady Komissii, 16,* 76–92.

Nathan, S. 1971. Geology and petrology of the Campbell-Aviator divide, northern Victoria Land, Antarctica. *N.Z. Jl Geol. Geophys., 14,* No. 3, 564–96.

Neethling, D. C. 1972. Age and correlation of the Ritscher Supergroup and other Precambrian rock units, Dronning Maud Land. (*In* Adie, R. J., *ed. Antarctic Geology and Geophysics.* Oslo, Universitetsforlaget, 547–56.)

Ravich, M. G., and Kamenev, E. N. 1972. *Kristallicheskii fundament Antarkticheskoi platformy* [Crystalline basement of the Antarctic platform]. Leningrad, Gidrometeoizdat, 658 pp. [English translation: New York, John Wiley, 1975, 582 pp.]

Ravich, M. G., Klimov, L. V., and Soloviev, D. S. 1965. *Dokembrii Vostochnoi Antarktidy* [The Precambrian of East Antarctica]. Moscow, Izdatel'stvo Nedra, 470 pp.

Rudiachonok, V. M. 1974. Rifeisko-rannepaleozoiskie skladchatye kompleksy Antarktidy [The Riphean-early Paleozoic fold complexes of Antarctica]. *Antarktika, Doklady Komissii, 13,* 61–84.

Stillwell, F. L. 1918. Metamorphic rocks of Adelie Land. *Scient. Rep. Aust. antarc. Exped.,* Ser. A, *3,* Pt. 1, 1–230.

53

Precambrian Geologic Relationships in the Vestfold Hills, Antarctica

R. L. OLIVER and P. R. JAMES
Department of Geology and Mineralogy, University of Adelaide, Adelaide, S. A. 5000, Australia

K. D. COLLERSON and A. B. RYAN
Department of Geology, Memorial University of Newfoundland, St. John's, Newfoundland, A1C 5S7, Canada

Abstract. Archean gneisses in the Vestfold Hills, East Antarctica, (lat. 68° 35′ S, long. 77° 85′ E) comprise four distinct rock types, each of which contains mineral assemblages indicative of crystallization or recrystallization under granulite facies conditions. The gneissic rock types appear to be structurally interleaved and tectonically repeated. Two of these rock types exhibit strong compositional layering and variably developed parallel schistosity that seems to have formed parallel to the axial planes of mesoscopic intrafolial folds (or traces thereof). This layering is displayed both by granitic to tonalitic (?) layered gray gneiss and by garnetiferous pelitic to semipelitic layered paragneiss. The garnetiferous paragneiss is associated with minor amounts of other metasedimentary rocks including metamorphic derivatives of banded ironstone, impure chert, and carbonate rocks, and with mafic layers or trains of mafic gneiss in the form of boudins and agmatite. Intrusive into these layered rocks are two groups of compositionally similar orthogneiss. The earlier of these intrusive groups exhibits a diffuse compositional (transposed?) layering and tectonic contacts with the layered gneiss. In contrast, the younger homogenous orthogneiss shows a simple fabric and distinct intrusive relationships.

The rock units form ENE-WSW–trending zones 1–2 km wide. A variably developed layering-schistosity ($S_0/S_1 + S_2$), manifested by mineral segregations, also trends ENE-WSW and dips northward. Mineral segregations also define a lineaton which plunges predominantly northward. Folds are mainly of two generations: D_1 isoclinal intrafolial mesofolds and macro- and meso-D_2 folds with axes plunging moderately steeply, predominantly to the north, parallel to the mineral lineation. The homogeneous orthogneiss is believed to have been intruded syntectonically with the D_2 ductile deformation. A later period of more brittle strain (D_3) resulted in the development of planar zones of shear which modified the trend and dip of $S_0/S_1 + S_2$ and the plunge of the lineation. Displacement of gneissic layering has occurred on many of the numerous vertical faults which trend north-south, northeast-southwest, northwest-southeast, and east-west. Swarms of basaltic and doleritic dikes were emplaced along many of the faults and also along parallel joints, probably during Proterozoic time. In the eastern end of the area, the gneiss is cut by a well-developed north-south–trending mylonite zone, as much as 1 km wide.

The Vestfold Hills, the hinterland to Davis Base, form a triangular region, approximately 400 km^2 in area, with a maximum altitude of 160 m above sea level. The area is almost entirely ice-free, and except for some small snowdrifts, it is also snow-free throughout much of the year.

The first geologic reconnaissance of the area was undertaken by P. W. Crohn of the Australian National Antarctic Research Expedition (ANARE) during 1955 (Crohn, 1959). He observed a variety of gneiss and commented on the prominent basic dikes which abound in the region. According to Ravich (1960) in a later study, the dominant rock types in the Vestfold Hills are migmatized "dipyroxenic plagiogneisses," "crystalline schists," and several varieties of "granitoids."

McLeod and others (1966) carried out the first detailed geologic investigation of the area, recognizing and mapping three distinct east-west–trending rock units which they interpreted as being structurally repeated throughout the terrane. Their divisions are as follows:

Unit 1: Variable lithology—dominantly garnetiferous gneiss;
Unit 2: Quartz-feldspar gneiss with minor pyroxene and poorly developed foliation; and
Unit 3: Pyroxene-biotite-quartz-feldspar gneiss.

They recognized the restricted occurrence of norite in the north and also referred to the abundant dolerite dikes.

Rb-Sr geochronology by Arriens (1975) has indicated ages of 2500 m.y. for gneiss and 1400 m.y. for dikes from the Vestfold Hills. Previously, however, Harding and McLeod (1967) had recorded a K-Ar age of 1000 m.y. for the same dikes.

The present paper is based predominantly on field work by Oliver and James in January 1976, following on preliminary field work by Collerson in January 1970, combined with laboratory study by Oliver and James at Adelaide University and Collerson and Ryan at Memorial University.

Geologic relationships and lithology

The Archean gneiss complex in the Vestfold Hills is a variably deformed, layered sequence of diverse rock types,

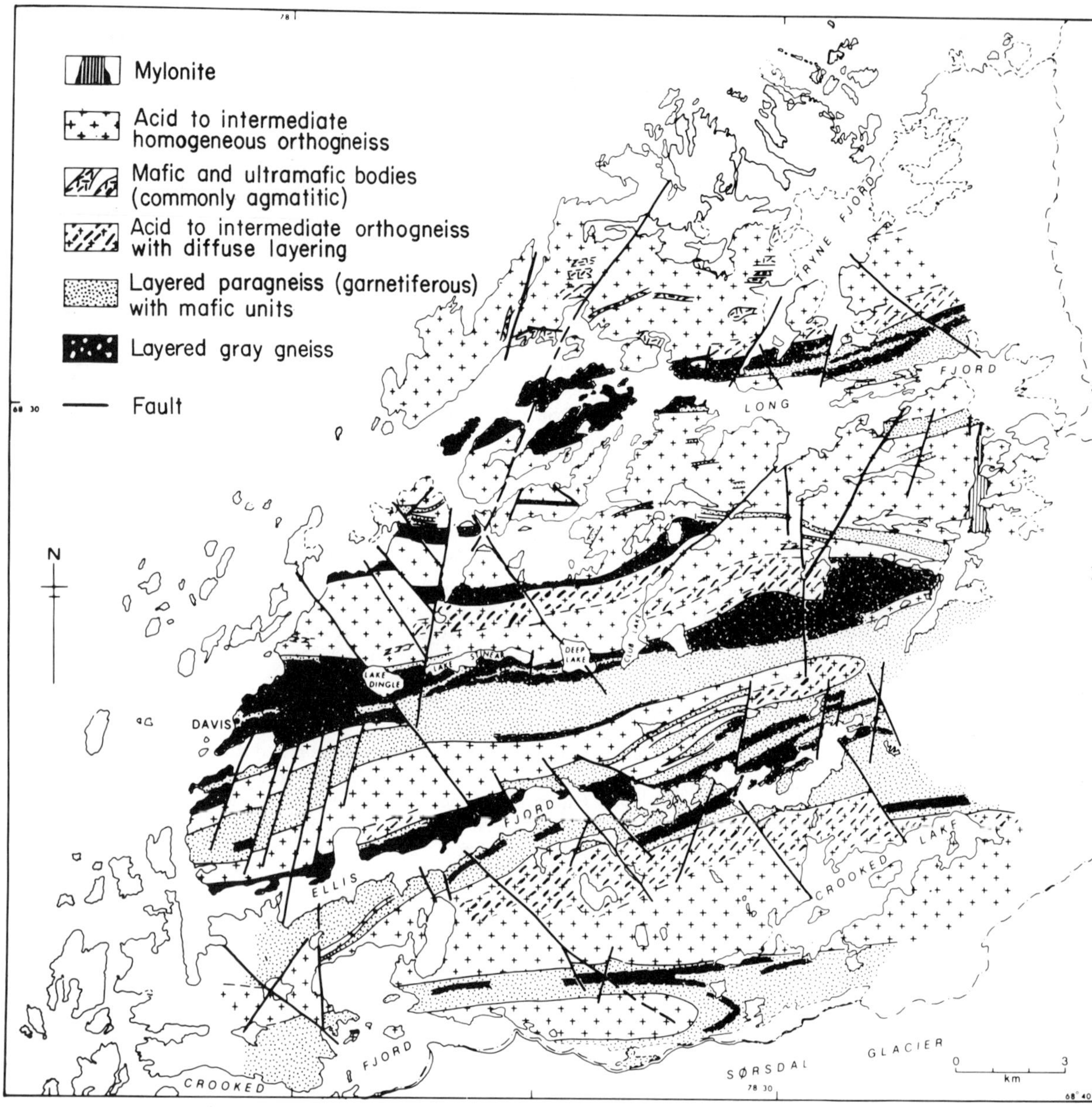

Figure 53.1. Geologic map of the Vestfold Hills.

comprising four principal map units (Figure 53.1). These units are (1) well-layered gray gneiss with intercalated, sharply bounded, mafic and felsic units and sporadic pods of metasedimentary rocks; (2) well-layered garnetiferous paragneiss with subsidiary mafic and submafic intercalated layers; (3) acid to intermediate orthogneiss with diffuse layering; and (4) acid to intermediate, relatively homogeneous orthogneiss, with simple fabric. Relationships between major layered units are interpreted as tectonic, thereby accounting for their repetition. Intrusive relationships between the homogeneous orthogneiss and other units are indicated by the presence of xenoliths and simpler fabric in the former. The mineralogy and texture of most gneissic rocks (Table 53.1) reflects crystallization and recrystallization under granulite facies conditions during several stages of the metamorphic and tectonic history.

Other rock types (Archean and Proterozoic) of lesser importance in the region are also present, including a broad zone of mylonite in the eastern edge of the area (Figure 53.1). The distribution of the Proterozoic basaltic dike swarm is shown in Figure 53.2.

Layered gray gneiss

The layered gray gneiss exhibits a well-developed compositional layering on a scale of 1-20 cm, caused by variation in the relative amounts of quartzo-feldspathic and mafic constituents (orthopyroxene and hornblende in equilibrium), and accentuated locally by the intercalation of discrete mafic units (Figure 53.3). Intrafolial folds of earlier formed layering and parallel schistosity are commonly preserved in the layered gray gneiss (Figure 53.3), and a strong schistosity, defined either by the presence of aggregates of platy quartz or by trains of prismatic orthopyroxene and hornblende, is developed parallel to the axial planes of these folds.

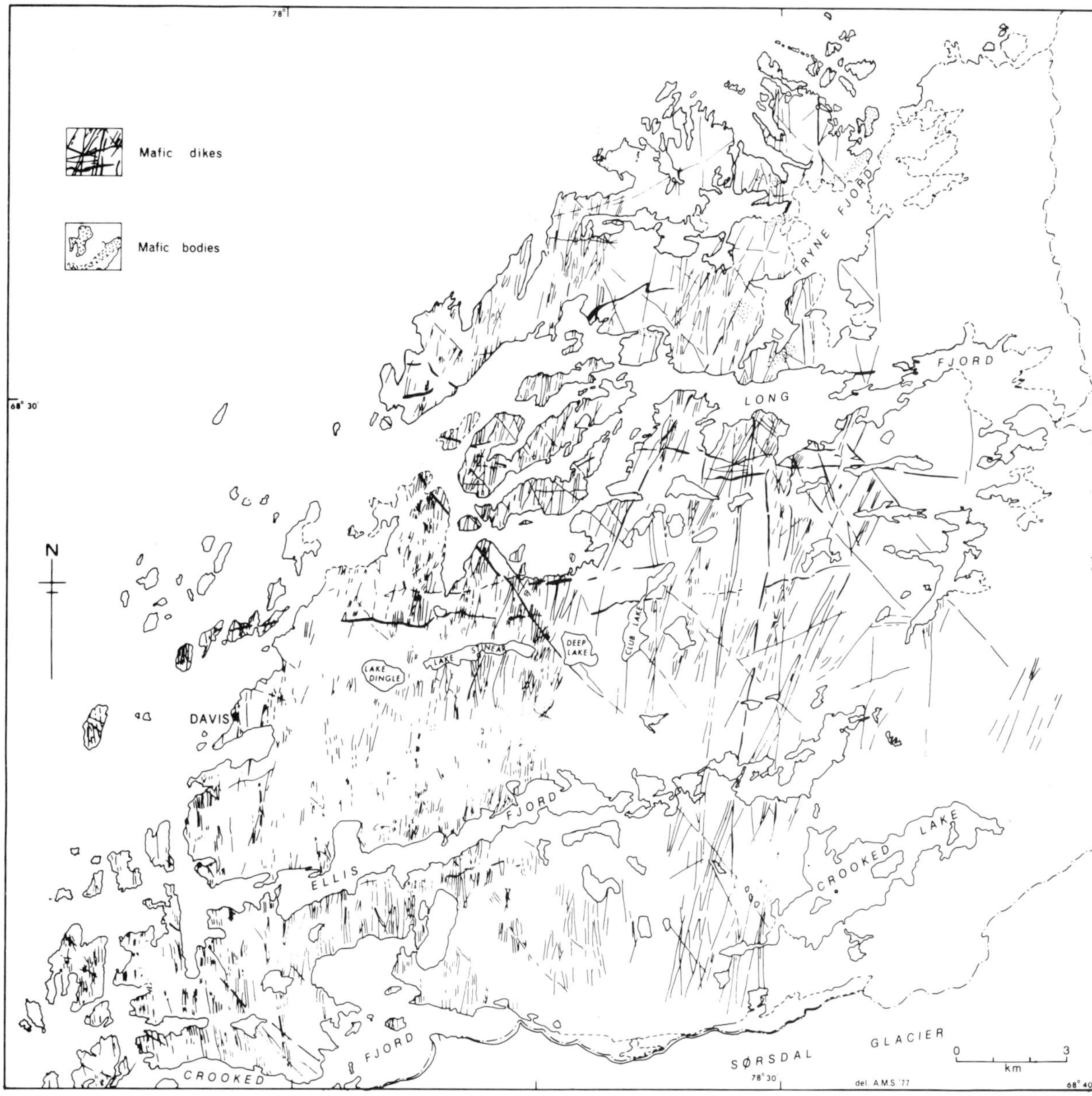

Figure 53.2. Map of the Vestfold Hills showing distribution of mafic dikes and bodies.

LAYERED PARAGNEISS

The most common of the metasedimentary rock is garnetiferous paragneiss. It typically consists of thin, elongate, lenticular layers of fine- to medium-grained quartz-feldspar-biotite-garnet-orthopyroxene gneiss intercalated with coarse-grained (average grain size about 1 cm) quartz-feldspar-garnet gneiss (Figure 53.4). This compositional layering is believed to be the result of metamorphic differentiation or local anatexis, and its present geometry is regarded as being the result of folding and transposition. Coarse-grained quartzo-feldspathic material also commonly transgresses the gneissic layering as a migmatitic

Table 53.1. Modal composition (in vol %) of typical Archean gneiss from the Vestfold Hills

	Homogeneous orthogneiss			Orthogneiss with diffuse layering				Layered paragneiss				Layered gray gneiss	Ultramafic rock		
	61A	14	119B	33	39	40	68	91	100	119E	188	86	87	167	181
Quartz	0	5	0	40	46	0	15	75	5	7	0	1	0	0	0
K-feldspar	0	0	0	0	46	0	30	5	0	0	0	0	0	0	0
Plagioclase	55	50	45	40	1	50	5	0	60	8	40	49	1	0	7
Orthopyroxene	30	20	45	15	0	0	25	10	22	0	5	25	80	85	0
Clinopyroxene	0		0	0	0	0	0	0	3	0	0	1			80
Hornblende	trace	0	0	0	0	16	0	0	3	0	40	15	15	0	7
Biotite	10	20	5	5	0	16	23	10	5	40	5	5		15	6
Opaque minerals	5	5	5		0	16	2	trace	2	0	0	4	4	0	0
Garnet	0	0	0	0	0	0	0	0	0	45	10	0	0	0	0

All specimens are prefixed by An 77 and are located at the Department of Geology, University of Adelaide, South Australia.

Figure 53.3. Layered gray gneiss folded into type-3 (Ramsay, 1962) interference pattern. South of Lake Stinear.

Figure 53.4. Garnetiferous layered paragneiss on left, with resistant quartzose layers. Massive garnet quartzite in center; pegmatite at top left. South of Lake Stinear near Ellis Fjord.

neosome and locally comprises an extensive matrix enclosing disconnected and randomly oriented gneiss fragments.

There is abundant textural evidence that garnet, biotite, and hypersthene coexist stably in both the paleosome and neosome components of this gneiss; the presence of hypersthene suggests that migmatization occurred under granulite facies conditions. In localized shear zones, however, hypersthene is replaced in part by fine-grained, red-brown biotite. Significant minerals of rare occurrence in the garnetiferous paragneiss are cordierite and sillimanite, commonly as poikiloblastic inclusions in garnet porphyroblasts and as spinel with coronas of corundum.

Other layered paragneiss rock types recognized include orthopyroxene-cordierite quartzite (originally Mg-Al chert?), quartz-diopside-scapolite rocks, diopside-anorthite rocks, forsterite marble (lime-rich metasedimentary rocks), and finely layered quartz and quartz-magnetite (banded-iron formation). As well as forming separate mappable units for detailed work, thin septa of these metasedimentary rocks are also structurally intercalated within the gray gneiss.

Layered and foliated medium- to fine-grained mafic gneiss containing variable amounts of orthopyroxene, clinopyroxene, amphibole, biotite, and plagioclase is commonly intercalated with the layered paragneiss as semicontinuous horizons or trains of boudins or agmatite. The common association of mafic units containing orthopyroxene, apparently in equilibrium with amphibole, together with amphibole-free mafic units, is regarded as reflecting local variations in water pressure during metamorphism.

By analogy with other Archean gneiss terranes (cf. Bridgwater and others, 1973; McGregor, 1973; Myers, 1976; Collerson and others, 1976), it is highly probable that the mafic units are metavolcanic or, alternatively, represent basic intrusions into the metasedimentary sequence.

Acid to intermediate orthogneiss with diffuse layering

The acid to intermediate diffusely layered orthogneiss is medium- to fine-grained and ranges in composition from granitic to dioritic. The characteristic diffuse, discontinuous layering (Figure 53.5) results from the elongation of mafic to submafic schlieren or variation in the texture of the quartzo-feldspathic component. More discrete, lenticular, metasedimentary xenoliths also occur. A relatively strong S-fabric defined by the alignment of orthopyroxene, hornblende, and biotite parallels the layering in some cases. Although intrafolial folds have not been recognized in the field, it is considered likely that the layering and schistosity are transposed earlier fabrics. Discrete, sharply bounded, mafic layers which occur in the layered gray gneiss are absent from this unit.

Acid to intermediate homogeneous orthogneiss

The dominant component of the homogeneous orthogneiss is a medium-grained, uniformly textured rock weathered to buff color, with "streaky" foliation and downdip lineation of variable intensity. The foliation and lineation, where well developed, are due to parallelism of ellipsoidal quartz grains, feldspar prisms, and ferromagnesian mineral clots 1–5 cm long (Figure 53.6). In places, lineation and foliation are virtually absent.

Xenoliths of both apparent metasedimentary and meta-igneous nature are relatively common within the orthogneiss, testifying to its igneous origin; invariably these xenoliths are in the form of elongated schlieren or early fold remnants (Figure 53.7). The size of such inclusions ranges from only a little larger than the ferromagnesian mineral clots, referred to above, to 20–30 m in length; intercalated layers may reach regional extent.

In thin section, the mineral content (Table 53.1) can be seen to consist essentially of 50% plagioclase and 50% pyroxene, with subsidiary biotite. In general, there is a weak fabric defined by the parallel alignment of orthopyroxene. Biotite, where present in significant quantity

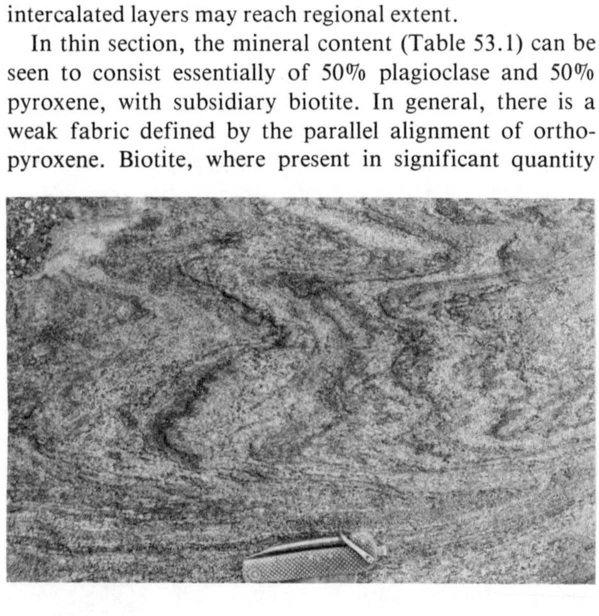

Figure 53.5. Acid to intermediate orthogneiss with diffuse layering and early schistosity folded during D_2, Near Deep Lake.

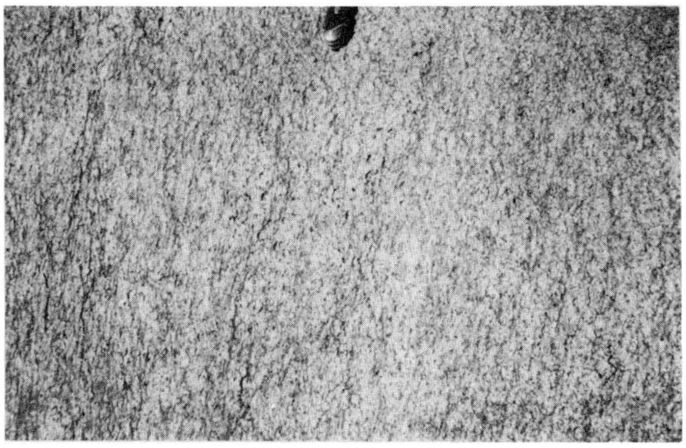

Figure 53.6. Homogeneous orthogneiss showing simple anastamosing D_2 fabric. North of Lake Dingle.

Figure 53.7. Homogeneous orthogneiss with mafic xenolithic remnants of early folding. Between Ellis Fjord and Lake Stinear.

(e.g., 20%), exhibits a stronger fabric, but in these cases there is no indication that the biotite and orthopyroxene have not crystallized synchronously and in equilibrium.

ULTRAMAFIC BODIES

There are a number of relatively large ultramafic bodies included within the gneiss. Many are linear features, as seen on air photos, or consist now of a line of boudins or pods, each 20-30 m in width and wrapped by the S_2 foliation in the surrounding gneiss; others appear to have been folded. Most have some foliation.

Within the metasedimentary rocks as well, a number of smaller bodies of the same ultramafic composition occur as boudins.

In Figure 53.1, a number of mafic bodies are shown north of Long Fjord. These may be genetically associated with those mentioned above, though one at least is known to be noritic in composition (McLeod and others, 1966). Many of these northern bodies are agmatites, which suggests that they predate the orthogneiss which surrounds them.

MYLONITE

A mappable zone of mylonitized rock occurs in the eastern edge of the area (Figure 53.1). It is characterized in the field by overall darker color, fine-grained texture, well-developed foliation, and westerly plunging lineation (see also below).

PEGMATITE

There are no large pegmatite bodies in the Vestfold Hills, but minor pegmatite development is widespread, mainly in the form of thin veins and patches. Gneissic rocks of a variety of compositions, from mafic to acid, serve as hosts for these pegmatites.

Structure

The structure of the Vestfold Hills metamorphic terrane is more complex than envisaged by McLeod and others (1966). The apparently simple pattern of ENE-WSW striking and north-dipping layering and schistosity is superimposed, in the belts of layered gray gneiss, layered paragneiss, mafic and submafic gneiss, and acid to intermediate, diffusely layered orthogneiss, on an earlier tectonothermal fabric which is largely absent in the homogeneous orthogneiss. The area also displays a complex pattern of large- and small-scale conjugate, simple shear zones, a major mylonite zone, and fault zones marked by pseudotachylite.

EARLY HISTORY

Evidence of an early tectonothermal and composite history of formation is provided by abundant isoclinal, attenuated, and commonly rootless, intrafolial folds, some of which show refolding of earlier folds to produce complex interference patterns (e.g., a fold near Deep Lake appears to be of the type-3 pattern discussed by Ramsay, 1962) (see Figure 53.3). These structures offer clear evidence for the tectonic origin of some of the layering within the gneiss complex. Elsewhere evidence of possible sedimentary origin is provided by fine layering of quartz and quartz-magnetite in rocks resembling banded iron formation. Parallel and discordant, coarse-grained, quartzo-feldspathic pegmatite layers are examples of the effects of metamorphic differentiation and anatectic remobilization in the formation of some layers.

Early folds within the layering have folded an even earlier tectonic layering and fabric (also at granulite grade?) and have also developed strong fabric parallel to the axial planes. These early folds show their complex origin in the variability of their plunge within the layering. A detailed analysis of the early deformational history in the different gneiss belts is obviously important to the understanding of the present geometry and early form of these belts. Such an analysis, however, is not possible at present, and more detailed observations are necessary.

For the sake of simplicity, all early deformational events are included in D_1, since for the present purpose it is sufficient to describe only the event which produced the main layering (and general parallelism of early features) in the gneiss.

D_2 STRUCTURES

No pre-D_2 structures have been discovered in the homogeneous orthogneiss bodies. These bodies, which display obvious intrusive relationships (see above), are therefore most likely to have been emplaced syntectonically with this major event, when the area was still undergoing granulite facies metamorphism. The preexisting anisotropy of the gneiss complex formed by the earlier events (D_1) and the developing D_2 compressional stress field may together explain the extremely elongate nature of these orthogneiss bodies.

Both the orthogneiss and the paragneiss of the Vestfold Hills contain a variably developed fabric formed by crystallization or recrystallization during D_2. Within the layered gneiss, this fabric is parallel to the axial planes of a set of tight folds of the $S_0 + S_1$ layering and schistosity; in the south near Crooked Fjord $S_0 + S_1$ forms a major closure. The more common F_2 minor folds have axial planes parallel to the D_2 fabric, are typically asymmetric, and plunge at moderate angles to the north-northeast (Figure 53.8). Within the orthogneiss, F_2 folds are not found because of the lack of a preexisting anisotropy.

Variations in morphology of the D_2 fabric in the orthogneiss and paragneiss are due mainly to differences in composition. The D_2 fabric is best displayed in the coarse homogeneous orthogneiss where both foliation (schistosity) and lineation are typically combined as an LS tectonite fabric. This S_2L_2 fabric is expressed mostly as a weak planar and linear elongation of quartz, feldspar, and ferromagnesian grain aggregates, as mentioned above. It is interpreted to have formed during the first deformation of the orthogneiss by a single, finite strain which possibly occurred during the intrusion of the igneous protoliths of the gneiss.

The D_2 fabric in the preexisting gneiss is clearly the product of D_2 plus all the earlier deformational events, namely D_1. As such, the shape of the fabric produced is variable, especially near D_2 folds where the superimposed strain has produced both purely planar and purely linear finite fabrics. The D_2 fabric on a large and small scale normally parallels the main layering in the gneiss and the boundaries of metasedimentary belts (except in the hinge regions of D_2 folds). This fabric appears to have formed with a regular ENE-WSW trend and moderate to steep northward dip (Figure 53.9). The morphology of the F_2 folds and distribution and style of the D_2 fabrics suggest deformation under highly ductile conditions, which is further evidence that this event occurred close to a peak of metamorphic conditions. The D_2 orientations, however, have also been much modified in some places by the later deformational events.

Modification of D_2 structures

As previously mentioned, the layering and schistosity produced during D_1 and D_2 generally trend ENE-WSW. However, some large planar zones can be delineated where the trend of the foliation is in a roughly ESE-WNW direction, e.g., in the south near Crooked Fjord, across Long Fjord, and (on a smaller scale) on Broad Peninsula (Figure 53.9). In addition to their variation in trend, the major layering and schistosity both appear to have markedly variable, but logical dip variations. These dip variations are contoured and shown in Figure 53.10. Most of the layering and fabric in the metasedimentary rocks ($S_0/S_1 + S_2$) and the fabric in the orthogneiss (S_2) dip at angles of 60°–80° N. However, in certain bands shown by the contours (Figure 53.10), the dip decreases markedly to 30°–60° (e.g., south of Long Fjord and north of Sørsdal Glacier). In the field these planar zones of gentler dip can be traced across strike, symmetrical about a central minimum. These zones vary from a few hundred meters to 1–2 km in width, and from the map they are seen to reach 5–10 km in strike length.

The elongation lineation (L_2) in the schistosity is shown in Figure 53.11. This lineation, which is mainly downdip,

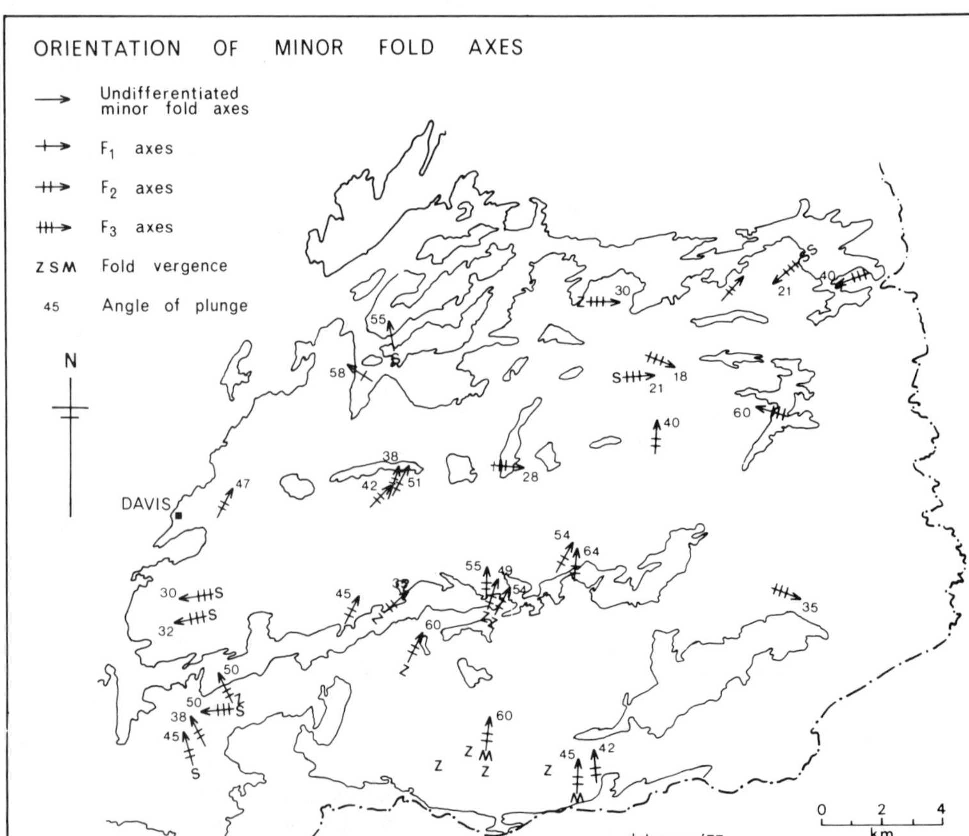

Figure 53.8. Map of the Vestfold Hills showing orientation of minor fold axes.

also changes its plunge angle. In the zones of gentle inclination, however, there is little variation in the trend of the lineation.

Together with measurements of the orientation of the finite strain fabric (S_2L_2) in the field, estimates of the visible shapes and axial ratios of the mineral aggregates defining the fabric were made. Ratios of the lengths of the three major axes of the subellipsoidal aggregates, which were assumed to be nearly equidimensional before deformation, were used to define the intensity of fabric development, shown in Figure 53.12, and fabric symmetry, shown in Figure 53.13. The intensity of fabric development is assumed to depend upon the amount of imposed strain. Variations of strain intensity (Figure 53.12) do not seem to correlate with variations in rock types, although fabric strength values are difficult to estimate in the coarser grained homogeneous orthogneiss. This supports the inference that the pre-D_2 metamorphism and modification of the area effectively produced almost equigranular granoblastic aggregates in most rock units; this interpretation, however, is difficult to reconcile with the existence of earlier layering and planar fabrics.

From Figures 53.10 and 53.12 it appears that the zones of gentler dip relate spatially to the zones of greatest intensity of strain and also, from Figure 53.9, that these zones broadly correlate with the planar zones of fabric strike inflection. This feature of the main fabric variation was also observed in the field, where gently north-dipping zones of simple shear were thought to cause rotation of the S_2 and L_2 fabric elements, with the development of later, more intense, finite strain fabrics (S_3L_3).

It appears that during the post-D_2 cooling history of the area, D_2 deformation was eventually unable to continue by wholesale ductile flow and production of a pervasive tectonothermal fabric throughout the area. Consequently, as a less ductile mode of response, D_3 deformation continued in more discrete shear zones on gently north-dipping shear planes formed by the northern area moving southward and upward relative to the southern area. As the plunge directions of the L_3 lineations do not vary consistently throughout the shear zones, it is probable that the movement direction for D_3 did not change noticeably from the D_2 principal shear direction.

Although the distortion of the fabrics caused by the D_3 shear zones is most evident on a regional scale (Figure 53.9), these are not the most prominent post-D_3 structures observed in the field. On the mesoscopic scale, the main foliations (both fabrics and layering) are commonly deformed either into steep, discrete, mylonitic shear zones ranging in width from 1 to 10 m (Figure 53.14) or into more gently dipping shear zones. The fabrics within the steep zones also increase in intensity symmetrically into the zones, and there is little variation in the trend of the linear component of the superimposed fabrics. The more discrete and more mylonitic nature of the steep zones suggests that they either formed because of higher values of resolved shear stress or formed slightly later than the shallow ones and reflect the decreasing temperature and ductility of the rocks. The gently and steeply inclined zones are therefore termed D_{3a} and D_{3b}, respectively. A possible late result of the ductility decrease is the multiplicity of narrow pseudotachylite zones which pervade the area. These brittle features transect all earlier structures and fabrics and are especially well developed in the homogeneous orthogneiss. They do not seem to be closely related to the major fault zones which also cut the area (Figure 53.1).

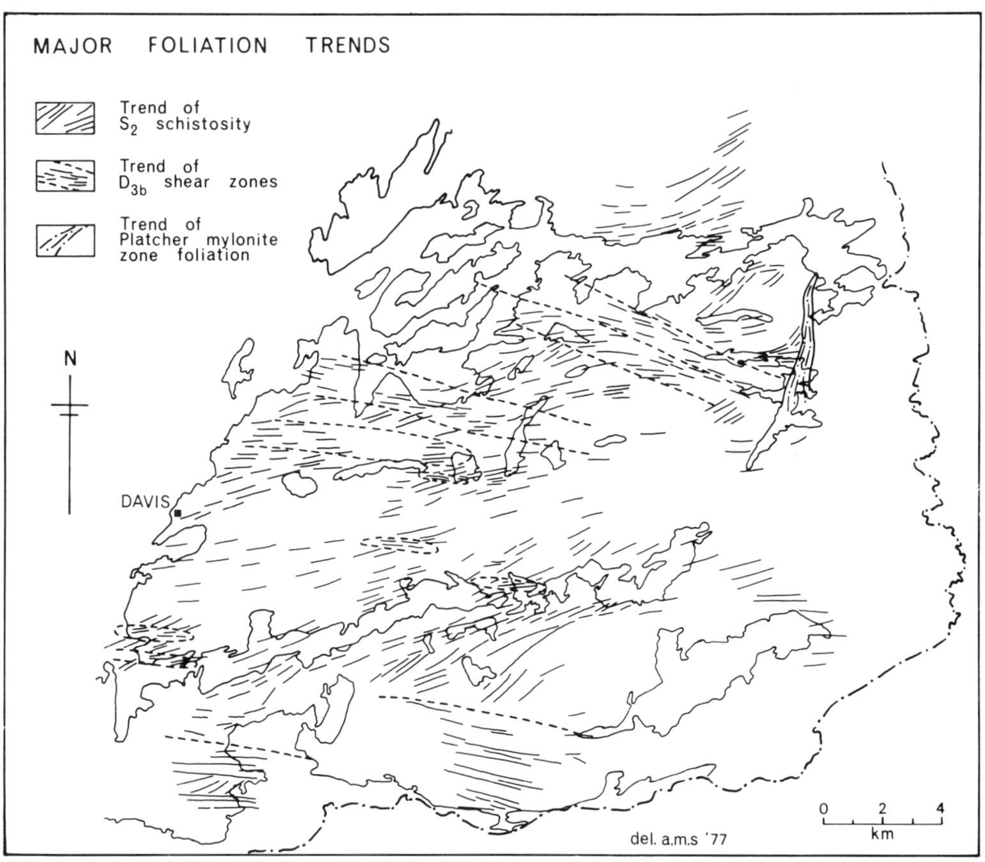

Figure 53.9. Map of the Vestfold Hills showing major foliation trends.

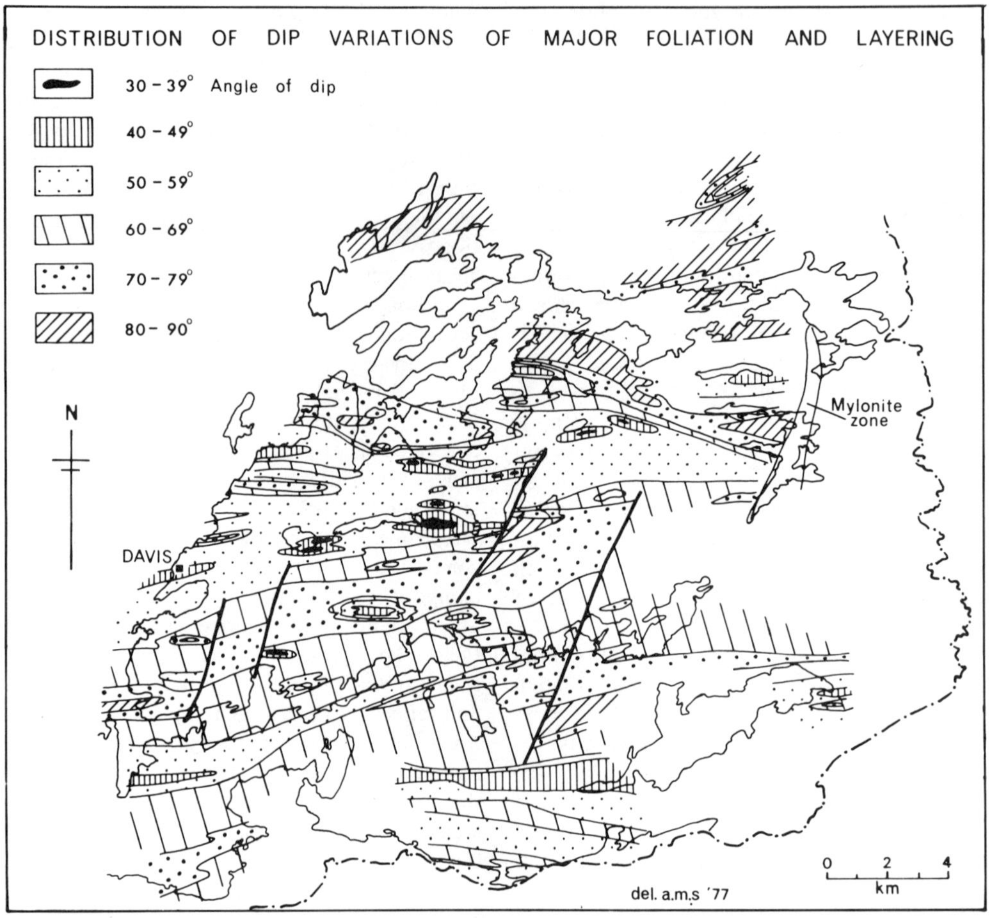

Figure 53.10. Map of the Vestfold Hills showing distribution of dip variations of major foliation and layering.

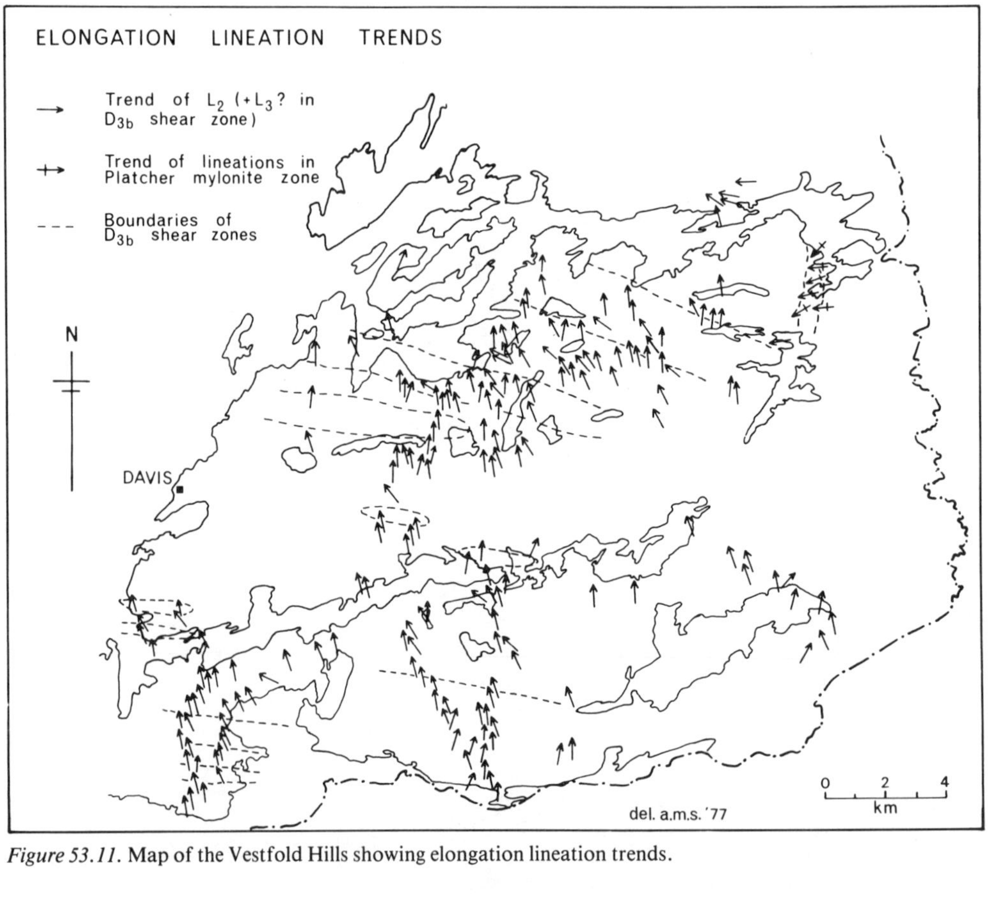

Figure 53.11. Map of the Vestfold Hills showing elongation lineation trends.

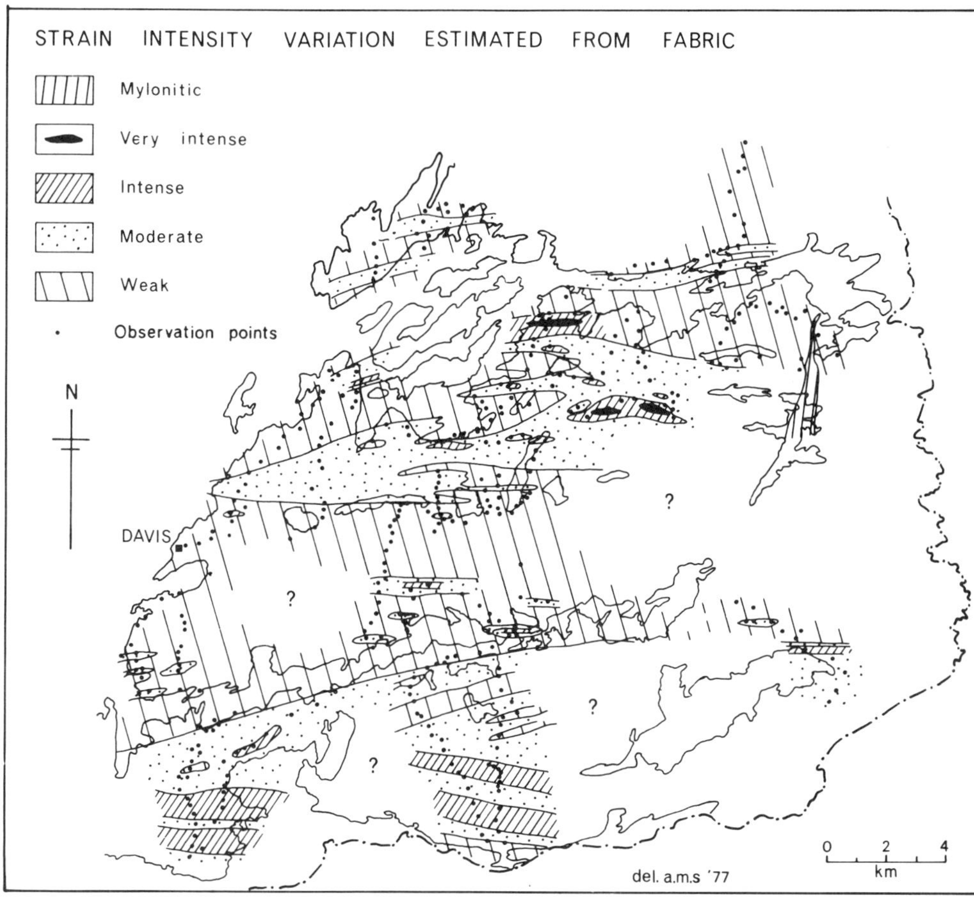

Figure 53.12. Map of the Vestfold Hills showing strain intensity variation estimated from fabric.

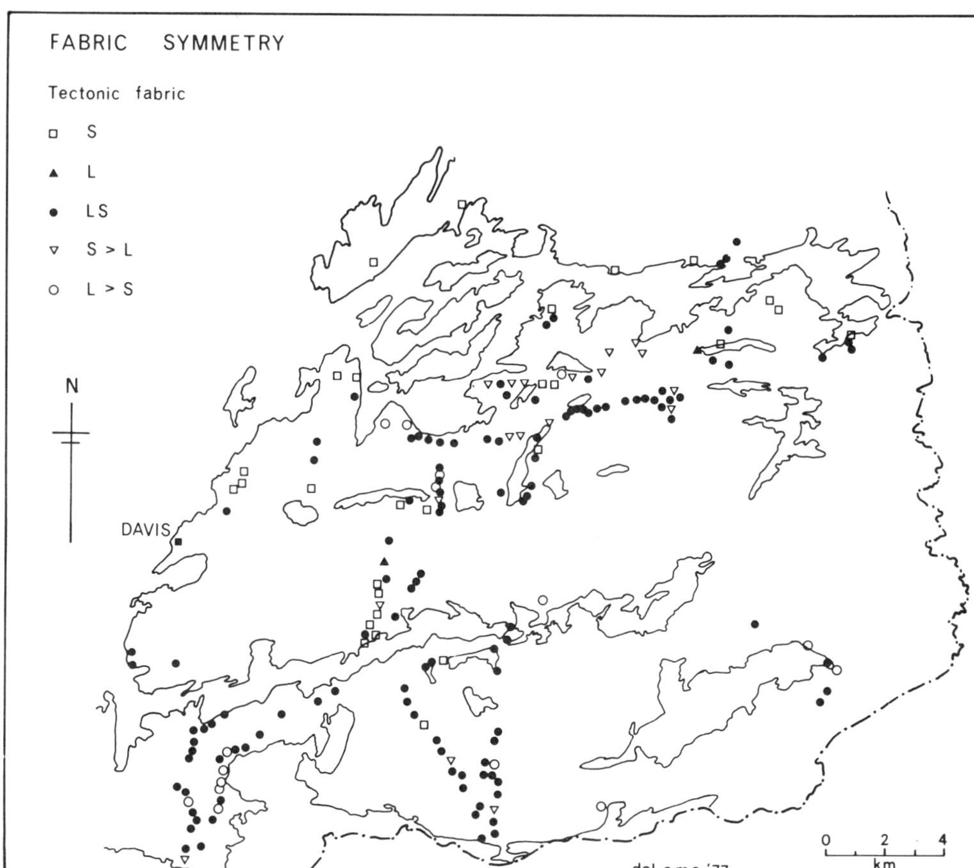

Figure 53.13. Map of the Vestfold Hills showing variation in fabric symmetry.

Figure 53.14. D_{3b} flexure, plunging approximately 10° to the east, in layered gray gneiss. Near Deep Lake.

PLATCHER MYLONITE ZONE

South of the head of Long Fjord, a major ductile mylonite zone crosses at a high angle to the regional structural trend. The textures in this zone vary from conchoidally fractured black phyllonite to porphyroclastic black mylonite and ultramylonite. The foliation dips at a gentle angle (20°–30°) to the west, trends parallel to the margins of the zone (generally due north-south), and contains a downdip elongation and mineral lineation. The zone appears to thicken markedly to the south (up to 1 km in outcrop width) and thins northward, almost dying out as it disappears into Long Fjord. At the margins of the zone, D_2 fabrics increase in intensity and curve into the zone, showing a sense of thrust movement with the west block moving eastward and upward over the eastern block.

The discordant nature of the zone and the obviously retrogressive nature of the biotite-rich mylonite fabric suggest that the Platcher Mylonite formed during a later, independent deformational event.

Discussion and summary

In the last decade considerable effort has been directed toward understanding the process responsible for the evolution of gneiss complexes. Research in the Archean craton of the North Atlantic region (McGregor, 1973; Bridgwater and others, 1973; Myers, 1976; Collerson and others, 1976; and Bridgwater and others, 1978) and in other areas (Ermanovics and Davison, 1976; Hunter, 1970) has demonstrated that despite the intense deformation and structural interleaving of gneiss rock types, it is possible in many cases to establish a logical sequence of events for a particular region. In many places it has also been possible to identify and interpret the original nature of different protoliths within a particular gneiss complex. For example, many Archean gneiss terranes are regarded as being composed of several generations of variably deformed gray gneiss derived from igneous (granitic to tonalitic) parents, which is interleaved with gneiss of metasedimentary (Al-rich mudstone, impure chert, ironstone, and carbonate rocks) and metavolcanic origin.

The assemblage of rock types recognized in the Vestfold Hills is remarkably similar to those found in other Archean cratons. By analogy with other Archean gray gneiss, it is suggested that the layered gray gneiss, which ranges in composition from granitic to dioritic, is in fact highly deformed orthogneiss within which paragneiss is structurally interleaved. The garnetiferous layered gneiss and other metasedimentary and metavolcanic rocks in the paragneiss belts may represent the structurally disrupted remnants of a cover sequence. Into these and into the interleaved layered gray gneiss, the protoliths of the acid to intermediate layered orthogneiss are thought to have been intruded, prior to or during D_1; this event was followed by the syntectonic emplacement of the parents to the homogeneous orthogneiss and their subsequent crystallization during D_2. Metamorphic and geochemical studies suggest that granulite facies metamorphic conditions prevailed during most of the long and structurally complicated tectonic and igneous history of the terrane.

Later deformation of a brittle nature, reflecting the cooling and stabilization of the gneiss terrane, resulted in the development of numerous vertical faults, many of which were subsequently occupied by basic dikes.

Acknowledgments. The authors wish to acknowledge logistical support from ANARE during the 1970–71 season (K. D. C.) and 1976–77 season (R. L. O. and P. R. J.). Laboratory work at Memorial University has been funded by a National Research Council of Canada Grant (#8694) to K. D. Collerson. The authors are grateful to D. F. Strong for critically reading the manuscript.

References cited

Arriens, P. A. 1975. Precambrian geochronology of Antarctica. 1st. Australian Geological Convention, Adelaide, 1975 (Abstr.). *Spec. Publ. geol. Soc. Aust.,* 97–98.

Bridgwater, D., Collerson, K. D., and Myers, J. S. 1978. The development of the Archean gneiss complex of the North Atlantic region. (*In* Tarling, D., *ed. Evolution of the Earth's Crust.* New York, Academic Press, 19–68.)

Bridgwater, D., Watson, J., and Windley, B. F. 1973. The Archaean craton of the North Atlantic region. *Phil. Trans. R. Soc. Lond.,* Ser. A, *273,* 493–512.

Collerson, K. D., Jesseau, C. W., and Bridgwater, C. 1976. Crustal development of the Archean gneiss complex, eastern Labrador. (*In* Windley, B. F., *ed. The Early History of the Earth.* London, Wiley, 237–53.)

Crohn, P. W. 1959. A contribution to the geology and glaciology of the western part of Australian Antarctic territory. *Bull. Bur. Miner. Resour. Geol. Geophys. Aust.,* No. 52, 43–44.

Ermanovics, I. F., and Davison, W. L. 1976. The Pikwitonei Granulites in relation to the north-western Superior Province of the Canadian Shield. (*In* Windley, B. F., *ed. The Early History of the Earth.* London, Wiley, 331–47.)

Harding, R. R., and McLeod, I. R. 1967. Age of dolerite dikes in the Vestfold Hills, Antarctica. *Nature, Lond., 215,* 149–51.

Hunter, D. R. 1970. The ancient gneiss complex in Swaziland. *Trans. Geol. Soc. S. Afr., 73,* 107–50.

McGregor, V. R. 1973. The early Precambrian gneisses of the Godthaab district, West Greenland. *Phil. Trans. R. Soc. Lond.,* Ser. A, *273,* 348–58.

McLeod, I. R., Trail, D. S., Cook, P. J., and Wallis, G. R. 1966. Geological work in Antarctica, January–March, 1965. *Rec. Bur. Min. Res. Geol. Geophys. Aust.,* 1966/9. [Unpublished] 71–78.

Myers, J. 1976. The early Precambrian gneiss complex of Greenland (*In* Windley, B. F., *ed. The Early History of the Earth.* London, Wiley, 165–75.)

Ramsay, J. G. 1962. Interference patterns produced by the superposition of folds of "similar" type. *J. Geol., 60,* 466–81.

Ravich, M. G. 1960. Gornye porody oazisa Vestfoll'. [Rocks of Vestfold Oasis]. *Nauch. -issled. inst. geol. Arktiki, Trudy, 113,* No. 2, 25–52.

54

Aspects of the Geologic History of the Windmill Islands, Antarctica

D. F. BLIGHT* and R. L. OLIVER
Department of Geology and Mineralogy, University of Adelaide, Adelaide, S. A. 5000, Australia

Abstract. Rocks of the Windmill Islands, Antarctica (lat. 66°, long. 110°E) consist of a layered sequence of schists, gneisses, and migmatites (the Windmill Metamorphics) intruded by a charnockite (the Ardery Charnockite) and a porphyritic granite (the Ford Granite) and cut by two swarms of east-west-trending dolerite dikes.

The rocks have undergone four deformations. The first two produced tight isoclinal folds, the third developed broader less appressed concentric folds, and the last deformation produced gentle warps which plunge steeply to the south.

The metamorphic grade of the Windmill Metamorphics ranges from (i) upper amphibolite facies (sillimanite-biotite-orthoclase) in the north, through (ii) biotite-cordierite-almandine granulite to (iii) hornblende-orthopyroxene-granulite in the south. The boundary between (i) and (ii) above is marked by the appearance of orthopyroxene and also by disappearance of sphene, and that between (ii) and (iii) by the disappearance of cordierite. Other metamorphic variations, apparent in the field, are the southward color changes of biotite (sepia to red brown) and of hornblende (blue green to brown green), and the greater abundance of migmatites and pegmatites in the northern part of the area.

Major element rock chemistry suggests that the protolith of the schists and gneisses probably consisted of acid and basic volcanic rocks interbedded with sedimentary rocks ranging from graywacke-type sandstones to shales. Partial melting of these rocks is thought to have produced some of the more acidic gneisses of the area.

The abundances of the elements K, Rb, and Th are lower than those of terranes of similar grade elsewhere; these abundances are regarded as representing the original rock composition. The ratios K/Rb, Th/K, and K/Rb/Sr, however, are comparable with those from similar terranes elsewhere, where K, Rb, and Th enrichment by crustal fractionation has been suggested. Hence, these ratios preclude the derivation of the Windmill Metamorphics from the upper granulite facies by retrogradation.

Use of the garnet-cordierite geothermometer and geobarometer (Currie, 1971), coupled with other experimentally determined mineral breakdown curves in *P-T* space, indicate that the northern part of the Windmill Islands recrystallized under conditions of 5.5–6.5 kbar load pressure, approximately 3.0 kbar water pressure, and temperatures around 760°C.

This paper summarizes the metamorphic geology of the Windmill Islands (lat. 66° 20′ S, long. 110° 30′ E) previously discussed by Blight and Oliver (1977) and, in addition, provides new information regarding the physical conditions operating during metamorphism. The rocks of the area constitute part of the East Antarctic "Basement Complex" (Harrington, 1965).

Rock formations

The Windmill Metamorphics

These comprise the majority of the rocks from the Windmill Islands (Figure 54.1) and yield Rb/Sr whole-rock isochrons between 1400 and 1100 m.y. (P. A. Arriens, personal communication). They have been subdivided into:

1. *acid gneisses* of several types with the modal assemblage: quartz + plagioclase (An_{20-45}) + microcline + biotite + opaques ± garnet, plus minor apatite and zircon;
2. *basic schists and gneisses* with the modal assemblage: plagioclase (An_{40-65}) + hornblende ± biotite ± orthopyroxene ± clinopyroxene ± garnet ± opaques;
3. *migmatite schists and gneisses* of both pelitic and psammitic compositions consisting of various combinations of quartz + plagioclase (An_{20-40}) + biotite + opaques ± microcline ± garent ± cordierite ± sillimanite ± spinel ± hornblende ± orthopyroxene;
4. *layered gneisses* with the modal assemblage: plagioclase (An_{45-56}) + orthopyroxene ± quartz ± biotite ± clinopyroxene ± microcline ± hornblende + opaques.

The Ardery Charnockite

As shown in Figure 54.1, this is a large body, superficially homogeneous, but on close inspection is seen to consist of a cluster of multiple intrusions. This rock intrudes the Windmill Metamorphics and yields a Rb/Sr whole-rock age of 1200 m.y. (P. A. Arriens, personal communication). The modal assemblage is quartz + plagioclase (An_{40}) + microcline + orthopyroxene + biotite + clinopyroxene + opaques ± hornblende, with minor zircon and apatite.

* Present address: Geological Survey of Western Australia, Department of Mines, Perth, W. A., Australia

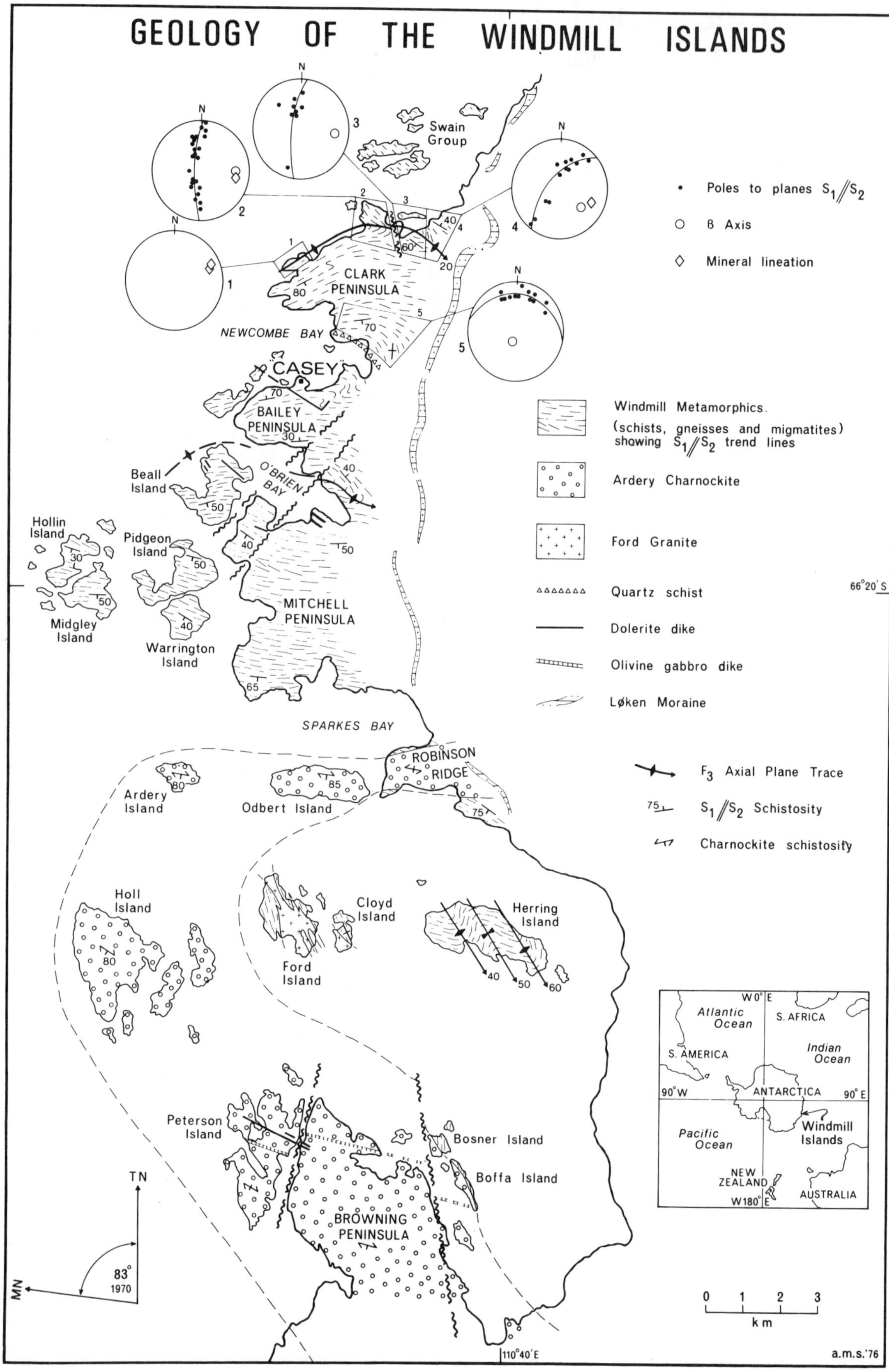

Figure 54.1. Geologic map of the Windmill Islands.

THE FORD GRANITE

This rock is intrusive into the Windmill Metamorphics and yields Rb/Sr whole-rock isochrons between 1100 and 1140 m.y. (P. A. Arriens, personal communication). It is a coarsely porphyritic granite with tabular potash feldspar megacrysts that tend to be aligned. A typical mineral assemblage is quartz + plagioclase (An_{30}) + microcline + biotite + hornblende + opaques ± hypersthene.

LATER INTRUSIVES

Pegmatites are abundant on Clark Peninsula and are also present on Pigeon, Ford (McLeod and Gregory, 1966), and Herring (Oliver, 1970) Islands. Thin aplite dikes occur extensively. Minor basic intrusives are represented by an olivine gabbro dike and two dolerite dike swarms (see Figure 54.1).

Metamorphic grade

The rocks of the mapped area can be subdivided into three metamorphic subfacies as defined by de Waard (1965), which are, from north to south: sillimanite-biotite-orthoclase subfacies of the almandine amphibolite facies, biotite-cordierite-almandine subfacies and hornblende-orthopyroxene-plagioclase subfacies of the granulite facies.

These areas are bounded by mineral isograds which, traversing southwards, are (Figure 54.2):

1. the appearance of hypersthene
2. the disappearance of sphene
3. the disappearance of cordierite.

Increasing metamorphic grade from north to south is also manifested (see Leake, 1965; Binns, 1969*a*) by (1) the southward increase of titanium in biotite and hornblende (reflected by the variation of biotite absorption color (*Y-Z*) from sepia in the north to rusty red in the south and hornblende absorption color from bluish green on Clark Peninsula through green (Bailey Peninsula) to greenish brown on Ford Island); and (2) the relative abundance of pegmatites and migmatites in the northern areas (north of Sparkes Bay) due to higher water content and consequently more extensive melting in the amphibolite facies terrane.

Variation of metamorphic grade within the mapped area is considered to be progressive. This is supported by trace element data (see below) and by the positive correlation of hornblende Mg/Fe ratios with those of the host rock (see Sen, 1970).

Structure

The deformational history of the mapped area is summarized in Table 54.1.

FOLDS

Four deformation episodes have been identified. D_1 is interpreted as being responsible for the development of the prevalent metamorphic layering—schistosity by transposition of an earlier layering. No F_1 folds have been recognized.

D_2 is manifested by tight mesoscopic folds, many of which are rootless. S_2 is recognizable only in the cores of F_2 folds; the latter are random in orientation.

D_3 produced broader, less appressed macroscopic (F_3) folds, and associated smaller folds, with vertical axial

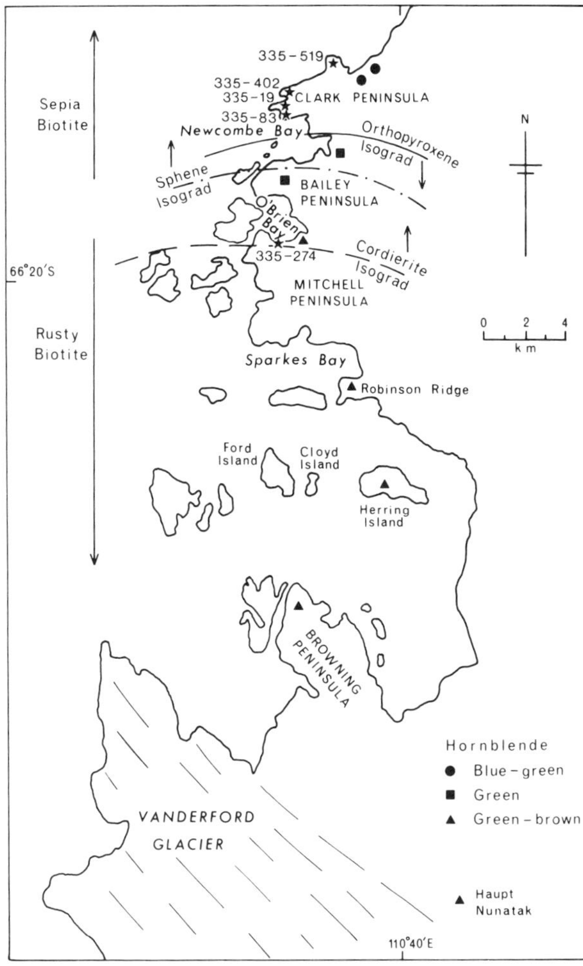

Figure 54.2. Isograd map of the Windmill Islands also showing variation in mineral color and the location of five analyzed garnet-cordierite pairs.

plane schistosity (S_3). Macroscopic F_3 antiforms have been delineated on the northern coast of Clark Peninsula and in the O'Brien Bay area, plunging generally eastward (Figure 54.1). The distribution of the large intrusive charnockite body in the southern part of the Windmill Islands is regarded as an expression of F_3. The general shape of this fold has been interpreted in part from aeromagnetics (Voronov and Karasik, 1963).

D_4 has produced broad macroscopic warps, the axes of which plunge steeply to the south. The warps can be seen to modify the axial surface trace of the F_3 antiform on Clark Peninsula and are presumably responsible for the variation in plunge direction of the F_3 axis (see Figure 54.1).

FAULTS

A major fault (or shear), manifest as a quartz schist, occupies the synformal area between the two large east-west-trending antiforms in the north of the area (Figure 54.1). Other faults cross-cutting the F_3 fold in O'Brien Bay, and displacing the olivine gabbro dike on Browning Peninsula, are shown in Figure 54.1.

Major element geochemistry

Chemical analyses of rocks from the Windmill Islands have been previously published (Blight and Oliver, 1977) and are not reproduced here. They have been grouped chemically into 29 acid rocks, 10 intermediate (60–70% SiO_2) and pelitic (> 3% normative corundum) rocks, and 17 basic rocks.

ACID ROCKS (>70% SiO_2)

An indication of the premetamorphic nature of these rocks has been sought by applying to them the discriminative function, D.F. (D.F. = 10.44 − $0.21SiO_2$ − $0.32Fe_2O_3$ (total Fe) − 0.98MgO + 0.55CaO + $1.46Na_2O$ + $0.54K_2O$), of Shaw (1972). All have positive D.F. values, which according to Shaw suggests an igneous parentage.

Field observation indicates that there are two types of acid rocks in the Windmill Islands, an obviously intrusive type—apparently crystallized from an almost completely liquid magma—and a not so obviously intrusive type.

Porphyritic granite from Ford and Cloyd Islands, and aplites and intrusive granite from Midgely, Warrington, Hollin, and Pigeon Islands constitute the first category. On a normative An-Ab-Or triangle (Figure 54.3) these intrusive types all plot in or close to the low-temperature trough of Kleeman (1965). Because of this, it is likely that the magmas from which these rocks crystallized were derived from the melting of country rock, which may or may not have been located close to the final emplacement position.

The second type comprises the rest of the acid rocks of the Windmill Islands. These are generally concordant with the layering of the surrounding gneisses. On the An-Ab-Or triangle they plot away from the low-temperature trough toward the An-Ab side. This suggests that unless temperatures were exceptionally high these rocks could not have crystallized from a complete melt. It is probable that they formed *in situ* with the development of only a small amount of liquid from partial melting.

PELITIC (> 3% NORMATIVE CORUNDUM) AND INTERMEDIATE (60–70% SiO_2) ROCKS

Shaw's discriminative function for these rocks yields negative values, indicating sedimentary parentage.

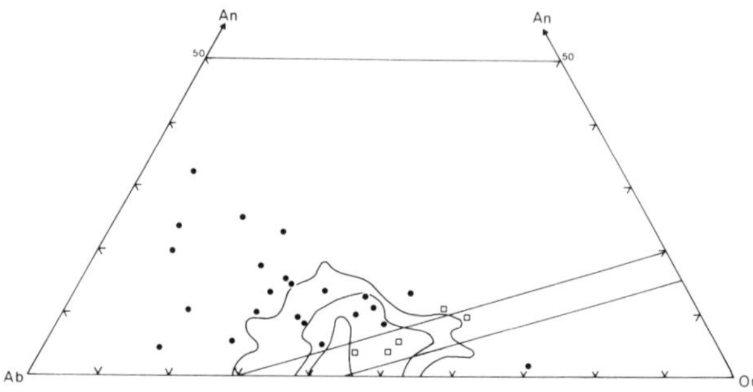

Figure 54.3. An-Ab-Or plot of acid rocks from the Windmill Islands. Square symbols are the "obviously intrusive" granitic rocks.

BASIC ROCKS (< 60% SiO_2)

Calculation of Niggli numbers from chemical analyses of the basic gneisses, and plotting of these on diagrams proposed by Leake (1964), indicates that the basic gneisses probably originated as igneous bodies.

Trace element geochemistry

The trace element compositions of rocks from the Windmill Islands are listed in Blight and Oliver (1977).

COMPARISON OF METAMORPHIC GRADES

Comparison of a limited number of rocks from the upper amphibolite facies (i.e., north of the orthopyroxene isograd on Figure 54.2) with those of similar composition in the lower granulite facies south of this isograd shows no significant trace element geochemical differences, thus supporting the similar findings of Lambert and Heier (1968*a,b*) pertaining to high-grade metamorphic rocks from Australia.

COMPARISON WITH OTHER TERRANES

As can be seen from Table 54.2, the Windmill Metamorphics have generally lower K, Th, and Rb values than similar rocks from elsewhere. However, the Th/K, Rb/Sr, K/Rb, and K/Pb ratios are compatible with these other

Table 54.1. Deformational history of the Windmill Islands terrane

Deformational or intrusive event	Fold style	Features	Metamorphic grade
D_1	F_1 ? tight isoclinal	S_1 dominant axial plane penetrative schistosity; sillimanite lineation	Upper amphibolite to lower granulite facies
D_2	F_2 tight isoclinal	Weak S_2 axial plane schistosity manifested in hinge zones by biotite	
Intrusion of charnockite			
D_3	F_3 broad open concentric folds	Weak S_3 axial plane schistosity and L_3, a lineation, developed in hinge zones; gentle NE to SE plunge Pegmatities developed in some hinge zones; S_m schistosity developed in quartz schist manifested by flattened quartz and muscovite; faulting of F_3 axial plane traces	Low to middle amphibolite facies
D_4	F_4 gentle warps	S_4 axial plane schistosity developed in quartz schist; steeply plunging to south	
Intrusion of olivine gabbro dike			
Faulting of olivine gabbro dike			?
Intrusion of dolerite dikes			

values. The low K, Th, and Rb values probably reflect the premetamorphic chemistry of the rocks.

Figure 54.4 shows log-log plots of K against Rb for each of the three chemical rock groups of the Windmill Metamorphics. A linear regression analysis was carried out on these data using the method described by Shaw (1968). The "main trend" of Shaw (1968) for igneous and quasi-igneous rocks ($\log_{10}Rb(ppm) = 1.115 \log_{10}K(\%) + 1.597$) and the "metamorphic trend" of Lewis and Spooner (1973) for upper amphibolite facies rocks ($\log_{10}Rb(ppm) = 1.136 \log_{10}K(\%) + 1.497$) are also plotted on this diagram for comparison.

Two observations are evident from this log-log plot. First, Rb is much richer with respect to K for the Windmill Islands rocks than for those that gave the metamorphic trend of Lewis and Spooner (1973), whose upper granulite facies rocks are characteristically depleted in Rb with respect to K (due to crustal fractionation). Secondly, the basic rocks are very much richer in Rb than the igneous "main trend" of Shaw (1968), suggesting that they are not showing their original igneous trace element (Rb) chemistry.

Plots of log Th against log K (Blight and Oliver, 1977) give values between 1×10^{-4} and 10×10^{-4}, corresponding to the values regarded by Heier and Adams (1965) as reflecting upper amphibolite-lower granulite facies conditions (namely, enrichment of Th with respect to K). In a plot of log K against log (Rb/Sr) (Blight and Oliver, 1977), the Windmill Metamorphics fall into the upper amphibolite-lower granulite field defined by Sighinolfi (1971), rather than the upper granulite field, thus confirming indications provided by Th and Rb.

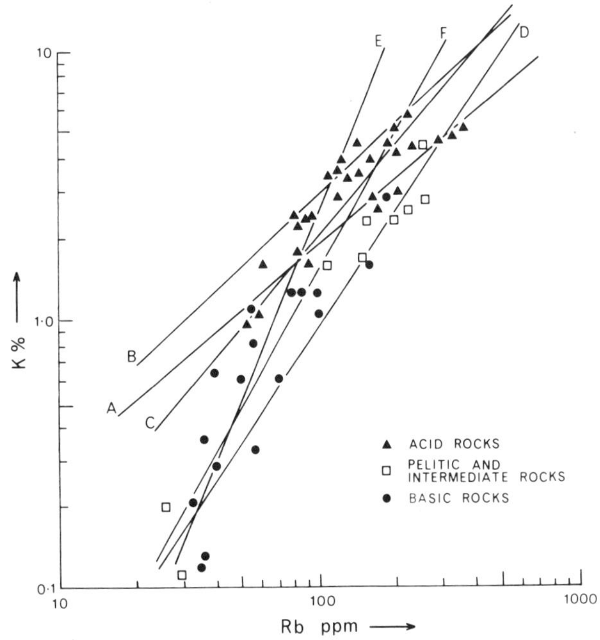

Figure 54.4. Plot of K % against Rb ppm for rocks from the Windmill Islands. A: Main trend of Shaw (1968) (see text for explanation). B: Metamorphic trend of Lewis and Spooner (1973). C: Linear regression analysis of Windmill Island acid rocks. D: Linear regression analysis of Windmill Island pelitic and intermediate rocks. E: Linear regression analysis of Windmill Island basic rocks. F: Linear regression analysis of all Windmill Island rocks.

Conclusions

The trace element ratios shown by the Windmill Metamorphics are characteristic of the upper amphibolite-lower granulite facies and are interpreted as being due to enrichment of the more lithophile elements by a crustal fractionation process. Biotite can concentrate Rb by a factor of three to four relative to feldspars (Lange and others, 1966), and as the basic rocks of the Windmill Metamorphics are generally fairly rich in modal biotite (up to 10%), they display this lithophile enrichment markedly.

This assumption of enrichment suggests that there has been differential migration of the more lithophile elements from a correspondingly depleted granulite facies source at great depth during "prograde" metamorphism of a section of the crust comprising these rocks.

The alternative interpretation, that development of the Windmill Islands upper amphibolite-lower granulite facies mineral assemblages occurred by retrogression from upper granulite facies rocks, a relationship postulated for some other amphibolite-granulite facies terranes (see Oliver, 1969), is not considered likely. During such retrogradation low Rb/K and Th/K ratios, representing depletion under former upper granulite facies conditions, are likely to be retained. The Windmill Islands ratios, indicating enrichment, cannot be reconciled with such a process.

Estimated pressure-temperature conditions of metamorphism

In the last fifteen years the stability field of cordierite in *P-T* space has been explored experimentally by many workers. Of particular interest has been the equilibrium position of coexisting garnet and cordierite of varying

Table 54.2. "Average" potassium and trace element abundances and ratios of upper amphibolite-lower granulite facies terranes

	Musgrave Block		Eyre Peninsula	Cape Naturaliste	North Norway	Windmill Islands (this study)		
	Lambert and Heier (1968*a*)	Collerson (1975)	Lambert and Heier (1968*a*)	Lambert and Heier (1968*a*)	Heier and Thorensen (1971)	4 Acid	1 Basic	Weighted average
K (wt %)	2.8	3.6	3.65	3.8	3.1	3.3	0.85	2.81
Rb (ppm)	140	163	185	185	155	152	64	134
Sr (ppm)	400	235	130	175	338	183	306	208
Th (ppm)	11	16	27	30	14	13	3	11
Pb (ppm)	25	—	30	32	24	51	14	44
Th/K(X10⁴)	2.1-5.0	6.4	2.8-9.5	3.2-9.6	4.5	6.5	8.7	6.9
K/Rb	144-688	310	175-232	182-686	198	274	132	246
K/Pb	701-1992	—	960-1785	861-1509	1300	911	726	874
Rb/Sr	0.01-1.39	1.08	0.21-3.16	0.03-4.55	0.46	0.84	0.35	0.74

compositions (Hensen and Green, 1969, 1970, 1971, 1972, 1973; Currie, 1971, 1973; Hutcheon, Froese, and Gordon, 1974). In all of these investigations it is the divariant reaction 3(Fe-Mg) Cordierite ⇋ 2(Mg-Fe) garnet + 4 sillimanite + 5 quartz that has been studied most closely. The three groups of workers mentioned above have produced three different calibrations of temperature and pressure against varying iron and magnesium in the garnet and cordierite.

For any of these calibrations to be valid in natural assemblages, these natural assemblages should fulfill the following conditions:

1. attainment of equilibrium, at least in regard to partitioning between iron and magnesium;
2. reasonable freedom of the cordierite and garnet from contaminants that would compete with Fe^{2+} and Mg^{2+} for lattice positions; and
3. besides coexisting garnet and cordierite, the investigated assemblage must contain sillimanite and quartz.

The assemblage quartz + garnet + sillimanite + cordierite is placed in the upper amphibolite to lower granulite facies (Binns, 1964; de Waard, 1965; Hietanen, 1967; Winkler, 1975). This assemblage is present in the Windmill Metamorphics (pelitic section of the migmatite gneiss) and, with the application of one or another of the calibrations mentioned above, is thus a potential geothermometer and geobarometer for the Windmill Island rocks. However, the three above-mentioned calibrations were first evaluated by applying each of them to naturally occurring garnet-cordierite pairs from other areas. The results of this literature survey are shown in Table 54.3 and Figure 54.5. Because of the paucity of information, it was not always possible to ascertain that the three conditions, outlined above, were always met.

As can be seen from Figure 54.5, the calibration of Hutcheon, Froese, and Gordon (1974) gives a very large temperature range (e.g., Saxena and Hollander's data in Table 54.3 show temperatures ranging from 350°C to 1200°C, implying an unrealistic geothermal gradient). In other words, the determined temperatures of equilibrium are sensitive to small changes in measured distribution coefficients. This is not surprising, for in order to reach their final equation, relating the temperature of equilibrium to the distribution coefficient, Hutcheon, Froese, and Gordon employ some rather tenuous assumptions. For example, to obtain a value for the change in the Gibbs Free Energy (ΔG) of a reaction, they estimate that this ΔG will be zero at 1000°C and 10 kbar pressure because ". . . it allows the rocks from the Daly Bay Complex to fall into the stability field of sillimanite"! The calibration of Hutcheon, Froese, and Gordon is thus rejected.

The calibration of Hensen and Green is the only one in which an attempt has been made to reverse the experimental reactions (Hensen, 1972). However, one serious problem associated with this calibration is obvious as soon as it

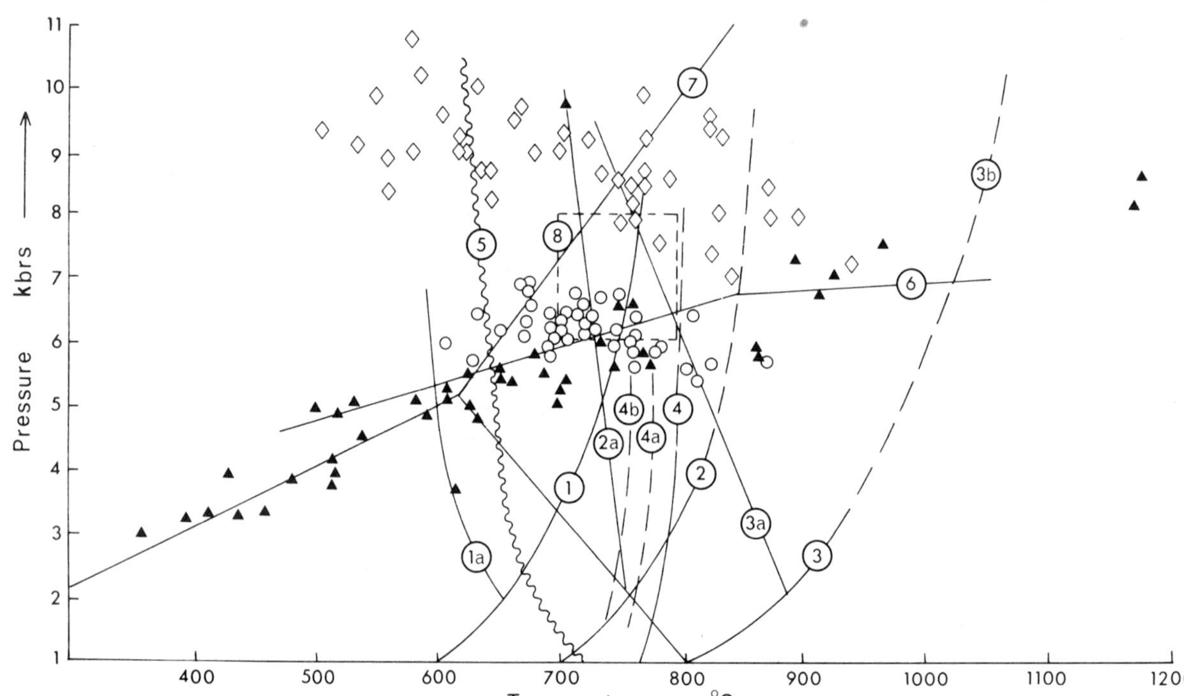

Figure 54.5. Petrogenetic grid showing estimated *P-T* conditions of coexisting garnet and cordierite pairs obtained from the literature. Open diamonds: calibration of Hensen and Green (1973). Open circles: calibration of Currie (1971). Solid triangles: calibration of Hutcheon, Froese, and Gordon (1974).
1. Muscovite + quartz ⇋ K feldspar + Al_2SiO_5 + H_2O: $P_{H_2O} = P_L$* (Winkler, 1967); 1a. Same as (1) except P_{H_2O} = 2 kbar; 2. Orthoamphibole ⇋ orthopyroxene $P_{H_2O} = P_L$ (Touret, 1971*a*); 2a. Same as (2) except P_{H_2O} = 2 kbar (Touret, 1971*a*); 3. Hornblende breakdown. $P_{H_2O} = P_L$ (Touret, 1971*a*); 3a. Same as (3) except P_{H_2O} = 2 kbar (Touret, 1971*a*); 3b. Same as (3) except at higher $P_{H_2O} = P_L$ (Lambert and Wyllie, 1972); 4. Hornblende + quartz ⇋ Hornblende + orthopyroxene + clinopyroxene + plagioclase, $P_{H_2O} = P_L$ (Binns, 1969*b*); 4a. Same as (4) except $P_{H_2O} = 0.5P_L$ (Binns, 1969*b*); 4b. Same as (4) except $P_{H_2O} = 0.3P_L$ (Binns, 1969*b*); 5. Beginning of melting of hydrous granite (Winkler, 1967); 6. Upper stability limit for anhydrous Mg-cordierite (Newton and others, 1974); 7. Al_2SiO_5 stability fields (Richardson and others, 1969); 8. Field of granulites from fluid inclusion studies (Touret, 1971*b*).

* P_{H_2O} ≡ partial pressure of water. P_L ≡ load pressure.

Table 54.3. Estimated *P-T* conditions for coexisting garnet and cordierite from various areas described in the literature

	Calibration of pressure and temperature					
	Currie (1973)		Hensen and Green (1969)		Hutcheon, Froese, and Gordon (1974)	
Source	*P*(kbr)	*T*°C	*P*(kbr)	*T*°C	*P*(kbr)	*T*°C
Reinhardt (1968)						
D-28	6.6	677	8.6	790	6.0	737
D-56	6.4	712	9.0	700	5.0	626
D-87	5.6	652	7.1	840	5.9	866
D-102	6.4	692	8.5	760	5.5	688
D-175	6.1(5)	641	8.0	870	6.8	918
R-114	6.1	711	7.9	750	4.7(5)	635
R-124	6.5	717	9.3	705	5.1	709
W-53	6.9	668	9.4	825	9.7	706
Hutcheon, Froese, and Gordon (1974)						
1	6.7(6)	713	10.0	770	5.5	625
2	6.4	704	9.2	725	5.3	650
3	6.8(6)	673	9.3	835	6.6	760
4	6.2(5)	694	9.2	770	5.8	680
5	6.9	675	9.5	825	6.6	750
Barker (1962)						
2	6.2	698	8.6	750	5.7	774
Gable and Sims (1969)						
EWT-90	5.9(6)	746	8.7	635	5.4	706
Wynne-Edwards and Hay (1963)						
H29	6.3(5)	673	8.7	770	7.1	926
H70	6.1	654	8.1	830	8.5	1182
Currie (1971)						
69-117	5.7	627	7.4	825	5.9	864
69-38	6.1(5)	652	8.0	760	5.6	745
69-77	6.2	722	8.7	735	5.4	653
69-6	6.4(5)	717	10.2	585	4.9	519
69-29	6.7	732	10.8	580	5.0	501
Saxena and Hollander (1969)						
6	6.4	810	9.9	550	3.9	428
1	5.6	870	9.3	480	3.0	357
12	5.6	823	9.1	510	3.3	412
7	5.4	812	8.4	560	3.2(5)	396
8	5.8	777	9.0	580	3.9	482
4	6.0	762	9.0	620	4.2	512
3	5.8	690	7.6	780	5.1	701
9	5.9	690	8.2	760	5.3	703
2	6.1	671	8.5	770	5.9	766
5	6.0	607	7.3	940	8.0	1175
10	6.4	632	8.5	870	7.5	965
Henry (1974)						
2	6.4	727	9.5	665	5.0(5)	582
3	6.7(5)	750	10.0	635	5.0(5)	534
8	6.4	720	9.7	670	5.1(5)	608
9	6.2	726	9.0	680	4.9	590
10	6.1	747	9.2	620	4.5	539
14	6.1	704	8.0	895	5.4	660
This study						
335/19	5.9	760	8.7	640	3.9	516
335/83	5.6	762	8.2	645	3.7	618
335/274	6.4	762	9.6	605	7.3	890
335/402	6.0	760	9.0	625	3.8	518
335/519	5.9	780	8.9	560	3.3	459

is applied to natural assemblages: the calibration curves of Hensen and Green (1973, p. 154) commonly need to be extrapolated into the kyanite field to accommodate natural assemblages. Since the reaction investigated involves sillimanite, and the natural assemblages examined contain sillimanite, this calibration is considered unsuitable.

The calibration developed by Currie appears to give the most acceptable answers. The coexisting garnet-cordierite pairs referred to in the literature generally plot in the stability field of sillimanite (Figure 54.5), above the breakdown of muscovite (at low P_{H_2O}), above the granite solidus (partial melt migmatites are common in these regions), below the breakdown of hornblende, around the appearance of hypersthene, and in the vicinity of the field—suggested by the fluid inclusion studies of Touret (1971*b*)—of lower granulites. The temperatures suggested by this calibration are thus of the order expected from rocks of the upper amphibolite to lower granulite facies implicit in the stability of the assemblage quartz + cordierite + sillimanite + garnet (see above).

Compositional data* and results based on the calibration developed by Currie, from rocks of the Windmill Metamorphics containing coexisting garnet and cordierite, are presented in Table 54.4. The results are also plotted in Figure 54.6, along with certain experimentally determined breakdown curves. The location of the various specimens is given in Figure 54.2. Specimens, 335–19, 335–83, 335–402, and 335–519 are from the upper amphibolite facies rocks north of the hypersthene isograd, and specimen 335–274 is from the north side of Mitchell Peninsula, in the lower granulite facies. All basic rocks from these areas contain hornblende. Thus these rocks namely, 335–19, 335–83, 335–274, 335–402, and 335–519) should plot close to the experimentally determined curve for the appearance of hypersthene (curves 2 and 4 of Figure 54.6) and on the low-temperature side of the hornblende breakdown curve. These conditions are satisfied if P_{H_2O} was around 3 kbar.

The *P-T* estimates of these rocks, obtained from the garnet-cordierite pair using the calibration developed by Currie, are compatible with other mineralogic evidence. None of the Windmill Metamorphics contain primary muscovite, but they contain instead sillimanite and K-feldspar; hence the rocks should plot on the high-temperature side of the muscovite breakdown curve and in the sillimanite stability field. All cordierite-bearing specimens were taken from regions abounding in migmatites, suggesting that the *P-T* conditions these rocks experienced must have been at least that necessary for the beginning of granite melting.

The rocks of the northern part of the Windmill Islands (north of Sparkes Bay) are thus believed to have been recrystallized under conditions of 5.5–6.5 kbar pressure and temperatures about 760°C with water pressure approximately 3.0 kbar.

* Mineral analyses were done on a Jeol Electron Microprobe X-ray Analyser, utilizing the technique developed by Dr. D. K. B. Sewell at Melbourne University. FeO and Fe_2O_3 were assumed to be similar to averages published in the literature. This Fe (total) to Fe^{3+} ratio is 20 for cordierite and garnet (i.e., little oxidation has taken place).

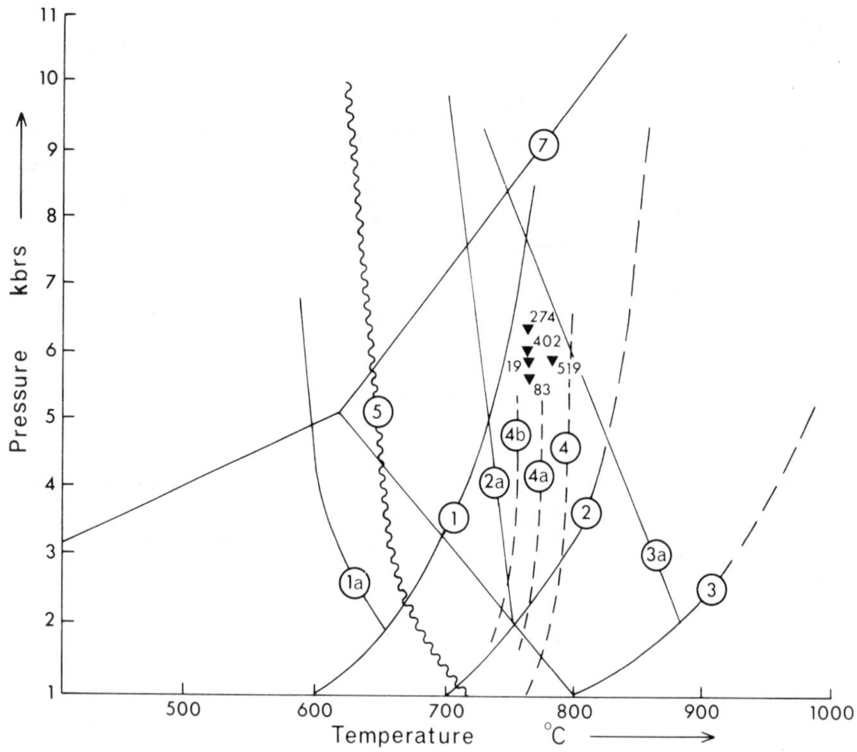

Figure 54.6. Petrogenetic grid showing estimated *P-T* conditions of coexisting garnet and cordierite pairs from the Windmill Metamorphics. Curve number as in Figure 54.5.

Table 54.4. Analyses of coexisting garnet and cordierite assemblages from the Windmill Islands, and derived pressure and temperature based on the calibration of Currie (1973). (Whole-rock and mineral analyses in wt %)

	335/19		335/83		335/274		335/402		335/519	
Whole-rock analyses										
SiO_2	70.46		62.07		64.87		64.44		63.32	
Al_2O_3	13.31		17.42		14.78		14.40		15.80	
Fe_2O_3	2.15		5.10		4.16		3.66		2.78	
FeO	4.57		7.77		6.18		5.11		5.85	
CaO	1.87		0.80		2.02		1.25		1.95	
Na_2O	2.02		1.12		1.81		1.56		2.52	
MgO	1.74		2.11		2.33		4.19		2.16	
K_2O	2.79		2.78		2.00		3.07		3.25	
TiO_2	0.82		1.66		1.08		1.12		0.99	
MnO	0.16		0.16		0.23		0.17		0.15	
P_2O_5	0.17		0.23		0.21		0.16		0.26	
H_2O+	0.50		0.72		1.05		1.81		1.07	
Total	100.56		100.94		100.72		100.94		100.10	
Modal percentages										
Quartz	38		33		29		22		31	
Plag	8		7		27		35		36	
Kspar	—		2		3		8		—	
Cordierite	24		17		13		11		9	
Biotite	27		29		12		20		20	
Sillimanite	trace		7		trace		1		1	
Garnet	1		2		13		1		trace	
Opaques	2		3		3		2		3	
Mineral analyses										
	Cord	Gnt	Cord	Gnt	Cord	Gnt	Cord	Gnt	Cord	Gnt
SiO_2	47.88	37.05	47.24	36.05	48.49	37.54	47.87	37.23	48.09	37.06
Al_2O_3	32.68	21.00	32.47	20.87	32.75	21.34	32.69	21.00	32.99	21.08
FeO	7.90	29.56	9.29	31.30	6.45	30.24	7.49	28.39	8.35	30.13
CaO	0.00	0.75	0.02	0.93	0.00	1.24	0.03	0.98	0.02	1.12
MgO	7.85	3.97	7.22	3.27	9.01	5.64	7.69	3.94	7.64	3.28
MnO	0.32	4.20	0.25	4.04	0.12	1.94	0.24	6.52	0.45	5.61
Estimated *P-T* conditions										
Currie *P* (kbr)	5.9		5.6		6.4		6.0		5.9	
Currie *T*°C	760		762		762		760		780	

Acknowledgments. We gratefully acknowledge the logistic support of the Australian National Antarctic Research Expeditions and in particular the 1970 Casey wintering personnel. Financial assistance came from a University of Adelaide Research Grant (D.F.B.) and the Australian Research Grants Committee (R.L.O.). We also acknowledge permission from the Editor of the *Journal of the Geological Society of Australia* to utilize certain illustrations and tables in this paper.

References cited

Barker, P. F. 1962. Cordierite-garnet gneiss and associated microcline rich pegmatite at Sturbridge, Massachusetts and Union, Connecticut. *Am. Miner., 47,* 907-18.

Binns, R. A. 1964. Zones of progressive metamorphism in the Willyama Complex, Broken Hill District, New South Wales, *J. geol. Soc. Aust., 11,* 283-330.

Binns, R. A. 1969*a*. Ferromagnesium minerals in high grade metamorphic rocks. *Spec. Publs. geol. Soc. Aust., 2,* 323-32.

Binns, R. A. 1969*b*. Hydrothermal investigations of the amphibolite-granulite facies boundary. *Spec. Publs. geol. Soc. Aust., 2,* 341-44.

Blight, D. F. 1975. *The Metamorphic Geology of the Windmill Islands and Adjacent Coastal Areas, Antarctica.* Ph.D. dissertation, University of Adelaide.

Blight, D. F., and Oliver, R. L. 1977. The metamorphic geology of the Windmill Islands, Antarctica, a preliminary account. *J. geol. Soc. Aust., 24,* 239-62.

Collerson, K. D. 1975. Contrasting patterns of K/Rb distribution in Precambrian high grade metamorphic rocks from central Australia. *J. geol. Soc. Aust., 22,* 145-58.

Currie, K. L. 1971. The reaction 3 cordierite ⇌ 2 garnet + sillimanite + 5 quartz as a geological thermometer in the Opinicon Lake Region, Ontario. *Contr. Miner. Petrol., 33,* 215-26.

Currie, K. L. 1973. A note on the calibration of the garnet-cordierite geothermometer and geobarometer. *Contr. Miner. Petrol., 44,* 35-44.

de Waard, D. 1965. A proposed subdivision of the granulite facies. *Am. J. Sci., 263,* 455-61.

Gable, D. J., and Sims, P. K. 1969. Geology and regional metamorphism of some high grade cordierite gneisses, Front Range, Colorado. *Spec. Pap. geol. Soc. Am., 128,* 87 pp.

Harrington, H. J. 1965. Geology and geomorphology of Antarctica. (*In* Van Mieghem, J., and Van Oye, P., *eds.* Biogeography and ecology in Antarctica. The Hague, Junk, 1-71.) [*Monographiae biol.,* 15.]

Heier, K. S., and Adams, J. A. S. 1965. Concentration of radioactive elements in deep crustal material. *Geochim. cosmochim. Acta, 29,* 53-61.

Heier, K. S., and Thorensen, K. 1971. Geochemistry of high grade metamorphic rocks, Lofoten-Vesteralen, North Norway. *Geochim. cosmochim. Acta., 35,* 89-100.

Henry, J. 1974. Garnet-cordierite gneisses near the Egersund-ogna anorthositic intrusion, southwestern Norway. *Lithos, 7,* 207-16.

Hensen, B. J. 1972. Cordierite-garnet equilibria as a function of pressure, temperature, and iron-magnesium ratio. Yb. *Carnegie Inst., 71,* 418-21.

Hensen, B. J., and Green, D. H. 1969. Experimental data on the stability of garnet and cordierite in high grade metamorphic rocks. *Spec. Publs. geol. Soc. Aust., 2,* 345-47.

Hensen, B. J., and Green, D. H. 1970. Experimental data on coexisting cordierite and garnet under high grade metamorphic conditions. *Phys. Earth Planet. Interiors, 3,* 431-40.

Hensen, B. J., and Green, D. H. 1971. Experimental study of cordierite and garnet in pelitic compositions at high pressures and temperatures: I. Compositions with excess alumino-silicate. *Contr. Miner. Petrol., 33,* 309-30.

Hensen, B. J., and Green, D. H. 1972. Experimental study of cordierite and garnet in pelitic compositions at high pressures and temperatures: II. Compositions without excess alumino-silicate. *Contr. Miner. Petrol. 35,* 331-54.

Hensen, B. J., and Green, D. H. 1973. Experimental study of cordierite and garnet in pelitic compositions at high pressures and temperatures: III. Synthesis of experimental data and geological applications. *Contr. Miner. Petrol., 38,* 151-66.

Hietanen, A. 1967. On the facies series in various types of metamorphism. *J. Geol., 75,* 187-214.

Hutcheon, I., Froese, E., and Gordon, T. M. 1974. The assemblage quartz-sillimanite-garnet-cordierite as an indicator of metamorphic conditions in the Daly Bay Complex, N.W.T. *Contr. Miner. Petrol., 44,* 29-34.

Kleeman, A. W. 1965. The origin of granitic magmas. *J. geol. Soc. Aust., 12,* 35-52.

Lambert, I. B., and Heier, K. S. 1968*a*. Geochemical investigations of deep-seated rocks in the Australian Shield. *Lithos, 1,* 30-53.

Lambert, I. B., and Heier, K. S. 1968*b*. Estimates of crustal abundance of Thorium, Uranium and Potassium. *Chem. Geol., 3,* 233-38.

Lambert, I. B., and Wyllie, P. J. 1972. Melting of gabbro (quartz eclogite) with excess water to 35 kilobars, with geological applications. *J. Geol., 80,* 693-708.

Lange, I. M., Reynolds, R. C., Jr., and Lyons, J. B. 1966. K/Rb ratios in coexisting K-feldspars and biotites from some New England granites and metasediments. *Chem. Geol., 1,* 317-22.

Leake, B. E. 1964. The chemical distinction between ortho- and para-amphibolites. *J. Petrology, 5,* 238-54.

Leake, B. E. 1965. The relationship between composition of calciferous amphibole and grade of metamorphism. (*In* Pitcher, W. S., and Flinn, G. W., *eds. Controls of Metamorphism.* Edinburgh, Oliver and Boyd, 299-318.)

Lewis, J. D., and Spooner, C. M. 1973. K/Rb ratios in Precambrian granulite terrains. *Geochim. cosmochim. Acta, 37,* 1111-18.

McLeod, I. R., and Gregory, C. M. 1966. Geological investigations along the Antarctic coast between longitude 108° E and 166° E. *A.N.A.R.E. Rep.,* Ser. A (111) Geology, Publ. 83, 53 pp.

Newton, R. C., Charlu, T. V., and Keppla, O. J. 1974. A calorimetric investigation of the stability of anhydrous magnesium cordierite with application to granulite facies metamorphism. *Contr. Miner. Petrol., 44,* 295-311.

Oliver, R. L. 1969. Some observations on the distribution and nature of granulite-facies terrains. *Spec. Publs. geol. Soc. Aust., 2,* 259-68.

Oliver, R. L. 1970. Some aspects of Antarctica-Australian geological relationships. (*In* Adie, R. J., *ed., Antarctic Geology and Geophysics.* Oslo, Universitetsforlaget, 859-64.)

Reinhardt, E. W. 1968. Phase relations in cordierite-bearing gneisses from the Gananoque area, Ontario. *Can. Jnl Earth Scis, 5,* 455-82.

Richardson, S. W., Gilbert, M. C., and Bell, P. M. 1969. Experimental determination of kyanite-andalusite and andalusite-sillimanite equilibria: the aluminosilicate triple point. *Am. J. Sci., 267,* 259-72.

Saxena, S. K., and Hollander, N. B. 1969. Distribution of iron and magnesium in coexisting biotite, garnet and cordierite. *Am. J. Sci., 267,* 210-16.

Sen, S. K. 1970. Magnesium-iron compositional variance in hornblende-pyroxene granulites. *Contr. Miner. Petrol., 29,* 76-88.

Shaw, D. M. 1968. A review of the K/Rb fractionation trends by co-variance analysis. *Geochim. cosmochim. Acta, 32,* 573-601.

Shaw, D. M. 1972. The origin of the Apsley Gneiss, Ontario. *Can. Jnl Earth Scis, 9,* 18-35.

Sighinolfi, G. P. 1971. Investigations into deep crustal levels: fractionating effects and geochemical trends related to high-grade metamorphism. *Geochim. cosmochim. Acta, 35,* 1005–21.

Touret, J. 1971*a*. Les facies granulite en Norvege Meridionale: I. Les associations mineralogiques. *Lithos, 4,* 239–49.

Touret, J. 1971*b*. Les facies granulite en Norvege Meridionale: II. Les fluides inclusions. *Lithos, 4,* 423–36.

Voronov, P. S., and Karasik, A. M. 1963. Brief geological and geophysical characteristics of Grearson Oasis in the Budd Coast area of East Antarctica. *Inf. Bull. Sov. Antarct. Exped., 42,* 5–11.

Winkler, H. G. F. 1967. *Petrogenesis of Metamorphic Rocks.* 2nd ed. Berlin, Springer-Verlag, 220 pp.

Winkler, H. G. F. 1975. *Petrogenesis of Metamorphic Rocks.* 4th ed. Berlin, Springer-Verlag, 334 pp.

Wynne-Edwards, H. R., and Hay, P. W. 1963. Coexisting cordierite and garnet in regionally metamorphosed rocks from the Westport area, Ontario. *Can. Mineralogist, 7,* 453–78.

55

The Geologic Evolution of the Prince Charles Mountains—An Antarctic Archean Cratonic Block

R. J. TINGEY
Bureau of Mineral Resources, P.O. Box 378, Canberra, A.C.T. 2601, Australia

Abstract. The Prince Charles Mountains consist mainly of a complex of Archean to upper Proterozoic metamorphic rocks, locally intruded by Cambrian granites and unconformably overlain by a small enclave of upper Permian sedimentary strata. Metamorphic grade ranges from greenschist to granulite facies, and at least two major metamorphisms—one Archean, the other late Proterozoic—have been distinguished despite geologic, metamorphic, and isotopic complexities caused by the overprinting of the later metamorphism. The wide range of $^{87}Sr/^{86}Sr$ initial ratios obtained in geochronologic studies is interpreted as evidence that both major metamorphisms involved crustal formation and reworking processes.

Archean rocks are exposed in the southern Prince Charles Mountains and comprise a 2800 m.y. old granitic basement complex unconformably overlain by metasedimentary beds characterized by prominent white fuchsite-bearing quartzites that are cut at one locality by a 2580 m.y. old muscovite pegmatite. Banded-iron formations in the area may also be Archean. These Archean rocks are intruded by now-amphibolitized tholeiitic basic dikes of probably middle Proterozoic age and are overlain by locally contorted upper Proterozoic greenschist facies metasedimentary beds. The greenschist facies metamorphism in the southern Prince Charles Mountains is thought to be a marginal manifestation of the major metamorphic episode that produced amphibolite and granulite facies metamorphic rocks in the northern Prince Charles Mountains between 800 and 1100 m.y. ago. Granitic stocks and pegmatites were intruded into the Precambrian rocks about 500 m.y. ago and were associated with a widespread thermal effect that reset mineral isotope systems. A few Phanerozoic alkaline basic intrusive rocks that intersect the basement are thought to be related to activity along the major rift structure now occupied by the Lambert Glacier–Amery Ice Shelf system.

The Prince Charles Mountains provide probably the best cross section of the East Antarctic metamorphic shield and contain the only Archean cratonic block as yet identified in Antarctica. Comparisons with other parts of Gondwanaland reveal important similarities with Africa and southwestern Australia but few correlations with peninsular India and Sri Lanka, the areas most commonly juxtaposed against the Prince Charles Mountains area in Gondwanaland reassemblies. The research described in this paper demonstrates yet again that the highest grade metamorphic rocks in an area are not necessarily the oldest, and it highlights the importance of geochronology in studies of the East Antarctic shield.

The Prince Charles Mountains (lat. 69°–74° S, long. 61°–70° E) were first sighted in 1947 by U.S. Navy aircraft participating in Operation Highjump, and they were named when a ground party from the Australian National Antarctic Research Expedition (ANARE) base at Mawson visited them in 1954. Since then they have been visited by several Australian and Soviet field parties, most of which included geologists (Crohn, 1959; Ruker, 1963; Trail, 1964*a,b*; McLeod, 1964; Trail and McLeod, 1969; Soloviev, 1972; Iltchenko, 1972). This paper is mainly the result of systematic 1:250,000 scale regional geologic mapping between 1968 and 1974 by Bureau of Mineral Resources (BMR) geologists working with helicopter-supported ANARE field parties and the result of associated regional geochronologic studies by P. A. Arriens at the Australian National University, Canberra. Recent publications include papers by Arriens (1975), Tingey (1974), Grikurov and Soloviev (1974), Halpern and Grikurov (1975), Grew (1976), Tingey and others (1976), and Wellman and Tingey (1976). In addition, the BMR issued 13 preliminary edition 1:250,000 scale geologic maps of the area between 1973 and 1976.

The Prince Charles Mountains constitute the best cross section through the East Antarctic shield and are exposed in a major 600 × 300 km trough in the East Antarctic ice sheet that is caused by ice drainage through the Lambert Glacier-Amery Ice Shelf system (Trail, 1964*a*). The Lambert Glacier, together with its tributaries, dominates and controls the topography and scenery of the Prince Charles Mountains; it seems to be located on a major crustal structure (Wellman and Tingey, 1976). The mountains consist of isolated nunataks and peaks near the edge of the trough and, near the major glaciers, very steep-sided, flat-topped massifs that stand up to 1000 m above the surrounding ice and range up to 500 km² in area. Apart from these massifs, the Mawson Escarpment (around lat. 73 ° S, long. 68°–69° E) and the mountains near and to the west of Beaver Lake (lat. 70° 45′ S, long. 68° E) are the only extensive ice-free areas. Most outcrops are covered with *felsenmeeren* of broken rock debris of mostly local origin, through which

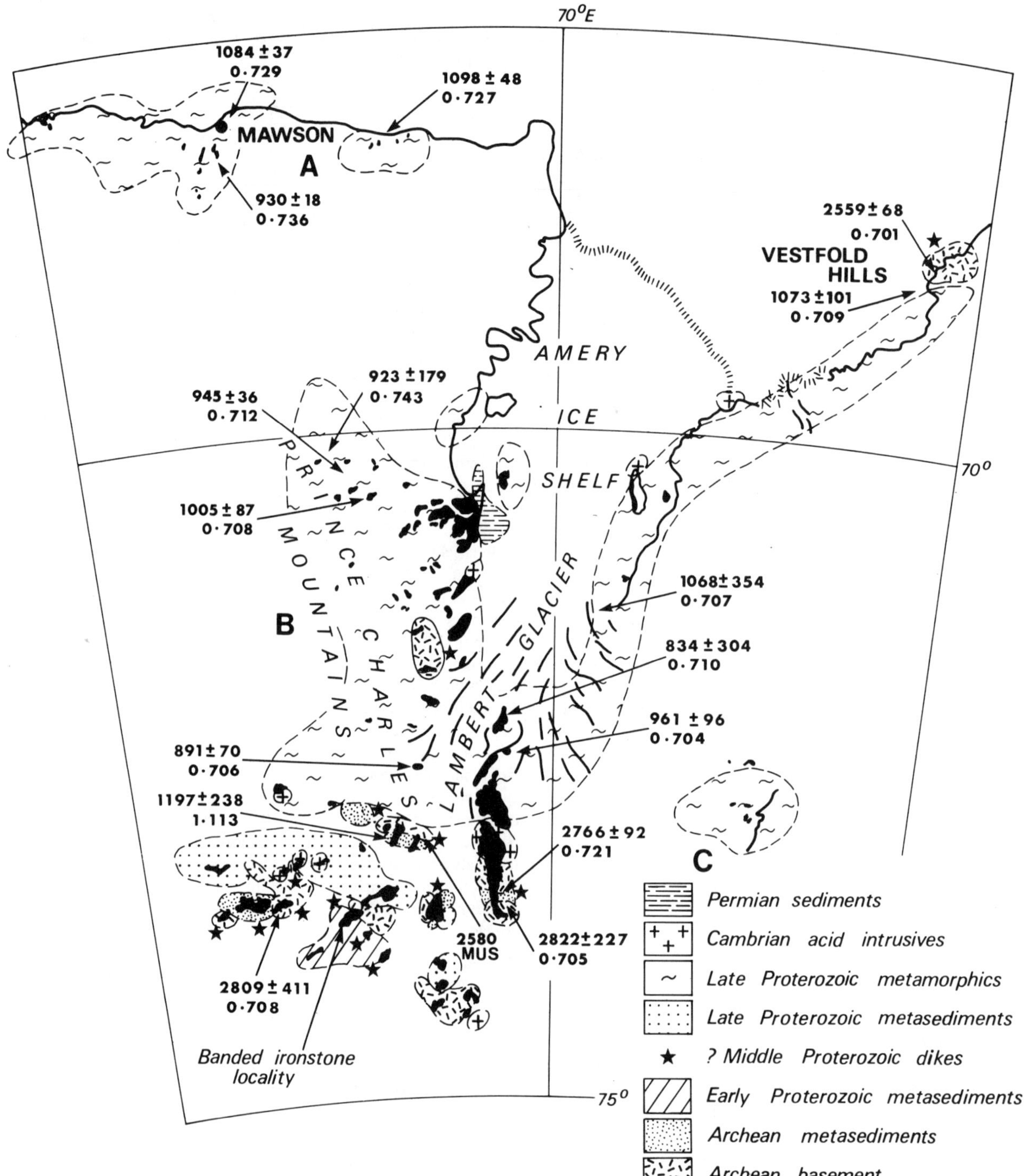

Figure 55.1. Schematic geologic map of the Prince Charles Mountains region with selected Rb-Sr geochronologic data. Solid black indicates outcrop; boundaries under glacial cover are diagrammatic only. Geochronologic data are provisional and subject to revision.

bedrock trends can be mapped only by photogeologic interpretation. Bedrock locally protrudes through this debris, but the best exposures are in cliff faces. Correlations between scattered outcrops are inferred from similarities in rock type, deformation, metamorphic grade, and stratigraphic succession.

This paper describes the stratigraphy and metamorphic geology of the Prince Charles Mountains and briefly discusses their significance to an understanding of the geologic evolution of the East Antarctic shield and Gonwanaland as a whole. In addition to Arriens' (1975) geochronologic data, it incorporates the results of geochemical, petrologic, and petrographic studies made in the BMR laboratories (see Sheraton, Chapter 59, this volume).

Bedrock geology

Trail and McLeod (1969) and Craddock (1972) show the Prince Charles Mountains as consisting of Precambrian metamorphic rocks whose grade decreases from granulite facies in the north to greenschist facies in the south. The metamorphic rocks are shown as intruded by Precambrian

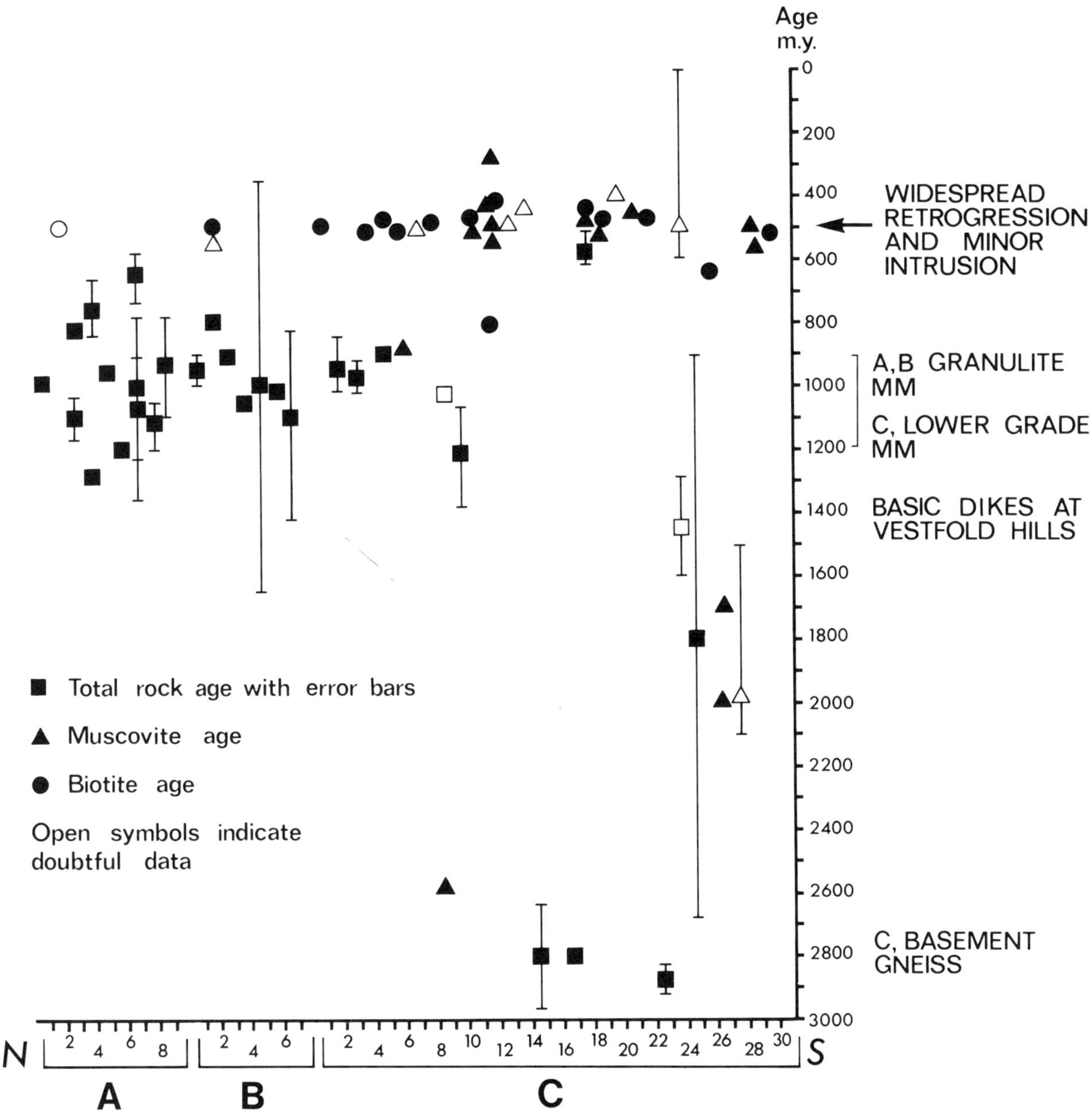

Figure 55.2. Rb-Sr geochronologic data from the Prince Charles Mountains region. A, B, and C refer to areas on Figure 55.1. Data are provisional and subject to revision.

charnockite and biotite granite, and overlain in the Beaver Lake area by a small enclave of Permian sedimentary beds. Craddock's interpretation was based directly on that of Trail and McLeod (1969) who summarized available results of reconnaissance mapping.

The recent Soviet geologic map of Antarctica (Ravich and Grikurov, 1976) refines Craddock's interpretation by assigning ages to various elements of the Prince Charles Mountains metamorphic complex. It depicts the granulite-facies rocks in the north as Archean and shows most of the lower grade rocks in the south as Proterozoic, with decrease in metamorphic grade generally correlated with decrease in age.

A very different interpretation is presented in this paper. The highest grade metamorphic rocks in the Prince Charles Mountains are attributed to late Proterozoic metamorphism, whereas rocks of lower metamorphic grade, exposed in southern outcrops, are Archean. Figure 55.1 is a geologic map of the area, and geochronologic results are plotted on Figure 55.2. Localities referred to in the text can be found on topographic maps of Antarctica or on maps in the various references cited, particularly Soloviev (1972), Halpern and Grikurov (1975), and Trail (1964*a, b*).

Archean

Two groups of rocks of Archean age are exposed in the southern Prince Charles Mountains in a 230 km × 80 km belt between the Mawson Escarpment in the east and Mount Mather in the west. They are, respectively, granitic basement dated at 2750–2900 m.y. and metasedimentary rocks characterized by prominent, thick, white, quartzite beds whose minimum age is defined by a single 2580 m.y. old pegmatite dike. In addition, other rocks nearby, and a small enclave of granitic rocks near Mount Collins in the northern Prince Charles Mountains, are inferred to be Archean (Figure 55.1). The Archean rocks are typically

Figure 55.3. Cambrian granite with very sharp contacts intruded into Archean gneisses, Mount Borland, southern Prince Charles Mountains.

intersected by tholeiitic intrusive rocks metamorphosed to lower amphibolite-facies mineral assemblages. In places there are also crosscutting acid intrusive rocks of Cambrian age (Figures 55.1, 55.2, and 55.3).

Granitic basement rocks

The Archean granitic basement rocks are mostly massive or poorly foliated and have amphibolite-facies mineral assemblages; they are composed of metamorphosed biotite granite, biotite adamellite, biotite granodiorite, augen gneiss, biotite-ferrohastingsite gneiss, and well-foliated biotite gneiss. Unpublished chemical analyses (J. W. Sheraton, personal communication) confirm field observations that most of these rocks are of igneous intrusive origin. For example, biotite-ferrohastingsite gneiss, a widely distributed, massive, dark brown rock of uniform appearance, is chemically homogenous and diopside normative.

A few small enclaves of metasedimentary rocks, which may be either xenoliths of older rocks or fault slivers of younger ones, are contained within the granitic basement rocks. No evidence of Archean greenstones has been found, although large metabastite masses in the northern Prince Charles Mountains (Soloviev, 1972; Tingey, 1974) require further study. The relationships of the granitic basement rocks with various younger formations are generally obscured by poor outcrop, faulting, folding, and intrusive activity. They are displayed at three localities, and there is no evidence that the granitic basement is intrusive into the various cover rocks that overlie it.

Specimens from three widely separate localities (see Figure 55.1) have given Rb-Sr isochron ages between 2750 and 2900 m.y. and are the basis for the Archean age assigned to the granitic basement as a whole. Other rocks, including granites which are mapped on the basis of field and chemical characteristics as part of the Archean basement, have given Proterozoic isochron ages (1000–1800 m.y.; Figure 55.2), with, in most cases, very high initial ratios (up to 1.113; Figure 55.1). The limited geochronologic data for the granitic basement rocks are generally consistent with field relationships and are thought to indicate a major episode of crustal formation between 2900 and 2750 m.y. ago, although some initial Sr-isotope ratios imply the existence of still older sialic crustal rocks.

Elsewhere in Antarctica, Archean ages are reported for single localities in western Queen Maud Land (Halpern, 1970) and Enderby Land (Sobotovich and others, 1976; Kamenev, 1977), but their regional significance awaits confirmation. The Archean amphibolite-facies granitic basement in the southern Prince Charles Mountains has by contrast been mapped over a considerable area and is thought to be part of an Archean cratonic block analogous to the ancient granitic nuclei now preserved in various parts of Africa (see Clifford, 1974; Kröner, 1977).

Metasedimentary rocks

The Archean granitic basement is overlain by younger sedimentary rocks characterized by thick (as much as 200 m) white to green beds of locally fuchsite-bearing quartzite, which also include conglomerate with mainly quartzite and granite clasts, pelitic and psammitic beds, calc-silicate rocks, iron-rich strata, and para-amphibolite. In most areas these rocks show lower amphibolite-facies assemblages, probably retrograde, with kyanite, muscovite, and staurolite in the pelitic rocks; but there are some signs of earlier higher grade assemblages, such as replacement of larger, more magnesian garnets by clusters of smaller garnet euchedra and replacement of sillimanite by staurolite. Disequilibrium assemblages, such as garnet + cordierite + kyanite + staurolite + biotite + chlorite + quartz, have been found at only very few localities. In the southernmost outcrops, sillimanite, probably relict from these early high-grade assemblages, is the dominant aluminum silicate, and its preservation suggests that the retrograde effects of the late Proterozoic metamorphism were less intense toward the south. The early high-grade assemblages are thought to be the products of an ill-defined early Proterozoic metamorphic episode that included the intrusion of granites and pegmatites in the southernmost Prince Charles Mountains (see Figures 55.1 and 55.2 and below).

Contacts between the Archean metasedimentary rocks and the underlying granitic basement are mostly faulted or not seen because of poor outcrops, but at one place the quartzite rests unconformably on an irregular basement surface. Elsewhere, although there is a lithologic contrast (but not an intrusive contact) between the granitic basement rocks and the overlying strata, later folding has imposed the same foliation on both units, and there is now no obvious structural discontinuity. The Archean strata were probably deposited upon an eroded granitic basement, and the local presence of conglomerate, together with the preservation of cross-lamination in quartzite, demonstrates that at least some of the sedimentation took place in a high-energy environment.

The Archean strata, like the underlying basement rocks, are intersected by amphibolitized tholeiitic intrusive rocks and locally by muscovite and biotite granites and pegmatites. Most of the latter are unmetamorphosed and almost certainly of early Paleozoic age, but one pegmatite, metamorphosed under low-amphibolite or greenschist-facies conditions with partial recrystallization of muscovite, has given a Rb-Sr muscovite age of 2580 m.y.; its field relationships with the basic dikes are unknown. The Archean age of the metasedimentary rocks is based on this single date. The pegmatite was probably metamorphosed during the same (late Proterozoic) event that produced retrograde assemblages in nearby pelites and converted the tholeiitic intrusive rocks to amphibolites.

Proterozoic

Metamorphic rocks ranging in age from early to late Proterozoic are exposed throughout the Prince Charles Mountains. The lower Proterozoic rocks, like the Archean rocks, are intersected by amphibolitized tholeiitic dikes of possibly middle Proterozoic age, but there are no such dikes in the upper Proterozoic rocks. In the northern Prince Charles Mountains late Proterozoic high-grade metamorphism has obliterated older structures, including any crosscutting basic intrusive rocks that might have existed. However, near the southern edge of this high-grade zone, and in a small area within it (Figure 55.1), amphibolitized tholeiitic intrusive rocks are preserved in original Archean granitic basement rocks that give late Proterozoic metamorphic ages (Figure 55.1). Upper Proterozoic sedimentary rocks deposited after the emplacement of the tholeiitic intrusive rocks were also metamorphosed at this time.

Lower Proterozoic

The earliest known Proterozoic igneous activity (1700–2000 m.y.) in the area is indicated by sparse geochronologic data (Figure 55.2) from pegmatites and granites that intrude migmatites and from pelitic and psammitic beds, at a few of the most southerly outcrops. Local high-grade assemblages in these rocks, and in others mapped as Archean, are attributed to metamorphism associated with this igneous activity. The white quartzite so typical of the upper Archean strata is absent from these particular outcrops, and the metasedimentary rocks there—which are, along with the granitic rocks that intrude them, cut by amphibolitized tholeiitic dikes—may be either early Proterozoic in age or remnants of a quartzite-free facies of the Archean sequence. Similarly the migmatites may be either Archean or early Proterozoic.

In places these supposed lower Proterozoic rocks are markedly retrogressed: metapelites have chlorite-chloritoid assemblages, and tholeiitic dikes are converted to finely matted actinolite, chlorite, and plagioclase. Pelitic metasedimentary beds and amphibolitized tholeiitic dikes within the granitic and pegmatitic intrusive rocks are, however, only slightly affected by the retrogression, which postdates the amphibolitization of the basic dikes and is attributed to the Cambrian episode of thermal metamorphism and intrusive activity.

Basement-cover relationships are not displayed by these metasedimentary rocks but are exposed at Mount Ruker, where low-grade metamorphosed psammite, slate, and shale, with minor quartzite and banded ironstone lenses (Soloviev, 1972) overlie an eroded Archean granitic basement (Halpern and Grikurov, 1975). Recent Australian research (P. A. Arriens, personal communication; J. W. Sheraton, personal communication) shows that these basement rocks are strongly fractionated and give a Rb-Sr isochron age of 1442 ± 152 m.y. (initial $^{87}Sr/^{86}Sr = 0.8902$); this is considered to be a metamorphic, rather than an intrusive, age. The exposed banded ironstones appear to be uneconomic, although aeromagnetic surveys by the Soviet Antarctic Expedition have traced anomalies attributed to them for some 120 km under nearby glaciers. Quite apart from their remote location, they are unimpressive in terms of outcrop area (10 km^2), thickness (100 m), and grade (about 35% Fe).

Achritarch studies by Iltchenko (1972) are cited by Soloviev (1972) as evidence that the ironstone-bearing sequence is early Proterozoic in age, although more recent work (Halpern and Grikurov, 1975) seems to cast doubt on some of Soloviev's interpretations (see below). The sequence is cut by amphibolitized tholeiitic dikes and by unmetamorphosed lower Paleozoic biotite granites and pegmatites, and it could be correlative with the Archean metasedimentary rocks. This is not apparent, however, from its metamorphic grade and mineral assemblages.

Middle Proterozoic(?)

Metamorphosed tholeiite dikes intersect all the Archean and Proterozoic rocks so far described, but none are known in either the high-grade upper Proterozoic metamorphic rocks in the north or the lower grade upper Proterozoic calcareous metasedimentary beds in the south. Related subconcordant metabasic rocks up to 200 m thick are exposed within the Archean metasedimentary strata at Mount Menzies (Trail, 1964*b*). Most of these basic rocks have retrograde lower amphibolite-facies assemblages, with characteristic blue-green hornblende, which are attributed to late Proterozoic metamorphism. Locally (e.g., in the southern part of the Mawson Escarpment) there are networks of basic dikes with complicated crosscutting relationships, but elsewhere (Cumpston Massif; also see Figure 55.4) shearing, probably associated with recumbent folding (Figure 55.5) in the latter stages of the late Proterozoic metamorphism, has caused the dikes to be streaked out into subconcordant amphibolite lenses and bands, somewhat in the manner illustrated on a larger scale from Greenland by Escher and others (1975).

There is no direct evidence for the age of the metamorphosed tholeiitic intrusive rocks, but they postdate the lower Proterozoic granitic intrusive rocks and predate upper Proterozoic metasedimentary rocks that are intruded by lower Paleozoic granites. In the Vestfold Hills, some 500 km northeast of the Prince Charles Mountains, Archean metamorphic rocks are intruded by unaltered dolerite dikes whose Rb-Sr whole-rock isochron age is quoted at about 1400 m.y. (Arriens, 1975). The chemical composition of these dikes is unknown, but the striking similarity between their stratigraphic relationships and

Figure 55.4. Tightly folded Archean gneiss at Mount Twigg, southern Prince Charles Mountains. A is a folded amphibolitized tholeiitic dike of possibly middle Proterozoic age. The whole structure is crosscut by veins of Cambrian pegmatite. Note helicopter for scale.

Figure 55.5. Large recumbent fold in Archean or lower Proterozoic gneiss, southern Blake Nunataks, southern Prince Charles Mountains. The cliff face is about 500 m high and 5 km long. View looking southeast.

those of the metamorphosed basic dikes in the southern Prince Charles Mountains suggest that both sets may be of the same age. A 1400 m.y. age for the latter dikes would fall between the maximum age of 2000–1700 m.y. imposed by the lower Proterozoic rocks they intrude and the minimum age of 1100–800 m.y. imposed by the younger upper Proterozoic high-grade metamorphic rocks (see Figure 55.2).

Upper Proterozoic metasedimentary rocks

Original sedimentary features such as ripple marks, mudcracks, and cross-laminae are preserved in low-grade, mostly calcareous metasedimentary rocks and phyllites which crop out in a broad southeasterly trending band across the center of the southern Prince Charles Mountains. These rocks generally have low to middle greenschist-facies prograde assemblages containing quartz and combinations of chlorite, sericite, biotite, epidote, actinolite, tremolite, and carbonate; slightly higher grade (lower amphibolite facies) assemblages, with kyanite and staurolite in pelites, are found near the southernmost outcrops. There is no evidence that any assemblage is retrograde. The calcareous beds are not cut by the amphibolitized tholeiitic intrusive rocks that typically intersect older formations at nearby outcrops, nor are there concordant metabasic rocks that might reasonably be interpreted as sheared-out remnants of such intrusive rocks. However, they are intruded by small bodies of Cambrian granite and pegmatite at outcrops along the north side of the Fischer Glacier and are thus younger than the tholeiitic dikes (1400 m.y.?) and older than the granites (about 500 m.y.).

These low-grade rocks were originally described from Mount Rubin by Soloviev (1972), who grouped them, together with jaspilite-bearing, low-grade metasedimentary rocks at nearby Mount Ruker, into the "Sodruzhestvo Series." Grikurov and Soloviev (1974) reported jaspilite clasts in intraformational conglomerate at Mount Rubin, where the base of the sequence is not exposed. The clasts could have been derived from concealed strata lower in the sequence, but the presence or absence of the metamorphosed basic dikes appears to be the critical field relationship. They are present at Mount Ruker and absent from Mount Rubin. The Sodruzhestvo Series grouping is thus probably invalid, despite the similarities in metamorphic grade; observations by Grikurov and Soloviev (1974) and Halpern and Grikurov (1975) support this.

Halpern and Grikurov (1975) indicate that the Mount Rubin beds were metamorphosed 800 m.y. ago and again at about 500 m.y. ago. Initial $^{87}Sr/^{86}Sr$ ratios (0.738 and 0.730, respectively) were interpreted as indicating that during the Cambrian intrusive episode, Rb-Sr isotopic re-equilibration occurred in the phyllites (500 m.y. age) but not in clasts (800 m.y. age) from an introformational conglomerate. The 800 m.y. age could be either that of an early metamorphism of the metasedimentary sequence or of the rocks from which the clasts were derived. The age of the metasedimentary beds is thus imprecisely known.

Late Proterozoic metamorphism

Granulite facies and upper amphibolite facies rocks of late Proterozoic age are exposed (Figure 55.1) in the northern Prince Charles Mountains (Crohn, 1959; Tingey, 1974), at the Mawson Coast (Crohn, 1959; Trail, 1970; Sheraton, Chapter 59, this volume) along the eastern flanks of the Amery Ice Shelf (Trail and McLeod, 1969; Craddock, 1972), and in and near the northern part of the Mawson Escarpment. They are not cut by metamorphosed tholeiitic intrusive rocks, but they locally contain concordant mafic lenses and streaks that may represent deformed remnants of such intrusive rocks.

Metamorphic grade generally decreases southward from the northern Prince Charles Mountains, where granulite facies rocks predominate, through a zone of upper amphibolite facies rocks to the southern outcrops where the main effects were (1) retrograde lower emphibolite facies metamorphism of older rocks and (2) low-grade prograde metamorphism of upper Proterozoic calcareous strata. Moderate metamorphic pressures are inferred for the northern area because high-grade pelitic rocks there contain sillimanite, cordierite, and garnet but lack andalusite and kyanite, and because metabasic rocks do not contain garnet. Farther south lower temperature metamorphism at relatively higher pressures is indicated by retrograde assemblages with kyanite, garnet, and staurolite. The contrast between the two areas is interpreted as evidence that the north-south decrease in metamorphic grade cannot be entirely attributed to the present exposure of different crustal levels. No equivalent gradation of metamorphic grade is seen on the northern edge of the high-grade zone where it is exposed near the Vestfold Hills (Figure 55.1), and a major but, at present, concealed geologic dislocation must be postulated for that area.

Figure 55.2 summarizes Rb-Sr geochronologic data from the Prince Charles Mountains area (Arriens, 1975, and personal communication) and shows that whole-rock isochron dates for the high-grade metamorphic rocks, especially those in the northern Prince Charles Mountains (B on Figures 55.1 and 55.2), cluster around 1100–800 m.y. Initial $^{87}Sr/^{86}Sr$ ratios (see Figure 55.1) range from 0.705 to 1.113 and are interpreted as evidence that the late Proterozoic metamorphism involved material only recently derived from the mantle and—as is evidenced by such obvious metasedimentary rocks as marble and metapelite—older sialic crustal rocks.

Crust formation and crustal reworking are thus both invoked for the major metamorphic episode that affected a 600-km-wide zone in this part of Antarctica between 1100 and 800 m.y. ago. Sparse geochronologic data presented by Maegoya and others (1968) and Grew (1975) suggest a westward extension of the zone through the Rayner Complex (Ravich and Grikurov, 1976) in Enderby Land and at least as far as the Lutzow-Holm Bay area of Dronning Maud Land. The 1100–800 m.y. metamorphism corresponds well to the Kibaran orogeny (Clifford, 1974) in Africa and to isotopic ages from the Vijayan Series in Sri Lanka (see Crawford, 1974) and southwestern Australia (Arriens, 1975), but similar ages are uncommon in peninsular India (Crawford, 1974).

PHANEROZOIC

The Phanerozoic geologic record in the Prince Charles Mountains is limited to Cambrian granitic intrusive rocks, Permian sedimentary beds, and alkaline basic intrusive and extrusive rocks of various ages. Cenozoic glacial phenomena are well displayed (Trail, 1964*a*; Tingey, 1974), but there is at present no direct means of dating them.

Granitic intrusive rocks

Thermal metamorphism and igneous intrusive activity took place in the area between 600 and 400 m.y. ago and reached their maximum intensity about 500 m.y. ago (Figure 55.2). They resulted in the emplacement of biotite- and muscovite-bearing granite and pegmatite (see Figure 55.1), local low-grade retrograde metamorphism in the southern Prince Charles Mountains, and regional resetting of Rb-Sr isotope systems in biotite (Figure 55.2). The intrusive rocks cut sharply across country-rock structures (Figures 55.3 and 55.4), and contact metamorphism is confined to a maximum width of only a few meters. In the northern Prince Charles Mountains the lower Paleozoic intrusive rocks occur in association with (and are locally hard to distinguish from) anatectic granite and pegmatite produced in the late Proterozoic metamorphism discussed above, but Rb-Sr biotite ages reflect the later intrusive activity.

Thermal metamorphism and intrusive activity were widespread in Gondwanaland in early Paleozoic time and are referred to as the Pan African orogeny in Africa (see Clifford, 1974; Kröner, 1977) and as the Ross orogeny elsewhere in Antarctica (Gunn and Warren, 1962). This widespread activity is poorly understood (Hurley, 1973), but Kröner (1977) implies that it might be due to substantial fragmentation of the crust in anticipation of the "continental drift" tectonic regime that became established later in Phanerozoic time.

Sedimentary rocks

Upper Permian clastic sedimentary rocks and coal measures exposed around Beaver Lake in the northern Prince Charles Mountains have been described by Crohn (1959), Mond (1972), and Bennett and Taylor (1972); the contained flora have been described by Kemp (1973) and White (1973). Mond (1972) also discussed possible correlation of these rocks with parts of the Beacon Supergroup of the Transantarctic Mountains. Fine-grained red siltstone debris with *Glossopteris* plant fossils was also discovered in moraine in the southern Prince Charles Mountains (Ruker, 1963; White, 1962), but no outcrops of Permian rocks have been found in that area despite an extensive search.

It is perhaps significant that the Permian beds are exposed at Beaver Lake, the lowest area of exposed rock in the Prince Charles Mountains, for the sedimentary sequence appears to be a valley-floor deposit. Beaver Lake is the only site in the Prince Charles Mountains where such a valley floor is exposed by modern ice retreat, but other enclaves of Permian strata are probably buried beneath the ice at low topographic levels. Such an enclave could be the source of the morainal debris discovered by Ruker (1963).

Basic intrusive and extrusive rocks

Basic igneous rocks with alkaline affinities have been mapped at a few widely scattered places in the Prince Charles Mountains. They range in age from 504 m.y. to 50 m.y. and include a magnophorite basalt dike (Sheraton, in prep.), alnöite sills (Walker and Mond, 1971), and olivine leucitite flows. The wide spread of K-Ar ages appears to rule out any common genesis for these rocks, but they may be related to major structural features such as the crustal rift postulated by Wellman and Tingey (1976).

Structure

Because of the scattered outcrop and the broad-ranging method of fieldwork adopted, little is known about the structural evolution of the area apart from its obvious complexity (Figures 55.4 and 55.5) and the large scale of some of its structures (Figure 55.5).

Large-scale recumbent folds (Figure 55.5) are exposed at several places in the southern Prince Charles Mountains. None of them are cut by amphibolitized tholeiitic intrusive rocks, although in some cases the intrusive rocks are themselves folded and streaked out (Figures 55.4 and 55.5). The folds thus postdate these intrusive rocks, but predate the 500 m.y. old acid intrusive rocks that crosscut all structures in upper Proterozoic strata along the north flanks of the Fisher Glacier. This, together with Rb-Sr geochronologic data from one locality, suggests that the folding occurred in later phases of the 1100–800 m.y. old metamorphism.

Faults have been mapped in the largest outcrops, notably near Beaver Lake, at Mount Stinear, and on the Mawson Escarpment, but scattered outcrop prevents proper analysis. The major glaciers are possibly located on major structural features, and Wellman and Tingey (1976) relate this to gravity anomaly evidence; there is little geologic evidence apart from the straight courses of the glaciers.

Conclusions

Geochronologic and field data from the Prince Charles Mountains demonstrate that this area has had a long and complex geologic history that broadly resembles the pre-Mesozoic metamorphic history of Africa as described by Clifford (1974) and Kröner (1977). Correlation with peninsular India and Sri Lanka (Crawford, 1969, 1974; Berger and Jayasinghe, 1976; Nagpaul and Mehta, 1975) is less satisfactory despite the juxtaposition in Gondwanaland reconstructions of that subcontinent against this part of Antarctica (Dietz and others, 1972).

Archean granitic basement rocks and metasedimentary beds exposed in the southern Prince Charles Mountains constitute an Archean cratonic block—the only one as yet known from Antarctica. It is bordered on its northern margin by a 600-km-wide zone marked by late Proterozoic high-grade metamorphism, probably equivalent to the Kibaran orogeny in Africa (Clifford, 1974) and to high-grade metamorphism in Madagascar (Vachette-Caen and Hottin, 1976), in Wilkes Land, Antarctica, and in the Esperance and Fraser Range areas of southwestern Australia (Arriens, 1975). Data from Sri Lanka also indicate a major metamorphism at this time (Berger and Jayasinghe, 1976), but no equivalent events are evident in data from peninsular India (Crawford, 1974), despite the common juxtaposition of this area to the Prince charles Mountains region in Gondwanaland reassemblies (Dietz and others, 1972). This high-grade metamorphism in the northern Prince Charles Mountains resulted in the reconstitution of older rocks, obliteration of original structures and rock relationships, and the re-equilibration of Rb-Sr isotope systems. Low $^{87}Sr/^{86}Sr$ initial ratios probably indicate that there was also large-scale generation of new crustal rocks from the mantle. Lower grade marginal effects of the high-grade metamorphism are also apparent in the southern Prince Charles Mountains.

Intrusion of granite and pegmatite and the widespread re-equilibration of biotite Rb-Sr systems resulted from a pervasive early Paleozoic intrusive and thermal metamorphic episode. This reached its maximum intensity about 500 m.y. ago and probably corresponds to—and is as poorly understood as—the Pan African (Clifford, 1974; Kröner, 1977) and Ross (Gunn and Warren, 1962) orogenies elsewhere in Gondwanaland. The later Phanerozoic record is very limited and probably reflects continued geologic stability similar to that displayed by other metamorphic shielf areas in Gondwanaland. It comprises coal-bearing Permian sedimentary sequences probably deposited in valleys in the eroded metamorphic basement, spasmodic and scattered alkaline basic intrusions possibly related to major crustal structures (Wellman and Tingey, 1976), and the Cenozoic glaciation.

The geology and geologic evolution of the Prince Charles Mountains described in this paper reveals yet another example of an area where the oldest rocks are by no means the most deformed and where the degree of deformation and metamorphism is no guide to the antiquity of the rocks concerned. The interpretation here differs markedly from that presented by Ravich and Grikurov (1976), but it has close parallels in Africa, Australia, and other parts of the world. The area also demonstrates the crucial importance of geochronology in attempts to unravel the complexities of metamorphic rocks in Antarctica.

Acknowledgments. This presentation relies greatly on the work, thoughts, and experience of my colleagues R. N. England and J. W. Sheraton and on the geochronologic data of our co-worker P. A. Arriens. The fieldwork that forms the basis for our work was made possible by the efforts, often under severely adverse conditions, of many members of the Australian National Antarctic Research Expeditions. Logistic support for these expeditions is provided by the Antarctic Division of the Australian Department of Science, and I gladly acknowledge the sympathetic and ready response they have made to our many requests.

Critical comment by BMR colleagues J. W. Sheraton, L. A. Offe, D. S. Trail, J. H. C. Bain, R. G. Dodson, W. B. Dallwitz, and J. F. Truswell is gratefully acknowledged. The paper is published by permission of the Director, Bureau of Mineral Resources, Canberra, Australia.

References cited

Arriens, P. A. 1975. Precambrian geochronology of Antarctica. *First Australian Geological Convention, Adelaide, 1975* (Abstract only). Sydney, Geological Society Australia, 97–98.

Bennett, A. J. R., and Taylor, G. H. 1972. Coals from the vicinity of the Prince Charles Mountains, Antarctica. (*In* Adie, R. J., *ed. Antarctic Geology and Geophysics.* Oslo, Universitetsforlaget, 591–98.)

Berger, A. R., and Jayasinghe, N. R. 1976. Precambrian structure and chronology in the Highland Series of Sri Lanka. *Precambr. Res.,* 3, 559–76.

Clifford, T. N. 1974. Review of African granulites and related rocks. *Spec. Pap. geol. Soc. Am., 156,* 49 pp.

Condie, K. C., and Hunter, D. R. 1976. Trace element geochemistry of Archaean granitic rocks from the Barberton region, South Africa. *Earth planet. Sci. Lett., 29,* 389–400.

Craddock, C. 1972. Geologic Map of Antarctica 1:5,000,000. New York, American Geographical Society.

Crawford, A. R. 1969. Reconnaissance Rb-Sr dating of Precambrian rocks of South Peninsula India. *J. Geol. Soc. India, 10,* 117–66.

Crawford, A. R. 1974. Indo-Antarctica, Gondwanaland, and the distortion of a granulite belt. *Tectonophysics, 22,* 141–57.

Crohn, P. W. 1959. A contribution to the geology and glaciology of the western part of Australian Antarctic Territory. *Bull. Bur. Miner. Resour. Geol. Geophys. Aust., 52,* 103 pp.

Dietz, R. S., Holden, J. C., and Sproll, W. P. 1972. Antarctica and continental drift. (*In* Adie, R. J., *ed. Antarctic Geology and Geophysics.* Oslo, Universitetsforlaget, 837–42.)

Escher, A., Escher, J. C., and Watterson, J. 1975. The reorientation of the Kangamiut dike swarm, West Greenland. *Can. Jnl Earth Scis, 12,* 158–73.

Grew, E. S. 1975. Geologic studies of Precambrian basement around Molodezhnaya Station, Enderby Land. *Antarct. Jnl U.S., 10,* No. 5, 245–48.

Grew, E. S. 1976. History of metamorphism and deformation of Precambrian rocks of Enderby Land and the Prince Charles Mountains, East Antarctica. *Abstr. 25th Int. Geol. Congr., Sydney, 1976, 1,* 122–23.

Grikurov, G. E., and Soloviev, D. S. 1974. Geologicheskoe stroenie gornogo obramlenia lednika Lamberta [Geological structure of the mountain fringe of Lambert Glacier]. *Inf. Byull. sov. antarkt. Eksped., 88,* 21–29. [English translation: *Inf. Bull. Sov. Antarct. Exped. 8,* No. 10, 543–48.]

Gunn, B. M., and Warren, G. 1962. Geology of Victoria Land between the Mawson and Mulock Glaciers, Antarctica. *Bull. geol. Surv. N.Z.,* No. 71, 157 pp.

Halpern, M. 1970. Rubidium-strontium date of possibly 3 billion years for a granitic rock from Antarctica. *Science, N.Y., 164,* No. 3949, 977–78.

Halpern, M., and Grikurov, G. E. 1975. Rubidium strontium data from the southern Prince Charles Mountains. *Antarct. Jnl U.S., 10,* No. 1, 9–15.

Hurley, P. M. 1973. On the Origin of 450 ± 200 m.y. Orogenic Belts. (*In* Tarling, D. H., and Runcorn, S. K., *eds. Implications of Continental Drift to the Earth Sciences. 2.* London, Academic Press, 1083–95.)

Iltchenko, L. N. 1972. Late Precambrian acritarchs of Antarctica. (*In* Adie, R. J., *ed. Antarctic Geology and Geophysics.* Oslo, Universitetsforlaget, 599–602.)

Kamenev, E. N. 1977. Nekotorye zakanomernosti raspredeleniya fatsiy regional'nogo metamorfizma v Antarktide [Some regularities of facies distribution of regional metamorphism in Antarctica]. *Antarktika, Doklady Komissii, 16,* 100–20.

Kemp, E. M. 1973. Permian flora from the Beaver Lake area, Prince Charles Mountains, Antarctica. 1. Palynological examination of samples. *Palaeont. Pap. 1969, Bull. Bur. Miner. Resour. Geol. Geophys. Aust., 126,* 7–12.

Kröner, A. 1977. The Precambrian geotectonic evolution of Africa: plate accretion versus plate destruction. *Precambr. Res., 4,* No. 2, 163–213.

McLeod, I. R. 1964. An outline of the geology of the sector from longitude 45° to 80° E, Antarctica. (*In* Adie, R. J., *ed. Antarctic Geology.* Amsterdam, North-Holland Publishing Company, 237–47.)

Maegoya, T., Nohda, S., and Hayase, I. 1968. Rb-Sr dating of the gneissic rocks from the east coast of Lutzow-Holm Bay, Antarctica. *Mem. Fac. Sci. Kyoto Univ.,* Ser. Geol. Mineral., No. 35, 131–38.

Mond, A. 1972. Permian sediments of the Beaver Lake area, Prince Charles Mountains. (*In* Adie, R. J., *ed. Antarctic Geology and Geophysics.* Oslo, Universitetsforlaget, 585–89.)

Nagpaul, K. K., and Mehta, P. P. 1975. Cooling history of South India as revealed by fission track studies. *Am. J. Sci., 275,* 753–62.

Ravich, M. G., and Grikurov, G. E. 1976. Geological Map of Antarctica 1:5,000,000. Leningrad, Kartfabrika ob'edineniia Aereogeologiia.

Ruker, R. A. 1963. Geological reconnaissance in Enderby Land and the southern Prince Charles Mountains, Antarctica. *Rec. Bur. Miner. Resour. Geol. Geophys. Aust.,* 1963/154 [Unpublished].

Sheraton, J. W., and England, R. N. In press. Highly potassic mafic dykes from Antarctica. *J. Geol. Soc. Australia.*

Sobotovich, E. V., Kamenev, E. N., Komaristyy, A. A., and Rudnik, V. A. 1976. Drevneishie porody Antarktidy (Zemlia Enderbi) [The oldest rocks of Antarctica (Enderby Land).]. *Izv. Akad. Nauk SSSR,* Ser. geol., No. 11, 30–50. [English translation: *Int. geol. rev., 18,* No. 4, 371–88.]

Soloviev, D. S. 1972. Geological structure of the mountain fringe of the Lambert Glacier and Amery Ice Shelf. (*In* Adie, R. J., *ed. Antarctic Geology and Geophysics.* Oslo, Universitetsforlaget, 573–77.)

Tingey, R. J. 1974. Australian geological mapping in the Prince Charles Mountains, 1968–1973. *Polar Rec., 17,* No. 107, 150–53.

Tingey, R. J., England, R. N., Sheraton, J. W., and Arriens, P. A. 1976. Geology, geochemistry, and geochronology of the Archaean rocks of the Prince Charles Mountains, Antarctica. *Abst. 25th Int. Geol. Congr., Sydney, 1976, 1,* 22–23.

Trail, D. S. 1964*a*. The glacial geology of the Prince Charles Mountains. (*In* Adie, R. J., *ed. Antarctic Geology.* Amsterdam, North-Holland Publishing Company, 143–51.)

Trail, D. S. 1964*b*. Schist and granite in the southern Prince Charles Mountains. (*In* Adie, R. J., *ed. Antarctic Geology.* Amsterdam, North-Holland Publishing Company, 492–97.)

Trail, D. S. 1970. ANARE 1961 geological traverses on the Mac-Robertson Land and Kemp Land coast. *Rep. Bur. Miner. Resour. Geol. Geophys. Aust., 135,* 32 pp.

Trail, D. S., and McLeod, I. R. 1969. Geology of the Lambert Glacier Region. (Sheet 12 Lambert Glacier Region) (*In* Bushnell, V. C., and Craddock, C., *eds.* Geologic Maps of Antarctica. *Antarct. Map Folio Ser.,* Folio 12, Pl. XI.)

Vachette-Caen, M., and Hottin, G. 1976. Geochronology of the Precambrian of Madagascar. *Abstr. 25th Int. Geol. Congr., Sydney, 1976, 1,* 23.

Walker, K. R., and Mond, A. 1971. Mica lamprohyre (alnöite) from Radok Lake, Prince Charles Mountains, Antarctica. *Rec. Bur. Miner. Resour. Geol. Geophys. Aust.,* 1971/108 [Unpublished], 6 pp.

Wellman, P., and Tingey, R. J. 1976. Gravity evidence for a major crustal fracture in eastern Antarctica. *Jnl Bur. Miner. Resour. Geol. Geophys. Aust., 1,* 105–8.

White, M. E. 1962. Permian plant remains from Mount Rymill, Antarctica. In Report on 1961 plant fossils collections. *Rec. Bur. Miner. Resour. Geol. Geophys. Aust.,* 1962/114 [Unpublished], 14 pp.

White, M. E. 1973. Permian flora from the Beaver Lake area, Prince Charles Mountains, Antarctica. 2. Plant fossils. *Palaeont. Pap. 1969, Bull. Bur. Miner. Resour. Geol. Geophys. Aust., 126,* 13–18.

56

Amphibolite Facies Rocks of the Southern Prince Charles Mountains, East Antarctica

B. G. LOPATIN
Department of Geologic Mapping, SEVMORGEO, Moyka 120, Leningrad 190121, U.S.S.R.

V. S. SEMENOV
Department of Antarctic Geology, SEVMORGEO, Moyka 120, Leningrad 190121, U.S.S.R.

Abstract. A detailed study of metamorphic mineral assemblages of the amphibolite facies, based on chemical and spectroscopic analysis of rocks and minerals, along with isotopic age and paleotemperature determinations, reveals a complex geologic history of the southern Prince Charles Mountains area.

The upper Archean basement gneisses, granite gneisses, and schists (Lambert Series) are strongly migmatized and belong to the high-temperature amphibolite facies (sillimanite-almandine-orthoclase subfacies). This complex is of polymetamorphic nature and includes primary granulite facies rocks with hypersthene relics. Migmatites of lit-par-lit type are cut by beryl-bearing pegmatitic granite very similar to the lower Paleozoic granite from Mount McCauley and probably of the same magmatic suite.

The Proterozoic complex of schists, quartzites, and amphibolites (Menzies Series) underwent at least three metamorphic episodes: (1) regional metamorphism (probably about 1000 m.y. ago), slightly zonal from its lowermost limit in the staurolite-almandine subfacies (Mount Menzies, Goodspeed Nunataks) to its uppermost limit; (2) subsequent metamorphic-metasomatic transformation associated with the lower Paleozoic granite intrusive rocks (Mounts Scherger and McCauley); and (3) greenschists facies diaphthoresis.

Typical stable mineral assemblages of regional metamorphism are: staurolite + biotite + muscovite + quartz ± plagioclase; staurolite + garnet + biotite + muscovite + quartz ± plagioclase; staurolite + kyanite + biotite + muscovite + quartz; and garnet + biotite ± muscovite + quartz. Mineral thermometers and barometers indicate ranges of $T = 485°–610°C$ and $P = 4.2–7.5$ kbar.

The post-kinematic nature of the subsequent metamorphic-metasomatic event is evidenced by the massive, coarse-grained structure and by the unstable mineral composition of the rocks. Two types of mineral assemblages can be distinguished (each having two varieties): (1) cummingtonite + anthophyllite + biotite + plagioclase; and gedrite (Al_2O_3-bearing anthophyllite) + garnet + quartz, cordierite + gedrite + plagioclase + quartz; and (2) kyanite + muscovite + quartz; and fibrolite (fibrous sillimanite) + quartz. The authors suggest an origin under the influence of postmagmatic alkaline solutions in the stage of acidic leaching. The estimated conditions are: $T = 570°C$, $P =$ 5–7 kbar.

Low-temperature metamorphic transformations of the basement rocks are mainly confined to fault zones which probably formed as a result of Ross orogenic movements.

This report is based on data obtained during the geologic survey of the Prince Charles Mountains by the Soviet Antarctic Expedition in 1972–73 and 1973–74. Information on general geologic and geomorphologic features of the area under study is presented in several publications (Soloviev, 1972; Grikurov and Soloviev, 1974; Lopatin and others, 1977; Ravich and others, 1978) and in papers presented at this symposium (and contained in this volume) by R. J. Tingey (Chapter 55), E. S. Grew (Chapter 57), L. V. Fedorov (Chapter 118), M. G. Ravich (Chapters 60 and 107), and J. Hofmann (Chapter 58).

The authors emphasize the complexity of the geology of the southern Prince Charles Mountains, an area composed of folded metamorphic rocks of different ages and facies, exposed in separate tectonic blocks. The authors distinguish four groups of metamorphic rocks: the upper Archean Lambert Series,* the lower–middle Proterozoic Ruker and Menzies Series, and the upper Proterozoic–lower Paleozoic Sodruzhestvo Series (Figure 56.1).

The Lambert Series is a part of the crystalline basement, and its upper Archean age is confirmed by isotopic determinations. The age of the uppermost Sodruzhestvo Series is proved by a problematic microfauna and by the occurrence of granite and jaspilite pebbles in conglomerates (Ravich and others, 1978). As to the time of origin, folding, and regional metamorphism of the Ruker and Menzies Series, the authors can at present define these

* Following Russian usage, *series* here means a lithostratigraphic unit formed during a period or longer.

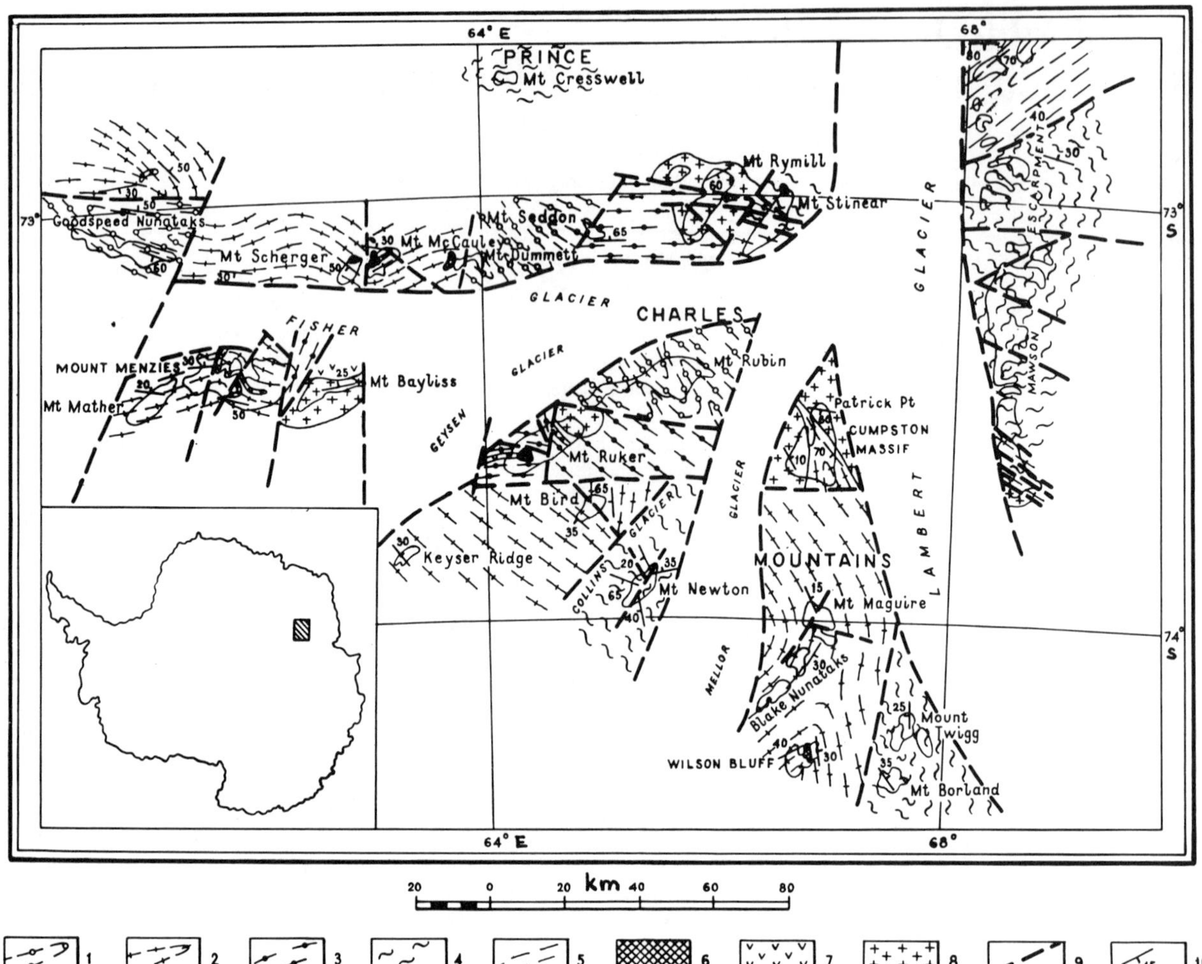

Figure 56.1. Geologic map of the southern Prince Charles Mountains, compiled by B. G. Lopatin and L. V. Fedorov. Fisher Glacier flows eastward. 1. Upper Proterozoic–lower Paleozoic greenschist complex (Sodruzhestvo Series); 2. Proterozoic complex of the lower temperature amphibolite facies (Menzies Series); 3. Proterozoic greenschist complex (Ruker Series); 4. Upper Archean complex of the higher temperature amphibolite facies (Lambert Series); 5. Archean complex of granulite facies; 6. Lower Paleozoic garnet-mica granites; 7. Upper Proterozoic metabasites; 8. Upper Archean gneissic granites; 9. Faults; and 10. Strike and dip of bedding.

events only within a broad interval from 2500 to 900 m.y. ago. The stratigraphic relationships of these series are also poorly known.

The metamorphic rocks of the Menzies Series are the main subject of our detailed study. Their complex mineralogy is evidence of heterogeneous transformations and of a polymetamorphic origin. On the whole, these rocks differ from those in the Lambert Series by the lower grade of metamorphism, although both are within the almandine-amphibolite facies. The authors do not describe in this report greenschist facies rocks of the Ruker and Sodruzhestvo Series.

Metamorphic complex of the almandine-amphibolite facies (Lambert Series)

This complex is composed of strongly migmatized and granitized biotite, garnet-biotite, and pyroxene-hornblende plagiogneiss, with minor biotite-garnet-sillimanite and biotite-garnet-cordierite plagiogneiss with amphibolite, and tremolite-diopside calciphyre. These rocks make up the southern part of Mawson Escarpment, Mount Newton, Mount Twigg, Mount Crosswell, and other peaks. The widespread migmatites are of the lit-par-lit type, with biotite-garnet and biotite-hornblende granite as a neosome. Mineral assemblages of paleosome include such index minerals as sillimanite, pyrope-almandine garnet, cordierite, and orthoclase, and in mafic rocks, brown-green hornblende, salite, and calcareous plagioclase (Table 56.1). These assemblages indicate sillimanite-almandine-orthoclase subfacies metamorphism under $T = 650°–700°C$ and $P = 4–5$ kbar. In some places hypersthene relics occur in plagiogneisses (Mount Newton), showing the presence of primary granulite facies rocks within the complex and consequently its polymetamorphic origin.

Lit-par-lit migmatites enclose mafic intrusive rocks, and both are cut by coarse-grained, garnet-muscovite, beryl-bearing, granitic veins, commonly forming an agmatite-like pattern. This granite is very similar to the lower Proterozoic granite from Mounts McCauley and Scherger and probably belongs to the same magmatic suite, formed as a result of the tectono-magmatic activization of the crystalline basement during the Ross orogeny.

Metamorphic complex of the lower temperature amphibolite facies (Menzies Series)

Structure and composition

Metamorphic rocks of the lower temperature amphibolite facies make up several major nunataks: Mount Menzies, Mount McCauley, Mount Scherger, Mount Stinear, Mount Maguire, Mount Wilson, and others (Figure 56.1). The entire thickness of the sequence is no less than 8 km. These metamorphic rocks have a distinctive, thin schistosity. Quartz (or plagioclase-quartz) schists with staurolite, muscovite, biotite, garnet, kyanite, and fibrolite (fibrous sillimanite) predominate. Commonly they grade into nearly mono-mineral quartzites. In subordinate amounts, biotite-amphibole schists, amphibole-clinopyroxene-plagioclase-quartz schists, and quartz-epidote-amphibole schists also occur. Phlogopite-tremolite marbles and calciphyres are scarce.

The northern group of Goodspeed Nunataks is composed of mica-quartz and epidote-mica-quartz-plagioclase schists and quartzites. The same rocks are common at Mount Menzies. Staurolite here is very scarce and occurs in fine grains. From Mount Menzies the Australian geologists (Tingey and England, 1973; England and Langworthy, 1975) reported assemblages of chloritoid with kyanite and garnet. It is very likely that in the area of the Goodspeed Nunataks and Mount Menzies the grade of regional metamorphism was lower and close to the lowermost limit of the almandine-amphibolite facies, if not the upper greenschist facies (Winkler, 1967).

The schists are commonly penetrated by orthoamphibolite sills and dikes. At Mount McCauley separate sills occur that are as much as 70 m in thickness. Both the orthoamphibolites and the surrounding rocks were folded and metamorphosed. Primary magmatic textures in the mafic rocks are nearly absent.

On Mounts McCauley and Scherger metamorphic rocks are intruded by garnet-mica granite stocks with numerous apophyses and swarms of aplite and pegmatite dikes and veins. The folding here is of the disharmonic type. Endomorphic and exocontact alterations are nearly absent, implying thermodynamic equilibrium between magma and host rocks. The peculiarity of the Mount McCauley and Mount Scherger area lies in the occurrence of metasomatic rocks superimposed on the regional metamorphics. These rocks include massive, coarse-grained, non-uniform plagioclase-cordierite-gedrite, quartz-gedrite-garnet, and plagioclase-quartz-cummingtonite-anthophyllite, forming large (as much as 12 m in thickness) lens-shaped bodies, mainly in the amphibolites near their contacts with quartzites and quartz-muscovite-kyanite and quartz-fibrolite pockets (commonly enriched in tourmaline) located mainly in the quartz schists. The crystals of kyanite, cummingtonite, and anthophyllite, as large as 2 to 10 cm, show no preferred orientation. The blue kyanite is common in radial vug fillings. The structure of the metasomatic rocks indicates their origin under post-kinematic conditions after the regional metamorphism.

Zones of diaphthoritic metamorphism, typically represented in Mount Stinear, are composed of strongly chloritized cataclasites with augen structure and distinct secondary schistosity. To a lesser degree these rocks also occur in Mount Menzies and Mount Maguire.

Apart from the above-mentioned metamorphic rocks, there also occur composite rocks with polymetamorphic mineral assemblages that include staurolite, kyanite,

Table 56.1. Mineral assemblages of metamorphic rocks of almandine-amphibolite facies

Staurolite-almandine subfacies (Menzies Series)	Sillimanite-almandine-orthoclase subfacies (Lambert Series)
Rocks CaO deficient and Al_2O_3 oversaturated	
Staurolite + biotite + muscovite + quartz + plagioclase (22–25% An)	Sillimanite + garnet + biotite + potassic feldspar + plagioclase + quartz
Staurolite + garnet + biotite + muscovite + quartz + plagioclase	Garnet + biotite + potassic feldspar + plagioclase + quartz
Staurolite + kyanite + biotite + muscovite + quartz + plagioclase	Garnet + cordierite + biotite + plagioclase ± potassic feldspar + quartz
Kyanite + biotite + muscovite + quartz	Biotite + potassic feldspar + plagioclase + quartz
Garnet + biotite ± muscovite + quartz ± plagioclase (35–45% An)	
Biotite + muscovite + quartz + plagioclase (28–40% An)	
Biotite + quartz + plagioclase (25–35% An)	
Rocks rich in calcium	
Hornblende + salite[a] + plagioclase (50–60% An) + quartz	Salite[a] + hornblende + plagioclase ± quartz
Hornblende + biotite + plagioclase (30–40% An) + quartz	Hornblende + plagioclase ± biotite ± quartz
Hornblende + garnet + biotite + plagioclase + quartz	
Hornblende + epidote + biotite + plagioclase + quartz	
Epidote + salite[a] + quartz ± plagioclase	
Epidote + biotite + plagioclase (70–80% An) + quartz	
Metasomatic rocks	
Kyanite + muscovite ± tourmaline + quartz	
Fibrolite[b] + quartz ± magnetite	
Cummingtonite + anthophyllite + biotite + plagioclase (30–50% An) + quartz	
Gedrite[c] + garnet + quartz	
Cordierite + gedrite[c] + plagioclase (23–37% An) ± quartz	

[a] Salite: diopside with small amount of iron. [b] Fibrolite: fibrous sillimanite. [c] Gedrite: Al_2O_3-bearing anthophyllite.

garnet, biotite, muscovite, plagioclase, and quartz, as well as some relics of cordierite and andalusite. Andalusite is reported from Mount Stinear and Mount Rymill; it is usually replaced by muscovite. At Mount Stinear occurrences of andalusite are confined to narrow tectonic blocks between uplifts of the granite-gneiss crystalline basement rocks (Lambert Series). Since andalusite and kyanite cannot belong to the same stable metamorphic assemblage, we suggest that andalusite (in association with cordierite) is a result of contact metamorphism that occurred before the regional metamorphism, possibly around basic intrusives; this interpretation, however, needs further study and confirmation. At present, the geologic position of andalusite and cordierite relict assemblages is not clear.

Mineral assemblages of the regional metamorphic rocks

To throw light upon the thermodynamic conditions of the regional metamorphism the authors first established stable mineral assemblages. To approach this problem, in addition to the study of mineral relationships in thin sections, the authors investigated the dependence of mineral composition upon the chemical composition of the rocks. Two common and useful rock groups were investigated—(1) CaO-deficient and Al_2O_3-oversaturated rocks and (2) CaO-rich rocks—using Thompson's and Winkler's diagrams (Winkler, 1967).

CaO-deficient and Al_2O_3-oversaturated rocks vary considerably in mineral composition (Table 56.2), commonly including biotite, muscovite, garnet, and staurolite, and rarely kyanite and sillimanite (fibrolite). The rocks are usually porphyroblastic with a fine-grained matrix. Schistosity folia are thin and distinct, due to parallel orientation of micas, quartz aggregates, and, to a lesser degree, staurolite and kyanite.

Staurolite grains vary from small skeletal individuals to well-shaped crystals with cruciform twins, usually containing numerous inclusions of quartz, plagioclase, and micas.

Garnet occurs as equant fissured grains. Quartz inclusions are common within them and commonly show S-shaped forms (snowball garnets), implying synkinematic crystallization of the garnet. The almandine component comprises 70% by volume of the garnet.

Kyanite forms isometric or tabular grains usually with quartz inclusions. Fibrolite (fibrous sillimanite) occurs in ordinary felty aggregates. From the relationships observed in thin sections the authors conclude that fibrolite developed after biotite, and they do not regard fibrolite as a mineral of the primary regional metamorphism.

Forms of biotite range from small thin flakes to large

Table 56.2. Summary of rock compositions (in wt %)

Rocks CaO deficient and Al_2O_3 oversaturated

Specimen number	1	2	3	4	5	6	7	8	9	10	11	12	13	15	16	17	18	19	20	21	22	23	24	25	26	27
Location	Mt Stinear	Mt Stinear	Mt Stinear	Mt Stinear	Mt Stinear	Mt Stinear	Mt Stinear	Mt Stinear	Mt Maguire	Mt Maguire	Mt Maguire	Mt Maguire	Seavers Ntk	Mt Rymill	Mt Stinear	Mt Stinear	Mt Stinear	Mt McCauley	Mt McCauley	Mt Stinear	Blake Ntks	Blake Ntks	Mt Maguire	Mt Stinear	Mt Stinear	Mt Stinear
Quartz	45	45	50	60	40	25	50	30	50	20	55	35	40	25	55	30	5	40	50	50	60	60	10	30	20	40
Plagioclase	5	1	7	5	1	5	19	3	14	20	15	10	3	20	15	5	50	30	5	0	5	5	10	10	5	5
Biotite	20	15	23	10	10	20	25	15	15	15	10	20	25	0	25	15	40	27	30	20	17	10	10	35	25	20
Muscovite	10	12	1	5	30	1	0	25	0	5	4	15	10	5	0	0	0	0	0	10	5	20	10	15	15	20
Staurolite	1	17	7	10	1	20	5	10	6	30	2	6	5	15	0	0	0	0	1	0	0	0	45	5	20	10
Garnet	1	7	9	5	0	10	0	2	5	5	10	5	2	30	5	0	4	0	0	5	5	5	5	1	0	0
Kyanite	10	0	0	0	10	0	0	0	8	0	0	0	15	0	0	0	0	0	0	0	0	0	5	0	0	0
Fibrolite	0	0	0	0	0	0	5	5	0	5	0	4	0	5	0	0	0	0	12	5	5	0	0	1	0	0
Opaques	0	0	0	0	3	0	0	0	0	0	0	0	0	0	0	0	1	3	2	0	0	0	3	0	0	0
Chlorite	8	3	3	5	5	14	5	10	2	0	4	5	0	0	0	50	0	0	0	10	3	0	2	3	15	5

Metamorphic rocks rich in Calcium

Specimen number	29	30	31	32	33	34	35	36	37	38	39	40	41	42	43	44
Location	Goodspeed Ntk.	Mt Stinear	Mt Stinear	Mt McCauley	Mt McCauley	Mt McCauley	Mt McCauley	Mt McCauley	Mt McCauley	Mt McCauley	Mt Rymill	Mt Rymill	Mt Stinear	Mt Stinear	Mt Stinear	Mt Stinear
Quartz	50	35	0	15	35	23	33	22	10	3	59	20	25	30	50	20
Plagioclase	10	20	0	43	25	0	7	20	22		0	0	20	20	10	30
Microcline	0	0	0	0	25	0	8	0	0	0	10	0	0	0	0	0
Biotite	25	15	20	20	7	5	5	0	5	0	20	0	20	15	25	5
Muscovite	5	0	0	0	0	0	2	0	0	0	0	0	0	0	0	0
Hornblende	3	20	80	22	8	42	17	58	53	95	10	60	30	20	0	5
Cummingtonite	0	0	0	0	0	0	0	0	0	0	0	0	0	10	5	25
Clynopyroxene	0	0	0	0	0	0	28	0	0	0	0	0	0	0	0	0
Garnet	7	0	0	0	0	0	0	0	1	0	0	0	0	5	0	0
Epidote	0	10	0	0	0	30	0	0	0	0	0	0	0	0	0	0
Opaques	0	0	0	0	0	0	0	0	4	2	1	3	0	0	0	3
Chlorite	0	0	0	0	0	0	0	0	0	0	0	17	5	—	10	12

Specimen numbers 14 and 28 are not representative and have been omitted on purpose.

porphyroblasts. Biotite is not uniform in composition, having a broad range of FeO-MgO isomorphism, and certainly belongs to several generations. Composition of plagioclase ranges from oligoclase (An_{25}) to andesine (An_{40}). The secondary low-temperature minerals chlorite and sericite are common everywhere except in the Mount McCauley and Mount Scherger area.

The authors plotted compositions of CaO-deficient and Al_2O_3-oversaturated rocks and ferromagnesia minerals from these rocks on the AFM diagram of Thompson (see Winkler, 1967) (Figure 56.2). This diagram considers Mg and Fe as independent components and therefore provides sufficient accuracy to establish mineral assemblages in equilibrium with a certain metamorphic subfacies. The analyzed rock specimens contain quartz and common muscovite and therefore suit the requirements of this diagram.

According to the diagram the most common equilibrium mineral assemblage in this case is staurolite + biotite + almandine + muscovite + quartz. The occurrence of kyanite in specimen numbers 1, 5, and 9 is in good agreement with their position on the diagram near the line connecting the biotite and staurolite points. In these specimens kyanite is associated with biotite, staurolite, and rare garnet. Kyanite in these assemblages reflects the chemical composition of the rocks and does not show the appearance of a different subfacies.

Study of assemblages with fibrolite revealed that in only three cases out of eight (specimen numbers 7, 8, and 15) can its occurrence be explained by favorable rock composition; in the other specimens (numbers 10, 12, 20, 21, and 22) fibrolite "breaks the rule" and apparently is a superimposed mineral that confirms the relationships with biotite observed in thin sections.

Mineral assemblages of the analyzed CaO-rich rocks are well illustrated on the ACF diagram for staurolite-almandine subfacies (Figure 56.3). The equilibrium of epidote in this subfacies in association with intermediate and calcic plagioclase can be explained by high pressure, favorable

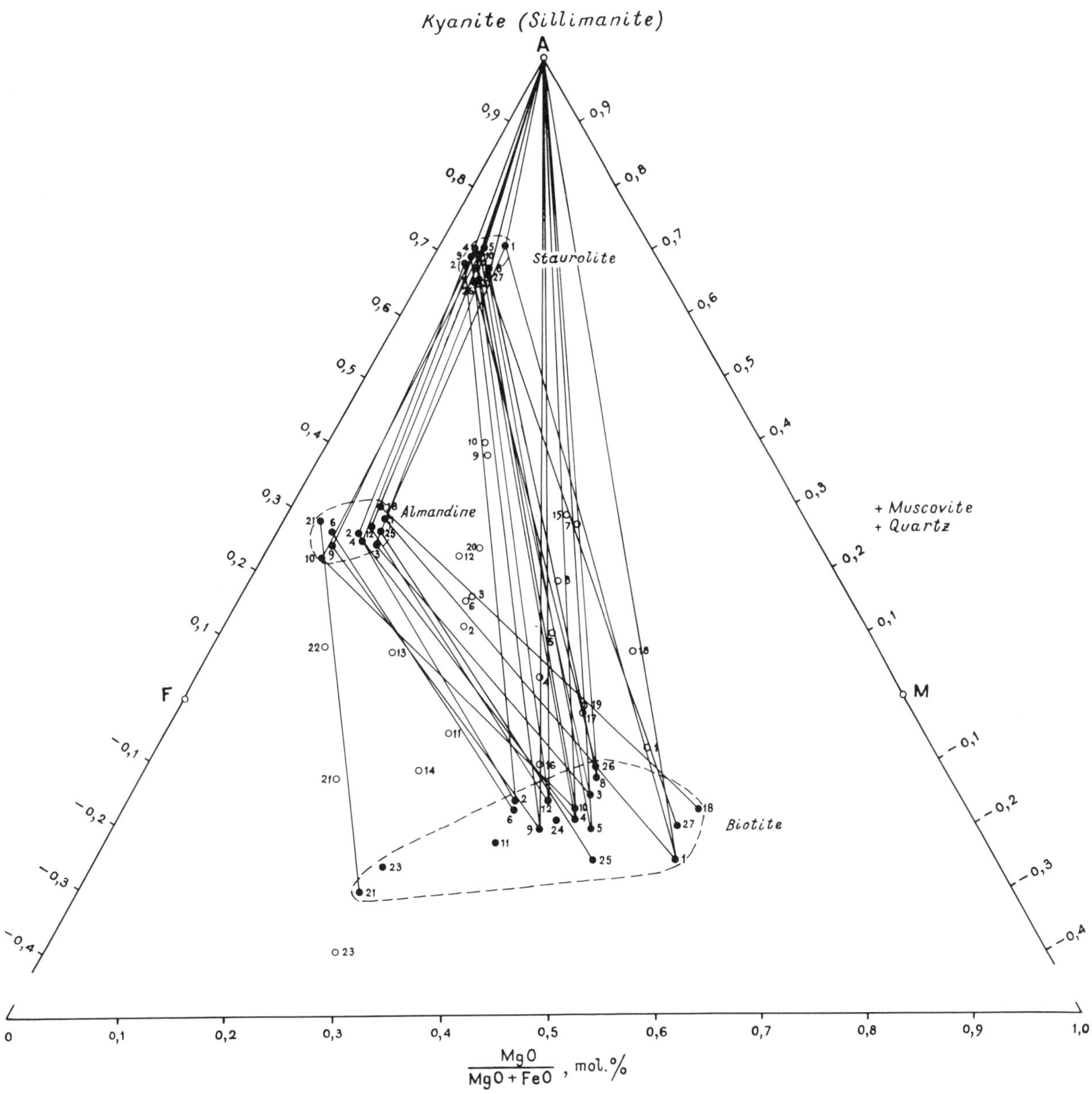

Figure 56.2. AFM diagram for CaO-deficient and Al_2O_3-oversaturated quartz-bearing schists of the staurolite-almandine subfacies. A = Al_2O_3; F = FeO; and M = MgO. Solid dots: compositional points of minerals. Open dots: compositional points of rocks.

for expansion of the stability field of epidote + quartz toward a higher temperature direction (Glebovitskiy, 1973). The ACF diagram confirms the equilibrium of hornblende with clinopyroxene. The crystallization of the latter depends on the chemical composition of the rocks.

The nature of cummingtonite seems complex. In only one case (specimen number 43) can crystallization of cummingtonite instead of hornblende be explained by higher ferruginous rock composition. In the other cases (specimen numbers 42 and 44) cummingtonite-bearing rocks do not differ chemically from common hornblende schists. This fact supports the secondary nature of cummingtonite, occurring after hornblende due to late metasomatic events. This is in good agreement with the reactional relationship between the amphiboles observed in thin sections.

Thus, the established stable mineral assemblages from regional metamorphism correspond to the single staurolite-almandine facies of higher pressure (Table 56.1).

Occurrences of fibrolite and potassic feldspar in schists from Mount McCauley and Mount Scherger, in addition to decreased amounts of staurolite, may suggest transitional conditions to higher temperature facies: staurolite + muscovite + biotite I + quartz → fibrolite + biotite II + H_2O. However, many facts make the authors believe that this transformation actually took place because of a subsequent metasomatic episode.

Mineral assemblages of the superimposed metamorphic-metasomatic transformation

Regional metamorphic rocks in the Mount McCauley and Mount Scherger area underwent transformations apparently associated with granitic magmatism, which is also typical of this area.

These transformations include development of bundles of fibrolite after biotite, cummingtonite after hornblende, reversed zones in plagioclase, and tourmaline-quartz-kyanite and quartz-fibrolite pockets in schists. Within amphibolites, near their contacts with quartzites, the secondary products are represented by massive coarse-grained rocks consisting of rhombic amphiboles, cummingtonite, cordierite, and garnet. These mineral assemblages are typical of acidic leaching zones where, under the influence of postmagmatic alkaline solutions, mobilization and transportation of strong cations (iron and magnesium) and concentration of silica and the weak cation aluminum take place. As a result biotites are replaced by fibrolite and feldspars by quartz with kyanite and fibrolite. In mafic rocks (amphibolites) contemporaneous magnesium-ferruginous metasomatic rocks were formed. In the initial stage cummingtonite replaced hornblende, followed by anthophyllite or gedrite (Al_2O_3-bearing anthophyllite) crystallization in large (as much as 8 cm) pole-like grains. In the late stage of the metasomatic transformation of basic rocks, gedrite formed with large nodules of cordierite and pyrope-grossular-almandine garnet in idiomorphic crystals as large as 3 cm in diameter. Plagioclases in these assemblages have distinct reverse zonation. Four measurements gave the following results: $An_{25,28,70}$ from the grain cores and $An_{45,65,82}$ from the grain margins, respectively.

Conditions of metamorphism

Regional metamorphism of the Menzies Series took place under high pressure within the entire temperature interval of the staurolite-almandine subfacies, as is evidenced by the stable mineral assemblages (Table 56.1) and indirect temperature and pressure determinations, using

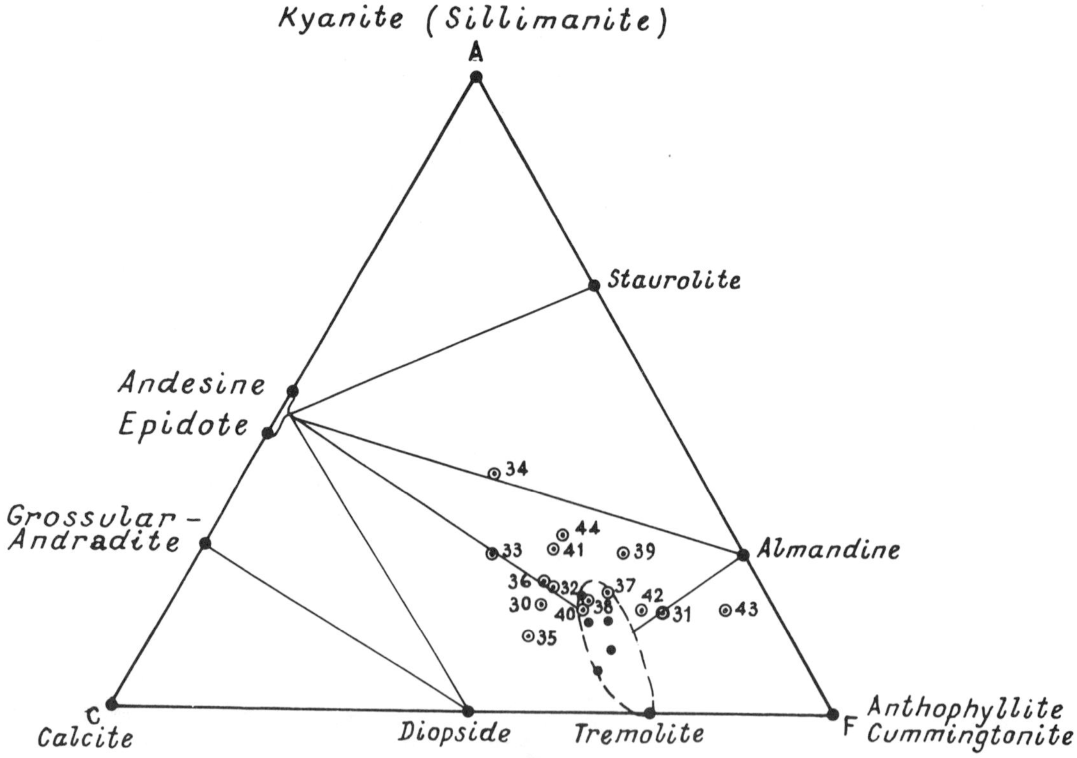

Figure 56.3. ACF diagram for CaO-rich rocks of the staurolite-almandine subfacies. A = $Al_2O_3 + Fe_2O_3 - (Na_2O + K_2O)$; C = $CaO - 3.3P_2O_5$; and F = FeO + MgO + MnO. Solid dots: compositional points of amphiboles. Circled dots: compositional points of rocks. The area inside the dashed line is the composition field of amphiboles. A + C + F = 100% (mol).

Table 56.3. Mineralogic thermometers and barometers

Specimen number	Mg/(Mg + Fe + Mn) (%)			T°C		P (kbar)	
	Biotite	Garnet	Staurolite	Biotite-garnet	Staurolite-garnet	Biotite-garnet	Staurolite-garnet
2	0.457	0.147	0.158	610	600	5.8	6.0
3	0.556	0.180	0.216	570	580	6.3	6.4
4	0.531	0.111	0.173	485	540	4.4	4.8
6	0.460	0.101	—	510	—	4.2	—
9	0.497	0.106	0.171	510	530	4.2	5.6
10	0.529	0.120	0.201	500	500	4.8	4.7
12	0.496	0.162	0.214	580	540	6.2	5.8
24	0.506	0.141	—	530	—	5.6	—
25	0.553	0.133	—	510	—	4.3	—
28	0.421	0.133	—	600	—	5.8	—
42	0.410	0.105	—	590	—	4.9	—
45	0.488	0.136	—	570	—	7.5	—

equilibrium pairs of magnesium-ferruginous minerals (Perchuk, 1970; Fed'kin, 1975). The authors obtained the most reliable results using well-studied biotite-garnet and staurolite-garnet pairs (Table 56.3). Inferred temperatures vary from 485° to 610°C, and pressure from 4.2 to 7.5 kbar.

Parameters of the subsequent metamorphic-metasomatic transformation were apparently near the upper limit of the staurolite-almandine subfacies, judging by widespread fibrolite and the appearance of potassic feldspar. One temperature determination using a biotite-garnet pair gives 570°C (Table 56.3, specimen number 45). Pressure in this stage (about 5–7 kbar) was high enough to produce such minerals as kyanite and gedrite, but probably of a hydrostatic nature since schistosity is practically absent in the metasomatic rocks. Hot alkaline solutions enriched in volatile components are thought to have been a major factor in the metamorphism. The granite magma intruded during the Ross orogeny, as a result of a general rise of the thermal front, was most likely the source of the solutions.

Summary

The upper Archean crystalline basement complex (Lambert Series) belongs to the sillimanite-almandine-orthoclase subfacies of the amphibolite facies. A few hypersthene relics suggest its polymetamorphic origin.

The Proterozoic metamorphic complex (Menzies Series) underwent zonal regional metamorphism of higher pressure within the staurolite-almandine subfacies of the amphibolite facies. The rocks of the lowermost limit of the subfacies are exposed in the Mount Menzies and Goodspeed Nunataks area. In other occurrences the metamorphic rocks of this complex are closer to the middle or upper limit of the subfacies. The paleotemperature and pressure values of the regional metamorphism, estimated by biotite-garnet and staurolite-garnet mineral pairs, range from about 500° to 600°C and from about 5 to 7 kbar, respectively. The age of the regional metamorphism is about 1000 m.y. (Halpern and Grikurov, 1975).

The metamorphic zonation was strengthened by the subsequent metamorphic-metasomatic episode during the Ross orogeny (about 550 m.y. ago) in connection with the granite magmatism, which was intensive in the Mount McCauley–Mount Scherger area. In other tectonic blocks, probably less deeply submerged, the Ross orogeny affected the basement complexes by forming zones of the broken and diaphthoritic rocks.

One of the problems of Prince Charles Mountains geology for future studying is the position and correlation of the Proterozoic greenschist facies, jaspilite-bearing Ruker Series.

Acknowledgments. The authors are grateful to G. E. Grikurov for critical review of the manuscript.

References cited

England, R. N., and Langworthy, A. P. 1975. Geological work in Antarctica—1974. *Publ. Bur. Miner. Resour. Geol. Geophys. Aust., 30,* 19 pp.

Fed'kin, V. V. 1975. *Stavrolit* [Staurolite]. Moscow, Izdatel'stvo Nauka, 272 pp.

Glebovitskiy, V. A. 1973. *Problemy evolutsii metamorficheskikh processov v podvizhnykh oblastiikh* [Problems of metamorphic processes evolution in mobile zones]. Leningrad, Izdatel'stvo Nauka, 126 pp.

Grikurov, G. E., and Soloviev, D. S. 1974. Geologicheskoe stroenie gornogo obramlenia lednika Lamberta [Geological structure of the mountain fringe of Lambert Glacier]. *Inf. Byull. sov. antarkt. Eksped.,* No. 88, 21–29.

Halpern, M., and Grikurov, G. E. 1975. Rubidium-strontium data from the southern Prince Charles Mountains. *Antarct. Jnl U.S., 10,* No. 1, 9–15.

Lopatin, B. G., Fedorov, L. V., and Polyakov, M. M. 1977. Geologiia i metamorfizm proterozoiskikh otlozhenii gor Mak-Koli i Sherger [Geology and metamorphism of the proterozoic rocks from Mount McCauley and Mount Scherger]. *Antarktika, Doklady Komissii, 16,* 76–92.

Perchuk, L. L. 1970. Ravnovesiie porodoobrazuyschikh mineralov [Equilibrium of rock-forming minerals]. Moscow, Izdatel'stvo Nauka, 390 pp.

Ravich, M. G., Soloviev, D. S., and Fedorov, L. V. 1978. Geologicheskoe stroenie Zemli Mak-Robertson [Geological structure of MacRobertson Land]. Leningrad, Hydrometeoizdat, 225 pp.

Soloviev, D. S. 1972. Geological structure of the mountain fringe of Lambert Glacier and the Amery Ice Shelf. (*In* Adie, R. J., *ed. Antarctic Geology and Geophysics.* Oslo, Universitetsforlaget, 573–77.)

Tingey, R. J., and England, R. N. 1973. Geological work in Antarctica—1972, *Publ. Bur. Miner. Resour. Geol. Geophys. Aust., 161,* 17 pp.

Winkler, H. G. F. 1967. *Die Genese der Metamorphen Gesteine.* 4th ed. Berlin, Springer-Verlag, 334 pp.

57

Geology of the Southern Prince Charles Mountains, East Antarctica

EDWARD S. GREW
Department of Earth and Space Sciences, University of California–Los Angeles, Los Angeles, California 90024, U.S.A.

Abstract. The southern Prince Charles Mountains (lat. 73°–74° 30′ S, long. 60°–70° E) consist of (1) a gneissic basement complex and (2) a superjacent series of metamorphosed sedimentary and volcanic rocks. The basement complex is largely granite and quartzo-feldspathic gneiss containing biotite, hornblende, or garnet. The most widespread superjacent metamorphic rocks are quartzite, slate, phyllite, mica schist, and amphibole schists, whereas metamorphosed banded iron formation, marble, gypsum, calc-silicate rocks, and metaconglomerate are of limited distribution. At least two sequences of superjacent metamorphic rocks may be present; one includes the metamorphosed iron formation, the second includes the metaconglomerate and a distinctive quartzite–mica schist–amphibole schist association.

Dikes, sills, and large tabular bodies of intrusive mafic rocks are widespread. These bodies commonly are discordant, have chilled contacts, and show evidence of having been recrystallized and deformed during regional metamorphism and folding. The mafic intrusive rocks are probably the intrusive equivalents of the metavolcanic rocks in the superjacent series. Dikes and small bodies of granite and pegmatite (commonly with muscovite or garnet) cut the metamorphic rocks, generally in the sillimanite zone of regional metamorphism.

At Mount Stinear, steeply dipping faults were active prior to the regional metamorphism. Mafic rocks were subsequently emplaced along some of these faults, notably those striking N 25° W to N 45° W. Postmetamorphic steeply dipping faults with displacements of as much as 300 m cut the rocks in many exposures. A gently southward dipping postmetamorphic thrust fault is exposed on Mount Maguire. The metamorphic grade of the superjacent metamorphic rocks ranges from the biotite zone to the sillimanite zone (greenschist and amphibolite facies). Kyanite is generally the only other Al_2SiO_5 mineral present, although andalusite and kyanite occur together on Mounts Stinear and Rymill.

Geochronologic data suggest that the basement complex is largely Archean in age. However, at Mount Stinear it has been affected by a late Precambrian event. The sequence of metamorphic rocks that includes banded iron formation is probably at least 1700 m.y. old, but the second sequence may be less than 1000 m.y. old. The regional metamorphism and folding of the superjacent rocks, and emplacement of granite and pegmatite, most likely occurred in early Paleozoic time.

Field work was carried out by the author as a U.S. exchange scientist with the 18th and 19th Soviet Antarctic Expeditions in 1973 and 1974 (Grew, 1975). Detailed mapping of five rock exposures was supplemented by reconnaissance study of about twenty other exposures, which the author examined in person or from which he obtained samples and field data. One hundred fifty rocks have been studied in thin section. The identity of a few minerals has been confirmed by X-ray diffraction and by wavelength scans on the electron microprobe.

Description of rock units

The following rock units have been recognized in the southern Prince Charles Mountains, listed in order of decreasing age: (1) basement complex; (2) superjacent metamorphic rocks, largely derived from sedimentary rocks and to a lesser extent from volcanic rocks; (3) mafic intrusive rocks, which may be contemporaneous with the volcanic rocks in unit 2; and (4) granite and pegmatite. The intensity of metamorphism ranges from the greenschist to the granulite facies (Figure 57.1), and some areas have been affected by more than one metamorphic event.

Basement complex

The basement complex is exposed at Mounts Ruker, Stinear, Rymill, McCauley, Scherger, Newton, and Bayliss (east of Menzies; Figure 57.1).

The basement complex at Mount Ruker is granite containing muscovite and biotite; at Mount Rymill, a pink-gray hornblende-biotite granite gneiss; and at Mounts McCauley and Scherger, a gray layered quartzo-feldspathic gneiss and subordinate amphibolite. The quartzo-feldspathic gneiss consists of biotite, hornblende, garnet, allanite, and locally clinopyroxene.

At Mount Stinear, the basement complex consists predominantly of pink or gray quartzo-feldspathic gneiss containing biotite, hornblende, or garnet, and locally fluorite. This gneiss seems to be in part plutonic in origin. Associated with the gneiss in the southernmost block (Figure 57.2) are quartz schist, mica schist with staurolite, garnet, and kyanite, clinopyroxene-bearing amphibolite, and migmatitic gneiss. Pegmatite is abundant in the basement complex at Mount Stinear.

Plagioclase-biotite-garnet gneiss, plagioclase-hornblende-cummingtonite-orthopyroxene gneiss, and garnet-

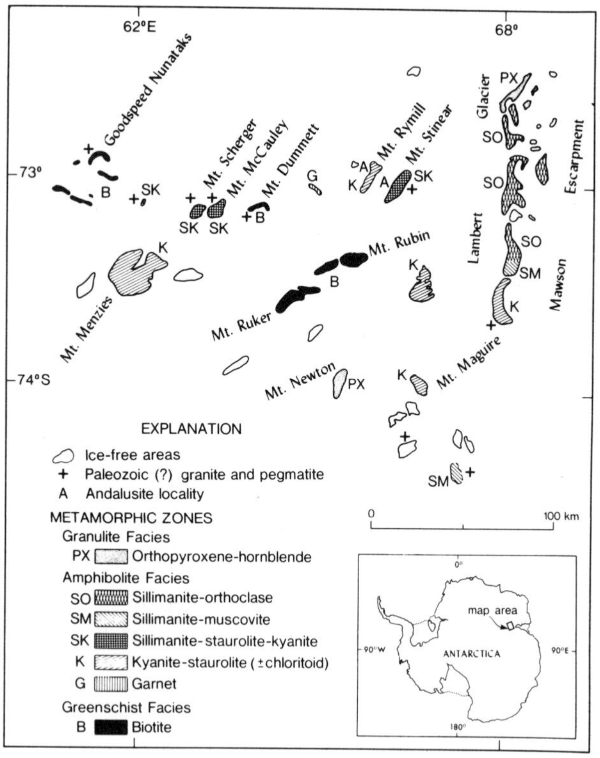

Figure 57.1. Map of metamorphic zones in the southern Prince Charles Mountains. This map is based on field work by the author, supplemented by information from Ravich and Kamenev (1972, pp. 596–98) on garnet-zone rocks; England (1975), England and Langworthy (1975, Figure 1), and Sheraton (1975) for Mawson Escarpment and for the exposure of kyanite-zone rocks near Mount Rubin; Tingey and England (1973) for Mount Menzies; and Halpern and Grikurov (1975) for Mount Rubin. Exposures for which the author has insufficient information have been left blank.

sillimanite-sericite gneiss containing rare chloritoid at Mount Newton are included in the basement complex.

Superjacent metamorphic rocks

Greenschist facies—Mount Ruker

About 2000 m of metasedimentary rocks are exposed in a homoclinal sequence in the central part of Mount Ruker (Figure 57.2). Truncations in cross-bedded quartzite imply that the section is right side up. The rock sequence can be divided into the following units, which appear to have conformable contacts between them (Grew, 1976): (1) metamorphosed banded iron formation (150 to 200 m thick, bottom not exposed); (2) green slate (1500 m thick); (3) quartzite, slate, and phyllite alternating in layers 5 to 50 m thick (300 to 400 m thick); (4) metamorphosed conglomeratic mudstone (10 to 35 m thick); and (5) green phyllite (top not exposed).

The metamorphosed banded iron formation (unit i, Figure 57.2) consists of alternating layers a few millimeters to 10 cm (rarely to 35 cm) thick of gray to black quartzite, dark biotitic or riebeckitic slate and schist, and magnetite. The quartzite is very fine grained (in places chertlike) and is commonly finely laminated; some laminae are red. The following minerals have been identified in the metamorphosed iron formation: magnetite, hematite, quartz, plagioclase, K-feldspar, biotite, stilpnomelane, riebeckite, aegirine, siderite, and ankerite. Biotite is characteristically yellow-green in thin section. Brown stilpnomelane occurs locally in small amounts. Yellow-green aegirine occurs as rosettes of prisms or as scattered megacrysts (as much as 2 cm long and 0.5 cm thick) that lie parallel to the bedding in thin layers. The aegirine is end-member $NaFeSi_2O_6$ in composition (microprobe analysis, R. Jones, personal communication, 1977). An emerald-green mica (celadonite?) and magnesioriebeckite are found with the aegirine. Associated with the metamorphosed iron formations are layers 0.5–2 m thick of brown phyllite containing ankerite, talc, biotite, quartz, and feldspars. The metamorphosed banded iron formation passes upward into the overlying green slate by a gradual decrease in the number of magnetite layers and by the appearance of chlorite and brown biotite. The green slate is a well-layered (a few millimeters to 20 cm thick) calcareous or ankeritic rock containing chlorite, biotite, epidote, sericite, pyrite, and quartz.

Quartzite, which constitutes one half to two-thirds of the overlying unit, is white to dark gray and in places cross-bedded or conglomeratic. The remainder of this unit consists in part of alternating layers of green phyllite and a poorly foliated ankerite-rich rock; individual layers are several centimeters to several tens of centimeters in thickness. Other rocks in this unit are sericitic and chloritic quartzose slates.

The metamorphosed conglomeratic mudstone (subunit c of unit sp, Figure 57.2) is a brown rock dominated by the matrix. The clasts are derived from sedimentary rocks, mostly quartzite. South of the vertical fault only one bed 10 m thick is present, but north of the fault are three beds totaling 35 m in thickness. The individual beds are unlayered.

The green phyllite at the top of the section consists of muscovite, biotite, chlorite, actinolite (locally), and epidote.

The metamorphosed conglomeratic mudstone resembles the metaconglomerate at Mount Rubin described by Grikurov and Soloviev (1974, p. 27), and the well-layered phyllite and slate found with the quartzite is not unlike these authors' (1974, pp. 27–28) "zebralike" unit at the top of the Mount Rubin section.

Amphibolite facies—Mount Maguire

The metasedimentary rocks (unit ms, Figure 57.2) are green-gray to gray mica schist and micaceous quartzite, apparently lying in a homoclinal sequence. Truncations in cross-beds at one locality indicate that the section is right side up. Staurolite, kyanite, garnet, magnetite, calcite, and epidote are common. Hornblende-bearing lenses a few centimeters thick are rare. Marble, calc-silicate rocks, and subordinate gypsum form beds about 4 to 5 m thick; these rocks constitute no more than 1% of the exposed section.

The amphibole rocks (unit a, Figure 57.2) are dominantly dark-colored schists consisting of garnet, hornblende, cummingtonite, biotite, plagioclase, and quartz. Black amphibolite and light-colored amphibole-garnet schists are subordinate. These rocks alternate in layers a few tens of centimeters to 5 m thick.

Amphibolite facies—Mounts McCauley, Scherger, Rymill, and Stinear

Four mappable units are recognized in the amphibolite facies rocks of this area: quartzite and mica schist (unit qs,

Figure 57.2); quartzite, mica schist, and amphibole rocks (subunit qa); metaconglomerate (subunit c); and mica schist (unit s). The rock types in units qs and qa alternate in layers a few meters to more than 100 m thick. A similar sequence is described from Mount Menzies (Tingey and England, 1973).

Cross-beds are preserved in the quartzite at Mount Stinear. Minerals in the quartzite are mica, garnet, staurolite, kyanite, sillimanite (mostly fibrolite), and hornblende. The association quartz-muscovite-chlorite-chloritoid-staurolite occurs at Mount Rymill. Green mica is found in seams and in some cross-beds on Mounts Rymill and Stinear.

The amphibole rocks at Mount Stinear are schists containing quartz, plagioclase, garnet, biotite, chlorite, hornblende, and cummingtonite. At Mounts McCauley and Scherger, the amphibole rocks are black amphibolite and coarse-grained rocks containing anthophyllite, garnet, cordierite, and minor staurolite. These rocks at Mount McCauley are associated with kyanite-sillimanite(fibrolitic and prismatic)-cordierite-staurolite-biotite schists lacking muscovite.

The mica schists at Mounts Rymill and Stinear are white, gray, or light-green rocks containing fibrolitic sillimanite (rare and at Mount Stinear only) and megacrysts of plagioclase, staurolite, garnet, biotite, kyanite, and andalusite (Figure 57.1). Andalusite crystals at Mount Rymill are subhedral to euhedral and have been partially replaced along the margins by chlorite and white mica. At Mount Stinear andalusite is anhedral. Paragonite has been identified (using X-ray diffraction and microprobe wavelength scans) in mica schist from Mount Rymill; it constitutes most of the white mica replacing andalusite but is rare in the matrix. Mica schist at Mount McCauley contains abundant sillimanite (largely fibrolitic) and garnet or staurolite.

Marble and calc-silicate rocks with diopside, grossular, and epidote, in layers as much as 10 m thick, are common at Mount McCauley, but rare at the other localities. Rocks of limited distribution include, at Mount Rymill, micaceous quartzite with garnet and actinolite and, at Mount Stinear, amphibole and mica schists with layers and small pods (as much as 3 cm thick) of pink to red "coticule," a garnet-rich rock, some of which contains considerable magnetite.

Metaconglomerate (subunit c, Figure 57.2) crops out at Mount Stinear in a band several hundred meters wide at the base of the mountain. This band consists of alternating layers, generally a few to 10 meters thick, of schist and metaconglomerate. Most of the schist is micaceous and contains garnet or staurolite; some contains cummingtonite. The clasts in the metaconglomerate are predominantly of micaceous (or less commonly, amphibolitic) quartzite. Mica schist, garnet-amphibole rock, amphibolite, and vein quartz are also found. The clasts range in size from a few centimeters to as much as 2 m across. The matrix and some clasts contain biotite, garnet, staurolite, kyanite, hornblende, or cummingtonite.

The abundance of clasts distinguishes this metaconglomerate from the metamorphosed conglomeratic mudstone of Mounts Rubin and Ruker. Metaconglomerate similar to the Stinear rock crops out at the south end of the Mawson

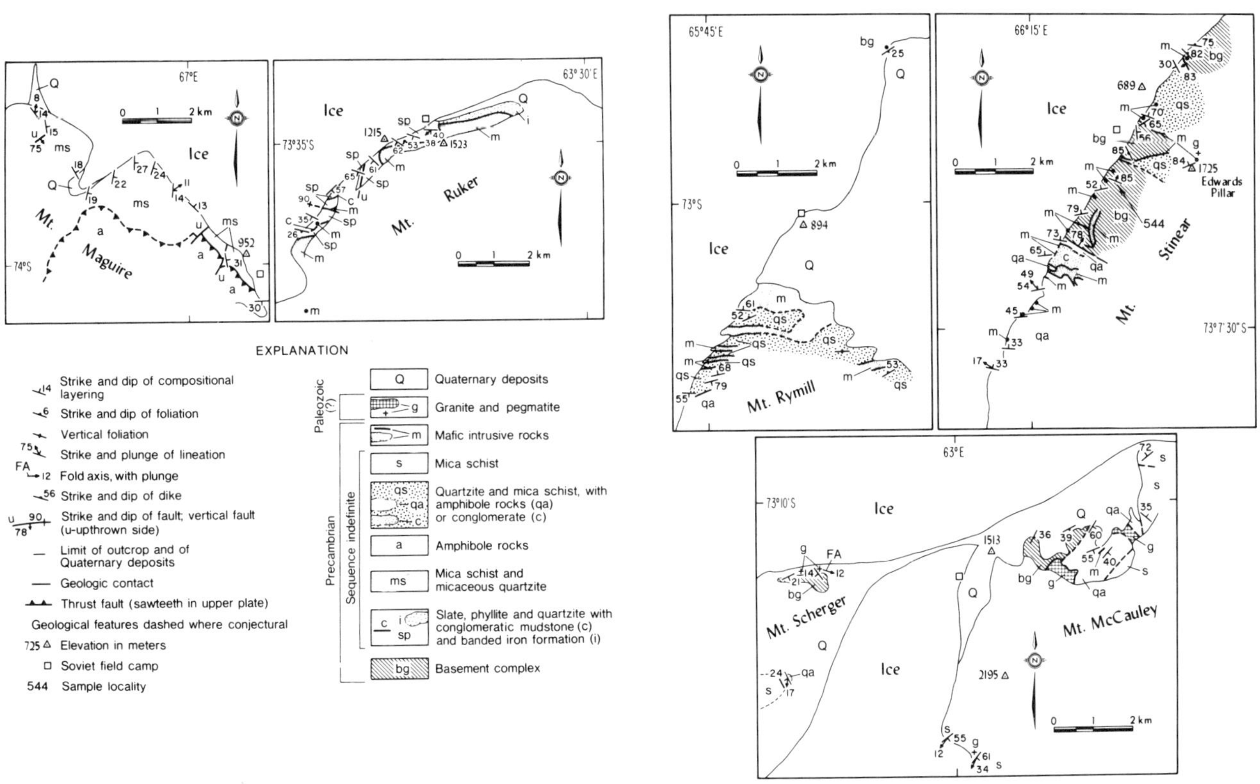

Figure 57.2. Geologic sketch maps of areas studied in detail in the southern Prince Charles Mountains. Field data have been plotted on aerial photographs taken in 1973 at an altitude of 6100 m by the Australian Division of National Mapping. Maps have not been corrected for distortion in the photographs. Scales, which are approximate, coordinates, and elevations are from Tingey's (1975*a*) Cumpston Massif sheet. Since the Cumpston Massif sheet was compiled on an uncontrolled base derived from LANDSAT imagery, the coordinates in the sketch maps are only approximate. In general, only the larger bodies of mafic intrusive rocks and of the granite and pegmatite are shown; some areas of Quaternary deposits have been omitted.

Escarpment (England and Langworthy, 1975), Mount Dummett, and in the Goodspeed Nunataks. The clasts at these localities are predominantly micaceous quartzite. Metaconglomerate with clasts of granitic rocks was found in float on the most northwesterly of the Goodspeed Nunataks.

The mica schist unit at Mounts McCauley and Scherger (unit s, Figure 57.2) consists largely of fibrolitic mica schist and micaceous quartzite; calc-silicate rocks are subordinate. Garnet is rare in the mica schist. Kyanite, sillimanite, and cordierite are found in quartz veins in mica schist on Mount McCauley.

Mafic intrusive rocks

Dikes, sills, and tabular bodies (1 m to about 100 m across and 1 km in extent) of diorite, amphibole schists, and greenstone are abundant at Mounts Stinear, Rymill, and Ruker, but rare at Mount McCauley (unit m, Figure 57.2). Similar mafic intrusive rocks are reported from many exposures in the southern Prince Charles Mountains (Tingey, 1975*a,b,* and Chapter 55, this volume).

Diorite (at Mount Rymill only) is a massive rock, locally foliated near its contacts. It consists largely of plagioclase and hornblende; biotite is minor. Intrusive amphibole rocks are found at Mounts Stinear, Rymill, and McCauley. They are commonly foliated, lineated, or layered, and contain quartz, plagioclase, olivine (Mount McCauley only), cummingtonite, hornblende, garnet, biotite, or chlorite. The greenstone (Mount Ruker only) is typically massive, locally schistose, and contains chlorite, sericite, epidote, biotite, calcite, plagioclase (mostly saussuritized), and minor hornblende.

The mafic intrusive rocks at Mounts Rymill, Stinear, and Ruker locally have discordant and chilled contacts and show evidence of recrystallization (Grew, 1976). The green slate at Mount Ruker has been converted to hornfels around the greenstone, and the country rocks at Mount Stinear are locally enriched in cummingtonite in the vicinity of the intrusive amphibole schists. The presence of andalusite and the fine-grained appearance of some of the schists at Mount Rymill may be due to contact metamorphism by the diorite.

Granite and pegmatite

Granite and pegmatite that are intrusive into the superjacent metamorphic rocks, as well as into the basement complex, are found in many parts of the southern Prince Charles Mountains (Figure 57.1; unit g, Figure 57.2). These rocks, typically white or gray, commonly contain muscovite or garnet. At Mounts McCauley and Scherger, veins of granite and pegmatite are abundant, and at Mounts McCauley and Dummett, larger bodies of granite are exposed. At Mount Stinear, on the other hand, only one pegmatite was found in the superjacent metamorphic rocks (in quartzite of unit qs near Edwards Pillar; Figure 57.2). As this pegmatite contains muscovite and garnet, the author tentatively correlates it with the pegmatite at Mounts McCauley and Scherger, and believes it is not related to the pegmatite that is so abundant in the basement complex at Mount Stinear.

In general, the granite and pegmatite are associated with sillimanite-zone metamorphism, a spatial relationship suggesting that the regional metamorphism of the superjacent rocks and the plutonic activity are related events and took place at roughly the same time.

Structure

Structures in the metamorphic rocks include premetamorphic steeply dipping faults, synmetamorphic folds, a postmetamorphic thrust, and, lastly, younger steeply dipping faults.

Steep faults along which mafic rocks were subsequently emplaced were active prior to the regional metamorphism on Mount Stinear (Grew, 1976). These trend N 25° W to N 45° W. Blastomylonite zones as much as 2 m wide are exposed along some of the faults. A thick zone of blastomylonitic gneiss in the basement complex at Mount Stinear (southernmost block) may be related to premetamorphic movement along the nearby fault dipping 78° N (Figure 57.2).

All exposures of the basement complex and of the overlying metamorphic rocks show evidence of folding. The deformation of the metamorphic rocks, at least in part, seems to have been contemporaneous with the regional metamorphism. Mica schistosity, alignment of amphibole prisms, and elongation and flattening of clasts in the metaconglomerate appear to be related to this deformation. Large-scale recumbent folding is indicated by minor folds with amplitudes of tens (Mount Scherger) to hundreds of meters (Blake Nunataks near Mount Maguire; Tingey and England, 1973, Figure 23).

A shallow, south-dipping thrust fault is well exposed on Mount Maguire (Figure 57.2). The metasedimentary rocks have been brecciated and locally mineralized with pyrite and hematite for distances as great as 100 m below the thrust plane; in contrast, the amphibole rocks above the thrust have been little affected. The thrust is postmetamorphic and is offset by steeply dipping faults (Figure 57.2). Northeast- and northwest-trending steep faults with vertical displacements of as much as 300 m were mapped at Mounts Ruker, McCauley, Stinear, and Maguire (Grew, 1976). A 2- to 3-m-thick zone mineralized with fluorite fills the most northerly fault on Mount Maguire.

Regional metamorphism

The basement rocks at Mounts Newton and Stinear seem to have been affected by an upper amphibolite to granulite facies metamorphic event prior to the deposition of the overlying metasedimentary rocks. Migmatization accompanied this early event at Mount Stinear.

The overlying sedimentary and volcanic rocks and the mafic intrusive rocks were regionally metamorphosed under conditions of the greenschist and amphibolite facies. With the possible exception of the Mawson Escarpment, the exposures of metamorphic rocks shown in Figure 57.1 were affected by a single event that was accompanied by folding and emplacement of granite and pegmatite. The metamorphism was of the Barrovian type over most of the area. Andalusite at Mount Stinear, unlike that at Mount Rymill, seems to have formed during the regional event and thus would indicate slightly lower metamorphic pressures in this area.

Correlation of the rock units

Rocks that have been mapped as basement complex include the following: (1) plutonic rocks that do not intrude the superjacent metamorphic rocks and appear to be overlain by them unconformably (for example, Mount Ruker; England and Langworthy, 1975, pp. 4–5); (2) high-grade gneisses that have been affected by later lower grade events (Mount Newton); and (3) at Mount Stinear, gneissic and migmatitic rocks that are cut by numerous pegmatites not found in the superjacent metamorphic rocks, with the single exception noted above.

In the exposures studied by the author, at least two sequences of superjacent metamorphic rocks seem to be present. One sequence is distinguished by the presence of a metamorphosed banded iron formation and was found only at Mount Ruker (Fedorov and Tarutin, 1977). The second sequence, which is roughly equivalent to the Menzies Series* as originally defined by Soloviev (1972), is characterized by metaconglomerate and by the quartzite–amphibole schist–mica schist association. The rocks at Goodspeed Nunataks and Mounts Scherger, McCauley, Dummett, Rymill, and Stinear belong to this sequence; the amphibole schists above the thrust at Mount Maguire may be correlative with it.

Origin and tectonic setting of the superjacent metamorphic rocks

Most of the superjacent metamorphic rocks are of sedimentary origin. On the other hand, the amphibole schists seem to be of volcanic origin, including both lava flows and volcanic debris deposited by water. The mineralogic similarity of the amphibole schists to the mafic intrusive rocks at Mount Stinear suggests, moreover, that the amphibole schists alternating with the metasedimentary rocks may be extrusive equivalents of the mafic intrusive rocks.

The metaconglomerate in the Menzies Series does not seem to be the basal unit in the sequence. The large size of some clasts, however, suggests that the conglomerate may have been deposited near active faults. Inasmuch as some of the steeply dipping faults at Mount Stinear were active prior to the regional metamorphism, it is not unreasonable to assume that the conglomerate deposition and the faults are related. Furthermore, the mafic intrusive rocks were emplaced during the faulting. Taking these observations together, the following picture emerges for the tectonic setting during accumulation of the Menzies Series: volcanism and block faulting, rapid denudation of uplifted blocks, and deposition of sediments in down-dropped blocks, including conglomerate in the vicinity of the faults.

In contrast, there is less evidence that the Ruker Series was deposited in a tectonically active environment. The conglomeratic mudstone was probably deposited by subaqueous slumping or flow (Pettijohn, 1975, p. 182) and thus is probably not related to tectonic activity. The banded iron formation may have been deposited in a tectonically quiet environment. It resembles more the miogeosynclinal Superior type than the eugeosynclinal Algoma type, as defined by Gross (1965), for it is associated with quartzite and black slate, and volcanic rocks are subordinate (see also England and Langworthy, 1975, p. 9).

* Following Russian usage, *series* here means a lithostratigraphic unit formed during a period or longer.

Ages of the rocks

Basement complex

Total-rock Rb-Sr data on granitic rocks from the basement complex at several localities indicate that the complex is at least 2500 m.y. old or Archean in age (England, 1975; Halpern and Grikurov, 1975; Tingey, 1975*b;* Tingey and others, 1976; Tingey, Chapter 55, this volume). At Mount Stinear, however, a Pb-Pb age of 1035 m.y. was obtained by W. I. Manton (unpublished data) on a mixture of yttrobritholite(?), xenotime(?), and other unidentified minerals in a nodule from pegmatite in the basement complex (sample 544, Figure 57.2). The basement complex at Mount Stinear may thus be a southward extension of the high-grade metamorphic terrane of the northern Prince Charles Mountains, which was affected by a late Precambrian event (Tingey and others, 1976; Grew, 1978).

Superjacent metamorphic rocks

Iltchenko (1972) describes an acritarch collection from Mounts Ruker, Rubin, and Seddon (located between Rymill and Dummett) that suggests a middle to late Proterozoic age for these rocks. Professor L. R. Wilson (personal communication, 1977) of the Oklahoma Geological Survey has examined quartzite, slate, and phyllite from the metamorphosed banded iron formation and overlying rocks at Mount Ruker. He found no organic remains in thin section or in acid-insoluble residues of these rocks.

The restricted distribution of banded iron formations in geologic time strongly suggests that the Ruker Series is at least 1700 m.y. old (Halpern and Grikurov, 1975, p. 9; Tingey, Chapter 55, this volume). The metamorphosed banded iron formation at Mount Ruker has several of the physical and chemical features of Proterozoic banded iron formations described by Lepp and Goldich (1964), Govett (1966), and Goldich (1973). Its bulk composition (Ravich and Kamenev, 1972, Table 126; Ravich and Tarutin, 1977) is close to the average for Proterozoic banded iron formations of the Canadian Shield and has the following chemical characteristics of these iron formations: high SiO_2 and low Al_2O_3, CaO, P_2O_5, and TiO_2 contents (Lepp and Goldich, 1964, Table 10).

Tingey and others (1976) and Tingey (Chapter 55, this volume) report a Rb-Sr age of 2580 m.y. on muscovite from the pegmatite cutting quartzite near Edwards Pillar (Figure 57.2), indicating an Archean age for the Menzies Series at Mount Stinear. This age assignment is not consistent with the observed field relationships between the superjacent metamorphic rocks and basement complex, and with the Pb-Pb age of 1035 m.y. on pegmatite from the basement complex, a pegmatite that seems to be older than the superjacent metamorphic rocks.

Rb-Sr total-rock and mica data show that the emplacement of granite and pegmatite (except that at Mount Stinear) and the regional metamorphism both occurred during early Paleozoic time, about 500 m.y. ago (Halpern and Grikurov, 1975).

Summary

The author's preferred interpretation of the geologic history of the southern Prince Charles Mountains can be summarized as follows:

1. Consolidation of basement during Archean time.
2. Deposition of sedimentary rocks, including banded iron formation during Proterozoic time.
3. Metamorphism, plutonism, and deformation in the northern part of the area (Mount Stinear) during late Precambrian time.
4. Deposition of the Menzies Series accompanied by block-faulting, volcanic activity, and contact metamorphism during late Precambrian time.
5. Regional metamorphism, folding, and plutonism during early Paleozoic time.
6. Formation of steeply dipping faults and thrust faults.

Acknowledgments. I thank D. S. Soloviev and other Soviet geologists for their overall assistance and for sharing samples and field data; R. J. Tingey for his assistance, particularly in keeping me informed of the results of Australian work; B. P. Lambert for sending me aerial photographs taken by the Australian Division of National Mapping; L. C. Noakes for permission to use the information in the 1:250,000 preliminary geologic maps and in the unpublished records of the Australian Bureau of Mineral Resources, Geology and Geophysics; and Martin Halpern for samples collected by G. E. Grikurov and O. A. Tarutin in 1972. W. G. Ernst and P. C. Grew critically reviewed the manuscript. The author is grateful for their comments and suggestions, but he accepts all responsibility for the interpretations presented in this paper.

The research was supported by National Science Foundation grants DPP 72-05797 to the University of Wisconsin-Madison and DPP 75-17390 to the University of California-Los Angeles.

References cited

England, R. N. 1975. Mawson Escarpment South, Sheet SS40-42/8, 1:250,000 Geological Series, Preliminary Edition. *Bur. Miner. Resour. Geol. Geophys. Aust.*

England, R. N., and Langworthy, A. P. 1975. Geological work in Antarctica—1974. *Rec. Bur. Miner. Resour. Geol. Geophys. Aust.* 1975/30, [Unpublished] 19 pp.

Federov, L. V., and Tarutin, O. A. 1977. Geologicheskoe stroenie gory Ruker (iuzhnaia chast'gor Prins Charl'z, Vostochnia Antarktida) [Geological structure of Mount Ruker (southern part of the Prince Charles Mountains, East Antarctica)]. *Antarktika, Doklady Komissii, 16,* 93-99.

Goldich, S. S. 1973. Ages of Precambrian iron-formations. *Econ. Geol, 68,* No. 7, 1126-34.

Govett, G. J. S. 1966. Origin of banded iron formations. *Geol. Soc. Am. Bull., 77,* No. 11, 1191-212.

Grew, E. S. 1975. With the Soviets in Antarctica. *Antarct. Jnl U.S., 10,* No. 1, 1-8.

Grew, E. S. 1976. Geologic studies in the southern Prince Charles Mountains. *Antarct. Jnl U.S., 11,* No. 4, 240-42.

Grew, E. S. 1978. Precambrian basement at Molodezhnaya Station, East Antarctica. *Geol. Soc. Am. Bull., 89,* 801-13.

Grikurov, G. E., and Soloviev, D. S. 1974. Geologicheskoe stroyenie gornogo obramlenia lednika Lamberta [Geologic structure of the mountainous fringe of the Lambert Glacier]. *Inf. Byull. sov. antarkt. Eksped.,* No. 88, 21-29.

Gross, G. A. 1965. Geology of iron deposits in Canada. 1. General geology and evaluation of iron deposits. *Econ. Geol. Rep. geol. Sur. Cana.,* No. 22, 181 pp.

Halpern, M., and Grikurov, G. E. 1975. Rubidium-strontium data from the southern Prince Charles Mountains. *Antarct. Jnl U.S., 10,* No. 1, 9-15.

Iltchenko, L. N. 1972. Late Precambrian acritarchs of Antarctica. (*In* Adie, R. J., *ed. Antarctic Geology and Geophysics.* Oslo, Universitetsforlaget, 599-602.)

Lepp, H., and Goldich, S. S. 1964. Origin of Precambrian iron formations. *Econ. Geol., 59,* No. 6, 1025-60.

Pettijohn, F. J. 1975. *Sedimentary Rocks.* 3rd ed. New York, Harper and Row, 628 pp.

Ravich, M. G., and Kamenev, E. N. 1972. *Kristallicheskii fundament Antarkticheskoi platformy* [Crystalline basement of the Antarctic platform]. Leningrad, Gidrometeoizdat, 658 pp. [English translation: New York, John Wiley, 1975, 582 pp.]

Ravich, M. G., and Tarutin, O. A. 1977. Zhelezorudnaia provintsia v Vostochnoi Antarktide [An iron-ore province in East Antarctica]. *Antarktika, Doklady Komissii, 16,* 54-61.

Sheraton, J. W. 1975. Mawson Escarpment North, Sheet SS40-42/4, 1:250,000 Geological Series, Preliminary Edition. *Bur. Miner. Resour. Geol. Geophys. Aust.*

Soloviev, D. S. 1972. Geological structure of the mountain fringe of Lambert Glacier and the Amery Ice Shelf. (*In* Adie, R. J., *ed. Antarctic Geology and Geophysics.* Oslo, Universitetsforlaget, 573-77.)

Tingey, R. J. 1975*a*. Cumpston Massif Sheet SS40-42/7, 1:250,000 Geological Series, Preliminary Edition. *Bur. Miner. Resour. Geol. Geophys. Aust.*

Tingey, R. J. 1975*b*. Mount Menzies Sheet SS40-42/6, 1:250,000 Geological Series, Preliminary Edition. *Bur. Miner. Resour. Geol. Geophys. Aust.*

Tingey, R. J., and England, R. N. 1973. Geological work in Antarctica—1972. *Rec. Bur. Miner. Resour. Geol. Geophys. Aust.,* 1973/161, [Unpublished] 17 pp.

Tingey, R. J., England, R. N., Sheraton, J. W., and Arriens, P. A. 1976. Geology, geochemistry, and geochronology of the Archaean rocks of the Prince Charles Mountains, Antarctica. *Abstr. 25th Int. Geol. Congr., 1,* 22-23.

58

Main Tectonic Features and Development of the Southern Prince Charles Mountains, East Antarctica*

J. HOFMANN
Section of Geosciences, Mining Academy, 92 Freiberg, German Democratic Republic

Abstract. This paper deals with results of tectonic field work done during the 19th Soviet Antarctic Expedition (SAE) in the 1973–74 austral summer on the middle–upper Proterozoic low-grade metamorphic complex of the southern Prince Charles Mountains. This complex, disconformably overlying the upper Archean–lower Proterozoic high-grade metamorphic basement, was deformed under the influence of basement block movements. Its east-west structural trends (fold axes) are parallel to the crystalline basement trends. Due to the protective influence of the basement, the intensity of folding decreases from the upper part (Sodruzestvo Series) to the lower part (Menzies and Ruker Series) of the low-grade complex. The lower, lesser folded unit is characterized, however, by an intensive cleavage which has served as an additional element of deformation and as a main factor of tectonic transport. In the regions of tectonic contacts between the low-grade complex and the basement gneisses, the greenschist facies metamorphism of the low-grade complex changes laterally into stress-accented almandine-amphibolite facies metamorphism with staurolite and kyanite. Magmatic activity within the low-grade complex is restricted to basic synkinematic intrusions connected with the deformation of the lower part of the complex (Menzies and Ruker Series). According to scarce absolute age determinations, the deformation of the Menzies and Ruker Series has an age of approximately 1000 to 900 m.y., and that of the Sodruzestvo Series of 600 to 500 m.y.

This paper deals with the results of tectonic field work done by the author during the 19th Soviet Antarctic Expedition (SAE) in the 1973–74 austral summer on the Precambrian low-grade metamorphic complex of the southern Prince Charles Mountains (Figure 58.1). The new observations were supplemented by field data from the 17th and 18th SAE and published by ANARE data.

Geology

The southern Prince Charles Mountains (lat. 69° 30′–74° 30′ S, long. 60° 30′–68° 30′ E) consist of a group of about 45 nunataks forming mountain fringes of the Fisher, Mellor, and Collings Glaciers, which are western and southern tributary glaciers to the Lambert and Kaliningrad Glaciers.

The folded low-grade complex, mainly metamorphosed under greenschist conditions and surrounded by outcrops of crystalline basement rocks, was first described by Trail (1963, 1964). A jaspilite-bearing sequence of chlorite-sericite and chlorite-biotite-sericite schists and quartzites was discovered by the late D. S. Soloviev in 1965. The first lithostratigraphic subdivision into a lower Menzies Series**, containing jaspilites, and an upper Sodruzestvo Series was given by Soloviev (1972*a*). A more detailed subdivision, based on 1971–74 SAE fieldwork, was published by Grikurov and Soloviev (1974). A lower sequence (Menzies and Ruker Series, >2500 m) is disconformably overlain by an upper group (Sodruzestvo Series, >3500 m). The Menzies and Ruker Series are thought to be approximately the same age, two facies of a middle Proterozoic lithostratigraphic unit. The Sodruzestvo Series is probably of late Proterozoic (Vendian) age (Iltchenko, 1972; Grikurov and Soloviev, 1974). In the present state of knowledge, the low-grade complex may be interpreted as a terrigenous, molasse-like sequence, formed in an intracratonic basin of the pre-middle Proterozoic East Antarctic craton (Kossygin and others, 1974). Correlations in lithostratigraphic, structural, metamorphic, and magmatic development with the Proterozoic Cuddapah Basin of the Indian peninsular shield are probable (Fedorov and others, 1977).

In this paper, written from a more structural point of view, the low-grade complex is divided into lower and upper structural stages, characterized by pronounced differences in deformation and metamorphism. The lower stage, which includes the Menzies and Ruker Series, is separated from the upper one (Sodruzestvo Series) by an angular unconformity. Pebbles of jaspilites, occurring in the basal conglomerates of the Sodruzestvo Series, prove the existence of a significant period of folding and metamorphism before the deposition of the Sodruzestvo sedimentary beds.

General tectonic features of the low-grade complex

Detailed field work at Mount Maguire, Mount McCauley, Mount Stinear, Mount Ruker, Mount Dummett, and Mount Rubin, which was supplemented by structural observations during helicopter traverses and by field data

* Publication No. 749, Sektion Geowissenschaften, Bergakademie Freiberg/GDR under permission of the Zentralinstitut für Physik der Erde der Akademie der Wissenschaften der DDR, Potsdam.

** Following Russian usage, *series* here means a lithostratigraphic unit formed during a period or longer.

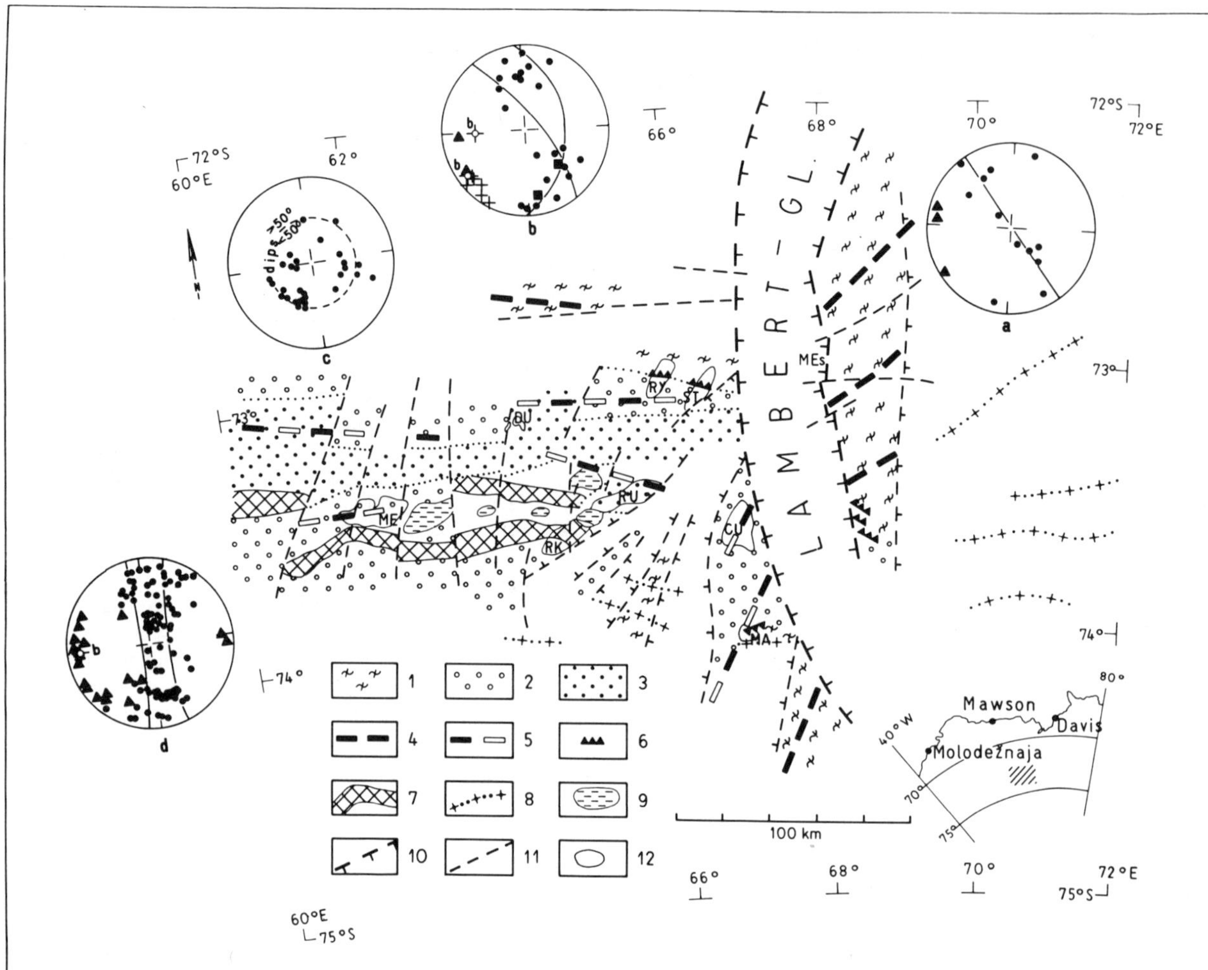

Figure 58.1. Schematic structural map of the southern Prince Charles Mountains. 1. Crystalline basement, undivided; 2. Low-grade complex, lower structural stage (Menzies and Ruker Series); 3. Low-grade complex, upper structural stage (Sodruzestvo Series); 4. Crystalline basement structural trends; 5. Low-grade complex structural trends; 6. Observed tectonic contacts between crystalline basement and low-grade complex, lower structural stage; 7. Positive magnetic anomaly of jaspilite sequence (Ruker Series); 8. Axes of positive magnetic anomalies; 9. Areas of negative magnetic anomalies (uplifts of crystalline basement in the low-grade complex?); 10. Marginal faults of Lambert graben and related grabens of lower order; 11. Faults, unclassified; 12. Contours of some important nunataks, mountains and escarpments. MEs: Mawson Escarpment. MA: Mount Maguire. CU: Mount Cumpstone. RK: Mount Ruker. RU: Mount Rubin. ST: Mount Stinear. RY: Mount Rymill. DU: Mount Dummett. ME: Mount Menzies.

Stereographic projections (lower hemisphere, equal-area) show planar and linear elements (gneissosity, foliation, bedding, axes of folds, b-axes, and lineations) of the basement complex and the low-grade complex. *a*. Structural data from Mawson Escarpment; *b*. From crystalline basement outcrops north of the low-grade complex; *c*. From low-grade complex, lower structural stage; *d*. From low-grade complex, upper structural stage. Dots: poles of planar elements (gneissosity, foliation, and bedding). Squares: poles of axial planes from mesoscopic folds. Triangles: poles of fold axes. Crosses: poles of lineations. Crossed circles: constructed b-axes. Compiled from structural data of the author and SAE geologists, from SAE magnetic data, and from published ANARE data.

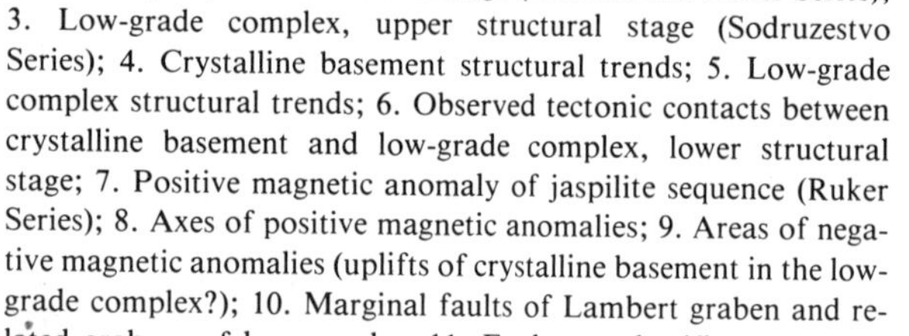

Figure 58.2. Profile from northern flank of Mount Dummett. Folded basal sequences of Sodruzestvo Series after Ravich and others (1978). 1. Greenish, mica-bearing quartzites; 2. Gray quartzites; 3. Metaconglomerates; 4. Faults; 5. Reconstructed parts of folds.

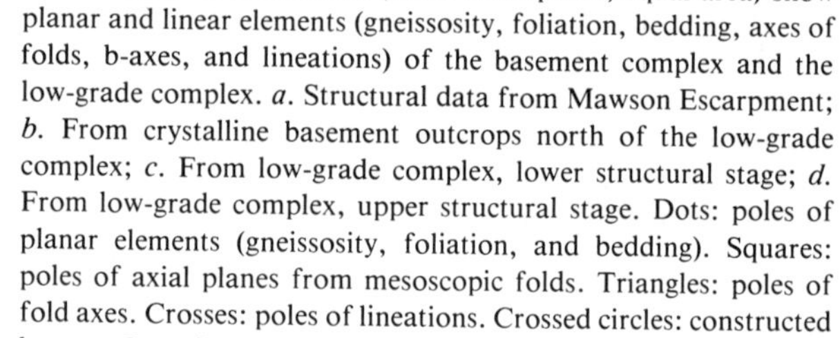

Figure 58.3. Block diagram from the northern flank of Mount Maguire, showing structural conditions of the Menzies Series. On the eastern side are upthrusted metabasites (see Figure 58.10). On the western side is a sequence of quartzites and staurolite-kyanite-garnet-mica-quartz schists (see Figure 58.4). Stippled planes indicate faults.

Figure 58.4. Homoclinal sequence of quartzites (dark) and staurolite-kyanite-garnet-mica-quartz schists (light) on the northern flank of Mount Maguire, western corner. The nunatak is about 400 m high.

from SAE geologists (D. S. Soloviev, V. V. Samsonov, V. S. Semenov, and O. A. Tarutin), led to the following general results about the structural, metamorphic, and magmatic development of the low-grade complex.

1. Simple, linear, and in most cases, overturned fold structures, locally complicated by syngenetic thrusts, are predominant. In some cases, however, rock sequences have been observed that are only gently folded, and yet they have cleavage and show evidence of strain (drag folds and secondary cleavage) within individual layers. Structural trends (fold axes and lineations) in both of the structural stages are east–west, parallel to the trends in the crystalline basements. Aberations of fold axes and lineations from the latitudinal trend into southwest-northeast or south-southwest–north-northeast directions, respectively, were observed in the vicinity of supposed faults or fault zones (Figure 58.1).

2. The intensity of folding increases form the lower to the upper part of the low-grade complex. The intensely folded sequences of the Sodruzestvo Series observed at Mount Dummett (Figure 58.2) and at Mount Rubin (Grikurov and Soloviev, 1974) are in striking contrast to the more gently folded sequences of the Ruker and Menzies Series at Mount Maguire (Figures 58.3 and 58.4), Mount McCauley, Mount Ruker, and Mount Bayliss(?), which are characterized by dips of $<50°$ (Figures 58.1, 58.5*c*, and 58.6*a, d*).

3. Locally, especially in the jaspilite sequences of Mount Ruker and in well-bedded metasiltstone and metacarbonate sequences of the Sodruzestvo Series at Mount Rubin, fold structures of a higher order are complicated by intensive disharmonic folds (drag folds) on a scale up to several hundred meters in wavelength and amplitude (Figure 58.7).

4. In more homogenous rock sequences such as chlorite-(biotite)-sericite quartzites of the Ruker Series, an intensive cleavage (Transversalschieferung of German authors), which is steeply dipping and oriented parallel or symmetric to axial planes of folds, is present (Figures 58.6*c* and *e* and 58.8). In distinct parts of the lower structural stage (less intensely folded areas), cleavage seems to be the main element of deformation.

5. At the southern and northern margins, low-grade complex sequences (lower structural stage) are overthrust by basement gneisses. This is so at Mount Stinear (Figure 58.9), at Mount Rymill, at Mount Maguire (Figures 58.3 and 58.10) and, according to England and Langworthy (1975), at the southern end of the Mawson Escarpment. Due to the thrusting of basement gneisses upon sequences of the Menzies Series, these latter rocks are intensely folded and partially overturned (Figure 58.11). Cleavage in these cases is replaced by foliation parallel to bedding.

6. The intensity of regional metamorphism in the lower structural stage does not exceed greenschist facies conditions (sericite-chlorite and biotite-chlorite subfacies) according to Soloviev (1972*b*). In the areas of tectonic contacts between basement gneisses and the low-grade complex, however, the appearance of staurolite, kyanite, and almandine proves the existence of stress-accented metamorphism under somewhat higher temperatures (stauro-

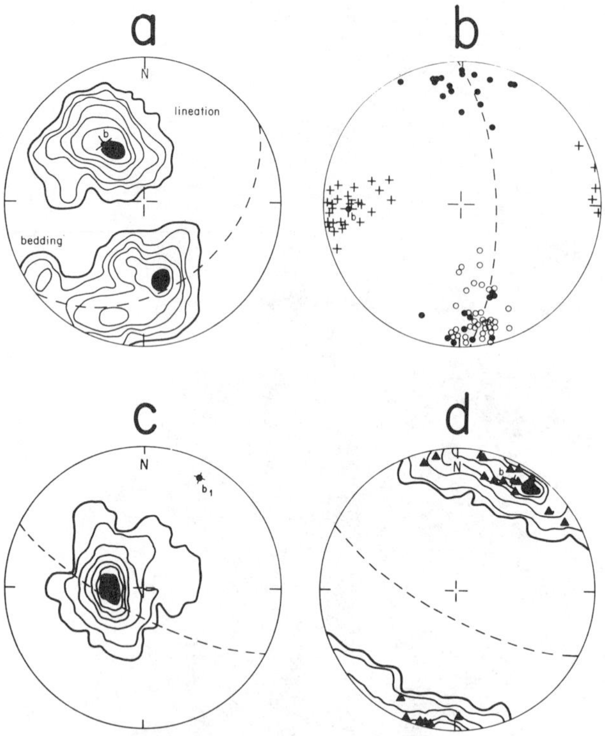

Figure 58.5. Diagrams of planar and linear structural elements (lower hemisphere, equal-area) in the low-grade complex and in basement gneisses. *a.* Bedding (104 poles) and lineation (68 poles) at southern part of Mount Stinear. Contours at 0, 1, 2, 3, 5, 10, and 15%; *b.* Gneissosity and lineation in basement gneisses at central part of Mount Stinear (see Figure 58.14). Light and dark circles indicate gneissosity, and crosses indicate lineation; *c.* Foliation in sericite-mica-quartz schists and quartzites at northern side of Mount Maguire (see Figure 58.3); 192 poles of foliation. Contours at 0, 1, 5, 10, 15, 20, and 25%; *d.* Lineation and fold axes (triangles) in sericite-mica-quartz schists and quartzites at northern side of Mount Maguire; 155 poles of lineation. Contours at 0, 1, 5, 10, 15, and 20%.

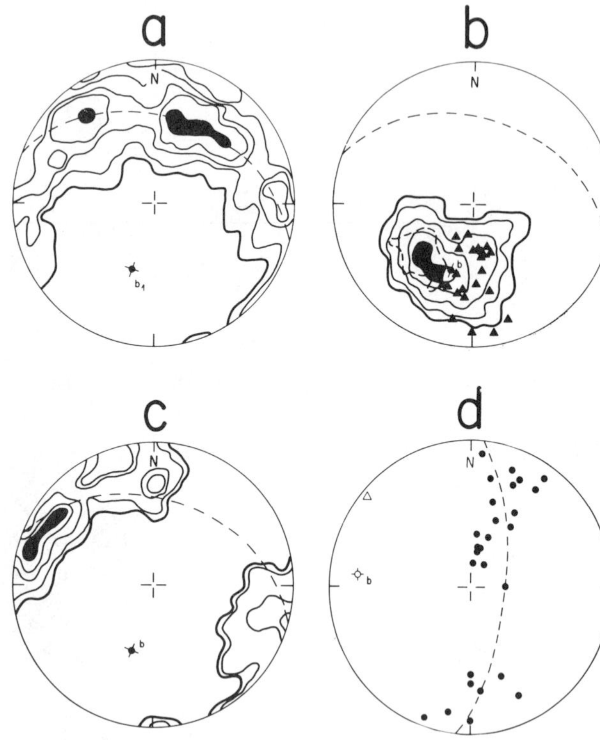

Figure 58.6. Diagram of planar and linear structural elements (lower hemisphere, equal area) in the low-grade complex. *a.* Bedding (218 poles) in chlorite-sericite schists and quartzites at western side of Mount Ruker. Contours at 0, 1, 3, 5, and 10%; *b.* Lineation (141 poles) and fold axes (triangles) in chlorite-sericite schists and quartzites at western side of Mount Ruker. Contours at 0, 1, 5, 10, and 15%. Dashed line indicates area of elongated pebble axes in metaconglomerates; *c.* Cleavage (84 poles) in chlorite-sericite schists and quartzites at western side of Mount Ruker. Contours at 0, 1, 5, 7, and 10%; *d.* Bedding in Sodruzestvo Series at northwestern side of Mount Rubin (see Grikurov and Soloviev, 1974, Figure 2).

lite-kyanite-muscovite subfacies of almandine-amphibolite facies). Examples of migmatization and granitization are completely absent. The upper structural stage, separated by an angular unconformity from the lower one, is characterized by a widely uninvestigated weak dynamometamorphism of pumpellyite-prehnite-quartz facies in connection with folding of the Sodruzestvo Series.

7. Magmatic activity seems to be restricted to the lower structural stage, in which basic dikes, sills, and small intrusive bodies of probably late synkinematic origin appear (Mount Ruker and Mount Stinear). The radiometric ages of 1040–980 m.y. (Halpern and Grikurov, 1975) give the first information about the deformation time of the lower structural stage. Small intrusive bodies of granite in the lower structural stage (Mount McCauley) of about 500 m.y. belong to a later period of platform magmatic activity (Soloviev, 1972*c*). The rather flat-lying sequences of the lower structural stage (Mount McCauley and Mount Bayliss) seem to have acted as a kind of structural trap for the intruding granites.

The deformation and metamorphism of the low-grade complex were probably caused by basement block movements, under the influence of which overlying sequences were folded and faulted, mainly by vertical movements. More intensive folding appears in marginal parts where basement gneisses are thrust upon older sequences of the low-grade complex. The structural picture in these areas is similar to that of the Owl Creek uplift in Wyoming, as described by Wise (1963).

In the present state of knowledge, the structure and deformation in the low-grade complex seem to be comparable to that of "saxonotype" or "germanotype" tectonics

Figure 58.7. Concentric, disharmonic fold in jaspilites of Ruker Series. Northwestern side of Mount Ruker, central part. Height of outcrop is about 12 m.

Figure 58.8. Cleavage and bedding in contact-metamorphosed sericite-biotite quartzites of Ruker Series. Bedding dips from right to left; cleavage runs parallel to observer. Intersection lines between bedding and cleavage dip steeply (40°) south (to the left). Northwestern side of Mount Ruker, central part.

Figure 58.10. Metabasite complex (dark) upon sericite-biotite-quartz schists (light gray) with folded metacarbonate layers. Light area is thrust breccia. Northern side of Mount Maguire, eastern corner (see Figure 58.3). Height of cliff is about 400 m.

according to Stille (1924) or that of "dictyogenese" according to Bubnoff (1954), characterized, however, by low-grade, partially stress-accented metamorphism and basic magmatic activity. A more detailed description of the subjects discussed above has been given by Hofmann (1978).

Structural map

A structural map based on all structural field data available is shown in Figure 58.1. Structural trends in the basement as well as in the low-grade complex were constructed giving consideration to paracrystaline linear elements (mesoscopic fold axes, *b*-axes, and lineations) and steeply dipping planar elements (gneissosity, foliation, and bedding) where rotation due to later faulting is unlikely. Planar and linear elements (single measurements and average outcrop data) are represented in synoptic stereographic projections (Figure 58.1).

Pertinent geomagnetic data, axes of positive anomalies, and contours of jaspilite anomalies in the Ruker Series were all taken from SAE maps. Fault patterns were constructed from geomagnetic and geomorphologic data.

In the following sections, short descriptions of some distinctive structural features of the low-grade complex are given.

Fold structures and folding

Single structural observations in the low-grade complex have been made by Trail (1963), Grikurov and Soloviev (1974), England and Langworthy (1975), and Ravich and others (1978).

Figure 58.9. Tectonic contact between basement gneisses (left) and low-grade complex (right). Migmatized muscovite-biotite basement gneisses (left), blastomylonites and augengneisses (center), and staurolite-kyanite metaquartzites of the Menzies Series (right). In the blastomylonites, basic dikes (small dark bodies) occur. Height of outcrop is 600 m. Western side of Mount Stinear, central part.

Figure 58.11. Anticline in metaconglomerates (light) and kyanite-garnet-biotite schists (dark) of the Menzies Series. Western flank of Mount Stinear, southern part. Height of outcrop is about 600 m.

In the lower structural stage, fold structures are somewhat varied in type and morphology. Intensely folded sequences, probably belonging to the Menzies Series, were observed or studied at Mount Stinear (southern part), Mount Rymill (southern part), and Mount Cumpstone These nunataks belong to the marginal northern part of the low-grade complex. In these areas, folding obviously was connected with block movements in the basement gneisses. At Mount Stinear (southern part) overturned folds were observed in air photos and later studied in the field (Figure 58.11). Repeated occurrence of metaconglomerates and quartzites between biotite-amphibole schists indicate fold amplitudes of about 1500 m and wavelengths up to 2 km. Folds observed at Mount Cumpstone (eastern side) have approximately the same dimensions.

At Mount Menzies, no structural field observations were made by SAE expeditions. Trail (1964), however, mentioned isoclinal folds with amplitudes as high as 1000 m and steeply dipping axial planes. This structure is similar to that at the southern part of Mount Stinear, so that Mount Menzies is thought to be also in a marginal position at the southern flank of the low-grade complex.

At Mount Ruker, the jaspilite sequence, regardless of its disharmonic folding, and the psammitic-pelitic rocks in the roof of metagabbro-diabase sills are more gently folded in somewhat smaller dimensions, with amplitudes as high as 300–500 m and wavelengths as much as 100–2000 m (see Hofmann, 1978, p. 109). The presence of intense cleavage (Figures 58.6*c* and 58.8) seems to be related genetically to smaller fold amplitudes and less intensive folding, which is probably effected at shallow depths by uplifts of the basement.

Gently inclined rock sequences (dips <45°) were also observed and partially studied at Mount McCauley, Mount Scherger, Seavers Nunatak, Goodspeed Nunataks (northern group), and the northern side of Mount Maguire. In general, the structure in these areas is characterized by rather flat-topped, cupolalike folds up to 10–12 km in diameter. In incompetent sequences evidences of intense internal deformation, mostly recumbent folds (~20 m in scale) of disharmonic character, occur (Figures 58.12 and 58.13*b, c,* and *d*). In places, their limbs are extremely

Figure 58.12. Intensely folded staurolite-kyanite-mica-quartz schists and quartzites between unfolded quartzites (above and below). Northern flank of Mount Maguire, western corner. Height of outcrop is 12 m. Detail from Figure 58.4.

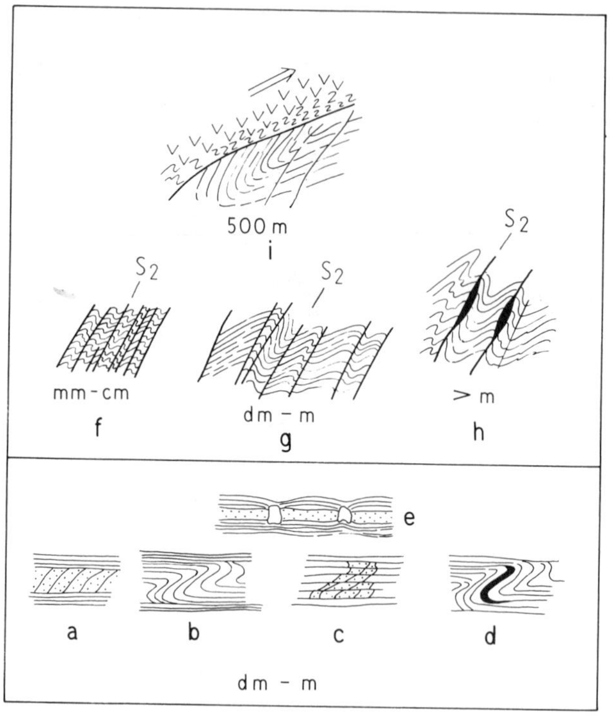

Figure 58.13. Mesocopic and megascopic structural elements in Menzies Series rocks at Mount Maguire, northern flank. *a.* Cleavage in competent layers; *b.* Drag folds in incompetent layers; *c.* Sheared drag folds in incompetent layers; *d.* Drag folds with syngenetic quartz veins in incompetent layers; *e.* Interboudin quartz veins in quartzites; *f–h.* Secondary foliation in chlorite-mica-quartz schists near upthrusted metabasites; *i.* Cross section through metabasites and chlorite-mica-quartz schists (see Figure 58.10).

thinned or replaced by shear planes. In competent layers (metaquartzites and amphibolites), boudinage and fracture cleavage are widely developed (Figure 58.13*a* and *e*). Interboudin veins are filled with syntectonic brecciated and recrystallized quartz.

The apparent homoclinal structure in the deeper parts of the low-grade complex may be interpreted as a less intensely folded, but highly internally deformed, sheared, and cleaved, structural unit resting directly on the basement.

At Mount Ruker (northern part), a slightly metamorphosed weathering crust disconformably overlies a basement granite (Ravich and others, 1978), the age of which was determined by Halpern and Grikurov (1975) as being about 2.8 b.y. A group of isolated negative magnetic anomalies, exactly situated along a latitudinal line on westernmost Mount Ruker, could be interpreted as basement uplifts within the low-grade complex.

Tectonic contacts between the low-grade and basement complexes

Tectonic contacts between the lower structural stage and basement gneisses have been observed at Mount Stinear (see Hofmann, 1978, p. 108), Mount Rymill, Mount Seddon(?), and Mount Maguire. At the southern part of the Mawson Escarpment, according to England and Langworthy (1975), a highly disturbed sequence of basement gneisses and staurolite-kyanite metaconglomerates and

Figure 58.14. Anticline in partially migmatized muscovite-biotite basement gneisses. Western flank of Mount Stinear, central part. The crest of the anticline is about 350 m high.

quartzites is present. It is comparable with that at the western side of Mount Stinear, where intensely folded staurolite-kyanite metaconglomerates and quartzites are separated from folded basement gneisses (Figures 58.5*b* and 58.14) and granites (central part) by blastomylonitic zones about 100–300 m in thickness.

The most interesting zone at Mount Stinear was observed approximately 5 km south of Edward's Pillar (Figure 58.9). In this zone, partially migmatized muscovite-biotite basement gneisses are transformed into banded and foliated blastomylonites and augen gneisses (Figure 58.15). Their foliation is parallel to the east-west striking, nearly vertically dipping fault planes. Primary banding in basement gneisses is completely extinguished by blastomylonitic foliation. In the nearly vertically dipping metaquartzites and metaconglomerates of the Menzies Series, the intensity of foliation and lineation and the degree of recrystallization increases from south-southwest to north-northeast (Figure 58.5*a*). The presence of staurolite and kyanite megablasts or megablastic aggregates up to 10 cm in diameter seems to be connected with their immediate proximity to the blastomylonitic zone. Kyanite and staurolite growth is obviously connected with a second substage of metamorphism. In the vicinity of thrust zones, biotite and quartz streaks have been observed in poikiloblastic relationship with kyanite and staurolite megablasts. The megablasts do not show linear orientation, but only belteroporic orientation, probably in connection with a final phase of increasing pressure after thrusting.

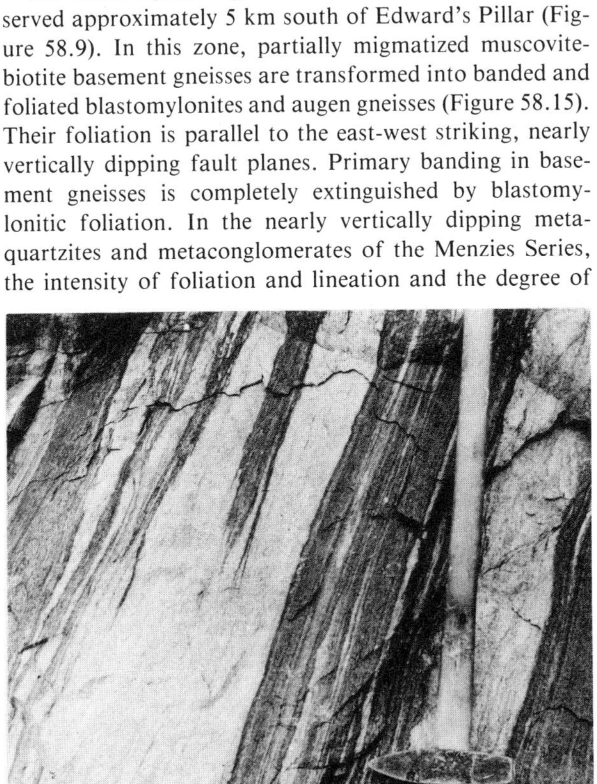

Figure 58.15. Blastomylonitic augen gneisses resulting from muscovite-biotite basement gneisses. Western side of Mount Stinear, central part. Detail from Figure 58.9, middle of cliff.

Figure 58.16. Zone of blastomylonites (banded rocks in foreground) at foot of Edward's Pillar, western flank of Mount Stinear (central part). In background, white quartzites are cut by basic dikes. Dark rocks on the left are amphibole-biotite granite, and light rocks on the right are muscovite-biotite basement gneisses.

A similar, but somewhat thinner, zone appears northward of Edward's Pillar bordering the northern flank of the central Mount Stinear block. At the footwall of Edward's Pillar, a blastomylonitic zone cuts into the basement gneisses and the Mount Stinear granites, transforming these rocks into banded blastomylonites (Figure 58.16). A well-developed linear growth of biotite and quartz was observed in all the mylonites. Their orientation coincides with the lineation in Menzies metaquartzites and conglomerates at the southern part of Mount Stinear. This fact proves a close temporal relationship between deformation of the low-grade complex and block movements in the crystalline basement.

At the northern side of Mount Maguire (eastern part), sericite-biotite schists and quartzites of Menzies Series are overthrust by an intensely deformed metagabbroidic rock complex (Figure 58.10). The thrust plane strikes east-west and dips 40°–50° to the south (Figure 58.13*j*). The petrographic composition, deformation, and metamorphism of the metagabbroid complex lead to the supposition that this belongs to the basement complex. Underlying sericite-biotite schists are dragged upward to form an east-west-

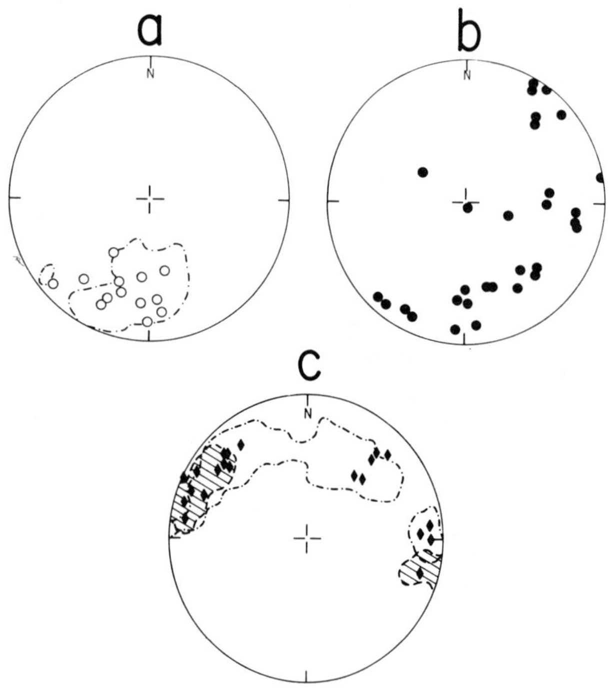

Figure 58.17. Structural diagrams of basic dikes in the low-grade complex and in the basement gneisses (lower hemisphere, equal-area). *a.* Basic dikes in the Menzies Series, southern part of Mount Stinear. Dashed line indicates 3% contour of poles to bedding (see Figure 58.5*a*); *b.* Basic dikes in basement gneisses, central part of Mount Stinear. Dots are poles to bedding; *c.* Basic dikes in the Ruker Series at western side of Mount Ruker. Dashed line indicates 3% contour of poles to bedding (see Figure 58.6*a*); hatched area indicates 3% area of poles to cleavage (see Figure 58.6*c*).

oriented synclinal structure. The thrust plane, displaced by southwest-northeast–striking faults, is marked by a light-colored blastomylonite (30–50 m) which contains tectonic inclusions of metagabbroidic composition. Banding in the metagabbroidic rocks is intensely folded in a zone up to a hundred meters above the thrust plane. The folds (as much as 30 or 40 m in wavelength) are inclined. Their axial planes are mainly parallel to the thrust plane. In the underlying sericite-biotite schists, a second system of east-west oriented, nearly vertically dipping foliation (Schubklüftung of German authors) appears (Figure 58.13*f*, *g*, and *h*). Its intensity increases toward the thrust plane. The parallel trends of the thrust plane, the second foliation, and the fold axes in the metagabbroidic rocks prove a close genetic and temporal relationship between basement thrusting and development of the second foliation.

In general, field observations at Mount Maguire and Mount Stinear demonstrate the existence of tectonic contacts and temporal relationships between folding of the low-grade complex and block movements in the basement.

Structural control of basic dikes

Basic dikes have been found widely distributed in the lower structural stage (see Trail, 1964). Their structural relationships with adjacent rocks at Mount Stinear and Mount Ruker are discussed here in more detail.

At the southwestern part of Mount Stinear slightly metamorphosed basic rocks have been found in the following positions:

1. Sills parallel to bedding or foliation of metaconglomerates, metaquartzites, and biotite-amphibole schists (Figure 58.17*a*);
2. Dikes parallel to *ac* joints;
3. Irregular or s-shaped dikes as fillings of feather joints and similar structures between parallel dikes;
4. Small intrustive bodies with irregular contacts but mostly parallel to general trends of schistosity.

Structural control of dikes in the Menzies Series is in sharp contrast to the much more variable orientation of basic dikes in basement gneisses and granites at the central part of Mount Stinear, which follow, in addition to gneissosity, several sets of preexistent joints (Figure 58.17*b*).

The relative temporal position of the dike intrusion event is evidenced by the presence of dikes in blastomylonite zones which separate the basement gneisses from the Menzies Series at southwestern and central Mount Stinear (Figure 58.9). Otherwise, thin dikes mostly show a weak schistosity connected with linear growth of neogenic amphiboles parallel to quartz and mica lineation of adjacent rocks. These facts prove a late kinematic position of the basic dikes in relation to the deformation and metamorphism of the Menzies Series.

At Mount Ruker (northwestern side) sills and dikes of metagabbro-diabase with a total thickness of more than several hundred meters, are intruded into jaspilites and biotite-sericite-quartz schists and quartzites of the Ruker Series. The lowest of the sills was emplaced along the discontinuity plane between the jaspilite squence and higher biotite-sericite-quartz schists. Sheets of contact-metamorphosed jaspilites occur locally within this sill. Dikes above and below the sill are controlled by well-developed bedding or cleavage in the host rocks (Figures 58.17*c* and 58.18). In the metagabbro-diabase, a rather well-developed joint system is present whose position coincides with that of the cleavage in the adjacent rocks of the Ruker Series, proving that the dike intruded during the final stages of cleavage formation.

At Mount Ruker, structural control of sills and dikes by bedding and cleavage, the presence of fracture-cleavage-like joints, and slight metamorphism place the intrusions of the metagabbro-diabase during the final stages of folding. Their late kinematic position coincides with that of the Mount Stinear basic dikes.

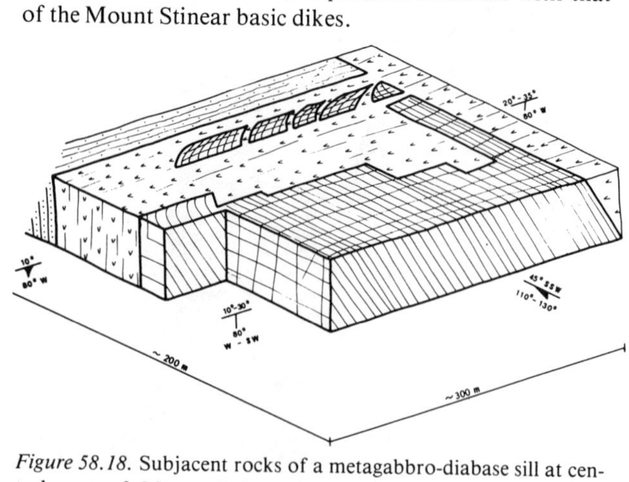

Figure 58.18. Subjacent rocks of a metagabbro-diabase sill at central part of Mount Ruker, northwestern side, 300 m above moraine. Control of sill contacts and dike borders by bedding (110°–130°/45°SSW) and cleavage (10°–30°/80°WSW).

Conclusions

From consideration of tectonic field data and radiometric age determinations (Halpern and Grikurov, 1975), the following stages in the geologic development of the low-grade complex in the southern Prince Charles Mountains are recognized:

1. Denudation of the crystalline basement of late Archean–early Proterozoic age and deposition of the Menzies and Ruker Series.
2. Folding and metamorphism of Menzies and Ruker sedimentary beds connected with basement block movements. During the final stages of folding and metamorphism, intrusion of basic dikes and sills into the basement and low-grade complex, and their subsequent slight metamorphism and deformation (1040–980 m.y.).
3. Denudation of the crystalline basement and the folded metamorphosed Menzies and Ruker sequences, associated with deposition of the Sodruzestvo sedimentary beds, probably restricted to a central, latitudinal area in the low-grade complex (grabensynclinorium according to Soloviev, 1972*a*).
4. Folding and metamorphism of the Sodruzestvo Series (600–500 m.y.).

Tectono-metamorphic and basic magmatic activity was accompanied by intrusions of platform granites 800 and 500 m.y. old. The periods of tectono-metamorphic activity mentioned above could be interpreted as events corresponding with Nimrod and Beardmore orogenesis in the Transantarctic Mountains.

These results allow a subdivision of the East Antarctic shield cover, from a structural point of view, into two types: (1) flat-lying, faulted, and unmetamorphosed sequences, such as the Ritscher Supergroup from Queen Maud Land (Neethling, 1972) and the Sandow Series from the Pravda Coast (Ravich and others, 1965); and (2) the block-folded complex of the southern Prince Charles Mountains metamorphosed under greenschist and stress-accented almandine-amphibolite conditions. These two types reflect different degrees of basement tectonic activity in the East Antarctic shield.

Acknowledgments. The author expresses his appreciation to the late M. G. Ravich for introduction to Antarctic geology, to the late D. S. Soloviev, leader of the 19th SAE field party, and to all of his Soviet colleagues during the 1973–74 field season in the southern Prince Charles Mountains. The Leningrad Research Institute of Arctic and Antarctic Geology provided me with use of some field materials from the 17th and 18th SAE. This research was carried out under the financial support of the Central Institute of Physics of the Earth, Academy of Sciences of the German Democratic Republic.

References cited

Bubnoff, v. S. 1954. *Grundprobleme der Geologie.* Berlin, Akademie-Verlag, 234 pp.

England, R. N., and Langworthy, A. P. 1975. Geological work in Antarctica. *Rec. Bur. Miner. Resour. Geol. Geophys. Aust.,* 1975/30, 19 pp.

Fedorov, L. V., Hofmann, J., and Ravich, M. G. 1977. Ein Vergleich des geologischen Baues Südostindiens einschließlich Sri Lankas und der Ostantarktis (35°–90°E). *Zeitschr. Geolog. Wiss., Berlin, 5,* No. 11–12, 1319–36.

Grikurov, G. E., and Soloviev, D. S. 1974. Geologicheskoe stroenie gornogo obramlenia lednika Lamberta [Geological structure of the mountain fringe of Lambert Glacier]. *Inf. Byull. sov. antarkt. Eksped.,* No. 88, 21–29.

Halpern, M., and Grikurov, G. E. 1975. Rubidium-strontium data from the southern Prince Charles Mountains. *Antarct. Jnl U.S., 10,* No. 1, 9–15.

Hofmann, J. 1978. Tektonische Beobachtungen im hoch- und schwach-metamorphen Präkambrium der Gebirgsumrandung des Lambert-Gletschers (Ostantarktis). *Freiberger Forsch. Hft.. 335,* 5–111.

Iltchenko, L. N. 1972. Late Precambrian acritarchs of Antarctica. (*In* Adie, R. J., *ed. Antarctic Geology and Geophysics.* Oslo, Universitetsforlaget, 599–602.)

Kossygin, A., Jr., Basarin, A. K., Berzin, N. A., and Barchatov, B. P. 1974. *Karta Tektoniki Kontinentov 1:15,000,000.* [Tectonic Map of Continents 1:15,000,000]. Moscow, Izdatel'stvo Nauka, 74 pp.

Neethling, D. C. 1972. Age and correlation of the Ritscher Supergroup and other Precambrian rock units, Dronning Maud Land. (*In* Adie, R. J., *ed. Antarctic Geology and Geophysics.* Oslo, Universitetsforlaget, 547–56.)

Ravich, M. G., Klimov, L. V., and Soloviev, D. S. 1965. *Dokembriia vostocnoi Antarktidy* [The Precambrian of East Antarctica]. Moscow, Izdatel'stvo Nedra, 470 pp.

Ravich, M. G., Soloviev, D. S., and Fedorov, L. V. 1978. *Geologicheskoe stroenie zemli Mak-Robertsona, Vostochnaia Antarktida* [Geology of MacRobertson Land, East Antarctica]. Leningrad, Gidrometeoizdat, 230 pp.

Soloviev, D. S. 1972*a*. Geological structure of the Mountain Fringe of Lambert Glacier and the Amery Ice Shelf. (*In* Adie, R. J., *ed. Antarctic Geology and Geophysics.* Oslo, Universitetsforlaget, 573–77.)

Soloviev, D. S. 1972*b*. Zelenoslancevaia faciia. [Greenschist facies]. (*In* Ravich, M. G., and Kamenev, E. N. *Kristallicheskii fundament Antarkticheskoi platformy* [Crystalline basement of the Antarctic platform]. Leningrad, Gidrometeoizdat, 587–608.)

Soloviev, D. S. 1972*c*. Platform magmatic formations of East Antarctica. (*In* Adie, R. J., *ed. Antarctic Geology and Geophysics.* Oslo, Universitetsforlaget, 531–38.)

Stille, H. 1924. *Grundfragen der vergleichenden Tektonik.* Berlin, Bornträger-Verlag, 443 pp.

Trail, D. S. 1963. Low-grade metamorphic rocks from the Prince Charles Mountains, Antarctica. *Nature, Lond., 197,* 548–50.

Trail, D. S. 1964. Schist and granite in the southern part of Prince Charles Mts. (*In* Adie, R. J., *ed. Antarctic Geology.* Amsterdam, North-Holland Publishing Company, 492–97.)

Wise, D. U. 1963. Keystone faulting and gravity sliding, driven by basement uplift of Owl Creek Mountains, Wyoming. *Bull. Am. Ass. Petrol. Geol., 47,* 586–98.

59

Origin of Charnockitic Rocks of MacRobertson Land

J. W. SHERATON
Bureau of Mineral Resources, P.O. Box 378, Canberra, A.C.T. 2601, Australia

Abstract. The hypersthene granites and associated granodiorites of the Mawson Coast (the Mawson Charnockite) are characterized by a trend leading to extreme K enrichment as shown by a normative Ab-Or-An diagram. They differ from more typical granitic rocks in having considerably higher mafic (hypersthene) contents and plagioclase of more calcic composition (andesine-labradorite). These unusual chemical features are rather typical of charnockitic intrusives, and a two-stage model is proposed for the genesis of such rocks. Initial melting of sialic crustal rocks under "wet" amphibolite-facies conditions would produce granitic liquids and a variety of quartz-syenitic or tonalitic residues, depending on source-rock composition. Further melting of quartz-syenitic residues under dry granulite facies conditions at greater depths in the crust would produce potassic melts. Different degrees of partial melting and retention of various proportions of the more refractory phases (hypersthene and calcic plagioclase) as xenocrysts in the magma could account for the observed chemical variations of the Mawson Charnockite. Hypersthene tonalites (enderbites) could be derived in a similar fashion by melting of more sodic residues.

Granulite-facies, quartzo-feldspathic gneisses from the northern Prince Charles Mountains and the Mawson Coast have apparently been depleted in granitic components by partial melting, and they are richer in more refractory residual materials than amphibolite-facies gneisses from the southern Prince Charles Mountains, which are generally granitic in composition. Rocks similar to the more potassic of these granulite-facies gneisses would be suitable source materials for the charnockitic magmas.

The charnockitic rocks of the Mawson Coast, MacRobertson Land, which crop out over an area of at least 2000 km^2 and intrude mainly quartzofeldspathic gneisses of the granulite facies, were named the Mawson Charnockite by Trail (1970). They include a variety of hypersthene-bearing felsic intrusives, predominantly granites, but also some granodiorites and tonalites (the classification of charnockitic rocks proposed by Streckeisen (1976) is used in this paper). These rocks have been divided by McLeod and others (1966) into porphyroblastic and even-grained charnockite, both of which occur over wide areas. Field relationships and petrology have been described in detail by Crohn (1959), McLeod and others (1966), and Trail (1970).

At Mawson Rock (the type area of the Mawson Charnockite and site of an Australian National Antarctic Research Expedition station), weakly foliated, coarse-grained porphyroblastic hypersthene granite, containing hypersthene (10–17%), quartz (20–25%), andesine-labradorite (25–30%), and porphyroblasts of orthoclase perthite (20–35%), is the predominant rock type. It intrudes even-grained hypersthene granodiorite of similar mineralogy, but with less orthoclase (10–15%). The granodiorite occurs as inclusions in the porphyroblastic granite, which also contains mafic pyroxene-granulite xenoliths and inclusions of pelitic metasedimentary rocks. Both the granite and the granodiorite contain minor clinopyroxene, biotite, and hornblende, and were recrystallized and foliated under granulite-facies conditions, probably soon after their emplacement (Kizaki, 1972). A Rb-Sr isochron age of 1084 ± 37 m.y. (initial $^{87}Sr/^{86}Sr$ ratio, 0.729) has been obtained for the Mawson Charnockite (P. A. Arriens, 1975, personal communication), although a younger thermal event is suggested by a K-Ar age of 535 m.y. (Ravich and Krylov, 1964).

Geochemistry

A total of nineteen samples, including three relatively mafic inclusions (a hypersthene tonalite and two mafic granulites), have been analyzed. Most are from Mawson Rock, but some from Evans and Macey Islands and the Framnes Mountains are also included (Figure 59.1). The analytical results are given in Table 59.1.

Most granite and granodiorite samples are relatively low in SiO_2 (60–67%) and correspondingly high in TiO_2, FeO, MgO, and CaO. The more mafic rocks, in particular, have much higher TiO_2 contents for a given SiO_2 content than typical calc-alkaline igneous rocks (cf. Tarney, 1976) (Figure 59.2). Contents of Zr, Ba, La, and Ce are high, and those of Th, U, and, in some samples, Rb are low compared with those for average granitic rocks (Turekian and Wedepohl, 1961; Taylor, 1968). K/Rb, Ba/Rb, and Th/U ratios are generally high, the latter due mainly to the very low U content which is possibly a result of high-grade metamorphism (see below). In most aspects—high contents of many trace elements, relatively low Na_2O, and $Al/(Na + K + [Ca/2]) > 1.1$—these rocks are comparable to the S-type granitoids of Chappell and White (1974), which were derived by melting of predominantly sedimentary (or metasedimentary) source materials. However, these are either diopside-normative or only slightly corundum-normative, a characteristic of I-type (igneous-derived) granitoids.

The hypersthene granites from Mawson Rock are remarkably similar in chemical composition to the granodiorites, with almost identical trace-element contents, except that the former are considerably higher in K_2O and Rb; the higher normative orthoclase contents are compensated for by slightly lower normative hypersthene and anorthite

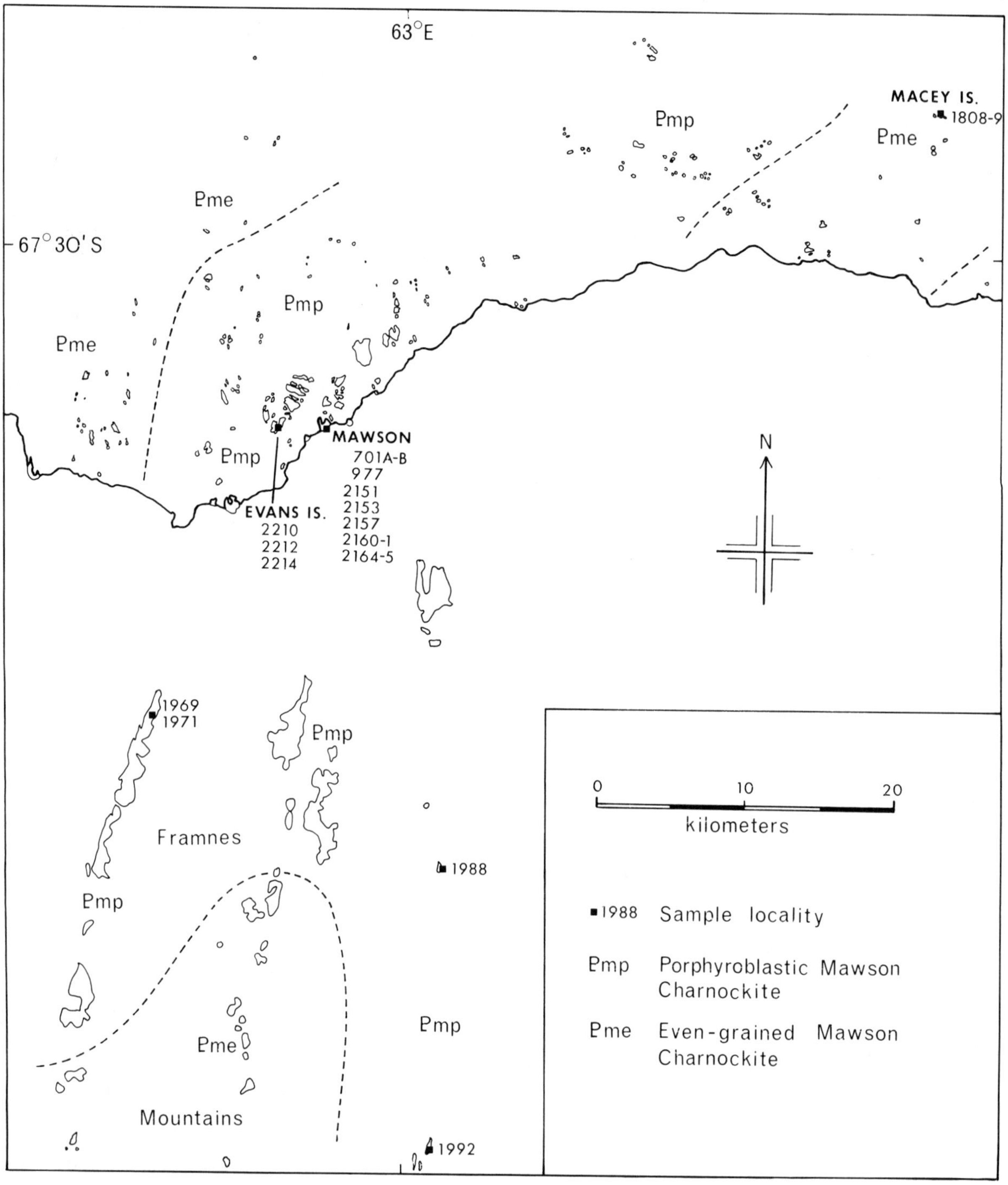

Figure 59.1. Map showing sample localities. Sample numbers below 1000 are prefixed "72280"; those greater than 1000 are prefixed "7328."

(Table 59.1*a* and *b*). Coarse-grained, mostly porphyroblastic hypersthene granites from elsewhere on the Mawson Coast are chemically very similar to those at Mawson Rock, although one specimen from Evans Island is particularly rich in K_2O, Rb, Ba, and Th (Table 59.1*e* and *f*). An even-grained hypersthene granite from Evans Island is even richer in K_2O and is very rich in Rb, La, Ce, Pb, and Th, whereas a fine-grained dike from the same locality is generally similar in composition to the porphyroblastic hypersthene granites from Mawson; Cr, Ni, La, Ce, and Th contents, however, are significantly higher (Table 59.1*g* and *h*). Even-grained hypersthene granodiorites from Macey Island are similar to those at Mawson, although Cr content is higher and Cu, Zr, Nb, Ba, La, and Ce contents are lower (Table 59.1*i*).

The general chemical homogeneity for both major and trace elements of the charnockitic rocks of the Mawson area support the field observations and leave little doubt that they are of magmatic origin, although they have been subsequently deformed and recrystallized. The hypersthene granites, with the exception of those from Evans Island, plot as smooth, igneous-type trends on SiO_2 variation diagrams, and the hypersthene granodiorites differ significantly only in having lower K_2O and Rb, and slightly

Table 59.1. Chemical analyses (in wt %) and C.I.P.W. norms (after Cross and others, 1902) of charnockitic rocks of the Mawson area

	1	2	3	4	5	6	7	8	9
SiO_2	62.7	61.5	57.6	48.8	63.8	63.4	66.0	66.3	65.1
TiO_2	1.67	1.66	2.37	1.33	1.43	0.86	0.91	1.05	0.65
Al_2O_3	14.77	14.83	15.54	15.68	14.50	16.05	14.97	13.57	15.30
Fe_2O_3	0.94	1.11	0.81	1.93	1.42	1.06	0.98	1.08	0.83
FeO	6.61	8.00	10.1	13.1	5.26	4.25	2.95	4.95	5.28
MnO	0.17	0.23	0.23	0.28	0.14	0.09	0.05	0.11	0.13
MgO	2.45	2.71	2.99	8.47	2.03	2.52	1.10	2.59	2.89
CaO	4.68	5.54	6.73	8.07	4.30	3.52	2.63	3.71	5.40
Na_2O	2.07	2.20	2.20	1.35	2.00	2.10	1.85	2.05	2.35
K_2O	3.42	1.83	1.03	0.58	4.01	5.59	6.82	3.40	1.81
P_2O_5	0.42	0.33	0.54	0.12	0.35	0.32	0.36	0.37	0.16
H_2O	0.30	0.33	0.34	0.34	0.52	0.36	0.40	0.40	0.22
Total	100.20	100.27	100.48	100.05	99.76	100.12	99.02	99.68	100.12
					Norms				
Q	21.0	21.5	17.0	0.6	22.7	16.8	21.4	27.6	25.4
C	0.2	—	—	—	—	0.9	0.6	0.7	0.04
Or	20.2	10.8	6.1	3.4	23.7	33.0	40.3	20.1	10.7
Ab	17.5	18.6	18.6	11.4	16.9	17.8	15.7	17.4	19.9
An	20.5	25.2	29.5	35.0	18.8	15.4	10.7	16.0	25.7
Di	—	0.1	0.3	3.5	0.3	—	—	—	—
Hy	15.0	18.2	21.7	40.2	11.3	12.0	5.9	13.1	15.4
Mt	1.4	1.6	1.2	2.8	2.1	1.5	1.4	1.6	1.2
Il	3.2	3.2	4.5	2.5	2.7	1.6	1.7	2.0	1.2
Ap	1.0	0.8	1.3	0.3	0.8	0.8	0.9	0.9	0.4
mg[a]	0.36	0.34	0.33	0.50	0.35	0.46	0.34	0.43	0.46
k[a]	0.52	0.35	0.24	0.22	0.57	0.64	0.71	0.52	0.34
% An in plag.	52	56	60	74	51	45	39	46	55
				Trace elements (ppm)					
V	82	96	127	337	80	95	52	98	94
Cr	31	34	25	366	33	97	15	101	95
Ni	9	10	10	106	12	24	10	29	14
Cu	19	25	40	22	16	20	13	35	4
Zn	84	97	111	143	73	62	41	91	68
Ga	21	22	23	22	20	20	17	21	19
Rb	110	48	22	33	139	246	240	131	77
Sr	164	173	208	92	177	231	225	146	212
Y	37	41	49	35	30	24	23	31	25
Zr	364	340	394	105	369	239	502	383	164
Nb	25	19	24	5	21	12	14	15	10
Ba	1040	918	1290	363	1060	1520	1180	685	579
La	63	63	74	23	64	89	213	114	32
Ce	126	124	156	49	128	167	416	223	69
Pb	19	16	15	6	18	27	47	27	12
Th	3	3	9	2	13	28	133	68	8
U	0.5	0.5	1.0	1.5	1.5	1.5	2.5	1.0	1.0
K/Rb	258	317	386	145	240	189	236	215	195
Rb/Sr	0.67	0.28	0.11	0.36	0.79	1.06	1.07	0.90	0.36
Ca/Y	910	970	980	1650	1020	1050	820	850	1500
Ba/Rb	9.5	19	58	11	7.6	6.2	4.9	5.2	7.5
Th/U	6	6	9	1.3	9	19	53	68	8

1. Average of five porphyroblastic hypersthene granites from Mawson Rock (72280701A, 977; 73282151, 2157, 2161[b]).
2. Average of two even-grained hypersthene granodiorites from Mawson Rock (72280701B, 73282153).
3. Hypersthene tonalite inclusion from Mawson Rock (73282164).
4. Average of two mafic granulite inclusions from Mawson Rock (73282160, 2165).
5. Average of four porphyroblastic hypersthene granites from the Framnes Mountains (73281969, 1971, 1988, 1992).
6. Porphyroblastic hypersthene granite from Evans Island (73282214).
7. Even-grained hypersthene granite from Evans Island (73282210).
8. Hypersthene granite dike from Evans Island (73282212).
9. Average of two even-grained hypersthene granodiorites from Macey Islands (73281808, 1809).

FeO and H_2O analyses by the Australian Mineral Development Laboratories, Adelaide; remaining elements by X-ray fluorescence in the Bureau of Mineral Resources laboratories.

[a] Atomic Mg/(Mg + total Fe) and K/(K + Na), respectively.

[b] Bureau of Mineral Resources registered sample numbers.

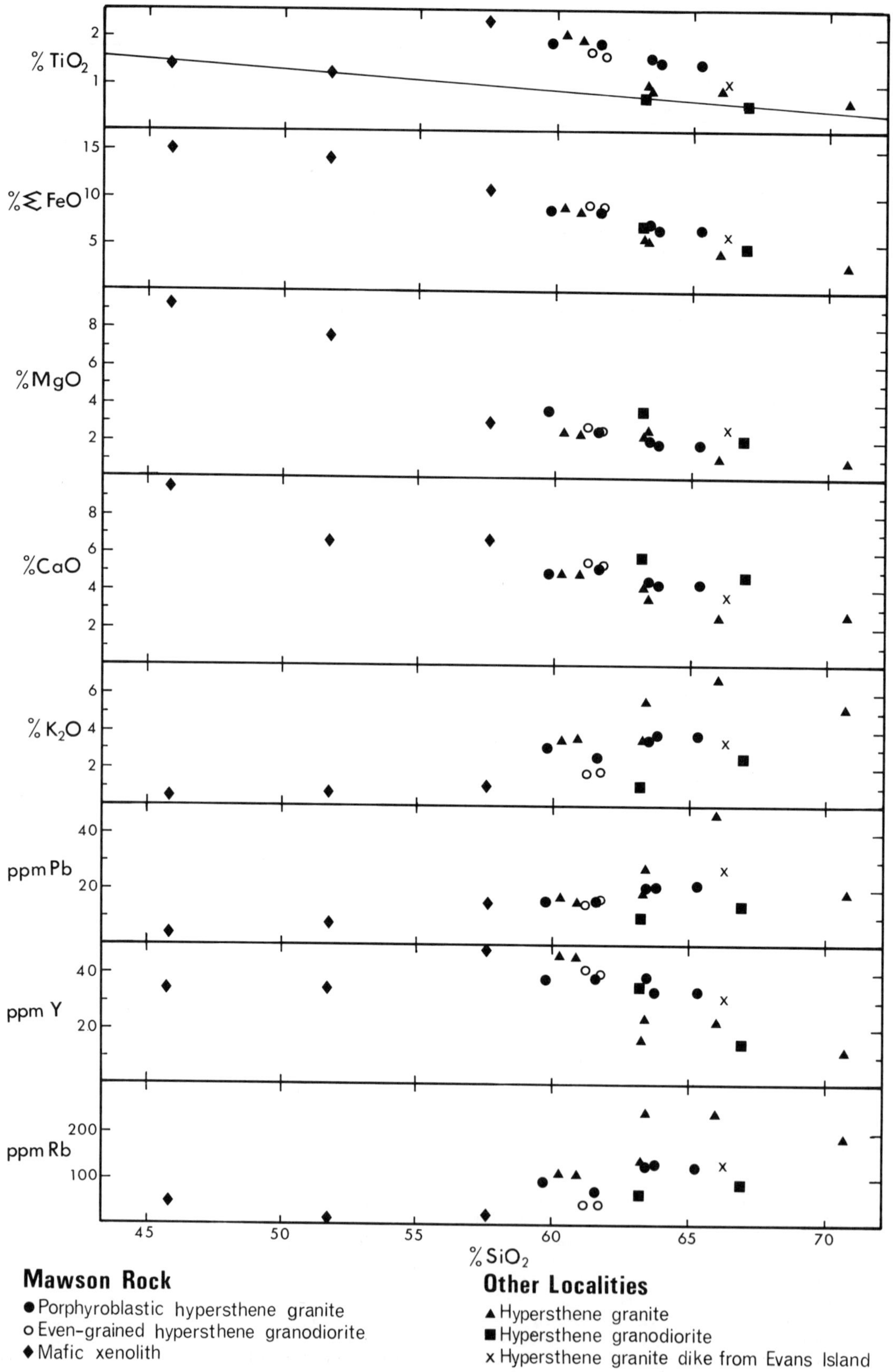

Figure 59.2. SiO_2 variation diagrams for charnockitic rocks. The line on the SiO_2-TiO_2 plot separates the field of calc-alkaline igneous rocks (below line) from that of sedimentary rocks (after Tarney, 1976).

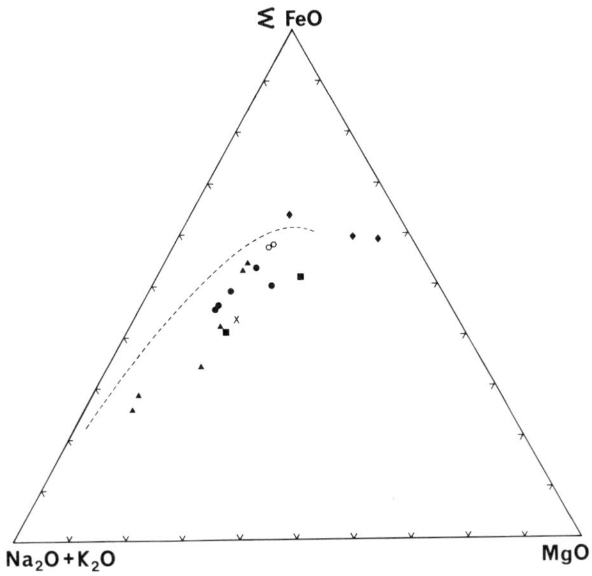

Figure 59.3. FMA (FeO, MgO, Na_2O + K_2O) diagram for charnockitic rocks. The dashed line shows the approximate trend of East Antarctic intrusive charnockitoids (Ravich and Kamenev, 1972). Symbols as in Figure 59.2.

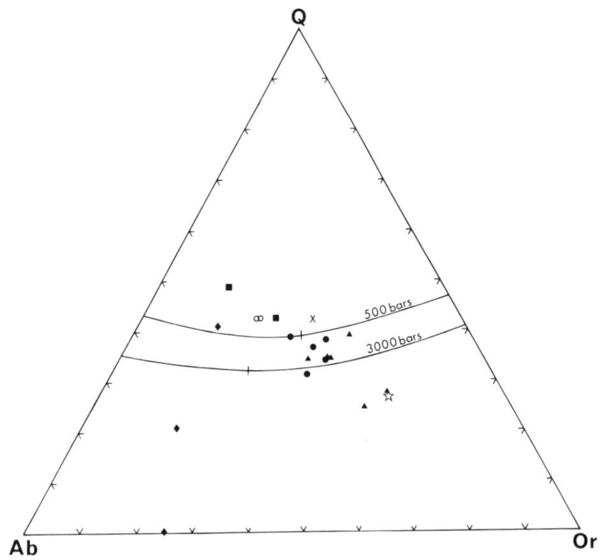

Figure 59.4. Normative Q-Ab-Or diagram for charnockitic rocks showing quartz-feldspar field boundaries at 500 and 3000 bars P_{H_2O}, and positions of quaternary isobaric minima (after Tuttle and Bowen, 1958). Star indicates composition of hypothetical granitic liquid; other symbols as in Figure 59.2.

higher CaO contents (Figure 59.2). The fine-grained, hypersthene tonalite inclusion and two mafic xenoliths also follow the trends for most elements (Table 59.1*c* and *d*). The decrease of Ca/Y with CaO content is rather small compared to typical igneous trends (cf. Lambert and Holland, 1974). Other elements not shown on the variation diagrams show little systematic variation. Cr, Ni, Ga, Sr, Zr, Nb, Ba, La, and Ce contents are remarkably constant, particularly among the specimens from Mawson Rock, although V, Cu, and Zn show a slight increase in the less siliceous rocks. Many of these features may be attributed to an unusual degree of equilibration (homogenization) of the various phases during metamorphism, which would also explain the virtually uniform mg values (atomic Mg/(Mg + Fe)) of the charnockitic rocks from Mawson, in spite of considerable differences in the proportions of hypersthene and other phases (Figure 59.3). Linear trends on FMA diagrams have been established for the granulite-facies gneisses of the northern Prince Charles Mountains and Mawson Coast (Sheraton, unpublished data), as well as for several other high-grade metamorphic complexes (e.g., Sheraton and others, 1973; Lambert, 1976), and these have been attributed to homogenization during metamorphism (Sheraton and others, 1973). The rather iron-rich character of the more mafic phases of the Mawson Charnockite, compared to typical calc-alkaline rocks (cf. Best, 1969), may be explained on this basis. The two mafic xenoliths have relatively high V, Cr, and Ni contents and low Zr, Nb, Ba, La, and Ce contents. These xenoliths, which consist mainly of labradorite/bytownite, hypersthene, and as much as 10% biotite, are not chemically similar to common mafic igneous rocks and possibly represent the more refractory parts of the charnockitic source rocks. The low K/Rb ratios presumably reflect the presence of significant amounts of biotite, which itself usually has a low K/Rb ratio (Lange and others, 1966; Ragland and others, 1968).

Most of the hypersthene granites plot near the quaternary minima at moderate water pressures in the Q-Ab-Or-H_2O system; the almost anhydrous assemblages, however, suggest crystallization under low P_{H_2O} conditions. The hypersthene granodiorites plot farther towards the Q-Ab side of the triangle, although they are unusual in being relatively quartz-rich (Figure 59.4). The trend on the Ab-Or-An diagram is remarkably linear (Figure 59.5) but differs from the normal igneous trend in pointing almost directly toward the Or apex (cf. Irvine and Baragar, 1971). The trend is also unusual in that the hypersthene granodio-

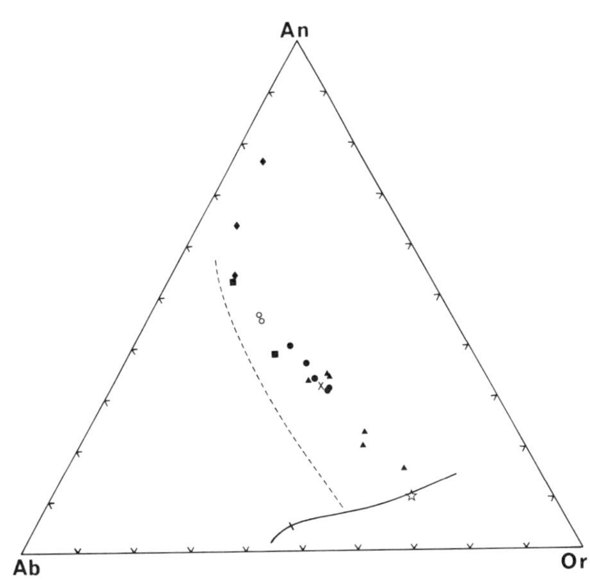

Figure 59.5. Normative Ab-Or-An diagram for charnockitic rocks showing the univariant boundary curve separating the fields of plagioclase and alkali feldspar (solid line) in the quartz-saturated ternary feldspar system at 1000 bars P_{H_2O}, projected onto the Ab-Or-An face of the tetrahedron (after James and Hamilton, 1969). The dashed line shows the approximate trend of East Antarctic intrusive charnockitoids (Ravich and Kamenev, 1972). Star indicates composition of hypothetical granitic liquid; other symbols as in Figure 59.2.

rites from Macey Island are relatively siliceous, but plot well toward the An apex.

The hypersthene granodiorites of the Mawson Coast are mostly similar in composition to some of the average tonalities and granodiorites of Nockolds (1954), but the hypersthene granites are difficult to match chemically with any common igneous rock type outside granulite-facies terranes. They differ from typical granites in the high mafic (hypersthene) and anorthite contents. There is some similarity with high-K diorites (Gulson and others, 1972), except that the Mawson rocks have an even higher K_2O content which suggests syenitic affinities. A high K content seems rather typical of relatively mafic charnockitic rocks in general, many of which may be classified as quartz-pyroxene syenites (Carmichael and others, 1974) or "granosyenites" (Ravich and Kamenev, 1972), although more sodic varieties (hypersthene granodiorites and tonalites) are also common. Rocks of composition comparable to that of the charnockitic rocks of the Mawson area are found among the Charnockite Series of Madras (Howie, 1955) and the mangerites of southern Quebec (Philpotts, 1966), as well as over wide areas of the East Antarctic basement (Ravich and Kamenev, 1972). Similar trends toward strong K enrichment, as shown by the Ab-Or-An diagram, are found in the Madras charnockites (Howie, 1955) and the charnockitic rocks of British Guiana (Singh, 1966). The former also show marked Fe enrichment, similar to that of the Mawson Charnockite. Data on other East Antarctic charnockitic rocks given by Ravich and Kamenev (1972) indicate a lesser degree of K enrichment, but stronger Fe enrichment, compared to the Mawson Charnockite (Figures 59.3 and 59.5).

Table 59.2. Effect of partial melting on hypersthene granite from Mawson Rock. (Minor components are omitted.)

	a	b	c	d	e
Q	22.3	22.8	21.4	18.3	20.8
Or	21.4	11.4	12.5	6.6	7.1
Ab	18.6	19.7	18.5	20.0	18.4
An	21.8	26.7	26.4	31.8	29.2
Hy	15.9	19.3	21.2	23.4	24.5

a. Average normative composition of five hypersthene granites from Mawson Rock.
b. Average normative composition of two hypersthene granodiorites from Mawson Rock.
c. Residue after subtraction of 25% of melt of composition $Q_{25}Or_{48}Ab_{19}An_8$ from the average composition given in column a.
d. Normative composition of hypersthene tonalite inclusion in granite from Mawson Rock.
e. Residue after subtraction of 35% of melt of composition $Q_{25}Or_{48}Ab_{19}An_8$ from column a.

Petrogenesis

Although the hypersthene granites and granodiorites have many chemical features in common, it is difficult to explain the higher K_2O content of the former in terms of a magmatic differentiation series which would normally lead to an accompanying increase in SiO_2. Trail (1970) has suggested that the porphyroblastic hypersthene granite may have been derived by K metasomatism of rocks similar to the even-grained hypersthene granodiorites. Although K metasomatism on a scale of a few meters may have occurred locally, the sharp contacts between the two types (at least at Mawson) and their general homogeneity effectively preclude large-scale, *in situ* metasomatism. Soviet authors have pointed out the objections to regarding intrusive charnockitic rocks as forming a magmatic differentiation series—the consistently high TiO_2 and FeO, and particularly high K_2O contents of the more alkaline types. They favor an origin by granitization and alkali metasomatism of more mafic intrusives, commonly, but not necessarily, accompanied by remobilization (Klimov and others, 1964; Ravich, 1972; Ravich and Kamenev, 1972). The source of the vast quantities of potassium necessary for large-scale metasomatism presents a formidable problem, and the preservation in charnockitic massifs of xenoliths which have undergone only minimal change of composition (Ravich and Kamenev, 1972 [English 1975 ed., p. 511]) is difficult to explain on such a basis. Consequently, alkali metasomatism is considered here to be confined to relatively minor redistribution of K and Na within metamorphic terranes, and thus not important in the formation of charnockitic rocks in general. Instead, a process involving anatexis of lower crustal rocks, which avoids some of these difficulties, is proposed.

Much of the chemical variation of the charnockitic rocks (both hypersthene granites and granodiorites) near Mawson Station can be explained on the basis of mixtures of various proportions of potassic, granitic liquid with crystals of hypersthene and calcic plagioclase. For example, a composition equivalent to the hypersthene granodiorite could be derived from the hypersthene granite (Table 59.2*a*) by the subtraction of 25% of a melt of composition Q_{25}-Or_{48}-Ab_{19}-An_8 (Table 59.2*b* and *c*). Similarly, extraction of 35% of such a liquid would give a composition close to that of the hypersthene tonalite inclusion (Table 59.2*d* and *e*). In reality, of course, the granodiorite and tonalite magmas would probably have been formed by more extensive partial melting than that for the granitic magmas. Magma of granodiorite composition (Table 59.2*b*) would be only about 25% melt if the melt was of the same composition as the hypothetical granitic liquid. Low degrees of partial melting of rocks with a wide range of compositions would tend to produce K-rich liquids, whereas more extensive melting would give relatively sodic magmas (e.g., James and Hamilton, 1969). However, the incorporation of significant proportions of probable xenocrystic minerals in even the most potassic phases of the Mawson Charnockite suggests that moderately extensive melting of K-rich (possibly quartz-syenitic) parent materials was involved. Early breakdown of any biotite in the source rocks during anatexis would also lead to some enrichment of K in the melt and would provide water (von Platen, 1965; Brown and Fyfe, 1972; Winkler, 1974). The liquid used in the calculations is necessarily rather potassic but is very similar in composition to the hypersthene granites from Evans Island (Table 59.3). Much of the normative hypersthene and anorthite in the Evans Island granites is thought to have been present in xenocrystic phases in the magma, and these components have therefore been omitted from the table.

Although these calculations are undoubtedly an oversimplification, they do indicate that a process involving different degrees of partial melting and separation of K-rich liquid from the more refractory phases (xenocrysts of calcic plagioclase and hypersthene) and possibly from quartz, perhaps by crystal settling or filter-press action,

Table 59.3. Comparison of normative quartz, orthoclase, and albite contents of hypothetical granitic liquid with those of hypersthene granites from Evans Island (recalculated to 100%)

	1	2	3
Q	27.1	27.6	24.8
Or	52.2	52.1	48.8
Ab	20.7	20.3	26.3

1. Hypothetical granitic liquid (calculated An-free).
2. Even-grained hypersthene granite from Evans Island (73282210).
3. Porphyroblastic hypersthene granite from Evans Island (73282214).

could explain many of the observed chemical trends of the Mawson Charnockite. In addition, variations of source-rock composition and some degree of fractionation were probably also important. The fairly uniform plagioclase compositions in both the hypersthene granites and granodiorites are compatible with this hypothesis, although recrystallization subsequent to intrusion would have effectively eliminated any zoning. The relatively uniform mg values may be a consequence of derivation from metamorphosed source rocks of fairly uniform mg number, and the relatively Fe-rich composition of the more mafic charnockitic rocks compared to typical calc-alkaline rocks may merely reflect this factor.

It remains to explain the rather unusual source-rock compositions which appear to be necessary for the large-scale production of highly potassic liquids. De Vore (1975) has shown that even fairly extensive partial melting of graywackes and other geosynclinal sedimentary rocks can leave residues with considerable amounts of alkali feldspar; some residues have syenitic compositions, although in his examples the graywackes were relatively sodic. Frith and Currie (1976) have demonstrated that partial melting of a wide range of rock types with an excess of normative Or over Ab will produce residues enriched in the Or component and trending toward monzonite or quartz syenite in composition. More sodic rocks would leave granodioritic or tonalitic residues after anatexis. Further melting of quartz-syenitic residues could form granitic magmas of compositions similar to those of the hypersthene granites of the Mawson area. Such a two-stage model is considered to explain the "residual" chemical features—relatively high K/Rb and Ba/Rb, and low Rb/Sr ratios—and the high normative hypersthene and anorthite contents of the charnockitic rocks of the Mawson area more adequately than a single episode of extensive partial melting. The hypersthene granites from Evans Island, although more fractionated than those from elsewhere in the Mawson area, still have relatively unfractionated ratios compared to more typical granitic rocks.

Some support for such a hypothesis comes from analytical data on quartzo-feldspathic gneisses from the Prince Charles Mountains and Mawson Coast (Sheraton, unpublished data). Archean to Proterozoic layered, amphibolite-facies gneisses from the southern Prince Charles Mountains have a relatively restricted range of composition and tend to group near the ternary minima in the granite system (Figure 59.6), whereas many of the Proterozoic granulite-facies gneisses from the northern Prince Charles Mountains and Mawson Coast are relatively quartz- or K-feldspar-rich, with granitic compositions

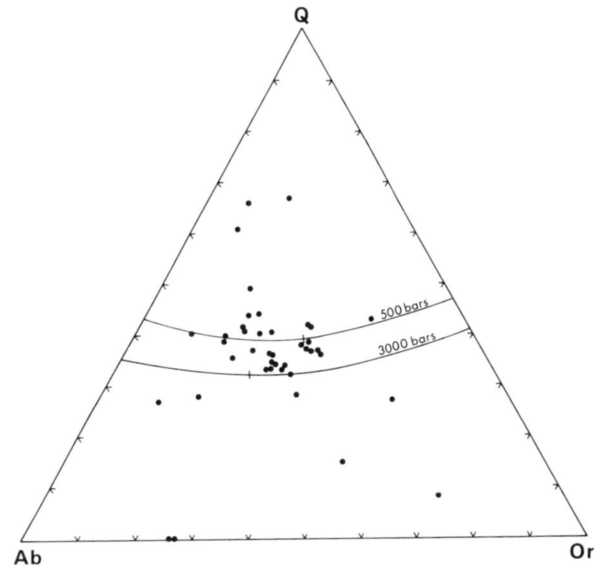

Figure 59.6. Normative Q-Ab-Or diagram for amphibolite-facies quartzo-feldspathic gneisses from the southern Prince Charles Mountains.

much less common (Figure 59.7). The higher grade terranes have presumably been depleted in a granitic component by partial melting and tend to be richer in more refractory rock types, although there may also have been original differences in composition. The more potassic of these "residual" rocks should yield K-rich liquids after anatexis.

In fact, the charnockitic rocks appear to have an even more refractory chemistry (i.e., higher normative anorthite and hypersthene) than the granulite-facies gneisses which they intrude. The field of the gneisses on an Ab-Or-An diagram is bounded on one side by the charnockitic rocks, with virtually all of the analyzed gneisses plotting on the Ab-Or side of the charnockitic trend (Figure 59.8). Table 59.4 compares Th and U contents and Th/U and K/Rb ratios of gneisses of the amphibolite and granulite facies

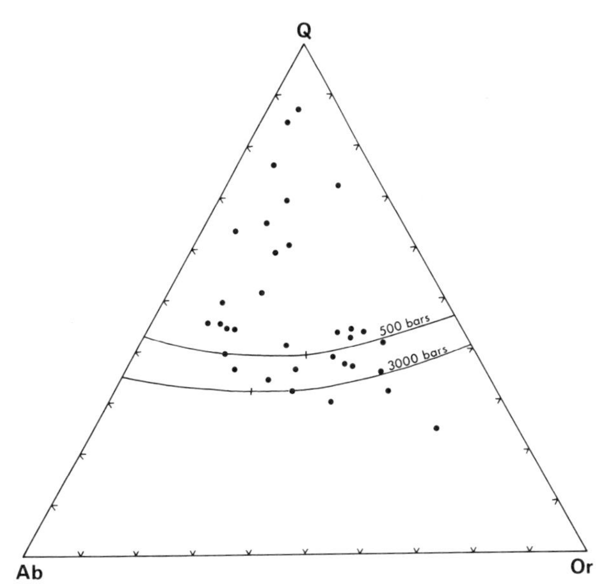

Figure 59.7. Normative Q-Ab-Or diagram for granulite-facies quartzo-feldspathic gneisses from the northern Prince Charles Mountains and Mawson Coast.

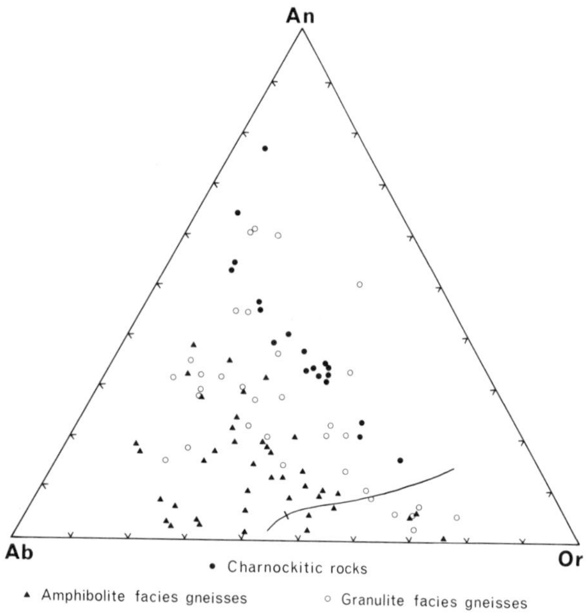

Figure 59.8. Normative Ab-Or-An diagram comparing charnockitic rocks with amphibolite- and granulite-facies gneisses. The univariant boundary curve separating the fields of plagioclase and alkali feldspar is shown by the solid line.

with those of the charnockitic rocks. The charnockitic rocks, with the exception of three samples from Evans Island with anomalously high Th contents, are low in both Th and U. The granulite-facies gneisses appear to have been depleted only in U relative to the amphibolite-facies rocks, and hence have higher Th/U ratios. Lambert and Heier (1968) found that U is depleted relative to Th and K in moderate grades of metamorphism, whereas both Th and U are depleted in medium- to high-pressure granulites; and Kalsbeek (1976) showed that high-grade Archean gneisses of West Greenland are depleted in U relative to Th. There is no evidence for depletion of Rb relative to K in the high-grade gneisses, although the charnockitic rocks have relatively high K/Rb ratios. Lambert and Heier (1968) and Sighinolfi (1971) found that such depletion is characteristic only of medium to high-pressure granulites. The relatively unfractionated chemistry of the charnockitic rocks is thus consistent with an origin in partial melting of depleted, granulite-facies gneisses near the base of the crust. There is abundant geologic and geophysical evidence that the lower continental crust is made up of a complex series of igneous and metamorphic rocks of roughly intermediate composition (Smithson and Brown, 1977).

Table 59.4. Comparison of charnockitic rocks with amphibolite and granulite-facies gneisses (Th and U contents in ppm)

	Th	U	Th/U	K/Rb
Average of 41 amphibolite-facies gneisses	21.9	3.8(2.7[a])	5.8(8.0[a])	228
Average of 32 granulite-facies gneisses	24.7	1.9	13.1	156
Average of 14 charnockitic rocks[b]	7.0	0.9	7.8	250

[a] Excluding two high-U values (22 and 27 ppm).
[b] Excluding three high-Th samples from Evans Island and two mafic granulite xenoliths from Mawson Rock.

Conclusions

A two-stage model is proposed for the formation of the Mawson Charnockite which may also be applicable to the genesis of charnockitic rocks in general. An initial phase of melting of sialic crustal rocks under "wet" amphibolite-facies conditions would produce liquids with low K/Rb and Ba/Rb, and high Rb/Sr ratios. The products of such anatexis—pegmatites and granitic rocks—are particularly common in amphibolite-facies terrane, whereas at greater depths in the crust, corresponding to the granulite facies, rather more refractory rocks, somewhat depleted in granitic components and partly of quartz-syenitic composition, are thought to predominate. A second phase of melting of these rocks would produce liquids with various degrees of K enrichment or depletion (charnockitic or enderbitic), depending on the initial composition and extent of melting. Temperatures adequate for the production of charnockitic melts may well be reached under the conditions of the granulite facies even when $P_{H_2O} \ll P_{LOAD}$ (Frith and Currie, 1976), particularly since liquidus temperatures would be significantly lower if much of the plagioclase and hypersthene were not dissolved in the magma during intrusion (Whitney, 1975; White and Chappell, 1976). Indeed, many of the observed chemical features of the Mawson Charnockite can best be explained if hypersthene and calcic plagioclase were present as xenocrysts in the magma because of only partial separation of the liquid from these more refractory phases.

Acknowledgments. I am indebted to J. G. Pyke for the X-ray fluorescence analyses and to W. B. Dallwitz, J. Ferguson, A. Y. Glikson, and R. J. Tingey for critically reading the manuscript. Logistic support in the field was provided by the Antarctic Division, Department of Science, Melbourne. This paper is published with the permission of the director of the Bureau of Mineral Resources, Australia.

References cited

Best, M. G. 1969. Differentiation of calc-alkaline magmas. (*In* McBirney, A. R., *ed.* Proceedings of Andesite Conference, Oregon, 1968. *Bull. St. Ore., 65,* 65–75.)

Brown, G. C., and Fyfe, W. S. 1972. The transition from metamorphism to melting: status of the granulite and eclogite facies. *24th Int. Geol. Congr., Montreal, 1972,* Sec. 2, 27–34.

Carmichael, I. S. W., Turner, F. J., and Verhoogen, J. 1974. *Igneous Petrology.* New York, McGraw-Hill, 739 pp.

Chappell, B. W., and White, A. J. R. 1974. Two contrasting granite types. *Pacif. Geol., 8,* 173–74.

Crohn, P. W. 1959. A contribution to the geology and glaciology of the western part of Australian Antarctic Territory. *Bull. Bur. Miner. Resour. Aust., 52,* 103 pp.

Cross, W., Iddings, J. P., Pirsson, L. V., and Washington, H. S. 1902. A quantitative chemico-mineralogical classification and nomenclature of igneous rocks. *J. Geol., 10,* No. 6, 555–690.

De Vore, G. W. 1975. The role of partial melting of metasediments in the formation of the anorthosite-norite-syenite complex, Laramie Range, Wyoming. *J. Geol., 83,* No. 6, 749–62.

Frith, R. A., and Currie, K. L. 1976. A model for the origin of the Lac St. Jean anorthosite massif. *Can. Jnl Earth Scis, 13,* No. 2, 389–99.

Gulson, B. L., Lovering, J. F., Taylor, S. R., and White, A. J. R. 1972. High-K diorites, their place in the calc-alkaline association and relationship to andesites. *Lithos, 5,* No. 3, 267–79.

Howie, R. A. 1955. The geochemistry of the charnockite series of Madras, India. *Trans. R. Soc. Edinb., 62,* Part 3, 725–68.

Irvine, T. N., and Baragar, W. R. 1971. A guide to the chemical classification of the common volcanic rocks. *Can. Jnl Earth Scis, 8,* No. 5, 523–48.

James, R., and Hamilton, D. 1969. Phase relations in the system $NaAlSi_3O_8$–$KAlSi_3O_8$–$CaAl_2Si_2O_8$–SiO_2 at 1 kilobar water vapour pressure. *Contr. Miner. Petrol., 21,* No. 2, 111–41.

Kalsbeek, F. 1976. Metamorphism of Archaean rocks of West Greenland. (*In* Windley, B. F., *ed. The Early History of the Earth.* London, Wiley, 225–35.)

Kizaki, K. 1972. Sequence of metamorphism and deformation in the Mawson Charnockite of East Antarctica. (*In* Adie, R. J., *ed. Antarctic Geology and Geophysics.* Oslo, Universitetsforlaget, 527–30.)

Klimov, L. V., Ravich, M. G., and Soloviev, D. S. 1964. Charnockites of East Antarctica. (*In* Adie, R. J., *ed. Antarctic Geology.* Amsterdam, North-Holland Publishing Company, 455–62.)

Lambert, I. B., and Heier, K. S. 1968. Geochemical investigations of deep-seated rocks in the Australian Shield. *Lithos, 1,* No. 1, 30–53.

Lambert, R. St. J. 1976. Amîtsoq gneiss geochemistry: preliminary observations. (*In* Windley, B. F., *ed. The Early History of the Earth.* London, Wiley, 191–201.)

Lambert, R. St. J., and Holland, J. G. 1974. Yttrium geochemistry applied to petrogenesis utilizing calcium-yttrium relationships in minerals and rocks. *Geochim. cosmochim. Acta, 38,* No. 9, 1393–1414.

Lange, I. M., Reynolds, R. C., and Lyons, J. B. 1966. K/Rb ratios in coexisting K-feldspars and biotites from some New England granites and metasediments. *Chem. Geol., 1,* No. 4, 317–22.

McLeod, I. R., Trail, D. S., Cook, P. J., and Wallis, G. R. 1966. Geological work in Antarctica, January–March, 1965. *Rec. Bur. Miner. Resour. Geol. Geophys. Aust.* 1966/9 [Unpublished] 71–78.

Nockolds, S. R. 1954. Average chemical compositions of some igneous rocks. *Bull. geol. Soc. Am., 65,* No. 10, 1007–32.

Philpotts, A. R. 1966. Origin of the anorthosite-mangerite rocks in southern Quebec. *J. Petrology, 7,* Part 1, 1–64.

Ragland, P. C., Billings, C. K., and Adams, J. A. S. 1968. Magmatic differentiation and autometasomatism in a zoned granitic batholith from central Texas, U.S.A. (*In* Ahrens, L. H., *ed.* Origin and distribution of the elements. *Int. Ser. Monogr. Earth Sci., 30,* 795–823.)

Ravich, M. G. 1972. The charnockite problem. (*In* Adie, R. J., *ed. Antarctic Geology and Geophysics.* Oslo, Universitetsforlaget, 523–26.)

Ravich, M. G., and Kamenev, E. N. 1972. *Kristallicheskii fundament Antarkticheskoi platformy* [Crystalline basement of the Antarctic platform]. Leningrad, Gidrometeoizdat, 658 pp. [English translation: New York, John Wiley, 1975, 582 pp.]

Ravich, M. G., and Krylov, A. Ya. 1964. Absolute ages of rocks from East Antarctica. (*In* Adie, R. J., *ed. Antarctic Geology.* Amsterdam, North-Holland Publishing Company, 579–89.)

Sheraton, J. W., Skinner, A. C., and Tarney, J. 1973. The geochemistry of the Scourian gneisses of the Assynt district. (*In* Park, R. G., and Tarney, J., *eds. The Early Precambrian of Scotland and Related Rocks of Greenland.* Newcastle, England, University of Keele, 13–30.)

Sighinolfi, G. P. 1971. Investigations into deep crustal levels: fractionating effects and geochemical trends related to high-grade metamorphism. *Geochim. cosmochim. Acta, 35,* No. 10, 1005–21.

Singh, S. 1966. Orthopyroxene-bearing rocks of charnockitic affinities in the South Savanna–Kanuku Complex of British Guiana. *J. Petrology, 7,* Part 2, 171–92.

Smithson, S. B., and Brown, S. K. 1977. A model for lower continental crust. *Earth planet. Sci. Lett., 35,* No. 1, 134–44.

Streckeisen, A. L. 1976. To each plutonic rock its proper name. *Earth Sci. Rev., 12,* No. 1, 1–33.

Tarney, J. 1976. Geochemistry of Archaean high-grade gneisses, with implications as to the origin and evolution of the Precambrian crust. (*In* Windley, B. F., *ed. The Early History of the Earth.* London, Wiley, 405–17.)

Taylor, S. R. 1968. Geochemistry of andesites. (*In* Ahrens, L. H., *ed.* Origin and distribution of the elements. *Int. Ser. Monogr. Earth Sci., 30,* 559–83.)

Trail, D. S. 1970. ANARE 1961 geological traverses on the MacRobertson Land and Kemp Land Coast. *Rep. Bur. Miner. Resour. Geol. Geophys. Aust., 135,* 32 pp.

Turekian, K. K., and Wedepohl, K. H. 1961. Distribution of the elements in some major units of the Earth's crust. *Geol. Soc. Am. Bull., 72,* No. 2, 175–92.

Tuttle, O. F., and Bowen, N. L. 1958. Origin of granite in the light of experimental studies in the system $NaAlSi_3O_8$-$KAlSi_3O_8$-SiO_2-H_2O. *Mem. geol. Soc. Am., 74,* 153 pp.

von Platen, H. 1965. Experimental anatexis and genesis of migmatites. (*In* Pitcher, W. S., and Flynn, G. W., *eds.* Controls of metamorphism. *Spec. Issue Geol. Jnl,* No. 1, 203–18.)

White, A. J. R., and Chappell, B. W. 1976. Ultrametamorphism and granitoid genesis. *Abstr. 25th Int. Geol. Congr., Sydney, 1976, 3,* 674–75.

Whitney, J. A. 1975. The effects of pressure, temperature, and X_{H_2O} on phase assemblage in four synthetic rock compositions. *J. Geol., 83,* No. 1, 1–31.

Winkler, H. G. F. 1974. *Petrogenesis of Metamorphic Rocks.* 3rd ed. Berlin, Springer-Verlag, 320 pp.

60

Geologic Structure of MacRobertson Land and Princess Elizabeth Land, East Antarctica

M. G. RAVICH† and L. V. FEDOROV
Department of Antarctic Geology, SEVMORGEO, Moyka 120, Leningrad 190121, U.S.S.R.

Abstract. During 1971–74 Soviet geologists and geophysicists carried out a reconnaissance and partly detailed survey within areas of MacRobertson Land and Princess Elizabeth Land, totaling about 500,000 km². As a result, a geologic map covering an area of 150,000 km² was compiled.

The surveyed area is 85% covered by ice, and only 15% consists of exposed rocks ranging in age from Archean to Permian. The crystalline basement of the East Antarctic craton dominating this region is composed of Archean metamorphic rocks, as much as 20 km thick, that were formed under granulite and superimposed amphibolite facies conditions. Ultrametamorphic enderbites, charnockites, and shadow granites are widespread, along with migmatites, gneisses, and amphibolites. Metamorphic rocks are cut by upper Archean to lower Proterozoic gneissic granites and rheomorphic charnockitoids (porphyroblastic granosyenites).

In an aulacogen-type structure superimposed upon the crystalline basement, a folded Proterozoic complex, not less than 8 km thick, was formed; it is thought to have been initially metamorphosed to greenschist facies and subsequently to epidote-amphibolite and amphibolite facies. As a result, zonal metamorphic rocks developed. A jaspilite-bearing unit occurs at the base of the sequence. Chlorite-mica schists, quartzites, and conglomerates predominate, and they are cut by numerous metabasite sills and dikes, and in some localities by mica granites and pegmatites.

A rift zone traverses the whole region from north to south. A coal-bearing Permian formation of Gondwanian type was found, with an observable thickness of 1300 m, which is very similar to the Beacon Supergroup and is probably associated with the initial stage of development of the rift zone in the Transantarctic Mountains. This formation is cut by dikes of Cretaceous alkaline-ultramafic lamprophyres.

The ice-free ranges within MacRobertson Land and Princess Elizabeth Land, including the Prince Charles Mountains, extend for more than 600 km southward from the mainland coast. The exposed part of the Antarctic craton has a very complex geology (Crohn, 1959; Soloviev, 1972; Grikurov and Soloviev, 1974). Archean crystalline basement rocks predominate, underlying an area of 130,000 km². Proterozoic metasedimentary rocks are subordinate; they occupy a graben-like synclinorium that formed on the site of an early Paleozoic aulacogen. The upper Paleozoic cover sequence is exposed only in the small Beaver Lake area, and the unexposed Mesozoic–Cenozoic unconsolidated cover is believed to fill a giant longitudinal rift graben extending for more than 500 km (Figure 60.1). Glacial moraines are scattered over the entire territory.

Archean crystalline basement

The studied areas of MacRobertson Land and Princess Elizabeth Land consist mainly of dome-shaped structures of polymetamorphic schists and shadow granitoids of the amphibolite facies. The cores of the dome-shaped structures are characterized by major brachiform folds complicated by a few disharmonic minor folds. Strong linear folds, commonly overturned and complicated by boudinage structures, are typical of the outer dome zones. The trends of main fold axes vary from sublatitudinal to northeasterly; the dip angles are from 20° to 45° in the cores, and from 70° to 80° on the flanks.

Two units are recognized in the dome-shaped structures: the Reinbolt Series* and the Larsemann Series. The Reinbolt Series consists mainly of ultrametamorphic enderbites and charnockites, interlayered with pyroxene-plagioclase schists. Biotite-garnet plagiogneisses with migmatites and shadow granites are subordinate, along with minor layers of quartzites and calciphyres. This series comprises mainly the dome cores, and apparently it is the most ancient in this region. The Larsemann Series is predominantly composed of biotite-garnet and sillimanite-cordierite-garnet plagiogneisses with migmatites and shadow granites; these gneisses include subordinate beds of pyroxene-plagioclase schist and a few enderbite and charnockite bands. Calciphyre beds (mainly phlogopite-forsterite) are more abundant than in the Reinbolt Series.

As the Larsemann Series usually makes up the outer parts of the domes, it is regarded as younger than the Reinbolt Series; however, the major metamorphic event, possibly of middle Archean age, was presumably the same one for both these series. Their total thickness is close to 10 km. The Reinbolt and Larsemann Series are correlative to the Humboldt Series (Complex) from Queen Maud Land, which is regarded as upper Archean (Ravich and Kamenev, 1972; Ravich, Chapter 51, this volume).

† Deceased.

* Following Russian usage, *series* here means a lithostratigraphic unit formed during a period or longer.

The youngest rocks of the Antarctic craton crystalline basement are represented by the Lambert Series. These rocks have been metamorphosed in an amphibolite facies environment, near the granulite facies boundary. They are concentrated mainly in the southern part of the region. The Lambert Series is 4–5 km thick; it is composed of biotite, garnet-biotite, and amphibole plagiogneisses (locally amphibolites) and shadow granites with migmatites (Tingey, 1972). Quartzites and calciphyres (tremolite and diopside) are subordinate, but they are more abundant here than in the older series. The fold pattern of the Lambert Series is highly variable owing to widespread minor folding, which complicated the isoclinal, linear folds. This pattern is caused by intense migmatization and granitization processes, and they are also responsible for the abundant flowage and drag folds. The Lambert Series in Prince Charles Mountains is similar to the Insel Series (Complex) in Queen Maud Land (Ravich and Kamenev, 1972). Like in

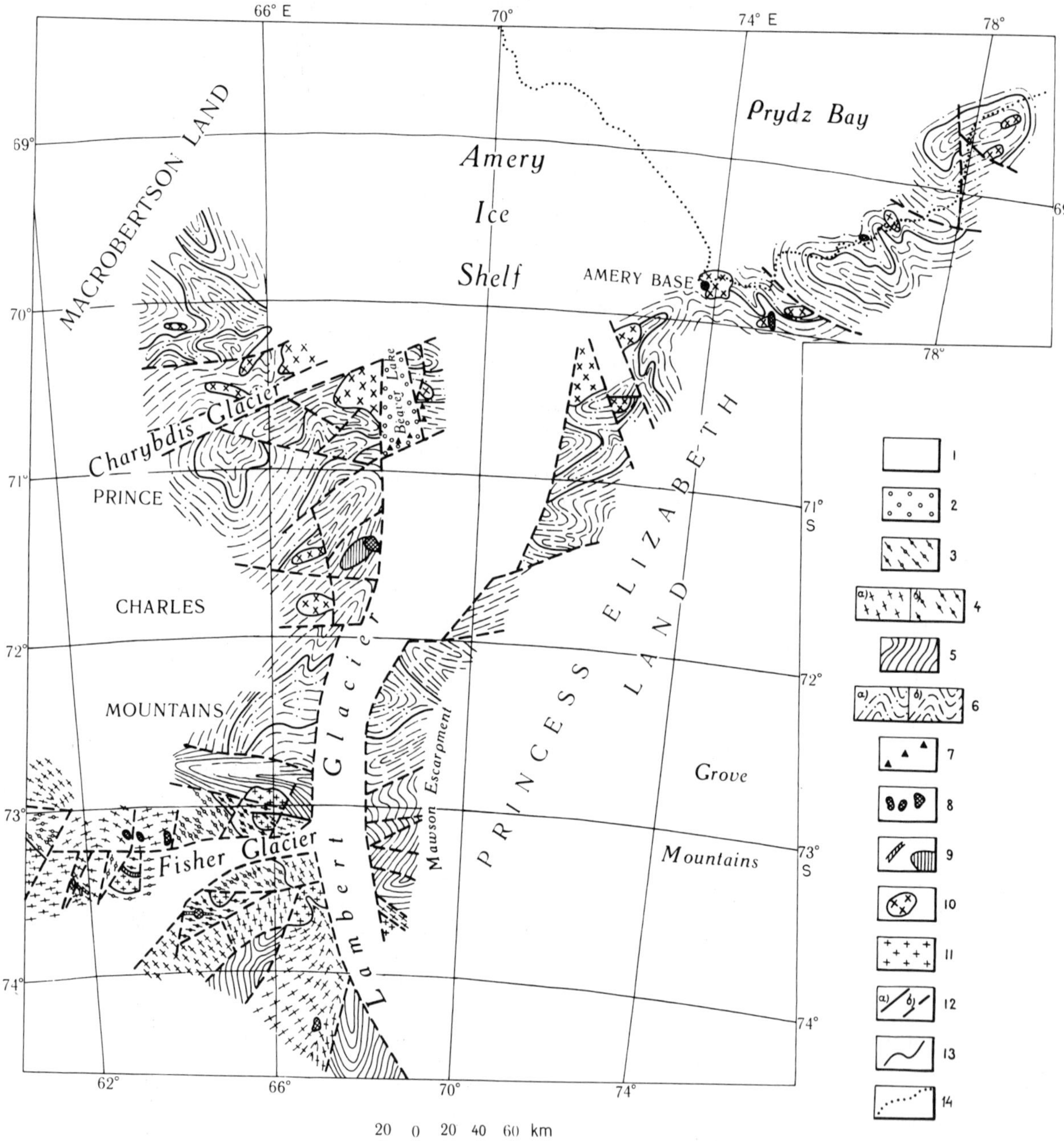

Figure 60.1. Geologic map of MacRobertson and Princess Elizabeth Lands.

1. Continental ice; 2. Permian conglomerates, sandstones, siltstones, claystones, and coals; 3. Upper Proterozoic–lower Paleozoic Sodruzhestvo Series; 4. Lower–middle Proterozoic: (a) Menzies Series and (b) Ruker Series; 5. Upper Archean Lambert Series; 6. Archean: (a) Larsemann Series and (b) Reinbolt Series; 7. Upper Mesozoic (mainly Lower Cretaceous) sills and dikes of alkaline–ultramafic lamprophyres; 8. Lower Paleozoic two-mica and biotite granites; 9. Upper Proterozoic metabasites; 10. Lower Proterozoic charnockitoids; 11. Upper Archean gneissic granites; 12. Faults: (a) observed and (b) inferred; 13. Geologic contacts; 14. Ice barrier.

the latter area, the rocks of this type of metamorphism indicate the most mobile zones within the crystalline basement.

The regional metamorphism responsible for formation of the Lambert Series was the final event in the consolidation of the Antarctic craton crystalline basement; it took place possibly at the end of Archean and/or beginning of Proterozoic time. From this time onwards, the crystalline basement behaved as a rigid structure, reacting to tectonic movements by fault and block tectonics. Subsiding block movements along the major faults caused the formation of Proterozoic grabens which were subsequently filled with terrigenous deposits derived from the disintegrated crystalline basement, mainly the Lambert Series. In connection with different stages of Proterozoic tectonic activity, various intrusive rocks were developed, of which rheomorphic charnockites within the Archean rocks were the earliest; later some granitic and mafic intrusions were formed. During Phanerozoic time, only minor granite stocks and mafic dikes intruded the crystalline basement in response to continued impulses of its activization.

Proterozoic to lower Paleozoic sequences

In the southern Prince Charles Mountains, folded metasedimentary rock units occupy an area of about 200 × 100 km (Soloviev, 1972; Tingey, 1972; Tingey and England, 1973). The tectonic setting of these units is controlled by a latitudinal graben-synclinorium limited by deep faults. The age of these rocks is tentatively considered to be Proterozoic, since they developed on the cratonized upper Archean crystalline basement and are capped by the uppermost Proterozoic to lower Paleozoic metasedimentary rocks containing Vendian microfossils that range in age from 500 to 700 m.y. (Halpern and Grikurov, 1975). The total thickness of the epicratonic folded sequences from geologic observations is about 7–8 km. This value is confirmed by geophysical data, which show an increase in the thickness of the earth's crust to 40–45 km beneath the Proterozoic formations, as compared to a thickness of 32–35 km for the crystalline basement terrane.

The lower part of the sequence is composed of lower–middle Proterozoic units named the Ruker Series and Menzies Series. Boundaries between these series are not exposed, and their age relationship is not clear. However, structurally these two series, though different in rock composition and metamorphic grade, make up a single complex with a total thickness of about 4 km. This complex is often referred to as the Ruker Complex (Ravich, Chapter 51, this volume), while the tectonic events and structures related to its formation are also termed Ruker (cycle, orogenies, fold system, etc.) (Grikurov, Chapter 98, this volume). The basal unit of the Ruker Series was observed to rest directly on the weathered surface of basement granites that have a Rb-Sr isochron age of 3200–2500 m.y. (Halpern and Grikurov, 1975). The Ruker Series consists of chlorite-mica and mica-chlorite carbonate schists, interlayered with mica-chlorite quartzites. Isolated beds are composed of chlorite-actinolite schists, formed after metabasites. In its lower part, a thick (up to 350 m) jaspilite bed occurs. All the rocks have been metamorphosed in a greenschist facies environment.

The Menzies Series (Figure 60.2) has a considerably more varied rock composition; it was metamorphosed in a greenschist facies environment and subsequently metamorphosed to epidote-amphibolite and amphibolite facies (Tingey and England, 1973). This resulted in transformation of chlorite-mica schists and quartzites (locally preserved at the base of the series) into garnet-mica, sillimanite-biotite, and garnet-staurolite-mica (with some kyanite) schists and garnet-mica quartzites. In addition, the Menzies Series includes cummingtonite-amphibole schists and amphibolites, marble, and tremolite calciphyres. The Ruker and Menzies Series are cut by sills and dikes of metabasites; their isotopic age ranges from 860 to 1100 m.y. according to Rb-Sr isochron determinations. The rocks of both series are linearly folded and locally complicated with minor faults and disharmonic folds. Secondary foliation and cleavage are present. The folds have sublatitudinal axial strike, changing to northwesternly (250–330°), and the dip angles of the fold limbs range from 30° to 50–60°.

At the northern and eastern margins of the graben-synclinorium occurs the upper Proterozoic to lower Paleozoic Sodruzhestvo Series, as much as 3.5 km thick (Figure 60.3). This series comprises the upper unit in the folded sequence (Grikurov and Soloviev, 1974). Its direct contact with the Ruker and Menzies Series was not observed, and the stratigraphic position is deduced from its composition, somewhat different folding pattern, presence of unidentifiable fossils, and isotopic age determinations. The Sodruzhestvo Series is composed of phyllites, metasandstones, quartzites and sericite-carbonate quartzites, schists, and metaconglomerates. Quartzitic metasandstones and quartzites with metaconglomerate layers predominate in the upper half of the section, while phyllites and schists are more common in the lower half. The rocks underwent metamorphism to the greenschist facies. They are deformed into symmetrical linear folds, usually overturned, with sublatitudinal axial strikes changing to northwesternly (270–315°) and steep limbs dipping 70–80°. In contrast to the underlying series, secondary cleavage and foliation are not developed in the Sodruzhestvo Series. Metabasites and jaspilites, also typical of the underlying series, are present only as clasts in conglomerates.

A characteristic feature of the Proterozoic to lower Paleozoic fold sequence is the total absence of synkinematic granitization phenomena, either in the form of migmatites or plutonic intrusions. This feature distinguishes the sequence from geosynclinal fold complexes and suggests that its formation was controlled by an epi-cratonic trough of aulacogen type. However, some small postkinematic stocks of lower Paleozoic granites occur both within the Ruker Complex and the underlying Archean basement. They are usually several hundred meters to 2 km across and are composed predominantly of two-mica granites, commonly garnet-bearing. The K-Ar ages of these rocks range from 490 to 550 m.y. These granite bodies are accompanied by numerous garnet-muscovite, muscovite, and commonly beryl-bearing pegmatite veins, occurring not only in the enclosing rocks but also within the granite bodies. The emplacement of granites and veins was probably associated with folding and metamorphism in the Sodruzhestvo Series rocks.

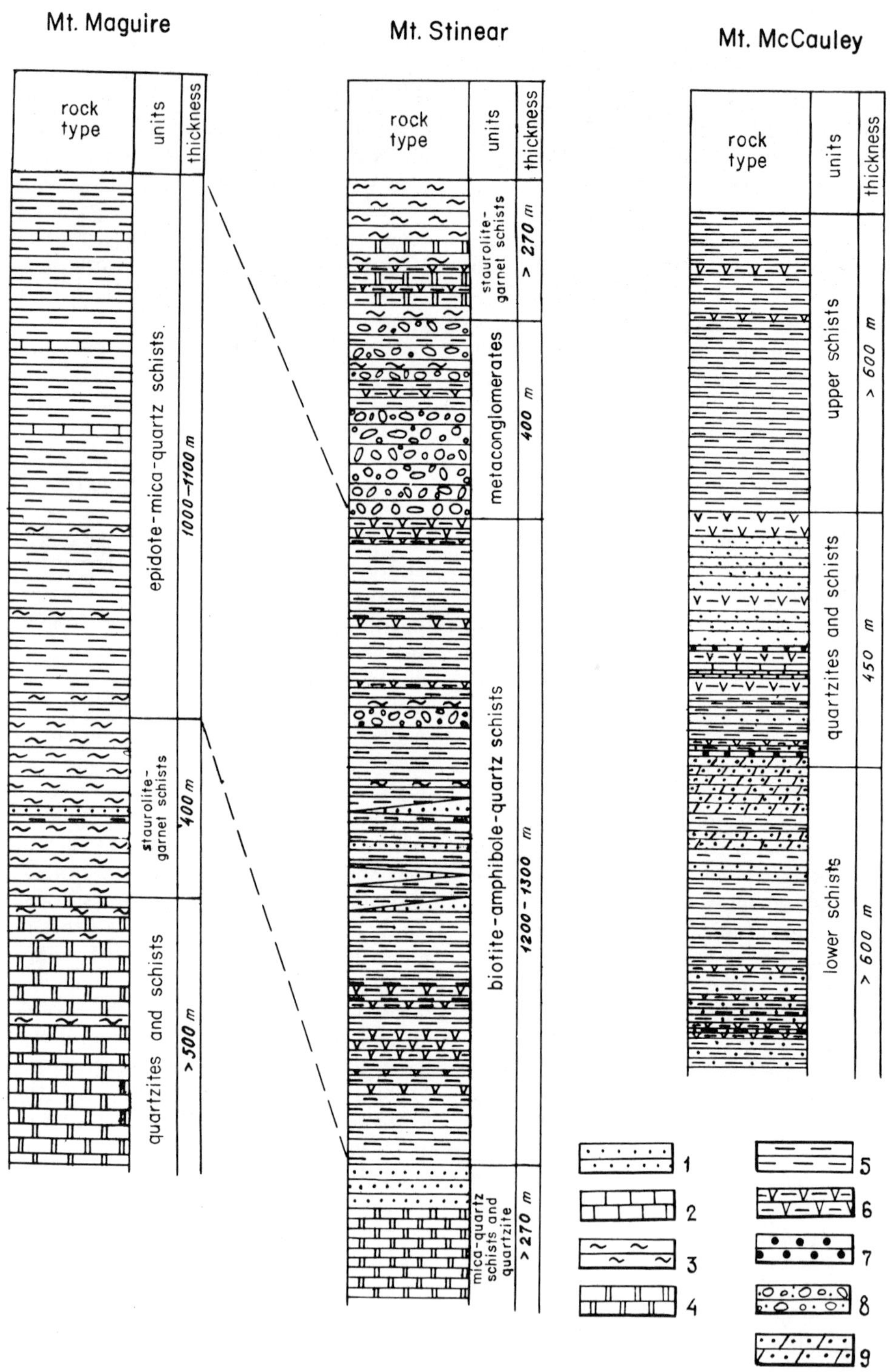

Figure 60.2. Generalized geologic sections of the Menzies Series. 1. Quartzites, micaceous and mica-garnet quartzites; 2. Marbles and calciphyre; 3. Garnet-staurolite-mica-quartz schists, locally with kyanites; 4. Mica-quartz and garnet-mica-quartz schists; 5. Biotite-epidote-quartz and biotite-epidote-amphibole-plagioclase-quartz schists; 6. Orthoamphibolites and metaultrabasites; 7. Gedrite-cummingtonite-garnet-cordierite schists; 8. Metaconglomerates; 9. Staurolite-fibrolite-mica-plagioclase-quartz schists.

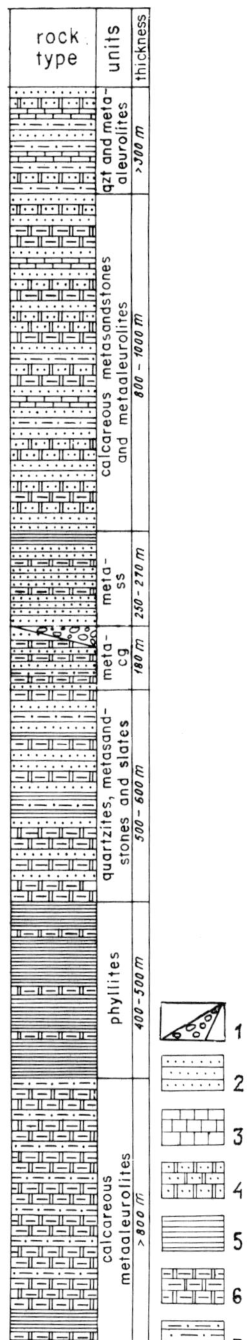

Figure 60.3. Generalized geologic section of the Sodruzhestvo Series at Mount Rubin. 1. Metaconglomerates; 2. Quartz metasandstones and metasiltstones; 3. Marbled siliceous limestones; 4. Calcareous metasandstones and metasiltstones; 5. Phyllites; 6. Mica-carbonate-quartz schists; 7. Chlorite-carbonate schists.

Upper Paleozoic, Mesozoic, and Cenozoic formations

After the early Paleozoic orogeny, the tectonic regime in the study area was dominated by crustal-block movements, resulting in the formation of a longitudinal system of horsts and narrow depressions. In one such depression, namely the Beaver Lake graben, coal-bearing, flat-lying, Permian deposits are preserved, which directly overlie the strongly eroded crystalline basement rocks (Crohn, 1959; Ravich, 1974). These strata, 1300 m in thickness, are identical to the upper Paleozoic Beacon Supergroup in the Transantarctic Mountains. The Permian deposits from the Beaver Lake graben are subdivided into three units. The basal unit consists of conglomerates, gritstones, and coal-siltstone shales, rhythmically interbedded, totaling as much as 250 m in thickness. Fragments of gneissic granitoid rocks predominate in the coarse clasts. The middle unit is 900 m thick and is composed of rhythmically interlayered sandstones, siltstones, claystones, and 2-m-thick coal seams. The uppermost unit consists of regularly interbedded arkose, coarse-grained sandstones, and various fine-grained sandstones and claystones totaling 200 m in thickness.

The Permian rocks are cut by Mesozoic (Lower Cretaceous?) alkaline ultramafic intrusive bodies (Walker and Mond, 1971; Grikurov and Soloviev, 1974). Polzenites predominate in sills and monchikites in dikes and pipe-like bodies. In the pipe-like bodies, monchikites appear as a matrix of eruptive breccia, the clasts of which consist of peridotites (abundant), shadow granites and migmatites of the crystalline basement, quartz porphyries (rare), and Permian sandstones and coals.

Emplacement of alkaline ultramafic rocks was probably related to the initiation of rifting in the area. Available geophysical evidence obtained from aeromagnetic, seismic, gravimetric, and radio-echo sounding observations indicates the existence of a major rift zone underlying the Lambert Glacier and the Amery Ice Shelf. This zone crosses the whole region from north to south for more than 500 km and is 100 km wide. It is likely to be filled with sedimentary deposits as much as 10 km thick. Large blocks of crystalline basement rocks form longitudinally oriented horsts along the strike of the rift zone.

The youngest geologic formations in the study region are the Cenozoic ice mantle and moraine deposits. The latter, with a maximum thickness of 50 km, consists of boulder conglomerate and boulder clay, concentrated mainly along the glacier's margins. Among the boulders, local rocks from the crystalline basement predominate, along with minor jaspilites and other Proterozoic rocks.

References cited

Crohn, P. W. 1959. A contribution to the geology and glaciology of the western part of Australian Antarctic Territory. *Bull. Bur. Miner. Resour. Geol. Geophys. Aust.*, No. 52, 103 pp.

Grikurov, G. E., and Soloviev, D. S. 1974. Geologicheskoe stroenie gornogo obramleniia lednika Lamberta [Geology of the mountain fringe of Lambert Glacier]. *Inf. Byull. sov. antarkt. Eksped.*, *88*, 21–29.

Halpern, M., and Grikurov, G. E. 1975. Rubidium-strontium data from the southern Prince Charles Mountains. *Antarct. Jnl U.S.*, *10*, No. 1, 9–15.

Ravich, G. M. 1974. Razrez permskih uglenosnyh otlozhenii raiona ozera Biver (gory Prins-Charl'z, Vostochnaia Antarktida) [Section through the Permian coal measures in the Beaver Lake area (Prince Charles Mountains, East Antarctica)]. *Antarktika, Doklady Komissii, 13,* 19–35.

Ravich, M. G., and Kamenev, E. N. 1972. *Kristallicheskii fundament Antarkticheskoi platformy* [Crystalline basement of the Antarctic platform]. Leningrad, Gidrometeoizdat, 658 pp. [English translation: New York, John Wiley, 1975, 582 pp.]

Soloviev, D. S. 1972. Geological structure of the mountain fringe of Lambert Glacier and the Amery Ice Shelf. (*In* Adie, R. J., *ed. Antarctic Geology and Geophysics.* Oslo, Universitetsforlaget, 573–77.)

Tingey, R. J. 1972. Geological work in Antarctica, 1971. *Rec. Bur. Miner. Resour. Geol. Geophys. Aust.,* 1972/132 [Unpublished] 49 pp.

Tingey, R. J., and England, R. N. 1973. Geological work in Antarctica, 1972. *Rec. Bur. Miner. Resour. Geol. Geophys. Aust.,* 1973/161 [Unpublished] 17 pp.

Walker, K. R., and Mond, A. 1971. Mica lamprophyre (alnoite) from Radok Lake, Prince Charles Mountains, Antarctica. *Rec. Bur. Miner. Resour. Geol. Geophys. Aust.,* 1971/108 [Unpublished] 6 pp.

61

Antarctica's Oldest Metamorphic Rocks in the Fyfe Hills, Enderby Land

E. N. KAMENEV
Department of Antarctic Geology, SEVMORGEO, Moyka 120, Leningrad 190121, U.S.S.R.

Abstract. Lead isotope determinations on enderbites and pyroxene-plagioclase schists from the Fyfe Hills, Enderby Land, show that they are the oldest rocks in Antarctica, formed about 4000 ± 200 m.y. ago (Sobotovich and others, 1974). An 800-m-thick succession at the southern nunatak of the Fyfe Hills belongs to a stratified complex metamorphosed in the two-pyroxene (granulite) facies. Seventy percent of the succession consists of enderbites and charnockites enclosing lenses and beds of pyroxene-plagioclase schist, pyroxenite, peridotite, garnet-pyroxene-magnetite-quartz schist and garnet gneiss, mainly mesoperthitic. The sequence includes 59 beds of pyroxene-plagioclase schist (15% of the total succession thickness), 28 beds and lenses of pyroxenite (6%), 11 beds and lenses of peridotite (2%), 17 beds of garnet-pyroxene-magnetite-quartz schist (3%), and 24 beds of garnet gneiss (4%). Enderbites, pyroxene-plagioclase schists, and peridotites may represent primary metamorphic rocks, but the other rocks may be their derivatives resulting from metamorphic differentiation and metasomatism. The overall composition of the succession is that of melanocratic enderbites. The primary (oldest) crust may have had a composition of enderbite or quartz diorite.

The determination of a reliable age for the oldest crustal rocks, along with their composition and conditions of formation, is of great theoretical importance not only for the reconstruction of peculiar features and the sequence of geologic events in the earliest Precambrian, but also for the construction of a model for the earth's generation as a planet.

Lead isotope determinations on enderbites and pyroxene-plagioclase schists from the Fyfe Hills, Enderby Land, show that they are the oldest rocks in Antarctica, formed about 4000 ± 200 m.y. ago (Sobotovich and others, 1974, 1976). In addition, it was established that lead isotope ratios in these rocks are similar to those of lead in meteorites. The isotope and geologic data suggest that a succession of metamorphic rocks of the Napier Complex in the Fyfe Hills, Enderby Land (Kamenev, 1975) is a relic of the primary (ancient) earth's crust, which formed by accretion during the late stages of planetary creation and which was metamorphosed later under conditions of the two-pyroxene (granulite) facies.

General geology of the Fyfe Hills

The Fyfe Hills are situated on the southern coast of Khmara Bay in central Enderby Land, at lat. 67° 19′–67° 23′ S and long. 49° 15′–49° 18′ E. They consist of several nunataks separated by grounded glaciers 2–4 km in width.

The metamorphic sequence forms a system of anticlinal folds with wavelengths on the order of several kilometers, sublatitudinal axes, and warped axial surfaces; fold hinges plunge to the west at an angle of 10°–12°. Dip angles of the fold limbs range from 25° to 75°. Gneissosity planes and layering in metamorphic rocks of the same section have various dip angles and slightly different strikes. All this implies gentle, small-scale folding of the layers, with fold axes parallel to dip and strike. Steeper small-scale folding is characteristic in places of migmatite development, but it seldom extends more than 5–10 m.

Tectono-metamorphic lens-like boudinage and attenuation should be emphasized. Lens-like boudinage is characteristic of pyroxene-plagioclase schist, pyroxenite, and peridotite. The ratio of lens length along strike to the greatest thickness is 4:1 on the average. As a rule, lenses are asymmetrical; some of their ends are elongate and curved, and opposite ones are obtuse and streamlined (Figure 61.1). Forms of tectono-metamorphic attenuation are typical of garnet and sillimanite-garnet gneiss, thinning out down dip along strike where underlying and overlying beds are bent. Such structures within the succession imply an important role for tectono-metamorphic differentiation in the creation of their appearance.

A large number of shear zones, as well as zones of blastomylonites, blastocataclasites, and much younger diaphthorites conformable with bedding or obliquely cutting the gneissosity of country rocks, confirm this inference. Shear zones have thicknesses from 0.6 m to 5–7 m and as a rule are related to layered and lens-like bodies of ultramafic rock and pyroxene-plagioclase schist. Zones of blastomylonites (garnet-magnetite-pyroxene-plagioclase or garnet-magnetite-pyroxene-quartz schist), commonly enriched by magnetite and garnet, are 0.2–5 m thick, and units of blastocataclasites of enderbite and charnockite (with garnet) composition reach 30 m in thickness.

In the central Fyfe Hills, submeridional plicated diaph-

Figure 61.1. Lenses of pyroxenites in enderbites. *a.* (top) General view of Enderbite Mount; *b.* (bottom) Detail of outcrop. Fyfe Hills.

thorite zones, 10–15 m wide, including cataclasized veins of muscovite pegmatites, cut across the foliation in mica schists of amphibolite facies. The largest cross-cutting zone of diaphthorites, about 400 m thick, is in the westernmost Fyfe Hills; it strikes 335°. Diaphthorites consist of quartz, plagioclase-quartz, and quartz-mica schists with kyanite, cordierite, staurolite, garnet, actinolite, hornblende, epidote, zoisite, and anthophyllite. Undoubtedly, the diaphthorites were formed from the enclosing interlayered garnet-bearing enderbites, pyroxene-plagioclase schists, garnet gneisses, and peridotites.

Several granite pegmatite veins and five dolerite and diabase dikes 1.8–4.6 m wide were also found in the Fyfe Hills.

Rock composition of the succession of Enderbite Mount

Some parts of Enderbite Mount in the southern Fyfe Hills (Figure 61.2), several kilometers from diaphthoresis zones, were studied in detail. Enderbites and charnockites comprise more than 70% of the total thickness of the succession; they are the enclosing rocks for lenses and beds of pyroxene-plagioclase schist, pyroxenite, peridotite, garnet-pyroxene-magnetite-quartz and plagioclase schist, and garnet (sillimanite-garnet) gneiss, mainly mesoperthitic. There are 59 beds of pyroxene-plagioclase schist 0.15–13.35 m thick (15% of total thickness of the succession), 28 beds and lenses of pyroxenite 0.15–12.25 m thick (6%), 11 beds and lenses of peridotite 0.15–3.25 m thick (2%), 17 beds of garnet-pyroxene-magnetite-quartz or plagioclase schist 0.25–5.10 m thick (3%), and 24 beds of garnet gneiss 0.36–2.50 m thick (4%).

Rather persistent combinations of metamorphic rocks occur, such as units of garnet gneiss and pyroxene-plagioclase schist (as a rule, bearing secondary garnet) and units of pyroxene-plagioclase schist, pyroxenite, and peridotite. Central parts of the units commonly consist of peridotite surrounded first by pyroxenite, then by pyroxene-plagioclase schist, and finally by enderbite. All these units are 10–30 m thick and are "drowned" in a huge mass of enderbite and charnockite, which also contains thick layers of other rocks. Because of the difficulty of distinguishing these host rocks in the field, the proportions of char-

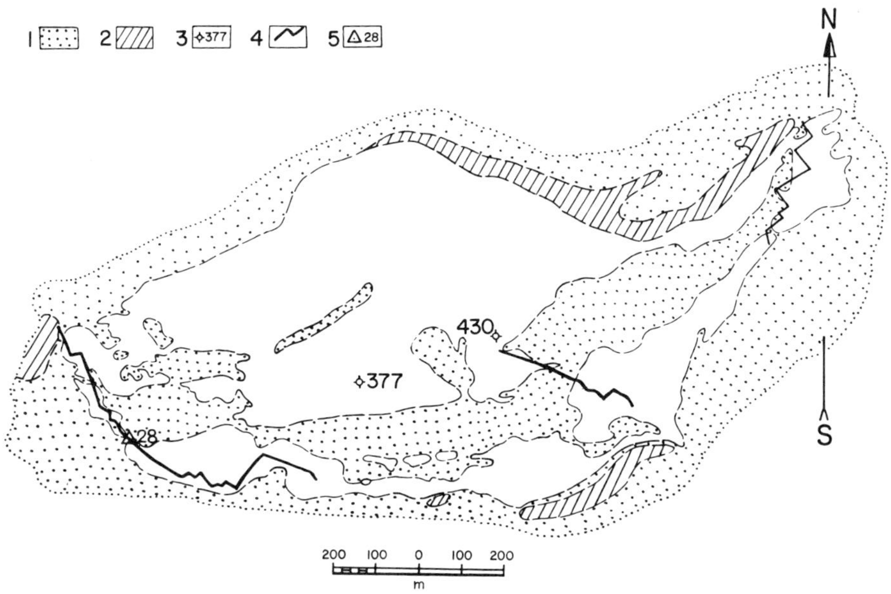

Figure 61.2. Sketch map of Enderbite Mount (lat. 67° 22′S, long. 49° 17′E), Fyfe Hills. 1. Snow; 2. Frozen lakes; 3. Altitude of peak, in m; 4. Line of detailed section; 5. Occurrence of enderbites with age of 4100 ± 100 m.y.

nockite and enderbite in the section were determined only approximately; enderbite comprises 40% and charnockite, 60%. The largest bed consisting completely of enderbite is 23.65 m thick. Some enderbites are thinly laminated and resemble migmatite.

Most pyroxene-plagioclase schists and their garnetiferous varieties have a massive structure, but there are also a few banded varieties. Lit-par-lit and veined migmatites in which the melanosome is represented by pyroxene-plagioclase schist and the leucosome by a coarse-grained enderbite also occur. Leucosome accounts for 5–20% of the migmatite. The thickness of migmatite units does not exceed 10 m. Massive leucocratic varieties with large but rare garnet porphyroblasts are most widespread among garnet gneisses. Thin-layered gneisses, a type of migmatite, are uncommon. Sillimanite-garnet gneiss occurs only in two beds, each 2 m thick.

Pyroxenite and peridotite are massive. Grains of rock-forming olivine and pyroxene are flattened, which implies that they formed under anisotropic stress conditions. There are two textural varieties of ultramafic rocks: fine and coarse grained. Fine-grained varieties occur in layers or elongate lenses. Coarse-grained varieties make up lenses not more than 3–4 m long but as much as 4.5 m wide. Lenses are separated by veinlets of coarse-grained pyroxene-plagioclase or plagioclase rocks 0.5 m thick; rims around lenses are made up of the same rocks. Smaller lenses of coarse-grained pyroxenite as much as 0.5 m thick and about 1 m long are rather common among pyroxene-plagioclase schists and enderbites; in some cases they amount to 15% of the layer volume. The mode of occurrence of coarse-grained pyroxenite suggests that it is the result of metamorphic differentiation on the concretional principle. Undoubtedly, mechanical differential movements between layers (shearing?) were of importance in their genesis.

Thin beds of magnetite-bearing schists do not differ from elongate lenses of other schists either in structure or mode of occurrence. Thicker beds, as mentioned above, form zones intersecting the gneissosity of enclosing enderbites and charnockites (as a rule, poor in migmatite) at an angle of not more than 10°. The magnetite content of magnetite-bearing schists varies from 20% to 50%, increasing considerably toward the central parts of the beds. Gneissose structures, blastomylonite textures, presence of bulges, and lens-like patterns all suggest that the formation of the magnetite-bearing schists occurred in a mobile environment.

Discussion

Mineral assemblages are shown in Table 61.1. It is necessary to emphasize the nature of the development of garnet and magnetite forming rims around pyroxene. Hornblende and biotite developed later than garnet, commonly along fissures. Garnet is formed mainly in strongly deformed rocks, in blastocataclasites and blastomylonites, and is itself undeformed. The garnet in garnet gneisses, and in some cases even in granulites, forms undeformed large porphyroblasts scattered unevenly across the rock. Small intergrowths of garnet and magnetite, very similar to those in magnetite-rich schists, enderbites, garnetiferous charnockites, garnetiferous pyroxene-plagioclase schists, and garnet gneisses, have developed along thin shear zones, but after the shearing. Thus, it is possible to conclude with some certainty that garnet gneisses could arise from charnockites and enderbites as a result of tectono-metamorphic differentiation. An excessive (tectonic) pressure and the redistribution of components of rocks are necessary conditions for such a differentiation. The approximate reaction may be described as:

$$4\,\mathrm{Opx} + \mathrm{An} \rightleftharpoons \mathrm{Gr} + \mathrm{Mgt} + 5\mathrm{Q} + \mathrm{Ca}\uparrow$$

The remaining rocks could also result from tectono-metamorphic differentiation. The mechanism of this differentiation may be briefly formulated as follows: local gradients of pressure and, consequently, gradients of

Table 61.1. Mineral assemblages in rocks from the Fyfe Hills

Mineral assemblages[a]	Secondary minerals
Enderbites	
Q + $Pl_{antiperthite}$ + Opx (+ Cpx)	Hrb, Mgt
Q + Pl + Opx	Bi, Mgt
Q + $Pl_{antiperthite}$ + Opx (+ Cpx)	KFsp, Gr, Mgt, Bi
Charnockites	
Q + Pl + $KFsp_{perthite}$ + Opx (+ Mgt)	Bi, Gr
Q + Pl + Mesoperthite + Opx	—
Q + Mesoperthite + Opx	—
Q + Mesoperthite + Opx	Bi, Gr
Pyroxene-plagioclase schists	
Pl + Opx + Cpx	Q, Bi, Hrb
Pl + Opx + Cpx	Gr, Sph, Mgt, Q, Hrb
$Pl_{antiperthite}$ + Opx + Cpx + Mgt	—
Pyroxenites	
Cpx + Pl	—
Opx + Cpx + Phlog	—
Opx + Cpx + Sp + Mgt	—
Opx + Cpx + Mgt	—
Opx + Cpx + Pl	Hrb, Gr, Mgt
Opx + Sp	Gr, Mgt
Opx (+ Cpx)	Amf, Bi, Pl
Peridotites	
Opx + Cpx + Ol + Pl	Hrb, Serp
Opx + Cpx + Ol + Sp	Phlog, Serp
Opx + Cpx + Ol + Mgt	Phlog, Amf, Serp
Opx + Ol + Mgt	Phlog, Serp
Garnet	
Q + $KFsp_{perthite}$ + Pl + Gr + Mgt + Opx (relics)	—
Q + Mesoperthite + Gr	Bi, Mu
Q + $KFsp_{perthite}$ + Gr (+ Pl)	—
Q + $KFsp_{perthite}$ + Gr + Sill (+ Sp)	—
Garnet-pyroxene-magnetite-quartz and plagioclase schists	
Opx + Cpx + Q	Gr, Mgt
Opx + Cpx + Pl	Gr, Mgt
Cpx + Q	Gr, Mgt
Opx + Cpx + Pl	Q, Gr, Mgt
Other rocks	
Q + KFsp + Pl	—
Q + Pl	Hrb, Bi
Pl + Opx	—
Opx	Sapp, Sill, Gr, Phlog
Opx + KFsp	Sill, Gr, Phlog

[a] Accessory minerals are apatite, rutile, and zircon.

chemical potentials of components originate from differential movements in a plastically flowing rock mass. Matter immediately diffuses into the areas of low pressure. The diffusion distance is apparently small, as evidenced by neighboring complementary products of metamorphic differentiation, such as accumulation of femic and salic minerals and enrichment in magnetite zones.

Chemical compositions of enderbites and pyroxene-plagioclase schists from the Fyfe Hills are shown in Table 61.2 and in the diagram by Zavaritskii (1950, 1960) (Figure 61.3). The varieties of rocks studied differ little from corresponding magmatic rocks and thus may be assigned to a normal series of rocks, saturated and oversaturated in silica, and poor in alkalis. A lower degree of iron oxidation than for normal igneous rocks (characteristic only of olivine-bearing gabbroids, ultramafic rocks, and some alkaline basaltoids) and low contents of water and other volatiles are rather common features. A study of the chemical compositions of enderbite and pyroxene-plagioclase schist using La Roche (1972) diagrams reveals an important feature of their chemistry, namely a low alumina content as compared to the average compositions of mafic igneous rocks (Figure 61.4*a* and *b*). In addition, the diagrams support the hypothesis that all the rocks studied are magmatogenic in origin since their analyses plot into narrow fields elongated along the curve of average compositions of igneous rocks after Daly (1933). The diagrams become even more emphatic when the chemical compositions of other metamorphic rocks (~50 analyses of enderbite, charnockite, pyroxene-plagioclase schist, garnet and sillimanite-garnet gneiss, peridotite, and pyroxenite) (Ravich and Kamenev, 1972) of the Napier Complex, Enderby Land—of which the metamorphic sequence of the Fyfe Hills is a part—are also plotted for comparison. The diagrams give convincing evidence that the metamorphic rocks listed above have been derived from magmatic rocks. Average chemical compositions of the major rock varieties of the Fyfe Hills and an average composition for the whole succession (Table 61.3) were estimated using chemical compositions of rocks from the Napier Complex. The average rock is a melanocratic enderbite in composition. Since the age of these rocks is quite ancient, it is suggested that the primary (oldest) earth's crust formed from a melt of enderbite (or quartz diorite) composition. Pyroxene-plagioclase schists and peridotites may also be considered as primary materials, while other rocks could arise from the primary ones as a result of tectono-metamorphic differentiation, partial melting, and metasomatism.

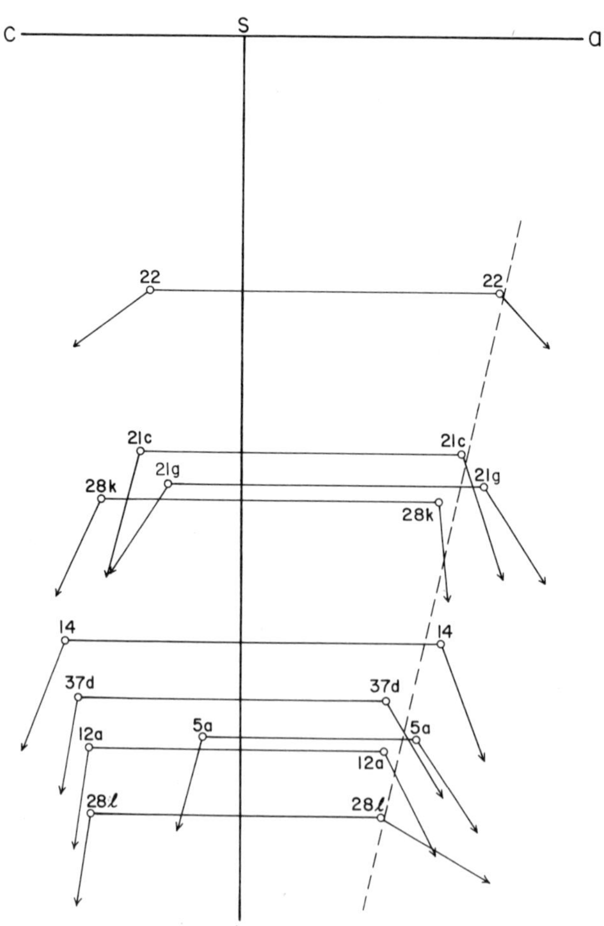

Figure 61.3. Zavaritskii's (1950, 1960) diagram for enderbites and pyroxene-plagioclase schists from the Fyfe Hills. Specimen numbers shown are also in Table 61.2.

Table 61.2. Chemical compositions (in wt %) of pyroxene-plagioclase schists and enderbites of the Fyfe Hills

	1	2	3	4	5	6	7	8	9
SiO_2	50.62	53.07	56.37	53.72	52.44	57.89	62.85	63.67	65.46
TiO_2	0.57	0.62	0.74	0.57	0.65	1.63	0.49	0.51	1.38
Al_2O_3	13.71	13.93	13.25	14.52	16.71	14.50	12.23	13.27	13.46
Fe_2O_3	2.47	1.65	2.29	1.62	1.22	1.52	1.84	1.10	1.30
FeO	6.85	9.65	8.77	9.06	8.16	9.59	5.44	4.49	4.76
MnO	0.74	0.22	0.62	0.23	0.17	0.15	0.23	0.18	0.19
MgO	4.75	6.91	6.09	5.97	6.28	4.19	4.04	5.07	1.29
CaO	16.36	10.34	7.09	10.56	9.33	5.18	6.40	5.97	4.84
Na_2O	2.32	2.42	2.60	2.32	3.42	2.50	2.82	3.32	2.00
K_2O	0.56	0.46	1.10	0.72	0.50	1.82	2.76	1.43	4.22
P_2O_5	0.32	0.10	0.13	0.12	0.13	0.16	0.19	0.15	0.25
Loss on ignition	0.42	0.72	0.85	0.65	0.69	0.53	0.44	0.69	0.43
Total	99.69	100.09	99.90	100.06	99.70	99.66	99.73	99.85	99.58
H_2O^-	0.12	0.03	0.10	0.01	0.07	0.05	0.14	0.03	0.10
Numeral characteristics, after Zavaritskii (1950, 1960)									
a	5.8	5.9	7.2	6.0	8.2	8.1	9.9	9.0	10.5
b	30.9	28.3	27.8	26.3	24.0	18.4	17.8	16.5	10.1
c	6.1	6.2	1.5	6.6	7.2	5.7	3.0	4.1	3.8
s	57.2	59.6	63.5	61.1	60.6	67.8	69.3	70.4	75.6
f^1	30.2	37.8	38.1	38.5	37.2	57.5	37.4	32.0	57.4
c^1	44.0	21.7	25.4	23.3	18.1	3.7	25.6	17.6	21.0
m^1	25.8	40.5	36.5	38.2	44.7	38.8	37.0	50.4	21.6
Q	−3.3	1.2	11.1	3.6	−2.4	13.7	15.8	18.7	26.4
n	86.0	88.6	77.8	84.1	91.7	67.8	60.0	77.9	41.6
t	0.8	0.9	1.0	0.8	0.9	2.0	0.6	0.6	1.6
φ	7.0	5.2	3.4	5.2	4.0	6.7	8.1	5.6	10.8

1. 28l: Clinopyroxene-plagioclase schist with secondary Gr, Mgt, and Sph.
2. 12a: Pyroxene-plagioclase schist with secondary Q, Gr, and Mgt.
3. 5a: Pyroxene-plagioclase schist with secondary Q, Bi, Hrb, and Mgt.
4. 37d: Pyroxene-plagioclase schist with secondary Q, Gr, and Mgt.
5. 14: Leucocratic pyroxene-plagioclase schist; also contains magnetite.
6. 28k: Enderbite with clinopyroxene and secondary perthite, Gr, Mgt, and Bi.
7. 21g: Enderbite with secondary perthite; also contains magnetite.
8. 21c: Enderbite with secondary Bi; also contains magnetite.
9. 22: Enderbite with secondary Gr, Ort, and Mgt; also contains zircon.

Apatite is a common accessory mineral.

Table 61.3. Average chemical compositions (in wt %) of metamorphic rocks from the Napier Complex of Enderby Land and of the metamorphic unit of the Fyfe Hills

	Peridotite (2)	Pyroxenite (4)	Pyroxene-plagioclase schist (11)	Enderbite (7)	Charnockite (17)	Garnet gneiss (6)	Enderbite + charnockite of the Fyfe Hills	Entire metamorphic unit of the Fyfe Hills
SiO_2	44.00	49.99	52.01	67.39	70.55	71.98	69.26	64.8
TiO_2	0.39	2.84	1.00	0.51	0.52	0.20	0.52	0.7
Al_2O_3	3.30	4.87	14.19	15.54	13.45	14.05	14.31	13.5
Fe_2O_3	1.72	2.66	2.79	1.12	1.28	0.46	1.21	1.5
FeO	14.00	12.75	8.48	3.59	3.29	2.97	3.41	4.9
MnO	0.21	0.27	0.20	0.06	0.07	0.06	0.07	0.1
MgO	31.85	19.72	7.17	2.02	1.36	0.92	1.63	4.5
CaO	2.80	4.96	8.96	4.14	2.13	1.44	2.95	4.0
Na_2O	0.38	0.52	2.53	3.99	2.52	2.81	3.12	2.8
K_2O	0.28	0.55	1.68	1.20	4.34	4.79	3.05	2.7
P_2O_5	0.02	0.26	0.27	0.16	0.09	0.09	0.12	0.1
Loss on ignition	1.05	0.61	0.72	0.28	0.40	0.23	0.35	0.4
Total	100.00	100.00	100.00	100.00	100.00	100.00	100.00	100.00

Numbers in parentheses indicate the number of analyses used for the calculation of average composition.

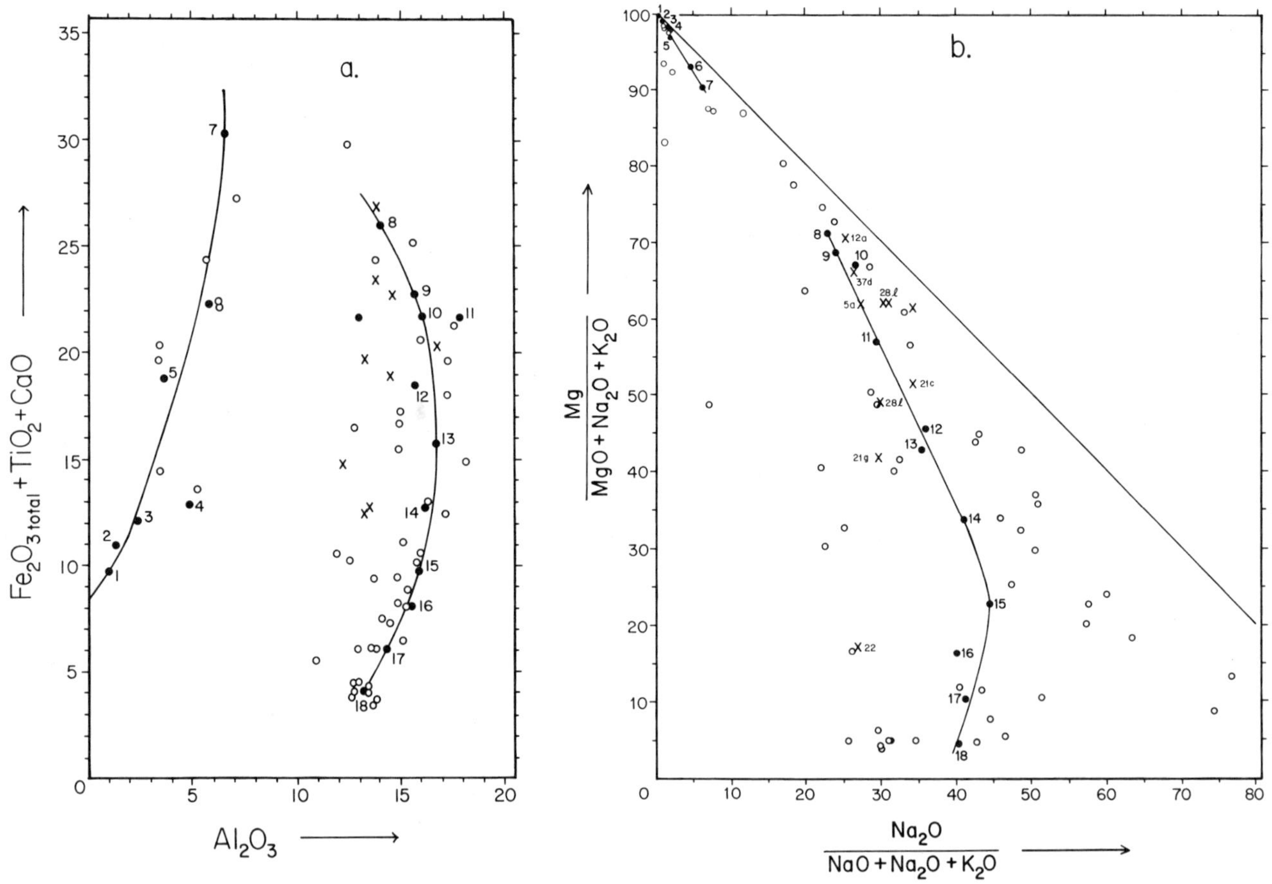

Figure 61.4 a and b. La Roche's (1972) diagrams for rocks from the Napier Complex of Enderby Land. X's: enderbites and pyroxene-plagioclase schists from the Fyfe Hills. Open circles: two-pyroxene facies rocks from the Napier Complex (Ravich and Kamenev, 1972). Dots: average igneous rocks after Daly (1933). 1. Dunite; 2. Harzburgite; 3. Bronzitite; 4. Lherzolite; 5. Websterite; 6. Wehrlite; 7. Diallagite; 8. Plateau basalt; 9. Basalt; 10. Norite; 11. Gabbro; 12. Quartz gabbro; 13. Diorite; 14. Quartz diorite; 15. Granodiorite; 16. Quartz monzonite; 17. Granite; 18. Rhyolite. Specimen numbers in diagram *b* are also in Table 61.2.

References cited

Daly, R. A. 1933. *Igneous Rocks and the Depths of the Earth.* McGraw-Hill, New York, 598 pp.

Kamenev, E. N. 1975. Geologiia Zemli Enderbi [Geology of Enderby Land]. *Antarktika, Doklady Komissii, 14,* 34–58.

La Roche, H. de. 1972. Geokhimicheskaia kharakteristika oblastei metamorfizma: priznaki i dokazatel'stva ikh dometamorficheskoi istorii [Geochemical characteristic of metamorphic regions: indications and evidence of their premetamorphic history]. (*In* Vinogradov, A. P. I. Mezhdunarodnii geokhimicheskii kongress, kniga I. Metamorfizm i metasomatoz. Moscow, 263–90.)

Ravich, M. G., and Kamenev, E. N. 1972. *Kristallicheskii fundament Antarkticheskoi platformy* [Crystalline basement of the Antarctic platform]. Leningrad, Gidrometeoizdat, 658 pp. [English translation: New York, John Wiley, 1975, 582 pp.]

Sobotovich, E. V., Kamenev, E. N., Komaristy, A. A., and Rudnik, V. A. 1974. Drevneishie porody Antarktidy (Zemlia Enderbi) [The oldest rocks of Antarctica (Enderby Land)]. *Izv. Akad. Nauk SSSR,* Ser. geol., No. 11, 30–50.

Sobotovich, E. V., Rudnik, V. A., Iskanderova, A. D., Kamenev, E. N., Komaristy, A. A., Korol'kov, V. G., and others. 1976. Ranneazoiskiy vozrast drevneishikh porod Zemli (po rezul'tatam svintsovo-izotopnykh issledovanii kristallicheskikh porod Antarkticheskoi platformy) [Early Azoic age of the oldest rocks of the Earth (on results of Pb^{207}/Pb^{206} determinations of the crystalline rocks of the Antarctic platform)]. (*In* Afanas'ev, G. D. Opredelenie absolutnogo vozrasta rudnykh mestorozhdenii i molodykh magmaticheskikh porod XVIII sessiia.) Moscow, Izdatel'stvo Nauka, 59–70.

Zavaritskii, A. N. 1950. Vvedenie v petrokhimiiu izverzhennykh gornykh porod [Introduction to petrochemistry of igneous rocks]. Leningrad Izdatel'stvo AN SSSR, 400 pp.

Zavaritskii, A. N. 1960. Pereschet khimicheskikh analisov izverzhennykh gornykh porod [Recalculation of chemical analyses of igneous rocks). Leningrad, Gosgeoltekhizdat, 156 pp.

62

Tectonics and Microstructure of Charnockites around Lützow-Holmbukta, East Antarctica

MASARU YOSHIDA
Department of Geosciences, Osaka City University, Sumiyoshi-ku, Osaka, Japan

KEIZO YANAI
National Institute of Polar Research, Itabashi, Tokyo, Japan

SHINICHI YOSHIKURA
Institute of Geology, Kochi University, Asakura, Kochi, Japan

TERUMI ISHIKAWA
Mitsui Kinzoku Engineering Company, Chuo-ku, Tokyo, Japan

SATOSHI KANISAWA
Department of Earth Science, Tohoku University, Kawauchi, Sendai, Japan

Abstract. The crystalline rocks occurring around Lützow-Holmbukta (Lützow-Holm Bay) consist of three lithostratigraphic groups: the Skallen Group composed of charnockites, siliceous gneiss, and marble; the Ongul Group composed of charnockites and garnet-biotite gneiss; and the Okuiwa Group composed of pink granites and granitic gneiss. These rocks all suffered superposed tectonic events. The first stage (D_1) formed recumbent folds followed by isoclinal folds, the trends of the axes of which vary from north-south to east-west, parallel to the coast line of Lützow-Holmbukta; the second stage (D_2) formed close to open folds with easterly trends; the third stage (D_3) formed gentle folds with northerly trends, commonly associated with faults; and the fourth stage (D_4) formed a diagonal system of fractures with the acute bisectrix in an easterly direction.

The tectonic and microstructural investigations of the charnockitic rocks reveal four microstructural episodes: the first episode formed fine-grained, granular texture, characterized by clinopyroxene, quartz, and biotite; the second episode formed small- to medium-grained polygonal texture, characterized by plagioclase and two pyroxenes; the third episode formed coarse-grained, coalescent texture, characterized by plagioclase, orthopyroxene, and hornblende; and the fourth episode formed inequigranular, lobate texture, characterized by fine granitic aggregates and porphyroblastic quartz, K-feldspar, and alteration biotite. The first microstructural episode is referred to the D_1 stage, the second and the third episodes between the D_1 and D_2 stages, and the last episode to the D_2 and D_3 stages of tectonic events.

Differences are recognized in conditions of metamorphism and in chemical compositions of certain minerals related to each microstructural episode, and correlations are suggested between macrostructures, mesostructures, microstructures, and the principal plutonic and metamorphic events.

The geology and tectonics of the region around Lützow-Holmbukta have recently been studied in some detail (e.g., Kizaki, 1964; Tatsumi and others, 1964; Tatsumi and Kizaki, 1969; Yoshida and Ando, 1971; Ishikawa, 1976; Yoshida, 1977; Yoshida and others, 1977), but petrologic studies of the metamorphic rocks done so far (e.g., Banno and others, 1964; Suwa, 1968) have been completely devoid of the microstructural viewpoint; hence the metamorphic history of the region was left unsurveyed. This article first gives a new summary of the tectonics of the region around Lützow-Holmbukta; secondly, investigates the petrology of the charnockitic rocks from both the tectonic and microstructural viewpoints; and finally, attempts to relate the charnockite petrology to regional tectonics.

Outline of the geology

The crystalline rocks occurring in the region around Lützow-Holmbukta can be divided into (1) the paragneisses and allied rocks, including various gneisses, quartzite, marble, and metabasites (these include both granulite and amphibolite facies rocks), (2) the charnockitic rocks (granulite facies rocks), and (3) the pink granites (amphibolite facies rocks). These rocks consist of three lithostratigraphic groups: the Skallen Group, distributed in the region around Botnneset and in the southern part of the Soya Coast, is mainly composed of charnockites, siliceous gneiss, and marble; the Ongul Group, distributed in the central part of the Soya Coast, is mainly composed of charnockites and garnet-biotite gneiss; and the Okuiwa Group, distributed in the northern region of the Soya Coast, is mainly composed of pink granites and granitic gneisses. These three groups comprise the Lützow-Holm Bay "System" of Tatsumi and Kizaki (1969). (This "system," however, may have to be redetermined in the future.) The distribution of these three groups is shown in Figure 62.1, along with some tectonic elements.

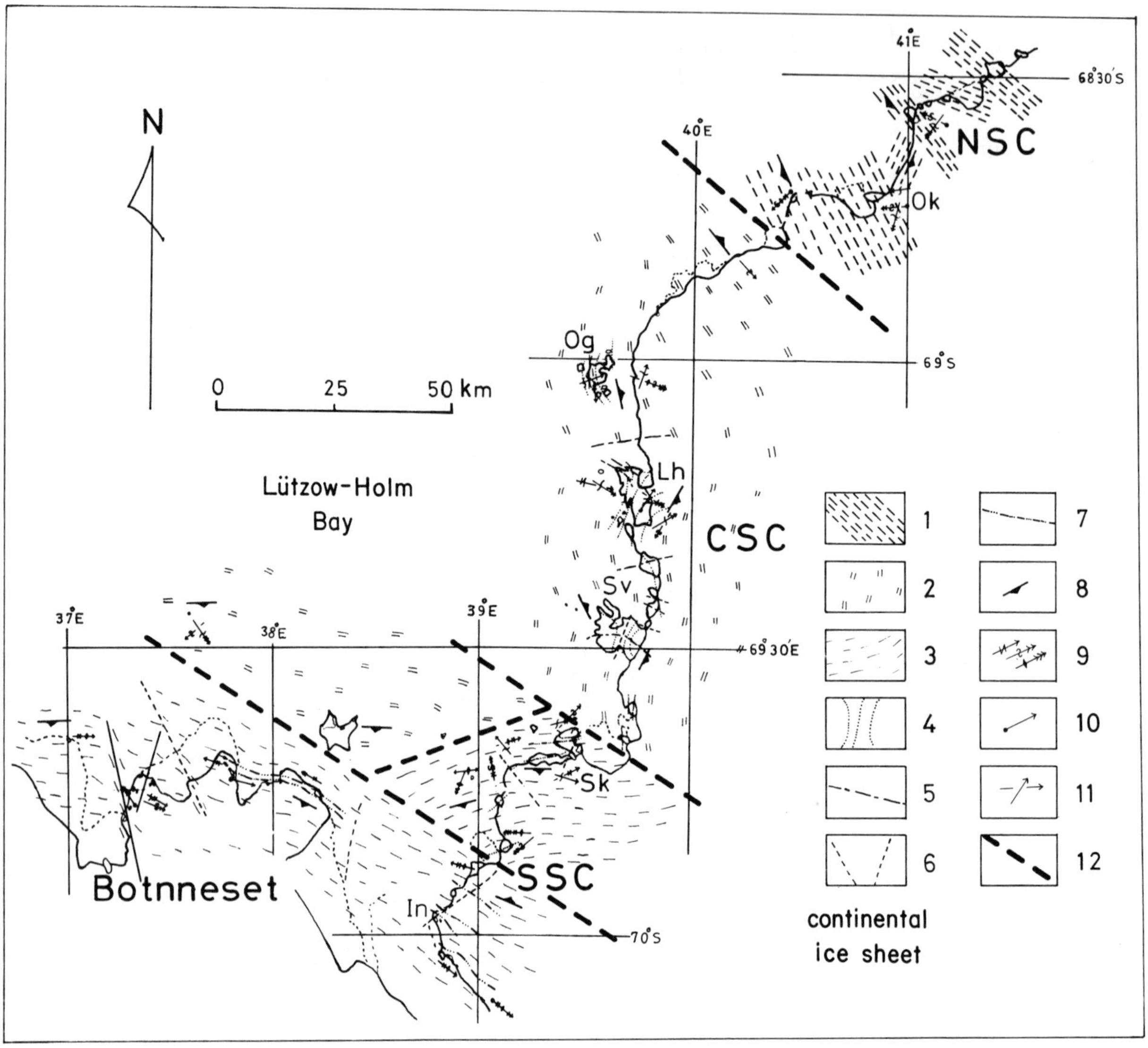

Figure 62.1. Distribution of Lützow-Holm Bay "System" and tectonic outline of the region around Lützow-Holmbukta. Lützow-Holm Bay "System"; 1. Okuiwa Group; 2. Ongul Group; and 3. Skallen Group. 4. Axial trace of the D_1 folds; 5. Axial trace of the D_2 folds; 6. Axial trace of the D_3 folds; 7. Axial trace of D_1 and D_2 folds, superposed; 8. General trend of banding and/or foliation; 9. Lineations (Upper: crenulation lineation, gently plunging. Middle: axis of minor fold, moderately plunging. Lower: mineral lineation, steeply plunging.); 10. Lineation or axis of minor fold, developed in or associated with the pink granites; 11. Two superposed lineations (in this example, the northeasterly plunging lineation cuts the older easterly plunging one); 12. Inferred fault. NSC: Northern Soya Coast. SSC: Southern Soya Coast. CSC: Central Soya Coast. Ok: Okuiwa. Og: Ongul Islands. Lh: Langhovde. Sv: Skarvsnes. Sk: Skallen. In: Instekleppane. This figure is from Yoshida (1977*b*).

The Lützow-Holm Bay "System" suffered superposed tectonic events. The first tectonic stage (D_1) formed early recumbent folds and later isoclinal folds, axial traces of both fold sets running south, southwest, or west, paralleling the coastline of Lützow-Holmbukta. The second tectonic stage (D_2) formed close to open folds with axes trending nearly east. The third stage (D_3) formed gentle folds with axes trending generally north, associated in some areas with faults running north. The fourth stage (D_4) is evidenced by a diagonal system of shear fractures with the acute bisectrix in an easterly direction. The details of the geologic outline are given fully in Yoshida (1977*b*; 1978) and partly in the previous Japanese works already mentioned.

The charnockitic rocks occupy some tens of percent of all the rocks of the Lützow-Holm Bay "System." They contribute many data important to the deciphering of the tectonic and metamorphic histories.

Classification of the charnockites

The charnockitic rocks are small- to coarse-grained holocrystalline, felsic to mafic rocks principally composed of equant grains of plagioclase, generally associated with one or several of the following: quartz, K-feldspar, orthopyroxene, clinopyroxene, hornblende, biotite, and opaque minerals. The color of these rocks is characteristically dark with a brownish or greenish tint to the unaided eye. They are geologically grouped into three classes: charnockitic gneisses, charnockitic plutonites, and charnockitic pegmatites. The charnockitic gneisses resemble the paragneisses in having a (relict) banded structure, and in that a layer of gneiss is thin. The charnockitic plutonites characteristi-

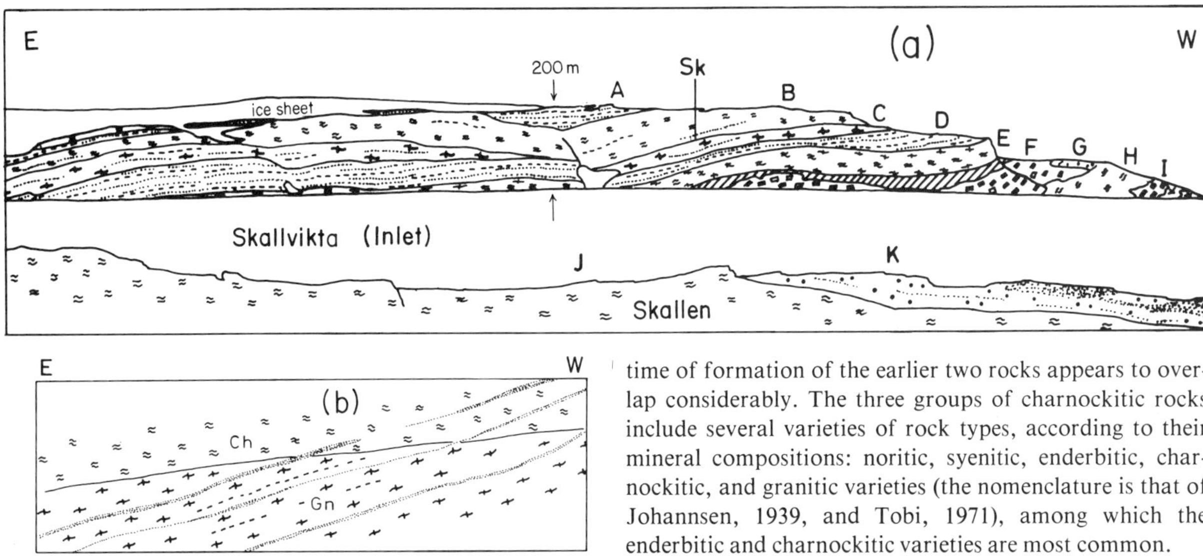

Figure 62.2 a. Occurrence of charnockitic rocks on the northeastern cliff of Skallevikhalsen, viewed from Skallen. A and D: Skallen siliceous formation. B,E,H, and J: Charnockites. C: Garnet gneissose granite. F and I: Marble bed. G and K: Paragneiss of the Skallen lower calcareous formation. Sk: Sketch point shown in Figure 62.2*b*. Base of cliff is at sea level. Outcrops of J and K are part of Skallen (a different locality). This figure is from Yoshida (1977*a*). *b*. Enlarged view of the Sk point in Figure 62.2*a*. The garnet gneissose granite with interlayered metabasite (Gn, lower half) is transformed *in situ* into the charnockitic mass (Ch, upper half). The metabasite layers (dotted, several meters in width) gradually become indistinct and disappear inward.

cally have many features of common plutonic rocks; they are somewhat homogeneous and monotonous in both petrography and tectonics over a considerable width of outcrop. The charnockitic pegmatites are coarsely crystalline rocks occurring only in minor amounts, generally as small dikes, veins, or pods within the other charnockitic rocks.

These charnockitic rocks are regarded as having developed in the order (1) charnockitic gneisses, (2) charnockitic plutonites, and (3) charnockitic pegmatites, although the time of formation of the earlier two rocks appears to overlap considerably. The three groups of charnockitic rocks include several varieties of rock types, according to their mineral compositions: noritic, syenitic, enderbitic, charnockitic, and granitic varieties (the nomenclature is that of Johannsen, 1939, and Tobi, 1971), among which the enderbitic and charnockitic varieties are most common.

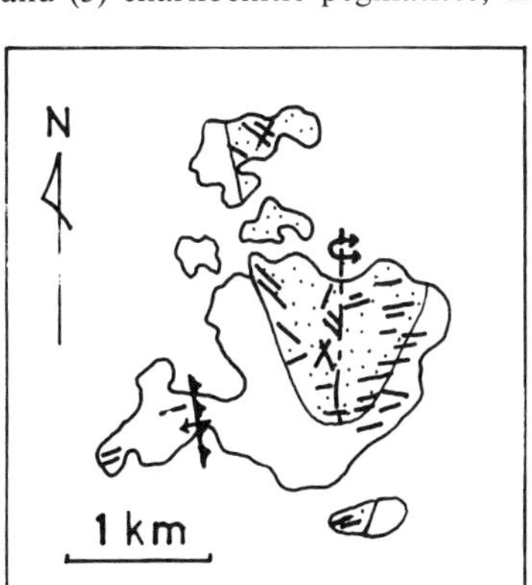

Figure 62.3. The charnockitic pegmatites (short thick bars) on East Ongul Island are scattered and trend about at right angles to the major structure. Dotted: garnet gneiss. Blank: Charnockitic rocks, including metabasites and hornblende gneiss. A southerly plunging isoclinal antiform with an easterly inclination of the axial plane is found in the northeastern part of the island (the axial trace is shown by a chain with a dot), and a thrust fault running northward and inclined eastward is found in the southwestern part (line with solid triangles). (Simplified after Kizaki, 1964; cited from Yoshida, 1977*b*.)

Tectonics of the charnockites

Occurrence

The charnockitic gneisses and charnockitic plutonites occur as sheetlike masses intercalated with the paragneisses and allied rocks; the intercalation is commonly concordant but locally discordant with the macroscopic banding (alternation structures) of the paragneisses and allied rocks. A typical mass of the charnockitic gneisses or plutonites has a thickness of some tens to hundreds of meters and a lateral extent of several kilometers to tens of kilometers.

In detail, interlayered paragneiss and metabasite grade *in situ* into the charnockitic gneisses, and the banded structure gradually becomes indistinct and disappears toward the central part of a mass of the charnockitic rocks, where the charnockitic gneisses pass into and become charnockitic plutonites (Figures 62.2*a* and *b*).

The charnockitic pegmatites occur either as crosscutting veins (Figure 62.3) or as veins along axial planes of tight minor folds.

The pink granites and the granitic gneisses either cut clearly or pass gradually into the charnockitic rocks; the gradation is more distinct with the granitic gneisses than with the pink granites. The schistosity in the intrusive body of the pink granite is generally parallel to the wall, and the intrusive body clearly cuts the tight minor folds of the D_1 tectonic stage but was subjected to gentle folds of the D_3 tectonic stage (Figure 62.4).

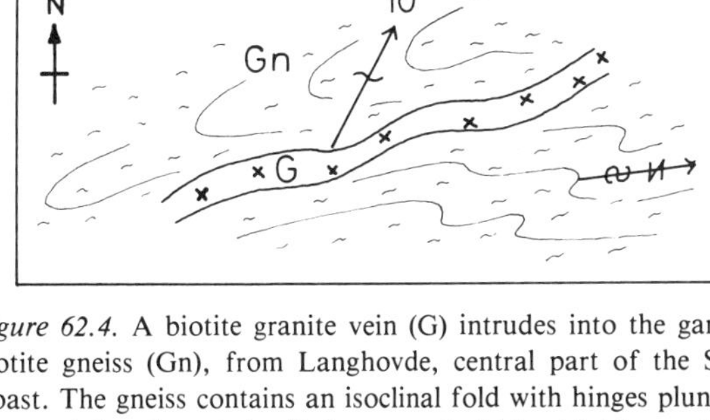

Figure 62.4. A biotite granite vein (G) intrudes into the garnet-biotite gneiss (Gn), from Langhovde, central part of the Soya Coast. The gneiss contains an isoclinal fold with hinges plunging eastward, but the vein itself shows gentle folds with hinges plunging north-northeastward. Structural symbols are as in Figure 62.1. Width of the vein is about 10 cm. This figure is from Yoshida, (1977*b*.)

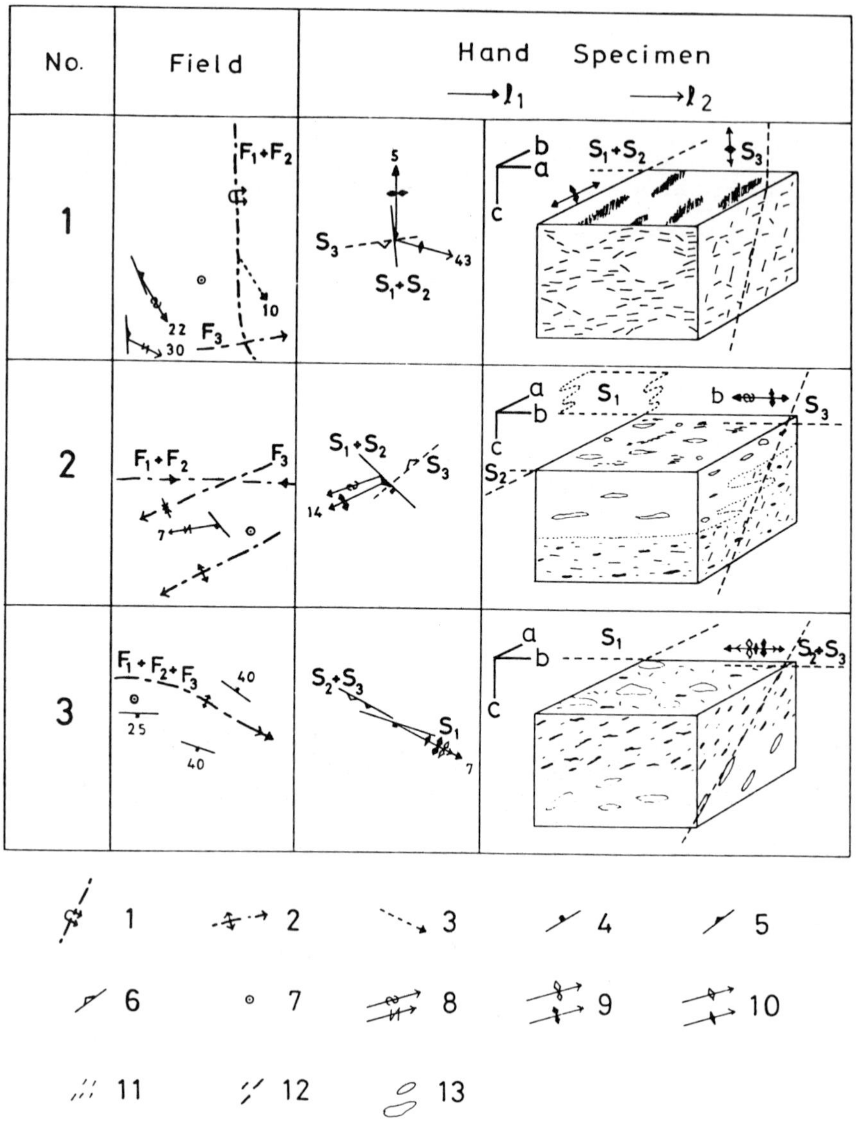

Figure 62.5. Hand specimen structures and related field structures. Specimen no. 1: Charnockitic plutonite collected from East Ongul Island, Y69042302. Specimen no. 2: Banded gneiss composed of quartz-feldspathic biotite gneiss and biotite-pyroxene-plagioclase rock collected from Skallen, A69020302. Specimen no. 3: Charnockitic gneiss collected from Instekleppane, southern part of the Soya Coast, A69100705. 1. Axial trace of recumbent antiform with axial surface inclined eastward (F_1 and F_2 are folds belonging to the D_1 tectonic stage, and F_3, to the D_2 tectonic stage); 2. Axial trace of plunging fold; 3. Estimated plunge of hinge line of fold; 4. Banding; 5. Foliation; 6. Schistosity (For 4–6, S_1, S_2, and S_3 indicate *s* planes related to the F_1, F_2, and F_3 folds, respectively); 7. Location of the specimen collected; 8. Axis of minor folds (upper symbol) and crenulation lineation (lower symbol); 9. Mineral distribution lineation either of mafic mineral (solid) or of salic mineral (open); 10. Mineral elongation lineation, including mineral orientation lineation (For 8–10, l_1 is undulation of S_1 by the D_1 folds or the intersection of S_1 and S_2, and l_2 is the intersection of S_1 (or S_2) and S_3); 11. Dimensional preferred orientation of a scattered mafic mineral; 12. Dimensional preferred orientation of mafic clots; 13. Dimensional preferred orientation of coarse salic crystals or their clots. Hand specimen structures were determined in laboratory from the oriented specimen. Although the field structures are written in an arbitrary scale, good conformity between the macroscopic structures and hand specimen structures exists in all specimens. The figure is modified from Yoshida (1977*b*), referring to his descriptions.

PLANAR AND LINEAR STRUCTURES

Alternation and banding

Alternation, banding, and thin banding structures are made of alternating bands of different compositions (Figures 62.2*a*, *b*, and 62.5). The alternations (meters thick) and the banding structures (centimeters to decimeters thick) are direct reflections of the primary sedimentary structures such as bedding, because the mineral compositions of layers commonly differ significantly, and because these structures are concordant with the marble and quartzite layers which are locally intercalated with the charnockitic rocks.

The thin banding structures (less than a centimeter thick) are generally parallel to the banding and alternation structures, but the compositions of the mafic laminae rarely differ significantly from each other; this fact is favorable to a metamorphic segregation origin for the thin banding. Some coarser minerals show elongated shapes parallel to the banding structures, but all minerals except some grains of biotite lack lattice preferred orientations. Fracture cleavage parallel to the bedding is a common forerunner of

Figure 62.6. Microscopic view of the foliation structure of the charnockitic plutonite. Foliation is made by rough elongation of mafic clots which appear to be folded in *ac* section; the foliation parallels the axial surface. (Specimen 68032502, collected from Teøya, Ongul Islands.) This figure is from Yoshida, 1977*b*.

metamorphic segregation along the bedding plane (e.g., Turner and Weiss, 1963, p. 98; Vernon, 1976, p. 217). Such cleavage may be formed by tectonic folding, such as the flexural slip type (Donath, 1962). Thus the formation of some of the thin compositional banding may be related to some of the folding, although Hobbs and others (1976, p. 264–6) have pointed out that metamorphic foliation parallel to the folded surface is not common.

The thin compositional banding is folded by all folds in the region and therefore may be synchronous with the oldest folding, i.e., the recumbent folds of the D_1 tectonic stage. The microfold lineation and some of the mineral distribution lineation, as shown in Figure 62.5, may also be referred to the tectonics of this stage. These lineations plunge variably and generally lie parallel to the hinges of the D_1 folds in each area.

Foliation

The foliation structure is made of either mafic or salic clots of leaflike shape (Harker, 1950, p. 203, 316). The microscopic characteristics are the same as those of the banding structures. The compositions of the mafic and salic folia are rather homogeneous as was the case with the thin compositional banding, and again this may indicate that the cause of the foliation was metamorphic segregation and/or modification of the thin banding.

The foliation is generally parallel to the thin banding, but in the crestal zones of the isoclinal folds of D_1 tectonics, the two *s* planes crosscut each other (Figures 62.5, 62.6, and 62.7). The discordancy can only be distinguished in a section cut nearly normal to the microfold axis and the mineral distribution lineations; the intersection of the foliation and the banding forms a distinct lineation of mineral distribution, which is generally parallel to the hinges of the isoclinal folds of the D_1 tectonic stage in each area. Thus the axial plane cleavage of the isoclinal folds is the preferred geometric explanation for the foliation in relation to the banding.

Schistosity

The schistosity is generally a result of lattice preferred orientation of biotite, and in some cases, of dimensional preferred orientations of other mafic or salic minerals (Harker, 1950, p. 195). The schistosity resulting from the dimensional preferred orientations of minerals is generally parallel to the foliation and therefore may be regarded as genetically grouped with this foliation. The biotite schistosity is locally parallel to, but otherwise discordant with the banding and/or foliation structures, and two kinds of biotite schistosity discordant with each other are sometimes detected together in one specimen (see Figure 62.8). The schistosity parallel to the banding structure is interesting, since minerals other than biotite are not as distinctly oriented, and furthermore, biotite is generally an alteration product after other mafic minerals. Since the microstructural facts, discussed below, indicate that some biotites of this kind predate the other mafic minerals, no explanation other than the crystal growth of the biotite on the preexistent biotite flakes which were preferentially oriented parallel to the banding structure is acceptable. Experimental study of the recrystallization of biotite by Etheridge and others (1974) pointed out the strong mimetic nature of this mineral in developing preferred orientation.

The biotite schistosity cutting cross the banding and/or foliation structures is a later stage product, developed parallel to the axial plane of folds of the D_2 tectonic stage.

In the rocks on the Ongul Islands where the hinges of the D_2 folds are nearly at right angles to those of D_1 folds, and therefore to the main lineations as well, a *bc* section (fabric axis *b* parallels the mineral distribution or microfold lineation related to the D_1 stage, *a* lies in the *s* surface and normal to *b*, and *c* is normal to both *a* and *b*) of specimens exhibits schistosity that crosscuts the banding and foliation structures. In the southern part of the Soya Coast, where the main lineation trends around to the east and is therefore parallel to the hinges of the D_2 folds, the crosscutting biotite schistosity is only detectable in an *ac* section of specimens (Figure 62.5).

Microstructures of the charnockites

Microstructural petrography

The charnockitic rocks around Lützow-Holmbukta have microstructural characteristics common to granulites everywhere else in the world (e.g., Behr and others, 1971). They are principally small- to medium-grained and granoblastic; the *s* surface, where present, is generally composed of a mafic or salic band or folium; and the constituent minerals generally lack lattice preferred orientations. The charnockitic gneisses differ from the charnockitic plutonites in that they are usually finer grained and commonly carry platy quartz. The general microstructural characteristics of the charnockitic gneisses and charnockitic plutonites are summarized in Table 62.1.

Microstructural development

The systematic method of microstructural analysis is not yet well developed, although some excellent speculations

Table 62.1. Microstructural characteristics of charnockitic plutonites and charnockitic gneisses

	Charnockitic plutonites	Charnockitic gneisses
Hand specimen and general texture	Holocrystalline, small- to medium-grained, granoblastic. Leucocratic but brownish to brown-greenish, and foliated or thinly banded by mafic-rich folia or laminae	Similar, but banded or thinly banded structures are commonly conspicuous, and platy quartz is common
Mineral association	Plagioclase ±, quartz, K-feldspar, orthopyroxene, clinopyroxene, biotite, hornblende, opaque mineral and garnet. Clinopyroxene is scarce, and garnet is rare	Similar, but commonly quartz-dominant
Plagioclase	Small granular—small/medium polygonal, commonly with later lobation of the grain boundary and coarsening in grain size	Analogous, but dominantly small granular
Quartz	a. Fine granular to polygonal xenocrysts in plagioclase and pyroxene, although this is absent in the basic variety b. Small lobate—embayed, at plagioclase-plagioclase boundary c. Infiltrated—porphyroblastics, especially in the acid varieties or the acid band	Analogous, but elongated shape is common
K-feldspar	a. Fine polygonal xenocrysts in one specimen of the syenitic variety b. Small polygonals in some specimens c. Antiperthitic patches in plagioclase d. Porphyroblastics are not rare, especially in the acid varieties or the acid band	Analogous, but slightly elongate shape is not rare
Orthopyroxene	Small to coarse, the periphery commonly poikiloblastic	Similar
Clinopyroxene	a. Fine granular xenocrysts in some specimens b. Small polygonals in some specimens c. Medium to coarse grains rarely occur in the charnockitic variety	Similar
Biotite	a. Fine xenocrysts b. Medium-grained, at grain boundary c. Medium to coarse grains, commonly with a wormy texture, replacing other mafics	Similar
Hornblende	a. Fine xenocrysts in some specimens b. Small polygonals in some specimens c. Poikiloblastic in some specimens	Analogous, but not common
Garnet	a. Fine to small irregular and rounded relict grains embedded in plagioclase-orthopyroxene domain in several specimens b. Medium to coarse, irregular, and poikiloblastic, associated with clinopyroxene and quartz in one specimen	Analogous, but *b*-type is not rare as in the plutonites

and petrologic works have been made from this viewpoint (e.g., Harker, 1950; Smith, 1964; Spry, 1969). Examination of the time relationships of the constituent minerals listed in Table 62.1 reveals four episodes of microstructural development, as follows.

First episode of microstructure

The oldest microstructure of the charnockitic rocks is considered to be fine-grained and equigranular, judging from the fine-grained, equigranular, and rather equant quartz and clinopyroxene, along with fine flaky biotite xenocrysts generally in small- to medium-grained polygonal plagioclase and locally in orthopyroxene (Figures 62.7 and 62.8). They occur in most of the charnockitic rocks, but are distinctly poor or absent in the granitic varieties. Very small amounts of micro- to fine-grained rounded garnet are found in some specimens, included within the small- to medium-grained polygonal plagioclase. Hornblende also occurs in limited amounts.

Second episode of microstructure

The second episode of microstructure is characterized by a small- to medium-grained foam texture, generally explained as a result of annealing recrystallization. This is exemplified by small- to medium-grained plagioclase in almost all the charnockitic rocks. K-feldspar, clinopyroxene, and hornblende occur locally and display similar foam texture associated with the plagioclase. They have smooth, locally straight grain boundaries, and near-120° triple-junction angles were found in places, although the grain boundary and triple junction are commonly disturbed by later events (Figures 62.7 and 62.9*b*). In Many specimens, quartz, K-feldspar, and orthopyroxene have lobate, amoeboid, or poikiloblastic peripheries. The central part of these minerals is considered to have participated in the construction of the foam texture of this episode, although their peripheries may belong to a later microstructure.

Third episode of microstructure

The third episode of microstructure is the coarsening of grain size. Poikiloblastic orthopyroxene and medium- to coarse-grained plagioclase are the most characteristic examples, and almost all the other major constituent minerals such as biotite, hornblende, and clinopyroxene also become coarser grained. However, in many cases, clinopyroxene decreases in amount, while in the same specimen, hornblende increases.

Fourth episode of microstructure

The microstructure of this episode is inequigranular lobate, characterized by porphyroblastic microclinic K-feldspar and quartz, along with alteration biotite (Figures

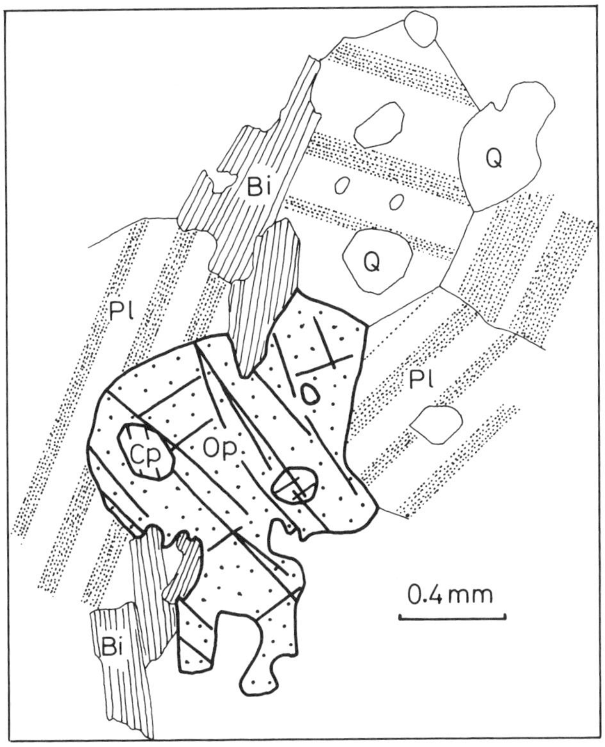

Figure 62.7. Fine xenocrystic clinopyroxene and quartz in the polygonal plagioclase and orthopyroxene. Pl: Plagioclase. Q: Quartz. Bi: Biotite. Cp: Clinopyroxene. Op: Orthopyroxene. (Charnockitic plutonite, collected from Teøya, Ongul Islands, specimen 68032502.) This figure is from Yoshida, 1977*b*.

62.9*a* and *b*). The formation of mesoperthite and the separation into two isolated feldspars may have occurred in this episode, probably near the beginning of it. Intergranular to network development of fine granitic aggregate is another characteristic of the microstructure of this episode. This may indicate a condition something like a combination of anatexis and recrystallization.

Metamorphic conditions

The microstructures of the four episodes described above are regarded as having formed by different metamorphisms or different phases of metamorphism. Preliminary considerations of the conditions of these metamorphisms are given below.

First metamorphism

Fine-grained biotite of the first episode of microstructure is arranged parallel to the thin banding and foliation, indicating predominance of a directed stress field during this microstructural episode. From mineral paragenesis, somewhat different physical conditions from the second episode are probable, judging from the relict occurrence of both clinopyroxene and garnet, which are included within orthopyroxene and plagioclase, and therefore suggest their transformation into these minerals. Such reactions are cited by Green and Ringwood (1967) as indicative of transition from the high-pressure granulites to intermediate-pressure granulites. The dominance of relict quartz and the scarcity of xenocrystic plagioclase are compatible with this interpretation.

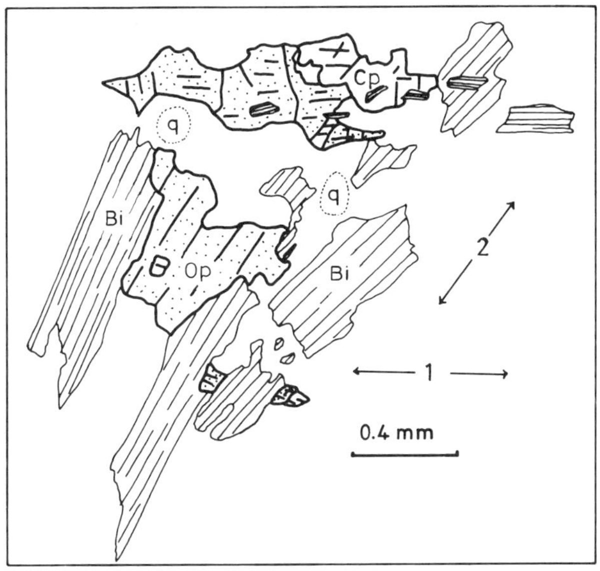

Figure 62.8. Two stages of biotites. Fine xenocrystic biotite (finely striated) is embedded in the poikiloblastic orthopyroxene (Op, high relief, stippled), clinopyroxene (Cp, high relief), and in coarse alteration biotite (Bi, roughly striated). Blank area is plagioclase, with fine granular quartz inclusions (q, circled by dotted line). Arrow 1 indicates the trend of the banding and foliation, and arrow 2 shows trend of the new schistosity. (Charnockitic plutonite, West Ongul Island, specimen 68022801.) This figure is from Yoshida, 1977*b*.

From preliminary microprobe analyses, xenocrystic clinopyroxene was found to be slightly but generally poorer in Al and Fe in comparison to the polygonal clinopyroxene of the second episode of microstructure occurring in the same thin section. This difference may be a reflection of the different genesis of the xenocrystic grains from the other grains of the same mineral.

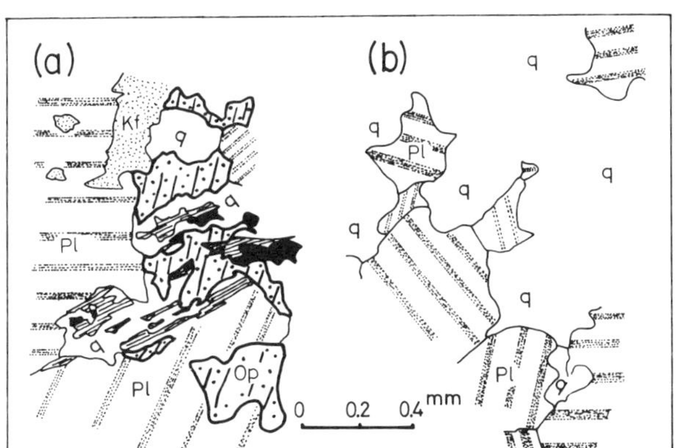

Figure 62.9 a. Alteration biotite on orthopyroxene. Striated: Biotite. High relief with dots (Op): Orthopyroxene. Solid: Opaque mineral. Blank (q): Quartz. Striped (Pl): Plagioclase. Dotted (Kf): K-feldspar. (Charnockitic plutonite, collected from Ongul Islands, specimen 68032504.) *b.* Porphyroblastic quartz. Very coarse porphyroblastic quartz (blank, with q) develops over the aggregates of small polygonal plagioclase (Pl, striped). The orientation of q indicates the crystallographic orientation. Thus, almost all quartz is continuous, forming large grains. Lobation of the originally straight grain boundary of the plagioclase is seen. (Charnockitic plutonite, Teøya, Ongul Islands, specimen 68032509.) This figure is from Yoshida, 1977*b*.

Table 62.2. Microprobe analyses (in wt %) of pyroxenes from a specimen of charnockitic plutonite (syenitic variety) collected from the Ongul Islands (No. 68030103)

	Host orthopyroxene[a]	Xenocrystic clinopyroxene[b]	Small isolated clinopyroxene	
SiO_2	50.44	51.06	50.92[a]	50.68[b]
TiO_2	0.11	0.19	0.20	0.23
Al_2O_3	0.81	1.89	2.14	2.09
FeO	32.32	13.86	14.32	14.88
MnO	0.74	0.28	0.33	0.31
MgO	14.02	10.16	10.03	10.12
CaO	0.73	21.24	21.13	20.75
Na_2	0.01	0.49	0.56	0.55
Total	99.18	99.17	99.63	99.61
Number of ions in formula on the basis of 6 oxygens				
Si	1.994	1.965	1.955	1.950
Ti	0.004	0.006	0.006	0.007
Al	0.037	0.085	0.097	0.095
Fe	1.068	0.446	0.460	0.479
Mn	0.025	0.009	0.011	0.010
Mg	0.826	0.583	0.574	0.580
Ca	0.031	0.876	0.869	0.855
Na	0.001	0.036	0.042	0.042
Molecular percent				
Mg	42.4	30.4	30.0	30.2
Fe	56.0	23.8	24.6	25.4
Ca	1.6	45.8	45.4	44.4
Total	100.0	100.0	100.0	100.0

Analyst, S. Yoshikura
[a] Average of 3 analyses on different parts of one single crystal.
[b] Average of 4 analyses on different parts of one single crystal.

Chemical analytical data of pyroxenes are listed in Table 62.2. The lower content of Al in the xenocrystic clinopyroxene may not appear consistent with the assumption of higher pressure condition (e.g., Wilson, 1976), but the composition of pyroxenes may also be affected by mineral paragenesis; if garnet coexisted with the xenocrystic clinopyroxene, Al would tend to enter preferentially into the garnet and to decrease in the clinopyroxene. Systematic analyses of the minerals of this microstructural episode are needed.

Second metamorphism

Foam texture, the principal feature of the second episode of microstructure, is characteristic of annealing recrystallization under static tectonic conditions. Granuloblastic plagioclase and two pyroxenes are the typomorphic minerals of this episode along with K-feldspar and hornblende. Biotite existed in a stable or metastable condition during this episode. The occurrence of a garnet–sillimanite–K-feldspar assemblage in paragneisses, and the absence of cordierite and olivine in all rocks of appropriate compositions (Yoshida, 1978), may also provide critical limits. These mineral assemblages indicate granulite facies conditions, suggestive of the intermediate–pressure granulites of Green and Ringwood (1967).

Metamorphism of the third episode of microstructure

Coalescence of plagioclase and enlargement of orthopyroxene are the dominant events of the third episode of microstructure along with the enlargements of K-feldspar and hornblende. This metamorphism may prove to be a late phase of the second metamorphism.

In some specimens of charnockites with low Mg:Mg + Fe^{2+} ratios, coarse clinopyroxene and garnet appear in association with quartz. This association here may indicate either different physical or chemical conditions than those of the main phase of the second metamorphism (e.g., Wagner and Crawford, 1975). These minerals came either in the slightly later phase of the third episode or in the next metamorphism.

Third metamorphism

Acidification of plagioclase, microclinization of perthite, and alteration of pyroxenes and some garnet into biotite are characteristics of the fourth episode of microstructure, indicating retrograde conditions. Porphyroblastic developments of quartz and K-feldspar in some specimens indicate metasomatic conditions. Network developments of fine granitic aggregate may indicate anatectic conditions combined with recrystallization. All these textures and minerals indicate intermediate metamorphic conditions with significant water pressure, that is, the amphibolite facies.

Development of charnockites in the tectonic-metamorphic history of the region around Lützow-Holmbukta

The second microstructural episode is characterized by a posttectonic annealing over both thin banding and foliation; the banding has been referred to the recumbent folds, and the foliation to the isoclinal folds, of the D_1 tectonic stage. Schistosity, formed mainly by dimensional preferred orientation of biotite parallel to the axial planes of D_2 folds, is referred to the fourth episode of microstructure, characterized by development of porphyroblastic salic minerals and alteration biotite. The second and third microstructural episodes, therefore, are placed between the D_1 and D_2 tectonic stages.

Table 62.3. Generalized tectonic and plutonometamorphic history of the region around Lützow-Holmbukta (Lützow-Holm Bay)

Marcroscopic structures	Mesoscopic structures	Microstructures	Metamorphism and plutonism
D_1 tectonic stage 1. Recumbent folds 2. Isoclinal folds	1. Bedding schistosity and thin banding 2. Axial plane foliation	(1st episode) Fine granular clinopyroxene, quartz, biotite (and garnet)	High-pressure granulites (?)
	Transversal joints filled with charnockitic pegmatites	(2nd and 3rd episodes) Plagioclase and two pyroxenes	Intermediate-pressure granulites and charnockitic rocks
D_2 tectonic stage Open to close folds	Schistosity parallel to the axial plane	(4th episode) Porphyroblastasis of salic minerals and alteration biotite	Amphibolite facies and granitic gneisses
D_3 tectonic stage Gentle folds	Schistosity in pink granites, parallel to the intrusive boundary, very sporadic	Fine granitic aggregates of the 4th episode may be about this age	Pink granites
D_4 tectonic stage Shear fractures	Nonpenetrative fractures	Not observed	Not observed

Field observations appear to demonstrate the concordance of charnockitic gneisses and plutonites with both the recumbent and isoclinal folds of the D_1 tectonic stage, but this may only indicate the position of the charnockite formation after the foldings, since the charnockitic rocks are considered, from their occurrence, to be formed from some distinct stratigraphic horizon of rocks. Development of some charnockitic pegmatite along axial planes of the minor folds of the D_1 stage only indicates the formation of some quartz-feldspathic rocks along with the folding, since the quartz-feldspathic rocks could have been charnockitized with the country rocks in some later period.

Those charnockitic pegmatites developed in thin veins trending at right angles to the axial planes of the D_1 folds may indicate the youngest limit of charnockite formation. These veins are considered to fill transverse joints of the D_1 folds.

Mutual relationships between macroscopic, mesoscopic, and microscopic tectonics and microstructures, supplemented with the inferred metamorphic sequence, are summarized in Table 62.3.

References cited

Banno, S., Tatsumi, T., Ogura, Y., and Katsura, T. 1964. Petrographic studies on the rocks from the area around Lützow-Holmbukta. (*In* Adie, R. J., *ed. Antarctic Geology.* Amsterdam, North-Holland Publishing Company, 405–14.)

Behr, H. J., den Tex, E., de Waard, E., Mehnert, K. R., Sharbet, H. G., and others. 1971. Granulites—Results of a discussion. *Neues Jb. Miner. Mh., 1971,* 97–123.

Donath, F. A. 1962. Role of layering in geologic deformation. *Trans. N. Y. Acad. Sci., 24,* 236–48.

Etheridge, M. A., Paterson, M. A., and Hobbs, B. E. 1974. Experimentally produced preferred orientation in synthetic mica aggregates. *Contr. Miner. Petrol., 44,* 275–94.

Green, D. H., and Ringwood, A. E. 1967. An experimental investigation of the gabbro to eclogite transformation and its petrological applications. *Geochim. cosmochim. Acta, 31,* 767–833.

Harker, A. 1950. *Metamorphism.* London, Methuen, 362 pp.

Hobbs, B. E., Means, W. D., and Williams, P. F. 1976. *An Outline of Structural Geology.* New York, John Wiley, 572 pp.

Ishikawa, T. 1976. Superimposed folding of the Precambrian metamorphic rocks of the Lützow-Holm Bay region, East Antarctica. *Mem. Natn Inst. Polar Res. Japan,* Ser. C, *9,* 41 pp.

Johannsen, A. 1939. *A Descriptive Petrography of the Igneous Rocks, 1.* 2nd ed. Chicago, University of Chicago Press, 360 pp.

Kizaki, K. 1964. Tectonics and petrography of the East Ongul Island, Lützow-Holmbukta. Antarctica. *JARE Sci. Rep.,* Ser. C, *2,* 24 pp.

Smith, C. S. 1964. Some elementary principles of polycrystalline microstructure. *Metall. Rev., 9,* 1–48.

Spry, A. 1969. *Metamorphic Textures.* Oxford, Pergamon Press, 350 pp.

Suwa, K. 1968. Petrological studies on the metamorphic rocks from the Lützow-Holmbukta area, East Antarctica. *23rd Int. Geol. Congr., Czechoslovakia, 1968, 4,* 171–87.

Tatsumi, T., and Kizaki, K. 1969. Geology of the Lützow–Holm Bay region and the "Yamato Mountains" (Queen Fabiola Mountains). (*In* Bushnell, V. C., and Craddock, C., *eds.* Geologic Maps of Antarctica. *Antarct. Map. Folio Ser.,* Folio 12, Pl. IX, X.)

Tatsumi, T., Kikuchi, T., and Kizaki, K. 1964. Geology of the region around Lützow-Holmbukta and the "Yamato Mountains" [Dronning Fabiolafjella]. (*In* Adie, R. J., *ed. Antarctic Geology.* Amsterdam, North-Holland Publishing Company, 293–303.)

Tobi, A. C. 1971. The nomenclature of the charnockitic rock suite. *Neues Jb. Miner. Mh., 1971,* 193–205.

Turner, F. J., and Weiss, L. E. 1963. *Structural Analysis of Metamorphic Tectonites.* New York, McGraw-Hill, 545 pp.

Vernon, R. H. 1976. *Metamorphic Processes.* London, Thomas Murby, 243 pp.

Wagner, M. E., and Crawford, M. L. 1975. Polymetamorphism of the Precambrian Baltimore gneiss in southeastern Pennsylvania. *Am. J. Sci., 16,* 323–38.

Wilson, A. F. 1976. Aluminium in coexisting pyroxenes as a sensitive indicator of changes in metamorphic grade within the mafic granulite terrane of the Fraser Range, Western Australia. *Contr. Miner. Petrol., 56,* 255–77.

Yoshida, M. 1977*a.* Geology of Skallen region, Lützow-Holmbukta, East Antarctica. *Mem. Natn Inst. Polar Res. Japan,* Ser. C, *11,* 38 pp.

Yoshida, M. 1977*b. Tectonics and Petrology of Charnockites around Lützow-Holmbukta, East Antarctica.* D.Sci. dissertation, Hokkaido University, 75 pp.

Yoshida, M. 1978. Tectonics and petrology of charnockites around Lützow-Holmbukta, East Antarctica. *J. Geoscis Osaka Cy Univ., 21,* 65–152.

Yoshida, M., and Ando, H. 1971. Geological surveys on the vicinity of Lützow-Holm Bay and the Yamato Mountains, East Antarctica. *Antarctic Rec., 38,* 46–54.

Yoshida, M., Yanai, K., Matsumoto, Y., Ishikawa, T., and Kizaki, K. 1977. Geologic structure of the region around Lützow-Holmbukta, East Antarctica (in Japanese). *Abstr. 84th Annual Meeting, Geol. Soc. Japan, Kōchi, 1977,* 90.

63

Geologic Structure of the Prince Charles Mountains, East Antarctica

Abstract

L. V. FEDOROV
Department of Antarctic Geology, SEVMORGEO, Moyka 120, Leningrad 190121, U.S.S.R.

During 1971–74 the Soviet Antarctic Expedition carried out a geologic-geophysical survey of the Prince Charles Mountains area totalling 300,000 km². As a result, geologic and tectonic maps on a scale of 1:2,000,000 were compiled.

The bedrock comprises about 15% of the surveyed area and ranges in age from Archean to Permian. The crystalline basement of the Antarctic platform substantially prevails over other rock units and consists of metamorphic sequences as much as 20 km thick; these sequences have been formed under granulite and superimposed high-temperature amphibolite facies with widespread migmatite and skialithic granitoid of ultrametamorphic origin. Ultrametagenic enderbite and charnockite are widespread along with aluminiferous gneiss and amphibolite. Metamorphic units are cut by rheomorphic intrusions of ancient charnockitoid (porphyroblastic granosyenite) and younger biotite granite.

Latitudinal aulacogens that formed on the crystalline basement surface were filled by thick (as much as 7 km) Proterozoic terrigenous deposits initially metamorphosed to greenschist facies and partially recrystallized in a high-pressure amphibolite facies environment, which resulted in development of a wide variety of zonal metamorphic rocks such as staurolite, disthene, and sillimanite schist. Thick horizons of jaspilite are found at the base of the Proterozoic sequence otherwise consisting of chlorite-micaceous schist, quartzite, intraformational conglomerate, and sills and dikes of metabasite; early Paleozoic intrusions of two-mica granite and veins of pegmatite are also present.

In post-Permian time, a longitudinal continental rift transversed the Precambrian complexes along the present site of the Lambert Glacier, and only relics of Permian coal measures of Gondwanian type reaching 1300 m in thickness are now preserved within the rift graben. This Permian formation is identical to the synchronous strata of the Beacon Group in the Transantarctic Mountains. Thin dikes and sills of Cretaceous alkaline-ultrabasic lamprophyre are present in the Permian sedimentary rocks.

64

Structural Development of Precambrian Rocks in the Mountain Fringe of Lambert Glacier and the Southern Part of the Prince Charles Mountains, East Antarctica

Abstract

L. V. FEDOROV
Department of Antarctic Geology, SEVMORGEO, Moyka 120, Leningrad 190121, U.S.S.R.

J. HOFMANN
Section of Geosciences, Mining Academy, 92 Freiberg, German Democratic Republic

Results of tectonic investigations in granulite, almandine-amphibolite, and greenschist facies rocks in the sector between long. 40° and 80° E are briefly discussed. An older early Proterozoic–late Archean complex with east-west and northeast-southwest oriented structural trends, widespread in the whole area, is overlain in the southern Prince Charles Mountains by a middle–late Proterozoic unit of greenschist facies and stress-accented, low-temperature, almandine-amphibolite facies. Main tectonic features and inner subdivisions of the basement complex are largely unknown. Fold structures and tectonic development of the younger complex are influenced by block movements in the basement complex. Locally, the basement complex is in tectonic contact with sequences of the younger unit. A division of the younger unit into two subcomplexes is possible.

IV Upper Precambrian–Paleozoic Rocks

65

Uppermost Proterozoic and Lower Paleozoic Geology of the Transantarctic Mountains

Review Paper

M. G. LAIRD
New Zealand Geological Survey, Department of Scientific and Industrial Research, Christchurch, New Zealand

J. D. BRADSHAW
Geology Department, University of Canterbury, Christchurch, New Zealand

Abstract. The Transantarctic Mountains are a linear zone of Cenozoic horsts which expose a number of upper Proterozoic and lower Paleozoic sedimentary sequences and tectonic belts. Early syntheses recognized a comprehensive Ross Supergroup, deformed by a Ross orogeny of early Paleozoic age and truncated by a sub-Beacon Supergroup erosion surface. Subsequent work has shown that the Ross "Supergroup" can be divided into separate Precambrian and lower Paleozoic sequences.

All fossiliferous lower Paleozoic rocks in the Transantarctic Mountains are Cambrian in age and constitute an unconformity-bounded sequence which in the central region exceeds 8 km in thickness. Lower Cambrian Archeocyatha limestones crop out in the Argentina Range and in the central Transantarctic Mountains, and erratics of similar rocks occur in the Whichaway Nunataks. Middle Cambrian clastic and carbonate rocks are widespread, occurring sporadically throughout much of the mountain chain. Fossiliferous Upper Cambrian strata are known only from northern Victoria Land, but unfossiliferous sequences in the Pensacola and Queen Maud Mountains may also be of this age. Diagnostic early Paleozoic fossils younger than Cambrian are not known, but coarse clastic sequences of probably Ordovician age occur in northern Victoria Land and perhaps also in the Neptune Range.

The base of the Cambrian sequence is generally unconformable on older rocks and is marked by an abrupt change in sedimentary regime, with a Cambrian carbonate-quartzite-mudstone assemblage overlying older rocks of "graywacke" type. The term *Ross Supergroup,* which in current usage includes rocks both above and below the unconformity, is inconvenient and should be abandoned. It is suggested that the term *Ross Sequence* be applied to the mainly Cambrian rocks which rest unconformably on the Beardmore or older orogens and which were folded during the Late Cambrian–Early Ordovician Ross orogeny. Units such as the Neptune Group, which postdate the Ross orogeny, are excluded. The scheme may not apply in northern Victoria Land, which appears to lie in a subparallel but separate sedimentary-tectonic belt.

Although sedimentary rocks of early Paleozoic age are known or inferred in many parts of Antarctica, they are best developed in the Transantarctic Mountains, which stretch for more than 3000 km across the continent. The presence of lower Paleozoic rocks in the Transantarctic Mountains (and in the whole of Antarctica) was first recorded by Taylor (1914)—from limestone erratics containing the Cambrian index fossil Archaeocyatha, collected from the Beardmore Glacier by members of Shackleton's 1908–9 expedition. Another report treated similarly fossiliferous specimens collected *in situ,* reportedly from Buckley Nunatak (Skeats, 1916). Archaeocyatha-bearing specimens were collected from moraines at the foot of the Beardmore Glacier by the pole party of Scott's last expedition, who dragged them for hundreds of kilometers until all in the party died. The later dredging of similar fossiliferous material from the floor of the Weddell Sea (Gordon, 1920) suggested that the distribution of Cambrian limestone could be widespread, but at that time its stratigraphic relationship to the flat-lying upper Paleozoic Beacon Supergroup and the strongly deformed metasedimentary rocks and gneisses of the Precambrian basement could not be determined.

It was not until the early 1960's that this problem was solved by New Zealand and United States field parties working respectively in the Ross Sea and Weddell Sea sectors of the Transantarctic Mountains; fossiliferous Cambrian rocks were found to lie unconformably on Precambrian metasedimentary beds and to underlie unconformably the flat-lying Beacon Supergroup. Since then the results of numerous geologic expeditions have shown that lower Paleozoic rocks are widespread throughout the length of the Transantarctic Mountains (Figure 65.1), locally attaining great thickness, as in the central part where more than 8,000 m are preserved.

Lower Cambrian Archaeocyatha limestones are known in outcrop from the Nimrod and Beardmore Glacier areas and from the Argentina Range, while erratics of similar rocks occur in the Whichaway Nunataks in Coats Land. Middle Cambrian rocks, both clastic and carbonate, are widespread, occurring sporadically throughout the entire mountain chain. Upper Cambrian strata are known with

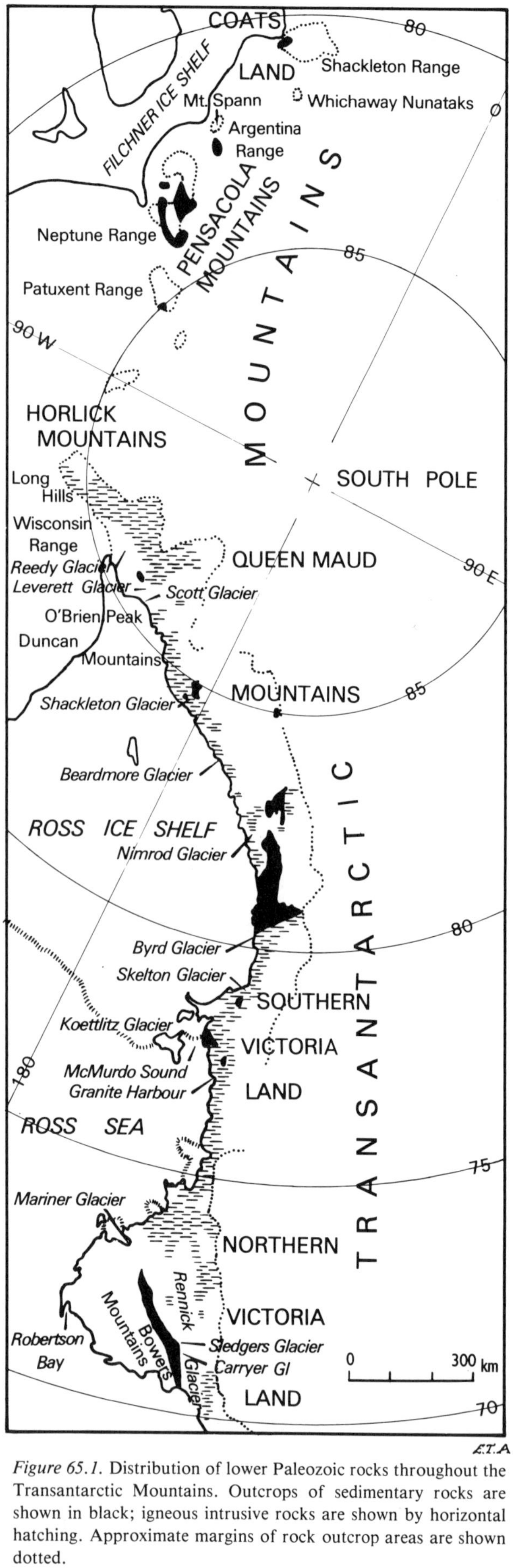

Figure 65.1. Distribution of lower Paleozoic rocks throughout the Transantarctic Mountains. Outcrops of sedimentary rocks are shown in black; igneous intrusive rocks are shown by horizontal hatching. Approximate margins of rock outcrop areas are shown dotted.

certainty only from northern Victoria Land, but unfossiliferous sequences lying above fossiliferous Middle Cambrian rocks in the Pensacola Mountains and in the Queen Maud Mountains may include Upper Cambrian beds. No diagnostic early Paleozoic fossils younger than Late Cambrian in age have yet been reported from anywhere in the region, although unfossiliferous formations overlying dated Upper Cambrian units in northern Victoria Land almost certainly include some Ordovician beds.

Figure 65.2 shows the correlation of lower Paleozoic units in the Transantarctic Mountains. The lower Paleozoic formations of the Ellsworth Mountains are also included for reference.

Regional descriptions

Coats Land

Although not included in the Transantarctic Mountains proper by some writers, the isolated ranges and nunataks extending along the eastern side of the Filchner Ice Shelf into Coats Land form a logical topographic and geologic extension of these mountains.

In the western part of the Shackleton Range, the basement rocks, which consist of the Precambrian Shackleton Range Metamorphic Complex, are overlain unconformably by the Blaiklock Glacier Group, a 6000-m-thick clastic sedimentary sequence (Clarkson, 1972). The Mount Provender Formation, the older of the two units into which the group is divided, is composed almost entirely of feldspathic sandstone, with breccia forming the basal 6 m. The sandstone is red or gray, mainly fine grained, and locally cross-bedded. Red-gray micaceous shales and slates make up a minor part of the unit. The top of the formation is not exposed, but the minimum thickness is 760 m.

The stratigraphically higher Otter Highlands Formation consists mainly of a uniform sequence of hard, gray-green, feldspathic, fine-grained conglomerate and sandstone. Cross-bedding and intraformational folding are common. The lowest exposed units of the formation include a compact even-bedded conglomerate with pebbles and small boulders of metamorphic rocks, thin beds of micaceous shale and sandstone, and, at the bottom of the sequence, thick beds of calcareous grit and sandstone. Neither the top nor the bottom of the formation is exposed, but the minimum thickness is 5300 m.

No fossils have been recorded from the Blaiklock Glacier Group, but a minimum age is provided by a crosscutting dolerite dike K-Ar dated at 297 ± 12 m.y. (Late Carboniferous) (Rex, 1972). Nearby fossiliferous erratics of silty shale containing inarticulate brachiopods are assumed to belong to an intermediate unit of the Blaiklock Glacier Group lying between the two exposed formations but concealed by glacier ice. The brachiopods have been identified as obolids (Thompson, 1972), which range from Lower Cambrian to Upper Ordovician; thus, the age of the Blaiklock Glacier Group is probably Cambro-Ordovician. Recently trilobites of probably Middle Cambrian age have been found in similar morainal material from the same area by the 1976–77 Soviet Antarctic Expedition (Soloviev and Grikurov, 1978).

No other outcrops of lower Paleozoic rocks have been recorded from Coats Land. However, morainal material collected from the Whichaway Nunataks, which lie 100 km

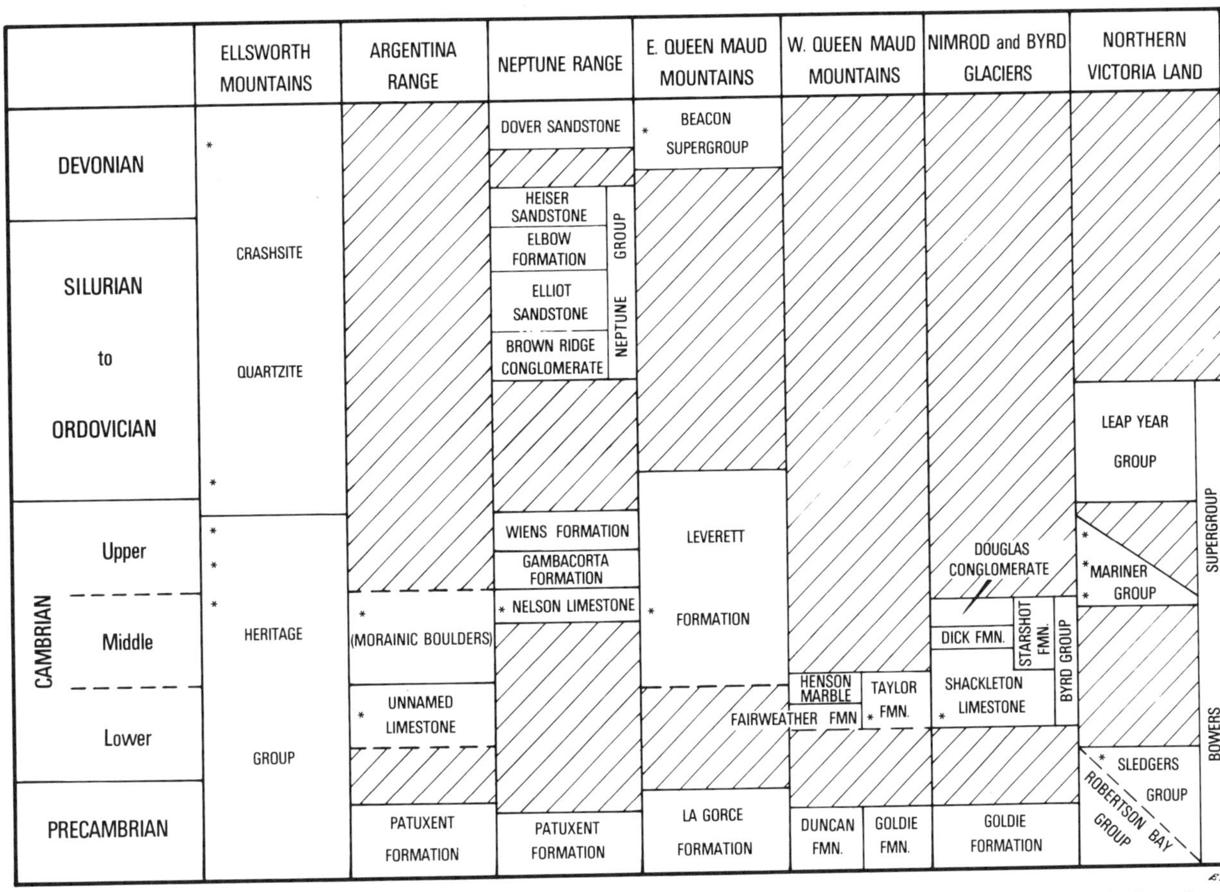

Figure 65.2. Correlation of lower Paleozoic stratigraphic units in the Transantarctic Mountains. The lower Paleozoic units of the Ellsworth Mountains are also included for reference. Correlation of Precambrian units is approximate only. Asterisks indicate fossil horizons.

south of the Shackleton Range and are themselves composed of Upper Paleozoic Beacon Supergroup sedimentary beds, contains numerous specimens of limestone, some of which contain Archaeocyatha dated by Hill (1965) as Early Cambrian in age. No outcrops of limestone occur in the area, and it is considered that Cambrian limestone lies nearby, buried under the ice.

Pensacola Mountains

In the southern Argentina Range occur extensive outcrops of an unnamed Archaeocyatha limestone unit, but the fauna has not yet been described. At Mount Spann, about 65 km to the northeast, morainal limestone boulders bear Archaeocyatha, mixed Archaeocyatha and trilobites, and trilobite-mollusk associations which indicate the presence nearby of limestone ranging in age from Early Cambrian to late Middle Cambrian. Almost all the trilobites examined are close relatives of Australian, Chinese, or Siberian forms (Palmer and Gatehouse, 1972).

Eleven sedimentary and volcanic rock units (of which at least three are lower Paleozoic) have been mapped in the Neptune and Patuxent Ranges. The oldest unit, the Patuxent Formation, consists of isoclinally folded sandstone, slate, and basaltic and rhyolitic flows of late Precambrian age. Overlying the Patuxent Formation with angular unconformity are three formations of Cambrian age: the Nelson Limestone, the Gambacorta Formation, and the Wiens Formation (Schmidt and others, 1965).

The Nelson Limestone is exposed in the southern part of the Patuxent Range and in the Neptune Range, where five members are recognized. The lowest two members are, respectively, a thin basal conglomerate as much as 6 m thick and an overlying red-bed clastic succession reaching 18 m in thickness; the upper three members are limestone, oolitic and pisolitic in part. The total thickness varies from 180 m to at least 550 m. The formation evidently consists of shallow-water deposits formed during marine transgression. Two different faunal assemblages of trilobites, indicating a middle or late Middle Cambrian age, were collected from the oldest limestone beds in the section (Palmer and Gatehouse, 1972).

The Gambacorta Formation, which conformably overlies the Nelson Limestone, crops out only in the Neptune Range. The base and lower part of the formation, about 90 m thick, consist of intensely altered sandstone and conglomerate composed of volcanic detritus. The rest of the formation consists of interlayered dark brown, red-brown, and light green rhyolite flows, volcanic breccias, pyroclastic deposits, and detrital sandstone and conglomerate composed mainly of volcanic clasts. The maximum thickness of this formation is 330 m. Fossils have not been found, but the isotopic age of interbedded rhyolite is 563 ± 35 m.y. (Eastin and others, 1969).

The Wiens Formation conformably overlies the Gambacorta Formation in the Neptune Range, reaching a maximum thickness of 330 m. In several places the lower part of the Wiens Formation intertongues with volcanigenic sedimentary rocks of the upper part of the Gambacorta

Formation. It consists of interlayered green and red-brown thin-bedded shale, siltstone, and fine-grained sandstone, with thin gray oolitic limestone beds at some horizons. Only trace fossils have been recorded; although locally abundant, they give no indication of age.

A marked angular unconformity, attributed to Late Cambrian uplift and folding (Schmidt and Ford, 1969), separates the Wiens Formation from the overlying Neptune Group. This group is a folded red-bed clastic sequence without diagnostic fossils and of indefinite age, although a minimum age of Late Devonian is given by plant fossils in disconformably overlying sandstones (Schmidt and Ford, 1969). The Neptune Group has been included within the Beacon Supergroup by Barrett and others (1972), although the age may be in part early Paleozoic.

Horlick and eastern Queen Maud Mountains

Lower Paleozoic rocks in this area are restricted to a few widely separated outcrops. North of the Leverett Glacier a sequence of limestone, shale, sandstone, and rhyolite, making up the Leverett Formation, crops out. It probably exceeds 2 km in thickness (Minshew, 1967), although neither the top nor bottom of the formation is exposed. The formation consists of seven units composed of, from oldest to youngest: 300 m of coarse-grained, red, cross-bedded, arkosic sandstone; poorly sorted, red, massive, pebbly mudstone 100 m thick; thin-bedded shale 125 m thick; interbedded limestone and shale 150–175 m thick; thin-bedded, dark gray, argillaceous limestone at least 175 m thick and containing trilobite fragments near the middle of the unit; massive, white limestone 100 m thick; and interbedded rhyolite and coarse-grained conglomeratic sandstone, probably 1000 m or more thick. All these sedimentary deposits accumulated largely in a shallow-marine environment. The strongly deformed and slightly metamorphosed trilobite fragments are considered to be of Middle Cambrian age (Minshew, 1966; Palmer, 1970). In the same nunatak, the rhyolite stratigraphically above the trilobite-bearing bed gives a Rb-Sr radiometric age of 489 ± 30 m.y.; an apparently equivalent rhyolite in the Long Hills, southeastern Wisconsin Range, gives an age of 498 ± 45 m.y. (Faure and others, 1968.)

Along the strike from the Leverett Glacier succession near O'Brien Peak, exposures of limestone, marble, and sandstone, although more highly metamorphosed than those north of the Leverett Glacier, may be correlative (Katz and Waterhouse, 1970). Similar rocks also occur in an isolated nunatak at the head of the Reedy Glacier in the southwestern Horlick Mountains (Mirsky, 1969).

Central Transantarctic Mountains

This region, stretching from the eastern Queen Maud Mountains to the Byrd Glacier, contains the thickest sequence of fossiliferous Cambrian sedimentary rocks known in Antarctica. In the Beardmore Glacier–Byrd Glacier area, all the Cambrian sedimentary rocks are included within the Byrd Group (Laird, 1963), which is divided into four formations. The Shackleton Limestone, which comprises the lowermost unit and rests unconformably on the upper Precambrian Beardmore Group, is also the most widespread and distinctive unit. Between the Nimrod and Byrd Glaciers, where almost continuous exposures occur, this formation attains a thickness of nearly 8300 m (Grindley and Laird, 1969; Laird and others, 1971). The Shackleton Limestone is a highly calcareous unit, consisting dominantly of pure limestone, but with local lenses of clastic sedimentary rocks, including conglomerate as much as 150 m thick. The basal deposits vary from locality to locality. South of the Nimrod Glacier they are quartzites, reaching a total thickness of 500 m; in the immediate vicinity of the glacier, limestone is in direct contact with Precambrian rocks. Fifty km north of the Nimrod Glacier as much as 1200 m of breccia derived from local Precambrian rocks underlie a great thickness of limestone (Laird and others, 1971). Archaeocyatha occur sporadically throughout most of the limestone sequence, but only those from immediately north of the Nimrod Glacier have been positively dated. Here a small collection from beds within 1000–1300 m of the base of the Shackleton Limestone has been assigned to the lower Lena Stage (upper Lower Cambrian) (Hill, 1964). The stratigraphically highest Archaeocyatha occur at least 5400 m above the base of the formation in beds that may be as young as early Middle Cambrian. The uppermost 2000 m of the formation is characterized by a thick limestone conglomerate unit and thin limestones.

East of the carbonate rocks is a narrow belt of clastic sedimentary rocks referred to as the Starshot and Dick Formations and the Douglas Conglomerate (Laird, 1963, 1964; Skinner, 1964, 1965; Laird and others, 1971). This succession consists of at least 3000 m of conglomerate, calcareous sandstone, mudstone, and several thin rhyolite flows. The formations are unfossiliferous, and their relationship to the Shackleton Limestone is not clear, but the presence in the conglomerate of limestone boulders very similar to the Shackleton Limestone suggests that it is at least in part younger than that formation.

In the Shackleton Glacier area, 100 km south of the southernmost mapped exposures of the Byrd Group in the Beardmore Glacier area, a thick sequence of quartzite, marble, and felsic volcanic rocks, referred to as the Taylor Formation, crops out (McGregor and Wade, 1969). Fossils identified as the Lower Cambrian form *Cloudina?* (Stump, 1974) have been collected from the sequence and suggest correlation of the Taylor Formation with the Byrd Group. Two formations regarded as the stratigraphic equivalents of the Taylor Formation, the Fairweather Formation and the Henson Marble, occur 60 km further south in the western Queen Maud Mountains. The Fairweather Formation, considered to be Lower Cambrian, (McGregor, 1965), is a homogeneous unit of recrystallized quartzitic rock between 1500 m and 3000 m thick. It is overlain gradationally by 500 m of coarsely crystalline calcareous rocks of the Henson Marble.

Southern Victoria Land

Many of the sedimentary rocks in this area are highly metamorphosed, commonly to amphibolite facies. In the south, where the rocks are less altered, they are included in the Skelton Group (Gunn and Warren, 1962), while the northern, more metamorphosed sequences are included in the Koettlitz Group (Grindley and Warren, 1964; Warren, 1969). Although these two groups have different metamorphic grades, it is likely that they are in part correlative.

In the Skelton Glacier area, the Skelton Group has been

divided into two formations, the Anthill Limestone and the Teall Greywacke. The Anthill Limestone consists primarily of white, gray, or green recrystallized limestone and marble, but it has been divided by Flory and others (1971) into five units, including graywacke-mudstone, quartzite, and slate, as well as limestone, which total perhaps 1000 m in thickness (D. Skinner, personal communication). Recent field work (Skinner, Chapter 69, this volume) has suggested that the Anthill Limestone includes as a lateral facies variant the type Teall Greywacke, and the definition of the Teall Greywacke is at present undergoing revision. The thickness of this unit at Teal Island (the type locality) is approximately 400 m (D. Skinner, personal communication). In the central part of the Skelton Glacier an unnamed unit of volcaniclastic metasedimentary rocks and mainly felsic metavolcanic rocks rests with sharp angular unconformity on the Anthill Limestone (Skinner, Chapter 69, this volume). No fossils have been recorded from the Skelton Group, but the calcareous nature of the Anthill Limestone suggests possible correlation with the Lower Cambrian Shackleton Limestone to the south. The unnamed felsic volcanic unit may represent extrusive activity associated with the emplacement of the Granite Harbour Intrusives during the Late Cambrian and Early Ordovician Ross orogeny (Skinner, Chapter 69, this volume).

In the Koettlitz Glacier–McMurdo Sound region, the Koettlitz Group (Grindley and Warren, 1964) consists of more than 6000 m of schist, marble, gneiss, and quartzite metamorphosed to amphibolite grade. Cylindrical siliceous objects resembling Archaeocyatha have been recovered from marble of the Koettlitz Group (Blank and others, 1963), but their origin is now considered to be inorganic (Findlay, 1978); hence the age of the group is not fixed, but it may be Cambrian in part.

Northern Victoria Land

All known lower Paleozoic sedimentary rocks of northern Victoria Land, which forms the Pacific terminus of the Transantarctic Mountains, are included within the Bowers Supergroup (Laird and others, 1976; Laird and others, Chapter 66, this volume). These sedimentary beds are preserved within a fault-bounded strip, 20–25 km wide, extending southeastward from the mouth of the Rennick Glacier for a distance of at least 270 km. They are in probable fault contact with the upper Precambrian Robertson Bay Group to the northeast, and they are in part faulted against the Precambrian Wilson Group to the southwest.

The Bowers Supergroup includes three major units, separated from each other by unconformities—the Sledgers Group, the Mariner Group, and the Leap Year Group—together with an older unit, the Husky Conglomerate. This latter unit, which consists mainly of amphibolitic conglomerate and breccia at least 370 m thick, is in fault contact with the Sledgers Formation and rests unconformably on amphibolite of the Wilson Group. It is of uncertain affiliation but may correlate with either the Robertson Bay Group or the base of the Sledgers Group. It is probably Vendian in age or older.

The Sledgers Group, whose base is not exposed, is at least 3500 m thick. It is composed in its northern part mainly of angular unsorted basaltic breccia with some thin flow units and interbedded thin units of shallow-water mudstone, sandstone, conglomerate, and limestone. The volcanic rocks thin to the south and interfinger with deeper water sedimentary beds consisting mainly of mudstone and well-sorted, wavy bedded, volcanigenic, fine-grained sandstone. Interspersed in these fine-grained sedimentary rocks are lenticular mass-movement deposits. Shelly fossils are absent from all but the youngest beds of the Sledgers Group, but acritarchs present suggest that the age of the group ranges from Vendian to earliest Cambrian (Cooper and others, Chapter 77, this volume).

The Mariner Group is separated from the underlying Sledgers Group by a hiatus which may include much of Early and Middle Cambrian time. The group, which is probably at least 2 km thick, represents a transgressive/regressive sequence dominated by shallow and marginal marine environments. The basal few meters consist of brown, muddy sandstone and mudstone showing dessication cracks and arthropod trails, overlain by approximately 100 m of quartzitic sandstone. This is followed by several hundred meters of calcareous mudstone with fossiliferous limestone lenses. The regressive part of the sequence is well exposed at the type locality at the head of the Mariner Glacier, where a continuous succession 1600 m thick occurs. Here fossiliferous calcareous mudstone passes upward in turn into wavy bedded, fine-grained sandstone, hyolithid limestone, and finally reddish-brown cross-bedded or wavy bedded sandstone with mud cracks, relatively abundant trace fossils, and rare shelly fossils (Andrews and Laird, 1976). Shelly Fossils (mainly brachiopods and trilobites) are common throughout much of the Mariner Group and give an age ranging from late Middle Cambrian to middle or late Late Cambrian (Cooper and others, 1976; Chapter 77, this volume).

A regional unconformity, which has resulted from the erosion of perhaps 2 km or more of Mariner Group strata in the north of the area (Bradshaw and others, Chapter 100, this volume; Laird and others, Chapter 66, this volume) separates the Mariner Group from the overlying Leap Year Group. This latter group is exposed in two separate strips restricted to the eastern and western margins of the Bowers Supergroup outcrop. In the eastern strip, at least 3 km, and perhaps as many as 7 km, of mainly fluvial, commonly cross-bedded, quartzose sandstone and minor quartzose conglomerate (Camp Ridge Quartzite) is exposed. No shelly fossils are known, but trace fossils have been collected from horizons near the base of the unit and may indicate local marine influence. In the western strip of outcrop, the Leap Year Group is represented by two fluvial formations. The older Carryer Conglomerate reaches a maximum thickness of 800 m in the lower Carryer Glacier, but thins southward to disappear near the head of the Sledgers Glacier. The Carryer Conglomerate is conformably overlain by the Reilly Conglomerate, which consists of quartzose conglomerate and minor quartzose sandstone at least 300 m thick, and which may be a stratigraphic equivalent of the Camp Ridge Quartzite (Laird and others, Chapter 66, this volume). No diagnostic fossils have been found in the Leap Year Group. However, a maximum age of middle to late Late Cambrian is given by the unconformably underlying Mariner Group, and a minimum age of Early Silurian is given by radiometric dates attributed to metamorphism of the Bowers Supergroup (Adams and others, Chapter 67, this volume). A late Late Cambrian to Ordovician age is considered probable for the Leap Year Group.

Paleogeography

Cambrian sedimentation along the trend of the present Transantarctic Mountains began on the site of an earlier sedimentary belt in which a great thickness of strata, mainly sandstone and mudstone, had accumulated. These older rocks, of late Precambrian age (Patuxent Formation, Beardmore Group, and their correlatives) were folded and the sedimentary belt everted during the late Precambrian Beardmore orogeny (Grindley and McDougall, 1969), and the resulting mountains were subsequently eroded.

Subsequent marine transgression, accompanied by basaltic volcanism, is likely to have occurred first in northern Victoria Land during late Vendian time in a graben or grabens formed during a period of block faulting; this activity continued into at least Early Cambrian time (Figure 65.3*a'*). Uplift here occurred sometime in Early or early Middle Cambrian time, bringing a halt to sedimentation. Although the site of the Pensacola Mountains appears to have been dry land in Early Cambrian time, a seaway evidently formed between these mountains and the edge of the East Antarctic shield, allowing deposition of shallow-water Archaeocyatha limestone on the site of the present Argentina Range and the Whichaway Nunataks. This seaway may have been continuous as far as the central Transantarctic Mountains, where mainly quartzitic sandstone was being deposited in the south, and Archaeocyatha limestone in the north. Early Cambrian seas may have extended northward into southern Victoria Land and beyond, but lack of age data there makes this uncertain (Figure 65.3*a*).

Indications of tectonic instability in late Early Cambrian time in the central Transantarctic Mountains are given by the occurrence of slumped masses of limestone and conglomerate derived from Archaeocyatha limestone. Facies relationships suggest biohermal development on a tectonically subsiding shelf with a shoreline to the west (toward the craton) (Laird and others, 1971).

By Middle Cambrian time, marine conditions prevailed along almost the entire length of the Transantarctic Mountains. While clastic sediments were being deposited close to the margin of the East Antarctic shield on the site of the Shackleton Range, calcareous or mixed calcareous and clastic deposits were being formed throughout most of the remainder of the region. A series of carbonate banks stretched from the Weddell Sea to the Ross Sea. These are represented by the limestone erratics in the Argentina Range, by the Nelson Limestone in the Pensacola Mountains, by limestone units of the Leverett Formation, and possibly by the upper parts of the Henson Marble, Shackleton Limestone, and perhaps the Anthill Limestone in the central Transantarctic Mountains and in southern Victoria Land. The oldest (late Middle Cambrian) mixed carbonate rocks and clastic units of the Mariner Group in northern Victoria Land may have been deposited in a separate basin from the remainder of the lower Paleozoic rocks. Continued tectonic activity in the central Transantarctic Mountains in early Middle Cambrian time may have caused the great thickness of conglomerate that overlies the Archaeocyatha limestone. Sedimentation offshore from this unstable shelf may be represented by the older part of the Starshot and other clastic formations (Figure 65.3*b*).

Late Cambrian events are less clear because of scant paleontologic control except in northern Victoria Land. On the site of the Pensacola Mountains the extrusion of the rhyolitic lavas and pyroclastic deposits of the Gambacorta Formation indicates felsic volcanism during late Middle Cambrian or Late Cambrian time. Other felsic volcanism, perhaps contemporaneous, also occurred in Late Cambrian or Early Ordovician time in the Queen Maud Mountains (rhyolites of the Leverett Formation). Rhyolite flows associated with the Starshot Formation in the Central Transantarctic Mountains and with unnamed units in southern Victoria Land, may also be correlative. In each case the volcanic rocks are included within thick clastic sedimentary rock sequences. In northern Victoria Land, Late Cambrian time was marked by regression of the seas and probable renewed uplift in latest Late Cambrian and Early Ordovician time to provide thick continental sequences of quartzose conglomerate and sandstone (Figures 65.3*c* and *c'*). These tectonic movements in northern Victoria Land, and the extrusion of felsic volcanics elsewhere in the Transantarctic Mountains, may have been associated with the onset of the Late Cambrian–Ordovician Ross orogeny, which in most places along the length of the modern mountain chain brought sedimentation to a halt.

Ross Supergroup or Ross Sequence?

The evolution of a satisfactory stratigraphic terminology for a region as large as the Transantarctic Mountains is necessarily slow and complex. To early investigators the most obvious subdivision was into a largely undeformed upper Paleozoic–Mesozoic cover (the Beacon Supergroup) and an underlying basement complex. The boundary was drawn at the sub-Devonian Kukri Peneplain or on younger surfaces of planation.

In the Ross Sea sector the basement complex was considered the product of the Ross orogeny (Gunn and Warren, 1962) which numerous subsequent isotopic studies have shown to be a mainly Late Cambrian to Early Ordovician event, although there are indications of activity—which include polyphase deformation, metamorphism, migmatization, and several phases of granitic injection—throughout a period of 50–80 m.y.

Some have defined the Ross orogeny in a very broad way. Craddock (1972) suggested that events as early as 1000 m.y. ago may have been "early pulses" of the Ross orogeny, and Grikurov and others (1972) include in the Ross orogeny all Riphean to Devonian tectonism throughout the Transantarctic Mountains. Others have taken a restricted view and distinguished separate early orogenies such as the Beardmore orogeny (Grindley and McDougall, 1969; Elliot, 1975) in late Precambrian time and a younger Borchgrevink orogeny in Devonian time (Craddock, 1970).

The rocks deformed by the Ross orogeny were regarded as the deposits of a major sedimentary belt, the Ross geosyncline (Gunn, 1963), and were united in a comprehensive Ross Supergroup. This supergroup has been used as a lithostratigraphic term from almost the north coast of Victoria Land to the central Transantarctic Mountains, and it embraces a wide variety of stratigraphic units which may range in age from Riphean to Ordovician. It has been used for rocks which predate the Beardmore orogeny (Grindley and others, 1964) to rocks which are not folded by the Ross orogeny (in the strict sense), which may be Ordovician in age (Laird and others, Chapter 66, this volume; Bradshaw

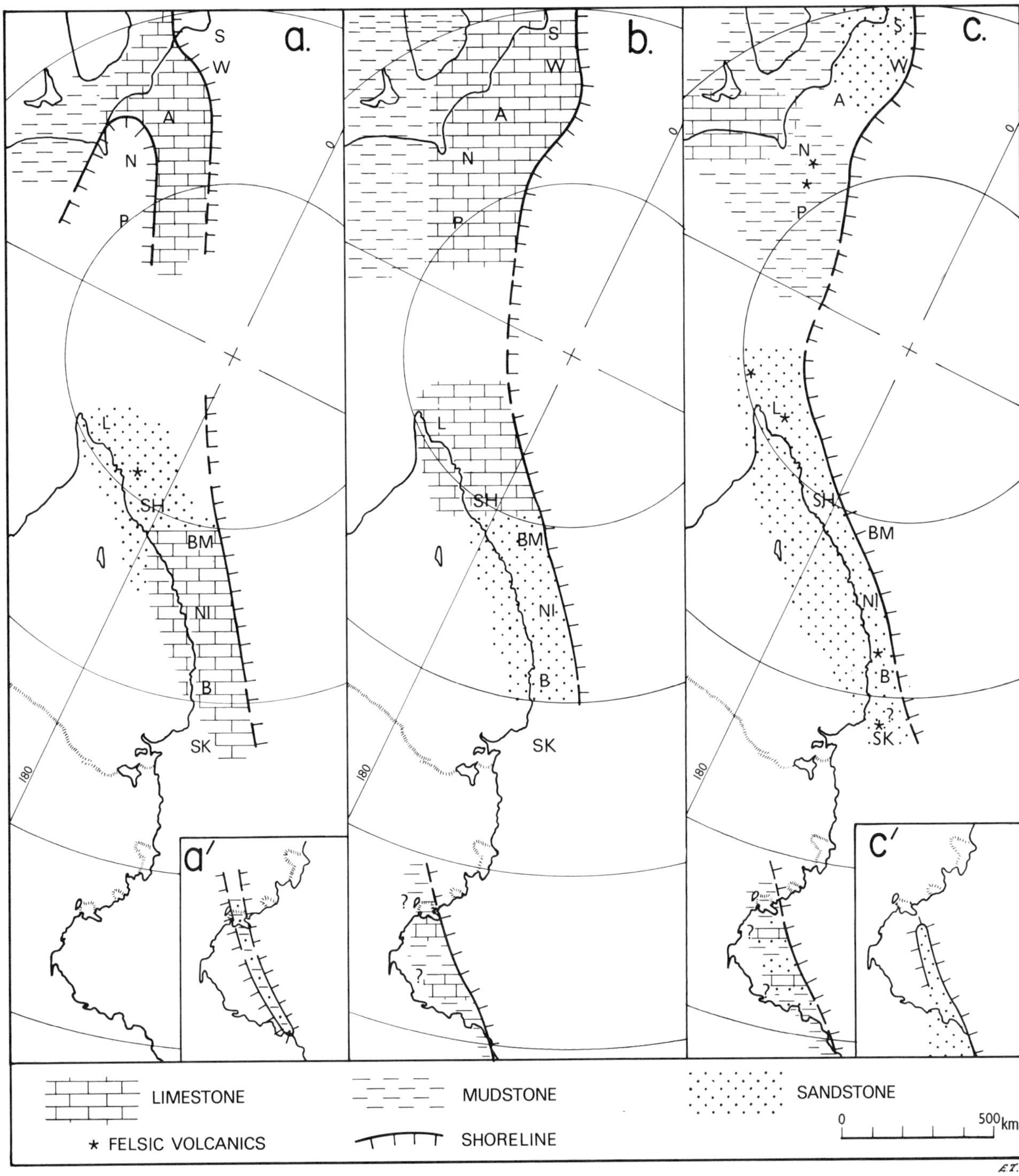

Figure 65.3. Early Paleozoic paleogeography of the Transantarctic Mountains, showing shorelines, the probable extent of depositional areas, and lithofacies distribution. Data for lithofacies shown northwest of the Neptune and Patuxent Ranges were inferred from Craddock and others (1964) and Craddock (1969). *a′*. Vendian to earliest Cambrian; *a*. Early Cambrian; *b*. Middle Cambrian; *c*. Late Cambrian; *c′*. Latest Cambrian to Ordovician. Geographic locations: S: Shackleton Range. W: Whichaway Nunataks. A: Argentina Range. N: Neptune Range. P: Patuxent Range. L: Leverett Glacier. SH: Shackleton Glacier. BM: Beardmore Glacier. NI: Nimrod Glacier. B: Byrd Glacier. SK: Skelton Glacier.

and others, Chapter 100, this volume) and which in large part may be detritus from Ross orogenic uplift.

A stratigraphic term so wide in scope and subject to such variable interpretation is not merely unhelpful but can be a definite impediment to geologic communication. We feel that recognition of more closely defined orogenies, such as the Beardmore and the Borchgrevink, requires a corresponding refinement of stratigraphic terms and the abandonment of the term *Ross Supergroup*.

The only lower Paleozoic rocks with diagnostic fossils are Cambrian in age, and in the central and Weddell Sea sectors of the Transantarctic Mountains, the Cambrian forms a major sequence dominated by a carbonate-quartzite sedimentary assemblage. These rocks formed in an intracratonic (or parageosynclinal) basin which stretched from the Shackleton Range at least as far as the Byrd Glacier, a distance of 2000 km. These sedimentary rocks stand in marked contrast to upper Precambrian "graywacke"-type clastic sequences and older gneisses on which they can be seen or reasonably inferred to rest unconformably.

Because the Cambrian was universally included in the old Ross Supergroup we suggest that the name *Ross* be retained in the form *Ross Sequence.* The term *sequence* is employed in the sense of Sloss and others (1949), Krumbein and Sloss (1963), and Sloss (1963), i.e., a major rock stratigraphic unit traceable over a very large area, bounded by unconformities and showing a uniform tectonic history. The new term "synthem" for unconformity-bounded units (Chang, 1975) was considered, but not adopted as Chang appears to include granites and other rocks of the succeeding orogeny in a synthem (Chang, 1975, p. 1546). Though this is perfectly logical, we did not wish to include the Granite Harbour Instrusives in the new unit and therefore preferred *sequence.*

The term *Ross Sequence* is proposed for mainly Cambrian rocks which rest unconformably on rocks deformed during the Vendian Beardmore or older orogenies, and which were in turn themselves deformed by the Late Cambrian–Early Ordovician Ross orogeny. It would include almost all of the lower Paleozoic succession in the Transantarctic Mountains, but would exclude such units as the Neptune Group of the Pensacola Mountains which clearly postdate the Ross orogeny.

It would also be unwise to extend the term Ross Sequence to northern Victoria Land where the lower Paleozoic sedimentary rocks may well have accumulated in a subparallel but completely independent basin (See Figure 65.3). Here fossiliferous Cambrian rocks (Mariner Group) form part of an essentially concordant succession extending from Vendian to Upper Cambrian or Ordovician. The basal units of the Bowers Supergroup may thus be contemporaneous with the Beardmore orogeny, as may parts of the Robertson Bay Group. Although there is a widespread thermal event corresponding to the Ross orogeny (Adams and others, Chapter 67, this volume) there was little or no deformation here at that time. The Leap Year Group, which may be contemporaneous with or postdate the Ross orogeny and could be a result of it, is essentially concordant with underlying fossiliferous Cambrian beds. In other words, correlatives of the Ross Sequence form parts of a discrete and distinct succession of rocks with a sedimentologic-tectonic framework different from that of the Ross Sequence. Unless it can be shown that the culminative deformation is the Ross orogeny and that the oldest units are younger than the Beardmore Orogeny, the use of Ross Sequence would not be justified in northern Victoria Land. Similar reservations apply to the clastic sedimentary rocks of the Blaiklock Glacier Group of Coats Land.

Conclusions

The quality of regional synthesis is improved by using well-defined lithostratigraphic and tectonic terms of restricted age range. The practice of defining stratigraphic units and tectonic events in such a broad way as to include a period equal to a third or more of Phanerozoic time can only obscure relationships. A more restricted Late Cambrian–Early Ordovician Ross orogeny and a largely Cambrian Ross Sequence are suggested. More detailed research on the Beardmore and Borchgrevink episodes is needed. It is interesting to note that in a similar way and for comparable reasons, Lambert and McKerrow (1976) have proposed subdivision of the Scottish Caledonian orogeny into separate Grampian and Caledonian (restricted) episodes. If a need is felt for a comprehensive term for all late Proterozoic and early Paleozoic sedimentary and orogenic belts in the Transantarctic Mountains, the term *Rossides* (Grikurov and others, 1972) is still available.

References cited

Andrews, P. B., and Laird, M. G. 1976. Sedimentology of a Late Cambrian regressive sequence (Bowers Group), northern Victoria Land, Antarctica. *Sedim. Geol., 16,* No. 1, 21–44.

Barrett, P. J., Grindley, G. W., and Webb, P. N. 1972. The Beacon Supergroup of East Antarctica. (*In* Adie, R. J., *ed. Antarctic Geology and Geophysics.* Oslo, Universitetsforlaget, 319–32.)

Blank, H. R., Cooper, R. A., Wheeler, R. H., and Willis, I. A. G. 1963. Geology of the Koettlitz–Blue Glacier region, southern Victoria Land, Antarctica. *Trans. R. Soc. N.Z., 2,* No. 5, 79–100.

Chang, K. H. 1975. Unconformity-bounded stratigraphic units. *Geol. Soc. Am. Bull., 86,* No. 11, 1544–52.

Clarkson, P. D. 1972. Geology of the Shackleton Range: a preliminary report. *Bull. Br. antarc. Surv.,* No. 31, 1–15.

Cooper, R. A., Jago, J. B., MacKinnon, D. I., Simes, J. E., and Braddock, P. E. 1976. Cambrian fossils from the Bowers Group, northern Victoria Land, Antarctica (preliminary note). *N.Z. Jl Geol. Geophys., 19,* No. 2, 283–88.

Craddock, C. 1969. Geology of the Ellsworth Mountains (Sheet 4). (*In* Bushnell, V. C., and Craddock, C., *eds.* Geologic Maps of Antarctica. *Antarct. Map Folio Ser.,* Folio 12, Pl. IV.)

Craddock, C. 1970. Tectonic map of Antarctica. (*In* Bushnell, V. C., and Craddock, C., *eds.* Geologic Maps of Antarctica. *Antarct. Map Folio Ser.,* Folio 12, Pl. XXI.)

Craddock, C. 1972. Antarctic tectonics. (*In* Adie, R. J., *ed. Antarctic Geology and Geophysics.* Oslo, Universitetsforlaget, 449–55.)

Craddock, C., Anderson, J. J., and Webers, G. F. 1964. Geologic outline of the Ellsworth Mountains. (*In* Adie, R. J., *ed. Antarctic Geology.* Amsterdam, North-Holland Publishing Company, 155–70.)

Eastin, R., Faure, G., Shultz, C. H., and Schmidt, D. L. 1969. Rb-Sr ages of the Littlewood Volcanics and of the acid volcanic rocks of the Neptune Range, Pensacola Mountains, Antarctica. *Abstr. geol. Soc. Am.,* Pt. 6, 13.

Elliot, D. H. 1975. Tectonics of Antarctica: a review. *Am. J. Sci., 275-A,* 45–106.

Faure, G., Murtaugh, J. G., and Montigny, R. J. E. 1968. The geology and geochronology of the basement complex of the central Transantarctic Mountains. *Can. Jnl Earth Scis, 5,* No. 3, Pt. 2, 555–60.

Findlay, R. H. 1978. Provisional report on the geology of the region between the Renegar and Blue Glaciers, Antarctica. *Antarctic Rec., 1,* No. 1, 39–44.

Flory, R. F., Murphy, D. J., Smithson, S. B., and Houston, R. S. 1971. Geologic studies of basement rocks in southern Victoria Land. *Antarct. Jnl U.S., 6,* No. 4, 119–20.

Gordon, W. T. 1920. Cambrian organic remains from dredgings in Weddell Sea (Scottish National Antarctic Expedition, 1902–4). *Trans. R. Soc. Edinb., 52,* 681–84.

Grikurov, G. E., Ravich, M. G., and Soloviev, D. S. 1972. Tectonics of Antarctica. (*In* Adie, R. J., *ed. Antarctic Geology and Geophysics.* Oslo, Universitetsforlaget, 457–68.)

Grindley, G. W., and Laird, M. G. 1969. Geology of the Shackleton Coast (Sheet 15). (*In* Bushnell, V. C., and Craddock, C., *eds.* Geologic Maps of Antarctica. *Antarct. Map Folio Ser.*, Folio 12, Pl. XIV.)

Grindley, G. W., and McDougall, I. 1969. Age and correlation of the Nimrod Group and other Precambrian rock units in the central Transantarctic Mountains, Antarctica. *N.Z. Jl Geol. Geophys.*, *12*, No. 2–3, 391–411.

Grindley, G. W., and Warren, G. 1964. Stratigraphic nomenclature and correlation in the western Ross Sea region. (*In* Adie, R. J., *ed.* *Antarctic Geology*. Amsterdam, North-Holland Publishing Company, 314–33.)

Grindley, G. W., McGregor, V. R., and Walcott, R. I. 1964. Outline of the geology of the Nimrod-Beardmore-Axel Heiberg Glaciers region, Ross Dependency. (*In* Adie, R. J., *ed.* *Antarctic Geology*. Amsterdam, North-Holland Publishing Company, 206–19.)

Gunn, B. M. 1963. Geological structure and stratigraphic correlation in Antarctica. *N.Z. Jl Geol. Geophys.*, *6*, 423–43.

Gunn, B. M., and Warren, G. 1962. Geology of Victoria Land between the Mawson and Mulock Glaciers, Antarctica. *Bull. geol. Surv.*, *N.Z.*, No. 71, 157 pp.

Hill, D. 1964. Archaeocyatha from the Shackleton Limestone of the Ross System, Nimrod Glacier area, Antarctica. *Trans. R. Soc. N.Z.*, *(Geology)*, *2*, No. 9, 137–46.

Hill, D. 1965. Geology 3. Archaeocyatha from Antarctica and a review of the phyllum. *Scient. Rep.*, *Transantarct. Exped. 1955–58*, No. 10, 151 pp.

Katz, H. R., and Waterhouse, B. C. 1970. Geological reconnaissance of the Scott Glacier area, south-eastern Queen Maud Range, Antarctica. *N.Z. Jl Geol. Geophys.*, *13*, No. 4, 1930–37.

Krumbein, W. C., and Sloss, L. L. 1963. *Stratigraphy and Sedimentation*. 2nd Ed. San Francisco, W.H. Freeman, 660 pp.

Laird, M. G. 1963. Geomorphology and stratigraphy of the Nimrod Glacier–Beaumont Bay region, southern Victoria Land, Antarctica. *N.Z. Jl Geol. Geophys.*, *6*, No. 3, 465–84.

Laird, M. G. 1964. Petrography of rocks from the Nimrod Glacier-Starshot Glacier region, Ross Dependency. (*In* Adie, R. J., *ed.* *Antarctic Geology*. Amsterdam, North-Holland Publishing Company, 463–72.)

Laird, M. G., Bradshaw, J. D., and Wodzicki, A. 1976. Re-examination of the Bowers Group (Cambrian), northern Victoria Land, Antarctica (Preliminary note). *N.Z. Jl Geol. Geophys.*, *19*, No. 2, 275–82.

Laird, M. G., Mansergh, G. D., and Chappell, J. M. A. 1971. Geology of the central Nimrod Glacier area, Antarctica. *N.Z. Jl Geol. Geophys.*. *14*, No. 3, 427–68.

Lambert, R. St. J., and McKerrow, W. S. 1976. The Grampian orogeny. *Scot. J. Geol.*, *12*, No. 4, 271–92.

McGregor, V. R. 1965. Geology of the area between the Axel Heiberg and Shackleton Glaciers, Queen Maud Range, Antarctica. Part 1—Basement Complex, structure, and glacial geology. *N.Z. Jl Geol. Geophys.*, *8*, No. 2, 314–43.

McGregor, V. R., and Wade, F. A. 1969. Geology of the western Queen Maud Mountains (Sheet 16). (*In* Bushnell, V. C., and Craddock, C., *eds.* Geologic Maps of Antarctica. *Antarct. Map Folio Ser.*, Folio 12, Pl. XV.)

Minshew, V. H. 1966. Stratigraphy of the Wisconsin Range, Horlick Mountains, Antarctica. *Science*, *N.Y.*, *152*, No. 3722, 637–38.

Minshew, V. H. 1967. *Geology of the Scott Glacier and Wisconsin Range Areas, Central Transantarctic Mountains, Antarctica.* Ph.D. dissertation, Ohio State University, 268 pp.

Mirsky, A. 1969. Geology of the Ohio Range–Liv Glacier area (Sheet 17). (*In* Bushnell, V. C., and Craddock, C., *eds.* Geologic Maps of Antarctica. *Antarct. Map Folio Ser.*, Folio 12, Pl. XVI.)

Palmer, A. R. 1970. Early and Middle Cambrian trilobites from Antarctica. *Antarct. Jnl U.S.*, *5*, No. 5, 162.

Palmer, A. R., and Gatehouse, C. G. 1972. Early and Middle Cambrian trilobites from Antarctica. *Prof. Pap. U.S. geol. Surv.*, 456-D, 37 pp.

Rex, D. C. 1972. K-Ar age determinations on volcanic and associated rocks from the Antarctic Peninsula and Dronning Maud Land. (*In* Adie, R. J., *ed.* *Antarctic Geology and Geophysics.* Oslo, Universitetsforlaget, 133–36.)

Schmidt, D. L., and Ford, A. B. 1969. Geology of the Pensacola and Thiel Mountains (Sheet 4). (*In* Bushnell, V. C., and Craddock, C., *eds.* Geologic Maps of Antarctica. *Antarct. Map Folio Ser.*, Folio 12, Pl. V.)

Schmidt, D. L., Williams, P. L., Nelson, W. H., and Ege, J. R. 1965. Upper Precambrian and Paleozoic stratigraphy and structure of the Neptune Range, Antarctica. *Prof. Pap. U.S. geol. Surv.*, 525-D, D112–19.

Skeats, E. W. 1916. Report on the petrology of some limestones from the Antarctic. *Rep. Br. antarc. Exped. 1907–09*, *Geol. 2*, No. 12, 189–200.

Skinner, D. N. B. 1964. A summary of the geology of the region between Byrd and Starshot Glaciers, south Victoria Land, Antarctica. (*In* Adie, R. J., *ed.* *Antarctic Geology*. Amsterdam, North-Holland Publishing Company, 284–92.)

Skinner, D. N. B., 1965. Petrographic criteria of the rock units between the Byrd and Starshot Glaciers, south Victoria Land, Antarctica. *N.Z. Jl Geol. Geophys.*, *8*, No. 2, 292–303.

Sloss, L. L. 1963. Sequences in the cratonic interior of North America. *Geol. Soc. Am. Bull.*, *74*, No. 2, 93–114.

Sloss, L. L., Krumbein, W. C., and Dapples, E. C. 1949. Integrated facies analysis. (*In* Longwell, C. R., *ed.* Sedimentary Facies in Geologic History. *Mem. geol. Soc. Am.*, *39*, 91–123.)

Soloviev, I. A., and Grikurov, G. E. 1978. Pervyye nakhodki srednekembriyskikh trilobitov v khrebte Sheklton (Antarktida) [First discovery of Middle Cambrian trilobites in the Shackleton Range (Antarctica)]. *Antarktika*, *Doklady Komissii*, *17*, Moscow, Izdatel'stvo Nauka, 187–98.

Stump, E. 1974. Volcanic rocks of the Early Cambrian Taylor Formation, central Transantarctic Mountains. *Antarct. Jnl U.S.*, *9*, No. 5, 228–29.

Taylor, T. G. 1914. *In* David, T. W. E., and Priestley, R. E., *eds.* Glaciology, physiography, stratigraphy, and tectonic geology of South Victoria Land. *Rep. Br. antarc. Exped. 1907–9*, *Geol. 1*, 399 pp.

Thomson, M. R. A. 1972. Inarticulate Brachiopoda from the Shackleton Range and their stratigraphical significance. *Bull. Br. antarc. Surv.*, No. 31, 17–20.

Warren, G. 1969. Geology of the Terra Nova Bay–McMurdo Sound area, Victoria Land (Sheet 14). (*In* Bushnell, V. C., and Craddock, C., *eds.* Geologic Maps of Antarctica. *Antarct. Map Folio Ser.*, Folio 12, Pl. XIII.)

66

Stratigraphy of the Upper Precambrian and Lower Paleozoic Bowers Supergroup, Northern Victoria Land, Antarctica

M. G. LAIRD
New Zealand Geological Survey, Department of Scientific and Industrial Research, Christchurch, New Zealand

J. D. BRADSHAW
Geology Department, University of Canterbury, Christchurch, New Zealand

A. WODZICKI
Geology Department, Western Washington University, Bellingham, Washington, 98225, U.S.A.

Abstract. The Bowers Supergroup is composed of more than 8 km of sedimentary and volcanic rocks exposed in the Bowers Mountains and Evans Neve areas. It consists of three major units (Sledgers Group, Mariner Group, and Leap Year Group) separated by unconformities. An additional unit, the Husky Conglomerate, consisting dominantly of amphibolitic conglomerate, is faulted against the Sledgers Group and rests unconformably on amphibolitic rocks of the Precambrian Wilson Group. It is of uncertain affiliation but is almost certainly older than the Sledgers Group.

The Sledgers Group (Vendian–Lower Cambrian) consists of more than 3 km of basic volcanic rocks interfingering to the southeast with clastic marine sedimentary rocks. The Mariner Group, which overlies the Sledgers Group with probable unconformity, contains Middle and Upper Cambrian shelly faunas. It is composed of more than 2 km of sandstone, mudstone, and limestone, forming a transgressive-regressive sequence deposited in dominantly shallow and marginal marine environments. The Mariner Group is truncated by the unconformably overlying, mainly fluvial Leap Year Group, which is exposed in two separate strips of outcrops with contrasting rock types. The eastern strip consists of at least 3 km of quartzose sandstone and minor conglomerate (Camp Ridge Quartzite). In the northwest, however, the deeply-channeled Mariner Group, in places, is overlain by as much as 800 m of polymictic conglomerate and minor sandstone (Carryer Conglomerate); this unit, in turn, is overlain by 300 m of quartzose conglomerate and minor sandstone (Reilly Conglomerate). No shelly fossils are known from the Leap Year Group, but it is considered to be upper Upper Cambrian or Ordovician.

Paleocurent data and facies relationships within the Bowers Supergroup suggest that the Sledgers and Leap Year Groups, and perhaps the Mariner Group, were deposited in an elongate northwest-trending basin (the Bowers Trough).

The Bowers Supergroup, which includes all known lower Paleozoic sedimentary rocks of northern Victoria Land, occurs within a fault-bounded strip 20–25 km wide, extending southeast from the lower Rennick Glacier across the southern Bowers Mountains to at least the upper Mariner Glacier, a distance of 270 km (Figure 66.1). It is flanked on the southwest by the Wilson Group (Precambrian), which consists of schists and gneisses of amphibolite metamorphic facies, and to the northeast by the metasedimentary Robertson Bay Group (probably upper Precambrian) (Cooper and others, Chapter 77, this volume). In its central part it is overlain by gently dipping rocks of the Beacon Supergroup and Ferrar Group of late Paleozoic and Mesozoic age. The Bowers sequence, which shows steep dips, is folded about northwest-trending axes into a complex anticline in the west and a major syncline in the east (Figure 66.1).

Stratigraphy

Previous descriptions of the succession have included it within the Bowers Group (Sturm and Carryer, 1970), which consisted of three conformable formations (Dow and Neall, 1972, 1974) (Table 66.1). However, subsequent field observations and paleontologic study have shown the need for comprehensive stratigraphic revision, including changes in the order of succession (Laird and others, 1976). The sequence includes two important unconformities, and the revised stratigraphy now accommodates four major units, the upper three forming part of a newly defined Bowers Supergroup (Table 66.1). These units are, from oldest to youngest: the Husky Conglomerate, which rests unconformably on the Wilson Group and is of uncertain affiliation; the Sledgers Group of Vendian to Early Cambrian age; the Mariner Group, of late Middle Cambrian to middle or late Late Cambrian age; and the Leap Year Group, of probably late Late Cambrian to Ordovician age. Their stratigraphic relationships are indicated in Figure 66.2.

Husky Conglomerate

The Husky Conglomerate is a sedimentary succession of dark green amphibolitic conglomerate and minor sandstone and mudstone cropping out on the eastern flanks of the Lanterman Range, to the north and south of Husky Pass

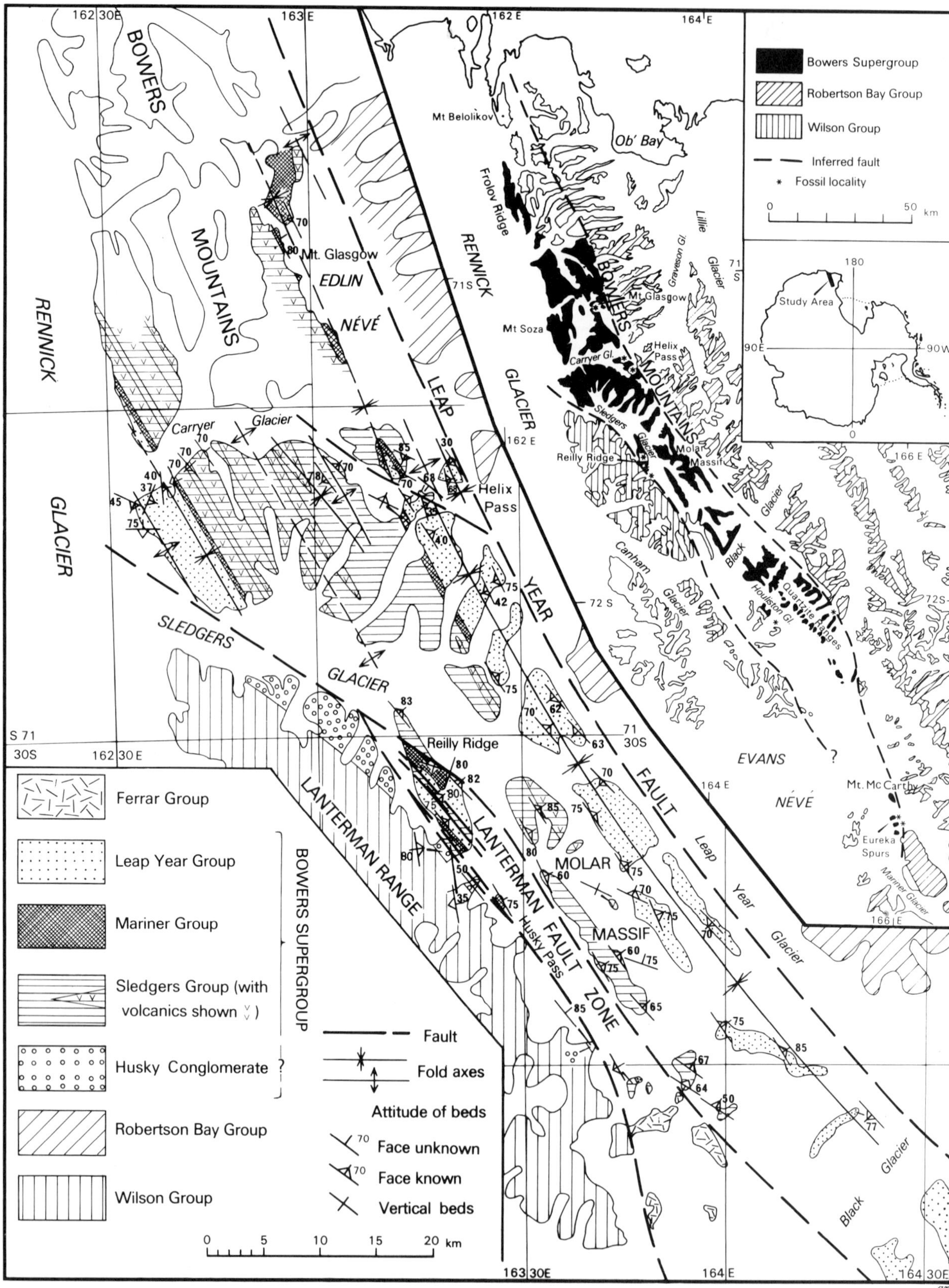

Figure 66.1. Geologic maps of part of northern Victoria Land showing the distribution of the Bowers Supergroup. The main map shows details of stratigraphic relationships and structure of the rocks in the southern Bowers Mountains.

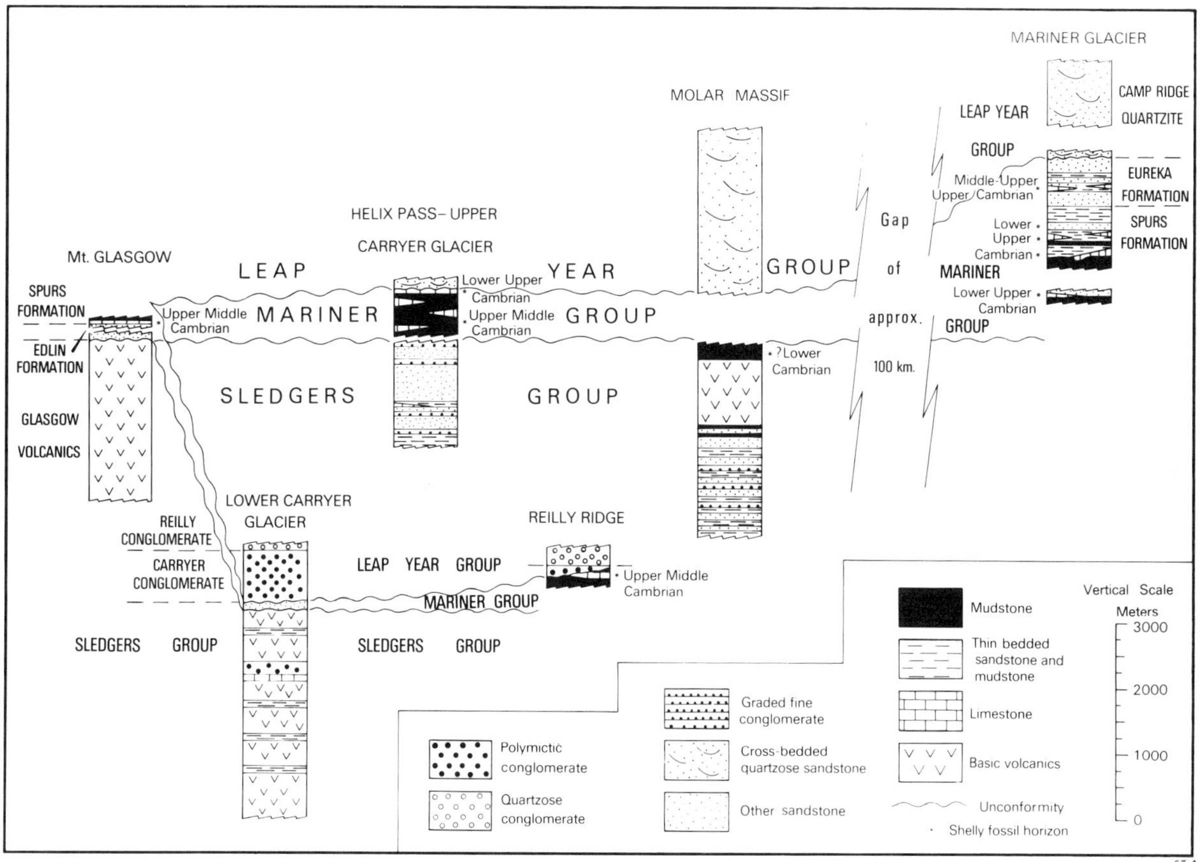

Figure 66.2. Correlation of main stratigraphic sections through the Bowers Supergroup, showing proposed nomenclature.

Table 66.1. Lower Paleozoic stratigraphic units

Dow and Neall, 1972, 1974			Laird, Bradshaw, and Wodzicki (this paper)			
Group	Formations		Groups	Formations	Thickness (m)	Age
Bowers Group	Camp Ridge Quartzite	Bowers Supergroup	Leap Year Group	Camp Ridge Quartzite / Reilly Conglomerate; Carryer Conglomerate	3000 +	Ordovician to late Late Cambrian(?)
	Sledgers Formation		Hiatus			
			Mariner Group	Eureka Formation	2000 +	middle or late Late Cambrian to late Middle Cambrian
				Spurs Formation		
				Edlin Formation		
			Hiatus			
			Sledgers Group	Glasgow Volcanics	3500 +	Early Cambrian to Vendian
				Molar Formation		
	Carryer Conglomerate	Husky Conglomerate			370 +	Vendian(?)

and to the west of Reilly Ridge. Binocular observation suggests that it also occurs on several northern spurs of the Lanterman Range extending into the lower Sledgers Glacier.

At the type locality west of Reilly Ridge, 370 m of vertical, eastward younging, dark green conglomerate and breccia, consisting dominantly of amphibolite clasts, with minor sandstone and mudstone, rest on amphibolite of the Wilson Group. Although the basal contact is obscured by snow, an unconformity is inferred. The top of the sequence is separated by a snow gap and probable fault from overturned, eastward-dipping, interbedded mudstone and volcanigenic sandstone correlated with the Sledgers Group.

Conglomerate units, some as thick as 200 m, make up the bulk of the Husky Conglomerate. The clasts are mainly subrounded to subangular, reaching 3 m in diameter. They are commonly tightly packed in a sparse matrix, but mudstone units as much as 95 m thick with scattered subrounded boulders also occur. Massive breccia units as thick as 40 m and massive mudstone and sandstone units as thick as 15 m form a smaller fraction of the sequence. The clasts at all localities visited consisted almost entirely of amphibolite indistinguishable from the Wilson Group amphibolite unit on which the Husky Conglomerate rests at localities north of Husky Pass. Rounded pebble-sized clasts of quartz and granite make up less than 1%.

Because its upper contact is faulted, and no fossils have been recorded, the stratigraphic relationship between the Husky Conglomerate and other units is uncertain. Its metamorphic rank (greenschist facies) suggests correlation with either the Robertson Bay Group or the basal part of the Sledgers Group.

Sledgers Group

Originally included in the Bowers Group as the Sledgers Formation (Crowder, 1968), the unit is here given separate group status within the Bowers Supergroup. It is well exposed in the northern Molar Massif (Figure 66.1), where more than 2500 m of nearly continuous outcrop occurs, and this is taken as the type area.

The group crops out throughout the Bowers Mountains at least as far north as Frolov Ridge. It has not been definitely identified south of the Black Glacier, but some of the rocks lying south and southeast of the Quartzite Ranges and previously included within the Robertson Bay Group (such as those in the Mount McCarthy area, northeast of Eureka Spurs), may form part of the Sledgers Group. Sedimentary rocks exposed in isolated nunataks lying west of the Quartzite Ranges in the Houliston Glacier area have previously been mapped as Sledgers Formation (now Group) (Dow and Neall, 1974; Cooper and others, 1976). However, their stratigraphic position is uncertain, and the discovery within the sequence by Cooper and others (1976) of Middle Cambrian trilobites suggests possible correlation with the Mariner Group (see below). The Sledgers Group is overlain with probable unconformity by the Mariner Group. Its base is not exposed, but the minimum thickness of the Sledgers Group is 3500 m. It is divided into two formations, the Glasgow Volcanics and the Molar Formation (Table 66.1).

Glasgow Volcanics

This formation is named for the massive basaltic breccia at its type section on Mount Glasgow, where the formation is at least 2500 m thick. The Glasgow Volcanics occur as sheets and lenses within the sedimentary facies (Molar Formation) of the Sledgers Group, occupying a high proportion (perhaps more than half) of the succession in the northern part of the mapped area but apparently wedging out to the south where the formation is unknown south of the Molar Massif. Its northern extent is undefined, but similar volcanic rocks which are almost certainly part of the same sequence have been reported from Frolov Ridge in the northern Bowers Mountains (Sturm and Carryer, 1970). The base of the succession at Mount Glasgow was not observed, but the top is marked by a sharp contact with the basal beds of the Mariner Group. The sequence here consists mainly of unsorted angular to subangular basaltic breccia, with minor flows, some containing pillows. In the Molar Massif, the formation occurs as one massive sheet-like body as much as 1000 m thick lying between sandstone and mudstone of the Molar Formation near the top of the Sledgers Group. It consists there entirely of unsorted angular to subangular basaltic breccia with a few fragments of pillows. In the lower Carryer Glacier area, the formation consists of several thick units composed entirely of volcanic breccia except for 100 m of reddish weathered lava near the top of the sequence. No pillows were observed in this area. The top surface of the uppermost breccia is notably weathered and overlain abruptly by quartzose sandstone and mudstone of the Mariner Group; the base of the succession was not observed.

Figure 66.3. Dark mudstone containing thin beds of well-sorted, rippled and wavy-bedded, fine-grained sandstone. Molar Formation, northeastern Molar Massif.

Molar Formation

This formation is named for the sequence of sandstone, mudstone, and minor conglomerate cropping out in the western part of the Molar Massif. The unit is also well exposed in the nunataks lying immediately south of the Molar Massif and in the mountains between the Sledgers and Carryer Glaciers. The base of the formation is not exposed, but at its type locality in the northwestern Molar Massif a minimum of 1350 m of sedimentary rocks lies below the sharp contact with the Glasgow Volcanics. The sequence here consists dominantly of dark mudstone containing thin beds of well-sorted, rippled or wavy-bedded sandstone (Figure 66.3). Intercalated within the mudstone are isolated units of massive or graded-bedded sandstone,

Figure 66.4. Calcareous mudstone containing a 20-m unit of ridge-forming fossiliferous limestone. Spurs Formation, northern Reilly Ridge.

lenses of graded breccia or coarse sandstone, and a 40-m-thick chaotic slide breccia containing angular blocks as much as 1 m in diameter. The sequence is similar throughout the Molar Massif and in the nunataks to the south, although rare thin conglomerate horizons with rounded boulders as large as 30 cm occur in exposures south of the type section. In the Molar Massif and the ridges to the north, as much as 350 m of sedimentary rocks, tentatively included within the Molar Formation, lie with apparent conformity above the Glasgow Volcanics (Figure 66.2). This succession consists of mudstone with thin, cross-laminated beds and lenses of calcareous sandstone or limestone, locally with burrows. The sequence is generally similar in the area between the Sledgers and Carryer Glaciers, except in the lower Carryer Glacier area where the rocks consist of thin (as much as 350 m) units of conglomerate, sandstone, mudstone, and one bed of limestone 10 m thick, interbedded within thick sequences belonging to the Glasgow Volcanics (Figure 66.2).

The graded, coarse sandstone, breccia, and conglomerate beds are poorly sorted; they contain angular to moderately well-rounded fragments, mainly of volcanic and intrusive rocks (about 70%), and lesser amounts of sandstone, siltstone, and limestone. Basaltic fragments, similar in rock type to the Glasgow Volcanics, comprise about half the igneous clasts; the remainder includes andesites, dacites, rhyolites, and intrusive rocks ranging from granodiorite to granite (Laird and others, 1976).

Acritarchs extracted from the Molar Formation indicate a Vendian age for most of the sequence (Cooper and others, Chapter 77, this volume), although inarticulate brachiopods from the highest beds in the Molar Massif suggest that it ranges up into the Lower Cambrian.

Mariner Group

This unit, originally defined as the Mariner Formation (Andrews and Laird, 1976), is here raised to group status. The Mariner Group is well exposed in the Mariner Glacier area at Eureka Spurs, where 1600 m occurs, but crops out intermittently to the north of this as far as Mount Glasgow. The base of the group has been seen only at Mount Glasgow and in the lower Carryer Glacier, where it rests sharply on Glasgow Volcanics. Although the contact is apparently concordant, the weathered nature of the upper surface of the volcanic rocks in the lower Carryer Glacier, the abrupt change in rock type from volcanic rocks and volcanigenic sedimentary rocks to dominantly quartzose and calcareous sedimentary rocks, and the lack of fossils older than late Middle Cambrian age from the Mariner Group all suggest that a hiatus possibly representing much of Early and Middle Cambrian time is present between the Mariner and Sledgers Groups.

No complete Mariner Group sequence has been seen, but the thickness of the group, though unknown, is likely to be in excess of 2 km. The group is subdivided into three units: the Edlin, Spurs, and Eureka Formations (Table 66.1).

Edlin Formation

The basal unit of the Mariner Group has been identified only in the lower Carryer Glacier and north of Mount Glasgow in the Edlin Névé, from which it takes its name. These sedimentary rocks are characteristically highly quartzose and represent marginal marine sands deposited transgressively on the underlying Sledgers Group.

In the type locality, 1 km north of Mount Glasgow, a poorly exposed composite sequence 110 m thick consists of a basal 8 m of red-brown, muddy sandstone which rests sharply on the Glasgow Volcanics, followed by thin-bedded or massive quartzose sandstone. This passes upward into limestone and calcareous mudstone of the Spurs Formation.

In the lower Carryer Glacier, 100 m of fine-grained, thin-bedded, quartzose sandstone sharply, but apparently concordantly, overlies weathered breccia of the Glasgow Volcanics and is itself overlain by the Carryer Conglomerate. The sandstone, which contains abundant ripple marks and low-angle cross-bedding, is muddy and notably red-brown colored in the basal 10 m, where desiccation cracks and arthropod trails occur.

No shelly fossils were seen in the Edlin Formation, but trilobites of late Middle Cambrian age (Cooper and others, 1976; Chapter 77, this volume) occur in limestone of the Spurs Formation immediately overlying quartzose sandstone of the Edlin Formation northeast of Mount Glasgow. Thus, a Middle Cambrian age is probable for the Edlin Formation.

Spurs Formation

Named for Eureka Spurs, at the head of the Mariner Glacier, this formation is the most widely represented unit of the Mariner Group (see Figure 66.2). At the type locality, 900 m of fissile mudstone with thin limestone lenses pass gradationally upward, with increasing sandy layers, into wavy-bedded sandstone forming the basal unit of the Eureka Formation. Elsewhere, the Spurs Formation is dominated by mudstone, commonly with thin lenses and layers of limestone. In the Helix Pass area, scattered outcrops of this rock type also include thin beds of quartzose or calcareous sandstone as well as limestone. Thicker limestones (as much as 20 m; Figure 66.4) occur on Reilly Ridge* and also at Mount Glasgow, where a limestone lens 30 m thick occurs. East of the Mount Glasgow sequence, at

* At this locality some horizons several meters thick consist of slumps and olistostromes composed of large (as much as 3 m) irregular blocks of limestone in mudstone.

Figure 66.5. Trough cross-bedded, medium-grained, quartzose sandstone. Camp Ridge Quartzite, central Molar Massif.

the head of Edlin Névé, several hundred meters of calcareous mudstone containing thick units of poorly sorted, quartzo-feldspathic, thin-bedded to massive, medium-grained sandstone exposed in low ridges can probably be correlated with the Spurs Formation.

In Houliston Glacier, mudstone, sandstone, and conglomerate occurring in isolated nunataks have been previously included within what is now the Sledgers Group (see above). The presence of Middle Cambrian fossils, however, shows that these beds are much younger than the youngest known beds of the Sledgers Group (Lower Cambrian) (Cooper and others, Chapter 77, this volume). Their lithostratigraphic correlation is speculative, but rock types similar to those in the Spurs Formation are common (R. A. Cooper, personal communication), and inclusion in this formation is possible.

As no continuous section through the Spurs Formation has been recorded, its total thickness is uncertain. The 900 m of nearly continuous outcrop exposed on Eureka Spurs is a minimum, and scattered outcrops of similar rocks lying apparently stratigraphically below this sequence suggest that at least 1500 m is present.

The Spurs Formation is notably fossiliferous, the main shelly fauna being dominated by trilobites and brachiopods with a range from upper Middle Cambrian to lower Upper Cambrian (Cooper and others, 1976; Chapter 77, this volume).

Eureka Formation

This formation is named for the upper part of the succession at Eureka Spurs. The unit does not occur farther north where it has probably been removed by erosion prior to deposition at the Leap Year Group.

At the type locality, 700 m of sedimentary rocks, dominated by quartzose sandstone, are well exposed above the Spurs Formation. The succession has been informally divided into: the basal Wavy-bedded Sandstone Unit (250 m), into which the Spurs Formation passes gradationally upward; the Limestone Unit (100 m), which contains lenses of limestone as much as 8 m thick in mudstone; and the Red Bed Unit (350 m), consisting mainly of brownish-gray, very fine- to fine-grained, quartzose sandstone and minor mudstone with mud cracks and arthropod tracks (Andrews and Laird, 1976). The formation seems to be of shallow-marine origin and to become regressive upward into intertidal deposits. The Eureka Formation is overlain sharply by basal conglomerate of the Camp Ridge Quartzite (Leap Year Group) which fills small scours eroded into the underlying rocks.

Shelly fossils occur at several horizons within the Eureka Formation and indicate a middle to late Late Cambrian age for the unit (Cooper and others, 1976).

LEAP YEAR GROUP

This newly defined group rests on an erosion surface which truncates the Mariner Group. Although locally the two groups appear to be concordant, the Leap Year Group rests on middle or upper Upper Cambrian beds in the upper Mariner Glacier region, and on lower Upper Cambrian and upper Middle Cambrian beds at Helix Pass and Reilly Ridge, suggesting that a considerable thickness of rock (perhaps in excess of 2 km; see Figure 66.2) has been eroded from the Mariner Group in the north.

The Leap Year Group is exposed in two separate outcrop strips of contrasting rock type occupying the eastern and western flanks of a complex anticline with Sledgers Group rocks in its core. It is divided into three dominantly fluvial formations: the Camp Ridge Quartzite, which is recognized only in the eastern strip of outcrop, and the Carryer and Reilly Conglomerates, which crop out in the western strip.

Camp Ridge Quartzite

This formation is the most widely represented unit of the Leap Year Group and occupies the core of a major syncline which has been mapped from the upper Mariner Glacier northwestward for 180 km to Helix Pass. At its type locality in the Quartzite Ranges (Le Couteur and Leitch, 1964), it consists of red-brown or yellow quartzose sandstone, quartzose conglomerate, and minor mudstone. Large-scale, unimodal cross-bedding is common in the coarser beds (Figure 66.5), and small channels also occur. Conglomerate pebbles are set in a sandy matrix; they are well rounded and as much as 8 cm in diameter. Rock type is relatively uniform throughout the Camp Ridge Quartzite outcrop, although in exposures north of the Molar Massif, conglomerate becomes rare.

The basal contact of the formation is known only at Eureka Spurs and 5 km north of Helix Pass. At the first locality it fills small scours eroded in the underlying Eureka Formation; at the second locality, quartzose sandstone rests sharply and possibly with slight angular discordance on the Spurs Formation (Cooper and others, 1976). The upper contact of the Camp Ridge Quartzite was not observed, so its true thickness is unknown. However, 2900 m has been measured at the type locality (Laird and others, 1974), and it has been suggested (Dow and Neall, 1972) that as much as 7000 m may be exposed 20 km to the north.

No shelly fossils have been collected from the Camp Ridge Quartzite, and its age is not established. At Helix Pass and in outcrops within and south of the Quartzite Ranges, *Scolithus* burrows occur, and at the type locality, the lower part of the Camp Ridge Quartzite contains a variety of trace fossils which are consistent with an early Paleozoic age (Laird and others, 1974; Cooper and others, Chapter 77, this volume). A maximum age is provided by the middle or late Late Cambrian age of the unconformably underlying Mariner Group at Eureka Spurs. A minimum age is suggested by a radiometric age of 421 m.y. (Early Silurian) determined from the formation by Adams and others (Chapter 67, this volume). The Camp Ridge Quartzite is thus probably no older than late Late Cambrian in age and is likely to range up into the Ordovician.

Carryer Conglomerate

This formation (Dow and Neall, 1972), which is the older of the two divisions of the Leap Year Group occurring in the western strip, is very well exposed along the lower Carryer Glacier and on Reilly Ridge, and it also appears to occupy the southwestern flanks of Mount Soza and much of the massif between the lower Sledgers and Carryer Glaciers.

The Carryer Conglomerate consists dominantly of polymictic, well-bedded conglomerate of well-rounded clasts, some reaching 3 m in diameter and consisting mainly of granite, quartz, sandstone, rare volcanic rocks, and locally, limestone. Lenses of cross-bedded, pebbly, coarse-grained sandstone are numerous (Figure 66.6), commonly filling small scours. The succession at Reilly Ridge, although dominated by polymictic conglomerate, also locally includes units of mudstone and fine-grained sandstone as much as 70 m thick.

The basal contact with the Mariner Group is obscured by scree along the lower Carryer Glacier but appears to be concordant. However, on Reilly Ridge the Carryer Conglomerate appears to fill channels as deep as 200 m cut into the underlying mudstone and limestone of the Helix Formation (see Figure 100.1 of Bradshaw and others, Chapter 100, this volume). At both Reilly Ridge and lower Carryer Glacier, the formation passes upward by gradation within a few meters into basal sandstone of the Reilly Conglomerate. It attains its greatest thickness of 800 m at the type locality. At Reilly Ridge it has a maximum thickness of approximately 300 m but thins steadily southeastward and is not present in outcrops 6 km southwest of the southern end of Reilly Ridge.

No body or trace fossils are known from the Carryer Conglomerate, but a maximum age is provided by the unconformably underlying Mariner Group, which contains late Middle Cambrian trilobites. Hence, a Late Cambrian to Ordovician age is likely.

Reilly Conglomerate

This formation is named for pebbly quartzose conglomerate cropping out on Reilly Ridge and on nunataks to the south. It also crops out in the lower Carryer Glacier where 20 m of reddish, medium- to coarse-grained, quartzose sandstone and quartzose conglomerate rest with apparent conformity on Carryer Conglomerate; the unit probably continues southward as a strip of outcrop to the northern flank of the lower Sledgers Glacier. Isolated outcrops of quartzose pebbly sandstone and conglomerate, tentatively correlated with the Reilly Conglomerate, also occur 15 km southeast of Husky Pass and in the eastern Evans Névé (Laird and others, 1974).

At the type locality at the south end of Reilly Ridge, mudstone and polymictic conglomerate forming the uppermost beds of the Carryer Conglomerate pass gradationally upward within 2 m into 10 m of parallel-laminated,

Figure 66.6. Pebbly sandstone and coarse polymictic conglomerate. Carryer Conglomerate, mouth of the Carryer Glacier.

thinly-bedded, fine-grained, quartzose sandstone forming the basal unit of the Reilly Conglomerate. This in turn passes upward into parallel- and cross-stratified quartzose conglomerate containing well-rounded quartz pebbles as much as 10 cm in diameter. The top of the sequence rests in inferred fault contact with interbedded mudstone and volcanigenic sandstone correlated with the Molar Formation. To the south of Reilly Ridge the underlying Carryer Conglomerate thins and ultimately wedges out, and the Reilly Conglomerate rests directly on limestone of the Mariner Group 6 km southeast of Reilly Ridge.

No upper stratigraphic limit to the Reilly Conglomerate was seen, but a minimum thickness of 300 m is present in outcrops of the formation north of the type locality.

No fossils or fossil traces have been recorded from the Reilly Conglomerate, but its age, like that of the conformably underlying Carryer Conglomerate, is probably late Cambrian or Ordovician. Although coarser grained, the Reilly Conglomerate is otherwise closely similar in composition to the Camp Ridge Quartzite and occupies a similar statigraphic position (at least south of Reilly Ridge), overlying rocks of the Mariner Group. The two formations are considered to be stratigraphic equivalents.

Depositional environments and provenance

The depositional environment of the Husky Conglomerate is somewhat difficult to determine because of its modest thickness and the limited number of outcrops visited. However, its generally coarse nature, and particularly the presence of angular, unsorted breccias, suggests that the unit was deposited very close to its source with little sorting. The composition leaves little doubt that the source lay to the southwest in the Lanterman Range, which probably represented a tectonically active basin margin. The bulk of the rocks of the Sledgers Group are marine in origin and consist of fine-grained deposits emplaced mainly by traction currents moving from northwest to southeast. By contrast, the subordinate slide breccias and conglomerates, and lenticular bodies of graded, coarse-grained sandstone and fine breccia, appear to have been emplaced by downslope mass-transport mechanisms. Sparse directional

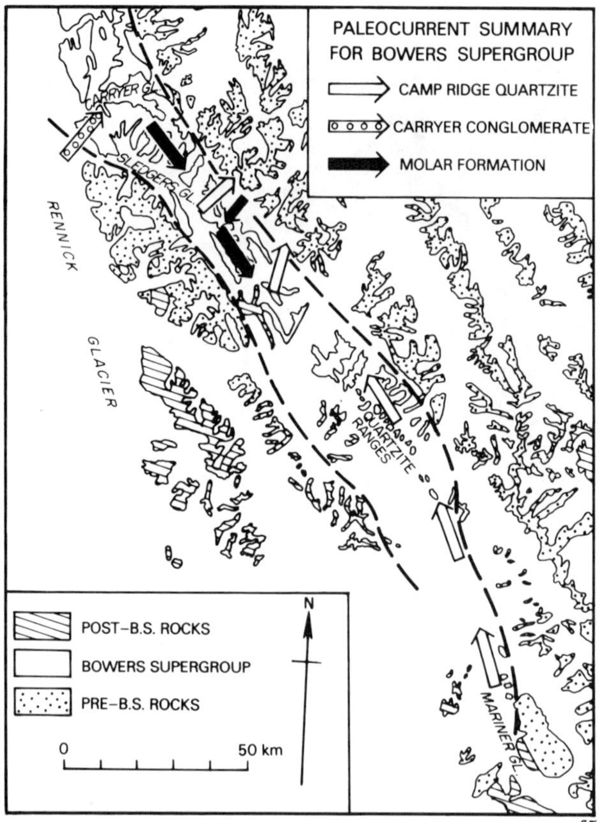

Figure 66.7. Sketch map of Bowers Supergroup outcrop showing main paleocurrent trends of the Molar Formation, Camp Ridge Quartzite, and Carryer Conglomerate.

data suggest that these deposits were derived from the northeastern flank of the main basin of deposition.

The deposits of the Mariner Group represent a shallow-marine environment, with transgressive and regressive phases represented by the relatively coarse-grained deposits of the Edlin and Eureka Formations, respectively. The generally fine-grained nature of this group indicates tectonic quiescence, although the occurrence of slump horizons and olistostromes at Reilly Ridge suggests the presence of a slope or the proximity of a tectonically active basin margin. Paleocurrent data are conflicting and no provenance direction or basin morphology could be determined. However, the high quartz content and absence of volcanigenic detritus indicate that the Sledgers Group was not an important source. Derivation from an acid plutonic or high-grade metamorphic terrane, such as the Precambrian Wilson Group now exposed to the south and west, is likely.

All three formations of the Leap Year Group show characteristics typical of fluvial deposits, although the presence of trace fossils near the base of the Camp Ridge Quartzite may indicate some local marine influence. Cross-bedding azimuths from the Camp Ridge Quartzite show a transport direction toward the northwest or north-northwest south of the Molar Massif; north of this locality there is a change in direction toward the northeast. The highly quartzose and quartzo-feldspathic nature of the detritus, coupled with the presence of rare pebbles of gneiss and granite, suggest a granitic and gneissic provenance, probably from the Wilson Group or similar rocks. The direction of transport of the Carryer Conglomerate at its type locality is toward the northeast; paleocurrent directions at Reilly Ridge are sparse and show wide dispersion, but the dominant clasts are compatible with a Wilson Group source. The few current directional readings obtained from the Reilly Conglomerate suggest a transport direction toward the northwest, and the petrography suggests a provenance similar to that of the Camp Ridge Quartzite. The main paleocurrent trends for the Sledgers and Leap Year Groups are summarized in Figure 66.7.

In summary, the sedimentary data suggest that most of the Bowers Supergroup was deposited in a structurally controlled troughlike basin (here called the Bowers Trough), elongated northwest-southeast, lying parallel to and immediately northeast of a crustal block of Precambrian crystalline rocks. Periodic tectonic uplift of the southwestern margin of the trough accompanied subsidence and filling of the trough from Vendian to Late Cambrian or Ordovician time. A northeastern margin of the Bowers Trough during Vendian and Early Cambrian time is implied by the presence of northeasterly derived mass-flow deposits in the Molar Formation, the coarseness and brecciated nature of which suggest accompanying tectonic activity and uplift along the margin.

Acknowledgments. Field work was supported by the Department of Scientific and Industrial Research under the New Zealand Antarctic Research Programme. We appreciate the cooperation of R. A. Cooper and his (1974–75) field party during the field work, and also thank him and J. B. Jago for helpful comments on the manuscript. The diagrams were drawn by E. T. Annear and the manuscript was typed by C. M. Johnstone.

References cited

Andrews, P. B., and Laird, M. G. 1976. Sedimentology of a Late Cambrian regressive sequence (Bowers Group), northern Victoria Land, Antarctica. *Sedim. Geol., 16,* No. 1, 21–44.

Cooper, R. A., Jago, J. B., MacKinnon, D. I., Simes, J. E., and Braddock, P. E. 1976. Cambrian fossils from the Bowers Group, northern Victoria Land, Antarctica (Preliminary note). *N.Z. Jl Geol. Geophys., 19,* No. 2, 283–88.

Crowder, D. F. 1968. Geology of a part of north Victoria Land, Antarctica. *Prof. Pap. U.S. geol. Surv.,* 600-D, D95–107.

Dow, J. A. S., and Neall, V. E. 1972. A summary of the geology of the lower Rennick Glacier, northern Victoria Land, Antarctica. (*In* Adie, R. J., *ed. Antarctic Geology and Geophysics.* Oslo, Universitetsforlaget, 339–44.)

Dow, J. A. S., and Neall, V. E. 1974. Geology of the lower Rennick Glacier, northern Victoria Land, Antarctica. *N.Z. Jl Geol. Geophys., 17,* No. 3, 659–714.

Laird, M. G., Andrews, P. B., and Kyle, P. R. 1974. Geology of northern Evans Névé, Victoria Land, Antarctica. *N.Z. Jl Geol. Geophys., 17,* No. 3, 587–601.

Laird, M. G., Bradshaw, J. D., and Wodzicki, A. 1976. Re-examination of the Bowers Group (Cambrian), northern Victoria Land, Antarctica (Preliminary note). *N.Z. Jl Geol. Geophys., 19,* No. 2, 275–82.

Le Couteur, P. C., and Leitch, E. C. 1964. Preliminary report on the geology of an area southwest of upper Tucker Glacier, northern Victoria Land. (*In* Adie, R. J., *ed. Antarctic Geology.* Amsterdam, North-Holland Publishing Company, 229–36.)

Sturm, A., and Carryer, S. J. 1970. Geology of the region between Matusevich and Tucker Glaciers, north Victoria Land, Antarctica. *N.Z. Jl Geol. Geophys., 13,* No. 2, 408–35.

67

Potassium–Argon Geochronology of the Precambrian–Cambrian Wilson and Robertson Bay Groups and Bowers Supergroup, Northern Victoria Land, Antarctica*

C. J. D. ADAMS and J. E. GABITES
Institute of Nuclear Sciences, Department of Scientific and Industrial Research, Lower Hutt, New Zealand

A. WODZICKI
Geology Department, Western Washington University, Bellingham, Washington 98225, U.S.A.

M. G. LAIRD
New Zealand Geological Survey, Department of Scientific and Industrial Research, Christchurch, New Zealand

J. D. BRADSHAW
Geology Department, University of Canterbury, Christchurch, New Zealand

Abstract. K-Ar age data are presented for Precambrian–Cambrian rocks in northern Victoria Land belonging to the Wilson and Robertson Bay Groups and the Bowers Supergroup.

K-Ar hornblende, biotite, and muscovite ages for the Wilson Group are in the range of 451–483 m.y. both for rocks of the amphibolite facies and for rocks that have undergone retrogressive metamorphism to epidote-amphibolite facies. Whole-rock ages for the Robertson Bay Group have a similar range of 447–502 m.y. irrespective of whether the rocks have been metamorphosed to prehnite-pumpellyite, pumpellyite-actinolite, or greenschist facies. In the Sledgers Group rocks, four whole-rock ages are in the same range as the stratigraphically older Wilson and Robertson Bay Group rocks (441–467 m.y.), but four others fall into a younger group from 384 to 421 m.y. with those of Mariner and Leap Year Group rocks. Two low ages of 354 and 361 m.y. are probably associated with the intrusion of Admiralty Intrusives (Devonian–Carboniferous).

These K-Ar ages reflect two main low-grade metamorphic events: a pre–Late Cambrian, post–Early Cambrian event in the Wilson, Robertson Bay, and Sledgers Groups; and a younger, Silurian–Devonian event recorded in the Bowers Supergroup. These Cambrian and Silurian–Devonian regional metamorphisms are probably associated with the Ross and Borchgrevinck orogenies, respectively.

The Precambrian–lower Paleozoic rocks of the Bowers Mountains, northern Victoria Land, provide one of the best areas in the Transantarctic Mountains to study the chronology of tectonism, metamorphism, and plutonism associated with the Paleozoic Ross and Borchgrevink orogenies and the less well-known Precambrian orogenic history. This paper presents some of the results of a broader potassium-argon (K-Ar) dating study of Precambrian–Paleozoic geochronology in the Transantarctic Mountains, and it follows an approach similar to that of a parallel project in the Nimrod Glacier region (see Adams, Gabites, and Grindley, Chapter 101, this volume). By using the sensitive K-Ar whole-rock technique on low-grade metamorphic rocks such as slates, one commonly can avoid the complicated isotopic histories encountered in mineral dating of higher grade metamorphic rocks from the deeper zones of orogenic belts.

Geologic outline

Since the earlier geologic surveys of Crowder (1968), Sturm and Carryer (1970), and Dow and Neall (1974), the Precambrian–Cambrian rocks of northern Victoria Land and their critical role in relation to the early Paleozoic history of Antarctica, Australia, and New Zealand have attracted intensive study (Laird and others, 1976, 1977; Cooper and others, 1976). The most recent geologic work is compiled and reviewed in this volume (Laird and others, Chapter 66; Bradshaw and others, Chapter 100; Wodzicki and others, Chapter 68) and only a very brief summary of the stratigraphy relevant to the K-Ar dating is given here.

1. *Wilson Group* (Precambrian?). Quartzo-feldspathic, pelitic, and amphibolitic gneisses, regionally metamorphosed to amphibolite facies and subsequently retrogressively metamorphosed to epidote-amphibolite facies.
2. *Robertson Bay Group* (Vendian). Slates and metasandstones of granitic provenance, metamorphosed to prehnite-pumpellyite, pumpellyite-actinolite, and greenschist facies.
3. *Bowers Supergroup* (Vendian–Upper Cambrian).
   a. *Sledgers Group* (Vendian–Lower Cambrian). Shallow-marine fossiliferous volcanic rocks and volcaniclastic breccias, sandstones, and metapelites, metamorphosed to prehnite-pumpellyite and pumpellyite-actinolite facies;

* INS contribution number 871.

Table 67.1. Potassium-argon age determinations on the Wilson and Robertson Bay Groups and Bowers Supergroup,

I.N.S. "R" number	Field number	Rock description	Location	Lat. S	Long. E	K (wt. %)	^{40}Ar (radiogenic) nl/g	^{40}Ar (radiogenic) % total	(m.y.)
Wilson Group (Precambrian ?)									
3898bi	W1	Quartzo-feldspathic garnet-biotite-muscovite granodiorite gneiss	Upper Zenith Glacier Lanterman Range	71° 47′	163° 41′	7.361	161.4	94.8	483 ± 3
3898mu						7.926	173.7	88.3	483 ± 3
3899bi	W2	Quartzo-feldspathic biotite-muscovite chlorite granodiorite gneiss	Upper Zenith Glacier Lanterman Range	71° 47′	163° 42′	6.858	144.3	97.6	466 ± 3
3900hb	W3	Hornblende (blue-green) albite-epidote amphibolitic gneiss	Upper Zenith Glacier Lanterman Range	71° 47′	163° 42′	0.142	2.91	61.3	455 ± 3 } 453[a]
							2.87	83.4	450 ± 3 }
3901mu	W4a	Quartzo-feldspathic biotite-muscovite-chlorite granodiorite gneiss	Upper Zenith Glacier Lanterman Range	71° 47′	163° 42′	7.745	167.1	82.3	476 ± 3
3902bi	W7	Quartzo-feldspathic biotite-muscovite-garnet granodiorite gneiss	S side, Husky Pass head of Sledgers Glacier	71° 44′	163° 40′	7.108	152.5	96.8	474 ± 3
3903hb	W11	Hornblende (blue-green) amphibolitic gneiss	S side, Husky Pass head of Sledgers Glacier	71° 43′	163° 36′	0.653	10.15	86.4	355 ± 3 } 355[a]
							10.16	84.0	355 ± 3 }
3904hb	W13	Hornblende (brown-green) biotite-garnet-oligoclase amphibolitic gneiss	Husky Pass, head of Sledgers Glacier	71° 40′	163° 27′	0.542	11.00	86.3	451 ± 3
3905bi 152-422μ	W15	Quartzo-feldspathic biotite-muscovite-hornblende (blue-green) gneiss	Husky Pass, head of Sledgers Glacier	71° 39′	163° 29′	6.511	141.5	96.6	479 ± 3
3962mu 152-853μ	W12	Quartzo-feldspathic biotite-muscovite granodiorite gneiss	S side, Husky Pass head of Sledgers Glacier	71° 43′	163° 36′	6.618	130.3	96.3	467 ± 3
Robertson Bay Group (Vendian)									
3208TR	MGL/RB2	Argillite, from sandstone/siltstone sequence	4 km NW of Mt McCarthy, Evans Névé	72° 35′	166° 08′	2.420	52.17	98.3	476 ± 3
3661TR	P13	Red siltstone from sandstone/siltstone sequence	4 km NNW of Pyramid Peak, Evans Névé	72° 14′	165° 32′	3.809	86.68	93.6	499 ± 3
3885TR	RB1	Slate, pumpellyite-actinolite facies, textural zone 2	Leap Year Glacier, at head of Champness	71° 30′	163° 49′	5.823	133.3	99.1	502 ± 4
3887TR	RB4	Slate, pumpellyite-actinolite facies, textural zone 2	N side of Leap Year Glacier, 4 km S of Mt Verhage	71° 28′	163° 40′	3.067	61.09	98.4	444 ± 3 } 447[a]
							62.08	77.7	450 ± 3 }
3888TR	RB5	Slate, pumpellyite-actinolite facies, textural zone 2	N side of Leap Year Glacier, 4 km S of Mt Verhage	71° 27′	163° 42′	2.989	67.40	97.5	495 ± 4
3889TR	RB6	Slate, pumpellyite-actinolite facies, textural zone 2	N side of Leap Year Glacier, 4 km S of Mt Verhage	71° 27′	163° 43′	2.229	47.14	97.0	468 ± 3
3890TR	RB7	Phyllite, greenschist facies, textural zone 4	N side, Ian Peak Champness Glacier	71° 31′	163° 57′	2.847	57.40	82.9	448 ± 3
3891TR	RB8	Slate, pumpellyite-actinolite greenschist facies, textural zone 3	N side, Head of Champness Glacier	71° 28′	163° 52′	2.809	62.56	97.7	490 ± 4
3892TR	RB9	Slate, pumpellyite-actinolite greenschist facies, textural zone 3	N side, at head of Champness Glacier	71° 28′	163° 51′	2.733	61.94	97.5	491 ± 4
3893TR	RB10	Slate, pumpellyite-actinolite greenschist facies, textural zone 3	2 km SE of Ian Peak Leap Year Glacier	71° 32′	164° 02′	2.608	56.25	84.2	476 ± 3
3895TR	RB13	Slate, pumpellyite-actinolite/greenschist facies, textural zone 3	Head of Leap Year and Graveson Glaciers	71° 17′	163° 25′	3.208	68.91	86.7	474 ± 3
3896TR	RB14	Slate, pumpellyite-actinolite/greenschist facies, textural zone 3	N side, Leap Year Glacier, 4 km S of Mt Verhage	71° 26′	163° 42′	4.008	87.58	90.3	482 ± 3
3897TR	RB15	Slate, pumpellyite-actinolite/greenschist facies, textural zone 3	N side, Leap Year Glacier, 4 km S of Mt Verhage	71° 27′	163° 42′	3.369	75.33	97.7	491 ± 4
6203TR	J9-17	Gray-green slate, greenschist facies	S of Austin Peak Mirabito Range	71° 45′	165° 30′	2.164	48.47	94.2	492 ± 4
6204TR	J10-7	Gray-green slate, greenschist facies	S of Austin Peak Mirabito Range	71° 45′	165° 30′	2.473	55.24	84.0	491 ± 4
6206TR	J13-8	Gray slate, slightly recrystallized to hornfels	NW end, Mirabito Range	71° 30′	163° 37′	3.202	88.70	93.1	361 ± 3

b. *Mariner Group* (upper Middle Cambrian–Upper Cambrian). Shallow-marine fossiliferous sedimentary rocks of granitic provenance, including calcareous siltstones and argillites, metamorphosed to prehnite-pumpellyite facies;

c. *Leap Year Group* (Upper Cambrian–?). Generally coarse-grained, shallow-marine and fluvial sedimentary rocks, including red siltstones and shales.

4. *Admiralty (and Granite Harbour?) Intrusives.* Several generations of unfoliated granites and granodiorites intrude the Wilson and Robertson Bay Groups. Of those dated, the Champness Granodiorite, Freyberg Adamellite, and Gallipoli Porphyries all give mineral ages characteristic of the Admiralty Intrusives, 300–380 m.y., but earlier generations of granites associated with the Granite Harbour Intrusives are probably also present.

Although direct relationships are not seen, the structural and metamorphic evidence suggests that the Wilson Group is the oldest unit in the region, and it is probably unconformably overlain by the Robertson Bay Group and the Bowers Supergroup. An intra-Vendian deformational event between the deposition of the Robertson Bay and Sledgers Groups is also probable. No angular unconformities exist within the Bowers Supergroup, although an important erosive break occurs between the Mariner and Leap Year Groups, and a probable hiatus occurs between the Sledgers and Mariner Groups.

Potassium-argon dating

Potassium-argon mineral and whole-rock ages have been determined for 9 Wilson Group pelitic and amphibolitic gneisses and 16 Robertson Bay Group and 17 Bowers Supergroup argillites, slates, and phyllites. Full analytical data and age results with brief petrographic notes and locality information are presented in Table 67.1. Sample locations are plotted on a simplified geologic map in Figure 67.1. Experimental methods used in this work have been described by Adams (1975).

northern Victoria Land, Antarctica

I.N.S. "R" number	Field number	Rock description	Location	Lat. S	Long. E	K (wt. %)	^{40}Ar (radiogenic) nl/g	^{40}Ar (radiogenic) % total	(m.y.)
Sledgers Group (Vendian–Lower Cambrian), Bowers Supergroup									
3868TR	S13	Slate, prehnite-pumpellyite facies, Vendian microfossil location	Molar Massif at Husky Pass	71° 38′	163° 37′	3.202	67.57 67.50	96.4 98.4	467 ± 3 467 ± 3 } 467[a]
3870TR	S19A	Slate, prehnite-pumpellyite facies (Sledgers Group?)	N side, Sledgers Glacier, W of icefall	71° 28′	163° 26′	2.798	49.85	96.7	401 ± 3
3871TR	S29	Slate, prehnite-pumpellyite facies	Mt Jamroga, head of Carryer Glacier	71° 20′	163° 05′	2.805	51.74	97.7	414 ± 3
3872TR	S30	Slate, prehnite-pumpellyite facies	Mt Jamroga, head of Carryer Glacier	71° 20′	163° 05′	2.784	51.13	97.9	413 ± 3
3873TR	S32	Slate, prehnite-pumpellyite facies	Mt. Jamroga, head of Carryer Glacier	71° 20′	163° 05′	2.924	54.24	96.0	416 ± 3
6187TR	2M83	Slate, prehnite-pumpellyite facies, Vendian microfossil location	SW ridge, Molar Massif at Husky Pass	71° 40′	163° 44′	2.116	42.80	98.1	450 ± 3
6207TR	2M85	Slate, prehnite-pumpellyite facies, Vendian microfossil location	Central section, Molar Massif	72° 35′	163° 39′	0.928	18.93	77.6	453 ± 3
6293TR	S2	Siltstone, prehnite-pumpellyite facies, Vendian microfossils and Cambrian fossils	Molar Massif, northern section	71° 35′	163° 36′	1.874	37.08	93.4	441 ± 3
Mariner Group (Middle–Upper Cambrian), Bowers Supergroup									
3875TR	S44	Slate, prehnite-pumpellyite facies (Mariner Group correlative)	S end, Reilly Ridge, Sledgers Glacier	71° 34′	163° 22′	3.636	66.81	55.2	413 ± 3
3877TR	S49	Slate, prehnite-pumpellyite facies (Mariner Group correlative)	S end, Reilly Ridge, Sledgers Glacier	71° 34′	163° 22′	3.392	60.52	90.7	402 ± 3
3878TR	S52	Slate, prehnite-pumpellyite facies (Mariner Group correlative)	Central section, Reilly Ridge, Sledgers Glacier	71° 32′	163° 17′	3.719	69.09	81.7	417 ± 3
3880TR	S55	Slate, prehnite-pumpellyite facies (Mariner Group correlative)	Central section, Reilly Ridge, Sledgers Glacier	71° 32′	163° 17′	2.058	35.31 34.49	90.5 97.9	388 ± 3 380 ± 3 } 384[a]
3882TR	S57	Slate, prehnite-pumpellyite facies (Mariner Group correlative)	Central section, Reilly Ridge, Sledgers Glacier	71° 32′	163° 17′	2.974	49.55	74.7	380 ± 3
3884TR	S59	Slate, prehnite-pumpellyite facies (Mariner Group correlative)	Central section, Reilly Ridge, Sledgers Glacier	71° 32′	163° 17′	2.946	52.24	98.2	400 ± 3
6294TR	H8	Siltstone, prehnite-pumpellyite facies (Mariner Group?)	Helix Pass, head of Leap Year Glacier	71° 20′	163° 15′	2.957	54.37	96.1	413 ± 3
6296TR	S43	Siltstone, prehnite-pumpellyite facies	4 km WSW of Helix Pass	71° 20′	163° 15′	1.369	23.87	93.9	394 ± 3
Leap Year Group (Upper Cambrian–), Bowers Supergroup									
6292TR	P5	Red siltstone, Camp Ridge Quartzite	6 km WNW Pyramid Rock Evans Névé	72° 16′	165° 22′	1.037	19.49	91.6	421 ± 3

TR: total (whole) rock. bi: biotite. mu: muscovite. hb: hornblende.
Decay constants, ^{40}K: λ_β = 0.472 × 10^{-9}/yr; λ_e = 0.584 × 10^{-10}/yr. ^{40}K/K = 0.0119 atomic %.
Errors are one standard deviation.
All 221–422μ fraction unless stated otherwise.
[a] Mean ages.

The range of apparent ages for all the samples dated is 355–502 m.y., but there are two important concentrations at 475–500 m.y. and 380–420 m.y. This age pattern, distributed by stratigraphic groups, can be seen in the histograms presented in Figure 67.2.

The Wilson Group

The seven biotite and muscovite ages from Wilson Group gneisses form a cluster at 466–483 m.y. The three hornblende ages from amphibolitic gneisses are younger, at 355, 451, and 453 m.y. In view of the usually high argon retentivity of hornblende (Hart, 1964), one might expect these to yield the oldest ages, but it seems likely that these hornblendes have been partially and preferentially affected by a young thermal event. Sample R3903hb, at 355 m.y., does in fact show some recrystallization into blue-green from brown-green hornblende during a late retrogressive metamorphism. The hornblende age data therefore do not confirm, as anticipated, the presumed Precambrian age of the Wilson Group. The only other age determination relevant to this problem is a single Rb-Sr analysis of the Rennick Group schists, a possible equivalent of the Wilson Group in the Upper Rennick Glacier area about 150 km to the south (Faure and Gair, 1970). A probable maximum age of 770 ± 20 m.y. was obtained, assuming a low initial $^{87}Sr/^{86}Sr$ ratio of 0.704. The uncertainty in the age quoted does not reflect the great uncertainty in the upper limit for the initial $^{87}Sr/^{86}Sr$ ratio; a higher value could reasonably reduce the apparent Precambrian age to 400–500 m.y. Thus the oldest event definitely recorded so far by the K-Ar and Rb-Sr measurements for the Wilson Group seems to be a metamorphism about 500 m.y. ago.

Robertson Bay Group and Bowers Supergroup

In general, the Robertson Bay Group and Bowers Supergroup yield a similar age pattern to the Wilson Group, having an age range of 361–502 m.y. with a sharp upper limit at about 500 m.y. (see Table 67.1). Since the prehnite-pumpellyite facies rocks are from a structurally higher level than the deep-seated amphibolite facies gneisses of the Wilson Group, their general age concordance implies rapid uplift and cooling after a metamorphic event.

In detail, the whole-rock slate ages of the Robertson Bay Group and constituent groups of the Bowers Supergroup show some important differences (see Figure 67.2). It can be seen that the Wilson, Robertson Bay, and Sledgers Groups all yield some ages in the range of 450–500 m.y., but the Sledgers Group alone also has four significantly younger ages, 401–416 m.y. In contrast, the Mariner Group yields only younger whole-rock ages in the range of 380–417 m.y. These younger ages must relate to a later thermal event in the Bowers Supergroup which is not recorded in the adjacent Wilson and Robertson Bay Groups.

The Borchgrevink and Ross Orogenies

The age data obtained from the Wilson Group, Robertson Bay Group, and Bowers Supergroup are best interpreted as reflecting two thermal events. These are most

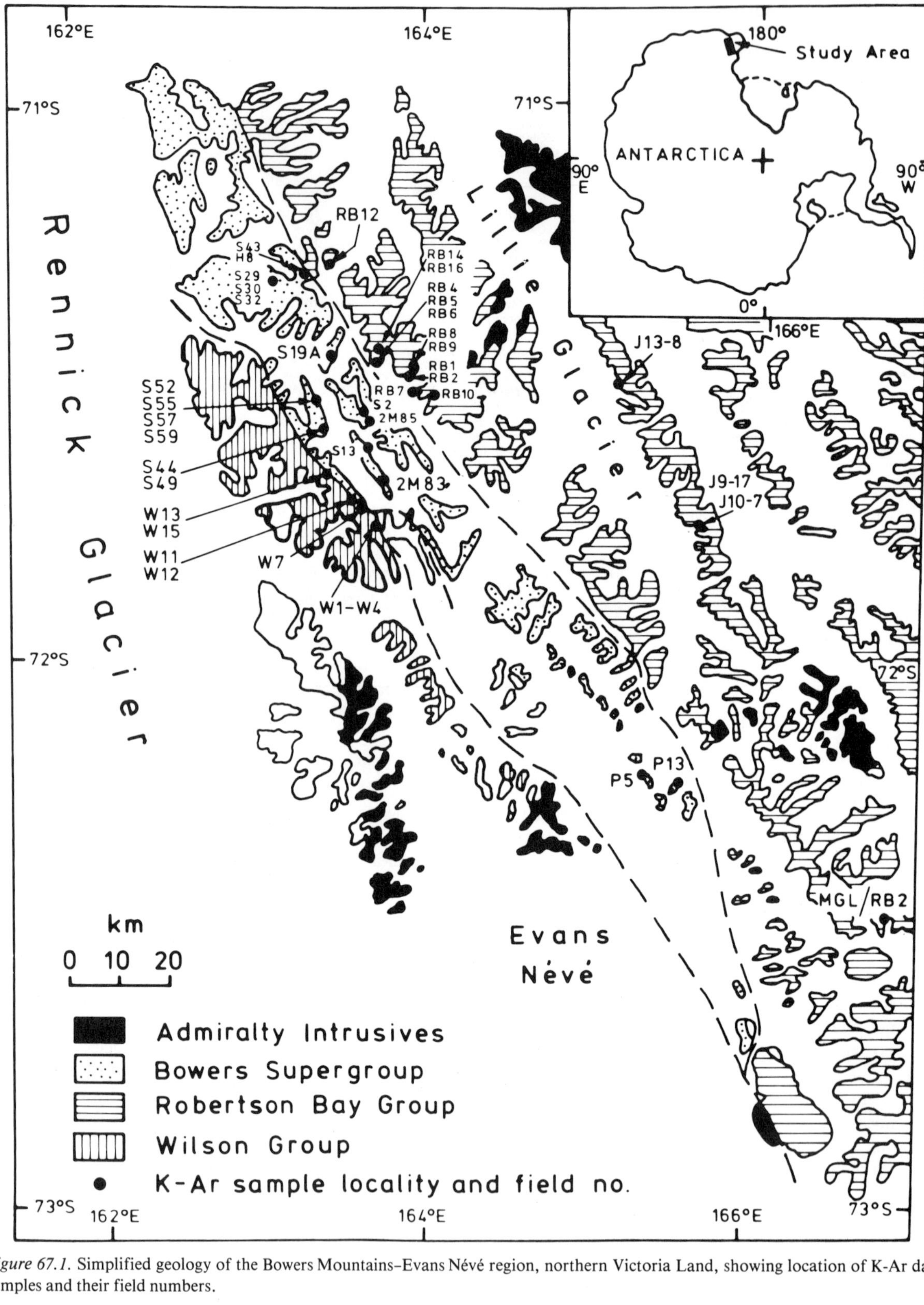

Figure 67.1. Simplified geology of the Bowers Mountains-Evans Névé region, northern Victoria Land, showing location of K-Ar dating samples and their field numbers.

likely associated with low-grade regional metamorphism during the two main orogenic episodes in northern Victoria Land, namely the Ross orogeny (Cambrian-Ordovician) and the Borchgrevink orogeny (Silurian-Devonian). K-Ar ages cannot really provide a maximum age for metamorphism, but ages on low-grade slates such as these usually provide a minimum value and, in many cases, approximate quite closely the actual time of an event. Thus, the maxima of each of the two groups of ages (380–420 m.y. and 450–500 m.y.) suggest minimum ages for the two events. Unfortunately, field evidence for the timing of these metamorphic events is poor, as a full, continuous section from Precambrian (Robertson Bay Group) to Upper Cambrian is not exposed at any one place. There is field evidence for

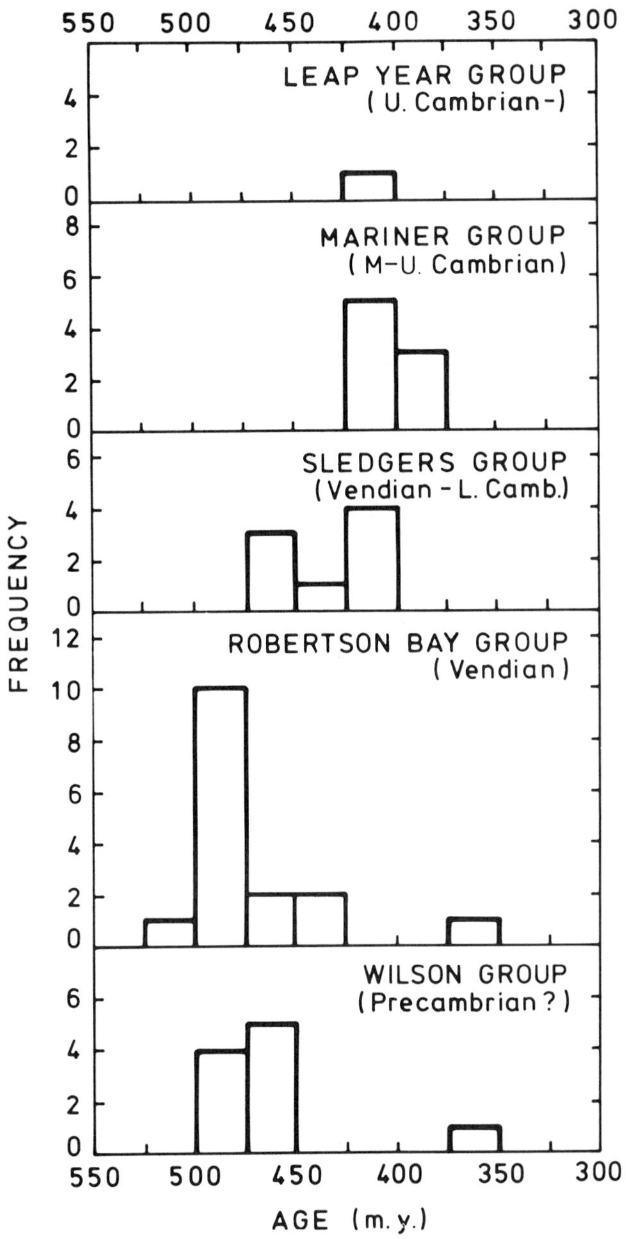

Figure 67.2. Histograms of K-Ar ages for the Wilson and Robertson Bay Groups and Bowers Supergroup (subdivided into constituent Sledgers, Mariner, and Leap Year Groups).

a metamorphic event which is post-Cambrian and in which the youngest rocks affected are Upper Cambrian. The Ross orogeny is generally considered to be Ordovician in age, but there are no Ordovician post-Ross orogeny rocks in the area. The isotopic evidence for the age of the orogeny is also tenuous. From the K-Ar ages, the event must have occurred here at or slightly before 500 m.y. ago.

In fitting the Ross orogeny into the local geologic sequence, the lack of old ages from Mariner Group rocks must be considered. The young ages of the Mariner Group slates are most probably related to a tectonic and low-grade metamorphic episode associated with the Borchgrevink orogeny, which is recorded elsewhere in northern Victoria Land (Gair and others, 1969). Whether this event obliterated older Ross orogeny metamorphic dates (450–500 m.y.) in the Mariner Group is not clear. Certainly none are preserved now. It is possible that the Mariner and Leap Year Groups were insufficiently buried during the Ross orogeny to be affected by either (a) regional metamorphism or (b) emplacement of Granite Harbour Intrusives at depth at that time. But in any case, there are no geochronologic data to confirm the presence of Granite Harbour Intrusives. It is significant that the older metamorphic ages have been largely preserved in the pre-Mariner rocks; thus it seems more likely that a metamorphic event occurred before rather than after Mariner Group sedimentation, i.e., before the late Middle Cambrian. Whether this event is a major or minor part of the Ross orogeny cannot be decided on the age data alone. The slate ages suggest a minimum age for the metamorphism of 500 m.y., i.e., latest Cambrian on the Phanerozoic timescale of Harland and others (1964). It should be stressed, however, that the relevant data points for the Cambrian timescale are notoriously weak. Recent revisions, additions, and deletions may alter the situation significantly, possibly reducing the Lower Ordovician and Upper and Middle Cambrian boundaries by 10–20 m.y. The 500 m.y. age maximum for the K-Ar ages presented here would thus provide a minimum age for a metamorphic event in early Late Cambrian or late Middle Cambrian time.

From a geochronologic viewpoint, a middle Cambrian age for the Ross orogeny therefore seems logical, but there is no evidence for any important metamorphic or structural discordance between the Sledgers and Mariner Groups. There may, however, be a major gap in the fossil record between the upper Middle Cambrian (basal Mariner) and the Lower Cambrian (uppermost Sledgers) which may indicate an important sedimentary break. Alternatively, if the metamorphism is considered to predate the Sledgers Group, then the Ross orogeny would merge with the late Precambrian Beardmore orogeny (Grindley and McDougall, 1969), and the low-grade metamorphic rocks would have to have remained at considerable depths and above their argon retention temperature for 70–100 m.y. before final uplift and cooling. In view of their low metamorphic rank, and thus high structural level, this seems improbable. However, if the Ross metamorphism was Middle Cambrian (i.e., pre-500 m.y.), then uplift and cooling (as suggested by ages up to 500 m.y.) must have occurred rapidly afterward. The individual ages between 500 and 450 m.y. could then be interpreted as continued uplift and cooling of the Wilson and Robertson Bay Groups throughout the Late Cambrian and Early Ordovician. Such uplift must imply erosion and consequent rapid (shallow-water?), and synchronous, sedimentation. This could be reflected in the changes of sedimentary style from the volcanigenic Sledgers Group (Lower Cambrian) at one extreme, to the coarse-grained, arenaceous Leap Year Group (Upper Cambrian–Lower Ordovician?) at the other extreme. Whether this process was terminated by emplacement of the Granite Harbour Intrusives is unclear; the answer must await new Rb-Sr whole-rock dating studies of these granites.

The maximum age of the younger group of K-Ar ages implies a minimum age of 420 m.y. for the second thermal event, that associated with the Borchgrevink orogeny. As mentioned previously, young ages are confined chiefly to the Leap Year and Mariner Groups and partly to the Sledgers Group. The most probable explanation for this is that the higher metamorphic grade Robertson Bay and

Wilson Groups (and part of the Sledgers Group) rendered them mineralogically and isotopically resistant to the younger and slightly lower grade metamorphic overprinting event. At the same time, it is possible that most of the Sledgers, Robertson Bay, and Wilson Group samples dated were geographically outside the area of influence of the younger event. In this respect it is significant that between lat. 71° S and lat. 72° S an approximately north-south line at about long. 163° 30′ E (see Figure 67.1) separates an area of younger, 350–420 m.y. ages on the western side from 450–500 m.y. ages to the east. However, there seems to be no obvious geologic explanation for this pattern and the cause is not clear. One possibility, a thermal metamorphic effect associated with the Admiralty Intrusives, seems unsatisfactory for three reasons:

1. such intrusives, e.g., the Gallipoli Porphyries (Faure and Gair, 1970) and the Champness Granodiorite (Dow and Neall, 1974) occur throughout northern Victoria Land and are not confined to the area of younger K-Ar slate ages (see Figure 67.1);
2. the ages of the Admiralty Intrusives, although ranging from 300 to 380 m.y. (Gair and others, 1969; Dow and Neall, 1974), seem to be younger than the metamorphic ages recorded in the Mariner and Leap Year Groups (378–421 m.y.); and
3. Robertson Bay Group samples R3890TR and R3893TR occur close to the Champness Granodiorite (350 m.y.), yet their total rock ages, 448 and 476 m.y. respectively, suggest that they are little if at all affected by any contact metamorphism (see Figure 67.1). On the other hand, elsewhere in the area of older ages there are two samples, R6206TR (Robertson Bay Group) and R3903hb (Wilson Group), whose younger ages (361 and 354 m.y.), falling within the age range of the Admiralty Intrusives, can be obviously related to local contact effects.

Summary

K-Ar mineral and whole-rock ages of the Wilson and Robertson Bay Groups and Bowers Supergroup, in the range of 355–502 m.y., reflect two low-grade regional metamorphic events:

1. a younger, Silurian–Devonian event recorded in the Bowers Supergroup associated with the Borchgrevink orogeny; and
2. an earlier, Cambrian (and possibly pre-Late Cambrian) event, recorded in the Wilson, Robertson Bay, and Sledgers Groups associated with the Ross orogeny.

The age data suggest that this earlier metamorphic event occurred in the time interval between the deposition of the uppermost Sledgers and the basal Mariner Groups. Ages of the Sledgers and Wilson Groups reflect their uplift after the metamorphism. This uplift was synchronous with renewed sedimentation of the Mariner and Leap Year Groups.

Very locally, there seems to be a further thermal metamorphic overprint associated with the emplacement of the Admiralty Intrusives (300–380 m.y.).

Acknowledgments. R. A. Cooper is thanked for providing additional specimens from the Bowers Supergroup. D. S. Nicholson is thanked for careful mineral separation of Wilson Group gneiss specimens.

References cited

Adams, C. J. D. 1975. New Zealand potassium-argon age list—2. *N.Z. Jl Geol. Geophys., 18,* No. 3, 443–67.

Cooper, R. A., Jago, J. B., MacKinnon, D. I., Simes, J. E., and Braddock, P. E. 1976. Cambrian fossils from the Bowers Group, northern Victoria Land, Antarctica. (Preliminary note) *N.Z. Jl Geol. Geophys., 19,* No. 2, 283–88.

Crowder, D. F. 1968. Geology of a part of north Victoria Land, Antarctica. *Prof. Pap. U.S. geol. Surv.,* No. 600-D, D95–107.

Dow, J. A. S., and Neall, V. E. 1974. Geology of the lower Rennick Glacier, northern Victoria Land, Antarctica. *N.Z. Jl Geol. Geophys., 17,* No. 3, 659–714.

Faure, G., and Gair, H. S. 1970. Age determinations of rocks from northern Victoria Land, Antarctica. *N.Z. Jl Geol. Geophys., 13,* No. 4, 1024–26.

Gair, H. S., Sturm, A., Carryer, S. J., and Grindley, G. W. 1969. The geology of northern Victoria Land (Sheet 13). (*In* Bushnell, V. C., and Craddock, C., *eds.* Geologic Maps of Antarctica. *Antarct. Map. Folio Ser.,* Folio 12, Pl. XIII.)

Grindley, G. W., and McDougall, I. 1969. Age and correlation of the Nimrod Group and other Precambrian rock units in the central Transantarctic Mountains, Antarctica. *N.Z. Jl Geol. Geophys., 12,* No. 2–3, 391–411.

Harland, W. B., Gilbert Smith, A., and Wilcock, B. 1964. *eds.* The Phanerozoic Time-scale. *Q. Jl geol. Soc. Lond., 120S,* 458 pp.

Hart, S. R. 1964. The petrology and isotopic-mineral age relations of a contact zone in the Front Range, Colorado. *J. Geol., 72,* 493–525.

Laird, M. G., Bradshaw, J. D., and Wodzicki, A. 1976. Re-examination of the Bowers Group (Cambrian), northern Victoria Land, Antarctica. (Preliminary note) *N.Z. Jl Geol. Geophys., 19,* No. 2, 275–82.

Laird, M. G., Cooper, R. A., and Jago, J. B. 1977. New data on the lower Palaeozoic sequence of northern Victoria Land, Antarctica, and its significance for Australian-Antarctic relations in the Palaeozoic. *Nature, Lond., 265,* 107–10.

Sturm, A., and Carryer, S. J. 1970. Geology of the region between Matusevitch and Tucker Glaciers, north Victoria Land, Antarctica. *N.Z. Jl Geol. Geophys., 13,* 408–35.

68

Petrology of the Wilson and Robertson Bay Groups and Bowers Supergroup, Northern Victoria Land, Antarctica

ANTONI WODZICKI
Geology Department, Western Washington University, Bellingham, Washington 98225, U.S.A.

J. D. BRADSHAW
Geology Department, University of Canterbury, Christchurch, New Zealand

M. G. LAIRD
New Zealand Geological Survey, Department of Scientific and Industrial Research, Christchurch, New Zealand

Abstract. In the Bowers Mountains, northern Victoria Land, metamorphic rocks of the Precambrian Wilson Group and the Late Precambrian Robertson Bay Group are separated by a northwesterly trending belt of metasedimentary and metavolcanic rocks of the Vendian to Lower Paleozoic Bowers Supergroup.

The Wilson Group includes quartzo-feldspathic and pelitic gneisses, amphibolites, and deformed conglomerates. Metamorphism to the amphibolite facies during early folding led to the formation of quartz, oligoclase, garnet, cordierite, biotite, and muscovite in pelitic rocks, and oligoclase, green hornblende, biotite, and garnet in mafic rocks. Retrogression to the epidote-amphibolite facies during subsequent folding resulted in the partial replacement of oligoclase by albite and clinozoisite or epidote, and in the formation of a blue-green amphibole.

The Robertson Bay Group consists of interbedded metapelites and metasandstones of granitic provenance. Early synmetamorphic folding produced schistosity and mineral lamination which were later folded into an antiform. On the flanks of the antiform only a weak schistosity is exposed; plagioclase is replaced by albite, K-feldspar is partly replaced by sericite, and the matrix has recrystallized to quartz, albite, sericite, chlorite ± pumpellyite ± calcite ± epidote ± prehnite ± sphene—assemblages of the prehnite-pumpellyite to pumpellyite-actinolite facies. Toward the core of the antiform prominent schistosity and mineral lamination are present, and, with the disappearance of prehnite and pumpellyite and the sporadic appearance of clinozoisite and biotite, the rocks grade into the greenschist facies.

The Bowers Supergroup has been folded into a northwest-striking anticlinorium flanked by major synclines. The Sledgers Group consists mainly of shallow-marine basaltic breccias which interfinger with volcaniclastic sedimentary beds derived from mafic and felsic volcanics and shallow intrusives. The overlying Mariner and Leap Year Groups are largely of granitic provenance. Folding of the Bowers Supergroup produced axial planar cleavage and metamorphism to the prehnite-pumpellyite and pumpellyite-actinolite facies. The facies boundary, marked by the disappearance of prehnite and the sporadic appearance of actinolite in the mafic rocks, lies near the contact of the Sledgers and Mariner Groups; the higher grade pumpellyite-actinolite facies occupies the core of the main anticlinorium.

Early accounts of the geology of the Bowers Mountains, northern Victoria Land, include the work of Crowder (1968), Sturm and Carryer (1970), and Dow and Neall (1974). During the 1974–75 field season the Bowers Mountain region was the subject of an intensive geologic study by two New Zealand expeditions. Preliminary results of that work were reported by Laird and others (1976) and Cooper and others (1976). Various aspects of their work are discussed in greater detail in this volume by Laird and others (Chapter 66), Bradshaw and others (Chapter 100), and Adams and others (Chapter 67) who deal respectively with stratigraphy, structure and tectonic history, and K-Ar dating of metamorphic events.

Briefly, the region consists of two major Precambrian units separated by a northwest-trending belt of Vendian to lower Paleozoic rocks of the Bowers Supergroup (Figure 68.1). The older of the two Precambrian units, the Wilson Group, lies to the southwest and the younger, the Robertson Bay Group, to the northeast of the Bowers Supergroup. Both Precambrian groups have undergone at least two, and possibly three, periods of folding about northwesterly axes, and both are in probable fault contact with the Bowers Supergroup.

The Bowers Supergroup is subdivided into the Husky Conglomerate (oldest), the Sledgers Group, the Mariner Group, and the Leap Year Group, which are separated by unconformities and which are, respectively, at least 0.4 km, 3 km, 2 km, and 3 km thick. The Sledgers Group con-

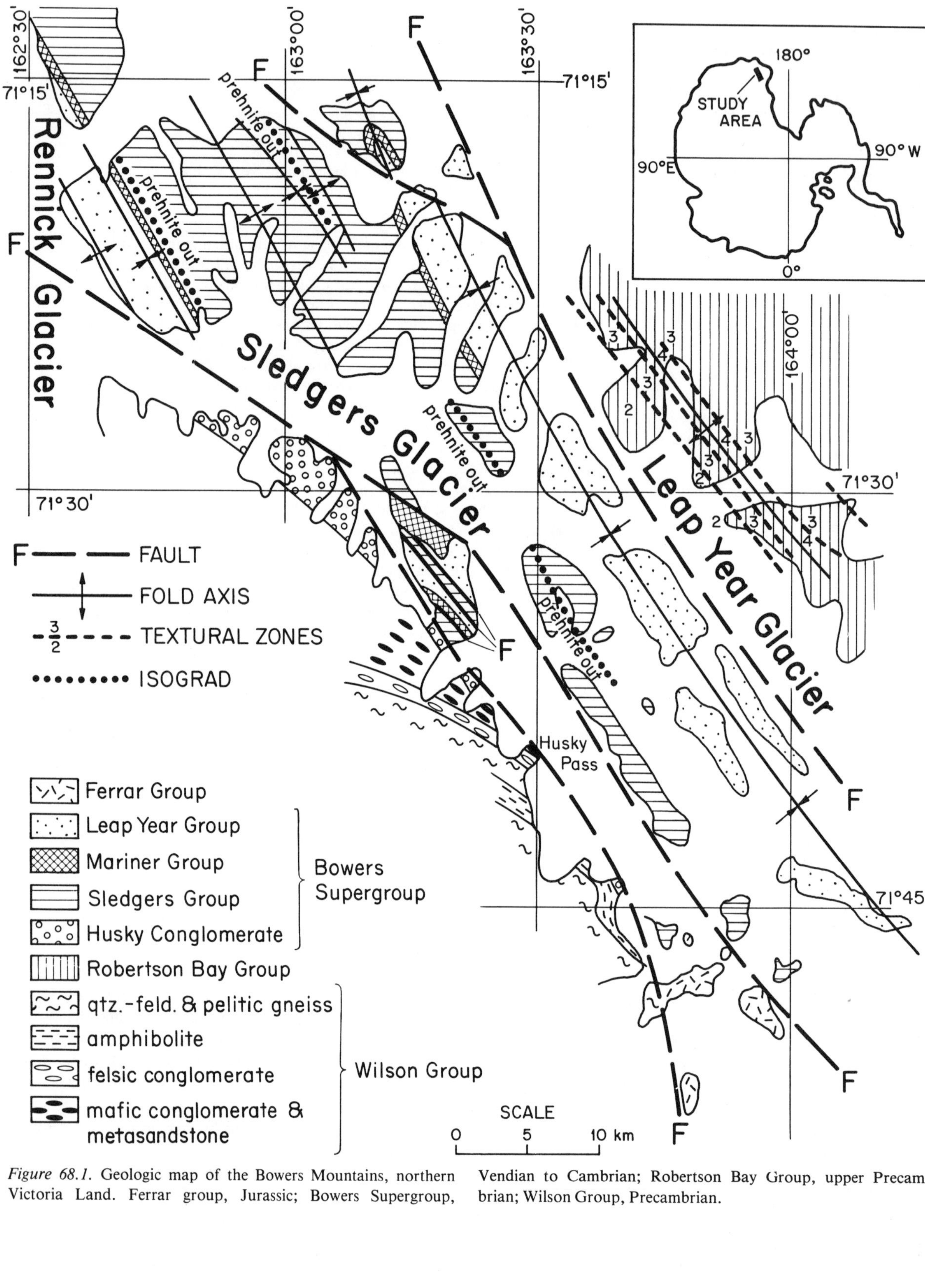

Figure 68.1. Geologic map of the Bowers Mountains, northern Victoria Land. Ferrar group, Jurassic; Bowers Supergroup, Vendian to Cambrian; Robertson Bay Group, upper Precambrian; Wilson Group, Precambrian.

sists of basaltic breccias which interfinger to the southeast with clastic sedimentary beds. The Mariner Group contains fossiliferous shallow-marine sandstones, mudstones, and limestones. The Leap Year Group consists of quartz sandstones and conglomerates in the northeast (Camp Ridge Quartzite) and polymictic conglomerate (Carryer Conglomerate) and quartzose conglomerate (Reilly Conglomerate) in the northwest. The rocks of the Bowers Supergroup have been folded about northwest-striking axes into a broad anticlinorium bounded on both sides by major synclines. They are unconformably overlain by rocks of the Ferrar Group, which will not be discussed further in this paper.

The purpose of this paper is to describe the petrology of this complex suite of rocks, and, where possible, to interpret it in terms of sediment provenance, tectonic history, and conditions of metamorphism.

The Wilson Group

The Wilson Group was examined along a 2–4 km wide strip on the southwest side of the Sledgers Glacier (Figure 68.1). It is unconformably overlain by the Husky Conglomerate and is in probable fault contact with the other units of the Bowers Supergroup.

Four major rock units are present in the mapped area: interlayered quartzo-feldspathic and pelitic gneisses, amphibolites—both fine and coarse grained, deformed felsic conglomerates, and deformed mafic conglomerates with interbedded metasandstones. A small metamorphosed ultramafic body is present 8 km south of Husky Pass. Near the confluence of the Rennick Glacier with the Sledgers Glacier, the Wilson Group gneisses contain pegmatitic segregations and are intruded by granitic rocks of unknown age.

The Wilson Group has probably undergone three periods of folding about northwesterly axes. The earliest recognizable folds (F_1) have steeply plunging axes, are isoclinal, and have resulted in the development of a strong foliation (S_1), a mineral lineation in the amphibolites, and the elongation of pebbles parallel to *a* in the conglomeratic facies. Metamorphism to the amphibolite facies has formed quartz, oligoclase, garnet, cordierite, biotite, and muscovite in pelitic rocks, and oligoclase, green hornblende, biotite, and garnet in mafic rocks.

Later folds (F_2) have variable but steeply plunging axes; they fold S_1 and locally are associated with the development of a younger, weak foliation (S_2). The F_2 folding is associated with retrograde metamorphism to the epidote-amphibolite facies. In the pelitic and quartzo-feldspathic gneisses this resulted in the replacement of oligoclase by albite and clinozoisite or epidote. In specimens that have only partly retrogressed, albite occurs as rims around oligoclase grains. In mafic rocks, retrogression also resulted in replacement of oligoclase by albite and clinozoisite or epidote and in the change from a green to a blue-green, finer grained hornblende.

The ultramafic rocks occur as a lens 40 m thick at the contact between quartzo-feldspathic gneisses and amphibolites; they have been affected by F_2 folding. The core of the ultramafic body consists of antigorite and magnesite, and the periphery of talc and magnesite, indicating CO_2 and H_2O metasomatism.

A third period of folding (F_3) is suggested by the local presence of a penetrative cleavage and the fact that F_2 folds have variable plunge. Partial retrogression to a greenschist facies assemblage, including alteration of garnet and biotite to chlorite and of hornblende to a fibrous actinolite, may be associated with this folding episode.

The Robertson Bay Group

The Robertson Bay Group was examined along a 6 km strip on the northeast side of Leap Year Glacier (Figure 68.1). It is in probable fault contact with rocks of the Bowers Supergroup. An early period of folding resulted in the development of schistosity and mineral segregation lamination which is generally discordant to bedding. During a later episode the rocks were folded about northwesterly axes into a broad antiform, exposing more intensely deformed, higher grade rocks in the core.

Three textural zones, similar to zones 2A and 2B, 3A and 3B, and 4 as defined by Bishop (1972), were recognized in the field (Figure 68.1). Textural zone 4 occupies the core of the antiform and is characterized by mineral segregation lamination >2 mm wide. It is flanked by zone 3 which is characterized by mineral segregation lamination <2 mm wide. These in turn are flanked by rocks of textural zone 2, which show penetrative foliation but no mineral segregation lamination.

Sandy rocks in textural zone 2 are poorly sorted and subangular. They contain clastic quartz, plagioclase, perthitic orthoclase, microcline, muscovite, minor lithic fragments of microgranite, quartzite and muscovite schist, and traces of tourmaline, sphene, and apatite. The detritus is dominantly of granitic provenance with a lesser contribution from a metamorphic terrane. The clastic texture is clearly preserved, but a weak foliation is present. Plagioclase is replaced by albite and sericite, K-feldspar is progressively replaced by sericite, and the matrix is recrystallized to the following assemblage: quartz, albite, sericite, chlorite ± calcite ± epidote ± sphene ± pumpellyite ± prehnite. Samples containing more than a few percent calcite do not contain other calcium-bearing minerals. Pumpellyite is present in about half of the samples and prehnite is present in only one sample, suggesting that the rocks of textural zone 2 include the prehnite-pumpellyite to pumpellyite-actinolite facies boundary (marked here by the disappearance of prehnite).

In textural zone 3, clastic texture is progressively obliterated, but mineral segregation lamination is clearly visible in this section. The following metamorphic minerals are present: quartz, albite, sericite, chlorite ± calcite ± epidote ± sphene ± pumpellyite. Prehnite is absent and pumpellyite is present in about half the samples, and the rocks are probably in the pumpellyite-actinolite facies.

In textural zone 4, no clastic textures are preserved and mineral segregation lamination and foliation are well developed. In some samples a second foliation, associated with the later episode of folding, is present. Metamorphic minerals present are: quartz, albite, muscovite, chlorite ± calcite ± epidote ± clinozoisite ± sphene ± biotite. With the disappearance of pumpellyite and the sporadic appearance of clinozoisite and biotite, the rocks grade into the greenschist facies.

The Bowers Supergroup

The Bowers Supergroup units are discussed first in terms of their primary mineralogy and provenance, and then in terms of their metamorphism and tectonics.

Primary mineralogy and provenance

Husky Conglomerate

This formation is present in a narrow strip between the Wilson and the Sledgers Groups. It lies unconformably on the Wilson Group and is in probable fault contact with the Sledgers Group. It consists mainly of conglomerate and breccia derived almost entirely from the underlying Wilson Group amphibolites and mafic conglomerates.

Sledgers Group

This group consists of the Glasgow Volcanics, which predominate in the north and northwest, and the Molar Formation, which occurs chiefly in the south and southwest of the region studied and which is probably contemporaneous with the Glasgow Volcanics. The Sledgers Group is in fault contact with the Wilson Group and the Husky Conglomerate and is unconformably overlain by the Mariner Group.

The Glasgow Volcanics consist dominantly of angular, unsorted, commonly vesicular basaltic and andesitic breccias, with pillow fragments in places, and locally (in the Lower Carryer Glacier) with lava flow units as much as 100 m thick and minor fine-grained limestone. Most of the volcanic rocks are porphyritic and contain various combinations of plagioclase, olivine (completely replaced with serpentine), orthopyroxene (almost completely replaced by chlorite), augite, and rare hornblende phenocrysts.

The Molar Formation consists mainly of interbedded mudstone and well-sorted sandstone which are intercalated with redeposited clastic rocks consisting of graded-bedded sandstones, conglomerates, and breccias. The redeposited rocks consist of angular to subrounded, poorly sorted lithic detritus. They contain fragments of basalt, andesite, and quartzo-feldspathic sandstone, lesser amounts of dacite, spherulitic rhyolite, fine-grained diorite, quartz diorite, trondhjemite, albite granite, graphic granite, limestone, quartzite, and muscovite schist fragments. The andesite, basalt, and limestone fragments are similar to rocks present in the Glasgow Volcanics; the clasts of quartzo-feldspathic sandstone are mineralogically and texturally identical to those that occur in nearby contemporaneous Sledgers Group sedimentary beds; and the metamorphic rock fragments are similar to those in the Robertson Bay Group. The source rocks for the remaining clast types, however, are not exposed in this region. Sparse sedimentologic data (Laird and others, Chapter 66, this volume) suggest that the redeposited sediments were derived from the northeast, an area where the Robertson Bay Group now crops out.

The Mariner Group

This group unconformably overlies the Sledgers Group and is unconformably overlain by the Leap Year Group. It consists of shallow-water marine sandstone, mudstone, and limestone. The sandstones contain clastic quartz, plagioclase, biotite, muscovite, and chlorite and are generally calcareous. They differ from the Sledgers Group in being poor in volcanigenic material and in being more micaceous. Quartzo-feldspathic and pelitic gneisses of the Wilson Group or Robertson Bay Group schists could be the sources of the sediments.

The Leap Year Group

This group consists of fluvial deposits which unconformably overlie the Mariner Group. Camp Ridge Quartzite crops out on the northeast side of the Bowers Supergroup outcrop area, and the Carryer Conglomerate and Reilly Conglomerate crop out on the southwest side.

Camp Ridge Quartzite consists mainly of poorly sorted, subrounded to subangular sandstones and quartz conglomerates which consist dominantly of quartz (75–95%) and contain variable amounts of plagioclase, perthitic orthoclase, microcline, muscovite, minor granite, and traces of biotite, epidote, clinozoisite, tourmaline, zircon, apatite, and garnet. The provenance is mainly granitic with possibly a small component originating from a metamorphic terrane. Sedimentologic data (Laird and others, Chapter 66, this volume) show that over most of the area studied, sediment transport was from the southwest, a region where the Wilson Group now crops out. Evidently at the time of Camp Ridge Quartzite sedimentation, granitic rocks must have been exposed extensively in this region.

The Carryer Conglomerate consists mainly of bedded, well-rounded conglomerate with a poorly sorted, angular, sandy matrix, interbedded with minor sandstone. The clasts consist of sandstone and some limestone; alaskite, biotite-quartz monzonite, biotite granodiorite, hornblende granodiorite, and pyroxene-quartz diorite; minor volcanic clasts which include hornblende dacite and pyroxene andesite; and minor metamorphic clasts which include quartzite and biotite-muscovite schists. Sparse sedimentologic data (Laird and others, Chapter 66, this volume) suggest derivation of the conglomerate from the southwest, an area where metamorphic rocks of the Wilson Group now crop out.

The Reilly Conglomerate consists of poorly sorted quartzose sandstones and conglomerates. The clasts are angular to subrounded and consist mainly of quartz and quartzite with minor muscovite and biotite schists and traces of plagioclase, perthitic orthoclase, and microcline. Sedimentologic data show derivation of these sediments from the southwest. The quartz-rich nature of both the Reilly Conglomerate and the Camp Ridge Quartzite suggests a possible correlation between the two formations.

Metamorphism of the Bowers Supergroup

Folding of the Bowers Supergroup was accompanied by development of an axial plane cleavage and metamorphism ranging from prehnite-pumpellyite to greenschist facies.

The Reilly Conglomerate and the Camp Ridge Quartzite contain quartz, albite, chlorite, sericite ± calcite assemblages. Plagioclase is entirely replaced by albite, sericite ± calcite; biotite is altered to chlorite; and K-feldspar remains unaltered. Quartz has recrystallized, especially near fold axes. The Carryer Conglomerate contains quartz, albite, sericite, chlorite, epidote ± calcite ± prehnite ± pumpellyite assemblages and is entirely in the prehnite-pumpellyite facies. Plagioclase is replaced by albite, sericite ± pumpellyite ± prehnite ± calcite; pyroxene by chlorite ± calcite; hornblende by chlorite, epidote ± calcite; biotite by chlorite; and K-feldspar remains unaltered. The Mariner Group contains quartz, albite, sericite, chlorite ± calcite. It has no diagnostic metamorphic mineral assemblages but is both underlain and overlain by rocks with prehnite-pumpellyite mineral assemblages.

The Sledgers Group includes the prehnite-pumpellyite facies and pumpellyite- actinolite facies boundary. On the southwest side of the Sledgers Group outcrop area the boundary lies within 100–200 m below the contact with the Mariner Group. Above the boundary the assemblages quartz, albite, sericite, chlorite, epidote ± calcite ± prehnite ± pumpellyite are present, and pyroxene and hornblende are partly replaced by chlorite; plagioclase is replaced by albite, sericite, prehnite, pumpellyite, and epidote; and the rocks are cut by prehnite-pumpellyite-albite and calcite-albite veinlets. Below the boundary prehnite is absent, but actinolite is present in the mafic rocks. On the northeast side of the Sledgers Group outcrop, prehnite is present down to 1000 m below the Mariner Group contact; however, no actinolite was observed down to 2000 m below the Mariner contact there, which was the deepest part of the section examined.

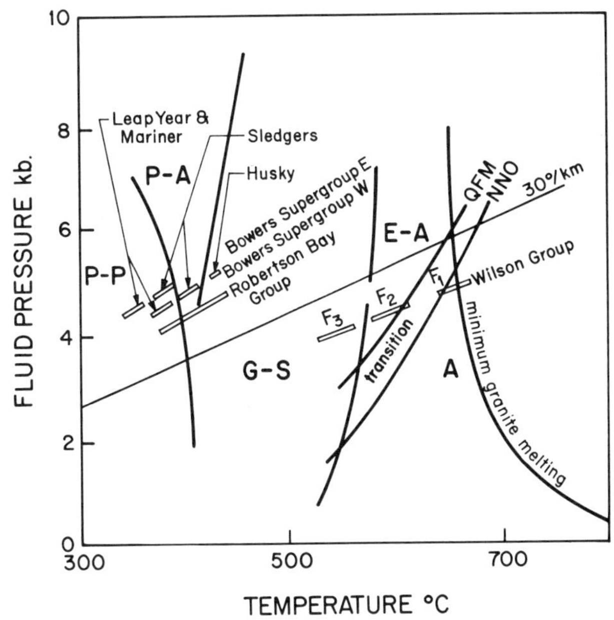

Figure 68.2. Estimated *P-T* fields of metamorphic facies redrawn from Liou (1971) and Liou and others (1974) and estimated *P-T* conditions prevailing during various metamorphic episodes in the Bowers Mountains, northern Victoria Land. Metamorphic facies: P-P: prehnite-pumpellyite. P-A: pumpellyite-actinolite. G-S: greenschist. E-A: epidote-amphibolite. A: amphibolite. QFM: quartz-fayalite-magnetite buffer. NNO: nickel-nickel oxide buffer.

Discussion

Provenance

The Wilson Group is the oldest unit that crops out in the area studied. The presence of both felsic and mafic conglomerates within the Wilson Group, however, shows that both granitic and basaltic crust were exposed at the time of Wilson Group sedimentation. The rocks have been too intensely deformed and metamorphosed to determine sediment transport direction.

The Robertson Bay Group has also been too intensely deformed and metamorphosed to determine the transport direction of the detritus, but the dominantly granitic nature of the sediments suggests derivation from a cratonic area, possibly lying to the south and west.

The redeposited facies of the Molar Formation are unique in the Bowers Mountain region because they are dominantly volcaniclastic and because they were transported from the northeast. Mafic volcanic and sedimentary clasts may have originated from the Glasgow Volcanics and the lower Molar Formation within the Bowers Trough. However, the felsic volcanic and hypabyssal clasts must have originated further to the northeast where the Robertson Bay group now crops out. Evidently, a wide variety of volcanic and plutonic rocks were exposed in this area during the time of Sledgers Group sedimentation.

The Leap Year Group sediments were derived from the southwest where Wilson Group gneisses now crop out. The detritus was mainly derived from granitic and quartz-rich metamorphic rocks. A minor volcanic component is restricted to the Carryer Conglomerate, and the ratio of granitic- to metamorphic-derived clasts is higher in the Carryer Conglomerate than in the Reilly Conglomerate. This could be explained by denudation of a terrane where volcanic rocks were underlain by granitic rocks, both in a structurally high position within the Wilson Group.

Metamorphic conditions and timing

The pressure-temperature conditions prevailing during the various metamorphic episodes are shown schematically in Figure 68.2. The three folding episodes that affected the Wilson Group took place under progressively lower pressure and temperature conditions. The presence of amphibolite facies rocks suggests that the geothermal gradient was greater than 30°/km. Fold phases F_1 and F_2 are Vendian or older because of the presence of epidote-amphibolite facies Wilson Group clasts in the Husky Conglomerate.

The Bowers Supergroup and the Robertson Bay Group include the boundaries between the prehnite-pumpellyite, pumpellyite-actinolite, and greenschist facies. The widespread development of pumpellyite-actinolite facies in the Robertson Bay Group and Bowers Supergroup suggests that the geothermal gradient during metamorphism was probably less than 30°/km in both terranes. Furthermore, in the Bowers terrane, the boundary between the prehnite-pumpellyite and pumpellyite-actinolite facies lies just below the Leap Year–Mariner contact in the west and 1000 to 2000 m below that contact in the east, suggesting that the geothermal gradient was steeper in the west than in the east. This may have resulted from different basement rocks underlying the two areas: the Wilson Group in the west and the Robertson Bay Group in the east.

Radiometric ages from northern Victoria Land fall into two main groups (Adams and others, Chapter 67, this volume). The older group is in the range of 440 to 500 m.y. and includes most Wilson Group and Robertson Bay Group, and about half of the Sledgers Group, samples. The younger group, in the range 380 to 420 m.y., is only recorded in the Mariner and Leap Year Groups and in the upper part of the Sledgers Group on the northeastern flank of the Sledgers Group outcrop, rocks that have only been metamorphosed to the prehnite-pumpellyite facies. Evidently, the K-Ar ages have been reset in the lower grade rocks by the younger metamorphic event.

Acknowledgments. The writers wish to thank Ian Curfy, whose assistance in the field was greatly appreciated. One of us (A. W.) started the project while a member of the New Zealand Geological Survey but completed it after resignation. We are grateful to Randall S. Babcock who read the manuscript and made several suggestions for its improvement.

References cited

Bishop, D. G. 1972. Progressive metamorphism from prehnite-pumpellyite to greenschist facies in the Dansey Pass area, Otago, New Zealand. *Geol. Soc. Am. Bull., 83,* 3177–98.

Cooper, R. A., Jago, J. B., MacKinnon, D. I., Simes, J. E., and Braddock, P. E. 1976. Cambrian fossils from the Bowers Group, northern Victoria Land, Antarctica (Preliminary note). *N.Z. Jl Geol. Geophys., 19,* 283–88.

Crowder, D. F. 1968. Geology of a part of north Victoria Land, Antarctica. *Prof. Pap. U.S. Geol. Surv.,* No. 600-D, D95–107.

Dow, J. A. S., and Neall, V. E. 1974. Geology of the lower Rennick Glacier, northern Victoria Land, Antarctica. *N.Z. Jl Geol. Geophys., 17,* 659–714.

Laird, M. G., Bradshaw, J. D., and Wodzicki, A. 1976. Re-examination of the Bowers Group (Cambrian), northern Victoria Land, Antarctica (Preliminary note). *N.Z. Jl Geol. Geophys., 19,* 275–82.

Liou, J. G., 1971. Synthesis and stability relations of prehnite, $Ca_2Al_2Si_3O_{10}(OH)_2$. *Am. Miner., 56,* 507–31.

Liou, J. G., Kuniyoshi, S., and Ito, K. 1974. Experimental studies of the phase relations between greenschist and amphibolite in a basaltic system. *Am. J. Sci., 274,* 613–32.

Sturm, A., and Carryer, S. J. 1970. Geology of the region between Matusevitch and Tucker Glaciers, north Victoria Land, Antarctica. *N.Z. Jl Geol. Geophys., 13,* 408–35.

69

Stratigraphy and Structure of Low-Grade Metasedimentary Rocks of the Skelton Group, Southern Victoria Land—Does Teall Greywacke Really Exist?

D. N. B. SKINNER
New Zealand Geological Survey, Department of Scientific and Industrial Research, Otara, Auckland, New Zealand

Abstract. Teall Greywacke at its type locality on the central bluff of Teall Island is not a true graywacke but is comparable to calcarenaceous beds within the Anthill Limestone; inferred current directions indicate derivation from the southwest. Elsewhere, Teall Greywacke is lower in carbonate and generally higher in quartz, with—at Red Dike Bluff—pilotaxitic felsic volcanic clasts in fine-grained, actinolite-rich, tuffaceous(?) beds; inferred current directions indicate derivation from the north.

Lying to the north, but separated from Red Dike Bluff by granites, are two belts of "greywacke-like" volcaniclastic metasedimentary and metavolcanic rocks. These have not previously been distinguished from the Anthill Limestone. In the western belt these rocks rest with sharp angular unconformity upon the Anthill Limestone, and the lowest beds are discontinuous lenses of conglomerate which rest on and fill channels cut into the Anthill Limestone. Intraformational conglomerates are a common feature of these metasedimentary rocks which are here named the Cocks Formation.

The Anthill Limestone (in the strict sense) and the "type" Teall Greywacke have been complexly folded during three phases of deformation, a history common to other basement rocks in the McMurdo Sound region. However, the Teall Greywacke at Red Dike Bluff and possibly at the northern and southern ends of Teall Island, along with the Cocks Formation have only been deformed during the younger two phases.

The type "Teall Greywacke" may be a facies variant of the Anthill Limestone. However, all other Teall Greywacke may be a distal facies of the Cocks Formation, a volcanic sedimentary sequence that formed late in the history of the Ross orogen, prior to the emplacement of late and postkinematic intrusions.

The Anthill Limestone and the Teall Greywacke of Gunn and Warren (1962) comprise the Skelton Group (Grindley and Warren, 1964) metasedimentary rocks generally of greenschist facies; higher grade rocks north of the Koettlitz Glacier are assigned to the Koettlitz Group. The 1975–76 NZARP New Zealand Geological Survey program initiated a continuing study of the relationships between the Skelton and Koettlitz Groups. During that season outcrops along the Skelton Glacier, the Upper Cocks Glacier, the Foster Glacier, and the Mount Cocks–Mount Moxley–Hooper Crags massif were examined (Figure 69.1). Since the brief reconnaissance study of Gunn and Warren (1962) there has been only two short field seasons by University of Wyoming parties north of the Cocks Glacier (Murphy and others, 1970; Flory and others, 1971), and a brief visit by Soviet exchange geologists (Lopatin, 1972).

Teall Greywacke

No direct contact is known between the Anthill Limestone and the Teall Greywacke, which has been correlated with the older Goldie Formation of the Beardmore Group because "both units are similar and consist of folded, massive quartzo-feldspathic to arkosic greywackes and cleaved argillites with rare limestone and quartzite" (Grindley and Warren, 1964, p. 323). Although unfossiliferous itself, the Anthill Limestone has generally been correlated with the Lower to Middle Cambrian Shackleton Limestone of the Byrd Group.

Only the rocks of northern and central Teall Island (the latter being the type locality) and Red Dike Bluff were mapped as Teall Greywacke (Figure 69.1); all other metasedimentary rocks along Skelton Glacier were assigned to the Anthill Limestone. However, Flory and others (1971) briefly note "greywacke-mudstone" within the Anthill Limestone sequence north of Cocks Glacier, and the writer has mapped a discontinuous belt of greywacke-like metasedimentary and metavolcanic rocks along the east side of Skelton Glacier and another belt along the Cocks-Koettlitz Glacier divide. These rocks are here named the Cocks Formation. The southern end of Teall Island is also metagreywacke, similar to that of the northern Teall Island outcrops.

What is Teall Greywacke?

Gunn and Warren's brief definition (1962, p. 70) of "steeply dipping, complexly folded and thermally altered greywackes and argillites exposed . . . on the central part of the eastern side of Teall Island" was modified by their comment that the rocks at the type locality are "rather

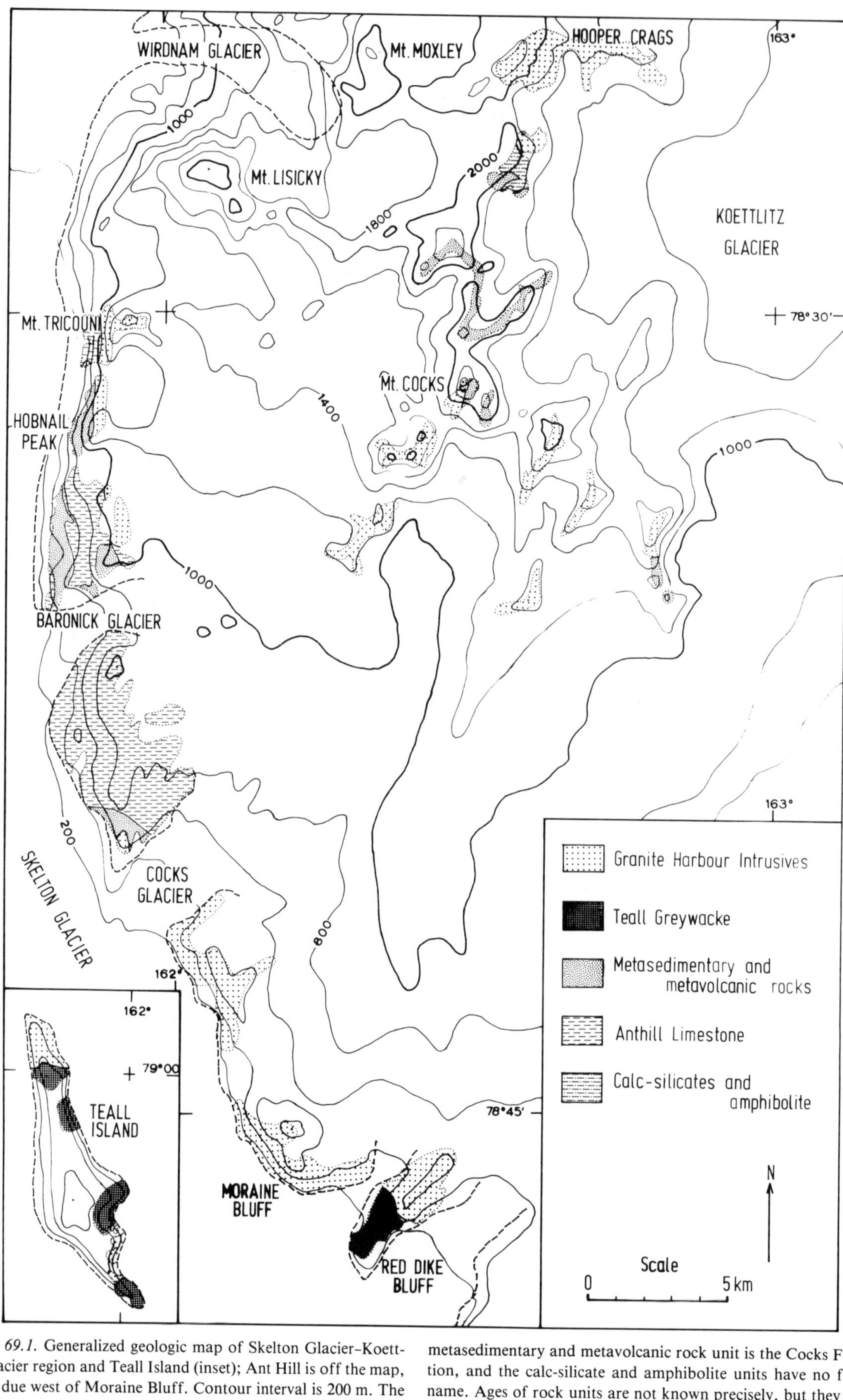

Figure 69.1. Generalized geologic map of Skelton Glacier–Koettlitz Glacier region and Teall Island (inset); Ant Hill is off the map, 16 km due west of Moraine Bluff. Contour interval is 200 m. The Teall Greywacke is as mapped by Gunn and Warren (1962); the metasedimentary and metavolcanic rock unit is the Cocks Formation, and the calc-silicate and amphibolite units have no formal name. Ages of rock units are not known precisely, but they probably range from Late Precambrian to Ordovician.

Figure 69.2. Folded, well-bedded calcarenites of Gunn and Warren's (1962) type Teall Greywacke on the central eastern bluff of Teall Island. The folds are F_1 phase and have a well-developed crenulation cleavage.

more calcareous'' than those at Red Dike Bluff. Grindley and Warren's (1964) description of the rocks as ''massive'' is hardly proper in view of their well-bedded, graded, current-laminated, and rippled nature (Figure 69.2), but are the rocks really quartzo-feldspathic to arkosic greywackes as they state?

The writer has re-examined Gunn and Warren's 14 thin sections, plus 44 more from 96 specimens he collected in 1975. If we accept a definition of greywacke as a sandstone with more than 15% matrix, which typically contains <50% quartz, as well as subordinate rock fragments and/or feldspar, insignificant cement, and only minor accessory carbonate (Pettijohn and others, 1972, p. 159 and 199), then it is evident that, even allowing for the effects of metamorphism, very little, if any, true greywacke is present at the type locality.

Most of the rock is siltstone or mudstone with relatively rare sandstone made more obvious by crystalloblastic clinozoisite, actinolite, hornblende, biotite, and diopside. Modal analyses of sandstone (Table 69.1) show how much more calcareous the rocks are from the type ''Teall Greywacke,'' and how similar they are to calcarenaceous sandstone typical of rocks interbedded with the Anthill Limestone. This calcite difference cannot be explained by incorporation of CaO in actinolite or diopside, since the growth of these minerals is predominantly related to the presence of volcanic detritus.

Petrography

Quartz occurs as angular to subrounded single grains and subrounded multiple grains with secondary overgrowths. Calcite is generally recrystallized but remains granular in low-grade rocks; the Anthill Limestone calcarenite is only slightly metamorphosed and retains calcite as fine-grained matrix, crystalline aggregates, and ?granular fragments. Feldspar is albitized plagioclase (andesine-labradorite) and perthite, with the latter more common in the calcarenites. Igneous rock fragments are mostly felsic to quartzo-feldspathic aphanitic, allotriomorphic, and porphyritic volcanic clasts, along with some granite. Only at Red Dike Bluff are there also pilotaxitic volcanic grains. It is commonly difficult to differentiate aphanitic felsite clasts from the very fine-grained cherty matrix; most volcanic clasts are obscured by crystalloblastic actinolite and diopside with minor biotite. Sedimentary rock fragments consist of polycrystalline quartzite, chert(?) (quartz siltstone), and siltstone-argillite flakes. Matrix, where separable from volcanic felsite, is very minor, but its original presence in some rocks is suggested by the enveloping and separation of detrital grains by crystalloblastic biotite that may represent recrystallized clay. At Red Dike Bluff a tuffaceous matrix is suggested by the texture and association of actinolite and minor biotite. Opaque minerals are predominantly pyrite cubes, but carbon granules occur in the Anthill calcarenite.

Table 69.1. Modal analyses (in %) of field specimens (collected in 1975) of Teall Greywacke as mapped by Gunn and Warren (1962)

	1[a]	2	3	4	5	6	7	8	9	10	11
Quartz	28	20	27	10	29	44	53	50	34	38	23
Calcite	46	63	35	59	14	8	0	0	15	4	61
Feldspar	4	9	9	<1	6	5	12	17	4	9	4
Igneous clasts	0	2	7	4	8	10	5	2	2	7	0
Sedimentary clasts	0	0	0	22	0	<1	0	0	0	0	0
Matrix	2	3	3	3	2	4	2	<1	0	1	7
Opaque minerals	0	<1	<1	1	0	<1	0	1	1	1	0
Biotite	20	3	7	0	<1	6	21	25	2	0	0
Actinolite	<1	<1	12	1	<1	23	7	5	42	40	6
Diopside	0	0	0	0	41	0	0	0	0	0	0

1. P22632: Central Teall Island; Gunn and Warren (1962).
2. 82B: Central Teall Island.
3. 82C: Central Teall Island.
4. 82D: Central Teall Island.
5. 84A: North Teall Island.
6. 86C: North Teall Island.
7. 88B: South Teall Island, previously unmapped.
8. 88E: South Teall Island, previously unmapped.
9. 74G: Red Dike Bluff.
10. 74K: Red Dike Bluff.
11. 3: North of Cocks Glacier, calcarenaceous sandstone from Anthill Limestone.

[a] Gunn and Warren (1962) specimen.

Figure 69.3. Unconformable contact of Cocks Formation (dark rock on the left) resting on Anthill Limestone (light gray) at Cocks Glacier; several F_1 folds are present in the limestone, and both limestone and Cocks Formation have been intruded by a felsic ?andesite dike which has been offset twice by right-lateral flexural slip on S_0/S_1. The basal conglomerate lies along the gully containing the 7-m-long snowdrift (in the foreground), and pillow lava is at its head between the dike and the limestone.

Figure 69.4. Cocks Formation basal conglomerate at Cocks Glacier with almost uniaxial ellipsoidal clasts, z-symmetry drag folds, and boudinage structures.

Cocks Formation—greywacke-like metasedimentary and metavolcanic rocks

The Western Belt

Gunn and Warren (1962) describe, as Anthill Limestone, a rock from the north side of Cocks Glacier (P22648), 1.6 km east of Skelton Glacier, consisting of granular diopside with relatively minor orthoclase, calcite, apatite, and pyrite. The locality, level with the top of the lower ice falls, is of some importance, for there is a well-exposed angular unconformity between the Anthill Limestone (in the strict sense) and the overlying dark greenish-gray to black metasedimentary and metavolcanic rocks (Figure 69.3). The succession above, i.e., west of, the unconformity is designated here as the type section for the Cocks Formation.

Channels as much as 2 m deep and 3 m wide eroded into the complexly folded limestone/marble are filled by ellipsoidal limestone/marble and metavolcanic clasts as large as 35 cm × 10 cm set in a calcareous, tuffaceous(?), quartz sandstone matrix. This channel-fill passes up into a lenticular basal conglomerate as much as 3 m thick, which contains fewer limestone fragments but numerous ellipsoidal clasts of felsic, porphyritic, and pilotaxitic meta-andesite and biotite-perthite-quartz granite similar to prekinematic granite intruding the limestone to the northeast (Figure 69.4). Thin lenses of coarse-grained gray and fine-grained dark marble are interbedded with the conglomerate and the overlying well-bedded volcaniclastic sandstone. Well-developed graded bedding and cross-bedding confirm the decrease in age away from the limestone. There has been soft-sediment deformation around 1-m-thick, felsic, porphyritic pillow lava along the contact with the limestone. Discontinuous lenses of basal conglomerate lie along this contact and can be traced west-northwestward to some 3 km north of the Cocks Glacier–Skelton Glacier confluence.

North of the Baronick Glacier there is another well-exposed unconformity above the Anthill Limestone. No major basal conglomerate is present, but the lower metasedimentary sequence contains abundant scattered quartz pebbles and pebble beds. Volcaniclastic conglomerate lenses are common in the lower 150 m of the sequence, culminating in thick beds of stretched ellipsoidal green-gray volcanic clasts in a yellowish matrix of quartz and lithic volcanic sandstone (Figure 69.5).

Below the granite bluffs of Mount Tricouni, a small outcrop of Anthill Limestone lies unconformably below schistose greywacke; no conglomerate was seen, but most of the area is inaccessible because of active icefalls.

Contact metamorphism to at least hornblende-hornfels facies has occurred adjacent to postkinematic granite, gabbro, and diorite bodies at Cocks Glacier, to the north of Baronick Glacier, and at Hobnail Peak. Most specimens contain crystalloblastic hornblende, diopside, biotite, and feldspar, but retain blastoporphyritic and blastophitic pla-

Figure 69.5. Ellipsoidal metavolcanic rock clasts in the Cocks Formation, 150 m above its base on the north bank of Baronick Glacier; the largest clast is 21 cm long.

Figure 69.6. Tight, attenuated F_1 folds, with S_0 transposed to S_1, in fine-grained marbleized limestone at Ant Hill.

Figure 69.7. Hinges of isoclinal F_1 folds in calcarenaceous Anthill Limestone, adjacent to Skelton Glacier, 4.5 km north of Cocks Glacier; visible ice axe shaft (lower left) is about 60 cm long.

gioclase in the volcanites. Along the Skelton Glacier the rocks are regionally metamorphosed to biotite or garnet zone (greenschist facies).

THE EASTERN BELT

At the southern end of the Cocks-Koettlitz divide, fine- to medium-grained, laminated, graded, and thin-bedded biotite-quartzo-feldspathic schist and fine-grained quartzose marble are interbedded with actinolite amphibolite and with blastoporphyritic felsic to quartzo-feldspathic ?andesite-dacite. The amphibolite has formed from volcaniclastic and finely porphyritic volcanic rocks in which the quartz may be a secondary premetamorphic mineral phase; calcite, except in the thin marble beds, is a minor accessory mineral.

The local metamorphic grade rises from albite-epidote-hornfels in the south to hornblende-hornfels on Mount Cocks; near granite bodies, the metasedimentary rocks have been migmatized to hybrid "diorites" but still retain sharp contacts. Sedimentary features are still preserved, and the rocks resemble those to the south except that amphibolites are hornblende-diopside-feldspar aggregates.

Further north near Mount Moxley, the regional metamorphic grade is amphibolite facies, and the metasedimentary and metavolcanic rocks are interfolded with older sillimanite-muscovite calc-silicate rocks. The metasedimentary rocks are thin-bedded to laminated, graded, slaty to schistose biotite-diopside-quartz siltstone and sandstone, hornblende-garnet amphibolite, and rare bands of diopside marble. Blastoporphyritic and blastophitic quartz-plagioclase amphibolite layers probably originated as volcanic flows, and sills of similar rock in the steeply dipping sillimanite calc-silicates may have been feeders to these volcanics.

Tectonic deformation of the Skelton Group

Early workers (Gunn and Warren, 1962; Blank and others, 1963) considered the tectonic history of both Skelton and Koettlitz Groups to be relatively simple, with only one period of deformation and accompanying pre-, syn-, and postkinematic granitoid intrusions; the latter authors also recognized multiple metamorphism. Folding of the Skelton Group beds was described as minor but commonly complex, and of the Koettlitz Group, as close but essentially simple.

In contrast, Williams and others (1971) recognized far greater complexity within the Koettlitz Group, describing interference structures denoting two separate intense deformation events. Similarly, two fold systems have been recognized in the Skelton Group in the Cocks Glacier region by Murphy and others (1970) and by Flory and others (1971). To the north in the McMurdo Sound region, in rocks usually correlated with the Koettlitz Group, Smithson and others (1970) have proposed a ten-fold tectono-magmatic history with three deformation events, the last of which was preceded, accompanied, and followed by increasing magmatic activity. Magmatism also preceded the second deformation (Smithson and others, 1971).

ANTHILL LIMESTONE

Three fold phases have deformed the Anthill Limestone (F_1, F_2, and F_3). In the Ant Hill block west of the Skelton Glacier, F_1 folds are preserved as tightly isoclinal, overturned (in the sense of Ramsay, 1967) folds in the limbs of F_2 folds and like them verge to the southeast; F_1 fold axes have gentle to moderate plunges to both the northeast and southwest, and bedding (S_0) has been transposed into axial plane cleavage (S_1) (Figure 69.6). F_1 amplitudes generally range from a few centimeters to a few tens of meters, but two major folds with recrystallized marble cores have amplitudes of a few hundred meters. A similar F_1 fold geometry prevails north of and along the south bank of the Baronick Glacier, but fold axes curve so as to plunge north-northwest in the north and south-southwest in the south, and vergence is to the west.

In the Cocks Glacier block, there is Ant Hill–style F_1 deformation adjacent to the Skelton Glacier (Figure 69.7) and the lower Cocks Glacier, but the vergence changes from north-northwestward to northeastward with westerly and southeasterly axial plunges. However, in the central

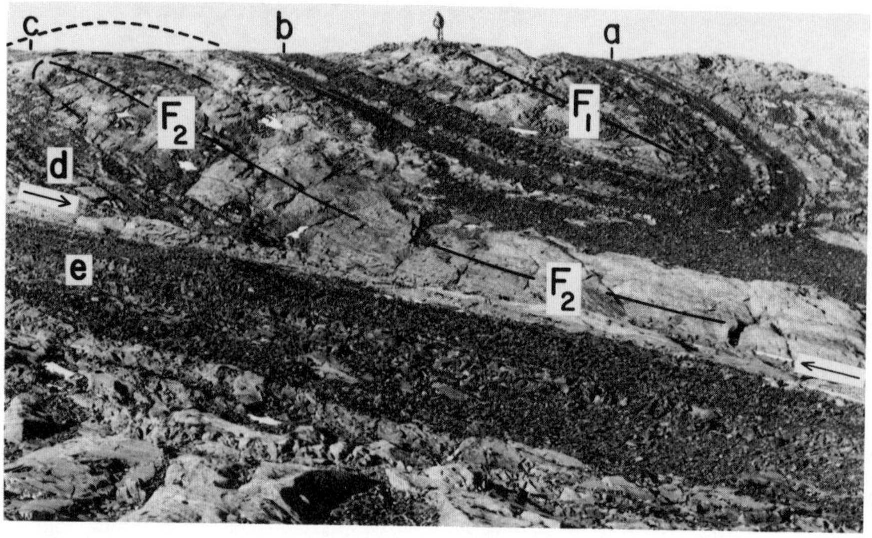

Figure 69.8. A large refolded structure in Anthill Limestone north of Cocks Glacier. The man stands on the axis of an F_1 fold which has been refolded into a large coplanar, almost isoclinal, F_2 hook structure. The F_2 axial surface dips away from the viewer and runs through the wide, light-colored rock band across the middle of the photo. A gully (arrowed) separates this band from the dark foreground beds (e), and to the left, separates beds (d) and (e); the gully lies along the repeated F_1 axial surface. All the dark beds are a single, refolded marker horizon; the F_1/F_2 limb (b) is continuous around F_2 nose (c) with F_2 limb (d); likewise, and parallel to (b-c-d), F_1 limb (a) is continuous out of sight beyond the man and to the left beyond F_2 nose (c), where it rejoins the dark foreground beds (e) out to the left of the photo.

and northeastern parts of the block, well-preserved F_1 folds are overturned but asymmetric to nearly isoclinal and are much less deformed by F_2, so that bedding (S_0) is preserved and no transposition into S_1 has occurred (Figure 69.8). Amplitudes range from 200 to 2000 m, and because of F_2 folding, vergences are variable in direction but never between south and east. Fold axes plunge at moderate angles to between northeast and northwest.

Swarms of blastophitic quartzo-feldspathic and blastoporphyritic amphibolite dikes and sills preceded F_2 deformation and act as markers for F_2 folds, which are the most obvious on a macroscale in most blocks, with amplitudes on the order of a few hundred meters. At Ant Hill and in the north and south Baronick Glacier blocks, the F_2 structures are overturned to nearly recumbent, asymmetric to nearly isoclinal folds with axial surfaces coplanar with those of F_1 folds. Fold axes plunge steeply to the northwest in the Ant Hill block, but southeastward and northeastward in the Baronick Glacier region. A crenulation cleavage has formed where S_1 has been folded about F_2 fold hinges, but otherwise, S_2 nearly parallels S_0/S_1. At Cocks Glacier, F_2 folds are generally open, upright to overturned and asymmetric to subisoclinal, and both bedding and nearly axial-planar fracture cleavage are present (Figure 69.8). Because of strong F_3 folding, vergence is variable between west and northeast; fold axes plunge steeply to between north and east.

Post-F_2 intrusions of gabbro, diorite, and gneissic granite occupy sheared-out flexural slip planes of the F_2 folding; along with the fracture cleavage in F_2 folds, they are folded by F_3 folds.

F_3 folds vary progressively from large nearly isoclinal folds to asymmetric warps northward from Ant Hill and, except for those at Ant Hill, are too large to see in the field; fold axes plunge moderately to steeply east to southeast. Similarly, north of the Baronick Glacier, F_3 is discernible only as broad warps with east-west axes. However, throughout the whole Cocks Glacier block, F_3 folds have deformed F_2 axial planes about steep, east- to southeast-plunging axes into open asymmetric folds with amplitudes of 5 km and northerly vergence.

Plutons of granite, granodiorite, quartz monzonite, and diorite were intruded during and after the F_3 folding, with synkinematic intrusions in particular causing the tightness of folding at Ant Hill and north of Baronick Glacier. Microgranites, aplites, pegmatites, and lamprophyres are younger than the major postkinematic intrusions.

Teall Greywacke

Central Teall Island

Three deformation phases are demonstrable here. Bedding (S_0) has been folded by F_1 into asymmetric, northwest-verging, upright folds with S_1 crenulation cleavage (Figure 69.2); axes plunge gently to moderately northeastward. The F_2 folds are almost coplanar with F_1 folds, but axial plunges are steeper and slaty cleavage (S_2) has developed; zones of intrafold décollement and flexural slip have been intruded by post-F_2 mafic sills. F_3 folding about northwesterly axes verges to the southwest and has caused fracture cleavage (S_3), rodding of siltstone, and pencil jointing of argillite. F_3 is associated with synkinematic granite, and F_3 axial planes have been deflected by postkinematic granite.

North and South Teall Island

Only two fold systems may be present here, the younger being deformed by postkinematic granites. The northern outcrops form a tight, nearly isoclinal, overturned, north-verging, gently to moderately east-plunging complex anticline, with well-developed secondary mesofolds as much as 3 m in amplitude and with axial plane schistosity. A younger fracture cleavage and cleavage mullions are associated with southeast-plunging, broad meso- and microfolds that parallel the contact with the postkinematic granite.

Figure 69.9. Typical second deformation mesofold and fan cleavage in ripple-bedded quartz litharenite of the Teall Greywacke at Red Dike Bluff. These beds are now considered to be a correlative of the Cocks Formation.

In southern Teall Island, the apparently simple, south-dipping, upright sequence may be part of a nearly recumbent structure. A gently east-dipping, semi-schistose cleavage is associated with small, isoclinal, recumbent mesofolds as much as 2 m in amplitude that have axial planes parallel to S_0 and axes plunging southward nearly down the dip of the bedding; these folds form a mullion-like lineation. Open, upright, asymmetric, east-plunging, broad folds less than 1 m in amplitude cross the recumbent set, and their axes trend parallel to the contact with the postkinematic granite.

Red Dike Bluff

At this locality the evidence clearly indicates only two deformations. The earlier structure is a north-northeast-trending, upright, asymmetric, west-verging, south-plunging anticline of at least 300 m amplitude, with both fan-fracture cleavage and intrabedding flexural slip–kink band cleavage. The Red Dike and other small tourmaline-rich granitoids postdate the main structure, but they have been folded with the metasedimentary beds during the intrusion of large, subgneissic synkinematic granitoids and diorites into upright, nearly symmetric, open folds with east-west axes (Figure 69.9). Postkinematic granite followed this second deformation.

COCKS FORMATION

The Western Belt

The Cocks Glacier rocks have been deformed on both the small (cm) and large (100 m) scale, particularly along their contact with the Anthill Limestone, into typical skewed mushroom-shaped type 2 interference patterns of Ramsay (1967), thus obscuring the angular unconformity. Early west- to northwest-verging, steeply northeast-plunging, tight, nearly isoclinal, overturned folds are crossed by younger northwest-trending, southwest-verging, asymmetric to overturned dextral (viewed down plunge) mesofolds. Flexural slip near and along the greywacke-limestone contact is indicated by dextral sill/dike dislocation (Figure 69.3), by nearly uniaxial attenuation, by dextral folding of conglomerate clasts (Figure 69.4), and by boudinage and extension fractures, thus suggesting a major synclinal hinge to the south. Igneous activity is pre-, syn-, and postkinematic with respect to the younger deformation.

Incomplete data from north of the Baronick Glacier suggest a type 1–2 interference pattern (Ramsay, 1967) of interfolding between the Cocks Formation and the Anthill Limestone. A series of overturned, west-verging, attenuated, curved, dome-like structures have been produced by the interaction of an earlier north-northwest-trending set of folds with a later west-northwest-trending set associated with pre-, syn-, and postkinematic diorite-granite intrusions. On Hobnail Peak, south-plunging, west-verging, nearly isoclinal, north-northeast-trending overturned folds are crossed by younger southeast-plunging, almost recumbent southwest-verging mesofolds below the contact with posttectonic granite.

The Eastern Belt

The small outcrop at the Cocks-Koettlitz saddle reveals two nearly isoclinal, slightly northwest-verging, southwest-plunging steep folds associated with pre- and postkinematic granitoids; evidence for a second fold phase has not been seen. On Mount Cocks, asymmetric, upright to overturned mesofolds plunge at moderate angles to the south-southwest, parallel to the regional bedding, and are warped along east-west axes into broad, open folds with variable vergence.

South of Mount Moxley, the Cocks Formation metagreywackes are interfolded with the older calc-silicate rocks in a type 3 interference pattern of Ramsay (1967). The calc-silicate rocks show three deformation phases, the earliest of which precedes deposition of the metagreywackes and intrusion of felsic and amphibolite sills. The F_1 folds of this early phase are preserved in the calc-silicate rocks as tight, isoclinal, moderately plunging hinge zones in the limbs of F_2 folds; the S_0 bedding has been transposed into S_1 axial plane cleavage. The F_2 folds in the calc-silicate rocks also involve the metagreywackes on steeply plunging, northeast- to east-trending axes, producing asymmetric, north-verging mesofolds with inverted stratigraphy. These mesofolds have been broadly warped about southeast-trending axes to produce a fracture cleavage into which postkinematic granite and lamprophyre have intruded.

Stratigraphic and tectonic summary

A summary of the geologic history of the Skelton Group is given in Table 69.2.

There are two mapped Teall Greywacke varieties, neither of which are commonly true greywacke: one is predominantly calcarenaceous and not unlike parts of the Anthill Limestone, the other a slightly calcareous, quartzose litharenite. Unfortunately, the calcarenaceous variety comprises the type locality of the Teall Greywacke. Although both varieties contain felsic igneous clasts, only the quartzose litharenite Teall Greywacke at Red Dike Bluff contains, in addition, obviously volcanic clasts (felsic pilotaxitic) and probably volcanic (tuffaceous?) groundmass, since converted to actinolite.

Table 69.2. Summary of the sedimentary, magmatic, and tectonic history of the Skelton Group

Time	Event
Middle Ordovician	Intrusion of microgranites and lamprophyres; Rb-Sr, 470 ± 7 m.y. Postkinematic granite, granodiorite, quartz monzonite, and diorite; Rb-Sr, 480 ± 44 m.y.
Early Ordovician	Third deformation phase F_3 and synkinematic granitoids; Rb-Sr, 500 ± 44 m.y. Post-F_2 intrusion of gabbro, diorite, and granite Second deformation F_2
Late Cambrian–Early Ordovician	Penecontemporaneous intermediate volcanism and sedimentation (Cocks Formation) and felsic to intermediate hypabyssal intrusion; sedimentary beds of north and south Teall Island and Red Dike Bluff deposited away from volcanic centers
Late Cambrian	First deformation F_1
Early to Middle Cambrian	Sedimentation of limestones and calcarenites of the Anthill Limestone and contemporaneous intermediate to basic volcanism (amphibolite); sedimentary beds of central Teall Island deposited penecontemporaneously

A previously undescribed greywacke-like metasedimentary-metavolcanic rock sequence (Cocks Formation) unconformably overlies the Anthill Limestone. Although the limestone has been deformed by three folding phases, the younger Cocks Formation postdates the first deformation and has only been deformed twice. Like the Anthill Limestone, the calcarenaceous "type" Teall Greywacke and amphibolite-grade calc-silicate metasedimentary rocks have been deformed three times. In contrast, the quartzose litharenite Teall Greywacke of northern and southern Teall Island and of Red Dike Bluff have only been folded twice. The youngest fold phase in each case has developed about east-west to northwest-southeast axes and is generally associated with the major synkinematic granites and diorites. The postkinematic intrusions have locally deformed the youngest fold structures.

Excluding penecontemporaneous volcanism in the Anthill Limestone, the earliest intrusive igneous activity, in the form of felsic sills and dikes, began after the first of the three folding phases; these felsic rocks are generally similar to the unconformably younger metavolcanic rocks of the Cocks Formation. More mafic rocks were intruded between the second and third folding phases.

It is argued that the rocks at the type locality of the "Teall Greywacke" (central Teall Island) are a facies variant of the Anthill Limestone, and that what is left of the unit (at northern and southern Teall Island and Red Dike Bluff) is correlative with the Cocks Formation, and hence unconformable upon and younger than the Anthill Limestone.

Regional significance

Throughout the Transantarctic Mountains, a sequence of Cambrian or presumed Cambrian calcareous units, overlain by noncalcareous, commonly conglomeratic units and felsic volcanic rocks, is not unusual, e.g., in the Pensacola Mountains (Schmidt and Ford, 1969), in the Wisconsin Range and southern Queen Maud Mountains (Mirsky, 1969), in the western Queen Maud Mountains (McGregor and Wade, 1969; McGregor, 1965; Stump, 1974; Wade, 1974), and along the Shackleton Coast (Laird, 1963, 1964; Skinner, 1964, 1965). Although most of the volcanic rocks can only be assigned a Middle Cambrian or younger age, those of the Wisconsin Range and southern Queen Maud Mountains (Mirsky, 1969) have been dated as Early to Middle Ordovician (Rb-Sr; 498 ± 45 m.y. and 472 ± 11 m.y.).

The question is, where does the Skelton Group fit into the regional picture? Although the limestones of the Shackleton and Anthill Formations are dissimilar, the latter being more carbonaceous, pyritic, and sulphurous, and containing actinolite layers perhaps derived from volcanic material, nevertheless, the calcarenaceous and argillaceous rocks of the two units are comparable despite the greater tectonism and metamorphism of the Anthill Limestone. This metamorphism has also obscured the original character of the metavolcanic rocks that postdate the Anthill Limestone, although relict blastoporphyritic and blastophitic plagioclase (andesine and labradorite) and perhaps the quartz suggest intermediate to silicic varieties. If the Skelton and Byrd Groups are correlatives, then nowhere else in the Ross orogen has there been reported an angular unconformity and a tectonic intensity difference between Cambrian calcareous rocks and younger volcanic units.

Only in northern Victoria Land, however, is there an apparently complete Middle Cambrian to Ordovician(?) succession (Laird and other, Chapter 66, this volume). In the Wisconsin Range and southern Queen Maud Mountains, there is an absence of Upper Cambrian rocks between the Middle Cambrian limestones and the dated Ordovician volcanic rocks, although the two units are apparently conformable (Mirsky, 1969). If all the Paleozoic volcanic rocks are indeed of comparable Ordovician age, then a regional Upper Cambrian disconformity may be present south of the Skelton Glacier. The volcanic rocks would be equivalent to the postkinematic Granite Harbour Intrusives with respect to the Late Cambrian Ross orogeny of Adams and others (Chapter 101, this volume).

If the volcanic rocks of the Cocks Formation are also Ordovician, then in the Skelton Glacier–McMurdo Sound region, the Ross orogeny must have continued beyond Cambrian time with a relatively brief tectonic pause during deposition of the Cocks Formation. Dated specimens (Rb-Sr) from the Wright and Victoria Intrusives provide time constraints for the extended orogeny, since the two units are representative of the regional syn- and postkinematic Granite Harbour Intrusives. The Wright Intrusives underwent isotopic closure at 500 ± 44 m.y. and the Victoria Intrusives at 480 ± 44 m.y. The Cocks Formation volcanic-sedimentary event therefore took place during Late Cambrian to Early Ordovician time.

Not only is the Skelton-McMurdo region the only part of the Ross orogen where three-phase deformation has been described, but it is also the only part where volcanism may have been andesitic rather than rhyolitic. Because rhyolitic volcanism is a feature of extensional, and andesitic of compressional, crustal regimes (Snyder and others, 1976), this interrelationship of magmatism and tectonism, unique within the Ross orogen, also supports a longer lived Ross orogeny in the McMurdo Sound region.

Acknowledgments. I am grateful to Sue West, Barry C. Waterhouse, Garry Brehaut, and Ken Sullivan for their support during the 1975-76 field season. Field logistic support by helicopters of U.S. Navy Squadron VXE-6 is gratefully acknowledged. Drafts of this paper were reviewed by K. B. Spörli, G. W. Grindley, M. G. Laird, B. C. Waterhouse, and F. E. Bowen.

References cited

Blank, H. R., Cooper, R. A., Wheeler, R. H., and Willis, I. A. G. 1963. Geology of the Koettlitz-Blue Glacier Region, southern Victoria Land, Antarctica. *Trans. R. Soc. N.Z., Geol., 2,* No. 5, 79-100.

Faure, G., and Jones, L. M. 1974. Isotopic composition of strontium and geologic history of the basement rocks of Wright Valley, southern Victoria Land, Antarctica. *N.Z. Jl Geol. Geophys., 17,* No. 3, 611-27.

Flory, R. F., Murphy, D. J., Smithson, S. B., and Houston, R. S. 1971. Geologic studies of basement rocks in southern Victoria Land. *Antarct. Jnl U.S., 6,* No. 4, 119-20.

Grindley, G. W., and Warren, G. 1964. Stratigraphic nomenclature and correlation in the western Ross Sea region. (*In* Adie, R. J., *ed. Antarctic Geology.* Amsterdam, North-Holland Publishing Company, 314-33.)

Gunn, B. M., and Warren, G. 1962. Geology of Victoria Land between the Mawson and Mulock Glaciers, Antarctica. *Bull. Geol. Surv. N.Z., 71,* 157 pp.

Laird, M. G. 1963. Geomorphology and stratigraphy of the Nimrod Glacier-Beaumont Bay region, southern Victoria Land, Antarctica. *N.Z. Jl Geol. Geophys., 6,* No. 3, 465-84.

Laird, M. G. 1964. Petrography of rocks from the Nimrod Glacier-Starshot Glacier region, Ross Dependency. (*In* Adie, R. J., *ed. Antarctic Geology.* Amsterdam, North-Holland Publishing Company, 463-72.)

Lopatin, B. G. 1972. Basement complex of McMurdo 'Oasis,' south Victoria Land. (*In* Adie, R. J., *ed. Antarctic Geology and Geophysics.* Oslo, Universitetsforlaget, 287-92.)

McGregor, V. R. 1965. Geology of the area between the Axel Heiberg and Shackleton Glaciers, Queen Maud range, Antarctica. Part 1—Basement complex, structure, and glacial geology. *N.Z. Jl Geol. Geophys., 8,* No. 2, 314-43.

McGregor, V. R., and Wade, F. A. 1969. Geology of the western Queen Maud Mountains (Sheet 16). (*In* Bushnell, V. C., and Craddock, C., *eds.* Geologic Maps of Antarctica. *Antarct. Map Folio Ser.,* Folio 12, Pl. VI.)

Mirsky, A. 1969. Geology of the Ohio Range-Liv Glacier Area (Sheet 17). (*In* Bushnell, V. C., and Craddock, C., *eds.* Geologic Maps of Antarctica. *Antarct. Map Folio Ser.,* Folio 12, Pl. XVI.)

Murphy, D. J., Flory, R. F., Houston, R. S., and Smithson, S. B. 1970. Geological studies of basement rocks in south Victoria Land. *Antarct. Jnl U.S., 5,* No. 4, 102.

Pettijohn, F. J., Potter, P. E., and Siever, R. 1972. *Sand and Sandstone.* New York, Springer-Verlag, 618 pp.

Ramsay, J. G. 1967. *Folding and Fracturing of Rocks.* New York, McGraw-Hill, 568 pp.

Schmidt, D. S., and Ford, A. B. 1969. Geology of the Pensacola and Thiel Mountains (Sheet 5). (*In* Bushnell, V. C., and Craddock, C., *eds.* Geologic Maps of Antarctica. *Antarct. Map Folio Ser.,* Folio 12, Pl. V.)

Skinner, D. N. B. 1964. A summary of the geology of the region between Byrd and Starshot Glaciers, south Victoria Land. (*In* Adie, R. J., *ed. Antarctic Geology.* Amsterdam, North-Holland Publishing Company, 284-92.)

Skinner, D. N. B. 1965. Petrographic criteria of the rock units between the Byrd and Starshot Glaciers, south Victoria Land, Antarctica. *N.Z. Jl Geol. Geophys., 8,* No. 2, 292-303.

Smithson, S. B., Fikkan, P. R., and Toogood, D. J. 1970. Early geologic events in the ice-free valleys, Antarctica. *Geol. Soc. Am. Bull., 81,* 207-10.

Smithson, S. B., Fikkan, P. R., Murphy, D. R., and Houston, R. S. 1971. Development of augen-gneiss in the ice-free valley area, south Victoria land. (*In* Adie, R. J., *ed. Antarctic Geology and Geophysics.* Oslo, Universitetsforlaget, 293-98.)

Snyder, W. S., Dickinson, W. R., and Silberman, M. L. 1976. Tectonic implications of space-time pattern of Cenozoic magmatism in western United States. *Earth planet Sci. Lett., 32,* No. 1, 91-106.

Stump, E. 1974. Volcanic rocks of the early Cambrian Taylor Formation, central Transantarctic Mountains. *Antarct. Jnl U.S., 9,* No. 5, 228-29.

Wade, F. A. 1974. Geological surveys of Marie Byrd Land and the central Queen Maud Range. *Antarct. Jnl U.S., 9,* No. 5, 241-42.

Williams, P. F., Hobbs, B. E., Vernon, R. H., and Anderson, D. E. 1971. The structural and metamorphic geology of basement rocks in the McMurdo Sound area, Antarctica. *J. geol. Soc. Aust., 18,* No. 2, 127-42.

70

The Ross Supergroup in the Queen Maud Mountains

EDMUND STUMP
Department of Geology, Arizona State University, Tempe, Arizona 85281, U.S.A.

Abstract. The apparently oldest rocks of the upper Precambrian–Cambrian Ross Supergroup in the Queen Maud Mountains are those of the Beardmore Group, which includes the Goldie, Duncan, and LaGorce Formations. These formations are composed of the metamorphic equivalents of graywacke, shale, and sandstone. In the Amundsen Glacier–Reedy Glacier area the LaGorce Formation is intruded and apparently overlain by hypabyssal and extrusive phases of the Wyatt Formation, a dark, usually massive, porphyritic, metafelsite of late Precambrian age. Metavolcanic and associated carbonate and clastic rocks of Cambrian age, herein named the Liv Group, are represented by the Taylor, Fairweather, and Leverett Formations, and tentatively the Greenlee Formation. The Taylor Formation of the Shackleton Glacier area is composed of porphyritic metarhyolites, with minor metabasalts in the lower part of the formation, and associated metamorphosed clastic, volcaniclastic, and carbonate rocks. Early Cambrian fossils from the middle part of the formation serve to date it. The Greenlee Formation, a sequence of fine-grained quartzite and phyllite, occurs conformably beneath the Taylor Formation. The Fairweather Formation of the Duncan Mountains area contains rock types and a stratigraphic succession similar to the Taylor Formation, but no positive evidence of its age has yet been found. Thick, white, coarsely crystalline marble units occurring in the vicinity of the Taylor and Fairweather Formations are referred to as the Henson Marble Member of each formation. The Leverett Formation, occurring on Mount Webster north of Leverett Glacier, contains slightly metamorphosed limestone, sandstone, and rhyolite; it has yielded fossils of Middle Cambrian age.

The Ross Supergroup, a suite of rocks of sedimentary and volcanic origin, is widely exposed throughout the Ross Sea sector of the Transantarctic Mountains (Figure 70.1). These rocks are late Precambrian and Cambrian in age. They are younger than the Nimrod Group in the upper Nimrod Glacier area, which had been dated by Rb-Sr at 1.98 b.y. (Gunner and Faure, 1972), and are older than the deformation, metamorphism, and intrusion of the Cambro-Ordovician Ross orogeny (520–450 m.y.) (Grindley and McDougall, 1969). The Ross Supergroup encompasses a diverse assemblage of continental margin deposits, including thick sequences of clastic, carbonate, and volcanic rocks. The extensive, fossiliferous limestones of the Byrd Group, found in the Nimrod Glacier–Byrd Glacier area to the west (Laird, 1963; Skinner, 1965; Laird and others, 1971), are not represented in the Queen Maud Mountains, which extend from the Beardmore Glacier to east of the Scott Glacier.

Beardmore Group

Sequences of rhythmically interbedded graywacke and shale, or their metamorphic equivalents, preserved in three separate areas in the Queen Maud Mountains, are referred to as the Beardmore Group (Figure 70.2). One formation has been identified in each of the three areas as follows: Goldie Formation, Ramsey Glacier–Beardmore Glacier area (Oliver, 1972; Wade and others, 1965); Duncan Formation, Duncan Mountains area (McGregor, 1965); and LaGorce Formation, Amundsen Glacier–Reedy Glacier area (Minshew, 1967; Murtaugh, 1969).

Only one conformable contact with the Beardmore Group is known throughout the Ross Sea sector—on Ackerman Ridge in the LaGorce Mountains, between the LaGorce and Wyatt Formations. Katz and Waterhouse (1970), who discovered this locality, have suggested that the Wyatt Formation is older. Based on observations of cross-bedding on Ackerman Ridge, the author has interpreted the Wyatt Formation as being stratigraphically younger than the LaGorce Formation (Stump, 1976*b*). Such an order is also followed in this paper.

The Ackerman Ridge locality is tremendously important to tectonic interpretations of the Ross orogen because it holds the answer to whether silicic volcansim preceded or only postdated deposition of the Beardmore Group. It is therefore with some hesitation that the author bases his conclusion on a few cross-beds. Further study is needed to settle the question.

Elsewhere neither the top nor the bottom of the Beardmore Group has been found. In the Beardmore Glacier–Ramsey Glacier area, the Goldie Formation is isolated from other formations of the Ross Supergroup by granitic intrusions and glacial cover. However, to the northwest in the Nimrod Glacier area, the Goldie Formation is unconformably overlain by the Shackleton Limestone. In the Duncan Mountains area the Duncan Formation is thrust over the Fairweather Formation along a high-angle, reverse fault (Stump, 1976*b*).

The age of the Beardmore Group is considered to be late Precambrian on the basis of both the various isotopic dates which imply that it was involved in tectonism during the late Precambrian Beardmore orogeny (Grindley and MacDougall, 1969; Faure and others, 1979) and also the unconformably overlying fossiliferous Shackleton Limestone which is dated as Early Cambrian (Laird and Waterhouse, 1962).

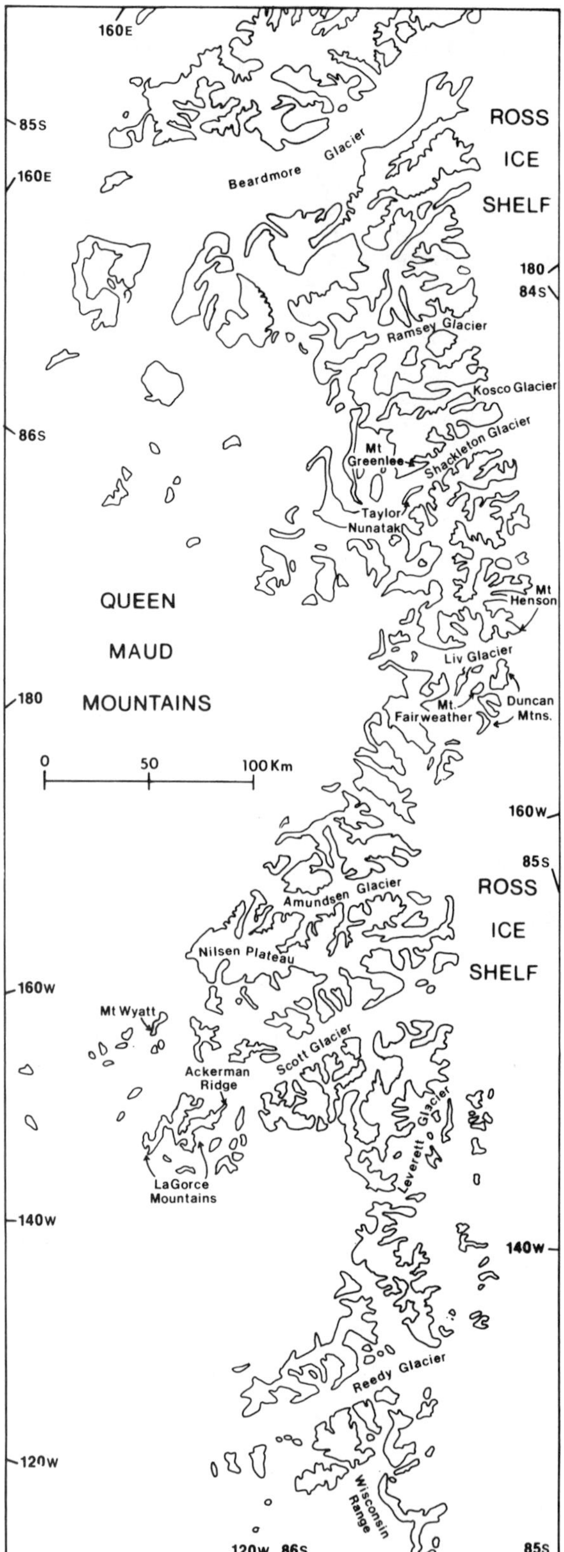

Figure 70.1. Location map, Queen Maud Mountains.

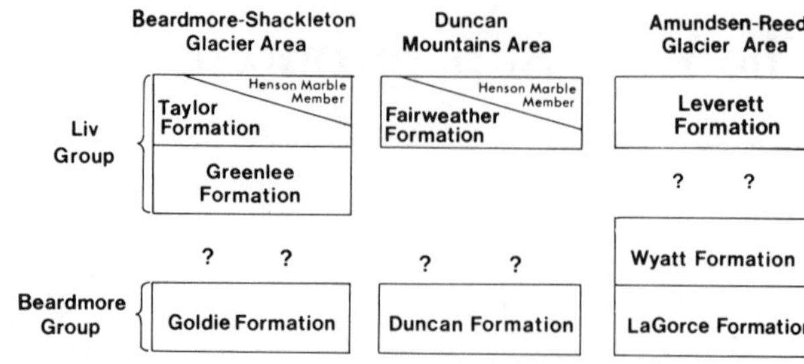

Figure 70.2. Formations of the Ross Supergroup, Queen Maud Mountains.

Thickness estimates on the Beardmore Group range from 2500–3500 m in the Ramsey Glacier area (Stump, 1976*b*; Wade and others, 1965), to 3500–4600 m in the Duncan Mountains area (McGregor, 1965; Stump, 1976*b*), to 2000 to several thousand meters in the Amundsen Glacier–Scott Glacier area (Stump, 1976*b*; Minshew, 1967). Because of the unknown boundary relationships, all of these figures are minimum estimates.

The coarser grained units of the Beardmore Group are metagraywackes or impure quartzites. The Goldie and LaGorce Formations are poorly sorted and contain angular grains of quartz, with some plagioclase, K-feldspar, and muscovite, which "float" in a fine-grained, recrystallized matrix of biotite, chlorite, or sericite. The Goldie Formation also is slightly calcareous in places. The Duncan Formation has been recrystallized to such a degree that coarse-grained quartz grains in some samples are the only vestiges of the original clasts. Metamorphism has produced both foliated and hornfels textures in different parts of the formation. Calc-silicate assemblages indicate that some layers of the Duncan Formation were also carbonate-bearing rocks. The finer grained units of the Beardmore Group are argillites, slates, and schists, depending on the degree of metamorphism that has affected a particular area.

The strata of the Beardmore Group are evenly bedded, rhythmic interlayerings of this coarser and finer grained clastic material, with bedding ranging from a few centimeters to several meters. Laminations are by far the most common sedimentary structure, but many beds lack any internal structure whatsoever. Although graded bedding has been reported from the LaGorce and Duncan Formations (Minshew, 1967; Murtaugh, 1969; McGregor, 1965; Stump, 1976*b*) and from the Goldie Formation northwest of the Queen Maud Mountains (Laird and others, 1971; Gunner, 1976), this feature is apparently rather rare. Other sedimentary structures include cross-bedding and sole marks in the LaGorce Formation (Minshew, 1967), dish structures in the Duncan Formation (Nilsen and others, 1977), and low-angle cross-bedding in the Goldie Formation (Stump, 1976*b*).

Lithologies, bedding characteristics, and sedimentary structures all indicate that the Beardmore Group was deposited in deep-sea fans through the action of turbidity currents (Minshew, 1967; Stump, 1976*b*). The paucity of graded bedding and the abundance of laminated and structureless beds probably indicates that at least in the Queen Maud Mountains the Beardmore Group is representative of the distal facies of a deep-sea fan system (Walker, 1976).

Wyatt Formation

The Wyatt Formation, exposed around the upper reaches of the Amundsen, Scott, and Reedy Glaciers, is composed of porphyritic silicic metafelsite (Minshew, 1967; Murtaugh, 1969; Katz and Waterhouse, 1970; Stump, 1976*b*). It remains the most poorly understood of the basement formations in the Queen Maud Mountains, and as mentioned in the previous section, its conformable contact with the LaGorce Formation on Ackerman Ridge is still controversial.

The age of the Wyatt Formation is thought to be late Precambrian. Faure and others (1979), using Rb-Sr geochronometry, suggest a date of 790 m.y. as a maximum limit on the age of the formation on the west side of Scott Glacier.

With the exception of rocks in the LaGorce Mountains, the Wyatt Formation is composed of massive metafelsites lacking any observable internal structure. A discordant contact between the Wyatt and LaGorce Formations at the head of Amundsen Glacier has been interpreted as an intrusive contact (Stump, 1976*b*). However, except for the rocks in the LaGorce Mountains, which are clearly extrusive, the question of whether the rest of the formation originated as extrusions or hypabyssal intrusions remains to be answered.

The massive metafelsites of the Wyatt Formation contain phenocrysts of quartz, plagioclase, biotite, and in some places K-feldspar. The presence of biotite may be a characteristic that distinguishes the Wyatt Formation from the silicic volcanic formations of the Liv Group, which apparently have only phenocrysts of quartz, plagioclase, and K-feldspar (Stump, 1976*b*). The groundmass of the massive metafelsites contains quartz, plagioclase and sericite, biotite, or chlorite. Coarseness of the groundmass varies according to the proximity of granitic intrusions.

Liv Group

The name Liv Group is proposed for scattered formations in the Queen Maud Mountains, known or thought to be of Cambrian age, and in which silicic volcanic rocks are prominent. These include the Taylor, Fairweather, and Leverett Formations, and tentatively the Greenlee Formation. The Taylor Formation contains Early Cambrian fossils (Yochelson and Stump, 1977) and the Leverett Formation, Middle Cambrian fossils (Palmer and Gatehouse, 1972). The Fairweather Formation is somewhat similar in lithology and stratigraphy to the Taylor Formation (Stump and others, 1976), but no fossils have been found to establish its age.

Although these formations occur in approximately the same stratigraphic position as the Byrd Group of the Byrd Glacier–Nimrod Glacier area, they have distinct lithologic differences. The Byrd Group is dominated by the thick Shackleton Limestone and contains only minor volcanic beds, whereas the proposed Liv Group is largely silicic volcanic rock with minor associated clastic and carbonate rocks.

As is customary in the naming of basement groups, the name Liv is selected from a local major valley glacier which transects the range. The selection of a type unit is arbitrary since facies change so rapidly in a silicic volcanic terrane. The section measured on Mount Fairweather (85° 04′ S lat., 166° 32′ W long.) and the ridge crest to the northeast of it is the longest and contains most of the lithologies represented in the group (Stump, 1976*b*); it is therefore chosen as the type section. However, reference should also be made to measured sections in the Leverett and Taylor Formations to compare differences within the group (Minshew, 1967; Stump, 1976*b*).

Taylor Formation

The Taylor Formation, cropping out on both sides of the lower Shackleton Glacier (Wade and others, 1965), contains the best preserved primary volcanic textures in the group. Wade's (1974) minimum thickness estimate of 2500 m as recorded on Mount Greenlee is fair since it is not known whether sections on the east and west sides of Shackleton Glacier are overlapping or not.

Porphyritic rhyolite with phenocrysts of quartz, plagioclase, and in places, K-feldspar is the most abundant rock type in the formation. Both ash-flow tuffs with welded-shard structure and lavas with flow banding are represented (Stump, 1974, 1976*b*). Spherulites occur in some of these units (Stump, 1976*a*). Also present are ash-fall tuffs. On Mount Greenlee, in what is thought to be the lower part of the formation, several units of amygdaloidal, basaltic lava crop out.

At the Taylor Nunatak locality, the silicic volcanic rocks are associated with carbonate and volcaniclastic rocks. The carbonates are micrites, clastic oosparites, and a breccia from which the Early Cambrian fossils have been recovered (Yochelson and Stump, 1977). Certain units contain only coarse-grained quartz and plagioclase, some of which is euhedral, and seem to be detrital concentrations of phenocrysts. Rocks which appear in the field to be chert can be seen in thin section to be deposits of fine ash. Microscopic graded bedding indicates that these were deposited subaqueously.

The rest of the Taylor Formation to the north on the east side of Shackleton Glacier is distinct from the Taylor Nunatak section. Judging from the dip of beds in both areas, the Taylor Nunatak rocks are stratigraphically beneath those to the north, unless some structural complication is concealed beneath the small tributary glacier north of Taylor Nunatak.

Although there appears to be stratigraphic continuity within the area to the north, only a few of the ridge crests have been traversed, so a coherent picture of the stratigraphic sequence is lacking there. Detailed mapping in this area would probably be fruitful for determining the nature of facies changes in this volcanic terrane.

Interbedded with the silicic volcanic rocks in this area are units of argillite and festoon cross-bedded quartzite, probably representing a fluvial or shallow-marine environment. Also found toward the northern part of the area are volcaniclastic conglomerates and chert-like deposits which most likely originated as water-laid ash.

The Taylor Formation on the eastern side of Shackleton Glacier has suffered only minor and local metamorphic and deformational effects. On the western side of the glacier, however, especially north and west of Mount Greenlee, the rocks are so badly sheared that much of their original character is lost.

The environment during deposition of the Taylor Formation was probably one of active volcanic islands, perhaps becoming more emergent through time, as indicated by the cross-bedded quartzite and argillite in the supposed upper part of the formation.

Greenlee Formation

Lying stratigraphically below the Taylor Formation on Mount Greenlee and Epidote Peak is a sequence of fine-grained quartzite and phyllite called the Greenlee Formation (Wade, 1974). The sedimentary and volcanic rocks are interbedded in a transition zone, and the Taylor Formation basal contact is marked at the lowest thick felsite unit.

The thickness of exposed Greenlee Formation is probably not in excess of 1000 m. Bedding is 5 cm to 1 m thick, and even. Laminations are the only common sedimentary structure. Unlike the Beardmore Group, these rocks lack abundant graywacke and do not contain interbedded coarser and finer grained layers. The author will not speculate on the environment of deposition more than to say that it was low energy.

No lithologic equivalents of the Greenlee Formation are known throughout the Queen Maud Mountains. It is quite distinct from the volcanic formations of the Liv Group and may well precede initiation of this period of volcanism. However, since there is no other place to put it, the Greenlee Formation is included in the Liv Group.

Fairweather Formation

The Fairweather Formation occurs at several localities in the vicinity of the Duncan Mountains (McGregor, 1965). The composite section measured on Mount Fairweather and along the ridge to the northeast of it is in excess of 3000 m. In many ways the Fairweather Formation is like the Taylor Formation, but it has been considerably more affected by metamorphism, so that groundmass and matrix have taken on hornfels or foliated textures.

The Fairweather Formation has thick units of porphyritic silicic volcanic rock throughout. Basaltic rocks occur in the lower part of the formation, and chert-like rock is fairly common in the lower and middle parts. The upper part of the formation contains cross-bedded quartzites, schists, metaconglomerates, and some thin calcareous beds (Stump, 1976*b*).

The similarity of the geologic histories of the Fairweather and Taylor Formations, as well as their geographic proximity, strongly suggests that they formed during the same volcanic episode.

Henson Marble Member

Thick, white, coarsely crystalline marble units are found overlying the Fairweather Formation in two places, as well as in isolated outcrops from west of Kosco Glacier to the Herbert Range. Though it is impossible to establish a strict equivalence of these occurrences, nor would one be likely in a volcanic terrane where facies changes are rapid, these marble units are grouped together for convenience in mapping. The Henson Marble was named by McGregor (1965) for the occurrence at Mount Henson. The author concurs with the suggestion of Wade and Cathey (in press) to extend the name Henson Marble Member to all these marble units, including them in the Fairweather or Taylor Formation, according to location.

Leverett Formation

The Leverett Formation has been measured and described from an isolated mountain north of Leverett Glacier (Minshew, 1967); the rest of the outcrop in the vicinity remains to be examined geologically. The formation is more than 1000 m thick. The lowest exposed unit consists of cross-bedded sandstone, which gives way upward to shale and then to thick limestone in which Middle Cambrian fossils have been found (Palmer and Gatehouse, 1972). Overlying the limestone is a unit of silicic volcanic rock interbedded with and overlain by conglomerate and sandstone.

Minshew (1967) has suggested that the Leverett Formation accumulated in a shallow-marine environment, with intermittent volcanism.

The Middle Cambrian fossils preclude correlation with the Taylor Formation, but the presence of silicic volcanic, clastic, and carbonate rocks warrants inclusion in the Liv Group.

Summary

The rocks of the Ross Supergroup record geosynclinal development in the Queen Maud Mountains area prior to the Cambro-Ordovician Ross orogeny. The initial phase of deposition, represented by the turbidite sequences of the Beardmore Group, appears to have been along a quiescent continental margin. The belt became activated during late Precambrian time, producing the magmatic rocks of the Wyatt Formation and parts of the Wisconsin Range Batholith (Murtaugh, 1969), as well as causing deformation (followed by profound erosion) in areas beyond the Queen Maud Mountains (Laird and others, 1971; Schmidt and others, 1965). This earlier activity, referred to as the Beardmore orogeny (Grindley and McDougall, 1969), probably represents the first tectonism resulting from subduction. This tectonism continued through Cambrian time, producing widespread silicic volcanism represented by the Liv Group, and culminated in the emplacement of the Queen Maud–Wisconsin Range Batholith, which had cooled through argon-retention temperatures by Late Ordovician time (Faure and others, 1979).

Acknowledgments. Research leading to this paper was supported by the National Science Foundation through grants GV-26652 and OPP74-04897 to the Ohio State University, Institute of Polar Studies.

References cited

Faure, G., Eastin, R., Ray, P. T., McLelland, D., and Schultz, C. H. 1979. Geochronology of igneous and metamorphic rocks, central Transantarctic Mountains. (*In* Laskar, B., and Raja Rao, C. S., *eds. Fourth International Gondwana Symposium Papers, Delhi,* Hindustan Publishing Corporation, 805–13.)

Grindley, G. W., and McDougall, I. 1969. Age and correlation of the Nimrod Group and other Precambrian rock units in the central Transantarctic Mountains, Antarctica. *N. Z. Jl Geol. Geophys., 12,* 391–411.

Gunner, J. 1976. Isotopic and geochemical studies of the pre-Devonian basement complex, Beardmore Glacier region, Antarctica. *Rep. Inst. polar Stud., Ohio State Univ., 41,* 126 pp.

Gunner, J., and Faure, G. 1972. Rubidium-strontium geochronology of the Nimrod Group, central Transantarctic Mountains. (*In* Adie, R. J., *ed. Antarctic Geology and Geophysics.* Oslo, Universitetsforlaget, 305–11.)

Katz, H. R., and Waterhouse, B. C. 1970. Geological reconnaissance of the Scott Glacier area, southeastern Queen Maud Range, Antarctica. *N. Z. Jl Geol. Geophys., 13,* 1030–37.

Laird, M. G. 1963. Geomorphology and stratigraphy of the Nimrod Glacier-Beaumont Bay region, southern Victoria Land, Antarctica. *N. Z. Jl Geol. Geophys., 6,* 465–84.

Laird, M. G., and Waterhouse, J. B. 1962. Archeocyathine limestones of Antarctica. *Nature, Lond., 194,* 861.

Laird, M. G., Mansergh, G. D., and Chappell, J. M. A. 1971. Geology of the central Nimrod Glacier area, Antarctica. *N. Z. Jl Geol. Geophys., 14,* 427–68.

McGregor, V. R. 1965. Geology of the area between the Axel Heiberg and Shackleton glaciers, Queen Maud Range, Antarctica. Part 1—Basement complex, structure and glacial geology. *N.Z. Jl Geol. Geophys., 8,* 314–43.

Minshew, V. H. 1967. *Geology of the Scott Glacier and Wisconsin Range Areas, Central Transantarctic Mountains, Antarctica.* Ph.D. dissertation, The Ohio State University, 268 pp.

Murtaugh, J. G. 1969. Geology of the Wisconsin Range batholith, Transantarctic Mountains. *N. Z. Jl Geol. Geophys., 12,* 526–50.

Nilsen, T. H., Bartow, J. A., Stump, E., and Link, M. H. 1977. New occurrences of dish structure in the stratigraphic record. *J. Sedim. Petrol., 47,* 1299–304.

Oliver, R. L. 1972. The geology of an area near the mouth of the Beardmore Glacier, Ross Dependency, Antarctica. (*In* Adie, R. J., *ed. Antarctic Geology and Geophysics.* Oslo, Universitetsforlaget, 379–85.)

Palmer, A. R., and Gatehouse, C. G. 1972. Early and Middle Cambrian trilobites from Antarctica. *Prof. Pap. U.S. geol. Surv.,* No. 456-D, 37 pp.

Schmidt, D. L., Williams, P. L., Nelson, W. H., and Ege, J. R. 1965. Upper Precambrian and Paleozoic stratigraphy and structure of the Neptune Range, Antarctica. *Prof. Pap. U.S. geol. Surv.,* No. 525-D, D112–19.

Skinner, D. N. B. 1965. Petrographic criteria of the rock units between the Byrd and Starshot Glaciers, south Victoria Land, Antarctica. *N. Z. Jl Geol. Geophys., 8,* 292–303.

Stump, E. 1974. Volcanic rocks of the Early Cambrian Taylor Formation, central Transantarctic Mountains. *Antarct. Jnl U.S., 9,* 228–29.

Stump, E. 1976*a*. Accretionary lapilli and lithophysal spherulites from the Taylor Formation, Queen Maud Mountains, Antarctica. *Antarct. Jnl U.S., 11,* 246–48.

Stump, E. 1976*b*. On the late Precambrian-early Paleozoic metavolcanic and metasedimentary rocks of the Queen Maud Mountains, Antarctica, and a comparison with rocks of similar age from southern Africa. *Rep. Inst. polar Stud. Ohio State Univ., 62,* 212 pp.

Stump, E., Corbató, C. E., and Browning, A. W. 1976. Stratigraphy of the Fairweather Formation, central Transantarctic Mountains. *Abst. with Program, Geol. Soc. Am.,* 511–12.

Wade, F. A. 1974. Geological surveys of Marie Byrd Land and the central Queen Maud Range. *Antarct. Jnl U.S., 9,* 241–42.

Wade, F. A., and Cathey, C. A. In press. Geology of the basement complex, western Queen Maud Mountains, Antarctica. (*In* Turner, M. D., and Splettstoesser, J. F., *eds. Geology of the Central Transantarctic Mountains.* Washington, American Geophysical Union) [Antarctic Research Series].

Wade, F. A., Yeats, V. C., Everett, J. R., Greenlee, D. W., La Prade, K. E., and Shenk, J. C. 1965. Geology of the central portion of the Queen Maud Range, Transantarctic Mountains. *Science, N. Y., 150,* 1808–9.

Walker, R. G. 1976. Facies Models 2. Turbidites and associated coarse clastic deposits. *Geosci. Can., 3,* 25–36.

Yochelson, E. L., and Stump, E. 1977. Discovery of Early Cambrian fossils at Taylor Nunatak, Antarctica. *J. Paleont., 51,* 872–75.

71

Sedimentology and Paleocurrent Analysis of the Basal Part of the Beacon Supergroup (Devonian [and Older?] to Triassic) in South Victoria Land, Antarctica

R. W. PLUME
Department of Geology, Victoria University of Wellington, Wellington, New Zealand

Abstract. The New Mountain Subgroup includes a basal sandstone about 40 m thick (the Windy Gully Sandstone) conformably overlain by a thinner dark siltstone (the Terra Cotta Siltstone) which is in turn overlain conformably by about 150 m of quartz sandstone (the New Mountain Sandstone). Locally the New Mountain Sandstone includes two members: a breccia-conglomerate unit at the base (the Boreas Subgreywacke Member) overlain by a shaly unit about 7 m thick (the Plane Table Member). The New Mountain Subgroup rests nonconformably on the Kukri Peneplain and is separated from overlying formations of the Taylor Group by the Heimdall Erosion Surface. A tentative age of Early Devonian for the Terra Cotta Siltstone is indicated on the basis of microfossil evidence.

Sedimentologic evidence indicates that the Windy Gully Sandstone and most of the New Mountain Sandstone were deposited by braided rivers, the Terra Cotta Siltstone and Plane Table Member in lakes, and the Boreas Subgreywacke Member by debris flows. Some eolian deposition is recorded in the upper part of the New Mountain Sandstone. Analysis of cross-bedding directions shows that the mean flow direction in river channels throughout deposition of the New Mountain Subgroup was to the west and northwest. Tectonic stability of the area during deposition of the New Mountain Sandstone is indicated by stable, unimodal flow patterns at three localities and by the consistent lithologic character of channel deposits. Eolian cross-bedding indicates a dominant paleowind direction to the southeast.

This paper treats the basal part of the Taylor Group (Beacon Supergroup) in the McMurdo Sound region, that is, those rocks resting on the Kukri Peneplain and separated from the rest of the Taylor Group by the Heimdall Erosion Surface. Known as the New Mountain Subgroup (Plume, 1978), these rocks comprise a basal sandstone (the Windy Gully Sandstone) conformably overlain by a thinner siltstone (the Terra Cotta Siltstone), which is in turn overlain conformably by a thick sandstone (the New Mountain Sandstone). Locally, two members (the Boreas Subgreywacke Member and the Plane Table Member) are included in the New Mountain Sandstone. Outcrops of the subgroup are now known from as far south as Mount Kempe in the Royal Society Range and as far north as Mount Suess near the MacKay Glacier.

Age

Recently, poorly preserved palynomorphs were extracted from a single sample of the Terra Cotta Siltstone collected at Table Mountain (lat. 77° 59′ S; long. 161° 57′ E). A tentative age of Early Devonian is indicated (Kyle, 1977). A well-established age of Middle to Late Devonian for the overlying Aztec Siltstone determines the minimum possible age for the New Mountain Subgroup.

Lithologic description

Windy Gully Sandstone

The Windy Gully Sandstone is composed largely of clean, white, light-tan or buff, medium- to coarse-grained, massive to blocky, well-bedded sandstone. Dark-green interbeds a few centimeters thick occur rarely.

Large-scale cross-bedding (Conybeare and Crook, 1968), in places as much as 3 m thick, is common. Medium-scale (20 cm) cross-bed sets were recorded at Mount Kempe. Mudcracks and the traces of burrowing and boring animals are also present. Ripple marks are rare and usually symmetric. The formation includes discontinuous basal conglomerate beds ranging in thickness from a few centimeters to about 3 m. Clasts range in size from granules and pebbles to boulders.

Terra Cotta Siltstone

The Terra Cotta Siltstone is composed of dark purple, chocolate brown or red, black, gray and yellowish green, thin-bedded siltstone, shale, and minor pink or white sandstone. The color of the siltstone varies from red to dark purple or black, depending on the local degree of heating by intrusion of the Ferrar Dolerite peneplain sill.

Sedimentary structures include symmetric ripple marks, mudcracks, and trace fossils. Mudstone-clast conglomerate, loading features, and, rarely, deformation of sandstone interbeds are also present. Carbonate concretions as much as 20 cm long and elongate parallel to bedding are common in the upper part of the formation. In the Ferrar Glacier–Taylor Glacier region the top of the formation is marked by a poorly developed vein network or mudcrack horizon.

NEW MOUNTAIN SANDSTONE

Medium- to coarse-grained, moderately well sorted and rounded, light tan or buff, quartzose sandstone characterizes this formation. Interbedded olive-green siltstone beds, a few centimeters thick and laterally persistent, are found both at the base and near the top of the formation.

Cross-bedding, as much as 7 m thick but more commonly about 2 m thick, is the most outstanding feature of the formation. Smaller scale cross-bed sets are also known but are relatively uncommon. Other primary features include both symmetric and asymmetric ripple marks. Mudcracks, locally superimposed on ripple marks, are also common on the olive-green siltstone beds described above. Primary current lineation is widespread, and some scour-and-fill channels were seen. Burrowing and boring trace fossils are very abundant within plane-laminated intervals, and a distinctive vertically burrowed interval occurs at the top of the formation. Walking-type trace fossils were also observed. Red beds were recorded at Table Mountain but not at any other locality.

Field work

During the 1974–75 field season, the writer, as a member of Victoria University of Wellington Antarctic Expedition (VUWAE) 19, visited nine outcrops of the New Mountain Subgroup in the McMurdo Sound region. Stratigraphic sections were described in detail and measured using a Jacob's staff and Abney level. Paleocurrent measurements, using a Brunton compass, were made on cross-bedding, ripple marks, and primary current lineation. Rock specimens were taken from regular intervals and these are now held in the Victoria University of Wellington rock collection (Nos. 26130–26300).

Features of the New Mountain Subgroup

RED BEDS

Red beds occur at two intervals within the New Mountain Subgroup. Red siltstones are common throughout the Terra Cotta Siltstone at Mount Handsley and Table Mountain. However, most outcrops of the Terra Cotta Siltstone are dark (purple, black, or chocolate brown) rather than red. Both X-ray and thin-section examination by the writer have shown that the color is nevertheless due to the presence of hematite (Plume, 1974). Its recrystallization into larger and hence opaque grains by the intrusion of dolerite causes the darker color.

In the New Mountain Sandstone red beds are known only at Table Mountain. They occur 200 m above the base of the formation in two intervals, each 0.2 m thick, separated by 4.6 m of section. The beds are continuous laterally for about 50 m but gradually thin and disappear. Both beds are associated with large-scale cross-bedding considered to be of eolian origin (see below). Because of the characteristic red color and the association with green color, the color of the red beds in the New Mountain Sandstone at Table Mountain is considered to be also due to hematite.

The presence of red beds in both the Terra Cotta Siltstone and New Mountain Sandstone favors a nonmarine paleoenvironment.

CARBONATE CONCRETIONS

Carbonate concretions are common in the Terra Cotta Siltstone, especially near the upper contact of the formation. Bedding of the host material conforms to the shape of concretions indicating that the carbonate crystallized before induration. Carbonate matrix also occurs in some fine-grained sandstone interbeds of the Terra Cotta Siltstone. Samples of carbonate from specimens collected at Table Mountain and Mount Handsley (where the red color of the Terra Cotta Siltstone shows that the sedimentary rocks have been relatively unaffected by dolerite) were analyzed for their $\delta^{13}C$ and $\delta^{18}O$ values (Table 71.1).

Keith and Weber (1964) suggested that specimens older than Jurassic can be best classified as marine or fresh water in origin on the basis of $\delta^{13}C$ values alone. Both $\delta^{13}C$ values from the Terra Cotta Siltstone lie within one standard deviation of the mean ($\bar{x} = -4.93$, $s = 2.75$) for Devonian and younger fresh-water carbonates. Hence the values of $\delta^{13}C$ for carbonates from the Terra Cotta Siltstone indicate that the formation is a fresh-water deposit.

TRACE FOSSILS

Trace fossils of many different types are very common throughout the subgroup. The most common form observed are complexly interwoven horizontal tubular structures thought to be the result of burrowing, worm-like animals (Vialov, 1962, Figures 7 and 8), possibly annelids (Webb, 1963, Figure 8). Walking traces, similar to those described by Gevers and others (1971) as *Arthropodichnus antarcticum* and *A. gouldi*, were also observed. In the Ferrar Glacier–Taylor Glacier region, specimens of *Skolithus* are very abundant in the upper part of the New Mountain

Table 71.1. Carbon and oxygen isotope values for two samples of carbonate from specimens of the Terra Cotta Siltstone

Locality	Specimen number	Lithology	Sample	$\delta^{13}C_{PDB}$[a]	$\delta^{18}O_{SMOW}$[b]
Mount Handsley	26200	Sandstone	Matrix	−5.8	8.0
Table Mountain	26176	Concretion	Whole-rock	−4.0	16.0

Analyst M.A. Cox, Institute of Nuclear Sciences, Gracefield, New Zealand.
Specimens are held in the Victoria University of Wellington rock collection.
[a] Pee Dee Belemnite.
[b] Standard mean ocean water.

Sandstone. Lobate structures very similar to *Cruziana* and *Isopodichnus* (Hantzschel, 1962, pp. W189 and W201, respectively) were found in the Terra Cotta Siltstone at Table Mountain.

In the New Mountain Sandstone at Plane Table an unusual trace, probably of biogenic origin, was found in a medium-grained sandstone. The traces are depressions, each containing a sand body which duplicates the shape of, but is slightly smaller than, the depression. The sand body may be cylindrical, wedge-shaped, or helical in structure. Complete structures range in length from 2 to 15 cm, are about 1 cm in depth, and appear to be randomly oriented on the bedding surface.

Trace fossils are commonly taken as evidence of marine deposition, but traces are also known from both modern environments (Coleman, 1969) and ancient deposits (Barrett and Kohn, 1975; MacPherson, 1975) interpreted as continental. Hence, the trace fossils in the New Mountain Subgroup do not give definite evidence of either marine or continental deposition, though their abundance and complexity indicate an aqueous environment.

MUDCRACKS

Mudcracks, formed by subaerial dessication of fine-grained sediment, are found in all formations of the subgroup. Their presence is strong evidence in favor of a nonmarine environment. Some cracks of probably subaqueous origin were also observed in the Terra Cotta Siltstone.

RIPPLE MARKS

Both symmetric and asymmetric ripple marks occur in all formations of the subgroup.

PRIMARY CURRENT LINEATION

In the New Mountain Subgroup, primary current lineation (Allen, 1964) is most common in the New Mountain Sandstone but is also present in the Windy Gully Sandstone. It is normally found on plane-laminated surfaces but may also be found on the face of foreset bedding.

Primary current lineation (or parting lineation) forms in the lower part of the upper flow regime when the Froude number is greater that about 0.75 (Allen, 1964). Dunes form on the bed during lower flow regime and are destroyed upon transition to upper flow regime conditions. Since dunes and primary current lineation form in different flow regimes they cannot coexist under conditions of constant flow. Thus, the presence of primary current lineation on foresets shows that a hydrodynamic disequilibrium existed frequently during deposition of the New Mountain Sandstone.

Tidal flow in the North Sea has a Froude number of 0.04 (McCave, 1971), far lower than required for transition to upper flow regime conditions. Since primary current lineation is therefore not likely to develop in submarine tidal environments, its presence is taken as evidence favoring a continental rather than marine interpretation.

PENECONTEMPORANEOUS DEFORMATION

In the New Mountain Sandstone deformation of medium- and large-scale cross-bedding is common. It is generally coherent, but in some places bedding is completely destroyed. Distortion on a scale of as much as 3 m is known. Where deformation occurs, it involves only one cross-bedding set while overlying and underlying sets are unaffected. Thus, the deformation was clearly penecontemporaneous.

CROSS-BEDDING

Description

Cross-bedding in the New Mountain Subgroup is most abundant in the New Mountain Sandstone, though it is found in the Windy Gully Sandstone and rarely in the Terra Cotta Siltstone. Virtually all is either medium- or large-scale (5 cm to 2 m and 2–8 m, respectively) (Conybeare and Crook, 1968, p. 32).

Sets in the Windy Gully Sandstone and New Mountain Sandstone are tabular and wedge planar, tangential and angular (McKee and Weir, 1953), and range in thickness from 0.1 m to a maximum of 7 m. They usually have erosional bases, and the average maximum dip of foresets is 15° (Barrett and Kohn, 1975) to 17° (Matz, 1968), but lower angle and oversteepened foresets do occur.

Types I and II cross-bedding

Three distinct size populations of cross-bedding can be identified within the New Mountain Sandstone. Near the base of the formation there is a population of small-scale cross-bedding (0.2 to 0.5 m thick), here designated Type I cross-bedding. In the Ferrar Glacier–Taylor Glacier region most of the cross-bedding is from 0.5 to 2 m thick, here designated Type II cross-bedding. A third type of cross-bedding is discussed below. Type II cross-bedding has a strong westerly trend (Figure 71.1), and Type I cross-bedding trends at a large angle to Type II cross-bedding (Figure 71.2).

The cross-bedding of the Windy Gully Sandstone is very similar in style, scale, and orientation to Types I and II cross-bedding of the New Mountain Sandstone. In addition, other features such as mudcracks and trace fossils occur in both formations. Because of the similarities in the two formations, the cross-bedding of both will be discussed together.

Occurrence

In the Ferrar Glacier–Taylor Glacier region the three types of cross-bedding found in the New Mountain Sandstone occur in a definite vertical sequence. The basal part of the formation is distinguished by the presence of Type I cross-bedding (as at Table Mountain) or by a thick plane-laminated interval (as at New Mountain). Wherever the basal part of the formation is exposed, laterally persistent beds of olive-green siltstone a few centimeters thick are

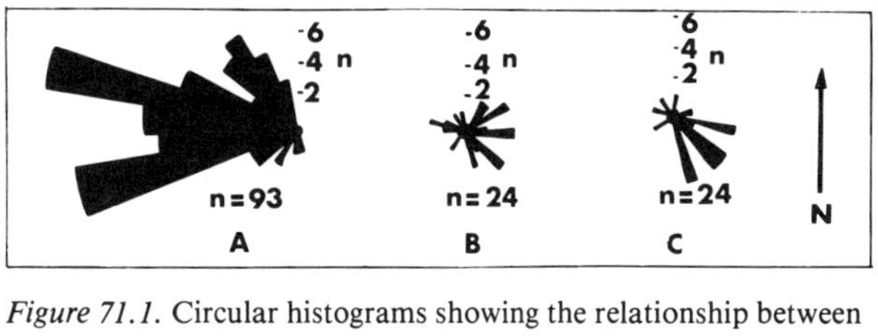

Figure 71.1. Circular histograms showing the relationship between cross-bedding thickness and paleocurrent direction for cross-bedding in the New Mountain Sandstone in the Ferrar Glacier–Taylor Glacier region. A. Cross-bedding 0.5 to 2 m thick (Type II cross-bedding); B. Cross-bedding 2 to 3 m thick; C. Cross-bedding greater than 4 m thick.

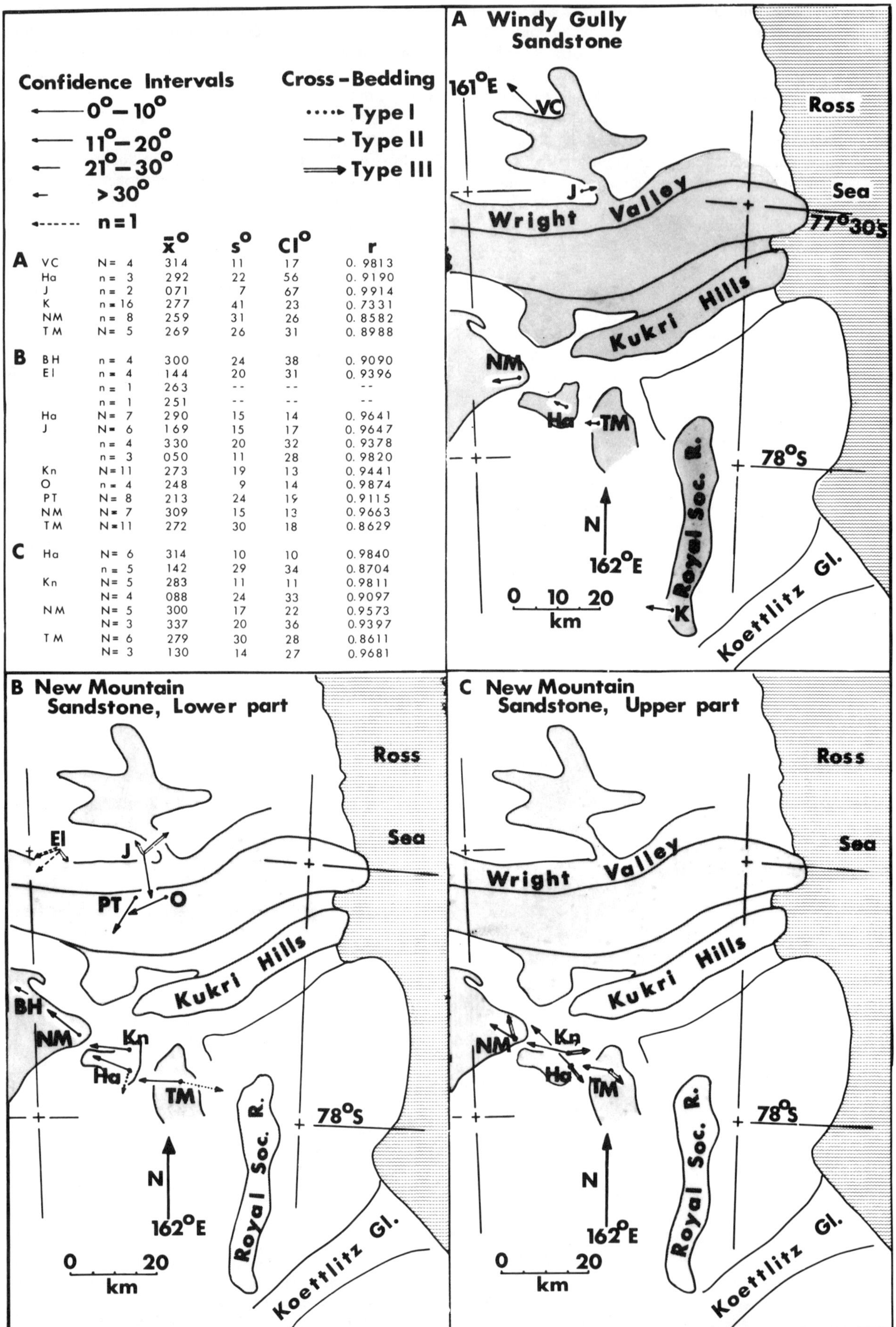

			$\bar{x}°$	s°	CI°	r
A	VC	N = 4	314	11	17	0.9813
	Ha	n = 3	292	22	56	0.9190
	J	n = 2	071	7	67	0.9914
	K	n = 16	277	41	23	0.7331
	NM	n = 8	259	31	26	0.8582
	TM	N = 5	269	26	31	0.8988
B	BH	n = 4	300	24	38	0.9090
	El	n = 4	144	20	31	0.9396
		n = 1	263	--	--	--
		n = 1	251	--	--	--
	Ha	N = 7	290	15	14	0.9641
	J	N = 6	169	15	17	0.9647
		n = 4	330	20	32	0.9378
		n = 3	050	11	28	0.9820
	Kn	N = 11	273	19	13	0.9441
	O	n = 4	248	9	14	0.9874
	PT	N = 8	213	24	19	0.9115
	NM	N = 7	309	15	13	0.9663
	TM	N = 11	272	30	18	0.8629
C	Ha	N = 6	314	10	10	0.9840
		n = 5	142	29	34	0.8704
	Kn	N = 5	283	11	11	0.9811
		N = 4	088	24	33	0.9097
	NM	N = 5	300	17	22	0.9573
		N = 3	337	20	36	0.9397
	TM	N = 6	279	30	28	0.8611
		N = 3	130	14	27	0.9681

common. These thin beds are most abundant at New Mountain, where mudcracks are common.

The middle part of the New Mountain Sandstone is an interval 50 to 100 m thick dominated almost entirely by Type II cross-bedding, i.e., cross-bedding 0.5 to 2 m thick, and with a westerly paleocurrent trend. Sets commonly persist laterally for tens of meters.

The upper part of the formation is distinguished by the presence of Type III cross-bedding (see below) and the reappearance of thin olive-green siltstone beds identical to those found in the basal part of the formation (see above). These beds commonly show symmetric ripple marks and mudcracks. The siltstone beds are separated by plane-laminated sandstone beds averaging 50 cm in thickness. A few of these sandstone interbeds have Type I cross-bedding, and a scour channel 25 cm deep and 50 cm wide was observed well exposed on the upper surface of one interbed.

Interpretation

The Ganges-Brahmaputra River of India is a large braided stream which carries fine sand as its dominant load. Dunes as high as 15 m form during flood cycle, and sections of large ones reveal a complex internal structure of cosets of cross-strata as much as 7 m thick. Individual sets are about 1 m thick, though thicker sets do occur (Coleman, 1969).

Type II cross-bedding of the New Mountain Sandstone (and Windy Gully Sandstone) averages about 1.5 m in thickness, slightly more than the average thickness in the Brahmaputra River. The average dip of Type II foresets is 15° (Barrett and Kohn, 1975) to 17° (Matz, 1968), closely comparable to the 15° dip of foreset bedding in the Brahmaputra River. In addition, both angular and tangential foresets occur in the New Mountain Subgroup and in the modern analogue. Thus, the form and internal structure of the cross-bed sets in the New Mountain Subgroup are consistent with the interpretation that they formed as migrating dunes in a large braided river.

Coleman (1969) also carried out a detailed analysis of the relationship between current-flow direction and the orientation of cross-bedding in preserved bars. He found that the average cross-bedding direction was in excellent agreement with the current-flow direction and that the standard deviation of cross-bedding averages about 28°. The mean angular deviation of Type II cross-bedding in the New Mountain Subgroup is 20° (equivalent to a standard deviation of about 25°), and values range from 10° to a maximum of 30° for grouped data.

P. J. Barrett (personal communication, 1977) has shown that the standard deviation of uniformly sampled cross-bedding directions is a measure of sinuosity. Furthermore, on the basis of empirical evidence he suggested that standard deviations of approximately 50° to 60° separate braided river deposits from meandering river deposits. Thus, values of standard deviation from the New Mountain Subgroup are consistent with a braided river interpretation.

In the Bengal Basin, crevasse splay and levee deposits from when sediment-laden water overtops the banks of the Brahmaputra River during flood stage. Sediment not carried in suspension is distributed by a radial network of small channels running over the surface of the splay deposit. Small-scale cross-bedding, which may trend at a large angle to the normal stage mean flow direction, is developed in these sandy deposits. These sandy deposits interfinger with thin silt beds deposited from suspension after the flood has subsided. Suspension deposits are usually only a few centimeters thick and are called flood basin deposits by Coleman (1969).

Sedimentary beds of the basal and upper parts of the New Mountain Sandstone are interpreted as flood basin deposits for a number of reasons. Interbedded thin siltstones and cross-bedded sandstones are associated with scour channels, primary current lineation, desiccation cracks, and symmetric ripple marks. The strongly divergent paleocurrent pattern of the Type I cross-bedding strongly suggests deposition in crevasse splay deposits. Subsequent suspension deposition apparently formed the interbedded green siltstone beds in ephemeral lakes where wind-induced wave activity and later desiccation resulted in symmetric ripple marks and mudcracks.

Type III cross-bedding

A third population, designated Type III cross-bedding, is composed of large-scale sets (2 to 8 m thick), formed by currents trending in an easterly direction (Figure 71.1). All Type III cross-bedding occurs in the upper part of the New Mountain Sandstone, to a limited extent interbedded with Type II cross-bedding.

Because large-scale submarine landforms have been unknown until relatively recently, there has been a tendency to interpret thick cross-bedding as eolian largely on the basis of scale. The discovery of submarine sand waves in the North Sea and other tidal areas of the world (e.g., Off, 1963; Houbolt, 1968) has led some workers to reexamine established interpretations of some cross-bedded formations (e.g., Freeman and Visher, 1975).

To deduce the depositional environment from cross-bedding in ancient sandstone, it is essential to know the internal structure of bedforms used as modern analogues. Sparker profiles across asymmetric sand waves in the North Sea (Houbolt, 1968) have revealed foreset beds which dip in the direction of steepest slope. However, oriented box cores from these sand waves also showed cross-bedding dipping in the opposite direction, and the profiles showed megaripples migrating over the surface of the sand waves. The bimodal current distribution and the migrating megaripples suggest that the portrayal of submarine sand waves as cross-bedded in a simple, regular, and unimodal fashion (Houbolt, 1968, Figure 17, p. 257) is unjustified.

In spite of the accessibility of eolian bedforms, little was known of their internal structure until McKee (1966) described sand dunes at White Sands National Monument in the United States. By cutting deep trenches in various

Figure 71.2. Means of cross-bedding in the Windy Gully Sandstone and New Mountain Sandstone at various localities. BH: Beacon Heights. El: Mount Electra. Ha: Mount Handsley. J: Mount Jason. K: Mount Kempe. Kn: Knobhead. NM: New Mountain. O: Mount Odin, PT: Plane Table. TM: Table Mountain. VC: Vashka Crag. Data for Beacon Heights, Mount Jason, and Vashka Crag taken from Barrett and Kohn (1975). N implies grouped data; n, ungrouped data. $\bar{x}$°: Mean. s°: Standard deviation. Cl°: Confidence limits. r: Vector sum.

types of sand dunes, McKee demonstrated that some types of eolian structures are cross-bedded in a simple and regular fashion similar to that observed in the New Mountain Sandstone.

Textural and mineralogic features are frequently cited as evidence of an eolian origin (e.g., sorting, frosting, roundness, and mean grain size). However, such features as good sorting and roundness may result from reworking, and frosting may result from diagenetic corrosion of grains. Similarly, other textural and mineralogic features are not necessarily diagnostic of eolian mechanisms and hence must be regarded as only a suggestion of eolian deposition. Sandstone in the upper part of the New Mountain Sandstone associated with Type III cross-bedding commonly shows textural features which suggest an eolian origin. Quartz grains are commonly frosted, well sorted, and rounded. Sieve analysis of a disaggregated sample gave a mean diameter of 0.83 ϕ and a standard deviation of 0.42 ϕ (measures of Folk and Ward, 1957).

The paleocurrent pattern of Types II and III cross-bedding also suggests an eolian paleoenvironment. Types II and III cross-bedding interbed in the upper part of the New Mountain Sandstone. Since Type II cross-bedding formed in a fluvial channel, its orientation is therefore closely related to the paleoslope direction. Type III cross-bedding is found in some places closely associated with Type II cross-bedding but trending in nearly the opposite direction. Therefore, the mechanism by which Type III cross-bedding was deposited must be independent of paleoslope. The only geologically reasonable mechanism capable of producing such large-scale features in a continental environment is the wind.

Paleocurrent analysis

Methods

Collection of data

Orientations of medium- and large-scale cross-bedding, primary current lineation, and the crest strike of ripple marks were measured using a Brunton compass. All readings were corrected by 152° for magnetic declination (true north being taken as long. 162° E). No correction was made for the gentle westerly dip of Beacon strata. Because of the scale of cross-bedding, sets of data were taken from intervals at least 10 m thick.

If the cross-bedding was well exposed in plan, the trough axis could be measured directly to give the paleocurrent direction. The direction parallel to the trough axis could be estimated if bedding could be seen on at least two intersecting vertical planes. Where trough axes were not exposed, they were assumed to parallel the direction of maximum dip of foresets.

A total of 564 (n) individual measurements were taken from the Windy Gully Sandstone, Terra Cotta Siltstone, and New Mountain Sandstone. Measurements were grouped stratigraphically and according to structure type into 130 (N) data sets.

Statistical treatment of data

Statistical measures used for vector resultant and angular deviation described by Reyment (1971) were used for analysis of paleocurrent data. Vector resultants for grouped data were compared at a 5% level of significance using the circular F-statistic described by Watson (1966).

Variance

In floodplain systems the variance of cross-bed orientations in channel sands arises from three main sources: between-channel variation, within-channel variation, and measurement error. The effects of the latter can be reduced by not measuring poorly exposed structures. The effects of within-channel variability and measurement error can be reduced by averaging orientation measurements over small stratigraphic intervals (about 10 m, see above). The variance calculated about the means of these averages is considered a reliable estimate of the variance resulting from between-channel differences.

Results

The results of the paleocurrent analysis are shown in Figure 71.2. The relevant statistics are given in terms of N for grouped data, and in terms of n for ungrouped data.

Windy Gully Sandstone

The dominant trend of Type II cross-bedding in the Windy Gully Sandstone is westerly to northwesterly, though a few cross-beds in the formation at Mount Jason trend in an easterly direction.

New Mountain Sandstone

Chi-square tests for normality of Type II cross-bedding at Table Mountain, Knobhead, and New Mountain show nonsignificant deviation from circular normality (Table 71.2). By contrast, the distributions in both the lower

Table 71.2. Results of χ^2 (chi-square) tests for normality of Type II cross-bedding in the New Mountain Sandstone

	Table Mountain			Mount Handsley			Knobhead		New Mountain	
	L	U	A	L	U	A	L	A	L	U
n	43	27	77	46	38	84	49	55	41	54
$\chi^2_{calculated}$	6.04	2.18	7.23	22.39	18.59	15.04	2.77	7.80	10.38	7.68
$\chi^2_{critical}$	9.24	6.25	12.02	11.07	9.49	15.51	9.24	9.24	9.24	12.02
P	0.50	0.50	0.50	0.01	0.01	0.06	0.50	0.18	0.08	0.35
Decision	NS	NS	NS	S	S	ND	NS	NS	ND	NS

L: Lower part of section.
U: Upper part of section.
A: All readings from entire section.
NS: Not significantly different from a normal distribution at the 10% level of significance.
S: Significantly different from a normal distribution at the 5% level of significance.
ND: No decision, implying that the probability (P) of $\chi^2_{calculated}$ lies between $P = 0.10$ and $P = 0.05$.
Data are fitted to a von Mises circular normal distribution; data are ungrouped (n).

Table 71.3. Results of circular *F*-tests comparing the mean direction of Type II cross-bedding in the lower and upper parts of the New Mountain Sandstone at four localities in the Ferrar Glacier-Taylor Glacier region

	Table Mountain	Mount Handsley	Knobhead	New Mountain
N (Total)	17	13	16	12
Upper mean (N_2)	279° (6)	314° (6)	283° (5)	300° (5)
Lower mean (N_1)	272° (11)	290° (7)	273° (11)	309° (7)
R_2	5.17	5.90	4.91	4.79
R_1	9.49	6.75	10.38	6.76
R	14.59	12.37	15.23	11.51
$F_{calculated}$	0.45	8.80	0.53	0.89
$F_{critical, 0.05}$	4.55	4.70	4.60	5.00
Decision	NS	S	NS	NS

NS: Not significantly different at the 5% level of significance.
S: Significantly different at the 5% level of significance.
Data are grouped (*N*).

(the basal and middle parts in the previous discussion) and upper parts at Mount Handsley differ significantly from normal. The χ^2_{calc} value for the entire distribution has a probability between 5% and 10%.

Comparison of the mean directions of the lower and upper parts shows that there is no significant change in paleocurrent direction at Table Mountain, Knobhead, and New Mountain. The change in direction at Mount Handsley, however, is significant (Table 71.3).

The normal distribution and unchanged paleocurrent direction at Table Mountain, Knobhead, and New Mountain indicate that the paleocurrent sample at each of these localities can be regarded as coming from a single, large, unimodal population. The significant change in current direction at Mount Handsley is unlikely to be the result of a paleoslope change that left the other areas unaffected. Hence, stability of the paleoslope is indicated by the paleocurrent patterns at Table Mountain, New Mountain, and Knobhead.

Table 71.4 shows that the mean paleocurrent directions from the entire distribution at Table Mountain and the lower part of Mount Handsley are not significantly different, nor are the means of the entire distribution at New Mountain and the upper part of Mount Handsley. However, the means of the entire distributions of Table Mountain and New Mountains *are* significantly different.

If the basin was tectonically stable, as suggested previously, then mixing of cross-bedding formed in two different major channel systems seems the most likely cause of the relationship described above. If two channel systems, each draining a different part of the source area, are responsible for the two current directions, then a regional migration of channels is indicated by the stratigraphic relationships of paleocurrents at Mount Handsley. The basal, middle, and upper parts of the New Mountain Sandstone have been interpreted above as flood basin, channel, and flood basin deposits, respectively. The vertical sequence of these facies changes also indicates a regional channel migration.

Schumm (1972) has shown that the type of sediment transport, regional gradient, and sinuosity are related in such a way that a decrease in gradient results in an increased sinuosity and a change in the mode of sediment transport. The character of Type II cross-bedding remains nearly constant throughout the New Mountain Subgroup, suggesting that the paleoslope was stable with respect to gradient as well as to slope direction. The Altar Mountain Formation, which overlies the New Mountain Subgroup, has a similar mean current direction (excepting the Odin Arkose Member) but a standard deviation of about 70° (Barrett and Kohn, 1975). In addition, when compared with the New Mountain Subgroup, the Altar Mountain Formation is markedly different in lithologic character. Fining-upward cycles with numerous silt beds, mudcracks, ripple marks, and cross-bedding suggest deposition by meandering streams. Hence, the transition from the New Mountain Sandstone to the Altar Mountain Formation is marked by a change from braided to meandering type streams with a consequent increase in sinuosity (as reflected by the increase in standard deviation). This relationship is probably the result of a decrease in paleoslope gradient.

Table 71.4. Results of F-tests comparing paleocurrent directions of groups of Type II cross-bedding in the New Mountain Sandstone

	All Table Mountain (TM) versus lower Mount Handsley (Ha)	All New Mountain (NM) versus upper Mount Handsley	All Table Mountain versus all New Mountain
N (Total)	24	18	29
Means (N_1)	TM 268° (17)	NM 305° (12)	TM 268° (17)
(N_2)	Ha 290° (7)	Ha 314° (6)	NM 305° (12)
R_1	14.59	11.51	14.59
R_2	6.75	5.90	11.51
R	21.01	17.37	23.69
$F_{calculated}$	2.71	0.27	22.43
$F_{critical, 0.05}$	4.30	4.50	4.30
Decision	NS	NS	S

NS: Not significantly different at the 5% level of significance.
S: Significantly different at the 5% level of significance.
Data are grouped (*N*).

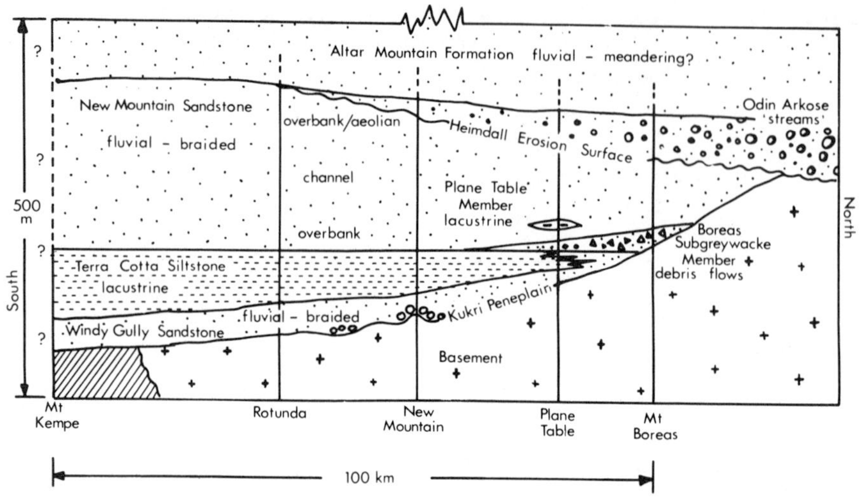

Figure 71.3. Restored section summarizing the relationships between units in the basal part of the Taylor Group. Columnar sections are projected to a line joining Mount Kempe and Mount Electra (K and El, Figure 71.2). Horizontal scale is approximate; there is 120 times vertical exaggeration. Upper contact of the Terra Cotta Siltstone is taken as datum.

Paleoenvironmental interpretation of the New Mountain Subgroup

Kukri Peneplain

The known relief on the Kukri Peneplain varies from a few meters to 32 m (at New Mountain). Relief can also be inferred from the Boreas Subgreywacke Member, which required prominent relief for deposition (see discussion below). South- and southwest-trending cross-bedding, and decreasing grain size in the same direction, in the Boreas Subgreywacke Member indicate that hills were exposed to the north of the present Olympus Range. Hence, the topography in the McMurdo Sound region immediately prior to Beacon Supergroup deposition had local irregularities, though hills were well worn (Figure 71.3).

Windy Gully Sandstone

By analogy with the New Mountain Sandstone, which it closely resembles, the Windy Gully Sandstone is interpreted as having been deposited in large braided streams. Some cross-bedding in the formation may have an eolian origin.

Terra Cotta Siltstone

Environment of deposition

The lithologic character of the Terra Cotta Siltstone is consistent with both lacustrine and marginal marine interpretations. Subaqueous and subaerial mudcracks are known from both environments and are thus not diagnostic (Picard and High, 1972). Fine, persistent lamination can also form in both environments.

Chemical evidence such as the presence of abundant hematite (which requires oxidation to form) and $\delta^{13}C$ values, which indicate deposition in fresh water, favors a nonmarine interpretation.

It has been suggested above that the Windy Gully Sandstone and New Mountain Sandstone represent deposition by braided rivers. If so, the Terra Cotta Siltstone is conformably bounded by fluvial deposits, a stratigraphic relationship which also favors a lacustrine interpretation.

The evidence discussed above suggests a continental interpretation, and hence the Terra Cotta Siltstone is interpreted as a lacustrine deposit.

Paleogeography

The limits of lacustrine deposition can be tentatively drawn based on facies changes and extrapolated thinning trends. During field work for this paper, the writer identified the Terra Cotta Siltstone on the south side of Mount Jason. Kohn and MacPherson (*in* Barrett and Webb, 1973, p. 63) report purple shales interbedded with grit on the north side of Mount Jason. Rapid thinning of the Terra Cotta Siltstone northward is indicated. The formation crops out at Mount Heimdall where it is mainly composed of various sandstones (Kohn and Webb, *in* Barrett and Webb, 1973, p. 46). On the east face of Plane Table there is a very clearly exposed lateral facies change from Terra Cotta Siltstone into sandstone. The formation thins from 24 m to less than 1 m in about 500 m laterally. Thin beds of siltstone which may represent the limits of Terra Cotta Siltstone deposition were noted at Mount Coates. In the Ferrar Glacier–Taylor Glacier region there is a westerly thinning trend, and by extrapolation of the trend in the northwest and southwest directions from Table Mountain, tentative boundaries may be drawn, as shown in Figure 71.4. The limits of deposition derived in this way suggest that the lake may have covered an area of 3000 to 4000 km^2.

New Mountain Sandstone

The New Mountain Sandstone is interpreted as having been deposited by braided streams for reasons discussed above. A sequence of vertical facies changes in the formation indicates a regional channel migration, possibly the result of basin filling. Eolian sand dunes migrated in a southeasterly direction across the flood plain. A few large cross-bedding sets, which may be interpreted as eolian on the basis of scale, trend in a westerly direction, possibly the result of seasonal or storm winds.

Boreas Subgreywacke Member

At Mount Boreas coarse beds of the Boreas Subgreywacke Member contain large angular fragments (as much

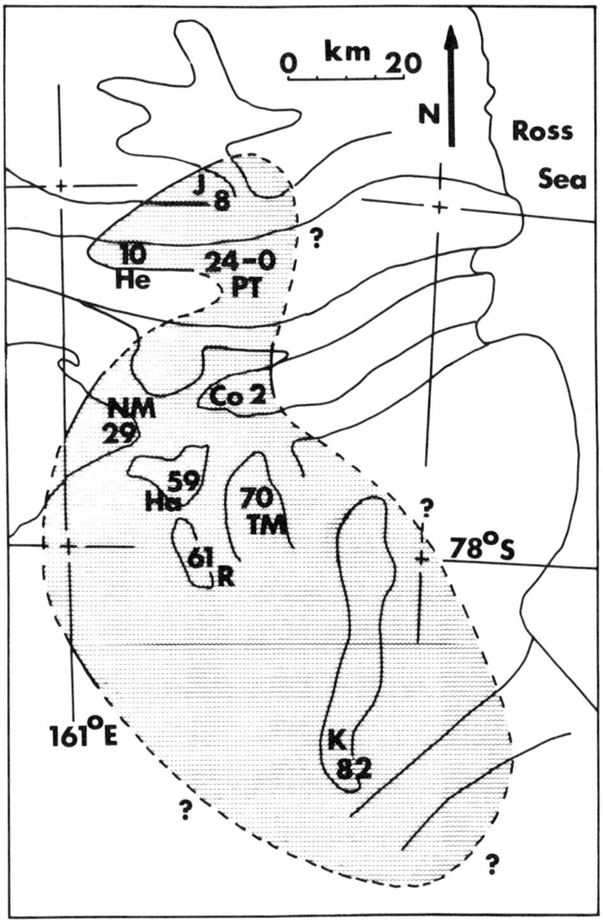

Figure 71.4. Inferred limits of Terra Cotta Siltstone deposition. Solid lines based on extrapolated thinning trends and facies changes. Thicknesses shown are in m.

as 4 cm) of quartz, feldspar, and granite supported by a matrix of fine-grained, well-sorted sand and recrystallized clay. Mudstone clasts were observed both in thin section and hand specimen, and tabular clasts locally show preferred orientation.

The angularity and coarseness of feldspar, granite, and mudstone fragments show that the material has been deposited very close to its source. The open framework fabric and extremely poor sorting suggest that the Boreas Subgreywacke was deposited by a gravity flow mechanism. Because of the large size of included clasts (Middleton and Hampton, 1973), the unit is considered to have been deposited by debris flows.

A granitic source area is implied by the clast lithology, and some topographic relief is indicated by the inferred depositional mechanism. The angularity and freshness of granite and feldspar clasts suggest physical rather than chemical weathering processes. This observation, in conjunction with the cyclic nature of the bedding, suggests that the member was deposited as an alluvial fan during periodic floods, possibly in an arid or semi-arid environment.

Plane Table Member

The Plane Table Member is known only from Plane Table where it is found near the base of the New Mountain Sandstone. It is dominantly a siltstone unit though lenses of sandstone do occur within it. By analogy with the Terra Cotta Siltstone, this member is interpreted as a lacustrine deposit.

Conclusions

The entire New Mountain Subgroup is interpreted as nonmarine for a number of reasons. Subaerial desiccation cracks were observed in all formations of the subgroup, strong evidence in favor of a continental interpretation. Chemical evidence—the presence of hematite and the $\delta^{13}C$ values—also indicates a nonmarine environment of deposition for at least part of the New Mountain Subgroup. Primary current lineation, a common feature of the subgroup, probably does not form in marine environments.

The Terra Cotta Siltstone is interpreted as lacustrine primarily because $\delta^{13}C$ values are strongly nonmarine in character. Sedimentologic (lamination and mudcracks) and biologic (lack of body fossils) characteristics are consistent with this interpretation. The Plane Table Member is considered lacustrine because of its similarity to the Terra Cotta Siltstone.

Parts of the New Mountain Sandstone and Windy Gully Sandstone are interpreted as having been deposited in braided rivers similar to the Brahmaputra River, primarily because the paleocurrent pattern of cross-bedding shows low standard deviation and unimodal distribution. The form of cross-bedding (its thickness, angle of dip, and lateral persistence) is consistent with such an interpretation. Finally, other features of the two formations, such as primary current lineation, scour-and-fill channels, and interbedded thin siltstone beds, are consistent with a braided river interpretation.

The Boreas Subgreywacke Member is interpreted as debris flow deposits because of its extremely poor sorting, closeness to source area, open framework fabric, and the large size of included clasts.

Eolian deposition in the upper part of the New Mountain Sandstone is indicated by the observation that the large-scale cross-bedding trends in a direction independent of the inferred paleoslope.

Acknowledgments. This paper represents part of a thesis submitted at Victoria University of Wellington, New Zealand. Many groups and individuals deserve recognition for their help; without their efforts this work would not have been possible. In particular, the New Zealand Antarctic Division and the U.S. Navy deserve special thanks, along with Peter Barrett for his invaluable assistance. Peter Blattner and M. A. Cox of the Institute of Nuclear Sciences, Gracefield, New Zealand are thanked for their assistance in the carbon and oxygen isotope analyses. The companionship and help of the members of VUWAE 18 and 19 are gratefully acknowledged.

References cited

Allen, J. R. L. 1964. Primary current lineation in the Lower Old Red Sandstone (Devonian), Anglo-Welsh Basin. *Sedimentology, 3,* 89–108.

Barrett, P. J., and Kohn, B. P. 1975. Changing sediment transport directions from Devonian to Triassic in the Beacon Supergroup of south Victoria Land, Antarctica. (*In* Campbell, K. S. W., *ed. Gondwana Geology.* Canberra, Australia National University Press, 15–35.)

Barrett, P. J., and Webb, P. N., *eds.* 1973. *Stratigraphic Sections of the Beacon Supergroup (Devonian and Older(?) to Jurassic) in South Victoria Land.* Antarc. Data Ser. Victoria Univ., Wellington, 3, 165 pp.

Coleman, J. M. 1969. Brahmaputra River: channel processes and sedimentation. *Sedim. Geol., 3,* 129–239.

Conybeare, C. E. B., and Crook, K. A. W. 1968. Manual of sedimentary structures, Canberra, Australia. *Bull. Miner. Resour. Geol. Geophys. Aust., 102,* 327 pp.

Folk, R. L., and Ward, W. C. 1957. Brazos River bar: a study in the significance of grain size parameters. *J. sedim. Petrol., 27,* 3–26.

Freeman, W. E., and Visher, G. S. 1975. Stratigraphic analysis of the Navajo Sandstone. *J. sedim. Petrol., 45,* 651–58.

Gevers, T. W., Frakes, L. A., Edwards, L. N., and Marzolf, J. E. 1971. Trace fossils in the Lower Beacon sediments (Devonian), Darwin Mountains, southern Victoria Land, Antarctica. *J. Paleont., 45,* 81–94.

Hantzschel, W. 1962. Trace fossils and problematica. (*In* Moore, R. C., *ed. Treatise on Invertebrate Paleontology, Part W, Miscellanea.* Boulder and Lawrence, Geological Society of America and University of Kansas, W177–W245.)

Houbolt, J. J. H. C. 1968. Recent sediments in the southern bight of the North Sea. *Geologie Mijnb., 47,* 245–73.

Keith, M. L., and Weber, J. N. 1964. Carbon and oxygen isotopic composition of selected limestones and fossils. *Geochim. cosmochim. Acta, 28,* 1787–1816.

Kyle, R. A. 1977. Devonian palynomorphs from the Basal Beacon Supergroup of South Victoria Land, Antarctica (note). *N.Z. Jl Geol. Geophys., 20,* No. 6, 1147–50.

Matz, D. B. 1968. *Stratigraphy and Petrology of the Beacon Sandstone, Southern Victoria Land, Antarctica.* Master's thesis, University of Massachusetts, 110 pp.

McCave, I. N. 1971. Sand waves in the North Sea off the coast. *Mar. Geol., 10,* 199–225.

McKee, E. D. 1966. Structures of dunes at White Sands National Monument, New Mexico (and comparison with structures of dunes from other selected areas). *Sedimentology, 7,* 1–69.

McKee, E. D., and Weir, G. W. 1953. Terminology for stratification and cross-stratification in sedimentary rocks. *Bull. geol. Soc. Am., 64,* 381.

MacPherson, J. G. 1975. *The Aztec Siltstone: An Upper Devonian Alluvial Plain Red Bed Sequence, Southern Victoria Land, Antarctica.* Ph.D. dissertation, Victoria University of Wellington, New Zealand.

Middleton, G. V., and Hampton, M. A. 1973. Sediment gravity flows: mechanics of flow and deposition. (*In* Middleton, G. V., and Bouma, A. H. [chairmen], Turbidites and Deep Water sedimentation. Short Course, *Soc. econ. Paleont. Mineral.,* 1–13.)

Off, T. 1963. Rhythmic linear sand bodies caused by tidal currents. *Bull. Am. Ass. Petrol. Geol., 47,* 324–41.

Picard, M. D., and High, L. R. 1972. Criteria for recognizing lacustrine rocks. (*In* Rigby, J. K., and Hamblin, W. K., *eds.* Recognition of Ancient Sedimentary Environments. *Spec. Publs Soc. econ. Paleont. Miner., Tulsa,* No. 16, 108–45.)

Plume, R. W. 1974. The stratigraphy, petrology and sedimentology of the New Mountain Sandstone at Plane Table, Asgard Range, South Victoria Land, Antarctica. Unpublished work, Victoria University of Wellington, New Zealand, 53 pp.

Plume, R. W. 1978. Proposed changes to the existing stratigraphy of the New Mountain Sandstone (Beacon Supergroup), South Victoria Land, Antarctica. *N.Z. Jl Geol. Geophys., 21,* No. 2, 167–73.

Reyment, R. 1981. *Introduction to Quantitative Paleoecology.* Amsterdam, Elsevier, 226 pp.

Schumm, S. A. 1972. Fluvial paleochannels. (*In* Rigby, J. K., and Hamblin, W. K., *eds.* Recognition of Ancient Sedimentary Environments. *Spec. Publs Soc. econ. Paleont. Miner., Tulsa,* No. 16, 98–107.)

Tanner, W. F. 1967. Ripple mark indices and their uses. *Sedimentology, 9,* 89–104.

Vialov, O. S. 1962. Problematica of the Beacon Sandstone at Beacon Heights West, Antarctica. *N.Z. Jl Geol. Geophys., 5,* 718–32.

Watson, G. S. 1966. The statistics of orientation data. *J. Geol., 74,* 786–97.

Webb, P. N. 1963. Geological investigations in southern Victoria Land, Antarctica. Part 4. Beacon Group of the Wright Valley and Taylor Glacier region. *N.Z. Jl Geol. Geophys., 6,* 361–87.

72

Petrology and Petrography of the Beacon Supergroup, Shackleton Glacier Area, Queen Maud Range, Transantarctic Mountains, Antarctica

KERBY E. LA PRADE
Department of Earth Sciences, East Texas State University, Commerce, Texas 75428, U.S.A.

Abstract. A total of 203 thin sections from specimens of Beacon Supergroup rocks, a 1425 m thick sequence in the Shackleton Glacier area (lat. 85°/S, long. 176°/W), were studied for composition and grain parameters by counting 500 points and measuring 500 grains per thin section. Seven rock specimens were friable enough to permit dis-association of grains for a typical sieve and binocular microscope analysis. The rocks are now graywacke varieties, having been diagenetically derived form parent arkoses. Regression analysis of composition and grain size shows that quartz content decreases with decreasing grain size, and feldspar increases with decreasing grain size in the Pagoda and Fairchild formations but decreases with decreasing grain size in the Mackellar, Buckley, and Fremouw-Falla Formations. Clay increases with decreasing grain sizes in all formations. Nearly all clay is secondary, derived from hydrothermal alteration of feldspars. The most common accessory minerals are biotite, muscovite, and zircon; the clay minerals are by far the most abundant secondary minerals. Beacon sedimentary rocks fall mainly within the fine-size grades, and they are nearly all moderately well sorted and slightly skewed toward the fine-grain sizes. The tails (above and below the 16 and 84 percentile levels on the cumulative frequency curves) of the sediments are better sorted than are the central parts. Contour maps of statistical parameters for the Fairchild and Fremouw-Falla were drawn from some grain size parameters in an effort to determine the paleoslopes. Results of this method do not show much agreement with paleoslopes inferred from sedimentary and glacial structures, but these maps are based on only a few control points in a small area, and thus their validity is considerably suspect.

Clastic sedimentary rocks of the Beacon Supergroup (Figure 72.1) are exposed in a 1425 m thick sequence along Shackleton Glacier and its tributaries (lat. 85° S, long. 176° W) in the Queen Maud Range of the Transantarctic Mountains. Geologic field work for this study was conducted in the Shackleton Glacier area by field parties from Texas Tech University during the austral summers of 1962-63, 1964-65, and 1968-69.

Synopsis of stratigraphy

Beacon Supergroup sedimentary rocks and their included diabase intrusive sheets crop out along the full length of the Transantarctic Mountains. These clastic sedimentary rocks are of mainly fluviatile origin, as shown by abundant land plant fossils, coal deposits, cross-bedding, current ripplemarks, mudcracks, discontinuity of beds, and the absence of marine fossils. Individual beds within the Beacon sequence are numerous, typically wedging out laterally in short distances and intertonguing complexly with other facies to give the unit a heterogeneous but distinctive character. There is no problem in recognizing Beacon rocks anywhere along the Transantarctic Mountains, but subdividing the sequence and correlating individual units of formations is difficult and subjective. Thus regional, and even local, correlations are difficult to establish.

The Beacon Supergroup includes many names for correlative formations, because each investigator working in isolated areas of the Transantarctic Mountains has assigned his own names to the local Beacon sequence. Table 72.1 shows the correlation of Beacon strata along the central Transantarctic Mountains.

Petrology and petrography

A total of 203 representative specimens of Beacon rocks were selected for thin section analysis of composition and grain parameters. Seven specimens were disaggregated and analyzed by the conventional sieving method.

Composition

Mineralogic composition of each specimen was determined by counting 500 points per thin section, and these compositional data were used to identify rock types. Classification of terrigenous rocks generally utilizes a triangular diagram. The writer has used the pole components of Pettijohn (in Krumbein and Sloss, 1963, p. 157), and the eight rock types with the percentage values selected for their boundaries of Folk (1954). Thus, the classification scheme used here is based on both texture and mineralogy rather than solely on mineralogy as in Folk's classification. Nearly all Beacon rocks can be classified as some sort of graywacke (Figure 72.2). Compositions of Beacon sedimentary rocks were determined for the present-day rock, which is generally very different than the original rock. Nearly all feldspars are highly altered to clay. The clay component is now large, but the rocks were probably clay poor. Thus, the graywackes of the Beacon Supergroup are of secondary or diagenetic origin; the original rocks were a

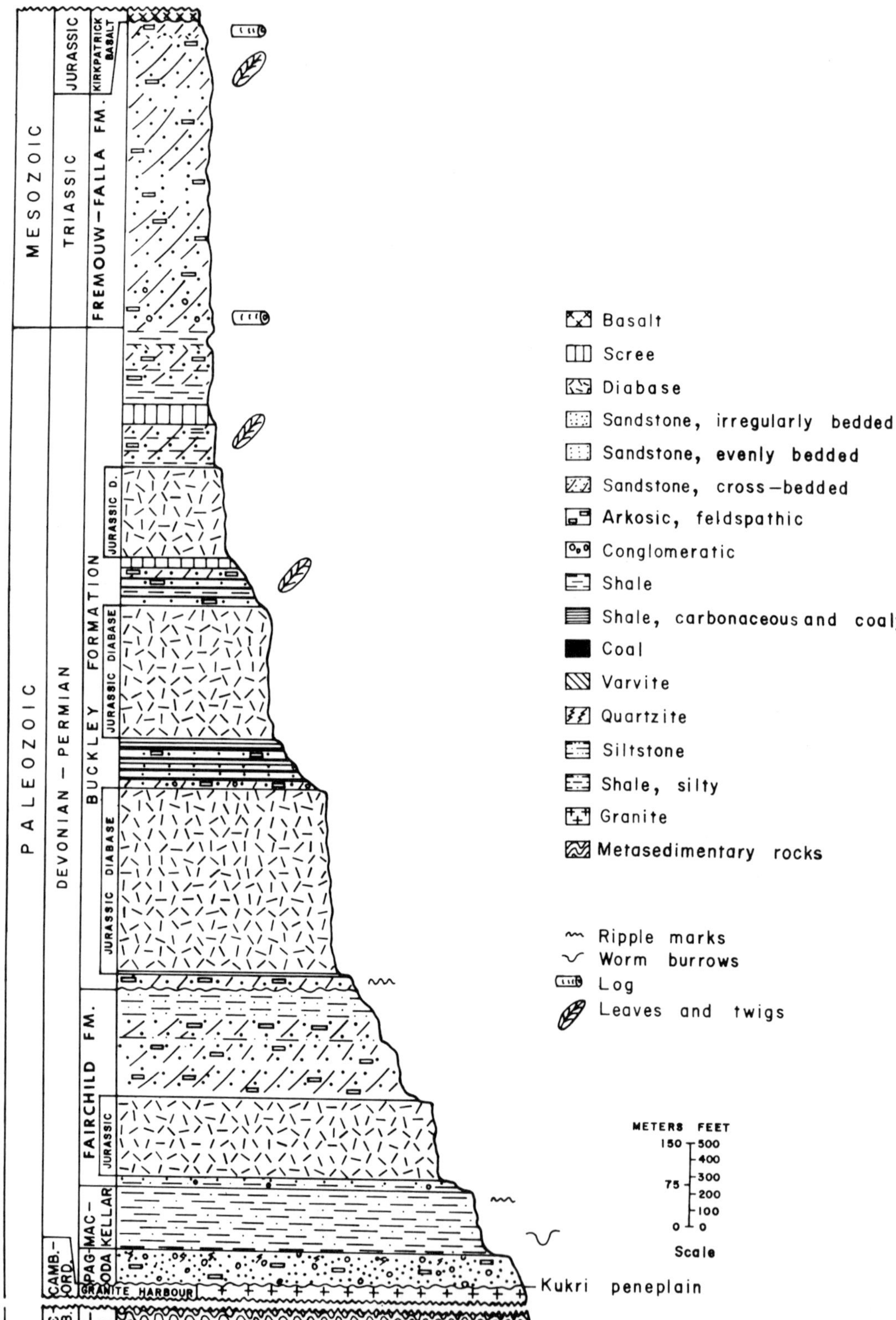

Figure 72.1. Generalized geologic column for the Shackleton Glacier area, Queen Maud Range, Antarctica.

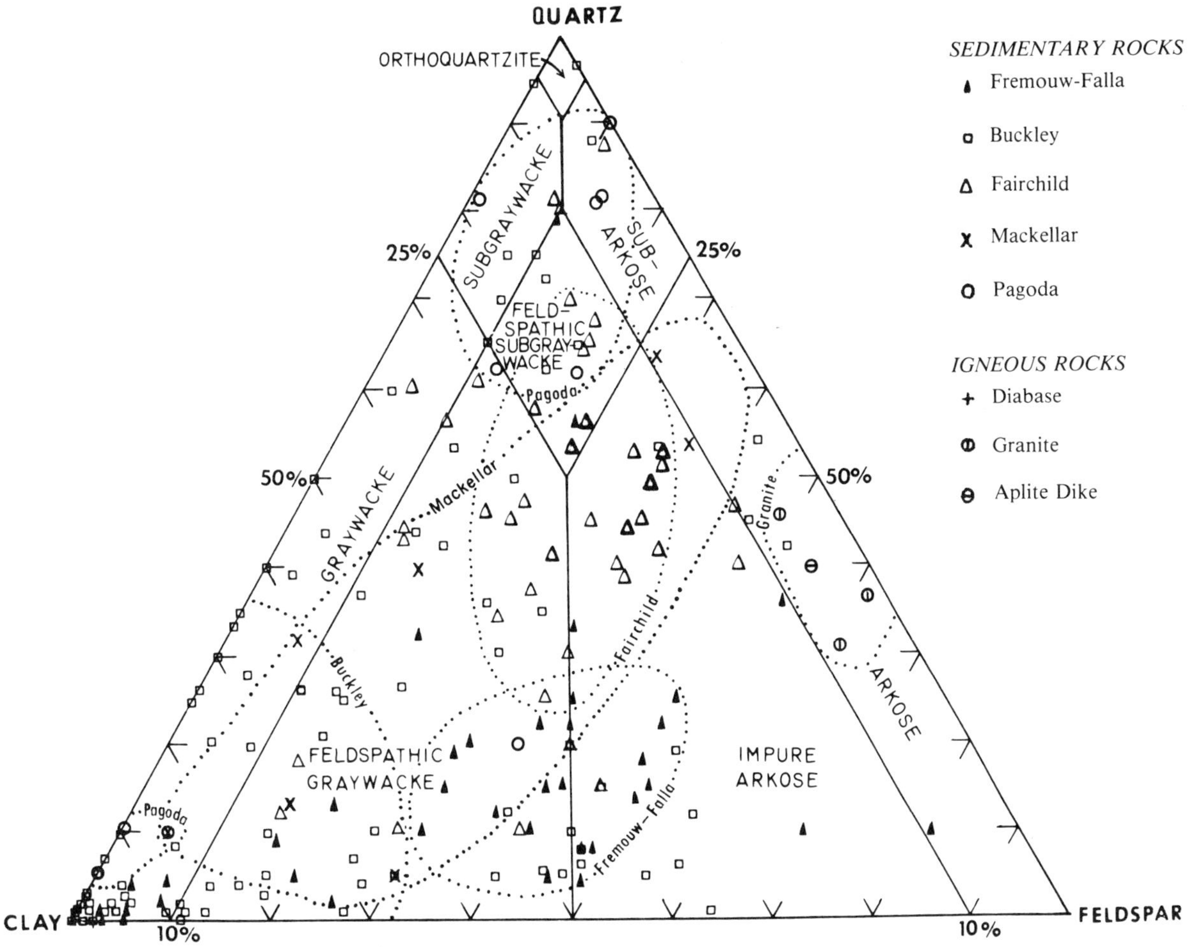

Figure 72.2. Triangular diagram showing compositions of Beacon and associated rocks in the Shackleton Glacier area. Rock classification after Folk (1954). Quartz, feldspar, and clay recalculated to 100% where other components are present.

Table 72.1. Correlation of Beacon Supergroup rocks in the Transantarctic Mountains

				Shackleton Glacier area			
	Ohio Range	Robert Scott Glacier	Mount Fridtjof Nansen	Mount Wade	Cape Surprise	Cumulus Hills	Beardmore Glacier
Triassic	Long, 1959, 1962*a*, *b*, 1964, 1965; Minshew, 1966, 1967	Minshew and Summerson, 1964; Doumani and Minshew, 1965; Minshew, 1966, 1967		Barrett, 1965, 1966		This paper; La Prade, 1970 1972	Grindley, 1963; Barrett and others, 1967; Barrett, 1969, 1972
Triassic						Fremouw Falla[a] Formation 563 + m	Falla Formation 160–530m Fremouw Formation ~ 650 m
Permian	Mount Glossopteris Formation: Upper Coal Measures 551 m	Queen Maud Formation (unit E) 433 + m	(Unit C) 259 + m	(Unit C) 253 + m	(Unit C) 259 + m	Buckley Formation 467 m	Buckley Formation ~750 m
Permian	Mount Glossopteris Formation: Lower Sandstone 150 m	Weaver Formation 185 m: Burrowed sandstone (unit D) Platy sandstone and shale (unit C)	(Unit B) 195 m	(Unit B) 253 m	(Unit B) 137 m	Fairchild[b] Formation 214 m	Fairchild Formation 130–220 m
Permian	Discovery Ridge Formation 166–195 m	Weaver Formation 185 m: Dark shale (unit B) 15 m	Mackellar Formation (unit A) 92 m	Mackellar Formation (unit A) 92 m	Mackellar Formation (unit A) 137 m	Mackellar Formation 117 m	Mackellar Formation 60–140 m
Permian	Buckeye Tillite 250–350 m	Scott Glacier Formation 0–18 m	Pagoda Formation 0–3 m	Pagoda Formation Basal pebbly beds 0–3 m	Pagoda Formation 0–15 m	Pagoda Formation 0–61.3 m	Pagoda Formation 125–295 m
Low Dev.	Horlick Formation 0–46 m		Metamorphic-granitic basement				Alexandra Formation 0–320 m
Camb.-Ord.	Granitic basement						Granitic basement

[a] Formerly Mount Kenyon Formation (La Prade, 1970, 1972).

[b] Formerly Mount Butters Formation (La Prade, 1970, 1972).

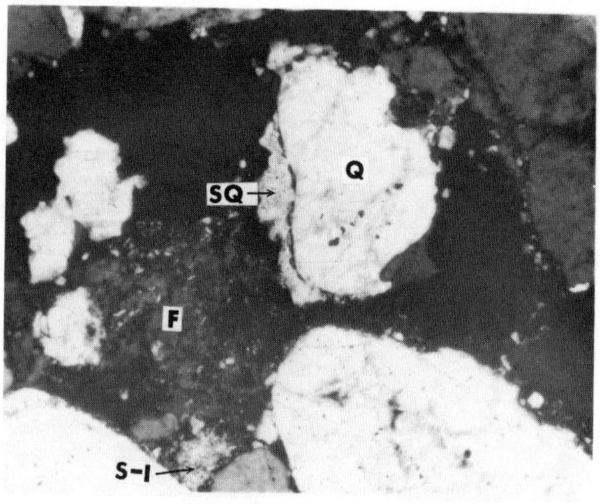

Figure 72.3. Photomicrograph of typical Fremouw-Falla Formation thin section of specimen 11-1 showing secondary quartz overgrowth (SQ) on primary quartz (Q). Sericite-illite (S-I) is produced by alteration of feldspar (F). Crossed nicols, ×75.

type of arkose which was low in primary matrix material (perhaps <5%).

Quartz

Throughout the Beacon Sequence, detrital quartz grains typically show trains of bubble inclusions and wavy extinctions. Much of the quartz is secondary, as indicated by relict outlines of grains (Figure 72.3), sutured contacts with local penetration of quartz by feldspar grains, and the occurrence in at least one specimen of hexagonal outlines of quartz crystals in secondary growth areas. Sutured contacts are commonly completely welded, as in quartzites. Considerable quartz has formed by *in situ* replacement of feldspar. In some stratigraphic sections, especially in the Buckley Formation, the original arkosic rocks have been nearly completely altered by replacement of most of the feldspar with chert, while sericite-illite-kaolinite has replaced the remaining feldspar.

Grain-to-grain contact varies, depending on the extent of the alteration of feldspar to clay. Where the quartz content is high (>20%), grain-to-grain contact is common, but where the feldspar content was high in the original rocks, its alteration to clay has been so pervasive that little grain-to-grain contact remains. Quartz regression lines (Figure 72.4) for Beacon formations show that the quartz content decreases with decreasing grain size throughout the Beacon Supergroup. This reflects the increasing dominance of clay in the progressively finer size grades. It should be noted that the statistical confidence level is low for those regression lines based on very few samples. The same qualifying statement applies also to the feldspar and clay regression lines discussed below.

Feldspar

The feldspars consist of both potassium and plagioclase varieties. The composition of the plagioclase feldspars was determined from maximum extinction angles to twinning. The maximum extinction angles, when plotted against Michel Levy's curve, indicate that the plagioclase feldspars are oligoclase and primarily andesine, having an anorthite content from 30 to 40%. X-ray diffraction traces were obtained from twelve specimens; they show the presence of oligoclase and orthoclase.

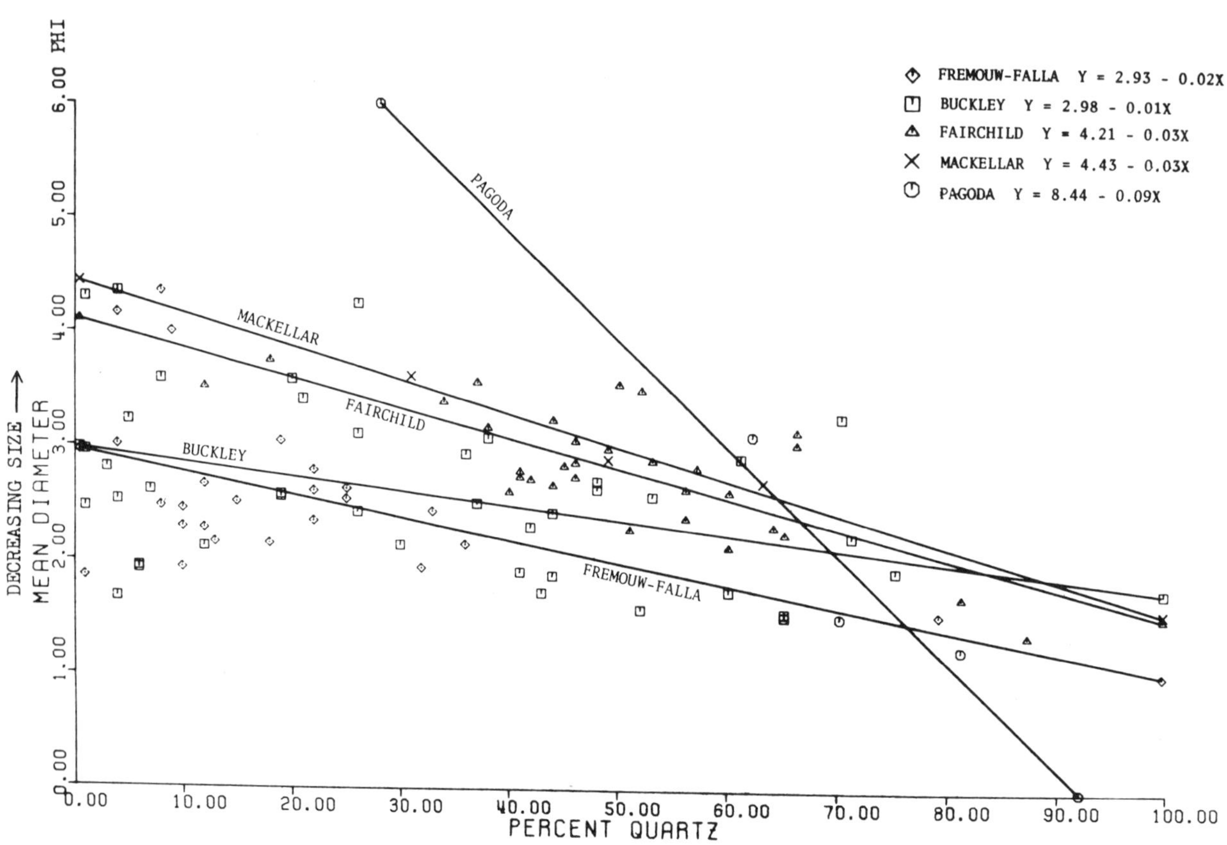

Figure 72.4. Computer drawn scatter plot of percent quartz against mean diameter (M_ϕ).

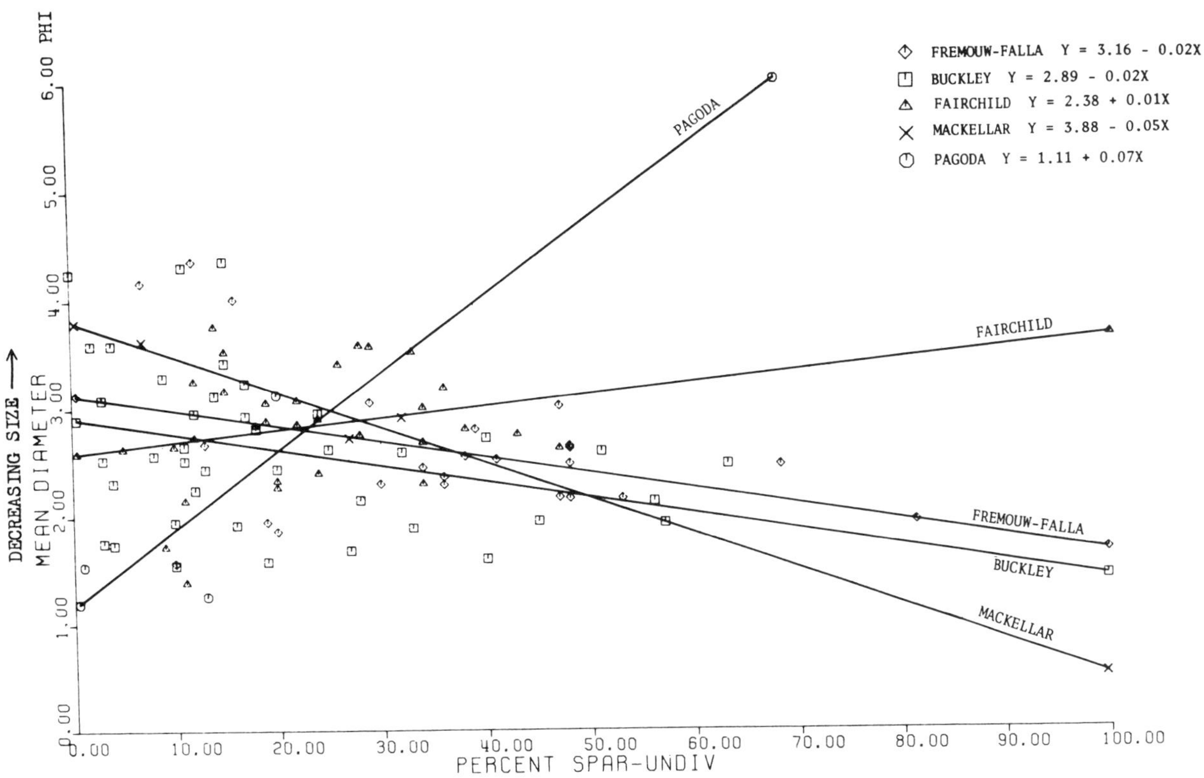

Figure 72.5. Computer drawn scatter plot of percent feldspar content against mean diameter (M_ϕ).

Feldspar regression lines (Figure 72.5) indicate that in the Pagoda and Fairchild Formations, the feldspar content increases with decreasing grain size. However, in the Mackellar, Buckley, and Fremouw-Falla Formations the opposite relationship exists, with the feldspar content decreasing with decreasing grain size. The increasing feldspar content with decreasing grain size may seem anomalous since it may be expected that as clastic sediments are cycled, feldspar would drop out because of decomposition; hence, finer and finer sediments, presumably the most recycled, should show a decrease in feldspar. Increasing feldspar in progressively finer grade sizes may result from very limited cycling and deposition under cool climatic conditions where decomposition was very slow. Widespread Buckeye Tillite, a Pagoda correlative (Table 72.1), followed by *Glossopteris* fossils throughout the Buckley Formation indicate that the climate ranged from polar in Early Permian time to temperate in Late Permian time. Continuation of temperate climate into Triassic time is indicated by petrified conifer-type trees. It is believed that decomposition of the feldspars is postdepositional, having occurred during and after emplacement of the Jurassic diabase sills. Because of the excellent cleavage of feldspar, it easily disintegrates, and thus it became increasingly concentrated in the finer and finer sediment fractions.

Clays

Clays form the matrix material of these clastic Beacon rocks and are commonly the dominant component, with increasing amounts occurring in progressively decreasing grade sizes (Figure 72.6). Clays in Beacon rocks, determined from X-ray diffraction, consist dominantly of sericite, illite, kaolinite, and chlorite. The diagenetic origin of much, perhaps nearly all, of the matrix clay material in Beacon rocks is obvious from many thin sections, where parts of feldspar grains can be seen to have been partly or entirely altered to clay minerals. It is common for sericitization to be more intense along alternate twinning planes of plagioclases. Most of the feldspar grains have been altered so much that grain boundaries are generally difficult to distinguish from matrix. Relict outlines of completely altered feldspar grains disclose the alteration that has occurred, as does relict twinning of plagioclase in the clay (Figure 72.7). Barrett (1966) studied secondary minerals in Beacon sedimentary rocks in the vicinity and likewise concluded that the clay (sericite) was postdepositional, based upon the radiating habit of sheaflike aggregates and upon the presence of sericite within grains of altered plagioclase; in one or two cases, the sericite extended from the plagioclases into the matrix. He identified some of the sericite in sandstones and coarse shales as detrital if the flakes were much longer than normal secondary sericite and were subparallel to the bedding, but found the distinction difficult. This writer encountered the same problem, except where the clay occurred on feldspar crystals, or relict outlines of feldspar crystals could be seen in the clay.

Typical graywacke sandstones are mainly unaltered rock debris, and thus are thought to have formed under cool climatic conditions near their source. This seems to have been the situation with the Beacon rocks. It is reasonable to infer that the rocks (originally arkoses) remained relatively unaltered until emplacement of diabase sills in Jurassic time. After these intrusions, circulation of warm ground water through the rocks caused rapid decay of the feldspars.

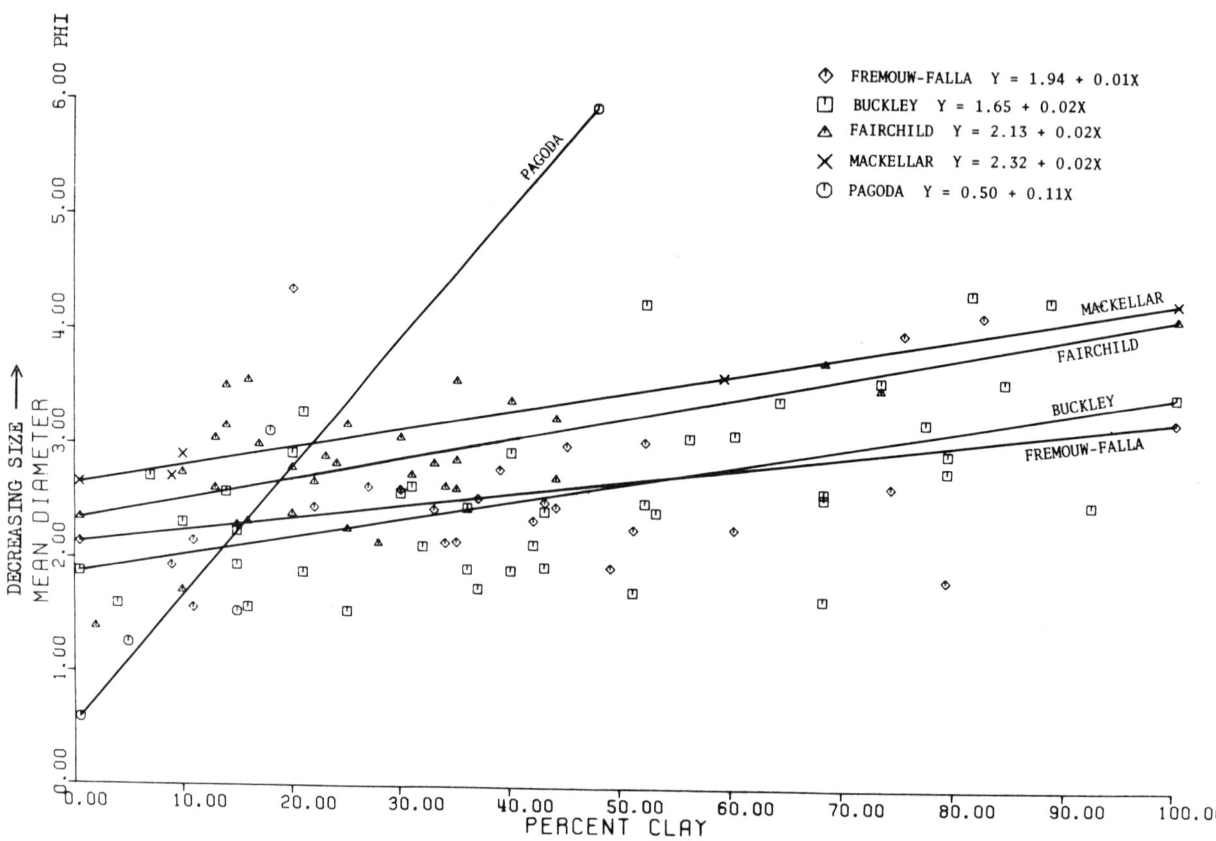

Figure 72.6. Computer drawn scatter plot of percent clay content against mean diameter (M_ϕ).

Accessory minerals

A cursory examination for accessory minerals in Beacon rocks was made. No accessory minerals at all were noted in many specimens, possibly because the low-power objective (×10) was used, plus many of the commercially prepared thin sections were too thick for detailed examination. The most common accessory minerals are biotite, muscovite, and zircon, but other minerals present in some sections include chlorite, garnet, hematite, hornblende, ilmenite, epidote, magnetite-chromite, sillimanite, tourmaline, and tremolite-actinolite. Zircon (white) occurs as rounded grains or as inclusions in quartz grains. Tourmaline and sillimanite (needlelike) occur as inclusions in quartz, the biotite is bleached, and the garnets include one or more of the red varieties.

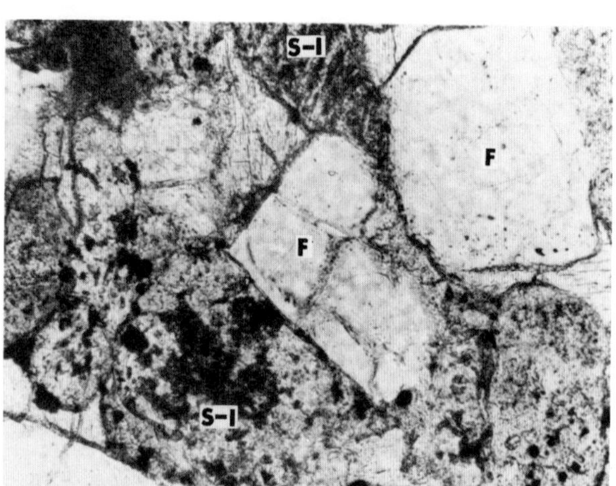

Figure 72.7. Photomicrograph of typical Fremouw-Falla Formation thin section of specimen 10-20 showing intense alteration of feldspar (F) to sericite-illite (S-I). Crossed nicols, ×75.

All Beacon formations have about the same suite of accessory minerals, suggesting the probability of a common detrital source. Barrett (1966) postulated a different source for Buckley sediments because here he found that potassium feldspar is rare and zircon is more common than in other Beacon formations. The writer did not note any general decrease in potassium feldspars upward in the section, except in those samples which were so highly altered that all the feldspar was reduced to clay; nor was any variance in the abundance of zircon noted.

This heavy mineral suite and the rounded zircon grains suggest that Beacon sediments were derived from metamorphic and sedimentary parent rocks (Krumbein and Sloss, 1963, p. 140). Though the zircon is rounded, it is easy to distinguish the double pyramids, indicating limited reworking of the sedimentary parent rocks, possibly only one cycle. The light detrital minerals, consisting dominantly of quartz and feldspar (potassic and plagioclase), suggest a nearby source rock of silicic to intermediate igneous type. Therefore, silicic to intermediate igneous, metamorphic, and sedimentary rocks, may have been exposed in the source area and may have all contributed simultaneously to the Beacon sediments.

Secondary minerals

Secondary minerals in Beacon sedimentary rocks include kaolinite, illite, sericite, calcite, chert, quartz, plagioclase, epidote, graphite, limonite, prehnite, zeolite, and zoisite-clinozoisite. The clay minerals, mainly sericite-illite, are by far the most abundant secondary minerals in Beacon rocks, being ubiquitous and representing a major component of the rocks.

The most unusual secondary mineral noted was plagioclase, which occurred as microscopic veinlets in a few specimens. Maximum extinction angle for the twinning is about 18°, which indicates andesine according to the Michel Levy curve.

Composition of individual formations

Pagoda Formation

Based on eight specimens, the present Pagoda rocks have a composition typical of feldspathic subgraywackes (Folk, 1954) (Figure 72.2): 41.1% quartz, 11.1% feldspar, and 39.8% clay, with a quartz/feldspar ratio of 1:0.27. These percentages are averages for the total grain size range, but the highest percentages of all components lie mainly between the 1.5ϕ and 3.5ϕ size grades. Quartz is the dominant component through all size grades; feldspar and clay are important constituents in size grades smaller than 1.0ϕ.

Mackellar Formation

Analyses of seven specimens, including two analyses (numbers 53 and 65) published by Barrett (1966), show the formation to consist of 30.1% quartz, 18.7% feldspar, and 49.0% clay, with a quartz/feldspar ratio of 1:0.62, making the rocks feldspathic graywackes (Folk, 1954) (Figure 72.2). The highest percentages of all major components are in the 2ϕ to 4ϕ size grades. Quartz is the dominant component through all size grades coarser than 3.5ϕ, but in grades finer than 3.5ϕ, feldspar and clay are dominant.

Fairchild Formation

Analyses of 37 thin sections of Fairchild rocks, plus published data by Barrett (1966) for 5 additional thin sections of Fairchild rocks (numbers 69, 71, 80, 82, and 84), show the formation to consist of 44.3% quartz, 25.8% feldspar, and 29.4% clay, and to have a quartz/feldspar ratio of 1:0.58. This composition identifies the rocks as feldspathic graywackes (Folk, 1954) (Figure 72.2). The highest percentages of all major components lie between 1.0ϕ and 4.0ϕ. Quartz is the dominant component in size grades coarser than about 4.0ϕ, and in grades finer than 4.0ϕ, feldspar and clay are dominant.

Buckley Formation

Analyses of 108 thin sections of Buckley rocks, plus published data on 3 specimens (numbers 58, 61, and 75) by Barrett (1966), show 18.3% quartz, 14.3% feldspar, and 53.3% clay, with a quartz/feldspar ratio of 1:0.78. These rocks are also feldspathic graywackes (Folk, 1954) (Figure 72.2). The highest percentages of major components lie between 1.0ϕ and 4.0ϕ. Quartz is dominant in grades coarser than about 0.5ϕ, but in grades finer than 0.5ϕ, clay is dominant over both quartz and feldspar. Feldspar is dominant over quartz in grades finer than about 3.5ϕ.

Fremouw-Falla Formation

Analyses of 42 thin sections of Fremouw-Falla rocks were made for composition, showing 14.3% quartz, 29.4% feldspar, and 54.2% clay. The quartz/feldspar ratio is 1:2.06. These rocks also fall into the feldspathic graywacke classification (Folk, 1954) (Figure 72.2). Most of the sediments lie within the 0.5ϕ to 4.5ϕ size grades. Quartz is not the dominant constituent as it is in other Beacon formations, being subordinate here to feldspar and clay. Feldspar is the dominant component between 0.5ϕ and about 2.0ϕ, and clay becomes dominant in grades finer than about 2.0ϕ.

Grain Analysis

Beacon rocks are mainly well indurated, requiring that they be studied in thin sections. Seven specimens were friable enough to permit disaggregation so that heavy and light fractions could be separated in the heavy liquid bromoform (density 2.89). Mechanical sieve analyses of these samples served to check the accuracy of the thin section studies. Grain size cumulative curves drawn from sieve data were compared with curves drawn from thin section data of the same rock, which had been adjusted in a manner described by Friedman (1962) so that thin section grain measurements are made comparable to sieve measurements. However, in this study, the cumulative curves drawn from unadjusted thin section data correlated more closely with the sieve data curves than did the adjusted curves. For this reason, the unadjusted thin section data were used in this study. The writer can offer no explanation for why the Friedman method did not work satisfactorily in this study.

Grain parameters were obtained by measuring diameters of 500 grains, using a counting stage graduated to move 0.1 mm intervals. Cumulative curves were drawn and various descriptive grain parameters were determined from them in a manner described by Inman (1952). Parameters include median diameter (Md_ϕ), mean diameter (M_ϕ), sorting coefficient (σ_ϕ), primary skewness (α), second skewness (α_2), and kurtosis (β); definitions of these terms are given by Inman (1952). Folk and Ward (1957) modified Inman's statistical equations to include more points on the curve; this is an improvement for working with bimodal and strongly skewed sediments. The Inman equations, however, seemed adequate for this study. However, Folk's (1974) so-called "verbal" or descriptive limits for statistical parameters (i.e., sorting coefficient 0.35–0.50ϕ = well sorted, etc.) have been used. All formations of the Beacon Supergroup show rather similar statistical parameters, and the average values are given in Table 72.2.

An attempt was made to determine the direction of sediment transport from grain parameters. Contour maps were drawn of various statistical parameters for the Fairchild and Fremouw-Falla Formations to determine if such maps would indicate a paleoslope that would agree or disagree with the paleoslope determined from sedimentary and glacial structures for the Shackleton area (Barrett, 1965) and for the Beardmore area (Barrett, 1970, 1972). It was not expected that completely reliable conclusions could be drawn from the maps because of the limited number of control points and the small area of study. These maps are not given here because of space limitations.

Sediments generally become finer and better sorted in the central part of the grain size distribution curve (between 16–84 percentiles) the farther they are transported from their source, most commonly in the downslope direc-

Table 72.2. Average statistical parameters with their "verbal" descriptions as defined by Folk (1974)

	Median diameter (Md_ϕ)	Mean diameter (M_ϕ)	Sorting coefficient (σ_ϕ)	Primary skewness (α)	Second skewness (α_2)	Kurtosis (peakedness (β)
Fremouw-Falla Formation	2.56ϕ fine sand	2.62ϕ fine sand	0.62ϕ moderately well sorted	0.09 nearly symmetrical	0.26	0.70 platykurtic
Buckley Formation	2.52ϕ fine sand	2.60ϕ fine sand	0.82ϕ moderately well sorted	0.09 nearly symmetrical	0.20	0.76 platykurtic
Fairchild Formation	2.70ϕ fine sand	2.83ϕ fine sand	0.69ϕ moderately well sorted	0.17 fine skewed	0.51	0.82 platykurtic
Mackellar Formation	2.87ϕ fine sand	3.08ϕ very fine sand	0.66ϕ moderately well sorted	0.29 fine skewed	0.69	0.96 mesokurtic
Pagoda Formation	1.92ϕ medium sand	1.97ϕ medium sand	0.57ϕ moderately well sorted	0.10 fine skewed	0.39	0.80 platykurtic

tion. Friedman (1961, 1967) points out that stream deposits are positively skewed because their agent of transport is unidirectional and has limited competency in transporting coarse fractions. But since no such limit affects transport of fine particles, a tail is lacking at the coarse end of the distribution curve. Indeed, the absence of the coarse tail may be more pronounced in a downstream direction as the suspended load becomes greater and traction load less with the decrease in the energy of transport. Thus, increasing positive skewness of fluvial deposits may be used to determine paleoslope direction. Kurtosis likewise is expected to become greater in the downstream direction as the sorting of the fine particles decreases relative to the sands in the central part of the grain size distribution curve. The composition of the sediments should also change in the downstream direction and thus reflect the paleoslope, with feldspar content decreasing and quartz increasing.

The paleoslope directions obtained by mapping of grain parameters and composition are summarized in Table 72.3 and are compared with published paleoslope determinations based on sedimentary and glacial structures. Five of the six grain and compositional paleoslope indicators for the Fairchild Formation suggest a slope to the northwest or west. However, cross-beds in the Shackleton Glacier area suggest a paleoslope to the southeast (Barrett, 1965) as do cross-bedding, parting lineations, distribution of plant debris, and rounded quartz pebbles in the Beardmore area (Barrett, 1972). Paleoslope indicators for the Fremouw-Falla Formation show too much variation to be considered definitive: three grain and compositional indicators suggest some component of westward paleoslope, the other three suggest some component of an eastward paleoslope. Barrett (1972) has determined from sedimentary structures that the paleoslope was to the northwest in Fremouw and Falla time in the Beardmore area, exactly opposite to his paleoslope in the older underlying formations.

Summary and conclusions

Beacon rocks are classified as varieties of graywacke, using a triangular diagram with quartz, feldspar, and clay at the poles and the subdivisions of Folk (1954). The rocks are of diagenetic origin, derived from original arkoses low in clay content. The average composition of the Pagoda, Mackellar, Fairchild, Buckley, and Fremouw-Falla Formations is that of feldspathic graywackes, though indi-

Table 72.3. Paleoslopes suggested by mapping of grain parameters and composition, compared with paleoslopes indicated by sedimentary and glacial structures

Formation	Mean Diameter (m_ϕ)	Sorting Coefficients (σ_ϕ)	Skewness (α)	Kurtosis (β)	Quartz Content	Feldspar Content	Sedimentary and Glacial Structures Shackleton Glacier Area (Barrett, 1965); Beardmore Glacier Area (Barrett, 1970, 1972; Lindsay, 1969)
Fremouw-Falla	↘	↑	←	↓	↖	↑	↖
Buckley[a]	—	—	—	—	—	—	↘
Fairchild	↓	←	↑	↖	↖	↖	↘
Mackellar[b]	—	—	—	—	—	—	↘
Pagoda[b]	—	—	—	—	—	—	↘

Arrows indicate compass direction of paleoslope.
[a] No maps were compiled.
[b] Insufficient data to compile maps.

vidual specimens may be other varieties of graywackes or arkoses (Figure 72.2). Quartz content decreases with decreasing grain sizes in Beacon sediments, and there is increasing dominance of clay in the finer grades. Potassic feldspars are dominant over plagioclase. The potassic feldspars are believed to be mainly microcline, and extinction angles of plagioclase twins indicate that they are andesine (mainly) and oligoclase, having an anorthite content from about 30 to 40%

The feldspar content increases with decreasing grain sizes in the Pagoda and Fairchild Formations, but feldspar decreases with decreasing grain size in the Mackellar, Buckley, and Fremouw-Falla Formations. The anomalous increase of feldspar content in decreasing grain sizes is believed to reflect limited cycling of sediments and perhaps deposition in cool climates where decomposition of feldspar was slow. The excellent cleavage of feldspars caused it to be increasingly concentrated in finer size grades. Clay content increases in the finer grain sizes in all Beacon formations. Clays are dominantly sericite-illite, but kaolinite and chlorite are present. Nearly all clay is secondary, produced by hydrothermal alteration of feldspars during and after emplacement of thick diabasic sills (Jurassic). The most common accessory minerals are biotite, muscovite, and zircon, but chlorite, sillimanite, tourmaline, and tremolite-actinolite are present in traces. The accessory minerals suggest parent rocks of metamorphic, sedimentary, and intermediate igneous types (Krumbein and Sloss, 1963, p. 140). The clay minerals, mainly sericite-illite, are the dominant secondary minerals, but calcite, chert, quartz, plagioclase, epidote, graphite, limonite, prehnite, zeolite, and zoisite-clinozoisite are present.

Most mean grain diameters for Beacon sediments are within the fine-grade sizes. Sorting coefficients indicate that the sediments are moderately well sorted. On the average, the sands are slightly skewed toward the fine-size grades and are platykurtic, with the tails of the sediments better sorted than the central parts. Maps of statistical parameters can suggest directions of paleoslopes if the transport direction is taken as the downslope direction. However, maps drawn of parameters for the Fairchild and Fremouw-Falla Formations do not completely agree with paleoslopes determined by other workers from sedimentary and glacial structures. This may only reflect the fact that the maps are drawn from only a few control points obtained in a small area. Nevertheless, it is interesting to note that the paleoslope determined here for the Fairchild Formation is nearly opposite to that obtained by earlier workers.

Acknowledgments. Field studies were made possible through National Science Foundation grants G-23806 and GA-122 to F. A. Wade of Texas Tech University.

References cited

Barrett, P. J. 1965. Geology of the area between the Axel Heiberg and Shackleton Glaciers, Queen Maud Range, Antarctica, Part 2—Beacon Group. *N.Z. Jl Geol. Geophys., 8,* No. 2, 344–70.

Barrett, P. J. 1966. Petrology of some Beacon rocks between the Axel Heiberg and Shackleton Glaciers, Queen Maud Range, Antarctica. *J. sedim. Petrol., 36,* No. 3, 794–805.

Barrett, P. J. 1969. Stratigraphy and petrology of the mainly fluviatile Permian and Triassic Beacon rocks, Beardmore Glacier area, Antarctica. *Rep. Inst. polar Stud., Ohio State Univ.,* No. 34, 132 pp.

Barrett, P. J. 1970. Paleocurrent analysis of the mainly fluviatile Permian and Triassic Beacon rocks, Beardmore Glacier area, Antarctica. *J. sedim. Petrol., 40,* No. 1, 395–411.

Barrett, P. J. 1972. Stratigraphy and petrology of the mainly fluviatile Permian and Triassic part of the Beacon Supergroup, Beardmore Glacier area. (*In* Adie, R. J., *ed. Antarctic Geology and Geophysics.* Oslo, Universitetsforlaget, 365–72.)

Barrett, P. J., Elliot, D. H., and Lindsay, J. F. 1967. Geology of the Beardmore Glacier area, Antarctica. *Antarct. Jnl U.S., 2,* No. 4, 110–12.

Doumani, G. A., and Minshew, V. H. 1965. General geology of the Mount Weaver area, Queen Maud Mountains, Antarctica. (*In* Hadley, J. B., *ed. Geology and Paleontology of the Antarctic.* Washington, American Geophysical Union, 127–39.) [Antarctic Research Series, Vol. 6.]

Folk, R. L. 1954. The distinction between grain size and mineral composition in sedimentary-rock nomenclature. *J. Geol., 62,* 344–59.

Folk, R. L. 1974. *Petrology of Sedimentary Rocks.* Austin, Texas, Hemphill Publishing Co., 182 pp.

Folk, R. L., and Ward, W. C. 1957. Brazos River bar: a study in the significance of grain size parameters. *J. sedim. Petrol., 27,* No. 1, 3–26.

Friedman, G. M. 1961. Distinction between dune, beach, and river sands from their textural characteristics. *J. sedim. Petrol., 31,* No. 4, 514–29.

Friedman, G. M. 1962. Comparison of moment measures for sieving and thin-section data in sedimentary petrological studies. *J. sedim. Petrol., 32,* No. 1, 15–25.

Friedman, G. M. 1967. Dynamic processes and statistical parameters compared for size frequency distribution of beach and river sands. *J. sedim. Petrol., 37,* No. 2, 327–54.

Grindley, G. W. 1963. The geology of the Queen Alexandra Range, Beardmore Glacier, Ross Dependency, Antarctica; with notes on the correlation of Gondwana sequences. *N.Z. Jl Geol. Geophys., 6,* No. 3, 307–47.

Inman, D. L. 1952. Measures for describing the size distribution of sediments. *J. sedim. Petrol., 22,* No. 3, 125–45.

Krumbein, W. C., and Sloss, L. L. 1963. Stratigraphy and sedimentation. (*In* Krumbein, W. C., and Sloss, L. L. *Classification and Description of Sedimentary Rocks.* 2nd ed. San Francisco, W. H. Freeman, 150–89.)

La Prade, K. E. 1970. Permian–Triassic Beacon Group of the Shackleton Glacier area, Queen Maud Range, Transantarctic Mountains, Antarctica. *Geol. Soc. Am. Bull., 81,* 1403–10.

La Prade, K. E. 1972. Permian–Triassic Beacon Supergroup of the Shackleton Glacier area, Queen Maud Range, Transantarctic Mountains. (*In* Adie, R. J., *ed. Antarctic Geology and Geophysics.* Oslo, Universitetsforlaget, 373–78.)

Lindsay, J. F. 1969. Stratigraphy and sedimentation of lower Beacon rocks in the central Transantarctic Mountains, Antarctica. *Rep. Inst. polar Stud., Ohio State Univ.,* No. 33, 58 pp.

Long, W. E. 1959. Preliminary report of the geology of the central range of the Horlick Mountains, Antarctica. *Rep. Ohio State Univ. Research Foundation,* 825-2, Pt. 7, 1–23.

Long, W. E. 1962*a*. Permo-Carboniferous glaciation in Antarctica (abstract). *Spec. Pap. geol. Soc. Am., 68,* 314.

Long, W. E. 1962*b*. Sedimentary rocks of the Buckeye Range, Horlick Mountains, Antarctica. *Science, N.Y., 136,* No. 3513, 319–21.

Long, W. E. 1964. Stratigraphy of the Horlick Mountains. (*In* Adie, R. J., *ed. Antarctic Geology.* Amsterdam, North-Holland Publishing Company, 352–63.)

Long, W. E. 1965. Stratigraphy of the Ohio Range, Antarctica. (*In* Hadley, J. B., *ed. Geology and Paleontology of the Antarctic.* Washington, American Geophysical Union, 71–116.) [Antarctic Research Series, Vol. 6.]

Minshew, V. H. 1966. Stratigraphy of the Wisconsin Range, Horlick Mountains, Antarctica. *Science, N.Y., 152,* No. 3722, 637–38.

Minshew, V. H. 1967. *Geology of the Scott Glacier and Wisconsin Range Areas, Central Transantarctic Mountains, Antarctica.* Ph.D. dissertation, The Ohio State University, 268 pp.

Minshew, V. H., and Summerson, C. H. 1964. Sedimentary section of the Robert Scott Glacier area, Queen Maud Range, Antarctica. *Spec. Pap. geol. Soc. Am., 76,* 312.

73

The Precambrian Sedimentary-Volcanic Sequence and Associated Intrusive Rocks of the Ahlmannryggen, Western Dronning Maud Land: A New Interpretation

J. H. BREDELL
Geological Survey, Private Bag X112, Pretoria, South Africa

Abstract. Nunataks between lat. 71°50′ and 72°35′ S and long. 1°30′ and 4°00′ W were mapped geologically on a scale of 1:25,000 during the summer of 1970–71. A new lithostratigraphic subdivision is proposed, based mainly on detailed investigation in strategic areas. The oldest rocks belong to the Pyramiden Formation (390 m) which consists mostly of feldspathic graywacke and siltstone. The overlying Grunehogna Formation (450 m), which has the widest distribution of all the sedimentary formations in this area, consists of interbedded persistent layers of light-colored arenite and dark-colored argillite. This succession is conformably overlain by the Högfonna Formation which is subdivided into a lower Borgmassivet Member (650 m), consisting of feldspathic quartzite, shale, and jasper-bearing conglomerate, and an upper Jekselen Member, consisting of calcareous, jasper-bearing, arenaceous sedimentary rocks. These three formations are assigned to the Ahlmannryggen Group; the Raudberget Formation, which overlies the Högfonne Formation conformably in the Borgmassivet to the south, forms a fourth unit in the group. The volcanic-sedimentary succession at Tindeklypa and Istind is believed to overlie the Högfonna Formation conformably and is called the Viddalen Formation, a time-equivalent of the Raudberget Formation. It is subdivided into two conformable members: a lower Tindeklypa Member (500 m), consisting of agglomerate with subordinate tuff and andesitic lava flows, and an upper Istind Member (340 m), consisting of interbedded feldspathic quartzite, agglomerate, tuff, and lava flows. A continuous upward gradation from plutonic through hypabyssal to volcanic rocks (near-surface intrusions) was detected in the 300 m, approximately 1700 m.y.-old mafic sill (Krylen Intrusions) at Jekselen. The lava flows in the Viddalen Formation are considered to be the volcanic equivalent of the Krylen Intrusions. The majority of the large mafic and associated minor felsic intrusions in the area are believed to have an age of about 1000 m.y. and are called the Ytstenut Intrusions.

During the summer of 1970–71 the 12th South African National Antarctic Expedition carried out geologic investigations in the Ahlmannryggen (Ahlmann Ridge) and the northeastern Borgmassivet (Borg Massif) between lat. 71° 50′ and 72° 35′ S and long. 1° 30′ and 4° 00′ W (Figure 73.1). Twenty separate nunatak groups within an area of 3400 km² were mapped on a scale of 1:25,000 (Bredell, 1976).

The mountain ranges of western Dronning Maud Land between lat. 71° and 73° S and long. 1° and 4° W comprise a mainly subhorizontal Precambrian sedimentary-volcanic sequence with a maximum thickness of approximately 3000 m. It was invaded at different times by predominantly mafic intrusions. This sequence was named the Ritscher Supergroup and was subdivided as follows (Neethling, 1972):

Trollkjellrygg Group
- Straumsnutane Formation
- Istind Formation
- Tindeklypa Formation

Jutul volcanics
- Fassett and Ütkikken lava

Ahlmannrygg Group
- Raudberg Formation
- Högfonna Formation
- Schumacher Formation
- Pyramiden Formation

Field observations by the author and his colleague, A. W. W. Paterson, indicated that Neethling's proposed subdivision requires some revision, especially regarding the relative and absolute ages of formations belonging to the upper part of the Ritscher Supergroup. In revising the stratigraphic nomenclature it was necessary to rename or redefine most of Neethling's unit terms, not only to incorporate new information, but also to comply with the new code of the IUGS Commission on Stratigraphy. Most of the stratigraphic units which were grouped under the Ritscher Supergroup have now been placed in a newly defined "Ahlmannryggen Group." The revised subdivision is presented in Figure 73.2.

The Pyramiden Formation

This formation was named (Neethling, 1972) after Pyramiden, a nunatak at the western margin of the Ahlmannryggen, where an approximately 200-m-thick succession of well-bedded feldspathic graywacke and intercalated dark shale with mud-pebble conglomerate is exposed. Similar rocks have also been reported from Sfinksen and Knallen to the immediate south and west of Pyramiden. The present investigation proved, however, that the regional distribution of the Pyramiden Formation is much more extensive. Correlative rocks were found at Nashornet and Viddalskollen at the confluence of the Viddalen and

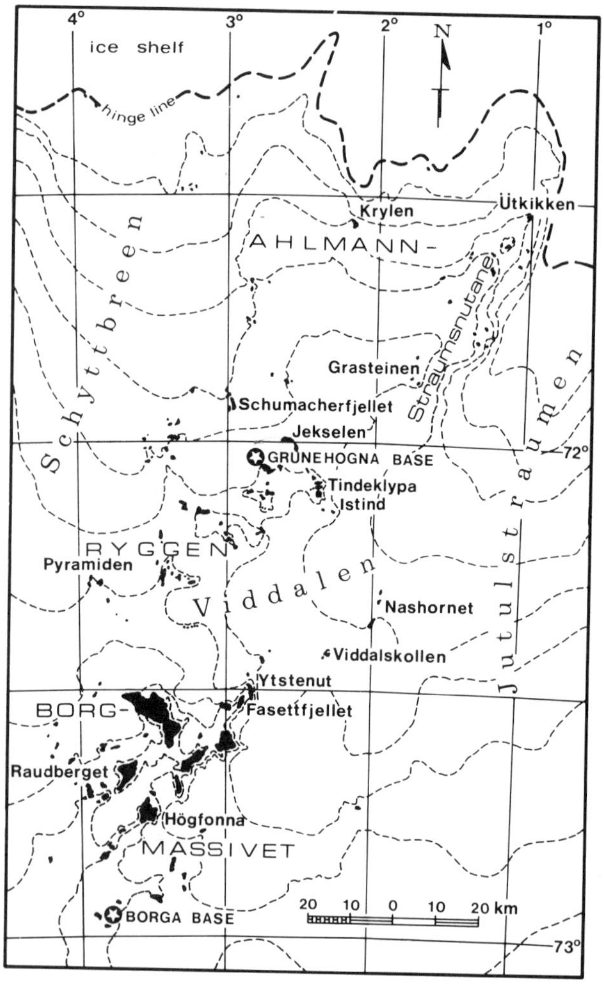

Figure 73.1. Map of the Ahlmannryggen (Ahlmann Ridge) and Borgmassivet (Borg Massif) showing localities referred to in the text. Darkened areas indicate rock outcrops. Contours (on the ice surface) are at 200 m intervals.

Jutulstraumen glaciers, as well as at Babordsranten and Kjölrabbane, 14 km east of the type locality (Bredell, 1973).

In the Nashornet-Viddalskollen area the Pyramiden Formation was lithologically subdivided into two conformable members: a lower member, some 140 m thick, consisting of feldspathic graywacke and siltstone which was deposited under low-energy, deep-water conditions, and an upper member, about 250 m thick, consisting of feldspathic graywacke, siltstone, and conglomerate which was deposited in shallower water under slightly higher energy conditions.

The Pyramiden Formation has not yet been found in contact with any of the other formations in the area, and its stratigraphic position within the Ahlmannryggen Group has not been established conclusively. It is tentatively believed to represent the lowermost sedimentary unit in the group.

The Grunehogna Formation

This formation was previously known as the Schumacher Formation, the name being derived from Schumacherfjellet, a prominent nunatak in the northwestern part of the Ahlmannryggen (Neethling, 1972). Since Neethling's visit to Schumacherfjellet in 1962, subsequent investigations have proved, however, that the most typical exposure is at the Grunehogna group of nunataks (lat. 72° 02′ S, long. 2° 48′ W). Although thicker successions are found at other localities in the Ahlmannryggen, the occurrences at Grunehogna are stratigraphically the most significant owing to a well-exposed conformable relationship with the overlying Borgmassivet Member of the Högfonna Formation. Because of this and the fact that "Schumacher" is not a geographic name, the author proposes the term "Grunehogna Formation" for all those sedimentary rocks in the Ahlmannryggen which have up to now been grouped under the Schumacher Formation.

The Grunehogna Formation is by far the most widely distributed sedimentary unit exposed in the area. A

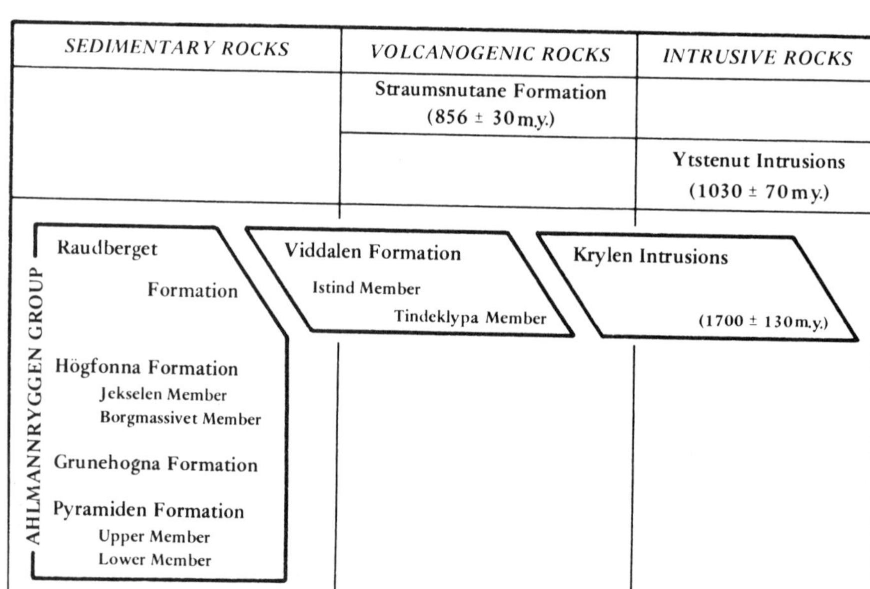

Figure 73.2. Revised stratigraphic subdivision of the rock units in the Ahlmannryggen and Borgmassivet.

minimum thickness of 450 m was estimated by Neethling (1972). In outcrop it is characterized by interlayered persistent beds of light-colored arenite and dark-colored argillite, which makes it the most easily identified unit in the Ahlmannryggen. Although conglomerates are usually conspicuously absent, some rare well-rounded mudstone pebbles were observed in the upper part of the sequence at Lyftingen (Van Zyl, personal communication). The base of the formation is not exposed, and its relationship with the supposedly underlying Pyramiden Formation has not yet been proven in outcrop.

The Högfonna Formation

General discussion and type locality

The Högfonna Formation takes its name from Högfonna, a nunatak in the northeastern part of the Borgmassivet. De Ridder (1970) defined this formation as being "the jasper-bearing conglomeratic sedimentary sequence exposed in the central part of the Borg Massif," and divided it into a lower and an upper member "on the basis of local red beds and tuff in the upper part." Subsequently Aucamp (1972) found that the Höfonna Formation also occurs at various localities in the Ahlmannryggen. He also described red beds and ash particles in the lower part of the formation at Grunehogna. It seems therefore that although De Ridder's subdivision might have been justified for the Borgmassivet, it does not apply for the area as a whole. Furthermore, a lithologically different correlative sequence, which is believed to constitute the upper part of the Högfonna Formation in the Ahlmannryggen, was found at Jekselen and Tindeklypa.

In view of these recent findings, a new two-fold subdivision for the Högfonna Formation is proposed:

1. A Borgmassivet Member, which includes all sedimentary rocks assigned by De Ridder (1970) to the originally defined Högfonna Formation, as well as all lithologically similar rocks found in the Ahlmannryggen; and
2. A Jekselen Member, which includes all dark gray to buff-colored arenaceous, jasper-bearing, calcareous sedimentary rocks occurring at Jekselen and Tindeklypa.

The Borgmassivet Member has been dealt with in detail by various authors (De Ridder, 1970; Neethling, 1970; Aucamp, 1972).

The Jekselen Member

At Jekselen, the type locality, sedimentary beds belonging to this member are confined to the southeasternmost part of the nunatak. They occur typically as isolated sections separated by amygdaloidal dikes, and bedding invariably dips at angles of 36° or more. The approximately 80-m-thick succession comprises mainly greenish-gray to buff-colored calcareous shale and sandstone with polymictic jasper-bearing conglomerate near the top.

At Tindeklypa, a 50-m-thick succession of similar sedimentary beds is exposed on the northwestern side of Nunatak 1390, about 5 km southeast of Jekselen.

Jekselen Member sedimentary beds are also abundantly present as xenoliths within the Tindeklypa Member of the Viddalen Formation.

The Viddalen Formation

Definition and type locality

The term Viddalen Formation is here introduced to replace the previously used stratigraphic units "Tindeklypa Formation" and "Istind Formation" as defined by Neethling (1970). It includes all the agglomerate, tuff, lava flows, and penecontemporaneous intrusions, as well as the shallow-water sedimentary rocks with intercalated agglomerate, tuff, and lava flows, which occur at the Tindeklypa-Istind range of nunataks at the confluence of the Viddalen and Jutulstraumen Glaciers. Exposures of similar rock types at Fasettfjellet (De Ridder, 1970) and Grasteinen (Van Zyl, 1974) are considered to be correlative.

On the basis of the presence of nonvolcanic sedimentary beds in its upper part, the Viddalen Formation is subdivided into a basal Tindeklypa Member and a conformably overlying Istind Member.

The Tindeklypa Member

This member consists predominantly of agglomerate with subordinate tuff, lava flows, and penecontemporaneous intrusions. Except for relatively minor occurrences at Fasettfjellet and Grasteinen, it is confined to the Tindeklypa nunataks south of Nunatak 1390 and to the base of Istind Peak. The base of the member is not exposed and the estimated total thickness of 500 m should be regarded as a minimum.

All the clasts in the agglomerate are angular and show no definite orientation. There is absolutely no sorting, and the clasts range in size from small microscopic fragments to large angular boulders several meters in diameter, occurring side by side. About 90% of the clasts are of sedimentary origin, and all pre-Viddalen formations are represented. A characteristic feature of the matrix is that microscopically, it closely resembles the macroscopic appearance of the agglomerate as a whole. It consists mainly of unsorted angular to subrounded quartz grains which in turn are embedded in a finer argillaceous matrix. No volcanic materials, such as glass shards or ash particles, were observed, indicating that the agglomerate was probably produced during the first stages prior to the establishment of an eruptive vent (Geikie, 1897).

The Istind Member

The Istind Member has a very limited distribution; it is confined to Istind Peak where a thickness of about 340 m of strata is exposed and to Nunatak 1320 where a succession, 215 m thick, has been recognized as correlative.

In contrast with the underlying Tindeklypa Member, the Istind Member also includes quartzite and shale, rocks of nonvolcanic origin. In general, the Istind Member is composed of interbedded red and gray agglomerate (identical to that of the Tindeklypa Member), lava flows, red, gray, and buff quartzites, and subordinate tuff and shale.

There is no unconformity between the Istind and Tindeklypa Members as suggested by Neethling (1970). Van Zyl (1974), who made a detailed study of the succession at Istind Peak, regards the "Istind Formation" as a "mere upward continuation of the Tindeklypa Formation" and suggests that the term Istind "Formation" should be abandoned.

THE ORIGIN OF THE VIDDALEN FORMATION AND ITS RELATION TO THE AHLMANNRYGGEN GROUP

The postulated volcaniclastic origin for the Tindeklypa agglomerate (Neethling, 1970) is accepted and supported. This idea should, however, also be extended to incorporate a large part of the Istind Member, which contains much more agglomerate than has been previously known. The agglomerate is believed to have formed during explosive eruptions which accompanied tectonic movements while the Jutulstraumen and Viddalen troughs were forming. These violent eruptions were periodically interrupted by emission of magma which in the case of the Tindeklypa Formation partly intruded still unconsolidated volcaniclastic material. The underlying sedimentary formations served as the main source of material for the formation of the agglomerates. Fragments of mafic intrusive rock resembling the Krylen Intrusions probably originated from deep-seated sills which had already solidified when the eruptions took place, while clasts of lava reported by Butt (1962) could have been included in the agglomerate by new eruptions which brought up material from already existing agglomerate and intercalated lava flows. Volcanic activity ceased for short periods during which shallow-water sediments were deposited (Istind Member).

Volcanic activity was most probably concentrated along fissures parallel to the Jutulstraumen, which explains the restricted, yet linear, distribution of volcaniclastic rocks at Fasettfjellet, Tindeklypa-Istind, and Grasteinen. The steeply dipping sedimentary beds of the Jekselen Member, which are associated with amygdaloidal intrusions at Jekselen, indicate the western boundary of the volcanic activity.

The Viddalen Formation at the type locality does not occur in contact with any other formation, and its stratigraphic position relative to the Ahlmannryggen Group can therefore only be derived by indirect means:

1. The most obvious feature of this formation is the variety of rock types represented as clasts in the agglomerate. It has already been mentioned that all the sedimentary formations of the Ahlmannryggen Group are represented as clasts in the agglomerate. This, however, only indicates that the Viddalen Formation is younger than the Högfonna Formation.

2. The high ratio of igneous to sedimentary rocks exposed in the Ahlmannryggen today is definitely not reflected in the agglomerates of the Viddalen Formation where clasts of sedimentary rock predominate. This suggests that the agglomerates were formed prior to the emplacement of the numerous mafic sheets exposed at present.

3. A lithologic connection exists between Jekselen and Tindeklypa Nunataks in the form of the hypabyssal and volcanic zones (see section below on Krylen Intrusions) of the Jekselen sill ("Borg Intrusions" of Neethling, 1970), which are exposed at the northern extremity of the Tindeklypa Range. To the south in this range at Nunatak 1599, the amygdaloidal volcanic zone is present as lava flows and penecontemporaneous intrusions in the agglomerate. Only the upper contact of the lower lava flow is exposed at the base of this nunatak and this exposure, which was first described by Butt (1962), has led to Neethling's postulation of a pre-Viddalen lava, the so-called "Jutul volcanics." The mode of occurrence of this lava and its compositional similarity to other lava flows in the agglomerate, however, suggests to the writer that all these lava flows are closely related and contemporaneous with the deposition of the agglomerate unit. This seems to indicate therefore that the 1700 m.y. "Borg intrusive episode" (Allsopp and Neethling, 1970) and the Viddalen Formation are contemporaneous, with the lava flows in the latter formation constituting the volcanic equivalent of the "Borg Intrusions," which have now been redefined by the writer and renamed the Krylen Intrusions.

4. The beds correlative with the Tindeklypa Member at Fassettfjellet overlie the Högfonna Formation conformably (Watters and Aucamp, 1969), while the "Raudberg" Formation is reported to have the same relationship with the Högfonna Formation at Högfonna Nunatak (De Ridder, 1970). It thus follows that the "Raudberg" and Viddalen Formations are probably contemporaneous. (Although the "Raudberg" Formation is not exposed within the area investigated, the writer wishes to suggest the name Raudberget Formation, as this is the proper geographic name of the type locality as described by De Ridder.)

CONCLUSIONS

Shortly after deposition of the Jekselen Member of the Högfonna Formation, the volcanic activity, of which signs already appeared in the lower part of the Högfonna Formation, reached its peak in the Ahlmannryggen area. In the north, igneous activity was restricted to the intrusion of thick mafic sheets, but at Jekselen signs of near-surface intrusions can be observed. At Tindeklypa and Istind eruption of agglomerate and lava took place, at the same time sediments of the Raudberget Formation were being deposited in the quieter areas to the south in the Borgmassivet.

Igneous rocks of Viddalen and post-Viddalen age

The sedimentary sequence of the Ahlmannryggen Group has been invaded on a large scale by predominantly mafic intrusions. With the exception of the Krylen Intrusions, all the igneous rocks in the Ahlmannryggen and Borgmassivet are believed to postdate the Viddalen Formation.

Field relationships supported by additional radiometric age determinations in strategic regions enabled a simplification of Neethling's (1970) subdivision of igneous rocks in the area. The following geochronologic subdivision is proposed:

1. *Krylen Intrusions (about 1700 m.y. old).* These intrusions have a volcanic equivalent (part of the Viddalen Formation), and include Neethling's "Jutul Group," part of his "Borg Intrusions," and the lavas in his "Tindeklypa" and "Istind" Formations.
2. *Ytstenut Intrusions (about 1000 m.y. old).* This unit incorporates the greater part of Neethling's "Borg Intrusions" as well as his "Jørgen Intrusions."
3. *Straumsnutane Formation (about 850 m.y. old).* This formation comprises 860 m of andesitic lava of uncertain correlation (Neethling, 1970), which is exposed in the Straumsnutane Range to the northeast of the investigated area.
4. *Post-Ytstenut Intrusions (about 190 m.y. old).* These are the youngest rocks in the investigated area and occur abundantly in the form of minor vertical and near-vertical dolerite dikes.

It should be noted that the term "volcanic" as used below is not regarded as a synonym for "extrusive," but refers to near-surface intrusions which are included in the volcanic class, as described by Turner and Verhoogen (1951) and as defined in the AGI Glossary of Geology (Gary and others, 1973).

The Krylen Intrusions

Definition

The Krylen Intrusions include all igneous rocks, plutonic as well as hypabyssal and volcanic, which were emplaced or extruded about 1700 m.y. ago. They are contemporaneous with the Raudberget and Viddalen Formations.

The Krylen Intrusions correspond in age to the "Borg Intrusions" as defined by Neethling (1970), but differ in regional distribution and relationship to the upper part of his "Ritscher" Supergroup. To avoid confusion, the term "Borg" has been abandoned in favor of "Krylen," which is derived from Krylen Nunatak (lat. 71° 32′ S, long. 2° 10′ W) where the first age of about 1700 m.y. was obtained in western Dronning Maud Land (McDougall, 1969). Because of the close lithologic and mineralogic resemblance of the Krylen and Ytstenut Intrusions, the exact distribution of these rocks is now known at present. The nature of the inclusions in the Tindeklypa Member of the Viddalen Formation suggests, however, that most of the mafic intrusions in the area are of post-Viddalen age, i.e., younger than 1700 m.y.

Jekselen Nunatak

The existence of approximately 1700-m.y.-old volcanic rocks here was recognized by Neethling (1970). The assigned age was based partly on an age of 1760 m.y. obtained from altered lava from Ütkikken (Eastin and others, 1970) and partly on the presence of lava "below" the Tindeklypa Formation at Tindeklypa. The relationship of these rocks to plutonic rocks of similar age was, however, not established until the writer and his colleague, Andrew Paterson, discovered a continuous upward gradation from plutonic through hypabyssal to volcanic rocks in the 300-m-thick mafic sill at Jekselen.

The following sequence has been established:

1. *Plutonic zone.* The greater part of Jekselen is composed of coarse- to medium-grained intrusive rock which, judging from its texture, must have crystallized under plutonic conditions. A striking feature is the extensive layering due to variations in composition (granodioritic to quartz monzodioritic).

2. *Plutonic-hypabyssal transition zone.* On top of the typical plutonic zone, but without any sharp boundary, there is a relatively thin transition zone (10–15 m) within which occur porphyritic crystals of pigeonite. Rocks of this zone also have a quartz monzodioritic composition, and except for the sparsely distributed bladed pigeonite crystals which are approximately 0.5 mm across and up to 10 mm long, they are identical to those of the plutonic zone.

3. *Hypabyssal zone.* The hypabyssal zone can be subdivided into a lower quartz monzodioritic subzone and an upper quartz dioritic subzone, each about 50 m thick, with a gradual transition between them. The lower subzone contains larger porphyritic crystals (8–10 mm long), while in the upper subzone these crystals are markedly smaller (2–4 mm long). Toward the top of this zone, the porphyritic texture gradually becomes more dominant until the bladed crystals of pigeonite occupy about 30% (by volume) of the rock. The matrix of rocks from the hypabyssal zone is fine-grained compared to the medium- to coarse-grained matrix of rocks from the plutonic zone.

4. *Hypabyssal-volcanic transition zone.* The characteristic feature of this transition zone (5–10 m thick) is the combined occurrence of textures typical of both hypabyssal and volcanic rocks. Small spheroidal quartz amygdules (1–8 mm in diameter) occur at the base of this zone, but the small porphyritic pigeonite crystals dominate the texture. Upward, the amygdules gradually become more abundant and the porphyritic crystals decrease in number until they disappear completely. Representative specimens indicate a quartz dioritic composition, i.e., the same as that of the upper subzone of the underlying hypabyssal zone. Because of their amygdaloidal texture, however, rocks from this zone should rather be termed porphyritic quartz andesite.

5. *Volcanic zone.* The greater part of this zone, which is 80–100 m thick, consists of rock similar in general appearance to that of the underlying transition zone. The porphyritic texture is, however, characteristically absent. Small spheroidal quartz amygdules in a fine-grained matrix are the dominant macroscopic textural feature of this part, which is here termed the main volcanic subzone. At the contacts with sedimentary beds of the overlying Jekselen Member, however, the vesicles are much larger (up to 70 mm in diameter) and filled with calcite. Here, where the volcanic rocks are usually present as transgressive bodies along fault zones in the sedimentary beds, the matrix is aphanitic and dark grayish-green in color. This part of the zone is termed the volcanic contact subzone. Rocks from both subzones can be classified as quartz andesites.

Tindeklypa-Istind area

Except for the plutonic and plutonic-hypabyssal transition zones, which are not exposed, the sequence as observed for the Krylen Intrusions in the sill at Jekselen is also developed along the Tindeklypa-Istind range of nunataks. The main volcanic and volcanic contact subzones are present as interbedded flows (or penecontemporaneous intrusions) within the agglomerate and are here considered part of the Tindeklypa Member of the Viddalen Formation. At Istind Peak and Nunatak 1320 the Krylen Intrusions are also present in the form of the hypabyssal and volcanic zones, found as concordant flows (or penecontemporaneous sills) within the sedimentary beds and agglomerate of the Istind Member.

Fasettfjellet

The amygdaloidal lava which caps this nunatak is considered to correlate with the volcanic zone of the Krylen Intrusions.

The age of the Krylen Intrusions

The first suggestion of an age on the order of 1700 m.y. for intrusive rocks in Dronning Maud Land came from K-Ar isotopic age measurements on hornblende from Krylen Nunatak (McDougall, 1969). An age this great, however, was considered to be somewhat doubtful until an age of

1700 ± 130 m.y. was established by the Rb-Sr method on four whole-rock samples from Jekselen Nunatak (Allsopp and Neethling, 1970). This age was interpreted as being the date of emplacement of the intrusions.

Bowman (1971) determined a Rb-Sr age of between 1190 and 1280 m.y. on one whole-rock sample of the lava at Fasettfjellet. As mentioned earlier, this lava can, according to field relationships, be correlated with the volcanic zone of the Krylen Intrusions, and Bowman's ages are therefore regarded by the writer as representing minimum values, probably resulting from overprinting during regional metamorphism.

A whole-rock sample from the main volcanic subzone was submitted to Geological Dating Services in Britain for determination of $^{40}Ar/^{39}Ar$ age spectra. Extensive replacement of the original minerals prevented the determination of the original date of intrusion. The three age components determined were: 1339 ± 55 m.y., 1140 ± 56 m.y., and 806 ± 28 m.y. It was concluded that crystallization must have taken place before 1339 m.y. ago. (Bredell, 1973). In the light of field evidence, the writer has no doubt that the real age of these volcanic rocks is the same as that established by Allsopp and Neethling (1970) for the plutonic zone, i.e., 1700 ± 130 m.y.

A specimen from the lower subhorizontal lava flow (forming part of the Istind Member of the Viddalen Formation) at Istind was also submitted to the same analysts for age determination. This specimen was even more altered than the one from Jekselen, and it was therefore expected that any age obtained would most probably relate to the major episode of alteration rather than to the age of original crystallization. Three major apparent age components could be recognized, each of which had the same apparent age of about 600 m.y. It was concluded that these rocks, although probably very much older than 600 m.y., were completely re-equilibrated during a major hydrothermal event around 600 m.y. In the writer's opinion, this lava belongs to the Krylen Intrusions and should therefore have a real age of about 1700 m.y.

The Ytstenut Intrusions

Definition

The Ytstenut Intrusions include all intrusive rocks which are younger than the whole of the Ahlmannryggen Group and the Viddalen Formation and which have an age of around 1000 m.y.

The Ytstenut Intrusions correspond in age to the Jørgen Intrusions as defined by Neethling (1970), but as the latter type of granodiorite intrusions were found to be only a local differentiation product of more widespread mafic intrusives, they are now incorporated in and replaced by the term Ytstenut Intrusions. The Ytstenut Intrusions also provisionally include, as far as distribution (but not necessarily age) is concerned, the greater part of Neethling's Borg Intrusions.

The name "Ytstenut" is derived from Ytstenut Nunatak (lat. 72° 30′ S, long. 2° 50′ W) where the first age of about 1000 m.y. was obtained from intrusive rocks in the Borgmassivet (Bredell, 1973). All the major intrusives of the Ahlmannryggen and Borgmassivet, with the exception of those at Ütkikken, Krylen, Jekselen, and Istind-Tindeklypa, are provisionally grouped together as Ytstenut Intrusions. The following line of reasoning supports the wide distribution of the Ytstenut Intrusions. The Raudberget Formation overlies the Högfonna Formation conformably at Högfonna Nunatak (De Ridder, 1970). Pyroclastic rocks and lavas, correlated with the Viddalen Formation, overlie the Högfonna Formation conformably at Fasettfjellet (Watters and Aucamp, 1969). The Raudberget and Viddalen Formations are thus regarded as contemporaneous. The Viddalen Formation is coeval with the Krylen Intrustions. The Raudberget Formation was thus deposited while the Krylen Intrusions were being emplaced. Large mafic sheets, however, are commonly found intrusive into both the Högfonna and Raudberget Formations in the Borgmassivet (De Ridder, 1970). The most logical conclusion is that these intrusives are of a younger age, i.e., Ytstenut Intrusions.

The age of the Ytstenut Intrusions

The first indication of an age of about 1000 m.y. for intrusive rocks in the area was from K-Ar isotopic age measurements on plagioclase (825 m.y.) and pyroxene (1010 m.y.) from Boreas Nunatak (lat. 71° 15′ S, long. 4° W) (McDougall, 1969). As argon loss due to regional metamorphism was expected, these datings were considered unreliable (Neethling, 1970). Subsequently an age of 1030 ± 70 m.y. (Rb-Sr whole-rock) was determined on syenodiorite from Grunehogna (Allsopp and Neethling, 1970). They considered this to be the age of a separate episode of felsic intrusions younger than the Borg Intrusions. The term "Jørgen Intrusions" was proposed for all these felsic intrusions with an age of about 1000 m.y.

The writer, however, came to the conclusion that there invariably exists a close relationship between the emplacement of the major mafic sheets and the minor felsic intrusions, and that the latter only represents late differentiates and in places a reaction product of the intrusion of the sedimentary rocks by the mafic magma. The age of 1030 m.y. determined for the felsic intrusions was therefore also expected to be the age of the larger sills in which they occur.

In order to settle this dispute, a whole-rock sample from the dioritic sill at Grunehogna was submitted to Geological Dating Services in Britain for determination of $^{40}Ar/^{39}Ar$ age spectra. A minimum age of 832 ± 2 m.y. was determined (Bredell, 1973). Due to the extensive alteration of the original minerals, this age reflects a major episode of reheating and not the age of emplacement. A second less well-defined age component of 716 ± 4 m.y. was also indicated. Although not very definite, the result is quite significant if seen in the light of the minimum age of 1339 ± 55 m.y. obtained on an equally extensively altered sample from Jekselen, and it certainly does not oppose the writer's interpretation of field observations.

For reasons mentioned above, the large mafic sheets in the Borgmassivet were expected to be younger than 1700 m.y. (the age of the Krylen Intrusions). Neethling (1970) grouped these intrusions with his 1700 m.y. Borg Intrusions, but it was never verified by isotopic age determinations. A whole-rock sample from Ytstenut in the northeastern Borgmassivet was submitted by the writer to the same analysts for $^{40}Ar/^{39}Ar$ age determinations. Again the extensive alteration of the sample prevented determination of a reliable date of intrusion. A minimum age of 924 ± 4 m.y. was established (Bredell, 1973). A second well-defined age component of 761 ± 11 m.y. was also indicated. Although not in contradiction with the writer's view, this result is also not very satisfactory. Therefore, until more accurate isotope ages can be obtained, an age of emplacement of 1030 ± 70 m.y. for the Ytstenut Intrusions is proposed.

Acknowledgments. I am indebted to the Geological Survey of South Africa, the South African Scientific Committee for Antarctic Research, and the Department of Transport for having made this investigation possible. In particular I would like to thank Andrew Paterson for his companionship and interest during and after the expedition, and Leon Wolmarans, L. E. Kent, and D. C. Neethling for critical reading of the manuscript.

References cited

Allsopp, A. L., and Neethling, D. C. 1970. Rb-Sr isotopic ages of Precambrian intrusives from Queen Maud Land, Antarctica. *Earth planet. Sci. Lett., 8,* No. 1, 66–70.

Aucamp, A. P. H. 1972. The geology of Grunehogna, Ahlmannryggen, western Dronning Maud Land, Antarctica. *S. Afr. J. Antarct. Res.,* No. 2, 16–22.

Bowman, J. R. 1971. *Use of the Isotopic Composition of Sr and SiO_2 Content in Determining the Origin of Mesozoic Basalt from Antarctica.* Master's thesis, The Ohio State University, 179 pp.

Bredell, J. H. 1973. The geology of the Nashornet-Viddalskollen area, western Dronning Maud Land, Antarctica. *S. Afr. J. Antarct. Res.,* No. 3, 2–10.

Bredell, J. H. 1976. *The Ahlmannryggen Group, the Viddalen Formation and the Associated Igneous Rocks in the Viddalen Area, Western Dronning Maud Land, Antarctica.* Master's thesis, University of Pretoria, 129 pp.

Butt, B. C. 1962. The geology of the Straumsnutane and Istind nunataks, western Queen Maud Land, Antarctica. *Rep. S. Afr. Nat. Antarct. Exped. Earth Sci., 9,* Pretoria, Geological Survey, 23 pp. (Unpublished).

De Ridder, E. 1970. *The Sedimentary Succession Exposed in Part of the Borg Massif, Western Queen Maud Land, Antarctica.* Master's thesis, University of Pretoria, 61 pp.

Eastin, R., Faure, G., and Neethling, D. C. 1970. The age of the Trollkjellrygg Volcanics of western Queen Maud Land, Antarctica. *Antarct. Jnl U.S., 5,* No. 5, 157–58.

Gary, M., McAfee, R., Jr., and Wolf, C. L., *eds.* 1973. *Glossary of Geology.* Washington, American Geological Institute, 805 pp.

Geikie, A. 1897. *The Ancient Volcanoes of Great Britain. 1.* London, Macmillan, 476 pp.

McDougall, I. 1969. Potassium-Argon dates on minerals from dolerites from western Queen Maud Land, Antarctica. Appendix to Neethling, D. C. 1969. Pre-Gondwana sedimentary rocks of Queen Maud Land, Antarctica. (*In Gondwana Stratigraphy. IUGS Symposium. Buenos Aires, 1-15 October 1969.* Paris, UNESCO, 1153–62.)

Neethling, D. C. 1970. *South African Earth Science Exploration of Western Queen Maud Land, Antarctica.* Ph.D. dissertation, University of Natal, 139 pp.

Neethling, D. C. 1972. Age and correlation of the Ritscher Supergroup and other Precambrian rock units, Dronning Maud Land. (*In* Adie, R. J., *ed. Antarctic Geology and Geophysics.* Oslo, Universitetsforlaget, 547–56.)

Turner, F. J., and Verhoogen, J. 1951. *Igneous and Metamorphic Petrology.* 2nd ed. New York, McGraw-Hill, 602 pp.

Van Zyl, C. Z. 1974. The geology of Istind, western Dronning Maud Land, and the relationship between the Istind and Tindeklypa Formations. *S. Afr. J. Antarct. Res.,* No. 4, 2–5.

Watters, B. R., and Aucamp, A. P. H. 1969. Notes on the geology of Trioen, Knallen, Tindegga and Fastettfjellet, Ahlmann Ridge. Pretoria, Geological Survey, *Rep. S. Afr. Nat. Antarct. Exped. Earth Sci., 154,* 7 pp. (Unpublished).

74

Geology and Petrology of the Southern Heritage Range, Ellsworth Mountains

AUDUN HJELLE, YOSHIHIDE OHTA, and THORE S. WINSNES
Norsk Polarinstitutt, P.O. Box 158, 1330 Oslo Lufthavn, Norway

Abstract. Sedimentary and magmatic rocks occur in the southern Heritage Range (lat. 80° S. long. 82° W), which comprises the southern half of the Ellsworth Mountains. Volcanic rocks are abundant in the lower part of the succession, in the Minaret Group and the lower part of the Heritage Group of Craddock (1969), but they were not observed in the upper part of the Heritage Group. On the basis of this lithostratigraphic contrast, a new classification is proposed: the Heritage Group is redefined to include the Middle Horseshoe Formation (530⁺ m, the upper part of the original Minaret Group) and the Edson Hills Formation (3900 m, the lower part of the original Heritage Group). A stratigraphic gap exists above the Edson Hills Formation, and a separate formation, the Dunbar Ridge Formation (1600 m, the upper part of the original Heritage Group) is proposed. The Dunbar Ridge Formation is Middle–Late Cambrian in age, with trilobite-bearing limestones.

The structure of this area is governed by a major anticline with a gentle northwesterly plunge; it includes many parasitic folds of chevron and isoclinal styles. Contrast of fold intensity and change of fold style below the Whiteout Conglomerate (Permian?) suggest a possible late paleozoic deformation phase.

Two phases of magmatism are distinguished, corresponding to the stratigraphic gaps. The pre–Middle Cambrian magmatism is dominantly K-alkaline and suggests the beginning of block subsidence of a continental crust. Late Paleozoic(?) magmatism is Na-alkaline and tholeiitic; this activity may be later than the main deformation, but the rocks have suffered regional metamorphism to the actinolite-greenschist facies. This younger activity is considered to represent an advanced stage of block tectonics.

Succeeding the three expeditions of the University of Minnesota (Craddock and others, 1964), a four-man Norwegian expedition mapped the southern part of the Heritage Range, Ellsworth Mountains, during two months in the 1974–75 southern summer (Figure 74.1). The main purpose of this work was to make a geologic traverse of the mountains.

Stratigraphy

The following stratigraphic units, described by Craddock (1969), were observed in this area: the Minaret Group, the Heritage Group, the Crashsite Quartzite, and the Whiteout Conglomerate, in ascending order. A major north-northwest–striking anticline in the western part of the area controls the distribution of these units. The descriptions in this chapter follow Craddock's classification.

The Minaret Group

This group, exposed in a belt about 5 km wide, was observed along the axial zone of the major anticline to the south of the Schanz Glacier (shown as the *Middle Horseshoe Formation* in Figure 74.1 according to the new classification). The base of the group has not been observed. A large strike fault limits the eastern border, while the western border is a supposedly conformable contact with the overlying Heritage Group.

The observed succession extends 530 m; the upper 480 m is a volcanic-calcareous sequence with dense lava, amygdaloidal agglomerate, and pillow lava; a 50-m-thick argillaceous succession occurs below. A polymictic conglomerate unit near the middle of the sequence has a black, coarse-grained matrix and boulders of granite, gabbro, mafic volcanic rocks, and white marble. A distinct oolitic and pisolitic marble member, 150–230 m in thickness, occurs at the top of the group. Craddock (1969) reported that farther south in the Minaret Hills, this group is totally composed of marble.

The Heritage Group

This group is divided here into two formations: the lower volcaniclastic *Edson Hills Formation* and the upper calc-argillaceous *Dunbar Ridge Formation.* The maximum calculated thickness is 5500 m.

Dense lavas with porphyritic texture and agglomerates are dominant in the lower part of the Edson Hills Formation (Figure 74.2*b*). Fine- to medium-grained sandstone layers of green and red tints are interbedded. In the Liberty Hills, interbedded fine- to medium-grained sandstone of gray to green color and conglomeratic layers make up the majority of the succession. The thickness of this formation is about 3900 m in the Edson Hills and 3200 m on the eastern slope of the Liberty Hills.

The Dunbar Ridge Formation is separated from the lower formation by a distinctive 150-m-thick white-yellow marble layer to the north of the Drake Icefall. The marble is overlain by a 300-m-thick black slate, the lower part of which has calcareous nodules with many trilobite fragments (Figure 74.1, fossil locality A). These fossils are considered to be probably Middle Cambrian in age by Webers

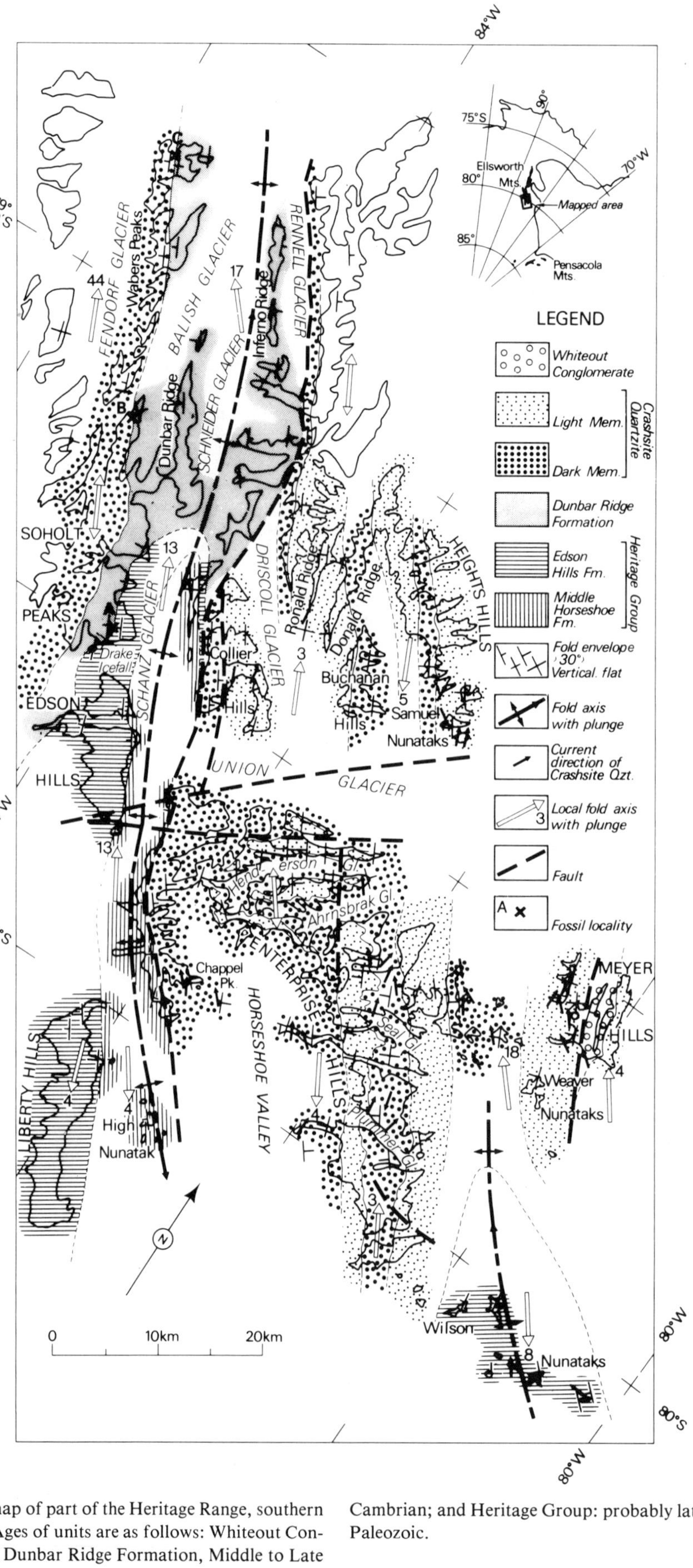

Figure 74.1. Geologic map of part of the Heritage Range, southern Ellsworth Mountains. Ages of units are as follows: Whiteout Conglomerate, Permian(?); Dunbar Ridge Formation, Middle to Late Cambrian; and Heritage Group: probably late Proterozoic to early Paleozoic.

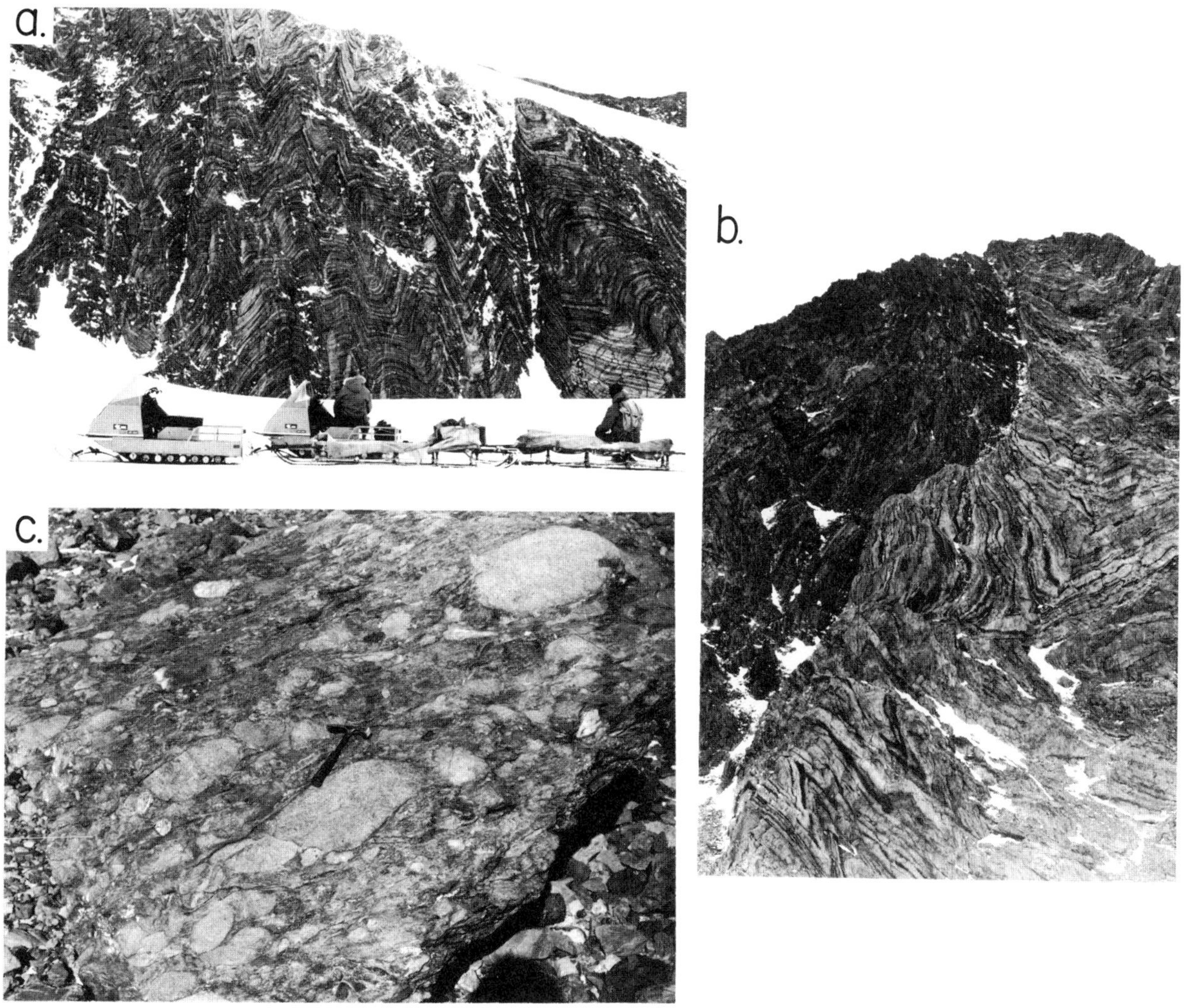

Figure 74.2. Typical exposures of rocks in the Heritage Range. *a.* Strongly folded Crashsite Quartzite, the Light Member on the ridge; north side of Henderson Glacier. *b.* Branch of the major strike fault separating the Dark and Light Members of the Crashsite Quartzite; Collier Hills, *c.* Agglomerate of the Edson Hills Formation; Edson Hills.

(personal communication). About 200 m of gray-green sandstone, shale, and thin calcareous beds occur above the slate.

The slaty facies of this formation crops out widely to the north along the major anticline axis. This slate is commonly calcareous, and sandstone and a few quartzite beds are intercalated in the lower part. About 100 m below the top of this formation, a fossil-rich gray limestone, 5–6 m thick, occurs (Figure 74.1, fossil locality C). These fossils have already been reported by Webers (1972) as probably Late Cambrian in age. This bed is traceable for 35 km along the eastern foothills of Webers Peaks, and the thickness increases southward to more than 70 m. A new fossil locality was found there (Figure 74.1, fossil locality B). The trilobites from here were examined by Webers (personal communication) and may be a little older than those from fossil locality C. The calculated thickness of the Dunbar Ridge Formation is 700–1600 m.

A succession of coarse clastic rocks crops out in scattered nunataks in the Wilson Nunataks near the southeastern end of the Heritage Range. The rocks are arkosic conglomerate, coarse-grained sandstone, gray limestone, black shale, and quartzite. Total observed thickness from eight separate exposures is less than 450 m, but the calculated thickness including unexposed areas may reach 1200 m. A small gabbro stock and some porphyrite dikes cut this succession and have caused weak thermal metamorphism of the shale. The conglomerate boulders are quartzite, conglomerate, sandstone, gray limestone, gneiss, and quartz diorite. Based on the rock types and the structural interpretation, this succession is considered to be part of the Edson Hills Formation.

Volcanic activity was continuous during the period from late Minaret Group time to Edson Hills Formation time. The volcanogenic rocks are thickest around Edson Hills, decreasing abruptly both northward and southward; clastic rocks dominate in other areas. The arkosic conglomerate in the Wilson Nunataks indicates that the source area was not far from this area and was composed of a granite-gneiss complex with some sedimentary cover. A high-energy sedimentary environment of shallow water is suggested by these coarse-grained sedimentary rocks.

The Dunbar Ridge Formation was deposited during Middle to Late Cambrian time, and the dominant black slate suggests a quiet, closed, shallow basin. Volcanism was totally terminated before this time.

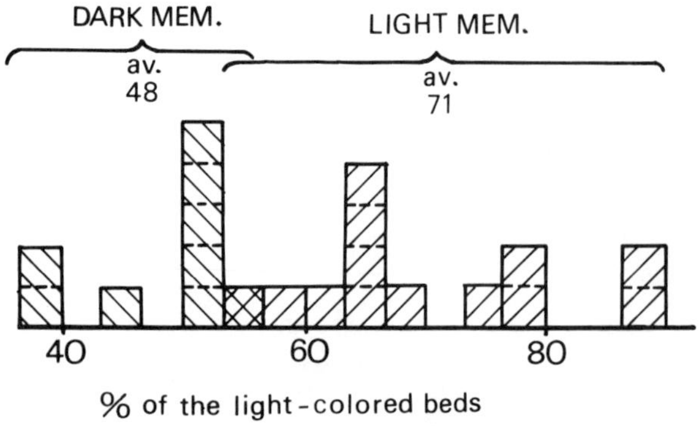

Figure 74.3. Percentages of light colored (quartzitic) beds at 22 localities in the Crashsite Quartzite. Measured on photographs of large exposures; each square represents one exposure. Counted at 3.33% intervals. av: average percentage of light colored beds.

THE CRASHSITE QUARTZITE

This formation consists mainly of white, gray, and green quartzite, commonly intercalated with dark-colored sandstone and, in a few places, with conglomerate and shale beds (Figures 74.2*a* and *c*). Two members are easily distinguished in each major exposure: the *lower Dark Member* and the *upper Light Member* (Figure 74.3); their thicknesses are 750–1000 m and 800–850 m, respectively.

The quartzites and sandstones have abundant cross-bedding, from which the paleocurrent direction has been determined. The current direction lies subparallel to the general structural trend and is from the south-southeast toward the north-northwest (Figure 74.1). This trend is roughly parallel to the distribution trend of the Heritage Group in the southern part of the mapped area, and it suggests that the major anticline might have been a submarine barrier which controlled the current direction. Ripple-marks and mudcracks are common in the silty beds.

Modal analyses and classifications of the quartzites and sandstones are summarized in Tables 74.1 and 74.2. The two members are generally similar, but the rocks of the Light Member are somewhat more mature than those of the Dark Member (Figure 74.4 and Table 74.1).

The sedimentary structures suggest a shallow-water, high-energy environment, and the modal data indicate a relatively low relief in the source area. No distinct volcanic fragments have been seen, and this suggests that the volcanic beds of the Minaret and Heritage Groups along the major anticline had not been exposed to erosion. Thus, the source land was not in the west, but possibly in the southeast.

Table 74.1. Classification of the quartzites and sandstones (by percentage of arenite and wacke) from the Crashsite Quartzite, following the system of Okada (1971). A total of 44 specimens were analyzed.

	Arenite (matrix < 15%)	Wacke (matrix ≥ 15%)
Dark Member		
Quartzite	4	23
Quartzitic sandstone	23	23
Feldspathic sandstone	0	4
Lithic sandstone	0	23
Total	27	73
Light Member		
Quartzite	11	11
Quarzitic sandstone	16	47
Feldspathic sandstone	0	5
Lithic sandstone	5	5
Total	32	68

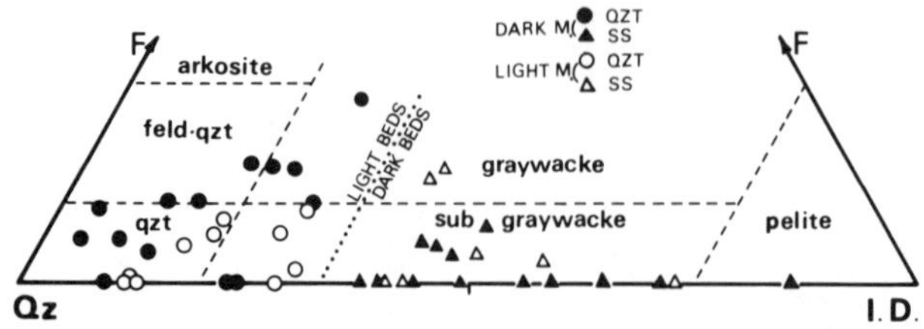

Figure 74.4. Rock classification of clastic rocks of the Crashsite Quartzite. Broken lines: after Pettijohn (1949). Dotted line: distinct border in the present rocks. Qzt: quartzite. Ss: sandstone. Feld: feldspathic. Corners of diagram: F: feldspar, Qz: quartz, and I.D.: inter-granular detritals.

Craddock (1969) distinguishes an upper Dark Member and describes the upper Crashsite Quartzite as including a significant amount of volcanic material. This unit has not been observed in the present study area.

THE WHITEOUT CONGLOMERATE

This is a dark brown, almost black formation with scattered pebbles of various sizes. This formation crops out here only in the Meyer Hills. The observed thickness is 400 m, but no upper contact was seen.

The matrix of this rock is coarse- to medium-grained sandstone and shale (modal matrix = 50–70%), very poorly sorted, and with nearly vertical cleavage strongly developed. The pebbles and boulders are K-feldspar porphyroblastic granite, hornblende-biotite granite, siliceous granitic migmatite, garnet-biotite gneiss, quartzite, various types of limestone, and gray and red sandstone. The maximum size of the boulders observed is 80 cm across.

The lower contact of this formation is conformable with the underlying Crashsite Quartzite in the northeast, but to the southwest the underlying quartzite shows a brown weathered zone for less than 1 m just below the conglomerate. This evidence, and the lack of an upper Dark Member, suggests a clino-unconformity at the base of the Whiteout Conglomerate.

STRATIGRAPHIC DISCONFORMITIES AND PROPOSAL OF NEW CLASSIFICATION

From rapid facies change and structural evidence, two possible disconformities are inferred within the sedimen-

Table 74.2. Modal proportions of the detrital minerals in the Crashsite Quartzite

	Percentage of rocks containing each mineral		Maximum modal percentages	
	Dark Member	Light Member	Dark Member	Light Member
Potash feldspar	65	71	15.7	11.2
Plagioclase	35	48	5.3	2.8
Muscovite	73	76	2.0	2.0
Biotite	27	19	<0.5	6.3
Garnet	58	67	5.9	3.1
Number of specimens[a]	26	21	26	21

[a] A total of 47 specimens were analyzed: 44 (as in Table 74.1) plus 3 calcareous shale specimens.

tary succession of this area: one at the base of the Dunbar Ridge Formation and the other at the base of the Whiteout Conglomerate.

The first may reflect a change of sedimentary conditions by the cessation of intense, local volcanism and associated block tectonics of the Edson Hills Formation time, with no accompanying folding.

In the second case, the Whiteout Conglomerate shows very gentle open folds of km-scale wavelength, while the Crashsite Quartzite below has many small-scale tight chevron and isoclinal folds of less than a few meters in wavelength. The fold styles are different in the two formations, suggesting a deformation phase before the deposition of the Whiteout Conglomerate. Although these structural contrasts may depend on the different physical natures of the two formations, the lack of an upper Dark Member of the Crashsite Quartzite indicates a clino-unconformity. If this is the case, the major deformation and metamorphism of the Ellsworth Mountains is older than the Permo-Carboniferous glaciation. This conforms with the K-Ar age of 298 m.y. reported by Craddock (1969) from a phyllitic rock.

On the basis of its exposure in the Marble Hills to the south, Craddock (1969) defined the Minaret Group as a succession of carbonate rocks. In the present study area, from the evidence of volcanism, this group, at least its upper part, is not distinguishable from the Heritage Group, and some marble beds are also intercalated in the latter. Moreover, the most distinct lithologic change occurs at the base of the Dunbar Ridge Formation. For these reasons a revised classification is proposed here as follows:

Dunbar Ridge Formation	
Heritage Group:	Edson Hills Formation: volcaniclastic sequence
	Middle Horseshoe Formation: volcanic-calcareous sequence
Minaret Group:	dolomitic succession in the Marble and Independence Hills

The description hereafter follows this new classification.

Igneous rocks

Igneous rocks occur in the Heritage Group in two separate areas: along the axial zone of the major anticline and in the Wilson Nunataks. In the former locality the rocks occur as concordant volcanic-stratigraphic piles, while in the latter a gabbro stock and some porphyrite dikes cut the folded Edson Hills Formation and are weakly metamorphosed. Thus, the igneous rocks of these two localities are of different ages: pre–Middle Cambrian and presumably late Paleozoic.

Pre–Middle Cambrian volcanism

This volcanic succession underlies the slate which includes Middle Cambrian trilobites and is therefore Middle Cambrian in age or older. In the Middle Horseshoe Formation various agglomerates are common, and deformed pillow lava occurs southwest of Chapell Peak, Horseshoe Valley. Volcanic rocks also occur as clasts in the conglomerate below the oolitic limestone, including a large number of mafic volcanic rock and plutonic rock pebbles and boulders.

In the Edson Hills Formation a mafic sill with baked country rocks and andesitic-basaltic lava flows occurs associated with a thick agglomerate and tuff breccia sequence in the Edson Hills and at Soholt Peaks. Boulders of gabbro, basalt, and dacite found on the ridge north of Drake Icefall are also assumed to be from the Edson Hills Formation. The rapid lateral facies changes suggest that the center of volcanic activity was nearby.

Twenty-four igneous rocks were analyzed for major and minor elements by X-ray fluorescence, titration, and flame-photometric methods. In chemical classification (Table 74.3), 71% of the analyzed rocks are alkaline according to MacDonald and Katsura (1964). Assuming a 1% addition of alkalis during diagenesis and low-grade metamorphism, 50% are still alkaline after correction. A total of 46% are K-alkaline (when 1% alkalies is subtracted, it is 30%), and the maximum K_2O is 6.75 wt % (No. 9 of Table 74.3). The pillow lava and the mafic rock pebbles of the conglomerate of the Middle Horseshoe Formation are all K-alkaline. This compositional evidence argues strongly for the existence of thick continental crust during this period.

The volcanic rocks near the contact between the Middle Horseshoe and Edson Hills Formations are mainly Na-alkaline, and a tholeiitic dacite also occurs in the uppermost marble of the Middle Horseshoe Formation (No. 12, Table 74.3). Mafic and dacitic rock boulders (Nos. 18–24, Table 74.3) have a tholeiitic nature.

The P_2O_5-TiO_2 and Zr-Sr diagrams (Figures 74.5*a* and *b*) indicate that most rocks are of a non-oceanic type. Low Zr suggests no contamination by crustal materials. Minor heavy metal elements show an almost flat or very gently negative relation to the FeO*/MgO (FeO* = FeO + $0.9Fe_2O_3$) ratio, being different from the oceanic rocks, e.g., in Figure 74.5*c*. The major elements SiO_2 and TiO_2 plotted against the FeO*/MgO ratio make a cluster on Figures 74.6*a* and *b* in the range of FeO*/MgO < 2.0, but some scattered plots faintly suggest that these rocks are of a continental alkaline type. This trend is evident in the K_2O-SiO_2 diagram (Figure 74.6*c*) by a more rapid increase of K_2O than in common oceanic alkaline rocks in a basaltic composition range. Rb, Ba, and Sr show the same trend.

Table 74.4 shows that the early eruptives in the Trap Series of the African Rift Valley (Mohr, 1960) tend to have a higher K-alkaline/Na-alkaline ratio and fewer calc-alkaline and tholeiitic rocks than the later Quaternary volcanic sequence. To have only K-alkaline rocks in the Middle Horseshoe Formation, associated with shallow-marine oolitic limestone, suggests a beginning of block subsidence on a continental crust. Lack of intermediate and silicic rocks argues against a back-arc origin.

The volcanic rock association of the Edson Hills Formation is Na-alkaline ≫ K-alkaline and has as many tholeiitic rocks as Na-alkaline ones. The tholeiites are not of an oceanic type, as shown in Figures 74.5*a* and *b*. This may indicate an advanced stage of block tectonics.

Late Paleozoic(?) Magmatism

A hornblende gabbro stock, measuring 100 m × 300 m, intrudes the presumptive Edson Hills Formation in the

Table 74.3. Classification and petrographic summary of 24 analyzed specimens of pre-Middle Cambrian volcanic rocks

Mafic Rocks

Specimen number	Symbol[a]	Primary minerals/Rock name	Norm[b] Q	Ne	Ol	Classification	Chemical classificiation: Macdonald and Katsura (1964)	Middlemost (1975)
Edson Hills Formation								
18	●	Cpx, pl/gabbro	0	0	10.76	Ol-th	Th-ol-bs	Flood bs
20	●	Pl/amygdaloidal dolerite	4.51	0	0	Th-bs	Quartz th-bs	Flood bs
19	○	Cpx, pl/porphyrite	0	0	12.06	Ol-th	Ol-bs	K-bs
13	○	Hb, pl/porphyrite	0	0.85	11.06	Alk-ol-bs	Alk-ol-bs	Hawaiite
14	○	Cpx, hb, pl/porphyrite	3.84	0	0	Th-bs	Quartz th-bs	High Al-bs/trachy bs
15	○	Cpx, pl/basalt	0	0	13.21	Ol-th	Ol-bs	Hawaiite
16	○	Pl/amygdaloidal basalt	3.81	0	0	Th-bs	Alk-bs	Hawaiite
17	○	Pl/andesite porphyrite	0	0	13.71	Alk-ol-bs	Alk-ol-bs	Hawaiite
Middle Horseshoe Formation								
1	×	Cpx, hb, pl/gabbro	0	0	14.32	Ol-th	Th-ol-bs/ol-bs	Hawaiite
2	×	Pl/schistose lava	0	0	1.24	Ol-th	Alk-bs	Hawaiite
3	×	Pl/rhomb porphyry	1.06	0	0	Th-bs	Quartz bs	Trachy bs
4	×	Cpx, hb, pl/rhomb porphyry	0	0	0.71	Ol-th	Th-bs/alk-bs	Flood bs/trachy bs
5	×	Ol, cpx, pl/gabbro	0	0	16.66	Ol-th	Th-ol-bs/ol-bs	Flood bs/K-bs
6	×	Ol, cpx, pl/porphyrite	0	0	9.46	Ol-th	Ol-bs	Trachy bs
7	×	Cpx, pl/amygaloidal basalt	0	5.79	7.63	Alk-ol-bs	Basanite	K-bs
8	×	Cpx, hb, bi, pl/diabase porphyrite	1.34	0	0	Th-bs	Quartz th-bs	High Al-bs/K-bs
9	×	Hb, pl, K-feldsp/porphyrite	0	0	3.30	Ol-th	Alk-bs	High K-bs
10	×	Ol, cpx, bi, pl, quartz/porphyrite	0	0	0.73	Ol-th	Alk-bs	Trachy bs
11	×	Ol, cpx, pl/porphyrite	0	1.20	12.93	Alk-ol-bs	Alk-ol-bs	Trachy bs

Silicic Rocks

Specimen number	Symbol[a]	Primary minerals/Rock name	Norm[b] Q	Or	Ab + An	Classification	Chemical classification: Kuno (1959)	Miyashiro (1975)
Edson Hills Formation								
21	▲	Pl/glassy dacite	34.32	9.57	50.47	Dacite	Subalkaline	Tholeiite
22	▲	Quartz/dacite	42.57	22.08	23.34	Rhyodacite	Subalkaline	Tholeiite
23	▲	Pl, quartz/parlite	40.39	16.77	35.80	Dacite	Subalkaline	Tholeiite
24	△	Pl/andesite (Cu-bearing)	19.89	4.37	71.18	Andesite	Alkaline	—
Middle Horseshoe Formation								
12	△	Pl/amygdaloidal lava	30.95	13.42	41.85	Dacite	Subalkaline	Tholeiite

Cpx: clinopyroxene. Pl: plagioclase. Hb: hornblende. Ol: olivine. Bi: biotite. K-feldsp: potash feldspar. Th: tholeiite. Bs: basalt. Alk: alkaline.

[a] Symbols keyed to Figures 74.5 and 74.6

[b] Norm classification of mafic rocks by Yoder and Tilley (1962) and of silicic rocks by Rittmann (1952).

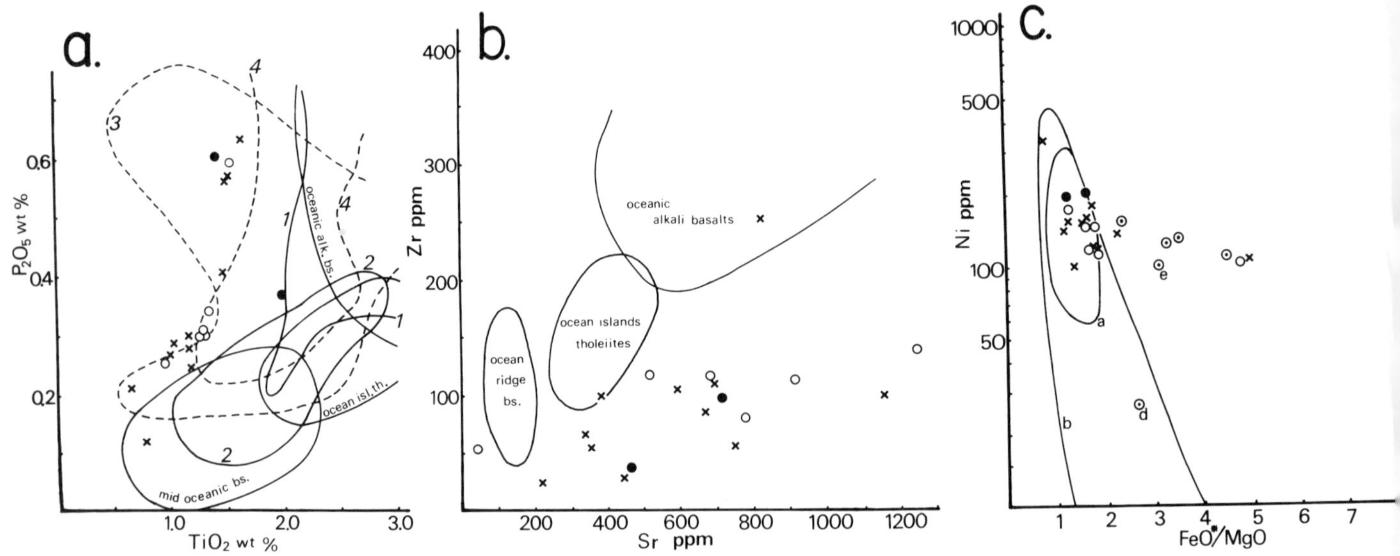

Figure 74.5. Minor elements of the pre-Middle Cambrian igneous rocks.

a. P_2O_5-TiO_2 diagram. 1. Hawaiian alkali rocks (Macdonald and Katsura, 1964); 2. Galapagos alkali rocks (McBirney and Williams, 1969); 3. Ethiopian alkali rock province (Mohr, 1960); 4. Circum-Japan Sea alkali rock province (Ono, 1962). Three curves of oceanic rocks are from Bass and others (1973). The symbols refer to the stratigraphic division in Table 74.3.

b. Zr-Sr diagram. The curves are from Bass and others (1973).

c. Ni-FeO* diagram, representing an example of differentiation of minor heavy elements (FeO* = FeO + 0.9 Fe_2O_3). Circle with dot in center: Late Paleozoic(?) igneous rocks of the Wilson Nunataks. d: a contaminated variety of the gabbro. e: porphyrite dike. All other points represent common varieties of the gabbro. The curves are from Miyashiro (1975); a: abyssal tholeiites. b: tholeiites from island arcs and active continental margins.

Table 74.4. Proportion of various rock series in volcanic rocks from the Heritage Group, Ellsworth Mountains, and from some other alkali rock provinces (in % analyzed specimens)

	Ellsworth Mountains		Ethiopia[a]		
Rock series	Middle Horseshoe Fm (n = 12)	Edson Hills Fm (n = 12)	Trap Series (n = 73)	Aden Series (n = 65)	Circum-Japan Sea[b] (n = 92)
Calc-alkaline	8	8	1	3	7
Tholeiitic	8	42	30	42	41
Na-alkaline	17	34	32	29	23
K-alkaline	67	16	37	26	26
Intermediate rocks ($55 < SiO_2 < 70$)	0	33	18	22	34
Silicic rocks ($SiO_2 > 70$)	8	0	29	30	30

n: number of analyses.
[a] Data from Mohr, 1960.
[b] Data from Ono, 1962.

middle of the Wilson Nunataks. Hornblende porphyrite dikes are associated and cut both the gabbro and the country rocks. The gabbro is a passive intrusion with a concordant contact along the steep bedding of the country rocks; it terminates abruptly along the strike.

All varieties of the gabbro have essentially a hornblende-plagioclase mineral assemblage, representing a relict ophitic texture, and are strongly decomposed by low-grade metamorphism.

Five varieties of gabbro and a porphyrite (e., in Figures 74.5, 74.6, and 74.7) were analyzed by the same methods as the pre-Middle Cambrian volcanic rocks. Figure 74.7 shows that these rocks have lower Fe contents and higher Na contents than the Mesozoic Ferrar Dolerites of the Transantarctic Mountains. All but one (d, in Figures 74.5, 74.6, and 74.7, a contaminated variety of the gabbro) are alkaline according to MacDonald and Katsura (1964). In Figure 74.6*c* rapid increase of K_2O occurs with increasing SiO_2, as in the older volcanic rocks, excluding d, but the trend is uncertain because of the limited SiO_2 range. The basic minor elements show a differentiation trend similar to that in the older rocks, for example as in Figure 74.5*c*. The TiO_2 and SiO_2 differentiation trends indicate a continental alkaline rock type.

Thus, from their chemical characteristics, these rocks cannot be correlated with the Ferrar Dolerites of the Transantarctic Mountains. If there occurred a regional deformation and metamorphism older than the Whiteout Conglomerate, then the emplacement of the gabbro and porphyrite may be late Paleozoic in age.

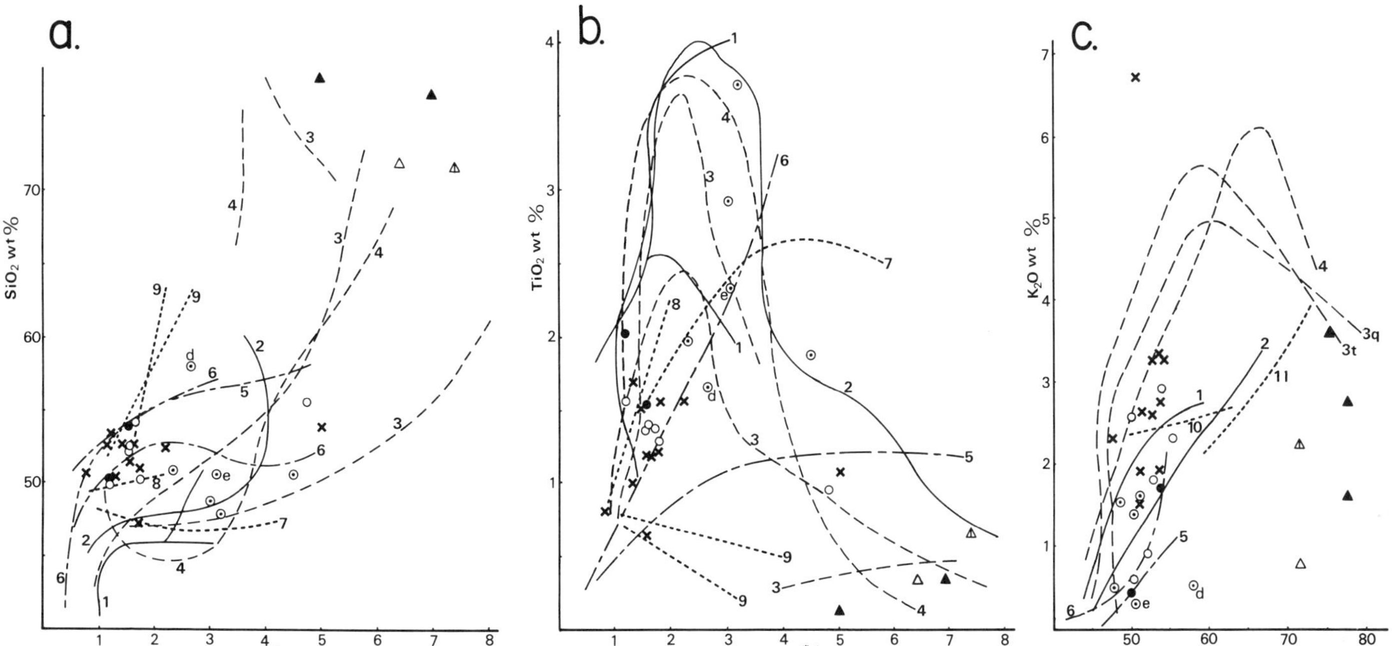

Figure 74.6. a. Differentiation trend of SIO_2; *b.* Differentiation trend of TiO_2; *c.* Differentiation trend of K_2O. Curves 1 to 4 are the same as in Figure 74.5*a*. 3t. The Trap Series; 3q. The Aden Series; both from Ethiopia (Mohr, 1960); 5. Ferrar Dolerites (Gunn, 1966); 6. Karroo Dolerites (Walker and Poldervaart, 1949); 7. Skaergaard Intrusion; 8. Abyssal basalts; 9. Island-arc calc-alkaline rocks; 10. New Guinea shoshonites; 11. Andes volcanic rocks. Data for curves 7, 8, and 9 are from Miyashiro (1975) and for 10 and 11 from Jakeš and White (1972). All curves show general trend of differentiation. The symbols refer to the stratigraphic division in Table 74.3.

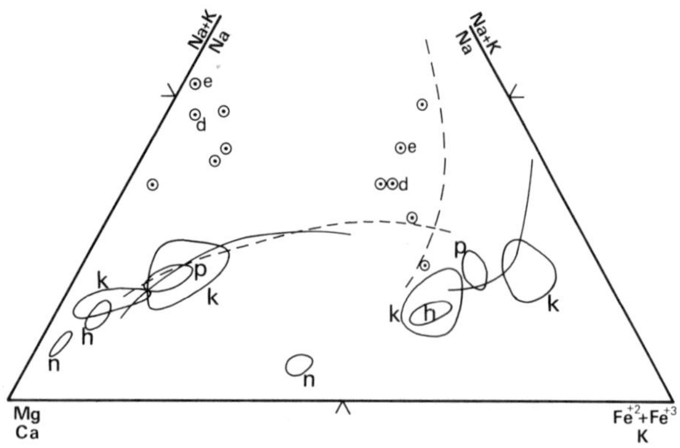

Figure 74.7. Atomic proportion of the upper Paleozoic(?) gabbro and porphyrite from the Wilson Nunataks (circles with dots in center), with reference to the Ferrar Dolerite and Andean intrusive rocks (curves). Two diagrams are superimposed: K-Na-Ca to the left and Fe^{2+} + Fe^{3+}-Na + K-Mg to the right. d: contaminated variety of the gabbro. e: porphyrite. All other points represent common varieties of the gabbro. n: norite zone, Lake Vanda Sill. h: hypersthenic tholeiites. p: pigeonitic tholeiites. Solid curve: New Mount Sill. (All four above from Gunn, 1966.) k: Kirkpatrick Basalt (Elliot, 1972). Broken curve: Andean intrusive rocks of eastern Ellsworth Land (Laudon and others, 1969).

Geologic structures

Folds

The major structure of this area is an anticline from the western Horseshoe Valley northward to the Schneider Glacier; the axis plunges 10–20° northwestward in the north and is almost horizontal in the south. The fold style is cylindrical in general, but the southern part shows a slightly conical nature. This anticline is not a single fold, but includes parasitic folds of some hundreds of meters in wavelength, and it has a slightly overturned eastern limb which is cut by the major strike fault (Figures 74.1 and 74.8).

Most of the area, except for the northwestern part, belongs to the eastern limb of the major anticline. Intense parasitic folds prevent the study of regional structures (Figures 74.2*a* and *c*); however, fold-envelopes were calculated from the general dip of some distinct beds and from local fold axes in many large exposures. Three anticline-syncline pairs are mapped in the area of the Enterprise and Meyer Hills; each has a 7–10 km wavelength and an amplitude of a few km (Figure 74.8). The axial trend is N 32° W with a gentle northwesterly plunge to the north of the Union Glacier and a very gentle southeasterly plunge in the

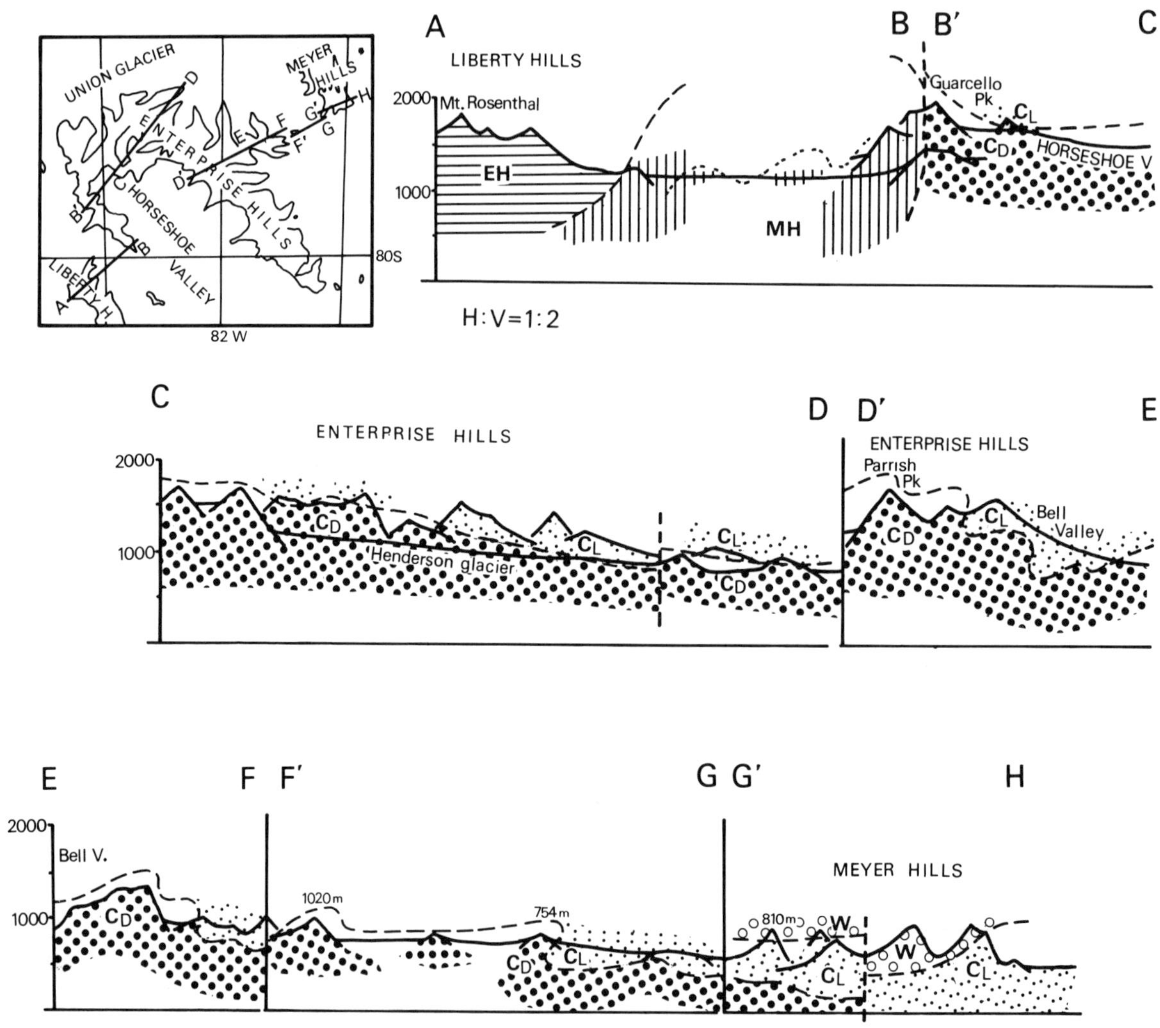

Figure 74.8. Composite profile across the Enterprise Hills, southern Heritage Range. MH: Middle Horseshoe Formation. EH: Edson Hills Formation. C_D and C_L: Dark and Light Members of the Crashsite Quartzite. W: Whiteout Conglomerate. Height in meters.

south. These folds are asymmetric with a steeper eastern limb, indicating tectonic transport from the west as in the major anticline.

The parasitic folds are mainly of an asymmetric chevron style with some isoclinal folds; many are overturned. Steep axial plane cleavages are very well developed in the shaly beds, and the competent beds show radial cleavages at the crests. The calculated fold axes conform well with the observed ones (Figure 74.9). A cross-joint system is developed in the area and has caused small disturbances of the bedding planes; many of the calculated subordinate fold axes lie parallel to the joint trends.

FAULTS

A distinct strike fault occurs along the eastern limb of the major anticline, following the oolitic limestone of the Middle Horseshoe Formation in the south. In the north, this fault trends slightly more northerly and separates the shale of the Dunbar Ridge Formation and the Crashsite Quartzite. This fault is nearly vertical, but has a slight reverse tendency consistent with the tectonic transport from the west. A branch of this fault occurs in the Collier Hills (Figure 74.2*c*). The throw of the major fault is 5–6 km in the south and less than 1 km in the north.

A strike fault which is observed in the Meyer Hills cuts the Whiteout Conglomerate. The throw is less than 300 m.

Two transverse faults are inferred along the Union Glacier, based on displacement of the oolitic limestone bed. The northern sides are thrown down by nearly vertical faults. These occur in the axial culmination zone of the major anticline, and the strike is consistent with the direction of the cross-joints and the subordinate fold axes.

Metamorphism

Quartz, phyllosilicates, and carbonates are well recrystallized in all the sedimentary rocks, but all plagioclase and garnet grains are detrital, and muscovite and biotite are commonly preserved as large fragmental flakes. Detrital garnet is commonly decomposed into chlorite along the margins and cracks.

The analyzed volcanic rocks, when plotted in the ACF diagram, are mostly in the field of epidote-chlorite-actinolite, and neither pumpellyite nor stilpnomelane have been found in any partial assemblage. Dolomite and calcite occur sparsely in the crushed matrix. Plagioclases have been converted into an albite-epidote (zoisite)-sericite mixture. The actinolite-chlorite tie line may be partially disturbed by the formation of biotite. All observed hornblende and pyroxene grains are relict from original magmatic associations. Typical metamorphic assemblages include actinolite, albite, epidote, sphene, sericite, and carbonates. The grade of metamorphism is the actinolite-greenschist facies, corresponding to zones III and IV of New Zealand and to the chlorite and biotite zones of Scotland. Temperature range can be limited by the reactions prehnite + calcite + quartz → epidote + actinolite + vapor on the lower side (330°–350° C, Nitsch, 1971) and on the upper side by the decomposition of Fe-chlorite (550–600° C, Turnock, 1960), both in the pressure range of 1–5 kb. No evidence for high pressure has been found. This metamorphic grade is common from the Heritage Group through the Whiteout Conglomerate, while the overlying Polarstar Formation is of the laumontite facies with dominant illite and carbonates in the matrix (Castle and Craddock, 1975).

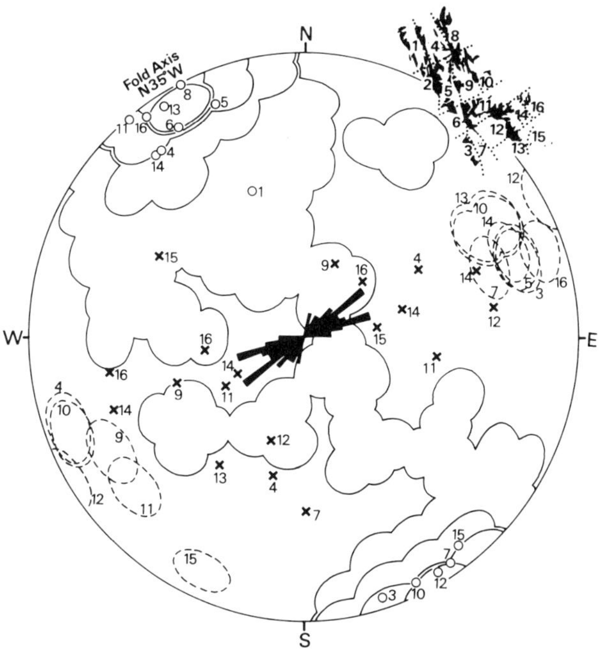

Figure 74.9. Composite fabric diagram. Solid contours: observed fold axes (1, 5, 10, 15, and 17.5% contours). Dashed ovoids: maxima of cleavages. Open circles: calculated main fold axes. Crosses: calculated subordinate fold axes. Rosette: observed joints. The numbers refer to subareas shown in the inset map of the Heritage Range in the upper right corner. Lower hemisphere equal-area projection.

Conclusions

The basis of stratigraphic division is mainly lithostratigraphic in the present study area. The distinct change in rock types at the base of the Dunbar Ridge Formation is considered to be the major disconformity below the Crashsite Quartzite. A new classification is proposed in the lower part of the sequence:

Dunbar Ridge Formation
Heritage Group { Edson Hills Formation / Middle Horseshoe Formation }
Minaret Group

A possible clino-unconformity is inferred at the base of the Whiteout Conglomerate, and the metamorphic grade is distinctly lower in the overlying Polarstar Formation (Permian). The lone phyllite K-Ar age (Craddock, 1969) also conforms with this postulated regional deformation and metamorphism during late Paleozoic time.

Two phases of magmatism have been distinguished, corresponding to the stratigraphic gaps: pre-Middle Cambrian and probably late Paleozoic. Chemical characteristics of the magmatic rocks are K-alkaline in the early activity, with a change to Na-alkaline + tholeiite in the later stage. These alkaline rocks are considered to be a non-oceanic suite, and their presence argues for block tectonics on a continental crust.

Acknowledgments. We are indebted to the Norsk Polarinstitutt of Norway and the National Science Foundation of the U.S.A. for making the expedition possible. We are grateful to C. Craddock, G. F. Webers, and A. B. Ford for their kind sharing of information and discussion.

References cited

Bass, M. N., Rhodes, J. M., Shih, Chi-Yu, and Church, S. E. 1973. Volcanic rocks cored in the central Pacific, Leg 17, Deep Sea Drilling Project. (*In* Winterer, E. L., Ewing, I. I., and others. *Initial Reports of the Deep Sea Drilling Project,* Vol. 17. Washington [U.S. Government Printing Office], 429–503.)

Castle, J. W., and Craddock, C. 1975. Deposition and metamorphism of the Polarstar Formation (Permian), Ellsworth Mountains. *Antarct. Jnl U.S., 10,* No. 5, 239–41.

Craddock, C. 1969. Geology of the Ellsworth Mountains (Sheet 4, Ellsworth Mountains). (*In* Bushnell, V. C., and Craddock, C., *eds.* Geologic Maps of Antarctica. *Antarct. Map Folio Ser.,* Folio 12, Pl. IV.)

Craddock, C., Anderson, J. J., and Webers, G. F. 1964. Geologic outline of the Ellsworth Mountains. (*In* Adie, R. J., *ed. Antarctic Geology.* Amsterdam, North-Holland Publishing Company, 155–70.)

Elliot, D. H. 1972. Major oxide chemistry of the Kirkpatrick Basalt, central Transantarctic Mountains. (*In* Adie, R. J., *ed. Antarctic Geology and Geophysics.* Oslo, Universitetsforlaget, 413–18.)

Gunn, B. M. 1966. Modal and element variation in Antarctic tholeiites. *Geochim. cosmochim. Acta, 30,* No. 9, 881–920.

Jakeš, P., and White, A. J. R. 1972. Major and trace element abundances in volcanic rocks of orogenic areas. *Bull. U.S. geol. Surv., 83,* 29–40.

Kuno, H. 1959. Origin of Cenozoic petrographic provinces of Japan and surrounding areas. *Bull. volcan.,* Śer. 2, *20,* 37–76.

Laudon, T. S., Lackey, L. L., Quilty, P. G., and Otway, P. M. 1969. Geology of eastern Ellsworth Land (Sheet 3, eastern Ellsworth Land). (*In* Bushnell, V. C., and Craddock, C., *eds.* Geologic Maps of Antarctica. *Antarct. Map Folio Ser.,* Folio 12, Pl. III.)

Macdonald, G. A., and Katsura, T. 1964. Chemical composition of Hawaiian lavas. *J. Petrol., 5,* 82–133.

McBirney, A. R., and Williams, H. 1969. Geology and petrology of the Galapagos Islands. *Mem. U.S. geol. Surv., 118,* 197 pp.

Middlemost, E. A. K. 1975. The basalt clan. *Earth Sci. Rev., 11,* 337–64.

Miyashiro, A. 1975. Classification, characteristics and origin of ophiolites. *J. Geol., 83,* 249–81.

Mohr, P. A. 1960. *The Geology of Ethiopia.* Adis Abeba, University-College, 268 pp.

Nitsch, K. H. 1971. Stabilitätsbeziehungen von Prehnit- und Pumpellyite-haltigen Paragenesen. *Contr. Miner. Petrol., 30,* 240–60.

Okada, H. 1971. Classification of sandstone: analysis and proposal. *J. Geol., 79,* 509–25.

Ono, K. 1962. *Chemical Composition of Volcanic Rocks in Japan.* Tokyo, Geological Survey of Japan, 441 pp.

Pettijohn, F. J. 1949. *Sedimentary Rocks.* New York, Harper and Brothers, 526 pp.

Rittmann, A. 1952. Nomenclature of volcanic rocks. *Bull. volcan., 12,* 75–102.

Turnock, A. C. 1960. The stability of iron chlorites. *Yearbook Carnegie Inst., 59,* 98–103.

Walker, F., and Poldervaart, A. 1949. Karroo dolerites of the Union of South Africa. *Bull. geol. Soc. Am., 60,* No. 4, 591–705.

Webers, G. F. 1972. Unusual Upper Cambrian fauna from West Antarctica. (*In* Adie, R. J., *ed. Antarctic Geology and Geophysics.* Oslo, Universitetsforlaget, 235–37.)

Yoder, H. S., and Tilley, C. E. 1962. Origin of basalt magmas: an experimental study of natural and synthetic rock systems. *J. Petrol., 3,* 342–52.

75

The Swanson Formation, Ford Ranges, Marie Byrd Land— Evidence For and Against a Direct Relationship with the Robertson Bay Group, Northern Victoria Land

F. ALTON WADE† and DONA R. COUCH*
Antarctic Research Center, The Museum, Texas Tech University, Lubbock, Texas 79409, U.S.A.

Abstract. A sequence of geosynclinal metasedimentary rocks constitutes the Swanson Form on which crops out in the central and southern parts of the Ford Ranges, Marie Byrd Land. Thickness is estimated to be in excess of 4300 m. Based upon limited evidence, the age is late Precambrian–Cambrian. Deformation of the Swanson Formation occurred during the Ford orogeny, resulting in broad, open to closed folds whose major axes trend roughly northwest-southeast. The orogeny culminated with the intrusion of granodioritic plutons during Devonian–Carboniferous time. Metamorphism was a low grade, only locally altering the rocks to higher than greenschist facies. Thermal metamorphism associated with middle Cretaceous granitic intrusions further modified the rocks. The metasedimentary sequence includes metagraywacke, hornfels, schist and semischist, slate and phyllite, and rare marble, metavolcanic rocks, and spilite. Deposition was in a geosynclinal basin of varying depth controlled by the rate of subsidence. The source rock of the sediments was probably of granitic composition. A suggested direct relationship between the Swanson Formation and the Robertson Bay Group in northern Victoria Land is based upon similarities in rock types, deformation patterns, and metamorphic histories. In order to determine the validity of this suggestion, the literature concerning the geology of northern Victoria Land was reviewed and the data were compared with those concerning the geology of the Ford Ranges. The similarities were confirmed. Differences, however, are just as striking. After weighing all evidence, we favor the suggestion that the two sectors are totally unrelated.

The Swanson Formation crops out in western Marie Byrd Land in the central and southern parts of the Ford Ranges and in the Alexandra and Rockefeller Mountains in Edward VII Peninsula. The Ford Ranges are a series of ridges whose trends range from sublatitudinal in the north to northwest-southeast in the south (Figure 75.1). The valleys between the ridges are partially filled with outlet glaciers that drain the plateau ice which is impounded to the east and south by these ranges. The Rockefeller Mountains are grouped in the south-central part of Edward VII Peninsula in no particular pattern. The Swanson Formation crops out only in the south-central sector of these mountains. The Alexandra Mountains are widely spaced nunataks aligned roughly along the northeast coast of the peninsula. Outcrops of the Swanson Formation occur only in the southern half of the mountains.

Maximum elevations of ridges and individual peaks in the region do not exceed 1300 m. Relief of the rock exposures ranges from a few meters to more than 800 m.

History of exploration

The northernmost of the Alexandra Mountains were seen by Scott (1905) in 1902 from aboard ship, and they were explored in 1911 by Prestrud of Amundsen's expedition (Amundsen, 1913). In 1929 the Rockefeller Mountains and the Ford Ranges were discovered by Byrd (1930) during exploratory flights from Little America. Washington Peak, southernmost of the Rockefeller Mountains, was briefly visited by Gould (Byrd, 1930) that same year.

Exploration of the Ford Ranges was begun in 1934 by a four-man party which included the senior writer as a member (Wade, 1937). A more detailed survey of the central part was carried out by Warner (1945) and Passel (1945) in 1940. During the same year Wade (1945) surveyed the Rockefeller Mountains. The Clark Mountains, easternmost of the Ford Ranges, were investigated briefly in 1960 by the oversnow traverse from Byrd Station (Boudette and others, 1966). The U.S. Antarctic Research Program geologic survey of Marie Byrd Land began in 1966 and continued for three years. During the first field season, 1966–67, investigations were mainly in the Ford Ranges (Beitzel and Bentley, 1967; Klimov, 1967; Scharon and Early, 1967; Wade, 1967, 1969). In 1967 Wilbanks led a three-man survey team which investigated the Fosdick Mountains in the Ford Ranges (Wilbanks, 1972). No field parties have worked in the area since that year.

Geologic setting

The Swanson Formation is a sequence of metasedimentary rocks which is associated with (1) an older, highly metamorphosed and migmatized sequence of metasedimentary rocks (the Fosdick Complex), (2) two generations of intrusive plutons, and (3) Cenozoic extrusive rocks. The

† Deceased.

* Present address: Oil Development Co., Box 12058, Amarillo, Texas 79100, U.S.A. Formerly Dona R. Long.

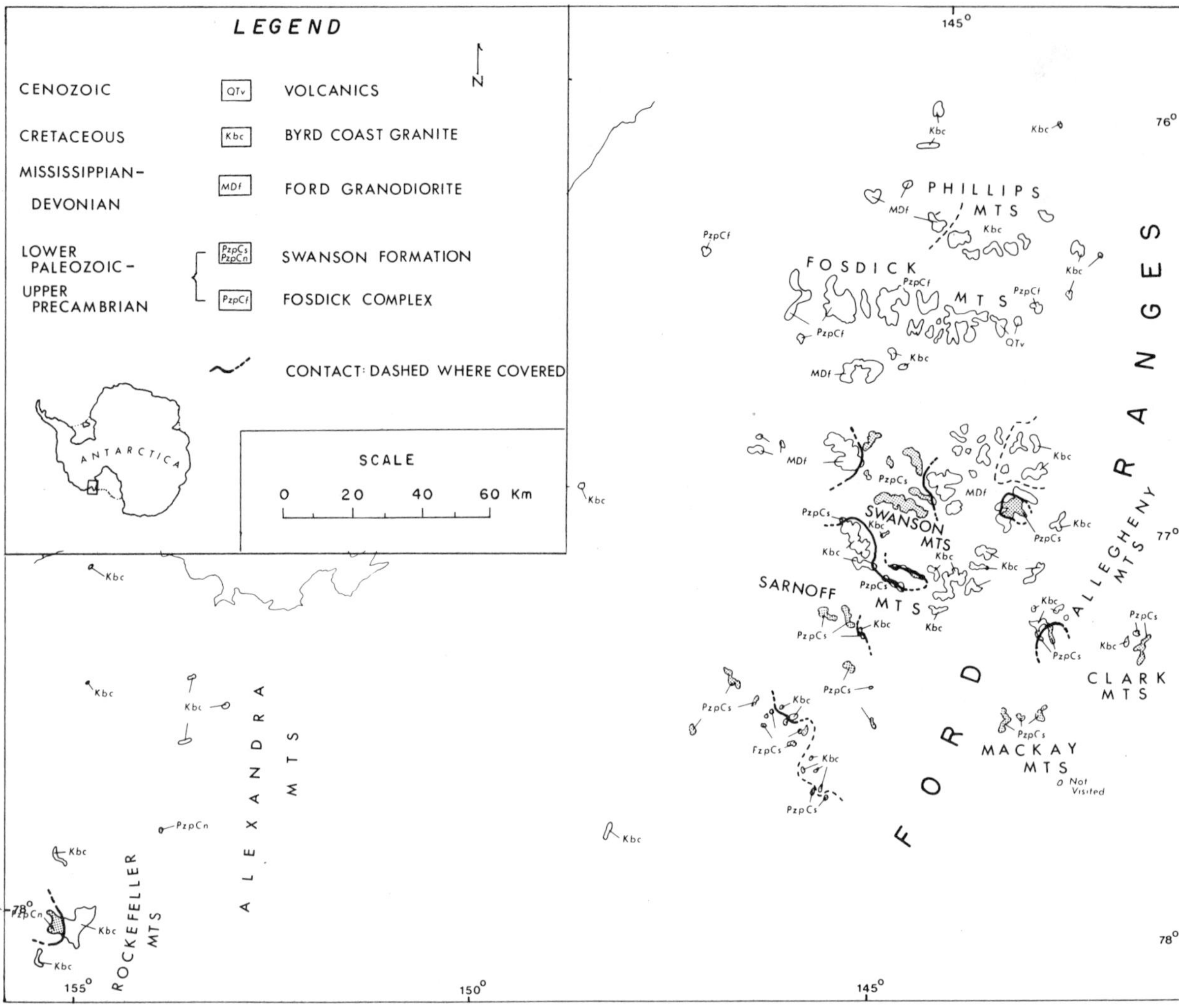

Figure 75.1. Geologic map of the part of western Marie Byrd Land covered by this report.

Fosdick Complex, Swanson Formation, and the older intrusive plutons constitute what is generally referred to as the "basement complex."

The thickness of the Swanson Formation cannot be measured because neither the lower contact nor the upper contact is exposed. Folds and faults also complicate the problem. Warner (1945) estimated the thickness to be in excess of 4300 m. Based upon the investigations of Klimov (1967), Lopatin and Orlenko (1972) suggested a thickness of 12 to 15 km.

The rocks of the Swanson Formation (Figure 75.2) are predominantly dark gray, greenish gray, and tan. Thin films of red and brown secondary iron minerals cover some surfaces.

The sedimentary rocks comprising the Swanson Formation were, with few exceptions, originally graywacke, sandstone, and argillite. Deformation and low-grade metamorphism have altered the sedimentary rocks to metagraywacke, hornfels, slate, phyllite, semischist, and schist. All these rock types are very similar in mineral composition. In most units, regardless of rock type, quartz, biotite, and muscovite constitute approximately 90% of the rock. Feldspars are present in nearly all rocks, but—except where present as phenoblasts in graywacke—they compose less than 5% of the total volume. Minor constituents include apatite, tourmaline, chlorite, zircon, rutile, magnetite, and ilmenite. The principal secondary constituents are calcite, sericite, kaolinite, epidote, limonite, hematite, and leucoxene.

Petrography

Metagraywacke

Metagraywacke dominates the formation. One section which seemed to be quite typical was measured; in this 415 m sequence, 60% of the rock was metagraywacke. In general the metagraywacke can be described as composed of matrix material and phenoblasts (Figure 75.3). In the matrix, quartz is the dominant mineral, constituting 60% of the total volume on the average. The matrix grains are subangular to subrounded. Biotite is generally the second most common mineral in the matrix material, comprising about 35%. In a few beds muscovite is more abundant than biotite. Minor and accessory constituents generally make up <5% of the matrix. Phenoblasts range upward in size from that of the matrix grains, which are usually <0.1 mm. In most of the metagraywacke units, maximum size of the phenoblasts is <1 mm, but in a few specimens it is as much as 5 mm. The phenoblasts are randomly distributed and are sharply angular to subrounded. Most phenoblasts are either quartz or potassium feldspar. Lithic frag-

Figure 75.2. The Swanson Mountains, type locality for the Swanson Formation.

ments of quartzite are common, and of older graywacke, quite rare. Graded bedding characterizes some units, and poorly developed cross-bedding has been noted.

HORNFELS

This type of rock constitutes approximately 24% of the measured section. The hornfels is essentially the same as the metagraywacke but without the phenoblasts. It is fine grained and usually lacks structural features. In some places, however, bedding, cross-bedding, and slump structures have been noted. As in the metagraywacke, quartz and biotite constitute 90%–95% of the rock in most units. In rare specimens muscovite exceeds 5%. The most common accessories are the opaque minerals.

SLATE AND PHYLLITES

The pelitic sedimentary rocks, through diagenetic and metamorphic processes, have been altered to slate and low-grade phyllite. Identifiable minerals include quartz, biotite, muscovite, chlorite, and opaques. Quartz consti-

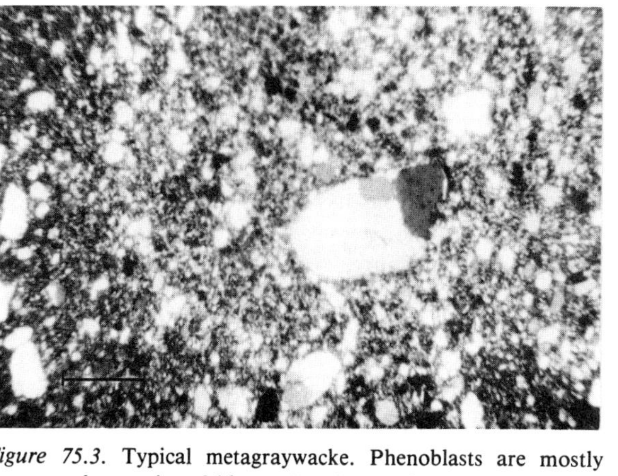

Figure 75.3. Typical metagraywacke. Phenoblasts are mostly quartz and potassium feldspar. Dominant minerals in the matrix are quartz and biotite. Length of bar is 0.5 mm.

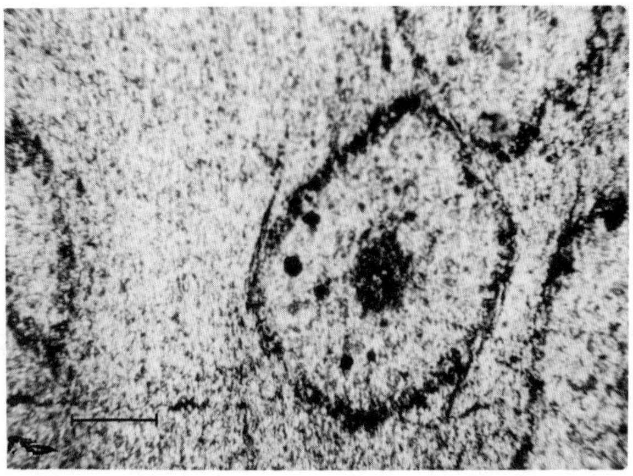

Figure 75.4. Biotite-deficient "spot" in a slate. Rim is composed of a concentration of biotite and hematite. Bar is 0.5 mm long.

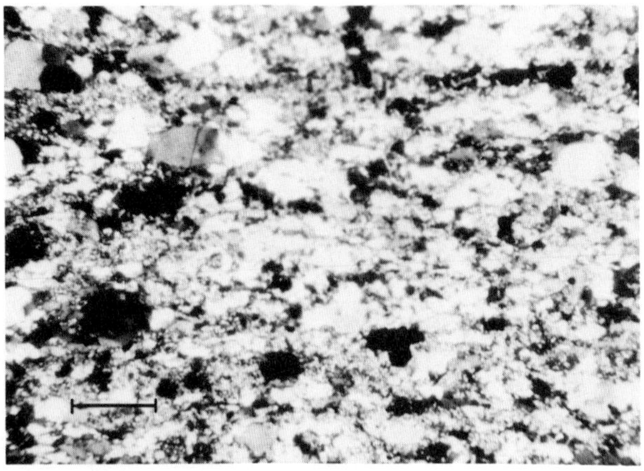

Figure 75.5. Semischist. Segregation of minerals into layers has proceeded far enough to produce fair schistosity. Length of bar is 0.5 mm.

tutes 35% to 40% of the total volume, and biotite and muscovite together about 50% to 55%. Chlorite is present in most specimens but only as a minor accessory mineral. Opaque minerals constitute as much as 10% of some specimens. Pyrite and an iron carbonate, probably ankerite, are scattered throughout some specimens. Good slaty cleavage has developed in most of these fine-grained rocks. The only basis for designating some of these rocks as phyllite is the sheen which is exhibited on some cleavage surfaces. Much of the slate and phyllite is spotted. The "spots" are actually parts of the rock which are biotite deficient. In many places the spots are rimmed with a zone in which biotite and hematite are concentrated (Figure 75.4). They probably result from metamorphic differentiation processes. The slate and phyllite constitute <4% of the measured section. In other areas, however, they dominate the section.

Semischist and schist

In the measured section, approximately 12% of the sequence is classified as semischist or schist. The principal mineral components are the same as in the rock types just described. Quartz constitutes about 40% of the total volume, and biotite and muscovite together approximately 50%. Minor constituents include chlorite, tourmaline, zircon, rutile, and opaque minerals. In the semischist the schistosity is fairly well developed. Segregation of minerals into bands is minimal (Figure 75.5). In the schist, however, mica-rich and mica-lean bands are characteristic, and the schistosity is well developed.

Marble

Beds of carbonate rock in the Swanson Formation are rare. Calcitic marble has been noted in several widely separated localities. Few beds exceed 10 cm in thickness. The marble is impure, being contaminated with clastic detritus.

Volcanic rocks

Two occurrences of volcanic rocks have been observed interbedded with the metasedimentary sequence. One, a rhyolite tuff that crops out at Bailey Ridge in the Sarnoff Mountains, has devitrified and includes secondary silica in the form of chalcedony (Figure 75.6). The second volcanic rock crops out at Mount Spencer in the Allegheny Mountains. It is an extremely fine-grained ash-flow tuff which has devitrified (Figure 75.7). The composition is probably rhyolitic. Two layers of spilite were also noted.

Structural features

Deformation during the first of two orogenies to which the Swanson Formation has been subjected resulted in a system of folds whose axes trend, in general, northwest-southeast and plunge to the northwest. The intensity of the folding seems to increase from southwest to northeast, from broad open folds to closed folds, some of which may be isoclinal. Synclinoria and anticlinoria have been identified. Local deviations from this general structure can largely be attributed to deformation caused by intrusion of plutons at two subsequent times. Faults of small displacement are common.

Primary structures, which developed during deposition and which are still identifiable, include bedding, cross-bedding, graded bedding, ripple marks, mud cracks, and slump structures.

Schistosity ranges macroscopically from absent in the metagraywacke and hornfels to well developed in the schist. However, microscopic *s*-surfaces are identifiable in nearly all specimens, and two generations of these are obvious. Many of the older *s*-surfaces parallel axial planes or bedding surfaces. The younger surfaces intersect the older ones, generally at angles between 30° and 40°.

Metamorphism

During the first of the two orogenic events which affected the Swanson Formation, regional metamorphism converted the sedimentary rocks to metagraywacke, hornfels, slate, phyllite, semischist, and schist. Except in localities where the sedimentary rocks are adjacent to large plutons, the degree of metamorphism did not exceed that necessary to produce the biotite stage of the greenschist facies. The metamorphic grade is remarkably uniform. The muscovite and biotite flakes are in fair to good alignment. Superimposed upon and partially obliterating this fabric are randomly oriented mica flakes, principally biotite, which developed from thermal metamorphism when younger plutons were intruded. The degree of this metamorphism was slightly lower than that attained during the

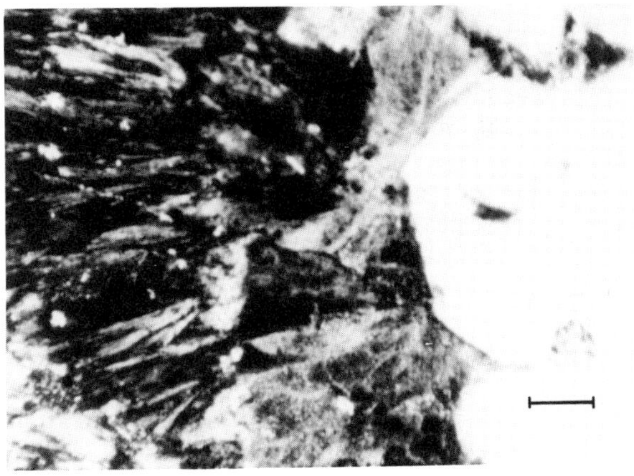

Figure 75.6. Metamorphosed rhyolite tuff. Phenocrysts of embayed potassium feldspar crystals on the right, secondary silica in form of chalcedony on the left. Bar is 0.1 mm long.

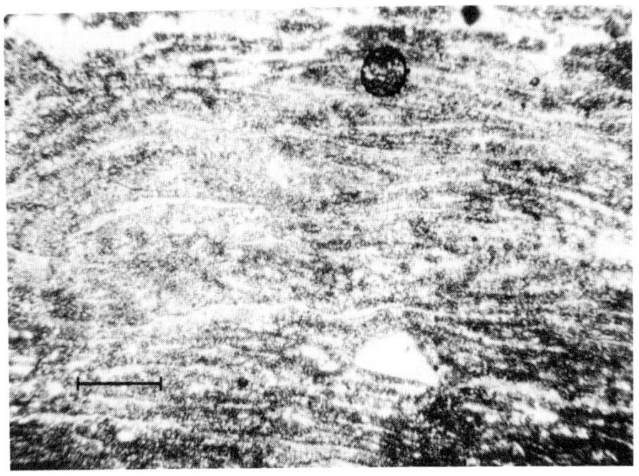

Figure 75.7. Ash-flow tuff, partially devitrified. Length of bar is 0.5 mm.

dynamothermal event. Small amounts of chlorite were formed by retrogressive metamorphism during this event.

Associated formations

Fosdick Complex

The Fosdick Mountains are located in the northern part of the Ford Ranges along lat. 76° 30′ S and between long. 143° 45′ W and 147° 30′ W. They are composed of rock types dissimilar to other rock units in the Ford Ranges. Wilbanks (1972) surveyed in detail the eastern and central parts of the Fosdick Mountains in 1967–68. In his report he states: "The Fosdick Mountains crystalline mass comprises an upper biotite-gneiss and granitoid injection complex passing laterally and vertically to some extent into a lower more 'granitized' and deformed migmatite and nebulite complex." He groups the rock types into three broad classes: banded biotite-gneiss, porphyroblastic biotite-gneiss, and quartzo-feldspathic nebulite. There is evidence indicating two deformational events. The older event resulted in folding along a northeast-striking axis, and the younger produced folds along a west-northwest–striking axis. In the central and western parts of the range the younger deformation virtually obliterated most evidence of the older one. A logical interpretation of the evidence led Wilbanks (1972) to suggest that the Fosdick Complex and the beds of the Swanson Formation had an infrastructure-superstructure relationship. The most reliable evidence for this interpretation was that the fold orientation of the Swanson Formation was identical with that of the younger folds in the Fosdick Complex. He discarded the suggestion that the Fosdick Mountains represent an uplift of older basement. We believe, however, that this concept should not be discarded; it should be considered a possible explanation, and the evidence reexamined. The Fosdick Mountains are bounded on the north by Block Bay and Balchen Glacier which together may occupy a graben valley. The terminations of the north-south ridges on the south side of the Fosdick Mountains, when connected, make nearly an east-west line. The Fosdick Mountains could well be an upthrown block. Fault surfaces, however, are not exposed to direct observation. To date no remnants of the Swanson Formation have been observed capping the Fosdick Complex; the nearest outcrop of the former is located over 30 km south of the latter. No outcrops of the Swanson Formation have been observed to the north of the Fosdick Mountains. The basin in which the Swanson sediments were deposited may have been located entirely south of the Fosdick Mountains. The source of the sediments composing the Swanson Formation could well have been the Fosdick Complex.

Age relationships of these two formations have not been satisfactorily established. There is general agreement that the Fosdick Complex is the older of the two. From specimens of the Swanson Formation, Iltchenko (1972) recovered acritarchs and other microfossils which indicate a late Precambrian–early Paleozoic age. Three specimens of the formation collected at Mount West yielded K-Ar ages ranging from 445 to 475 m.y. (Krylov and others, 1970). Radiometric ages for rocks of the Fosdick Complex, however, indicate an apparent Cretaceous age (Halpern, 1972; Krylov and others, 1970). Halpern, explaining this anomalous situation, suggested that the Fosdick Mountains had been greatly affected during Cretaceous time by deformation and granitic intrusions, possibly resetting the Rb-Sr ages.

Middle Paleozoic intrusive rocks

Large plutons of granodiorite were intruded into the Swanson Formation and older rocks of the Ford Ranges during middle Paleozoic time. K-Ar ages for specimens from these plutons range from 301 ± 11 m.y. to 355 ± 12 m.y. (Wade, 1972). Rb-Sr isotopic analyses (three whole-rock and five on biotite concentrates) on specimens collected in the Phillips Mountains indicate an age of 352 ± 21 m.y. (Halpern, 1968). The relative ages of the Swanson Formation and the granodiorite have been determined at Mount Swan (Swanson Range) where the contact relationships are observable.

Middle Cretaceous intrusive rocks

Granite plutons of Cretaceous age are widely distributed throughout the Ford Ranges and Edward VII Peninsula (Boudette and others, 1966; Wade, 1969, 1972; Halpern, 1968). Contacts of the granite with rocks of the Swanson Formation are exposed in many localities. These contacts are sharp (Figure 75.8) except where interfingering has occurred. Low-grade thermal metamorphism of the Swanson rocks resulted during and subsequent to the intrusive

Figure 75.8. Contact of Cretaceous granite (above) with metagraywacke (below). Contact is sharp, and no contact metamorphism is evident. Bar is 0.5 mm long.

event. Only in narrow zones adjacent to contacts was the grade of metamorphism raised above greenschist facies.

CENOZOIC VOLCANIC ROCKS

Volcanic rocks, principally olivine basalt and hyaloclastite, are present only in the eastern Ford Ranges and, there, only in association with the Fosdick Complex. They have not been observed in contact with rocks of the Swanson Formation.

Geologic history

From the information summarized above, a reconstruction of the geologic events leading to the present state of the Swanson Formation is attempted. A paucity of critical data makes interpretation of the early part of the history highly speculative.

We visualize this sector of Marie Byrd Land during late Precambrian time as ancient continental crust, possibly greater in area than at present. During a Precambrian orogenic event, the "Ancestral Fosdick Mountains" were developed; the trend of the fold axes was to the northeast. In late Precambrian–early Paleozoic time a geosyncline developed in the area now occupied by the central and southern Ford Ranges and in Edward VII Peninsula. This event marked the beginning of the Ford orogenic cycle. The basin, or basins, was located mainly south of the "Ancestral Fosdick Mountains." The trend of the basin was probably northwest-southeast. The eroding Fosdick highlands supplied the detritus which was deposited in the geosyncline to form the Swanson Formation. Intense folding of the sedimentary strata along northwest-trending axes occurred during Ordovician–Silurian time. The adjacent mountain remnant and roots of the ancient Fosdick Mountains were involved in the deformation. This event resulted in the development of the younger system of folds in the Fosdick Complex.

In Late Devonian time large plutons of granodiorite intruded the Precambrian and lower Paleozoic rocks. They crop out in an east-west zone in the northern part of the Swanson Mountains and in the westernmost Phillips Mountains located to the north of the Fosdick Mountains.

There is a hiatus in the geologic record from Carboniferous to middle Cretaceous time. We can only speculate that it was primarily a time of erosion, but where the products of the erosive processes were deposited is not known. About 95 to 105 m.y. ago, throughout West Antarctica huge granite plutons were intruded, preceded by only minor or no deformation of the country rock. Such deformation as occurred was caused by the intruding magma and was minimal. It appears likely that the part of the Fosdick Complex which is exposed today was deep seated at the time of the intrusive activity and was locally migmatized and granitized. This intrusive activity resulted in a reset of the radiometric time clocks, causing the radiometric analyses to indicate the Fosdick Complex as Cretaceous in age.

Block faulting probably occurred during early Cenozoic time, and the Fosdick Mountain block was elevated to its present position, thus exposing what had been a deep-seated part of the crust. Extrusive volcanism began about the same time, restricted areally to the eastern Fosdick Mountains and nearby areas. Vents appear to have developed along the two major fault zones bounding the Fosdick horst. Glaciation in this area began about 50 m.y. ago (LeMasurier, 1972), and present geomorphic features are the result of glacial erosion.

Northern Victoria Land

A direct relationship between western Marie Byrd Land and northern Victoria Land has been suggested by several investigators. The evidence upon which the suggestions were based, however, is limited. In essence it consists of: (1) the presence in each sector of thick sequences of similar basin deposits, primarily graywacke, sandstone, and argillite; (2) the rocks in both areas have been intensely folded and altered by low-grade metamorphic processes; (3) major structural trends are similar and, if projected, would be continuous; (4) the rocks in both areas, based on fossil evidence, were deposited during the same time interval; and (5) both sequences were intruded by granodioritic magma in middle Paleozoic time.

In order to provide background to pursue this question further, we present brief descriptions of the Wilson Group and the Robertson Bay Group of northern Victoria Land.

WILSON GROUP

The Wilson Group is comprised of two units: (1) the Rennick Schist, which is biotite and garnet schist, and (2) quartz-feldspar-biotite-garnet gneiss and migmatite (Gair and others, 1969). These two units grade into one another, and intermediate rock types are also present. They were derived from a thick sequence of sedimentary rocks containing a few beds of tuff or mafic volcanic rocks (Dow and Neall, 1974). Deposition probably took place in late Precambrian time. Regional metamorphism which accompanied deformation and which developed the schist has been assigned a minimum age of 770 ± 20 m.y. (Dow and Neall, 1974) based upon a Rb-Sr whole-rock date (Faure and Gair, 1970). Local migmatization of the Wilson Group occurred during the Ross orogeny with the intrusion of the Granite Harbour plutons. K-Ar ages for this event, as reported by Sturm and Carryer (1970), range from 434 to 565 m.y.

The dominant structural feature displayed by the Wilson Group is a northwest-trending regional schistosity (Dow and Neall, 1974). This trend parallels that developed in the adjacent Robertson Bay and Bowers Groups during Ordovician time, and it probably developed at the same time in the Wilson Group. Older structural features, developed during an earlier deformational event, have not been reported; either they were obliterated during the second deformation (Ordovician), or the two sets are parallel in trend.

Outcrops of the Wilson Group and the associated Granite Harbour Intrusives are isolated from the younger Robertson Bay and Bowers Groups by major faults.

Ross Supergroup

Robertson Bay Group

The Victory Mountains, Admiralty Mountains, Anare Mountains, and the northeastern Bowers Mountains are composed mostly of rocks of the Robertson Bay Group. Scattered outcrops of Admiralty Intrusives and McMurdo volcanic rocks also occur along the Borchgrevink Coast.

The Robertson Bay Group is a thick sequence consisting principally of metagraywacke and argillite. Probably deposited and indurated in a marginal basin as graywacke, sandstone, and shale, these sedimentary rocks were subsequently altered by deformation and metamorphism to metagraywacke, hornfels, phyllite, and semischist. In some areas sedimentary structures such as bedding, cross-bedding, and graded bedding are preserved. Sole markings and flute casts have been observed.

There is a distinct range in the degree of metamorphism exhibited by rocks in this group. In the Robertson Bay area and elsewhere, removed from the intrusive rocks, the quartz-albite-muscovite-chlorite subfacies is represented. Much of the rock is quartzose metagraywacke, siliceous slate, and spotted slate. In the vicinity of the Admiralty Intrusives, contact metamorphism has produced biotite, hornblende, and cordierite hornfels. Most of the lithic fragments in the metagraywacke and coarse hornfels are quartz, quartzite, slate, and mafic volcanic rocks. Layers of spilite are present in some parts of the sequence.

The principal structural feature in the Robertson Bay Group is the northwest trend of the axes of the intensely folded beds and the axial plane cleavage. Although few major folds have been mapped, anticlinoria and synclinoria have been identified.

The age of the Robertson Bay Group is tentatively considered to be late Precambrian to Cambrian based upon the presence of spores.

Bowers Group

A sequence of shallow-water, slightly metamorphosed sedimentary rocks—the Bowers Group—is believed to overlie the Robertson Bay Group. Cambrian fossils are present in some units. As there is no known counterpart of this group exposed in the Ford Ranges, further consideration of this group is unnecessary.

Summary and speculations

In considering a possible close relationship of the Swanson Formation with the Robertson Bay Group, it is necessary to include discussions of the associated rocks and their histories. To facilitate the discussion, the geologic events are presented in chronologic order in Table 75.1 for both the Ford Ranges and northern Victoria Land.

There are sufficient similarities in the early histories of the two sectors to justify the suggestion that they represent segments of a single, continuous geologic province. The 1000 km gap across the Ross Sea between the two sectors could be the result of depression of that central sector, possibly by block faulting. The oldest known rocks are the Fosdick Complex in the Ford Ranges and the Wilson Group in northern Victoria Land. Both are believed to be Precambrian in age, although for each the actual age is poorly documented. If we tentatively accept the suggestion that an orogen extended across that part of Antarctica, we can assume that the ancient mountains which were developed included the rocks of the Wilson Group and of the Fosdick Complex. In late Precambrian–early Paleozoic time, a new geosyncline developed adjacent to the mountain ranges. In the basin were deposited the sediments of the Robertson Bay Group and the Swanson Formation.

Table 75.1. Timetable for principal geologic events in the Ford Ranges of Marie Byrd Land and in northern Victoria Land

Geologic time scale	Ford Ranges	Northern Victoria Land
Cenozoic	Block faulting Extrusion of basic volcanic rocks and hyaloclastite	Block faulting Extrusion of McMurdo volcanic rocks
Cretaceous	Andean orogeny Intrusion of granite (88 – 107 m.y.) Migmatization of Fosdick Complex	
Permian – Jurassic		Deposition of Beacon Group Emplacement of FerrarGroup
Devonian – Carboniferous	Ford Intrusives (300 – 355 m.y.)	Admiralty Intrusives (313 – 385 m.y.)
Cambrian – ordovician	Deformation of Swanson Formation (445 – 475 m.y.) and of Fosdick Complex	Ross orogeny Deformation of Robertson Bay Group (420 m.y.) and Bowers Group Granite Harbour Intrusives and local migmatization of Wilson Group
Cambrian		Deposition of Bowers Group
Late Precambrian – early Paleozoic	Deposition of Swanson Formation	Deposition of Robertson Bay Group
Precambrian	Deposition, deformation, and metamorphism of Fosdick Complex	Deposition, deformation, and metamorphism of Wilson Group

The source of the former was the mountains and highlands of the Wilson sector, and of the latter, those of the Fosdick sector. During Cambrian–Ordovician time, an orogeny resulted in the intense folding and low-grade metamorphism of both sedimentary sequences. Major structural features and patterns are similar in both areas. In Devonian–Carboniferous time, magmatic activity resulted in the emplacement of granodiorite plutons in both sectors. This comparison briefly summarizes the similar historical events in both sectors.

Other factors, however, cast doubt upon the validity of the concept of close relationship. The schist of the Wilson Group, on the basis of one radiometric date, was metamorphosed about 770 m.y. ago. The Middle–Upper Cambrian Bowers Group is present in northern Victoria Land but has no counterpart in the Ford Ranges. These two facts point to a possible extension of the Beardmore orogen to include northern Victoria Land, and the Beardmore orogeny thus could have been responsible for the deformation and metamorphism of the Robertson Bay and Bowers Groups. The Granite Harbour Intrusives were emplaced in the Wilson Group during the Ross orogeny and were responsible for local migmatization and the development of gneiss. Similar processes affected the Fosdick Complex but seemingly not until Cretaceous time, during the Andean orogeny.

If Marie Byrd Land and East Antarctica were separate geologic provinces during their early histories, the model suggested by Elliot (1975) may be valid: that Marie Byrd Land was part of a separate block that was welded onto Antarctica by plate convergence during late Precambrian or early Paleozoic time. We believe that such a convergence might have occurred in late Paleozoic or even early Mesozoic time.

Before we can resolve the difficult question of the relationship of East and West Antarctica, much additional knowledge is required. In particular we must considerably expand our knowledge of West Antarctica. At the moment, however, we favor the idea that Marie Byrd Land is totally unrelated to East Antarctica, and that it and adjacent western Ellsworth Land were not related to the Antarctic Peninsula until Mesozoic time.

Acknowledgments. Critical reviewers of this paper were John P. Brand and Carl A. Cathey; we are grateful for their helpful criticisms and suggestions. We wish to thank members of the field parties during the 1934, 1945, and 1966 seasons for their assistance. This work was supported by National Science Foundation Grant No. OPP75-19130.

References cited

Amundsen, R. 1913. *The South Pole. 2.* London, Murray, 449 pp.

Beitzel, J. E., and Bentley, C. R. 1967. Geophysical investigations in Marie Byrd Land. *Antarct. Jnl U.S., 2,* No. 4, 95–96.

Boudette, E. L., Marvin, R. F., and Hedge, C. E. 1966. Biotite, potassium-feldspar, and whole-rock ages of adamellite, Clark Mountains, West Antarctica. *Prof. Pap. U.S. geol. Surv.,* No. 550-D, D190–94.

Byrd, R. E. 1930. *Little America.* New York, Putnam's, 167–87.

Dow, J. A. S., and Neall, V. E. 1974. Geology of the lower Rennick Glacier, northern Victoria Land, Antarctica. *N.Z. Jl Geol. Geophys., 17,* No. 3, 659–714.

Elliot, D. H. 1975. Tectonics of Antarctica: a review. *Am. J. Sci., 275-A,* 45–106.

Faure, G., and Gair, H. S. 1970. Age determinations of rocks from northern Victoria Land. *N.Z. Jl Geol. Geophys., 13,* No. 4, 1024–26.

Gair, H. S., Sturm, A., Carryer, S. J., and Grindley, G. W. 1969. Geology of northern Victoria Land (Sheet 13, northern Victoria Land.) (*In* Bushnell, V. C., and Craddock, C., *eds.* Geologic Maps of Antarctica. *Antarct. Map Folio Ser.,* Folio 12, Pl. XII.)

Halpern, M. 1968. Ages of Antarctic and Argentine rocks bearing on continental drift. *Earth planet. Sci. Lett., 5,* No. 3, 159–67.

Halpern, M. 1972. Rb-Sr total rock and mineral ages from the Marguerite Bay area, Kohler Range and Fosdick Mountains. (*In* Adie, R. J., *ed. Antarctic Geology and Geophysics.* Oslo, Universitetsforlaget, 197–204.)

Iltchenko, L. N. 1972. Late Precambrian acritarchs of Antarctica. (*In* Adie, R. J., *ed. Antarctic Geology and Geophysics.* Oslo, Universitetsforlaget, 599–602.)

Klimov, L. V. 1967. Nektorye rezul'taty geologicheskikh issledovanii na Zemle Meri Berd v 1966–1967 gg. [Some results of geological observations in Marie Byrd Land, 1966–1967.] *Inf. Byull. sov. antarkt. Eksped.,* No. 66, 18–25. [English translation: *Inf. Bull. Sov. Antarct. Exped., 6,* No. 6, 555–59.]

Krylov, A. Ya. 1972. Antarctic geochronology. (*In* Adie, R. J., *ed. Antarctic Geology and Geophysics.* Oslo, Universitetsforlaget, 491–94.)

Krylov, A. Ya, Lopatin, B. G., and Mazina, T. I. 1970. The age of the rocks from the Ford Mountains and Ruppert Coast (the western part of Marie Byrd Land). *Inf. Bull. Sov. Antarct. Exped., 80,* 34–37.

LeMasurier, W. E. 1972. Volcanic record of Cenozoic glacial history of Marie Byrd Land. (*In* Adie, R. J., *ed. Antarctic Geology and Geophysics.* Oslo, Universitetsforlaget, 251–59.)

Lopatin, B. G., and Orlenko, E. M. 1972. Outline of the geology of Marie Byrd Land and the Eights Coast. (*In* Adie, R. J., *ed. Antarctic Geology and Geophysics.* Oslo, Universitetsforlaget, 245–50.)

Passel, C. F. 1945. Sedimentary rocks of the southern Edsel Ford Ranges, Marie Byrd Land, Antarctica. *Proc. Am. phil. Soc., 89,* No. 1, 123–31.

Scharon, L., and Early, R. 1967. Paleomagnetic measurements in Marie Byrd Land. *Antarct. Jnl U.S., 2,* No. 4, 94–95.

Scott, R. F. 1905. *The Voyage of the "Discovery."* London, Scribners, 183.

Sturm, A., and Carryer, S. J. 1970. Geology of the region between the Matusevich and Tucker Glaciers, north Victoria Land, Antarctica. *N.Z. Jl Geol. Geophys., 13,* No. 2, 408–35.

Wade, F. A. 1937. Petrologic and structural relations of the Edsel Ford Range, Marie Byrd Land, to other Antarctic mountains. *Bull. geol. Soc. Am., 48,* No. 9, 1387–96.

Wade, F. A. 1945. The geology of the Rockefeller Mountains, King Edward VII Land, Antarctica. *Proc. Am. phil. Soc., 89,* No. 1, 67–77.

Wade, F. A. 1967. Geology of the Marie Byrd Land coastal sector of West Antarctica. *Antarct. Jnl U.S., 2,* No. 4, 93–94.

Wade, F. A. 1969. Geology of Marie Byrd Land (Sheet 18, Marie Byrd Land). (*In* Bushnell, V. C., and Craddock, C., *eds.* Geologic Maps of Antarctica. *Antarct. Map Folio Ser.,* Folio 12, Pl. XVII.)

Wade, F. A. 1972. Geologic survey of Marie Byrd Land. *Antarct. Jnl U.S., 7,* No. 5, 144–45.

Wade, F. A., and Wilbanks, J. R. 1972. Geology of Marie Byrd and Ellsworth lands. (*In* Adie, R. J., *ed. Antarctic Geology and Geophysics.* Oslo, Universitetsforlaget, 207–14.)

Warner, L. A. 1945. Structure and petrography of the southern Edsel Ford Ranges, West Antarctica. *Proc. Am. phil. Soc., 89,* No. 1, 78–122.

Wilbanks, J. R. 1972. Geology of the Fosdick Mountains, Marie Byrd Land. (*In* Adie, R. J., *ed. Antarctic Geology and Geophysics.* Oslo, Universitetsforlaget, 277–84.)

V Paleontology

76

Mesozoic Vertebrates of Antarctica

Review Paper

EDWIN H. COLBERT
Curator Emeritus, American Museum of Natural History, New York. Professor Emeritus, Columbia University, New York.
Curator of Vertebrate Paleontology, Museum of Northern Arizona, Flagstaff, Arizona 86001, U.S.A.

Abstract. Our knowledge of Mesozoic vertebrates from Antarctica is presently confined to a Lower Triassic assemblage of amphibians and reptiles collected from the Fremouw Formation of the Beacon Supergroup and to some Lower Jurassic lacustrine fishes collected from interbeds within the Kirkpatrick Basalt of the Ferrar Group. All of these fossils come from the central Transantarctic Mountains. The amphibians from the Fremouw Formation are labyrinthodonts belonging to two genera, *Austrobrachyops* and *Cryobatrachus,* related to Lower Triassic labyrinthodonts found in the *Lystrosaurus* zone of South Africa and the Blina Formation of Australia. Fremouw reptiles are numerous and varied. The most prevalent form is the dicynodont *Lystrosaurus,* of which two species typical of the South African Triassic *Lystrosaurus* zone are recognized, *L. murrayi* and *L. curvatus.* Also present are theriodonts, including *Thrinaxodon liorhinus* and *Scaloposaurus;* all are *Lystrosaurus* zone forms. A small cotylosaur, *Procolophon trigoniceps,* is well represented in the Fremouw assemblage and in Africa. In addition, there are fossils of a prolacertilian from the Fremouw Formation, closely related to *Prolacerta* of South Africa. In essence, the Fremouw fauna is remarkably similar to the *Lystrosaurus* fauna of South Africa, an indication of the contiguity of these continental blocks at the beginning of Mesozoic time.

The Jurassic fishes belong to the genus and species *Oreochima ellioti,* related to the freshwater holostean fishes *Aetheolepis* and *Aphnelepis* of Australia. The implications of these relationships, as well as those of the amphibians mentioned above, to the connections between Antarctic and Australia in Early Triassic time need no elaboration. Although Cretaceous vertebrates are as yet unknown in Antarctica, there are good reasons to expect that dinosaurs and marsupials will eventually be discovered, probably on the Antarctic Peninsula.

Our knowledge of the position of Antarctica within Gondwanaland and of the relationship of the present South Polar continent to the other great blocks that once formed the vast supercontinent of the Southern Hemisphere depends upon studies carried out through various scientific disciplines, not the least of which is paleontology. Within the field of paleontology the evidence supplied by fossils of land-living animals is particularly important because such animals were in the past, as they are today, intimately associated with the continents on which they live or have lived. Terrestrial animals, unless they are adapted for flight, are confined to the firm lands across which they wander. Therefore, the presence of identical or closely related fossils in separated regions, either contiguous to or far distant from each other, implies dry-land connections by which genera and species may have become distributed across areas of varying dimensions. Such considerations are basic to the interpretation of terrestrial fossils found in Antarctica.

The possibility of finding definitive fossils that would reveal the relationship of Antarctica to other continents has long been in the minds of many geologists. In 1931 L. M. Gould, today the doyen of Antarctic geologists, stated the problem in the following manner: "Great stories about the past climates and life of the Antarctic are still waiting to be read in the rocks. Nowhere in the world is the hunt for fossils filled with prospects of greater promise. From fossils we shall learn whether the present isolation of the Antarctic continent is a recent condition, or whether it has enjoyed such isolation throughout much of recorded geological time. It is in the study of extinct life that we must seek the solution of one of the greatest problems in animal migration" (Gould, 1931, pp. 262–63).

Of course terrestrial fossils had been found in Antarctica years before Gould made this statement, these being the remains of fossil plants, particularly of the genus *Glossopteris,* discovered during the course of the Scott *Terra Nova* expedition of 1910–13. Reference to these fossils by F. Debenham, geologist on the Scott expedition, was published in 1913: "The 35 lbs. of specimens brought back by the Polar Party from Mt. Buckley contain impressions of fossil plants of late Palaeozoic age, some of which a cursory inspection identifies as occurring in other parts of the world. When fully examined, they will assuredly prove to be of the highest geological importance" (Debenham, 1913, p. 300). Since the time of Scott, *Glossopteris* has been found abundantly preserved in the Paleozoic sedimentary rocks of Antarctica.

Glossopteris has long been recognized as a characteristic late Paleozoic Gondwanaland plant, but there has been considerable disagreement as to the manner in which it may have spread from one part of the ancient supercontinent to another, or whether, for that matter, its presence today in the Southern Hemisphere and in peninsular India can be accepted as good evidence for the existence of Gondwanaland. Is its wide distribution proof positive of the former unity of the postulated single continent, or might the spread of this ancient plant have been facilitated across water barriers by the wind-borne distribution of spores? This latter possibility is slight at best, yet it does introduce a small element of doubt into the problem.

In the austral summer of 1966–67 the presence of Mesozoic vertebrates in Antarctica was established by the discovery by D. H. Elliot of Jurassic fishes, presumably of freshwater type, in the Transantarctic Mountains. These fossils were not described until 1972; thus their significance could not be appreciated before then. (They are discussed below.)

In the meantime, the discovery of ancient terrestrial vertebrates in Antarctica—the remains of tetrapods fully adapted for life on the land and thus unable to cross wide water barriers, either by their own efforts or by natural rafts driven by currents and vagrant winds, and thus necessarily present in Antarctica because of former land connections—was something much to be desired.

The hoped-for discovery took place in December 1967 when a field party including P. J. Barrett and R. J. Baillie, then of the Institute of Polar Studies, Ohio State University, found by chance a small portion of the lower jaw of a labyrinthodont amphibian in the Lower Triassic Fremouw Formation, on the slopes of Graphite Peak in the Transantarctic Mountains. The fossil, fragmentary though it was, emphasized the potential value of careful and detailed paleontologic field work in search of more Triassic fossils (Barrett and others, 1968).

Consequently, an expedition designed specifically to prospect for and to excavate Lower Triassic fossil vertebrates went into the field during the austral summer of 1969–70. The expedition collected many specimens, so it was followed by a second effort in 1970–71. The expeditions, under the supervision of D. H. Elliot of the Institute of Polar Studies, included E. H. Colbert of the American Museum of Natural History, J. A. Jensen of Brigham Young University, W. J. Breed of the Museum of Northern Arizona, and J. S. Powell of the University of Arizona during the first field season, and J. W. Kitching of the Bernard Price Institute for Paleontological Research of the University of the Witwatersrand, Johannesburg, South Africa, J. Ruben of the University of California at Berkeley, and briefly T. Rich of the American Museum of Natural History during the second field season.

A rich fauna of early Triassic age was obtained and is now being studied (Elliot and others, 1970).

Triassic and Jurassic stratigraphy in the Transantarctic Mountains

The search for Triassic tetrapods in the Transantarctic Mountains, although initiated upon evidence that was indeed slender, consisting of a single jaw fragment, was not really a case of paleontologic wildcatting. The stratigraphic setting in the central Transantarctic Mountains bears a very strong resemblance to the stratigraphy of the Karroo Series in South Africa, a sequence that has long been famous for its wealth of fossil amphibians and reptiles. In both regions the sedimentary rocks of late Paleozoic and early Mesozoic age are for the most part essentially horizontal and are intruded and capped by Jurassic volcanic rocks.

These Antarctic rocks were first described in 1907 by H. T. Ferrar, the geologist on the Scott *Discovery Expedition.* He named the sedimentary rocks the Beacon Sandstone on the basis of observations made in southern Victoria Land. In 1972 Barrett elevated the Beacon to the status of a supergroup, containing formations ranging in age from Devonian to Jurassic. In the central Transantarctic Mountains the Triassic formations within this supergroup are, from bottom to top, the Fremouw, Falla, and Prebble. Intruding and capping the Beacon sequence are the Jurassic Ferrar Dolerite and the Kirkpatrick Basalt, both in the Ferrar Group.

The Fremouw Formation, some 750 m in thickness, rests unconformably upon the Permian Buckley Formation which contains dark shale and coal with an abundance of *Glossopteris* leaves. The Fremouw itself consists of sandstone and mudstone with igneous intrusions. The sedimentary units show cyclic rhythms, with basal coarse-grained channel sand deposits grading upward into fine-grained floodplain deposits. Fossil remains are found principally within the lower part of the formation. In the upper part of the sequence are fossil logs and coal with *Dicroidium,* a very characteristic Gondwana plant, near the top. At the present time fossil vertebrates are not known from the overlying Falla and Prebble Formations, nor have they been found in the underlying Buckley Formation, so rich in *Glossopteris.*

Thus our knowledge of Triassic fossil vertebrates in Antarctica is restricted to the base of the Triassic—to the lower part of the Fremouw Formation. This is in distinct contrast to the situation in South Africa, where fossil amphibians and reptiles occur abundantly in the Lower Beaufort beds of Late Permian age, as well as in the Early Triassic Middle Beaufort beds and the Late Triassic Upper Beaufort beds. It must be remembered, however, that paleontologic explorations in Triassic rocks of Antarctica are in their initial stages, with a total of only six or seven months of field work having been accomplished—and this includes a considerable amount of lost time resulting from Antarctic storms. Thus there remains much to be done, and in view of the similarities of geologic expression in Triassic rocks of the two continents, it seems likely that more Triassic faunas will eventually be discovered in the Transantarctic Mountains.

Our knowledge of Jurassic vertebrates is presently confined to the fossil fishes found in lacustrine interbeds within the Kirkpatrick Basalt. They are discussed below.

The succession of Triassic and Jurassic rocks in the Transantarctic Mountains is presented in Table 76.1, which is adapted from Colbert (1974).

The Fremouw fauna

Localities and occurrences

The fragmentary amphibian jaw, the discovery of which initiated paleontologic exploration of the Fremouw Formation, was collected at Graphite Peak (lat. 85° 03.3′ S, long. 172° 19′ E). Subsequently, other fossils have been found at this locality. The main thrust of the first season of collecting, however, was carried on at Coalsack Bluff,

Table 76.1. Triassic and Jurassic rocks of the Transantarctic Mountains

	Group	Formation	Description
Jurassic	Ferrar	Kirkpatrick Basalt 600+ m	Tholeiitic flows; interbeds with holostean fish and conchostracans
		Ferrar Dolerite 1000 m (intruded into Beacon sedimentary beds)	Sills and a few dikes
		Disconformity	
Triassic	Beacon	Prebble Formation 3–460+ m	Mudflows, pyroclastic breccia, tuff, tuffaceous sandstone
		Falla Formation 160–530 m	Sandstone, shale, tuff, and *Dicroidium* in upper part
		Fremouw Formation 650 m	Subarkose and arkosic volcanic sandstone, shale, and greenish gray mudstone. Logs, coal, and *Dicroidium* near top. *Lystrosaurus* fauna in lower beds
		Disconformity	
Permian		Buckley	Coal measures

about 125 km to the east of Graphite Peak, at lat. 84° 15′ S, long. 162° W. During the second season, work was centered around the junction of the Shackleton and McGregor Glaciers, at about lat. 85° 08′ S, long. 174° 50′ W. The discovery of fragments at some other localities, together with the extensive exposures of the Fremouw Formation in the Transantarctic Mountains, indicates that fossils will almost surely be found elsewhere in this section of Antarctica. The relative abundance of fossils at the localities where they were collected points to the probability that fossils in the Fremouw Formation occur in numbers that may be comparable to their prolific presence in the Karroo strata of South Africa.

The fossils collected at Coalsack Bluff were found in stream-channel deposits. For this reason, the specimens consist largely of individual bones—the remains of carcasses that disintegrated as a result of transportation along stream beds. At McGregor Glacier the fossils were found in fine-grained sedimentary beds, so that several articulated or partially articulated skeletons were recovered. Thus the two sites complement each other, providing a record of fossils buried under different sedimentary conditions.

Amphibia

Amphibians in the Fremouw Formation are at present not so abundantly represented as are reptiles, which may be an indication of the "upland" nature of the fauna. Nevertheless, several taxa of Labyrinthodontia (upper Paleozoic–lower Mesozoic amphibians with heavily roofed skulls) are now known, showing definite relationships with Lower Triassic amphibians in South Africa and Australia.

The first fragmentary jaw, discovered at Graphite Peak, represents a large, short-faced or brachyopid labyrinthodont of the type that is quite characteristic of Lower Triassic rocks on other continents. This specimen is supplemented by other material from Coalsack Bluff, and the fossils have been designated as a new genus and species, *Austrobrachyops jenseni.* These fossils show resemblance to *Batrachosuchus,* from the *Lystrosaurus* zone of South Africa, and to *Blinasaurus,* from the Blina Formation of Western Australia.

Of particular interest among the Fremouw labyrinthodonts are the bones of a very small form which has been named *Cryobatrachus kitchingi.* This is quite obviously a lydekkerinid—closely related to *Lydekkerina,* a genus very characteristic of the South African *Lystrosaurus* zone. In addition, *Cryobatrachus* may be compared with *Chomabatrachus,* a lydekkerinid from the Lower Triassic Knocklofty Formation of Tasmania (Figure 76.1).

These are the two named amphibians now known from the Lower Triassic sequence of Antarctica (Colbert and Cosgriff, 1974). It should be pointed out that amphibians as we know them today are unable to tolerate salt water. Therefore the presence of such freshwater tetrapods in Antarctica, Africa, and Australia points to former connec-

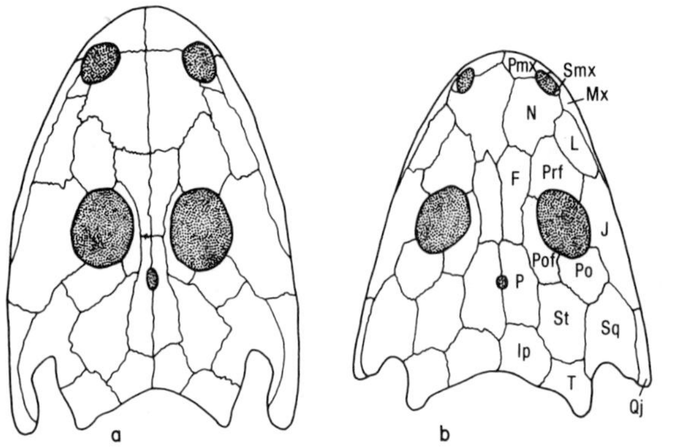

Figure 76.1. a. Cryobatrachus kitchingi, a labyrinthodont amphibian from the Lower Triassic Fremouw Formation of Antarctica. Dorsal view of skull, ×1. *b. Lydekkerina huxleyi,* a labyrinthodont amphibian from the Lower Triassic *Lystrosaurus* zone of South Africa. Dorsal view of skull (after Parrington, 1948), ×1/2.

tions between these now separate continents, by which amphibians could move from one drainage system to another, thereby ensuring their eventual distribution across a broad continental region.

Other amphibian bones, as yet too incomplete for definite identification, are known from the Fremouw Formation. Therefore it can be expected that the amphibians in the Fremouw fauna should increase in numbers and variety in future years.

Reptilia

The fossil remains of reptiles are much more numerous and varied than are those of amphibians in the Fremouw Formation, a situation that nicely parallels the occurrences of fossils in the *Lystrosaurus* zone of South Africa. Conversely, Lower Triassic rocks of Australia have yielded amphibians in some numbers, but very few reptilian remains. Since Antarctic Lower Triassic amphibians show relationships to similar fossils found both in Africa and Australia, the contrast in the occurrences of reptiles in Antarctica and Australia may be attributed to the probable differences in the environments represented by the sedimentary sequences of the two continents.

The most abundant reptilian fossils by far from the Fremouw Formation belong to the genus *Lystrosaurus,* the reptile so very characteristic of the *Lystrosaurus* zone in South Africa. In South Africa about 85% of all fossils found in the *Lystrosaurus* zone belong to this genus. The collections so far obtained from Antarctica are not sufficiently large for a meaningful census to be made, but the proportion of *Lystrosaurus* to other reptiles in the Fremouw Formation may be comparable to what is present in the African *Lystrosaurus* zone.

During the years, numerous species of *Lystrosaurus* have been described from South Africa, but recent careful studies have shown that these can be reduced to about six valid species. Two of these African forms, *Lystrosaurus murrayi* and *Lystrosaurus curvatus,* seem to be present in the Fremouw fauna. The presence of identical species of *Lystrosaurus* in Africa and in Antarctica certainly is significant when considering past continental relationships, for *Lystrosaurus* undoubtedly was a land-living reptile incapable of crossing wide oceanic barriers (Colbert, 1972, 1974) (Figure 76.2).

Lystrosaurus is a dicynodont reptile, a highly specialized therapsid or mammal-like reptile, which has a rather bizarre looking skull characterized by a pair of tusks set in otherwise edentulous jaws. (Not all dicynodonts possess tusks.) The beak-like jaws evidently were covered with horn, like the jaws of turtles. The eyes and the nostrils are located high on the skull, an indication that *Lystrosaurus* probably was semiaquatic, living along rivers and in lakes. The body is heavy and capacious, indicating that *Lystrosaurus* probably was a herbivore, feeding upon bulky food. The limbs are short and stout, the feet are broad, and the tail is very short. In its general adaptations, *Lystrosaurus* may be compared to a modern hippopotamus, although it is much smaller; the largest *Lystrosaurus* is about the size of a sheep (Figure 76.3).

This reptile probably inhabited ancient Antarctica in large herds, feeding upon abundant vegetation, walking across the land, and wading and swimming in lakes and streams. Likewise, it existed in large numbers in what is now southern Africa, where it typifies the *Lystrosaurus* zone, and in peninsular India, where it is found in the Lower Triassic Panchet Formation. Furthermore, *Lystrosaurus* has been found in southern China as well as in Sinkiang in western China, which gives some credence to the view, derived from geophysical studies, that eastern Asia may have been a part of Gondwanaland.

There are some other dicynodont reptiles in the Fremouw Formation, as yet represented only by fragmentary material. One very tantalizing fossil is a huge dicynodont stapes bone, from the ear region, found at Coalsack Bluff. It shows that there are some very large dicynodonts waiting to be found in Antarctica.

The *Lystrosaurus* zone of South Africa, although replete with the fossils of *Lystrosaurus,* also contains therapsid or mammal-like reptiles of other kinds, notably the very advanced forms which are known as theriodonts. The same is true for the Fremouw Formation. One of the theriodonts particularly characteristic of the *Lystrosaurus*

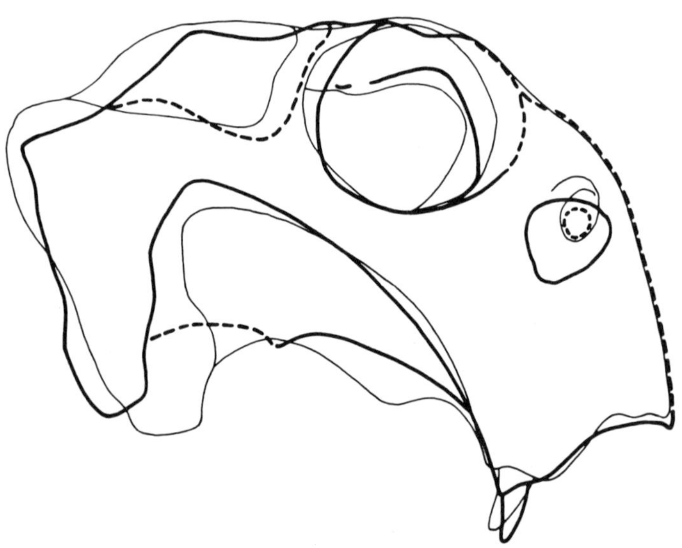

Figure 76.2. Lystrosaurus curvatus. Outline of skull (heavy lines) from the Lower Triassic Fremouw Formation of Antarctica superimposed over outline of skull (light lines) from the Lower Triassic *Lystrosaurus* zone of South Africa. Right lateral views.

Figure 76.3. Restoration of *Lystrosaurus murrayi,* by M. Colbert. (Approximately 1/3 the size of a small individual.)

zone in Africa is *Thrinaxodon liorhinus,* and this same form is well represented in the Fremouw Formation (Colbert and Kitching, 1977). *Thrinaxodon,* as now known from Antarctica, is restricted to the beds at the McGregor Glacier site, but in the future it probably will be found elsewhere in the Transantarctic Mountains. An interesting feature concerning the fossilization of *Thrinaxodon* at McGregor Glacier is that many of the specimens are preserved in the form of natural molds in the rocks. Consequently our knowledge of *Thrinaxodon* in Antarctica is derived to a considerable degree from latex casts made from these natural molds. Yet the information thus obtained is good. Altogether we know *Trinaxodon* in the Fremouw fauna from skulls and skeletons and parts of skeletons (Figure 76.4).

Thrinaxodon was a small carnivore that anticipated in

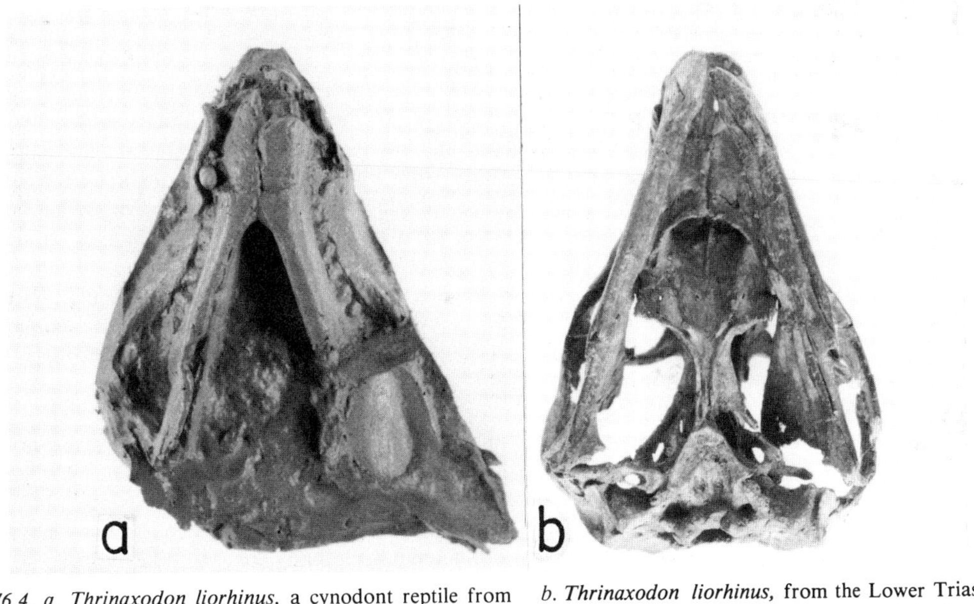

Figure 76.4. a. Thrinaxodon liorhinus, a cynodont reptile from the Lower Triassic Fremouw Formation of Antarctica. Ventral view of portion of skull with lower jaw in articulation, ×1.

b. Thrinaxodon liorhinus, from the Lower Triassic *Lystrosaurus* zone of South Africa. Ventral view of skull with lower jaw in articulation, ×1.

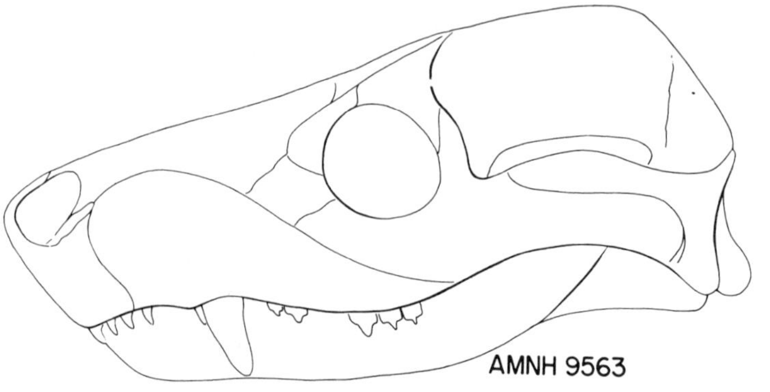

Figure 76.5. Thrinaxodon liorhinus, reconstruction of skull in left lateral view, ×1.

reptilian terms such animals as weasels or foxes in our modern world. The skull is very "dog-like"—elongated, with a secondary palate that increased the efficiency of breathing and eating, and with differentiated teeth, including nipping incisors, piercing canines, and cutting cheek teeth for reducing the food to small pieces. This indicates a quick assimulation and transformation of food into energy, which in turn suggests that *Thrinaxodon* was very active, with a relatively high rate of metabolism. Perhaps it was "warm-blooded" and had an external protective covering of hair. This is further suggested by the form of the body, because the trunk is clearly divided into neck or cervical vertebrae with small ribs, thoracic or dorsal vertebrae with widely expanded ribs, and lumbar vertebrae with peculiar short, wide ribs that look more like small plates than ribs. Perhaps this interesting reptile possessed a diaphragm. The limbs are slender and carried well beneath the body, and the feet are compact, with sharp claws. All in all, *Thrinaxodon* must have been an efficient little predator (Figure 76.5).

There are still other theriodonts in the Fremouw fauna, including scaloposaurians, very small mammal-like reptiles that perhaps preyed upon insects or other such small fare. They are also characteristic of the *Lystrosaurus* zone.

Lystrosaurus and *Thrinaxodon* in Africa are associated with a small reptile known as *Procolophon,* and the same is true for Antarctica. Again there is an identity to the species level, for *Procolophon trigoniceps* characterizes the Fremouw fauna as it does the *Lystrosaurus* zone fauna in Africa (Colbert and Kitching, 1975).

Procolophon is a small reptile of primitive relationships belonging to the Order Cotylosauria, among which are the earliest and most primitive of the reptiles, first appearing in strata of Pennsylvanian age. The cotylosaurs were, however, a persistent group, and one branch, the Procolophonia, continued through Triassic time. *Procolophon* itself was ecologically lizard-like and perhaps filled a niche at the beginning of Mesozoic time that later was to be occupied by some of the true lizards. It was a robust little reptile, not unlike a modern horned lizard *(Phrynosoma)* in its proportions. The skull is broad and the jaws are short. The teeth are limited in number and those in the sides of the jaws are chisel-shaped, evidently for cutting food into small pieces. Perhaps this reptile was a herbivore. Close relatives of *Procolophon* are found in Lower Triassic rocks of China (Figure 76.6).

Even though *Procolophon* may have been an ecological "lizard," there are also forerunners of the true lizards in the Lower Triassic beds of South Africa and Antarctica. These are the prolacertilians, which although probably not on the direct ancestral line to the lizards of later times, are nevertheless not far removed from that line, so that they anticipate the lizards in many ways. In the African *Lystrosaurus* zone is the genus *Prolacerta,* which occurs in the Fremouw Formation as well. The prolacertilians were quite different from *Procolophon* in their adaptation, being slender and evidently swift-running reptiles, as indicated by their long, slender limbs. The skull in these reptiles is delicately constructed, the jaws are long, and the sharp-pointed teeth are numerous. In short, the prolacertilians have the appearance of insect-eaters; for this reason there probably was no real competition between these active little reptiles and the rather stodgy *Procolophon.*

Prolacerta is rather rare in the *Lystrosaurus* zone; however, it occurs in considerable abundance in the Fremouw Formation at McGregor Glacier.

Finally there should be mentioned the thecodont reptiles, as yet poorly represented in the Fremouw fauna. In South Africa these reptiles are represented in the *Lystrosaurus* zone by the single genus *Proterosuchus,* often designated as *Chasmatosaurus.* This was an elongated reptile reaching as much as 1 m in length, with short limbs and long jaws, armed with numerous sharp teeth. The material from the Fremouw Formation may represent a similar type of reptile.

The thecodonts are noteworthy in being the Triassic ancestors of the Mesozoic ruling reptiles, namely the crocodilians, the flying reptiles, and the dinosaurs, collectively known as archosaurs. From the scanty material so far available it would appear that Antarctica, as well as other parts of Pangaea, was the habitat of these forerunners of the dominant Mesozoic reptiles.

Such is the composition of the Fremouw fauna, as it is now known. It is in essence a *Lystrosaurus* fauna identical in many respects to the Lower Triassic *Lystrosaurus* fauna of South Africa, yet showing some differences as well. The geographic range of this fauna, which was centered in southern Africa, extends not only to Antarctica, but also to peninsular India and eastern Asia. Thus it would appear that this fauna was once a continuous assemblage of amphibians and reptiles that stretched across several thousand kilometers latitudinally. It is to be expected that a fauna with such a wide range would show some local differences among its constituent members, even though it was united

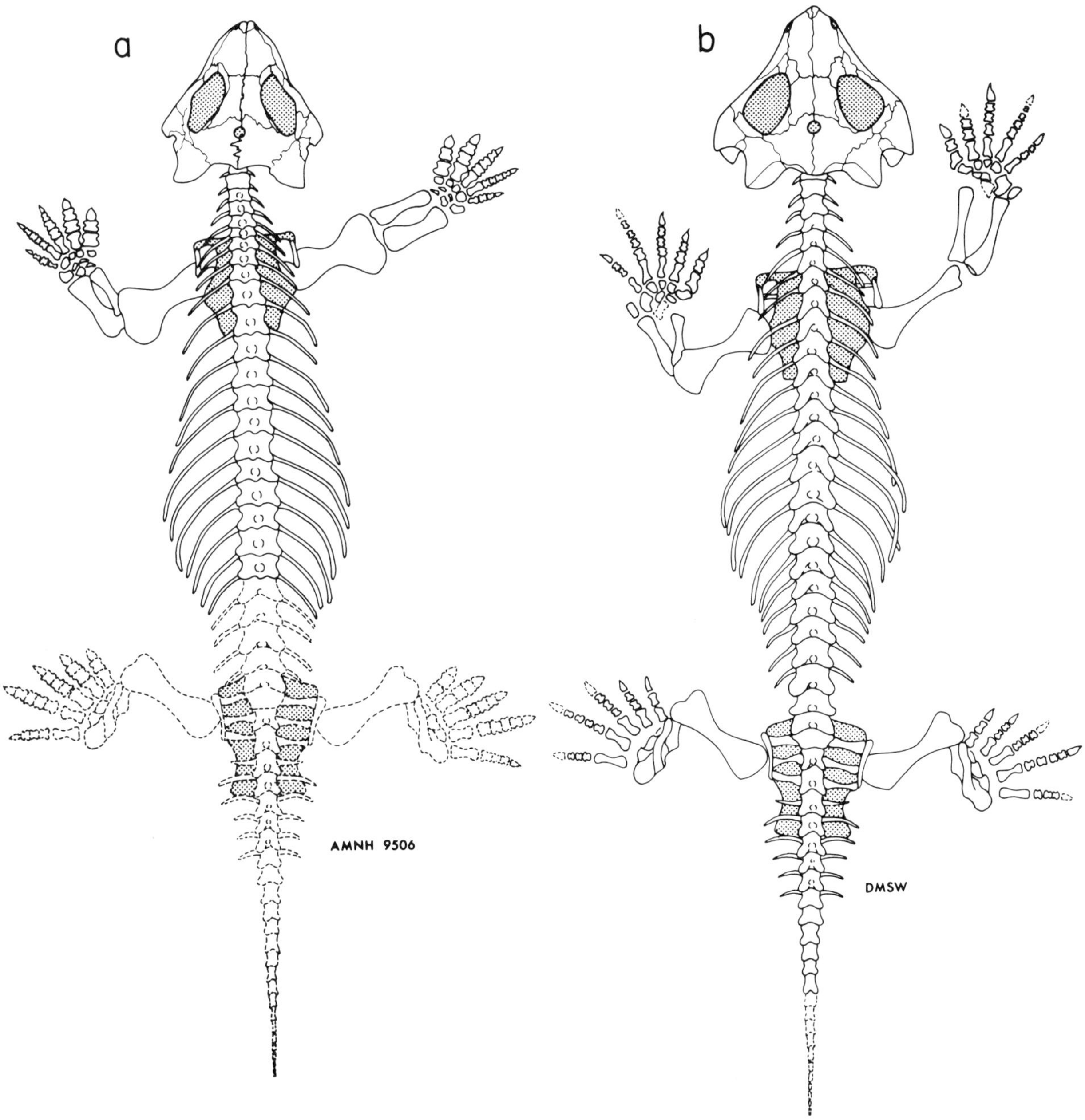

Figure 76.6. a. Procolophon trigoniceps, a cotylosaurian reptile from the Lower Triassic Fremouw Formation of Antarctica. Restoration of skeleton in dorsal view, approximately $\times 1/2$. *b. Procolophon trigoniceps,* from the Lower Triassic *Lystrosaurus* zone of South Africa. Restoration of skeleton in dorsal view, approximately $\times 1/2$.

by many identical genera and species. With the break-up of Pangaea, the formerly continuous distribution of the *Lystrosaurus* fauna was fragmented by the separation of several parts of Gondwanaland from one another and the drift to their present mutually distant positions, extending across the Southern Hemisphere and into the Northern Hemisphere as well. Yet the essential unity of this Lower Triassic fauna indicates the probable reality of a once-continuous Gondwanaland, where active terrestrial amphibians and reptiles could roam from one region to another (Colbert, 1975) (Figure 76.7).

The Jurassic fishes

We now come to the other Mesozoic vertebrates known from Antarctica, the Jurassic fishes mentioned above. These fishes, described by B. Schaeffer, were collected from lacustrine interbeds within the Lower Jurassic Kirkpatrick Basalt, which occurs in the Transantarctic Mountains above the Triassic sequence (Schaeffer, 1972).

The fossil material, which is well represented, has been described as a new genus and species, *Oreochima ellioti,* a holostean fish belonging to the Order Pholidophoriformes. These were freshwater fishes of rather primitive organization, related in a general way to the modern garpikes and bowfins of North America. However, the closer relationships of *Oreochima* are with genera from Jurassic rocks of Australia, namely *Aetheolepis* and *Aphnelepis.* In one detail, the nature of the suborbital bone in the skull, the Antarctic fish is more primitive than the Australian genera (Figure 76.8).

Oreochima is a small fusiform fish covered with heavy

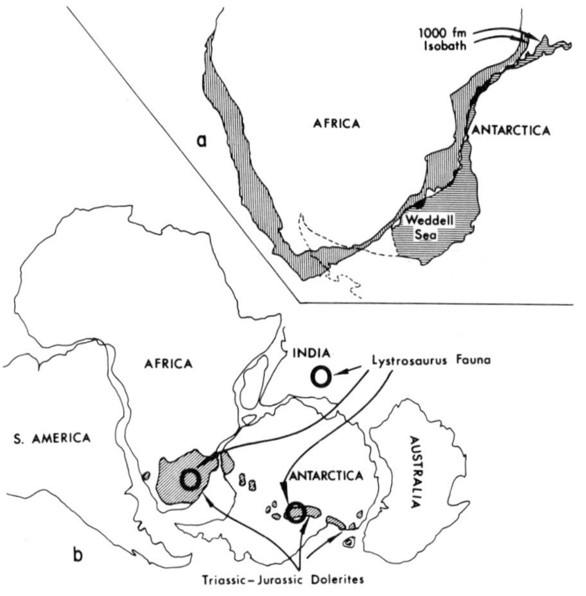

Figure 76.7. a. The best fit between Antarctica and Africa, based upon the 1000 fathom isobath. Overlaps are shown by black areas, underlaps by clear areas. Map adapted from Dietz and Sproll (1970). *b.* Restoration of Gondwanaland in Early Triassic time, showing the occurrences of the *Lystrosaurus* fauna and dolerites (oblique lines). Map adapted from Smith and Hallam (1970).

rhombic scales. The skull is generally primitive and the eye is large. The pectoral and pelvic fins are in the primitive positions, and behind the pelvics are the dorsal and anal fins. The caudal fin is structurally heterocercal, as is characteristic of primitive fishes, but morphologically it is almost homocercal, with the upper and lower lobes essentially equal in development. As Schaeffer has remarked, "If we assume that the Archaeomaenidae [a family of the Pholidophoriformes], as here delimited, were entirely freshwater, their presence in both Australia and Antarctica during the Jurassic is not surprising, as these two continents were in contact throughout the Mesozoic" (Schaeffer, 1972, p. 15).

Cretaceous possibilities

The Triassic amphibians and reptiles of the Fremouw fauna and the Jurassic freshwater fishes from the interbeds within the Kirkpatrick Basalt are at present the only known vertebrates from Mesozoic rocks of Antarctica. Our paleontologic knowledge of the continent is, however, manifestly incomplete, so there is reason to think that other fossil faunas remain to be discovered. Is it possible that Cretaceous vertebrates will eventually be found in Antarctica?

Probably so, and probably on the Antarctic Peninsula. In this part of the continent there are Upper Jurassic and Cretaceous sedimentary sequences from which, as a result of diligent search, some exciting and very significant fossils may be unearthed. What may be expected?

Two groups of vertebrates can be mentioned; one is the dinosaurs, the other the marsupials. Geologic and geophysical evidence suggest that Antarctica was still an integral part of Gondwanaland during Cretaceous time, possibly being joined on the west to South America and on the east to Australia. In view of these continental relationships, what may be predicted as to future discoveries in the Cretaceous beds of Antarctica, specifically in the exposures along the Antarctic Peninsula?

Cretaceous dinosaurs are known from South America, India, Africa, and Australia. In the Upper Cretaceous sedimentary rocks of South America there have been found large carnivorous dinosaurs, various sauropods (these being the dinosaurian giants so well exemplified by such North American genera as *Apatosaurus* (*Brontosaurus*) and *Diplodocus*), and armored dinosaurs. Some of the South American sauropods, notably *Antarctosaurus, Laplatasaurus,* and *Titanosaurus* are recorded from peninsular India and to a lesser extent from Madagascar. The fossil evidence from Australia is presently scanty at best, but trackways of a large carnosaur are present in Lower Cretaceous rocks of Western Australia, while a skeleton of an iguanodont-like dinosaur has been found in Queensland (Colbert and Merrilees; 1967).

Is it not possible, therefore, that there were migrations of Cretaceous dinosaurs from South America, Africa, and India through Antarctica to Australia at a time when these continental blocks were still connected? The entrance of dinosaurs into Australia might even have occurred from South America by way of the ancestral Antarctic Peninsula, at a time in the Cretaceous when Africa and India had become separated from Antarctica by drift, as shown in the reconstructions published by Elliot (1972). Thus the Antarctic Peninsula is a logical place in which to search for the remains of Cretaceous dinosaurs.

While the discovery of dinosaurs in Antarctica would be spectacular, the finding of Cretaceous marsupials would be particularly important. For many decades it was a tenet of paleontologic faith that the marsupials had entered Australia from the northwest—down through Asia and the

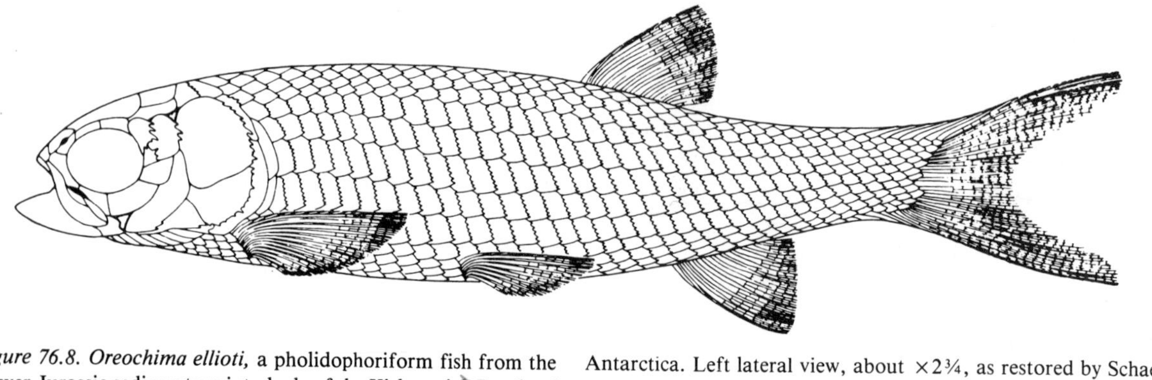

Figure 76.8. Oreochima ellioti, a pholidophoriform fish from the Lower Jurassic sedimentary interbeds of the Kirkpatrick Basalt of Antarctica. Left lateral view, about ×2¾, as restored by Schaeffer (1972).

East Indies and into the Australian island continent. But fossil marsupials are quite unknown in Asia.

In South America, however, a diverse marsupial fauna of Cretaceous age has recently been found. Australia is of course noted for its dominant marsupial fauna. Is it not possible, therefore, that the marsupials entered Australia from South America by way of an Antarctic connection? The possibility has been succinctly set forward by Cox: "In brief, the simplest explanation is that marsupials crossed from South America to the Antarctica-Australia landmass before this broke away, and that this separation took place before placentals arrived in South America. [presumably from North America]. As a result, the placentals were unable to reach Australia" (Cox, 1974, p. 89). Hence, the Antarctic Peninsula also appears to be a most promising locale in which to search for ancient marsupials.

Conclusions

This has been a brief survey of our knowledge concerning the presence of Mesozoic vertebrates in Antarctica. As we have seen, a Lower Triassic *Lystrosaurus* fauna containing amphibians and reptiles is present in the Transantarctic Mountains, while in Lower Jurassic rocks in this same region lacustrine fishes have been found. But these discoveries represent only the beginning of exploration in search of vertebrate fossils on the South Polar continent, which shows that Antarctica was indeed faunally an integral part of Gondwanaland. One may expect that additional Mesozoic vertebrates, or even vertebrate faunas, will be found in the future. Perhaps the Antarctic Peninsula will also yield the remains of Cretaceous dinosaurs and marsupials. Why not? Almost a half century ago, L. M. Gould (1931) wrote "And I had rather go back to the Antarctic and find a fossil marsupial than three gold mines!"

Acknowledgments. The studies of Triassic tetrapods upon which this article is largely based were supported in part by grant GV-25341 from the National Science Foundation. Continuing research on Triassic tetrapods from Antarctica is being supported in part by grant DPP75-23126 from the National Science Foundation. The study of Antarctic Jurassic fishes by B. Schaeffer was supported by the American Museum of Natural History, New York. The Antarctic specimens discussed in this paper will be deposited at the American Museum of Natural History, New York.

References cited

Barrett, P. J., Baillie, R. J., and Colbert, E. H. 1968. Triassic amphibian from Antarctica. *Science, N.Y., 161,* No. 3840, 460–62.

Colbert, E. H. 1972. Antarctic Triassic tetrapods. (*In* Adie, R. J., *ed. Antarctic Geology and Geophysics.* Oslo, Universitetsforlaget, 393–401.)

Colbert, E. H. 1974. *Lystrosaurus* from Antarctica. *Am. Mus. Novit.*, No. 2535, 1–44.

Colbert, E. H. 1975. Early Triassic tetrapods and Gondwanaland. *Mems Mus. nat. Hist. Nat, N.S.*, Ser. A, *88,* 202–15.

Colbert, E. H., and Cosgriff, J. W. 1974. Labyrinthodont amphibians from Antarctica. *Am. Mus. Novit.*, No. 2552, 1–20.

Colbert, E. H., and Kitching, J. W. 1975. The Triassic reptile *Procolophon* in Antarctica. *Am. Mus. Novit.*, No. 2566, 1–24.

Colbert, E. H., and Kitching, J. W. 1977. Triassic cynodont reptiles from Antarctica. *Am. Mus. Novit.*, No. 2611, 1–30.

Colbert, E. H., and Merrilees, D. 1967. Cretaceous dinosaur footprints from Western Australia. *J.R. Soc. West. Aust., 50,* 21–25.

Cox, C. B. 1974. Vertebrate palaeodistributional patterns and continental drift. *J. Biogeography, 1,* 75–94.

Debenham, F. 1913. The geological theory of South Victoria Land. (*In* Huxley, L., *arr. Scott's Last Expedition. 2.* New York, Dodd, Mead, 295–300.)

Dietz, R. S., and Sproll, W. P. 1970. Fit between Africa and Antarctica: a continental drift reconstruction. *Science, N.Y., 176,* 1612–14.

Elliot, D. H. 1972. Aspects of Antarctic geology and drift reconstructions. (*In* Adie, R. J., *ed. Antarctic Geology and Geophysics.* Oslo, Universitetsforlaget, 849–58.)

Elliot, D. H., Colbert, E. H., Breed, W. J., Jensen, J. A., and Powell, J. S. 1970. Triassic tetrapods from Antarctica: evidence for continental drift. *Science, N.Y., 169,* No. 3951, 1197–201.

Gould, L. M. 1931. *Cold. The Record of an Antarctic Sledge Journey.* New York, Brewer, Warren and Putnam, 275 pp.

Parrington, F. R. 1948. Labyrinthodonts from South Africa. *Proc. Zool. Soc.. London, 118,* 426–45.

Schaeffer, B. 1972. A Jurassic fish from Antarctica. *Am. Mus. Novit.*, No. 2495, 1–17.

Smith, A. G., and Hallam, A. 1970. The fit of the southern continents. *Nature, Lond., 225,* 139–44.

77

Late Precambrian and Cambrian Fossils from Northern Victoria Land and their Stratigraphic Implications

R. A. COOPER
New Zealand Geological Survey, Department of Scientific and Industrial Research, Lower Hutt, New Zealand

J. B. JAGO
South Australian Institute of Technology—Applied Geology, Adelaide, S.A. 5098, Australia

D. I. MACKINNON
Geology Department, University of Canterbury, Christchurch, New Zealand

J. H. SHERGOLD*
Bureau of Mineral Resources, P.O. Box 378, Canberra, A.C.T. 2601, Australia

G. VIDAL
Department of Historical Geology and Paleontology, Solvegatan 13, Lund, Sweden

Abstract. Fossils recently discovered in northern Victoria Land include trilobites, brachiopods, molluscs, acritarchs, and trace fossils; they range in age from Vendian (latest Precambrian) or Early Cambrian to Late Cambrian or Ordovician. Acritarchs from three localities in rocks mapped as Robertson Bay Group indicate an Early Cambrian to possibly Vendian age. Acritarchs from several localities in the Molar Formation (Sledgers Group), together with acrotretid and obolid inarticulate brachiopods from the highest beds, indicate an Early Cambrian (Atdabanian) to possibly Vendian age for the formation.

The concordantly overlying Spurs Formation (Mariner Group) has yielded trilobites, brachiopods, and molluscs from several horizons, and they range from late Middle Cambrian to middle Late Cambrian in age. The trilobite faunas bear strong affinity with Australian-Asian Cambrian faunas, which is consistent with previous interpretations of Antarctic Cambrian trilobites. A previously unsuspected hiatus representing late Early and early Middle Cambrian time is inferred to be between the Sledgers and Mariner Groups. The faunal data and stratigraphic relationships also suggest that the Mariner Group was uplifted, slightly tilted, and partially eroded prior to deposition of the overlying Camp Ridge Quartzite (Leap Year Group). Trace fossil assemblages from low in the Camp Ridge Quartzite are consistent with a Late Cambrian to Ordovician age and indicate that, in places, deposition was marine.

Prior to 1971, body fossils were unknown from the pre-Devonian basement complex of northern Victoria Land. The first report of fossils was that of Laird and others (1972)—trilobites, brachiopods, and molluscs at Eureka Spurs, Mariner Glacier. The fossils were described by Shergold and others (1976) and assigned a Late Cambrian age. Trace fossils were reported from Eureka Spurs and East Quartzite Range by Laird and others (1974). Additional fossils were collected at each of these localities as well as at several other new localities in northern Victoria Land in 1974–75 (Cooper and others, 1976). All the above fossil localities lie within rocks of the Bowers Supergroup (formerly Bowers Group). In addition, samples collected by Cooper and others, and by earlier expeditions, from the Molar Formation (Bowers Supergroup) and from the Robertson Bay Group have been processed for organic-walled microfossils by one of us (Vidal). Acritarchs have been recovered that have important implications for the age and stratigraphic relationships of both stratigraphic units. Fossil localities are shown in Figure 77.1; lithostratigraphic units and fossiliferous horizons are correlated in Figure 77.2.

All fossils are currently under study by the authors, and a full description of them will be published later. The purpose of this note is to summarize the age and stratigraphic significance of the fossils. The terms Lower and Middle Cambrian are used here as generally applied in Australia (Öpik, 1968).

The stratigraphic terminology of the Bowers Supergroup has been revised by Laird and others (Chapter 66, this volume), and their revision is followed here:

- Bowers Supergroup
  - Leap Year Group
    - Camp Ridge Quartzite
    - Reilly Conglomerate
    - Carryer Conglomerate
  - Mariner Group
    - Eureka Formation
    - Spurs Formation
    - Edlin Formation
  - Sledgers Group
    - Glasgow Volcanics
    - Molar Formation

* J. H. Shergold publishes with approval of the Director, Bureau of Mineral Resources, Canberra, Australia.

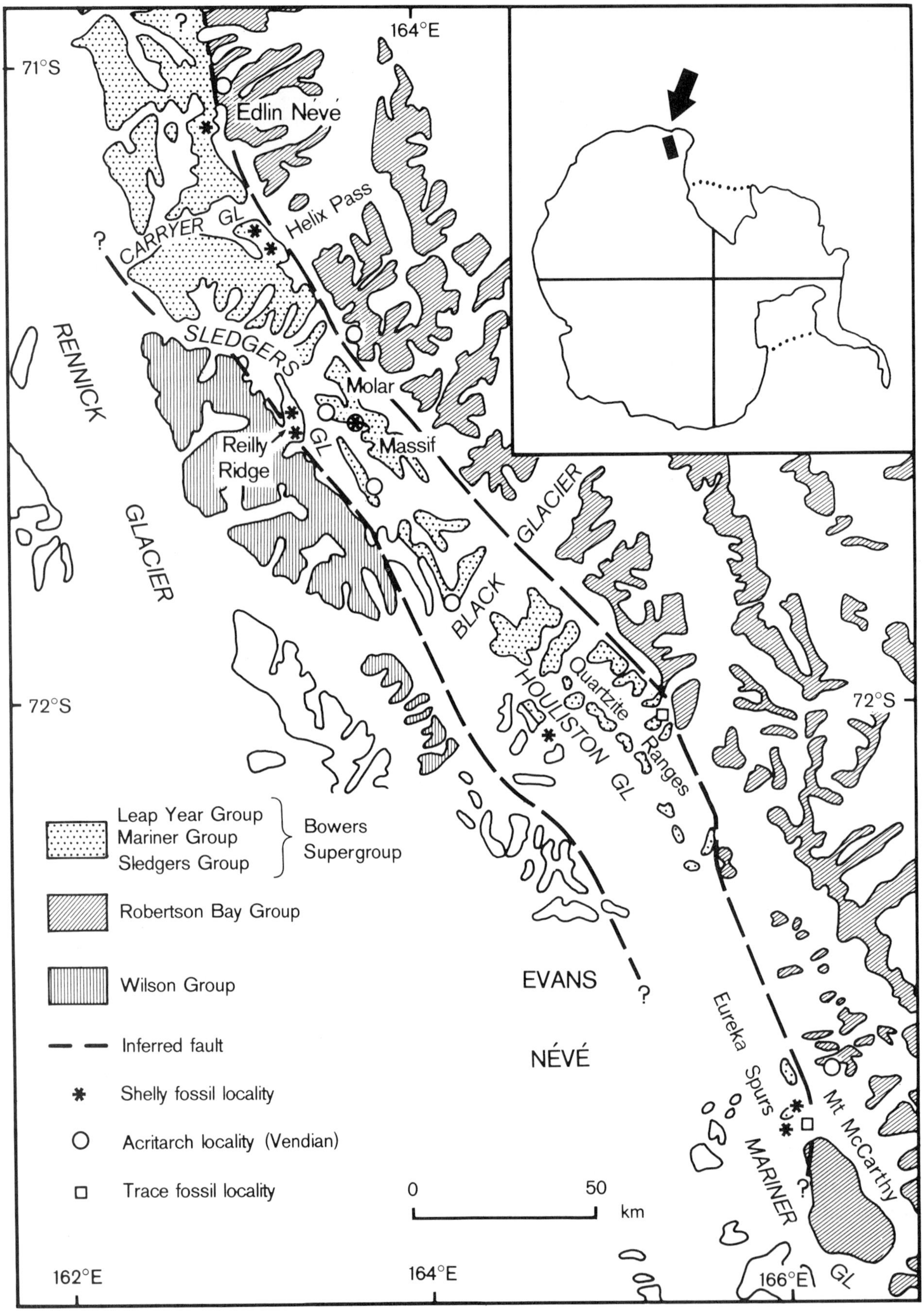

Figure 77.1. Simplified basement geology of Evans Névé–Bowers Mountains area, northern Victoria Land, showing distribution of fossil localities. Wilson Group is Precambrian; ages of other units are shown in Figure 77.2.

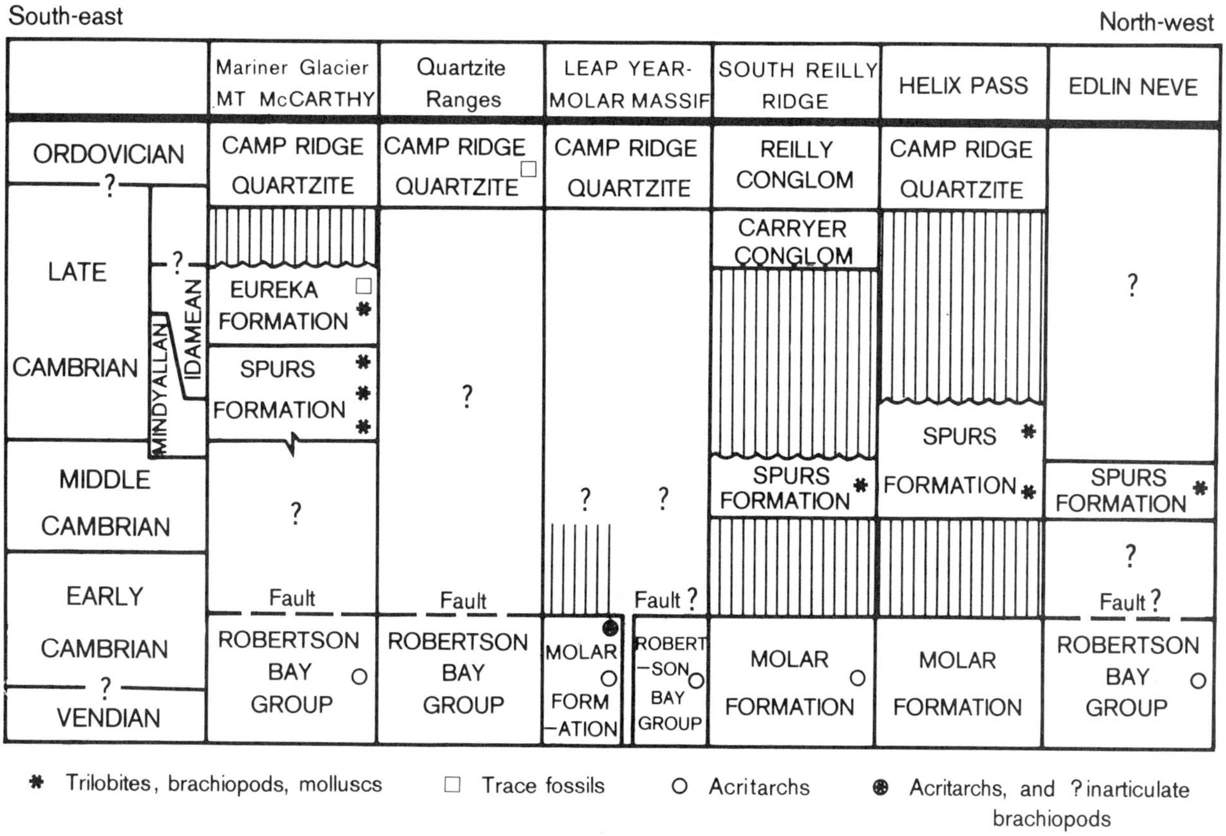

Figure 77.2. Correlation table showing lithostratigraphic units and fossiliferous horizons in northern Victoria Land. Vertical hatching: beds excluded by unconformity.

Robertson Bay Group

Acritarchs and other microfossils have been recovered from several samples of mudstone and muddy sandstone mapped as Robertson Bay Group (Gair and others, 1969) at three main localities: Mount McCarthy in the southeast; Leap Year Glacier near Molar Massif; and Mount Sturm near Edin Névé, 330 km to the northwest of Mount McCarthy. The richest assemblage was obtained from Mount McCarthy where the following taxa are present:

Bavlinella faveolata (Shepeleva) Vidal
Trachysphaeridium levis (Lopukhin) Vidal
T. cf. *timofeevi* Vidal
Eosphaera cf. *tylery* Barghoorn
Protosphaeridium cf. *tuberculiferum*
Trematosphaeridium holtedahli Timofeev
cf. *Stictosphaeridium* sp. (*sensu* Vidal, 1976)
Chuaria circularis Walcott ?
Vendotaenia sp.
Micrhystridium sp.

Chuaria circularis is restricted to the Late Proterozoic (Vidal, 1976), but its presence in the Mount McCarthy assemblage is uncertain because of the poor preservation of the specimens.

Bavlinella faveolata and *Trachysphaeridium levis* are known to range from late Proterozoic time into Early Cambrian time (as high as the Scandinavian zone with *Schmidtiellus mickwitzi* and *Holmia mobergi;* Vidal, 1979). The presence of scattered micrhystrids and acanthomorphs (most probably baltisphaerids), on the other hand, suggests a Cambrian (or possibly younger) age. Taken together, the last two groups suggest an early Cambrian age.

From Leap Year Glacier and Mount Sturm only sparse assemblages are present, but they include *Bavlinella faveolata,* which is also in the Mount McCarthy assemblage, and are consistent with an Early Cambrian age.

That part of the Robertson Bay Group sampled here is therefore best regarded as of Early Cambrian age, possibly ranging down into the Vendian. The only other indication of the age of Robertson Bay Group rocks is that of Iltchenko (1972). Acritarchs from the Oates Coast were assigned a Riphean (pre-Vendian) age, older than that reported here. However, it should be noted that all sample localities of the present study lie along strike from each other and near the contact with the Bowers Supergroup. The age range of the extensive belt of rocks mapped as Robertson Bay Group (Gair and others, 1969) may well be greater than that represented by our samples.

Bowers Supergroup—Molar Formation

Samples of mudstone, sandstone, limestone, and conglomerate from several localities in the Molar Formation between Black Glacier and upper Sledgers Glacier have yielded assemblages of acritarchs and other microfossils. The following taxa are present:

Bavlinella faveolata (Shepeleva) Vidal
Trachysphaeridium levis (Lopukhin) Vidal
cf. *Stictosphaeridium* sp. (sensu Vidal 1976)
Chuaria circularis Walcott?
Vendotaenia sp.
Leiosphaeridia sp.
Pterospermopsimorpha cf. *densicoronata* Vidal
Lacunosphaera cf. *simplex* Pychora

The assemblage is characteristically a Vendian one, but it could be as young as Early Cambrian (see discussion of Robertson Bay Group above).

At one locality in Molar Massif above a unit of volcanics and in the highest exposed beds of the Molar Formation, poorly preserved inarticulate brachiopods were found in calcareous mudstone. The assemblage includes acrotretids and an obolid and indicates an age of no older than middle Atdabanian in terms of the Russian sequence (or early Early Cambrian; A. B. Rowell, personal communication). Acritarchs recovered from the same beds include *Trachysphaeridium levis* (Lopukhin) Vidal.

The age of the youngest exposed beds of the Molar Formation is therefore taken as Early Cambrian (about Atdabanian); the lower part of the formation may be Vendian.

Houliston Glacier beds

In Houliston Glacier area, isolated nunataks are composed of alternating mudstones, sandstones, and conglomerates that have previously been mapped as Molar Formation (formerly Sledgers Formation) (Dow and Neall, 1974; Cooper and others, 1976). Very poorly preserved ptychagnostids (trilobites) were found in cleaved mudstones by Cooper and others (1976), suggesting a Middle Cambrian age. The Houliston Glacier beds are thus much younger than the youngest known beds of the Molar Formation, and their lithostratigraphic correlation is here regarded as uncertain. Sandstone and mustone clasts in conglomerate contain the Early Cambrian to Vendian acritarchs *Trematosphaeridium holtedahli, Bavlinella faveolata,* and *Chuaria circularis* (?) and may well have been derived from the Molar Formation or the Robertson Bay Group.

Mariner Group—Spurs Formation

The Spurs Formation is the most richly fossiliferous unit in northern Victoria Land. A diverse assemblage of trilobites, brachiopods, and molluscs from a single horizon in the upper part of the formation at Eureka Spurs has been described by Shergold and others (1976) and was assigned a late Idamean (possibly *Erixanium sentum* Zone) age, equivalent to the late Dresbachian of North America. The following taxa were recorded:

Brachiopoda:
- *Schizambon reticulatus* MacKinnon
- acrotretid sp. (previously listed as *Prototreta* sp.)
- *Billingsella* sp.

Mollusca:
- *Contitheca webersi* Yochelson
- *Hyolithes* sp.
- *Pelagiella* sp.
- *Scaevogyra* sp.

Trilobita:
- *Pseudagnostus* sp.
- *Stigmatoa* sp.
- *Olentella* cf. *olentensis* Ivshin
- *Irvingella?* sp.
- *Pedinocephalus* sp. of *P. bublichenkoi* Ivshin
- Aphelaspids (3 species)
- *Talbotinella?* sp.
- Olenid (1 species)
- *Prochuangia* sp. aff. *P. granulosa* Lu
- *Proceratopyge* cf. *P. lata* Whitehouse

The trilobites bear marked affinity with trilobites of Australia, China, and central Kazakhstan, in agreement with affinities shown by Cambrian trilobites from other Antarctic localities (Palmer and Gatehouse, 1972; Webers, 1972).

Trilobites, brachiopods, and molluscs were also collected from several lower horizons in the formation at Eureka Spurs by Cooper and others (1976). The trilobites include *Pseudagnostus, Proceratopyge,* and *Paleodotes* cf. *italops* Opik and indicate a late Mindyallan (early Dresbachian, Late Cambrian) age.

At Helix Pass, trilobites have been found in the formation a few meters below the unconformable contact with the overlying Camp Ridge Quartzite. They include *Glyptagnostus stolidotus* and *Aspidagnostus,* and indicate a late Mindyallan age. Somewhat lower in the sequence, *Centropleura, Pianaspis,* and *Ptychagnostus* have been collected, indicating a late Middle Cambrian age.

At Edlin Névé, a limestone band in the lower slopes of Mount Glasgow contains poorly preserved and deformed trilobites including *Dorypyge, "Amphoton,"* and *Centropleura*(?), suggesting a late Middle Cambrian age.

At Reilly Ridge a rich trilobite assemblage includes *"Amphoton," Doropyge, Corynexochus, "Solenopleura," Leiopyge, Hypagnostus,* and *Ptychagnostus,* and indicates a late Middle Cambrian age.

The age range of fossils from the Spurs Formation is thus late Middle Cambrian to Idamean (*E. sentum* Zone ?).

Mariner Group—Eureka Formation

At Eureka Spurs (see Figure 77.1 for location), the Spurs Formation is conformably overlain by the Eureka Formation. Trilobites were found by Cooper and others (1976) in a thin limestone bed in the middle part of the Eureka Formation 500 m below the contact with the overlying Camp Ridge Quartzite. Provisional identifications show the presence of *Pseudagnostus (Pseudagnostus), Homagnostus, Amorphella* sp., *Kujandaspis* sp., *Proceratopyge, Kazelia*(?) sp., and *Olentella*(?) sp., and the age is probably early post-Idamean. Trace fossils from the uppermost beds of the Eureka Formation (Red Bed unit of Andrews and Laird, 1976) include *Rusophycus, Cruziana,* and *Diplichnites*(?).

Leap Year Group—Camp Ridge Quartzite

At Camp Ridge, in the east Quartzite Range, a variety of trace fossils are found in beds which probably represent the lower part of the predominantly fluvial Camp Ridge Quartzite. They include *Scolithus, Daedalus, Arthrophycus, Monocraterion* (?), and one of unknown affinity, listed by Laird and others (1974, pp. 593–94) as "probably *Diplocraterion.*" The assemblage is consistent with a late Cambrian or Ordovician age and indicates that, in places, deposition was marine.

Stratigraphic implications

In addition to their importance for dating the rocks, the fossils also have important implications for stratigraphic relationships. Acritarch assemblages in the Robertson Bay Group at Mount McCarthy have many taxa in common with those from the Molar Formation, and the two units appear, at least in part, to be of similar (Early Cambrian) age. The range of lithotypes of the sediments sampled in each group is similar, leading to the possibility that the two represent the same stratigraphic unit. This suggestion has important implications for the interpretation of regional stratigraphy and structure, and certainly merits further investigation.

The difference in age now known between the generally concordant Molar and Spurs Formations indicates the presence of a considerable hiatus between them, apparently involving late Early Cambrian and early Middle Cambrian time.

Beds immediately underlying the Camp Ridge Quartzite at Helix Pass (Spurs Formation) are appreciably older than beds underlying it at Eureka Spurs (Eureka Formation). The entire Eureka Formation and the upper part of the Spurs Formation have apparently been eroded at Helix Pass prior to deposition of the Camp Ridge Quartzite. The lack of marked angular discordance between the Camp Ridge Quartzite and the older units suggests that erosion resulted from uplift without strong folding or tilting (Cooper and others, 1976). Evidence for deeper erosion with strong discordance, however, is reported from Reilly Ridge by Laird and others (1976) and by Bradshaw and others (Chapter 100 this volume).

Acknowledgments. This work forms part of International Geological Correlation Project 7, "Southwest Pacific Basement Correlation." The acritarch studies were undertaken by Vidal during the tenure of a New Zealand National Advisory Research Council postdoctoral fellowship at Lower Hutt, New Zealand.

Field work was supported by the Department of Scientific and Industrial Research under the New Zealand Antarctic Research Program.

We wish to acknowledge the cooperation of M. G. Laird and his (1974–75) field party during the field work, and we also thank him for reading the manuscript.

References cited

Andrews, P. B., and Laird, M. G. 1976. Sedimentology of a Late Cambrian regressive sequence (Bowers Group), northern Victoria Land, Antarctica. *Sedim. Geol., 16,* 21–44.

Cooper, R. A., Jago, J. B., MacKinnon, D. I., Simes, J. E., and Braddock, P. E. 1976. Cambrian fossils from the Bowers Group, northern Victoria Land, Antarctica (preliminary note). *N.Z. Jl Geol. Geophys., 19,* No. 2, 283–88.

Dow, J. A. S., and Neall, V. E. 1974. Geology of the Lower Rennick Glacier, northern Victoria Land, Antarctica. *N.Z. Jl Geol. Geophys., 17,* No. 3, 659–714.

Gair, H. S., Sturm, A., Carryer, S. J., and Grindley, G. W. 1969. The Geology of northern Victoria Land (Sheet 13). (*In* Bushnell, V. C., and Craddock, C., *eds.* Geologic Maps of Antarctica. *Antarct. Map Folio Ser.,* Folio 12, Pl. XII.)

Iltchenko, L. N. 1972. Late Precambrian acritarchs of Antarctica. (*In* Adie, R. J., *ed. Antarctic Geology and Geophysics.* Oslo, Universitetsforlaget, 599–602.)

Laird, M. G., Andrews, P. B., and Kyle, P. R. 1974. Geology of northern Evans Névé, Victoria Land, Antarctica. *N.Z. Jl Geol. Geophys., 17,* No. 3, 587–601.

Laird, M. G., Bradshaw, J. D., and Wodzicki, A. 1976. Reexamination of the Bowers Group (Cambrian), northern Victoria Land, Antarctica (preliminary note). *N.Z. Jl Geol. Geophys., 10,* No. 2, 275–82.

Laird, M. G., Andrews, P. B., Kyle, P., and Jennings, P. 1972. Late Cambrian fossils and the age of the Ross Orogeny, Antarctica. *Nature, Lond., 238,* 5358.

Öpik, A. A. 1968. The Ordian Stage of the Cambrian and its Australian Metadoxididae. *Bull. Miner. Resour. Geol. Geophys. Aust., 92,* 133–69.

Palmer, A. R., and Gatehouse, C. G. 1972. Early and Middle Cambrian trilobites from Antarctica. *Prof. Pap. U.S. geol. Surv.,* No. 456-D, D1–37, pls. 1–6.

Shergold, J. H., Cooper, R. A., MacKinnon, D. I., and Yochelson, E. L. 1976. Late Cambrian Brachiopoda, Mollusca, and Trilobita from northern Victoria Land, Antarctica. *Palaeontology, 19,* Pt. 2, 38–42.

Vidal, G. 1976. Late Precambrian microfossils from the Visingso Beds in southern Sweden. *Fossils and Strata, 9,* 1–57.

Vidal, G. 1979. Acritarchs from the Upper Proterozoic and Lower Cambrian of East Greenland. *Grønl. Geol. Unders., 134,* 19, 24.

Webers, G. F. 1972. Unusual Upper Cambrian fauna from West Antarctica. (*In* Adie, R. J., *ed. Antarctic Geology and Geophysics.* Oslo, Universitetsforlaget, 235–38.)

78

Upper Cambrian Molluscs from the Ellsworth Mountains

GERALD F. WEBERS
Department of Geology, Macalester College, St. Paul, Minnesota 55105, U.S.A.

Abstract. Twenty species of molluscs have been recovered in a single fauna in the north-central Heritage Range of the Ellsworth Mountains, West Antarctica. This fauna is preserved in a thin (6 m thick), highly fossiliferous limestone in the upper part of the Heritage Group. Of these species, ten are monoplacophorans, four are gastropods, four are hyolithids, and two are rostroconchians. Eight new genera and a number of new species are illustrated but not formally described. The dominant monoplacophorans show a wide variety of shell forms, indicating an evolutionary expansion of the Monoplacophora into a variety of niches in the shallow-marine environment. One form, *Knightoconus,* is associated with the origin of the cephalopods. Protoconchs and early growth stages indicate that the monoplacophoran forms evolved from low, cap-shaped forms. The four species of gastropods exhibit rapidly expanding hyperstrophic shell forms that are intermediate in their expansion rates between *Pelagiella* and younger Late Cambrian and Early Ordovician forms. There are three genera of hyolithids present, two of which have been reported recently from the Bowers Group of northern Victoria Land (Shergold and others, 1976).

The molluscs are part of a larger fauna which includes twelve species of trilobites, articulate and inarticulate brachiopods, pelmatozoans, and archaeocyathids. This great faunal diversity indicates a mild climate, perhaps warm temperate or tropical. The fauna is considered to be a shoreline population, and this probably represents the main ecologic niche of the monoplacophorans during Late Cambrian time. The molluscan fauna most closely resembles Franconian and younger molluscan faunas of North America, but it is considered to be of Dunderbergian or earliest Franconian age on the basis of the trilobites.

Webers (1972) has reported an unusual Upper Cambrian fauna from the Heritage Range of the Ellsworth Mountains. The general location of the fauna is shown in Figure 78.1. This fauna is dominated by trilobites and molluscs, but it also contains archaeocyathids, articulate and inarticulate brachiopods, and pelmatozoans. Subsequent more detailed study of the molluscs has revealed an unusually diverse and well-preserved fauna, which includes eight new genera and twenty species. Space does not permit formal description of these forms here, but a specimen of each is illustrated (Figure 78.2) to convey a sense of the diversity and preservation. The molluscs are probably among the best preserved Upper Cambrian forms from anywhere in the world. Shell material is preserved in most specimens. The thin (6 m thick) limestone in which the fauna is preserved is folded and recrystallized, but fossil specimens have suffered only minor distortion of their original shapes.

Molluscan faunal composition

Monoplacophorans dominate the fauna and exhibit a wide variety of shell forms (Figure 78.2). Ten species are present, nine of them new. Genera include *Knightoconus, Proplina,* and a number of new genera. One of the new genera is similar to *Ozarkoconus,* and a second genus is close to *Archinacella.* Shells range from low, cap-shaped forms to high, hypseloconidlike forms. Cross sections reveal round, oval, rectangular, triangular, and rhombic forms. A series of growth forms from protoconch to adult is present for the more common species and records some dramatic changes in overall appearance. Most early growth stages are low and flat in comparison to the adult stages. One species shows a complete reversal of shell curvature during ontogeny.

The four species of gastropods present are all hyperstrophic. All of the shell forms expand rapidly and are intermediate in their expansion rates between pelagiellids and younger gastropods. It does not appear, however, that pelagiellids were ancestral to the Heritage Range forms (at least not directly) since the overall coiling characteristics differ markedly. A small species of *Scaevogyra* is the most

Figure 78.1. Map of Antarctica, showing the generalized location of the Transantarctic Mountains and Ellsworth Mountains. Approximate position of the Heritage Range fauna in the Ellsworth Mountains is shown by the arrow.

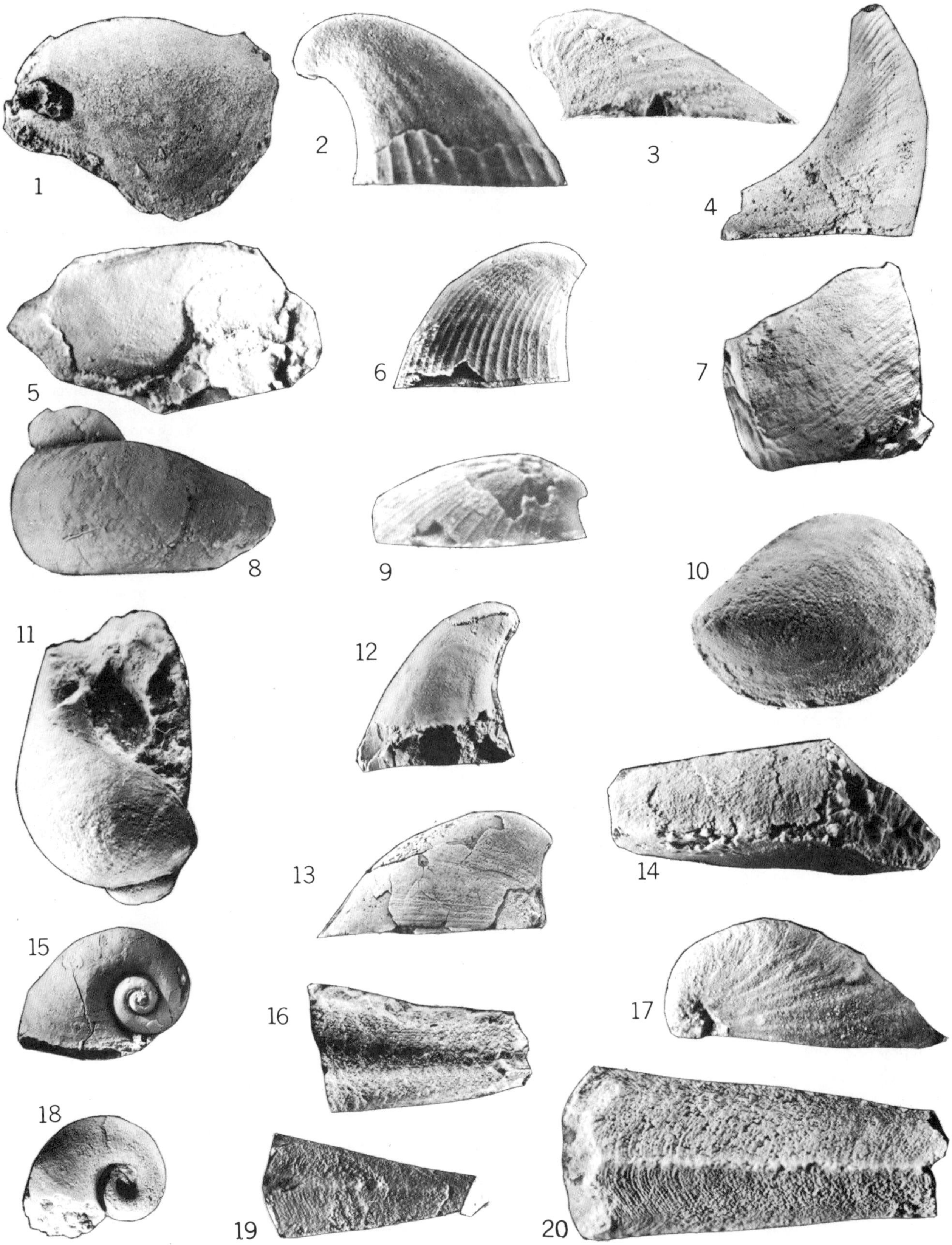
1
2
3
4
5
6
7
8
9
10
11
12
13
14
15
16
17
18
19
20

common gastropod (Figure 78.2, no. 8). Two of the gastropods (nos. 15 and 18) are macluritellids and are possibly ancestral to Ordovician forms in North America.

Hyolithids are present in four species and are rather common in the fauna. Two species of *Hyolithus* occur (Figure 78.2, nos. 19 and 20), and one of them is common. *Orthotheca* (no. 16) and *Contitheca* (no. 14) are uncommon, and each is represented by a single species.

Two rostroconchs are present and mark the first appearance of this class of molluscs in Antarctica. One species is referable to *Ribeiria* (no. 5), and a second (no. 1) represents a new genus and species.

Correlation of the molluscan fauna

The molluscs of the Heritage Range correlate most closely with those of Late Cambrian (Franconian and Trempealeauan) and Early Ordovician (Canadian) ages in the north-central United States. *Scaevogyra* is known from Franconian sedimentary rocks of Minnesota and Wisconsin. *Knightoconus* is very similar to a septate *Hypseloconus* species described by Berkey (1897, 1898) from Taylors Falls, Minnesota. *Ozarkoconus* is a Lower Ordovician monoplacophoran from the central and north-central United States with an unusual ornament consisting of sharp costae extending from apex to margin at right angles to the growth lines. A similar ornament is found in four species in the Heritage fauna (Figure 78.2, nos. 2, 6, 9, and 12). There are also general similarities in both shell diversity and form between the Heritage Range fauna and the fauna described by Berkey (1897, 1898) from Franconian strata of Minnesota and Wisconsin. Surprisingly, agnostid and aphelaspid trilobites in the Heritage fauna indicate a late Dresbachian (Dunderbergian) age.

Within Antarctica, the only Upper Cambrian molluscan fauna reported is from northern Victoria Land (Shergold and others, 1976). These four genera and species were reported: *Scaevogyra sp., Contitheca webersi, Hyolithus sp.,* and *"Pelagiella" sp.* Of these, *Hyolithus, Contitheca,* and *Scaevogyra* are also found in the Heritage fauna, and the two faunas are thus considered to be close in age. The small species of *Scaevogyra* in the Victoria Land fauna may be conspecific with that of the Heritage fauna. Yochelson (in Shergold and others, 1976) considered the molluscs to be representative of Trempealeau or younger faunas. There also, agnostid and aphelaspid trilobites give an age of late Dresbachian (Dunderbergian) for the Victoria Land fauna. Dresbachian molluscan faunas are not well known, and it appears that Franconian and younger molluscan faunas were already present in general shell form and diversity in late Dresbachian time.

Figure 78.2. Representative mollusc specimens, Heritage Range fauna. Monoplacophorans: numbers 2, 3, 4, 6, 7, 9, 10, 12, 13, and 17. Gastropods: numbers 8, 11, 15, and 18. Hyolithids: numbers 14, 16, 19, and 20. Rostroconchs: numbers 1 and 5.

1. New genus, new species, side view, steinkern, ×4.
2. New genus, new species, side view, ×5.
3. New genus, new species, side view, ×2.5.
4. *Knightoconus antarcticus* Yochelson, Flower & Webers, side view, ×2.
5. *Ribeiria sp.,* side view, ×6.
6. New genus, new species, side view, ×3.
7. New genus, new species, side view, ×1.5.
8. *Scaevogyra* new species, side view, ×6.
9. New genus, new species, side view, ×10.
10. New genus, new species, top view, slightly deformed, ×6.
11. New genus(?), new species, side view, ×4.
12. New genus, new species, side view, steinkern, ×3.
13. *Proplina,* new species, side view, ×1.2.
14. *Orthotheca* species, side view, ×5.
15. New genus, new species, bottom view, ×1.1.
16. *Contitheca sp.,* bottom view, ×6.
17. New genus, new species, side view, ×4.5.
18. New genus, new species, bottom view, ×2.5.
19. *Hyolithus,* species A. top view, ×5.
20. *Hyolithus,* species B, top view, ×6.

Diversity of the Heritage Range fauna

The Heritage Range fauna reveals a great degree of diversity of species, especially in comparison to the typical low-diversity faunas of Cambrian sequences. Virtually every major animal group of Cambrian time is represented—archaeocyathids, articulate and inarticulate brachiopods, pelmatozoans, trilobites, and molluscs. The fauna is dominated by trilobites and molluscs. Twelve species of trilobites have been recognized, and the presence of twenty species of molluscs probably ranks the collection among the most diverse molluscan faunas yet found in Upper Cambrian rocks. This faunal diversity suggests mild, favorable conditions, here interpreted as tropical. Faunal diversity of this general level is also found in an Upper Cambrian fauna from the north-central United States. As indicated above, these two faunas show marked similarity in the kinds of molluscs present, but they are not the same age. Sedimentary beds containing the north-central United States faunas are also interpreted as originating under tropical conditions (Dott, 1974).

The inferred tropical nature of the Heritage Range fauna is supported by paleomagnetic investigations in the Southern Hemisphere. Utilizing paleomagnetic data and the distribution of upper Precambrian glacial deposits, McElhinny and others (1974) have postulated that Gondwanaland existed at least as long ago as 750 m.y., and they have plotted a polar wandering path for the supercontinent from late Precambrian to Middle Ordovician time. This path shows considerable movement of the pole during this time interval and indicates a subtropical to tropical latitude for the Heritage Range in Late Cambrian time. A slightly modified reconstruction by Hargraves (1976, Figure 2, p. 364) places the Heritage Range almost on the equator in Late Cambrian time.

McElhinny and others (1974, p. 558) list two pole positions for Antarctica: one for Late Cambrian–Early Ordovician time at lat. 2° S, long. 20° E, and a second for Early–Middle Ordovician time at lat. 28° S, long. 10° E. Independent of reconstructions of Gondwanaland, Antarctic paleomagnetic data thus support a tropical latitude for the Heritage Range during Late Cambrian time. Figure 78.3 shows the late Precambrian–early Paleozoic polar wandering curve for Gondwanaland.

Paleoecologic and evolutionary considerations

The paleoenvironment in which the Heritage Range fauna lived is considered to have been shallow, marine, nearshore, and tropical. Shallow-water sedimentary struc-

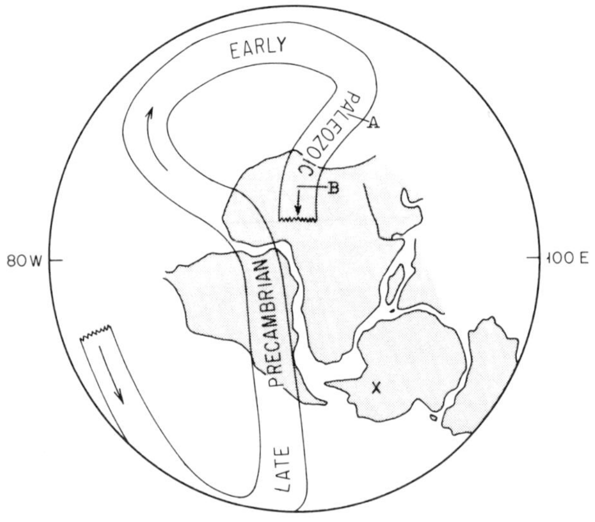

Figure 78.3. Late Precambrian to early Paleozoic polar wandering path for Gondwanaland (modified after McElhinny and others, 1974 and Hargraves, 1976). A: Late Cambrian–Early Ordovician Pole for Antarctica. B: Early–Middle Ordovician Pole for Antarctica. X: approximate position of the Heritage Range.

tures in the overlying quartzite include abundant cross-bedding and rare mud cracks. The fossil-bearing limestone is a coquina representing an accumulation of wave- or current-washed fossil debris. Many of the fossils show evidence of transport. Archaeocyathids are cylindrical in shape and invariably broken; the lack of attachments argues for transportation. Trilobites are disarticulated and commonly broken. The molluscs are well preserved and present in all growth stages, indicating that they probably occupied the site of accumulation, perhaps with a sprinkling of other faunal elements. Conditions for the molluscs must have been close to optimum, as indicated by their numbers and diversity. There appears to have been a period of adaptive radiation of the molluscs in late Dresbachian time in Antarctica. Molluscs were apparently occupying and perhaps dominating the ecologic niches in the nearshore, shallow-marine environment. Two representatives of the new molluscan class Rostroconchia (Pojeta and others, 1972) are indicative of adaptation to the infaunal environment. *Knightoconus antarcticus* is ancestral to the cephalopods (Yochelson and others, 1973) and marks a divergence toward the pelagic mode of life. The hyolithids, contithecids, and orthothecids are possibly nektonic, but their mode of life is still open to question. Most of the Monoplacophora in the Heritage Range fauna were probably epifaunal detritus feeders, but some may have been rock clingers in the manner of modern limpets. The four species of gastropods are probably epifaunal.

Once established, the environmental adaptations of the molluscs to the tropical nearshore, shallow-marine environment apparently persisted through Franconian time and possibly later. As mentioned above, the closest correlative fauna to the Heritage Range fauna is found in the Franconian and younger rocks of the central and north-central United States, where the molluscan fauna is found in nearshore tropical paleoenvironments which surrounded islands in the Cambrian sea.

Acknowledgments. The author would like to thank Ellis Yochelson and John Pojeta of the U.S. National Museum for their help throughout the study. The support of the National Science Foundation is gratefully acknowledged.

References cited

Berkey, C. P. 1897. Geology of the St. Croix Dalles. *Am. Geol.*, *20*, 345–83.

Berkey, C. P. 1898. Geology of the St. Croix Dalles. *Am. Geol.*, *21*, 270–74.

Dott, R. H., Jr. 1974. Cambrian tropical storm waves in Wisconsin. *Geology*, *2*, 243–46.

Hargraves, R. B. 1976. Precambrian geologic history. *Science, N.Y.*, *193*, 363–71.

McElhinny, M. W., Giddings, J. W., and Embleton, B. J. J. 1974. Palaeomagnetic results and late Precambrian glaciations. *Nature, Lond.*, *248*, 557–61.

Pojeta, J., Jr., Runnegar, B., Morris, N. J., and Newell, N. D. 1972. Rostroconchia: a new class of bivalved mollusks. *Science, N.Y.*, *177*, 264–66.

Shergold, J. H., Cooper, R. A., Mackinnon, D. I., and Yochelson, E. L. 1976. Late Cambrian Brachiopoda, Mollusca, and Trilobita from northern Victoria Land, Antarctica. *Palaeontology*, *19*, 247–91.

Webers, G. F. 1972. Unusual Upper Cambrian fauna from West Antarctica. (*In* Adie, R. J., *ed. Antarctic Geology and Geophysics.* Oslo, Universitetsforlaget, 235–37.)

Yochelson, E. L., Flower, R. H., and Webers, G. F. 1973. The bearing of the new Late Cambrian monoplacophoran genus *Knightoconus* upon the origin of the Cephalopoda. *Lethaia*, *6*, 275–309.

79

Trace Fossils and their Environment in Devonian (Silurian?) Lower Beacon Strata in the Asgard Range, Victoria Land, Antarctica

T. W. GEVERS
Geology Department, University of the Witwatersrand, Johannesburg, South Africa

A. TWOMEY
Department of Geology and Geophysics, University of Wisconsin, Madison, Wisconsin 53706, U.S.A.

Abstract. A broad platform, overlooking Upper Wright Valley, in the Asgard Range is underlain by trough (festoon) cross-bedded, fine-grained sandstones, some 25 m thick, overlying even-bedded sandstones and forming the uppermost part of the New Mountain Sandstone of the basal Taylor Group (Beacon Supergroup). The deeper and more steep-sided troughs are generally truncated by other less steeply inclined thin-bedded layers, on whose surfaces an abundance of 8–30 mm wide arthropod trails (*Beaconichnus gouldi*) occur. These trails match walking patterns of certain trilobites better than those of any other group of arthropods. On surfaces of the underlying even-bedded sandstones are found large trails (*Beaconichnus giganteum*), varying from 80 to 200 mm in width. Their origin is less clear. They do not fit the walking pattern of eurypterids, of which large fossil specimens are known elsewhere from the Silurian and Devonian. The pattern of these trails, however, does not rule out trilobites, of which two Devonian genera are known to have been large enough to produce such trails. Both types of sandstone are extremely mature, more than 99% of the rounded and well-sorted grains being quartz. These sand grains must have been reworked many times before final deposition in a shallow-water marine environment.

The trough cross-bedded sandstones are truncated by the Heimdall erosion surface, above which lies a cyclic sequence of subarkosic grits, sandstones, siltstones, and shales, part of the basal Altar Mountain Formation. Grit and sandstone surfaces below shale-siltstone layers exhibit very abundant intersecting grooves and ridges closely resembling tracks of some living gastropods. Estuarine and lagoonal conditions of deposition are indicated. Dessication cracks demonstrate occasional emergence.

The conditions of track and trail preservation are discussed in relation to sedimentologic features and actual observations in present-day shore environments. The clear detail observed in such pure sandstones requires the former presence of a binding agent; in water, fine organic matter, such as algal slime, is excellent for the retention of detail; subaerially, impregnation with saline spray serves this purpose.

Locality

An abundance of trace fossils have been found by the authors and J. Forsythe, in December of 1967, on a wide sandstone ledge flanking a hanging valley that extends down from the glaciated massif of St. Pauls (about 2410 m) and Round Mountains in the central Asgard Range. The ledge terminates in a broad promontory overlooking the deeply entrenched trough of Upper Wright Valley. The locality is approximately lat. 77° 34′ S and long. 161° 22′ E.

Stratigraphy

A basement complex of metasedimentary, metavolcanic, and intrusive rocks (Ross Supergroup) is truncated by the Kukri peneplain, above which lie some 1200–1300 m of Taylor Group (Lower Beacon Supergroup) sedimentary beds. The basal unit is the New Mountain Sandstone, which also includes siltstones and is about 130–160 m thick. Above a disconformity (Heimdall erosion surface) there follow upwards in succession the Altar Mountain Formation (about 160 m), the Arena Sandstone (385 m), the high, cliff-forming Beacon Heights Orthoquartzite (335 m), and the Aztec Siltstone (0–140 m). Above a very marked disconformity (Maya erosion surface) lie the sedimentary strata of the Victoria Group (Upper Beacon Supergroup), beginning with Permo-Carboniferous glacial deposits, overlain by Permian coal-bearing beds and Triassic sedimentary beds, and ending with Jurassic basalts (McKelvey and others, 1972).

Age

The trace fossils to be described occur in the uppermost New Mountain Sandstone (arthropod trails) immediately below the Heimdall erosion surface, and in the basal layers (Odin Arkose Member) of the Altar Mountain Formation (worm burrows and gastropod? tracks).

Since the Aztec Siltstone at the top of the Taylor Group has yielded an abundance of freshwater fish remains held to be (as is also a microflora) of Middle to Upper Devonian age (McKelvey and others, 1970, 1972), and since lycopod stems, dated as Lower–Middle Devonian (Plumstead, 1964), have been found in the upper part of the Beacon Heights Orthoquartzite, the underlying 1000–1100 m of

Figure 79.1. Promontory platform showing trough cross-bedded sandstones truncated by Heimdall erosion surface. Mountains of Olympus Range in background.

Figure 79.2. Truncation of trough (festoon) cross-beds by other sets. Same locality as Figure 79.1. Note thin-bedded flaggy nature of sandstones. Height of exposure is about 2 m.

dominantly arenaceous sedimentary rocks are generally also assigned to the Devonian. Poorly preserved palynomorphs from the lower units indicate an Early Devonian age; the lowermost may hence be still older, namely Silurian (Kyle quoted by Plume, Chapter 71, this volume).

Lithology

TROUGH CROSS-BEDDED UNIT

The Promontory platform is underlain by well-exposed fine- to medium-grained sandstones exhibiting strongly developed trough (festoon) cross-stratification (Figures 79.1 and 79.2). This appears to be a widespread feature elsewhere in the uppermost part of the New Mountain Sandstone (McKelvey and others, 1972; Plume, 1978). The sandstones are mostly grayish-white in color and rather friable; in places, surficial silicification has resulted in hardening and oxidation to a rust-red tint. These beds extend down to about 20–25 m below the marginal cliff, where they are replaced by even-bedded sandstones in which the thicker layers commonly show planar cross-bedding. Arthropod trails occur in both types of sandstone.

Troughs are up to 2 m deep. Fillings throughout are flaggy (Figure 79.2) and generally concave upwards; both concordant and discordant layering is present. Inclinations along the margins of troughs may be as high as 25°–30°. Troughs are truncated in an irregular fashion by other troughs or along master stratification planes by laterally more extensive beds with cross-beds of lesser inclinations (Figure 79.2).

The grains of both cross- and even-bedded sandstones are subrounded to rounded, with a sphericity of 0.6–0.8. More than 99% of the grains are quartz. Size analyses provide statistical parameters and cumulative frequency curves (omitted for lack of space) that corroborate a high degree of sorting. These sand grains must have been reworked many times before final deposition.

CYCLIC UNIT

At the inner margin of the Promontory platform and above the Heimdall erosion surface, there reposes on the truncated trough cross-bedded sandstones a 45–60 cm thick layer of yellowish-green shales and argillaceous siltstones which weather light yellowish-brown. These fine-textured sedimentary rocks are directly overlain by coarse, somewhat arkosic grits containing scattered small pebbles, some conglomeratic zones, and lenticular layers of coarse sandstone. The coarse-grained sedimentary rocks show great irregularity in deposition, with rapid lateral variation. Above them lie a number of similar sequences in which the shale layers become more silty and the grits progressively less coarse, with fewer conglomerate layers and smaller pebbles. Some of the coarse-grained sedimentary units show planar cross-bedding. In places, "rip-up" mudflake breccias are present in the basal portions of coarse grits overlying shale-silt layers.

Dark ferruginous concretions, up to 105 cm in diameter, are present, particularly in the lowermost coarse grits of the cyclic assemblage.

The fact that the light-colored, somewhat arkosic grits possess an admixture of macroscopically rather inconspicuous clayey material (probably whitish kaolin) is shown by the occurrence of polygonal dessication cracks on bedding surfaces below layers of shale and argillaceous siltstone. The infilling of the broad cracks, however, is not of these overlying materials, but of light-colored grit identical with the dessicated rock. Similarly, the many track-grooves on such surfaces, probably made by gastropods, are filled with grit, not clay or silt.

According to the classification of McKelvey and others (1972), this cyclic unit is a part of the basal Odin Arkose Member of the Altar Mountain Formation.

Trace fossils

McKelvey and Webb (1962), Allen (1962), and Webb (1963) have referred to "worm casts," to "worm borings and burrows," and to "track and trail markings by invertebrate animals" in the lower Beacon beds of the Wright Valley region. Webby (1968) supplied a more general account of trace fossils then known in these sedimentary layers. The first detailed description of arthropod trails was given by Gevers and others (1971) from the upper Hatherton Sandstone in the Darwin Mountains. Due to prior use in India (Chiplonkar and Badwe, 1970), the term *Arthropodichnus* for three types of arthropod walking trails was replaced (Gevers, 1973) by the generic name *Beaconichnus,* as listed in the new (1975) edition of Part W, Supplement I, of the *Treatise on Invertebrate Palaeontology.*

Arthropod trails (Arthropodichnia)

BEACONICHNUS GOULDI (GEVERS AND OTHERS, 1971)

Distinctive features: paired parallel rows of closely spaced footprints 8–30 mm apart; no medium "tail" drag-mark (Figure 79.3).

These trails are similar to those described in detail by Gevers and others (1971) from the even-bedded upper Hatherton Sandstone in the Darwin Mountains. On the Promontory platform in the Asgard Range they occur in great profusion on the surfaces of sandstone slabs belonging to less steeply inclined cross-bedding sets which truncate troughs of more highly inclined cross-beds. The degree of crowding of criss-crossing trails in some localities can be gauged from Figure 79.3. The longest trail seen, 168 cm, is broadly sinuous.

The trails consist of double rows of well-marked footprints in regular linear progression. Imprint depths in the wider trails are around 2 mm and diameters are 2–3 mm. Their spacing is, in general, remarkably regular; but at trail bends, the *outer* stemming side may have fewer imprints than the inner side.

BEACONICHNUS GIGANTEUM SP. NOV.

Distinctive features: large trails, widths 95–200 mm; biserial footprints progress linearly and are usually arranged in short transverse rows of two to four imprints or grouped in clusters; no median "tail" drag-mark (Figures 79.4 and 79.5).

Below the low cliffs on the western side of the Promontory platform, loose slabs from an even-bedded zone with some planar cross-bedded layers that immediately underlie

Figure 79.3. Sandstone slab from Promontory platform, with numerous criss-crossing narrow arthropod walking trails *(Beaconichnus gouldi)*. Length of pencil is about 14 cm.

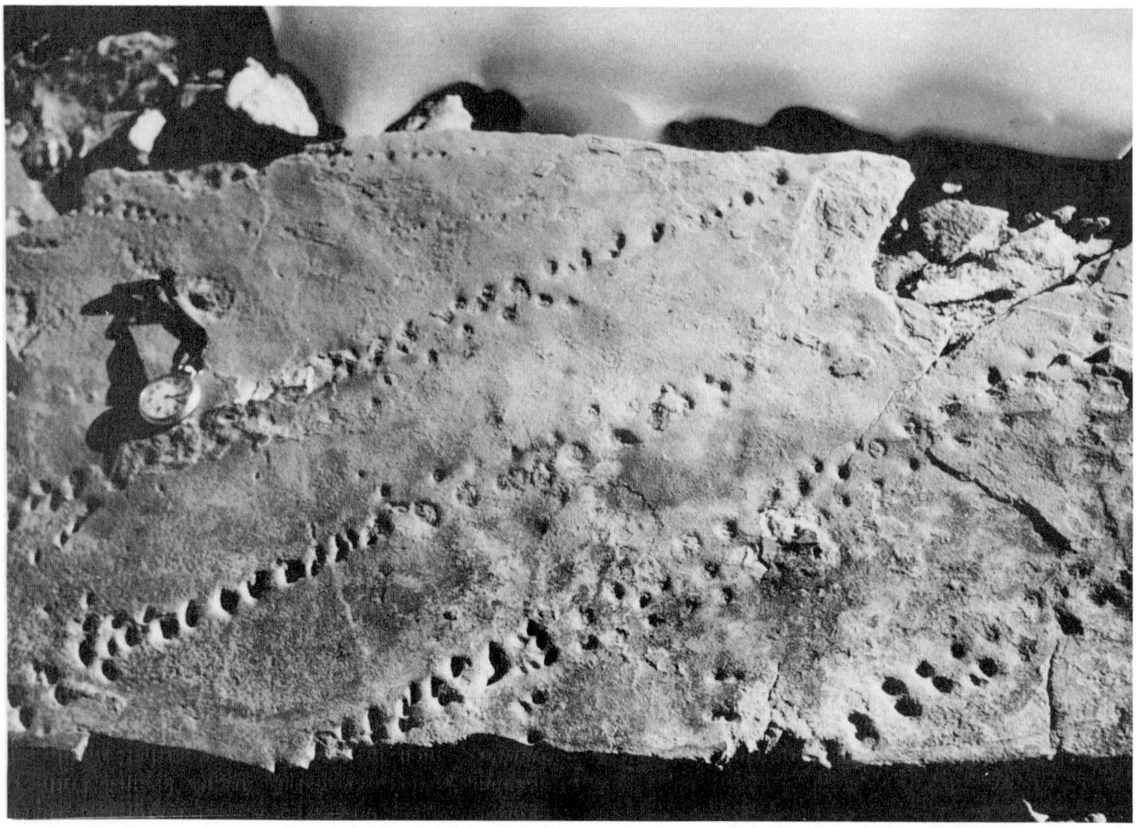

Figure 79.4. Two giant parallel arthropod walking trails *(Beaconichnus giganteum)* on sandstone slab below Promontory. Note irregular pattern of footprints in each tread line.

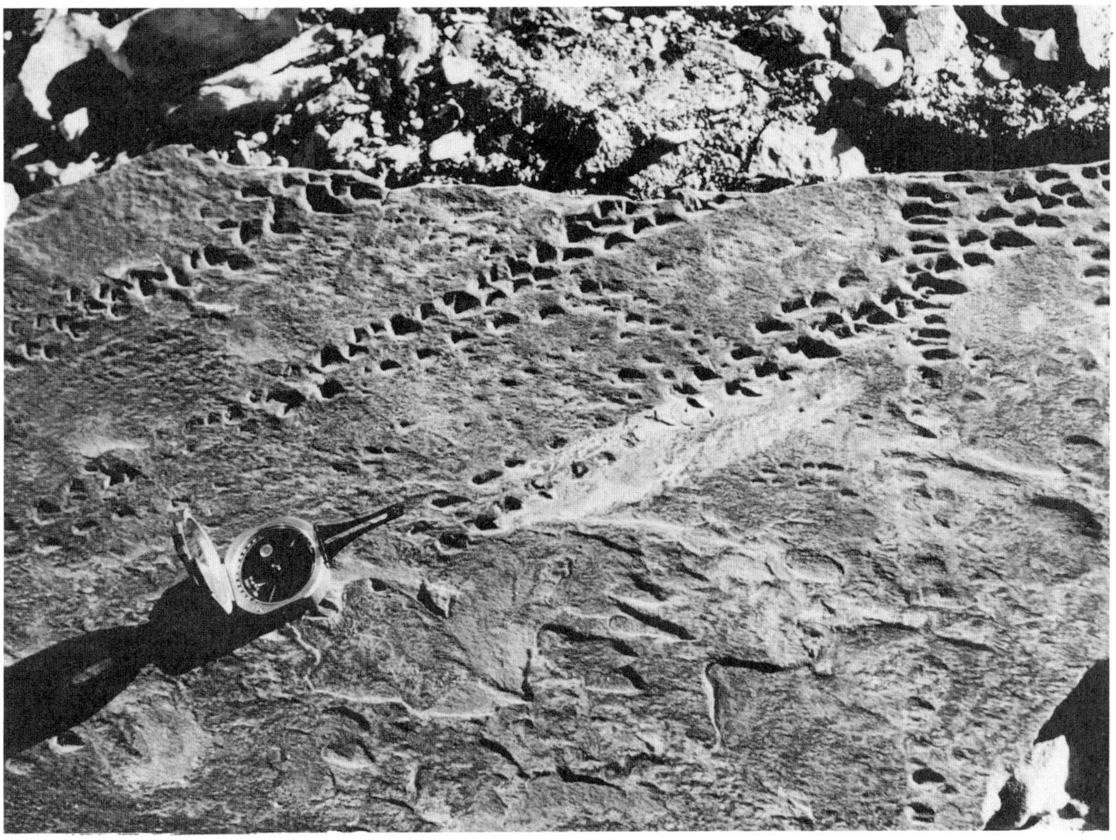

Figure 79.5. Intersecting giant arthropod walking trails *(Beaconichnus giganteum).* Same locality as in Figure 79.4. Footprint pits are strongly wind-eroded.

the strongly trough cross-bedded flaggy sandstones, frequently show trails of large arthropods.

The largest trail seen extends over more than 120 cm. Widths of individual tread-lines (trackways) are highly variable, depending on the number and spacing of pits. Those measured range between 15 and 30 mm, the greater widths belonging to wider trails. Generally there are more than just one line of pits in each tread line. Usually there is considerable irregularity in the spacing and arrangement of pits, commonly differing on opposite sides of the trail (Figure 79.4).

Goldring and Seilacher (1971) have drawn attention to the fact that many arthropod trails are *undertracks* and thus their details depend on the depth of erosion below the original surface. Walking limbs penetrate to differing depths and also at different angles, causing the relative position of imprints, as well as the width of the track-ways and trails, to vary with depth. The more heteropodus the animal, the greater the degree of undertrack diversity.

The extraordinary parallelism of the two adjacent giant trails in Figure 79.4 is rather striking. The two, no doubt adult, animals must have been of very nearly equal size and appear to have marched along "shoulder to shoulder."

Beaconichnus wrighti, sp. nov.

Distinctive feature: chevron-shaped footprints.

On the surface of a single small slab at the same locality and same stratigraphic horizon (see above) a short trail of triangular footprints was found. These are aligned in two parallel rows; the spacing between individual footprints in each row is almost equidistant. Two grooves diverge from the apex, making an angle of about 25°-30° with one another and deepening rearwards. A detailed description awaits further finds.

Identification of arthropods responsible for trails

Since the age of the sedimentary beds is at least Devonian, possibly older, the groups of aquatic arthropods to be considered are: *Trilobita* (Cambrian-Permian); *Merostomata,* both *Xiphosurida* (Limulus—King Crab) (Cambrian-Recent) and *Eurypterida* (Ordovician-Permian); and *Crustacea* (Cambrian-Recent).

By a process of elimination, the narrow trails *(Beaconichnus gouldi)* can with some confidence be attributed to Trilobita, as set out by Gevers and others (1971) for the similar trails in the upper Hatherton Sandtone of the Darwin Mountains. The nature of the originators of the large trails *(Beachonichnus giganteum)* is more debatable. They are probably *undertracks* and therfore are most likely incomplete with regard to actual numbers of appendages; but with such a large number of footprints still showing, even in undertracks, V-shaped divergences should be recognizable. The total absence of such features, as well as of median telson drag-marks, argues against both *Xiphosurida,* which in any case were rather small, and *Eurypterida.* Some trails of present-day *Crustacea,* investigated by the senior author on the Indian Ocean coast of Natal, somewhat resemble on a smaller scale those of *B. giganteum;* the footprints, however, are scattered over a wider zone, the individual tread-lines being less compact.

Figure 79.6. Maze of intersecting grooves probably made by gastropods, in part filled with grit, on surface of a grit layer below shales, in cyclic unit above Heimdall erosion surface. Majority of grooves are 13–15 mm wide.

The giant *B. giganteum* trails must have been made by large animals. Among *Eurypterida* there existed many large forms, even giant ones; *Pterygotus* found in the Devonian (Old Red Sandstone) of Scotland, with a length of 180 cm, was the largest arthropod known. But the trail pattern of *B. giganteum* does not seem to fit Eurypterids; there is no evidence of rearward divergence of footprints and there is no sign of rear paddles. While the majority of the great diversified class of *Trilobita* were comparatively small creatures, well within the range of the narrow *B. gouldi* trails, large forms capable of producing the *B. giganteum* trails, while rather rare, did exist among the *Homalonotida* of Devonian age and among the truly giant forms of the *Lichnida,* e.g., *Terataspis grandis* from the Devonian of North America. Size of these trails, therefore, does not rule out *Trilobita,* nor does the trail pattern.

All trails are purely of the walking type; no "raking" tracks, produced in feeding activities and commonly associated with trilobite trails (Seilacher, 1970), have been found.

Grooves and ridges

Figure 79.6 depicts an example of the mazes of intersecting grooves on the upper surface of grit beds below shale and silt layers of the cyclic unit. The majority of the grooves are 13–15 mm wide. In most cases their depth is considerably less than the width. Many well-preserved grooves exhibit a low and narrow, commonly crenulated median ridge; in others, the latter is accentuated by flanking depressions. Some structures are raised above their surroundings and possess a slight median depression.

In many respects these structures are strongly reminiscent of gastropod tracks. Locomotion is accomplished by rhythmic waves along the broad "foot." Many gastropod trails show a median ridge or groove. The former are due to the lateral parts of the foot being pressed deeper into the substrate than the inner parts. Grooves can be caused by downward projection of the dragged-along shell. The forms pictured in Figure 79.6 are practically identical with tracks made on sandy beaches and lagoonal estuaries of the Indian Ocean in South Africa by the common gastropod *Bullia rhodostoma*.

Environmental conditions

The mode of deposition of the predominantly sandy sedimentary beds of the lower (Taylor) group of the Beacon Supergroup has been debated by many authors. Space forbids their enumeration. Formerly most authors favored a non-marine environment. Thus Gunn and Warren (1962) considered "wandering dunes and shallow impermanent lakes, possibly grading into estuarine conditions near the coast." They emphasized the freshwater nature of the fish remains in the Aztec Siltstone at the top of the group and the "general absence of marine organisms."

As stressed by Seilacher (1970), many sandstones of proven marine origin do not provide favorable media for the preservation of body fossils—especially of chitinous arthropods—but nevertheless may be crowded with *traces* of life activities. Such beds have now been found in many horizons of the Taylor Group, and proposals for a shallow-water marine origin of at least certain layers have consequently been put forward (Vialov, 1962; Allen, 1962;

Webby, 1968; Barrett and others 1972; Matz and others, 1972). In the Darwin Mountains, Gevers and others (1971) deduce a shallow-marine environment from the intercalation of highly bioturbated layers, from the presence of many worm burrows, and from abundant trails on the surface of even-bedded sandstones, for which evidence of trilobite origin is strong.

The abundant presence of the same *Beaconichnus gouldi* on sandstone slabs of the trough cross-bedded unit of the Promontory platform in the Asgard Range leads to a similar conclusion. Suggestions that rather steep cross-bedding within the Taylor Group indicates eolian dune origin cannot apply to these sandstones. All troughs are repeatedly truncated by sharp master stratification planes on which lie more extensive layers cross-bedded at much lesser angles, in some cases approaching planar style, indicative of flowing water rather than wind. Horizontal or near-horizontal beds (Figure 79.2) are normally absent in dune deposits.

The high compositional and textural maturity of sandstones is no definitive criterion. Windblown sands round faster than those transported in water; of the latter, beach sands round faster than river sands (Folk, 1965, p. 15). Furthermore, frosting and pitting are not reliably diagnostic, as pointed out by Pettijohn (1957, p. 70). An investigation of windblown sands of the Kalahari Desert, South Africa, by Ritchie (unpublished) has shown frosting and pitting to be mainly post-depositional, being particularly noticeable in and around "pans" and other depressions with at least temporarily more abundant ground moisture, in regions where eolian sands have been stabilized for some time.

The uniformly thin-bedded (flaggy) nature of the trough fillings, generally concave upwards and mostly concordant with the floor, is best explained by deposition in relatively quiet trough pools under a thin sheet of water transporting sand at low velocities.

The sedimentary beds of the cyclic unit are obviously water-laid. The abundance of worm burrows and tracks probably made by gastropods, together with sedimentologic features, indicate estuarine and lagoonal deposits. The common occurrence of pyrite crystals within ferruginous concretions in grit beds immediately below shales indicates reducing conditions and most likely also the presence of organic matter.

From his recent detailed sedimentologic investigation and paleocurrent analysis of the basal part of the Taylor Group, Plume (Chapter 71, this volume) concludes that the New Mountain Sandstone had been deposited by braided rivers, with some eolian deposition in the upper layers. Cross-bedding is dealt with in great detail in his paper. Medium (.05–2m) and large-scale (2–8 m thick) sets are considered to be of eolian origin. The trace fossil–bearing, trough cross-bedded sandstones of the Promontary platform would fall into these categories. In briefly referring to trace fossils, Plume states that these "do not give incontrovertible evidence of either marine or continental deposition though their abundance and complexity does indicate an aqueous environment." The present authors do not as yet consider the trilobite origin of *Beaconichnus gouldi* as incontrovertibly proved; the admission by Plume of the subaqueous nature of these trace fossils, however, is at variance with the previously deduced eolian deposition of the sedimentary beds in which they occur.

Plume's model is the present-day deposits in the delta of the Ganges-Brahmaputra River in India, as described by Coleman (1969), where low sand dunes can be seen to migrate across the floodplain. It is well known, however, that the very low elevation of the latter renders it particularly liable to marine invasions during monsoon storms in the Bay of Bengal. Only slight tectonic movements would submerge a large area.

The recent detailed work by Plume (Chapter 71, this volume) greatly adds to the knowledge of the nature, succession, and origin of the sediments of the lower Taylor Group. The present authors, however, place greater emphasis on the significance of trace fossils. They are aware of the difficulties entailed in identifying the originators of the many trace fossils. However, they consider the deductions presented in this paper to be the most plausible. Further work on the numerous and varied trace fossils in certain horizons of these sediments is essential.

The authors envisage a low coastal environment involving, under rather stable tectonic conditions, only such mild epeirogenic movements as are necessary for slow and gradual transgressions and regressions of the sea, thus enabling long-continued reworking and redeposition of sediment. If low eolian dunes accumulated in near-shore, deltaic, or fringing-estuary environments, they existed only to be swept away again from time to time and to have their materials redistributed by tidal wash, estuarine floods, and shallow invasions of the sea.

The coarse and somewhat pebbly grits and subarkosic sandstones of the cyclic unit at the base of the Altar Mountain Formation indicate renewed instability inland; coarser material again became available to be spread by floods from shifting water courses over an advancing estuary. High-velocity channel turbulence tailed off laterally into more quiescent waters of bordering floodplains and coastal lagoons. Each cyclic unit marks a lateral shift, and also an upward recession, of transport capacity.

The authors strongly support Harrington's (1965) view that the lower Taylor succession represents an interdigitation of marine and non-marine sedimentary beds, with probably more marine horizons, however, than formerly envisaged.

The rich marine fauna of the Horlick Formation, forming the lowermost unit of the Beacon Supergroup in the Ohio Range, Antartica, shows close affinity with that of the Bokkeveld Beds of the Cape Supergroup in South Africa. The fossils include abundant trilobites. The Table Mountain Sandstone, underlying these Lower Devonian shallow-water marine sedimentary rocks, was long held to have been deposited in a fluvial-deltaic environment. Numerous arthropod trails, attributed to trilobites by Taljaard and Rust (Rust, 1967) (Seilacher concurring), but viewed noncommittally by Anderson (1975), have been found in certain horizons. This was followed by the discovery of brachiopod and trilobite body fossils in an interbedded siltstone-shale layer (Cocks and others, 1970). Similar discoveries have been made in the Witteberg Beds, overlying the Bokkeveld, also previously held to be freshwater sedimentary deposits. So far, body fossils of brachiopods, pelecypods, and trilobites have been located in recent years only in fine-grained sediments of the lower portion of this stratigraphic correlative of the Taylor Group. Among the numerous trace fossils found in coarser sedimentary beds are arthropod trails akin to those in the Table Mountain Sandstone (J.N. Theron, personal communication, 1969), as well as *Cruziana* specimens similar to those attributed to trilobites by Seilacher (1970).

Plume (Chapter 71, this volume) dismisses the evidence of a marine environment provided by the presence of lobate structures very similar to *Cruziana* and *Rusophycus* in the Terra Cotta Siltstone underlying the New Mountain Sandstone, in favor of a lacustrine environment, mainly because of the presence of red beds and ^{13}C values in limestone concretions that suggest fresh-water as the medium of deposition. The absence of body fossils, noted by Plume, is poor evidence against marine deposition in porous permeable sandstones deposited under well-aerated conditions, in which the chitinous exoskeleton of trilobites has practically no chance of survival. In less permeable siltstones the absence of body fossils carries somewhat more weight, but is not a strong argument. To throw further light on this problem, all fine-grained siltstone and shale layers interbedded with sandstones in the Lower Taylor Group should be carefully searched for body fossils.

Trace preservation

How can animal traces, especially trails of comparatively slender arthropod walking limbs, be preserved in such pure sands devoid of apparent binding material? The senior author (T. W. G.) has carried out extensive investigations on sands of similar grain size (fine to medium) and purity along the shores of the Indian Ocean in South Africa, both subaerially and by diving in large rock pools, in estuarine lagoons, and in the sea.

The presence of well-preserved traces on mud bottoms needs no comment; compacted muds are remarkably scour resistant. On sandy bottoms, in the absence of scour, retention and detail depend on the degree of compaction of the sand, the weight of the animal in water, and above all, the presence of a binding substance. In the absence of clay, fine organic material of colloidal dimensions, such as algal slime, was found to be an excellent binding agent. Pure sand, freshly swept by spring tides and storms into pools and depressions, had but a poor capacity for trace preservation, except for those of some gastropods whose foot exudes a slimy mucus. As algal slime accumulated and infiltrated into the surficial pore space, however, animal traces became progressively better developed and trails of crustaceans (crabs and crayfish) became more and more distinct until finally of excellent detail.

Bedding surfaces of the very mature and clean upper Hatherton Sandstone (over 99% quartz, Haskell and others, 1965) in the Darwin Mountains are commonly crowded with feeding burrows of worms *(Beaconites antarcticus);* i.e., they are concentrated in what would have been the uppermost portion of the substrate close to the water-sediment boundary, and yet the sandstone shows no readily discernible organic content. Quite obviously, however, the burrowing sand-ingesting worms must have had food to live on. The abundant narrow arthropod trails *(Beaconichnus gouldi)* in the same sandstones exhibit good detail, indicating a former binding agent. The present absence of organic matter may be due to diffusion by downward elutriation in the originally highly permeable sediment and to subsequent oxidation.

In discussing the optimum conditions for the creation of trails, Ager (1963, p. 108) mentions experiments carried out by McKee (1947), who is said to have found that the sharpest outlines were made by lizards, scorpions, wood lice, and other animals on *loose dry sand.* The observations by Gevers on the shores of the Indian Ocean, where strong winds cause eolian modification of beach sands and the piling up, in places, of low sand dunes, do not substantiate this claim. By far the clearest trails, showing minute detail, were made by crabs, lizards, and large beetles on sand wetted by dew or surf-spray, maintainance of good relief downwards depending on the depth of wetting. On absolutely dry sand, some distance from the surf, the relief is nowhere near as good; in imprints of sufficient depth the dry sand "runs" to the angle of repose (30–34°) for loose material of this nature. When trails of excellent detail on sand wetted by dew or very light rain dry out under the sun, detail is progressively lost in the same way. When wetted by surf-spray, however, sufficient saline cementation of sand grains is achieved for excellent relief to be retained on drying, resistant enough to be subsequently filled with sand by winds not strong enough for actual scour. As indicated, however, the authors consider the arthropod trails described in this paper to have been made subaqueously.

Acknowledgments. The authors are indebted to the U.S. National Science Foundation and its Division of Polar Programs, as well as to the pilot-officers and crewmen of the U.S. Naval Support Force, Antarctica. The authors also wish to express their great indebtedness to the late Professor W. Haentzschel, compiler of Volume W, Supplement I (Trace Fossils) in the *Treatise on Invertebrate Palaeontology,* for his interest in the original, paleontologically more extensive version of this paper.

References cited

Ager, D. V. 1963. *Paleoecology.* New York, McGraw-Hill, 108 pp.

Allen, A. D. 1962. Formations of the Beacon Group in the Victoria Valley Region. *N.Z. Jl Geol. Geophys., 5,* No. 2, 278–94.

Anderson, A. M. 1975. The "Trilobite" trackways in the Table Mountain Group of South Africa. *Palaeont. afr., 18,* 35–45.

Barrett, P. J., Grindley, G. W., and Webb, P. N. 1972. The Beacon Supergroup of East Antarctica. (*In* Adie, R. J., *ed. Antarctic Geology and Geophysics.* Olso, Universitetsforlaget, 319–29.)

Chiplonkar, G. W., and Badwe, R. M. 1970. Trace fossils from the Bagh Beds. *J. palaeont. Soc. India, 14,* 4–10.

Cocks, L. R. M., Brunton, C. H. C., Rowell, A. J., and Rust, I. C. 1970. The first lower Palaeozoic fauna proved from South Africa. *Q. Jl geol. Soc. Lond., 125,* 583–603.

Coleman, J. M. 1969. Brahmaputra River: channel processes and sedimentation. *Sedim. Geol., 3,* 129–239.

Folk, R. L. 1965. *Petrology of Sedimentary Rocks.* Hemphills, Austin, Texas, 15, 102.

Gevers, T. W. 1973. A new name for the ichnogenus *Arthropodichnus. J. Paleont., 47,* No. 5, 1002.

Gevers, T. W., Frakes, L. A., Edwards, L. N., and Marzolt, J. E. 1971. Trace fossils in lower Beacon sediments (Devonian), Darwin Mountains, southern Victoria Land, Antarctica. *J. Paleont., 45,* No. 1, 81–94.

Goldring, R., and Seilacher, A. 1971. Limulid undertracks and their sedimentological implications. *Neues Jb. Geol. Palaeont. Abh., 137,* No. 3, 422–42.

Gunn, B. M., and Warren, G. 1962. Geology of Victoria Land between the Mawson and Mulock Glaciers, Ross Dependency, Antarctica, *Bull. geol. Surv. N.Z.,* No. 71, 157 pp.

Haentzschel, W. 1962. Trace fossils and problematica. (*In* Moore, R. C., and Teichert, C., *eds. Treatise on Invertebrate Paleontology, Part W, Miscellanea.* Boulder, Colo. and Lawrence, Kans., Geological Society of America and University of Kansas, W177–W245.)

Haentzschel, W. 1975. Trace fossils and problematica. (*In* Moore, R. C., and Teichert, C., *eds. Treatise on Invertebrate Paleontology, Part W, Miscellanea, Supplement 1.* Boulder, Colo. and Lawrence, Kans., Geological Society of America and University of Kansas, W269 pp.)

Harrington, H. J. 1965. Geology and morphology of Antarctica. (*In* Van Mieghem, J., and Van Oye, P., *eds.* Biogeography and ecology in Antarctica. The Hague, Junk, 1–71.) [*Monographiae biol.* 15.]

Haskell, T. R., Kennett, J. P., and Prebble, W. M. 1965. Geology of the Brown Hills and Darwin Mountains, southern Victoria Land, Antarctica. *Trans. R. Soc. N.Z., 2,* 231–48.

McKee, E. D. 1947. Experiments on the development of tracks in fine cross-bedded sand. *J. sedim. Petrol., 17,* 23–28.

McKelvey, B. C., and Webb, P. N. 1962. Geological investigations in southern Victoria Land. Pt. 3. Geology of Wright Valley. *N.Z. Jl Geol. Geophys., 5,* No. 1, 143–62.

McKelvey, B. C., Webb, P. N., Gorton, M. P., and Kohn, B. P. 1970. Recent discoveries of Devonian fish in Antarctica. *Polar Rec., 15,* No. 95, 216–17.

McKelvey, B. C., Webb, P. N., Gorton, M. P., and Kohn, B. P. 1972. Stratigraphy of the Beacon Supergroup between the Olympus and Boomerang Ranges, Victoria Land. (*In* Adie, R. J., *ed. Antarctic Geology and Geophysics.* Oslo, Universitetsforlaget, 345–52.)

Matz, D. B., Pinet, P. R., and Hayes, M. O. 1972. Stratigraphy and petrology of the Beacon Supergroup, southern Victoria Land. (*In* Adie, R. J., *ed. Antarctic Geology and Geophysics.* Oslo, Universitetsforlaget, 355–58.)

Pettijohn, F. J. 1957. *Sedimentary Rocks.* New York, Harper, 70.

Plume, R. W. 1978. Proposed changes to the existing stratigraphy of the New Mountain Sandstone (Beacon Supergroup), South Victoria Land, Antarctica. *N.Z. Jl Geol. Geophys., 21,* No. 2, 167–73.

Plumstead, E. P. 1964. Palaeobotany of Antarctica. (*In* Adie, R. J., *ed. Antarctic Geology.* Amsterdam, North-Holland Publishing Company, 637–54.)

Rust, I. C. 1967. *On the Sedimentation of the Table Mountain Group in the Western Cape Province.* Ph.D. dissertation, University of Stellenbosch.

Seilacher, A. 1959. Vom Leben der Trilobiten. *Die Naturwiss, 46,* No. 12, 398–93.

Seilacher, A. 1964. Biogenic sedimentary structures. (*In* Imbrie, J., and Newell, N., *eds. Approaches to Paleoecology.* New York, John Wiley, 296–310.)

Seilacher, A. 1970. Cruziana stratigraphy of "non-fossiliferous" Palaeozoic sandstones. (*In* Crimes, T. P., and Harper, J. C., *eds.* Trace fossils. *Geol. Jnl Spec. Iss.,* No. 3, 447–76.)

Vialov, O. S. 1962. Problematica of the Beacon sandstone at Beacon Heights West, Antarctica. *N.Z. Jl Geol. Geophys., 5,* No. 5, 718–32.

Webb, P. N. 1963. Geological investigations in southern Victoria Land, Antarctica, 4. Beacon Group of the Wright Valley and Taylor Glacier Region. *N.Z. Jl Geol. Geophys., 6,* No. 3, 361–87.

Webby, B. D. 1968. Devonian trace fossils from the Beacon Group of Antarctica. *N.Z. Jl Geol. Geophys., 11,* No. 4, 1001–8.

80

Permian and Triassic Palynostratigraphy of the Victoria Group, Transantarctic Mountains

ROSEMARY A. KYLE*
Institute of Polar Studies, The Ohio State University, Columbus, Ohio 43210, U.S.A.

JAMES M. SCHOPF†
Department of Geology and Mineralogy, The Ohio State University, Columbus, Ohio 43210, U.S.A.

Abstract. Plant microfossil assemblages from the Victoria Group (upper Beacon Supergroup) of southern Victoria Land, the Beardmore Glacier area, the Queen Maud Mountains, and the Ohio and Wisconsin Ranges reflect a succession of three distinctive floras of Permian and Triassic age. Monosaccate pollen dominate assemblages associated with the upper Paleozoic glacial deposits. Rare taeniate bisaccate pollen, some probably allied with the early glossopterids, also occur. This flora is of Early Permian or possibly Late Carboniferous age by comparison with similar Australian assemblages. It is recognized in the Darwin Tillite of the Darwin Mountains, in the Buckeye Formation of the Ohio and Wisconsin Ranges, and in the postglacial Roaring Formation shale of the Nilsen Plateau.

Plant microfossil assemblages associated with *Glossopteris* in the overlying coal measures contain an abundance of taeniate and non-taeniate bisaccate pollen. Early Permian assemblages are recognized in the Weller Coal Measures, southern Victoria Land. Late Permian assemblages are present in the Queen Maud Formation, Nilsen Plateau, and the upper part of the Mount Glossopteris Formation, Ohio Range, which are stratigraphically equivalent. Plant megafossils provide general confirmation of these correlations and age assignments.

Assemblages associated with *Dicroidium* contain abundant non-taeniate bisaccate pollen and a variety of distinctive trilete and monolete spores. Four biostratigraphic units of late Early Triassic to Late Triassic age can be recognized in the uppermost Feather Conglomerate and Lashly Formation of southern Victoria Land. In the Beardmore and Shackleton Glacier areas, the upper three biostratigraphic units of Middle to Late Triassic age occur in the upper Fremouw and lower Falla Formations.

Flat-lying continental sedimentary rocks of the Victoria Group (upper Beacon Supergroup) crop out along the length of the Transantarctic Mountains. The basal Victoria Group formation disconformably overlies Devonian Taylor Group sandstone and consists of tillite and associated glaciofluvial deposits (Buckeye Formation in the Ohio and Wisconsin Ranges, Scott Glacier Formation in the Nilsen Plateau, Pagoda Tillite in the Beardmore Glacier area, Darwin Tillite in the Darwin Mountains, and Metschel Tillite in southern Victoria Land). These geographic localities and local formation names are given in Figures 80.1 and 80.2.

In the central Transantarctic Mountains, the glacial beds are overlain by dark lacustrine shale with thin sandstone and limestone beds (Discovery Ridge Formation, Roaring Formation, and Mackellar Formation) and a feldspathic sandstone unit (Mount Glossopteris Formation [lower part], Amundsen Formation, and Fairchild Formation). *Glossopteris*-bearing coal measures (Mount Glossopteris Formation [upper part], Queen Maud Formation, and Buckley Formation) overlie this sequence.

In southern Victoria Land (Mulock-Mackay Glacier area) the *Glossopteris*-bearing Weller Coal Measures consist of three members and overlie the Metschel Tillite. The essentially noncarbonaceous and unfossiliferous Feather Conglomerate overlies the Weller beds and consists of lower conglomerate and coarse sandstone and an upper cyclic succession of finer sandstone and green siltstone (Fleming Member). The overlying *Dicroidium*-bearing Lashly Formation is composed of four members. Member A (lower) is a cyclic sequence of fine-grained sandstone and root-bearing green siltstone; member B, predominantly medium-grained sandstone; member C, interbedded sandstone and carbonaceous siltstone; and member D, predominantly medium-grained sandstone.

In the Beardmore and Shackleton Glacier areas, central Transantarctic Mountains, the Buckley Formation is overlain by the Fremouw Formation, which consists of three members. The lower member is a cyclic sequence of sandstone and green siltstone which contains fossil vertebrate bones of the *Lystrosaurus* fauna. The middle member is root-bearing green mudstone with fine-grained sandstone interbeds. Redbeds are locally developed in this member in the Shackleton Glacier area. The upper member is predominantly medium-grained sandstone with carbonaceous siltstone interbeds containing *Dicroidium*. The overlying Falla Formation consists of cyclic fine-grained sandstone and siltstone, overlain by greenish-gray fine-grained sandstone and carbonaceous shale with *Dicroidium*, and an upper part of tuff and tuffaceous sandstone. Paraconglomerate and tuffs of the Prebble Formation make up the youngest Victoria Group formation in this area.

Correlations of these sequences have been proposed on

* Now Rosemary A. Askin. Present address: Department of Geology, Colorado School of Mines, Golden, Colorado 80401, U.S.A.

† Deceased.

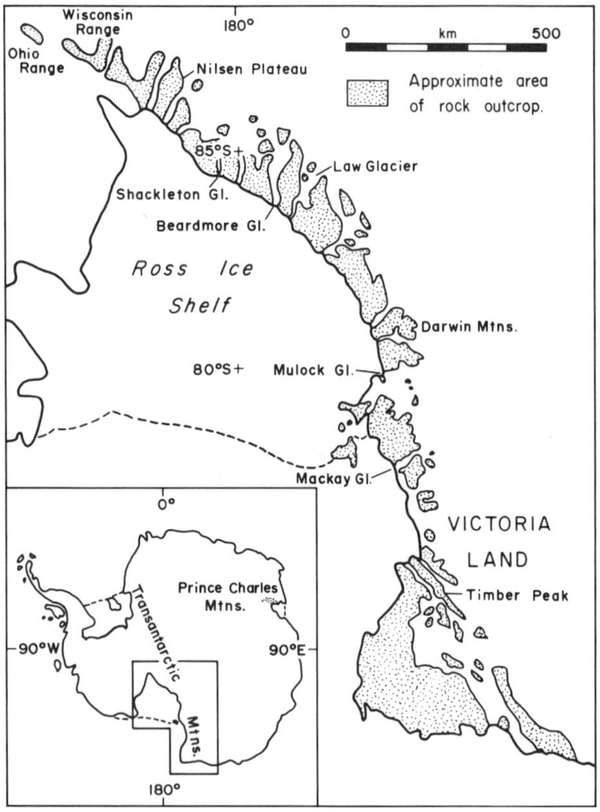

Figure 80.1. Locality map of the Transantarctic Mountains.

the basis of lithologic similarities and plant megafossils (e.g., Barrett, 1971), although relationship of the southern Victoria Land Weller Coal Measures and lower Feather Conglomerate with the central Transantarctic Mountains Mackellar, Fairchild, and Buckley Formations is not clearly established. Barrett and Kohn (1975) correlated the Fleming Member of the Feather Conglomerate with the lower Fremouw Formation and members A and B of the Lashly Formation with the middle and upper members of the Fremouw Formation. They placed the Permian-Triassic boundary in southern Victoria Land between the lower and upper parts of the Feather Conglomerate by correlating an abrupt change in paleocurrent direction at this horizon with a paleocurrent reversal in the Beardmore Glacier area between the Upper Permian *Glossopteris*-bearing Buckley Formation and Lower Triassic *Lystrosaurus*-bearing lower Fremouw Formation.

Upper Permian conchostracans in the upper Mount Glossopteris Formation of the Ohio Range (Doumani and Tasch, 1965) are the only invertebrate fossils in the Victoria Group of the Transantarctic Mountains. Vertebrate fossil remains are found in the lower member of the Fremouw Formation in the Beardmore and Shackleton Glacier areas; these include reptile and amphibian bones included in the *Lystrosaurus* Zone of Early Triassic age (Elliot and others, 1970; Kitching and others, 1972).

Abundant plant megafossils in much of the Victoria Group have been used to date the rocks within a broad range. Rigby and Schopf (1969) concluded that most of the *Glossopteris* collections are assignable to a middle and late Permian age, while Kyle (1976) regarded the plant fossils from the Weller Coal Measures of southern Victoria Land as early and middle Permian.

Plant megafossils dominated by *Dicroidium* from the upper Victoria Group have generally been regarded as Middle or Upper Triassic in age (Townrow, 1967). A mixed *Glossopteris-Taeniopteris-Dicroidium* assemblage from near the base of the Lashly Formation at Allan Nunatak has been correlated by Rigby and Schopf (1969) with floras from the Panchet beds of India, considered to be Early Triassic in age.

Plant microfossils are widespread in a variety of lithologies in much of the Victoria Group and provide the potential for making relatively precise correlations within Antarctica and with comparable sequences in other Gondwana continents. Palynomorph preservation is generally poor, and many samples are nonproductive due to thermal metamorphism during emplacement of the Jurassic Ferrar Group basalt flows and dolerite sills (Kyle and Schopf, 1977).

The plant microfossils of the Victoria Group fall into three broad groups: those associated with the glacial and postglacial strata, those associated with the Permian *Glossopteris* flora, and those associated with the Triassic *Dicroidium* flora. Early Permian or possibly Late Carboniferous palynomorphs have been described from the Ohio and Wisconsin Ranges (Schopf, *in* Long, 1964; Rigby and Schopf, 1969; Kemp, 1975) and Darwin Mountains (Bar-

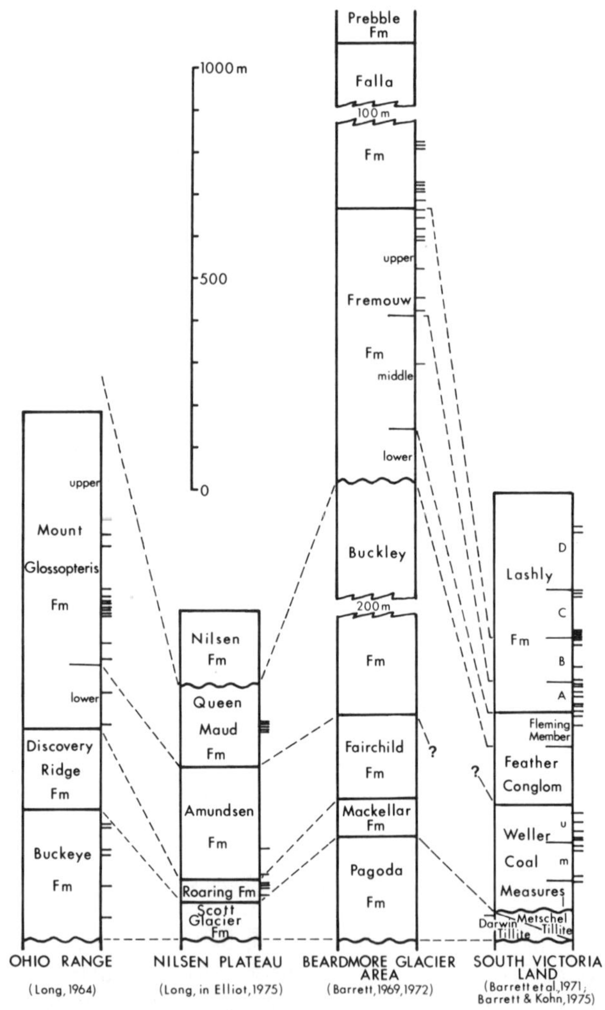

Figure 80.2. Generalized columns of the Victoria Group in the Transantarctic Mountains showing lithostratigraphic correlation of formations and approximate stratigraphic horizon of samples containing identifiable plant microfossils. Samples containing very poorly preserved palynomorphs are not shown.

rett and Kyle, 1975); Permian palynomorphs from the Ohio Range (Schopf, 1962) and Prince Charles Mountains (Balme, *in* Crohn, 1959; Balme and Playford, 1967; Kemp, 1973); and Triassic palynomorphs from northern Victoria Land (Norris, 1965), southern Victoria Land (Helby and McElroy, 1969), and the Beardmore Glacier area (Fasola, 1974; Kyle and Fasola, 1978). Kyle (1976; 1977) described Permo-Carboniferous to Late Triassic palynomorphs from southern Victoria Land and proposed a preliminary informal biostratigraphic scheme for the Victoria Group. Additional material from the Ohio Range, the Nilsen Plateau area of the Queen Maud Mountains, and the Shackleton and Beardmore Glacier areas has now been examined. This paper reviews the palynologic data from the Victoria Group of the Transantarctic Mountains.

Early Permian or possibly Late Carboniferous flora

Distribution

Plant microfossils representing a periglacial flora and included in the informal *Parasaccites* zone (Kyle, 1977) have been extracted from the upper shale beds of the Darwin Tillite of the Darwin Mountains (Barrett and Kyle, 1975), from the Buckeye Formation of the Ohio and Wisconsin Ranges (Schopf, *in* Long, 1964; Kemp, 1975), and from the postglacial Roaring Formation of the Nilsen Plateau.

The sandstone sequence overlying the postglacial shale has yielded a few poor palynomorph assemblages from the lower Amundsen Formation of the Nilsen Plateau and from the lower Mount Glossopteris Formation of the Ohio Range.

Composition

Monosaccate pollen dominate assemblages from the glacial beds and postglacial shale. These are mainly radially symmetric forms (*Parasaccites*), and some assignable to *Potonieisporites* also occur. Taeniate bisaccate pollen, some probably allied with the early glossopterids, and nontaeniate bisaccate pollen are both rare. Monocolpate pollen assigned to the species *Cycadopites cymbatus* are common to abundant. Characteristic species include the following (also see Figure 80.3):

Leiotriletes directus Balme & Hennelly, 1956
Punctatisporites gretensis var. *minor* Hart, 1965
Phyllothecotriletes nigritellus (Luber) Luber, 1955
Granulatisporites sp. cf. *G. micronodosus* Balme & Hennelly, 1956
Apiculatisporis levis (Balme & Hennelly) Segroves, 1970
Microbaculispora tentula Tiwari, 1965
Verrucosisporites spp.
Latosporites colliensis (Balme & Hennelly) Bharadwaj, 1962
Parasaccites mehtae (Lele)
Parasaccites gondwanensis (Balme & Hennelly) Segroves, 1969
Parasaccites diffusus Tiwari, 1965
Potonieisporites spp.
Alisporites spp.
Limitisporites moersensis (Grebe) Klaus, 1963
Protohaploxypinus spp.
Striatoabieites multistriatus (Balme & Hennelly) Hart, 1964
Vittatina spp.
Cycadopites cymbatus (Balme & Hennelly) Segroves, 1970
Quadrisporites horridus Hennelly, 1958

The assemblages from the lower Amundsen Formation and the base of the Mount Glossopteris Formation are of similar composition. Monosaccates are dominant, except in one sample where *Cycadopites* spp. predominate, and taeniate bisaccates are rare. One sample from higher in the lower part of the Mount Glossopteris Formation contains 18% bisaccate pollen, nearly half of which are taeniate.

Correlation and age

In the Late Carboniferous and Permian palynologic succession of eastern Australia, a series of five zones designated "Stages 1-5" was initially proposed by Evans (1969), with subsequent refinements and subdivisions by several workers (Paten, 1969; Norvick, 1971, 1974; Kemp and others, 1977). The assemblages from the tillite and the Roaring Formation in the Transantarctic Mountains are correlated with Upper Stage 2 of eastern Australia; in Western Australia they are correlated with the upper *Microbaculispora* Assemblage from the Nangetty Formation of Perth Basin (Segroves, 1971) and with Unit II from the Grant Formation of Canning Basin (Balme, *in* Kemp and others, 1977) (Figure 80.4).

The age of rocks containing Stage 2 plant microfossils in Australia is still far from certain in terms of the standard Russian sequence. Late Sakmarian ammonoids (Glenister and Furnish, 1961) associated with Stage 3 palynomorphs (*Quadrisporites* Assemblage; Segroves, 1971) in the Holmwood Shale, Perth Basin, impose an upper age limit on the underlying Nangetty Formation. Other faunal evidence is meager, and radiometric age determinations are few; thus the position of the Carboniferous-Permian boundary in Australia remains controversial. Stage 2, which includes the first occurrence of taeniate bisaccate pollen associated with the appearance of the *Glossopteris* flora, has been traditionally accepted as earliest Permian in age, but some workers (Balme, 1973; Kemp and others, 1977) now regard this flora as possibly Late Carboniferous.

The assemblages from the lower Amundsen Formation and basal part of the Mount Glossopteris Formation also appear to be comparable to Upper Stage 2. The assemblage with the greater taeniate bisaccate pollen content from higher in the lower Mount Glossopteris Formation may be of Stage 3 or younger age. No index fossils diagnostic of Stage 3 or younger zones occur in this sample, and its age will remain in doubt until better preserved material is available.

Permian flora

Distribution

Palynomorphs from the *Glossopteris*-bearing coal measures have been recovered from the middle and upper members of the Weller Coal Measures in southern Victoria Land (Kyle, 1977), from the Queen Maud Formation in the Nilsen Plateau, and from the upper part of the Mount Glossopteris Formation in the Ohio Range.

Composition

Nine plant microfossil assemblages from the middle and upper Weller Coal Measures have been described by Kyle

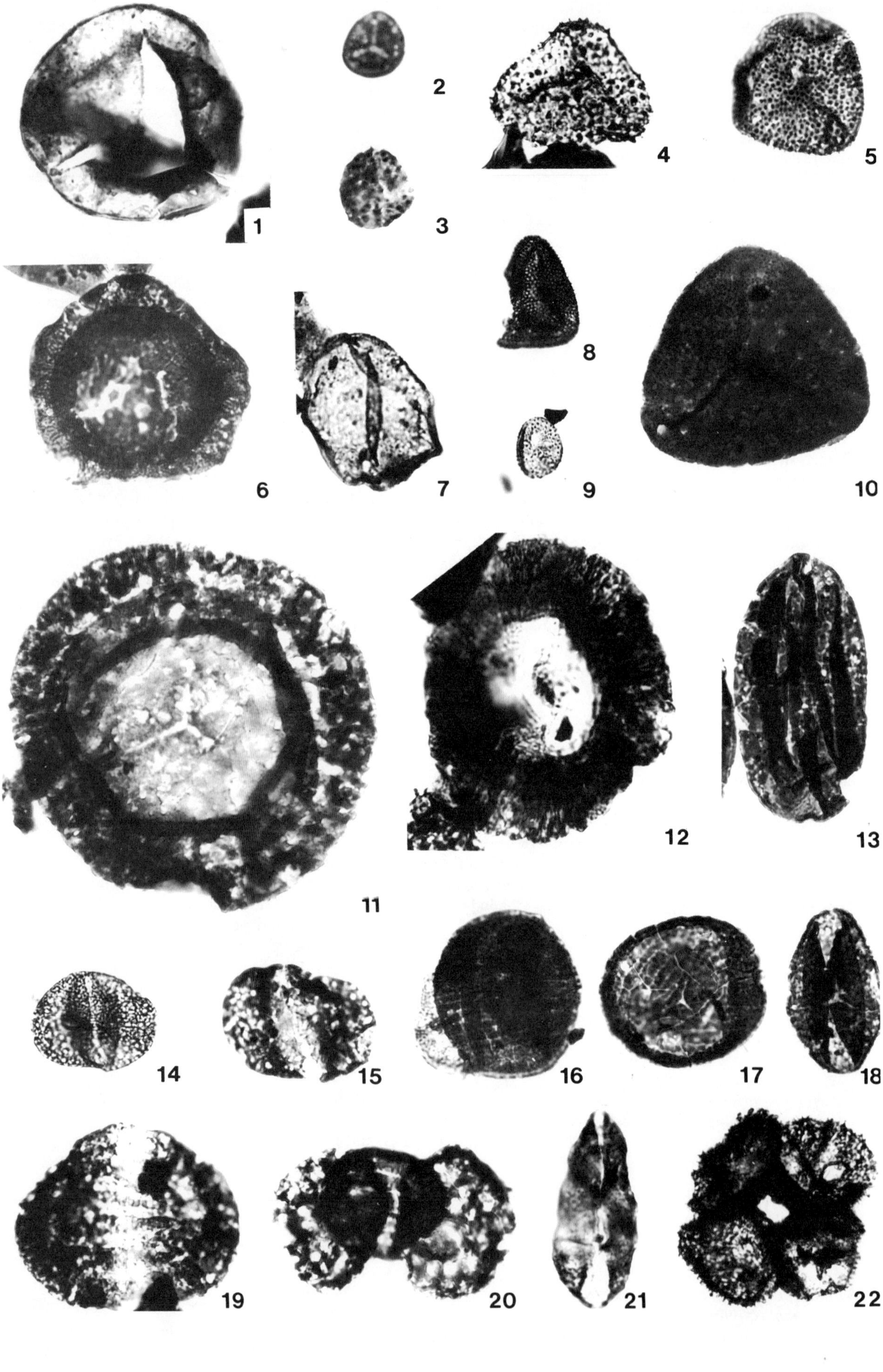

(1977) in an informal *Protohaploxypinus* zone. These contain 47–55% bisaccate pollen, over half of which are taeniate. The majority of the taeniate forms are assigned to the genus *Protohaploxypinus*. They closely resemble pollen extracted from *Arberiella*-type sporangia (e.g., Pant, 1958; Pant and Nautiyal, 1960; of glossopterid staminate fertiligers (Schopf, 1976).

Characteristic plant microfossil species from the Weller Coal Measures include the following (see also Figure 80.3):

Leiotriletes directus Balme & Hennelly, 1956
Phyllothecotriletes nigritellus (Luber) Luber, 1955
Granulatisporites trisinus Balme & Hennelly, 1956
Granulatisporites sp. cf. *G. micronodosus* Balme & Hennelly, 1956
Acanthotriletes tereteangulatus Balme & Hennelly, 1956
Apiculatisporis levis (Balme & Hennelly) Segroves, 1970
Didecitriletes sp. cf. *D. ericianus* (Balme & Hennelly) Venkatachala & Kar, 1965
Microbaculispora tentula Tiwari, 1965
Laevigatosporites minimus (Wilson & Coe) Schopf, Wilson & Bentall, 1944
Polypodiisporites leopardus (Balme & Hennelly) Foster, 1976
Tuberculatosporites modicus Balme & Hennelly, 1956
Parasaccites spp.
Alisporites tenuicorpus Balme, 1970
Sulcatisporites ovatus (Balme & Hennelly) Bharadwaj, 1962
Limitisporites moersensis (Grebe) Klaus, 1963
Hamiapollenites dettmannae Segroves, 1969
Protohaploxypinus amplus (Balme & Hennelly) Hart, 1964
Protohaploxypinus limpidus (Balme & Hennelly) Balme & Playford, 1967
Protohaploxypinus sp. cf. *P. rugatus* Segroves, 1969
Striatoabieites multistriatus (Balme & Hennelly) Hart, 1964
Striatopodocarpites cancellatus (Balme & Hennelly) Hart, 1964
Striatopodocarpites spp.
Vittatina spp.
Cycadopites cymbatus (Balme & Hennelly) Segroves, 1970

Figure 80.3. Permian palynomorphs from the Transantarctic Mountains, magnification ×600. 1. *Punctatisporites gretensis* Balme & Hennelly, 1956; 2. *Phyllothecotriletes nigritellus* (Luber) Luber, 1955; 3. *Apiculatisporis levis* (Balme & Hennelly) Segroves, 1970; 4. *Acanthotriletes tereteangulatus* Balme & Hennelly, 1956; 5. *Granulatisporites* sp. cf. *G. micronodosus* Balme & Hennelly, 1956; 6. *Bascanisporites undosus* Balme & Hennelly, 1956; 7. *Latosporites colliensis* (Balme & Hennelly) Bharadwaj, 1962; 8. *Microbaculispora tentula* Tiwari, 1965; 9. *Tuberculatosporites modicus* Balme & Hennelly, 1956; 10. *Granulatisporites trisinus* Balme & Hennelly, 1956; 11. *Parasaccites gondwanensis* (Balme & Hennelly) Segroves, 1969; 12. *Parasaccites mehtae* (Lele); 13. *Praecolpatites sinuosus* (Balme & Hennelly) Bharadwaj & Srivastava, 1969; 14. *Sulcatisporites ovatus* (Balme & Hennelly) Bharadwaj, 1962; 15. *Alisporites tenuicorpus* Balme, 1970; 16. *Striatoabieites multistriatus* (Balme & Hennelly) Hart, 1964; 17. *Marsupipollenites triradiatus* Balme & Hennelly, 1956, striate form; 18. *Marsupipollenites triradiatus* Balme & Hennelly, 1956; 19. *Protohaploxypinus limpidus* (Balme & Hennelly) Balme & Playford, 1967; 20. *Striatopodocarpites* sp.; 21. *Cycadopites cymbatus* (Balme & Hennelly) Segroves, 1970; 22. *Quadrisporites horridus* Hennelly, 1958.

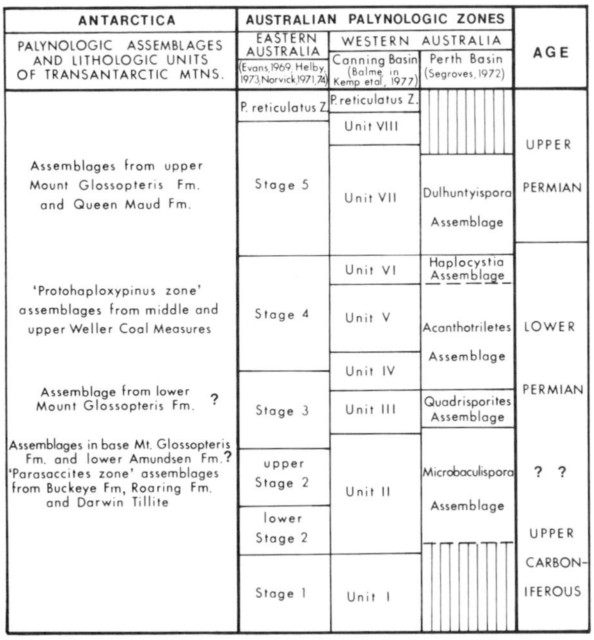

Figure 80.4. Suggested correlation of Australian Upper Carboniferous and Permian palynologic zones with assemblages from the Transantarctic Mountains. Correlation between the Australian zones is based on Kemp and others (1977).

Cycadopites follicularis Wilson & Webster, 1946
Marsupipollenites triradiatus Balme & Hennelly, 1956 (including striate forms)
Circulisporites parvus de Jersey *emend.* Norris, 1965
Quadrisporites horridus Hennelly, 1958

Bisaccate pollen (60–70%) dominate assemblages from the Queen Maud Formation of the Nilsen Plateau. Taeniate bisaccate forms (mainly *Protohaploxypinus*) make up nearly 50% of the total assemblage in some samples. *Marsupipollenites* pollen are a common component in many samples. Most of the species listed above for the Weller Coal Measures also occur in these assemblages, along with a variety of other species including the stratigraphically significant *Bascanisporites undosus* Balme & Hennelly, 1956, and *Praecolpatites sinuosus* (Balme & Hennelly) Bharadwaj & Srivastava, 1969.

Assemblages from the upper part of the Mount Glossopteris Formation in the Ohio Range also contain abundant taeniate and non-taeniate bisaccate pollen (as much as 80%) and have a species composition similar to that listed above.

Correlation and Age

The Weller Coal Measures assemblages are broadly comparable with Stage 4 of eastern Australia. For Western Australia they are comparable with the *Vittatina* Assemblage from the Byro Group, Carnarvon Basin (Balme, 1964); with *Acanthotriletes* and *Haplocystia* Assemblages from the Irwin River Coal Measures and Carynginia Formation, Perth Basin (Segroves, 1971); and with Units V and VI(?) from the Noonkanbah Formation, Canning Basin (Balme, *in* Kemp and others, 1977) (Figure 80.4).

In eastern Australia, Stage 4 palynomorphs occur in beds containing Fauna II invertebrates in the Sydney Basin (Evans, 1969; Dickins, 1968) and Fauna II and III in the Denison Trough (Dickins and others, 1964; Dickins and Malone, 1968; Dear, 1972). Ammonoids associated with these faunas in the northern Bowen Basin are of Artinskian age (Armstrong and others, 1967), and in the Carnarvon Basin of Western Australia, late Artinskian (Baigendzhinian) ammonoids occur in the Byro Group (Glenister and Furnish, 1961). An Early Permian (Artinskian) age is therefore inferred for the Weller Coal Measures.

The greater abundance of taeniate and non-taeniate bisaccate pollen in the Queen Maud Formation and upper Mount Glossopteris Formation samples implies that these sequences are younger than the Weller Coal Measures. Correlation with Stage 5 of eastern Australia is indicated by the high proportion of bisaccates, particularly taeniate forms, and by the presence of *Praecolpatites sinuosus* and *Bascanisporites undosus*. Comparable assemblages in Western Australia include the *Dulhuntyispora* Assemblage from the Wagina Sandstone, Perth Basin (Segroves, 1971) and Unit VII from the Liveringa Formation, Canning Basin (Balme, *in* Kemp and others, 1977) (Figure 80.4). Both *Praecolpatites sinuosus* and *Bascanisporites undosus* characterize Stage 5 assemblages and their Western Australian equivalents, although they first appear in older sequences (Paten, 1969; Segroves, 1971; Foster, 1976; Rigby and Hekel, 1977; Kemp and others, 1977).

Comparable assemblages from the Radok Conglomerate and Bainmedart Coal Measures of the Amery Group, Prince Charles Mountains (Antarctica), contain high proportions of taeniate bisaccate pollen, *Praecolpatites sinuosus* and *Bascanisporites undosus,* plus other Stage 5 index species (Balme and Playford, 1967; Kemp, 1973). *Dulhuntyispora,* the main Stage 5 index fossil in Australia, has not yet been found in Antarctica.

In the Springsure area of Queensland, Stage 5 (Evans, 1969; Rigby and Hekel, 1977) extends down into rocks with Fauna III (Mollan and others, 1969) of late Artinskian (Baigendzhinian) age (Armstrong and others, 1967; Runnegar, 1969). Most of Stage 5, however, is of Late Permian age. Stage 5 palynomorphs (Evans, 1969) are associated with Fauna IV in the Sydney Basin and in Queensland (Dickins, 1968; Mollan and others, 1969) of possibly early Late Permian age (Runnegar, 1969). In the Canning Basin, Western Australia, Stage 5 equivalents occur with faunas of Late Permian age and extend back to late Artinskian age (Glenister and Furnish, 1961). A Late Permian age is thus suggested for the Queen Maud Formation and the upper Mount Glossopteris Formation.

Plant megafossils provide general confirmation of these correlations and age assignments. Kyle (1976) has considered the Weller Coal Measures *Glossopteris* assemblages to be assignable to early and middle Permian ages, while Rigby and Schopf (1969) have concluded that the majority of Antarctic *Glossopteris* collections are assignable to middle and late Permian ages. Collections from the upper Mount Glossopteris and Queen Maud Formations are dominated by *Glossopteris* spp. (Cridland, 1963; Rigby and Schopf, 1969) and imply a Late Permian age. A similar age was suggested by Lacey (*in* Lambrecht and others, 1973) for the *Glossopteris* collections from the Buckley Formation of the Law Glacier area.

Triassic flora

Distribution

Triassic palynomorphs occur in the uppermost Feather Conglomerate (upper Fleming Member) and throughout much of the Lashly Formation in southern Victoria Land. They are described by Kyle (1977) and grouped in the informal *Alisporites* zone. Comparable assemblages have also been recovered from the middle and upper members of the Fremouw Formation in the Beardmore and Shackleton Glacier areas and from the lower part of the Falla Formation in the Beardmore Glacier area (Fasola, 1974; Kyle and Fasola, 1978). Norris (1965) has described an assemblage from Timber Peak, northern Victoria Land.

Composition

These assemblages are dominated by colpate bisaccate pollen included in the genus *Alisporites* Daugherty. They resemble bisaccate pollen grains extracted from *Pteruchus* sporangia (Townrow, 1962), allied with the corystosperm *Dicroidium* which dominates plant megafossil assemblages in the Lashly, Fremouw, and Falla Formations. Other palynomorphs occurring throughout the *Alisporites* zone assemblages include the following (see also Figure 80.5):

Dictyophyllidites mortoni (de Jersey) Playford & Dettmann, 1965
Calamospora tener (Leschik) de Jersey, 1962
Leiotriletes directus Balme & Hennelly, 1956
Stereisporites antiquasporites (Wilson & Webster) Dettmann, 1963
Osmundacidites wellmanii Couper, 1953
Neoraistrickia taylorii Playford & Dettmann, 1965
Uvaesporites verrucosus (de Jersey) Helby *ex* de Jersey, 1971
Uvaesporites sp. A
Guthoerlisporites cancellosus Playford & Dettmann, 1965
Aratrisporites spp.
Platysaccus queenslandi de Jersey, 1962
Cycadopites follicularis Wilson & Webster, 1946

The following species are rare to uncommon throughout the sequence:

Lophotriletes novicus Singh, 1964
Densoisporites playfordi (Balme) Dettmann, 1963
Nevesisporites limatulus Playford, 1965
Punctatosporites walkomi de Jersey, 1962
Circulisporites parvus de Jersey *emend.* Norris, 1965

Kyle (1977) has divided the *Alisporites* zone into four

Figure 80.5. Triassic palynomorphs from the Transantarctic Mountains, magnification ×600. 1. *Dictophyllidites mortoni* (de Jersey) Playford & Dettmann, 1965; 2. *Osmundacidites wellmanii* Couper, 1953; 3. *Anapiculatisporites* sp. A; 4. *Neoraistrickia taylorii* Playford & Dettmann, 1965; 5. *Cadargasporites baculatus* de Jersey & Paten, *emend.* Reiser & Williams, 1969; 6. *Cadargasporites granulatus* de Jersey & Paten *emend.* Reiser & Williams, 1969; 7. *Cadargasporites senectus* de Jersey & Hamilton, 1967; 8. *Uvaesporites* sp. A; 9. *Duplexisporites problematicus* (Couper) Playford & Dettmann, 1965; 10. *Rugulatisporites* sp. A; 11. *Nevesisporites limatulus* Playford, 1965; 12. *Uvaesporites verrucosus* (de Jersey) Helby *ex* de Jersey, 1971; 13. *Annulispora microannulata* de Jersey, 1962; 14. *Annulispora folliculosa* (Rogalska) de Jersey, 1959; 15. *Densosporites raceviewensis* de Jersey, 1971; 16. *Lundbladispora brevicula* Balme, 1963; 17. *Polypodiisporites ipsviciensis* (de Jersey) Playford & Dettmann, 1965; 18. *Aratrisporites wollariensis* Helby, 1967; 19. *Aratrisporites parvispinosus* Leschik *emend.* Playford, 1965; 20. *Guthoerlisporites cancellosus* Playford & Dettmann, 1965; 21. *Alisporites* sp.; 22. *Cycadopites follicularis* Wilson & Webster, 1946; 23. *Equisetosporites steevesi* (Jansonius) de Jersey, 1968; 24. *Grebespora concentrica* Jansonius, 1962.

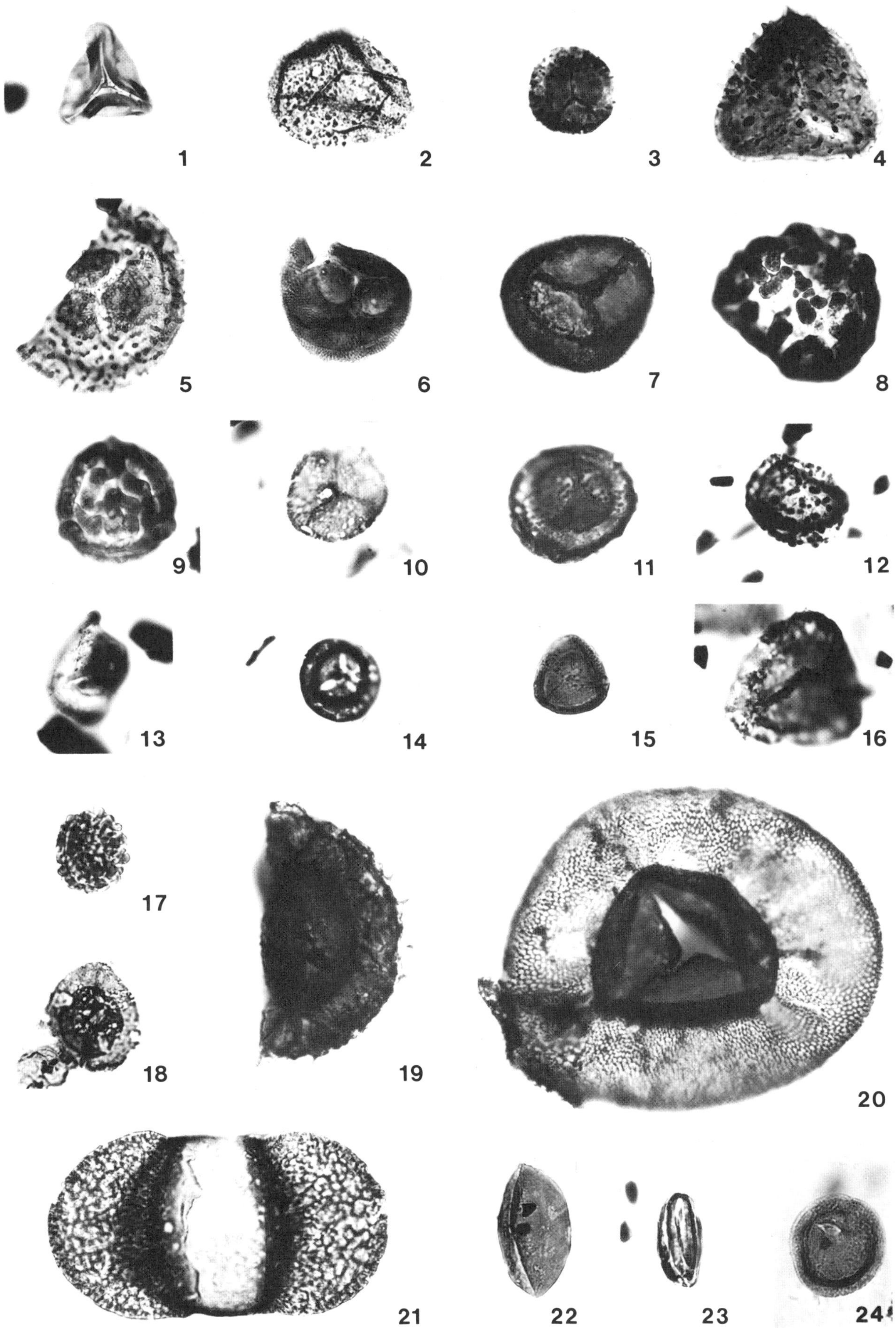
1
2
3
4
5
6
7
8
9
10
11
12
13
14
15
16
17
18
19
20
21
22
23
24

subzones based on the proportions of *Aratrisporites* and *Alisporites* and the presence of certain trilete and monolete spores. Common to abundant *Aratrisporites* (as much as 55%), particularly *A. wollariensis,* characterize subzone A assemblages. *Aratrisporites* spores decrease in abundance and become rare to uncommon components in younger samples. *Alisporites* spp. are common to abundant in older samples included in subzones A and B and become dominant in subzones C and D samples, with some assemblages consisting almost entirely of these bisaccate forms. Taeniate bisaccate pollen are rare to uncommon in most of the southern Victoria Land samples. In the Beardmore Glacier area they are slightly more common (as much as 8%; Fasola, 1974) in subzone B assemblages.

The ranges of species other than those listed above as occurring throughout the *Alisporites* zone are indicated in Table 80.1. Some of the listed species have greater ranges in Australian Triassic rocks than observed to date in Antarctic rocks. The stratigraphic significance of restricted species is discussed in Kyle (1977).

In southern Victoria Land, subzone A assemblages occur in the uppermost Feather Conglomerate and in the lower part of member A of the Lashly Formation, subzone B assemblages in upper member A and possibly in lower member B, and subzone C assemblages in uppermost member B and in member C. Two subzone D assemblages occur in the middle part of member D of the Lashly Formation. The assemblage described by Helby and McElroy (1969) from Mount Feather, southern Victoria Land, is included in subzone B.

The assemblage from Timber Peak, northern Victoria Land (Norris, 1965), is included in the upper part of the *Alisporites* zone (subzones C and D) because of the high proportion of bisaccate pollen.

The recognizable palynomorphs in one poorly preserved sample from the middle member of the Fremouw Formation, Beardmore Glacier area, are mostly *Alisporites* and taeniate bisaccate pollen. This assemblage is equated with the lower part (subzones A and B) of the *Alisporites* zone. Fasola (1974) and Kyle and Fasola (1978) have described assemblages from the upper member of the Fremouw Formation, which are included in subzone B. Subzone C assemblages occur at the top of the Fremouw Formation and in the lower part of the Falla Formation in the Beardmore and Shackleton Glacier areas. Possible subzone D assemblages, marked by the appearance of *Cadargasporites baculatus* and *C. granulatus,* occur in the lower part of the Falla Formation.

Table 80.1. Species distribution in the Triassic *Alisporites* zone

Species	Subzone A	B	C	D
Cyathidites breviradiatus Helby, 1967	X			
Osmundacidites senectus Balme, 1963	X			
Anapiculatisporites sp. A	X			
Lundbladispora brevicula Balme, 1963	X			
Lundbladispora sp. cf. *L. willmotti* Balme, 1963	X			
Aratrisporites wollariensis Helby, 1967	X	X		
Apiculatisporis levis (Balme & Hennelly) Segroves, 1970	X	X		
Apiculatisporis ?clematisi de Jersey, 1968	X	X		
Grebespora concentrica Jansonius, 1962	X	X		
Chomotriletes sp.	X	X	X	
Retusotriletes junior de Jersey & Hamilton, 1967		X		
Cadargasporites senectus de Jersey & Hamilton, 1967		X		
Verrucosisporites carnarvonensis de Jersey & Hamilton, 1967		X		
Tigrisporites playfordi de Jersey & Hamilton, 1967		X		
Kraeuselisporites verrucifer de Jersey & Hamilton, 1967		X		
Equisetosporites steevesi (Jansonius) de Jersey, 1968		X		
Rugulatisporites sp. A		X	X	
Duplexisporites problematicus (Couper) Playford & Dettmann, 1965		X	X	X
Baculatisporites comaumensis (Cookson) Potonié, 1956		X	X	X
Vitreisporites pallidus (Reissinger) Nilsson, 1958		X	X	X
Converrucosisporites cameroni (de Jersey) Playford & Dettmann, 1965		X	X	X
Polypodiisporites ipsviciensis (de Jersey) Playford & Dettman, 1965		X	X	X
Chordasporites australiensis de Jersey, 1962		X	X	X
Annulispora folliculosa (Rogalska) de Jersey, 1959			X	X
Annulispora microannulata de Jersey, 1962			X	X
Densosporites raceviewensis de Jersey, 1971			X	X
Semiretisporis denmeadi (de Jersey) de Jersey, 1970			X	X
Tuberculatosporites sp. cf. *T. aberdarensis* de Jersey, 1964			X	X
Cyathidites minor Couper, 1953				X
Cadargasporites baculatus de Jersey & Paten *emend.* Reiser & Williams, 1969				X
Cadargasporites granulatus de Jersey & Paten *emend.* Reiser & Williams, 1969				X
Anapiculatisporites sp. cf. *A. pristidentatus* Reiser & Williams, 1969				X
?Polycingulatisporites crenulatus Playford & Dettman, 1965				X
Hamiapollenites ?insculptus Playford & Dettman, 1965				X

Correlation and Age

Subzone A

Assemblages comparable to those of subzone A in eastern Australia have been described from the upper Rewan Formation (de Jersey, 1970*a*) and overlying Clematis Sandstone (de Jersey, 1968) of the Bowen Basin, from the upper Narrabeen Group of the Sydney Basin (included in the *Aratrisporites tenuispinosus* Zone; Helby, 1973), and from the Tiers Formation of the Poatina area, Tasmania (Playford, 1965) (Figure 80.6).

In Western Australia, subzone A is correlated with the *Tigrisporites playfordi* Zone (Dolby and Balme, 1976) from the Locker Shale of the Carnarvon Basin. Associated

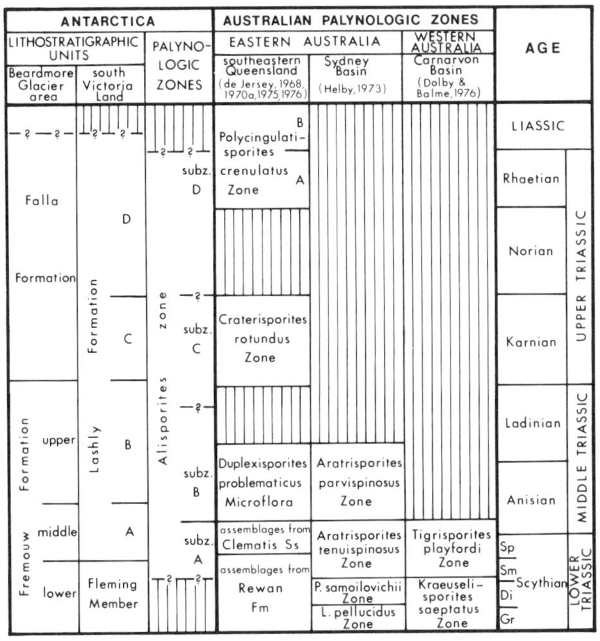

Figure 80.6. Suggested correlation of Australian Triassic palynologic zones with those of southern Victoria Land and the Beardmore Glacier area. Ages are based on de Jersey (1970*b*, 1971, 1972, 1975) and the subdivisions of the Lower Triassic on Dolby and Balme (1976). Gr, Di, Sm, and Sp refer to Griesbachian, Dienerian, Smithian, and Spathian time, respectively.

conodont faunas are of Smithian, Spathian, and earliest Anisian age (McTavish, 1973).

Subzone B

Subzone B is correlated with the *Duplexisporites problematicus* Microflora of southeastern Queensland, an informal zone proposed by de Jersey (1975) which includes assemblages from the Moolayember Formation of the Bowen Basin (de Jersey and Hamilton, 1967) and the Esk Formation of the Esk Trough (de Jersey, 1972). In the Sydney Basin the *Aratrisporites parvispinosus* Zone from the Hawkesbury Sandstone and Wianamatta Group (Helby, 1973) is comparable to subzone B.

Broad age inferences for these and younger sequences in eastern Australia have been made by comparison with European and North American palynomorph assemblages, as direct faunal evidence is lacking. By comparison with northern assemblages de Jersey (1972) assigned a Middle Triassic age to the Esk Formation assemblages. A Middle Triassic age is therefore suggested for subzone B in upper member A and lower member B of the Lashly Formation and in the upper member of the Fremouw Formation.

Subzone C

Subzone C is correlated with the *Craterisporites rotundus* Zone (de Jersey, 1975) which includes assemblages from the Ipswich Coal Measures of the Moreton Basin area (de Jersey, 1962, 1970*b*, 1971), believed to be of early Late Triassic (Karnian) age (de Jersey, 1971). Similar assemblages have been described from the Brady Formation and the "Feldspathic Sandstone" of Poatina, Tasmania (Playford, 1965) and from the Leigh Creek Coal Measures (older assemblages) of South Australia (Playford and Dettmann, 1965). A Late Triassic age is suggested for subzone C and thus for uppermost member B and member C of the Lashly Formation, the uppermost Fremouw Formation, and the lower Falla Formation.

Subzone D

Assemblages comparable to subzone D are included in the lower part (Assemblage A; de Jersey, 1976) of the *Polycingulatisporites crenulatus* Zone (de Jersey, 1975) from the Aberdare Conglomerate and Raceview Formation of the basal Bundamba Group, Moreton Basin (de Jersey, 1970*b*). De Jersey (1970*b*, 1975) has suggested a Late Triassic (Rhaetian?) age for these assemblages, which implies a Late Triassic age for subzone D assemblages in member D of the Lashly Formation and in the lower part (above subzone C) of the Falla Formation.

Summary

The plant microfossil assemblages described here reflect a succession of three distinctive floras ranging in age from possibly Late Carboniferous through Permian and Triassic. Subdivisions within these floras enable correlations between sequences in different areas of the Transantarctic Mountains and also with the divisions of the Australian palynologic successions (Figures 80.4 and 80.6).

Palynomorphs from the glacial beds and the postglacial shale unit suggest that these strata were deposited during a relatively short period of time throughout the Transantarctic Mountains, as was at least part of the feldspathic sandstone unit in the central Transantarctic Mountains. No palynomorphs were extracted from the lower Weller Coal Measures samples; thus palynologic correlation of these strata with rocks in the areas to the south is not possible. Plant megafossils from the lower Weller member are considered to be Early Permian in age, similar to the Indian Talchir-Karharbari floras (Kyle, 1976). Plant microfossils from the middle and upper Weller Coal Measures are of Early Permian age and are older than those from the Queen Maud Formation and the upper Mount Glossopteris Formation. Plant microfossil and plant megafossil evidence both suggest that the Weller Coal Measures of southern Victoria Land were deposited before the *Glossopteris*-bearing coal measures of the central Transantarctic Mountains, as indicated by Barrett and Kohn (1975).

Middle Triassic subzone B palynomorphs from the upper Fremouw Formation and lower member B of the Lashly Formation support the lithologic correlation of these members by Barrett and Kohn (1975). Subzone C palynomorphs suggest correlation of the top of Fremouw Formation and lower Falla Formation with the top of member B and member C of the Lashly Formation. The range of subzone D is unknown, but the few subzone D assemblages indicate correlation of at least the middle of member D of the Lashly Formation with part of the lower Falla Formation. Subzone D is regarded as Late Triassic (Rhaetian) in age. A Late Triassic age for the upper Falla Formation is suggested by a whole-rock K-Ar age of 198 ± 3 m.y. for a trachyte pebble (Barrett and Elliot, 1972) and by a whole-rock Rb-Sr isochron age of 203 ± 12 m.y. for a tuff (Hill, 1969).

Acknowledgments. We are grateful to the National Science Foundation for support under Grant DPP76-83030 and to Stephen R. Jacobson for critical review of this manuscript.

References cited

Armstrong, J. D., Dear, J. F., and Runnegar, B. 1967. Permian ammonoids from eastern Australia. *J. geol. Soc. Aust., 14,* No. 1, 87–97.

Balme, B. E. 1963. Plant microfossils from the Lower Triassic of Western Australia. *Palaeontology, 6,* No. 1, 12–40.

Balme, B. E. 1964. The palynological record of Australian pre-Tertiary floras. (*In* Cranwell, L. M., *ed. Ancient Pacific Floras.* Honolulu, University of Hawaii Press, 48–80.)

Balme, B. E. 1970. Palynology of Permian and Triassic strata in the Salt Range and Surghar Range, Western Pakistan. (*In* Kummel, B., and Teichert, C., *eds. Stratigraphic Boundary Problems: Permian and Triassic of West Pakistan.* Lawrence, The University Press of Kansas, 306–453.)

Balme, B. E. 1973. Age of a mixed *Cardiopteris-Glossopteris* flora. *J. geol. Soc. Aust., 20,* 103–4.

Balme, B. E., and Hennelly, J. P. F. 1956*a*. Monolete, monocolpate and alete sporomorphs from Australia Permian sediments. *Aust. Jnl Bot., 4,* 54–67.

Balme, B. E., and Hennelly, J. P. F. 1956*b*. Trilete sporomorphs from Australian Permian sediments. *Aust. Jnl Bot., 4,* 240–50.

Balme, B. E., and Playford, G. 1967. Late Permian plant microfossils from the Prince Charles Mountains, Antarctica. *Revue Micropaleont., 10,* No. 3, 179–92.

Barrett, P. J. 1969. Stratigraphy and petrology of the mainly fluviatile Permian and Triassic Beacon rocks, Beardmore Glacier area, Antarctica. *Rep. Inst. polar Stud., Ohio State Univ.,* No. 34, 132 pp.

Barrett, P. J. 1971. Stratigraphy and paleogeography of the Beacon Supergroup in the Transantarctic Mountains, Antarctica. *Second Symposium on Gondwana Stratigraphy. Cape Town and Johannesburg, July, 1970,* 249–56.

Barrett, P. J. 1972. Stratigraphy and petrology of the mainly fluviatile Permian and Triassic part of the Beacon Supergroup, Beardmore Glacier area. (*In* Adie, R. J., *ed. Antarctic Geology and Geophysics.* Oslo, Universitetsforlaget, 365–72.)

Barrett, P. J., and Elliot, D. H. 1972. The early Mesozoic volcaniclastic Prebble Formation, Beardmore Glacier area. (*In* Adie, R. J., *ed. Antarctic Geology and Geophysics.* Oslo, Universitetsforlaget, 403–9.)

Barrett, P. J., and Kohn, B. P. 1975. Changing sediment transport directions from Devonian to Triassic in the Beacon Supergroup of south Victoria Land, Antarctica. (*In* Campbell, K. S. W., *ed. Gondwana Geology.* Canberra, Australia National University Press, 15–35.)

Barrett, P. J., and Kyle, R. A. 1975. The Early Permian glacial beds of south Victoria Land and the Darwin Mountains, Antarctica. (*In* Campbell, K. S. W., *ed. Gondwana Geology.* Canberra, Australia National University Press, 333–46.)

Barrett, P. J., Kohn, B. P., Askin, R. A., and McPherson, J. G. 1971. Preliminary report on Beacon Supergroup studies between the Hatherton and Mackay Glaciers, Antarctica. *N.Z. Jl Geol. Geophys., 14,* No. 3, 605–14.

Bharadwaj, D. C. 1962. The miospore genera in the coals of Raniganj stage (Upper Permian), India. *Palaeobotanist, 9,* 68–106.

Bharadwaj, D. C., and Srivastava, S. C. 1969. A Triassic mioflora from India. *Palaeontographica, 125B,* 119–49.

Couper, R. A. 1953. Upper Mesozoic and Cainozoic spores and pollen grains from New Zealand. *Palaeont. Bull. geol. Surv. N.Z., 22,* 77 pp.

Cridland, A. A. 1963. A *Glossopteris* flora from the Ohio Range, Antarctica. *Am. J. Bot., 50,* No. 2, 186–95.

Crohn, P. W. 1959. A contribution to the geology and glaciology of the western part of Australian Antarctic Territory. *Bull. Bur. Miner. Resour. Geol. Geophys. Aust., 52,* 1–103.

Dear, J. F. 1972. Preliminary biostratigraphic subdivision of the Permian brachiopod faunas in the northern Bowen Basin and their distribution throughout the basin. *Rep. Geol. Surv. Qd., 49,* 19 pp.

de Jersey, N. J. 1959. Jurassic spores and pollen grains from the Rosewood coalfield. *Qd. Govt Min. J., 60,* 346–66.

de Jersey, N. J. 1962. Triassic spores and pollen grains from the Ipswich coalfield. *Publns Geol. Surv. Qd., 307,* 18 pp.

de Jersey, N. J. 1964. Triassic spores and pollen grains from the Bundama Group. *Publns Geol. Surv. Qd., 321,* 21 pp.

de Jersey, N. J. 1968. Triassic spores and pollen grains from the Clematis Sandstone. *Publns Geol. Surv. Qd., 338, Palaeont. Pap., 14,* 44 pp.

de Jersey, N. J. 1970*a*. Early Triassic miospores from the Rewan Formation. *Publns Geol. Surv. Qd., 345, Palaeont. Pap., 19,* 29 pp.

de Jersey, N. J. 1970*b*. Triassic miospores from the Blackstone Formation, Aberdare Conglomerate and Raceview Formation. *Publns Geol. Surv. Qd., 348, Palaeont. Pap., 22,* 41 pp.

de Jersey, N. J. 1971. Triassic miospores from the Tivoli Formation and Kholo Sub-group. *Publns Geol. Surv. Qd., 353, Palaeont. Pap., 28,* 40 pp.

de Jersey, N. J. 1972. Triassic miospores from the Esk Beds. *Publns Geol. Surv. Qd., 357, Palaeont. Pap., 32,* 40 pp.

de Jersey, N. J. 1975. Miospore zones in the Lower Mesozoic of southeastern Queensland. (*In* Campbell, K. S. W., *ed. Gondwana Geology.* Canberra, Australia National University Press, 159–72.)

de Jersey, N. J. 1976. Palynology and time relationships in the lower Bundamba Group (Moreton Basin). *Qd. Govt Min. J., 77,* 460–65.

de Jersey, N. J., and Hamilton, M. 1967. Triassic spores and pollen grains from the Moolayember Formation. *Publns Geol. Surv. Qd., 336, Palaeont. Pap., 10,* 61 pp.

Dettmann, M. E. 1963. Upper Mesozoic microfloras from southeastern Australia. *Proc. R. Soc. Vict., 77,* 1–148.

Dickins, J. M. 1968. Correlation of the Permian of the Hunter Valley, New South Wales, and the Bowen Basin, Queensland. *Palaeont. Pap. 1965, Bull. Bur. Miner. Resour. Geol. Geophys. Aust., 80,* 27–44.

Dickins, J. M., and Malone, E. J. 1968. Regional subdivision of the Back Creek Group (Middle Bowen Beds), Bowen Basin, Queensland. *Qd. Govt Min. J., 69,* 292–301.

Dickins, J. M., Malone, E. J., and Jensen, A. R. 1964. Subdivision and correlation of the Permian Middle Bowen Beds. *Rep. Bur. Miner. Resour. Geol. Geophys. Aust., 70,* 1–12.

Dolby, J. H., and Balme, B. E. 1976. Triassic palynology of the Carnarvon Basin, Western Australia. *Rev. Palaeobot. Palynol, 22,* 105–68.

Doumani, G. A., and Tasch, P. 1965. A leaiid conchostracan zone (Permian) in the Ohio Range, Horlick Mountains, Antarctica. (*In* Hadley, J. B., *ed. Geology and Paleontology of the Antarctic.* Washington, American Geophysical Union, 229–39.) [Antarctic Research Series, Vol. 6.]

Elliot, D. H. 1975. Gondwana basins of Antarctica. (*In* Campbell, K. S. W., *ed. Gondwana Geology.* Canberra, Australia National University Press, 493–536.)

Elliot, D. H., Colbert, E. H., Breed, W. J., Jensen, J. A., and Powell, J. S. 1970. Triassic tetrapods from Antarctica: evidence for continental drift. *Science, N.Y., 169,* No. 3951, 1197–201.

Evans, P. R. 1969. Upper Carboniferous and Permian palynological stages and their distribution in eastern Australia. (*In* Amos, A. J., *ed. Gondwana Stratigraphy.* Paris, UNESCO, 41–54.)

Fasola, A. 1974. *Palynological Study of Triassic Samples from the Beardmore Glacier Area in Antarctica.* Master's thesis, The Ohio State University, 163 pp.

Foster, C. B. 1976. Permian plant microfossils from the Blair Athol Coal Measures, central Queensland, Australia. *Palaeontographica, 154B,* No. 5–6, 121–71.

Glenister, B. F., and Furnish, W. M. 1961. The Permian ammonoids of Australia. *J. Paleont., 35,* No. 4, 673–736.

Hart, G. F. 1964. A review of the classification and distribution of the Permian miospore: Disaccate Striatiti. *C.r. 5th Cong. Int. Strat. Geol. Carbonif. Paris,* 1171–99.

Hart, G. F. 1965. *The Systematics and Distribution of Permian Miospores.* Johannesburg, Witwatersrand University Press, 253 pp.

Helby, R. J. 1967. Triassic plant microfossils from a shale within the Wollar Sandstone, N.S.W. *Proc. R. Soc. N.S.W., 100,* 61–73.

Helby, R. J. 1973. Review of Late Permian and Triassic palynology of New South Wales. *Spec. Publs. geol. Soc. Aust., 4,* 141–55.

Helby, R. J., and McElroy, C. T. 1969. Microfloras from the Devonian and Triassic of the Beacon Group, Antarctica. *N.Z. Jl Geol. Geophys., 12,* No. 2–3, 376–82.

Hennelly, J. P. F. 1958. Spores and pollen from a Permian–Triassic transition, N.S.W. *Proc. Linn. Soc. N.S.W., 83,* 363–69.

Hill, R. L. 1969. *Strontium Isotope Composition of Basaltic Rocks of the Transantarctic Mountains, Antarctica.* Master's thesis, The Ohio State University, 87 pp.

Jansonius, J. 1962. Palynology of Permian and Triassic sediments, Peace River area, western Canada. *Palaeontographica, 110B,* 35–98.

Kemp, E. M. 1973. Permian flora from the Beaver Lake area, Prince Charles Mountains, Antarctica. 1. Palynological examination of samples. *Palaeont. Pap. 1969. Bull. Bur. Miner. Resour. Geol. Geophys. Aust., 126,* 7–12.

Kemp, E. M. 1975. The palynology of Late Palaeozoic glacial deposits of Gondwanaland. (*In* Campbell, K. S. W., *ed. Gondwana Geology.* Canberra, Australia National University Press, 397–413.)

Kemp, E. M., Balme, B. E., Helby, R. J., Kyle, R. A., Playford, G., and Price, P. L. 1977. Carboniferous and Permian palynostratigraphy in Australia and Antarctica: a review. *J. Bur. Miner. Resour. Geol. Geophys. Aust., 2,* No. 3, 177–208.

Kitching, J. W., Collinson, J. W., Elliot, D. H., and Colbert, E. H. 1972. *Lystrosaurus* Zone (Triassic) fauna from Antarctica. *Science, N.Y., 175,* No. 4021, 524–27.

Klaus, W. 1963. Sporen aus dem Sudalpinen Perm. *Jb. geol. Bundesanst, Wien., 106,* 229–63.

Kyle, R. A. 1976. *Palaeobotanical Studies of the Permian and Triassic Victoria Group (Beacon Supergroup) of South Victoria Land, Antarctica.* Ph.D. dissertation, Victoria University of Wellington, 306 pp.

Kyle, R. A. 1977. Palynostratigraphy of the Victoria Group of south Victoria Land, Antarctica. *N.Z. Jl Geol. Geophys., 20,* No. 6, 1081–102.

Kyle, R. A., and Fasola, A. 1978. Triassic palynology of the Beardmore Glacier area of Antarctica. *Palinologia, num. extraord. 1,* 313–19.

Kyle, R. A., and Schopf, J. M. 1977. Palynomorph preservation in the Beacon Supergroup of the Transantarctic Mountains. *Antarct. Jnl U.S., 12,* No. 4, 121–22.

Lambrecht, L. L., Lacey, W. S., and Smith, C. S. 1973. Observations on the Permian flora of the Law Glacier area, central Transantarctic Mountains. *Bull. Soc. belge Géol., Paléont., Hydrol., 81,* No. 394, 161–67.

Long, W. E. 1964. The stratigraphy of the Horlick Mountains. (*In* Adie, R. J., *ed. Antarctic Geology.* Amsterdam, North-Holland Publishing Company, 352–63.)

Luber, A. A. 1955. Atlas spor i pyl'tsy paleozoyskikh otlozheniy Kazakhstana [Atlas of spores and pollen of the Paleozoic deposits of Kazakhstan]. *Trudy Vses. Nauchno-issled. Geol. Inst., 139,* 107 pp.

McTavish, R. A. 1973. Triassic conodont faunas from Western Australia. *Abh. Neues Jb. Geol. Palaont., 143,* No. 3, 275–303.

Mollan, R. G., Dickins, J. M., Exon, N. F., and Kirkegaard, A. G. 1969. Geology of the Springsure 1:250,000 Sheet area, Queensland. *Rep. Bur. Miner. Resour. Geol. Geophys. Aust., 123,* 1–119.

Nilsson, T. 1958. Uber das Vorkommen eines Mesozoischen Sapropelgesteins in Schonen. *Acta Univ. Lund., Arsskr. N.F. Adv. 2, 54,* No. 10, 112 pp.

Norris, G. 1965. Triassic and Jurassic miospores and acritarchs from the Beacon and Ferrar Groups, Victoria Land, Antarctica. *N.Z. Jl Geol. Geophys., 8,* No. 2, 236–77.

Norvick, M. 1971. Some palynological observations on Amerada Thunderbolt No. 1 Well, Galilee Basin, Queensland. *Rec. Bur. Miner. Resour. Geol. Geophys. Aust., 1971/53* [Unpublished].

Norvick, M. 1974. Permian and Late Carboniferous palynostratigraphy of the Galilee Basin, Queensland. *Rec. Bur. Miner. Resour. Geol. Geophys. Aust., 1974/141* [Unpublished].

Pant, D. D. 1958. The structure of some leaves and fructifications of the *Glossopteris* flora of Tanganyika. *Bull. Br. Mus. Nat. Hist., 3,* 125–75.

Pant, D. D., and Nautiyal, D. D. 1960. Some seeds and sporangia of *Glossopteris* flora from Raniganj coalfield, India. *Palaeontographica, 107B,* 1–64.

Paten, R. J. 1969. Palynologic contributions to petroleum exploration in the Permian formations of the Cooper Basin, Australia. *J. Aust. Petrol. Expl. Ass., 9,* 79–87.

Playford, G. 1965. Plant microfossils from Triassic sediments near Poatina, Tasmania. *J. geol. Soc. Aust., 12,* No. 2, 173–207.

Playford, G., and Dettmann, M. E. 1965. Rhaeto-Liassic plant microfossils from the Leigh Creek Coal Measures, South Australia. *Senckenberg. leth., 46,* Nos. 2–3, 127–81.

Potonié, R. 1956. Synopsis der Gattungen der Sporae dispersae. 1. Teil: Sporites. *Jb. Beih. geol., 23,* 103 pp.

Reiser, R. F., and Williams, A. J. 1969. Palynology of the Lower Jurassic sediments of the northern Surat Basin, Queensland. *Publns Geol. Surv. Qd., 339, Palaeont. Pap., 15,* 24 pp.

Rigby, J. F., and Hekel, H. 1977. Palynology of the Permian sequence in the Springsure Anticline, central Queensland. *Publns Geol. Surv. Qd., No. 363, Palaeont. Pap., 37,* 76 pp.

Rigby, J. F., and Schopf, J. M. 1969. Stratigraphic implications of Antarctic paleobotanical studies. (*In* Amos, A. J., *ed. Gondwana Stratigraphy.* Paris, UNESCO, 91–106.)

Runnegar, B. N. 1969. The Permian faunal succession in eastern Australia. *Spec. Publs. geol. Soc. Aust., 2,* 73–98.

Schopf, J. M. 1962. A preliminary report on plant remains and coal of the sedimentary section in the central range of the Horlick Mountains, Antarctica. *Rep. Inst. polar Stud., Ohio State Univ., 2,* 61 pp.

Schopf, J. M. 1976. Morphologic interpretation of fertile structures in glossopterid gymnosperms. *Rev. Palaeobot. Palynol., 21,* 25–64.

Schopf, J. M., Wilson, L. R., and Bentall, R. 1944. An annotated synopsis of Paleozoic fossil spores and the definition of generic groups. *Rep. Invest. Illinois State Geol. Surv., 91,* 1–73.

Segroves, K. L. 1969. Saccate plant microfossils from the Permian of Western Australia. *Grana palynol., 9,* 174–227.

Segroves, K. L. 1970. Permian spores and pollen grains from the Perth Basin, Western Australia. *Grana, 10,* No. 1, 43–73.

Segroves, K. L. 1971. The sequence of palynological assemblages in the Permian of the Perth Basin, Western Australia. *Second Symposium on Gondwana Stratigraphy. Cape Town and Johannesburg, July 1970,* 511–29.

Singh, H. P. 1964. A miospore assemblage from the Permian of Iraq. *Paleontology, 7,* No. 2, 240–65.

Townrow, J. A. 1962. On *Pteruchus,* a microsporophyll of the Corystospermaceae. *Bull. Br. Mus. Nat. Hist., 6,* No. 2, 287–320.

Townrow, J. A. 1967. Fossil plants from Allan and Carapace Nunataks, and from the upper Mill and Shackleton Glaciers, Antarctica. *N.Z. Jl Geol. Geophys., 10,* No. 2, 456–73.

Tiwari, R. S. 1965. Miospore assemblage in some coals of Barakar Stage (Lower Gondwana) of India. *Palaeobotanist, 13,* No. 2, 168–214.

Venkatachala, B. S., and Kar, R. K. 1965. Two new trilete spore genera from the Permian of India. *Palaeobotanist, 13,* No. 3, 337–40.

Wilson, L. R., and Webster, R. M. 1946. Plant microfossils from a Fort Union coal of Montana. *Am. J. Bot., 33,* 271–78.

81

Experimental Valve Geothermometry Applied to Fossil Conchostracan Valves, Blizzard Heights, Antarctica

PAUL TASCH
Department of Geology, Wichita State University, Wichita, Kansas 67208, U.S.A.

Abstract. Conchostracan-bearing samples collected from Jurassic sedimentary beds between two basalt flows (Blizzard Heights, Marshall Mountains) display various valve conditions (partially jet black, white enameloid, and others). These variations are apparently related to the thermal history caused by the overlying basalt flow. Experimental data were obtained from furnace heating the valves of a modern conchostracan (*Cyzicus* sp.) that has Paleozoic and Mesozoic equivalents. Dried cyziciid valves, in either the naked state or embedded in tri-layered clay briquettes, were exposed to temperatures of 200°C–1200°C.

Thermally induced color and other valve changes were observed, and seven classes were established. Experimental temperature ranges for each class are: (1) 300°–400°C, 100% char (carbonization) (C); (2) 450°–600°C, brown crust (Br); (3) 600°C, 50% char (C-1/2) plus other valve changes; (4) 600°–900°C, blue green (Bl); (5) 600°–1100°C, white enameloid (W); (6) 800°–1100°C, white enameloid plus brown crustose condition (WBr); and (7) 1200°C, valves glassy (G). Occurrence of more than one class of color or other valve change can be attributed to differential thermal effects on the surficial cuticle, as compared to the underlying cuticle(s).

Experimentally determined color classes were assigned to fossil valves from measured sections at two stations at Blizzard Heights. X-ray diffraction data on these two stations conformed with other thermal data in showing recurrent high sanidine. Other minerals present are microcline, quartz, calcite, and probably grossularite at one station, and calcite, quartz, montmorillonite, and a single occurrence of grossularite at the other. High sanidine represents a temperature of 1000 + °C, which falls in the white enameloid range—the dominant color class throughout both Blizzard Heights stations. Accordingly, "white enameloid" is a valid independent indicator of that temperature. At Brimstone Peak (all C) and Carapace Nunatak (Br and C) low temperatures are indicated. X-ray diffraction for Carapace Nunatak Tasch Station 2, beds 1 and 2, indicates alpha-quartz and wollastonite. Mauger Nunatak valves represent the Blizzard Heights color/temperature class.

Jurassic branchiopod conchostracans were collected from sedimentary beds between two successive basalt flows at Blizzard Heights (lat. 84° 37′ S, long. 163° 53′ E), Marshall Mountains, Antarctica, during the austral summer of 1969–70. The interbed was approximately 72 m below the upper dolerite. Station 0 (see Figure 81.4*d*) was at the east end and some 33 m to the west was Station 1, the direction in which the bed thickened. (For another view of this interbed see Elliot and Tasch, 1967, Figure 2.) A study of the material disclosed some puzzling valve characteristics: there were many variably charred valves, others white or blue enameloid in appearance, some with a brown crustlike exterior, and still others displaying a combination of white enameloid and brown crust conditions. Some of these valve-color class types were confined to a particular bed in the thin shale between the two basalt flows (Figures 81.1 and 81.2).

The observed variations seem to be related to heating by the overlying basalt flow. If so, does variation in valve appearance have value as a geothermometer? An experimental study has been undertaken to probe this question since there is no available literature pertinent to the conchostracans.

Pre-experiment considerations

Several factors have been taken into account in the experimental program in order to correlate field and laboratory observations. These include: (1) choice of experimental material, (2) nature of enclosing sedimentary materials, (3) thickness and size of synthetic substrate, and (4) temperature range.

The first factor led to the choice of the conchostracan genus *Cyzicus,* an ubiquitous modern form, for which hundreds of dried valves were easily available. It is also the dominant genus represented by the fossils at Blizzard Heights. The second factor led to the use of clay mud from an aquarium, originally taken from freshwater Kansas ponds. The Blizzard Heights sedimentary unit likewise, in its original state, had been lacustrine mudstone or indurated clay mud. The third factor required varying clay briquette thickness and size to simulate greater or lesser depths below the upper basalt flow. The final requirement was met with the use of a Lindblad Hevi-Duty Furnace with readout and a temperature range of 1200°C. This spread was enough to include the temperature of the basalt flow in its original molten condition.

Arthropod integuments, in general, are chitinous or

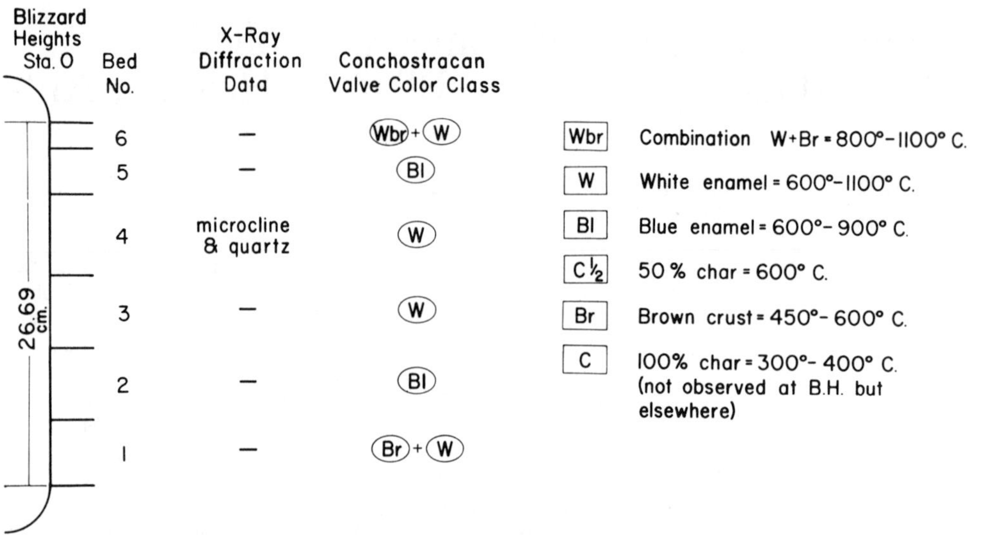

Figure 81.1. Blizzard Heights sedimentary unit (Tasch Station 0, between two basalt flows). First column: measured section and bed numbers. Second column: X-ray diffraction data for bed 4 (by Dah Cheng Wu; see text for supplementary data by Sue Erlewein). Third column: valve-color class. Fourth column: color class and temperature range (maximum-minimum).

chitino-proteinaceous (Richards, 1951*b*). Branchiopod cuticles that have been studied (including notostracans, anostracans, and cladocerans) give a positive chemical reaction for chitin (Richards, 1951*a*). The flexibility of the conchostracan valve in modern forms, as well as its close affinities with the cladocerans, indicates a typical arthropod integument composition for conchostracans. A study of living forms in the author's laboratory has shown that the valve consists of an indeterminate number of cuticles or layers, each about 100μ thick and all alike in composition. Valves used in the experiments described below were thoroughly cleaned and dried, and only complete, undamaged valves were included.

Experimental procedure

Mud in solution was poured into aluminum foil molds in each preparation of a briquette and allowed to dry into a homogeneous solid. Three different sizes of clay briquettes were prepared as matrix for valves. Each briquette consisted of three equal layers. Dried right or left valves were placed on the surfaces of the upper, middle, and lower layers. Valves were placed in three orientations, evenly spaced on each surface: concave upward, convex upward, and one valve (midway between the others) perpendicular to the surface and half-buried in the mud.

Briquettes of three sizes (in cm) were used: 10 × 3 × 2.5, with valves arranged as described above with the top

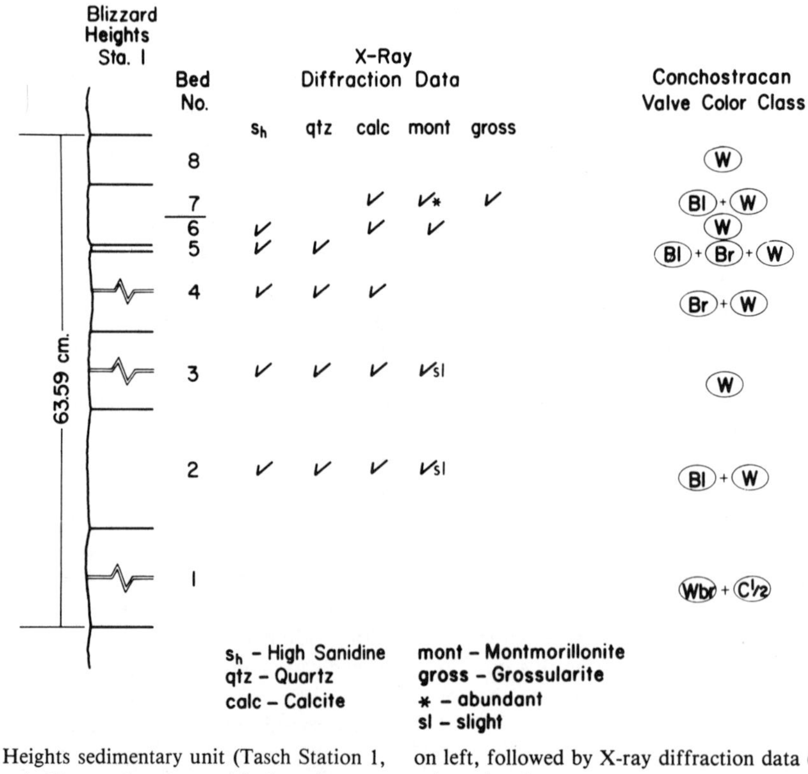

Figure 81.2. Blizzard Heights sedimentary unit (Tasch Station 1, between two basalt flows). Measured section and bed numbers are on left, followed by X-ray diffraction data (Dah Cheng Wu) and valve color classes.

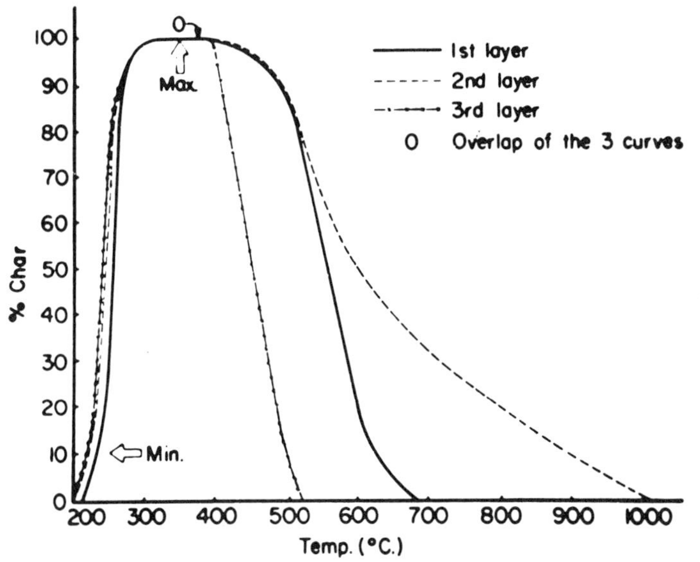

Figure 81.3. Experimental temperature and related valve char data for each of three briquette layers containing conchostracan valves. First layer: uppermost layer of a given briquette. Second and third layers: middle and basal layers, respectively, of the same briquette. Number of briquettes = 6. Number of valves on each layer = 3, or 9 per briquette. Total valves on which curves are based, n = 54. Temperature is indicated in 100°C increments; percentage of char (carbonization) is in 10% increments. Valves have similar char percentage regardless of briquette layer on which they occur, up to the point of overlap. This char affected only the surficial cuticle of the valve wall. Past the temperature of overlap, curves diverge, denoting that higher temperature increments are required to char the nonsurficial cuticles and that these temperatures vary in the different briquette layers. Actual readout temperatures are given. Corrected furnace temperatures (°C) applicable to all temperature figures are: 200° = 211.6°, 300° = 311.9°, 400° = 407.7°, 500° = 501.5°, 600° = 595°, 700° = 689.2°, 800° = 786.4°, 900° = 887.2°, 1000° = 992.8°, 1100° = 1104°, and 1200° = 1222.4°.

of the middle layer 2 cm below the surface; 10 × 3 × 5, with the top of the middle layer more than 2.5 cm below the surface; and 10 × 3 × 7.5, with the top of the middle layer 5.0 cm below the surface. Duration of a standard run was one hour at the given temperature. The furnace was not preheated on these runs.

Experimental runs

Naked valves

As a control, single naked valves were placed one at a time in porcelain dishes; the sample size was $n = 11$. The furnace was then turned on and set to the desired temperature. When attained, that temperature was continued for one hour. Therafter the furnace was shut down and the specimen was not removed until the furnace readout was zero.

The valve-color class of naked valves changed with rise of temperature to 500°C (Figure 81.1). By contrast, briquette-embedded valves, less directly exposed to furnace heat, required temperatures above 500°C to achieve the same color changes.

Briquette-embedded valves

2.5 cm, 5.0 cm, and 7.5 cm thick briquettes were either placed in a porcelain dish or, for larger briquettes, encased in aluminum foil, and inserted in the furnace. There was a separate run for each of three thicknesses. Conditions of heating and duration of runs were the same as for naked valves. Results are given in Figure 81.3.

Special runs

Three oversized briquettes, 20 × 10 × 5 cm, were run to determine how greater areal spread might affect embedded valves. The large briquettes were submitted to temperatures of 300°C, 500°C, and 1000°C, respectively, for one hour.

Most valve changes corresponded to those attained from the standard briquette runs. However, there were some variations, as discussed below.

Runs of naked valves at the same temperature used for embedded valves provided the definitive temperature for a specific resultant valve condition. This allowed comparison of temperatures required to cause a valve-color change in unembedded valves with those needed to achieve the same color change in embedded valves.

To test for effects from longer temperature duration, five runs were made with naked valves: runs at temperatures of 200°C, 300°C, 500°C, and 1000°C for two hours each and one run at 200°C for four hours. There was no detectable difference in valves due to the extended duration at a given temperature.

To test for possible effects from differences in preheating, two types of preheat were made at different temperatures (300°C, 400°C, and 500°C), one with slow cooling and one with rapid cooling. Neither of these runs had any observable effect on temperature-induced valve color.

Conchostracan valve geothermometry

Recent valves

Experimental runs with briquette-embedded valves create seven megascopically recognizable valve changes that are temperature-induced (Figures 81.4 and 81.5). Since the conchostracan valve consists of several chitinous or chitino-proteinaceous cuticles, at lower temperatures a change in the exposed valve-wall layer does not correspond

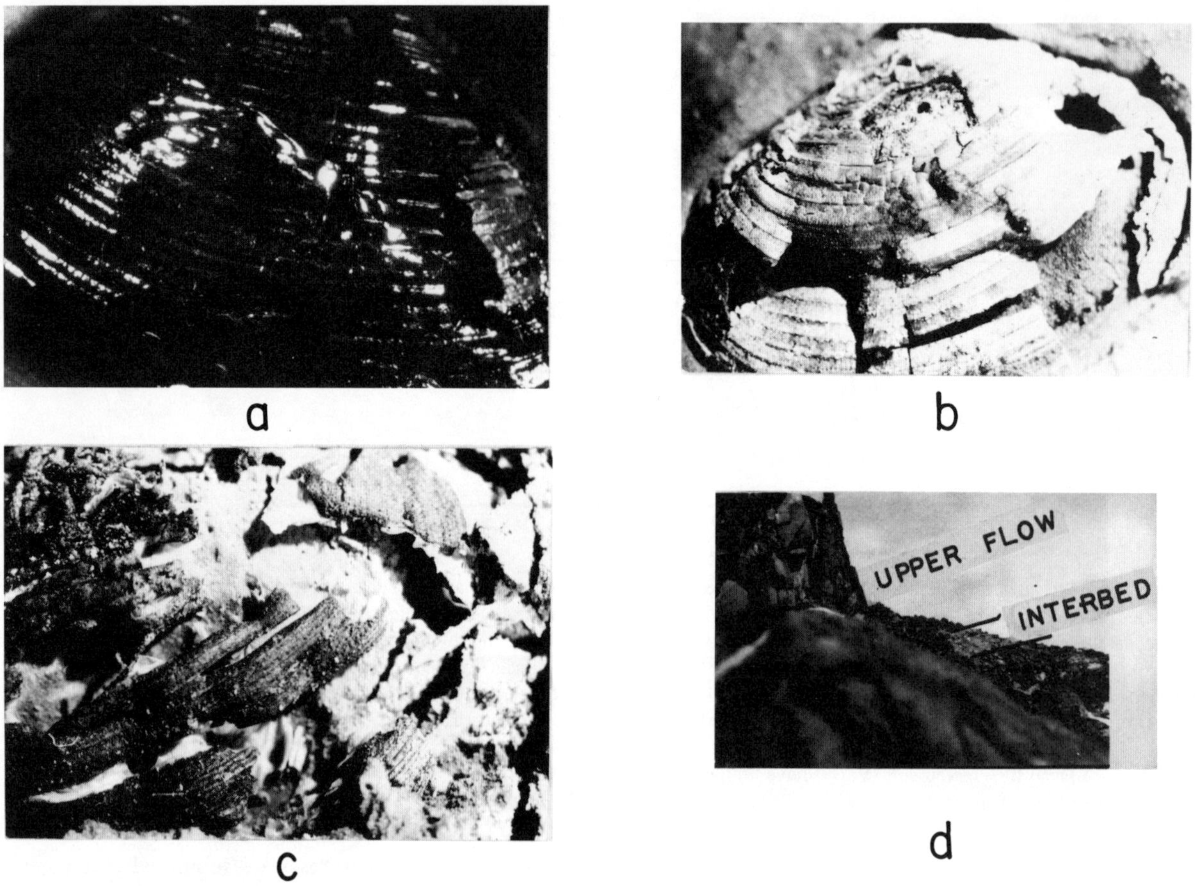

Figure 81.4. Experimental valve changes. *a.* C condition, 100% char, temperature range 300°–400°C, in a 5.1-cm-thick briquette and heated for one hour at 300°C (see Figure 81.5*d*). Valves appear glossy black. *b.* Br condition, brown crust, temperature range 450°–600°C, in a 2.54-cm-thick briquette and heated at 450°C for one hour. *c.* WBr condition, white enamel and brown crust, temperature range 800°–1100°C; valve embedded in a 5.1 cm briquette and heated at 1000°C for one hour. White enameloid layer (W) represented by white area; brown crust (Br) overlies the white patches (see Figure 81.5*a*). *d.* Blizzard Heights, Antarctica; interbed and upper flow. Tasch Station 0 around the bend.

to equivalent changes in underlying layers. However, as temperature increases, color changes also ensue in successive underlying cuticles. Hence, the surficial changes monitored commonly reflect a combination of outer and inner cuticle changes.

The experimentally determined valve geothermometry is outlined below.

1. 300°–400°C (color class "C"). In this range valves are completed charred (Figure 81.4*a*) regardless of valve depth below the exposed briquette surface. Equivalent valve effects are not obtained at any temperature lower than 300°C or higher than 400°C.
2. 450°–600°C (color class "Br"). Valves in this range display a surficial light-brown crust (Figure 81.4*b*). Deformation of the shell begins, and the valve is often cracked and split and begins to curl. Deformation, splitting, and curling continue to 1100°C. Below 450°C no value is deformed by heating. The brown color itself is not created by heating since unheated valves are a brownish hue. Usually what transpires is swelling and displacement of growth bands creating a brown, crustose appearance.
3. About 600°C (color class "C-1/2"). Fifty % char accompanies other valve changes; that is, an underlying cuticle is now charred.
4. 600°C–900°C (color class "Bl"). Valves turn bluish-green (Figure 81.5*c*).
5. 600°–1100°C (color class "W"). Within this temperature range valves turn white and take on a white enameloid or porcelanous appearance (Figure 81.4*c* (W) and Figure 81.5*b*).
6. 800°–1100°C (color class "WBr"). This temperature range yields a combined change in valve appearance: white enameloid plus brown crust (Figure 81.4*c* (WBr) and Figure 81.5*a*).
7. 1200°C (color class "G"). Valves acquire a transparent, glassy aspect, with numerous trapped bubbles visible. Valve configuration is barely decipherable due to flow.

All valve changes except the C-1/2 condition were found within a specific range of temperatures. This signifies that the valve changes occurred at the lowest end of the temperature range in experimental runs. Thus, for example, in the 600°–900°C range, valve change occurred at 600°C, and no additional major change transpired at elevated temperatures up to 900°C. Any temperature within the cited range would create the same valve effect.

However, for fossil valves, the situation was reversed, i.e., the highest part of a given temperature range would have been initially effective on sediment-embedded valves. Lava that flowed over the lacustrine sediments cooled from its maximum temperature. The resultant cycle of rapid heating and slow cooling would not have altered the valve color change induced by the initial maximum temper-

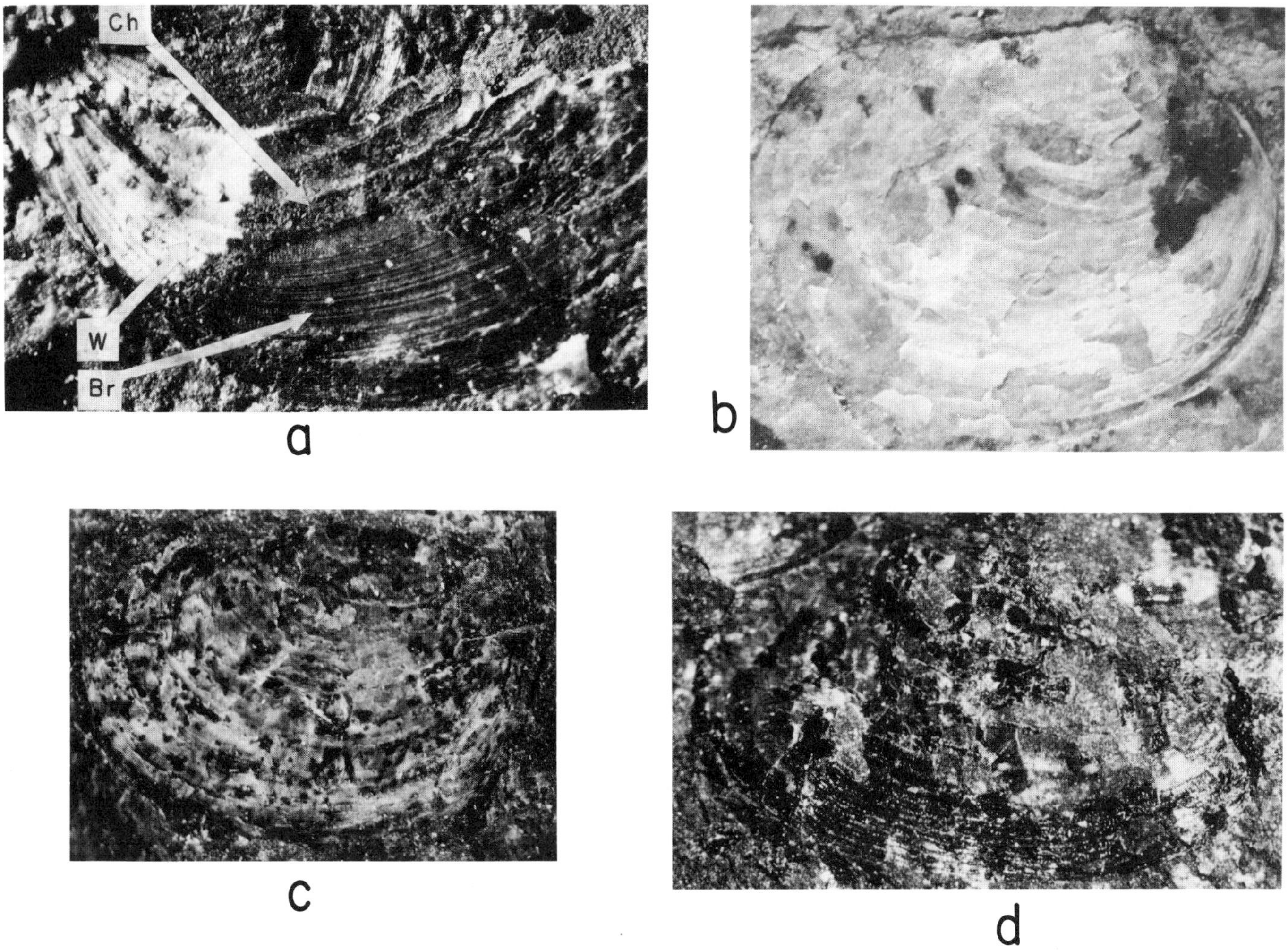

Figure 81.5. Fossil valve changes. *a.* WBr condition, white enameloid plus brown crust, temperature range 800°–1100°C; Tasch Station 0, bed 6, Blizzard Heights, in direct contact with upper flow. W: white enameloid. Br: brown crust. Ch: char. *b.* W condition, white enameloid, temperature range 600°–1100°C; Tasch Station 0, bed 2, Blizzard Heights. Upper part of this bed has Bl condition dominant. Note slight area of char in upper right-hand corner. *c.* Bl condition, blue enameloid, temperature range 600°–900°C; Tasch Station 0, bed 5, Blizzard Heights. *d.* British Museum specimen M. S. 116, L3412, Brimstone Peak, southern Victoria Land. Note glossy, jet black condition of valve (see Figure 81.4*a*).

ature. (See discussion below on geochemistry and recurrent high sanidine in the conchostracan-bearing sediments.) As discussed below, there are also other ways to discriminate the upper and lower ends of certain temperature ranges.

Fossil valves

In the Queen Alexandra Range, the thin sedimentary beds (baked fossiliferous mudstone) representing lacustrine deposits were originally explored by D. H. Elliot (1970) and subsequently by the writer (austral season 1969–70). The sections shown in Figures 81.1 and 81.2 are highly generalized; further details are given here.

At Blizzard Heights (Tasch Station 0), the uppermost flow of the Kirkpatrick Basalt (Figure 81.4*d*) forms a cliff that extends approximately 72 m below the summit to the top of the fossiliferous sedimentary interbed (Elliot and Tasch, 1967). In the field, beds were numbered downward from the basalt flow through the sedimentary unit. In the measured section below, however, numbers are given to sedimentary beds only, and numbers are assigned from the bottom to the top of the sedimentary unit.

Measured section at Blizzard Heights, Tasch Station 0

Bed Number	*Thickness (cm)*
6. Just below base of the overlying basalt. (For detailed description of flow, see Elliot, 1970.) Sharp contact between basalt and laminated, baked calcareous argillite, containing fossil conchostracan valves. The valves occur both at the contact surface and at 1.35 cm below that surface. Valves are classified as WBr and W on the experimental scale (Figure 81.1, right column). X-ray diffraction data indicate strong calcite and grossularite(?). (Field bed number 4.)	1.35
5. Baked argillite with pseudobanding, containing conchostracan fossils in lower 1.3 cm. These valves display a blue enameloid appearance (Bl condition). X-ray diffraction data indicate alpha-quartz and high sanidine. (Field bed number 5.)	3.19
4. Massive and blocky, weathered argillite, containing fossil conchostracans in upper 1.3 cm. Fossil valves have a white enameloid appearance (W condition). X-ray diffraction data indicate microcline and quartz. (Field bed number 6.)	6.30

3. Blocky aspect of argillite continues downward, weathered a dark brown and containing fossil conchostracans. Valves have the W condition. X-ray diffraction data indicate alpha-quartz, high sanidine, and calcite. (Field bed number 7.) 5.05
2. Fissile shale, conchostracan-bearing. Bl-type valves are dominant; upper part has Bl- and W-type valves, lower and middle parts have Bl-types. X-ray diffraction data indicate calcite and high sanidine. (Field bed number 8.) 5.05–5.75
1. Generally a baked, dark, blocky argillite, with upper 1.9 cm laminated and containing fossil conchostracans. These valves are Br- and W-types. X-ray diffraction data indicate alpha-quartz and high sanidine. (Field bed number 9.) 5.05

Approximate total thickness 26.69

Due to talus at Station 0, the nature of the contact with the lower basalt flow had to be determined a short distance (2.7 m) away from Station 0, toward the east. The somewhat gradational contact with the lower flow observed at this offset site can be attributed to a thin weathered zone (soil?) on the underlying basalt.

In an easterly direction about 33 m from Station 0, at Tasch Station 1, the sedimentary unit (shown below) is thicker by a factor of more than two (Figure 81.2).

Measured section at Blizzard Heights, Tasch Station 1

Bed Number		*Thickness (cm)*
10.	Tuff, light, buff to brown, massive, jointed; overlain by agglomerate. Base of tuff forms an abrupt contact with the underlying argilliate bed. (Field bed number 1.)	—
9.	Baked fossiliferous argillite overlain by buff clay and then by tuff; conchostracan-bearing. Valves are classified as W. (Field bed number 2.)	3.8
8.	White weathered, fossiliferous argillite; single bed; readily crumbles; conchostracan-bearing. Valves are classified as W. (Field bed number 3.)	3.2
7.	Blocky, white weathered, conchostracan-bearing argillite; coprolite with fish scales and spines. X-ray diffraction data indicate calcite, abundant montmorillonite, and grossularite. Valves are classified as Bl. (Field bed number 4.) (See bed 6 for thickness.)	—
6.	Same as bed 7 but yellow brown, thin bedded; conchostracan-bearing. X-ray diffraction data indicate calcite, high sanidine, and montmorillonite. Valves are classified as W. (Field bed number 5.) (Thickness of beds 6 and 7 is taken together.)	3.9
5.	Argillite, conchostracan-bearing; fish scales. X-ray diffraction data indicate high sanidine and quartz. Valves are classified as Br, Bl, and W (minor char on underlying cuticle). (Field bed number 6.)	1.3
4.	Medium-bedded, ligh-gray argillite. X-ray diffraction data indicate high sanidine, quartz, and calcite. Valves are classified as Br and W. (Field bed number 7.)	9.0
3.	Light-gray, blocky argillite; conchostracan-bearing; spoor (burrows); single bed. X-ray diffraction data indicate high sanidine, quartz, calcite, and some montmorillonite. Valves are classified as W. (Field bed number 8.)	9.0
2.	Conchostracan-bearing argillite. X-ray diffraction data are same as for bed 3 above. Valves are classified as Bl and W. (Field bed number 9.)	7.7
1.	Tan to khaki argillite; sharp conformable contact with overlying conchostracan-bearing beds; contact with underlying amygdaloidal basalt is a low relief surface (~2.5 cm). Valves are classified as WBr and C-1/2. (Field bed number 10.)	28.0
	Total thickness	65.65

X-ray diffraction data for both sections at Blizzard Heights show recurrent high sanidine. Blizzard Heights Station 1, alpha-quartz also occurs. At both stations, the youngest bed sampled contains calcite and grossularite. The W-valve condition is found throughout both sections. (Cf. Figures 81.1 and 81.2.)

Valve-color classes

The Bl and W enameloid conditions are phases of the same category. They represent "flow" of one or more layers of the previously solid cuticular valves that in places wholly or partly overlap or conceal growth bands. The Bl condition was not observed in naked valves and hence is probably due to some pigment added by the baked sediments. Such pigments can account for fossil occurrences of this color class.

An exploration was undertaken of the possibility that more than one effective temperature acted on a layer (bedding plane) containing valves. This was done with the experimental briquette runs, by reviewing all valve-color classes on a given briquette layer. A further check was made on fossil valves on a given bedding plane (Storm Peak sections, lat. 84° 35′ S, long 163° 55′ E). The conclusion reached is that while on every briquette layer and on every bedding plane (fossil scan) there was a dominant tendency in color class for all valves, a secondary component of some valves belonging to other color classes was present. The significance of this observation involves the theory of heat conduction and loss and is beyond the scope of this exploratory study. However, since both experimental and fossil valve data are consistent in this regard, the subject merits further study by specialists.

There is another factor to consider. To achieve the *same* valve change above 500°C required higher temperatures for briquette-embedded valves than for naked valves. Thus, to attain the C-1/2 condition (50% char), briquette-embedded valves required an increase of 100°C over naked valves, while the WBr condition required an increase of 300°C over naked valves. It follows that an increase from 100°C to 300°C furnace heat did not induce any new valve-color changes in briquette-embedded valves over those attained in naked valves at lower temperatures.

Valve flow and plastic deformation

The higher temperature accounting for valve conditions W and Wbr can cause valve flow and plastic deformation—particularly in the upper limits of the temperature ranges. Valve flow can occur without destruction of conchostracan growth lines or with only partial destruction of some growth lines (as observed in a specimen from Storm Peak) where flow continues longer than usual. Still higher temperatures cause plastic deformation of the valve so that the configuration is wholly or partially destroyed along with most growth bands or growth band ornamentation. The presence of flow and deformation can therefore further delimit the effective temperature within a range.

Chitin structure and thermal change

Richards (1951*a*) has observed that in arthropods the cuticle is a chitin-protein complex, with 50% or less of the dry weight being chitin. Crustacean branchiopod cuticular chitin, as detailed above, underwent change in response to varying degrees of increased temperature caused in nature by igneous heat and simulated in this study by furnace heat.

Data set forth here must correspond to structural and other changes in the cuticular chitin-protein complex due to *loss of volatiles* and/or *flow*. At 200°C there is no char effect, i.e., the chitin-protein structure presumably remains intact. Above this temperature (up to 400°C), varying degrees of valve char, as great as 100% char, suggest loss of volatiles and first-level changes in the complex—probably effects on the protein component (Figure 81.3).

The brown crustose condition (450°-600°C) is due to swelling of the cuticular complex. Thus, volatiles continue to escape from the cuticle, expanding and breaking growth bands to form the crustose condition. Some protein is also charred. Where more than one valve condition coexisted, this may be attributed to response by both surficial and underlying cuticles, with different effects on each.

The thermal yield point of cuticular chitin, embedded in a clay briquette, appears at 600°C (100°C less when naked valves are exposed). At the temperature several valve changes occur, chiefly enameloid condition and char. At the higher end of the enameloid temperature range, the cuticular chitin itself becomes plastic, and deformation and warping or flow result. Usually some protein is charred in the underlying cuticle.

At 1200°C, volatiles are still escaping (trapped bubbles are visible). Valve chitin presents a transparent glassy appearance. Complete absence of char effect, even slight, indicates total or almost total loss of the protein component, while the hyaline aspect denotes a second structural change in cuticular chitin.

Further delineation of the detailed structural and chemical changes corresponding to the thermal events discussed above is beyond the scope of this investigation but these changes deserve more detailed examination.

Thermal record in other fossiliferous beds of Antarctica

At Mauger Nunatak (lat. 85° 44′ S, long. 176° 40′ E), near the Beardmore Glacier area, a very thin (20 cm) sedimentary bed of Jurassic age occurs. It is overlain by approximately 30 m of basalt and tuff. D. H. Elliot collected outcrop samples here which were used in this study.

At Elliot Station 62.10.1, valves with the W condition occur 3.6 cm below the top of a slab in basal bed A. All rock slabs of the entire section are oxidized white (see Tasch and Gafford, 1969). Nevertheless, the white enameloid appearance of the valves cannot be attributed to oxidation; the same effect is obtained in experimental runs. Valves glisten under light and are glossy and brittle compared to the dull white oxidized matrix in which they are embedded. Bed B, the next younger bed, also has valves of W type. Beds C and D respectively, near to and in contact with the overlying Kirkpatrick Basalt flow, also have only W type. Based on the 1000+ °C temperature indicated for the dominant W condition at Blizzard Heights, the same temperature is applicable at Mauger Nunatak.

At Brimstone Peak (lat. 75° 38′ S, long. 158° 35′ E), Ricker's locality in southern Victoria Land, some 151 m or more above the base of the exposed section and above the eighth basalt flow, conchostracan-bearing sedimentary beds of Jurassic age occur. They show a single valve condition, C (100% char). Appearance of the valves, whole or fragmentary, is glossy jet black. The British Museum collection of Ricker's specimens could denote a rather low effective temperature (see Figure 81.1). However, larger collections and more X-ray diffraction data, as at Blizzard Heights (which lacks C types), are needed to confirm this inference.

At Carapace Nunatak (lat. 76° 54′ S, long. 159° 27′ E) Tasch Stations, C and Br conditions are the only color classes observed, which suggests a rather low effective temperature (see Figure 81.1). Borns and Hall (1969) have reported pillow lavas from Carapace Nunatak indicative of rapid quenching in lake water. Examination of conchostracans from their collections show only C and Br conditions. Two other bits of evidence support the low temperature inference: (1) palynomorphs are preserved in excellent condition and with unaltered color, which signifies a very low maximum effective temperature (Tasch and Lammons, 1978); and (2) additional X-ray diffraction analyses by S. Erlenwein on two samples from Carapace Nunatak, Tasch Station 2, beds 1 and 2, show alpha-quartz and a previously unidentified calcium mineral, wollastonite (low-temperature species). This mineral assemblage fits a temperature range of 573°-600°C at low pressure (Harker and Tuttle, 1956), which fits the Br valve color/temperature class of this paper (Figure 81.1).

Summary

An exploratory study has been undertaken. A previously unavailable thermally induced color scale has been experimentally derived (temperature range 200°-1200°C), using valves of a modern conchostracan with Paleozoic and Mesozoic equivalents. The same valve color class was found at Blizzard Heights and at Mauger Nunatak, and based on this, both sites were assigned a temperature maximum of 1000+ °C. X-ray diffraction data confirm this assignment. A much lower effective temperature is indicated at Brimstone Peak (300°-400°C) and at Carapace Nunatak (about 600°C). For the latter site, the inferred valve-color class temperature has several types of independent support; for the former, X-ray diffraction data are needed.

Acknowledgments. This research was carried out under National Science Foundation Grant GA-1709, sponsored by the Office of Polar Programs. Air support was provided during the 1969–70 austral season by the U.S. Navy. Field assistants during the 1969–70 season were Dietmar Schumacher and William Lokey. Laboratory assistants for this study were Rod Anderson and Dennis Carlton. Conchostracan photographs were taken by Bill Hart. X-ray diffraction data were developed originally by Dah Cheng Wu and supplemented in further studies by Sue Erlenwein. David Elliot offered constructive criticism of an earlier version of this paper.

References cited

Borns, H. W., Jr., and Hall, B. A. 1969. Mawson "Tillite" in Antarctica: preliminary report of a volcanic deposit of Jurassic age. *Science, N.Y., 166,* No. 3907, 807–72.

Elliot, D. H. 1970. Major oxide chemistry of the Kirkpatrick Basalt, central Transantarctic Mountains. (*In* Adie, R. J., *ed. Antarctic Geology and Geophysics.* Oslo, Universitetsforlaget, 413–18.)

Elliot, D. H., and Tasch, P. 1967. Lioestheriid conchostracan: a new Jurassic locality and regional and Gondwana correlations. *J. paleont., 41,* No. 6, 1561–63.

Harker, R. I., and Tuttle, O. F. 1956. Experimental data on the P_{CO_2}-T curve for the reaction: Calcite + quartz ⇌ wollastonite + carbon dioxide. *Am. J. Sci., 254,* 239–56.

Richards, A. G. 1951*a*. *The Integument of Arthropods, the Chemical Components and their Properties, the Anatomy and Development, and the Permeability.* Minneapolis, University of Minnesota Press, 9–58.

Richards, A. G. 1951*b*. Structure and development of the integument. (*In* Roeder, K. D., *ed. Insect Physiology.* New York, John Wiley, 29.)

Tasch, P., and Gafford, E. L. 1969. Weathering of an Antarctic argillite: field, geochemical, and mineralogical observations. *J. sedim. Petrol., 39,* No. 1, 369–77.

Tasch, P., and Lammons, J. M. 1978. Palynology of some lacustrine beds of the Antarctic Jurassic. *Palinologia, Núm. extraord. I,* 455–60, Pl. I.

82

Tectonic and Other Implications of Middle–Upper Jurassic Rocks and Marine Faunas from Ellsworth Land, Antarctica

PATRICK G. QUILTY*
School of Earth Sciences, Macquarie University, North Ryde, N.S.W. 2113, Australia

Abstract. Studies of Middle and Late Jurassic faunas, particularly bivalves, from Ellsworth Land, Antarctica, have led to a reappraisal of tectonic hypotheses about the history of the Antarctic Peninsula. The Middle Jurassic faunas show European or cosmopolitan affinities, a reflection of uniform global conditions. These faunas may have accumulated in the area under much more stable tectonic conditions than later faunas.

By Late Jurassic time, a south Pacific bioprovince seems to have developed, marked particularly by the presence of *Inoceramus galoi* Böhm. The affinities of the Ellsworth Land bivalve faunas of this age are overwhelmingly with coeval New Zealand forms, and in the absence of enough ammonite control, New Zealand stage terminology is followed. Late Jurassic faunas seem to have accumulated under tectonically unstable conditions, probably in an island-arc environment.

The Antarctic Peninsula was most likely part of the same tectonic system as New Zealand and shared the same coastline; similar conditions may also have encompassed what is now New Caledonia. The Antarctic Peninsula is envisaged as having been on the western side of the southern tip of South America, more in keeping with the Tarling Gondwanaland fit rather than the Smith-Hallam fit. There are two possible origins for the Weddell Sea with the data presently available. One hypothesis is that it arose as early as Middle Jurassic time when a spreading ridge separated Africa–South America from India-Australia-Antarctica. The other hypothesis is that it began in early Tertiary time as the western end of the spreading ridge that developed between Australia and Antarctica.

An area at the southern end of the Antarctic Peninsula (Figure 82.1) was mapped by the University of Wisconsin Ellsworth Land Nunataks Geology Expedition during the Southern Hemisphere summer of 1965–66, and fossils were recovered from 22 sites. Study of the faunas by the writer has proceeded since they were collected; several papers have appeared and others are in press. Of the 22 localities, only 15 have been dated by the fossils studied so far. Poor preservation of the fossils suggests that it is unlikely that many of the remaining localities can be dated. Most important species mentioned in the text are illustrated in Figure 82.2.

The first mention of the Ellsworth Land faunas is by Behrendt and Laudon (1964) who recorded belemnites and worm tubes from Lyon Nunatak (then Johnson Nunatak) and assigned a Cretaceous age on the basis of the worm tube *Rotularia*. Stevens (1967) subsequently examined the belemnites and bivalves from Lyon Nunatak and revised the age to Jurassic.

The first major summary of the new faunas and floras is contained in Laudon and others (1969), and this is still the only paper to refer in any detail to the poorly preserved flora. Quilty (1970) reported on the ammonite faunas, which are Bajocian, Oxfordian, and probably Callovian in age. Subsequently, Bajocian brachiopods (Quilty, 1972*a*) and crinoids (Quilty, 1972*b*) and Late Jurassic bivalves (Quilty, 1977*b*) have also been described.

* Present address: Antarctic Division, Department of Science and the Environment, Melbourne, Victoria 3004, Australia.

The major faunal element still to be studied consists of Middle Jurassic bivalves, and work on these is in progress. This paper reiterates some of the significant features reported by Quilty (1977*b*) but also contains the results of other studies not yet published.

Palynomorphs were destroyed during the Cretaceous plutonic phase. Foraminifera have been seen in thin section but they have not been recovered by disaggregation techniques.

Geologic history

The area under consideration consists of Middle and Upper Jurassic sedimentary and volcanic rocks, dominantly arkosic and graywacke sandstone (Laudon and others, 1969; Laudon, 1971) with some shale and common intraformational conglomerate. The sediment source seems to have been a nearby volcanic arc, probably related to contemporaneous subduction activity. There is some meager faunal evidence to suggest that conditions were less active in the Middle than in the Late Jurassic.

The entire sequence was intensely folded soon after deposition and has been intruded by a series of mafic dikes, in turn intruded by granodioritic plutons during Early Cretaceous time (102–109 m.y., Halpern, 1967). The age of the mafic dikes is not accurately known but could be either Late Jurassic (post-Kimmeridgian) or Early Cretaceous.

Since middle Cretaceous time the area has been peneplaned, uparched, and block-faulted, with minor extrusion of Pliocene olivine basalt (Halpern, 1971).

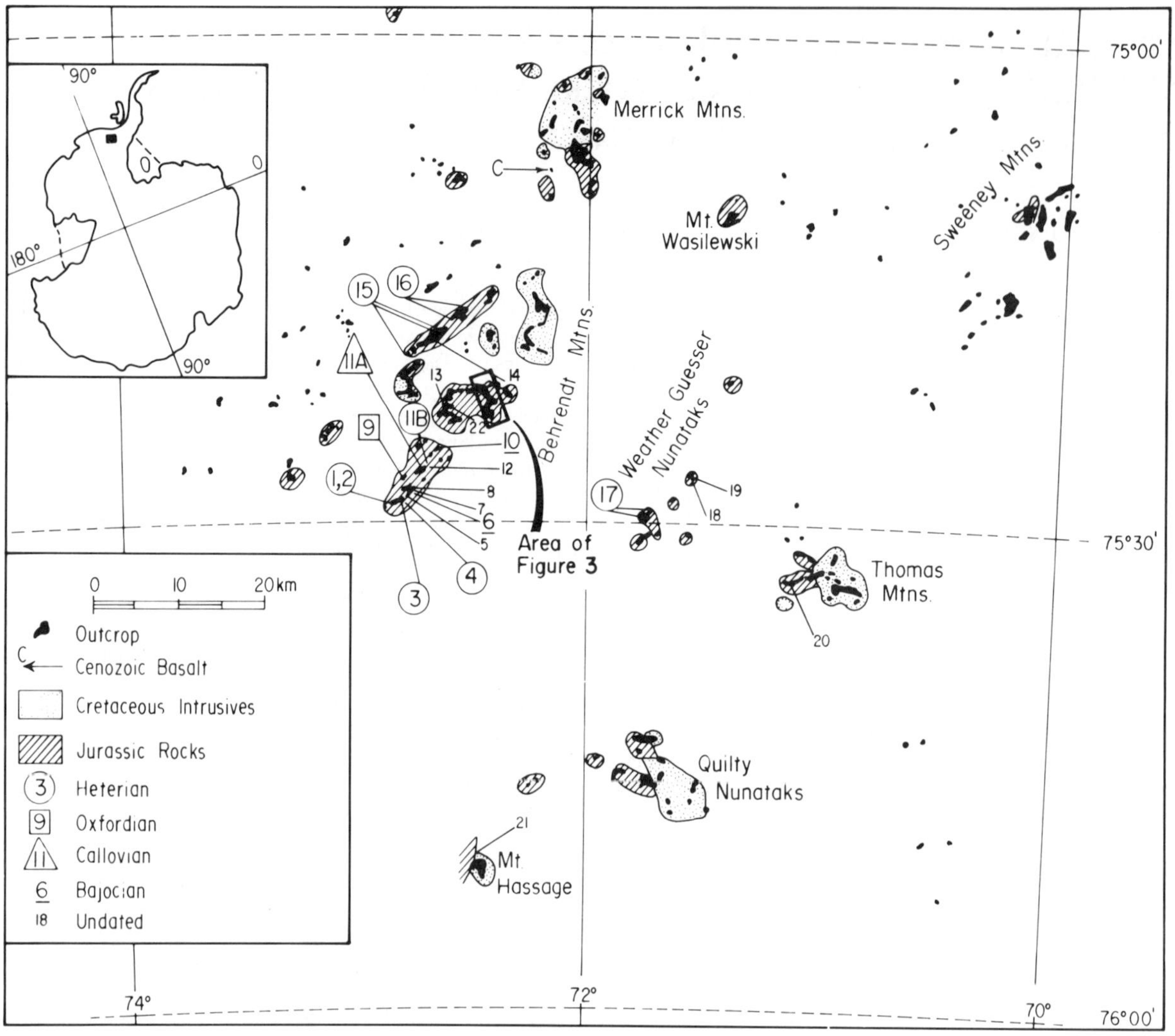

Figure 82.1. Map of eastern Ellsworth Land, showing fossil localities.

Jurassic faunas

Preservation of fossils

Preservation of the fossils is generally poor. Plant macrofossils have been heated intensely, probably during granodiorite emplacement. Hence these fossils are not identifiable except in a general way. Palynomorphs were also destroyed at this time.

Marine macrofaunas are composed almost entirely of what were originally calcareous forms. The bivalves, ammonites, and brachiopods are now decalcified and consist only of external and internal molds, most of which show little evidence of distortion. In Callovian shales, those fossils incorporated early in pyritic concretions have preserved their shape and are potentially identifiable, but those in the surrounding shale are now merely flattened, unidentifiable impressions. Belemnites are abundant locally, and some still consist of calcite; however, it is recrystallized to the extent that potentially valuable isotope studies are no longer feasible. The most abundant fossils in the area are undoubtedly the coiled, calcareous worm tube *Rotularia,* which occurs in most faunas. Here also the calcite has usually been removed. Soft-bodied organisms are represented by various tracks and trails; study of these has not commenced.

Stratigraphic nomenclature—The Heterian Stage

The ammonite study (Quilty, 1970) revealed Bajocian, probably Callovian, and Oxfordian faunas which can be integrated into the internationally accepted zonation of the Jurassic based on European stratotypes. Consequently, most fossils associated with the ammonites can also be so

Figure 82.2. Important fossils from Ellsworth Land. 1. *Teloceras* n. sp., ×0.8, UTGD 87285b, locality 10; 2. *Normannites* cf. *vulgaricostatus* Westerman, ×0.8, UTGD 87344d, locality 10; 3. *Megasphaeroceras* cf. *rotundum* Imlay, ×1.2, UTGD 87268c, locality 10; 4 and 5. *Flabellirhynchia australensis* Quilty. (4) ×2.0, UTGD 87247a. (5) ×2.5, UTGD 87255c, locality 10; 6. *Pholadomya* n. sp., ×0.8, UTGD 87367c and d, locality 10; 7. *Pleuromya* cf. *milleformis* Marwick, ×0.8, UTGD 87285b (IV), locality 10; 8 and 9. *Nothocephalites*? sp., ×0.8, UTGD 87114a, locality 11A; 10. *Perisphinctes (Discosphinctes)* cf. *antillarum* Jaworski, ×0.8, UTGD 87358b, locality 9; 11. Perisphinctid sp. indet., ×0.8, UTGD 87358d, locality 9; 12. *Vaugonia kawhiana* (Marwick), ×1.0, UTGD 87171c, locality 2; 13. *Grammatodon (Indogrammatodon) sparsilineatus* Marwick, ×1.6, UTGD 87359c, locality 9; 14. Representative fauna from locality 2, ×0.8: *a. Vaugonia kawhiana* Marwick; *b. Entolium* sp.; *c. Rotularia* sp.; and *d. Ditrupa* sp.; 15. *Inoceramus galoi* Böhm, ×0.4, UTGD 87033, locality 17.

Specimens shown in this figure are housed in the Geology Department, University of Tasmania (UTGD), Hobart.

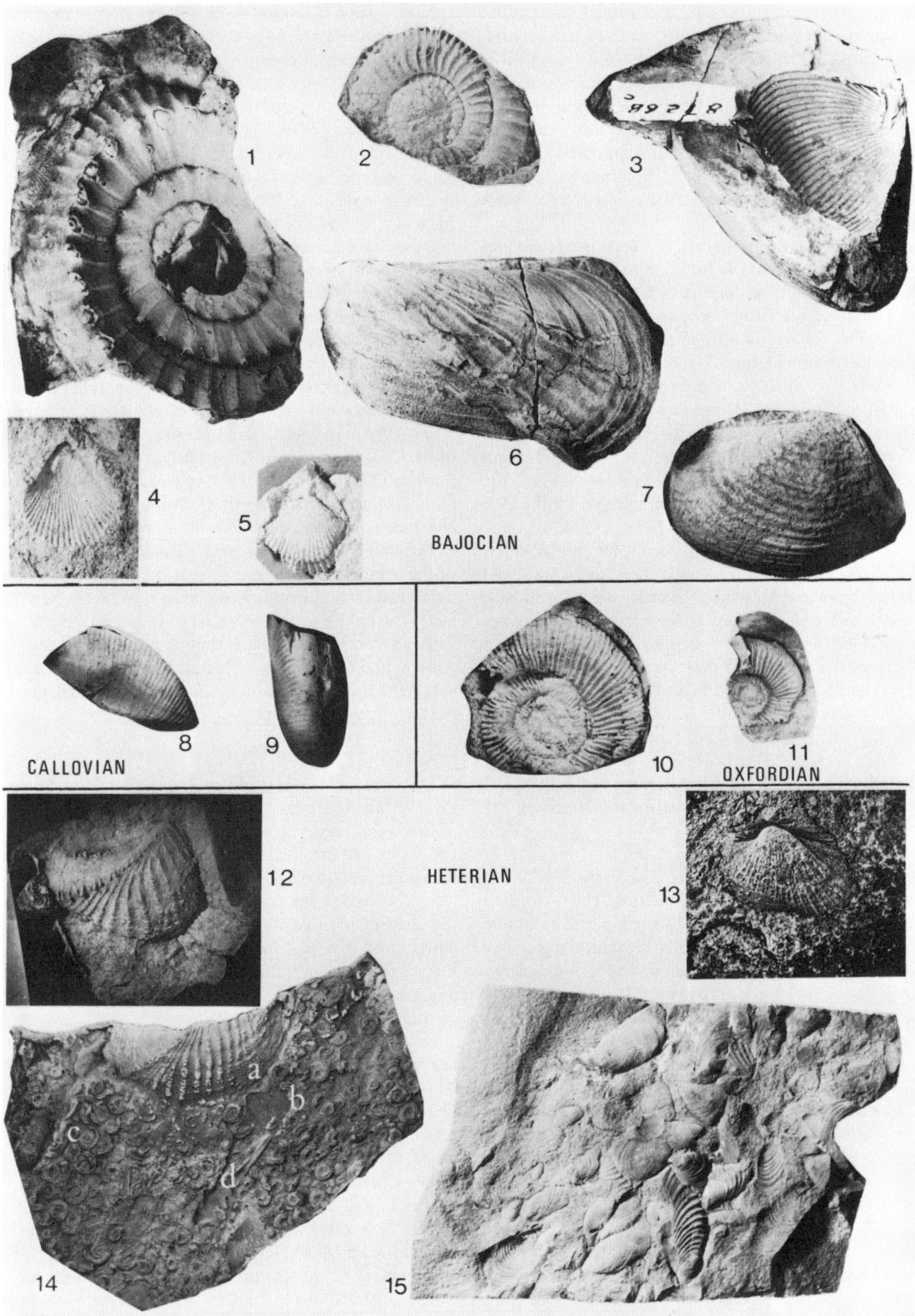
1
2
3
4
5
6
7
BAJOCIAN
CALLOVIAN
8
9
10
11
OXFORDIAN
12
HETERIAN
13
14
a
b
c
d
15

integrated. There is, however, one faunal association which cannot, but which is very important. Figure 82.1 is a location map for the collections discussed here, and it is noteworthy that 8 of the 22 collections are recorded as Heterian, a New Zealand Late Jurassic stage term, whereas all others are either undated or referred to European stage terms. This is because of the lack of ammonites in many collections and of the presence, commonly in large numbers, of the characteristic bivalve *Inoceramus galoi* Böhm (Figure 82.2).

Thus, a dual stage terminology has been used here and by Quilty (1977*b*) because *I. galoi*, an index fossil in New Zealand for the Heterian, occurs in Ellsworth Land associated with other species also known from the Heterian in New Zealand. The relationship of the Heterian to the conventional international stages is unknown. Until recently it was regarded in New Zealand as early to middle Kimmeridgian, but Enay (1973) has indicated that it could be anywhere during Oxfordian–early Kimmeridgian time. The Antarctic ammonite faunas include a few Oxfordian species (Quilty, 1970), but their relationship to the Heterian is unknown, a problem complicated by the lack of proven Oxfordian in New Zealand.

An Oxfordian fauna occurs at locality 9 and includes, among the few bivalves, *Vaugonia kawhiana* (Marwick) and *Pinna kawhiana* Marwick, the latter regarded in New Zealand as a Heterian species. However, *Inoceramus galoi* is absent. This apparent absence may be due to the fact that locality 9 was sampled only very briefly. Two other observations should be noted here: *I. galoi* was described from the Oxfordian (actually post-Oxfordian) by Böhm (1907), and when recorded from New Zealand (Marwick, 1951, 1953), it was taken as Heterian which was in turn correlated with the Oxfordian. Although the evidence, I believe, supports an Oxfordian age for the Heterian, the mixed nomenclature is used here.

Environment of deposition

The faunas discussed here are all marine. The Oxfordian and Heterian faunas occur in well-sorted quartz and lithic wackes in which valves are usually separated and commonly broken, indicating a shallow-water, high-energy environment. This is in marked contrast to the Bajocian and Callovian faunas which occur, respectively, in fine-grained sandstones and concretionary black shales. In these beds the fossils are not broken or disarticulated, suggesting quiet-water, low-energy environments. This contrast may indicate a general change in tectonic activity in the area after Callovian time, but there are rather few samples on which to base any conclusion.

Affinities of the faunas

Bajocian

Middle Bajocian faunas occur at localities 6 and 10, the latter having by far the more diverse fauna. So far, the only forms described fully are the ammonites, brachiopods, and crinoids (Quilty, 1970, 1972*a,b*); a comprehensive work on the bivalves remains to be completed. The bivalves are the dominant part of the fauna and include both epifaunal and infaunal species. Preservation is as molds only, but quality is good. Many of the bivalves, both epi- and infaunal, are identical with New Zealand species, many of which have not yet been described in detail, but there are many Antarctic forms which are new. However, many are very similar to described European forms. There is very little in common with other Bajocian localities, such as Western Australia (Arkell and Playford, 1954; Skwarko, 1974). The whole fauna suggests deposition in relatively quiet water, as all elements of the infauna seem to be *in situ* and the valves are intact. This is in strong contrast to the Late Jurassic faunas which seem to be of a shallower water, higher energy type.

Quilty (1970) commented on the European affinities of the Bajocian ammonites, the only Ellsworth Land ammonite fauna which is diverse enough to allow any meaningful comparison. Species of *Teloceras* and *Stephanoceras* are clearly similar to European forms and entirely distinct from any of the Canadian forms. However, *Megasphaeroceras* cf. *rotundum* Imlay is similar to the Alaskan form. This apparent anomaly is removed if Stone (1980) is correct in suggesting that certain rocks of the Cook Inlet area of Alaska originated far south in the Pacific Ocean off Central America. The brachiopod *Flabellirhynchia australensis* Quilty is very similar to *F. concinna* Crickmay of the coeval Californian Jurassic.

The Bajocian faunas, then, are European in the affinity of the nektonic element and also of the benthonic element. Other benthonic forms are similar to those from New Zealand. The linkup with European faunas seems to have been through Central America, including what is now southern Alaska, and it may be that similarities over such a distance are due to the last phase of cosmopolitanism which marked the earlier Jurassic faunas.

Callovian

The only locality to yield probable Callovian fossils is locality 11A where macrocephalitid ammonites occur with a few very poorly preserved bivalves, belemnites, fish vertebrae, and the crustacean *Cycleryon*. The fauna occurs in concretion-rich black shale.

Thierry (personal communication) has suggested that the identification (Quilty, 1970) of *Nothocephalites* is incorrect and that this fossil is really the Bathonian genus *Eurycephalites*. Until this change is documented, however, a Callovian age for these fossils will be tentatively adopted.

Not enough evidence is available from this sparse fauna to suggest faunal affinities.

Oxfordian

An Oxfordian fauna has been identified at locality 9 based on perisphinctid ammonites. Bivalves present include *Vaugonia kawhiana* (Marwick), *Pinna kawhiana* Marwick, and *Tancredia* aff. *T. allani* Marwick, all suggesting New Zealand affinities. Ammonites of Oxfordian age are not known in New Zealand, and *Perisphinctes (Discosphinctes)* cf. *antillarum* Jaworski is the sole species which suggests any affinities, in this case with South American forms.

Heterian

The Heterian (probably Oxfordian) is the time interval best represented in the Antarctic faunas. No ammonites are known, however, unless the above-mentioned Oxfordian fauna can eventually be shown to have definitive

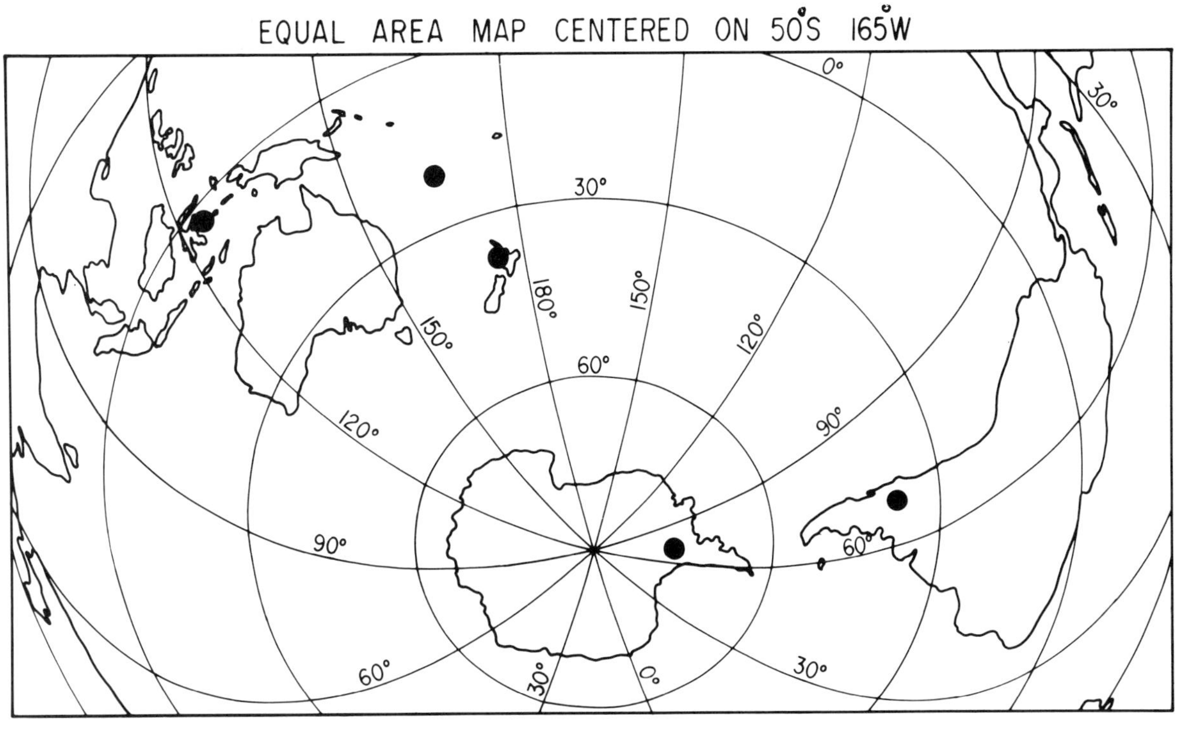

Figure 82.3. Distribution of *Inoceramus galoi* Böhm, a possible biogeoprovince index fossil.

Heterian components, and the faunas consist dominantly of species of robust bivalves indicative of shallow-water, high-energy marine conditions.

The most important species to be considered here is *Inoceramus galoi* Böhm, originally described from Wai Galo in Indonesia (Böhm, 1907). This species is now known from New Caledonia (Avias, 1953), New Zealand (Marwick, 1953, and many later records, especially Speden, 1970), the Antarctic Peninsula (Quilty, 1977*b*), and Argentina (Stehn, 1923; Cox and others, 1969) (Figure 82.3). There is here a sound basis for recognition of a Heterian faunal province based on the occurrence of *I. galoi*.

The affinities of the benthonic faunas with New Zealand equivalents, and the lack of affinity with coeval faunas of East Africa (Cox, 1965) and Western Australia (Skwarko, 1974; Brunnschweiler, 1960), suggest that no seaway connected the Antarctic Peninsula area to these latter localities. All the faunal evidence suggests that the Antarctic Peninsula lay very close to New Zealand and probably on the western side of the southern part of South America, in marked contrast to the widely used Smith-Hallam fit and more in keeping with the Tarling fit or, even better, those proposed by Powell and others (1980) or by Norton and Molnar (1977).

Geologic interpretation

Geologic structure

A feature that has been mentioned previously (e.g., Laudon and others, 1969) but not documented is the intensity of the folding of the sedimentary beds. Figure 82.4, based on a northerly directed panorama in the northern Behrendt Mountains, gives an idea of the structural complexity that has impeded understanding of stratigraphic relations in the area.

Relationship of sedimentation to possible sea-level fluctuation

There have been several recent attempts to relate episodes of sedimentation to sea-level fluctuations (Vail and others, 1978; Quilty, 1977*a*, and 1980). Vail and others (1978) envisage that sea level rose generally throughout Jurassic time and that several minor rises and falls were superimposed on the general rise. Their detailed curve is perhaps less useful than that of Hallam (1975, p. 173), which is generally compatible with that of Vail and others, except at the very end of the Jurassic, when Vail and others would have sea level rising but Hallam has a sharp fall. This discrepancy does not affect the present case.

Figure 82.5 includes Hallam's curve with a generally rising, long-term eustatic component of Vail and others superimposed. It is clear from this that the ages assigned to the marine faunas from Antarctica coincide very closely with sea-level highs and that time intervals not recognized as yet in faunas coincide closely with expected sea-level lows. From these lines of evidence it is suggested that unconformities representing sea-level lows may be identified by further field work.

The coincidence of the faunal ages with times of high sea level suggests that the curve can be validly applied in accounting for sedimentation in the area. It also suggests that sedimentation was complete before major folding occurred; if this were not true, the major tectonism should have caused discrepancies between the Vail-Hallam curve and the field evidence. The coincidence of the faunas and the curve also supports the earlier comment that the Heterian may well be Oxfordian.

The relationship of the Antarctic Peninsula to other continental masses

Tectonic factors have an important role in the affinities of faunas, migration routes, paleolatitude determinations,

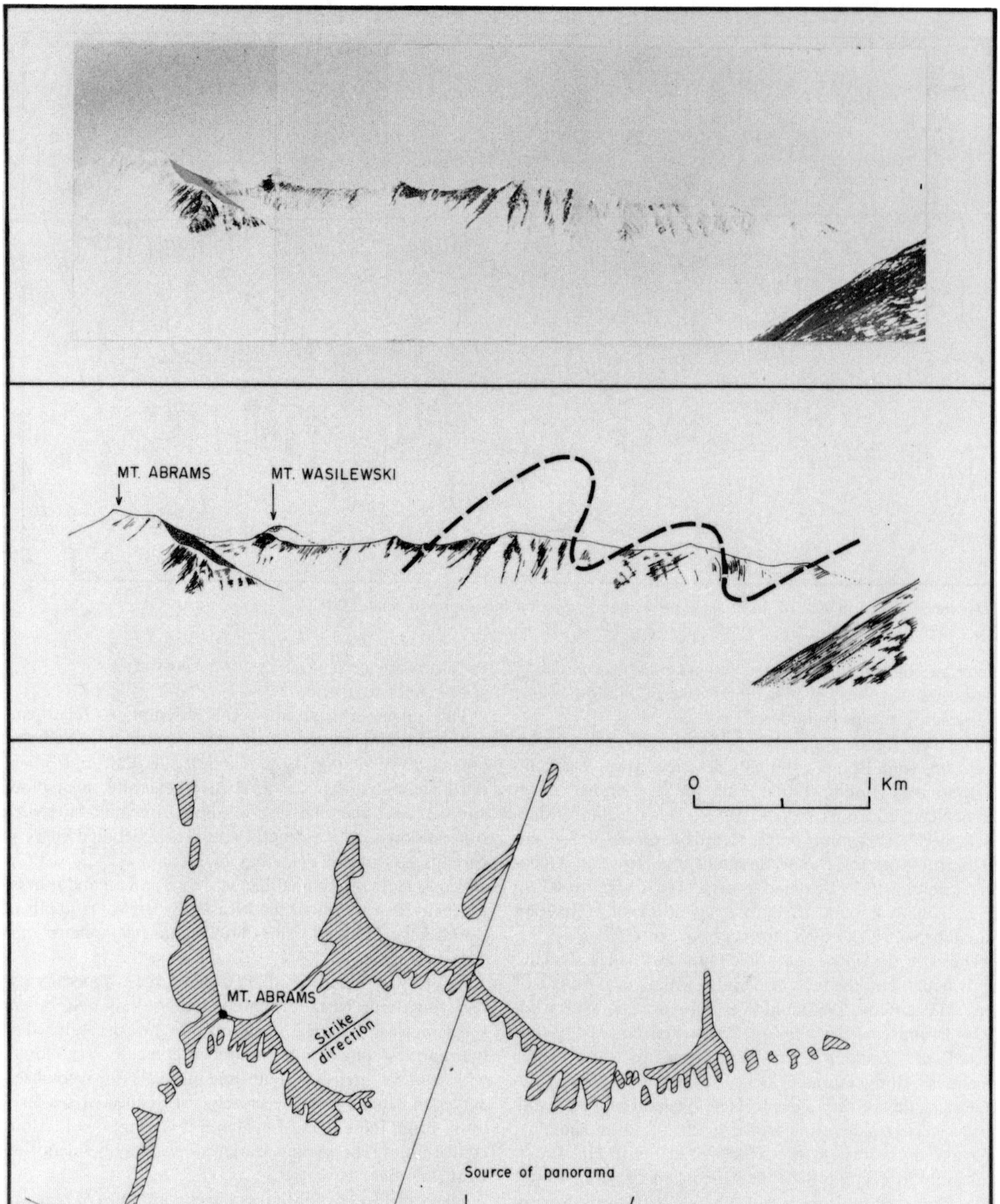

Figure 82.4. Structural complexity, eastern part of the northern Behrendt Mountains.

and related matters. In turn, study of the faunas can often place some constraints on tectonic hypotheses.

The tectonic history of the region, and especially relationships among the Antarctic Peninsula, southern South America, and New Zealand, have not been well documented. Figure 82.6 is an attempt to show the positions of relevant areas but in keeping with present knowledge of the Jurassic faunas and before the major phase of sea-floor spreading led to the present configuration. The map is based mainly on previous reconstructions, particularly those of Carey (1958), Griffiths (1971), Dietz and others (1972), and also on discussion with colleagues at Macquarie University (Powell and others, 1980).

Relationships to be rejected

The following hypotheses also have been presented to show the relationship of the present Antarctic Peninsula with the nearby landmasses.

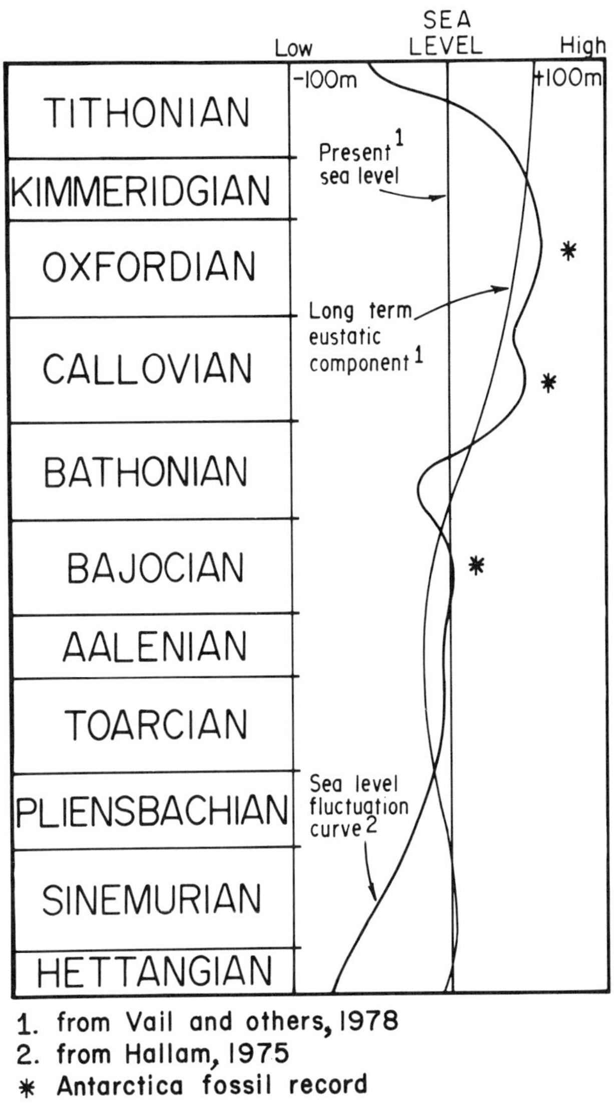

Figure 82.5. Jurassic sea levels and their relationship to Ellsworth Land sedimentation.

1. The Antarctic Peninsula rotated counter-clockwise from an original position near the Coats Coast (Caird Coast) of East Antarctica.

This hypothesis (e.g., Cullen, 1970) must be rejected because the Antarctic Peninsula Jurassic faunas have such strong New Zealand affinities and also because similar sedimentary rock types and tectonic history suggest that the Antarctic Peninsula and New Zealand must have been part of the same tectonic regime. The comments of Katz (1972) are relevant as they indicate that no continental crust existed immediately west of the Antarctic Peninsula. He emphasized contrasts between southern South American and Antarctic Peninsular tectonic histories. His interpretation is strengthened by the paleomagnetic striping patterns documented by Molnar and others (1975) which indicate that west of the Antarctic Peninsula, in the Amundsen Sea area, most sea floor formed since Jurassic times is still preserved, whereas along the west coast of southern South America, much old sea floor has been subducted.

2. The Antarctic Peninsula originally was situated between Africa and South America or directly south of Africa.

Several authors (e.g., Dietz and others, 1972; Smith and others, 1973) have shown reconstructions which vary a great deal in detail but which fall in this category. These also must be rejected because of the New Zealand affinities mentioned above, because most sea floor to the west of the peninsula which formed since middle Cretaceous time is preserved and because the marine sedimentary beds of the peninsula were being formed at the same time that the area was supposedly the site of continental crust.

3. The Antarctic Peninsula has maintained its present position with respect to South America.

Elliot (1972), Suarez (1976), and several papers by Dalziel and collaborators (e.g., Dalziel and Elliot, 1971) have implied that the Antarctic Peninsula and South America have long formed a more or less straight southeastern margin to the Pacific Ocean. This hypothesis is

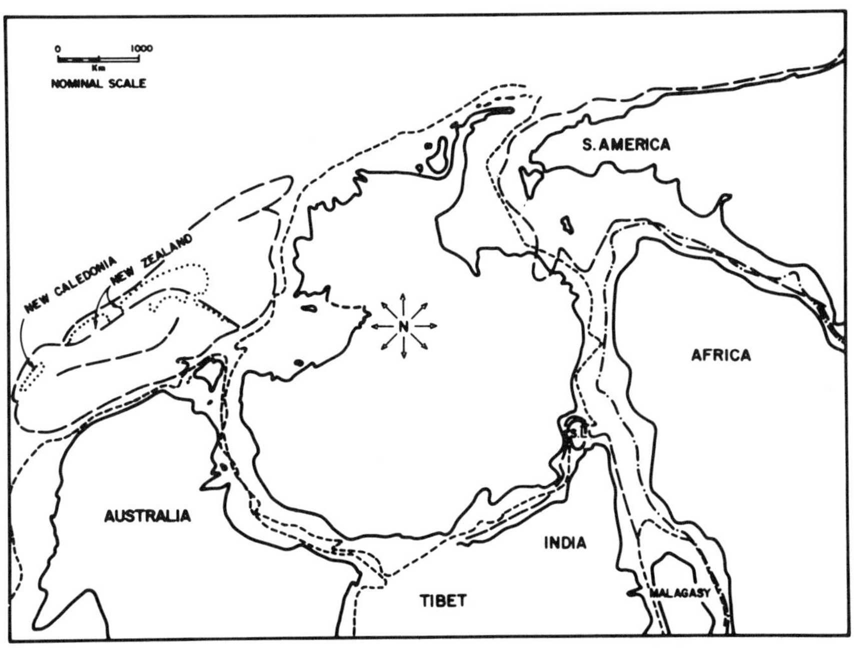

Figure 82.6. Reconstruction of the southern continents in Middle Jurassic time.

more in accord with the facts than the earlier mentioned ideas, but it still suffers geometrically.

Since Jurassic times, and especially during the last 55 m.y., Australia has moved *almost* due north (in fact, slightly west as well) from Antarctica. During the same time interval, the Atlantic and Indian Oceans have formed, with no known subduction zones in any meridional sense. Thus Africa has moved west relative to Australia, but this has been achieved largely by meridional movement; in fact, relative to its original position against India and Antarctica, it has moved eastward some 50°. This is about as would be expected from evolution of Gondwanaland according to the reconstruction used in Figure 82.6.

South America has moved west and north, relative to its original position, about 30°–40° of longitude for a point at the latitude of Cape Town. Therefore, it is impossible that its position relative to the Antarctic Peninsula has been maintained. In fact, it has moved relatively westward to its present meridional position. The arguments presented earlier regarding faunal affinities, Katz's (1972) interpretations, and those of Molnar and others (1975) on margin structure and magnetic striping also support the conclusion reached here that South America has moved west relative to the Antarctic Peninsula.

Recently, Suarez (1976) has discussed the tectonic history of the Antarctic Peninsula, recognizing Late Jurassic–Early Cretaceous island-arc provinces. His remarks are very interesting as they suggest possible late Paleozoic subduction on the western side of the peninsula. The other feature he noted is the difference in tectonic style and time and type of sedimentation between the area covered in this paper and the area of the peninsula to the north (including Alexander Island), the boundary being at about lat. 72° S, coincident with the change of trend of the peninsula from north-south to roughly east-west. The significance of this change must be fundamental, but its nature is not yet understood. The boundary at lat. 72° S differentiates a southern area of Middle–Late Jurassic island-arc tectonics from a northern one in which the island-arc activity continued until Early Cretaceous time. Another interpretation of Suarez (1976) that is consistent with the hypothesis presented here is his suggestion that the shape of the peninsula is not a geologically young feature but rather a relic of Late Jurassic–Early Cretaceous history.

Possible relationships

Having shown why he does not believe that any of the above three hypotheses is correct, the writer advances two possible histories for the Antarctic Peninsula, neither of which can be proved with presently published information. Both begin with the continental reconstruction shown in Figure 82.6. The biggest difficulty in such analysis is the lack of information on the area east of the Antarctic Peninsula, in the Weddell Sea. Bregman and Frakes (1970) recognized poorly defined magnetic striping anomalies there which seem to trend roughly east-west, thus suggesting an east-west–trending spreading ridge and a system of north-south transform faults. How old are these anomalies, and how do they relate to the geology of the Antarctic Peninsula? Two possibilities present themselves.

1. The first possibility is that the ridge indicated by these anomalies may be the weakly seismic ridge in the southwest Indian Ocean. The northeastern extension of this ridge is now inactive but has produced part of the magnetic striping pattern documented in the Indian Ocean by Sclater and Fisher (1974), Larson (1975), and others. The results of Deep Sea Drilling Project Leg 27 (Veevers and Heirtzler, 1974) have shown that the history of this spreading system began in Late Jurassic time. Marine sedimentary rocks of the Antarctic Peninsula (Laudon and others 1969; Quilty, 1970; Berger and Winterer, 1974) may suggest a history extending further back, probably into Middle Jurassic time. It may well be that the magnetic striping pattern in the Weddell Sea predates that of anywhere else and that this ocean basin is as old as Middle Jurassic in age. This possibility is hinted at by Berger and Winterer (1974, Figure 6). During Middle Jurassic time, the Antarctic Peninsula probably was part of southern South America, as suggested in Figure 82.6. The reconstruction shown in Figure 82.6 is identical in this area with that proposed by Tarling (1972).

Early movement on the Jurassic spreading ridge would have led to separation of South America–Africa from Australia-India-Antarctica, forming the east-west magnetic striping pattern and separating the Antarctic Peninsula from South America, probably along a transform fault immediately to the east of the Antarctic Peninsula. The next phase of movement was the change of the spreading ridge to a northward orientation, eventually separating Africa from South America, the latter moving west to its position north and slightly west of the Antarctic Peninsula. Slightly later than this, Australia and New Zealand separated, as described by Hayes and Ringis (1973). The last phase relevant to this history is the separation of Australia from Antarctica, which took place in early Tertiary time (Weissel and Hayes, 1972; Johnstone and others, 1973).

Water depths in the Weddell Sea, however, are shallower than would be expected for an ocean basin of Jurassic or Cretaceous age (Sclater and Detrick, 1973). This problem would be removed if future work shows that the Weddell Sea has a significant sediment layer. Should this be the case, then this hypothesis would seem preferable to the next.

2. The other possibility is that the magnetic lineations are related to a more recent history (approximately Paleocene–Eocene) and to the spreading system which separated Australia from Antarctica. This system separated Antarctica from New Zealand in Cretaceous time and made an attempt to separate Australia and New Zealand, an attempt which ceased about early Paleocene time (Hayes and Ringis, 1973). The ridge then passed westward between Australia and Antarctica, possibly extending as far west as the Weddell Sea and forming the magnetic anomalies recorded by Bregman and Frakes (1970). This phase also was destined to be short-lived; it ended when the spreading system changed direction and moved into the northwest Indian Ocean, initiating the present phase of sea-floor spreading. A late Eocene–Oligocene age for the cessation of the Weddell Sea formation would be compatible with the Sclater and Detrick (1973) age/depth hypothesis (see also Morgan, 1975).

Deep sea drilling in the Weddell Sea area, as suggested by Kennett and others (1976), would be the most useful means of testing the hypotheses discussed above.

Summary

Middle Jurassic marine faunas from Ellsworth Land have cosmopolitan features, but by Late Jurassic time, a southern Pacific bioprovince had differentiated. Thus, international stage terminology can be used for the Middle Jurassic, but for at least part of the Late Jurassic, a New Zealand stage—the Heterian—is more appropriate.

Sedimentation in Middle and Late Jurassic time may have been related in part to sea-level fluctuation. Soon after deposition, the sedimentary beds were severely folded; later, during Cretaceous time, they were intruded by granodiorites.

The New Zealand affinities of many elements of the faunas suggest that the Antarctic Peninsula and New Zealand were along the coast of the same landmass and that the Antarctic Peninsula lay to the west of the southern tip of South America.

The Weddell Sea may have sea floor as old as Middle Jurassic.

Acknowledgments. Field work in Antarctica was supported by National Science Foundation grants GA-199 and GA-14355 to the University of Wisconsin. Field support was provided by the United States Navy.

Study of the faunas has been supported at various times by the University of Tasmania Geology Department, West Australian Petroleum Pty. Limited (WAPET), and the Macquarie University School of Earth Sciences.

This paper has been helped by discussions with I. Speden, M. R. Banks, B. D. Johnson, M. H. Johnstone, and J. J. Veevers, and their comments are gratefully acknowledged.

References cited

Arkell, W. J., and Playford, P. E. 1954. The Bajocian ammonites of western Australia. *Phil. Trans. R. Soc. Lond.*, Ser. A, *237*, No. 651, 547-605.

Avias, J. 1953. Contribution a l'etude stratigraphique et paleontologique de la Nouvelle-Caledonia Centrale. *Sciences Terre, 1,* No. 1-2, 1-276.

Behrendt, J. C., and Laudon, T. S. 1964. Cretaceous fossils collected at Johnson Nunatak, Antarctica. *Science, N.Y., 143,* No. 3604, 353-54.

Berger, W. H., and Winterer, E. L. 1974. Plate stratigraphy and the fluctuating carbonate line. (*In* Hsu, K. J., and Jenkins, H. C., *eds.* Pelagic Sediments: On Land and Under the Sea. *Spec. Publs Int. Ass. Sedim.*, 11-48.)

Böhm, G. 1907. Beitrage zur Geologie von Niederlandische—Indien. 1. Abt. die Sudküsten der Sula—Inseln Taliabu and Mangoli. 3. Abschn. Oxford des Wai Galo. *Palaeontogr. Supp. IV. 1,* No. 3, 59-120.

Bregman, M. L., and Frakes, L. A. 1970. Magnetic anomalies in the Weddell Sea and vicinity. *Earth planet. Sci. Lett., 9,* 322-26.

Brunnschweiler, R. O. 1960. Marine fossils from the Upper Jurassic and the Lower Cretaceous of Dampier Peninsula, western Australia. *Bull. Bur. Miner. Resour. Geol. Geophys. Aust., 59,* 1-53.

Carey, S. W. 1958. The tectonic approach to continental drift. (*In* Carey, S. W., *ed. Continental Drift: a Symposium.* Hobart, Univerity of Tasmania, 177-355.)

Cox, L. R. 1965. Jurassic Bivalvia and Gastropoda from Tanganyika and Kenya. *Brit. Mus. (Nat. Hist.) Geol. Supp.*, No. 1, 1-213.

Cox, L. R., Newell, N. D., and others. 1969. *Bivalvia.* (*In* Moore, R. C., *ed. Treatise on Invertebrate Paleontology.* Lawrence, University of Kansas Press, and Boulder, Geological Society of America, Part N, 3 vols., N954-N1224.)

Cullen, D. J. 1970. A tectonic analysis of the south-west Pacific. *N.Z. Jl Geol. Geophys., 13,* 7-20.

Dalziel, I. W. D., and Elliot, D. H. 1971. The evolution of the Scotia arc. *Nature, Lond., 233,* 246-52.

Dietz, R. S., Holden, J. C., and Sproll, W. P. 1972. Antarctica and continental drift. (*In* Adie, R. J., *ed. Antarctic Geology and Geophysics.* Oslo, Universitetsforlaget, 837-42.)

Elliot, D. H. 1972. Aspects of Antarctic geology and drift reconstructions (*In* Adie, R. J., *ed. Antarctic Geology and Geophysics.* Oslo, Universitetsforlaget, 849-58.)

Enay, R. 1973. Upper Jurassic (Tithonian) ammonites (*In* Hallam, A., *ed. Atlas of Palaeobiogeography.* Amsterdam, Elsevier, 297-307.)

Griffiths, J. R. 1971. Reconstruction of the south-west Pacific margin of Gondwanaland. *Nature, Lond., 234,* No. 5326, 203-7.

Hallam, A. 1975. *Jurassic Environments.* Cambridge, Cambridge University Press, 269 pp.

Halpern, M. 1967. Rubidium-strontium isotopic age measurements of plutonic igneous rocks in eastern Ellsworth Land and northern Antarctic Peninsula, Antarctica. *J. geophys. Res., 72,* No. 20, 5133-42.

Halpern, M. 1971. Evidence for Gondwanaland from a review of West Antarctica radiometric ages. (*In* Quam, L. O., *ed.* Research in the Antarctic. *Am. Ass. Adv. Sci.*, Publn No. 93, 717-30.)

Hayes, D. E., and Ringis, J. 1973. Seafloor spreading in the Tasman Sea. *Nature, Lond., 243,* 454-58.

Johnstone, M. H., Lowry, D. C., and Quilty, P. G. 1973. The geology of southwestern Australia—a review. *J. R. Soc. West. Aust., 56,* 5-15.

Katz, H. R. 1972. Plate tectonics and orogenic belts in the southeast Pacific. *Nature, Lond., 237,* 331-32.

Kennett, J. P., Sclater, J. G., and van Andel, T. H. 1976. For JOIDES Southern Ocean targeted. *Geotimes, 21,* No. 3, 21-24.

Larson, R. L. 1975. Late Jurassic sea-floor spreading in the eastern Indian Ocean. *Geology, 3,* 69-71.

Laudon, T. S. 1971. Stratigraphy of eastern Ellsworth Land. (*In* Adie, R. J., *ed. Antarctic Geology and Geophysics.* Oslo, Universitetsforlaget, 215-23.)

Laudon, T. S., Lackey, L. L., Quilty, P. G., and Otway, P. M. 1969. Geology of eastern Ellsworth Land. (*In* Bushnell, V. C., and Craddock, C., *eds.* Geologic Maps of Antarctica. *Antarct. Map Folio Ser.*, Folio 12, Pl. III.)

Marwick, J. 1951. Series and stage division of N.Z. Triassic and Jurassic rocks. *N.Z. Jl Sci. Technol.*, Ser. B, *32,* No. 3, 8-10.

Marwick, J. 1953. Divisions and faunas of the Hokonui System. *Bull. geol. Surv. N.Z.*, No. 21, 1-141.

Molnar, P., Atwater, T., Mammerickx, J., and Smith, S. W. 1975. Magnetic anomalies, bathymetry and the tectonic evolution of the South Pacific since the Late Cretaceous. *Geophys. J. R. astr. Soc., 40,* 383-420.

Morgan, W. J. 1975. Heat flow and vertical movements of the crust. (*In* Fischer, A. G., and Judson, S., *eds. Petroleum and Global Tectonics.* Princeton, Princeton University Press, 23-43.)

Norton, I., and Molnar, P. 1977. Implications of a revised fit between Australia and Antarctica for the evolution of the eastern Indian Ocean. *Nature, Lond., 267,* 338-40.

Powell, C. McA., Johnson, B. D., and Veevers, J. J. 1980. A revised fit of East and West Gondwanaland. *Tectonophysics, 63,* 13-29.

Quilty, P. G. 1970. Jurassic ammonites from Ellsworth Land, Antarctica. *J. Paleont., 44,* No. 1, 110-16.

Quilty, P. G. 1972*a*. Middle Jurassic brachiopods from Ellsworth Land, Antarctica. *N.Z. Jl Geol. Geophys., 15,* No. 1, 140-47.

Quilty, P. G. 1972*b*. *Pentacrinites* and (?)*Apiocrinus* from the Jurassic of Ellsworth Land, Antarctica. *Neues Jb. Geol. Paläont. Mh., 8,* 484-89.

Quilty, P. G. 1977*a*. Cenozoic sedimentation cycles in Western Australia. *Geology, 5,* 336–40.

Quilty, P. G. 1977*b*. Late Jurassic bivalves from Ellsworth Land, Antarctica: their systematics and paleogeographic implications. *N.Z. Jl Geol. Geophys., 20,* 1033–80.

Quilty, P. G. 1980. Sedimentation cycles in the Cretaceous and Cenozoic of Western Australia. *Tectonophysics, 63,* 349–66.

Sclater, J. G., and Detrick, R. 1973. Elevation of midocean ridges and the basement age of JOIDES deep sea drilling sites. *Geol. Soc. Am. Bull., 84,* 1547–54.

Sclater, J. G., and Fisher, R. L. 1974. Evolution of the east Central Indian ocean, with emphasis on the tectonic setting of the NinetyEast Ridge. *Geol. Soc. Am. Bull., 85,* 683–702.

Skwarko, S. K. 1974. Jurassic fossils of Western Australia, 1: Bajocian Bivalvia of the Newmarracarra Limestone and the Kojarena Sandstone. *Bull. Bur. Miner. Resour. Geol. Geophys. Aust.,* No. 150, 1–54.

Smith, A. G., and Hallam, A. 1970. The fit of the southern continents. *Nature, Lond., 225,* No. 5228, 139–44.

Smith, A. G., Briden, J. C., and Drewry, G. E. 1973. Phanerozoic world maps. (*In* Hughes, N. F., *ed.* Organisms and Continents through Time. *Spec. Pap. Palaeont., 12,* 1–39).

Speden, I. G. 1970. Three new inoceramid species from the Jurassic of New Zealand. *N.Z. Jl Geol. Geophys., 13,* 825–51.

Stehn, F. 1923. Beitrage zur Kenntniss des Bathonien und Calloven in Südamerika. *Neues Jb. Min. Geol. Paläont., 49,* 52–158.

Stevens, G. R. 1967. Upper Jurassic fossils from Ellsworth Land, West Antarctica, and notes on Upper Jurassic biogeography of the South Pacific region. *N.Z. Jl Geol. Geophys., 10,* No. 2, 345–93.

Stone, D. B. 1980. The Alaska Orocline, paleomagnetism and the paleogeography of the Alaskan oil fields. *Tectonophysics, 63,* 63–73.

Suarez, M. 1976. Plate-tectonic model for southern Antarctic Peninsula and its relation to southern Andes. *Geology, 4,* 211–14.

Tarling, D. H. 1972. Another Gonwanaland. *Nature, Lond., 238,* 92–93.

Vail, P. R., Mitchum, R. M., Jr., and Thompson, S., III. 1978. Global cycles of relative changes of sea level. (*In* Payton, C. E., *ed.* Seismic Stratigraphy—Applications to Hydrocarbon Exploration. *Mem. Am. Ass. Petrol. Geol., 26,* 83–98.)

Veevers, J. J., and Heirtzler, J. R. 1974. Tectonic and paleogeographic synthesis. (*In* Veevers, J. J., Heirtzler, J. R. and others. *Initial Reports of the Deep Sea Drilling Project,* Vol. 27. Washington [U.S. Government Printing Office], 1049–52.)

Weissel, J. K., and Hayes, D. E. 1972. Magnetic anomalies in the southeast Indian Ocean. (*In* Hayes, D. E., *ed. Antarctic Oceanology. 2. The Australian-New Zealand Sector.* Washington, American Geophysical Union, 165–96.) [Antarctic Research Series, Vol. 19.]

83

Antarctic Fossil Echinoids: Review and Current Research

FREDERICK H. C. HOTCHKISS*
Academy of Natural Sciences, Philadelphia, Pennsylvania 19103, U.S.A.

Abstract. Antarctic Peninsula region fossil echinoids occur in Aptian rocks of Alexander Island, Campanian rocks of Seymour and Snow Hill Islands, Paleogene rocks of Seymour and Cockburn Islands, and Neogene strata of King George Island. In the McMurdo Sound region, fossil echinoids occur in erratic boulders of Eocene(?) rocks at Black Island, in the Plio-Pleistocene Scallop Hill Formation at White Island, and in the Quaternary Taylor Formation at Taylor Valley, on Ross Island, and near Backstairs Passage. Cidaroid spines occur in Quaternary terraces of the Vestfold Hills area, Princess Elizabeth Land. An unidentified Mesozoic echinoid spine has been reported from the South Orkney Islands.

Fossil echinoids from Black Island in East Antarctica and from the James Ross Island Group in the Antarctic Peninsula region are represented in Tierra del Fuego and in Patagonia by the same or closely related species. This distribution is in general agreement with the concept of a Weddellian Province. However, family-level differences between the echinoids of Australasia and those of South America and Antarctica seem to require recognition of an early Paleogene local provincialism not reflected by molluscs.

Current research on undescribed collections from Cretaceous and Tertiary rocks of Seymour Island, Tertiary rocks of Patagonia, and Neogene strata of King George Island is directed toward better understanding of the content and relationships of the echinoid faunas of the Weddellian Province and is expected to have a bearing on the early history of the circum-Antarctic West Wind Drift.

In an earlier paper on the zoogeographic implications of a Paleogene echinoid from East Antarctica, Hotchkiss and Fell (1972) have demonstrated a zoogeographic connection between South America and East Antarctica during Eocene time and have argued against an early Tertiary shallow-water benthic migration route between Australasia and South America via Antarctica. It is important to recognized that the conflict this presents with respect to geophysical evidence on the timing of the breakup of Gondwanaland derives from the view that echinoderms evolve slowly. The principal result of the review presented here is the suggestion that the unique Australasian echinoid stocks arose in a burst of adaptive radiation in middle to late Paleogene time from unknown ancestors derived from the western part of the Late Cretaceous Austral Province.

Review of Antarctic fossil echinoids

The 1901–03 Swedish South Polar Expedition brought back the first collection of fossil echinoids from Antarctica (Lambert, 1910; Wilckens, 1924). *Cyathocidaris nordenskjoeldi* Lambert and *C. erebus* Lambert from Seymour Island, *C. patera* Lambert from Seymour Island and Snow Hill Island, and *Holaster lorioli* Lambert, *Nordenskjoeldaster antarcticus* Lambert, *Vomeraster vomer* (Lambert), and *Vomeraster* sp. from Snow Hill Island were assigned a Cretaceous age. (See Figure 83.1 for locations.) *Cassidulus andersoni* Lambert and *Schizaster antarcticus* Lambert from Seymour Island and *Cidaris* sp. from Cockburn Island have been assigned a Tertiary age. Lambert noted that *C. nordenskjoeldi* and *Schizaster antarcticus* bear some resemblance to the Patagonian *Cidaris antarctica* Ortmann and *Schizaster patagonensis* Lambert. The Cretaceous sequence of the James Ross Island Group has been more precisely dated as Campanian by Howarth (1966), and the Tertiary sequence of Seymour Island has been shown to range from Paleocene to possibly lower Oligocene, with lower and middle Eocene strata missing (Zinsmeister, 1977). Both Croft (*in* Marples, 1953) and Zinsmeister (1977) have indicated that fossil echinoid remains are locally abundant in the Seymour Island sequence.

Taylor (1966) has reported *Hemipedina*(?) sp., *Rhabdocidaris* sp., *Epiaster*(?) sp. (possibly two species), and *Anorthopygus*(?) sp. from Aptian strata of the east coast of Alexander Island (Figure 83.1). His main conclusion was that the Aptian sequence of Alexander Island includes several stratigraphically important echinoderm horizons. Elliott (1974) has reported additional rhabdocidaroid spines from the Ablation Point area of Alexander Island.

The 1964–65 Victoria University of Wellington Antarctic Expedition No. 9 discovered echinoid remains in erratic sandstone boulders at Black Island, McMurdo Sound (Vella, 1969) (Figure 83.1). These included a well-preserved ambulacral petal of a clypeasteroid echinoid (Hotchkiss and Fell, 1972). The petal morphology placed the specimen in the suborder Scutellina and excluded the family Astriclypeidae, thus precluding African or Australasian affinities. A probable Eocene age, indicated by the associated microflora and microplankton (McIntyre and Wilson, 1966; Wilson, 1967), suggested the Eocene families Monophorasteridae and Protoscutellidae; comparison of the Black Island fossil with specimens of the Patagonian monophorasterid genera *Monophoraster* and *Iheringiella* strongly suggested that the specimen belongs to *Iheringi-*

* Present address: 153 Dunham Corner Road, East Brunswick, New Jersey 08816, U.S.A.

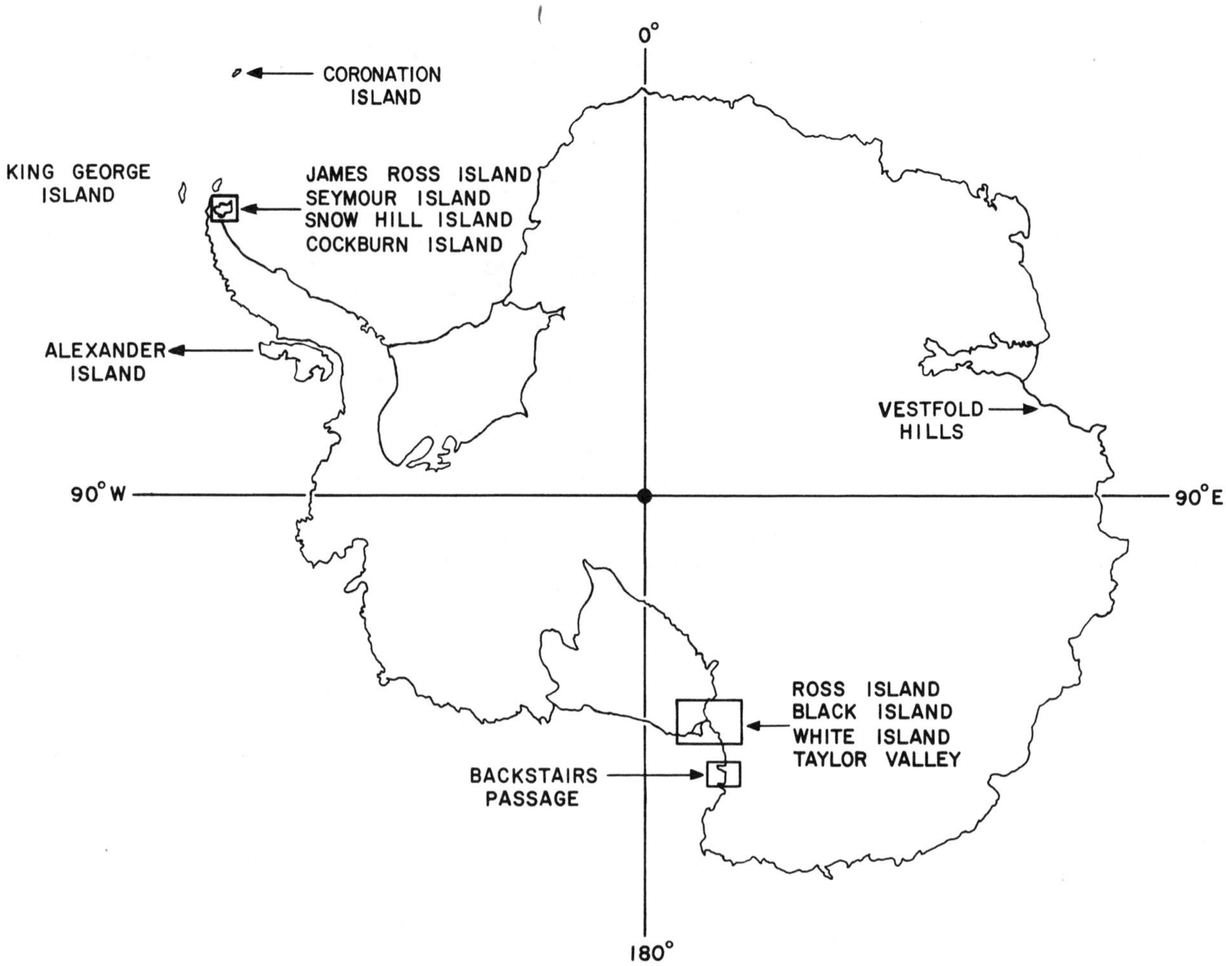

Figure 83.1. Location map showing fossil echinoid collecting localities discussed in text.

ella. Hotchkiss and Fell have inferred an early Tertiary zoogeographic connection between South America and the McMurdo Sound sector of Antarctica and a zoogeographic barrier to shelf-fauna migration between Antarctica and the New Zealand Plateau.

Shackleton's 1907–09 *Nimrod* Expedition collected unidentified Quaternary echinoid test fragments and spines near Backstairs Passage, Terra Nova Bay region, and at two localities on Ross Island, McMurdo Sound (David and Priestly, 1914; Chapman, 1916; Debenham, 1920) (Figure 83.1). Speden (1962) has referred these deposits to the Taylor Formation of postglacial age (except at Ross Island where they may be older than postglacial). Speden (1962) has reported the echinoid *Sterechinus* cf. *S. neumayeri* in a collection from the mouth of Taylor Valley. The caption to his Figure 13 refers this collection to the Taylor Formation, but the text refers it to a miscellaneous deposit not certainly identified as to formation. Speden (1962) has also reported unidentified echinoid spines from the Scallop Hill Formation at White Island. He has assigned this formation an early Pleistocene interglacial or possibly pre-Pleistocene age and suggested a probable general correlation with the Pecten Conglomerate on Cockburn Island. Wrenn and Webb (1976) have reported unidentified fragmentary echinoid spines in Pleistocene beds of Dry Valley Drilling Project holes 8, 9, and 10 in Taylor Valley. Crespin (1960) has reported delicate cidaroid spines in Quaternary terraces in the Vestfold Hills area of Princess Elizabeth Land.

The only fossil echinoid described from the Scotia Arc is an unidentified echinoid spine from Mesozoic rocks of Coronation Island, South Orkney Islands (Thomson and Willey, 1975). Unidentified echinoderm plates, not necessarily echinoid, have been reported from Mesozoic beds of South Georgia (Skidmore, 1972).

Relationships of the echinoids

There are clear taxonomic relationships among the echinoids of Antarctica, the Magellanic region, and southern Chile and Argentina. Furque and Camacho (1949) have reported *Cyathocidaris patera* and *Nordenskjoeldaster* sp. from the Campanian Policarpo Group and *Holaster* sp. from the Maestrichtian Leticia Group, Rio Clara Series, of the Atlantic coast of the southern part of Argentine Tierra del Fuego. These authors have noted relationships with the Cretaceous echinoids of Snow Hill Island and Seymour Island. Bernasconi (1954) has described *Holaster mortenseni* from Cretaceous strata of Rio Grande on the Atlantic coast of Tierra del Fuego and compared it with *H. lorioli* from Snow Hill Island and with *H. feruglioi* Melinossi from Lago Argentina, Patagonia. Most recently, the genus *Iheringiella,* originally described from Paleogene beds of Patagonia, has been found in Tierra del Fuego and in East Antarctica (Bernasconi, 1959; Hotchkiss and Fell, 1972).

There are significant family-level differences between the fossil echinoids of Australasia and those of southern

South America (Fell, 1953; Hotchkiss and Fell, 1972). The characteristic Australasian and Indo-Pacific family Temnopleuridae, present in Eocene rocks of southeastern Australia, is completely unknown in lower and middle Tertiary rocks of Antarctica and southern South America. In contrast, temnopleurids are well represented in the Eocene beds of southeastern Australia and include four marsupiate species of *Paradoxechinus* and the monotypic marsupiate genus *Pentechinus* (Philip and Foster, 1971). The characteristic Australasian clypeasteroid family Arachnoidae, which first appears in the Oligocene strata of New Zealand, and the equally characteristic family Fossulasteridae, which has as its type genus the marsupiate *Fossulaster* from Miocene rocks of Australia, are both unknown in Antarctica and South America. Reciprocally, the southern South American Paleogene clypeasteroid family Monophorasteridae is unknown in Australasia, although it occurs in the Paleogene rocks of East Antarctica. It is important to note that the Australasian and South American clypeasteroid families did not diverge from a single stock but are members of entirely different clypeasteroid suborders. As a further distinction between Australasian and South American fossil echinoids, it must be noted that marsupiate echinoids have yet to be found in the Tertiary sequence of southern South America.

Three cases of established or putative relationships between the fossil echinoids of Australasia or the Indo-West Pacific and those of South America exist. The Southern Ocean cidaroid subfamily Ctenocidarinae is represented in the Paleogene rocks of Patagonia by *Austrocidaris jorgensis* Loriol, in the Paleogene strata of southeastern Australia by *Austrocidaris operta* Philip, and in the Pliocene beds of New Zealand by *Notocidaris vellai* Fell. It has not yet been found as a fossil in Antarctica. The temnopleurid genus *Pseudechinus,* present in Australasia from late Paleogene to Holocene time, is represented in southern South America from Pliocene to Holocene time by *P. magellanicus.* Finally, the Indo-Pacific clypeasteroid family Laganidae may be represented in the Pliocene sequence of Chile by *Jacksonaster chilensis* Lambert and Thiery.

Discussion

The distribution of Paleogene echinoids suggests that there was an Eocene zoogeographic connection between East and West Antarctica and South America. Because it is inconceivable that echinoids would have failed to cross a shallow-water benthic migration route had it existed, the significant family-level differences between the fossil echinoids of Australasia and those of southern South America imply a major barrier to exchange between East Antarctica and Australasia. The generally acknowledged slow rate of evolution of echinoderms implies that this zoogeographic barrier must substantially predate the Eocene Epoch to account for the evolution of such distinctive families as the South American Monophorasteridae and the Australasian Arachniidae and Fossulasteridae, and to account for the diversity of the distinctly Australasian temnopleurid genera. This result is, of course, at variance with the claim that shallow-marine conditions existed between New Zealand and Antarctica until about the end of early Paleocene time and between Australia and Antarctica until about late Eocene or early Oligocene time (Zinsmeister, 1979).

Two of the echinoids in common between Australasia and South America do not affect the interpretation of shallow-water migration routes. *Pseudechinus* seems to have a common ancestry with the fossil Australasian genus *Brochopleurus* (Fell, 1962*a*), and its presence from Pliocene to Holocene time in Chile and Argentina can be explained either by larval planktonic or postlarval epiplanktonic dispersal on the West Wind Drift (Fell, 1962*b;* Larrain, 1975). Similarly, *Jacksonaster chilensis* may be a transpacific laganid migrant that did not permanently establish itself. However, as noted by Durham (1954, 1955), it may instead by a juvenile *Clypeaster* or an unusual and geologically young neolaganid, and therefore not of zoogeographic significance.

It is the Ctenocidarinae that suggest a compromise solution. Cidaroids arose in Paleozoic time and seem to have had an extremely slow rate of evolution. Thus the evolution of a distinctive Antarctic cidaroid subfamily has been considered as evidence that the Antarctic echinoid fauna evolved over a long period in isolation (Fell, 1961). Mounting evidence argues against this view. *Austrocidaris* has now been found as a fossil in southeastern Australia (Philip, 1964) and has been dredged from the sea floor near the Antipodes Islands on the New Zealand Plateau (Pawson, 1968), *Notocidaris* has been found as a fossil in New Zealand (Fell, 1954), and the monotypic living genus *Ogmocidaris* is endemic to New Zealand waters. Furthermore, Philip (1964) has concluded from a revision of the southeastern Australian Tertiary cidaroids that the Ctenocidarinae differentiated from a goniocidarine-stereocidarine plexus in Australasia during late Paleogene time. It is therefore necessary to change our views on the origin of the Ctenocidarinae and the rate of evolution of cidaroids.

According to Philip and Foster (1977), the presence of ctenocidarine and marsupiate echinoids in southeastern Australia refutes the traditional view of an Indo-Pacific origin for the Australasian Cenozoic echinoid fauna and indicates an Austral origin. They have correlated a progressive endemicity of Australian temnopleurid and clypeasteroid echinoids from Eocene into Oligocene time with isolation due to rifting of Australia from Antarctica in the Cretaceous Period, followed by sea-floor spreading initiated in Eocene time. With respect to the Ctenocidarinae, their conclusion is consistent with Zinsmeister's (1979) interpretation that shallow-marine conditions existed between Australia and Antarctica until about late Eocene or early Oligocene time. However, with respect to temnopleurid and clypeasteroid echinoids, there still remains the question of why Australasian stocks did not reach South America via Antarctica and why South American clypeasteroids did not enter Australasia. Because the Ctenocidarinae are more eurybathic than temnopleurids and clypeasteroids, one answer may be that a narrow, deep-water barrier, not traversed by currents favorable for planktonic larval dispersal, separated Australia from Antarctica.

It is probably correct to interpret the Australian marsupiate fossil echinoids as a record of cold water and infer a high southern paleolatitude, but without a substantiating fossil record in Antarctica or South America this does not necessarily imply an Austral origin for the Australasian Cenozoic echinoid fauna. The Late Cretaceous Madagascan marsupiate echinoid *Tripylus pseudoviviparus* Lambert proves this and, in fact, suggests an alternative direction of origin for the Australasian stocks. Thus the question of the ages and origins of the distinctly Australasian temnopleurid and clypeasteroid stocks is still unresolved, and a reconciliation of the early Tertiary biogeography with geophysical evidence on the timing of the breakup of Gondwanaland is needed. Because the conflict presented by existing data is between the estimated time of separation of Australasia from Gondwanaland and the inferred rates of evolution of echinoderms, the best available hypothesis is that the unique Australasian temnopleurid and clypeasteroid stocks arose in a burst of adaptive radiation during middle to late Paleogene time from unknown ancestors derived from the western part of the Late Cretaceous Austral Province of Kauffman (1973), and not from the south and east.

The Weddellian Province

The Weddellian Province (Zinsmeister, 1979) includes the southern part of South America, Antarctica from the Antarctic Peninsula to northern Victoria Land, New Zealand, and the southeastern corner of Australia. Zinsmeister constructed this province on the basis of the paleobiogeography of *Struthiolaria, Struthiolarella, Lahillia* (subgenus *Lahilleona*), and *Struthioptera* molluscan lineages from Late Cretaceous through Eocene time. These lineages made their first appearance during the Cretaceous Period and are known or inferred to be present in New Zealand, Australia, Antarctica, and South America and are not known outside the Southern Hemisphere. They demonstrate an early Tertiary trend toward local provincialism which Zinsmeister has correlated with the breaking away of Australasia from Antarctica. These lineages disappear from New Zealand in rocks younger than early Paleocene in age presumably because of latitudinal climatic changes associated with sea-floor spreading.

The Weddellian Province is seen as one of the products of the fragmentation of the initially larger Austral Province of Kauffman (1973); by middle Tertiary time, the Weddellian Province itself had undergone further fragmentation. Zinsmeister (1979) has reviewed several sources of evidence and concluded that the Cretaceous shallow-marine conditions between New Zealand and Antarctica ceased to exist by about the end of early Paleocene time. According to Zinsmeister, deep-marine conditions began to develop between Antarctica and Australia by late Paleocene time, but shallow seas and highlands continued to link Antarctica's northern Victoria Land with southeastern Australia through the Tasman Rise until about late Eocene or early Oligocene time. This shallow-marine connection had disappeared by middle Oligocene time because there is evidence for the deep-flowing Antarctic Circumpolar Current south of the South Tasman Rise at that time. Warm-water Indo-Pacific molluscan groups did not appear in New Zealand until late Eocene time and in southeastern Australia until late Eocene to early Oligocene time.

The zoogeography of Antarctic and related fossil echinoids generally supports Zinsmeister's thesis. The major unresolved problem is the general absence of faunal elements in common between Australasia in the west and South America and Seymour Island in the east. I agree with Zinsmeister (personal communication) that the absence of these taxa may be the result of incomplete knowledge of the fossil record in Australasia for the uppermost Cretaceous and lowermost Eocene sequences. However, the family-level differences between the echinoids of Australasia and those of South America and Antarctica seem to require recognition of local provincialism of echinoid stocks not reflected by the molluscs.

Current research

Current research is directed toward the study of undescribed collections of Antarctic, Patagonian, and Chilean fossil echinoids of Cretaceous, Tertiary, and Plio-Pleistocene age, and toward the restudy of type and published specimens. The objectives are to document the faunas, to attempt to classify species and genera now *incertae sedis,* to develop echinoids as a tool for dating Antarctic formations, to test working hypotheses of zoogeographic relationships implied by continental-drift reconstructions, and to analyze the data for information on the rifting apart of Gondwanaland and on the rates of evolution of echinoderms. Prominent taxonomic problems of evolutionary or zoogeographic interest include the following: the systematic position of *Nordenskjoeldaster* from the Cretaceous beds of Snow Hill Island (Lambert, 1910) and Tierra del Fuego (Furque and Camacho, 1949), classified *incertae sedis* in the Holasteroida by Wagner and Durham (1966); the systematic position of *Platipygus posthumus* from the Patagonian beds, classified *incertae sedis* in the Cassiduloida by Kier (1962); and whether the Patagonian fossil *Isechinus praecursor* belongs in Echinidae or Echinometridae—Mortensen (1943) has rejected the latter possibility only on zoogeographic grounds.

Materials now under study include: (1) An echinoid test and spines collected by G. J. Hobbs of the Falkland Islands Dependencies Survey from the "*Pecten* conglomerate" of Lion's Rump, King George Island, South Shetland Islands, made available by M. R. A. Thomson of the British Antarctic Survey. Hobbs' collections were briefly referred to by Barton (1965), but the echinoid remains were not specifically mentioned; (2) Echinoid spines from the Cretaceous of Seymour Island collected and made available by W. J. Zinsmeister (see Zinsmeister, 1977); (3) Tertiary echinoids from Seymour Island collected by Croft (see Marples, 1953) and made available by R. P. S. Jefferies of the British Museum (Natural History); and (4) echinoids collected by the 1911 Eighth Amherst Expedition from Patagonian beds, mostly in the vicinity of Mazaredo on the Gulf of St. George, made available by Gerald P. Brophy of Amherst College and Karl Waage of Yale University. The Amherst Expedition invertebrate fossils were only briefly discussed by Loomis (1913, 1914) because the principal expedition objective was to collect fossil vertebrates.

The King George Island echinoids are the first to come from the South Shetland Islands and are of great interest with respect to both the history of the Antarctic fauna and the age and correlations of the "*Pecten* conglomerate." Similar conglomerates occur at Cockburn Island, Heard Island, and in the McMurdo Sound region. Preliminary study suggests that the King George Island echinoid test may belong to the extant South African genus *Polyechinus* (Echinidae). Preliminary examination of the Seymour Island echinoids indicates that Zinsmeister's Cretaceous echinoid spines probably represent *Cyathocidaris* spp. and that Croft's Tertiary echinoids probably represent *Cassidulus andersoni* and *Schizaster antarcticus;* it is too early to know whether these specimens will provide any data not noted in Lambert's (1910) report on the Swedish Antarctic Expedition materials. Despite the work of Ortmann (1902), de Loriol (1901, 1902), Lambert (1903), Bernasconi (1959), and others, the full content and the higher systematic relationships of the Patagonian echinoid fauna are still not known. It is hoped that the Amherst Expedition materials will lead to an improved systematic and paleobiogeographic analysis of the Patagonian faunas.

Acknowledgments. In addition to those acknowledged in the text, I thank Krister Brood, Swedish Museum of Natural History; Horacio H. Camacho, Universidad de Buenos Aires; J. Wyatt Durham, University of California, Berkeley; David H. Elliot, Institute of Polar Studies, Ohio State University; Phyllis Hasson, Princeton University; Roger L. Kaesler and Larry Martin, University of Kansas, Lawrence; J. S. Lawless, Yale University; David N. Lewis, British Museum (Natural History); A. C. Riccardi, Universidad Nacional de La Plata; Eugene S. Richardson, Jr., Field Museum of Natural History; and Mort D. Turner and Guy Guthridge, National Science Foundation. I thank Anita Pruzan for reading the manuscript and John L. Boonshaft and Erenow Books for timely practical assistance.

References cited

Barton, C. M. 1965. The geology of the South Shetland Islands III. The stratigraphy of King George Island. *Scient. Rep. Br. antarct. Surv.*, No. 44, 33 pp.

Bernasconi, I. 1954. Nota sobre una nueva especie de Equinoideo fósil de Tierra del Fuego. *Physis, 20,* No. 59, 397–400.

Bernasconi, I. 1959. Equinoideos fósiles de la colección del Museo Argentino de Ciencias Naturales. *Physis, 21,* No. 61, 137–76.

Chapman, F. 1916. Report on the Foraminifera and Ostracoda from elevated deposits on the shores of the Ross Sea. *Rep. Scient. Invest. Br. Antarct. Exped. 1907–09,* Geology 2, 27–52.

Crespin, I. 1960. Some Recent Foraminifera from Vestfold Hills, Antarctica. *Hanzawa Mem., Spec. Vol. Scient. Rep. Tohoku Univ., 4,* 19–31.

David, T. W. E., and Priestly, R. E. 1914. Glaciology, physiography, stratigraphy, and tectonic geology of south Victoria Land. *Rep. Scient. Invest. Br. Antarct. Exped. 1907–09,* Geology 1, 1–319.

Debenham, F. 1920. A new mode of transportation by ice: the raised marine muds of south Victoria Land. *Q. Jl geol. Soc. Lond., 75,* Pt. 2, 51–76, Pl. 4.

de Loriol, P. 1901. *Notes pour servir à l'étude des Echinodermes.* Ser. 1, No. 9. Basil, Georg, 1–45, Pl. I–III.

de Loriol, P. 1902. *Notes pour servir à l'étude des Echinodermes.* Ser. 2, No. 1. Basil, Georg, 1–51, Pl. I–III.

Durham, J. W. 1954. A new family of clypeasteroid echinoids. *J. Paleont., 28,* No. 5, 677–84.

Durham, J. W. 1955. Classification of clypeasteroid echinoids. *Univ. Calif. Publ. geol. Sci., 31,* No. 4, 73–198, Pls. 3–4.

Elliott, M. H. 1974. Stratigraphy and sedimentary petrology of the Ablation Point area, Alexander Island. *Bull. Br. antarct. Surv.*, No. 39, 87–113.

Fell, H. B. 1953. The origin and migrations of Australasian echinoderm faunas since the Mesozoic. *Trans. R. Soc. N. Z., 81,* No. 2, 245–55.

Fell, H. B. 1954. Tertiary and Recent Echinoidea of New Zealand. Cidaridae. *Paleont. Bull. geol. Surv. N.Z.*, No. 23, 62 pp., 15 pls.

Fell, H. B. 1961. The Fauna of the Ross Sea. Part I. Ophiuroidea. *Bull. N. Z. Dep. scient. ind. Res.*, No. 142, 79 pp., 19 plates.

Fell, H. B. 1962*a*. Embryological evidence of evolutionary trends in some temnopleurid echinoids. (*In* Leeper, G. W., *ed. The Evolution of Living Organisms, A Symposium of the Royal Society of Victoria, Melbourne, December 1959,* 307–10.)

Fell, H. B. 1962*b*. West-wind-drift dispersal of echinoderms in the Southern Hemisphere. *Nature, Lond., 193,* No. 17, 759–61.

Furque, G., and Camacho, H. H. 1949. El Cretacico de la Costa Atlantica de Tierra del Fuego. *Revta. Soc. geol. argent., 4,* No. 4,263–98.

Hotchkiss, F. H. C., and Fell, H. B. 1972. Zoogeographical implications of a Paleogene echinoid from East Antarctica. *J. R. Soc. N. Z., 2,* No. 3, 369–72.

Howarth, M. K. 1966. Ammonites from the Upper Cretaceous of the James Ross Island Group. *Bull. Br. antarct. Surv.*, No. 10, 55–69.

Kauffman, E. G. 1973. Cretaceous Bivalvia. (*In* Hallum, A., *ed. Atlas of Paleobiogeography.* Amsterdam, Elsevier, 353–83.)

Kier, P. M. 1962. Revision of the cassiduloid echinoids. *Smithson. misc. Collns, 144,* No. 3, 262 pp., 44 pls.

Lambert, J. 1903. Notes sur les Echinides recueillis par M. A. Tournouër en Patagonie. *Bull. Soc. geol. Fr.*, Ser. 4, *3,* No. 4, 474–84, Pl. 15.

Lambert, J. 1910. Les Echinides fossiles des îles Snow-Hill et Seymour. *Wiss. Ergebn. Schwed. Südpolarexped. 1900–1903, 3,* No. 11, 15 pp.

Larrain, A. P. 1975. Los equinoideos regulares fósiles y recientes de Chile. *Gayana, Zoologia, 35,* 189 pp.

Loomis, F. B. 1913. *Hunting Extinct Animals in the Patagonion Pampas.* Concord, N.H., Rumford Press, 141 pp.

Loomis, F. B. 1914. *The Deseado Formation of Patagonia.* Concord, N.H., Rumford Press, 232 pp.

Marples, B. J. 1953. Fossil penguins from the mid-Tertiary of Seymour Island. *Scient. Rep. Falkld Isl. Depend. Surv.*, No. 5, 15 pp.

McIntyre, D. J., and Wilson, G. J. 1966. Preliminary palynology of some Antarctic Tertiary erratics. *N.Z. Jnl Bot., 4,* No. 3, 315–21.

Mortensen, T. 1943. *A Monograph of the Echinoidea, 3,* Pt. 2. Copenhagen, Reitzel, 533 pp., 56 pls.

Ortmann, A. E. 1902. Tertiary invertebrates. *Rep. Princeton Univ. Expds. Patagonia 1896–1899, 4,* Paleontology 1, Pt. 2, 47–332, Pls. 11–39.

Pawson, D. L. 1968. The echinozoan fauna of the New Zealand Subantarctic Islands, Macquarie Island, and the Chatham Rise. *Bull. N. Z. Dep. scient. ind. Res.*, No. 187, 35 pp.

Philip, G. M. 1964. The Tertiary echinoids of South-eastern Australia II. Cidaridae (2). *Proc. R. Soc. Victoria,* New Ser., *77,* No. 2, 433–77, Pls. 58–67.

Philip, G. M., and Foster, R. J. 1971. Marsupiate Tertiary echinoids from South-Eastern Australia and their zoogeographic significance. *Palaeontology, 14,* No. 4, 666–95, Pls. 124–34.

Philip, G. M., and Foster, R. J. 1977. The origins and relationship of the Australasian Cainozoic echinoid fauna. (Abstr.). *J. Paleont., 51,* Sup. to No. 2, Pt. 3, 21.

Skidmore, M. J. 1972. The geology of South Georgia. III. Prince Olav Harbour and Stromness Bay areas. *Scient. Rep. Br. antarct. Surv.*, No. 73, 50 pp.

Speden, I. G. 1962. Fossiliferous Quaternary marine deposits in the McMurdo Sound Region, Antarctica. *N. Z. Jl Geol. Geophys., 5,* No. 5, 746–77.

Taylor, B. J. 1966. Taxonomy and morphology of Echinodermata from the Aptian of Alexander Island. *Bull. Br. antarct. Surv.*, No. 8, 1–18.

Thomson, M. R. A., and Willey, L. E. 1975. Fossils from the South Orkney Islands, I. Coronation Island. *Bull. Br. antarct. Surv.*, No. 40, 15–22.

Vella, P. 1969. Surficial geological sequence, Black Island and Brown Peninsula, McMurdo Sound, Antarctica. *N. Z. Jl Geol. Geophys., 12,* No. 4, 761–70.

Wagner, C. D., and Durham, J. W. 1966. Holasteroids. (*In* Moore, R. C., *ed. Treatise on Invertebrate Paleontology.* Boulder and Lawrence, Geological Society of America and University of Kansas, Pt. U., U523–43.)

Wilckens, O. 1924. Die tertiäre Fauna der Cockburn-Insel (Westantarktica). *Further zool. res. Swed. Antarct. Expd. 1901–1903, 1,* No. 5, 1–17.

Wilson, G. J. 1967. Some new species of Lower Tertiary dinoflagellates from McMurdo Sound, Antarctica. *N.Z. Jnl Bot., 5,* No. 1, 57–83.

Wrenn, J. H., and Webb, P. N. 1976. Microfossils from Taylor Valley. *Antarct. Jnl. U. S., 11,* No. 2, 82–85.

Zinsmeister, W. J. 1977. Note on a new occurrence of the Southern Hemisphere aporrhaid gastropod *Struthioptera* Finlay & Marwick on Seymour Island, Antarctica. *J. Paleont., 51,* No. 2, 399–404.

Zinsmeister, W. J. 1979. Biogeographic significance of the late Mesozoic and early Paleogene molluscan faunas of Seymour Island (Antarctic Peninsula) to the final breakup of Gondwanaland. (*In* Gray, J., and Boucot, A. J., *eds.* Historical biogeography, plate tectonics, and the changing environment. *37th Biological Colloquium of the Oregon State University.* Corvallis, University of Oregon Press, 347–55.)

VI *Igneous Rocks*

84

The Antarctic Cenozoic Volcanic Provinces and their Implications in Plate Tectonic Processes

Review Paper

OSCAR GONZÁLEZ-FERRÁN
Departamento de Geofísica, Universidad de Chile, Casilla 2777, Santiago, Chile

Abstract. Two main Cenozoic volcanic regions in the Antarctic area are distinguished: the subantarctic volcanic islands located on the mid-oceanic ridge around the Antarctic continent, and the continental margin volcanism in apparently aseismic West Antarctica. A great deal of information related to the volcanic rocks and tectonic environment of Antarctica has been obtained in recent years. This paper reviews and analyzes the actual data on West Antarctica Cenozoic volcanism.

At least five volcanic provinces have been recognized in the West Antarctic region: (1) The South Sandwich Islands in the Scotia Arc, which form an intraoceanic arc between the South Sandwich Trench and a Holocene westward spreading line. Late Pliocene to Holocene tholeiitic to calc-alkaline volcanic rocks are found in this province. (2) The South Shetland Islands–Antarctic Peninsula province. Eocene to Pliocene, calc-alkaline, high-alumina basalts predominate in this zone. The Shetland Islands are related to the subduction of the Drake plate and to the late Pliocene–Holocene active volcanic line, which includes Deception, Penguin, and Bridgeman Islands. The latest volcanic line is composed of olivine-bearing subalkali-basalts and probably represents the spreading axis of the active rift located in the Bransfield Strait. (3) The James Ross volcanic island. This province consists of Miocene to Pleistocene middle-alkaline volcanic rocks related to a continental margin extension zone located toward the northwest. Associated with this extension zone is a younger volcanism (Seal Nunataks–Paulet Island) of olivine tholeiites to more alkaline nepheline-normative hawaiites. (4) Marie Byrd Land and neighboring regions. This province is characterized by Eocene to Holocene nepheline-normative alkali-basalts, peralkaline trachytes, phonolites, and alkali-rhyolites, with relatively recent volcanic activity. (5) The Hallett and McMurdo province (including Mount Weaver), which consists of Miocene to Holocene alkali-basaltic and trachytic associations and active volcanic centers (Mounts Erebus and Melbourne).

In each province, the Cenozoic volcanism is characterized by a basal subhorizontal sequence of pyroclastic rocks and lavas, which locally form real plateaus; in the alkaline provinces palagonites and tuff breccias are the predominant rocks. These plateaus have been subjected to differential block movements, followed by the development of central vent-type volcanism on the uplifted plateau. K-Ar dating, strontium isotope ratios, major and trace element data, and petrographic features of the volcanic rocks of West Antarctica are compared.

This paper reviews and synthesizes the available information on Antarctic Cenozoic volcanism. The Symposium on Antarctic Geology and Geophysics held in Oslo in 1970 revealed notable progress in field studies and surveys of the Cenozoic volcanic areas, especially of Marie Byrd Land, the Antarctic Peninsula, the Scotia Arc, and some subantarctic islands. Later investigations have been centered on petrologic, geochemical, and radiometric studies which, in conjucntion with geophysical advances, have generated petrologic and tectonic models for these regions. part of the progress related to these studies was presented to the IAVCEI Symposium on Andean and Antarctic Volcanology Problems held in Santiago in 1974.

Deception Island has been very active during the past decade. This has attracted the attention of many investigators who have contributed through several papers to the knowledge of the petrology, chemistry, volcanic evolution, and morphology of the island. Especially important are the contributions of Baker and others (1975) and Schultz (1976).

Age determinations, strontium isotope rations, and REE data are being used to model the genesis of the Antarctic Cenozoic volcanic rocks. However, more field work is necessary to clarify the geologic structures of the volcanic areas.

Volcanic activity, which is tied to tectonic processes, has been intermittent through the Antarctic continental history. The late Cenozoic volcanic cycles are responsible for the actual morphologic features of parts of West Antarctica.

The Antarctic continent represents the major polar tectonic element that limits the austral oceanic plates. Its location and plate relationship during Eocene–Miocene time have been discussed by Herron and Tucholke (1976). Their model is important since Antarctic volcanic activity reached its maximum intensity during that interval. The

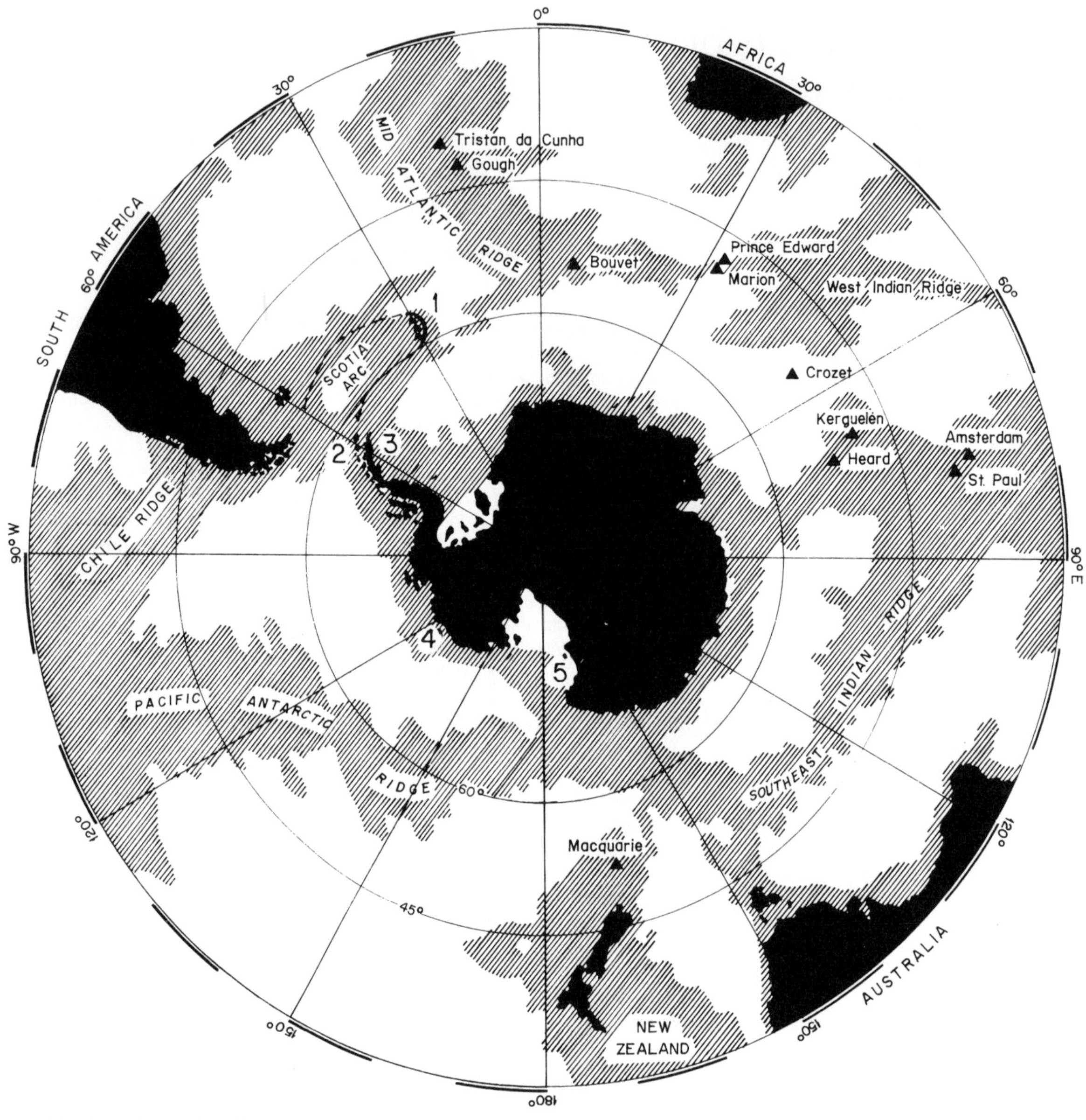

Figure 84.1. Location of the principal Cenozoic volcanic provinces of Antarctica (circled numbers), including subantarctic Cenozoic islands. 1. South Sandwich Islands; 2. South Shetland Islands and Antarctic Peninsula; 3. James Ross Island; 4. Marie Byrd Land and Ellsworth Land; 5. Hallet and McMurdo.

actual plate relationships around the Antarctic continent have been discussed by Hayes and Pitman (1972).

Two main regional units characterize Antarctic Cenozoic volcanism: (1) the subantarctic volcanism and (2) the West Antarctic or circum-Pacific volcanism. The first unit can, in turn, be divided into two groups: (a) the volcanism associated with active seismic ridges such as that of Tristan da Cunha Island, Gough Island, and Bouvet Island; and (b) the volcanism related to aseismic areas such as Crozet, Kerguelen, and Heard Islands. The West Antarctic or circum-Pacific unit also includes the Scotia Arc. Antarctic Cenozoic volcanic evolution is complex, and five geographic provinces have been recognized (Figure 84.1) They differ in tectonic environment and in petrologic composition, which ranges from the tholeiitic series through the calc-alkaline to the alkaline series (Figure 84.2).

In general, two main cycles of volcanic activity separated by intense differential block tectonism are recognized for Cenozoic time. The first eruptive cycle formed extensive plateau-like volcanic deposits, while the second is fundamentally a central-type volcanism represented by stratovolcanoes, calderas, and parasitic cones. At least 13 stratovolcanoes are still active (Figure 84.3). Cenozoic volcanism has not been synchronous throughout West Antarctica. Thus, in the alkaline provinces of Marie Byrd Land, Hallett, and McMurdo, the hyaloclastites and palagonitic breccias are Eocene–Pliocene in age, while at James Ross Island the palagonitic breccias appeared in Pliocene time.

Cenozoic volcanic provinces

The South Sandwich Islands volcanic province

The South Sandwich Islands are late Pliocene to Holocene volcanic islands (Baker, 1978). They are the youngest feature around the Soctia Sea and form an intraoceanic arc between the South Sandwich Trench and the Pliocene–

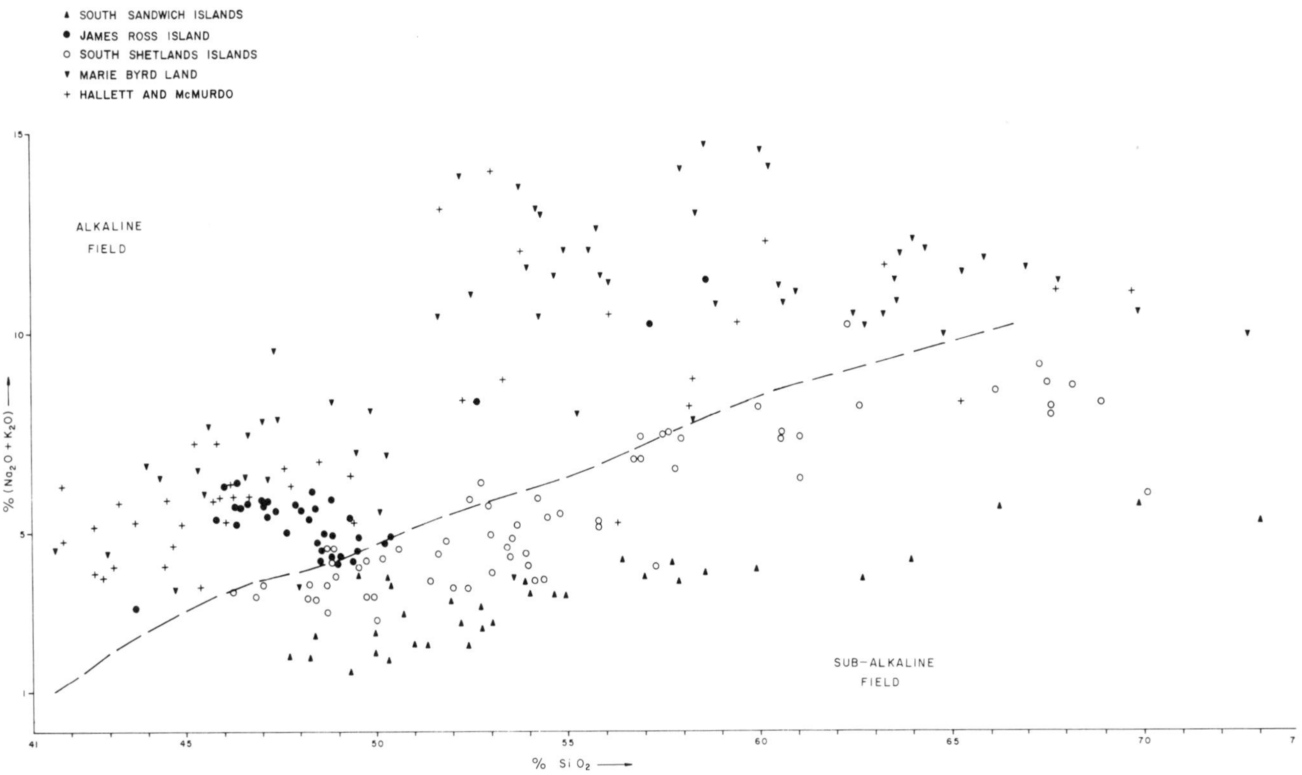

Figure 84.2. Comparative total alkalies versus silica for the South Sandwich Islands, James Ross Island, South Shetland Islands, Marie Byrd Land, and Hallett volcanic series of Antarctica. References for data include the following: Tyrrell (1945), Gass and others (1963), Baker (1972, 1976, 1978), and Tarney and others (1977), for South Sandwich Islands; Nelson (1966), Baker and others (1976), and González-Ferrán and Baker (1973), for James Ross Island; González-Ferrán and others (1970, 1971) and Baker and others (1975), for South Shetland Islands; González-Ferrán (1972) and LeMasurier and Wade (1976) for Marie Byrd Land; and Hamilton (1972), Cole and Ewart (1968), Nathan and Schulte (1967), and Treves (1967), for Hallett and McMurdo.

Holocene westward spreading line (Baker, 1976, 1978; Barker and Griffiths, 1972). At least 8 of these islands display recent fumarolic activity (Baker, 1978). Gass and others (1963) discussed a submarine eruption which occurred in 1962 on the axis of the South Sandwich Arc. Geologic, chemical, and petrologic studies have been carried out by Tyrrell (1945) and Baker (1972, 1976, 1978). According to Baker (1976), the islands are dominated by basalts and basaltic andesites of island-arc tholeiite composition. Most of these lavas are poor in K_2O (less than 0.5%), and the series as a whole is characterized by iron enrichment and intermediate silica composition. This type of low-K tholeiite is apparently characteristic of those island arcs situated in an oceanic environment and is thought to represent a relatively early phase of volcanism, generally preceeding a calc-alkaline phase. Some of the South Sandwich lavas may be regarded as transitional between the island-arc tholeiite and the calc-alkaline types, but true calc-alkaline two-pyroxene andesites occur only on Leskov Island. It is significant that this island is located some 50 km west of the main axis of the arc, suggesting that the magma is perhaps derived from a deeper part of the Benioff zone. The only known rhyolite rock in the group is the pumice that was the product of the submarine eruption north of Zavadovski Island (Gass and others, 1963). The total alkali versus silica variations diagram (Figure 84.2) shows that most of the samples fall in the tholeiitic basalt field of Kuno (1960); just a few of them fall within the field of the high-alumina basalts.

K-Ar age determinations show that the oldest exposed rocks have ages varying from 4 m.y. to 0.7 m.y. (Baker, 1972), suggesting that the volcanism may be contemporaneous with the active spreading ridge located about 280 km west of the South Sandwich Islands. This spreading ridge was identified in 1972 by Baker who assigned to it an age of at least 8 m.y. Tarney and others (1977) described some pillow basalt fragments dredged by the R.R.S. *Shackleton* from various points along the axis of the spreading center and also along the charted magnetic anomaly tracks. These basalts range in composition from quartz-normative to olivine-normative tholeiite, and their $^{87}Sr/^{86}Sr$ ratios range from 0.7028 to 0.7032. These values are higher than those observed in mid-oceanic ridge basalts. Baker (1976) reports a $^{87}Sr/^{86}Sr$ ratio of 0.7038 for a basalt from Bristol Island.

South Shetland Islands volcanic province

The South Shetland Islands are 15 major islands and many small islands which form an island arc separated from the Antarctic Peninsula by the Bransfield Strait. Their basement consists of Paleozoic and Mesozoic igneous and sedimentary rocks. The Cenozoic volcanic rocks lie uncomformably on the Paleozoic–Mesozoic basement and are characterized by the predominance of calc-alkaline olivine basalts and augite-olivine–bearing andesites (Adie, 1972; González-Ferrán and Vergara, 1972; González-Ferrán and Katsui, 1970; Baker, 1976; Shultz, 1976). Two cycles of volcanic activity are dis-

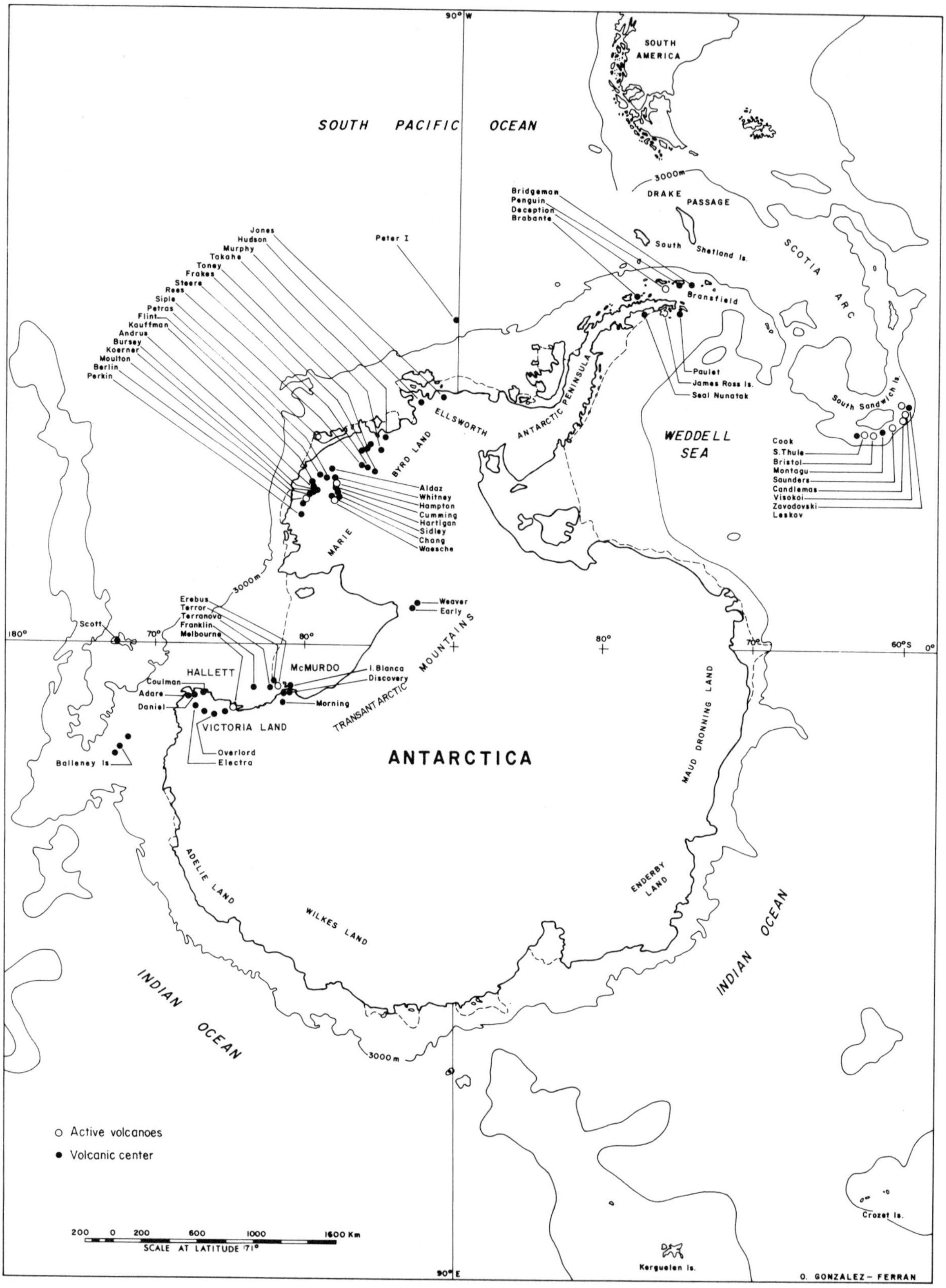

Figure 84.3. Location of the late Cenozoic volcanic centers and active volcanoes in Antarctica.

tinguished in this province: one was developed from Eocene to Miocene time and the other from late Pliocene to Holocene time.

The first cycle is represented by volcanic necks, lava flows, and aggolmerates of andesitic to basaltic composition deposited in association with Eocene–Miocene sedimentary rocks containing fossil plants (Barton, 1965; Orlando, 1964; Covacevich and Lamperein, 1972). This volcanism extended regionally from the South Shetland Islands to the west coast of the Antarctic Peninsula and southwest of the Bransfield Strait into the Palmer Archipelago. K-Ar dating of basalts from Two Hummock and Tower Island have been reported, and ages of 35 m.y. and 54 m.y., respectively, are given by Rex (1972). This dating corroborates the Eocene–Miocene age assigned to this volcanic series.

The geology, petrography, and chemical composition of this province have been studied by Adie (1972), Barton (1965), Hawkes (1961), González-Ferrán and others (1970), and Dalziel and Elliot (1973). Based on geophysical data, Barker and Griffiths (1972), Barker (1976), and Herron and Tucholke (1976) concluded that the cenozoic volcanism in this region was associated with subduction processes which began during mesozoic time and ceased in late Cenozoic time. This could explain the calc-alkaline character of this volcanic rock series.

The second cycle of volcanic activity began during late Pliocene time and is still active, as revealed by the recent eruptions on Deception Island occurring in 1967, 1969, and 1970 (Valenzuela and others, 1968; Baker and others, 1969; Clapperton, 1969; González-Ferrán and others, 1971; Baker and McReath, 1971; Schultz, 1976; Roobal, 1973; Viramonte and others, 1976; Orheim, 1972). This volcanism extends along a fracture located on the southeast edge of the South Shetland Islands, and it is represented by Deception, Penguin, and Bridgeman Islands and probably by some seamounts. According to geophysical data provided by the authors mentioned above, this fracture and subsequent volcanism represent a spreading line produced by the extension of Bransfield Strait (Karig, 1971; Barker and Griffiths, 1972; Barker, 1976). This process is related to the ending of subduction beneath the South Shetland area.

Penguin Island volcanic rocks are characterized by subalkaline olivine basalts with phenocrysts of olivine and clinopyroxene. High-alumina basalt and basaltic andesite lava flows with phenocrysts of plagioclase, clinopyroxene, and olivine predominate on Bridgeman Island (González-Ferrán and Katsui, 1970). Deception Island has a longer history of crystal fractionation from basalt (50% SiO_2) to rhyodacites (70% SiO_2); basalt and andesite basalt may be aphyric, but most of the lavas have plagioclase phenocrysts with additional olivine and clinopyroxene in the more basic lavas, hypersthene in the intermediate lavas, and fayalitic olivine in the salic lavas (Tarney and others, 1977). Chemical analyses plotted on the total alkali-versus-silica diagram (Figure 84.2) are within the field of the calc-alkaline series as was previously reported by González-Ferrán and Vergara (1972). Baker and others (1975) have given the most complete report about the evolution, chemistry, and petrology of Deception Island, and they point out that the volcanic rock suite differs in several essential aspects from the typical calc-alkali series. Schultz (1976) noted the difficulty of assigning these rocks to a magmatic series, especially of showing that they are nearer to the tholeiitic series than to the alkaline, as was previously reported.

Baker (1976) pointed out that the Deception Island lavas constitute a virtually unique suite in which unusually high Na_2O contents are associated with comparatively low K_2O values. The precise affinities of the Deception Island volcanic rocks have long been a matter of discussion. Analyses of initial $^{87}Sr/^{86}Sr$ ratios of these volcanic rocks give values of 0.7030 to 0.7039 (Halpern and González-Ferrán, 1973) and 0.7036 for rocks from the eruption of 1969 (Faure and others, 1972), indicating a probable upper mantle source for the magma.

James Ross Island volcanic province

The volcanic sequence and the chemical character of this province have been described by Nelson (1966), Bibby (1965), and Adie (1972), as well as by Baker and others (1976) and González-Ferrán and Baker (1973), who investigated part of this province in the Prince Gustav Channel region.

The James Ross Island volcanic province is exposed on James Ross Island and on numerous small islands and peninsulas along the Prince Gustav Channel near the northeastern part of the Antarctic Peninsula. This province consists of a sequence (approximately 1,500 m thick) of lavas and palagonite breccias overlying Cretaceous and lower Miocene sedimentary rocks. Adie (1972) inferred a middle to late Miocene age for this volcanic group since it rests on lower Miocene sedimentary rocks and is overlain by the Pliocene Pecten Conglomerate on Cockburn Island. However, K-Ar age determinations ranging from 6 m.y. to 1 m.y. (Rex, 1972) place the entire group within the upper Pliocene; some of the associated volcanic rocks, particularly those of Paulet Island and Seal Nunataks, are still younger (less than 1 m.y.). The ages obtained by Rex suggest a migration of the Pliocene volcanic activity from James Ross Island toward the north, where a line of volcanic activity developed from Seal Nunataks to Paulet Island during Pleistocene (0.4 m.y.) to Holocene time (Baker and others, 1976).

The present scattered outcrops of the James Ross Island volcanic province probably represent a relic of an extensive basaltic plateau that reached the southeast side of the Antarctic Peninsula and the adjacent islands. However, there is no evidence that it extended farther across the Antarctic Peninsula.

Nelson (1966) distinguished five separate volcanic phases of alternating palagonite breccias and olivine basalt flows. Recent investigations (Baker and others, 1973, 1976) have recognized a general sequence from deep-water hyaloclastites and palagonite tuffs to shallow-water pyroclastic rocks accompanied by subaerial components, redistributed shallow-water tuff, and later subaerial lava flows and red scoria. The origin of the hyaloclastite formation is almost certainly submarine rather than subglacial, suggesting that the area must have experienced an uplift of at least 400 m since the end of Pliocene time.

Paulet Island appears to be a younger volcanic structure and, in this respect, is perhaps comparable to the island of Seal Nunataks located farther south. Both islands represent a very young and probably still active volcanic line between the southeastern edge of the Antarctic Peninsula and James Ross Island.

Petrographically and geochemically, the lavas from this province are hawaiites rather than basalts since the normative feldspar is in the andesine compositional range rather than the labradorite range. The southerly occurrences at Seal Nunataks and James Ross Island are essentially olivine tholeiite or slightly nepheline-normative hawaiite. The alkalinity of the lavas increases toward the northeast; thus, the Paulet Island lavas and the lavas at Purvis Point on Dundee Island are richer in alkalis than those at Seal Nunataks. Baker and others (1976) have suggested that the magmatism of this province is related to a thermal event within the upper mantle, which would have also been responsible for the inferred major post-Pliocene uplift in the northern part of the peninsula. The tholeiitic composition of the lavas agrees with a high degree of mantle partial fusion combined with shallow fractionation and controlled by a relatively steep geothermal gradient. The increasingly alkalic composition found to the northeast presumably reflects a diminishing thermal gradient away from the main focus of activity.

Marie Byrd Land volcanic province

The geology of Marie Byrd Land and neighboring Ellsworth Land has been investigated and described by Wade and Wilbanks (1972), Wilbanks (1972), Rutford and others (1972), LeMasurier (1972), LeMasurier and others (1976), LeMasurier and Wade (1976), Laudon (1972), Halpern and González-Ferrán (1973), González-Ferrán and González-Bonorino (1972), González-Ferrán and Vergara (1972), and Andrews and LeMasurier (1973). This region is one of the major Cenozoic volcanic provinces in Antarctica. Here, two volcanic units are distinguished; (1) a basal unit composed of a large, thick succession of palagonite breccias and lava flows of alkali basalt composition that probably represent the products of fissure eruptions, and (2) stratovolcanoes superimposed on the basal unit (González-Ferrán and González-Bonorino, 1972; González-Ferrán and Vergara, 1972; LeMasurier, 1972, 1976).

Outcrops with subhorizontal stratification observed in many places in Marie Byrd Land probably represent relics of an extensive basaltic plateau. This basal sequence rests on eroded Mesozoic rocks. Some central vent volcanoes, like Mount Murphy and Mount Takahe are related to this sequence. The occurrence of palagonite breccias has been attributed to the presence of a thick ice sheet throughout most of Cenozoic time (LeMasurier, 1976).

The stratovolcanoes occur in linear chains oriented north-south or east-west and are controlled by a fault system. The lavas of these volcanoes are predominantly nepheline-normative alkaline basalts and peralkaline trachytes, phonolites, and rhyolites. They belong to the highly sodic basalt mugearite-benmorite-trachyte series (LeMasurier and Wade, 1976). In the total alkali versus silica diagram they fall in the alkaline field of Kuno (1960) and Irvine and Baragar (1971). Their Peacock's index is 46.5 (González-Ferrán and Vergara, 1972) (Figure 84.2).

Cenozoic lavas exhibiting initial $^{87}Sr/^{86}Sr$ ratios varying from 0.7025 to 0.7036 are characteristic of this province (LeMasurier and Wade, 1976; Halpern and González-Ferrán, 1973). These data suggest that the Cenozoic lavas are probably derived from the upper mantle and that they have undergone a negligile degree of crustal contamination.

More than 30 radiometric ages from Marie Byrd Land have been published (Halpern, 1970; Halpern and González-Ferrán, 1973; LeMasurier, 1976) These data suggest that Cenozoic volcanism began during Eocene time and that the stratovolcano activity started about 13 m.y. ago and has continued practically without interruption up to the present (LeMasurier and Wade, 1976).

Hallett and McMurdo volcanic province

The Hallett and McMurdo volcanic province of late Miocene to Holocene age lies along the Ross Sea coast of northeastern Victoria Land and the neighboring islands from Balleny to McMurdo; it also includes the most southern Cenozoic volcanic locality represented by Mount Early near Mount Weaver in the central Transantarctic Mountains. Hamilton (1972) divided this province into three major subprovinces: the McMurdo subprovince (Mount Morning to Ross Island), which is characterized by high subaerial volcanoes; the Hallett subprovince (Coulman Island to Adare Peninsula), which consists of long coastal domes of subglacial breccias surmounted by low subaerial volcanoes; and the Balleny Islands, which also form long domes but rise from the deep ocean. Treves (1967) described the Cenozoic volcanic rocks from the Mount Weaver area, which is located at about lat. 87° S and long. 152° W. This region was previously studied by Doumani and Minshew (1965) who suggested that the volcanic rocks may be late Tertiary or Quaternary in age. Hamilton (1972) described subglacial palagonitic breccias and pillow-breccia complexes in the Hallett province, which had erupted during late Miocene, Pliocene, and Quaternary time under the Antarctic ice sheet. This was followed by subaerial lavas and scorias formed after the volcanoes had penetrated the ice. The volcanic rocks of Hallett constitute a very sodic assemblage containing approximately 70% alkaline olivine basalt and subordinate olivine trachybasalt, 15% trachyte, and no more than a few percent each of basanite, andesite, basalt, latite, phonolite, and quartz trachyte. The series is bimodal: only a small amount of rock is intermediate between basalt and trachyte. This province is characterized by sodic basalts with subordinate felsic rocks that range from extremely sodic phonolite to relatively potassic quartz trachyte. The highly alkaline character is shown in Figure 84.2. The volcanic rocks from the McMurdo region and the Mount Weaver area are chemically and petrographically similar to those of the Hallett region.

The volcanic rocks are late Miocene, Pliocene, Pleistocene, and Holocene in age. Mounts Erebus and Melbourne are still active (Kyle and others, 1976; Giggenbach and others, 1973). The K-Ar ages for pillows in the palagonite breccias range from 7.3 m.y. to 2.6 m.y. (Hamilton, 1972). Forbes and others (1974) gave a K-Ar age of 1.18 m.y. for the trachyte from Observation Hill on Ross Island.

An offshore volcanic island

The Peter I Island lies in the southeastern Pacific Ocean (lat. 68° 50′ S; long. 90° 40′ W) facing the Eights Coast of West Antarctica. Bastien and Craddock (1976) describe this island as a volcanic island composed of interbedded basalt flows and more siliceous lavas. Basalts, andesites, and trachyandesites have been identified, with the oceanic-type olivine basalts predominating. All these rocks may represent a single volcanic series somewhat alkaline in character, but lying mainly between the tholeiitic basalt series and the alkali olivine basalt series. The authors have found a K-Ar age of 12 m.y. for one of the lavas of the island.

Conclusions

Cenozoic volcanism along the Antarctic Peninsula, Ellsworth Land, and Marie Byrd Land suggests a large scale continuity in circum-Pacific processes throughout Cenozoic time. However, Antarctic volcanism differs in many aspects from volcanism in other circum-Pacific regions (Baker, 1972; González-Ferrán, 1972; LeMasurier and Wade, 1976). Geophysical investigations (Herron and Tucholke, 1976) show that the tectonic processes are not synchronous along the Antarctic volcanic belt.

The Cenozoic suite of alkali-rich volcanic rocks in the Marie Byrd Land, Hallett, and McMurdo Provinces suggest that an extensional tectonic environment has affected continental West Antarctica for a large part of Cenozoic time. The oldest alkaline rocks are Eocene, but marine geophysical studies suggest that extension associated with spreading in the Pacific Antarctic Ridge began about 80 m.y. ago. Block faulting seems to be the major mechanism of deformation (LeMasurier and Wade, 1976).

A tholeiitic series is present in the South Sandwich Island, but it is probably being replaced by a more calc-alkaline composition as an expression of the subduction of the oceanic crust at the South Sandwich Trench, a process which probably just started during Pliocene time (Baker, 1976).

Cenozoic calc-alkaline volcanism is restricted to the South Shetland Islands and islands close to the Antarctic Peninsula. The volcanic activity has migrated from the South Shetlands southeast into the Bransfield Strait during Pliocene to Holocene time and has developed the active volcanic line represented by Deception, Penguin, and Bridgeman Islands.

Acknowledgments. I am grateful to W. LeMasurier, Natural and Physical Sciences, University of Colorado, and to L. Lopez, Department of Geology of the University of Chile, who read the manuscript and made some suggestions. This paper was supported by grants from the Inter-Union Commission on Geodynamics, Instituto Antártico Chileno and Servicio de Desarrollo Científico de la Universidad de Chile.

References cited

Adie, R. J. 1972. Evolution of volcanism in the Antarctic Peninsula. (*In* Adie, R. J., *ed. Antarctic Geology and Geophysics.* Oslo, Universitetsforlaget, 137-41.)

Andrews, J. T., and LeMasurier, W. E. 1973. Rates of Quaternary glacial erosion and Corrie Formation, Marie Byrd Land, Antarctica. *Geology, 1,* No. 2, 75-80.

Baker, P. E. 1972. Recent volcanism and magmatic variation in the Scotia arc. (*In* Adie, R. J., *ed. Antarctic Geology and Geophysics.* Oslo, Universitetsforlaget, 57-60.)

Baker, P. E. 1976. Volcanism and plate tectonics in the Antarctic Peninsula and Scotia arc. (*In* Gonzáles-Ferrán, O., *ed. Andean and Antarctic Volcanology Problems.* Rome, International Association of Volcanology and Chemistry of the Earth's Interior, 347-56.)

Baker, P. E. 1978. The South Sandwich Islands: II. Petrology and Geochemistry. *Scient. Rep. Br. antarc. Surv.,* No. 93, 1-34.

Baker, P. E., and McReath, I. 1971. 1970 volcanic eruption at Deception Island. *Nature Phys. Sci., 231,* 5-9.

Baker, P. E., Davies, T. G., and Roobal, M. J. 1969. Volcanic activity at Deception Island in 1967 and 1969. *Nature, Lond., 224,* 553-60.

Baker, P. E., Gonzáles-Ferrán, O., and Vergara, M. 1973. Paulet Island and the James Ross Island Volcanic Group. *Bull. Br. antarc. Surv.,* No. 32, 89-95.

Baker, P. E., Buckley, F., and Rex, D. C. 1977. Cenozoic volcanism in the Antarctic. *Phil. Trans. R. Soc. Lond.,* Ser. B, *279,* 131-42.

Baker, P. E., González-Ferrán, O., and Vergara, M. 1976. Geology and geochemistry of Paulet Island and the James Ross Island Volcanic Group. (*In* González-Ferrán, O., *ed. Andean and Antarctic Volcanology Problems.* Rome, International Association of Volcanology and Chemistry of the Earth's Interior, 39-47.)

Baker, P. E., McReath, I., Harvey, M. R., Roobol, M. J., and Davies, T. G. 1975. Geology of the South Shetland Islands: V. Volcanic evolution of Deception Island. *Scient. Rep. Br. antarc. Surv.,* No. 78, 1-81.

Barker, P. F. 1976. The tectonic framework of Cenozoic volcanism in the Scotia Sea Region: a Review. (*In* Gonzalez-Ferrán, O., *ed. Andean and Antarctic Volcanology Problems.* Rome, International Association of Volcanology and Chemistry of the Earth's Interior, 330-46.)

Barker, P. F., and Griffiths, D. H. 1972. The evolution of the Scotia Ridge and Scotia Sea. *Phil. Trans. R. Soc. Lond.,* Ser. A, *271,* 151-83.

Barton, C. M. 1965. The geology of the South Shetland Islands: III. The stratigraphy of King George Island. *Scient. Rep. Br. antarct. Surv.,* No. 44, 1-33.

Bastien, T. W., and Craddock, C. 1976. Igneous rocks of Peter I Island. *Antarct. Jnl U.S., 11,* 267-69.

Bibby, J. S. 1965. Some observations on sea level changes in the James Ross Island Group. *Bull. Br. antarc. Surv.,* No. 6, 67-75.

Clapperton, C. M. 1969. The volcanic eruption at Deception Island, December 1967. *Bull. Br. antarc. Surv.,* No. 22, 83-90.

Cole, J. W., and Ewart, A. 1968. Contributions to the volcanic geology of the Black Island, Brown Peninsula, and Cape Bird areas, McMurdo Sound, Antarctica. *N. Z. Jl Geol. Geophys., 2,* No. 4, 793-828.

Covacevich, V., and Lamperein, C. 1972. Ichnites from Fildes Peninsula, King George Island, South Shetland Islands. (*In* Adie, R. J., *ed. Antarctic Geology and Geophysics.* Oslo, Universitetsforlaget, 71-74.)

Craddock, C. 1972. *Geologic Map of Antarctica, 1:5,000,000.* New York, American Geographical Society.

Dalziel, I. W. D., and Elliot, D. H. 1973. The Scotia arc and the Antarctic margin. (*In* Nairn, A. E. M., and Stehli, F. G., *eds. The Ocean Basins and Margins. 1. The South Atlantic.* New York, Plenum Press, 171-246.)

Dort, W., Jr. 1972. Late Cenozoic volcanism in Antarctica. (*In* Adie, R. J., *ed. Antarctic Geology and Geophysics.* Oslo, Universitetsforlaget, 645-52.)

Doumani, G. A., and Minshew, V. H. 1965. General geology of the Mount Weaver area, Queen Maud Mountains, Antarctica. (*In* Hadley, J. B., *ed. Geology and Paleontology of the Antarctic.* Washington, American Geophysical Union, 127-39.) [Antarctic Research Series, Vol. 6.]

Faure, G., Hill, R. L., Jones, L. M., and Elliot, D. H. 1972. Isotope composition of strontium and silica content of Mesozoic basalt and dolerite from Antarctica. (*In* Adie, R. J., *ed. Antarctic Geology and Geophysics.* Oslo, Universitetsforlaget, 617-22.)

Forbes, R. B., Turner, D. L., and Carden, J. R. 1974. Age of trachyte from Ross Island, Antarctica. *Geology, 2,* 297-98.

Gass, I. G., Harris, P. G., and Holdgate, M. W. 1963. Pumice eruption in the area of the South Sandwich Islands. *Geol. Mag., 100,* No. 4, 321-30.

Giggenbach, W. F., Kyle, P. R., and Lyon, G. L. 1973. Present volcanic activity on Mount Erebus, Ross Island, Antarctica. *Geology, 1,* 135-36.

Girod, M., and Nougier, J. 1972. Volcanism of the Sub-Antarctic Islands. (*In* Adie, R. J., *ed. Antarctic Geology and Geophysics.* Oslo, Universitetsforlaget, 777-91.)

Goldich, S. S., Treves, S. B., Suhr, N. H., and Stuckless, J. S. 1975. Geochemistry of the Cenozoic volcanic rocks of Ross Island and vicinity, Antarctica. *J. Geol., 83,* 415-35.

González-Ferrán, O. 1972. Distribution, migration and tectonic control of Upper Cenozoic volcanism in West Antarctica and South America. (*In* Adie, R. J., *ed. Antarctic Geology and Geophysics.* Oslo, Universitetsforlaget, 173–79.)

González-Ferrán, O., and Baker, P. E. 1973. Isla Paulet y el volcanismo reciente en las Islas del Weddell Noroccidental, Antártica. Terra Australis. *Revta geogr. Chile,* No. 22–23, 25–43.

González-Ferrán, O., and González-Bonorino, F. 1972. The volcanic ranges of Marie Byrd Land between long. 100° and 140° W. (*In* Adie, R. J., *ed. Antarctic Geology and Geophysics.* Oslo, Universitetsforlaget, 261–75.)

González-Ferrán, O., and Katsui, Y. 1970. Estudio integral del volcanismo cenozoico superior de las Islas Shetland del Sur, Antártica. *Ser. cient. Inst. antart. chil., 1,* No. 2, 123–74.

González-Ferrán, O., and Vergara, M. 1972. Post-Miocene volcanic petrographic provinces of West Antarctica and their relation to the southern Andes of South America. (*In* Adie, R. J., *ed. Antarctic Geology and Geophysics.* Oslo, Universitetsforlaget, 187–94.)

González-Ferrán, O., Katsui, Y., and Tavera, J. 1970. Contribución al conocimiento geológico de Península Byers, Isla Livingston, Islas Shetland del Sur, Antárctica. *Ser. cient. Inst. antart. chil., 1,* No. 1, 41–54.

González-Ferrán, O., Munizaga, F., and Moreno, H. 1971. 1970 eruption at Deception Island: distribution and chemical features of ejected material. *Antarct. Jnl U.S., 6,* No. 4, 97–98.

Halpern, M. 1970. Rubidium-strontium dates and Sr^{87}/Sr^{86} initial ratios of rocks from Antarctic and South America: a progress report. *Antarct. Jnl U.S., 5,* 159–61.

Halpern, M., and González-Ferrán, O. 1973. Edad del Volcán Andrus y relación inicial Sr^{87}/Sr^{86} de rocas volcánicas de Tierra María Byrd e Islas Shetland del Sur, Antártica. *Revta bras. Geoc.,* No. 3, 141–48.

Hamilton, W. 1972. The Hallett volcanic province, Antarctica, *Prof. Pap. U.S. geol. Surv.,* 456-C, C1–62.

Hawkes, D. D. 1961. The geology of the south Shetland Islands. II. Geology and petrology of Deception Island. *Scient. Rep. Falkld Isl. Depend. Surv.,* 27, 1–43.

Hayes, D. E., and Pitman, W. C., III. 1972. Review of marine geophysical observations in the Southern Ocean. (*In* Adie, R. J., *ed. Antarctic Geology and Geophysics.* Oslo, Universitetsforlaget, 725–38.)

Herron, E. M., and Tucholke, B. E. 1976. Sea-floor magnetic patterns and basement structure in the southeastern Pacific. (*In* Hollister, C. D., Craddock, C., and others. *Initial Reports of the Deep Sea Drilling Project,* Vol. 35. Washington [U.S. Government Printing Office], 263–78.)

Irvine, T. N., and Baragar, W. R. A. 1971. A guide to the chemical classification of the common volcanic rocks. *Can. Jnl Earth. Scis, 8,* 523–48.

Jones, L. M., and Walker, R. L. 1972. Geochemistry of the McMurdo Volcanics, Victoria Land. I. Strontium isotope composition. *Antarct. Jnl U.S.,* 7, 142–44.

Karig, D. E. 1971. Origin and development of marginal basins in the western Pacific. *J. geophys. Res., 76,* 2542–61.

Kuno, H. 1960. High-alumina basalt. *J. Petrology, 1,* 121–45.

Kyle, P. R., Giggenbach, W. F., and Keys, H. J. R. 1976. Volcanic activity of Mount Erebus, Ross Island. *Antarct. Jnl U.S., 11,* No. 4, 270–71.

Laudon, T. S. 1972. Stratigraphy of eastern Ellsworth Land. (*In* Adie, R. J., *ed. Antarctic Geology and Geophysics.* Oslo, Universitetsforlaget, 215–23.)

LeMasurier, W. E. 1972. Volcanic record of Cenozoic glacial history of Marie Byrd Land. (*In* Adie, R. J., *ed. Antarctic Geology and Geophysics.* Oslo, Universitetsforlaget, 251–59.)

LeMasurier, W. E. 1976. Interglacial volcanoes in Marie Byrd Land. *Antarct. Jnl U.S., 11,* 269–70.

LeMasurier, W. E., and Wade, F. A. 1976. Volcanic history in Marie Byrd Land: implications with regard to Southern Hemisphere tectonic reconstructions. (*In* Gonzáles-Ferrán, O., *ed. Andean and Antarctic Volcanology Problems.* Rome, International Association of Volcanology and Chemistry of the Earth's Interior, 398–424.)

LeMasurier, W. E., Kyle, P. R., and Raukin, P. C. 1976. Rare-earth element geochemistry of volcanic rocks from the Executive Committee Range, Marie Byrd Land. *Antarct. Jnl U.S., 11,* 263–67.

Nathan, S., and Schulte, F. J. 1967. Recent thermal and volcanic activity on Mount Melbourne, northern Victoria Land, Antarctica. *N.Z. Jnl Geol. Geophys., 10,* No. 2, 422–30.

Nelson, P. H. H. 1966. The James Ross Island volcanic group of northeast Graham Land. *Scient. Rep. Br. antarct. Surv.,* No. 54, 1–62.

Orheim, O. 1972. Volcanic activity on Deception Island, South Shetland Islands. (*In* Adie, R. J., *ed. Antarctic Geology and Geophysics.* Oslo, Universitetsforlaget, 117–20.)

Orlando, H. A. 1964. The fossil flora of the surroundings of Ardley Peninsula, King George Island, South Shetland Islands. (*In* Adie, R. J., *ed. Antarctic Geology.* Amsterdam, North-Holland Publishing Company, 629–36.)

Rex, D. C. 1972. K-Ar age determinations on volcanic and associated rocks from the Antarctic Peninsula and Dronning Maud Land. (*In* Adie, R. J., *ed. Antarctic Geology and Geophysics.* Oslo, Universitetsforlaget, 133–36.)

Roobal, M. J. 1973. Historic volcanic activity at Deception Island. *Bull. Br. antarc. Surv.,* No. 32, 23–30.

Rutford, R. H., Craddock, C., White, C. M., and Armstrong, R. L. 1972. Tertiary glaciation in the Jones Mountains. (*In* Adie, R. J., *ed. Antarctic Geology and Geophysics.* Oslo, Universitetsforlaget, 239–43.)

Schultz, C. H. 1976. Petrology and petrochemistry of Deception Island, Antarctica. (*In* González-Ferrán, O., *ed. Andean and Antarctic Volcanology Problems.* Rome, International Association of Volcanology and Chemistry of the Earth's Interior, 498–517.)

Sun, S. S., and Hanson, G. N. 1975. Origin of Ross Island basanitoids and limitations upon the heterogeneity of mantle sources for alkali basalts and nephelinites. *Contr. Miner. Petrol., 52,* 77–106.

Sun, S. S., and Hanson, G. N. 1976. Rare earth element evidence for differentiation of McMurdo volcanics, Ross Island, Antarctica. *Contr. Miner. Petrol., 54,* 139–55.

Tarney, J., Saunders, A. D., and Weaver, S. D. 1977. Geochemistry of volcanic rocks from the island arcs and marginal basins of the Scotia arc region. (*In* Talwani, M., and Pitman, W. C., III, *eds. Maurice Ewing Series, 1: Island Arcs, Deep Sea Trenches, and Back-Arc Basins.* Washington, American Geophysical Union, 367–77.)

Treves, S. B. 1967. Volcanic rocks from the Ross Island, Marguerite Bay and Mt. Weaver areas, Antarctica. (*In* Nagata, T., *ed.* Proceedings the Symposium on Pacific-Antarctic Sciences. *JARE Sci. Rep., Special Issue No. 1,* 126–49.)

Tyrrell, G. W. 1945. Report on rocks from West Antarctica and the Scotia arc. *'Discovery' Rep., 23,* 37–102.

Valenzuela, E., Chavez, L., and Munizaga, F. 1968. Preliminary report on the eruption of Deception Island that occurred in December 1967. *Boln. Inst. antárt. chil., 3,* 5–16.

Viramonte, J. G., Sureda, R. J., Bossi, G. E., Fourcade, N. H., and Omarini, R. H. 1976. Geochemical and minerological study of the high temperature fumaroles from Deception Island, South Shetland, Antarctica. (*In* Gonzáles-Ferrán, O., *ed. Andean and Antarctic Volcanology Problems.* Rome, International Association of Volcanology and Chemistry of the Earth's Interior, 543–61.)

Wade, A. F., and Wilbanks, J. R. 1972. Geology of Marie Byrd and Ellsworth Lands. (*In* Adie, R. J., *ed. Antarctic Geology and Geophysics.* Oslo, Universitetsforlaget, 207–14.)

Wilbanks, J. R. 1972. Geology of the Fosdick Mountains, Marie Byrd Land. (*In* Adie, R. J., *ed. Antarctic Geology and Geophysics.* Oslo, Universitetsforlaget, 277–84.)

85

Granitoid Complexes in Antarctica

G. E. GRIKUROV, E. N. KAMENEV, and G. I. KAMENEVA
Department of Antarctic Geology, SEVMORGEO, Moyka 120, Leningrad 190121, U.S.S.R.

Abstract. Granitoid rocks varying in composition and origin are widely exposed in Antarctica and comprise more than half of the bedrock outcrops in mountain areas. Archean and lower Proterozoic granitoids as protocrustal, ultrametamorphic, and intrusive rocks make up the bulk of the crystalline basement of the East Antarctic craton. Granitic terranes of an early Precambrian age are also believed to underlie the Ross and West Antarctic fold systems. Upper Proterozoic to lower Paleozoic granitoids were generated mainly in connection with development of the Ross geosyncline and related tectono-thermal activizations within the East Antarctic craton. Middle-upper Paleozoic, Mesozoic, and Cenozoic granitoids are characteristic of the Pacific margin of the continent. Maximums of granitoid generation and resulting complexes of intrusive rocks can be recognized in each major structural province on the basis of geochronologic evidence.

The principal trends of evolution of granitoid complexes in Antarctica are characterized by an increase with time of intrusive-magmatic processes, structural control of intrusions, geochemical differentiation, and alkalinity of final products of individual intrusive phases. Other factors such as the duration of emplacement, the total volume of granitoid masses, the silica content, and the activity of volatiles seem to decrease from older to younger complexes. These trends are consistent with progressive thickening of the earth's crust in the course of geologic history.

Granitoid rocks varying in composition and origin comprise more than half of the bedrock outcrops in mountain areas of Antarctica. As protocrustal and ultrametamorphic formations, they form a substantial part of the basement of the East Antarctic craton and the Ross fold system. In the latter, as well as in the West Antarctic fold system, they are also the most widespread plutonic intrusions.

This paper attempts to determine the evolution of Antarctic granitoids within the entire continent and in each separate structural province. The lack of review papers on Antarctic granitoids forced the authors to use a great number of publications elucidating granitoid features and geochronology in separate areas or even in nunatak groups. These publications exceed 500 in number. In this report, only those sources which contained descriptions of granitoids permitting their assignment to a particular group, or which summarized geochronologic data, are cited.

Archean and lower Proterozoic granitoids

Ultrametamorphic rocks

The cores of shields and massifs of the East Antarctic craton consist of lower Archean protocrustal and ultrametamorphic granitoids of the two-pyroxene facies. The region of their most typical development is Enderby Land, but they also occur extensively along the entire coast of East Antarctica from central Queen Maud Land into Adelie Land (Ravich, 1974; Ravich and Kamenev, 1972; Ravich and Grikurov, 1976). Granitoids constitute 25–80% of the volume of cratonic outcrops and 90% of the volume of the oldest Raggatt Series* (Kamenev, 1975*a*). Lower Archean granitoids should be considered as products of reworking of the primary sialic crust due to juvenile solutions and fusions. Enderbites are closest to the primary sialic crust in composition; their isotopic dates (4100 ± 100 m.y.) on specimens from the Fyfe Hills, from Enderby Land, and from a geologic position at the beginning of a series of metasomatic granitoids within the crystalline basement, show them to be among the oldest rocks of the earth's crust (Sobotovich and others, 1974). Rocks such as diorite, quartz diorite, plagiogranite, and tonalite occur among the enderbites, locally even in the same outcrops. Charnockites and granite gneisses are highly variable in composition, as well, and include granodiorites, granites, quartz syenites, quartz monzonites, and alkali-feldspar granites. All these rocks grade into one another both in composition and in their field interrelationships; this suggests that granitoid masses have undergone little differentiation. The rock bodies are folded layers, sheets, lenses, or amorphous masses. Thicknesses of homogeneous granitoid units reach several kilometers. In the course of evolution, the alkali and silica contents of these rocks increase, but the activity of alkalis remains generally low, having an effect which is reflected in the wide distribution of garnetiferous rock varieties.

In marginal parts of the craton's shields and massifs and in the basements of the Ross and West Antarctic fold systems, migmatites and granite gneisses of the high-grade amphibolite facies are developed (Ravich and others, 1965; Ravich and Soloviev, 1966; Grikurov, 1973; Ravich and Kamenev, 1972; Lopatin and Polyakov, 1976). These rocks formed from lengthy transformation (metasomatism, selective melting, rheomorphism, and mobilization) of the older two-pyroxene facies (crustal) granitoids and, in a few places, of the earliest sedimentary-volcanic (supracrustal) sequences. Maximums in histogram plots of age determinations (2900 m.y., 2500 m.y., 2100 m.y., and 1700 m.y.) imply, beyond any doubt, the existence of

* Following Russian usage, *series* here means a lithostratigraphic unit formed during a period or longer.

cyclic peaks in granite generation. However, no particular type of granitoids can be associated with these maximums. Numerous isotopic dates (1450 m.y., 1050 m.y., 650 m.y., and 470 m.y.) show multiple "rejuvenation" ages of the amphibolite facies granitoids, synchronous with stages of intrusive magmatism in mobile belts of Antarctica. In general, the regional amphibolite facies metamorphism is known to terminate in early Proterozoic time. Thus, the generation of migmatites and granite gneisses of the amphibolite facies must have occurred during Archean or early Proterozoic time. The amphibolite facies granitoids differ from those of the two-pyroxene facies by having a lower temperature of formation and by having a higher silica content in the final products (granites and plagiogranites prevail). Granitoid neosome of migmatites, and stratiform, fusiform, and more complicated rootless bodies of granite gneisses, form no less than 30% of all exposed cratonic areas, while in the regions composed exclusively of high-temperature amphibolite facies rocks, granitoids constitute no less than two-thirds of the exposed terranes. In the craton these uppermost lower Precambrian granitoids occur in Queen Maud Land (Insel Series), on the Prince Olav Coast (Polkanov Series), in the Hansen Mountains, in the Prince Charles Mountains (Lambert Series), and on the Vincennes Bay coast. In the Ross fold system basement, migmatites and granite gneisses form a small part of the Nimrod Series on the Shackleton Coast and the Wilson Series in the Wilson Hills and along the upper reaches of Rennick Glacier. Similar granitoids make up a part of the basement of the West Antarctic mobile belt in Marie Byrd Land (Fosdick Series) and in an axial part of the Antarctic Peninsula (the "basement complex"), but their Precambrian age has not been proven conclusively.

Intrusive rocks

Granitoids with some features of intrusive rocks (i.e., porphyritic and relatively isotropic in composition) seem to have existed in the crust of Antarctica from late Archean time onward. The oldest and least altered of them are represented by porphyroblastic charnockites localized in Archean metamorphic rocks of the two-pyroxene facies in Enderby Land and MacRobertson Land. Porphyroblastic charnockites that occur among polymetamorphic complexes in Queen Maud Land are only partly preserved from replacement by later "charnockitoids" (Ravich and Kamenev, 1972).

The largest massifs of porphyroblastic charnockites in the Napier and Framnes Mountains reach 1000 and 1200 km^2 in size, respectively. The massifs are of irregular oval or round configuration on the present erosion surface; some stratiform bodies occur as well. The most homogeneous porphyroblastic charnockites are close to quartz syenites in composition, but they can vary to quartz monzonites and syenites. Porphyroblastic textures in charnockites originated by alkaline metasomatism of different metamorphic rocks of the two-pyroxene facies, chiefly of protocrustal and ultrametamorphic charnockites and enderbites, rather than by crystallization from magma. However, occasional sharp contacts of these bodies with enclosing rocks of the two-pyroxene facies suggest that anatexis played some part in their formation. The Archean to lower Proterozoic age of porphyroblastic charnockites is so far based only on geologic evidence such as their development within the lower Archean sequences of the two-pyroxene facies, preservation in the polymetamorphic upper Archean complexes, and replacement by upper Proterozoic to lower Paleozoic granitoids.

Upper Archean gneissic granites occur in the Sør Rondane Mountains, in the Shackleton Range, in the southern Prince Charles Mountains, in the Grove Nunataks (Princess Elizabeth Land), and in erosional inliers of the ancient cover in the Ritscher Upland (Clarkson, 1972; Ravich and Kamenev, 1972; Soloviev and Halpern, 1975; Fedorov and Tarutin, 1977). They are surrounded by schists, gneisses, and migmatites of the high-temperature amphibolite facies in cratonic terranes, or they form basement projections (gneissic granite domes) among less metamorphosed lower to middle Proterozoic sequences of intracratonic fold systems. Small gneissic granite domes also occur in areas of Proterozoic greenschist sequences of the Ross and, possibly, Andean fold systems (Kamenev, 1975*b*). Areas of gneissic granite outcrops vary from several to some hundreds of square kilometers. Central parts of the massifs have the structure of igneous rocks and a composition varying from granites to quartz diorites, while marginal parts are gneissic and bear traces of metamorphism similar to that of the enclosing metamorphic rocks. Xenoliths of surrounding gneisses and schists are found locally in marginal parts of large massifs; intrusive contacts, obscured by metamorphic processes, are noted in a few places. Gneissic granite massifs are intersected by bodies of coarse biotite granites, granitic pegmatite, and aplite dikes. Some highly deformed gneissic granites, transformed into blastomylonites, also occur. The isotopic age of a granodiorite dike that cuts gneissic granites in the Shackleton Range is about 1500 m.y. In western Queen Maud Land and in the Prince Charles Mountains, Rb-Sr dates have been obtained directly on gneissic granites from several massifs, and they range from 2500 to 3200 m.y.

Upper Proterozoic to lower Paleozoic granitoids

Charnockitoids

The rocks termed "charnockitoids" occur widely in East Antarctica in numerous massifs of variable size, usually not greater than a few hundred square kilometers. Their structural setting is not always clear; generally, they are believed to represent orogenic intrusions related to tectono-thermal activizations of the craton (Klimov and others, 1964; Ravich and others, 1965; Ravich and Soloviev, 1966; Soloviev, 1972; Ravich and Kamenev, 1972).

Charnockitoids intrude the Archean metamorphic rocks of the two-pyroxene and amphibolite facies, as well as massifs of anorthosites and gabbro-peridotites. The largest intrusions are believed to be tabular or lens-like, and small intrusions are stocks. Wall rocks of the intrusions are intersected by apophyses. In some cases, zones of skarns and hornfels have been observed at contacts with enclosing carbonate rocks; elsewhere, mylonites and cataclasites occur at contacts with gneisses. Fayalite–two-pyroxene quartz monzonites, locally transitional to quartz monzodiorites and monzogabbros, occur in marginal parts of the massifs, while their bulk is composed of porphyroblastic eulite-clinopyroxene and biotite-hornblende quartz syenites, partially transitional to biotite-hornblende granites and granodiorites in central parts of the intrusions. The most common geochemical features of charnockitoids are: (1) relatively high concentration of alkalis; (2) high total ferruginosity of rocks and mafic minerals; (3) a low oxidation coefficient resulting in the predominance of ferrous iron over ferric; and (4) an affinity with charnockites by other petrochemical features.

The geologic and geochemical features of charnockitoids allow their comparison with the rapakivi rocks of the Baltic shield, Ukrainian Massif, and East Siberia (Sudovikov, 1967; Velikoslavinsky, 1973; Rudnik, 1975). Charnockitoids have been formed as a result of crystallization of granitoid magma, rheomorphically generated in the deep zone of the crust, and intruded and protruded into overlying horizons through deep fault zones. The analogy with the rapakivi rocks suggests that the generation of charnockitoids commenced near the early–late Proterozoic time boundary, about 1700 m.y. ago. However, complete homogenization of radioactive isotopes in these rocks took place only after Ross activization of the craton, which is reflected in the wide range of abundant isotopic dates from 470 m.y. to 1050 m.y.

Syenites and granites

Small (50 to 150 km^2) stock-like and lens-like massifs of augite and fayalite-augite syenites, hornblende quartz syenites, and biotite, biotite-hornblende, and two-mica granites are typical of zones of Ross activization of the craton (Ravich and Soloviev, 1966; Soloviev, 1972; Lopatin and others, 1977). The vein facies is rather abundant and is represented by aplites, leucogranites, plagiogranites, quartz syenites, syenite-porphyres, and albitites. Rocks of this group occur in western and central Queen Maud Land, the Sør Rondane and Yamato Mountains, the Prince Charles Mountains, and the more eastern parts of the East Antarctic coast; single massifs are known in Enderby Land. Vast pegmatitic fields, widespread in all mountain regions of the craton, probably also belong to this group. Intrusions and their vein series have an active contact effect on enclosing rocks through their supply of volatiles. Enclosing rocks consist mainly of high-grade amphibolite facies schists of the craton basement and lower grade sequences of the intracratonic troughs. From ratios of alkaline, alkaline-earth elements, and aluminium, the granitoids under discussion are rocks of a normal series. Fayalite syenites grade into alkaline types with distinctive sodic enrichment. Their index of iron oxidation is about 1. A majority of massifs have formed by means of successive intrusions of palingenetic melts and their crystallization at intermediate depths. Isotopic age determinations of granites and syenites vary from 360 to 800 m.y.

Diorites and granodiorites

Rocks of this composition are extremely abundant in the Transantarctic Mountains, where a complex intrusive suite was formed in latest Precambrian and early Paleozoic time in connection with orogenic events in the Ross geosyncline. Intrusions of the latest Precambrian phase are prevalent in Victoria Land but also occur in the Queen Maud Range and the Horlick and Thiel Mountains (Gunn and Warren, 1962; Murtaugh, 1969; Lopatin, 1972; and others). Areas of massifs commonly reach 100 km^2; they are elongate along the strike of large folds. For as much as 3 to 7 km inside the contact, the intrusive rocks are intensely gneissified and altered to augen gneiss. Small vein- and layer-like bodies of granitoids are folded and gneissified together with the enclosing rocks. Porphyritic biotite and biotite-hornblende quartz monzonites and quartz leucomonzodiorites dominate the massifs, but granodiorite varieties occur in the central parts of large massifs. Veins consist of leucogranites and leucogranodiorites. At the periphery of the massifs there are orbicular gabbro-diorites, large xenoliths of biotite-hornblende-clinopyroxene diorites and quartz diorites, as well as abundant xenoliths of country rocks such as hornfels (after schist) and skarn (after marble). Endo- and exocontact zones of plutons are obscured, as a rule, by intense blastocataclasis and blastomylonitization. Small intersecting bodies of lower Paleozoic granitoids generally occur within large intrusions, causing most rock varieties to yield ages of 440–500 m.y. Features of palingenesis were also noted in the uppermost Precambrian granitoids. Nevertheless, at several localities relict isotopic dates of 600–630 m.y. were obtained on gneissified quartz monzonites.

Intrusions of the early Paleozoic phase are represented by highly discordant, stock-like, subvertical, laminar and dike bodies. Areas occupied by these intrusions are not known to exceed 150 km^2. Two groups of intrusions can be distinguished. Earlier intrustions consist of biotite-hornblende-clinopyroxene diorites and quartz diorites; their marginal parts are enriched in alkalis and quartz and contain abundant xenoliths of biotitized and uralitized gabbros. Later intrusions are represented by biotite, biotite-hornblende, and two-mica granites and granodiorites, locally as small stocks, veins, and dikes cutting diorites. Veins are abundant and include a variety of compositions: leucogranites, alaskites, albitites, pegmatites, lamprophyres, diorite porphyrites, granite, granodiorite, and syenite porphyres. The formation of hornfels and skarn are usual alterations in exocontact aureoles of these intrusions. Lower Paleozoic granitoids occur throughout the Transantarctic Mountains from Victoria Land to the Pensacola Mountains. In this direction, they decrease in number and their ages become somewhat older within the range from 425 to 520 m.y.

Middle to upper Paleozoic granitoids

Middle to upper Paleozoic intrusions of granitoids are localized mainly along the Pacific margin of the continent. They intrude the upper Proterozoic to lower Paleozoic metasedimentary sequences in the extreme northern part of Victoria Land and in western Marie Byrd Land (Sturm and Carryer, 1970; Nathan, 1971; Lopatin and Polyakov, 1976). In northern Victoria Land, granitoids make up small stocks, elongate massifs, and dike-like bodies which are probably the apophyses of batholith-like intrusions not exposed by erosion. Prevailing rock varieties are biotite tonalites, two-mica granites, and quartz syenites. The borders of massifs are impregnated with xenoliths of enclosing metasedimentary rocks, as well as of gneisses, schists, and quartz diorites. A vein series, intersecting both the source granitoids and surrounding rocks, occurs locally. The K-Ar ages for granitoids in northern Victoria Land range from 355 to 325 m.y.

In western Marie Byrd Land, intrusions of this group include hornblende-biotite granodiorites, biotite quartz monzonites, and two-mica granites. They form massifs, a few hundred square kilometers in size, elongated parallel to the fold axes in the enclosing metasedimentary sequence. Petrochemical features of the intrusions are (1) relatively low ferruginosity and (2) oversaturation with alumina. An isochron Rb-Sr age for granodiorites in the Ford Range is 352 m.y., while K-Ar dates on biotite and two-mica granites range from 290 to 325 m.y.; thus, the latter are likely to belong to the final phase of the middle–late Paleozoic plutonic activity. In eastern Marie Byrd Land and on Thurston Island, a few intrusions of middle to upper Paleozoic quartz diorites and granodiorites are present, but their size, shape, and relationship to enclosing rocks remain unknown. On Thurston Island, where granitoids are closely associated with gabbroids, outcrops of intrusive rocks constitute about 2000 km². Biotite-hornblende quartz diorites prevail; they are distinctly cataclastic and gneissified at some localites and massive at others. Granodiorites and quartz monzodiorites occur rarely and are considered to be the late phase of the same intrusive complex. The middle to late Paleozoic age of these rocks is indicated by available isotopic determinations ranging from 265 to 300 m.y.

Mesozoic and Cenozoic granitoids

Lower Mesozoic intrusions of granitoids occur mainly in the Antarctic Peninsula (a massif on Detroit Plateau and smaller intrusions on the eastern Nordenskjöld Coast, on the Loubet Coast, and on the Oscar II Coast) and in eastern Marie Byrd Land (Eights Coast and Jones and Isherwood Mountains). Their formation is believed to be associated with the major early Mesozoic (Gondwanide) phase of folding in the Pacific geosyncline (Grikurov, 1973). The most widespread rock varieties are quartz diorites, granodiorites, tonalites, and subalkaline granites; quartz monzonites are rarely encountered. These intrusions are heavily cataclased and contain abundant large xenoliths of enclosing rocks in endomorphic zones; active assimilation of the xenoliths is responsible for hybridism phenomena. Isotopic dates range from 130 to 230 m.y., the younger dates probably being reset in Cretaceous time.

In nunataks between the Ellsworth and the Transantarctic Mountains, the deformed Paleozoic(?) sequences are pierced by abundant small granitoid intrusions with isotopic ages ranging from 170 to 190 m.y.

Upper Mesozoic intrusions occur exclusively along the Pacific continental margin where their emplacement accompanied the most intense Andean orogeny in middle Cretaceous time. Upper Mesozoic granitoids in the Antarctic Peninsula compose at least 40% of all bedrock exposures (Adie, 1955). The most widespread rocks are quartz diorites and associated varieties. Gabbros, gabbrodolerites, and quartz gabbros are also present; their formation predates emplacement of quartz monzonites, granodiorites, and granites.

On the Loubet Coast the maximum size of the massifs reaches 20 km across. Gabbroids, quartz diorites, and granitoids are distributed in equal amounts. On the Oscar II Coast the outcrops of the intrusions reach 400 km²; granites prevail over plagiogranites, diorites, and quartz diorites; some olivine gabbro also occurs. At the northern tip of the Antarctic Peninsula a tabular body about 1000 m thick is composed of banded gabbroids. In eastern Ellsworth Land the largest massifs do not exceed 100 km². They consist mainly of diorites, quartz diorites, and leucocratic gabbros. Smaller stocks, ranging from 5 to 10 km², are composed of quartz monzonites and granodiorites. A more felsic composition typifies batholiths on the Lassiter Coast, where granodiorites and granites are very abundant, and is also common for small plutons in the northern Antarctic Peninsula and South Shetland Islands.

In eastern Marie Byrd Land and on the Eights Coast (Lopatin and Polyakov, 1976) the Cretaceous intrusions are represented mainly by quartz monzonites with subordinate diorites and granites. Outcrops of quartz monzonites occur in the Kohler Range area, where they probably belong to a single sublatitudinal massif. Intrusions of quartz monzonites reaching 600 km² in size are exposed on Thurston Island. Granites in western Marie Byrd Land are characterized by higher alkalinity. On the Saunders Coast leucocratic subalkaline biotite granites are most widespread; they form a series of subvolcanic intrusions reaching several hundred square kilometers in size. On the Ruppert and Hobbs Coasts the intrusions consist mainly of porphyritic quartz syenites and associated leucocratic granites. They intrude older gneissic granodiorites, diorites, and gabbros, as well as the middle–upper Paleozoic metavolcanic rocks. The intrusions of quartz syenites and subalkaline granites probably reach several hundred square kilometers in area.

The Cretaceous age of the intrusions is supported by geologic evidence and by numerous isotopic determinations, ranging from 90 to 130 m.y. All the Cretaceous intrusions produce a pronounced effect on enclosing sedimentary-volcanic sequences, turning them into hornfels of an andalusite grade over vast areas.

Upper Cenozoic granitoids in the Antarctic Peninsula area have been found quite recently, after discovery of a K-Ar age value of 9.5 m.y. on biotite from biotite granodiorite on Cornwallis Island in the South Shetland archipelago (Rex and Baker, 1973). Enclosing rocks are greenschists of unknown age. The presence of Pliocene granite suggests that intrusive granitic magmatism did not terminate during the Cretaceous stage of orogenesis but continued, at least in the southern Scotia Arc, until as recently as in the Andes.

Geochronologic data

All the available information up to 1976 on the geochronology of Antarctica has been obtained for this study to distinguish eras of granite generation (e.g., Craddock and others, 1964; Ravich and Krylov, 1964; Picciotto and Coppez, 1964; Webb and Warren, 1965; Faure and others, 1970; Craddock, 1972; Lopatin and Polyakov, 1976; and others). The total of 1135 available isotopic age determinations on Antarctic granitoids are distributed among the major structural provinces in the following way: 301 determinations were carried out on granitoids of the West Antarctic System, 242 on granitoids of the Ross System, and 592 on granitoids of the East Antarctic craton. Over 70% of the determinations were made using the K-Ar method, about 25% using the Rb-Sr method, and less than 5% using lead isotopes.

Naturally, there is not complete agreement between the

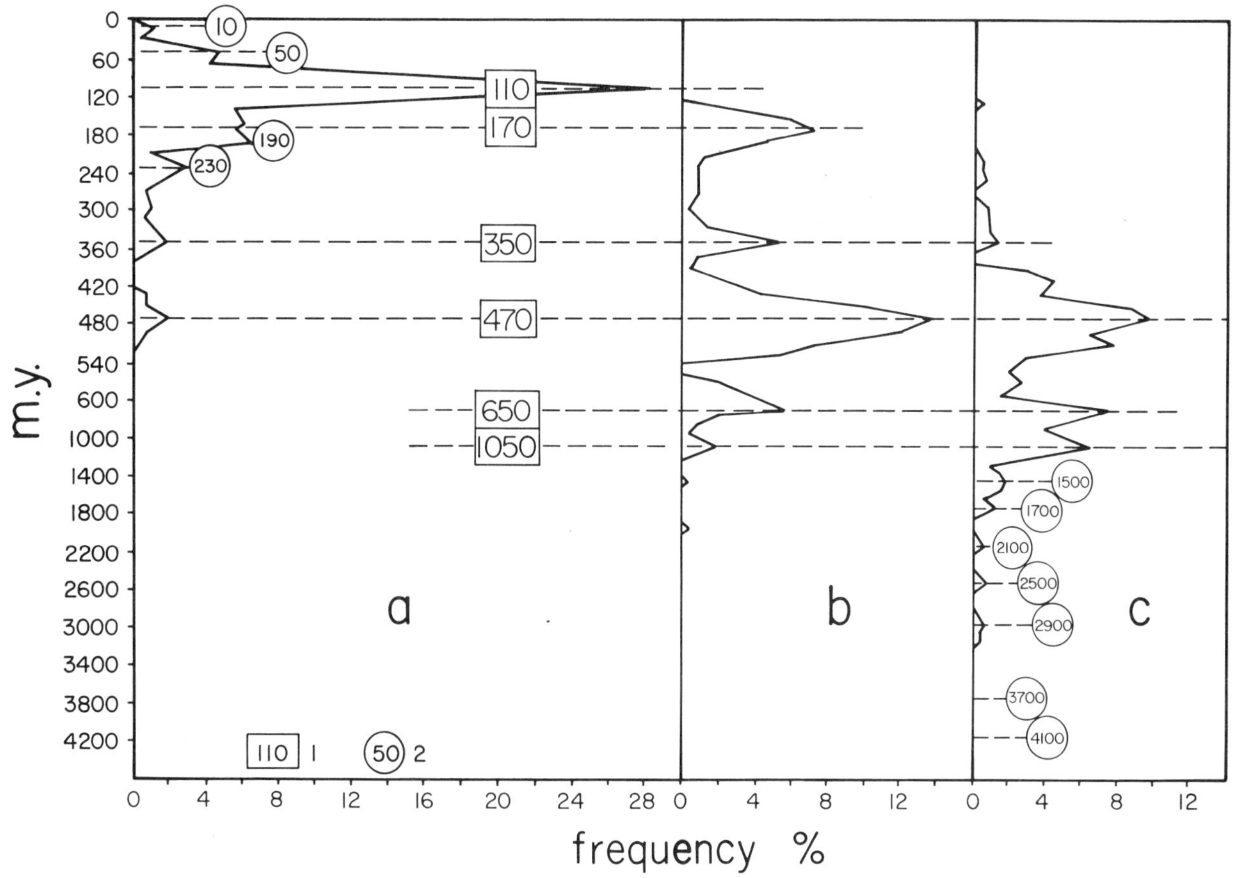

Figure 85.1. Generalized histograms of isotopic age determinations, by K-Ar, Rb-Sr, and Pb-isotope methods, of plutonic activity in major structural provinces of Antarctica. A: West Antarctic fold system, total of 301 determinations. B: Ross fold system, total of 242 determinations. C: East Antarctic craton, total of 592 determinations. Ages in boxes indicate maximums of granitoid generation characteristic of two or more structural provinces, and ages in circles indicate miscellaneous maximums. The scale is 60 m.y. per interval for the Phanerozoic and 400 m.y. per interval for the Precambrian. Compiled by G. I. Kameneva from all available data published before 1977.

values obtained by different methods, nor is there agreement in all cases between the values and the geologic evidence. The authors applied a statistical treatment in isotopic determinations, plotting histograms of their distribution according to frequency of occurrence (Figure 85.1). The maximums of isotopic dates obtained in this manner are expected to indicate ages of the true geologic events, apparently represented or, at least, accompanied by granite generation. Because of the predominance of K-Ar determinations, which usually indicate only the last alterations in minerals and rocks, these events appear more reliably established for the more recent part of the geochronologic scale, while older, chiefly early Precambrian events, require much additional investigation. However, on the whole, diverse isotopic determinations showed surprisingly good agreement and consistent results, revealing the most widespread maximums of granite generation (Figure 85.1).

In the East Antarctic craton, evident maximums correspond to age values of 1050, 650, 470, and 350 m.y. Seven less emphatic maximums of granite generation are older than 1050 m.y.; they are 4100, 3700, 2900, 2500, 2100, 1700, and 1500 m.y. ago. As granite generation is known to have taken place extensively in Precambrian time, it can be foreseen that for this period of time more maximums will emerge with more isotopic dates, and that continuous development of some ancient granitoid complexes will be established. In the Ross fold system, five distinct maximums of granite generation are established: 1050, 650, 470, 350, and 170 m.y. ago. Six maximums are distinguished for the West Antarctic system: 470, 350, 230, 170, 110, and 50 m.y. ago; the most recent data on augen gneisses of granitoid composition from the Orville Coast area (Clarkson and Brook, 1977) suggest two more maximums of granite generation: about 1000 and 650 m.y. ago. Thus, in the Ross system and in the craton, four maximums of granite generation coincide, while five maximums are probably common for the Ross and Andean systems and four maximums—1050, 650, 470, and 350 m.y.—are typical of the entire Antarctic continent. However, each structural province of Antarctica is characterized by specific granitoid complexes, the sequence of generation and composition of which are associated with peculiarities of the geologic evolution of the crust.

Six granitoid complexes are recognized in the East Antarctic craton: (1) a lower Archean complex of protocrustal and ultrametamorphic granitoids of the granulite facies, consisting of enderbites, charnockites, and granite gneisses (3000–4200 m.y.); (2) an Archean to lower Proterozoic complex of ultrametamorphic granite gneisses of the amphibolite facies (1800–3500 m.y.); (3) an Archean to lower Proterozoic complex of rheomorphic porpyroblastic charnockites (1800–3500 m.y.); (4) an upper Archean to lower-middle Proterozoic remobilized complex of gneissic granite domes (1500–3200 m.y.), with rock compositions ranging from quartz diorite to normal granite; (5) an upper Proterozoic to lower Paleozoic complex of intrusive and metasomatic charnockitoids (500–1700 m.y.) represented by rocks of quartz-syenitic, quartz-monzonitic, granodioritic, and monzo-gabbroic composition; and (6) an uppermost Precambrian to lower-middle Paleozoic complex of intrusive granites, quartz syenites, and syenites (360–800 m.y.). The first four complexes are associated with formation and cratonization of the crystalline basement in early Precambrian time, and the last two are related to repeated activizations of this basement during late Precambrian and Phanerozoic time.

In the Ross fold system there are one ultrametamorphic and three intrusive granitoid complexes: (1) a lower(?) Precambrian complex of ultrametamorphic granite gneisses (older than 1050 m.y.) that is part of the pre-Ross basement; (2) an uppermost Precambrian batholithic complex (600–1050 m.y.) of quartz monzonites, monzodiorites, granodiorites, and granites; (3) a lower Paleozoic complex of minor intrusions (450–500 m.y.) of granitic, dioritic, and quartz-dioritic composition; and (4) a middle–upper Paleozoic complex (325–355 m.y.) of granodiorites, tonalites, granites, and quartz syenites.

In the Pacific fold belt of West Antarctica there are one ultrametamorphic and four intrusive granitoid complexes, each of them characterized by a close association with earlier gabbroids: (1) a lower(?) Precambrian complex (older than 1050 m.y.) of ultrametamorphic granite gneisses and associated meta-igneous rocks of more mafic composition, which constitutes the basement of the Mesozoic fold structures; (2) a middle–upper Paleozoic complex of batholiths and minor intrusions of quartz diorites and granodiorites (265–300 m.y.); (3) a lower Mesozoic complex of "synkinematic" granitoids (150–230 m.y.) consisting of quartz diorites, granodiorites, tonalites, and subalkaline granites; (4) the most widespread upper Mesozoic complex of "postkinematic" gabbro-granite batholiths intruded into the axial zone of Mesozoic uplift during the Cretaceous orogeny (90–130 m.y.) and composed mainly of quartz diorites and granodiorites with subordinate final alkaline rocks of the potassic series—monzonites and quartz syenites; and (5) an upper Cenozoic complex as yet represented by a single stock of granodiorite in the South Shetland archipelago (9.5 m.y.).

Conclusions

The principal trends of evolution of granitoid complexes in Antarctica are the following:

1. An increase of intrusive-magmatic factors and decrease of metasomatic-anatexical factors in the course of geologic time; this trend can equally be applied to individual structural provinces and the whole of the Antarctic continent;
2. A shorter duration of emplacement of the younger complexes as compared to the older ones, as well as a greater structural control, a more pronounced deep-level differentiation, and a higher degree of geochemical polarization of the final intrusive products (within basic-acid and/or alkaline-aluminiferous series). Within the craton, this trend is apparently associated with an overall decrease with time of a generated volume of granitoid rocks. Within geosynclinal terranes, such a decrease can positively be demonstrated only for the orogenic and post-orogenic stages of their development;
3. An increase in alkalinity in the course of magmatic and/or metasomatic processes during every epoch of granitoid generation irrespectively of its age and structural environment, and a general increase in alkalinity of final and/or subsequent activization differentiates of granitoid complexes, from the older to the younger ones, within individual structural provinces. However, if the rocks formed during major intrusive phases are compared, they would appear progressively less alkaline and silicic in the younger complexes;
4. A general decrease in the amount and activity of volatiles in the younger complexes as opposed to a relative increase of the role of volatiles during every individual epoch of granitoid formation;
5. A broad synchronism of the major stages of granitoid generation throughout the continent as opposed to formation of specific granitic products within individual structural provinces.

These trends may be explained by permanent submergence with time of source of excitation of regional plutonic processes. Such submergence is consistent with a concept of progressive thickening of the earth's crust through the course of geologic history.

References cited

Adie, R. J. 1955. The petrology of Graham Land. II. The Andean granite-gabbro intrusive suite. *Scient. Rep. Falkld Isl. Depend. Surv.*, No. 12, 39.

Clarkson, P. D. 1972. Geology of the Shackleton Range: a preliminary report. *Bull. Br. antarc. Surv.*, No. 31, 1–15.

Clarkson, P. D., and Brook, M. 1977. Age and position of the Ellsworth Mountains crustal fragment, Antarctica. *Nature, Lond.*, *265*, No. 5595, 615–16.

Craddock, C. 1972. *Geologic Map of Antarctica, 1:5,000,000.* New York, American Geographical Society.

Craddock, C., Gast, P. W., Hanson, G. N., and Linder, H. 1964. Rubidium-strontium ages from Antarctica. *Geol. Soc. Am. Bull.*, *75*, No. 3, 237–40.

Faure, G., Eastin, R., Gunner, J., Hill, R., Jones, L., and Elliot, D. 1970. Studies in the geochronology and geochemistry of the Transantarctic Mountains. *Rep. Inst. polar Stud., Ohio State Univ.*, No. 3, 180 pp.

Fedorov, L. V., and Tarutin, O. A. 1977. Geologicheskoe stroenie gory Ruker (iuzhnaia chast' gor Prins-Charl'z, Vostochnaia Antarktida) [The geological structure of Mount Ruker (southern Prince Charles Mountains, East Antarctica)]. *Antarktika, Doklady Komissii*, *16*, 93–99.

Grikurov, G. E. 1973. *Geologiia Antarkticheskogo poluostrova* [Geology of the Antarctic Peninsula]. Moscow, Izdatel'stvo Nauka, 119 pp.

Gunn, B. M., and Warren, G. 1962. Geology of Victoria Land between the Mawson and Mulock Glaciers, Antarctica. *Bull. geol. Surv. N.Z.*, No. 71, 157.

Kamenev, E. N. 1975*a*. Geologiia Zemli Enderbi [Geology of Enderby Land]. *Antarktika, Doklady Komissii*, *14*, 34–58.

Kamenev, E. N. 1975*b*. O geologicheskom stroenii dobikonskogo fundamenta v raione sukhikh dolin, Transantarkticheskie gory [Geological structure of the pre-Beacon basement in the dry valleys, Transantarctic Mountains]. *Trudy. sov. antarkt. Eksped.*, *65*, 204–14.

Klimov, L. V., Ravich, M. G., and Soloviev, D. S. 1964. Charnokity Vostochnoi Antarktidy [Charnockites of East Antarctica]. *Doklady sov. geol. MGK XXII sessiya*, 45–53.

Lopatin, B. G. 1972. Metamorfizm i magmatizm rossid iuzhnoi chasti Zemli Viktorii [Metamorphism and magmatism of rossides in southern Victoria Land]. *Trudy. sov. antarkt. Eksped., 60,* 6-35.

Lopatin, B. G., and Polyakov, M. M. 1976. *Geologiia i petrologiia Zemli Meri Berd i Berega Eitsa* [Geology and petrology of Marie Byrd Land and Eights Coast]. Moscow, Izdatel'stvo Nauka, 175 pp.

Lopatin, B. G., Fedorov, L. V., and Polyakov, M. M. 1977. Geologiia i metamorfizm proterozoiskikh otlozhenii gor Mak-Koli i Sherger [Geology and metamorphism of the Proterozoic deposits of Mount McCauley and Mount Scherger]. *Antarktika, Doklady Komissii, 16,* 76-92.

Murtaugh, J. G. 1969. Geology of the Wisconsin Range Batholith, Transantarctic Mountains, *N.Z. Jl Geol. Geophys., 12,* No. 2-3, 526-50.

Nathan, S. 1971. Geology and petrology of the Campbell-Aviator Divide, northern Victoria Land, Antarctica. *N.Z. Jl Geol. Geophys., 14,* No. 3, 564-96.

Picciotto, E., and Coppez, A. 1964. Bibliography of absolute age determinations in Antarctica (addendum). (*In* Adie, R. J., *ed. Antarctic Geology.* Amsterdam, North-Holland Publishing Company, 563-69.)

Ravich, M. G. 1974. Formatsiya charnokitoidov (na primere Gondwanskikh platform) [Charnockitoid formation (evidence from Gondwana platforms)] (*In* Shurkin, K. A., *ed. Problemy dokembriyskogo magmatizma.* Leningrad, Izdatel'stvo Nauka, 119-24.)

Ravich, M. G., and Grikurov, G. E., *eds.* 1976. *Explanatory notes to the Geologic Map of Antarctica 1:5,000,000.* Leningrad, Research Institute of Arctic Geology, 83 pp.

Ravich, M. G., and Kamenev, E. N. 1972. *Kristallicheskii fundament Antarkticheskoi platformy* [The crystalline basement of the Antarctic platform]. Leningrad, Gidrometeoizdat, 658 pp. [English translation: New York, John Wiley, 1975, 582 pp.]

Ravich, M. G., and Krylov, A. Ya. 1964. Absolutny vozrast gornykh porod Vostochnoi Antarktidy [Absolute age determinations of the East Antarctic rocks]. *Antarktika, Doklady Komissii,* 33-44.

Ravich, M. G., and Soloviev, D. S. 1966. *Geologiia i petrologiia tsentral'noi chasti gor Zemli Korolevy Mod* [Geology and petrology of the central part of the Queen Maud Land Mountains]. *Trudy instituta geologii Arktiki, 141,* 290 pp.

Ravich, M. G., Klimov, L. V., and Soloviev, D. S. 1965. *Dokembrii vostochnoi Antarktidy* [Precambrian of East Antarctica]. Moscow, Izdatel'stvo Nedra, 470 pp.

Rex, D. C., and Baker, P. E. 1973. Age and petrology of the Cornwallis Island granodiorite. *Bull. Br. antarc. Surv.,* No. 32, 55-61.

Rudnik, V. A. 1975. *Granitoobrazovanie i formirovanie zemnoi kory v kokermbrii* [Granite and crust formation in the Precambrian]. Leningrad, Izdatel'stvo Nedra, 415 pp.

Sobotovich, E. V., Kamenev, E. N., Komaristy, A. A., and Rudnik, V. A. 1974. Drevneishie porody Antarktidy (Zemlia Enderbi) [The oldest rocks of Antarctica (Enderby Land)]. *Izv. Akad. Nauk SSSR,* Ser. geol., No. 11, 30-50.

Soloviev, D. S. 1972. Platformennye magmaticheskie formatsii vostochnoi Antarktidy [Platform magmatic formations of East Antarctica]. *Antarktika, Doklady Komissii, 11,* 59-72.

Soloviev, D. S., and Halpern, M. 1975. Pervye arkheiskie izotopnye vozrasty, poluchennye v Antarktide po porodam kristallicheskogo fundamenta [The first Archean isotopic ages determined on the crystalline basement rocks of Antarctica]. *Inf. Byull. sov. antarkt. Eksped.,* No. 90, 23-25.

Sturm, A., and Carryer, S. J. 1970. Geology of the region between the matusevich and Tucker Glaciers, north Victoria Land, Antarctica. *N.Z. Jl Geol. Geophys., 13,* 408-35.

Sudovikov, N. G. 1967. *Problema rapakivi i pozdneorogennykh intruzii* [Rapakivi problem and late orogenic intrusions]. Moscow, Izdatel'stvo Nauka, 117 pp.

Velikoslavinsky, D. A. 1973. K voprosy o formatsiiakh sobstvenno granitov rapakivi i rapakiviobraznykh granitov na Baltiiskom shchite [To the problem of rapakivi granite and rapakivi-like granite formations in the Baltic shield]. *Izv. Akad. Nauk SSSR,* Ser. geol., No. 8, 46-54.

Webb, P. N., and Warren, G. 1965. Isotope dating of Antarctic rocks: a summary—2. *N.Z. Jl Geol. Geophys., 8,* No. 2, 221-30.

86

Subdivision of Granitoids of Southern Victoria Land by Means of Feldspar Structural Diffractometry

E. N. KAMENEV
Department of Antarctic Geology, SEVMORGEO, Moyka 120, Leningrad 190121, U.S.S.R.

S. S. KUMEEV
Kalmyk State University, Elista, Kalmyk ASSR, 358000, U.S.S.R.

Abstract. Data on the geology and petrology of granitoids of southern Victoria Land, obtained by New Zealand, U.S., and Soviet investigators, indicate the presence of (1) pre-Ross System metamorphic, partly rheomorphosed, granite gneisses and (2) Ross magmatic early orogenic, orogenic, and late orogenic granitoids. Microscopic and X-ray examinations of granitoid feldspars have lent strong support to the above subdivision. Feldspars of metamorphic granite gneisses are characterized by the lowest and most heterogeneous structural state, i.e., disordered varieties are present in rocks along with ordered ones. The same heterogeneity with predominant disorder is traced into the stage of mobilization of the basement rocks in the early orogenic granitoids. Feldspars of the initial orogenic granites do not inherit the mode of origin of phase and structural state of granite-gneiss feldspars. The situation is indicative of a separate initial magmatic melt, whose higher temperature evolution resulted in formation of the final phase of the orogenic complex where most disordered feldspars developed. The definite similarity in change of composition and order observed in feldspars of late orogenic granitoids suggests that these rocks are also a product of subsequent evolution of a single magmatic chamber under thermal conditions with rising temperature. Thus, structural diffractometry of granitoid feldspars of southern Victoria Land allows us not only to identify granitoids as belonging to certain structural complexes, but also to tackle the problem of their genesis during the evolution of the Ross orogen.

Feldspars of granitoids from southern Victoria Land have been investigated in material collected by one of us (E. N. K.) in 1971-73 when he participated as an exchange scientist in the U.S. Antarctic Research Program (Kamenev, 1973; Kamenev, 1975). The region under study is situated in the Transantarctic Mountains on the McMurdo Sound coast between Skelton and Mackay Glaciers.

General geology

Data on the geology and petrology of granitoids from southern Victoria Land obtained by New Zealand, American, and Soviet geologists (Webb and McKelvey, 1959*a*; Allen and Gibson, 1962; Gunn and Warren, 1962; McKelvey and Webb, 1962; Blank and others, 1963; Webb, 1963; Grindley and Warren, 1964; Haskell and others, 1965; Palmer and others, 1967; Smithson and others, 1968; Ghent, 1970; William and others, 1971; Smithson and others, 1972; Lopatin, 1972; and others) indicate the presence of (1) pre-Ross system metamorphic, partly rheomorphosed granite gneisses and (2) Ross orogenic intrusive granitoids, subdivided into three groups: preorogenic or early orogenic, proper orogenic, and late or postorogenic granitoids. The enclosing folded sequence of upper Proterozoic carbonate rocks and schists has been irregularly metamorphosed under conditions of thermal (contact) and dynamothermal (near-fault) metamorphism and transformed into diverse hornfels, skarns, marbles, and schists. To the north of Ferrar Glacier and to the west of Blue Glacier, platform-type terrigenous sedimentary rocks of the Beacon Supergroup (Devonian–Jurassic) rest subhorizontally on an eroded surface cut on igneous and folded metamorphic rocks (the so-called Kukri peneplain). To the northwest of Blue Glacier, along gently inclined shear fractures, the basement and the sedimentary cover are intruded by sills of the Jurassic trappean complex of the Ferrar Dolerite (Allen, 1962; Hamilton, 1965; Grapes and Reid, 1971). To the southeast of Blue Glacier, the metamorphic rocks and granitoids are pierced and overlain by the large piles of the Cenozoic Discovery and Morning Volcanoes and are locally overlain by Cenozoic basalt and kenyte lavas. Small volcanic vents and remnants of lava flows of the same age occur in glacial valleys between the Blue and Koettlitz Glaciers and in the western part of the Taylor Valley (McIver and Gevers, 1970).

Geology of granitoid complexes

A pre-upper Proterozoic granite-gneiss complex is localized in small (as much as 13 km across) oval and rounded dome-shaped inliers between Koettlitz Glacier and Miers Valley. They are cut by younger intrusions and overlapped by upper Proterozoic metasedimentary sequences along a surface of structural unconformity. These granite gneisses consist of shadow and rheomorphic biotite, biotite-garnet, and biotite-hornblende granites, granodiorites, tonalites, and quartz diorites; they contain thin intercalations, lenses, and boudins of amphibolites, and biotite-clinopyroxene-plagioclase and biotite-hornblende-feldspar schists and migmatites. Evidence suggesting selective melting of granitoid material of the domes has been observed near the boundary of the granite-gneiss domes at the present erosion level.

Upper Proterozoic, early orogenic granitoids (Olympus granite gneiss) are represented by plutonic masses of gneissic porphyritic biotite and biotite-hornblende quartz monzonites and quartz monzodiorites. The intrusions are conformable to the folded structures of the enclosing sequence. At the present erosional level, early orogenic intrusions look like bands, 3–7 km wide and about 50 km long, with a northwesterly strike. Smaller bodies are deformed together with enclosing rocks and squeezed into fold hinges. In the marginal parts of the granitoid bands, occur numerous layers and lenses of tectonized hornfels, skarn, and migmatite of dendritic and lit-par-lit type, probably xenoliths of the enclosing sequence. In the authors' opinion, early orogenic granitoid intrusions are likely to be mobilized gneisses of the pre-Ross System basement, i.e., mobilized analogues of the granite-gneiss domes.

Upper Proterozoic orogenic granitoids proper are represented by stocks and batholiths of massive porphyritic coarse-grained biotite-hornblende granodiorites (Larsen Granodiorite and Dais Granite?). They are the largest plutons in the region. They are only partially conformable with enclosing metamorphic rocks and early orogenic granitoids. Locally in the marginal parts of granodiorite massifs, there occur as xenoliths orbicular gabbro-diorites, biotite-hornblende-clinopyroxene diorites, and quartz diorites—all xenoliths of enclosing metamorphosed sedimentary rocks. Granodiorites have both sharp contacts and gradual transitions with enclosing early orogenic granitoids. In some places early orogenic quartz monzonites represent gneissified endocontact parts of orogenic granodiorites. In early orogenic granitoids there is evidence of selective melting, expressed in the formation of veins and spherical segregations of granites and plagiogranites. Therefore, orogenic granodiorites are believed to have been formed, at least partly if not completely, during the process of palingenesis of early orogenic granitoids.

Late orogenic lower Paleozoic granitoids (Theseus Granodiorites?) include stocks or steeply dipping sheets of coarse-grained biotite-hornblende-clinopyroxene diorites and quartz diorites containing abundant xenoliths of biotitized and uralitized gabbros in the marginal parts of bodies enriched in alkali and quartz. At contacts with enclosing carbonate-schist sequences, on which the plutons have a pronounced metamorphic effect, the marginal parts of the latter consist of monzonites and syenites. Later biotite, biotite-hornblende, and two-mica granites and granodiorites (Vida Granite and Irizar Granite) locally intrude diorites as small stocks, veins, and dikes, or form separate laccoliths or sill-like bodies. The vein phase of late orogenic intrusions is rather abundant and diverse: leucogranite, alaskite, albitite, pegmatite, lamprophyre, diorite porphyrite, granite, granodiorite, and syenite porphyre.

Methods of feldspar investigations

Feldspars of granitoids from the area under study have been investigated microscopically (121 thin sections) and radiographically (79 specimens). The aim of microscopic examination was to determine (1) the distribution in the rocks of certain morphologic types, such as one- and two-feldspar varieties, (2) the presence of dynamic effects (3) the nature and composition of plagioclases and K-feldspars, and (4) the presence of perthites and antiperthites and other unusual forms.

X-ray examination of feldspars, performed on the copper-radioactivated YPE-50 M diffractometer, involved calculation of elementary crystal structures and subsequent interpretation of composition and ordering of feldspars by cell sizes (Kumeev, 1973). Six reflections on diffractograms have been used for the calculation of cell sizes (Kumeev, 1975). The relationships of unit-cell parameters are shown in diagrams of α–γ (international symbols for interaxial angles of unit-cells) (Figure 86.1) and b–c (axial dimensions) (Figure 86.2). Parameters of the end-members of alkali feldspars of low and high structural states are plotted (straight lines) in the diagrams for comparison (Stewart and Wright, 1974), because the data for the potassic or sodic end-members are used to estimate structural state and Or content and to recognize anomalous cell dimensions in natural alkali feldspars. From the relationships of the given parameters, plagioclases of granite gneisses and of late orogenic granitoids are characterized by a certain stability of unit-cell parameters. Plagioclases of orogenic granitoids in the final phase, in contrast, occupy a wide field in the diagram. In the K-feldspar field of the diagrams, analyses of late orogenic formations are relatively most closely grouped.

Discussion

Results of microscopic study

The results of microscopic study are summarized in Figure 86.3. The frequency of occurrence (percentage of specimens within entire population) of plagioclase is seen to be rather high: from 75% (orogenic granitoids) to 89% (granite gneisses). In magmatic granitoids, the frequency increases in younger complexes. The frequency of occurrence of rock varieties with only plagioclase and of normally twinned, unchanged plagioclase crystals also increases. Plagioclases of orogenic granitoids are generally characterized by antiperthitic structure, and plagioclases of late orogenic granitoids, by zoned structure. Dynamic effects in feldspar crystals are most distinctly manifested in the basement granite gneisses. Early orogenic granitoids are represented by only a few thin sections, but unaltered plagioclase and untwinned K-feldspar are present in each of them.

Younger granitoids are characterized, as a rule, by a decrease in K-feldspar, but pure K-feldspar rock varieties more commonly occur in orogenic granitoids. Magmatic formations generally contain untwinned K-feldspar, and only in late orogenic granitoids does gridiron-twinned K-feldspar occur along with the untwinned variety. Perthitic structure is rather common in K-feldspar of the orogenic granitoids but rather uncommon in K-feldspars of the late orogenic rocks. K-feldspars of granite gneisses differ considerably from K-feldspars of magmatic granitoids. Though K-feldspar is rather abundant in granite gneisses, it occurs almost everywhere with plagioclase. Untwinned varieties prevail over gridiron-twinned ones, but both K-feldspars have been observed together very rarely. Per-

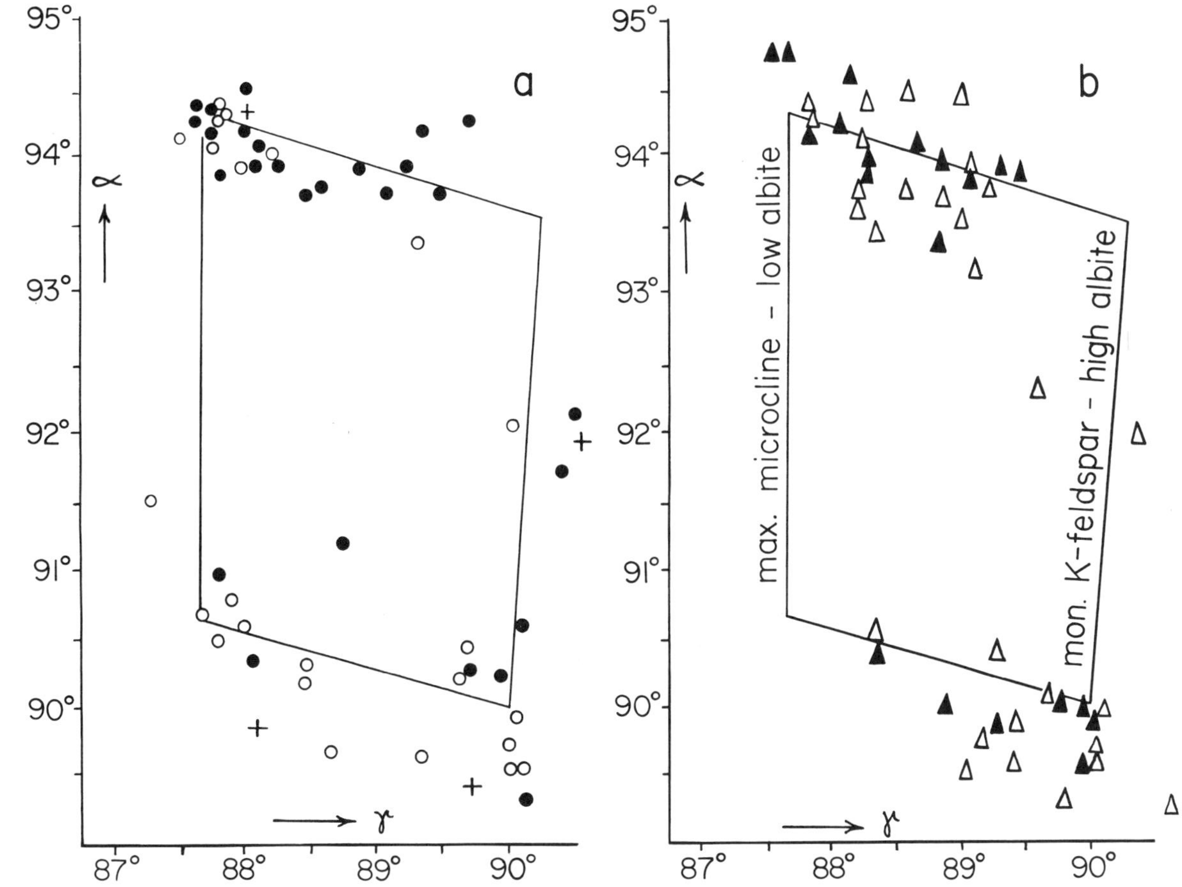

Figure 86.1. Diagram of the interaxial angles α–γ (international symbols) for the feldspars of granitoids from southern Victoria Land. *a.* Dots: granite gneisses. Crosses: early orogenic granitoids. Open circles: orogenic granitoids. *b.* Open triangles: orogenic granitoids of final phase. Solid triangles: late orogenic granitoids. Straight lines are alkali-exchange paths for end-members of alkali feldspars of low and high structural states. The parameters for the end-members are from Stewart and Wright (1974).

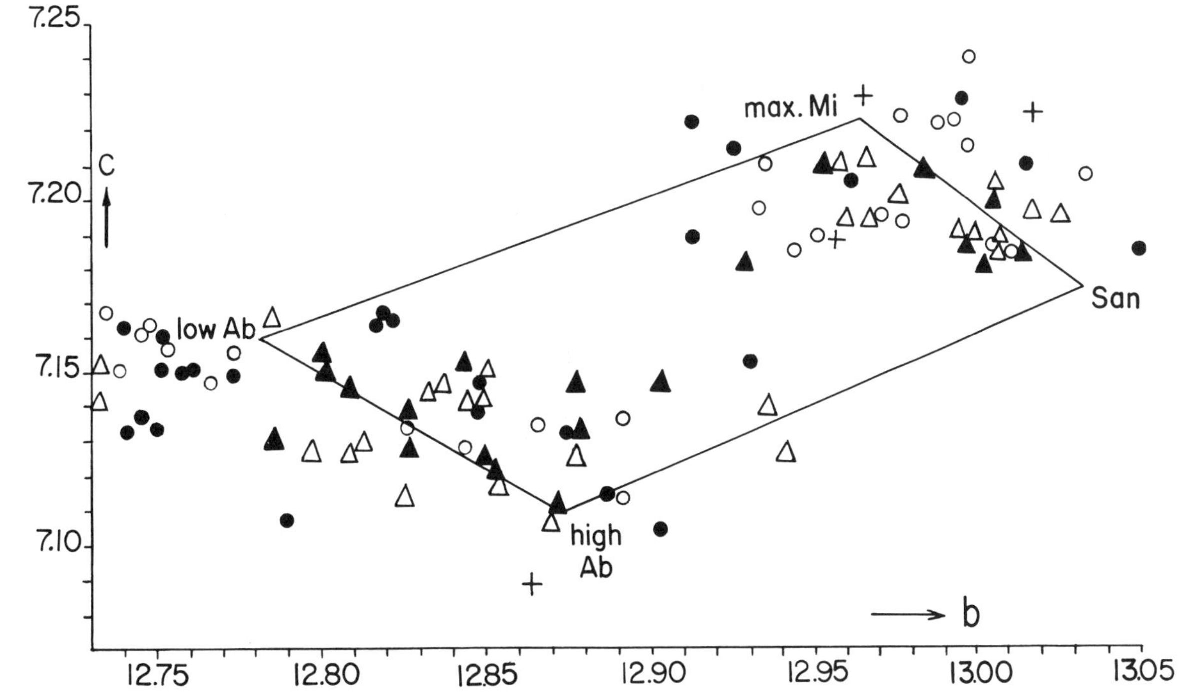

Figure 86.2. Diagram of the axial dimensions b–c for the feldspars of granitoids from southern Victoria Land. Dots: granite gneisses. Crosses: early orogenic granitoids. Open circles: orogenic granitoids. Open triangles: orogenic granitoids of final phase. Solid triangles: late orogenic granitoids. Straight lines are alkali-exchange paths for end-members of alkali feldspars of low and high structural states. The parameters for the end members are from Stewart and Wright (1974).

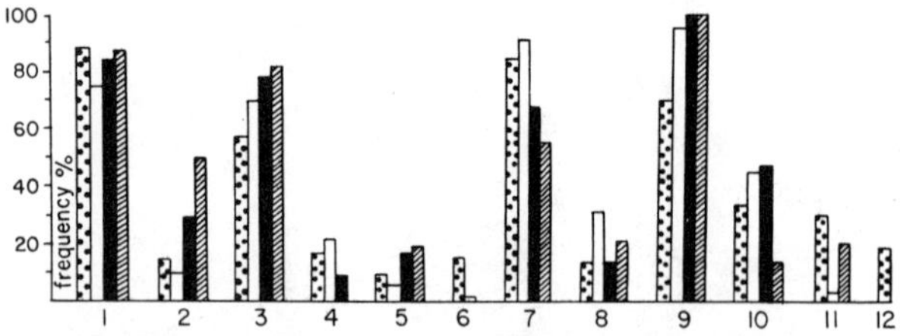

Figure 86.3. Frequency of occurrence of optical-morphologic feldspar types in granitoids from southern Victoria Land: 1. Plagioclase; 2. Plagioclase only (without KFsp); 3. unaltered plagioclase; 4. Antiperthite; 5. Zoned plagioclase; 6. Altered crushed plagioclase. (Number 1 represents the percentage of specimens with plagioclase within the entire population, whereas numbers 2–6 represent the percentage of specimens that show this characteristic within the population of specimens with plagioclase.) 7. K-feldspar; 8. K-feldspar only (without Pl); 9. Untwinned K-feldspar; 10. Perthitic untwinned K-feldspar; 11. Gridiron-twinned K-feldspar; 12. Perthitic gridiron-twinned K-feldspar (Number 7 represents the percentage of specimens with K-feldspar within the entire population, whereas the numbers 8–12 represent the percentage of specimens that show this characteristic within the population of specimens with K-feldspar; two varieties of K-feldspar may occur in the same specimen.) Dotted symbol: granite gneisses. White symbol: orogenic granitoids. Black symbol: orogenic granitoids of final phase. Lined symbol: late orogenic granitoids. (There are 20–30 specimens in each category.)

thitic structure in K-feldspars of granite gneisses has been recorded in more than half of the investigated specimens, but more commonly in gridiron-twinned varieites.

Thus, microscopic study of feldspars shows that changes in their properties are systematic in magmatic granitoids and that these variations are in agreement with the chronologic pattern of magmatism. Properties of feldspars in the metamorphic granite gneisses do not follow these same patterns of variation.

RESULTS OF X-RAY STUDY

As mentioned above, structural and phase states of feldspars have been interpreted by their unit-cell sizes. The orthoclase content of K-feldspar was calculated by the equation $N_{Or} = 2.1099a + 0.0207\,(\Delta bc) - 17.1975$ (Luth, 1974). An content of plagioclase was determined from the function "γ – % An" (Nissen, 1969). Order-disorder of plagioclases and K-feldspars was calculated by the concentration of aluminum in tetrahedral position (T_1O); this allows us to consider the thermal history of feldspars from a single position. According to Luth (1974) and Stewart and Ribbe (1969),

$$\mathrm{Al}\,(T_1O) = \frac{\Delta bc + \Delta\alpha^*\gamma^*}{2}$$

$$\text{where } \Delta bc = \frac{c - 0.4864b - 0.7711}{1.8388 - 0.1306b} \quad \text{and}$$

$$\Delta\alpha^*\gamma^* = \frac{\alpha^* - 2.0066 \times \gamma^* + 90.594}{0.208245\,\gamma^* - 23.2978}$$

The computer program of cell calculation used allows simultaneous calculations of order-disorder (Kumeev, 1975). It should be noted that for plagioclase of the non-albitic structural type, the Al (T_1O) values are not only a function of the thermal history of the rock in the specimen but a function of anorthite content as well. Consequently, for the specimens under study, order-disorder has been calculated as a function of plagioclase composition only and expressed in terms of Al (T_1O) as well (Viswanathan, 1972). For K-feldspars, besides Al (T_1O), the degree of triclinicity has been determined from the function "γ – triclinicity" (Laves and Viswanathan, 1967).

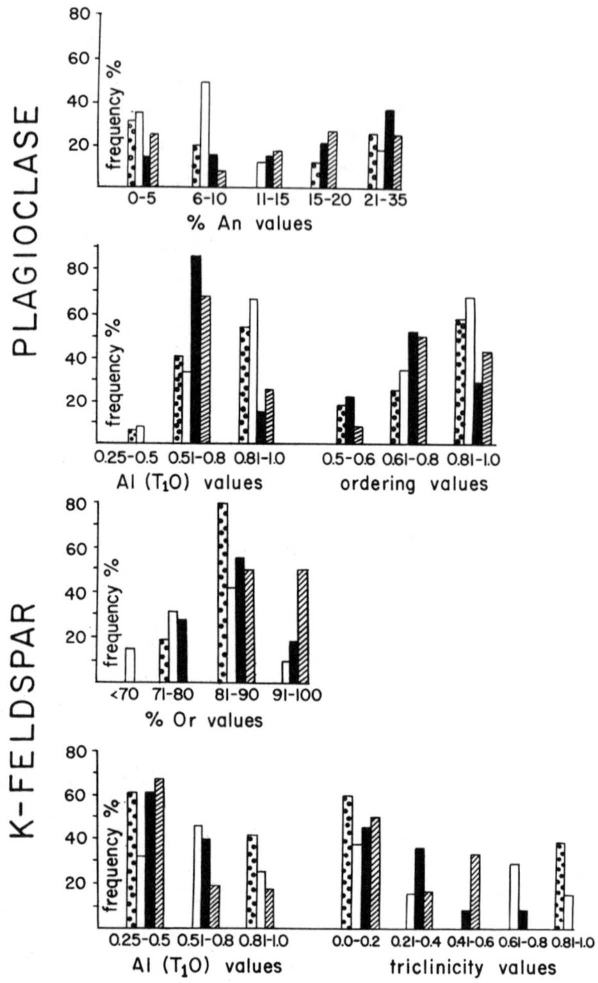

Figure 86.4. Frequency of occurrence (percentage of specimens within entire population) of different compositional and structural varieties of plagioclase and K-feldspar in granitoids from southern Victoria Land. Dotted symbol: granite gneisses. White symbol: orogenic granitoids. Black symbol: orogenic granitoids of final phase. Lined symbol: late orogenic granitoids.

Figure 86.4 illustrates the frequency of occurrence of plagioclase and K-feldspars of variable composition in granite gneisses of the basement and in magmatic granitoids, as well as their distribution of order-disorder. Compositional variations of plagioclase in granite gneisses and in late orogenic granitoids are uniform, and maximums fall in the albite and oligoclase-andesine categories. In orogenic granitoids of the early phase, the maximums fall in the albite (An_{6-10}) composition, and their final phase is characterized by plagioclase of a wide variety of An content with maximums in the oligoclase-andesine range. In K-feldspars, the maximums of orthoclase content commonly fall in the 80–90% range, but the distribution of values of orthoclase content is more specific here for each complex. Thus, K-feldspars with 71–80% of the orthoclase component occur in granite gneisses, varieties with the widest range of orthoclase content are present in orogenic granitoids of the early phase, variations of orthoclase content narrow in orogenic granitoids of the final phase, and K-feldspars of late orogenic granitoids are characterized by the highest values of orthoclase content (80–100%).

Al (T_1O) values are subdivided into three groups: 0.25–0.50 (high, disordered varieties of feldspars), 0.51–0.80 (intermediate), and 0.81–1.00 (low, ordered varieties). K-feldspars are characterized by a greater number of subdivisions: Al (T_1O) = 0.25 is typical of sanidine; 0.35, orthoclase; 0.45–0.55, monoclinic-triclinic transition; 0.55–0.80, intermediate microcline; and 0.80–1.00, maximum microcline (Stewart and Wright, 1974). Plagioclase order-disorder, being a function both of the thermal history and composition [Al (T_1O)], corresponds mainly to a low structural state for granite gneisses and orogenic granitoids of the initial stage, and to an intermediate structural state for orogenic granitoids of the final phase and for late orogenic granitoids. In granite gneisses and in the latest granitoids occur high plagioclases. Aluminum distribution, which is only a function of plagioclase composition, generally repeats the maximums of undifferentiated ordering. Some details also indicate partial thermal disordering of plagioclases from granite gneisses and from late orogenic granites (values of ordering, which are a function of composition, are higher than actual calculated values).

As for K-feldspars, Al (T_1O) maximums fall in an interval from 0.25 to 0.50 for granite gneisses, orogenic granitoids of the final phase, and late orogenic granitoids. In addition, some K-feldspars of granite gneisses are ordered varieties, and some K-feldspars of orogenic granitoids are intermediate varieties. The maximum occurrence for K-feldspars of the initial stage of the orogenic complex falls in the same interval as the values for aluminum concentrations.

Triclinicity of K-feldspars varies from 0.0 to 1.0, with values up to 0.2 usually being considered as monoclinic varieties; 0.21–0.40, monoclinic-triclinic inversion; 0.4–0.8, intermediate microcline; and 0.81–1.00, maximum microcline. (Triclinicity is a conventional measure of the ordering of K-feldspars, indicative of the degree of deviation of angles from 90°, i.e., from the monoclinic cell.) The triclinicity distribution in feldspars of the granitoids under study is straightforward: maximums occur in the range 0.0–0.2 for all four rock types. In other ranges of triclinicity values, K-feldspars of each rock type have individual, more common occurrences: in the interval 0.81–1.00, granite gneisses; 0.61–0.80, orogenic granitoids of the initial phase; 0.21–0.40, orogenic granitoids of the final phase; and 0.41–0.60, late orogenic granitoids.

Figure 86.5 shows mean values of the phase states and order-disorder of feldspars from the granitoids. Comparison of orogenic granitoids of two phases reveals an increase in An content of plagioclase and in orthoclase content of K-feldspars in the final phase and a decrease in order of both feldspars. The late orogenic granitoids are characterized by some decrease in An content of plagioclase and by the highest orthoclase content of the K-feldspars. As a whole, ordering of both plagioclases and K-feldspars does not vary as compared to the previous stage of granite generation (the presence of separate disordered feldspars in rocks of this complex was stated above). As for granite gneisses, because of known non-uniformity

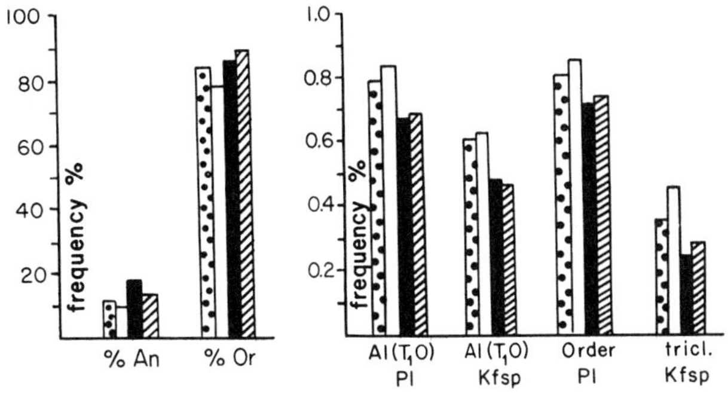

Figure 86.5. Mean compositions and mean values of nonperthitic feldspar ordering in granitoids from southern Victoria Land. Dotted symbol: granite gneisses. White symbol: orogenic granitoids. Black symbol: orogenic granitoids of final phase. Lined symbol: late orogenic granitoids.

both of composition and ordering, mean values of these constants are rather high. It can be emphasized, for example, that both anorthite and orthoclase contents increases but order somewhat decreases in feldspars of granite gneisses as compared to feldspars of orogenic granitoids. Mean Al (T_1O) values allow the assignment of plagioclases of orogenic granitoids of the final phase and those of late orogenic granitoids to an intermediate type, and plagioclases of granite gneisses and those of granitoids of the initial phase of the orogenic complex to the ordered category. K-feldspars can also be assigned to a field of monoclinic-triclinic inversion and intermediate microcline, respectively, and this differentiation can be traced both by aluminum concentration [Al (T_1O)] and by triclinicity degree.

All the data mentioned above concern plagioclases and K-feldspars occurring in one specimen or in different specimens, but they do not deal with perthites and antiperthites. Co-existing in one grain, but not intermixing, these plagioclase and K-feldspar phases (perthites and antiperthites) generally reveal some anomaly in cell sizes (Smith, 1974), and, therefore, it is not always possible to calculate order-disorder from parameters used for this purpose. Calculations of mean values of composition and ordering of K-feldspars, carried out with Na and K phases of perthites taken into account, show minor decreases in An content of plagioclase and orthoclase content of K-feldspar in granitoids of all genetic groups. Order-disorder of K-feldspars does not vary in granite gneisses (only triclinicity of K-feldspars decreases somewhat), but increases slightly in magmatic granitoids (triclinicity of K-feldspars remains unchanged). Some decrease in the degree of ordering, depending on composition, can be noted in orogenic granitoids of the final phase. The decrease in An content of plagioclases of the given rock group found during simultaneous consideration of uniform varieties and perthites, and the decrease in the degree of ordering, may indicate higher temperature conditions of perthitic decay as compared to rocks of other complexes.

Study of feldspar states can be supplemented with some evidence from the early orogenic granitoids. Data on this rock group are rather heterogeneous: ordered and disordered K-feldspars, intermediate plagioclases, and anorthoclases are all observed. It is quite possible that the relatively high-temperature states of these feldspars are associated with conditions of mobilization of the ancient basement.

Conclusions

The general conditions of granite generation can be roughly determined on the basis of structure-phase states of feldspars. Feldspars in the metamorphic granite gneisses are characterized on the whole by low, but heterogeneous, structural states, i.e., disordered varieties, in particular anorthoclases, which are present in rocks along with ordered ones. The same heterogeneity with predominant disorder is also traced into the stage of mobilization of the basement rocks in the early orogenic granitoids. The initial stage of orogenic granite generation, however, does not inherit the same phase and structural states of granite-gneiss feldspars. The situation is probably indicative of an independence of the initial magmatic melt, whose higher temperature evolution results in formation of the final phases of the orogenic complex, where more disordered feldspars are observed. The definite similarity in change of composition and order-disorder observed in late orogenic granitoids suggests that these rocks are also products of subsequent evolution of a single magmatic chamber under a continuing thermal regime with rising temperatures.

Thus, structural diffractometry of granitoid feldspars from southern Victoria Land allows us not only to identify granitoids as belonging to certain rock groups, but also to tackle the problem of their genesis associated with the evolution of the Ross orogen.

References

Allen, A. D. 1962. Geological investigations in southern Victoria Land, Antarctica. Pt. 7. Formations of the Beacon Group in the Victoria Valley region. *N.Z. Jl Geol. Geophys., 5,* No. 2, 278–94.

Allen, A. D., and Gibson, G. W. 1962. Geological investigations in southern Victoria Land, Antarctica. Pt. 6. Outline of the geology of the Victoria Valley region. *N.Z. Jl Geol. Geophys., 5,* No. 2, 234–42.

Blank, H. R., Cooper, R. A., Wheeler, R. H., and Willis, I. A. G. 1963. Geology of the Koettlitz-Blue Glacier region, southern Victoria Land, Antarctica. *Trans. R. Soc. N.Z.,* Ser. geol., *2,* No. 5, 79–100.

Ghent, E. D. 1970. Chemistry and mineralogy of the Mt. Falconer pluton and associated rocks, lower Taylor Valley, southern Victoria Land, Antarctica. *Trans. R. Soc. N.Z.,* Ser. geol., *7,* No. 9, 107–32.

Grapes, R. H., and Reid, D. L. 1971. Rhythmic layering and multiple intrusion in the Ferrar Dolerite of south Victoria Land, Antarctica. *N.Z. Jl Geol. Geophys., 14,* No. 3, 600–604.

Grindley, G. W., and Warren, G. 1964. Stratigraphic nomenclature and correlation in the western Ross region. (*In* Adie, R. J., *ed. Antarctic Geology.* Amsterdam, North-Holland Publishing Company, 314–33.)

Gunn, B. M., and Warren, G. 1962. Geology of Victoria Land between the Mawson and Mulock Glaciers, Antarctica. *Bull. geol. Surv. N.Z., n.s.,* No. 71, 157.

Hamilton, W. 1965. Diabase sheets of the Taylor Glacier region, Victoria Land, Antarctica. *Prof. Pap. U.S. geol. Surv.,* 456-B, 70 pp.

Haskell, T. R., Kennet, J. P., Prebble, W. M., Smith, G., and Willis, I. A. G. 1965. The geology of the middle and lower Taylor Valley of south Victoria Land, Antarctica. *Trans. R. Soc. N.Z.,* Ser. geol., *2,* No. 12, 169–86.

Kamenev, E. N. 1973. Geological investigations at McMurdo station, 1972–1973. *Antarct. Jnl U.S., 8,* No. 6, 338–39.

Kamenev, E. N. 1975. O geologicheskom stroenii dobikonskogo fundamenta v raione Sukhikh Dolin (Transantarkticheskie gory) [About geological structure of the Pre-Beacon basement in the Dry Valley region (Transantarctic Mountains)]. *Trudy. sov. antarkt. Eksped., 65,* 204–14.

Kumeev, S. S. 1973. *Strukturnaia difraktometriia polevykh shpatov* [Structural diffractometry of feldspars]. Elista, Kalmytskoe knizhnoe izdatelstvo, 158 pp.

Kumeev, S. S. 1975. K raschetam polevoshpatovych elementarnych yacheek [To the calculations of feldspar elementary cells]. (*In* Kumeev, S. S., *ed. Geologiia i poleznye iskopaemye Kalmytskoy ASSR. 3,* Elista, Kalmytskii universitet, 133–92.)

Laves, F., and Viswanathan, K. 1967. Relations between the optic axial and triclinicity of potash feldspars and their significance for the definition of stable-unstable states of alkali feldspars. *Schweiz. miner. petrogr. Mitt., 47,* No. 1, 147–61.

Lopatin, B. G. 1972. Metamorfizm i magmatizm rossid yuzhnoi chasti Zemli Viktorii [Metamorphism and magmatism of rossides in southern Victoria Land]. *Trudy. sov. antarkt. Eksped., 60,* 5–35

Luth, W. C. 1974. Analysis of experimental data on alkali feldspars: unit cell parameters and solvi. (*In* Mackenzie, W. S., and Zussman, J., *eds. Feldspars.* Manchester, England, Manchester University Press, 249–96.)

McIver, J. R., and Gevers, T. W. 1970. Volcanic vents below the Royal Society Range, central Victoria Land, Antarctica. *Trans. Geol. Soc. S. Afr., 73,* 65–90.

McKelvey, B. C., and Webb, P. N. 1962. Geological investigations in southern Victoria Land, Antarctica. Pt. 3. Geology of Wright Valley. *N.Z. Jl Geol. Geophys., 5,* No. 1, 143–62.

Nissen, H. U. 1969. Lattice changes in the low plagioclase series. *Schweiz. miner. petrogr. Mitt., 49,* No. 3, 491–508.

Palmer, D. F., Bradley, J., and Prebble, W. M. 1967. Orbicular granodiorite from Taylor Valley, South Victoria Land, Antarctica. *Geol. Soc. Am. Bull., 78,* No. 11, 1423–28.

Smith, J. V. 1974. *Feldspar Minerals. 1. Crystal Structure and Physical Properties.* Berlin, Springer-Verlag, 627 pp.

Smithson, S. B., Fikkan, P. R., and Toogood, D. 1968. Geological studies in the Dry Valley area of Victoria Land. *Antarct. Jnl U.S., 3,* No. 4, 107.

Smithson, S. B., Fikkan, P. R., Murphy, D. J., and Houston, R. S. 1972. Development of augen gneiss in ice-free valley area, south Victoria Land. (*In* Adie, R. J., *ed. Antarctic Geology and Geophysics.* Oslo, Universitetsforlaget, 293–98.)

Stewart, D. B., and Ribbe, P. H. 1969. Structural explanation for variations in cell parameters of alkali feldspar with Al/Si ordering. *Am. J. Sci., 267-A,* 444–62.

Stewart, D. B., and Wright, T. L. 1974. Al/Si order and symmetry of natural potassic feldspars, and the relationship of strained cell parameter to bulk composition. *Bull. Soc. fr. Miner. Cristallogr., 97,* No. 2–5, 353–77.

Viswanathan, K. 1972. Kationenaustausch an Plagioklasen. *Contr. Miner. Petrol., 37,* No. 4, 277–90.

Webb, P. N. 1963. Geological investigations in southern Victoria Land, Antarctica. Pt. 4. Beacon Group of the Wright Valley and Taylor Glacier region. *N.Z. Jl Geol. Geophys., 6,* No. 3, 361–87.

Webb, P. N., and McKelvey, B. C. 1959*a*. Geological investigations in south Victoria Land, Antarctica. Pt. 1. Geology of Victoria Dry Valley. *N.Z Jl Geol. Geophys., 2,* No. 1. 120–36.

Webb, P. N., and McKelvey, B. C. 1959*b*. Geological investigations in south Victoria Land, Antarctica. Pt. 2. Geology of upper Taylor Glacier region. *N.Z. Jl Geol. Geophys., 2,* No. 4, 718–28.

William, P. F., Hobbs, B. E., Vernon, R. H., and Anderson, D. E. 1971. The structural and metamorphic geology of basement rocks in the McMurdo Sound area, Antarctica. *J. geol. Soc. Aust., 18,* Pt. 2, 127–42.

87

The Inception and Duration of Mesozoic Volcanism in the Allan Hills–Carapace Nunatak Area, Victoria Land, Antarctica

B. A. HALL
Department of Geological Sciences, University of Maine, Orono, Maine 04469, U.S.A.

J. F. SUTTER
Department of Geology and Mineralogy, The Ohio State University, Columbus, Ohio 43210, U.S.A.

H. W. BORNS, JR.
Department of Geological Sciences, University of Maine, Orono, Maine 04469, U.S.A.

Abstract. In the Allan Hills–Carapace Nunatak area igneous activity younger than the Beacon Sandstone (Triassic and older) is recorded by laharic breccia and volcanogenic sandstone of the Mawson Diamictite, basalt flows, vent breccia, hyaloclastite, and intrusive dikes and sills of basalt and dolerite. Radiometric ages obtained from a basalt flow (specimen H-31) and a basalt pillow (specimen H-32) at Carapace Nunatak provide an age for the upper part of the volcanic section. Ages obtained from a basalt dike (specimen H-8) at Allan Hills provide a minimum, and probably approximate, age for the inception of post-Beacon volcanism. This dike intrudes the basal part of the Mawson Diamictite which rests unconformably on the Beacon Sandstone. The dike has chilled margins but was plastically distended by subsequent slight movement of the diamictite. Also at Allan Hills, massive dolerite (specimen H-5), which is intrusive into the Beacon Sandstone, is interpreted to predate emplacement of the Mawson Diamictite.

Averaged conventional K-Ar ages are 184.3 m.y. for H-5, 187.6 m.y. for H-8, 162.7 m.y. for H-31, and 151.6 m.y. for H-32. These ages suggest a time span of several tens of millions of years between the inception of volcanism and the emplacement of the uppermost volcanic flows. More recently, $^{40}Ar/^{39}Ar$ age spectrum data have been obtained for H-8 and H-32. $^{40}Ar/^{39}Ar$ plateau ages are 185.7 ± 3.4 m.y. for H-8 and 182.8 ± 3.5 m.y. for H-32. The combined K-Ar and $^{40}Ar/^{39}Ar$ data suggest that the post-Beacon Supergroup volcanic and volcanogenic rocks (more than 500 m thick) exposed in the Allan Hills–Carapace Nunatak area is middle Jurassic in age and was extruded during a relatively short geologic time interval (3–4 m.y.) about 184 m.y. ago.

Numerous papers have dealt with the post-Beacon Supergroup Mesozoic rocks of the Allan Hills–Carapace Nunatak and Battlements Nunatak area (Figure 87.1) since the initial report of Gunn and Warren (1962). General geologic relationships and stratigraphy have been covered by Borns and Hall (1969*a,b*), Ballance and Watters (1971), and Ball and others (1979). The origin and paleoenvironment of rocks exclusive of basalt flows and dikes have been considered by Ballance and others (1965), Borns and Hall 1969*a,b*), Ballance and Watters (1971), Hall and Borns (1972), and Ball and others (1979). Fossil plants and invertebrates have been described or discussed by Plumstead (1962), Townrow (1967), Carpenter (1969), Tasch (1970, 1973, 1974), Borns and others (1972), and Ball and others (1979).

Two of us (Hall and Borns) have obtained whole-rock K-Ar ages from one dolerite and three basalt localities at Allan Hills and Carapace Nunatak (Jones and others, 1973; Ball and others, 1979). The apparent ages of these rocks and other pertinent data are summarized in Table 87.1. These ages are important inasmuch as they provide estimates for the inception and duration of Mesozoic volcanism in this part of Victoria Land. Age determinations for several of these samples were subsequently checked by the $^{40}Ar/^{39}Ar$ age spectrum technique (Table 87.2). These new data confirm the K-Ar estimate of the age for inception of volcanism, but they provide a considerably different estimate for the age of the uppermost part of the section.

The purposes of this paper are: (1) to describe the geologic relationships that permit meaningful estimates of the volcanic chronology from radiometric dating of sampled basalt and dolerite units, (2) to present radiometric age data for selected rock specimens, (3) to discuss discrepancies between whole-rock K-Ar and whole-rock $^{40}Ar/^{39}Ar$ age spectrum data, and (4) to provide estimates for the time of inception and duration of post-Beacon Supergroup mesozoic volcanism in the area being considered.

Geology

Stratigraphy

The post-Beacon Supergroup stratigraphy of the area under consideration is relatively simple. The lowest part of the sequence is exposed at Allan Hills and has been described by Gunn and Warren (1962), Ballance and others (1965), Borns and Hall (1969*a*), and Ballance and Watters

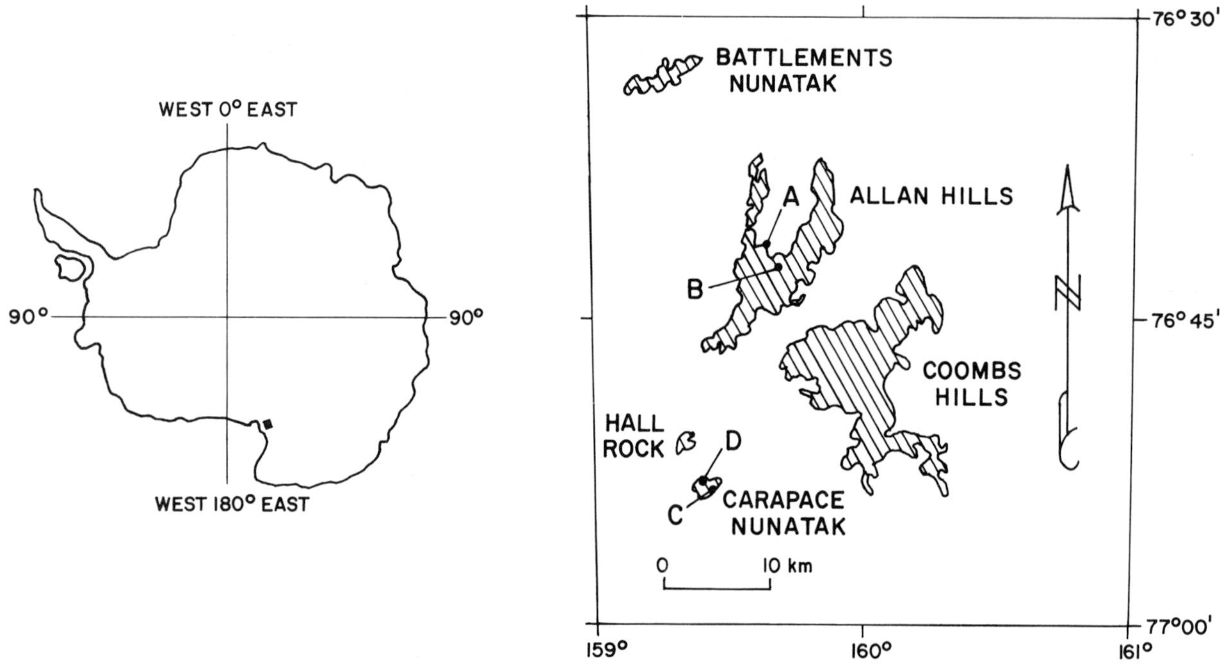

Figure 87.1. Location of field area in Antarctica. A–D of inset map are basalt localities referred to in text.

(1971). An erosion surface with minimum relief of 50 m is cut into Beacon rocks at least as young as Permian. This surface is overlain in post-Beacon erosional valleys by sandstone and conglomerate, as much as a few meters thick, that were derived directly through erosion of Beacon rocks and are probably fluvial in origin (Borns and Hall, 1969*a*). This basal unit is overlain by up to about 490 m (Gunn and Warren, 1962) of largely laharic breccia referred to as the Mawson Diamictite (Ballance and Watters, 1971). Graded beds from several meters to tens of meters thick are common (see Figure 2 of Borns and Hall, 1969*a*). The upper, finer grained parts of many such beds consist of cross-bedded or massive volcanogenic sandstone.

Approximately 15 km south of Allan Hills a second pertinent section ($\sim$400 m thick) is exposed at Carapace Nunatak. The lower 85 m of this exposed section consists of brownish, cross-bedded, volcanogenic sandstone initially placed in the Beacon Supergroup by Gunn and Warren (1962). Borns and Hall (1969*a*) first suggested that these rocks are related to the "Mawson" sequence rather than to the Beacon. Ballance and Watters (1971) subsequently assigned the sandstone and conglomerate to a new formation, the Carapace Sandstone, within the Ferrar Group. The precise stratigraphic relationship of these rocks to the Mawson Diamictite of Allan Hills is not clear because of ice cover between the two areas and the possibility of unexposed fault relationships. Nevertheless, the composition of the sandstone at Carapace Nunatak is similar to the Mawson Diamictite at Allan Hills, and two of us (Hall and Borns) interpret the Carapace Nunatak sandstone as a slightly higher (topographically) or more distal current-sorted facies of the diamictite at Allan Hills.

Above the sandstone at Carapace Nunatak are several hundred meters of interbedded basalt flows, fossiliferous Middle Jurassic lake deposits (Ball and others, 1979), hyaloclastite units, and a single interbed of laharic breccia similar to the Mawson Diamictite (Ballance and Watters, 1971).

Stratigraphic relationships of radiometrically dated rocks

Extrusive and intrusive basalt and dolerite are common at Allan Hills. As noted above, extrusive basalt is common at Carapace Nunatak. The basalt and dolerite are related

Table 87.1. K-Ar whole-rock data for basaltic rocks from Allan Hills–Carapace Nunatak area, Victoria Land, Antarctica

Specimen number	K (wt. %)	$^{40}Ar_{(radiogenic)}$/gm (cc STP x 10^{-6})	$^{40}Ar_{(radiogenic)}$ (% of total)	Apparent age (m.y.)
H-5	1.23	9.311	69	185.0 ± 4
H-5	1.25	9.391	64	183.7 ± 4
H-8	0.557	4.308	86	188.8 ± 4
H-8	0.556	4.321	87	189.7 ± 4
H-8	0.602	4.540	58	184.3 ± 4
H-31	0.650	4.345	61	164.3 ± 3
H-31	0.650	4.406	62	166.5 ± 3
H-31	0.710	4.540	50	157.5 ± 3
H-32	1.12	6.855	91	151.0 ± 3
H-32	1.12	6.924	91	152.3 ± 3

Decay constants (calculated following Steiger and Jäger, 1978):
$\lambda^{40}K_{\beta-} = 4.962 \times 10^{-10}$/yr; $\lambda^{40}K_e + \lambda'^{40}K_e = 0.581 \times 10^{-10}$/yr;
$^{40}K/K = 1.167 \times 10^{-4}$ atom/atom; atomic ratio $^{40}Ar/^{36}Ar$ atmospheric = 295.5. Analytical uncertainties calculated following Cox and Dalrymple (1967).
Analyses done by Isotopes, Inc., and by R. L. Armstrong.

to the stratigraphic section in at least five different modes. These are discussed below and their relationships are described. Specimen numbers (see Table 87.1) are identified where appropriate.

1. A basalt dike (locality A, Figure 87.1), intrusive into the Beacon rocks, clearly predates the pre-Mawson Diamictite erosion surface. The relative age of nearby massive dolerite (H-5) is interpreted to be the same.
2. At locality B (Figure 87.1) a basalt dike (H-8) clearly intrudes the basal part of the Mawson Diamictite. However, the dike and its chilled margins are distended and broken (Figure 87.2), suggesting some remaining mobility of the Mawson Diamictite after dike intrusion. Specimen H-8 is interpreted to be roughly contemporaneous with the onset of Mawson laharic breccia emplacement.
3. Specimen H-31 (locality C, Figure 87.1) is from a basalt flow interbedded with fossiliferous lacustrine deposits. The flow lies about 10 m above the top of the basal sandstone-conglomerate unit of the south spur of Carapace Nunatak.
4. Specimen H-32 (locality D, Figure 87.1) is from a basalt pillow in a hyaloclastite delta deposit on the north spur of Carapace Nunatak. The base of the hyaloclastite unit is estimated to be 30 m above the basal sandstone-conglomerate unit. Some pillows have foundered in underlying lake deposits (Figure 87.3), demonstrating contemporaneity or near contemporaneity with the lake deposits. These relationships are discussed in detail by Ball and others (1979).

In summary, specimen H-5 is believed to predate Mawson volcanic activity at Allan Hills, and specimen H-8 roughly dates the onset of the same volcanic activity.

Figure 87.2. Distended basalt dike intrusive into Mawson Diamictite at Allan Hills (locality B of Figure 87.1 and specimen H-8 of text and Tables 87.1 and 87.2). (See hammer for scale.)

Table 87.2. $^{40}Ar/^{39}Ar$ whole-rock age spectra data for basaltic rocks from Allan Hills-Carapace Nunatak area, Victoria Land, Antarctica

Temperature (°C)	$^{40}Ar/^{39}Ar$ (measured)	$^{37}Ar/^{39}Ar^a$ (corrected)	$^{36}Ar/^{39}Ar$ (measured)	$^{40}Ar_R/^{39}Ar_K$	$^{39}Ar_K$ (% of total)	$^{40}Ar_R$ (% of total)	$^{36}Ar_{Ca}$ (% of total)	Apparent age (m.y.)
Specimen H-8 (J = 0.006509)								
450	76.25	0.8758	0.1982	17.76	16.9	23.31	0.12	197.3 ± 9.9
650	28.11	1.788	0.3878	16.81	25.7	59.56	1.25	187.3 ± 4.1
800	29.08	3.342	0.4330	16.58	21.1	56.56	2.10	184.9 ± 4.6
900	28.60	4.185	0.04247	16.42	10.7	56.82	2.68	183.2 ± 7.4
1,000	26.47	5.580	0.03715	15.99	8.6	59.51	4.09	178.6 ± 6.7
Fusion	25.15	24.50	0.04089	15.27	17.0	56.40	16.30	170.9 ± 4.8
Total								
Gas	—	—	—	—	—	—	—	184.5
650-1,000	—	—	—	—	66.1	—	—	185.7 ± 3.4
Specimen H-32 (J = 0.004978)								
600	42.46	1.691	0.07231	21.25	34.0	49.91	0.64	181.4 ± 4.0
800	31.18	2.803	0.03361	21.51	18.2	68.59	2.27	183.5 ± 4.2
900	29.91	2.455	0.02858	21.69	17.1	72.17	2.34	185.0 ± 3.8
1,025	26.28	5.565	0.02733	18.71	21.4	70.25	5.54	160.7 ± 3.5
1,100	29.39	11.47	0.04129	18.24	6.6	60.29	7.56	156.8 ± 6.7
Fusion	34.83	17.07	0.06294	17.79	2.7	48.87	7.38	153.1 ± 14.8
Total								
Gas	—	—	—	—	—	—	—	175.6
600-900	—	—	—	—	69.3	—	—	182.8 ± 3.5

$^{40}Ar_R$: radiogenic ^{40}Ar. $^{39}Ar_K$: potassium-derived ^{39}Ar. $^{36}Ar_{Ca}$: calcium-derived ^{36}Ar.

The value of J is given by $J = \frac{e^{t_m/T} - 1}{(^{40}Ar_R/^{39}Ar_K)m}$, where t_m = known age of monitor mineral, T = mean life of $^{40}K = 1/\lambda^{40}K = 1.804 \times 10^9$ yr, and

$(^{40}Ar_R/^{39}Ar_K)m$ = measured ratio of $^{40}Ar_R$ to $^{39}Ar_K$ in monitor mineral irradiated along with samples of unknown age.

Decay constants are listed in notes to Table 87.1. Corrections for interfering argon isotope were made following Dalrymple and Lanphere (1971); analytical uncertainties were calculated following Dalrymple and Lanphere (1971).

The reader is referred to Dalrymple and Lanphere (1971), Lanphere and Dalrymple (1971), and Fleck and others (1977) for discussion of the $^{40}Ar/^{39}Ar$ age spectrum method of K/Ar dating.

^{a} Corrected for decay of ^{37}Ar.

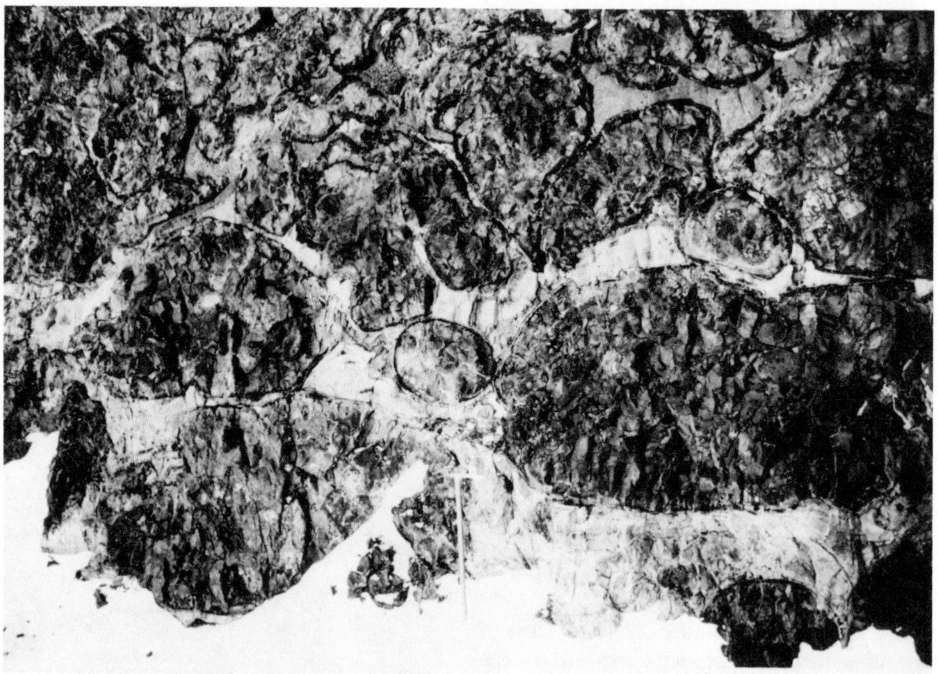

Figure 87.3. Basalt pillows foundered in Jurassic lake sediment at Carapace Nunatak (locality D of Figure 87.1, similar to specimen H-32 of text and Tables 87.1 and 87.2). (See ice axe for scale.)

Specimens H-31 and H-32 are from basalt units in the upper part of the Allan Hills–Carapace Nunatak volcanogenic sequence. Among them, their ages should provide estimates for the inception and duration of this volcanic activity.

Interpretation of ages

Our interpretation of the inception and duration of volcanism in the Allan Hills–Carapace Nunatak area, Victoria Land, Antarctica is based on the plateau (constant-age) portions of the $^{40}Ar/^{39}Ar$ age spectra of specimens H-8 and H-32 (Table 87.2). Specimen H-8 is from a distended basaltic dike which is interpreted as being contemporaneous with the inception of volcanism in this area, and the $^{40}Ar/^{39}Ar$ plateau age of 185.7 ± 3.4 m.y. dates this inception. Specimen H-32 is from a pillow basalt which overlies and deforms Jurassic lake beds, and it represents a late stage of Mesozoic volcanism in this area. Its $^{40}Ar/^{39}Ar$ plateau age is 182.8 ± 3.5 m.y., which is the same (within analytical uncertainty) as that of H-8, indicating that the duration of volcanism was quite short in this area, probably not more than 3 or 4 m.y.

The conventional K-Ar ages (Table 87.1) of specimens H-31 and H-32 are interpreted as showing effects of argon loss or inhomogeneity of the K-bearing phases. This interpretation is strengthened by comparing the K-Ar ages with the $^{40}Ar/^{39}Ar$ age spectrum for H-32. Although the plateau portion of the age spectrum of H-32 yields an age in agreement with H-8, the rest of the spectrum of H-32 indicates the presence of a nonretentive K-phase which must have influenced the value of nonretentive K-phase which must have influenced the value of the K/Ar ages. This is evidence suporting the usefulness of the $^{40}Ar/^{39}Ar$ age-spectrum technique in analyzing Mesozoic basaltic whole rocks, especially where tight stratigraphic and paleontologic control is lacking.

Acknowledgments. Fieldwork for this study was carried out during the 1968–69 and 1970–71 Antarctic summers, supported by National Science Foundation grant GA-1148 to Borns and Hall.

References cited

Ball, H. W., Borns, H. W., Jr., Brooks, H. K., Carpenter, F. M., Delevoryas, T., and Hall, B. A. 1979. Biota, age and significance of lake deposits, Carapace Nunatak, Victoria Land, Antarctica. *4th Int. Gondwana Symposium, Papers,* 1, 166–75.

Ballance, P. F., and Watters, W. A. 1971. The Mawson Diamictite and the Carapace Sandstone, formations of the Ferrar Group at Allan Hills and Carapace Nunatak, Victoria Land, Antarctica. *N.Z. Jl Geol. Geophys., 14,* No. 3, 512–27.

Ballance, P. R., Warren, G., and Watters, W. A. 1965. A reinterpretation of the Mawson Tillite, Antarctica. *Abstr., Aust. N.Z. Ass. Advmnt Sci.,* Hobart, 1965.

Borns, H. W., Jr., and Hall, B. A. 1969*a*. Mawson "Tillite" in Antarctica: preliminary report of a volcanic deposit of Jurassic age. *Science, N.Y., 166,* 870–2.

Borns, H. W., Jr., and Hall, B. A. 1969*b*. A reinvestigation of the Mawson Tillite, Victoria Land, Antarctica. *Antarct. Jnl U.S., 4,* No. 4, 133–34.

Borns, H. W., Jr., Hall, B. A., Ball, H. W., and Brooks, H. K. 1972. Mawson Tillite, Victoria Land, East Antarctica: reinvestigation continued. *Antarct. Jnl U.S., 7,* No. 4, 106–107.

Carpenter, F. M. 1969. Fossil insects from Antarctica. *Psyche, 76,* 418–25.

Cox, A., and Dalrymple, G. B. 1967. Statistical analysis of geomagnetic reversal data and the precision of Potassium-Argon dating. *J. geophys. Res., 72,* No. 10, 2603–14.

Dalrymple, G. B., and Lanphere, M. A. 1971. $^{40}Ar/^{39}Ar$ technique of K-Ar dating: a comparison with the conventional technique. *Earth planet. Sci. Lett., 12,* 300–308.

Fleck, R. J., Sutter, J. F., and Elliot, D. H. 1977. Interpretation of discordant $^{40}Ar/^{39}Ar$ age spectra of Mesozoic tholeiites from Antarctica. *Geochim. cosmochim. Acta, 41,* 15–32.

Gunn, B. M., and Warren, G. 1962. Geology of Victoria Land between the Mawson and Mulock Glaciers, Antarctica. *Bull. geol. Surv. N.Z., n.s., 71,* 157 pp.

Hall, B. A., and Borns, H. W., Jr. 1972. Jurassic geology of the Allan Battlements-Carapace nunataks area, Victoria Land. (*In* Adie, R. J., *ed. Antarctic Geology*. Amsterdam, North-Holland Publishing Company, 411.)

Jones, L. M., Walker, R. L., Hall, B. A., and Borns, H. W., Jr. 1973. Origin of the Jurassic dolerites and basalts of southern Victoria Land. *Antarct. Jnl U.S., 8,* No. 5, 268–70.

Lanphere, M. A., and Dalrymple, G. B. 1971. A test of the $^{40}Ar/^{39}Ar$ age spectrum technique on some terrestrial materials. *Earth planet. Sci. Lett., 12,* 359–72.

Plumstead, E. P. 1962. Fossil floras of Antarctica. *Scient. Rep. transantarct. Exped., 9,* 1-132.

Steiger, R. H., and Jäger, E. 1978. Subcommission on geochronology: convention on the use of decay constants in geochronology and cosmochronology. (*In* Cohee, G. V., Glaessner, M. F., and Hedberg, H. D., *eds.* Contributions to the Geologic Time Scale. *Stud. Geol. Am. Ass. Petrol. Geol.,* No. 6, 67–71.)

Tasch, P. 1970. Antarctic and other Gondwana conchostracans and insects: new data: significance for drift theory. *Proc. Pap. 2nd Gondwana Symposium, Pretoria, 1970,* 589–92.

Tasch, P. 1973. Jurassic beetle from southern Victoria Land, Antarctica. *J. Paleont., 47,* No. 3, 509–91.

Tasch, P. 1974. Food chain relationships in ancient fresh water ecosystems of Antarctica. *Antarct. Jnl U.S., 9,* No. 5, 238–39.

Townrow, J. A. 1967. Fossil plants from Allan and Carapace Nunataks, and from the Upper Mill and Shackleton Glaciers, Antarctica. *N.Z. Jl Geol. Geophys., 10,* No. 2, 456–73.

88

Systematic Variations of $^{87}Sr/^{86}Sr$ Ratios and Major Element Concentrations in the Kirkpatrick Basalt of Mount Falla, Queen Alexandra Range, Transantarctic Mountains

GUNTER FAURE, KAREN K. PACE, and DAVID H. ELLIOT
Institute of Polar Studies, The Ohio State University, Columbus, Ohio 43210, U.S.A.

Abstract. Mount Falla (lat. 84°22′ S, long. 164°55′ E) is capped by 14 flows of the Kirkpatrick Basalt. The rocks are subalkaline tholeiite basalts with unusually high concentrations of SiO_2 (55.82%) and total iron as FeO (11.51%) but with low concentrations of Al_2O_3 (13.55%), MgO (3.50%), and CaO (8.12%). The major elements vary systematically through the sequence, with a major chemical discontinuity between flows 10 and 11 and a lesser one between 6 and 7. The initial $^{87}Sr/^{86}Sr$ ratios are anomalously high (0.7118), vary systematically in a stratigraphic sense, and have discontinuities between the same flows listed above. Flows 1 through 6 have a constant initial ratio and permit dating by the Rb-Sr isochron method. The age of the flows is 173 ± 6 m.y. The chemical and isotopic properties are not readily explainable in terms of magma generation in a heterogeneous mantle. An alternate model based on progressive contamination of basalt magma with material derived from the granitic rocks of the continental crust is preferred. This model permits derivation of a mixing equation from which the compositions of the parent basalt and the hypothetical contaminant are estimated. The basalt parent was a tholeiite with a low Sr content of 92 ppm. The contaminant contained radiogenic ^{87}Sr, silica, iron, and alkali metals but little alumina and even less MgO and CaO. The abundance of the contaminant ranges form 18 to 32% by weight. The flow sequence on Mount Falla can be correlated with the flows on nearby Storm Peak by matching discontinuities in the chemical and isotopic parameters.

Mount Falla is located at lat. 84° 22′ S, long. 164° 55′ E in the Queen Alexandra Range of the Transantarctic Mountains. It is capped by 14 flat-lying lava flows of basaltic composition belonging to the Kirkpatrick Basalt. These flows are very similar to lava flows on nearby Storm Peak studied by Faure and others (1974). One of the unusual features of these rocks is their high initial $^{87}Sr/^{86}Sr$ ratios(0.7094 to 0.7133). High ratios have also been reported from other exposures of the Kirkpatrick Basalt and the Ferrar Dolerite by Compston and others (1968), Faure and others (1972), and Jones and others (1973). High initial $^{87}Sr/^{86}Sr$ ratios in basaltic rocks of Mesozoic age have also been observed in Tasmania by Heier and others (1965) and in the Parana Basin of Brazil by Halpern and others (1974).

The previous study of the lava flows on Storm Peak by Faure and others (1974) revealed systematic variations of the initial $^{87}Sr/^{86}Sr$ ratios and the concentrations of major oxide components in a stratigraphic sense. These variations were interpreted as progressive assimilation by basalt magma of a contaminant derived from the granitic rocks of the underlying continental crust. The present study was undertaken to test the applicability of this model to lava flows preserved on Mount Falla, which is located about 24 km northeast of Storm Peak.

Chemical compositions

The lava flows that cap Mount Falla have a total thickness of about 450 m and consist of 14 flows, numbered consecutively from the bottom up, as shown in Figure 88.1. The flows range in thickness from about 100 m for flow 1 to only 2 m for flow 7. One specimen from each of the 14 flows has been analyzed in this study.

Complete chemical analyses for flows 1 and 10 are shown in Table 88.1. The average concentrations of the major oxide components of the two flows from Mount Falla are similar to those of the Storm Peak suite (Faure and others, 1974) but differ in several respects from normal tholeiite basalts (Manson, 1967). The rocks on Mount Falla and Storm Peak have higher concentrations of SiO_2, FeO + Fe_2O_3, and K_2O than average tholeiite basalt but are depleted in TiO_2, Al_2O_3, MgO, and CaO. According to the chemical classification of Irvine and Baragar (1971), the volcanic rocks on Mount Falla are subalkaline tholeiite basalts. They plot entirely within the tholeiite field on the AFM diagram but do not show any evidence of the enrichment in alkalis that commonly occurs in the tholeiite series after initial enrichment in iron. Both rocks contain quartz, orthopyroxene, and clinopyroxene in their norms without olivine or nepheline.

The concentrations of SiO_2, Al_2O_3, total iron as FeO, MgO, Na_2O, and K_2O of the 14 flows are given in Table 88.2. The concentrations of FeO (total iron) and Na_2O + K_2O rise with increasing SiO_2, while those of Al_2O_3 and MgO decrease; these relationships are not unusual. In Figure 88.2 the concentrations of SiO_2, Al_2O_3, FeO (Total Fe), MgO, CaO, and Na_2O have been plotted versus their flow number. This diagram reveals that the concentrations of these components tend to change systematically upward through the section and thus as a function of time. The concentrations of SiO_2 and FeO (total Fe) decrease sequen-

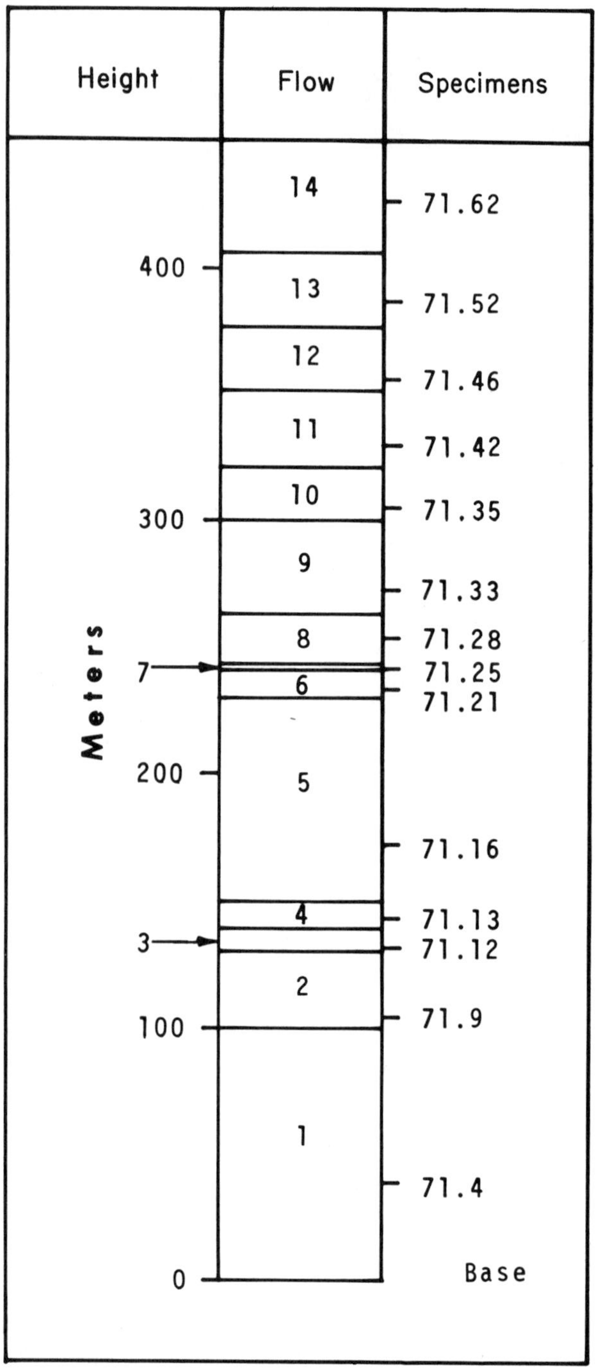

Figure 88.1. Sequence of basalt flows on Mount Falla in the Queen Alexandra Range, Transantarctic Mountains.

tially from flows 1 through 10, while those of Al_2O_3 and MgO (and CaO) increase, and Na_2O remains nearly constant. A discontinuity in these trends occurs between flows 10 and 11. However, the sense of variation in flows 11, 12, 13, and 14 is similar to that of the lower suite of flows (1 to 10). Evidently, the petrogenesis of these lava flows proceeded initially in some regular fashion which was interrupted after the extrusion of flow 10 and was later repeated. Figure 88.2 also reveals that the first flow to be extruded in each suite (flows 1 and 11) is more highly evolved chemically than flows extruded later. For example, flow 1 has a higher content of SiO_2 and FeO (total Fe) than succeeding flows. Similarly, its concentrations of Al_2O_3,

Table 88.1. Chemical composition (in wt. %) of Kirkpatrick Basalt specimens from Mount Falla and Storm Peak, Queen Alexandra Range

	71.4 Flow 1	71.35 Flow 10	Average Storm Peak	Average tholeiite
SiO_2	58.20	54.74	56.60	51.1
TiO_2	1.65	1.05	1.28	1.6
Al_2O_3	12.29	13.54	12.92	16.2
Fe_2O_3	6.35	3.65	4.50	3.1
FeO	6.57	6.81	7.52	7.6
MnO	0.16	0.20	0.22	0.17
MgO	2.53	5.14	3.44	6.2
CaO	6.72	9.52	7.91	9.9
Na_2O	2.34	2.13	2.29	2.5
K_2O	1.72	1.16	1.29	0.7
P_2O_5	0.183	0.121	0.16	0.22
H_2O+	1.48	0.97	1.62	0.7
H_2O-	0.49	0.43	0	0
CO_2	<0.05	0.75	0	0
Total	100.68	100.21	99.75	99.99
		C.I.P.W. norms[a]		
Q	17.9	10.0	14.0	
Or	10.6	7.2	7.8	
Ab	19.9	18.3	19.9	
An	18.1	24.2	21.4	
Opx	12.0	14.8	12.6	
Cpx	13.4	19.7	17.9	
Mt	4.6	3.7	4.2	
Il	3.2	2.0	2.4	
Norm plagioclase	47.6	56.9	51.8	
Norm color index	33.2	40.1	37.4	

Specimen analyst, A. S. McCreath and Son, Inc., Harrisburg.
Storm Peak data are from Faure and others (1974); tholeiite data from Manson (1967).
[a] Fe_2O_3 was reduced to less than TiO_2 + 1.5, as suggested by Irvine and Baragar (1971).

MgO, and CaO are lower. Apparently, the most highly differentiated fraction of the magma, which presumably occurred near the top of the chamber, was extruded first followed by flows which may have tapped deeper and less differentiated levels of the magma chamber.

Isotopic composition of strontium

The $^{87}Sr/^{86}Sr$ ratios of all of the specimens were measured on a mass spectrometer (Nuclide Corp. 6-60-S, single filament, Ta), and the concentrations of Rb and Sr were determined by X-ray fluorescence spectrometry (Diano Corp., Model XRD-6, LiF (220), Mo-target X-ray tube). The Eimer and Amend $SrCO_3$ standard was analyzed 16 times during this investigation and yielded an average $^{87}Sr/^{86}Sr$ ratio of 0.70800 ± 0.00005 (one standard deviation of the mean). All measured $^{87}Sr/^{86}Sr$ ratios were corrected for isotope fractionation to $^{86}Sr/^{88}Sr$ = 0.1194. The Rb and Sr determinations are based on calibrations obtained from the United States Geological Survey rock standards (G-2, GSP-1, BCR-1, AGV-1, and W-1). Replicate analyses of all specimens indicate reproducibilities of ± 1.2 ppm for Sr and ± 0.8 ppm for Rb. The initial $^{87}Sr/^{86}Sr$ ratios for each flow were calculated by means of the measured Rb/Sr ratios, assuming an age of 170 m.y. for the flows and setting λ (^{87}Rb) = 1.39×10^{-11}/hr. The age of the flows is known from K-Ar whole-rock dates, reviewed by Faure and others (1974), and from $^{40}Ar/^{39}Ar$ partial-release dates reported by Fleck and others (1977).

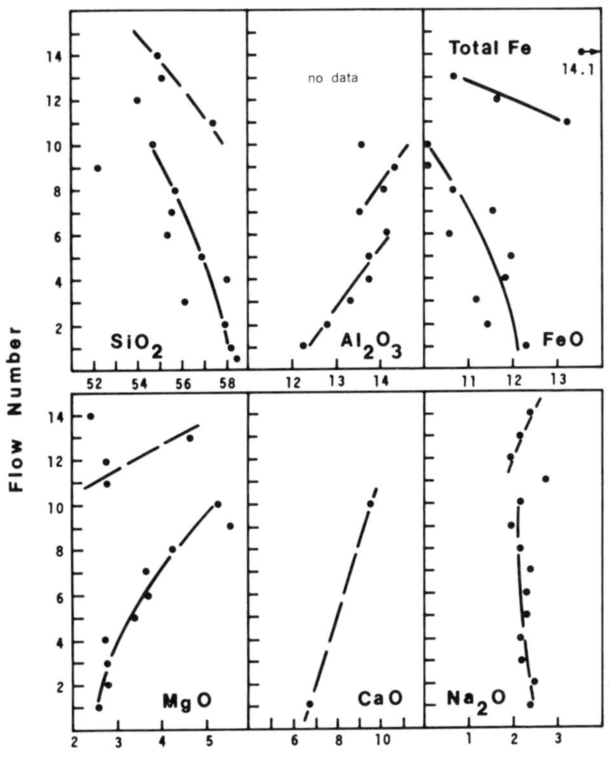

Figure 88.2. Stratigraphic variation of the major oxide components (in wt. %). A prominent discontinuity occurs between flows 10 and 11.

The concentrations of Rb and Sr listed in Table 88.3 range from 13.7 to 102.9 ppm and from 86.3 to 166.9 ppm, respectively. The average concentrations are: Rb = 54.5 ppm and Sr = 127.6 ppm. These results are very similar to those from the Storm Peak suite reported by Faure and others (1974) but differ from those of average tholeiites compiled by Prinz (1967). He reported Rb = 30 ppm and Sr = 400 ppm which yields a Rb/Sr ratio of 0.075. The rocks from the Mount Falla suite have a much higher Rb/Sr ratio of 0.45 ± 0.22, primarily because of their unusually low Sr concentrations. In this respect, the Mount Falla suite resembles oceanic tholeiites whose average Sr content is 124 ppm (Faure, 1978). However, oceanic tholeiites have low Rb concentrations of 1.11 ppm (Hart, 1971), and as a consequence, their average Rb/Sr ratio is only 0.009.

The initial $^{87}Sr/^{86}Sr$ ratios range from 0.7100 to 0.7131 and have a mean of 0.7118. This value is significantly higher than the initial $^{87}Sr/^{86}Sr$ ratios of most basaltic rocks, which range from 0.702 to 0.706 (Faure and Powell, 1972). However, the average initial $^{87}Sr/^{86}Sr$ ratio of the Mount Falla suite is very similar to that of the Storm Peak section for which Faure and others (1974) have reported a value of 0.7112. These high initial ratios are characteristic of the Kirkpatrick Basalt and of the Ferrar Dolerite in the Transantarctic Mountains from northern Victoria Land to the Dufek Massif of the Pensacola Mountains (Faure and others, 1972). However, the Jurassic basalts of Queen Maud Land have more normal initial $^{87}Sr/^{86}Sr$ ratios clustered about two means of 0.7037 ± 0.0004 and 0.7072 ± 0.0003 (Faure and Elliot, 1971).

The initial ratios of flows 1 to 6 are constant at 0.7129, whereas those of flows 7 to 10 decrease sequentially to 0.7101, as suggested in Figure 88.3. The initial ratios then increase abruptly to 0.7121 in flow 11 and subsequently decline to 0.7102 in flow 14. Evidently, the discontinuity between flows 10 and 11, recognized in Figure 88.2 on the basis of the concentrations of major elements, is also expressed in the initial $^{87}Sr/^{86}Sr$ ratios of these flows.

In contrast with certain elements of low atomic number (e.g., carbon, oxygen, and sulfur), the isotopes of Sr are not fractionated in nature. Moreover, the effects of instrumental fractionation are removed by a correction that is routinely applied to all measured $^{87}Sr/^{86}Sr$ ratios. Therefore, the systematic variation of the initial $^{87}Sr/^{86}Sr$ ratios and the change in this ratio between flows 10 and 11 must have a petrogenetic cause.

Before considering alternative petrogenetic models to explain the variation in the initial $^{87}Sr/^{86}Sr$ ratios, we examine the possibility that the initial $^{87}Sr/^{86}Sr$ ratios of the Mount Falla suite, as calculated from the data in Table 88.3, were adversely affected by changes in the Rb/Sr ratios after crystallization. If the Rb/Sr ratio of a specimen was altered recently, then the calculated initial $^{87}Sr/$

Table 88.2. Major element concentrations (in wt %) of flows from Mount Falla, Queen Alexandra Range

Specimen	71.62	71.52	71.46	71.42	71.35	71.33	71.28
Flow	14	13	12	11	10	9	8
SiO_2	54.94	55.12	54.02	57.39	54.74	52.12	55.74
Al_2O_3	—	—	—	—	13.54	14.32	14.06
FeO (total Fe)	14.12[a]	10.64[a]	11.65[a]	13.22[a]	10.09	10.08	10.67
MgO	2.39	4.66	2.71	2.74	5.25	5.58	4.21
CaO	—	—	—	—	9.52	—	—
Na_2O	2.38	2.14	1.94	2.70	2.13	1.97	2.12
K_2O	1.82	1.18	1.78	0.97	1.16	1.47	1.16
Specimen	71.25	71.21	71.16	71.13	71.12	71.9	71.4
Flow	7	6	5	4	3	2	1
SiO_2	55.54	55.32	56.90	58.04	56.10	57.90	58.20
Al_2O_3	13.53	14.13	13.72	13.79	13.34	12.81	12.29
FeO (total Fe)	11.55	10.53	11.99	11.84	11.12	11.41	12.28
MgO	3.62	3.67	3.37	2.72	2.73	2.76	2.53
CaO	—	—	—	—	—	—	6.72
Na_2O	2.33	2.27	2.23	2.14	2.17	2.45	2.34
K_2O	1.03	1.53	1.53	1.52	0.88	2.21	1.72

—: not determined.
Analysts, A. S. McCreath and Son, Inc., Harrisburg (SiO_2, Al_2O_3, FeO, Fe_2O_3, CaO, Na_2O, and K_2O) and L. M. Jones (MgO, Na_2O, and K_2O).
[a]Determined by XRF.

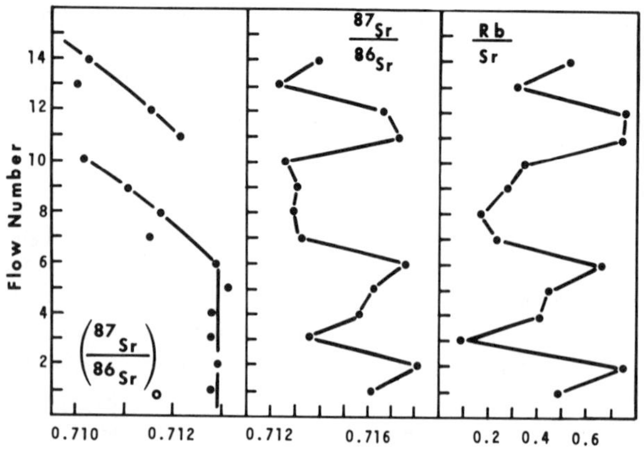

Figure 88.3. Stratigraphic variation of the initial $^{87}Sr/^{86}Sr$ ratios (left) and of the present $^{87}Sr/^{86}Sr$ and Rb/Sr ratios (center and right). Note the discontinuity of the initial $^{87}Sr/^{86}Sr$ ratio between flows 10 and 11.

^{86}Sr ratio of that specimen would be in error. However, if the alteration occurred soon after crystallization and if the $^{87}Sr/^{86}Sr$ ratio remained essentially unchanged by the alteration, then the calculated initial $^{87}Sr/^{86}Sr$ ratio will be correct. The accuracy of the initial ratios as calculated can be evaluated by consideration of the measured $^{87}Sr/^{86}Sr$ and Rb/Sr ratios plotted stratigraphically in Figure 88.3. The resulting profiles have remarkably similar shapes, which indicates that the present $^{87}Sr/^{86}Sr$ ratios of these specimens are controlled by their Rb/Sr ratios. This conclusion implies that any changes in the concentrations of Rb and Sr that may have occurred took place within a few million years after crystallization and therefore do not affect the calculated values of the initial $^{87}Sr/^{86}Sr$ ratios.

The variability of the initial $^{87}Sr/^{86}Sr$ ratios of volcanic rocks on Mount Falla and elsewhere makes them unsuitable for dating by the Rb-Sr method. The isochron method of dating is based in part on the assumption that different specimens of comagmatic plutonic or volcanic igneous rocks had the same initial $^{87}Sr/^{86}Sr$ ratio. The basaltic rocks on Mount Falla and Storm Peak, as well as suites of igneous rocks in island arcs and subduction zones elsewhere, do not satisfy this assumption. Such rocks should scatter about the isochron whose slope, and hence the date, may be biased as a result. Fortunately, flows 1 to 6 on Mount Falla seem to have constant initial $^{87}Sr/^{86}Sr$ ratios and therefore should be datable by the Rb-Sr isochron method. Figure 88.4 is the isochron based on these specimens; it yields a date of 173 ± 6 m.y. and an initial $^{87}Sr/^{86}Sr$ ratio of 0.7128 ± 0.0001. The date confirms the results of previous K-Ar age determinations on basalts from this area and is dependent only on the assumptions that these rocks had the same initial $^{87}Sr/^{86}Sr$ ratios and that their Rb/Sr ratios have not changed since crystallization except for decay of ^{87}Rb to ^{87}Sr.

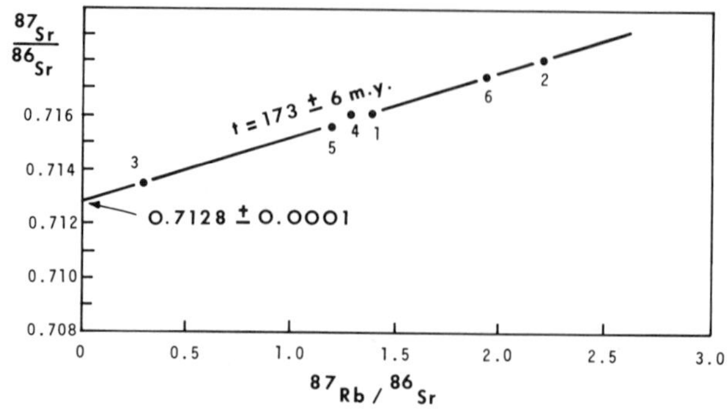

Figure 88.4. Rb-Sr whole-rock isochron for flows 1 through 6. The errors of the age and the initial $^{87}Sr/^{86}Sr$ ratio are one standard deviation.

Petrogenesis

The stratigraphic variations of the initial $^{87}Sr/^{86}Sr$ ratios and major oxide components of the lava flows on Mount Falla restrict the choices of possible petrogenetic models for these rocks. These rocks could not have formed by fractional crystallization of a single batch of basaltic magma. Such an origin may be compatible with the chemical characteristics of the flows, but does not account for the variation of the initial $^{87}Sr/^{86}Sr$ ratios. The chemical and isotopic properties of these flows may have resulted either from fractional melting of heterogeneous source rocks in the upper mantle or from systematic interactions of a basalt magma with derivatives of the continental crust.

Heterogeneous-Mantle Model

It is possible that the lithospheric part of the mantle under continents consists of a heterogeneous assemblage of rocks that separated from the asthenosphere or mesosphere at some time in the past. Such segregations may

Table 88.3. Initial $^{87}Sr/^{86}Sr$ ratios and concentrations of Rb and Sr, Kirkpatrick Basalt, Mount Falla

Specimen number	Flow	Rb (ppm)	Sr (ppm)	Rb/Sr	$^{87}Sr/^{86}Sr$	$^{87}Sr/^{86}Sr_0$[a]
71.62	14	68.7	130.7	0.525	0.7138	0.7102
71.52	13	42.0	132.3	0.318	0.7122	0.7100
71.46	12	65.9	86.3	0.763	0.7167	0.7115
71.42	11	102.9	136.5	0.754	0.7173	0.7121
71.35	10	41.9	120.7	0.347	0.7125	0.7101
71.33	9	46.5	166.9	0.279	0.7130	0.7111
71.28	8	22.2	132.1	0.168	0.7129	0.7117
71.25	7	24.9	109.6	0.227	0.7131	0.7115
71.21	6	65.7	98.7	0.666	0.7175	0.7129
71.16	5	66.5	149.6	0.444	0.7161	0.7131
71.13	4	57.6	140.1	0.411	0.7156	0.7128
71.12	3	13.7	142.3	0.096	0.7135	0.7128
71.9	2	77.8	102.7	0.757	0.7181	0.7129
71.4	1	66.4	137.6	0.483	0.7161	0.7128

[a] Initial $^{87}Sr/^{86}Sr$ ratio calculated with the assumptions that the age of the rocks is 170 m.y. and that $\lambda(^{87}Rb) = 1.39 \times 10^{-11}/yr$.

have elevated Rb/Sr ratios and may therefore result in Sr that is variously enriched in ^{87}Sr. If melting occurs in such a heterogeneous mantle, the $^{87}Sr/^{86}Sr$ ratio of the liquid will depend on the age and Rb/Sr ratio of the solid, and perhaps also on the degree of melting. If melting occurs in several different places, and if the Rb/Sr ratios of the resulting liquids are the same as those of the solids that melted, then the age of the mantle rocks could be determined from an isochron obtained by plotting initial $^{87}Sr/^{86}Sr$ ratios of the resulting plutonic or volcanic igneous rocks versus their $^{87}Rb/^{86}Sr$ ratios. This approach was taken by Brooks and others (1976) who calculated an "apparent age" of 470 ± 80 m.y. from data published by Compston and others (1968) and Faure and others (1972) for flows and sills of Mesozoic age from northern Victoria Land to the Dufek Massif, a distance of 2700 km. The significance of this date with respect to the age of the upper mantle is unclear, not only because of the widely scattered collecting sites of the specimens, but also because it seems unlikely that the rocks produced in the process would have the same Rb/Sr ratios as the source rocks from which magma was generated. The effect of changes of Rb/Sr ratios on "mantle isochrons" was recently discussed by Duncan and Compston (1976).

According to the heterogeneous-mantle model, the initial $^{87}Sr/^{86}Sr$ ratios of the flows on Mount Falla are attributable to magma formation in a heterogeneous mantle in which differences in the Rb/Sr ratio had given rise to corresponding differences in $^{87}Sr/^{86}Sr$ ratios. Presumably, flows 1 through 6 formed from the same batch of magma, whereas later flows were derived from separate magmas that originated from source rocks having different Rb/Sr ratios. The entire process of magma generation and subsequent extrusion must have occurred in an orderly sequence as indicated by the stratigraphic profile of initial $^{87}Sr/^{86}Sr$ ratios in Figure 88.3 and by the systematic chemical changes displayed in Figure 88.2.

If the flows on Mount Falla originated from several discrete sources in the lithospheric mantle, then they may define a line in coordinates of initial $^{87}Sr/^{86}Sr$ and $^{87}Rb/^{86}Sr$ ratios from which the age of the mantle source can be estimated. The relevant data in Figure 88.5*a* do not form the expected isochron. Flows 1 through 6 define a straight line of zero slope whereas the other specimens scatter widely. The colinearity of the first six flows arises from the fact that they have the same initial $^{87}Sr/^{86}Sr$ ratio but have widely varying Rb/Sr ratios. A second group of flows (7, 8, 9, 10, and 13) actually scatter about a line having a negative slope. Such lines can have no time significance but may be generated by mixing of two components having differing $^{87}Sr/^{86}Sr$ and Rb/Sr ratios.

In general, the attempt to account for the observed chemical and isotopic properties of the Mount Falla suite on the basis of selective melting in a heterogeneous mantle is unsatisfactory. The systematic variations that are observed require a degree of orderliness that seems improbable for such a process.

CRUSTAL-CONTAMINATION MODEL

The observed variations of the initial $^{87}Sr/^{86}Sr$ ratios of the lava flows on Mount Falla may be due to additions of radiogenic ^{87}Sr to the basalt magma. It is probable that the radiogenic ^{87}Sr originated from Rb-rich phases in the

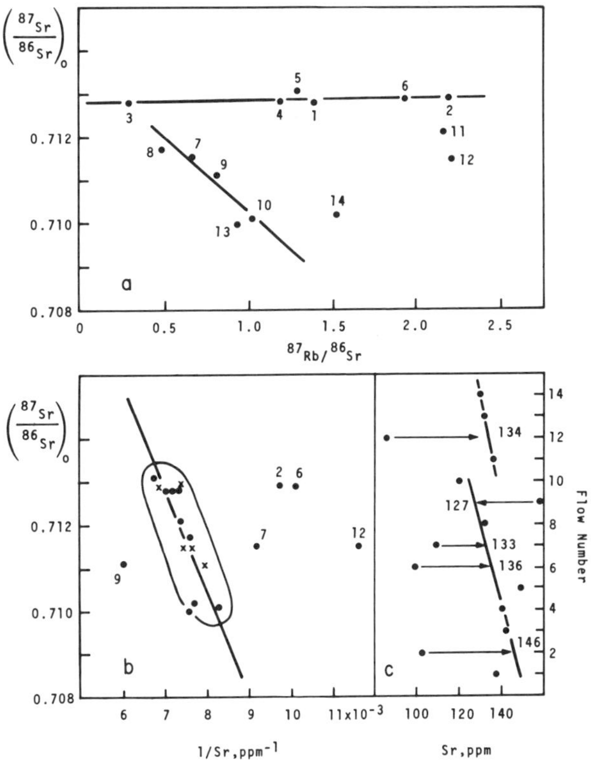

Figure 88.5. a. Plot of initial $^{87}Sr/^{86}Sr$ ratios versus $^{87}Rb/^{86}Sr$ to test for the existence of a "mantle isochron." No isochron is present. *b.* Mixing plot for basalt flows on Mount Falla. The numbered specimens scatter off the mixing line. *c.* Stratigraphic plot which shows that the deviant specimens have abnormal Sr concentrations. When Sr is adjusted to the trend line shown here, the data points on the mixing diagram move inside the envelope (X's shown in diagram *b*).

granitic rocks of the underlying continental crust. However, the details of the process by which it was transferred from the country rock into the magma are not immediately apparent. Comparisons between the stratigraphic profiles shown in Figures 88.2 and 88.3 suggest that increases in the initial $^{87}Sr/^{86}Sr$ ratio correlate with increasing concentration of silica, iron, and alkali metals and with decreasing concentrations of Al_2O_3, CaO, and MgO. These tendencies may indicate that the radiogenic ^{87}Sr was incorporated into the magma together with silica, iron, and alkali metals, but without much Al_2O_3, CaO, or MgO.

When two components having differing $^{87}Sr/^{86}Sr$ ratios and Sr concentrations are mixed in varying proportions, the resulting mixtures satisfy an equation of the form

$$\left(\frac{^{87}Sr}{^{86}Sr}\right) = A + \frac{B}{Sr_m}, \tag{1}$$

where A and B are constants and the subscript m identifies measured values of the mixtures (Faure and others, 1974; Boger and Faure, 1976; Shaffer and Faure, 1976). The equation is a hyperbola that is transformable into a straight line when $1/Sr_m$ is chosen as the abscissa. Therefore, two-component mixtures must fit a straight line in coordinates of $^{87}Sr/^{86}Sr$ and 1/Sr provided that (1) their Rb and Sr concentrations remained unchanged after mixing, (2) radiogenic ^{87}Sr produced in the specimens is subtracted, and (3) the $^{87}Sr/^{86}Sr$ ratios and Sr concentrations of the end members were constant while a given series of mixtures was produced. If these conditions are satisfied, then the mixing equation can be derived by fitting a straight line to data points in coordinates of initial $^{87}Sr/^{86}Sr$ ratios and 1/Sr.

The initial $^{87}Sr/^{86}Sr$ ratios and the reciprocals of the Sr concentrations of the flows on Mount Falla have been plotted on the mixing diagram in Figure 88.5*b*. Nine of the specimens (solid circles) cluster about a straight line, but the others scatter widely. The Sr concentrations of the deviant specimens may have been altered either by fractional crystallization of the magma after mixing or after extrusion and crystallization. This explanation seems to be confirmed by a stratigraphic plot of Sr concentrations in Figure 88.5*c*. The deviant specimens (2, 6, 7, 9, and 12) also diverge from the stratigraphic trend established by the other specimens. When the Sr concentrations of the deviant specimens are changed to the values indicated in Figure 88.5*c*, they then move *inside* the envelope drawn around the regular specimens in Figure 88.5*b* (shown by X's). The behavior of the deviant specimens on the mixing diagram is compatible with the conclusion that their Sr concentrations were changed after contamination of the magma from which they formed. However, the alteration cannot be due to recent weathering or to processes that occurred some time after crystallization because, in that case, specimens 2 and 6 would not fit on the isochron in Figure 88.4. Four of the deviant specimens (2, 6, 7, and 12) have apparently *lost* Sr, which means that their $^{87}Sr/^{86}Sr$ ratios probably remained unchanged. Specimen 9 is the only one that seems to have gained Sr, but its initial $^{87}Sr/^{86}Sr$ ratio fits into the stratigraphic profile in Figure 88.3.

Following the restoration of the deviant specimens, the mixing equation was derived by means of a least-squares fit to all of the data points. The result is:

$$\left(\frac{^{87}Sr}{^{86}Sr}\right) = 0.7286 - \frac{2.271}{Sr_m}. \qquad (2)$$

The coefficient of correlation is 0.67, which is highly significant for 14 pairs of variables. The Sr concentrations of the parent basalt and of the hypothetical contaminant were calculated from equation 2 for assumed values of their respective $^{87}Sr/^{86}Sr$ ratios. The isotopic composition of Sr in basaltic rocks varies only within narrow limits and is represented by a $^{87}Sr/^{86}Sr$ ratio of 0.704 ± 0.002 (Faure and Powell, 1972). The corresponding value of the hypothetical contaminant was taken as 0.720 ± 0.005, following Faure and others (1974). The resulting estimate of the Sr concentration of the basalt magma is 92 ppm, with a range of 85 to 100 ppm. The Sr content of the hypothetical contaminant is 264 ppm, ranging from 167 to 631 ppm. The upper limit seems to be excessive, which suggests that the $^{87}Sr/^{86}Sr$ ratio of the contaminant is probably less than 0.725.

The concentrations of the major oxides in the basalt magma and the hypothetical contaminant were estimated from the linear relationships between Sr and other elements. The results are listed in Table 88.4. The restored values of Sr concentrations (Figure 88.5*c*) were used throughout. Straight lines were fitted by eye, and the best estimates and ranges of the concentrations were determined both for the basalt magma and the contaminant. The estimated composition of the basalt magma is more reliable than that of the contaminant because: (1) its $^{87}Sr/^{86}Sr$ ratio is less uncertain, (2) the shape of the mixing hyperbola is such that its Sr content is insensitive to the assumed $^{87}Sr/^{86}Sr$ ratio, and (3) the extrapolations are shorter. In general, the inferred major element concentrations of the parent basalt are compatible with those of average tholeiite listed in Table 88.1.

The hypothetical contaminant has high concentrations of SiO_2 (68.5%), total iron as FeO (22.4%), and alkali metals ($Na_2O + K_2O = 6.2\%$). The concentrations of Al_2O_3 (5.8%), MgO, and CaO are low or close to zero. The upper limit of the Sr concentration of this component (631 ppm) results in nonsensical estimates of several major element concentrations: $SiO_2 = 104\%$, $Al_2O_3 = -17.0\%$, $FeO = 55.4\%$, and $MgO = -43.5\%$. The results thus confirm that the $^{87}Sr/^{86}Sr$ ratio of the hypothetical contaminant is less than 0.725.

The high concentrations of silica, iron, and alkalis of the contaminant may be explained in terms of a partial melt of a biotite-bearing granitic rock. It may also be the result of removal of selected elements from the granitic country rock by aqueous solutions at elevated temperatures. In this case, the isotopic composition of oxygen of the hybridized magma may have been changed as well, which may be reflected by anomalous (low?) $\delta^{18}O$ values of the rocks. No systematic efforts have yet been made to test this hypothesis. An example of partial melting of quartz monzonite xenoliths in basalt magma has been reported by

Table 88.4. Estimated chemical compositions of basalt magma and contaminant

	Basalt		Contaminant	
	Best value	Range	Best value	Range
$^{87}Sr/^{86}Sr$	0.704	0.702-0.706	0.720	0.715-0.725
Sr (ppm)	92	85-100	264	167-631
SiO_2(%)	51.9	51.2-52.7	68.5	59.1-104
Al_2O_3	16.5	16.9-16.0	5.8	11.8-(-17.0)
FeO (total Fe)	6.9	6.3-7.6	22.4	13.7-55.4
MgO	7.7	8.4-7.0	(-8.6)	0.6-(-43.5)
CaO	<14.2?	—	—	—
Na_2O	1.7	1.6-1.8	3.5	2.5-7.3
K_2O	1.0	0.9-1.0	2.7	1.7-6.4
f (max)	—	—	0.32	0.72-0.10
f (min)	0.17	0.21-0.13	—	—

—: not determined.
f (max): weight fraction of contaminant, based on specimen 71.16, assuming Sr_b = 92 ppm.
f (min): weight fraction of contaminant, based on specimen 71.52, assuming Sr_c = 264 ppm.

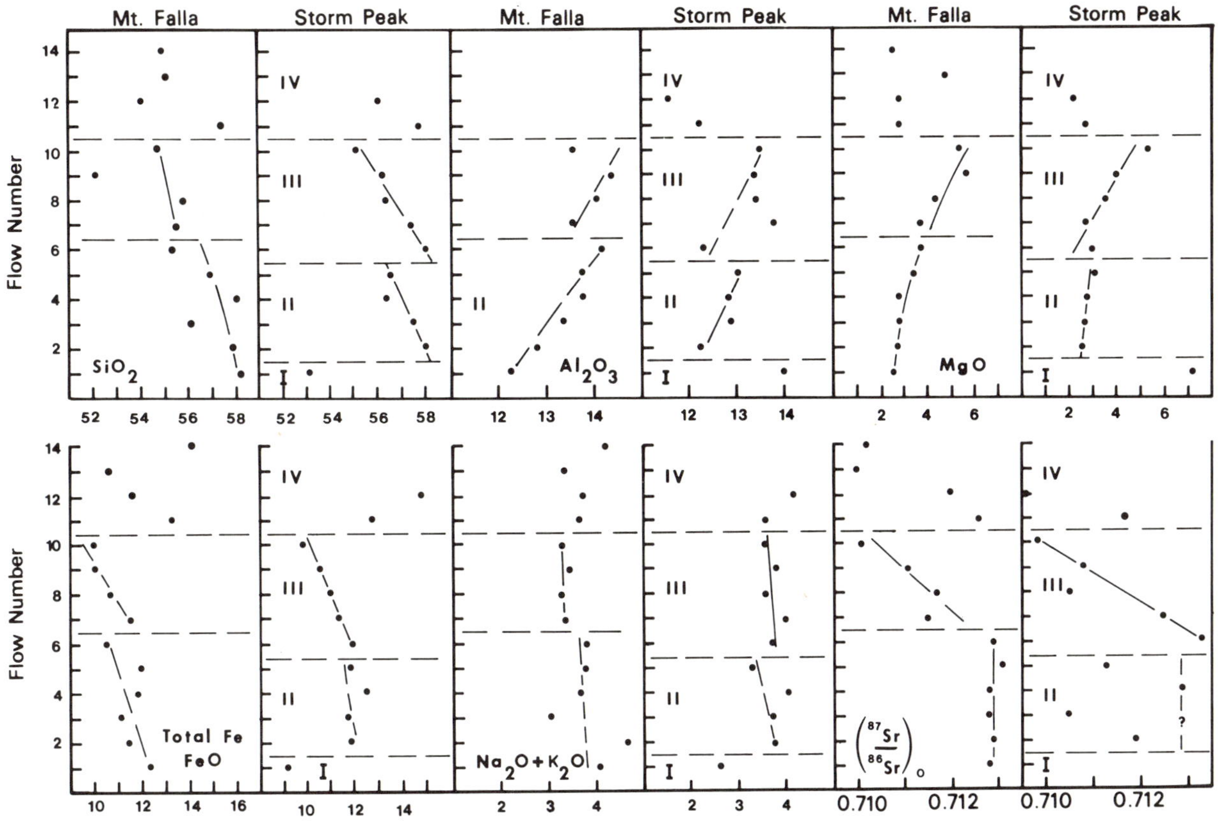

Figure 88.6. Comparison of the stratigraphic variation of major elements and initial $^{87}Sr/^{86}Sr$ ratios on Mount Falla and Storm Peak. The correlation is based on the chemical discontinuities recognized at the two sites.

Pushkar and Stoeser (1975). They found evidence that biotite had melted preferentially to form a brown (Fe-rich?) glass enriched in radiogenic ^{87}Sr.

The initial $^{87}Sr/^{86}Sr$ ratios of the Mount Falla flows, which range from 0.7100 to 0.7131, can be used to set limits on the extent of contamination. The calculation is based on the equation

$$Sr_m = Sr_c f + Sr_b (1 - f), \tag{3}$$

where Sr_m, Sr_c, and Sr_b are the concentrations of Sr in the specimen (*m*), in the contaminant (*c*), and in the basalt magma (*b*), respectively, and where *f* is the weight fraction of contaminant in a given specimen. Since $Sr_c = 264$ ppm and $Sr_b = 92$ ppm, equation 3 can be solved for *f*, given a value for Sr_m. The lowest level of contamination is suggested for flow 13 because its initial $^{87}Sr/^{86}Sr$ ratio is 0.7100. According to the mixing model (equation 2), its Sr content should be 122 ppm. With these assumptions, flow 13 contains 17.5% of the contaminant. Flow 5 is most highly contaminated having an initial $^{87}Sr/^{86}Sr$ ratio of 0.7131. The mixing model indicates a Sr content of 146.5 ppm for this flow and a contamination level of 31.5%. These estimates are listed in Table 88.4, together with possible ranges for these values based on the uncertainties of the Sr concentrations of the basalt magma and the contaminant. The range of *f* values for flow 13 (least contaminated) was calculated from the high and low Sr concentrations of the basalt magma, while holding Sr_c constant at 264 ppm. The analogous calculation for flow 5 (most contaminated) was done by varying Sr_c while holding Sr_b constant at 92 ppm. The results of these calculations indicate that the estimates of the level of contamination are *sensitive* to the Sr concentrations used in equation 3. The most reasonable set of assumptions indicates a range from 17.5 to 31.5%. The highest value ($f = 0.72$) arises from the assumption that the $^{87}Sr/^{86}Sr$ ratio of the contaminant is 0.715 and hence $Sr_c = 167$ ppm. The unreasonably large value of *f* indicates that the $^{87}Sr/^{86}Sr$ ratio of the contaminant is very probably greater than 0.715. These and earlier tests confirm the validity of the initial assumption that the $^{87}Sr/^{86}Sr$ ratio of the contaminant was greater than 0.715 but less than 0.725 and that 0.720 is probably a good approximation.

Correlation with Storm Peak

Although Mount Falla and Storm Peak are separated by a distance of only about 24 kilometers, a flow-by-flow correlation has not been possible in the field. The only correlations that seem reasonable based on thickness and stratigraphic position are flow 1 on Mount Falla with flow 2 on Storm Peak and flow 14 on Mount Falla with flow 12 on Storm Peak. These tentative correlations can be tested by matching discontinuities in the chemical and isotopic parameters of the flows at the two locations. The absolute values of these parameters are not necessarily reliable because both the chemical compositions and the initial $^{87}Sr/^{86}Sr$ ratios may vary throughout a given flow (Brookins and others, 1975). The reasons for this variation include (1) the initial heterogeneity of the magma, (2) differences in cooling rates and hence in fractional crystallization, and (3) local contamination of flows during extrusion or after crystallization.

Examination of Figure 88.6 makes clear that the discon-

Table 88.5. Correlation of flow sequence on Storm Peak and Mount Falla

Sequence	Storm Peak	Mount Falla
IV	Flows 11 and 12	Flows 11, 12, 13, and 14
III	Flows 6, 7, 8, 9, and 10	Flows 7, 8, 9, and 10
II	Flows 2, 3, 4, and 5	Flows 1, 2, 3, 4, 5, and 6
I	Flow 1	Absent

tinuity between flows 10 and 11 on Mount Falla corresponds to the break between flow series III and IV on Storm Peak. In addition, it is apparent that flow 1 on Storm Peak is not present on Mount Falla. The correlation of flow 12 (Storm Peak) with 14 (Mount Falla) and of flow 11 (Storm Peak) with 11 (Mount Falla) implies that flows 12 and 13 of Mount Falla are missing on Storm Peak. The discontinuity between series II and III (between flows 5 and 6) on Storm Peak may be represented by the break in the profiles of Al_2O_3 and the initial $^{87}Sr/^{86}Sr$ ratios between flows 6 and 7 on Mount Falla. The resulting correlations between flow sequences are listed in Table 88.5.

Summary

The chemical compositions and initial $^{87}Sr/^{86}Sr$ ratios of the lava flows on Mount Falla support the following conclusions. The rocks are subalkaline tholeiite basalts and are characterized by high concentrations of silica and iron and low concentrations of magnesium and calcium. The concentrations of SiO_2, FeO, and alkalis decrease upward through the section, while those of Al_2O_3, MgO, and CaO increase. All chemical parameters have a sharp discontinuity between flows 10 and 11. A lesser discontinuity is indicated by Al_2O_3 between flows 6 and 7. The rocks have low Sr concentrations (127.6 ppm) and high Rb concentrations (54.5 ppm) compared to average tholeiite basalts. Their average Rb/Sr ratio is 0.45 ± 0.22, which is higher than in other basaltic rocks.

The initial $^{87}Sr/^{86}Sr$ ratios of the flows range from 0.7100 to 0.7131 and vary stratigraphically. A well-defined discontinuity occurs between flows 10 and 11, and a lesser one between flows 6 and 7. Flows 1 through 6 have a constant initial $^{87}Sr/^{86}Sr$ ratio and are therefore datable by the Rb-Sr method. An isochron indicates an age of 173 ± 6 m.y. This result confirms previous K-Ar dates that suggested an age of about 170 m.y. The close positive correlation between $^{87}Sr/^{86}Sr$ and Rb/Sr ratios, as well as the success in dating flows 1 through 6, indicate that the Rb and Sr concentrations of the flows have not been altered since crystallization and cooling.

The high initial $^{87}Sr/^{86}Sr$ ratios and the systematic variation of this parameter are explainable either by heterogeneous magma sources in the upper (lithospheric) mantle or by variable contamination of basalt magma with radiogenic ^{87}Sr derived from the continental crust. The heterogeneous-mantle hypothesis is not supported because the data points do not form a "mantle isochron" and because batch melting in a heterogeneous mantle is not likely to produce the highly systematic stratigraphic variations of both chemical and isotopic parameters. The contamination hypothesis permits derivation of a mixing equation from the observed linear relationship between the initial $^{87}Sr/^{86}Sr$ ratios and the reciprocal of Sr concentrations. This model also permits estimates of the chemical composition of the basalt magma and the contaminant. The parent basalt magma was a tholeiite basalt with a low Sr content of about 92 ppm. The contaminant contained high proportions of silica, iron, and alkalis with lesser or near zero content of Al_2O_3, MgO, and CaO. The proportion of the contaminant ranges from 17.5 to 31.5% by weight. The flow sequence on Mount Falla was correlated with similar flows on nearby Storm Peak by matching discontinuities of the chemical and isotopic parameters. Hybridization of basalt magma by interaction with old granitic basement rocks has played an important role in determining both the isotopic composition of Sr and the bulk chemical composition of the Kirkpatrick Basalt on Mount Falla and on Storm Peak.

Acknowledgments. We thank Lois M. Jones for chemical analyses and Philip R. Kyle for discussions regarding alternate petrogenetic models. However, the conclusions expressed here are our responsibility. This research was supported by the Division of Polar Programs of the National Science Foundation through Grant GA-12315 to D. H. Elliot and by Grants OPP72-00459-AO2 and OPP76-11871 to G. Faure. This is Contribution no. 40 of the Laboratory for Isotope Geology and Geochemistry, Department of Geology and Mineralogy, Ohio State University, and contribution IPS C-323 of the Institute of Polar Studies, Ohio State University.

References cited

Boger, P. D., and Faure, G. 1976. Systematic variations of sialic and volcanic detritus in piston cores from the Red Sea. *Geochim. cosmochim. Acta, 40,* 731–42.

Brookins, D. G., Carden, J. R., and Laughlin, A. W. 1975. Additional note on the isotopic composition of strontium in McCartys flow, Valencia County, New Mexico. *Earth planet. Sci. Lett., 25,* 327–30.

Brooks, C., James, D. E., and Hart, S. R. 1976. Ancient lithosphere: its role in young continental volcanism. *Science, N.Y., 193,* 1086–94.

Compston, W., McDougall, I., and Heier, S. K. 1968. Geochemical comparison of the Mesozoic basaltic rocks of Antarctica, South Africa, South America and Tasmania. *Geochim. cosmochim. Acta, 32,* 129–49.

Duncan, R. A.. and Compston, W. 1976. Sr-isotopic evidence for an old mantle source region for French Polynesian volcanism. *Geology, 4.* 728–32.

Faure, G., 1978. Concentration of strontium in igneous rocks. *(In* Wedepohl, K. H., *ed. Handbook of Geochemistry,* Section 38-E. Heidelberg, Springer-Verlag, E1–14.)

Faure, G., and Elliot, D. H. 1971. Isotope composition of strontium in Mesozoic basalt and dolerite from Dronning Maud Land. *Bull. Br. antarct. Surv.,* No. 25, 23–27.

Faure, G., and Powell, J. L. 1972. *Strontium Isotope Geology.* Heidelberg, Springer-Verlag, 188 pp.

Faure, G., Bowman, J. R., Elliot, D. H., and Jones, L. M. 1974. Strontium isotope composition and petrogenesis of the Kirkpatrick Basalt, Queen Alexandra Range, Antarctica. *Contr. mineral. Petrol., 48,* 153–69.

Faure, G., Hill, R. L., Jones, L. M., and Elliot, D. H. 1972. Isotope composition of strontium and silica content of Mesozoic basalt and dolerite from Antarctica. *(In* Adie, R. J., *ed. Antarctic Geology and Geophysics.* Oslo, Universitetsforlaget, 617–24.)

Fleck, R. J., Sutter, J. F., and Elliot, D. H. 1977. Interpretation of discordant $^{40}Ar/^{39}Ar$ age-spectra of Mesozoic tholeiites from Antarctica. *Geochim. cosmochim. Acta, 41,* 15–33.

Halpern, M., Cordani, J. G., and Berenholc, M. 1974. Variations in strontium isotopic composition of Parana Basin volcanic rocks of Brazil. *Revta bras. Geosc., 4,* 223–27.

Hart, S. R. 1971. K, Rb, Cs, Sr and Ba contents and Sr isotopic ratios of ocean floor basalts. *Phil. Trans. R. Soc. Lond.,* Ser. A *268,* 573–87.

Heier, K. S., Compston, W., and McDougall, I. 1965. Thorium and uranium concentrations, and the isotopic composition of strontium in the differentiated Tasmanian dolerites. *Geochim. cosmochim. Acta, 29,* 643–59.

Irvine, T. N., and Baragar, W. R. A. 1971. A guide to the chemical classification of the common volcanic rocks. *Can. Jnl Earth Scis, 8,* 523–48.

Jones, L. M., Walker, R. L., Hall, B. A., and Borns, H. W., Jr. 1973. Origin of the Jurassic dolerites and basalts of southern Victoria Land. *Antarct. Jnl U.S., 8,* No. 5, 268–70.

Manson, V. 1967. Geochemistry of basaltic rocks: major elements. (*In* Hess, H. H., *ed. Basalts.* 1. New York, Wiley Interscience, 213–69.)

Prinz, M. 1967. Geochemistry of basaltic rocks: trace elements. (*In* Hess, H. H., *ed. Basalts.* 1. New York, Wiley Interscience, 271–324.)

Pushkar, P., and Stoeser, D. B. 1975. $^{87}Sr/^{86}Sr$ ratios in some volcanic rocks and some semifused inclusions of the San Francisco volcanic field. *Geology, 3,* 669–71.

Shaffer, N. R., and Faure, G. 1976. Regional variation of $^{87}Sr/^{86}Sr$ ratios and mineral compositions of sediment from the Ross Sea, Antarctica. *Geol. Soc. Am. Bull., 87,* 1491–1500.

89

Volcanic Record of Cenozoic Glacial History in Marie Byrd Land and Western Ellsworth Land: Revised Chronology and Evaluation of Tectonic Factors

W. E. LEMASURIER
Geology Department, University of Colorado, Denver, Colorado 80202, U.S.A.

D. C. REX
Department of Earth Sciences, University of Leeds, Leeds LS2 9JT, U.K.

Abstract. The volcanic record of glacial history previously reported is now geographically extended to cover the area from the Fosdick Mountains (lat. 145° W) to the Hudson Mountains (lat. 100° W). A total of 17 new and repeated K-Ar ages on hyaloclastite sequences are presented, and preliminary results of a comparative study of submarine and intraglacial volcanic deposits from several other regions are reported.

Hyaloclastite sequences are found in a coastal sector 1500 km long and extending 400 km inland, within the area covered by the 1966–69 Marie Byrd Land Survey. Thicknesses are commonly in the range of 100–2000 m, and the ages of these deposits now appear to extend from late Oligocene to late Pleistocene. A marine origin for these deposits seems unlikely, particularly in comparison with known submarine volcanic sequences, because no Cenozoic marine sedimentary beds have been found in the entire region. The absence of pillow lavas in all localities except the Hudson Mountains seems to be related to characteristics of the lava and its mode of eruption, rather than to some aspect of glacial or tectonic history. The hyaloclastite in this region seems to be predominantly or entirely the product of eruptions beneath a continental ice sheet that was present during each eruptive episode. Because both volcanism and the formation of deep subglacial grabens seem to postdate the formation of the ice sheet, it seems likely that the original West Antarctic ice sheet formed on a subaerial erosion surface—perhaps concurrent with a period of regional doming—and that the base of the ice sheet has been tectonically depressed below sea level since that time. Hughes has suggested that the Pleistocene hyaloclastite in this region may record major fluctuations in the level of the West Antarctic ice sheet. Moraine data acquired during the 1967–68 Marie Byrd Land Survey are presented which may support his suggestion, though not necessarily on the same time scale.

Within roughly the past decade the number of recognized occurrences of hyaloclastite and related rocks in Antarctica has grown from three widely separated localities (Harrington, 1958; Craddock and others, 1964; Doumani, 1964) to include nearly every volcanic locality along the Pacific margin of the continent (Figure 89.1). In the Antarctic Peninsula region, hyaloclastite has been described from James Ross Island (Nelson, 1966), Deception Island (Baker and others, 1975), and Alexander Island (Bell, 1973). In Ellsworth Land they have been reported from the Jones Mountains (Craddock and others, 1964; Rutford and others, 1968, 1972), and the Hudson Mountains (Lopatin and Polyakov, 1974; K. E. LaPrade, personal communication, 1972). Thick, extensive hyaloclastite sequences occur in Marie Byrd Land (LeMasurier, 1972*a,b*), on the coast of northern Victoria Land (Hamilton, 1972), and on Ross Island and neighboring Beaufort Island (Harrington, 1958; Luckman, 1974; Lyon, 1974; Kyle and Treves, 1974; Treves, 1975). Altogether, they occur in a discontinuous coastal belt, roughly 5000 km long, that is one of the most extensive and voluminous continental occurrences of hyaloclastite in the world.

Hyaloclastite may form during volcanic eruptions either under water or under ice. In the latter case, the growth of the volcano is generally visualized as taking place within a large vault, or lake, formed by melting within the glacier. It is difficult, however, to distinguish hyaloclastite of submarine origin from that of subglacial origin because the two types appear to be petrographically identical. A submarine eruption environment for the Deception Island hyaloclastite has never been questioned because of its recent activity and its location in Bransfield Strait (Baker and others, 1975). James Ross Island is a Plio-Pleistocene volcanic center (Rex, 1972), with good exposures of palagonitized hyaloclastite breccia up to elevations of about 900 m and an ice-covered summit elevation of 1628 m. It probably had a submarine origin as well, but the evidence consists only of the presence of ophiuroids in tuffs at one locality and the fact that it is an island (Nelson, 1966). Deuterium/hydrogen and $^{18}O/^{16}O$ analyses provide the basis for interpretations by Lyon (1974) and Luckman (1974) that the Ross Island hyaloclastite was produced by eruptions that rose from the floor of the Ross Sea and broke through a floating ice shelf. The discovery of dia-

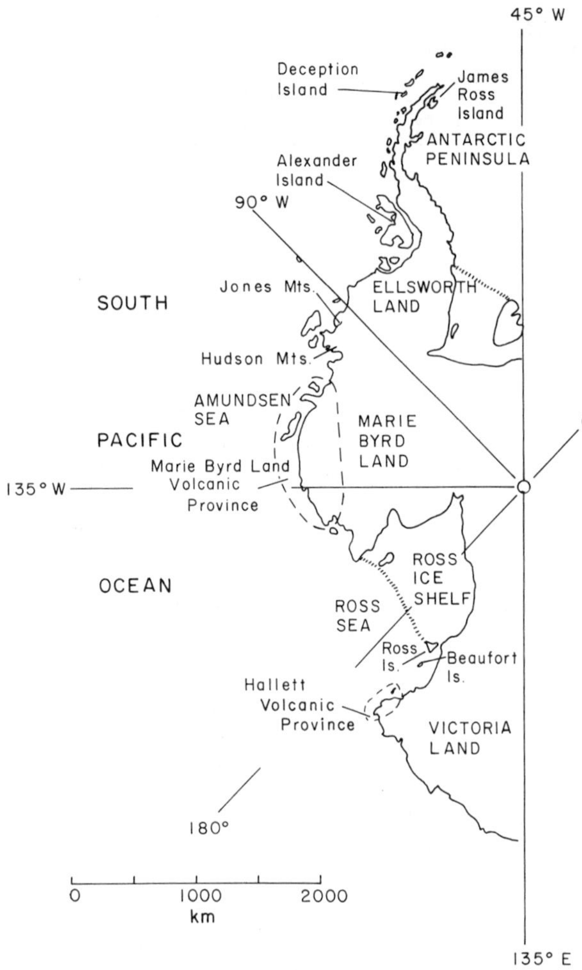

Figure 89.1. Antarctic location map, showing reported hyaloclastite localities.

toms in the Hut Point hyaloclastite (DVDP hole 3) at depths between 120 and 381 m below sea level supports this interpretation (Kyle and Treves, 1974).

The Jones Mountains hyaloclastite deposits overlie lenses of tillite and a glacially striated bedrock pavement (Rutford and others, 1968). This is the only direct field evidence of subglacial eruption reported from an Antarctic hyaloclastite locality. Marie Byrd Land hyaloclastite sequences have been interpreted as subglacial in origin because of their thicknesses, elevations, and inland extent (LeMasurier, 1972*a,b*). Similar reasoning applies to the Hallett Coast exposures (Hamilton, 1972). A preliminary study of the oxygen isotope composition of some Marie Byrd Land hyaloclastite produced inconclusive results (D. B. Wenner, written communication, 1973).

In summary, past experience suggests that field studies and stable isotope analyses may provide direct information about the origins of Antarctic hyaloclastite under favorable circumstances, but that most interpretations must rely on indirect reasoning, involving a knowledge of glacial and tectonic history. It nevertheless appears that few of the Antarctic deposits can be satisfactorily interpreted without postulating an overlying layer of glacial ice. The K-Ar ages of these deposits have suggested that the ice sheet has been in existence throughout a large part of Cenozoic time (Craddock and others, 1964; Rutford and others, 1968; Hamilton, 1972; LeMasurier, 1972*a,b*).

Scope of paper

This paper focuses on the intraglacial* volcanic deposits of Marie Byrd Land and western Ellsworth Land that were visited during the 1966–67, 1967–68, and 1968–69 reconnaissance surveys. Seventeen new K-Ar age determinations are presented. Seven of these are from localities in the Hudson and Fosdick Mountains, which have not been dated previously. The remainder are new age determinations of specimens that have been dated previously (LeMasurier, 1972*a*) but which needed reevaluation because of the potential problems of dating hyaloclastite by K-Ar techniques. The ages of hyaloclastite sections at 10 localities have now been checked either by analysis of the same specimens in two different laboratories or by analysis of several specimens from the same collection site.

The new data support most of the interpretations presented earlier (LeMasurier, 1972*a,b*), but enlarge the geographic region that must be considered in interpreting these deposits. This paper also reevaluates these deposits in the light of newly published geophysical and glaciologic data that bear on tectonic and glacial history and that provide new insights on the significance of the hyaloclastite sequences.

Chronology of hyaloclastite deposits

The ages of hyaloclastite specimens collected from the area between the Fosdick Mountains (lat. 76° 30′ S, long. 144° 00′ W) and the Jones Mountains (lat. 73° 30′ S, long. 95° 00′ W) are summarized in Table 89.1 (see Figure 89.2 for locations). Because basaltic glasses commonly leak argon, material selected for dating was, for the most part, holocrystalline basalt from cogenetic lenses or nodules in hyaloclastite. Whole-rock analysis of this basalt was the main method of age determination, but a few feldspar separates were dated. We avoided specimens that contained coarse-grained, complexly zoned feldspars, or abundant ultramafic nodules, because these materials tend to retain argon from earlier magmatic episodes and give anomalously old dates. Multiple specimens from the same collection site were dated at Mount Takahe (specimens 65B and 65C) and Turtle Peak (specimens 61A and 61B). If the work of Rutford and others (1972) and of Laudon is included, then specimens from different stratigraphic levels in the same section have now been dated from Mount Perkins in the Fosdick Mountains, Mount Manthe in the Hudson Mountains, and the Jones Mountains (Table 89.1).

All specimens from Mount Manthe in the Hudson Mountains seem to be Pliocene in age, and the 200 m section seems to be nearly the same age from bottom to top, suggesting that it formed during one eruptive episode. The Fosdick Mountains (Mount Perkins and North Flank) rocks were difficult to analyze because they contained glass and weathering products. The large discrepancy between the two whole-rock and the two feldspar ages suggests either argon loss from the glass or argon retention in the feldspars. We favor the Pliocene ages for these rocks because the fresh morphology of the volcanic cones sug-

* The term intraglacial is used for hyaloclastite and pillow lava formed by eruptions into or beneath glaciers, as opposed to subglacial deposits which could refer to any deposit found beneath a glacier, regardless of origin.

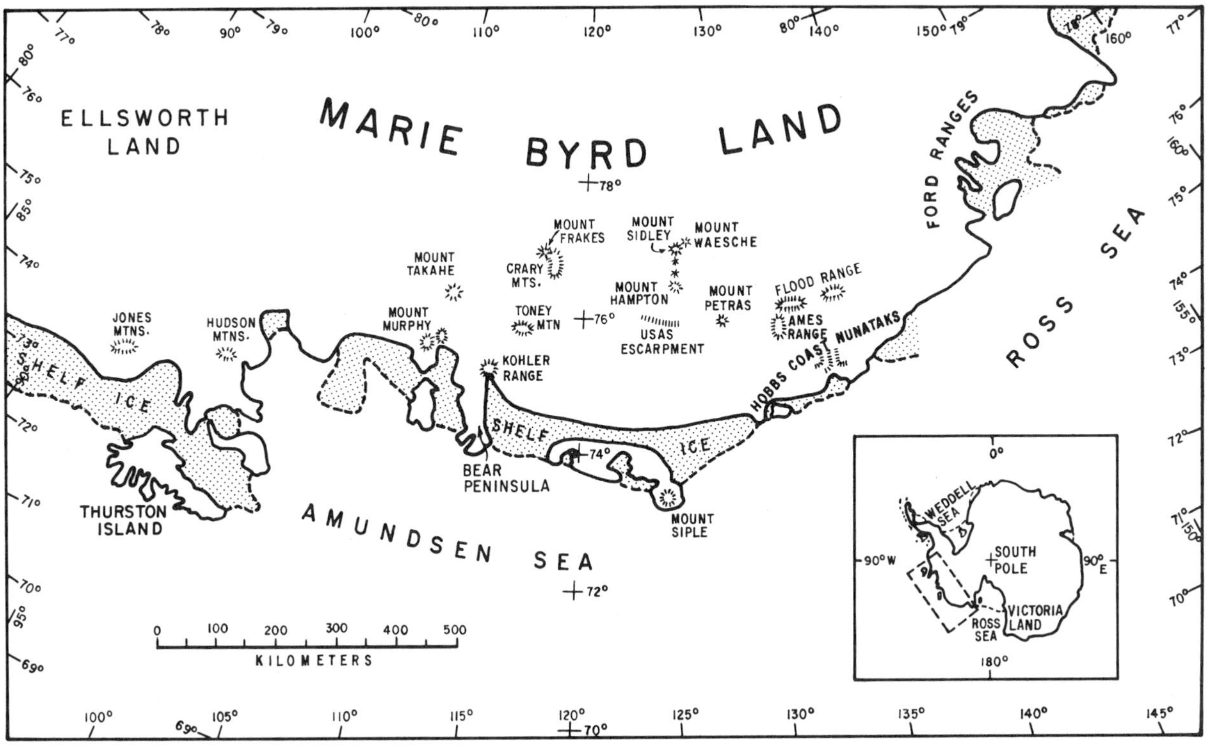

Figure 89.2. Location map for Marie Byrd Land and western Ellsworth Land.

gests that they are very young (Warner, 1945), but the dating of the two Fosdick Mountains localities must be regarded as unsatisfactory.

Four of the central Marie Byrd Land specimens (6c, 13b, 58b, and 62A) presented no analytical problems and gave reasonably good confirmation of the dates reported earlier (LeMasurier, 1972*a*). Reanalysis of the Bowyer Butte specimen (5) suggests that it is middle Miocene rather than late Oligocene in age. The Mount Steere specimen (73) is somewhat altered by oxidation, but gave a date only slightly outside the limits of error reported earlier, after oxidation products were separated from the rock. The Turtle Peak specimens (61A and 61B) were difficult to analyze, but this section is unusually interesting because of its thickness (800–1000 m) and the fact that it seems to be one of the oldest hyaloclastite sections in the region. The analytical range of 10–20 m.y. (Miocene) adequately represents our best estimate of the age of this section. It is certainly not Eocene, as previously reported (LeMasurier, 1972*a*).

The revised age data show the following major features. (1) The oldest hyaloclastite is 27 ± 1 m.y. in age. These are the oldest Cenozoic extrusive rocks that have been found in Marie Byrd Land or Ellsworth Land. The age of the Dorrel Rock alkaline gabbro (42 ± 2, 44 ± 2 m.y.) suggests that there may also have been Eocene volcanic activity in this region, but the apparently cogenetic dikes in the gabbro give an age of 28 ± 2 m.y. (Table 89.1, specimen 60D). We suspect, therefore, that the gabbro-syenite complex at Dorrel Rock formed about 28 m.y. ago, perhaps contemporaneously with the volcanic activity noted above (Table 89.1, specimen 58b), and that the age obtained on the gabbro is a result of excess argon in the feldspars. It seems likely that there was volcanic activity in this region in middle Oligocene time, but we are uncertain whether there was any activity before that time. (2) The new and revised hyaloclastite dates do not fall into groups that support a model with periods of glaciation and deglaciation. Combining these data with all available data on subaerial volcanoes (LeMasurier and Wade, 1976) does not reveal any intervals of volcanism that were predominantly subaerial or predominantly subaquatic, unless they were of such short duration that they could not generally be identified by K-Ar dating.

Lithologic character and spatial relationships of hyaloclastite deposits

The general nature and distribution of Marie Byrd Land hyaloclastite deposits have been described previously (LeMasurier, 1972*a,b*). The major features of these deposits are: (1) they are found nearly everywhere that the base of the volcanic sequence is exposed; (2) in some localities very thick sections are entirely hyaloclastite; (3) they are predominantly basaltic in composition; and (4) they are not associated with pillow lavas. Typical hyaloclastite deposits are subhorizontally stratified accumulations of ash and lapilli-sized sideromelane and tachylyte clasts. They range in thickness from more than 2000 m to less than 100 m. Discontinuous cogenetic lenses and nodules of holocrystalline lava are commonly interbedded with hyaloclastite. Varying degrees of oxidation and palagonitization of vitric clasts produce brown, yellow, rust, and black colors in individual sections. Most hyaloclastite is extremely friable, but the nature of glacial erosion is such that it is commonly exposed in cliffs. In spite of the generally protective role of cold ice sheets, glacial erosion of these deposits has been unusually rapid (Andrews and LeMasurier, 1973).

The Hudson Mountains hyaloclastite that we have examined differs from that in Marie Byrd Land in some petrographic details and also in the fact that it is associated with well-developed pillow lava at Mount Nickens (Lopatin and Polyakov, 1974; K. R. LaPrade, personal

Table 89.1. Potassium-argon ages of Cenozoic intraglacial volcanic deposits in Marie Byrd Land and western Ellsworth Land

				New data				
Specimen number	Location	Material[a] dated	Previously reported age (m.y.)	K (wt %)	Vol^{40}Ar rad. scc/g x 10^{-5}	%^{40}Ar rad.	Apparent age (m.y.)	Remarks
65C	Mt. Takahe	WR	<0.240					LeMasurier, 1972*a*
	(76°00′S, 112°00′W)							
65B	Mt. Takahe	WR	0.2 ± 0.2					LeMasurier, 1972*c*
67A	Mt. Takahe	WR	0.3 ± 0.3					LeMasurier, 1972*c*
62A	Mt. Murphy			0.70	0.0023	4.6	0.90 ± 0.14	
	(75°30′S, 116°00′W)	WR	0.82 ± 0.14		0.0028	8.5		LeMasurier, 1972*a*
66-D-28	Mt. Perkins, Fosdick Mtns	WR		0.57	0.0077	9.4	3.4 ± 0.3	Stratigraphically below 66-D-30
	(76°30′S, 144°00′W)							
28-3A	Velie Nuntak, Hudson Mtns	WR		0.542	0.0079	15.6	3.6 ± 0.2	
	(74°22′S, 99°00′W)				0.0077	15.1		
6C	Shibuya Peak, Hobbs Coast	WR	6.6 ± 0.7	1.00	0.0173	22.9	4.4 ± 0.2	
	(75°40′S, 133°30′W)				0.0176	25.0		LeMasurier, 1972*a*
66-D-30	Mt. Perkins, Fosdick Mtns	WR		0.59	0.0107	2.3	4.5 ± 0.5	Upper part of hyaloclastite section roughly 40 m thick
42-6A	Mt. Manthe, Hudson Mtns	WR		0.686	0.0121	13.4	4.8 ± 0.3	Top of 200 m hyaloclastite section
	(74°50′S, 99°30′W)				0.0141	11.0		
42-5A	Mt. Manthe	WR		0.733	0.0129	31.8	4.5 ± 0.2	Below 42-6A
						33.2		
42-4A	Mt. Manthe	WR		0.650	0.0135	11.4	4.9 ± 0.3	Below 42-5A
					0.0122	17.1		
H-6	Mt. Manthe	WR					5.5 ± 1.9	Subaerial cap of 200 m hyaloclastite section (T. S. Laudon, unpub. ms.)
H-4	Mt. Manthe	WR					8.3 ± 1.0	Nodule within hyaloclastite roughly 50 m below top of 200 m section (Laudon, unpub. ms.)
H-2	Mt. Manthe	WR					4.8 ± 1.6	Middle of hyaloclastite section (Laudon, unpub. ms.)
	Jones Mtns	WR					7–10;	500 m section dated 7–10 m.y. by average of 10 K-Ar dates (Rutford and others, 1972). One 12.1 m.y. fission track date (Rutford and others, 1973)
	(73°30′S, 95°00′W)						12.1 ± 1.7	
73	Mt. Steere, Crary Mtns	WR	7.0 ± 1.1	0.97	0.0331	47.0	8.3 ± 0.3	
	(76°30′S, 118°00′W)				0.0321	38.0		LeMasurier, 1972*a*
61A	Turtle Peak	WR		1.54	0.0795	6.1	13 ± 3	Base of 800–1000 m (est. from air photos) section of hyaloclastite
	(75°15′S, 111° 30′W)							
61B	Turtle Peak	WR	42 ± 9	1.35	0.0891	1.4	15 ± 5	Same locality as 61A
					0.0793	1.2		
5	Bowyer Butte	WR	23.2 ± 2.1	0.71	0.0405	3.5	13 ± 2	LeMasurier, 1972*a*
	(75°00′S, 135°00′W)				0.0405	3.5		
66-D-27	Mt. Perkins, Fosdick Mtns	Feldspar		0.137	0.0104	7.5	19 ± 2	Stratigraphically below 66-D-28
56b	Mt. Aldaz, USAS Escarpment	WR	19.4					LeMasurier, 1972*a*. Material remaining from previous analysis unsuitable for dating because of abundant ultramafic nodules
	(76°03′S, 124°30′W)							
13b	Mt. Petras	WR	22.2 ± 1.6	1.35	0.1158	69.8	22 ± 1	LeMasurier, 1972*a*
	(75°50′S, 128°30′W)				0.1187	75.7		
58b	USAS Escarpment	WR	31.3 ± 2.01	1.36	0.1476	60.2	27 ± 1	LeMasurier, 1972*a*
	(75°58′S, 125°15′W)				0.1476	6.15		
66-D-91	North Flank, Fosdick Mtns	Feldspar		0.11	0.0173	13.8	39 ± 4	Scoriaceous olivine basalt
	(76°28′S, 144°50′W)							
60A	Dorrel Rock	WR	53.1 ± 4.2	0.44	0.0785	55.1	44 ± 2	Coarse-grained alkaline gabbro
	(75°27′S, 111°20′W)			0.35	0.0605	53.2	42 ± 2	
					0.0593	53.1		
60D	Dorrel Rock	WR		4.83	0.5058	81.7	28 ± 2	Aegerine syenite dike perpendicular to flow structure in the gabbro
					0.5817	81.8		

Constants: $\lambda_\beta = 4.72 \times 10^{-10} yr^{-1}$; $\lambda_e = 0.58 \times 10^{-10} yr^{-1}$; $^{40}K/K = 0.0119$ atomic %.
WR: whole rock.
[a] All dated materials are cogenetic basalt lenses, or nodules in hyaloclastite, or feldspars separated from these nodules, unless otherwise noted under remarks.

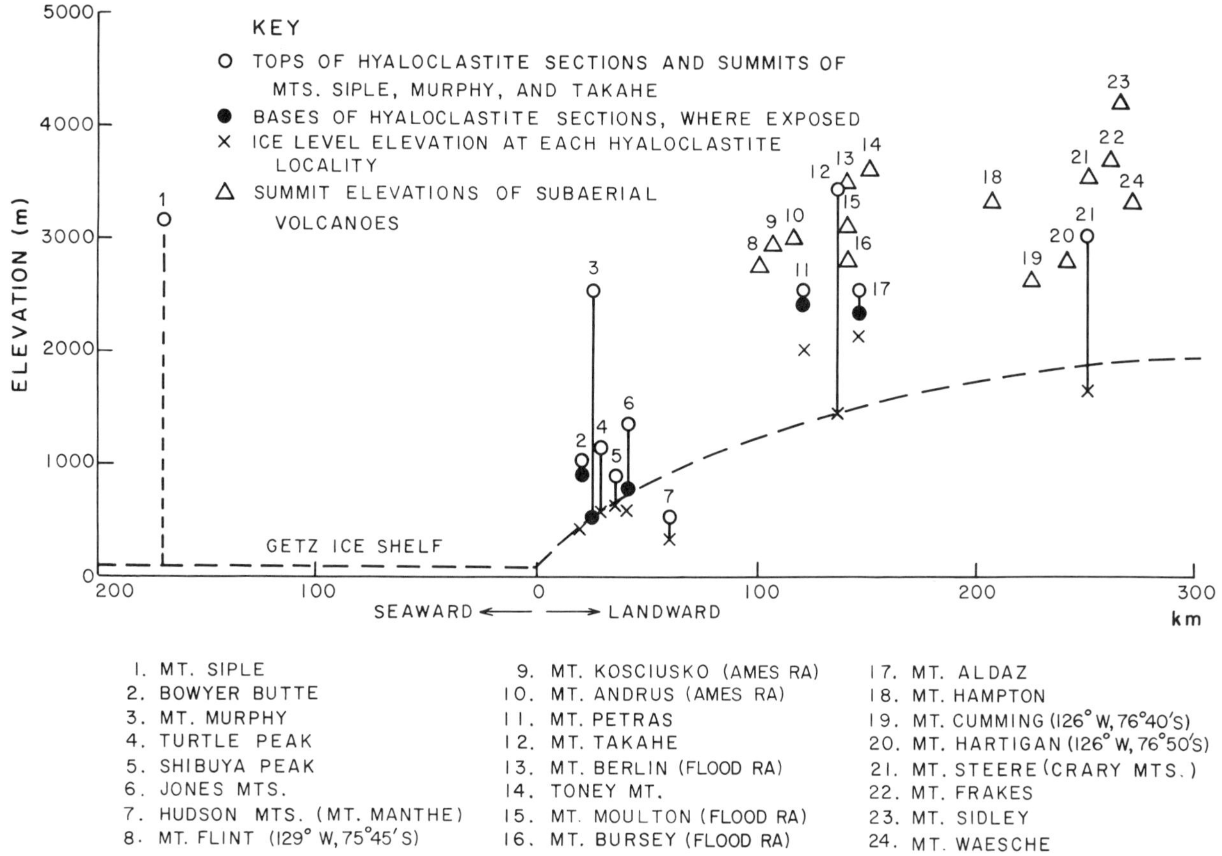

Figure 89.3. Diagramatic representation of the thicknesses and elevations of major hyaloclastite occurrences and of volcano summit elevations in Marie Byrd Land and western Ellsworth Land. The abscissa represents a composite topographic profile, parallel to longitude lines, from the Amundsen Sea southward. The zero point represents the landward edge of floating ice shelves (Getz Ice Shelf, Dotson Ice Shelf, and Thwaites Glacier Tongue). Distances from the ice shelves to each locality shown were measured along meridians. The dashed line represents a generalized profile of the ice sheet, derived from widely scattered elevation data. Elevations of the tops and bases of hyaloclastite sections, and of ice levels at each locality, are taken from the following sources: U.S. Geological Survey Antarctica Sketch Maps, 1:500,000, Hobbs Coast—Byrd Land and Bakutis Coast—Byrd Land sheets; and U.S.G.S., 1:250,000 Reconnaissance Series, Mount Sidley, Mount Hampton, Mount Galla, Crary Mountains, Mount Takahe, Mount Kosciusko, Grant Island, and McCuddin Mountains quadrangles.

communication, 1972). The hyaloclastite in the Hudson Mountains is composed almost entirely of sideromelane shards, in contrast to the blocky, weakly vesicular or nonvesicular clasts of both sideromelane and tachylyte that are characteristic of Marie Byrd Land hyaloclastite. On the other hand, the Hudson Mountains hyaloclastite is similar to that in Iceland and some Ross Island localities. The significance of this observation seems to be that the Marie Byrd Land type of hyaloclastite may form in lieu of pillow lava.

Figure 89.3 illustrates the thicknesses and relationships to present ice level of the major hyaloclastite sections in Marie Byrd Land and western Ellsworth Land. All sections shown have been dated, except that at Mount Siple. Mount Siple, though it has never been visited, is tentatively shown as if it were composed of hyaloclastite because its shape and degree of dissection are both suggestive of an intraglacial volcano (LeMasurier, 1972*b*; Andrews and LeMasurier, 1973). A total thickness of 3000 m of hyaloclastite may be present at this volcano. Even without Mount Siple, however, it is evident from the diagram that hyaloclastite sections in this region are commonly very thick and stand much higher than the present ice level.

Figure 89.3 also shows the summit elevations of all the subaerial stratovolcanoes and of three intraglacial volcanoes. Thorarinsson (1967) used a similar plot to compare the elevations of Icelandic intraglacial volcanoes and postglacial basalt shield volcanoes to their distances from the coast. He was able to show that (1) both types of volcanoes become higher inland, and (2) the intraglacial volcanoes (table mountains) are higher than the postglacial shield volcanoes in every geographic position. He interpreted these relationships to be an effect of the influence of crustal and glacial loading on the height to which lava would rise in a column. The comparison of Marie Byrd Land intraglacial and subaerial volcanoes in Figure 89.3 shows no consistent pattern to their distributions, elevations, or thicknesses. The intraglacial volcanoes do not seem to be consistently higher, as in Iceland, or consistently lower, as one would expect if their elevations were controlled by a stable ice level in a tectonically stable terrain. It may, however, be significant that the highest hyaloclastite units are the upper parts of the thickest sections.

In further comparison of subaerial lava and hyaloclastite, there seem to be few differences in stratigraphic level or structural position. The only stratigraphic difference seems to be that hyaloclastite deposits are almost invariably found at the base of each volcanic section; but they are not restricted to that occurrence. Neither subaerial nor intraglacial volcano elevations correlate with basement contact elevations, which are represented in Figure 89.3 as the exposed bases of volcanic sections. The basement contact, as discussed below, is an erosion surface of very low relief that has been offset by faulting, and it seems to be the best structural and stratigraphic datum in the region. In addition to the surface exposures of the basement represented in Figure 89.3, the seismic work of Bentley and Clough (1972) indicates that the basement beneath Toney Mountain lies about 3000 m below sea level. The total relief on the basement is therefore very large (more than 5000 m), compared to the differences in constructional summit elevations (approximately 1500 m). In view of the fact that the hyaloclastite formed below ice level and the subaerial lava formed above, it is surprising that so few differences can be found between their present elevations, thicknesses, stratigraphic levels, and structural positions.

Glacial geology

Moraines and striated bedrock pavements are uncommon in Marie Byrd Land, and they are difficult to date. Nevertheless, their present elevations seem to provide a record of deglaciation that may be helpful in interpreting the history recorded by hyaloclastite.

Moraines composed of looping ridges of bouldery materials at ice level are easy to recognize and were observed around the bases of Executive Committee Range volcanoes, along the margins of outlet glaciers in the Kohler Range, and at several other localities. Older, elevated moraines are more difficult to recognize, except where they are composed of a mixture of volcanic and basement rock types that is clearly too heterogeneous to have been locally derived. The most unambiguous example of elevated moraine of this type was found on and around the summit of Chang Peak, 8 km north of Mount Waesche, in the Executive Committee Range. The moraine rests on 1.6-m.y.-old pumice (LeMasurier, 1972*c*) at an elevation of 2900 m, roughly 800–1000 m above present ice level. Similar deposits were found resting on the upper Miocene lavas of Mount Hartigan, 500–700 m above ice level. A less heterogeneous bouldery deposit was found just below and southwest of the summit cone of Toney Mountain, at an elevation of 3000 m above sea level and 1400 m above ice level. Lava in place at this locality has been dated at 0.5 m.y. old (LeMasurier, 1972*a*).

Interpretation of striated and polished rock pavements is sometimes difficult in this region, because rocks long exposed at the surface have been highly polished by wind-driven snow. Loose boulders have been faceted and grooved by the same process to form excellent ventifacts. The gabbro at the summit of Dorrel Rock is highly polished and striated, and little evidence exists for control of striation direction by bedrock structure. It is perhaps the least ambiguous example of glacial polish that was observed during the 1967–68 Marie Byrd Land Survey. Topographic data have not yet been published for this locality (Mount Murphy quadrangle), but the polished surface is estimated to be 400–500 m above ice level.

Doumani (1964) reported glacial grooves from the southern flank of Mount Sidley that are wider and much deeper than the striae on Dorrel Rock. They parallel flow structure within the lava and are overlain by exceedingly well developed ventifacts. They may have originated as glacial grooves, later modified by wind, or they could be entirely the product of wind action on the crenulated surface of a flow. A glacial influence is supported by the fact that the grooves lie roughly 800 m below the level of the Chang Peak moraines, 15 km to the west.

Discussion

Environment of hyaloclastite eruption

Many of the arguments bearing on an intraglacial as opposed to a submarine mode of eruption for the hyaloclastite in Marie Byrd Land and western Ellsworth Land have been presented earlier (LeMasurier, 1972*a,b*; Rutford and others, 1973) and need only be briefly reviewed. The major difficulties with a submarine interpretation are: (1) the great thicknesses of hyaloclastite to be explained, (2) the complete absence of interbedded marine sediment, (3) the wide geographic extent of these deposits and length of the time interval during which they formed, and (4) the extraordinary amount of recent uplift that would be required to raise the hyaloclastite deposits above sea level to their present elevations.

The great thicknesses of the hyaloclastite sequences, illustrated in Figure 89.3, are one of the major factors to be explained by any interpretation. A 2000 m thickness of hyaloclastite is well exposed at Mount Murphy (LeMasurier, 1972*a*). Mount Takahe, which rises 2100 m above ice level but which is less deeply dissected than Mount Murphy, is believed to be composed entirely of hyaloclastite because all outcrops on the volcano are composed of this material and because its surface slopes are only 7°–8° (LeMasurier, 1972*b*). Subaerial volcanoes in this region have surface slopes of 12°–16°. Similarly, Mount Siple, which rises 3000 m above the Getz Ice Shelf, is inferred to be predominantly hyaloclastite because of its 7°–8° surface slopes and because its degree of dissection is greater that that of any subaerial volcano in the region, although it is less than that at Mount Murphy (Andrews and LeMasurier, 1973). The incompletely exposed Miocene hyaloclastite sequences at Turtle Peak and Mount Steere are roughly 800 m and 1200 m thick, respectively. These thicknesses all imply great thicknesses of ice or great depths of water during the growth of each volcano (LeMasurier, 1972*a,b*).

Were these hyaloclastite sequences formed at their present elevations, beneath a much thicker ice sheet, or were they tectonically elevated from a much lower subglacial or submarine environment of eruption? One could imagine the USAS Escarpment deposits being uplifted roughly 2400 m from a hypothetical submarine environment within the past 27 m.y., but as one considers successively younger and thicker deposits, the likelihood of a predominantly submarine eruption environment greatly diminishes, simply from considerations of water depths and the required magnitudes and rates of uplift.

The apparent absence of Cenozoic marine sedimentary deposits anywhere in the region also argues against significant marine involvement in the origin of these deposits. A submarine origin for just the Marie Byrd Land–western Ellsworth Land hyaloclastite would require repeated periods of submergence throughout the past 27–28 m.y. to depths well over 1000 m in some areas, of a coastal region

1500 km long and 250 km wide, without leaving a trace of marine sediment. New Zealand is a region of comparable length and width that contains an abundance of hyaloclastite and pillow lava. In virtually all New Zealand localities, however, the presence of interbedded marine sedimentary layers clearly shows that these are submarine volcanic deposits. In the Northland and East Cape localities of North Island, the strata are composed of graywacke and chert. At the Oamaru submarine volcano on South Island, hyaloclastite and pillow lava are interbeded with limestone that is fossiliferous in places (Coombs and Dickey, 1965). On the other hand, undisputed intraglacial volcanic deposits such as those in Iceland rarely contain tills (LeMasurier, 1972*b*), and the basal tillite in the Jones Mountains provides direct evidence of an intraglacial origin for the Miocene hyaloclastite in that area (Rutford and others, 1972).

Glacial history implied by the hyaloclastite

The ages and thicknesses of intraglacial hyaloclastite sequences provide a record of the antiquity of the ice sheet in coastal West Antarctica, and the elevations of these deposits may possibly provide a record of fluctuations of the ice surface (LeMasurier, 1972*a,b*). The West Antarctic ice sheet is believed to be unstable because it is grounded below sea level over large areas. Consequently, it is viewed by some glacial geologists and glaciologists as a younger feature than the East Antarctic ice sheet (Mercer, 1975) and also more subject to large-scale changes in ice level (Hughes, 1973). However, it is likely that both subglacial bedrock elevations and the elevations of hyaloclastite sequences in this region have been affected by some degree of vertical tectonic displacement (LeMasurier, in press), and this uncertainty must be weighed in any interpretation of glacial history.

Duration of glaciation

Cores from DSDP Legs 28 and 29 indicate that there have been glaciers in Antarctica continuously for the past 25–35 m.y. and possibly longer (Allis and others, 1975; Shackleton and Kennett, 1975). Volcanic data from Marie Byrd Land and western Ellsworth Land (Rutford and others, 1968; LeMasurier, 1972*a*) and the new and revised dates summarized in Table 89.1 conform very well to this idea. It has been suggested, however, that the ice sheet "as we know it today" has existed only a little more than 10 m.y. (Shackleton and Kennett, 1975) and that the presumably unstable West Antarctic part of the continental ice sheet is only 4–6 m.y. old (Mercer, 1975). This point seems to represent a major area of dispute, because the alternatives of a submarine origin for the hyaloclastite, or major errors in all but the youngest dates, are unacceptable to us.

It has been suggested that the Jones Mountains hyaloclastite was produced by eruption beneath a Miocene mountain glacier (Mercer, 1975). However, the thicknesses of the Crary Mountain and Turtle Peak hyaloclastite sections, both of which are incompletely exposed, imply thicknesses of Miocene ice well over 1000 m. These are exceptional thicknesses for mountain glaciers and would surely require very large mountain ranges. However, the only existing large mountains in central Marie Byrd Land are volcanoes, most of which were not in existence when these older hyaloclastite sequences were formed, and the prevolcanic landscape appears to have been exceptionally flat, at least in coastal exposures (Rutford and others, 1968; LeMasurier and Wade, 1976). It is much more likely that the older hyaloclastite in Marie Byrd Land and western Ellsworth Land formed beneath a large continental ice sheet that was grounded above sea level, and that the present subglacial topography in West Antarctica developed subsequently. This requires block faulting and subsidence of a previously higher continental crust in Marie Byrd Land and Ellsworth Land. Evidence for both block faulting and subsequent subsidence seems reasonably well recorded in this region, as discussed below.

Tectonic history and its influences on glacial history

The tectonic history of Marie Byrd Land and Ellsworth Land has a fundamental bearing on both the nature of the ancestral West Antarctic ice sheet and on the significance to glacial history of the present elevations of hyaloclastite deposits. The existence of tectonic activity in this region is suggested by the extent and magnitude of Cenozoic volcanism and by the occurrence of a rectangular system of block faults in Marie Byrd Land. The fault system can be recognized on the basis of: (1) the rectilinear arrangement of chains of Cenozoic volcanoes, (2) the sharply rectilinear shapes of basement nunataks, especially notable along the Hobbs Coast, and (3) differential vertical displacements of a very flat, prevolcanic erosion surface (LeMasurier, 1972*a*; LeMasurier, in press). The bimodal and alkaline character of Cenozoic volcanism suggests that these structures were produced in an extensional tectonic environment that has existed as long as the volcanic activity. Evidence for block faulting and extensional tectonics has also been described in the Ross Sea region (Hamilton, 1972; Hayes and Davey, 1975), and this, together with a similar volcanic history, suggests a tectonic unity of the entire Ross Sea–Marie Byrd Land region. The apparent tectonic unity of the region and Cenozoic sea-floor spreading relationships in the southern ocean (Molnar and others, 1975) suggest that the Byrd Subglacial Basin and the Bentley Subglacial Trench represent the axial region of a West Antarctic rift system and that the Marie Byrd Land volcanic province is tectonically related to the Byrd Basin as one of the flanks of the rift (LeMasurier, in press). If this postulated relationship is true, then tectonic relationships determined in coastal Marie Byrd Land and the Ross Sea region should be broadly applicable to a large part of the region that underlies the West Antarctic ice sheet.

The presence of alkaline volcanism suggests that rifting has been active in West Antarctica for at least the last 27–28 m.y., but independent geologic evidence for the timing and sense of displacement of Marie Byrd Land faults has been difficult to acquire. As there are no Cenozoic sedimentary rocks in this region, it is necessary to consider the stratigraphic significance of three less precise "units": (1) the prevolcanic erosion surface, (2) the ice sheet, and (3) the dated volcanic rocks.

The main virtue of the erosion surface is that it is exceedingly flat in most places and apparently is of regional extent (Rutford and others, 1968; Laudon, 1972; LeMasurier, 1972*a*). It is the only continuous datum plane available. It appears to have developed in Late Cretaceous or early Tertiary time. Using this surface as a datum, one can infer that Cenozoic block faulting has produced a vertical displacement of 3000–4000 m between the prevolcanic basement in the Kohler Range and the seismic basement beneath Toney Mountain, 75 km to the south (Bentley and Clough, 1972). Between the subsurface basement at Toney Mountain and the basement exposures 250 km eastward on the USAS Escarpment, there is a total relief of 5000 m. However, at Mount Petras, 70 km beyond the western end of the USAS Escarpment, there is erosional relief of roughly 400 m on the contact between an 80-m.y.-old pluton and the overlying 22-m.y.-old hyaloclastite. This erosional relief is much greater than has been observed on the prevolcanic erosion surface at any other locality, and it indicates the scale of uncertainty involved in using the erosion surface to measure structural relief.

The stratigraphic value of the ice sheet lies in the fact that there can be no significant sedimentation in an area covered by a grounded ice sheet. From this point of view, the deep subglacial troughs in Marie Byrd Land are of great interest. The Bentley Trench, which extends more than 2500 m below sea level, is well known. However, the linear subglacial trough that extends WNW-ESE between the Crary Mountains and Toney Mountain is perhaps even more remarkable because it is only 40–50 km wide and extends nearly 2000 m below sea level (Bentley and Chang, 1971). Its dimensions and configuration are like many of the Tertiary basins in the Rocky Mountains (U.S.A.), but it is filled with ice rather than sediment. Another possible analogue is the Imperial Valley–Gulf of California rift, which has been filled with 6 km of sedimentary beds and lava flows since Miocene time, or possibly within only the last few million years (Elders and others, 1972). Because these examples suggest that intracontinental rifts are rapidly filled with sediment under normal circumstances, it seems likely that the deep subglacial troughs in West Antarctica, including the entire Byrd Basin, have formed largely or entirely since the development of the continental ice sheet. This interpretation is consistent with tectonic inferences based on the geochemical character of volcanism.

The stratigraphic value of the volcanic rocks seems more limited than that of the ice sheet or the erosion surface, because each mountain massif is a separate volcanic center. Therefore, one cannot demonstrate magnitude or timing of offset between the different ranges or between ranges and adjacent basins by offset of volcanic units. The best potential indicator of fault displacement would seem to be the elevations of hyaloclastite units with respect to ice level, assuming that ice level is a reasonably stable datum. If this were so, then Quaternary vertical displacements on the order of 2000 m would be recorded by Mount Murphy and Mount Takahe (Figure 89.3), which is extraordinary for an aseismic region. Some glaciologists believe ice levels in West Antarctica are very unstable, however, and may have fluctuated through a range as large as 2000 m in some coastal localities (Hughes, 1973). If ice-level fluctuations have actually been this large, then no faulting at all would be required to explain the present elevations of any Marie Byrd Land hyaloclastite (Hughes, 1973), except those suspected at Mount Siple. One could argue then that the summit elevations of both subaerial and intraglacial volcanoes were controlled by the hydrostatic head of the magma column (Bradley, 1965; Thorarinsson, 1967; Walker, 1975) and that the present elevations of the hyaloclastite were governed by the level of ice at the time of eruption. This does not seem to be an entirely satisfactory model either, however, because the intense and nearly continuous volcanic activity that has occurred in Marie Byrd Land through the past 18 m.y. strongly implies that fault displacements have occurred during this time, by analogy with more thoroughly studied rift regions such as the Gulf of California (mentioned above), the Basin and Range province in the western U.S.A., and the African rift valleys. Furthermore, if Mount Siple is in fact composed of hyaloclastite, its present elevation would require major fault displacement, probably in Quaternary time. It stands at least 1000 m higher than the highest ice level anyone has suggested for a glacial maximum at this locality, and its degree of dissection suggests that its age lies between that of Mount Murphy (0.8 m.y.) and Mount Takahe (<0.3 m.y.) (Andrews and LeMasurier, 1973).

In summary, it seems clear that block faulting has displaced the flat prevolcanic erosion surface in Marie Byrd Land, that the maximum vertical offsets are between 3000 m and 5000 m, and that the time of displacement must be at least within the past 60–80 m.y. (post-erosion surface). The existence of deep, ice-filled grabens in Marie Byrd Land suggests that a significant amount of this displacement has taken place since the formation of the West Antarctic ice sheet, or within at least the past 27–28 m.y. Further details of the timing and magnitude of these displacements cannot be reconstructed with confidence from the relationships of the exposed volcanic sections.

There are important implications that arise from the possibility that the Byrd Basin has formed largely since the development of the ice sheet. First of all, the volcanic evidence for thick and extensive Miocene ice in Marie Byrd Land and western Ellsworth Land can be reconciled with the problems that seem inherent in forming an ice sheet over West Antarctica as it presently exists. The West Antarctic ice sheet may have evolved from a warm continental ice sheet grounded above sea level, to a cold continental ice sheet, to the present cold marine ice sheet, with the gradual accompanying deterioration of Cenozoic climate and structural collapse of the Byrd Basin.

Because the oldest volcanic rock is hyaloclastite, it seems likely that the West Antarctic ice sheet developed on the prevolcanic erosion surface. This surface is now exposed at 500–900 m elevations from the Hobbs Coast to the Jones Mountains, where it seems to be undissected, and inland at 2300–2700 m elevations along the USAS Escarpment–Mount Petras line, where at least 400 m of erosional relief is evident. The flatness of the erosion surface at most localities suggests that it formed near sea level and has since been uplifted to its present elevations. It is possible that there was regional doming over the entire West Antarctic rift system prior to the formation of major graben structures and that the ice sheet developed on this high upland surface. Early regional doming has been well documented in the East African rift system, which bears many similarities to the rift zone in West Antarctica (Baker and others, 1972; LeMasurier, in press).

The past 7 m.y., characterized by Hughes (1973) as a period of glacial instability in West Antarctica, was a period of intense volcanic activity in Marie Byrd Land and western Ellsworth Land and may therefore have been a period of continuing tectonic instability. The use of present hyaloclastite elevations to measure former ice levels, as

Hughes has suggested, cannot be rigorously evaluated until some reliable datum is found to estimate fault displacements during this time interval and until the nature of Mount Siple has been determined and evaluated. However, the moraines in this region suggest a recent large-scale lowering of ice levels, unless they too have been elevated by faulting, and the continental-scale aseismicity of West Antarctica suggests a tectonic stability that is paradoxical in view of the wide extent of volcanic activity in the region. The present elevations of hyaloclastite should be interpreted with caution, but they clearly indicate instability in Marie Byrd Land within the recent geologic past. Is the instability glacial, or tectonic, or both?

Summary

Hyaloclastite is a common rock type throughout the coastal volcanic terrain that extends from the Fosdick Mountains eastward to the Jones Mountains and inland to the Crary Mountains. It appears to range from Oligocene (27–28 m.y.) to Quaternary (<0.2 m.y.) in age. No associated sedimentary rocks have been found anywhere in this region, with the exception of the tillite at the base of the Jones Mountains sequence. Miocene hyaloclastite sections 500–1500 m thick are found in the Crary Mountains, at Turtle Peak (near Mount Murphy), and in the Jones Mountains. Quaternary hyaloclastite at Mount Murphy and Mount Takahe is at least 2000 m thick, and it extends 2000 m or more above the present level of the ice sheet.

Geologic history in West Antarctica has evidently involved a complex interplay of volcanic, tectonic, and glacial phenomena, all of which must be viewed together in order to understand any one of the three. If an ice sheet has existed in Marie Byrd Land since Oligocene time, as suggested by the volcanic record, then it would seem that the ice sheet is not as unstable as many glaciologists believe or else that the deep sub-sea-level troughs in Marie Byrd Land formed after the ice sheet. The latter interpretation seems consistent with other geologic relationships in this region and implies that the ancestral West Antarctic ice sheet was originally grounded above sea level.

The present elevations of hyaloclastite sequences and moraines above ice level are so great that they suggest either massive fluctuations of ice level or major tectonic displacements, or a combination of ice level and tectonic displacements on the order of 1000 m each. At the present time, it is difficult to suggest an interpretation that eliminates all conflicts. The interpretation of glacial history that seems to cause the fewest contradictions with existing data and that requires the fewest special circumstances involves the following steps: (1) Formation of the ice sheet in middle Oliogcene time or earlier, on a broad subaerial surface, and perhaps concurrent with a period of regional doming; (2) Structureal subsidence of the Byrd Basin and other sub-sea-level grabens in Marie Byrd Land, following the formation of the ice sheet and accompanied by volcanism. The basement rock–ice sheet interface in Marie Byrd Land may have been depressed below sea level for the first time in middle to late Miocene time, based on the earliest period of unstable behavior described by Hughes (1973) and on the ages of thick hyaloclastite deposits at Turtle Peak (13–15 m.y.) and Mount Steere (8.3 m.y.); (3) Present glaciologic characteristics of the West Antarctic ice sheet presumably date from the time the base of the ice sheet was tectonically depressed below sea level. From this time until the present, the geologic history has certainly been characterized by intense subglacial and subaerial volcanic activity. Some combination of major fluctuations in ice level and vertical displacement of fault-bounded crustal blocks also occurred during this interval, but the time scale and relative magnitudes of these phenomena remain to be deciphered.

Acknowledgments. This work has been supported by National Science Foundation Grant OPP70-02980, administered by the Division of Polar Programs. We are grateful to C. R. Bentley, A. L. Bloom, P. D. Rowley, and D. L. Schmidt for reviews and discussions of the manuscript. Although several of these friends are responsible for many words of caution in the manuscript, they are in no way responsible for errors of presentation or interpretation.

References cited

Allis, R. G., Barret, P. J., and Christoffel, D. A. 1975. A paleomagnetic stratigraphy for Oligocene and early Miocene marine glacial sediments at Site 270, Ross Sea, Antarctica. (*In* Hayes, D. E., Frakes, L. A., and others. *Initial Reports of the Deep Sea Drilling Project,* Vol. 28. Washington [U.S. Government Printing Office], 879–84.)

Andrews, J. T., and LeMasurier, W. E. 1973. Rates of Quaternary glacial erosion and corrie formation, Marie Byrd Land, Antarctica. *Geology, 1,* 75–80.

Baker, B. H., Mohr, P. A., and Williams, L. A. J. 1972. Geology of the eastern rift system of Africa. *Spec. Pap. geol. Soc. Am., 136,* 67 pp.

Baker, P. E., McReath, I., Harvey, M. R., Roobol, M. J., and Davies, T. G. 1975. The geology of the South Shetland Islands: V. Volcanic evolution of Deception Island. *Scient. Rep. Br. antarc. Surv.,* No. 78, 81 pp.

Bell, C. M. 1973. The geology of Beethoven Peninsula, southwestern Alexander Island. *Bull. Br. antarc. Surv.,* No. 32, 75–83.

Bentley, C. R., and Chang, F. K. 1971. Geophysical exploration in Marie Byrd Land, Antarctica. (*In* Crary, A. P., *ed. Antarctic Snow and Ice Studies 2.* Washington, American Geophysical Union, 1–38.) [Antarctic Research Series, Vol. 16.]

Bentley, C. R., and Clough, J. W. 1972. Antarctic subglacial structure from seismic refraction measurements. (*In* Adie, R. J., *ed. Antarctic Geology and Geophysics.* Oslo, Universitetsforlaget, 683–91.)

Bradley, J. 1965. Intrusion of major dolerite sills. *Trans. R. Soc. N.Z., 3,* 27–55.

Coombs, D. S., and Dickey, J. S. 1965. The early Tertiary petrographic province of north-east Otago: Waiareka and Deborah volcanic formations. (*In* Thompson, B. N., and Kermode, L. O., *eds.* New Zealand Volcanology, South Island. *Inf. ser. N.Z. Dep. scient. ind. Res.,* No. 51, 38–54.)

Craddock, C., Bastien, T. W., and Rutford, R. H. 1964. Geology of the Jones Mountains area. (*In* Adie, R. J., *ed. Antarctic Geology.* Amsterdam, North-Holland Publishing Company, 171–87.)

Doumani, G. A. 1964. Volcanoes of the Executive Committee Range, Byrd Land. (*In* Adie, R. J., *ed. Antarctic Geology.* Amsterdam, North-Holland Publishing Company, 666–75.)

Elders, W. A., Rex, R. W., Meidav, T., Robinson, P. T., and Biehler, S. 1972. Crustal spreading in southern California. *Science, N.Y., 178,* 15–24.

Fillon, R. H. 1972. Extensive late Cenozoic disconformity in the Ross Sea, Antarctica (abstract). *Trans. Am. Geophys. Un., 53,* 422.

Hamilton, W. 1972. The Hallett volcanic province, Antarctica. *Prof. Pap. U.S. geol. Surv.*, 456-C, 62 pp.

Harrington, H. J., 1958. Beaufort Island, remnant Quaternary volcano in Ross Sea, Antarctica. *N.Z. Jl Geol. Geophys., 1,* 595–603.

Hayes, D. E., and Davey, F. J. 1975. A geophysical study of the Ross Sea, Antarctica. (*In* Hayes, D. E., Frakes, L. A., and others. *Initial Reports of the Deep Sea Drilling Project,* Vol. 28. Washington [U.S. Government Printing Office], 887–907.)

Hughes, T., 1973. Is the West Antarctic ice sheet disintegrating? *J. geophys. Res., 78,* 7884–7910.

Kyle, P. R., and Treves, S. B. 1974. Geology of DVDP 3, Hut Point Peninsula, Ross Island, Antarctica. *Bull. Dry Valley Drilling Project,* No. 3, 13–48.

Laudon, T. S. 1972. Stratigraphy of eastern Ellsworth Land. (*In* Adie, R. J., *ed. Antarctic Geology and Geophysics.* Oslo, Universitetsforlaget, 215–24.)

LeMasurier, W. E. 1972*a*. Volcanic record of Cenozoic glacial history of Marie Byrd Land. (*In* Adie, R. J., *ed. Antarctic Geology and Geophysics.* Oslo, Universitetsforlaget, 251–60.)

LeMasurier, W. E. 1972*b*. Volcanic record of Antarctic glacial history: implications with regard to Cenozoic sea levels. (*In* Price, R. J., and Dugden, D. E., comp. Polar Geomorphology. *Spec. Publs Inst. Brit. Geogr.,* No. 4, 59–74.)

LeMasurier, W. E. 1972*c*. Marie Byrd Land Quaternary volcanism: Byrd ice core correlations and possible climatic influences. *Antarct. Jnl U.S., 7,* 139–41.

LeMasurier, W. E. In press. Rift valley volcanism in Marie Byrd Land, West Antarctica. (*In* Green, J., *ed. Volcanoes and Volcanology.* Dowden, Hutchinson and Ross.)

LeMasurier, W. E., and Wade, F. A. 1976. Volcanic history in Marie Byrd Land: implications with regard to Southern Hemisphere tectonic reconstructions. (*In* Golzález-Ferrán, O., *ed. Andean and Antarctic Volcanology Problems.* Rome, International Association of Volcanology and Chemistry of the Earth's Interior, 398–424.)

Lopatin, B. G., and Polyakov, M. M. 1974. Geologiia vulkanicheskikh gor Khadson (Bereg Uolgrina, Zapadnaia Antarktida) [Geology of the volcanic Hudson Mountains (Walgreen Coast, West Antarctica)]. *Antarktika, Doklady Komissii, 13,* 52–60.

Luckman, P. G. 1974. *Products of Submarine and Subglacial Volcanism in the McMurdo Sound region, Ross Island, Antarctica.* B.Sc. Hons. Proj., Victoria University of Wellington, 115 pp.

Lyon, G. L. 1974. Stable isotope analyses of ice from DVDP 3. *Bull. Dry Valley Drilling Project,* No. 3, 160–70.

Mercer, J. H. 1975. Cainozoic temperature trends in the Southern Hemisphere: Antarctic and Andean glacial evidence. (*In* van Zinderen Bakker, E. M., *ed. Paleoecology of Africa and of the Surrounding Islands and Antarctica. 8.* Cape Town, Balkema, 85–114.)

Molnar, P., Atwater, T., Mammerickx, J., and Smith, S. M. 1975. Magnetic anomalies, bathymetry and the tectonic evolution of the South Pacific since the late Cretaceous. *Geophys. J.R. astr. Soc., 40,* 383–420.

Nelson, P. H. H. 1966. The James Ross Island volcanic group of north-east Graham Land. *Scient. Rep. Br. antarc. Surv.,* No. 54, 62 pp.

Rex, D. C. 1972. K-Ar age determinations on volcanic and associated rocks from the Antarctic Peninsula and Dronning Maud Land. (*In* Adie, R. J., *ed. Antarctic Geology and Geophysics.* Oslo, Universitetsforlaget, 133–36.)

Rutford, R. H., Craddock, C., and Bastien, T. W. 1968. Late Tertiary glaciation and sea level changes in Antarctica. (*In* Tanner, W. F., *ed.* Tertiary sea level fluctuations. *Paleogeogr., Paleoclim., Paleoecol., 5,* No. 1, 15–39.)

Rutford, R. H., Craddock, C., White, C. M., and Armstrong, R. L. 1972. Tertiary glaciation in the Jones Mountains. (*In* Adie, R. J., *ed. Antarctic Geology and Geophysics.* Oslo, Universitetsforlaget, 239–43.)

Rutford, R. H., LeMasurier, W. E., LaPrade, K. E., Laudon, T. S., and Boellstorff, J. D. 1973. A summary of the terrestrial record of Antarctic glacial history in Ellsworth Land and Marie Byrd Land, West Antarctica. *Abstr. 9th Congr. Int. Un. Quat. Res.,* 309.

Shackleton, N. J., and Kennett, J. P. 1975. Paleotemperature history of the Cenozoic and the initiation of Antarctic glaciation: oxygen and carbon isotope analyses in DSDP Sites 277, 279, and 281. (*In* Kennett, J. P., Houtz, R. E., and others. *Initial Reports of the Deep Sea Drilling Project,* Vol. 29. Washington [U.S. Government Printing Office], 743–55.)

Thorarinsson, S. 1967. Some problems of volcanism in Iceland. *Geol. Rdsch., 57,* 1–20.

Treves, S. B. 1975. Hyaloclastite of DVDP 3, Hut Point Peninsula, Ross Island, Antarctica. *Bull. Dry Valley Drilling Project,* No. 6, 32.

Walker, G. P. L. 1975. Intrusive sheet swarms and the identity of Crustal Layer 3 in Iceland. *Q. Jl geol. Soc. Lond., 131,* 143–61.

Warner, L. A. 1945. Structure and petrography of the southern Edsel Ford Ranges, Antarctica. *Proc. Am. phil. Soc., 89,* 78–122.

90

Volcanic Activity Associated with the Anorthoclase Phonolite Lava Lake, Mount Erebus, Antarctica

PHILIP R. KYLE
Institute of Polar Studies, The Ohio State University, Columbus, Ohio 43210, U.S.A.

RAYMOND R. DIBBLE
Department of Geology, Victoria University of Wellington, Wellington, New Zealand

WERNER F. GIGGENBACH
Chemistry Division, Department of Scientific and Industrial Research, Lower Hutt, New Zealand

J. (HARRY) KEYS
Department of Chemistry, Victoria University of Wellington, Wellington, New Zealand

Abstract. Mount Erebus sustains a persistent anorthoclase phonolite lava lake which has gradually expanded in size over a four year period from 1972 and 1976. Small patches of molten lava were first observed in the northern half of the Inner Crater of Mount Erebus during a period of observation from 24 December 1972 to 6 January 1973. Strombolian eruptions occurred at a rate of 2–3 per day. Observations during November 1973 indicated an increase in the area of exposed molten lava; a small lava flow issued from a vent in the northeastern corner of the Inner Crater and terminated in a small lava lake (~40 m diameter). Excellent observations of the Inner Crater were made from the floor of the Main Crater during the 1974–75 field season. Lava filled much of the northern half of the Inner Crater and formed an irregular semicircular lake. The lava welled up from two sources at the northeastern end of the lake and flowed slowly along a curved path before disappearing into a tunnellike opening at the southern end. Mild degassing explosions in the lava lake ejected bombs over the southern half of the Inner Crater and occasionally onto the floor of the Main Crater. Stronger strombolian eruptions occurred from a new vent (the Active Vent) on the southern half of the Inner Crater, ejecting bombs as high as 900 m and about 700 m laterally.

Observations during December 1974 using an array of seismometers about 1 km from the lava lake recorded about 1000 volcanic earthquakes per week with energy exceeding 10 J; no seismic tremor was recorded. Two earthquakes associated with the largest eruptions were recorded on the World Wide Standardized Seismograms at Scott Base, 37 km away. The largest explosive earthquake had a seismic energy of 10^5 J and a peak seismic power of 10^4 W.

During November–December 1975 and December 1976, lava occupied the entire northern half of the Inner Crater, with the 1974 convective movements replaced by a series of randomly distributed overlapping areas where lava was welling up. An average of two explosions per day were heard in the camp area during 1975; no explosions occurred during 1976.

Preliminary data indicate no major variations in the chemistry of the anorthoclase phonolite bombs erupted in 1972–73, 1974, and 1975. Visual observations suggest that the magma viscosity has decreased, possibly as a result of an increase in the magma temperature. The increased size of the lava lake may be due to thermal expansion of the magma column.

Molten anorthoclase phonolite lava has been present in the crater of Mount Erebus (3794 m), Ross Island, Antarctica, for at least four years. The first such observations of lava were made in December 1972 and were reported by Giggenbach and others (1973). Initially, fluid lava was exposed only in a few small patches. These have since developed into a large single convecting lava lake about 140 m long. Subsequent observations of the activity in November 1973, November–December 1974, December 1975, and December 1976 have documented the gradual growth of the lava lake. During 1974 clear conditions and access to the floor of the Main Crater allowed close observations of the lava lake from the rim of the Inner Crater, about 120 m above the actual lake. It was at this time that the first direct observations were made of strombolian eruptions in the lava lake and from adjacent vents. This paper summarizes observations on the growth of the lava lake and describes the known history of volcanic activity at Mount Erebus.

Mount Erebus is one of two active volcanoes in the Ross Sea area of Antarctica; the other, Mount Melbourne (2733 m) on the coast of northern Victoria Land, shows only weak thermal activity (Nathan and Schulte, 1968; Lyon and Giggenbach, 1974). Mount Erebus is the largest of three major volcanoes which form Ross Island. Unlike the predominantly basaltic stratovolcanoes at Mount Bird and Mount Terror, Mount Erebus is composed predominantly of anorthoclase phonolite lavas. Seven K-Ar age determinations (Treves, 1967; Armstrong, 1978) on lavas from

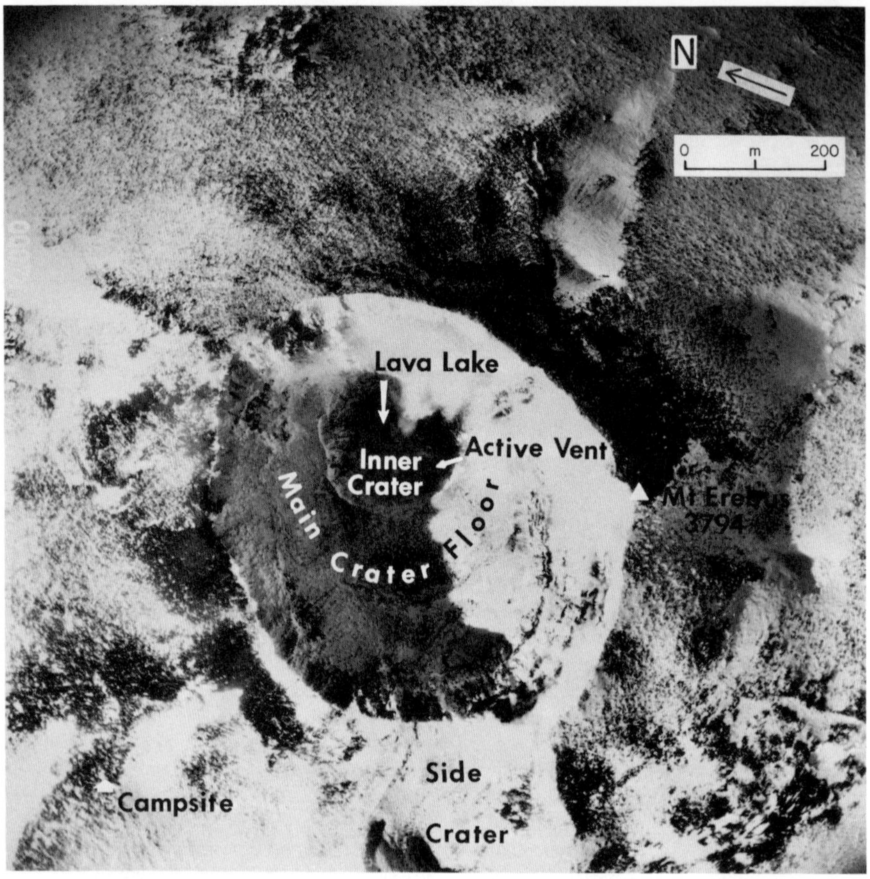

Figure 90.1. Vertical air photograph of the crater of Mount Erebus, U.S. Navy photograph (TMA 1214, F32, no. 74) taken on 4 November 1963.

Mount Erebus range from 0.93 to 0.20 m.y. It is the youngest of the major volcanic vents in the McMurdo Sound area.

The elliptical summit crater of Mount Erebus measures ~600 m north-south and ~500 m east-west (Figure 90.1); the nearly flat floor of the Main Crater is ~160 m below the summit and ~100 m below the lowest part of the crater rim. A circular pit (~250 m in diameter and 100 m deep), comprising the Inner Crater, is situated at the north end of the Main Crater (Figure 90.1). An east-west ridge divides the Inner Crater. The southern half is snow covered and is the site of several noisy high-pressure fumaroles and a few weak fumaroles. Molten lava and most of the explosive activity are confined to the northern half of the Inner Crater.

Volcanic activity

Previous reports

Previous reports of volcanic activity at Mount Erebus are summarized in Table 90.1. Large eruptions were occurring when James Ross discovered the mountain on 27 January 1841. From 1841 to 1956 observations of the mountain were infrequent (Table 90.1), but it is apparent that throughout historic time Mount Erebus has had a large vapor plume. Although the plume has been absent for brief periods, this is probably related to climatic factors such as low humidity, rather than a cessation of activity.

A few expeditions climbed to the crater of Mount Erebus during the late 1950's and the 1960's; none reported the presence of molten lava. However, a U.S. Navy air photograph (TMA 1214, F32, no. 74) taken on 4 November 1963 from directly above the Inner Crater indicates the presence of lava (Figure 90.1). In the photo, snow covers much of the Inner Crater; however, the northeast quadrant of the crater shows an extensive area of bare ground with a vapor cloud rising from it. Bombs litter the snow-covered areas of the Inner Crater and skid marks appear to radiate out from the area of bare ground. It is thus likely that molten lava was present at the time the photograph was taken. The topography of the Inner Crater at that time was very similar to the form observed in recent years. Lava may have been present on numerous other occasions, but the first direct observations were not made until 1972 (Giggenbach and others, 1973). The present state of activity is probably not exceptional, but it appears to have persisted for considerable time.

Eruptive activity from 1972 to 1975

During observations between 1972 and 1975, a record has been kept of all explosive eruptions audible at the campsite about 100 m outside the Main Crater rim. In December 1972–January 1973 and November 1973 the length of each reputation was estimated in seconds, while in 1974 and 1975 the size of an eruption was referred to an arbitrary scale of 1 to 5, 1 being only just audible at the camp while 5 was the most intense (Figure 90.2). In 1973 the eruptions varied in duration from 1 to 25 seconds, while the intensity of the noise accompanying the eruption appeared to be similar no matter what the duration. In 1974, the eruptions were generally of similar short dura-

Table 90.1. Summary of previous reports of volcanic activity of Mount Erebus

Date	Observations	Reference
28 January 1841	"... emitting flame and smoke in great profusion" "At 4 p.m. Mount Erebus was observed to emit smoke and flame in unusual quantities ... the bright red flame that filled the mouth of the crater was clearly perceptible; and some of the officers believed they could see streams of lava pouring down its sides ..."	Ross (1847, p. 216, 220-221)
February 1900	"From the crater of Mt Erebus a smoke cloud was from time to time shot up into the frosty air"	Scholes (1953, p. 67) Borchgrevink's observations.
1901-04	"... only steam was ever seen to be erupted"	Ferrar (1907, p. 8).
1901-04	"On the *Discovery* expedition we saw a glow once or twice during the winter month ..."	Shackleton (1909, p. 174).
10 March 1908	David and party on making the first ascent of Erebus, noted "After a continuous loud hissing sound, lasting for some minutes, there would come from below a big dull boom, and immediately great globular masses of steam would rush upwards .... There were at least three well-defined openings at the bottom of the cauldron, and it was from these that the steam explosions proceeded."	Shackleton (1909, p. 187)
14 June 1908	8:45 a.m. "the steam cloud over its summit was broader and taller than usual ..." 11:30 a.m. "Immense masses of steam rushed upwards to at least 2000 ft. above the summit ..." 3:15 p.m. "... a bright glow was seen ..." 3:25 p.m. "... a remarkably bright glow suddenly lit up all the lower part of the steam column ..." 3:50 p.m. "... there was another bright glow."	David and Priestley (1909, p. 308-309)
	"... we were unable to observe, any trace of lava actually flowing down the side of the cone. Apparently on this occasion the crater was only partly filled with molten lava."	David and Priestley (1914, p. 216).
1908	"It became quite an ordinary thing to hear reports from men who had been outside during the winter that there was a strong glow on Erebus.... when the barometer showed a period of extreme depression, the glow was much more active, waxing and waning at intervals of a quarter of an hour..., and at other times we have seen great bursts of flame crowning the crater."	Shackleton (1909, p. 175).
14 June and 25 June 1908	"On several occasions, when the voluminous smoke cloud showed unusual activity, there appeared to be red-hot or incandescent matter thrown out to a considerable height ..."	Murray (1909, p. 313)
12 December 1912	"Priestley and party on making the second ascent of Erebus, noted there was a loud explosion, and amongst the smoke I could see large blocks of pumice hurled aloft."	Priestley (1913, p. 278)
22 March 1915	"Erebus is emitting large volumes of smoke, ... and a red glare is also discernible."	Shackleton (1919, p. 259)
16 January 1935	"... a plume of vapour floating above its cone."	Byrd (1935, p. 372)
1956-1976	Frequent observations of Mount Erebus made from Scott Base (New Zealand) and McMurdo Station (U.S.A.) have not recorded any significant changes in the activity; exceptions are given below	
July-August 1974	Red glow seen above Mount Erebus	Scott Base personnel
4 September 1974	Large ash eruption which lasted six hours	S. Treves, Univ. of Nebraska

tion; however, the intensity of the explosions varied, thus necessitating the change in the recording scale. No explosions were heard during December 1976.

1973

Giggenbach and others (1973) reported hearing strombolian eruptions in December 1972–January 1973, and they recorded their frequency and duration (Figure 90.2). Similar eruptions were continuing in November 1973, but the duration of the explosions had noticeably increased, while the frequency was variable (Figure 90.2). It should be noted that during November 1973 only two men were on the mountain, and it is very likely that shorter eruptions went undetected, especially during hours of sleep. Both in 1972–73 and in November 1973 no direct observations of an eruption were made, although Giggenbach and others (1973) reported hearing ejected material hit the ground ~6 seconds after an eruption, indicating an eruptive height of ~50 m; dense condensation in the crater precluded any view. Fresh volcanic ejecta were collected on the Main Crater rim in 1972–73, but none was found in November 1973. No signs of bomb marks or ejected material were found on the snow-covered Main Crater floor.

A heavy snowfall 13–15 November 1973 covered the Main Crater and southern Inner Crater floors and in places the crater walls with fresh snow. No volcanic ejecta were noted on the snow during observations on 18 November, except around a small fumarole vent on the southwest sector of the Inner Crater where a 20 × 10 m patch of ash was seen on the crater wall above the vent. It was from this vent that a small ash eruption of 45 seconds duration occurred on 4 February 1973 (Treves and Kyle, 1973). It is likely that the large 25 second eruption recorded on 17 November 1973 (Figure 90.2) was from this vent.

In summary, the eruptions of November 1973 lasted longer than those of the previous year, and there was no

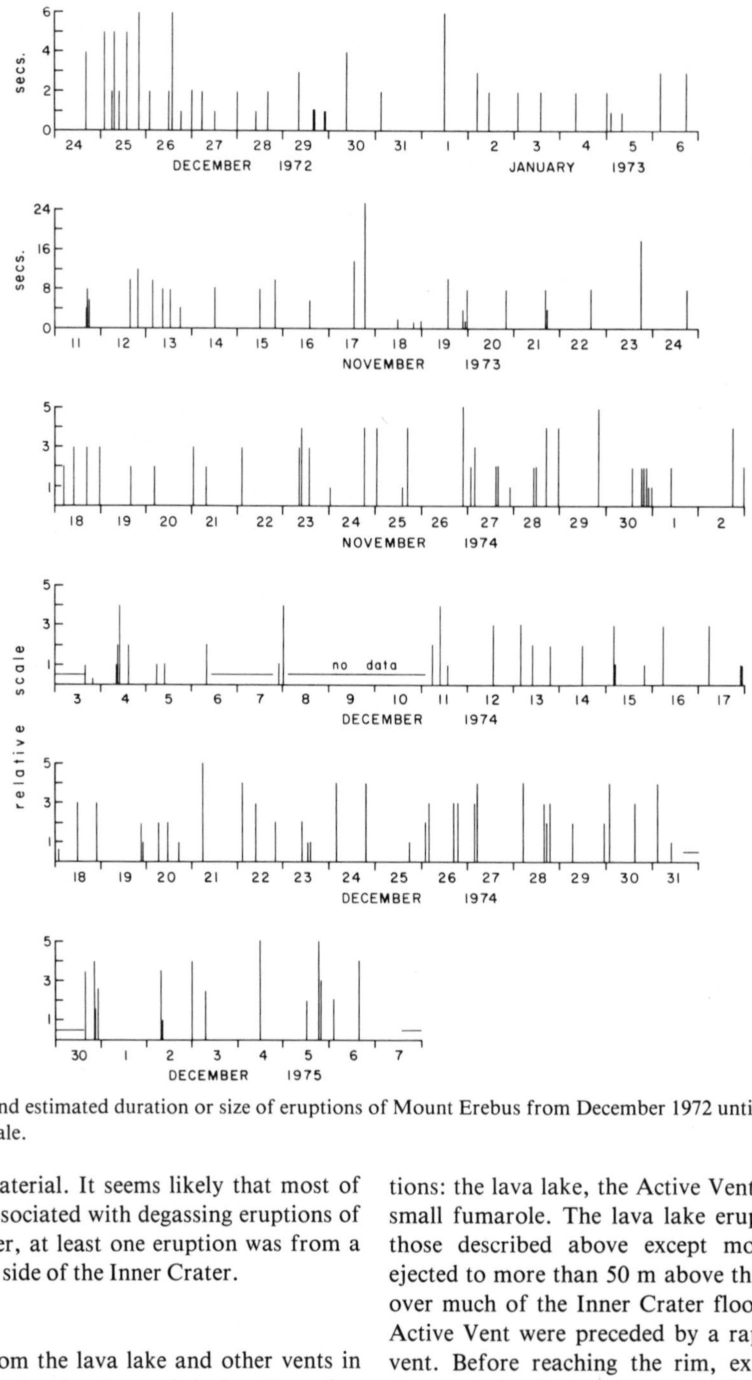

Figure 90.2. Frequency and estimated duration or size of eruptions of Mount Erebus from December 1972 until December 1975. See text for explanation of size scale.

evidence of ejected material. It seems likely that most of the explosions were associated with degassing eruptions of the lava lake. However, at least one eruption was from a fumarole on the south side of the Inner Crater.

1974

Eruptive activity from the lava lake and other vents in the Inner Crater were closely observed during December 1974. The eruptions can be subdivided rather arbitrarily into weak, medium, and strong. It is apparent from the seismic observations, discussed below, that there is a complete gradation among the types.

Weak eruptions occurred in the lava lake and consisted of small degassing explosions. Orange-red glowing hemispherical bubbles about 2–6 m in diameter would pierce the gray crust on the lake, burst, and splatter a few incandescent blobs of lava only a short distance. Lava surrounding the bubble could be seen streaming toward the depression left by the escaping gases. Eruption frequency was variable, with intervals from 10 minutes to several hours. The eruptions occurred over much of the lava lake. There was, however, a marked preference for the northern and eastern areas of the lake.

Medium-strength eruptions occurred from three locations: the lava lake, the Active Vent (Figure 90.3*d*), and a small fumarole. The lava lake eruptions were similar to those described above except more violent: lava was ejected to more than 50 m above the lake, sending bombs over much of the Inner Crater floor. Eruptions from the Active Vent were preceded by a rapid rise of lava in the vent. Before reaching the rim, explosions sent twirling domes of glowing lava some 20–50 m into the air. Most of the bombs landed near the vent and formed an elevated rim of coarse spatter. Only one eruption from a fumarole at the base of the Inner Crater south wall (Figure 90.3*d*) was observed directly. Following a sharp explosion, a cloud of fine debris containing some larger, probably cool fragments was ejected to a level just above the Main Crater floor. The build-up of a rim of debris around this vent suggests that such lower temperature gas blasts have been quite common.

Eruptions audible in the camp area (Figure 90.1), outside the Main Crater, belong to the strong group. During these eruptions, which all apparently occurred from the Active Vent, bombs and other ejecta were usually thrown onto the Main Crater floor and occasionally onto the Main Crater rim. Several strong eruptions were witnessed and are discussed below with the seismic observations.

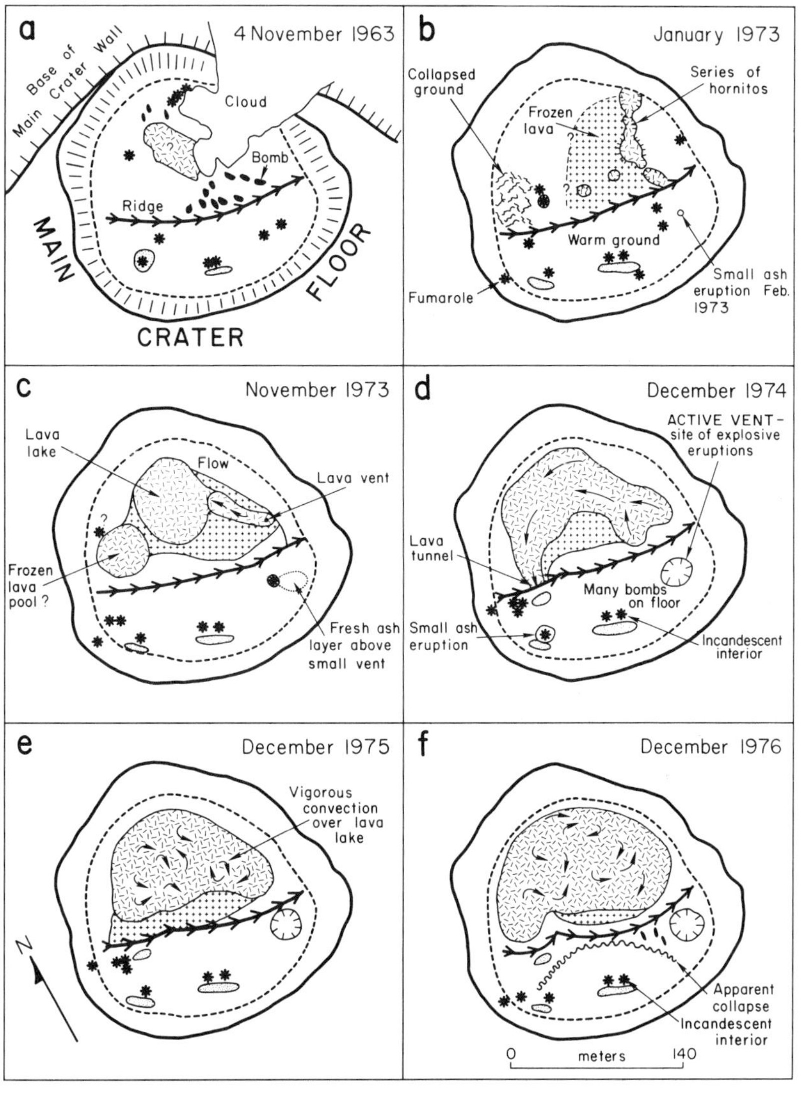

Figure 90.3. Development of the anorthoclase phonolite lava lake within the Inner Crater of Mount Erebus from January 1973 until December 1976.

1975

Eruptions audible at the camp occurred with the same frequency as previously, averaging 2 or 3 a day (Figure 90.2). Their character, however, had changed from the short, sharp bangs heard before to more drawn and whooshing eruptions. The small size (0.2–0.4 m) of fresh bombs found on the northeastern Main Crater rim indicate that the eruptions were probably of a less violent type than those witnessed in December 1974. Small bursting bubble-type eruptions were common in the lava lake.

1976

No large eruptions were recorded. Convective upwellings and small degassing eruptions occurred within the lava lake. The Active Vent was quiescent, although whooshing noises indicated that it had become a fumarolic vent. A small puff of tephra was seen to be ejected from the Active Vent.

Development of the lava lake

Small patches of incandescent lava first observed at Mount Erebus in December 1972 were described by Giggenbach and others (1973). A small active lava lake, ~40 m in diameter, had formed by November 1973 and this has continued to increase in size. The development of the lake is summarized in Figure 90.3.

On 11 November 1973 viscous lava was observed flowing from a small vent about 0.5 to 1 m in diameter (Figure 90.3*c*). A small flow about 35 m long with a maximum width of 5 m moved slowly (0.01 to 0.03 m/sec) west for 5 m, then turned northwest until it entered a lava lake 40 m in diameter.

Observations during December 1974 indicated an increase in the size of the lava lake: it was about 150 m long and 20–30 m wide (Figure 90.3*d*). The overall shape of the partly incandescent, moving surface area of the lake varied from a semicircle to S-shaped, as solidifying folds of lava accumulated on the shore, redirecting the surface flow. The lava welled up from two main sources at the northeastern end of the lake and flowed slowly, with an estimated speed of about 0.05 m/sec, along a curved path before disappearing into a tunnellike opening at the southern end. The surface of the flow was covered by a highly viscous greenish-gray layer of partially solidified lava, which would occasionally break along transverse rifts and reveal the orange glow of the underlying melt. The deeper layers appeared to be moving somewhat faster than the more viscous surface layer.

In December 1975 the lava lake had expanded to a semicircular outline with a maximum width of about 140 m. Vigorous random convection occurred over most of the lake surface. Numerous disturbances would break the thin, partly solidified crust, revealing incandescent material.

In December 1976 the lava lake showed behavior similar to the previous year. The size, however, had further increased by eating into the east-west dividing ridge (Figure 90.3*f*).

Development of the lake has involved only a minor increase in the volume of lava present in the crater. A slight rise of perhaps 10–20 m is likely to be responsible for some increase in the area of exposed lava. Minor lateral stoping of older consolidated material on the floor and walls may also have contributed to the areal increase. Development in December 1976 of a semicircular scarp on the south side of the Inner Crater indicates that future expansion of the lava will probably involve collapse and lateral stoping. The lava lake apparently represents the top of a highly stable column of magma.

Volcanic ejecta

Volcanic ejecta from the lava lake and Active Vent consist largely of coarse, initially molten material. Ash and lapilli appear to be produced only during eruptions from fumarolic vents.

Fresh volcanic ejecta are easily recognized by their black color and lack of sulphur staining. Through interaction with gases released from the lake surface and fumaroles, bombs acquire a faint yellow sulphur stain within a few weeks after eruption onto the Main Crater floor or rim. After a year, ejecta develop a stain indistinguishable from that of much older bombs and lava. No fresh ejecta were found in January 1973 and December 1976, in contrast to 1974 and 1975. The largest bomb observed erupted in December 1974 and was estimated to be about 6 m in length and 1 m in diameter, with a weight of 500–1000 kg. It landed on the Main Crater floor near the eastern rim of the Inner Crater. Bombs were found on the Main Crater rim; in a large eruption, one bomb weighing about 100 kg bounded several times on the crater slopes before coming to rest about 100 m below the Main Crater rim.

Most of the bombs were erupted as partly molten material and were drawn out to the typical spindle-shape (fusiform) during flight through the air. When these bombs hit the snow on the Main Crater floor they invariably melted a large cavity. Some bombs were apparently cold when ejected: they had iregular shapes and did not melt the snow. These probably represent partly solidified material from the surface skin of the lava lake, or those erupted from the Active Vent could represent older material from the vent walls.

Table 90.2. Preliminary major element analyses (in wt. %) of volcanic ejecta erupted from Mount Erebus in 1972, 1974, and 1975

	1972	1974[a]	1975
SiO_2	56.13	56.32	55.96
TiO_2	0.95	0.93	0.98
Al_2O_3	20.04	20.18	19.92
Fe_2O_3 (total Fe)	5.36	5.17	5.49
MnO	0.21	0.21	0.22
MgO	0.84	0.83	0.89
CaO	2.68	2.67	2.65
Na_2O	8.41	8.54	8.48
K_2O	4.50	4.54	4.57
P_2O_5	0.39	0.38	0.42

Analyzed by P. R. Kyle at Victoria University of Wellington using X-ray fluorescence analysis.

[a] Mean of three analyses.

All bombs consist of highly vesicular dark olive-green to black phonolitic glass with ubiquitous anorthoclase phenocrysts, as long as 20 mm. Bombs erupted in 1972 contain phenocrysts of anorthoclase and microphenocrysts of olivine ($Fa_{47}Fo_{53}$), clinopyroxene ($Wo_{48}Fs_{15}En_{37}$), titanomagnetite ($Usp_{70}Mt_{30}$), pyrrhotite, and apatite (Kyle, 1977). Preliminary studies indicate that bombs erupted in 1974 and 1975 are chemically (Table 90.2) and mineralogically identical to the 1972 eruptives. The magma from 1972 until 1975 was therefore part of a closed system and was not modified by addition of new batches of magma. Phenocryst contents of ejecta from the three years are also similar and suggest that little or no additional crystallization occurred during this period. Similarly no increase in the size of anorthoclase phenocrysts has been noted in bombs sampled during the three years.

It is concluded from examination of the ejecta that the lava and phenocryst-microphenocryst phases are in equilibrium. The lack of any apparent continued crystallization of anorthoclase suggests the present ~1000°C temperature* is not within the supercooled region in which rapid crystallization and growth are expected to occur. Growth of these crystals is likely therefore to have occurred before attainment of the presently observed steady-state conditions.

Volcanogenic salt deposits

One of the major present-day effects of volcanic activity in the summit region of Mount Erebus is the deposition of salts through interaction of the phonolite lava with sulfuric acid derived from the acidic volcanic gases emitted from the crater. The most common salts are gypsum and mixed sodium, magnesium, potassium, aluminum, calcium, and iron hydrated sulphates, silicates, and chlorides. They accumulate mainly as white to yellow encrustations, needles, massive deposits, and efflorescences under rocks and blocks of frozen lava, in joints, on the surfaces of rocks and old bombs, around fumaroles, and in certain soils. Halite, sylvite, thenardite, and mirabilite also occur in the summit region though not to the same extent. They are generally found as efflorescences downslope of and under pebbles and cobbles and most commonly on north-facing slopes. Such efflorescences are similar to those found in some areas of southern Victoria Land. Hydrothermal deposits of calcite occur around the Main Crater rim and at many other places inside and outside the Main Crater. Inside the cave system near the camp (Giggenbach, 1976), gypsum and mirabilite were found. Within the Inner Crater, volcanic sublimates and other salts have been noted around large fumarole vents and near the western end of the lava lake. These have not been sampled to date.

Outside the Main Crater, the alum deposits generally increase in amount toward the rim of the summit plateau. Individual accumulations on the order of a gram occur near the Main Crater rim, while deposits on the order of

* Temperature was estimated using the olivine-clinopyroxene geothermometer (Powell and Powell, 1974) and also by direct observation using an optical pyrometer.

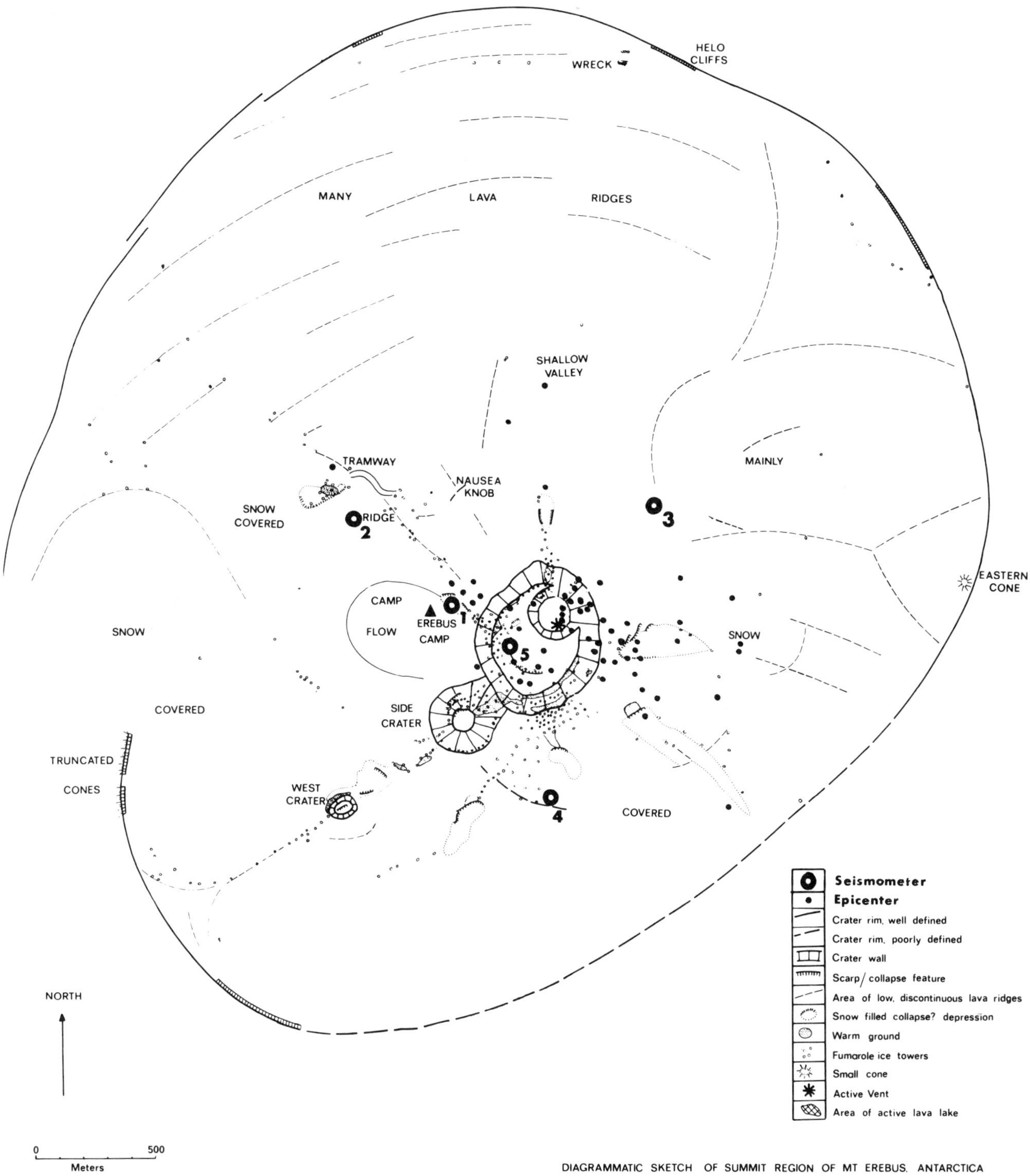

Figure 90.4. Sketch map of Erebus summit showing the geophone sites 1 to 5 and epicenters during the period 24 to 29 December 1974. The names of topographic features are provisionally approved by the N.Z. Geographic Board and the N.Z. Antarctic Place Names Committee.

100 g occur in suitable places on the plateau rim. One sampling traverse showed a steady reduction in the amount of salt down the outer mountain slopes below this rim, although deposits of 1–2 g were found in some places on the Fang Ridge 4 km outside and 1000 m below the main crater rim. At a distance of about 500 m from the Main Crater rim, the greatest accumulation of colored salts occurs to the northwest between Camp Flow and Nausea Knob (Figure 90.4), and the least accumulation to the south and southeast. This is consistent with the dominant wind direction being from the south to southeast as indicated by sastrugi orientations and by observations of wind at the summit. Inside the Main Crater, alum accumulations of as much as 10 g are relatively common, especially in fumarole areas.

Although most of the salt in the summit area is due to the geothermal and volcanic activity, some of the halite, thenardite, mirabilite, and probably sylvite is likely to be derived from marine aerosols. Sampling at sea level and at various places up the mountain to the summit cone showed that although the amounts of these salts decrease with altitude, a partial marine source for them cannot be discounted in the summit area.

Redistribution and accumulation of salts are processes currently active in the summit area. Leaching is occurring in some very moist soils around fumaroles, and on warm summer days evaporating melt from snow and ice brings salts to the ground surface on north-facing slopes. On the Main Crater floor a yellow weathering product forms on bombs not more than a year old. Older bombs in the same area have thicker encrustations. The rate of encrustation decreases with increasing distance from the Inner Crater.

Seismologic observations

Seismograph array

Five vertical Willmore (Mark 1) seismometers of 1-second period were installed within 1 km of the lava lake during December 1974 (Figure 90.4). Sites 1 to 4 recorded by wire telemetry on two slow-motion tape seismographs (Dibble, 1964) in the geothermally heated ice cave described by Giggenbach (1976). Seismic power (Dibble, 1969) was monitored by a chart recorder. Site 5 recorded on a third slow-motion tape seismograph partially buried in warm ground in the Main Crater.

Volcano seismic activity

There were numerous volcanic earthquakes of six types: explosion type, B-type, and A-type of Minakami (1963), high-frequency harmonic type, long emergent type, and microearthquake swarm type; these were recorded against a low background (~100 nm/s) of the sea microseisms. Volcanic tremor was not recorded, even inside the crater.

The larger earthquakes were of explosion type and accompanies the 60 or so audible explosions. The largest of all—at 0350 NZST, 23 December 1974—had a seismic energy of 200 kJ (Richter magnitude, $M_L = 1$) and was weakly recorded on the World Wide Standardized Seismograms at Scott Base, 37 km away. The eruption threw bombs more than 700 m from the crater. Explosion earthquakes had their dominant spectral peak between 3 and 7 Hz. On high-gain seismograms, all were emergent (with consequent errors in times and epicenters), and about 20 had variable but definite forerunning vibrations commencing as much as 30 seconds before the main onset. A similar number (but not all the same events) were followed by as many as six small aftershocks within 5 minutes.

The B-type earthquakes were the most persistently numerous (about 50 per day $\leq$ 10 J) and were distinguishable from explosion earthquakes only by absence of audible or visible explosions, by greater variability in vibration frequency and depth of occurrence, and by an apparent limit of about 1 kJ in seismic energy. A-type earthquakes were infrequent and could be double events misinterpreted as A-types. Five out of six had apparent S-P intervals in the range of 3.5 to 4.5 seconds, indicating distances of at least 15 km, but none were recorded at Scott Base. The high-frequency harmonic type had simple impulsive onsets and a comparatively long regular tail of about 10 Hz. The swarms of microearthquakes were mostly recorded at sites 1 and 2. There were as many as four events per minute, with a dominant wave frequency of 7–10 Hz, duration of 2–4 seconds, and indefinite form. In the largest and longest swarm, 0200 to 0900 hours NZST, 28 December 1974, many were also recorded at site 3.

The Gutenberg parameter b (relating rate of occurrence of all earthquakes at Mount Erebus to their magnitude, taken as log velocity-amplitude) had the value $b = 1.2$ (Figure 90.5), which is intermediate to the values considered normal for volcanic and tectonic earthquakes. The poor fit for the largest earthquakes is not adequately explained by overloading of the seismographs, which was found to be minor by measurements of earthquake tails and which was absent during the seven days when the recording gain was 12 dB lower than usual. Instead it appears that there was a preferred size between 32 and 100 kJ for explosion earthquakes.

A diurnal periodicity in activity appeared to exist in December 1974 (Figure 90.6), with a maximum at 0200 to 0600 which was several times greater than the minimum at 1400 to 1800 hours local solar time (0.8 hour behind NZST). The microearthquake swarms were present most nights during 0100 and 0300 and never occurred outside 2000 to 0700 local solar time. In local lunar time the maximum activity was at 20–22 hours. If the seismic maximums were to correlate with the maximums of the gravity tide, which at Mount Erebus (lat. 77.5° S) is strongly diurnal with the lunar component, 71% of the solar component, the monthly maximum should have been at new moon on 14 December. Figure 90.7 shows a peak on that day, but the greatest activity was on 23 and 24 December, when the average rate of energy release (seismic power) was 2.5 watts.

Epicenters and velocities

Five detectors are the minimum needed to determine the latitude, longitude, depth, and origin time, as well as the P velocity in a homogeneous medium, for each earthquake. However, the percentage errors in timing emergent low-frequency earthquakes, and in positioning detectors on a mountain by uncontrolled aerial photography, are both large, and the medium is not homogeneous. To permit averaging, or rejection of discordant data, in each epicenter determination, zero focal depth and an apparent P velocity of 2.1 km/s were initially assumed. Epicenters clustered near the northeast wall of the Main Crater, but there was a weak east-west trend outside the crater (Figure 90.4). The group of earthquakes near site 1 were of the high-frequency harmonic type. The arrival times at site 1, however, were commonly late (0.1 sec), suggesting a lower than average velocity between that site and the crater.

Apparent velocities for individual earthquakes were determined from the epicenters, and the distribution (Figure 90.8) shows a clear peak at 1.5 km/s, a minimum of 1 km/s (probably an indication of the errors), and a maximum of 4 km/s. The six best recorded and observed explosions in the vent, assumed to have occurred at zero depth, gave an apparent P-velocity of 1.4 ± 0.2 km/s. Earthquakes with apparent velocities >1.6 km/s probably have appreciable depth, and a scale of nominal depth in Figure 90.8 indicates that about 25% were less than 300 m deep and the remainder were evenly distributed down to 1.5 km depth.

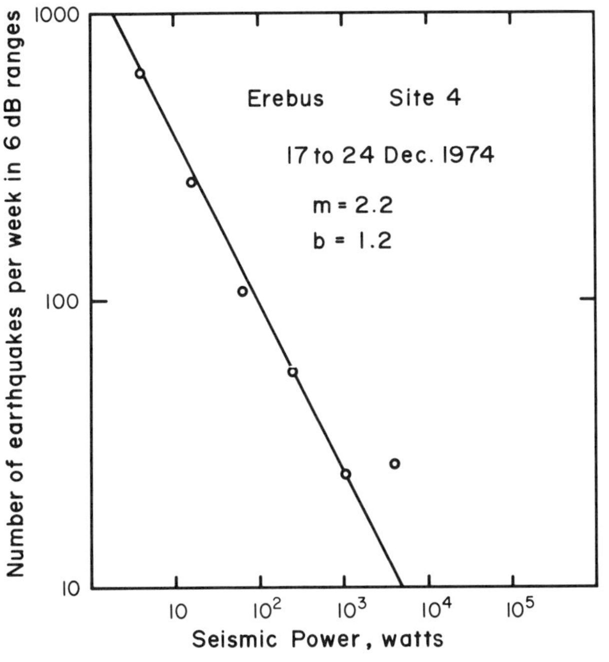

Figure 90.5. Frequency of occurrence per 6 dB range versus seismic power, for Mount Erebus earthquakes of all types. A factor of 10^2 in seismic power is 1 unit of Richter magnitude. A 6 dB range is 0.3 magnitude units. The Ishimoto-Iida parameter *m* relates occurrence with earthquake amplitude, and the Gutenberg-Richter parameter *b* relates occurrence with magnitude, and are derived from the slope of the graph.

CORRELATIONS BETWEEN HUMAN AND SEISMIC OBSERVATIONS

Explosions large enough to be heard on the outer slopes of the volcano between 10 and 31 December 1974 all had seismic energies between 1 and 200 kJ. Three explosions were especially well observed.

The explosion from the Active Vent at 0653 NZST, 21 December 1974, occurred after 20 seconds of forerunning vibration, beginning with a B-type earthquake of energy one joule, and continuing as a vibration of seismic power one watt until the explosion earthquake of 70 kJ. Concurrently, lava rose in the vent and was about to overflow when it exploded violently, throwing two observers, 200 m away, to the ground, and ejecting numerous bombs, some with low trajectory angles. Aftershocks followed but the observers had departed.

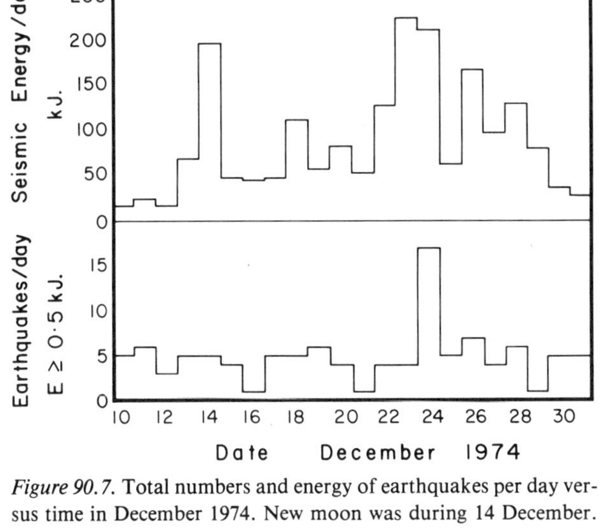

Figure 90.7. Total numbers and energy of earthquakes per day versus time in December 1974. New moon was during 14 December. 225 kJ/day is an average seismic power of 2.6 W.

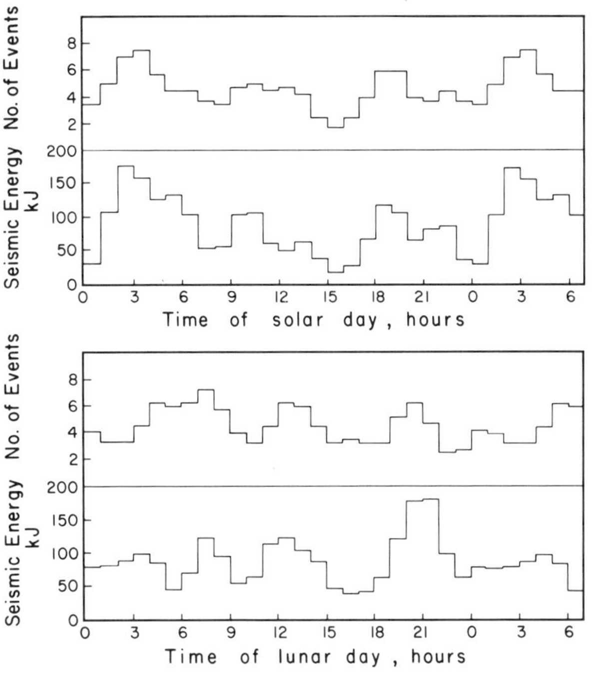

Figure 90.6. Total number and energy of earthquakes at Mount Erebus versus local solar time (top) and lunar time (bottom) in the day over a 22 day period in December 1974. Values are running means of the type $0.25\ (N_{-1} + 2N + N_{+1})$.

The explosion from the eastern part of the lava lake at 0400 NZST, 24 December, threw lava 100 m above the normally calm slow-moving lake. The explosion earthquake of 16 kJ had a lower frequency onset (2.7 Hz), a shorter tail, and relatively less energy at high frequencies (consistent with lower elastic quality factor *Q*) than for Active Vent explosions, but the dominant frequency of 5 Hz was the same. Any forerunning vibration to the lava lake explosion was obscured by a small explosion in the Active Vent 34 seconds earlier.

The third explosion observed in detail, a brief event at 1918 NZST, 28 December, threw bombs totalling a few

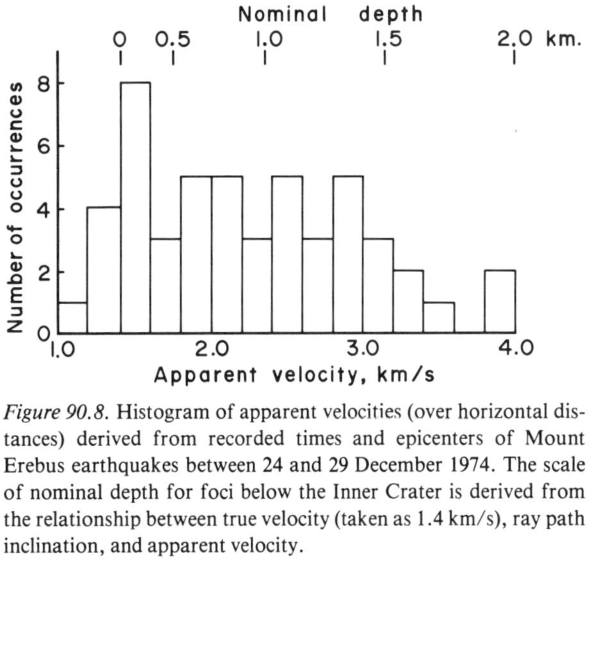

Figure 90.8. Histogram of apparent velocities (over horizontal distances) derived from recorded times and epicenters of Mount Erebus earthquakes between 24 and 29 December 1974. The scale of nominal depth for foci below the Inner Crater is derived from the relationship between true velocity (taken as 1.4 km/s), ray path inclination, and apparent velocity.

hundred kilograms to a height of about 450 m. The kinetic energy was 1 MJ. The explosion earthquake of 25 kJ had no forerunning vibration and was timed as occurring only 1 ± 0.5 second before the explosion; therefore the earthquake had a shallow origin.

No events with seismic energy between 0.3 and 1 kJ were seen or heard by humans, and those less than 0.3 kJ were acoustically weak or silent. Twenty-one of the forty-one small events which were timed during the 40 hours total of documented human observation of the crater could not be detected on the seismograms. They included small explosions unaccompanied by visible ejecta, commencement or cessation of fumarole discharge, bursting of bubbles several meters in diameter in the lava lake, and even eruptions of lava to heights of 15 m from the Active Vent and 45 m from the lake (unless the reported times are in error by at least 18 minutes). The largest of the ten B-type earthquakes without audio-visual expression during the observation period had an energy of 900 J.

Eruptive mechanism

Dynamics of eruptions

In previous investigations into the pressures required to bring about the observed distribution of material ejected during volcanic eruptions, straightforward application of relationships based on the Bernoulli equation led to extremely high preeruption pressures and the problem of how to confine these pressures (McBirney, 1973). More realistic approaches (Blackburn and others, 1976), however, show that much lower pressures are sufficient to explain the often extremely violent behavior of erupting volcanoes.

The typical course of events for medium to minor eruptions (as opposed to violent ones) from the lava lake on Mount Erebus was frequently observed in detail. Shortly before an eruption, a bulge would form on the lake surface which would then rapidly expand, and the initially subspherical bubble would break into incandescent blobs. Mathematically the events may be described in a simplified way by assuming that a bubble with initial diamter s_P and pressure P expands to atmospheric pressure B, causing the acceleration of a disc of lava with cross section area $A = \pi r^2$ and mass $m = \pi r^2 t \varrho$, where t is the thickness and ϱ the density of the lava disc. The energy imparted to the lava disc then is given by:

$$E_e = \int_{s_P}^{s_B} AB(s)ds = mgh.$$

After integration and rearrangement and employing SI-units, an expression valid for $P > (B + 0.05)$ is obtained:

$$P = 1.16\,(B + 2ght\varrho/s_P),$$

allowing the initial pressure P (in MPa, or megapascals) to be calculated from measured or estimated values for B, h, t, ϱ, and s_P. Direct observations suggest that the ratio of the height h reached by the ejecta to the initial diameter of the bubbles has values around 5–10. For an average thickness of the bombs ejected of about 0.5 m, a density of 1.5 Mg/m³, and with a value of 0.065 MPa for the atmospheric pressure in the crater area of Mount Erebus, the pressure within the bubble before eruption is calculated to be 0.16 to 0.25 MPa or approximately 1 to 1.8 bar above atmospheric pressure. These pressures are of the same magnitude as those derived by Blackburn and others (1976) on the basis of observations made during eruptions at Heimaey and Stromboli.

For the eruption from the Active Vent on 0653 NZST, 21 December 1974, the gas pressure causing the explosion can be estimated from the report that the observers were thrown down. If it is assumed that gas in an initial bubble of radius $r = 15$ m (the same as the vent) expanded isothermally as a shock wave with velocity $c = 330$ m/s to radius $R = 200$ m in a cone of angle $2\theta = 90°$ (diameter of Inner Crater twice its depth) where it impulse-accelerated the observers of bulk density $\varrho = 0.7$ Mg/m³ to velocity $V = 0.5$ m/s or more, causing them to fall over, then the initial pressure, given by

$$P = \frac{(1 - \cos\theta)\varrho c V R^2}{2r^2}$$

was 3 MPa or more, equivalent to a depth of 150 m or more in vesicular magma of nominal density 2 Mg/m³.

The two pressures calculated above represent limiting extremes. Small bubblelike eruptions in the lava lake probably occur due to a slight build-up of gas pressure beneath the partly solidified crust. Such eruptions were generally quiet and were not recorded on the seismographs. The larger eruptions were extremely loud and overloaded all the seismographs; they all originated from the Active Vent. Such eruptions probably resulted from a rapid transfer of gas pressure from depth. This is in agreement with the seismic observations which showed forerunning vibrations prior to the eruptions.

Possible eruption mechanism

In December 1975 the character of the eruptions had changed from the short sharp bangs observed earlier to more drawn out whooshing eruptions, while during December 1976 no eruptions audible outside the Main Crater were observed at all. This gradual change in eruptive behavior was possibly caused by a decrease in the viscosity of the lava reaching the surface. This change in viscosity may be inferred from the variations in lava flow patterns observed. During 1972–73 most of the area presently covered by the lava lake appeared to be frozen over. The generally gray, nonincandescent surface layer would only occasionally break open in a few areas to reveal the underlying melt. During the following years, highly viscous lava was observed to well up from distinct sources and to move slowly away with estimated speeds of about 0.05 to 0.1 m/s. In 1975 these unidirectional, convective flow patterns were replaced by random, considerably faster movements from several upwelling areas—with incandescent material exposed over about half the lava lake. In 1976 gas was released from the now quite fluid lava lake in the form of bursting bubbles.

From these findings it would appear that eruptive behavior was largely controlled by the viscosity of the upper parts of the magma column. With the decrease in viscosity, convection within the magma column apparently became more vigorous, resulting in an easier and more rapid transfer of exsolved gases from depth. Gas bubbles may have had less time to grow and hence would have been smaller, thus accounting for the decrease in the magnitude of eruptions from 1973 to 1976.

The observed changes in general appearance, size, viscosity, and eruptive behavior of the lava lake suggest a steady increase in the temperature of the magma column. The heat is probably derived from a crystallizing magma chamber at depth and is likely to be transferred from depth largely through convective movements, possibly assisted by the formation of gas bubbles in the rising parts of the magma column. The 10 to 20 m increase in the lava lake level is probably the result of thermal expansion, as analyses of volcanic ejecta show no evidence for changes in the bulk composition of the magma.

Acknowledgments. Special thanks are due to the Antarctic Division, Department of Scientific and Industrial Research, and to the U.S. Navy VXE-6 Antarctic Support Squadron for field and logistic support. Assistance and companionship in the field of many colleagues is gratefully acknowledged. The manuscript was reviewed by C. Craddock, D. H. Elliot, and W. E. LeMasurier. Financial support was provided by the New Zealand University Grants Committee and Victoria University of Wellington. P. R. K. acknowledges support from the Institute of Polar Studies and National Science Foundation Grant DPP76-23440.

References cited

Armstrong, R. L. 1978. K-Ar dating: Late Cenozoic McMurdo Volcanic Group and dry valley glacial history, Victoria Land, Antarctica. *N.Z. Jl Geol. Geophys., 21,* No. 6, 685–98.

Blackburn, E. A., Wilson, L., and Sparks, R. S. J. 1976. Mechanisms and dynamics of strombolian activity. *J. Geol. Soc. Lond., 132,* 429–40.

Byrd, R. E. 1935. *Discovery.* New York, Putnam's, 405 pp.

David, T. W. E., and Priestley, R. E. 1909. Notes in regard to Mount Erebus. (*In* Shackleton, E. H. *The Heart of the Antarctic, Being The Story of the British Antarctic Expedition, 1907–09,* 2. London, Heinemann, 308–309.)

David, T. W. E., and Priestley, R. E. 1914. Glaciology, physiography, stratigraphy and tectonic geology of south Victoria Land (with appendices by T. G. Taylor and E. J. Goddard). *Rep. Br antarc. Exped., 1907–9, Geol., 1,* 319 pp.

Dibble, R. R. 1964. A portable slow motion tape recorder for geophysical purposes. *N.Z. Jl Geol. Geophys., 7,* No. 3, 445–65.

Dibble, R. R. 1969. Seismic power recordings during hydrothermal eruptions from Ruapehu Crater Lake in April 1968. *J. geophys. Res., 74,* 6545–51.

Ferrar, H. T. 1907. Report on the field geology of the region explored during the "Discovery" Antarctic Expedition, 1901–4. *Nat. Antarc. Exped., 1901–1904, Nat. Hist., 1,* 1–100.

Giggenbach, W. F. 1976. Geothermal ice caves on Mt. Erebus, Ross Island, Antarctica. *N.Z. Jl Geol. Geophys., 19,* No. 3, 365–72.

Giggenbach, W. F., Kyle, P. R., and Lyon, G. L. 1973. Present volcanic activity on Mount Erebus, Ross Island, Antarctica. *Geology, 1,* 135–36.

Kyle, P. R. 1977. Mineralogy and glass chemistry of recent volcanic ejecta from Mt. Erebus, Ross Island, Antarctica. *N.Z. Jl Geol. Geophys., 20,* 1123–46.

Lyon, G. L., and Giggenbach, W. F. 1974. Geothermal activity in Victoria Land, Antarctica. *N.Z. Jl Geol. Geophys., 17,* No. 3, 511–21.

McBirney, A. R. 1973. Factors governing the intensity of explosive andesitic eruptions. *Bull. volcan., 37,* 444–53.

Minakami, T. 1963. A note on earthquakes originating from volcanoes. (*In Geophysical Papers Dedicated to Professor Kenzo Sassa.* Kyoto, Kyoto University Geophysical Institute, 277–83.)

Murray, J. 1909. Additional notes on eruptions. (*In* Shackleton, E. H. *The Heart of the Antarctic, Being The Story of the British Antarctic Expedition, 1907–09.* 2. London, Heinemann, 310–15.)

Nathan, S., and Schulte, F. J. 1968. Geology and petrology of the Campbell-Aviator divide, northern Victoria Land, Antarctica. Part 1. Post-Paleozoic rocks. *N.Z. Jl Geol. Geophys., 11,* 940–75.

Powell, M., and Powell, R. 1974. An olivine-clinopyroxene geothermometer. *Contr. Miner. Petrol., 48,* 249–63.

Priestley, R. E. 1913. The ascent of Erebus, December 1912 (*In* Scott, R. F. *Scott's Last Expedition.* 2. London, Smith, Elder, 278.)

Ross, J. 1847. *Voyage to the Southern Seas.* 1. London, John Murray, 366 pp.

Scholes, A. 1953. *Seventh Continent.* London, George Allen and Unwin, 226 pp.

Shackleton, E. H. 1909. *The Heart of the Antarctic, Being the Story of the British Antarctic Expedition, 1907–09.* 1. London, Heinemann, 372 pp.

Shackleton, E. H. 1919. *South, The Story of Shackleton's 1914–1917 Expedition.* London, Heinemann. 376 pp.

Treves, S. B. 1967. Volcanic rocks from the Ross Island, Marguerite Bay, and Mt. Weaver areas, Antarctica. (*In* Nagata, T., *ed. Proceedings of the Symposium on Pacific-Antarctic Sciences. JARE Sci. Rep., Special Issue,* No. 1, 136–49.)

Treves, S. B., and Kyle, P. R. 1973. Renewed volcanic activity of Mt. Erebus, Antarctica. *Antarct. Jnl U.S., 8,* 156.

91

Volcanic Geology of The Pleiades, Northern Victoria Land, Antarctica*

PHILIP R. KYLE
Institute of Polar Studies, The Ohio State University, Columbus, Ohio 43210, U.S.A.

Abstract. The Pleiades are situated near the crest of the Transantarctic Mountains in northern Victoria Land and comprise part of the Melbourne volcanic province of the upper Cenozoic McMurdo Volcanic Group. The Pleiades consist of a series of scoria cones, endogenous domes, lava flows, and associated pyroclastic rocks which extend in a north-northeast direction for more than 13 km. K-Ar age determinations on four lava flow specimens range from 3,000 ± 14,000 to 40,000 ± 50,000 years. The large analytical uncertainties in the age determinations all overlap, and therefore the ages cannot be used to define stratigraphic relationships. Mount Atlas, a stratovolcano at the southern end of The Pleiades, is the largest eruptive center and consists of three main cones. The lavas there are mildly potassic, weakly undersaturated trachyandesite, tristanite, and K-trachyte. North of Mount Atlas, the lavas are mainly tristanite, K-trachyte, and peralkaline K-trachyte, overlain by small quantities of sodic and strongly undersaturated basanite, Ne-hawaiite, and Ne-benmoreite eruptives. The Pleiades lavas are all undersaturated alkali volcanic rocks belonging to (1) a weakly undersaturated, mildly potassic lava lineage, and (2) a strongly undersaturated sodic lava lineage. Each lineage represents a simple differentiation series which formed by fractional crystallization.

The Pleiades comprise part of the Melbourne volcanic province (Kyle and Cole, 1974) of the alkaline upper Cenozoic (older than 15 m.y. to present day) McMurdo Volcanic Group (Harrington, 1958; Nathan and Schulte, 1968; Kyle and others, 1979). The Melbourne volcanic province includes all upper Cenozoic volcanic rocks in northern Victoria Land that extend inland from Mount Melbourne to The Pleiades (Figure 91.1). At its eastern extremity, the province borders the Hallett volcanic province (Hamilton, 1972; Kyle and Cole, 1974). The field occurrences of Melbourne volcanic province rocks have been described by Nathan and Schulte (1968), who divided them into a Local Suite and a Central Suite. The Central Suite is the more extensive and petrologically diverse and consists of the large stratovolcanoes of Mount Melbourne (2733 m) and Mount Overlord (3396 m) and the cones at The Pleiades and Surprise Hills. The Local Suite includes all other occurrences not assigned to the Central Suite and, with one exception, consists of olivine-bearing basanites, which generally occur as cinder cones and other small volcanic centers (Figure 91.1). By definition, lavas at The Pleiades belong to the Central Suite, but the final eruptive products show a close chemical affinity to rocks of the Local Suite.

The distribution of eruptive centers within the Melbourne volcanic province forms a curvilinear trend (Figure 91.1), which Nathan and Schulte (1968) suggested could be related to a "deep-seated crustal rift or tension zone." Kyle and Cole (1974) considered the distribution of the Central Suite volcanoes to be controlled primarily by north- to northwest-trending graben and fault structures.

A southwestward extension of the alignment of eruptive centers at The Pleiades intersects Mount Overlord (Figure 91.1) and defines a lineament approximately parallel to the northern Victoria Land coastline and the general trend of the Transantarctic Mountains. Intersection of a deep crustal fracture with the fault and graben structures noted by Kyle and Cole (1974) could account for the distribution of Mount Overlord and The Pleiades.

The location and range of K-Ar ages (Armstrong, 1978) for eruptive centers within the Melbourne volcanic province are shown in Figure 91.1. Mount Overlord has a mean age of 7 m.y. (based on three determinations: two whole-rock and one feldspar separate). The upper cone of the active (Nathan and Schulte, 1967; Lyon and Giggenbach, 1974) Mount Melbourne has ages ranging from 0.01 to 0.26 m.y., whereas specimens from areas at the base of the volcano yield ages varying from 0.19 to 2.4 m.y. Two Local Suite basaltic lavas have ages of 4.28 m.y. (Hades Terrace) and 18.0 m.y. (Nathan Hills). The cone at Nathan Hills has a very youthful appearance, however, and the lavas are also extremely fresh (S. Nathan, personal communication, 1972). The 18.0 m.y. age is therefore believed to be erroneous.

The Pleiades (lat. 72° 40′ S, long. 165° 30′ E) are situated 120 km inland from the northern Victoria Land coast at the head of the Mariner Glacier, at an elevation of 3000 m in the Transantarctic Mountains (Figure 91.1).

* Contribution no. 326 of the Institute of Polar Studies, The Ohio State University.

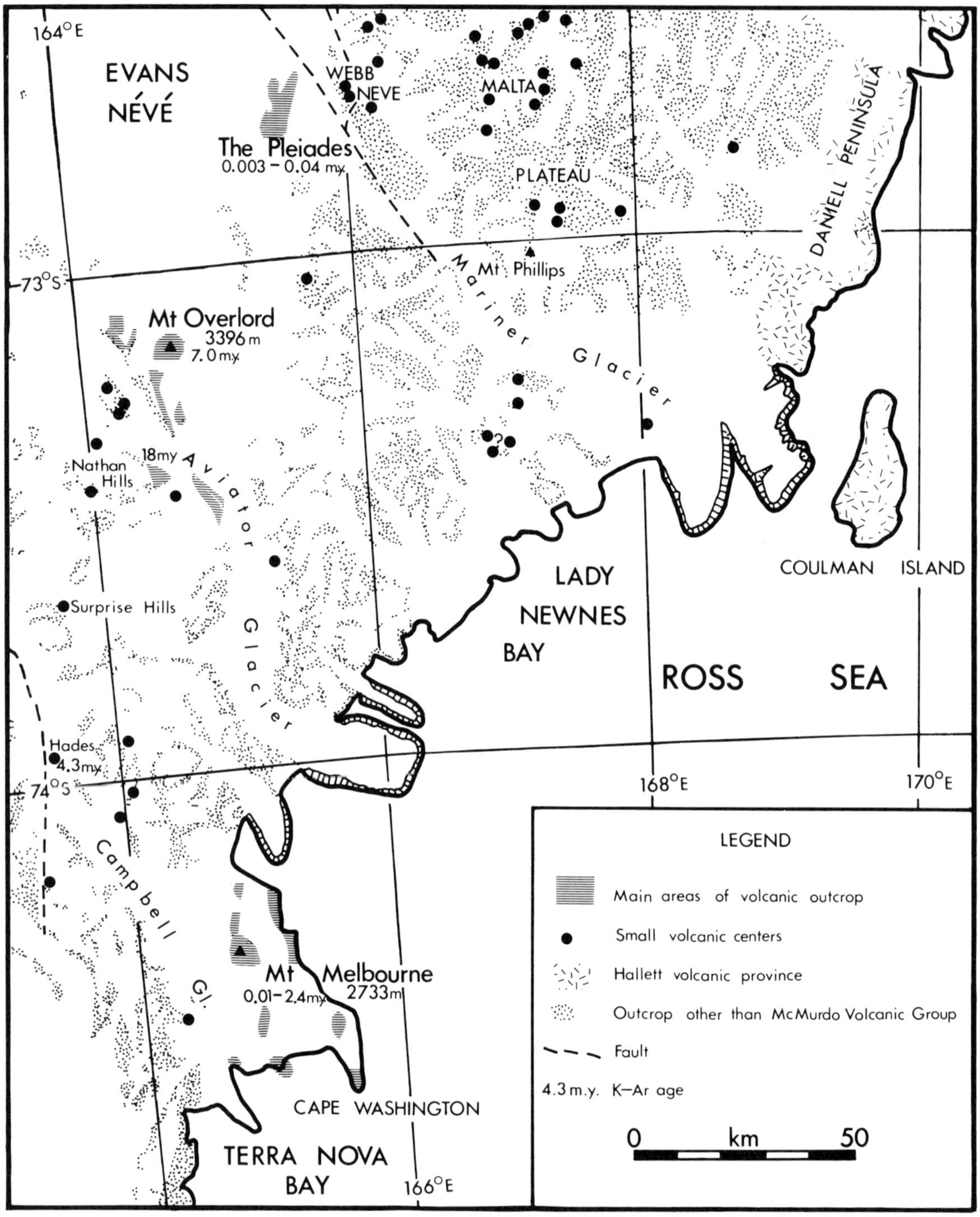

Figure 91.1. Location map of the Melbourne volcanic province, showing distribution of eruptive centers. See text for a discussion of the K-Ar age determinations.

Sedimentary rocks of the Robertson Bay Group and Bowers Supergroup of Precambrian and Paleozoic age occur to the north and east of The Pleiades, and Paleozoic granite and granodiorite intrusions of the Granite Harbour and Admiralty Intrusives predominate to the south (Gair and others, 1969). Granodiorite xenoliths in lavas from The Pleiades suggest that intrusive rocks underlie the volcanic center.

The volcanic rocks at The Pleiades represent some of the most varied compositions presently known in the McMurdo Volcanic Group. The close spatial distribution of this varied suite of rocks offers an ideal opportunity to study their origin, petrogenesis, and genetic relationships. The Pleiades, along with Mount Overlord, are the only major suites of intermediate and trachytic rocks erupted in the uplands of the Transantarctic Mountains; the other major volcanic piles of the Hallett and Erebus volcanic provinces are situated along the foot of the Transantarctic Mountains.

This paper describes the volcanic geology of The Pleiades and briefly discusses the main chemical characteristics of the eruptive rocks. A detailed description of the petrography, mineralogy, geochemistry, and petrogenesis of the rocks is in preparation.

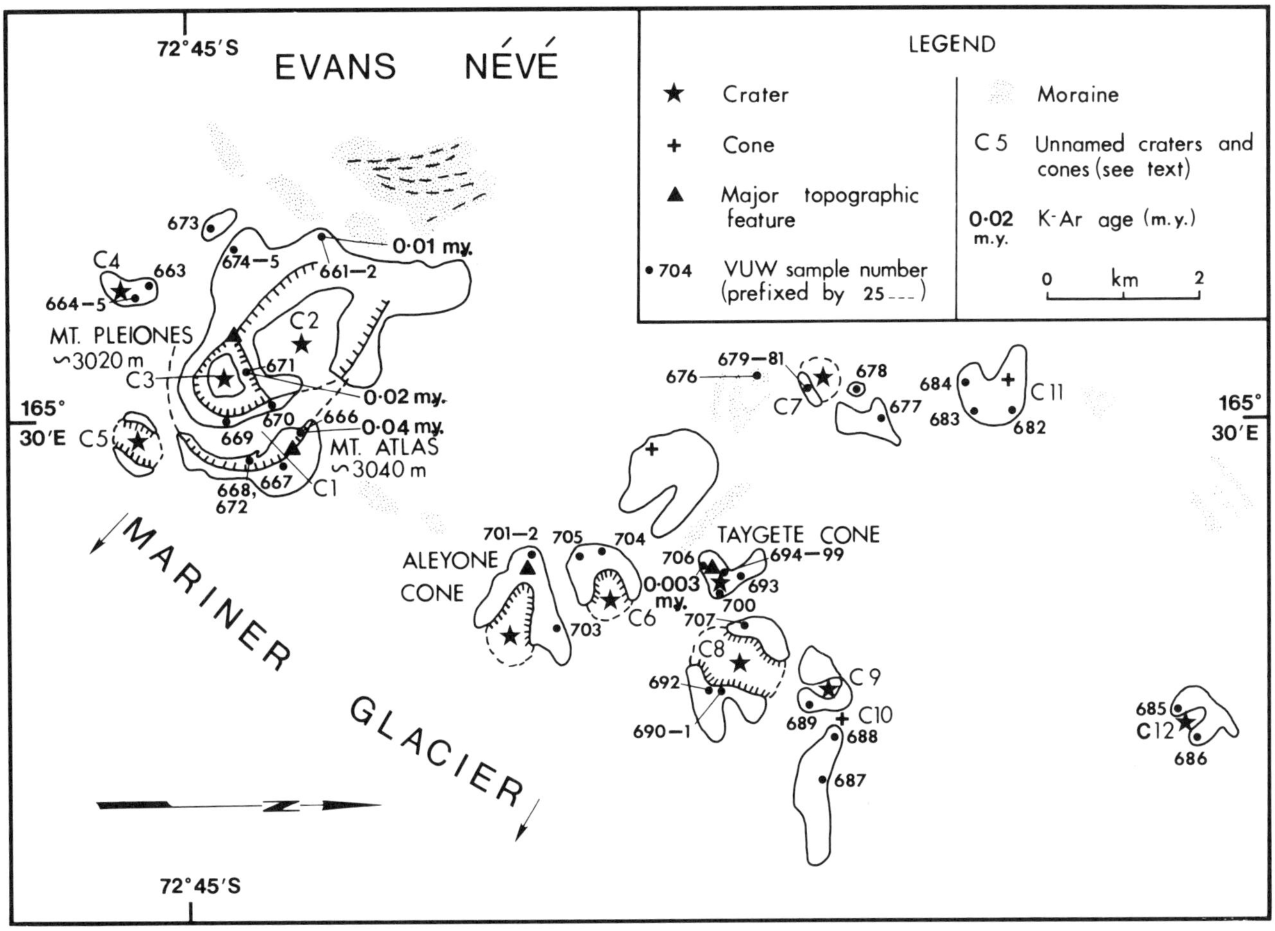

Figure 91.2. Sketch map of The Pleiades, showing the distribution of eruptive centers and locations of specimens.

Previous work

Prior to this study, only two brief visits had been made to The Pleiades. Riddolls and Hancox (1968) have described thin sections of a basalt and two trachytes, but no locations for the specimens were given. They have suggested that The Pleiades were formed by an early trachytic episode followed by a basaltic episode of volcanism. Nathan and Schulte (1968) have discussed in detail the geology of volcanic rocks cropping out between Mount Melbourne and The Pleiades. Several thin section descriptions, a geologic map, and a discussion of the geology of Mount Atlas, the largest cone at the south end of The Pleiades, were presented. Kyle and Rankin (1976) have discussed the rare-earth-element chemistry of eight representative specimens from The Pleiades.

Volcanic geology

The Pleiades comprise a series of small cones which extend 13 km in a north-northeast direction. Mount Atlas* in the south is the dominant feature, with smaller, less prominent cones to the north (Figure 91.2). In the following discussion, unnamed features (cones and craters) are referred to by numbers (i.e., C7; see Figure 91.2).

Four K-Ar age determinations on specimens from The Pleiades (Table 91.1) indicate that this volcanic center is extremely young. The oldest dated specimen, from Mount Atlas, is 40,000 ± 50,000 years old and the youngest, from Taygete Cone, is 3,000 ± 14,000 years old. The large analytical uncertainties in the four age determinations all overlap, and therefore the ages cannot be used to define stratigraphic relationships. Although young K-Ar ages are commonly unreliable, the general agreement of all the determinations suggests that they are meaningful. A lower

* Nathan and Schulte (1968) referred to a point on Mount Atlas as Mount Electra. However, the name was apparently unacceptable, because on the Freyberg Sheet of the U.S. Geological Survey 1:250,000 map series, the point was named Mount Pleiones. This is not the summit of the volcano, so to avoid confusion, the name Mount Atlas has been submitted for the true summit to the N.Z. Geographic Board. In the following discussion Mount Atlas refers to the whole volcano unless otherwise stated.

Table 91.1. K-Ar age determinations for volcanic rocks from the Pleiades (after Armstrong, 1978)

Specimen number	Specimen description and location	Mean age (m.y.)
25706	Peralkaline K-trachyte, south side Taygete Cone	0.003 ± 0.014
P37081[a]	Ne-tristanite, cone C2, 1.4 km west of Mount Atlas	0.012 ± 0.04
25671	Trachyandesite dike on northwest side of cone C3, Mount Atlas	0.02 ± 0.04
25666	Tristanite, cone C1, 300 m west of Mount Atlas	0.04 ± 0.05

[a] Collected by S. Nathan; equivalent to 25661.

age limit (at 95% probability) for the dated specimens is given by the age plus twice the analytical uncertainty (the analytical uncertainty is quoted as one standard deviation); thus the bulk of Mount Atlas is considered to be younger than 140,000 years. The extremely young ages show that the specimens do not contain excess radiogenic argon. Complete degassing of the lavas, therefore, occurred prior to or during eruption.

Mount Atlas

Mount Atlas (~3040 m) (Figure 91.2) rises 500 m above the Evans Névé. It is a stratovolcano, consisting of three main cones, each of which is represented by a crater (Nathan and Schulte, 1968). The oldest and largest cone (C1, Figure 92.1), 40,000 ± 50,000 years old (Table 91.1), is composed of tristanite (666)* and K-trachyte (667). The K-trachyte is exposed on the north side of the crater as oxidized scoriaceous flows. Cognate xenoliths (663 and 668) of essexite as large as 0.4 m but averaging 0.1 to 0.2 m in diameter are scattered around the rim of the center.

Cone 2 (C2, Figure 91.2) is less well defined, but exposures along the western side indicate an interlayered tristanite and K-trachyte sequence. Tristanite (675) is overlain by K-trachyte, and this in turn by more tristanite (661, 662, and 673), and finally K-trachyte (674). Nathan and Schulte (1968) have described a Ne-tristanite** flow (P37081 equivalent to 661) from cone 2 with a "streaked" texture due to flowage of semiconsolidated lava. A K-Ar age determination on the Ne-tristanite gave 12,000 ± 40,000 years (Table 91.1).

The youngest cone of Mount Atlas (C3, Figure 91.2) is composed of tristanite (670) and has a subcircular crater. It is cut on the western side by two nearly vertical trachyandesite (671) dikes, which intrude through oxidized, poorly bedded, tristanite agglomerate. One of the dikes gave a K-Ar age of 20,000 ± 40,000 years (Table 91.1).

Nathan and Schulte (1968) suggested that the two small undissected cones C4 and C5 are the same age as C3 of Mount Atlas. Cone C4 has a crater 200 m in diameter and is composed mainly of K-trachyte (665). The crater rim, however, is capped by red oxidized and rare black unoxidized trachyandesite (664) which shows a ropy texture. A few trachyandesite bombs are scattered over the cone.

At the north-northwest base of Mount Atlas, Nathan and Schulte (1968) described a pattern of semicircular ridges which they stated "appear to be the remnant of an old trachytic stratovolcano which was presumably eroded down to a low level and later planed off by moving ice." The ridges, however, are composed of moraine, similar in appearance to that found in association with volcanic rocks in the McMurdo Sound area. The surface of the ridges consist of slightly rounded, well-sorted pebbles and cobbles 0.01 to 0.2 m in diameter. A few large boulders with very fine glacial striae are also present. Rock types are diverse, but all are volcanic and cannot be considered to be in place. The young age of The Pleiades allows little time for erosion to have occurred. Processes are also very slow judging from the youthful appearance of 7-m.y.-old Mount Overlord. Several boulders and pebbles of tristanite in the moraine are salt-weathered, and many are coated with carbonate on their underside.

* Specimen numbers refer to the petrologic collection, Department of Geology, Victoria University of Wellington, New Zealand. All numbers are prefixed by 25; i.e., 666 is 25666. Specimen numbers prefixed by P refer to the petrologic collection, New Zealand Geological Survey, Lower Hutt.

** Nomenclature is summarized in Table 91.3 and discussed later in text.

Aleyone Cone

Aleyone Cone,‡ 2 km northeast of Mount Atlas, is only slightly lower in elevation than Mount Atlas but much smaller in areal extent (Figure 91.2). On the east side, Aleyone Cone is nearly covered in snow, while to the north, there is a large slaggy scree composed of large blocks and bombs of scoriaceous trachyandesite (703). This scree is only a thin cover, however, and is underlain by K-trachyte (701 and 702). The trachyandesite was apparently erupted from a poorly defined crater to the east of Aleyone Cone. A thin (~1 m) trachyandesite flow mantles the trachyte near the summit.

About 100 m north of Aleyone Cone is a low, horseshoe-shaped crater (C6, Figure 91.2) breached to the northeast. It is composed of K-trachyte (705) similar in composition and probably age to K-trachyte (701) from Aleyone Cone. The trachyte is overlain by blocks and bombs, and to the west, by a 1-m-thick scoriaceous flow of Ne-benmoreite (704).

Taygete Cone

Taygete Cone‡ is an endogenous dome of peralkaline K-trachyte (696, 698, 699, and 706). It was probably emplaced in part as a consolidated or semiconsolidated plug, as slickensides are well developed along the south flanks. Slickensides are also common within the cone, and this, together with the irregular structure and weakly developed flow banding, indicates an endogenous origin. A K-Ar age determination (Table 91.1) on the trachyte (706) gives an age of 3,000 ± 14,000 years, indicating that the cone is very young, possibly postdating Mount Atlas. Five areas of hydrothermal alteration near the summit of the cone are 3 to 6 m in diameter and appear to represent the most recent activity. Each area shows a zonal pattern of alteration with white, strongly bleached, silicified sulfur-coated trachyte surrounded by a zone of less altered, hematite-stained trachyte (probably a result of alteration of ferromagnesium minerals) and around this, light gray, weakly altered trachyte. The zonal arrangement and the presence of a few areas of strong alteration aligned with the slope suggest that the alteration resulted from a hot spring.

A small semicircular crater 100 m in diameter is superimposed on the eastern side of Taygete Cone. The crater is mantled on the south and east rim by yellow Ne-mugearite (700) agglomerate. Ne-mugearite lava flows on the northern end of Taygete Cone are probably the same age as the agglomerate. The areas of hydrothermal alteration discussed above may also be contemporaneous with the crater development and associated agglomerate formation.

Northeast of Taygete Cone there is a larger, mainly snow-covered crater (C8, Figure 91.2). Only the east and west side of the crater rim are exposed, with outcrops rare.

‡ Name submitted to the N.Z. Geographic Board for approval.

Slope material suggests that the crater is composed of a light greenish-gray trachyte (692). The eastern outcrop is cut on the west side by a small Ne-benmoreite (690) dike, which may feed a few thin Ne-benmoreite (691) flows. Xenoliths of comagmatic(?) syenite and basement granite are abundant (1–3%). The Ne-benmoreite is overlain by a dark trachyandesite(?). On the northern rim there is another small crater (C9, Figure 91.2) composed mainly of trachyte which is again overlain by trachyandesite(?).

To the north and west there are four more eruptive centers. C7 (Figure 91.2) is a small crater, mainly snow covered, with only the east rim exposed. Although no outcrop occurs, slope deposits indicate that the crater is mainly light-colored pumiceous trachyte (681) and peralkaline trachyte (680) which is overlain by basanite (679). A small trachyte flow (678) occurs to the north of crater 7.

The cone C10 (Figure 91.2) and the slopes leading down to the Mariner Glacier are strongly eroded K-trachyte (687) flows(?) that show strong cavernous salt weathering. The exposure consists mainly of rubbly trachytic slope material containing fragments of dark basaltic rock. Tristanite pyroclastic rocks (688) from an unknown source cover much of the upper part of cone C10.

C11 (Figure 91.2), an endogenous dome, is composed mainly of phonolite (683) which, as on most of the other cones and craters, is mantled by later basanite lava and pyroclastic rocks. The basanite (682) forms a semicircular crater on the east side of C11.

The nearly circular crater C12 (Figure 91.2) is the most northern outcrop of The Pleiades. It is breached to the south and appears to be composed mainly of Ne-hawaiite (685).

Stratigraphy

The limited exposure, lack of crosscutting relationships, and the wide range of diverse rock types mask the time relationships between the various cones and craters of The Pleiades. Mount Atlas developed in three stages and, throughout most of its history, erupted in alternating order tristanite and K-trachyte lavas. Trachyandesite was the final eruptive product. A similar eruptive sequence is observed at Aleyone Cone and cone C4.

Vents north of Mount Atlas are composed mainly of tristanite, K-trachyte, and peralkaline K-trachyte. Many of the vents are overlain by strongly undersaturated sodic lavas which vary in composition from basanite to Ne-benmoreite.

In conclusion, the apparent eruptive sequence at The Pleiades was alternating eruptions of tristanite and K-trachyte, followed by trachyandesite and minor peralkaline K-trachyte. Strongly undersaturated basanite to Ne-benmoreite lavas were the last eruptive products, but in terms of volume, they amount to less than 5%.

Petrology

A detailed discussion of the petrology and geochemistry of volcanic rocks from The Pleiades is outside the scope of this paper. However as the classification of the rocks and the inferred eruptive sequence are based on their chemical compositions, a brief summary of their chemical characteristics is included. Analyses of representative specimens are listed in Table 91.2.

Nomenclature

Alkalic volcanic rocks are divided into sodic and potassic series, the former being classically represented by Hawaiian and Hebridean lavas and the latter by Tristan de Cunha and Gough Island lavas. Division of a volcanic rock suite into the two series is subjective. Irvine and Baragar (1971) used the normative orthoclase content whereas Macdonald (1960) and Macdonald and Katsura (1964) used a Na_2O/K_2O ratio of 2. In strongly undersaturated lavas, Coombs and Wilkinson (1969) noted several difficulties with the Na_2O/K_2O ratio, such as the tendency of lava series to show a decreasing Na_2O/K_2O ratio with differentiation. In such a situation, basaltic lavas generally belong to the sodic series, whereas rocks with high differentiation indexes (benmoreites, tristanites, and trachytes) are potassic. Notwithstanding this and other difficulties mentioned by Coombs and Wilkinson (1969), the division based on a Na_2O/K_2O ratio has been found convenient in this study. It has been possible to assign The Pleiades rocks to both the potassic and sodic series. Once a rock is assigned to either the sodic or potassic series, the appropriate nomenclature (Table 91.3) may be applied.

Geochemistry

Lavas from The Pleiades are undersaturated and range from basanite and trachyandesite to Ne-benmoreite and peralkaline K-trachyte. Two lava series or lineages can be recognized on the basis of Na_2O/K_2O ratios (Figure 91.3) and plots of differentiation index (D.I.) against normative nepheline (Figure 91.4). The bulk of the lavas have $Na_2O/K_2O < 2$ (Figure 91.3) and consist of a mildly potassic trachyandesite to tristanite to K-trachyte to peralkaline K-trachyte lineage. Lavas with $Na_2O/K_2O > 2$ (Figure 91.3) and a trend of increasing Ne with increasing D. I. (Figure 91.4) belong to a sodic basanite to Ne-hawaiite to Ne-benmoreite lineage.

Lavas of the sodic lineage are the final eruptive products at The Pleiades and constitute about 1% by volume of the lavas. At all outcrops they intrude older potassic lineage lavas. As all the sodic lineage lavas seem to be closely related in time, it is suggested that their eruption represents outpourings from the same magma chamber. The degree of alkalinity (or degree of fractionation) decreases toward the west and southwest. Baker (1968) described a similar systematic compositional variation in parasitic bodies and dikes at Saint Helena, from phonolite in the central area, outward through trachyte to trachybasalt. He suggested that a magma chamber with a domed roof was tapped at varying depths and horizontal distances from its center and that this accounted for the compositional variations. This situation may occur under the northern Pleiades, but with only six data points it is not possible to define the shape of any underlying magma chamber.

The two lava lineages from The Pleiades are believed to represent simple differentiation series which resulted from fractional crystallization. Kyle and Rankin (1976) used

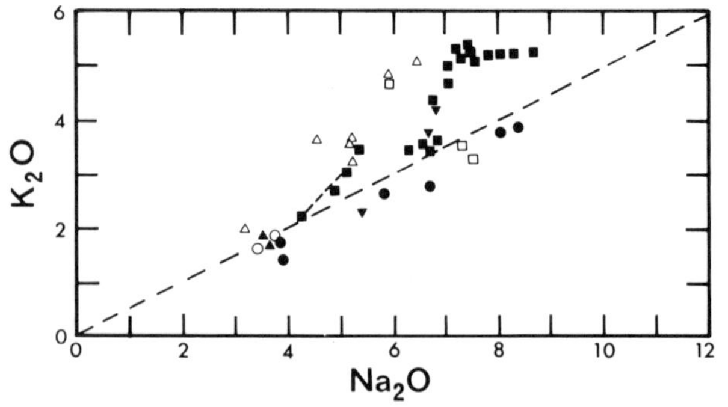

Figure 91.3. Plot of K_2O against Na_2O (in wt %) for Melbourne volcanic province rocks. Dashed line indicates a Na_2O/K_2O ratio of 2. Small dashed line joins whole-rock sample to its groundmass. Symbols for data from The Pleiades (this paper): solid squares: potassic lineage; solid circles: sodic lineage; open circles: essexite inclusions; open squares: transitional. Symbols for other Melbourne volcanic province rocks (Kyle, 1976; Nathan and Schulte, 1968): inverted triangles: Mount Overlord; open triangles: Mount Melbourne; solid triangles: Local Suite basanites.

rare-earth–element data to reconstruct the evolution of the lineages.

Comparison with other Melbourne volcanic province rocks

Potassic lineage lavas from The Pleiades are similar to three analyzed specimens from Mount Overlord (Kyle, 1976), but they have considerably higher alkali contents than the oversaturated Q-trachyandesite, Q-tristanite, and Q–K-trachyte lavas from Mount Melbourne (Nathan and Schulte, 1968; Kyle, 1976) (Figures 91.3 and 91.5). Mount Melbourne lavas are strongly enriched in K_2O (Figure 91.3) compared to The Pleiades. Sodic lineage basanites from The Pleiades are similar to two analyzed Local Suite basanites (Figures 91.3, 91.4, and 91.5).

Discussion and conclusions

The Pleiades are young eruptive centers, perhaps less than 140,000 years old, and represent one of the youngest phases of activity in the 7 m.y. history of the Melbourne volcanic province. Chemical analyses indicate that the lavas form at least two differentiation trends (lava lineages) which are believed to have evolved by fractional crystallization. Lavas at The Pleiades belong mainly to a mildly potassic trachyandesite to peralkaline K-trachyte lineage; it is the most complete lineage among McMurdo Volcanic Group rocks from northern Victoria Land.

Local Suite basanites are commonly found as small scoria cones and lava flows throughout the Melbourne volcanic province (Nathan and Schulte, 1968). In some places they have been slightly modified by differentiation from a more primitive mantle-derived basanite parent. It is only at The Pleiades that extensive fractionation of the basanites has resulted in the basanite to Ne-benmoreite lava lineage. No other lavas comparable to this lineage are found in the Melbourne volcanic province; however, similar lavas occur at Hut Point Peninsula in the Erebus volcanic province (Kyle and Rankin, 1976).

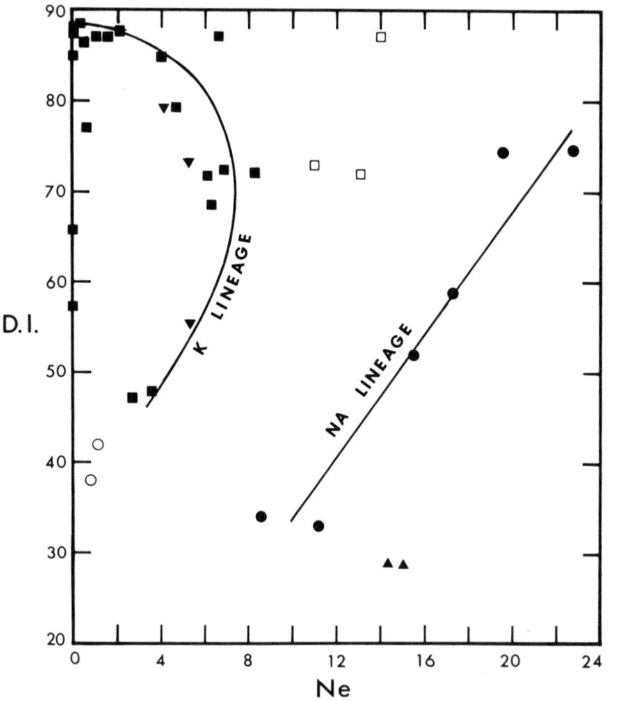

Figure 91.4. Plot of the Thornton and Tuttle (1960) differentiation index (D.I.) against normative nepheline (Ne) (in wt %) for rocks from the Melbourne volcanic province, excluding Mount Melbourne. The three potassic lineage lavas with D.I. < 80 and Ne $<$ 1 are oxidized and would have a higher Ne content if a correction were made to the Fe^{3+}/Fe^{2+} ratio, to account for oxidation. Symbols are as in Figure 91.3.

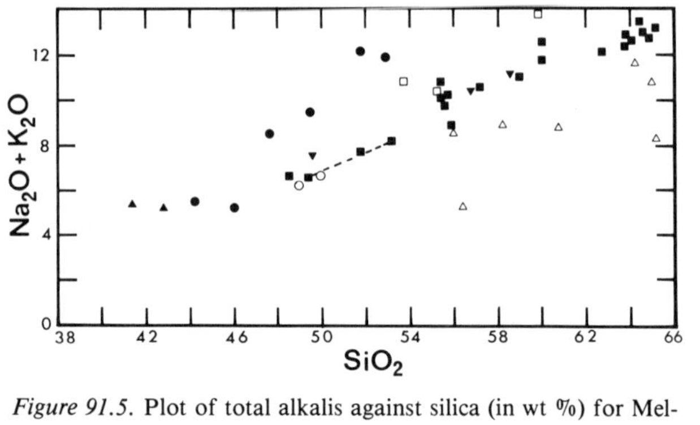

Figure 91.5. Plot of total alkalis against silica (in wt %) for Melbourne volcanic province rocks. Symbols are as in Figure 91.3.

The undersaturated alkalic nature of the volcanism is typical of that found within rifted and uplifted intraplate environments. The location of major volcanic features adjacent to and within the Transantarctic Mountains is probably related to deep crustal fractures developed during the Cenozoic uplift of the mountain range. Why volcanism was initiated is unknown; however, it may not be a coincidence that the bulk of the activity in the McMurdo Volcanic Group has occurred during the last 10 m.y. (Armstrong, 1978), about the same period of time that the East Antarctic Ice Sheet has been in a fully developed state. Perhaps isostatic depression of the crust due to glacial loading has caused thermal perturbations in the mantle which initiated partial melting.

Table 91.2. Chemical analyses (in wt. %) of representative rocks from The Pleiades, Melbourne volcanic province, Antarctica

	1	2	3	4	5	6	7	8
VUW number	25679	25682	25685	25664	25663	25703	G mass[b]	25700
SiO_2	44.26	46.11	47.58	48.44	48.93	49.26	53.1	49.46
TiO_2	3.66	2.41	2.96	2.85	1.90	2.63	2.75	2.01
Al_2O_3	14.82	14.35	16.97	16.19	13.55	16.10	17.0	18.01
Fe_2O_3	4.01	3.30	3.53	5.11	2.51	4.67	3.79	3.90
FeO	8.70	8.35	8.08	6.33	8.86	6.49	5.27	6.36
MnO	0.20	0.21	0.24	0.20	0.19	0.20	0.21	0.24
MgO	7.21	8.94	3.51	4.72	8.66	5.09	3.05	2.81
CaO	10.98	10.04	8.15	7.82	8.16	8.14	7.13	7.05
Na_2O	3.80	3.84	5.80	4.41	3.42	4.24	5.14	6.72
K_2O	1.74	1.48	2.67	2.33	1.71	2.25	3.11	2.80
P_2O_5	0.78	0.59	0.99	0.89	0.35	0.73	—	0.76
H_2O+	—	—	—	0.20[a]	0.95[a]	0.19	—	—
H_2O-	—	—	—	—	—	—	—	—
Total	100.16	99.62	100.48	99.49	99.19	99.99	100.55[c]	100.12

	9	10	11	12	13	14	15	16
VUW number	25668	25671	25691	25690	25704	25661	25675	25662
SiO_2	49.83	51.72	51.76	52.77	53.74	55.12	55.26	55.36
TiO_2	2.00	2.23	1.15	1.23	1.17	1.29	1.26	1.31
Al_2O_3	14.43	16.97	18.96	19.10	18.23	17.39	17.57	17.57
Fe_2O_3	2.56	7.17	4.39	5.15	5.16	3.07	3.24	3.24
FeO	8.40	3.26	3.86	3.45	4.24	5.66	5.50	5.55
MnO	0.19	0.20	0.25	0.25	0.28	0.23	0.23	0.23
MgO	7.68	3.50	1.22	1.29	1.13	1.65	1.60	1.63
CaO	7.81	5.97	4.58	4.72	4.82	4.11	3.96	4.14
Na_2O	3.75	4.86	8.34	8.03	7.47	7.31	6.88	6.69
K_2O	1.91	2.78	3.85	3.90	3.35	3.52	3.61	3.53
P_2O_5	0.38	0.79	0.39	0.41	0.48	0.59	0.57	0.60
H_2O+	—	—	—	—	—	0.02	—	0.00
H_2O-	—	—	—	—	—	0.02	—	0.03
Total	98.94	99.45	98.75	100.30	100.07	99.96	99.68	99.88

	17	18	19	20	21	22	23	24
VUW number	25684	25670	25666	25667	25674	25683	25687	25665
SiO_2	55.50	55.58	55.81	57.10	59.04	59.79	59.94	59.98
TiO_2	1.58	1.29	1.66	1.22	0.88	0.21	0.60	0.70
Al_2O_3	16.97	17.61	16.58	18.52	17.83	18.83	18.19	17.67
Fe_2O_3	2.71	3.41	3.77	5.65	2.15	3.54	4.03	5.57
FeO	5.38	5.44	4.18	2.51	4.76	1.89	1.34	0.98
MnO	0.18	0.23	0.16	0.11	0.18	0.20	0.18	0.17
MgO	2.66	1.65	2.98	0.77	1.06	0.16	0.57	0.75
CaO	4.98	4.09	5.04	3.20	3.02	1.34	2.13	2.47
Na_2O	6.28	6.61	5.35	5.89	6.79	8.68	7.29	7.09
K_2O	3.50	3.56	3.52	4.70	4.38	5.26	5.22	4.74
P_2O_5	0.55	0.59	0.53	0.67	0.35	0.09	0.17	0.24
H_2O+	—	—	—	—	—	—	—	—
H_2O-	—	—	—	—	—	—	—	—
Total	100.29	100.06	99.58	100.34	100.44	99.99	99.66	100.36

	25	26	27	28	29	30	31	32
VUW number	25702	25701	25705	25696	25706	25699	25680	25698
SiO_2	62.71	63.84	63.86	63.90	64.36	64.56	64.83	65.16
TiO_2	0.46	0.29	0.28	0.15	0.14	0.15	0.15	0.17
Al_2O_3	17.56	16.52	16.59	16.77	16.85	16.95	15.99	16.70
Fe_2O_3	1.55	2.00	3.12	2.82	2.34	3.28	2.33	4.06
FeO	3.21	2.88	1.65	1.12	1.52	0.74	2.38	0.00
MnO	0.15	0.17	0.17	0.15	0.15	0.15	0.23	0.15
MgO	0.51	0.14	0.20	0.05	0.08	0.07	0.06	0.04
CaO	1.94	1.30	1.27	0.95	0.93	0.94	1.08	0.89
Na_2O	7.02	7.21	7.46	7.57	8.26	7.81	7.53	7.96
K_2O	5.07	5.32	5.36	5.12	5.24	5.18	5.26	5.21
P_2O_5	0.17	0.07	0.07	0.02	0.03	0.02	0.05	0.03
H_2O+	—	—	—	0.45	—	—	—	—
H_2O-	—	—	—	—	—	—	—	—
Total	100.35	99.74	100.03	99.07	99.90	99.85	99.89	100.37

1 and 2, Basanite; 3, Ne-hawaiite; 4, Trachyandesite; 5, Essexite inclusion; 6 and 7, Trachyandesite; 8, Ne-mugearite; 9, Essexite inclusion; 10, Trachyandesite; 11-13, Ne-benmoreite; 14, Ne-tristanite; 15-19, Tristanite; 20-21, K-trachyte; 22, Phonolite; 23-27, K-trachyte; 28-32, Peralkaline K-trachyte.

Analyses done by P.R. Kyle at Victoria University of Wellington using X-ray fluorescence analysis for all elements except Na_2O, FeO, and H_2O, which were determined using atomic absorption spectroscopy and standard wet-chemical techniques.

[a] Loss on ignition.

[b] Broad beam electron microprobe analysis of glassy groundmass of specimen 25703.

[c] Total iron as FeO = 8.68, Fe_2O_3 calculated using the same Fe_2O_3/FeO as the whole rock.

Table 91.3. Nomenclature, based on normative mineralogy, used here to describe rocks of the McMurdo Volcanic Group

Sodic series	Potassic series	Features and comments
Alkali basalt	Alkali K-basalt	An > 50, Ne < 5
Basanite	K-basanite	An > 50, Ne > 5
Basanitoid	—	Used synonymously with basanite
Hawaiite	Trachyandesite	An = 30–50, Ne < 10
Ne-hawaiite	Ne-trachyandesite	An = 30–50, Ne > 10
	Q-trachyandesite	An = 30–50, Q > O, D.I. < 65
Mugearite	Trachyandesite	An = 10–30, Ne < 10, D.I. < 65
Ne-mugearite	Ne-trachyandesite	An = 10–30, Ne > 10, D.I. < 65
Benmoreite	Tristanite	D.I. = 65–75, Ne < 10
Ne-benmoreite	Ne-tristanite	D.I. = 65–75, Ne > 10
	Q-tristanite	D.I. = 65–75, Q > 0
Trachyte	K-trachyte	D.I. > 75, Ne < 10
Phonolite	Phonolite	D.I. > 75, Ne > 10
—	Q-K-trachyte	D.I. > 75, Q > O, Q > Ac
—	Peralkaline K-trachyte	D.I. > 75, Ac > O, Ac > Q, Ac > Ne

An: normative $\frac{100An}{An + Ab}$.
Ne: normative nepheline.
D.I.: Differentiation index (Thornton and Tuttle, 1960).
Q: normative quartz.
Ac: normative acmite.
Mineralogic prefixes are used to distinguish some rock types: the phonolites in particular have been classified as anorthoclase phonolite (equivalent to the kenyte lavas of the Mount Erebus center), pyroxene phonolite (for aegirine-augite-bearing phonolites), and kaersutite phonolite.

Acknowledgments. This study was undertaken during 1971–72 as part of Victoria University of Wellington Antarctic Expedition 16 (VUWAE 16). Special thanks are due to the Antarctic Division, Department of Scientific and Industrial Research, and to the U.S. Navy VXE-6 Antarctic Support Squadron for field and logistic support. Malcolm Laird, Peter Andrews, and the late Peter Jennings provided invaluable advice, assistance, and companionship in the field. Financial support for field and laboratory work was provided by the New Zealand University Grants Committee and Victoria University of Wellington. Preparation of this paper was made possible by support from the Institute of Polar Studies and National Science Foundation Grant DPP76-23440. D. H. Elliot and W. E. LeMasurier are thanked for reviewing the manuscript. P. J. Barrett and J. W. Cole commented on a preliminary version of this manuscript which formed part of a Ph.D. dissertation submitted to Victoria University of Wellington.

References cited

Armstrong, R. L. 1978. K-Ar dating: Late Cenozoic McMurdo Volcanic Group and dry valley glacial history, Victoria Land, Antarctica. *N.Z. Jl Geol. Geophys., 21,* No. 6, 685–98.

Baker, I. 1968. Compositional variation of minor intrusions and the form of a volcanic magma chamber. *Q. Jl geol. Soc. Lond., 124,* 67–79.

Coombs, D. S., and Wilkinson, J. F. G. 1969. Lineages and fractionation trends in undersaturated volcanic rocks from East Otago Volcanic Province (New Zealand) and related rocks. *J. Petrol., 10,* 440–501.

Gair, H. S., Sturm, A., Carryer, S. J., and Grindley, G. W. 1969. The geology of northern Victoria Land (Sheet 13, northern Victoria Land). (*In* Bushnell, V. C., and Craddock, C., *eds.* Geologic Maps of Antarctica. *Antarct. Map Folio Ser.* Folio 12, Pl. XII.)

Hamilton, W. 1972. The Hallett Volcanic Province. *Prof. Pap. U.S. geol. Surv.,* 456-B, 62 pp.

Harrington, H. J. 1958. Nomenclature of rock units in the Ross Sea region, Antarctica. *Nature, Lond., 182,* 290.

Irvine, T. N., and Baragar, W. R. A. 1971. A guide to the chemical classification of the common volcanic rocks. *Can. Jnl Earth Scis, 8,* 523–48.

Kyle, P. R. 1976. *Geology, Mineralogy and Geochemistry of the Late Cenozoic McMurdo Volcanic Group, Victoria Land, Antarctica.* Ph.D. dissertation, Victoria University of Wellington, New Zealand, 444 pp.

Kyle, P. R., and Cole, J. W. 1974. Structural control of volcanism in the McMurdo Volcanic Group, Antarctica. *Bull. volcan., 38,* 16–25.

Kyle, P. R., and Rankin, P. C. 1976. Rare earth element geochemistry of Late Cenozoic alkaline lavas of the McMurdo Volcanic Group, Antarctica. *Geochim. cosmochim. Acta, 40,* 1497–1507.

Kyle, P. R., Adams, J., and Rankin, P. C. 1979. Geology and petrology of the McMurdo Volcanic Group at Rainbow Ridge, Brown Peninsula, Antarctica. *Geol. Soc. Am. Bull., 90,* 676–88.

Lyon, G. L., and Giggenbach, W. F. 1974. Geothermal activity in Victoria Land, Antarctica. *N.Z. Jl Geol. Geophys., 17,* 511–21.

Macdonald, G. A. 1960. Dissimilarity of continental and oceanic rock types. *J. Petrol., 1,* 172–77.

Macdonald, G. A., and Katsura, T. 1964. Chemical composition of Hawaiian lavas. *J. Petrol., 5,* 82–133.

Nathan, S., and Schulte, F. J. 1967. Recent thermal and volcanic activity on Mt. Melbourne, northern Victoria Land, Antarctica. *N.Z. Jl Geol. Geophys., 10,* 422–30.

Nathan, S., and Schulte, F. J. 1968. Geology and petrology of the Campbell-Aviator divide, northern Victoria Land, Antarctica. Part 1—Post Paleozoic rocks. *N.Z. Jl Geol. Geophys., 11,* 940–75.

Riddolls, B. W., and Hancox, G. T. 1968. The geology of the upper Mariner Glacier region, north Victoria Land, Antarctica. *N.Z. Jl Geol. Geophys., 11,* 881–99.

Thornton, C. P., and Tuttle, O. F. 1960. Chemistry of igneous rocks. I. Differentiation index. *Am. J. Sci., 258,* 664–84.

92

Volcanism of Saint Paul and Amsterdam Islands (TAAF): Some Aspects of Volcanism along Plate Margins

JACQUES NOUGIER
Laboratoire de Géologie, Faculté des Sciences, 84000 Avignon, France

Abstract. The structural position of Saint Paul and Amsterdam Islands is important because they border the active rift of the East Indian Ridge. The study of the eruptive dynamism of the two volcanoes shows a similar evolution during the Brunhes epoch:

1. settling of a very wide paleo-volcano with a caldera at Amsterdam and eruption of basaltic tuffs at Saint Paul;
2. migration of the eruptive centers at distance of 0.7 to 2 km;
3. edification of a neo-volcano which overlays the paleo-volcano almost entirely, having a center crater in Saint Paul and a complex caldera in Amsterdam;
4. noticeable collapse phase due to faulting; and
5. recurrence of the volcanism (lateral cones).

Petrographic and geochemical studies show that these volcanoes belong to the same unit of tholeiitic basalt containing hypersthene, the composition of which is governed by the plagioclase crystallization. The chronology of the lava sequences, as well as the study of their variations, allows for identification of the magmatic parameters involved. Comparisons are made with other lava units belonging to similar magmatic and structural provinces, such as the Mid Indian Ocean Ridge, the Afar Rift, and the Red Sea islands (Jebel at Taïr Island).

The district of Saint Paul and Amsterdam Islands (Terres Australes et Antarctiques Françaises or TAAF) is made up of two volcanoes lying very close to the axis of the East Indian Ocean Ridge, between lat. 35° 52′ S and 38° 43′ S and about long. 77° 30′ E. This exceptional position at an accreting plate boundary, between the Antarctic and Indian plates (Le Pichon and others, 1973), has only two equivalents in the southern hemisphere: Bouvet Island, located on the crest of the South Atlantic Ridge, and Ascension Island, less than 100 km away from the ridge.

From bathymetric charts (Fisher, 1966; Berthois and others, 1969) and analyses of the magnetic anomalies (Schlich and Patriat, 1968, 1971; Schlich, 1975), the two volcanoes, Saint Paul and Amsterdam, are seen to lie on the western border of the ridge, which is dislocated along a probable northeast-southwest transform fault (Figure 92.1) that most likely passes between these islands. The northwest-southeast axial depression could correspond to the active rift, whereas the −645 m shallow which faces Amsterdam, as well as the −450 m one which faces Saint Paul on the east, could represent some parts of the volcanic edifice moved aside during the spreading of the ridge (Fisher, 1966). It can be noted, concerning Saint Paul (Figure 92.2), that if the −450 m shallow represents a fragment of the bedrock of the island, then its estimated age is about 0.3 m.y. based on a 3.4 cm/yr half-spreading rate of the ridge. This is in agreement with the isotopic measurements made on the subaerial neo-volcano, which give a 0.1 m.y. age (Watkins and others, 1975). We do not as yet have very detailed paleomagnetic information for this area.

In regard to the emergent volcanoes, the previous studies consist of petrographic reconnaissances by Hochstetter (1866), Philippi (1909), and Reinisch (1909), as well as an accurate volcanologic study of Saint Paul by Vélain (1878). Recently, within the framework of the scientific mission to the TAAF, some detailed studies have been carried out: petrologic (Girod and others, 1971; Gunn and others 1971, 1975), paleomagnetic (Watkins and Nougier, 1973), and isotopic (Hedge and others, 1973; Watkins and others, 1975). This new work has provided a good knowledge of these islands.

The purpose of this paper is to relate the volcanologic evolution of the islands, and their petrographic and geochemical characteristics, to the historical and regional context of their setting on the active margin of an oceanic plate.

Amsterdam Island

The Amsterdam Island stratovolcano (Figure 92.3) is elliptically shaped. The axes are respectively 9.8 and 7.4 km in length, which is an area of about 58 km². It is fringed with a 30–60 m cliff made by marine abrasion, which prevents any landing except at La Cale in the northern part of the island. On the west, two large faults trending 20° and 330° have breached the island, causing steep cliffs as high as 600–700 m. As is the case for the Piton de la Fournaise on Réunion Island, the summit here lies on the border of the caldera of the Volcan de la Dives (881 m).

Chronology of the Eruptions

The first generalized description of the volcano was made by Vélain (1878), and it has been completed by Gunn and others (1971). A paleo-volcano (Mont du Fernand, le Grand Balcon, le Pignon) is overlapped by a neo-volcano (Mont de la Dives) which presently makes up the major part of the island.

The Fernand paleo-volcano

Originally, the Fernand paleo-volcano took up an area of about 44 km² in the southwestern part of the island;

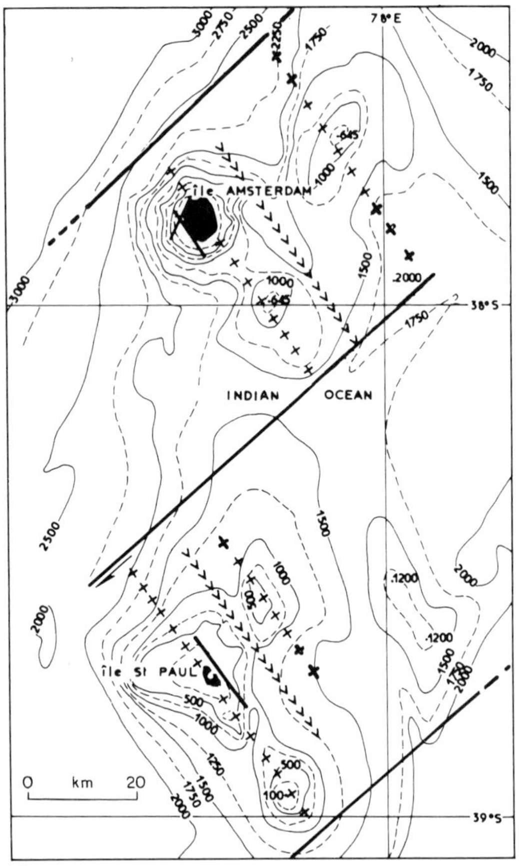

Figure 92.1. Structural interpretation of Saint Paul and Amsterdam Islands inspired from the bathymetric map of Berthois and others (1969) and from magnetic anomalies (Schlich, 1975) (contours in m). Line of crosses: ridge crest with volcanoes and seamounts. Chevron line: axis of the rift. Solid line with arrows: transform fault.

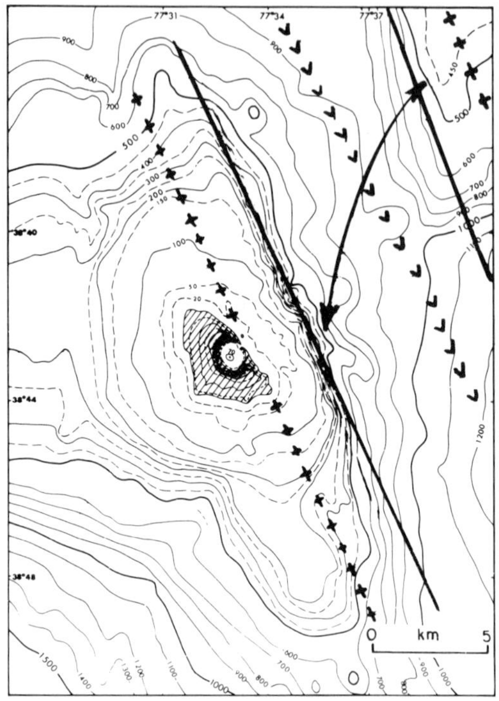

Figure 92.2. Surroundings of Saint Paul Island, with new bathymetric data after Berthois and others (1969) and possible lateral offset of the northeastern part of the island inspired by Fisher (1966). (Contours in m.) Line of crosses: ridge crest with volcanoes and seamounts. Chevron line: axis of the rift. Solid line: transform fault.

today less than 3 km² of it are visible, representing 5% of the island. The remnants of the paleo-caldera, whose diameter was about 2 km, are the Rempart du Pignon and the Môle du Fernand. This ancient caldera has been filled by the outpourings of a fissure set on the flank of Mont de la Dives. A good section of the edifice can be seen on the western wall of the island, but unfortunately it is inaccessible. A basal unit is intruded by many diversely trending dikes which break the surrounding lavas into strongly altered panels. A breccia marks the top of this unit as well as its contact with the upper unit, which is lacking in feeder dikes. The petrographic likeness between the older unit and the needle of the Pointe d'Entrecasteaux shows the latter to be a feeder vent. The strata of the upper unit are intersected at the base of les Grandes Ravines by a strongly eroded fault plane; these strata occur also in the paleo-caldera at le Fernand and le Pignon, where they are deeply altered.

The composition of the whole paleo-volcano indicates, as will be detailed below, the evolution of a very feldspathic lava toward an aluminobasalt. The paleomagnetic measurements (Watkins and Nougier, 1973) have shown the whole volcano to belong to the Brunhes epoch, and an age of less than 0.1 m.y has been proposed for the paleo-volcano. This cannot be more accurate because of the lack of isotopic measurements.

The Dives neo-volcano

The Dives neo-volcano results from a migration of the magmatic activity 2 km towards the east-northeast. The slopes that lead to the summits are gently inclined (15°) near the base, but the slope reaches 30° in the upper parts. An elliptical caldera with a maximum length of 1.5 km crowns the volcano; morphologically it looks very much like the Fournaise at la Réunion or the Erta Ale "sink" in Ethiopia (Varet 1971). This caldera first poured out westward (level of the upper step), then northward, where two intermediary steps remain. The flows have spread out into mingled sheets, and some individual ones reach 3 km in length. They are on the whole pahoehoe-type, and lava tubes are common. The most remarkable flows are those of le Plateau des Tourbières where the flows have sealed a fault-bounded coastal compartment at the northwestern tip of the island. The emptying of the caldera must have occurred through this pit, which shows many analogies with the pits that can be observed at le Karthala (Grande Comore) and at the Cratère Commerson of la Réunion. The hypothesis of an explosive crater being built up during a paroxysmal phase (Vélain, 1878) does not seem to be corroborated.

On the whole, this volcanism is characterized by aluminous and plagioclase-rich lavas trending toward an enrichment in ferromagnesians at the end of the eruptions, according to a scheme which approaches that observed in the paleo-volcano. The age of the Dives volcano has not been determined; it can be estimated, however, at some tens of thousands of years at the most, by comparing it to the condition of the Piton de la Fournaise, which formed under equivalent climatic conditions. The large western faults are

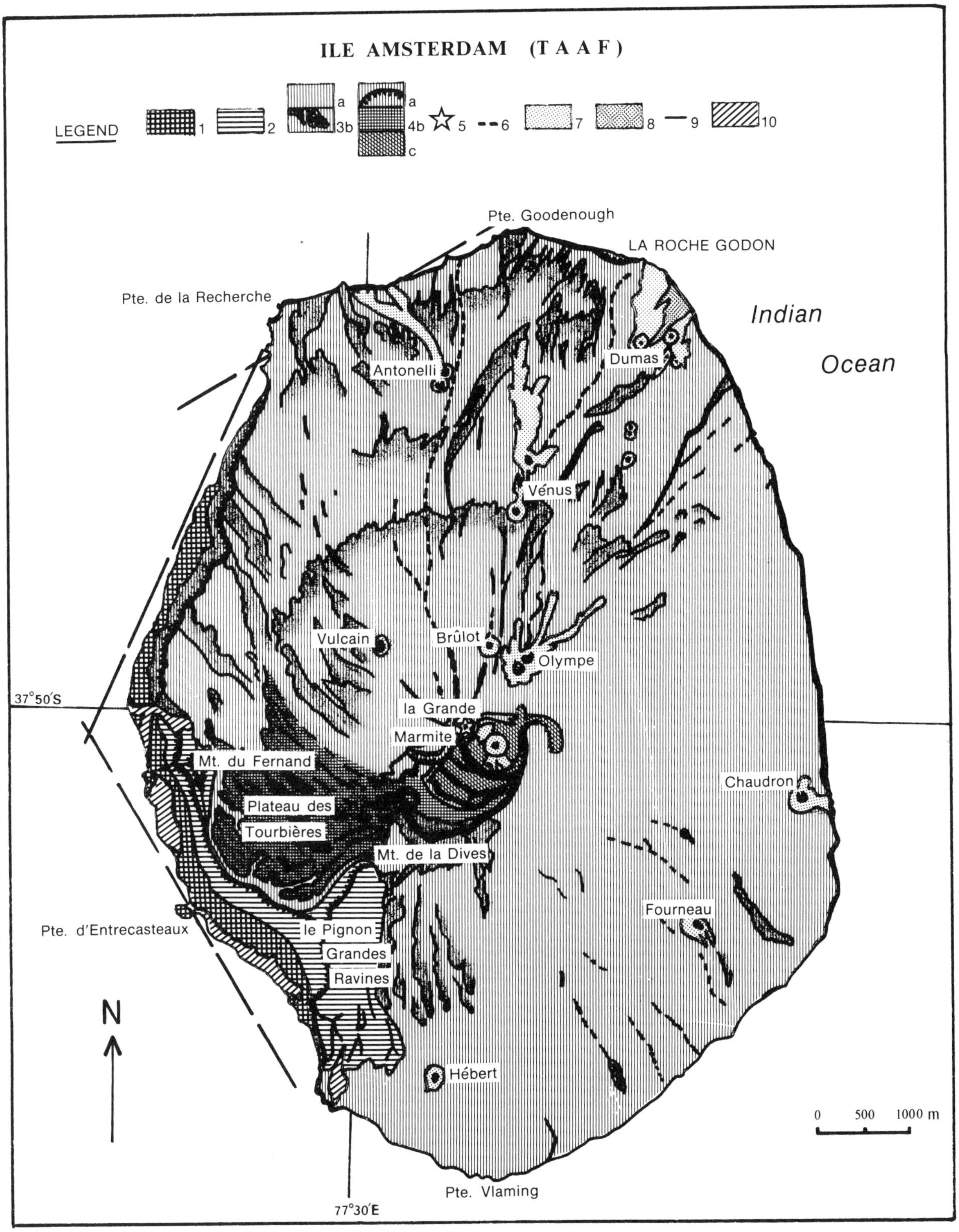

Figure 92.3. Schematic geologic map of Amsterdam Island. *The Fernand paleo-volcano:* 1. Lower units cut by veins; 2. Upper units. *The Dives neo-volcano:* 3*a*. Non-individual superficial flows; 3*b*. Individual flows and front of lava flows; 4*a*. Steps of the lava lake; 4*b*. Westward draining of the lava lake and of the top fissure; 4*c*. Successive levels of the lava lake and its eastward overflowing; 5. The Grande Marmite pit; 6. Radial fissures and lavas tubes; 7. Scoriaceous lateral cones; 8. Present eruptions of Dumas East Crater; 9. Fault; 10. Scree.

presumably more recent, except for the Grandes Ravines fault which is overlapped by the basal flows of la Dives.

Many short, late-stage eruptions formed scoriaceous cones, small aa-type flows, and lava fountains; 15 main cones and about 10 secondary edifices, localized either at the top or on the outer flank, have been discovered. The most remarkable ones are arranged along a meridian direction. The latest volcanism is represented by the crater Dumas, rising less than 100 m above sea level. The scoriaceous lava looks remarkably fresh, and the gently eroded flanks of the cones show that these lavas have formed in historical time, probably about 100 years ago at the most. The scoria are olivine basalts, which corrobates the magmatic compositional trend initiated during the earlier eruptions.

Amsterdam Island must be regarded as a potentially active volcano, although secondary phenomena such as fumaroles are absent today.

PETROGRAPHIC AND GEOCHEMICAL EVOLUTION

The Fernand paleo-volcano

The basal unit of the paleo-volcano has a porphyritic tendency, with labradorite phenocrysts (An_{60-65}) that can reach 7 mm in length; yellowish clinopyroxenes and titanomagnetite are constant elements of the groundmass as is labradorite (An_{55}). Olivine appears only at the top of the unit and in a small amount. In the chemical analysis of specimen NA20 (Table 92.1), the relatively high alumina percentage caused the formation of normative anorthite, which is confirmed in thin section by the abundance of calcic plagioclase. This analysis resembles that of a dolerite collected at Saint Paul Island by Girod and others (1971), but unfortunately its precise sampling locality was not given. Within the silica/alkali diagram of Kuno (1959) (Figure 92.4), the lavas belong to the field of the hypersthene tholeiites. The contact between the basal and the

Table 92.1. Chemical analyses of rocks from Amsterdam volcano, classified in approximate chronologic order of emission (major elements in %)

	Paleo-volcano					Dives neo-volcano					
	1	2	3	4	5	6	7	8	9	10	11
SiO_2	49.50	50.10	46.90	48.70	49.00	47.70	48.47	46.00	48.60	45.50	44.70
Al_2O_3	19.20	18.40	23.00	15.00	16.70	25.00	15.69	19.85	17.10	15.70	15.50
Fe_2O_3	3.51	5.80	4.28	4.61	2.25	0.84	4.35	2.39	2.03	3.46	10.62[a]
FeO	3.81	4.15	3.34	6.88	9.52	5.38	6.55	6.44	9.07	8.53	1.23[b]
MnO	0.12	0.13	0.11	0.17	0.27	0.10	0.19	0.11	0.20	0.21	0.19
MgO	4.37	2.98	3.40	5.00	4.90	3.40	6.46	4.13	4.90	5.77	6.14
CaO	11.23	9.72	11.67	9.83	11.01	12.72	12.71	12.45	11.49	11.10	10.93
Na_2O	2.73	2.93	2.49	2.54	2.88	2.36	2.48	2.43	2.78	2.28	2.26
K_2O	0.68	0.73	0.53	0.53	0.85	0.55	0.65	0.68	0.75	0.76	0.72
TiO_2	1.78	1.98	1.40	1.96	2.22	1.04	1.75	1.75	2.08	2.08	2.03
H_2O	3.21	2.83	2.71	3.61	0.09	0.15	0.68	2.72	0.03	4.27	4.10
Total	100.14	99.75	99.83	98.83	99.69	99.24	99.98	98.95	99.03	99.66	98.42
					Trace elements (ppm)						
Ba	106	447	87	164	70	130	179	165	200	210	182
Co	29	282	34	34	40	30	< 10	24	40	43	38
Cr	147	558	85	104	140	52	85	82	86	86	134
Cu	101	87	53	87	100	40	50	78	60	115	120
Ni	58	554	58	67	60	50	55	48	60	67	77
Sr	266	204	311	217	260	310	333	321	270	220	206
V	300	1146	214	352	320	160	288	< 10	215	316	288
					Norms						
Qz	3.84	8.22	2.10	5.46	0.0	0.0	0.0	0.0	0.0	0.0	0.0
Or	3.89	3.89	2.78	2.78	5.00	3.34	3.34	3.89	4.45	4.45	3.89
Ab	23.06	24.63	20.96	21.48	24.63	19.91	20.96	20.44	23.58	19.39	19.39
An	38.09	35.03	50.04	28.08	30.02	55.88	30.02	41.14	31.97	30.30	30.02
Di	13.77	10.15	6.08	16.59	20.23	5.94	26.50	16.78	20.66	20.18	19.65
Hy	5.63	2.80	6.06	10.30	5.70	4.18	7.48	.99	1.03	2.56	5.15
Ol	0.0	0.0	0.0	0.0	6.55	6.69	1.24	6.01	9.52	8.57	7.06
Mt	5.10	8.12	6.26	6.73	3.25	1.16	6.26	3.48	3.02	5.10	4.64
Il	3.56	3.80	2.74	3.65	4.26	1.98	3.34	3.34	3.80	3.80	3.80

New data from C.R.P.G., Nancy, analyst.

1. NA21A: Basalt, high Al, basal unit. Pointe d'Entrecasteaux.
2. NA21C: Basalt, feldsparphyric, upper unit. Pointe d'Entrecasteaux. (Add in norm, He = 0.16.)
3. NA20: Basalt, coarsely feldsparphyric. Feeder intrusion of the upper unit (cf. NA21C). Pointe d'Entrecasteaux.
4. NA22: Basalt, low-Al, finely granular, flow from Fernand volcano. Pointe del Cano cliff.
5. NA25: Olivine basalt, basement of Mont du Fernand.
6. NA31: Olivine basalt, feldsparphyric, inner flank of the Dives neo-volcano.
7. NA13: Basalt, feldsparphyric, basement of Chaussée des Otaries. (Pointe de la Recherche.)
8. NA16: Basalt, high-Al, upper flow from Dives lava lake, Chaussée des Otaries.
9. NA30: Olivine basalt, upper level of the lava lake of the Dives volcano.
10. NA40: Olivine basalt, upper flow from Dives, near Grande Marmite pit.
11. NA36: Scoriaceous weathered basalt. Lava fountain on the southeast flank of the island.

[a] Rectified value of Fe_2O_3 = 3.20.
[b] Rectified value of FeO = 7.90.

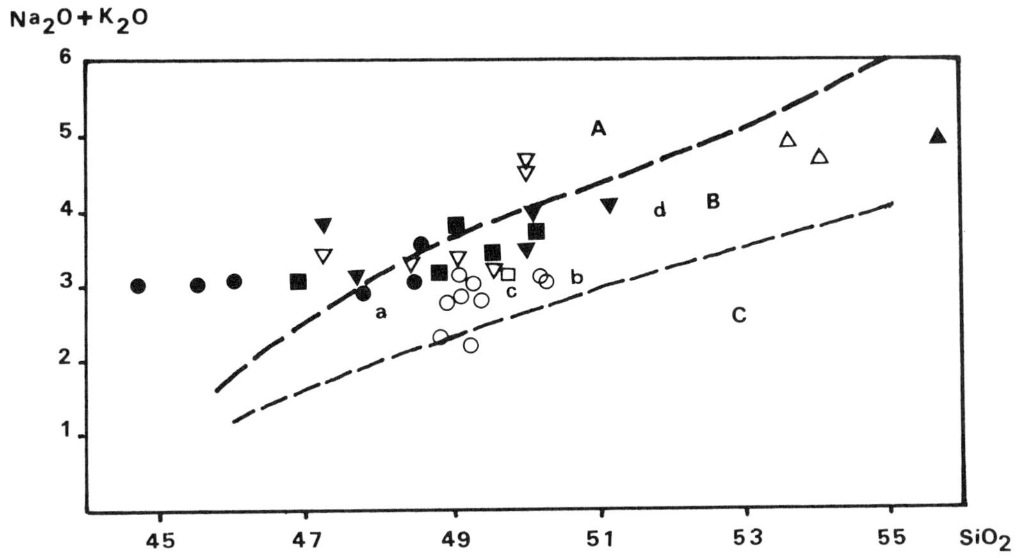

Figure 92.4. Variation diagram for alkalis (Na_2O + K_2O) with respect to SiO_2 (both in wt %) in the lavas of Amsterdam and Saint Paul volcanoes and the mid-ocean ridges. In black: new data. In white: literature data from Gunn and others, 1971, 1975; Girod and others, 1971.
Square: paleo-volcano on Amsterdam. Circle: Dives volcano on Amsterdam. Triangle, right-side up: tuffs from paleo-volcano on Saint Paul Island. Triangle, up-side down: younger volcano on Saint Paul Island. a. basalt, feldsparphyric, from East Indian Ridge, from Hekinian, 1978; b. basalt, tholeiitic, average of 4 analyses, Mid-Indian Ocean Ridge, from Engel and others, 1965; c. basalt, tholeiitic, average of 7 analyses, Mid-Atlantic Ridge, from Engel and others, 1965; and d. basalt, tholeiitic, average of 4 analyses, Carlsberg Ridge, from Wiseman, 1937. A. field of alkali basalts; B. field of hypersthene tholeiites; and C. field of pigeonite tholeiites, after Kuno, 1959.

upper units is characterized by a residual concentration of Co, Cr, Ni, and V on the bedrock by weathering of the overlying peridotite under a very wet, temperate climate.

The paleo-volcano unit is followed, at the level of les Grandes Ravines, by an aphyric sequence, vesicular and homogenous, with plagioclase microlites, minor clinopyroxene, and abundant titanomagnetite; the presence of minor olivine is to be noticed even though a small amount of quartz appears in the norm (specimen NA22). The lack of plagioclase phenocrysts in the rock finds expression in the analysis as relatively low alumina content (15%). Higher in the Fernand sequence, the porphyritic tendency (labradorite, An_{60-70}) recurs, but the olivine becomes more and more commonly associated with the constant elements of the groundmass. In addition, this is the lowest unit for which the chemical analysis gives some amount of normative olivine. The last Fernand emissions are characterized by feldsparphyric lavas, with olivine forming small phenocrysts as well as small prisms in the groundmass. In the analysis, the whole volcano shows relatively constant amounts of the major elements; the most significant variation occurs in the level of alumina, and it controls the plagioclase fractionation. The appearance and persistence of olivine at the end of the eruptive cycle indicate the rise of a deeper magma after a probable depletion of the superficial magma by eruption from the magmatic chamber.

The Dives neo-volcano

This complex has formed the subject of a relatively detailed study (Gunn and others, 1971, 1975) which related to a global examination of lavas. The revival of volcanism at the Dives volcano is marked first by very porphyritic lavas, with labradorite-bytownite phenocrysts, titanomagnetite, and matrix plagioclases less calcic than the phenocrysts. As in some of the older rocks of the paleo-volcano, the rocks of the neo-volcano show a high percentage of normative anorthite, as much as 56% (specimen NA31). These plagioclases have been considered as cumulates.

The same kind of lavas also occur at the front of the Chaussée des Otaries flow (specimen NA13), thus proving the similarity of outpourings more than 5 km apart. At this place, the composition of the lavas is close to the upper Dives lavas (specimens NA30–NA40).

Laterally, these rocks become aphyric (Pointe Goodenough) with a modal composition as follows: plagioclase, 54%; clinopyroxene, 37%; dark elements, 9%; and a trace of olivine. In the whole sequence, good correlations are found between Al_2O_3 and other major elements, and are in agreement with the plagioclase fractionation.

The lateral cones, which represent the latest volcanism, are characterized by nearly isotropic lavas with plagioclase phenocrysts. In specimen NA36, an obvious decrease of the silica content is related to a MgO increase, whereas the alkalis remain in constant ratio. In fact, at the Dives neo-volcano, an evolution towards an alkaline series with olivine exists, as discussed above for the Fernand paleo-volcano. In both cases, the eruption of a magma controlled by plagioclase fractionation was followed by the issuing of a deeper magma characterized by the appearance of olivine.

Saint Paul Island

This island (Figure 92.5) has been explored by the expeditions of the *Novara* (1857), the *Valdivia* (1876), and the *Gauss* (1901); these visits yielded short publications (Hochstetter, 1866; Philippi, 1909), but the main monography has been Vélain's work (1878). The geologic map presented in this paper is the first document effected since that date.

This island is triangular in shape; its northeastern part has collapsed, owing to major northwest-southeast faults. Originally the island had an area of about 12 km², and it

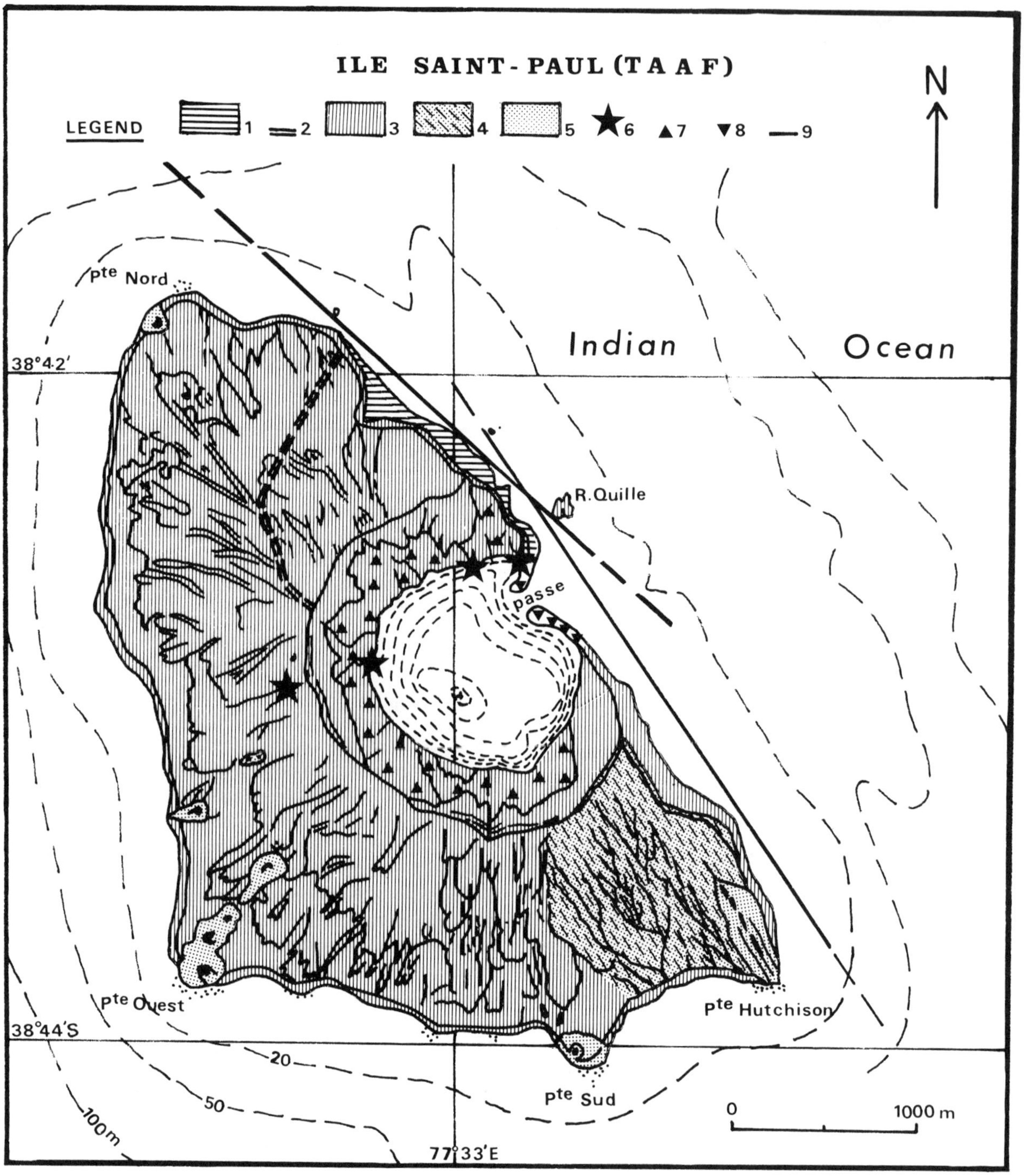

Figure 92.5. Schematic geologic map of Saint Paul Island. *Paleo-volcano:* 1. Tuffs, volcanic ashes, and pyroclastic beds cut by veins; 2. Border of the paleo-volcano, overlain by younger outpourings. *Neo-volcano:* 3. Basaltic volcano and individual flows; 4. Superficial volcanic ashes of the southeastern flank; 5. Lateral cones, scoriaceous lava fountains; 6. Hot areas and fumaroles; 7. Slope scree; 8. Coastal spit of the crater pass; 9. Fault. Contour interval of the crater isobaths is 10 m.

exhibited a perfectly circular central crater, 1800 m in diameter and about 300 m in height. Because of the collapse of the northeastern part, however, the island does not take up more than 6.5 km², and the crater communicates with the sea through a shallow pass (2.5 m deep during high tides, 0.8 m deep during low tides). The internal basin reaches a depth of 80 m (Bogdanov and others, 1967) at a point which represents the central vent. A characteristic asymmetry is presented by the slopes of the volcano: 7° slopes on the external flank and 45° slopes on the internal one. The volcano is surrounded by a cliff formed by coastal abrasion which reaches 30 to 50 m in height. Landing is impossible everywhere but inside the crater, after entering through the pass by means of shallow-draft boats. The ruins of an old crayfish canning factory are visible on the north side of the pass, near the site of a fumarolic hot area.

Chronology of the Eruptions

As for Amsterdam Island, two main eruptive phases can be recognized, culminated by late scoriaceous lateral volcanism.

The paleo-volcano is visible only on the outer northeastern cliff. It is wholly overlapped disconformably by the flows of the neo-volcano, about 60 m in thickness. This overlap is marked on the outer northern flank by a slope break at the superficial flows, which allows an estimation of the area of the paleo-volcano at about one-seventh of the present area of the island. This paleo-volcano is made of strata of friable, yellow, pumiceous tuffs overlain by agglomerates of pumice and tephra arranged as graded beds. Dip of bedding is variable and can reach 30° in places. Changes in the thickness of the deposits show that a major fault intersects the side of a tuffaceous edifice whose center lies underwater north of the Roche Quille. The basaltic character of these tuffs has been pointed out by Gunn and others (1975). Vélain (1878) mentioned the presence of rhyolites at the Ilôt du Milieu occurring as dikes and sills within the tuffs. These rhyolites could not be closely observed because of their difficult access, but they appear to be limited to the paleo-volcano.

The neo-volcano results from a 700-m migration of the magmatic activity towards the south. Rhythmic activity, mainly above sea level, is indicated by the piling up of basaltic flows interlayered with beds of cinders and scoria, as well as by the lack of pillow structures. Dikes from a second generation intersect the paleo-volcano and served as feeders for the present cone, which must have been built up from the overflowing of a lava lake of the Amsterdam type. This origin is emphasized by the outpourings of late basalts which make up the present topographic surface of the volcano. The extensional fissures by which the lavas of the peripheral scoriaceous edifices have risen are perfectly radial from the central vent of the crater. Hence, this crater can be considered as an undersea pit, comparable with the Grande Marmite of Amsterdam.

At Saint Paul, the volcanism has ended with the emission of 0.4-m-thick cinder beds restricted to the southeast flank, and with the building of scoriaceous cones or hornitos lined up along the fissure axis. Formed at very low altitudes, these cones are breached by the cliff of coastal abrasion. Some cones, such as at Pointe Nord, have been obliterated by the basaltic flows that issued from the main crater; others are obviously superimposed, like the Dumas of Amsterdam Island. Some steamy emanations were reported near these cones at the end of the eighteenth century (McCartney, 1 July 1793; cited by Vélain, 1878).

There are still some hot springs and fumarolic areas within the crater and near sea level. The temperature of the water in the northwestern basin was 44° C in 1857, 42° C in 1874, and 37° C in January 1972. The fumaroles to the west of the crater were 104° C in 1874, and they contained more than 10% carbonic gas. We measured them again in 1972; their temperature was between 80° and 85° C, and they were made of only steam. Some other hot spots are visible at higher altitudes in a westerly direction; their temperature is relatively low.

The Saint Paul volcanism has been dated by K-Ar isotopic measurements made on the basalts of the neo-volcano (Watkins and others, 1975). The age obtained is less than 0.1 m.y. Consequently the big fault and collapse on the northeastern side of the island (possibly contemporaneous with the Amsterdam Island collapses) can be regarded as having occurred some thousands or tens of thousands of years ago at the most.

PETROGRAPHIC AND GEOCHEMICAL EVOLUTION

The paleo-volcano

The paleo-volcano is built from a thick stratified sequence of yellow palagonitic tuffs (Vélain, 1878), including some scarce isolated crystals of calcic plagioclase and clinopyroxene. The whole sequence is strongly oxidized and weathered. Recent chemical analyses (Gunn and others, 1975, Table 6; and this paper, Table 92.2, specimen AV1) show these rocks to be altered basaltic tuffs, rich in silica (55.6%) and in ferric iron and poor in CaO, a consequence of the oxidation and leaching of this pulverized material. Thus, the high content of normative quartz (10.4%) does not express the true original composition of this material, which must be considered as a basalt whose composition approaches that of the oldest lavas.

The basaltic lavas are associated with a few intrusions of light-colored rhyolite, bluish-gray where altered. According to Vélain (1878), the rhyolitic lava occurs as an amorphous mass spangled with small quartz feldspathic microlites, sanidine in particular, and iron oxide grains. Chemical analyses (Table 92.2, specimen AV3) indicate alkaline rhyolite of the comendite type as described by MacDonald (1976). These lavas must be compared with those from the north of the Kerguelen archipelago (Table 92.2, specimen AV4) (Nougier, 1970), which are associated with a primitive volcanism, presumably tholeiitic (Watkins and others, 1974). They must also be compared with the comendite (Table 92.2, specimen BT) from Bouvet Island, associated with tholeiites (Table 92.2, specimen AV5) (Broch, 1947), which are very similar to certain lavas of Saint Paul, such as specimen SP2.

The dikes which intersect these basal units, and thus feed the neo-volcano, show various structures: aphyric with flow structures, glassy and vesicular, and doleritic with an ophitic tendency. Labradorite phenocrysts (An_{55}) are associated with andesine microlites, clinopyroxene, and titanomagnetite. The brown interstitial glass is partly devitrified, and the vesicles are locally zeolitized.

The neo-volcano

Good sampling in the interior of the crater shows that the neo-volcano began with basalts lacking in olivine but with subidiomorphic phenocrysts of labradorite (An_{55-60}) within a very fine groundmass of microlites and small laths of andesine, titanomagnetite, and yellowish clinopyroxene. Isotropic brown glass is commonly abundant. Specimen SP2 has the composition of an aluminous tholeiitic basalt with 2% normative quartz and a noticeable titanium dioxide content (2.35%). The low value for normative hypersthene is due to oxidation of Fe^{+2}. This analysis approaches the average values obtained from 14 specimens by Hekinian (1974) on the Ninety East Ridge.

Above 30 m from the base, the section shows thick porphyritic basalt flows with olivine. Several modal analyses indicate the following mineral percentages: plagioclase, 50–52%; clinopyroxene, 33–36%; olivine, 3–4%; titanomagnetite, 8–10%; and glass, 1–3%. Such a structure and mineralogy have also been found at la Roche Quille, which allows the latter to be considered as a remnant of the collapsed eastern part of the volcano.

These widespread younger units have been accurately described and analyzed in some previous works (Girod and others, 1971; Gunn and others, 1975). The plagioclase phenocrysts have an An_{82} core and an An_{57} periphery, whereas the microlite compositions vary from An_{58} to An_{25}. The crystallization of the clinopyroxene was late, the pheocrysts being enriched in Al_2O_3 and CaO, compared to those of the former unit; the common olivine is a fayalitic type (Fo_{35}). The discontinuity between the composition of the plagioclase phenocrysts and microlites suggests a gap during the crystallization (Gunn and others, 1975). Alkali feldspar microlites have been observed by Girod and others (1971) as 3–11% of the modal analysis. In the new chemical analysis (Table 92.2, specimen SP6), similar to specimen NA31 (Table 92.1) of Amsterdam Island, Al_2O_3–plagioclase controls (Thornton and Tuttle Differentiation Index, D.I. = 70.7) are found in these very porphyritic types, with normative olivine contents as great as 3.4%.

The uppermost unit is a very fine aphyric basaltic lava (specimen SP30) in which no olivine can be seen. The lack of feldspar phenocrysts leads to a chemical composition which approaches the average calculated for the whole volcano (Table 92.2, specimen AV2). The latter, in turn, is comparable to some tholeiites of the Ninety East Ridge described by Hekinian (1974), with a D.I. close to 30 (DSDP site 216).

The peripheral cones include various facies, from porphyritic basalts with labradorite phenocrysts within a vesicular brown glass that includes andesine microlites and small olivine crystals (cone of the Pointe Nord and western fountain), to very glassy, dark lavas, without any visible crystals (Les Quatre Collines). Analysis of this glassy lava (Table 92.2, specimen SP33), rich in water (5.22%) and containing 2.42% CO_2, shows the maintenance of the same type of magma, enriched in titomagnetite.

The Saint Paul–Amsterdam volcanic province

In summary, the tholeiitic character (Figure 92.6) of the magmas of Saint Paul and Amsterdam is confirmed by these studies. The average SiO_2 contents are 47.7% at Amsterdam and 49.7% at Saint Paul, and the sum of the

Table 92.2. Chemical analyses of rocks from Saint Paul volcano, classified in chronologic order of emission, and analyses of rocks from Kerguelen Islands and Bouvet Island, included for comparison (major elements in %)

	1	2	3	4	5	6	7	8	9	10	11
SiO_2	55.66	48.20	50.10	49.10	49.90	51.09	49.98	72.46	71.09	70.47	49.15
Al_2O_3	15.08	16.70	14.90	21.15	14.00	12.24	15.54	14.28	12.99	12.28	17.69
Fe_2O_3	8.78	5.78	6.23	1.11	4.32	7.54	12.92	2.55	1.10	2.96	4.28
FeO	2.90	5.76	5.51	6.87	11.07	9.04	—	2.10	1.10	1.99	5.70
MnO	0.21	0.18	0.20	0.13	0.22	0.19	0.19	—	0.05	0.06	0.16
MgO	4.12	4.22	4.87	3.49	4.00	2.90	3.92	0.15	0.30	0.18	4.21
CaO	4.04	10.80	10.35	11.47	9.20	6.04	10.41	0.84	0.50	0.44	10.18
Na_2O	3.70	3.18	3.14	2.99	3.38	3.08	3.58	2.67	3.72	4.91	3.11
K_2O	1.14	0.64	0.74	0.55	0.90	0.94	0.64	4.29	5.00	4.85	0.85
TiO_2	1.84	2.35	2.23	1.63	2.62	4.00	2.45	—	0.12	0.29	2.50
H_2O	1.90	1.81	0.47	0.27	0.02	2.91	0.36	1.52	3.82	0.92	1.51
Total	99.37	99.62	98.70	98.76	99.63	99.97	99.99	100.86	100.04	99.35	99.34
Trace elements (ppm)											
Ba	204	137	144	60	180	—	—	—	—	—	—
Co	35	39	40	35	45	—	—	—	—	—	—
Cr	66	88	109	80	80	—	—	—	—	—	—
Cu	26	93	55	50	55	—	—	—	—	—	—
Ni	60	54	60	60	50	—	26	—	—	—	—
Sr	140	245	264	304	230	—	—	—	—	—	—
V	135	324	283	219	380	—	—	—	—	—	—
Norms											
Qz	10.45	2.04	4.50	0.0	0.66	17.64	0.50	37.56	29.16	22.92	2.40
Or	6.56	3.34	4.45	2.78	5.56	5.56	3.76	25.58	29.47	28.91	5.00
Ab	30.96	27.25	26.72	25.15	28.82	25.68	54.75	22.33	31.44	36.15	26.20
An	19.71	29.47	24.19	42.81	20.02	13.07	54.75	4.17	0.83	0.0	31.97
Di	15.31	19.17	21.46	11.76	21.31	0.0	20.76	0.0	1.76	1.67	14.64
Hy	3.05	3.66	3.69	7.53	11.77	11.29	9.07	2.12	0.0	1.52	6.68
Mt	4.56	8.35	9.05	1.62	6.26	10.90	4.68	3.71	1.62	2.09	6.26
Il	3.44	4.41	4.26	3.04	5.02	7.60	4.66	0.0	0.15	0.61	4.71

New data from C.R.P.G., Nancy, analyst.

1. AV1: Average of two altered yellow tuffs from the paleo-volcano. (Add in norm, He = 5.51; Cor = 0.39.)
2. SP2: Basalt, high-Al, feldsparphyric, from the neo-volcano; sea level at the southeast rim of the crater.
3. SP12B: Basalt tholeiitic, feldsparphyric, from the neo-volcano; basement of la Roche Quille.
4. SP6: Basalt high-Al, feldsparphyric, from the neo-volcano; altitude 180 m, rim of the crater. (Add in norm, Ol = 3.44.)
5. SP30: Basalt low-Al, aphyric, from the neo-volcano; top of the southeast crater rim.
6. SP33: Glassy lava, low-Al, from the neo-volcano; scoriaceous lateral cone. Les Quatre Collines. In the original analysis, water = 5.22; CO_2 = 2.42. In this table, the analysis is calculated free of water. Add P_2O_5 = 2.91, and in norm, Ap = 6.72, Cor = 1.43.
7. AV2: Average of 14 analyses from Saint Paul, all basalt, tholeiitic; from Gunn and others (1975). Calculated free of water, and total iron as Fe_2O_3. Add P_2O_5 = 0.35, and in norm, Ap = 0.77.
8. AV3: Comendite from Saint Paul, average of 3 analyses; from Velain (1878). Add in norm, Cor = 3.67.
9. AV4: Comendite from Kerguelen Islands, average of 10 analyses from Nougier (1970). Add in norm, Cor = 1.12.
10. BT: Comendite from Bouvet Island (Lars Islet); from Broch (1947). Add in norm, Aeg = 4.62.
11. AV5: Basalt, tholeiitic, from Bouvet Island, average of 2 analyses from Broch (1947). Compare with SP2.

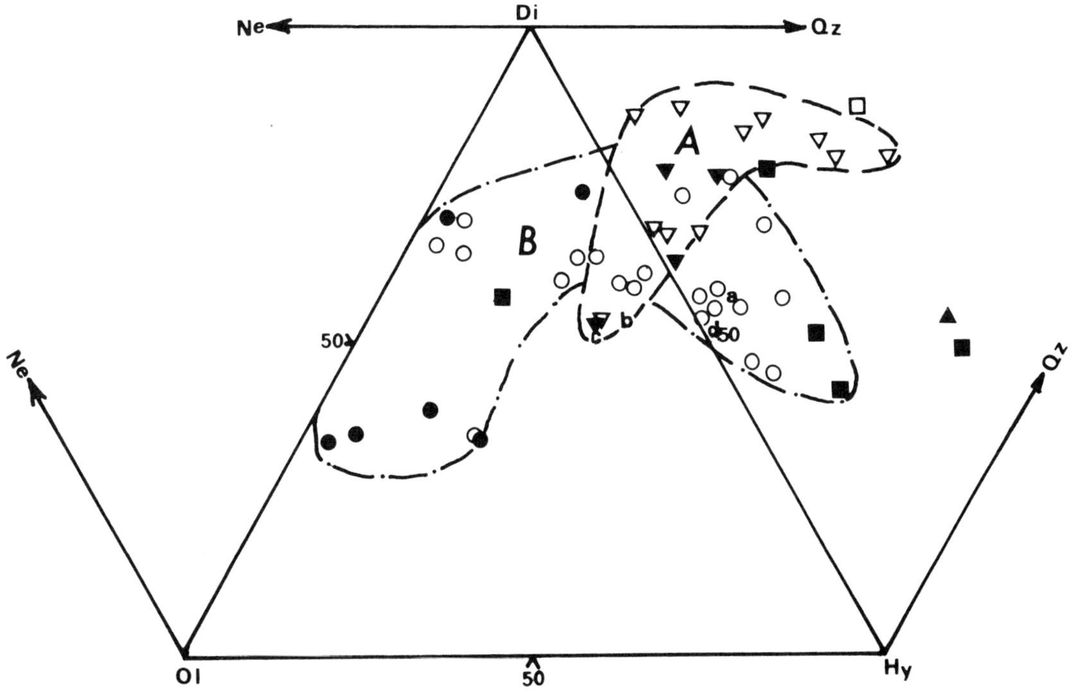

Figure 92.6. Open tetrahedron, based on normative Qz-Di-Ol-Hy-Ne, for Saint Paul and Amsterdam Islands. A. Field of Saint Paul basaltic lavas; B. Field of Amsterdam lavas. Same legend as for Figure 92.4.

alkalis is 3.2% at Amsterdam and 3.9% at Saint Paul (Figure 92.4). Potash remains less than 0.9%, and the Na_2O/K_2O ratio approaches 4 at Amsterdam and 5.6 at Saint Paul. As has been noted, alumina content varies according to the porphyritic character of the lavas and of the observed plagioclase cumulates. The residual lavas have an "andesitic" composition (Gunn and others, 1975), although not equivalent to that of the calc-alkaline lavas commonly found in island arcs (Miyashiro, 1975) or to that of the oceanic andesite of Hekinian (1974). The $^{87}Sr/^{86}Sr$ ratio, studied by Hedge and others (1973), is 0.7039 at Amsterdam and 0.7037 at Saint Paul. These values are higher than those which have been noted elsewhere on the Mid-Indian Ocean Ridges (M.I.O.R.), namely 0.7034 (Subbarao and Hedge, 1973), but close to the value obtained at Réunion Island (0.7040). As for the trace element values (Co, Ni, and Rb), they are comparable for the two islands and approach the averages proposed by Engel and others (1965) for oceanic tholeiitic basalts.

Perceptible variations occur in the magmatic evolution of the two islands, which are both characterized by two volcanic phases (paleo- and neo-volcanoes) formed during the Brunhes epoch:

1. The Amsterdam paleo-volcano is mainly tholeiitic. The emissions began with feldsparphyric lavas, and they ended with an olivine-enriched basalt, following the depletion of the more superficial feldspathic magma.
2. The Saint Paul paleo-volcano is made of altered basaltic tuffs (probably of the tholeiitic kind) associated with rhyolitic injections of the comendite type; no intermediate types were observed. This compositional association, which seems to characterize the crestal zones of oceanic ridges, has been observed at Bouvet, Easter, and Ascension Islands (Baker, 1975).
3. The Amsterdam neo-volcano shows a revival of the feldspathic magma, but it also trends toward olivine enrichment, according to a process that has been described from the paleo-volcano.
4. The Saint Paul neo-volcano does not vary much from an average composition of aluminous tholeiite (Figure 92.6), characterized by the appearance near the end of the cycle of magma enriched in mafic elements, as was also observed on Amsterdam Island.

Comparisons with mid-ocean ridge volcanism of the Indian Ocean region

Plate boundary igneous activity in a seismically active zone has been studied in the Indian Ocean from rocks dredged from the M.I.O.R., and also from volcanoes built up on the central rift, like the Jebel at Taïr Island in the Red Sea and the Assal rift formations in the Djibouti Republic. From the outset, the large part played by weathering and alteration must be emphasized: as the Fe_2O_3/FeO ratio increases, this leads to a decreasing percentage of normative olivine and even to its disappearance and the appearance of normative quartz (Miyashiro and others, 1969; Girod and others, 1971).

The M.I.O.R.

The fresh lavas of the southeastern branch of the M.I.O.R. are traditional, K-poor, olivine tholeiites that are common on the ocean floor (Cann and Vine, 1966; Hekinian, 1968). They differ from the Carlsberg Ridge lavas, which are enriched in mafic elements. This variation has been explained by Hekinian (1968) as the consequence of different spreading rates, with low rates favoring the concentration of mafic elements.

The Ninety East Ridge

The Ninety East Ridge is seismically inactive and overlain by sediments of Cretaceous to Miocene age. According to Morgan (1972), this ridge is the surface expression of plates moving away from fixed hot spots in the mantle and is secondarily associated with transform faults (McKenzie and Sclater, 1971). Hekinian (1974) classified the lavas as intermediate tholeiites and andesites; they differ from the M.I.O.R. lavas by having higher amounts of whole iron (>11%) and titanium dioxide (2–3%), and variable amounts of K_2O (0.2–1.5%). They are thus appreciably different from typical mid-ocean ridge lavas (Vinogradov and others, 1969), and their variations from the Saint Paul and Amsterdam lavas do not suggest a similar magmatic evolution.

The Jebel at Taïr Island

Located at the Red Sea, the Jebel at Taïr Island can be regarded as an expression of the ocean floor (Chase, 1969). It is made of K-poor tholeiites enriched in light rare-earth elements (Gass and others, 1973), which suggests formation from partial melting of the mantle at less than 15 km depths (5 kb), the melting being possible at shallow depth if the thermal gradient is high. It can be thought of as a fractionation of the olivine alone from a primary tholeiitic magma and a rapid rising up of tholeiitic main magma, the radial fissures favoring (as in Saint Paul and Amsterdam) the rapid evacuation of the products.

The Afar rift

The Afar rift has an axial zone built of K-poor tholeiites (Treuil and Varet, 1973) that evolve towards slightly more alkaline compositions; this is explained by a more-or less intense partial melting. When the spreading rate is decreasing, the fractionation may lead to rhyolitic magmas (Barberi and Varet, 1970). As for the Ethiopian Plateaus, they are made of silica-saturated tholeiites with hypersthene and quartz or olivine, and show tendencies toward the alkaline basalts (Jones, 1976).

Mineral paragenesis (Bizouard and others, 1976) corroborates the important role played by fractionation effected under low fugacity of oxygen, low water pressure, and at temperatures from 1240° to 1150° C., in the formation of basalts, according to probable models for magmatic chambers in oceanic accretional zones.

The comparison between the mid-Indian Ocean floor basalts and the mid-Indian Ocean ridge basalts is complicated by $^{87}Sr/^{86}Sr$ measurements. For the M.I.O.R., Subbarao and Hedge (1973) obtained values of about 0.7032–0.7035, whereas for Saint Paul and Amsterdam Hedge and others (1973) reported slightly higher values of 0.7037–0.7039, as well as a higher content of Rb and a higher Rb/Sr ratio. This suggests either isotopic heterogeneity of primary magmas or perhaps distinct origins from the mantle.

In conclusion, the magmatism of the Saint Paul and Amsterdam province is similar to that produced in other active accretional zones. The differences noted follow the peculiar evolution of each individual volcano. They are influenced by (1) the degree of partial melting in the upper mantle; (2) the fractionation of olivine and the residual liquid, which occur at relatively shallow depths; (3) the variation in the spreading rate of the rift; and (4) the source level of each eruption with regard to the subvolcanic magmatic chamber.

Acknowledgments. This work was carried out within the framework of the scientific program of the Terres Australes et Antarctiques Françaises (Administrateur Supérieur R. Barberot; Director of the Scientific Laboratories J-P. Bloch), with the support of the *M.S. Gallieni* and the helicopter crews of the French Air Force during January 1972. Field work was done in cooperation with the late N. Watkins and J-P. Bloch, whose help is very much appreciated.

References cited

Baker, P. E. 1975. Peralkaline acid volcanic rocks of oceanic islands. *Bull. volcan., 38,* No. 3, 736–54.

Barberi, F., and Varet, J. 1970. The Erta Ale volcanic range (Danakil depression, northern Afar, Ethiopia). *Bull. volcan., 34,* 848–917.

Berthois, L., Schlich, R., and Patriat, P. 1969. Carte bathymétrique du Sud de l'Océan Indien. *Inst. Phys. Globe, Paris,* No. 26, 8 pp., 4 maps.

Bizouard, H., Barberi, F., and Varet, J. 1976. Minéralogie de séries volcaniques axiales de l'Afar: précisions sur les processus et les conditions de fractionnement des magmas basaltiques dans les zones en accrétion. *Bull. Soc. Géol. Fr.,* Ser. 7, *18,* No. 4, 837–40.

Bogdanov, Y., Kuksa, V., Pasternak, F., and Romankevich, Y. 1967. Recent sedimentation conditions in the St. Paul lagoon. *Okeanologia, 7,* No. 4, 636–44.

Broch, O. 1947. Lavas of Bouvet island. *Scient. Res. Norw. Antarct. Exped., 2,* No. 25, 1–26.

Cann, J. R., and Vine, F. J. 1966. An area on the crest of the Carlsberg Ridge: petrology and magnetic survey. *Phil. Trans. R. Soc., Lond.,* Ser. A, *259,* 198–208.

Chase, R. 1969. Basalt from the axial trough of the Red sea. (*In* Degens, E. T., and Ross, D. A., *eds. Hot Brines and Recent Heavy Metal Deposits in the Red Sea.* New York, Springer Verlag, 122–28.)

Engel, A. E. J., Engel, C. G., and Havens, R. G. 1965. Chemical characteristics of oceanic basalts and the upper mantle. *Geol. Soc. Am. Bull., 76,* No. 7, 719–34.

Fisher, R. 1966. The median ridge in the south central Indian Ocean. (*In* Irvine, T. N., *ed. The World Rift System.* Pap. Geol. Surv. Can., 66–14, 135–47.)

Gass, I., Mallick, D., and Cox, K. 1973. Volcanic islands of the Red Sea. *J. Geol. Soc. Lond., 129,* Pt. 3, 275–309.

Girod, M., Camus, G., and Vialette, Y. 1971. Sur la présence de tholeiites à l'île St. Paul (Océan Indien). *Contr. Miner. Petrol., 33,* No. 2, 108–17.

Gunn, B., Watkins, N., Trzcienski, W., and Nougier, J. 1975. The Amsterdam–St Paul volcanic province and the formation of low-Al tholeiitic andesites. *Lithos, 8,* 137–49.

Gunn, B., Abranson, C., Nougier, J., Watkins, N., and Hajash, A. 1971. Amsterdam island, an isolated volcano in the southern Indian Ocean. *Contr. Miner. Petrol., 32,* No. 2, 79–92.

Hedge, C., Watkins, N., Hildreth, R., and Doering, W. 1973. $^{87}Sr/^{86}Sr$ ratios in basalts from islands in the Indian Ocean. *Earth planet. Sci. Lett., 21,* 29–34.

Hekinian, R., 1968. Rocks from the mid ocean ridge in the Indian Ocean. *Deep Sea Res., 15,* No. 2, 195–213.

Hekinian, R. 1974. Petrology of the Ninety East Ridge (Indian Ocean) compared to other aseismic ridges. *Contr. Miner. Petrol., 43,* 125–47.

Hochstetter, F. Von. 1866. *Geologische Betrachtungen während der Reise der Osterreichischen Fregatte "Novara" um die Erde. 2,* Vienna, Gerold, 58 pp.

Jones, P. 1976. Le Magmatisme au stade initial de la fragmentation des plaques arabique, numbienne et somalienne. *Bull. Soc. Géol. Fr.,* Ser. 7, *18,* No. 4, 829–30.

Kuno, H. 1959. Origin of Cenozoic petrographic provinces of Japan and surrounding areas. *Bull. volcan., 20,* 37–76.

Le Pichon, X., Francheteau, J., and Bonnin, J. 1973. *Plate Tectonics.* Amsterdam, Elsevier, 300 pp. [Developments in Geotectonics, Vol. 6.]

MacDonald, R. 1976. Nomenclature and petrochemistry of the peralkaline oversaturated extrusive rocks. *Bull. volcan., 38,* No. 3, 498–516.

McKenzie, D. P., and Sclater, J. G. 1971. The evolution of the Indian Ocean since the Late Cretaceous. *Geophys. J. R. astr. Soc., 25,* 437–528.

Miyashiro, A. 1975. Island arc volcanic rock series: a critical review. *Pétrologie, 1,* No. 3, 177–87.

Miyashiro, A., Shido, F., and Ewing, M. 1969. Diversity and origin of abyssal tholeiite from the Mid-Atlantic Ridge near 24° and 30° North Lat. *Contr. Miner. Petrol., 23,* 38–52.

Morgan, W. 1972. Deep convection plumes and plate motions. *Bull. Am. Ass. Petrol. Geol., 56,* No. 2, 203–13.

Nougier, J. 1970. Contribution of l'étude géologique et géomorphologique des îles Kerguelen (TAAF). *Paris, Comité National Français des Recherches Antarctiques,* No. 27, *1,* 440 pp., *2,* 256 pp.

Philippi, E. 1909. Geologie von St. Paul. *Deutsche Südpolar-expéd. 2,* No. 5, 370–76.

Reinisch, R. 1909. Gesteine von St. Paul. *Deutsche Südpolar-expéd., 2,* No. 5, 385–98.

Schlich, R. 1975. Structure et âge de l'océan Indien occidental. *Mem. Soc. Géol. Fr., 6,* 103 pp.

Schlich, R., and Patriat, P. 1968. Interprétation possible de données géophysiques recueillies sur la dorsale médio-indienne entre 20° et 40° Sud. *C. r. Acad. Sci., Paris, 266,* 820–22.

Schlich, R., and Patriat, P. 1971. Anomalies magnétiques de la branche Est de la dorsale médio-indienne entre les îles Amsterdam et Kerguelen. *C. r. Acad. Sci., Paris, 272,* 773–76.

Subbarao, K., and Hedge, C. 1973. K, Rb, Sr, and $^{87}Sr/^{86}Sr$ in the rocks from the mid-Indian oceanic Ridge. *Earth planet. Sci. Lett., 18,* No. 2, 223–28.

Treuil, M., and Varet, J. 1973. Critères volcanologiques, pétrologiques et géochimiques de la genèse et de la différenciation des magmas basaltiques: exemple de l'Afar. *Bull. Soc. Géol. Fr., 15,* 506–40.

Varet, J. 1971. Sur l'activité récente de l'Erta Ale (Dankalie, Ethiopie). *C. r. Acad. Sci., Paris, 272,* No. 15, 1964–67.

Vélain, C. 1878. *Description Géologique de la Presqu'île d'Aden, de L'île de la Réunion, des Îles St. Paul et Amsterdam.* Paris, Hennuyer, 356 pp.

Vinogradov, A., Udintsev, G., Dmitriev, L., Kanaev, V., Neprochnov, Y., Petrova, G., and Rikunov, L. 1969. The structure of the mid-oceanic rift zone of the Indian Ocean and its place in the world rift system. *Tectonophysics, 8,* No. 4–6, 377–401.

Watkins, N., and Nougier, J. 1973. Excursion and secular variation of the Brunhes epoch geomagnetic field in the Indian Ocean region. *J. geophys. Res., 78,* No. 26, 6060–68.

Watkins, N., Macdougall, I., and Nougier, J. 1975. Paleomagnetism and potassium-argon age of St. Paul Island, Southeastern Indian Ocean; contrasts in geomagnetic secular variation during the Brunhes epoch. *Earth planet. Sci. Lett., 24,* No. 3, 377–84.

Watkins, N., Gunn, B., Nougier, J., and Baksi, A. 1974. Kerguelen: continental fragment or oceanic island? *Geol. Soc. Am. Bull., 85,* No. 2, 201–12.

Wiseman, J. 1937. Geological and mineralogical investigations. 1. Basalts from the Carlsberg Ridge, Indian Ocean. *Rep. John Murray Exp. Brit. Mus. N.H., 3,* No. 1, 1–28.

93

Geology of Ile de l'Est, Crozet Archipelago (TAAF)

JEAN LAMEYRE
Laboratoire de Pétrologie, Université Pierre et Marie Curie, 75230 Paris, Cedex 05, France

JACQUES NOUGIER
Laboratoire de Géologie, Faculté des Sciences, 84000 Avignon, France

Abstract. Preliminary studies and geologic mapping of Ile de l'Est (Crozet Archipelago) indicate that the island is composed of four main units:

Unit I, a basal complex, consists of stratified coarse-grained plutonic rocks emplaced concordantly in a volcano-detrital complex and dissected by a dense dike swarm; the unit is epidotized and zeolitized;

Unit II contains a basal conglomerate made up of blocks of unit I, overlain by interlayered agglomerates and flows;

Unit III consists of the same materials resting on an erosional surface developed on Unit I; and

Unit IV contains scoria cones.

The basal unit comprises the foundation of the island. Its study can contribute to an understanding of certain atypical ophiolitic complexes and provide insight into the evolution of suites of oceanic alkaline igneous rocks.

During the last decade, the western Indian Ocean and the Crozet Plateau and islands have been the subject of important oceanographic, geophysical, and geologic studies. They have been carried out at sea by the Terres Australes et Antarctiques Françaises (TAAF) vessels *Gallieni* and *Marion Dufresne.* On land, they have consisted of reconnaissance geologic missions effected at the Ile de l'Est by N. Watkins and E. Abranson in 1969 and by J. Lameyre and J. Nougier in 1976. Both the exploration and the mapping of the island group will be pursued in 1978.

The Crozet Basin

From previously published observations, it appears that the Crozet Plateau, which lies east of the West Indian Ridge (Figure 93.1) has a surface area of about 4500 km^2. To the west it is flanked by a 1500-m-deep submarine plateau (Berthois and others, 1969). The whole region has been thought by Laughton and others (1970) to be a continental fragment; the aseismic Kerguelen Plateau has also been added by Dietz and Holden (1970). However, McKenzie and Sclater (1971), and then Schlich (1975), have concluded in favor of the oceanic character of this structure. The latter worker has supported his arguments by the analysis of the magnetic anomalies of the neighboring oceanic basins. Anomalies 33 to 31 (Schlich, 1975) (Late Cretaceous) trending 126°, have been recorded to the south of the Crozet Plateau, and anomalies 30 to 23 (Late Cretaceous–Paleocene), trending 118°, have been identified north-northeast of the plateau. The changes of the trends and spacing of these anomalies correspond to distinct spreading phases, with spreading (half) rates amounting respectively to 5.4 cm/yr, 13.3 cm/yr, and 7.8 cm/yr in the Crozet Basin (Schlich, 1975). From these data, added to those obtained in the Madagascar and Mascarene Plateau, an oceanic origin for the Crozet Plateau is inferred. A Campanian age for its oldest bedrock is implied by the magnetic anomalies.

The Crozet Plateau

The Crozet Plateau, which is defined by the 500-m isobath, is oriented along a generally WNW–ESE direction. It consists of a western plateau on which the Ile aux Cochons (60 km^2), the Ilot des Pingouins, and the Ilot des Apôtres are located (Girod and Nougier, 1972). The eastern plateau has a smaller surface area, but the two islands are larger in

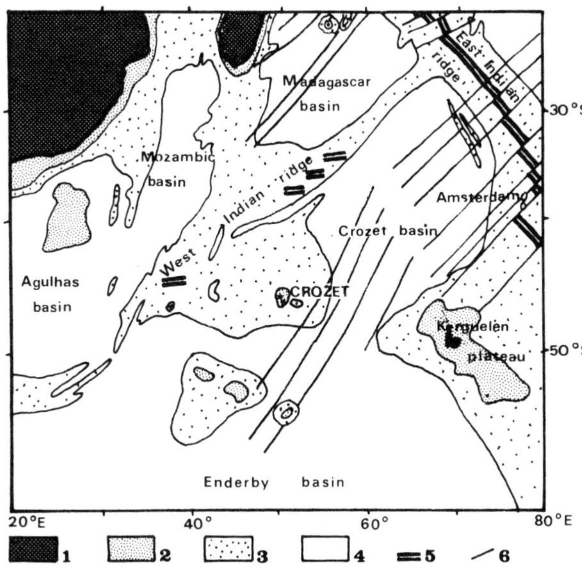

Figure 93.1. Bathymetric and structural schematic map of the southwestern Indian Ocean (after Schlich, 1975) 1. Emerged areas; 2. Depths between 0 and 1000 m; oceanic plateaus; 3. Depths between 1000 and 4000 m, including mid-ocean ridges; 4. Depths below 4000 m; ocean basins; 5. Mid-ocean ridges; rift structures; 6. Main fractures perpendicular to magnetic anomalies.

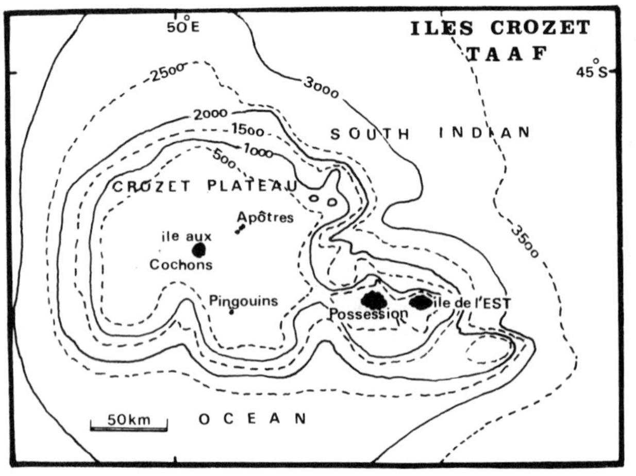

Figure 93.2. The Crozet Plateau; depth in meters.

size. The Ile de la Possession takes up about 150 km², and the Ile de l'Est—the easternmost one—is 16 km away from the other and is 170 km² in area (Figure 93.2).

Crozet Islands

Among all the islands, only the Ile de la Possession had been geologically investigated before 1969. During the Watkins and Abranson mission to the Ile de l'Est, 43 lavas flows were sampled, and they have subsequently carried out petrologic and geochemical studies. Gunn and others (1970) have analyzed a continuous series of oceanites-ankaramites-olivine basalts-feldsparphyric basalts. The extension of this sequence on the Ile de la Possession (Gunn and others, 1972) has shown the latter to form a distinct petrologic province of alkaline basalt characterized by ankaramitic fractionation.

The aim of our 1976 mission was the identification and mapping of the main volcanic units of the Ile de l'Est. Such is the subject of this preliminary paper. The petrologic, geochemical, and geochronologic studies will be presented later.

The Ile de l'Est: Summary description

The Ile de l'Est (Figure 93.3) is oriented along a main WNW-ESE axis, that is, according to the same axis as defined for the Crozet Plateau. This axial ridge on the island, whose altitude varies between 800 and 900 m, drops in three passes to less than 400 m in height. Two other mountainous axes, less than 1000 m in height and perpendicular to the former, partition the island so that communications between diverse parts of the island are difficult. Between these aretes, in the central part, the Vallée du Naufrage to the north and de l'Abondance to the south, form wide, flat, but shallow, tectonic depressions. Two other similar valleys, symmetrical with regard to the east-west axis, and also some secondary sunken valleys which do not exceed 4 km in length, diverge from the main axial ridge. Thus, they complete the drainage network which has locally retained the marks of former glaciations. The coast near the main valley is protected by nearly vertical cliffs which prevent any access except in some bays.

Our geologic exploration has allowed us to recognize four characteristic and superimposed rock units.

Unit I, the basal series

Outcrops of this unit in the center of the island are peripherally overlain by more recent formations. Westward, the basal series is in fault contact with the overlying unit, so Unit I is not visible at the western end of the island. If the younger formations, by which Unit I is locally overlapped, are ignored, it can be estimated that this series forms two-thirds of the island's surface and that its outcrops cover about 40 km². Its massifs do not exceed 600 m in height; they are characterized by greenish landscapes with scree-covered, gentle slopes which are reinforced with a dense network of dikes.

The series is composed of a volcano-detrital matrix. A stratified plutonic complex has been emplaced within this matrix. Its base, visible on the northern side of the island, is made of pyroxenites. Southward, the plutonic series contains stratified gabbros and monzodiorites. In the uppermost parts of the complex, some monzonitic veins can be observed. This plutonic complex is explained as a sill dipping slightly southward and about 300 m in thickness.

The volcano-sedimentary beds, as well as the plutonic complex, are cut by a very dense network of feeder dikes forming up to 90% of the outcrops, according to Gunn and others (1970). From a systematic study of the trends of these dikes, they can be shown to belong to three distinct families. Dikes with highly variable trends are by far the most numerous ones. They are truncated at the level of the erosion surface that marks the top of the Unit I. Two sets of dikes which trend 100°-110° and 170°-180° served as feeders to the upper units and, consequently, cross the upper boundary of Unit I.

All of Unit I, including the major part of the first generation dikes, have been subjected to an appreciable metamorphism characterized by the presence of epidote in the central and eastern parts of the island and by the presence of zeolites in the west central part. This unit was not sampled by Watkins (Gunn and others, 1970).

The stratified complex is characterized by a potassic tendency, marked in some gabbros by the presence of biotite comprising asmuch as 10% of the rock. It can be noted that Schlich's (1975) assertion of a strictly oceanic origin for the Crozet Plateau is strengthened by this hypothesis.

Unit II

This unit lies on an erosion surface which obliquely intersects the bedding of Unit I. The base is characterized by conglomerates, localized in restricted channels, which exhibit graded-bedding and cross-bedding. They give evidence for emergence of the underlying unit, accompanied by strong erosion. In fact, most of the pebbles of this conglomerate correspond to the dikes of the first generation; pebbles from the stratified massifs are lacking. The eroding surface which supplied the conglomerate with clastic materials probably did not include outcrops of the stratified plutonic massifs.

A volcanic sequence about 300 to 700 m thick lies on these conglomerates. It is made of an interlayering of not very differentiated lavas (alkaline-olivine basalts) and yellow agglomerates. Among the latter, very thin-bedded units, resembling palagonitic tuffs, were observed at their base.

This unit is crossed by the second family of dikes, which

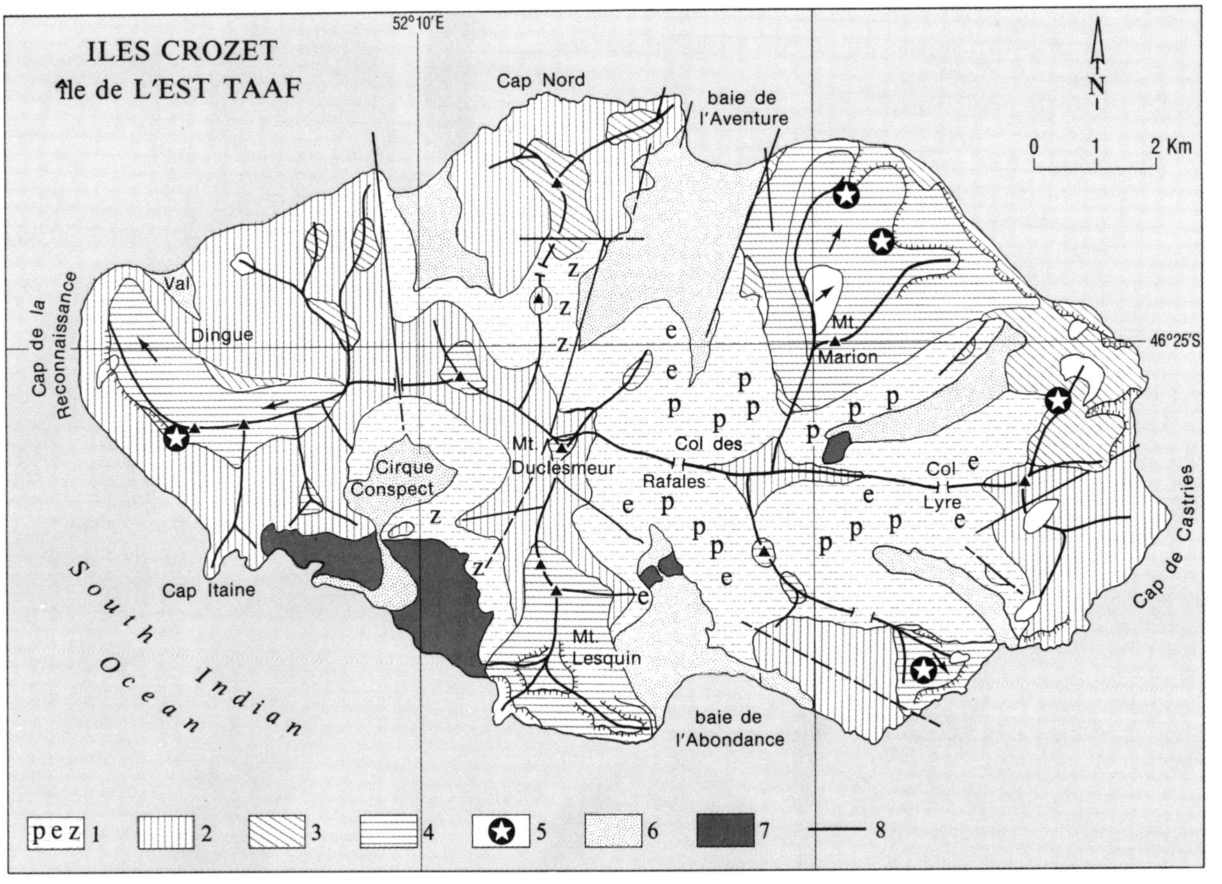

Figure 93.3. Schematic preliminary geologic map of Ile de l'Est. 1. Basal series or Unit I: p = plutonic stratified rocks, e = epidote zone, and z = zeolite zone; 2. Unit II: volcanic agglomerates with some lava flows; 3. Unit III (lower): volcanic agglomerates with hawaiite lava flow at the top; 4. Unit III (upper): volcanic agglomerates with lava flows (basalts, oceanites, and ankaramites); 5. Unit IV: scoriaceous cone (star) or red scoriaceous recent massifs (white areas); 6. Valleys between 0 and 100 m altitude. Scree and fluvio-glacial sediments of the valleys; 7. Morainal deposits; 8. Observed faults.

trend 100°–110° (i.e., parallel to the main axis of the island), as well as by the dikes of the third family, which trend 170°–180°.

Unit II is widely present in the western and the northwestern parts of the island, where it forms the major part of the cliffs. It can also be found as a cap in the center of the island, and in the periphery of the southeastern part where it has a gentle dip which seems to diverge from both sides of a north-south median axis. The unit is locally affected by translation faults. These faults die out (eastward in one case) against the base of overlying unit III. The 100°–110° dike network indicates that a period of extension affected the island between the deposition of unit II and unit III.

Unit III

This unit also begins with agglomerates of reddish pyroclastites with yellowish tuffaceous cement interlayered with lava flows. The real contact between Units II and III is of a structural kind; Unit III lies unconformably over Unit II, and even directly on Unit I (in the northeastern and southeastern parts of the island).

Unit III has been subdivided into two groups. The lower group, about 150 m in thickness, is composed of lower agglomerates on the unconformity; the overlying group is a distinct, thick flow of columnar-jointed hawaiite, rich in peridotite nodules.

The upper group is widely present on the most noticeable summits of the island (Mont Lesquin, Pic Marion, Mont Duclesmeur, and Table de l'Ouest), and it can be very thick (400 m). Just above the reference flow of hawaiite, the group is built of agglomerates, and in the upper part it contains thick flows of alkaline basalts rich in olivine and clinopyroxene. The Mont Lesquin section described in detail by Gunn and others (1970) corresponds to this sequence, and the Cascade section corresponds to the top of the lower group in Unit III. Some scarce dikes, oriented 170°–180°, intersect this group, whose bedding dips diverge from the central part of the island. The lower group is apparently affected by few faults (massif between the Cactus and the Mont de l'Accolade), but we will have to wait for aerial photographs in order to affirm this conclusion.

Unit IV

This unit is a very limited one, the result of the most recent volcanism. It consists of: (1) five very reddened scoriaceous cones, localized in the periphery of the island; (2) a larger massif, set in a paleo-valley at the top of Mont Lesquin, built of a 50-m accumulation of flows of glassy lava interlayered with pyroclastic beds; and (3) some scraps or mingled trails of red scoria which are the final expression of these last eruptions. The lava is a basalt with olivine and clinopyroxene.

Conclusions

It can be asserted that the Ile de l'Est shows these obvious geologic characteristics:

1. Four unconformable volcanic units, each fed by its own distinctive network of dikes;
2. The presence of an interesting basal unit with a volcano-detrital matrix in which a stratified plutonic complex was emplaced.
3. A noticeable abundance of agglomerates intercalated with lava flows, which show an evolution towards ankaramitic composition at the end of the eruptive cycle.

Studies now underway will allow a detailed analysis of these observations, as well as an investigation of the extent to which these characteristics can be found in the neighboring islands. Thus, a new dimension could be added to understanding the genesis and the evolution of the Crozet oceanic plateau.

Acknowledgments. Thanks are due to the Administration of the Terres Australes et Antarctiques Françaises (R. Barberot, Administrateur Supérieur; J. P. Bloch, Director of Scientific Laboratories) who permitted the realization of this field work, with the active cooperation of the officers and crew of M.S. *Marion Dufresne* (Captain Burdin) and of helicopters *Alouette II* of the French Air Force (Lt. De Valicourt, Adj. Siffert, Sgt. Doucet and Huré).

References cited

Berthois, L., Schlich, R., and Patriat, P. 1969. Carte bathymétrique du Sud de l'Océan Indien. *Inst. Phys. Globe. Ann. Rep.*, Paris, No. 26, 7 pp.

Dietz, R., and Holden, J. 1970. Reconstruction of Pangaa: break up and dispersion of continents, Permian to Present. *J.geophys. Res., 75,* 4939–50.

Girod, M., and Nougier, J. 1972. Volcanism of sub-antarctic islands. (*In* Adie, R. J., *ed. Antarctic Geology and Geophysics.* Oslo, Universitetsforlaget, 777–88.)

Gunn, B., Abranson, C., Watkins, N., and Nougier, J. 1972. Petrology and geochemistry of Iles Crozet: a summary. (*In* Adie, R. J., *ed. Antarctic Geology and Geophysics.* Oslo, Universitetsforlaget, 825–29.)

Gunn, B., Coy-Yll, R., Watkins, N., Abranson, C., and Nougier, J. 1970. Geochemistry of an oceanite-ankaramite-basalt suite from East Island, Crozet archipelago. *Contr. Miner. Petrol., 28,* No. 4, 319–39.

Laughton, A., Matthews, D., and Fischer, R. 1970. The structure of the Indian Ocean. (*In* Maxwell, A., *ed. The Sea.* New York, Wiley Interscience, *4,* 543–86.)

McKenzie, D. P., and Sclater, J. 1971. The evolution of the Indian Ocean since the Late Cretaceous. *Geophys. J. R. astr. Soc., 25,* 437–528.

Schlich, R. 1975. Structure et âge de l'Océan Indien Occidental. *Mem. Soc.Géol.Fr., 6,* 103 pp.

94

Trace Element Geochemistry of Volcanic Rocks from Bouvetøya, South Atlantic

TORE PRESTVIK
Department of Geology, The University of Trondheim, 7034 Trondheim, Norway

Abstract. Analyses of 11 trace elements, including six rare earth elements (REE), have been determined for representative volcanic rock specimens from Bouvetøya. The whole series, which includes transitional basalts, trachytic icelandites, and comendites, displays a chondrite-normalized pattern characterized by light REE enrichment. The most "primitive" basalts have negative Eu anomalies, whereas the more evolved transitional basalts and the intermediate trachytic icelandites do not. The strong negative Eu anomaly of the comendites suggests strong feldspar fractionation under low P_{O_2} at a late stage in the evolution of the series. The generally parallel REE patterns are consistent with a fractional crystallization model for the series, but the different transitional basalt groups are not interrelated through *simple* crystal fractionation. The transitional character of the whole series is also emphasized by the generally high concentrations of the incompatible elements (Hf, Ba, U, and Th) when compared to tholeiitic and alkaline series. The Sc data suggest that clinopyroxene was a fractionation phase in the development of the transitional basalts.

Bouvetøya is situated in the South Atlantic Ocean, close to the Bouvet triple junction (Figure 94.1; Sclater and others, 1976) and is thought to be the surface manifestation of a mantle plume (Morgan, 1971, 1973). The geology, geochemistry, and petrology of the island have recently been described by Verwoerd and others (1976) and Imsland and others (1977).

The island is built up of a series of volcanic rocks comprising basic, intermediate, and silicic types. The whole series displays a transitional character, and the terms "transitional basalt," "trachytic icelandite," and "comendite" used by Imsland and others (1977) are adopted here. Furthermore, the series is characterized by a high Fe/Mg ratio in both the basic and intermediate rocks. Verwoerd and others (1976) concluded that the transitional basalts (hawaiites) were derived from more primitive magmas mainly by olivine fractionation, possibly at high pressure. These authors also concluded that the rock types of Bouvetøya were interrelated by simple crystal fractionation. On the other hand, Imsland and others (1977), on the basis of model calculations and trace element considerations, concluded that the rock types included in the series were *not* interrelated by *simple* crystal fractionation. These last authors proposed the following model for the evolution of the series: fractional crystallization under varying conditions of *P, T* and P_{H_2O} *either* from slightly different magmas *or* from one parental magma type in an open system with loss of Rb.

This paper presents new analytical data on 11 trace elements (La, Ce, Sm, Eu, Yb, Lu, Sc, Ba, Hf, U, and Th) obtained from some of the same samples that were analyzed by Imsland and others (1977). These rock specimens include four transitional basalts, three trachytic icelandites, and three comendites. The analytical work was carried out by the Institutt for Atomenergi, Lillestrøm, Norway, using instrumental neutron activation analysis (Gordon and others, 1968).

The analytical results are presented in Table 94.1.

Discussion of data

The Rare Earth Elements (REE)

The chondrite-normalized REE patterns (Figure 94.2) show that the whole series is enriched in the light REE. The REE pattern of the transitional basalts is similar to that of tholeiites from oceanic islands and also to the basalt dredged nearby from Discovery Tablemount, which is thought to overlie a small mantle plume (Kempe and Schilling, 1974). Similar REE patterns are also found in some transitional basalts from Deccan, India (Alexander and Gibson, 1977) and in the transitional basalt-pantellerite suite of Boina, Afar (Barberi and others, 1975).

The least fractionated transitional basalts (BOV 2 and 10) both show negative Eu anomalies. The two more evolved transitional basalts (BOV 8 and 9), however, do not exhibit significant Eu anomalies. This seems to exclude

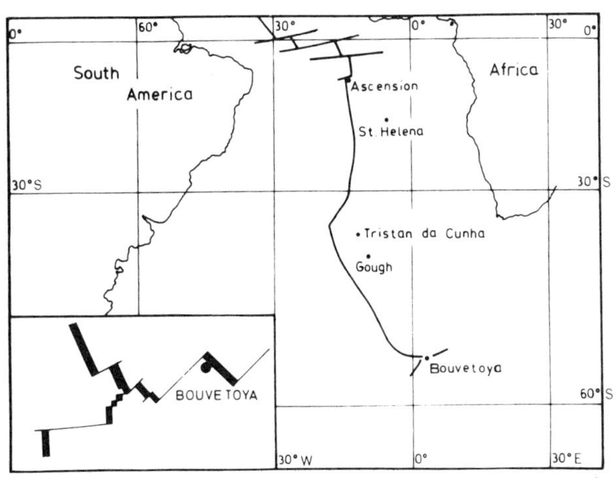

Figure 94.1. Location map of Bouvetøya. The tectonic relations of the Bouvet triple junction, as interpreted by Sclater and others (1976) are inserted; heavy lines are spreading ridges, light lines are transform faults.

Table 94.1. Trace element concentrations (ppm) in volcanic rocks from Bouvetøya

Rock Specimen	La	Ce	Sm	Eu	Yb	Lu	Sc	Hf	Ba	U	Th	Th/U	Rb[a]	Zr[a]
Transitional basalts														
BOV 2	22.5	53	7.2	1.64	1.83	0.40	26.9	6.2	222	1.0	2.13	2.13	24	185
BOV 8	29.6	71	9.3	2.6	2.5	0.60	28.5	7.8	249	1.1	2.34	2.13	16	231
BOV 9	26.8	65	8.7	3.0	2.6	0.51	26.4	6.2	274	1.0	2.57	2.57	26	217
BOV 10	23.8	45	7.5	1.62	2.9	0.47	26.1	5.6	154	0.9	2.06	2.29	20	174
Trachytic icelandites														
BOV 17	56.6	125	14.3	4.5	4.7	1.04	12.0	14.3	523	2.7	6.14	2.27	58	598
BOV 18	55.4	116	13.6	4.3	5.2	0.92	11.9	13.5	587	2.7	6.25	2.31	47	570
BOV 19	57.4	124	15.0	4.5	5.4	1.13	13.4	14.3	558	2.4	6.31	2.63	56	578
Comendites														
BOV 20	56.2	123	12.5	1.29	9.3	1.67	0.7	30.7	245	4.6	12.6	2.74	99	1,259
BOV 28	100.0	215	20.0	1.81	10.6	1.90	0.4	29.2	216	5.3	13.5	2.55	121	1,470
BOV 30	100.0	228	20.5	1.86	12.0	1.92	0.5	32.8	183	5.5	14.8	2.69	127	1,420

[a] Taken from Imsland and others (1977)
Analyst: Institutt for Atomenergi, Lillestrøm, Norway.

the possibility that the most evolved transitional basalts formed by crystal fractionation of liquids, corresponding to the samples BOV 2 and 10. Imsland and others (1977) also concluded that the two groups of transitional basalts were not interrelated by simple crystal fractionation. The observed Eu anomalies in BOV 2 and 10 are probably due to earlier plagioclase fractionation. The question is then whether or not the most fractionated basalts evolved by fractionation of plagioclase. There has been some dispute in the literature about the origin of Eu anomalies in basaltic rocks, and it seems evident that the redox conditions (Eu^{2+}/Eu^{3+}) and the composition of the fractionating feldspar are the most important factors (Towell and others, 1969; Schnetzler and Philpotts, 1970; Philpotts, 1970; Weill and Drake, 1973; Ferrara and Treuil, 1975; Barberi and others, 1975). The Sr and Ca abundances of the most evolved basalts (Imsland and others, 1977) indicate that plagioclase was a principal fractionating phase in their development. Thus, one explanation is that the lack of significant Eu anomalies in these rocks is due to higher oxygen fugacity since the compositional difference of fractionating plagioclase for the two groups of basalt is expected to be small. Another possibility is that the parental liquids for BOV 8 and 9 were primary magmas with positive Eu anomalies and that plagioclase fractionation fortuitously produced a nearly flat pattern.

The normalized REE patterns of the trachytic icelandites are nearly parallel to those of the most evolved transitional basalts, and they show no Eu anomalies. Imsland and others (1977) concluded that these rocks might be derived

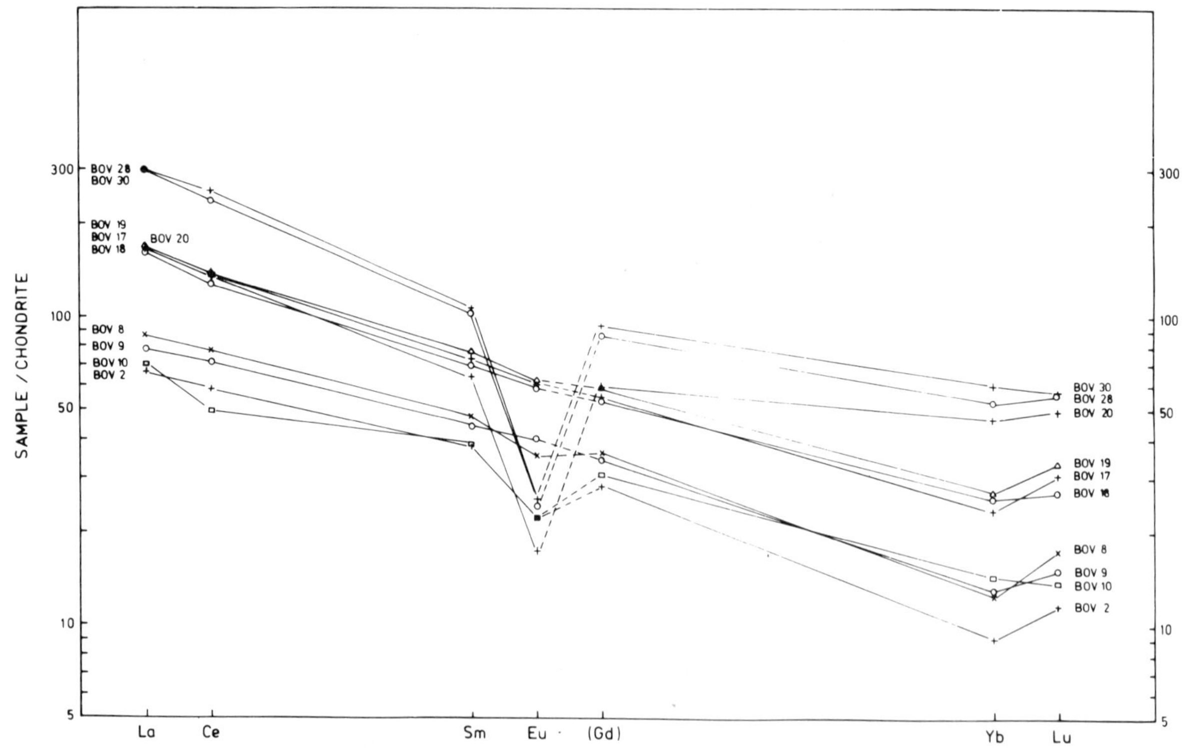

Figure 94.2. Chondrite-normalized REE patterns of volcanic rocks from Bouvetøya. In order to emphasize the Eu anomalies, gadolinium (Gd) plots are indicated as points on a straight line between the Sm and Yb-Lu plots. Rock specimens include: transitional basalts (BOV 2, 8, 9, and 10), trachytic icelandites (BOV 17, 18, and 19), and comendites (BOV 20, 28, and 30). Gadolinium plots were obtained by interpolation on graph between Sm and an averaged normalized value of Yb and Lu.

by fractionation of plagioclase (34%), clinopyroxene (16%), olivine (4%), and oxides (8%) from magma with a composition like that of the most evolved basalts. Because this high percentage of plagioclase fractionation did not result in depletion of Eu, the oxygen fugacity in the magma was probably high during this stage in the evolution of the series. On the other hand, the large negative Eu anomalies of all comendites reflect low oxygen fugacity at this stage of fractionation. Low oxygen fugacity is also indicated by the coexistence of oxide and olivine (QFM-buffer; Carmichael, 1967) in comendite modes (Imsland and others, 1977). A similar situation is found in the Boina suite where negative Eu anomalies first occur in rocks with $SiO_2 >$ 65%, even though plagioclase is thought to have been one of the main fractionating phases. Barberi and others (1975) found that this relative depletion of Eu began at the same time that the oxygen fugacity passed through a minimum value.

The two obsidians (BOV 28 and 30) have a higher content of REE than the crystalline comendite. This is especially conspicuous for the light REE and corresponds to a general depletion in Na_2O, Rb, Y, and Zr for the crystalline comendites compared to the glassy equivalents (Imsland and others, 1977). Similar REE depletion in crystalline comendite has recently been described from Kenya (Baker and Henage, 1977).

O'Nions and others (1977) published a few data on Nd and Sm in rocks from Bouvetøya. Except for one specimen (BV 5), all their results are similar to the results presented in this paper as far as Sm is concerned. Their specimen BV 5 (O'Nions and Pankhurst, 1974) is very high in both Sm and Nd even though it is much lower in K_2O (O.22%) and Rb (3.3 ppm) than any of the rocks analyzed by Verwoerd and others (1976) and Imsland and others (1977).

Scandium

The concentration of Sc in the transitional basalts ranges from 26.1 to 28.5 ppm. This is lower than in South Atlantic ocean floor tholeiites (Frey and others, 1974), but similar to basalts from Terceira Island, Azores (Allègre and others, 1977). If the transitional basalts of Bouvetøya were derived by olivine fractionation (Verwoerd and others, 1976), the Sc content of the primary magma was probably lower than in the transitional basalts. Data on Sc abundances of primary basaltic magmas are scarce, but Helmke and Haskin (1973) report initial Sc abundances of 27.2 ppm for Steens Mountain basalt, Oregon, U.S.A. Furthermore, Allègre and others (1977) infer initial Sc concentration of Terceira basalts to be 28 ppm. If the primary Bouvetøya magma had a similar Sc concentration, then clinopyroxene was probably also a fractionating phase during the evolution of the transitional basalts.

Sc decreases during the Bouvetøya rock series and is barely present in the comendites. This is qualitatively in agreement with the evolutionary model proposed by Imsland and others (1977) where clinopyroxene is a dominant fractionating phase.

Hafnium

The transitional basalts of Bouvetøya are higher in Hf (5.6–7.8 ppm) than basalts of corresponding K_2O values from Terceira (Allégre and others, 1977) and have about twice the concentration reported for South Atlantic ocean floor tholeiites (Frey and others, 1974). The Boina basalts have similar Hf concentrations, whereas the supposed primary basalts from Steens Mountain, Oregon (Helmke and Haskin, 1973) are lower in Hf. Basalts from Deccan (Alexander and Gibson, 1977) range from 4.2 to 6.5 ppm in Hf while the picrites from that area, believed to be close in composition to the parental liquids, have Hf concentrations from 2.3 to 4.7 ppm. The fact that transitional basalts from several other areas have lower Hf concentrations than corresponding rocks from Bouvetøya further substantiates the evolved character of the latter.

Hf increases throughout the compositional series of Bouvetøya, reaching about 30 ppm in the comendites. This concentration is lower than in comendites from the continental Kenya rift valley (Baker and Henage, 1977).

Barium

As with other large-ion lithophile elements, Ba content is high in the transitional basalts of Bouvetøya compared to ocean floor basalts (Frey and others, 1974) but significantly lower than in the otherwise similar Boina suite (Barberi and others, 1975). The K/Ba ratio of basic and intermediate rocks from Bouvetøya is about 40 compared to 20–25 for both Boina basalts (Barberi and others, 1975) and Terceira primary magma, and 100 for ocean floor basalts (Hart, 1971). The South Atlantic Discovery Tablemount basalt also has a K/Ba ratio of about 20 (Kempe and Schilling, 1974). This indicates that the source region for Bouvetøya magmas is to some degree depleted in Ba relative to K, but this depletion is much less than in ocean floor tholeiites.

Ba concentration decreases markedly from trachytic icelandite to comendite, and even considerably among the comendites. This effect is probably due to fractionation of sodic plagioclase and anorthoclase (Imsland and others, 1977).

The Ba data presented here are in good agreement with the data of Verwoerd and others (1976).

Uranium and Thorium

Verwoerd and others (1976) also reported analyses of U and Th in rocks from Bouvetøya. As there are discrepancies between their data and those presented here, two samples (BOV 10 and 30) were reanalyzed together with the USGS-standard BCR-1. The result obtained for BCR-1 (Table 94.2) is about 11% *above* the *average* value for U and 15.8% *below* the *recommended* value for Th (Flanagan, 1973). If this is taken as consistent for all U and Th analyses of this work, then the presented Th/U ratios are about 25% too low.

Table 94.2. Uranium and thorium concentrations obtained by contemporaneous analysis of BOV 10, BOV 30, and U.S.G.S. standard BCR-1

Rock Specimen	U	Th	Th/U
BOV 10	0.9	2.17	2.41
BOV 30	4.84	14.2	2.93
BCR-1	1.96[a]	5.05[a]	2.58
BCR-1	1.74[b]	6.0[c]	3.45[d]

[a] Values obtained in this work
[b] Average value (Flanagan, 1973)
[c] Recommended value (Flanagan, 1973)
[d] Ratio calculated from average and recommended values

The U and Th contents of the transitional Bouvetøya basalts are significantly higher than in oceanic tholeiites (Frey and others, 1974) and in Icelandic tholeiites (Heier and others, 1966; Polyakov and Sobornov, 1975). Compared to the Boina suite, the Bouvetøya basalts are slightly higher in U and lower in Th. According to Polyakov and Sobornov (1975), transitional and mildly alkaline basalts from Iceland have U and Th contents similar to those reported here for the Bouvetøya basalts.

Even if we correct for the analytical uncertainty mentioned above, the Th/U ratios for the Bouvetøya basalts are lower than the Th/U ratios Heier and Rodgers (1963) found for basaltic rocks. Variations in Th/U ratios are usually interpreted as due to oxidation (Watkins and others, 1970), but there is no positive correlation between Th/U and Fe_2O_3/FeO ratios for the Bouvetøya basalts. It should be mentioned, however, that the Th/U ratios of Bouvetøya are well within the range of values presented by Polyakov and Sobornov (1975) for 78 volcanic rocks from Iceland. In particular, the transitional and alkaline basalts from Iceland with K_2O values similar to the Bouvetøya basalts have about the same abundances of both U and Th as those reported here.

Conclusions

The new analytical data presented in this paper further substantiate the transitional character of the volcanic series of Bouvetøya. When compared to other transitional suites, the evolved character of the Bouvetøya basalts is emphasized. The data suggest that in addition to olivinc, clinopyroxene and plagioclase were also fractionating phases in the development of the transitional basalts.

Acknowledgments. The author is indebted to Elvis Steinnes for the analytical work. This work was supported by The Norwegian Research Council for Science and the Humanities (NAVF), grant D. 42.31-11. The author would like to thank his colleagues, Páll Imsland, Jørgen G. Larsen, and Ellen Sigmond, for the cooperation on an earlier paper on the geology and petrology of Bouvetøya from which this work benefited greatly.

References cited

Alexander, P. O., and Gibson, I. L. 1977. Rare earth abundances in Deccan trap basalts. *Lithos, 10,* 143–47.

Allègre, C. J., Treuil, M., Minster, J. F., Minster, B., and Albarède, F. 1977. Systematic use of trace element in igneous process. Part I: fractional crystallization processes in volcanic suites. *Contr. Miner. Petrol., 60,* 57–75.

Baker, B. H., and Henage, L. F. 1977. Compositional changes during crystallization of some peralkaline silicic lavas of the Kenya rift valley. *J. Volcan. Geotherm. Res., 2,* 17–28.

Barberi, F., Ferrara, G., Santacroce, R., Treuil, M., and Varet, J. 1975. A transitional basalt-pantellerite sequence of fractional crystallization, the Boina Centre (Afar rift, Ethiopia). *J. Petrology, 16,* 22–56.

Carmichael, I. S. E. 1967. The iron-titanium oxides of salic volcanic rocks and their associated ferromagnesian silicates. *Contr. Miner. Petrol., 14,* 36–64.

Ferrara, G., and Treuil, M. 1975. Petrological implications of trace element and Sr isotope distribution in basalt-pantellerite series. *Bull. volcan., 38,* 548–74.

Flanagan, F. J. 1973. 1972 values for international geochemical reference samples. *Geochim. cosmochim. Acta, 37,* 1189–1200.

Frey, F. A., Bryan, W. B., and Thompson, G. 1974. Atlantic ocean floor: geochemistry and petrology of basalts from Legs 2 and 3 of the Deep-Sea Drilling Project. *J. geophys. Res., 79,* 5507–27.

Gordon, G. E., Randle, K., Goles, G. G., Corliss, J., Beeson, M., and Oxley, S. 1968. Instrumental activation analyses of standard rocks with high resolution gamma ray detectors. *Geochim. cosmochim. Acta, 32,* 369–96.

Hart, S. R. 1971. K, Rb, Cs, Sr, and Ba content and Sr isotope ratios of ocean floor basalts. *Phil. Trans. R. Soc. Lond.,* Ser. A., *268,* 573–88.

Heier, K. S., and Rodgers, J. J. W. 1963. Radiometric determination of thorium, uranium, and potassium in basalts and in two magmatic differentiation series. *Geochim. cosmochim. Acta, 27,* 137–54.

Heier, K. S., Chappell, B. W., Arriens, P. A., and Morgan, J. W. 1966. The geochemistry of four Icelandic basalts. *Norsk geol. Tidsskr., 46,* 427–37.

Helmke, P. A., and Haskin, L. A. 1973. Rare-earth elements, Co, Sc, and Hf in Steens Mountain basalts. *Geochim. cosmochim. Acta, 37,* 1513–29.

Imsland, P., Larsen, J. G., Prestvik, T., and Sigmond, E. 1977. The geology and petrology of Bouvetøya, South Atlantic Ocean. *Lithos, 10,* 213–34.

Kempe, D. R. C., and Schilling, J.-G. 1974. Discovery Tablemount basalt: petrology and geochemistry. *Contr. Miner. Petrol., 44,* 101–15.

Morgan, W. J. 1971. Convection plumes in the lower mantle. *Nature, Lond., 230,* 42–43.

Morgan, W. J. 1973. Plate motions and deep mantle convection. *Mem. geol. Soc. Am., 132,* 7–22.

O'Nions, R. K., and Pankhurst, R. J. 1974. Petrogenetic significance of isotope and trace element variations in volcanic rocks from the Mid-Atlantic. *J. Petrology, 15,* 603–34.

O'Nions, R. K., Hamilton, P. J., and Evensen, N. M. 1977. Variations in $^{143}Nd/^{144}Nd$ and $^{87}Sr/^{86}Sr$ rations in oceanic basalts. *Earth planet. Sci. Lett., 34,* 13–22.

Philpotts, J. A. 1970. Redox estimation from a calculation of Eu^{2+} and Eu^{3+} concentrations in natural phases. *Earth planet. Sci. Lett., 9,* 257–68.

Polyakov, A. I., and Sobornov, O. P. 1975. Uranium, thorium, and potassium in the volcanic rocks of Iceland. *Geochem. Int., 12,* Pt. 5, 20–27. (Translated from *Geokhimiya, 9,* 1309–16).

Schnetzler, C. C., and Philpotts, J. A. 1970. Partition coefficients of rare-earth elements between igneous matrix material and rock-forming mineral phenocrysts—II. *Geochim. cosmochim. Acta, 34,* 331–40.

Sclater, J. G., Bowin, C., Hey, R., Hoskins, H., Peirce, J., Phillips, J., and Tapscott, C. 1976. The Bouvet triple junction. *J. geophys. Res., 81,* 1857–69.

Towell, D. G., Spirn, R. V., and Winchester, J. W. 1969. Europium anomalies and the genesis of basalt: a discussion. *Chem. Geol., 4,* 461–65.

Verwoerd, W. J., Erlank, A. J., and Kable, E. J. D. 1976. Geology and geochemistry of Bouvet Island. (*In* Gonzáles-Ferrán, O., *ed. Andean and Antarctic Volcanology Problems.* Rome, International Association of Volcanology and Chemistry of the Earth's Interior, 201–37.)

Watkins, N. D., Gunn, B. M., and Coy-Yll, R. 1970. Major and trace element variations during the initial cooling of an Icelandic lava. *Am. J. Sci., 268,* 24–49.

Weill, D. F., and Drake, M. J. 1973. Europium anomaly in plagioclase feldspar: experimental results and semiquantitative model. *Science, N.Y., 180,* 1059–60.

95

Geochemistry of Mesozoic and Cenozoic Igneous Rocks, Eastern Ellsworth Land

THOMAS S. LAUDON
Department of Geology, University of Wisconsin–Oshkosh, Oshkosh, Wisconsin 54901, U.S.A.

Abstract. Rock units mapped in eastern Ellsworth Land include clastic marine sedimentary rocks of Middle and Late Jurassic age which are correlated with the Latady Formation; andesite flow rocks and felsic volcanic rocks of Jurassic(?) age which are correlated with the Antarctic Peninsula Volcanic Group; mafic and intermediate dikes of Cretaceous(?) age, plutonic intrusive rocks of Cretaceous age, and felsic dikes of Cretaceous(?) age, all of which are correlated with the Andean Intrusive Suite; and olivine basalt breccias and flows of Pliocene age.

Variation diagrams based on major oxide analyses display similar calc-alkalic trends for Jurassic(?) volcanic rocks, Cretaceous plutonic rocks, and Cretaceous(?) felsic dikes. Cretaceous(?) mafic and intermediate dikes appear to be more alkalic, but variation trends are less definite. Pliocene olivine basalt is strongly alkalic and bears chemical similarities to Pliocene olivine basalts from the Hudson Mountains of western Ellsworth Land.

Eastern Ellsworth Land is the southwesternmost part of the Antarctic Peninsula and is physiographically separated from western Ellsworth Land by areas in which the bedrock surface lies below sea level (Behrendt and Parks, 1962; Behrendt, 1964). All rock units mapped within a 15,000 km^2 area (Figure 95.1), as well as rocks collected by the 1962 oversnow traverse from three localities in the Sweeney Mountains to the east and from Mount Rex to the west, can be correlated with rock units which are widespread elsewhere in the Antarctic Peninsula (Laudon and others, 1969; Laudon, 1972).

The most abundant rocks in eastern Ellsworth Land are intensely deformed volcaniclastic marine sedimentary rocks of Middle to Late Jurassic age (Laudon and others, 1969; Laudon, 1972; Quilty, 1970, 1972*a,b,* 1977, and Chapter 82, this volume; Stevens, 1967). They are in part correlative with the lithologically similar Latady Formation of Late Jurassic age on the Lassiter Coast (Williams and others, 1972; Rowley and Williams, Chapter 38, this volume).

Igneous rocks have been assigned to the following units (Laudon and others, 1969): (1) andesite flow rocks of Jurassic(?) age; (2) felsic volcanic rocks of Jurassic(?) age; (3) mafic and intermediate dikes of Cretaceous(?) age; (4) plutonic rocks of Cretaceous age (96 m.y.; Halpern, 1971); (5) felsic dikes of Cretaceous(?) age; and (6) olivine basalt breccias and flow rocks of Pliocene age (6 m.y.; Halpern, 1971).

Units 1 and 2 are correlated with the Upper Jurassic Volcanic Group (Adie, 1964*a,b*) for which Thomson (Chapter 37, this volume) proposes the new name *Antarctic Peninsula Volcanic Group.* Units 3, 4, and 5 are correlated with the Andean Intrusive Suite (Adie, 1954, 1955). Unit 6 is correlated with other upper Tertiary basalts which are widespread in West Antarctica.

Antarctic Peninsula Volcanic Group

Deformed, highly altered, amygdaloidal dark-green and black flow rocks of andesitic and basaltic composition occur interbedded with black shale and siltstone at the southern end of Mount Neuner in the Behrendt Mountains. Thicknesses of individual flows range from less than 10 cm to greater than 1 m. Light-colored amygdules are prominent in most beds. Phenocrysts of pyroxene and amphibole occur in some flows.

Most thin sections show porphyritic textures with phenocrysts of plagioclase and pyroxene, commonly exhibiting flow alignment. Olivine and amphibole are present as phenocrysts in some sections. Groundmass textures are pilotaxitic. All specimens examined have been extensively altered, many to the extent that little of their original character is revealed. Amphibole, chlorite, biotite, serpentine, epidote, calcite, and quartz are abundant secondary minerals, occurring both as replacements and in amygdules.

These rocks are interbedded with shale and siltstone correlated with the Jurassic Latady Formation.

Masses of deformed, altered porphyries of dacitic and rhyodactic compositions form several prominent nunataks including Mount Rex, Cheeks Nunatak, and Mount Wasilewski north of lat. 75° 05′ S in eastern Ellsworth Land. Most outcrops are colored red by hematite streaks and coats on joint faces. Colors in hand specimens are either pale green, dark green, gray, or black. Quartz phenocrysts are common in many specimens. Altered feldspar grains are white and have euhedral outlines.

Thin sections show porphyritic textures with rounded to angular phenocrysts of plagioclase (An_{26-42}), quartz, and sanidine. Sericitization, saussuritization, and carbonatization of plagioclase phenocrysts are common. Groundmass material is usually cryptocrystalline, and commonly only secondary minerals can be discerned. Chlorite, epidote, sphene, magnetite, glass, and quartz are common groundmass constituents. Pyroclastic textures noted in a few thin sections, and flow structures in others, indicate extrusive origin of the dacites.

Indirect structural and sedimentologic evidence strongly suggests that these rocks are contemporaneous with Middle and Upper Jurassic sedimentary rocks assigned to the

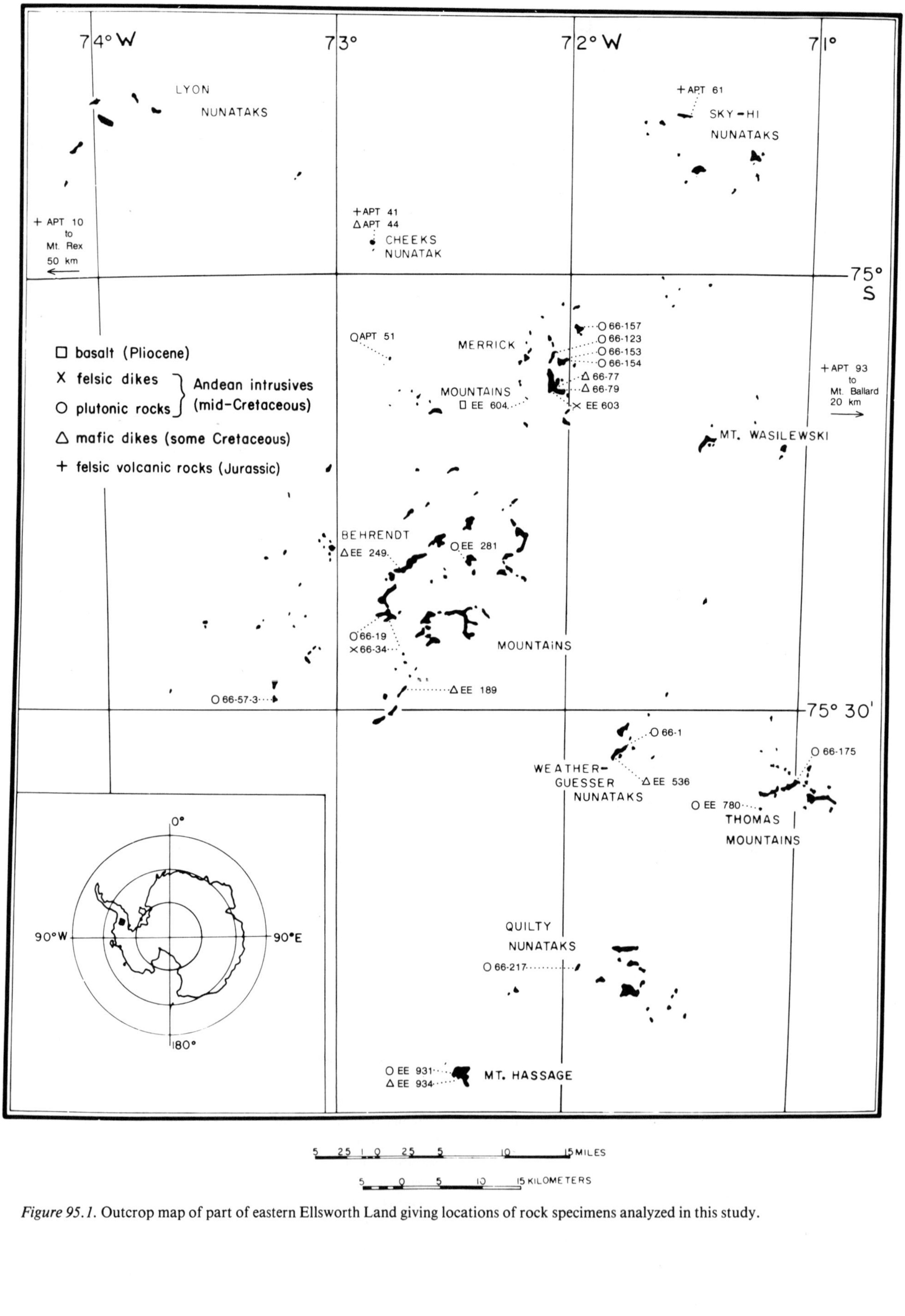

Figure 95.1. Outcrop map of part of eastern Ellsworth Land giving locations of rock specimens analyzed in this study.

Latady Formation (Laudon and others, 1969; Laudon, 1972).

Table 95.1 presents major oxide analyses and C.I.P.W. norms for four felsic volcanic rocks. Because of their highly altered condition, none of the amygdaloidal flow rocks from Mount Neuner have been analyzed.

Mafic dikes

Dikes of mafic and intermediate composition are common throughout eastern Ellsworth Land. Some dikes intrude deformed sedimentary rocks of Middle and Late Jurassic age, others intrude deformed felsic volcanic rocks of Jurassic(?) age, and still others intrude plutonic rocks of Cretaceous age. East of Mount Boyer, in the Merrick Mountains, gabbro dikes, intrusive into deformed Jurassic sedimentary rocks are truncated by plutonic diorite. Near the summit of Mount Hassage, basalt dikes intrusive into deformed Jurassic sedimentary rocks are truncated by plutonic granodiorite.

Andesite textures are usually porphyritic and intergranular. Euhedral zoned phenocrysts of plagioclase (An_{45-58}) predominate. Plagioclase, quartz, chlorite, epidote, magnetite, and hornblende are common groundmass minerals, with hyperstheme, biotite, magnetite, apatite, and sphene less common. Alteration of both phenocrysts and groundmass is common.

Table 95.1. Chemical analyses of Mesozoic volcanic rocks

	1	2	3	4
SiO_2	67.5	67.1	71.35	70.46
Al_2O_3	13.9	14.1	12.82	13.10
Fe_2O_3	1.6	2.1	1.08	2.71
FeO	2.9	2.4	2.40	1.83
MgO	1.0	1.0	0.87	2.08
CaO	1.5	1.7	2.05	2.25
Na_2O	2.7	2.2	1.96	2.19
K_2O	3.5	4.5	4.66	1.96
MnO	0.05	0.05	0.04	0.07
TiO_2	0.41	0.38	0.53	0.55
P_2O_5	0.17	0.17	0.22	0.25
CO_2	—	—	—	—
H_2O	—	—	1.83	2.48
Loss on ignition	1.24	0.73	—	—
Total	96.5	96.4	99.81	99.92
		Norms		
Q	32.62	31.45	35.81	42.82
Or	20.68	26.59	27.54	11.58
Ab	22.85	18.62	16.59	18.53
An	6.33	7.32	8.73	9.60
Ne	—	—	—	—
Di	—	—	—	—
Hy	5.91	4.63	4.88	5.52
Ol	—	—	—	—
Mt	2.32	3.04	1.57	3.93
Il	0.78	0.72	1.01	1.04
Ap	0.39	0.39	0.51	0.56
Cc	—	—	—	—
C	3.35	2.93	1.35	3.86

—: not reported

1. APT 61: dacite, Sky-Hi Nunataks; new analysis.
2. APT 93: dacite, north of Mount Ballard, Sweeney Mountains; new analysis.
3. APT 10: dacite, Mount Rex; from Laudon (1972, Table 3).
4. APT 41: dacite, Cheeks Nunatak; from Laudon (1972, Table 3).

Specimens collected by J.C. Behrendt in 1962. Analyses 1 and 2 by Rocky Mountain Geochemical Corporation, Midvale, Utah. Analyses 3 and 4 by Japan Analytical Chemistry Research Institute, Tokyo.

Basalt textures are usually intergranular. Phenocrysts of plagioclase (An_{58-67}) and hornblende predominate, with phenocrysts of augite usually present. Common matrix minerals are plagioclase (An_{40-47}), hornblende, and augite. Magnetite, epidote, biotite, and sphene are the most common accessory minerals. Alteration of both phenocrysts and groundmass is common.

Many mineralogic and textural varieties of diorite and gabbro occur, commonly varying even within the same intrusive body. Quartz-bearing varieties, quartz- and olivine-free varieties, olivine gabbros, and ultramafic varieties are present. Textures are usually hypidiomorphic-granular or allotriomorphic-granular. Plagioclase (strongly zoned) and augite are present in all specimens examined. Hornblende, olivine, actinolite, hypersthene, biotite, chlorite, and magnetite are abundant in many specimens.

Similar dikes are associated with Mesozoic plutonic rocks in many parts of the Antarctic Peninsula. On the Lassiter Coast and southern Black Coast, both preplutonic and postplutonic dikes are present (Williams and others, 1972; Rowley and Williams, 1974 and Chapter 38, this volume). Farrar and others (Chapter 39, this volume) conclude that plutonism in the southern Antarctic Peninsula, including the emplacement of all or most dikes, occurred during a short interval of Late Cretaceous time. Field relationships and chemistry of the dikes suggest that this is also true in eastern Ellsworth Land.

Table 95.2 presents major oxide analyses and C.I.P.W. norms for seven mafic dikes.

Plutonic rocks and felsic dikes

Cretaceous plutonic rocks and associated apophyses intrude deformed sedimentary rocks of Jurassic age in the Merrick Mountains, in the Behrendt Mountains, in the Thomas Mountains, in the Weatherguesser Nunataks, in the Quilty Nunataks, at Mount Glowa, and at Mount Hassage. Magnetometer surveys by P. J. Wasilewski (unpublished, personal communication) suggest that snow-covered central parts of the Merrick Mountains and the Behrendt Mountains are underlain by plutonic rocks. Emplacement of plutonic rocks was followed by intrusion of felsic dikes of rhyolite porphyry, aplite, and pegmatite. Exposed portions of plutons are generally smaller than elsewhere in the Antarctic Peninsula, consisting of stocks or cupolas. This situation seems to represent the continuation of a trend noted by Rowley and Williams (1974) on the Lassiter Coast where plutonic rocks underlie more than a third of the area; most plutonic bodies in the southern third of the region are stocks, while batholiths predominate farther north. This pattern is attributed in part to progressively smaller amounts of postplutonic uplift and erosion in the southern and western parts of the Antarctic Peninsula.

Exposed plutonic contact zones tend to be steeply dipping and are characterized by abundant apophyses and inclusions of wall rock. Contact breccias occur in places, and there is a tendency for xenoliths to be less abundant and more rounded at increasing distances from the contact. In the Thomas Mountains and at the northeast end of Mount Boyer, sedimentary rocks adjacent to plutons have been tightly refolded by plutonic emplacement. Contact aureoles of andalusite-hornfels grade occur adjacent to the pluton boundaries. Similar hornfelses in many parts of the Behrendt Moutnains are believed to reflect the presence of plutonic rocks which have not yet been exposed by erosion.

Diorite was found only in the Merrick Mountains where it is the most abundant plutonic rock type. It is associated with hybrid rocks of more mafic compositions east of Mount Becker and at Mount Matheson and grades into quartz diorite in some places. Mafic inclusions are common in diorites and occur locally as mottled patches of leucogabbro.

Diorites and quartz diorites from the Merrick Mountains are hypidiomorphic-granular and contain plagioclase, amphibole, quartz, biotite, pyroxene, and K-feldspar. Those specimens classified as diorite contain less than 10% quartz and only traces of K-feldspar; those classified as quartz diorite contain greater than 10% quartz and greater than 5% K-feldspar. Chlorite, epidote, zircon, and magnetite are common accessory minerals.

Plagioclase (An_{42-60}) is euhedral to subhedral, tends to be strongly zoned parallel to crystal boundaries, and is commonly sericitized. Hornblende occurs as euhedral to anhedral grains and as reaction rims on pyroxene; it is commonly altered to biotite or chlorite and less commonly to epidote. The pyroxene is diopsidic augite, and it invariably has reaction rims of hornblende. Many pyroxene grains are completely altered in hornblende. Biotite is present as euhedral grains and as an alteration product of hornblende, and it is commonly altered to fibrous masses of penninite. Quartz is interstitial.

A small pluton at Mount Glowa is classified as a quartz diorite on the basis of modal composition, but it is considerably more felsic than the quartz diorites from the Merrick Mountains.

Plutons and associated dikes composed of quartz monzonite and granodiorite occur south of lat. 75° 15′ S, in the Behrendt Mountains, in the Thomas Mountains, in the Weatherguesser Nunataks, in the Quilty Nunataks, at Mount Hassage, and in a small nunatak located approximately 7 km northwest of Mount Matheson in the Merrick Mountains. Field observations suggest that granodiorite is more abundant than quartz monzonite; however, only slight compositional differences exist between some rocks classified as granodiorite and others classified as quartz monzonite. Fresh specimens of these rocks are gray or pink; weathered surfaces are gray or tan. Discoloration of weathered surfaces by limonite streaks is common. Joint surfaces are commonly coated with epidote.

These rocks are medium to coarse grained with hypidiomorphic-granular textures, except at Mount Caywood and Luck Nunatak in the Behrendt Mountains where quartz

Table 95.2. Chemical analyses of mafic dikes

	1	2	3	4	5	6	7
SiO_2	45.97	46.62	58.48	52.10	44.1	45.8	55.4
Al_2O_3	15.82	16.52	16.90	19.24	13.7	12.3	14.2
Fe_2O_3	4.77	2.10	2.00	2.28	3.2	5.8	1.0
FeO	4.36	10.07	6.12	6.03	9.2	7.8	8.5
MgO	6.27	6.88	2.47	2.17	6.4	11.4	3.3
CaO	8.86	10.26	4.34	6.12	5.4	9.2	7.3
Na_2O	2.40	2.27	3.45	3.13	2.7	2.1	3.5
K_2O	1.45	1.84	2.70	4.22	0.9	1.8	3.6
MnO	0.12	0.14	0.08	0.14	0.14	0.2	0.15
TiO_2	1.30	1.13	0.42	1.04	1.13	0.99	0.61
P_2O_5	0.29	0.19	0.14	0.42	0.45	0.19	0.38
Co_2	6.01	—	—	—	—	—	—
H_2O	2.75	2.75	3.45	3.85	—	—	—
Loss on ignition	—	—	—	—	9.71	2.57	0.15
Total	99.78	100.77	100.56	100.74	97.0	100.2	98.1
			Norms				
Q	14.16	—	11.38	—	—	—	—
Or	8.57	10.87	15.96	24.94	5.32	10.64	21.28
Ab	20.31	15.99	29.19	26.49	22.85	17.55	29.62
An	4.07	29.45	20.62	25.99	22.61	18.82	12.40
Ne	—	1.74	—	—	—	0.12	—
Di	—	16.65	—	1.37	1.02	20.44	17.97
Hy	18.25	—	15.19	10.15	26.65	—	13.12
Ol	—	17.68	—	1.71	1.04	19.28	0.07
Mt	6.92	3.04	2.90	3.31	4.64	8.41	1.45
Il	1.90	2.15	0.80	1.98	2.15	1.88	1.16
Ap	0.67	0.44	0.32	0.97	1.04	0.44	0.88
Cc	13.67	—	—	—	—	—	—
C	8.81	—	0.75	—	—	—	—

—: not reported.

1. APT 44: andesite, Cheeks Nunatak; from Laudon (1972, Table 3).
2. 66-77: gabbro, Mount Boyer; 2–7 are new analyses.
3. 66-79: diorite, Mount Boyer.
4. EE 249: andesite, Mount Neuner.
5. EE 189: basalt, Mount Hirman.
6. EE 536: basalt, Weatherguesser Nunataks.
7. EE 934: andesite, Mount Hassage.

Specimen APT 44 collected by J. C. Behrendt in 1962. Analyses 1-4 by Japan Analytical Chemistry Research Institute, Tokyo. Analyses 5–7 by Rocky Mountain Geochemical Corporation, Midvale, Utah.

monzonite is porphyritic with pink orthoclase phenocrysts as long as 5 cm. Essential minerals include quartz, orthoclase, microcline, biotite, and hornblende. Apatite, zircon, sphene, and magnetite are accessory minerals. Epidote, chlorite, and calcite are secondary minerals.

Plagioclase is euhedral and well twinned, commonly exhibiting strong normal zoning (An_{20-47}). Sericitization and less common saussuritization are concentrated in the more calcic cores. K-feldspar is anhedral and generally less altered than plagioclase. Quartz is anhedral and interstitial, exhibiting moderate to strong undulatory extinction. Hornblende is euhedral, pleochroic, and, where altered, replaced by biotite. Euhedral brown biotite has dark pleochroic halos and alters to chlorite.

Table 95.3 presents chemical analyses and C.I.P.W. norms for 13 plutonic rocks.

Felsic dikes intrude Jurassic sedimentary rocks and cretaceous plutonic and hypabyssal rocks. Rhyolite porphyry dikes occur in the Merrick Mountains. The largest dikes exceed 10 m in thickness and 1 km in length. Contacts are sharp with a narrow chill zone present. Inclusions of host rock are absent.

Texture is porphyritic, with 15–20% quartz phenocrysts and 7–10% feldspar phenocrysts in a 70–75% cryptocrystalline felsic groundmass. Quartz "eyes" up to 3.5 mm in diameter are generally rounded and deeply embayed, through a few are bipyramidal with only slight rounding; almost all are fractured. Sanidine is the principal feldspar; orthoclase is also present. Feldspar phenocrysts are usually altered to sericite, less commonly to calcite. The cryptocrystalline groundmass contains abundant sericite and may, in part, represent devitrified glass.

In the Behrendt Mountains, in the Thomas Mountains, and at Mount Hassage, randomly oriented aplite dikes as wide as 10 m intrude Jurassic sedimentary rocks and Cretaceous plutons. Contacts are sharp with a fine-grained chill zone usually present. Banding parallel to dike margins, caused by concentrations of feldspar, occurs in smaller dikes. Some granitic rocks adjacent to aplite dikes have granoblastic textures with pronounced schistosity due to alignment of biotite grains.

Aplites are fine to medium grained, but some dikes contain local pockets of pegmatite. Textures are allotriomorphic. Perthite, microcline, and quartz are the most abundant minerals. Sodic plagioclase is subordinate. Biotite and magnetite are accessory minerals. Alteration is limited to sericitization of feldspars.

Table 95.4 gives chemical analyses and C.I.P.W. norms for a rhyolite porphyry from Mount Boyer and an aplite from Mount Trimpi.

Table 95.3. Chemical analyses of plutonic rocks

	1	2	3	4	5	6	7	8	9	10	11	12	13
SiO_2	51.32	53.30	53.40	54.96	58.23	59.10	60.48	62.4	65.62	66.21	67.5	68.43	71.50
Al_2O_3	18.32	18.85	19.07	17.42	17.68	16.04	16.09	15.6	15.83	16.17	14.7	15.19	14.55
Fe_2O_3	1.41	3.63	3.23	3.17	1.04	3.71	3.45	1.2	2.00	1.58	1.0	0.77	0.22
FeO	6.57	5.65	5.79	4.74	4.90	3.29	3.67	3.6	2.79	1.86	2.0	2.63	1.48
MgO	5.38	3.36	3.22	3.70	2.36	2.68	2.36	1.6	1.69	1.23	0.8	1.08	0.95
CaO	9.92	8.29	8.19	7.57	6.65	5.56	5.50	3.9	3.63	3.89	2.5	2.65	2.15
Na_2O	2.48	3.34	3.67	3.16	3.86	3.10	3.27	2.7	3.04	3.59	2.7	3.05	2.79
K_2O	1.15	1.30	1.13	2.29	2.77	3.90	2.68	2.5	3.86	4.12	3.2	4.59	4.19
MnO	0.16	0.13	0.13	0.14	0.14	0.12	0.13	0.06	0.10	0.08	0.06	0.07	0.01
TiO_2	0.67	0.85	0.84	0.80	0.63	0.92	0.89	0.40	0.50	0.31	0.18	0.38	0.25
P_2O_5	0.15	0.33	0.32	0.27	0.36	0.45	0.35	—	0.27	0.26	—	0.24	0.19
CO_2	—	—	—	—	—	—	—	—	1.32	—	—	1.20	1.65
H_2O	2.70	1.78	1.80	1.67	1.40	1.24	1.06	—	—	1.23	—	—	—
Loss on ignition	—	—	—	—	—	—	—	0.53	—	—	0.32	—	—
Total	100.23	100.81	100.79	99.89	100.05	99.39	99.93	94.5	100.65	100.53	95.0	100.28	99.93
Norms													
Q	1.80	5.60	4.54	6.54	6.78	12.10	16.47	24.07	22.34	19.71	31.78	24.83	32.77
Or	6.80	7.68	6.68	13.53	16.37	23.05	15.84	14.77	22.81	24.35	18.91	27.13	24.76
Ab	20.99	28.26	31.06	26.74	32.66	26.23	27.67	22.85	25.72	30.38	22.85	25.81	23.61
An	35.46	32.60	32.22	26.59	22.74	18.33	21.31	19.35	16.25	15.84	12.40	11.58	9.43
Ne	—	—	—	—	—	—	—	—	—	—	—	—	—
Di	10.47	5.22	5.20	7.47	6.59	5.03	2.98	—	—	1.43	—	—	—
Hy	18.35	12.02	12.28	10.61	9.95	5.96	7.10	9.05	7.04	4.11	4.65	6.39	4.51
Ol	—	—	—	—	—	—	—	—	—	—	—	—	—
Mt	2.04	5.26	4.68	4.60	1.51	5.38	5.00	1.74	2.90	2.29	1.45	1.12	0.32
Il	1.27	1.61	1.60	1.52	1.20	1.75	1.69	0.76	0.95	0.59	0.34	0.72	0.47
Ap	0.35	0.76	0.74	0.63	0.83	1.04	0.81	—	0.63	0.60	—	0.56	0.44
Cc	—	—	—	—	—	—	—	—	—	—	—	—	—
C	—	—	—	—	—	—	—	1.36	0.70	—	2.25	0.96	1.97

– : not reported.
1. 66-157: diorite, Mount Berger.
2. 66-123: diorite, Mount Becker.
3. 66-153: diorite, Mount Becker.
4. 66-154: quartz-diorite, Mount Becker.
5. EE 931: granodiorite, Mount Hassage.
6. 66-175: quartz-monzonite, Thomas Mountains.
7. 66-217: granodiorite, Quilty Nunataks.
8. 66-57.3: quartz-diorite, Mount Glowa.
9. EE 780: quartz-monzonite, Thomas Mountains.
10. 66-1: granodiorite, Weatherguesser Nunataks.
11. EE 281: quartz-monzonite, Mount Caywood.
12. 66-19: quartz-monzonite, Mount Trimpi.
13. APT 51: quartz-monzonite, northwest Merrick Mountains.

Specimen APT 51 collected by J. C. Behrendt, 1962. Analyses 8 and 11 by Rocky Mountain Geochemical Corporation, Midvale, Utah. Other analyses by Japan Analytical Chemical Research Institute, Tokyo. Analyses 2, 3, 8, and 11 are from Laudon (1972, Table 3). The remainder are new analyses.

Olivine basalt

One small (~100 m long), low (~30 m high) nunatak, located approximately 2.1 km south-southwest of Mount Matheson in the Merrick Mountains, is composed entirely of olivine basalt breccias and flow rocks of Pliocene age (6 m.y.; Halpern, 1971). Most of the surface is covered with rubble. A few small, discontinuous outcrops occur mainly on the southern slope. Exposed beds strike approximately east-west and dip gently to the north (approximately 20°). On the southern third of the nunatak, only breccias were observed in place, and rubble consists of both breccias and flow rocks. On the northern third, including the summit and higher parts of the southern slope, only flow rocks occur in place, and breccia does not occur in the rubble. Thus, exposed flows lie stratigraphically above exposed breccias. However, it was not possible to determine either thicknesses of individual beds or whether the breccias are interbedded with flow rocks.

Breccias consist of angular fragments of black vesicular olivine basalt as long as 5 cm, embedded in tan, fine-grained, probably glassy matrix; they may be hyaloclastic.

Exposed flow rocks are of pahoehoe type with 20-25% vesicles. Much of the basaltic rubble is scoriaceous.

Texture in thin section is porphyritic with 11% olivine (Fo_{65-75}) phenocrysts in a pilotaxitic groundmass. The groundmass is composed predominantly of plagioclase (An_{45-56}) microlites and brown glass, with accessory olivine, apatite, and magnetite, and some cryptocrystalline material. Neither pyroxene nor amphibole is present.

Table 95.5 presents major oxide analyses and C.I.P.W. norms for one specimen of basalt from the Merrick Mountains and three specimens of Pliocene olivine basalt from Mount Manthe in the Hudson Moundains, western Ellsworth Land.

Table 95.4. Chemical analyses of felsic dikes

	1	2
SiO_2	72.81	76.47
Al_2O_3	13.83	11.98
Fe_2O_3	0.41	0.80
FeO	0.34	1.15
MgO	0.17	0.56
CaO	1.25	0.92
Na_2O	2.66	3.19
K_2O	4.36	3.54
MnO	0.14	0.01
TiO_2	0.06	0.33
P_2O_5	0.15	0.10
CO_2	2.65	1.00
H_2O	1.47	—
Loss on ignition	—	—
Total	100.30	100.05
	Norms	
Q	40.19	41.42
Or	25.77	20.92
Ab	22.51	26.99
An	—	3.91
Ne	—	—
Di	—	—
Hy	0.87	2.32
Ol	—	—
Mt	0.59	1.16
Il	0.11	0.63
Ap	0.35	0.23
Cc	1.88	—
C	4.74	1.47

—: not reported.
1. EE 603: rhyolite-porphyry, Mount Becker; from Laudon (1972, Table 3).
2. 66-34: aplite, Mount Trimpi; from Laudon (1972, Table 3).

Analyses by Japan Analytical Chemistry Research Institute, Tokyo.

Geochemistry

Major oxide analyses and C.I.P.W. norms for four specimen of dacite, seven mafic dikes, thirteen specimens of plutonic rocks, two felsic dikes, and one specimen of olivine basalt from eastern Ellsworth Land are presented in Tables 95.1 through 95.5, respectively. Table 95.5 also includes analyses of three Pliocene basalts from Mount Manthe in the Hudson Mountains of western Ellsworth Land. Figures 95.2, 95.3, 95.4, and 95.5 are variation diagrams based on these analyses.

Silica variation diagrams (Figure 95.2) for plutonic rocks and felsic dikes display definite, expectable trends. As SiO_2 increases, K_2O increases and Na_2O shows only slight variation, while Al_2O_3, CaO, MgO, and total iron oxides decrease, AFM (Figure 95.3) and KCN (Figure 95.4) diagrams of these data show typical calc-alkaline trends

Table 95.5. Chemical analyses of Pliocene basalts from Merrick Mountains and Mount Manthe, Hudson Mountains, western Ellsworth Land

	1	2	3	4
SiO_2	44.64	46.7	46.5	46.2
Al_2O_3	15.11	14.0	14.1	14.8
Fe_2O_3	2.84	2.3	2.3	3.0
FeO	7.49	13.0	12.6	9.8
MgO	8.13	8.9	6.8	4.9
CaO	8.13	9.2	8.7	9.1
Na_2O	4.24	2.8	2.5	2.6
K_2O	2.54	1.1	1.0	1.0
MnO	0.15	0.18	0.15	0.13
TiO_2	2.24	1.71	1.69	2.04
P_2O_5	0.79	0.23	0.17	0.19
CO_2	—	—	—	—
H_2O	3.54	—	—	—
Loss on ignition	—	0.52	0.69	1.03
Total	99.84	100.6	97.2	94.8
		Norms		
Q	—	—	—	—
Or	15.01	6.50	5.91	5.91
Ab	12.74	20.26	21.16	22.00
An	14.70	22.38	24.30	25.76
Ne	12.54	1.86	—	—
Di	16.48	17.92	14.74	15.02
Hy	—	—	9.75	14.32
Ol	14.64	24.06	13.72	2.08
Mt	4.12	3.33	3.33	4.35
Il	4.25	3.25	3.21	3.87
Ap	1.83	0.53	0.39	.44
Cc	—	—	—	—
C	—	—	—	—

—: not reported.
1. EE 604: olivine-basalt, southwest Merrick Mountains; from Laudon (1972, Table 3).
2. H2: olivine-basalt, Mount Manthe, Hudson Mountains; new analysis.
3. H5: olivine-basalt, Mount Manthe, Hudson Mountains; new analysis.
4. H6: olivine-basalt, Mount Manthe, Hudson Mountains; new analysis.

Analysis 1 by Japan Analytical Chemistry Research Institute, Tokyo. Analyses 2-4 by Rocky Mountain Geochemical Corporation, Midvale, Utah.

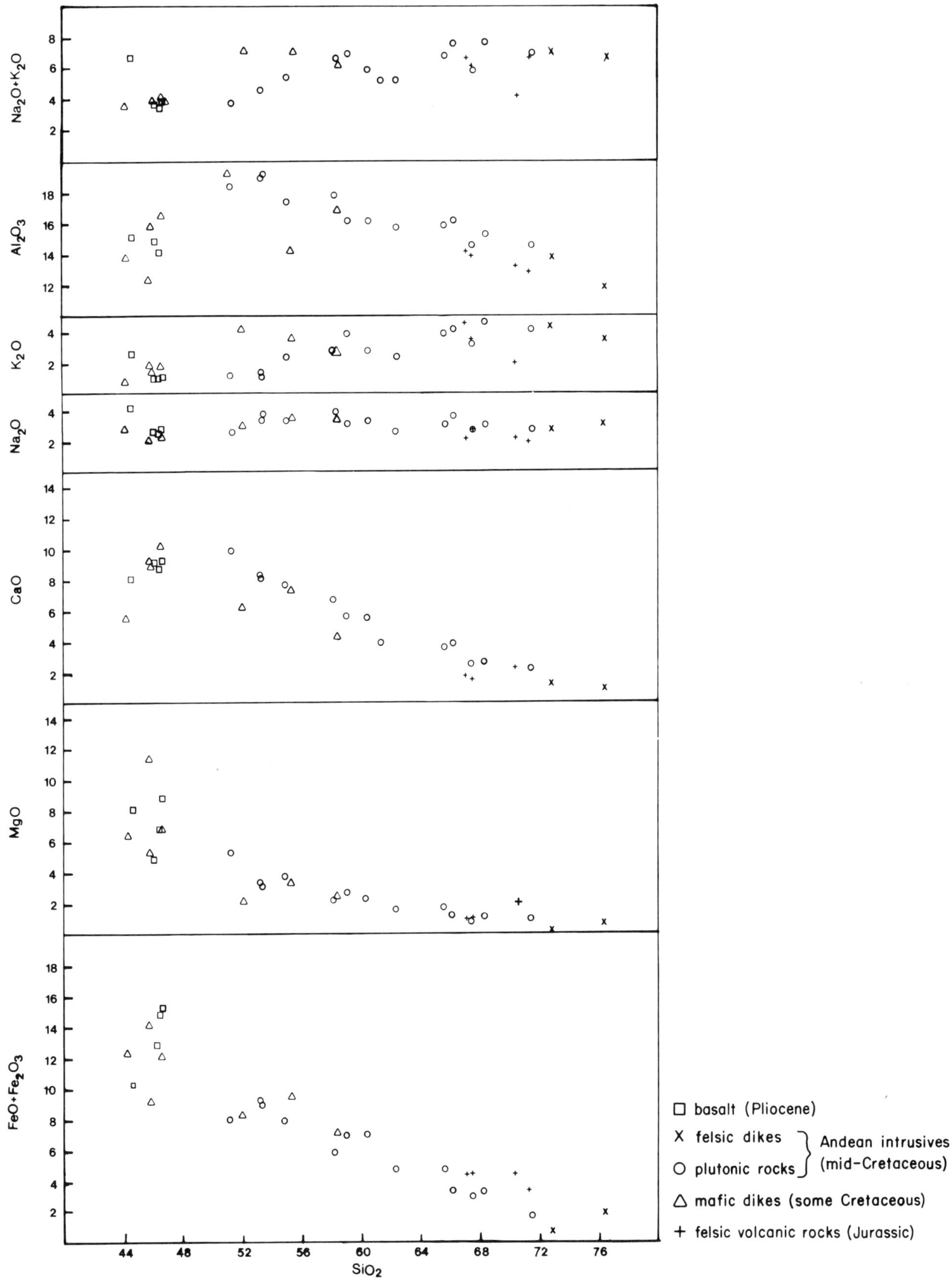

Figure 95.2. Silica variation diagrams for igneous rocks from Ellsworth Land.

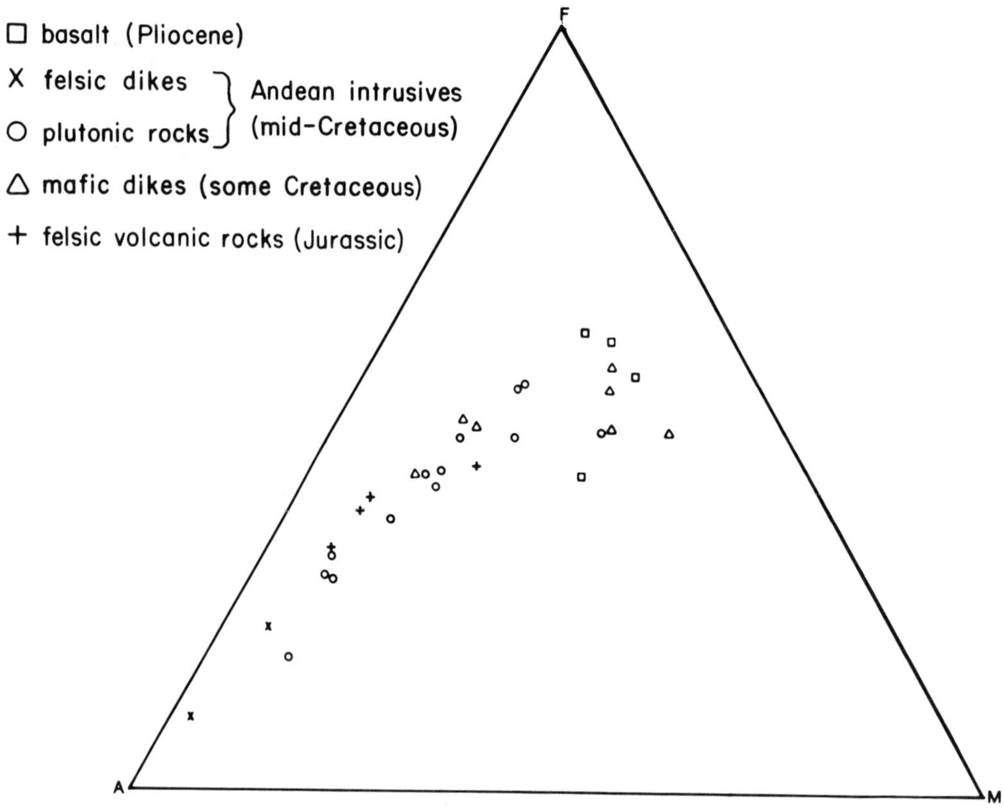

Figure 95.3. Ternary diagram showing distribution of Na_2O + K_2O, FeO + Fe_2O_3, and MgO (AFM) for igneous rocks from Ellsworth Land.

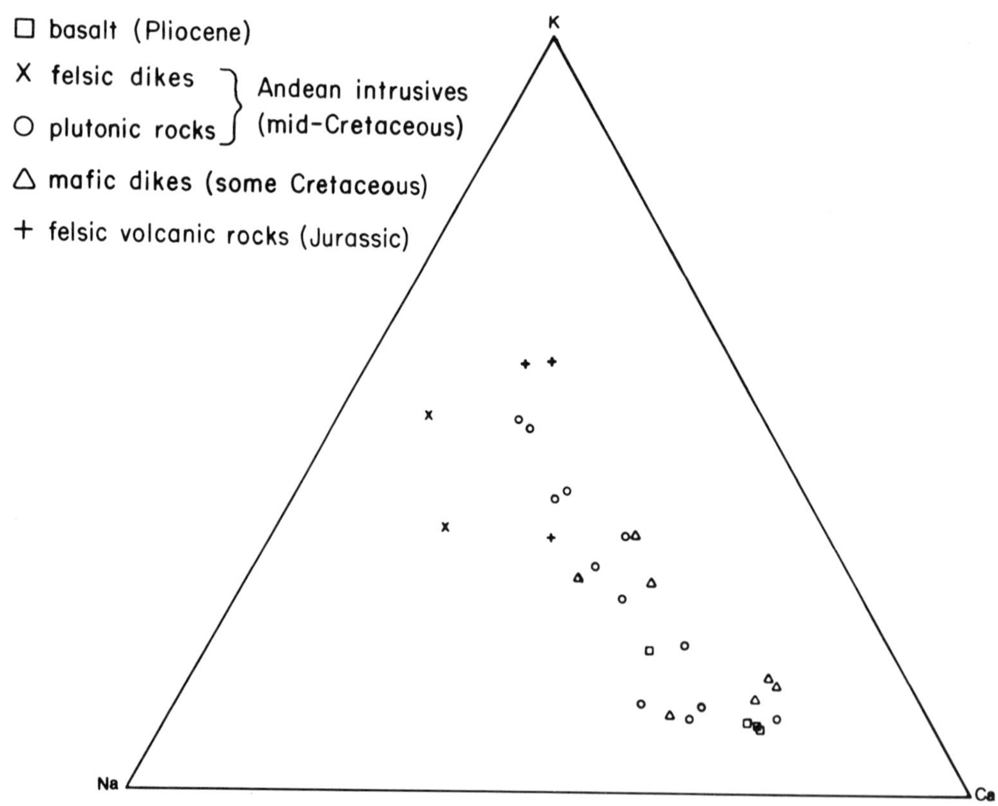

Figure 95.4. Ternary diagram showing distribution of K_2O, CaO, and Na_2O (KCN) for igneous rocks from Ellsworth Land.

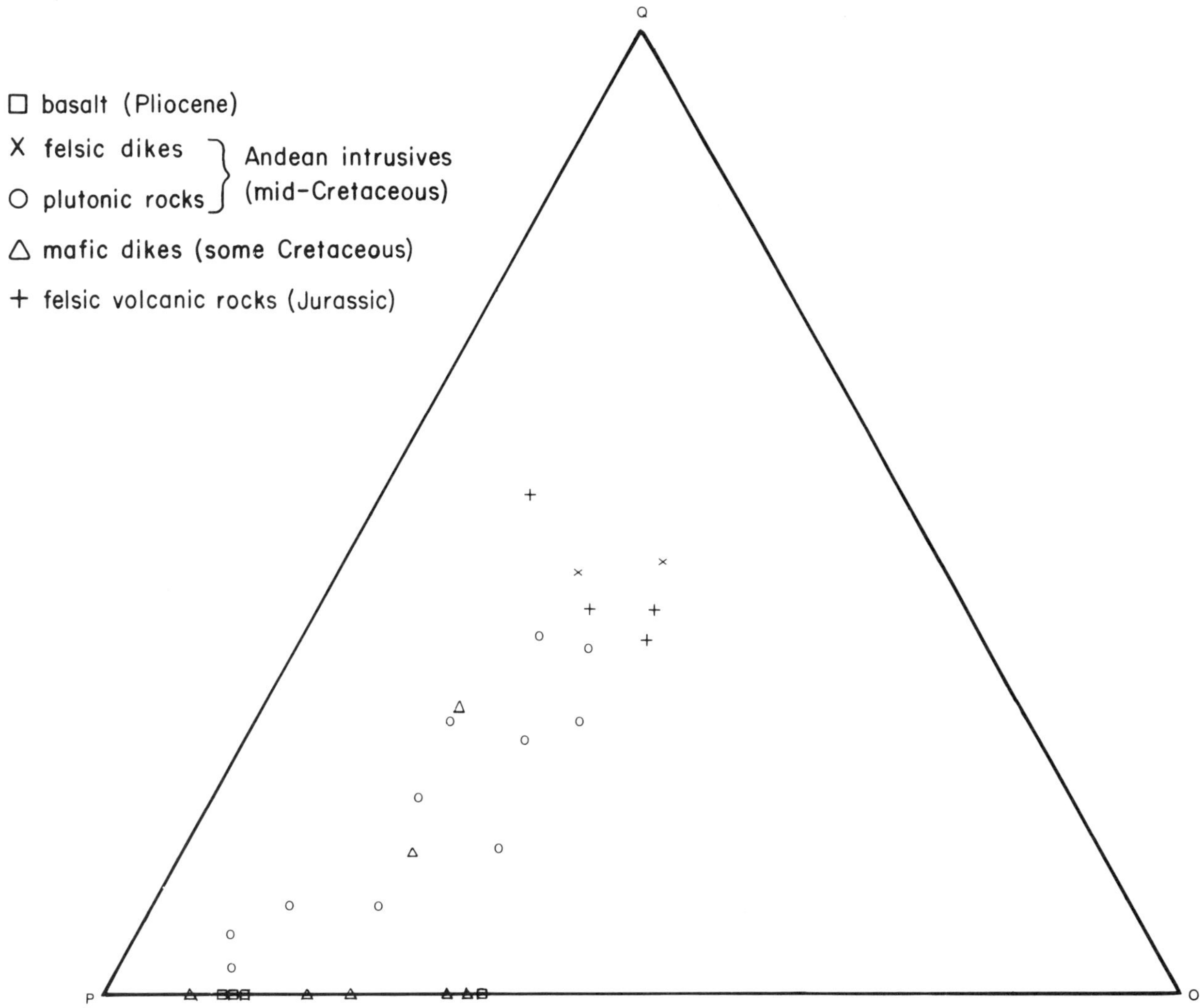

Figure 95.5. Ternary diagram showing distribution of normative plagioclase, orthoclase, and quartz for igneous rocks from Ellsworth Land.

and on alkali-lime index of about 59, clearly indicating these rocks belong to a calc-alkalic magma series. They are correlated with the Andean Intrusive Suite of the northern Antarctic Peninsula (Adie, 1955, 1964*a,b*) and with similar rocks from the Lassiter Coast and southern Black Coast (Williams and others, 1972; Rowley and Williams, 1974; Rowley and Williams, Chapter 38, this volume) on the basis of age, field relationships, and chemical similarity.

Plots of Mesozoic volcanic rocks fall within or very near the fields defined by the plutonic rocks and felsic dikes, suggesting that the volcanic rocks represent a chemically similar calc-alkalic magma series.

Mesozoic volcanic rocks in eastern Ellsworth Land are correlated with calc-alkalic volcanic rocks that are widespread in the Antarctic Peninsula, which have previously been called the Upper Jurassic Volcanic Group (Adie, 1964*a*), but for which Thomson (Chapter 37, this volume) has proposed the new name *Antarctic Peninsula Volcanic Group*. All these rocks are chemically similar to Mesozoic plutonic rocks of the Andean Intrusive Suite (Adie, 1964*a*, 1972; Farrar and others, chapter 39, this volume; Weaver and others, Chapter 29, this volume).

Analyzed specimens of Mesozoic volcanic rocks, Andean felsic dikes, and most of the more felsic plutons (five of six having $SiO_2 > 62\%$) from eastern Ellsworth Land are corundum normative. Vennum (1978) noted that three of four felsic plutonic rocks and an aplite dike from the Werner Batholith of the southwestern Dana Mountains are also corundum normative and pointed out that this is an uncommon feature of rocks of the Andean Intrusive Suite from elsewhere in the Antarctic Peninsula (Adie, 1955, 1964*a*).

When plotted on variation diagrams, analyses of mafic and intermediate dikes show considerably more scatter than do analyses of Mesozoic volcanic rocks and of Andean plutonic rocks and felsic dikes. In general, the mafic dikes are more alkalic (potassic) and less calcic than projected variation trends for the generally more felsic plutonic rocks. These characteristics probably reflect the altered condition of many of the mafic dikes; however, Vennum (1978) noted similar chemical characteristics in a late hornblende andesite dike from the Warner Batholith and suggested that such dikes formed by remobilization of mafic rocks during intrusion of plutonic granodiorite. Two of the analyzed dikes (specimen 66-77, gabbro from Mount Boyer and specimen EE 536, basalt from Weatherguesser Nunataks; Table 95.2) are nepheline normative.

Table 95.6. Whole-rock K-Ar ages, volcanic sequence, Mount Manthe, Hudson Mountains, western Ellsworth Land

Specimen Number	Rock type	Occurrence	K (wt %)	^{40}Ar radiogenic (ppm)	Age (m.y.)
H6	Basalt	Subaerial flow-rocks[a] overlying hyaloclastite sequence	0.935	0.000300	5.5 ± 1.9
H4	Basalt	Nodule within hyaloclastite[b]; ~50 m below top of 200 m section	0.730	0.000432	8.3 ± 1.0
H2	Basalt	Flow rock within hyaloclastite sequence[b]; about middle of 200 m section	1.068	0.000302	4.8 ± 1.6

Constants used: $\lambda_\beta = 4.72 \times 10^{-10}$/year; $\lambda_e = 0.585 \times 10^{-10}$/year; $K^{40}/K = 1.22 \times 10^{-4}$ g/g;

$$\text{Age} = \frac{1}{\lambda_e + \lambda_\beta} \ln \left[\frac{\lambda_\beta + \lambda_e}{\lambda_e} \times \frac{^{40}\text{Ar rad.}}{^{40}\text{K}} + 1 \right].$$

Analyses by Geochron Laboratories Division, Krueger Enterprises, Inc.

[a] Upper lava formation of Lopatin and Polyakov (1974).

[b] Middle tuff formation of Lopatin and Polyakov (1974).

The analyzed specimen of Pliocene olivine basalt from the Merrick Mountains is strongly alkalic, with 12.54% normative nepheline, and is classified as a basanite (Green and Ringwood, 1967). The chemical affinities of this rock are closer to Pliocene alkaline basalts from the Jones Mountains (Craddock and others, 1964) and the Hudson Mountains of western Ellsworth Land, than to Cenozoic basalts from the northern Antarctic Peninsula and the Scotia Arc (Baker, 1972; Weaver and others, Chapter 29, this volume). Three analyses of olivine basalts from Mount Manthe in the Hudson Mountains are included in Table 95.5. Specimens H2 and H5 are from the hyaloclastic middle tuff formations, and specimen H6 is from the upper lava formation (Lopatin and Polyakov, 1974). Table 95.6 gives K-Ar ages for three basalts from Mount Manthe.

Conclusions and speculations

Mesozoic volcanic rocks in eastern Ellsworth Land are mainly of Middle and Late Jurassic age and are correlative with the Antarctic Peninsula Volcanic Group. Intrusive rocks are of Cretaceous age and are correlative with the Andean Intrusive Suite. The volcanic rocks, the plutonic rocks and the postplutonic felsic dikes all possess similar calc-alkalic chemical characteristics. Mafic and intermediate dikes appear to be more alkalic. All of these rocks are representatives of a magmatic arc associated with Mesozoic subduction of the Pacific sea floor beneath West Antarctica.

Pliocene basalt from a single nunatak in eastern Ellsworth Land is strongly alkalic and is correlative with Pliocene basalts from western Ellsworth Land.

In western Ellsworth Land and Marie Byrd Land, alkalic volcanic rocks formed under conditions of rifting that have been active for the last 26–46 m.y. Subglacial basins there are believed to have formed since the development of the continental ice sheet (LeMasurier and Rex, Chapter 89, this volume; LeMasurier and Wade, 1976).

Laudon (1972) has speculated that extrusion of alkali basalt in eastern Ellsworth Land was associated with late Cenozoic arching and block faulting of eastern Ellsworth Land, although no such faults have been identified in the field. However, bedrock elevation profiles based on reflection seismology and free-air gravity anomalies (Behrendt, 1964) reveal steep, narrow depressions extending hundreds of meters below sea level immediately north of the Sweeney Mountains, between Sky-Hi Nunataks and Cheeks Nunatak, and between Cheeks Nunatak and Mount Rex. At Behrendt's station 940, located 5 km east of the summit of Mount Hirman (elevation 1071 m), between the Behrendt Mountains and the Weatherguesser Nunataks, a bedrock reflection was recorded from an elevation of −1150 ± 70 m. It is suggested that these sub-sea-level depressions are grabens, formed since the development of the ice sheet, as postulated for western Ellsworth Land and Marie Byrd Land by LeMasurier and Rex (Chapter 89, this volume) and by LeMasurier and Wade (1976).

Acknowledgments. This work was supported by grants GA-199 and GA-14355 from the National Science Foundation and by grant R-107 from the University of Wisconsin—Oshkosh. I gratefully acknowledge the contributions of the other members of the 1965–66 field party in eastern Ellsworth Land: L. L. Lackey, P. G. Quilty, P. M. Otway, M. Halpern, P. J. Wasilewski, and R. White; and of A. B. Ford and J. M. Aaron who accompanied me to the Hudson Mountains on 11 February 1961. Petrographic analyses were done by L. L. Lackey, D. J. DeWeese, and T. W. Bastien. Logistic support was provided by the U.S. Antarctic Research Program and by the U.S. Navy Operation Deep Freeze.

References cited

Adie, R. J. 1954. The petrology of Graham Land—I. The basement complex: early Palaeozoic plutonic and volcanic rocks. *Scient. Rep. Falkld Isl. Depend. Surv.*, No. 11, 22 pp.

Adie, R. J. 1955. The petrology of Graham Land—II. The Andean granite-gabbro intrusive suite. *Scient. Rep. Falkld Isl. Depend. Surv.*, No. 12, 39 pp.

Adie, R. J. 1964*a*. Stratigraphic correlation in West Antarctica. (*In* Adie, R. J., *ed. Antarctic Geology*. Amsterdam, North-Holland Publishing Company, 308–13.)

Adie, R. J. 1964*b*. The geochemistry of Graham Land. (*In* Adie, R. J., *ed. Antarctic Geology*. Amsterdam, North-Holland Publishing Company, 541–47.)

Adie, R. J. 1972. Evolution of volcanism in the Antarctic Peninsula. (*In* Adie, R. J., *ed. Antarctic Geology and Geophysics*. Oslo, Universitetsforlaget, 137–41.)

Baker, P. E. 1972. Recent volcanism and magmatic variation in the Scotia arc. (*In* Adie, R. J., *ed. Antarctic Geology and Geophysics*. Oslo, Universitetsforlaget, 57–60.)

Behrendt, J. C. 1964. Antarctic Peninsula traverse geophysical results relating to glaciological and geological studies. *Res. Rep. Ser. Geophys. Polar Res. Center, Univ. Wisconsin,* No. 64-1, 112 pp.

Behrendt, J. C., and Parks, P. E. 1962. Antarctic Peninsula traverse. *Science, N.Y., 137,* No. 3530, 601–2.

Craddock, C., Bastien, T. W., and Rutford, R. H. 1964. Geology of the Jones Mountains area. (*In* Adie, R. J., *ed. Antarctic Geology*. Amsterdam, North-Holland Publishing Company, 171–87.)

Green, D. H., and Ringwood, A. E. 1967. The genesis of basaltic magmas. *Contr. Miner. Petrol.,* No. 15, 103–90.

Halpern, M. 1971. Evidence for Gondwanaland from a review of West Antarctic radiometric ages. (*In* Quam, L. O., *ed.* Research in the Antarctic. *Am. Ass. Adv. Sci.,* Publn No. 93, 717–30.)

Laudon, T. S. 1972. Stratigraphy of eastern Ellsworth Land. (*In* Adie, R. J., *ed. Antarctic Geology and Geophysics.* Oslo, Universitetsforlaget, 215–23.)

Laudon, T. S., Lackey, L. L., Quilty, P. G., and Otway, P. M. 1969. Geology of eastern Ellsworth Land (Sheet 3, eastern Ellsworth Land). (*In* Bushnell, V. C., and Craddock, C., *eds.* Geologic Maps of Antarctica. *Antarct. Map Folio Ser.,* Folio 12, Pl. III.)

LeMasurier, W. E., and Wade, F. A. 1976. Volcanic history in Marie Byrd Land: implications with regard to Southern Hemisphere tectonic reconstructions. (*In* González-Ferrán, O., *ed. Andean and Antarctic Volcanology Problems.* Rome, International Association of Volcanology and Chemistry of Earth's Interior, 398–424.)

Lopatin, B. G., and Polyakov, M. M. 1974. Geology of the volcanic Hudson Mountains, Walgreen Coast, West Antarctica (in Russian). *Antarktika, Doklady Komissii, 13,* 52–60.

Quilty, P. G. 1970. Jurassic ammonites from Ellsworth Land, Antarctica. *J. Paleont., 44,* No. 1, 110–16.

Quilty, P. G. 1972*a*. Middle Jurassic brachiopods from Ellsworth Land, Antarctica. *N.Z. Jl Geol. Geophys., 15,* No. 1, 140–47.

Quilty, P. G. 1972*b*. *Pentacrinites* and (?) *Apiocrinus* from the Jurassic of Ellsworth Land, Antarctica. *Neues Jb. Geol. Palaont. Mh., 8,* 484–89.

Quilty, P. G. 1977. Late Jurassic bivalves from Ellsworth Land, Antarctica: their systematics and paleogeographic implications. *N.Z. Jl Geol. Geophys., 20,* No. 6, 1033–80.

Rowley, P. D., and Williams, P. L. 1974. Plutonic rocks of the Lassiter Coast. *Antarct. Jnl U.S., 9,* No. 5, 225–26.

Stevens, G. R. 1967. Upper Jurassic fossils from Ellsworth Land, West Antarctica, and notes on Upper Jurassic biogeography of the South Pacific region. *N.Z. Jl Geol. Geophys., 10,* No. 2, 345–93.

Vennum, W. R. 1978. Igneous and metamorphic petrology of the southwestern Dana Mountains, Lassiter Coast, Antarctic Peninsula. *J. Res. U.S. geol. Surv., 6,* No. 1, 95–106.

Williams, P. L., Schmidt, D. L., Plummer, C. C., and Brown, L. E. 1972. Geology of the Lassiter Coast area, Antarctic Peninsula: preliminary report. (*In* Adie, R. J., *ed. Antarctic Geology and Geophysics.* Oslo, Universitetsforlaget, 143–48.)

96

Explosive Volcanic Eruptions in the Vicinity of the Balleny Islands Recorded in the Abyssal Sediments of the Southern Oceans

Abstract

T. C. HUANG and N. D. WATKINS†
Graduate School of Oceanography, University of Rhode Island, Kingston, Rhode Island 02881, U.S.A.

Analysis of the fine volcanic glass distribution in nine dated Eltanin cores, across a 4000-km-long traverse downwind from the Balleny Islands, reveals a series of seven separate periods of volcanic activity over the past 2 m.y. Correlation of specific ash horizons across the traverse was initially made using paleomagnetic, micropaleontologic, and sedimentologic methods, but verification was later carried out by micropit-population and chemical analyses using both the scanning electron microscope and the electron microprobe. The glass shards in the ash layers are rhyolitic and chemically homogeneous, consistent with a single source.

The period of the most explosive volcanism occurred between $t = 1.8$ and 1.6 m.y. The average volcanic dust production rate for the 88 to 11 μm size fraction was on the order of 1.7×10^8 kg/yr over an area of 4×10^6 km^2 for a period of as long as 2×10^5 years. Analysis of the ash particle size distribution across the traverse shows that volcanic clouds reached as far as 33 km at that time, consistent with minimum explosivity of 30 megatons TNT equivalent (10^{17} joules).

† Deceased.

97

Syenite and Granite Ring Complexes in the Kerguelen Islands: Petrologic and Geochronologic Data*

Abstract

J. M. CANTAGREL
Laboratoire Associé No. 10, Centre National de la Recherche Scientifique et Université, 63038 Clermont-Ferrand Cedex, France

L. DOSSO and P. VIDAL
Institut de Géologie, 35031 Rennes Cedex, France

J. LAMEYRE and S. ZIMINE
Laboratoire de Pétrologie, Université Pierre et Marie Curie, 75230 Paris, Cedex 05, France

Alkaline plutonic rocks from the Rallier du Baty peninsula in the southwestern part of the Kerguelen Islands have intruded and metamorphosed a thick series of tholeiitic and alkaline basalts. Detailed field mapping completed between 1973 and 1976 indicates five distinct plutonic centers. Each center consists of a sequence of ring dikes; chilled margins and enclaves show that the sequence of injection is centripetal.

Alkaline granite appears only in the core of the large southwestern center. Nordmarkite is the most abundant rock type. The occurrence and composition of fayalite, clinopyroxene, perthite, amphibole, biotite, quartz, chevkinite, and other minerals indicates a temperature of intrusion higher than 900°C and a pressure lower than 1 kb.

K-Ar and Rb-Sr radiometric dates for the southern center give ages ranging from 12 m.y. (outer ring) to 7 m.y. (core). The time interval between each intrusion is about 1 m.y. Ages for the whole group of centers range from more than 12 m.y. in the south to less than 5 m.y. in the north, suggesting a very slow rate of migration of the magmatic activity. Furthermore, two layered basic and intermediate intrusions have been discovered, the syenitic differentiates of which are very similar to the syenitic rocks of the ring complexes.

* Published: Dosso, L., Vidal, P., Cantagrel, J. M., Lameyre, J., Marot, A., and Zimine, S. 1979. Kerguelen: continental fragment or oceanic island? Petrology and isotopic geochemistry evidence. *Earth planet. Sci. Lett., 43,* 46–60.

VII Structural Geology and Tectonics

98

Structure of Antarctica and Outline of its Evolution

Review Paper

G. E. GRIKUROV
Department of Antarctic Geology, SEVMORGEO, Moyka 120, Leningrad 190121, U.S.S.R.

Abstract. The following structural units are recognized in Antarctica: (1) the East Antarctic craton, (2) the Weddell Sea submerged platform, (3) the Ross fold system of the Transantarctic Mountains, (4) the West Antarctic fold system of the circum-Pacific mobile belt, (5) transitional areas, and (6) rift zones.

The East Antarctic craton is characterized by a predominance of Archean high-grade schists and various intrusive rocks in the remnant hills of the pre-glacial erosional surface, but vast subglacial plains and bedrock depressions may be underlain by layered rocks of the platform cover. The Ross fold system is no longer included in the Antarctic platform (as in previous publications by Soviet authors) but is now considered to be a relic of early stages of development of the Pacific geosynclinal belt, which is typically represented by the West Antarctic orogenic fold system. The transitional areas include the Pensacola and Ellsworth Mountains, northern Victoria Land and western Marie Byrd Land, the Byrd Subglacial Basin, and the Ross Ice Shelf. Rift zones essentially modify the pre-Cenozoic structural pattern.

Four cycles of tectonic evolution (Napier, Rayner, Humboldt, and Insel) are suggested for the Archean, during which the Panantarctic crystalline basement was formed by granitization of the enderbitic protocrustal shell. The two subsequent cycles (2600–1700 m.y. and 1700–1000 m.y. ago) have been named early and late Ruker; they were responsible for formation of ancient epicratonic platform covers and miogeosynclinal fold systems, as well as for initiation of pericratonic Pacific geosynclines. The latter were fully developed during the Ross cycle (1000–500 m.y. ago), after which the geosynclinal process continued only at the Pacific margin, while in the rest of the continent the Beacon cover was formed (Gondwanide cycle, 500–200 m.y. ago). The latest stages of evolution (200–100 m.y. ago and since 100 m.y. ago) have been named early and late Andean and are characterized mainly by orogenic and riftogenic processes, respectively.

The ensialic evolution of the crust is evidenced not only by Antarctic data but is also confirmed by recent geologic studies on other continents. The ensialic model is incompatible with the horizontal tectonics philosophy.

At the second Symposium on Antarctic Geology and Geophysics held in 1970 in Oslo, papers and maps dealing with Antarctic tectonics were presented (Craddock, 1970, 1972; Ford, 1972*a*; Grikurov, 1972; Grikurov and others, 1972) that summarized all research in this field prior to 1970 and gave comprehensive analyses of the problem updated to the time of the meeting. Since then, more studies have been published (Craddock, 1975; Dalziel and Elliot, 1973; Elliot, 1975*a,b*; Ford, 1972*b*; Grikurov, 1973; Grikurov and Lopatin, 1975; Stump, 1973) discussing various aspects of the tectonic evolution of Antarctica; the authors of those works have largely maintained their previous views. These views and the principal controversies are briefly outlined below, but no detailed summary of early publications on the subject is attempted, as this has been done in many of the papers mentioned above.

The most valuable recent contributions to Antarctic tectonics have resulted from geophysical and geochronologic studies. The former yielded a much better knowledge of bedrock topography due to extensive airborne radio-echo sounding (Robin, 1972; Scott Polar Research Institute, 1974; Drewry, 1972, 1976; Fedorov, 1973; Korotkevich and others, 1977); consequently, it became possible to make structural extrapolations through vast subglacial terrains. Some regions of lesser areal extent but of utmost tectonic importance, such as the Lambert Glacier rift zone, were studied in sufficient detail to gain deep crustal information (Fedorov and others, Chapter 118, this volume). Geochronology based on improved radiogenic isotope methods has finally succeeded in obtaining early Precambrian (Archean) ages for the East Antarctic metamorphic terrane (Halpern and Grikurov, 1975; Australia Bureau of Mineral Resources, 1975; R. J. Tingey, written communication), and even an unprecedented figure of 4000 m.y. is now available for Antarctic enderbites (Sobotovich and others, 1974). An extremely prolonged crustal evolution for Antarctica is thus established beyond doubt and must be taken into account in paleotectonic reconstructions.

Structural stages, individual complexes, and intrusive suites will not be specifically described in the present paper in terms of age, lithology (petrology), thickness, metamorphism, stratigraphic relationships, and deformational characteristics. This information can be found in previous publications on the subject, the most important of which are listed above (see also Ravich and Grikurov, 1976*a,b*); it is believed that the tectonic complexes identified in these works are commonly recognized. Instead, the emphasis in this paper is on the historical and genetic aspects responsible for typical tectonic combinations of major structural features, and on some debatable problems. Units and events of dubious nature are discussed at greater length, but some apparent situations are only briefly characterized or merely mentioned. The author admits the vulnerability of subglacial extrapolations and realizes that subglacial structural patterns thus conceived will inevitably be reviewed in the course of future investigations.

Summary of previous work

The majority of Soviet authors have long maintained a two-fold subdivision of the continent; the Antarctic platform of Gondwana type and the West Antarctic fold belt of Pacific affinity (Antarctandes) were usually distinguished as major geotectonic elements. An important, although admittedly debatable, point involved extending the Antarctic platform across the Transantarctic Mountains and well into West Antarctica, so that only the Antarctic Peninsula, the Eights Coast, and possibly Marie Byrd Land were recognized as a typical Pacific area. The general idea behind this approach was that, up to early Paleozoic time, geotectonic evolution proceeded more or less uniformly over nearly the whole of Antarctica and resulted in formation of the Precambrian Panantarctic craton, dissected into large crustal blocks by the late Precambrian to early Paleozoic epigeosynclinal fold belts which were cratonized during the Ross orogeny. Such structural pattern was considered typical of the heterogeneous basements of Gondwana platforms. The West Antarctic part of the circum-Pacific mobile belt was thought to have evolved by continuation of geosynclinal development (or by its regeneration on incompletely cratonized Ross structures) throughout middle-late Paleozoic and Mesozoic time, while either erosion or Beacon-type sedimentation on the epi-Ross peneplain took place outside the Pacific domain. Both the Weddell Sea and Ross Sea basins were believed to represent ancient platform plates of pre-Ross stabilization. As noted by Elliot (1975*a*), a "stabilist" philosophy has always been quite obvious in this classification.

In contrast to the vertical tectonics postulate maintained by the Russian authors, a vast majority of non-Russian publications on the subject (mostly by American workers) proceeded from the horizontal tectonics philosophy. Their views were clearly "mobilist" with respect to global geotectonic development and especially the Gondwanaland problem, but as far as the Antarctic continent is concerned, the main emphasis was placed on lateral accretion to its Pacific margin, rather than on movement of the continent as a whole. The term "Antarctic platform" was usually avoided, but instead the East Antarctic Shield was distinguished—an ancient (Precambrian) cratonic nucleus onto which successively younger geosynclinal belts welded a Pacific-ward series of orogens—Ross (early Paleozoic), Borchgrevink (middle Paleozoic), Ellsworth or Weddell (early Mesozoic), and Andean (late Mesozoic to Cenozoic). This accretional development was thought to have occurred along the Antarctic part of the Pacific continental margin of Gondwanaland, and (after the Gondwanide orogeny in early Mesozoic time and subsequent Gondwanaland breakup) along the edge of the isolated Antarctic segment of the circum-Pacific mobile belt (Craddock, 1975).

However, Craddock (1975) admits that simple accretion cannot account for the presence of ancient crystalline basement rocks within the Phanerozoic orogenic belts, and Ford (1972*a*) considers the concept of progressive migration of younger mobile zones across West Antarctica, from the Transantarctic Mountains toward the Pacific Ocean, to be "oversimplified." Both Ford (1972*a*) and Elliot (1975*a*) mention that the orogenic belts listed above are at least partly epicratonic and might have developed by recycling of older terranes. It is therefore clear that these authors do not imply that the boundary of the East Antarctic shield delineates the limits of areal extent of the Precambrian continental crust.

Thus "mobilists" appear to agree with "stabilists" in that the evolution of the Transantarctic Mountains and at least some of West Antarctica was essentially epicratonic, and that new continental crust, if ever formed, developed mainly by transformation of the pre-existing sialic crust rather than by lateral growth of the continent at the expense of adjacent oceanic domain. Furthermore, there seems to be a close similarity between Ford's (1972*b*) assertion that the evolution of Antarctica followed a course of increasing duration of mobility toward the Pacific margin, and the concept of West Antarctic development suggested by Grikurov and Lopatin (1975). On the other hand, the latter authors are no longer inclined to include any part of the Ross geosynclinal terranes in the Antarctic platform; this opens opportunity for a better correlation between the structural units recognized in the present paper and those distinguished earlier by Craddock (1970, 1972, 1975), Elliot (1975*a*), Ford (1972*b*), and other workers.

Major structural units and typical tectonic features

The following major structural elements are distinguished (Figure 98.1): (1) the East Antarctic craton, (2) the Weddell Sea submerged plate, (3) the Ross fold system of the Transantarctic Mountains, (4) the West Antarctic fold system of the circum-Pacific mobile belt, (5) areas where a convergence of tectonic characteristics precludes positive determination, and (6) rift zones superimposed onto all the other structures and probably responsible for distortion of the Antarctic landmass during Cenozoic time.

The East Antarctic craton

This is the largest geotectonic unit of the continent. It occupies the whole of East Antarctica (with the exception of the Transantarctic Mountains) and extends well beyond the Greenwich meridian into the Weddell Sea area. Its preglacial erosional surface is characterized by a predominance of Archean schists penetrated by various intrusive rocks. These rocks compose mountains in Queen Maud Land and Enderby Land, and compose the greater part of the Prince Charles Mountains, as well as numerous small outcrops along the Indian Ocean coast; they are also believed to make up the subglacial Gamburtsev Mountains and subglacial hills in southern Wilkes Land. Other characteristic geomorphic features in East Antarctica are the subglacial plains and the bedrock surface depressions which separate and surround projections of the crystalline basement and are probably underlain by a platform cover. The latter crops out extensively only in the Ritscher Upland (western Queen Maud Land) where undeformed lower to middle Proterozoic strata overlap upper Archean granitized basement. Upper Precambrian to lower Paleozoic undisturbed volcanic rocks and Permian sedimentary rocks occur farther to the southwest along the Weddell Sea coast, and this region of the oldest platform stabilization probably extends to the Shackleton Range where mature quartz sandstone and stromatolite limestone rest on deeply weathered ancient granites.

The East Antarctic craton was affected by intense post-Archean differential movements which resulted in development of diverse tectonic forms on a heterogeneous Archean foundation. Among those forms the most characteristic are: (1) areas of persistent but slow and quiet uplift-shields *sensu stricto* (Enderby Land, southern subglacial part of Wilkes Land); (2) zones of multiple tectono-thermal activizations of variable intensity—recurrent epicratonic orogens *sensu stricto* (Queen Maud Land, the northern Prince Charles Mountains, possibly the Gamburtsev Mountains); (3) intracratonic fold systems developed from aulacogen-like troughs (the southern Prince Charles Mountains); and (4) regions of active subsidence that took place mostly in Mesozoic and Cenozoic time—intraplatform basins, partially filled by sediment and concealed under the thickest parts of the ice sheet (bedrock depressions between long. 100° and 150° E).

All these tectonic forms within the East Antarctic craton have an important feature in common: they show no evidence of geosynclinal development, either during the Archean eon of crystalline basement creation, or during the course of the craton's subsequent evolution.

The Weddell Sea plate

This region can so far be distinguished only provisionally, as dependable geologic evidence for its existence is restricted to the Weddell Sea eastern coast in the uplifted terrain of East Antarctica. The greater part of this hypothetical ancient platform is either submarine or concealed under extensive shelf ice. It is characterized by a predominance of submerged plains and related features of low relief apparently associated with intense depositional processes that buried and smoothed out the smaller tectonic forms. The presence of a thick supracrustal cover is confirmed by geophysical data indicating a magnetic basement 7–8 km below sea level at a distance of 200–250 km off the Filchner Ice Shelf front.

It is reasonable to suspect that the geologic structure of the southeastern Weddell Sea is similar to that of the adjacent land (Shackleton Range–Theron Mountains–Coats Land–Ritscher Upland area), i.e., that the Archean basement is overlain there by a cover of wide stratigraphic range (Proterozoic to lower Mesozoic sedimentary and volcanic rocks, most likely supplemented by a thick upper Mesozoic–Cenozoic pile). The central and western sectors of the Weddell Sea may have a similar combination of thick platform complexes overlying ancient basement, but nearer the Antarctic Peninsula the latter may also include post-Archean fold complexes.

The Ross fold system

As mentioned above, the Transantarctic Mountains were commonly classified by Soviet authors as the youngest (epi-Ross or epi-early Paleozoic) element of the Antarctic platform. Such classification was nominally justified by the distinct two-stage structure of this extended belt, consisting of the Precambrian to lower Paleozoic basement (Ross complex with inliers of underlying older crystalline rocks) overlain by the middle Paleozoic–Mesozoic undeformed cover (Beacon "complex"). The close similarity between this cover and the "Gondwana System" elsewhere, both in stratigraphic range and in rock composition, implied an additional tie to the Gondwana-type platforms. The latter are commonly characterized by a highly heterogeneous basement which contains, along with the lower Precambrian crystalline rocks, the upper Precambrian to lower Paleozoic fold complexes (for instance, the Brazilides outside the Andean orogen in South America and the Katanga and Damara belts of South Africa). However, the post-lower Precambrian fold complexes found in the basement of most typical Gondwana platforms—such as Africa or peninsular India—are formed in intracratonic miogeosynclinal troughs and not in true geosynclines. Such formations are now known in the Prince Charles Mountains within the East Antarctic craton (Ravich and Fedorov, Chapter 60, this volume). They comprise a Proterozoic to lower Paleozoic fold system which, in fact, is very similar to the Ross belt in the time of development of the parental volcanic-sedimentary trough and in the time of its tectonic reversal accompanied by folding and greenschist facies metamorphism. Nevertheless, there is a notable distinction from a typically geosynclinal Ross area, namely, the absence or negligible degree of reworking of the pre-geosynclinal basement. Such reworking, in the author's opinion, is the most important feature in recognizing a genuine geosynclinal development, even if the composition of the pre-geosynclinal substratum and, consequently, the general nature of the geosynclinal process, remain open to debate.

Hence it appears reasonable to regard the Ross geosynclinal fold complexes of the Transantarctic Mountains as relics of an ancient ("preparatory") stage of development of the future Pacific geosynclinal belt. They are preserved in a relatively unaltered state due to their consolidation with the cratonized area of East Antarctica and due to the conserving effect of the overlying Beacon cover. Elsewhere in the circum-Pacific belt, the synchronous fold structures (with the possible exception of the Adelaide geosyncline in Australia) served as a basement for younger geosynclinal complexes, and therefore appear totally reworked. If the Transantarctic Mountains are not considered a part of the

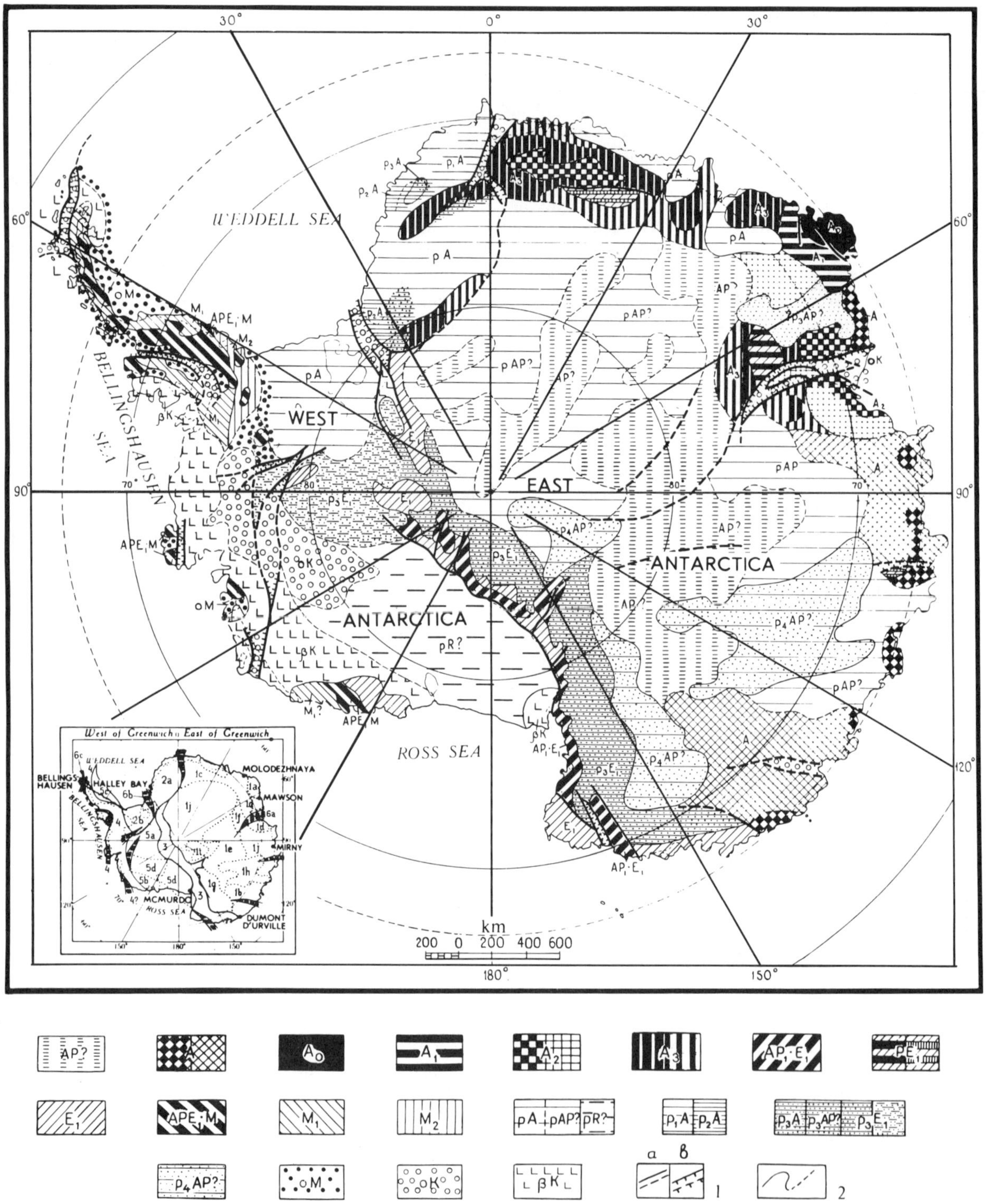
30°
0°
30°
60°
60°
90°
90°
120°
180°
150°
WEDDELL SEA
BELLINGSHAUSEN SEA
WEST
EAST
ANTARCTICA
ANTARCTICA
ROSS SEA
km
200 0 200 400 600
West of Greenwich East of Greenwich
WEDDELL SEA
MOLODEZHNAYA
BELLINGSHAUSEN
HALLEY BAY
MAWSON
MIRNY
MCMURDO
ROSS SEA
DUMONT D'URVILLE
AP?
A₀
A₁
A₂
A₃
AP₁E₁
E₁
APE₁M
M₁
M₂
pA pAP? pR?
p₁A p₂A
p₃A p₃AP? p₃E₁
p₄AP?
oM
oK
βK
a
в
1
2

Figure 98.1. Tectonic scheme of Antarctica. *Fold complexes.* AP?: inferred Precambrian in subice mountains of East Antarctica. A: undifferentiated Archean, identified and inferred. A_0: relics of protocrustal ("enderbitic") shell formed about 4000 m.y. ago (Napier orogeny). A_1: formed during initial stage of metamorphism and granitization about 3500 m.y. ago (Rayner orogeny). A_2: formed during mature stage of metamorphism and granitization about 3000 m.y. ago (Humboldt orogeny), identified and inferred. A_3: formed during the culmination of metamorphic and granitization processes about 2600 m.y. ago (Insel orogeny), identified and inferred. AP_1E_1: undifferentiated lower Precambrian, reworked in the infrastructure of Ross geosynclines as basement projections and anatexitic plutons. PE_1: undifferentiated Proterozoic to lower Paleozoic of a miogeosynclinal type, deformed and metamorphosed at 1800–1700, 1000, and 500 m.y. age levels (Ruker and Ross orogenies), identified and inferred. E_1: upper Proterozoic to lower Paleozoic geosynclinal rocks deformed 650–450 m.y. ago (Ross orogeny). APE_1M: undifferentiated Precambrian to lower Paleozoic, reworked in the infrastructure of Pacific geosynclines as basement projections (rejuvenated to different degrees) and anatexitic plutons. M_1: middle(?)-upper Paleozoic geosynclinal rocks deformed about 200 m.y. ago (Gondwanide orogeny). M_2: Mesozoic geosynclinal(?) rocks deformed about 100 m.y. ago (early Andean orogeny).
Platform cover complexes. pA: undifferentiated in age of accumulation, with Archean fold basement; pAP?: same, with Precambrian(?) fold basement; pR?: same, with late Precambrian(?) fold basement deformed during Ross orogeny. p_1A: lower-middle Proterozoic, with Archean fold basement; p_2A: upper Proterozoic to lower Paleozoic with Archean fold basement. p_3A: middle Paleozoic to lower Mesozoic (Beacon), with Archean fold basement; p_3AP?: same, with Precambrian(?) fold basement; p_3E_1: same, both undeformed, and deformed during Gondwanide orogeny, with lower Paleozoic (Ross) fold basement. p_4AP?: upper Mesozoic to Cenozoic with Precambrian(?) fold basement.
Orogenic and riftogenic complexes. oM: Mesozoic molasse. oK: Cenozoic molasse. βK: Cenozoic volcanic rocks and Mesozoic Dufek Intrusion.
Structure. 1. Major disjunctive lineaments, (*a*) fractures and fracture zones, identified and inferred and (*b*) faults bounding rift grabens, identified and inferred; 2. Boundaries of structural complexes, identified and inferred.
Major structural units of Antarctica (inset). 1. *East Antarctic craton.* Shields: *a.* Enderby Land and *b.* Wilkes Land. Recurrent epicratonic orogens: *c.* Queen Maud Land, *d.* northern Lambert Glacier area, *e.* Gamburtsev Mountains, and *f.* intracratonic fold system of the southern Prince Charles Mountains. Mesozoic–Cenozoic sedimentary basins related to large basement depressions: *g.* Victoria Basin or Wilkes Basin of Drewry (1976), *h.* West Wilkes Basin or Aurora basin of Drewry (1976), *i.* South Pole Basin, and *j.* subglacial lowlands of unknown tectonic nature. 2. *Weddell Sea ancient platform (plate): a.* exposed in the uplifted part of western Queen Maud Land, and *b.* submerged beneath the Ronne and Filchner Ice Shelves. 3. *Ross fold system of the Transantarctic Mountains.* 4. *West Antarctic (Pacific) orogenic-fold system.* 5. *Transitional areas: a.* Pensacola-Ellsworth Mountains block, *b.* northern Victoria Land–western Marie Byrd Land block, *c.* Larsen Ice Shelf area, and *d.* Ross Ice Shelf–Byrd Subglacial Basin massif (median?). 6. *Rift zones confirmed by geophysical evidence: a.* Lambert Glacier Zone, *b.* Filchner Ice Shelf zone, and *c.* Bransfield Strait zone. Other rifts are inferred from geologic and geomorphologic evidence.

Antarctic platform, the latter term becomes a synonym for the Antarctic craton and conveys a certain historic-genetic meaning. In this usage it denotes only that part of the Antarctic continental crust which was formed during early Precambrian time, without participation of a geosynclinal process.

The West Antarctic orogenic-fold system

This region embraces the Antarctic Peninsula with adjacent islands, Ellsworth Land, Thurston Island, and probably much of Marie Byrd Land. The most typical tectonic features of the circum-Pacific mobile belt are present within this system, namely, the pronounced and widespread Mesozoic to Cenozoic igneous activity (both intrusive and effusive) and the dramatic orogenic movements of a highly heterogeneous substratum. The latter consists of reworked lower Precambrian(?) crystalline basement and younger complexes formed during the late Precambrian and Phanerozoic geosynclinal cycles and/or corresponding terminal orogenies. Evidently, a polycyclic geosynclinal process played the leader role in the tectonic evolution of the Pacific margin of Antarctica.

Transitional areas

These tectonic elements are recognized mainly in West Antarctica where they border and separate other major geostructural units. They are characterized by a convergence of typical tectonic features and include the following regions:

1. The Pensacola and Ellsworth Mountains with adjacent nunataks in the West Antarctic interior, where rocks equivalent to those in the Ross fold belt contain very few granitic intrusive rocks as a result of a lesser degree of geosynclinal reworking of the underlying crystalline basement. The overlying Beacon-equivalent formations lack their typical platform appearance—they have increased thickness, include molasse deposits, and contain minor granitic stocks and fold deformations developed during the early Mesozoic orogeny.

2. Northernmost Victoria Land and western Marie Byrd Land, which differ from other areas of Ross tectogenesis, first, in a deficiency of Ross synkinematic granites and, secondly, in the apparent absence of the epi-Ross stratified formations of either Gondwana (platform) or Pacific (geosynclinal) affinity. Nevertheless, it is quite obvious that the pre-Ross crystalline basement of these two areas has been strongly reworked and remobilized into a higher crustal level. This fact is clearly demonstrated by the abundance of the post-Ross granitoid intrusions.

3. Western shelf of the Weddell Sea along the Antarctic Peninsula, where a deep post-Early Cretaceous molasse basin developed directly on the marginal part of an ancient stable platform.

4. An extensive submerged part of West Antarctica occupied by the Byrd Subglacial Basin and the adjoining Ross Ice Shelf, provisionally considered a transitional type. The difficulties in tectonic classification of this region are caused by an acute lack of information rather than by uncertainties in the available data. Ambiguity may, in fact, be quite characteristic of evidence in that area, which may have experienced considerable extensional deformation in the process of Cenozoic rifting.

Sporadic seismic soundings in the Byrd Subglacial Basin (Bentley and Clough, 1972) seem to show a submerged high-velocity (crystalline?) basement overlain directly by a thick (as much as 3 km) layer of stratified rocks; the layered rocks include some low-velocity varieties consistent with an undisturbed Cenozoic volcanic-sedimentary cover. There is no reliable indication of an intermediate middle-velocity layer comparable to the folded epicratonic formations. On the other hand, carbonate rocks similar to the uppermost horizons of the Ross complex are present in the Ross Sea bottom, as shown by the seismic and drill core data summarized by Elliot (1975*a*). The submarine geology of the Ross Sea is apparently dominated by longitudinal structures, both in the basement and in the platform cover, which accumulated in south-trending basins during Cenozoic and, possibly, latest Mesozoic time. Remnants of the older (Beacon) cover are probably also present locally, although in the Devonian Period the Ross Sea (or some parts of it) might have been an uplifted gneissic terrane which supplied sediments to the early Beacon basins (Elliot, 1975*a*).

Whatever little is known at present about the Byrd Subglacial Basin and the Ross Sea, it seems improbable that this area participated directly in the Ross geosynclinal development. More likely it behaved as a relatively rigid block—median massif(?) (Grikurov and Lopatin, 1975)—within the Ross geosynclinal domain and separated the latter into individual belts. Consequently, the Ross complex here, if present, can be expected to have a geanticlinal, or even subplatformal, rather than a geosynclinal appearance.

Rift Zones

Rift zones are very important in the Antarctic structure as indicators of the most dramatic neotectonic movements, responsible for much of the present-day landscape and, possibly, even the configuration of the continent, particularly its western part. Another critical function of rifts is the control of recent violent volcanic activity, which is especially widespread in West Antarctica.

The largest known and best studied rift in East Antarctica coincides with the Lambert Glacier and the Amery Ice Shelf (Fedorov and others, Chapter 118, this volume). Beneath it there is a prominent rise of the Moho discontinuity and a corresponding decrease of the crustal thickness down to 22 km in the axial zone of the rift in contrast to 30–33 km in its "shoulders." This zone is also characterized by a disappearance of the Conrad interface, a slight increase in average crustal seismic velocities, and a decrease of velocities in the upper mantle to 7.7–7.9 km/sec. Other big outlet glaciers in East Antarctica with apparent graben-like appearance of valleys—such as the Denman, Rennick, and Jutulstraumen Glaciers—are likely to be associated with rifts as well, but this is purely geomorphologic speculation, as yet unconfirmed by geologic and/or geophysical data. The bedrock relief gives evidence of a subglacial rift on the Wilkes Coast at long. 130° E.

The Weddell Sea plate is also split by a major rift, as indicated by the Soviet geophysical work carried out from the Druzhnaya Base in 1977. This rift is positively established beneath the northern Filchner Ice Shelf and can be continued into the open sea on the basis of bathymetric information. Its extension southward into the Pensacola Mountains (Dufek Massif?) is very likely, although so far speculative. If this is the case, initiation of the rift may be as old as middle Mesozoic.

Rifts at the Pacific margin of Antarctica are accompanied by vast areas of upper Cenozoic basalts in Marie Byrd Land and on the Bellinsghausen Sea coast; in the Antarctic Peninsula and Victoria Land these volcanic rocks are less extensive. Geologic and geomorphic data suggest the presence of rift zones between the South Shetland Islands and Graham Land (the Bransfield Strait), between Alexander Island and Palmer Land (George VI Sound), immediately south of Thurston Island (the Abbot Ice Shelf), and in some localities in Marie Byrd Land (e.g., on the Bakutis Coast and between Mount Crary and Mount Toney). Available geomorphologic evidence from the Ellsworth Mountains area also seems to indicate the presence of deep narrow grabens bounding the Sentinel Range. Geophysical information on the Pacific rift zones is obtained only for the Bransfield Strait, where the existence of a rift is confirmed by thinning of the crust to 15 km and by its suboceanic nature (absence of the "granitic layer"), as well as by a lower density of the upper mantle.

As yet there are insufficient data to determine the nature of the most prominent geomorphic feature of Antarctica, namely, a great escarpment that brings the uplifted Transantarctic Mountains in juxtaposition with the submerged Ross Sea shelf. A conspicuous gradient in crustal thickness and/or in other properties is almost certain and is most likely associated with a deep fault. The presence of the latter is supported by outpourings of alkaline basalts along the chain of the Transantarctic Mountains as far south as the head of Scott Glacier. However, at present it is impossible to tell whether this deep fault delineates a one-sided horst or whether it delineates an asymmetric rift graben adjacent to the Transantarctic Mountains, with the lower Pacific-ward "shoulder" concealed under the Ross Sea, and a considerable accumulation of fill in the axial part. If such a rift zone (or a series of rifts) exists, it may be responsible for some lateral extension within the Ross Sea basin.

Principal stages of structural evolution

Data on this subject are compiled in Table 98.1 and illustrated in Figure 98.2. The table also shows the correlation between tectonic events (and/or epochs) recognized in the present paper and those in earlier publications.

Discussion

The tectonic evolution of Antarctica as outlined above postulates a sialic nature of the primitive crust and ensialic development of all subsequent structures of the continent. The early Precambrian (Archean) eon of evolution appears in many respects unique in crustal history. In the author's opinion, such a point of view is strongly corroborated by recent studies of the ancient crystalline basement in Antarctica (Ravich, Chapter 51, this volume; Ravich and Kamenev, 1972; Ravich and Grikurov, 1976*a,b*) and on other continents (Kröner, 1977; Vajner, 1976; Spinivasan and others, 1976; Baadsgaard, 1976). There is also a steady increase in evidence for the presence of ancient crystalline rocks in the basement of orogenic fold belts along the Pacific margins of the Gondwana continents of Antarctica (Clarkson and Brook, 1977), South America (Cobbing and others, 1977), and Australia (Cooper and others, 1976).

The suggested model contradicts the classical geosynclinal theory, which postulates a constructive role for the geosynclinal process, i.e., creation of a continental-type sialic crust at the expense of an oceanic-type substratum by lateral accretion onto the ancient sialic nuclei of successively younger geosynclinal orogenic belts. However, at present almost all Antarctic geologists agree that simple accretion cannot account for many structural features of the continent.

The major structural units of Antarctica recognized in this paper are in better agreement with the tectonic subdivisions of American authors than before, mainly because of a revised understanding of the nature and boundaries of the Antarctic platform (craton). Some essential discrepancies still remain in a number of fields, but they are certainly subject to critical discussion from both sides. For instance, Craddock (1970, 1972) shows the Ross orogen to continue as far as the Weddell Sea eastern coast, while in the present work it is believed to terminate in the Shackleton Range area; where Craddock (1970, 1972) distinguishes independent structural units—Ellsworth and Borchgrevink orogens—the author perceives a transition between the Ross and West Antarctic fold systems. Lack of evidence is certainly the principal cause of these disagreements, which mainly arise in the least exposed and/or least studied regions. In such cases the shortage of information compels these general speculations, which in turn reveals different philosophies about fundamental geotectonic concepts.

The ensialic model of tectonic evolution is incompatible with the "mobilist" philosophy, as it implies the primitive nature of a continental-type crust and, consequently, the secondary (superimposed) character of oceanic-type structures. The accretionary model of development does not contradict that philosophy, and probably retains popularity largely for that reason. However, the idea of accretionary tectonics is not an offspring of the continental drift hypotheses: it is much "older" than the drift concept and had been extensively pursued by proponents of vertical tectonics long before the development of "mobilist" ideas; advocates of the latter hypothesis have employed accretion as a "ready" mechanism, which "works" equally successfully both prior to and subsequent to Gondwanaland break-up, i.e., apparently quite independent from the drift itself.

Right there the "mobilist" philosophy seems the most vulnerable. Indeed, if the Andean orogen of Antarctica is *younger* than the Gondwana separation and developed on the *isolated* Antarctic fragment of the supercontinent (see Craddock, 1975), then its similarity to other late Mesozoic and Cenozoic provinces along the Pacific margins of the Gondwana continents implies that identical geologic processes and resulting structures can develop synchronously on different continental blocks. If so, the latter need not be reassembled into a single unit in order to explain the likeness of geologic features. But if a reconstruction is nevertheless attempted, it must not include the structures that simply did not exist in the pre-drift Gondwanaland. Therefore, "matching" of the Andean orogen in Antarctica and South America (see also Ford, 1972*b*) is, to say the least, illogical if these orogens post-date the breakup.

The difficulties in application of the plate tectonics model to Antarctica result from the apparent absence of subduction zones around the continent (Ford, 1972*b*). At the same time, the geosynclinal orogenic and riftogenic processes which have been operating in West Antarctica and the Transantarctic Mountains since late Precambrian time require the classification of this area as an active margin related to the consuming plate boundary. Such a definition is hardly applicable to a terrane measuring many hundreds of kilometers across, and hence the early Paleozoic plate convergence is introduced into the tectonic model (Elliot, 1975*a*, p. 90), presumably as the reason for the welding of Marie Byrd Land (hitherto remote) onto the Antarctic continent; in many other works this effect is assigned to the Andean orogeny. Likewise, the plate tectonics interpretation of the Beardmore-Ross pre-orogenic and orogenic events (Elliot, 1975*a*, p. 57) is not entirely consistent with a distinctive epicontinental character of deposition, which preceded both the Beardmore and Ross orogenies. Such deposition implies the tectonic regime of an extensive, passive (Atlantic-type) margin throughout most of late Precambrian and early Paleozoic time, and it is difficult to visualize the sudden change of this regime into an active Pacific-type (compressional) margin. Apparent absence in the Transantarctic Mountains of suture-type features and/or indications of oceanic crust of about 500–600 m.y. in age provides further restrictions for plate tectonic interpretations, unless subduction is postulated to result from the collision of continental plates.

Conclusions

Inconsistencies in the structural evolution of Antarctica as viewed from the "mobilist" philosophy are at least as numerous as the questions unanswered by the "stabilist" interpretation. Nevertheless, the latter seems advantageous to the author as offering better opportunities for analysis of geotectonic evolution. Twenty years ago, when the geologic similarity of Antarctica to other Gondwana continents was documented by reconnaissance study of exposed areas, it was only natural to accept this information as confirmation of Gondwanaland and the continental drift concept. Since then, however, a wealth of new evidence has accumulated showing that striking similarities in crustal history exist throughout the Southern and Northern Hemispheres, and that the older the rocks being considered, the greater is the resemblance revealed. The magnitude of those similarities and their nature are such that a global uniformity of the crust is required to explain the greater part of its history. Gondwanaland (or Pangea) reconstructions provide, in fact, but quasiuniformity, because they canonize the primary subdivision of the crust into irreconcilable and virtually independent types.

As it happens, evidence long collected and commonly employed to sustain horizontal tectonics has now piled up to a "critical mass" capable of undermining the original idea by offering a broader concept of ensialic crustal history and the destructive (i.e., leading to oceanization) trend of geostructural evolution.

Table 98.1. Principal stages of structural evolution in Antarctica. Heavy vertical lines correspond to orogenic events,

Geologic time-scale		Stages of evolution (tectonic cycles)	Tectonic events				
			East Antarctic craton	Weddell Sea plate	Ross fold system	West Antarctic fold system	Transitional areas
ARCHEAN	Early I	Napier Cycle	Formation of a protocrustal shell from the primitive (meteorite?) compounds of the earth; accretional heating of the near-surface zone of the planet to a high-grade granulite facies level and consequent transformation of the primitive substance into rocks of enderbitic composition				
			4000 ± 200 m.y(?) Napier orogeny?				
	Early II	Rayner Cycle	Thickening of the protocrustal "enderbitic layer" and its charnockitization under increasing pressure and decreasing temperature (to a moderate-grade granulite facies); beginning of differentiation of the crust by metamorphic development of the monomineral rocks (quartzites and marbles); appearance of granitization and migmatization				
			3500 m.y.(?) Rayner orogeny?				
	Late I	Humboldt Cycle	Extensive reworking of older enderbite-charnockitic crust, wide development of granitization and migmatization; formation of high-pressure polymetamorphic complexes under temperature decreased to the granulite/amphibolite facies transition. Irregular growth of the crustal volume and development of highly tectonized zones resulting from active generation of migmatites and shadow granites				
			3000 m.y.(?) Humboldt orogeny?				
	Late II	Insel Cycle	Culmination of the granitization process in a high-grade amphibolite facies environment; formation of vast migmatite and shadow-granite terranes consisting entirely of amphibolite facies rocks derived from completely reworked older crustal blocks and/or the oldest (epi-Humboldt?) supracrustal sequences				
		2650 ± 150 m.y.	**Insel Orogeny**: cratonization of the Panantarctic crystalline basement, or the granitic-metamorphic crust; formation of massifs of gneissic granites in the zones of most intense granitization by selective melting and rheomorphic mobilization; rheomorphic generation of porphyroblastic charnockitoids at deeper levels of the same zones, and/or in adjacent blocks of the pre-Insel crust, preserving a higher temperature environment; intrusive mobilization of anorthosites				
PROTEROZOIC	Middle? to Early	Early Ruker Cycle	Development of intra-cratonic troughs of a miogeosynclinal type with thick volcanic(?) sedimentary fill (southern Prince Charles Mountains); probably igneous (mostly mafic?) activity elsewhere	Formation of the ancient sedimentary platform cover (Ritscher Upland; probably also Coats Land, Shackleton Range area, and Weddell Sea Shelves	No evidence of supracrustal formations—probably continuation of a predominantly metamorphic evolution of the crust, with granitization processes gradually focused in extended linear mobile zones instead of operating regionally; resulting crustal anisotropy is responsible, to a large extent, for structural inhomogeneity of subsequent geosynclinal terranes		
			1700 m.y.(?) **Early Ruker orogeny** (perhaps continuous with an older event at 2000 m.y. recognized by Gunner and Faure, 1972)				
			Deformation and greenschist facies metamorphism in the Ruker fold system (southern Prince Charles Mountains); dolerite dikes elsewhere	Trap intrusions in the ancient platform cover (Börg metamafic rocks)	First impulse of plutonic (zonal) metamorphism and/or granitization in the early Precambrian basement of actively evolving mobile belts and consequent appearance of geanticlinal embryos; terminal episode of regional granitization in relatively stable crustal blocks separating the mobile belts and consequent delineation of future median massifs		
	Late		Mostly erosion and planation(?)	Regression(?) of the sea and subaerial erosion, local red beds (Ritscher Upland), possibly accompanied by volcanic eruptions; local (Shackleton Range) sedimentation in a stable platform environment— mature quartz sands, carbonate beds with stromatolites	Initial submarine volcanic eruptions and sedimentation related to the early stage of subsidence in the proto-Pacific geosnclinal terrane; increase in intensity of subsidence and final differentiation of the Ross geosynclinal troughs by the end of the cycle; possibly volcanic activity alternating with subaerial erosion in positive tectonic forms related to the early Ruker granitization event		
			1000 ± 150 m.y. **Late Ruker orogeny** or Nimrod orogeny of Grindley and McDougall (1969)				
		Late Ruker Cycle	Activization of the crystalline basement; numerous minor trap intrusions (dikes and subvolcanic stocks) locally differentiated to granitoids; isotopic rejuvenation of the Archean rocks; emplacement of metabasites, zonal superposition of the epidote-amphibolite facies assemblages on the initially greenschist facies rocks, and partial erosion in the early Ruker fold system (southern Prince Charles Mountains)	Igneous activity of both intrusive and extrusive facies (Ritscher Upland and Coats Land)	Final emergence of positive tectonic forms in the proto-Pacific geosynclinal terrane Development of geanticlinal zones within the Ross geosynclinal belts at the expense of plutonic-metamorphic reworking and infrastructural elevation of the crystalline basement A Rb-Sr superimposed metamorphic event in the Nimrod Group rocks (Gunner and Faure, 1972)	A K-Ar superimposed (?) metamorphic event in the Haag Nunatak "Basement Complex" rocks (Clarkson and Brook, 1977)	Development of the hypothetical Ross Ice Shelf–Byrd Subglacial Basin median massif between two major mobile areas—the Transantarctic Mtns and the Marie Byrd Land–Antarctic Peninsula Ross geosynclinal belts; felsic volcanic rocks in the Pensacola Mtns early geosynclinal complex (Elliot, 1975a)

and light vertical lines to times of anorogenic development

Geologic time-scale		Stages of evolution (tectonic cycles)	Tectonic events: East Antarctic craton	Weddell Sea plate	Ross fold system	West Antarctic fold system	Transitional areas
PROTEROZOIC	Late	Ross Cycle	Scarce evidence of granitic and charnockitoid intrusions possibly associated with spasmodic orogenic episodes throughout the cycle; carbonate-terrigenous sedimentation in the epi-Ruker miogeosynclinal trough (southern Prince Charles Mountains), in the Sør-Rondane Mountains and the Denman Glacier area	Episodic volcanic activity alternating with sedimentation ?	Major stage of the proto-Pacific geosynclinal development: Thick sediments (slate-graywacke and carbonate) carbonate-terrigenous sequences locally interbedded with volcanic rocks in rapidly subsiding troughs; reduced sedimentary-volcanic successions in geanticlinal zones	Terrigenous sedimentation in Marie Byrd Land and northern Victoria Land. Carbonate-terrigenous sediments and mafic volcanic rocks in the Antarctic Peninsula ?	Carbonate-terrigenous sediments in the Ellsworth Mountains, continuation(?) of rapid graywacke deposition in the Pensacola Mountains; possibly accumulation of the lowermost platform cover formations of the Ross Ice Shelf-Byrd Subglacial Basin median massif
			650 ± 50 m.y. **Early Ross orogeny** or Beardmore orogeny of Grindley and McDougall (1969) and Elliot (1975*a*)				
			Activization of the crystalline basement: isotopic rejuvenation of older metamorphic rocks and intrusive charnockitoids; uplift and intraformational erosion in the epi-Ruker trough recorded in deposition of coarse tillite-like deposits; deformation and low-grade greenschist facies metamorphism in the Denman Glacier area sedimentation basin	No record except dolerite dikes in western Queen Maud Land ?	Reversal of subsidence and major phase of epi-geosynclinal linear folding; vigorous synkinematic mobilization of the pre-Ross basement; development of rheomorphic diorite-granodiorite batholiths and related extrusive rocks; expansion of infrastructural cores in geanticlinal rises due to the amphibolite facies plutonic (zonal) metamorphism of the Ross superstructural geosynclinal pile, which is only slightly metamorphosed (low-grade greenschist facies) away from the plutonic bodies	No record ?	No record except in the Pensacola Mountains and northern Victoria Land where a marked change in rock type, degree of deformation, and metamorphism occurs at this stratigraphic level of the section
PALEOZOIC	Cambrian		Tectonic stabilization; local continuation of sedimentation in residual depositional troughs (southern Prince Charles Mountains)	Continuation(?) of predominant sedimentation in extensive epi-continental sea which periodically flooded surrounding land (Shackleton Range, Ellsworth Mountains, and Pensacola Mountains)	Short period of stabilization succeeded by total epi-geosynclinal orogenic uplift; discontinuous shallow-water carbonate sediments are mixed with molassic coarse clastic and volcanic rocks. Complete regression of the sea by the time of the Cambrian-Ordovician boundary	Continuation(?) of subsidence and further thickening of geosynclinal pile; complete filling of geosynclinal troughs by the end of the cycle	Mainly carbonate, shallow-water terrigenous, and coarse clastic sedimentation, in the Ellsworth Mountains apparently continuous since the preceding cycle, and in northern Victoria Land associated with prominent mafic volcanic activity; volcanic rocks also occur in the Pensacola Mountains
			480 ± 30 m.y. **Late Ross orogeny** or Ross orogeny of Gunn and Warren (1962), Craddock (1972, 1975), Elliot (1975*a*), and others				
	Ordovician		Vigorous tectono-thermal activization; emplacement of numerous intrusions of pegmatoid mica granites all over East Antarctica, and of minor syenite bodies in Queen Maud Land; widespread development of apogabbroic charnockitoids; total orogenic uplift and related isotopic "rejuvenation" of older rocks to 450–500 m.y.; folding and low-grade greenschist facies metamorphism in the epi-Ruker miogeosynclinal complex in the southern Prince Charles Mountains. Onset of predominant erosion	Uplift(?) and subaerial outpourings of basaltic lavas (Kraul Mountains) probably associated with dolerite dikes in the Shackleton Range and Coats Land	Final episode of epi-geosynclinal orogenic uplift accompanied by deformation of the Cambrian sequences into broad open folds and faulted monoclines and by post-kinematic granitic intrusions. Onset of total erosion and planation	Reversal (probably partial) of tectonic regime and major(?) phase of folding in the Ross geosynclinal superstructure, accompanied by a greenschist facies metamorphism; infrastructural rheomorphic mobilization of the pre-geosynclinal amphibolite-facies basement; possible volcanic activity and molassic sedimentation in residual troughs geosynclinally recycled at subsequent stages	No record in the Ellsworth Mountains; deformation and minor granitic intrusion in the Pensacola Mountains; a K-Ar event in western Marie Byrd Land apparently associated with folding and greenschist facies metamorphism in the Ross geosynclinal complex, and possible final deformational episode in the same complex in northern Victoria Land; uplift and beginning of erosion of the hypothetical Ross Ice Shelf-Byrd Subglacial Basin median massif

(continued on following pages)

Table 98.1 (continued)

Geologic time-scale	Stages of evolution (tectonic cycles)	Tectonic events: East Antarctic craton	Weddell Sea plate	Ross fold system	West Antarctic fold system	Transitional areas
PALEOZOIC: Silurian, Devonian	Gondwanian Cycle (Beacon Cycle of Grikurov and Lopatin, 1975)	No record; probably predominant subaerial erosion	No record; probably predominant shallow-water sedimentation interrupted by intraformational erosional and volcanic episodes ?	Planation of the epi-Ross cordillera during the Silurian down to the Kukri Peneplain; subsequent Devonian shallow-marine sedimentation with the source area for the lowermost beds in West Antarctica and in the East Antarctica craton for higher in the section	No record; probable existence of a highly differentiated terrane with numerous geanticlinal rises which focused granitic and felsic volcanic activity and supplied sediments to adjacent depositional troughs, residual and/or geosynclinally recycled	Accumulation of thick terrigenous sequences in the Ellsworth and Pensacola Mountains and adjacent areas of the West Antarctica interior; no record for northern Victoria Land and western Marie Byrd Land; probable denudation of a crystalline terrane in the Ross Ice Shelf–Byrd Subglacial Basin area
		325 ± 25 m.y. **Early Gondwanide orogeny** or Borchgrevink orogeny of Craddock (1972, 1975) and Elliott (1975*a*)				
PALEOZOIC: Carboniferous		Minor intrusions of nepheline syenites in western Queen Maud Land; uplift and local glaciation(?) elsewhere	No record ?	Faulting responsible for block distortion of the Kukri Peneplain and partial erosion of the Devonian basal platform formations	Igneous activity in geosynclines evolving along the Pacific margin; mafic to felsic volcanic rocks in subsiding troughs and metadiorite intrusions in geanticlinal zones and/or other source areas; K-Ar and Rb-Sr events in the South Orkney Islands, Thurston Island, and Marie Byrd Land metamorphic rocks	Orogenic uplift in northern Victoria Land and western Marie Byrd Land accompanied by post-kinematic granites, associated felsic volcanic rocks and inferred refolding of the Ross complex; post-Devonian deformation in the Pensacola Mountains (and Ellsworth Mountains); uplift and subsequent glaciation of the West Antarctica interior (possibly including much of the hypothetical Ross Ice Shelf–Byrd Subglacial Basin median massif)
PALEOZOIC: Permian; MESOZOIC: Triassic		Permian (to Triassic?) shallow-marine and nonmarine sedimentation and coal measures so far established only in the Prince Charles Mountains (local character of the present-day distribution of Permo-(?)Triassic beds is likely to reflect their limited accumulation restricted to isolated source areas, but may also be caused by post-Paleozoic denudation of originally more widespread successions)	Permian sedimentation (shallow-marine strata and coal measures in the Kraul and Theron Mountains, Shackleton Range, and western Queen Maud Land) probably followed by Triassic regression	Major stage of formation of the Beacon platform cover: accumulation of the post-early Gondwanide glacial deposits presumably derived from the West Antarctica interior (Elliot, 1975*b*); shallow-water sediments and floodplain coal measures of Permo-Triassic age, locally interbedded and/or capped by felsic volcaniclastic rocks	Major stage of geosynclinal development along the Pacific margin: rapid subsidence and accumulation of thick Carboniferous slate-graywacke and volcanic sequences in the Antarctic Peninsula; no record of Permian events; indirect evidence of basin compensation in Early-Middle Triassic; late Paleozoic to Early Triassic volcanic activity in eastern Marie Byrd Land	Glacial deposits in the Ellsworth and Pensacola Mountains succeeded by the Permian coal measures, but no direct Triassic record; on the contrary, exclusively Triassic nonmarine sediments in northern Victoria Land; no record in western Marie Byrd Land and the Ross Ice Shelf–Byrd Subglacial Basin area (erosion?)
		190 ± 20 m.y. **Late Gondwanide orogeny** or Ellsworth orogeny of Craddock (1972, 1975); Weddell orogeny of Ford (1972*a*); Gondwanian orogeny of Elliot (1975*a*) and Dalziel and Elliot (1973); early Mesozoic (early Kimmerian) folding of Grikurov (1972, 1973), Grikurov and Lopatin (1975), and Grikurov and others (1972); or early Kimmerian (Weddell) orogeny of Ravich and Grikurov (1976*a*)				
MESOZOIC: Jurassic		Abundant dolerite dikes; probable general uplift	Gabbro intrusion in the Kraul Mountains; basaltic outpourings in western Queen Maud Land; numerous dolerite sills and dikes in the Theron Mountains–Shackleton Range area; probably more widespread trap activity associated with uplift and initiation of rifting northwards from the Pensacola Mountains	Faulting, block movements, and areal trap magmatism of both intrusive and effusive types	Reversal of tectonic regime and major phase of linear folding in the Pacific-margin geosynclines; refolding and isotopic rejuvenation of older metamorphic basement complexes; granitic intrusions in geanticlinal zones	Intense folding in the Ellsworth Mountains and the West Antarctica interior, associated uplift and granitoid intrusions; emplacement of a giant stratiform Dufek Intrusion in the Pensacola Mountains probably accompanied by uplift and refolding of the Paleozoic strata; dolerite intrusions in northern Victoria Land; no record in western Marie Byrd Land and the Ross Ice Shelf–Byrd Subglacial Basin Area (uplift?)

Table 98.1 (continued)

Geologic time-scale		Stages of evolution (tectonic cycles)	Tectonic events				
			East Antarctic craton	Weddell Sea plate	Ross fold system	West Antarctic fold system	Transitional areas
MESOZOIC	Jurassic / Early Cretaceous	Early Andean Cycle	No record; presumably mostly elevation and denudation; local shallow-marine sedimentation is indicated by boulders of Lower(?) Cretaceous fossiliferous sandstone dredged off the East Antarctica coast	No direct record; indirect data suggest a rising sialic terrane with volcanic activity in the western and southwestern Weddell Sea which supplied sediments to the Jurassic basin at the base of the Antarctic Peninsula	No sedimentary record; possibly local continuation of basaltic volcanic activity into Middle-Late Jurassic time	Accumulation of lacustrine and shallow-marine sediments and thick subaerial volcanics on an erosional surface cut into the pre-Mesozoic complexes; local continuation (or regeneration) of subgeosynclinal downwarping responsible for graywacke-shale sedimentation and submarine volcanic activity (at the base of the Antarctic Peninsula and on the Ruppert Coast in Marie Byrd Land?)	No record
	Late Cretaceous	Early Andean Cycle	100 ± 15 m.y. **Early Andean orogeny** or older events of Andean orogeny of Craddock (1972, 1975) and Elliot (1975*a*); late Mesozoic (late Kimmerian) folding of Grikurov (1973), Grikurov and Lopatin (1975), and Grikurov and others (1972); or late Kimmerian (Andean) orogeny of Ravich and Grikurov (1976*a*)				
			Crustal up-arching in the crystalline basement terranes and initiation of rifting with associated alkaline-ultramafic intrusions (Prince Charles Mountains, probably also western Queen Maud Land, Denman Glacier, etc); development of compensatory downwarpings—intracratonic molasse basins	No record, except in the Larsen Ice Shelf area which reversed to subsiding and rapid shallow-marine accumulation of Late Cretaceous molassic deposits supplied from the Antarctic Peninsula cordillera; probably less active sedimentation elsewhere	No record	Vigorous orogenic uplift over the whole of the Pacific margin; associated deformation of the pre-orogenic Mesozoic strata—negligible in molassic sequences (open folding and block faulting and drag folds adjacent to major faults) and linear isoclinal folding and thrusting in the graywacke-shale formations. Culmination in reworking of the ancient crystalline basement: upward mobilization and related isotopic rejuvenation of spacious infrastructural ("Basement Complex") projections and plutonic development of intrusive (synkinematic) gabbro-granite association	No record, except in western Marie Byrd Land (granitic intrusions apparently accompanying orogenic uplift) and western Weddell Sea (see column 4); presumably denudation of still elevated Ellsworth Mountains and adjacent West Antarctica interior, and beginning of submergence of the Ross Ice Shelf–Byrd Subglacial Basin area
CENOZOIC	Paleogene	Late Andean Cycle	Rapid infilling (during latest Cretaceous? time) of molasse basins adjacent to the early Andean uplifts; gradual decrease in intensity of erosional-accumulative processes due to their increasing balance. Beginning (in Paleogene time) of general planation under a relatively warm climate, with sedimentation rate approximately equal to that of submergence of depositional basins; probable development of more active linear graben-like depressions with greater thickness of Paleogene fill (buried rifts?) in the Ross Ice Shelf area. Local minor orogenic and/or igneous activity related to later events of Andean orogeny of Craddock (1972, 1975) and Elliot (1975*a*)—the latest Cretaceous to earliest Paleocene intrusive and extrusive rocks in the Antarctic Peninsula and on Alexander Island, the Eocene uplift accompanied by riftogenic gabbro intrusions and basaltic eruptions in the Bransfield Strait area, contemporaneous molassic sedimentation in northern Larsen Ice Shelf, alkaline basaltoids in the Prince Charles Mountains, etc. Probable beginning of glaciation in isolated, relatively uplifted inland centers, but no catastrophic spreading of the ice sheet until climatic deterioration and neotectonic activization in Miocene.				
	Quaternary and Neogene	Late Andean Cycle	20 ± 5 m.y. (to present) **Late Andean orogeny** or Victoria orogeny(?) of Gunn and Warren (1962)				
			Neotectonic activization in the form of dissection of a peneplained terrain cut into the present-day mountains (revived Mesozoic orogens) and areas of intensified submergence: the renewed post-early Andean (late Mesozoic to early Cenozoic) inland depressions and marginal shelves; the lack of sediment supply caused by glaciation is compensated in the inland depressions by continental ice sheet, but the marginal shelf of Antarctica reveals an apparent deficiency in the late Cenozoic sediment cover and hence is characterized by increased depth and relatively rugged bottom topography. Wide development of rifting and associated volcanicity are responsible for formation of vast basaltic plateaus and a number of stratovolcanoes; enlargement of apparent lateral dimensions of the glaciated Antarctic landmass (especially in West Antarctica) is possibly due to the extensional effect of rift openings, as well as simultaneous oceanization phenomena within the rift zones and/or beneath the deepest crustal depressions.				

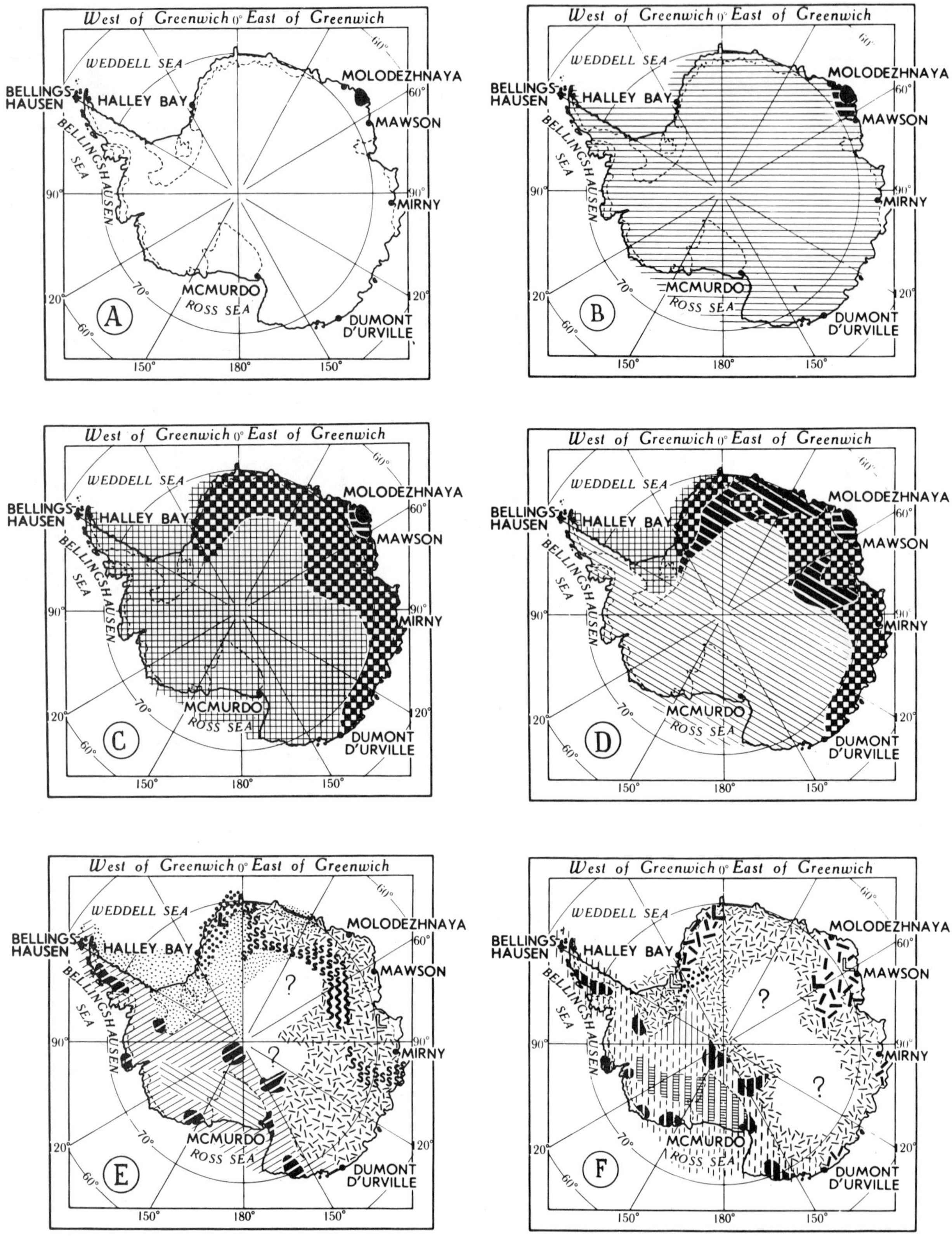

Figure 98.2. Schematic presentation of paleotectonic evolution of Antarctica, showing the structural environment developed by the closure of each tectonic cycle (stage) in the course of terminal orogeny rather than pre-orogenic events (unless specifically indicated otherwise). A through K represent orogenic epochs. A. Napier orogeny, about 4000 m.y. ago; B. Rayner orogeny, 3500(?) m.y.; C. Humboldt orogeny, 3000(?) m.y.; D. Insel orogeny, 2650 ± 150 m.y.; E. Early Ruker orogeny, 2000(?)–1700 m.y.; F. Late Ruker orogeny, 1000 ± 150 m.y.; G. Ross cycle, pre-orogenic (the latest Precambrian); H. Ross orogeny, 650 ± 50 to 480 ± 30 m.y.; I. Gondwanide orogeny, 325 ± 25 to 190 ± 20 m.y.; J. Early Andean orogeny, 100 ± 15 m.y.; K. Late Andean orogeny, 20 ± 5 m.y. ago to present.

Tectonic events and resulting structures (1–14) are shown in explanation as (*a*) established and (*b*) inferred. *Regional metamorphism and resulting crustal complexes of the crystalline basement.* 1. Protocrust of enderbitic composition; 2. Regional charnockitization and resulting granulite facies complexes; 3. Moderate regional

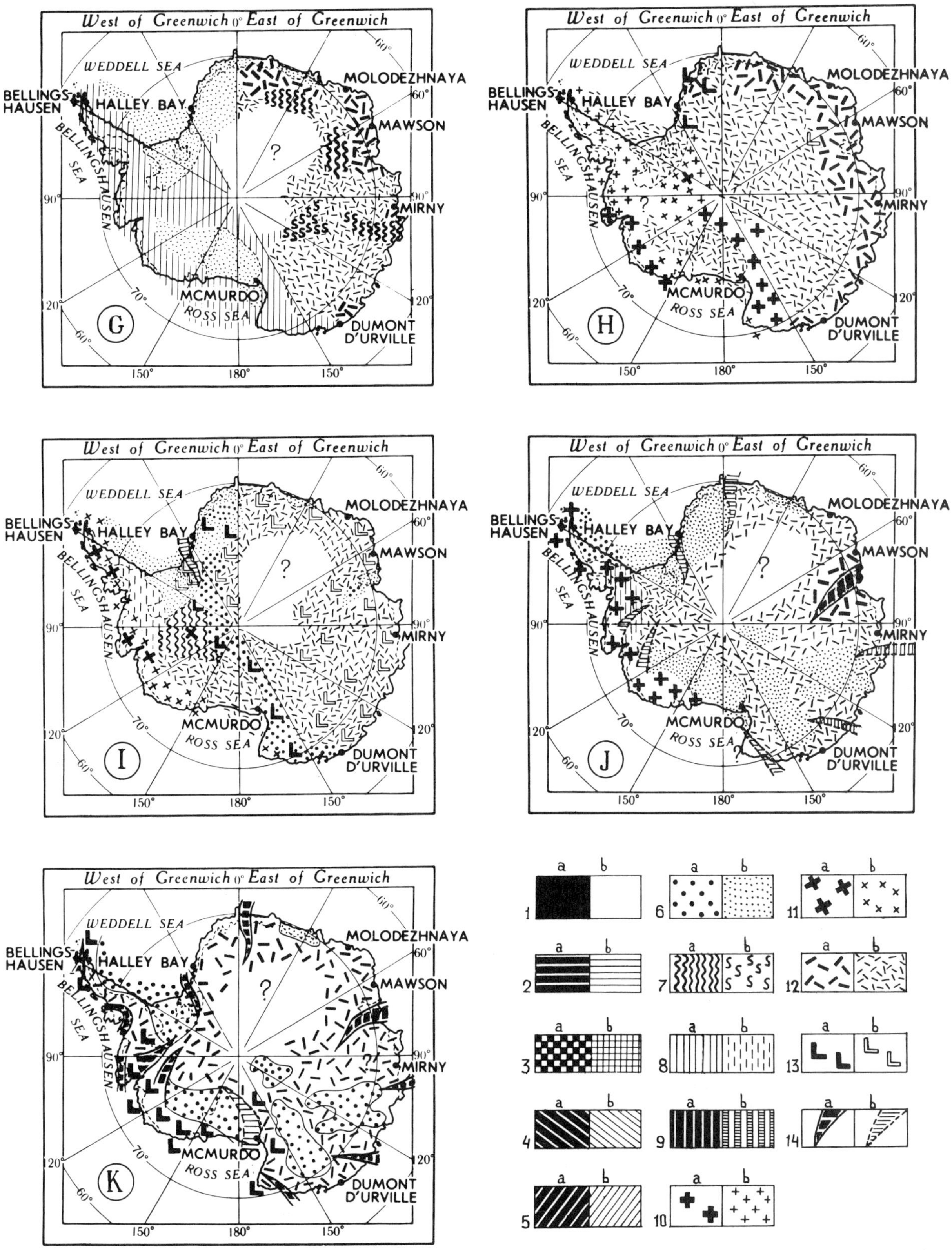

granitization and resulting polymetamorphic complexes; 4. Vigorous regional granitization and migmatization and resulting high-grade amphibolite facies complexes; and 5. Plutono-metamorphic granitization localized in linear mobile belts. *Supracrustal sedimentation and volcanicity and resulting sequences.* 6. Undeformed platform covers; 7. Folded miogeosynclinal (intracratonic) rocks; and 8. Geosynclinal rocks. *Degradation and/or reversal of geosynclinal regime and resulting structures.* 9. Geanticlines and median massifs; 10. Epigeosynclinal and postgeosynclinal orogens with vigorous reworking and synkinematic mobilization of the pregeosynclinal basement, and resulting zonal metamorphic complexes; 11. Same, with moderate to slight basement mobilization and the consequent absence of zonal metamorphism; 12. Elevated areas (such as recurrent orogens, crustal up-archings, and source areas) not preceded by geosynclinal subsidence, and associated with various activization events, commonly very intensive; 13. Mafic igneous activity of areal extent; and 14. Rifting.

References cited

Australia. Bureau of Mineral Resources, Geology and Geophysics. 1975. Preliminary edition, 1:250,000 Geological Series. Mawson Escarpment South (Sheet 40-42/8).

Baadsgaard, H. 1976. Further U-Pb dates on zircon from the early Precambrian rocks of the Godthaabsfjord area, West Greenland. *Earth planet. Sci. Lett., 33,* No. 2, 261–67.

Bentley, C. R., and Clough, J. W. 1972. Antarctic subglacial structure from seismic refraction measurements. (*In* Adie, R. J., *ed. Antarctic Geology and Geophysics.* Oslo, Universitetsforlaget, 683–91.)

Clarkson, P. D., and Brook, M. 1977. Age and position of the Ellsworth Mountains crustal fragment, Antarctica. *Nature, Lond., 265,* No. 5595, 615–16.

Cobbing, E. J., Ozard, J. M., and Snelling, N. J. 1977. Reconnaissance geochronology of the crystalline basement rocks of the Coastal Cordillera of southern Peru. *Geol. Soc. Am. Bull., 88,* No. 2, 241–46.

Cooper, J. A., Fanning, C. M., Flook, M. M., and Oliver, R. L. 1976. Archaean and Proterozoic metamorphic rocks on southern Eyre Peninsula, South Australia. *J. geol. Soc. Aust., 23,* Pt. 3, 287–92.

Craddock, C. 1970. Tectonic map of Antarctica. (*In* Bushnell, V. C., and Craddock, C., *eds.* Geologic Maps of Antarctica. *Antarct. Map Folio Ser.,* Folio 12, Pl. XXI.)

Craddock, C. 1972. Antarctic tectonics. (*In* Adie, R. J., *ed. Antarctic Geology and Geophysics.* Oslo, Universitetsforlaget, 449–55.)

Craddock, C. 1975. Tectonic evolution of the Pacific margin of Gondwanaland. (*In* Campbell, K. S. W., *ed. Gondwana Geology.* Canberra, Australia National University Press, 609–18.)

Dalziel, I. W. D., and Elliot, D. H. 1973. The Scotia arc and Antarctic margin. (*In* Nairn, A. E. M., and Stehli, F. G., *eds. The Ocean Basins and Margins. 1. The South Atlantic.* New York, Plenum Press, 171–246.)

Drewry, D. J. 1972. Subglacial morphology between the Transantarctic Mountains and the South Pole. (*In* Adie, R. J., *ed. Antarctic Geology and Geophysics.* Oslo, Universitetsforlaget, 693–704.)

Drewry, D. J. 1976. Sedimentary basins of the east Antarctic craton from geophysical evidence. *Tectonophysics, 36,* No. 1–3, 301–14.

Elliot, D. H. 1975*a*. Tectonics of Antarctica: a review. *Am. J. Sci.,* 275-A, 45–106.

Elliot, D. H. 1975*b*. Gondwana basins of Antarctica. (*In* Campbell, K. S. W., *ed. Gondwana Geology.* Canberra, Australia National University Press, 493–536.)

Fedorov, B. A. 1973. Podlednyi antarkticheskii rel'ef v raione Stantsii Molodezhnoi [Subglacial Antarctic relief in the Molodezhnaya Station area]. *Trudy. sov. antarkt. Eksped., 56,* 121–24.

Ford, A. B. 1972*a*. Weddell Orogeny—Latest Permian to Early Mesozoic deformation at the Weddell Sea margin of the Transantarctic Mountains. (*In* Adie, R. J., *ed. Antarctic Geology and Geophysics.* Oslo, Universitetsforlaget, 419–25.)

Ford, A. B. 1972*b*. Fit of Gondwana continents—drift reconstruction from the Antarctic continental viewpoint. *24th Int. Geol. Congr., Montreal, 1972, 3,* 113–21.

Grikurov, G. E. 1972. Tectonics of the Antarctandes. (*In* Adie, R. J., *ed. Antarctic Geology and Geophysics.* Oslo, Universitetsforlaget, 163–68.)

Grikurov, G. E. 1973. *Geologiia Antarkticheskogo poluostrova* [Geology of the Antarctic Peninsula]. Moscow, Izdatel'stvo Nauka, 119 pp.

Grikurov, G. E., and Lopatin, B. G. 1975. Structure and evolution of the west Antarctic part of the circum-Pacific mobile belt. (*In* Campbell, K. S. W., *ed. Gondwana Geology.* Canberra, Australia National University Press, 639–50.)

Grikurov, G. E., Ravich, M. G., and Soloviev, D. S. 1972. Tectonics of Antarctica. (*In* Adie, R. J., *ed. Antarctic Geology and Geophysics.* Oslo, Universitetsforlaget, 457–68.)

Grindley, G. W., and McDougall, I. 1969. Age and correlation of the Nimrod Group and other Precambrian rock units in the central Transantarctic Mountains, Antarctica. *N.Z. Jl Geol. Geophys., 12,* No. 2–3, 391–411.

Gunn, B. M., and Warren, G. 1962. Geology of Victoria land between the Mawson and Mulock Glaciers, Antarctica. *Bull. geol. Surv. N.Z., n.s.,* No. 71, 157 pp.

Gunner, J., and Faure, G. 1972. Rb-Sr geochronology of the Nimrod Group, central Transantarctic Mountains. (*In* Adie, R. J., *ed. Antarctic Geology and Geophysics.* Oslo, Universitetsforlaget, 305–11.)

Halpern, M., and Grikurov, G. E. 1975. Rubidium-strontium data from the southern Prince Charles Mountains. *Antarct. Jnl U.S., 10,* No. 1, 9–15.

Korotkevich, E. S., Koblentz, Ya. P., and Kosenko, N. G. 1977. Karta korennogo rel'efa Antarktidy masshtaba 1:10,000,000 [Map of bedrock relief of Antarctica, scale 1:10,000,000]. Moscow, Soyuzmorniiproekt.

Kröner, A. 1977. The Precambrian geotectonic evolution of Africa: plate accretion versus plate destruction. *Precambr. Res., 4,* No. 2, 163–213.

Ravich, M. G., and Grikurov, G. E. *eds.* 1976*a*. *Explanatory Notes to the Geologic Map of Antarctica (scale 1:5,000,000).* Leningrad, Research Institute of Arctic Geology, 83 pp.

Ravich, M. G., and Grikurov, G. E. *eds.* 1976*b*. *Geologic Map of Antarctica, scale 1:5,000,000.* Leningrad, Kartfabrika ob'edineniia Aerogeologiya.

Ravich, M. G., and Kamenev, E. N. 1972. *Kristallicheskii fundament Antarkticheskoi platformy* [Crystalline basement of the Antarctic platform]. Leningrad, Gidrometeoizdat, 658 pp. [English Translation: New York, John Wiley, 1975, 582 pp.]

Robin, G. de Q. 1972. Radio-echo sounding applied to the investigation of the ice thickness and sub-ice relief of Antarctica. (*In* Adie, R. J., *ed. Antarctic Geology and Geophysics.* Oslo, Universitetsforlaget, 675–82.)

Scott Polar Research Institute. 1974. *Antarctica Radio-echo Sounding Map Series A, scale 1:5,000,000.* Sheet 3. Ice sheet surface and sub-ice relief 90° E-180°. Cambridge.

Sobotovich, E. V., Kamenev, E. N., Komaristy, A. A., and Rudnik, V. A. 1974. Drevneishie porody Antarktidy (Zemlia Enderbil) [The oldest rocks of Antarctica (Enderby Land)]. *Izv. Akad. Nauk SSSR,* Ser. geol., No. 11, 30–50.

Spinivasan, R., Sreenivas, B. L., and Krishna, R. B. 1976. Evolution of peninsular gneiss (restraints, reasons and reflections). *J. geol. Soc. India, 17,* No. 1, 94–96.

Stump, E. 1973. Earth evolution in the Transantarctic Mountains and West Antarctica. (*In* Tarling, D. H., and Runcorn, S. K., *eds. Implications of Continental Drift to the Earth Sciences. 2.* New York, Academic Press, 909–24.)

Vajner, V. 1976. Epeirogeny, orogeny and the tectonic development of the Kaapvaal basin, southern Africa. *Geol. Rdsch., 1,* No. 65, 333–60.

99

On the Significance of Oceanization in the Process of Crustal Evolution

G. E. GRIKUROV and E. N. KAMENEV
Department of Antarctic Geology, SEVMORGEO, Moyka 120, Leningrad, 190121, U.S.S.R.

Abstract. The study of Antarctic geology, the East Antarctic ancient crystalline terrane in particular, has led the authors to some global speculations presented in this paper. In Antarctica, as well as on all other continents, the existence of a granitic-gneissic basement underneath all volcanic-sedimentary supracrustal sequences is reliably established. In the authors' opinion, this basement is the result of early Precambrian (mostly Archean) pregeosynclinal reworking of the primitive protocrustal (sialic) shell, which originally had an essentially enderbitic composition but became highly granitized in the course of regional metamorphism. The formation of greenstone belts which accompanied this early stage of crustal evolution may be considered as embryonic oceanization phenomena. The next stage (geosynclinal) commenced in late Precambrian time and was characterized by a more intense oceanization trend manifested by the splitting of the early Precambrian sialic crust and by thick basaltic outpourings within linear mobile belts. However, this trend still failed to overcome the continuing enlargement of continental crust, new parts of which were eventually formed in the late Precambrian mobile zones in the course of termination of their geosynclinal development during Paleozoic and early Mesozoic time. With the sealing of thermally active zones, there remained no path for the energy discharge of the earth's interior other than a global explosion of basaltic magmatism in late Mesozoic time. Accompanied by subsidence of global-size crustal segments, this basaltic flood led to the formation of the oceanic basins and the beginning of the third stage (oceanic) of crustal evolution, which is still in progress. Oceanization is thus, in general, a global process of formation of oceanic-type crustal structures in place of continental-type ones. Its mechanism is enigmatic and can probably be best assessed by the expanding earth hypothesis. This process is geologically quite young; therefore the principle of uniformity must not be extrapolated back into geologic history beyond a limiting point, which is likely to lie within the last 200–500 m.y.

Early stages of crustal evolution

Composition and derivation of the protocrust

The existence of a granitic-gneissic basement underneath all volcanic-sedimentary sequences is now reliably established (Kosygin and Parfenov, 1977; Kröner, 1977; Vajner, 1976). Even the so-called "greenstone belts," long considered as primitive igneous products, are found to be secondary formations underlain by a sialic basement complex of variable composition. The most ancient structural complexes, with a 4000 m.y. old nucleus composed of enderbites, have so far been encountered widely only in Antarctica (Sobotovich and othes, 1974). Similar enderbites are locally recorded in the crystalline basements of other continents as well (Gorlov, 1977), but the charnockitic shield complexes with isotopic ages in excess of 3000 m.y. are more common. These charnockitic complexes, in turn, are extensively replaced by granitic-gneissic rocks which may also be as old as 3000 m.y., although more often are dated at 2500–2600 m.y. or younger (Shurkin and Mitrofanov, 1977).

Recent study of the Antarctic crystalline basement has shown that charnockitic and granite-gneissic complexes represent successive stages of reworking of the initially enderbitic crust. Such reworking was achieved through multiple tectono-metamorphic differentiation, selective melting, metasomatism (granitization in particular), and rheomorphism, i.e., through the processes commonly referred to as regional metamorphism, which lead to spatial redistribution of crustal matter in the course of geologic history. A retrograde trend of those processes was determined from a general decrease of temperature under highly variable pressure (Kamenev, Chapter 52, this volume; Grikurov and others, Chapter 85, this volume; Ravich and Grikurov, 1976).

The protocrustal, enderbitic shell might have developed in the course of slow cooling (probably during a few hundred million years) of primitive surface accumulates that were melted by heating caused by the last major accretion of the earth's mass (Rudnik and Sobotovich, 1973). The textural and chemical characteristics of the oldest enderbites indicate their close affinities with plutonic magmatic rocks of tonalitic composition. Similar characteristics are commonly observed in later charnockitic and granite-gneissic complexes as well; this suggests that some parts of these lower grade complexes could have developed from similar parental melts and not by reworking (charnockitization or granitization) of initially enderbitic rocks. In such areas, the charnockitic and granite-gneissic complexes lack any relics of pre-existing rocks, and hence, can locally be considered as the oldest surviving crustal rocks.

Additional arguments for the primitive nature of these early Precambrian crystalline complexes, in comparison with the younger igneous and sedimentary formations, are provided by the extremely low initial $^{87}Sr/^{86}Sr$ ratios of ancient granite-gneisses (0.7010–0.7044), which indicate a negligible degree of geochemical differentiation. There is also some evidence (Sobotovich and others, 1974) for a close similarity in Pb-isotope compositions between the oldest enderbites and meteorites.

THE OLDEST SUPRACRUSTAL FORMATIONS

The presence of eroded, early Precambrian granitic-metamorphic basement underneath all known successions of unaltered or low-grade metamorphic rocks of volcanic or sedimentary derivation strongly suggests that sedimentation and volcanism became global geologic processes only subsequent to the formation of a sialic crustal layer over a greater part of the planet. Volcanic and sedimentary processes are conventionally called "supracrustal," "superficial," etc.; it seems, however, that current usage of these terms does not recognize the fact that the actual *surface* separating the solid earth from its outer liquid and/or gaseous spheres could hardly have developed prior to the appearance of a consolidated crust which acted as the most effective factor in thermal and density differentiation.

It can further be argued that the considerable amount of free water required for widespread denudation and/or sedimentation could have appeared on the surface of consolidated crust only *after* this crust had cooled sufficiently to allow precipitation. The process of slow cooling of the initial protocrustal shell is best illustrated by the above-mentioned increase with time in the amount of amphibolite-facies granitization products, regressive with respect to the primitive high-grade (two-pyroxene) granulite-facies rocks. The granitization itself must have also been responsible for significant transport and liberation of water. A number of examples are available to show that the oldest supracrustal sequences (platform covers, greenstone belts, aulacogen fills) overlie sialic nuclei with the most ancient age of granitization, while in some younger sedimentary-volcanic basins of geosynclinal affinity, the granitization age of the basement seems to extend into latest early Precambrian time. In addition, the supracrustal formations never directly overlie the primitive crustal complexes of enderbitic-charnockitic composition, and wherever the sialic basement projects through younger supracrustal layered rocks, it has in every case a highly granitized appearance.

The separation of water from granitized crust into a hydrosphere could have contributed to the deep-seated generation of "dry" basaltic melts and their supracrustal emplacement as outpourings of the oldest basaltic magmas. These eruptions and the earliest sedimentation thus represent the embryonic manifestations of oceanization.

Therefore, the development of water-filled basins and accumulation of the oldest supracrustal volcanic-sedimentary sequences depended to a great extent on the time of completion of granitization in the underlying sialic crust. Because the granitization process occurred irregularly in both space and time, the oldest supracrustal sequences in various regions differ considerably in age and, consequently, in structural position. Some of these sequences may well have been involved in the granitization process too, and thus would become indistinguishable from their crystalline basement.

The available evidence seems to indicate that the primitive protocrust of the earth had a "granitic" rather than a "basaltic" composition. If so, many current theoretical concepts concerning the early stages of geologic history must be revised.

Intermediate stages of crustal evolution

Intermediate stages of crustal evolution were characterized by a predominance of geosynclinal processes and, consequently, by intensification of granitization-oceanization competition. Indeed, the geosynclinal process has a number of transitional features. On one hand, it seems to contribute to crustal strengthening by creation of extensive plutonic-metamorphic complexes reminiscent of the early Precambrian granitic-metamorphic terranes. On the other hand, geosynclinal belts delineate zones of splitting of the preexisting continental crust, and geosynclinal accumulation usually has begun with thick basaltic extrusions accompanied by rapid subsidence, similar to the events of the oceanization process. Important features of the geosynclinal process are folding of layered rocks and *progressive* zonal greenschist- to amphibolite-facies metamorphism in the supracrustal sequences or in the superstructure; these events are closely related to the synchronous reworking of the pregeosynclinal basement or of the infrastructure.

If the oldest supracrustal sequences discussed above (the greenstone belts in particular) are now considered in the light of these observations, they can probably be regarded as prototypes of geosynclinal structures. These belts apparently had not achieved full development in accordance with a normal "geosynclinal cycle" and hence degenerated into structures intermediate between complete geosynclinal systems and simple platform covers.

In spite of the obvious intensification of an oceanization trend in crustal evolution through the global development of geosynclinal processes, this trend still failed to overcome the continuing growth of continental crust, new parts of which were formed in the majority of geosynclinal belts, welding together the fragments of the old Precambrian granitic-metamorphic layer (the ancient platforms). Therefore, the geosynclinal process may be viewed as an unsuccessful attempt at oceanization which began by destroying continental crust but led eventually to its transformation and enlargement.

Present stage of crustal evolution

With the continued growth of the sialic crust, originally by regional granitization, and later through sealing of thermally active zones by geosynclinal reworking, there remained no escape for the energy discharge of the earth's interior other than a global explosion of basaltic magmatism and the associated extension and subsidence of oceanic basins. This revolutionary process probably began in early Phanerozoic time in the Pacific area, but culminated in latest Paleozoic and Mesozoic time when all other oceans were formed (Belousov, 1975). Thus, huge volumes of basaltic rocks appeared on both the ocean floors and the continents, and the long conflict between oceanization and the growth of continental crust that dominated the greater part of the earth's geologic history was finally resolved. At present, oceanization represents the leading geologic process, having almost completely replaced the geosynclinal means of evolution. Only in some still-active mobile zones can oceanization phenomena, such as the development of inner and marginal seas, riftogenesis, and basaltic volcanism, be equally well interpreted as geosynclinal features.

The validity of these speculations depends to a large extent on the similarities between oceanic and continental basaltic provinces. No definite conclusions can be reached at the present stage of geologic and geophysical knowledge of both environments, but in the authors' opinion, these crustal regions will, with further exploration, display more common features than dissimilarities. Indeed, just a brief scanning of the available evidence shows that oceanic crust is much more heterogeneous than previously thought and, in fact, is comparable in this respect to continental terrane. Recent data indicate that oceanic tholeiites include alkaline varieties long believed to be typical exclusively of continental volcanic rocks. Even mafic and ultramafic intrusive rocks are now known among oceanic basaltic sequences, which illustrates their similarity to the continental trap complexes; like the latter, the oceanic basalts seem to be interlayered with sedimentary beds. In addition, dredging on the oceanic rises has provided specimens of granite-gneisses, charnockites, greenstones and other metamorphic, sedimentary, and intrusive rocks that yield a Paleozoic or even Precambrian age (Peive, personal communication). Similar discoveries among the continental trap complexes are commonly explained by the presence of older sialic rocks below the trap formations. These complexes on continents commonly have great areal extent (over a million square kilometers) and thickness (3-4 km) and appear quite comparable to some oceanic basaltic terranes. Finally, it is well known that some of the greatest continental trap rock provinces were formed in latest Paleozoic and Mesozoic time, i.e., synchronous with the younger oceanic basins.

Unfortunately, the deep crustal structure of large continental trap provinces is too poorly known to afford reliable correlation with the oceanic areas. However, the above arguments suggest that the formation of trap and plateau-basaltic complexes was the major manifestation of the oceanization process both in the terrestrial and submarine realms.

Conclusions

The history of crustal evolution appears to fall into three principal megastages: (1) pregeosynclinal creation of the global crystalline granitic-metamorphic crust; (2) geosynclinal reworking and transformation of this crust; and (3) development of basaltic oceanic crust. Each of these megastages was characterized by a major geologic process which determined the nature and general trend of development during the corresponding cycles of crustal evolution. Other processes were also active simultaneously, but they were subdued by the dominating events; these same processes, however, were the major ones in other times.

During the first (early Precambrian) megastage, evolution was dominated by formation of the crystalline crust of continental (sialic) type by the processes of regional metamorphism and granitization. The predecessors of future oceanic structures during this megastage were represented by metamorphic differentiates of mafic and ultramafic composition within the enderbitic-charnockitic protocrust; the later greenstone belts may be considered embryonic geosynclines. During the second megastage (Proterozoic to Paleozoic), the prevailing process of geocynclinal reworking of the early Precambrian crust coexisted with both decreasing granitization-metamorphic phenomena and widespread basaltic volcanism which nearly represented the oceanization process; true oceanization was probably realized by the end of this megastage in the Pacific area. This started the third, still continuing, megastage of crustal history—the oceanic revolution which finally succeeded by late Mesozoic time. The process of oceanization has not yet completely replaced the geosynclinal means of development which can still be recognized in some mobile zones.

The authors realize that their arguments and conclusions are highly speculative, but one point seems to emerge quite clearly from this discussion: the principle of uniformity must not be extrapolated back into geologic history beyond a limiting point, which probably lies within the last 200-500 m.y. The processes of geologic evolution are irreversible and changing, and therefore they cannot be considered purely in terms of uniformitarianism.

References cited

Belousov, V. V. 1975. *Osnovy geotektoniki* [Fundamental geotectonics]. Moscow, Izdatel'stvo Nedra, 262 pp.

Gorlov, N. V. 1977. O "tonalitovom" sloe kontinental'noi kory [On the "tonalitic" layer of the continental crust]. (*In* Dagelaysky, V. B., and Sokolov, Yu. M., *eds. Problemy geologii rannego dokembriia* [Problems of early Precambrian geology]. Leningrad, Izdatel'stvo Nauka, 58-64.)

Kosygin, Yu. A., and Parfenov, L. M. 1977. Nekotorye problemy tektoniki dokembriia kontinentov [Some problems of Precambrian tectonics of continents]. (*In* Dagelaysky, V. B., and Sokolov, Yu. M., *eds. Problemy geologii rannego dokembriia* [Problems of early Precambrian geology]. Leningrad, Izdatel'stvo Nauka, 16-29.)

Kröner, A. 1977. The Precambrian geotectonic evolution of Africa: plate accretion versus plate destruction. *Precambr. Res., 4,* No. 2, 163-213.

Ravich, M. G., and Grikurov, G. E., *eds.* 1976. *Explanatory Notes to the Geologic Map of Antarctica (scale 1:5,000,000.)* Leningrad, Research Institute of Arctic Geology, 83 pp.

Rudnik, V. A., and Sobotovich, E. V. 1973. *Ranniaia istoriia zemli* [The early history of the earth]. Leningrad, Izdatel'stvo Nedra, 22 pp.

Shurkin, K. A., and Mitrofanov, F. P. 1977. Osnovnye napravleniia i zadachi issledovanii dokembriiskogo magmatisma [General lines and problems of the study of the Precambrian magmatism]. (*In* Dagelaysky, V. B., and Sokolov, Yu. M., *eds. Problemy geologii rannego dokembriia* [Problems of early Precambrian geology]. Leningrad, Izdatel'stvo Nauka, 72-79.)

Sobotovich, E. V., Kamenev, E. N., Komaristy, A. A., and Rudnik, V. A. 1974. Drevneischie porody Antarktidy (Zemlia Enderbi) [The oldest rocks of Antarctica (Enderby Land)]. *Izv. Akad. Nauk SSSR,* Ser. geol., No. 11, 30-50.

Vajner, V. 1976. Epeirogeny, orogeny and the tectonic development of the Kaapvaal basin, southern Africa. *Geol. Rdsch., 65,* No. 1, 333-60.

100

Structural Style and Tectonic History in Northern Victoria Land

J. D. BRADSHAW
Geology Department, University of Canterbury, Christchurch, New Zealand

M. G. LAIRD
New Zealand Geological Survey, Department of Scientific and Industrial Research, Christchurch, New Zealand

A. WODZICKI
Geology Department, Western Washington University, Bellingham, Washington 98225, U.S.A.

Abstract. In the Bowers Mountains there are three major structural domains, each with a distinctive history. They correspond to the northwest-trending belts of the Wilson Group (west), Bowers Supergroup (central), and Robertson Bay Group (east). The Wilson Group shows polyphase deformation and metamorphism which produced a strong northwest-trending fabric in schist and paragneiss. Later open folding about gently plunging northwest axes was accompanied by retrograde metamorphism and the local development of crenulation schistosity. The structure of the Bowers Supergroup comprises a major syncline flanked to the west by a complex, faulted anticline-syncline couplet. These structures are upright or steeply inclined and have a very gentle southerly plunge. The Robertson Bay Group to the east shows a northwest-trending, steeply inclined, plunging antiform which refolds weakly schistose rocks.

Stratigraphic, structural, and metamorphic interrelationships and available K-Ar dates suggest a minimum of five main events:

1. Early deformation (wF_1 and wF_2) and metamorphism of the Wilson Group in a pre-Vendian orogeny.
2. Possible first folding of the Vendian Roberston Bay Group.
3. Intrusion of the "Granite Harbour"-type granites and regional uplift and tilting of rocks up to and including the Middle–Upper Cambrian Mariner Group. Rejuvenation of source areas of Leap Year Group and widespread K-Ar dates of 447–499 m.y. (Ordovician).
4. Folding of Bowers Supergroup, refolding of Robertson Bay Group. Uplift and cooling with widespread K-Ar dates of 386–417 m.y. (Silurian–Lower Devonian), prior to the emplacement of the Admiralty Intrusives (300–385 m.y., Devonian–Lower Carboniferous).
5. Post-Jurassic folding and faulting of Beacon Supergroup and Ferrar Group.

The history of northern Victoria Land suggests that the concept of distinct Ross and Borchgrevink orogens needs careful reconsideration.

Structural relationships

The structure and structural evolution of the Bowers Mountains area is particularly interesting because it was thought to contain the deformed rocks of two discrete orogens, the lower Paleozoic Ross orogen and the middle Paleozoic Borchgrevin orogen. The structure of the central area between the Evans Névé, the Rennick Glacier, and the Lillie Glacier was reexamined in 1974 in conjunction with studies of stratigraphy, sedimentology, metamorphism, paleontology, and geochronology reported elsewhere in this volume (Laird and others, Chapter 66; Wodzicki and others, Chapter 68; Cooper and others, Chapter 77; Adams and others, Chapter 67). Earlier work is summarized by Dow and Neall (1974) and by Nathan and Skinner (1972).

The Bowers Mountains can be divided into three northwest-trending belts of the Wilson Group (west), the Bowers Supergroup (central), and the Robertson Bay Group (east), each with a distinctive structural history. The boundaries of the central belt lie just west of the Lanterman Fault zone and along the Leap Year Fault (Figure 100.1).

Fold descriptions throughout the paper follow the suggestions of Fleuty (1964) and Ramsay (1967, Chapter 7).

Wilson Group

The Wilson Group consists of foliated orthogneiss, paragneiss, and schist which were formed from basic and ultrabasic igneous rocks, quartzose sandstone, conglomerate, and mudstone. The degree of textural reconstitution varies, and in the coarsest grained rocks original bedding, channel structures, and cross-bedding are preserved. However, in finer grained rocks original layering can only be inferred from compositional variation.

The most striking deformation structure is a schistosity or mineral lamination which strikes 120°–140° and dips at more than 65°. This fabric has two components; one, an early schistosity (wS_1) developed in synmetamorphic isoclinal folds (wF_1) which is tightly refolded to form crenulation, gleittbrett structure, rodding in the hinges, and a mild crumpling of S_1 on the limbs. A second schistosity (wS_2) is variably developed with further growth of micas (Figures 100.2 and 100.3).

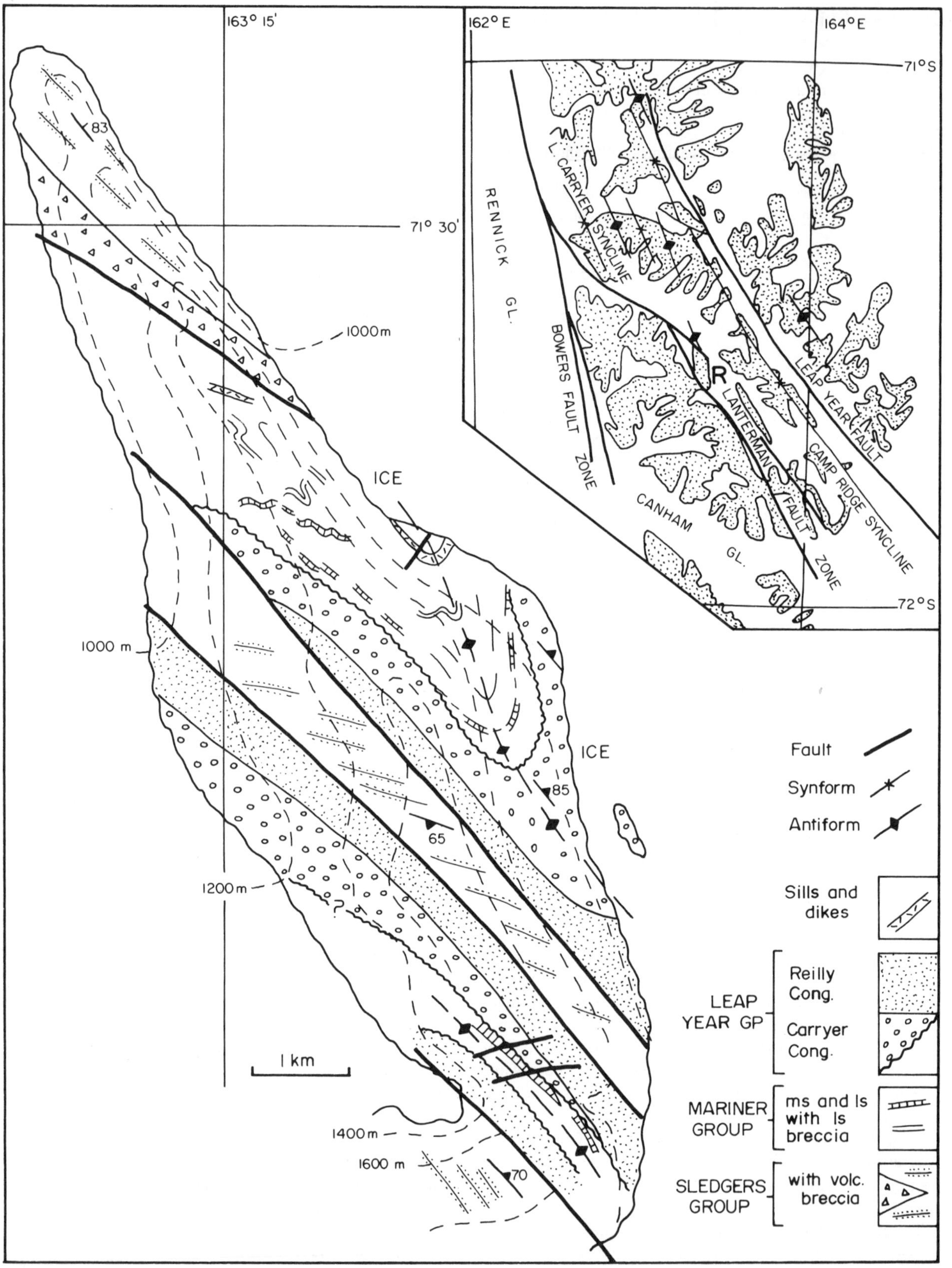

Figure 100.1. Detailed geologic map of Reilly Ridge, Sledgers Glacier, Bowers Mountains. The inset shows the position of Reilly Ridge (R) and the major structural features of the region.

Figure 100.2. Small fold of wF_1 generation, Wilson Group, Husky Pass. The fold deforms original compositional layering.

First fold hinges and S_0-S_1 lineations plunge steeply from 40° northwest to 40° southeast, with the majority between 65° and 90°. The second folds are upright or steeply inclined and axes plunge at moderate angles to the southeast. The primary orientation of wF_1 axes is uncertain but was oblique to F_2. The amount of structural data collected does not permit more detailed analysis.

Particularly towards the eastern edge of the range, a third set of folds is developed. These are open, steplike chevron folds, which trend and gently plunge toward the northwest. A northeast-dipping crenulation schistosity is developed in these folds.

Between Reilly Ridge and Husky Pass a spectacular deformed conglomerate is exposed in which very oblate pebbles, flattened in wS_1 ($=S_0$), can be seen bent around wF_3 hinges and cut by wS_3 crenulation cleavage (Figure 100.4). In the wS_1 plane the oblate spheroids have long axes which pitch 45°–55° southeast on steep limbs and trend north-northeast on the flat limbs. It is possible that this elongation is a relic of extension perpendicular to wF_1. No wF_1 axes were detected in the conglomerates. Sandstone pebbles are more deformed than volcanic (now amphibolite) clasts.

On the western side of the Lanterman Range, open folds occur above the Canham Glacier. On the Rennick Glacier face of the range, the Wilson Group is a gray quartz-feldspar-biotite gneiss net-veined by granitic sheets and cut by discordant plutons of porphyritic "Granite Harbour"-

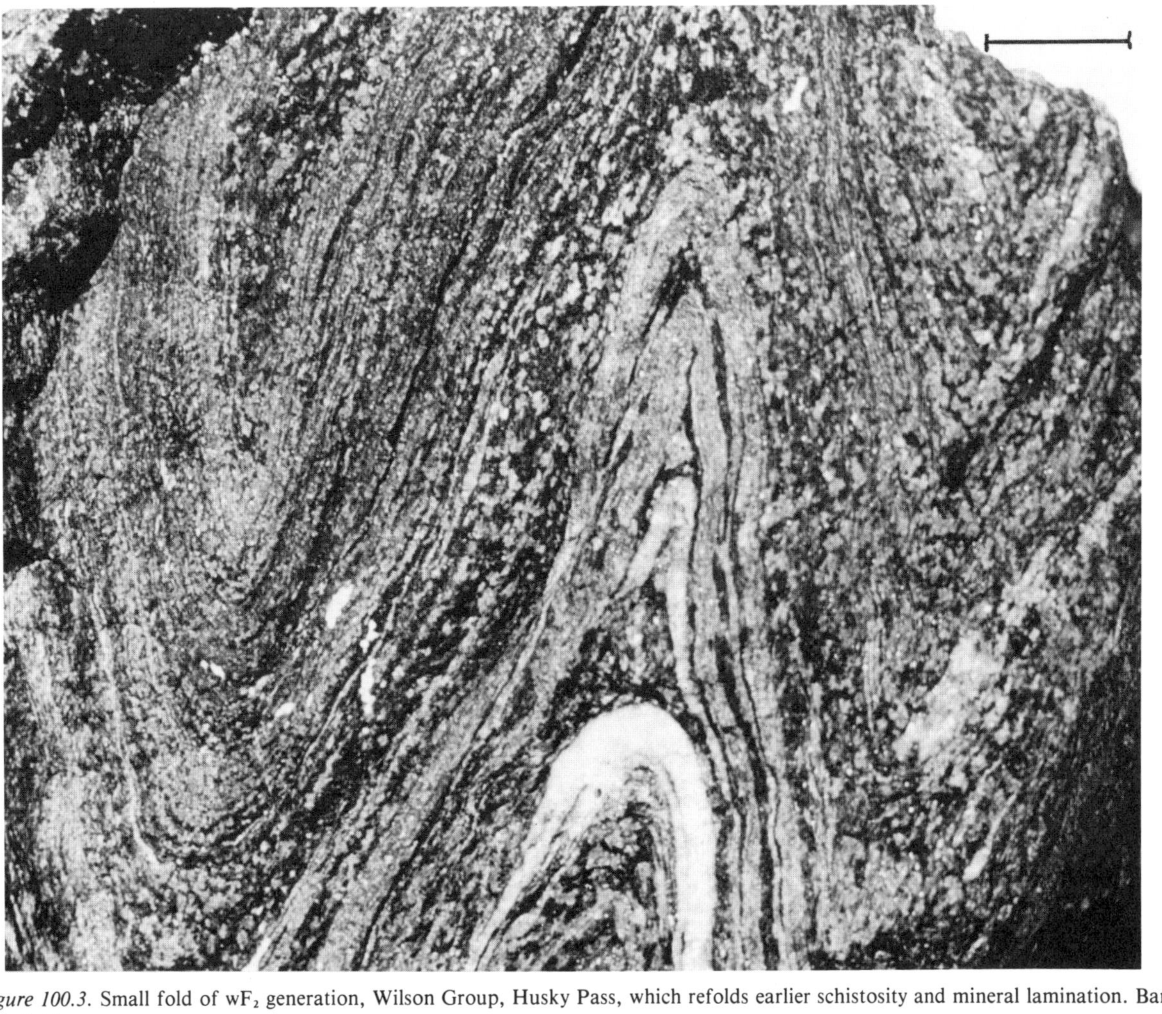

Figure 100.3. Small fold of wF_2 generation, Wilson Group, Husky Pass, which refolds earlier schistosity and mineral lamination. Bar scale is 10 cm.

Figure 100.4. Conglomerate with pebbles strongly flattened in wS_1 refolded by open chevron-style wF_3. A new schistosity parallel to the chevron hinge surface is visible within some pebbles. Wilson Group, Sledgers glacier (at lat. 71° 37′ S, long. 163° 50′ E). Bar scale is 10 cm.

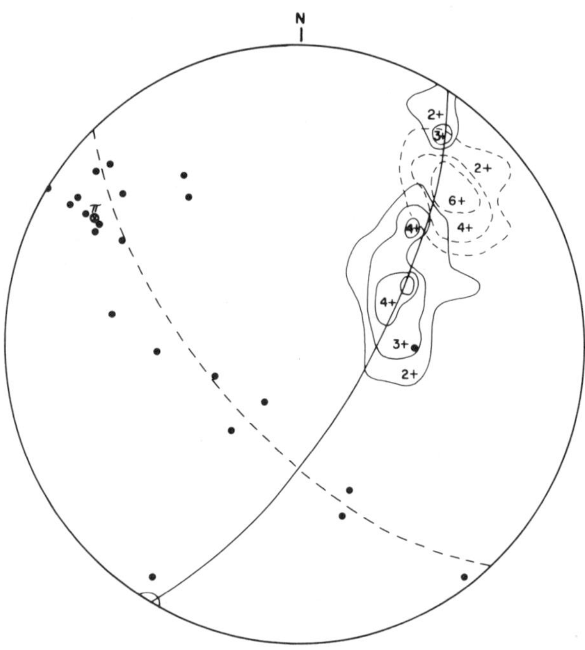

Figure 100.5. Equal-area plot of poles to bedding (solid contours), poles to cleavage (dashed contours), and RrS_0/rS_2 intersection lineation (dots) from the Robertson Bay Group east of the Leap Year Glacier. The dashed great circle is the mean orientation of rS_2. Numbers indicate poles per 1% area.

type granite near the mouth of the Sledgers Glacier. Similar granite is reported cutting the Wilson Group and Robertson Bay Group in the Morozumi Range to the west (Dow and Neall, 1974).

Robertson Bay Group

The Robertson Bay Group rocks east of the Leap Year and upper Graveson Glaciers are gray, low-grade metasedimentary rocks, almost entirely interbedded sandstone and mudstone. Rare graded bedding, low-angle cross-stratification, and convolute lamination can be detected at some localities.

The most widely developed structure is a refracted fracture cleavage/slaty cleavage that dips steeply to the southwest and is related to a northwest-trending and steeply inclined, gently plunging antiform (Figure 100.5). Despite the simplicity of Figure 100.5, the cleavage and antiform are considered second generation structures. At several places on the southwest limb the sedimentary structures indicate that the rocks were inverted before the development of the antiform. At widely scattered localities there is clear evidence of early schistosity and mineral lamination which predate the dominant cleavage. Although most cleavage-bedding intersections plot near the π pole, a significant number are distributed at varied angles within the cleavage, suggesting that local bedding had varying orientation prior to this folding. Finally, the core of the anti-

Figure 100.6. Sedimentary carbonate breccia interbedded with Mariner Group calcareous mudstone. South end of Reilly Ridge, Sledgers Glacier. Ice axe is 85 cm long.

form contains higher grade rocks, suggesting that isograds may be folded. The core may correspond to the "island of medium-grade gneiss" recorded by Crowder (1968) in this area.

The widespread northwest-trending cleavage and the antiform are considered rS_2 and rF_2, respectively. Little can be said of rF_1 except that it was probably coaxial with rF_2. Northwest-trending folds, similar in form to those here termed rF_2, are widespread throughout northern Victoria Land (Gair and others, 1969).

Bowers Supergroup

With the exception of some in the Reilly Ridge area discussed below, all the units in the supergroup are concordant despite indications of significant stratigraphic breaks (Laird and others, Chapter 66, this volume).

All folds are open with gentle plunge. Cleavage is well developed in mudstone and visible in poorly sorted, fine-grained sandstone, but it decreases with improved sorting and increased grain size. The oolites of a limestone in the Mariner Group are distinctly ellipsoidal.

The northeastern side of the belt is dominated by a major syncline, which from Helix Pass (lat. 71° 18′ S, long. 163° 18′ E) southeastward for 175 km has a core of Camp Ridge Quartzite (Leap Year Group, Upper Cambrian–Lower Ordovician). North of Helix Pass in the Edlin Névé area (lat. 71° .07′ S, long. 163° .05′ E) the core is marked by the Older Mariner Group (Middle–Upper Cambrian). The fold is generally upright with an axial surface that dips very steeply to the northeast or southwest. The axis undulates but has an overall very gentle southerly plunge. In the Leap Year Glacier area the eastern limb seems to be faulted against the Robertson Bay Group, but to the north this limb extends into a small anticline.

The main outcrops of the Sledgers Group are exposed to the west in the core of a complex anticline. In the Lower Carryer Glacier, Leap Year Group rocks reappear in a second major syncline with a moderately inclined axial surface dipping to the northeast.

The outlier of Mariner and Leap Year Groups at Reilly Ridge is a much faulted anticline, and it may owe its existence to the oblique convergence of the Lanterman Fault zone and the subsidiary anticlines between the lower Carryer and Leap Year Synclines.

The base of the Leap Year Group is an important unconformity, but except at Reilly Ridge no discordance in dip can be seen. If the Eureka Formation (Upper Cambrian) (Laird and others, Chapter 66, this volume) was uniformly developed, then about 2 km of sedimentary beds must have been removed in the Edlin Névé–Reilly Ridge area before Leap Year sedimentation started.

The geology of the Reilly Ridge area suggests differential movement during and after Mariner Group sedimentation (Figure 100.1). The Mariner rocks at the south end of the ridge include sedimentary carbonate breccias with some blocks more than 2 m in diameter (Figure 100.6). The blocks seem to be the same age as the matrix. The breccias are very similar to those associated with syndepositional faulting in other areas (e.g., Scotland—Bailey and Wier, 1932; Crowell, 1961; and Swiss Alps—Schindler, 1959) and they indicate local strong bathymetric relief. There is no evidence of subaerial exposure. Emplacement was probably by debris flows. These sedimentary rocks also show mesoscopic slump folds, and anomalous cleavage-bedding intersections may relate to contemporaneous slump folding. In general, strikes are more meridional and the southerly plunge is more marked here than elsewhere.

The base of the Leap Year Group on Reilly Ridge is locally strongly discordant, and a minimum of 200 m of Mariner beds has been removed in places. Whether pre–Leap Year Group folding or syndepositional faulting and large-scale slumping was responsible could not be established. Some pre–Leap Year Group erosion is obvious, and greater instability is indicated here than elsewhere in the Cambrian of northern Victoria Land.

Late Mesozoic or Cenozoic structures

Late Mesozoic or Cenozoic (post-Jurassic) faulting is indicated by displacement of Ferrar Dolerite sills near the head of the Johnstone Glacier and by major faults along the Rennick Glacier. Ferrar Dolerites and Beacon sedimentary rocks dip gently westwards in the Canham Glacier area, but small outliers along the west margin of the Lanterman Range are faulted and dip at angles up to 50° southwestwards. Crowder (1968) mentions other areas of deformed Beacon rocks high in the Lanterman Range which suggest that the deformation is more than small-scale bending adjacent to faults.

Tectonic history

The structural relationships outlined above, in conjunction with new stratigraphic and paleontologic data, provide some constraints on theories of the structural evolution of northern Victoria Land.

Two phases of strong deformation and metamorphism in the Wilson Group predate the unconformably overlying Husky Conglomerate and the succeeding Sledgers Group (Vendian). The age of these phases is uncertain: the single Rb-Sr age (Faure and Gair, 1970) of 770 m.y. from Wilson Group rocks is not an adequate guide, and a number of K-Ar ages obtained recently (Adams and others, Chapter 67, this volume) relate to a post-Vendian event, perhaps to the F_3 deformation, or more probably to the intrusion of granites. The relationship to Vendian strata suggests that the Wilson events may be older than the Beardmore orogeny, for which ages of 580–650 m.y. (i.e., equivalent to the Vendian) have been reported (Adams and others, Chapter 101, this volume).

The history of the Bowers Supergroup and the Robertson Bay Group is more problematic, and its unraveling relies in part on parallel studies of sedimentology, paleontology, and isotopic ages.

Only Vendian fossils have been found in typical Robertson Bay Group rocks. Some rocks previously mapped as Robertson Bay Group (Gair and others, 1969; Dow and Neall, 1974) in the Mount McCarthy area differ in composition and fossil content (Cooper and others, Chapter 77, this volume), and may prove to be Sledgers Group emerging on the east limb of the Camp Ridge Syncline. The Vendian–Early Cambrian Sledgers Group is probably younger than the Robertson Bay Group of the Bowers Mountains. However, with a unit that covers a vast area like the Robertson Bay Group, it is unwise to generalize about age.

The Robertson Bay Group shows two deformations, the second similar in style and intensity to that of the Bowers Supergroup. As the rF_2 antiform is adjacent and parallel to the Camp Ridge Syncline, it is tempting to regard them as components of the same fold system (see also Nathan and Skinner, 1972; Craddock, 1972). However, the Bowers Supergroup shows no sign of the rF_1 folding and accompanying metamorphism, and it therefore seems likely that rF_1 is an intra-Vendian deformation which predates the Sledgers Group. This interpretation is consistent with paleocurrent and provenance data which suggest that the Robertson Bay Group was desposited in a major basin east of the Wilson Group basement block, and that the younger northwest-trending volcanic "Bowers Trough" (Laird and others, Chapter 66, this volume) subsided between the Wilson basement block and the Robertson Bay block (already deformed) to the east.

The salient isotopic, paleontologic, and structural data for the early Paleozoic are brought together in Figure 100.7). The two clusters of K-Ar ages suggest two main periods of isotopic closure, probably representing uplift and cooling episodes. Cluster A is from older rocks: Wilson Group, Robertson Bay Group, and the lower part of the Sledgers Group. The indicated closure age, 470–500 m.y. (Early Ordovician) is younger than the fossiliferous Middle to Late Cambrian Mariner Group. However, the Mariner Group and the upper part of the Sledgers Group give the younger 400–420 m.y. (Silurian) dates of cluster B.

The isotopic dates and their stratigraphic distribution are puzzling. The three rock groups (Sledgers, Mariner, and Leap Year) are concordant and show a single major deformation and low-grade metamorphism. The most probable explanation is that the younger dates relate to this deformation in which Mariner rocks were cleaved, and that the older event was a post-Mariner, pre–Leap Year thermal episode probably associated with "Granite Harbour"-type granites at depth, and marked by slight tilting, uplift (2 km), and paleogeographic changes at the surface prior to Leap Year Group deposition. The Mariner rocks are thought to show no sign of this episode because they were a poorly lithified, near-surface unit at that time.

The events of cluster A fall into the period allotted to the Ross orogeny and syntectonic and post-tectonic granites of the Granite Harbour suite. Though the Ross orogeny is widely recognized in Victoria Land and in the Transantarctic Mountains, the Bowers Mountains area was at best peripheral to the Ross orogenic belt and was probably not significantly deformed at that time. The younger mid-Paleozoic Borchgrevink orogeny was the more significant event for Cambro-Ordovician rocks.

In addition, two alternative hypotheses have been considered: 1) that the early event was pre-Mariner Group, and 2) that the early event post-dated Leap Year Group sedimentation. The first alternative hypothesis leads to even greater dissonance between paleontologic and isotopic ages and still fails to provide an accompanying phase of deformation. The second fails either to explain the young ages or to provide sufficient time for erosion followed by deposition of up to 7 km of Leap Year Group

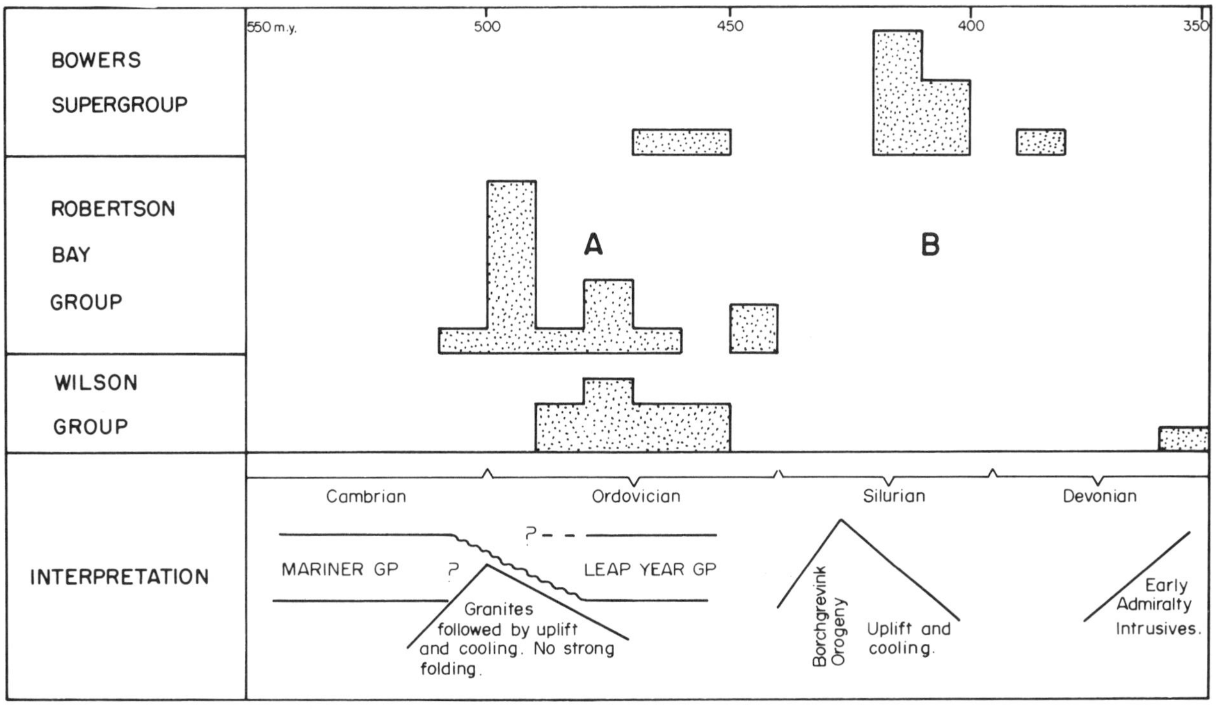

Figure 100.7. Diagram showing the relationship of isotopic ages (K-Ar) and geologic events in the Bowers Mountains. The column width of the stippled areas represents 10 m.y. The minimum column height is one date. The tall column for Robertson Bay Group represents seven dates. The placing of Leap Year Group sedimentation in the Ordovician is interpretive and based on the preferred hypothesis (see text). Isotopic ages from Adams and others (Chapter 67, this volume).

strata (Dow and Neall, 1974), though it does provide a convenient "Ross orogeny." The interpretation advanced in Figure 100.7 based on the preferred hypothesis, agrees well with the geologic data. The alternatives are further discussed in Adams and others (Chapter 67, this volume).

Regional implication

In northern Victoria Land, two orogenies have been recognized (Nathan and Skinner, 1972): the Cambro-Ordovician Ross orogeny, and the mid-Paleozoic Borchgrevink orogeny (Craddock, 1970) seen mainly in the Robertson Bay Group and Bowers Supergroup. Two corresponding orogens have been delineated with a junction along the line of the Rennick Glacier. The Ross orogeny is usually dated by reference to copious granites (Granite Harbour Intrusives) and metamorphic rocks, and the Borchgrevink orogeny is linked to the granitic Admiralty Intrusives. Recently it has become clear that the Robertson Bay Group is much older than previously believed and that its first deformation was at least of Cambro-Ordovician age (i.e., "Ross orogeny") (Adams and others, Chapter 67, this volume), or more probably of late Precambrian age. The Borchgrevink episode is also somewhat older than previously thought, and it is probably the second deformational event in the Robertson Bay terrane and the principal deformation of the Bowers Supergroup (see also Craddock, 1972, p. 452).

In addition, the "Ross orogeny" here is not marked by significant compression but only by a thermal event and modest vertical tectonics. Thus the extension of the Ross orogen north of 74° is open to question.

Depending on the interpretation, the Robertson Bay Group could be included in the Beardmore orogen on the basis of a pre-Vendian or intra-Vendian rF_1, in the Ross orogen because of widespread K-Ar ages similar to those of the Ross orogen, or in the Borchgrevink orogen because of rF_2 and later Admiralty Intrusives. The third course results in a simple accretive pattern of orogens in the Transantarctic Mountains sector of East Antarctica. However, there is risk that a geologist unfamiliar with the Antarctic might find the pattern deceptively simple.

Acknowledgments. The field party was supported by the Antarctic Division of D.S.I.R. as part of the New Zealand Antarctic Research Program. The help of the Scott Base staff and U.S. Navy *Hercules* aircrews is greatly appreciated.

References cited

Bailey, E. B., and Wier, A. J. 1932. Submarine faulting in Kimmeridgian times: East Sutherland. *Trans. R. Soc. Edin., 47,* 431-67.

Craddock, C. 1970. Tectonic map of Antarctica. (*In* Bushnell, V. C., and Craddock, C., *eds.* Geologic Maps of Antarctica. *Antarct. Map Folio Ser.,* Folio 12, Pl. XXI.)

Craddock, C. 1972. Antarctic Tectonics. (*In* Adie, R. J., *ed. Antarctic Geology and Geophysics.* Oslo. Universitetsforlaget, 449-55.)

Crowder, D. F. 1968. Geology of a part of north Victoria Land, Antarctica. *Prof. Pap. U.S. geol. Surv.,* 600-D, D95-107.

Crowell, J. C. 1961. Depositional structures from Jurassic boulder beds, East Sutherland. *Trans. Edinb. geol. Soc., 18,* 202-20.

Dow, J. A. S., and Neall, V. E. 1974. Geology of the Lower Rennick Glacier, northern Victoria Land, Antarctica. *N.Z. Jl Geol. Geophys., 17,* No. 3, 659-714.

Faure, G., and Gair, H. S. 1970. Age determinations of rocks from northern Victoria Land, Antarctica. *N.Z. Jl Geol. Geophys., 13,* No. 4, 1024-26.

Fleuty, M. J. 1964. The description of folds. *Proc. Geol. Ass. Lond., 75,* 461-92.

Gair, H. S., Sturm, A., Carryer, S. J., and Grindley, G. W. 1969. The geology of northern Victoria Land (Sheet 13). (*In* Bushnell, V. C., and Craddock, C., *eds.* Geologic Maps of Antarctica. *Antarct. Map Folio Ser.,* Folio 12, Pl. XII.)

Nathan, S., and Skinner, D. N. B. 1972. Recent advances in the pre-Permian geology of northern Victoria Land. (*In* Adie, R. J., *ed. Antarctic Geology and Geophysics.* Oslo, Universitets-forlaget, 333–37.)

Ramsay, J. G. 1967. *Folding and Fracturing of Rocks.* New York, McGraw-Hill, 568 pp.

Schindler, C. M. 1959. Zur Geologie des Glarnischs. *Beitr. geol. Karte Schweiz, 107,* 136 pp.

101

Orogenic History of the Central Transantarctic Mountains: New K-Ar Age Data on the Precambrian–Lower Paleozoic Basement*

C. J. D. ADAMS and J. E. GABITES
Institute of Nuclear Sciences, Department of Scientific and Industrial Research, Lower Hutt, New Zealand

G. W. GRINDLEY
New Zealand Geological Survey, Department of Scientific and Industrial Research, Lower Hutt, New Zealand

Abstract. Recent K-Ar dating on amphiboles and micas has cast doubt on the previously reported dating of the Nimrod orogeny at either 1000 m.y. (K-Ar on amphiboles) or 2000 m.y. (Rb-Sr whole-rock isochron), but it has provided further evidence for isotopic closures of K-Ar and Rb-Sr radiogenic systems during the Beardmore (latest Precambrian) and Ross (Cambro-Ordovician) orogenies. Published Rb-Sr isochron ages on the Nimrod Group, ranging from 600 to >3750 m.y., and old and new K-Ar hornblende ages, ranging back to 1150 m.y., were obtained from severely tectonized augen gneiss, amphibolite, skarn, and migmatitic rocks from the Aurora, basal Miller, and upper Argosy Formations that formed along and beneath a major synmetamorphic thrust fault—the Endurance Thrust—at depths of 15–20 km. Incorporation of excess radiogenic argon and strontium released during metamorphism into crystal lattices of minerals such as amphibole, plagioclase, and pyroxene, may account for these old ages. From all present age data, the only well-determined events that can be inferred are the following:

1. Archean: U-Pb dates of 2800 m.y. on zircons provide minimum age for cratonic source area and/or maximum age for Nimrod sedimentation;
2. Late Precambrian: Beardmore orogeny at 580–650 m.y. Deformation, overthrusting, and metamorphism of Nimrod Group. Uplift and gravity sliding of Beardmore Group. Minimum age provided by isotopic closure of whole rocks (Rb-Sr) and some amphiboles (K-Ar) in Nimrod Group;
3. Late Cambrian: Ross orogeny at 500–570 m.y. Deformation and metamorphism of Beardmore (Upper Precambrian) and Byrd (Lower–Middle Cambrian) Groups. Isotopic closure of whole rocks in Beardmore Group and of micas and amphiboles in Nimrod Group away from granite contacts;
4. Ordovician: Granite Harbour Intrusives at 450–500 m.y. Intrusion of Hope Granite and related plutons. Isotopic closure of all micas in Nimrod and Beardmore Groups. Isotopic closure of whole rocks in Hope Granite;
5. Silurian–Early Devonian. Uplift of Ross orogenic belt. Isotopic closure of fine-grained micas in some Byrd Group slates from eastern part of orogenic belt.

In the central Transantarctic Mountains, lower Paleozoic and Precambrian sedimentary rocks, metasedimentary rocks and gneisses underlie Beacon Supergroup (Devonian–Triassic) sedimentary rocks over a large region, providing one of the best opportunities to study the chronology of several pre-Devonian phases of tectonic, metamorphic, and igneous activity. In this paper, the basement metamorphic rocks of the Nimrod Group, the lower grade sedimentary rocks of the Beardmore and Byrd Groups, and the Granite Harbour Intrusives are re-investigated by the potassium-argon (K-Ar) dating method. These new age determinations and all previous K-Ar and rubidium-strontium (Rb-Sr) datings are discussed in relation to the orogenic history of the region.

* I.N.S. Contribution number 872.

Rock units

Since the early surveys in the region (Gunn and Walcott, 1962), the geology of the central section of the Transantarctic Mountains between 80° S and 84° S has been extensively studied (Grindley, 1963, 1972; Grindley and others, 1964; Laird, 1963, 1964; Laird and others, 1971; Skinner, 1964, 1965; Gunner, 1969, 1976; Oliver, 1972), providing an excellent basis for geochronologic studies. The lower Precambrian–Paleozoic stratigraphy is summarized below.

Nimrod Group (Precambrian)

Because of its intense polyphase deformation, the Nimrod Group has been considered older than the adjoining lower grade, and more simply deformed, Beardmore Group (Grindley and others, 1964). The original Nimrod strata consisted of shallow-water quartzose sandstone, siltstone, limestone, dolomite, and interbedded mafic volcanic rocks; all are now metamorphosed to the amphibolite facies. Three concordant units are known—the Worsley, Argosy, and Aurora Formations—separated by a major overthrust from a fourth, tectonically higher unit, the Miller Formation. Along this major overthrust (Endurance Thrust), small lenses of gneissic diorite and gabbro (Argo Gneiss) containing eclogitic inclusions have been emplaced during an early phase of intense penetrative deformation (F_1 of Grindley, 1972). Augen gneisses produced during this deformation were metasomatized to K-feldspar-rich migmatites, and the whole sequence was refolded during a second deformational episode (F_2) coaxial with the first. A third folding episode (F_3), with axes plunging southwest down the dip of the gneisses, is also recognized locally (Grindley, 1972). These three deformational episodes, with accompanying metamorphism to the amphibolite facies, define the Nimrod orogeny (Grindley and McDougall, 1969). Metamorphism commonly was in the staurolite-almandine subfacies but locally attained kyanite grade close to the Endurance Thrust.

Beardmore Group (Upper Precambrian)

The coastal ranges south of 81°S (Figure 101.1) are underlain by a very thick sequence of terrigenous sedimentary rocks of a quartzose to arkosic graywacke–green slate facies termed the Goldie Formation (Gunn and Walcott, 1962). Field relations with the Nimrod Group to the west are not exposed. The sedimentary rocks are metamorphosed to the low greenschist facies, with development of slaty cleavage parallel to the axial planes of numerous closely spaced isoclinal folds. Only rare carbonate, chert, and conglomerate beds interrupt the thick, monotonous, flyschlike sequence in which graded bedding and soft-sediment deformation structures indicate deposition in a major geosynclinal trough. No fossils are known. In the Cobham Range, north of the Nimrod Glacier, an older sequence of carbonate rocks, schist, and quartzite (Cobham Formation) underlies the Goldie Formation conformably (Laird and others, 1971).

Byrd Group (Cambrian)

This group, found mainly in the ranges between the Byrd and Nimrod Glaciers, is faulted against the Beardmore Group on the east and overlies it unconformably on the west. The unconformity resulted from the late Precambrian Beardmore orogeny of Grindley and McDougall (1969). The Byrd Group forms synclinal belts unconformable on the Goldie Formation in the Queen Elizabeth Range and crops out as a small inlier beneath Beacon sedimentary rocks at the head of Beardmore Glacier (Young and Ryburn, 1968). Archeocyathid-bearing limestone is the typical and dominant rock type of the Shackleton Limestone (Laird, 1963), but interbedded phyllite and conglomerate with abundant carbonate clasts are prominent locally, especially to the west. A thick quartzite unit underlies the limestone in sections south of Nimrod Glacier (Laird and others, 1971). Coarse-grained clastic formations adjoin the Shackleton Limestone belt in the east and are described variously as the Starshot (Laird, 1963, 1964), Douglas, and Dick Formations (Skinner, 1964, 1965). Arkosic and calcareous flysch contain roundstone conglomerates with abundant carbonate clasts, suggesting that these units are equivalent to or younger than the limestone. Spilitic, trachytic, and rhyolitic volcanic rocks and dikes are also present.

Granite Harbour Intrusives (Upper Cambrian–Lower Ordovician)

Folding of both the Beardmore and Byrd Groups in the Ross orogeny was succeeded by the medium- to high-level emplacement of numerous post-tectonic plutons ranging from biotite-muscovite adamellite to granodiorite, with minor quartz diorite (Oliver, 1972), collectively termed the Hope Granite (Gunn and Walcott, 1962). An older pyroxene-biotite diorite is known from the coast near the Starshot Glacier (Laird, 1964), and a younger granite is known from the mouth of the Beardmore Glacier (Ida Granite of Gunner, 1974). Hope Granite plutons also intruded the Nimrod Group in the Miller Range after considerable post-metamorphic uplift (Grindley, 1972). Isotopic and other geochemical work supports an anatectic origin for the Hope Granite, with lower $^{87}Sr/^{86}Sr$ ratios characterizing the plutons intruded into the Beardmore Group, compared with those intruding the Nimrod Group (Gunner and Mattinson, 1975).

Previous geochronologic studies

K-Ar dating

In the earliest dating work, McDougall and Grindley (1965) obtained K-Ar biotite and muscovite ages for the Hope Granite and one lamprophyre dike in the range 450–478 m.y., similar to the ages obtained for the Granite Harbour Intrusives of North Victoria Land, which concentrate in the range 460–500 m.y. (see compilations by Deutsch and Webb, 1964; Faure and Jones, 1974). A similar range, with two exceptions at 520 and 633 m.y., was obtained for K-Ar biotite and muscovite ages from Nimrod Group gneisses. The general similarity between ages from the granites and from the gneisses suggests that these date the crustal cooling during orogenic uplift following the deformation and regional metamorphism of the Ross orogeny, and also date the emplacement of the Hope Granite. Grindley and McDougall (1969) continued their study by dating hornblendes from Nimrod Group gneisses; they found that two ages, 504 and 528 m.y., were reasonably concordant with earlier biotite ages, but that three others were much older—1006, 1011, and 1043 m.y. While recognizing the possibility that the older ages might be anomalous, resulting from the incorporation of excess argon in the hornblendes, they tentatively inferred that the gneisses were at least 1000 m.y. old and that the ages reflected an early tectonic-metamorphic event (Nimrod orogeny).

Rb-Sr dating

Rb-Sr dating studies of the Nimrod and Beardmore Groups and the Hope Granite by Gunner and Faure (1972) and Gunner and Mattinson (1975) largely confirmed the K-Ar ages of micas from the Hope Granite, but they further complicated the metamorphic history of the Nimrod Group. Rb-Sr whole-rock isochrons of (i) several Hope Granite plutons in the Beardmore-Nimrod Glaciers area, and (ii) the equivalent, Martin Dome pluton in the Miller Range, gave ages of 463 ± 6 m.y. and 488 ± 17 m.y.,

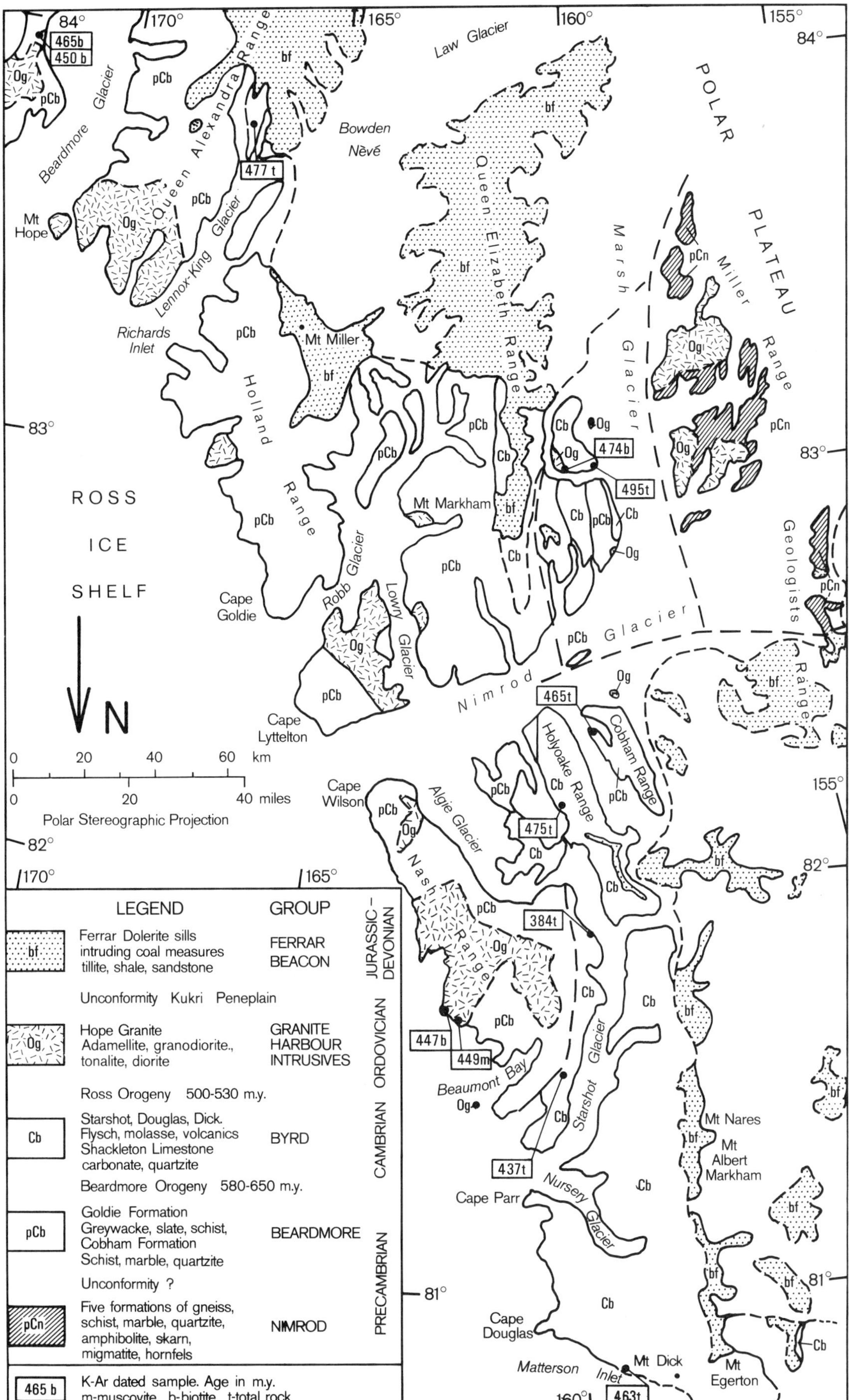

Figure 101.1. Geologic map of central Transantarctic Mountains showing locations of new K-Ar age determinations on Beardmore and Byrd Group phyllites and Hope Granite (after Grindley and Laird, 1969).

respectively (Gunner and Mattinson, 1975; Gunner, 1976), which are similar to maximum K-Ar mineral ages and indicate intrusion during early Ordovician time. A concordant U-Pb date on sphene of 476 m.y. was also obtained from the Martin Dome pluton.

In an attempt to establish the depositional age of the Beardmore Group, Gunner (1976) obtained Rb-Sr whole-rock data from several selected composite sedimentary rock samples which defined an imprecise isochron age of 538 ± 28 m.y. (Cambrian). However, this is younger than its depositional age (Precambrian) and reflects isotopic rehomogenization during the Ross orogeny.

Rb-Sr whole-rock data from the Nimrod Group (Gunner and Faure, 1972) were even more confusing and widely scattered on an isochron diagram, and despite careful assembly of composite sample data, they failed to establish a clear age pattern. However, it was shown that while practically all the data could be accommodated within general maximum (3700 m.y.) and minimum (600 ± 39 m.y.) isochrons, with the use of three exclusion criteria (relating mainly to the absence of a molten history and to metasomatism by younger granites), the remaining data could be resolved into subparallel isochrons giving ages of about 1800–2000 m.y. Gunner and Faure (1972) regarded these ages, somewhat tentatively, as the time of sedimentation, diagenesis, or metamorphism of the paragneisses. Discordant U-Pb zircon dates (Gunner and Mattinson, 1975) of relict detrital zircons in two paragneisses in the Aurora Formation indicate that the ancestry of the gneisses may go back to at least 2800 m.y.

New K-Ar age determinations

The interpretation by Gunner and Faure (1972) of the scattered Rb-Sr whole-rock isochron data from the Nimrod Group gneisses suggested that the main complicating influence was the adjacent Hope Granite plutons. It was thus decided to re-examine the Nimrod Group, using the K-Ar method and dating mainly hornblendes. Nineteen amphibolite facies gneisses from the Miller Range were dated, representing the Miller Formation calc-schists, pelitic schists, and amphibolites (seven), the Aurora Formation paragneisses and migmatites (seven), the Argosy Formation pelitic schists (three), and the Worsley Formation calcareous and pelitic schists (two). The specimen locations and simplified geology are shown in Figures 101.1 and 101.2.

To compare the age patterns of the lower metamorphic grade Beardmore and Byrd Groups, two Goldie Formation phyllites, two phyllites within the Shackleton Limestone, two Starshot Formation phyllites, and one phyllite from the Dick Formation of Skinner (1964) were dated by the whole-rock method.

Eleven specimens of Hope Granite and related adjacent hornfels (Martin Dome and Astro Glacier plutons) were dated as well as a quartz diorite intrusive into the Shackleton Limestone.

All analytical data, age results and errors (one standard deviation), and petrographic and locality information for the Hope Granite, Beardmore/Byrd Groups, and the Nimrod Group are listed in Tables 101.1, 101.2, and 101.3, respectively. Experimental methods used in this study have been fully described by Adams (1975). Our new K-Ar dates for the Nimrod Group and Hope Granite have been combined with those of McDougall and Grindley (1965) and Grindley and McDougall (1969) on cumulative frequency diagrams (Figure 101.3*a, b*). Detailed discussions of each group follow.

Hope Granite

Nine new ages for the Hope Granite and associated xenoliths range from 474 to 495 m.y. forming a closer group than other published data (Figure 101.3*a*). Two hornblende-biotite age pairs are concordant (Table 101.1, specimens 3669 and 3671). This suggests that the Hope Granite had a rapid uplift and cooling history after being intruded into the Nimrod Group about 490 ± 10 m.y. ago in Early Ordovician time (line H, Figure 101.3*a*). The related plutons (Hope Granite?) intruding the Goldie Formation in the coastal ranges could be somewhat younger, at about 450 m.y. (Table 101.1, specimens 6297 and 6301).

Beardmore and Byrd Groups

Fine-grained cleaved slates in the Goldie Formation of the Beardmore Group gave whole-rock ages of 465–477 m.y. (Table 101.2), similar to the age range of the Hope Granite, and thus probably reflect the same period of uplift and cooling after granite intrusion.

Younger ages (384–463 m.y.) for some Starshot Formation slates (Byrd Group) seem to indicate a separate and younger event not recorded in the older rocks, perhaps related to continued uplift of the orogenic belt prior to deposition of the Beacon Supergroup. The rocks concerned are secondarily deformed slates with boudinage structure, containing fine-grained micas that might be susceptible to low-temperature argon loss during late deformation and/or uplift. The youngest age of 384 m.y. (late Early Devonian) provides a maximum age of Middle Devonian for the unconformably overlying basal Beacon strata; this age agrees well with fossil evidence from the McMurdo Sound region (Barrett and others, 1972, Table 101.1).

Nimrod Group

Twenty-six biotite and hornblende ages from nineteen Nimrod Group specimens (Table 101.3) fall in the range of 499–1153 m.y. with most biotite ages in a close group (502–530 m.y.) but with hornblende ages widely scattered (499–1153 m.y.) (Figure 101.3*b*). In all these data, a single biotite age of 633 m.y. (GA770, McDougall and Grindley, 1965) appears anomalous, since biotite from an adjacent, similar specimen is dated at 515 ± 4 m.y. (Table 101.1, 3673bi) and coexisting hornblende at 521 ± 4 m.y. (Table 101.1, 3673hb). As is usual, most of the biotite ages are younger than the hornblende ages (see Figure 101.3*b*) and group in the range of 470–500 m.y., probably reflecting the thermal overprint of the Hope Granite (line H in Figure 101.3). Some biotite and hornblende pairs in the range of 500–560 m.y. are concordant and seem to date a period of rapid postmetamorphic uplift. The concentration of concordant hornblende-biotite ages in the range of 500–560 m.y., with a marked limit at 570 m.y. (line M, Figure 101.3), reflect this uplift and cooling of the complex in the Ross orogeny.

Most of the remaining hornblende ages in Figure 101.3*b* range, with no grouping, from about 560 to 1153 m.y., and

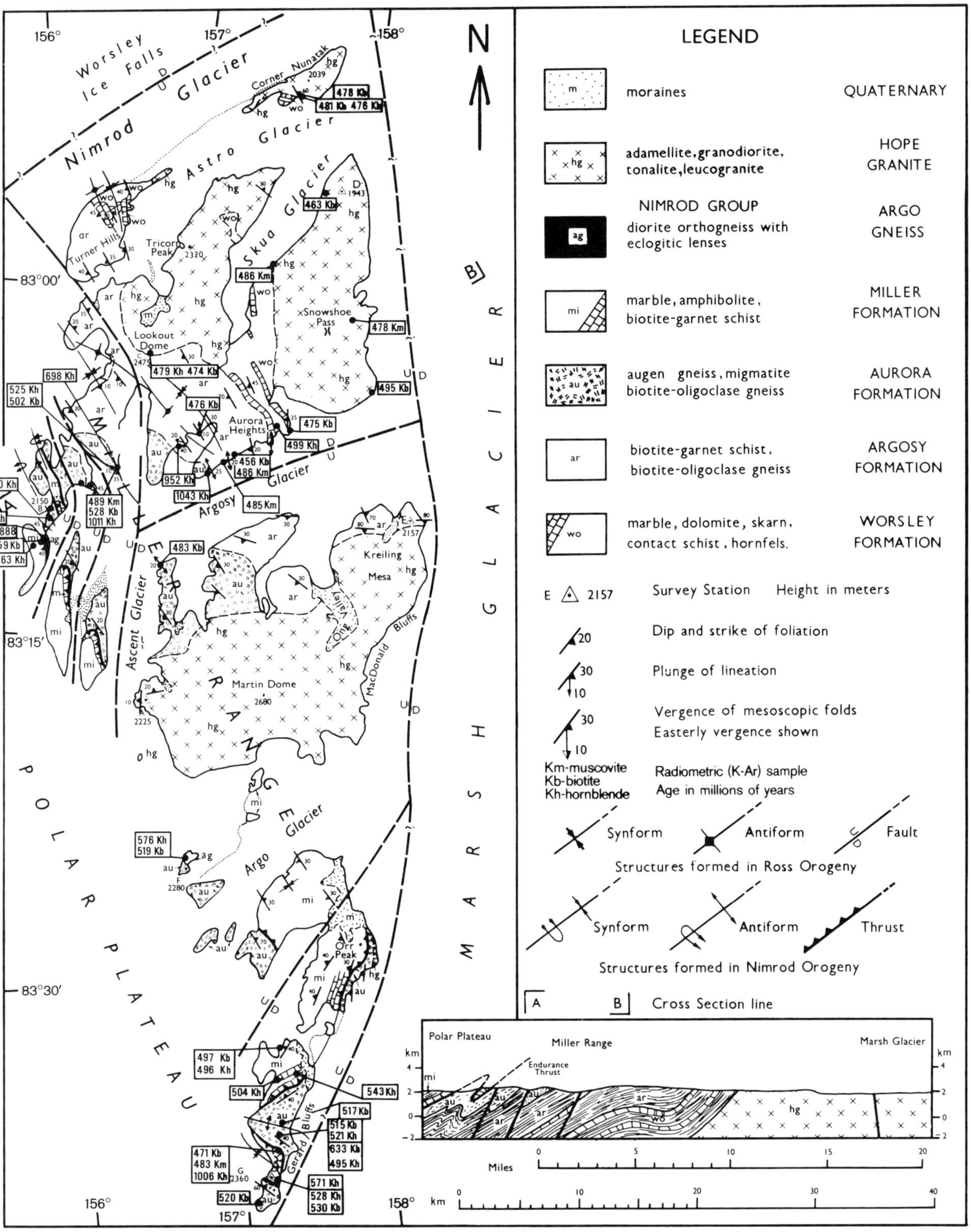

Figure 101.2. Geologic map of Miller Range, showing locations of K-Ar age determinations on Nimrod Group metasedimentary rocks and amphibolites and Hope Granite plutons (modified from Grindley, 1972, Figure 101.1).

a clear interpretation is not possible. The suggestion of an older metamorphic event at about 1000–1100 m.y. (Grindley and McDougall, 1969) can be neither confirmed nor rejected, but there is a strong possibility that the older ages are anomalous and arise from "initial" or "excess" argon introduced into these hornblendes during crystallization or later metamorphism. The following points are important to such an interpretation:

1. the biotite and hornblende ages are widely and variably discordant;
2. the old hornblende ages all come from low-potassium skarns and amphibolites within potassium-rich granitic migmatites of the Aurora and Argosy Formations, or from close to the margins of the formations;
3. the amphiboles from such rocks have variable potassium contents which are inversely proportional to their

Table 101.1. Potassium-argon age determinations on the Hope Granite and related intrusive rocks, southern Victoria Land, Antarctica

I.N.S. "R" number	N.Z.G.S. "P" number	Rock type	Lat. S	Long. E	K (wt %)	^{40}Ar (radiogenic) nl/g	^{40}Ar (radiogenic) % total	Age (m.y.)
Hope Granite (hornfels and contact schist included)								
3674bi	26769	Biotite-quartz-diorite; xenolith in Hope Granite, Martins Dome pluton	83° 18′	156°28′	7.418	164.9	96.4	489 ± 4
3675bi	26771	Biotite-oligoclase granite, Martins Dome pluton	83° 18′	156° 28′	6.502	143.1	96.6	485 ± 4
3665bi	26842A	Biotite-muscovite-oligoclase granite; northern pluton close to contact	83° 05′	157° 53′	6.298	141.2 142.1	92.8 98.2	494 ± 4 495 ± 4 } 495[a]
3664mu	26831D	Quartz-muscovite-biotite-almandine-staurolite schist close to granite	83° 01′	157° 15′	5.973	131.7	93.4	486 ± 4
3669hb 3669bi	26807A	Quartz-oligoclase-biotite-hornblende hornfels, contact with northern granite pluton	83° 03′	156° 27′	1.247 5.642	27.1 121.1	91.7 95.5	479 ± 4 474 ± 3
3671hb 3671bi	26827A	Quartz-diorite; at granite contact with skarn marbles	82° 53′	157° 24′	1.278 6.498	27.9 140.3	95.6 93.5	481 ± 3 476 ± 3
3663bi	26826	Biotite-hornblende granodiorite; sill at contact of northern pluton with marbles	82° 53′	157° 24′	6.223	132.4 134.5	92.7 93.4	470 ± 3 477 ± 4 } 474[a]
Related intrusives								
6295bi	31554	Biotite-hornblende tonalite intruding Shackleton Limestone (Byrd Group)	83° 01′	160° 00′	6.198	133.0	97.9	474 ± 3
6297mu	27065	Tourmaline-muscovite pegmatite intruding Goldie Formation (Beardmore Group)	81° 37′	161° 40′	7.109	143.5	97.8	449 ± 3
6301bi	27037	Biotite-hornblende adamellite intruding Goldie Formation (Beardmore Group)	81° 42′	162° 24′	5.399	108.5	97.5	447 ± 3
6298bi	26854A	Biotite-muscovite adamellite, granitic apophyse intruding Aurora gneiss	83° 35′	157° 13′	6.751	160.0	98.6	517 ± 4

hb: hornblende. bi: biotite. mu: muscovite.
Errors are one standard deviation.
Decay constants ^{40}K: $\lambda_\beta = 0.472 \times 10^{-9}/yr$; $\lambda_e = 0.584 \times 10^{-10}/yr$; and $^{40}K/K = 0.0119$ atomic %.
[a] Mean ages.

Table 101.2. Potassium-argon age determinations on the Beardmore and Byrd Groups, southern Victoria Land

I.N.S. "R" number	N.Z.G.S. "P" number	Rock type	Lat. S	Long. E	K (wt %)	^{40}Ar (radiogenic) nl/g	^{40}Ar (radiogenic) % total	Age (m.y.)
Goldie Formation (Beardmore Group)								
3996TR	26865	Green-gray slate, greenschist facies	83° 50′	167° 00′	2.683	58.04	90.3	477 ± 3
6208TR	31600 (E′4)	Green-gray slate, greenschist facies	82° 23′	159° 43′	4.194	88.11	96.6	465 ± 3
Shackleton Limestone (Byrd Group)								
6186TR	N4	Green-gray slate, greenschist facies interbedded with limestone and quartzite	83° 02′	159° 30′	3.660	82.54	94.6	495 ± 4
6304TR	27121	Siliceous siltstone interbedded with Cambrian limestone	82° 13′	160° 24′	1.288	27.70	97.9	475 ± 3
Starshot Formation (Byrd Group)								
6209TR	J4	Blue-gray slate, micaceous sandstone	81° 53′	159° 40′	4.016	68.03	96.6	384 ± 3
6303TR	27099	Black lineated argillite interbedded with sandstone and conglomerate	81° 30′	162° 32′	3.134	61.40	96.4	437 ± 3
6291TR	27536	Green-gray siliceous siltstone (Dick Formation)	80° 51′	159° 32′	1.118	23.35	98.0	463 ± 3

TR: total (whole) rock.
Errors are one standard deviation.
Decay constants ^{40}K: $\lambda_\beta = 0.472 \times 10^{-9}/yr$; $\lambda_e = 0.584 \times 10^{-10}/yr$; and $^{40}K/K = 0.0119$ atomic %.

Table 101.3. Potassium-argon age determinations on the Nimrod Group, southern Victoria Land, Antarctica

I.N.S. "R" number	N.Z.G.S. "P" number	Rock type	Lat. S	Long. E	K (wt %)	^{40}Ar (radiogenic) nl/g	^{40}Ar (radiogenic) % total	Age (m.y.)	
Miller Formation (Nimrod Group)									
3660hb	21763	Biotite-hornblende schist; metavolcanic band in pelitic schists and marbles	83° 38′	157° 09′	0.352	8.55	79.9	528 ± 4	
3660bi					5.431	132.4	93.6	530 ± 4	530[a]
						132.1	98.3	529 ± 4	
3679hb	26850	Hornblende schist; metavolcanic band in pelitic schists and marbles	83° 38′	157° 09′	0.710	18.81	93.1	570 ± 4	571[a]
						18.84	94.4	571 ± 4	
3678hb	26797	Diopside-hornblende lineated schist, metadolomite or marble in pelitic schist	83° 12′	155° 52′	0.582	15.27	93.3	563 ± 4	
3677hb	26794	Diopside-scapolite skarn pegmatite with coarse minor hornblende; intruding marble	83° 12′	155° 52′	0.619	28.38	94.1	899 ± 6	888[a]
						27.52	93.1	877 ± 6	
6308hb	26795	Garnet-biotite-muscovite-hornblende schist 30 m above Endurance Thrust	83° 12′	155° 52′	1.839	46.86	98.1	551 ± 4	
6308bi					6.149	159.4	97.8	559 ± 4	
6311hb	26859	Diopside-hornblende-biotite schist in marble sequence	83° 34′	157° 17′	0.971	24.32	92.7	543 ± 4	
6312hb	26861	Quartzo-feldspathic biotite-hornblende-garnet schist 200 m above Endurance Thrust	83° 33′	157° 18′	1.470	33.2	96.5	496 ± 4	
6312bi					6.902	156.2	97.4	497 ± 4	
Aurora Formation (Nimrod Group)									
3673hb	26856	Biotite-hornblende pegmatitic augen-gneiss	83° 36′	157° 15′	1.402	33.49	96.0	521 ± 4	
3673bi					6.737	159.5	98.5	517 ± 4	515[a]
						158.2	97.3	513 ± 4	
3672hb	26844	Biotite-hornblende coarse quartz-diorite gneiss; pretectonic intrusive rock (Argo Gneiss)	83° 25′	156° 40′	0.424	11.38	92.0	576 ± 4	
3672bi					6.897	164.6	98.5	520 ± 4	519
						163.9	94.2	518 ± 4	
3661bi	26774	Quartz-oligoclase-biotite-muscovite gneiss	83° 14′	156° 32′	7.759	170.1	94.1	483 ± 4	
3666hb	26788	Amphibolite facies biotite-hornblende schist; metavolcanic horizon in migmatites	83° 10′	155° 54′	0.494	20.83	92.9	840 ± 6	
3666hb′					0.695	25.84	92.2	758 ± 5	755[a]
						25.54	94.1	751 ± 5	
3676hb	26790	Diopside-hornblende skarn pegmatite; reaction zone at marble in migmatites	83° 11′	155° 52′	0.068	4.30	74.1	1152 ± 8	1153[a]
						4.31	72.3	1154 ± 8	
3662hb	26801B	Biotite-hornblende granitic gneiss, pegmatitic veins in paragneiss sequence	83° 08′	156° 00′	1.341	32.32	93.7	525 ± 4	
3662bi					7.540	173.3	95.4	503 ± 4	502[a]
						173.5	98.5	504 ± 4	
						171.1	98.0	498 ± 4	
6299hb	26855	Quartz-oligoclase-biotite-garnet-hornblende schist	83° 36′	157° 12′	1.646	37.13	98.0	485 ± 3	
Argosy Formation (Nimrod Group)									
3668hb	26800	Biotite-hornblende schist band (1 m) overlying fuchsite-quartzite	83° 08′	156° 00′	0.712	23.86	93.7	696 ± 5	698[a]
						24.03	97.6	700 ± 5	
3667hb	26776A	Quartz-muscovite-biotite-hornblende-garnet schist in pelitic sequence	83° 07′	156° 45′	0.365	18.01	89.8	952 ± 7	
6309mu	26809	Quartz-muscovite-garnet-epidote schist; pelitic schist with quartzite beds	83° 09′	156° 56′	6.734	148.3	98.0	485 ± 3	
Worsley Formation (Nimrod Group)									
3670hb	26813A	Quartz-muscovite-biotite-hornblende schist in pelite and marble sequence	83° 07′	157° 15′	0.279	6.35	84.7	499 ± 4	
6310bi	26817	Quartz-oligoclase-biotite-clinozoisite schist	83° 07′	157° 20′	7.294	157.0	98.4	475 ± 3	

hb: hornblende. hb′: amphibole. bi: biotite. mu: muscovite.
Errors are one standard deviation.
Decay constants ^{40}K: $\lambda_\beta = 0.472 \times 10^{-9}$/yr; $\lambda_e = 0.584 \times 10^{-10}$/yr; and $^{40}K/K = 0.0119$ atomic %.
[a] Mean ages.

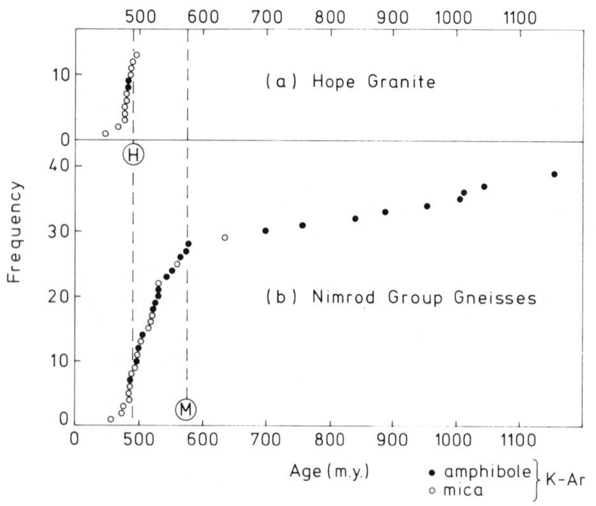

Figure 101.3. Cumulative frequency diagrams of K-Ar mineral age determinations from, a) Hope Granite, and b) Nimrod Group; includes published ages by McDougall and Grindley (1965) and Grindley and McDougall (1969).

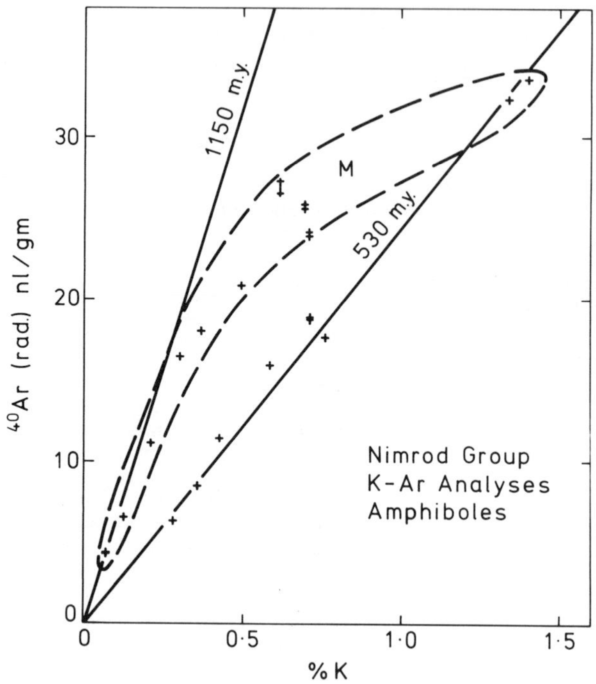

Figure 101.4. K-Ar isochron diagram for hornblende analyses from Nimrod Group, including published data by Grindley and McDougall (1969). Reference isochrons shown for 1150 and 530 m.y.

apparent age. This is shown on an isochron diagram, Figure 101.4, where amphiboles from the migmatic sequence (enclosed in envelope M, Figure 101.4) vary from 0.07 to 1.40% K with ages ranging widely from about 1150 to about 530 m.y., as indicated by reference isochrons of this age range;

4. those hornblendes from the Miller and Worsley Formation (outside the envelope M, Figure 101.4) show the more usual pattern of consistent ages (530–576 m.y.) despite variable potassium content (0.25–0.75% K).

The presence of excess argon in amphibolites has been previously argued for metamorphic rocks of this type, especially where low-potassium amphiboles (as in metavolcanics) are closely related to potassic mica-schists or K-feldspathic gneisses at medium to high grades of regional metamorphism (Pearson and others, 1966; Wilson, 1972). In similar geologic situations, particularly where high-grade gneisses are close to major thrust planes, excess argon may also be present in biotites as well as in hornblende, pyroxene, and plagioclase (Brewer, 1969; Lambert, 1970; Wanless and others, 1970; Wilson, 1972). Thus the association of anomalously old hornblende ages in the Aurora and Argosy migmatites in locations near the Endurance Thrust, which separates the Miller and Aurora Formations, could be another example of such a situation. Similarly, the oldest (633 m.y.) biotite (GA707) mentioned above might also contain excess argon. If the true age of the hornblendes is 530–580 m.y., then the amount of excess argon indicated could be in the range 2–8 nl/gm ^{40}Ar.

The older hornblende ages can be correlated with the anomalous Rb-Sr data from the Nimrod Group gneisses (Gunner and Faure, 1972) obtained on rock specimens chiefly from the basal Miller, Aurora, and upper Argosy migmatitic sequences. One specimen (518) mentioned only briefly by Gunner and Faure (1972, Figure 101.3, p. 309) yielded an apparent age of about 20,000 m.y., several times older than the age of the earth, clearly demonstrating that the radiogenic strontium in this rock could not have originated by *in situ* decay of rubidium (Figure 101.5). Thus, the rock has behaved as an open system with respect to rubidium and radiogenic strontium; once this situation is recognized, it must be assumed to be at least possible for the other Rb-Sr samples. Hence, the Rb-Sr data can, at best, only date the final closure of the isotopic systems. Since the Aurora gneisses commonly show metasomatic introduction of K-feldspar in porphyroblastic augen, in veins, and in migmatitic layering, the migration of alkalis, particularly K and Rb, is well established; associated with these alkalis would be ^{87}Sr and ^{40}Ar liberated from the older source of the alkalis during metasomatism. High-grade regional metamorphic conditions might thus produce a correlation of high initial $^{87}Sr/^{86}Sr$ ratios in granitic gneisses with excess argon in associated hornblendes and pyroxenes in amphibolites and skarns. Within such a metasomatic gneissic sequence, only high-potassium minerals would minimize the excess or initial argon problem, and even these minerals would probably only date the time of metasomatism. Distant from the migmatite sequence of the Aurora Formation, in the Miller and Worsley Formations, biotite and hornblende approach concordance, and two hornblende ages of 571 and 565 m.y. provide a minimum age for the Nimrod Group metamorphism. This is supported by the younger limiting isochron age of 600 ± 38 m.y. from the Rb-Sr whole-rock data (Gunner and Faure, 1972), which probably reflects the time at which these rocks became closed isotopic systems with respect to Rb and Sr.

Gunner and Faure (1972), by eliminating specimens within 12 km of the Hope Granite and those migmatitic and granitic rocks which had been molten or strongly metasomatized, produced a best-fit isochron for the remaining specimens of 1980 ± 77 m.y., which they considered the most likely age for the metasedimentary part of the Nimrod Group (Figure 101.5). They were unable to determine, however, whether the date referred to sedimentation, diagenesis, or metamorphism. If this isochron does

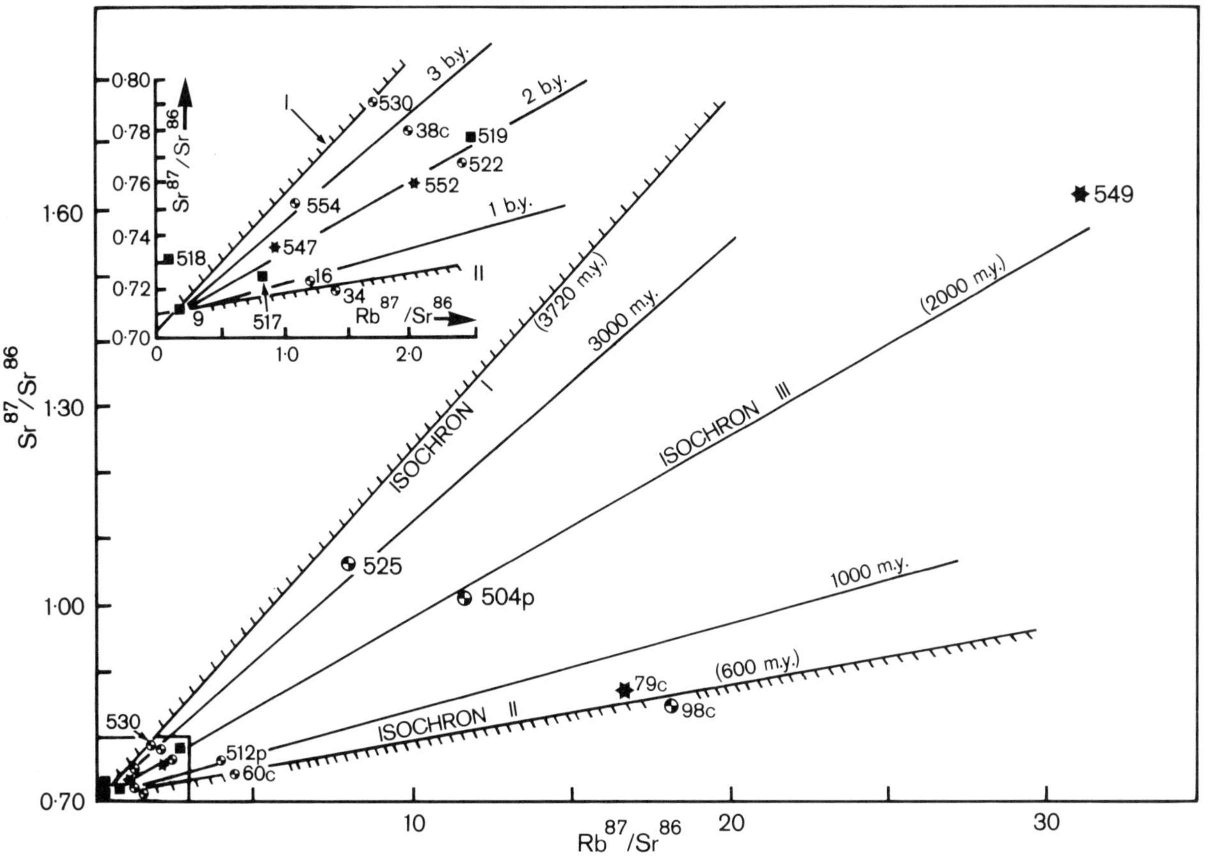

Figure 101.5. Limiting Rb-Sr isochrons for whole-rock determinations from the Nimrod Group. Isochron I (3700 m.y.) refers to the apparently oldest migmatite while Isochron II refers to contact schists in the aureole of the Hope Granite. Isochron III (2000 m.y.) is a best-fit isochron for selected paragneisses and may indicate the age of sedimentation (modified from Gunner and Faure, 1972, Figure 101.3). Circles—Aurora Formation; squares—Miller Formation; stars—Argosy Formation; p—pegmatites; c—contact schists.

have geologic significance, it might apply to sedimentation, diagenesis, or early low-grade burial metamorphism, but it cannot date the high-grade metamorphism and migmatization associated with the overthrusting and penetrative three-phase deformation of the Nimrod orogeny (Grindley, 1972). It may be significant that the specimens used in determining the 1980 m.y. isochron (Isochron III of Figure 101.5) are all from paragneisses, schists, and micaceous quartzites of the Miller and Argosy Formations and lack signs of metasomatism. New U-Pb dates of 2800 m.y. for detrital zircons in paragneisses (Gunner and Mattinson, 1975) indicate a late Archean age for the source area. Therefore, an early Proterozoic age for Nimrod Group sedimentation remains a possibility.

Conclusions

The known elements of the pre-Devonian geochronologic history of the central Transantarctic Mountains can now be summarized:

1. U-Pb data indicate a minimum age of 2800 m.y. for part of the source area of the Nimrod Group sedimentary rocks and also indicate a maximum age for sedimentation. This cratonic source area presumably now underlies the East Antarctic ice sheet to the west.
2. Rb-Sr data on selected nonmigmatized paragneisses and other metasedimentary rocks indicate a possible early Proterozoic age (1800–2100 m.y.) for Nimrod Group sedimentation (Gunner and Faure, 1972).
3. The K-Ar and Rb-Sr data from Nimrod Group migmatites, skarns, and amphibolites relate to the late Precambrian Beardmore orogeny (about 600–630 m.y. ago) when migration of K (+Rb) + Na and SiO_2 along a major overthrust zone occurred, leading to excess radiogenic argon in amphiboles, pyroxenes, and some biotites and to excess radiogenic strontium in migmatites, giving spurious old ages ranging back to 1150 m.y. (hornblendes) and >3700 m.y. (migmatites).
4. Except for anomalous hornblende ages, excluded because of the possibility of excess argon, the remaining Nimrod Group hornblendes and one biotite provide a minimum age of 580 m.y. for cooling and uplift following the Beardmore orogeny.
5. Concordant biotite-hornblende ages for the Nimrod Group as great as 560 m.y. date cooling and uplift of the gneisses during the Ross orogeny.
6. Much of the Rb-Sr isotopic evidence indicates open-system behavior with respect to Rb and Sr about 600 m.y. ago, and this is consistent with the K-Ar age anomalies. Both the K-Ar and Rb-Sr data appear to indicate a late Precambrian metamorphism and migmatization associated with the Beardmore orogeny. Little direct evidence now exists for a separate Nimrod Orogeny (either 1 b.y. or 2 b.y. ago) in the Transantarctic Mountains.

7. K-Ar whole-rock ages from the Beardmore Group provide a minimum age of 495 m.y. for the low-grade regional metamorphism of the Ross orogeny, and the Rb-Sr whole-rock isochron age of 538 m.y. is almost certainly associated with this same event.
8. The Hope Granite was intruded about 490 ± 10 m.y. ago (Upper Cambrian–Lower Ordovician), but K-Ar mineral ages show that uplift and cooling continued until 450 m.y. ago.
9. The Starshot Formation and other Byrd Group coarse-grained clastic rocks may have been deposited during the Ross orogeny prior to the intrusion of the Hope Granite. They were folded 490–500 m.y. ago. Uplift followed during Ordovician–Silurian time, and some rocks remained open to argon loss until Early Devonian time.

Acknowledgments. N. Orr and D. S. Nicholson are thanked for their very careful mineral separation work. W. A. Watters and P. Blattner are thanked for their comments on the manuscript. M. G. Laird and D. N. B. Skinner collected the rock specimens from the Byrd-Nimrod region. K. Baty typed the manuscript and R. Phillips supervised the drafting at D.S.I.R. Cartographic Section.

References cited

Adams, C. J. D. 1975. New Zealand potassium-argon age list-2. *N.Z. Jl Geol. Geophys., 18,* No. 3, 443–67.

Barrett, P. J., Grindley, G. W., and Webb, P. N. 1972. The Beacon Supergroup of East Antarctica. (*In* Adie, R. J., *ed. Antarctic Geology and Geophysics.* Oslo, Universitetsforlaget, 319–32.)

Brewer, M. S. 1969. Excess radiogenic argon in metamorphic micas from the eastern Alps, Austria. *Earth planet. Sci. Lett., 6,* 321–31.

Deutsch, S., and Webb, P. N. 1964. Sr/Rb dating on basement rocks from Victoria Land—evidence for a 1000 million year old event. (*In* Adie, R. J., *ed. Antarctic Geology.* Amsterdam, North-Holland Publishing Company, 557–62.)

Faure, G., and Jones, L. M. 1974. Isotopic composition of strontium and geological history of the basement rocks of Wright Valley, southern Victoria Land, Antarctica. *N.Z. Jl Geol. Geophys., 17,* No. 3, 611–27.

Grindley, G. W. 1963. The geology of the Queen Alexandra Range, Beardmore Glacier, Ross Dependency, Antarctica, with notes on the correlation of Gondwana sequences. *N.Z. Jl Geol. Geophys., 6,* No. 3, 307–47.

Grindley, G. W. 1972. Polyphase deformation of the Precambrian Nimrod Group, central Transantarctic Mountains. (*In* Adie, R. J., *ed. Antarctic Geology and Geophysics.* Oslo, Universitetsforlaget, 313–18.)

Grindley, G. W., and Laird, M. G. 1969. Geology of the Shackleton Coast (Sheet 15). (*In* Bushnell, V. C., and Craddock, C., *eds.* Geologic maps of Antarctica. *Antarct. Map Folio Ser.,* Folio 12, Pl. XIV.)

Grindley, G. W., and McDougall, I. 1969. Age and correlation of the Nimrod Group and other Precambrian rock units in the central Transantarctic Mountains, Antarctica. *N.Z. Jl Geol. Geophys., 12,* No. 2–3, 391–411.

Grindley, G. W., McGregor, V. R., and Walcott, R. I. 1964. Outline of the geology of the Nimrod-Beardmore-Axel Heiberg Glaciers region, Ross Dependency, Antarctica. (*In* Adie, J. R., *ed. Antarctic Geology.* Amsterdam, North-Holland Publishing Company, 206–19.)

Gunn, B. M., and Walcott, R. I. 1962. The geology of the Mount Markham region, Ross Dependency, Antarctica. *N.Z. Jl Geol. Geophys., 5,* No. 3, 407–26.

Gunner, J. D. 1969. Petrography of metamorphic rocks from the Miller Range, Antarctica. *Rep. Inst. polar Stud. Ohio State Univ.,* No. 32, 44 pp.

Gunner, J. 1974. Investigations of lower Paleozoic granites in the Beardmore Glacier region. *Antarct. Jnl U.S., 9,* No. 3, 76–81.

Gunner, J. 1976. Isotopic and geochemical studies of the pre-Devonian basement complex, Beardmore Glacier region, Antarctica. *Rep. Inst. polar Stud. Ohio State Univ., No. 41,* 126 pp.

Gunner, J., and Faure, G. 1972. Rb-Sr geochronology of the Nimrod Group, central Transantarctic Mountains. (*In* Adie, R. J., *ed. Antarctic Geology and Geophysics.* Oslo, Universitetsforlaget, 305–11.)

Gunner, J., and Mattinson, J. M. 1975. Rb-Sr and U-Pb isotopic ages of granites in the central Transantarctic Mountains. *Geol. Mag., 112,* No. 1, 25–31.

Laird, M. G. 1963. Geomorphology and stratigraphy of the Nimrod Glacier–Beaumount Bay region, southern Victoria Land, Antarctica. *N.Z. Jl Geol. Geophys., 6,* No. 3, 465–84.

Laird, M. G. 1964. Petrography of rocks from the Nimrod Glacier–Starshot Glacier region, Ross Dependency. (*In* Adie, R. J., *ed. Antarctic Geology.* Amsterdam, North-Holland Publishing Company, 463–72.)

Laird, M. G., Mansergh, G. D., and Chappell, J. M. A. 1971. Geology of the central Nimrod Glacier area, Antarctica. *N.Z. Jl Geol. Geophys., 14,* No. 3, 427–68.

Lambert, R. St.J. 1970. A potassium-argon study of the margin of the Tauernfenster at Dollach, Austria. *Eclog. geol. Helv., 63,* No. 1, 197–205.

McDougall, I., and Grindley, G. W. 1965. Potassium-argon dates on micas from the Nimrod-Beardmore-Axel Heiberg region, Ross Dependency, Antarctica. *N.Z. Jl Geol. Geophys., 8,* No. 2, 304–13.

Oliver, R. L. 1972. The geology of an area near the mouth of the Beardmore Glacier, Ross Dependency, Antarctica. (*In* Adie, R. J., *ed. Antarctic Geology and Geophysics.* Oslo, Universitetsforlaget, 379–85.)

Pearson, R. C., Hedge, C. E., Thomas, H. H., and Stern, T. W. 1966. Geochronology of the St. Kevin granite and neighboring Precambrian rocks, northern Sawatch Range, Colorado. *Geol. Soc. Am. Bull., 77,* 1109–20.

Skinner, D. N. B. 1964. A summary of the geology of the region between Byrd and Starshot Glaciers, south Victoria Land, Antarctica. (*In* Adie, R. J., *ed. Antarctic Geology.* Amsterdam, North-Holland Publishing Company, 284–92.)

Skinner, D. N. B. 1965. Petrographic criteria of the rock units between the Byrd and Starshot Glaciers, south Victoria Land, Antarctica. *N.Z. Jl Geol. Geophys., 8,* No. 2, 292–303.

Wanless, R. K., Stevens, R. D., and Loveridge, W. D. 1970. Anomalous parent-daughter isotopic relationships in rocks adjacent to the Grenville Front near Chibougaman, Quebec. *Eclog. geol. Helv., 63,* No. 1, 345–64.

Wilson, M. R. 1972. Excess radiogenic argon in metamorphic amphiboles and biotites from the Sulitjelma region, central Norwegian Caledonides. *Earth planet. Sci. Lett., 14,* 403–12.

Young, D. J., and Ryburn, R. J. 1968. The geology of Buckley and Darwin Nunataks, Beardmore Glacier, Ross Dependency, Antarctica. *N.Z. Jl Geol. Geophys., 11,* No. 4, 922–39.

102

Post-Beacon Tectonics in the Region of Amundsen and Scott Glaciers, Queen Maud Range, Transantarctic Mountains

H. R. KATZ
New Zealand Geological Survey, Department of Scientific and Industrial Research, Lower Hutt, New Zealand

Abstract. A structural map of the base of the Beacon Supergroup has been compiled from field and air-photo studies. It covers some 20,000 km^2 in the southernmost part of the Transantarctic Mountains between about long. 140° and 160° W. From rock type, the small thickness variations in the basal Beacon sedimentary beds, and observation of the contact with the underlying basement (well exposed over great distances), it is concluded that there was virtually no relief when Beacon sedimentation commenced. The present relief on the underlying peneplain, which is exposed through a vertical range of 1500 m (from 2200 to 3700 m above sea level), is therefore the result of post-Beacon (i.e., post-Permian in this area) tectonic deformation. Relative vertical displacements of several thousands of meters are inferred for the mountain range as a whole, in relation to the adjacent Ross Sea and continental interior depressions.

The structure of the mountain belt consists of a mosaic of differentially uplifted fault blocks which are locally tilted and warped. Faults are mainly normal but apparently include horizontal components; they are predominantly parallel to the trend of the mountains, but cross faulting is widespread. True folds appear to be absent, nor is there evidence for reverse or thrust faults. It is concluded that the tectonic process was principally one of vertical movements with principal horizontal stress more or less along the axis of the mountain belt. Evidence for crustal extension across the range, such as longitudinal graben structures, is supported by the occurrence of late Cenozoic alkali basalt volcanism on both sides of the mountains.

From direct evidence in the Transantarctic Mountains, the age of their uplift is post-Jurassic, but regional considerations and indirect evidence from Ross Sea sediments suggest that this deformation took place mainly during Tertiary time. Initial movements may have begun as early as Late Cretaceous time, while the final stages of uplift may have occurred as recently as Plio-Pleistocene time.

Excellent outcrops of the basement peneplain underlying the Beacon Supergroup sedimentary beds (Permian) are widespread in the southernmost part of the Transantarctic Mountains, between long. 140° and 160° W. Around the Nilsen Plateau and along the Watson Escarpment, exposures are virtually continuous for many tens of kilometers and provide clear evidence for the structure of the Beacon Supergroup. From both field observation (Katz and Waterhouse, 1970*a*) and air-photo interpretation, these outcrops have been mapped and plotted onto the U.S. Geological Survey (1962–) topographic maps (Antarctica 1:250,000 Reconnaissance Series, sheets "D'Angelo Bluff," "Mount Blackburn," "Nilsen Plateau," "Mount Goodale," and "Leverett Glacier").

Detailed observations along the eroded basement surface suggest that it represents a very flat and originally near-horizonal peneplain. The Beacon sedimentary beds overlying this peneplain are a rather uniform sequence, with little change in thickness and rock type throughout the region. They consist of basal tillites overlain by a *Glossopteris*-bearing sandstone-shale sequence with carbonaceous intercalations and coal beds, which locally include dolerite sills (Minshew, 1967; Katz and Waterhouse, 1970*a*). Apart from scattered Quaternary glacial and volcanic deposits, the Beacon Supergroup represents the youngest rock unit in the region; consequently, its total original thickness is not known. The greatest remaining thickness is on the Nilsen Plateau (Figures 102.1 and 102.2), where it reaches up to 1400 m (this amount also includes an unknown thickness of present-day ice cover). Local differences in the basal sequences, which are mainly glacial but also lacustrine to fluviatile, are primarily due to action of the Permian ice sheet, particularly its irregular and differnetial waning and melting; meltwater streams and rivers must have created some erosional relief on the depositional surface. There is no clear evidence that topographic relief was generated by tectonic movements during the later stages of Beacon sedimentation. Certainly no significant original relief in the basement peneplain itself is apparent. Minshew (1967) reports local relief of as much as 10 m, negligible for the purpose of this study. Thus, an originally horizontal basement surface has been assumed and taken as the datum plane for the present structural map (Figure 102.1). It is suggested that the pattern exhibited by this map and the cross section (Figure 102.2) reflects post-Beacon Supergroup tectonic deformation of the region.

The construction of structural contours is based on detailed mapping from field and air-photo study of all Beacon rock occurrences and observed faults in the region, together with data on the elevation and attitude of the base of the Beacon sequence (Figures 102.3 and 102.4). Also, the exhumed peneplain surface can still be recognized at a number of localities where erosion has left no Beacon

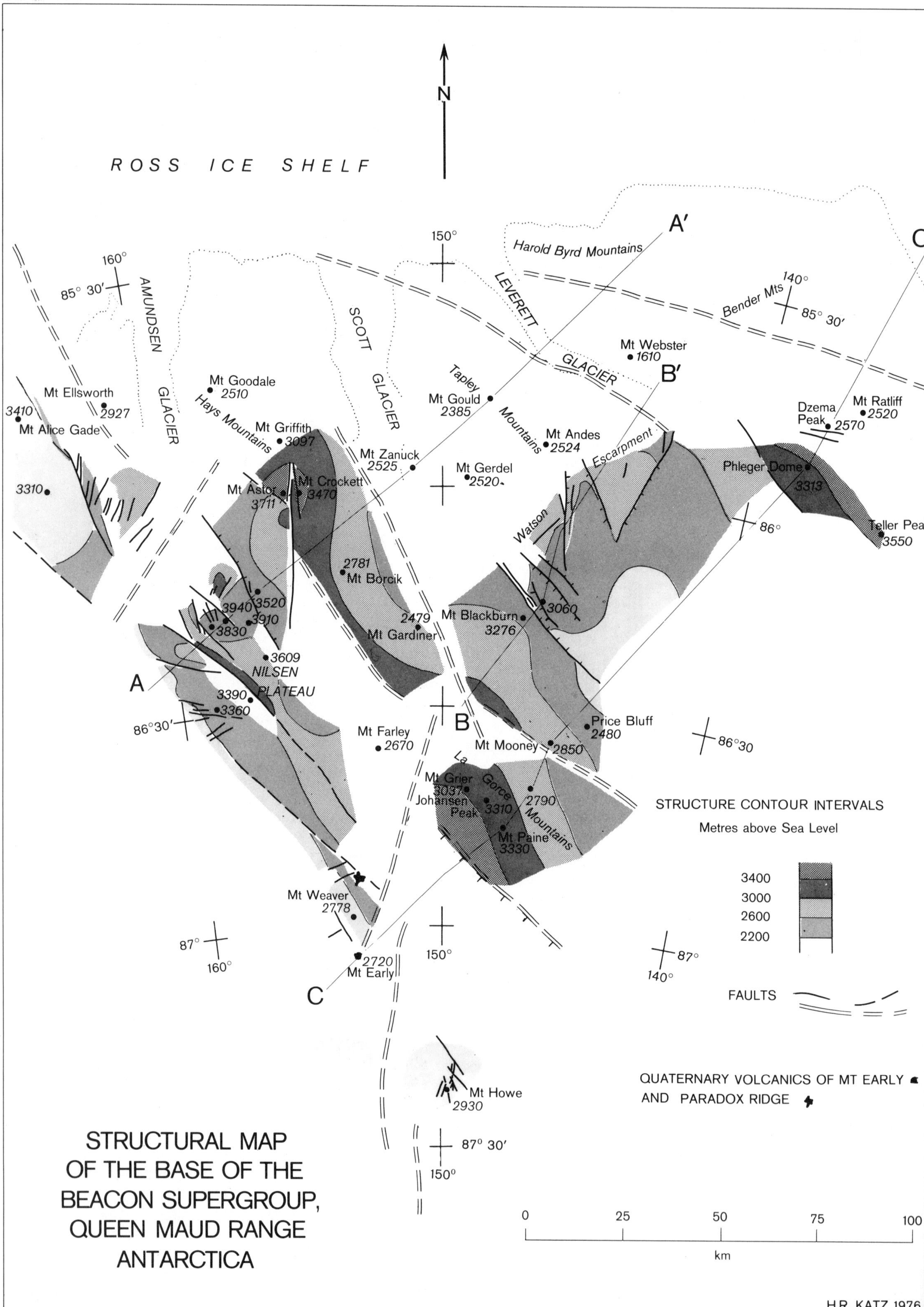
ROSS ICE SHELF
N
A'
B'
C'
A
B
C
Harold Byrd Mountains
Bender Mts
AMUNDSEN GLACIER
SCOTT GLACIER
LEVERETT GLACIER
Watson Escarpment
Tapley Mountains
Hays Mountains
NILSEN PLATEAU
La Gorce Mountains
Mt Ellsworth 2927
Mt Alice Gade 3410
3310
Mt Goodale 2510
Mt Griffith 3097
Mt Astor 3711
Mt Crockett 3470
Mt Zanuck 2525
Mt Gould 2385
Mt Gerdel 2520
Mt Andes 2524
Mt Webster 1610
Dzema Peak 2570
Mt Ratliff 2520
Phleger Dome 3313
Teller Peak 3550
2781 Mt Borcik
2479 Mt Gardiner
Mt Blackburn 3276
3060
3940
3520
3830
3910
3609
3390
3360
Mt Farley 2670
Mt Mooney 2850
Price Bluff 2480
Mt Grier 3037
Johansen Peak 3310
2790
Mt Paine 3330
Mt Weaver 2778
2720 Mt Early
Mt Howe 2930
160°
85° 30'
150°
140°
85° 30'
86°
86°30'
86°30
87°
160°
150°
87°
140°
87° 30'
150°
STRUCTURE CONTOUR INTERVALS
Metres above Sea Level
3400
3000
2600
2200
FAULTS
QUATERNARY VOLCANICS OF MT EARLY AND PARADOX RIDGE
0 25 50 75 100
km
STRUCTURAL MAP OF THE BASE OF THE BEACON SUPERGROUP, QUEEN MAUD RANGE ANTARCTICA
H.R. KATZ 1976

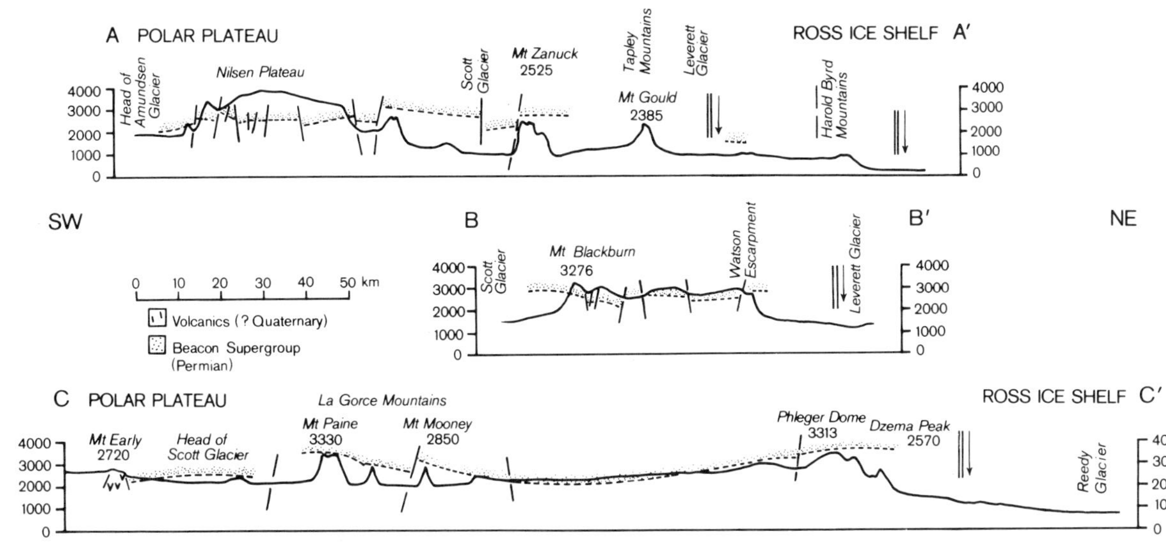

Figure 102.2. Structural cross sections showing post-Beacon deformation in the region of Amundsen and Scott Glaciers, Transantarctic Mountains. For location of sections, see Figure 102.1. Dashed line denotes inferred (or projected) base of Beacon Supergroup where eroded away above present-day ground level.

rocks overlying it, particularly around the head of Scott Glacier in the southeastern La Gorce Mountains and on Szabo and Price Bluffs to the north of the head of Robison Glacier. In addition, minimum elevations of this peneplain surface are provided by mountain peaks formed of basement rock (Figure 102.2). This is particularly important in the area toward the edge of the Ross Ice Shelf, where Beacon strata are generally missing. A combination of all these data provides a fairly good basis for the construction of a regional contour map of the base of the Beacon Supergroup, i.e., the basement peneplain.

This map (Figure 102.1) includes approximately 20,000 km^2 of the southernmost part of the Transantarctic Mountains, lying roughly between lat. 85°30′ and 87°30′ S and long. 140° to 160° W. More importantly, it covers virtually the whole width (about 150 km) of the exposed mountain range; this is apparently only little less than the entire width of the uplifted belt (including its ice-covered part) between the structurally depressed areas of the Polar Plateau and the Ross Sea–Marie Byrd Land region (see Drewry, 1972).

Discussion of structure

Although cursory field observation suggests a virtually horizontal attitude of the Beacon sedimentary beds and the pre-Beacon basement peneplain over wide areas, contouring of the peneplain surface reveals considerable present-day relief, reflecting deformation to a much greater extent and magnitude than expected. At Mount Astor (lat. 86° S, long. 155° 30′ W) (Figure 102.1) the base of the Beacon sequence is 3700 m high while less than 60 km to the west it is at 2200 m. Numerous faults, clearly visible in the high cliffs of Nilsen Plateau and Watson Escarpment, give rise to tilted blocks and wide, gently warped, complex synclinal and anticlinal structures. Where the nature of the faults is evident, they are invariably normal; there are no reverse faults, but in some cases a strike-slip component is suggested. The strike of the faults is predominantly parallel to the trend of the mountain range, but a secondary conjugate set of cross faults intersects the main trend at an angle of generally 40°–45°; this is particularly apparent to the southwest and north of Nilsen Plateau and at Mount Howe. A number of main regional faults (so-called master faults) are proposed (Figure 102.1); their positions are inferred mainly from topography, but partly from the structural attitude and relative position of the base of the Beacon sequence in adjacent areas. If correctly located, it seems that these inferred master faults control to a considerable extent the position of the main glaciers flowing from the Polar Plateau to the Ross Ice Shelf.

Figure 102.1. Structural map of the base of the Beacon Supergroup, Queen Maud Range. Proposed main regional faults (master faults) are indicated with broken double line. A-A′, B-B′, and C-C′: trace of profiles shown in Figure 102.2.

Overall, the structure of the mountain range is dominated by fault blocks or horsts. Graben-and-horst structures, both longitudinal and transverse, are clearly visible in the Watson Escarpment north of Mount Blackburn (Long Valley; Figure 7 of Katz and Waterhouse, 1970*a*) (Figure 102.4) and also on Nilsen Plateau and its northward continuation towards Mount Astor and Mount Crockett (Figures 102.1 and 102.2, profile A-A′). The summit of Mount Astor is capped by a remnant of Beacon rocks with its base at about 3700 m, while air-photo interpretation suggests that another Beacon remnant is downfaulted into the saddle between Mount Astor and Mount Crockett, with its base at little more than 3000 m. In general, however, the observed throw of individual faults is considerably less, being only a few hundred, and more often a few tens, of meters. At Mount Weaver at the head of Scott Glacier, Minshew (1967) reports Beacon rocks disrupted by numerous faults striking mostly east-west, with an average displacement of less than 10 meters.

The La Gorce Mountains, some distance to the east, appear to represent a large horst block with a consistent tilt to the northeast (Figure 102.2, profile C-C′); this interpretation is based on the summit heights and on an extensive, exhumed pre-Beacon peneplain east of Mount Paine, where according to Minshew (1967) some small isolated Beacon outliers still exist. Mount Mooney, however, is too high to form part of this tilted horst block; a fault along its steep southwest face is therefore inferred, presumably con-

Figure 102.3. Post-Beacon fault displacements on the north side of Mount Blackburn (3276 m), Scott Glacier. Note that there is no reverse displacement on these faults, although it may appear so from this angle of view. Vertical height of central summit over granite basement is about 600 m.

Figure 102.4. Northwest-trending graben structure in Beacon sedimentary rocks and underlying granite basement, Watson Escarpment north of Mount Blackburn, Scott Glacier. Height of escarpment, situated in the back of Long Valley, is about 1100 m. View to the southeast. The middle and lower reaches of Scott Glacier run nearly parallel to the fault-and-graben trend for more than 120 km and are most probably similarly fault controlled (see Figure 102.1).

tinuing down Robison Glacier and along the remarkably straight northeast side of Ackerman Ridge, which is in line with this trend. (See U.S. Geological Survey, 1962–, sheets D'Angelo Bluff and Mount Blackburn, for locations not shown on figure 102.1.) Thus the location of Robison Glacier, which descends from the Polar Plateau to join Scott Glacier, may be structurally controlled. On their southwestern side, the La Gorce Mountains, which here terminate in steep cliffs, are inferred to be uplifted by a fault roughly parallel to that along Robison Glacier. This fault may have a throw of about 1000 m (Figure 102.2, profile C-C′). Although the base of the Beacon sequence at Mount Paine is 3330 m or higher, it is 2400 m and less between Mount Wilbur and the innermost nunataks around the head of Scott Glacier (D'Angelo Bluff and Mount Howe, which are the southernmost rock exposures in Antarctica, a little less than 300 km from the South Pole). From the stratigraphic level of the Beacon strata exposed in Mount Howe, including thick dolerite sills, the Beacon base at this place is inferred to be less than 2200 m above sea level. Mount Howe is dissected by several faults trending in a southeasterly direction; a predominantly strike-slip displacement is inferred from its marked topographic features which strongly suggest fault offsets. Cross faults strike north to northeast, parallel to inferred main regional faults which have been traced (Figure 102.1) farther west along the south-trending eastern cliff of D'Angelo Bluff and along another very long and straight ice cliff between D'Angelo Bluff and Mount Howe, which continues far to the south into the Polar Plateau. This fault system, components of which are visible at Mount Howe itself, may be responsible for the regional ice outlet and the location of the head of Scott Glacier.

While a full cross section of the mountain range shows a number of fault blocks with differential tilt and warping, the landward edge towards the Polar Plateau is generally the structurally lowest part. This feature is probably of regional significance. Thus the overall structure is to some degree asymmetric, the entire belt being tilted to the southwest, i.e., dipping underneath the Polar Ice Cap. Yet the abrupt and rectilinear margin leaves little doubt that some fault control is involved here, too. This is also suggested by the subglacial morphology, as seen from radio-echo soundings, which clearly indicates a general fault-block character along this inland flank of the Transantarctic Mountains (Drewry, 1972).

The edge facing the Ross Ice Shelf, in contrast, generally is structurally high, if not *the* most uplifted part of the whole mountain range. This is so in the northeastern corner of the area mapped here and apparently also further east along the Wisconsin Range beyond Reedy Glacier. In the central part, between Leverett and Amundsen Glaciers, such frontal ranges as the Tapley Mountains, Mount Zanuck, and the Hays Mountains (Figures 102.1 and 102.2) are all entirely in basement rocks but have elevations between 2500 and 3000 m; here also the mountain belt is uplifted to a greater height than along the Polar Plateau. Farther along the Transantarctic Mountains, northwest of Beardmore Glacier, Lindsay and others (1973) have mapped a large synclinal structure with its strongly uplifted eastern flank immediately opposite the Ross Ice Shelf. In general, the Transantarctic Mountains break off very abruptly toward the deeply depressed area covered by the Ross Ice Shelf, in a way that suggests major faulting. In the region mapped here, the longitudinal arrangement of such ranges as the Hays and Tapley Mountains, and also of the markedly lower ones to the north such as O'Brien Peak (Katz and Waterhouse, 1970*b*), Harold Byrd Mountains, and Bender Mountains (Figures 102.1 and 102.2) at the very edge of the ice shelf, is also strongly suggestive of faulting. Unfortunately, here as in most places along the edge of the Ross Ice Shelf, there are no Beacon rocks preserved, and faults cannot be directly shown. However, at Cape Surprise (hence the name, I suppose) east of the mouth of Shackleton Glacier, Beacon rocks are indeed found at low elevations, apparently downdropped along a fault parallel to the edge of the ice shelf; the fault throw is at least 3000 m and possibly as much as 5000 m (Barrett, 1965). There is probably additional downfaulting somewhere along the inner edge of the Ross Ice Shelf itself; evidence from seismic surveys (Houtz and Davey, 1973) and DSDP drilling (Hayes and others, 1975) in the Ross Sea farther north suggests that the basement rocks not far away from this coast may be depressed as much as 2000 m below sea level. In the McMurdo Sound area, however, seismic investigations (Northey and others, 1975) have failed to detect evidence of faults in the sediments overlying the basement, which are believed to date back to at least late Miocene time (P. J. Barrett, personal communication).

In conclusion, the Transantarctic Mountains in the region mapped represent a belt of differentially uplifted horsts comprising a mosaic of fault blocks which are locally tilted and warped. Faults are mainly normal, but some probably have strike-slip components; there is no evidence for regional reverse or thrust faults. Folding, as opposed to tilting and warping, appears not to have occurred.

Comparison with previous interpretations and the general structure of the Transantarctic Mountains

Although lacking detailed maps, Gould as far back as 1935 interpreted the general structure of the Transantarctic Mountains in this region as an extensive fault-block mountain range; he was convinced also of the importance of cross faults, calling attention to the major outlet glacier valleys which in his opinion were undoubtedly controlled tectonically. From the Wisconsin Range to the east, post-Beacon cross faults in a north-south and east-west direction were reported by Murtaugh (1969), the faults being normal and with a throw of as much as 300 m. To the northwest, however, Grindley (1963) and Grindley and others (1964) were unable to find evidence for major faults between Axel Heiberg and Nimrod Glaciers, although previously Gunn and Walcott (1962) had suggested fault-block structure for these ranges and pointed out that the altitude of the sub-Beacon peneplain varies between 1000 m and 3000 m within a relatively limited area south of Nimrod Glacier. Along the western edge of this area, Barrett and others (1970) have mapped a large longitudinal fault with a westward downthrow of 400–500 m toward Marsh Glacier, which itself seems to be a regional graben structure parallel to the inland flank of the mountain range. Between here and the Nimrod Glacier the total vertical displacement increases to 1200 m, being taken up by two parallel faults with westerly downthrows of 750 m and 450 m, respectively (Laird and others, 1971). Down the Nimrod Glacier itself, Laird and others (1971), on strong circumstantial evidence, have postulated a major cross fault which they infer to be downthrown to the north by some 2000 m.

North of Nimrod Glacier, Laird (1963) and Skinner (1964) have interpreted the present topography as strongly affected by extensive Cenozoic block faulting, although in fact they did not present evidence of actual displacement of geologic elements. Similarly Gunn and Warren (1962), who studied part of Victoria Land opposite McMurdo Sound and coined the term "Victoria orogeny" for what they believe were block-faulting movements that probably occurred in late Tertiary and Quaternary time and that form the present-day mountains of Victoria Land (see Gunn and Warren, 1962, p. 56), have mentioned and shown on their map a large number of regional faults. However, these faults have been traced only on morphologic grounds; the exception is apparently one single verifiable stratal dislocation near Mount Davidson, which is the main basis for their "MacKay Fault" (Gunn and Warren, 1962, p. 58).

Because of this ambiguity and general lack of real evidence, King (1965) challenged the entire horst block structural interpretation of the Transantarctic Mountains, maintaining that they were due to a simple homoclinal dip eastward underneath the Ross Sea and Ross Ice Shelf, although heavily dissected by erosion. Hamilton (1963) also dismissed the idea of the mountain range being a horst, saying—of the Royal Society Range in Victoria Land—that it "is actually uplifted largely in an anticlinal structure" (Hamilton, 1963, p. 7). He did admit, though, that within the mountains there are many normal faults and gently tilted fault blocks, "but the broad structure, as defined by the gently deformed strata overlying the basement rocks, is anticlinal rather than horst-like." But this, I contend, is a question of semantics rather than basic principle. From Hamilton's descriptions the reader is left with little doubt that the mountain range as a whole is composed of differentially uplifted fault blocks, whether or not it is bounded by major fault scarps on either side. Thus, regardless of whether the overall structure may locally appear slightly anticlinal (or domal), there is no question that normal faults do play a major role.

Based on the relatively detailed evidence now available from the Amundsen and Scott Glaciers region, it is concluded that the overall structure throughout the greater part of the Transantarctic Mountains is a complex mosaic of fault blocks; however tilted and warped individual segments may be, the main structure is really horst-and-graben type. The original concept of a great Antarctic horst obviously is too generalized and simple, but in principle the old idea of Priestley and David (1912) has now been amply confirmed, in spite of rather heavy and elaborate attacks by Hamilton (1963) and King (1965).

Tectonic interpretation

The present data are still insufficient for a detailed analysis of the history of fault movements and the orientation of stress fields. However, the predominance of normal faults and the formation of graben-and-horst structures in a longitudinal direction indicate a mainly extensional regime across the mountains, with the principal horizonal stress roughly coinciding with the axis of the mountain belt itself. If this is correct, then the cross faults, which in many places on the map enclose an angle of a little less than 45° with the longitudinal faults, must have a considerable strike-slip component; field observation indeed suggests such displacement in several places. No evidence exists on the dip of fault planes at the proposed main regional faults (Figures 102.1 and 102.2), nor is it established that there has been only one stress situation during post-Beacon time. Different sets of faults could conceivably have arisen at different times under different stress fields.

This notion is particularly important with regard to the main boundary fault (or fault system) along the Ross Ice Shelf. Under the stress conditions visualized above, and from direct observation of faults within the mountain belt, a reverse fault in this region, as suggested by Barrett (1965), seems impossible; there can only be a normal fault, perhaps with a horizontal component. However, as this zone forms the boundary between two major tectonic provinces (the Transantarctic Mountains and the Ross Sea depression), there may have been more complex conditions; local changes in stress fields may have modified a fault zone that was basically normal. The special character of this zone—which in fact is the boundary between the two different and separate crustal units of East and West Antarctica—is also born out by the occurrence of a large gravity gradient and negative Bouguer anomaly following the eastern boundary of the mountains at McMurdo Sound. This gravity anomaly has been interpreted by Smithson (1972) as being due to a crust, east of the Transantarctic Mountains, that is not only thinner but more mafic and "fundamentally" different (i.e., different in structure, composition, density, etc.).

A regime of regional extension in a direction approximately perpendicular to the mountain belt is also indicated from the most recent geologic history in northern Victoria Land, where late Cenozoic undersaturated, alkaline volcanic rocks (the McMurdo Volcanic Group) are scattered along the eastern margin of the mountains and are related to the occurrence here of longitudinal faults and graben structures (Kyle and Cole, 1974). It is interesting to note that in the Scott Glacier area similar volcanic rocks (olivine basalt flows and pyroclastics) at Mount Early and Paradox Ridge near Mount Saltonstall (Minshew, 1967; Treves, 1967) are similarly related to faults but, in this case, situated at the inner, southwestern margin of the mountain range (Figure 102.1, both south and north of Mount Weaver at about lat. 87° S and long. 153° 30′ W). Along the fault zone of Mount Saltonstall to the north of Mount Weaver, Minshew (1967, p. 22) reports brecciated and epidotized basement rocks.

On the whole, the very limited data available at present do not allow a more detailed interpretation of the tectonic history in post-Beacon time; in particular, the precise nature of the northeastern margin of the mountain belt is still a matter of speculation, although I consider it beyond doubt that this margin is mainly fault controlled. The tectonic process involved in the formation of the mountains is basically one of differential vertical uplift of horst blocks. From both the fault pattern and the distribution of

Quaternary volcanism, it appears that principal horizontal stress lies roughly in a longitudinal direction, which causes extension in a direction across the mountain belt. This is a tectonic situation quite contrary to the one that could be deduced from Hamilton's (1963) anticlinal theory.

Superimposed on this tectonic uplift may be an isostatic adjustment caused by loading from the thick inland ice sheet (G. Lensen, personal communication). This ice sheet from the Polar Plateau has been spreading for some distance over and across part of the mountain range and is undoubtedly depressing the crust in the west and south of our region. The conspicuous regional tilt landward may thus be due to isostatic rather than tectonic causes.

Age of the uplift

The faulting has generally affected the whole Beacon Supergroup, including the dolerite sills within it; the latter therefore are not the cause of such faulting or doming. Radiometric ages of the dolerites are Jurassic (McDougall, 1963) so that the deformation is certainly post-Jurassic. However, regional considerations suggest that the uplift is concomitant with, and related to, the foundering of the Ross Sea depression immediately adjacent to it. This subsidence may have started in Late Cretaceous or early Tertiary time, as deduced from the presence near McMurdo Sound and in the Ross Sea of fossil-bearing glacial erratics of this age (Wilson, 1968; Webb and Neall, 1972). On the other hand, from direct observation of *in situ* sediments in the Ross Sea, it is evident that subsidence has mainly occurred during Neogene time, but beginning in late Oligocene time (Hayes and others, 1975). The pebble composition in the sediment sequence of DSDP hole 273 in the western Ross Sea (Barrett, 1975) suggests that the Transantarctic Mountains have been under active erosion since at least middle Miocene time. The marked change in sediments in the post-Miocene sequence, compared to those of the Miocene, probably implies continued uplift of the mountain range during that time, with erosion by then having cut through the Beacon Supergroup and into basement rocks.

It is thus concluded that the uplift of the Transantarctic Mountains is mainly a Tertiary event; initial movements may have begun as early as Late Cretaceous time, while the final stages probably occurred in Plio-Pleistocene time.

Acknowledgments. I am indebted to P. J. Barrett, R. A. Cooper, and M. G. Laird for valuable suggestions regarding improvements of the text. Margaret Phelan typed the manuscript and the Cartographic Section of the Department of Scientific and Industrial Research, under the supervision of R. Phillips, drew the map and sections.

References cited

Barrett, P. J. 1965. Geology of the area between the Axel Heiberg and Shackleton Glaciers, Queen Maud Range, Antarctica. Part 2—Beacon Group. *N.Z. Jl Geol. Geophys., 8,* No. 2, 344–70.

Barrett, P. J. 1975. Characteristics of pebbles from Cenozoic marine glacial sediments in the Ross Sea (DSDP sites 270–274) and the South Indian Ocean (site 268). (*In* Hayes, D. E., Frakes, L. A., and others. *Initial Reports of the Deep Sea Drilling Project,* Vol. 28. Washington [U.S. Government Printing Office], 769–73.)

Barrett, P. J., Lindsay, J. R., and Gunner, J. 1970. Reconnaissance geologic map of the Mount Rabot quadrangle, Transantarctic Mountains, Antarctica. U.S. Geological Survey, *U.S. Antarctic Research Program,* Antarctic Map No. 1.

Drewry, D. J. 1972. Subglacial morphology between the Transantarctic Mountains and the South Pole. (*In* Adie, R. J., *ed. Antarctic Geology and Geophysics.* Oslo, Universitetsforlaget, 693–703.)

Gould, L. M. 1935. Structure of the Queen Maud Mountains, Antarctica. *Bull. geol. Soc. Am., 46,* 973–84.

Grindley, G. W. 1963. The geology of the Queen Alexandra Range, Beardmore Glacier, Ross Dependency, Antarctica; with notes on the correlation of Gondwâna sequences. *N.Z. Jl Geol. Geophys., 6,* No. 3, 307–47.

Grindley, G. W., McGregor, V. R., and Walcott, R. I. 1964. Outline of the geology of the Nimrod-Beardmore-Axel Heiberg Glaciers region, Ross Dependency, Antarctica. (*In* Adie, R. J., *ed. Antarctic Geology.* Amsterdam, North-Holland Publishing Company, 206–19.)

Gunn, B. M., and Walcott, R. I. 1962. The geology of the Mt. Markham region, Ross Dependency, Antarctica. *N.Z. Jl Geol. Geophys., 5,* No. 3, 407–26.

Gunn, B. M., and Warren, G. 1962. Geology of Victoria Land between the Mawson and Mulock glaciers, Antarctica. *Bull. geol. Surv. N.Z., n.s., 71,* 157 pp.

Hamilton, W. 1963. Tectonics of Antarctica. (*In* Childs, O. E., and Beebe, B. W., *eds.* Backbone of the Americas. *Mem. Am. Ass. Petrol. Geol., 2,* 4–15.)

Hayes, D. E., Frakes, L. A., and others. 1975. *Initial Reports of the Deep Sea Drilling Project,* Vol. 28. Washington (U.S. Government Printing Office), 1017 pp.

Houtz, R. E., and Davey, F. J. 1973. Seismic profiler and sonobuoy measurements in Ross Sea, Antarctica. *J. geophys. Res., 78,* 3448–68.

Katz, H. R., and Waterhouse, B. C. 1970*a*. Geological reconnaissance of the Scott Glacier area, south-eastern Queen Maud Range, Antarctica. *N.Z. Jl Geol. Geophys., 13,* No. 4, 1030–37.

Katz, H. R., and Waterhouse, B. C. 1970*b*. Geologic situation at O'Brien Peak, Queen Maud Range, Antarctica. *N.Z. Jl Geol. Geophys., 13,* No. 4, 1038–49.

King, L. C. 1965. Geological relationships between South Africa and Antarctica. *Geol. Soc. South Africa, Alex. L. du Toit Memorial Lectures,* No. 9, 32 pp.

Kyle, P. R., and Cole, J. W. 1974. Structural control of volcanism in the McMurdo Volcanic Group, Antarctica. *Bull. volcan., 38,* No. 1, 16–25.

Laird, M. G. 1963. Geomorphology and stratigraphy of the Nimrod Glacier-Beaumont Bay region, southern Victoria Land, Antarctica. *N.Z. Jl Geol. Geophys., 6,* No. 3, 465–84.

Laird, M. G., Mansergh, G. D., and Chappell, J. M. A. 1971. Geology of the central Nimrod Glacier area, Antarctica. *N.Z. Jl Geol. Geophys., 14,* No. 3, 427–68.

Lindsay, J. F., Gunner, J., and Barrett, P. J. 1973. Reconnaissance geologic map of the Mount Elizabeth and Mount Kathleen quadrangles, Transantarctic Mountains, Antarctica. U.S. Geological Survey, *U.S. Antarctic Research Program,* Antarctic Map A-2.

McDougall, I. 1963. Potassium-argon age measurements on dolerites form Antarctica and South Africa. *J. geophys. Res., 68,* No. 5, 1535–46.

Minshew, V. H. 1967. *Geology of the Scott Glacier and Wisconsin Range Areas, Central Transantarctic Mountains, Antarctica.* Ph.D. dissertation, The Ohio State University, 268 pp.

Murtaugh, J. G. 1969. Geology of the Wisconsin Range batholith, Transantarctic Mountains. *N.Z. Jl Geol. Geophys., 12,* No. 2-3, 526–50.

Northey, D. J., Brown, C., Christoffel, D. A., Wong, H. K., and Barrett, P. J. 1975. Continuous seismic profiling survey in McMurdo Sound, Antarctica—1975. *Bull. Dry Valley Drilling Project,* No. 5, 167–78.

Priestley, R. E., and David, T. W. E. 1912. Geological notes of the British Antarctica expedition, 1907–09. *C. r. 11th Int. Geol. Congr., Stockholm, 1910,* 767–811.

Skinner, D. N. B. 1964. A summary of the geology of the region between Byrd and Starshot Glaciers, south Victoria Land. (*In* Adie, R. J., *ed. Antarctic Geology.* Amsterdam, North-Holland Publishing Company, 284–92.)

Smithson, S. B. 1972. Gravity interpretation in the Transantarctic Mountains near McMurdo Sound, Antarctica. *Geol. Soc. Am. Bull., 83,* No. 11, 3437–42.

Treves, S. B. 1967. Volcanic rocks from the Ross Island, Marguerite Bay and Mount Weaver areas, Antarctica. (*In* Nagata, T., *ed.* Proceedings of the Symposium on Pacific-Antarctic Sciences. *JARE Scient. Rep., Special Issue No. 1,* 136–49.)

U.S. Geological Survey. 1962- . Antarctica 1:250,000 Reconnaissance Series [topographic maps]. Sheets D'Angelo Bluff, Mount Blackburn, Nilsen Plateau, Mount Goodale, and Leverett Glacier.

Webb, P. N., and Neall, V. E. 1972. Cretaceous Foraminifera in Quaternary deposits from Taylor Valley, Victoria Land. (*In* Adie, R. J., *ed. Antarctic Geology and Geophysics.* Oslo, Universitetsforlaget, 653–57.)

Wilson, G. J. 1968. On the occurrence of fossill microspores, pollen grains, and microplankton in bottom sediments of the Ross Sea, Antarctica. *N.Z. Jl Mar. Freshwat. Res., 2,* No. 3, 381–89.

103

Tectonic Significance of the Shackleton Range

P. D. CLARKSON
British Antarctic Survey, Natural Environment Research Council, Madingley Road, Cambridge, CB3 0ET, U.K.

Abstract. Two major structural trends have been recognized in the Shackleton Range: an east-west trend that originated in Precambrian time and a north-south trend of Cambro-Ordovician age. In the Precambrian metamorphic basement an overall east-west trend can be detected, but it shows many local variations. In the upper Precambrian Turnpike Bluff Group, the east-west fold trend is predominant with associated axial-plane cleavage in the slates. The present northern and southern margins of the range are east-west fault zones occupied by two major westward-flowing glaciers. A recent report of seismic activity from this area may indicate continued fault movement within the northern zone. This trend, approximately perpendicular to the local trend of the Ross orogen, is apparently unique in Antarctica.

The north-south trend occurs as a broad anticline in the Cambro-Ordovician sedimentary rocks and is also detectable in some of the Precambrian rocks at the western end of the range. It represents the eastern margin of the Ross orogen and possibly its northernmost exposure on the Atlantic coast of the continent.

It is suggested that an upper Precambrian orogen with an orthogonal trend to the Transantarctic Mountains probably predates the Beardmore orogen, and that any equivalent trends recognized elsewhere may be significant in Gondwanaland reconstructions.

The Shackleton Range (lat. 80° 07′–80° 50′ S, long. 31°–19° W) lies east of the Filchner Ice Shelf at the head of the Weddell Sea. The greater part of the range is formed of basement rocks of the Shackleton Range Metamorphic Complex, unconformably overlain by the Turnpike Bluff Group in the south and by the Cambro-Ordovician Blaiklock Glacier Group in the northwest (Clarkson, 1972). The range is bounded to the north and south by Slessor and Recovery Glaciers flowing westward along the major fault zones between which the range has been uplifted. The western end of the range is a discontinuous erosion scarp and the eastern end gradually disappears beneath the Antarctic ice sheet.

The stratigraphic succession is shown in Table 103.1, and the distribution of outcrops is shown in Figure 103.1. The basement complex has been broadly divided into (1) the older gneisses which have been migmatized and intruded by granite stocks, particularly in the Read Mountains, and (2) the younger schists, marbles, and amphibolites, representing a metamorphosed sedimentary sequence. A granodiorite, which intrudes one of the granite stocks, represents the last event in the migmatization; it has been dated (K-Ar) at 1446 ± 60 m.y. (Rex, 1972), providing the only radiometric age for the basement complex. The Turnpike Bluff Group, a sequence of slates and quartzites with minor conglomerates, rests unconformably on the basement complex in the Read Mountains. It has been divided into four formations on lithologic and structural evidence and is younger toward the west. In the northern Haskard Highlands, the Blaiklock Glacier Group also rests uncomformably on the basement complex, but it is nowhere seen in contact with the Turnpike Bluff Group. It is divided into two formations and is considered to be Cambro-Ordovician from the evidence of fossiliferous erratics (Thomson, 1972) assumed to be derived from intermediate strata hidden beneath the Blaiklock and Stratton Glaciers. Dolerite dikes were intruded during Ordovician time (457 ± 18 m.y. [K-Ar], according to Rex, 1972) into schists of the basement complex, and during Carboniferous time (297 ± 12 m.y. [K-Ar], according to Rex, 1972) into the Blaiklock Glacier Group. A number of dikes have chemical affinities with the dolerite intrusions of western Dronning Maud Land and the Transantarctic Mountains, suggesting a third period of intrusion during Jurassic time (Clarkson, in press).

Structure

In the western Shackleton Range the older basement rocks appear to be folded about northwest-plunging axes; the few observed fold axes are in close agreement with the axis calculated from the plot of the poles to foliation planes (Figure 103.2*a*). The younger basement rocks in this area have a north-south fold trend which is discussed below. In the Read Mountains the gneisses and migmatites show a more complicated fold pattern; the plot of poles to foliation planes (Figure 103.2*b*) does not indicate an obvious fold trend. One interpretation of the diagram suggests two fold axes plunging very steeply to the north and south, but the distribution could also imply an east-plunging fold with steeply dipping limbs. In the field very few fold axes were observed, but these plunged either northwestward or, less commonly, westward. In certain areas, particularly in the western Read Mountains, the foliation dips very steeply north or south, or is subvertical. These data are regarded as indicating that the rocks were originally folded about northwest-plunging axes, as in the western part of the range, but that subsequent migmatization slightly modified the structure and the intrusion of the granite stocks probably steepened the fold limbs. The predominantly east-west folds in the younger basement rocks may also have affected the older gneisses.

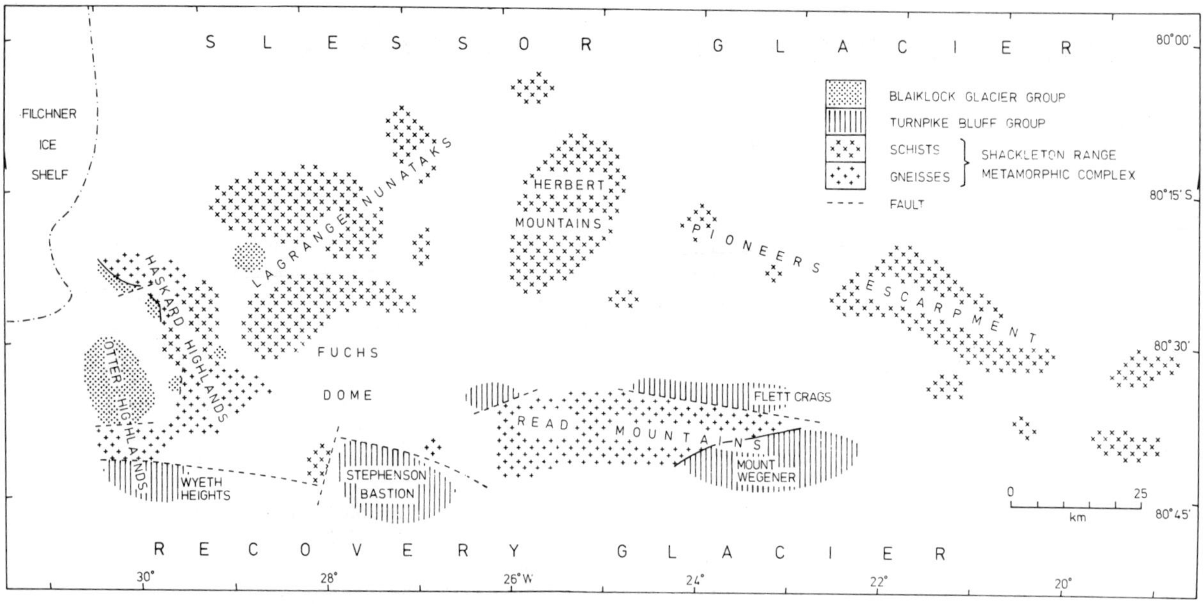

Figure 103.1. Diagrammatic geologic sketch map of the Shackleton Range showing the major areas of outcrop of the middle Precambrian Shackleton Range Metamorphic complex, the upper Precambrian Turnpike Bluff Group, and the Cambro-Ordovician Blaiklock Glacier Group.

Elsewhere in the range the younger schists, marbles, and amphibolites of the basement complex are much more regularly folded about east-west axes. The diagram of the poles to foliation (Figure 103.2*c*) shows a well-defined girdle indicating an axis plunging gently in a direction just north of east; this axis lies at the edge of the maximum concentration contour of the observed fold axes for the same areas (Figure 103.2*d*). In the field, the Herbert Mountains straddle the axis of a major east-west synform. A few observed axes trend approximately north-south; they are discussed below.

Poles to original bedding in the slates and quartzites of the Turnpike Bluff Group plot as a girdle, indicating folding about subhorizontal axes trending generally east-west (Figure 103.2*e*). This diagram also shows the orientations of observed fold axes which, although mostly gently plunging, have widely scattered bearings. The plot of poles to cleavage planes (Figure 103.2*f*) in the slates indicates northward-dipping axial planes and overfolding to the south. The data are exaggerated by the inclusion of the cleavage plane measurements from the Flett Crags Formation where the cleavage is a bedding cleavage. The reason for the different style of cleavage development in this formation is not clear, but its present position to the north of the line joining the other formations of the group and its higher level of outcrop above the escarpment of the Read Mountains, may imply a slightly different tectonic regime.

The structures in the Precambrian rocks described above

Table 103.1. Stratigraphic sequence in the Shackleton Range

	Southern Shackleton Range	Western Shackleton Range	Northern Shackleton Range	
Tertiary to Cretaceous	————	Peneplanation	————	
Jurassic		Dolerite Dikes	Dolerite Dike	
Carboniferous		Dolerite Dike		
Ordovician			Dolerite Dike	
	— — ———	Deformation (Ross orogeny)	——— — —	
to Cambrian		Otter Highlands Formation Mount Provender Formation		Blaiklock Glacier Group
	— — — — ———	Erosion	——— — — — —	
Upper Precambrian	Wyeth Heights Formation Stephenson Bastion Formation Flett Crags Formation Mount Wegener Formation			Turnpike Bluff Group
	———	Erosion	——— — — — — —	
Middle Precambrian		———— Schists, marbles, and amphibolites	————	Shackleton Range Metamorphic Complex
	———— Gneisses, migmatites, and granite stocks	————		

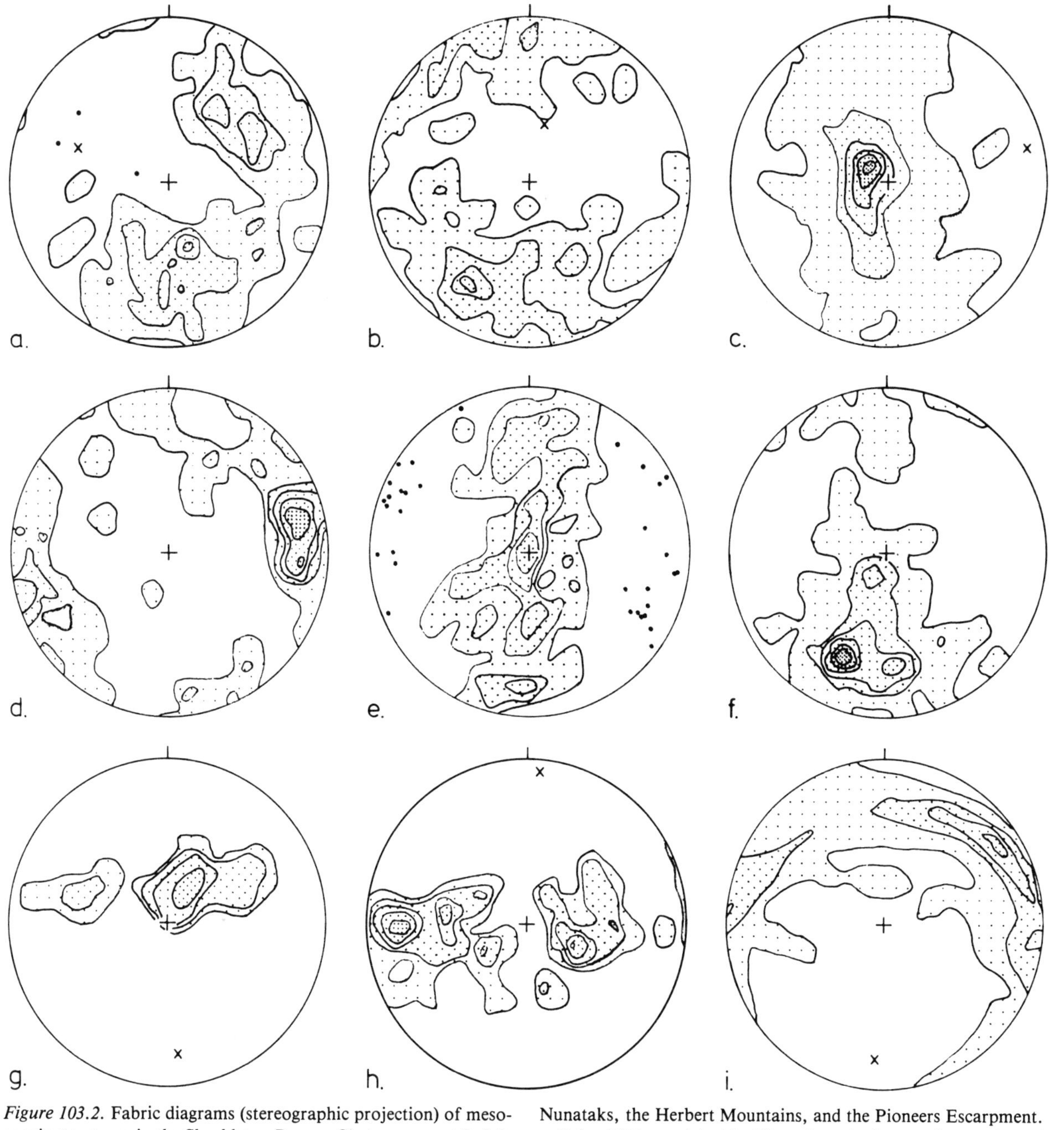

Figure 103.2. Fabric diagrams (stereographic projection) of mesoscopic structures in the Shackleton Range. Contours are at 0, 2.5, 5, 7.5, 10, and 12.5% per 1% unit area except in Figure 103.2*g* where they are at 0, 5, 10, and 20% per 1% unit area. Large dot: plunge of observed fold axis. x: plunge of calculated fold axis.
a. Poles (160) to foliation planes in older Precambrian basement rocks from Otter and Haskard Highlands. *b.* Poles (200) to foliation planes in older Precambrian basement rocks from the Read Mountains. *c.* Poles (604) to foliation planes in younger Precambrian basement rocks from the Lagrange Nunataks, the Herbert Mountains, and the Pioneers Escarpment. *d.* Plunges (97) of fold axes in younger Precambrian basement rocks from the Lagrange Nunataks, the Herbert Mountains, and the Pioneers Escarpment. *e.* Poles (129) to original bedding planes in slates and quartzites of the Turnpike Bluff Group, excluding bedding-cleavage planes in the Flett Crags Formation. *f.* Poles (198) to cleavage planes in slates of the Turnpike Bluff Group, including bedding-cleavage planes in the Flett Crags Formation. *g.* Poles (67) to bedding planes in sandstones of the Blaiklock Glacier Group. *h.* Poles (66) to foliation planes in younger Precambrian basement rocks from the central Haskard Highlands. *i.* β-diagram of 607 bedding-cleavage intersections in slates of the Wyeth Heights Formation (Turnpike Bluff Group).

all show an overall east-west trend although there are variations, some of which may be significant (e.g., the complicated fold pattern of the Read Mountains). In the Phanerozoic rocks one fold trend has been recognized—a north-south trend affecting the lower Paleozoic sedimentary rocks and some of the Precambrian rocks at the western end of the range.

No folds have been observed in the Blaiklock Glacier Group in the field, but the dip of the bedding (southwest in the west and southeast in the east) is indicative of an anticline plunging southward; this is confirmed by the stereographic plot of the poles to bedding planes (Figure 103.2*g*). A similar north-south trend can also be seen in some of the older rocks. The younger basement rocks which crop out in the central part of Haskard Highlands seem to have been folded about a horizontal north-south axis (Figure

103.2*h*). Some fold axes observed in other rocks of the same age farther to the east also show a north-south trend (Figure 103.2*d*) although it is not so pronounced. A β-diagram (Figure 103.2*i*) of a few observed and many constructed bedding-cleavage intersections in the Wyeth Heights Formation, the most westerly formation of the Turnpike Bluff Group, suggests a north-south fold trend. Although this last diagram grossly magnifies the data and may not be reliable (Turner and Weiss, 1963), it does illustrate an expected trend and one which is not found farther east, except to a limited extent in the Stephenson Bastion Formation.

The Shackleton Range had a long history of folding around predominantly east-west axes during Precambrian time, probably beginning as a more northwesterly trend but later becoming east-west, particularly by the time of deformation of the Turnpike Bluff Group. During Phanerozoic time, however, the only fold trend known to have affected the range was north-south. This trend is considered to be of Cambro-Ordovician age and representative of the Ross orogen; it affected the sandstones of the Blaiklock Glacier Group and some of the younger Precambrian basement rocks at the western end of the range. Although no younger folds are seen in the range, the important east-west trend affects the shape of the range, elongated and elevated between the major westward-flowing Slessor and Recovery Glaciers which flow along major fault zones to the north and south of the range. This suggests that the east-west trend is probably a very deep-seated crustal weakness which prevailed through and after the Ross orogeny and may even be active today, if the report of an earthquake at approximately lat. 80° S, long. 20° W (Kaminuma and Ishida, 1971) is related to the northern boundary fault zone beneath Slessor Glacier.

Regional correlation

The east-west fold trend of the Precambrian rocks is apparently unique in the Transantarctic Mountains. No outcrops are known to the west of the range, except in West Antarctica, and there is no outcrop to the east for more than 1500 km, at which distance structural correlation has not been attempted. The nearest outcrops of similar rocks occur in Heimefrontfjella, but the structural data are very limited and suggest a north-south trend (Juckes, 1972). The Shackleton Range Metamorphic Complex is regarded as part of the Precambrian shield of East Antarctica, although no precise correlation has been made.

The Turnpike Bluff Group is difficult to correlate with other areas because the true age of the group is unknown; the only constraints on the age of the group are that it rests unconformably on the older part of the basement complex, which has a minimum age (K-Ar) of 1446 ± 60 m.y. (Rex, 1972), and that it is presumed to be older than the Cambro-Ordovician Blaiklock Glacier Group from its contrasting rock types, structural trend, and fold style. The closest equivalent of the group, both geographically and lithologically, is the upper Precambrian Patuxent Formation in the Pensacola Mountains (Nelson and others, 1968). Similar rocks have been described from many parts of the Transantarctic Mountains (e.g., Elliot, 1975), but the structural trends of these deformed Precambrian rocks are parallel to the trends formed by the Ross orogeny.

The Blaiklock Glacier Group is also difficult to correlate owing to its limited extent and to the local origin of its sediment. However, the group consists of sandstones deposited in near-shore or estuarine conditions near the inward margin of the Ross geosyncline. The nearest compositionally equivalent rocks appear to be the early Paleozoic(?) Urfjell Group in Kirwanveggen (Aucamp and others, 1972); the Nelson Limestone in the Pensacola Mountains (Williams, 1969) may represent a distal equivalent.

Discussion

The east-west structural trend of the Precambrian rocks in the Shackleton Range is approximately perpendicular to the structural trend of the Transantarctic Mountains, the Ross orogen, and possibly some older and younger orogenic belts subparallel to the Ross orogen. The Beardmore orogen (formed about 620–680 m.y. ago; Grindley and McDougall, 1969) was probably about parallel to the Ross orogen. This Precambrian event was probably responsible for the main folding of the Patuxent Formation. It seems likely that the folding of the Turnpike Bluff Group represents a regeneration of earlier Precambrian fold trends in the Shackleton Range and therefore predates the Beardmore orogeny, rather than forming an orthogonal trend to, and intermediate in time between, the Beardmore and Ross orogens. Aughenbaugh and others (1965) dated (K-Ar) some rhyolites from the Littlewood Nunataks (lat. 77° 53′ S, long. 34° 08′ W) at 840 ± 30 m.y. and suggested that they may have formed an island arc associated with the Beardmore orogeny. Although this age is greater than the age for the Beardmore orogeny suggested by Grindley and McDougall (1969), it is not as great as the Rb-Sr age (1044 m.y.) obtained by Faure and others (1968) for the same rocks from the Littlewood Nunataks. According to the alternative Rb-Sr dates given by Gunner and Faure (1972) for the Nimrod orogeny (1980 or 1000 m.y.), the rocks at Littlewood Nunataks could be equivalent to or could postdate the Nimrod orogeny. Whatever is the true age of the Nimrod orogeny, there may be a pre-Beardmore orogeny isotopic event related to the folding of the Turnpike Bluff Group, at either 840 m.y. or 1044 m.y. ago. In either case the Littlewood Nunataks could still represent an ancient island arc, but one related to an orogeny which folded the Turnpike Bluff Group. The Shackleton Range might then have represented a geanticline, and flysch-type sedimentary rocks of this age may form the unexposed basement rocks of the Theron Mountains 150 km north of the Shackleton Range. These arguments imply that a new orogenic episode may have to be defined, but present data are considered insufficient to make such a definition. It is hoped that isotopic work on the slates of the Turnpike Bluff Group and radio-echo profiling of the region planned for the near future will provide some of the data necessary to resolve these problems of the Turnpike Bluff Group. Similar questions arise when the metamorphic history of the Shackleton Range Metamorphic Complex is considered, but planned geochronologic studies may offer solutions to these problems.

The north-south fold trend cutting across the western end of the Shackleton Range is considered to be the northward continuation of the Ross orogen. This early Paleozoic orogenic event also folded some of the Precambrian rocks at the western end of the range; its effects are not seen farther east in the same rock types, however, so the western end of the Shackleton Range probably forms the eastern margin of the Ross orogen in this area. The only exception to this may be the dolerite dike near the eastern end of Pioneers Escarpment (Clarkson, 1972), which has been dated at 457 ± 18 m.y. (Rex, 1972) and is presumably related to a late phase of the Ross orogeny. North of the Shackleton Range the continuation of the Ross orogen is not known, unless it swings to the northeast and is present in Kirwanveggen, which Neethling (1972) has described as the continuation of the Transantarctic Mountains, possibly synonymous with the Ross orogen. Likewise, Aucamp and others (1972) described fold trends in the Urfjell Group, Kirwanveggen, which would also parallel the trend of the Ross orogen if it extended into that area.

At least once during the Phanerozoic the Shackleton Range was uplifted between the east-west fault zones which form the northern and southern boundaries of the range. The time of uplift is unknown, but the presence of probable Mesozoic dikes in the range and the absence of any Mesozoic lavas and sills and any Beacon Supergroup sedimentary rocks may imply that the uplift and peneplanation took place during Cretaceous or early Tertiary time.

Schopf (1969) suggested that the Ellsworth Mountains have moved to their present position by transform faulting from somewhere near the coast of Coats Land. More recently, Clarkson and Brook (1977) have suggested a possible relationship in time and space between some gneisses at Haag Nunataks (lat. 77° 02′ S, long. 78° 20′ W) and the Littlewood Nunataks. The deep-seated crustal weakness in the fault zones forming the northern and southern margins of the Shackleton Range may have been an important factor in determining the positions of crustal fractures during the fragmentation of the western margin of East Antarctica. Finally, any Precambrian structural trends which are approximately perpendicular to known late Precambrian or early Paleozoic trends may well be related to the anomalous east-west structural trend of the Shackleton Range. Such structural trends in a group of Precambrian rocks could indicate a pre-breakup position to the west of and aligned with the Shackleton Range.

Conclusions

There are two major structural trends in the Shackleton Range: a predominatly east-west trend which formed in Precambrian time when the basement rocks and the younger slates and quartzites were folded about subhorizontal axes, and which was subsequently regenerated during Phanerozoic time when the range was uplifted between two major fault zones; and a north-south fold trend of Cambro-Ordovician age which is the local manifestation of the Ross orogen. The younger fold trend also affects some of the Precambrian rocks at the western end of the range.

The folding of the Turnpike Bluff Group took place during late Precambrian time, probably before the Beardmore orogeny; its folds trend perpendicular to the axis of the Beardmore orogen. The time and trend of the Turnpike Bluff Group folding may represent a separate orogeny, but as yet there are insufficient data to define such an orogeny. The east-west Precambrian fold trend seems related to a deep-seated crustal weakness which may have been an important factor in determining the position of crustal fractures during the fragmentation of the western margin of East Antarctica.

Acknowledgments. I wish to thank my field companions for their assistance, especially R. B. Wyeth and M. J. Skidmore who made their field notes and collections available to me. I am grateful to my colleagues in the Earth Sciences Division of the British Antarctic Survey for much useful discussion and criticism. Finally I wish to thank the officers and men of the U.S. Navy Squadron VXE-6 without whose transport facilities this work would not have been done.

References cited

Aucamp, A. P. H., Wolmarans, L. G., and Neethling, D. C. 1972. The Urfjell Group, a deformed(?) early Palaeozoic sedimentary sequence, Kirwanveggen, western Dronning Maud Land. (*In* Adie, R. J., *ed. Antarctic Geology and Geophysics.* Oslo, Universitetsforlaget, 557–62.)

Aughenbaugh, N. B., Lounsbury, R. W., and Behrendt, J. C. 1965. The Littlewood Nunataks, Antarctica. *J. Geol., 73,* No. 6, 889–94.

Clarkson, P. D. 1972. Geology of the Shackleton Range: a preliminary report. *Bull. Br. antarc. Surv.,* No. 31, 1–15.

Clarkson, P. D. In press. Geology of the Shackleton Range: IV. The dolerite dykes. *Bull. Br. antarc. Surv.*

Clarkson, P. D., and Brook, M. 1977. Age and position of the Ellsworth Mountains crustal fragment, Antarctica. *Nature, Lond., 265,* No. 5595, 615–16.

Elliot, D. H. 1975. Tectonics of Antarctica: a review. *Am. J. Sci., 275-A,* 45–106.

Faure, G., Hill, R. L., Eastin, R., and Montigny, R. J. E. 1968. Age determinations of rocks and minerals from the Transantarctic Mountains. *Antarct. Jnl U.S., 3,* No. 5, 173–75.

Grindley, G. W., and McDougall, I. 1969. Age and correlation of the Nimrod Group and other Precambrian rock units in the central Transantarctic Mountains, Antarctica. *N.Z. Jl Geol. Geophys., 12,* No. 2/3, 391–411.

Gunner, J., and Faure, G. 1972. Rb-Sr geochronology of the Nimrod Group, central Transantarctic Mountains. (*In* Adie, R. J., *ed. Antarctic Geology and Geophysics.* Olso, Universitetsforlaget, 305–11.)

Juckes, L. M. 1972. The geology of north-eastern Heimefrontfjella, Dronning Maud Land. *Scient. Rep. Br. antarc. Surv.,* No. 65, 44 pp.

Kaminuma, K., and Ishida, M. 1971. Earthquake activity in Antarctica. *Antarctic Rec.,* No. 42, 53–60.

Neethling, D. C. 1972. Submarine and subglacial morphology, Kronprinsesse Martha Kyst, Dronning Maud Land. (*In* Adie, R. J., *ed. Antarctic Geology and Geophysics.* Oslo, Universitetsforlaget, 705–11.)

Nelson, W. H., Schmidt, D. L., and Schopf, J. M. 1968. Structure and stratigraphy of the Pensacola Mountains, Antarctica. *Spec. Pap. geol. Soc. Am.,* No. 115, 344.

Rex, D. C. 1972. K-Ar age determinations on volcanic and associated rocks from the Antarctic Peninsula and Dronning Maud Land. (*In* Adie, R. J., *ed. Antarctic Geology and Geophysics.* Oslo, Universitetsforlaget, 133–36.)

Schopf, J. M. 1969. Ellsworth Mountains: position in West Antarctica due to sea-floor spreading. *Science, N.Y., 164,* No. 3875, 63–66.

Thomson, M. R. A. 1972. Inarticulate Brachiopoda from the Shackleton Range and their stratigraphical significance. *Bull. Br. antarc. Surv.,* No. 31, 17–20.

Turner, F. J., and Weiss, L. E. 1963. *Structural Analysis of Metamorphic Tectonites.* New York, McGraw-Hill, 545 pp.

Williams, P. L. 1969. Petrology of upper Precambrian and Paleozoic sandstones in the Pensacola Mountains, Antarctica. *J. sedim. Petrol., 39,* No. 4, 1455–65.

104

Geology of the Whitmore Mountains

GERALD F. WEBERS
Department of Geology, Macalester College, St. Paul, Minnesota 55105, U.S.A.

CAMPBELL CRADDOCK
Department of Geology and Geophysics, University of Wisconsin, Madison, Wisconsin 53706, U.S.A.

M. ALAN ROGERS
2019 Thousand Pines, Humble, Texas 77338, U.S.A.

JOHN J. ANDERSON
Department of Geology, Kent State University, Kent, Ohio 44242, U.S.A.

Abstract. The Whitmore Mountains (lat. 82° 30′ S, long. 104° 30′ W) consist of a series of small mountains and nunataks arranged in a roughly semicircular pattern of about 16 km radius. Bedrock consists of predominant igneous rocks and minor metasedimentary rocks. The Mount Seelig Granite, a coarsely crystalline porphyritic granite, comprises about 95% of the exposed bedrock. A second granite, the Linck Nunataks Granite, is a finely crystalline equigranular rock that crosscuts the Mount Seelig Granite. The Linck Nunataks Granite is considered to be a late-phase differentiate of the Mount Seelig Granite. Radiometric apparent ages (K-Ar) of the two granites support this conclusion. The Mount Seelig Granite is Late Triassic in age (190 ± 8 m.y.) and the Linck Nunataks Granite is Early Jurassic in age (176 ± 5 m.y.). Metasedimentary rocks make up a small proportion of the bedrock and consist primarily of roof pendants and blocks within the Mount Seelig Granite. The thickest exposed sequence of metasedimentary rocks is estimated to be 800 m thick. Rock types include phyllite, slate, graywacke, subarkose, and impure quartzites. Because of insufficient knowledge of the stratigraphy, the metasedimentary rocks have not been given formational status. If the surrounding ice melted, the Whitmore Mountains would stand as a southwest-trending peninsula attached to the narrow landmass extending southward from the Ellsworth Mountains. Radiometric ages obtained on the Mount Seelig and the Linck Nunataks Granites indicate that they belong to the West Antarctic tectonic province (Ellsworth orogen) formed by the Ellsworth Mountains and certain nunataks to the south.

The Whitmore Mountains (centered at about lat. 82° 30′ S, long. 104° 30′ W) consist of a series of small mountains and nunataks arranged in a roughly semicircular pattern of 16 km radius. The highest mountain, Mount Seelig, is 3022 m in height and rises approximately 610 m above the snow surface.

These mountains were first visited during the 1957–58 austral summer by a traverse party under the direction of C. R. Bentley (Bentley and others, 1960; Bentley and Ostenso, 1961); this traverse party passed through the southernmost part of the Whitmore Mountains and made geologic and geophysical investigations. The Whitmore Mountains were again briefly visited during the 1962–63 austral summer when a series of ice flowage markers was established by a group from Ohio State University between New Byrd Station and Mount Chapman, the westernmost peak in the Whitmore Mountains (Bull, 1963; Brandenberger, 1964). During the 1964–65 austral summer, a group from the University of Minnesota conducted a geologic reconnaissance of the Whitmore Mountains as part of a larger program of exploration in West Antarctica (Anderson and others, 1962; Craddock, 1966, 1978; Craddock and others, 1964). Members of this expedition were Gerald F. Webers and M. Alan Rogers (geologists, University of Minnesota) and Roger Buchanan (entomologist, Bishop Museum). All accessible rock exposures were studied for geologic and biologic information. No flora or fauna were observed.

Bedrock geology

The bedrock consists of both igneous and metasedimentary rocks with the former greatly predominating (Figure 104.1).

Mount Seelig Granite

A gray, coarsely crystalline, porphyritic granite, here called the Mount Seelig Granite, is found throughout the Whitmore Mountains and accounts for more than 95% of the exposed bedrock. Twinned orthoclase phenocrysts as much as 4 cm in length comprise as much as 20% of this rock (Figures 104.2 and 104.3). Mineralogic composition and appearance are quite uniform throughout the Whitmore Mountains; the coarseness of the Mount Seelig Gran-

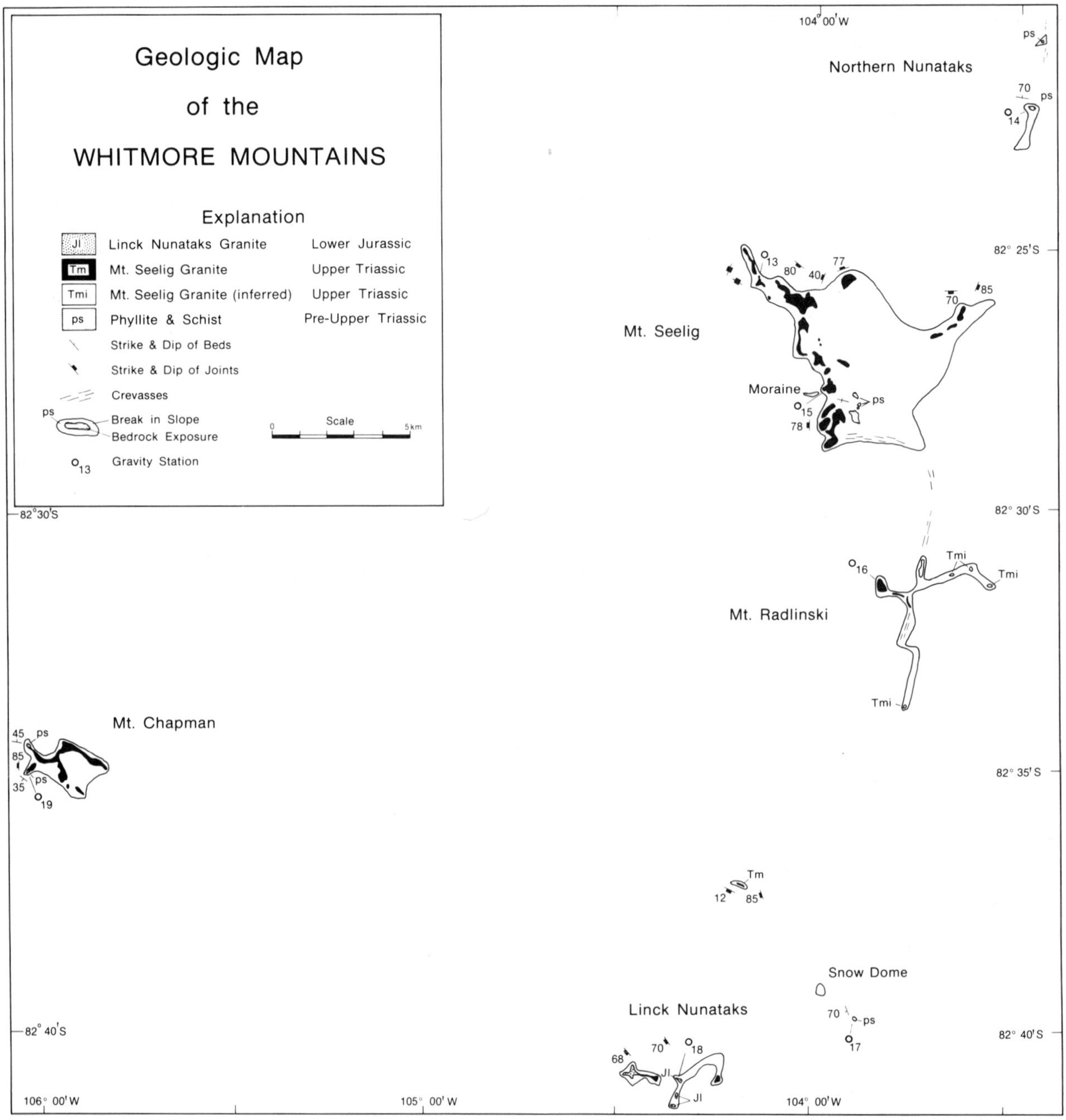

Figure 104.1. Geologic map of the Whitmore Mountains. Cartographic Lab, University of Wisconsin-Madison.

Figure 104.2. Mount Seelig Granite (W-65-77), crossed nicols, 20×. Q: quartz. B: biotite. Or: orthoclase. Ol: oligoclase.

Figure 104.3. Hand specimen of Mount Seelig Granite, 1/2×.

Table 104.1. Point-count percentage estimates of mineral abundances in igneous rocks of the Whitmore Mountains

	Mount Seelig Granite										
	W-65-44	W-65-45	W-65-48	W-65-54	W-65-57	W-65-58	W-65-64	W-65-65	W-65-66	W-65-77	Average
Quartz	42	8	35	39	11	28	28	22	38	14	27
Orthoclase	16	29	19	35	24	27	10	15	33	27	24
Microcline	3	42	p	p	p	14	18	30	4	22	13
Oligoclase	14	15	32	17	47	29	25	28	17	23	25
Biotite	21	5	11	9	14	3	19	5	7	10	10
Hornblende	3	p	2	0	2	p	0	0	1	4	1
Chlorite	p	1	p	p	p	0	p	p	p	0	p
Muscovite	0	p	0	p	p	p	0	p	p	0	p
Apatite	p	p	p	p	p	p	p	p	p	p	p
Zircon	1	p	1	p	1	p	p	p	p	p	p
Magnetite	p	p	p	p	p	p	p	p	p	p	p
Hematite	p	0	p	p	p	0	p	p	p	0	p
Sericite	p	p	p	p	p	p	p	p	p	p	p
Rutile	0	0	0	0	p	p	0	0	0	0	0

p: present but less than 1%.
J. J. Anderson, petrographer.

Location of specimens:
Mount Seelig: W-65-44, 45, 48, 54, 55, 57, 58, and 59
Mount Radlinski: W-65-64
Nunatak north of Linck Nunataks: W-65-65 and 66
Linck Nunataks: W-65-73, 75, 76, 78, and 80

ite, however, causes the petrographic point count of the minerals (Table 104.1) to seem more variable than it actually is. Chemical analyses (Table 104.2) and mineralogic compositions indicate that this intrusive rock is close to the boundary between granite and quartz monzonite in composition. Assignment to the granite category is because modal analyses of the majority of specimens of this intrusive rock fall on the granite side of the boundary.

The rock appears fresh in both thin section and field exposure, and it shows no postemplacement shearing or alteration. It weathers to rounded pink blocks.

The Mount Seelig Granite intrudes metasedimentary rocks in three locations and is in turn cut by a younger granite in the Linck Nunataks. Radiometric dating (K-Ar) of a granite specimen (W-65-45) from Mount Seelig has yielded a Late Triassic age of 190 ± 8 m.y. (Table 104.3).

Linck Nunataks Granite

A gray, equigranular, finely crystalline granite, here called the Linck Nunataks Granite, is exposed in the Linck Nunataks. Texture is hypautomorphic granular (Figure 104.4). In two outcrops the granite is present as dikes as much as 7 m wide crosscutting and containing inclusions of the Mount Seelig Granite. In a third outcrop the granite occupies the entire exposure. Aplite dike swarms in Mount Seelig and elsewhere are similar in color and composition

Table 104.2. Chemical analyses (in wt %) of igneous rocks from Mount Seelig and Linck Nunataks

	Mount Seelig Granite W-65-44[a]	Linck Nunataks Granite W-65-75[a]
SiO_2	65.83	71.90
Al_2O_3	16.29	15.74
Fe_2O_3	2.12	0.82
FeO	2.86	1.30
CaO	3.42	1.76
MgO	0.92	0.40
MnO	0.06	0.04
TiO_2	0.74	0.34
Na_2O	2.49	2.67
K_2O	3.90	3.69
CO_2	0.17	0.05
P_2O_5	0.28	0.18
H_2O^-	0.11	0.05
H_2O^+	0.33	0.54
Total	99.52	99.48

[a] Reported values are averages of standard duplicate analyses performed by Technical Services, Incorporated, Toronto.

Table 104.3. K-Ar age data on biotites from granites, Whitmore Mountains

Specimen number	Location	K (wt %)	^{40}Ar (radiogenic) ppm	^{40}Ar (radiogenic) % total	Age (m.y.)
W-65-76	Linck Nunataks	6.68	0.0876	85.5	176 ± 5
W-65-45	Mount Seelig	5.57	0.0795	74.9	190 ± 8
W-65-4	Pagano Nunatak	6.48	0.0845	77.3	175 ± 4

Constants: $\lambda_\beta = 4.72 \times 10^{-10}$/yr; $\lambda_e = 0.585 \times 10^{-10}$/yr; $^{40}K/K = 1.22 \times 10^{-4}$ g/g.

Table 104.1 (continued)

	Linck Nunataks Granite (and aplite dikes)							
	W-65-55[a]	W-65-59[a]	W-65-73	W-65-75	W-65-76	W-65-78	W-65-80	Average
Quartz	25	35	28	41	30	29	40	33
Orthoclase	47	46	24	24	36	39	32	36
Microcline	9	6	24	5	5	11	8	9
Oligoclase	17	12	18	18	18	19	16	17
Biotite	2	1	6	6	8	2	4	4
Hornblende	0	0	0	0	0	0	0	0
Chlorite	p	0	p	p	p	p	p	p
Muscovite	p	p	0	6	3	p	p	1
Apatite	p	p	p	p	p	p	p	p
Zircon	p	p	p	p	p	p	p	p
Magnetite	p	p	p	p	p	p	p	p
Hematite	p	p	p	0	0	p	p	p
Sericite	p	p	0	p	p	p	p	p
Rutile	0	0	p	p	p	p	p	p

[a] Aplite dikes.

and are considered to be a part of the same later intrusion. If so, the Linck Nunataks Granite is probably a late phase of the same intrusion that formed the Mount Seelig Granite. The dike swarms follow only one set of joints, presumably cooling joints that had formed by that time in the Mount Seelig Granite. Radiometric dating (K-Ar) of the Linck Nunataks Granite has yielded an Early Jurassic apparent age of 176 ± 5 m.y. (Table 104.3), compatible with its inferred slightly younger age.

The Linck Nunataks Granite appears fresh in both hand specimen and thin section, and it displays no postemplacement shearing or alteration. The rock weathers to a rubble of pink cobbles.

Metasedimentary rocks

Four exposures of metasedimentary rocks were found in the Whitmore Mountains. The thickest stratigraphic sequence, estimated to be 800 m thick, is exposed on a south-trending spur of the highest peak of Mount Seelig. Although this section proved inaccessible, a sampling of rock types was obtained in a talus deposit at the base of the exposure. Other exposures are small and generally consist of large blocks or roof pendants within the Mount Seelig Granite.

The metasedimentary rocks consist of a variety of altered clastic rocks, including phyllites, schists, graywackes, subarkoses, and impure quartzites. Feldspar con-

Figure 104.4. Linck Nunataks Granite (W-65-75), crossed nicols, 40×. Q: quartz. B: biotite. Or: orthoclase.

Table 104.4. Point-count estimates of mineral abundances in metasedimentary rocks of the Whitmore Mountains

	W-65-52	W-65-53	W-65-60	W-65-61	W-65-67	W-65-63	W-65-69	W-65-70	W-65-79
	Quartz-mica phyllite	Siliceous graywacke	Quartz-chlorite schist	Quartz-chlorite phyllite	Quartz-mica phyllite	Quartz-mica phyllite	Quartz-biotite phyllite	Quartz-biotite phyllite	Biotite schist
Quartz	42	28	41	55	69	67	39	40	9
Quartz-mica groundmass	41	46	0	0	0	0	0	0	0
Orthoclase	6	5	0	0	0	0	0	0	2
Orthoclase & albite	0	0	21	11	13	9	19	17	0
Plagioclase	1	3	0	0	0	0	0	0	38
Biotite	0	0	0	23	6	15	29	34	39
Chlorite	p	p	3	9	1	3	1	1	4
Muscovite	p	p	0	0	5	2	11	7	0
Apatite	p	0	0	p	p	p	p	p	p
Zircon	p	p	p	p	p	p	0	p	p
Sphene	0	0	2	1	1	1	p	p	2
Pyrite	p	0	1	1	0	0	0	0	0
Magnetite	p	p	p	p	p	p	p	p	p
Ilmenite	p	p	1	p	p	p	0	0	p
Hematite	p	p	p	p	p	p	p	p	p
Limonite	p	p	0	p	p	p	p	p	0
Calcite	9	2	3	p	5	3	1	1	0
Sericite	0	p	0	0	p	p	p	p	p
Metamorphic rock fragments	1	16	0	0	0	0	0	0	0
Actinolite	0	0	26	0	0	0	0	0	6
Epidote	0	0	2	0	0	0	0	0	0
Diopside (?)	0	0	0	0	0	0	0	0	0

p: present but less than 1%
J.J. Anderson, petrographer.

Location of specimens:
Mount Seelig: W-65-60 and 61
Northern Nunataks: W-65-52 and 53
Nunatak northeast of Linck Nunataks: W-65-67, 68, 69, and 70
Linck Nunataks: W-65-79
Mount Chapman: W-65-81, 82, 83a, 83b, 84, and 85

tent (Table 104.4) is high and averages more than 19%. Alkali feldspars constitute a high percentage of the feldspars. This feldspar component and the presence of metamorphic rock fragments in the metasedimentary rocks is interpreted as an indication that they were derived from a nearby granitic and metamorphic terrane. As the granites in nunataks south of the Ellsworth Mountains and at Pagano Nunatak are similar in age to the Whitmore Mountains granites which intrude these metasedimentary rocks, the source of this detritus is considered to have lain in the nearby Transantarctic Mountains area where older granites of the Ross orogen (Cambro-Ordovician) crop out extensively (Angino and Turner, 1963; Barrett, 1965; Ford and Aaron, 1962; Schmidt, 1964; Schmidt and others, 1964; Schmidt and others, 1965).

Because of very limited knowledge of the stratigraphy and the extent of the metasedimentary rocks in the Whitmore Mountains, they have not been given formational status.

Geomorphology

The Whitmore Mountains consist of several low mountains and nunataks. Slopes are sheer only on Mount Chapman and the west flank of Mount Seelig. Mount Seelig, Mount Radlinski, and the intervening ridges form a north-south impediment to the westward flow of the ice. Ice elevations on the east side of this ridge are as much as 130 m above those of the west side. The strongest crevassing in the Whitmore Mountains is along this ridge.

Several small ice-cored moraines are present in the blue-ice areas on the west side of Mount Seelig. All of the morainic material appears to be locally derived.

Rubble-covered surfaces are found in the Linck Nunataks at several locations. The thickest covers (as much as 25 cm) include clasts ranging from boulders to clay-size particles. No patterned ground was observed.

Regional implications

Bentley and others (1960), on an early map of West Antarctic rock surface elevations, showed the Whitmore Mountains as a positive area surrounded by ice grounded below sea level. If, as they suggested, the −500 m contour represents the approximate preglacial sea level, the Whitmore Mountains would then have been part of a southwest-trending peninsula attached to the narrow landmass extending southward from the Ellsworth Mountains.

Radiometric apparent ages of 196 ± 8 m.y. and 176 ± 5 m.y. on the Mount Seelig Granite and the Linck Nunataks Granite, respectively, indicate that the Whitmore Mountains belong to the West Antarctic tectonic province (Ellsworth orogen) formed by the Ellsworth Mountains and some nunataks and ranges to the south (Craddock, 1978).

The structural grain of the Whitmore Mountains is difficult to establish. Attitudes of bedding in the metasedimentary rocks from widely separated exposures most commonly show WNW-ESE orientations, but a total of only five strikes and dips of this bedding were obtained.

Table 104.4 (continued)

W-65-81	W-65-82	W-65-83a	W-65-83b	W-65-84	W-65-85
Felds-pathic quartzite	Granulite	Biotite schist	Granulite	Granite gneiss	Granite gneiss
54	53	44	43	25	30
0	0	0	0	0	0
8	8	27	28	36	35
0	0	0	0	0	0
1	4	0	p	12	13
0	0	26	0	27	22
p	1	2	2	p	p
0	0	0	0	0	0
0	p	p	p	p	p
p	1	p	1	p	p
1	p	p	p	p	0
0	0	0	0	0	0
0	p	p	0	p	p
0	0	0	0	0	0
0	p	p	0	p	p
0	0	p	0	p	p
17	8	1	4	0	0
p	0	p	p	p	p
0	0	0	0	0	0
0	5	0	3	0	0
16	2	0	0	0	0
18	18	0	19	0	0

Acknowledgments. This work was conducted as part of the United States Antarctic Research Program under grants from the National Science Foundation to the University of Minnesota (Campbell Craddock, principal investigator); this assistance is gratefully acknowledged. Much credit is also due to Roger Buchanan of the Bishop Museum, Honolulu, for assistance throughout the field season, and to the U.S. Navy for transportation and logistic support.

References cited

Anderson, J. J., Bastien, T. W., Schmidt, P. G., Splettstoesser, J. F., and Craddock, C. 1962. Antarctica: geology of the Ellsworth Mountains. *Science, N.Y., 138,* No. 3542, 824-25.

Angino, E. E., and Turner, M. D. 1963. Antarctic orogenic belts as delineated by absolute age dates. (*In* Adie, R. J., *Antarctic Geology*. Amsterdam, North-Holland Publishing Company, 551-56.)

Barrett, P. J. 1965. Geology of the area between the Axel Heiberg and Shackleton Glaciers, Queen Maud Range, Antarctica. Part 2—Beacon Group. *N.Z. Jl Geol. Geophys., 8,* No. 2. 344-70.

Bentley, C. R., and Ostenso, N. A. 1961. Glacial and subglacial topography of West Antarctica. *J. Glaciol.. 3,* No. 29, 882-910.

Bentley, C. R., Crary, A. P., Ostenso, N. A., and Thiel, E. C. 1960. Structure of West Antarctica. *Science, N.Y., 131,* No. 3394, 131-36.

Brandenberger, A. J. 1964. Aerial triangulation in the Antarctic. *Photogram. Engin., 30,* 197-201.

Bull, C. 1963. Measurement of the surface velocity of inland parts of the Antarctic ice sheet by an aerial triangulation method. *General Assembly of Berkeley, Int. Ass. Sci. Hydrol.,* No. 61, 144-46.

Craddock, C. 1966. The Ellsworth Mountains fold belt—a link between East and West Antarctica. *Spec. Pap. geol. Soc. Am., 87,* 37-38.

Craddock, C. 1978. Antarctica and Gondwanaland. (*In* McWhinnie, M. A., *ed. Polar Research—To the Present and the Future.* Boulder, Westview Press, 63-96.)

Craddock, C., Anderson, J. J., and Webers, G. F. 1964. Geologic outline of the Ellsworth Mountains. (*In* Adie, R. J., *ed. Antarctic Geology*. Amsterdam, North-Holland Publishing Company, 155-70.)

Ford, A. B., and Aaron, J. M. 1962. Bedrock geology of the Thiel Mountains, Antarctica. *Science, N.Y., 137,* 751-52.

Schmidt, D. L. 1964. Geology of the Pensacola Mountains. *Bull. U.S. Antarct. Projects Officer, 5,* No. 10, 98-101.

Schmidt, D. L., Dover, J. H., Ford, A. B., and Brown, R. D. 1964. Geology of the Patuxent Mountains. (*In* Adie, R. J., *ed. Antarctic Geology*. Amsterdam, North-Holland Publishing Company, 276-83.)

Schmidt, D. L., Williams, P. L., Nelson, W. H., and Ege, J. R. 1965. Upper Precambrian and Paleozoic stratigraphy and structure of the Neptune Range, Antarctica. *Prof. Pap. U.S. geol. Surv.,* 525-D, D112-19.

105

Geology of the Ellsworth Mountains–Thiel Mountains Ridge

Abstract

CAMPBELL CRADDOCK
Department of Geology and Geophysics, University of Wisconsin, Madison, Wisconsin 53706, U.S.A.

GERALD F. WEBERS
Department of Geology, Macalester College, St. Paul, Minnesota 55105, U.S.A.

JOHN J. ANDERSON
Department of Geology, Kent State University, Kent, Ohio 44242, U.S.A.

An area of scattered nunataks and small mountain ranges between the Ellsworth Mountains (lat. 79° S, long. 85° W) and the Thiel Mountains (lat. 85° S, long 88° W) comprises the only presently exposed geologic tie between West Antarctica and East Antarctica. Within this area bedrock exposures occur, in roughly southward order, in the Pirrit Hills, the Martin Hills, the Nash Hills, the Whitmore Mountains, Pagano Nunatak, the Hart Hills, and the Stewart Hills. In places between these outcrops, the ice sheet-bedrock contact commonly lies near or below sea level.

The Ellsworth Mountains consist of a strongly deformed, slightly metamorphosed sequence of upper Precambrian(?)-Paleozoic sedimentary rocks and minor volcanic rocks. The section is at least 13,000 m thick and contains fossils ranging from Middle Cambrian to Permian in age. The structural grain in this 350 km mountain chain trends northwest at its southern end but changes gradually to a north trend at its northern end.

Although outcrops are limited, geologic similarities exist among the Pirrit Hills, Martin Hills, Nash Hills, Whitmore Mountains, and Pagano Nunatak. Each of these five areas consists wholly or partly of intrusive plutons of mainly felsic composition. Except for Pagano Nunatak, there are some exposures in each area of the older country rock, which consists of deformed metasedimentary rocks. A total of 11 radiometric ages have been obtained on rocks from this region, including one or more from each outcrop area except the Martin Hills, and all of these ages fall in the 163–190 m.y. range.

The Hart Hills consist of deformed metasedimentary rocks and probably younger altered gabbroic rocks. Tightly folded metasedimentary rocks in the Stewart Hills have yielded a radiometric age of 508 m.y. In the Thiel Mountains, upper Precambrian metasedimentary rocks, including one thin marble bed with stromatolites, are tilted and cut by plutons with ages in the 470–720 m.y. range.

On the basis of composition, age, and structure the rocks of Pagano Nuntak, the Whitmore Mountains, and areas northward are considered part of the early Mesozoic Ellsworth orogen. The rocks of the Stewart Hills and the Thiel Mountains seem to preserve also the record of an older late Precambrian intrusive event. The tectonic position of the Hart Hills is problematic.

106

The Structure of Central East Antarctica from Recent Geophysical Investigations

Abstract

DAVID J. DREWRY
Scott Polar Research Institute, University of Cambridge, Cambridge CB2 1ER, U.K.

The results of radio-echo soundings obtained from a 50–100 km square grid in East Antarctica have been used to investigate terrain characteristics and gross geophysical properties of bedrock. The principal techniques employed have been: (1) isolation of large scale terrain units from detailed subice mapping (including both present-day and isostatically compensated elevations); (2) numerical parameterization of surface roughness using autocorrelation and other statistical methods; (3) computation of "terrain fabrics" (by consideration of surface derivatives in two or more different directions) providing data on lineation or structural "grain"; and (4) radio wave reflection characteristics including calculation of power reflection coefficients which are related, in part, to electrical properties of subice materials.

In central East Antarctica such studies indicate the presence and nature of distinctive geo-units such as the Gamburtsev Mountains (differentiated into several subprovinces). In the Dome "C" region, two large block-faulted massifs have been detected (lat. 72° S, long. 132° E and lat. 76.5° S, long. 128° E). A pronounced trench, possibly a rifted structure, flanks the northwest margin of the latter highland area. It extends for ~200 km, is 30–40 km wide, attains a mean depth of 750 m below sea level (locally 1000 m below sea level), and possesses scarps as high as 1700 m. Two major intracratonic basinal structures are examined—the Wilkes subglacial basin and the "Aurora" Basin. The former lies inland and subparallel to the Transantarctic Mountains; the latter forms a complicated branching system trending southeast from the Wilkes Land coast. Detailed analysis of radar and other geophysical data suggest that these basins contain a smoothing cover of sediments of as much as 3 km in thickness and are partially fault controlled.

VIII *Mineral Deposits*

107

Precambrian Iron Deposits of the Prince Charles Mountains

Review Paper

M. G. RAVICH†, L. V. FEDOROV, and O. A. TARUTIN
Department of Antarctic Geology, SEVMORGEO, Moyka 120, Leningrad 190121, U.S.S.R.

Abstract. In the course of Soviet geologic and geophysical exploration of the Lambert Glacier area, three iron ore occurrences were discovered in the Precambrian terrane. Two of them are bedrock outcrops at Mount Ruker and at Mount Stinear, located 80 km apart in the southern Prince Charles Mountains. The third occurrence is represented by glacial erratics in the southern Vestfold Hills, at a distance of almost 800 km from the first two localities.

At Mount Ruker, jaspilites are most abundant. They form about ten horizons of limited thickness, interbedded with metamorphosed mafic igneous rocks, ferruginous quartzites, and other greenschist facies rocks; the major jaspilite unit is nearly 400 m thick. This sequence (Ruker Series) belongs to a supracrustal complex of lower to middle Proterozoic age resting on the eroded surface of the upper Archean, highly granitized basement. A detailed section of the major jaspilite body is described, including its geologic relationships with the metabasites. The jaspilites have a distinctly banded structure with magnetite and subordinate hematite as major ore minerals, and quartz as the principal non-metallic component; carbonate, alkaline amphibole (riebeckite), albite, microcline, and brown and green mica are also present. The silica content in the Mount Ruker jaspilites varies from 35.84% to more than 60.58%, and the total amount of iron oxides is in the range of 33.72% to 57.35%. The Mount Ruker jaspilites have been traced westward under the ice sheet for 120 km by aeromagnetic survey.

Ferruginous quartzites at Mount Stinear form horizons of limited thickness among the Menzies Series schists (of staurolite-almandine facies). They possess a much lower content of iron oxides (as much as 25.95%); a considerable increase in magnesium and manganese occurs in garnet-bearing beds. The bedrock sources of the Vestfold Hills jaspilites are not known.

Analyses of initial compositions and existent mineral assemblages of the Antarctic jaspilites permit their correlation with similar jaspilite-bearing sequences on other continents formed in epicratonic environments under specific climatic and chemical conditions.

History of exploration and general geologic setting of the Precambrian Iron Formation

In 1956 Soviet geologists recorded large (1.5–2.0 m) subrounded boulders of ferruginous quartzites resting on high-grade metamorphic rocks of the crystalline basement exposed in the northeastern Vestfold Hills (Ravich and Voronov, 1958). The sources of these boulders was unknown. In 1966 jaspilites and ferruginous quartzites were found at Mount Ruker (lat. 73° 38′ S, long. 68° 40′ E) in the southern Prince Charles Mountains in a bedrock outcrop; the jaspilites were observed to form a horizon 70 m thick in a greenschist facies metasedimentary sequence (Soloviev and others, 1967).

During 1971–74 the Soviet Antarctic Expedition carried out a systematic geologic-geophysical survey of the Prince Charles Mountains area. In the course of this work, led by D. S. Soloviev†, a detailed study of Mount Ruker was undertaken which resulted in further discoveries of much thicker jaspilite units and in a better understanding of their geologic position. Two major aeromagnetic anomalies were discovered and were found to extend west of Mount Ruker in a sublatitudinal direction for 120–180 km, each of them 5 to 10 km wide with positive amplitudes ranging from 600 to 3000 gamma. On the basis of the apparent spatial association of these anomalies with the Mount Ruker bedrock exposures, as well as the high relief of the anomalies despite the considerable height of the aircraft above the potential magnetic source, these anomalies are almost certainly indicative of subglacial continuation of the Mount Ruker jaspilite bodies.

The presence of iron formations elsewhere in the Prince Charles Mountains area is indicated by discoveries of tectonic fragments of jaspilite units in some outcrops (Mount Stinear). These most likely represent tectonic slices of jaspilite-bearing sequences of unknown thickness and lateral extent, which may form the subglacial source of several other intense magnetic anomalies recorded in the surveyed area. So far, however, only the southern Prince Charles Mountains (Fisher Glacier area) can be identified as having real potential for iron deposits.

The larger northern part of the territory under consideration (between the Mawson Escarpment and the coast) is a relatively uniform high-grade crystalline base-

† Deceased.

ment terrane composed of polymetamorphic assemblages (granulite facies and high-grade amphibolite facies of regional metamorphism) typical of the Archean complexes. The rocks include two-pyroxene, biotite-hypersthene-garnet, and cordierite-sillimanite schists, migmatites and shadow granites, and porphyroblastic charnockites (Ravich and Kamenev, 1972). In the southern part of the Mawson Escarpment and adjacent mountains separating the upper reaches of Lambert Glacier, two major structural complexes are exposed. The crystalline basement is represented there by its youngest—uppermost Archean to lower Proterozoic(?)—complex composed of intensely migmatized and granitized biotite-amphibole, biotite-garnet, and amphibole plagiogneisses, and schists of the upper grade amphibolite facies of regional metamorphism. The jaspilite-bearing supracrustal rocks (considered to be lower to middle Proterozoic in age) differ from the crystalline basement in having a considerably lesser degree of metamorphism and in having a different zonal character (greenschist and lower to middle-grade amphibolite facies), as well as in having a total absence of migmatization and granitization.

In the eastern part of Mount Ruker the supracrustal complex was observed to rest with a structural unconformity on the eroded surface of basement granites with a minimum age of 2500 m.y. Elsewhere in the southern Prince Charles Mountains the original relationship between the crystalline basement and the overlying supracrustal rocks is obscured by severe faulting, responsible for the "broken-plate" appearance of the whole area. Nevertheless, a major structure of a graben-synclinorium type can be recognized along Fisher Glacier; the axial zone of this synclinorium appears to be occupied by the greenschist facies Ruker Series—a major iron formation consisting of metabasites, various metasedimentary rocks, ferruginous quartzites, and jaspilites. Along the flanks of the synclinorium the somewhat higher grade rocks of the Menzies Series occur; these rocks are interpreted as remetamorphosed equivalents of the Ruker Series.

The Mount Ruker site

Stratigraphy of the iron formation

The Ruker Series (Figure 107.1) crops out through an area of 10,000 km^2 near the Fisher Glacier. Its best exposures are located in the central part of the northern face of Mount Ruker. The section is tectonically much disturbed, especially in its lower ferruginous part rich in jaspilites and metabasites, so that a reliable stratigraphic succession can be worked out only for the upper part of the section, which contains no ferruginous beds. It is, however, quite clear that the very base of the section is composed of interlayered schists (quartz, quartz-chlorite, and mica-chlorite), microquartzites, metabasites, and jaspilites in horizons as much as 30–50 m thick, as a sequence of that composition was observed to rest with structural unconformity on the basement granites in the southeastern part of Mount Ruker. This sequence apparently strikes in a northwesterly direction under moraine and snow, and reappears in bedrock outcrops at the eastern end (at the base?) of the section exposed along the northern slope of Mount Ruker. There, however, much thicker (as much as 300–400 m)

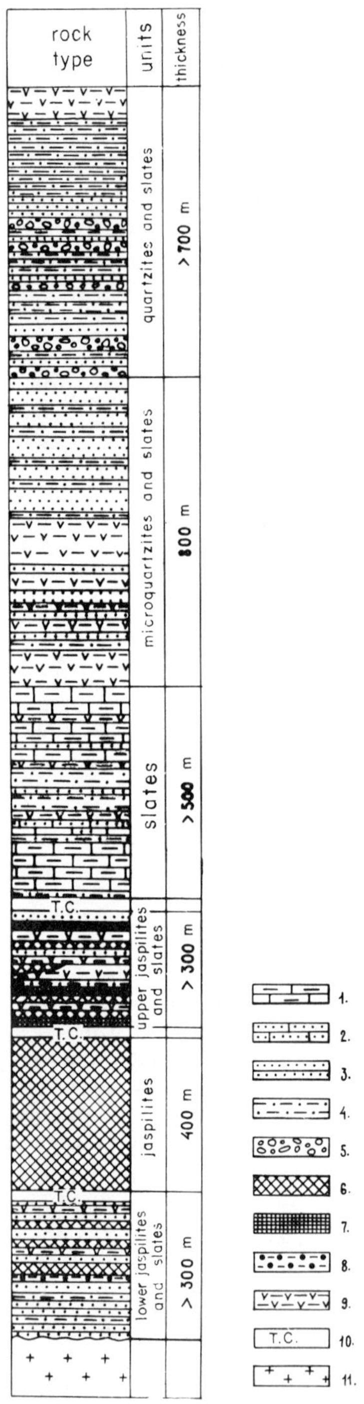

Figure 107.1. Generalized geologic section of the Ruker Series at Mount Ruker. 1. Sericite-carbonate-quartz schists; 2. Chlorite-carbonate-quartz schists; 3. Quartzites, micaceous quartzites, and quartz metasandstones and metasiltstones; 4. Calcareous quartzites, metasandstones, and metasiltstones; 5. Metaconglomerates and metaconglomerate breccias; 6. Jaspilites; 7. Ferruginous chlorite-mica-quartz schists; 8. Spessartite-bearing mica-chlorite-quartz schists; 9. Metabasites; 10. T.C.: tectonic contact; 11. Gneissic granites.

bodies of jaspilites and metabasites occur, along with interlaying of the type mentioned above. A detailed measured section of the major jaspilite body at Mount Ruker is given below, from base to top:

1. Gray laminated slaty jaspilites with about 20% of quartzite intercalations, 26 m.
2. Dark gray and black laminated jaspilites, impregnated with needle-like metacrysts of alkaline amphibole, with about 15–20% of quartzite intercalations, 33 m.
3. Intensely altered metaporphyrite, 3 m.
4. Jaspilites similar to those of unit 1, 30 m.
5. Interlayered purplish and gray crinkled jaspilites in beds 0.5–5.0 m thick, 28 m.
6. Dark gray slaty laminated jaspilites with rare intercalations of chlorite-quartz schists, 0.3–0.5 m thick, and light brown quartzites, 0.1–0.3 m thick, 50 m.
7. Jaspilites similar to those of unit 1, 10 m.
8. Gray slaty jaspilites intensely impregnated with alkaline amphibole (riebeckite) in glomeroblasts as large as 3 cm, 10 m.
9. Gray schistose crinkled jaspilites with quartz-carbonate and quartz-actinolite veinlets, 10 m.
10. Dark gray and brown-gray crinkled jaspilites with rare thin intercalations of quartzites, 45 m.
11. Dark gray and black massive jaspilites, 5 m.
12. Light brown quartzites, intensely cleaved, with numerous quartz-siderite and quartz-hematite veinlets, 5 m.
13. Rhythmically bedded gray slaty laminated jaspilites and quartzites, 25 m.
14. Interlayered purplish and black microfolded jaspilites, with abundant quartz-siderite veinlets and metacrysts of alkaline amphibole at the top of the unit, 50 m.
15. Gray laminated jaspilites with rare intercalations (as thick as 0.5 m) of light gray quartzites, 20 m.
16. Jaspilites similar to those of unit 1, 10 m.
17. Black laminated slaty jaspilites; at the top of the unit, near the tectonic contact with a sill of metagabbro, they are intensely brecciated and sheared and impregnated by quartz veinlets with albite, pyrite, siderite, and rare large metacrysts of alkaline amphibole, 20 m.

The jaspilite unit described above forms a large outcrop (about 1000 m in length and 100–150 m in height) at the eastern end of a continuous cliff exposing the Ruker Series section in northern Mount Ruker. The jaspilites within this outcrop are rather regular, thinly laminated rocks, highly distorted by minor disharmonic plications ranging in wavelength from a few centimeters to a few dozen meters; the microfolds are commonly isoclinal, with steep (60°–70°) axial planes striking in a northwest-southeast direction.

The jaspilite unit has an upper tectonic contact with a large body of metabasite (metagabbro?) composing adjacent cliffs for 2.0–2.5 km westward. The configuration and geologic position of this body are obscure; most likely it is a semiconcordant intrusion of lensoid or sheet form with a number of feeders, but it may be a steep tabular body with numerous satellite sills and dikes. A few spurs at the base of the cliff are composed of ferruginous rocks, one of those being a 70-m-thick jaspilite unit first sighted by D. S. Soloviev in 1966. At these spurs, ferruginous rocks and metagabbros can be observed to have both sharply discordant and conformable relationships. The former are certainly intrusive contacts, as along them metagabbros are chilled against the jaspilites for 0.3–0.5 m. In the latter case both concordant intrusive and tectonic contacts can be argued, as metabasites in contact zones are highly sheared and turned into typical "greenschist" with schistosity parallel to that in the adjacent ferruginous rocks. At the western termination the metabasite body seems to thin out and terminate against a rhythmic ferruginous sequence similar to that at the very base of the Ruker Series section. Only a limited thickness of this sequence is exposed before it is cut by a steep fault, beyond which, farther to the west, the upper part of the Ruker Series is exposed.

As was mentioned above, the upper part of the section contains no ferruginous rocks, although some of the sedimentary beds are certainly iron-enriched. It begins with approximately 200 m of essentially carbonate rocks: carbonate-quartz-mica schists with layers of chloritic marbles, marbled (partly recrystallized) limestones, and intercalations of micaceous microquartzites and quartz-mica schists. Much of the carbonate is in the form of siderite. The next 500–600 m is composed of various phyllites interbedded with essentially quartzose sedimentary rocks, and the uppermost 700 m of the exposed section consists of interbedded chlorite-mica schists, quartzites, metasandstones, and metaconglomerates. Sheets of metabasite are abundant; in places they reach a few dozen meters in thickness, and they are especially numerous in the second unit.

Composition and geochemistry of Mount Ruker Jaspilites

The Mount Ruker jaspilites are predominantly dark gray and purplish-black, banded, crinkled, or microfolded rocks. Coarse banding is caused by rhythmic intercalations of opaque mineral layers as much as 1–2 cm in thickness with quartzites, 2–3 cm thick, which have little or no opaque mineral content. Intercalations of both types are further complicated by microlamination of a similar nature (Figures 107.2*a* and *b*). The opaque intercalations are composed predominantly of polyhedral aggregates of granoblastic magnetite and hematite which usually occur in a 3:2 ratio; the grain size is 0.05 mm to 0.8–1.0 mm. Small granoblastic quartz grains do not exceed 20% of the opaque bands. Within the opaque intercalations there are lenticular leucocratic microlaminae not exceeding 2–3 mm and consisting of quartz with some admixture of tiny subidiomorphic opaque grains. Essentially the quartz intercalations contain magnetite and hematite (up to 15–20%) in grains grouped into thin chain-like aggregates and lenses, or irregularly spaced as micro-inclusions in granoblastic quartz. In some opaque intercalations hematite is replaced by magnetite (muschketowite); magnetite and hematite are commonly replaced by limonite. These processes are usually associated with contact metamorphism caused by sills of metagabbroid rock. The reverse process of martitization is also recorded.

In addition to the above minerals, jaspilites contain various amounts of carbonate, alkaline amphibole (riebeckite), albite and microcline, brown biotite, and green mica. Riebeckite is encountered in a majority of thin sections as a product of alkaline metasomatism, along with microcline, albite, and green mica. The riebeckite forms needle-shaped, elongate prismatic, and rhombic isolated grains, or their glomeroblastic aggregates. The alkaline minerals usually develop along the opaque mineral layer–quartzite contacts in thinly banded varieties. The amount of secondary carbonate reaches 18–20%; it is mostly developed along the bedding planes. Carbonate is represented by siderite, which forms the largest crystals, and by smaller grains of calcite and dolomite. In some places siderite grains group around magnetite aggregates. In zones of brecciation the jaspilite fragments are cemented with mosaic aggregates of quartz, calcite, and siderite. The matrix of breccias is commonly also impregnated by pyrite and recrystallized hematite.

A number of jaspilite specimens have been subjected to chemical analyses (Table 107.1). The results show a rather uniform composition for the Mount Ruker jaspilites. The greatest variations are observed in silica content (from 35.84% to 60.58%) and in total iron oxide content (from 33.72% to 57.35%), which reflect variations in the ratio of quartz to opaque constituents in analyzed jaspilites. The total amount of other oxides does not exceed 6–7%, while ignition losses reach 3–4% in some specimens due to the high content of constitutional water in chlorite and micas. Subordinate elements are mainly calcium and magnesium, which are present as a constant admixture of carbonates in jaspilites. An increased amount of alkalis in some specimens is related to the presence of such products of superimposed alkaline metasomatism as alkaline amphibole, albite, and microcline. The low content of sulfur and phosphorus must be specifically noted. Quantitative spectral analyses additionally show the presence in jaspilites of as much as 0.2% nickel; copper, chrome, cobalt, and zirconium contents range from 0.001% to 0.01%.

Mount Stinear site

At Mount Stinear, ferruginous quartzites contain garnet and are thought to belong to the Menzies Series of the staurolite-almandine facies. They occur as distinct units of variable thickness (5–6 m to 15–20 m) among quartzites and kyanite-staurolite-garnet schists, which are preserved as down-faulted wedges in the granitic-gneissic complex of the crystalline basement. Ferruginous quartzites have an indistinctly banded structure caused by intercalation of opaque mineral lenses with essentially quartzitic ones. Dark gray and black opaque laminae are 0.5–1.0 cm thick and consist of fine granules of magnetite, hematite, and quartz. Light intercalations are 3–5 cm thick; they consist of quartz and garnet with an admixture of magnetite, and have a pink color in hand specimen. In some specimens the amount of opaque material reaches 25%.

Figure 107.2. a. Microlamination of jaspilites. *b.* Banding of jaspilites. Light laminae: quartz. Black laminae: magnetite.

Chemical analyses of ferruginous quartzites from Mount Stinear (Table 107.1) show that the total content of ferric and ferrous iron is relatively low (21.64%–25.95%). A sharp increase in the amount of magnesium oxide (up to 4.15%) and manganese oxide (up to 4.60%) is associated with the presence of garnet.

Vestfold Hills site

In the Vestfold Hills area, jaspilites were found not only as great isolated lumps scattered over the ice-polished surface of the Archean rocks, but also as smaller and more numerous boulders, concentrated near the coast in terminal moraine deposits together with clasts of the greenschist facies metasedimentary rocks. The source of these ferruginous deposits is not clear. It is tempting to look for the subglacial source in an extensive zone of positive magnetic anomalies south of the Vestfold Hills. This zone extends in a latitudinal direction for about 120 km and consists of a number of anomalies ranging from 10 to 40 km in length, from 2 to 4 km in width, and from 800 to 5900 gamma in positive amplitudes. Both the configuration and intensity of some anomalies are quite comparable to those recorded over the Mount Ruker jaspilites, under similar conditions of aircraft height and ice thickness. On the other hand, in contrast to the Fisher Glacier area, the

Table 107.1. Chemical composition (in wt %) of jaspilites and ferruginous quartzites

	Jaspilites, Mount Ruker													
	707-V	707-d	707-i	707-c	707-t	707-u	507-7	507-8	507-9	212-a	212-b	1007-21	1007-26	1007-31
SiO_2	51.46	53.12	60.58	49.52	42.15	60.32	40.70	48.68	49.58	57.30	35.84	42.34	43.64	43.87
TiO_2	0	0	trace	0.01	0	0	0	0	trace	trace	0.04	0.07	0.03	0.07
Al_2O_3	1.18	0.54	0.32	1.60	0.65	0.61	0.37	0.53	0.67	1.15	1.35	0.66	0.36	0.36
Fe_2O_3	33.38	34.05	26.72	27.18	40.06	25.59	38.48	31.70	34.04	26.03	40.96	39.81	34.85	39.78
FeO	8.80	8.15	7.00	13.34	12.35	8.58	14.04	11.35	10.68	10.97	16.39	9.88	14.77	10.52
CaO	0.78	0.48	1.26	1.94	0.56	0.75	1.29	0.68	0.76	0.80	0.56	1.52	1.29	0.96
MgO	1.30	1.32	1.58	1.55	1.96	2.07	1.76	2.29	1.51	1.50	2.23	1.71	1.57	1.10
MnO	0.13	0.12	trace	0.17	trace	trace	0.31	trace	trace	0.16	0.16	0.09	0.11	0.12
K_2O	0.11	0.11	0.04	0.84	0.15	0.25	0.25	0.55	0.42	0.18	0.58	0.75	0.32	0.25
Na_2O	0.16	0.83	0.07	0.19	1.24	1.40	0.64	2.78	0.89	0.60	0.65	0.90	0.52	0.42
P_2O_5	0.08	0.11	0.10	0.62	0.08	0.08	0.11	0.44	0.08	0.08	0.06	0.48	0.13	0.14
Loss on ignition	2.20	0.75	2.34	3.21	0.78	0.64	2.33	0.76	1.43	1.54	1.29	1.68	2.50	2.10
Total	99.58	99.58	100.01	100.17	99.98	100.29	100.28	99.76	100.06	100.31	100.11	99.89	100.09	99.69
H_2O at 105°-110°C	0.16	0.20	0.30	0.12	0.12	0.24	0.04	0.20	0.14	trace	0.14	0.10	0.14	0.14
Sulfur (total)	0	0	0	0	0	0	0	0	0	0	0	trace	trace	trace

	Jaspilites, Mount Ruker							Ferruginous quartzites, Mount Stinear		
	1007-7	1007-12	1007-16	1008-A	1008-11A	1009-3A	1007-34	1018-21B	1018-22	1018-23
SiO_2	47.31	53.99	41.68	46.08	42.92	44.70	36.93	53.05	52.25	50.25
TiO_2	0.03	trace	0.05	0.07	0.03	0.10	0.03	0.50	0.46	0.37
Al_2O_3	0.54	0.24	0.43	0.38	0.35	1.22	0.46	14.08	11.75	10.70
Fe_2O_3	29.01	27.56	39.81	36.90	35.08	29.98	45.77	7.54	9.59	11.29
FeO	14.65	11.33	13.59	12.33	14.48	14.66	11.31	14.48	12.05	14.66
CaO	1.07	1.18	0.59	0.73	1.34	1.55	1.30	1.28	1.83	1.80
MgO	1.69	1.62	1.32	1.54	1.44	2.04	1.70	3.93	4.15	3.46
MnO	0.25	0.15	0.09	0.18	0.28	0.35	0.11	1.67	3.81	4.60
K_2O	0.54	0.34	0.50	0.30	0.47	0.95	0.45	1.70	2.07	1.34
Na_2O	0.31	0.72	0.48	0.70	0.34	0.21	0.60	0.42	0.55	0.14
P_2O_5	0.26	0.17	0.12	0.14	0.32	0.19	0.17	0.17	0.12	0.17
Loss on ignition	4.43	2.38	1.19	0.91	2.68	3.83	1.20	1.04	1.37	0.93
Total	100.09	99.68	99.85	100.26	99.73	99.78	100.03	99.86	100.00	99.71
H_2O at 105°-110°C	0.12	0.22	0.24	0.23	0.15	0.16	0.16	0.15	0.25	0.16
Sulfur (total)	0.02	0.01	trace	0.08	trace	0.06	trace	0.01	trace	0.01

	Jaspilites, Vestfold Hills boulders				
	1349-20	1349-16	1348-43	1350-20	1350-19
SiO_2	50.20	60.63	49.69	47.54	51.94
TiO_2	0.15	trace	0.29	trace	trace
Al_2O_3	0.18	0.48	6.90	35.35	trace
Fe_2O_3	37.84	25.10	8.74	13.54	32.67
FeO	10.59	11.96	25.38	1.25	10.24
CaO	0.31	0.17	5.42	0.94	2.76
MgO	0.24	0.87	2.26	0.09	1.41
MnO	0.06	0.19	0.13	0.05	0.06
K_2O	0.05	0.07	0.06	0.05	0.06
Na_2O	0.04	0.04	0.14	0.05	0.07
P_2O_5	0.14	0.16	0.17	0.14	0.12
Loss on ignition	0.24	0.53	0.82	1.13	0.64
Total	100.04	100.20	100.00	100.13	99.97
H_2O at 105°-110°C	0.04	0.20	0.19	0.26	0.22
Sulfur (total)	trace	trace	0.29	trace	trace

axis of this anomalous zone is apparently related to a narrow depression in the bedrock surface, which could hardly have developed over a considerable jaspilite body. Furthermore, the jaspilite-bearing epicratonic complexes tend to accumulate on highly granitized metamorphic terranes of amphibolite facies rather than on the older, less granitized granulite and/or polymetamorphic complexes which dominate the Vestfold Hills area. Finally, it is difficult to visualize the means of transportation of the rock debris for even a short distance in the absence of large outlet glaciers.

An alternative explanation would require transport of jaspilites into the Vestfold Hills area from the southern Prince Charles Mountains by the Lambert Glacier stream. However, even if such transport were feasible, the Vestfold Hills jaspilites display certain differences in appearance, mineralogy, and chemistry from both the Mount Ruker and Mount Stinear jaspilites. Hence, they cannot as yet be correlated with any known bedrock sources of ferruginous deposits.

Discussion

Of the three iron formation occurrences described above—Mount Ruker, Mount Stinear, and the Vestfold Hills—only the first has been studied in sufficient detail to attempt genetic correlation with better known jaspilite provinces elsewhere in the world. By analogy with such provinces of the Russian Platform, the metasedimentary rocks of the Ruker Series can be subdivided into three groups according to their initial composition and ultimate mineral assemblages: (1) clastic sands, silts, and muds, turned into quartzites and phyllites, now characterized by the association of quartz with sericite and chlorite and an admixture of magnetite (hematite), calcite, and epidote; (2) clastic sediments of type 1 mixed with chemically formed carbonates, turned into various carbonate-bearing schists and marbled limestones, with a typical mineral assemblage of calcite, dolomite, quartz, epidote, and chlorite; (3) siliceous and ferruginous chemical precipitates, turned into jaspilites, with an association of magnetite, hematite, quartz, dolomite, and siderite. All of the above mineral assemblages are typical of the sericite-chlorite subfacies of the greenschist facies.

Similar conditions of sedimentation in the jaspilite-bearing sequences of the Russian platform are believed to indicate periodic changes of environment in shallow-water basins typical neither of geosynclinal nor of classical platform development. Such basins can best be described as aulacogens, i.e., slowly deepening elongate troughs, semicompensated by sedimentation, and formed in connection with steady subsidence of a cratonized crystalline basement in linear zones of crustal weakening. The considerable thickness of supracrustal complexes in such basins is related to their exceedingly prolonged subsidence, so that even extremely low rates of accumulation provide great piles of well-sorted clastic and chemical sediments. The orogeny which caused folding and metamorphism of the supracrustal sedimentary fill was the first major tectonic event since cratonization of the crystalline basement, at about the end of Archean time.

Intense chemical weathering of highly ferruginous older rocks must have occurred under a warm and humid climate to give rise to jaspilite formation. Such a terrane was probably represented by a near-peneplane underlain by the lower Precambrian crystalline complexes rich in pyroxene, amphibole, biotite, garnet, etc. A stable tectonic environment in the source area stimulated perfect sorting of detritus, which was reduced to only a few minerals resistant to chemical and mechanical weathering (quartz, rutile, tourmaline, zircon, etc.), while rapid changes of pH value facilitated alternate precipitation of silica and iron oxide layers.

There is no direct evidence for the time of accumulation of the ferruginous sedimentary sequence of the Prince Charles Mountains, and the age of its metamorphism is also poorly known. It is beyond any doubt, however, that the metasedimentary sequence is younger than the east Antarctic crystalline basement cratonized during latest Archean to earliest Proterozoic time. Hence, the iron formation of East Antarctica can provisionally be assigned to the early to middle Proterozoic. A similar conclusion is obtained by analogy with the Singareni Series of the lower part of the Cuddapah System in southern India (Crawford, 1973), and with the Krivorozhsky Basin of the U.S.S.R. (Kalganov, 1965).

References cited

Crawford, A. R. 1973. The age of the Cuddapah and Karnool systems, southern India. *J. geol. Soc. Aust., 19,* 453–64.

Kalganov, M. I. 1965. Zhelezorudnye formatsii v dokembriyskom fundamente Russkoi platformy [Iron deposits in the Precambrian basement of the Russian platform]. (*In* Pustovalov, L. V., *ed. Rudonosnost' Russkoi Platformy* [Ore Deposits in the Russian Platform]. Moscow, Izdatel'stvo Nauka, 46–53.)

Ravich, M. G., and Kamenev, E. N. 1972. *Kristallicheskii fundament Antarkticheskoi platformy* [Crystalline basement of the Antarctic platform]. Leningrad, Gidrometeoizdat, 658 pp. [English translation: New York, John Wiley, 1975, 582 pp.]

Ravich, M. G., and Voronov, P. S. 1958. Geologicheskoe stroenie poberezhiia Vostochnoi Antarktidy (mezhdu 55 i 110° v.d.) [Geological structure of the East Antarctic coast (between 55° and 110° E)]. *Sovet. Geol.,* No. 2, 3–26.

Soloviev, D. S., Kamenev, E. N., and Ravich, G. M. 1967. Geologicheskie issledovaniia v 1965/66 g. [Geological activities in 1965/66]. *Inf. Byull. sov. antarkt. Eksped.,* No. 62, 10–18.

108

Metallic Mineral Resources of the Antarctic Peninsula

Review Paper

PETER D. ROWLEY
U.S. Geological Survey, Denver, Colorado 80225, U.S.A.

DOUGLAS E. PRIDE
Department of Geology and Mineralogy, The Ohio State University,
Columbus, Ohio 43210, U.S.A.

Abstract. Geologic similarities between the Andes of South America and the Antarctic Peninsula, coupled with Gondwana reconstructions, indicate that the Antarctic Peninsula is part of the circum-Pacific Andean (Mesozoic and Cenozoic) orogenic belt. However, predicitons of great mineral riches in the peninsula, made by analogy with the mineral-rich central Andes, may be premature because the two regions have somewhat different Cenozoic tectonic histories. Many metallic mineral occurrences are known in the Antarctic Peninsula, but none contain ore-grade material, and they are not likely to be mined now or in the foreseeable future. These occurrences, however, indicate widespread mineralization; thus it is possible that further mapping will reveal mineable mineral deposits.

Metallic minerals are found in trace amounts at many places, and some occurrences of mineralized and hydrothermally altered rocks in the Antarctic Peninsula cannot be appraised adquately because published reports are too sketchy. Occurrences that are interesting, though not presently economic, include: (1) widespread hydrothermal(?) pyrite and quartz veins and disseminated pyrite on King George Island; (2) a hydrothermal copper-lead vein and disseminated sulfide deposit on Livingston Island; (3) scattered veinlets and disseminated crystals of copper, iron, and molybdenum sulfides and oxides, as well as nearby high-temperature alteration zones—possibly reflecting porphyry copper systems—in the Anvers Island area and on Adelaide Island, Brabant Island, and the Melchior Islands; (4) magnetite-rich lava flows on Brabant Island; (5) magmatic magnetite and copper minerals in layered gabbro plutons in the Argentine and Terra Firma Islands; and (6) a low-grade porphyry-type deposit containing veinlets and disseminated sulfides and oxides of copper, iron, lead, and molybdenum adjacent to sheared high-temperature alteration zones in the Lassiter Coast of the southeastern Antarctic Peninsula. Virtually all mineralized and hydrothermally altered rocks in the Antarctic Peninsula are localized near Mesozoic and Cenozoic plutons of the Andean Intrusive Suite. Most of the more promising occurrences are hydrothermal deposits related to Tertiary trondhjemitic plutons in westward islands that are separated by faulting from mainland parts of the northern and central Antarctic Peninsula.

The enormous climatic, logistic, technologic, and political problems confronting any endeavor to recover minerals or hydrocarbons from Antarctica have been discussed by numerous workers, among them Schnellman (1955), Kosack (1955), Anonymous (1956, 1974), Chalmers (1957), McLeod (1965), Potter (1969), Smith (1972), Wade (1976), Piper (1976), Splettstoesser (1976), Institute of Polar Studies (1977), Mitchell (1977), and Rowley and Williams (1979). The Antarctic Peninsula itself is largely covered by an ice sheet and is fringed by ice shelves as much as 200 km wide. Approaches to the peninsula from the sea are marked by frequent heavy storms and widespread pack ice. Access by ship to land is limited to the northern and northeastern parts of the peninsula during calm days of the summer season. Elevations in the interior parts of the peninsula are as high as 4000 m; access to the interior by land is difficult because of crevassed glaciers. The geologic constraints likewise are severe. With the possible exceptions of iron in East Antarctica and coal on the margins of East Antarctica (Wright and Williams, 1974; Ravich and others, Chapter 107, this volume), no mineral deposit is yet known in Antarctica that could be exploited economically, even if it were located in a more favorable climate and nearer to markets.

These drawbacks notwithstanding, recent demands upon world supplies of minerals and energy materials have focused considerable attention on the prospects of finding exploitable mineral deposits in Antarctica. This attention has been accompanied by legitimate concern over the effects that mineral exploitation might have on the environment (Cousteau, 1976; Schofield, 1976). Certainly the spectacular and unusual beauty of Antarctica and its living resources deserve continued protection by all nations. However, if proper decisions are to be made concerning the possible future use of its resources, it is important that knowledge of the region be actively sought. This paper reviews the known occurrences of metallic minerals in the Antarctic Peninsula and attempts to interpret them in terms of modern theories of ore genesis.

Regional geology

Most rocks exposed in the Antarctic Peninsula are stocks and batholiths of the Mesozoic and Tertiary Andean orogen. These plutons, named the Andean Intrusive Suite by Adie (1955), intrude a variety of rocks of different ages: (1) Paleozoic(?) basement rocks; (2) a thick folded sedimentary sequence of Carboniferous(?) to Triassic age, in most places named the Trinity Peninsula Series; (3) thick sequences of folded Jurassic and Lower Cretaceous sedimentary and volcanic rocks; and (4) scattered patches of horizontal or slightly deformed Upper Cretaceous and Tertiary sedimentary and volcanic rocks. These stratified and plutonic units are locally overlain unconformably by mostly undeformed sedimentary and volcanic rocks ranging in age from Late Cretaceous to Quaternary (Adie, 1964, 1969*a,b*; Craddock, 1972*a*; Dalziel and Elliot, 1973; Grikurov, 1973).

The rocks of the Antarctic Peninsula clearly belong within the circum-Pacific orogenic system. Numerous workers have noted the strong geologic similarities between the South American Andes and the Antarctic Peninsula, most recently Adie (1964, 1972*a*), Craddock (1972*b* and Chapter 1, this volume), Dalziel and Elliot (1973), Grikurov (1973), Wright and Williams (1974), Wade (1976), and Ericksen (1976). The two regions probably belonged to a single magmatic arc system in Mesozoic time (Suárez, 1976) and were fragmented and separated from each other by subduction and transform faulting during development of the Scotia arc in Cenozoic time (Dalziel and Elliot, 1973; Herron and Tucholke, 1976; Craddock and Hollister, 1976; de Wit, 1977). South of the peninsula, the Andean orogen probably passes westward through western Ellsworth Land and Marie Byrd Land, and from there curves northward through New Zealand and the western Pacific region (Craddock, 1969, 1972*b*, 1977, and Chapter 1, this volume; Wright and Williams, 1974; Molnar and others, 1975; Craddock and Hollister, 1976; LeMasurier and Wade, 1976).

The presence of a variety of mineral deposits in Gondwana continents formerly adjacent to Antarctica has been used by many workers (e.g., Runnels, 1970; Wright and Williams, 1974) to suggest that parts of Antarctica may also contain significant mineral deposits. In particular, the geologic similarities between the Andes and the Antarctic Peninsula have led many persons to assume that mineral deposits like those in the central Andes (northern Peru to northern Chile), one of the world's richest metallogenic areas, may be found farther south in the poorly known areas of the southern Andes and Antarctic Peninsula. In a recent review paper, Ericksen (1976) has projected metal subprovinces the length of the Andean orogen of South America, but noted that deposits in the southern Andes and Antarctic Peninsula seem sparse and of lower grade than those of the central Andes. In the central Andes, mineralized rocks are related to the emplacement of calc-alkaline plutons from Triassic through Cenozoic time (Farrar and others, 1970). The intrusive activity and associated mineralization are apparently related to rapid subduction of oceanic crust beneath the South American continent (James, 1971; Herron and Tucholke, 1976). Subduction is continuing along much of the western margin of South America, as indicated by the presence there of a spectacular ocean trench and intense seismicity.

At about lat. 45° S (southern Chile), however, the southeast-trending Chile Ridge of the ocean basin comes in contact with, and is subducted beneath, South America (Figure 108.1 inset). Beginning in Late Cretaceous or early Tertiary time, the tectonic history of the area south of the Chile Ridge intersection changed markedly from that north of the Chile Ridge. Except for the oceanic trench in and near the Sandwich Islands in the eastern Scotia arc, southernmost South America has low seismicity, and Antarctica is essentially aseismic; neither southernmost South America nor Antarctica exhibit definite evidence of present-day subduction. Sea-floor magnetic anomalies south of the Chile Ridge give evidence of an extremely complex subduction history since Late Cretaceous time and perhaps before (Herron and Tucholke, 1976). Interpretation of these anomalies (Herron and Tucholke, 1976; Craddock and Hollister, 1976) and of geologic information from land areas (Farrar and others, Chapter 39, this volume; Rowley and Williams, Chapter 38, this volume; LeMasurier and Wade, 1976; LeMasurier and Rex, Chapter 89, this volume) suggests that subduction beneath northern Marie Byrd Land and western Ellsworth Land probably ceased by early or middle Tertiary time, whereas it continued at a relatively slow rate and in different directions until late Tertiary time under most parts of southernmost South America and the Antarctic Peninsula. In Marie Byrd Land and western Ellsworth Land, crustal extension and alkaline igneous activity started in early to middle Tertiary time and have continued up to the present (LeMasurier and Wade, 1976; LeMasurier and Rex, Chapter 89, this volume), whereas extensional faulting with mildly alkaline volcanism and transform faulting with tholeiitic volcanism were dominant in the Scotia arc and along the coastline of the Antarctic Peninsula, starting locally no later than middle Tertiary time. Ericksen (1976, p. 537) has suggested that differences in the "position, inclination, and subduction rate of the oceanic lithospheric plate" may have produced differences in ages, types, and numbers of mineral deposits. Thus the postulated subduction histories north and south of the Chile Ridge may account in part for the relatively fewer mineral deposits south of the Chile Ridge and must surely temper predictions of great mineral riches in West Antarctica. Metallic minerals seem to be even less abundant in Marie Byrd Land than in the Antarctic Peninsula (Wade, 1976; Wade, personal communication, 1977); perhaps this is because subduction ceased earlier in Marie Byrd Land than in the peninsula.

These ideas notwithstanding, the many geologic similarities between the South American Andes and the Antarctic Peninsula suggest that an understanding of the mode of occurrence of mineral deposits in western South America may be useful in the search for mineralized rocks in the peninsula. Many of the significant mineral deposits in western South America—the porphyry copper deposits—occur within and adjacent to high-level Andean-age stocks. Those porphyry deposits of Cenozoic age are the most significant economically; many of these are related to adjacent strike-slip faults (Hollister, 1974). In addition to their large low-grade copper mineralization, porphyry copper or similar igneous systems may spawn a variety of associated hydrothermal mineral deposits, such as massive replacement deposits and base- and precious-metal veins. Contact metasomatic deposits associated with plutons, and massive sulfide deposits associated with island-arc volcanism, might also occur in the Antarctic Peninsula.

Because porphyry copper systems are likely to be present in the Antarctic Peninsula, the following characteristics of porphyry copper deposits (from Lowell and Guilbert, 1970) may help in interpreting the mineral "shows" recorded to date in the peninsula. Porphyry copper deposits generally occur in small, high-level (subvolcanic), calc-alkaline plutons commonly associated with multiple intrusive events. Metallization occurs within faulted areas or fracture networks and/or in breccia pipes. Pervasive hydrothermal alteration of the stock and enclosing rocks accompanies metallization. Hydrothermal alteration occurs in distinctive concentric patterns, commonly cylindrical in three-dimensional development, with the intensity of alteration increasing inward toward the center of the stock. The common alteration assemblages are, from most to least intense, potassic, phyllic, argillic, and propylitic. Ore minerals in a porphyry copper system commonly include disseminated and stockwork pyrite, chalcopyrite, bornite, and molybdenite in the inner alteration zones, and veins and veinlets with galena, sphalerite, gold, and silver within the peripheral alteration facies. Some porphyry deposits ("plutonic porphyry deposits") have formed at greater depth or in stocks and batholiths of hypidiomorphic-granular (granitoid) texture; mineralization of these deposits is related to a regular vein system, and alteration is weakly developed and also related to the veins (Sutherland Brown, 1969; Sutherland Brown and others, 1971; Soregaroli, 1974; Kimura and others, 1976). Supergene enrichment is found in many porphyry deposits, but may not be significant in the peninsula because of the harsh climate and the effects of glaciation since middle Tertiary time. Thus the recognition of porphyry deposits in the peninsula may well depend on recognizing the associated hydrothermal alteration patterns and peripheral vein-type minerals.

Mineral occurrences in the Antarctic Peninsula region

Most of the known metallic mineral occurrences reported from the Antarctic Peninsula region are of hydrothermal origin and are associated with Mesozoic and Tertiary plutons of the Andean Intrusive Suite. Some occurrences are more promising than others, but, with few exceptions (e.g., the Lassiter Coast copper deposit; Rowley and others, 1975, 1977), they have not been studied in detail or assessed in terms of modern concepts of ore genesis. As a general rule, the mineralogy, geochemistry, and distribution of altered and mineralized rocks in the peninsula have not been investigated. This review attempts to assemble the data and to interpret the significance of those few occurrences that have been described.

The reported occurrences of metallic minerals in the Antarctic Peninsula region are summarized in the following section. Where known, the geologic setting is given briefly. For completeness, the mineral occurrences of the Scotia arc are also mentioned. Possible resources of hydrocarbons, manganese nodules on the ocean floor, and nonmetallic mineral deposits have been dealt with elsewhere (Wright and Williams, 1974; Anonymous, 1974; Ericksen, 1976; Glasby, 1976; Splettstoesser, 1976, 1979; Institute of Polar Studies, 1977; Behrendt, 1979) and thus will not be considered here. Metallic occurrences in the peninsula are described from north to south and are shown as numbered localities on Figure 108.1.

Islands of the Scotia Arc

Falkland Islands (Locality 1)

Pegmatite and camptonite dikes that intrude the Cape Meredith metamorphic complex (Precambrian) may contain metallic minerals, but assays for gold and silver have proved negative and so far only sparse xenotime crystals from pegmatites have been found (Adie, 1972*b*).

South Sandwich Islands (Locality 2)

Native sulfur in an "important quantity" has been reported from "Savadovski" Island (Kosack, 1955), the site of active volcanism (Gass and others, 1963) in a volcanic island arc.

South Orkney Islands (Locality 3)

Small amounts of disseminated sulfide minerals occur at several outcrops in Paleozoic(?) schist (Dalziel, 1972) and in an Andean dolerite dike on southeastern Coronation Island (Thomson, 1974).

South Shetland Islands

Aspland and Gibbs Islands (Locality 4)

Nickel, cobalt, and chromium, perhaps associated with graphite and asbestos, were reported by Kosack (1955) in dunite on Aspland Island. The *San Francisco Sunday Examiner* (6 February 1977) suggested that "chromates [*sic*] probably in commercial quantities" occur on Gibbs Island. Gibbs Island and apparently Aspland Island are underlain by a metamorphic complex into which a layered body of dunite and serpentinized dunite was emplaced; all these rocks are probably of late Paleozoic or early Mesozoic age (Tyrrell, 1945; Matthews, 1959; Cox, 1964; Dalziel and Elliot, 1973; Dalziel, 1972, 1975; de Wit and others, 1977). The chromite on Gibbs Island occurs in disseminated form and in thin cumulate or tectonic layers in the ultramafic body, but—contrary to the newspaper report mentioned above—major concentrations of chromite have not been found (I. W. D. Dalziel, personal communication, 1977; de Wit and others, 1977).

King George Island (Locality 5)

The most extensive mineralized area yet known in the Antarctic Peninsula area is on King George Island. Most of the island, which is nearly 80 km long and 30 km wide, is underlain by folded and faulted andesitic volcanic rocks and subordinate sedimentary rocks of Late Jurassic and Tertiary age; these rocks are intruded by at least three Andean plutons of mafic to intermediate composition (Hawkes, 1961; Barton, 1965). Ferguson's (1921) mapping in the Admiralty Bay area of the south-central part of the island has revealed quartz veins containing abundant

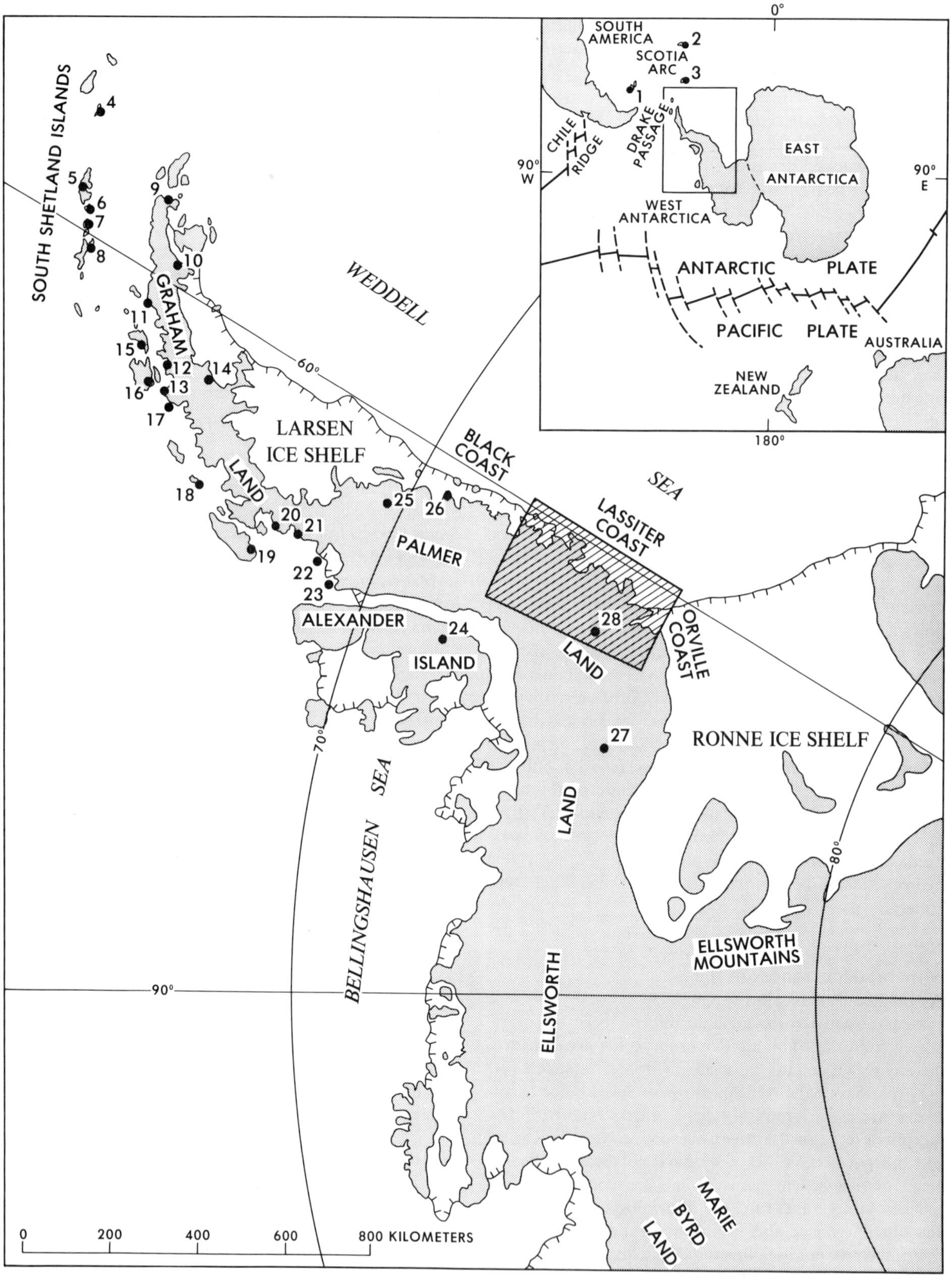

Figure 108.1. Map showing numbered localities of mineral occurrences in the Antarctic Peninsula region (Scotia Arc, Graham Land, Palmer Land, and eastern Ellsworth Land). Cross-hatched area shows Lassiter Coast project area.

pyrite and subordinate secondary hematite and limonite in several places. Barton has described two quartz-pyrite replacement bodies on Keller Peninsula in the northern Admiralty Bay area: one more than 2 km long (Ferguson, 1921) and as much as 155 m wide, and another that is a mineralized breccia occurring along a "shatter zone." Mineralized and altered rocks occur scattered over a known area of more than 100 km^2 along western Admiralty Bay. A quartz diorite Andean pluton, flecked with pyrite and about 1 km in diameter, occurs near the center of this area, and an Andean granodiorite pluton having a diameter of at least 7 km occurs east of Admiralty Bay (Barton, 1965). These igneous bodies, of medium-grained hypidiomorphic-granular texture, may represent the upper parts of a large batholith that underlies the Admiralty Bay area (Hawkes, 1961; Barton, 1965). Petrographic descriptions by Tyrrell (1921) and Hawkes (1961) have suggested that the rocks in the Admiralty Bay area have been hydrothermally altered to grades having alteration intensities as high as phyllic and perhaps potassic grades in present usage (Lowell and Guilbert, 1970); alteration minerals include quartz, adularia, albite, sericite, chlorite, epidote, and calcite.

Ferguson (1921) has mapped a large pyrite-quartz replacement deposit in folded Upper Jurassic volcanic rocks on the northeastern side of the island and on nearby offshore islands. He has reported that the deposit is as much as 30 m wide and 1.5 km long and estimated that it contains "many thousands of tons" of pyrite. He has predicted that "large deposits of similar character are concealed under the ice-fields" that cover most of the interior of the island. The volcanic country rocks here are strongly feldspathized and silicified. Hawkes (1961) has mapped a quartz diorite pluton, having a diameter of at least 3 km, near this locality, but it was not noted by Barton (1965).

Quartz veins, pyrite, and hydrothermally altered rocks also occur at the western end of the island (Ferguson, 1921; Tyrrell, 1921). On Fildes Peninsula, for example, Barton has mapped faulted Upper Jurassic and Tertiary volcanic rocks that are intensely hydrothermally altered and contain disseminated pyrite, calcite, and quartz. Grikurov and Polyakov (1971*a,b*) have described hydrothermally altered, silicified, and mineralized (pyrite, hematite?, and limonite?) rocks near Bellingshausen Station (U.S.S.R.) on Fildes Peninsula and in areas farther south. The altered and mineralized rocks occur along faults and fractures in Tertiary volcanic rocks and in subordinate sedimentary rocks that occur near, and are intruded by, Tertiary andesitic volcanic necks and dikes. On Barton Peninsula, south of Fildes Peninsula, Barton mapped mineralized, feldspathized, and silicified Upper Jurassic volcanic rocks adjacent to a quartz diorite pluton having a diameter of less than 1 km. The pluton has yielded whole-rock K-Ar ages of 50–46 m.y. (Watts, Chapter 28, this volume). At least three quartz-pyrite replacement bodies, the largest of which is more than 2 km long and as much as 0.3 km wide, occur in brecciated rock, apparently along fault zones (Barton, 1965). Disseminated pyrite occurs in adjacent volcanic rocks over an area of several square kilometers.

The genesis of the mineralized and altered rocks of King George Island is unclear. The large volume and widespread distribution of the pyrite within volcanic rocks locally resemble a volcanogenic massive sulfide deposit affected by subsequent faulting and plutonism. Some disseminated pyrite at Fildes Peninsula and Admiralty Bay, in fact, is localized in pyroclastic beds and at contacts between lava flows (Barton, 1965). Yet Hawkes (1961) and Barton (1965) have concluded that altered, silicified, and mineralized rocks are derived from hydrothermal solutions within the faulted roof of a batholith complex. Recently Littlefair (1978) has studied the mineral deposits and alteration facies and has concluded instead that the pyrite had a volcanogenic origin and that both it and the altered rocks may be due to epithermal-sulfataric low-temperature hydrothermal systems.

Robert Island (Locality 6)

Despite its suggestive name, Coppermine Cove on Robert Island contains only red and "copper colored" stains (apparently from iron oxides, chlorite, and zeolite) on Tertiary volcanic rocks (Thomas, 1921; Mueller, 1964; Caballero and Fourcade, 1958).

Greenwich Island (Locality 7)

Greenwich Island consists of volcanic rocks, apparently of andesitic composition (Mueller, 1964), of Tertiary age (Craddock, 1972*a*). Two veins, 1–2.5 m wide, on the island contain chalcopyrite and "its oxidation products," as well as ankerite, barite, calcite, magnetite, and quartz; grade does not reach more than 0.8% copper (Mueller, 1964).

Livingston Island (Locality 8)

One or more mineral deposits may occur in the southeastern part of Livingston Island. This area is underlain by clastic rocks of the Miers Bluff Formation of late Paleozoic(?) or early Mesozoic age (Schopf, 1973) and andesitic volcanic rocks of Late Jurassic and Tertiary age. These rocks have been intruded by two Andean plutons (Hobbs, 1968). The plutons are called tonalites by Hobbs, but his modal analyses suggest that the plutons are entirely or at least partly of trondhjemitic composition. One of the plutons, about 1 km in diameter, is on Hurd Peninsula; it has been extensively hydrothermally altered (sericite, epidote, calcite, and prehnite), apparently to grades as high as phyllic alteration. The other, about 6 km to the south across False Bay, has a diameter of at least 10 km; it is mostly fresh to slightly hydrothermally altered, but in places is intensely altered, with formation of K-feldspar, sericite, chlorite, and epidote. It has a Rb-Sr age of 40 ± 10 m.y. (Dalziel and others, 1973). Both plutons have a fine- to medium-grained hypidiomorphic-granular texture.

Detailed geologic mapping of Hurd Peninsula by Caminos and others (1973) has disclosed hydrothermally altered rocks (albite, sericite, epidote, clinozoisite, chlorite, and calcite) and quartz and calcite veins containing chalcopyrite, covellite, galena, pyrite, sphalerite, malachite, azurite, hematite, and limonite. These deposits are localized along a fault zone of Late Cretaceous or early Tertiary age in folded sedimentary rocks of the Miers Bluff Formation. Altered and mineralized rocks extend for a distance of 1.5 km along the fault zone and are locally as much as 10 m wide. Del Valle and others (1974) studied this deposit and identified, in addition to the minerals listed above, tetrahedrite, tennantite, chalcocite, bornite, linarite, antlerite, cerussite, and witherite. They concluded that mineralization was due to hydrothermal-mesothermal solutions controlled by faults and that copper and lead are the main metal additions. Caminos and others (1973) and del Valle and others (1974) have concluded that alteration and mineralization are genetically related to the larger of the two plutons mapped by Hobbs (1968).

The area east and south of False Bay, mostly underlain by the larger of the two plutons, may contain a second mineral deposit. Malachite stains and veins of molybdenite in granodiorite boulders occur in moraines along the eastern shore of False Bay (J. E. Curl, personal communication, 1975). The material was probably derived from small cirques northeast of Charity Glacier and east and southeast of False Bay. The plutonic rock near this probable source is intensely altered to sericite, chlorite, epidote, and calcite; quartz is intensely deformed (Hobbs, 1968).

The mineralized and altered rocks of southeastern Livingston Island clearly define one or more hydrothermal mineral deposits, but the type of deposits remains unclear. A hydrothermal polymetallic base-metal vein or porphyry deposit seems most likely. The region should be subjected to additional field and laboratory investigations.

Graham Land

Trinity Peninsula

Trinity Peninsula consists of folded sedimentary rocks of the Trinity Peninsula Series that are intruded by Andean plutons. Adie (1957, 1964) has reported masses and numerous small veins of pyrite, chalcopyrite, and tourmaline in the Trinity Peninsula Series at the contacts of Andean quartz diorite and diorite plutons that are 1–4 km in diameter (Adie, 1969*a*) in northeastern Trinity Peninsula (Locality 9). Aitkenhead (1975) has mapped small veins containing quartz, pyrite, and green copper stains in rock of the Trinity Peninsula Series near the contact of the Mount Reece Granite (Locality 10), an Andean pluton about 170 m.y. old (Rex, 1976) and about 20 km in diameter.

Northern Danco Coast

Scattered pyrite and minor amounts of chalcopyrite and other copper minerals (including paratacamite), in veins and as disseminations, as well as low-grade hydrothermally altered rocks, have been reported in the Spring Point area (Locality 11) by Vieira and others (Chapter 109, this volume). The minerals, which are ascribed to hydrothermal copper vein mineralization, occur in small amounts in folded and faulted Trinity Peninsula Series beds and volcanic rocks that are intruded by Andean granite bodies, some of which are porphyritic (Vieira and others, Chapter 109, this volume).

Central and southern Danco Coast (Locality 12)

Veins and disseminations of pyrite, limonite, and minor amounts of chalcopyrite, sphalerite, galena, silver minerals, copper oxides, and marcasite, as well as low-grade hydrothermally altered rocks, have been noted by Vieira and others (Chapter 109, this volume) in numerous places near Wilhelmina Bay, central Danco Coast. The occurrences represent hydrothermal polymetallic vein mineralization related to Andean plutons; they are not economic (Vieira and others, Chapter 109, this volume). Country rocks consist of folded and faulted Trinity Peninsula Series and Jurassic volcanic rocks that are intruded by Andean granitic plutons (West, 1974; Vieira and others, Chapter 109, this volume).

Ferguson (1921) has observed chalcopyrite in folded rocks of the Trinity Peninsula Series (West, 1974) near andesitic dikes on "Coughtrey Island" (now Coughtrey Peninsula), in the Paradise Harbor area of the southern Danco Coast; the area is about 1 km from an Andean pluton (West, 1974). Vieira and others (Chapter 109, this volume) have observed pyrite and copper minerals in the same locality, and they have recorded other sightings of low-grade disseminated pyrite and lesser amounts of chalcopyrite in scattered nearby areas within Jurassic volcanic rocks and adjacent to Andean plutons (West, 1974). Several small veins of copper oxides on Bryde Island and along Andvord Bay occur in or adjacent to Andean plutons (West, 1974; Vieira and others, Chapter 109, this volume). Vieira and others have attributed the occurrences to hydrothermal copper vein mineralization.

Northern Graham Coast (Locality 13)

Hydrothermal copper vein mineralization, represented by minor scattered disseminations and veins of pyrite, chalcopyrite, and pyrrhotite and by low-grade hydrothermal alteration, took place in the LeMaire Channel area, northern Graham Coast (Vieira and others, Chapter 109, this volume). The country rocks are largely Jurassic volcanic rocks, adjacent to Andean plutons (Adie, 1969*a*).

Oscar II Coast

According to Fleet (1968), sparse metallic minerals are widespread in and adjacent to Upper Cretaceous (Rex, 1976) granitic Andean plutons along the Oscar II Coast. These plutons intrude folded sedimentary rocks of the Trinity Peninsula Series and folded sedimentary and silicic to intermediate volcanic rocks of late Jurassic age. Fleet found pyrite and fluorite in a pegmatite dike near a granitic pluton about 9 km in diameter on the southern Oscar II Coast (Locality 14). A fragment from a copper-bearing vein that contains chalcopyrite, bornite, malachite, fluorite, quartz, and epidote was observed in scree near this locality (Fleet, 1968).

Islands just west of Graham Land

Brabant Island (Locality 15)

Although not presently economic, magmatic iron of apparently significant volume is exposed in widely scattered places on Brabant Island (Vieira and others, Chapter 109, this volume). Most iron occurs within magnetite-rich lava flows—locally altered to martite, hematite, and limonite—that occur interbedded with subhorizontal andesitic lava flows and minor sedimentary rocks. Some hematite and limonite occurs in the sedimentary rocks. All these rocks are considered to be Pleistocene. Significant amounts of chalcopyrite, locally associated with native copper, occur on southern Brabant island and the nearby Melchior Islands; they may represent porphyry copper occurrences (Vieira and others, Chapter 109, this volume).

Anvers Island and adjacent islands (Locality 16)

Ferguson (1921) has noted copper carbonate and brown altered rock on Green Spur, east of Copper Peak on the southeastern side of Anvers Island, which is about 70 km long and 45 km wide. The copper mineral was confirmed to be malachite by Hooper (1962), who has mapped the island. Vieira and others have also observed leached zones and abundant disseminated copper minerals in this area that they attributed to porphyry copper mineralization. The copper minerals occur in a large (at least 70 km in greatest diameter) batholith complex of gabbroic to granodioritic composition that underlies Anvers Island and vicinity (Hooper, 1962). The complex intrudes folded andesitic volcanic rocks of Late Jurassic age. K-Ar ages of 53–48 m.y. were determined for specimens of the complex from Wiencke and Doumer Islands (Scott, 1965; Rex, 1976), which are just southeast of Anvers Island. Plutonic rocks in the Palmer Station (U.S.A.) area, on the southwestern part of the island, have yielded isotopic ages of about 20 m.y. (Gledhill and others, Chapter 35, this volume). Some rocks of the batholith complex are typical of Andean magmatic rocks and have a medium-grained hypidiomorphic-granular texture. Many plutonic rocks of the complex, however, have undergone high-temperature metasomatism that involved introduction of silicon, sodium, and potassium, and resulted in formation of quartz, K-feldspar, biotite, plagioclase, and subordinate amounts of other minerals. The altered plutonic rocks show varied fine- to medium-grained textures (including subhedral crystal form, quartz and plagioclase porphyroblasts, and replacement textures) and varied compositions; rock types include diorite, trondhjemite, and plagioclase-rich "feldspar rock" (Hooper, 1962). Such high-grade alteration, coupled with numerous examples of igneous intrusive breccia, suggests that rocks on Anvers Island represent the roots of an igneous complex characterized by several major periods of emplacement.

Hydrothermally altered rocks, ranging from propylitic to potassic in grade, are abundant and widespread within the plutonic rocks and country rocks on Anvers Island. Hooper has noted that the Upper Jurassic rocks on Anvers Island are considerably more altered than is common adjacent to most Andean plutons. Numerous dikes that cut the plutonic rocks also are intensely altered (Hooper, 1966). Alteration minerals include quartz, K-feldspar, albite, biotite, sericite, chlorite, and calcite (Hooper, 1962). Hooper also described "dike-like bodies of recrystallized mylonite" affected by high-grade alteration (albite, quartz, K-feldspar, biotite, chlorite, and epidote) about 8 km west of Copper Peak. These features resemble, but show a higher alteration grade than, the sheared phyllic-argillic zones of the Lassiter Coast copper deposit (see below).

A boulder of botryoidal malachite and azurite and "blue-green vitreous-lustered copper silicate" was found on a beach at the foot of a glacier near Port Lockroy on Wiencke Island (Thomas, 1921). The area drained by this glacier is underlain by trondhjemitic rock of the batholith complex of Anvers Island (Scott, 1965; Hooper, 1962). Ferguson (1921) has noted thin films of chalcopyrite along joints in the same pluton on Isle Casabianca, near Wiencke Island. On Doumer Island, deeply altered (argillic- and phyllic-grade) mafic dikes intrude the batholith (Scott, 1965). Major high-angle faults are common on Doumer and Wiencke Islands and separate these islands from the mainland. High-grade hydrothermally altered rocks (sericite, quartz, epidote, calcite, and chlorite), considered by Hooper (1962) to be associated with the last stages of Andean plutonism, are concentrated along the fault zones. The Wauwermans Islands, just south of Anvers Island, also are underlain by the batholith complex (Hooper, 1962); Anderson (1965) has cited a report that quartz, including amethyst, and "a great quantity of copper ore" occur in these islands. Significant amounts of disseminated and vein-type copper minerals and pyrite, as well as leached zones and hydrothermally altered rocks, occur within plutonic rocks at several places in northeastern Anvers Island (Vieira and others, Chapter 109, this volume).

Multiple plutonism is also evident in the Arthur Harbor area of southwestern Anvers Island, where intense rock alteration, numerous intrusive contacts, and several areas of igneous breccia occur (Hooper, 1962). Some breccia is stained by iron oxides. Other evidence of mineralization in this area has been reported by C. C. Plummer (*in* Potter, 1969), who has found a quartz-chalcopyrite vein (bearing about 10% chalcopyrite) exposed below the high-tide mark on Bonaparte Point near Palmer Station. The vein, in hydrothermally altered diorite that was intruded by quartz diorite and aplite, also includes bornite, pyrite, magnetite, and malachite (C. C. Plummer, written communication, 1977). Small amounts of pyrite, chalcopyrite, bornite, and molybdenite have been observed just inland of Palmer Station (W. J. Zinsmeister, personal communication, 1977). On the Outcast Islands, just south of Arthur Harbor, Hooper (1962) has mapped volcanic country rocks that are locally altered to quartz, biotite, sericite, epidote, and chlorite, and contain disseminated pyrite. Where they have been weathered, the rocks are stained by iron oxide. The Joubin Islands west of the Outcast Islands is another area of multiple plutonism, igneous brecciation, intense alteration (albite, quartz, biotite, sericite, chlorite, and epidote), and local disseminated pyritization of volcanic and plutonic rocks (Hooper, 1962). A huge xenolith about 50 m long in one of the plutons contains abundant pyrrhotite and sparse pyrite and chalcopyrite. Breccia dikes occur locally within the plutons.

Anvers Island and vicinity is a geologically complex area. It contains widely scattered disseminated metallic minerals and quartz veinlets containing minor copper and iron sulfides and oxides, apparently significant amounts of high-temperature metasomatic altered rock, conventional potassic to propylitic hydrothermally altered rock, multiple intrusions, and faults. Taken all together, the numerous reports of mineral occurrences indicate that the exposed rocks on Anvers Island were at the level of primary hydrothermal mineralization, but no mineral deposits have any economic potential now. Such deposits, perhaps of the "plutonic porphyry" type or as hydrothermal veins, might be discovered by detailed mapping; but it is also possible that this area represents the roots of a porphyry system and that any significant mineral deposits were removed by erosion. Vieira and others (Chapter 109, this volume) have also suggested that porphyry copper deposits occur on Anvers Island and at several other places along nearby Gerlache Strait.

Argentine Islands (Locality 17)

The Argentine Islands are underlain by a composite Andean pluton, at least 5 km in diameter, that has a medium-grained hypidiomorphic-granular texture, ranges in composition from gabbro to trondhjemitic granodiorite, and intrudes Upper Jurassic andesitic and dacitic volcanic rocks (Fraser, 1964; Elliot, 1964). Some parts of the pluton have yielded K-Ar ages of 57–54 m.y. (Rex, 1976). In the Anagram Islands of the Argentine chain, widespread late-magmatic magnetite and subordinate pyrite, chalcopyrite, malachite, and limonite occur in quartz veins, pegmatite lenses, and miarolitic cavities in an Andean banded gabbro; alteration of the gabbro has produced talc, hornblende, serpentine, epidote, and calcite (Fraser, 1964). Fraser has hypothesized that the iron, related metals, and altered rocks were due to injection of residual liquids from the magma chamber, apparently following cumulate crystallization of most of the pluton. Elsewhere in the Argentine Islands, more typical hydrothermal metals and alteration products are exposed. Local hydrothermal alteration of the volcanic rocks has produced albite, K-feldspar, sericite, quartz, epidote, and chlorite; the most intensely altered rocks are centered in the area of Winter, Galindez, and Indicator Islands in the southern Argentine Islands, near extensive areas of disseminated pyrite and subordinate pyrrhotite and magnetite (Elliot, 1964). Trace amounts of disseminated pyrite and local concentrations of pyrite, magnetite, molybdenite, hematite, and quartz veins occur at other localities in the Argentine Islands (Elliot, 1964). Although widespread, no metallic minerals in the Argentine Islands occur in significant volumes (D. H. Elliot, personal communication, 1977).

Biscoe Islands

Pyritic hornfels within Upper Jurassic volcanic rocks has been described by Goldring (1962) near Upper Jurassic(?) porphyritic microdiorite sills in the southern Biscoe Islands (Locality 18).

Adelaide Island and adjacent islands (Locality 19)

Southeastern Adelaide Island and adjacent small islands are underlain by Andean plutons, from 1 to 25 km in diameter, that intrude flat-lying to gently warped Upper Jurassic sedimentary rocks, tuffs, and silicic- to intermediate-composition lava flows and volcanic breccias (Dewar, 1970). The largest pluton, which is located on southeastern Adelaide Island, is of heterogeneous composition, ranging from gabbro to granodiorite; it is also trondhjemitic, based on modal analyses given by Dewar. The textures of most rocks of the pluton are fine- to coarse-grained hypidiomorphic-granular or porphyritic. This body is characterized by intrusive breccias and is intruded and brecciated by a younger granodiorite pluton of medium- to coarse-grained hypidiomorphic-granular and xenomorphic-granular texture that is at least 13 km across (Dewar, 1970); the central part of the pluton contains sparse disseminated pyrite and is slightly altered (antigorite, chlorite, and epidote). Several mineral occurrences are known on southeastern Adelaide Island and nearby islands. At one locality, minor molybdenite and iron minerals occur in quartz veins along shear fractures in tuff at the contact with the large heterogeneous pluton (Dewar, 1970). At another locality nearby, magnetite-limonite masses, disseminated pyrite, numerous quartz-pyrite veinlets, thick quartz-magnetite veins, and minor quartz-specularite veins occur in the sedimentary wall rocks less than 1.5 km from the intrusive contact of the largest pluton and less than 1 km from where this pluton is intruded by the granodiorite pluton (Dewar, 1970). Elsewhere on Adelaide Island, scattered veins containing quartz, aragonite, pyrite, magnetite, molybdenite, and "copper ores" are associated with Andean granodiorite and tonalite plutons; these veins occur in the plutons themselves or in volcanic country rocks (Dewar, 1970). Nichols (1955) has noted chalcopyrite, malachite, and limonite from rocks of two islands about 10 km southwest of Adelaide Island.

As on Anvers Island, the presence on Adelaide Island of scattered disseminated and vein metal occurrences and of features of multiple plutonism suggests an origin by a "plutonic porphyry" copper system or hydrothermal veins. Future detailed mapping might disclose mineral deposits. It is possible, however, that any significant mineral deposits have been removed by deep erosion of the batholith complex.

MARGUERITE BAY AREA

Numerous reports of minor base- and precious-metal occurrences have been made throughout the Marguerite Bay area, yet their meaning is unclear because few modern, detailed geologic maps exist. The geologic history is complex and includes several periods of plutonism, dike intrusions, and faulting. Most rocks apparently are Andean plutons (Adie, 1969*b*), but some are Upper Jurassic volcanic rocks and others may belong to a Paleozoic(?) basement complex (Adie, 1955; Nichols, 1955; Fraser, 1965; Grimley, 1966).

Stonington Island (Locality 20)

Stonington Island consists largely of basement complex into which some Andean dikes were emplaced (Fraser, 1965). Knowles (1945) recorded manganese stains, crystals of molybdenite, and a "brecciated pyrite mineralized 8-foot zone," near the British base on Stonington Island. Several assays from this area yielded gold values of as much as 0.05 oz/ton and silver values of as much as 0.37 oz/ton. Nichols (1955) has noted limonite stains due to the oxidation of sulfides on the island. Fraser (1965) and Grimley (1966) have found minor magnetite and pyrite in dikes, as well as traces of chalcopyrite, malachite, and azurite.

The coast of central Marguerite Bay and nearby islands (Locality 21)

Most rocks along the coast of central Marguerite Bay and on nearby islands consist of Andean plutons and basement complex. From north to south, the following minor occurrences of metals have been noted: (1) limonite and pyrite that show assay traces of gold and silver at Neny Island (Knowles, 1945; Nichols, 1955); (2) magnetite float and malachite at Roman IV Mountain (Knowles, 1945; Nichols, 1955); (3) malachite on an island in Neny Fjord (Nichols, 1955); (4) limonite "gossan" zones and malachite at Neny Fjord Thumb (Nichols, 1955); (5) limonite stains and molybdenite crystals at Red Rock Ridge (Nichols, 1955; Adie, 1955); (6) limonite and pyrite at the Refuge Islands (Nichols, 1955); (7) malachite at "Skua

Gull Islands'' (Nichols, 1955); (8) malachite, sulfide minerals, and limonite at Black Thumb Mountain (Nichols, 1955); (9) pyrite at ''Chasm Island'' (Nichols, 1955); and (10) pyrite-quartz veins and limonite stains at ''Moraine Point'' (Nichols, 1955).

Terra Firma Islands (Locality 22)

Malachite, limonite, and manganese oxide were observed in an Andean layered gabbro in the Terra Firma Islands by Nichols (1955). Adie (1955) has found that the limonite is associated with magnetite and subordinate hematite in a prominent zone about 6 m thick near the base of the layered pluton; the malachite was found with chalcopyrite in a second prominent zone of similar thickness that is about 45–60 m stratigraphically above the limonite zone. These mineral layers represent cumulate deposits; other metallic minerals, such as those found in many other layered mafic complexes, might also be present here.

Islands in southern Marguerite Bay (Locality 23)

Nichols (1955) has reported the occurrence of small veins of pyrite and calcite at Mushroom Island. Knowles (1945) has observed malachite stains on an unnamed island north of the mouth of Fleming Glacier.

ALEXANDER ISLAND (LOCALITY 24)

Minor pyrite, hematite, magnetite, and cassiterite occur in fractures near Andean plutons and dikes on southern Alexander Island (Bell, 1973).

PALMER LAND

An assay of a pegmatite in the Eternity Range (Locality 25), sampled by Knowles (1945) during a traverse of the Wilkins Coast, shows traces of gold and silver. Knowles has also reported chalcopyrite and copper stains in hornblendite at ''Cape Eielson'' (now Cape Boggs, Locality 26); an assay of pyrite from this locality shows minor amounts of gold and silver.

Reconnaissance mapping along the southern Black Coast, Lassiter Coast, and northern Orville Coast (cross-hatched area on Figure 108.1) documented about 30 metal occurrences and an equal number of areas of hydrothermally altered rocks (Rowley and others, 1975, 1977). Minerals identified include magnetite, hematite, limonite, chalcopyrite, pyrite, malachite, azurite, chrysocolla(?), and molybdenite. Spectrographic analyses of material from some of the localities revealed anomalous amounts of Cu, Pb, Zn, Mo, As, Bi, W, and Ag. Hydrothermally altered rocks attain grades as high as phyllic facies. Mineralized and altered rocks are related to Andean plutons ranging in age from 120 to 95 m.y. (Farrar and others, Chapter 39, this volume). Except for the Lassiter Coast and Sky-Hi Nunataks copper deposits, the metal occurrences are insignificant and were not investigated in detail.

Sky-Hi Nunataks copper deposit (Locality 27)

Recent geologic mapping on the Orville Coast and in eastern Ellsworth Land disclosed a small granodiorite stock containing abundant disseminated pyrite, pervasive hydrothermally altered rocks, and a stockwork system of shear zones (Rowley, 1978). The exposures in the Sky-Hi Nunataks may represent the barren outer pyrite zone of a porphyry copper occurrence, inward from which may be a subsurface body of disseminated chalcopyrite. Peripheral quartz-chalcopyrite veins occur in Jurassic volcanic rocks elsewhere in the nunatak cluster. Minor traces of malachite and copper and iron sulfides were also observed at numerous widely scattered parts of eastern Ellsworth Land (T. S. Laudon, personal communication, 1977), as well as elsewhere on the northern Orville Coast.

Lassiter Coast copper deposit (Locality 28)

The Copper Nunataks, near the crest of the peninsula ice plateau of the western Lassiter Coast, was the site of two periods of plutonism, hydrothermal alteration, and mineralization during Late Cretaceous time (Rowley and others, 1975, 1977). The first occurred about 105 m.y. ago, when emplacement of a medium-grained, porphyritic and hypidiomorphic-granular granodiorite batholith was followed by stockwork-type shearing, minor propylitic and argillic hydrothermal alteration, and deposition of minor amounts of disseminated pyrite and of pyrite veins. The second period of activity occurred about 95 m.y. ago, when emplacement of a concentrically zoned, fine- to coarse-grained, porphyritic to hypidiomorphic-granular quartz monzonite stock was followed by intrusion of granodiorite porphyry dikes; this, in turn, was followed by shearing and by hydrothermal alteration (argillic and phyllic grades) along northwest-striking zones and by related mineralization that resulted in the Lassiter Coast copper deposit. The main metallic minerals are chalcopyrite, pyrite, and magnetite, which occur as disseminated crystals or in quartz veinlets; molybdenite, malachite, hematite-limonite, chrysocolla(?), and chalcocite(?) are less abundant. Mineralized rocks are exposed over an area of about 1 km^2, and snow and ice undoubtedly cover several more square kilometers of mineralized rock. The deposit demonstrates affinities with porphyry-type copper deposits but, insofar as can be ascertained, does not approach commercial grade. The most highly mineralized nunatak has an average grade of about 200 ppm Cu, 100 ppm Pb, and 50 ppm Mo. Although greater amounts of metals may occur either laterally beneath the ice or at depth, the inaccessibility and high altitude (1500–2000 m above sea level) indicate that this deposit will never be economic.

Conclusions

Although metallic minerals are widespread in the Antarctic Peninsula, nearly all occur as isolated crystals or small masses, and, as far as preliminary reports reveal, they have no economic significance. In several places, however, descriptions are suggestive of unusual metal concentrations. All are associated with subduction-generated Andean plutonism, or perhaps with postsubduction rifting. Most of the localities need closer investigation. The particularly interesting localities are summarized below.

Large quartz-pyrite veins, widespread disseminated pyrite, and high-temperature hydrothermally altered rocks on King George Island appear to be the result of volcanogenic massive sulfide deposition or major hydrothermal sulfide mineralization. The occurrences on southeastern Livingston Island, in veins and in disseminated form, of copper, lead, and related metals, as well as of high-temperature hydrothermally altered rocks, suggest that one or more polymetallic hydrothermal vein deposits or porphyry deposits may exist in this area. In the Anvers Island area, southeastern Adelaide Island, Brabant Island, and the Melchior Islands, scattered minor amounts of copper and related metals, high-temperature hydrothermally altered rocks, and features of intrusive brecciation and multiple plutonism are suggestive of porphyry-type systems. So far, however, no large-scale deposits have been found at these localities; such deposits may have been removed by deep erosion. Magnetite-rich lava flows occur on Brabant Island. A low-grade porphyry-type copper deposit occurs on the Lassiter Coast. Additional areas of potential interest include minor late magmatic enrichment of magnetite in a banded gabbro in the Argentine Islands, and cumulate zones of copper and iron minerals in the layered gabbro of the Terra Firma Islands.

The areas exhibiting the greatest effects of mineralization and hydrothermal alteration occur on islands separated by faulting from the western coast of the Antarctic Peninsula. Most of the intrusions associated with the mineralized and altered rocks on these islands are Tertiary in age. The metallization and alteration commonly accompany the youngest pluton in an area of multiple intrusive activity. Many plutons on the islands are trondhjemitic, and thus are petrologically different from the Mesozoic plutons exposed elsewhere in West Antarctica. Tertiary trondhjemitic plutons may have formed under conditions of slower subduction and more pronounced extension and transform faulting than the Mesozoic plutons (Rowley and Williams, Chapter 38, this volume).

Through careful future geologic and geochemical study, it seems probable that larger deposits may be found in the Antarctic Peninsula. Yet tectonic differences between West Antarctica and the mineral-rich central Andes suggest that past predictions of mineral riches in the Antarctic Peninsula may have been premature. Plutons in the Antarctic Peninsula may have been emplaced at deep levels in the crust where temperature and pressure conditions did not favor accumulation of significant mineral deposits; or the plutons of the peninsula may have contained only small volumes of hydrothermal fluids and associated metals. Also, the depth of erosion in the peninsula may have been such that most high-level mineral deposits were removed. No deposit presently known in the Antarctic Peninsula can be exploited economically now or in the foreseeable future.

Acknowledgments. P. L. Williams made significant additions to this study by contributing unpublished data, through discussions that helped crystallize ideas, and through technical review. Discussions with D. L. Schmidt, R. Gordon Schmidt, and D. P. Cox also helped formulate concepts. We appreciate the assistance of Campbell Craddock, University of Wisconsin—Madison, and D. C. Rex, University of Leeds, for preprints of papers. We thank C. C. Plummer, California State University, Sacramento; J. E. Curl, Shell Oil Company, Houston; T. S. Laudon, University of Wisconsin—Oshkosh; D. H. Elliot, Ohio State University; W. J. Zinsmeister, Ohio State University; I. W. D. Dalziel, Columbia University; and F. A. Wade, Texas Tech. University for unpublished information on mineral occurrences in the Antarctic Peninsula. Technical reviews by G. E. Ericksen, W. B. Myers, C. H. Thorman, and D. L. Schmidt greatly improved the manuscript. The study was financed by National Science Foundation Grant AG-187.

References cited

Adie, R. J. 1955. The petrology of Graham Land—II. The Andean granite-gabbro intrusive suite. *Scient. Rep. Falkld Isl. Depend. Surv.*, No. 12, 39 pp.

Adie, R. J. 1957. The petrology of Graham Land—III. Metamorphic rocks of the Trinity Peninsula Series. *Scient. Rep. Falkld Isl. Depend. Surv.*, No. 20, 26 pp.

Adie, R. J. 1964. Geological history. (*In* Priestley, R. E., Adie, R. J., and Robin, G. de Q., *eds. Antarctic Research, A Review of British Scientific Achievement in Antarctica.* London, Butterworths, 118–62.)

Adie, R. J. 1969*a*. Northern Antarctic Peninsula (sheet 1, northern Antarctic Peninsula). (*In* Bushnell, V. C., and Craddock, C., *eds.* Geologic Maps of Antarctica. *Antarct. Map Folio Ser.*, Folio 12, Pl. I.)

Adie, R. J. 1969*b*. Southern Antarctic Peninsula (sheet 2, southern Antarctic Peninsula). (*In* Bushnell, V. C., and Craddock, C., *eds.* Geologic Maps of Antarctica. *Antarct. Map Folio Ser.*, Folio 12, Pl. II.)

Adie, R. J. 1972*a*. Recent advances in the geology of the Antarctic Peninsula. (*In* Adie, R. J., *ed. Antarctic Geology and Geophysics.* Oslo, Universitetsforlaget, 121–24.)

Adie, R. J. 1972*b*. Appendix A—economic geology. (*In* Greenway, M. E., *ed.* The Geology of the Falkland Islands. *Scient. Rep. Br. antarc. Surv.*, No. 76, 31–37.)

Aitkenhead, N. 1975. The geology of the Duse Bay–Larsen Inlet area, northeast Graham Land (with particular reference to the Trinity Peninsula Series). *Scient. Rep. Br. antarc. Surv.*, No. 51, 62 pp.

Anderson, J. J. 1965. Bedrock geology of Antarctica—a summary of exploration, 1831–1962. (*In* Hadley, J. B., *ed. Geology and Paleontology of the Antarctic.* American Geophysical Union, 1–70.) [Antarctic Research Series, Vol. 6.]

Anonymous. 1956. Minerals of the Antarctic. *Min. J. Lond., 246,* No. 6285, 149–50.

Anonymous. 1974. Antarctica—world hunger for oil spurs Security Council review. *Science, N.Y., 184,* No. 4138, 776–79.

Barton, C. M. 1965. The geology of the south Shetland Islands—III. The stratigraphy of King George Island. *Scient. Rep. Br. antarc. Surv.*, No. 44, 33 pp.

Behrendt, J. C. 1979. Speculations on petroleum resource potential of Antarctica (abstract). *Bull. Am. Ass. Petrol. Geol., 63,* No. 3, 418.

Bell, C. M. 1973. The geology of southern Alexander Island. *Bull. Br. antarc. Surv.*, No. 33–34, 1–16.

Caballero, M. A., and Fourcade, N. H. 1958. Observaciones geologicas en Caleta Copper Mine, Islas Shetland del Sur, Antartida Argentina. *Contr. Inst. antart. argent.*, No. 23, 39 pp.

Caminos, R., Marchese, H. G., Massabie, A. C., Morelli, J. R., Rinalde, C. A., and Spikermann, J. P. 1973. Geologia del Sector noroccidental de la Peninsula Hurd, Isla Livingston, Shetland del Sur, Antartida Argentina. *Contr. Inst. antart. argent.*, No. 162, 32 pp.

Chalmers, R. O. 1957. Mineral possibilities of the Antarctic. *Aust. Mus. Mag., 12,* No. 8, 252–55.

Cousteau, J. 1976. Reasonable utopias. *Bull. Sierra Club, 61,* No. 6, 25–27.

Cox, M. J. G. 1964. Seismic refraction measurements in Bransfield Strait. *Bull. Br. antarc. Surv.*, No. 4, 1–12.

Craddock, C. 1969. Map of Gondwanaland. (*In* Bushnell, V. C., and Craddock, C., *eds.* Geologic Maps of Antarctica. *Antarct. Map Folio Ser.*, Folio 12, Pl. XXIII.)

Craddock, C. 1972*a*. *Geologic Map of Antarctica 1:5,000,000.* New York, American Geographical Society.

Craddock, C. 1972*b*. Antarctic tectonics. (*In* Adie, R. J., *ed. Antarctic Geology and Geophysics.* Oslo, Universitetsforlaget, 449–55.)

Craddock, C. 1977. The evolution and fragmentation of Gondwanaland. (*In Proceedings of Fourth International Gondwana Symposium.* Calcutta, Geological Survey of India, 17 pp.)

Craddock, C., and Hollister, C. D. 1976. Geologic evolution of the southeast Pacific basin. (*In* Hollister, C. D., Craddock, C., and others. *Initial Reports of the Deep Sea Drilling Project,* Vol. 35. Washington [U.S. Government Printing Office], 723–43.)

Dalziel, I. W. D. 1972. K-Ar dating of rocks from Elephant Island, South Scotia Ridge. *Geol. Soc. Am. Bull., 83,* No. 6, 1887–94.

Dalziel, I. W. D. 1975. The Scotia Arc tectonics projects, 1969–1975. *Antarct. Jnl U.S., 10,* No. 3, 70–81.

Dalziel, I. W. D., and Elliot, D. H. 1973. The Scotia Arc and the Antarctic margin. (*In* Nairn, A. E. M., and Stehli, F. G., *eds. The Ocean Basins and Margins. 1. The South Atlantic.* New York, Plenum Press, 171–246.)

Dalziel, I. W. D., Kligfield, R., Lowrie, W., and Opdyke, N. D. 1973. Paleomagnetic data from the southernmost Andes and the Antarctandes. (*In* Tarling, D. H., and Runcorn, S. K., *eds. Implications of Continental Drift to the Earth Sciences. 1.* New York, Academic Press, 87–101.)

del Valle, R., Morelli, J., and Rinaldi, C. 1974. Manifestacion cupro-plumbifera "Don Bernabe," Isla Livingston, Islas Shetland del Sur, Antartida Argentina. *Contr. Inst. antart. argent.,* No. 175, 35 pp.

Dewar, G. J. 1970. The geology of Adelaide Island. *Scient. Rep. Br. antarc. Surv.,* No. 57, 66 pp.

de Wit, M. J. 1977. The evolution of the Scotia Arc as a key to the reconstruction of southwestern Gondwanaland. *Tectonophysics, 37,* No. 1–3, 53–81.

de Wit, M. J., Dutch, S., Kligfield, R., Allen, R., and Stern, C. 1977. Deformation, serpentinization and emplacement of a dunite complex, Gibbs Island, South Shetland Islands: possible fracture zone tectonics. *J. Geol., 85,* No. 6, 745–62.

Elliot, D. H. 1964. The petrology of the Argentine Islands. *Scient. Rep. Br. antarc. Surv.,* No. 41, 31 pp.

Ericksen, G. E. 1976. Metallogenic provinces of southeastern Pacific region. (*In* Halbouty, M. T., Maher, J. C., and Lian, H. M., *eds.* Circum-Pacific Energy and Mineral Resources. *Mem. Am. Ass. Petrol. Geol.,* No. 25, 527–38.)

Farrar, E., Clark, A. H., Haynes, S. J., Quirt, G. S., Conn, H., and Zentilli, M. 1970. K-Ar evidence for the post-Paleozoic migration of granitic intrusion foci in the Andes of northern Chile. *Earth planet. Sci. Lett., 10,* No. 1, 60–66.

Ferguson, D. 1921. Geological observations in the South Shetlands, the Palmer archipelago, and Graham Land, Antarctica. *Trans. R. Soc. Edinb., 53,* Pt. 1, 29–56.

Fleet, M. 1968. The geology of the Oscar II Coast, Graham Land. *Scient. Rep. Br. antarc. Surv.,* No. 59, 46 pp.

Fraser, A. G. 1964. Banded gabbros of the Anagram Islands, Graham Land. *Bull. Br. antarc. Surv.,* No. 4, 23–38.

Fraser, A. G. 1965. The petrology of Stonington and Trepassey Islands, Marguerite Bay. *Scient. Rep. Br. antarc. Surv.,* No. 52, 51 pp.

Gass, I. G., Harris, P. G., and Holdgate, M. W. 1963. Pumice eruption in the area of the South Sandwich Islands. *Geol. Mag., 100,* No. 4, 321–30.

Glasby, G. P. 1976. Manganese nodules in the South Pacific: a review. *N.Z. Jl Geol. Geophys., 19,* No. 5, 707–36.

Goldring, D. C. 1962. The geology of the Loubet Coast, Graham Land. *Scient. Rep. Br. antarc. Surv.,* No. 36, 50 pp.

Grikurov, G. E. 1973. *Geologiia Antarkticheskogo poluostrova* [Geology of the Antarctic Peninsula]. Moscow, Izdatel'stvo Nauka, 119 pp.

Grikurov, G. E., and Polyakov, M. M. 1971*a.* Geologic structure of Fildes peninsula, southeastern tip of King George Island (Waterloo). *Inf. Bull. Sov. Antarct. Exped., 7,* No. 3, 189–92.

Grikurov, G. E., and Polyakov, M. M. 1971*b.* New data on the geology of the South Shetland Islands (from field work by the Thirteenth Soviet Antarctic expedition in January–February, 1960). *Inf. Bull. Sov. Antarct. Exped., 7,* No. 3, 193–96.

Grimley, P. H. 1966. Geology of part of Stonington Island, Marguerite Bay. *Scient. Rep. Br. antarc. Surv.,* No. 9, 71–74.

Hawkes, D. D. 1961. The geology of the South Shetland Islands—I. The petrology of King George Island. *Scient. Rep. Falkld Isl. Depend. Surv.,* No. 26, 28 pp.

Herron, E. M., and Tucholke, B. E. 1976. Sea-floor magnetic patterns and basement structure in the southeastern Pacific. (*In* Hollister, C. D., Craddock, C., and others. *Initial Reports of the Deep Sea Drilling Project,* Vol. 35. Washington [U.S. Government Printing Office], 263–78.)

Hobbs, G. J. 1968. The geology of the South Shetland Islands—IV. The geology of Livingston Island. *Scient. Rep. Br. antarc. Surv.,* No. 47, 34 pp.

Hollister, V. F. 1974. Regional characteristics of porphyry copper deposits of South America. *Trans. Soc. Mining Engrs., A.I.M.E., 256,* No. 1, 45–53.

Hooper, P. R. 1962. The petrology of Anvers Island and adjacent islands. *Scient. Rep. Falkld Isl. Depend. Surv.,* No. 34, 69 pp.

Hooper, P. R. 1966. The dykes of Anvers Island and adjacent islands. *Bull. Br. antarc. Surv.,* No. 9, 75–85.

Institute of Polar Studies. 1977. *A Framework for Assessing Environmental Impacts of Possible Antarctic Mineral Development.* Columbus, Ohio State University, 620 pp.

James, D. E. 1971. Plate tectonic model for the evolution of the central Andes. *Geol. Soc. Am. Bull., 82,* No. 12, 3325–46.

Kimura, E. T., Bysouth, G. D., and Drummond, A. D. 1976. Endako. (*In* Sutherland Brown, A., *ed.* Porphyry Deposits of the Canadian Cordillera. *Spec. Vol. Can. Inst. Min. Metall., 15,* 444–54.)

Knowles, P. H. 1945. Geology of southern Palmer Peninsula, Antarctica. *Proc. Am. phil. Soc., 89,* No. 1, 132–45.

Kosack, H. P. 1955. La explotacion de los yacimientos minerales de al Antarctica. *Revta Geogr. Chile, 13,* 83–90.

LeMasurier, W. E., and Wade, F. A. 1976. Volcanic history in Marie Byrd Land: implications with regard to Southern Hemisphere tectonic reconstructions. (*In* González-Ferrán, O., *ed. Andean and Antarctic Volcanology Problems.* Rome, International Association of Volcanology and Chemistry of the Earth's Interior, 398–424.)

Littlefair, M. J. 1978. The "quartz-pyrite" rocks of the South Shetland Islands, western Antarctic Peninsula. *Econ. Geol., 73,* No. 6, 1184–89.

Lowell, J. D., and Guilbert, J. M. 1970. Lateral and vertical alteration-mineralization zoning in porphyry ore deposits. *Econ. Geol., 65,* No. 4, 373–408.

Matthews, D. H. 1959. Aspects of the geology of the Scotia arc. *Geol. Mag., 96,* No. 6, 425–41.

McLeod, I. R. 1965. Antarctica—geology and mineral occurrences. (*In* Dew, J. M., *ed. Handbook, Australia and New Zealand, Eighth Commonwealth Mining and Metallurgical Congress.* Melbourne, Australasian Institute of Mineralogy and Metallurgy, 165–66.)

Mitchell, B. 1977. Antarctica—a special case? *New Scient., 73,* No. 1034, 64–66.

Molnar, P., Atwater, T., Mammerickx, J., and Smith, S. M. 1975. Magnetic anomalies, bathymetry and the tectonic evolution of the South Pacific since the Late Cretaceous. *Geophys. J.R. astr. Soc., 40,* 383–420.

Mueller, G. 1964. Some notes on mineralization in Antarctica. (*In* Adie, R. J., *ed. Antarctic Geology.* Amsterdam, North-Holland Publishing Company, 393–94.)

Nichols, R. L. 1955. Bedrock geology of Marguerite Bay area, Palmer Peninsula, Antarctica. *Tech. Rep. Ronne Antarct. Res. Exped.,* No. 13, 60 pp.

Piper, A. 1976. Frozen mineral waters. *Nature, Lond., 260,* No. 5548, 182–83.

Potter, N. 1969. *Natural Resource Potentials of the Antarctic.* New York, American Geographical Society, 97 pp.

Rex, D. C. 1976. Geochronology in relation to the stratigraphy of the Antarctic Peninsula. *Bull. Br. antarc. Surv.,* No. 43, 49–58.

Rowley, P. D. 1978. Geologic studies in Orville Coast and eastern Ellsworth Land, Antarctic Peninsula. *Antarct. Jnl U.S., 13,* No. 4, 7–9.

Rowley, P. D., and Williams, P. L. 1979. Metallic mineral resources of Antarctica—development far in future (abstract). *Bull. Am. Ass. Petrol. Geol., 63,* No. 3, 518.

Rowley, P. D., Williams, P. L., and Schmidt, D. L. 1977. Geology of an Upper Cretaceous copper deposit in the Andean Province, Lassiter Coast, Antarctic Peninsula. *Prof. Pap. U.S. geol. Surv.,* No. 984, 36 pp.

Rowley, P. D., Williams, P. L., Schmidt, D. L., Reynolds, R. L., Ford, A. B., Clark, A. H., Farrar, E., and McBride, S. L. 1975. Copper mineralization along the Lassiter Coast of the Antarctic Peninsula. *Econ. Geol., 70,* No. 5, 982–92.

Runnels, D. D. 1970. Continental drift and economic minerals in Antarctica. *Earth planet. Sci. Lett., 8,* 400–402.

Schnellman, G. A. 1955. Mineral exploration. *A. Rev. Min. J. Lond., 1955,* 79–85.

Schofield, E. A. 1976. Antarctica up for grabs. *Bull. Sierra Club, 61,* No. 6, 17–19.

Schopf, J. M. 1973. Plant material from the Miers Bluff Formation of the South Shetland Islands. *Rep. Inst. polar Stud. Ohio State Univ.,* No. 45, 45 pp.

Scott, K. M. 1965. Geology of the southern Gerlache Strait region, Antarctica. *J. Geol., 73,* No. 3, 518–27.

Smith, P. J. 1972. The resource potential of Antarctica. *Comments Earth Sci.—geophys., 3,* No. 1, 23–28.

Soregaroli, A. E. 1974. Geology of the Brenda copper-molybdenum deposit in British Columbia. *Bull. Can. Min. Metall., 67,* No. 750, 76–83.

Splettstoesser, J. F. 1976. Mining in Antarctica—survey of mineral resources and possible exploitation methods. (*In Proceedings of the Third International Conference on Port and Ocean Engineering under Arctic Conditions. 2.* Fairbanks, University of Alaska, 1137–55.)

Splettstoesser, J. F. 1979. Coal in Antarctica (abstract). *Bull. Am. Ass. Petrol. Geol., 63,* No. 3, 532.

Suárez, M. 1976. Plate-tectonic model for southern Antarctic Peninsula and its relation to southern Andes. *Geology, 4,* No. 4, 211–14.

Sutherland Brown, A. 1969. Mineralization in British Columbia and the copper and molybdenum deposits. *Trans. Can. Inst. Min. Metall., 72,* 1–15.

Sutherland Brown, A., Cathro, R. J., Panteleyev, A., and Ney, C. S. 1971. Metallogeny of the Canadian cordillera. *Bull. Can. Min. Metall. 64,* No. 709, 37–61.

Thomas, H. H. 1921. On the Innes Wilson collection of rocks and minerals from the South Shetland Islands and Trinity Island. *Trans. R. Soc. Edinb., 53,* Pt. 1, 81–90.

Thomson, J. W. 1974. The geology of the South Orkney Islands—III. Coronation island. *Scient. Rep. Br. antarc. Surv.,* No. 86, 39 pp.

Tyrrell, G. W. 1921. A contribution to the petrography of the South Shetland Islands, the Palmer archipelago, and the Danco Land Coast, Graham Land, Antarctica. *Trans. R. Soc. Edinb., 53,* No. 4, 57–79.

Tyrrell, G. W. 1945. Report on rocks from West Antarctica and the Scotia Arc. *Discovery Rep., 23,* 37–102.

Wade, F. A. 1976. Antarctica: an unprospected, unexploited continent—summary. (*In* Halbouty, M. T., Maher, J. C., and Lian, H. M, *eds.* Circum-Pacific Energy and Mineral Resources. *Mem. Am. Ass. Petrol. Geol., 25,* 74–79.)

West, S. M. 1974. The geology of the Danco Coast, Graham Land. *Scient. Rep. Br. antarc. Surv.,* No. 84, 58 pp.

Wright, N. A., and Williams, P. L. 1974. Mineral resources of Antarctica. *Circ., U.S. geol. Surv.,* No. 705, 29 pp.

109

Metallic Mineralization in the Gerlache Strait Region, Antarctica

CAMILO VIEIRA
Instituto Antártico Chileno, Correo 21, Santiago, Chile, and
Instituto de Investigaciones Geológicas, Casilla 10465, Santiago, Chile

BORIS ALARCÓN
Instituto Antártico Chileno, Correo 21, Santiago, Chile, and
Departamento de Geología, Universidad de Chile, Santiago, Chile

JOSÉ AMBRUS
Instituto Antártico Chileno, Correo 21, Santiago, Chile, and
Codelco Chile—División Chuquicamata, Calama, Chile

LUIS OLCAY
Instituto Antártico Chileno, Correo 21, Santiago, Chile, and
Departamento de Geología, Universidad de Chile, Santiago, Chile

Abstract. Areas showing evidence of different types of metallic mineralization have been detected in the Gerlache Strait region. These mineralized zones are comparable to ore deposits well known along the Pacific margin of South America, and they include the following four types: (1) polymetallics (copper, lead, and zinc), (2) vein-type copper, (3) porphyry copper, and (4) magmatic iron. None of these mineralized areas can be considered economically workable, at least on the basis of available evidence. There are some features common to all mineralized zones in Antarctica:

1. Most ore deposits are localized in altered rocks, and therefore they are more easily removed by glacial erosion than are unaltered rocks.
2. The nature of meteoric waters, neutral and slightly oxidizing, hinders the leaching of primary minerals, and therefore the development of rich concentrations of secondary sulfides is unlikely.

Six areas which show evidence of metallic mineralization have been detected in the Gerlache Strait region. From a genetic point of view, these occurrences include four separate types of ore deposits, all of which are well known in the Pacific margin of South America (Figure 109.1). These four types are: (1) polymetallics (copper, lead, and zinc) in the Wilhelmine Bay area; (2) vein-type copper in the Spring Point, Paradise Harbour, and Lemaire Channel areas; (3) porphyry copper in the Palmer Archipelago area; and (4) magmatic iron in the Brabant Island area. The available evidence suggests that none of these zones can be considered economically workable at the present time.

Due to strong glacial erosion and to the fact that most ore deposits are localized in altered rocks which are more easily removed than unaltered country rocks, the majority of Antarctic ore deposits are most likely to occur in topographic lows. Because of this, they will be mostly covered by thick layers of ice, and the only evidence of the existence of an ore deposit to be found on the surface will be related to its peripheral zones, such as propylitic rings, secondary vein systems, and silicified zones.

With regard to the possibility of finding supergene minerals, there is no evidence for the existence of a subtropical climate and peneplanation in the Gerlache Strait region since Cretaceous time that would make strong leaching possible and keep a stable groundwater level. Accordingly, the primary minerals (except iron) are scarcely affected by leaching, and secondary sulfides are rare. Surface mineralization is represented mainly by primary minerals and some copper oxides. Iron is easily mobilized from primary magnetite and redeposited as hematite. It does not seem possible, however, to expect rich concentrations of secondary sulfides in this region (Alarcón and others, 1976).

Mineralized zones and economic possibilities

Wilhelmine Bay Area

The main rocks of this area are metagraywackes and quartz sandstones intruded by Cretaceous granites, but micaceous slates, pelitic metasedimentary rocks, quartzites, brecciated quartzites, and other metasedimentary rocks are also observed. All of these bedded rocks are strongly folded, and locally they are significantly faulted as well (Table 109.1).

Patches with surficial limonitic crust are common in this

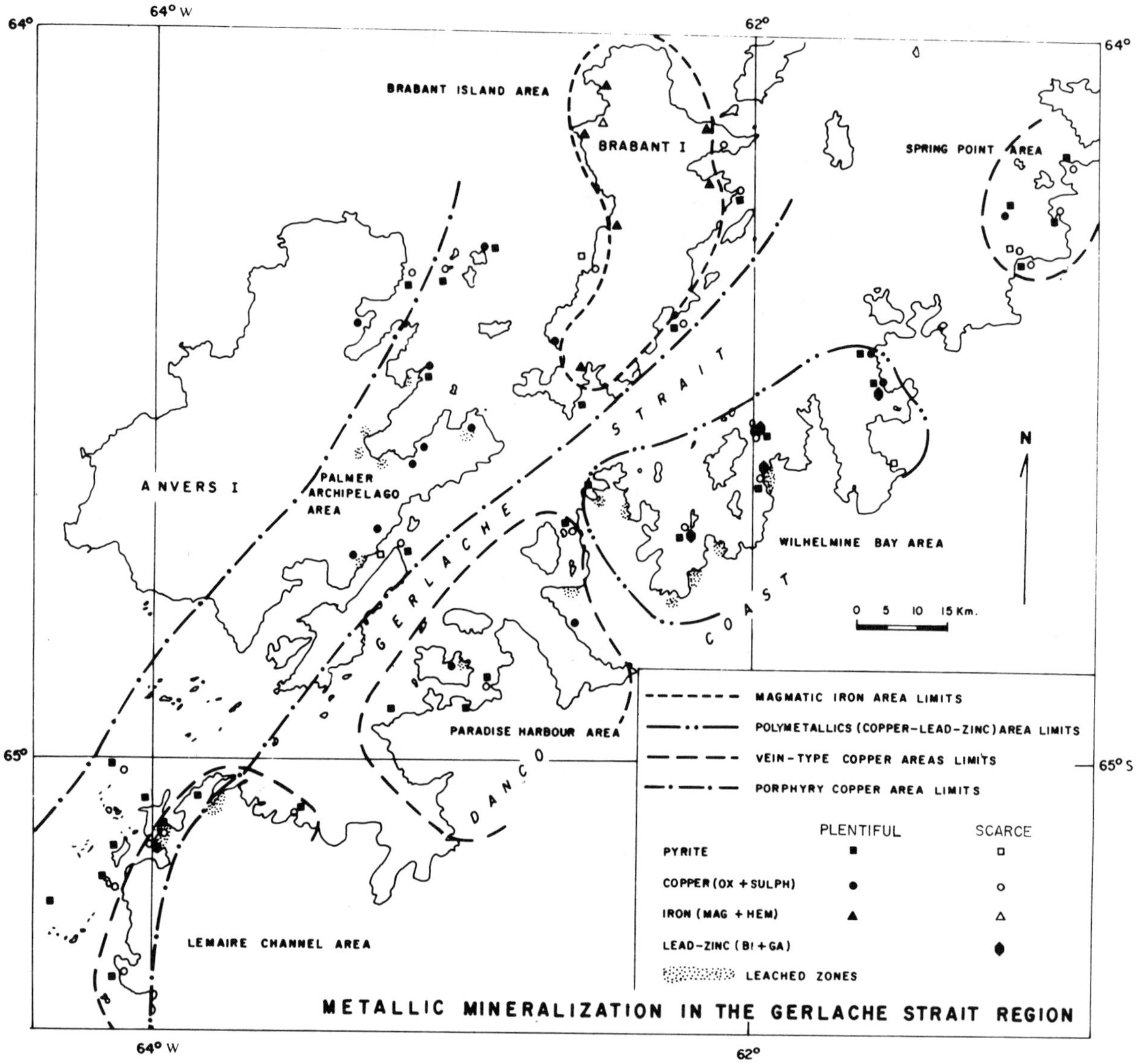

Figure 109.1. Map of Gerlache Strait region showing areas with evidence of different types of metallic mineralization.

area. These occurrences are related to silicification and to disseminated euhedral pyrite crystals. Minor amounts of chalcopyrite, sphalerite, galena, silver minerals, and scarce marcasite have also been found. The origin of this zone seems to be related to a low-temperature hydrothermal alteration of Triassic or upper Paleozoic sedimentary rocks generated by the intrusion of Cretaceous granites. It is possible that the mineralization may be restricted only to certain sedimentary layers.

The limonitic crusts are the best prospecting targets in this area, but geochemical studies are necessary to define the nature and extent of subsurface mineralization. Generally, polymetallic mineralization is observed in bodies not associated with well-defined tabular shapes.

Table 109.1. Mineralization in the Wilhelmine Bay area

Rock type	Alteration products	Mineralization	Occurrence
Micaceous slate	Quartz veins	Pyrite, chalcopyrite (abundant), copper oxides (scarce)	In veins and disseminated
Pelitic metasedimentary rocks	Silica, chlorite (abundant),	Pyrite, chalcopyrite (abundant), sphalerite, silver (?) (scarce)	In veins and disseminated
Metasedimentary rocks	Chlorite, silica (abundant)	Pyrite (abundant)	Disseminated
Quartzite	Silica (abundant)	Pyrite (abundant), sphalerite, marcasite, galena, chalcopyrite (scarce)	Disseminated
Brecciated quartzite	Silica (abundant)	Pyrite (abundant), chalcopyrite, sphalerite (scarce)	Disseminated
Metaandesite	Thermal metamorphism	Pyrite (abundant)	Disseminated

Table 109.2. Mineralization in the Spring Point area

Rock type	Alteration products	Mineralization	Occurrence
Granite	Argillite, silica (abundant), sericite, chlorite (scarce)	Pyrite, chalcopyrite (abundant)	Disseminated
Porphyritic andesite	Chlorite (abundant), epidote (scarce), K-feldspar (scarce)	Pyrite, chalcopyrite, paratacamite (abundant)	In copper veins and disseminated in the wall rock
Brecciated lavas	Chlorite (abundant)	Pyrite (abundant), chalcopyrite (scarce)	In small veins and disseminated
Metasedimentary rocks	Silica, chlorite, calcite (scarce), argillite (scarce)	Pyrite (abundant), chalcopyrite (scarce)	Disseminated
Metagraywacke and metaquartzite	Silica (abundant), quartz veins (scarce)	Pyrite, chalcopyrite (scarce)	Disseminated

Spring Point area

Chemically intermediate to basic volcanic rocks, volcanic breccias, and metasedimentary rocks make up most of the bedrock in this area. All of these rocks have been intruded by Cretaceous granites, some of which are porphyritic in texture. The stratified rocks are folded and faulted, in some places as much as the underlying beds, which presumably belong to the Trinity Peninsula Series (Table 109.2).

Pyrite with minor amounts of chalcopyrite, and locally copper oxides such as paraatacamite, are found in veins. However, these minerals are also found disseminated within the country rock. The mineralization occurs in zones of the country rock, which exhibits weak hydrothermal alteration and some degree of thermal metamorphism. The alteration and the metamorphism have been caused by the Cretaceous intrusions (Vergara and others, in press).

In future prospecting, vein-type copper ores should be the most likely type of deposit in this area.

Paradise Harbour area

This area is underlain by folded Upper Jurassic volcanic rocks that are intruded by granites, olivine monzodiorites, and monzogabbros of the Danco Coast (Table 109.3).

Several localities show disseminated pyrite and lesser amounts of chalcopyrite. A copper vein was found at Bryde Island, while another was observed at Andvord Bay. The mineralization seems to be related to contact metamorphism produced by chemically basic intrusive rocks, but it is also influenced by some form of tectonic control.

Vein-type mineralization, similar to that of the Spring Point area, seems to be the most widespread type of mineralization in this area. However, it is possible that other types of mineralization, such as nickel and chromium deposits, could be found associated with certain ultrabasic intrusive rocks.

Lemaire Channel area

This area is characterized by gently folded intermediate to basic volcanic rocks, breccias, and low-grade metamorphosed sedimentary rocks intruded by Eocene granodiorites and monzogabbros (Table 109.4).

Pyrite, chalcopyrite, and pyrrhotite comprise the most common mineralization in this area. The minerals are widely disseminated in the volcanic rocks and in the plutonic intrusive rocks, but the major concentrations of these minerals are found within veins (Figure 109.2). The origin of this mineralization seems to be related to the Eocene intrusions, which produced some alteration and thermal metamorphism in the country rock. Vein-type deposits,

Table 109.3. Mineralization in the Paradise Harbour area

Rock type	Alteration products	Mineralization	Occurrence
Porphyritic granodiorite	Argillite, chlorite (scarce)	Pyrite (abundant), chalcopyrite (scarce)	Disseminated
Granite	Argillite, sericite (scarce)	Copper oxides	In veins
Metamorphic rocks		Pyrite (abundant)	Disseminated
Basic andesite	Chlorite, argillite (abundant)	Pyrite (abundant), chalcopyrite (scarce)	Disseminated
Andesitic lavas intruded by basic and gabbroic rocks	None	Copper oxides	In veins

Table 109.4. Mineralization in the Lemaire Channel area

Rock type	Alteration products	Mineralization	Occurrence
Porphyritic andesite	Chlorite (abundant), K-feldspar (scarce), epidote (scarce)	Pyrite, chalcopyrite (abundant)	Disseminated and in copper-bearing veins
Granodiorite	Argillite, silica (abundant), sericite, chlorite (scarce)	Pyrite (abundant), chalcopyrite (scarce)	Disseminated
Andesite	Chlorite, argillite (abundant)	Pyrite (abundant), chalcopyrite (scarce)	Disseminated and in small veins
Brecciated lavas	Chlorite (abundant)	Pyrite (abundant), chalcopyrite (scarce)	In small veins and disseminated

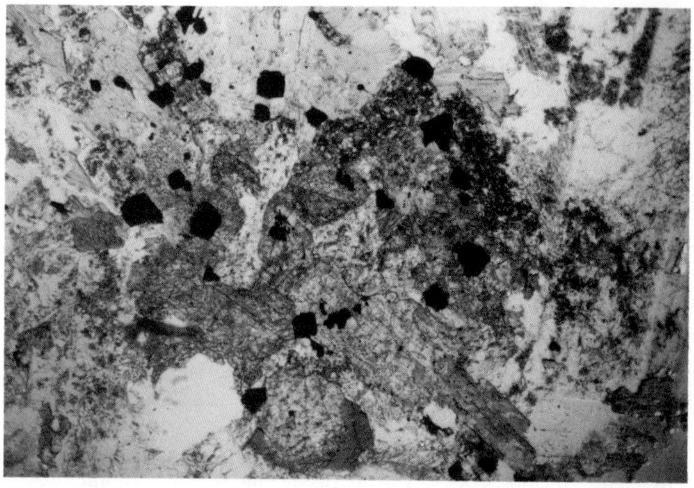

Figure 109.2. Disseminated pyrite and chalcopyrite (small black grains) in a granodiorite from the Lemaire Channel area (parallel nicols, ×10).

especially of copper, seem to be the only type of mineralization likely to be found in this area.

PALMER ARCHIPELAGO AREA

This area is underlain by gently folded Cretaceous lavas intruded by Eocene quartz diorites and granodiorites. In some places, these andesitic lavas are intruded by plutons of granitic and dioritic porphyries. Previous to the intrusive event, the volcanic rocks acquired a north-northeast tectonic grain (Table 109.5).

The mineralization is related to propylitic phases in some very large areas of hydrothermal alteration. These areas contain large amounts of disseminated pyrite and subordinate chalcopyrite (Figure 109.3). In other places, mineralization occurs as veins (Figure 109.4). Along the east coast of Anvers Island, on Copper and High Peaks, are several outcrops with porphyry-copper type mineralization, copper oxides, and huge altered zones. In places such as the Melchior Islands and Brabant Island, major amounts of chalcopyrite, and locally native copper, are present.

This mineralization is probably related to hydrothermal alteration associated with chemically intermediate mesothermal intrusions. Propylitization, sericitization (Neumayer Channel), and biotization of mafic minerals (Thompson Peninsula) are the most common hydrothermal alteration phases. Due to the similarities in the type of intrusion, in the type of country rock, in the age of the intrusion, and in its tectonic control, this area can be compared to other porphyry-copper districts in the Pacific margin of South America, such as Chuquicamata in northern Chile (Vieira and Cepeda, in press).

Because of the small possibility of secondary enrichment, the porphyry-copper ore deposits in this area do not have economic importance at present. However, it would be worthwhile to continue the investigation for the presence of economically important primary minerals such as gold or silver.

BRABANT ISLAND AREA

This area is characterized by the presence of subhorizontal Pleistocene volcanic rocks overlying a paleotopography cut on continental sedimentary and volcanic rocks of Cretaceous age. The volcanic rocks are andesites and olivine

Table 109.5. Mineralization in the Palmer Archipelago area

Rock type	Alteration products	Mineralization	Occurrence
Aphanitic andesite	Quartz, chlorite (abundant)	Pyrite (abundant), chalcopyrite (scarce)	Disseminated
Porphyritic hornblende andesite	Chlorite (scarce)	Pyrite (scarce)	Disseminated
Diorite	Chlorite, sericite (scarce)	Copper oxides (abundant)	In large areas of copper oxides related to altered zones
Diorite in zone of contact with andesite	Chlorite, sericite (scarce)	Copper oxides (abundant)	In large areas of copper oxides
Granodiorite	Chlorite (abundant), argillite, epidote, quartz (scarce), sericite (abundant)	Chalcopyrite, pyrite, bornite, copper oxides (abundant)	In veins, disseminated, and in areas with copper oxides. Locally related to altered zones.
Contact hornfels	Chlorite, sericite (abundant), argillite (scarce)	Pyrite (abundant), chalcopyrite (scarce)	In veins and disseminated
Porphyritic granodiorite	Chlorite, argillite (scarce)	Pyrite, chalcopyrite, copper oxides (abundant), native copper (scarce)	Disseminated

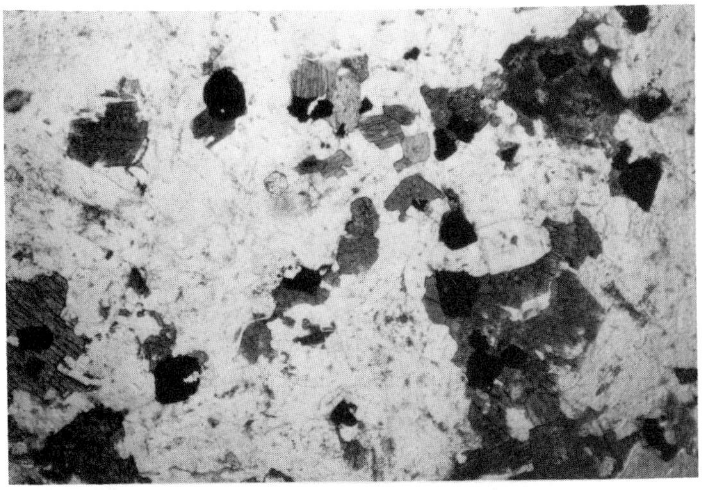

Figure 109.3. Disseminated pyrite and subordinate chalcopyrite (black grains) in a diorite from Neumayer Channel, Palmer Archipelago area (parallel nicols, ×10).

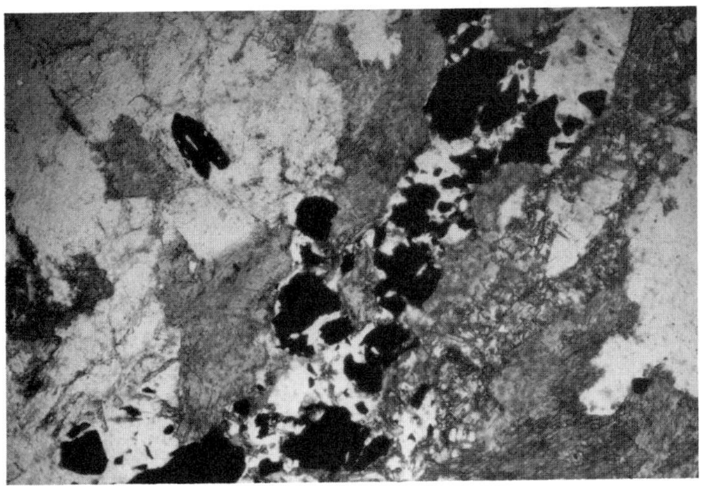

Figure 109.4. Chalcopyrite (black areas) in a vein in a hornblende diorite from the Melchior Islands, Palmer Archipelago area (parallel nicols, ×10).

basalts with remarkable columnar structure. Several volcanic necks of the same material have been observed in the northeastern part of Brabant Island (Table 109.6).

Large amounts of magnetite associated with the Pleistocene lavas have been detected in this area (Figures 109.5 and 109.6). The magnetite is partially altered to martite, and it has locally been transported into the underlying sedimentary rocks. In such localities the strata are fully impregnated with hematite and limonite. The presence of magnetite in the Pleistocene lavas of Brabant Island marks one of the few cases of iron oxide-rich volcanic flows in the world. The classic locality for this type of volcanic ore deposit is El Laco, in the highlands of Antofagasta, northern Chile. The comparison between the two deposits is remarkable, because there are similarities in mineralogy, age, and texture. However, the variety of different types of volcanic structures described in El Laco has not been recognized in Brabant Island.

Table 109.6. Mineralization in the Brabant Island area

Rock type	Alteration products	Mineralization	Occurrences
Basic lavas	Hematite (abundant)	Magnetite (abundant), hematite (abundant)	Associated with lava flows
Olivine basalt	Martite	Magnetite (abundant), martite (scarce)	Associated with lavas with columnar structures
Vesicular olivine andesite	Groundmass altered to hematite	Hematite (scarce)	Associated with lava flows
Conglomerate with andesitic pebbles	Sandy groundmass strongly altered to hematite	Hematite (abundant)	Associated with continental sedimentary rocks
Columnar basalt	Hematite (abundant)	Magnetite (scarce), hematite (abundant)	Volcanic necks (?)

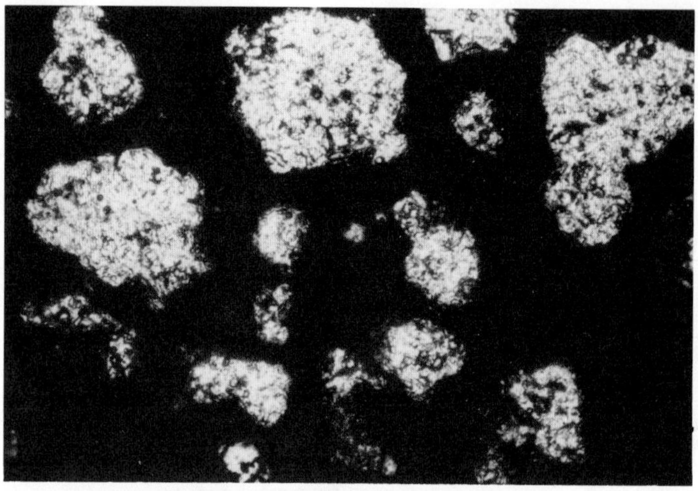

Figure 109.5. Magnetite groundmass (black) of a vesicular lava from Brabant Island (parallel nicols, ×10).

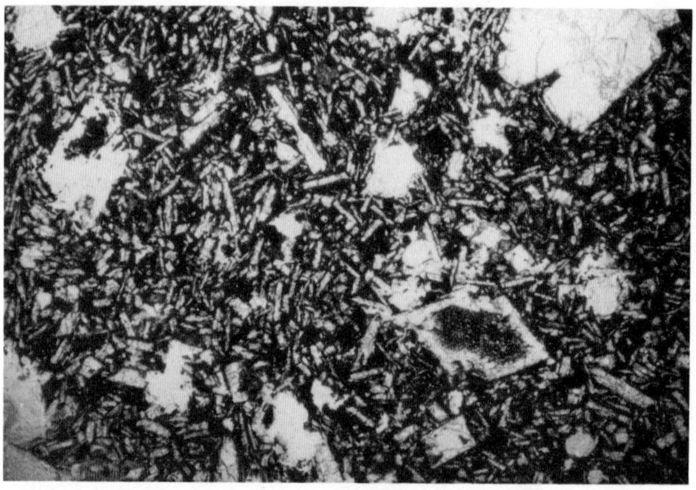

Figure 109.6. Large amount of magnetite (black areas) in the groundmass of a vesicular lava from Brabant Island (parallel nicols, ×10).

Acknowledgments. The authors wish to express thanks for the financial support during field work given by the Instituto Antártico Chileno and the logistic support given by the Armada de-Chile. Codelco Chile—División Chuquicamata, Departamento de Geología of the Universidad de Chile, and Instituto de Investigaciones Geológicas all contributed during the laboratory studies. We also want to thank in a special way Lieutenants De Bonnafos and Thornton for their enthusiasm during the extensive reconnaissance flights.

References cited

Alarcón, B., Ambrus, J., Olcay, L., and Vieira, C. 1976. Geología del Estrecho de Gerlache entre los paralelos 64° y 65° latitud sur, Antártica Chilena. *Ser. Cient. Inst. antárt. Chil., 4,* No. 1, 7–51.

Vergara, M., Hoces de la G., G., and Cepeda, A. In press. Geología del Cabo Tisné, Antártica Chilena. *Ser. Cient. Inst. antárt. Chil.*

Vieira, C., and Cepeda, A. In press. Geología de las Costa de Danco y de la Costa de Graham entre el Cabo Willems y el Cabo Tuxen, y de las islas e islotes adyacentes. *Ser. Cient. Inst. antárt. Chil.*

110

Resource and Radioactivity Survey in Antarctica by Airborne Gamma-Ray Spectrometry

E. J. ZELLER, G. DRESCHHOFF, and K. CRISLER
Department of Geology, University of Kansas, Lawrence, Kansas 66045, U.S.A.

F. TESSENSOHN
Bundesanstalt für Geowissenschaften und Rohstoffe, 3 Hannover, Federal Republic of Germany

Abstract. This study constitutes the first systematic attempt at a gamma-ray survey in a part of Antarctica. The primary effort was directed toward an examination of the radioactivity of the exposed rocks in the Dry Valley region of southern Victoria Land. The principal objective of this phase of the work was an evaluation of the uranium resource potential of this extensive ice-free area. Prior to the airborne survey, a calibration test was made in which high-precision ground measurements were compared with results from a helicopter-borne detector. Subsequent to this test, the survey was accomplished by airborne gamma-ray measurement techniques, but ground checks were made in places where significant anomalies were detected. In the course of the survey, uranium-bearing pegmatites were found, and other potential mineral resources were discovered.

The geology of the Dry Valley region of southern Victoria Land has been studied extensively, and only a brief review of the types of rocks cropping out in the study area is needed here.

The oldest metasedimentary rocks in the area that was studied are 5000 to 8000 m thick and consist mainly of steeply dipping crystalline argillite, quartzite, and marble. The marble is widespread over the area and is at least 3000 m thick.

Granite, granodiorite, and diorite which were grouped together by Gunn and Warren (1962) as the Granite Harbor Intrusives are clearly closely associated with the metamorphic rocks of the Ross Supergroup. A posttectonic phase with true intrusive rocks can be readily separated from a syntectonic phase which consists mainly of gneiss. The Larsen Granodiorite is actually a gneiss, and the foliated texture is clearly developed almost everywhere. The fact that the Larsen Granodiorite and the augen gneiss associated with it lie in the boundary zone between high-grade metamorphic rocks from partial anatexis and plutonic igneous rocks from complete melting makes uranium concentrations unlikely because the original sediments were generally low in uranium content.

There are a number of small bodies of discordant granite which intrude the metamorphic sequence. These are posttectonic and are usually hornblende granites, though the composition varies widely. The lit-par-lit intrusive complex at Mount Dromedary, which consists of granite and pegmatite containing radioactive minerals, apparently should also be classified as one of the posttectonic intrusive rocks.

The youngest geologic events which affect the basement rocks are the intrusions of swarms of lamprophyres and porphyries. Only the Vanda Porphyries (McKelvey and Webb, 1962) are of interest in evaluating uranium resources. These porphyry dikes cut all of the older Precambrian and lower Paleozoic formations. They are usually reddish with dark chill margins and contain phenocrysts of quartz and feldspar or of feldspar alone.

After an erosional phase, most probably during Silurian time, the folded metamorphic sequence from the Ross orogeny was covered by Karroo-type continental sandstone and related rocks. This essentially undeformed flat-lying sequence exhibits very similar rock types throughout most of East Antractica. Its age ranges from Devonian to Jurassic.

At present, the Beacon Supergroup (Barrett, 1971) is divided into two stratigraphically different sequences, the older Taylor Group and the younger Victoria Group. The Taylor Group, includes the lower strata, which are essentially Devonian in age. There appears to be a hiatus above the Taylor Group, and no Carboniferous beds are present in the sequence. The Victoria Group includes mainly Permian and Triassic beds, and it is overlain by widespread Jurassic volcanic and sedimentary rocks.

In the Devonin part of the sequence, the strata are well bedded, yellowish quartz sandstones with conglomeratic and argillaceous zones near the base; a persistent series of green to maroon siltstones is present at the top. The Taylor and Victoria Groups are separated by a tillite of variable thickness which is thought to be of Early Permian age. The Permian–Triassic sedimentary rocks are much less well sorted than those in the underlying Devonian beds. They contain sandstone with abundant carbonaceous material, arkose, and many lenses of shale and siltstone. Permian coal beds are present, and in some places individual units attain a thickness of more than 3 m. Volcanic activity during Jurassic time caused widespread intrusions of massive sills in the underlying strata (Ferrar Dolerites) and effusive basalts at the surface (Kirkpatrick Basalts). Jurassic sedimentary rocks are not extensive and are only locally developed.

The last major magmatic event in the area was the eruption of a series of highly alkaline volcanic rocks. This event was probably associated with the block faulting at the Ross Sea margin. The volcanism probably began in late Tertiary time and continues to the present as the activity of Mount

Erebus. The extrusive rocks include olivine basalt lava flows, along with phonolites, trachytes, and kenytes which commonly occur as stocklike shallow intrusive bodies with flow textures. Recent descriptions of these rocks have been provided by Treves (1962), Cole and Ewart (1968), Cole and others (1971), and Kyle and Treves (1973).

Radiation counting apparatus

The radiation counting apparatus used was designed to furnish total count as well as to provide spectral data of bismuth-214 from the U-238 decay series, thallium-208 from the Th-232 decay series, and potassium-40. The instrument could be modified either to serve as a portable ground-survey instrument or to function as an airborne gamma-ray detector with an analog record output.

Ground survey apparatus

For ground surveys on foot, the portable gamma-ray spectrometer system consisted of the GPX-21 detector and the GR-410 analyzer, both manufactured by GeoMetrics. The detector is a NaI(Tl) crystal of 347 cm^3 (21 in^3) volume and was carried on a pack frame in a foam-insulated, thin-walled aluminum cylinder. The analyzer was carried in a pouch which had been insulated by 2-cm layers of foam. The standard "D" cell alkaline batteries were replaced by lithium cells to reduce voltage drop caused by low-temperature operation. The gamma-ray spectrum analyzer provides simultaneous measurement of uranium, thorium, potassium, and total count by sequential digital readout of the four channels. The total count is determined within the range of 0.4 to 3.0 MeV. The channel width for potassium (K-40) is 1.46 MeV ± 200 keV, for uranium (Bi-214) 1.76 MeV ± 200 keV, and for thorium (Tl-208) 2.62 MeV ± 400 keV.

The overall effectiveness of the portable spectrometer with the 7.6 cm × 7.6 cm (3 in × 3 in) crystal detector is judged to be optimal for ground surveys. A crystal of this size is the best possible compromise between improved counting statistics, which would require increased weight, and smaller crystal volume, which would reduce weight but would seriously degrade counting statistics. For example, when the total count mode with the 347 cm^3 detector was used, it was found that a 1 minute counting time resulted in an accuracy of better than ±3%. However, to achieve a similar accuracy for uranium and thorium separately would require 10 minute counting times. An internal Ba-133 source which allows a continuous check of energy calibration of the system is especially advantageous for field work. Gain drift of more than ±2% is indicated automatically so that gain stabilization can be achieved by immediate recalibration.

The counter response can be expressed in radioelement concentration units "ur" as suggested by the International Atomic Energy Agency (I.A.E.A.) (Løvborg and others, 1976), which is the instrument response in the total count channel to a source containing 1 ppm eU (equivalent uranium). According to this definition, the calibration of the portable scintillation counter was found to be 21 cps/ppm eU or 21 cps/ur, assuming that the geologic source was in equilibrium. The value is obtained by placing the counter in contact with a source of known uranium concentration that is in radioactive equilibrium.

Airborne survey apparatus

For airborne operations, the gamma-ray detector GPX-112 manufactured by GeoMetrics was used. This unit contains a 1835 cm^3 NaI(Tl) crystal and was enclosed in a foam-insulated aluminum cylinder to reduce temperature fluctuations during operation. The GR-410 analyzer was used in the analog mode for total counts during all of the helicopter surveys. The total counts were recorded on a single-channel strip chart recorder, model 7155B, manufactured by Hewlett Packard. The equipment selected for the 1976–77 survey was deliberately chosen for minimum cost because no previous airborne surveys had ever been conducted and there was no way of determining what levels of radioactivity might be encountered in Antarctica.

The term "figure of merit" has been applied to make comparisons between various airborne surveys. As defined by Darnley (1971), it involves three parameters: detector volume, mean ground clearance, and mean ground speed. Although our apparatus made use of only 10.2 × 10.2 × 17.8 cm^3 detector volume, its figure of merit in the total count mode was exceptionally high because the helicopter operation was conducted at low ground clearance and low ground speed.

Flight operations

Aircraft

U.S. Navy UH-1/N helicopters, which are identical to the civilian Bell model 212, were used for all of the helicopter survey operations. This aircraft with twin turbine engines is a high-performance helicopter particularly well suited to the high relief terrain. Maximum service ceiling is 5000 m above mean sea level, and maximum airspeed is 130 knots at sea level. Total flight time can be extended to nearly 4 hours with an auxiliary fuel bladder; this corresponds to an operational radius of about 250 km.

Counter location in the aircraft

The detector crystal was located in the center of the cabin directly behind the avionics console. The aircraft structure at this point provides the least attenuation for gamma radiation. This location is forward of the main fuel tanks and transmission, and only two thin sheets of aluminum alloy separate the detector package from the outer surface of the aircraft.

Airborne survey crew operation

All airborne surveys were performed by two-person teams. In the aircraft, one person was positioned directly over the detector crystal, between and behind the pilots where good forward visibility is available. This individual was responsible for directing and recording the flight path, as well as for observing changes in rock units, airspeed, and altitude.

The second crew member was seated next to a side window and monitored the spectrometer and analog recorder controls and output. This individual was responsible for noting position and rock unit changes on the recorder chart and for keeping the other survey and flight crew members informed of the count rate from the terrain below. Upon encountering an anomaly, the spectrometer operator could request a turn and a repeat run over the flight path or could recommend landing if a sufficiently strong anomaly was encountered.

Flight operations in high relief areas

Owing to the high relief which is characteristic of the terrain of southern Victoria Land, high performance helicopters are clearly the most practical vehicle for making radiation surveys. By experimentation it was found that the helicopters could be operated at forward speeds of 30 to 40 knots and at ground clearances of 15 to 30 m with maximum accuracy and ease. This height was maintained unless turbulence required a greater clearance for reasons of safety. In calm air, cliff faces and steep slopes can be approached without difficulty. It rapidly became apparent that it was most efficient to delineate the extent of each anomaly by circling the site where it was detected and then landing for a direct ground check. This practice had the advantage of reducing the need for repeat flights into remote areas which might become inaccessible because of rapid weather changes, making resurvey of anomalies impossible. In addition, this mode of operation was necessary because of the lack of high-precision navigation or position recovery equipment.

Radiometric survey

Geologic basis for survey target selection

For the first systematic airborne radiometric survey in an area, two possible modes of operation can be considered. A completely unprejudiced and essentially random search can be initiated, with an attempt made to scan all rock outcrops in an area. This mode was impractical when operating from McMurdo Station because of its intense and competitive utilization of helicopter support. Accordingly, the other alternative of a selective survey was chosen, in which an attempt was made to focus on targets selected for their known geologic characteristics.

Early in the field operation, the Beacon Supergroup was chosen as the prime survey objective because of the character of these sedimentary rocks. The sequence is thought to be part of the extensive blanket of continental strata that once covered the old Gondwana continent. Subsequent breakup of the supercontinent resulted in segments of this blanket now being in the land masses of South America, Africa, Australia, Antarctica, and India. The Permian–Triassic sequence of the Beacon Supergroup, with its continental depositional environment, channel sands, and abundant carbonaceous material, fulfills several important conditions for uranium enrichment. Moreover, there are two known occurrences of uranium deposits in lower Triassic beds of Karroo type—one in India (Udas and Mahadevan, 1974) and one in South Africa (Von Blackstroem, 1974). The survey area is shown in Figure 110.1.

Calibration flight path

In compliance with I.A.E.A. recommendations, a calibrated flight path (Figure 110.2) was established within the study area. This permits recalibration of airborne apparatus at any time, particularly in the event of modifications or alterations to the equipment. The area of the Nussbaum Riegel in Taylor Dry Valley was selected as the site for the establishment of a calibrated ground track which could serve as a flight path for airbone instrument testing. The track consists of a gently curving line of 127 ground stations distributed uniformly along a path 4.1 km long. The 347 cm^3 crystal was mounted on a pack frame so that it could be carried 1.8 m above the ground. At each station, 1 minute measurements were made using the accumulate mode, and the output of all four channels was recorded. Flags were set up at 100 m intervals to mark the calibrated path and make it readily visible from the air.

The first part of the flight path passes over the top of the Nussbaum Riegel and was selected because it crosses a sequence of vertical metamorphic beds at 90° to the strike. Tests show that there are only minor variations in radiation background along the strike of the beds; therefore, small lateral deviations from the marked flight path do not cause major variations in the radiation level from the ground directly below. Furthermore, the top of the riegel is almost free of moraine material, and the exposures of bedrock are excellent. The segment of the track across the bedrock exposures is approximately 2.2 km long and descends roughly 200 m to the moraine west of the peak of the riegel. The flight path then crosses 1.1 km of ground moraine which is composed mainly of slightly weathered blocks of granodiorite. Beyond the moraine the path traverses about 300 m of well-exposed granodiorite, followed by 400 m of basalt scoria which has been erupted from a small volcanic vent on the south slope of the valley. The track terminates in an outcrop of augen gneiss just before reaching the steep south wall of the valley. Nearly vertical basalt dikes are present in a number of places along the flight path over the exposed bedrock. These dikes are sufficiently wide or numerous to cause substantial variations in the radiation background at four locations along the path.

The total relief on the entire track is about 250 m, but there are no steep slopes and the terrain can be followed by a helicopter at nearly constant ground clearance without undue difficulty. Clearly, a level calibration flight path would have been more advantageous than one which involved altitude changes: a constant forward velocity cannot be maintained easily when the aircraft is required to change altitude. In selecting the site, however, it was felt that substantial variation in background radiation from highly varied rock types was more important than strict maintenance of forward velocity. Furthermore, alterations in wind speed and direction cannot be controlled, and they also produce differential "shrinkage" or "stretching" of the output which appears on the chart of the analog recorder.

The profile for the total-count channel is shown in Figure 110.2. there is an almost twofold increase in radiation background from the marbles, which have the lowest content of radioactive impurities, to the augen gneiss, which has the highest. This variability provides a real indication of the sensitivity of the airborne unit. By comparing the ground profile with the airborne measurements, it is also possible to determine the ability of the counter to resolve small-scale radiation anomalies.

The larger volume airborne counter was found to be approximately three times higher in sensitivity in the total-count channel when compared to the ground survey instrument. The total-count response of the 1835 cm^3 detector

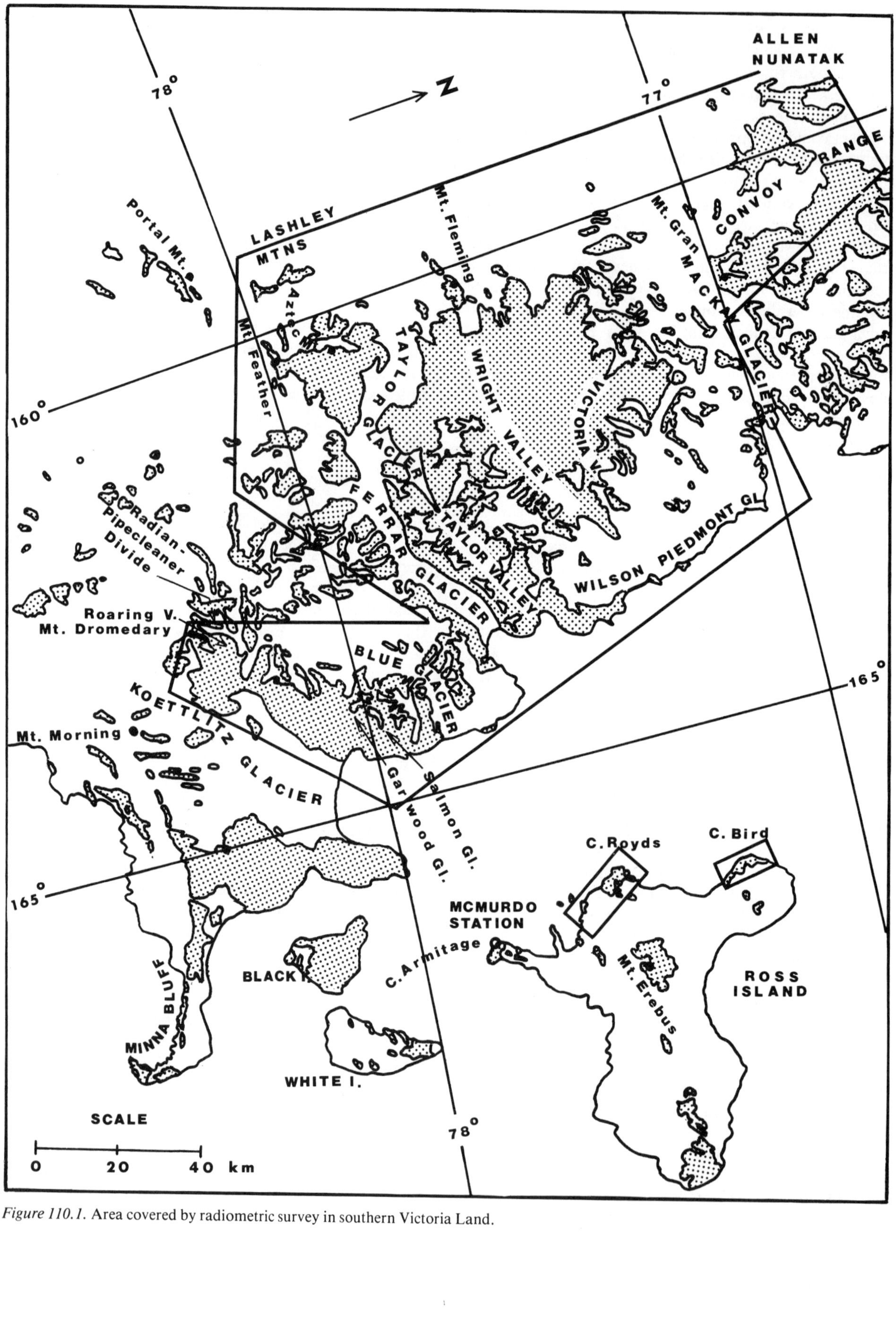

Figure 110.1. Area covered by radiometric survey in southern Victoria Land.

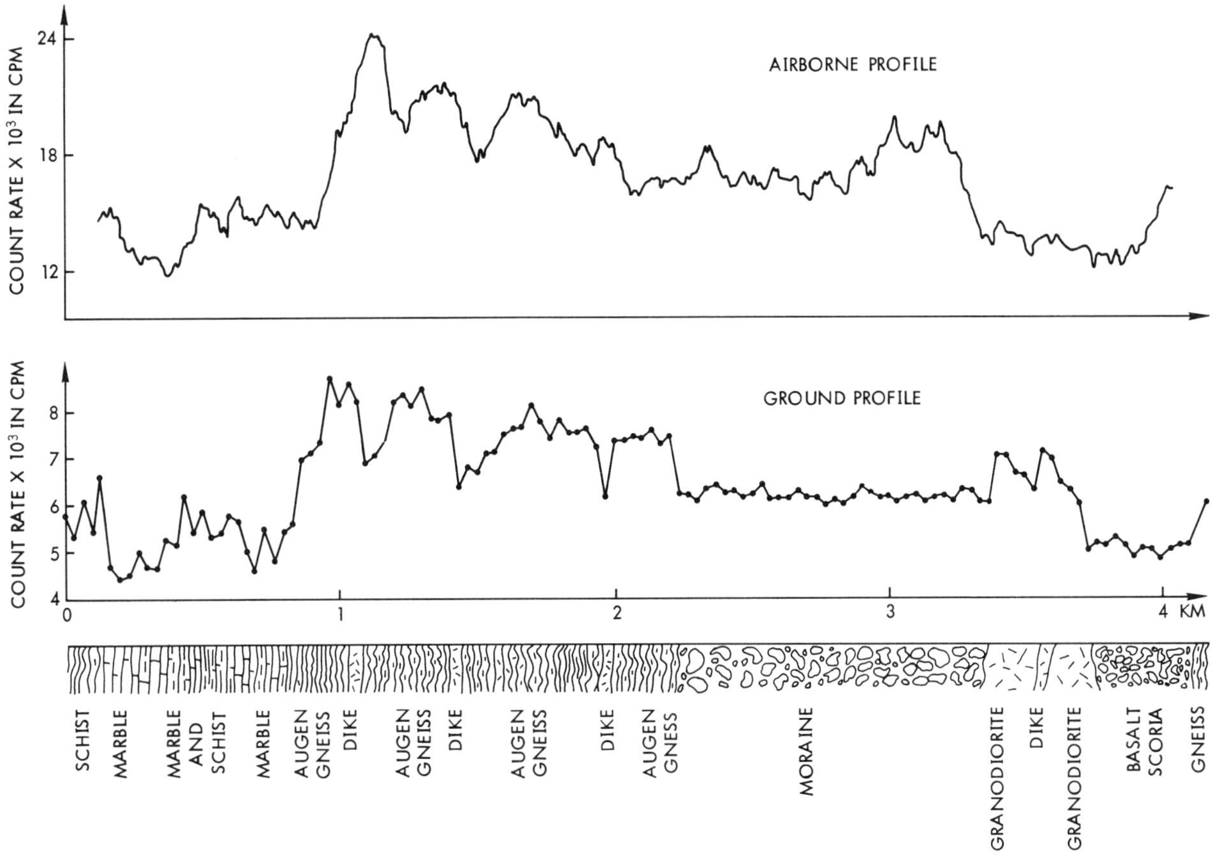

Figure 110.2. Calibration flight path, Nussbaum Riegel, Taylor Dry Valley. See text for explanation of profile orientation.

was found to be 300 cps at 100 ft (about 30 m) ground clearance over augen gneiss or granite. The counting error is ± 6%.

The influence of forward speed becomes especially apparent when attempting to detect anomalies. The ground speed at 40 knots will be 20 m/sec, and a flight line of 300 m will be covered in 15 seconds. In the example of an anomaly 300 m in diameter, with the radiation intensity from augen gneiss or granite, a total of 4500 counts would be detected in 15 seconds, with a standard deviation of better than ± 3% for the 1835 cm^3 crystal.

Correction of data

One of the principal advantages found in making radiation surveys in Antarctica is the ready availability of massive thicknesses of high purity ice which permits frequent rechecking of background contributions to the total count. The background in all four channels of the portable counter was measured 1.8 m above the surface of an ice-covered lake in Taylor Dry Valley. Measurements were made at the center of the lake, which was approximately 100 m in diameter. The contribution of background radiation to the count rate of the airborne survey instrument could be determined frequently during the flight. Since more than 100 km of sea ice had to be crossed between McMurdo Station and the survey area, background was routinely measured both before and after each survey flight. Over the survey routes, subsequent determinations of background could be made each time the aircraft flew over clean glacial ice.

Although we tried to maintain uniform ground clearance, in some cases measurements had to be normalized to a standard height above ground level. To accomplish this, counts were made for 20 seconds using the accumulate mode with the helicopter on the ground (350 ft above sea level) and then by hovering at intervals above the same point up to 300 ft above ground level. The area chosen was located in Taylor Dry Valley, 2 km south of Lake Fryxell. The ground surface is covered by a uniform moraine deposit and slopes upward to the south at approximately 3°; slope effects were considered negligible (Wormald and Clayton, 1976). The experiment shows that a reduction of approximately 20% in total count rate results for every 30 m increase in elevation above the surface.

Terrain effects which significantly alter the solid angle between the detector and the source can be very large in high-relief areas. Compared to data obtained over flat areas, a reduction of approximately one-half can be expected when flying along the edge of a cliff; an increase of almost 1.5 can be expected when flying close to the foot of a cliff in an ice-free valley (Wormald and Clayton, 1976). The ability of the counter survey crew to observe the terrain over which they were flying, together with the fact that the data could be evaluated directly during the flight, permitted false anomalies caused by terrain effects to be filtered out.

Results of the study

Sedimentary rocks

Although the investigation was too limited in scope and in areal coverage to make any general statement about ura-

Table 110.1. Relative[a] radioactivity of rock types in southern Victoria Land

Rock types	Average airborne radiation values[a]
Basalt	2.0
Trachyte	3.7
Dolerite	1.9
Triassic Beacon beds	2.3
Permian coal measures	2.1
Aztec Siltstone and fishbeds	2.6
Devonian Beacon beds	1.8
Basal Beacon beds	2.5
Lit-par-lit granite and pegmatite	4.1
Vanda Porphyry	3.7
Irizar Granite	2.6
Metasedimentary rocks	2.4
Marble	2.2
Augen gneiss	3.0

[a] Average airborne radiation values are determined by comparison with an ice background value of 1.0.

nium resources in Antarctica, it has provided some highly valuable information, which is summarized in Table 110.1. First, within the study area it was discovered that the Beacon Supergroup rocks, which seem to be the most promising host beds, show no significant concentrations of uranium. The sensitivity of the airborne counter was sufficient, however, to detect specific stratigraphic zones within the Beacon Supergroup with ease. The Aztec Mountain Siltstone and some of the shaly members of the Mount Fleming Formation always show substantially higher radiation levels than did most of the sandstone units. Thin beds of volcanic ash tend to show increased radioactivity, and the fish beds in the Devonian part of the Beacon Supergroup apparently contain enough phosphate to concentrate uranium and produce a consistent anomaly.

Strong anomalies were not found in the channel sandstone and carbonaceous zones in the upper part of the Beacon sequence. The high permeability of the sandstone beds, the dissemination of carbonaceous fragments, and the existence of an extensive channel system would seem to make these beds ideal host rocks for uranium. Clearly, the geochemical environment should have been such that uranium would have been deposited if uranium-bearing solutions had flowed through the strata. It seems that the lack or uranium in the sedimentary rocks in the study area may reflect an unfavorable position in the sedimentary basin. Unfortunately, the extensive ice cover precludes any detailed knowledge of the extent and configuration of the basin.

Igneous rocks

Within the area investigated, the igneous rocks are both highest in overall radioactivity and show the most promise as potential sources of uranium. This is true both for the Tertiary volcanic rocks and for the igneous plutons which intrude the metamorphic complex of the basement. Since it is unlikely that there can be any direct connection between the magma sources for the older and younger igneous rocks, they will be considered separately.

Volcanic rocks (Tertiary to Holocene)

The Tertiary and Holocene volcanic rocks in the study area consist mainly of olivine basalt, phonolite, and trachyte. The trachyte, especially the pyroxene trachyte, shows an unusually high uranium content. The uranium concentration of the pyroxene trachyte body on Observation Hill at Cape Armitage was determined by X-ray fluorescence and found to be 10 ppm. Hornblende trachyte from Trachyte Hill on Cape Bird shows a uranium concentration of only 5 ppm.

Intrusive rocks

Among all of the intrusive rocks which were examined, only the posttectonic granite and the pegmatite complex in the Mount Dromedary Massif show any substantial radioactivity. The posttectonic granite and the associated pink porphyry dikes, which are found in abundance in Victoria and Wright Valleys and to a lesser extent to the south in Taylor Valley, usually show total-count values which are two to three times normal. Ground examinations of outcrops were made in several places, but no significant local concentrations of radioactive minerals were found. Both the dikes and the granite masses are easily detectable by airborne surveys, and efforts were made to determine the existence of local concentrations of radioactive minerals by flying directly above some of the pink porphyry dikes along the length of the outcrop. In almost all cases, the dikes showed very uniform radioactivity. This was also true of the more massive intrusive granite bodies. They too were substantially higher in total radioactivity than the average of the other bedrock types, but like the pink porphyry dikes, they showed no localized concentrations.

The pegmatite dike complex which occurs in the lit-par-lit series at Mount Dromedary shows local concentrations of radioactivity that are higher than at any other place in the entire study area. This complex is similar in character to the uranium-bearing pegmatite series at Rössing, South-west Africa (Moreau, 1977). Differential counts made with the gamma-ray spectrometers indicate that both discrete uranium and thorium minerals are present. The radioactive rocks commonly exhibit the typical yellow-green fluorescence which is associated with some of the secondary minerals that develop as alteration products in the zone of weathering. The pegmatite veins commonly contain large masses of very dark smoky quartz. Rock specimens containing these radioactive minerals were tested by X-ray fluoresence and show values as high as 32 ppm uranium and 82 ppm thorium.

Pegmatites and small hydrothermal veins are present at a number of places in the upper part of Garwood Valley in the area between the Garwood and Salmon Glaciers, and these bodies have also been found to contain uranium minerals. These minerals seem to be similar in character to those in the Mount Dromedary Massif and are associated with large smoky quartz crystals. Thin encrustations showing strong yellow-green fluorescence are also present, and these too may be associated with secondary uranium minerals.

Weathering, uranium transport and secondary uranium mineral development

The presence of coatings of fluorescent secondary uranium minerals on many of the surfaces of specimens collected from the radioactive pegmatites suggests strongly that there is a high mobility of uranium in the weathering environment of the study area. This conclusion is supported by two additional pieces of evidence. Gamma-ray logs were made on both the hole and on the core taken from the Dry Valley Drilling Project hole at the Lake Vida drilling site. This hole, which was drilled into the Vida Granite, shows definite evidence of leaching and removal of radioactive elements from the upper 160 m. This is particularly noticeable in those segments of the core with the most fractures.

Potential fluorite resources

In the course of the airborne radioactivity survey, an area of fluorite-bearing marble was discovered on the divide between the Radian and Pipecleaner Glaciers. The occurrence of fluorite at this locality had been reported by Blank and others (1963) as follows: "On the Pipecleaner-Radian divide minute octahedra of purple fluorite and pegmatitic books of black biotite occur in marble adjacent to a thick, coarse-grained granitic dike." A field check of the dike showed that it had small areas of localized high radioactivity as much as eight times normal background, but the minute purple fluorite crystals could not be found. Instead, two parallel veins of almost pure botryoidal fluorite—each nearly 1 m wide and traceable for over 100 m—were discovered. The fluorite is white to pale green at the surface where it is exposed to the sun, but blue and purple crystals were found by digging. In view of this discovery and the previous report, this area merits a further examination to determine the extent of the fluorite deposits.

Acknowledgments. This research was supported in part by Grant DPP 75-08075 from the Division of Polar Programs of the National Science Foundation and by the Geological Survey of the Federal Republic of Germany.

References cited

Barrett, P. J. 1971. Stratigraphy and paleogeography of the Beacon Supergroup in the Transantarctic Mountains, Antarctica. (*In* Adie, R. J., *ed. Antarctic Geology and Geophysics.* Amsterdam, North-Holland, 249-56.)

Blank, H. R., Cooper, R. A., Wheeler, R. H., and Willis, I. A. G. 1963. Geology of the Koettlitz-Blue Glacier region, southern Victoria Land, Antarctica. *Trans. R. Soc. N. Z., Geol., 2,* 79-100.

Cole, J. W., and Ewart, A. 1968. Contributions to the volcanic geology of the Black Island, Brown Peninsula and Cape Bird areas, McMurdo Sound, Antarctica. *N.Z. Jl Geol. Geophys., 11,* 793-828.

Cole, J. W., Kyle, P. R., and Neall, V. E. 1971. Contributions to Quaternary geology of Cape Crozier, White Island and Hut Point Peninsula, McMurdo Sound region, Antarctica. *N.Z. Jl Geol. Geophys., 14,* 528-46.

Darnley, A. G. 1971. *Airborne Gamma Ray Survey Techniques.* Ottawa, Geological Survey of Canada, 35 pp.

Gunn, B. M., and Warren, G. 1962. Geology of Victoria Land between the Mawson and the Mulock Glaciers, Antarctica. *Bull. geol. Surv. N.Z., 71,* 157 pp.

Kyle, P. R., and Treves, S. B. 1973. Review of the geology of Hut Point Peninsula, Ross Island, Antarctica. *Bull. Dry Valley Drilling Project,* No. 2, 1-10.

Løvborg, L., Kirkegaard, P., and Mose Christiansen, E. 1976. Design of NaI(Tl) scintillation detectors for use in gamma-ray surveys of geological sources. (*In Exploration for Uranium Ore Deposits Symposium Proceedings.* Vienna, IAEA Proceedings Series, 127-48.)

McKelvey, B. C., and Webb, P. N. 1962. Geological investigations in southern Victoria Land, Antarctica. Pt. 3. Geology of Wright Valley, *N.Z. Jl Geol. Geophys. 5,* No. 1, 143-62.

Moreau, M. 1977. L'uranium et les granitoides: essai d'interpretation. (*In* Jones, M. J., *ed. Geology, Mining and Extractive Processing of Uranium.* London, Institution of Mining and Metallurgy, 83-102.)

Treves, S. B. 1962. The geology of Cape Evans and Cape Royds, Ross Island, Antarctica. (*In* Wexler, H., Rubin, M. J., and Caskey, J. E., *eds. Antarctic Research: The Matthew Fontaine Maury Memorial Symposium.* Washington, American Geophysical Union, 40-46.) [Geophysics Monograph No. 7.]

Udas, G. R., and Mahadevan, T. M. 1974. Controls and genesis of uranium mineralization in some geological environments in India. (*In Proceedings of a Symposium on the Formation of Uranium Ore Deposits, Athens, 1974.* International Atomic Energy Agency, SM-183/39, 425-36.)

Von Blackstroem, J. W. 1974. Uranium deposits in the Karroo Supergroup near Beaufort West, Cape Province, South Africa. (*In Proceedings of a Symposium on the Formation of Uranium Ore Deposits, Athens, 1974.* International Atomic Energy Agency, SM-183/48, 419-24.)

Wormald, M. R., and Clayton, C. G. 1976. Observations of the accuracy of gamma spectrometry in uranium prospecting. (*In Proceedings of a Symposium on the Exploration for Uranium Ore Deposits, Vienna, 1976.* International Atomic Energy Agency, 149-71.)

IX Crustal Structure

111

Geothermal Studies in the Ross Island–Dry Valley Region

Review Paper

EDWARD R. DECKER* and GERALD J. BUCHER
Department of Geology, University of Wyoming, Laramie, Wyoming 82071, U.S.A.

Abstract. Measured temperatures in Dry Valley Drilling Project (DVDP) holes in Lake Vanda and McMurdo Sound range between 3.0° and 24.0°C, and −2.0° and 0.5°C, respectively. Elsewhere the temperatures in DVDP holes are between −28.0° and −3.0°C, with different minimum and maximum temperatures at each locality. When the data from holes drilled on land are extrapolated downward, the depths to the 0°C isotherm in the Ross Island–dry valley region are inferred to lie between 230 and 980 m.

Three uncorrected heat flows in the Ross Island–dry valley region range from 2.0 to 3.4 HFU (1 HFU = 1 × 10^{-6}cal/cm^2•sec). The heat flow on Ross Island is 1.5–1.6 HFU, on the assumption that a correction is needed for a steady-state temperature difference (~16°C) between the surface of the island and the waters in McMurdo Sound. The flux near Lake Vida is 1.9 HFU, if an adjustment is made for the effects of indefinitely persisting terrain in Victoria Valley. The radiogenic heat production of core samples from DVDP hole 3 on Ross Island is calculated to be 3 HGU (1 HGU = 1 × 10^{-13}cal/cm^3•sec), and that of core samples from hole 6 at Lake Vida ranges from 2.2 to 3.2 HGU. The radioactive heat generations of surface samples of Irizar granite and Larson granodiorite from the dry valleys are calculated to be 2.6 and 3.7 HGU, respectively.

The combined heat flow and radioactivity data suggest that the Ross Island–dry valley region is a zone of high terrestrial heat flow. A zone of high heat flow in this part of Antarctica would imply near-melting conditions in the underlying lower crust and/or upper mantle.

Short discussions of geothermal measurements in Dry Valley Drilling Project (DVDP) drill holes have been given by Pruss and others (1974), Decker (1974), Decker and others (1975), and Bucher and Decker (1976*a,b*). These writers cite evidence for thick permafrost in the Ross Island–dry valley region, and they suggest that the geothermal gradients measured in DVDP drill holes that had been abandoned for short periods of time are reliable. Additional field and laboratory data for DVDP holes have now been collected. In addition to confirming earlier speculations, the present results imply that both Ross Island and the dry valleys are characterized by high heat flow. This paper briefly reviews all the existing geothermal data for DVDP drill holes.

The investigated drill sites are plotted in Figure 111.1. Surveyed locations for the holes are listed below.

Hole number	Locality	Latitude	Longitude
2	Ross Island	77°51′ S	166°40′29″ E
3	Ross Island	77°51′ S	166°40′29″ E
4	Lake Vanda	77°31′48″ S	161°32′10″ E
6	Lake Vida	77°22′48″ S	161°48′37″ E
8 and 10	New Harbor	77°34′43″ S	163°30′42″ E
11	Commonwealth Glacier	77°35′24″ S	163°24′40″ E
12	Taylor Valley (Lake Leon)	77°38′2″ S	162°51′13″ E
13	Don Juan Pond	77°33′20″ S	161°09′53″ E
14	Wright Valley (North Fork)	77°32′23″ S	161°24′22″ E
15	McMurdo Sound	77°28′24″ S	164°25′30″ E

* Present address: Geophysics Group, MS 977, University of California, Los Alamos Scientific Laboratory, Los Alamos, New Mexico, 87544.

Several sets of temperature measurements were made at sites 6, 10, 11, and 12 because each of these holes is cased to total logging depth with 4.10-cm-diameter iron pipe that is filled with diesel fuel (DFA). The hole at site 8 was also logged more than once, as were the accessible parts of holes 2, 3, and 13. Site 14 at North Fork was visited during the 1975–76 field season to collect one set of temperature measurements and to preserve the hole by "freezing in" DFA-filled, 4.10 cm iron pipe. Holes 4 and 15 were not cased, and our measurements at these sites were made after drilling had been stopped for about 2 hours and 15 hours, respectively.

Throughout the text, temperatures (T) and temperature differences (ΔT) are in °C, gradients of temperature (Γ) are in °C/km, and depths and distances are in meters or

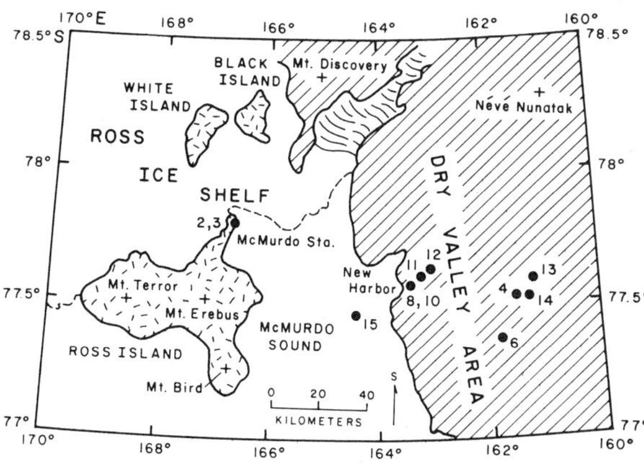

Figure 111.1. Map showing locations of investigated drill holes (numbered).

Table 111.1. Temperatures and temperature differences in selected stable drill holes

Depth (m)	Hole 3, 5-6 Nov. 1975, Probe 4, T°C	ΔT°C	Hole 3, 31 Dec. 1975–3 Jan. 1976, Probe 6, T°C	ΔT°C
50	−15.23		−15.22	
		0.66		0.65
70	−14.57		−14.57	
		0.69		0.70
90	−13.88		−13.87	
		0.67		0.66
110	−13.21		−13.21	
		0.58		0.59
130	−12.63		−12.62	
		0.61		0.60
150	−12.02		−12.02	
		0.32		0.31
160	−11.70		−11.71	
		0.71		0.72
180	−10.99		−10.99	
		0.74		0.75
200	−10.25		−10.24	
		0.82		0.81
220	− 9.43		− 9.43	
		0.45		0.46
230	− 8.98		− 8.97	
		0.46		0.45
240	− 8.52		− 8.52	
		0.45		0.45
250	− 8.07		− 8.07	
		0.41		0.41
260	− 7.66		− 7.66	

Depth (m)	Hole 11, 13–14 Nov. 1975, Probe 2, T°C	ΔT°C	Hole 11, 8–9 Dec. 1975, Probe 4, T°C	ΔT°C
40	−17.41		−17.42	
		0.81		0.83
60	−16.60		−16.59	
		0.96		0.93
80	−15.64		−15.66	
		0.94		0.94
100	−14.70		−14.72	
		0.91		0.92
120	−13.79		−13.80	
		0.86		0.86
140	−12.93		−12.94	
		0.93		0.92
160	−13.00		−13.02	
		0.89		0.91
180	−11.11		−11.11	
		0.91		0.90
200	−10.20		−10.21	
		1.09		1.07
220	− 9.11		− 9.14	
		1.00		1.02
240	− 8.11		− 8.12	
		1.04		1.04
260	− 7.07		− 7.08	
		0.91		0.92
280	− 6.16		− 6.16	
		1.05		1.03
300	− 5.11		− 5.13	

kilometers. The units of thermal conductivity, heat flow, and radiogenic heat production are abbreviated to CDU, HFU, and HGU, respectively, where 1 CDU = 1 × 10^{-3}cal/cm•sec°C, 1 HFU = 1 × 10^{-6}cal/cm^2•sec, and 1 HGU = 1 × 10^{-13}cal/cm^3•sec.

Field and laboratory procedures

Temperature measurements

Temperatures were measured at discrete points in the holes with thermistor probes in combination with a polypropylene-insulated three-lead cable, modified Wheatstone-type bridges, and electronic null detectors. Decker (1973) describes the temperature measuring equipment and the procedures used for calibration of the thermistor probes.

Reproducibility of absolute temperatures and temperature gradients in the most stable drill holes is illustrated by the selected examples listed in Table 111.1. Although the absolute temperatures in the relogged holes varied by as much as ±0.03°C, the mean gradient for depth intervals of 10 and 20 m was usually precise to within 5–10% (Table 111.1). The determination of the average gradient over a large depth range in the stable holes was therefore a very accurate part of our field measurements.

We have not been able to duplicate all of the temperatures in hole 6 at Lake Vida. Here the most recent measurements above 260 m were not stable, and the recordings at some points differed by as much as 0.11°C (Table 111.2). Below 260 m, however, the recordings were more stable, and the originally measured temperatures could be reproduced to within 0.03°C (Table 111.2).

Thermal conductivity measurements

Thermal conductivities were measured at the University of Wyoming with the divided-bar systems described by Decker (1973). The final conductivity used for each rock sample was the mean of two values determined relative to crystalline quartz and silica glass. Ratcliffe's (1959) curves were used for the conductivities of quartz (heat flow normal to the optic axis) and silica glass. The conductivities of remeasured samples were reproducible to within 1.5%.

Samples and standards were right-circular disks with faces ground flat and parallel to within 0.0033 cm. Disk thicknesses were within a few millimeters of either 1.64 cm or 2.46 cm. The diameters of the disks were within a few

Table 111.2. Precision of temperature measurements in hole 6

Depth (m)	11 Nov. 1975 T°C	19 Dec. 1975 T°C	ΔT°C
60	−23.06	−23.04	0.02
80	−22.52	−22.52	0.00
100	−21.96	−21.96	0.00
120	−20.81	−20.88	0.07
140	−20.39	−20.42	0.03
160	−19.84	−19.77	0.07
180	−19.19	−19.13	0.06
200	−18.64	−18.57	0.07
210	−18.24	−18.31	0.07
220	−18.14	−18.08	0.06
230	−17.89	−18.00	0.11
240	−17.61	−17.66	0.05
250	−17.35	−17.45	0.09
260	−17.11	−17.13	0.02
270	−16.84	−16.87	0.03
280	−16.59	−16.61	0.02
290	−16.35	−16.36	0.01
300	−16.11	−16.11	0.00

millimeters of either 3.21 cm or 4.17 cm. Samples and standards were measured under a minimum uniaxial stress of 200 bars.

The water-saturated samples of granite and gneiss from hole 6 were measured at laboratory temperatures in the range 18–22°C. To estimate the conductivity of these low-porosity units at mean *in situ* temperature, the room-temperature measurements were extrapolated to −20°C using the slopes of the Birch and Clark (1940) curves for the conductivities of igneous and metamorphic rocks.

The samples from holes 3 and 14 were saturated with water before measurements were made. Each sample was then placed in the divided bar and maintained at subzero temperatures to accomplish saturation with ice. Thin plastic retaining rings were placed around the porous samples of the volcanic rocks from hole 3 to prevent loss of water during freezing. The plastic rings were not needed for measurements of the low-porosity samples of granodiorite from hole 14. The "ice saturated" samples from hole 3 and hole 14 were measured at temperatures between −13°C and −17°C. The median down-hole temperature in hole 3 is about −12.5°C, and that in hole 14 is about −17.5°C.

Radiogenic heat production determinations

Radiogenic heat production values were calculated using gamma-ray spectrometric analyses for uranium (U), thorium (Th), and potassium (K) in surface and core samples. Each sample of crushed rock had a mass of 600 ± 1 g. The conversion factors used to calculate the heat generated by U, Th, and K were taken from Birch (1954). The heat production values are considered accurate to within 10%, based on comparisons between the University of Wyoming's results for the Lake Vida core samples and those obtained by the U.S. Geological Survey in Denver, Colorado.

Implications of temperature data

Permafrost thicknesses and circulating(?) ground fluids

Selected temperature-depth measurements in DVDP holes 3, 6, 8, and 10–14 are listed in Tables 111.3 and 111.4. Temperature-depth profiles for these holes are shown in Figure 111.2. If perennially frozen ground is considered to extend downward to the 0.0°C isotherm, then the data may be used to estimate approximate permafrost thicknesses. These estimates are compiled in Table 111.5, where the inferred permafrost thicknesses are shown to range from about 240 m at New Harbor to perhaps 970 m at Lake Vida.

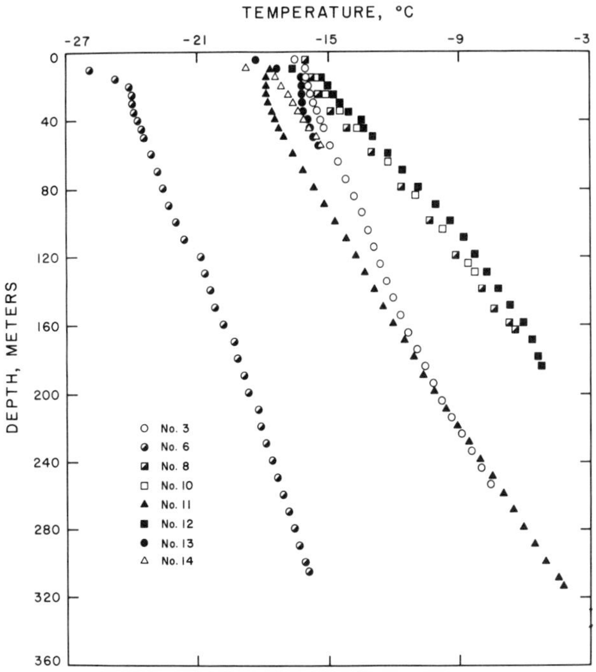

Figure 111.2. Temperatures as functions of depth in holes 3, 6, 8, and 10–14.

The arbitrary decisions embodied in the compilation in Table 111.5 should be clearly stated. Since there is no evidence for long depth intervals with significantly different gradients in holes 11, 12, and 14, the local permafrost thicknesses were calculated by downward continuation of the bottom-hole temperatures by projecting the least-squares temperature gradient at each site. In other holes there is evidence for two significantly different least-squares temperature gradients at different depths, and

Table 111.3. Selected temperatures in holes 8, 10, 12, 13, and 14

Depth (m)	Hole 13 *T*°C	Hole 14 *T*°C	Hole 8 *T*°C	Hole 10 *T*°C	Hole 12 *T*°C
20	−16.21	−17.38	−15.58	—	−14.98
40	−15.97	−16.31	−14.48	−14.55	−13.53
60	−15.44	−15.37	−12.98	−13.03	−12.26
61		−15.24[a]	—	—	—
80			−11.67	−11.61	−10.86
100			−10.43	−10.38	−9.49
120			−9.25	−9.21	−8.33
135			−8.37[a]	—	—
140				−8.05	−7.29
160				−6.81	−6.16
164				−6.55[a]	—
180					−5.48
186					−5.33[a]

[a] Deepest measurement.

Table 111.4. Selected temperatures in holes 3, 6, and 11

Depth (m)	Hole 3 *T*°C	Hole 6 *T*°C	Hole 11 *T*°C
20	−16.00	—	−17.84
40	−15.52	−23.65	−17.42
60	−14.90	−23.04	−16.59
80	−14.23	−22.52	−15.66
100	−13.50	−21.96	−14.72
120	−12.92	−20.88	−13.81
140	−12.33	−20.42	−12.94
160	−11.71	−19.77	−12.02
180	−10.99	−19.33	−11.11
200	−10.24	−18.57	−10.21
220	−9.43	−18.07	−9.14
240	−8.52	−17.66	−8.12
260	−7.66[a]	−17.13	−7.08
280		−16.61	−6.16
300		−16.11	−5.13
305		−15.97[a]	—
310			−4.55
315			−4.32[a]

[a] Deepest measurement.

Table 111.5. Estimated permafrost thicknesses near DVDP drill sites

Site	Γ°C/km	Permafrost thickness (m)
Hole 3 (Ross Island)	42, 32	440-500
Hole 6 (Lake Vida)	33, 25	800-970
Holes 8 and 10 (New Harbor)	80, 50	240-310
Hole 11 (Commonwealth Glacier)	48	405
Hole 12 (Lake Leon)	30	360
Hole 14 (North Fork)	51	350-360

both gradients were used in downward projections to estimate the limits of probable thicknesses. The projections of the observed gradient data at each site could represent oversimplifications because either the conductivity or the heat flow could differ at greater depth below the bottom of each hole. Without deeper measurements at each locality, speculation about conditions at greater depth must be confined to approximations based on extrapolation of in-hole measurements.

The temperatures in hole 13 near Don Juan Pond determine low gradients (1-4°C/km) between 15 and 35 m, and very different positive gradients (4-26°C/km) between 35 and 60 m (Figure 111.2). Those in hole 6 near Lake Vida similarly determine very diverse gradients (10-75°C/km) in the 20-220 m interval, and as discussed above, many of the measurements in this part of the hole could not be duplicated (Figure 111.2 and Table 111.2). For both holes, laboratory and field tests show that the low and/or diverse gradients are not due to vertical changes of conductivity or measurement errors. Consequently, we believe that these parts of both holes are disturbed by circulating ground fluids. Circulation of ground fluids in the 15-60 m part of the hole near Don Juan Pond is consistent with the fact that the brine level in the hole changes and is often above ground level in the surface casing. The fluid in the Lake Vida hole is presently below the surface of the ground, however, and we do not have independent evidence for circulating fluids in the 20-220 m interval, except to note that similar irregular temperature gradients are often detected in hydrologically disturbed drill holes (see Roy and others, 1972, Figure 1).

Equilibrium drill-hole gradients

Previous reports imply that temperature measurements taken shortly after drilling (one to two weeks) yielded reliable estimates for the equilibrium gradients in the lower parts of most of the DVDP holes (Pruss and others, 1974; Decker, 1974; Decker and others, 1975; Bucher and Decker, 1976*a*). This speculation is well documented by the summaries of data for selected holes listed in Tables 111.6 and 111.7. The best example is provided by the data for hole 3 on Ross Island. Here all of the measurements between 1 November 1973 and 6 November 1975 yield least-squares gradients in the 170-260 m depth range that agree to within 5% (Table 111.6). A similar conclusion can be inferred from the data for hole 8 and holes 10-12. At these sites, all of the deeper measurements at different times yield least-squares gradients that do not differ by more than 10% (Tables 111.6 and 111.7).

From analytical and field studies, Jaeger (1961) strongly suggests that reliable geothermal gradients can be measured in the central sections of diamond core holes that have been abandoned for short periods of time (one day to one week). Our determinations of gradients at various times in DVDP drill holes agree with Jaeger's conclusion and imply that his theory also applies to core holes in polar regions with permafrost conditions. Hence the recent studies at DVDP sites indicate for the first time that reliable heat flow data could be obtained from temperature measurements in recently abandoned, uncased core holes in Antarctica and the Arctic. Such measurements, in turn, could save considerable expense and time because it is costly and logistically difficult to permanently case drill holes in polar regions.

Limited studies of DVDP holes 2 and 4

Hole 2 on Ross Island and hole 4 in Lake Vanda were not studied in detail because they cannot be used for reliable heat flow calculations. The limited data for both sites can be used to substantiate other measurements, however, and it is for this reason that these holes are singled out and briefly discussed here.

Hole 2 is about 1 m from hole 3. When the most recent measurements were made in the 1975-76 austral summer, the observed temperatures ranged from −15.5°C to −12.5°C over the 40-130 m depth interval. These data agree with those from hole 3; for example, the observed temperatures ranged from −15.5°C to −12.6°C when the 40 to 130 m interval of this hole was logged during the same field season.

Table 111.6. Gradients as functions of time in holes 3 and 12

Date	Depth range (m)	Γ°C/km[a]	Depth range (m)	Γ°C/km[a]	Depth range (m)	Γ°C/km[a]
Hole 3; drilling completion date: 19 October 1973						
1 Nov. 1973	20-90	—	90-170	29.9 ± 0.5	170-260	41.0 ± 0.6
15 Nov. 1973	20-90	28.6 ± 0.8	90-170	30.0 ± 0.2	170-260	40.9 ± 0.6
6 Jan. 1974	20-90	30.3 ± 0.7	90-170	31.0 ± 0.3	170-260	41.2 ± 0.8
4-5 Feb. 1974	20-90	30.5 ± 0.7	90-170	30.9 ± 0.3	170-260	41.6 ± 0.5
1 Feb. 1975	20-90	32.1 ± 0.8	90-170	31.8 ± 0.2	170-260	41.5 ± 0.6
5-6 Nov. 1975	20-90	29.9 ± 0.9	90-170	30.8 ± 0.3	170-260	41.5 ± 0.6

Date	Depth range (m)	Γ°C/km[a]
Hole 12; drilling completion date: 2 January 1975		
6 Jan. 1975	40-180	53.0 ± 1.6
31 Jan. 1975	40-180	58.8 ± 1.7
19 Nov. 1975	40-180	58.4 ± 1.8

[a] Least-squares gradient with standard error.

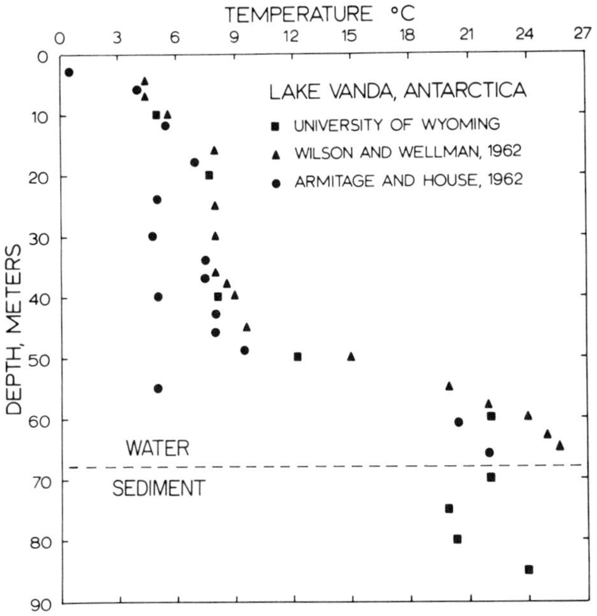

Figure 111.3. Temperatures in DVDP hole 4 (cased) and previously published measurements in waters of Lake Vanda.

Hole 4 was logged once, immediately after drilling was completed and just before the stabilizer casing was removed from the lake and underlying sediments. Because measurements were made at 5 m intervals, our temperature profile for the hole is not as detailed as those that other investigators obtained in earlier studies of other parts of the lake. Nevertheless, our widely spaced measurements within the lake waters agree best with those obtained by Wilson and Wellman (1962) (Figure 111.3). Like Wilson and Wellman, we also measured a negative gradient in the units just below lake bottom (68 m). However, our measurements below 75 m determine variable and positive gradients (Figure 111.3). Although these nonuniform and positive gradients could mean that some heat is being conducted into the lake, another explanation for them and the negative gradient just above could be that our measurements were made when a transient temperature regime existed in the shallow sublake units.

Table 111.7. Gradients as functions of time in holes 8, 10, and 11

Date	Depth range (m)	Γ°C/km[a]
Hole 8; drilling completion date: 23 January 1974		
25 Jan. 1974	30-136.5	59.2 ± 0.6
27 Jan. 1974	30-135	60.1 ± 0.7
3 Feb. 1974	30-135	62.8 ± 0.8
22 Nov. 1974	30-135	61.4 ± 0.9
30 Nov. 1974	30-135	63.1 ± 1.0
23 Jan. 1975	30-135	66.2 ± 0.9
16 Nov. 1975	30-135	64.2 ± 0.9
Hole 10; drilling completion date: 15 November 1974		
30 Nov. 1974	60-160	60.6 ± 0.8
23 Jan. 1975	60-160	62.2 ± 0.6
10 Dec. 1975	60-160	60.8 ± 0.6
Hole 11; drilling completion date: 8 December 1974		
11 Dec. 1974	60-300	44.3 ± 0.5
3 Jan. 1975	60-300	47.1 ± 0.4
24 Jan. 1975	60-300	47.4 ± 0.4
14 Nov. 1975	60-300	47.5 ± 0.3

[a] Least-squares gradient with standard error.

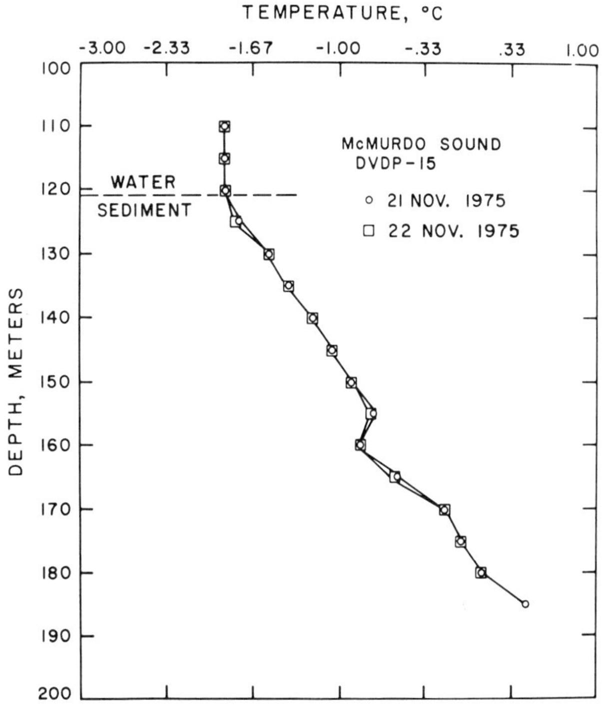

Figure 111.4. Temperatures in DVDP hole 15 in McMurdo Sound.

TEMPERATURES IN DVDP HOLE 15

Hole 15 in the western part of MuMurdo Sound was logged three times in November 1975. On 17 November a temperature of about −1.1°C was measured in the bottom of the hole at a depth of 184 m (62 m subsea). The 21 November log also reached a depth of 184 m, whereas the 22 November one did not go below 180 m because sediment had flowed up the hole after drilling. Measurements on the last two dates ranged from −1.9°C in the waters just above the sea floor to 0.1°C at a depth of 180 m (58 m subsea). The measured temperature was 0.4°C at 184 m on 21 November. Although the 21 and 22 November logs are characterized by variable and high gradients in the lower parts of the hole (Figure 111.4), we do not discuss their possible significance because these measurements so soon after drilling (6 to 8 hours) may not reflect equilibrium temperatures and gradients in the subsea sediments.

Heat flow and heat production

Preliminary basic heat-flow data for DVDP holes 3, 6, and 14 are listed in Table 111.8. The "best" heat-flow values for sites 3 and 6 are listed in Table 111.9, as are preliminary radioactivity and radiogenic heat production data for these localities. The heat-flow and radioactivity data are discussed below.

ROSS ISLAND (HOLE 3)

There are three intervals with uniform gradients in hole 3: 29.9°C/km between 100 and 160 m, 37.4°C/km between 160 and 220 m, and 45.0°C/km between 220 and 260 m. The mean of the conductivity measurements for these intervals ranges from 5.9 to 4.8 CDU, and the calculated interval heat flows are 1.79, 1.99, and 2.17 HFU, respectively (Table 111.8). A preferred observed flux cannot be selected from the interval values for two reasons: (1) the actual conductivity contrast could be greater than that

Table 111.8. Preliminary basic heat flow data for sites 3, 6, and 14

Site	Depth range (m)	Number of conductivity samples	Mean conductivity[a] (CDU)	Γ°C/km	Heat Flow (HFU) Uncorrected	Corrected[b]
Hole 3 (Ross Island)	100–160	5	5.9	29.9	1.79	1.4
	160–220	3	5.4	37.4	1.99	1.5
	220–260	3	4.8	45.0	2.17	1.6
			(Best values)		1.98	1.5–1.6
Hole 6 (Lake Vida)	60–190	19	6.3	33.4	2.10	1.90
	190–305	23	8.1	25.1	2.03	1.85
	260–305	23	8.1	25.2	2.04	1.85
			(Best values)		2.04	1.9
Hole 14 (North Fork)	30–60	12	7.0	48.0	3.4	—

[a] At *in situ* temperatures.
[b] See text for explanation of corrected data.

indicated by the small number (11) of conductivity measurements; and (2) increased gradients and flux at greater depth in the hole could be due to undetected disturbances caused by the environment near this drill site. Therefore, the 1.98 HFU value for the entire 100 to 260 m interval is considered the best value for the uncorrected flux on Ross Island.

The ambient temperature is about -1.9 ± 0.1°C in the waters below a depth of 3 m in McMurdo Sound (Littlepage, 1965) (see also Figure 111.4). In contrast, upward continuation of the temperatures and gradients in hole 3 implies that the local ground surface temperature here is about -18.5°C. Hence, the observed flux in hole 3 could be higher than the "undisturbed" flux at greater depth because heat will tend to flow away from the warmer sound bottom toward the cooler surface of Ross Island.

To obtain an estimate for the effect of McMurdo Sound at this site, we used Lachenbruch's (1957) method for correcting for the ground-temperature disturbances that may occur near juxtaposed land and oceans. Briefly, the earth's surface near hole 3 was taken to be a plane composed of land surface and sound bottom, and the shoreline was assumed to have shifted to its present position an infinitely long time ago. Hole 3 was positioned 200 m from the edge of the sound, and the temperature contrast between the land surface and the sound waters was taken to be 16°C. For these assumptions, application of Lachenbruch's steady-state correction decreased the gradient at site 3 by about 25%. Thus, our estimate for the corrected flux on Ross Island is 1.5–1.6 HFU (Table 111.8).

Lake Vida (hole 6)

The temperature measurements in the 260 to 305 m interval of hole 6 define a gradient of 25.2°C/km. When this gradient is multiplied by the 8.1 CDU conductivity of the samples of granite, the best "statistical" estimate for the flux in hole 6 is 2.04 HFU (Table 111.8).

On the basis of the other data here, the least-squares gradient and mean conductivity in the 60 to 190 m depth range are 33.4°C/km and 6.3 CDU, respectively. From 190 to 305 m the least-squares gradient is calculated to be 25.1°C/km, and the calculated conductivity of the core samples is 8.1 CDU. The uncorrected heat flows of 2.10 and 2.03 HFU between 60–190 m and 190–305 m, respectively, agree well with the "statistically" more reliable flux (2.04) in the 260–305 m interval (Table 111.8). This agreement, in turn, indicates vertically uniform heat flow near Lake Vida.

Our preliminary estimate for the corrected flux near Lake Vida is 1.9 HFU (Table 111.8). This value was calculated after the observed temperatures were corrected for the effect of a two-dimensional terrain model that closely approximates the topography in Victoria Valley. The steady-state terrain correction directly followed the theory outlined by Birch (1950) and was made for topography out to lateral distances equal to 40 times the depth of hole 6.

North Fork (hole 14)

The uncorrected flux at North Fork is a very high value of 3.4 (Table 111.8). However, the actual heat flow in the Wright Valley region remains unknown because the thermal data for very shallow hole 14 have not been corrected for nearby terrain and other possible disturbances.

Combined heat flow and heat production data

The heat flow and heat production data in Table 111.9 are plotted in Figure 111.5. Two heat production values (Table 111.9) are plotted for Ross Island because inclu-

Table 111.9. Preliminary radioactivity and heat flow data

Site and samples	Number of radioactivity samples	U (ppm)	Th (ppm)	K (wt. %)	Heat productivity (HGU)	Heat flow (HFU)
Hole 3 (Ross Island)[a]						
Core samples	5	1.9	6.7	1.4	3.0	1.5–1.6
Basement samples (?)						
Irizar granite	20	0.5	8.6	4.09	2.6	1.5–1.6
Larson granodiorite	6	0.9	15.5	3.12	3.7	1.5–1.6
Hole 6 (Lake Vida)[b]						
Core samples						
Granite	7	1.2	8.6	4.3	3.2	1.9
Gneiss	8	1.0	6.1	2.5	2.2	1.9

[a] RI in Figure 111.5.
[b] LV in Figure 111.5.

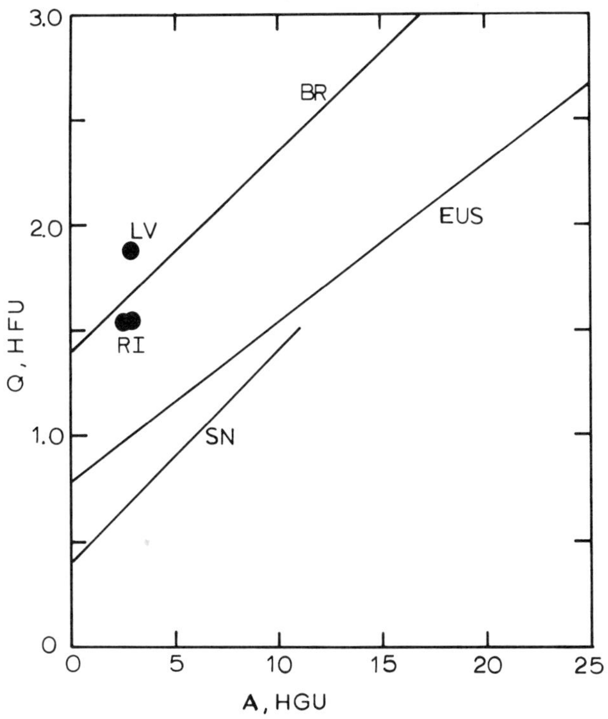

Figure 111.5. Combined heat flow (Q) and radiogenic heat production (A) data for hole 3 on Ross Island (RI) and hole 6 near Lake Vida (LV). Linear heat flow–heat production relationships in the United States are after Roy and others (1968). BR: Basin and Range province. EUS: Eastern United States. SN: Sierra Nevada.

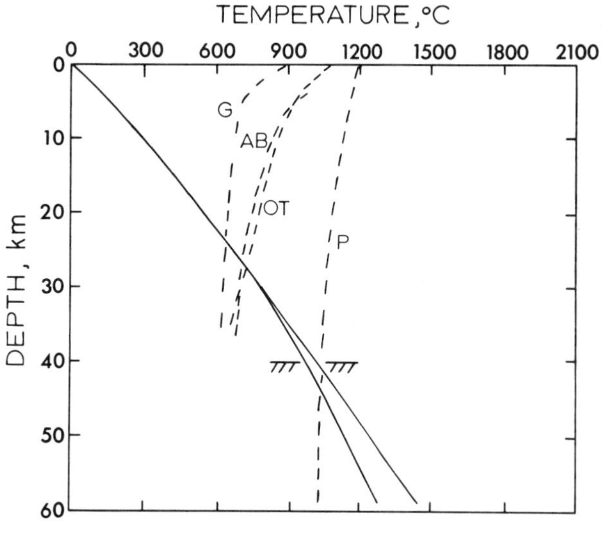

Figure 111.6. Preliminary steady-state temperature-depth model for the Lake Vida area. Temperature profiles (solid lines) based on commonly accepted models for uniform and exponentially decreasing crustal radioactivity (after Roy and others, 1968, 1972; Lachenbruch, 1968). The base of the crust (Robinson, 1964) is indicated by hatching. Solidus curves (dashed lines) in the presence of excess water are after Wyllie (1971), Lambert and Wyllie (1972), and Yoder and Tilley (1962). G: Granodiorite. P: Peridotite. AB: Alkaline basalt. OT: Olivine tholeiite.

sions in core samples from holes 2 and 3 and nearby outcrops suggest that the island is underlain by units similar to the basement rocks in the dry valleys (Treves and Kyle, 1973; Kyle and Treves, 1974; S. B. Treves, personal communication, 1977).

As is evident in Figure 111.5, the data for Ross Island and Lake Vida plot near the heat flow–heat production line for the Basin and Range province in the United States. This suggests that the Ross Island–dry valley region lies in a zone of high terrestrial heat flow. One interpretation of high flux in this part of Antarctica is that near-melting conditions exist near the crust-mantle boundary; for example, equilibrium temperature profiles based on the Lake Vida heat flow data suggest extensive partial melting of a "wet" lower crust and upper mantle (Figure 111.6). High temperatures in the crust and upper mantle, in turn, are consistent with Quaternary basaltic volcanism in the McMurdo Sound area and with the late Tertiary uplift of the Transantarctic Mountains. Although it is tempting to draw other inferences from the evidence for high subsurface temperatures in the Ross Island dry valley region, we hesitate to be more specific without more reliable heat flow values for DVDP drill sites.

Concluding remarks

The bottom-hole temperatures and observed gradients at DVDP drill sites indicate that permafrost is 440–500 m thick on Ross Island and 240–970 m thick in the dry valleys. The field work since 1973 also suggests that geothermal gradients obtained at most of the sites were reliable when the temperatures were measured shortly after drilling was completed (two days to two weeks). Thus, reliable heat flow studies may be conducted in recently abandoned core holes in polar regions without expensive and logistically difficult casing procedures.

Combined preliminary heat flow and radiogenic heat production values at DVDP sites 3 and 6 imply that the Ross Island–dry valley region is in a zone of high flux, comparable to the Basin and Range province in the United States. Although such high flux would be consistent with melting conditions in the mantle, tectonic uplift, and other thermal phenomena in this sector of Antarctica, more corrections must be applied to the existing thermal data for DVDP holes before reliable interpretations and modeling can be attempted.

Finally, the geothermal data for DVDP holes provide new insight into the thermal processes acting in the neighborhood of Ross Island. From temperature recordings over a period of 2.4 years in a 7.6 m hole, Risk and Hochstein (1974) obtained a 4.0 HFU heat-flow value for Hut Point Peninsula. In contrast, we obtain a significantly lower flux of 1.5–2.0 HFU at greater depths (100–260 m) in DVDP hole 3. The lower flux determined in hole 3 thus suggests that climatic changes, magmatic bodies, or other anomalous heat sources in the crust may not be needed to explain the actual heat flow regime in this part of Ross Island.

Acknowledgments. E. Pruss, K. Baker, and H. Harris generously helped with field measurements. The research was financially supported by U.S. National Science Foundation grants GV-37517, OPP75-20991, and DPP77-00207. The University of Wyoming also provided financial support for field and laboratory studies and provided free computer time for data processing.

References cited

Armitage, K. B., and House, H. B. 1962. A limnological reconnaissance in the area of McMurdo Sound, Antarctica. *Limn. Ocean., 7,* 36–41.

Birch, F. 1950. Flow of heat in the Front Range, Colorado. *Bull. geol. Soc. Am., 61,* 567–630.

Birch, F. 1954. Heat from radioactivity. (*In* Faul, H., *ed. Nuclear Geology.* New York, John Wiley, 148–74.)

Birch, F., and Clark, H. 1940. The thermal conductivity of rocks and its dependence upon temperature and composition. *Am. J. Sci., 238,* 529–58.

Bucher, G. J., and Decker, E. R. 1976*a*. Geothermal studies in the McMurdo Sound region. *Antarct. Jnl U.S., 11,* No. 2, 88–89.

Bucher, G. J., and Decker, E. R. 1976*b*. Temperature measurements in DVDP #15, McMurdo Sound, Antarctica. *Bull. Dry Valley Drilling Project, 7,* 111–12.

Decker, E. R. 1973. Geothermal measurements by the University of Wyoming. *Univ. Wyo. Contrib. Geol.,* No. 12, 21–24.

Decker, E. R. 1974. Preliminary geothermal studies of the Dry Valley Drilling Project holes at McMurdo Station, Lake Vanda, Lake Vida, and New Harbor, Antarctica. *Bull. Dry Valley Drilling Project, 4,* 22–23.

Decker, E. R., Baker, K. H., and Harris, H. H. 1975. Geothermal studies in the dry valleys and on Ross Island. *Antarct. Jnl U.S., 10,* No. 4, 176.

Jaeger, J. C. 1961. The effect of drilling fluid on temperatures in bare holes. *J. geophys. Res., 66,* No. 2, 563–69.

Kyle, P. R., and Treves, S. B. 1974. Geology of DVDP 3, Hut Point Peninsula, Ross Island, Antarctica. *Bull. Dry Valley Drilling Project, 3,* 13–48.

Lachenbruch, A. H. 1957. Thermal effects of the ocean on permafrost. *Bull. geol. Soc. Am., 68,* 1515–30.

Lachenbruch, A. H. 1968. Preliminary geothermal model of the Sierra Nevada. *J. geophys. Res., 73,* 6977–89.

Lambert, I. B., and Wyllie, P. J. 1972. Melting of gabbro (quartz eclogite) with excess water to 35 kilobars, with geological applications. *J. Geol., 80,* 693–708.

Littlepage, J. L. 1965. Oceanographic investigations in McMurdo Sound, Antarctica. (*In* Llano, G., *ed. Biology of the Antarctic Seals II.* American Geophysical Union, 1–37.) [Antarctic Research Series, Vol. 5.]

Pruss, E. F., Decker, E. R., and Smithson, S. B. 1974. Preliminary temperature measurements at DVDP holes 3, 4, 6 and 8. *Antarct. Jnl U.S., 9,* No. 4, 133–34.

Ratcliffe, E. H. 1959. Thermal conductivities of fused and crystalline quartz. *Br. J. Appl. Phys., 10,* 22–28.

Risk, G. F., and Hochstein, M. P. 1974. Heat flow at Arrival Heights, Ross Island, Antarctica. *N.Z. Jl Geol. Geophys., 17,* No. 3, 629–44.

Robinson, E. S. 1964. *Geologic Structure of the Transantarctic Mountains and Adjacent Ice-Covered Areas.* Ph.D. dissertation, University of Wisconsin.

Roy, R. F., Blackwell, D. D., and Birch, F. 1968. Heat generation of plutonic rocks and continental heat flow provinces. *Earth planet. Sci. Lett., 5,* 1–12.

Roy, R. F., Blackwell, D. D., and Decker, E. R. 1972. Continental heat flow. (*In* Robertson, E. C., *ed. The Nature of the Solid Earth,* New York, McGraw-Hill, 506–43.)

Treves, S. B., and Kyle, P. R. 1973. Geology of DVDP 1 and 2, Hut Point Peninsula, Ross Island, Antarctica. *Bull. Dry Valley Drilling Project, 2,* 11–82.

Wilson, A. T., and Wellman, H. W. 1962. Lake Vanda, an Antarctic lake, a solar energy trap. *Nature, Lond., 196,* 1171–73.

Wyllie, P. J. 1971. Experimental limits for melting in the earth's crust and upper mantle. (*In* Heacock, J. G., *ed. The Structure and Physical Properties of the Earth's Crust.* Washington, American Geophysical Union, 279–301.) [Geophysics Monograph No. 14.]

Yoder, H. S., and Tilley, C. E. 1962. Origin of basalt magmas: An experimental study of natural and synthetic rock systems. *J. Petrology, 3,* 342–532.

112

Crustal Structure of Part of East Antarctica from Geophysical Data

R. G. KURININ and G. E. GRIKUROV
Department of Antarctic Geology, SEVMORGEO, Moyka 120, Leningrad 190121, U.S.S.R.

Abstract. Deep seismic soundings carried out by the Soviet Antarctic Expedition in 1968 (Queen Maud Land) and 1973 (MacRobertson and Princess Elizabeth Lands) provided the first information about the deep crustal structure of these ancient metamorphic terranes. In both regions the crust reveals a typically continental block-layered structure and a thickness ranging from 30 to 40 km. The "granitic" layer is about 20 km thick and appears to have unusually high density and velocity values; these high values may reflect the presence in the crystalline basement of the oldest enderbite-charnockite crustal complexes which have a more mafic composition and a greater density than common granitic terranes. A similar interpretation is suggested for the Enderby Land positive isostatic gravity anomalies. The crustal structure of western Enderby Land and the adjacent submarine region is discussed in the light of available magnetic evidence and found to be comparable with other zones transitional from continents to oceans. A prominent crustal feature of the MacRobertson–Princess Elizabeth Lands area is the Lambert Glacier rift zone, which is characterized by large-scale normal faulting and the resulting development of a deep graben filled with low-density rocks. Beneath the graben there is a prominent compensatory rise of the mantle, and in the whole area, the seismic velocities just below the Mohorovičić discontinuity are 7.7–7.9 km/sec.

Until the late 1960's crustal structure studies in Antarctica were confined mainly to determination of crustal thickness. Some authors employed seismologic data for this purpose (Dewart, 1961; Evison and others, 1960; Kovach and Press, 1961; Dewart and Toksöz, 1965; Adams, 1971) and other workers used empirical correlations between crustal thickness and Bouger anomalies, or between crustal thickness and bedrock relief (Demenitskaya, 1960; Demenitskaya and Rosenberg, 1966). Later the so-called "effective relief" elevations were utilized, which introduced ice thickness values into the calculations (Rosenberg, 1971). In many publications isostatic anomalies were also taken into account (Frolov, 1964, 1971; Groushinsky and others, 1967; Stroev and others, 1967).

As the result of these studies, some general regularities in the behavior of the upper and lower boundaries of the crust were detected. However, the internal structure of the crust remained unknown until deep seismic soundings (DSS) were carried out in two parts of East Antarctica: Queen Maud Land in 1968 and the Lambert Glacier area in 1973.

The DSS in Queen Maud Land was accomplished by the 14th Soviet Antarctic Expedition (SAE) in the Novolazarevskaya Station area, 100 km north of the Humboldt Mountains (Figure 112.1). A sublatitudinal profile 425 km in length was obtained (Kogan, 1972). The second DSS, completed by the 18th SAE in MacRobertson and Princess Elizabeth Lands, consists of two profiles. The basic profile is 450 km long and crosses the northern Prince Charles Mountains, Lambert Glacier, and Amery Ice Shelf, and follows the Ingrid Christensen Coast; the supplementary 200-km-long profile lies along the axial zone of the northern Lambert Glacier (Kolmakov and others, 1975). In both regions it appears possible to determine the crustal thickness, its internal layered structure, and the position of major deep faults.

During 1971–74 (17th–19th SAE) an extensive airborne

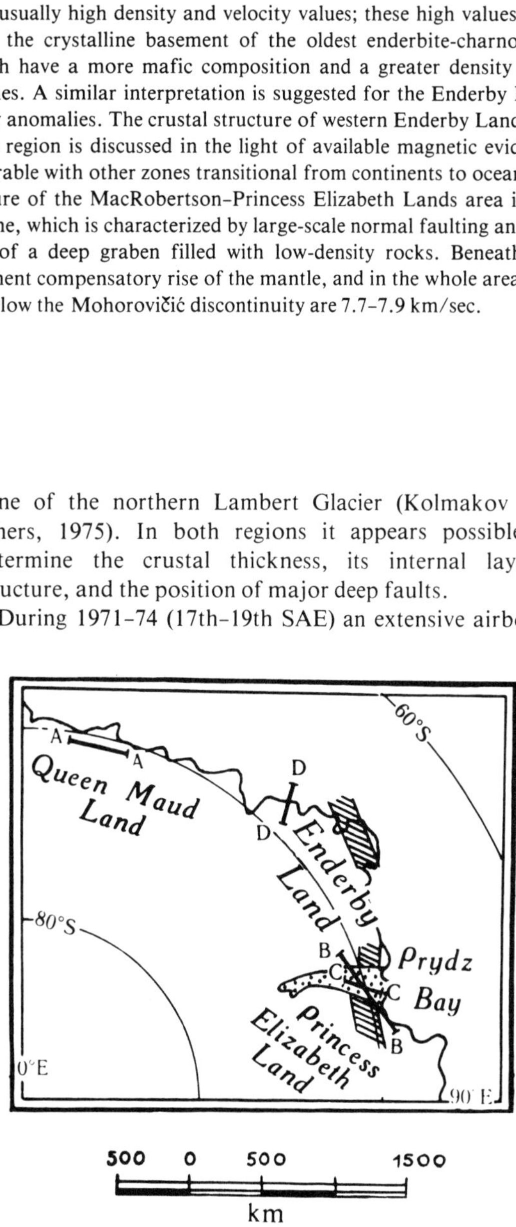

Figure 112.1. Position of the crustal sections. *Profile lines.* A-A: deep seismic sounding profile in the vicinity of Novolazarevskaya Station (Kogan, 1972); B-B and C-C: deep seismic sounding profiles across and along the Lambert Glacier (Kolmakov and others, 1975); D-D: magnetic profile across the Prince Olav Coast. Slashed pattern: blocks of high-density crust consisting of granulite facies rocks.

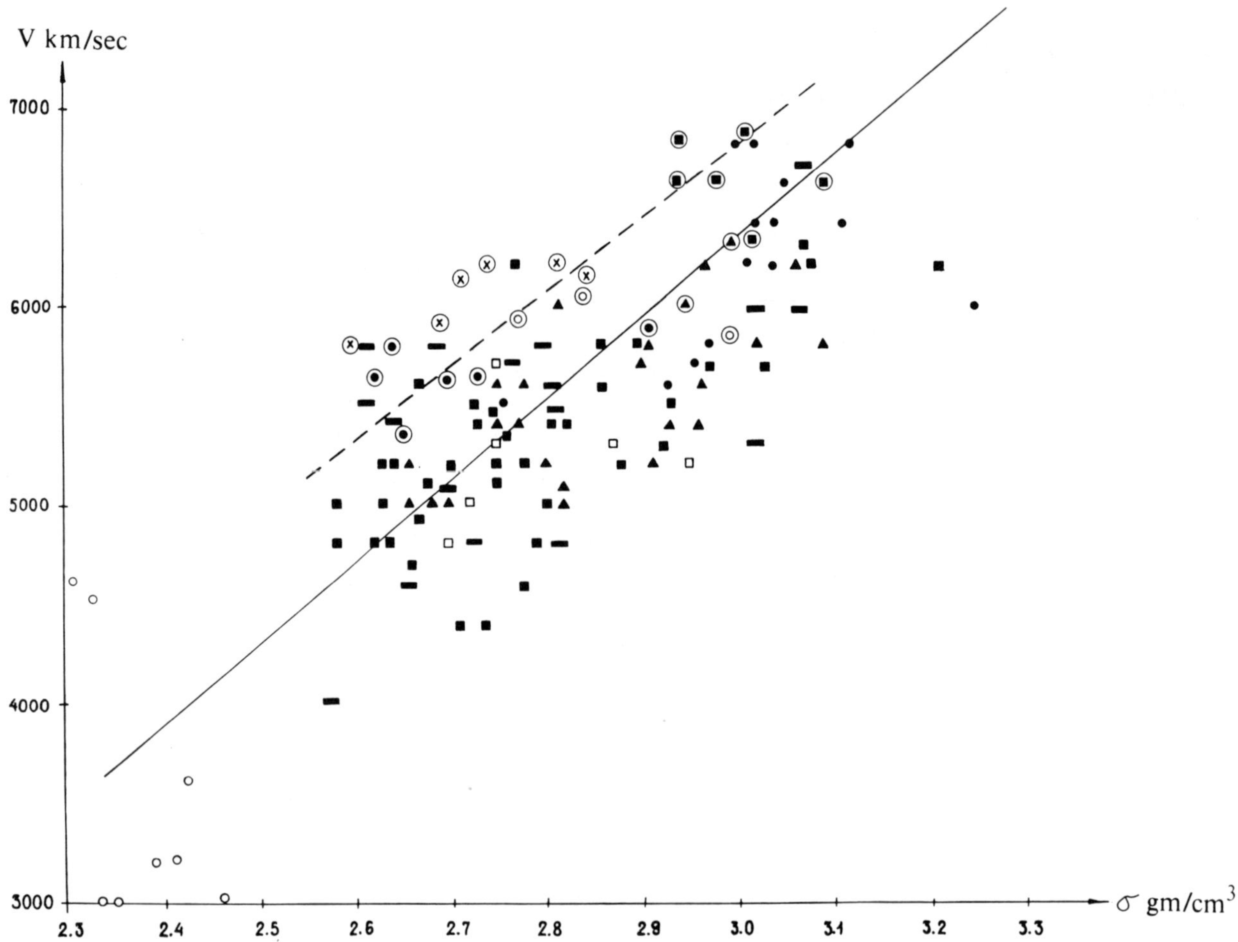

Figure 112.2. Correlation between density and velocity determined on the East Antarctic rocks. *MacRobertson and Princess Elizabeth Lands* (105 specimens). Rectangles: granulite facies rocks and polymetamorphic schists of the Larsemann Series. Solid boxes: granulite facies and polymetamorphic rocks of the Reinbolt Series. Solid triangles: greenschist facies rocks of the Ruker Series. Open boxes: low-grade greenschist facies metasedimentary rocks of the Sodruzhestvo Series. Solid dots: alkaline-ultramafic rocks. Open dots: Permian sandstones from the Beaver Lake area. *Queen Maud Land* (23 specimens). Circled solid dots: garnetiferous gneisses of the Humboldt Series. Circled open dots: garnetiferous plagiogneisses of the Insel Series. Circled X's: charnockites and anorthosites. Circled triangles: polymetamorphic schists of the Humboldt Series, undifferentiated. Circled boxes: dikes of dolerite or basalt. Solid line: regression line for the MacRobertson and Princess Elizabeth Land rocks. Dashed line: regression line for the Queen Maud Land rocks. Measurements were carried out on water-saturated slices at atmospheric pressure for the MacRobertson and Princess Elizabeth Land specimens, and at pressures of 600 bars on dry specimens for the Queen Maud Land rocks.

magnetic survey (1:2,000,000 scale) was carried out in the coastal zone of East Antarctica between long. 40° and 77° E. The total area surveyed was about 1,000,000 km². The regular flight lines within this area were spaced at 20 km; their total length, along with control traverses and a few long-distance flights, exceeded 100,000 km. An interpretation of the results of this survey is also used in the present paper for the recognition of certain crustal features.

Crustal sections

Correlation between density and seismic velocities in rocks

In order to obtain the density of individual crustal layers recognized in the DSS profiles, a graph of "density versus velocity" was plotted for specimens from Queen Maud Land and the Prince Charles area (Figure 112.2). Some discrepancy between the regression lines may reflect different physical properties of rocks from various regions, but is more likely to be caused by a difference in the methods of measurement of seismic velocities (see Figure 112.2 caption). However, the regression lines are sufficiently alike for regional correlations.

Queen Maud Land section

The DSS results in the Novolazarevskaya area had been reported at the previous SCAR meeting (Kogan, 1972), and therefore only a brief description of the Queen Maud Land crustal section is given below (Figure 112.3). The crustal characteristics were obtained at 16 sites. The crust is split into large blocks, each about 100 km in size. Beneath the mountains the crust is about 40 km thick. Four boundaries are recognized within the crust, the principal one lying at a depth of 17–20 km; it separates the two most prominent crustal layers. The upper layer has a velocity of 6.0–6.2 km/sec and consists of rocks with a density of 2.90–2.95 g/cm³. The lower layer has a considerably higher velocity (6.67–6.83 km/sec) and density (3.04–3.07 g/cm³). The boundary velocity along the Mohorovičić discontinuity is 7.9–8.0 km/sec; the upper mantle density is at least 3.3 g/cm³.

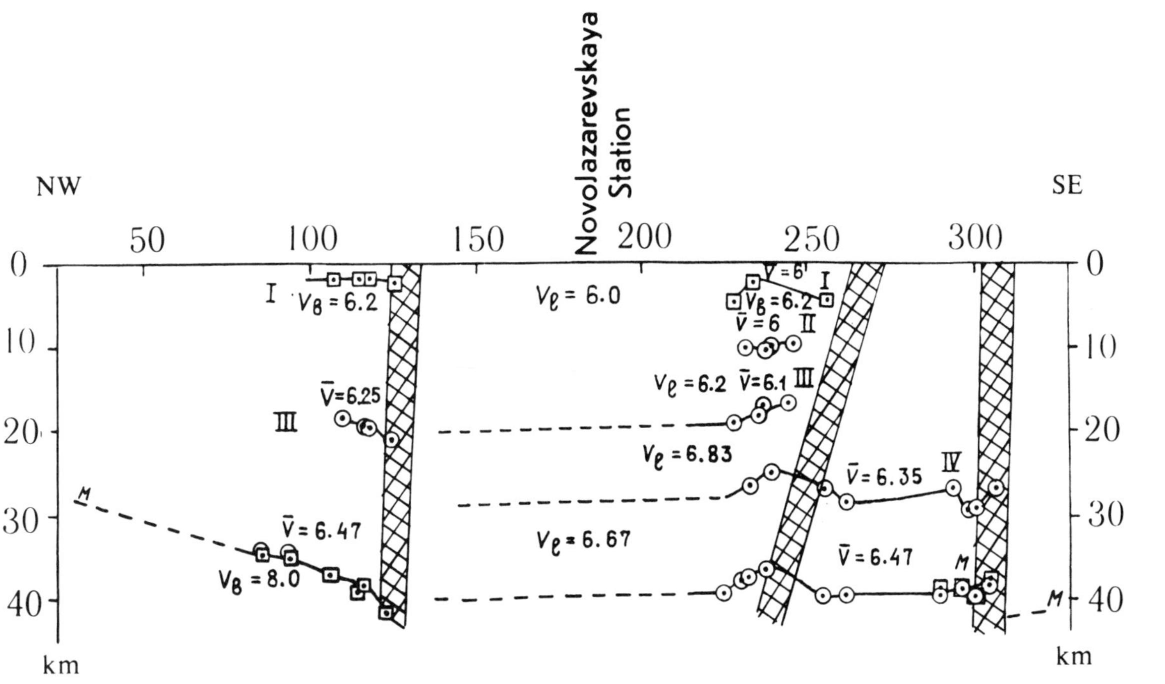

Figure 112.3. Crustal section in Queen Maud Land (modified from Kogan, 1972). Solid lines: seismic boundaries, identified. Dashed lines: seismic boundaries, inferred. Circled dots: depths from seismic reflection data. Boxed dots: depths from seismic refraction data. Grid pattern: crustal fault zones. V_g: boundary velocities, in km/sec. V_l layer velocities, in km/sec. $\bar{V}$: average velocities from the surface, in km/sec.

The density of the upper layer exceeds that of a common "granitic" layer. Hence it was suggested that this layer is composed here mostly of charnockites and therefore may be named "charnockitic" (Ravich and Kamenev, 1972). Both density and seismic velocity in the lower layer are similar to those of the usual "basaltic" layer.

Lambert Glacier area section

Methods

In order to obtain structural data throughout the whole thickness of the crust, two independent systems of discrete soundings were employed. The first system was intended to record the seismic waves returning from interfaces in the upper crust; the distances between explosion sites and recorders were 40 km and 60 km. The head wave was observed to refract at the boundary lying at a depth of 3-4 km. Subsequent arrivals registered the wave reflected from the intercrustal boundary, 18-22 km deep. This system was used only in the vicinity of Amery Base and Beaver Lake; it could not be employed in the central part of the profile due to the absence of water pools suitable for explosions. A total of thirty soundings were obtained for the upper crust.

The second system of soundings was designed to register the waves from the Moho discontinuity; consequently, the shot-recorder distance was increased to 160-220 km. Both the head refracted wave and a subsequent reflected wave were recorded with confidence. When the shot-recorder distance was decreased to about 80-100 km, a wave reflected from the Moho was still recognizable. Waves from the Moho were also traced along the axial part of the Lambert Glacier and across the southern Amery Ice Shelf, which helped to document the areal characteristics of the Moho depth and velocities. These data were obtained at 21 sites.

Four principal groups of waves were distinguished. The dynamic and kinematic features of these groups enabled the correlation of individual seismograms obtained at sites spaced 15-20 km apart.

The first group of waves is represented by oscillations registered in first arrivals at a shot-recorder distance of 35-100 km. These waves were identified as having been refracted at boundaries in the upper crust. The second group of waves are recognized in subsequent arrivals on seismograms recorded at a distance of 35-70 km from the explosion site; oscillations of this category represent waves reflected from interfaces in the middle part of the crust. The last two groups include the waves associated with the Moho discontinuity. The third group is recognized as subsequent arrivals of intense multiple waves at a shot-recorder distance of 80-100 km; these waves can be positively identified as having been reflected from the Moho discontinuity. The fourth group is distinguished in first arrivals when the distance from the shot site ranges from 160 to 220 km; the waves of this group produce a low-intensity recording, but their dynamic and kinematic patterns leave no doubt that their refracted nature is related to the Moho discontinuity. The accuracy of results obtained in this manner is 3-5%.

Description of the crustal section

The crustal section obtained in the Lambert Glacier area consists of three large blocks (Figure 112.4). Each block is characterized by relatively stable average and boundary velocities, by generally conformable behavior of seismic interfaces, and by roughly uniform depths to the Moho discontinuity. Changes in thickness of individual crustal layers and of the crust as a whole are associated with block boundaries and have a sharp, step-like nature. The western block corresponds to the Prince Charles Mountains, the

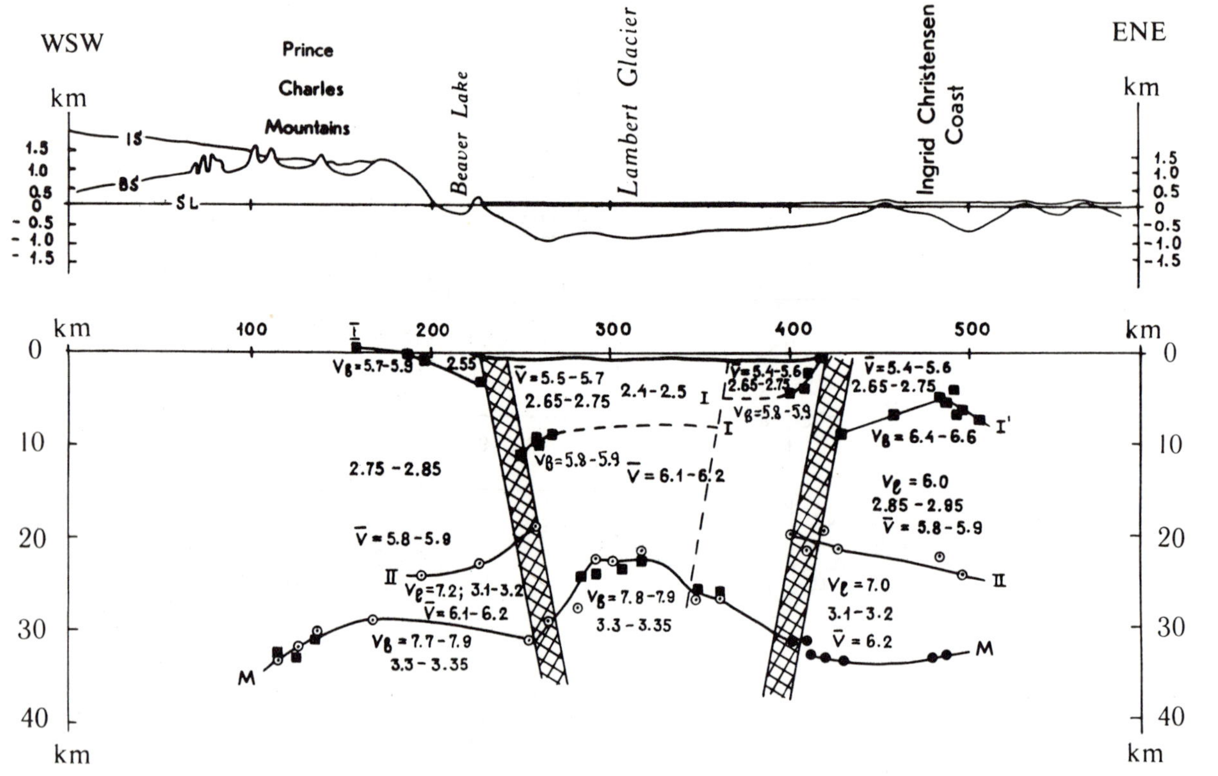

Figure 112.4. Crustal section across the Lambert Glacier (modified from Kolmakov and others, 1975). SL: sea level. BS: bedrock surface, IS: ice surface. Boxes: depths from seismic refraction data. Circled dots: depths from seismic reflection data. Solid dots: depths to the Moho discontinuity from seismic reflection data in the critical-angle area. V_g: boundary velocities, in km/sec. $\bar{V}$: average velocities from the surface, in km/sec. V_l: layer velocities, in km/sec. Grid pattern: crustal fault zones. Dashed line: projection of the eastern boundary of the Lambert Glacier graben on the section plane. Densities are given in g/cm³.

central block is represented by a depression in the bedrock surface occupied by the Lambert Glacier and the Amery Ice Shelf, and the eastern block has a relatively flat bedrock surface (subglacial plain of the Ingrid Christensen Coast). Basic differences in crustal features and present-day topography indicate that the blocks are bounded by deep faults penetrating down to the mantle, as suggested by the presence of high-density rocks (eruptive breccias, peridotites, and dunites) in the deep-fault zone bounding the Lambert Glacier graben on the west.

In the Prince Charles Mountains block, the crustal thickness gradually increases westward from 30 km to 33 km; the boundary velocity along the Moho discontinuity is 7.7–7.9 km/sec. Two intercrustal boundaries are recognized within this block. The lower boundary (Figure 112.4, II, presumably the Conrad interface) rises toward the graben from a depth of 24 km to a depth of 20 km; below it the layer velocity (7.2 km/sec) and estimated density (3.1-3.2 g/cm³) correspond to a normal "basaltic layer," which appears there to be 6–10 km thick. The upper boundary is identified as the surface of the crystalline basement; the latter has an estimated density of about 2.75–2.85 g/cm³. This boundary slopes down eastward from the bedrock surface to 3 km below sea level near the graben. Above it are the Permian sedimentary rocks of the Beaver Lake area with a measured density of 2.55 g/cm³.

In the central down-faulted block, the crustal thickness is decreased to 22–24 km. The gradual descent of the Moho discontinuity—with boundary velocities of 7.8–7.9 km/sec—in the eastern part of that block down to depths of 30–31 km, as shown in Figure 112.4, is caused by the fact that the profile line here comes nearer to the bedrock exposures bounding the Amery Ice Shelf on the southeast. The same reason probably accounts for the relatively shallow depth (only 2–3 km) of the upper refracting boundary, with a velocity 5.8–5.9 km/sec (the crystalline basement surface), in the eastern part of the central block. In contrast, a similar boundary underneath the axial and western parts of the Lambert Glacier is recorded at the depth of 8–10 km. The crystalline basement is overlain by low-velocity rocks with a density ranging from 2.4–2.5 g/cm³, in the axial zone of the block, to 2.65–2.75 g/cm³ on its flanks.

Although the crustal thickness of the eastern block is similar to that of the Prince Charles Mountains (32–34 km), the crustal structure here has some peculiar features. The upper refracting interface at a depth of 5–8 km has an anomalously high boundary velocity reaching 6.4–6.6 km/sec. Below this interface there is an intracrustal layer 15–19 km thick having a high density (2.85–2.95 g/cm³) and velocity (6.0 km/sec) compared to a normal "granitic" layer. Above this boundary the average velocity is unexpectedly low (5.4–5.6 km/sec), i.e., similar to that in the unconsolidated upper layer of the central block, in spite of the fact that the upper layer of the eastern block consists of crystalline basement rocks actually exposed in numerous outcrops along the Ingrid Christensen Coast. This apparent mass deficiency in the upper crust of the eastern block is most likely caused by the crushing of rocks in the fault zone associated with the Prydz Bay coast. The lowest crustal layer is about 10 km thick; its other parameters are

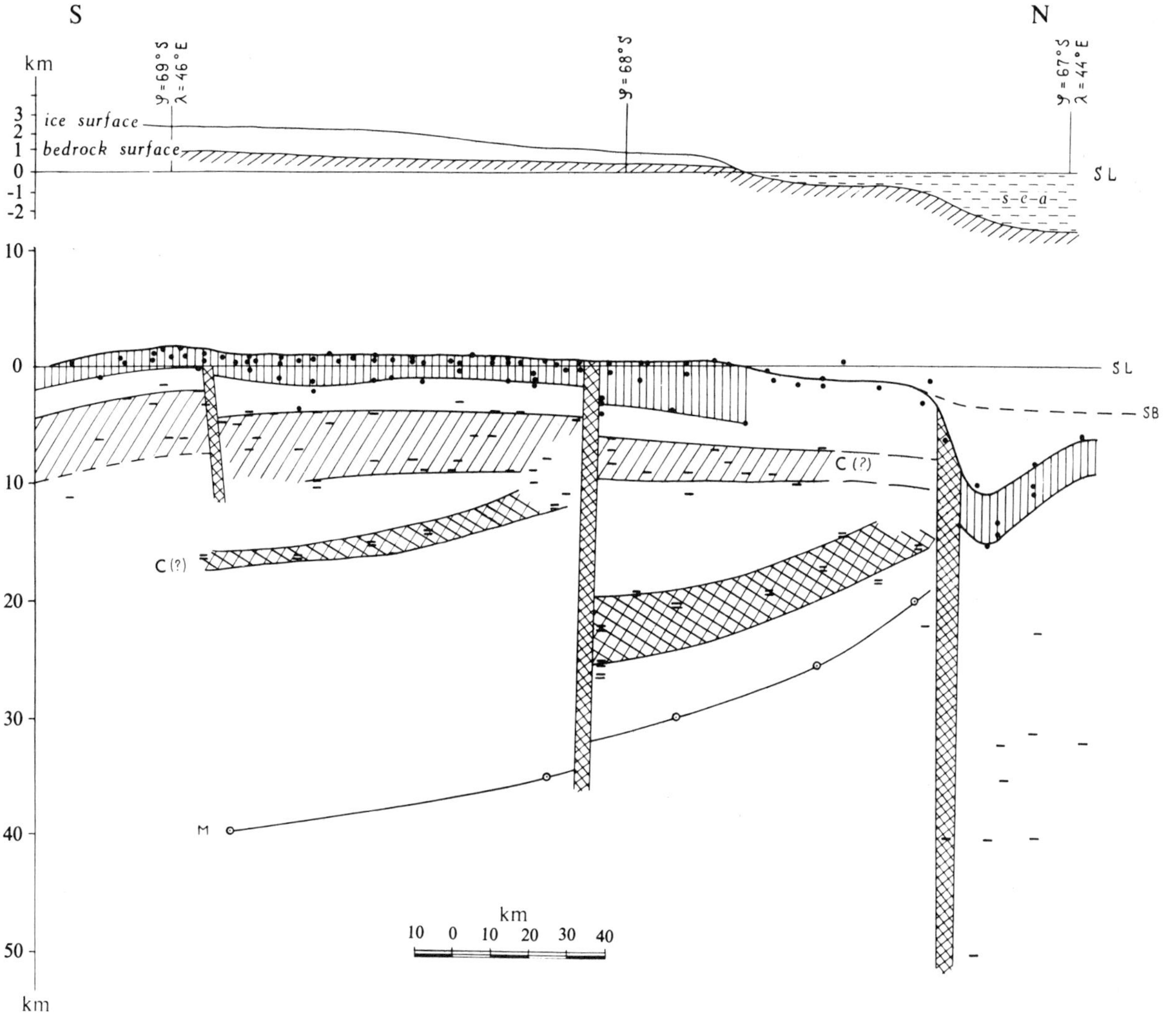

Figure 112.5. Crustal section in western Enderby Land from magnetic data. Vertical shading with dots: vertical range for position of the upper surface of the "magnetic layer" in the crust determined from the distribution of the upper boundaries (dots) of imaginary magnetic bodies used for model calculations. Oblique shading with dashes: vertical range for position of the lower surface of the "magnetic layer" determined from prevailing distribution of the lower boundaries (dashes) of the same bodies at moderate depths. Cross shading with double dashes: vertical range for position of the lower surface of the "magnetic layer" inferred from the distribution of some lower boundaries (double dashes) at greater depths. Line "M": Moho discontinuity from the "effective relief" data (based upon correlation between the mean "effective relief" elevations and the Moho depths obtained from the Lambert Glacier area DSS profiles). C(?): hypothetical Conrad interface. Grid pattern: crustal fault zone. SL: sea level. SB: sea bottom.

similar to those of the "basaltic" layer in the western block.

The crustal thickness and upper mantle characteristics in the central part of the Lambert Glacier graben remain uniform for about 200 km along the axial zone of the Lambert Glacier and Amery Ice Shelf depressions, as indicated by the supplementary DDS profile (see line C-C in Figure 112.1).

The crustal structure described above includes typical riftogenic features, such as:

1. considerable down-faulting of a large block of consolidated crust and the resulting development of a well-delineated graben,
2. growth of a prominent mantle rise compensating the graben structure and its low-density fill, and
3. unusually low seismic velocities along the Moho discontinuity.

Western Enderby Land section

Methods

The western Enderby Land section was compiled with the special purpose of observing the changes in crustal structure within a transition zone from continent to ocean. To compile this section, the depths to upper and lower boundaries of magnetic bodies were calculated for a coastal zone about 250 km wide, including the Prince Olav Coast, the adjacent continental shelf, and the continental rise. The calculated depths were then plotted on an arbitrary vertical plane perpendicular to the coast line (Figure 112.5.) Such an operation is appropriate because of the sublatitudinal orientation (parallel to the coast) of the major magnetic provinces, one of which corresponds to the continental part of the surveyed area, and another to the submarine territory. Both provinces are characterized by distinctive magnetic patterns, with clearly delineated anomalies ranging from −200 gammas over the continent to +600–800 gammas over the continental slope.

Description of the crustal section

Depths to lower and upper limits of magnetic bodies are plotted in Figure 112.5; their distribution suggests the presence of three major crustal blocks within the section under consideration. The blocks are separated by deep-seated faults, the northernmost of which is clearly marked in the magnetic field. The position of the Moho discontinuity is determined from "effective relief" heights (see Rosenberg, 1971).

The southern block has a typical continental crustal thickness, reaching 40 km. Most magnetic bodies appear to be concentrated within the upper 10 km of the crust. At about lat. 69° S there is a step-like change in the depth of the lower edges of magnetic bodies, suggesting the presence of a fault in the upper crust. A band of deeper boundaries of magnetic bodies, at approximately 12–17 km depth, may be related to an unusually shallow position of the Conrad interface.

In the central block there is a distinct oceanward rise of the Moho discontinuity from 35 km to 20 km. The lowest boundaries of magnetic bodies recorded in this block seem to concentrate in a band parallel to the Moho but 7–8 km higher; nevertheless, they are likely to be related to the latter within the accuracy of the method. An apparently greater vertical range for the shallow edges of magnetic bodies in the central block, as compared to the southern one, suggests a relatively elevated position for the latter. A similar conclusion can be reached if the second level of lower boundaries of magnetic bodies in the central block (at about 10 km below sea level) is identified with the Conrad interface. In this case the isostatic nature of the Conrad interface (and the Moho discontinuity) becomes quite evident, and the general thinning of the continental crust towards the continental rise acquires a more typical pattern, i.e., appears to occur mainly at the expense of the "basaltic" rather than the "granitic" layer.

The northern block has an oceanic-type crustal structure. The lower limits of magnetic sources are scattered at considerable depths (30–50 km) in the mantle. The upper boundaries of magnetic bodies delineate a conspicuous trough in the uppermost crustal layer, which is most likely represented here by a thin "basaltic" layer. However, the trough is not reflected in the sea-bottom topography, and therefore must be completely filled by a thick pile (as much as 7–8 km) of nonmagnetic rocks.

The character of the distribution of magnetic masses in the crust and upper mantle described above for the Prince Olav Coast and the adjacent submarine region agrees with the general pattern found in other zones of transition from continent to ocean (Kunin, 1972). The increase in depth of magnetic bodies oceanward, down to the level of some dozens of kilometers in the mantle, is particularly typical.

Discussion

Each of the three East Antarctic regions discussed above in terms of available geophysical information has certain specific features of crustal structure. Of these features the most prominent one is obviously the Lambert Glacier rift zone and the associated transformation of the whole crustal pattern from that typical of the area—total decrease in the crustal thickness by 6–8 km, sharply reduced thickness of the "basaltic" layer, low density in the upper mantle, and some other phenomena not encountered, for instance, in the Queen Maud Land area. No less obvious is the fact that the disappearance of typical continental features of the crust northward along the western Enderby Land section is caused by transition to the oceanic crust. If these local differences and their apparent explanations are disregarded, however, a crustal feature emerges which appears common to the whole East Antarctica coastal sector under consideration, namely, the presence of a certain mass excess within the "granitic" layer of the crust.

It has already been mentioned that both in the Queen Maud Land crustal section and in the eastern block of the Lambert Glacier area profile, the density of rocks in the upper crust exceeds that of a common "granitic" layer by 0.10–0.15 g/cm³. A similar "heavier" intracrustal layer seems to be present west of the Amery Ice Shelf, immediately to the north of the Prince Charles Mountains. The crustal section of western Enderby Land is compiled from magnetic data and hence gives no information on the masses involved, but some evidence concerning this region between two available DSS profiles can be obtained from earlier publications (Stroev and others, 1967; Stroev, 1971; Frolov, 1971).

The Tula and Napier Mountains in Enderby Land are marked by a large positive isostatic anomaly which can be traced westward along the Cosmonauts Sea shelf. This isostatic instability was first believed to be caused by the absence of the "root" below the Enderby Land mountains; however, more detailed study shows that the configuration of the anomaly does not depend on the "effective relief" topography. It is therefore likely that the anomaly is caused not by the lack of isostatic compensation of the surface masses, but rather by an extensive high-density inhomogeneity within the crust.

The Enderby Land mountains are composed of the oldest known protocrustal rocks, with an age reaching 4000 m.y. (Sobotovich and others, 1974). These rocks—enderbites, charnockites, and pyroxene-plagioclase schists—have a more mafic composition than common granites, and their measured density reaches 2.95 g/cm³. Recent studies of the early Precambrian evolution in Antarctica (Ravich, Chapter 51, this volume; Grikurov, Chapter 98, this volume) indicate that similar rocks once composed the Queen Maud Land and the Lambert Glacier areas as well, but later they were partly reworked there by superimposed tectono-metamorphic events. Hence, both geophysical and geologic evidence seems to indicate that the East Antarctica coastal zone (approximately 100 km wide) between Queen Maud Land and Princess Elizabeth Land may, in fact, consist of uncommonly high-density crystalline rocks with an exceedingly prolonged crustal history. However, more extensive crustal studies are needed before alternative explanations, such as a general density increase of the "granitic" layer in the coastal zone due to its basification adjacent to the oceanic basin, can be ruled out.

Conclusions

The crustal structure in the coastal sector of East Antarctica embracing Queen Maud, Enderby, MacRobertson, and Princess Elizabeth Lands is typically continental both in terms of the crustal thickness (30–40 km) and its block-layered nature. In all regions studied by geophysical methods (including deep seismic soundings) the thickness of a "granitic" layer is on the order of 20 km; however, its density in Queen Maud and Princess Elizabeth Lands is anomalously high (about 2.9 g/cm^3). Such a mass excess in the upper crust agrees with geologic evidence indicating that this part of East Antarctica is composed of the oldest known crystalline basement rocks, which have a more mafic composition and a higher measured density than common granitic terranes.

Deviations from a typically continental structure of the crust occur either in a zone transitional to the ocean or within a continental area in connection with riftogenic processes. In both cases the most characteristic change in the crustal structure is its overall thinning by about 10 km (mainly at the expense of the "basaltic" layer) due to a prominent mantle rise compensating the down-faulted crustal block and the low-density supracrustal rocks overlying it.

The Queen Maud Land and Lambert Glacier area crustal sections are closely comparable to those obtained in other Gondwanian ancient platforms, South Africa in particular (Willmore and others, 1952; Bonini and others, 1961). Hence the oldest crystalline complexes of the earth appear similar not only in superficial geologic pattern and in chronology of the earliest tectono-metamorphic events, but in the "geophysical" structure of the crust as well.

Integrated geophysical study of crustal structure in East Antarctica must be continued before further geologic conclusions may be reached. Of particular interest would be deep seismic sounding in Enderby Land where a unique protocrustal complex (enderbites 4000 m.y. old) is exposed. Both Enderby Land and Queen Maud Land offer excellent sites for deep seismic soundings along longitudinal profiles designed to reveal crustal structure in the continent to ocean transitional zone.

References cited

Adams, R. D. 1971. Reflections from discontinuities beneath Antarctica. *Bull. seism. Soc. Am., 61,* No. 5, 1441–51.

Bonini, W. E., Smith, T. J., and Woollard, G. P. 1961. Explosion studies of continental structure. *Publ. Carnegie Inst.,* No. 622.

Demenitskaya, R. M. 1960. Glavnye cherty stroeniia zemnoi kory v Antarktike [Main features of the earth's crust structure in the Antarctic]. *Inf. Byull. sov. antarkt. Eksped.,* No. 23, 10–14.

Demenitskaya, R. M., and Rosenberg, V. I. 1966. Karta tolshchiny zemnoi kory v Antarktide [Map of the thickness of the earth's crust in Antarctica]. (*In* Bakayev, V. G., Shcherbakov, D. I., and Tolstikov, E. I., *eds. Atlas Antarktiki, I* [Antarctic Atlas, 1]. Moscow, Glavnoe Upravlenie Geodezii i Kartografii MG SSSR, 60.)

Dewart, G. 1961. Issledovanie zemnoi kory v Antarktike [Study of the earth's crust in the Antarctic]. *Inf. Byull. sov. antarkt. Eksped.,* No. 25, 11–13.

Dewart, G., and Toksöz, M. N. 1965. Crustal structure in east Antarctica from surface wave dispersion. *Geophys. J. R. astr. Soc., 10,* No. 2, 127–39.

Evison, F. F., Ingham, C. E., Orr, R. H., and LeFort, J. H. 1960. Thickness of the earth's crust in Antarctica and the surrounding oceans. *Geophys. J. R. astr. Soc., 3,* No. 3, 289–306.

Frolov, A. I. 1964. Gravitatsionnoe pole i nekotorye cherty stroeniia zemnoi kory Antarktidy [Gravity field and some features of crustal structure of Antarctica]. *Izv. Akad. Nauk SSSR,* Ser. geol., No. 10, 1449–61.

Frolov, A. I. 1971. *Materik Antarktida po dannym geofizicheskikh issledovanii* [Antarctic continent as it appears from geophysical data]. Moscow, Izdatel'stvo Nauka, 82 pp.

Groushinsky, N. P., Demenitskaya, R. M., Karyakin, E. D., Stroev, P. A., Ushakov, S. A., and Frolov, A. I. 1967. Osnovnye cherty stroenia Antarktidy po gravimetricheskim dannym [Main features of Antarctic structure from gravity data]. (*In* Bugaev, V. A., *ed. Osnovnye itogi izucheniya Antarktiki za 10 let* [Principal results of research in the Antarctic during 10 years]. Moscow, Izdatel'stvo Nauka, 97–108.)

Kogan, A. L. 1972. Results of deep seismic soundings of the earth's crust in Antarctica. (*In* Adie, R. J., *ed. Antarctic Geology and Geophysics.* Oslo, Universitetsforlaget, 485–89.)

Kolmakov, A. F., Mishen'kin, B. P., and Soloviev, D. S. 1975. Glubinnye seismicheskie issledovaniia v Vostochnoi Antarktide [Deep seismic soundings in East Antarctica]. *Inf. Byull. sov. antarkt. Eksped.,* No. 91, 5–15.

Kovach, R. L., and Press, F. 1961. Surface wave dispersion and crustal structure in Antarctica and the surrounding oceans. *Annali Geofis., 14,* No. 2, 211–24.

Kunin, N. Ya. 1972. *Kompleksirovanie geofizicheskikh metodov pri geologicheskikh issledovaniyakh* [Complex of geophysical methods in the course of geological explorations]. Moscow, Izdatel'stvo Nedra, 272 pp.

Ravich, M. G., and Kamenev, E. N. 1972. *Kristallicheskii fundament Antarkticheskoi platformy* [Crystalline basement of the Antarctic platform]. Leningrad, Gidrometeoizdat, 658 pp. [English translation: New York, John Wiley, 1975, 582 pp.]

Rosenberg, V. I. 1971. Nekotorye rezultaty interpretatsii geofizicheskikh issledovanii Antarktidy [Some results of interpretation of geophysical research in Antarctica]. *Antarktika, Doklady Komissii, 1968,* 75–88.

Sobotovich, E. V., Kamenev, E. N., Komaristyi, A. A., and Rudnik, V. A. 1974. Drevneishie porody Antarktidy (Zemlia Enderbi) [The oldest rocks of Antarctica (Enderby Land)]. *Izv. Akad. Nauk SSSR,* Ser. geol., No. 11, 30–50.

Stroev, P. A. 1971. Moshchnost' i nekotorye cherty stroenia zemnoi kory v yuzhnom okeane i v pribrezhnoi zone Antarktidy [Thickness and some features of crustal structure in the Southern Ocean and the coastal zone of Antarctica]. *Antarktika, Doklady Komissii, 1969,* 42–59.

Stroev, P. A., Ushakov, S. A., and Frolov, A. I. 1967. *Zemnaia kora kraevoi zony Antarktidy* [The earth's crust in the marginal zone of Antarctica]. Moscow, Izdatel'stvo Nauka, 64 pp.

Willmore, P. L., Hales, A. L., and Gane, P. G. 1952. A seismic investigation of crustal structure in the western Transvaal. *Bull. seism. Soc. Am., 42,* No. 1, 53–80.

113

Some Problems of Crustal Geodynamics in Antarctica

R. M. DEMENITSKAYA
Department of Geophysical Investigations of Oceans, SEVMORGEO, Moyka 120, Leningrad 190121, U.S.S.R.

Abstract. Deep seismic soundings and aeromagnetic data obtained recently by Soviet investigators in Antarctica on the Atlantic and Indian Ocean coasts indicate that the estimates of crustal thickness in Antarctica made in 1960 from a theoretical correlation between the surface topography and the depth to the Moho discontinuity are generally correct and reveal broad structural features with considerable accuracy. On the basis of these structural features and the subglacial bedrock topography, a geodynamic analysis of the Antarctic continent is presented, and a classification of its crust in terms of plate tectonics is proposed.

The first general speculations about the thickness of the earth's crust in Antarctica resulted from analysis of the dispersion of P_g- and L_g-waves recorded at Mirny, Scott, and Hallett Stations. This method provides an average thickness of the crust along the path of wave propagation with an uncertainty of about 10%. However, interpretation of the data depends on the inferred crustal structure model. Because of this fact, some workers (Evison and others, 1959) have estimated average crustal thicknesses of 35 and 25 km for East and West Antarctica, respectively, while others have obtained 40 and 30 km (Bentley and others, 1960) based on the same initial data. On the whole, however, analyses of the dispersion of P_g- and L_g-waves crossing the continent in different directions did suggest a continental crust for Antarctica. The author also arrived at the same conclusion (Demenitskaya, 1960) utilizing data on the topography of the continent. Using mean values for height and gravity anomalies (the so-called "average plots method"), the author calculated the crust's thickness from the following empirical formulas:

$$H = 35\,(1 - th\,0.0037\Delta g), \tag{1}$$

$$H = 33\, th\,(0.38\Delta h - 0.18) + 38, \tag{2}$$

where H is the thickness of the earth's crust in km, Δg is the Bouguer anomaly in milligals, Δh is the surface elevation in km (positive on land and negative under the sea), and th is the sign of the hyperbolic tangent

$$(th\,x = \frac{e^x - e^{-x}}{e^x + e^{-x}}).$$

However, the presence of the ice layer must be allowed for in Antarctica, so we introduce $\Delta h = \Delta h_1 + 0.29\Delta h_2$ into equation (2), where Δh_1 is the elevation of the bed (also positive on land and negative under the sea); Δh_2 is the thickness of the ice, from bedrock in most places on land but from sea level in areas where the bedrock surface is below sea level; and 0.29 is the correction for ice density. Thus, equation (2) for the earth's crustal thickness in Antarctica becomes:

$$H = 33\, th\,[0.38\,(\Delta h_1 + 0.29\Delta h_2) - 0.18] + 38 \tag{3}$$

Using equation (3) the author estimated the thickness of the crust for different points in Antarctica and compiled the first contour map (Demenitskaya, 1960). More recently, on the basis of additional data on subglacial bottom topography, other improved versions have been published; one of them is given in the *Atlas of Antarctica* (Demenitskaya and Rosenberg, 1966). There are no important differences from the first map. Furthermore, recent geophysical studies carried out by the Soviet Antarctic Expedition in East Antarctica (including deep seismic soundings and extensive aeromagnetic surveys) confirmed the above crustal thickness estimations and demonstrated the usefulness of the suggested formulas (Kurinin and Grikurov, Chapter 112, this volume).

Crustal types

From the postulate that the Antarctic continent is similar to other continents of the earth in its crustal structure, the author has formulated an outline of the genesis of separate parts of the continental crust and the adjacent oceanic crust. The previously published genetic classification of the earth's crust (Demenitskaya, 1975) based on speculations about crustal development in terms of plate tectonics is used here. Unlike the parametric model (based on the physical properties of rocks), according to which the earth's crust is subdivided into continental, oceanic, and transitional provinces on the basis of seismic velocities and thickness of layers, this genetic model is based on the geodynamics of the lithosphere. Reasoning from the nature of crustal generation at the boundaries of two oceanic plates within the rift zones of mid-oceanic ridges and from the dynamic pattern of processes operating at the boundaries of continental and oceanic or two continental plates, the predominant stresses and the direction of flow of geologic masses can be deduced. These suggest the following four types of crust (see Table 113.1):

1. The constructive type. This is new, fresh crust formed at mid-oceanic ridges by emplacement of mantle material; it acquires a sedimentary layer and forms oceanic crust by sea-floor spreading and lateral migration.

2. The destructive-constructive type. This crust is generated at some ocean margins by sinking and subsidence of the lithosphere (subduction), followed by its destruction,

later reworking, and eventual secondary vertical injection; this setting is favorable for the creation of young continental crust of an island-arc type. This type of crust results from the destruction of oceanic crust and creation of young continental crust.

3. The passive-constructive type. This crust is generated at the boundary of two continental plates undergoing collision; it forms continental crust of greater thickness and also involves oceanic crust in the form of ophiolite belts. A new crust is not created here, but the old crust is "mechanically" transformed, becoming thicker as if it were "folded."

4. The conservative or exconstructive type. This crust consists of the last two types at the stage of their consolidation, stabilization, and denudation. Based on this classification, the following geodynamic explanation of the crustal structure of Antarctica is proposed.

Geodynamic analysis of Antarctic crust

Let us consider the subglacial topography of the Antarctic continent (Znachko-Yavorsky, 1977) and identify the major architectural features of the solid earth in the South Polar region. At the surface the ice sheet of Antarctica appears dome-like, but actually it covers a complex bedrock topography marked by certain regularities. Antarctica is divided by the Transantarctic Mountains into a rather stable eastern and a more mobile western part. The antipodal analogy inevitably comes to mind: the Arctic Ocean is similarly subdivided by the Lomonosov Ridge into two parts, i.e., the mobile Eurasian basin, and the stable Amerasian basin (Demenitskaya and Rosenberg, 1972).

The map of the bedrock surface in Antarctica shows that most of the land above sea level occupies the area from long. 20° W eastward to 100° E along the Atlantic and In-

Table 113.1. Genetic classification and geodynamics of the earth's crust, showing four crustal types.

Major parameters	Constructive (type 1)	Destructive-constructive (type 2)	Passive-constructive (type 3)	Conservative (exconstructive) (type 4)
Trends of initiation	Creation of young oceanic crust in a narrow zone of failing parts of plates (in rifts) due to supply of mantle material (spreading)	Creation of young continental crust due to assimilation and mantle reworking at high temperature and stresses of a subsiding and disintegrating oceanic crust in a narrow zone at the distal edge of an oceanic plate	Generation of a thickened continental crust in certain fold belts due to thrusting, folding, and partial overriding of plates	Stablization with time of destructive-constructive and passive-constructive types of crust. Evolutionary reconstruction of the crust.
Characteristics of initiation or transformation zone	Active zone of divergence (spreading) of two oceanic plates: (a) monorift oceans (b) polyrift oceans	Active zone of convergence (underthrusting) of two oceanic or oceanic and continental plates	Zone of plate collapse, at least one of them continental	Ubiquitous passive reconstruction with subordinate divergence, possible presence of weakened zones associated with paleo-divergent boundaries
Primary appearance	Mantle injections, submarine volcanism and hydrovolcanism	Granitization. Dehydration at depth. Active, discrete volcanism	Folding with mass overthrust and melange distortion (shrinkage)	Basification
Final transformation	Thrusting of an oceanic crust under continental crust (subduction) by active subsidence into the mantle and degradation of oceanic crust. Formation of relict regions, with oceanic crust as a subordinate process possible at the plate edge	Evolutionary, with surface erosion	Erosion: a transition to crust of the conservative type	Thinning (extension?) and breakup with formation of continental rifts or micro-oceanic basins and oceans where there has been a "failure"
Rate of formation	Fast, one-stage lateral rate of crustal accretion is 0 to 20 cm/yr	Fast or average, multistage	Fast in terms of stages, with the rate of vertical thickening of the crust up to 2 cm/100 yr	Gradual, slow, with inversions
Duration of development	Existence of one system of spreading pattern, not more than about 170 m.y.	Time of termination of continental crust generation at the stage (orogenic stage) of orogeny, about 15 m.y.	Orogenic stage, 1–10 m.y.	The age of the oldest conservative crust is 3700 m.y. However, the process of crustal initiation older than 600 m.y. is not clear and probably followed a slightly different pattern
Nature of principal horizontal stress	Tension	General compression, local zones of tension	Compression	Regional and local compression and tension (complex potential nodes)
Predominant movements	Active vertical movements in crestal zone; passive horizontal movements at flanks resulting from lateral spreading of vertical flow, convection of asthenospheric currents; extensive, rather drastic, vertical movements (sinking) of sea floor caused by cooling of plate and compensatory load	Vertical movements as a consequence of oblique subsidence of oceanic lithosphere and rise of crustal and mantle material	Vertical movements as a consequence of lateral compression, resulting in folding of sedimentary cover, metamorphism, and formation of mountain roots	Oscillatory movements as a consequence of isostasy due to changes in composition and density; processes related to differentiation in the upper mantle
Type of volcanism	Basaltic	Andesite	Absent at first stage	Basaltic, alkaline; felsic
Type of metamorphism	Glaucophane and greenschist facies. Serpentinization	Granulite, eclogite, amphibolite, greenschist, and glaucophane. Deserpentinization of oceanic crust	Granitization. Relics of all types of facies	All types of metamorphism from eclogite to granulite and greenschist facies
Parametric characteristics based on seismic data	Oceanic crust, thickness 5–6 km. Suboceanic crust, relics of the ocean, 6–15 m.	Continental crust, thickness, 35–50 km. Subcontinental crust, 15–30 km	Continental, thickened crust with thick granite and basalt layers; total thickness 45–75 km	Normal and thinned continental and subcontinental crust; thickness 25–45 km
Morphostructural characteristics illustrated by specific regions	Oceanic basins, some seas (Sea of Japan)	Transitional zone from continent to ocean of the Pacific type (island arcs, the Andes)	Fold systems of the Alpine-Himalayan belt type	Platforms, shields, blocky uplifts, continental rifts (Siberian platform, Colorado Plateau, western Europe.)
Manifestation within physical fields	Symmetric magnetic field, high heat flow, linear shallow-focus seismicity	Magnetic field characteristic of volcanic structures. Regionally high heat flow. Presence of seismic loci over the entire range of depth forming a single, oblique focal surface; intense gravity anomalies	Magnetic anomalies somewhat linear; sources of anomalies at great depths. High values of heat flow. Regional seismicity at different depths along sliding planes of plates or subsidence of their relics	Magnetic field diverse, with sources at different depths. Normal and low values of heat flow. Gravity field with local and regional anomalies. Ubiquitous aseismicity of some regions; microseismicity along faults

dian Ocean coasts and striking toward the pole. Elevations of 500 m, and rather large areas with elevations of 1000 m and more, such as the Gamburtsev Mountains, dominate this part of Antarctica. This area is marked by the presence of an elevated, arcuate plateau with a predominant elevation of 1000 m and numerous separate mountains as high as 3000 m, stretching eastward along the coast from the Weddell Sea. The topography of this part of Antarctica undoubtedly suggests a typical continental crust.

In a sector from long. 100° E farther east to 150° E, the bedrock topography of Antarctica is considerably lower: mainly about 500 m below sea level with some small sites above sea level. On the basis of its crustal structure, this part may be compared with shelf areas of other continents.

The Transantarctic Mountains, crossing Antarctica from the Coates Coast to the Weddell Sea, separate East and West Antarctica. The thickness of the earth's crust in this highland cannot be interpreted unambiguously. Previously it was believed that the crustal thickness increases in accordance with the height of the mountains, but now it seems likely that this mountain system may have no "roots."

West Antarctica is an archipelago situated near a large, considerably submerged shelf, which joins the continent of East Antarctica along the line of the Transantarctic Mountains; the shelf has an average depth of 500 m or more. Within the archipelago, the three largest tectonic components are the Marie Byrd Land plateau, the Ellsworth Mountains, and the Antarctic Peninsula. Some peaks have summits well above the ice sheet, their height locally reaching 5000 m. The highly dissected subglacial topography of this part of the Antarctic continent has no equivalent in other regions of the earth and implies a complex geodynamic environment, probably predating the formation of the ice sheet. It seems unlikely that this topography resulted from isostatic subsidence effected by glaciers, as has been suggested by some authors.

Crustal development

On the basis of these data and the evidence concerning the crustal thickness and distribution of magmatic units of different ages, the author suggests the following major periods when crustal and subcrustal processes were more active.

In early Paleozoic time, Antarctica was subjected to compression along its outer boundary, resulting in the formation of a marginal rise. The arcuate plateau of Queen Maud Land, with crustal thickness ranging from 25 km along the edge of the arc to 40 km in the center, is probably a relic of that rise.

In middle and late Paleozoic time, the Transantarctic Mountains area was receiving sedimentary deposits, probably close to sea level. During this same time, the continent was subdivided into two parts by the ancient Transantarctic Mountains along the suture separating the Archean and Proterozoic foundations of the continent. As mentioned above, these mountains may have no roots. East Antarctica became a zone of conservative crust 32 to 38 km thick, and apparently has not been activated since that time. In contrast, West Antarctica was subjected during Mesozoic and Cenozoic time to a number of crustal transformations, with tension leading to the decrease in thickness of ancient parts to 18–30 km and compression resulting in a series of folded arcs with thickness increasing to 40–50 km. It seems possible to infer a gradual retreat of tension zones with time from the ancient Transantarctic Mountains to the Pacific coast. The Jurassic–Cretaceous crustal transformation involved (or created?) the Pacific-ward, younger Transantarctic Mountains which embrace the Ross Sea shelf and Marie Byrd Land (inner arc). The Cenozoic crustal transformation manifested itself along the Pacific coast (outer arc of mountains). These arcs probably mark successive retreats of the Benioff zones which consumed a part of the oceanic crust generated at the Pacific mid-oceanic ridge during the late Mesozoic and Cenozoic stages of sea-floor spreading, i.e., crust of the constructive type, which surrounds the Antarctic continent at present.

Hence, East Antarctica, as there is no evidence of crustal transformation, has apparently been a single, rather passive, lithospheric plate at least since the end of early Paleozoic time. It is undoubtedly a conservative or fourth type of crust.

West Antarctica differs from East Antarctica not only in morphology but also in its kinematic and dynamic patterns. Since the end of early Paleozoic time, it has been subjected to active geodynamic transformations during middle to late Paleozoic, Late Triassic to Early Jurassic, Cretaceous, and Cenozoic times. The middle to late Paleozoic activization of the plate was concentrated within a zone of the ancient Transantarctic superregional suture, stretching from the Weddell Sea to the Ross Sea. The Late Triassic to Early Jurassic transformation migrated to another part of the plate, with manifestations in the Antarctic Peninsula and southwestward along two branches, one in central West Antarctica (Whitmore Mountains) and another following the coasts of the Bellingshausen and Amundsen Seas (Marie Byrd Land). The Early Jurassic activization of short duration involved the crust of the ancient Transantarctic Mountains. The Cretaceous and Cenozoic activizations were restricted by the Pacific slope of the Transantarctic Mountains and involved only West Antarctica. Evidence for the Cretaceous activization of crust is known almost exclusively from Marie Byrd Land.

Thus, West Antarctica, unlike East Antarctica, is underlain not only by crust of the fourth type—conservative—but also by some crust of the second and third types: destructive-constructive and passive-constructive. The crust of the conservative type in East and West Antarctica is different in thickness, being 30–40 km and 18–30 km thick in East and West Antarctica, respectively.

The Cenozoic Era was the most dynamic stage of crustal reconstruction of the Antarctic continent, resulting in the formation on the surface of West Antarctica of extensive lava sheets occupying most of the Pacific coastal zone and the formation of granitoids and gabbroids within the South Shetland Islands, the Antarctic Peninsula, and Marie Byrd Land. Manifestations of recent volcanic activity are known, for example, from McMurdo Sound and near the Antarctic Peninsula, where active volcanoes are present.

This preliminary and very schematic analysis of the crustal structure of the Antarctic continent implies that the subdivision into East Antarctica and West Antarctica by the first investigators was correct and reflects the basic geodynamic regime of the whole Antarctic.

References cited

Bentley, C. R., Crary, A. P., Ostenso, N. A., and Thiel, E. C. 1960. Structure of West Antarctica. *Science, N.Y., 131,* No. 3394, 131–36.

Demenitskaya, R. M. 1960. Glavnye cherty stroeniia zemnoi kory v Antarktike [Main features of the earth's crust structure in Antarctica]. *Inf. Byull. sov. antarkt. Eksped.,* No. 23, 10–14.

Demenitskaya, R. M. 1975. *Kora i mantiya Zemli* [Crust and mantle of the Earth]. Moscow, Izdatel'stvo Nedra, 255 pp.

Demenitskaya, R. M., and Rosenberg, V. I. 1966. Karta tolshchiny zemnoi kory v Antarktike, 1:40,000,000 [Map of thickness of the Earth's crust in Antarctica, 1:40,000,000]. (*In* Bakayev, V. G., *ed. Atlas Antarktiki* [Atlas of the Antarctic]. Moscow, Glavnoe Upravlenie Geodezii i Kartografii, 1, 60.)

Demenitskaya, R. M., and Rosenberg, V. I. 1972. Crustal structure of antipodal regions of the earth: Antarctic and Arctic. (*In* Adie, R. J., *ed. Antarctic Geology and Geophysics.* Oslo, Universitetsforlaget, 481–83.)

Evison, F. F., Ingham, C. E., and Orr, R. H. 1959. Thickness of the earth's crust in Antarctica. *Nature, Lond., 183,* No. 4657, 306–8.

Znachko-Yavorsky, G. A. 1977. Rel'ef Antarktidy [The Antarctic relief]. *Antarktika, Doklady Komissii, 16,* 45–53.

114

Some Features of Antarctic Crustal Structure

N. P. GROUSHINSKY
Sternberg Astronomical Institute, Moscow University, Moscow, U.S.S.R.

N. B. SAZHINA
VNII Zarubezhgeologiya, Moscow, U.S.S.R.

Abstract. Crustal thickness in Antarctica attains 60 km under the Vernadsky Mountains in East Antarctica. The crust under the Transantarctic Mountains and under central Antarctica is also thick, in some places as much as 40–50 km. Continental-type crust is typical for the entire Antarctic. The geoid reaches positive values of 20–30 m in the Vernadsky Mountains region where the thickness of the crust is the greatest. The elevation of the geoid in other parts of the continent has a negative value as great as 30–40 m. This emphasizes the asymmetry of the earth.

The gravimetric method is both important and convenient for geophysical investigations of crustal structure in Antarctica. It is of importance first of all because the nearly continuous ice cover greatly impedes direct investigation and measurement on bedrock, whereas the gravimetric method "X-rays" the ice cover and, because of the considerable difference in density between ice and bedrock ($2.8 - 0.96 = 1.84$ g/cm^3), permits rather accurate determination of the ice thickness, subice relief, and density anomalies. Secondly, this method is comparatively cheap and demands only light-weight instruments. Therefore, even in the early stages of development of gravimetric studies in Antarctica, some attempts were made to reconstruct the subice relief. Combined use of seismic, gravimetric, magnetic, and radio sounding methods have now made it possible to construct rather confidently a model of subice relief in Antarctica. As data have accumulated, maps of this kind have been compiled by a number of different authors. Among the latest, most complete models are the works of Kapitza (1968) and Bentley (1972). The most recent generalization of hypsometric data is that of Znachko-Yavorsky (1977).

Gravimetric data have accumulated for 20 years, and these have been reduced and a gravimetric map of Antarctica compiled. We consider it desirable now, after summarizing the data procured by other methods, to construct a new map of Antarctic subice relief. The generalization of gravimetric data in the form of a 1:5,000,000 scale map of free-air and Bouguer anomalies allows us to revise both the estimates of the earth's crustal thickness and the type of geoid in the Antarctic area.

The thickness of the crust

Early works on the structure of the earth's crust in Antarctica had emerged even by 1959. Studies by Evison and others (1959, 1960), however, were based on the few seismologic data available. The reliability of their calculations could not be determined. Evison and his colleagues determined the thickness of the crust using Love and Rayleigh wave propagation data from eight earthquakes, as recorded at Hallett, Scott, and Mirny stations. They obtained the following crustal thicknesses: 35 km for East Antarctica, 25 km for Marie Byrd Land, 30 km for the Ross Sea region, and 36 km for the eastern part of Victoria Land. They estimated a root-mean-square error of 3–4 km. The work by Demenitskaya and associates concerning the structure of the earth's crust in Antarctica was also published at about that time. Demenitskaya and Ushakov (1966) made use of some stochastic relationships between the thickness of the crust, obtained by deep seismic sounding, and surface elevation for other parts of the earth, and extrapolated them to Antarctica. Dewart and Toksöz (1965), using observations at Mirny Station of the velocities of Love and Rayleigh waves from distant earthquakes, obtained a thickness of 27 km for Marie Byrd Land and 35 km for the central part of Antarctica and the Antarctic Peninsula. Woollard (1962) made an attempt to determine the thicknesses of the antarctic crust on the basis of seismologic and gravimetric data. He obtained a crustal thickness of 30–35 km for the coastal area and 48 km for the center of the continent.

A map of Antarctic crustal thickness was published by Groushinsky and Frolov in 1967. Frolov then calculated the statistical relationship between crustal thicknesses, as obtained from the propagation of seismic waves from distant earthquakes, and Bouguer gravity anomalies in order to derive an interpolation formula for computing thicknesses of the crust from Bouguer anomalies. However, having very few data available, he had to determine hypothetical Bouguer anomalies, extrapolating them at a number of points in the internal part of the continent by surface elevations. Frolov (1971) obtained a crustal thickness of 50 km for the central part of East Antarctica and the Transantarctic Mountains, gradually decreasing to 25–20 km at the periphery of the continent. Many plots have since been made for thicknesses of the crust in individual areas of Antarctica. For example, Stroev (1972) calculated the thickness of the crust in the transitional zone from the Antarctic continent to the Indian Ocean on the basis of gravity data and interpolation formulas associating Bouguer anomalies with crustal thickness.

In 1969 geophysicists from the Soviet Antarctic Expedition carried out for the first time a deep seismic sounding on the Princess Astrid Coast near Novolazarevskaya Station. Depths to the Mohorovičić discontinuity of 28–38 km were obtained on a profile extending about 400 km along the coast; depths of 26–42 km were obtained from a profile extending 200 km northeast-southwest across the strike of

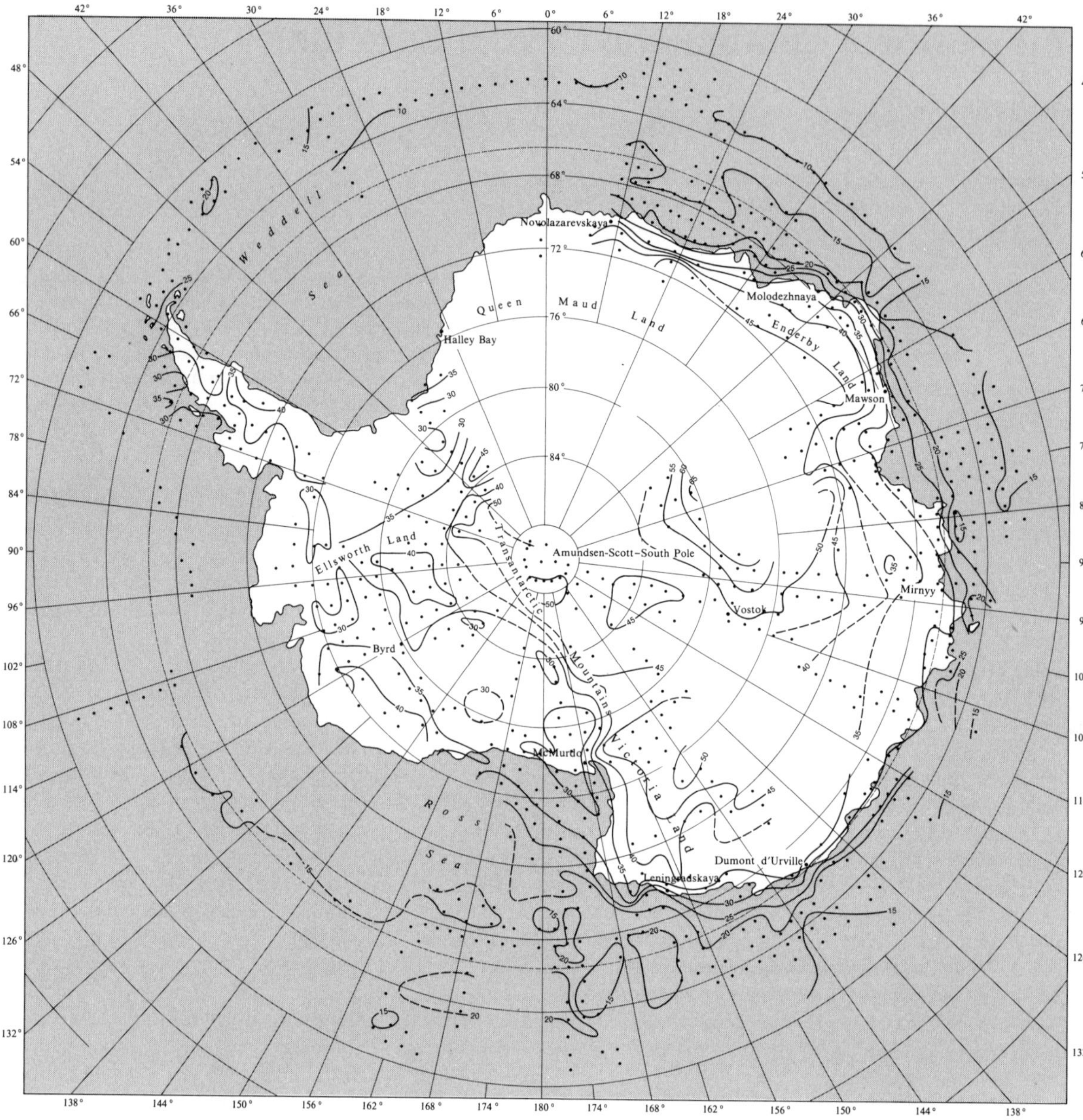

Figure 114.1. Map of Antarctic crustal thicknesses, showing contours (at 5 km intervals) of equal crustal thickness. Dashed contours indicate insufficient data. The dots are points where the thickness of the crust was calculated. Cartographic Lab, University of Wisconsin–Madison.

the coastline. In 1973 a new sounding was carried out by the Soviet Antarctic Expedition in the area of the Prince Charles Mountains and on the Amery Ice Shelf. As a result of these investigations, it was found that the crust in these areas has a complex block-faulted structure. The major crustal blocks are separated by deep faults, the greatest of which penetrate the upper mantle of the earth. At the boundaries between blocks, the thickness of the seismic layers, the velocities of elastic wave propagation, and the thickness of the earth's crust all change. The thickness varies from 22 to 34 km in the Prince Charles Mountains region. Gravity determinations carried out in these same areas led to a correlation between the depth to the Mohorovičić discontinuity (M) and the gravity anomalies. The relationship was determined as a linear function:

$$M = M_0 + kg$$

where M_0 is the depth in km corresponding to the zero anomaly, g is the Bouguer anomaly in mgal, and k is a constant. The gravitational anomalies were averaged over trapezoids equal in area to a trapezoic 1° on a side at the equator. As a result of the calculations, the following coefficients were obtained:

$$M_0 = 33.7 \pm 0.7$$
$$k = -0.11 \pm 0.02.$$

M values were calculated according to this empirical relationship for all regions of Antarctica for which there are gravimetric data available. For the adjacent ocean, the formula obtained earlier by Stroev (1972) for the Southern Ocean was used:

$$M = 32 - 0.057\,g$$

The calculations were made on the basis of the gravimetric map of Antarctica compiled at the Sternberg Astronomic Institute (SAI) of Moscow University and at the Scientific Research Laboratory of Foreign Geology, Ministry of Geology, U.S.S.R. (see Groushinsky and

Sazhina, Chapter 115, this volume). The map of crustal thicknesses obtained is shown in Figure 114.1. The contour interval is 5 km, prescribed by the accuracy of the coefficients in the linear equation.

The earth's crust attains its maximum thickness of 62–63 km in the area of the subglacial Gamburtzev massif. Across the Transantarctic Mountains from the Ross Ice Shelf to the polar plateau, the thickness is 45–50 km. The crust attains the same thickness in East Antarctica in the area of the Wegener Plateau (Queen Maud Land), in the Princess Elizabeth Mountains, and in Enderby Land.

In West Antarctica the thickness of the crust at the base of the Antarctic Peninsula, in the mountains of Marie Byrd Land, and in the region of the Ellsworth Mountains is as much as 40 km. Beneath the continental shelf, however, the thickness of the crust abruptly decreases to 10–15 km. In all other parts of the continent, the thickness of the crust is on the order of 30–35 km.

On the basis of these data, the crust of both East and West Antarctica is considered as typical continental crust. The continent as a unit is a structure that was depressed under the glacial load and has, on the whole, reached isostatic equilibrium.

In the transitional zone along the shores of all of East Antarctica, the Mohorovičić surface rises very steeply. In the first 100 km of the transitional zone from continent to ocean, it rises 10 km. At a distance of 300–500 km from the shoreline, the depth of the Moho surface decreases to 10–15 km. It was found that in some parts of the seas around Antarctica, the crust thickens (for example, north of the Shackleton Ice Shelf, in the region of Budd Coast, and around the Balleny Islands and Scott Islands). In these regions the thickness of the crust is 5–6 km greater than in the adjacent oceanic regions. An abrupt change in crustal thickness was observed in the region of Peter the First Island and in the Drake Passage. In this area, the earth's crust has a stratified-block structure. The Drake Passage may preserve old, buried structures that in the past could have connected the South America cordillera with Antarctica.

Heights of the geoid

As gravimetric data on Antarctica accumulated, study of the deviation of the geoid from an ellipsoid began. As early as 1967 a map of geoid altitudes for Antarctica, compiled by Frolov on the basis of limited gravimetric data and hypothetical gravitational anomalies, was published (Groushinsky and Frolov, 1967). Development of dynamic methods of satellite geodesy permitted the construction of generalized models of the geoid for all the earth, including Antarctica. According to a diagram of geoid altitudes in relation to an ellipsoid with flattening of 1:298.256 (data from Standard Earth 3, Smithsonian Observatory), the geoid in Antarctica is almost everywhere 10–40 m lower than the ellipsoid level. Only part of East Antarctica, between long. 20° W and 90° E, is characterized by positive geoid altitudes. A geoid diagram, based on harmonic analysis of the gravitational potential, reveals a pyramidal shape for the earth: in the North Pole region the geoid passes approximately 20 m above the ellipsoid, but in the south, it passes about 25–30 m below it (Figure 114.2). Planetary cross sections along meridians, exhibiting distinct longitudinal asymmetry, are shown in Figure 114.3.

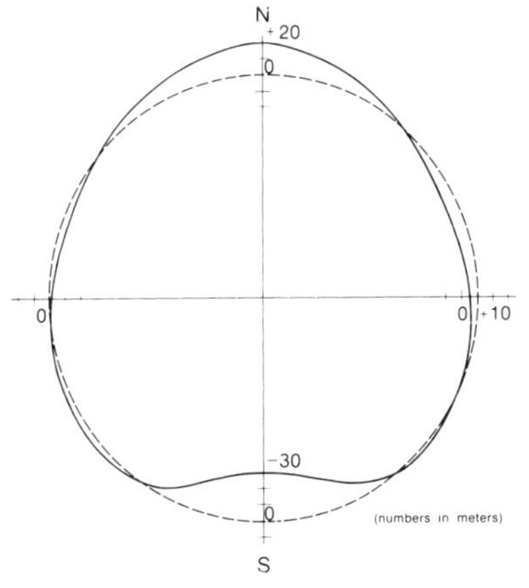

Figure 114.2. A meridional planetary cross section based only on zonal harmonics. The geoid is shown as a solid line and the ellipsoid as a dashed line. (Numbers in meters) Cartographic Lab, University of Wisconsin–Madison.

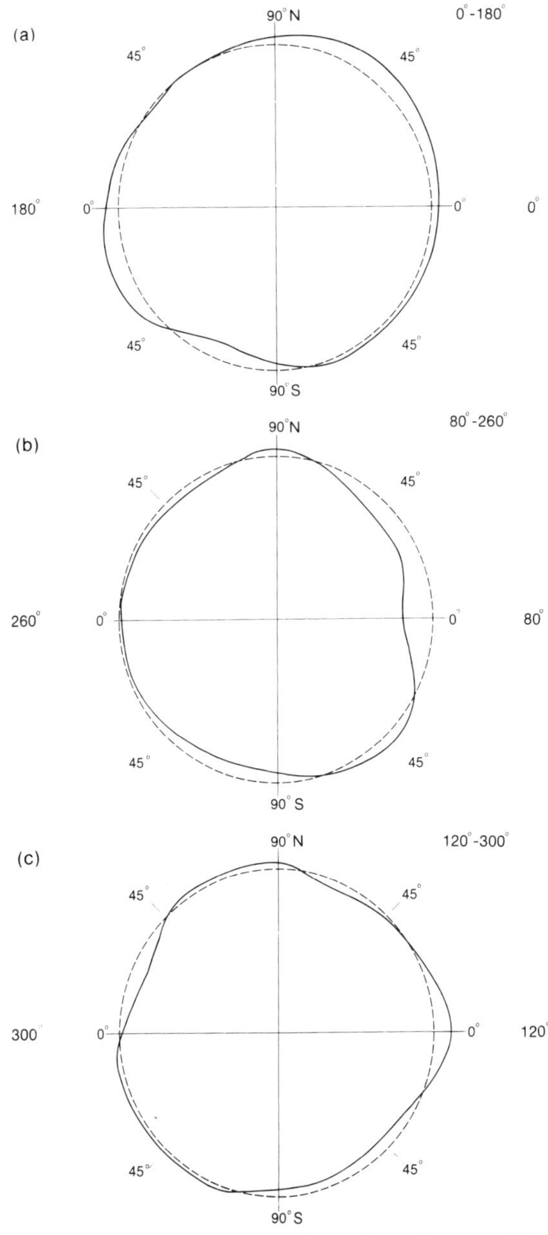

Figure 114.3. Planetary cross sections along three meridians, exhibiting distinct longitudinal asymmetry. The geoid is shown as a solid line and the ellipsoid as a dashed line. Cartographic Lab, University of Wisconsin–Madison.

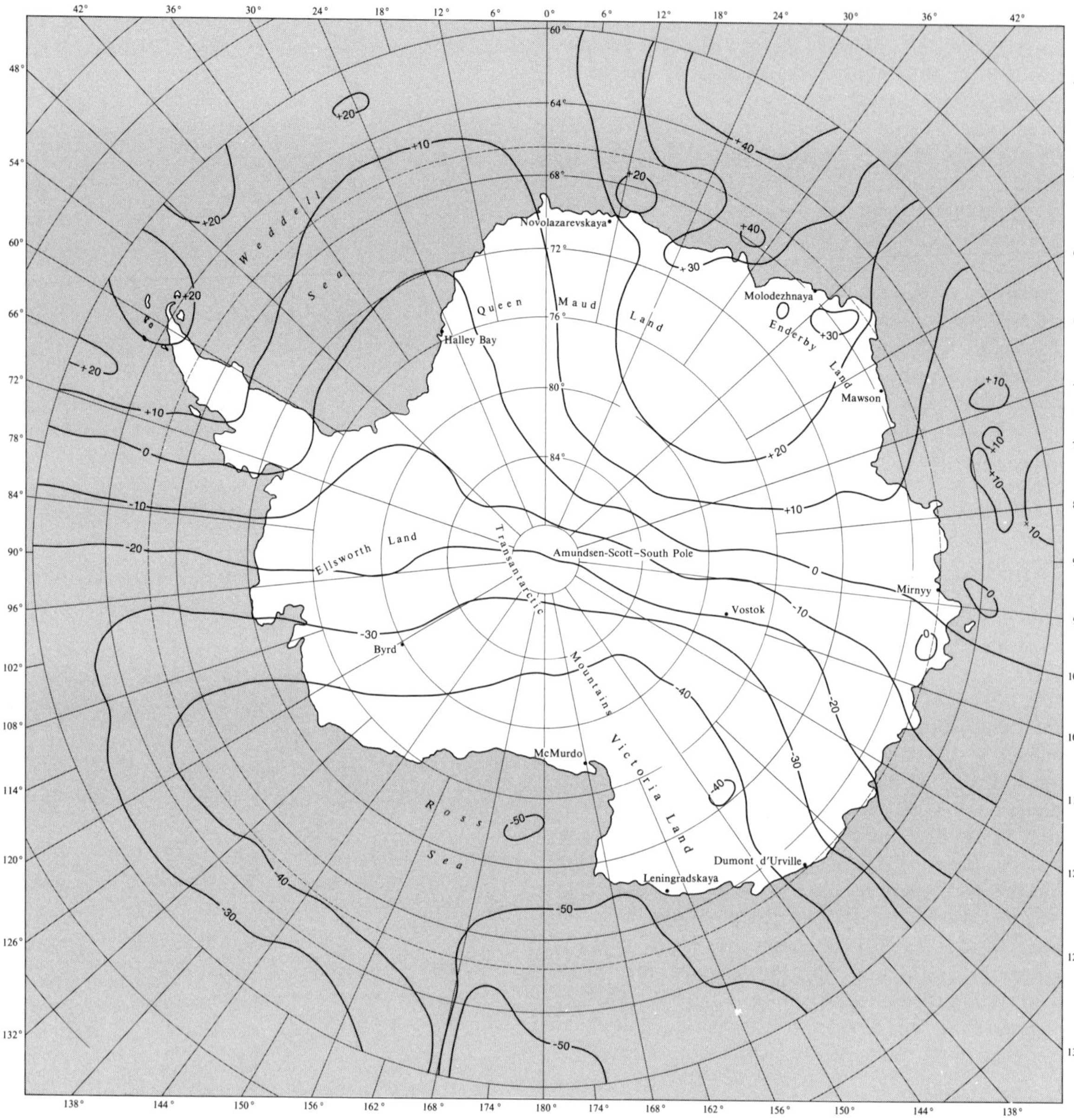

Figure 114.4. Map of the Antarctic geoid, showing contours of equal heights of the geoid surface relative to the reference ellipsoid. Contour interval is 10 m. Cartographic Lab, University of Wisconsin-Madison.

The available gravimetric data and the gravimetric map of Antarctica allow examination of the geoid shape in the Antarctic area to a greater extent and reveal certain details obscured in the satellite data alone. Averaged (by 1° trapezoids) free-air anomalies and a harmonic analysis of the gravity field of order 20 were used to compile the Antarctic geoidal height map. The harmonic analysis of the gravity field was used to interpolate the gravity field in remote areas and in areas of Antarctica where gravimetric observations have not yet been undertaken. The calculation was based on the Molodensky (1960) formulas. The area of integration was subdivided into zones, and the influence of the nearest zone was calculated directly from the average free-air anomalies while the influence of the remote zones was included as a correction factor to the satellite data. The map of the Antarctic geoid thus calculated is shown in Figure 114.4.

This new map supplies more details than other maps previously compiled from gravimetric and satellite data. It once again demonstrates that negative geoid altitudes are typical for Antarctica. The geoid surface over the major part of the continent passes below the reference ellipsoid, and deviations in West Antarctica reach 50 m, and locally 60 m. A considerable part of East Antarctica, which seems to be larger than first inferred from the satellite data, is characterized by positive values of about 20–30 m. However, the minimum value for East Antarctica reaches −50 m. Thus here we see another asymmetric feature in the geoid surface. The earth's shape should not be considered as a triaxial ellipsoid nor as a rounded pyramidal form because the whole southern part of the earth is not flattened. Indeed, slight bulging is found in the sectors between long. 0–15° W in West Antarctica and long. 0–90° E in East Antarctica (Queen Maud Land and Enderby Land).

References cited

Bentley, C. R. 1972. Subglacial rock surface topography. (*In* Heezen, B. C., Tharp, M., and Bentley, C. R., *eds.* Morphology of the Earth in the Antarctic and Subantarctic. *Antarct. Map Folio Ser.,* Folio 16, Pl. VII.)

Demenitskaya, R. M., and Ushakov, S. A. 1966. Glubinnoe stroeniia Antarktika. *Antarktika, Doklady Komissii.*

Dewart, G., and Toksöz, M. N. 1965. Crustal structure in East Antarctica from surface wave dispersion. *Geophys. J., 10,* 127–39.

Evison, F. F., Ingham, C. E., and Orr, R. H. 1959. Thickness of the earth's crust in Antarctica. *Nature, Lond., 183,* 306–8.

Evison, F. F., Ingham, C. E., Orr, R. H., and Le Forte, J. R. 1960. Thickness of the earth's crust in Antarctica and surrounding oceans. *Geophys. J., R. astr. Soc., 3,* 289–306.

Frolov, A. I. 1971. *Materik Antarktida po dannym geofizicheskikh issledovanii* [Antarctic continent as it appears from geophysical data]. Moscow, Izdatel'stvo Nauka, 82 pp.

Groushinsky, N. P., and Frolov, A. I. 1967. Some conclusions on the structure of Antarctica. *J. geol. Soc. Aust., 14,* Pt. 2, 215–24.

Groushinsky, N. P., and Stroev, P. A. 1975. Die Antarktic ist ein zerklüfteter Kontinent. *Umschau Wiss. technik, Oktob.,* No. 20, 638–40.

Kapitza, A. P. 1968. *Podlednyy rel'yef Antarktidy* [Subglacial relief of Antarctica]. *Akad. Nauk SSSR,* Mezhduved. Geofiz. Kom., Moscow, 100 pp.

Molodensky, M. S. 1960. The method of investigation outside the gravity field and the figure of the earth. *Trudi ZNIIGAIK,* No. 131, 1-250.

Stroev, P. A. 1972. *Uzhnii Okean i Antarctica po pravimetricheskim dannym.* Moscow, Nauka, 95 pp.

Woollard, G. P. 1962. Crustal structure in Antarctica. (*In* Wexler, H., Rubin, M. J., and Caskey, J. E., *eds. Antarctic Research: The Matthew Fontaine Maury Memorial Symposium.* Washington, American Geophysical Union, 53–73.) [Geophysical Monograph No. 7]

Znachko-Yavorsky, G. A. 1977. Rel'yef Antarktidy [The Antarctic Relief]. *Antarktika, Doklady Komissii, 16,* 45–53.

115

Gravitational Field of Antarctica

N. P. GROUSHINSKY
Sternberg Astronomical Institute, Moscow University, Moscow, U.S.S.R.

N. B. SAZHINA
VNII Zarubezhgeologiya, Moscow, U.S.S.R.

Abstract. The most representative gravity map of Antarctica at a scale of 1:5,000,000, which shows free-air and Bouguer anomalies, has been compiled and published. The map reduced to a scale of 1:20,000,000 is given in this paper. Large positive anomalies occur in regions of mountains and high plateaus. Negative anomalies are usually located over ice shelves and plains. Comparatively small amplitudes of anomalies demonstrate the isostatic equilibrium of the continent.

More information about Antarctica has been gathered over the last twenty years than was gathered over the whole prior period since its discovery. The gravimetric method has turned out to be very effective under Antarctic conditions. Because of the big difference in density between ice (0.9 g/cm^3) and the subglacial bedrock (mean value 2.7 g/cm^3), this method has been used for determining the thickness of the ice sheet. The first models of subglacial relief were constructed on the basis of gravimetric data supplemented by seismic sounding. The gravimetric method is one of the major methods used for studying the deep structure of the earth's crust in the Antarctic area and for acquiring knowledge of the regional geologic structure as well. It would be an even more effective method if Antarctica were covered with even a sparse network of gravimetric stations.

At present a large amount of gravimetric information requiring synthesis and generalization, which could be helpful for all uses, has accumulated. Therefore, as early as the second Symposium on Antarctic Geology and Geophysics in 1970 in Norway, work aimed at compiling a gravimetric map of Antarctica was recommended and subsequently begun. This work has been carried out by the authors in the Sternberg Astronomical Institute of Moscow University and in the Scientific Research Laboratory of Foreign Geology under the Ministry of Geology.

The data from more than 11,000 gravimetric stations on the continent and continental shelf, as well as continuous gravimeter recordings from more than 23,000 km of marine ship routes have been systematized and generalized for this work. In addition, this study has made use of data from more than 35 publications, some of them summaries, and data transmitted privately by national representatives, for which the authors express their sincere appreciation. The following are the sources which contain the most data: the catalog of gravimetric stations by P. Morgan (unpublished), compiled at Ohio State University and including more than 6000 stations; the catalog of the International Gravimetric Bureau in Paris, edited by Tardi (1961), including 2200 stations; the GAIS Catalog edited by Groushinsky and others (1972), including 5493 stations; and the album of the research ship *Eltanin,* edited by Talwani and Meijer (1972) and containing 13,700 km of continuous gravimetric recording. The number of gravimetric measurements carried out by different countries and used in the compilation of our gravimetric map of Antarctica are given in Table 115.1.

Gravimetric determinations in Antarctica are very heterogeneous in accuracy and distribution. Up to now there has not been a unified, reliable, base network. The surveys in Antarctica by various countries are linked with their national base stations through main base stations on Antarctica, which are not properly connected among themselves. The majority of gravimetric base stations in Antarctica have only a single connection with the world network, e.g., through the stations of Ellsworth, Shackleton, Oasis, Molodezhnaya, Esperanza, and TA-1. Mirny and McMurdo Stations have multiple connections with the world network.

In the process of increasing the extent of gravimetric surveys in Antarctica, one goal has been to create a system of base stations having reliable connections with the world gravimetric network. At present there are several systems of this type, which are linked to these stations: Ellsworth, McMurdo, Mirny, Wilkes, and Punta Arenas. There is little information on the precision of these base networks; it is probably low. Specific work on their control and correlation has not been carried out. The senior author drew attention to the unsatisfactory state of the base gravimetric network of Antarctica during the 1970 Symposium on Ant-

Table 115.1. Data on the gravimetric measurements carried out by various countries and used in the compilation of the gravimetric map of Antarctica

Country	Dates	Number of stations: On the continent	Number of stations: On the surrounding sea	Length of continuous marine survey record (km)	Type of survey	Type of instrument	Mean square error in free-air anomalies (mgal)
Australia	1953–69	1734	26	—	Traverse	Worden	± 1– ± 10
Great Britain	1959–72	257	810	9100	Areal survey and traverses	Worden	1–8
Argentina	1957–65	97	—	—	Traverses, sparse stations	Worden, Frost	—
Belgium	1958–67	59	—	—	Sparse stations	Worden	3–6
New Zealand	1957–62	272	—	—	Areal survey, short traverses, sparse stations	Worden	2–8
U.S.S.R.	1957–75	2092	1468	7400	Areal survey, traverses	SN-3, GAK, ANG-M, sea damped gravimeter pendulum apparatus	1–12
U.S.A.	1957–75	5355	—	13700	Areal survey, traverses, sparse stations	Worden, Frost La Coste-Romberg	1–17
France	1958–59	216	—	—	Traverses, sparse stations	Worden	—
Chile	1961–62	10	—	—	Sparse stations	Worden	—
Union of South Africa	1962–73	360	—	—	Areal survey	Worden	—
Japan	1958–73	1542	721	5200	Traverse, areal survey	Worden, La Coste–Romberg, string gravimeter	—

—: Information not given.

arctic Geology and Geophysics. The same point of view was reflected in a decision of the International Association of Geodesy adopted in 1971 at the General Assembly in Moscow. It was decided that it will be necessary in the near future to reduce all the main reference gravimetric determinations to one fixed system of Antarctic stations and to establish better gravity coverage in the interior of Antarctica with widely spaced gravity reference stations.

The accuracy of gravity anomaly determinations in Antarctica is characterized by mean square errors of from 0.5 to 20 mgal. The main sources of error for Bouguer anomalies are unreliable determinations of geographic coordinates, elevation, and ice thicknesses. By comparison of observation results along intersecting and closely spaced traverses in several areas, differences in gravity anomaly values have been revealed that enable us to introduce corrections for gravity anomalies. Thus, for instance, a correction of +30 mgal was introduced to the data obtained by the Japanese expedition in 1965 for gravity stations north of lat. 71° S. Only part of the results of the Ninth Soviet Antarctic Expedition were used (only in those areas where they were confirmed by traverses of other expeditions that intersect them). The values of American determinations made during 1957–60 in the area of McMurdo–Hillary Coast were reduced by −20 mgal. In spite of all these complications, the recently accumulated data contain a great deal of information and enable us to revise and reinterpret conclusions which had been drawn earlier on the basis of poor and biased data.

The gravity map of Antarctica with Bouguer and free-air reduction, with a scale of 1:5,000,000, was compiled on the basis of all these data and the data published in 1978. The intensity of the anomalies on the map was given by the contours drawn over 10 and 20 mgal and by the intensity of colors (Groushinsky and Sazhina, 1978). Figures 115.1 and 115.2 show these free-air and Bouguer anomaly maps decreased in scale to 1:20,000,000. In the Bouguer reduction the density was taken as 2.67 g/cm^3 and the submarine-topographic correction as 0.0419 $\Delta\sigma \cdot P$ mgal, where P is the depth of the water and $\Delta\sigma$ is the difference in density between rock and water.

In addition to these contour maps, diagrams of averaged Antarctic anomalies have been compiled. The averaging was done according to the chart system adopted by the International Gravimetry Bureau in Paris, based on trapezoids 1° on a side at the equator (Groushinsky and Sazhina, 1977). Maps showing these averaged free-air and Bouguer anomalies have been published in a paper by Groushinsky and Sazhina (1977). The extreme and average values of the free-air and Bouguer anomalies determined in East and West Antarctica are summarized here in Tables 115.2 and 115.3.

It is evident from the Antarctic free-air gravity map that the gravity field in general is rather complex. In East Antarctica, positive gravity anomalies prevail. Positive anomalies as great as 100 mgal are found over the subglacial Gamburtsev Mountains, where the rock surface rises to 3400 m and the ice surface is as high as 4000 m in altitude. Major areas of positive anomalies as high as 180 mgal are found along the Kemp Coast and in the Prince Charles Mountains, with altitudes of the subglacial rock surface as much as 2000 m and ice surface elevation as much as 3300 m. Positive anomalies of 170 mgal accompany the Gunnerus underwater ridge north of Princess Ragnhild Coast, which rises to a depth of 400 m, with average depths of about 1500 m.

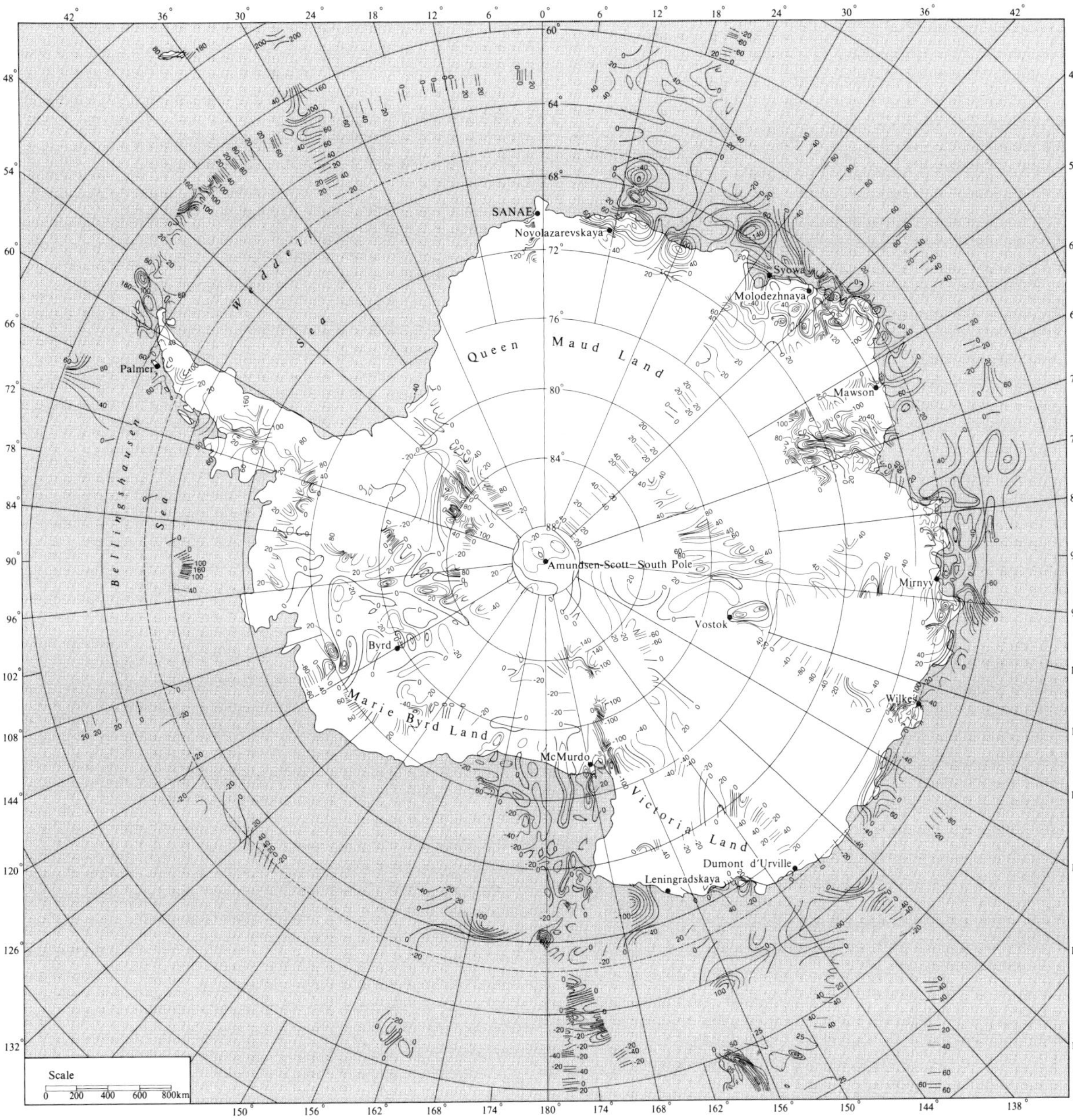

Figure 115.1. Gravity field of Antarctica showing contours of free-air anomalies. Cartographic lab, University of Wisconsin-Madison.

An area of strongly negative anomalies, down to −19 mgal, is confined to the wide Schmidt Plateau in southern Wilkes Land, with subglacial topographic highs of 1500 m and ice surface elevations of 3000 m. Negative minimum anomalies of considerable size are associated with basins. In Enderby Land, where the rock surface drops to −500 m and ice surface elevations are as much as 1500 m, anomalies decrease to −70 mgal. On the Lambert Glacier, with a rock level of −500 m and ice surface elevations of 2070 m, a minimum value of −70 mgal is found. Similarly, in the eastern plain close to Vostok Station, anomalies reach their minimum of −80 mgal where the rock elevation is about −20 m and the maximum ice surface elevation is 3500 m. In the Banzare Coast region, where the bedrock margin is also submerged to 200 m and the ice surface is as high as 1000 m, a minimum of −60 mgal is found. Minimums of −30 and −70 mgal were registered in the areas of Wilhelm II Coast; along with that, there is another widespread minimum of −80 mgal in the Davis Sea.

West Antarctica, limited by the Pacific edge of the Transantarctic Mountains, is generally characterized by mixed gravity anomaly fields, with low values prevailing. Major minimums are confined to the Ronne and Filchner Ice Shelves where the anomaly values decrease to −40 mgal at bottom depths of −1200 m or more. A wide area with minimum values as low as −100 mgal follows the edge of the Ross Ice Shelf along the Transantarctic Mountains.

A chain of positive anomalies as great as +130 mgal extends along the Antarctic Peninsula and passes into an island arc in the Scotia Sea. On the continent, this chain of high values reaches into the Pensacola Mountains, with surface altitudes of 2000 m and ice thicknesses of 1500–2000 m. In general, the anomalies follow the major land-

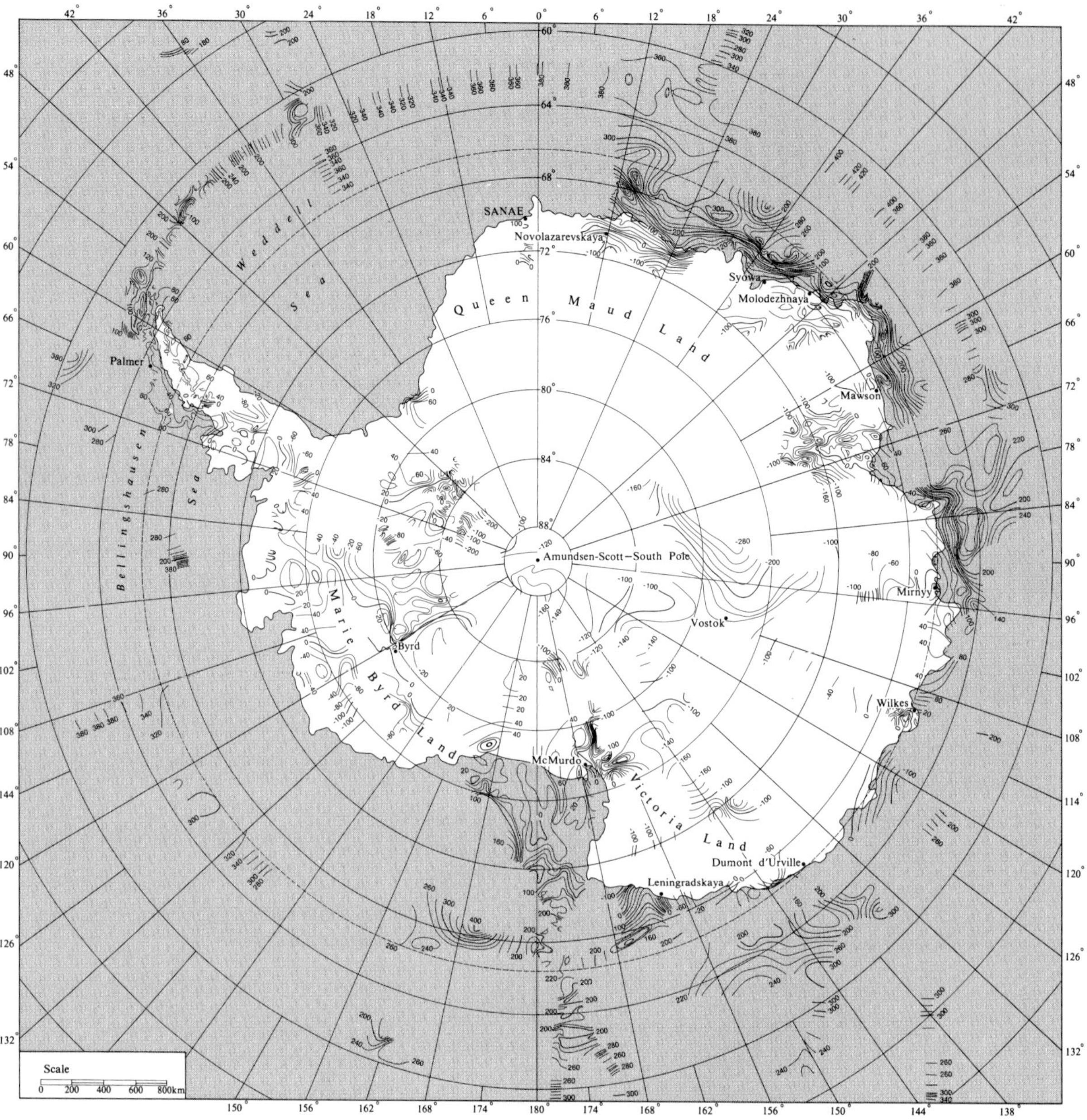

Figure 115.2. Gravity field of Antarctica showing contours of Bouguer anomalies. Cartographic Lab, University of Wisconsin-Madison.

Table 115.2. Free-air anomaly extremes and averages (in mgal)

	West Antarctica		East Antarctica	
	Actual	Averaged	Actual	Averaged
Amplitude	341	236	305	211
Maximum value	220	190	175	130
Minimum value	– 121	– 46	– 130	– 81
Average	50	72	22	25

Table 115.3. Bouguer anomaly extremes and averages (in mgal)

	West Antarctica		East Antarctica	
	Actual	Averaged	Actual	Averaged
Amplitude	370	283	418	211
Maximum value	130	90	109	80
Minimum value	– 240	– 193	– 309	– 280
Average	– 48	– 51	– 100	– 100

scape features, as was anticipated. Yet their range is not great—from about – 90 mgal in the eastern plain area up to + 170 mgal in the Prince Charles Mountains and Kemp Coast in East Antarctica, and from – 100 to – 140 mgal at the border between West and East Antarctica up to + 130 mgal on the Antarctic Peninsula. This limited range of values suggests that conditions in this area of the earth are close to isostatic equilibrium.

References cited

Groushinsky, N. P., and Sazhina, N. B. 1977. O gravimetricheskoi zuchennosti Antarctida i charaktere anomalii, sbornik [Gravimetric research in Antarctica and the nature of anomalies]. *Antarktika, Doklady Komissii, 16,* 130–34.

Groushinsky, N. P., and Sazhina, N. B. 1978. Gravity map of Antarctica, 1:5,000,000. Moscow, Sternberg Astronomical Institute and NII Zarubezhgeology.

Groushinsky, N. P., Koryakin, E. D., Stroev, P. A., Lazarev, G. E., Sidorov, D. V., and Virskaja, N. F. 1972. Katalog gravimetricheskikh punktov Antarctiki [The catalogue of the gravity stations in Antartica]. *Trudy Gosudarstvennogo Astronomickeskogo Instituta, 42,* 115–311.

Morgan, P. Catalogue of gravity stations in Antarctica 1957–64. *Inst. polar Stud., Ohio State Univ.* (Unpublished).

Talwani, M., and Meijer, R. R. *eds.* 1972. Preliminary report of USNS *Eltanin* Cruises, March 1967–March 1968. (*In* Lamont-Doherty Survey of the World Ocean. Gravity Measurements. *22,* 1–232.)

Tardi, P. *ed.* 1961. Rapport sur les travaux gravimetrique Antarctique, part 1, Bureau Gravimetrique International. *Annals IGY,* 1–108.

116

Seismicity in Antarctica

KATSUTADA KAMINUMA
National Institute of Polar Research, Itabashi, Tokyo, Japan

Abstract. No earthquakes have been located in the Antarctic area by the World Wide Seismological Station Network (WWSSN), with the exception of volcanic eruptions on Deception Island. However, the *Antarctic Seismological Bulletin* has been published on the basis of phase readings from Antarctic seismologic stations. Found in the bulletin were the events for which the phase readings of P-wave arrival times had been reported from more than four stations, but for which hypocenter determinations had not been made.

Seismograms at the Japanese Antarctic Station, Syowa, in Lützow-Holm Bay, East Antarctica, have indicated the occurrence of some local micro-earthquakes. This fact points toward the possibility of the occurrence of small and/or micro-earthquakes in other parts of Antarctica.

The present paper attempts to confirm the existence of earthquakes in Antarctica on the basis of data from the WWSSN and from occasional seismologic observations at some stations in Antarctica. The results can be summarized as follows: (1) Small earthquakes occur at least in the western part of Antarctica, though the activity is very low; these earthquakes are not, however, generally detectable by the insufficient seismologic network in Antarctica. (2) The seismicity around Syowa Station in East Antarctica is less than one micro-earthquake per month; however, earthquakes do occur in the marginal zone of the East Antarctic shield. (3) The seismicity around the dry valleys in Victoria Land is one micro- or ultramicro-earthquake every two days, and that around McMurdo Station on Ross Island is one earthquake per day. (4) The seismic activity in the modern volcanic region in the Antarctic is relatively higher than in the surrounding region. (5) The seismicity in the McMurdo Sound area is some ten times higher than that in East Antarctica.

The Antarctic continent and the surrounding ocean, which is called the Antarctic plate, although enclosed by marginal seismic belts, is one of the largest aseismic areas on the earth. More than ten seismologic stations have been operating in Antarctica since the International Geophysical Year (IGY) of 1957. Nevertheless, no earthquakes have been located in the Antarctic area by the World Wide Seismological Station Network (WWSSN), which can detect almost all earthquakes in the world with magnitude greater than five.

The geologic and geographic features of East Antarctica have remained as an essentially stable continental shield since before Devonian time. The orogenic belts on the Pacific Ocean side of the Transantarctic Mountains, which cross the Antarctic continent, are mainly composed of Mesozoic and Cenozoic rocks capped by Cenozoic volcanoes (Craddock, 1972). Two active volcanoes, Deception Island and Mount Erebus, are located near the tip of the Antarctic Peninsula in West Antarctica and on Ross Island in the Ross Sea, respectively.

With the exception of the volcanic eruptions of Deception Island, no significant earthquake has ever been located in Antarctica; however, the occurrence there of small-magnitude events has been recognized for some time. Since the IGY, the seismic activity of this region has been studied by many workers (e.g., Hatherton, 1961; Hatherton and Evison, 1962; Kogan and others, 1965; Lazareva and Sytinskiy, 1965; Browne-Cooper and others, 1967; Evison, 1967; Kaminuma, 1971 and 1976*b*). Some of them found strong evidence that many shocks recorded at the Antarctic stations originated from the calving of icebergs or cracking of the ice sheet. Although others have reported some shocks in the Antarctic region, these shocks were not confirmed clearly by other investigators. In any case, with the exception of the discoveries of Adams (1969, and Chapter 122, this volume), no earthquakes have been located in the Antarctic continent.

The present paper attempts to confirm the existence of earthquakes in Antarctica on the basis of data from WWSSN and from occasional seismologic observations at some stations in Antarctica.

Small earthquakes in the continent

The USNOAA (formerly USCGS) published the *Antarctic Seismological Bulletin* based on phase readings received telegraphically from Antarctic seismologic stations. Some of these readings of P-wave arrival times were not used by NOAA for hypocenter location, probably because no corresponding phase readings were available from other seismologic stations outside Antarctica. Those events for which the phase readings of P-wave arrival times were reported in the bulletin from more than four stations, but for which the hypocenter determination had not been made, were used by Kaminuma and Ishida (1971*b*) for trial location. The period from 1966 through 1969 was examined.

Only 24 events out of the 72 events which were studied could be located, as shown in Figure 116.1. The standard error of the calculation of the hypocenter locations is within 100 km of the epicenter coordinates, and O-C residual (the observed minus the computed travel time) is less

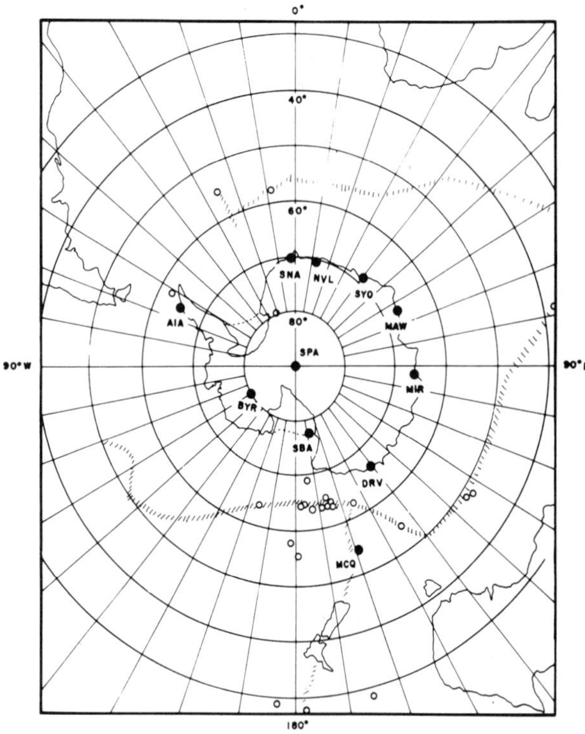

Figure 116.1. Earthquakes located in and around Antarctica for the period 1966 through 1969. Open circles indicate the epicenters. Solid circles indicate the seismologic stations. The hatched area indicates the seismic belt (Kaminuma and Ishida, 1971*b*).

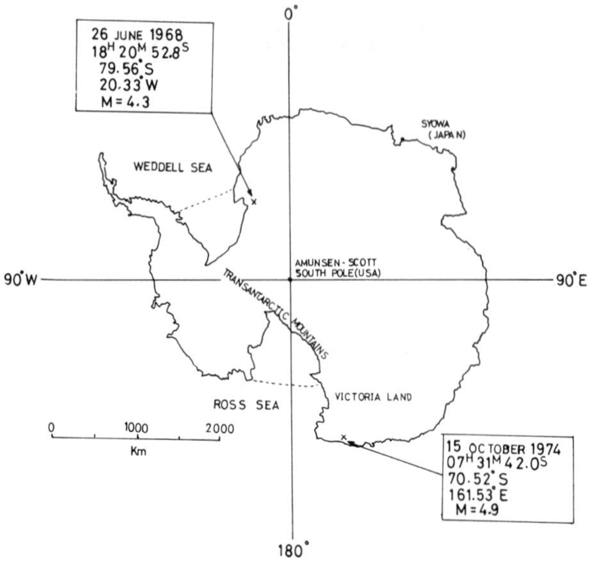

Figure 116.2. Two small earthquakes located in the Antarctic continent. One earthquake, which occurred on 26 June 1968, was determined by Kaminuma and Ishida (1971*b*). The other earthquake, which occurred on 15 October 1974, was determined by NOAA.

than 3 seconds. The epicenters of most of these events are located in or near the existing seismic belts that mark plate boundaries. The open circles in Figure 116.1 are the hypocenters that were located using P-wave arrival times from the seismologic stations in Antarctica, shown by the closed circles.

Only one event, which occurred on 26 June 1968, was located in the Antarctic continent, near lat. 80° S and long. 20° W. The stations used for the hypocenter locations were SANAE (SNA), South Pole (SPA), Byrd (BYR), one of the branch stations of the Stanford Network at Byrd (BY1), and Mawson (MAW). After examination of the location accuracy for this event, the hypocenter of the event is determined as follows:

Origin time	18h 20m 52.8s ± 1.1s
Latitude	79.56° S ± 15.11 km
Longitude	20.33° W ± 17.60 km
Depth	1 km (below sea level)
Magnitude	4.3

The altitude of the area where this event took place is between 1000 m and 1500 m, and the ice thickness is estimated to be less than 1000 m. Thus the earthquake occurred in the basement rock, as its depth was 1 km below sea level. Therefore, the event must have been a true earthquake, and this may be the first instrumentally recorded evidence of such an earthquake occurrence in Antarctica.

The epicenter is located along the northern edge of the Shackleton Range, which is east of the Filchner Ice Shelf at the head of the Weddell Sea. Geologic surveys of this range have been carried out by field parties of the Transantarctic Expedition and the British Antarctic Survey (Stephenson, 1966; Clarkson, 1972). It is pointed out by Clarkson (personal communication) that the event may be related to the major fault zone beneath Slessor Glacier which separates the Precambrian and lower Paleozoic rocks of the Shackleton Range from the upper Paleozoic sandstones and Jurassic intrusions of the Theron Mountains. The event may be reasonably considered as an earthquake from the geologic structure of the area where it occurred.

Another earthquake, which was located in the northern part of Victoria Land, occurred on 15 October 1974 (see Adams, Chapter 122, this volume). The location of the hypocenter was determined by NOAA using data from 12 stations. The results are as follows:

Origin time	07h 31m 42.0s ± 0.48s
Latitude	70.518° S ± 7.3 km
Longitude	161.538° E ± 10.8 km
Depth	33 km (below sea level)
Magnitude	4.9

The epicenters of both these earthquakes are shown in Figure 116.2. The two earthquakes occurred in the marginal zone of the Transantarctic Mountains, which is thought to be a tectonically active area. These two events show that minor earthquake activity does occur in Antarctica, though it is not generally detectable by the insufficient seismologic network presently available.

Earthquakes around Syowa Station

Seismologic observations at the Japanese Antarctic Station, Syowa, have been carried out since 1957 except for a period of four years, from 1962 to 1965 (Kaminuma, 1969). Syowa Station is located on Ongul Island, Lützow-Holm Bay, which is in the marginal zone of the East Antarctic shield. Sometimes small or micro-earthquakes are recorded by a three component Hagiwara's Electromagnetic Seismograph (HES) at Syowa Station. Many ice shocks are also recorded, but it is not difficult to distinguish between earthquakes and ice shocks because of their wave forms (Kaminuma, 1971; Kaminuma and Ishida,

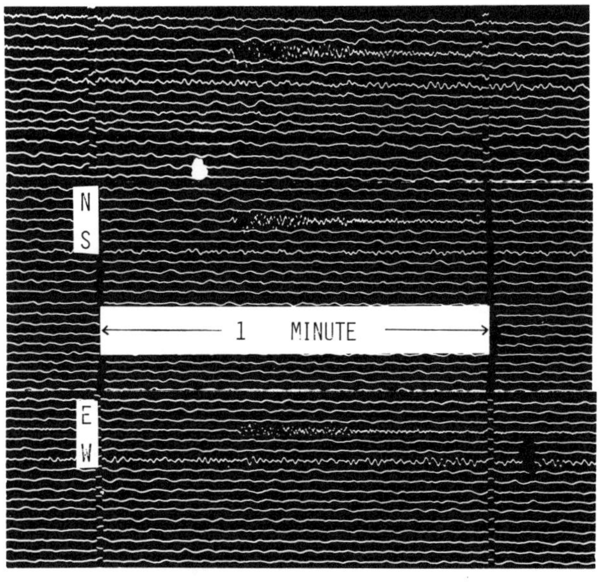

Figure 116.3. A typical seismogram of micro-earthquakes recorded by the HES seismograph at Syowa Station. This shock occurred on 29 January 1970.

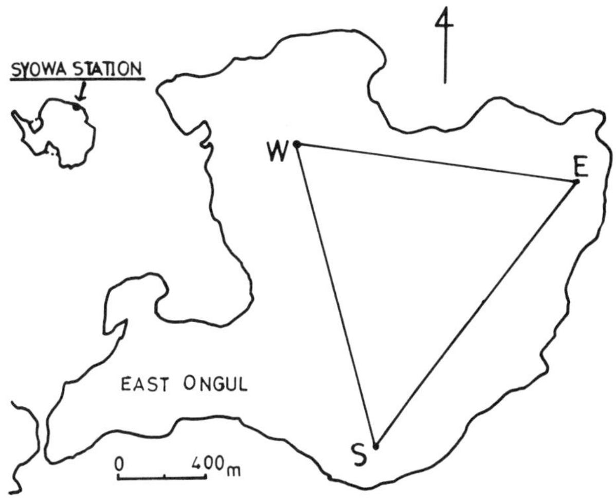

Figure 116.4. The tripartite network at Syowa Station installed on east Ongul Island, Antarctica, in 1973–74 and 1976–77.

1971*a*). A typical record of these small earthquakes is given in Figure 116.3.

For the purpose of determining the location of micro-earthquakes, seismologic observations using a tripartite network were made at Syowa Station in the winter seasons of 1973 (Kaminuma, 1976*b*) and 1976. Three seismographs with 1.0 second periods were installed on east Ongul Island. The location of the seismographs is shown in Figure 116.4. A magnetic tape recorder and three drums were used as the data recording system. The observations were carried out with a maximum magnification of 50,000.

Eight earthquakes were recorded during the 10 month period in 1973. Six out of the eight had S-P times of less than 10 seconds. The locations of two of the eight earthquakes were determined, but those of the other six earthquakes were not because of instrumental malfunction and/or much background noise.

One of the two earthquakes which were located had an epicenter about 20 km west-northwest of Syowa Station, and the other was 50 km to the west. The depth of both was about 10 km below mean sea level.

Two of the six earthquakes which were not located have the same wave form and nearly the same S-P times as the earthquake originating 20 km west-northwest of Syowa Station. Therefore, these two earthquakes were also estimated to be located about 20 km west of Syowa Station.

It is an interesting fact that the earthquakes occurred on the oceanic side of Syowa Station on Ongul Island, even though only a few hypocenters could be located precisely. The surface features of the island indicate that it was covered with glacial ice several thousand years ago and is now going through a rebounding movement associated with the retreat of the glacier. The micro-earthquakes which occurred around Syowa Station were probably caused by a release of strain due to this isostatic rebound.

Seven earthquakes were recorded by drum recorders from February 1976 to January 1977. S-P times of the seven earthquakes are less than 15 seconds; however, the locations of the earthquakes were not determined because the difference in P-wave arrival times between two observation points is less than 0.2 seconds and the paper speed of the drum recorder was 4 mm/sec. The frequency of S-P times is shown in Figure 116.5. The magnitudes of the earthquakes in Figure 116.5 were estimated to be between 0 and 1.

The seismicity around Syowa station is less than one earthquake per month. This fact shows that ultramicro- and/or micro-earthquakes do occur at least in the marginal zone of East Antarctica, but the activity is very low.

Seismicity around McMurdo Sound

For the purpose of observing micro-earthquakes around the McMurdo Sound area and volcanic earthquakes around Mount Erebus, seismologic observations were carried out in the summer seasons of 1974–75 and 1975–76, as one of the projects of the Dry Valley Drilling Project (DVDP) (Kaminuma, 1976*a,b*).

The McMurdo Sound area is located between the Antarctic continent and Ross Island, in the middle of the

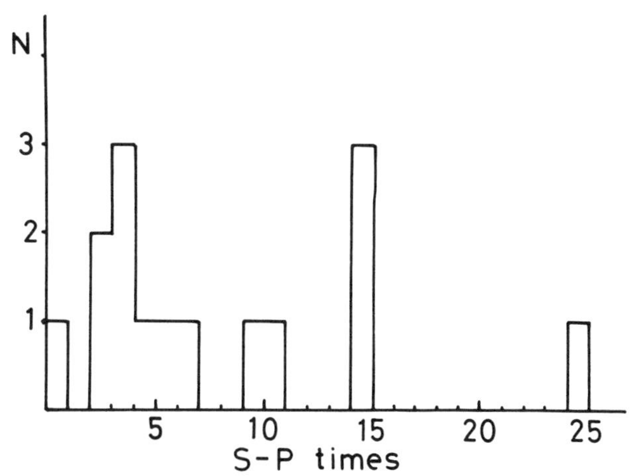

Figure 116.5. Frequency distribution of S-P times of earthquakes recorded by the tripartite network at Syowa Station in 1973–74 and 1976–77.

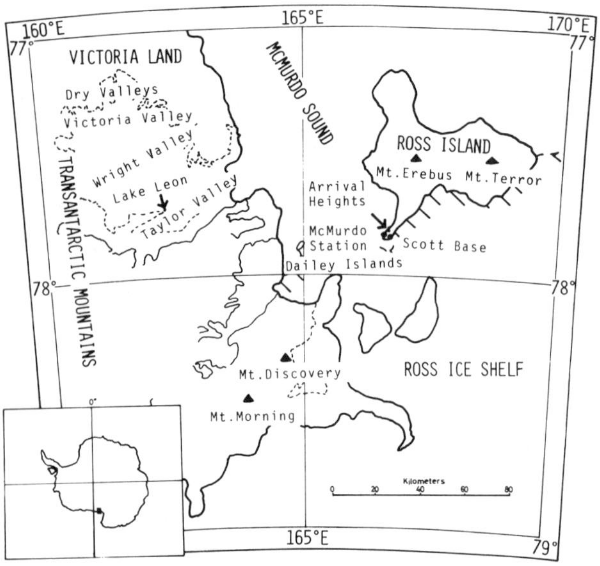

Figure 116.6. Map of McMurdo Sound area. Lake Leon is a small lake in the middle of Taylor Valley. Arrival Heights is located in the northeastern part of the McMurdo Station area, Ross Island. Arrows show positions of Lake Leon and Arrival Heights.

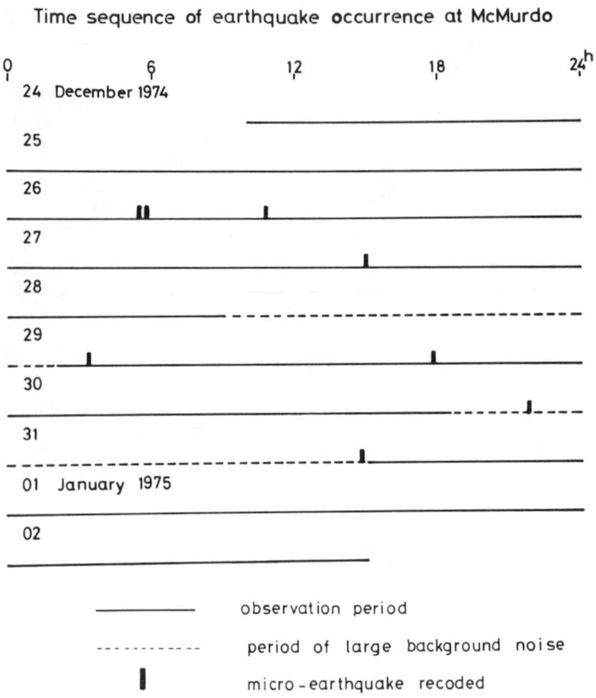

Figure 116.7. The time sequence of earthquake occurrence observed at Arrival Heights, McMurdo Sound, from 24 December 1974, to 2 January 1975. Solid columns show the occurrences of micro-earthquakes.

western part of the Ross Sea. A local map of McMurdo Sound is shown in Figure 116.6. Ross Island is of volcanic origin, with one active volcano, Mount Erebus. The Transantarctic Mountains trend from north to south on the west side of McMurdo Sound. On the eastern side of the Transantarctic Mountains are several snow-free valleys, called Dry Valleys.

The McMurdo Station area is not suitable for seismic observation because it is covered with thick layers of volcanic ejecta, mainly lapilli. A tripartite seismologic observation was made at Arrival Heights near McMurdo Station during three weeks in 1974–75. Arrival Heights consists of two small craters, First and Second Craters. A wide area for the tripartite observation is not available at Arrival Heights. Even so, it is the best site for making observations, considering the geologic and geographic limitations of any McMurdo Station site.

Three seismographs were installed, one near First Crater, another near Second Crater, and the third near the aeronomy observation hut of Scott Base on Arrival Heights.

A cassette data recorder with a magnification of 10^5 was used for the observation. The observation was disturbed frequently by high ground noise because of windy weather conditions and the high magnification of the recording system. Hence, data for only a total of one week are available. Hypocenter determination of earthquakes was not successful because the seismogram at the observation point near First Crater recorded many microtremors through all of the observation period.

The time sequence of earthquake occurrence is shown in Figure 116.7. Eight micro-earthquakes with S-P times of less than 10 seconds were identified during a nine day period. The real observation period was eight days because periods of much background noise activity were excluded from the record. The magnitudes of the earthquakes were around 0. Some ice shocks were also recorded by the observation network, but were easily recognized. No small or minor earthquakes were recorded. The seismicity around the McMurdo Station area is therefore considered to be about 1.0 micro-earthquake per day.

A seismologic observation at Lake Leon, in Taylor Valley, was carried out from 19 to 28 December 1974. A vertical seismograph and a cassette data recorder with three different magnifications (5000, 20,000, and 50,000) were used for the recording system. The observation point was near the drill hole of DVDP No. 12. Four micro-earthquakes with magnitudes between 0 and 1 were recorded during a period of 9.5 days. No minor or small earthquakes were recorded. The seismicity around the Taylor Valley is thus approximately one micro-earthquake every two days.

To confirm the results obtained in the summer season of 1974–75, seismologic observations were carried out again at McMurdo Station in 1975–76. Three seismographs were installed on the northern foot of Observation Hill.

To avoid artificial daytime noise, observations were carried out only at night. A pen recording system with a magnification of 120,000 was used. Thirteen micro-earthquakes with S-P times of less than 10 seconds were recorded between 15 November and 18 December 1975. The total time of the observation was 264 hours, or 11 days. Even though their amplitudes were very small, most of these earthquakes observed at McMurdo Station were also recognized on a seismogram at Scott Base, which is an Antarctic station of New Zealand and is located about 4 km east of McMurdo Station (Figure 116.6). The seismicity during this observation period was 0.9 micro-earthquakes per day. This seismicity is almost the same as that observed in the previous season. Therefore, we may conclude that the steady state of seismicity around McMurdo Station is about one micro-earthquake per day.

Table 116.1. Seismicities of volcanic and non-volcanic areas in Antarctica.

Station	Seismicity	
	Magnification of instrument	Number of earthquakes
McMurdo (volcanic) (74-75)	100,000	1.0 per day
(75-76)	120,000	0.9 per day
Lake Leon (non-volcanic)	50,000	0.5 per day
Syowa (non-volcanic)	50,000	7-10 per year

Discussion and conclusions

The seismicity around McMurdo Sound is some ten times that around Syowa Station. This fact may indicate that, generally, seismicity in the western part of the Antarctic continent is higher than that in East Antarctica (Table 116.1).

The seismicity around McMurdo Station is about one micro-earthquake per day and that in the Taylor Valley is 0.5 per day. The magnification of the instrument used at McMurdo Station is about two times greater than that used at Lake Leon. However, because of the higher ground noise around McMurdo Station, the observations at both sites are thought to be detecting earthquakes of the same scale. Micro- and/or ultra-micro-earthquake activity around the McMurdo Station area is therefore higher than that around the Dry Valleys. The ordinary earthquake activity around the McMurdo Sound area is estimated from these observations to be one micro-earthquake every two days. In addition to this ordinary earthquake activity, one micro-earthquake occurs every two days in the volcanic area around McMurdo Station.

Though small steam eruptions occurred repeatedly at the summit crater of Mount Erebus through the period of seismologic observation, no volcanic earthquakes corresponding to these eruptions were recorded. There is no close relationship between the high seismicity around McMurdo Station and the volcanic activity of Mount Erebus volcano, though the seismologic observation points at McMurdo Station are about 30 km from the active crater of Mount Erebus. The higher seismicity in the volcanic area is caused by the structural weakness of the upper crust which in turn causes the volcanic activity.

The main results are summarized as follows:

1. Small earthquakes occur at least in the Transantarctic Mountains, West Antarctica, but the activity is very low. However, the earthquakes are not generally detectable by the insufficient seismologic network in Antarctica.

2. The seismicity around Syowa Station in East Antarctica is less than one micro-earthquake per month. However, earthquakes do occur in the marginal zone of the East Antarctic shield.

3. The seismicity around the Dry Valleys in Victoria Land is one micro- or ultramicro-earthquake every two days, and that around McMurdo Station on Ross Island is one micro- and/or ultramicro-earthquake per day.

4. The seismic activity in the volcanic region in Antarctica is relatively higher than that in the surrounding region.

5. The seismicity in West Antarctica is some ten times higher than that in East Antarctica.

Small, micro- and ultramicro-earthquakes were detected in the Antarctic continent. The existence of such seismic activity, however, does not alter the fact that Antarctica is the most aseismic continent. Antarctica still has no large earthquakes and has the lowest seismicity of any continent. Why no large earthquakes occur in Antarctica is still one of the very important problems of geophysics.

References cited

Adams, R. D. 1969. Small earthquakes in Victoria Land, Antarctica. *Nature, Lond., 224,* 255-56.

Browne-Cooper, P. J., Small, G. R., and Whitworth, R. 1967. Probable local seismicity at Wilks, Antarctica. *N.Z. Jl Geol. Geophys., 10,* 443-45.

Clarkson, P. D. 1972. Geology of the Shackleton Range: a preliminary report. *Bull. Br. antarct. Surv.,* No. 31, 1-15.

Craddock, C. 1972. Antarctic tectonics. (*In* Adie, R. J., *ed. Antarctic Geology and Geophysics.* Oslo, Universitetsforlaget, 449-55.)

Evison, F. F. 1967. Note on the aseismicity of Antarctica. *N.Z. Jl Geol. Geophys., 10,* No. 2, 479-83.

Hatherton, T. 1961. A note on the seismicity of the Ross Sea region. *Geophys. J.R. astr. Soc., 5,* 252-53.

Hatherton, T., and Evison, F. F. 1962. A special mechanism for some Antarctic earthquakes. *N.Z. Jl Geol. Geophys., 5,* 864-73.

Kaminuma, K. 1969. The seismological observation and earthquake detection capability of Syowa Station, Antarctica. *Bull. Earthq. Res. Inst. Tokyo Univ., 47,* 453-66.

Kaminuma, K. 1971. Syowa-kichi de kansoku shita Bisho-zishin [Micro-earthquakes observed at Syowa Station, Antarctica]. *Antarctic Rec., 40,* 65-73.

Kaminuma, K. 1976*a*. Seismic activity around McMurdo Sound in Antarctica. *Antarctic Rec., 56,* 22-30.

Kaminuma, K. 1976*b*. Seismicity in Antarctica. *J. Phys. Earth, 24,* 381-95.

Kaminuma, K., and Ishida, M. 1971*a*. Earthquake interpretations at Syowa Station, Antarctica. *JARE Data Rep.,* No. 12, 76 pp.

Kaminuma, K., and Ishida, M. 1971*b*. Earthquake activity in Antarctica. *Antarctic Rec., 42,* 53-60.

Kogan, S. D., Pasechik, I. P., and Sultanov, D. D. 1965. Seismicheskaia karta Antarktiki [A seismic map of Antarctica]. *Izv. Akad. Nauk SSSR,* Ser. geophys., No. 2, 81-83.

Lazareva, A. P., and Sytinskiy, A. D. 1965. Seismicheskie nabliudeniia v Mirnom v 1961g. [Seismic observations at Mirny 1961]. *Inf. Bull. Sov. Antarct. Exped., 5,* No. 52, 355-60.

Stephenson, P. J. 1966. Geology. 1. Theron Mountains, Shackleton Range, and Whichaway Nunataks (with a section on palaeomagnetism of the dolerite intrusions, by D. J. Blundell). *Scient. Rep. transantarct. Exped.,* No. 88, 79 pp.

117

Modes of Gravity Anomaly Distributions in Relation to Crustal Structure in the Antarctic Continent

KATSUTADA KAMINUMA
National Institute of Polar Research, Itabashi, Tokyo, Japan

MEGUMI MIZOUE
Earthquake Research Institute, The University of Tokyo, Bunkyo, Tokyo, Japan

Abstract. An analytical method is introduced for applying gravity data to the study of crustal structure by using a reduced gravity anomaly where the thickness of the ice sheet is unknown. The reduced gravity anomaly, designated Δg_h (mgal), is defined by the equation $\Delta g_h = \Delta g_0 - 2\pi G\varrho(H - h)$, where Δg_0 (mgal) is the free-air anomaly, H (m) is the elevation of the measurement point, and h (m) is the elevation of the point for which the reduced gravity anomaly Δg_h is calculated. The density of the ice sheet and the universal constant of gravitation are designated as $\varrho(\mathrm{gr/cm^3})$ and G, respectively. For practical purposes, the value of h should be taken as constant and greater than or equal to the maximum value of H. The space bounded by the two levels, h and H, is assumed to be filled with an ice sheet of density ϱ in the reduction procedure. The reduced gravity anomaly Δg_h thus calculated is physically equivalent to the free-air anomaly corrected to a point on a surface above the ice sheet at a constant elevation h throughout the region.

Predominant modes of the spatial distribution of reduced gravity anomalies Δg_{4000} are found to occur in two wavelength ranges: a shorter wavelength range of 100–200 km and a longer wavelength range of more than 1000 km. The mode of the shorter wavelength range can possibly be correlated with topographic features of the rock and ice-sheet interface. On the other hand, the mode of the longer wavelength range is characterized by a decrease of Δg_{4000} toward the inland part of the continent. The results of the gravity data analysis lead to the conclusion that the crust of the Antarctic continent is in regional isostatic equilibrium similar to other continental regions on the earth. The thickness of the crust in West Antarctica, however, is less than that in East Antarctica.

Scientific surveys in Antarctica have been carried out extensively since the International Geophysical Year (IGY), which began in 1957. As one of the main subjects, studies of the crustal structure in the continent (Evison and others, 1960; Woollard, 1962) have been made by various geophysical means, overcoming great difficulties caused by the unavoidable, severe field conditions in the continent.

Structural features of the ice sheet and upper crust have been gradually clarified by seismic explosion studies in some limited parts of the continent, despite the fact that most of the seismic explosion experiments performed in Antarctica have been comparatively small in explosive power and were consequently limited in the length of survey lines (Bentley, 1964, 1965). One exception is the deep seismic sounding carried out by Russian scientists in the vicinity of Novolazarevskaya Station (Kogan, 1972).

Soon after the installation of long-period seismographs in Antarctica in the early stages of the IGY, preliminary estimates of the thickness of the crust in East and West Antarctica were given by Evison and others (1960) and Kovach and Press (1961), using the group velocity of surface waves. The crustal thickness was estimated at 35–40 km beneath East Antarctica, 30–35 km beneath West Antarctica, and 5–10 km in the oceanic regions surrounding the continent. No other reliable estimates from surface-wave group and/or phase velocities have ever been provided, mainly due to the insufficient number of long-period seismographs operated in Antarctica.

The gravimetric survey of Antarctica was started in the early stages of the many expeditions during the 20th century. A remarkable increase in the number of gravity stations in the inland part of the Antarctic continent has been achieved since the International Geophysical Year in 1957–58 due to cooperative efforts of traverse parties from various nations.

Gravity data thus accumulated up to the present are available for the study of regional modes of gravity anomaly distribution with special reference to crustal structure. However, neither the free-air nor the Bouguer reduction can provide gravity anomalies which permit direct interpretation in terms of underground structures in such a region as Antarctica where the thickness of the ice sheet is not known with sufficient accuracy.

A simple gravity reduction system is proposed in order to obtain standardized gravity anomalies which take into account existence of a thick ice-sheet layer and the undulatory ice sheet-bedrock interface. The gravity reduction introduced consists of the following two procedures: to reduce gravity values to a surface at the level greater than or equal to the maximum height of the ice-sheet surface, and to add the gravity effect produced by filling up the space between the two surfaces with material having ice-sheet density.

Analytical method

The area and the mean elevation of the Antarctic continent are 12.5 × 10^6 km^2 and 2300 m, respectively. An ice sheet with a mean thickness of 1880 m (Bentley, 1964) covers the continent, only 2% of which is found to be free of snow or ice. Information on the thickness of the ice sheet is essential for the reduction of gravity data and for its use in the study of the internal structure of the earth.

When gravity data are used for the study of crustal structure, however, a great difficulty arises from the fact that detailed knowledge of the thickness of the ice sheet has only rarely been obtained under the gravimetric observation points. To avoid this troublesome situation for gravity data analysis, a reduced gravity anomaly is introduced in this report.

The reduced gravity anomaly designated as Δg_h (mgal) is defined by the following equation:

$$\begin{aligned}\Delta g_h &= \{g_{ob} - c(h-H) + 2\pi G\varrho(h-H)\} - \{\gamma - ch\}\\ &= g_{ob} - (\gamma - ch) + 2\pi G\varrho(h-H)\\ &= \Delta g_0 - 2\pi G\varrho(H-h)\\ &= \Delta g_0 - 0.419(H-h)\end{aligned}$$

where g_{ob} (mgal) = the gravity measured at the point of the elevation H
γ (mgal) = the normal gravity
c (mgal/m) = the constant for free-air reduction
Δg_0 (mgal) = the free air anomaly
H (mg) = the elevation of the measurement point
h (m) = the elevation of the point for which the reduced gravity anomaly Δg_h is being calculated
ϱ (gr/cm^3) = the density of the ice sheet
G (cm^3/g•sec^2) = the universal constant of gravitation.

The relation of h and H in the above equation is shown in Figure 117.1. The value of h should be taken as constant, with a value greater than or equal to the maximum value of H, for practical purposes. The space bounded by the two levels of h and H is assumed to be filled with an ice sheet of density ϱ in the reduction procedure. The reduced gravity anomaly Δg_h thus introduced is physically equivalent to the free-air anomaly reduced, or adjusted, to a point on a surface above the ice sheet at a constant elevation h.

Reduced gravity anomaly along Syowa–South Pole traverse route

The reduced gravity anomaly has been calculated from gravity data obtained by the Japanese party which traversed from Syowa Station to the South Pole in 1968–69. The measurements were carried out with a La Coste–Romberg gravimeter at observation points every 4 km along the traverse route (Yanai and Kakinuma, 1971). More than 600 gravimetric observation points were established along

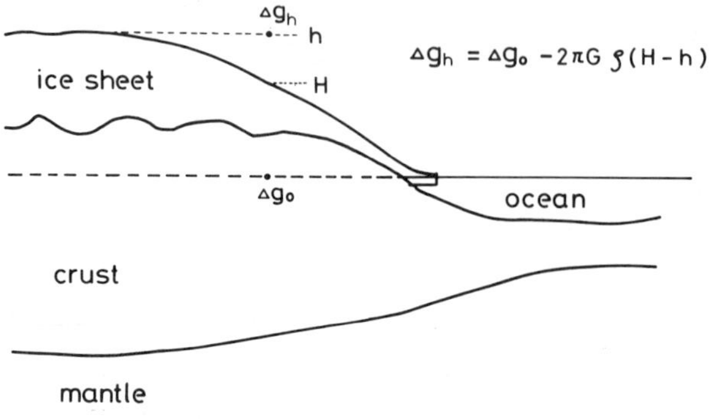

Figure 117.1. The concept of reduced gravity anomaly Δg_h is illustrated. Δg_0 is the free-air anomaly calculated from a measurement at the point of elevation H.

the traverse route, which is shown by the dashed line in Figure 117.2. The position (latitude and longitude), elevation, and gravity value, including free-air anomaly, were observed for each point. However, the thickness of the ice sheet is available for only a few of the points.

The elevation of the highest observation point along the traverse route is more than 3700 m. Therefore, 4000 m is taken as the value of h in the equation introduced in the previous section, and ϱ is taken as 0.9 gr/cm^3.

The reduced gravity anomalies along the survey routes are given in Figure 117.3. The maximum value of the reduced gravity anomalies Δg_{4000} is + 160 mgal at stations around lat. 70° S, and the minimum value is − 15 mgal at lat. 89° S.

Predominant modes of spatial distribution of the reduced gravity anomalies Δg_{4000} are found to occur two wavelength ranges, as shown in Figure 117.3: the shorter wavelength range of 100–200 km and the longer wavelength range of more than 1000 km. The mode of the shorter wave length range can possibly be correlated with topographic features of the Antarctic bedrock and ice-sheet interface. On the other hand, the mode of the longer wavelength is characterized by a decrease of Δg_{4000} values toward the inland part of the continent where the altitude of the ice-sheet surface is higher than in the marginal parts.

A remarkable decrease of Δg_{4000} values around lat. 89° S is interpreted as an indication of the topographic variation of basement rocks beneath the thick ice sheet.

The mean reduced gravity anomalies average over every 1° in latitude are shown with solid circles in Figure 117.4. The mean elevations corresponding to the gravity data are also given with open circles. The predominant mode of the longer wavelength range, clearly represented in Figure 117.4, is considered to reflect the regional variation of crustal thickness in the continent. In this connection, the lower values of the mean reduced gravity anomaly found in the central part of East Antarctica and the Precambrian shield can be correlated with an increase in thickness of the crust.

Regional variation in the crustal thickness of the Antarctic continent

Graviy data measured in and around Antarctica have been compiled by Groushinsky and others (1972). More than 6000 gravity stations have been established since the

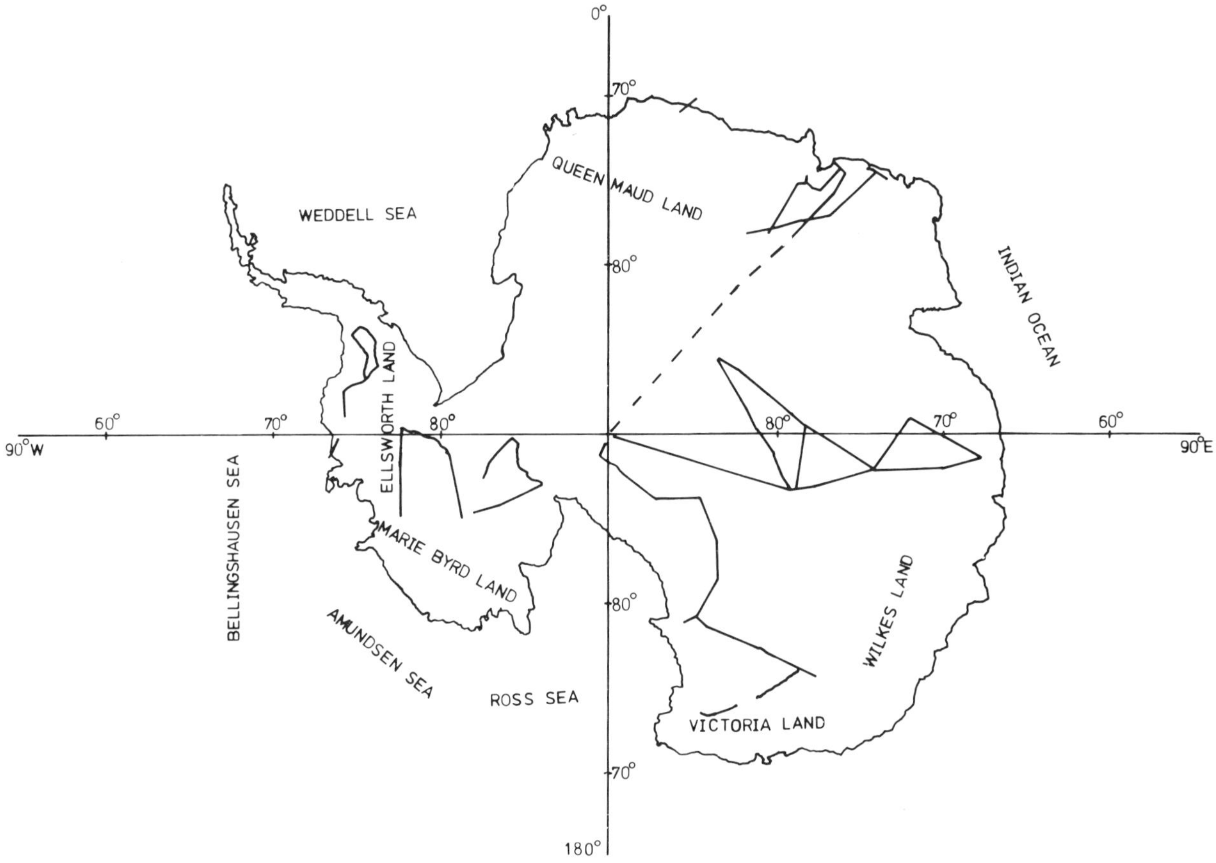

Figure 117.2. The gravimetric observation points in the inland parts of Antarctica are established along the traverses which are shown by solid and dashed lines. The dashed line is the traverse route of the Japanese party from Syowa Station to the South Pole in 1968–69.

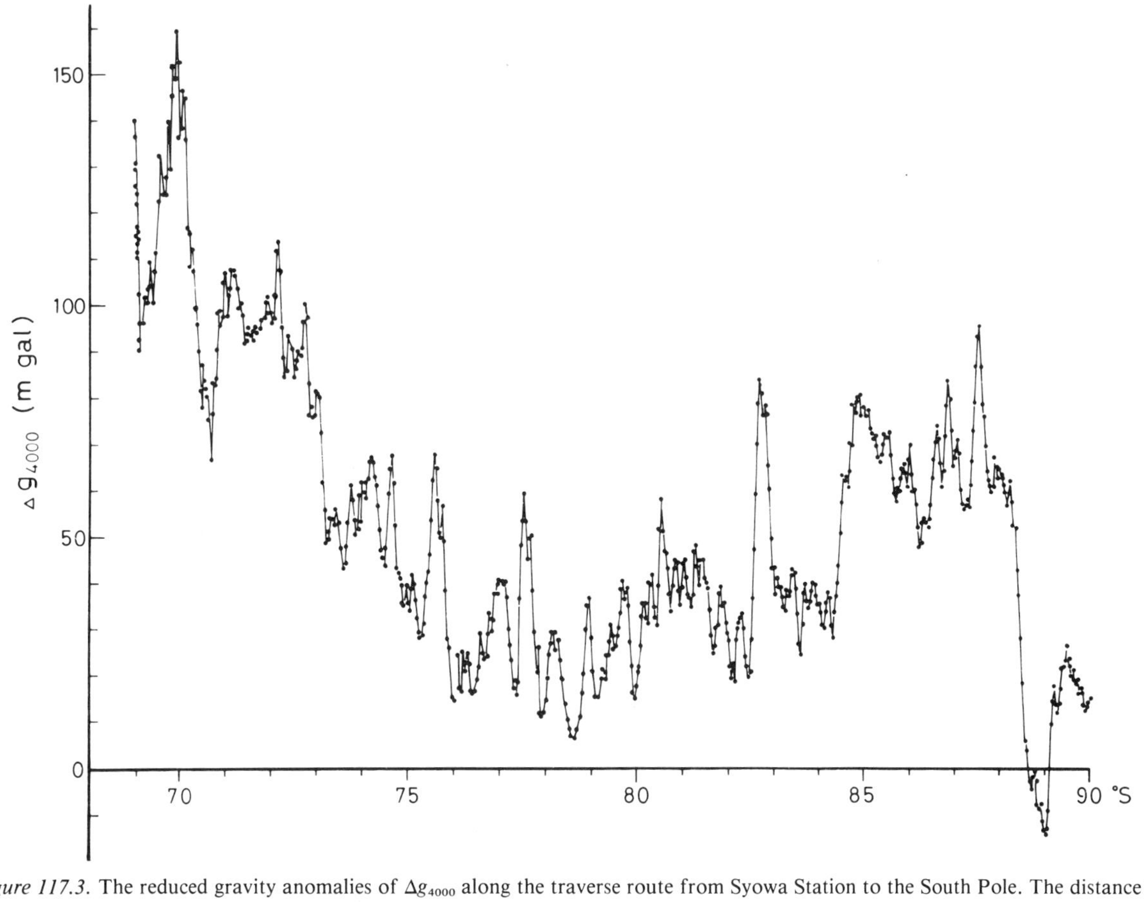

Figure 117.3. The reduced gravity anomalies of Δg_{4000} along the traverse route from Syowa Station to the South Pole. The distance between the Syowa Station and the South Pole is 2300 km.

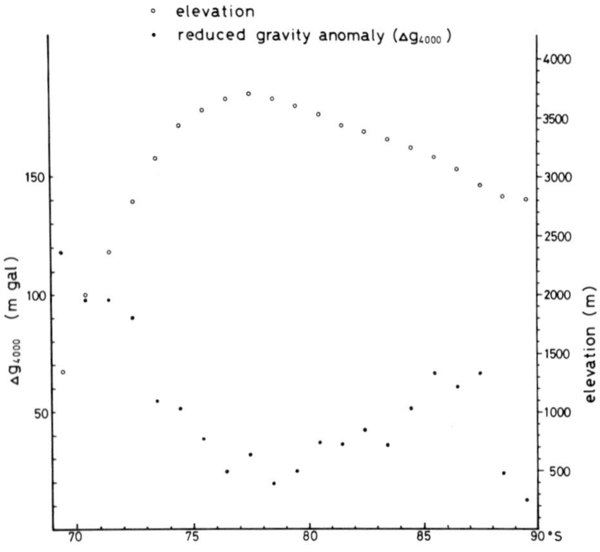

Figure 117.4. Ice-sheet elevations and the reduced gravity anomalies, averaged every 1° of latitude along the traverse route from Syowa Station to the South Pole.

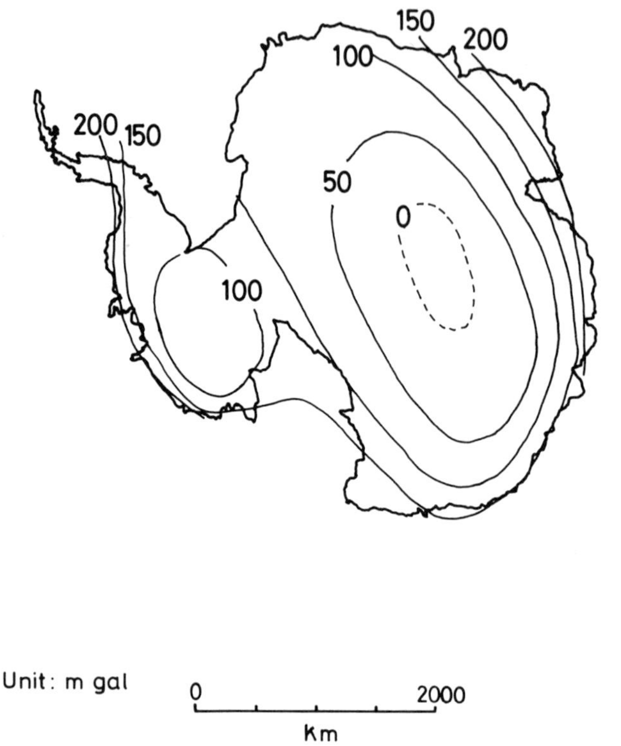

Figure 117.5. A contour map of the reduced gravity anomaly Δg_{4000}.

IGY. Most of these observations were carried out by the traverse parties of scientific expeditions to the inland area of Antarctica. The Japanese party participated through observations along the traverse route from Syowa Station to the South Pole, as described in the previous section. The traverse routes for which a gravity survey was accomplished are shown in Figure 117.2.

Reduced gravity anomalies have been calculated using the data presented in various publications (Yanai and Kakinuma, 1971; Groushinsky and others, 1972). Results are shown in Figure 117.5.

Most of the Δg_{4000} values in West Antarctica are larger than 100 mgal, as shown in Figure 117.5. However, some of Δg_{4000} values in East Antarctica show negative values, and those in the inland area are less than 30 mgal. It is of considerable interest that the values of Δg_{4000} in West Antarctica are, in general, larger than those in East Antarctica; it is highly probable that the crustal structure in East Antarctica is different from that in West Antarctica, which is in good agreement with geographic and geologic data (Craddock, 1972). The thickness of the crust has been estimated to be 35–40 km in East Antarctica but only 30–35 km in West Antarctica, from the surface wave studies (Evison and others, 1960).

The reduced gravity anomalies in West Antarctica and the marginal area of East Antarctica are relatively large compared to those of the inland part of East Antractica. This mode of gravity anomaly distributions is one of the geophysical characteristics of East Antarctica, an ancient and tectonically stable Precambrian shield.

The results of the gravity data analysis lead to the conclusion that the crust of the Antarctic continent is possibly in regional isostatic equilibrium, similar to other continental regions. The thickness of the crust in West Antarctica, however, is significantly less than that in East Antarctica—by about 3 km. This difference corresponds to a gravity anomaly difference as large as 80 mgal when a mean density contrast of $\Delta\varrho = 0.6$ gr/cm^3 between the crust and upper mantle is assumed.

Conclusions

An analytical method is introduced for applying gravity data to the study of crustal structure of Antarctica by using a reduced gravity anomaly where the thickness of the ice sheet is unknown. The reduced gravity anomalies Δg_h are calculated here from the available gravity observations in Antarctica, taking h as 4000 m, defined as the elevation of the point to which the free-air anomaly Δg_0 should be reduced. Modes of the distribution of reduced gravity anomalies Δg_{4000} have two predominant wavelength ranges: a shorter wavelength range of 100–200 km and a longer wavelength range of more than 1000 km. The first mode seems to be correlated with topographic features of the Antarctic bedrock and ice-sheet interface, and the second seems to reflect the regional variation of the thickness of the crust beneath the continent.

A remarkable decrease in Δg_{4000} values toward the inland part of East Antarctica indicates that the regional isostatic equilibrium condition is satisfied in the continent.

References cited

Bentley, C. R. 1964. The structure of Antarctica and its ice cover. (*In* Odishaw, H., *ed. Research in Geophysics. 2. Solid Earth and Interface Phenomena.* Cambridge, Mass. The MIT Press, 335–89.)

Bentley, C. R. 1965. The land beneath the ice. (*In* Hatherton, T., *ed. Antarctica.* London, Methnen, 259–77.)

Craddock, C. 1972. Antarctic tectonics. (*In* Adie, R. J., *ed. Antarctic Geology and Geophysics. Oslo, Universitetsforlaget,* 449–55.)

Evison, F. F., Ingham, C. E., Orr, R. H., and LeFort, J. H. 1960. Thickness of earth's crust in Antarctica and the surrounding oceans. *Geophys. J. R. astr. Soc., 3,* 289–306.

Groushinsky, N. P., Koryakin, E. D., Stroev, P. A., Lazarev, G. E., Sidorov, D. V., and Virskaja, N. F. 1972. Katalog gravimetricheskikh punktov Antarktiki [The catalogue of the gravity stations in Antarctica]. *Trudy Gosudarstvennogo Astronomickeskogo Instituta, 42,* 115–313.

Kogan, A. L. 1972. Results of deep seismic sounding of the earth's crust in East Antarctica. (*In* Adie, R. J., *ed. Antarctic Geology and Geophysics.* Oslo, Universitetsforlaget, 485–89.)

Kovach, R. L., and Press, F. 1961. Surface wave dispersion and crustal structure in Antarctica and the surrounding oceans. *Annali. Geofis., 14,* 211–24.

Woollard, G. P. 1962. Crustal structure in Antarctica. (*In* Wexler, H., Rubin, M. J., and Caskey, J. E., *eds. Antarctic Research: The Matthew Fontaine Maury Memorial Symposium.* Washington, American Geophysical Union, 53–73.) [Geophysics Monograph No. 7.]

Yanai, K., and Kakinuma, S. 1971. Measurement of gravity along the traverse route Syowa-South Pole. (*In* Murayama, M., *ed. Report of the Japanese Traverse Syowa-South Pole.* Tokyo, National Science Museum, 131–50.)

118

Crustal Structure of the Lambert Glacier Area from Geophysical Data

L. V. FEDOROV, G. E. GRIKUROV, R. G. KURININ, and V. N. MASOLOV
Department of Antarctic Geology, SEVMORGEO, Moyka 120, Leningrad 190121, U.S.S.R.

Abstract. The Lambert Glacier graben extends southward from the coast for almost 700 km, and the amplitude of bedrock relief across the graben is as high as 3000–3500 m. Magnetic data, however, reveal bedrock depressions 5 km or more in depth, filled with nonmagnetic rocks. The rift zone is sharply discordant relative to structural grain in the older rocks, mainly lower Precambrian complexes. Deep seismic-sounding results indicate deep faults bounding the rift graben. Beneath the graben, crustal thickness is as low as 22–24 km, while on both sides of the graben, the crust is 30–34 km thick. The beginning of rifting is provisionally dated as late Mesozoic, and it was probably associated with extensive crustal upwarping. Available evidence indicates a decrease in age of tectonic activity in a westward direction. The western side of the rift seems more heterogeneous than the eastern side, which may indicate that the rift zone inherited some ancient major lineament in the crust, perhaps as old as late Archean.

The Lambert Glacier is the largest outlet ice stream of the continent. Its unique geomorphic features have long attracted the interest of Antarctic geologists as the only place along the coast of East Antarctica where the mountain ridges pierce the ice sheet for almost 700 km southward, thus revealing the inland geologic structure. Until recently, geologic information from this region was confined to results from a few reconnaissance visits by the Australian and Soviet geologists (Crohn, 1959; McLeod, 1964; Trail, 1963, 1964; Soloviev and others, 1967; Soloviev, 1972). A systematic exploration of the Lambert Glacier area by the Soviet Antarctic Expedition was carried out during 1971–74 when, under the leadership of D. S. Soloviev, a field base was established on the Amery Ice Shelf; it served during three successive austral seasons as a logistic support center for extensive geographic, topographic, geologic, and geophysical investigations (including deep seismic sounding). The simultaneous field work of the Australian National Antarctic Research Expedition has greatly contributed to geologic understanding of the territory (Tingey, 1972; Tingey and England, 1973; England and Langworthy, 1975; Wellman and Tingey, 1976).

This paper presents a synthesis and a structural interpretation of all available geologic and geophysical data. The crustal structure of the area is shown to be dominated by the Lambert Glacier rift zone.

Geomorphology

The graben-like valley of the Lambert Glacier is the most prominent landform of the region (Figure 118.1). It stretches for more than 650 km from north to south, narrowing southward from 100 km to only 20 km. The bottom of the subglacial valley is locally more than 1500 m below sea level. Radio-echo sounding shows that the valley continues south of lat. 75° S (Boyarskii and Shalygin, 1976).

The Lambert graben is bounded on the west by the Prince Charles Mountains, with peaks from 1500 to 2000 m above sea level; the southern part of the graben is bounded on the east by the Mawson Escarpment, which is more than 100 km in length with absolute elevations of about 1500 m and relative elevations above the glacier surface of 100–800 m. The eastern boundary of the Lambert Glacier in its northern part, and of the Amery Ice Shelf, is marked by relatively low (less than 200 m) nunataks and cliffs extending longitudinally to the northeast between lat. 70° and 71° S.

The northern Prince Charles Mountains are dissected by narrow (10–20 km), rather deep (bedrock surface locally below sea level) sublatitudinal grabens, the largest of which is occupied by the Charybdis Glacier. The Fisher Glacier and some others represent similar structures in the southern Prince Charles Mountains.

The amplitude of bedrock relief alongside the Lambert Glacier is thus about 1500–2000 m; directly within the glacier valley, however, it reaches 3000–3500 m. Mount Menzies is the most striking landform (more than 3000 m) of distincltly alpinotype appearance. The strongly dissected Grove Mountains rise above 2000 m in the southeastern part of the region.

A large depression occurs between the northern and southern parts of the Prince Charles Mountains approximately along lat. 72° S (the West depression, which is 600 m or more below sea level). East of the Mawson Escarpment (approximately along long. 72° E) is the longitudinal Central depression with depth averaging 300 m. Bedrock topography within these low-lying areas is generally subdued.

Geophysical evidence

Aeromagnetic data

In the interpretation of magnetic data, the upper surface of the magnetic source was considered to be the upper surface of the consolidated crust, which includes the Archean crystalline basement and the Proterozoic to lower Paleozoic folded supracrustal complex (Figure 118.2). The bedrock surface of positive landforms generally coincides with the level of the upper edges of magnetic bodies, but in grabens and depressions there is, as a rule, a considerable

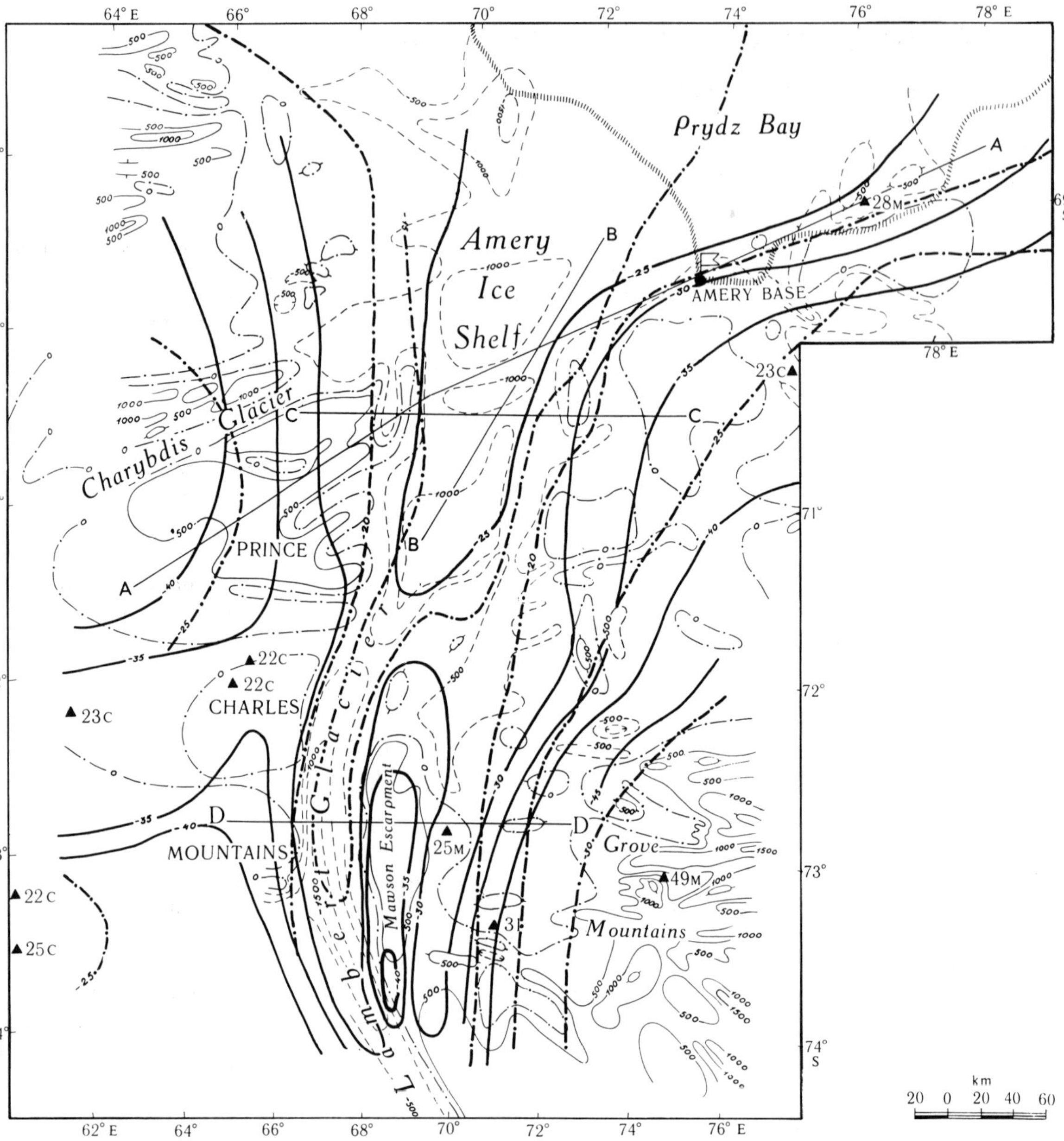

Figure 118.1. Relief of the Mohorovičić discontinuity and the Conrad interface in relationship to the subglacial topography. Hachured line: ice front. Solid contours: subglacial relief above sea level (500 m intervals). Dot-dash contours: subglacial relief at sea level. Dashed contours: subglacial relief below sea level. Solid heavy lines: depth to the Moho (5 km intervals). Heavy dot-dash lines: depth to the Conrad interface (5 km intervals). "C" numbers: depth to the seismologic boundary (from Adams, 1971) identified in the present paper as the Conrad interface. "M" numbers: depth to the seismologic boundary (from Adams, 1971) identified in the present paper as the Moho. Profile A-A: major line of deep seismic sounding shots. Profile B-B: supplementary deep seismic sounding line. Profile C-C: line of crustal section shown in Figure 118.3. Profile D-D: line of crustal section shown in Figure 118.4.

difference between the position of the upper surface of consolidated crust and that of the subglacial bedrock surface because of the presence of sedimentary graben-fill deposits.

The basement is deepest (more than 5 km) under the northern Lambert Glacier and beneath Prydz Bay. Deep but smaller depressions also occur west and east of the Mawson Escarpment and on the southern extension of the Beaver Lake graben. A considerable subsidence of the consolidated crust is also recorded within the West depression. Shallower basement depressions (about 3 km) occur within the Charybdis Glacier graben and also west and southeast of the Amery Ice Shelf. The differences in the hypsometric level of the bedrock topography and the inferred upper surface of the magnetic consolidated crust suggest that these structurally depressed areas are filled with nonmagnetic material.

Some large negative areas in the subglacial landscape are not reflected in corresponding lows of the magnetic basement, although gravity measurements indicate that these areas are also underlain by a low-density material. This may be explained by the presence of young magnetic (igneous?) rocks within the low-density sedimentary fill of depressions in the consolidated crust.

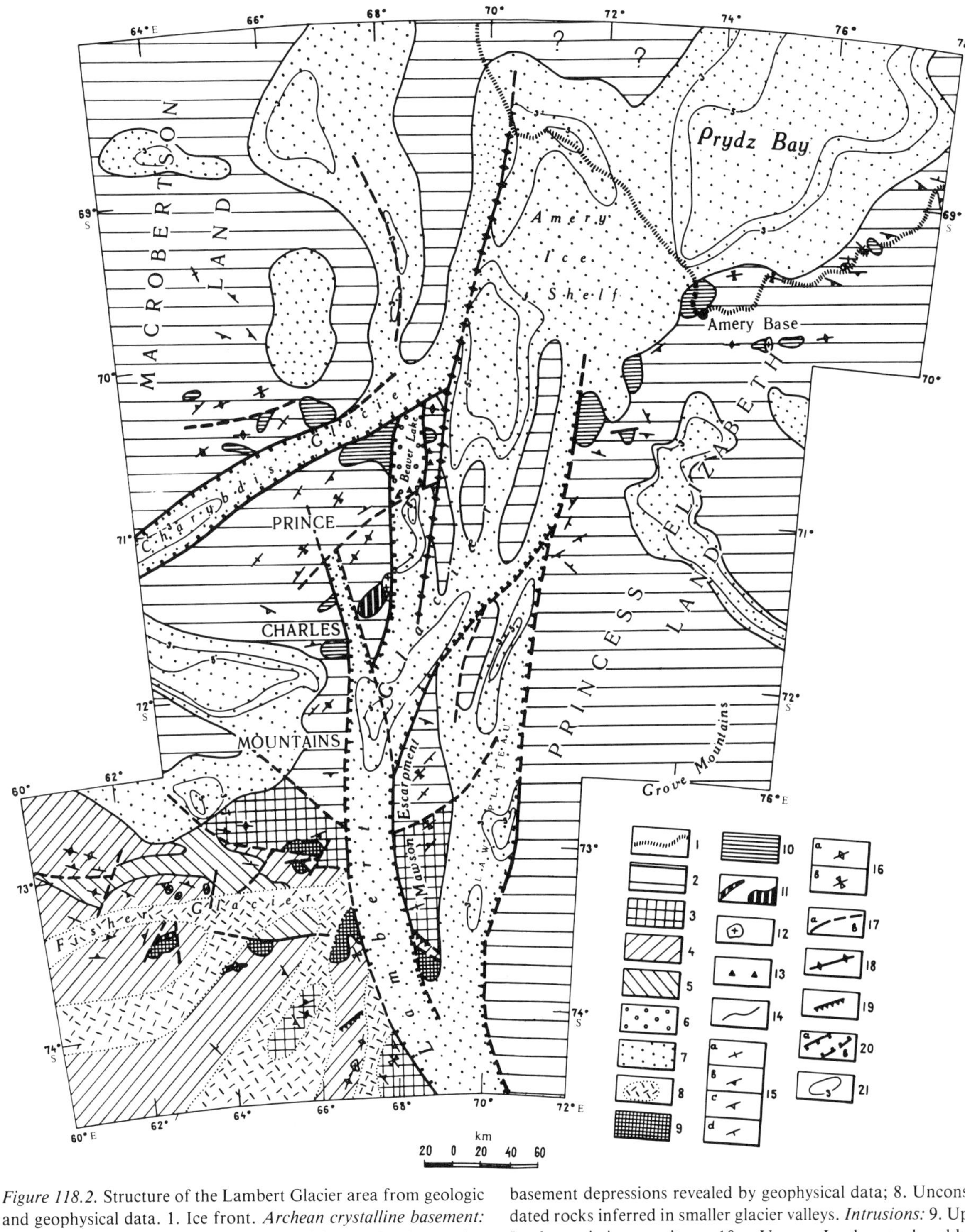

Figure 118.2. Structure of the Lambert Glacier area from geologic and geophysical data. 1. Ice front. *Archean crystalline basement:* 2. Polymetamorphic rocks of the Humboldt complex (upper Archean I), possibly with some Rayner complex granulite facies rocks (lower Archean); 3. High-grade amphibolite facies rocks of the Insel complex (upper Archean II). *Proterozoic to lower Paleozoic supracrustal formations in the intracratonic fold system of the southern Prince Charles Mountains:* 4. Greenschist and low-grade amphibolite facies sequences of the Ruker complex (lower middle Proterozoic); 5. Low-grade greenschist facies sequences of the Ross complex (upper Proterozoic to lower Paleozoic). *Upper Paleozoic to Recent undeformed formations:* 6. Permian sedimentary rocks of the Beacon Supergroup; 7. Upper Mesozoic(?) to Cenozoic molasse and/or riftogenic complex in major crystalline basement depressions revealed by geophysical data; 8. Unconsolidated rocks inferred in smaller glacier valleys. *Intrusions:* 9. Upper Insel gneissic granites; 10. Upper Insel porphyroblastic granosyenites (charnockitoids); 11. Upper Ruker metabasites; 12. Ross mica granites; 13. Lower Andean alkaline-ultramafic rocks. *Miscellaneous:* 14. Boundaries of structural complexes and individual formations; 15. Gneissosity, schistosity, banding, or bedding with dips (a) more than 80°, (b) 80–60°, (c) 60–30°, and (d) less than 30°; 16. Axes of folds (a) anticlinal and (b) synclinal; 17. Disjunctive dislocations (a) identified and (b) inferred; 18. Deep crustal fault (from geophysical evidence); 19. Thrust zones; 20. Faults bounding grabens (a) identified and (b) inferred; 21. Depth to magnetic basement (in km).

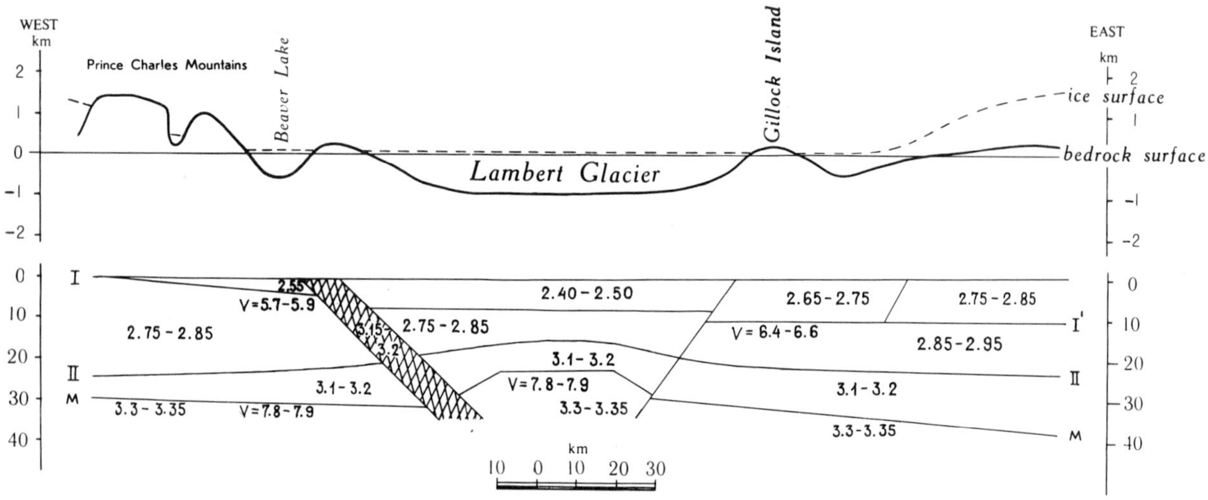

Figure 118.3. Crustal section across the principal structural features in the northern Lambert Glacier area inferred from geophysical data, especially deep seismic sounding. (For location see Figure 118.1, profile C-C). I: the upper surface of the crystalline basement. I′: the upper surface of heavier rocks within the "granitic layer." II: the Conrad interface. M: the Moho discontinuity. Grid pattern: zone of crustal weakness (deep fault?) with high-density rocks. Densities are given in g/cm³, and boundary velocities (V) in km/sec.

SEISMIC DATA

The detailed description of the deep seismic sounding (DSS) carried out during the 1972–73 season is given in another paper (Kurinin and Grikurov, Chapter 112, this volume) and in earlier publications (Kolmakov and others, 1975). The following description is based on a projection of the DSS profile, which was actually shot in a northeasterly direction, into a latitudinal crustal section directly across the northern Lambert Glacier graben (Figure 118.1, profile C-C, and Figure 118.3).

Deep fractures, marked in the bedrock topography as escarpments bounding the Lambert Glacier graben, are the major elements of the crustal structure. The most prominent is the western deep fault which, according to geophysical data, differs from the eastern one by the presence of high-density (σ = 3.2 g/cm³) plutonic rocks. In the Beaver Lake area such rocks have, in fact, been observed in bedrock exposures as sills, dikes, and stocks of alkaline-ultramafic rocks (Ravich and Fedorov, Chapter 60, this volume). The deep faults are responsible for offset of the main seismic boundaries.

The upper seismic boundary, correlated with the surface of Archean crystalline basement in the Lambert Glacier graben, lies 8–10 km below sea level; overlying it are the low-density rocks (σ = 2.4–2.5 g/cm³). Immediately west of the main western fault (beneath Beaver Lake) this boundary rises gently from −3 km to the surface, and rock exposures show the Permian sandstone (σ = 2.55 g/cm³) overlying it. Thus, the Lambert Glacier graben has relatively low bedrock surface relief (about 3000 m), but a considerably greater amplitude of vertical displacement (as much as 10 km) in the upper surface of ancient consolidated crust.

In the topography of the lower boundary of the crust, the Lambert graben is reflected in a step-like rise of the Moho discontinuity for 7–9 km. The upper mantle reveals anomalously low values of density and boundary velocity (3.3 g/cm³ and 7.8–7.9 km/sec, respectively) underneath the whole region. The total thickness of the crust beneath the Lambert graben is 22–24 km, but it increases to 33–35 km away from the graben. Hence, the zone under consideration is not only clearly expressed in the topography of the crystalline basement surface, but it also reveals a distinct antiform in the Mohorovičić surface.

The intracrustal boundaries are recorded west and east of the deep faults restricting the Lambert graben. The Conrad interface lies at a depth of 22–24 km. The density of a "basaltic layer" below this is 3.1–3.2 g/cm³. The geophysical data west of the Lambert Glacier imply the existence of a "granitic layer" within the upper crust, with a density of 2.75–2.85 g/cm³ and seismic velocities of 5.7–5.9 km/sec. To the east of Lambert Glacier, there is a layer above the Conrad interface that is about 13 km thick, with a higher density of 2.85–2.95 g/cm³; the boundary velocity along the top of that layer is 6.4–6.6 km/sec. A similar(?) intracrustal layer with higher density can also be distinguished immediately west of the Amery Ice Shelf.

Figure 118.4 shows a hypothetical crustal section along lat. 73° S (Figure 118.1, profile D-D) inferred from detailed interpretation of all geophysical data. The rugged topography of the upper and lower boundaries of consolidated crust is the main feature of the area. This is caused by southward virgation of the Lambert graben into two branches around the tectonic block of the Mawson Escarpment (see Figure 118.2). The western branch is very similar to the northern part of the Lambert Glacier valley described above. The eastern branch has little reflection in the topography, which obviously implies almost complete filling of the graben by low-density (σ = 2.30 g/cm³) material.

The principal features of the Mohorovičić and Conrad discontinuities are shown in Figure 118.1. The position of the Moho discontinuity was determined along the DSS profile in relation to the "effective relief"*; the correlation thus obtained was extrapolated for the whole area. The position of the Conrad interface was inferred from the areal magnetic data as the depth of lower limits of magnetic bodies, which were found to roughly coincide with the major intracrustal boundary along the DSS profile. On

* For explanation of "effective relief," see Chapter 112, this volume.

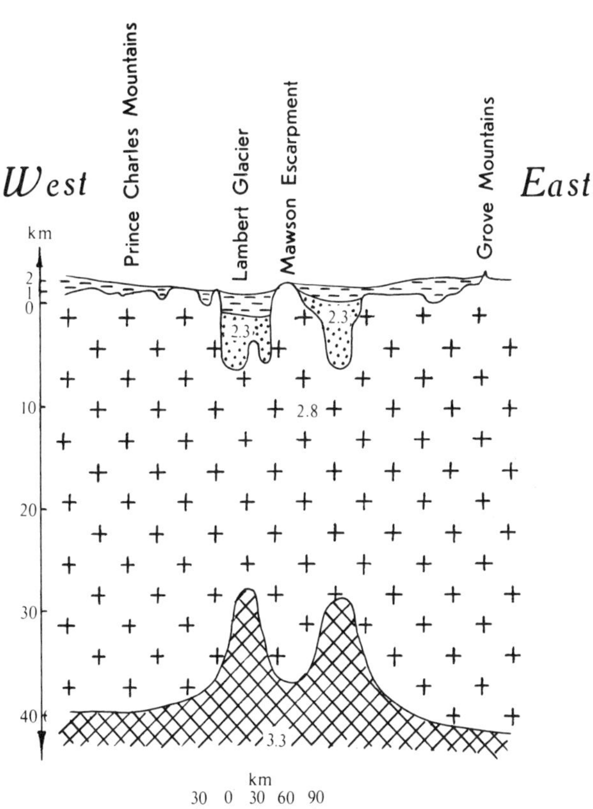

Figure 118.4. Crustal section across the principal structural features in the southern Lambert Glacier area, inferred from geophysical data. (For location see Figure 118.1, profile D-D.) Dashed pattern: ice. Dotted pattern: unconsolidated rocks. Plus pattern: consolidated crust. Grid pattern: mantle. Densities are given in g/cm³.

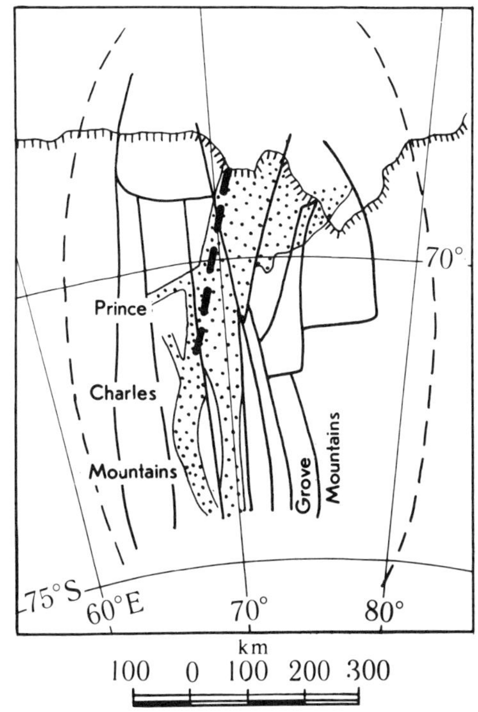

Figure 118.5. Reconstruction of the hypothetical late Mesozoic arch in the Lambert Glacier area. Hachured line: ice front. Dotted pattern: present configuration of the rift grabens filled with unconsolidated rocks. Heavy dashed line: position of deep fault. Solid lines: fault lines inferred from magnetic data, presumably responsible for splitting of the late Mesozoic arch. Light dashed lines: inferred boundaries of the late Mesozoic arch.

the average, the Conrad interface lies 10 km above the Moho discontinuity and repeats the configuration of the latter, i.e., it is also characterized by relatively higher elevation beneath grabens and downwarpings under mountains. Other seismologic data gathered from various parts of Antarctica (Adams, 1971) confirm the reality of these two boundaries at deep levels, as determined during our investigations.

Geologic interpretation of the presented data

The above geomorphologic and geophysical data positively indicate the rift nature of the Lambert Glacier graben, which possesses a number of features typical of continental rift zones (Milanovskii, 1976). The most important of these features are (1) thinning of the crust beneath an axial graben zone and (2) a local decrease in mantle density and boundary velocities beneath the graben. Continental rift zones are usually characterized by high heat flow and seismicity as well, but evidence for these phenomena is not as yet available.

The rift zone is distinctly discordant with respect to the prerift structural pattern, determined mainly by the trends in the basement complex (see Figure 118.2). The relationship between Permian sedimentation and the crustal structures which controlled initiation of the rift is not clear. It is quite possible that the position and contours of the Permian basin were to a certain degree predetermined by the same structural factors which later controlled more emphatically the configuration of the rift zone. On the other hand, the Permian strata might once have had a much greater areal extent, being preserved only locally (in the Beaver Lake area) due to their position within a graben.

The end of the Mesozoic Era may be suggested with greater assurance as the time of initiation of rifting. Alkaline-ultramafic magmatic activity, which took place at that time in the Beaver Lake area, indicates the development of extensive permeability zones in the crust which allowed the rising of deep-seated magmas to the surface during middle Cretaceous time. Significant late Mesozoic crustal events here are also suggested by numerous geologic data which imply widespread manifestations of extensive orogenic processes underway in many parts of Antarctica at that time (Grikurov, Chapter 98, this volume). Finally, such global phenomena as the Gondwanaland breakup and the formation of ocean basins are also believed to have occurred at the close of the Mesozoic Era.

Reconstruction of the initial stages of rifting may be attempted by analysis of magnetic anomalies which allow the recognition of extensive sublatitudinal faults in the basement, dissecting it into several narrow blocks (Figure 118.5). These faults are clearly more concentrated along long. 70° E and are not so numerous to the east or west, which gives a basis to the interpretation that their formation is due to the growth and splitting of a large (about 1000 km by 700 km) basement arch, the axis of which was just east of the present Mawson Escarpment. In the light of the above geologic observations and discussion, this arch was most likely formed during late Mesozoic time.

The present-day boundaries of rift grabens, viewed with respect to the system of faults concentrated in the axial part of the hypothetical paleoarch (Figure 118.5), show that the initial rifting is likely to have occurred along the longitudinal zone now occupied by the northern Lambert Glacier and its eastern branch near the Mawson Escarpment. The older age of the eastern branch of the graben is confirmed by the fact that it has no reflection in the subglacial topography, apparently due to complete compensation by a thick sequence of sedimentary deposits. On the other hand, the younger age of the western rift branch is evidenced both by its clear-cut expression in the bedrock topography and by a westward decrease in age of alkaline-ultramafic volcanicity which, in the Manning Massif (immediately west of Beaver Lake), is dated as Paleogene. The sharp increase in amplitude of relief and degree of bedrock dissection west of the Lambert Glacier also implies migration of tectonic activity in this direction, especially during late Cenozoic time.

The Prydz Bay Basin is of special interest since geophysical data seem to indicate here a rather deep subsidence of the basement (certainly more than 5 km, possibly 10–12 km). However, a genetic relation between this structure and the rifting is obscure. There is quite probably a correlation between the time of most intense subsidence of negative tectonic forms (e.g., the Prydz Bay Basin and West depression) and the development of rift grabens. Hence one can expect a similar age for the resulting sedimentary deposits, at least for their youngest part. Nevertheless, such synchronism, even if present, does not necessarily imply existence of a causal connection between rifting and the nearby tectonic subsidences. To clarify this problem, more detailed investigations are needed.

Conclusions

The Lambert Glacier graben and the Amery Ice Shelf have long attracted geologic interest as the largest disjunctive structures of East Antarctica (Voronov, 1964). The rift nature of these prominent geomorphic features is now established. A hitherto unknown (completely filled with sediment and hence not reflected in bedrock topography) branch of the rift zone east of the Mawson Escarpment is revealed. Extensive areas of relatively shallow bedrock depressions in the Prydz Bay and between the northern and southern Prince Charles Mountains seem to be underlain by much deeper basement subsidence; these areas are believed to represent the brachiform, syneclise-type structures in the platform cover.

Study of rift zones on other continents shows that their orientations are to a great extent predetermined by the geologic history of an older terrane. In the Lambert Glacier region, such a relationship may also exist. Available data indicate that the Archean terrane is more or less symmetric with respect to the rift zone. The symmetry is characterized by the homogeneous nature of the deeply metamorphosed strata on either side of the rift, by the similar composition and metamorphic grade of the rocks, and by the persistent strike and similar morphology of fold structures. On the other hand, the supracrustal Proterozoic to lower Paleozoic complex in the extreme southwestern part of the area seems to have no extension east of the rift zone. This lack of symmetry is also marked in the younger structures by the disposition of the largest tectonic depressions, which are related to the western flank of the rift zone in the south but attached to its eastern side in the northeast. On the whole, the western side of the rift appears more heterogeneous than the eastern side, the heterogeneity being apparently associated with epicratonic structures, i.e., with the post-Archean stages of tectonic evolution. If this observation is true and not merely a consequence of better exposure on the western side due to its more intensive neotectonic revival, then it is reasonable to assume that the Lambert Glacier rift zone has inherited some ancient inhomogeneity in the crust. This inhomogeneity might be as old as late Archean and may have grown more and more pronounced during Proterozoic and Paleozoic time. This presumed crustal discontinuity, and the rifting phenomena now associated with it, are likely to extend far beyond the studied territory both oceanward and inland.

The studies carried out in the Lambert Glacier area demonstrate the great value of integrated geologic and geophysical methods for investigation of the subglacial structure of Antarctica. Further development of geophysical methods is of great importance for progress in geologic investigation of the inland regions.

References cited

Adams, R. D. 1971. Reflections from discontinuities beneath Antarctica. *Bull. seism. Soc. Am., 61,* No. 5, 1441–51.

Boyarskii, V. I., and Shalygin, A. M. 1976. Radiolokatsionnaia s'iemka podlednogo rel'iefa Deviatnadtsatoi Sovetskoi antarkticheskoi expeditsiei [Radio-echo sounding investigations carried out by the Nineteenth Soviet Antarctic Expedition]. *Inf. Byull. sov. antarkt. Eksped.,* No. 92, 18–23.

Crohn, P. W. 1959. A contribution to the geology and glaciology of the western part of Australian Antarctic Territory. *Bull. Bur. Miner. Resour. Geol. Geophys. Aust.,* No. 52, 103 pp.

England, R. N., and Langworthy, A. P. 1975. Geological work in Antarctica—1974. *Rec. Bur. Miner. Resour. Geol. Geophys. Aust.,* 1975/30 [Unpublished], 19 pp.

Kolmakov, A. F., Mishen'kin, B. P., and Soloviev, D. S. 1975. Glubinnye seismicheskie issledovaniia v vostochnoi Antarktide [Deep seismic investigations in East Antarctica]. *Inf. Byull. sov. antarkt. Eksped.,* No. 91, 5–15.

McLeod, I. R. 1964. An outline of the geology of the sector from longitude 45° to 80° E., Antarctica. (*In* Adie, R. J., *ed. Antarctic Geology.* Amsterdam, North-Holland Publishing Company, 237–47.)

Milanovskii, E. E. 1976. *Riftovye zony kontinentov* [Rift zones of continents]. Moscow, Izdatel'stvo Nedra, 279 pp.

Soloviev, D. S. 1972. Geological structure of the mountain fringe of Lambert Glacier and the Amery Ice Shelf. (*In* Adie, R. J., *ed. Antarctic Geology and Geophysics.* Oslo, Universitetsforlaget, 573–77.)

Soloviev, D. S., Kamenev, E. N., and Ravich, G. M. 1967. Geologicheskie issledovaniia v 1965/66 g. [Geological activities in 1965/66]. *Inf. Byull. sov. antarkt. Eksped.,* No. 62, 10–18.

Tingey, R. J. 1972. Geological work in Antarctica, 1971. *Rec. Bur. Miner. Resour. Geol. Geophys. Aust.,* 1972/132 [Unpublished], 49 pp.

Tingey, R. J., and England, R. N. 1973. Geological work in Antarctica, 1972. *Rec. Bur. Miner. Resour. Geol. Geophys. Aust.,* 1973/161 [Unpublished], 17 pp.

Trail, D. S. 1963. The 1961 geological reconnaissance in the southern Prince Charles Mountains, Antarctica. *Rec. Bur. Miner. Resour. Geol. Geophys. Aust.,* 1963/155 [Unpublished].

Trail, D. S. 1964. Schist and granite in the southern Prince Charles Mountains. (*In* Adie, R. J., *ed. Antarctic Geology.* Amsterdam, North-Holland Publishing Company, 492–97.)

Voronov, P. S. 1964. Tectonics and neotectonics of Antarctica. (*In* Adie, R. J., *ed. Antarctic Geology.* Amsterdam, North-Holland Publishing Company, 692–700.)

Wellman, P., and Tingey, R. J. 1976. Gravity evidence for a major crustal fracture in eastern Antarctica. *Jnl Bur. Miner. Resour. Geol. Geophys. Aust., 1,* No. 2, 105–8.

119

A Gravity Survey of Enderby and Kemp Lands, Antarctica

P. WELLMAN and R. J. TINGEY
Bureau of Mineral Resources, P.O. Box 378, Canberra, A.C.T. 2601, Australia

Abstract. In 1977 the Australian National Antarctic Research Expedition carried out a helicopter-supported gravity survey of the 150-km-wide coastal strip of Enderby and Kemp Lands. The area is underlain by granulite-facies Precambrian gneisses. Bouguer anomalies (B) for a density of 2.67 metric ton m^{-3} range from +70 to −90 mgal at the 191 stations. If C is the attraction of an infinite slab with the same mass per unit area as the average mass above sea level in the surrounding 100 km × 100 km area, then the correlation of B with C is

$$B = 19.5 - (0.79 \pm 0.04) \times C$$

The mean deviation of the actual Bouguer anomalies from the values obtained from this equation is 22.5 mgal. This relation is consistent with the crustal and ice load in the area being isostatically compensated at the base of the crust.

Modified free-air anomalies were calculated by adding the attraction C to the Bouguer anomaly; these anomalies are essentially terrain-corrected, free-air anomalies. They range from +90 to −30 mgal and have a mean of +27.4 ± 1.5 mgal. An isolated area of high anomaly (maximum +90 mgal) over the Napier Mountains is surrounded by low anomalies; the high is interpreted as an area of relatively high-density upper crust with regional isostatic compensation. Other areas of high anomaly occur east of the Sandercock Nunataks and south of the Hansen Mountains. Areas of granite correspond to an area of low anomaly (+10 mgal) in the northwestern Napier Mountains and to a more extensive low (−20 mgal) west of the Nye Mountains. Many of the other zones of high gravity gradients correspond approximately to mapped granulite subfacies boundaries.

East Antarctica is composed mainly of a Precambrian shield with some sedimentary rock cover; it once formed the central part of the Gondwana continent. Now East Antarctica is almost completely buried under the Antarctic ice sheet. In the coastal belt of Enderby and Kemp Lands, however, there is an extensive area where small nunataks of rock project up through the ice sheet. Because of its extent, this outcrop area is important for geologic studies of the East Antarctic shield. Gravity mapping in this area provides information on lateral variations in crustal densities and on whether these crustal density variations and the ice thickness variations are isostatically compensated.

Geologic mapping of outcrop areas has been carried out at a 1:250,000 scale—in Kemp Land by previous Australian expeditions (McLeod, 1964) and in Enderby Land mainly during the 1977 season. It has also been mapped at a 1:1,000,000 scale by Soviet expeditions (Kamenev, 1972). The area is underlain by a Precambrian basement complex composed mainly of granulite-facies gneiss. The gneiss varies from massive rocks of possibly igneous origin to strongly layered rocks of possibly sedimentary origin. The structure is complex. Metamorphic grade varies regionally between granulite subfacies, with some retrogression, to amphibolite-facies mineralogy. In coastal Enderby Land, rocks containing quartz-sapphirine are widely distributed and indicate that metamorphic temperatures were exceptionally high, even for granulite-facies rocks. The gneisses are intruded by younger basic dikes and pegmatites and by granite.

The altitude of the ice surface (Figure 119.1) increases comparatively steeply from sea level at the coast, to 1000 m about 50 km inland, and then more gradually to 2000 m at 200 km inland. However, the Napier Mountains in northeast Enderby Land form an isolated dome with a maximum height of 2000 m, separated from the main Antarctic ice sheet by a saddle 1100 m high. The variation in ice surface level is very gradual except for narrow troughs occupied by major glaciers flowing into Edward VIII Gulf and Amundsen and Casey Bays. The thickness of ice in Enderby and Kemp Lands was mapped by ice radar measurements during the 1977 season; thicknesses are generally between 500 and 1500 m in the area of the gravity survey (I. Allison, personal communication).

Previous gravity observations in the area consist of Australian reconnaissance observations along the Kemp and MacRobertson Land coasts (Langron, 1966), scattered Soviet marine observations and a traverse south of Molodezhnaya (Groushinsky and others, 1972), a Soviet reconnaissance survey of Enderby Land and the adjacent continental shelf (but with only a few observations on rock) (Koriakine and others, 1970), and a Japanese glaciologic traverse from Syowa Base eastward to Sandercock Nunataks (H in Figure 119.1) (Watanabe and Yoshimura, 1972).

Survey procedures

This paper is based mainly on observations made during an Australian National Antarctic Research Expedition (ANARE) in the period 4 January to 8 February 1977. The survey was carried out by daily 6- to 10-hour flights by Hughes-500 helicopters between sites; nights were spent at a base camp on snow near Mount King (J in Figure 119.1). Gravity was measured with two LaCoste and Romberg

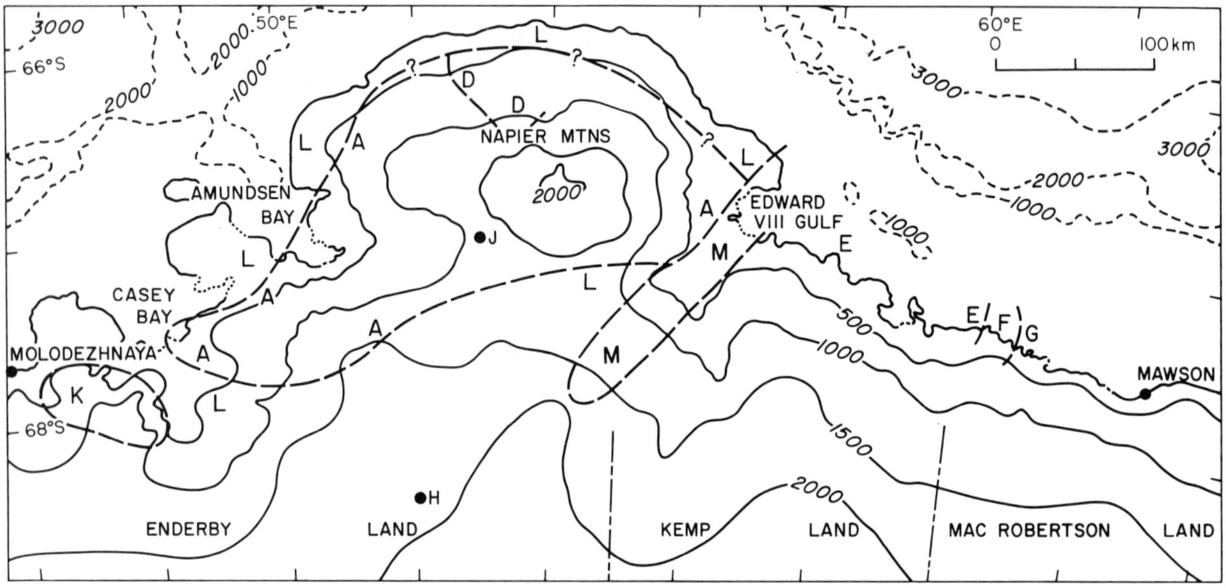

Figure 119.1. Enderby and Kemp Lands showing the configuration of the ice surface and bathymetry (contours in m). Rocks are granulite facies; subfacies boundaries are shown dashed. H: Sandercock Nunataks. J: Mount King. Other letters indicate metamorphic subfacies cited in the text.

geodetic gravity meters. Gravity values are based on an adopted value of 982, 153.14 mgal at Mount King camp, which is based in turn on Mawson A = 982,481.80 mgal in the Potsdam system. All gravity values within the survey area have a relative accuracy of 0.2 mgal and an absolute accuracy estimated to be 0.5 mgal. At Molodezhnaya base observed gravity values are between 0.5 and 1.0 mgal higher than reported by Koriakine and others (1970).

The altitude of the gravity stations was measured using "Mechanism" microbarometers—one at the base camp and two with each gravity meter. Heights were measured relative to 1114 m at Mount King camp, which is based on 1486.7 m at Mount King trigonometric station. From field stations close to sea level, and those at which repeat measurements were made on different days, the adopted heights are calculated to have a standard error of 10 m. Station positions in Enderby Land were identified on vertical, color aerial photographs (taken mainly during 1976); positions in Kemp Land were identified on oblique air photographs. Latitudes and longitudes were obtained from a 1:1,000,000 map of the Molodezhnaya-Mawson area produced from Landsat images by the Division of National Mapping, Canberra, in August 1976. Positions are thought to be accurate to 1 km.

For stations on rock the greatest error in the Bouguer anomalies is in the estimate of terrain correction; for stations near the tops of some mountains this correction reaches 30 mgal. Terrain corrections were estimated from field observations and air photographs; adopted values may have errors up to 10 mgal owing to restricted views from the gravity stations and to the terrain effect of the ice-rock interface.

The coverage given by gravity stations on rock was extended to the south using eight glaciologic stations that form a traverse from Mawson to Sandercock Nunataks (H in Figure 119.1). Estimates of ice thickness at these stations were derived from ground and aircraft-based ice radar measurements; ice thicknesses have an uncertainty of as much as 100 m, giving Bouguer anomalies corrected for ice thickness an uncertainty of 10 mgal.

Bouguer anomalies

The Bouguer anomalies for stations on rock are roughly corrected for local terrain and for the attraction of a slab of rock between the station and sea level. Stations on ice are corrected for the attraciton of rock and ice between the station and sea level. Densities assumed are 2.67 metric ton m^{-3} for rock and 0.917 metric ton m^{-3} for ice. The Bouguer anomaly map (Figure 119.2) shows the Enderby and Kemp Land stations and stations in adjacent MacRobertson Land (Wellman and Tingey, 1976). Bouguer anomalies are generally about +50 mgal at the coast line; they decrease comparatively steeply to −10 mgal at about the 700 m ice-surface elevation, then more gradually to about −80 mgal at the 2000 m ice-surface elevation in the southern part of the area, and to about −15 mgal at 2000 m in the Napier Mountains.

Bouguer anomalies reflect the regional free-air anomaly, the crustal root, and any crustal density variations. These effects can be roughly separated by regressing the bouguer anomaly (B) with the attraction of an infinite slab with the same mass per unit area as the average mass above sea level in the surrounding 100 km × 100 km area (C) (Figure 119.3). The same densities were used as above. The least squares relation, giving each gravity station equal weight, is

$$B = 19.5 - (0.79 \pm 0.04) \times C.$$

(Uncertainties are given as standard deviations.) The crustal root effect given by the correlation of b and C accounts for 60% of the gravity anomaly variation; upper crustal density variations and any other effects account for the remaining 40% of the gravity anomaly variation.

If the crust (including ice load) is in isostatic equilibrium with compensation at the base of the crust, then the expected relation between B and C is $B = k - 0.75 \times C$, where k is a constant. The number 0.75 is less than 1.00 because of the effects of earth curvature and because the gravity observation points are closer to the masses above sea level than to the compensating root at the base of the crust (Woollard, 1970, p. 491). The least-squares relation

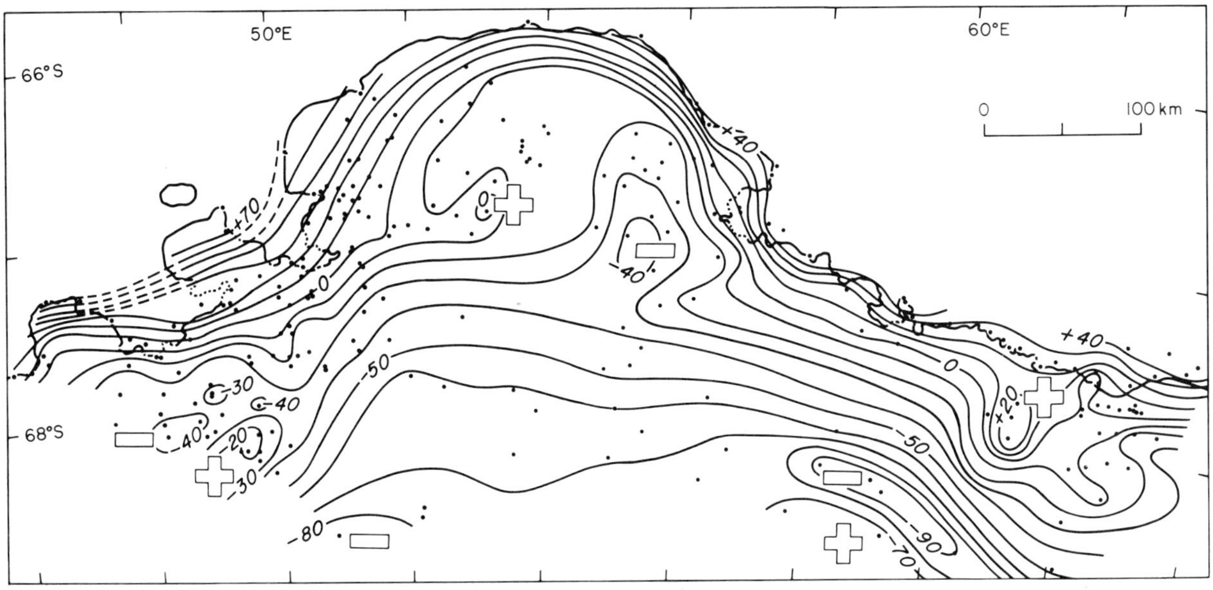

Figure 119.2. Bouguer anomaly map. Contour interval, 10 mgal. Dots show gravity station positions.

for Enderby and Kemp Lands given above is consistent within experimental error with the masses above sea level being isostatically compensated at the base of the crust. It is similar to the relation $B = (6 \pm 2) - (0.84 \pm 0.05) \times C$ found for the adjacent Prince Charles Mountains area to the southeast (Wellman and Tingey, 1976).

At the transition between the continental and oceanic crust there is a major change in crustal thickness and depth to the mantle. This change is generally reflected in the gravity anomalies—a free-air-anomaly high over the continental shelf and a low over the continental rise. Relative to the average free-air anomaly of the region these anomalies are generally about +40 and −40 mgal, respectively.

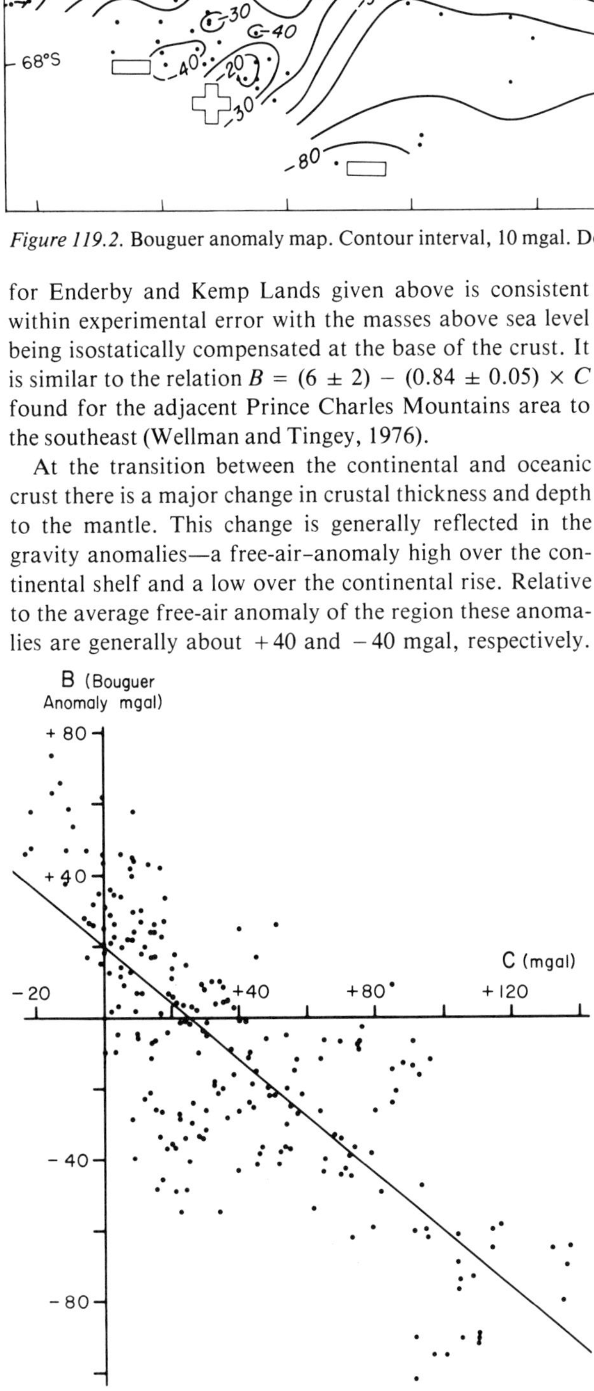

Figure 119.3. Relation between observed Bouguer anomaly at stations *(B)* and the atraction of an infinite slab with the same mass per unit area as the average mass above sea level in the surrounding 100 km × 100 km area *(C)*.

The edge of the high over the continental shelf is reflected in Figure 119.3, the plotted points generally being positive relative to the mean line for values of C less than +20 mgal.

Modified free-air anomalies

Modified free-air anomalies were also obtained; they are essentially terrain-corrected free-air anomalies. They were calculated by adding C to the Bouguer anomaly at each gravity station. The resulting map (Figure 119.4) has anomalies ranging from +90 to −20 mgal.

The largest anomaly is a positive one, reaching +90 mgal over the Napier Mountains in northern Enderby Land. This high gives a residual anomaly of +60 mgal relative to the average relation between B and C (Figure 119.3). In the Napier Mountains the ice thickness averages about 500 m thinner than elsewhere inland; however, the additional rock mass close to the surface explains only 10 of the +60 mgal. The remaining +50 mgal anomaly is interpreted as reflecting abnormally dense upper crust, even though there is no surface geologic expression of this. The high is surrounded by relative lows in a 250 km × 250 km area, so the excess mass is compensated in that area. This positive anomaly over the Napier Mountains was recognized by Koriakine and others (1970) on the basis of the free-air anomaly pattern. They attributed it to a lack of crustal thickening under the mountains, meaning that the area is not in local isostatic balance.

Relative lows in the northwestern part of the Napier Mountains and south of Casey Bay are thought to be due to granitic rocks of relatively low density. In the Napier Mountains, granite crops out in the area D of Figure 119.1; the lowest anomaly is +10 mgal. South of Casey Bay, massive gneiss intruded by granite sheets crops out over the area K of Figure 119.1; the lowest anomalies are −20 mgal.

The trend of many of the remaining zones of high gravity gradients correlate with the mapped boundaries of metamorphic subfacies in these granulite-facies rocks. The gravity gradients are unlikely to be directly caused by differences in density within the uppermost crust where all

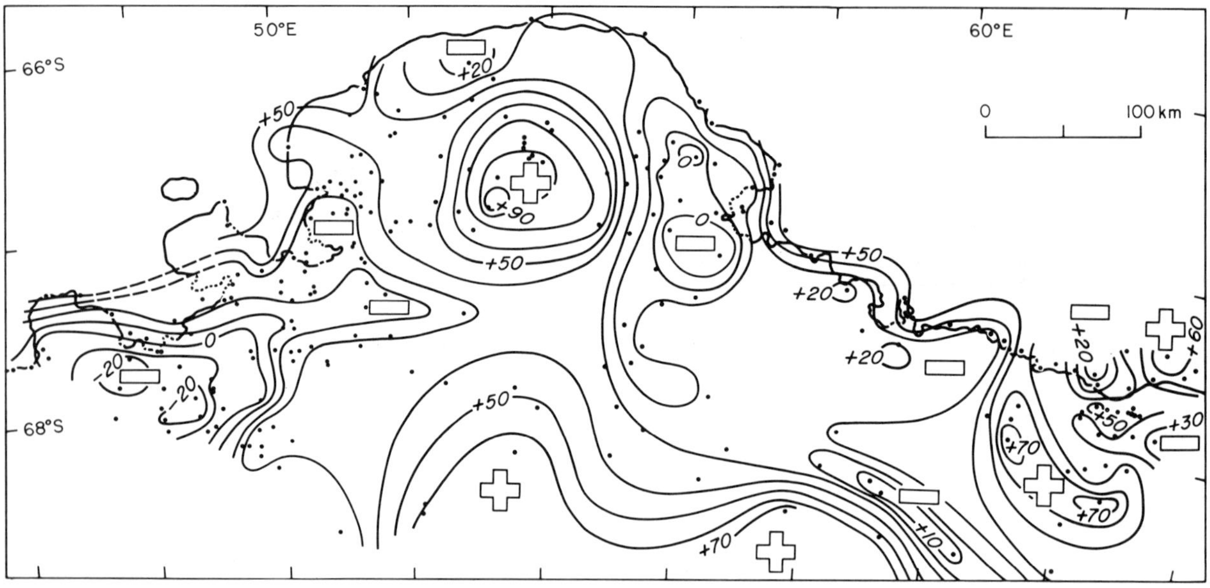

Figure 119.4. Map of modified free-air anomalies. Contour interval, 10 mgal.

rocks are granulite facies, but the mapped changes in metamorphic facies could reflect the boundaries of deeper structures in the middle and lower crust that cause the gravity anomalies. Figure 119.1 shows the metamorphic-facies boundaries in Enderby and Kemp Lands from Kamenev (1972) and Ravich and Kamenev (1972) and facies boundaries in MacRobertson Land from Trail (1970). In Enderby Land the boundary between the pyroxene granulite (A in Figure 119.1) and the hornblende granulite subfacies (L in Figure 119.1) corresponds well with the boundaries of the gravity low south of Amundsen Bay and the gravity high over the Napier Mountains. The reason for the lateral change from a gravity low to a gravity high within pyroxene granulite facies rocks is not known. In Kemp Land the area of amphibole-garnet-clinopyroxene granulite subfacies (M in Figure 119.1) corresponds roughly with the gravity low southwest of Edward VIII Gulf. In MacRobertson Land the change from banded gneiss (E) to garnet-bearing gneiss (F) to hypersthene charnockite (G) at long. 61° E corresponds to an increase in the anomalies.

Along the coastline there is a seaward increase in gravity anomalies; this is the edge of the zone of positive anomalies associated with the transition between continental and oceanic crust.

Acknowledgments. We thank the Antarctic Division, Department of Science, for providing logistic support and I. F. Allison and N. Young of the Antarctic Division for providing unpublished radar ice thicknesses and glaciologic traverse observations. We are grateful to W. Anfiloff for translating papers from Russian, B. C. Barlow for critically reading the draft paper, and the gravity drafting group, Geophysical Drawing Office, BMR, for drafting the figures. This paper is published with the permission of the Director, Bureau of Mineral Resources, Geology and Geophysics, Canberra.

References cited

Groushinsky, N. P., Koryakin, E. D., Stroev, P. A., Lazarev, G. E., Sidorov, D. V., and Virskaja, N. F. 1972. Katalog gravimetricheskikh punktov Antarktiki [The catalogue of the gravity stations in Antarctica]. *Trudy Gosudarstvennogo Astronomickeskogo Instituta, 42,* 115–313.

Kamenev, E. N. 1972. Geological structure of Enderby Land. (*In* Adie, R. J., *ed. Antarctic Geology and Geophysics.* Oslo, Universitetsforlaget, 579–83.)

Koriakine, E. D., Stroev, P. A., and Frolov, A. I. 1970. Resulyaty gravimetricheskikh issledovanii na zemle Enderbi (vostochnai Antarktida) [Results of gravity measurements in Enderby Land (East Antarctica)]. (*In* Fedinski, V. V., *ed. Moeskie Gravimetricheskie Issledovanii—sbornik statei, vbipusk 5* [*Gravity Measurements at Sea—collection of articles, No. 5,* Moscow, University of Moscow, 83–95].)

Langron, W. R. 1966. Gravity ties to Australian Antarctica, 1953–1963. *Rec. Bur. Miner. Resour. Geol. Geophys. Aust., 1966/24* [Unpublished], 14 pp.

McLeod, I. R. 1964. An outline of the geology of the sector from longitude 45° to 80°E., Antarctica. (*In* Adie, R. J., *ed. Antarctic Geology.* Amsterdam, North-Holland Publishing Company, 237–47.)

Ravich, M. G., and Kamenev, E. N. 1972. *Kristallicheskii fundament Antarkticheskoi platformy* [Crystalline basement of the Antarctic platform]. Leningrad, Gidrometeoizat, 658 pp. [English translation: New York, John Wiley, 1975, 582 pp.]

Trail, D. S. 1970. ANARE 1961 geological traverses on the MacRobertson Land and Kemp Land Coast. *Rep. Bur. Miner. Resour. Geol. Geophys. Aust., 135,* 32 pp.

Watanabe, O., and Yoshimura, A. 1972. Preliminary geoscientific researches on Sandercock Nunataks, Enderby Land, East Antarctica 1970. *Antarctic Rec., 45,* 47–65.

Wellman, P., and Tingey, R. J. 1976. Gravity evidence for a major crustal fracture in eastern Antarctica. *Jnl Bur. Miner. Resour. Geol. Geophys. Aust., 1,* 105–108.

Woollard, G. P. 1970. Evaluation of the isostatic mechanism and the role of mineralogic transformations from seismic and gravity data. *Phys. Earth Planet. Interiors, 3,* 484–98.

120

Gravity and Magnetic Anomalies across Jutulstraumen, a Major Geologic Feature in Western Dronning Maud Land

H. DECLEIR
Observatorium Rijksuniversiteit Gent, B-9000, Gent, Belgium

T. VAN AUTENBOER
Limburgs Universitair Centrum, Universitaire Campus, B-3610, Diepenbeek, Belgium

Abstract. Geophysical measurements have been carried out both over Jutulstraumen, a 50-km-wide ice stream forming the geographic continuation of the Pencksökket (Penck Trough) to which it is probably structurally linked, and over Viddalen, a tributary. The main gravity and magnetic profile runs between Nashornet Mountain, where the Pyramiden Formation (slightly metamorphosed sedimentary formations intruded by mafic sills) crops out, and Jutulrøra Mountain, composed of the metamorphic and intrusive rocks of Sverdrupfjella (Sverdrup Mountains).

The gravity anomalies, corrected for a marked regional gradient, are related to the existence of the ice streams. Jutulstraumen occupies an asymmetric subglacial valley with its deepest part 1500 m below sea level near the confluence with Viddalen. The considerable mass transport is related to the Trolltunga ice tongue and, compared to other discharge values, confirms the importance of the removal of ice by ice streams in the total mass balance of the Antarctic. Ice thicknesses were obtained by fitting anomalies computed for theoretical cross sections to the observed values. The adopted cross section is also used to calculate the effects of subglacial bedrock and magnetized bodies in the (total field) magnetic anomalies. Use was made of Talwani's method in which the cross section of the body is approximated by an n-sided polygon. The anomalies are then tentatively interpreted as a function of measured magnetic properties of the rocks on both sides of the glacier, while the known geology is extrapolated beneath the ice cover.

The results of this study are correlated with the general structural and geophysical information of the area.

Jutulstraumen and its northward extension, the Trolltunga ice tongue, are major morphologic features in western Dronning Maud Land. This important ice stream conceals an equally important geologic contact, that between the Precambrian basement complex to the east of Jutulstraumen and south of Pencksökket (Penck Trough), the southwest continuation of Jutulstraumen, and the Precambrian–Paleozoic(?) stable platform deposits to the west and north of these glaciers. The former consists mainly of different types of gneisses, migmatites, amphibolites with minor quartzites, and marbles, all belonging to the Sverdrupfjella Group (Roots, 1969). This high-grade metamorphic group is cut by posttectonic intrusions (syenite, monzonite, and granite) and younger dolerite dikes. The latter consist mainly of slightly metamorphosed, flat-lying or gently tilted, detrital sedimentary rocks (Ahlmannryggen Group) and associated volcanic rocks (Trollkjell Group) (Roots, 1969).

Similar platform deposits overlie the basement complex in several places in Dronning Maud Land (Craddock, 1972). Sedimentary formations in contact with the basement complex in the southeast Kirwanveggen (Kirwan Escarpment) were assigned to the Ahlmannryggen Group by Roots (1969) but have been interpreted differently by Aucamp and others (1972). Roots (1969) considered the Pencksökket depression separating the two rock terranes to be a down-faulted block or a series of faulted slices and, on the basis of morphologic evidence, suggested a major fault bounding the lower part of Jutulstraumen on its western side.

Theoretical geology thus seems well ahead of the basic geophysical data needed to interpret the geologic and structural significance of Jutulstraumen. The geophysical data reported here are also expected to furnish much-needed information on a major feature of the drainage pattern and on the mass balance of the ice cover in Dronning Maud Land.

Field work

The original plan of the 1968 Belgian expedition included gravity, geomagnetic, and ice movement measurements across Jutulstraumen and Viddalen, supplemented in the following season by radio-echo sounding. This would allow the isolation of "geologic" gravity anomalies from glaciologic ones. However, Antarctic meteorology and technology decided otherwise. The traverse ended halfway across Viddalen, the velocity measurements were abandoned, and aircraft failure limited the radio-echo sounding survey to the ice shelf north of lat. 71° 30′ S (Van Autenboer and Decleir, 1969, 1972, 1974). Fortunately, a Norwegian expedition subsequently carried out velocity measurements and repeated the gravity measurements (Gjessing, 1972). No radio-echo sounding has yet been carried out across Jutulstraumen as the oversnow geophysical work by SANAE parties concentrated on Ahl-

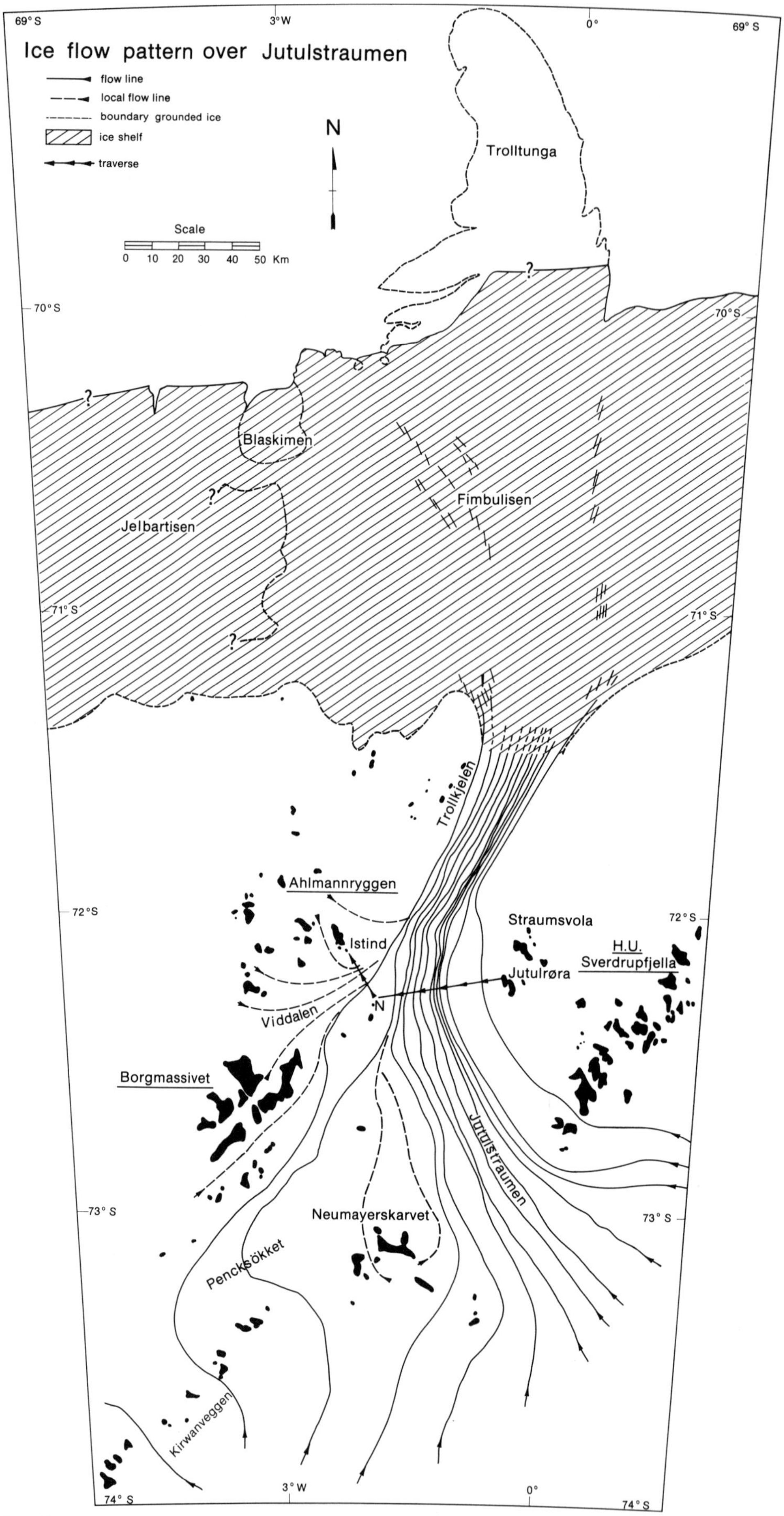

Figure 120.1. Location map and ice flow pattern for Jutulstraumen area. The traverse is indicated by arrows. Note the change in course at Nashornkalvane (N) from east-southeast to northwest (this change in course is indicated on Figure 120.3 by the line marked "bend in profile"). Scale is approximate.

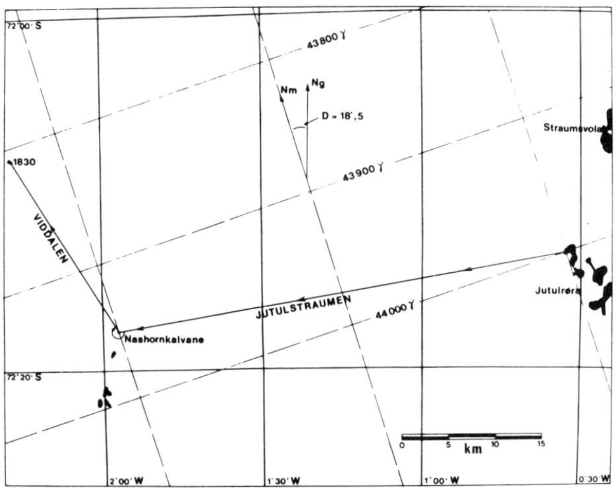

Figure 120.2. Geographic grid and adopted geomagnetic grid (February 1968) as calculated from traverse data, observed geomagnetic components at SANAE, and magnetic maps (Behrendt and Bentley, 1968). The observed magnetic values were reduced by subtracting the main earth field as shown in Figure 120.2. For the calculation of the components J_x (along the traverse) and J_z (vertical) of the intensity of magnetization J, the following values were adopted as being constant over the area: total magnetic field intensity, F = 44012 gamma; inclination, I = -64°; declination, D = −18.5°; vertical component, Z = −39558 gamma; horizontal component, H = 19294 gamma.

mannryggen (Ahlmann Ridge) and Borgmassivet (Borg Massif). Radioglaciology over Viddalen was successful only over the shallower ice as the limit of the equipment was quickly reached (Barnard, 1975).

The traverse originated on the west side of Jutulrøra Mountain, reached the northeastern extremity of Nashornkalvane (Nashornet Mountain) on a course of 263.5°, then turned northward (course 326°) toward Istind Peak (Figures 120.1 and 120.2). Distances were measured with a sledgewheel and adapted to the 1:250,000 maps of Queen Maud Land published by Norsk Polarinstitutt. The altitudes of the stations were measured with three aneroid barometers and reduced with the aid of the temperature and pressure records of a temporary meteorologic station located at Jutulrøra. The gravity measurements (Worden geodetic gravimeter) included those of 27 stations across Jutulstraumen and 9 across Viddalen, supplemented by five readings kindly communicated later by Gjessing. The total magnetic field intensity was measured every 500 m (a total of 124 stations) with an Elsec proton precession magnetometer.

Interpretation

Gravity profile

Due to the lack of radio-glaciologic ice-depth determinations, the gravity profile could only be interpreted as a function of the ice thickness. The data in Table 120.1 are therefore included to allow further interpretation in the future. The computation of the ice thickness follows essentially the method outlined by Van Autenboer and Blaiklock (1966) and further refined by Van Autenboer and Decleir (1974, 1978) for the glaciers of the Sør Rondane Mountains.

An initial model is calculated assuming that the gravity anomalies result from the replacement of an infinite slab of rock (a density contrast of 1.7 was used) by a slab of ice having identical dimensions. This model was then improved by computing the gravity effect of the calculated cross section on each of the stations using the line-integral method (Hubbert, 1948; Talwani and others, 1959). The difference between the observed anomaly and the computed effect was again converted into ice thickness by the infinite slab method, and the original calculated model was modified accordingly.

The regional gravity gradient is estimated from the difference between the two gravity values on rock, reduced to the same elevation and latitude. A correction for the

Table 120.1. Gravity data for Jutulstraumen and Viddalen

Station number	ΔX	Δh	Δg_m	Δg_{corr}	Δg_{calc}	ΔZ
			Jutulstraumen			
J1	0.0	0	0.0	0.0	—	0
J2	0.6	−10	−20.92	−23.60	−23.60	673
J3	2.6	20	−47.15	−42.00	−42.00	966
J4	4.5	49	−54.24	−41.21	−41.21	939
J5	6.6	46	−41.37	−29.44	−29.44	609
J6	8.6	48	−39.28	−26.68	−26.68	651
J7	10.5	24	−30.27	−24.18	−24.28	599
J8	12.5	38	−34.48	−24.95	−24.95	625
J9	14.4	23	−32.80	−27.30	−27.30	719
J10	16.4	13	−27.58	−24.96	−24.96	624
J11	18.4	16	−29.52	−26.09	−26.09	683
J12	20.3	15	−29.50	−26.53	−26.53	619
J13	22.3	38	−41.85	−32.91	−32.91	781
J13′	22.8	37	−42.92	−34.17	−34.17	581
J14	24.3	−2	−47.00	−49.81	−49.81	930
J15	26.2	−12	−71.50	−76.09	−76.09	1478
J16	28.2	−2	−96.23	−98.43	−98.44	1912
J17	30.2	32	−120.37	−113.34	−113.32	2717
J18	32.1	38	−122.60	−114.20	−114.22	2483
J19	34.1	34	−111.50	−104.10	−104.09	2238
J20	36.1	32	−91.34	−84.81	−84.81	1719
J21	38.0	23	−69.05	−64.91	−64.91	1256
J22	40.0	46	−58.45	−48.24	−48.24	1193
J23	42.0	−6	−24.37	−28.24	−28.24	661
J24	43.9	5	−23.65	−24.69	−24.69	866
J25	45.9	10	−14.08	−13.74	−13.74	687
J26	47.9	20	12.06	15.00	14.99	101
J27	48.25	45	12.97	19.32	—	0
			Viddalen			
V1	48.25	45	0.0	0.0	—	0
V2	48.4	25	1.11	−4.25	−4.25	132
V3	49.25	11	−8.93	−17.71	−17.71	277
V4	51.25	−39	−32.46	−53.95	−53.95	984
V5	53.25	−37	−59.97	−80.13	−80.14	1491
V6	55.25	−65	−71.56	−98.49	−98.43	2013
V7	57.25	−80	−75.00	−105.19	−105.48	2376
V7′	57.4	−77	−75.85	−105.19	−105.41	2558
V7″	58.25	−60	−78.42	−102.82	−102.30	1939
N21[a]	59.4	−78	−64.18	−92.97	−93.13	2075
N22[a]	63.1	−38	−43.54	−60.07	−60.04	1236
N23[a]	65.8	−18	−24.22	−34.28	−34.29	827
N24[a]	67.8	7	−14.11	−16.64	−16.64	727
N25[a]	69.0	50	7.02	16.22	—	0

ΔX: Distance from reference station at Jutulrøra in km.
Δh: Difference in elevation in m.
Δg_m: Measured gravity difference in mgal.
Δg_{corr}: Observed anomaly in mgal reduced to altitude and latitude of reference station.
Δg_{calc}: Calculated anomaly in mgal.
ΔZ: Total ice thickness in m.
[a] Gravity values from Gjessing (personal communication).

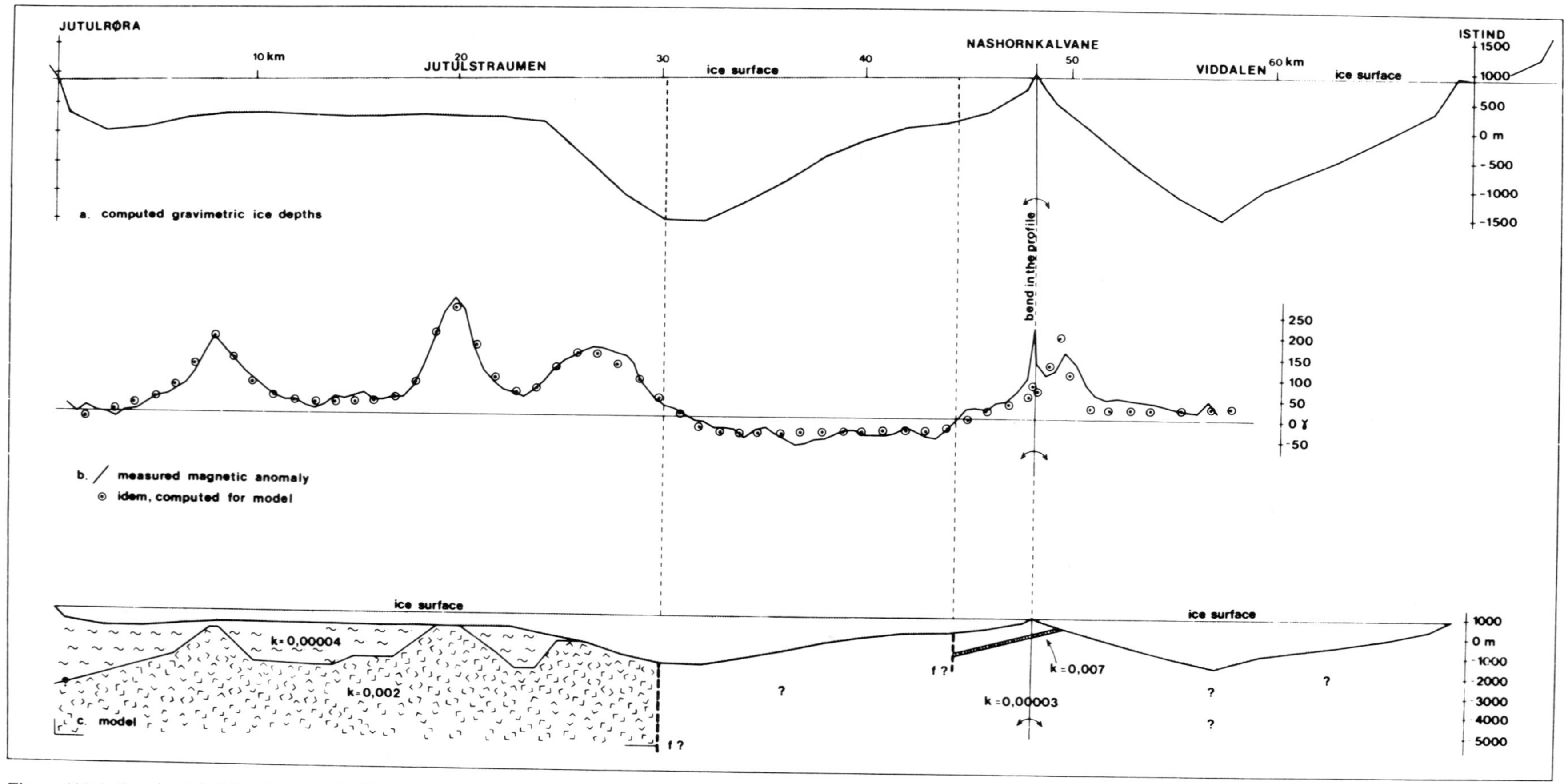

Figure 120.3. Gravimetric (*a*) and magnetic (*b*) models for Jutulstraumen and Viddalen. Line of traverse shown in Figure 120.1.

regional effect is applied for each station on the glacier by linear interpolation of this difference. Since the very existence of the glacier reduces the gravity value at the reference station, the original value for each reference station is too low. This change in the two reference values is also reflected in the regional gradient and consequently in the anomalies for each station. A third model was computed in which, at each step, a correction was applied to the observed values at the reference stations, allowing for the negative effect of the glacier model computed during the previous step. This iterative process converges rather rapidly towards a constant negative value at the reference stations. As can be expected, this gave increased depth readings for the glacier valley.

It is clear that at the reference stations, the mass deficit, representing all the surrounding ice, will be felt. As the gravity profiles here were measured over two neighboring glaciers, it was possible to take their mutual influence into account by combining both glaciers in one model, neglecting, however, the difference in orientation. This refinement further deepened the glacier valley. The total effect of the mass deficiency in the finally adopted model equals −15.80 mgal for Jutulrøra, −12.06 mgal for Istind "1320," and −19.83 mgal for Nashornkalvane, where the effect of the two glaciers is felt. These values indicate the error involved when omitting an "ice correction," which is similar to the normal terrain or topographic correction, when computing regional gravity fields over ice-covered mountain areas. The resulting regional gravity gradient equals 0.41 mgal/km increasing from south to north and 0.48 mgal/km increasing from east to west.

The polygonal approximation of the model makes the ice depths very sensitive to small irregular variations of the gravity values at adjacent stations and results in a rather angular subglacial relief. The adopted glacier cross section was therefore smoothed (Figure 120.3*a*). The computed values differ only by a few tenths of a milligal from the observed anomalies (Table 120.1), which indicates an accurate mathematical fit. This does not, however, represent the accuracy of the ice depths, as these can be affected by unknown factors such as geology, subglacial topography, deep-seated structures, errors in the assumed density contrast, and the two-dimensional representation of a three-dimensional structure.

The computed ice depths over Jutulstraumen exceed, as expected, those calculated by Gjessing (1972). Using that author's ice movement measurements, we recalculated the discharge across Jutulstraumen. The mass transport through this cross section equals 12.48 km^3/yr, of which 87% flows through the western side of the glacier. This is, of course, a maximum estimate, as no attempt was made to take the downward reduction of horizontal movement rate into account. This calculated discharge corresponds to a mass flux of 0.233×10^{15} g/km • yr for Jutulstraumen as a whole, with 0.406×10^{15} g/km•yr over the deep western half and 0.062×10^{15} g/km•yr over the eastern part of the section.

TOTAL MAGNETIC ANOMALIES

The measured total magnetic intensities were corrected for daily and substorm variations using the records of SANAE, the nearest geomagnetic observatory. This correction allowed the rather small anomalies to be identified and indicated that the northernmost measurements over Viddalen were not reliable because of large magnetic disturbances at the time these stations were occupied. In order to reduce the measured values to the same reference point, a correction for the geographic variation of the normal magnetic field was applied. This gradient correction (1.425 gamma/km over Jutulstraumen and 6.64 gamma/km over Viddalen; was calculated from the field data and the direction and magnitude of the geomagnetic field at SANAE (the geomagnetic meridian of SANAE passes along the eastern side of Jutulstraumen) (Figure 120.2).

The reduced profiles (Figure 120.3*b*) show three short-wave anomalies over the eastern side of Jutulstraumen with maxima at 8, 20, and 26 km, respectively. In contrast with the first two, the anomaly at 26 km is somewhat smoother and broader. The western part of the glacier is characterized by a relatively flat and negative anomaly curve (between 32 and 45 km). At Nashornkalvane (48 km) another more irregular anomaly, complicated by the rock outcrop, occurs.

In the interpretation of the magnetic anomalies, the subglacial topography as computed above was first introduced. In designing the different models, only the measured or calculated susceptibilities of the main rock types occurring west and east of Jutulstraumen were used (Table 120.2). The northwest side of Jutulrøra consists mainly of biotite and biotite-hornblende gneisses containing granite and pegmatite lenses, and amphibolite lenses and boudins. Numerous pegmatite dikes cut the gneisses as well as rectilinear, north-trending, 1-2-m-wide, vertical or subvertical basic dikes (Van Autenboer and Loy, 1972). At Tvora (a small nunatak situated 14 km northeast of Jutulrøra) an intrusive syenite massif is present, and it contains a nepheline syenite plug. Nepheline syenite also forms a large massif on Straumsvolane (Hjelle, 1974).

At Nashornkalvane feldspathic graywackes with intercalated siltstones were sampled. They belong to the lower

Table 120.2. Measured susceptibilities of the main rock types occurring east and west of Jutulstraumen

Specimen number	Description and locality	Susceptibility (e.m.u./cm³)
TVA 68 JU4	Small, folded, banded, biotite-hornblende gneiss containing thin quartz-feldspathic layers. Jutulrøra, northwest side	$k = 50.5 \times 10^{-6}$
TVA 68 JU14	Coarse (4-5 cm) pegmatite dike. Jutulrøra, northwest side	$k = 4.3 \times 10^{-6}$
TVA 68 JU11	Dark, homogeneous, fine-grained, ophitic mafic dike. Jutulrøra, northwest side	$k = 5139 \times 10^{-6}$
TVA 68 TV1	Coarse-grained, homogeneous syenite. Tvora, northside	$k = 2219 \times 10^{-6}$
HD 68 Nas1	Dark green siltstone. Nashornkalvane, north side	$k = 32.8 \times 10^{-6}$
HD 68 Nas2	Light gray graywacke. Nashornkalvane, north side	$k = 28.2 \times 10^{-6}$
—	Mafic sill. Nashornet. Susceptibility calculated from vol. % of opaques given by Bredell (1973)	$k = 0.007$

part of a sedimentary succession some 400 m thick, cut concordantly by a 150-m-thick mafic sill, which is not exposed at this nunatak but which occurs in a stratigraphically higher position farther to the south (Bredell, 1973). The strike (30°–42°) and dip (18°–20° E) used in the model are based upon field measurements by that author.

In the interpretation, the magnetic effect of the different models was computed in a manner similar to that of the gravity survey, with the cross section of the anomaly-producing body represented by an *n*-sided polygon in the direction of the traverse and extending to infinity at right angles to it (Talwani and Heirtzler, 1964). The magnetic anomalies were considered to be caused by induction only; the remanent magnetizations, as measured on the samples, could be neglected. Figure 120.2 provides the necessary data for the calculation of the components J_x (in the direction of the traverse) and J_z (vertical) of the intensity of magnetization J.

The introduction of several simple models showed that the general shape of the magnetic anomaly profile cannot be interpreted as a simple lateral contact between the gneissic complex and the sedimentary terrane. Nor can the high-susceptibility dolerite dikes—unless of unusual width—account for the peaks in the profile at 8, 20, and 26 km.

The gentle rise from a relatively flat negative low centered near 38 km to the positive peak at 26 km is suggestive of an anomaly caused by a vertical fault underneath the smooth transition area near 30 km that separates two contrasting susceptibilities. The magnitude of the anomaly further suggests that it could be caused by a fault bounding a broad structure with intermediate susceptibility to the east and low susceptibility to the west. Therefore, an intrusive massif with k = 0.002 corresponding to the susceptibility of the syenite at Tvora was introduced. The local anomaly profile peaks that are farther to the east (8 and 20 km) can then be explained by the geometric configuration of the top of the massif. Differentiation within the massif has not been attempted.

The Nashornkalvane anomaly is surprising, considering the low susceptibilities of the rock specimens collected there. The nearness of the rocks, both underneath and to the side of the station, situated on a small ridge of the nunatak at 48.25 km, can explain the very narrow peak, but the local topographic effect is very difficult to calculate with the two-dimensional model. The picture is further complicated by the change in course of the traverse.

The pronounced rise at 44 km and the peak at 50 km are tentatively explained by the existence within the sedimentary formations of a second (lower) mafic sill having the same geometry (thickness 150 m, dip 18° E, and strike 30°) and magnetic characteristics (k = 0.007 e.m.u./cm^3) as the one exposed on the other nunataks to the south on the ridge between Jutulstraumen and Viddalen. Truncation of the sill, which cuts the subglacial topography near 50 km, by a fault near 44 km may generate the observed anomalies.

The effect of the adopted model was calculated by taking into account the magnetization of all rocks including the low-susceptibility gneisses and sedimentary rocks. A comparison between the effect of the adopted model and the measured anomalies (Figure 120.3*b*) indicates a satisfactory fit and shows that the extrapolation of known surface geology allows the interpretation of the observed anomalies.

Flow line analysis

A glacial flow line analysis was attempted in order to interpret the drainage basin and its subglacial relief. This proved to be more complicated than a similar study for the Sør Rondane Mountains (Van Autenboer and Decleir, 1974, 1978). Analysis on the 1:250,000 topographic maps, which are restricted to the mountain area, resulted in flow lines which are not equivalent, i.e., the number of flow lines per unit area is not related to the mass flux.

The general flow pattern of the northwestern part of the Queen Maud Land ice dome had to be considered. The American Geographical Society 1:5,000,000 map suggests that the 3500-m-thick ice dome lying south of the Wohlthat Mountains causes a radially diverging flow, part of which is channelled directly into the main southeastern part of Jutulstraumen. The southeast-northwest alignment of the Kirwanveggen and Neumayerskarvet (30 km northeast of Kirwanveggen), locally parallel to this main flow pattern, can only result in a reduced flow through this escarpment. This likely difference in mass flux coupled with the dam effect of the escarpment itself was taken into account when drawing the flow lines.

The resulting pattern (Figure 120.1), which is compatible with a preliminary examination of ERTS imagery, confirms the importance of the measured mass flux through Jutulstraumen and its concentration in the narrow western part of the measured section. Some of the flow lines through Pencksökket are diverted at the northeastern boundary of Borgmassivet into Viddalen. This is confirmed not only by the considerable depths measured over the Viddalen section of the traverse but also by a subglacial relief map by Wolmarans (Chapter 124, this volume). The flow line map also indicates the main flux from the southeast that feeds Jutulstraumen, the divergence of the flow pattern through Kirwanveggen, and the marked lateral spreading that occurs when the ice stream reaches the ice shelf.

Discussion

The ice-covered topography in the mountain area between long. 0° and 3° W is dominated by Jutulstraumen, its tributary Viddalen, and Pencksökket. The floor of each glacier lies well below sea level: −1500 m for Jutulstraumen and Viddalen and −800 m for Pencksökket (Robin, 1958). Such depths are not exceptional, as shown by Wilson and Crary (1961) on the Skelton Glacier (−1300 m), by Morgan and Budd (1975) on the Lambert glacier (−2000 m), and by the 1308 m depth measured in Søgnefjord, Norway. The ice thicknesses on Viddalen are in general agreement with a radio-echo sounding map of the subglacial relief of the Ahlmannryggen and Borgmassivet further to the west (Wolmarans, Chapter 124, this volume).

The strongly asymmetric bedrock floor of Jutulstraumen reaches its maximum depth on the west side. In the regional picture it is clear that this deep subglacial depression forms the northward extension of the northwest side of the Pencksökket and that its further continuation can be traced northward to the Trollkjell escarpment.

The geomagnetic profile across Jutulstraumen was tentatively interpreted by an extrapolation of the known geology. The proposed model indicates the probable presence of at least two major discontinuities or faults. The most important one seems to be associated with the deepest part of Jutulstraumen and suggests a tectonic or intrusive contact between the basement complex and the sedimentary sequence. A large intrusive body with a susceptibility similar to that of the syenite at Tvora could explain the measured anomalies. K-Ar and Rb-Sr ages of a nepheline syenite formation, which seems to be the same as the one at Tvora, fall in the range of 165–270 m.y. (Soloviev, 1972). That author considers nepheline syenite bodies to be generally restricted to deep regional faults and that, in this case, they characterize the tectonic contact zone between basement and platform deposits. Geophysical evidence therefore supports Roots' (1969) contention that the Pencksökket is a down-faulted block and that the lower Jutulstraumen is limited by major faults.

The flow line analysis confirms this idea by showing the deviation of the flow and the existence of a deep trough (Pencksökket) in an area where the ice flow is not expected to be very high. The flow line analysis is in turn supported by the mass flux value of Jutulstraumen (0.233×10^{15} g/km•yr), which compares well with the mass flux of glaciers with large drainage basins (0.25×10^{15} g/km•yr) in the Transantarctic Mountains (Giovinetto and others, 1966). It is, however, very large when compared to the Sør Rondane area where this value equals 0.0067×10^{15} g/km•yr along a 220 km section and where the largest flux through an individual glacier only reaches 0.029×10^{15} g/km•yr (Van Autenboer and Decleir, 1974, 1978). The computed mass flux of Jutulstraumen is therefore eight times larger than that of the most active glacier in the Sør Rondane Mountains.

This confirms that ice streams are responsible for the removal of more ice than the sheet flow that occurs over the much larger part of the drainage periphery (Mellor, 1959). It seems that the main northerly flow from the plateau in Dronning Maud Land is channelled into two ice streams (Jutulstraumen and the glacier east of Novolazarevskaya) while the other outlet glaciers drain only a marginal section of the ice sheet.

Until recently, Jutulstraumen nourished Trolltunga, a major and characteristic ice tongue believed to have been observed as early as 1820. Satellite imagery and aerial observations show that the major part of the ice tongue calved off shortly before 1967 and the southern part during 1968. The area of the glacier that calved off in a short span of time was roughly estimated at 4500 km^2, which, by extrapolation of the 250 m ice thickness measured in its marginal parts by radio-echo sounding (Van Autenboer and Decleir, 1972), represents a total discharge of 1125 km^3. This is significant when compared to the estimated ablation by calving for the whole of Antarctica: 1450 km^3 (icebergs from all sources) and 520 km^3 (calving of glaciers only) (Loewe, 1967). The Trolltunga iceberg has been followed on satellite imagery during its westward drift into the Weddell Sea over the last nine years (Vinje, 1977).

Simple arithmetic gives that the minimum time necessary for the formation of the ice tongue is on the order of 100 years. This indicates that calving of glacier tongues is essentially a discontinuous process with long periods corresponding to the buildup of the glacier tongue and very short periods of major calving.

The scale of the mass wasting of Trolltunga and of the mass flux through Jutulstraumen and the discontinuous process of glacier wastage in the coastal areas show that a realistic estimate of mass wasting in Dronning Maud Land cannot be made at present. This supports Bull (1971) who noted that the possible errors in the ablation terms of the total mass balance in Antarctica are far larger than the magnitude of the assessed mass balance, so that at present no conclusion can be reached on whether the ice sheet is decreasing or increasing in size.

Acknowledgments. The authors thank all organizations and persons who were helpful: Baron G. de Gerlache, chairman of the Belgian Antarctic Committee, P. Dingens, director of the Observatory of the Gent State University; the South African National Antarctic Expeditions; J. J. Derwael; G. Kühn, who provided the geomagnetic records of SANAE; and J. Hus, who measured the susceptibilities of the specimens.

References cited

Aucamp, A. P. H., Wolmarans, L. G., and Neethling, D. 1972. The Urfjell group: a deformed(?) early Paleozoic sedimentary sequence, Kirwanveggen, western Dronning Maud Land. (*In* Adie, R. J., *ed. Antarctic Geology and Geophysics.* Oslo, Universitetsforlaget, 547–56.)

Barnard, H. P. 1975. Radio echo sounding in western Dronning Maud Land, 1974. *S. Afr. J. Antarct. Res.,* No. 5, 37–42.

Behrendt, J. C., and Bentley, C. R. 1968. Magnetic and gravity maps of the Antarctic. *Antarc. Map Folio Ser.,* Folio 9, Pl. IV.

Bredell, J. R. 1973. The geology of the Nashornet-Viddalskollen area, western Dronning Maud Land, Antarctica. *S. Afr. J. Antarct. Res.,* No. 3, 2–10.

Bull, C. 1971. Snow accumulation in Antarctica. (*In* Quam, L. O., *ed.* Research in the Antarctic. *Am. Ass. Adv. Sci.* Publn No. 93, 367–421.)

Craddock, C. 1972. Antarctic tectonics. (*In* Adie, R. J., *ed. Antarctic Geology and Geophysics.* Oslo, Universitetsforlaget, 449–55.)

Giovinetto, M. B., Robinson, E. S., and Swithinbank, C. W. M. 1966. The regime of the western part of the Ross Ice Shelf drainage system. *J. Glaciol., 6,* No. 43, 55–68.

Gjessing, Y. T. 1972. Mass transport of Jutulstraumen ice stream in Dronning Maud Land. *Arbok, Norsk Polarinst., 1970,* 227–32.

Hjelle, A. 1974. Some observations on the geology of H. U. Sverdrupfjella, Dronning Maud Land. *Arbok, Norsk Polarinst., 1972,* 7–22.

Hubbert, M. K. 1948. A line integral method of computing the gravimetric effects of two-dimensional masses. *Geophysics, 13,* No. 2, 115–25.

Loewe, F. 1967. The water budget in Antarctica. *JARE Sci. Rep.,* Special Issue No. 1, 101–10.

Mellor, M. 1959. Ice flow in Antarctica. *J. Glaciol., 3,* No. 25, 337–87.

Morgan, V. I., and Budd, W. F. 1975. Radio-echo sounding of the Lambert glacier basin. *J. Glaciol., 15,* No. 73, 103–11.

Robin, G. de Q. 1958. Seismic shooting and related investigations. *Scient. Res. Norw.-Br.-Swed. Antarct. Exped., 1949–1952, 5,* 134.

Roots, E. F. 1969. Geology of western Queen Maud Land. (*In* Bushnell, V. C., and Craddock, C., *eds.* Geologic Maps of Antarctica. *Antarct. Map Folio Ser.,* Folio 12, Pl. VI.)

Soloviev, D. S. 1972. Platform magmatic formations of East Antarctica. (*In* Adie, R. J., *ed. Antarctic Geology and Geophysics.* Oslo, Universitetsforlaget, 531–38.)

Talwani, M., and Heirtzler, J. R. 1964. Computation of magnetic anomalies caused by two-dimensional structures of arbitrary shape. (*In* Parks, G. A., *ed. Computers in the Mineral Industries.* Stanford, Stanford University, 464–80.)

Talwani, M., Worzel, J. L., and Landisman, M. 1959. Rapid gravity computations for two-dimensional bodies with application to the Mendocino submarine fracture zone. *J. geophys. Res., 64,* No. 1, 49–60.

Van Autenboer, T., and Blaiklock, K. V. 1966. Ice flow and thickness measurements in the Sør Rondane, Dronning Maud Land, Antarctica. *J. Glaciol., 6,* No. 43, 69–81.

Van Autenboer, T., and Decleir, H. 1969. Airborne radio-glaciological investigations during the 1969 Belgian Antarctic Expeditions. *Bull. Belg. Ver. Geol. Paleont. Hydrol., 78,* No. 2, 87–100.

Van Autenboer, T., and Decleir, H. 1972. Ice thickness and subglacial relief of the Jelbartisen-Trolltunga area, Dronning Maud Land. (*In* Adie, R. J., *ed. Antarctic Geology and Geophysics.* Oslo, Universitetsforlaget, 713–22.)

Van Autenboer, T., and Decleir, H. 1974. Jelbartisen-Trolltunga, Dronning Maud Land, Antarctica. Radioglaciological survey, data report. *Prof. Paper. Belg. geol. Surv., 1974,* No. 1, 14.

Van Autenboer, T., and Decleir, H. 1978. Glacier discharge in the Sør Rondane, a contribution to the mass balance of Dronning Maud Land, Antarctica. *Z. Gletsch. Glazialgeol., 14,* No. 1, 1–16.

Van Autenboer, T., and Loy, W. 1972. Recent geological investigations in the Sør Rondane Mountains, Belgicafjella and Sverdrupfjella, Dronning Maud Land. (*In* Adie, R. J., *ed. Antarctic Geology and Geophysics.* Oslo, Universitetsforlaget, 563–71.)

Vinje, T. E. 1977. Drift av Trolltunga i Weddellhavet. *Arbok, Norsk Polarinst., 1975,* 213.

Wilson, C. R., and Crary, A. P. 1961. Ice movement studies on the Skelton Glacier. *J. Glaciol., 3,* No. 29, 873–78.

121

Isostatic Gravity Anomalies in West Antarctica

C. R. BENTLEY and J. D. ROBERTSON*
Geophysical and Polar Research Center, University of Wisconsin, Madison, Wisconsin 53706, U.S.A.

Abstract. Through use of a three-dimensional computer gravity-modeling procedure, a correction for the finite size and lateral distribution of compensating masses at depth is applied to a 1° × 1° mean free-air gravity map of West Antarctica to produce an isostatic anomaly map. Extensive regions of West Antarctica, including the Byrd subglacial basin and the basin between the Ellsworth and Pensacola Mountains, are not detectably out of isostatic equilibrium. There are six significant isostatic anomalies on the map: (1) a positive anomaly of about 35 mgal at the base of the Antarctic Peninsula; (2) a positive anomaly of about 25 mgal on the northwestern flank of the Ellsworth Mountains; (3) a negative anomaly of about −20 mgal across the northeastern side of the Whitmore Mountains; (4) a negative anomaly of about −50 mgal on the Rockefeller Plateau; (5) a saddle-shaped positive anomaly ranging from about 40 mgal near the Hobbs Coast to about 25 mgal in an area immediately south of the Crary Mountains; and (6) a negative anomaly of about −50 mgal on the Amundsen Sea coast. The first and second anomalies are tentatively attributed to uncompensated crustal thinning, the third and fourth to downward extension of upper crustal rocks into the lower crust, the fifth to crustal rifting, and the sixth to a combination of recent retreat of ice and unknown tectonic factors.

Since the beginning of the International Geophysical Year, gravity observations have been made routinely on over-snow traverses in West Antarctica. Individual free-air gravity values on ice sheets depend primarily on subglacial rock elevations. Since this buried-terrain dependence is as likely to be negative as positive, Bentley (1968) used the technique of averaging gravity values for many adjacent stations to minimize the influence of topography on a free-air anomaly map constructed for West Antarctica. He computed means of free-air values both for squares 111 km (1° of latitude) on a side and for squares 222 km on a side and identified significant regional free-air anomalies in western Marie Byrd Land, near the Amundsen Sea coast, just north of the Horlick Mountains, and at the base of the Antarctic Peninsula.

A mean free-air anomaly map on an ice sheet is a good first approximation to an isostatic anomaly map even over a region of rough subglacial terrain, since the terrain effect will average close to zero. Even if isostatic compensation for the subglacial terrain is perfect, however, a correlation will still exist between mean elevation and mean free-air gravity owing to the finite size of the surficial and compensating rock masses. Correction for this effect, which reaches tens of milligals in some squares, produces an isostatic anomaly map from which both the state of isostatic balance in West Antarctica and the size and shape of anomalies rising from geologic structure can be assessed more readily than from the mean free-air anomaly map itself.

* Present address: Atlantic Richfield Company, Box 2819, Dallas, Texas 75221, U.S.A.

Method of analysis

Average free-air anomalies previously published (Bentley, 1968) were modified to incorporate more recent data (Kososki, 1972; Beitzel, 1972). The results for the squares shown in Figure 121.1 are tabulated in Table 121.1. In most cases, a 1° × 1° square mean is simply a linear average along a single traverse route and may be unrepresentative of the true areal mean in the square. However, the traverse routes were generally chosen without regard to subglacial terrain so that they should tend to sample a random section of the topography of each square.

From seismic soundings and the gravity observations themselves, a mean equivalent rock elevation, defined as the mean surface elevation if all ice in a square were compressed to the average density of 2.67 g/cm^3, was calculated for each 1° × 1° square (Table 121.1). With the assumption of perfect Airy isostatic compensation for the mean equivalent rock elevation, the gravitational attraction of topography and isostatic roots was calculated using the three-dimensional gravity-modeling procedure of Talwani and Ewing (1960). Subtraction of the model gravity field from the free-air anomaly field produces the isostatic anomaly map shown in Figure 121.2.

To carry out the modeling, it is necessary to adopt a "standard" crustal column for West Antarctica. Estimates of crustal thickness are not numerous—those that do exist, based on seismic and gravity data, suggest a standard sea-level thickness of about 30 km (Bentley, 1964, 1973). This value was therefore adopted.

The densities chosen were 2.67 g/cm^3 for the topography-forming rock, 2.95 g/cm^3 for the lower crust, and 3.32 g/cm^3 for the upper mantle. The existence of rocks in

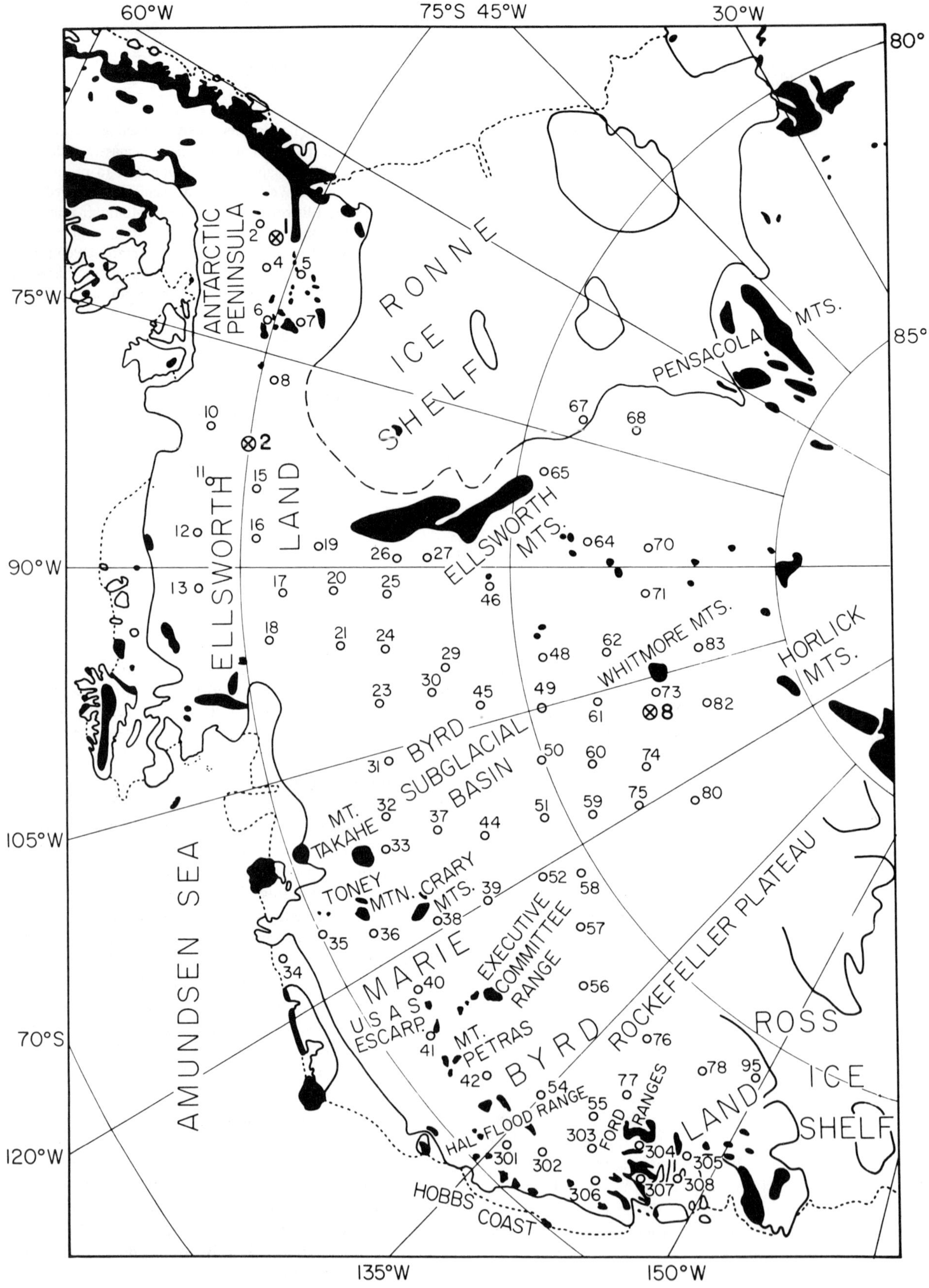

Figure 121.1. Location of centers of 1° × 1° squares in West Antarctica. Circled crosses show locations of the seismic velocity columns from Bentley and Clough (1972) that are referred to in the text.

Table 121.1. Gravity anomalies and equivalent rock elevations, means for 1° x 1° squares

Square number	Free-air gravity anomaly (mgal)	Mean equivalent rock elevation (m)	Isostatic gravity anomaly (mgal)	Number of stations	Square number	Free-air gravity anomaly (mgal)	Mean equivalent rock elevation (m)	Isostatic gravity anomaly (mgal)	Number of stations
2	76.3	1080	3	30	48	34.8	750	1	31
4	33.7	480	10	27	49	0.1	30	-5	20
5	82.1	1120	17	16	50	-38.1	-570	-13	37
6	47.4	490	36	31	51	-7.0	-100	-3	35
7	13.9	-90	34	26	52	-12.5	-180	-9	25
8	-2.5	-520	27	40	54	32.2	1075	9	24
10	30.3	220	24	26	55	18.8	881	-15	23
11	17.0	60	7	39	56	-23.6	-60	-18	17
12	23.0	160	14	25	57	5.0	160	-13	13
13	38.4	390	13	63	58	-9.3	-240	-1	22
15	1.7	-220	13	28	59	0.3	-70	-3	124
16	6.0	-80	6	15	60	-18.4	-150	-8	31
17	3.1	-340	13	38	61	-12.5	340	-23	21
18	-9.0	-550	11	20	62	-0.2	510	-11	22
19	7.2	-280	21	35	64	5.4	260	6	32
20	0.4	-80	-4	18	65	-4.9	-170	-3	44
21	12.4	30	-2	32	67	-11.7	-860	2	41
23	-27.4	-590	-15	62	68	-30.4	-710	-10	24
24	-38.8	-780	-9	20	70	6.7	330	-17	30
25	16.8	-160	25	20	71	-2.9	140	4	51
26	38.9	620	7	19	73	44.0	1160	-5	27
27	45.4	640	26	16	74	6.2	130	-5	59
29	-7.7	-50	-2	16	75	-21.3	-360	-11	39
30	-6.9	-240	-1	16	76	-63.0	-160	-53	35
31	-28.6	-630	-14	21	77	-39.8	127	-49	29
32	-26.4	-450	-19	19	78	-7.2	-120	-5	14
33	2.0	-200	4	22	80	-23.2	-500	-10	21
34	-74.9	-660	-54	10	82	9.5	220	17	54
35	-38.1	-80	-41	31	83	32.9	990	-21	52
36	2.0	450	-12	43	95	0.7	-200	1	16
37	-5.8	-240	-3	22	96	-18.8	-470	-10	49
38	-1.0	620	-21	24	301	42.0	440	41	11
39	7.4	90	26	21	302	26.9	596	22	10
40	23.7	780	-6	18	303	40.4	779	3	15
41	30.4	1050	-5	27	304	4.4	475	-6	48
42	0.5	810	-23	19	305	-1.1	120	-14	21
44	16.0	300	2	21	306	9.1	609	-11	11
45	-8.5	-50	-3	28	307	-37.0	112	-39	37
46	44.5	920	11	31	308	-5.9	7	2	11

the lower crust of distinctly higher density than that of the upper crust is to be expected not only from seismic and gravity studies elsewhere in the world (e.g., Woollard, 1968), but also from a few seismic refraction profiles in Antarctica (Bentley and Clough, 1972) that reveal a velocity of about 7 km/sec or a little less. The density of 2.95 g/cm^3 corresponds to that velocity, according to various empirical and experimental models (Woollard, 1959; Birch, 1961; Christensen and Fountain, 1975). The other densities are those commonly assumed for gravity modeling.

It should be emphasized that the isostatic corrections calculated are very insensitive to the choice of density, not only for the usual reason that they represent the difference between effects of identical masses, but also because, in this work, we are concerned with means over 1° squares. Furthermore, the gravity observations were made, on the average, more than 2 km above the rock topography, on the surface of the ice sheet.

The "standard" crustal thickness chosen is similarly not very important. Average isostatic anomalies in regional studies elsewhere (e.g., Erola, 1941; Tanni, 1942) generally vary by only one or two milligals as the crustal thickness varies between 20 and 40 km. Specifically for West Antarctica, Kososki (1972) examined the effect of different crustal models by calculating both Pratt-Hayford and Airy-Heiskanen isostatic anomalies. Differences between individual anomalies rarely exceeded 2 mgal. Such variations are negligible in comparison with other errors in Antarctic gravity studies.

Bentley (1964) estimated that the standard error in regional mean free-air gravity anomalies from Antarctic traverses on the grounded ice sheet is ± 10 mgal. To be conservative, we do not consider mean isostatic anomalies to differ significantly from zero unless they exceed 20 mgal in magnitude.

The interpretation of the overall mean requires a look at the basic datum level. The free-air anomalies used in this analysis were based on the International Gravity Formula (IGF). Satellite studies have shown, however, that the earth differs in shape significantly from that assumed in the development of the IGF. In 1971, the International Association of Geodesy adopted the Geodetic Reference System 1967, which has the effect of increasing anomalies by 4 mgal at latitudes greater than 75°. At the same time, the Potsdam datum was diminished by 14 mgal. The net result is that all gravity anomalies in the high polar regions should be diminished (algebraically) by 10 mgal.

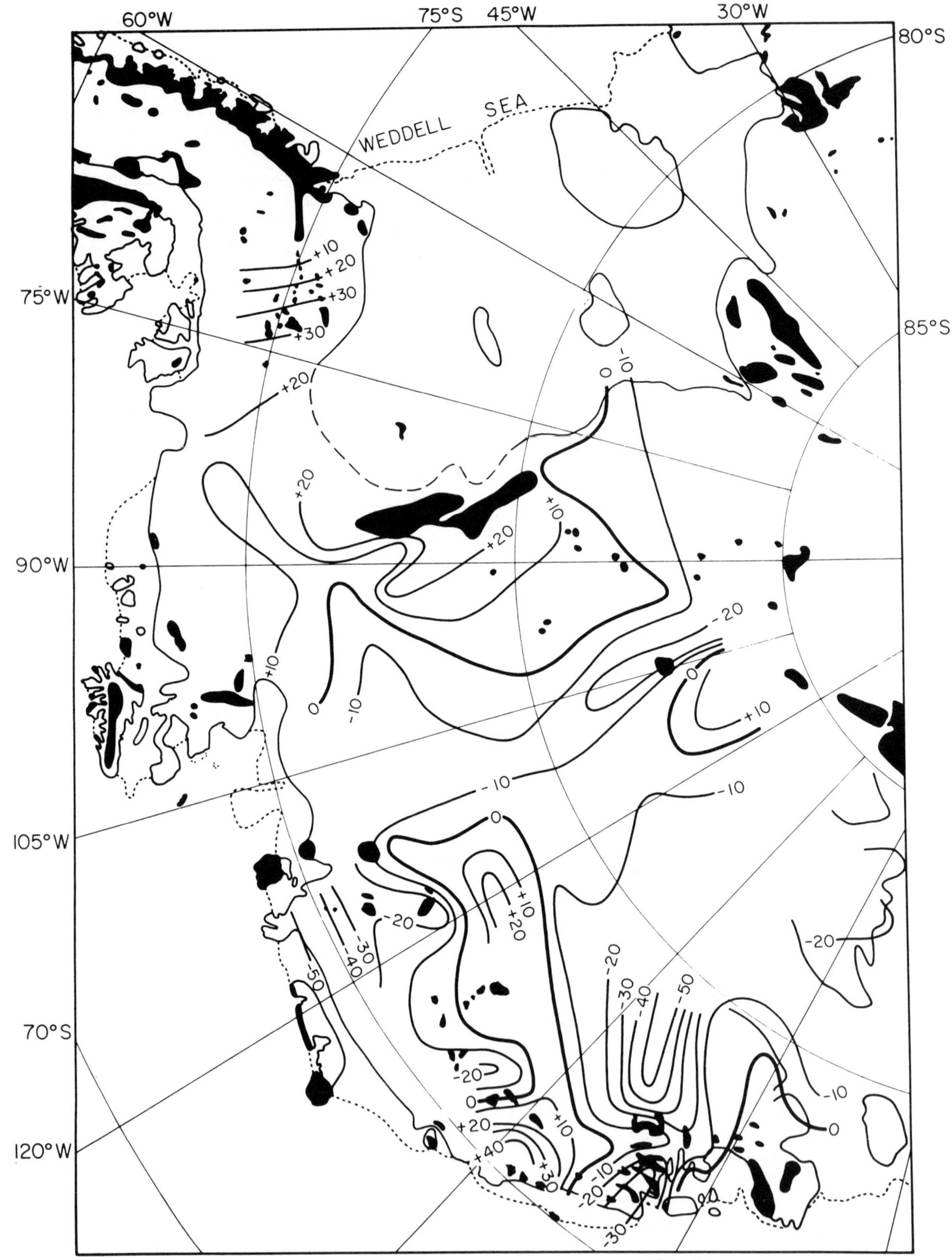

Figure 121.2. 1° × 1° isostatic anomaly map of West Antarctica. Anomalies are in mgals.

The effect of this on isostatic studies, however, is unclear. Regional gravity anomalies of large extent and tens of milligals in amplitude cover most of the Earth (for a recent compilation see Wagner and others, 1977). The cause of these anomalies must be deep-seated—they certainly are not related simply to crustal geologic features. Their existence makes it difficult to assess any regional West Antarctic glacio-isostatic imbalance. Without knowing what the mean isostatic anomaly would be in the absence of an ice sheet, we cannot tell from the present mean anomaly whether the ice sheet is compensated or not. We can only say that there is no evidence of a gross, regional imbalance (e.g., more than 500 m of uncompensated ice, corresponding to an anomaly of 20 mgal).

A recent satellite model shows a regional West Antarctic negative free-air anomaly of about −10 mgal (Wagner and others, 1977). Whether we attribute this to glacial or nonglacial causes, a West Antarctic area showing the same degree of isostatic compensation as West Antarctica on the whole should exhibit an isostatic anomaly of about −10 mgal. Coincidentally, this is the same magnitude as the corrections needed to the IGF, as described above. Thus, the maps based on the IGF can be used as a first approximation to a proper isostatic map, in the sense that zero isostatic anomaly implies an isostatic state similar to the regional state of West Antarctica.

Isostatic anomaly map

From Figure 121.2 it appears that large areas of West Antarctica are not detectably out of isostatic balance. In particular, the Byrd subglacial basin and the basin between the Ellsworth and Pensacola mountains seem to be reasonably well compensated. There are, however, six areas of anomalous isostatic gravity.

1. A positive anomaly reaching more than 35 mgal trends across the base of the Antarctic Peninsula. The anomaly lies on the western flank of a region of high topography and of a large positive free-air anomaly. To the east of the isostatic anomaly, a seismic velocity column (Column 1, Bentley and Clough, 1972; see Figure 121.1 for location) indicated a few kilometers of granitic intrusive rocks typical of the Andean Intrusive Suite underlain by crystalline basement and overlain by a thin layer of either frozen till or Mesozoic sedimentary rocks such as those found in the mountains of eastern Ellsworth Land. A second seismic column (Column 2, Bentley and Clough, 1972) to the west of the anomaly is thought to lie on a faulted boundary between the Antarctic Peninsula and the probable preglacial edge of the Weddell Sea. From both the broad extent of the isostatic anomaly and the absence of rock of density (as estimated from the seismic velocity) greater than 2.67 g/cm^3 in Columns 1 and 2, a likely cause of the Antarctic Peninsula positive anomaly is a widespread, abnormally shallow depth either to the lower crust or to the Moho, perhaps associated with an upthrown crustal block northwest of the Weddell Sea-Antarctic Peninsula boundary. From the same modeling program used to compute the isostatic map, it was found that the anomaly could be removed by raising the Moho beneath the anomaly by a few kilometers.

2. A positive isostatic anomaly of about 25 mgal lies on the northwestern side of the Ellsworth Mountains. Like the Antarctic Peninsula anomaly, the "Sentinel" anomaly sits on the flank of a major subglacial ridge. Modeling calculations indicate that the anomaly could be explained by an uncompensated decrease in depth to Moho of a few kilometers. There are no data in the northwestern section of the Ronne Ice Shelf to indicate whether the Antarctic Peninsula and Ellsworth highs are two spurs of the same feature.

3. The isostatic gravity anomaly falls below −20 mgal in two squares forming an elongated southeast-northwest low, trending across the Whitmore Mountains. Barely significant by themselves, the two values are substantially different from a positive isostatic value of 15 mgal in a square to the southwest over the trench between the Whitmore Mountains and the Horlick Mountains. A seismic velocity column (Column 8, Bentley and Clough, 1972) indicates that rock with a velocity of 7 km/sec lies within 4 km of sea level in this region. The Whitmore Mountains themselves are formed from felsic intrusive rocks and from the same Paleozoic sequence found in the Ellsworth Mountains, whereas the Horlick Mountains are composed of rocks from the Beacon Supergroup and the Granite Harbour Intrusive Suite of the Transantarctic Mountains (Craddock, 1972). Lying therefore on the boundary between the Ross and Ellsworth orogens, the relative isostatic gravity anomaly probably results from a mismatch of densities across the boundary. A model featuring intrusion of lower crustal mafic rocks along a crustal rift to a depth of 4 km below sea level in the Whitmore-Horlick trough substantially diminishes the one-point positive part of the anomaly. The inclusion in the model of felsic intrusive rocks in the lower crust in the vicinity of the Whitmore Mountains increases the negative part of the gravity anomaly to gravity values near zero. There are, of course, other models which would fit this rather complex anomaly equally well.

4. A negative isostatic anomaly of more than 50 mgal crosses the northern Rockefeller Plateau in a northwest-southeast direction. The anomaly coincides with a $1° \times 1°$ mean free-air anomaly of −60 mgal, but it is not coincident with any significant topographic feature. It does trend into the Ford Ranges, which are the site of a geosyncline that developed in late Precambrian-early Paleozoic time and in which were deposited more than 4600 m of highly quartzose sedimentary beds (Wade, 1969, 1972). The sedimentary rocks were moderately to intensely folded along nearly east-west axes and were intruded by granodiorite plutons in middle Paleozoic time. It appears likely that the Rockefeller Plateau negative anomaly lies along a southeastward extension of the axis of this geosyncline. A three-dimensional gravity model in which upper crustal rocks extend 5 km into the lower crust beneath the anomaly completely removes the isostatic feature.

5. A saddle-shaped positive anomaly stretches southeastward from Hobbs Coast around the southern side of the Executive Committee Range to an area south of the Crary Mountains. The anomaly exceeds 40 mgal in the west and 25 mgal in the east. In eastern Marie Byrd Land, it generally corresponds with elevated subglacial topography, although the anomaly is narrower than the mountainous region and is centered on its southern part. The region of high relief to the north, which extends from Mount Petras across the USAS Escarpment to Toney Mountain, has no positive isostatic expression. It is apparent from the

geologic map of Antarctica (Craddock, 1972) that, to the west, the isostatic anomaly coincides with the dividing line between the Cenozoic volcanic rocks of the Hal Flood Range and the Cretaceous volcanic rocks and upper Precambrian–lower Paleozoic sedimentary and metasedimentary rocks of the Ford Ranges. To the east, the anomaly separates the Crary Mountains and Mount Takahe on the north from the Byrd subglacial basin on the south. The Crary Mountains and Mount Takahe are composed of Cenozoic volcanic rocks as are most of the Executive Committee Range, the USAS Escarpment, Toney Mountain, and Mount Petras. The composition of the rocks of the Byrd subglacial basin is not directly known, but, from seismic refraction evidence, the rocks are probably either Cenozoic volcanic rocks or unmetamorphosed Paleozoic–Mesozoic strata of almost any kind (Bentley and Clough, 1972). Based on the geology of the area, a speculative determination of the cause of the positive isostatic anomaly is that it delineates a major crustal fissure along which Cenozoic volcanic rocks have penetrated Marie Byrd Land. Modeling calculations indicate that the anomaly could be essentially explained by an uncompensated 4 km extension of lower crustal rocks into the upper crust.

6. The last significant isostatic feature in Figure 121.2 is a negative anomaly exceeding −50 mgal over the Amundsen Sea coast. The coincident free-air anomaly of −40 mgal was attributed by Bentley and Chang (1971) partially to recent retreat of ice along the coast and partially to tectonic factors. Conversion of the free-air map to an isostatic map does not significantly affect that explanation. The extent of the anomaly is poorly defined.

Acknowledgments. This work was supported by the National Science Foundation under grants GA1706 and GV36963.

References cited

Beitzel, J. E. 1972. *Geophysical Investigations in Marie Byrd Land, Antarctica.* Ph.D. dissertation, University of Wisconsin–Madison, 111 pp.

Bentley, C. R. 1964. The structure of Antarctica and its ice cover. (*In* Odishaw, H., *ed. Research in Geophysics, 2. Solid Earth and Interface Phenomena.* Cambridge, Mass., MIT Press, 225–89.)

Bentley, C. R. 1968. Gravity maps. (*In* Behrendt, J. C., and Bentley, C. R., *eds.* Magnetic and Gravity Maps of the Antarctic. *Antarct. Map Folio Ser.,* Folio 9, PL. VIII–IX.)

Bentley, C. R. 1973. Crustal structure of Antarctica. *Tectonophysics, 20,* 229–40.

Bentley, C. R., and Chang, F. K. 1971. Geophysical exploration in Marie Byrd Land, Antarctica. (*In* Crary, A. P., *ed. Antarctic Snow and Ice Studies 2.* Washington, American Geophysical Union, 1–38.) [Antarctic Research Series, Vol. 16.]

Bentley, C. R., and Clough, J. W. 1972. Antarctic subglacial structure from seismic refraction measurements. (*In* Adie, R. J., *ed. Antarctic Geology and Geophysics.* Oslo, Universitetsforlaget, 683–91.)

Birch, F. 1961. The velocity of compressional waves in rocks to 10 kilobars, Part 2. *J. geophys. Res., 77,* 6385–91.

Christensen, N. I., and Fountain, D. M. 1975. Constitution of the lower continental crust based on experimental studies of seismic velocities in granite. *Geol. Soc. Am. Bull., 86,* 227–36.

Craddock, C. 1972. *Geologic Map of Antarctica. 1:5,000,000.* New York, American Geographical Society.

Erola, V. 1941. On the structure of the earth's crust in the neighborhood of the Ferghana Basin. *Ann. Acad. Sci., Fennicae, Helsingfors,* Ser. A3, *3,* 77 pp. (Isost. Inst. Int. Ass. Geodesy, No. 10.)

Kososki, B. A. 1972. *A Gravity Study of West Antarctica.* Master's thesis, University of Wisconsin–Madison, 78 pp.

Talwani, M., and Ewing, M. 1960. Rapid computation of gravitational attraction of three-dimensional bodies of arbitrary shape. *Geophysics, 25,* 203–25.

Tanni, L. 1942. On the isostatic structure of the earth's crust in the Carpathian countries and the related phenomena. *Ann. Acad. Sci., Fennicae, Helsingfors,* Ser. A3, *4,* 100 pp. (Isost. Inst. Int. Ass. Geodesy, No. 11.)

Wade, F. A. 1969. Geology of Marie Byrd Land. (*In* Bushnell, V. C., and Craddock, C., *eds.* Geologic Maps of Antarctica. *Antarct. Map Folio Ser.,* Folio 12, Pl. XVIII.)

Wade, F. A. 1972. Geologic survey of Marie Byrd Land. *Antarct. Jnl U.S., 7,* 144–45.

Wagner, C. A., Lerch, F. J., Brownd, J. E., and Richardson, J. A. 1977. Improvement in the geopotential derived from satellite and surface data (Gem 7 and 8). *J. geophys. Res., 82,* 901–14.

Woollard, G. P. 1959. Crustal structure from gravity and seismic measurements. *J. geophys. Res., 64,* 1521–44.

Woollard, G. P. 1968. The interrelationship of the crust, the upper mantle, and isostatic gravity anomalies in the United States. (*In* Knopoff, L., Drake, C. L., and Hart, P. J., *eds. The Crust and Upper Mantle of the Pacific Area.* American Geophysical Union, Washington, 312–41.) [Geophysical Monograph No. 12.]

122

Source Properties of the Oates Land Earthquake, October 1974

R. D. ADAMS*
Seismological Observatory, Geophysics Division, Department of Scientific and Industrial Research, Wellington, New Zealand

Abstract. An earthquake near the coast of Oates Land (lat. 70.5° S, long. 161.3° E) on 15 October 1974 is the only shock to have been located on the Antarctic continent by international agencies. The earthquake has the appearance of a normal shallow earthquake, with crustal phases recorded at Scott Base, the nearest station. The local magnitude (M_L) is about 5, and the seismic moment (M_o) determined from long-period records at Scott Base is 5×10^{16} newton-meters. This relation between magnitude and moment is near the middle of the range found for tectonic earthquakes in New Zealand and, if the source were in normal crustal rock, would imply a source radius of about 6 km, a stress drop of about 1 bar, and a source dislocation of about 1.5 cm. Although there is nothing unusual in these source parameters, the earthquake's location near the mouth of the Rennick Glacier suggests that, like smaller events previously located near glaciers in southern Victoria Land, it might be associated with ice movement or cracking in the glacier.

Earthquake locations in the Antarctic regions have in general been confined to the oceanic ridges surrounding the continent, but small local activity has been reported near most seismograph stations, including those at Mirny (Lazareva and Sytinskiy, 1965), Wilkes (Browne-Cooper and others, 1967), and Syowa (Kaminuma, 1971). Small disturbances have been reported from Scott Base that are consistent with calving of icebergs from the Ross Ice Shelf (Hatherton and Evison, 1962), and Adams (1971) has identified local disturbances recorded at Scott Base and Vanda as originating from areas near the Drygalski Ice Tongue and Harbord Glacier. More recently, Kyle and others (Chapter 90, this volume) have studied small volcanic shocks on Mount Erebus. Apart from an isolated shock suggested by Kaminuma and Ishida (1971) as occurring near lat. 80° S, long. 20° W in 1968, all these events are located in areas where they could be explained either by ice movement or volcanism, and none have been large enough to be recorded at seismograph stations outside Antarctica.

On 15 October 1974, the largest earthquake to have definitely originated on the Antarctic continent occurred near the coast of Oates Land. This was the first Antarctic earthquake to have been recorded elsewhere and to have been assigned locations by the Preliminary Determination of Epicenters Service of the U.S. Geological Survey (PDE) and by the International Seismological Centre (ISC). The earthquake was given body-wave magnitudes (m_b) of 4.9 by USGS and 4.7 by ISC, and by virtue of its size and unique location merits closer study.

* Present address: International Seismological Centre, Newbury and Department of Geology, University of Reading, U.K.

Location

Table 122.1 gives the formal solutions for the origin of the earthquake of 15 October 1974, as determined by PDE and ISC. The ISC solution is based on *P* readings at fourteen stations, of which four are in Antarctica, and the rest distributed in New Zealand, Australia, Africa, and South America. In addition, seven stations in the distance range from 146° to 156° reported *PKP* phases.

The epicenter is shown in Figure 122.1 to be near the coast of Oates Land. Also shown are the paths to the nearest two stations, Scott Base (SBA) and Dumont d'Urville (DRV). The inset shows the epicentral region on a larger scale, with the ISC epicenter and its standard errors. The epicenter lies at the western side of the Rennick Glacier, near its mouth. The PDE epicenter, not shown, is about 8 km to the east, closer to the center of the glacier, which is about 25 km wide at this point. Each epicenter lies within the area of confidence of the other, and in view of the errors involved in location, they must be considered identical.

The seaward extent of the glacier seems uncertain and may vary with time. Most maps, as drawn here, show the glacier extending to the vicinity of Znamenskiy Island, about 30 km north of the epicenter, but others, including the Geologic Map of Antarctica published by the American Geographical Society (Craddock, 1972), show the shoreline further south, nearer the position assigned to the epicenter.

The position of the earthquake is reasonably well determined, with four Antarctic stations reporting readings. The two closest, Scott Base and Dumont d'Urville, together with Milford Sound in New Zealand, give good azi-

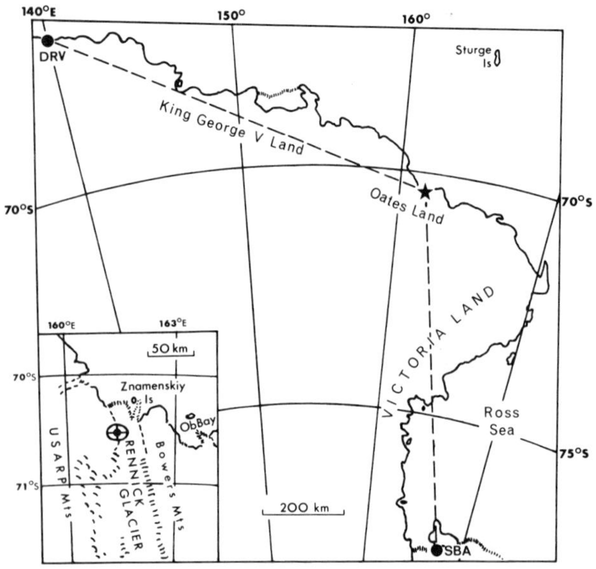

Figure 122.1. Epicenter of Oates Land earthquake of 15 October 1974, shown by the star, and paths to the nearest two stations, Scott Base (SBA) and Dumont d'Urville (DRV). Inset shows detail of epicentral area, with epicenter and standard errors determined by International Seismological Centre.

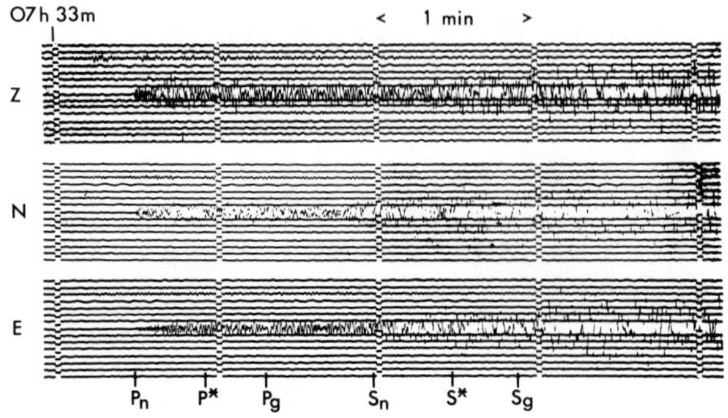

Figure 122.2. Short-period records at Scott Base of the earthquake of 15 October 1974 (distance 7.49°). Expected times of standard crustal phases are shown. (Seismometer period, 1.0 second; galvanometer period, 0.75 second).

muthal coverage, but there is no station close enough to enable the depth to be determined reliably. A standard depth of 33 km has been assumed by both agencies in default of other information and must not be considered well established. The only reported readings of depth phases are from Josvafo in Hungary, where the main *PKP* phase is followed by arrivals 4.2 seconds and 7.2 seconds later. If these are interpreted as *pPKP* and *sPKP* phases arising in normal crustal structure, they would correspond to a focal depth of about 10 km. Because of complexity in *PKP* phases, however, these interpretations cannot be given great weight.

Recordings at Scott Base

The records of this earthquake written at the World Wide Standard Seismograph station at Scott Base are shown in Figures 122.2 and 122.3. The short-period records (Figure 122.2) have the appearance of a typical shallow earthquake with much complexity due to crustal propagation, which favors a focus substantially shallower than that formally assigned. On the figure are marked the approximate arrival times at which crustal phases would be expected for a standard earth model. Although later arrivals are evident in the *P* group, they do not correspond clearly to the expected times of arrival of P^* and P_g, even though there is a large pulse on the north-south component about the time of P_g which would correspond to a wave velocity of 5.6 km/s, such as is expected for upper crustal rocks, including those of granitic type. The presence of this phase would add extra weight to the presumption of a focus shallower than about 10 km.

The *S* phases are not clearly distinguishable, although there is a general increase in amplitude corresponding to the time of S^*, and a lengthening of the period of the east-west recording at the expected time for S_n. The complexity of the short-period records at Scott Base is similar to that evident in records of closer shocks in Victoria Land reported by Adams (1971) and may be a feature, as suggested there, of transmission paths across the Ross Sea where there is likely to be substantial thickness of sediment and moraine.

Short-period records at Dumont d'Urville are much simpler in character and show clear *P* and *S* phases. This suggests that the crustal structure along the King George V Coast is simpler and more uniform than that between the earthquake and Scott Base.

Figure 122.3 shows the long-period recordings at Scott Base (seismometer period, 15 seconds; galvanometer period, 100 seconds). On the Z and north-south components the earthquake is superposed on surface waves from a larger distant earthquake. *P* waves are not recorded, and the first energy recorded on all components corresponds to the phase S*, traveling with a velocity of about 3.7 km/s in lower crustal rocks.

Waves with a period of up to 20 s are recorded on the Z and north-south components from 07h 36m. The polarization of these waves confirms that they are Rayleigh waves approaching from the north. The distance, about 7.5°, is too short to warrant a full surface-wave dispersion analysis, but the recorded velocities of 3.2 to 2.8 km/s are consistent with those found for continental paths. A further feature of these surface waves, indicative of a continental path, is their sudden cessation, particularly on the north-south component at about 07h 37.5m.

Table 122.1 Origin solutions for earthquake of 15 October 1974

Agency	Origin time h	m	s	Epicenter Lat.	Long.	Adopted Depth (km)	Standard deviation s	Number of stations	Magnitude (m_b)
PDE	07	31	42.1	70.52°S	161.53°E	33	1.3	12	4.9
			±0.5	±7km	±11 km				
ISC	07	31	42.9	70.55°S	161.3°E	33	1.7	14	4.7
			±0.5	±0.07	±0.3				

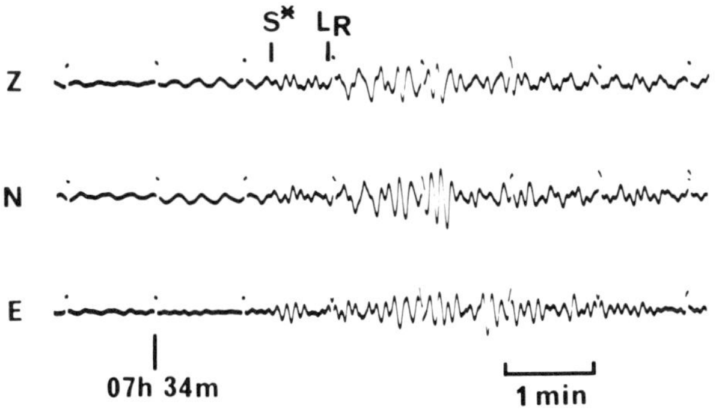

Figure 122.3. Long-period records at Scott Base of the earthquake of 15 October 1974. (Seismometer period, 15 seconds; galvanometer period, 100 seconds).

Source characteristics

The position of this earthquake near the mouth of the largest glacier in the area raises the question of whether, like previously reported seismic events in Antarctica, it could be associated with ice breakage or movement, rather than with the release of tectonic strain in the underlying rock.

To investigate this further, the source properties of the earthquake were analyzed according to the spectral theory of Brune (1970) as developed by Gibowicz (1975) for New Zealand earthquakes. The theory assumes that an earthquake arises from a simple dislocation over a circular area of a fault surface. The earthquake's magnitude (M_L) and seismic moment (M_o) can be determined from the short-period and long-period records, respectively. Once magnitude and moment are found, the radius of the equivalent circular source, the average displacement of the rupture, the stress drop, and the seismic energy can all be determined.

The recordings at Scott Base were used to derive these source properties. The equivalent local magnitude (M_L) was obtained from the maximum trace amplitude of the horizontal short-period instruments, adjusted for the difference in magnification between these instruments and the standard Wood-Anderson torsion seismograph. The magnitude thus obtained was $M_L = 5.1$, somewhat higher than the teleseismic magnitudes (m_b) found by the international agencies.

Seismic moment was derived by the method of Brune and others (1963) of summing the areas of the envelopes of the surface waves recorded on the three components of the long-period seismographs. The moment so found was 5×10^{16} newton-meters, or 6.5 on Gibowicz's (1975) scale of normalized moment.

From these values of M_L and M_o, the derived source parameters are:

Radius of equivalent source area	6 km
Average displacement of rupture	1.5 cm
Stress drop	1 bar
Seismic energy	7×10^{10} joules.

These parameters are in no way unusual for tectonic earthquakes recorded elsewhere. Gibowicz and Hatherton (1975), for instance, studied the relationship between seismic moment and average displacement for 270 shallow New Zealand earthquakes. For a normalized moment of 6.5, corresponding to the Oates Land earthquake, the displacement varied between about 0.2 and 10 cm. The value found here, 1.5 cm, is near the middle of the range of values on a logarithmic plot. Similarly, the value of 6 km found here for the source radius is well within the range of 1 to 15 km found by Adams (1975) for New Zealand earthquakes of similar magnitude.

This analysis assumes that the source is a simple dislocation occurring in rock of normal elastic properties, and there is nothing in the results to suggest otherwise. It is not clear, however, to what extent these source properties would be expected to differ if the source were in the glacial ice. The postulated source radius of 6 km could easily be accommodated in the Rennick Glacier in this area, where it is about 25 km wide.

Conclusions

The earthquake of 15 October 1974 is the only earthquake known to have originated in continental Antarctica and to have been large enough to be recorded elsewhere. It occurred near the mouth of the Rennick Glacier, an extensive ice system that extends 300 km south into the mountains of Victoria Land. The earthquake exhibits the seismic characteristics of an ordinary shallow earthquake, and an analysis of its source properties shows nothing that is inconsistent with its being an ordinary tectonic earthquake. Nevertheless, it may be significant that its location is at the mouth of the largest glacier to flow northward from Victoria Land to the Oates Coast, and that the only other nonvolcanic shocks in the area have been similarly located near the mouths of glaciers flowing into the Ross Sea. Although it cannot be unambiguously established, it seems most likely that this event is also associated with large-scale movement or cracking in the glacial ice.*

Acknowledgments. I thank Jean-Luc Debroux, Institut de Physique du Globe, Strasbourg, for making available copies of the Dumont d'Urville records.

* Since I completed this writing, E. S. Grew has drawn my attention to an earthquake listed by the International Seismological Summary as occurring in the same general area on 20 April 1952. *P* waves from this event were recorded at only six stations (two in Australia, three in South Africa, and one in South America, the nearest at a distance of 35°). A computer relocation using these readings gives the epicenter as lat. 69.1 ± 0.37° S, long. 158 ± 2.3° E. This position is about 200 km to the northwest of that of the 1974 earthquake, but it is not well constrained; the formal error in longitude is more than 100 km in either direction.

The occurrence of this event may be taken to confirm earthquake activity in the Oates Land, but it may also be significant that the preferred solution is once again on the coast near a prominent glacier and ice tongue, in this instance the Matusevich Glacier Tongue.

References cited

Adams, R. D. 1971. Local earthquakes in Victoria Land. (*In* Adie, R. J., *ed. Antarctic Geology and Geophysics.* Oslo, Universitetsforlaget, 495–99.)

Adams, R. D. 1975. Developments in studies of earthquake risk. *Bull. N.Z. Nat. Soc. Earthq. Engin., 8,* 1–11.

Browne-Cooper, P. J., Small, G. R., and Whitworth, R. 1967. Probable local seismicity at Wilkes, Antarctica. *N.Z. Jl Geol. Geophys., 10,* 443–45.

Brune, J. N. 1970. Tectonic stress and the spectra of seismic shear waves from earthquakes. *J. geophys. Res., 75,* 4997–5009.

Brune, J. N., Espinosa, A., and Oliver, J. 1963. Relative excitation of surface waves by earthquakes and underground explosions in the California-Nevada region. *J. geophys. Res., 68,* 3501–13.

Craddock, C. 1972. *Geologic Map of Antarctica 1:5,000,000.* New York, American Geographical Society.

Gibowicz, S. J. 1975. Variation of source properties: the Inangahua, New Zealand, aftershocks of 1968. *Bull. seism. Soc. Am., 65,* 261–76.

Gibowicz, S. J., and Hatherton, T. 1975. Source properties of shallow earthquakes in New Zealand and their tectonic associations. *Geophys. J.R. astr. Soc., 43,* 589–605.

Hatherton, T., and Evison, F. F. 1962. A special mechanism for some Antarctic earthquakes. *N.Z. Jl Geol. Geophys., 5,* 864–73.

Kaminuma, K. 1971. Micro-earthquakes observed at Syowa Station, Antarctica. *Antarctic Rec., 40,* 65–73.

Kaminuma, K., and Ishida, M. 1971. Earthquake activity in Antarctica. *Antarctic Rec., 42,* 53–60.

Lazareva, A. P., and Sytinskiy, A. D. 1965. Seismic observations at Mirny 1961. *Inf. Bull. Sov. Antarct. Exped., 5,* 355–60.

123

Crustal Structure and Seismic Activity around Dumont d'Urville Base

Abstract

M. A. CHOUDHURY, ROOPA GIR, and S. M. GIR SUBHASH
Laboratory of Global Seismology, Institut de Physique du Globe, 67084 Strasbourg, France

The crustal structure around Dumont d'Urville (DRV) base has been studied by the spectral ratio method. After a detailed examination of the crustal transfer function from theoretical models, a model for DRV was derived by utilizing deep focus earthquakes recorded on magnetic tape at distances between 40° and 60°. The frequency range to be considered for obtaining a detailed picture of the crust is discussed.

Seismic activity within about 500 km of the base is scarce. A large number of events, mostly of ice-cracking origin, were recorded. The component rotation technique was used to identify tectonic shocks.

X Subglacial Morphology

124

Subglacial Morphology of the Ahlmannryggen and Borgmassivet, Western Dronning Maud Land

L. G. WOLMARANS
Geological Survey, P.O. Box 900, Pietermaritzburg, South Africa

Abstract. During the summer seasons of 1971–72 to 1974–75, a series of closely spaced oversnow radio-echo sounding traverses, totalling 2400 km, was made across the Fimbulisen ice shelf, the mountain ranges of the Ahlmannryggen (Ahlmann Ridge) and Borgmassivet (Borg Massif), and the Pencksökket (Penck Trough) ice stream. A Scott Polar Research Institute Mk II radio-echo sounder was used.

The radio-echo sounding was concentrated mainly in the Ahlmannryggen, a highly dissected northeast-trending range consisting of small and scattered nunataks which are evidently in a late stage of their morphologic evolution. A contour map of the subglacial topography prepared from the radio-echo sounding data indicated that, contrary to previous views, the dissection of the Ahlmannryggen is systematic. A vector analysis of the subglacial valleys shows four predominant trends striking approximately 25°, 95°, 125°, and 165°, as well as two subsidiary trends at 45° and 145°. These directions coincide remarkably well with trends in the joint pattern of the area as well as with two previously recognized major structural directions, indicating the importance of tectonic control in the dissection of the landscape.

During the four austral summer seasons of 1971–72 to 1974–75, geophysical/glaciologic parties of the South African National Antarctic Expedition made a series of oversnow traverses across the Fimbulisen ice shelf, the mountain ranges of the Ahlmannryggen (Ahlmann Ridge) and Borgmassivet (Borg Massif), and the Pencksökket (Penck Trough) ice stream in western Dronning Maud Land (Figure 124.1). This project was planned and directed by the author. It included oversnow radio-echo sounding, and a total of 2400 km of almost continuous ice-thickness profiles was obtained on these traverses (Schaefer, 1972, 1973; Van Zyl, 1973; Schönfeldt and Van Zyl, 1974; Barnard, 1975).

The traverse lines

Because of difficulties in obtaining echoes from deep bedrock, as well as severe crevassing along the boundaries of the Jutulstraumen and Schyttbreen ice streams (Figure 124.1), traverses were mostly confined to internunatak areas where thinner and safer ice could be expected. Unlike the Borgmassivet, the sparseness of outcrops in the Ahlmannryggen makes it difficult to nearly impossible to recognize structural or morphologic lineaments from the 1:250,000 topographic maps published by Norsk Polarinstitutt, Oslo or even from LANDSAT imagery of the area. For this reason the majority of the traverses was concentrated in the Ahlmannryggen in order to attempt the distinction of possible subglacial lineaments by the use of a contour map of the subglacial topography prepared from radio-echo sounding data.

The navigational procedures on these traverses have been discussed by Schaefer (1973) and Van Zyl (1973); basically they consisted of resection to prominent mountain peaks and barometric altitude determinations, usually at 3-km intervals.

Instrumentation

A 35 MHz Scott Polar Research Institute Mk II radio-echo sounder was used; detailed descriptions of the equipment and its installation in the vehicle have been given by Schaefer (1972). He also (Schaefer, 1973) estimated that by using apropagation velocity for radio waves through ice of 169 m/μs, the ice thicknesses obtained were correct to within 5%.

Despite the fact that the sounder was functioning according to specifications, echoes from bedrock deeper than 1700 m were received only in very few cases; the maximum reflection attained was from 2100 m.

Reduction methods

Film records were scaled by feeding them through an enlarger fitted with a sprocket drive and projecting the image onto a grid so that the scale of the grid and that of the image of the calibration marks on the film were the same. Ice thicknesses could then be read directly from the grid at the required intervals, usually 500 m, and plotted on the maps and profiles. At the same time, the relief between these points was sketched in on the profiles.

During the contouring of the bedrock topography, due care was taken that the inferred subglacial landforms were consistent with surface topography as shown on the 1:250,000 topographic maps and on LANDSAT imagery. In this regard, the LANDSAT imagery was especially useful, as it often showed surface relief not resolved by the 100-m-interval surface contours on the topographic maps.

Contouring of bedrock topography was done at an interval of 100 m (Figure 124.2) even though this caused cluttering of contours in places. This smaller interval accentuated the subglacial relief as compared to a 200 m or larger interval.

No attempt has been made to deconvolute the echoes

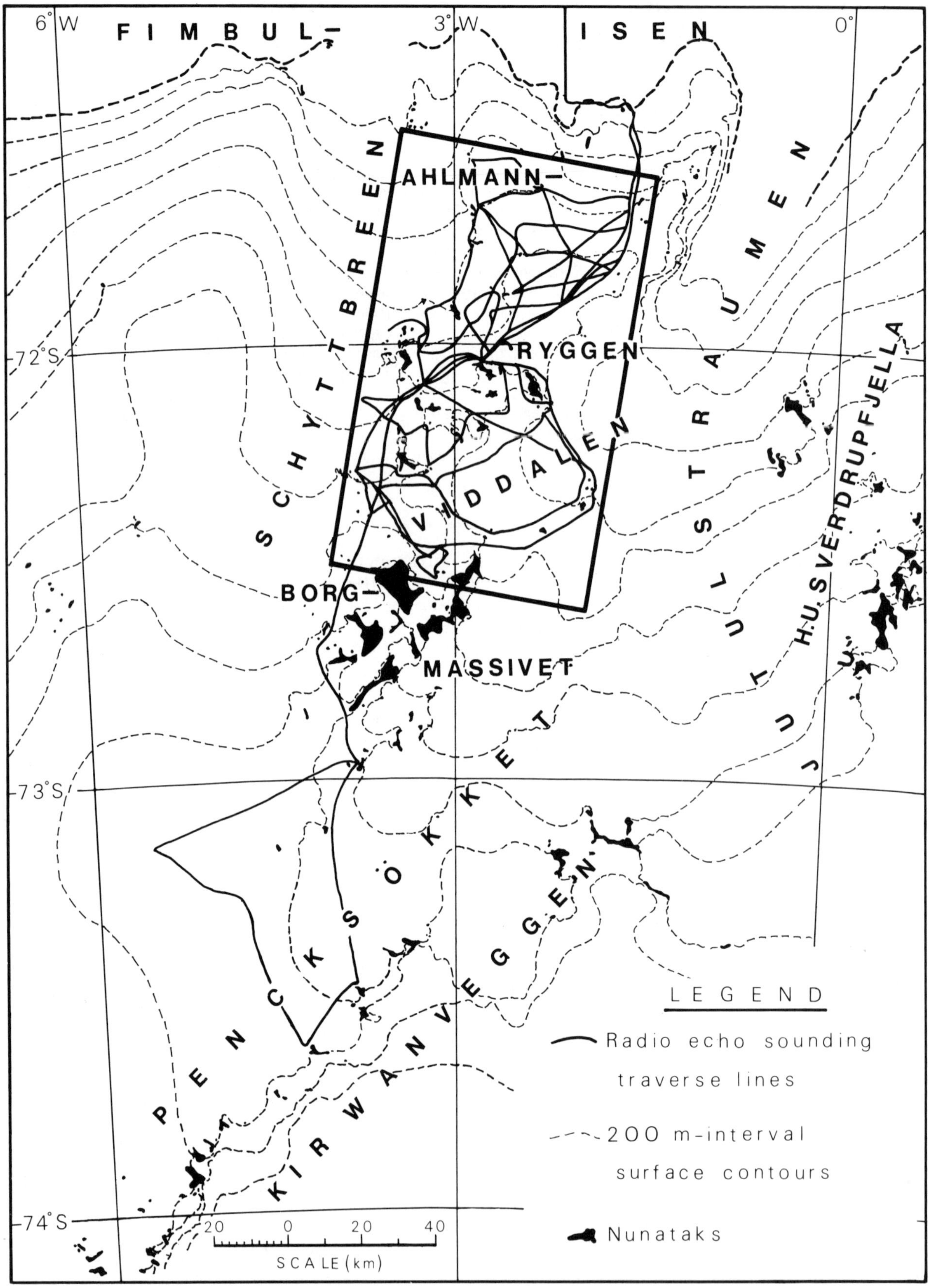

Figure 124.1. Map of western Dronning Maud Land with radio-echo sounding traverses indicated. The rectangle indicates the area covered by Figure 124.2.

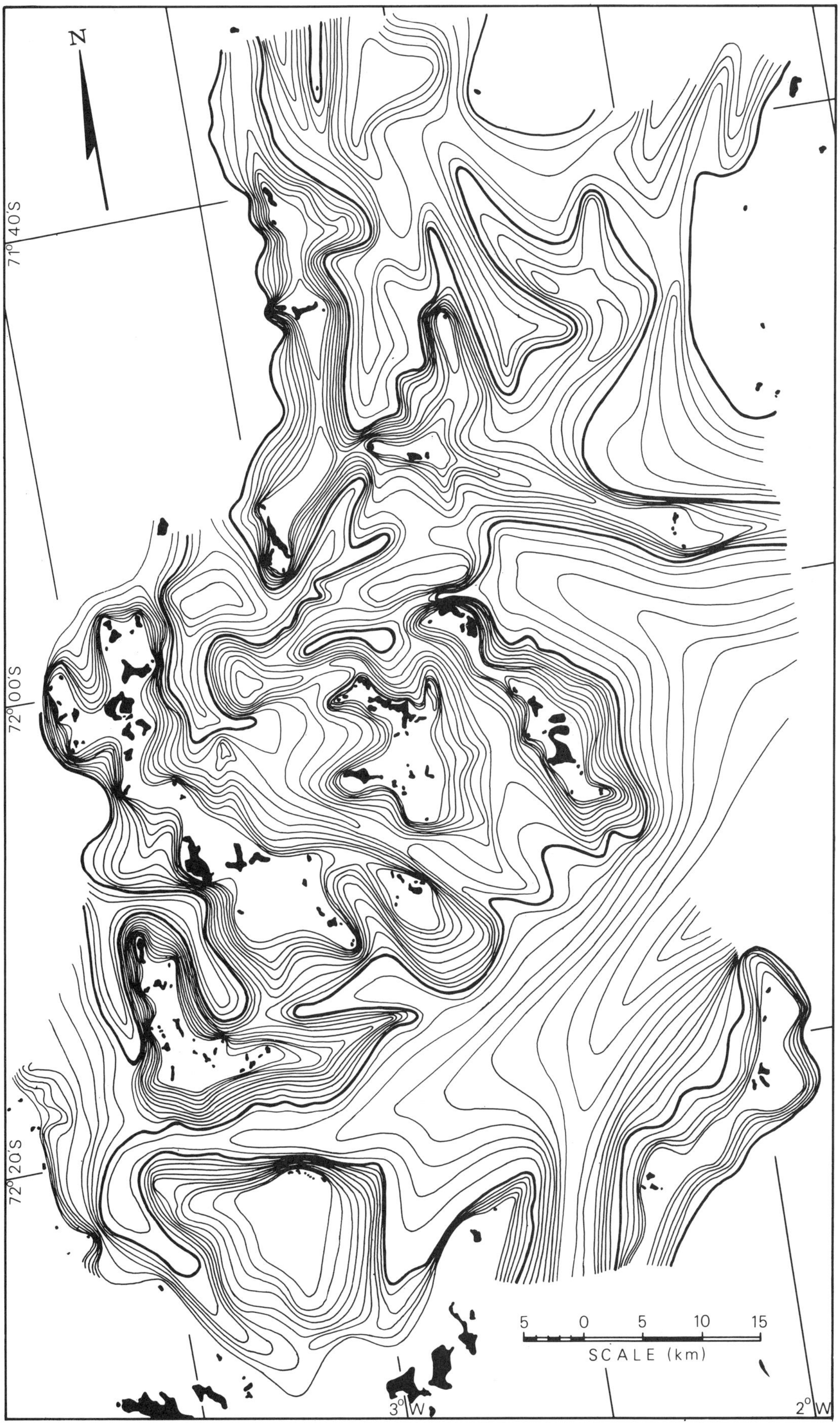

Figure 124.2. Contour map of bedrock elevation in the Ahlmannryggen. The heavier line is the sea-level contour. Contour interval is 100 m. Dark areas indicate exposed bedrock.

since at this stage only a map of subglacial topography for the tracing of subglacial lineaments was required. This map will also be used for the planning of future traverses and the placing of stake-lines to measure ice flow.

Surface morphology and geology

The first and by far the most comprehensive account of the morphology of western Dronning Maud Land is that by Swithinbank (1959). According to him, three distinct types of ice cover occur. These are the ice shelves, the inland ice sheet between the ice shelf and the plateau, and the ice of the main Antarctic, or polar, plateau. Their boundaries are particularly well defined. That between the polar plateau and the inland ice is formed by the Kirwanveggen, a 200-m- to 500-m-high northeasterly trending escarpment. Drainage from the plateau is mainly by the southern part of the Jutulstraumen and to a lesser extent by northwesterly trending breaches in the escarpment. The Jutulstraumen also separates the Kirwanveggen from the Sverdrupfjella (Sverdrup Mountains) to the northeast of it.

Geologically, the Sverdrupfjella and most of the Kirwanveggen consist of high-grade metamorphic rocks and minor intrusives (Roots, 1969). In the extreme southwestern part of the Kirwanveggen sedimentary rocks of probably Paleozoic age occur in tectonic contact with the metamorphic rocks. Both are unconformably overlain by Mesozoic sedimentary beds and lavas (Aucamp and others, 1972).

The Borgmassivet is located to the northwest of the Kirwanveggen and is separated from it by the Pencksökket, a northeastward flowing ice stream. The Borgmassivet is a range of tabular mountain blocks with mainly sheer rock walls, in places rising to 1000 m above the general ice level. Farther to the north, in the Ahlmannryggen, the mountains become progressively more dissected and pass into narrow ridges and sharp horns and ultimately into disconnected lines of nunataks. Most nunataks here are small and evidently in a late stage of their morphologic evolution (Swithinbank, 1959). The snow surface slopes roughly to the north in broad steps, descending from 2100 m in altitude in the southern part of the Borgmassivet to only 100 m in the northern parts of the Ahlmannryggen. Mountain summit levels also fall from a maximum of 2700 m in the Borgmassivet to about 200 m in the north. The summit levels are mainly controlled by flat-lying sills which are more resistant to weathering and ice excavation than the sedimentary rocks.

The Ahlmannryggen and Borgmassivet comprise a subhorizontal sequence of sedimentary and volcanic rocks with associated intrusive rocks. Neethling (1972*a*) considered the sequence to be the remnant Precambrian platform cover of a fault-margined block which has been subjected mainly to epeirogenesis since stabilization.

The Ahlmannryggen is separated from the metamorphic rocks to the east and west of it by the ice streams of the Jutulstraumen and Schyttbreen, respectively, and from the Borgmassivet by the Viddalen, a tributary of the Jutulstraumen. To the north the Ahlmannryggen is bordered by the Fimbulisen ice shelf.

Swithinbank (1959) described the present drainage pattern of the area as unsystematic but tending to a radial distribution from the center. He suggested that it might have been caused by a landscape dissected prior to its inundation by ice, but also emphasized that no real judgment on the origin of the subglacial topography could be made without a greater knowledge of the subglacial contours.

Subglacial morphology and valley directions

The major subglacial features in the area covered by Figure 124.1 are the valleys of the Jutulstraumen, Pencksökket, Viddalen, and Schyttbreen ice streams. They are clearly defined and can easily be recognized from surface topographic maps and especially from LANDSAT imagery. Seismic, gravimetric, and radio-echo sounding work done by Robin (1958), Van Autenboer and Decleir (1972), Schaefer (1972, 1973), Gjessing (1972), Van Zyl (1973), and Barnard (1975) indicates that these ice streams occupy valleys with floors well below sea level. They tend to parallel each other or smaller valleys and breaches in the Borgmassivet and the Kirwanveggen.

In its southern part the Jutulstraumen has a northwesterly direction, but at approximately lat. 72° 30′ S its direction changes markedly so that where it borders the Ahlmannryggen it follows a well-defined north-northeasterly direction. The Schyttbreen parallels the north-northeasterly part of the Jutulstraumen. The most prominent valley direction is that of the northeasterly striking Pencksökket which, with its extensions to the southwest, can be traced over some 250 km. This direction is also reflected by a series of ice-filled troughs in the Borgmassivet. The northwesterly direction of the upper reaches of the Jutulstraumen is also paralleled by glacier-filled breaches in the Kirwanveggen and to a lesser extent in the Borgmassivet.

In the Ahlmannryggen, glaciation has proceeded to the stage where, due to the extensive cover by the ice, very little about the subglacial morphology can be deduced from the surface morphology. Very few linear features can be discerned. The contour map of the subglacial topography (Figure 124.2), however, shows the Ahlmannryggen to be a highly dissected northeast trending ridge, characterized by numerous fjord-like valleys extending into it from the surrounding ice streams. These valleys have floors that are mainly below sea level. Separating them are sharp and narrow ridges, only the highest points of which protrude above the ice cover. The total relief is rugged and generally of the order of 500 m to 1000 m. Variations in bedrock elevation of more than 1500 m over relatively short distances are, however, commonly found. The deepest bedrock measured was 1200 m below sea level at the mouth of the Viddalen ice stream.

At first glance the subglacial topography of the Ahlmannryggen seems to confirm the view of Swithinbank (1959) that the drainage is unsystematic. Valleys appear to have random strike directions, but on analysis of these directions, definite trends emerge. A vector diagram was constructed by measuring the lengths and directions of the median lines of all the subglacial valleys and grouping these in ten-degree classes. The Jutulstraumen was included in this analysis because of its prominence and its position bordering the Ahlmannryggen.

The vector diagram is given in Figure 124.3 and shows the relative frequency of the subglacial valley directions for the ten-degree classes. From this, four predominant trends can be distinguished. These strike at approximately 25°,

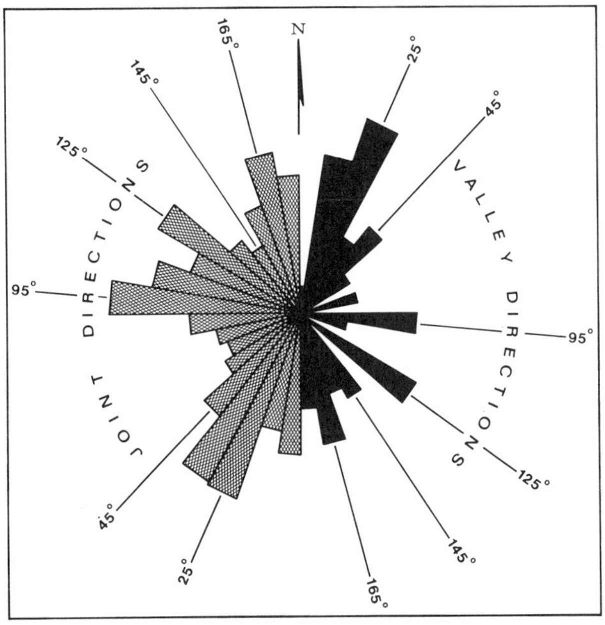

Figure 124.3. Composite vector diagram of subglacial valley directions and rose diagram of joint directions. Data are plotted in 10° classes.

95°, 125°, and 165°. Subsidiary trends striking approximately 45° and 145° are also present.

Comparison of valley and bedrock structural directions

Tectonic control over the forming of glacial landforms has often been suggested. Gregory (1913) held the view that fjords are of primary tectonic origin with negligible glacial excavation, but this received much opposition. O. Holtedahl (1929) suggested tectonic control over fluvial, and subsequent glacial, erosion. Such control over fjord direcions by the geologic structure has recently also been described by Randall (1961) and H. Holtedahl (1967) who compared fracture and fjord directions. Nicholson (1963), however, argued that such control is of local significance only and does not have universal validity.

Tectonic control over the landforms in the Borgmassivet-Pencksökket area has been suggested by Swithinbank (1959), Roots (1969), and Neethling (1970, 1972*a*,*b*). They ascribed the crudely rectangular pattern of the valleys in the Borgmassivet to block faulting along northeasterly and northwesterly lines. They also recognized the Pencksökket as a zone of faulting. Neethling (1970, 1972*a*) identified two major structural trends in this area. The first of these follows the Pencksökket and the northeasterly valleys in the Borgmassivet. He (1970) described this trend, which he called the Penck-Jutul rift, as a fault zone of fundamental significance, representing lines of recurrent movement between the platform and the repeatedly rejuvenated basement to the east. The other trend is the northwesterly one indicated mainly by the breaches in the Kirwanveggen and the Borgmassivet, as well as by the upper reaches of the Jutulstraumen.

Due to the paucity of outcrops in the Ahlmannryggen, structural trends are difficult to determine. No systematic surveys of minor faults or other minor structures have as yet been made. Over the years, however, several hundred joint measurments from the Ahlmannryggen and Borgmassivet have been accumulated. These data have been grouped in ten-degree classes according to their strike directions and plotted on a rose diagram. This shows four fairly strong trends of joint directions striking approximately 30°, 95°, 125°, and 165°. On comparison of this diagram with the vector diagram of the subglacial valleys, a close agreement between the two is immediately apparent (Figure 124.3). Both the subglacial valleys and the joint directions show strong trends striking north-northeast, east, east-southeast and south-southeast. None of these trends have previously been identified. The two major structural directions identified by Neethling (1970, 1972*a*) are not reflected by the joint patterns, except possibly for the northeasterly one. When the joint measurements from the Borgmassivet alone are considered, however, the northeasterly and northwesterly trends emerge very strongly. It would thus seem that those directions are characteristic of the Borgmassivet rather than of the Ahlmannryggen, and because a relatively small proportion of the joint measurements in Figure 124.3 are from the Borgmassivet, they are masked by the different and locally stronger trends in the Ahlmannryggen.

Conclusions

The major subglacial features in the area described are clearly defined and can easily be recognized without the use of radio-echo sounding. In the Borgmassivet-Pencksökket area these valleys follow definite trends, one striking northeast and the other northwest. These are also the major structural directions in the area. The valleys of the two major ice streams bordering the Ahlmannryggen strike north-northeast.

In the Ahlmannryggen, however, the detailed nature of the subglacial topography could be distinguished only through the use of radio-echo sounding and the preparation of a contour map of the subglacial topography. By this means it becomes evident that the Ahlmannryggen is deeply dissected by a number of fjord-like valleys. A vector analysis shows that their directions are systematic and that they have four major, and two subsidiary, trends.

An analysis of bedrock joint directions shows that the joint pattern in the Ahlmannryggen has four major trends. In the Bormassivet only two major trends are distinguished; these differ from those in the Ahlmannryggen but coincide with the two major structural directions in the Borgmassivet. Comparison of the trends in the joint pattern with those in the directions of the subglacial valleys shows a close agreement. This indicates a very large measure of tectonic control over the dissection of the landscape. This control was probably by block faulting and subsequent fluvial and/or glacial erosion. A primary tectonic origin for the subglacial valleys is doubtful. Exceptions to this might be the valleys of the Pencksökket, Jutulstraumen, and Schyttbreen, as well as the northeasterly striking valleys in the Borgmassivet; but, even if this is the case, glacial erosion probably played the major role.

Acknowledgments. The earth science investigations in western Dronning Maud Land are directed by the South African Geological Survey under the aegis of the South African Scientific Committee for Antarctic Research. Logistic and financial support was provided by the Department of Transport. I am indebted to these organizations for making this research possible.

I am also grateful to D. C. Neethling, L. E. Kent, and J. H. Bredell for reviewing the manuscript and for critical comments. Unpublished joint measurements were provided by B. R. Watters, A. P. H. Aucamp, J. H. Bredell, and A. W. W. Paterson.

References cited

Aucamp, A. P. H., Wolmarans, L. G., and Neethling, D. C. 1972. The Urfjell Group, a deformed (?) early Paleozoic sedimentary sequence, Kirwanveggen, western Dronning Maud Land. (*In* Adie, R. J., *ed. Antarctic Geology and Geophysics.* Oslo, Universitetsforlaget, 557–62.)

Barnard, H. P. 1975. Radio echo sounding in western Dronning Maud Land, 1974. *S. Afr. J. Antarct. Res.,* No. 5, 37–41.

Gjessing, Y. T. 1972. Mass transport of Jutulstraumen Ice Stream in Dronning Maud Land. *Arbok Norsk Polarinst., 1970,* 227–32.

Gregory, J. W. 1913. *The Nature and Origin of Fjords.* London, J. Murray, 542 pp.

Holtedahl, H. 1967. Notes on the formation of fjords and fjord-valleys. *Geogr. Annlr., 49A,* 188–203.

Holtedahl, O. 1929. On the geology and physiography of some Antarctic and Subantarctic islands. *Scient. Res. Norw. Antarct. Exped.,* 3, 172 pp.

Neethling, D. C. 1970. *South African Earth Science Exploration of Western Queen Maud Land, Antarctica.* Ph.D. dissertation, University of Natal, 139 pp.

Neethling, D. C. 1972*a*. Age and correlation of the Ritscher Supergroup and other Precambrian rock units, Dronning Maud Land. (*In* Adie, R. J., *ed. Antarctic Geology and Geophysics.* Oslo, Universitetsforlaget, 547–56.)

Neethling, D. C. 1972*b*. Submarine and subglacial morphology, Kronprinsesse Märtha Kyst, Dronning Maud Land. (*In* Adie, R. J., *ed. Antarctic Geology and Geophysics.* Oslo, Universitetsforlaget, 705–11.)

Nicholson, R. 1963. A note on the relation of rock fracture and fjord direction. *Geogr. Annlr., 45,* 303–5.

Randall, B. A. O. 1961. On the relationship of valley and fjord directions to the fracture pattern of Lyngen, Troms, N. Norway. *Geogr. Annlr., 43,* 336–38.

Robin, G. de Q. 1958. Glaciology. III. Seismic shooting and related investigations. *Scient. Res. Norw.-Br.-Swed. Antarct. Exped., 5,* 3–134.

Roots, E. F. 1969. Geology of western Queen Maud Land. (*In* Bushnell, V. C., and Craddock, C., *ed.* Geologic Maps of Antarctica. *Antarct. Map Folio Ser.,* Folio 12, Pl. VI.)

Schaefer, T. G. 1972. Radio-echo sounding in western Dronning Maud Land —a preview. *S. Afr. J. Antarct. Res.,* No. 2, 53–56.

Schaefer, T. G. 1973. Radio-echo sounding in western Dronning Maud Land, 1971. *S. Afr. J. Antarct. Res.,* No. 3, 45–52.

Schönfeldt, S. G., and Van Zyl, C. Z. 1974. Geophysical traverses in the Ahlmannryggen, western Dronning Maud Land, 1973. *S. Afr. J. Antarct. Res.,* No. 4, 50–57.

Swithinbank, C. 1959. Glaciology. I. The morphology of the inland ice sheet and nunatak areas of western Dronning Maud Land. *Scient. Res. Norw.-Br.-Swed. Antarct. Exped., 3D,* 97–117.

Van Autenboer, T., and Decleir, H. 1972. Ice thickness and subglacial relief of the Jelbartisen-Trolltunga area, Dronning Maud Land. (*In* Adie, R. J., *ed. Antarctic Geology and Geophysics.* Oslo, Universitetsforlaget, 713–22.)

Van Zyl, R. B. 1973. Radio-echo sounding in western Dronning Maud Land, 1972. *S. Afr. J. Antarct. Res.,* No. 3, 53–59.

125

Radio-Echo Sounding Investigations of Wilkes Land, Antarctica

R. H. N. STEED and D. J. DREWRY
Scott Polar Research Institute, University of Cambridge, CB2 1ER Cambridge, U.K.

Abstract. Radio-echo soundings along 23 tracks, spaced 50–100 km apart and forming an orthogonal grid in Wilkes Land, have been used to determine ice-sheet surface and bedrock surface configurations. Three principal, and several subsidiary, bedrock terrain zones are recognized. From east to west these are: the exposed and subice Transantarctic Mountains, the Wilkes subglacial basin, and a region of complex highlands and trenches. The smooth Wilkes Basin is shown to possess a major swell inland of the George V Coast; this swell divides the region into two linear troughs which reach the coast at the Cook Ice Shelf and the Ninnis Glacier. The Mertz basin forms a separate feature. Offshore dredge hauls, the composition of moraine clasts, and other geophysical studies suggest that these basins contain Mesozoic–lower Cenozoic sedimentary sequences. Beacon Supergroup strata may also underlie the basins as far west as long. ~143° E. Precambrian basement appears to comprise the bedrock inland for as far as 200 km from the Adélie Coast. The complex highland area on the west is subdivided into three key regions. A very rugged province inland of the Mertz Glacier is attributed to a granite batholith which crops out at Cape Bage and Cape Webb. A prominent trench and escarpment have been detected along long. ~135° E, and the structure is interpreted as sediment-filled half-graben. Ice more than 4.67 km thick fills the trough which reaches 2230 m below sea level. The existence of a postulated large meteorite crater in Wilkes Land is questioned. Eleven radio-echo sounding flights conducted during a joint National Science Foundation–Scott Polar Research Institute (NSR-SPRI) field program in 1971–72, along with twelve flights conducted during the NSF-SPRI–Technical University of Denmark (TUD) in 1974–75, provide data for compilation of ice-surface and bedrock-surface contour maps of Wilkes Land. The maps cover an area of 900,000 km^2.

Data collection and reduction

The radio-echo systems operate at 60 MHz (SPRI and TUD) and 300 MHz (TUD). The primary recording mode is the Z-scope display, which produces a photographic record with the flight time in the X-direction and echo delay time in the Y-direction (proportional to reflection distance). The brightness of any spot on the recording may be one of three discrete levels corresponding to predetermined values of the echo power gradient. This reduces the effect of noise and enhances echo resolution. Aircraft terrain clearance and ice thickness were later digitized from the Z-scope film recordings at 20 second intervals, corresponding to about 2.5 km on the ground. In some areas measurements were taken from the records twice as frequently.

Every 20 seconds, latitude and longitude values from one or two Intertial Navigation Systems (INS) were logged automatically. In addition, the output of a pressure transducer (connected to the aircraft static pipe), outside air temperature, and air speed were recorded. Errors in the INS track were reduced by fixing the aircraft's location from photographs of features of known position.

The altitude of the aircraft is the largest source of error. Pressure transducer output was converted to pressure altitude and corrected for temperature effects and the geostrophic wind. Aircraft "porpoising" effects were removed by smoothing the terrain clearance over a period of 40 seconds (~5 km). Absolute altitude above sea level was obtained by accurately determining the aircraft's altitude from radio-echo sounding over the sea or ground points of known elevation.

During the 1971–72 season, the aircraft operated at a constant pressure altitude, thereby reducing the systematic errors associated with elevation changes but increasing the chance of losing the bedrock echo through increased geometric loss associated with greater terrain clearance. Surface elevation data, along with measurements obtained during the 1958–59 Victoria Land Traverse-2 (VLT-2) and the 1958–59 Expédition Polaires Françaises Traverse, were combined to form a grid; discrepancies between crossing lines were removed by least squares adn random walk methods (Drewry, 1975*a*). The 1974–75 data, with more effective bedrock coverage but more frequent altitude changes, were then incorporated into the grid and additional interflight discrepancies removed. This process resulted in an estimated random error of less than 10 m and an overall systematic error of less than 50 m in the determination of ice surface elevations.

Map compilation

The flight-line grid has a separation of 50 to 100 km. The ice surface (Figure 125.1) is contoured at 100 m intervals, and this map also indicates the flight lines. The bedrock surface possesses greater roughness and has been contoured at 250 m intervals (Figure 125.2). Although flight-line spacing is sufficiently close to identify any large-scale

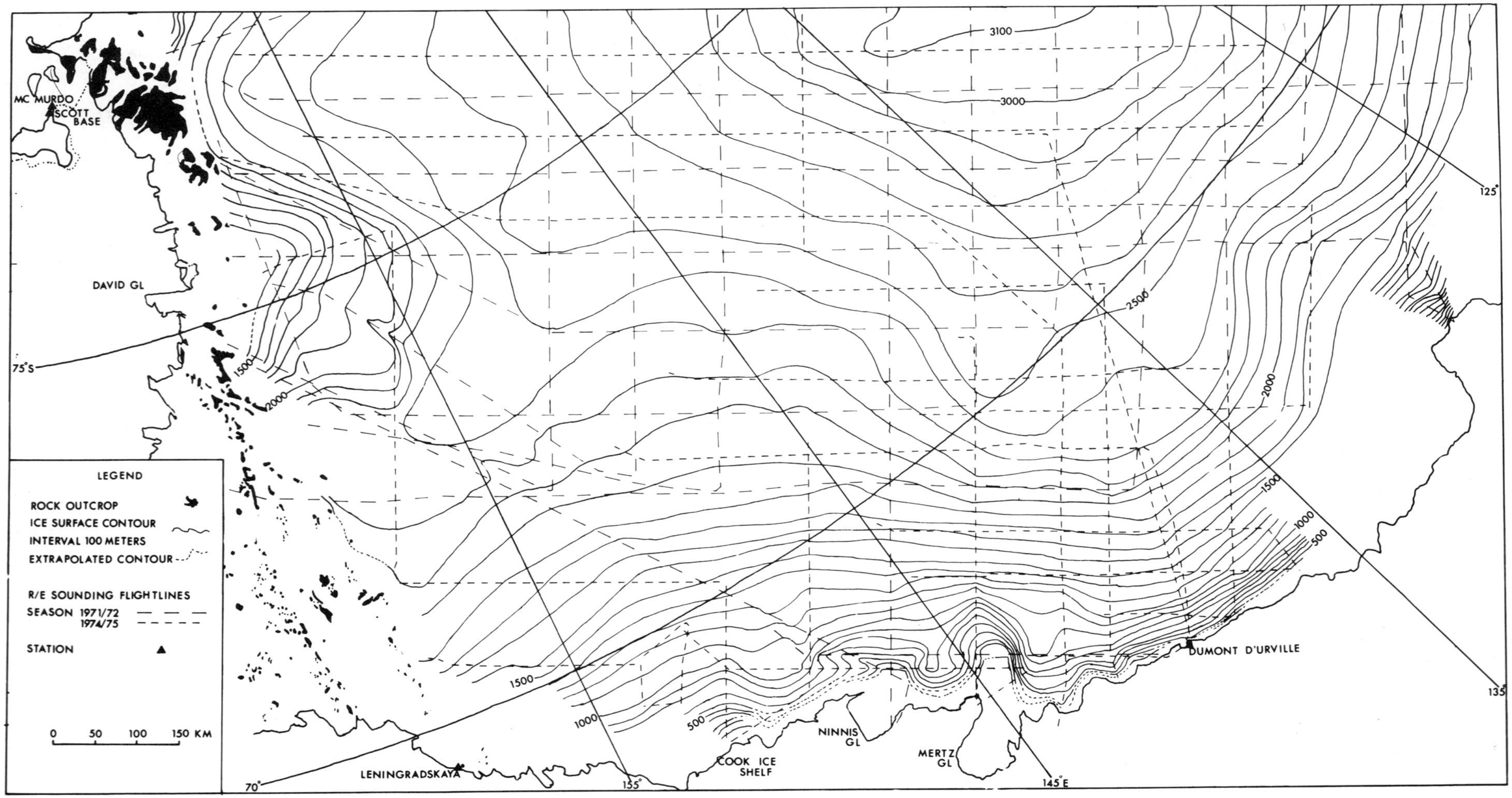

Figure 125.1. Ice-sheet surface configuration and radio-echo sounding flight lines, Wilkes Land.

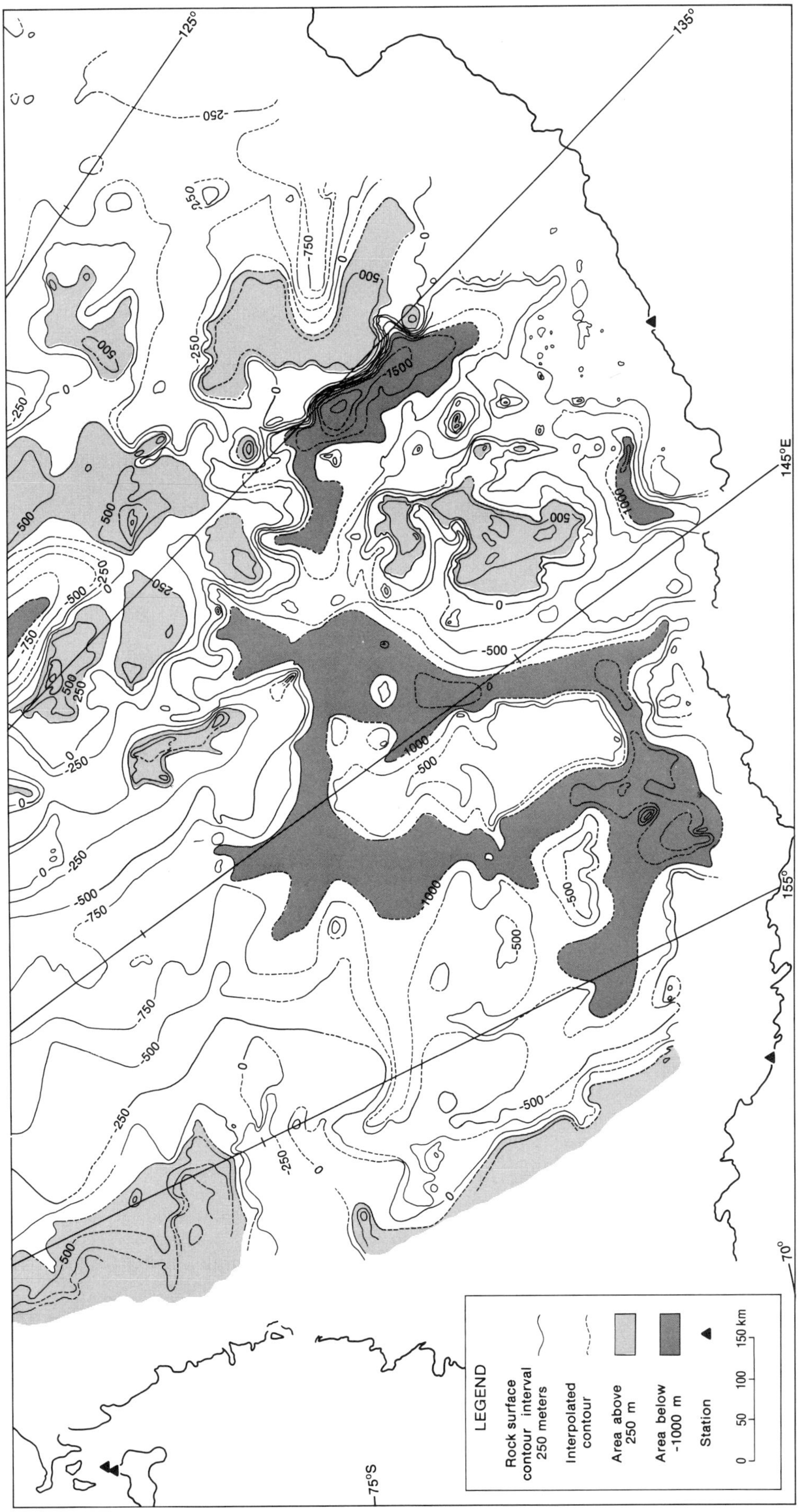

Figure 125.2. Bedrock surface configuration, Wilkes Land.

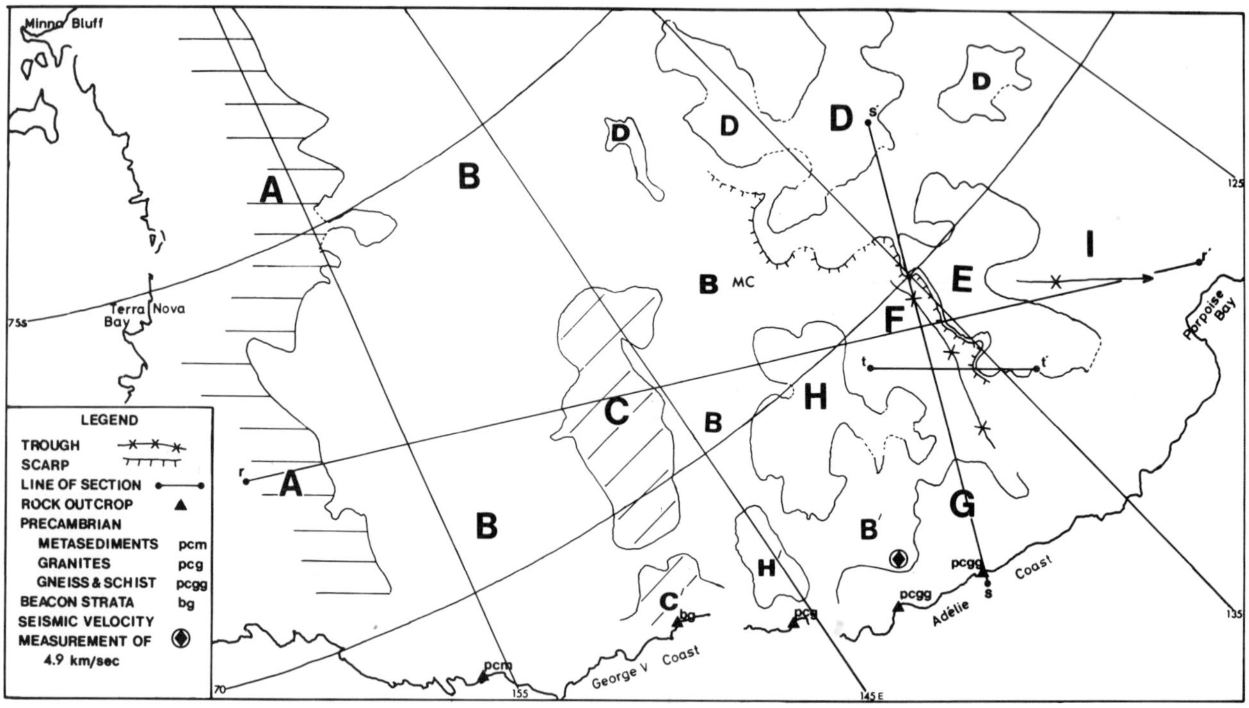

Figure 125.3. Wilkes Land, showing location of bedrock terrain regions (indicated by letters) discussed in text.

topographic elements, small-scale features may well remain undetected.

At times the bedrock echo was not detected because of increased ice thickness and aircraft terrain clearance. Minimum values, however, can be assigned to ice thickness in these areas, and any uncertainty is shown by dashed contours.

Ice surface

The ice surface map (Figure 125.1) confirms the convex-upward shape of the East Antarctic ice sheet, rising southward toward Dome C (off map). Surface gradients increase toward the coast and in some areas are affected by fast-flowing ice streams associated with large outlet glaciers. The David Glacier provides an eastern outlet into the Ross Sea and is associated with a major drainage basin at lat. 75° S. The bedrock surface data show a possible breach in this segment of the Transantarctic Mountains. The Ninnis and Mertz outlet glaciers on George V Coast make similar impressions on the ice surface for about 150 km inland and also correlate with troughs in the underlying bedrock topography.

A long ridge runs eastward between the David Glacier depression and the George V coastal region toward the mountainous area in northern Victoria Land.

Bedrock terrain

To facilitate discussion, the mapped bedrock area has been divided into nine principal regions (Figure 125.3). The topography (Figure 125.2) shows a general north-south trend. The Transantarctic Mountains (marked A in Figure 125.3) slope westward into the large, parallel Wilkes subglacial basin (B), which deepens toward the north. The basin is essentially low lying and smooth, but contains a high ridge (C) (500 m below sea level) in the northwest. A more complex area to the west of the Wilkes Basin contains mountainous regions extending inland from the Adélie Coast. This area also exhibits a deep, asymmetric trench (F) bounded by a massive, east-facing escarpment. Other highland regions rise between the Wilkes Basin and the "Aurora Basin" (not fully shown here; see Drewry, 1975*a*) in the southern part of the mapped area, gradually passing into a plain to the northwest.

It is interesting to note that many valleys and reentrants within the massifs show glacial characteristics. It is possible that they date from an earlier phase of upland glaciation prior to the establishment of the ice sheet and that they amplify ideas of early ice-sheet growth (Drewry, 1975*b*). The Mertz Glacier and Ninnis Glacier channels constitute major fjords.

Transantarctic Mountains and Wilkes subglacial basin

General description

The Transantarctic Mountains (Figure 125.3, region A) provide the eastern limit to the investigated area. The range shows a gentle slope westward into the Wilkes Basin on the order of 1:90 in the south and 1:40 in the north. The range is partially broken at lat. 75° S by a wide basin trending eastward to the Ross Sea. This feature may have been enhanced by glacial erosion due to the high rate of ice flow through David Glacier.

The extensive Wilkes Basin (Figure 125.3, region B) runs parallel to the Transantarctic Mountains for more than 1300 km, widening from about 100 km at lat. 81° S to about 300 km at lat. 70° S. The basin floor is smooth, becoming deeper to the north, and reaches a maximum measured depth of 1670 meters below sea level. The basin extends to the George V Coast in two locations—at the Cook Ice Shelf and at the Ninnis Glacier.

The basin contains several relatively high areas of small areal extent and a larger swell at its northwestern extremity (region C, bounded by the − 1000 m contour). This second feature maintains the smooth surface typical of the Wilkes Basin, but it rises to more than 500 m above the deeper part of the basin, reaches a maximum elevation of 250 m below sea level at its northern end, and is flanked by relatively steep slopes. It has a general north-south trend, with the crest line gently plunging to the south. Although separated from region C by a wide valley, region C′ may represent a continuation of this feature to the coast.

Geologic interpretation

The western flank of the northern Transantarctic Mountains consists primarily of a Precambrian basement complex which slopes steeply to the west (Bushnell and Craddock, 1969–70). Metasedimentary rocks of similar age crop out on George V Coast. South of lat. 72° 30′ S, basement rocks are overlain by the upper Paleozoic–lower Mesozoic Beacon Supergroup. The only occurrence of these strata farther west is at Horn Bluff (Mawson, 1940) (coastal region C′).

From an assessment of subice bedrock terrain characteristics, we suggest that Beacon strata overlie the basement (see Figure 125.6) from the Transantarctic Mountains at least as far as the western margin of the Wilkes Basin. The outcrop of Beacon rocks at Horn Bluff raises the possibility that regions C and C′ may represent a gentle swell or anticline within the basin, causing older Beacon strata to be elevated with respect to bordering younger sedimentary deposits.

Erosion during late Mesozoic and early Cenozoic time has probably resulted in further sedimentation in the Wilkes Basin. The Transantarctic Mountains and other highland areas in East Antarctica were likely sources of clastic material. Indeed, on the basis of interpretation of radio-echo reflection strengths and other geophysical data (Drewry, 1976), it seems that the southern part of the basin contains a sedimentary sequence. Offshore dredging west of Mertz Glacier Tongue (Mawson, 1940) has yielded traces of lignite (brown coal) and may suggest the presence of a basin inland of this coastal area with a young(?) sedimentary sequence.

Sedimentary rocks collected from the moraines near Cape Denison (125 km east of Dumont d'Urville) show an abundance of arkose, commonly red-stained (Mawson, 1940). Based on flow lines constructed from the surface map (Figure 125.1), a likely source area for these rocks can be defined in the vicinity of the southern reaches of the Mertz Basin. This may further strengthen the suggestion that a sedimentary sequence is preserved in the area.

Wilkes Land meteorite crater

Schmidt (1962) has suggested the presence of a meteorite crater of substantial size in Wilkes Land (lat. 71.5° S, 140° E; shown as MC in Figure 125.3) from examination of a prominent negative gravity anomaly recorded along the Victoria Land Traverse-2 (VLT-2). A number of radio-echo sounding flights cover the region, and although bottom echoes were not recorded on several 1971–72 missions (see above), the 1974–75 data enable reasonable interpretation of the bedrock surface configuration (Figure 125.2). The VLT-2 and adjoining Expédition Polaires Françaises (EPF) Traverses cross three major terrain provinces but reveal little resemblance to an impact crater. South of lat. 72.3° S, a broad highland area (D in Figure 125.3) culminates in peaks of 475 m above sea level before descending steeply into a very deep basin. This basin, an offshoot of the Wilkes Basin, reaches a depth of 1100 m below sea level. On the northern flank of the trough, the bedrock surface rises to the broad, dissected highland (area H) which was crossed by the French traverse.

The French seismic results and our radio-echo values are in good agreement. VLT-2 ice thickness determinations, however, show substantial differences, as is true elsewhere along this traverse (Drewry, 1975*c*). At station 524, for example, the seismic results (Weihaupt, 1961, 1976) significantly underestimate ice thickness. Interpretation of seismograms recorded at station 527 located above the trough yielded a rock elevation of 415 m below sea level; radio-echo sounding indicates 1120 m below sea level. It is interesting to note that, although the seismograms showed considerable noise, their reappraisal by Crary (1963) suggested to him a rock elevation of 1220 m below sea level. Thus the elevation difference between uplands and trough is at least 1600 m. This fact significantly reduces the importance of the 147 mGal anomaly by accounting for 110–120 mGal. The remainder of the anomaly could be explained by a sediment layer <1 km thick infilling the basin.

NORTHERN WILKES LAND HIGHLANDS

General description

To the west of the Wilkes Basin is a region of complex, mountainous terrain (region H, Figure 125.3). It shows a rugged relief with well-separated highs, reentrants, and isolated peripheral peaks, and it possesses a maximum measured elevation of 900 m above sea level. A similar area rises from the surrounding low basin to the northeast. Although this second area (H′) is generally lower and less rugged, it may well represent a northern continuation of region H that has suffered more severe glacial erosion closer to the coast. A shallow, 25-km-wide saddle separates H′ from H and provides a pass between the deeply eroded troughs of the Mertz and Ninnis Glaciers.

A very deep trench lies to the west of region H (region F). An east-west section (Figure 125.4, t–t′) has an asymmetric form, deepening gradually from the east but rising in a steep scarp on its western flank. The southern end terminates at a broad pass into the western margin of the Wilkes Basin. The trench flattens into a coastal plain to the north. The deepest measured point in the trench is 2230 m below sea level, and the ice thickness is 4670 m—the thickest yet recorded. (The Bentley subglacial trough in Marie Byrd Land is only 300 m less.) Figure 125.5 is a digitized profile running longitudinally along the trench. Isostatic readjustment indicates that the basin still would form a lake at least 200 km long and 1500 m deep. This is similar to Lake Baikal in the U.S.S.R., where the asymmetric structure is attributed to block faulting (Lamakin, 1968) and where the lake bottom is reported to be underlain by ~ 3 km of sediments.

The steepest part of the escarpment runs north-south for 260 km, eshibiting several reentrants and buttresses. It can be traced for a further 260 km southward in a more gentle form, providing the boundary of a southern high area (region D; Figure 125.3). The escarpment dies out to the north into a flat coastal plain and exhibits signs of intense glacial erosion (region G).

A minor, steep-sided basin lies between regions H and

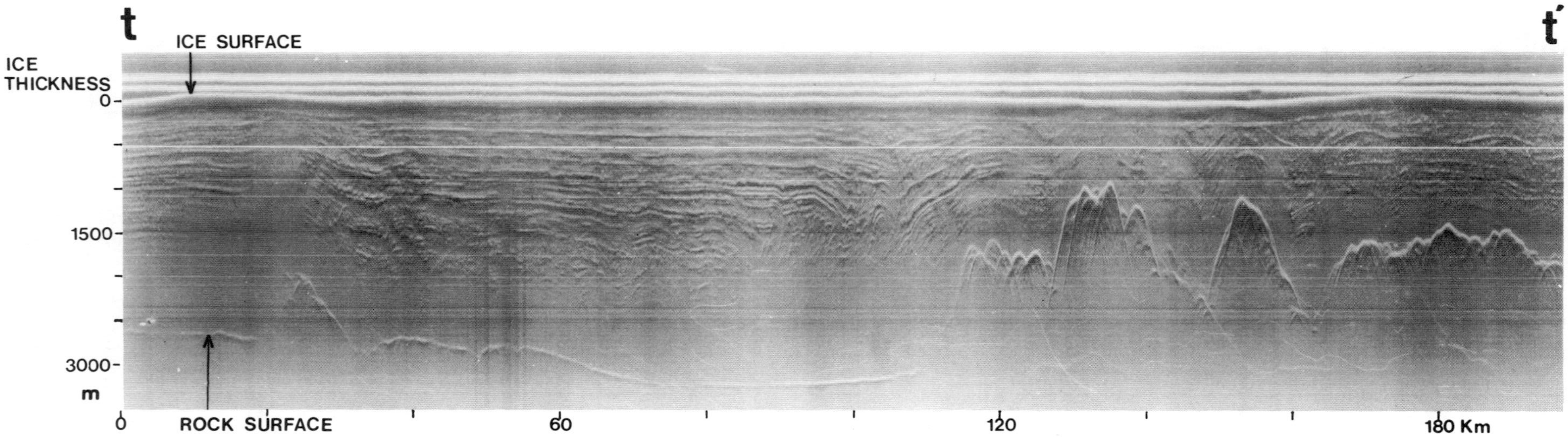

Figure 125.4. Section of Z-scope radio-echo sounding record along line t–t′ (see Figure 125.3. for location), showing ice surface and bedrock surface returns across trench and escarpment (region F).

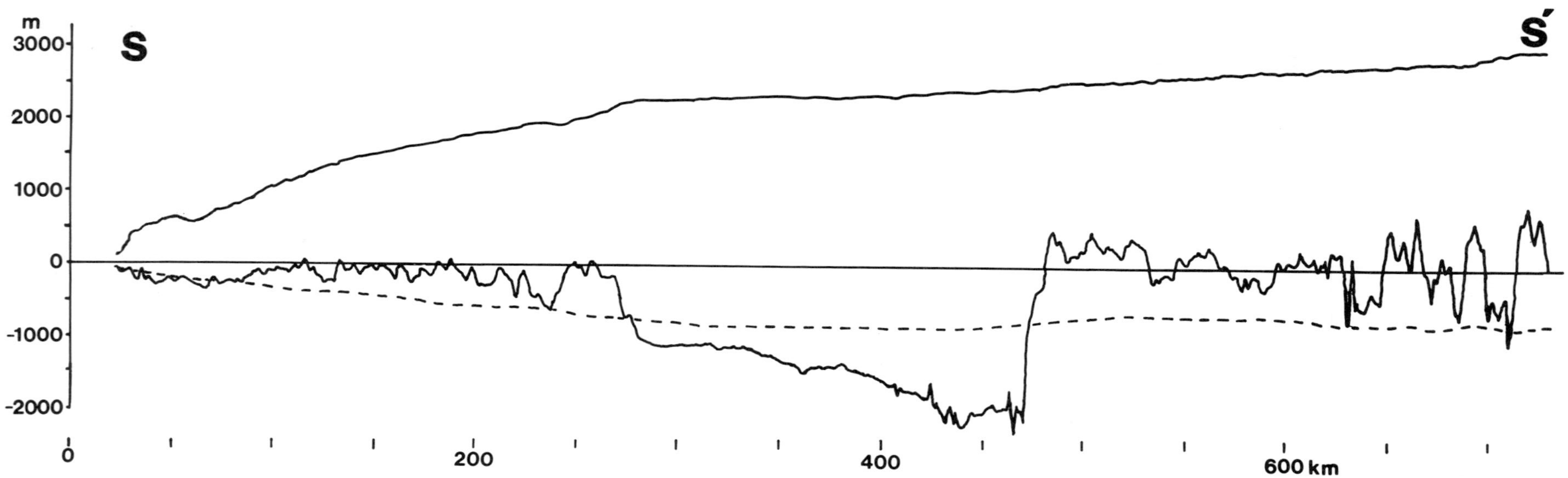

Figure 125.5. Digitized radio-echo sounding profile along the line s–s′ (see Figure 125.3 for location), across coastal plain (region G), trench, and escarpment (region F). Dashed line shows isostatically corrected sea level.

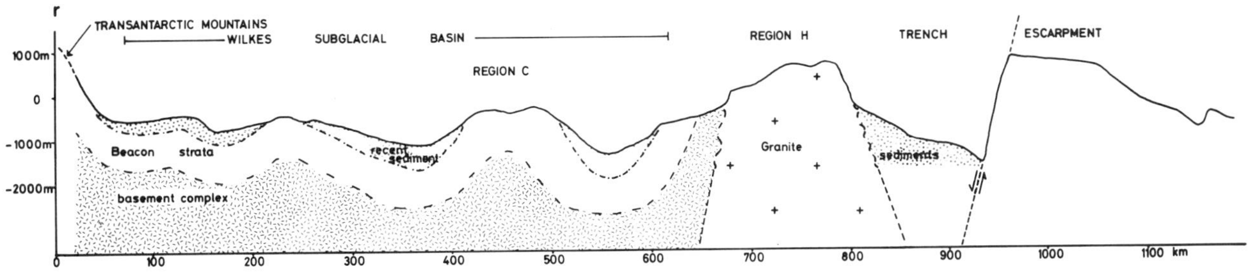

Figure 125.6. Speculative geologic structure across Wilkes Land along section r–r′ (see Figure 125.3 for location). Vertical exaggeration is 50:1. Thickness of stratified rock is not to scale.

G, running northeastward to the sea and constituting in its coastal terminus fjord beneath the Mertz Glacier (region B′) with a maximum measured depth of 1495 below sea level.

Geologic interpretation

Precambrian basement gneiss and schist crop out on the Adélie Coast and have been inferred 80 km farther inland from a seismic refraction velocity of 4.9 km/s (Imbert, 1953). It seems clear that the low-lying region labeled G, extending as far as 200 km inland, represents the continuation of this deeply eroded basement complex.

It is suggested that the elevated regions H and H′ may be associated with granitic plutons along a linear belt extending to the coast east of Mertz Glacier where Precambrian granite is exposed (Summers and Edwards, 1940; Bushnell and Craddock, 1969–70). Subsequent erosion has removed the original cover to produce the present ragged appearance, leaving the massifs upstanding and flanked by younger strata (Figure 125.6).

The asymmetric nature of trough F and the westerly sloping plateau suggest a fault-block structure or a half-graben, with a major normal fault running along the escarpment. The large throw could be explained by the structure being recent (as suggested by its high relief and fresh, angular appearance). The absence of any associated seismicity may be attributable to the "damping" effects of the ice sheet (Evison, 1967).

THE SOUTHERN HIGHLANDS

The highland region in the southwestern part of the mapped area (region D, outlined by the +250 m contour) is composed of four separate positive features. It is bounded to the east by the Wilkes Basin, to the south by the Aurora Basin (largely off this map) (see Drewry, 1975*a*), and to the west by a low plain. Steeper slopes occur in the northeast sector; they may be related to possible rifting activity more clearly shown in the trench-scarp zone (F) to the north.

Conclusions

The Wilkes subglacial basin and its salients constitute a major intercratonic zone of sedimentation with a history possibly extending back into Paleozoic time. The eastern edge of the basin probably provides a western limit to the area of orogenic activity identified in the Transantarctic Mountains. As a result, we consider that the Ross orogenic belt must narrow substantially to the north of Terra Nova Bay. Granites east of the Mertz Glacier have been radiometrically dated with an age range of 350 to 750 m.y. (Bushnell and Craddock, 1969–70; Craddock, 1972), but any relationships to orogenic events farther east are unclear.

Figure 125.6 is a speculative geologic cross section based on the closest rock outcrops and the subglacial terrain characteristics. It summarizes the general structures that have been discussed above.

Acknowledgments. Logistic support was generously provided by the U.S. National Science Foundation and by the U.S. Navy Antarctic Support Force and Air Development Squadron VXE-6. The SPRI project was funded under a U.K. Natural Environment Research Council grant. We wish to thank the aircrew of U.S. Navy C-130F (Buno 148320) and members of the radio-echo research groups. In particular we wish to acknowledge the assistance of K. E. Rose in providing computer programs and P. J. Barrett for useful discussions on Antarctic geologic interpretations.

References cited

Bushnell, V. C., and Craddock, C. *eds.* 1969–70. Geologic maps of Antarctica. *Antarct. Map Folio Ser.*, Folio 12.

Craddock, C. 1972. Antarctic tectonics (review). (*In* Adie, R. J., *ed. Antarctic Geology and Geophysics.* Oslo, Universitetsforlaget, 449–56.)

Crary, A. P. 1963. Results of the United States traverses in East Antarctica, 1958–61. *IGY Glaciol. Rep.*, No. 7, 144 pp.

Drewry, D. J. 1975*a*. Radio echo sounding map of Antarctica (N90°E–180°). *Polar Rec.*, *17*, No. 109, 359–74.

Drewry, D. J. 1975*b*. Initiation and growth of the east Antarctic ice sheet. *J. Geol. Soc. Lond.*, *131*, Pt. 3, 255–73.

Drewry, D. J. 1975*c*. Comparison of electromagnetic and seismic-gravity ice thickness measurements in East Antarctica. *J. Glaciol.*, *15*, No. 73, 137–50.

Drewry, D. J. 1976. Sedimentary basins of the east Antarctic craton from geophysical evidence. *Tectonophysics*, *36*, 301–14.

Evison, F. F. 1967. Note on the aseismicity of Antarctica. *N.Z. Jl Geol. Geophys.*, *10*, No. 2, 479–83.

Imbert, B. 1953. Sondages séismiques en Terre Adélie. *Annls Géophys.*, *9*, No. 1, 85–92.

Lamakin, V. V. 1968. The origin of the Baikal depression. *Priroda*, *4*, 48–59.

Mawson, D. 1940. Sedimentary rocks. *Rep. Australasian Antarc. Exped.*, Ser. A, *4*, Pt. 2, 347–67.

Schmidt, R. A. 1962. Australites and Antarctica. *Science, N.Y.*, *138*, No. 3538, 443–44.

Summers, H. S., and Edwards, A. B. 1940. Granites. *Rep. Australasian Antarc. Exped.*, Ser. A, *4*, Pt. 31, 87–113.

Weihaupt, J. G. 1961. Geophysical studies in Victoria Land, Antarctica. *Res. Rept. Ser.*, Geophys. Polar Res. Center, Univ. Wisconsin, No. 1, 123 pp.

Weihaupt, J. G. 1976. The Wilkes Land anomaly: evidence for a possible hypervelocity impact crater. *J. geophys. Res.*, *81*, No. 32, 5651–63.

126

Ice Flow, Bedrock, and Geothermal Studies from Radio-Echo Sounding Inland of McMurdo Sound, Antarctica

DAVID J. DREWRY
Scott Polar Research Institute, University of Cambridge, Cambridge CB2 1ER, U.K.

Abstract. Systematic airborne radio-echo sounding has been conducted, as part of a joint National Science Foundation–Scott Polar Research Institute–Technical University of Denmark program, along 10 subparallel lines 10–15 km apart, inland of the McMurdo Sound ice-free valley region to long. 156° E and between lat. 76.25° and 78.25° S. Information on ice surface and bedrock surface configurations, presence or absence of subice water bodies, and bedrock reflectivity characteristics has been obtained continuously along all flight tracks. Careful analysis of profiling data has enabled the compilation of an ice-surface contour map at an interval of 50 m. This clearly defines the pattern of ice flow, which is dominated by drainage into Mulock and Mawson Glaciers. An ice dome inland of Taylor Glacier restricts the catchment of Taylor and several other glaciers. This conclusion, supported by stable-isotope measurements, has important implications for glacial geologic studies because it casts doubt on the validity of the use of such glaciers in investigations of fluctuations of the East Antarctic ice-sheet.

A contour map of subice bedrock, at a 200 m interval, depicts the continuation of mountainous topography, declining in elevation westward. Reevaluation of seismic refraction profiles, combined with reflection and roughness characteristics from radio-echo sounding, indicates the presence of Beacon-type rocks beneath the ice sheet, although deep valleys seem to expose basement rocks.

Echo strength measurements have allowed estimates of the returned power from the ice-bedrock interface. Areas with strong, persistent returns are interpreted as having subice water bodies. A morainal layer may also be present at these sites. Such conclusions have implications for the geohydrologic regime of neighboring ice-free valleys. Based upon a boundary condition of pressure melting above subice lakes and upon surface glaciologic data, the basal temperature gradient of the ice sheet has been computed to be 50 ± 20°C/km. The heat flux through the ice sheet base is estimated to be 2.7 HFU. The technique described may be useful for obtaining this quantity elsewhere in ice-covered Antarctica.

The McMurdo Sound region is one of the most intensively studied areas of Antarctica. It has, for instance, been the subject of a recent major international earth science program—the Dry Valley Drilling Project (DVDP). In contrast, oversnow and airborne surveys, extending geophysical investigations beneath the ice sheet immediately inland of the Southern Victoria Land mountains and providing a valuable extra dimension, have been few and their results commonly ambiguous. There remains a need to determine the nature of bedrock materials and structures lying inland of the exposed mountains and to understand the controls exercised by bottom relief on past and present advances of glaciers into the valleys. The possibility that groundwater emerging in the ice-free areas has its origin at the base of the ice may also be studied by geophysical techniques, particularly radio-echo sounding.

Airborne radio-echo sounding missions have been flown into central East Antarctica during six field seasons since 1967, as part of a joint National Science Foundation–Scott Polar Research Institute (NSF–SPRI) program and later, a NSF–SPRI–Technical University of Denmark (TUD) program. Data obtained in 1967–68 and 1969–70 in the vicinity of southern Victoria Land have been discussed by Calkin (1974*a,b*). Although navigation, altimetry, and radar systems suffered from several occasionally severe limitations, the resulting information has been sufficient to outline the broad pattern of bedrock topography. Precision was increased by installation of inertial navigation during 1971–72, and the subsequent data from the McMurdo Sound area were incorporated into the SPRI radio-echo sounding map published in 1974 (Drewry, 1975*a*). In January 1975, a systematic airborne survey was conducted along 10 subparallel lines 10–15 km apart, inland of McMurdo Sound to long. 156° E and between lat. 76.25° and 78.25° S (Figure 126.1). Some of the results and implications of this investigation are presented below.

Data processing

NAVIGATION

Navigational information was recorded on punched paper tape from an Inertial System (LTN-51) every 20 seconds. The reconstructed flight track was fixed to known surface features, principally nunataks and mountains, by aerial photographs. Fix positions were obtained from the U.S. Geological Survey 1:250,000 Reconnaissance Series maps. Vertical air photographs allow position accuracy to

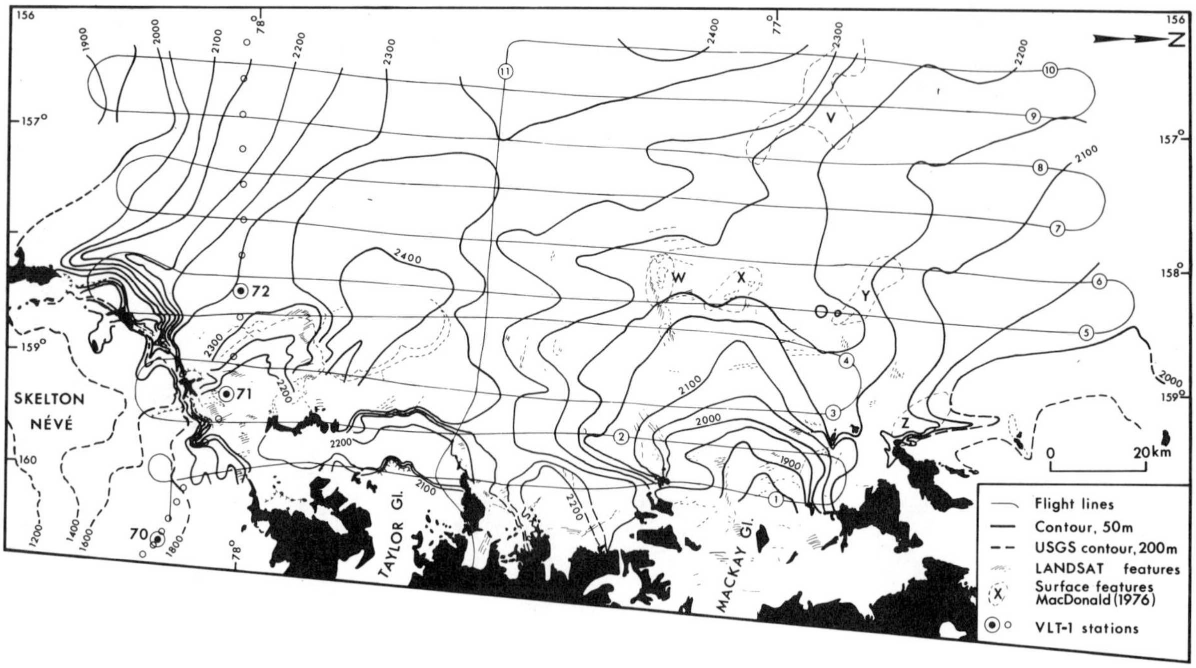

Figure 126.1. Ice-sheet surface configuration inland of McMurdo Sound from radio-echo sounding (flight lines shown numbered 1–11). Surface lineaments from LANDSAT images have been added, and exposed bedrock is shown in black. Major (numbered 70, 71, and 72) and minor stations occupied by Victoria Land Traverse—1 (VLT-1) are illustrated. Note the ice dome (>2400 m) inland of Taylor Glacier and the drainage basins of Mackay Glacier and other glaciers.

better than 100 m, while resections from oblique air photos are probably accurate to 0.5 km for points within 30–40 km of high-relief terrain. Close to fix positions, flight lines should be located to within 1 km and elsewhere to better than 2 km.

Radio-echo measurements

Radio-echo data were recorded simultaneously on two systems designed by the Technical University of Denmark, which operated at 60 and 300 MHz. The new 300 MHz equipment possesses a much narrower beamwidth (32°), and this has allowed the resolution of extremely rough subice terrain with increased precision. Most of the complex and troublesome hyperbolas and cusps generated by the 60 MHz radar are eliminated at 300 MHz (Robin and others, 1977). Increased absorption in ice of radio waves at this higher frequency, however, restricts the depth penetration of the present system. Nevertheless, because the ice sheet is relatively thin (<2,500 m) inland of McMurdo Sound, bedrock returns were obtained at 300 MHz along all the track flown. Terrain clearance and ice-thickness data were measured off the records at intervals of 10 seconds of flight time, corresponding to an average overland distance of 1.1 km. Accuracy of the latter measurements is 10 m or 1.5% ice thickness, whichever is the greater. An electromagnetic wave velocity in ice of 168 m/μs was used along with a +10 m firn correction.

Altimetry

Ice surface and bedrock surface elevations were determined by a combination of corrected aircraft pressure altimetry and radio-echo sounding data, tied to any known surface elevations. Corrections were made for outside air temperature effects and for the slope of the isobaric surface (cross-wind component). The resulting ice surface altitudes were checked where the last flight track (11 in Figure 126.1) crossed earlier tracks, and a mean difference of 15 m was observed. The data were subsequently refixed at these common points, and measurement of *relative* elevation is likely to be better than 10 m. Radio-echo sounding ice surface heights were then tied to ground control: altitudes were determined at points along the Victoria Land Traverse 1 (VLT-1; Crary, 1963), at points within the Transantarctic Mountains and at the McMurdo Station skiway airstrip. *Absolute* accuracy of the airborne determinations is probably better than 30 m. The new ice-sheet surface elevations are therefore considered to be more accurate than those presented from previous radio-echo sounding studies in this area that were based upon dead-reckoning navigation and crude altimetry (Calkin, 1974*b*).

Ice-sheet surface

Figure 126.1 presents an ice surface contour map with an interval of 50 m. Ice surface topography is dominated by the drainage basins of the Mulock, Mackay, and Mawson Glaciers. An ice ridge and dome system lies inland of Taylor Glacier, separating flow to the north and to the south. The ice surface consequently drops by 100 m or more before rising to the west to join the main East Antarctic plateau.

Surface features identified from LANDSAT satellite photographs have been added to Figure 126.1 and include mainly corrugations and lineations. Many of these elements show a high degree of correspondence to the independently constructed contours. New rock outcrops in this area, suggested by MacDonald (1976) from LANDSAT imagery, were not detected along radio-echo sounding tracks.

Principal ice divides and selected flow lines have been

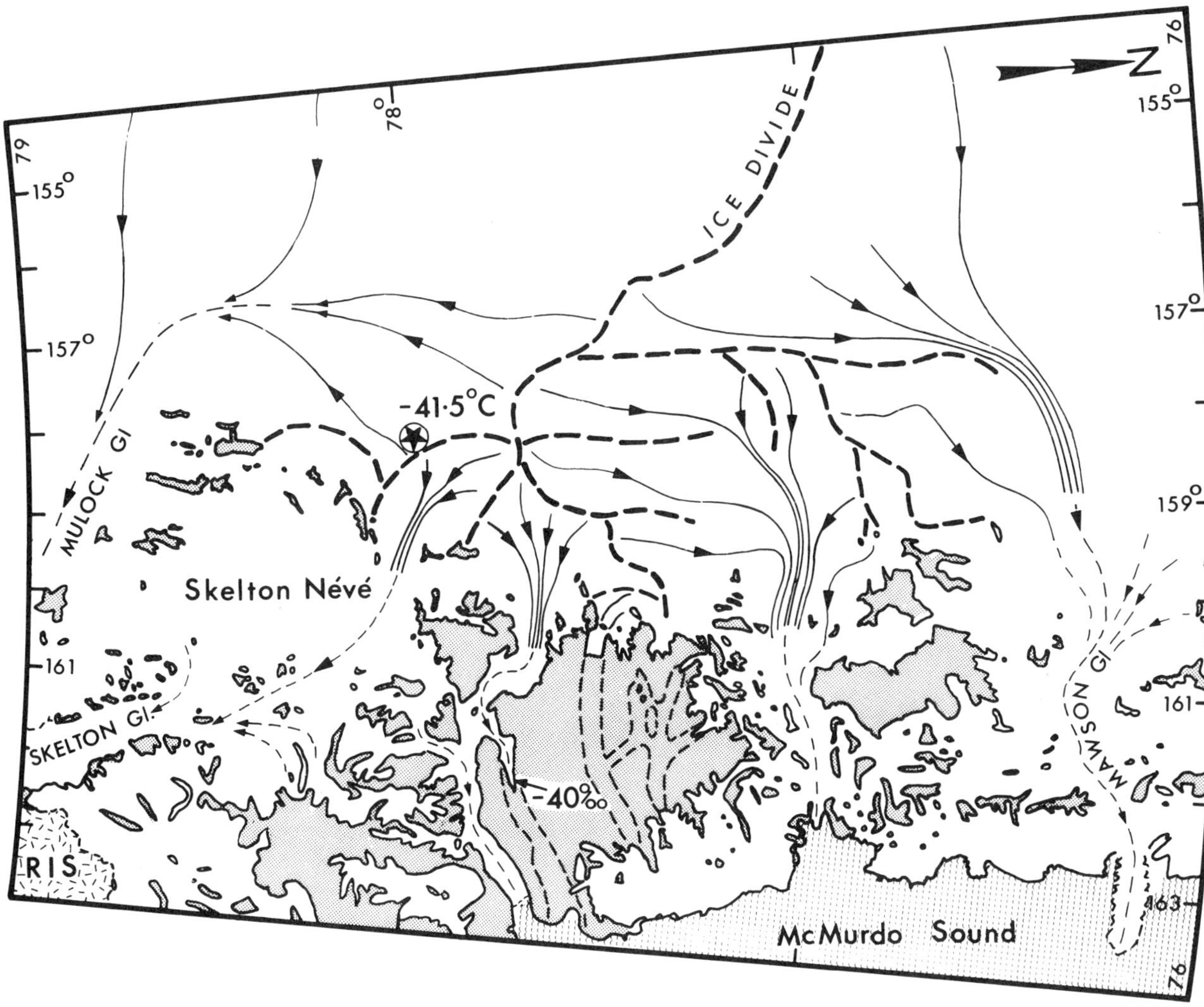

Figure 126.2. Ice divides and selected flow lines inland of the McMurdo Sound region, constructed from surface contour map. Station 72 mean annual ice-surface temperature (star) and mean stable-isotope ($\delta^{18}O$) value for Taylor Glacier snout ice are given. RIS: Ross Ice Shelf.

constructed from the surface contour map (Figure 126.2). The dome inland of Taylor Glacier acts as a locat center of outflow and is connected to a broad nose of the East Antarctic ice sheet, as depicted on the SPRI radio-echo sounding map (Drewry, 1975*a*) and by Steed and Drewry (Chapter 125, this volume). Figure 126.2 demonstrates that only the Mawson and Mulock Glaciers discharge any ice with a source more than 150 km from the coast. The smaller intervening glaciers (e.g., Skelton, Ferrar, Taylor, Wright Upper, Victoria Upper, and Mackay) are nourished by local ice originating within a few tens of kilometers of the Transantarctic Mountains. Stable-isotope determinations on ice from the snout of Taylor Glacier corroborate the notion of a nearby source. δ D values of between -327 ‰ (parts per thousand) and -341 ‰ were obtained with a mean value of -332‰. The deuterium ratio corresponds to a mean $\delta^{18}O$ value of -40.3 ‰ using:

$$\delta^{18}O = (\delta D - 10‰)/8$$

These values are isotopically much heavier than would be anticipated for ice derived from the central regions of the ice sheet ($\delta^{18}O$ of between -45 and -55‰, see Dansgaard and others, 1973). Minimum δ values for Taylor Glacier, predicted from elevation-temperature considerations (Dansgaard and others, 1973) at the ice shed and supported by the measured mean annual surface temperature at VLT-1 station 72 (Crary, 1963) of -41.5°C, are also of the same order (~ -40 to -41‰). The glaciologic régimes of these minor glaciers seem to be controlled by local, near-mountain climatic effects and not by the dynamics of the main ice sheet. This conclusion has important implications for glacial geologic studies within the McMurdo Sound region.

Advances and retreats of Taylor Glacier, for instance, have been considered to reflect fluctuations of the East Antarctic ice sheet (e.g., Denton and others, 1971). Only very large scale ice sheet events (i.e., several hundred meters of thickening) would be of sufficient magnitude to greatly modify or reverse local surface slopes, allow penetration of additional plateau ice into the mountain range, and thus be manifested by movement of the Taylor Glacier terminus. Vertical crustal movements, which in part control discharge over bedrock thresholds at the upper end of these glaciers, are probably small-magnitude effects (perhaps 100 m/m.y.) (Drewry, 1975*b*; Drewry and Robin, in press). Minor glaciers like Taylor within southern Vic-

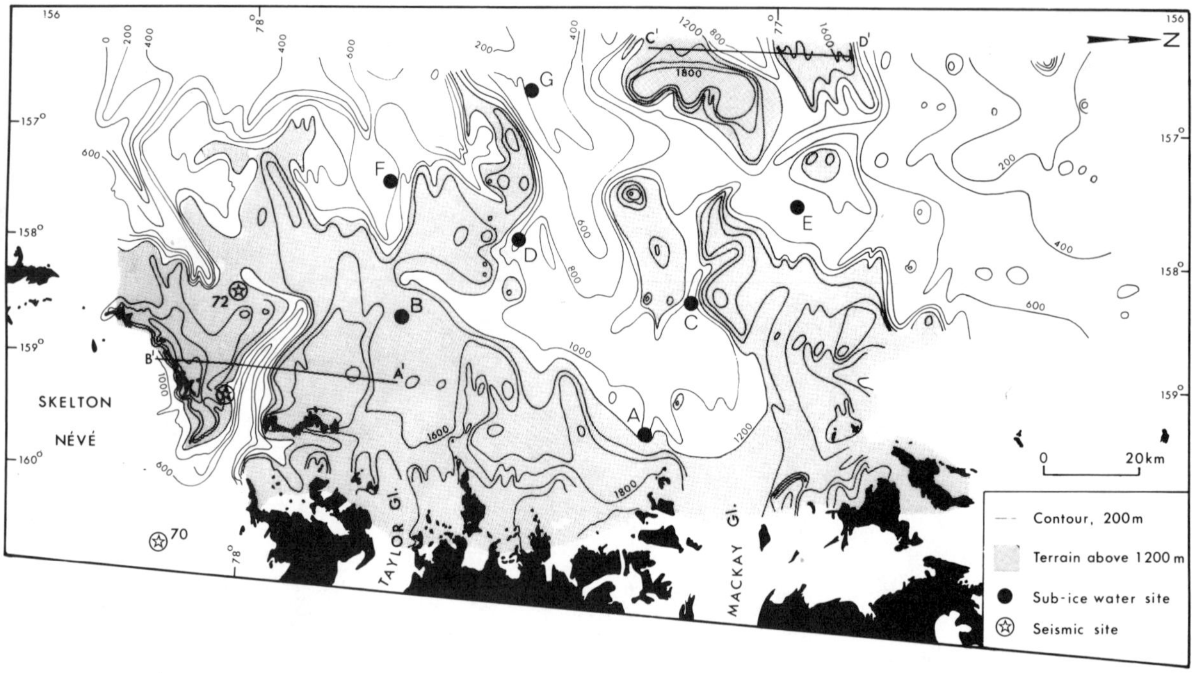

Figure 126.3. Bedrock configuration inland of McMurdo Sound from radio-echo sounding. Profiles along lines A′–B′ and C′–D′ are illustrated in Figure 126.4. Locations of subice water bodies, as detected by radio-echo sounding, are shown as black dots. Exposed bedrock is shown in black.

toria Land, fed principally by local ice, seem insensitive gauges of small- to medium-scale fluctuations of the East Antarctic ice sheet.

Bedrock surface

Calkin (1974*a*) has discussed geomorphologic results of some early radio-echo sounding profiles inland of McMurdo Sound. The subice bedrock surface derived from 1974–75 flight-line data is depicted in Figure 126.3, contoured at 200 m intervals. While the area is smaller than that discussed by Calkin, it is shown in substantially more detail.

The bedrock terrain lies entirely above sea level, although a deep bedrock channel entering the Mulock Glacier falls to 0 m at the edge of the area under consideration. The highest elevation is ~2200 m above sea level.

Table 126.1 summarizes salient statistical information on subice topography, computed for each parallel flight line. A principal feature is the continuation of the mountainous terrain exposed in southern Victoria Land. Nevertheless the subice surface is extremely uniform. Line 11 possesses properties similar to those of the other orthogonal north-south tracks, suggesting a significant degree of terrain isotropy. The statistics in Table 126.1 provide a more representative descriptive sample of Transantarctic Mountain terrain than previously available (Drewry, 1975*c*). They indicate a large degree of vertical roughness, although the moderately high autocorrelation distance* is probably indicative of structural control by subhorizonal Beacon Supergroup strata.

These data, averaged along each line, demonstrate a uniform decrease in elevation inland from >1500 m at the ice sheet edge to ~700 m at long. 156° E. The best-fit gradient suggests a surface sloping westward at 1°–2°. This value is close to the dip of the Beacon Supergroup, as reported for the Asgard Ranges (Warren, 1969).

Several deep valleys are cut into this surface, separating upland massifs. Four broad valleys trend directly inland while another feeds into the Mulock Basin. An exception-

* A distance over which roughness shows little or no statistical variation within given limits (see Table 126.1).

Table 126.1. Selected subice terrain statistics inland of McMurdo Sound

Flight track[a]	A	B	C	D
1	0.5	1517	459	14
2	10.5	1475	395	6
3	19	1417	383	6
4	34	1326	345	16
5	38.5	1295	313	5
6	49.5	1180	364	6
7	60	1070	322	4
8	70	828	326	5
9	79.5	888	524	23
10	90	735	504	12
	(Mean)	1173	393	10
11		1110	431	8

A: Average distance (in km) of line from western edge of mountains.
B: Mean elevation (in m) above sea level.
C: Root-mean-square of elevations (in m).
D: Autocorrelation distance (in km) calculated as the lag to l/e for the normalized autocorrelation function (with zero mean):

$$D_f = (n/n-k)\left[\left(\sum_{x=1}^{n-k} V_x \cdot V_{x+k}\right) / \sum_{x=1}^{n} V_x^2\right]$$

where V_x = discrete bedrock elevation
k = lag of unit 1.1 km (sample interval)
n = total number of sample elevations.

[a] See Figure 126.1 for locations.

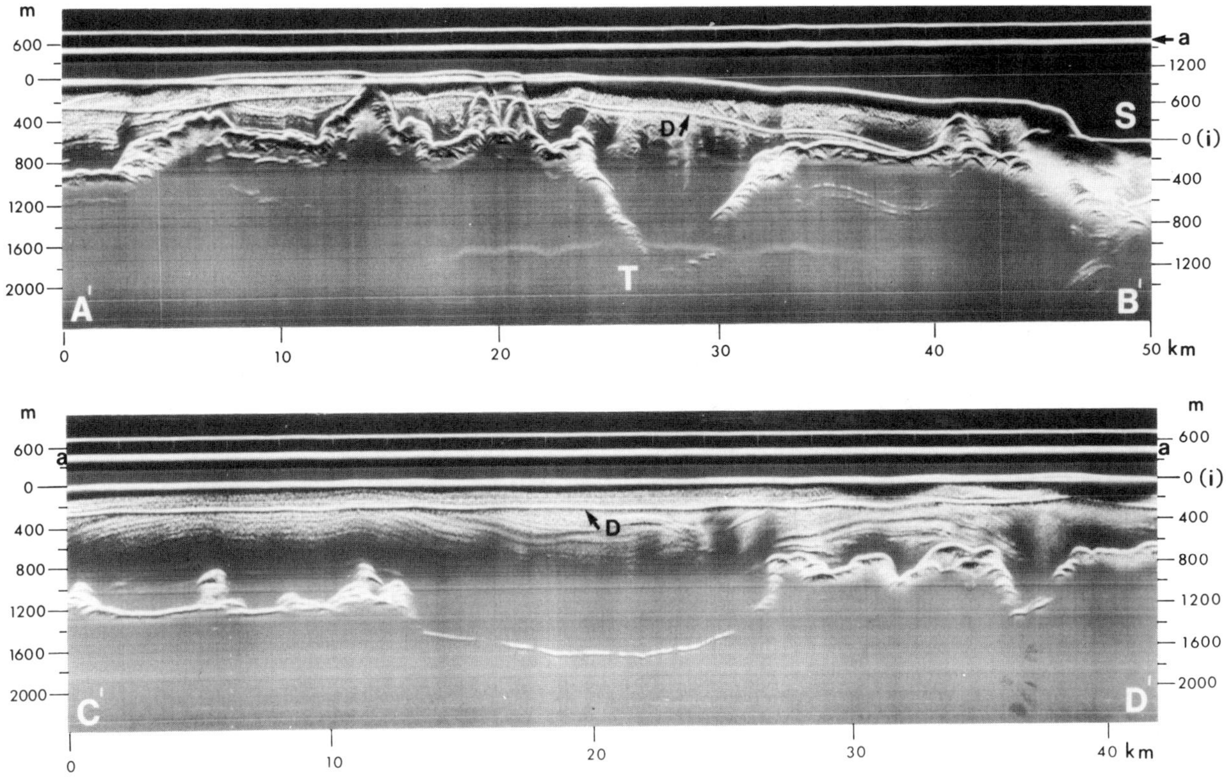

Figure 126.4. 300 MHz radio-echo sounding profiles. Note the different scales of elevation in air and ice depth due to different velocities of radio waves in these two media (300 and 168 m/μs, respectively). a: Transmitter pulse. D: Double surface echo. (i): Ice surface. A′–B′ is a section from Skelton Névé (S) northward across a deep bedrock channel (T) which drains through the "Portal" into Skelton Névé. Flanking terrain is probably developed in Beacon Supergroup strata. C′–D′ lies at the western edge of the investigated area and illustrates two high massifs (Beacon?) separated by a valley exhibiting different reflecting characteristics (presumed basement).

ally deep bedrock trough extends for 25 km inland eastward through the "Portal" and hence into Skelton Névé (T in profile A′–B′ of Figure 126.4). It seems probable that erosion by ice has been important in the formation of this valley. In the northwest corner of the mapped area, the bedrock surface is significantly less rough and at much lower elevation. This may be the result of older(?) erosion associated with extensive ice drainage into Mawson Glacier.

Figure 126.3 locates three seismic long-refraction profiles accomplished during VLT-1 (Crary, 1963). Station 70 is positioned within Skelton Névé and demonstrates a high-velocity layer dipping eastward, overlain by a wedge of 4.4 km/s material. Crary interpreted this structure as Beacon sedimentary rocks underlain by an intrusion of dolerite. Station 71 is located where a tributary valley enters the southern flank of the deep bedrock trough feeding Skelton Névé. This complex topography is probably responsible for the large westerly apparent dip of the lower 5.5 km/s layer and the thin, discontinuous nature of the uppermost sedimentary layer (4.4 km/s). Station 72, about 20 km farther west and located on a northwest-sloping bedrock highland, yielded an upper, 500 ± 100 m thick layer of Beacon-type sedimentary rock (4.3 km/s) overlying 1.2 km of typical granitic basement (5.4 km/s). At about 400 m below sea level, high-velocity (6.5 km/s) horizon was detected.

The results of radio-echo sounding in the vicinity of Crary's refraction stations clarify his interpretations. The two inland sites were located on high bedrock of undoubted Beacon-type terrain. Exposed rock on the western flank of the mountains consists, almost without exception, of Beacon Supergroup strata (including Ferrar Dolerite); locally the outcrops possess a capping of tholeiitic basalt. Radio-echo sounding and seismic results provide few grounds for rejecting the continuation of these units inland, particularly within high-standing tracts. Erosion has probably been responsible for isolating some of the discrete massifs depicted in Figure 126.3, although one cannot exclude the possible effects of faulting. It is unfortunate that no seismic refraction profile was shot within the deeper valleys of the region; these have probably cut down below Beacon rocks into granitic basement. Such a conclusion is supported by the marked contrast in radar reflection characteristics and surface roughness across valleys and adjacent highs (Figure 126.4) and is in keeping with a Beacon section here of < 1000 m thickness, as suggested by the seismic results.

Subice water

Substantial bodies of water have been detected beneath the ice sheet in East Antarctica from radio-echo sounding (Oswald and Robin, 1973; Robin and others, 1977). In general such "lakes" are located within basins on the bedrock surface in areas of low ice surface slopes and low ice velocity. These conditions are fulfilled inland of the McMurdo ice-free valleys where seven possible subice lakes of minor lateral extent (< 5 km) have been detected (Figure 126.3). The lakes are recognized on the basis of (1) in-

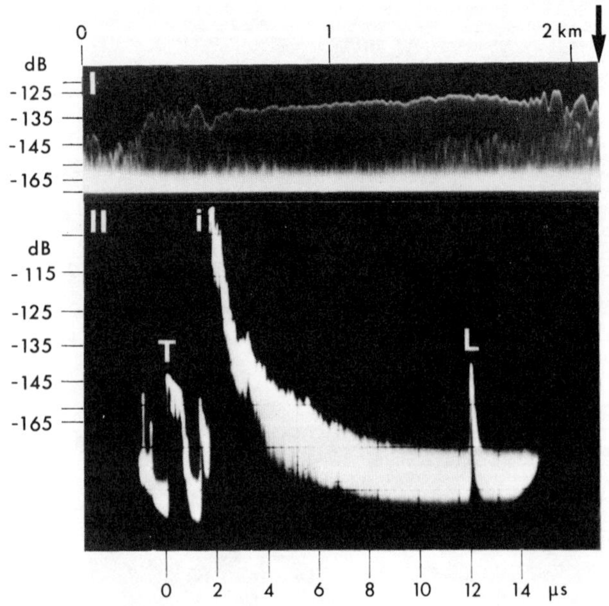

Figure 126.5. Radio-echo reflection characteristics over lake *F* (lat. 77.72° S, long. 157.42° E) shown in Figure 126.3. The upper print (I) shows a 2 km section of continuous basal echo strength measurements (ESM) in decibel (dB). Note the persistent high returned power. The arrow marks the location of the A-scope frame (displaying returned signal strength vs. echo delay time) below (II). T: Transmitter pulse. i: Ice surface return. L: Return from ice-water interface.

creased echo strength, (2) increased fading length, and (3) an almost horizontal upper reflecting surface. Figure 126.5 illustrates the A-scope (returned signal strength vs. echo delay time) and ESM (echo strength measurement; Neal, 1976) records over one such lake (F in Figure 126.3). The reflection coefficient over the lake lies in the range −5 to −10 decibel (dB). A typical ice-rock boundary possesses a reflection coefficient of −14 to −25 dB. While the coefficient at site F is much higher, indicating the presence of water, it does not attain that of a pure ice-water interface (about −3 dB). This suggests additional power loss at the sole from a surface somewhat rougher than specular—possibly caused by the presence of moraine.

McGinnis and Jensen (1971) have argued that saline waters in Taylor and Wright Valleys may result from the lateral movement of groundwater from beneath the East Antarctic ice sheet. Their hypothesis demands basal pressure melting inland of the mountains—a condition which is confirmed by the presence of subice lakes.

Estimates of the geothermal heat flux

Recognition of water at the bed from radio-echo sounding provides a limiting condition to the temperature profile through an ice sheet. This observation allows calculation of the basal temperature gradient and estimation of heat flux through the sole. Over most of Antarctica, with no rock exposed and few holes drilled to bedrock, the technique outlined below may provide a method of estimating this geophysical quantity.

At each "lake" depicted in Figure 126.3, the vertical temperature profile through the ice sheet has been computed based upon the steady-state solution (Robin, 1955):

$$k\,(d^2T/dx^2) - v(dT/dx) = 0$$

where T = temperature
k = thermal diffusivity of ice
v = vertical ice velocity
x = distance below surface.

The horizontal advection term has been neglected since the lakes lie close to a center of ice outflow. Frictional heating will also be minimal over an ice-water interface. The profile has been calculated to meet the pressure melting condition at the bed, in accord with the presence of water. The heat flux through the ice base can be computed by knowing the thermal conductivity of ice (K_i) and basal melting (θ):

$$F_b = (dT/dx)_i K_i + \theta$$

where $(dT/dx)_i$ = basal ice temperature gradient, calculated for the basal 50 m of the ice column.

INPUT DATA

Surface glaciologic data (accumulation rate and temperature) are necessary for accurate computation of vertical temperature profiles. Such data are meager inland of the southern Victoria Land mountains and result from three traverses that gained access to the East Antarctic ice sheet via Skelton Glacier. Unfortunately all three duplicated the route to station 84 (lat. 78° 02′ S, long. 154° 01′ E), which lies beyond the area considered here; consequently, the data have poor areal distribution. Close proximity to the Transantarctic Mountains must induce significant local variations in the patterns of mean annual surface temperature (MAST) and especially mean annual accumulation rate (A). Therefore, the values selected from traverse results have been taken only as indications of the true values, and additional computations were undertaken using a wider range of values to assess likely errors.

Mean annual surface temperature (MAST)

Between Skelton Névé and long. ~145° E, MAST as reported by Crary (1963) shows a mean change with elevation of ~4°C/100 m, which may reflect local mountain influences. As this gradient is four times higher than the average dry adiabatic lapse rate in air, temperatures have been computed using 1°C/100 m. The resulting MAST values given in Table 126.2 are probably accurate to ±5°C.

Table 126.2. Calculated basal temperature gradient and heat flux in ice above subice lakes

Site	Ice thickness (m)	MAST	A	dT/dx	F_b
A	1190	−35	14	−50	2.7
B	1160	−44	14	−64	3.5
C	1360	−41.5	14	−55	3.0
D	1480	−43	14	−54	2.9
E	1340	−42	14	−56	3.0
F	1740	−43.5	14	−51	2.7
G	2355	−44	14	−44	2.4
		(Rounded mean)		−50 ± 20	2.7

MAST: Mean annual surface temperature. (°C).
A: Mean annual accumulation rate. (gm cm^{-2}a^{-1})
dT/dx: Basal temperature gradient (°C/km) in lowermost 50 m.
F_b: Basal heat flux (HFU = μcal cm^{-2}s^{-1} = 41.86 mW m^{-2}).

Mean annual accumulation rate (A)

Accumulation is notoriously variable in mountainous areas. The only measurements available inland of McMurdo Sound are at stations 72 and 84 yielding 140 and 20 kg/m^2, respectively, which confirms wide variability. The second, low value is attributed to the effects of a topographic depression (Crary, 1963; Bull, 1971). With no *a priori* knowledge of local variations in A, 140 kg/m^2 has been used for all sites given in Table 126.2. The likely error of A may, however, be in the region of 50%.

We estimate the heat exchange from melting of freezing, assuming the rate is some reasonable proportion of accumulation at about 2 mm per year melting.

RESULTS

Values calculated for basal ice temperature gradient and heat flux are given in Table 126.2. They have not been corrected for nearby terrain or other effects. In view of uncertainties in MAST and A, it is suggested that the accepted mean values of 50°C/km and 2.7 HFU are open to considerable error (40–50%) and probably represent order of magnitude estimates of the upper limit.

Since the global, average continental geothermal heat flux is 1.65 HFU with a gradient of ~30°C/km, and despite uncertainty in the temperature calculations, the derived values for basal ice temperature gradient and heat flux inland of McMurdo Sound appear significantly high. Decker and Bucher (Chapter 111, this volume) have reported preliminary geothermal measurements in this area taken during the Dry Valley Drilling Project (DVDP). On Ross Island, gradients of 30–40°C/km were observed in DVDP-3, 25–33°C/km at Lake Vida (DVDP-6), and 48°C/km in the North Fork of Wright Valley (DVDP-14). At New Harbour, despite possible disequilibrium, very high gradients between 50 and 80°C/km were recorded. Uncorrected heat flows (i.e., compatible with data in Table 129.2) given by Decker and Bucker range from 1.8 to 3.4 HFU. These few results may not necessarily be representative of conditions inland of the mountains, nevertheless, they indicate that the high heat flow suggested from temperature calculations at subice lake sites may not be unreasonable.

Although it is not practical to discuss possible causes in detail here, the apparently high heat flow in Victoria Land could result from a variety of factors such as recent tectonic activity, crustal thickening, or nearby mantle convection. Lateral heat flushing induced by the Australia-Antarctica separation is also a possible mechanism; it is interesting to note that in southeast Australia abnormally high heat fluxes, between 2.8 and 2.9 HFU, have been reported (Sass, 1964).

Acknowledgments. The radio-echo sounding results discussed above were collected during a joint U.S. NSF-SPRI-TUD program in Antarctica during 1974–75. Logistic support was generously provided by the U.S. National Science Foundation and the U.S. Navy Antarctic Support Force and Air Development Squadron VXE-6. The SPRI project is funded under a U.K. Natural Environment Research Council grant (UR3/2291). I wish to thank Lt Cdr A. L. Herr, Jr., and the crew of U.S. Navy C-130F (Buno 148320) and members of the radio-echo sounding team for enthusiastic field support. I. Holyer assisted with the data reduction. G. de Q. Robin kindly read and commented on a draft of this paper.

References cited

Bull, C. B. B. 1971. Snow accumulation in Antarctica. (*In* Quam, L., *ed. Research in Antarctica. Am. Ass. Adv. Sci,* Publn No. 93, 367–421.)

Calkin, P. E. 1974*a*. Subglacial geomorphology surrounding the ice-free valleys of southern Victoria Land, Antarctica. *J. Glaciol., 13,* No. 69, 415–30.

Calkin, P. E. 1974*b*. Ice surface topography and ice-free area, southern Victoria Land. *Antarct. Jnl U.S., 9,* No. 5, 236–38.

Crary, A. P. 1963. Results of the United States traverses in east Antarctica, 1958–61. *IGY Glaciol. Rep.,* No. 7, 144 pp.

Dansgaard, W., Johnsen, S. J., Clausen, H. B., and Gunderstrup, N. 1973. Stable isotope glaciology. *Meddelelser om Grønland, 197,* No. 2, 53 pp.

Denton, G. H., Armstrong, R. L., and Stuiver, M. 1971. The late Cenozoic glacial history of Antarctica. (*In* Turekian, K. K., *ed. Late Cenozoic Glacial Ages,* New Haven, Yale University Press, 267–306.)

Drewry, D. J. 1975*a*. Radio echo sounding map of Antarctica (~90°E–180°). *Polar Rec., 17,* No. 109, 359–74.

Drewry, D. J. 1975*b*. Initiation and growth of the East Antarctic ice sheet. *J. Geol. Soc. Lond., 131,* No. 3, 255–73.

Drewry, D. J. 1975*c*. Terrain units in eastern Antarctica. *Nature, Lond., 256,* No. 5514, 194–95.

Drewry, D. J., and Robin, G. de Q. In press. Configuration and flow of the Antarctic ice sheet during the last 10^6 a. (*In* Robin, G. de Q., *ed. Isotopic and Temperature Profiles in Ice Sheets.* Cambridge, Cambridge University Press.)

MacDonald, W. R. 1976. Antarctic cartography. (*In* Williams, R. S., Jr., and Carter, W. D., *eds.* ERTS-1. A New Window on Our Planet. *Prof. Pap. U.S. geol. Surv.,* 929, 37–43.)

McGinnis, L. D., and Jensen, T. E. 1971. Permafrost-hydrologic regimen in two ice-free valleys, Antarctica, from electrical depth sounding. *Quaternary Res., 1,* No. 3, 389–409.

Neal, C. S. 1976. Radio-echo power profiling. *J. Glaciol., 17,* No. 77, 527–30.

Oswald, G. K. A., and Robin, G. de Q. 1973. Lakes beneath the Antarctic Ice Sheet. *Nature, Lond., 245,* No. 5423, 251–54.

Robin, G. de Q. 1955. Ice movement and temperature distribution in glaciers and ice sheets. *J. Glaciol., 2,* No. 18, 523–32.

Robin, G. de Q., Drewry, D. J., and Meldrum, D. T. 1977. International studies of ice sheet and bedrock. *Phil. Trans. R. Soc. Lond.,* Ser. B, *279,* No. 963, 185–96.

Sass, J. H. 1964. Heat-flow values from eastern Australia. *J. geophys. Res., 69,* 3889–93.

Warren, G. 1969. Geology of the Terra Nova Bay-McMurdo Sound area, Victoria Land. (*In* Bushnell, V. C., and Craddock, C., *eds.* Geologic Maps of Antarctica. *Antarct. Map Folio Ser.,* Folio 12, Pl. XIII.)

127

Radio-Echo Studies of Bedrock in Southern Marie Byrd Land, West Antarctica

K. E. ROSE
Scott Polar Research Institute, University of Cambridge, Cambridge CB2 1ER, U.K.

Abstract. Maps depicting the ice-sheet surface and the bedrock topography of 500,000 km² of Marie Byrd Land between the Ross Ice Shelf, Byrd Station, and the Transantarctic Mountains have been compiled from 15,000 km of radio-echo profiles obtained during the 1974–75 season. Five major ice streams are defined, and they are separated by ridges and domes in the ice surface near the Ross Ice Shelf. Further inland the ice sheet shows the typical convex-upward form.

The bedrock surface is generally below sea level, being smooth near the ice shelf and becoming rougher inland. A deep channel, aligned with the Shimizu Ice Stream, is evidence of erosion by ice flowing from East Antarctica prior to formation of the West Antarctic Ice Sheet. On the basis of topography the area is subdivided into five parts, the geology of which is discussed with the help of existing seismic, gravity, and magnetic data. One region in the west, bordering the Ross Ice Shelf, is believed to be part of a large sedimentary basin extending northwestward beneath the Ross Ice Shelf to the Ross Sea. In these areas resistant granitic massifs seem to provide tectonic control of the ice streams. Encompassing the Whitmore Mountains and extending 100 km south of the Thiel Mountains and westward to the edge of the Ohio Range, is an area of extreme roughness; dissected mountain blocks are separated by steep-sided valleys about 20 km wide and 2 km deep. A map of the isostatically compensated bedrock surface shows several new islands in a predominantly marine region.

During the 1974–75 Antarctic field season a joint NSF-SPRI-TUD* project allowed coverage of 500,000 km² of southern Marie Byrd Land to be obtained with a 50 km grid of radio-echo sounding flight lines.Other geophysical measurements, obtained during several earlier studies, have also been used in the interpretations given here. These data constitute mainly (1) seismic, gravity, and magnetic determinations on oversnow traverses in Little America—Byrd, the Sentinel Mountains, Marie Byrd Land, the Horlick Mountains (Bentley and Ostenso, 1961; Bentley and Chang, 1971), and the Wisconsin Range (C. R. Bentley, personal communication; (2) the airborne magnetic results of Project Magnet (Behrendt and Wold, 1963; Ostenso and Thiel, 1964; Behrendt, 1968); and (3) the seismic and radio-echo work undertaken by the Ross Ice Shelf Project (RISP) (Robertson, 1975).

SPRI has conducted radio-echo sounding flights in Antarctica since 1967. The early flights with dead reckoning navigation were mainly exploratory (Robin, Swithinbank, and Smith, 1970; Robin, Evans, and others, 1970). Inertial navigation was introduced during the 1971–72 season, and in 1974–75, TUD provided improved equipment and antennae capable of penetrating the warm, thick ice of Marie Byrd Land.

Data collection and reduction

The 1974–75 flights in Marie Byrd Land were planned on a 50 km square grid (Figure 127.1), aligned parallel to the 135° W meridian and passing over Byrd Station, so that the two sets of lines would approximate ice-flow lines and ice-surface contours. This paper discusses the results from 19 separate flights, including some over the Ross Ice Shelf, which are important for defining the grounding line. One flight studied the Byrd Station Strain Network (BSSN) of Whillans (1976).

Latitude, longitude, and ground speed from the inertial navigation system (INS), and altitude were recorded every 20 seconds on punched paper tape. Navigation was adjusted for terminal misclosure and to agree with additional fixes from trimetrogon photographs of mountains, nunataks, ground stations (Byrd, Roosevelt Island, and RISP Station J9), and the Ross Ice Shelf front (from Earth Resources Technology Satellite [ERTS] imagery). The presence of the Schuler frequency (Kayton, 1969) in the errors of the INS, having a maximum observed amplitude of 10 km, was not corrected for because of the impossibility of determining the magnitude and phase of this error at all times. Adjustments to the navigation were thus by linear interpolation. The mean shift of flight lines at 32 fixes was 5.9 km, with a standard deviation of 3.3 km. Navigational errors may also be assessed by comparison of values of measured ice thickness at the grid crossing points. The mean error at 104 intersections was 42 m, and the magnitude of the navigational shift required to eliminate errors greater than 30 m was 2.4 km.

Since ice-surface and bedrock elevations are computed by subtracting terrain clearance and ice thickness from the aircraft altitude, an accurate measurement of the altitude is essential. A pressure transducer, connected to the static line in the aircraft, was logged every 20 seconds and calibrated with the aircraft altimeter. Corrections were made for both the slope of the isobaric pressure surface, assuming all winds to be geostrophic, and for the temperature of the Antarctic atmosphere, which is colder than the

* National Science Foundation–Scott Polar Research Institute–Technical University of Denmark.

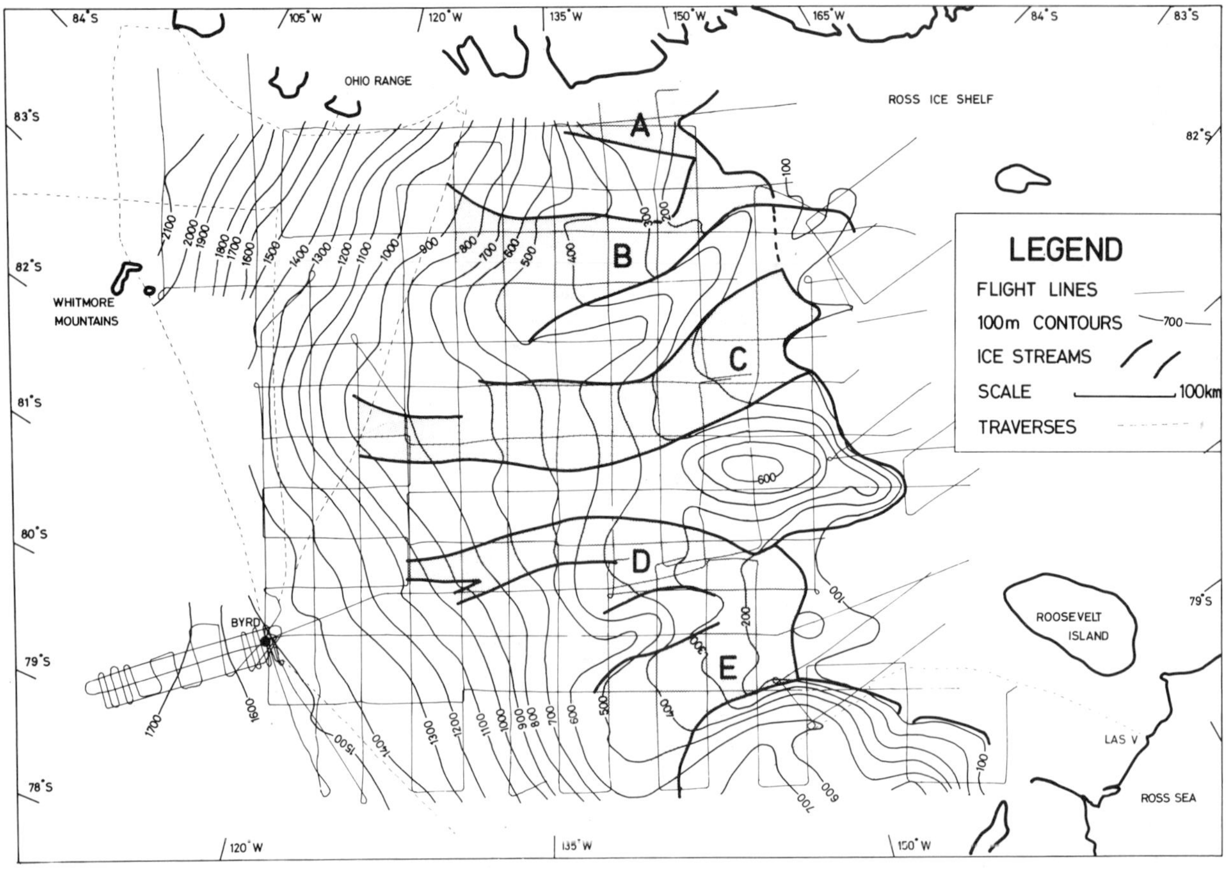

Figure 127.1. Southern Marie Byrd Land, showing ice-surface contours, ice streams A–E, and edge of the Ross Ice Shelf. LASV: Little America Station V.

International Standard Atmosphere. Altitudes were fixed by Byrd Station, elevation 1545 m (W. MacDonald, personal communication; King-Hele, 1975) and to floating parts of the Ross Ice Shelf, assuming hydrostatic equilibrium:

$$\text{RIS elevation (m)} = 0.108 \times \text{radio-echo ice thickness (m)} + 15.6 \quad (1)$$

(from Crary and others, 1962, Figure 14). Errors in surface elevations at flight grid intersections were distributed by random walk and least-squares techniques, disregarding those legs where the aircraft had changed elevation by more than 100 m. The flights were not fixed to ground traverse elevations except outside the area of the random walk, i.e., near the Whitmore Mountains and the Ohio Range. Overall accuracy should be comparable with that during the East Antarctic work, thought to be within about 30 m (Drewry, 1975*a*).

Radio-echo sounding information was recorded continuously on 35-mm film and later digitized semi-automatically on a back-projected trace reader, also with a 20 second interval. The resulting maps have been compiled from more than 10,000 data points. Radio-wave velocities of 300 m/μs and 168 m/μs for air and ice were used, plus a 10 m firn correction. Gaps in the flight lines in Figures 127.1 and 127.2 result from instrumental problems, inadequate system performance in areas of warm, thick ice when flying high, and scattering by surface crevassing. Random digitizing errors were removed by smoothing the terrain clearance values over a window of 5 km in length.

Maps of the flight lines, drawn and labeled with ice-surface and bedrock elevations by computer, were contoured by hand.

Ice-sheet surface

The ice-surface map (Figure 127.1) is much more complex than previously envisaged. In general, it conforms to a concave-up pattern below 1000 m elevation but changes to a convex-up profile at higher elevations. The lower region is dominated by five ice streams, as suggested by Robin, Evans, and others (1970). Intersections of flight lines and ice-stream boundaries are clearly defined by the strong hyperbolic echoes from crevasses. The boundaries were interpolated between these points with the extensive aid of trimetrogon photographs, taken during 1969–70 and 1974–75 radio-echo sounding flights, and interpolation for streams D and E was aided by ERTS images. The grounding line of the Ross Ice Shelf has been determined by comparing actual ice-surface elvations with the values predicted for a free-floating ice shelf of the same thickness, using equation 1. Between ice streams B and C, the position of the grounding line is indistinct, as confirmed by RISP measurements (Robertson, 1975).

Between ice streams, the ice-surface elevation rises. The most prominent such feature is the dome on the Siple Coast, between ice streams C and D; it rises to at least +657 m elevation, more than 300 m above the saddle to the east. Streams B and C and streams D and E are separated by ridges; the former has a small dome which rises to

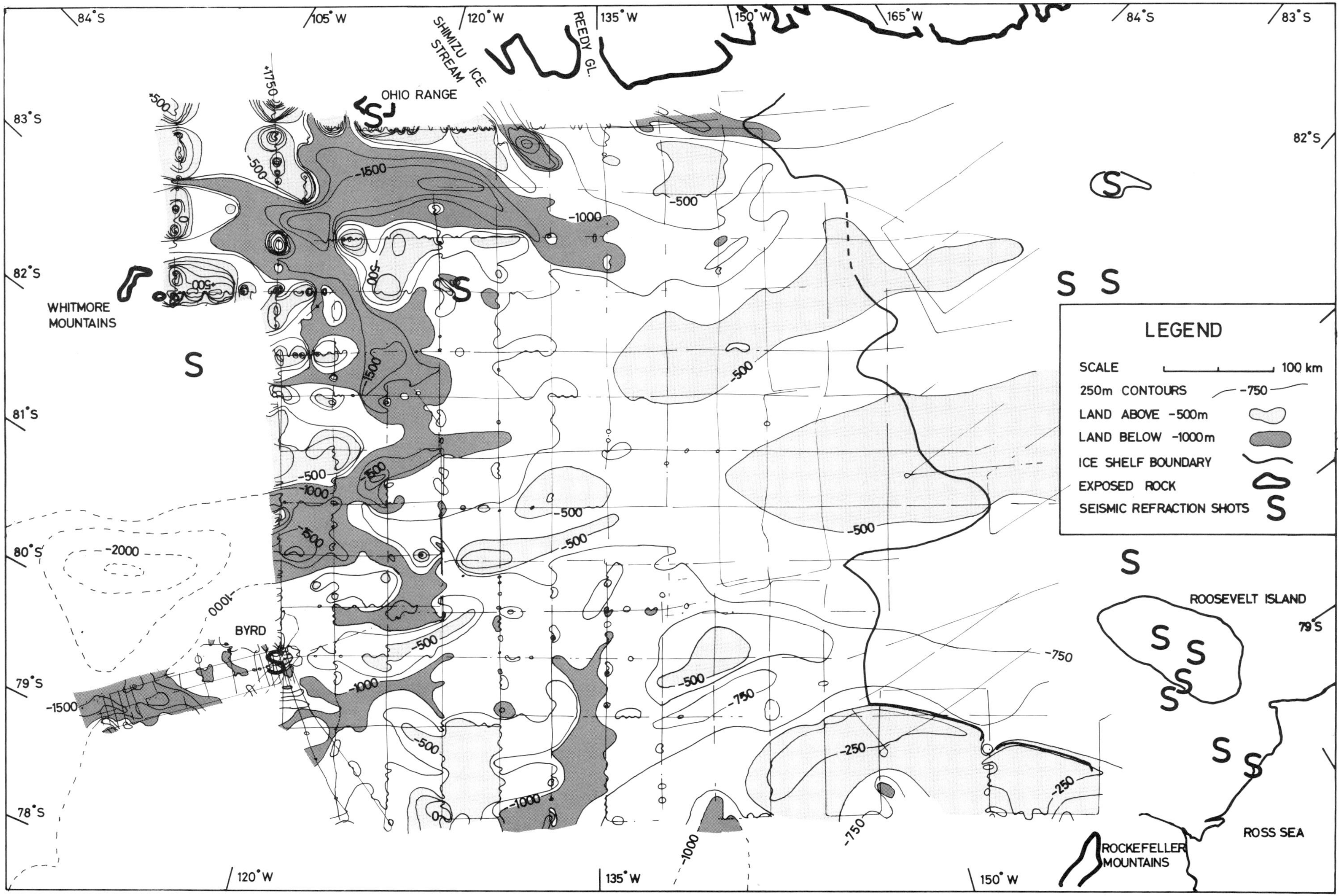

Figure 127.2. Southern Marie Byrd Land, showing bedrock surface contours.

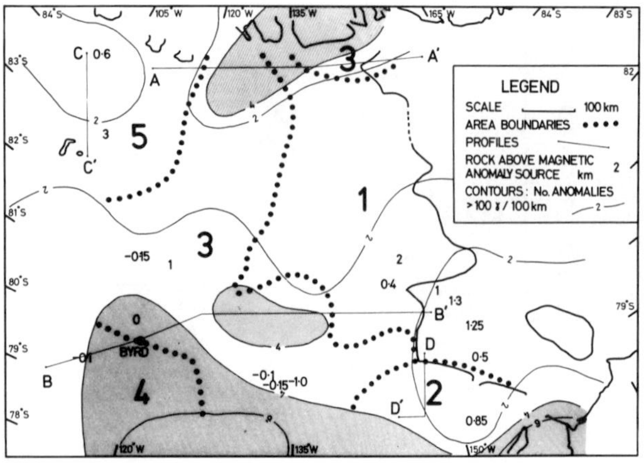

Figure 127.3. Southern Marie Byrd Land, showing location of areas 1–5 (see text) and additional aeromagnetic results. Numbers indicate depths of magnetic anomaly source below bedrock. Negative values imply that the source is calculated to be within the ice; these are interpreted as zeros, i.e., the source is at the bedrock surface.

+595 m above sea level at lat. 83° S, long. 139° W. South of ice stream E the surface raises sharply.

The boundaries of stream D and the surface contours around Byrd Station imply that the ice flowing past that base enters stream D, and not stream C as previously proposed (Hughes, 1973). Additional glaciologic implications of the radio-echo sounding work will be presented in Rose (1979).

Bedrock surface

The bedrock map (Figure 127.2) has been extended west of the grounding line with RISP data (Robertson, 1975). The dashed contours outside the flight grid have been taken from the work of Bentley and Chang (1971). Because of the roughness of the subglacial topography, it must be emphasized that the contour pattern depicted is but one of several possible interpretations. The rock surface changes in character from smooth near the Ross Ice Shelf, where the ice streams follow well-defined channels in the bedrock, to very rough in the southeastern part of the area. A major channel, 50 km wide and deeper than 1750 m below sea level beneath ice stream B, winds towards Byrd Station and connects with the pronounced low at lat. 80° S, long. 110° W. Two other deep channels are aligned with the Shimizu Ice Stream (depth 2061 m below sea level) and Reedy Glacier. The Shimizu subice channel connects with ice stream B, but today the Shimizu ice flows into ice stream A, suggesting that these deep channels were eroded by the action of East Antarctic ice before the development of the West Antarctic ice sheet.

For purposes of bedrock analysis, the data were reorganized into 346 sample groups, each containing 25 consecutive determinations of bedrock surface elevation and representing the terrain beneath about 55 km of flight line. Mean elevation, standard deviation, elevation relief ratio, skewness, and autocorrelation function (Drewry, 1975*b*) were computed for each group. The mean elevation and the standard deviation were found to be the best distinguishing criteria. To assist discussion, five areas have been defined. These areas are shown in Figure 127.3, together with contours of occurrences of short wavelength magnetic anomalies (Behrendt, 1968) and the inferred thickness of the rock column above several magnetic anomalies. These thicknesses have been calculated by combining source depths (Behrendt and Wold, 1963; Ostenso and Thiel, 1964) with radio-echo bedrock elevations. Negative values, implying sources shallower than the bedrock, are probably caused by errors in dead reckoning navigation, and are interpreted as meaning that the anomalies occur at the rock surface.

Area 1

This area, which borders the Ross Ice Shelf and extends inland for about 300 km (Figure 127.3) is a remarkable for its smooth terrain, as shown by the small range of mean elevations and the low standard deviations (Figure 127.4). Ice streams B, C, and D give rice to extensive areas of crevassing and consequent loss of the bedrock echo; otherwise a generally wet-based ice sheet is indicated. Situated beneath the ridges and domes of the ice surface are prominent bedrock plateaus possessing very smooth surfaces, generally rising about 300 m above the intervening channels which are less smooth. Data collected by the Ross Ice Shelf Geophysical and Glaciological Survey (RIGGS) from seismic reflection shooting indicate that these features continue west of the grounding line (Robertson, 1975).

Behrendt (1968) and Ostenso and Thiel (1964) report aeromagnetic flights within the region; these profiles show scarcely any anomalies except near the Siple dome (lat. 81° S, long. 150° W), which is weakly susceptible (Figure 127.3). On the Ross Ice Shelf only two significant anomalies are recorded, both at about lat. 81° S, long. 170° W, which indicate 3 km of overlying strata.

Three series of seismic refraction profiles have been obtained for the Ross Ice Shelf. At Little America V, Crary (1963) reports a 1.3-km-thick layer of sediment (velocity 2.4 km/sec) overlying 600 m of rock (velocity 4.2 km/sec), which rests on basement (velocity 6.4 km/sec). These two upper layers were also reported from station 49 (lat. 78.4° S, long. 160.3° W) on the Little America to Byrd Station trail (Crary, 1963). Robertson (1975) reports on Hochstein's four shots at Roosevelt Island; thin layers of low-velocity material overlie the basement (velocity 5.5 km/sec) at 500 m below sea level, but to the east 700 m of material (velocity 4.5 km/sec) cover basement which lies 1.6 km below sea level. Four profiles were shot by RIGGS

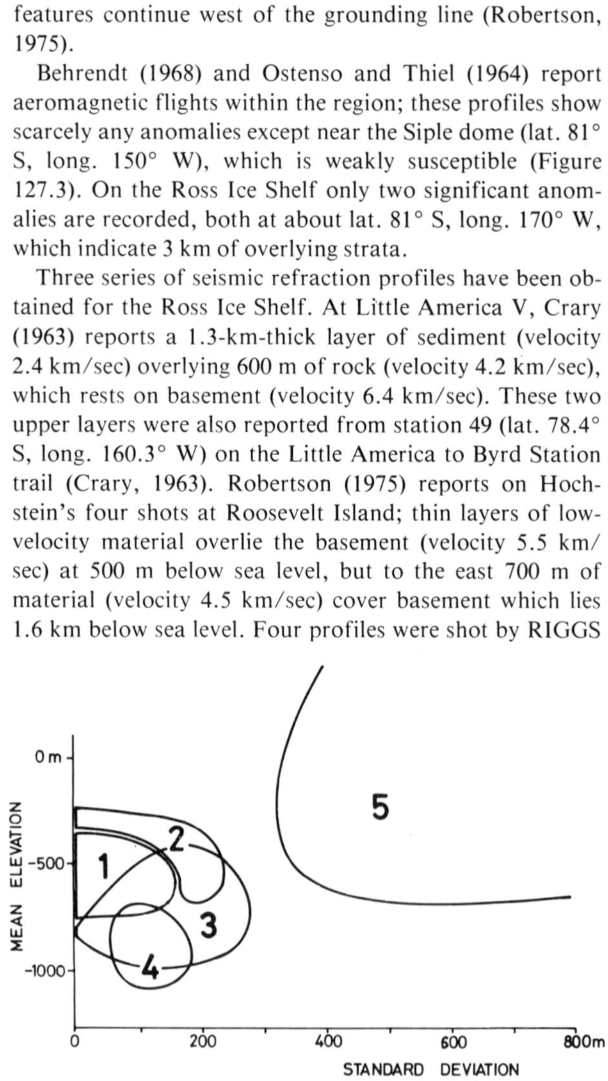

Figure 127.4. Bedrock surface characteristics of areas 1–5 in Figure 127.3.

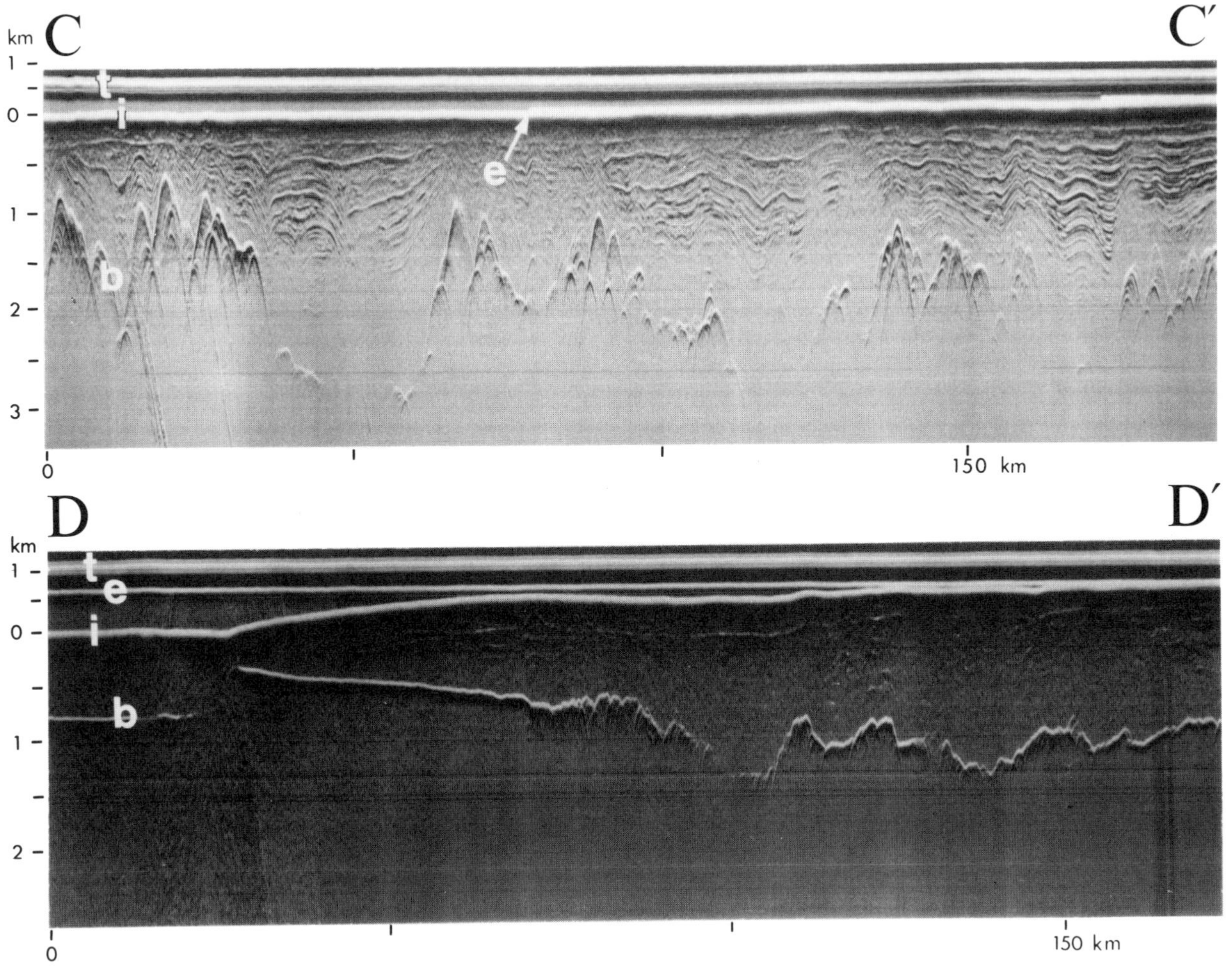

Figure 127.5. Radio-echo records. C-C′: Whitmore Mountains area. D-D′: to the north of ice stream E. t: transmitter pulse, e: end of suppression artifact, i: ice-surface return echo, and b: bedrock echo.

to the south of Roosevelt Island. Robertson (1975) reports two alternative interpretations for the northernmost of these, situated over a topographic low: either 700 m of 2.4 km/sec rock overlying 5.5 km/sec basement, or 100 m of 2.4 km/sec rock overlying 4.9 km/sec basement. Beneath the ice rise of lat. 83° S, long. 170° W, basement (5.7 km/sec) is found at 1.1 km below sea level, while just to the north it is at 1.9 km below sea level; at both sites basement is covered by 2.4 km/sec material. It thus seems that the grounded parts of the ice shelf are situated over basement highs and that the zone east of Roosevelt Island is anomalous, possessing 700 m of 4.5 km/sec material above the basement.

The pattern of isostatic gravity anomalies on the Ross Ice Shelf (Robertson, 1975) displays elongate features aligned with the topography, lows above lows. One area of significant positive isostatic gravity anomalies corresponds to the two magnetic anomalies at lat. 81° S, long, 170° W, over the north-facing slope of a ridge. Presumably this could be due to an isolated mafic igneous body.

Beneath the eastern Ross Sea continental shelf, a major basin has been identified from profiler and sonobuoy data (Houtz and Davey, 1973). This basin preserves as much as 4 km of sedimentary beds, which display a major unconformity. Hayes and Davey (1975), however, find no significant local or regional gravity anomalies and only local magnetic anomalies associated with the basin.

It seems likely that this eastern Ross Sea sedimentary basin extends inland under the ice shelf and beneath the grounded ice of area 1. During the late Mesozoic–early Cenozoic disintegration of Gondwanaland, this large marine basin may have been compressed between West Antarctic crustal blocks and the Transantarctic Mountains, resulting in folding and faulting with a strike parallel to the latter range, and being more severe in the east. These weaknesses may have allowed intrusion of a series of granitic bodies, occurring now as basement in the seismic profiles and in the intrusives of the Saunders Coast (northern Marie Byrd Land) and adjacent continental shelf.

Mesozoic and Cenozoic erosion in Marie Byrd Land must have been accompanied by complimentary deposition in adjacent basins, although various episodes of erosion also occurred there and produced unconformities. The present flat surfaces of the subglacial plateaus in area 1 are undoubtedly the result of the planing off of irregularities by the base of a partly floating ice sheet. The steep sides demonstrate erosion by the thicker, faster moving ice of the ice streams. The remaining thicknesses of sedimentary beds vary considerably across the basin, as the source-depth calculations and the seismic work indicate.

Area 2

North of ice stream E where the ice surface rises steeply, the bedrock surface exhibits an even more abrupt change, rising to a high, very smooth tableland (Figure 127.5, profile D-D′). This degenerates to the north as indicated by the curved plot of the statistics (Figure 127.4). The high plateau area is associated with wet-based ice, and the area is magnetically quiet—unlike the adjacent area around the Rockefeller Mountains. Because of the similarities, it is suggested that area 2 may have had a geologic history similar to that of area 1.

AREA 3

This is an intermediate zone, both in location and characteristics, separating the rough terrain of the Whitmore Mountains from area 1. Area 3 is approximately 200 km wide, and it contains the upper reaches of all the ice streams. Because of its size, its boundaries with all other areas, and its apparent lack of subdivisions, statistical parameters for this area have a greater range of values and overlap those of three of the four other regions. That part of profile B-B′ (Figure 127.6) which is between Byrd Station and 570 km from the start of the profile illustrates the topography in area 3.

One seismic refraction profile (Horlick Mountains traverse, station 222, Robinson, 1964) reported 1.1 km of 5.3 km/sec strata overlying basement (velocity 6.7 km/sec) in an area which appears generally smooth from radio-echo sounding. The upper layer is not correlated here with the Whitmore Mountains (Bentley and Clough, 1972) because of the intervening deep trench. Both Paleozoic sedimentary rocks and Mesozoic granitic intrusives could have the same velocity as this layer. The area near the seismic profile shows little magnetic activity, so perhaps this section represents an extension of area 1 where it has been eroded down to the granitic intrusion. To the south a "tongue" of magnetic activity is shown extending northwestward from the Transantarctic Mountains (Figure 127.3).

North of lat. 80° S the influence of Cenozoic volcanic activity is very strong. Calculated source depths suggest that the numerous magnetic anomalies may lie at the bedrock surface.

AREA 4

Area 4 is located immediately to the north of Byrd Station, near the limits of the radio-echo sounding grid. The bedrock surface in the area is generally low and relatively smooth (Figure 127.4). The first 200 km of profile B-B′ (Figure 127.6) are along the Byrd Station Strain Network (BSSN).

The wide-angle reflection seismic work at Byrd (Bentley and Ostenso, 1961) suggests that 1.4 km of 4.3 km/sec rock overlies basement (velocity 5.9 km/sec). As profile B-B′ shows, Byrd is located beside a bedrock high, and within 7 km the rock surface shows relief of 800 m. Such terrain may well hinder interpretation of the seismograms.

The magnetic character of the area indicates that the scarp in the center of the BSSN is also the edge of a province of more susceptible rock; the other edge may be the valley at kilometer 300 on profile B-B′. This magnetic province extends to the Cenozoic volcanic province in northern Marie Byrd Land.

One possible interpretation of the geology in area 4 is a granitic basement overlain by a sequence of metasedimentary rocks. Cenozoic igneous rocks may determine the magnetic character of the area.

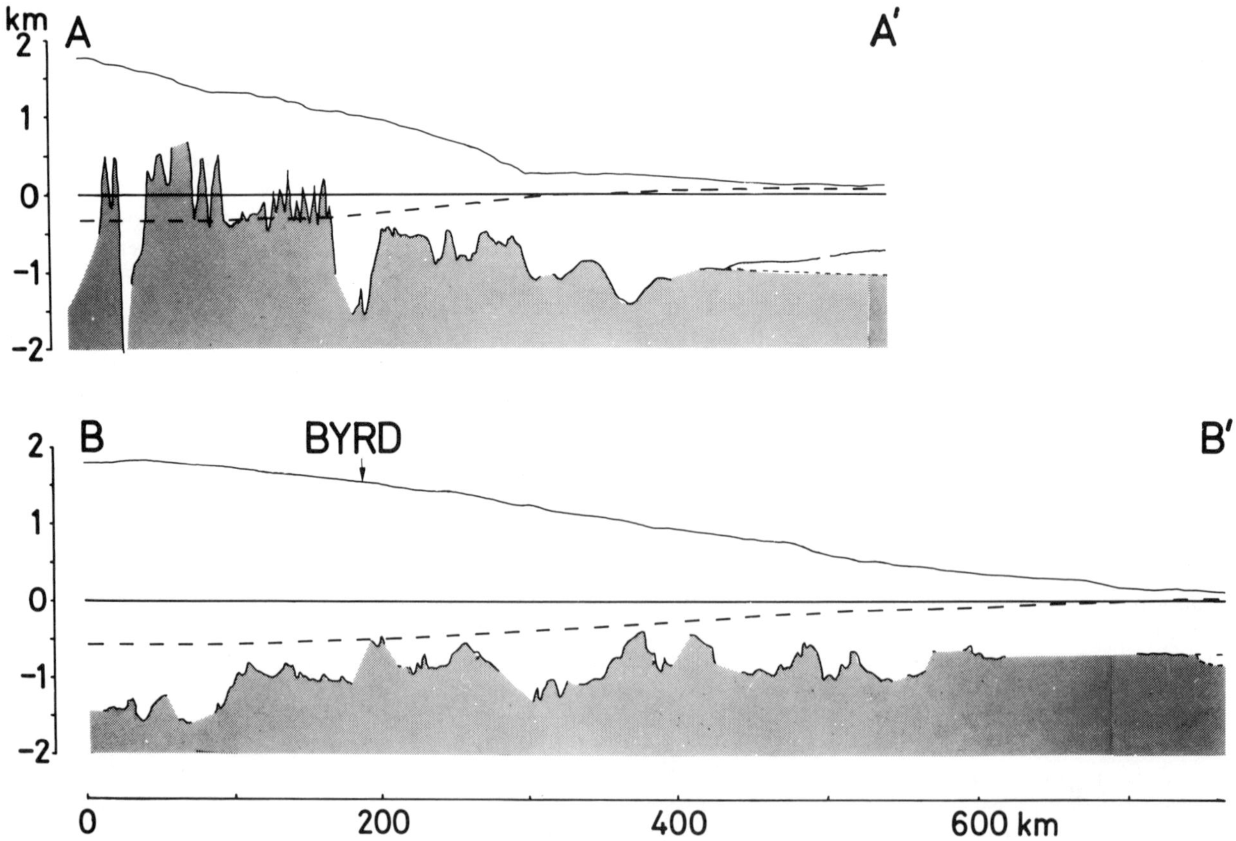

Figure 127.6. Radio-echo profiles. A–A′: from near the Ohio Range to the Ross Ice Shelf. B–B′: BSSN to the Ross Ice Shelf. Isostatically compensated sea level is shown by the broken line.

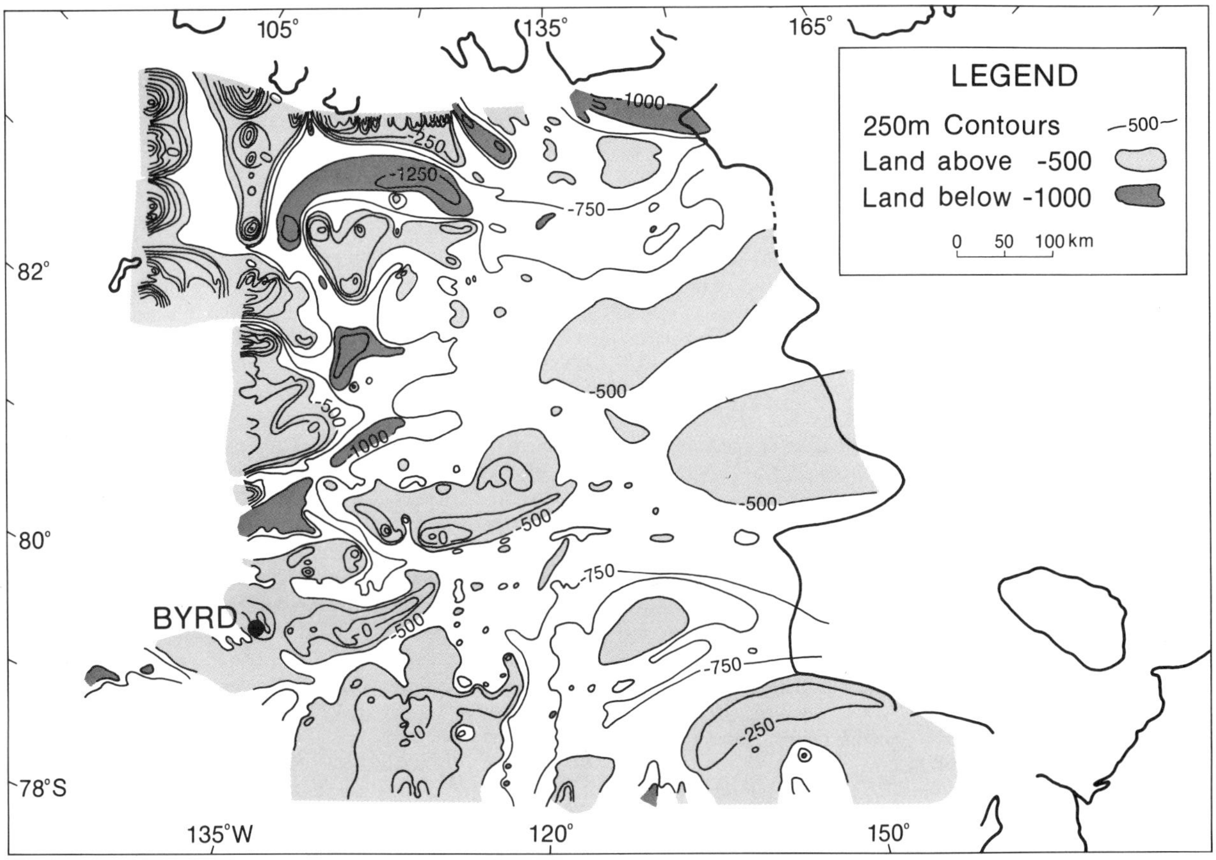

Figure 127.7. Southern Marie Byrd Land, showing contours of the isostatically compensated bedrock surface, allowing for a 65 m global rise in sea level.

AREA 5

Although visited by only three flights, this area contains the most distinctive terrain studied. Linking the Whitmore Mountains with the Ohio Range and the Thiel Mountains are groups of very rough mountains separated by deep, steep-sided valleys that are typically 20 km wide and 2 km deep. The statistics of this area plot apart from those of the other regions, and they show wider ranges of values (Figure 127.4). Profiles A-A′ (initial 110 km) and C-C′ are characteristic of this region.

The two seismic refraction profiles in the area, from the Horlick Mountains traverse, were both difficult to interpret (Robinson, 1964). The very rough configuration of bedrock is clearly responsible, and it is considered that the published results may be unreliable.

Two unterminated flight lines in area 5 extended beyond the Thiel Mountains to the South Pole. No change in bedrock morphology was recorded between the Whitmore and the Thiel Mountains; the latter extend 100 km toward the pole but are immediately terminated to the east by a deep trough. Drewry (1972) defined their western limit.

It would seem that the Thiel Mountains and the Whitmore Mountains are part of a large upland tract extending as far southwest as the Ohio Range but not including it. It is possible that the Thiel Mountains are the erosional remnants of a stratiform intrusive body overlying a subhorizontal sequence of Precambrian metasedimentary rocks. The Thiel Mountains area has experienced considerable faulting although apparently little folding. The high elevations of the mountian peaks in area 5 might have initiated early glaciation, which developed the valley systems detected by radio-echo sounding.

Isostatic bedrock surface map

Figure 127.7 shows the bedrock surface after allowing the overlying ice to melt, global sea level to rise by 65 m due to melting of the Antarctic and Greenland ice sheets, and the rock surface to regain isostatic equilibrium. A window of 150 km in length was used to adjust the central point of the window, taking densities of 900, 1030, and 3200 kg/m^3 for ice, ocean, and mantle respectively. Profiles A-A′ and B-B′ indicate the new sea level.

Where the ice loading is least, near the Ross Ice Shelf grounding line, the contours have not been altered much. Well inland, however, the rock surface has risen about 500 m, creating several new islands and enlarging those already existing. The possibility of a connection between the Ross Sea and the Amundsen Sea still exists.

Acknowledgments. Support for these investigations was provided by the Natural Environment Research Council in the U.K. and in the Antarctic by the U.S. National Science Foundation, U.S. Navy Antarctic Support Force and Antarctic Air Development Squadron VXE-6. Their assistance, by providing the aircraft and crews, especially after 15 January 1975, enabled us to substantially complete the West Antarctica grid. Radio-echo equipment was built by the Technical University of Denmark. Encouragement from and discussion with D. J. Drewry are acknowledged.

References cited

Behrendt, J. C. 1968. Magnetic anomalies of short wavelength. (*In* Behrendt, J. C., and Bentley, C. R., *eds.* Magnetic and Gravity Maps of the Antarctic. *Antarct. Map Folio Ser.,* Folio 9, Pl. VII.)

Behrendt, J. C., and Wold, R. J. 1963. Aeromagnetic survey in West Antarctica, 1963. *Res. Rep. Ser., Polar Res. Center, Univ. Wisconsin,* No. 63-1, 49 pp.

Bentley, C. R., and Chang, F. K. 1971. Geophysical exploration in Marie Byrd Land, Antarctica. (*In* Crary, A. P., *ed. Antarctic Snow and Ice Studies II.* Washington, American Geophysical Union, 1–38.) [Antarctic Research Series, Vol. 16.]

Bentley, C. R., and Clough, J. W. 1972. Antarctic subglacial structure from seismic refraction measurements. (*In* Adie, R. J., *ed. Antarctic Geology and Geophysics.* Olso, Universitetsforlaget, 683–91.)

Bentley, C. R., and Ostenso, N. A. 1961. Glacial and subglacial topography of West Antarctica. *J. Glaciol., 3,* No. 29, 882–911.

Crary, A. P. 1963. Marine-sediment thickness in the eastern Ross Sea area, Antarctica. *Geol. Soc. Am. Bull., 72,* No. 5, 787–90.

Crary, A. P., Robinson, E. S., Bennett, H. F., and Boyd, W. W. 1962. Glaciological studies of the Ross Ice Shelf, Antarctica, 1957–60. *IGY Glaciol. Rep., 6,* 193 pp.

Drewry, D. J. 1972. Subglacial morphology between the Transantarctic Mountains and the South Pole. (*In* Adie, R. J., *ed. Antarctic Geology and Geophysics.* Oslo, Universitetsforlaget, 693–703.)

Drewry, D. J. 1975*a*. Radio echo sounding map of Antarctica (90° E–180°). *Polar Rec., 17,* 359–74.

Drewry, D. J. 1975*b*. Terrain units in eastern Antarctica. *Nature, Lond., 256,* No. 5514, 194–95.

Hayes, D. E., and Davey, F. J. 1975. A geophysical study of the Ross Sea, Antarctica. (*In* Hayes, D. E., Frakes, L. A., and others. *Initial Reports of the Deep Sea Drilling Project,* Vol. 28. Washington [U.S. Government Printing Office], 887–907.)

Houtz, R., and Davey, F. J. 1973. Scientific profiler and sonobuoy measurements in Ross Sea, Antarctica. *J. geophys. Res., 78,* No. 17, 3448–68.

Hughes, T. 1973. Is the West Antarctic ice sheet disintegrating? *J. geophys. Res., 78,* No. 33, 7884–910.

Kayton, M. 1969. Inertial navigation. (*In* Kayton, M., and Fried, W. R., *eds. Avionics Navigation Systems.* New York, John Wiley, 281–341.)

King-Hele, D. G. 1975. A view of earth and air. *Phil. Trans. R. Soc. Lond., 278,* 67–109.

Ostenso, N. A., and Thiel, E. C. 1964. Aeromagnetic reconnaissance of Antarctica between Byrd and Wilkes stations. *Res. Rep. Ser., Polar Res. Center, Univ. Wisconsin,* No. 64–6, 33 pp.

Robertson, J. D. 1975. *Geophysical Studies on the Ross Ice Shelf, Antarctica.* Ph.D. dissertation, University of Wisconsin, 214 pp.

Robin, G. de Q., Swithinbank, C. W. M., and Smith, B. M. E. 1970. Radio echo exploration of the Antarctic ice sheet. (*In* Gow, A. J., Keeler, C., Langway, C. C., and Weeks, W. F., *eds. International Symposium on Antarctic Glaciological Exploration (ISAGE), Gentbrugge, 1968,* 97–115.) [International Association of Scientific Hydrology Publication, No. 86.]

Robin, G. de Q., Evans, S., Drewry, D. J., Harrison, C. H., and Petrie, D. L. 1970. Radio echo sounding of the Antarctic ice sheet. *Antarct. Jnl U.S., 5,* No. 6, 229–32.

Robinson, E. S. 1964. *Geologic Structure of the Transantarctic Mountains and Adjacent Ice Covered Areas.* Ph.D. dissertation, University of Wisconsin.

Rose, K. E. 1979. Characteristics of Ice Flow in Marie Byrd Land, Antarctica. *J. Glaciol., 24,* No. 90, 63–75.

Whillans, I. M. 1976. Radio echo layers and the recent stability of the West Antarctic ice sheet. *Nature, Lond., 264,* No. 5582, 152–55.

XI *Marine Geology*

128

The Antarctic Continental Shelf

Review Paper

G. L. JOHNSON
Office of Naval Research, Arlington, Virginia 22217, U.S.A.

J. R. VANNEY
Laboratoire de Géologie dynamique, Université Pierre et Marie Curie, 75230 Paris, Cedex 05, France

D. HAYES
Lamont-Doherty Geological Observatory of Columbia University, Palisades, New York 19064, U.S.A.

Abstract. The Antarctic continental shelf exhibits typical high-latitude morphology influenced both by tectonic history and climate. Although Arctic glaciation began about 5 m.y. ago in late Miocene time, Antarctic continental glaciation, as revealed by Deep Sea Drilling Project data, was initiated much earlier, in late Oligocene time. The contrasting morphology of northern and southern polar shelf areas reflects the differing durations of glacial influence.

The continental shelf of Antarctica has been deeply incised by glacial activity and by canyons, and well-developed submarine fans have been constructed on the continental rise. In contrast, Arctic shelves have a less deeply eroded network of channels and canyons and smaller associated continental rises and plains. During the growth stages and the time of maximum extension of the glaciers in Antarctica, erosion occurred as far as the outer edge of the continental shelf. Normal prolongation of inland valley glaciers and the flow of terrigenous debris resulted in an overdeepening and enlarging of the transverse valleys. Fractures parallel to the coast and related to initial continental rifting were also incised by the glaciers. The outer shelf "banks" were probably constructed by terminal moraines of piedmont glaciers. The greater depth of the Antarctic outer shelf (500–900 m), as compared to that of the Arctic (400–500 m), is perhaps due to depression by inland ice and incomplete postglacial isostatic rebound reflecting the persistence of thick ice sheets on the continent.

Arctic glacial conditions, as evidenced by glacial debris in ocean sediments, began about 5 m.y. ago in late Miocene time (Clark, 1975); in contrast, Antarctic continental glaciation, as revealed by Deep Sea Drilling Project (DSDP) data was initiated in late Oligocene time about 25 m.y. ago (Hayes and others, 1975*a*). Using the Arctic shelves as a basis for comparison, we should therefore expect the morphology of the Antarctic continental shelf to reflect a much greater level of glacial activity.

A "type section" for the Arctic glaciated coast is found in the exposed continental shelves of Greenland and Labrador, which have been extensively surveyed and studied (Henderson, 1975; Van der Linden and others, 1976). The origin of the gross morphology, as postulated by Johnson, Sommerhoff, and Egloff (1975) and Johnson, McMillan, and others (1975) is shown schematically in Figure 128.1. The sequence of events was: (1) rifting with antithetic faults to form an early Mesozoic basin in the Labrador Sea; (2) progradation of sediments across the pre-Pliocene Labrador Sea margin after initiation of sea-floor spreading; and (3) Pleistocene glacial alteration of the continental margin.

In the Labrador Sea, the inner shelf (*strandflat* of Nansen, 1922) is generally a submerged continuation of the Paleozoic and Precambrian continental basement. The outer banks and strandflat are usually separated by relatively long, narrow depressions several hundred meters deep. These marginal channels are generally parallel to the coastline. Holtedahl and Holtedahl (1961) of Norway did pioneering work delineating the structure and origin of these shelf features.

We suggest that, as in the case of Greenland and Labrador, the channels parallel to the Antarctic coast mark a fault or hinge line which separates rocks of the continental basement from the less resistant prograded Mesozoic and Tertiary sequences deposited after initial continental rifting. It is likely that these channels represent the most landward curved step fault (with block tilting) formed during the early uplift associated with initial rifting. This would account for the parallelism of the marginal channels with the coastline. During glacial periods the ice flows tended to drift along these shallow channels, locally breaking seaward in the transverse canyons. The advancing glacial ice abraded the channels, which originally were quite shallow, as well as the shallower outer parts of the shelf.

The forward motion of the ice results in a terminal moraine wherever the glacier either grounds on basement highs or otherwise reaches an equilibrium state. There may also be recessional and lateral morainal components within the outer banks. Due to the soft nature of the shelf strata, erosion and subsequent moraine construction are extensive. Recent seismic evidence (Grant, 1972) reveals that parts of the Labrador marginal trough are bordered by faults, with some faulting still occurring (Grant, 1970). The absence of faults in many areas, however, implies that

LABRADOR SEA

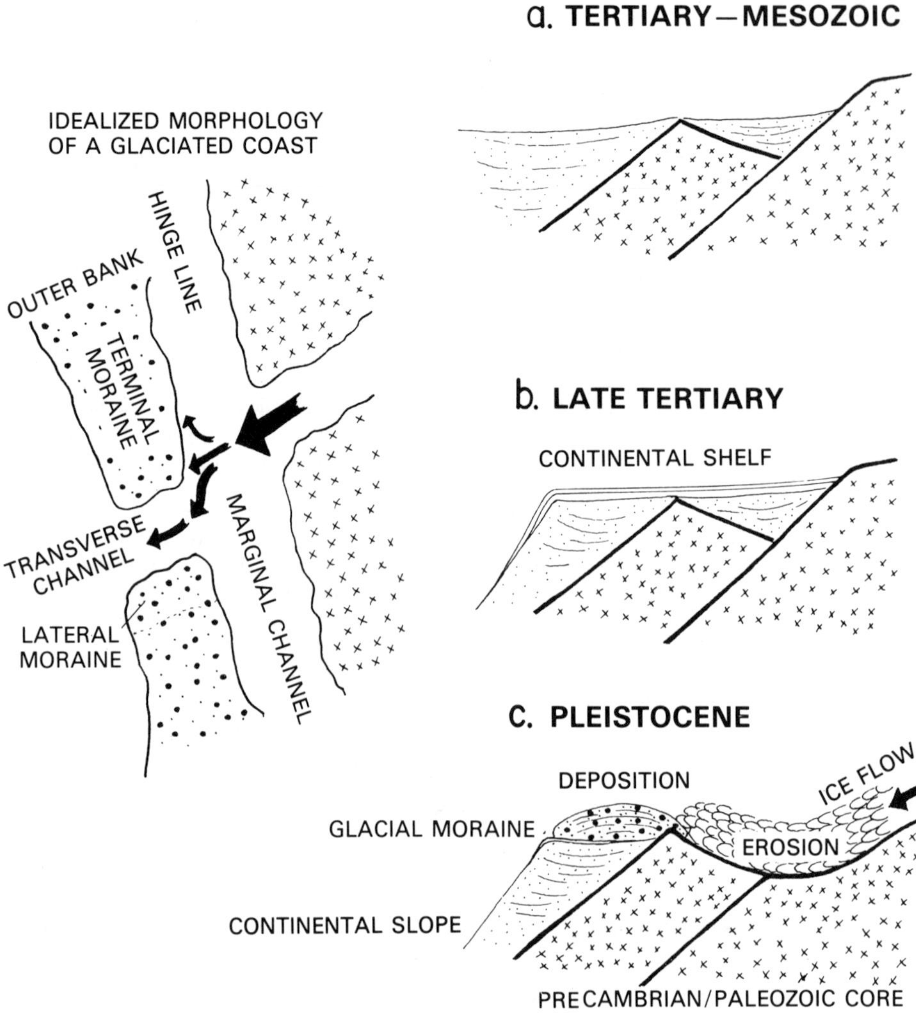

Figure 128.1. Idealized origin of glaciated continental shelf morphology, as shown along the coast of the Labrador Sea. (After Johnson, McMillan, and others, 1975.)

other factors play a significant role in developing the present Labrador shelf morphology.

Conversely, Zhivago and Lisitzin (1957) have suggested that the high-latitude morphology is the result of (1) differential isostatic movements due to differences in ice load between the glaciated and nonglaciated areas and (2) overdeepening of channels by erosion from glaciers and from the running water beneath them and by ice movement seaward (grounding and floating ice) during lower stands of sea level.

Nonglaciated Arctic coasts do not exhibit the shelf-channel system and, except for a somewhat deeper shelf break, generally resemble continental shelves from the temperate regions (Grantz and others, 1976).

Morphology

Labrador Sea

A well-studied example of a glaciated continental shelf complete with longitudinal and transverse canyons is the Hamilton Bank area (Figure 128.2). Hamilton Bank consists of a thick wedge of Mesozoic and Cenozoic strata that was built over a block-faulted, subsiding Precambrian crystalline basement in response to rifting of the Labrador Sea. The bank is underlain by gently seaward-dipping coastal plain strata that are covered unconformably by semistratified glacial drift (Van der Linden and others, 1976). The coastal plain deposits, in turn, unconformably overlie folded sedimentary rocks probably of Jurassic age (Grant, 1972). Beneath the inner Labrador shelf, landward of Hamilton Bank, the folded sedimentary rocks are missing, and undeformed coastal plain deposits directly overlie a rough basement which is the offshore extension of the Precambrian Labrador Shield. In the nearshore region the basement is widely exposed in a rough, rocky zone marked by thousands of islets and shoals. A more detailed treatment of the morphologic evolution of the Labrador continental margin may be found in Van der Linden and others (1976).

The continental rise is dissected by an unusually small number of submarine canyons, and the Labrador Sea

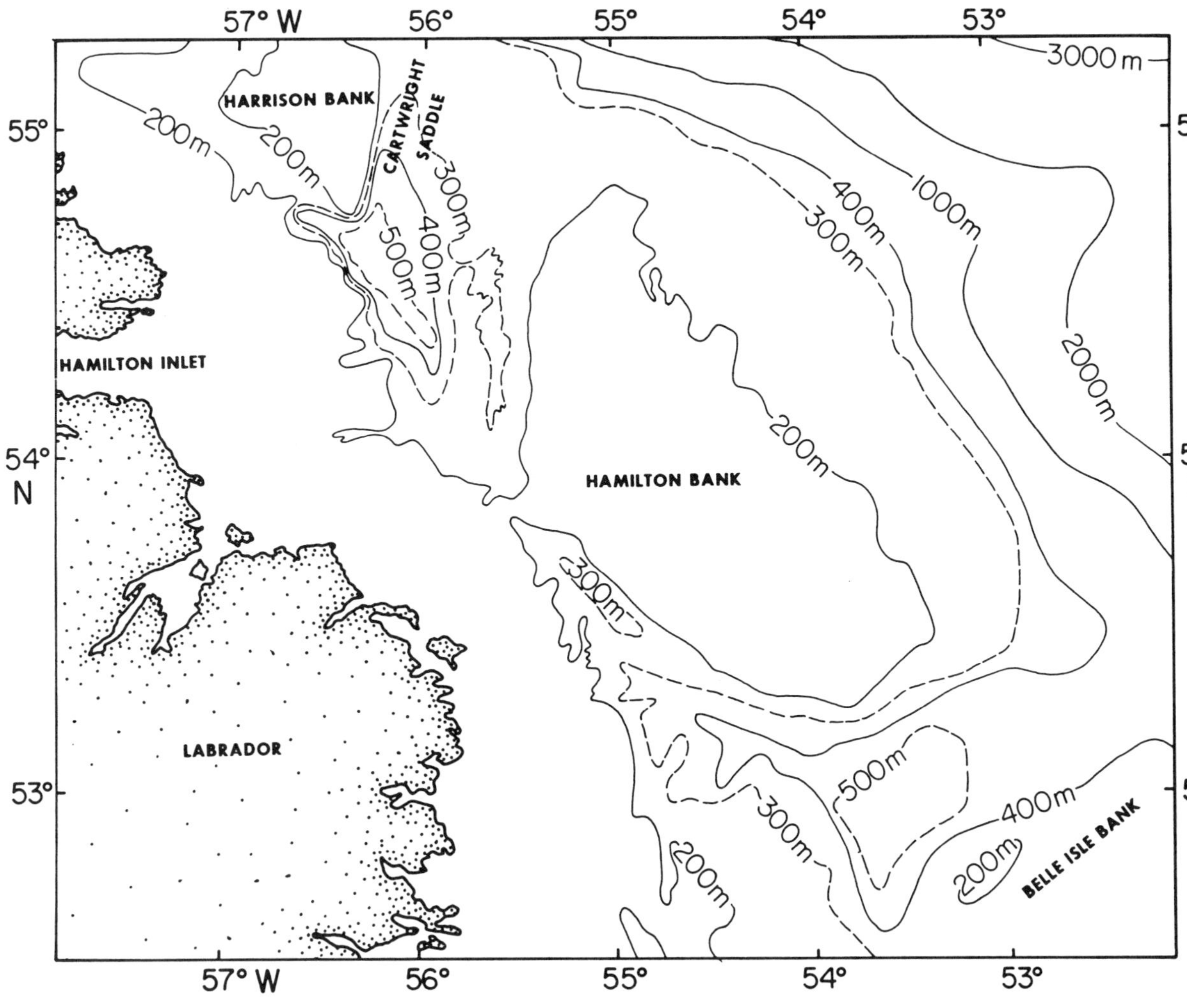

Figure 128.2. Bathymetry of the Hamilton Bank area seaward of the coast of Labrador. For scale, 1° of latitude equals 108 km. (After Van der Linden and others, 1976.)

basin is not floored by an abyssal plain (Johnson, McMillan, and others, 1975). The sea floor to the southeast and southwest of Greenland exhibits similar morphology (Johnson, Sommerhoff, and Egloff, 1975, and Johnson, McMillan, and others, 1975) except that the shelf here is narrower than that of Labrador.

Antarctic continental shelf

Around Antarctica, the continental shelf typically extends outward from the shore to depths of 500 to 900 m where a prominent change in slope occurs that serves to define the seaward edge of the shelf or the shelf break. The Antarctic continental shelf is unusual in that its average depth is as much as four times greater than average shelf depths around other continents. It is about twice as deep as the Arctic continental shelf, which generally has a shelf break at 400–500 m depth. It has been suggested by many workers that the depressed shelf depths of Antarctica have been caused by isostatic subsidence in response to loading by the massive Antarctic ice sheet (24×10^6 km^3 of ice) (Denton and others, 1971). The morphology of the shelf surface is known in detail in only a few selected regions, specifically the Ross continental shelf (Hayes and Davey, 1975; Vanney and Johnson, 1976*c*; Vanney and others, in press) and parts of the northern and western shelf of the Antarctic peninsula (Vanney and Johnson, 1976*a,b*).

Bellingshausen-Amundsen Sea

The coast here is bounded by high, rocky, glaciated cliffs or blocked by small ice shelves which bridge the bays and the interisland straits. The continental shelf is characterized by typical glacial morphology. The outer banks and strandflat are usually separated by relatively long, narrow depressions several hundred meters deep. These marginal channels are generally parallel to the coastline (Figure 128.3).

The rocky coastal belt is separated into segments by "transverse channels" which extend seaward between the banks of the outer shelf. They commonly begin abruptly landward (in the fjords), but commonly the channels' heads are hidden under the floating ice shelf. Seaward they almost invariably end in a counterslope. Marginal channels are present parallel to the main direction of the coastline, between the islands and between the inner and outer zones (Holtedahl, 1929). Some are locally more than 1000 m deep (1376 m in the Bismarck Strait) (see Figure 5 of Vanney and Johnson, 1976*a*). The western marginal channels are shallower than the eastern ones, suggesting greater tectonic activity or more vigorous glacial erosion adjacent to the Antarctic Peninsula.

The edge of the shelf is usually 400–500 m deep, but in some places it is more than 600 m deep. The depth at the shelf break decreases northward to the end of the peninsula

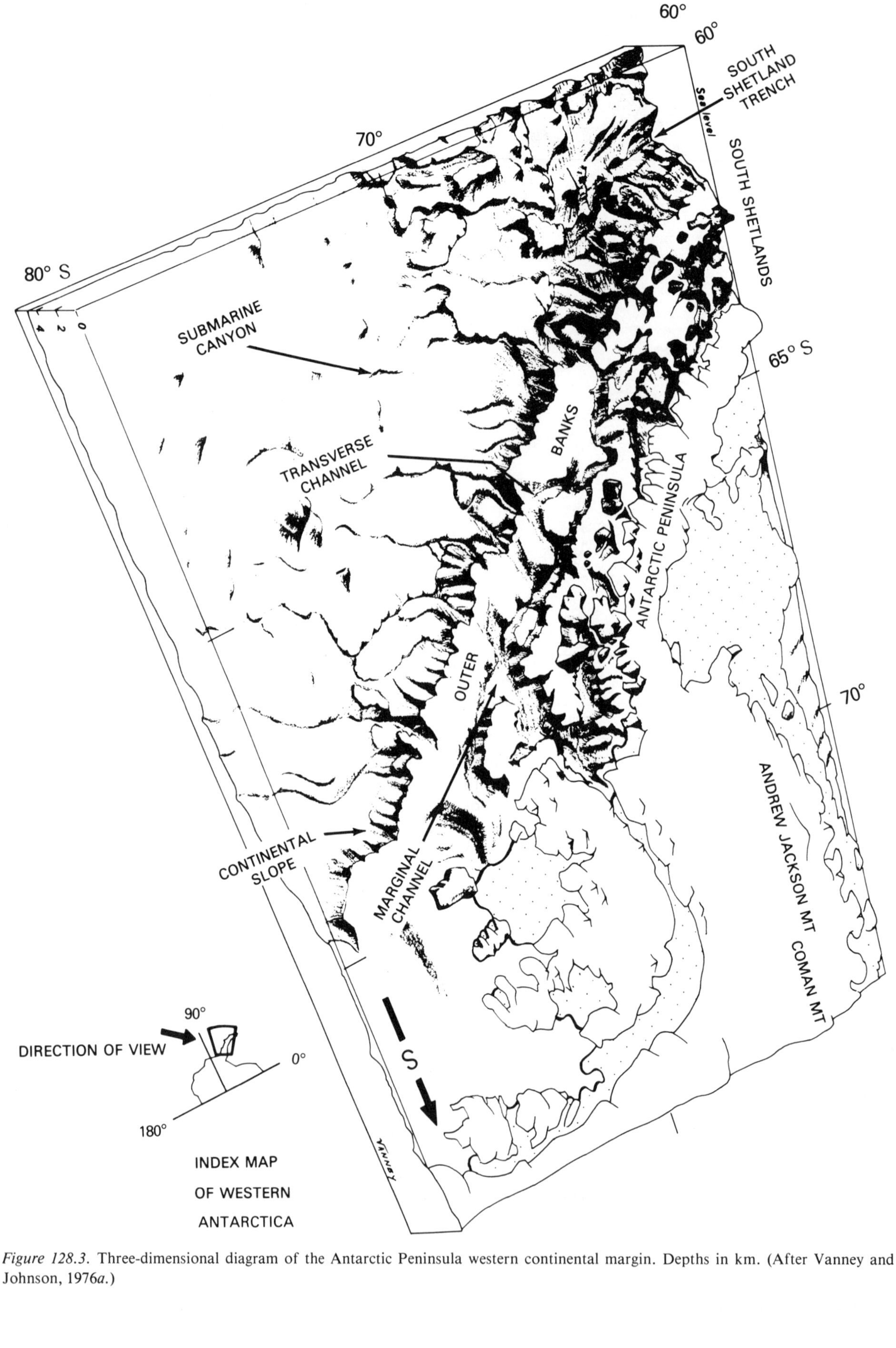

Figure 128.3. Three-dimensional diagram of the Antarctic Peninsula western continental margin. Depths in km. (After Vanney and Johnson, 1976*a*.)

where it usually lies between 200 and 300 m. In this outer shelf zone are found undulating plains with banks ("border banks" in the Soviet literature). Cenozoic subduction has occurred along the continental margin of the Antarctic Peninsula (Herron and Tucholke, 1976). Complex continental accretion and subduction processes interacting with the continental margin have led to a confused distribution of arc and trench facies. Whether this segment of shelf was consumed or enlarged as the crust was subducted is not clear. Subduction at fast rates may well erode the shelf, as, for example, the Peru-Chile trench where Paleozoic rocks occur on islands in the outer shelf and where slower rates might, by the plastering of sediments, build out the shelf. The subduction process along the Antarctic Peninsula died out northward, with the South Shetland trench ceasing activity about 5 m.y. ago (Herron and Tucholke, 1976).

If the dashed line in Figure 128.4 represents the innermost fault lineament, shelf progradation of 100 km has occurred. Prograding in the Labrador Sea extends as far as 120 km (Figure 128.2); however, seaward of southwest Greenland it is typically 20–30 km (Johnson, McMillan, and others, 1975).

The remainder of the continental shelf in the Bellingshausen and Amundsen Seas is poorly surveyed because of extensive ice cover. What little is known about the shelf there has been presented by Vanney and Johnson (1976*b*) and Heezen and others (1972).

Among the striking features of peri-Antarctic submarine morphology, the most conspicuous is the well-developed continental rise. This detrital apron attests to the importance of the erosional denudation of Antarctica. There are two basic morphologic characteristics of the peri-Antarctic submarine province: (1) extensive progradation of the shelf and (2) a well-developed slope and rise canyon system with resultant abyssal plains (Figure 128.3) (Vanney and Johnson, 1976*a,b*).

Ross Sea

The continental shelf here is broad and deep, and its southern half forms a great embayment occupied by the largest ice shelf in the world. Hayes and Davey (1975) have extensively studied the morphology and structure of the continental shelf. The Ross Sea shelf lies at the northern terminus of two subcontinents, divided roughly along the 180° meridian. The Transantarctic Mountains of Victoria Land lie to the west and Marie Byrd Land to the east. Bathymetric data reveal that the mean depth of the outer shelf is 500 m, with the eastern half deeper than the western. A positive linear gravity anomaly stirkes north-south across the western Ross Sea shelf at about long. 175° E and has been interpreted by Hayes and Davey (1975) to represent an ancient rift.

In the Ross Sea area, the continental shelf is characterized by a valley and swale relief believed to be the relict erosion surface of a grounded Ross ice sheet that previously (probably in latest Miocene or earliest Pliocene time) extended well north of its present position (Figure 128.5). The only clear evidence of grounded ice in the Ross Sea, and thus major continental glaciation, is the (500 m) erosional surface recorded over much of the continental shelf. The alternate progression and regression of the ice and of erosional and depositional conditions are the likely cause of the truncation of the undulating, northeast-dipping sedimentary sequence (Houtz and Meijer, 1970; Houtz and Davey, 1973; Hayes and Davey, 1975). The detailed topography of the banks is undulating, with small hummocks showing topographic relief of less than 100 m.

In contrast, however, to the areas discussed above, the banks and valleys of the Ross Sea embayment are not parallel to the coast, but instead the morphology consists of a series of northeast-trending banks and fjordlike valleys in the western sector. The trend of these features in the eastern Ross Sea is more northerly. This unusual morphology may be a reflection of the tectonic fabric, indicating that the Ross Sea, and by inference the Weddell Sea, are not simple passive continental margins but instead may reflect either shearing (LaBreque, Chapter 44, this volume) or ancient rifting active during the separation of New Zealand from Australia (Hayes and Davey, 1975).

The stratigraphy of the four DSDP drill sites on the Ross Sea continental shelf is presented in Hayes and others (1975*a,b*) and will not be repeated here.

As is apparent in Figure 128.5, both the slope and rise are dissected by an extensive network of submarine canyons.

Wilkes and Enderby Seas

The shelf in this region generally varies little in width. The narrowest part, approximately 50 km across, is off the Budd Coast, while the widest part, about 260 km across, is seaward of the Queen Mary Coast. In the second case, the width of the shelf is more than four times the world average of 78 km (Emery, 1969). Based on the structure of the Cape Adelie region, one of the few areas where sufficient data are available, the continental shelf is divided into two major morphology units (Vanney and Johnson, 1979*a,b*).

The inner shelf is heavily dissected, with the extreme contrast between rocky ridges and ledges and intervening depressions with depths from 100 m to about 1500 m. The narrow nearshore rocky zone consists of a succession of beaches and linear ridges and many glaciated low islands. Beyond this rocky zone, the shelf is deeply carved by a series of large, deep marginal basins and banks that present a roughly orthogonal pattern (Figure 128.6) (Vanney and Johnson, 1979*a,b*). The larger outer part (2/3) of the shelf is characterized by undulating sea floor. It is divided into triangular plateaus by the wide, open valleys (400–500 m deep, several tens of km wide) at the edges of the inner basins. The "outer banks" are oblique to the main orientation of the shelf edge.

The depth of the shelf edge varies only slightly from one sector to another. Generally, the shelf break is at about 600–700 m; it is relatively abrupt in the sectors where the shelf is bordered by shallow banks. Except for the detailed sinuosities, the shelf edge is comprised of a succession of linear segments of diverse orientation.

If deep shelf troughs parallel to the coast represent the most landward antithetic fault, progradation of more than 100 km by postrifting sedimentary fill is common in this part of the Antarctic shelf.

As is typical for Antarctica (Figure 128.6), the slope and rise are cut by an extensive submarine canyon system feeding into the Wilkes Abyssal Plain.

The western part of the area shown in Figure 128.6 con-

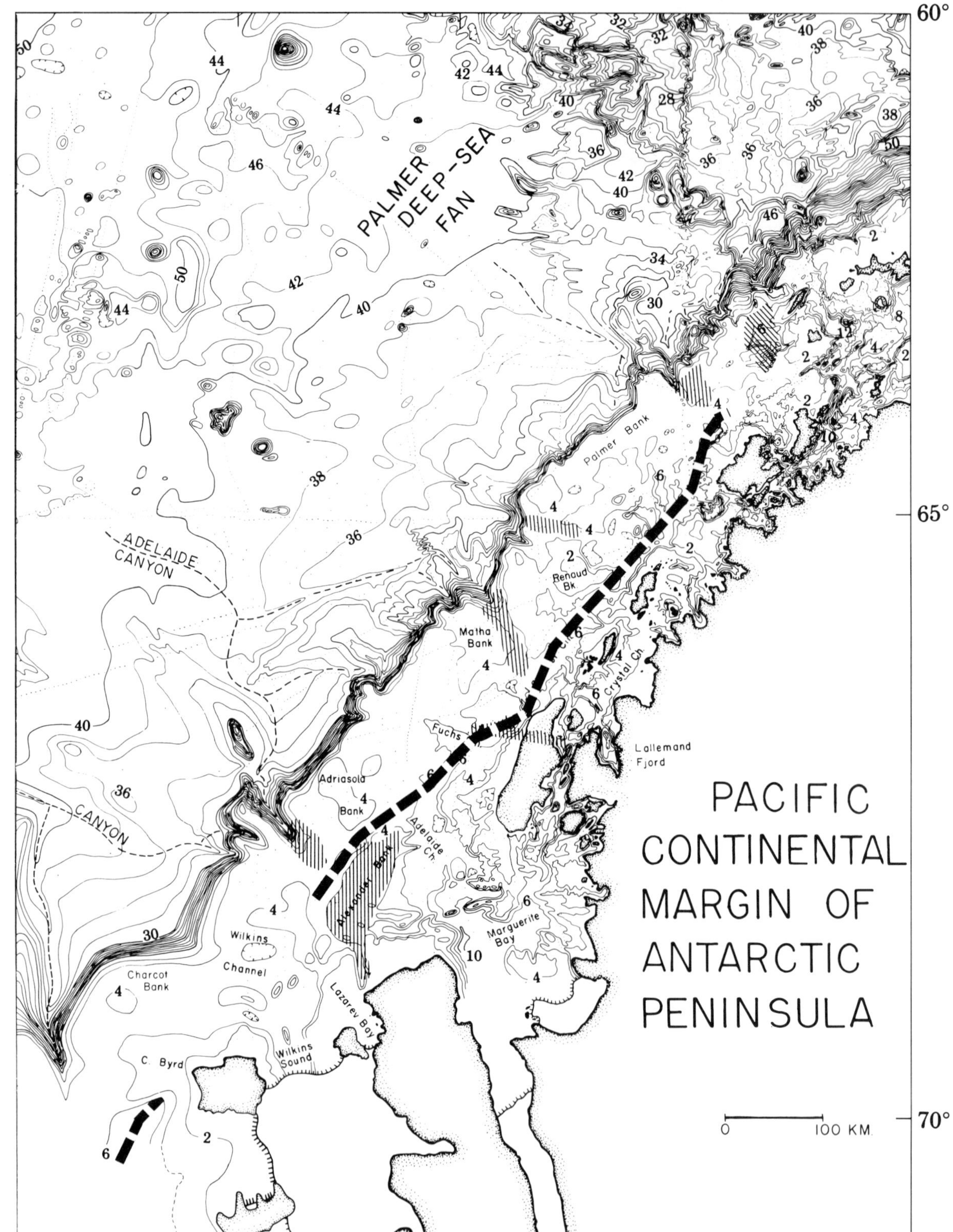

Figure 128.4. Bathymetric contour chart of part of the Antarctic Peninsula and adjacent sea floor. Lined pattern: shelf canyons. Heavy dashed line: trough parallel to the coast. Light dotted lines: ships' tracks. (After Vanney and Johnson, 1976*a*.)

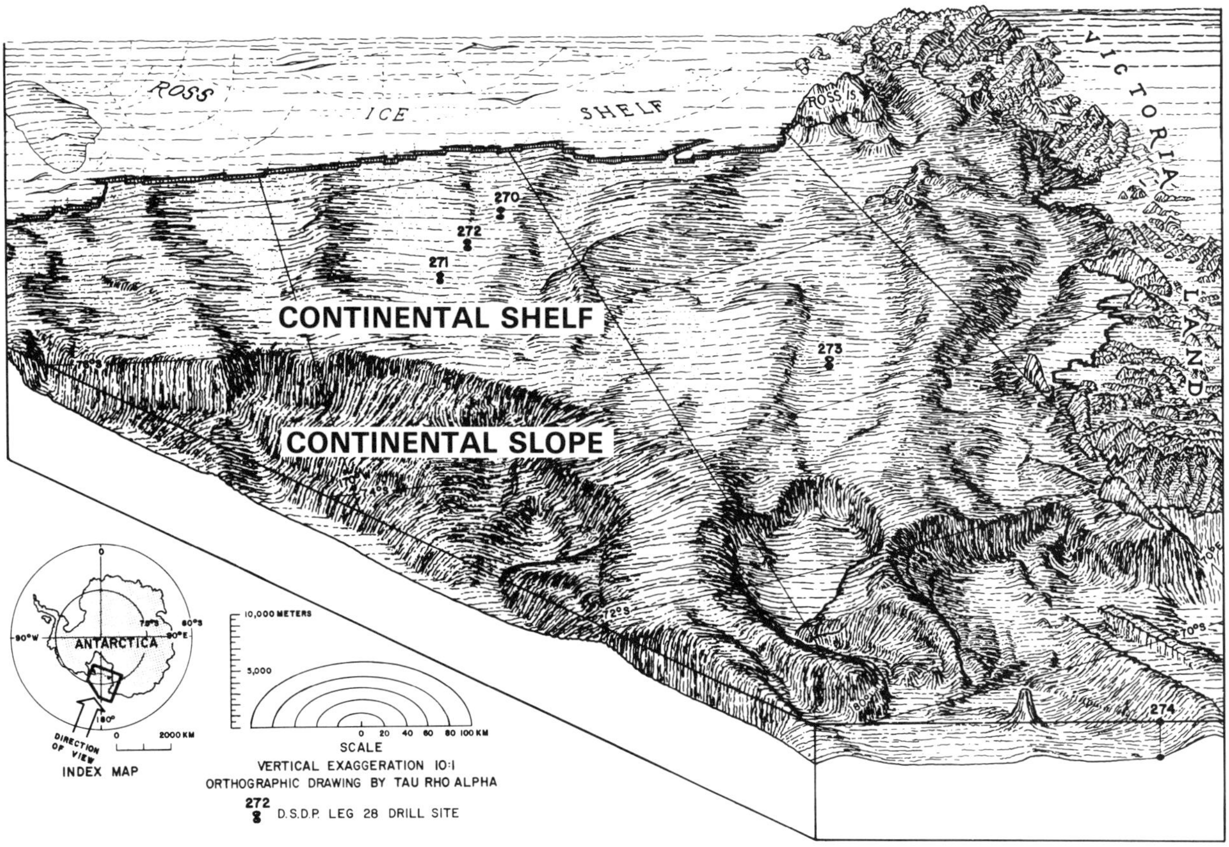

ROSS SEA CONTINENTAL SHELF

Figure 128.5. Three-dimensional diagram of the Ross Sea area. (After Hayes and Davey, 1975.)

tains relatively old sea floor formed when India departed on its northern journey in Late Jurassic or Early Cretaceous time (Sclater and Fisher, 1974; Sclater and others, 1976). The separation of Australia and East Antarctica, however, occurred mainly during early Tertiary time (Weissel and Hayes, 1972). The contrasts between the coastal segments of different ages should be apparent in the amount of shelf progradation. West of Bruce Rise (Figure 128.6), the shelf is somewhat wider than to the east; the difference, however, is not striking.

Weddell Sea

Although the morphology of the Weddell continental shelf is much less known than that of the Ross Sea (because of inaccessibility), its proximity to the other major ice shelf of the continent leads one to expect that the continental shelf morphology there is similar to that of the Ross shelf area. Like the Ross Sea, the Weddell Sea continental shelf may reflect a rift zone between East and West Antarctica (Craddock, Chapter 1, this volume). Seismic data of Fossum and others (Chapter 45, this volume) indicate as much as 100 km of progradation of the continental shelf.

Conclusions

With the exception of the Ross and Weddell Sea areas, the Antarctic continental shelf appears to be a typical passive margin. Troughs paralleling the coast probably mark the most landward antithetic fault created during the rifting and breakup of Gondwanaland. Progradation seaward

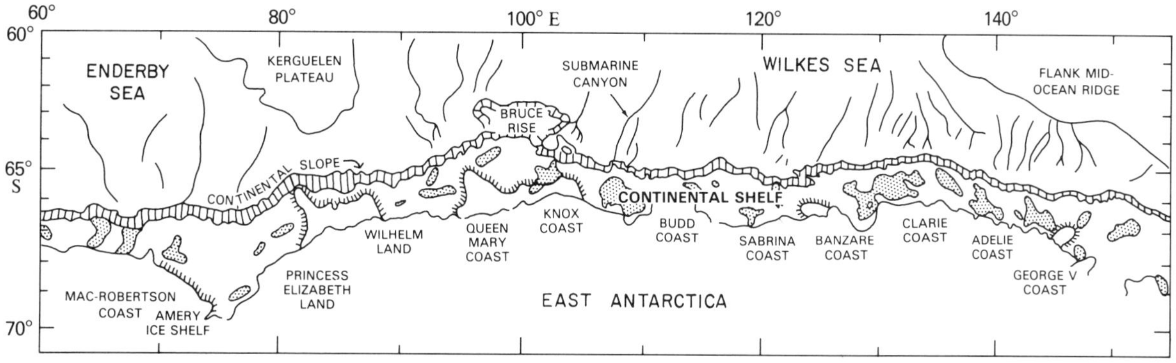

Figure 128.6. Physiographic provinces of the Wilkes and Enderby Sea areas. Dotted pattern: valleys on continental shelf. (After Vanney and Johnson, 1979*a*, *b*.)

of this "hinge line" is extensive, commonly exceeding 100 km.

The longer exposure of the shelf to glacial action in the Antarctic, as compared to the Arctic, has resulted in: (1) a more extensive continental rise, (2) a well-developed submarine canyon system, (3) widespread abyssal plain development, and (4) a deeper system of marginal channels and transverse canyons on the continental shelf.

The greater depth of the Antarctic outer shelf is perhaps due to depression by inland ice and incomplete postglacial isostatic rebound reflecting the persistence of thick ice sheets on the continent.

References cited

Clark, D. L. 1975. Geological history of the Arctic Basin. (*In* Yorath, C., Parker, E. R., and Glass, D. J., *eds.* Canadian Continental Margins and Offshore Petroleum Exploration. *Mem. Can. Soc. Petrol. Geol., 4,* 501–24.)

Denton, G. H., Armstrong, R. L., and Stuiver, M. 1971. The late Cenozoic glacial history of Antarctica. (*In* Turekian, K. K., *ed. Late Cenozoic Glacial Ages.* New Haven, Yale University Press, 267-306.)

Emery, K. O. 1969. The continental shelves. *Scient. Am., 221,* No. 3, 107–22.

Grant, A. C. 1970. Recent crustal movements on the Labrador Shelf. *Can. Jnl Earth Scis, 7,* 571–75.

Grant, A. C. 1972. The continental margin off Labrador and eastern Newfoundland—morphology and geology. *Can. Jnl Earth Scis, 9,* 1394–1430.

Grantz, A., Barnes, P. W., Eittreim, S. L., Reimnitz, E., Scott, E. W., Smith, R. A., Stewart, G., and Toimil, L. J. 1976. Summary report of the sediments, structural framework, petroleum potential, environmental conditions, and operational considerations of the United States Beaufort Sea, Alaska area. USGS Open-file report 76-830.

Hayes, D. E., and Davey, F. 1975. A geophysical study of the Ross Sea, Antarctica. (*In* Hayes, D. E., Frakes, L. A., and others. *Initial Reports of the Deep Sea Drilling Project,* Vol. 28, Washington [U.S. Government Printing Office], 881–907.)

Hayes, D. E., Frakes, L. A., and others. 1975*a*. *Initial Reports of the Deep Sea Drilling Project,* Vol. 28, Washington (U.S. Government Printing Office), 1017 pp.

Hayes, D. E., Frakes, L. A., and others. 1975*b*. Sites 270, 271, 272, The shipboard scientific party. (*In* Hayes, D. E., Frakes, L. A., and others. *Initial Reports of the Deep Sea Drilling Project,* Vol. 28, Washington [U.S. Government Printing Office], 211-334.)

Heezen, B. C., Tharp, M., and Bentley, C. R. 1972. Morphology of the earth in the Antarctic and subantarctic. *Antarct. Map Folio Ser.,* Folio 16, Pl. VIII.

Henderson, G. 1975. New bathymetric maps covering offshore west Greenland 59°–59° 30′ N. *Offshore Technology Conference, Dallas, OTC-2223,* 761–72.

Herron, E. M., and Tucholke, B. E. 1976. Sea-floor magnetic patterns and basement structure in the southeastern Pacific. (*In* Hollister, C. D., Craddock, C., and others. *Initial Reports of the Deep Sea Drilling Project,* Vol. 35, Washington [U.S. Government Printing Office], 263–78.)

Holtedahl, O. 1929. On the geology and physiography of some Antarctic and subantarctic islands. *Scient. Res. Norw. Antarct. Exped., 3,* 172 pp.

Holtedahl, O., and Holtedahl, H. 1961. On marginal channels along continental borders and the problems of their origins. *Bull. geol. Instn Univ. Upsala, 40,* 183–87.

Houtz, R., and Davey, F. J. 1973. Seismic profiler and sonobuoy measurements in Ross Sea, Antarctica. *J. geophys. Res., 78,* 3448–68.

Houtz, R., and Meijer, R. 1970. Structure of the Ross Sea shelf from profiler data. *J. geophys. Res., 75,* 6592–97.

Johnson, G. L., Sommerhoff, G., and Egloff, J. 1975. Structure and morphology of the west Reykjanes and the southeast Greenland and continental margin. *Mar. Geol., 18,* No. 3, 175–96.

Johnson, G. L., McMillan, N. J., Rasmussen, M., Campsie, U., and Dittmar, F. 1975. Sedimentary rocks dredged from the southwest Greenland continental margin. (*In* Yorath, C., Parker, E. R., and Glass, D. J., *eds.* Canadian Continental Margins and Offshore Petroleum Exploration. *Mem. Can. Soc. Petrol. Geol., 4,* 391–410.)

Nansen, F. 1922. The strandflat and isostasy. *Vidensk. Skr., I. Math., Nature, Lond., 2,* 313 pp.

Sclater, J. G., and Fisher, R. L. 1974. Evolution of the east central Indian Ocean. *Geol. Soc. Am. Bull., 85,* 5, 683–702.

Sclater, J. G., Luyendyk, B. P., and Meinke, L. 1976. Magnetic lineations in the southern part of the Central Indian Basin. *Geol. Soc. Am. Bull., 87,* 3, 371–78.

Van der Linden, W. J., Fillon, R. H., and Monahan, D. 1976. Hamilton Bank, Labrador margin: origin and evolution of a glaciated shelf. *Marine Science Pap. 14, Geol. Surv. Can. Pap., 75-40,* 31 pp.

Vanney, J. R., and Johnson, G. L. 1976*a*. Geomorphology of the Pacific Continental margin of the Antarctic Peninsula. (*In* Hollister, C. D., Craddock, C., and others. *Initial Reports of the Deep Sea Drilling Project,* Vol. 35. Washington [U.S. Government Printing Office], 279–89.)

Vanney, J. R., and Johnson, G. L. 1976*b*. The Bellingshausen-Amundsen Basins (southeastern Pacific): major sea floor units and problems. *Mar. Geol., 22,* No. 2, 71–101.

Vanney, J. R., and Johnson, G. L. 1976*c*. The floor of the Ross Sea and adjacent oceanic provinces. *Antarct. Jnl U.S., 11,* No. 4, 231–33.

Vanney, J. R., and Johnson, G. L. 1979*a*. The sea floor morphology seaward of Terre Adelie (Antarctica). *Dtsch. Hydrogr. Zeitschr., 32,* No. 2, 77–87.

Vanney, J. R., and Johnson, G. L. 1979*b*. Outlines of Wilkes Land continental margin physiography (East Antarctica). *Polarforshung., 49,* No. 1, 20–29.

Vanney, J. R., Falconer, R. K. H., and Johnson, G. L. In press. Geomorphology of the Ross Sea and adjacent oceanic provinces. *Mar. Geol.*

Weissel, J. K., and Hayes, D. E. 1972. Magnetic anomalies in the south eastern Indian ocean. (*In* Hayes, D. E., *ed. Antarctic Oceanology 2. The Australian-New Zealand Sector.* Washington, American Geophysical Union, 234–49.) [Antarctic Research Series, Vol. 19.]

Zhivago, A. V., and Lisitzin, A. P. 1957. New data on the bottom relief and the submarine deposits of eastern Antarctica. *Izv. Akad. Nauk SSSR,* Ser. Geogr. Geofiz., *1,* 19–35.

129

Sedimentation on the West Antarctic Continental Margin

JOHN B. ANDERSON and DENNIS KURTZ
Department of Geology, Rice University, Houston, Texas 77001, U.S.A.

FRED WEAVER
Exxon Production Research Company, Houston, Texas 77001, U.S.A.

MARIANNE WEAVER
Department of Geology, Florida State University, Tallahassee, Florida 32306, U.S.A.

Abstract. Terrigenous deposits of the West Antarctic continental margin reflect deposition from both grounded and floating ice shelves. These glaciogenic sediments are differentiated on the basis of size frequency data, stratification, pebble fabric, and fossil content. Glacial sedimentation probably occurs over relatively short intervals of geologic time and is followed by long periods of extensive reworking and redeposition of sediments by normal marine agents. The reworking of glacial marine deposits is largely controlled by the position of the floating ice shelf and perennial pack-ice front, which affects shelf circulation. In those regions where floating ice covers the inner continental shelf, such as in the southern Ross and Weddell Seas, bottom current activity is limited to thermohaline circulation, which is only moderately effective at reworking bottom sediments. Thick, laminated muds of the continental slope and rise may result from variations in contour-current intensity. These variations are due to changes in thermohaline circulation, which is a function of sea ice production on the continental shelf. Where the volume of sea-ice produced is sufficient to increase the density of shelf water above that of circumpolar deep waters, mixing occurs at the continental slope and contour currents are diminished. Variations in current intensity may occur quite frequently and could even be seasonal. Ice-rafting and bottom-current deposition on the continental slope is periodically interrupted by mass flow deposition. The mode of mass flow deposition on the continental slope is largely controlled by the textural properties of upper slope deposits. Textural properties of upper slope deposits are, in turn, controlled by shelf and slope circulation, and consequently by the extent of floating ice cover and the intensity of sea ice production on the shelf. Continental shelves and slopes with poor circulation (those subjected to thermohaline convection) are floored by muddy bottom sediments, and down-slope mass movement is largely by debris flows. The southwestern Weddell Sea and southern Ross Sea margins are examples. Continental shelves and slopes have good circulation where sea ice production is not intense enough to disrupt normal wind and geostrophic currents, and they are floored by reworked glacial marine deposits. Mass movement on these margins is primarily by turbidity currents. The eastern Weddell Sea and northwestern Ross Sea continental margins are examples.

The Antarctic continental margin is a unique sedimentary environment because terrigenous sediment deposition is almost entirely by ice. Initially, these glaciogenic sediments are unsorted. The extent to which they become sorted and redistributed on the sea floor reflects the importance and relative competence of various marine agents far better than in other marine environments where the initial textural properties of terrigenous sediments are unknown. This paper provides a general synthesis of results obtained from a sedimentologic investigation of those piston cores taken to date on the Ross, Weddell, and Bellingshausen continental margins and considers the sedimentary, glacial, and oceanographic processes that affected them. More detailed descriptions of our data are provided in recently published papers (Anderson and others, 1979; Anderson and others, 1980; Kurtz and Anderson, 1979.)

Methods

The locations of sediment cores studied to date are shown in Figure 129.1. Sediment cores were X-radiographed, and samples were collected from sedimentary units that could be identified on the basis of sedimentary structures, pebble fabric, and visual textural descriptions. Samples were wet-sieved at 63μ and decanted at 16μ. Larger-than-63 μ samples and 16 μ-to-63 μ samples were analyzed separately on 1.4-m-long and 30-cm-long rapid sediment analyzers, respectively. The less-than-16 μ fraction was analyzed by a hydrophotometer.

The apparent dips of the long axes of 250 pebbles in three cores were measured from X-radiographs to determine the pebble fabric. Orientations of all pebbles with a long axis at least twice the length of the short axis were measured. Pebbles situated near the margins of the core

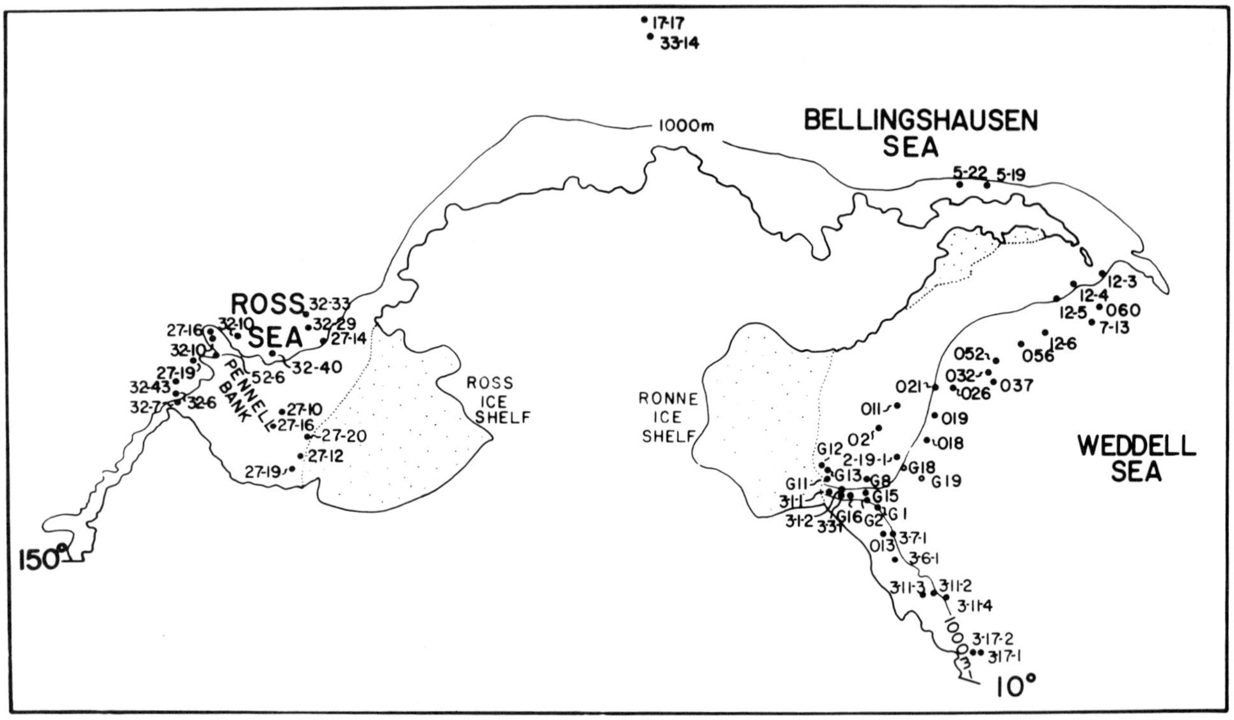

Figure 129.1. Locations of cores used in this study. The 1968 USCGC *Glacier* stations are labeled 02 through 060; 1969 *Glacier* stations, G1 through G19; and 1970 *Glacier* stations, 2-19-1 through 3-18-1. USNS *Eltanin* stations are labeled according to cruise number (7, 12, 27, and 32) and station number.

liner were not measured. Statistics were obtained using the von Mises distribution (Till, 1974).

Glacial sedimentation

Deposition from grounded ice shelves

Vast quantities of sediment may be transported in the basal zone of large ice sheets. Deposition of ice-transported sediments on land has been thoroughly examined; however, the fate of glacially transported debris upon entering the marine environment is poorly understood.

Sediment cores acquired on continental shelves contiguous to the Ross and Ronne-Filchner Ice Shelves consist primarily of diamictons (unstratified, unsorted, clastic debris consisting of sand and larger particles in a muddy matrix). Of 42 diamicton samples analyzed, the average gravel, sand, and mud percentages were found to be 7.8%, 33.8%, and 58.4%, respectively. These sediments are dark gray to black and generally devoid of fossils. They have a low water content and are typically overcompacted. Elongate pebbles tend to lie with their long axis parallel to bedding; however, this pebble fabric is crudely developed and may be inherited from the sediment-laden ice or may be due to postdepositional compaction. Pebble shape is highly variable and hence is of little diagnostic value. Representative cumulative curves and size frequency data for the -0.50 to $9.00\ \phi$ size fraction of representative diamicton samples from the Ross Sea and Weddell Sea continental shelves are given in Figure 129.2, along with an X-radiograph of a typical Antarctic diamicton.

Harland and others (1966) use the term *orthotill* to describe marine sediments derived directly from glaciers, which have undergone little or no reworking by bottom currents. That some Ross Sea and Weddell Sea diamictons were deposited directly beneath disintegrating grounded ice shelves is supported by the following lines of evidence.

1. The total absence of sorting and stratification (Figure 129.2) in these deposits and an absence of marine fossils reflects rapid deposition directly from ice, with essentially no reworking by bottom currents.
2. Supposed orthotills have not been cored beyond the shelf margin in either the Weddell or Ross Sea. We have examined several diamictons from the continental slopes in these regions and have concluded that all were derived either from floating ice shelves or from mass flow deposits (Kurtz and Anderson, 1979).
3. Orthotills are presently exposed on the sea floor in only a few scattered locations, and these may represent eroded surfaces. Hence, these deposits are apparently not being formed today seaward of the ice-shelf front.
4. Supposed orthotills appear to be confined to those continental shelves which adjoin large ice shelves. However, cores are sparsely distributed on the continental shelf.

Deposition from floating ice shelves

Sediments shed from floating ice shelves are likely to be influenced by marine currents, both during descent and after reaching the sea floor. Accordingly, such sediments should be distinguishable from those deposited by grounded ice shelves.

Harland and others (1966) propose the name *paratill* for marine sediments derived through a combination of glacial and normal marine processes. In the Weddell and Ross Seas we recognize at least two distinct types of paratill. Those paratills whose muddy matrix has been enriched in current-derived silt and clay relative to orthotills of the

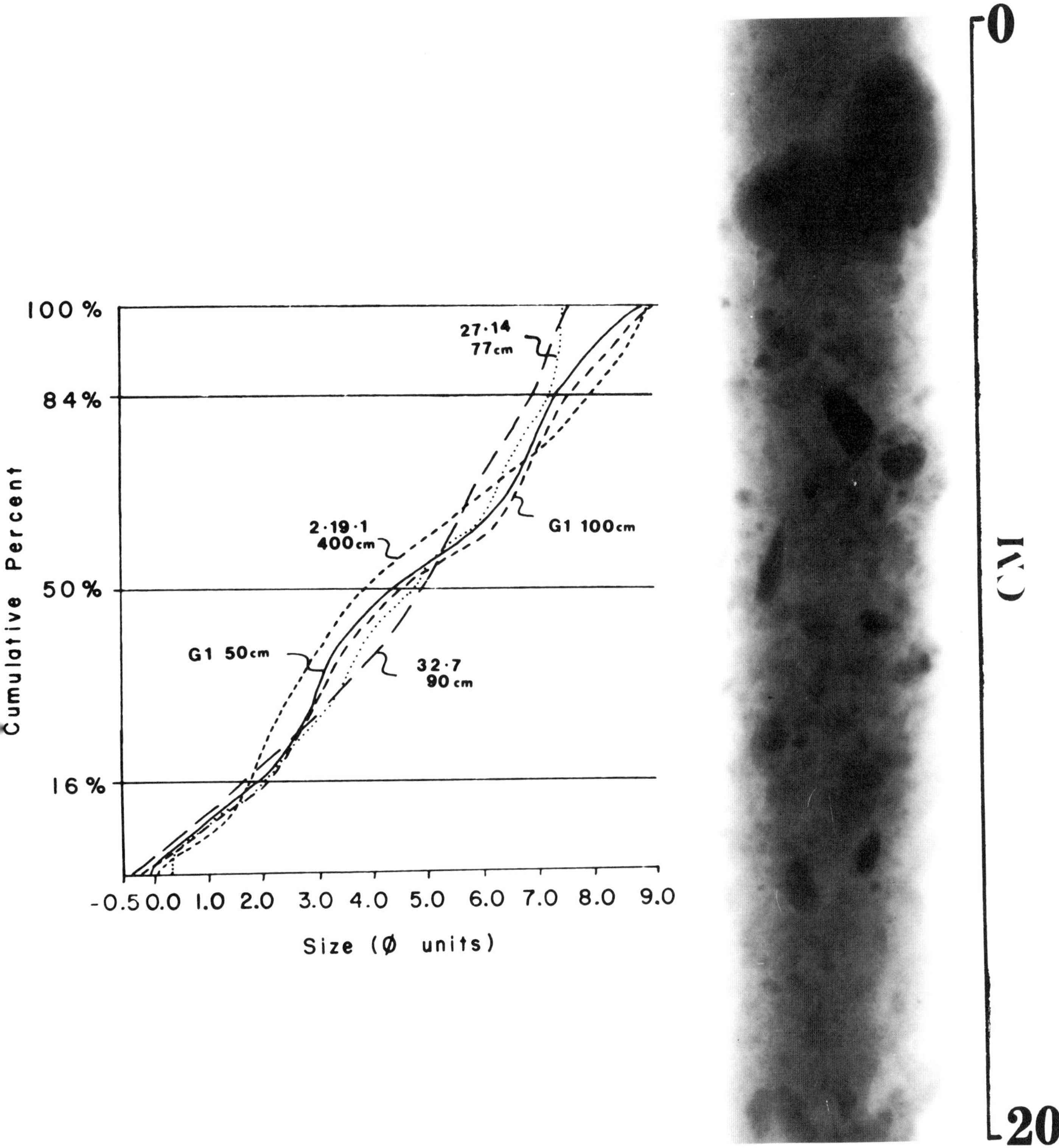

Figure 129.2. X-radiograph of typical diamicton (orthotill) (core G1), along with cumulative curves for the -0.50 to $+9.00$ ϕ size fraction of representative samples. See Figure 129.1 for locations of cores.

same area are called *compound paratills* (Anderson and others, 1977). Those which have had their silt-clay content depleted through current winnowing are called *residual paratills* (Anderson and others, 1977). Table 129.1 outlines the characteristics of orthotills, compound paratills, and residual paratills.

In the vicinity of the Ross and Ronne-Filchner Ice Shelves, orthotills are commonly overlain by, and grade seaward into, pebbly muds. These sediments are easily distinguished from orthotills in X-radiographs, as they exhibit crude stratification (Figure 129.3) and well-developed pebble fabric in which elongate clasts align themselves parallel to bedding. Size analyses of pebbly muds reflect a fair degree of sorting relative to orthotills, which is evident when representative cumulative curves of orthotills (Figure 129.2) and pebbly muds (Figure 129.3) are compared. The greater degree of sorting in pebbly muds is attributed to an additional current-derived silt-clay component (Chriss and Frakes, 1972) and, hence, the term compound paratill is used to describe these sediments. Compound paratills commonly contain abundant foraminiferal tests, in contrast to the essentially unfossiliferous orthotills.

In the Ross and Weddell Seas, compound paratills seem to be confined to those aras where large shelves cover the inner continental shelf and to bathymetric depressions on the continental shelf. Their presence reflects sluggish circulation. In contrast, those continental shelves whose coastal waters are ice free during part of the year, such as the eastern Weddell Sea and northwestern Ross Sea, are floored by coarse sediments which exhibit varying degrees of reworking. In these regions, the innermost shelf is commonly floored by gravelly, muddy sands. These deposits are texturally similar to orthotills, except that their muddy matrix has been partially or wholly removed. They are unstratified and generally contain abundant foraminiferal tests, as well as other fossil remains. Because they are believed to represent winnowed debris shed from floating ice, the name residual paratill seems appropriate.

Figure 129.4 shows the cumulative curves for the -0.50 to 9.00 ϕ size fraction of representative Ross and Weddell residual paratills. In both regions the degree of winnowing has obviously varied through time, probably in response to variations in the rate of both ice-rafting and in bottom-current intensity. Winnowing has mainly involved the 2.5 ϕ and finer fraction (Figure 129.4). As noted by Chriss and Frakes (1972), currents with velocities of only a few centimeters per second could easily account for this effect. Yet, beneath an ice shelf or permanent sea ice cover even such gentle currents may not exist. The fate of the debris winnowed from ice-rafted sediments is discussed below.

Ice-rafting beyond the shelf break

The concentration of ice-rafted debris (IRD) in Antarctic marine sediments decreases abruptly beyond the shelf break, and continental rise deposits are frequently devoid of ice-rafted debris. This is because the vast majority of sediment rafted out to sea by ice shelves is carried at or near the ice-rock interface. This material is deposited during basal melting seaward of, but near to, the buoyancy line (Boulton, 1972). The buoyancy line should, in a normal floating ice shelf, be located well inland of the calving line. Tabular bergs calved from large floating ice shelves, therefore, contain only minor amounts of debris, probably concentrated mainly within shear zones. The quantity of material transported in this manner is controlled by complex and poorly understood ice-sheet processes.

Icebergs calved from valley and piedmont glaciers are capable of transporting large quantities of sediment (Ovenshine, 1970), and the rate of supply of glacial sediments by these agents is probably greater than large ice shelves. Hence, it is possible that larger quantities of IRD could be transported to the deep-sea floor during glacial minima, when valley and piedmont glaciers play a greater role in supplying sediment to the sea. Apart from this, the parameters which control the rate of IRD accumulation on the sea floor are numerous and complex (see Weaver and Anderson, Chapter 149, this volume). For example, aside from changes in glacial intensity, IRD accumulation rates will vary (1) as the rate of terrigenous sediment supply from the ice shelf changes, (2) as bottom winnowing increases, and (3) as the rate of supply or dissolution of biogenic material changes. We doubt that attempts to utilize IRD as a climatic indicator can produce meaningful results until such time as these complexities are recognized and resolved.

Table 129.1. Glacial marine sediments of the Antarctic continental margin

	Orthotill		Residual paratill		Compound paratill	
Stratification	None		None		Present	
Gravel, sand and mud concentrations (in %)	*Average*	*Range*	*Average*	*Range*	*Average*	*Range*
Gravel	7.8	0.7–25.7	16.0	8-0–90.0	4.5	0.2–21.7
Sand	33.8	11.5–54.3	81.0	10.0–98.8	16.0	6.1–37.0
Mud	58.4	32.7–70.4	3.0	0.2– 8.0	79.5	55.0–92.7
Mean grain size (M_ϕ)	4.19 to 4.64 ϕ		2.25 to 3.15 ϕ		4.47 to 6.38 ϕ	
Sorting (σ_ϕ)	None		Poorly sorted (σ_I = 1.12 to 1.89 ϕ)		Poorly to very poorly sorted (σ_I = 1.54 to 2.27 ϕ)	
Skewness (SK_I)	−0.21 to −0.12		+0.07 to +0.64		−0.48 to −0.15	
Kurtosis (K_G)	−1.33 to −1.00		−0.81 to +1.94		−1.20 to +0.95	
Pebble fabric	Crude horizontal parallelism of elongate pebbles or none		None		Well-developed horizontal parallelism of elongate pebbles	
Benthic foraminifera	Extremely rare to absent		Abundant, low diversity, calcareous fauna (ice-shelf facies of Anderson, 1972)		Abundant calcareous or arenaceous fauna (ice-shelf facies or euryhaline facies of Anderson, 1972)	

Grain-size parameters are computed for the −0.50 to +9.00 ϕ size fraction of samples.

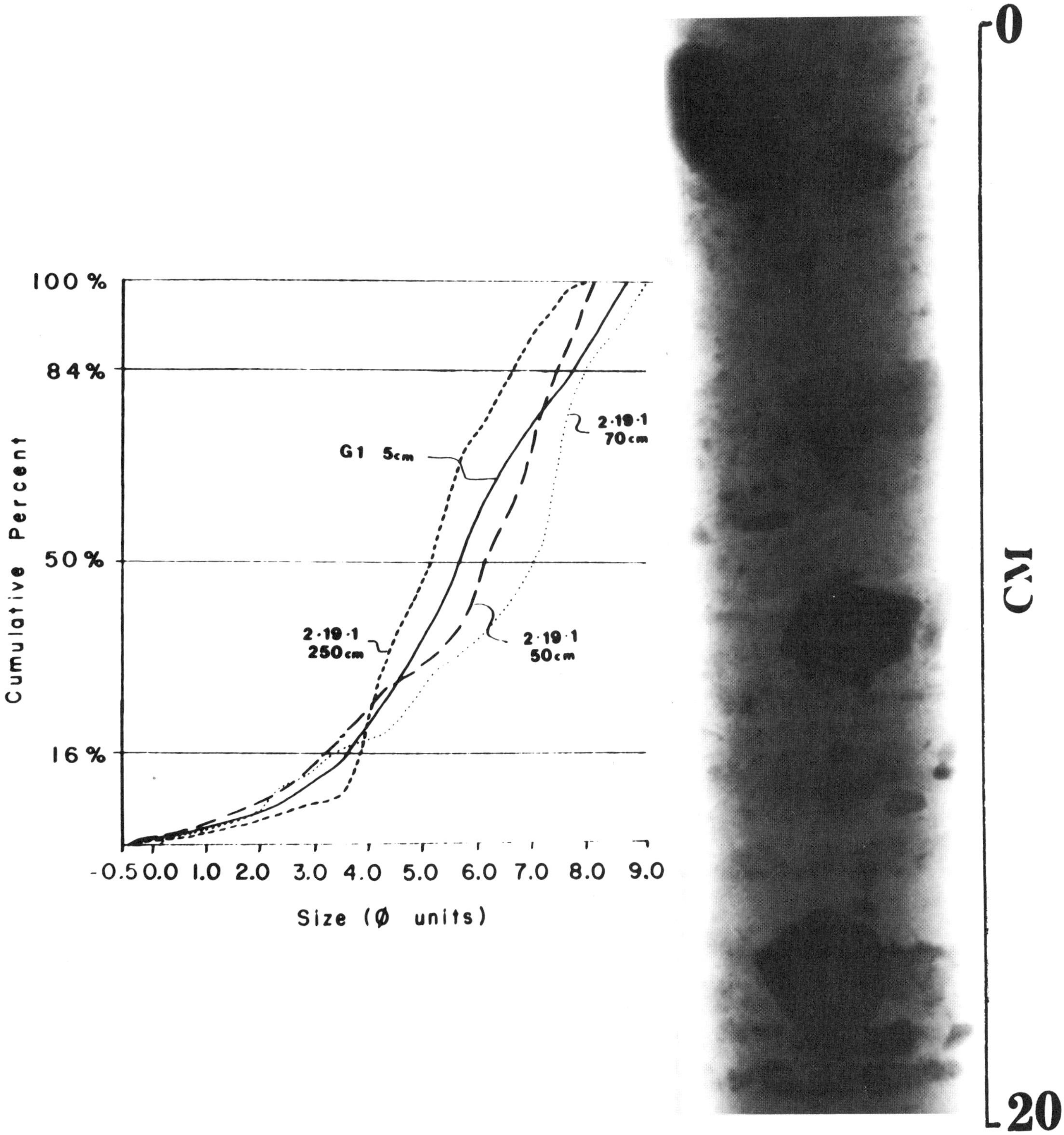

Figure 129.3. X-radiograph of compound paratill showing stratification and horizontal pebble alignment, along with cumulative curves for the -0.50 to $+9.00$ ϕ size fraction of representative samples.

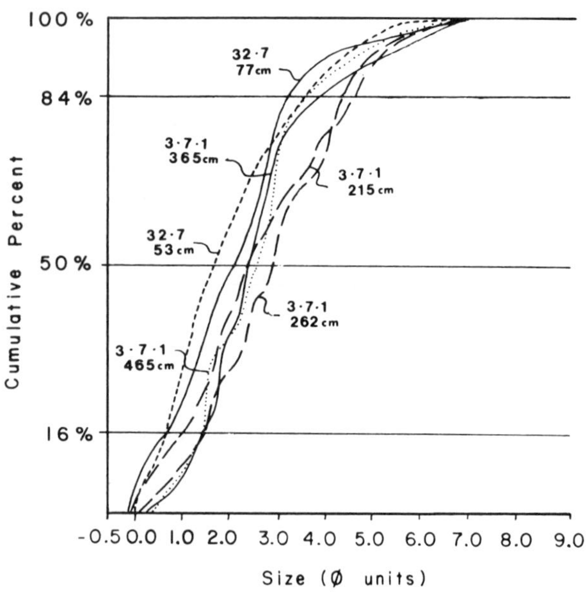

Figure 129.4. Cumulative curves for the −0.50 to +9.00 ϕ size fraction of representative residual paratill samples.

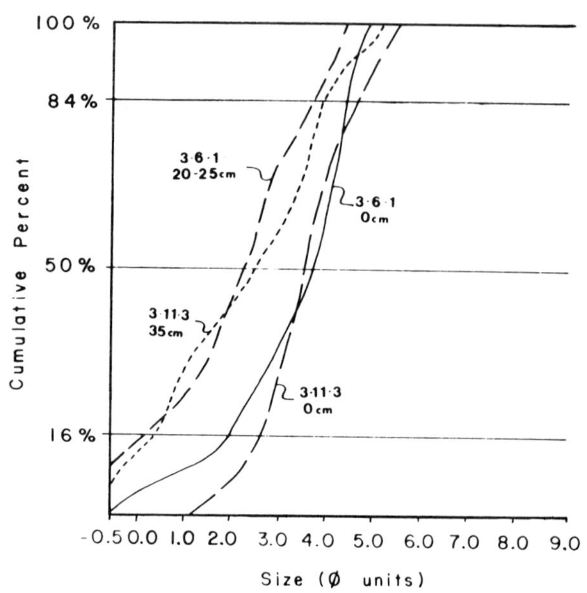

Figure 129.5. Cumulative curves for current-deposited sands from the eastern Weddell Sea continental shelf.

Nonglacial sedimentation

Redeposition of glacial sediments by bottom currents and mass flow mechanisms is actively occurring over much of the West Antarctic continental margin today. As expected, there is a strong correlation between the duration of seasonal ice production, which controls shelf circulation, and the degree of resedimentation by bottom currents.

Current deposits

Circulation on the Antarctic continental margin is dominated by westward-flowing Circumpolar Deep Water (Jacobs and others, 1970). This is a geostrophic current, and hence its intensity is largely controlled by the physiography of the margin. Maximum velocities appear to be associated with relatively straight segments of the margin, as in the eastern Weddell Sea where measured velocities range from 15 to 30 cm/sec. (Gill, 1973). Here the strongest velocities occur on the continental shelf and upper slope and are believed to diminish to a few centimeters per second below 800 to 1000 m (Foster and Carmack, 1976). The continental shelf in the eastern Weddell Sea is floored by residual paratills and traction deposits with well-defined 2.5 to 3.5 ϕ sand size modes (Figure 129.5).

Bottom current velocities on the eastern Ross Sea continental slope exceed 15 cm/sec (Jacobs and others, 1970). In both the Ross and Weddell Seas these strong contour currents are associated with gravelly lag deposits on the upper slope. Bottom photographs of the Weddell Sea slope (Hollister and Elder, 1969) and nephelometer measurements on the Ross Sea slope (Jacobs and others, 1970) show sediment-laden bottom waters. At the high-flow velocities measured, fine silts and clays could remain in suspension almost indefinitely.

Disruption of geostrophic flow can occur in two ways. First, physiographic irregularities in the continental margin can divert flow away from the continental slope. Second, flow can be diminished by frontal zone mixing, which both slows the flow and causes the contour current to be detached from the continental slope. This process is regulated by sea ice formation on the continental shelf and hence has important paleoclimatic implications.

As sea ice freezes, brines are formed which mix with shelf water, resulting in a density increase. If enough brine is produced to neutralize the density profile on the continental shelf, dense shelf waters will flow onto the continental slope and mix with less dense circumpolar deep waters (Foster and Carmack, 1976). The resultant mixture is Antarctic Bottom Water, which has a density greater than that of either mixing component. It flows down-slope, displacing circumpolar water. Direct measurements of Antarctic Bottom Water flow on the Weddell Sea slope range from 1 to 7 cm/sec and have a northward or down-slope component (Foldvik and Kringe, 1974). Mixing on the continental slope thus results in both a reduction in bottom-current velocities and a general increase in the down-slope component of flow.

The continental shelves of the southwestern Weddell Sea and the southeastern Ross Sea, where thermohaline circulation occurs, are floored by compound paratills. The very fine matrix of these sediments attests to sluggish bottom circulation in these areas. Sediments consisting of ice-rafted debris, biogenic material, and coarse, current-derived silts occur in a narrow zone restricted to the adjacent continental slope (Anderson, 1972; Anderson and others, 1979). Silts are deposited from suspension by waning contour currents. Biogenic sediments are associated with high biological productivity which results from upwelling of warm, nutrient-rich circumpolar water. Piston cores taken in these areas show major down-core changes in sediment type which have been attributed to changes in circulation (Anderson and others, 1979).

The most prominent sediment type on the Antarctic continental rise and lower slope is laminated mud (Figure 129.6). This sediment typically lacks microfossils and contains only sparse ice-rafted debris, which occurs in thin zones (Anderson and others, 1979). Figure 129.7 shows representative size frequency curves for closely spaced samples from a laminated mud unit in core 17-17 taken at 4508 m on the Bellingshausen rise (Figure 129.1). Samples at 106 and 99 cm were taken from thin layers containing ice-rafted debris.

The repeated occurrence of distinct silt modes within a given unit is characteristic of those laminated mud units studied in piston cores from the Bellingshausen, Ross, and Weddell Seas (Anderson and others, 1979). This is suggestive of more or less regular variations in contour-current intensity at a given level on the slope and rise.

MASS FLOW DEPOSITS

Laminated muds on the continental slopes in the eastern Ross Sea and the southern Weddell Sea are interbedded with diamictons which are texturally indistinguishable from orthotills (Kurtz and Anderson, 1979). If derived from floating ice, then the absence of bedding and fossils and the high silt and clay content of these deposits suggest extremely rapid deposition under unusually quiescent bottom conditions. On the other hand, the diamictons may have been derived through mass flow of orthotills from the continental shelf. Four lines of evidence support this second theory of origin.

1. Diamictons are in sharp contact with laminated muds (Figure 129.8). These interbedded muds show no significant increase in ice-rafted debris below the diamicton contact. It is not likely that deposition from floating ice would have commenced so abruptly. Contacts between diamictons and overlying laminated muds are also sharp and show little evidence of extensive erosion (i.e., lag deposits). Again, such rapid changes from glacial to nonglacial sedimentation are not easily explained by floating ice.
2. Benthic foraminifera are extremely rare within these diamictons, but those present include both shelf and slope forms. Similar displaced faunas were also observed in this area by Fillon (1974).
3. Mass flow deposits might well reveal differences in pebble fabric from ice-deposited orthotills and paratills, owing to much different transport mechanisms for these different types of deposits. Boulton (1971) sug-

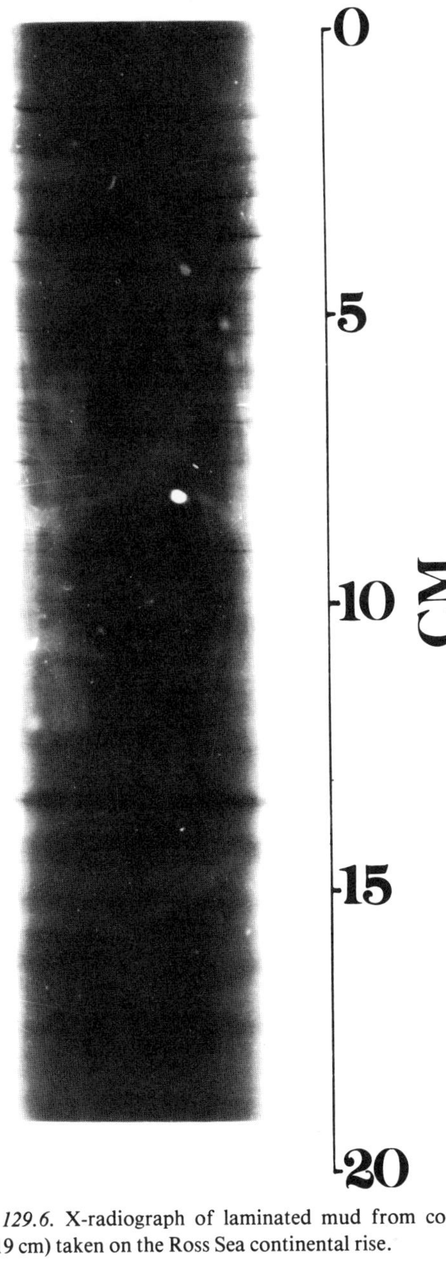

Figure 129.6. X-radiograph of laminated mud from core 32-40 (200–219 cm) taken on the Ross Sea continental rise.

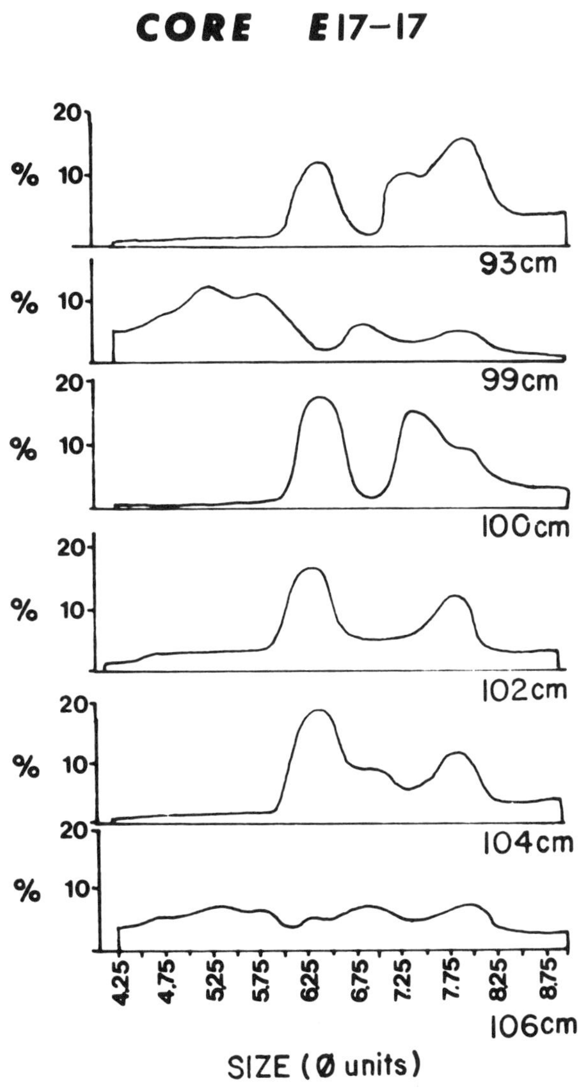

Figure 129.7. Frequency curves for samples taken at close intervals from a laminated mud unit in core E 17-17 from the Bellingshausen continental rise. Note the persistent occurrence of silt modes at 6.25 and 7.25 ϕ. Samples at 99 and 106 cm contain large ice-rafted components.

gests that prolate stones tend to be imbricated in an upslope direction in flow tills. Pebble-fabric analyses of a suspected mass flow deposit were conducted, and the results were compared with those of similar analyses of orthotills and compound paratills (Figure 129.9). Whereas pebbles in the orthotills and paratills exhibit a preferred horizontal orientation with low angular standard deviations, the long axes of clasts within a suspected mass flow deposit (core 32–10; Figure 129.1) show a preferred orientation of approximately 25° plunge from the horizontal. This crude imbrication is further support for mass flow deposition.

4. Another striking feature of supposed Antarctic debris flows is that they display remarkable textural homogeneity within a single unit (Kurtz and Anderson, 1979). Deposits from floating ice would not be expected to be homogeneous because variations in marine current intensity would result in textural variations in paratills.

In the southeastern Weddell Sea, near the front of the Filchner Ice Shelf, orthotills and residual paratills are overlain by normally and inversely graded gravels and sands. Cores G11 and G13 (Figure 129.1), taken at 585 and 659 m, respectively, penetrate this sequence. Figure 129.10 shows the cumulative curves for a graded interval in core G13. Note that sorting increases upward in the section and that sands finer than approximately 1.5ϕ have been removed. The fate of these finer sands can be determined from comparisons of cumulative curves from core G13 with those of samples taken from cores G15 and G16, which were acquired down slope at 859 and 1033 m, respectively, and 80 to 100 km from sites G11 and G13 (Figure 129.1). Here the ability of mass flow mechanisms to sort and redistribute even the most poorly sorted marine deposits is evident.

In the northwestern Ross Sea, west of Pennell Bank, residual paratills cover the bottom near the ice front and atop knolls, and turbidites fill many of the depressions and valleys on the continental shelf (Figure 129.11). On the adjacent continental slope and rise, turbidites cover vast areas of the sea floor. Southeast of Pennell Bank, the con-

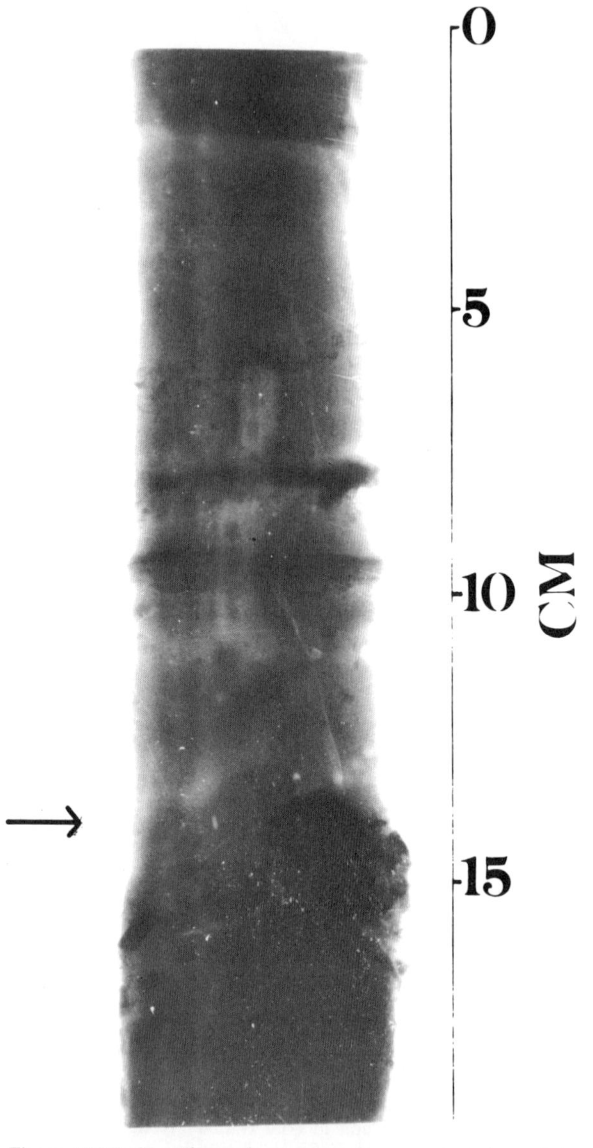

Figure 129.8. X-radiograph showing sharp contact (arrow) between a laminated mud unit and a mud flow deposit, in core E 32-33 from the Ross Sea continental slope.

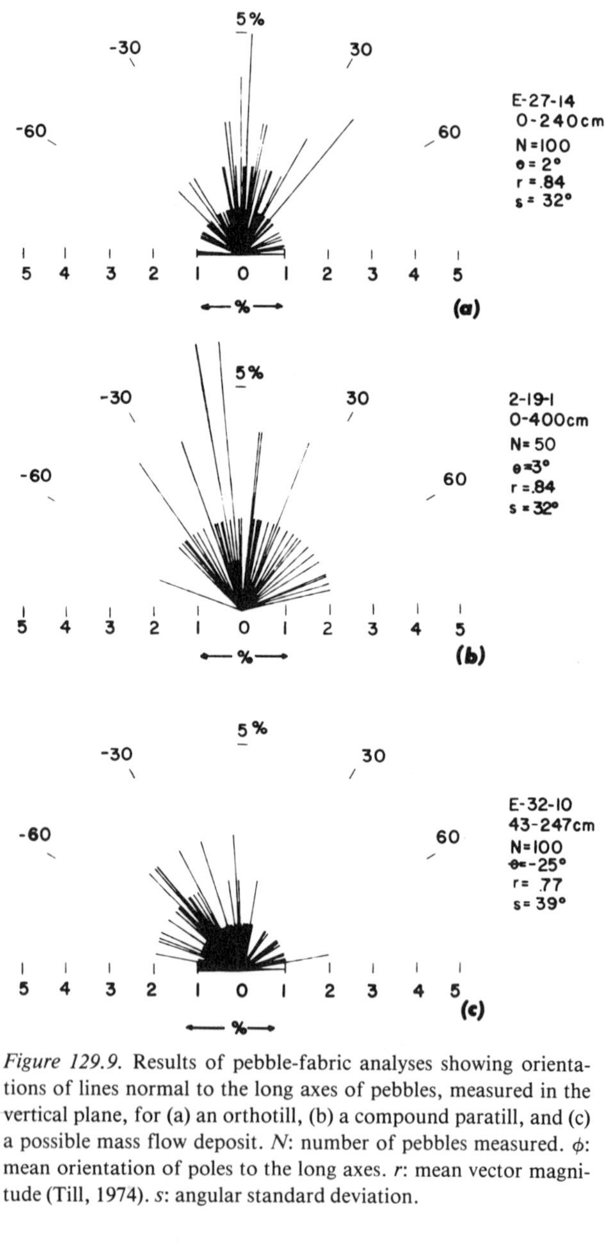

Figure 129.9. Results of pebble-fabric analyses showing orientations of lines normal to the long axes of pebbles, measured in the vertical plane, for (a) an orthotill, (b) a compound paratill, and (c) a possible mass flow deposit. *N*: number of pebbles measured. ϕ: mean orientation of poles to the long axes. *r*: mean vector magnitude (Till, 1974). *s*: angular standard deviation.

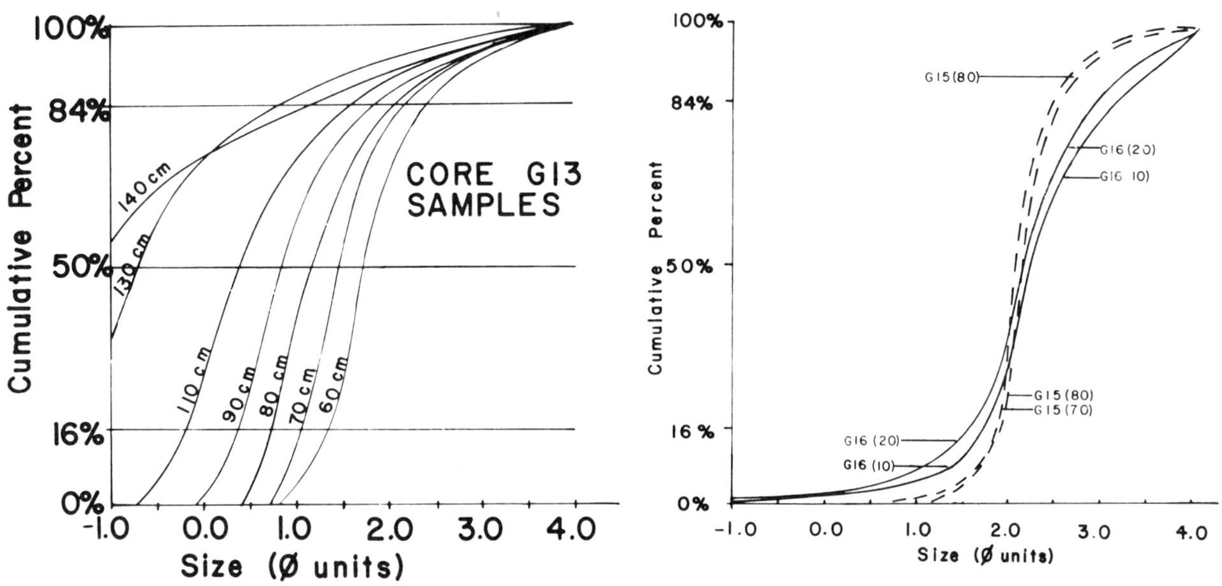

Figure 129.10. Cumulative curves for a proximal turbidite (core G13) and its distal counterparts (cores G15 and G16) (numbers in parentheses are depths, in cm).

tinental shelf slopes toward the continent, and bottom currents flow in that direction. Sediment supply to the outer continental shelf is largely confined to ice-rafting, and compound paratills are the most widespread deposits on the outer shelf. The down-slope movement of these poorly sorted, muddy deposits by debris flows appears to be a common process (Kurtz and Anderson, 1979).

Similar contrasts can be drawn with regard to continental slope sedimentation in the eastern, as opposed to the western, Weddell Sea. In the eastern Weddell Sea, glacial marine sediments are being actively reworked by strong contour currents flowing across the continental shelf. The continental shelf is floored by residual paratills and current-traction deposits. Down-slope transport is largely by turbidity currents which have shed well-sorted sand turbidites far out onto the continental rise and abyssal floor (Anderson, 1972). In contrast, the broad southwestern continental shelf of the Weddell Sea is almost entirely covered by perennial pack ice, and bottom currents along the outer continental shelf are more sluggish than along the eastern continental shelf (Gill, 1973). Compound paratills floor the outer continental shelf in the southwestern Weddell Sea. As in the southern Ross Sea, down-slope movement in the southwestern Weddell Sea is largely in the form of debris flows (Kurtz and Anderson, 1979).

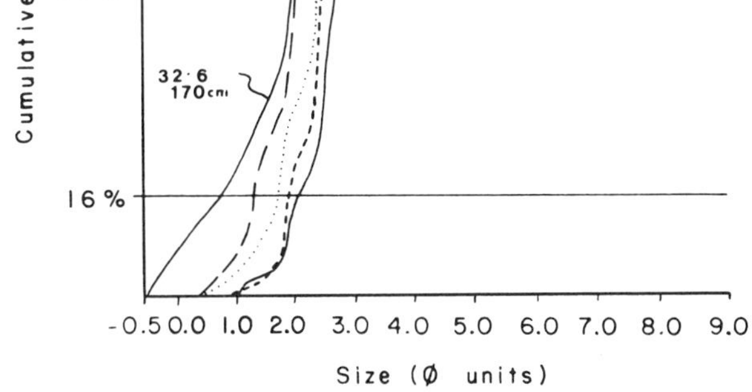

Figure 129.11. Cumulative curve for turbidite samples from the northeastern Ross Sea shelf. Note grading and improved sorting upward in the section.

Conclusions

Three distinct types of glacial marine deposits are recognized on the West Antarctic continental margin. They are distinguished on the basis of size frequency data, stratification, pebble fabric, and fossil content. They include:

1. orthotills—unsorted, unstratified, clastic sediments derived directly from the disintegration of grounded ice shelves;
2. residual paratills—unstratified, poorly sorted, gravelly, muddy sands believed to consist of the residual fraction of winnowed sediment derived from floating ice shelves; and
3. compound paratills—stratified, pebbly muds derived through a combination of ice rafting from floating ice (both ice shelves and icebergs) and current deposition.

Following a glacial maximum, glacial sedimentation on the continental shelf is probably rapid and short-lived and is confined to the initial disintegration of ice shelves. This burst of deposition is followed by a relatively long period of reworking of these glacially derived sediments. Reworking is largely by wind-generated and geostrophic currents. The intensity of these currents is closely tied to the extent of perennial ice cover over the continental shelf. The effects of sea ice cover are to retard wind-sea interaction, and the formation of sea ice causes thermohaline mixing which can, if pronounced, disrupt circulation. In those regions where ice shelves and perennial pack ice covers much of the shallow shelf, such as in the southern Ross and Weddell Seas, bottom-current activity is limited to themohaline circulation. Because these regions are floored by muddy sediments, such currents appear to be relatively ineffective sediment transporting agents.

Reworking of glacial sediments is most effective where the flow of wind-generated and geostrophic currents is uninhibited by thermohaline circulation (as in the eastern Weddell Sea and northwestern Ross Sea continental shelves). These areas are floored by sediments which have been either winnowed or redeposited by bottom currents.

Laminated muds are widespread on the Antarctic continental slope and rise. These deposits are believed to reflect variations in the relative intensity of contour currents and more sluggish down-slope flow of Antarctic Bottom Water. The flow regime over the continental slope is determined by the degree of brine production by sea ice formation over the continental shelf. This process also influences shelf circulation and hence the transport of sediment across the shelf.

Mass flow deposits are probably widespread on the Antarctic continental slope and rise and in depressions of the continental shelf. Debris flow deposits, which are texturally quite similar to glaciogenic deposits, may be distinguished on the basis of their sharp contacts with interbedded laminated muds, displaced benthic foraminifera, and pebble fabric that shows crude upslope imbrication. This mode of deposition is most prevalent on those continental slopes where muddy sediments mantle the adjacent shelf.

In contrast, turbidites, and probably grain flow deposits, are more common on those parts of the continental slope and rise where contiguous continental shelf sediments are winnowed by bottom currents. For example, turbidites are widespread in the eastern Weddell Sea and northwestern Ross Sea. No sandy turbidites have yet been cored on the continental slope or rise of the southwestern Weddell Sea and southeastern Ross Sea. Debris flows are apparently the most active mass flow mechanism in these areas. Hence, the importance of mass flow mechanisms in sorting some marine sediments may be documented by studies of Antarctic continental margin deposits.

Acknowledgments. Financial support for this research was provided through grants from the American Chemical Society Petroleum Research Fund (Grant PRF-9472-AC2) and from the National Science Foundation Division of Polar Programs (Grant DP-77-26407). We owe special thanks to Dennis Cassidy of the U.S. Antarctic Marine Geology Research Facility for his assistance in obtaining samples. We would also like to express our gratitude to Jenna Anderson and Susan Davis for their assistance in sample preparation and analyses.

References cited

Anderson, J. B. 1972. The marine geology of the Weddell Sea. *Florida State Univ. Sedim. Res. Lab. Rep.,* No. 35, 222 pp.

Anderson, J. B., Clark, H. C., and Weaver, F. M. 1977. Sediments and sediment processes on high latitude continental shelves. *Proc. Offshore Tech. Conf.,* 91–95.

Anderson, J. B., Kurtz, D. D., and Weaver, F. M. 1979. Sedimentation on the Antarctic Continental slope. (*In* Pilkey, O., and Doyle, L., *eds.* Continental Slopes. *Spec. Publs. Soc. Econ. Paleont. Miner., Tulsa,* No. 27, 265–83.

Anderson, J. B., Kurtz, D. D., Domack, E. W., and Balshaw, K. M. 1980. Glacial and glacial marine sediments of the Antarctic Continental Shelf. *J. Geol., 88,* No. 4, 399–414.

Boulton, G. S. 1971. Till genesis and fabric in Svalbard, Spitsbergen. (*In* Goldthwait, R. P., *ed. Till.* Columbus, Ohio State University Press, 41–72.)

Boulton, G. S. 1972. Role of thermal regime in glacial sedimentation. *Spec. Publs. Inst. Br. Geogr.,* No. 4, 1–19.

Chriss, T., and Frakes, L. A. 1972. Glacial marine sedimentation in the Ross Sea. (*In* Adie, R. J., *ed. Antarctic Geology and Geophysics.* Oslo, Universitetsforlaget, 747–62.)

Fillon, R. H. 1974. Late Cenozoic foraminiferal paleoecology of the Ross Sea, Antarctica. *Micropaleontology, 20,* No. 2, 129–51.

Foldvik, A., and Kvinge, T. 1974. Bottom currents in the Weddell Sea. *Geophys. Inst. Univ. Bergen Rep.,* No. 37, 38 pp.

Foster, T. D., and Carmack, E. C., 1976. Frontal zone mixing and Antarctic Bottom Water formation in the southern Weddell Sea. *Deep Sea Res., 23,* 301–17.

Gill, A. E. 1973. Circulation and bottom water production in the Weddell Sea. *Deep Sea Res., 20,* 111–40.

Harland, W. B., Herod, K. N., and Krinsley, D. H. 1966. The definition and identification of tills and tillites. *Earth Sci. Rev., 2,* 225–56.

Hollister, C. D., and Elder, R. B. 1969. Contour currents in the Weddell Sea. *Deep Sea Res., 16,* 99–102.

Jacobs, S. S., Amos, A. F., and Bruchhausen, P. M. 1970. Ross Sea oceanography and Antarctic Bottom Water formation. *Deep Sea Res., 17,* 935–62.

Kurtz, D. D., and Anderson, J. B. 1979. Recognition and sedimentologic description of recent debris flow deposits from the Ross and Weddell Seas, Antarctica. *J. sedim. Petrol., 49,* No. 4, 1159–70.

Ovenshine, A. T. 1970. Observations of iceberg rafting in Glacier Bay, Alaska, and the identification of ancient ice-rafted deposits. *Geol. Soc. Am. Bull., 81,* No. 3, 891–94.

Till, R. 1974. *Statistical Methods for the Earth Scientist (an Introduction).* New York, John Wiley, 154 pp.

130

Antarctic Glacial History Using Spatial and Temporal Variations of Ice-Rafted Debris in Abyssal Sediments of the Southern Ocean

N.D. WATKINS†, M. T. LEDBETTER*, and T. C. HUANG
Graduate School of Oceanography, University of Rhode Island, Kingston, Rhode Island 02881, U.S.A.

Abstract. Using piston cores collected during cruises of the USNS *Eltanin,* we have attempted to use ice-rafted debris (IRD) deposited on the floor of the Southern Ocean as an index of long-term Antarctic glacial history. A series of south-to-north core traverses between lat. 65° and 40° S in the southcentral Pacific have been examined and compared to an earlier study in the southeast Indian Ocean. The sedimentation rate of all cores was determined using micropaleontologic and paleomagnetic methods. The IRD between 62 and 150 μm was separated and counted, so that the apparent accumulation rate of IRD in mg/1000 yr/cm^2 could be measured at 10 cm core intervals for layers deposited during the Brunhes Epoch (0.73 m.y. ago to present).

The latitudinal gradient of decreasing IRD accumulation rates with increasing distance from Antarctica expected from the earlier study has been obscured in this region by active bottom currents which produced lag deposits of IRD that are not true indices of Antarctic glacial activity. We have corrected for the effect of lag deposits of IRD by using the accumulation rate of manganese micronodules as an index of bottom-current winnowing capacity. The latitudinal distribution of the resulting primary IRD shows a general decrease in IRD accumulation rates with distance from the source, but upon this is superimposed a maximum that corresponds to the Antarctic Convergence. The northern limit of icebergs—as reconstructed from the zone of IRD deposition—was, on the average, no greater during the past 0.74 m.y. than at present.

Increases in the accumulation rate of ice-rafted debris (IRD) in the Southern Ocean have generally been interpreted as an index of Antarctic glacial activity (Conolly and Ewing, 1965; Opdyke and others, 1966; Goodell and others, 1968; Margolis and Kennett, 1971). We have recently presented a model (Watkins and others, 1974; Keany and others, 1976) which suggests that deposition of IRD in the southeastern Indian Ocean is diachronous from south to north, with IRD being preferentially deposited in higher latitudes during interglacials and in lower latitudes during glacial periods (Figure 130.1). In an attempt to expand our study into the southeastern Pacific Ocean (Figure 130.2 and Table 130.1), we encountered complexities in the expected latitudinal distribution of IRD, including an observation that IRD maxima are commonly associated with manganese micronodule (MMN) maxima which, in turn, reflect increased bottom-current velocity (Goodell and others, 1971; Watkins and Kennett, 1971, 1972, 1977; Kennett and Watkins, 1975, 1976). Kennett and Brunner (1973) have mentioned that the dual occurrence of IRD and MMN is possible evidence for synchronous increases in bottom-water activity and in Antarctic glaciation. The increased bottom-current velocity, however, removes finer, lighter material and results in residual deposits of the more dense IRD. Therefore, these residual deposits are not solely a function of Antarctic glacial activity. This process was involved to explain apparently anomalous accumulation rates of IRD in two cores in the Indian Ocean study (Keany and others, 1976), as shown in Figure 130.3. In this paper, we examine the possibility that the accumulation rate of MMN may be used to calculate that portion of the total IRD accumulation rate which is the result of bottom-current winnowing, so that the primary IRD accumulation rate, representing true iceberg melting, can be obtained.

† Deceased.

* Present address: Department of Georgia, University of Georgia, Athens, Georgia 30602, U.S.A.

Methods and data analyses

Twelve cores from the *Eltanin* core collection have been selected from the Bellingshausen Basin in the southeastern Pacific Ocean (Figure 130.2). The cores were selected on the basis of age (as determined from paleomagnetic polarity), location, and length, as well as minimal distortion and turbidite content. The sediment types in the cores range from sandy, radiolarian-rich diatom oozes to sandy foraminiferal oozes. The general micropaleontology, sedimentology, age ranges, and sedimentation rates, and the methods employed in their analyses, are included in a large-scale study of the area (Hays, 1965; Goodell and Watkins, 1968; Huang and Watkins, 1977). Samples 6 cm^3 in volume were taken at 10 cm intervals in all cores. The coarse fraction (>62 μm) was retained after wet-sieving for sediment analysis. Composition of this coarse fraction was determined for the size range 62 to 250 μm. Fractions >250 μm were not measured in this study, since the few very large grains of ice-rafted debris that are found in deep-sea cores would tend to distort the within-core variation of any parameter derived from the mass of the debris. The lower size limit was subjectively chosen in order to minimize the inclusion of atmospherically transported terrigenous material or material that might have been sub-

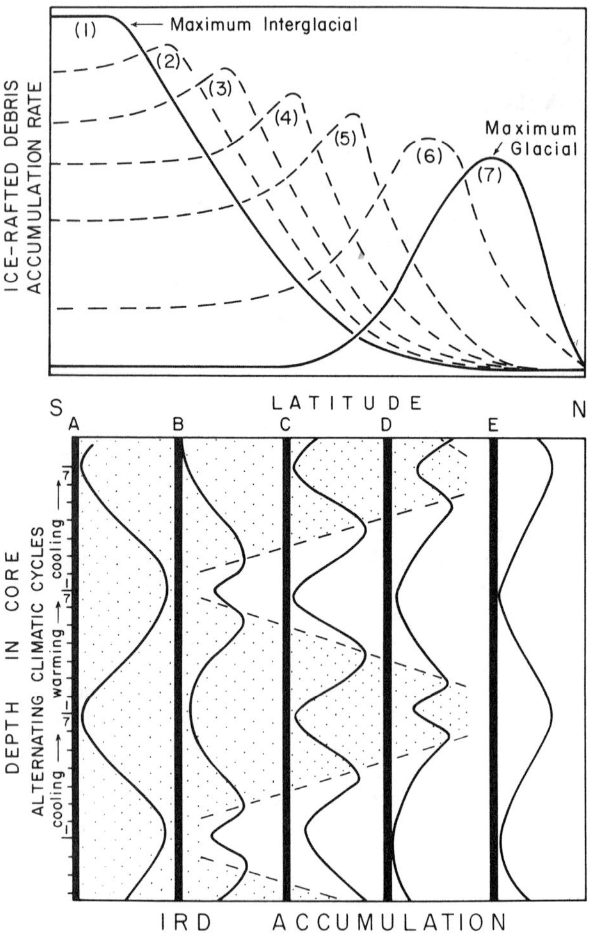

Figure 130.1. Upper diagram: extension of a simple model relating debris accumulation rate to latitude during glacial and interglacial periods (Watkins and others, 1974) to include periods of transition (curves 2 to 6) between extreme glacial (7) and interglacial (1) periods. Curves are spaced at arbitrary intervals and represent distribution of ice-rafted debris at different times (1 to 7). Lower diagram: adaptation of the seven curves in the upper diagram into debris accumulation rates (IRD) that would occur at five separate latitudes (A to E), assuming oscillations between glacial (7) and interglacial (1) conditions, without any sustained glacial or interglacial periods but rather simple reversals of climatic trend. Sequence 1 to 7 indicated on left corresponds to curves in upper diagram. Dotted areas represent those periods during which strong positive correlation can be expected between temperature and debris accumulation rate; unpatterned area represents those periods when correlation would be negative. Note increasing bifurcation of debris maxima in core A with decreasing latitude and virtual impossibility of establishing simple correlations between each core (from Keany and others, 1976).

Table 130.1. Core location, water depth, and core lengths

Core number	Lat.	Long.	Water depth (m)	Length (m)
E17-18	67°54′S	110°46′W	4120	6.5
E11-12	65°52′S	115°05′W	4760	18.8
E11-11	64°51′S	114°28′W	4900	12.1
E11-9	62°50′S	115°04′W	5055	17.0
E11-7	60°55′S	114°47′W	5075	9.7
E11-5	58°57′S	114°43′W	5110	11.4
E11-4	57°50′S	115°12.5′W	4815	11.2
E20-14	53°41′S	102°56′W	4260	11.6
E17-8	62°01′S	140°17′W	4590	9.9
E15-28	56°01′S	149°49′W	3360	12.6
E20-3	51°07′S	145°03′W	4130	8.1
E20-2	49°00′S	144°50′W	4560	8.6

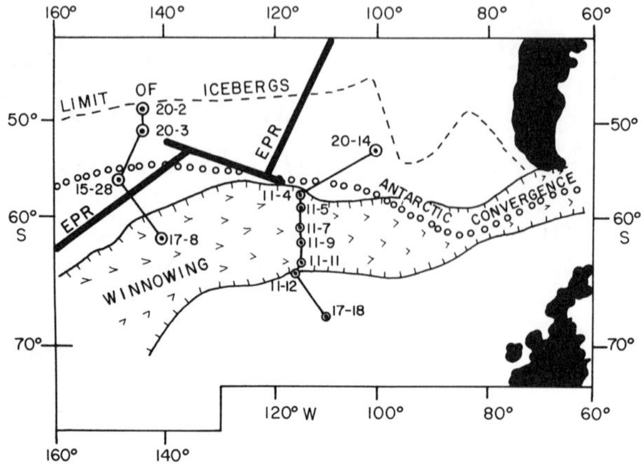

Figure 130.2. Map showing location of *Eltanin* cores used in this study. (For scale, 10° of latitude equal 1080 km.) The winnowing zone is taken from Hollister and Heezen (1967) who have described it on the basis of bottom photographs as an "erosional zone." Our data suggest it is the site of relatively high bottom-current velocities, which have caused the selective removal of finer, lighter sedimentary components, leaving a coarse lag deposit. The East Pacific Rise (EPR) is shown by a heavy line and is offset by the *Eltanin Fracture Zone.* The Antarctic Convergence (open circles) and the present limit of icebergs (dashed line) are taken from Goodell (1973). Exact locations, water depths, and lengths of cores are given in Table 130.1.

jected to abrasion to the extent that diagnosis of the ice-rafted debris would become difficult. We believe that both upper and lower limits on the size range of the debris are essential in order to recover signals reflecting net long-term fluctuations in the passage of melting icebergs. The debris was identified by morphology, composition, and the surface texture criteria of Margolis and Krinsley (1974). After the classification of debris grains in a sample, the total mass of the debris was calculated and used to compute the apparent ice-rafted debris accumulation rate (for a complete discussion of methods, see Keany and others, 1976).

The total accumulation rate of IRD is the sum of the lag deposits associated with increased bottom-current velocity and the real IRD deposited by the melting of passing icebergs. In an earlier study (Kennett and Brunner, 1973) the surface features on individual grains were examined using a scanning electron microscope in order to separate the two fractions. Reliable physical separation of the two components, however, is subject to many difficulties. We have therefore attempted to use the relationship between MMN and the total IRD to infer the lag deposit fraction of the total IRD so that the latitudinal distribution of primary IRD may be examined. A polynomial regression between IRD and MMN accumulation rates for the Brunhes age section of each core revealed that a linear regression best fit the data. Therefore, we have assumed that an increase in MMN proportion is matched by a linear increase in IRD due to the inclusion of residual IRD In the total IRD measured. The mean accumulation rate of primary IRD (IRD_{Real}) may be determined using the linear regression at the zero value of MMN which, in principle, represents the condition of minimum bottom-water activity. Samples with zero IRD were not used in the regression because of our interest in removing the effect of IRD lag deposits, which are formed only during periods of increased bottom-current velocity (i.e., when MMN were being formed).

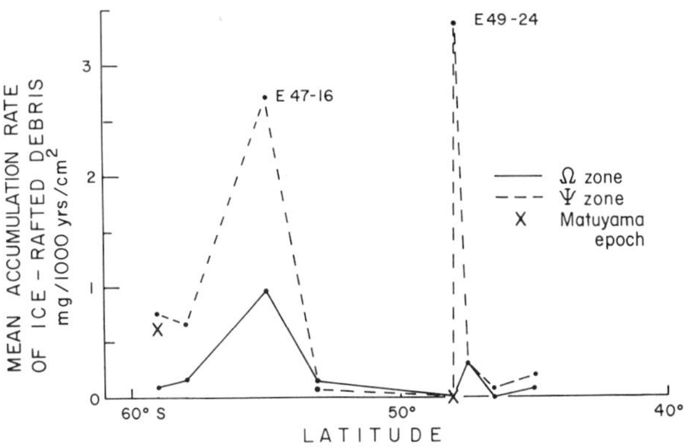

Figure 130.3. The mean accumulation rate of ice-rafted debris is plotted versus latitude for three periods for cores used in an earlier study of the southeastern Indian Ocean (Keany and others, 1976). The expected systematic decrease in accumulation rate with increasing distance from the source is obscured by anomalously high apparent rates in two cores. This was interpreted as a result of intense bottom-current winnowing, which has left a residual deposit of heavier ice-rafted debris (Keany and others, 1976). Radiolarian zones Ω and Ψ are from Hays (1965), and the Matuyama paleomagnetic epoch is from Keany and others (1976).

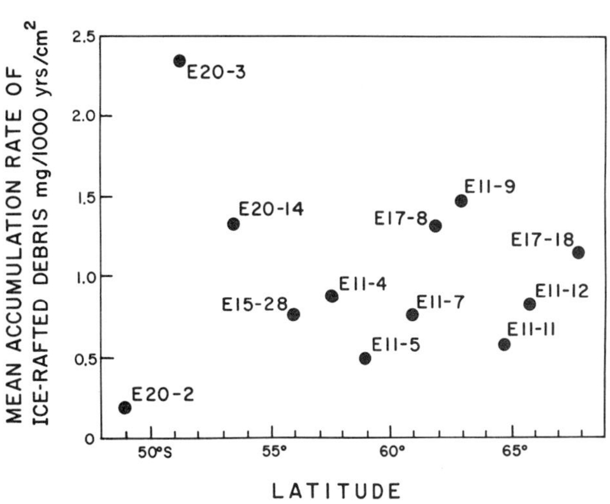

Figure 130.4. The mean accumulation rate during the Brunhes epoch of observed ice-rafted debris as a function of site latitude. No systematic relationship between accumulation rate and distance from the source is apparent. See Figure 130.2 and Table 130.1 for exact core locations.

Results

The mean total IRD accumulation rate was calculated for the Brunhes (0–0.73 m.y. old) section of each core and is shown as a function of latitude in Figure 130.4. The linear regression between total IRD and MMN was calculated for the Brunhes age section of each core. The intercept of the regression was then used as the mean accumulation rate of primary IRD during periods of no bottom-current-produced lag deposits of IRD. This empirically derived mean "primary" IRD accumulation rate (IRD_{Real}) has been plotted with latitude in Figure 130.5.

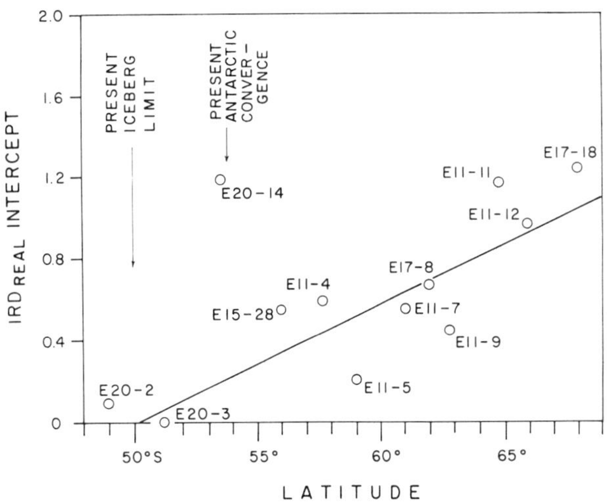

Figure 130.5. The intercept of the regression between ice-rafted debris (IRD) accumulation rates and manganese micronodules (see text) for each core plotted versus latitude. The intercept is the mean accumulation rate during the Brunhes epoch of primary IRD (IRD_{Real}) during periods when manganese micronodule production (and by inference, residual deposits of IRD) is zero. The least squares fit (solid line) shows a systematic decrease in IRD_{Real} with increasing distance from the source in Antarctica (compare with Figure 130.4). The position of the Antarctic Convergence and the present limit of icebergs (Figure 130.2) is taken from Goodell (1973). The latitudinal relationship suggests that the northern limit of icebergs for the last 730,000 years is not significantly different from that of the present and that the position of the Antarctic Convergence (where high melting rates would occur) is similarly unchanged.

Discussion

In Figure 130.4 we have shown that the total IRD has no systematic latitudinal variation. In Figure 130.5 we have plotted the same data for each core with the lag IRD removed by using the linear regression model. A systematic decrease in IRD with distance from the source had been expected in the previous study (Keany and others, 1976, Figure 3), but the relationship was apparently obscured by anomalously large apparent accumulation rates of IRD in areas of high bottom-current activity. As Figure 130.5 clearly shows, our technique has substantially suppressed the effects of winnowing on the latitude dependence of IRD, so that a systematic decrease in apparent accumulation rate with increased distance from the source materializes. The linear regression between IRD_{Real} and latitude (Figure 130.5) predicts a zero accumulation rate of IRD_{Real} very close to the position of the present average northern limit of icebergs in the South Pacific Ocean (Goodell, 1973). This relationship suggests that the northern limit of iceberg melting has not changed appreciably during the last 730,000 years. An anomalously high apparent accumulation rate of primary ice-rafted debris occurs at the site of core E20-14, which is located at the present Antarctic Convergence. This may be explained by an accelerated melting rate of debris-laden icebergs at their first encounter with warmer water north of the convergence.

Conclusions

Our earlier study of a series of *Eltanin* cores from the southeast Indian Ocean combined with the present study of 12 cores from the South Pacific Ocean has provided the following conclusions.

1. Increases in IRD cannot be simply associated with increased glacial activity on Antarctica. The origin of IRD maxima is far more complex than previously described. In-

creases in IRD may also be the results of increases in bottom-current velocity which actively winnow the lighter sedimentary components, leaving a residual lag deposit. A linear regression between the accumulation rates of MMN and IRD in a core may be used to derive a mean accumulation rate of primary IRD at that site. We have demonstrated that a change from erratic to systematic latitudinal variation of IRD with distance from source materializes if this technique is applied.

2. The northern limit of IRD accumulation for the Brunhes epoch is the same as for the present, which suggests that the factors controlling the melting rate of debris-laden icebergs have not drastically changed during the last 730,000 years. Similarly, the position of the Antarctic Convergence, at which icebergs experience an increase in melting rates, has been unchanged on the average during the same time interval.

Acknowledgments. We would like to thank R. Dow and G. Jones for their conscientious help in the sedimentology laboratory. B. B. Ellwood provided valuable computer programming assistance. Work supported by National Science Foundation Grants OPP75-19222-AO1 and DPP78-08511 from the Office of Polar Programs.

References cited

Conolly, J., and Ewing, M. 1965. Ice-rafted detritus as a climatic indicator in Antarctic deep-sea cores. *Nature, Lond., 238,* 40–42.

Goodell, H. G. 1973. The sediments. (*In* Goodell, H. G., Houtz, R., Ewing, M., and others, *eds.* Marine Sediments of the Southern Oceans. *Antarct. Map Folio Ser.,* Folio 17, 1-9.)

Goodell, H. G., and Watkins, N. D. 1968. Paleomagnetic stratigraphy of the Southern Ocean: 20°W to 160°E longitude. *Deep Sea Res., 15,* 89–112.

Goodell, H. G., Meylan, M. A., and Grant, B. 1971. Ferromanganese deposits of the South Pacific Ocean, Drake Passage, and Scotia Sea. (*In* Reid, J. L., *ed. Antarctic Oceanology 1.* Washington, American Geophysical Union, 27–92.) [Antarctic Research Series, Vol. 15.]

Goodell, H. G., Watkins, N. D., Mather, T. T., and Koster, S. 1968. The Antarctic glacial history recorded in sediments of the Southern Ocean. *Paleogeogr., Paleoclim., Paleoecol., 5,* 41–62.

Hays, J. D. 1965. Radiolaria and late Tertiary and Quaternary history of Antarctic seas. (*In* Llano, G. A., *ed. Biology of the Antarctic Seas 2.* Washington, American Geophysical Union, 125–84.) [Antarctic Research Series, Vol. 15.]

Hollister, C. D., and Heezen, B. D. 1967. The floor of the Bellingshausen Sea. (*In* Hersey, J. B., *ed. Deep-Sea Photography.* Baltimore, Johns Hopkins Press, 177–89.) [Johns Hopkins Oceanographic studies 3.]

Huang, T. C., and Watkins, N. D. 1977. Contrasts between the Brunhes and Matuyama sedimentary records of Bottom Water activity in the South Pacific. *Mar. Geol., 23,* 113–32.

Keany, J., Ledbetter, M. T., Watkins, N. D., and Huang, T. C. 1976. Diachronous deposition of ice-rafted debris in sub-Antarctic deep-sea sediments. *Geol. Soc. Am. Bull., 87,* 873–82.

Kennett, J. P., and Brunner, C. A. 1973. Antarctic late Cenozoic glaciation: evidence for initiation of ice-rafting and inferred increased bottom-water activity. *Geol. Soc. Am. Bull., 84,* 2043–53.

Kennett, J. P., and Watkins, N. D., 1975. Deep-sea erosion and manganese nodule development in the southeast Indian Ocean. *Science, N.Y., 188,* 1011–13.

Kennett, J. P., and Watkins, N. D. 1976. Regional deep-sea dynamic processes recorded by late Cenozoic sediments of the southeastern Indian Ocean. *Geol. Soc. Am. Bull., 87,* 321–39.

Margolis, S. V., and Kennett, J. P. 1971. Cenozoic paleoglacial history of Antarctica recorded in sub-Antarctic deep-sea cores. *Am. J. Sci., 271,* 1–36.

Margolis, S. V., and Krinsley, D. H. 1974. Process of formation and environmental occurrence of microfeatures on detrital quartz grains. *Am. J. Sci., 274,* 449–64.

Opdyke, N. D., Glass, B., Hays, J. D., and Foster, J. 1966. Paleomagnetic study of Antarctic deep-sea cores. *Science, N.Y., 154,* 349–57.

Watkins, N. D., and Kennett, J. P. 1971. Antarctic Bottom Water: major change in velocity during the late Cenozoic between Australia and Antarctica. *Science, N.Y., 173,* 813–18.

Watkins, N. D., and Kennett, J. P. 1972. Regional sedimentary disconformities and upper Cenozoic changes in bottom water velocities between Australia and Antarctica. (*In* Hayes, D. E., *ed. Antarctic Oceanology 2.* Washington, American Geophysical Union, 273–94.) [Antarctic Research Series, Vol. 19.]

Watkins, N. D., and Kennett, J. P. 1977. Erosion of deep-sea sediments in the Southern Ocean between longitudes 70°E and 190°E and contrasts in manganese nodule development. *Mar. Geol., 23,* 103–11.

Watkins, N. D., Keany, J., Ledbetter, M. T., and Huang, T. C. 1974. Antarctic glacial history from analysis of ice-rafted deposits in marine sediments: new model and initial tests. *Science, N.Y., 186,* 533–36.

131

Estimates of Antarctic Ocean Seasonal Sea-Ice Cover during Glacial Intervals

DAVID W. COOKE* and JAMES D. HAYS
Department of Geological Sciences and Lamont-Doherty Geological Observatory of Columbia University, Palisades, New York 10964, U.S.A.

Abstract. The past seasonal extent of Antarctic sea ice is estimated through the use of lithologic changes, sedimentation rate changes, and the distribution of ice-rafted detritus in Antarctic deep-sea sediments. Summer sea ice 18,000 years ago was greatly expanded (25×10^6 km^2) compared with today's summer coverage (2.5×10^6 km^2). Winter ice cover 18,000 Years ago (40×10^6 km^2) was probably double present winter ice cover (20×10^6 km^2). Changes in sea-ice cover are reflected in Antarctic deep-sea sediments by changes from relatively slowly accumulating diatomaceous clay with large amounts of volcanic detritus (expanded summer and winter ice cover) to rapidly accumulating diatomaceous ooze with small amounts of ice-rafted detritus (sea-ice conditions like today). Detailed biostratigraphic analysis indicates that the lithologic changes are synchronous throughout the Antarctic. Study of the sedimentary contacts in cores with accumulation rates of ~30 cm/1000 years indicates that seasonal ice cover changes within 200 years from relatively ice-free summers like today to summers characterized by approximately as much ice as today's winter. The disappearance of large amounts of summer sea ice was similarly rapid. The last onset of extensive sea ice began 80,000 years ago and ended 14,000 years ago. These changes in Southern Hemisphere sea ice either just precede or occur at the beginning stages of important changes of Northern Hemisphere ice volume.

The sediments blanketing the floor of the Antarctic Ocean form zonal patterns like the overlying ocean currents. South of lat. 60° S the sediments are dominantly silts and clays derived from Antarctica. Between lat. 60° and 50° S the sediments contain high percentages of siliceous microfossils, dominantly diatoms and Radiolaria, reflecting the high productivity of the surface waters between the Antarctic divergence and the Antarctic Polar Front. In the Atlantic and western Indian Oceans, these diatomaceous oozes are commonly very pure and, when dry, resemble glass wool. The boundary between the terrigenous sediments and the siliceous ooze is at about lat. 60° S, approximately the mean position of Antarctic sea ice (Hays, Lozano, and others, 1976), while the northern boundary of siliceous ooze closely follows the Antarctic Polar Front.

North of the polar front, deeper areas of the ocean floor are covered with clay, while shallower areas (<4000 m) accumulate calcareous ooze. These patterns were first noted by H.M.S. *Challenger* scientists (Murray and Renard, 1891), and they have been confirmed and the boundaries refined by subsequent workers (Philippi, 1912; Schott, 1939; Hough, 1956; Lisitzin, 1960; Goodell, 1973; Lozano and Hays, 1976).

Philippi (1912) was the first to observe that these sedimentary patterns change with time. He found calcareous ooze overlying siliceous ooze in cores from north of the polar front and siliceous ooze overlying "glacial marine" sediments in cores from south of the polar front. He interpreted this apparent northward displacement of the sediment zones to have been influenced by the glacial period. Schott (1939) argued that the subantarctic upward change from diatom ooze to Globigerina ooze was due to a southward movement of the polar front, while the upward change from the glacial marine to diatom ooze was due to a decrease in iceberg rafting of terrigenous material from Antarctica. Hays, Lozano, and others (1976) were able to show through new biostratigraphic techniques that the uppermost diatomaceous ooze layer comprises the Holocene sequence, while the underlying diatomaceous clay (glacial marine sediments of Philippi) was deposited during oxygen isotope stages 2, 3, 4, and 5a, or 2, 3, 4, and the upper two thirds of 5 (late Pleistocene time). They also showed that the diatomaceous clays accumulated much more slowly than the diatomaceous oozes. They argued that the diatomaceous clays were deposited during periods of greater sea-ice cover that inhibited the production of diatoms.

Through the use of improved stratigraphic (Hays, Lozano, and others, 1976) and chronostratigraphic techniques (Shackleton and Opdyke, 1973; Hays, Imbrie, and Shackleton, 1976), we here place these Antarctic lithologic changes in a time-stratigraphic framework. We will test their regional synchroneity and compare their timing with Northern Hemisphere ice-volume changes.

Stratigraphy

Shackleton and Opdyke (1973, 1976) have shown that late Pleistocene $\delta^{18}O$ changes in the shells of foraminifera preserved in deep-sea sediments are globally synchronous because these changes are caused by variations in global ice volume. In cores from south of the Antarctic Polar Front, continuous records containing $CaCO_3$ are rare; thus useful oxygen isotope data are not available.

* Present address: U.S. Geological Survey, P.O. Box 7944, Metairie, Louisiana 70010, U.S.A.

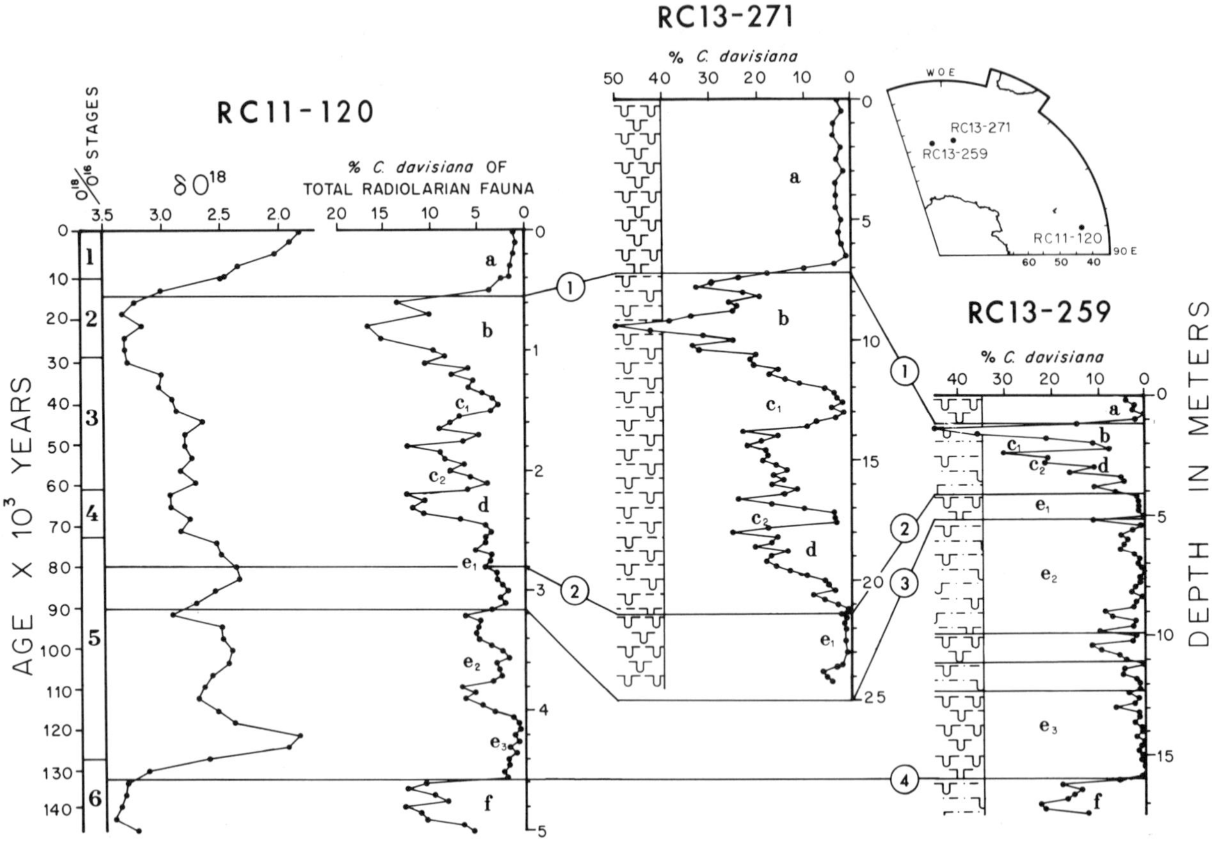

Figure 131.1. Left-hand diagram shows correlation of stratigraphic units determined by the oxygen isotope method and *Cycladophora davisiana* from the top 500 cm of core RC11-120. Oxygen isotopic composition is expressed as a deviation per mil from the PDB standard (modified from Hays, Lozano, and others, 1976). Middle and right-hand diagrams show correlation of *C. davisiana* abundance and lithostratigraphy of cores from south of the Antarctic Polar Front, RC13-271 and RC13-259, with core RC11-120 from north of the Antarctic Polar Front. Chronology of RC11-120 after Hays, Imbrie, and Shackleton (1976). From these correlations four lithologic changes are dated (see Table 131.1). For location of cores, see Table 131.4.

Hays, Lozano, and others (1976), have shown that one radiolarian species, *Cycladophora davisiana,* undergoes substantial relative abundance fluctuations in Antarctic and subantarctic sediments. They have shown—by comparing the oxygen isotope record with the record of *C. davisiana*—that the faunal abundance changes are synchronous over a broad area of the subantarctic. In Figure 131.1, it can be seen that the variations in $\delta^{18}O$ and relative *C. davisiana* abundances are nearly parallel, and thus the record of changes in *C. davisiana* abundance contains as much stratigraphic information as does the $\delta^{18}O$ curve over the last 140,000 years. This correlation has been extended further back in time by Hays, Imbrie, and Shackleton (1976).

Although the amplitude of the *C. davisiana* signal varies regionally, generally being maximal south of lat. 50° S, the shape of the curves are similar throughout the Antarctic and subantarctic, providing a high resolution and probably synchronous stratigraphy for this region.

Lithologic changes

In general, the sediments south of the polar front in the Atlantic and Indian Oceans are interbedded sequences of diatomaceous ooze and diatomaceous clay, the latter containing more silt and sand in the Atlantic and the southwest Indian Ocean. The contacts between these sediment types are sharp (usually within 5 cm) except where disturbed by bioturbation. In general, Pacific sector cores do not show these lithologic changes; however, one core from this sector (E17-9) does.

North of the polar front, cores taken within a few degrees of this important boundary and above the calcium carbonate compensation depth show a different sequence of sediment types. An upper layer of foraminiferal ooze or foraminiferal clay grades downward through diatomaceous ooze to diatomaceous clay containing both open marine and near-ice diatoms (L. H. Burckle, personal communication, 1977). This compositional change, from carbonate-rich to silica-rich sediments, appears to be synchronous with the boundary between overlying diatomaceous ooze and underlying diatomaceous clay in cores from south of the polar front at the transition between *C. davisiana* zones a and b.

The diatomaceous ooze south of the polar front consists primarily of open-marine diatoms (Burckle, 1972), with lesser numbers of polar Radiolaria (Lozano and Hays, 1976). In general, the carbonate content is less than 10% (Hays, Lozano, and others, 1976; Morley, 1977), and the nonbiogenic coarse fraction is negligible.

By contrast, the diatomaceous clay is usually gray or olive gray and smooth-textured; it contains open-marine and near-ice diatoms (Burckle, 1972; Cooke and others, 1977) and polar Radiolaria (Lozano and Hays, 1976). In

Table 131.1. Dating lithologic changes

Lithologic change number	Core number	Depth to lithologic change (cm)	*C. davisiana* zone	Depth of correlation in RC11-120 (cm)	$\delta^{18}O$ stage	Estimated age of lithologic change (years)
1	RC13-271	720	a/b boundary	55	2	14,000
2	RC13-271	2140	Top of e_1	280	5	80,000
3	RC13-259	515	Base of e_1	315	5	90,000
4	RC13-259	1595	Base of e_3	456	6	132,000

general, the carbonate content is negligible, and the terrigenous coarse fraction is as high as 25% by weight in the Atlantic sector but decreases toward the east. This terrigenous component—concentrated in and disseminated through the silty diatomaceous clay—is primarily volcanogenic and is found in cores between the South Sandwich Islands and Kerguelen Island. The volcanic particles consist of brownish glass shards, many of which contain inclusions of plagioclase and pyroxene, and vesicular rock fragments. They are similar to particles within ash layers, and volcanic particles like these are disseminated throughout cores just east of the South Sandwich Islands, as described by Ninkovich and others (1964) and Federman (1977). This similarity suggests that these islands are the primary source for the detritus. Most of the islands have been in some stage of activity throughout historical time. The oldest rocks are about 4 m.y. old (Baker, 1968, 1972), and geophysical data suggest that the island arc is about 8 m.y. old (Barker, 1972).

This material differs from airborne volcanic detritus found in deep-sea sediments in other areas because (1) the volcanic component of the diatomaceous clay is not in discrete layers but is disseminated throughout the unit; (2) the particle size ranges from fine-grained silt to pebbles several millimeters in diameter; and (3) the volcanic material is found as far as 5000 km from its source, while airborne volcanic material is ordinarily not found more than about 1000 km from the source (Ninkovich, 1968). The mode of transport of the volcanic material is probably a combination of air transport and ice-rafting by sea ice and small bergs.

Dating of lithologic changes

Correlation of the *Cycladophora davisiana* stratigraphies of Antarctic cores with the *C. davisiana* record of core RC11-120 permits the dating of Antarctic lithologic changes (Figure 131.1). To make these correlations as precise as possible, we have chosen two Antarctic cores with high sedimentation rates through the intervals to be correlated with RC11-120 (RC13-271 *C. davisiana* zones a, b, c, d, and most of e_1, and RC13-259 *C. davisiana* zones e_1, e_2, e_3, and f). To facilitate the comparison with RC11-120, we have sampled this core at 5 cm intervals to compare the *C. davisiana* pattern with the higher sedimentation rate cores RC13-271 and RC13-259, sampled at 20 cm intervals.

In Figure 131.1, we have numbered the contacts of four lithologic changes recognizable in numerous cores south of the polar front; the intersection of these contacts with the *C. davisiana* curve permits correlation with RC11-120 and estimates of their ages (Table 131.1). There are three lithologic changes within the e_2–e_3 interval of core RC13-259 that have not been numbered or dated because the sedimentation rate in this interval is so high that comparison with RC11-120 is difficult.

Sedimentation rates

Generalizations about regional sedimentation rates through time must be based on a large number of cores since local factors strongly influence the sedimentation rates of individual cores. We have examined 37 cores in the Atlantic and western Indian Ocean sectors of the Antarctic and subantarctic that penetrate to *C. davisiana* zone e_1, and from these observations we can make the following generalizations (Table 131.2). In cores south of the polar front, the sedimentation rates of the diatom oozes are greater than 10 cm/1000 years (e.g., *C. davisiana* zone a). The diatomaceous clays, on the other hand, have accumulation rates of about 4 cm/1000 years (e.g., *C. davisiana* zones b, c, and d).

In cores a few degrees north of the polar front the pattern is the opposite: low sedimentation rates are characteristic of *C. davisiana* zone a, and high sedimentation rates are characteristic of *C. davisiana* zones b, c, and d.

In the southeast Indian and Pacific Ocean sectors, on the other hand, average sedimentation rates are rarely more than 2 cm/1000 years, and gross changes in sedimentation rate have not been observed.

Since individual lithologic changes occur at specific segments on all Antarctic *C. davisiana* curves (see Figures 131.1 and 131.2), there is good evidence that these lithologic changes are synchronous or nearly synchronous throughout the Antarctic. In Figure 131.2, we have assumed their synchroneity and plotted these changes as time planes. In these six cores we have determined the sedimentation rates between lithologic changes. They further support the general pattern of sedimentation rates described above (Table 131.3).

Discussion

The lithologic changes with their accompanying sedimentation rate changes all suggest that conditions in the Antarctic during *C. davisiana* zones b, c, and d times ($\delta^{18}O$

Table 131.2. Sedimentation rates

Location	Core number	Sedimentation rates, zone *a* (cm/1000 yr)	Sedimentation rates, zones *b*, *c*, and *d* (cm/1000 yr)
Atlantic Ocean			
South of polar front	RC13-263	10.7	4.6
Near polar front	RC13-255	8.0–40.0	54.7
North of polar front	RC15-94	2.7	6.8
Southwest Indian Ocean			
South of polar front	RC17-61	12.3	8.2
Near polar front	V29-86	4.6	13.0
North of polar front	RC8-39	2.3	6.9

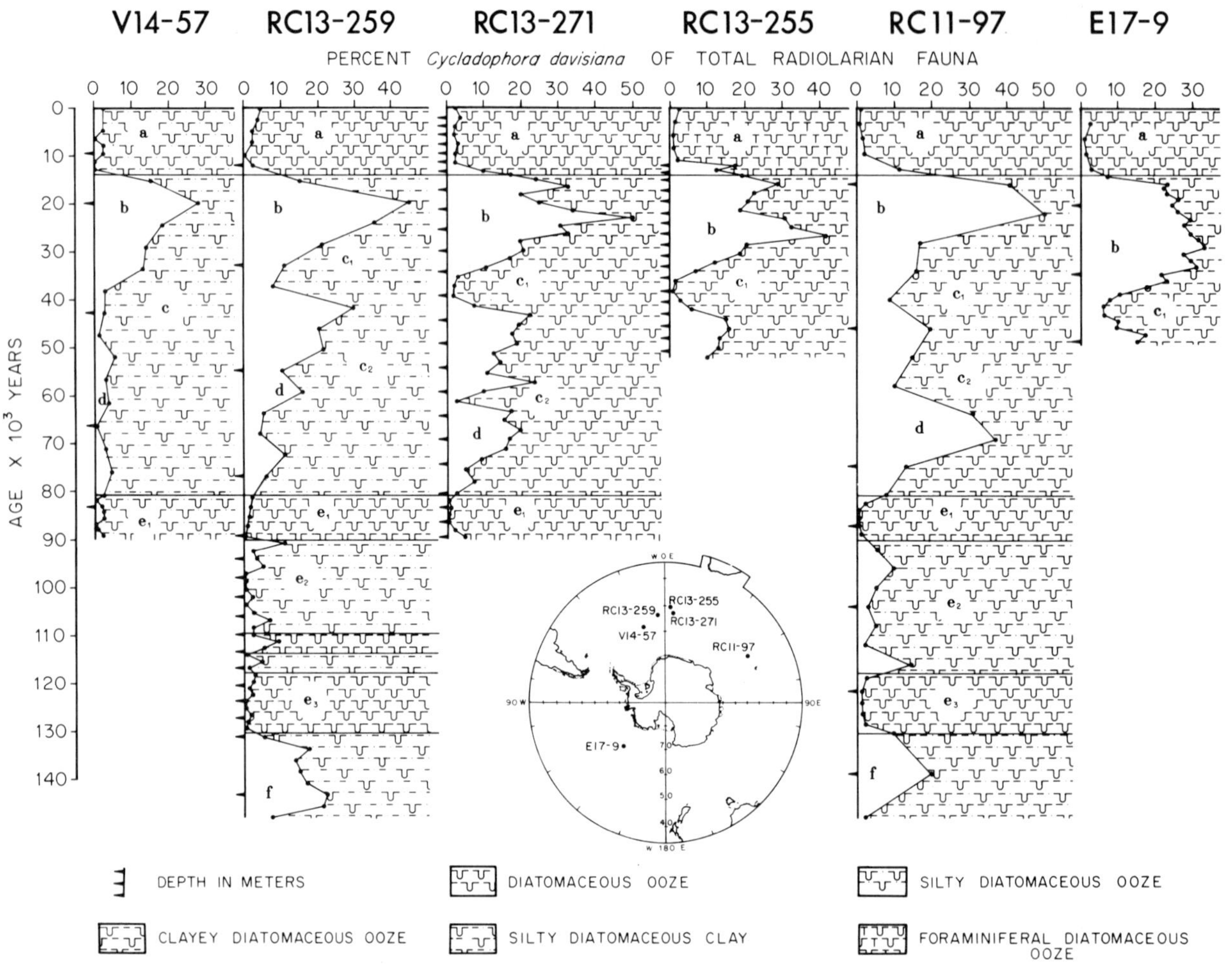

Figure 131.2. Correlation of biostratigraphic and lithostratigraphic units for six cores with variable sedimentation rates from the Antarctic Ocean. The four lithologic changes are drawn synchronous between cores while the *C. davisiana* units are defined assuming uniform sedimentation rates between time planes (see Table 131.3). Chronology after Hays, Imbrie, and Shackleton (1976).

stages 2, 3, and 4) were markedly different from those of today. Hays, Lozano, and others (1976) have suggested that these lithologic changes reflect changes in sea-ice cover, arguing that the diatomaceous oozes represent times like today when the Antarctic Ocean was relatively ice free in summer and that the diatomaceous clays with their lower sedimentation rates represent times when the Antarctic Ocean was covered or partially covered by sea ice through the summer, limiting diatom productivity and resulting in lowered accumulation rates.

This study provides many additional data beyond those of earlier work, including stratigraphic and lithologic analyses of 67 cores (Table 131.4), as well as detailed analyses of the ice-rafting component.

The presence of sea ice through much of the year during the deposition of the diatomaceous clay is supported by the additional data and by the following arguments.

1. The lithologic changes are synchronous over a huge area which argues against only local changes in the sedimentary regime.
2. Lower sedimentation rates in the diatomaceous clay, in spite of larger amounts of ice-rafted detritus, indicate reduced diatom and probably radiolarian productivity. Sea-ice cover, although not the only possible inhibiting factor, is a likely one in these high southern latitudes. To significantly inhibit diatom production, large amounts of sea ice must be present during the critical spring, summer, and fall months. It might be argued that the lithologic changes reflect a change in the bottom-water regime during glacial times (*C. davisiana*

Table 131.3. Sedimentation rates of cores shown in Figure 131.2

Core	Depth to lithologic change 1 (cm)	Sedimentation rate: top to contact 1 (cm/1000 yr)	Depth to lithologic change 2 (cm)	Sedimentation rate: contacts 1–2 (cm/1000 yr)	Depth to lithologic change 3 (cm)	Sedimentation rate: contacts 2–3 (cm/1000 yr)	Depth to lithologic change 4 (cm)	Sedimentati rate: contacts 3– (cm/1000 yr
V14-57	150	10.7	460	4.7	—	—	—	—
RC13-259	120	8.6	410	4.4	515	10.5	1595	25.7
RC13-271	720	51.4	2140	21.5	—	—	—	—
RC13-255	210	15.0	—	—	—	—	—	—
RC11-97	95	6.8	320	3.4	430	11.0	680	6.0
E17-9	45	3.2	—	—	—	—	—	—

Table 131.4. Location, depth of water, and depth in core of the 18,000 year level for 67 cores

Core number	Latitude (S)	Longitude	Water depth (m)	18,000 year level depth in core (cm)
E11-2	56°04.4′	115°05.6′W	3108	140
E11-3	56°54.2′	115°14.6′W	4023	60
E15-4	59°01.0′	99°45.5′W	4910	60
E17-9	63°05.0′	135°07.0′W	2651	160
E39-21[a]	48°51.7′	126°01.0′E	4078	80
E45-63	53°26.2′	114°15.4′E	3917	140
E45-64	52°29.0′	114°05.4′E	3822	140
E49-6	51°00.4′	109°59.3′E	3327	100
E49-7	53°02.2′	110°02.8′E	3592	60
E49-8	55°04.2′	110°01.1′E	3692	100
E49-10	59°00.8′	110°08.0′E	4371	60
E49-23	47°07.7′	95°04.8′E	3257	40
E49-25	49°22.7′	94°49.9′E	3336	80
E49-29	57°05.7′	94°57.3′E	4237	40
E49-33	57°45.8′	100°02.5′E	4040	140
E50-8	50°56.0′	104°54.5′E	3226	160
E50-11	55°56.7′	104°56.7′E	3923	100
E50-13	59°59.8′	105°00.0′E	4210	40
E50-17	63°00.1′	120°03.0′E	4082	90
RC8-39	42°53.0′	42°21.0′E	4330	90
RC8-43	48°41.0′	57°22.0′E	4319	40
RC8-61	46°32.0′	125°34.0′E	42.54	50
RC8-63	51°05′	129°58′ E	3442	20
RC9-139	47°46.2′	123°06.2′E	4158	50
RC11-76	54°23′	22°08′ W	5229	400
RC11-77	53°03′	16°27′ W	4098	200
RC11-78	50°52′	09°52′ W	3115	180
RC11-80	46°45′	00°03′ W	3656	80
RC11-83	41°36′	09°43′ E	4718	540
RC11-91	56°34′	34°11′ E	5373	20
RC11-94	54°29′	53°03′ E	4303	480
RC11-96	50°28′	59°35′ E	4839	100
RC11-97	50°19′	61°12′ E	4638	120
RC11-120[a]	43°31′	79°52′ E	3193	80
RC12-225	53°39.5′	123°08.0′W	2964	40
RC12-267	38°41′	25°47′ W	4144	60
RC12-289	47°54.2′	23°41.5′W	4484	60
RC13-243	36°54.0′	01°19.8′E	4790	40
RC13-251	42°30.7′	11°40.3′E	4341	20
RC13-254	48°34.2′	05°07.6′E	3636	280
RC13-255	50°34.5′	02°53.7′E	3332	810
RC13-256	53°10.9′	00°21.3′W	2525	400
RC13-257	55°00.2′	03°00.1′W	2836	20
RC13-259	53°52.8′	04°55.8′W	2677	140
RC13-261	56°07.3′	08°41.2′W	4221	80
RC13-263	53°48.3′	08°13.0′W	3389	180
RC13-269	52°37.6′	00°07.5′E	2591	200
RC13-271	51°59.1′	04°31.3′E	3634	950
RC13-275	50°43.0′	13°26.0′E	1984	60
RC15-93	46°05.8′	13°13.5′W	2714	100
RC15-94	42°53.9′	20°51.3′W	37.62	110
RC15-98	42°56.2′	29°46.6′W	4416	400
RC17-61	52°12.2′	54°27.7′E	3947	260
V14-57	57°33.5′	17°06.0′W	4978	200
V16-65	45°00′	45°46′ E	1618	40
V16-114	49°36′	138°13′ E	3116	50
V16-115	55°41′	141°17′ E	3147	100
V18-35	36°24′	41°07′ W	49.42	120
V22-86	41°36′	46°27′ W	5110	60
V22-108	43°11′	03°15′ W	4171	110
V29-84	43°51′	27°36′ E	5451	120
V29-86	49°34′	30°01′ E	5614	180
V29-87	49°06′	27°23′ E	5314	20
V29-89	45°44′	25°39′ E	5945	120
V29-90	43°42′	25°44′ E	5148	60
V29-104	50°02′	16°34′ E	4076	210
V29-105	48°05′	17°41′ E	4350	110

[a] Not sampled for ice-rafted detritus.

zones b, c, and d) characterized by strong bottom currents that winnowed the sediments, removing diatoms and concentrating coarse ice-rafted material. Although this argument is plausible, the fact that these lithologic changes are independent of topographic setting strongly suggests another cause.

3. The transport of volcanic detritus thousands of kilometers from its source is one of the most compelling arguments for increased sea ice. Discrete ash layers have only been observed in this region in the diatomaceous ooze layers (Ninkovich and others, 1964; Federman, 1977). Since volcanic eruptions can be assumed to be random with respect to season, the absence of discrete ash layers in the diatomaceous clay and their presence in the diatomaceous oozes suggests that ice persisted for much of the year during the deposition of the diatomaceous clays.
4. The presence of open-ocean diatoms as well as near-ice-dwelling forms in the diatomaceous clays (L. H. Burckle, personal communication, 1977) suggests that the sea ice was not a permanent pack. Either large leads opened annually, allowing intense local or regional diatom production, or the extent of summer meltback was highly variable from year to year.

Patterns of past sea-ice cover

These various lithologic changes allow us to estimate the past seasonal distribution of sea ice. Since the ice-rafted component is critical to these estimates in the Atlantic and western Indian Ocean sectors, we will first compare its distribution today with that of 18,000 years ago.

In this study the pattern of ice-rafted detritus from surface samples from the tops of 33 piston cores, known to be Holocene in age, from the Antarctic Ocean (Figure 131.3), are compared with the pattern of ice-rafted detritus from 65 piston core samples taken at the 18,000 year horizon (Figure 131.4). This level was chosen by the faunal method of Hays, Lozano, and others (1976). Table 131.4 lists the 65 cores used. The 18,000 year horizon had been determined for 23 cores by Hays, Lozano, and others (1976). Since the study, five of these horizons have been revised, and the 18,000 year level has been determined for the remaining cores by J. D. Hays and others (personal communication, 1977).

The ice-rafted detritus in this study is defined as the nonbiogenic coarse fraction $>63\mu$ and $<1000\ \mu$. The accumulation rates were calculated from the weight percent of coarse fraction, the sedimentation rates, and the dry weight, wet volume, bulk density (determined by weighing sections of dry core of known wet volume) (the results compare favorably with values used by Hays, Lozano, and others, 1976, and R. W. Embley, personal communication, 1977).

Figure 131.3 shows that north of lat. 60° S only a small amount of detritus is being ice-rafted today. The highest value is from the core closest to the South Sandwich Islands (V14-57). From satellite data it was determined that this core site is covered by winter sea ice for a few months each year (G. Kukla, personal communication, 1977).

Figure 131.4 shows a major change in the deposition of

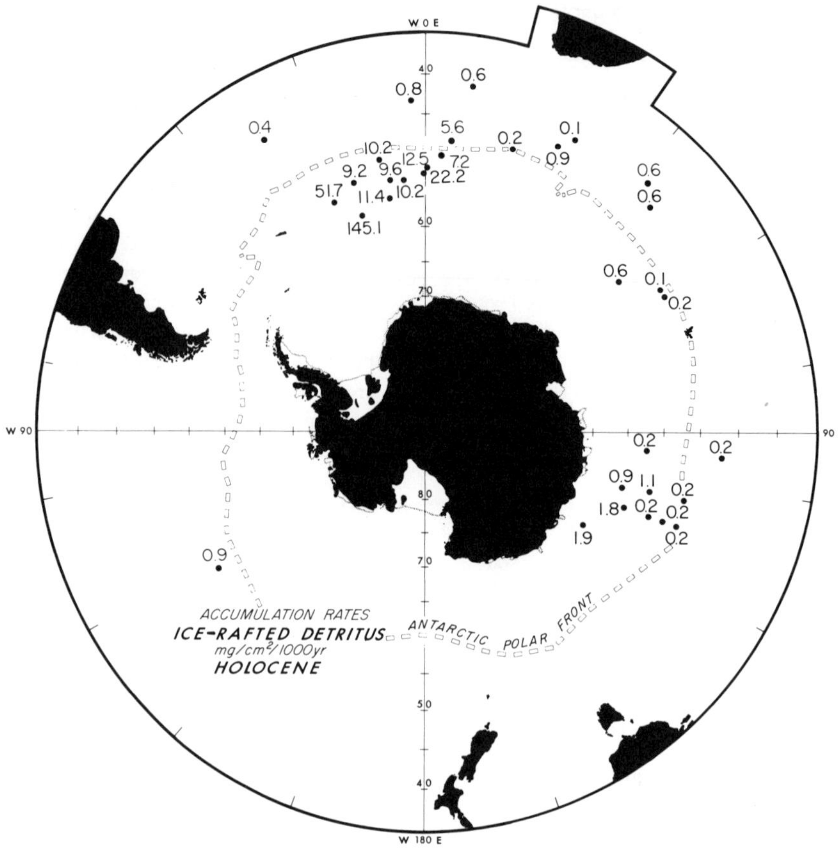

Figure 131.3. The accumulation rates of ice-rafted detritus in surface sediments, known to be Holocene in age, of 33 cores from the Antarctic Ocean.

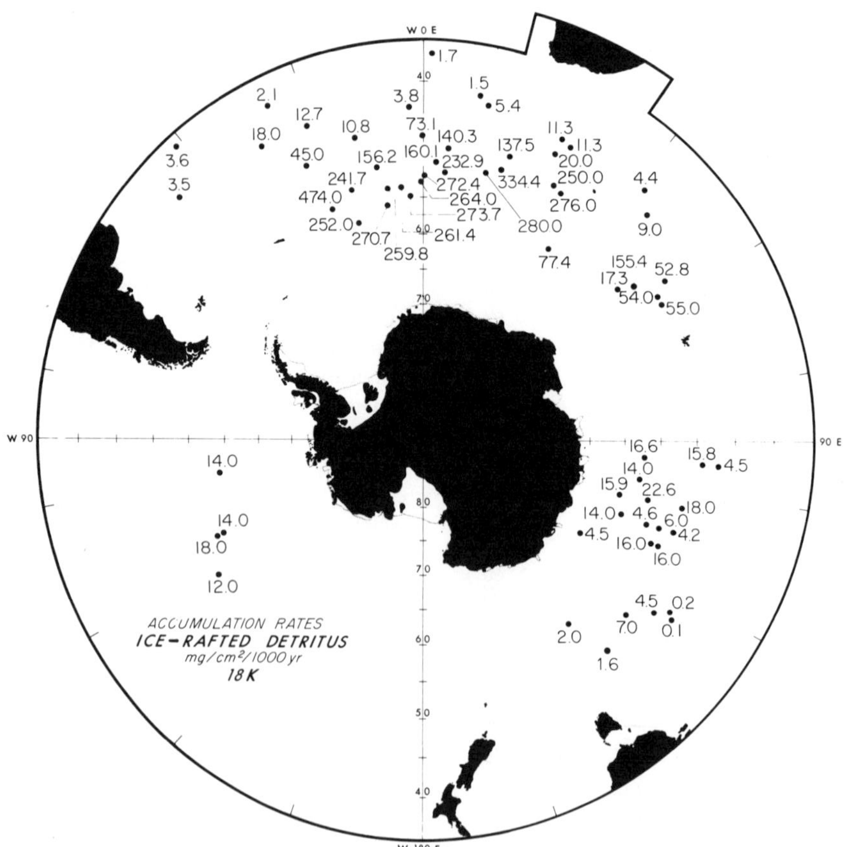

Figure 131.4. The accumulation rates of ice-rafted detritus 18,000 years ago (18 K) in 65 cores from the Antarctic Ocean (see Table 131.4).

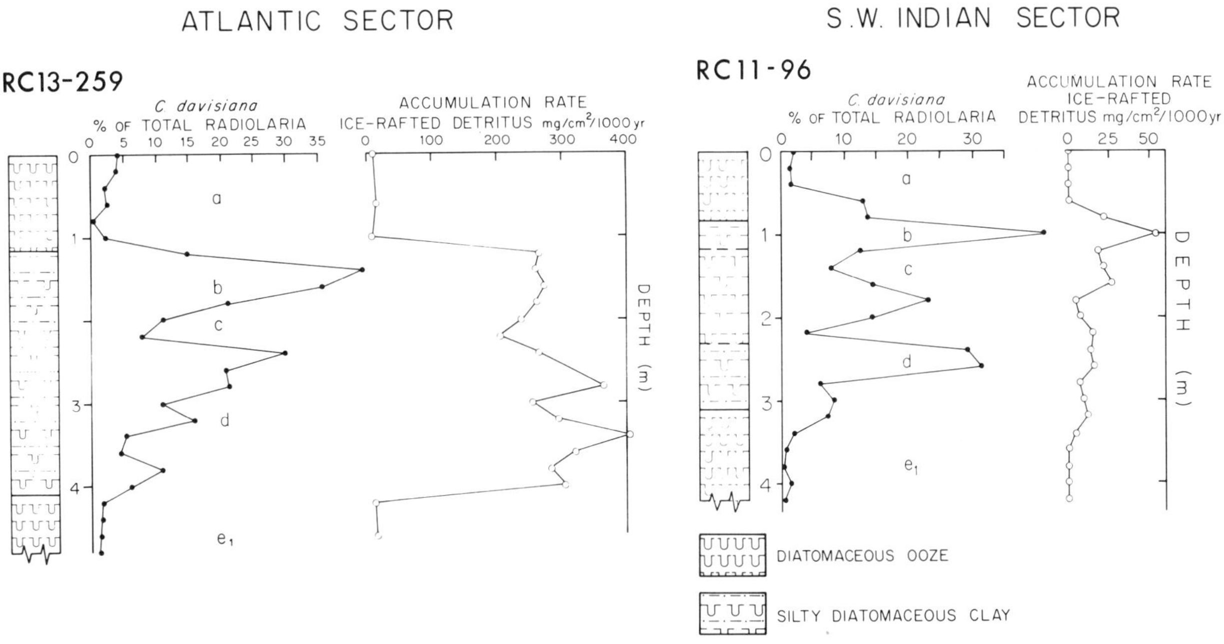

Figure 131.5. The accumulation rates of ice-rafted detritus in core RC13-259 from the Atlantic Ocean sector compared with the accumulation rates in core RC11-97 from the southwest Indian Ocean sector. See Table 131.4 for core locations.

ice-rafted detritus. A large tongue of ice-rafted detritus originating near the South Sandwich Islands stretches toward Kerguelen Island. The detritus is 80–90% volcanic, and thus the South Sandwich Islands are a logical primary source. Secondary sources could include active regions of West Antarctica or mid-ocean islands such as Bouvet Island in the Atlantic sector and Marion Island in the southwest Indian sector.

The accumulation rate of ice-rafted detritus decreases abruptly north of lat. 46° S in the Atlantic and southwest Indian Oceans, providing an estimate of the maximum (winter) extent of sea ice.

The distance from the South Sandwich Islands to sites near Kerguelen Island is about 5000 km. Little is known about the velocity of the circumpolar current, the West Wind Drift. Deacon (1960) has noted the eastward drift of bottles at 8 nautical miles per day (14.6 km/day). Coombs and Landis (1966) have reported the drift of pumice from an eruption in the South Sandwich Islands to New Zealand at rates of 8 to 13 nautical miles per day (14.6–23.8 km/day), while Sissala (1969) has noted that an iceberg tracked by satellite over a period of 105 days averaged 8.4 nautical miles per day (15.4 km/day). The time necessary to transport debris from the South Sandwich Islands to core sites near Kerguelen Island by ice could therefore be anywhere from 7 to 12 months.

Subsurface data (see Figure 131.5), from sites in the Atlantic and southwest Indian Oceans, indicate that the Atlantic Ocean received large amounts of detritus through *C. davisiana* zones b, c, and d, while the southwest Indian Ocean received its greatest influx during b time (18,000 years ago). At this time (18,000 years ago), volcanic material was spread from the South Sandwich Islands to Kerguelen Island. The time necessary to transport detritus 5000 km suggests sea ice may then have been present for most of the year and possibly year round. Open-marine and near-ice diatoms are dominant at this level in the cores (L. H. Burckle, personal communication, 1977). This suggests that open-water conditions existed at times, but not necessarily every year.

In the southeast Indian and Pacific Oceans, the accumulation rates are significantly less. The ice-rafted component is primarily quartz, feldspar, and heavy minerals and was probably transported by berg ice originating on the continent.

In Figure 131.6, we have estimated the extent of winter sea ice 18,000 years ago. Since sea ice can raft volcanic particles in summer or winter, the northern limit of high concentrations of particles disseminated through the diatomaceous clay can provide an estimate of maximum winter ice for the Atlantic and southwest Indian sectors. The extent of winter sea ice in the southeast Indian and Pacific Ocean sectors is difficult to determine. However, using the sedimentary parameters observed throughout the Antarctic Ocean (lithologic changes and sedimentation rate changes), and the relative position of winter ice with respect to summer ice in the Atlantic and southwest Indian sectors 18,000 years ago, we have mapped the winter ice extent in the southeast Indian and Pacific Oceans at about 5° north of the summer ice limit.

In Figure 131.7, we have mapped the extent of summer sea ice 18,000 years ago. South of the estimated summer-ice limit the glacial sedimentation rates (diatomaceous clay) are reduced with respect to the present rates (diatomaceous ooze). North of this ice limit, although sharp lithologic changes are observed, the glacial sedimentation rates are increased with respect to the present rates.

The contact between diatomaceous ooze and diatomaceous clay according to our interpretation, then, would represent a change in Antarctic waters from relatively ice-free summers like those of today to summers with extensive ice cover.

Our cores suggest that this change was rapid. The sedimentation rates exhibited in cores RC13-271 and RC13-255 across the upper lithologic change (see Figure 131.2), average greater than 30 cm/1000 years, and the lithologic

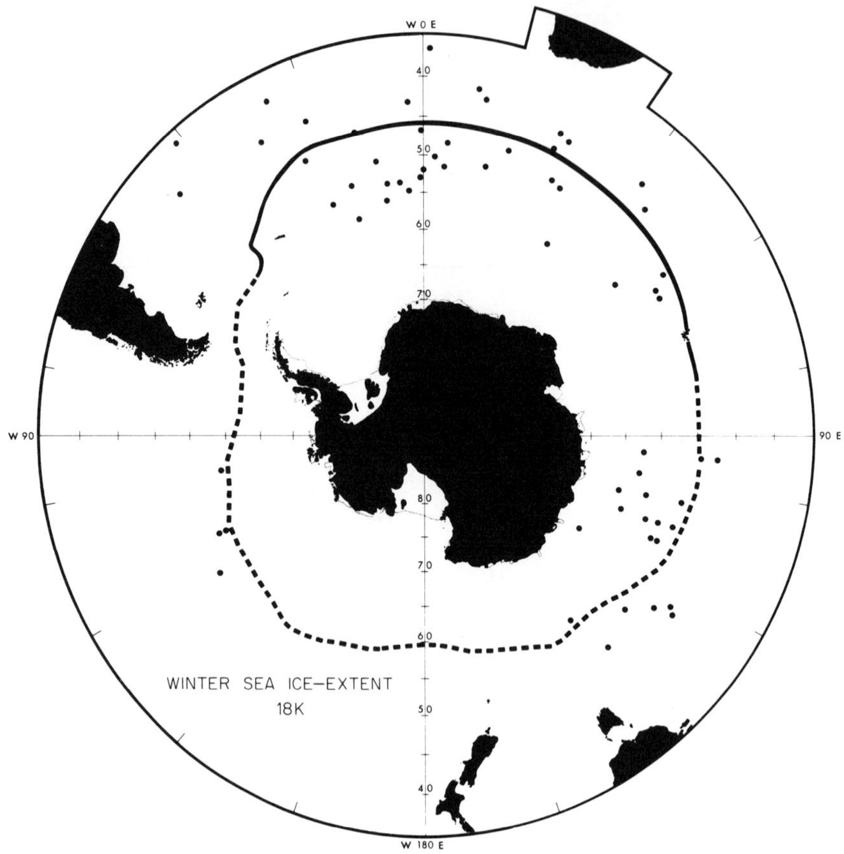

Figure 131.6. Map of estimated winter sea-ice extent 18,000 years ago (18 K). See text for discussion. The solid line indicates region where data control is strongest.

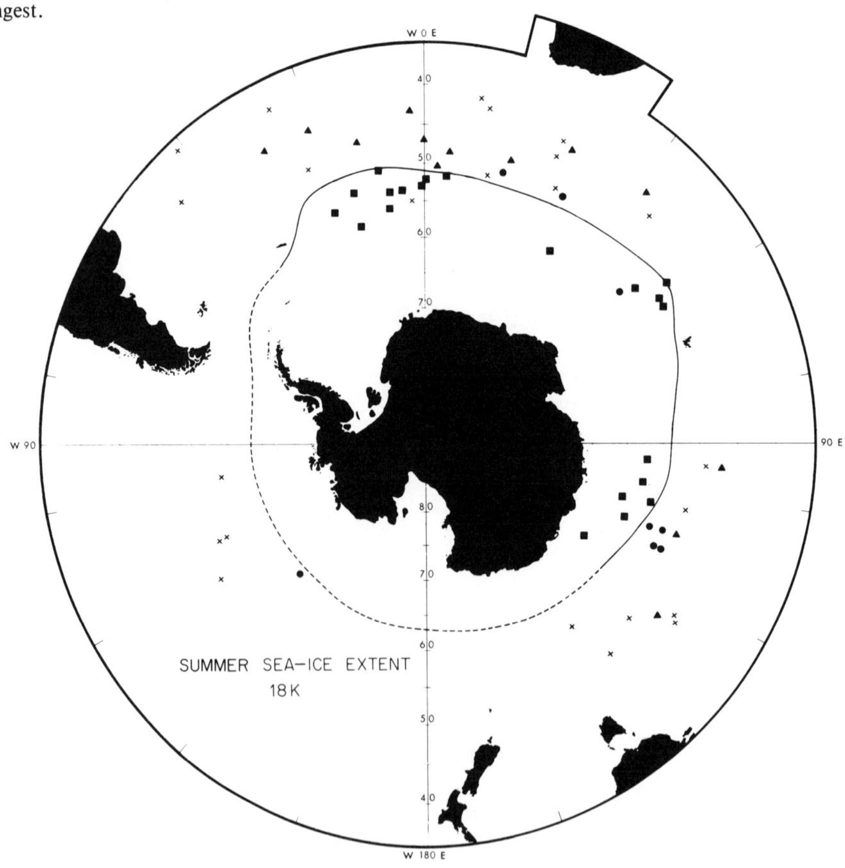

Figure 131.7. Map of estimated summer sea-ice extent 18,000 years ago (18 K). Symbols indicate core data used to determine summer sea-ice extent. Square: sharp lithologic change, diatomaceous ooze overlying silty diatomaceous clay. Triangle: sedimentation rate significantly higher during *C. davisiana* zone b compared with zone a. Circle: sharp lithologic change and high sedimentation rate observed. X: other cores used in this study. The solid line indicates region where data control is strongest.

change occurs within 5 cm. Consequently, we infer that within 200 years, the Antarctic Ocean changed from being dominantly covered by ice in the summer to its present state—ice-free in summer. This last change occurred about 14,000 years ago.

Core RC13-271 penetrates the boundary between underlying diatomaceous ooze and overlying diatomaceous clay in the upper part of zone e_1. This lithologic change also occurs within 5 cm, and although the sedimentation rate in this part of the core is somewhat less than 30 cm/1000 years, this change probably occurred as rapidly as the later one.

It is clear from Figure 131.1 that lithologic changes 1, 3, and 4, from diatomaceous clay to diatomaceous ooze, occurred just as Northern Hemisphere ice was beginning to melt, while lithologic change 2 from diatomaceous ooze to diatomaceous clay occurred just as a change toward increased Northern Hemisphere ice was beginning. How major changes in Southern Hemisphere sea ice would affect Northern Hemisphere glaciers is not clear. However, summer sea ice of the extent we propose would have made an important change in the earth's albedo and may have helped to accelerate the growth of Northern Hemisphere ice.

Acknowledgments. We thank L. H. Burckle, D. Cassidy, R. M. Cline, D. S. Cooke, R. W. Embley, E. Free, G. Irving, G. Kukla, R. Lynde, F. W. McCoy, J. J. Morley, D. Ninkovich, W. F. Ruddiman, and N. J. Shackleton for their scientific and editorial aid in the preparation of this paper. This research was funded by National Science Foundation Grant OCE75-19627 under the International Decade of Ocean Exploration project CLIMAP. Core curating and repository services at Lamont-Doherty Geological Observatory were funded by Office of Naval Research Grant N00014-75-C-0210 (Scope E) and National Science Foundation Grant OCE76-18049, and at Florida State University by the National Science Foundation Office of Polar Programs.

References cited

Baker, P. E. 1968. Comparative volcanology and petrology of the Atlantic island-arcs. *Bull. volcan.*, Ser. 2, *32*, No. 1, 189–206.

Baker, P. E. 1972. Recent volcanism and magmatic variation in the Scotia arc. (*In* Adie, R. J., *ed. Antarctic Geology and Geophysics.* Oslo, Universitetsforlaget, 57–60.)

Barker, P. F. 1972. Magnetic lineations in the Scotia Sea. (*In* Adie, R. J., *ed. Antarctic Geology and Geophysics.* Oslo, Universitetsforlaget, 17–26.)

Burckle, L. H. 1972. Diatom evidence bearing on the Holocene in the South Atlantic. *Quat. Res.*, *2*, No. 3, 323–26.

Cooke, D. W., Burckle, L. H., and Hays, J. D. 1977. Winter sea-ice cover during the Late Pleistocene in the Antarctic Ocean. *Trans. Am. geophys. Un.*, *58*, No. 6, 416.

Coombs, D. S., and Landis, C. A. 1966. Pumice from the South Sandwich eruption of March 1962 reaches New Zealand. *Nature, Lond.*, *209*, No. 5020, 289–90.

Deacon, G. E. R. 1960. The southern cold temperature zone. *Proc. R. Soc. Lond.*, Ser. B., *152*, 441–47.

Federman. A. N. 1977. The chemical characterization of abyssal tephras from the South Sandwich Islands. *Trans. Am. geophys. Un.*, *58*, No. 6, 528.

Goodell, H. G. 1973. The sediments. (*In* Goodell, H. G., Houtz, R., Ewing, M., and others, *eds.* Marine Sediments of the Southern Oceans. *Antarct. Map Folio Ser.*, Folio 17.)

Hays, J. D., Imbrie, J., and Shackleton, N. J. 1976. Variations in the earth's orbit: Pacemaker of the ice ages. *Science, N.Y.*, *194*, No. 4270, 1121–32.

Hays, J. D., Lozano, J. A., Shackleton, N. J., and Irving, G. 1976. Reconstruction of the Atlantic and western Indian Ocean sectors of the 18,000 B.P. Antarctic Ocean. (*In* Cline, R. M., and Hays, J. D., *eds.* Investigation of Late Quaternary Paleoceanography and Paleoclimatology. *Mem. geol. Soc. Am.*, *145*, 337–72.)

Hough, J. L. 1956. Sediment distribution in the southern oceans around Antarctica. *J. sedim. Petrol.*, *26*, No. 4, 301–6.

Lisitzin, A. P. 1960. Bottom sediments of the eastern Antarctic and the southern Indian Ocean. *Deep Sea Res.*, *7*, No. 2, 89–99.

Lozano, J. A., and Hays, J. D. 1976. Relationship of radiolarian assemblages to sediment types and physical oceanography in the Atlantic and western Indian Ocean sectors of the Antarctic Ocean. (*In* Cline, R. M., and Hays, J. D., *eds.* Investigation of Late Quaternary Paleoceanography and Paleoclimatology. *Mem. geol. Soc. Am.*, *145*, 303–36.)

Morley, J. J. 1977. *Upper Pleistocene Climate Variations in the South Atlantic Derived from a Quantitative Radiolarian Analysis.* Ph.D. dissertation, Columbia University, 344 pp.

Murray, J., and Renard, A. F. 1891. Report on deep-sea deposits based on the specimens collected in the years 1872–1876. (*In Report on the Scientific Results of the Voyage of the H.M.S. Challenger during the Years 1872–1876,* Vol. 18. London, Majesty's Stationery Office.)

Ninkovich, D. 1968. Pleistocene volcanic eruptions in New Zealand recorded in deep-sea sediments. *Earth planet. Sci. Lett.*, *4*, No. 2, 89–102.

Ninkovich, D., Heezen, B. C., Conolly, J. R., and Burckle, L. H. 1964. South Sandwich tephra in deep-sea sediments. *Deep Sea Res.*, *11*, No. 4, 605–19.

Philippi, E. 1912. Die Grundproben der deutschen Südpolar-Expedition 1901–1903. (*In Deutsche Südpolar-Expedition 1901–1903,* Vol. 2, Berlin, Geographie und Geologie, 415–616.)

Schott, W. 1939. Deep-sea sediments of the Indian Ocean. (*In* Trask, P., *ed. Recent Marine Sediments.* Tulsa, American Association of Petroleum Geologists, 396–408.)

Shackleton, N. J., and Opdyke, N. D. 1973. Oxygen isotope and paleomagnetic stratigraphy of equatorial Pacific core V28–238: oxygen isotope temperatures and ice volume on a 10^5 and 10^6 year scale. *Quat. Res.*, *3*, No. 1, 39–55.

Shackleton, N. J., and Opdyke, N. D. 1976. Oxygen-isotope and paleomagnetic stratigraphy of Pacific core V28–239: late Pliocene to Latest Pleistocene. (*In* Cline, R. M., and Hays, J. D., *eds.* Investigation of Late Quaternary Paleoceanography and Paleoclimatology. *Mem. geol. Soc. Am.* *145*, 449–64.)

Sissala, J. 1969. Observation of an Antarctic Ocean tabular iceberg from the Nimbus II satellite. *Nature, Lond.*, *224*, No. 5226, 1285–87.

132

Contrasts in Antarctic Bottom-Water Velocity between the Brunhes and Matuyama Epochs in the South Pacific

T. C. HUANG, M. T. LEDBETTER*, and N. D. WATKINS†
Graduate School of Oceanography, University of Rhode Island, Kingston, Rhode Island 02881, U.S.A.

Abstract. Textural and compositional variations in seven dated *Eltanin* sediment cores reveal that the velocity of Antarctic bottom waters in the South Pacific has fluctuated much less during the past 0.69 m.y. than in the preceding 1.7 m.y., when as many as three widespread disconformities or periods of substantially diminished sedimentation rates resulted from pulses of increased bottom-current activity. Polynomial regression analysis between mineral and biogenic constituents and skewness of the sediments shows that a skewness of −0.2 separates two depositional environments: those dominated by strong bottom currents (skewness values more positive than −0.2) and those influenced mainly by pelagic settling (skewness values more negative than −0.2). The temporal and spatial variations of the manganese micronodules in 23 cores from the same region confirm these interpretations and also demonstrate the much more widespread area of bottom-current activity during Matuyama time than during Bruhnes time.

Using various techniques, marine geologists have attempted for more than two decades to relate deep-sea sediment morphology and sediment textures to net bottom-water velocities (Menard, 1952; Heezen and others, 1959; Hurley and Fink, 1963; Heezen and Hollister, 1964; Heezen, Hollister, and Ruddiman, 1966; Heezen, Schneider, and Pilkey, 1966; Fox and Heezen, 1968; Laughton, 1968; Heezen and Johnson, 1969; Harms, 1969; Neumann and Ball, 1970; Allen, 1970; Johnson, 1972; Heezen and Tharp, 1972; Flood and Hollister, 1974; Kolla and others, 1976). The inferences which can be drawn from sea-floor photographs are, at best, semiquantitative. Recent recognition of evidence of bottom-water erosion and substantial erosional disconformities in both piston and Deep Sea Drilling Project cores (Heezen, Schneider, and Pilkey, 1966; Watkins and Kennett, 1971, 1972; Griggs and Kulm, 1970; Kennett and others, 1972, 1975; Davies and others, 1975) has sparked new interest in the significance of changes in long-term bottom-water velocities, especially at certain stages in continental separation (Watkins and Kennett, 1973).

Long-term changes in bottom-water velocity have been recognized in many sectors of the Southern Ocean (Watkins and Kennett, 1971, 1972; Kennett and Watkins, 1976). Late Pliocene time, including the Matuyama epoch, was characterized by high bottom-water velocity and erosion of abyssal sediment, while the Brunhes epoch was marked by lower velocities (Watkins and Kennett, 1971; Fillon, 1975; Huang and Watkins, 1977). Identification of sedimentary hiatuses was based primarily on paleontologic and paleomagnetic studies of the cores; the recognition of hiatuses by these methods is limited to those exceeding 0.2 m.y. (Watkins and Kennett, 1971, 1973). In this paper we describe our application of textural and compositional analyses to the detection of evidence for shorter periods of nondeposition and erosion in deep-sea sediments and the inferences which can be drawn concerning the associated bottom-water history. We also describe the areal distribution of manganese-micronodule accumulation rates, which we use to infer spatial changes in bottom-water activity.

Methods

The cores examined (Figure 132.1 and Table 132.1) have been dated using paleomagnetic and micropaleontologic methods, and the results have been published elsewhere (Hays, 1965, 1967; Watkins and Goodell, 1967, Goodell and Watkins, 1968; Kennett and Watkins, 1970; Watkins and Kennett, 1971, 1972). The tephra in some of the cores was also used to refine the stratigraphy (Huang and others, 1975*a*). The laboratory and statistical methods involved in textural and compositional analyses have been described by Huang and others (1975*b*), Keany and others (1976), and Huang and Watkins (1977).

Results and discussion

Sediment textures

The stratigraphy of selected *Eltanin* cores from the high latitudes of the South Pacific (Figure 132.1 and Table 132.1) has been established using three independent methods: paleomagnetic dating (Goodell and Watkins, 1968), radiolarian zonation, and tephra distribution (Huang and others, 1975*a*; Huang and Watkins, 1977). The hiatuses detected by these methods are mostly confined to sediments older than the Brunhes–Matuyama boundary at 0.69 m.y. ago (Figure 132.2). Average net

* Present address: Department of Geology, University of Georgia, Athens, Georgia 30602, U.S.A.

† Deceased.

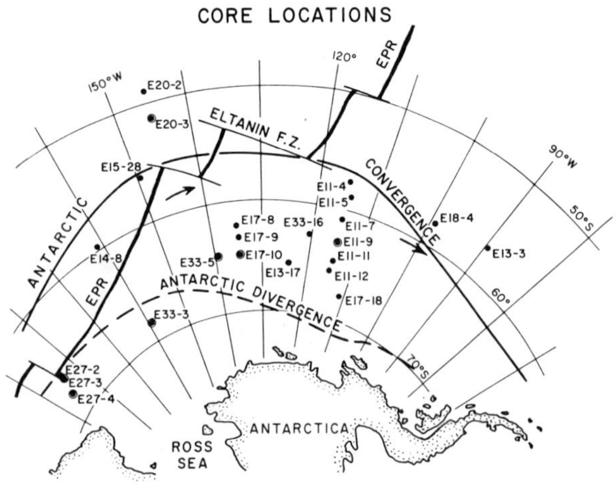

Figure 132.1. Map showing location of the 23 cores examined. The present circum-Antarctic surface-water flow directions are shown as solid arrows. The present position of the Antarctic Convergence, Antarctic Divergence, the *Eltanin* Fracture Zone, the East Pacific Rise (EPR), and other major tectonic elements are from Heezen and Tharp (1972) and Goodell (1973). Seven cores used for particle size analysis are indicated by circled dots.

sedimentation rates during the Brunhes epoch are as high as eight times those during the Matuyama epoch (Table 132.1). Hitherto, variations in sedimentation rates in this region have been attributed mainly to productivity and ice-rafted debris input variations (Goodell and Watkins, 1968; Goodell, 1973; Warnke, 1970; Gram and Warnke, 1974). To identify unambiguously the primary processes which caused variations in the observed sedimentation rates, statistical analyses of selected textural and compositional parameters have been made of samples taken at 10 cm intervals in the selected cores.

The textural parameters used here include mean grain size, sorting, skewness, and kurtosis. The moment functions given by Blatt and others (1972) were selected for this study. In the time since Krumbein and Aberdeen (1937) related textural parameters to changes in bottom-water velocity, well over 100 papers have been published on this topic. The consensus is that currents winnow fine-grained sediment and change the grain size distribution as bottom-water turbulence increases. In a given sample, such an increase in turbulence will cause the main grain size to shift from fine to coarse, sorting to change from poor to good, and skewness to change from negative to positive. These ideas have been applied extensively to interpret the sediments from various shallow-water environments (Visher, 1969), but only a few recent attempts have been made to apply these concepts to abyssal sediments (Ledbetter and Johnson, 1976; Huang and Watkins, 1977; Johnson and others, 1977; Ellwood and Ledbetter, 1977).

The fluctuation of skewness in four cores (Figure 132.2) is typical of the variations in textural parameters in all the cores examined (Figure 132.1). Brunhes epoch sediments almost invariably show negative skewness that fluctuates much less than that of the Matuyama sediments. A sharp positive skewness corresponding to coarser mean grain size and better sorting occurs within the 2.0 to 1.8 m.y. old segment of the three cores (Figure 132.2). Extreme positive skewness is to be expected where the finer components have been removed by bottom currents, and therefore the hiatuses in the cores are associated with intervals of positive skewness. In Figure 132.2 it can be seen that pronounced changes from negative to strong positive skewness occur across disconformities. This suggests that as bottom-current velocity increases to a critical point, erosion begins (probably removing positively skewed sediments previously formed during a period of diminished deposition rate) and is followed by a return to slower current velocities masked by negative skew. This sort of analysis could provide a method for estimating the amount of sediment removed during the hiatus. Examination of

Table 132.1. Location, water depth, length, and net sedimentation rates of *Eltanin* cores analyzed

Core number	Latitude	Longitude	Water depth (m)	Core length (cm)	Net sedimentation rates[a] (cm/1000 yr) Brunhes	Matuyama
E 11-4	57°49.7′ S	115°12.5′ W	4815	1123	1.10	0.65
E 11-5	58°56.6′ S	114°43.2′ W	5111	1038	0.65	0.23
E 11-7	60°55.1′ S	114°46.6′ W	5074	971	1.29	—
E 11-9	62°49.8′ S	115°04.0′ W	5055	2270	1.39	0.93
E 11-11	64°50.6′ S	114°29.5′ W	4908	1432	0.58	0.31
E 11-12	65°52.2′ S	115°05.0′W	4760	1880	2.23	—
E 13-3	57°00.3′ S	89°29.0′ W	5138	1603	0.35	0.37
E 13-17	65°41.0′ S	124°06.8′ W	4766	2642	0.66	0.44
E 14-8	59°40.0′ S	160°17.4′ W	3911	1830	0.77	0.30
E 15-28	56°01.0′ S	149°49.0′ W	3358	1342	0.72	0.41
E 17-8	62°01.0′ S	135°17.0′ W	4589	1289	1.11	0.14
E 17-9	63°05.0′ S	135°07.0′ W	4891	2018	2.23	—
E 17-10	65°01.0′ S	134°52.0′ W	4533	1585	0.60	0.57
E 17-18	67°54.0′ S	110°46.0′ W	4120	1935	1.74	1.29
E 18-4	58°00.0′ S	99°19.0′ W	4815	1374	0.53	0.40
E 20-2	49°00.0′ S	144°50.0′ W	4557	867	0.77	—
E 20-3	51°07.0′ S	145°03.0′ W	4133	820	0.80	—
E 27-2	66°01.6′ S	176°26.0′ E	3528	1760	0.80	0.63
E 27-3	66°03.7′ S	176°30.3′ E	3568	2173	0.71	0.52
E 27-4	68°02.9′ S	174°35.5′ E	3480	1991	0.97	0.91
E 33-3	68°04.6′ S	159°53.9′ W	3579	943	1.15	—
E 33-5	65°01.3′ S	139°43.1′ W	4373	915	0.49	0.26
E 33-16	63°14.2′ S	120°01.1′ W	4958	1512	0.69	0.43

[a] Brunhes is the normal magnetic polarity epoch (0.69 m.y. ago to present); Matuyama is the reversed magnetic polarity epoch (2.43 to 0.69 m.y. ago). The net sedimentation rates are based primarily on the paleomagnetic data given by Goodell and Watkins (1968).

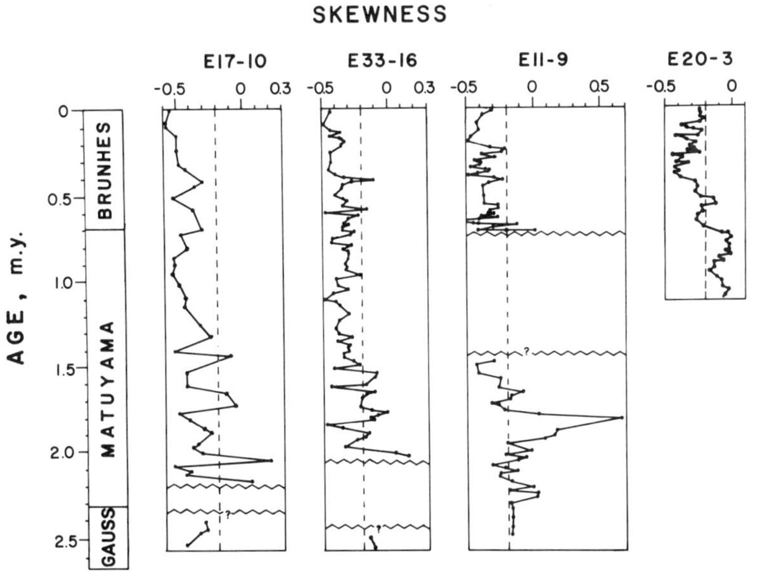

Figure 132.2. Sediment skewness variations in four selected cores (locations shown in Figure 132.1). Note the larger skewness fluctuations associated with the hiatuses (given as data gaps) and the general lack of positive skewness, or lack of inferred strong bottom-water activity, in the Brunhes age sediments.

X-ray radiographs of the sediment characterized by high positive skewness just above the observed disconformities indicates the presence of lag or residual deposits. In most of these residual deposits, thin current-laminar microstructure shows that lateral traction transport has occurred during deposition.

This simple interpretation is subject to doubt, however, since other factors can readily affect textural parameters. These factors include compositional variations, such as those which can result from ice-rafted debris or microfaunal productivity variations. Therefore, in theory, differential pelagic contributions alone might create a similar textural variation.

All textural parameters were computed using the total grain-size distribution, which is the sum of the different individual distributions of the subpopulations or components. The major components in the cores include siliceous and carbonate planktonic fossils, ice-rafted debris, and micronodules, all with specific and different genetic connotations.

To evaluate which component dominates in the total distribution or in the skewness variations, polynomial regression analyses of skewness versus six selected major components in the cores were made. The skewness data were arbitrarily divided into two groups: those with values more positive than -0.2 and those more negative than -0.2. The results (Huang and Watkins, 1977) show that during the deposition of sediments with skewness more positive than -0.2, particularly those with positive skewness (values above zero), manganese micronodules caused the most variation. In contrast, during the periods of deposition of sediments with skewness more negative than -0.2, siliceous plankton skeletons accounted for most of the skewness variation in the cores. Manganese micronodules are believed to form authigenically under high-oxidation conditions, favored by extremely slow ambient sediment deposition. This condition commonly occurs in areas with strong bottom currents (Watkins and Kennett, 1971, 1972) so that a high concentration of manganese micronodules is usually associated with residual deposits. Such environments have been reported beneath the Antarctic Convergence (Goodell, 1973) and on the floor of the southeast Indian Ocean (Kennett and Watkins, 1975). Thus, the skewness value of -0.2 seems to mark a useful division between those depositional environments influenced mainly by pelagic settling (skewness values more negative than -0.2) and those dominated by bottom currents (skewness values more positive than -0.2). The positive skewness values may ideally record the real long-term fluctuations of past bottom-current velocities in the area under examination. It is stressed that the relative current-velocity changes inferred from variations in skewness are net variations over periods of 1,800 to 14,000 years, since all the samples analyzed are about 2 cm thick, with sedimentation rates ranging from 0.14 to 1.11 cm/1000 yr. We conclude from Figure 132.2 that at least one and possibly three major discrete pulses of high bottom-water velocity occurred in the Pacific segment of the Southern Ocean during the early Matuyama epoch. In contrast, bottom currents during the last 0.7 m.y. seem to have been relatively weak in the region examined. The high-velocity pulses have caused disconformities in some cores, but have only reduced the depositional rates in others.

Manganese micronodules

The present distribution of higher Antarctic bottom-water velocity coincides roughly with the location of the present Antarctic Convergence and also correlates with the present distribution of manganese nodules (Hollister and Heezen, 1967; Goodell, 1973). Our inferred bottom-current pulses during the Matuyama epoch suggest either that the Antarctic Convergence may have been 5° of latitude farther south during early Matuyama time or that the region of strong bottom-current activity simply may have been of greater north-south extent. We choose to study the contrasting regional extent of increased bottom-water activity in the Brunhes and Matuyama epochs in an area surrounding the Bellingshausen Basin where there is evi-

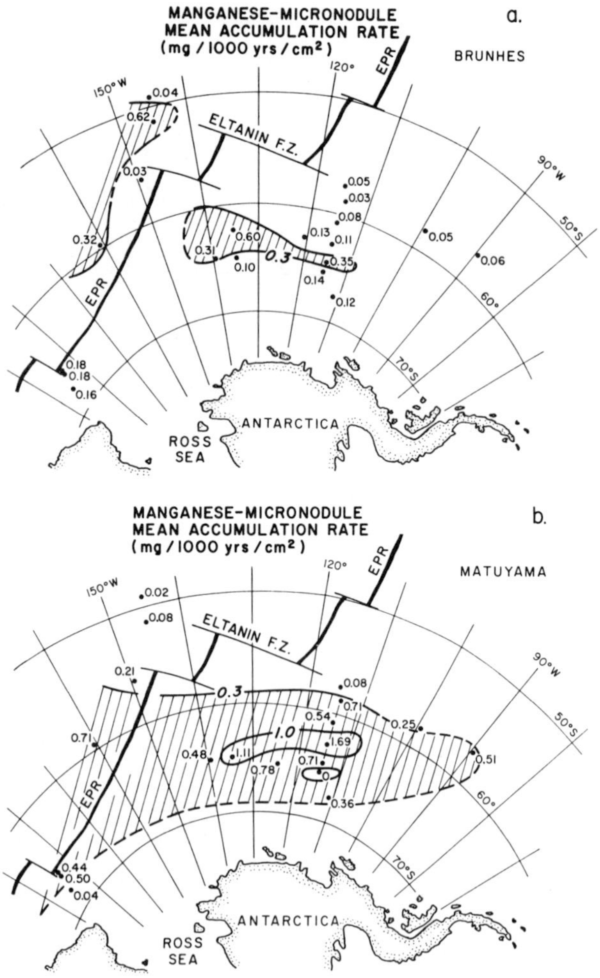

Figure 132.3. a. Mean manganese micronodule accumulation rates (in mg/1000 yr/cm²) in the Brunhes age sediments (0.69 m.y. ago to present) of 23 cores. The shaded area indicates the localities having a mean rate greater than 0.3 mg/1000 yr/cm², which implies sediments deposited under the effect of relatively strong bottom-current activity. *b.* As in *a*, but for Matuyama age sediments (2.43 to 0.69 m.y. ago). Note the contrast in shaded area size between the two different epochs.

dence of strong modern and ancient bottom currents (Hollister and Heezen, 1967; Heezen and others, 1968; Goodell and Watkins, 1968; Huang and Watkins, 1977).

The accumulation rate of manganese micronodules may be a function of bottom-current activity. The absence of the nodules does not always indicate reduced bottom-current activity (Watkins and Kennett, 1977), but an increase in the nodule accumulation rate almost certainly indicates a reduced sedimentation rate, which may in turn be due to increased current velocity. The mean accumulation rate of manganese micronodules in the Brunhes section of 23 cores shows that the area of relatively high-velocity bottom currents is restricted to a narrow zone between lat. 62° and 65° S (Figure 132.3*a*). The mean manganese micronodule accumulation rate of 0.3 mg/1000 yr/cm² was arbitrarily chosen as a dividing value in order to distinguish between sediments deposited under high and low bottom-water activity (Figure 132.3*a*). The sediments with mean manganese micronodule accumulation rates larger than 0.3 mg/1000 yr/cm² usually have skewness values more positive than −0.2.

In contrast, the mean accumulation rate in the Matuyama section of 18 cores show that the zone of high bottom-current velocity is a wide area between lat. 58° and 67° S (Figure 132.3*b*). Simple comparison of the distribution of the mean manganese micronodule accumulation rate indicates that the zone of relatively high bottom-current velocity had a much greater north-south extent in Matuyama time than in Brunhes time (Figure 132.3).

The more abundant bottom water may have originated in the Ross Sea and spread north and northeast (Figure 132.3*b*), as is the case for the present bottom currents in this area (Jacobs and others, 1970). The bottom-current activity was most vigorous in the area near the Ross Sea during Matuyama time, as shown in core E 33-3, which was taken to the north of the Ross Sea; the Matuyama sediments were almost entirely eroded by bottom currents (Huang and others, 1975*a*; Huang and Watkins, 1977), as also demonstrated in the Ross Sea area by Fillon (1975). Although a detailed study of the relative intensity as well as the timing of paleoclimatic change during Matuyama time is necessary, there is some evidence that the inferred increase in bottom-current activity may be a function of climatic change. Intense glaciation during early Matuyama time, associated with greater amounts of ice-rafted debris in the cores near the Antarctic continental margin, has been reported (Goodell and Watkins, 1968). The paleoclimatic evidence derived from foraminiferal assemblages also indicates that average temperatures during Matuyama time were, in general, cooler than those during Brunhes time (Keany and Kennett, 1972).

Conclusions

We suggest that past Antarctic bottom-current velocities can be inferred semiquantitatively from skewness variations in sediments. Short-period disconformities can be detected by analysis of skewness variation in conjunction with compositional parameters. At least one major and several minor Antarctic bottom-current pulses have been recognized in association with hiatuses and residual deposits of early Matuyama age. Subsequent reduction in bottom-water activity in the region is expressed by a greatly diminished area of inferred high bottom-water activity for the epoch beginning about 0.7 m.y. ago.

Acknowledgments. This research was supported by the National Science Foundation through Division of Polar Programs Grant DPP77-09491 and Submarine Geology and Geophysics Program Grant DES74-22347.

References cited

Allen, D. W. 1970. *Sedimentary Texture—a Key to Interpret Deep-Marine Dynamics.* Ph.D. dissertation, Oregon State University, 152 pp.

Blatt, H. G., Middleton, G. V., and Murray, R. 1972. *Origin of Sedimentary Rocks.* Englewood Cliffs, N.J., Prentice-Hall, 634 pp.

Davies, T. A., Weser, O. E., Luyendyk, B. P., and Kidd, R. B. 1975. Unconformities in the sediments of the Indian Ocean. *Nature, Lond., 253,* 15–19.

Ellwood, B. B., and Ledbetter, M. T. 1977. Antarctic bottom water fluctuations in the Vema Channel: effects of velocity changes on particle alignment and size. *Earth planet. Sci. Lett., 35,* 189–98.

Fillon, R. H. 1975. Cenozoic paleo-oceanography of the Ross Sea, Antarctica. *Geol. Soc. Am. Bull., 86,* 839–45.

Flood, R. D., and Hollister, C. D. 1974. Current-controlled topography on the continental margin off the eastern United States. (*In* Burk, C. A., and Drake, C. L., *eds. The Geology of Continental Margins.* New York, Springer-Verlag, 197–205.)

Fox, P. J., and Heezen, B. C. 1968. Abyssal antidunes. *Nature, Lond. 220,* 470–72.

Goodell, H. G. 1973. The sediments. (*In* Goodell, H. G., Houtz, R., Ewing, M., and others, *eds.* Marine Sediments of the Southern Oceans. *Antarct. Map Folio Ser.,* Folio 17, 1–5.)

Goodell, H. G., and Watkins, N. D. 1968. Paleomagnetic stratigraphy of the Southern Ocean: 20° W to 160° E longitude. *Deep Sea Res., 15,* 89–112.

Gram, R., and Warnke, D. A. 1974. New evidence for changing rates of biogenic sedimentation in the Southern Ocean. *EOS, 55,* 311.

Griggs, G. B., and Kulm, L. D. 1970. Sedimentation in Cascadia deep-sea channel. *Geol. Soc. Am. Bull., 81,* 1362–84.

Harms, J. G. 1969. Hydraulic significance of some sand ripples. *Geol. Soc. Am. Bull., 80,* 363–96.

Hays, J. D. 1965. Radiolaria and late Tertiary and Quaternary history of Antarctic seas. (*In* Llano, G. A., *ed. Biology of the Antarctic Seas 2.* Washington, American Geophysical Union, 125–84.)[Antarctic Research Series, Vol. 15.]

Hays, J. D. 1967. Quaternary sediments of the Antarctic Ocean. (*In* Sears, M., *ed. Progress in Oceanography.* New York, Pergamon Press, 117–31.)

Heezen, B. C., and Hollister, C. D. 1964. Deep-sea current evidence from abyssal sediments. *Mar. Geol., 1,* 141–74.

Heezen, B. C., and Johnson, G. L. 1969. Mediterranean undercurrent and microphysiography west of Gibraltar. *Bull. Inst. Oceans, Monaco, 67,* 95 pp.

Heezen, B. C., and Tharp, M. 1972. Morphology of the sea floor. (*In* Heezen, B. C., Tharp, M., and Bentley, C. R., *eds.* Morphology of the Earth in the Antarctic and Subantarctic. *Antarct. Map Folio Ser.,* Folio 16, 10–14.)

Heezen, B. C., Hollister, C. D., and Ruddiman, W. F. 1966. Shaping of the continental rise by deep geostrophic contour currents. *Science, N.Y., 152,* 502–8.

Heezen, B. C., Schneider, E. D., and Pilkey, O. H. 1966. Sediment transported by the Antarctic bottom current in Bermuda Rise. *Nature, Lond., 211,* 611–13.

Heezen, B. C., Tharp, M., and Ewing, M. 1959. The floors of the ocean 1: the North Atlantic. *Spec. Pap. geol. Soc. Am., 65,* 122 pp.

Heezen, B. C., Tharp, M., and Hollister, C. D. 1968. Illustrations of the marine geology of the Southern Ocean. *Symposium on Antarctic Oceanography, Santiago, Chile, 1968,* 13–16.

Hollister, C. D., and Heezen, B. C. 1967. The floor of the Bellingshausen Sea. (*In* Hersey, J. B., *ed. Deep-sea Photography.* Baltimore, Johns Hopkins Press, 177–89.) [Johns Hopkins Oceanographic studies 3.]

Huang, T. C., and Watkins, N. D. 1977. Contrasts between the Brunhes and Matuyama sedimentary records of bottom water activity in the South Pacific. *Mar. Geol., 23,* 113–32.

Huang, T. C., Watkins, N. D., and Shaw, D. M. 1975*a*. Atmospherically transported volcanic glass in deep-sea sediments: volcanism in Sub-Antarctic latitudes of the South Pacific during late Miocene and Pleistocene time. *Geol. Soc. Am. Bull., 86,* 1305–15.

Huang, T. C., Watkins, N. D., and Shaw, D. M. 1975*b*. Atmospherically transported volcanic glass in deep-sea sediments: development of a separation and counting technique. *Deep Sea Res., 22,* 185–96.

Hurley, R. J., and Fink, L. K. 1963. Ripple marks show that countercurrent exists in Florida Straits. *Science, N.Y., 139,* 603–5.

Jacobs, S. S., Amos, A. F., and Bruchhausen, P. M. 1970. Ross Sea oceanography and Antarctic Bottom Water formation. *Deep Sea Res., 17,* 935–62.

Johnson, D. A. 1972. Ocean-floor erosion in the equatorial Pacific. *Geol. Soc. Am. Bull., 83,* 3121–44.

Johnson, D. A., Ledbetter, M. T., and Burckle, L. H. 1977. Vema Channel paleo-oceanography: Pleistocene dissolution cycles and episodic bottom water flow. *Mar. Geol., 23,* 1–33.

Keany, J., and Kennett, J. P. 1972. Pliocene–early Pleistocene paleoclimatic history recorded in Antarctic-Subantarctic deep-sea cores. *Deep Sea Res., 19,* 529–48.

Keany, J., Ledbetter, M., Watkins, N. D., and Huang, T. C. 1976. Diachronous deposition of ice-rafted debris in sub-Antarctic deep-sea sediments. *Geol. Soc. Am. Bull., 87,* 873–82.

Kennett, J. P., and Watkins, N. D. 1970. Geomagnetic polarity change, volcanic maxima and faunal extinction in the South Pacific. *Nature, Lond., 227,* 930–34.

Kennett, J. P., and Watkins, N. D. 1975. Deep-sea erosion and manganese nodule development in the southeast Indian Ocean. *Science, N.Y., 188,* 1011–13.

Kennett, J. P., and Watkins, N. D. 1976. Regional deep-sea dynamic processes recorded by late Cenozoic sediments of the southeastern Indian Ocean. *Geol. Soc. Am. Bull., 87,* 321–39.

Kennett, J. P., Burns, R. E., Andrews, J. E., Churkin, M., Davies, T. A., Domitrica, P., Edwards, A. R. 1972. Australia-Antarctic continental drift, paleocirculation changes and Oligocene deep-sea erosion. *Nature. Phys. Sci., 239,* 51–55.

Kennett, J. P., Houtz, R. E., and others. 1975. Cenozoic paleoceanography in the southwest Pacific ocean, Antarctic glaciation, and the development of the circum-Pacific current. *Initial Reports of the Deep-Sea Drilling Project,* Vol. 26. Washington [U.S. Government Printing Office], 1155–69.

Kolla, V., Moore, D. G., and Curry, J. R. 1976. Recent bottom-current activity in deep western Bay of Bengal. *Mar. Geophys., 21,* 255–70.

Krumbein, W. C., and Aberdeen, E. J. 1937. The sediments of Baratoria Bay (La.). *J. sedim. Petrol., 7,* 3–17.

Laughton, A. S. 1968. New evidence of erosion on the deep ocean floor. *Deep Sea Res., 15,* 21–29.

Ledbetter, M. T., and Johnson, D. A. 1976. Increased transport of Antarctic bottom water in the Vema Channel during the last ice age. *Science, N.Y., 194,* 837–39.

Menard, H. W. 1952. Deep ripple marks in the sea. *J. sedim. Petrol., 22,* 3–9.

Neuman, A. C., and Ball, M. M. 1970. Submersible observations in the Strait of Florida: geology and bottom currents. *Geol. Soc. Am. Bull., 81,* 2861–74.

Visher, G. S. 1969. Grain size distributions and depositional processes. *J. sedim. Petrol., 39,* 1074–1106.

Warnke, D. A. 1970. Glacial erosion, ice rafting, and glacial marine sediments—Antarctica and the Southern Ocean. *Am. J. Sci., 269,* 276–94.

Watkins, N. D., and Goodell, H. G. 1967. Geomagnetic polarity change and faunal extinctions in the Southern Ocean. *Science, N.Y., 156,* 1083–87.

Watkins, N. D., and Kennett, J. P. 1971. Antarctic Bottom Water: major change in velocity during the late Cenozoic between Australia and Antarctica. *Science, N.Y., 73,* 813–18.

Watkins, N. D., and Kennett, J. P. 1972. Regional sedimentary disconformities and upper Cenozoic changes in Bottom Water velocities between Australia and Antarctica. (*In* Hayes, D. E., *ed. Antarctic Oceanology 2.* Washington, American Geophysical Union, 273–94.) [Antarctic Research Series, Vol. 19.]

Watkins, N. D., and Kennett, J. P. 1973. Response of deep-sea sediments to changes in physical oceanography resulting from separation of Australasia and Antarctica. (*In* Tarling, D. H., and Runcorn, S. K., *eds. Continental Drift, Sea Floor Spreading and Plate Tectonics—Implication for the Earth Sciences.* New York, Academic Press, 787–97.)

Watkins, N. D., and Kennett, J. P. 1977. Erosion of deep-sea sediments in the Southern Ocean between longitudes 70° E and 190° E and contrasts in manganese nodule development. *Mar. Geol., 23,* 103–111.

133

Variations in Particle Alignment and Size in Sediments of the Vema Channel Record Antarctic Bottom-Water Velocity Changes during the Last 400,000 Years

M. T. LEDBETTER* and B. B. ELLWOOD*
Graduate School of Oceanography, University of Rhode Island, Kingston, Rhode Island 02881, U.S.A.

Abstract. The anisotropy of magnetic susceptibility (AMS) and mean particle size in the silt fraction has been measured in surface sediments and selected cores from the Vema Channel (lat. 30° 20′ S, long. 39° 05′ W). Results show that a function which represents magnetic grain, long-axis alignment is highly correlated with variation in mean size of the carbonate-free silt fraction. An increase in alignment with increase in particle size reflects an increase in bottom-water velocity. Strongest alignment and coarsest particle size are found in the axis of the Vema Channel under high-velocity Antarctic Bottom Water (AABW). In contrast, poor particle alignment and fine silt mean size characterize the shallower region east of the channel axis, consistent with lower current velocities for North Atlantic Deep Water (NADW). Core *Chain* 115-61 was collected near the boundary between northward-flowing AABW and southward-flowing NADW. The high correlation between silt size and AMS in the core has been used to infer fluctuation in AABW velocity during the last 400,000 years. The increases in AABW velocity do not fit patterns predicted in previously proposed models of climate and abyssal circulation. Geographic orientation of the maximum AMS axes in core *Chain* 115-61 was accomplished using the declination of remanent magnetism and a geocentric axial dipole as the reference azimuth. The data show that the long axes of magnetic grains in these sediments are aligned normal to the bathymetric contours, which suggests an alignment normal to the flow direction as predicted by a traction transport mode of sedimentation. The result indicates that AABW in the Vema Channel has flowed along an azimuth of approximately N 20° E for the past 400,000 years and has experienced several periods of substantial increases in velocity. The period of highest inferred bottom-water velocity resulted in a late Pliocene hiatus marked by a buried manganese pavement in core CH 115-62.

High-velocity bottom currents of both Antarctic Bottom Water (AABW) and the Antarctic Circumpolar Current have created vast areas of erosion in the Southern Ocean floor (Watkins and Kennett, 1971, 1972, 1977; Kennett and Watkins, 1976). The erosion of abyssal sediment by AABW has also occurred in areas far removed from the Southern Ocean (Johnson, 1972, 1974; Davies and others, 1975; van Andel and others, 1975). Since the erosion of abyssal sediment by high-velocity AABW removes the sedimentary record, the best locations for studying temporal patterns of AABW paleovelocity are not within the scour zones. At the margins or far downstream from the bottom-water source, however, the erosional power of the bottom current is sufficiently reduced to allow sedimentation, but still high enough to produce winnowing zones which are preserved in the sedimentary record (Huang and Watkins, 1977). In order to investigate temporal fluctuations in the velocity of AABW, a site was chosen in the southwest Atlantic where AABW flows as a western boundary undercurrent through the constriction of the Vema Channel.

The Vema Channel (Figure 133.1) is the principal deep passage through the Rio Grande Rise permitting AABW to

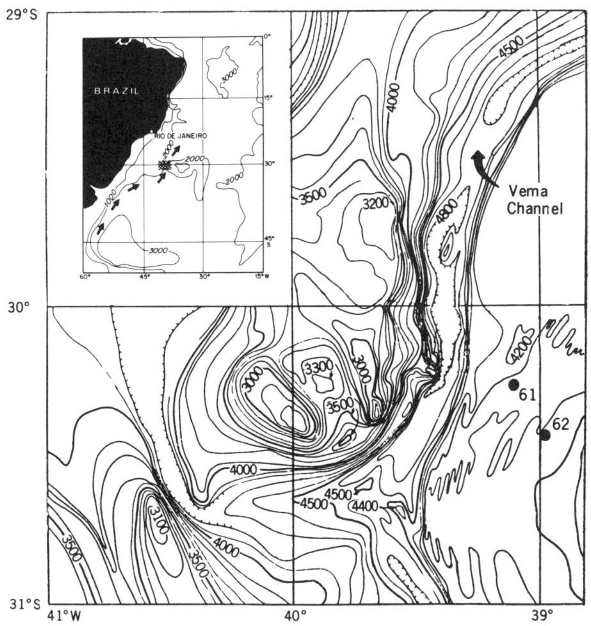

Figure 133.1. Bathymetry of the Vema Channel with location of *Chain* cores 115-61 and 115-62. Bathymetry in meters, from Johnson and others (1976). Inset shows location of the Vema Channel in the southwest Atlantic with the approximate path of Antarctic Bottom Water (arrows). For scale, 1° of lat. equals 108 km.

* Present address: Department of Geology, University of Georgia, Athens, Georgia 30602, U.S.A.

pass from the Argentine to the Brazilian Basin (Le Pichon and others, 1971; Johnson and others, 1976). The channel accommodates northward-flowing AABW beneath southward-flowing North Atlantic Deep Water (NADW) (Le Pichon and others, 1971; Melguen and Thiede, 1974; Johnson and others, 1976). It has been suggested that at the boundary between these two opposing water masses there exists a level with a minimum advective component (Ellwood and Ledbetter, 1977). This level has been identified from geologic evidence, and the transition between high-velocity AABW and low-velocity NADW corresponds to the midpoint in the benthic thermocline (Ledbetter and Johnson, 1976; Johnson and others, 1977). A vertical change in the position of this boundary has been used to infer a minor shallowing of the top of AABW during the last ice age (Ledbetter and Johnson, 1976). Similarly, textural sedimentary parameters in a core near the boundary have been used to infer fluctuations in bottom-water velocity during the last 160,000 years (Ellwood and Ledbetter, 1977). This paper extends our earlier AABW relative paleovelocity curve to approximately 400,000 years and examines the relationship of paleoclimate and inferred fluctuations in the velocity of AABW.

Anisotropy of magnetic susceptibility

For many years, the measurement of the anisotropy of magnetic susceptibility (AMS) has been used as a petrofabric indicator in natural sediment (Ising, 1942; Kent and Lowrie, 1975). Furthermore, AMS determinations of laboratory-deposited sediments have aided in defining conditions under which natural sediments are deposited (e.g., Hamilton and others, 1968).

Susceptibility (χ) is a proportionality constant relating an inducing magnetic field to an induced intensity of magnetization and is a tensor of the second rank which is usually represented by an ellipsoid. In a sediment sample which contains magnetic particles, this representation quadric is composed of the integrated shape effect of all the grains, each acting as individual ellipsoids. The magnitude and direction of the three orthogonal principal axes of the resultant ellipsoid determine the AMS. The final result from each sample is a maximum, intermediate, and minimum magnitude, χ_a, χ_b, and χ_c, respectively, and corresponding directions.

The two major grain fabrics in sediments are planar, due to particle settling, and linear, due to the action of forces tangential to the bed such as the drag of depositing currents. In the analysis of preferential bottom-current alignments, a parameter which emphasizes the linear fabric (or the maximum AMS magnitude, χ_a) is desirable. Such a function has been previously used (Ellwood and Ledbetter, 1977) in diagnosing the efficiency of magnetic grain long-axis alignment as a method for inferring fluctuations in relative bottom-current velocity. The function employed is F_s, where:

$$F_s = \frac{\chi'_a}{\sqrt{\chi'_b \chi'_c}} \qquad (1)$$

and χ'_a, χ'_b, and χ'_c are the maximum, intermediate, and minimum AMS magnitudes recalculated for a standard initial χ of 1×10^{-5} in c.g.s. mass units (representative of abyssal sediments), so that between-sample variation in F_s is not affected by simple variations in χ.

Particle size analysis

Sedimentation rates and particle size distributions are partially dependent on current-velocity fluctuations. Previous work (Hollister and Heezen, 1972) has summarized laboratory studies examining the relationship between current velocity and particle sizes during deposition, transportation, and erosion.

Flume studies have shown that calcareous oozes may be eroded by current velocities of 7 to 20 cm/sec, which are comparable to those commonly measured in the deep sea (Southard and others, 1971). Direct measurement of near-bottom current velocities in the Vema Channel yields maximum velocities of 15 to 20 cm/sec (Johnson and others, 1977; Reid and others, 1977). Changes in bottom-current velocity below the critical velocity required for scour can be inferred from changes in the mean particle size (Huang and Watkins, 1977). Therefore, changes in bottom-current velocity can be detected in winnowing zones which may be marginal to the more intense scour zones along the axis of the high-velocity flow.

Particle size variations in the silt fraction of carbonate-free samples from trigger core tops, representing surface sediment from the eastern flank of the Vema Channel, have been examined (Ledbetter and Johnson, 1976; Ellwood and Ledbetter, 1977). The silt particle size range of 4 μ to 62 μ (8 ϕ to 4 ϕ, where $\phi = -\log_2$, particle size in millimeters) was chosen for analysis because these particles should be sensitive to fluctuations in bottom-current velocity over the range of velocities suggested for the deep sea; it has been suggested that silt-size particles are kept in suspension by flow velocities of 1 cm/sec and are eroded by 6 to 15 cm/sec velocities (Hollister and Heezen, 1972). Therefore, the silt-size range should be a sensitive monitor of bottom-current velocity fluctuations in the Vema Channel. Earlier work has shown that a zone of coarse silt particles corresponds to a central stream of fast-flowing AABW in the axis of the Vema Channel, while finer grained silt-size sediment characterizes lower velocity NADW on the shallower flanks of the channel (Ledbetter and Johnson, 1976).

Methods

Cylindrical samples of approximately 11 cc (length to diameter ratio = 0.890) were taken at 10 cm intervals in CH 115-61 (790 cm length), which was taken near the present transition between the two bottom water masses. Our motive for studying this particular core in dtail is based on the argument that variations in AABW velocity should cause corresponding changes in this water mass boundary, which we have called the *Level of Least Motion* (Ellwood and Ledbetter, 1977). We expect, then, that the Level of Least Motion swept back and forth across the core site as a function of AABW velocity. The age of the bottom of core CH 115-61 was estimated from biostratigraphic evidence, which suggested that the basal carbonate peak corresponds to the base of isotopic stage 11 (Williams and Ledbetter,

1979) with an age of 410,000 years (Hays and Shackleton, 1976). The calculated sedimentation rate is therefore 1.8 cm/1000 yr.

Cylindrical samples were also taken continuously from a 50 cm section immediately above and below a manganese pavement between 563 and 572 cm depth in core CH 115-62 (Figure 133.3). A closely spaced sample interval was used in order to examine the sedimentary fabric associated with the hiatus.

The remanent magnetism was measured on a Superconducting Technology three-component cryogenic magnetometer, after alternating magnetic field partial demagnetization at a peak field of 200 oe, to minimize unstable components. The AMS for the same samples was then determined using a torsion balance, where the anisotropy in the plane of the observed section is detected by differences in the magnetic couple between the magnetic field applied across the sample and that induced in the sample (Stone, 1963). Initial susceptibility was measured on a Bison susceptibility bridge.

Each sample was then washed in distilled water, disaggregated in Calgon solution, and wet-sieved through a 62 μ sieve. Silt-size particles were separated from the fraction finer than 62 μ by repeated settling of the fine fraction in order to remove the bulk of the clay-size particles. The particle size analysis was performed on the silt fraction after removing the carbonate portion with 10% buffered glacial acetic acid. Analysis of carbonate-free sediment is preferred because it removes the effect of carbonate dissolution, which changes the particle size distribution of the total sediment in a manner that does not reflect bottom-current processes. Particles 4 ϕ to 8 ϕ (silt size) were analyzed on a Model ZB Coulter Counter using a 200 μ aperture, which allowed a resolution of the particle size distribution to 1/3 ϕ size class.

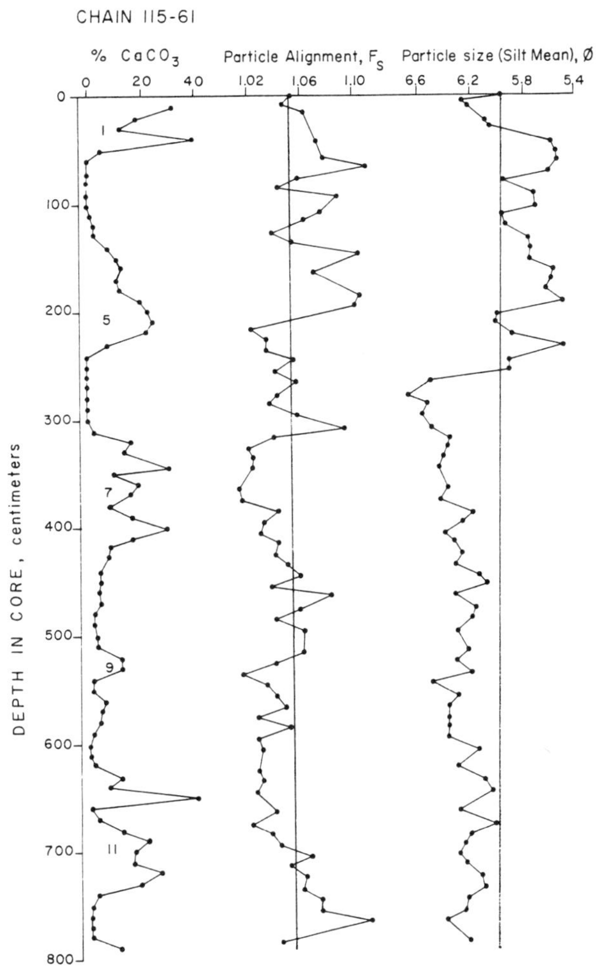

Figure 133.2. The percent total carbonate, the anisotropy of magnetic susceptibility parameter F_s (a function of preferred alignment of the long axis of magnetic particles), and the mean particle size of the carbonate-free silt fraction (in ϕ units) are plotted as functions of depth in core CH 115-61. The vertical line in the F_s and mean silt data separates values of inferred Antarctic Bottom Water paleovelocity which are greater than (to the right) or less than (to the left) values for the present sea floor. Numbers on the carbonate peaks refer to interglacial periods from the oxygen-isotopic stages which have been correlated with percent total carbonate in this area (Williams and Ledbetter, 1979).

Results

Figure 133.2 shows the percentage of calcium carbonate, the particle alignment (F_s), and the silt mean as functions of depth in the core *Chain* 115-61. The measurement of F_s and silt mean was performed on separate samples above 250 cm (Ellwood and Ledbetter, 1977) and on the same samples below. Measurement of both F_s and silt mean was also performed on the same samples in core CH 115-62. The silt mean and F_s are plotted with depth in the core for CH 115-62 in Figure 133.3.

Table 133.1 gives the mean remanent magnetism and mean AMS results for core CH 115-61, as well as the result of reorientation of the AMS principal axes using a zero declination remanent magnetism reference azimuth.

Discussion

AABW paleovelocity

Examination of Figure 133.2 shows down-core fluctuations in both the particle alignment and particle size parameter for core CH 115-61. The two parameters are positively correlated at the 99.9% significance level. The correspondence between the totally independent variables of particle alignment and silt mean suggests a clear dependence on a third variable. In Figure 133.2, the lines drawn down-core from the present sea-floor values of silt mean and F_s separate inferred fields of greater and reduced AABW velocity. In the data shown, particle alignment and silt-mean particle size determinations are clearly sensitive monitors of changing paleocurrent velocities. We have no other explanation for the parallel increase of particle alignment with increased particle size.

The fluctuations in inferred paleovelocity of AABW in core CH 115-61 have been compared to a dated paleoclimate curve derived from the correlation of the percent total carbonate curve with an ice-volume curve derived from oxygen isotopic studies of cores in the same area (Williams and Ledbetter, 1979). The base of the lowest carbonate peak (Figure 133.2) has been correlated with the isotopic stages 11/12 boundary (Williams and Ledbetter, 1979) at approximately 410,000 years ago (Hays and Shackleton, 1976). The paleovelocity fluctuations of

AABW do not correlate with the paleoclimate in any simple way, which suggests that models predicting a simple response of abyssal circulation to the glacial-interglacial cycles of the Pleistocene may be in need of refinement. For example, Arrhenius (1952) suggested that *increased* production of AABW during glacial periods was the result of increased upwelling; this was inferred from increased calcium carbonate preservation in equatorial Pacific cores. Weyl (1968), however, predicted *decreased* production of AABW during glacial periods because of decreased salinity at the source that resulted from a decrease in Norwegian Sea overflow water due to a proposed increased North Atlantic sea-ice cover. Newell (1974) proposed that AABW production increased at the inception and cessation of glacial periods because of changes in the oceanic poleward energy flux as it affects Antarctic sea-ice cover. In each model, a regular pattern of climate and abyssal circulation response is predicted, but no such regular pattern is apparent in the paleovelocity curve (Figure 133.2). The relationship of paleoclimate and AABW paleovelocity therefore seems to be more complex than predicted by a simple model involving only one feedback mechanism between paleoclimate and bottom-water production.

Paleovelocity fluctuations of AABW recorded in core CH 115-61 are inferred to be below the critical erosional velocity. We have also performed the same particle alignment and size measurements on samples from a known high-velocity environment in core CH 115-62 (Figure 133.2). The 8-cm-thick manganese pavement recovered at 560 cm in core CH 115-62 represents a hiatus in sedimentation caused by a late Pliocene high-velocity bottom-water event which occurred prior to 2.9 m.y. ago (Williams and Ledbetter, 1979). The values of F_s and silt mean record a gradual increase in paleovelocity below the hiatus until the critical erosional velocity was reached, at which time sedimentation ceased and the manganese pavement formed (Figure 133.3). Later, as the bottom-water velocity lessened, sedimentation was resumed. The high values measured for F_s and silt mean indicate that the paleovelocity was still relatively high but decreased rapidly with small, intermittent increases (Figure 133.3).

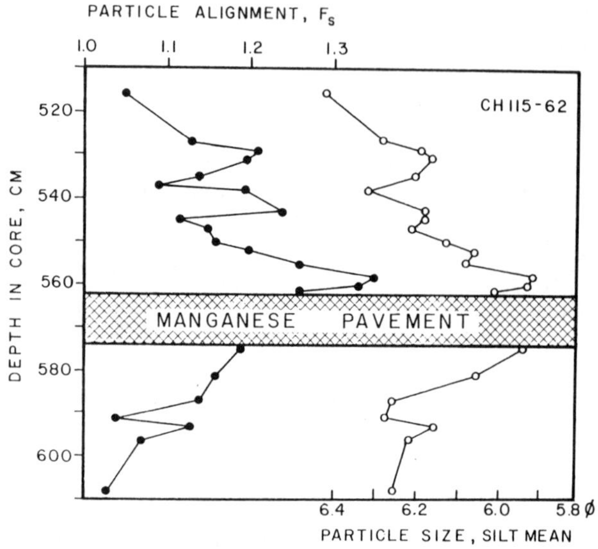

Figure 133.3. The alignment parameter F_s and the mean silt particle size (in ϕ units) are plotted as a function of depth in the core CH 115-62. The highest values of F_s and silt mean are associated with the 8-cm-thick manganese pavement which was formed during a period of high-velocity bottom-water flow in late Pliocene time (Williams and Ledbetter, 1979).

AABW DIRECTION

Core CH 115-61 was oriented during recovery with respect to top and bottom only, but the very stable Brunhes remanent magnetic (RM) directions recorded (with median destructive fields >100 oe) provide the possibility of rotating the mean of these vectors until it coincides with a declination of 0°, corresponding to a direction represented by an axial dipole field. Since each sample represents a magnetic recording which has lasted for more than 1000 years (calculated from an average sedimentation rate of 1.8 cm/1000 yr in the 2.3-cm-long samples), the axial dipole should approximate the earth's geomagnetic field during this length of time. The geographically oriented mean RM direction can then be used to reorient the mean AMS direction (Table 133.1).

This reorientation, using the mean RM declination, yields a long axis (χ_a) AMS azimuth approximately normal to the bathymetric contours and thus normal to the inferred current direction in the axis of the channel (compare Figure 133.1 to rotated D in TAble 133.1). Since grain sizes in these cores are small and expected current velocities large, we suggest that long AMS axes have been aligned normal to AABW flow direction in the Vema Channel, as

Table 133.1. Reorientation data for core CH 115-61

Remanent magnetism (RM)[a]				Long-axis anisotropy of magnetic susceptibility (AMS)[b]				
N	Mean *D*	Mean *I*	α_{95}	*N*	Mean *D*	Mean *I*	Rotated *D*	α_{95}
44	56.7°	−37.0°	3.2°	54	349.5° (169.5)	8.7° (−8.7)	292.8° (112.8)	5.0°

N: number of samples.
D: declination in degrees east of arbitrary reference point.
I: inclination in degrees above (or if negative, below) horizontal.
α_{95}: semi-vertical angle of 95% confidence cone (Fisher, 1953).
Rotated *D*: mean *D* of the maximum direction of AMS rotated into geograhic coordinates using the mean *D* of the RM which is assumed to be 0° (see text).

Numbers in parentheses are possible alternate AMS directions, since AMS determinations can only yield an azimuth, not a true direction (see text).

[a] RM results are after alternating field cleaning in a 200 oe field; mean RM is calculated for 0–500 cm in the core; below 500 cm, the RM is inconsistent with the top of the core.
[b] AMS data are for the entire core.

might be expected for slowly deposited long grains moving by a traction transport process. Reorientation of the AMS data then yields a mean flow azimuth for AABW of ~23° east of north (if one assumes that AABW is flowing northward), which is in good agreement with the known bathymetric trend at the site from which the cores were recovered (Figure 133.1). A net long-axis alignment normal to the current-flow direction, while not surprising from traction transport considerations, might at first examination appear to contradict the axial dipole assumption ($D = 0°$) used for the geographic reorientation. Since the RM resides mainly in the very finest fraction (which is aligned more readily by the geomagnetic field), we propose that the coarser fraction is the dominant contributor to the AMS signal and is controlled by current flow. The nearly normal orientation of the mean RM direction to the azimuth of the maximum AMS axis (Table 133.1) does not contradict this assumption.

Conclusions

Anisotropy of magnetic susceptibility (AMS) and mean silt particle size for samples in core CH 115-61, recovered from near the boundary of the two bottom-water masses, vary substantially and systematically. High AMS values, indicating efficient alignment of magnetic grain long axes, correspond to coarse particle sizes within the silt-size fraction. This strong positive association of preferred alignment and coarse particle sizes is therefore interpreted to be the direct effect of variations in AABW current velocity. Increases in AABW velocity result in better magnetic grain long-axis alignment and the deposition of a coarse lag deposit, due to the winnowing of the finer particles. Fluctuations of inferred AABW velocity do not show a simple pattern with respect to paleoclimate. The response of abyssal circulation to climatic fluctuations may be more complex than existing models have predicted, possibly because of complex feedback mechanisms.

The high values of particle aligment and size near the manganese pavement in core CH 115-62 suggest a bottom-current velocity which slowly increased until deposition was inhibited at the time of the formation of the pavement. Sedimentation resumed above the pavement as the inferred velocity decreased in a complex fashion, with small intermittent increases, until a low velocity was reached.

Acknowledgments. We would like to thank N. D. Watkins for his advice and constructive criticisms of the manuscript. We thank David Johnson for providing *Chain* core samples used in this study. The University of Rhode Island computer facility provided computer time. B. B. Ellwood was supported under National Science Foundation Grant GA37170 to N. D. Watkins, and M. T. Ledbetter was supported under a National Science Foundation Doctoral Dissertation Research grant and by the National Science Foundation Office of Polar Programs.

References cited

Arrhenius, G. 1952. Sediment cores from the east Pacific. *Swed. Deep-sea Exped. (1947-1948) Rep., 5,* Part 1, 89 pp.

Davies, T. A., Weser, O. E., Luyendyk, B. P., and Kidd, R. B. 1975. Unconformities in the sediments of the Indian Ocean. *Nature, Lond., 253,* 15-19.

Ellwood, B. B., and Ledbetter, M. T. 1977. Antarctic Bottom Water fluctuations in the Vema Channel: effects of velocity changes on particle alignment and size. *Earth planet. Sci. Lett., 35,* 189-98.

Fisher, R. A. 1953. Dispersion on a sphere. *Proc. R. Soc. Lond.,* Ser. A, *217,* 295-305.

Hamilton, N., Owens, W. H., and Rees, A. I. 1968. Laboratory experiments on the production of grain orientation in shearing sand. *J. Geol., 76,* 465-72.

Hays, J. D., and Shackleton, N. J. 1976. Globally synchronous extinction of the radiolarian *Stylotractus universus. Geology, 4,* 649-52.

Hollister, C. D., and Heezen, B. C. 1972. Geological effects of ocean bottom currents. (*In* Gordon, A. L., *ed. Studies in Physical Oceanography. 2.* New York, Gordon & Breach, 37-66.)

Huang, T. C., and Watkins, N. D. 1977. Contrasts between the Brunhes and Matuyama sedimentary records of bottom water activity in the South Pacific. *Mar. Geol., 23,* 113-32.

Ising, G. 1942. On the magnetic properties of varved clays. *Ark. Mat. Astr. Fys., A29,* 1-37.

Johnson, D. A. 1972. Ocean-floor erosion in the equatorial Pacific. *Geol. Soc. Am. Bull., 83,* 3121-44.

Johnson, D. A. 1974. Deep Pacific circulation: intensification during the early Cenozoic. *Mar. Geol., 17,* 71-78.

Johnson, D. A., Ledbetter, M. T., and Burckle, L. H. 1977. Vema Channel paleooceanography: Pleistocene dissolution cycles and episodic bottom water flow. *Mar. Geol., 23,* 1-33.

Johnson, D. A., McDowell, S. E., Sullivan, L. G., and Biscaye, P. E. 1976. Abyssal hydrography, nephelometry, currents and benthic boundary layer structure in the Vema Channel. *J. geophys. Res., 81,* 5771-86.

Kennett, J. P., and Watkins, N. D. 1976. Regional deep-sea dynamic processes recorded by late Cenozoic sediments of the southeastern Indian Ocean. *Geol. Soc. Am. Bull., 87,* 321-39.

Kent, D. V., and Lowrie, W. 1975. On the magnetic susceptibility anisotropy of deep sea sediment. *Earth planet. Sci. Lett., 28,* 1-12.

Ledbetter, M. T., and Johnson, D. A. 1976. Increased transport of Antarctic Bottom Water in the Vema Channel during the last ice age. *Science, N. Y., 194,* 837-39.

Le Pichon, X., Ewing, M., and Truchan, M. 1971. Sediment transport and distribution in the Argentine Basin, 2. Antarctic bottom current passage into the Brazil Basin. (*In* Ahrens, L. H., Press, F., Runcorn, S. K., and Urey, H. C., *eds. Physics and Chemistry of the Earth, 8.* New York, Pergamon Press, 31-48.)

Melguen, M., and Thiede, J. 1974. Facies distribution and dissolution depths of surface sediment components from the Vema Channel and the Rio Grande Rise (southwest Atlantic Ocean). *Mar. Geol., 17,* 341-53.

Newell, R. E. 1974. Changes in the poleward energy flux by the atmosphere and ocean as a possible cause for ice ages. *Quat. Res., 4,* 117-27.

Reid, J. L., Nowlin, W. D., Jr., and Patzert, W. C. 1977. On the characteristics and circulation of the southwestern Atlantic Ocean. *J. Phys. Ocean, 7,* 62-91.

Southard, J. B., Young, R. A., and Hollister, C. D. 1971. Experimental erosion of fine abyssal sediment. *J. geophys. Res., 76,* 5903-09.

Stone, D. B. 1963. Anisotropic magnetic susceptibility measurements on a phonolite and on a folded metamorphic rock. *Geophys. J., 7,* 375-90.

van Andel, T. H., Heath, G. R., and Moore, T. C. 1975. Cenozoic history and paleoceanography of the central equatorial Pacific. *Mem. geol. Soc. Am., 143,* 134 pp.

Watkins, N. D., and Kennett, J. P. 1971. Antarctic Bottom Water: major change in velocity during the late Cenozoic between Australia and Antarctica. *Science, N.Y., 173,* 813–18.

Watkins, N. D., and Kennett, J. P. 1972. Regional sedimentary disconformities and Upper Cenozoic changes in bottom water velocities between Australia and Antarctica. (*In* Hayes, D. E., *ed. Antarctic Oceanology* 2. Washington, American Geophysical Union, 273–94.) [Antarctic Research Series, Vol. 19.]

Watkins, N. D., and Kennett, J. P. 1977. Erosion of deep-sea sediments in the Southern Ocean between longitudes 70° E and 190° E and contrasts in manganese nodule development. *Mar. Geol., 23,* 103–11.

Weyl, P. K. 1968. The role of the oceans in climatic change: a theory of the ice ages. *Meteorol. Monogr., 8,* 37–62.

Williams, D. F., and Ledbetter, M. T. 1979. Chronology of late Brunhes biostratigraphy and late Pliocene disconformities in the Vema Channel (SW Atlantic). *Mar. Micropaleo., 4,* 125–36.

134

Rates of Particle Mixing in Antarctic Deep-Sea Sediments Using ^{210}Pb Measurements

DAVID J. DEMASTER*
Department of Geology and Geophysics, Yale University,
New Haven, Connecticut 06520, U.S.A.

Abstract. The distribution of excess ^{210}Pb activity (i.e., ^{210}Pb activity in excess of its parent, ^{226}Ra) in four Antarctic deep-sea cores has been examined to establish rates of particle mixing. ^{210}Pb ($t_{1/2}$ = 22 yr) should be confined to the top few millimeters of the deep-sea Antarctic sediment column, if sediment accumulation was the only process controlling its distribution. ^{210}Pb, however, is commonly found to a depth of 8 cm in Antarctic sediments, and in one siliceous core, it was found to a depth of 15 cm. The deep penetration of excess ^{210}Pb activity in Antarctic deep-sea sediments is attributed to particle mixing. The particle mixing coefficient, D_b, determined from the exponential decrease of excess ^{210}Pb activity with depth in four Antarctic cores, ranges between 40 and 160 cm^2/ky.

Particle mixing imposes important constraints on the interpretation of the sedimentary record. First, the distribution of any geologic or isotopic tracer associated with particles is influenced by sediment mixing. In a siliceous sediment accumulating at 1 cm/1000 yr, 15 cm of particle mixing means that fresh surface material has been intimately mixed with sediments as old as 15,000 years. Second, particle mixing coupled with sediment accumulation and dissolution can affect the siliceous species preservation pattern and the particle size distribution in the sedimentary record. These effects occur in siliceous sediments because dissolution acts preferentially on siliceous particles with high surface areas (small and delicate tests), whereas particle mixing on the deep-sea floor presumably acts on the bulk sediment. Particle mixing may increase the chance of survival in the sedimentary record of otherwise easily dissolvable diatom and radiolarian tests.

Biological and physical mixing of deep-sea sediments has been recognized for more than 35 years (Bramlette and Bradley, 1941; Arrhenius, 1963; Goldberg and Koide, 1962; Berger and Heath, 1968; Guinasso and Schink, 1975). Recent success in modeling the vertical distribution of plutonium and microtektites in marine sediments has produced quantitative estimates of the rate of particle mixing on the deep-sea floor (Guinasso and Schink, 1975). Naturally occurring radionuclides can also be used to establish the rate of particle mixing. This is a general technique which does not require knowledge of the mode and duration of introduction of the tracer into the sediment column, as is required when modeling the distribution of plutonium or microtektites.

Nozaki and others (1977) used naturally occurring ^{210}Pb to determine the rate of particle mixing for a calcareous core from the North Atlantic Ocean. ^{210}Pb ($t_{1/2}$ = 22 yr) is produced from the decay of ^{226}Ra in ocean water. Although ^{210}Pb can also be supplied to the water column directly from the atmosphere, this supply has been shown to be negligible in the Antarctic (Picciotto and others, 1968; DeMaster, 1979). ^{210}Pb is scavenged from the water column by settling particles and arrives at the sediment-water interface in excess of its parent, ^{226}Ra. In the near interface sediments, excess ^{210}Pb activity can be used as a particle tracer to establish the rate of sediment mixing. In the absence of sediment mixing, excess ^{210}Pb would be confined to the top few millimeters of the sediment column because of its short half-life and the slow rate of sediment accumulation in typical deep-sea deposits (about 1 cm/1000 yr). Excess ^{210}Pb profiles from deep-sea sediments supply a description of the sediment mixing which has occurred during the last 100 years. Radionuclides with longer half-lives (i.e., ^{32}Si, $t_{1/2}$ = 276 yr; ^{231}Pa, $t_{1/2}$ = 33,000 yr; and ^{230}Th, $t_{1/2}$ = 75,000 yr) must be used in addition to ^{210}Pb to provide a complete description of the long-term effects of mixing on the sedimentary record. This paper describes the use of naturally occurring radionuclides in establishing rates of particle mixing in Antarctic sediments and the consequences which arise from this mixing.

Methods

Four South Atlantic gravity cores taken aboard the *Atlantis II* during January 1976 were analyzed in this study (see Table 134.1). The gravity corer was lowered as slowly as possible near the ocean floor, in order to preserve the sediment-water interface.

Table 134.1. Core locations and descriptions

Core number	Location	Water depth (m)	Length of core (cm)	Sediment type
AII76-12	66°47′S, 30°08′E	4080	113	Clay
AII76-16	53°00′S, 35°38′E	4340	30	Siliceous ooze
AII76-17	52°59′S, 35°43′E	4730	19	Siliceous ooze
AII76-18	53°16′S, 35°54′E	4540	45	Siliceous ooze

* Present address: Department of Marine Science and Engineering, North Carolina State University, Raleigh, North Carolina 27650, U.S.A.

Pb carrier, ^{233}Pa, and ^{234}Th spike were added to 1 to 2 g of dried sediment before the sample underwent total dissolution. If uranium was measured, a ^{232}U-^{228}Th spike was also added. ^{226}Ra activity was measured by ^{222}Rn extraction followed by scintillation counting. Uranium, thorium, protactinium, and lead were isolated by anion exchange techniques (Ku, 1966; Benninger, 1976; DeMaster, 1979). ^{210}Pb was measured by beta counting the ingrowth of its daughter, ^{210}Bi ($t_{1/2}$ = 5 days). Thorium, protactinium, and uranium isotopes were measured by alpha spectrometry. ^{231}Pa activity was measured directly, as well as by counting its granddaughter, ^{227}Th. The agreement between these two analytical methods for determining ^{231}Pa activity is very good (Mangini and Sonntag, 1977; DeMaster, 1979).

Results

The radiochemical data are given in Table 134.2. The 1σ error given for each analysis was determined solely from counting statistics.

The accumulation rate and the depth of mixing were established for core AII76-16 using the excess $^{230}Th/^{231}Pa$ ratio (see Figure 134.1). ^{231}Pa ($t_{1/2}$ = 33,000 yr) and ^{230}Th ($t_{1/2}$ = 75,000 yr) are produced in sea water from the decay of ^{235}U and ^{234}U, respectively. Like ^{210}Pb they are scavenged from the water column by settling particles. The ratio of excess ^{230}Th to excess ^{231}Pa as a function of time is given by the following equation (Rosholt and others, 1961):

$$\frac{A'^{230}\mathrm{Th}}{A'^{231}\mathrm{Pa}}(t) = \left(\frac{A'^{230}\mathrm{Th}}{A'^{231}\mathrm{Pa}}\right)_{t=0} \cdot \exp\{(\lambda_{^{231}\mathrm{Pa}} - \lambda_{^{230}\mathrm{Th}})t\}, \quad (1)$$

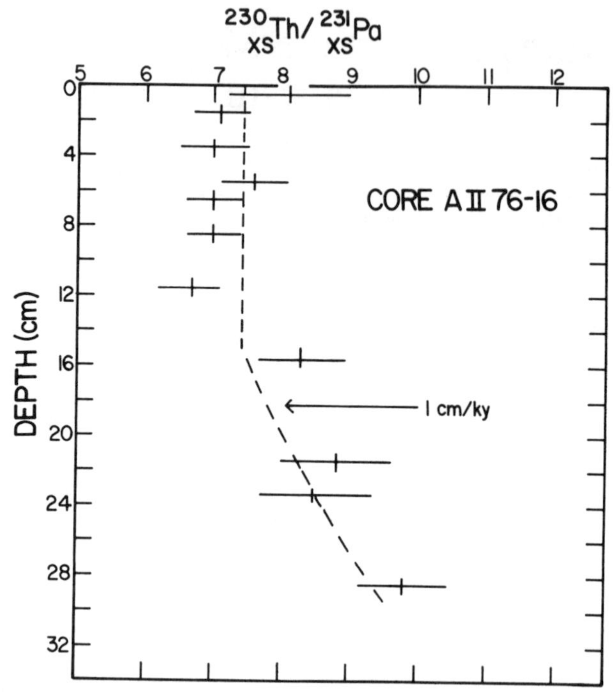

Figure 134.1. Excess ^{230}Th/excess^{231}Pa profile for core AII76-16. Excess ^{230}Th and excess ^{231}Pa activity were calculated by subtracting the activity of the appropriate uranium parent from the total radionuclide activity.

where A' is the unsupported activity of the radionuclide, t is time, and λ is the appropriate decay constant. In the top 12 cm of core AII76-16, the $^{230}Th/^{231}Pa$ ratio is about 7.3. This is significantly different from the dissolved uranium sea-water production ratio of 10.7 and is attributed to dif-

Table 134.2. Radiochemical data from Antarctic cores AII76-12, -16, -17, and -18[a]

Core number	Depth (cm)	^{226}Ra (dpm/g)	Total ^{210}Pb (dpm/g)	Excess ^{210}Pb (dpm/g)	Excess ^{230}Th / Excess ^{231}Pa
AII76-12	0–0.8	4.78 ± 0.16	24.18 ± 0.36	19.40 ± 0.39	—
	1.9–2.4	5.47 ± 0.18	10.03 ± 0.19	4.56 ± 0.27	—
	4.0–4.6	6.04 ± 0.21	6.77 ± 0.16	0.73 ± 0.26	—
	7.5–8.0	6.42 ± 0.28	6.83 ± 0.25	0.41 ± 0.32	—
	10.8–11.3	7.33 ± 0.31	6.98 ± 0.19	−0.35 ± 0.36	—
AII76-16	0–1	4.58 ± 0.18	32.04 ± 0.71	27.46 ± 0.74	8.1 ± 0.9
	1–2	4.00 ± 0.15	16.10 ± 0.44	12.10 ± 0.48	7.1 ± 0.4
	2–3	3.44 ± 0.12	7.74 ± 0.26	4.30 ± 0.29	—
	3–4	3.86 ± 0.14	5.73 ± 0.20	1.87 ± 0.24	7.0 ± 0.5
	5–6	2.70 ± 0.10	3.09 ± 0.11	0.39 ± 0.15	7.6 ± 0.5
	6–7	—	—	—	7.0 ± 0.4
	8–9	3.13 ± 0.12	3.85 ± 0.16	0.72 ± 0.20	7.0 ± 0.4
	11–12	2.90 ± 0.10	7.23 ± 0.21	4.33 ± 0.27	6.7 ± 0.5
	14–15	3.10 ± 0.13	4.46 ± 0.18	1.36 ± 0.22	—
	15–16	—	—	—	8.3 ± 0.7
	21–22	12.29 ± 0.60	11.47 ± 0.51	−0.82 ± 0.67	8.8 ± 0.7
	23–24	—	—	—	8.5 ± 0.7
	28–29	—	—	—	9.8 ± 0.6
AI76-17	0–1	3.49 ± 0.13	21.19 ± 0.42	17.70 ± 0.43	—
	1–2	3.79 ± 0.15	16.43 ± 0.38	12.64 ± 0.40	—
	3–4	3.22 ± 0.13	7.84 ± 0.20	4.62 ± 0.24	—
	7–8	3.04 ± 0.13	3.70 ± 0.19	0.66 ± 0.21	—
	11–12	4.38 ± 0.18	6.12 ± 0.13	1.74 ± 0.23	8.5 ± 0.6
AII76-18	0–2	5.33 ± 0.22	49.04 ± 0.8	43.71 ± 1.80	—
	1–2	4.70 ± 0.19	25.32 ± 0.56	20.62 ± 0.59	—
	3–4	4.28 ± 0.16	11.75 ± 0.24	7.47 ± 0.29	—
	7–8	3.39 ± 0.14	3.89 ± 0.14	0.50 ± 0.18	—
	11–12	4.58 ± 0.16	5.56 ± 0.26	0.98 ± 0.29	8.0 ± 0.5

[a] All radiochemical data are expressed in activity units of disintegrations per minute per gram of dried sediment (dpm/g).

ferences in the chemical behavior of thorium and protactinium during scavenging from the water column (Sackett, 1966).

Assuming that the chemical fractionation between thorium and protactinium has been constant with time, the bottom samples in core AII76-16 yield an accumulation rate of 1 cm/1000 yr. Sediment accumulation rates were not measured for nearby cores AII76-17 and AII76-18. For the purposes of this paper, it is assumed that the rates of accumulation for these two cores are equal to the rate of accumulation found for core AII76-16. Excess ^{230}Th measurements in core AII76-12 indicate that the sediment accumulation rate has been less than 3 cm/ky (DeMaster, 1979). Even if the accumulation rates were in error by a factor of two, the ^{210}Pb mixing coefficients would remain unchanged because of the short half-life of ^{210}Pb and the slow rates of deep-sea sediment accumulation.

As shown in Figure 134.2, all four cores show a decrease in excess ^{210}Pb activity in the top 8 cm of the sediment column. Core AII76-16, however, shows a second maximum in excess ^{210}Pb at approximately 12 cm.

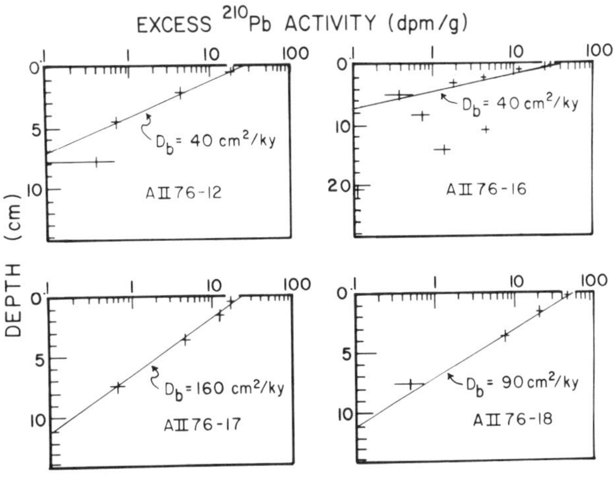

Figure 134.2. Excess ^{210}Pb profiles for Antarctic cores AII76-12, -16, -17, and -18. The best-fit line through the data is shown with the corresponding mixing coefficient, D_b.

Discussion

The sediment accumulation rates for the South Atlantic siliceous and clay-rich cores range between 1 and 3 cm/1000 yr. Excess ^{210}Pb should be confined to the top few millimeters of these cores if sediment accumulation was the only process controlling its distribution. As shown in Figure 134.2, excess ^{210}Pb is commonly found to a depth of 8 cm in Antarctic sediments. This deep penetration of excess ^{210}Pb activity is attributed to particle mixing.

A steady-state, one-dimensional equation commonly used to describe particle mixing (Goldberg and Koide, 1962; Guinasso and Schink, 1975) is shown below:

$$D_b \frac{\partial^2 N}{\partial x^2} - S\frac{\partial N}{\partial x} + P - \lambda N = 0, \qquad (2)$$

where D_b is the sediment mixing coefficient in cm²/ky, S is the sediment accumulation rate in cm/ky, N is the concentration of ^{210}Pb in atoms per gram, P is the *in situ* production rate of ^{210}Pb from ^{226}Ra in atoms per gram per ky, λ is the decay constant for ^{210}Pb in ky^{-1}, and x is the depth in the sediment column in cm. D_b, S, P, and the bulk sediment density are assumed constant within the mixed layer. Dissolution of silica is assumed to have a negligible effect on the distribution of excess ^{210}Pb and on the sediment accumulation rate. The solution to equation (2) using the boundary conditions, $A = A°$ at $x = 0$ and $A_{^{210}Pb} = A_{^{226}Ra}$ at $x = \infty$, is shown below (where $A = \lambda N$).

$$A(x) \text{ excess } ^{210}\text{Pb} = (A°_{^{210}\text{Pb}} - A_{^{226}\text{Ra}}) \cdot \exp\left(\frac{S - \sqrt{S^2 + 4D_b\lambda}}{2D_b} \cdot x\right). \qquad (3)$$

The mixing coefficient for each core was calculated using equation (3) by incorporating the excess ^{210}Pb data and the appropriate sediment accumulation rate. In core AII76-16 only the ^{210}Pb data from the upper 8 cm were used in the calculation of D_b.

The mixing coefficient for each core is given in Figure 134.2. Differences in these mixing coefficients do not seem to be related to sediment type. For example, the mixing coefficient for the Antarctic clay-rich core AII76-12 (40 cm²/ky) is comparable to the mixing coefficient for siliceous core AII76-16 (40 cm²/ky). There is good agreement between the ^{210}Pb mixing coefficients found in this Antarctic study (40 to 160 cm²/ky) and the ^{210}Pb mixing coefficient found by Nozaki and others (1977) for a calcareous Atlantic Ocean sediment (190 cm²/ky). Mixing coefficients (Guinasso and Schink, 1975) based on the distribution of plutonium in several cores from the Atlantic Ocean (Noshkin and Bowen, 1973) yield similar values of D_b (90 to 400 cm ²/ky).

Depth of Mixing

In core AII76-16, the maximum depth of mixing can be determined from the variation in the $^{230}Th/^{231}Pa$ ratio, as well as from the excess ^{210}Pb profile. The constant $^{230}Th/^{231}Pa$ ratio (Figure 134.1) near the surface of core AII76-16 indicates that mixing has occurred to a depth of at least 12 cm, whereas the higher $^{230}Th/^{231}Pa$ ratios found at depths of 22, 24, and 29 cm show that rapid mixing (D_b = 40 cm²/ky) has not occurred in the lower region of the sediment column. The $^{230}Th/^{231}Pa$ gradient in the lower 8 cm of core AII76-16, however, could result from slow episodic mixing rather than from sediment accumulation.

Particle mixing generally reduces the gradient in a profile of a radiochemical tracer that is added at a constant rate to a sediment accumulating at a constant velocity. Hence, mixing will cause the inferred accumulation rate to be higher than the true accumulation rate. If the $^{230}Th/^{231}Pa$ ratios near the bottom of core AII76-16 have been affected by particle mixing, then the true accumulation rate for this core would be even slower than 1 cm/1000 yr. In order to establish if slow episodic mixing or sediment accumulation has caused the observed $^{230}Th/^{231}Pa$ profile in core AII76-16, the accumulation rate must be determined deep in the core where all effects of particle mixing can virtually be ruled out. Siliceous core AII76-16 is too short for this purpose; however, ^{230}Th measurements made by the writer on a nearby 11 m piston core, RC17-58 (lat. 53° 31′S; long. 36° 38′ E), yield an accumulation rate of 1.8 cm/1000 yr (DeMaster, 1979). The agreement between this value and the value determined from the $^{230}Th/^{231}Pa$ gradient deep in core AII76-16 suggest that the high

$^{230}Th/^{231}Pa$ ratios near the bottom of core AII76-16 are a result of sediment accumulation and not a result of episodic particle mixing.

The excess ^{210}Pb activity found between 14 and 15 cm in core AII76-16 shows that episodic mixing (on a 100 year time scale) has occurred to a depth of 15 cm. With a sediment accumulation rate of 1 cm/1000 yr, this 15 cm of mixing means that fresh surface material has been mixed into and has contaminated 15,000 year old sediment. Conversely, a 15,000 year old siliceous test has a finite probability of occurring in the top centimeter of the sediment column. Such a disturbance in the last 15,000 years of the sedimentary record is an important consideration when interpreting the sedimentary record of the last transition from a glacial to a postglacial regime.

Non-steady state mixing

The mixing coefficients for the four cores examined in this study have been calculated assuming that the mixing process occurs in a diffusion-like manner (an infinite number of small random steps). This appears to be a good assumption in the upper 8 cm of these Antarctic cores. However, the second maximum in excess ^{210}Pb activity in core AII76-16 suggests that not all mixing is the result of a steady-state diffusion-like process acting on interface material. Such a maximum could result from organisms feeding on surface material but burrowing or defecating at a depth of 12 to 15 cm. This has been shown to occur in North Pacific deep-sea sediments (Donahue, 1971).

As sediment mixing may penetrate to different depths on different time scales, it is important to measure radionuclides with varying half-lives. The ^{210}Pb data in this paper demonstrate that most of the mixing in the upper 8 cm of the sediment column can be treated as a steady-state, diffusion-like process which occurs on a 100 year time scale. At depths greater than 8 cm, the assumption of steady-state mixing on a ^{210}Pb time scale breaks down because the events which mix sediments to these depths occur at an average time interval which is equal to or greater than the half-life of ^{210}Pb.

As shown in the section above, $^{230}Th/^{231}Pa$ ratios can provide additional information on the long-term mixing processes. Although mixing events between 8 and 15 cm in core AII76-16 seem episodic on a ^{210}Pb time scale, these events seem to approach steady-state conditions on a longer time scale, as indicated by the roughly constant $^{230}Th/^{231}Pa$ ratio throughout the upper 15 cm of the sediment column.

Further implications of mixing

In core AII76-16, the pore-water dissolved silica profile reaches its maximum value at a depth of 5 cm (K. Fanning, personal communciation, 1977). Below this depth, the dissolved silica concentration is constant, which means that there is no net dissolution of silica.

Particle mixing can affect the size distribution pattern of siliceous tests in a sediment column because dissolution preferentially acts on siliceous particles with high surface area (small and delicate tests), whereas sediment mixing on the deep-sea floor presumably acts on the bulk sediment. For example, if a fragile, high surface-area test and a robust siliceous test are deposited on a nonmixed, sediment-water interface, the robust test will have a higher probability of being incorporated into the long-term sedimentary record because it is less susceptible to dissolution. If these same two siliceous tests are deposited on a sediment column in which the near interface sediments are mixed, there is a chance that both tests will be quickly transferred to the base of the mixed layer (such as in core AII76-16, where there is excess ^{210}Pb at 15 cm).

In core AII76-16, the dissolution of siliceous tests below 5 cm is greatly retarded by the high dissolved silica concentrations in the pore water. If the time interval between the deep episodic mixing events is sufficiently long relative to the rate of sediment accumulation, both the delicate and robust tests can move downward out of the mixed layer by way of sedimentation. Once out of the mixed layer, these tests can not be remixed upward into the surface zone of rapid silica dissolution. In this way, sediment mixing provides a potential mechanism whereby small or delicate siliceous tests can bypass the region of intense dissolution (between 0 and 5 cm). Therefore, mixing may increase the chance of survival in the sedimentary record for otherwise easily dissolvable diatom or radiolarian tests.

Conclusions

The following main conclusions can be drawn from this study:

1. ^{210}Pb has been shown to be a useful tracer in establishing rates of particle mixing in Antarctic deep-sea sediments.
2. ^{210}Pb particle mixing rates from four Antarctic sediment cores range between 40 and 160 cm^2/ky. Differences in these mixing coefficients do not seem to be related to sediment type.
3. Sediment mixing has disturbed the last 15,000 years of the sedimentary record in a South Atlantic Antarctic siliceous sediment accumulating at 1 cm/1000 yr.
4. The intensity of particle mixing may affect the species preservation pattern and the particle size distribution in a deep-sea siliceous sediment because dissolution preferentially acts on siliceous tests with a high surface area, whereas particle mixing acts on the bulk sediment.

Acknowledgments. I thank K. K. Turekian, G. R. Holdren, Y. Nozaki, and S. Krishnaswami for their helpful reviews. Kent Fanning supplied some of the Antarctic sediment samples as well as information concerning the dissolved silicon profile for core AII76-16. This research was supported by the National Science Foundation through Grant OCE76-02-39 and its precursor (Karl K. Turekian, principal investigator).

References cited

Arrhenius, G. O. S. 1963. Pelagic sedimentation. (*In* Hill, M., *ed. The Sea, 3.* New York, John Wiley, 655–727.)

Benninger, L. K. 1976. *The Uranium-series Radionuclides as Tracers of Geochemical Processes in Long Island Sound.* Ph.D. dissertation, Yale University, 151 pp.

Berger, W. H., and Heath, G. R. 1968. Vertical mixing in sediments. *J. Mar. Res., 26,* No. 2, 134–43.

Bramlette, M. N., and Bradley, W. H. 1941. Geology and biology of North Atlantic deep-sea cores between Newfoundland and Ireland, 1. Lithology and geologic interpretations. *Prof. Pap. U.S. geol. Surv.,* No. 196A, 1–32.

DeMaster, D. J. 1979. *The Marine Budgets of Silica and* ^{32}*Si.* Ph.D. dissertation, Yale University, 307 pp.

Donahue, J. 1971. Burrow morphologies in north-central Pacific sediments. *Mar. Geol., 11,* M1–M7.

Goldberg, E. D., and Koide, M. 1962. Geochronological studies of deep-sea sediments using the ionium/thorium method. *Geochim. cosmochim. Acta, 26,* 417–50.

Guinasso, N. L., Jr., and Schink, D. R. 1975. Quantitative estimates of biological mixing rates in abyssal sediments. *J. geophys. Res., 80,* No. 21, 3032–43.

Ku, T. L. 1966. *Uranium Series Disequilibrium in Deep-sea Sediments.* Ph.D. dissertation, Columbia University, 167 pp.

Mangini, A., and Sonntag, C. 1977. ^{231}Pa dating of deep-sea cores via ^{227}Th counting. *Earth planet. Sci. Lett., 37,* 251–56.

Noshkin, V. E., and Bowen, V. T. 1973. Concentration and distributions of long-lived fallout radionuclides in open ocean sediments. (*In Radioactive Contamination of the Marine Environment.* Vienna, International Atomic Energy Agency, 671–86.)

Nozaki, Y., Cochran, J. K., Turekian, K. K., and Keller, G. 1977. Radiocarbon and ^{210}Pb distribution in submersible-taken deep-sea cores from Project Famous. *Earth planet. Sci. Lett., 34,* 167–73.

Picciotto, E., Cameron, R., Crozaz, R., Deutsch, G., and Wilgain, S. 1968. Determination of the rate of snow accumulation at the pole of relative inaccessibility, eastern Antarctica: a comparison of glacial and isotopic methods. *J. Glaciol., 7,* 273–87.

Rosholt, J. N. C., Emiliani, C., Geiss, J., Koczy, F., and Wangersky, P. 1961. Absolute dating of deep-sea cores by the ^{231}Pa/^{230}Th method. *J. Geol., 69,* 162–85.

Sackett, W. M. 1966. Manganese nodules: ^{230}Th/^{231}Pa ratios. *Science, N.Y., 154,* 646–47.

Turekian, K. K., Cochran, J. K., and DeMaster, D. J. 1978. Bioturbation in deep-sea deposits: rates and consequences. *Oceanus, 21,* 34–41.

135

Erosion of Deep Sea Sediments in the Southern Ocean between Longitude 70° E and 190° E, and Contrasts in Manganese Nodule Development

Abstract

N. D. WATKINS† and J. P. KENNETT
Graduate School of Oceanography, University of Rhode Island, Kingston, Rhode Island 02881, U.S.A.

Using our previously published results, we present trend surface maps of surface sediment age and the thickness of sediment deposited during the last 0.7 m.y. for the area outlined by the Kerguelen Plateau, Broken Ridge, Australia, New Zealand, and Antarctica. Four separate centers of widespread deep-sea sediment erosion have developed in response to high bottom-water velocities. Two of these, centered on the South Tasman Basin and the western end of the South Australian Basin, respectively, are associated with extensive manganese nodule development. The manganese nodule fields are almost entirely developed on sediment which averages at least 0.5 m.y. in age. In contrast, the southern and eastern flanks of the Kerguelen Plateau, the Ross Sea and the Campbell Plateau have very limited manganese nodule deposits, despite abundant evidence of widespread active erosion. It is concluded that sediment erosion is a necessary but insufficient requirement for extensive manganese nodule development in the area, and that a role is also played by submarine volcanism, with differences in the oxygenation of bottom water throughout the region also being possibly relevant.

† Deceased.

XII *Cenozoic History*

136

Search for Antarctic Meteorites

Review paper

TAKESI NAGATA
National Institute of Polar Research, Itabashi, Tokyo, Japan

Abstract. Until 1969 only four meteorites had been collected in Antarctica. From 1969 to 1975, however, 991 pieces of meteorites (including fragments) were recovered from the Meteorite Ice Field (71.5°–72.0° S, 35° E–36° E) near the Yamato Mountains. The Yamato meteorite collection consists of two irons, a pallasite, and all others, either chondrites or achondrites. From the preliminary analyses, 56 large Yamato stony meteorites have been identified as 1 E-chondrite, 29 H-chondrites, 14 L-chondrites, 2 LL-chondrites, 1 C-chondrite, and 9 achondrites.

Detailed analyses of eight Yamato stony meteorites have shown considerably varying cosmic-ray exposure ages and oxygen isotope ratios. All these preliminary data suggest that the Yamato meteorite collection is a fairly good partial assemblage of the known meteorites of the world.

In 1976–77, a field search for meteorites was carried out in the blue-ice areas of Victoria Land, and 11 meteorites (1 iron, 1 achondrite, and 9 chondrites) have been successfully recovered.

The frequency spectrum, $N(m)$, of the mass (m) distribution of Yamato stony meteorites is well represented by $N(m) = Am^{-5/3}$ for $m \gtrsim 5$ g. This spectral form is in good agreement with that of other registered stony meteorites of the world and with theoretical expectation.

The most plausible mechanism for accumulation of many meteorites within a small blue-ice area is as follows: Meteorites falling on the accumulation area of the ice sheet during about 10^6 to 10^7 years have been transported by the ice sheet with a velocity of about 1–10^2 m/year to the ablation area in the lower foot zone of the ice sheet. The mountain range standing against the ice sheet prevents its further flow, resulting in a horizontal convergence and an upward motion (about 5 cm/year on the average) of the ice flow carrying the meteorites. A strong ablation effect in the ablation area lowers the ice-sheet surface (by 5 cm/year on the average), continuously maintaining the flat blue-ice surface and leaving meteorites to collect on the surface.

In December of 1969, a Japanese field party conducting geologic and geomorphologic surveys of the Yamato Mountains in East Antarctica found nine pieces of meteorite within a small blue-ice area, between lat. 71° 48′ S and 71° 52′ S, and between long. 36° 10′ and 36° 30′ E, at the southeastern foot of the Yamato Mountains (Yoshida and others, 1971; Kusunoki, 1975). Eight pieces of these meteorites are chondrites and one is an achondrite. Four large pieces (greater than 60 g in weight) among these stony meteorites have been chemically, petrographically, mineralogically, and physically analyzed in fair detail (Nagata, 1975). In 1973, a field party for similar geologic and geomorphologic surveys of the Yamato Mountains made some effort to find more meteorites in the same blue-ice area and collected 12 new pieces of meteorite: 11 chondrites and 1 achondrite (Shiraishi and others, 1976).

Encouraged by these discoveries of meteorites in East Antarctica, a special effort to find and collect as many meteorites as possible was made by the 1974 field party in the same blue-ice area (Yanai, 1976). As a result, the 1974 field party collected 663 pieces of meteorite. Except for one stony-iron meteorite (pallasite), all meteorite specimens collected by the party are stony meteorites, either chondrites or achondrites.

In 1975, searches for meteorites were extended to a little wider area, including the blue-ice areas in the neighborhood of the northern Yamato Mountains (Matsumoto, 1978). Among 307 meteorite pieces collected by the 1975 field party, two are irons and all the others (305 in total) are chondrites or achondrites. Thus, a total of 991 meteorite pieces had been collected from a limited blue-ice area (about 4×10^3 km^2 in area) around the Yamato Mountains by the end of 1975 (Figure 136.1). Up until 1969, only four meteorites had been collected from the whole Antarctic continent (Hey, 1966). These four meteorites are from Adelie Land (olivine-hypersthene chondrite of about 1.0 kg in weight, found at lat. 67° S, long. 142° E, in 1912); from Lazarev Station (iron, comprising two fragments, 8 kg and 2 kg, found at lat. 72° S, long. 11.5° E, in 1961); from the Neptune Mountains (iron, octahedrite of 1.07 kg found at lat. 83° S, long. 55° W, in 1964); and from the Thiel Mountains (stony-iron pallasite, comprising two fragments, 22.7 kg and 9.0 kg, found at lat. 85° S, long. 87° W in 1961). The locations of the four places where the meteorites were found in the past and the location of the blue-ice field where a large number of Yamato meteorites have recently been found are illustrated in Figure 136.2 along with the elevation contours of the ice-sheet surface.

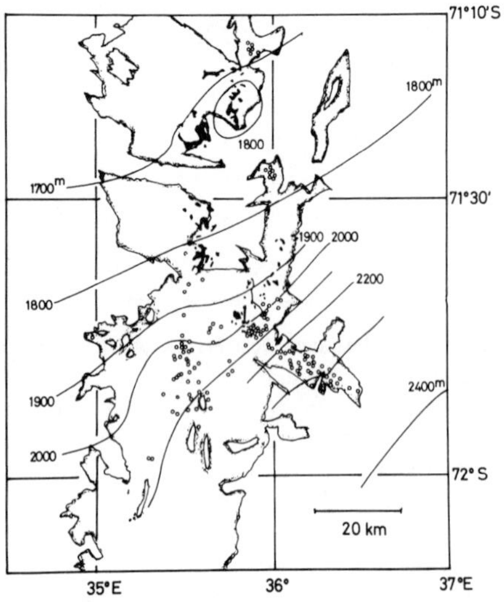

Figure 136.1. The Meteorite Ice Field near the Yamato Mountains, with equal elevation contours. Small circles: discovery site of large meteorite. Black areas: mountains or nunataks surrounded by ice. Irregular line indicates the boundary of blue-ice area (shaded side).

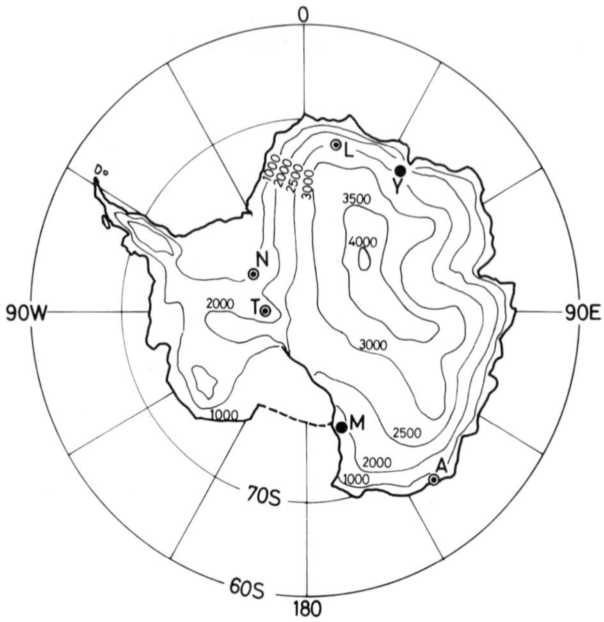

Figure 136.2. Locations of meteorite discoveries in Antarctica, with ice-sheet surface elevation contours (in meters). Y: Yamato (Meteorite Ice Field); L: Lazarev; N: Neptune Mountains; T: Thiel Mountains; M: McMurdo (Allan Nunatak and Mount Baldr); A: Adelie Land.

It can be seen in Figure 136.2 that all of these places where meteorites had been found are located in the lower foot zone of the East Antarctic ice sheet, namely, in the ablation zone of the large ice sheet. Although the Yamato meteorite collection of 991 pieces includes a considerable fraction of meteorite fragments which might have been produced by the collision of the original meteorites with the earth's surface, at least 200 pieces among the Yamato meteorite collection are almost entirely covered by fusion crust, suggesting that they have kept the original shapes of meteorites falling onto the earth's surface from interplanetary space. It seems that attempts at geometrical restoration of meteorite fragments to their original single body by reference to their partial fusion crusts are frequently successful if those fragments are scattered within a very limited area of the blue-ice surface. It seems possible to hope, therefore, that after the restoration work is completed in the near future, the true number of individual original meteorites in the Yamato meteorite collection can be approximately estimated. At the present stage of analysis, however, it can be argued that the Yamato meteorite collection comprises at least 300 individual original meteorites.

Since the Yamato meteorite collection includes almost all the different kinds of meteorites, i.e., irons, stony-irons, various types of chondrites, and achondrites, as will be described in the next section, the simple interpretation that a single or a few meteorite showers might have resulted in the large accumulation of meteorite pieces within the particular blue-ice area near Yamato Mountains is hardly acceptable. Then, the question of why so many pieces of meteorite could accumulate on the surface of a limited small area of blue-ice at a particular locality arises. A possible solution to this problem, based on theories of ice-sheet flow as well as observed glaciologic data, will be discussed later in this paper. As suggested in Figure 136.2, however, it seems very likely that those meteorites found to date in Antarctica might have been transported by the ice-sheet flow from a very large source area of the East Antarctic interior (about 6×10^6 km^2) to the ablation zone of the East Antarctic ice sheet, where the transported meteorites could have risen to the ice-sheet surface from its interior. In the Meteorite Ice Field near the Yamato Mountains, in particular, all meteorites were found only on the blue-ice surface, as shown in Figure 136.1. This observed fact suggests that a strong ablation effect on the ice-sheet surface, which removes the surface ice, results in maintaining a flat blue-ice surface and in keeping meteorites on the blue-ice surface.

Therefore, when a research program of field searches for meteorites in West Antarctica was jointly planned in 1976 by the National Institute of Polar Research of Japan and by W. A. Cassidy and Edward Olsen of the United States, blue-ice areas along the outer edge of the polar plateau ice sheet in Victoria Land were specifically selected for the field search. As illustrated in Figure 136.3, the meteorite search areas in Victoria Land are located very close to the eastern boundary line of the polar ice plateau, the seaward flow of which is severely impeded by the Transantarctic Mountain range. From the two blue-ice areas shown in Figure 136.3, 11 meteorites were collected in December, 1976 and January, 1977 (Yanai, 1978). The two Mount Baldr meteorites are chondrites, while the nine Allan Nunatak meteorites have been classified as one iron, seven chondrites, and one achondrite; thus, a good variety of meteoritic types is present in this new Antarctic meteorite collection, as well. From the locations of the meteorite discoveries in Victoria Land plotted in Figure 136.2, it may be concluded that all the meteorite sites in Antarctica so far are distributed in the ablation zone near the lower boundary of the great East Antarctic ice sheet.

In the following sections, the characteristics of these Antarctic meteorites known to date, the frequency distribution of their masses, and a plausible mechanism for

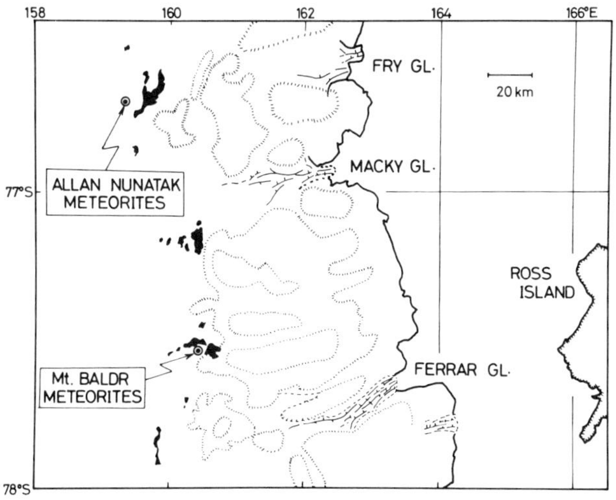

Figure 136.3. Approximate topography near Allan Nunatak and Mount Baldr in Victoria Land, near McMurdo Sound. Plateau ice flows eastward into the Ross Sea. Black areas: mountains or nunataks surrounded by ice.

accumulation of meteorites in the Meteorite Ice Field near the Yamato Mountains and in the blue-ice areas in Victoria Land will be summarized.

Chemical and petrographic characteristics of Antarctic meteorites

Adelie Land, Lazarev Station, Neptune Mountains, and Thiel Mountains

The Adelie Land meteorite (1.0 kg in weight) has been identified as in intermediate olivine-hypersthene chondrite (L_5), in which the fayalite component of the olivine is 24 mol % (Wasson, 1974). The Lazarev meteorite (10 kg in weight) was originally described by Ravich and Revnov (1963) as a coarse octahedrite containing 10.05% Ni, 0.67% Co, and 0.23% P, but Buchwald (1975) has suggested, after examining the Ravich and Revnov data, that the Lazarev meteorite may be the metal-rich fragment of a pallasite. The Neptune Mountains meteorite (1.07 kg in weight), containing 7.1% Ni and 0.20% P, has been identified as a coarse octahedrite, which is a typical Group I iron meteorite (Buchwald, 1975). The Thiel Mountains meteorite (31.7 kg in weight), containing 10.1% Ni, has been identified as a pallasite (Wasson, 1974).

Yamato meteorites

The 1969 collection of Yamato meteorites has been named Yamato (a), (b), . . . (i), and the 1973 collection, Yamato (j), (k), . . . (u). The weight of the specimens in the Yamato group (a)–(i) ranges from 715 g (Yamato [a]) to 10 g (Yamato [h] and [i]); that of Yamato group (j)–(u) ranges from 900 g (Yamato [k]) to 4 g (Yamato [n]). All 21 of these pieces of meteorite are stony meteorites, i.e., 19 chondrites and 2 achondrites, and all of them are almost completely covered by the fusion crust (Kusunoki, 1975; Shiraishi and others, 1976). Because the 1974 and 1975 collections of Yamato meteorites are so large (663 pieces in 1974 and 307 pieces in 1975), these meteorites have been provisionally named Yamato-74001 to Yamato-74663 for the 1974 collection and Yamato-75001 to Yamato-75307 for the 1975 collection. Among these Yamato-74 and Yamato-75 meteorites, 970 in total, Yamato-74044 (51.8 g) is a pallasite, and Yamato-75031 (60.2 g) and Yamato-75105 (19.6 g) are irons; all of the other 967 are either chondrites or achondrites, the weights of which range from 11.0 kg (Yamato-75102) to 0.1 g (Yamato-75040 and others).

Table 136.1. Classification of 56 pieces of Yamato stony meteorites, based on the Fe content of olivines and orthopyroxenes

Types	Number (total = 56)	Percentage
E	1	1.8
H	29	51.8
L	14	25.0
LL	2	3.6
C	1	1.8
Achondrite	9	16.1

Fifty-six pieces of Yamato stony meteorites heavier than 100 g have been petrographically examined to date for the purpose of provisional identification. The classification of chondrites in these examinations is based on the Fe content in olivines and orthopyroxenes. Results are presented in Table 136.1. The relative abundances of these chemically different types among the selected Yamato stony meteorites are not essentially different from those among stony meteorites registered worldwide (Hey, 1966), though the percentages of H-types and achondrites in the former are a little higher than in the latter. This problem will have to be re-examined in detail when the identification of all Yamato meteorites is completed. The preliminary results summarized above suggest, however, that the partial assemblage of the Yamato stony meteorites, selected by a criterion of mass greater than 100 g, could be considered as a nearly representative partial assemblage of the world's stony meteorites.

Systematic chemical, petrographic, and physical analyses of the Yamato meteorites are now under way. However, several large pieces of the 1969, 1973, and 1974 stony meteorite collections, along with two Yamato iron meteorites and one Yamato pallasite, have already been analyzed in fair detail. Table 136.2 gives the chemical compositions of eight Yamato stony meteorites (Shima and Shima, 1975; Yagi and others, 1978), and Table 136.3 gives the main compositions in the base of iron for two Yamato irons and one Yamato pallasite (Nagata and others, 1976; Fisher and others, 1978). The petrographic structures of the chemically analyzed stony meteorites have also been studied in detail (Okada, 1975; Yagi and others, 1978). They are an enstatite chondrite (E), two olivine-bronzite chondrites (H), two olivine-hypersthene chondrites (L), one carbonaceous chondrite (C), and two achondrites, as noted in Table 136.2. The cosmic-ray exposure ages of seven specimens among these stony meteorites have also been estimated from the analyses of rare-gas element isotopes such as (^{3}He, ^{4}He), (^{20}Ne, ^{21}Ne, ^{22}Ne), and (^{36}Ar, ^{38}Ar, ^{40}Ar) (Shima and Shima, 1975; Takaoka and Nagao, 1978). As summarized in Table 136.4, results of the rare-gas analyses have given significantly varying values for the cosmic-ray exposure ages of individual stony meteorites, ranging from 1.3 m.y. to 35 m.y. All these data on the bulk chemical compositions, the petrographic structures, and the cosmic-ray exposure ages of the Yamato stony meteorites strongly

Table 136.2. Chemical composition of some Yamato stony meteorites (in wt. %)

	Yamato specimens and classification							
	(a) E	(b) Diogenite	(c) C_3	(d) H	(j) H_4	(k) L_5	(l) Howardite	(m) L_5
	Silicate phase							
SiO_2	37.98	55.17	33.26	38.90	36.56	39.15	51.06	39.38
MgO	19.28	26.22	24.42	24.08	23.68	24.52	22.02	24.60
FeO	0.48	12.58	27.84	12.02	19.22	13.10	16.00	13.02
Al_2O_3	1.55	0.70	2.44	1.93	2.08	3.64	2.08	2.16
CaO	0.45	1.21	2.37	1.68	1.69	1.83	3.50	1.79
Na_2O	0.86	0.012	0.46	0.92	1.03	1.05	0.16	1.00
K_2O	0.085	0.006	0.039	0.104	0.12	0.13	0.06	0.12
Cr_2O_3	0.42	1.38	0.56	0.65	0.61	0.31	0.68	0.41
MnO	0.25	0.52	0.17	0.30	0.35	0.38	0.57	0.38
TiO_2	0.075	0.072	0.13	0.082	0.10	0.11	0.22	0.10
P_2O_5	0.46	0.009	0.22	0.25	0.25	0.27	0.01	0.27
NiO	—	—	—	—	0.89	—	—	—
CaO	—	—	—	—	0.05	—	—	—
	Metal phase							
Fe	22.18	0.66	0.15	12.69	7.33	7.64	0.39	7.50
Ni	1.86	< 0.004	1.32	1.52	0.78	0.96	0.012	0.83
Co	0.09	0.003	0.075	0.081	0.05	0.06	0.007	0.06
	Sulphide phase							
FeS	11.91	1.34	3.62	5.32	5.13	7.57	0.75	8.44
TOTAL	98.65	99.88	97.08	100.52	100.04	99.22	100.10	99.09

Analyst: M. Shima

Table 136.3. Chemical composition of Yamato iron and stony-iron meteorites (in the base of Fe; all values in wt. %)

	Pallasite (metal phase) 74044	Yamato Plessitic Octahedrite 75031	Hexahedrite 75105
Al	0.5	<0.01	0.25
Si	3.3	0.06	0.35
P	0.1	1.0	1.0
S	0.1	0.025	0.1
Ti	<0.01	<0.01	0.12
V	<0.01	<0.01	<0.01
Cr	<0.01	<0.01	<0.01
Mn	<0.01	<0.01	<0.01
Co	0.75	0.76	0.52
Ni	10.6	15.3	5.65
Cu	<0.01	0.035	—
Nb	0.02	<0.01	—
Mo	<0.01	<0.01	—

suggest that these stony meteorites are significantly different from one another and have come from different parts of some different meteorite-parent planet(s), probably falling on the earth's surface in different epochs. It has been concluded (Onuma and others, 1978) on the basis of oxygen isotope ratios, $^{18}O/^{16}O$ and $^{17}O/^{16}O$, in Yamato stony meteorites (including Yamato (l) and (m)), that Yamato (m), Yamato-74190 (L_5 chondrite), and Yamato-74191 (L_3 chondrite) are derived from one parent body, whereas Yamato (l) and Yamato-74013 (diogenite) are from another parent body.

The chemical compositions and metallographic properties of the Yamato irons and the metal phase of the Yamato pallasite are also considerably different from one another. The Ni content of the three specimens falls into three quite different ranges, i.e., about 5.7% and 15.3% for the two iron specimens and about 10.6% for the stony iron specimen. The P content in the two iron specimens is anomalously high (about 1.0%) due to the fact that both irons contain a number of coarse grains of schreibersite, $(Fe, Ni)_3P$. On the other hand, the metal phase of Yamato-74044 contains several grains of dark plessite enveloped by a taenite band (27–45% Ni) in a kamacite matrix (7% Ni), whereas the silicate phase is comprised of mostly olivines containing Si, Fe, Mg, Al, P, and S as the major elements.

These chemical and metallographic data indicate the slow cooling conditions of these specimens within their parent planets. The lowest limit of the slow cooling process is estimated to be 400°C, 500°C, and 650°C respectively for Yamato-74044, -75031, and -75105.

Allan Nunatak and Mount Baldr meteorites

The sizes and weights of 11 meteorites collected at the Allan Nunatak and the Mount Baldr blue-ice areas in the Victoria Land polar plateau are summarized, together with their provisional classifictions, in Table 136.5. As indicated in the table, Allan Nunatak (c) and (i) consist of 3 and 31 fragments, respectively. Since these fragments are distributed very close to one another in each group *in situ* and can certainly be restored to single meteorites well covered by the fusion crust, they have been counted as belonging to two single chondrites, i.e., Allan Nunatak (c) and (j). From mineralogic studies of olivines and orthopyroxenes, Mount Baldr (a) and (b) chondrites have been identified as H_4 and H_5 respectively; analyses of the other chondritic, achondritic, and iron meteorites are now under way.

Table 136.4. Cosmic-ray exposure ages (in m.y.) of Yamato meteorites

	Yamato specimens						
Rare gas	(a)	(b)	(c)	(d)	(j)	(k)	(m)
He	1.3	35	23	5.5	8.2	17	13
Ne	1.7	31	25	4.3	13	22	18
Ar	—	—	—	—	13	28	17

Frequency spectrum of mass distribution of Yamato stony meteorites

Since this is the first time that a large number of meteorites of different types have been found within a comparatively small area (about 4 × 10^3 km^2), a statistical examination of the mass distribution of these meteorites would be of a certain significance in relation to the mechanism of asteroid fragmentation.

A histogram of the frequency distribution of mass (m) of the 988 Yamato stony meteorite specimens (with respect to $\log_{10} m$) is illustrated in Figure 136.4 where the occurrence frequency, $N(\log_{10}m)$, is normalized by $N_o \equiv \Sigma N(\log_{10} m)$. In the range m ≥ 20 g in this figure, $N(\log m)/N_o$ is statistically well represented by the empirical formula

$$\log [N(\log m)/N_o] = a - 2/3 \log m. \quad (1)$$

The occurrence-frequency spectrum, $N(\log m)d(\log m)$, represented by equation (1) can be transformed to a corresponding mass-frequency spectrum, $N(m)dm$, as expressed by

$$N(m) = Am^{-5/3}, \quad (2)$$

for m ≥ 20 g. With respect to the whole range of m of the Yamato stony meteorite collection, the differential-frequency specum, $N(\log m)/N_o$, shown in Figure 136.3 can be approximated by

$$N(\log m)/N_o = Am^{-2/3}[1 - \exp(-m/m_c)]^a, \quad (3)$$

where α and m_c are numerically given by $\alpha \cong 1.7$ and $m_c \cong$ 5 g. The decrease of N with a decrease of m below the critical mass value, m_c, represented by the factor $[1 - \exp(-m/m_c)]_\beta$ may be mostly due to the effect of increasing difficulty in finding smaller meteorite pieces, depending on the conditions around the recovered meteorites. Hence, the factor, $[1 - \exp(-m/m_c)]^\alpha$ may be called the masking-effect factor.

A similar statistical examination is made on the known initial weights of the world's registered stony meteorites (Hey, 1966). As shown in Figure 136.4, the frequency spectrum of mass distribution for $m \geq 10$ kg can also be approximately represented by equations (1) or (2) (in this case $N_o = 1064$), and the whole spectrum can be reasonably well approximated by equation (3), where $\alpha \cong 1.35$ and $m_c \cong 2.5$ kg. These results suggest that it is not difficult to find stony meteorites heavier than 5 g on a flat blue-ice surface in Antarctica, but it is difficult to find stony meteorites less than 2.5 kg in weight on the ordinary land surface. It may then be concluded that the differential-frequency spectrum of mass distribution of stony meteorites, larger than 5 g in weight, can be generally represented by equation (2), namely, $N(m)/N_o = Am^{-5/3}$. By also taking into consideration the mass-frequency spectrum of the world's registered stony meteorites, we may expect that the relationship $N(m) \propto m^{-5/3}$ generally holds for a range of 5 g to 400 kg for stony meteorite mass.

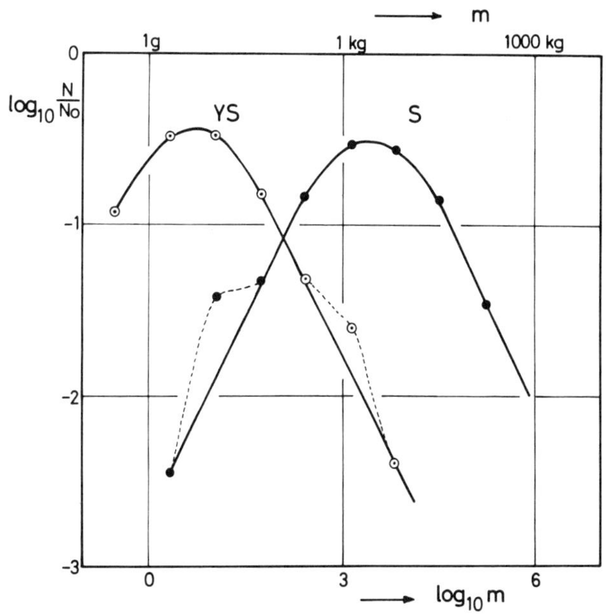

Figure 136.4. Differential frequency spectrum of mass distribution of the Yamato stony meteorites (YS) and the world's registered stony meteorites (S). Solid lines represent equation (3), which is the most easily fitted to observed data.

Hawkins (1960) has examined the frequency distributions of mass of the stone falls, stone finds, iron falls, and iron finds described in the Prior and Hey (1953) catalogue of meteorites in a similar way. He has obtained from the old data a somewhat different power law for the cumulative spectrum of mass distribution of the world registered stony meteorites; namely

$$\int_m^\infty N dm = \mathrm{B}\, m^{-s}, \qquad s = 1.0 \pm 0.3, \quad (4)$$

which corresponds to a differential spectrum expressed by $N(m) \propto m^{-(2\pm0.6)}$. Although the frequency spectrum of mass distribution of the world's registered stony meteorites

Table 136.5. Allan Nunatak and Mount Baldr meteorites

Meteorite	Weight (gm)	Size (cm x cm x cm)	Fragments	Type
Mount Baldr				
(a)	4,108	12 x 15 x 12		chondrite, H_4
(b)	13,773	33 x 17 x 15		chondrite, H_5
Allan Nunatak				
(a)	20,151	31 x 20 x 16		chondrite
(b)	1,510	11 x 10 x 7		iron
(c)	10,495	30 x 14 x 15	3	chondrite
(d)	305	10 x 7 x 2		chondrite
(e)	1,425	19 x 9 x 7		achondrite
(f)	1,137	9 x 9 x 8		chondrite
(g)	410	7 x 8 x 6		chondrite
(h)	1,151	16 x 8 x 6		chondrite
(i)	407,040		31	chondrite

may have some ambiguity, the frequency spectrum of Yamato stony meteorite mass itself represented by equation (2) is much more reliable, with errors of only ± 10% at most.

The frequency spectrum of mass distribution of small asteroids and meteorites has been theoretically estimated on the basis of a fragmentation model (e.g., Alfvén and Arrhenuis, 1976). On the assumption that each asteroidal body of mass m is broken and splits into n smaller bodies of equal weight whenever it collides with another body where the collision cross section is proportional to $m^{2/3}$, the steady-state frequency spectrum, $N(m)$, of fragment mass between m and $m + dm$ is represented by a complete compensation of the loss of bodies of mass m in collision by the production of new bodies of the same mass in collisions of larger bodies, namely

$$-N(m)m^{2/3}dm + N(nm)(nm)^{2/3}n\,dm = 0.$$

Expressing then $N(m)$ in a power form of m as

$$N(m) = Am^{-\gamma}, \tag{5}$$

we get

$$m^{-\gamma+2/3} = (nm)^{-\gamma+2/3}n,$$

which can be satisfied only by setting $\gamma = 5/3$. Equation (5) with $\gamma = 5/3$ is in exact agreement with the empirical expression given by equation (2).

In the present case of the Yamato stony meteorite collection, however, the effect of fragmentation of meteorite bodies caused by their collision with the earth's surface must also be taken into consideration. It appears, for example, that a number of groups in the 305 pieces of Yamato stony meteorites collected in 1975 appear likely to be broken fragments of single meteorite falls, while the others appear to be the original meteorites themselves which are almost entirely covered in the fusion crust. Thus the total number of original stony meteorite falls is estimated to be 102 (Matsumoto, 1978). Let us assume then that a fraction ($\beta < 1$) of all falling meteorites split into n' smaller bodies of equal mass when they hit the earth's surface where the collision cross section should be unity regardless of mass m. Then, the gain of the number ($\Delta N(m)$) of bodies of mass between m and $m + dm$ after collision with the earth's surface is given by

$$\Delta N(m) = \beta[-N(m) + n'N(n'm)]$$

$$= -\beta Am^{-5/3}(1 - (n')^{-2/3}).$$

Hence, the frequency spectrum of mass distribution ($N^*(m)$) of meteorite bodies after the collision with the earth's surface is expressed by

$$N^*(m) = A[1 - \beta(1 - (n')^{-2/3})]m^{-5/3}$$
$$= A^*(\beta, n')m^{-5/3}. \tag{6}$$

As the revised coefficient, $A^*(\beta, n')$ is dependent only on β and n', and both β and n' could be considered constant, the modified spectrum can still be represented by equation (2).

The agreement between the empirical spectrum $N(m)$ and the theoretical one may suggest first that the Yamato stony meteorite collection is a fairly good partial assemblage of all stony meteorites and, secondly, that the simple fragmentation model for the frequency spectrum of mass distribution of small asteroidal bodies and meteorites seems to be acceptable.

Probable mechanism of accumulation of meteorites within the Meteorite Ice Field in Antarctica

As briefly discussed in the introduction, the distribution of the locations of meteorite discoveries in Antarctica, illustrated in Figure 136.2, suggests that those meteorites found in Antarctica to date may have been transported by ice-sheet flow from the very large ice-sheet surface in the East Antarctic interior (an area of about 6×10^6 km^2) to the ablation zone of the ice sheet, where the transported meteorites can flow out to the ice-sheet surface from its interior. It may be further remarked that the horizontal ice-sheet flow in the neighborhood of meteorite discovery sites is prevented or at least strongly resisted by mountain ranges standing against the ice flow, so that the upward flowing of ice in the ice-sheet ablation zone is much more dominant there. It can be clearly seen in Figure 136.1 that the ice-sheet flow, which is approximately perpendicular to the elevation contours of the ice-sheet surface, is impeded by the Yamato Mountains. Similarly in Figure 136.3, the ice-sheet flow is generally prevented by the Transantarctic Mountains, though there are several narrow ice streams which nourish outlet glaciers near the sea coast. The circumstances are about the same for the Lazarev meteorite, which was found on sandy gravel on the inner (southern) side of the Humboldt Mountains, and for the Neptune Mountains meteorite, which was discovered on glacial moraines around a nunatak in the central part of the Neptune Mountains.

A third important point in considering the mechanism for accumulation of a number of meteorites of different types, such as is observed in the Meteorite Ice Field near the Yamato Mountains and in the blue-ice areas west of the Transantarctic Mountains, is the strong ablation effect which results in a continuous net ablation of the ice-sheet surface. This ablation maintains the permanent blue-ice surface and leaves the meteorites sitting on the blue-ice surface where they accumulate as a residue.

Figures 136.5 and 136.6 illustrate examples of a Yamato meteorite and a Mount Baldr meteorite as found *in situ* on the blue-ice surface. It has been reported (Yanai, 1976) that all meteorites collected in 1974, 663 pieces in total, were sitting on the blue-ice surface with only one exception, and that one exceptional meteorite was buried in the blue-ice at about 10 cm in depth.

The three key points mentioned above will have to be taken into account in considering a plausible mechanism of accumulation of meteorites within the blue-ice area. They are: (a) the ice-sheet flow from the accumulation area over the large central region to the ablation area on the lower zone, (b) the prevention of the ice sheet flow by the barrier of mountains in the ablation zone, and (c) a strong ablation effect which continuously lowers the ice-sheet surface and keeps transported meteorites on the blue-ice surface. These three points will be examined as quantitatively as possible in the remaining sections.

Figure 136.5. Yamato-74104 meteorite as found on the blue-ice surface (photo by K. Yanai).

Figure 136.6. Mount Baldr (a) meteorite as found on the blue-ice surface (photo by K. Yanai).

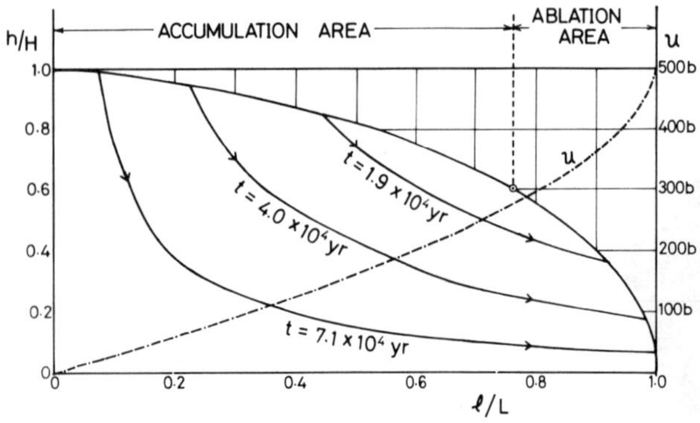

Figure 136.7. Steady-state profile of a two-dimensional ice-sheet model with stream lines of ice flow. h: height of ice-sheet surface. l: horizontal distance from the central crest point.
H: height of the central crest point
L: horizontal distance between the central crest point and the edge.
u: horizontal velocity of ice sheet flow in unit of b. b: snow accumulation rate.
t: time duration of ice-sheet flow from the accumulation area surface to the ablation area surface along a stream line (b = 10 cm/yr is assumed).

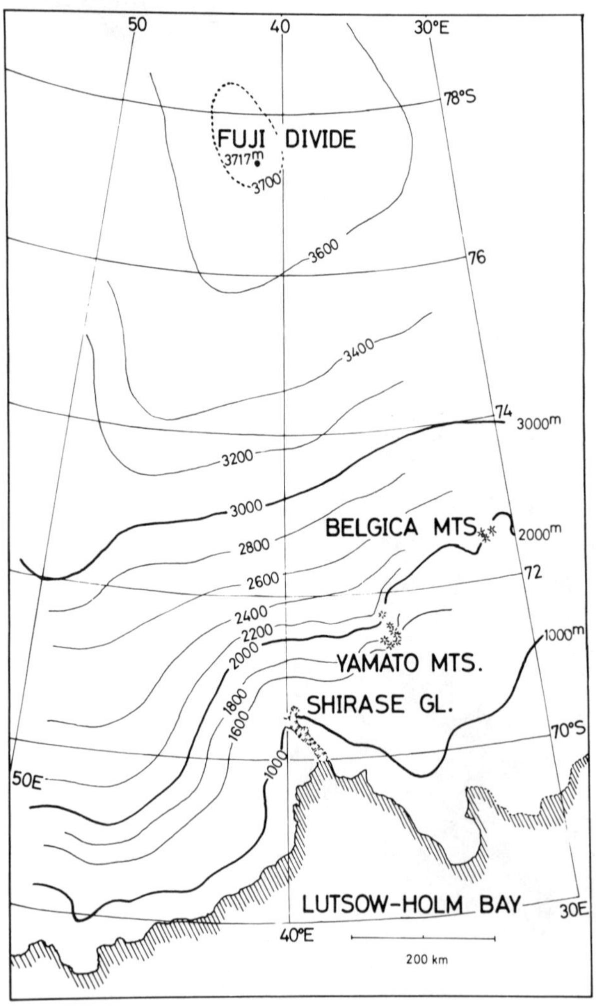

Figure 136.8. Ice-sheet surface topography in the vicinity of Yamato Mountains and Fuji Divide in East Antarctica.

The ice flow within an ice sheet

It is well known that the snow mass accumulated over an ice sheet becomes glacial ice at a certain depth and the ice thus formed gradually flows from the accumulation area to the ablation area (e.g., Nye, 1959; Haefeli, 1961). In Figure 136.7, a steady-state profile of a two-dimensional ice sheet and the stream lines of ice flow within it are illustrated according to a self-consistent model proposed recently by Nagata (1977). Figure 136.8, on the other hand, shows the ice-sheet surface topography along a line through the Yamato Mountains area and the crest point (Fuji Divide) of the ice sheet covering the Yamato Mountains area. The north-south profile of the ice sheet along the 40°E meridian in Figure 136.8 closely approximates the theoretical ice-sheet profile in Figure 136.7, provided that the values $L = 900$ km and $H = 3000$ m are used. The dotted curve u in Figure 136.7 shows the horizontal velocity of ice-sheet flow in units of the snow accumulation rate (b). In East Antarctica, $b = \sim 5$–20 cm water/year, the median value being $\bar{b} = 10$ cm water/year; hence, the horizontal velocity of ice-sheet flow at the outer edge amounts to about 50 m/year, which is generally in agreement with observed rates.

As shown by the solid curves in Figure 136.7, the snow or ice mass formed on the accumulation area surface gradually flows through the ice-sheet interior to the ablation area surface, where the velocity of ice flow depends on both b and the dimensions of the ice sheet. In Figure 136.7, travel times of small ice volumes from the accumulation area to the ablation area along three typical stream lines are given, based on appropriate parameters for the East Antarctic ice sheet, such as $b = 10$ cm water/year, $L = 900$ km, and $H = 3000$ m. This result suggests that the ice mass in the accumulation area surface on the northern slope of Fuji Divide must be transported to the ablation area surface during about 10^4–10^5 years. If meteorites falling on the accumulation area are perfectly trapped by the surrounding ice mass, the meteorites must also be transported to the ablation area together with the ice mass. In this regard, possible relative motion of a meteorite piece through a plastic medium of ice under the effect of gravity has been examined (Nagata, 1978). Results of that study have shown that the relative downward velocity of the largest Yamato meteorite (11.0 kg in weight) never exceeds 5×10^{-6} cm/year, which is negligibly small compared to the downward velocity of the ice flow itself (~ 10 cm/year). It may be concluded therefore that all meteorites falling on the accumulation area surface are transported to the ablation area surface within approximately 10^4–10^5 years.

Obstruction of ice-sheet flow by mountains

It may be obvious that mountain ranges aligned perpendicularly across the ice-sheet flow tend not only to prevent the horizontal ice flow but also to cause an upward component of ice flow. Actually, a number of stagnant areas of ice-sheet surface flow have been discovered in the neighborhood of the Yamato Mountains. Since a theoretical approach to this problem is extremely complicated and may be quantitatively unreliable, a special field experiment was carried out in the Meteorite Ice Field. Figure 136.9 shows a triangulation network set up in 1969 to measure the horizontal and vertical movements of the ice-sheet surface in the southeastern part of the Meteorite Ice Field where a large number of Yamato meteorites are particularly densely distributed. The base points, A1 and A2, were set up on nunataks so that both the horizontal and vertical movements of each triangulation point would be exactly determined. The horizontal and vertical velocities of the triangulation points obtained during the four years from 1969 to 1973 (Naruse, 1975) are illustrated in Figure 136.10. As shown in this figure, the velocity vectors of the horizontal ice flow are approximately perpendicular to the elevation contour lines, and the horizontal speed gradually decreases towards zero as the ice-sheet surface approaches the base points on the nunataks. The figure shows that all triangulation points on the blue-ice make upward movements from 0 to about 8 cm/year, the average being 4.6 cm/year.

For comparison, similar surveys, carried out during the same period around lat. 72° S and long. 40° E in the Mizuho Plateau, have revealed that the approximately northward horizontal flow velocity of the ice-sheet surface amounts to about 10–20 m/year, while the vertical movement is downward at an average rate of about 1 m/year (Naruse, 1975). This flow pattern, roughly in agreement with the ice-sheet surface shape, should represent the general ice-sheet flow pattern in the ablation area. It is concluded then that the ice-sheet surface flow pattern shown in Figure 136.10 is highly anomalous, clearly indicating an upward and horizontally convergent movement of the ice-sheet surface in the Meteorite Ice Field.

Ablation effect

Measurements of the snow accumulation or the ablation of blue-ice were carried out by the snow-stake method in the Meteorite Ice Field during 1969–1973 (Yokoyama, 1975). As shown in Figure 136.11, all points on the blue-ice surface are subject to the ablation effect (negative) amounting to about − 2 to − 7 cm/year, whereas the snow accumulation rate is positive at some points in the adjacent snow-covered area. The average annual rate of ablation at 10 points within the blue-ice area amounts to 5.4 cm/year. Over the blue-ice area, therefore, the ablation effect is large enough to remove completely any accumulating snow and also to lower the surface of the rising blue-ice.

In summary of all these results, the most plausible mechanism for accumulating a large number of meteorites ($\sim 10^3$ pieces) within a limited area of blue-ice surface ($\sim 4 \times 10^3$ km^2) is as follows: meteorite pieces falling on a large area of the northern slope of Fuji Divide ($\sim 2.5 \times 10^5$ km^2) during a long time (T) have been transported by convergent ice-sheet flow (about 1–10^2 m/year) toward the Meteorite Ice Field in the ablation area, where an upward movement of ice caused by the bedrock topography forming the Yamato Mountains further accelerates the rise of the transported meteorites to the ice-sheet surface (~ 5 cm/year in upward velocity), and the strong ablation effect (~ -5 cm/year) results in continuous ablation of the ice surface, leaving the transported meteorites to remain as a residue on the blue-ice surface.

According to Hawkins (1960), the influx rate of meteorite falls on the earth's surface is approximately 4.5×10^{-11} meteorites/km^2 per hour, or 4×10^{-7} meteorites/km^2 per year. If these figures for the influx rate of meteorites and

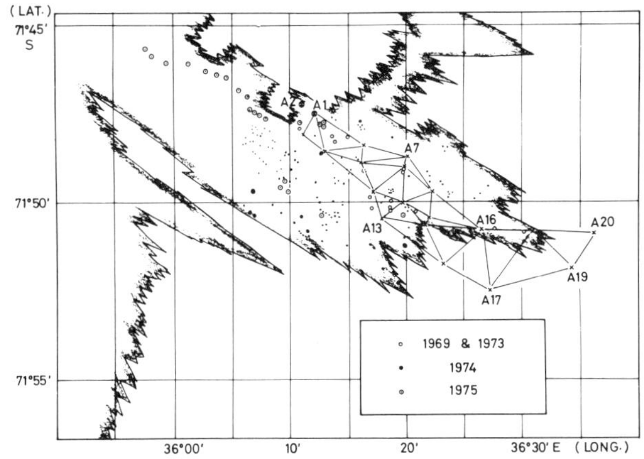

Figure 136.9. Distribution of Yamato meteorites (symbols indicate year of discovery) in the southeastern corner of the Meteorite Ice Field and the triangulation network established for measuring the ice-sheet surface movement. Irregular lines indicate the boundary of blue-ice area (shaded side).

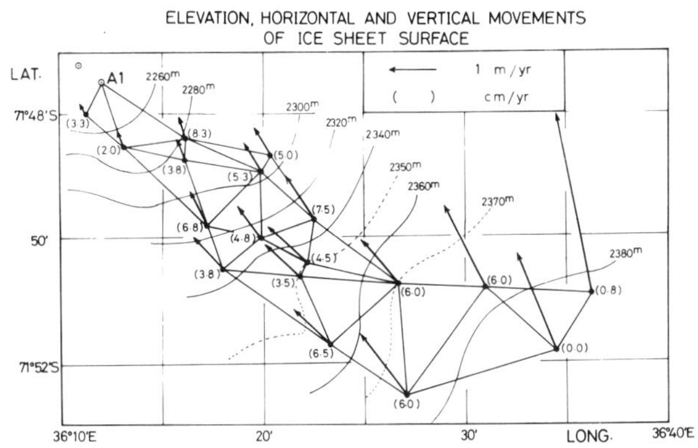

Figure 136.10. Annual rates of horizontal (represented by arrows) and upward vertical (numerals in parentheses) movements of the ice-sheet surface during 1969–73, with elevation contours of the ice-sheet surface.

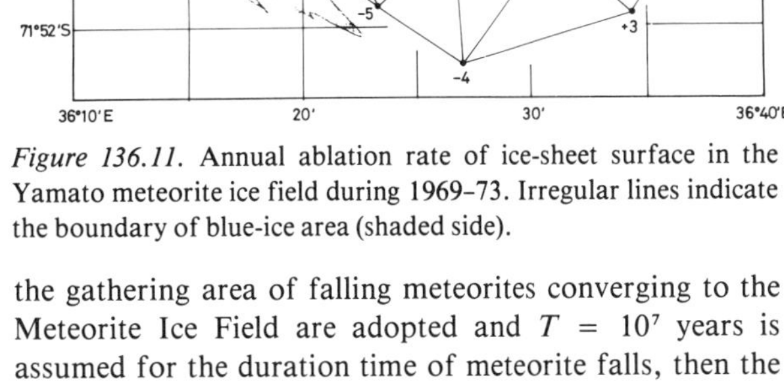

Figure 136.11. Annual ablation rate of ice-sheet surface in the Yamato meteorite ice field during 1969–73. Irregular lines indicate the boundary of blue-ice area (shaded side).

the gathering area of falling meteorites converging to the Meteorite Ice Field are adopted and $T = 10^7$ years is assumed for the duration time of meteorite falls, then the total number of meteorites within the Meteorite Ice Field amounts to as many as 10^6. Thus, the accumulation of approximately 10^3–10^4 meteorites in the Meteorite Ice Field seems reasonable, even if the great majority of the meteorites have been transported elsewhere.

Concluding remarks

Chemical, petrographic, metallographic, and physical analyses of Yamato, Mount Baldr, and Allan Nunatak meteorites have just started recently, and it will take a long time to complete the analyses of these meteorites in detail. One significant point about these Antarctic meteorites is that they have been kept in the Antarctic ice field, which has never been artificially contaminated, and these meteorite specimens have been collected and brought back to laboratories by scientists with extreme care. The Antarctic meteorites can therefore be considered the least contaminated specimens, and this uncontaminated condition is extremely important in various aspects of meteorite research, such as the analyses of rare-gas or organic composition and remanent magnetization.

The field searches for meteorites have not yet been completed in either the Meteorite Ice Field in East Antarctica or the Victoria Land blue-ice areas in the Transantarctic Mountains. This work will be systematically continued in both areas in the future. Since the hypothesis of meteorite transportation by ice-sheet flow appears reasonable for explaining the mechanism of meteorite accumulations in both the Meteorite Ice Field and the Mount Baldr–Allan Nunatak blue-ice area, this hypothesis could be applied to searching for meteorites in other places in Antarctica. For example, MacRobertson Land (~60° E–70° E) in Figure 136.2 is likely to be a good candidate for future meteorite search if wide blue-ice areas are present there, because the ice-sheet flow is converging toward this land, and the flow is obstructed by several mountains there.

References cited

Alfvén, H., and Arrhenius, G. 1976. *Evolution of the Solar System.* Washington, D.C., NASA Publication SP-345, 599 pp., 109–16.

Buchwald, V. F. 1975. *Handbook of Iron Meteorites.* Vol. 1–3. Berkeley and Los Angeles, University of California, 1418 pp.

Fisher, R. M., Spangler, E. C., Jr., and Nagata, T. 1978. Metallographic properties of Yamato iron meteorite, Yamato-75031 and stony iron meteorite, Yamato-74044. *Mem. Natn Inst. Polar Res.,* Special Issue No. 8, 248–59.

Haefeli, R. 1961. Contribution to the movement and the form of ice sheets in the Arctic and Antarctic. *J. Glaciol., 3,* 1133–50.

Hawkins, G. S. 1960. Asteroidal fragments. *Astr. J., 65,* 318–22.

Hey, M. H. 1966. *Catalogue of Meteorites.* London, British Museum, 637 pp.

Kusunoki, K. 1975. A note on the Yamato meteorites collected in December, 1969. *Mem. Natn Inst. Polar Res.,* Special Issue No. 5, 1–8.

Matsumoto, Y. 1978. Collection of Yamato meteorites, East Antarctica, in November and December, 1975 and January, 1976. *Mem. Natn Inst. Polar Res.,* Special Issue No. 8, 38–50.

Nagata, T., *ed.* 1975. Yamato meteorites collected in Antarctica in 1969. *Mem. Natn Inst. Polar Res.,* Special Issue No. 5, 110 pp.

Nagata, T. 1977. A theoretical steady-state profile of ice sheet. *Antarctic Rec., 60,* 13–27.

Nagata, T. 1978. A possible mechanism of meteorites within the Meteorite Ice-Field in Antarctica. *Mem. Natn Inst. Polar Res.,* Special Issue No. 8, 70–92.

Nagata, T., Fisher, R. M., and Sugiura, N. 1976. Metallographic and magnetic properties of a Yamato iron meteorite—Yamato-75105. *Mem. Natn Inst. Polar Res.,* Series C, No. 10, 1–11.

Naruse, R. 1975. Movement of the ice sheet observed by a triangulation chain. *JARE Data Rep.,* No. 28 (Glaciology), 48–61.

Nye, F. F. 1959. The motion of ice sheets and glaciers. *J. Glaciol., 3,* 493–507.

Okada, A. 1975. Petrological studies of Yamato meteorites. *Mem. Natn Inst. Polar Res.,* Special Issue No. 5, 14–66.

Onuma, N., Clayton, R. M., Mayeda, T. K., and Yanai, K. 1978. Oxygen isotopes in several Yamato meteorites. *Mem. Natn Inst. Polar Res.,* Special Issue No. 8, 220–24.

Prior, G. T., and Hey, M. H. 1953. *Catalogue of Meteorites.* London, British Museum, 432 pp.

Ravich, M. G., and Revnov, B. I. 1963. The iron meteorite Lazarev from the Antarctic. *Meteoritika, 23,* 30–35.

Shima, M., and Shima, M. 1975. Cosmochemical studies on the Yamato meteorites—a summary of chemical studies on Yamato (a), (b), (c), and (d) meteorites. *Mem. Natn Inst. Polar Res.,* Special Issue No. 5, 9–13.

Shiraishi, K., Naruse, R., and Kusunoki, K. 1976. Collection of Yamato meteorites, Antarctica, in December 1973. *Antarctic Rec., 55,* 44–60.

Takaoka, N., and Nagao, K. 1978. Rare-gas studies of Yamato-7331 (j), 7305 (k), and 7303 (m). *Mem. Natn Inst. Polar Res.,* Special Issue No. 8, 198–208.

Wasson, J. T. 1974. *Meteorites.* Berlin, Springer-Verlag, 316 pp.

Yagi, K., Lovering, J. F., Shima, M., and Okada, A. 1978. Mineralogical and petrographical studies of the Yamato meteorites (j), (k), (l), and (m) from Antarctica. *Mem. Natn Inst. Polar Res.,* Special Issue No. 8, 121–41.

Yanai, K. 1976. Search and collection of Yamato meteorites, Antarctica, in October and November, 1974. *Antarctic Rec., 56,* 70–81.

Yanai, K. 1978. Report of the US–Japan Antarctic program titled "Search for meteorites, 1976–77." *Mem. Natn Inst. Polar Res.,* Special Issue No. 8, 51–69.

Yokoyama, K. 1975. Net accumulation (of snow) obtained by stake measurement. *JARE Data Rep.,* No. 28 (Glaciology), 62–82.

Yoshida, M., Ando, H., Omoto, K., Naruse, R., and Ageta, Y. 1971. Discovery of meteorites near Yamato Mountains, East Antarctica. *Antarctic Rec., 39,* 62–65.

137

Middle Cenozoic Glacial Beds at Table Mountain, Southern Victoria Land

P. J. BARRETT and R. D. POWELL*
Antarctic Research Centre and Department of Geology, Victoria University of Wellington, Wellington, New Zealand

Abstract. The northwest slope of Table Mountain has a 15-m-thick sequence of diamictite, conglomerate, and sandstone exposed on a broad bench (elevation 1700 m above sea level) 700 m above the Ferrar Glacier. The morphology of the bench and the character of the sediments on it suggest that it is the remnant of a broad northeast-trending glacial valley now largely destroyed by the formation of the present Ferrar Valley.

The diamictite consists of muddy, fine- to medium-grained sandstone with scattered pebbles, cobbles, and rare boulders; it generally lacks stratification. The sandy matrix is dominated by rounded quartz grains from adjacent gently dipping Devonian quartzose sandstones. The clasts, which form about 20% of the rock, are more than 90% diabase from Jurassic sheets intruding the sandstone, and are mainly subrounded. A number of clasts are striated, some in several directions.

The sandstone forms beds up to 7 m thick and is mainly fine grained and moderately sorted. The conglomerate beds, which range from 1 to 5 m in thickness, consist largely of rounded diabase clasts. No basement rocks or minerals were found, although the present Ferrar Valley is cut 1000 m into basement granite, suggesting that the glacial beds described here were deposited before the valley was cut to its present depth.

The glacial beds comprise at least two, and possibly three diamictite beds deposited directly from ice, separated by current-sorted beds that may be ice-marginal or proglacial deposits. At least two advances of the ice are inferred. The age cannot be obtained directly, but it clearly predates the cutting of the present Ferrar Valley and is therefore probably older than late Miocene, the minimum age for the cutting of nearby Wright and Taylor Valleys. The sequence is unlikely to be older than 25 m.y., the probable age of the oldest glacial sediment in Ross Sea drill core (DSDP site 270).

Compact till preserved at high elevations in the Transantarctic Mountains was reported by Oliver (1964) and McGregor (1965) from the heads of the Beardmore and Shackleton Glaciers, but its significance was established by Mercer (1968, 1972) from observations in the Reedy and Beardmore Glacier areas. He recognized a "thick lodgement till that predates profound modification of the landscape by glacial erosion" (Mercer, 1972, p. 427) and named it the Sirius Formation, attributing its formation to local temperate glaciers.

Since then, Mayewski (1975) has reported another 22 sites with compact till like that at Mount Sirius spread along the Transantarctic Mountains from the Scott Glacier to the Coombs Hills, a distance of 1200 km. The sites range in elevation from 1600 to 3000 m, except for one at Mount Roth at 800 m. At most localities the lodgement till overlies pre-Tertiary rocks unconformably and consists "primarily of semi-lithified till and sparsely scattered stratified lenses" (Mayewski, 1975, p. 11), but at several sites a stratified sequence of gravel, sand, and mud was found overlying the till. Mayewski included all of this sediment in the Sirius Formation, which he believed represented debris from the ice sheet that at its maximum was almost twice the volume of the present East Antarctic Ice Sheet (Queen Maud Glaciation, Mayewski, 1975, p. 143). He argued that it was the Queen Maud Glaciation that most deeply sculpted the Transantarctic Mountains (Mayewski, 1975, p. 57) and that the compact till was deposited during the glacial maximum when the ice was thick enough to be at the pressure-melting point at the base. The stratified sediment was deposited after the ice retreated.

We describe newly located exposures of both tillite and stratified sedimentary strata on the northwest slopes of Table Mountain above the Ferrar Glacier (Figure 137.1). The deposit can be considered part of the Sirius Formation in the broad sense of Mayewski; however, deposition of the beds appears to predate deepest cutting of modern Ferrar Valley. The sequence represents initial wet-based sub-

* Present address: Department of Geology, Northern Illinois University, De Kalb, Illinois 60115, U.S.A.

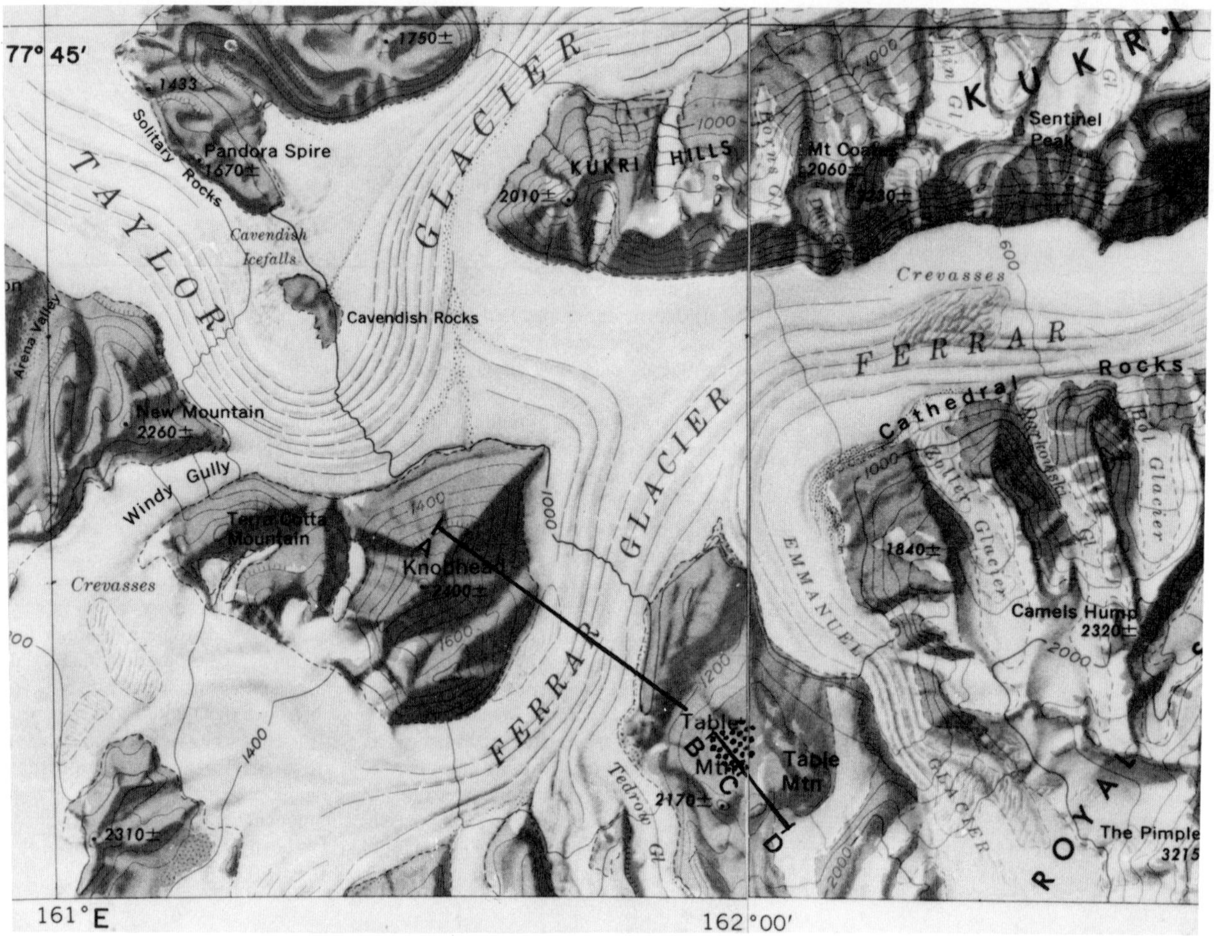

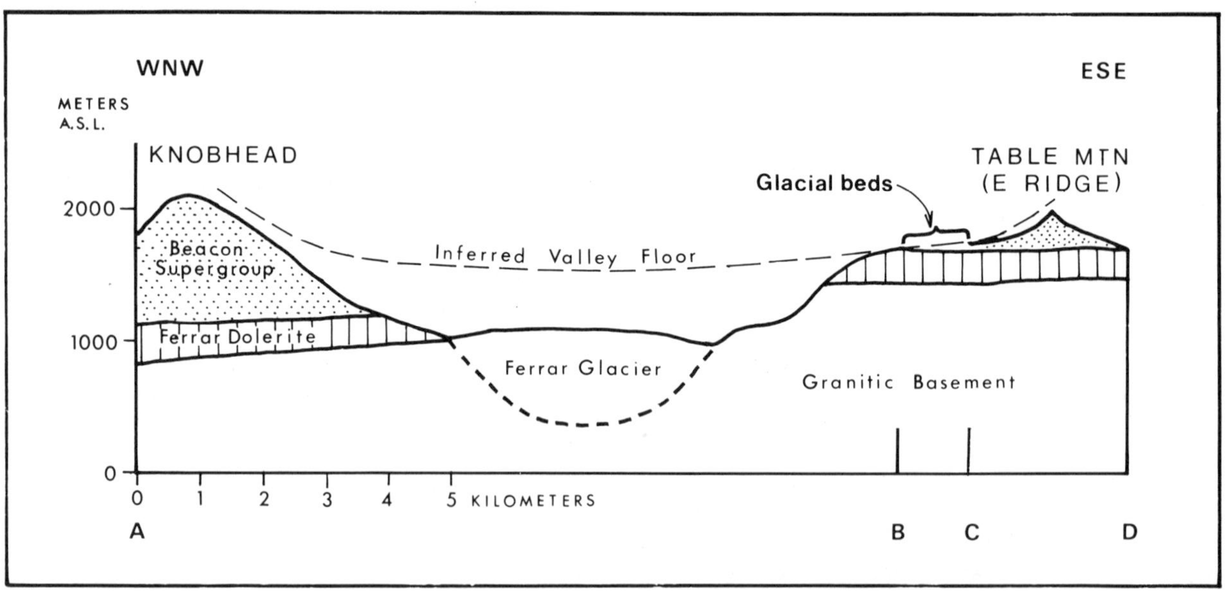

Figure 137.1. Map and cross section showing the glacial beds on Table Mountain. *a*. Topographic map (from U.S.G.S. Reconnaissance Series 1:250,000 Taylor Glacier sheet), showing the extent of the beds (black dots) and the location of the cross section. *b*. Geologic cross section from Knobhead to Table Mountain. Base of traverse is about 1800 m above sea level.

glacial deposition, followed by sedimentation in ice-marginal or proglacial streams; a subsequent advance of the ice front is also recorded. Thus our interpretation of the sequence requires a more complex local history than that proposed for the Sirius Formation of the region by Mayewski (1975).

Situation of the glacial beds at Table Mountain

The oldest rock on Table Mountain (Figure 137.1*b*) is basement granite of early Paleozoic age (Warren, 1970). This is nonconformably overlain by the Devonian Taylor Group (Beacon Supergroup), a continental sequence of quartzose sandstone with a 40 to 80 m thick shaly siltstone

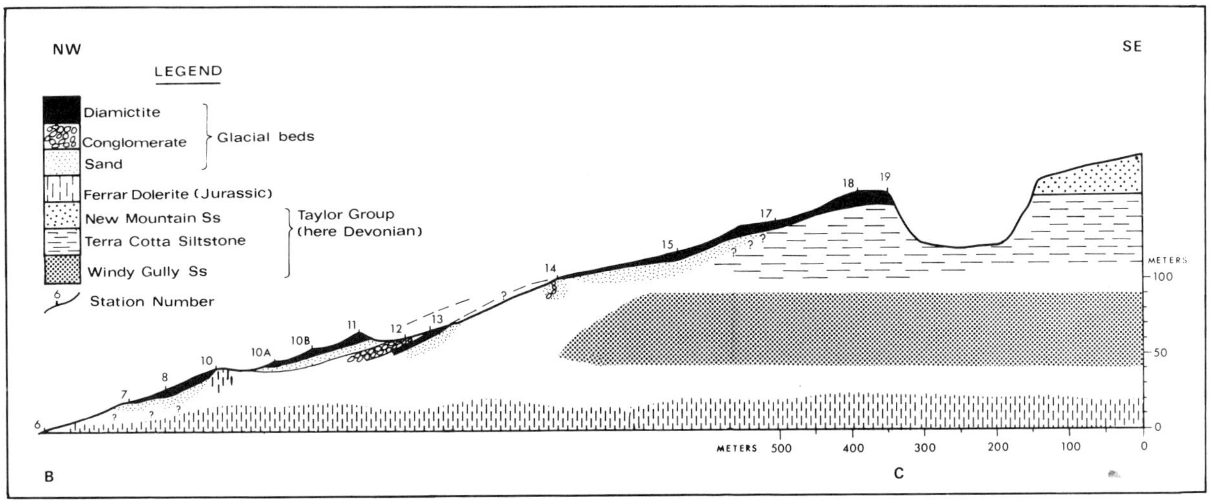

Figure 137.2. Section through the glacial beds on Table Mountain.

unit near the base (McKelvey and others, 1977). The sandstone is fine to coarse grained and well sorted; most of the grains are rounded. The region has been intruded by extensive sills of Ferrar Dolerite, tholeiitic diabase of Jurassic age. The mineralogy of the glacial beds shows them to have been derived entirely from the Taylor Group and the Ferrar Dolerite (see below).

The glacial beds on Table Mountain rest on a broad bench 1200 m wide and 1700 m above sea level, 700 m above the ice surface of the Ferrar Glacier (Figure 137.2). The bench slopes to the northwest at about 8° but coincides approximately with the top of a diabase sill 300 m thick that is intruded close to the base of the Beacon Supergroup (the Peneplain Sill of Gunn and Warren, 1962). Structural control of the bench is probable, but the apparent absence of basement-derived material in sediments on the bench (see below) indicates that the ice at the time had not cut through the Peneplain Sill upstream from this locality.

Ferrar Valley has another broad bench only 200 m above the present glacier surface. This level must have been entirely controlled by the erosional agent, as there is no rock contact here to cause the bench. It is therefore attributed to a period of glacial erosion by a much enlarged Ferrar Glacier. The cutting of the present valley, now filled with glacial ice, may represent a third period of glacial erosion. The floor of Ferrar Valley is now more than 1000 m below the valley floor inferred for the deposition of the glacial beds described here (Figure 137.1*b*). If these beds can be counted as part of the Sirius Formation, their situation seems inconsistent with Mayewski's contention that it was the Queen Maud Glaciation that most deeply sculpted the Transantarctic Mountains.

Description of the glacial beds

The glacial beds at Table Mountain cover an area of about 2 km² and reach a thickness of at least 15 m. However, they are in most places exposed only in scattered inliers surrounded by a veneer of diabase gravel, debris from a more recent glaciation. The hummocky topography has an average slope similar to the dip of the sediments themselves, and consequently the thickness of the beds exposed at any one place is rarely more than a few meters. The three main sediment types are pebbly, muddy sandstone (diamictite), sandstone, and conglomerate.

The basal contact is well exposed along the upslope (southeast) edge of the outcrop. Elsewhere it can be located only approximately because it is gravel covered. The upslope exposures show the diamictite in razor-sharp contact with fresh greenish-gray Devonian siltstone, which is also the dominant clast rock type, forming as much as 30% of the lower 0.3 m of the formation. The siltstone does not survive above that level, however, and is replaced by diabase. The basal contact has irregular local relief of about 0.2 m, evidently from the plucking out of siltstone blocks.

Diamictite

The most striking lithology of the glacial beds is a semilithified, very light gray, muddy sandstone with scattered pebbles and cobbles and rare boulders (Figure 137.3)—a diamictite (Flint and others, 1960). Generally it lacks stratification apart from a few thin sandstone lenses but has a faint crude fissility that parallels the regional dip of the formation and the slope of the bench (8° to the northwest). This description conforms to that of a typical till (Goldthwait, 1971), though no striated bedrock surface was found, and fabric data were not collected.

There are at least two, and probably three separate beds of diamictite. The sequence exposed between stations 11 and 14 shows two diamictite beds separated by several meters of current-sorted sediment. The basal part of the formation does not crop out there, but it could include another diamictite bed, for diamictite forms the base of the formation farther up the valley wall (stations 17–19).

The diamictite has a uniform appearance over most of its area of outcrop. Clasts larger than 2 mm form about 20% of the rock. Most are of pebble size, but they commonly measure up to 0.5 m, and one was found to be 1.4 m long. Diabase is by far the most common clast lithology (Table 137.1), apart from the siltstone in the basal 0.3 m where the diamictite rests directly on Terra Cotta Siltstone (Figure 137.2). The clasts are on the average subrounded,

Figure 137.3. The glacial beds at station 15 (Figure 137.2.) *a.* The 3 m thick body of diamictite conformably overlies friable sandstone at the base of the outcrop. Sandstone is also exposed on the flat, but may be weathered diamictite. Circled figure indicates scale. Arrow indicates area of Figure 137.3*b*. *b.* A closer view of the diamictite at the large arrow in Figure 137.3*a*, showing the scattered diabase clasts that form about 20% of the rock, the lack of stratification, and the crude fissility. Ice axe is 0.80 m long. *c.* Diamictite grades downward into stratified, poorly sorted, medium-grained sandstone (small arrow at far right in Figure 137.3*a*; sample 20, Table 137.3).

and a substantial proportion (a third of the finely crystalline diabase clasts) are striated (Table 137.2, Figure 137.4).

The matrix of the diamictite is poorly sorted, and between 8% and 30% is of mud size (Table 137.3). This texture, along with the moderate degree of rounding and the high proportion of striated clasts, indicates that the sediment was deposited directly from basal glacier ice. These criteria have been previously used by Goldthwait (1971) to infer subglacial deposition. Faint stratification in the youngest diamictite indicates a little sorting by water currents.

One feature of the size distributions in all of the diamictite samples is the broad but clear mode in the fine to medium (3.0 to 1.0 phi) sand range (see Figure 137.5). On its own, this could be taken as a sign of current sorting during deposition, but it is much more likely a result of the

Table 137.1. Percentages of pebble types (8–64 mm) in the diamictite at Table Mountain (100 pebbles per count)

	Coarse diabase	Fine diabase	Quartzose sandstone	Vein quartz	Other
Station 10B	86	9	4	1	1
Station 15	74	19	2	4	1

provenance of these sediments. The lower part of the Taylor Group, on which the glacial beds rest, consists in the Ferrar-Taylor Glacier area largely of well-sorted, fine- to medium-grained sandstone (Angino and Owen, 1961; Korsch, 1974) (see Table 137.4), apart from the 50-m-thick Terra Cotta Siltstone. About 70% of the diamictite sand fraction is rounded quartz grains (Table 137.3), a characteristic of Taylor Group sandstone. Quartz sand wears very slowly—subangular quartz sand loses less than 0.1% of its weight by abrasion in 1000 km of water transport (Kuenen *in* Blatt and others, 1972). The roundness of the grains in the diamictite and the sandstone is therefore believed to be inherited from the sandstone source. The Terra Cotta Siltstone also seems to have had its influence on the texture of the diamictite. Mud forms about 30% of the matrix in samples within 2 m of the basal contact with the siltstone (samples 10 and 27), but is substantially lower (and more variable) in samples from higher in the sequence where the influence of the siltstone is less immediate and dilution from the far more extensive sandstone has had an effect.

None of the diamictite examined was derived from the basement, whose upper surface is within 200 m of the base

Figure 137.4. A well-striated diabase boulder from diamictite at station 18. The striae, which become fainter upward, were acquired before deposition at the base of the glacier, but have now been lost entirely from the upper left corner due to weathering.

of the deposit and dips gently west toward the South Polar Plateau. Evidently the upstream ice at the time of deposition had not yet eroded through the Peneplain Sill (Figure 137.1*b*). No basement material was seen as clasts in the outcrop or in grain mounts from this unit, although a few granite boulders were noted in more recent moraines about 100 m lower on the west spur of Table Mountain. The dominance of sandstone-derived material in the matrix contrasts with the dominance of diabase in the gravel fraction and simply reflects the low resistance of the poorly cemented sandstone to glacial erosion and transport. The difference in behavior of the two rock types has already been described by Souchez (1966) from moraines at the edge of the upper Ferrar Glacier. The relative contributions from Taylor Group sediment and Ferrar Group diabase can be estimated by combining grain size and compositional data, and are in a ratio of 3:1.

CONGLOMERATE

Conglomerate crops out along the floor of a broad northeast-trending valley (stations 11-14, Figure 137.2), thickening from 1 to 5 m over a distance of 400 m to the

Table 137.2. Pebble characteristics, mean roundness, and number striated for different size classes and the two main rock types for diamictite at two localities at Table Mountain

	Krnss	Frnss	*N*	Striated	Size	Rock type
Station 15	0.41	0.12	10	1	32-64 mm	coarse diabase
	0.50	0.14	10	1	16-32 mm	
	0.38	0.09	10	—	8-16 mm	
				2/30		
	0.38	0.09	10	2	32-64 mm	fine diabase
	0.39	0.12	10	6	16-32 mm	
	0.38	0.11	10	3	8-16 mm	
	$\bar{x}$ 0.41	0.11		11/30		
	s 0.05	0.02				
Station 15A	0.52	0.20	10	—	32-64 mm	coarse diabase
	0.46	0.11	10	—	16-32 mm	
	0.51	0.15	10	—	8-16 mm	
				0/30		
	0.47	0.15	10	6	32-64 mm	fine diabase
	0.44	0.11	11	5	16-32 mm	
	0.47	0.14	10	6	8-16 mm	
	$\bar{x}$ 0.48	0.14		17/30		
	s 0.03	0.03				

Krnss: roundness after Krumbein (1941).
Frnss: roundness after Dobkins and Folk (1970).
N: number of pebbles.
$\bar{x}$: mean.
s: standard deviation.

Table 137.3. Grain size distribution (in weight percent), mean ($\bar{x}$), and standard deviation(s) for samples of diamictite matrix (finer than − 1.0 phi) and sandstone from the glacial beds at Table Mountain

	Percentage coarser than − 1.0	Class Limits in Phi Units												
		− 1.0 to − 0.5	− 0.5 to 0.0	0.0 to 0.5	0.5 to 1.0	1.0 to 1.5	1.5 to 2.0	2.0 to 2.5	2.5 to 3.0	3.0 to 3.5	3.5 to 4.0	Rest	$\bar{x}$	s
Windy Gully Ss (from Angino and Owen, 1961)		—	2.5	5.8	11.4	52.0	16.8	6.6	2.3	1.1	0.5	1.0	1.33	0.58
New Mountain Ss		—	—	0.3	0.8	6.5	21.3	53.8	14.9	1.2	0.6	0.6	2.15	0.42
Averages for three overlying sandstone formations of the Taylor Group (from Korsch, 1974)													2.2	0.7
Diamictite resting on bedrock (Terra Cotta Siltstone)														
10—along strike from sta. 18, 200 m to southwest	1.7	0.8	1.2	1.6	7.0	12.0	16.5	14.0	6.8	5.6	4.8	29.8	3.1	2.3
27—along strike from sta. 18, 1 km to northeast	6.6	0.6	1.0	1.4	7.5	11.1	14.4	13.7	10.2	6.3	4.1	29.7	3.3	2.4
Diamictite resting on penecontemporaneous sediment														
23—above sandstone and thin conglomerate at sta. 8	4.7	1.9	2.5	2.2	10.6	13.4	16.9	14.7	10.4	6.7	4.2	16.5	2.4	1.7
25—above conglomerate near sta. 13	16.6	3.0	3.5	3.5	8.6	15.9	17.6	15.9	11.6	7.2	4.6	8.5	2.0	1.3
28—bed above diamictite sample 27	0.4	0.3	0.6	0.8	6.1	10.7	17.5	16.6	13.1	6.9	4.0	23.4	3.0	1.9
Sandstone														
18—below deflation gravel in basin adjacent to sta. 15		0.1	0.2	0.5	2.3	10.3	13.8	18.6	14.0	6.2	5.1	23.9	3.0	1.7
19—probably weathered diamictite			0.1	0.1	1.0	5.2	13.7	19.7	17.0	11.3	8.3	23.6	3.1	1.3
20—below diamictite at sta. 15		0.2	0.5	1.0	6.5	15.1	21.6	17.7	11.6	5.6	4.6	15.6	2.5	1.4
22—below diamictite and thin conglomerate at sta. 8			0.1	0.1	0.1	0.3	1.6	19.8	39.1	16.4	11.9	10.6	3.0	0.7

Grain size data from the Taylor Group are included for comparison. The diamictite samples show a wide range in percentage coarser than − 1.0 phi due to small sample size (~ 20 gm); the gravel fraction in fact is typically in the range of 15 to 20%, judging from measurements over outcrop surfaces 0.2m x 0.6 m.

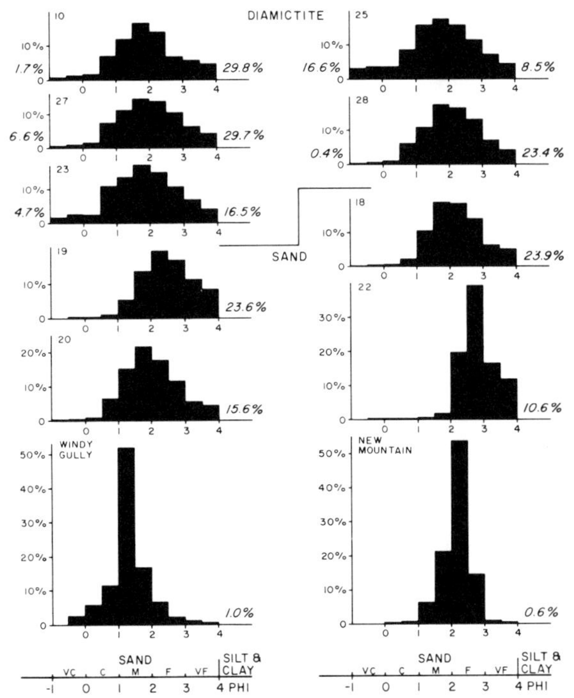

Figure 137.5. Histograms illustrating the grain size distributions (in wt. %) in the glacial sediments (five diamictites and four sands) and in two bedrock units (see Table 137.3 for sample descriptions) at Table Mountain. Percentages in italics to the left and right of each graph are proportions of the samples coarser than − 1.0 phi and finer than 4.0 phi, respectively.

southwest. It disconformably overlies a diamictite bed (Figure 137.6); the contact with the overlying sandstone is not exposed. The conglomerate must pinch out to the northwest because it is not exposed between stations 10 and 11. It probably has the form of an elongate lens paralleling the strike of the small valley.

Figure 137.6. Conglomerate overlying diamictite at station 13 (Figure 137.2). The contact is shown by arrows.

The conglomerate consists of a framework of moderately sorted and rounded diabase clasts, averaging about 30 mm in size; the matrix is light colored and sandy. These features suggest it was reworked from the diamictite or from debris-laden ice.

Conglomerate interbedded with sandstone crops out over a few square meters at station 14 (Figure 137.2), but here it has a rather different character. Thin beds of sandy pebble and cobble conglomerate are intercalated with very light gray fine-grained sandstone in a sequence that is folded; dips reach almost 90° and beds strike northeast, parallel to the valley side. This attitude and the style of folding indicate downslope movement of water-saturated sediment.

SANDSTONE

Sandstone forms bodies typically about 5 m thick; it is best exposed in the upper meter, where it is finely laminated and grades upward into unstratified or poorly stratified diamictite. The gradation zone in some places includes conglomeratic and sandy lenses (station 11), but other

Table 137.4. Grain composition (in %) for sediment from the glacial beds at Table Mountain

Sample No. and size	Quartz rounded	Quartz angular	Plagioclase (calcic)	Clinopyroxene (altered)	Hypersthene	Diabase grains	Other	Number of grains
Diamictite								
ZS-10								
1-2 phi	90	5	1	2	1	1	1	214
2-3 phi	82	12	1	1	trace	2	1	227
ZS-25								
1-2 phi	66	13	3	5	2	10	3	200
2-3 phi	55	19	2	9	5	9	1	246
ZS-27								
1-2 phi	75	15	2	3	2	2	1	224
2-3 phi	55	33	3	3	1	3	1	224
ZS-28								
1-2 phi	75	16	3	3	1	1	0	203
2-3 phi	60	29	3	2	2	2	1	205
Sandstone								
ZS-18								
1-2 phi	81	14	1	2	1	1	1	216
2-3 phi	71	22	2	2	1	1	1	212
ZS-19								
1-2 phi	84	10	2	2	trace	2	trace	245
2-3 phi	71	22	3	1	trace	2	1	232
ZS-20								
1-2 phi	77	13	2	4	2	2	1	204
2-3 phi	60	28	4	3	1	4	1	217
ZS-22								
1-2 phi	76	13	1	5	1	3	1	149
2-3 phi	76	15	1	3	1	3	1	214

Figure 137.7. This 2.7 m long diabase boulder caused flow deformation of the laminated fine-grained sandstone beneath by dropping from melting ice.

places lack these lenses (station 15). The sandstone is fine grained and moderately sorted and has a composition indistinguishable from the sand fraction of the diamictite (Tables 137.3 and 137.4). It is believed to have been deposited by ice-marginal or proglacial streams reworking debris from basal glacier ice.

The upper meter of sandstone near station 11 contains a giant diabase dropstone that has clearly bowed down and plastically deformed the sand laminae beneath it (Figure 137.7). Sandstone close to the right side of the boulder has no trace of stratification and is believed to have flowed into the moat that the boulder created by depressing the laminated sand. This situation suggests to us that the boulder was dropped from melting ice. Such a boulder would require a block of ice of at least 110 m^3 to float it. Therefore, it appears that deposition was in water at least several meters deep. Another boulder like this was seen in a similar situation near station 10A.

Sediment near station 15, poorly exposed beneath a veneer of diabase gravel (Figure 137.3*a*), has the appearance of sand, though size analyses (Table 137.4, samples 18 and 19) show that it has a texture of diamictite matrix. The "sand" may result from freezing and thawing above the permafrost (here about 0.2 m below the surface), gradually breaking down the underlying diamictite and segregating the gravel-size debris.

Discussion

The texture of the basal diamictite and the high proportion of striated clasts indicate that it is a basal till. However, diamictite overlying current-sorted sediment appears slightly current sorted. The current-sorted diamictite is still regarded as till because of the poor sorting and scattered gravel-size debris, but it may have been washed by water currents during rapid melting of the debris-charged basal ice.

The conglomerate and sandstone bodies within the sequence have no very distinctive features. Based on their textures and the glacial context, they seem to have been deposited from running water at high to moderate flow intensities and were eroded from the same source as the diamictite, or were reworked from the diamictite itself. They lack a number of features characteristic of ice-contact deposits, such as abrupt changes in grain size, sharp lateral discontinuities of bedding, and marked local deformation (Flint, 1975). Only one patch of disturbed bedding (by station 14) was seen in the area, and this most likely resulted from downslope mass movement (see discussion of conglomerate, above). However, the large dropstones described earlier indicate active calving of debris-laden ice in water of moderate depth. We conclude that the conglomerate and sandstone bodies were probably deposited close to the ice margin, but there is no evidence that they were in contact with the ice.

The upper diamictite (between stations 11 and 14, Figure 137.2) is similar to the basal diamictite in composition and texture, and is also regarded as a basal till.

The local climate during the deposition of the Table Mountain glacial beds was much warmer than at present. Stream deposits like the current-sorted beds are presently forming only in the lowest parts of the Dry Valley region (Powell, 1976), where the temperature is about 15°C higher than on the upper slopes of Table Mountain (assuming an adiabatic lapse rate of 1°C/100 m). The warmer local climate when these beds were deposited may have been due to either a lower elevation or a warmer regional climate, or both.

Age of the glacial beds

It is not possible to date the glacial beds directly, as they seem to be unfossiliferous. Instead their age must be estimated from other known and dated events. There seems to have been little more erosion in Ferrar than in Taylor Valley since the latter was cut to its present form. The glaciers have cut down to about the same level on either side of Knobhead (100 to 300 m above sea level, Calkin, 1974), but Ferrar Valley is a little deeper in its lower part than is lower Taylor Valley. Valley cutting at present must be minimal as ice in the valleys is moving very slowly (5 to 10 m/year for the Taylor Glacier, Robinson, 1979). Therefore, the last period of significant glacial erosion in Ferrar Valley is believed to be of the same age as that for the adjacent Taylor Valley. Late Miocene paleontologic ages from sediment fill in lower Taylor Valley (Webb and Wrenn, Chapter 144, this volume) give a minimal age for this last major period of valley cutting and thus for the deposition of the glacial beds on Table Mountain. Brady (Chapter 145, this volume) has found even older sediment in Wright Valley. Marine diatoms from sediments beneath Lake Vanda in Wright Valley indicate that the Dry Valleys had formed by the middle Miocene.

A maximum age for the glacial beds is suggested by the time of earliest ice-rafting in the Ross Sea. This has been well dated at 25 m.y. from a core taken 400 km east of McMurdo Sound (site 270, Deep Sea Drilling Project; see Hayes and others, 1975). The oldest glacial beds there have palynomorphs indicating a late Oligocene parent vegetation dominated by *Nothofagus,* but pollen abundance drops off markedly upward before Miocene strata are reached (Kemp and Barrett, 1975). The palynomorph assemblage in these oldest beds is similar to that in Eocene mudstone erratics from McMurdo Sound, from which McIntyre and Wilson (1966) deduced a vegetation and climate like that of the southern beech forest in South America and New Zealand today. The high rate of glacial deposition in the south-central Ross Sea throughout the late Oligocene and Miocene (41 m/m.y.) is itself an indication of extensive, wet-based glacial erosion around the Ross Embayment (Barrett, 1975).

We believe that the glacial beds at Table Mountain represent glacial and glacially derived sediment deposited on land at some time in the late Oligocene or early Miocene before the outlet glaciers had cut deeply into the Transantarctic Mountains. However, the age of these deposits can be determined more precisely only through some means of correlation with marine sequences beneath the Ross Sea.

Acknowledgments. We would like to thank the Antarctic Division of the Department of Scientific and Industrial Research and the U.S. Navy VXE-6 Antarctic Support Squadron for field and logistic support. Financial support for field work was provided by the New Zealand University Grants Committee and Victoria University of Wellington. We also gratefully acknowledge use of facilities at the Scott Polar Research Institute, Cambridge (P. J. B.), and the Institute of Polar Studies and Department of Geology and Mineralogy, The Ohio State University, Columbus, Ohio (R. D. P.).

References cited

Angino, E. E., and Owen, D. E. 1961. Sedimentologic study of two members of the Beacon Formation, Windy Gully, Victoria Land, Antarctica. *Trans. Kans. Acad. Sci., 65,* No. 1, 61–69.

Barrett, P. J. 1975. Textural characteristics of Cenozoic preglacial and glacial sediments at Site 270, Ross Sea, Antarctica. (*In* Hayes, D. E., Frakes, L. A., and others. *Initial Reports of the Deep Sea Drilling Project,* Vol. 28. Washington [U.S. Government Printing Office], 757–67.)

Blatt, H. G., Middleton, G. V., and Murray, R. 1972. *Origin of Sedimentary Rocks.* Englewood Cliffs, N.J., Prentice-Hall, 634 pp.

Calkin, P. E. 1974. Subglacial geomorphology surrounding the ice-free valleys of southern Victoria Land, Antarctica. *J. Glaciol., 13,* No. 69, 415–29.

Dobkins, J. E., Jr., and Folk, R. L. 1970. Shape development on Tahitinui. *J. sedim. Petrol., 40,* No. 3, 1167–203.

Fleck, R. J., Jones, L. M., and Behling, R. E. 1972. K-Ar dates of the McMurdo volcanics and their relation to the glacial history of Wright Valley. *Antarct. Jnl U.S., 7,* No. 6, 245–46.

Flint, R. F. 1975. Features other than diamicts as evidence of ancient glaciations. (*In* Wright, A. E., and Moseley, F., *eds. Ice Ages: Ancient and Modern.* Liverpool, See House Press, 121–36.)

Flint, R. F., Sanders, J. E., and Rodgers, J. 1960. Diamictite, a substitute term for symmictite. *Bull. geol. Soc. Am., 71,* 1809–10.

Goldthwait, R. P. 1971. Introduction to till, today. (*In* Goldthwait, R. P., *ed. Till/A Symposium.* Columbus, The Ohio State University Press, 3–26.)

Gunn, B. M., and Warren, G. 1962. Geology of Victoria Land between the Mawson and Mulock Glaciers, Ross Dependency, Antarctica. *Bull. geol. Surv. N.Z., 71,* 157 pp.

Hayes, D. E., Frakes, L. A., and others. 1975. *Initial Reports of the Deep Sea Drilling Project,* Vol. 28. Washington (U.S. Government Printing Office), 1017 pp.

Kemp, E. M., and Barrett, P. J. 1975. Antarctic glaciation and early Tertiary vegetation. *Nature, Lond., 258,* 257–58.

Korsch, R. J. 1974. Petrographic comparison of the Taylor and Victoria Groups (Devonian to Triassic) in south Victoria Land, Antarctica. *N.Z. Jl Geol. Geophys., 17,* No. 3, 523–41.

Krumbein, W. C. 1941. Measurement and geological significance of shape and roundness of sedimentary particles. *J. sedim. Petrol., 11,* No. 1, 64–72.

Mayewski, P. A. 1975. Glacial geology and late Cenozoic history of the Transantarctic Mountains, Antarctica. *Rep. Inst. polar Stud. Ohio State Univ.,* No. 56, 168 pp.

Mercer, J. H. 1968. Glacial geology at the Reedy Glacier area, Antarctica. *Geol. Soc. Am. Bull., 79,* No. 4, 471–85.

Mercer, J. H. 1972. Some observations on the glacial geology of the Beardmore Glacier area. (*In* Adie, R. J., *ed. Antarctic Geology and Geophysics.* Oslo, Universitetsforlaget, 427–33.)

McGregor, V. R. 1965. Notes on the geology of the area between the heads of the Beardmore and Shackleton Glaciers, Antarctica. *N.Z. Jl Geol. Geophys., 8,* No. 2, 278–91.

McIntyre, D. J., and Wilson, G. J. 1966. Preliminary palynology of some Antarctic Tertiary erratics. *N.Z. Jnl Bot., 4,* No. 3, 315–31.

McKelvey, B. C., Webb, P. N., and Kohn, B. P. 1977. The Taylor Group (Devonian) and the lower Victoria Group (Permian) in southern Victoria Land. *N.Z. Jnl Geol. Geophys., 20,* No. 5, 813–63.

Oliver, R. L. 1964. Geological observations at Plunket Point, Beardmore Glacier. (*In* Adie, R. J., *ed. Antarctic Geology.* Amsterdam, North-Holland Publishing Company, 138–42.)

Powell, R. D. 1976. *Textural Characteristics of some Glacial Sediments in Taylor Valley, Antarctica.* Master's thesis, Victoria University of Wellington N.Z., 316 pp.

Robinson, P. H. 1979. Taylor Glacier: glaciological and sedimentological data tables. *Victoria University of Wellington Antarctic Data Series No. 6,* 33 pp.

Souchez, R. A. 1966. The formation of shear moraines: an example from south Victoria Land. *J. Glaciol., 6,* No. 44, 837–43.

Warren, G. 1969. Terra Nova Bay–McMurdo Sound area. (*In* Bushnell, V. C., and Craddock, C., *eds.* Geologic Maps of Antarctica. *Antarct. Map Folio Ser.,* Folio 12, Pl. XIII.)

138

Composition of East Antarctic Moraines and Some Problems of Cenozoic History

V. I. BARDIN
Soviet Committee on Antarctic Research, Moscow, U.S.S.R.

Abstract. New results of multi-disciplinary studies of continental moraines in different areas of East Antarctica, of interest for Cenozoic paleogeography, are given. This paper mainly considers tills of different ages in the Prince Charles Mountains and Queen Maud Land, studied by the author in the field and later analyzed by a group in the laboratory. The samples of till were studied by a variety of traditional methods (granulometric, mineralogic, petrographic, and chemical), along with use of the X-ray method and the electron microscope (for clay minerals). Special methods were used in studies of a set of authigenic minerals, which are sensitive indicators of climatic conditions. Five distinct morainal complexes, which formed during Miocene–Pleistocene time, are defined and characterized. This study makes it possible to envision the succession, the character, and the time range of the processes of glacial sedimentation in East Antarctica.

Glacial deposits—moraines—have been studied by the author in a number of mountainous areas of East Antarctica: Queen Maud Land, MacRobertson Land, and Coats Land. These areas, which are characterized by rather large expanses of mountains free of ice, are of great interest for widely based investigations of continental moraines. To a large extent, these exposures can be regarded as key areas since most of the remainder of East Antarctica is covered with smooth, permanent ice.

Glacial deposits

The main results of moraine studies in Queen Maud Land have been published (Bardin, 1969; Bardin and Ravich, 1968; Bardin and Sudakova, 1972) and were also reported at the Second Symposium on Antarctic Geology and Geophysics in Oslo in 1970 (Bardin, 1972). Materials collected in the 1976–77 field season during investigations in the Shackleton Range in Coats Land are just now being studied. The results of investigations in the Prince Charles Mountains in MacRobertson Land, obtained during 1972–74 by Soviet expeditions collecting geologic, geophysical, and geographic material, are the essence of this report. In this region we have succeeded in discovering the most complete series of glacial deposits, extending from modern moraines to the most ancient ones, possibly formed near the beginning of Antarctic glaciation (Figure 138.1).

Moraines are rather abundant in the Prince Charles Mountains (Bardin, 1975). Ancient glacial deposits are typically found on gentle slopes of high mountains. Quite commonly they can be traced as heterogeneous beds elongated down the slope, with one above the other. They differ both in peculiarities of microrelief and in overall color of the pebble to boulder material.

Aerial observations of the slopes of mountain massifs (Mount Meredith, Mount Fisher, Mount Collins, and others) facing the Lambert Glacier valley and its tributaries clearly reveal the upper limit of loose material spread along the slope. Even more apparent are two lower moraines which mark the later stages of ancient continental glaciation.

Ground investigations on the northern slopes of Mount Collins confirmed the ideas about a three-unit structure of the surface morainal mantle. From the foot of the slope at the level of the modern glacier up to the mountain peak, the morainal material is grouped into three strips, one above the other, which differ greatly both in morphology and in color (Figure 138.2*a*).

The lowest strip consists of moraine beds distinctly expressed in the relief and which have still preserved the ice core. Boulders here are almost totally unrounded, i.e., "fresh," and they are light gray in color. The overlying strip of morainal material is of a gray-brown color; there is no ice preserved under the boulder material, which does not create sharp bed forms in the relief but rather a gentle, hilly bank. Weathering processes have partly influenced the surface of boulders. Still higher on the slope the third moraine is represented by a thin interrupted cover which has completely lost its original form. This upper, most ancient moraine is light brown in color.

Distinct differences and clear borders between the three strips of glacial deposits on Mount Collins indicate that they are moraines of three ages representing three successive stages (periods) of glacial development (Figure 138.2*b*). Such a situation is also characteristic of the remaining part of the region, judging from direct aerial observation and from the study of aerial photographs. Three stages of moraine formation have also been established in the mountains of Queen Maud Land (Bardin, 1969) and Victoria Land (Mercer, 1972), proof of an analogous sequence of changes (successive reduction of the glacial cover).

Our speculations about the history of glaciation in the Prince Charles Mountains would have ended with the model of three descending stages of glacial cover but for the discovery of thick ancient moraine in the northern area

Figure 138.1. The northern part of the Prince Charles Mountains near Lake Beaver and Lake Radok. Here, contemporary and ancient moraines are especially widely preserved.

of the Amery Oasis (Bardin, 1974). The study of this moraine has yielded important additional paleoglaciologic information.

These glacial deposits (with a visible thickness of about 60 m) are located along the northwestern side of Pagodrome Gorge, at the bottom of which is Lake Beaver Bay—an epishelf reservoir lying at sea level (Bardin, 1974). The full moraine thickness appears in vertical cliffs which crown the steep slope of the gorge eroded in Permian sandstones (Figures 138.3 and 138.4).

Upon first investigation of this moraine, its most conspicuous characteristic is the difference in color between the lower (gray) and upper (brown) parts. The lower gray layer of the moraine is 15–20 m thick. The gray moraine consists mainly of angular clasts of Permian sandstone,

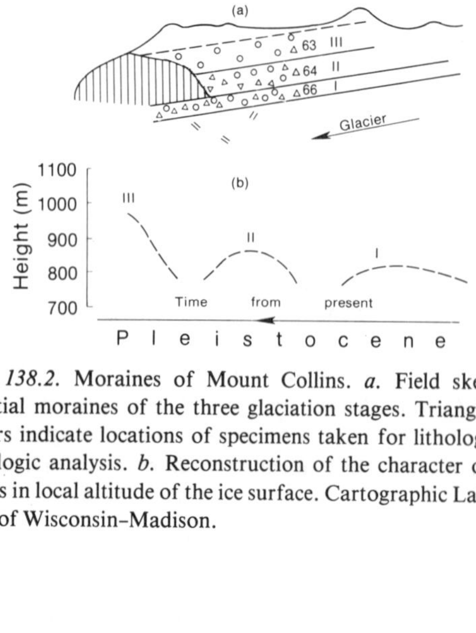

Figure 138.2. Moraines of Mount Collins. *a.* Field sketch of sequential moraines of the three glaciation stages. Triangles and numbers indicate locations of specimens taken for lithologic and mineralogic analysis. *b.* Reconstruction of the character of fluctuations in local altitude of the ice surface. Cartographic Lab, University of Wisconsin–Madison.

but it also contains pebbles, cobbles, and small boulders of crystalline rock.

The gray moraine layer passes upward into a 25–40-m-thick brown moraine layer, which can be traced up to the peak. In the contact zone there is sand and conglomerate. There are many pebbles, cobbles, and boulders of crystalline rocks. Alongside clasts of sandstone are large rounded boulders of crystalline rocks in the brown moraine layer.

The presence within the stratigraphic section of two moraines which are different in color and which are separated by conglomerate and rubblestone pointing to an interruption in glacial deposition, leads to the interpretation that traces of at least two different stages of glaciation have been preserved here.

A number of questions have arisen in regard to the discovery in the Prince Charles Mountains of ancient glacial deposits of the various types characterized above. The most important of these are the determination of ages and the correlation of the Mount Collins moraines with the glacial deposits which appear in the vertical cliffs along the Pagodrome Gorge.

Many genetic peculiarities of the glacial deposits in the Amery Oasis sections remain unexplained, and the reason for the difference in color of the moraine material (gray for the more ancient moraine, brown for the younger one) is particularly unclear. The situation is further complicated by the fact that, on the surface of the moraines at Mount Collins, a reverse correlation of colors has been observed. An explanation of these differences is necessary in any theory of moraine origin.

Unfortunately, other sections through similar moraines near the Antarctic coast are not known. Glacial deposits whose cross sections are similar in thickness and in texture

Figure 138.3. General view of section of moraine deposits, the uppermost unit in the wall of Pagodrome Gorge, Amery Oasis.

have been discovered in the mountains of Victoria Land (Mercer, 1972). However, they are far from the coast and lie at considerable elevations, not only well above the contemporary surface of the glacier, but even higher than the level of its ancient maximum (as recorded in the three stages of development mentioned above). In contrast, sections at Amery Oasis are still lower than the present-day surface of continental ice flowing around the oasis. Therefore, special arguments are needed to draw any analogies and correlations.

The complex character of this problem is aggravated by the absence of even approximate age dates on the moraines. It was not possible to obtain them through methods commonly applicable on other continents (biostratigraphic or radiometric dating of organic remains). In this connection, the author has tried to investigate the

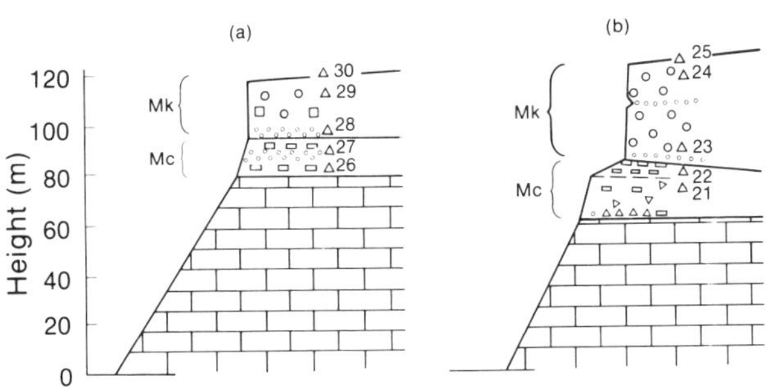

Figure 138.4. Field sketch of schematic profiles of investigated outcrops in Pagodrome Gorge, Amery Oasis. *a.* Outcrop 2. *b.* Outcrop 1. Mk: brown moraine. Mc: gray moraine. Triangles and numbers indicate locations of specimens taken for lithologic and mineralogic analysis. Cartographic Lab, University of Wisconsin-Madison.

age interrelationships among moraine thicknesses, as well as the genetic conditions for their formation, based on analysis of their composition.

Attempts at the use of lithologic peculiarities to establish the age of moraines in Antarctica has been undertaken before. Morpho-lithologic differences among various moraines in the mountains of Queen Maud Land have been determined (Bardin, 1969). Geochemical composition and some peculiarities of mineralogy and chemistry of several moraines in the same region were also studied (Bardin and Ravich, 1968; Bardin and Konoplyova, 1971; Bardin and Sudakova, 1972).

Although lithologic investigations carried out through a variety of traditional methods did give an extensive profile of the composition of glacial deposits (Bardin and others, 1976) (Figure 138.5), they did not provide sufficiently clear-cut criteria for distinguishing moraines of different ages.

Determination of the age and succession of ancient Antarctic glacial deposits containing no organic material is a task of great complexity that remains unsolved to the present day. Data on the absolute age of volcanogenic rocks in glacial deposits in the dry valleys of Victoria Land and the Jones Mountains (Denton and others, 1970; Fleck and others, 1972; Rutford and others, 1968, 1972), which allow the dating of these moraines, are so far unique, and at present they cannot be used for correlation of other distant moraines.

In the present paper an attempt has been made to use the results of mineralogic and petrographic studies of glacial deposits as indications of age and, in a broader sense, for paleogeographic interpretation. These results include: (1) data on the mineralogic composition of authigenic and non-authigenic components of heavy mineral fractions

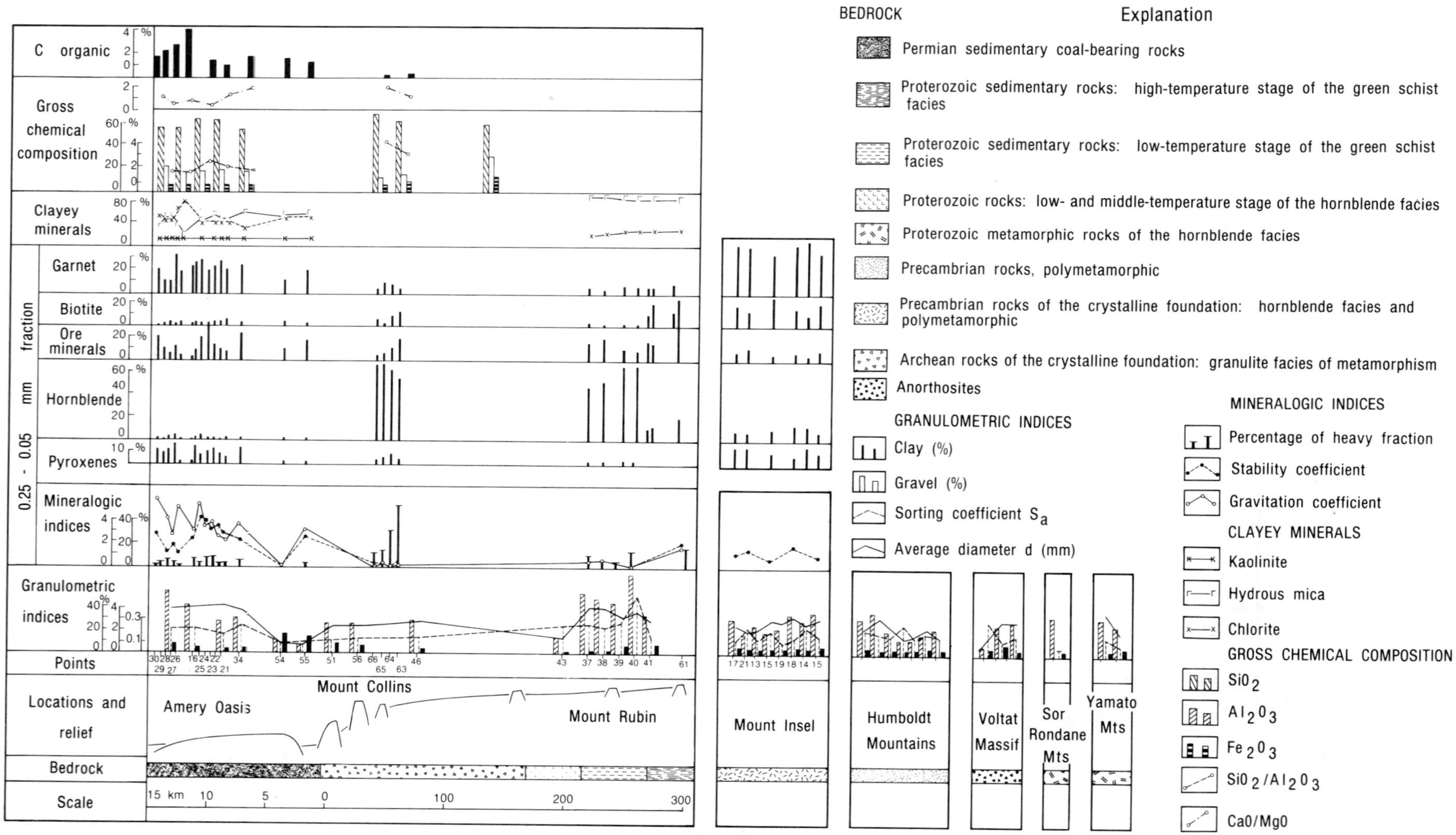

Figure 138.5. Material composition of moraines in Antarctica; after Bardin and others (1976). Cartographic Lab, University of Wisconsin–Madison.

(0.10–0.25 mm size); and (2) data on the petrographic composition of clayey matrix material of various moraines.

Application of a specific investigational method based on the study of authigenic minerals—sensitive indicators of the landscape and the climatic environment of deposition—was a distinctive element in this approach. Specimens collected by the author were processed, in keeping with this method, by Moscow geologists-analysts G. N. Kolosova and V. I. Konoplyova, in cooperation with whom the main conclusions were made.

In order to differentiate moraine layers of Antarctica, and because of the absence of sections of dated moraines within the South Polar continent, we were forced to use experience gained in work on moraines in northeastern U.S.S.R. (upper reaches of the Kolyma and Indigirka Rivers). These moraines are fairly well studied and subdivided stratigraphically according to mineralogic and palynologic analyses (Vas'kovsky, 1959; Alexandrova and Boyarskaya, 1973; Kolosova and Ananyeva, 1974).

Comparison of results of the study of moraines in regions located geographically far apart is rather unconventional. We have elected, however, to attempt such a correlation, taking into consideration the similar features of climate that exist in these comparable regions (continentality and inclemency) and their comparable subpolar locations, and introducing corrections into our models for the difference in geographic conditions of deposition both at the present time and during the whole of the Cenozoic Era.

As is well known, climatic conditions determine the character and intensity of weathering processes. Inclemency of climate, characteristic both of northeastern U.S.S.R. and the Antarctic coast, on the whole results in poor manifestation of these processes.

Through studies in the U.S.S.R. it was established that the authigenic process of mineral formation there takes place extremely slowly because of weak leaching, slowing the speed of cation exchange processes. Investigations in subpolar regions of northeastern U.S.S.R. (the Verkhne-Kolymsk Range) have shown that age distinctions in deposits are recorded in the quantity, composition, and morphology of authigenic minerals. Mineralogic spectra of adolescent moraines of late Pleistocene to Holocene in age are characterized by the complete absence of authigenic minerals. In a middle Pleistocene moraine, the content of authigenic minerals does not exceed 1–3%, but a lower Pleistocene moraine contains 8–23%.

The still more severe climatic conditions of Antarctica (low moisture together with low temperatures) evidently cause a reduction in the rate of authigenic mineral formation as compared with conditions in northeastern U.S.S.R. The highest content of authigenic minerals, however, will still be characteristic of the most ancient deposits, and younger Pleistocene moraines should not contain new minerals in any appreciable quantity.

Age and correlation of moraines

With the aim of classification and correlation of the glacial deposits of East Antarctica, the three moraines of Mount Collins and the exposed sections of moraine layers at Amery Oasis were lithologically studied.

Investigation of moraines on Mount Collins has shown that these are the youngest of the deposits. Weathering of moraines on Mount Collins is very limited; all minerals have a fresh appearance, and the usual group of authigenic minerals is missing (Table 138.1). Similar conditions of weak weathering processes within moraine deposits and the absence of an authigenic mineral complex are typical of younger moraines of the upper Kolyma (late Pleistocene to Holocene in age).

However, taking into account the considerably slower rate of weathering processes in the mountains of Antarctica as compared to that in northeastern U.S.S.R., the age of these moraines should be much older than the age of those in the U.S.S.R., which are dated as late Pleistocene and are similar in mineralogic characteristics. We are not able at the moment to offer a realistic transformation coefficient. It is likely that the rates of weathering processes in the regions in question differed from one another by no more than one order of magnitude. Therefore, formation of the three moraines or Mount Collins can apparently be framed within the whole of Pleistocene time, keeping in mind that the age of the most ancient of these moraines could be no older than early Pleistocene. The author is aware of the conditionality of both this procedure and the resulting conclusion.

Analysis of the lithologic and mineralogic composition of moraine layers at Amery Oasis has enabled us to distinguish three different stages of deposition. Maximum enrichment with the group of authigenic minerals (up to 31%) is a characteristic feature of the mineralogic spectrum of the lower moraine of gray color (Figure 138.4, specimens 21 and 26). New minerals include limonite, calcite, marcasite, and iron-rich carbonates. In terrigenous components of the mineralogic spectrum, iron oxides, iron-hydromica combinations of hydrothermal genesis (16–28.5%), and garnet-almandite (15–33%) prevail, the percentage of hornblende and pyroxenes being insignificant. The composition of accompanying minerals is rather diverse (Table 138.2 and Figure 138.6).

Active alteration by weathering processes and consider-

Table 138.1. Mineralogic composition of the heavy mineral fraction (0.10–0.25 mm size) from moraine deposits of Mount Collins (in wt. %)

Specimen number	63	64	66
Total heavy fraction (wt. %)	37.0	27.0	17.0
Stable minerals			
Ilmenite	0.0	1.5	0.5
Iron oxides	0.0	0.0	1.5
Sillimanite	0.0	0.0	0.0
Sphene	0.0	0.0	0.5
Zircon	0.5	0.0	0.0
Garnet-almandite	3.5	6.0	8.0
Unstable minerals			
Apatite	0.0	1.0	2.0
Epidote	0.0	0.5	3.0
Hornblende	87.0	78.0	27.5
Pyroxenes	1.0	6.0	11.5
Biotite	4.5	5.5	42.0
In-growths together	3.5	0.0	0.5
Hornblende grown with quartz	0.0	0.5	2.5
Ilmenite grown with quartz	0.0	0.0	0.5
Mineral changed by clayey film	15.5	3.0	2.5

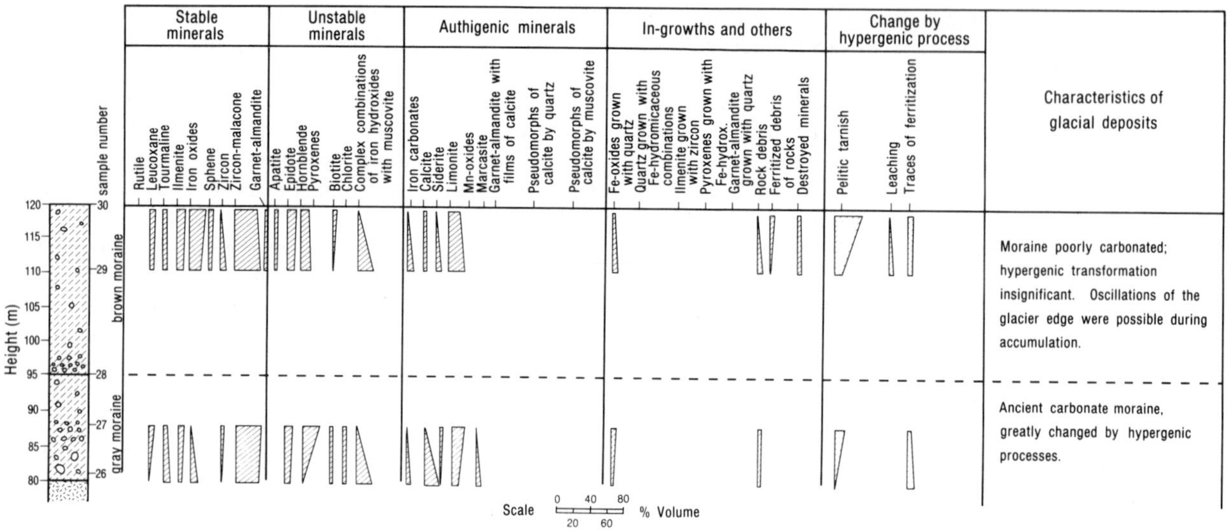

Figure 138.6. Diagram of mineralogic composition of the heavy mineral fractions of moraines from Pagodrome Gorge, Amery Oasis (outcrop 2); after Bardin and others (1977). The moraines consist of loamy soils, shown by dashed lines, and pebbles and boulders (scattered and in layers), shown by circles. Below 80 m is bedrock. Scale, in vol. % of each constituent, refers to width of shaded areas. Cartographic Lab, University of Wisconsin-Madison.

able carbonate content are characteristic of the gray moraine. Such phenomena are possible in conditions of active migration of microelements, leaching, and cation exchange; that is, in conditions of sufficiently moist, "warm" climates (much warmer than those of the present time).

Petrographic study of the clay matrix material of the moraine has shown that the moraine clay in the lower gray layer (specimen 26) contains sand (the content of the sand fraction fluctuates from 20 to 30%. In the sand fraction there occur the following grain types: quartz—36%, feldspar—42%, granite—6%, sandstone—3%, garnet—1%,

Table 138.2. Lithologic and mineralogic composition (in %) of moraines at Amery Oasis and Mount Collins and their estimated ages.

	Mount Collins moraines 1, 2, and 3	Pagodrome Gorge, Amery Oasis		
		Brown moraine	Intermediate conglomerate	Gray moraine
Sand grain composition				
Quartz	—	68 – 72	30	36
Feldspar	—	17 – 26	35	42
Granite	—	6 – 7	5	6
Sandstone	—	0 – 3	5	3
Coal	—	0	2	12
Calcite	—	1 – 4	0	0
Ore minerals	—	0	3	0
Garnet	—	0	5	0
Mineralogic composition of heavy mineral fractions (0.10 – 0.25 mm)				
In-growths together	0.5 – 3.5	4.7	2 – 3	0.7 – 1.5
Hornblende	27 – 87	2.2 – 8.5	4.0 – 8.5	1.4 – 2.5
Pyroxenes	1.0 – 11.5	4.5 – 15.4	9.0 – 19.0	1.4 – 3.5
Authigenic minerals				
Mn-oxides	nf	0.5	0.0	0.0
Fe-hydroxides, limonite	nf	9 – 25	12.5 – 15.0	5.6 – 17.0
Calcite	nf	0 – 2	3.0	17.0
Fe-carbonates	nf	1 – 3.5	0.0	1.4
Siderite	nf	0.5 – 1.6	0.5 – 4.5	0.0
Marcasite	nf	0.0	0.0	1.4
Total	—	14.0	17.5	26.0
Clayey minerals				
Kaolinite	trace	10	10	10
Montmorillonite	trace	trace	trace	trace
Hydrous mica	100	40 – 50	45 – 30	15
Chlorite	trace	40 – 50	45 – 60	85
Mixed layer	0.0	trace	trace	trace
Texture (fabric)	—	flow evidence	wash	disorderly
Carbonate as CO_2	0.02 – 0.16	0.17 – 0.35	0.09	0.3 – 0.42
Organic carbon	0.08 – 0.12	0.95 – 2.14	2.73	6.2 – 3.89
Estimated age of deposition	Pleistocene	late Pliocene	early Pliocene	Miocene-Pliocene

—: no data obtained. nf: not found to occur.

and coal—12%. The clay contains large quantities of thinly dispersed, dust-like calcite and a coal-like substance which, together with intensive iron-enrichment, give it a dense grayish and brown color.

Characteristic features of the petrographic composition of the clay matrix material within the gray moraine are the following: (1) extensive iron-enrichment of the clay mass; (2) the presence of large quantities of coal dust and uniformly dispersed calcite; (3) the presence of sandstone debris among the coarse grains; (4) approximately equal amounts of feldspar and quartz in the sand fraction; (5) appearance of grains of ore minerals in the section all along the upper unit; (6) an increase in granitoid content in the composition of the sand fraction; (7) a decrease in density of the rocks; and (8) evidence of the influence of melted-ice waters.

Mineralogic spectra of heavy mineral fractions from the overlying conglomeratic deposits (Figure 138.4, specimens 22 and 27) are similar to those of the gray moraine, both in the range of minerals and in the total heavy fraction. They differ in the following: (1) the percentage of iron oxides and complex iron-hydromica combinations of hydrothermal genesis is two to four times less in the conglomerate, whereas the content of garnet-almandite is 1.5 times greater. The content of hornblende and pyroxenes is also much greater, which testifies to the fact that granite massifs were drawn into transit. (2) The authigenic mineral content, mainly calcite and limonite, is less in the conglomerate, which indicates oxidizing conditions in the environment of deposition. Marcasite disappears; instead, there appears siderite, which forms under conditions of a weak oxygen deficit (Perel'man, 1968).

The presence of authigenic minerals—calcite and limonite, indicators of oxidizing conditions—testifies to an unstable geochemical regime. The decrease in content of authigenic minerals in the conglomeratic deposit (down to 17.5%, Table 138.2) and the lesser degree of weathering are surely connected with an interruption in the process of deposition. In addition, a maximum enrichment by accompanying minerals has been observed, which is possibly caused by the washing up of material under the influence of melted-ice waters.

Deposition of the overlying brown moraine took place under geographic conditions different from those above. Weathering of this material, as compared with the gray moraine, has been greatly suppressed. The total amount of authigenic minerals already comprises 14%; the average percentage of calcite and siderite is just 3–4%. Proportions of mineral groups change from layer to layer within the moraine. The lower layer of the moraine (Figure 138.4, specimens 23 and 29) is poorly calciferous; the upper layer is void of carbonates (specimens 25 and 30). The terrigenous components in the mineralogic spectrum of the brown moraine are similar to the spectrum of the conglomeratic deposits and testify to a constant source area feeding the glacier throughout the whole period of accumulation.

The petrographic composition of the sand fraction of the matrix in the brown moraine is rather monotonous: quartz—68–72%, feldspar—17–20%, granite—6–7%, sandstone—0–3%, and calcite—1–4%.

The clay matrix of the brown moraine is characterized by: (1) extensive iron-enrichment; (2) clear-cut traces of substance differentiation; (3) prevalence of products of granitoid destruction in the sand fraction, along with small quantities of Permian sandstone and some calcite; (4) presence of coal dust in considerably smaller amounts than in the gray moraine; and (5) the absence of coal in the grains from the sand and silt fraction.

Evidently, the periods of accumulation of the lower and upper moraines were separated by a lengthy interruption since the lower moraine is characterized by a more ancient appearance—a greater density and more corrosion of the debris material. The existence of evidence for current reworking of the layers capping the lower moraine bears witness to the same phenomenon.

Thus, available data testify to the following characteristics of the gray moraine, as compared to the brown moraines—the early Pleistocene moraines and subaqueous deposits of Miocene–Pliocene time. During the entire Cenozoic Era, climatic conditions in Antarctica had been more larger total quantity of authigenic minerals.

Such a high content of authigenic minerals in northeastern U.S.S.R. is typical only of the most ancient moraines—the early Pleistocene moraines and subaqueous deposits of Miocene–Pliocene time. During the entire Cenozoic Era, climatic conditions in Antarctica had been severe than in the northern polar regions, and consequently the rate of weathering processes had been slower. Thus, we can infer that the gray moraine at Amery Oasis is far more ancient, probably of Miocene–Pliocene age. The degree of epigenic change in the rocks supports this conclusion (i.e., corrosion of sand grains and alteration of clay minerals). The Permian sedimentary coal-bearing rocks served as the main source material for that moraine. Apparently, the gray color of the moraine can be explained by the presence of large quantities of coal dust. The presence of kaolinite, chlorite, and mixed-layer clay minerals also speaks for the Permian rocks playing a large role in the deposition of the gray moraine (Table 138.2).

Conclusions

The change of basic lithologic characteristics takes place quite regularly upward through the section of the gray moraine. From the base to the top, the total amount of authigenic minerals decreases, the composition of authigenic materials changes, and there occurs an enrichment in accompanying minerals. In contrast, the brown moraine differs both in variety and quantity of authigenic minerals and in the petrographic composition of rocks. These differences, and the presence of an intermediate conglomerate layer that formed under the influence of water flow, all point to a lengthy interruption in the moraine deposition process. Thus, the conglomeratic deposit indicates a prolonged stage of glacial melting and retreat. The contrasting lithologic and mineralogic composition of the brown moraine suggests that it may be much younger than the gray one and perhaps of late Pliocene age. In the spectrum and quantity of its authigenic minerals, this moraine is similar to the lower Pleistocene moraines of the Kolyma.

Continuing the age series of moraines at Amery Oasis are the glacial deposits on Mount Collins, which are characterized by the complete absence of authigenic minerals and by well-preserved mineral grains. As previously mentioned, these peculiarities enable us to regard them as Pleistocene deposits (see Table 138.2).

Despite certain unconventional procedures followed in this study, application of these methods to the moraines in MacRobertson Land (fully studied in the field) has provided us with a better idea of the succession, character, and time range of processes involved in the origin of glacial sediments in Antarctica. It would seem that conclusions derived from investigations of lithologic composition of moraines in the Prince Charles Mountains may have broader regional implications beyond their narrow local significance.

References cited

Alexandrova, A. N., and Boyarskaya, T. D. 1973. Amplituda izmenchivosti prirodnykh uslovii pleisotsena v rayonakh s kontinental'nym i morskim klimatom [Amplitude of changeability of natural conditions of the Pleistocene in regions with continental and sea climate]. *Izv. AN SSSR,* Ser. geogr., *1,* 213.

Bardin, V. I. 1969. Paleoglatsiologicheskiie issledovaniia na Zemle Korolevy Mod [Paleoglaciologic investigations on Queen Maud Land]. *Inf. Byull. sov. antarkt. Eksped.,* No. 75, 5-11.

Bardin, V. I. 1972. Moraines of Antarctica (*In* Adie, R. J., *ed. Antarctic Geology and Geophysics.* Oslo, Universitetsforlaget, 663-67.)

Bardin, V. I., 1974. Geograficheskiie issledovaniia na Zemle MakRobertsona [Geographical investigations on MacRobertson Land]. *Inf. Byull. sov. antarkt. Eksped., 88,* 15-20.

Bardin, V. I. 1975. Gliatsiogeomorfologiia lednika Lamberta i yego gornogo obramleniia [Glacial geomorphology of Lambert Glacier and of its mountainous frame]. *Antarktika, Doklady Komissii, 14,* 110-27.

Bardin, V. I., and Konoplyeva, V. I. 1971. O protsessakh vyvetrivaniia i probleme geokhronologii lednikovogo perioda Antarktidy [On weathering processes and on the problem of geochronology of the glaciation period of Antarctica]. *Antarktika, Doklady Komissii, 1969,* 102-12.

Bardin, V. I. and Ravich, M. G. 1968. Petrograficheskii sostav morennykh otlozhenii v gorakh tsentral'noi chasti Zemli Korolevy Mod [Petrographic composition of moraine deposits in mountains of the central part of the Queen Maud Land]. *Inf. Byull. sov. antarkt. Eksped.,* No. 68, 28-35.

Bardin, V. I., and Sudakova, N. G. 1972. Litologicheskiie osobennosti raznovozrastnykh moren v gorakh Zemli Korolevy Mod [Lithological peculiarities of different-age moraines in mountains of the Queen Maud Land]. *Inf. Byull. sov. antarkt. Eksped., 83,* 25-29.

Bardin, V. I., Glushankova, N. I., and Sudakova, N. G. 1976. Litologiia moren Vostochnoi Antarktidy [Lithology of moraines in East Antarctica]. *Antarktika, Doklady Komissii, 15,* 53-71.

Bardin, V. I., Kolosova, G. N., and Konoplyeva, V. I. 1977. Paleogliatsiologicheskie aspekty izucheniia litologii moren v gorakh Prins-Charl'z (Vostochnaia Antarktida) [Paleoglaciological aspects of moraine lithology in the Prince Charles Mountains (East Antarctica)]. *Antarktika, Doklady Komissii, 16,* 135-53.

Denton, G. H., Armstrong, R. L., and Stuiver, M. 1970. Late Cenozoic glaciation in Antarctica: the record in the McMurdo Sound region. *Antarct. Jnl U.S., 5,* No. 1, 15-21.

Fleck, R. J., Jones, L. M., and Behling, R. E. 1972. K-Ar dates of the McMurdo volcanics and their relation to the glacial history of Wright Valley. *Antarct. Jnl. U.S., 7,* No. 6, 245-46.

Kolosova, G. N., and Ananyeva, E. G. 1974. Metodika mineralogicheskogo analiza rykhlykh otlozhenii pri paleogeomorfologicheskikh postroyeniyakh (na primere Severo-Vostoka SSSR) [Method of mineralogical analysis of loose deposits when making paleogeomorphological constructions (taking as an example the Northeast of the USSR)]. *J. Geomorf.,* No. 4, 26-35.

Mercer, J. H. 1972. Some observations on the glacial geology of the Beardmore Glacial Area. (*In* Adie, R. J., *ed. Antarctic Geology and Geophysics.* Oslo, Universitetsforlaget, 427-33.)

Perel'man, A. I. 1968. *Geokhimiia epigeneticheskikh protsessov (zona gipergeneza)* [Geochemistry of epigenetic processes (the zone of hypergenesis)]. Moscow, Izdatel'stvo Nedra, 18-20.

Rutford, R. H., Craddock, C., and Bastien, T. W. 1968. Late Tertiary glaciation and sea-level changes in Antarctica. (*In* Tanner, W. F., *ed.* Tertiary sea level fluctuations. *Paleogeogr. Paleoclimatol., Paleoecol., 5,* No. 1, 15-39.)

Rutford, R. H., Craddock, C., Armstrong, R. L., and White, C. M. 1972. Tertiary glaciation in the Jones Mountains. (*In* Adie, R. J., *ed. Antarctic Geology and Geophysics.* Oslo, Universitetsforlaget, 239-43.)

Vas'kovsky, A. P., 1969. Kratikii ocherk rastitel'nosti, klimata i khronologii chetvertichnogo perioda v verkhovyakh rek Kolymy i Indigirki i na severnom poberezhye Okhotskogo morya [A brief review of vegetation, climate and chronology of the Quaternary in the upper flows of the Kolyma and Indigirka rivers and on the northern coast of the Okhotsk Sea]. In: sb. "Lednikovy period na territorii Evropeyskoy chasti SSSR i Sibiri," Izdatel'stvo Moscow State University.

139

Isostatic Gravity Anomalies on the Ross Ice Shelf

CHARLES R. BENTLEY, JAMES D. ROBERTSON*, and LARRY L. GREISCHAR
Geophysical and Polar Research Center, University of Wisconsin, Madison, Wisconsin 53706, U.S.A.

Abstract. Gravity observations covering the Ross Ice Shelf have been reduced to isostatic anomalies assuming perfect local Airy compensation. Roosevelt Island, Crary Ice Rise, and Discovery Deep seem to be isostatically in balance with the surrounding regions. The main ridge-trough submarine topography characteristic of the grid-western two-thirds of the ice shelf, however, does not seem balanced, suggesting that tectonic structure, not glaciation, is the fundamental cause of that topography.

Anomalies have been smoothed by averaging over 2° × 2° squares. The resulting map shows relative highs in the grid southeastern and southwestern corners of the shelf, with a minimum along the Siple Coast. After correction to the new Potsdam datum and the Geodetic Reference System 1967, all anomalies are moderately to strongly negative, but not so negative as the 12th-order harmonic field of the earth. If the latter approximates the "baseline" regional gravity field, then positive "adjusted" isostatic anomalies occur on the Ross Ice Shelf. On the other hand, if the 4th-order harmonic field is taken as the "baseline," then the anomalies are negative. This ambiguity demonstrates the difficulty in ascertaining the current state of glacial depression from gravity data in the presence of regional anomalies with deep-seated sources. The observed gradient is consistent with a plausible model of glacial retreat, but cannot be taken as strong supporting evidence for such a model.

During the past several summers, gravity values have been measured at more than 150 locations on the Ross Ice Shelf as part of the airlifted Ross Ice Shelf Geophysical and Glaciological Survey (RIGGS). These measurements, on a 55 km grid system, have been supplemented by data from surface traverses on the ice shelf (Crary and others, 1962; Bennett, 1964) and on Roosevelt Island (M. P. Hochstein, personal communication, 1966) and reduced to maps of free-air, Bouguer, and isostatic anomalies over the Ross Ice Shelf. The station locations are shown and the free-air and Bouguer anomaly maps discussed in a companion paper (Robertson and others, Chapter 140, this volume); we are concerned here with isostatic anomalies. In accordance with other measurements in Antarctica, gravity values have been calculated on the old Potsdam datum, and anomalies have been calculated relative to the International Gravity Formula. This procedure suffices so long as one is concerned only with anomaly differences from point to point on the ice shelf. However, before the significance of absolute values of gravity anomalies can be considered, corrections to a more modern datum and reference ellipsoid must be made.

Isostatic anomaly calculations

The three-dimensional gravity modeling procedure of Talwani and Ewing (1960) has been used to calculate the isostatic gravity anomalies, assuming that the sea-bottom topography is perfectly locally compensated on the Airy model. For reasons described more fully in another companion paper (Bentley and Robertson, Chapter 121, this volume), the standard crustal column was taken to be 30 km thick with densities of 2.67 Mg m^{-3} for the topography-forming rock, 2.95 Mg m^{-3} for the lower crust, and 3.32 Mg m^{-3} for the upper mantle. The estimated standard error in the gravity anomalies is ± 5 mgal, the major source of error being uncertainty in the surface elevations (Bennett, 1964; Whiting, 1975).

Isostatic anomaly map

The map of isostatic anomalies (Figure 139.1) is generally similar to the free-air anomaly map (Robertson and others, Chapter 140, this volume). This is not surprising since most of the topographic variations are rather broad, so that the gravity effect of the calculated isostatic compensation is nearly the same as that of the topography. A few of the features on the free-air anomaly map, however, are nearly or completely eliminated. The free-air gravity highs over Roosevelt Island and Crary Ice Rise have disappeared on the isostatic map, showing that these shoal areas on the sea floor are isostatically compensated. Similarly, the strong negative anomalies associated with Discovery Deep (around 10° S, 2.7° E grid) have been completely eliminated; in fact, the whole grid-southeastern part of the ice shelf, characterized by topographic variations exceeding 800 m, shows very subdued isostatic anomalies. (For an explanation of grid directions, used throughout this paper, see Robertson and others, Chapter 140, this volume.) This suggests that the topography beneath that part of the ice shelf does not reflect a deeper structural control. It seems quite likely that the features have been eroded by ice associated with the very active Byrd Glacier

* Present address: Atlantic Richfield Company, P.O. Box 2819, Dallas, Texas 75221, U.S.A.

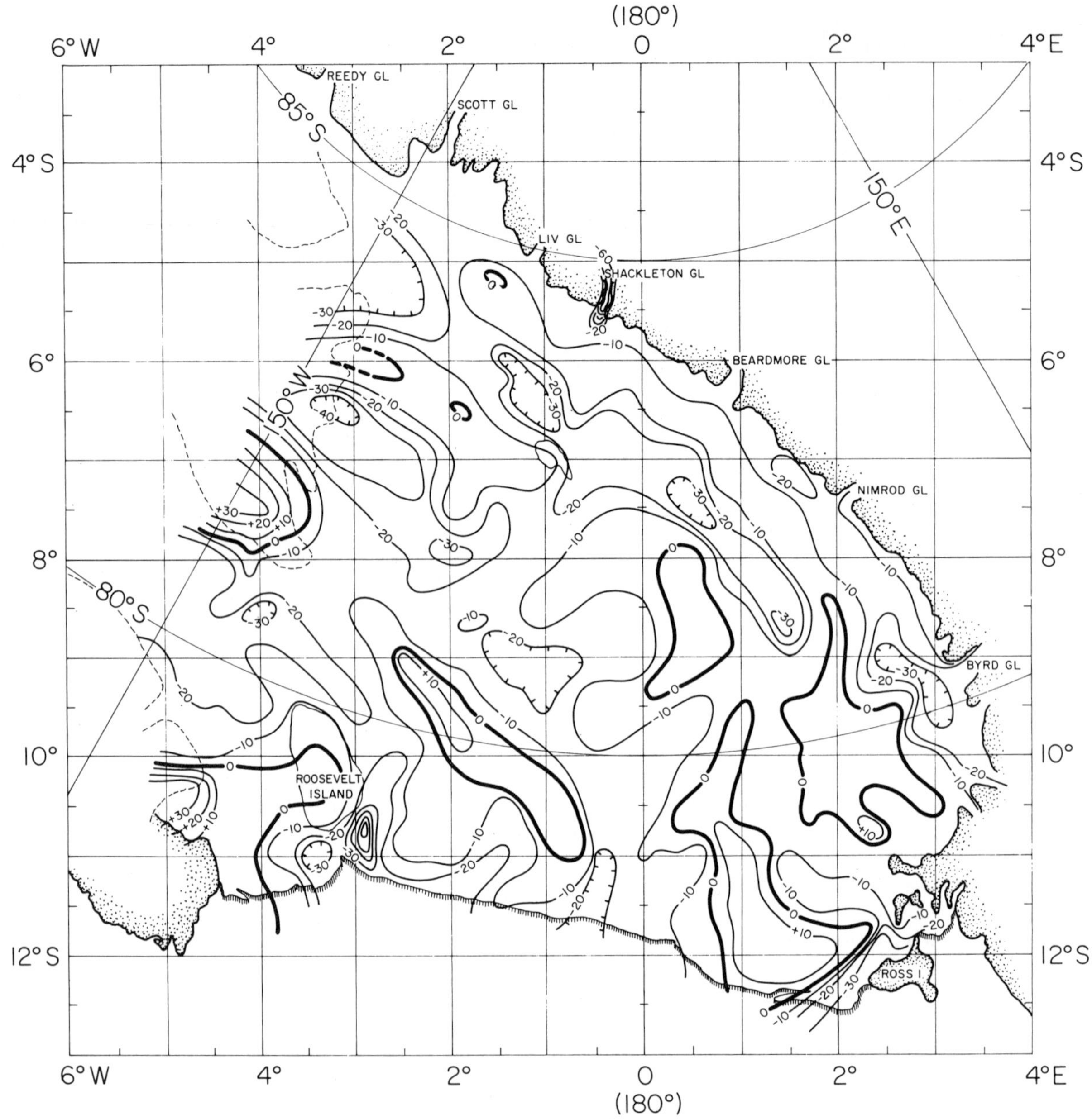

Figure 139.1. Isostatic anomalies on the Ross Ice Shelf, referred to the International Gravity Formula and the old Potsdam datum. Contour interval is 10 mgals. Grid coordinates are given along the margins; a few geographic coordinate lines are also shown for reference.

at an earlier time when the ice in the region of the present Ross Ice Shelf was grounded.

On the other hand, with the exception of the diminished anomalies over the ice rises, the isostatic anomalies on the western two-thirds of the ice shelf look very similar to the free-air anomalies, particularly in the association between anomalies and the ridge-trough submarine topography. The anomalies are too large to be explained by a process of glacial scouring and partial filling of the troughs with glacial marine sediments (Robertson, 1975); furthermore, the anomalies are not correlated with variations in sediment thickness where the latter has been measured (Robertson and others, Chapter 140, this volume). This suggests that, with the exception of the ice rises, tectonic structure and not glaciation is the fundamental determinant of the sea-bottom topography beneath the grid-western part of the Ross Ice Shelf.

Smoothed isostatic anomalies

To expose the large-scale features of the isostatic anomalies over the ice shelf, individual station values have been averaged over $1° \times 1°$ squares. Since variations in the submarine topography are generally 100 km or less across, this averaging largely removes the effect of that topography. The anomaly map has been still further smoothed by averaging the $1° \times 1°$ squares over squares 2° on a side, with the results shown in Figure 139.2. All correlation with the submarine topography has been removed by this process. The map shows that the most negative anomalies (< -15 mgal) lie along the Siple Coast and that the highest values (>0 mgal) lie in the grid-southeastern and southwestern extremes of the ice shelf. Anomalies are more negative in the grid-northern half of the ice shelf than in the grid-southern half, with a total range, away from the Siple Coast, of about 15 mgal.

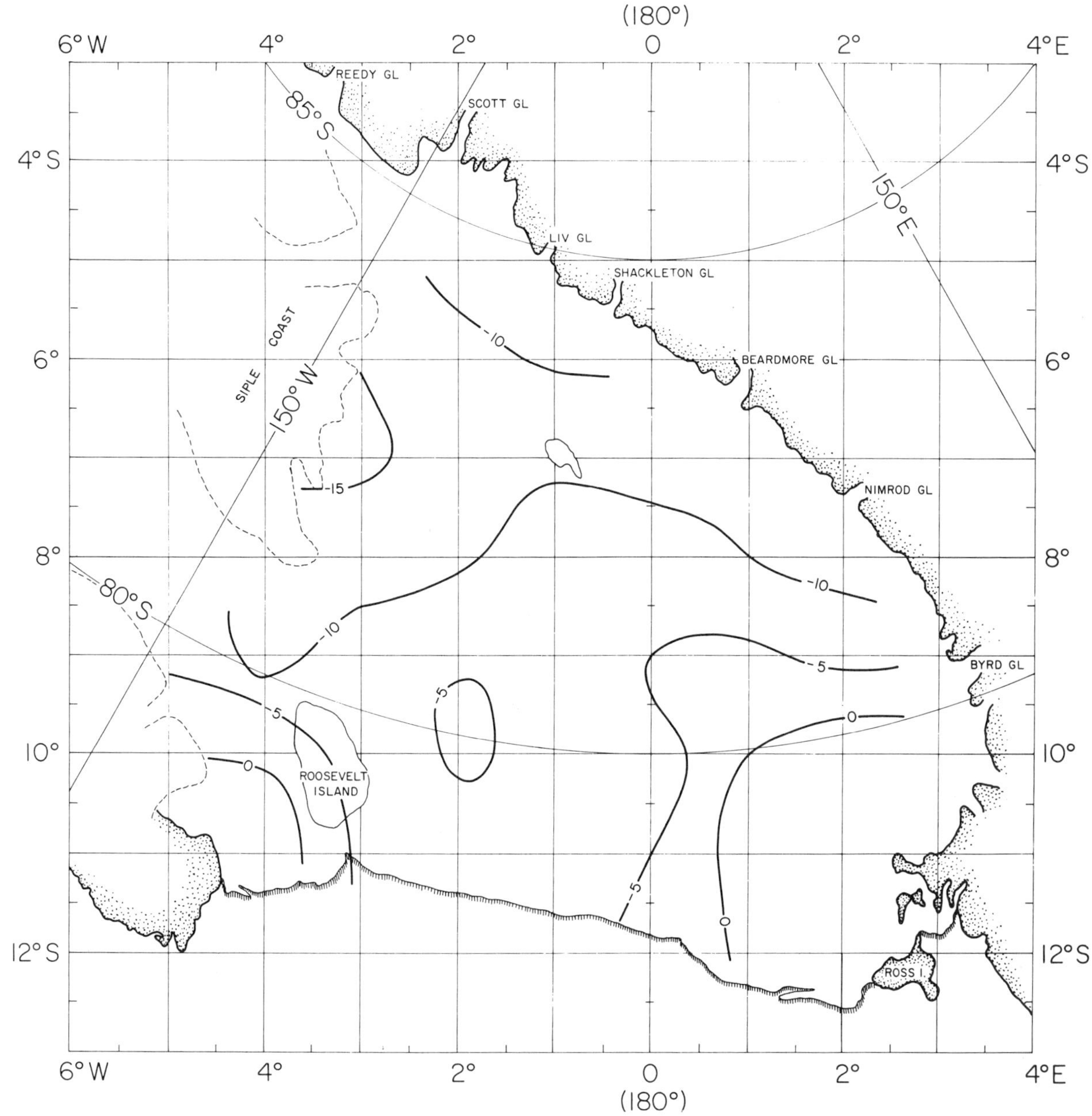

Figure 139.2. Isostatic anomalies from Figure 139.1 averaged over 2° × 2° squares. Contour interval is 5 mgal.

Discussion

As mentioned in the introduction, the anomalies have been calculated on the basis of the International Gravity Formula and the old Potsdam datum. We now wish to consider the absolute value of the anomalies in looking for evidence of isostatic imbalance, so it is important to bring the anomalies to the proper datum and refer them to the proper ellipsoid. Application of the − 14 mgal correction for the new Potsdam datum and the + 4 mgal correction to the Geodetic Reference System 1967 (Bentley and Robertson, Chapter 121, this volume) results in all anomalies being decreased by 10 mgal. (In recent studies of world-wide gravity anomalies based on satellite measurements, various "best fitting" ellipsoids, as well as the Geodetic Reference System 1967, have been used, but for our purposes, none of these ellipsoids varies significantly from the 1967 ellipsoid.)

Application of this − 10 mgal correction to Figure 139.2 would yield a map showing large negative isostatic anomalies decreasing from − 10 mgal in the grid-southeastern and southwestern parts of the shelf to − 25 mgal on the Siple Coast. It is not valid, however, to interpret these isostatic anomalies directly in terms of crustal isostatic imbalance. It is first necessary to consider what the isostatic anomalies would be in the Ross Shelf region in the absence of any crustal imbalance. It is by no means certain that the anomalies would be zero, since regional anomalies tens of milligals in magnitude, presumably arising from sources deep within the mantle, abound on the surface of the earth (see, e. g., Wagner and others, 1977). Only by estimating a "base line" configuration and subtracting from the present anomalies can we get "adjusted" anomalies to interpret in terms of isostasy.

To get an idea of the "base line," we turn to the satellite data. The most recent published determination of the free-air gravity anomalies on the surface of the earth from a

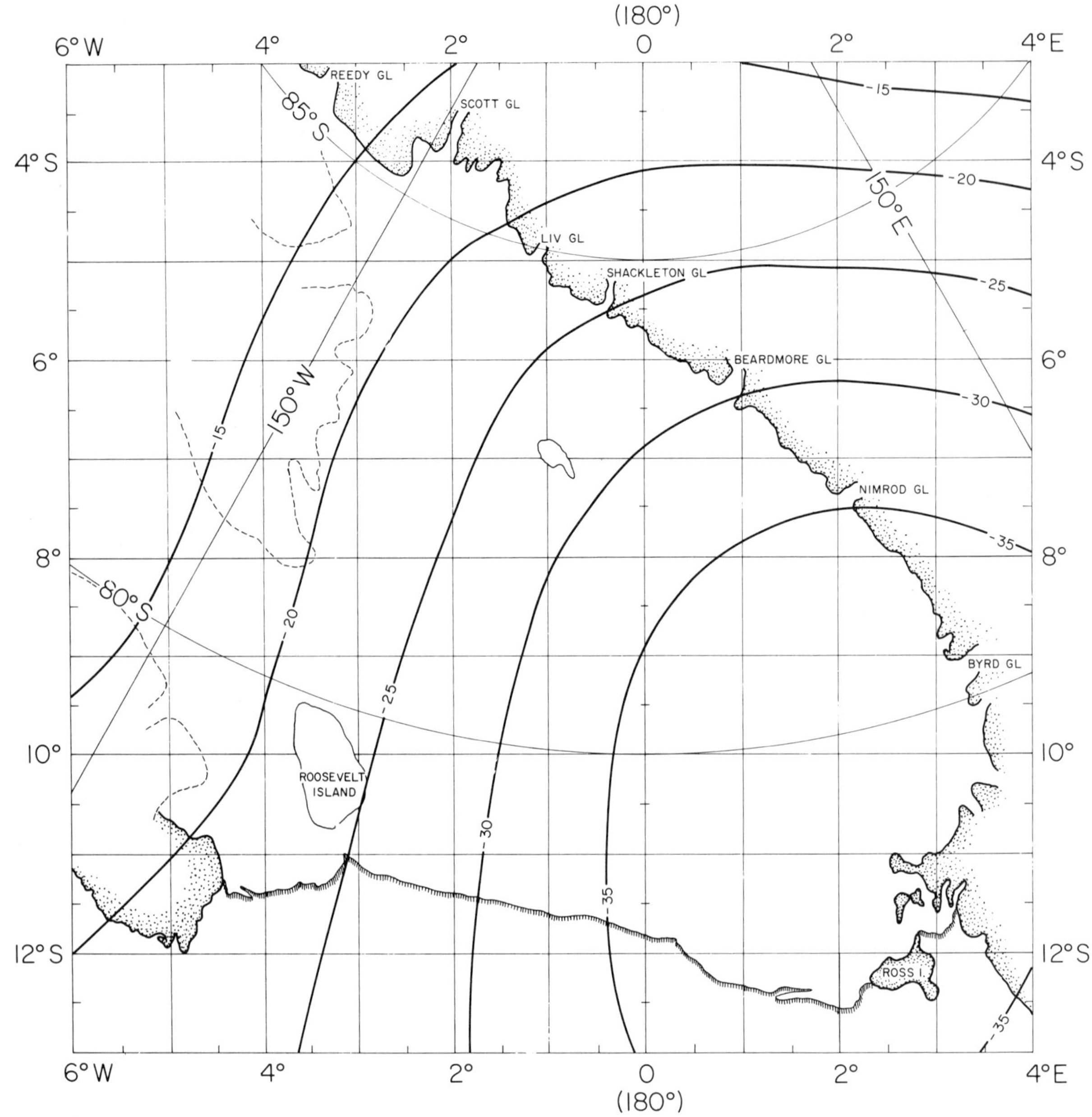

Figure 139.3. Free-air gravity anomalies from Goddard Earth Model 8 (GEM 8). Contour interval is 5 mgal.

combination of satellite data and earth-bound gravity measurements is called GEM 8 (for Goddard Earth Model 8) (Wagner and others, 1977). GEM 8, which includes all spherical harmonic coefficients up through degree 22, shows the strongly negative gravity values in the Ross embayment that are typical there for all satellite measurements (Figure 139.3). Not only are these anomalies negative in the extreme grid-southeast corner of the ice shelf (the minimum is very close to −40 mgal), but they show a strong gradient, the anomalies becoming more negative from the Siple Coast toward Ross Island. If this model were to be used as the "base line," the "adjusted" anomalies (observed values minus base-line values) would be strongly positive over almost the whole ice shelf and would exhibit a 35 mgal increase from the Siple Coast to Ross Island.

GEM 8, however, includes harmonics of high enough order to be affected by any broad, negative isostatic anomalies in the Ross Sea and vicinity. In fact, large negative gravity anomalies are characteristic of East Antarctica from the South Pole to Victoria Land along the East Antarctic side of the Transantarctic Mountains (Bentley, 1968). Isostatic anomalies are also generally negative in West Antarctica (Bentley and Robertson, Chapter 121, this volume). It is thus natural that a smoothed harmonic fit should show a minimum in the region between, as is seen in Figure 139.3.

Wagner and others (1977) have also calculated the gravity anomalies for harmonic orders 13 to 22 alone. We subtracted orders 13 to 22 from the full GEM 8 model to yield a 12th-order harmonic surface, hoping to minimize the effect of the Antarctic regional anomalies, but found that the 12th-order surface does not differ greatly from the full GEM 8 model.

In order to determine the extent of the negative gravity feature and to examine to what extent it is a truly deep-seated phenomenon, we next considered the very lowest order harmonics. Terms through the second-order zonal harmonic are used to describe the ellipsoid itself. Introduction of just the third-order and fourth-order terms yields a gravity anomaly of about −10 mgal over the central ice shelf. Since the half wavelength corresponding to the fourth harmonic is 5000 km, it is clear that at least 10 mgal of the large negative anomaly in the Ross Ice Shelf area is part of a fundamental and deep-seated feature of a major part of the earth, extending over an area 100 times that of the Ross Ice Shelf. We can then be reasonably certain that even in the absence of any crustal imbalance, there would be a regional negative isostatic anomaly of at least −10 mgal. (Over areas of these dimensions, the free-air and isostatic anomalies may be considered essentially synonymous; for example, the mean difference between free-air and isostatic anomalies from the RIGGS measurements is only 2 mgal.) For these low-order harmonics, the variation in the anomaly over the Ross Ice Shelf is small (about 3 mgal difference between the Siple Coast and Ross Island); for simplicity we will simply adopt the value of −10 mgal for the whole area. This "baseline" value just equals the correction for the datum changes discussed above, yielding no overall correction to the mean isostatic anomalies. This is a minimum correction, however, since it is based on the implicit assumption that all higher order harmonics refer to crustal imbalance. If that is not the case, a positive correction would be implied.

With this minimum correction, the adjusted anomalies are negative over most of the ice shelf (Figure 139.2). The strongest negative adjusted anomaly is about −15 mgal right at the Siple Coast, decreasing in magnitude to zero in the grid-southwestern and southeastern corners of the ice shelf. The pattern then does suggest isostatic depression from a recently retreated ice sheet.

The adjusted isostatic anomaly would not necessarily be near 0 in the absence of glacial isostatic depression. Since, as already noted, regional anomalies of 10 or 20 mgal are commonplace on the surface of the earth, it is likely that the natural, undisturbed anomaly picture of the Ross Shelf area would show an adjusted anomaly field, δg, that varied through the region. In that case, any present adjusted anomalies algebraically less than δg could represent glacial isostatic depression.

This suggests a modified approach to the question of imbalance. Although anomalies of 10 mgal are common on the earth's surface, regional gradients of 15 mgal in 500 km, such as is seen on the Ross Ice Shelf, are much less common in regions away from lithospheric plate boundaries. We might thus assume that before glaciation there was a nearly constant gravity anomaly throughout the Ross Ice Shelf region and that the current northwest-southeast slope (Figure 139.2) results from isostatic depression.

As a simple model, let us assume that the grid-southeast corner of the ice shelf shows no remanent depression and that before glaciation the whole ice shelf exhibited an adjusted mean anomaly of +10 mgal. There is, then, a 15 mgal decrease to the Siple Coast. Assuming that this value represents incomplete rebound after glacial depression and assuming a simple exponential recovery rate with a time constant of 4400 years (Cathles, 1975), we can, following Bennett (1964), calculate the approximate time of ice removal as a function of ice thickness removed. The 15 mgal negative anomaly would correspond to the removal of 3000 m of ice about 6500 years ago. This is in reasonable agreement with a model involving rapid deglaciation about 6000 years ago. It implies a residual depression of about 150 m, or about 25% of the maximum equilibrium depression under 3000 m of ice.

Without improved knowledge of the "base line" gravity anomalies, there is no good way to make definitive quantitative estimates of deglaciation history. This conclusion is in contrast with an earlier one (Bentley, 1976) that was in error because the datum corrections were not applied, and the "base line" field was not taken into account. The model just described, however, involves the *ad hoc* assumption of a rather large, positive "base line" anomaly field. Models requiring more recent ice-sheet retreat would involve even larger positive "base line" anomalies and are thus, *a priori,* decreasingly likely. On the basis of glacial geologic evidence, some model of ice-sheet retreat is certainly to be preferred, but reconciliation with the gravity data would be difficult for any model implying ice thicknesses substantially greater than at present within the last 5000 years. Perhaps a better solution to the problem will be possible when there has been extensive gravity surveying in the surrounding regions of West Antarctica and the Ross Sea.

Acknowledgments. This work was supported by National Science Foundation Grant DPP72-05802.

References cited

Bennett, H. F. 1964. A gravity and magnetic survey of the Ross Ice Shelf area, Antarctica. *Res. Rep. Ser., Geophys. Polar Res. Center, Univ. Wisconsin,* No. 64-3, 97 pp.

Bentley, C. R. 1968. Gravity maps. (*In* Behrendt, J. C., and Bentley, C. R., *eds.* Magnetic and Gravity Maps of the Antarctic. *Antarct. Map Folio Ser.,* Folio 9, Pl. VIII–IX.)

Bentley, C. R. 1976. Isostatic imbalance and the past extent of the grounded ice sheet in the Ross Ice Shelf region, Antarctica. *Abstr. with Programs geol. Soc. Am., 8,* 773.

Cathles, L. M. 1975. *The Viscosity of the Earth's Mantle.* Princeton, N.J., Princeton University Press, 386 pp.

Crary, A. P.. Robinson, E. S., Bennett, H. F., and Boyd, W. W., Jr. 1962. Glaciological studies of the Ross Ice Shelf, Antarctica, 1957–60. *I.G.Y. Glaciol. Rep., 6,* 193 pp.

Robertson, J. D. 1975. *Geophysical Studies on the Ross Ice Shelf, Antarctica.* Ph.D. dissertation, University of Wisconsin, 214 pp.

Talwani, M., and Ewing, M. 1960. Rapid computation of gravitational attraction of three-dimensional bodies of arbitrary shape. *Geophysics, 25,* 203–25.

Wagner, C. A., Lerch, F. J., Brownd, J. E., and Richardson, J. A. 1977. Improvement in the geopotential derived from satellite and surface data (Gem 7 and 8). *J. geophys. Res., 82,* 901–14.

Whiting, L. R. 1975. *A Gravity Survey of the Eastern Portion of the Ross Ice Shelf, Antarctica.* Master's thesis, University of Wisconsin, 97 pp.

140

Sea-Bottom Topography and Crustal Structure below the Ross Ice Shelf, Antarctica

JAMES D. ROBERTSON* and CHARLES R. BENTLEY
Geophysical and Polar Research Center, University of Wisconsin, Madison, Wisconsin 53706, U.S.A.

JOHN W. CLOUGH**
Ross Ice Shelf Project, University of Nebraska, Lincoln, Nebraska 68588, U.S.A.

LARRY L. GREISCHAR
Geophysical and Polar Research Center, University of Wisconsin, Madison, Wisconsin 53706, U.S.A.

Abstract. Seismic and gravity measurements were made at more than 150 stations on the Ross Ice Shelf during the austral summers of 1973–74, 1974–75, and 1976–77 as part of the Ross Ice Shelf Geophysical and Glaciological Survey (RIGGS). The data have been used to construct maps of sea-bottom topography and gravity anomalies and to calculate seismic velocity columns. Sea-bottom topography is characterized by a series of troughs (depths 700–1300 m below sea level) and ridges (depths between 200 and 500 m) which cross the ocean floor parallel to the Transantarctic Mountains. The deepest trough lies along the mountain front throughout the front's entire boundary with the ice shelf. The ridge and trough topography is clearly the dominant influence on the free-air gravity anomaly map. Anomalies on the Bouguer gravity map also show correlations with topography, particularly next to the Transantarctic Mountains and between the center of the ice shelf and Discovery Deep, that are not removed by isostatic corrections. The gravity anomalies are inconsistent with seismically derived thicknesses of glacial till, suggesting that tectonic structure and not glaciation is the fundamental determinant of sea-bottom topography.

Interval seismic velocities and acoustic impedances in the layer of sediment at the sea floor match those expected for glacial marine till. When RIGGS and previous seismic refraction measurements are considered, it seems most likely that "basement" beneath the ice shelf, found at a depth of 1.5 to 2 km below sea level, consists of crystalline rocks typical of the Ross or Andean orogenic provinces. On and near Roosevelt Island, an intermediate layer 200–700 m thick, interpreted as either Cenozoic volcanic strata or Paleozoic–Mesozoic sedimentary/metasedimentary rocks, lies between glacial till and crystalline basement. The layer may or may not be present under the main body of the ice shelf.

An airlifted geophysical survey was carried out on the Ross Ice Shelf, Antarctica, during the austral summers of 1973–74, 1974–75, and 1976–77 as part of the Ross Ice Shelf Geophysical and Glaciological Survey (RIGGS). Included in the geophysical program were seismic reflection recordings at 157 locations, spot radio-echo soundings at 171 locations, long seismic refraction profiles at 6 locations, and spot gravity measurements at 172 locations (Bentley and others, 1974; Clough and Robertson, 1975; Robertson, 1975; Whiting, 1975; Bentley, 1976; Bentley and Jezek, 1977). The data, supplemented by information from Crary and others (1962), Bennett (1964), and Hochstein (personal communication, 1966), have been converted to maps of sea-bottom topography, free-air and Bouguer gravity anomalies, and seismic velocity columns. The entire ice shelf, except for a small sector around Ross Island, has been covered on an approximately 50 km grid (Figure 140.1), and the remainder will be surveyed during the 1977–78 Antarctic summer field season. Geophysical stations on Figure 140.1 are located on the rectangular system of *grid* coordinates centered on the South Pole, which is normally used for air navigation in Antarctica and which has been adopted by RIGGS. Directions and locations in this paper will be given relative to this grid system, according to which north is always at the top of the map (Figure 140.1), and the other directions in their accustomed positions.

Sea-bottom topography

The elevation of the sea bottom below the ice shelf (Figure 140.2) has been calculated from the travel times of radio-echo and seismic reflections. At an individual survey station, the seismic technique generally produces a recognizable ocean-bottom echo, whereas the radio-echo sounding technique produces a clear, high-amplitude reflection from the bottom of the ice. The combination of the two methods satisfactorily produces a reliable estimate of depth to the ocean floor. Sea-bottom elevations at individ-

* Present address: Atlantic Richfield Company, P.O. Box 2819, Dallas, Texas 75221, U.S.A.

** Present address: Department of Geosciences, North Carolina State University, Raleigh, North Carolina 27650, U.S.A.

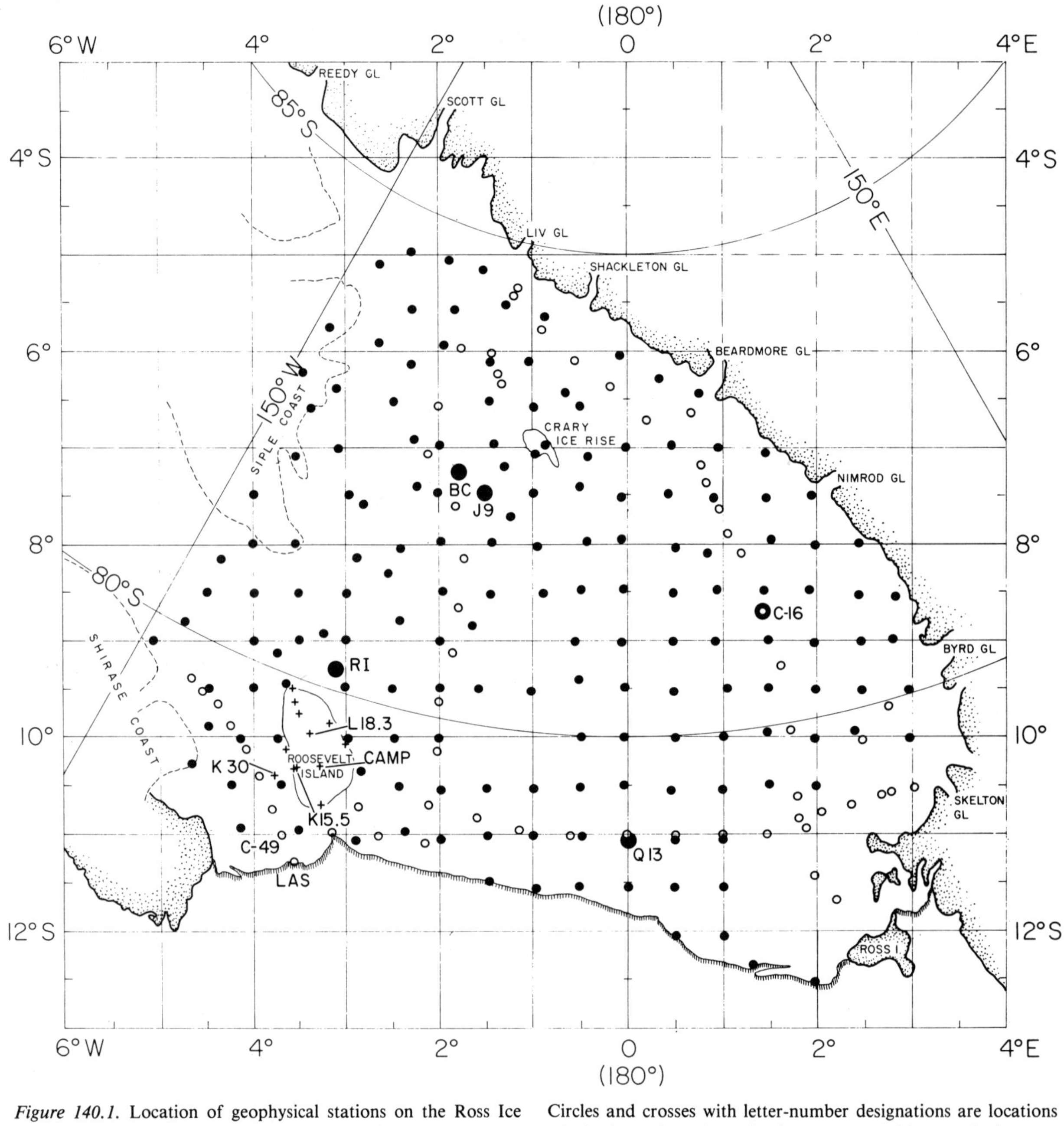

Figure 140.1. Location of geophysical stations on the Ross Ice Shelf. Small dots: RIGGS stations. Large dots: RIGGS base camps. Circles: stations, from Crary and others (1962). Crosses: stations, from Hochstein (personal communication, 1966). Circles and crosses with letter-number designations are locations of seismic sections shown in Figure 140.6. Grid system is that normally used for air navigation in Antarctica and which has been adopted by RIGGS, as explained in text.

ual stations are estimated to be accurate to within ± 15 m.

Sea-bottom topography is characterized by a series of troughs and ridges which cross the ocean floor parallel to the Transantarctic Mountains. The deepest trough lies immediately adjacent to the mountains and descends to more than 700 m below sea level along its entire length. Local depressions more than 900 m below sea level are superimposed on the trough off the mouths of the Liv, Beardmore, Byrd, and Skelton Glaciers. The last two of these depressions trend into Discovery Deep (10.5° S, 2.75° E grid), an oval-shaped trench reaching more than 1300 m below sea level beneath the extreme grid-eastern part of the ice shelf.

A pair of ridges extends southeastward from the Siple Coast. The northern ridge rises to within 300 m of sea level at Crary Ice Rise (7° S, 1° W) and extends to a termination at 10° S, 0°. A trough south of the ridge bifurcates around it. The eastern branch of this trough, which is separated from Discovery Deep by a ridge ascending to less than 500 m below sea level, intersects the trench along the Transantarctic Mountains in the vicinity of the Nimrod Glacier. The western branch of the trough extends from 11° S, 0° north and northwest to the ice-shelf grounding line at 6.5° S, 3° W and separates the two ridges stretching southeastward from the Siple Coast. The more southerly of these ridges crosses the grounding line at 7.5° S, 3.5° W and rises to within 400 m of sea level. The ridge is continuous with a submarine bank intersecting the ice-shelf front at 11.5° S, 1.5° W.

A trough locally descending more than 800 m below sea

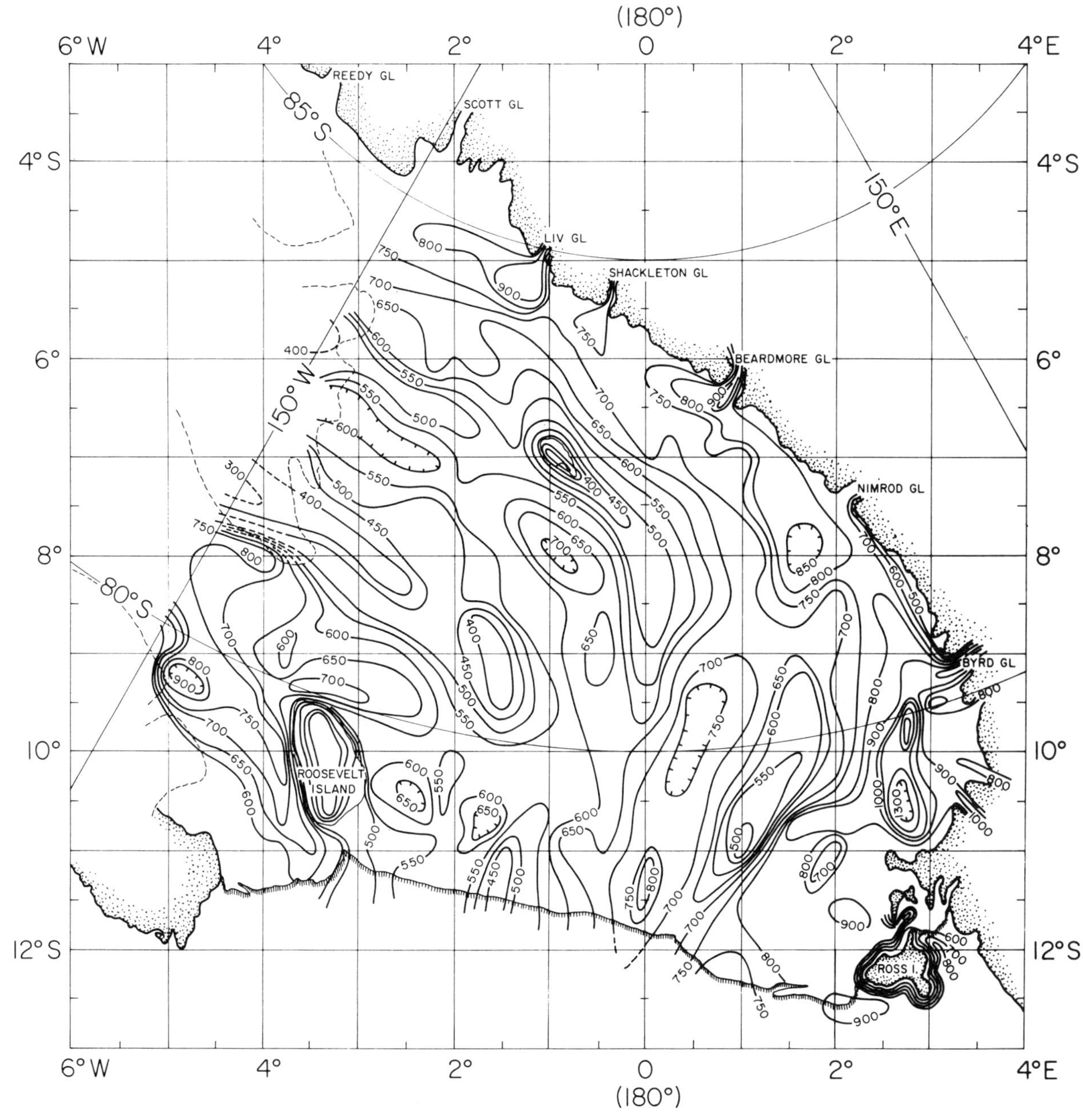

Figure 140.2. Sea-bottom topography below the Ross Ice Shelf. Contour interval is 50 m, but changes to 100 m for depths greater than 800 m (except where otherwise indicated) and for Roosevelt Island and the area surrounding Ross Island.

level lies between the Siple and Shirase Coasts. This trough bifurcates to pass on both sides of Roosevelt Island, so that the northern and western sides of the island are characterized by large gradients in submarine elevation. Roosevelt Island itself, with a subglacial elevation shallower than −200 m, is the northern extremity of a submarine bank that crosses the ice-shelf front at 11° S, 3° W. There is no ridge structure extending northwest from Roosevelt Island into Marie Byrd Land.

Free-air gravity anomalies

The ridge and trough topography of the sea bottom is clearly the dominant influence on the free-air gravity anomalies (Figure 140.3). Linear negative anomalies of −20 to −30 mgal or more are associated with the troughs, and linear positive anomalies of about 0 to +10 mgal with the ridges. However, the correlation is not perfect. A strong negative anomaly of more than −60 mgal is associated with the submarine bank southeast of Roosevelt Island. A weaker but nevertheless negative anomaly of at least −20 mgal coincides with the submarine bank at 11.5° S, 1.5° W. To the east of the island, in the vicinity of 10° S, 2° W, an elongated positive anomaly of more than +10 mgal lies between a trough and a ridge instead of over the crest of the associated ridge. Between the Liv and Nimrod Glaciers, three small, oval, negative anomalies of more than −30 mgal dot the southwest flank of the trough adjacent to the Transantarctic Mountains. Though two of the features are defined by only a few points, the anomaly at 8.5° S, 1.5° E was outlined in detail by surface traverses from station C-16 (Figure 140.1) during the austral summer of 1976–77. The anomalies are not associated with local topographic depressions along

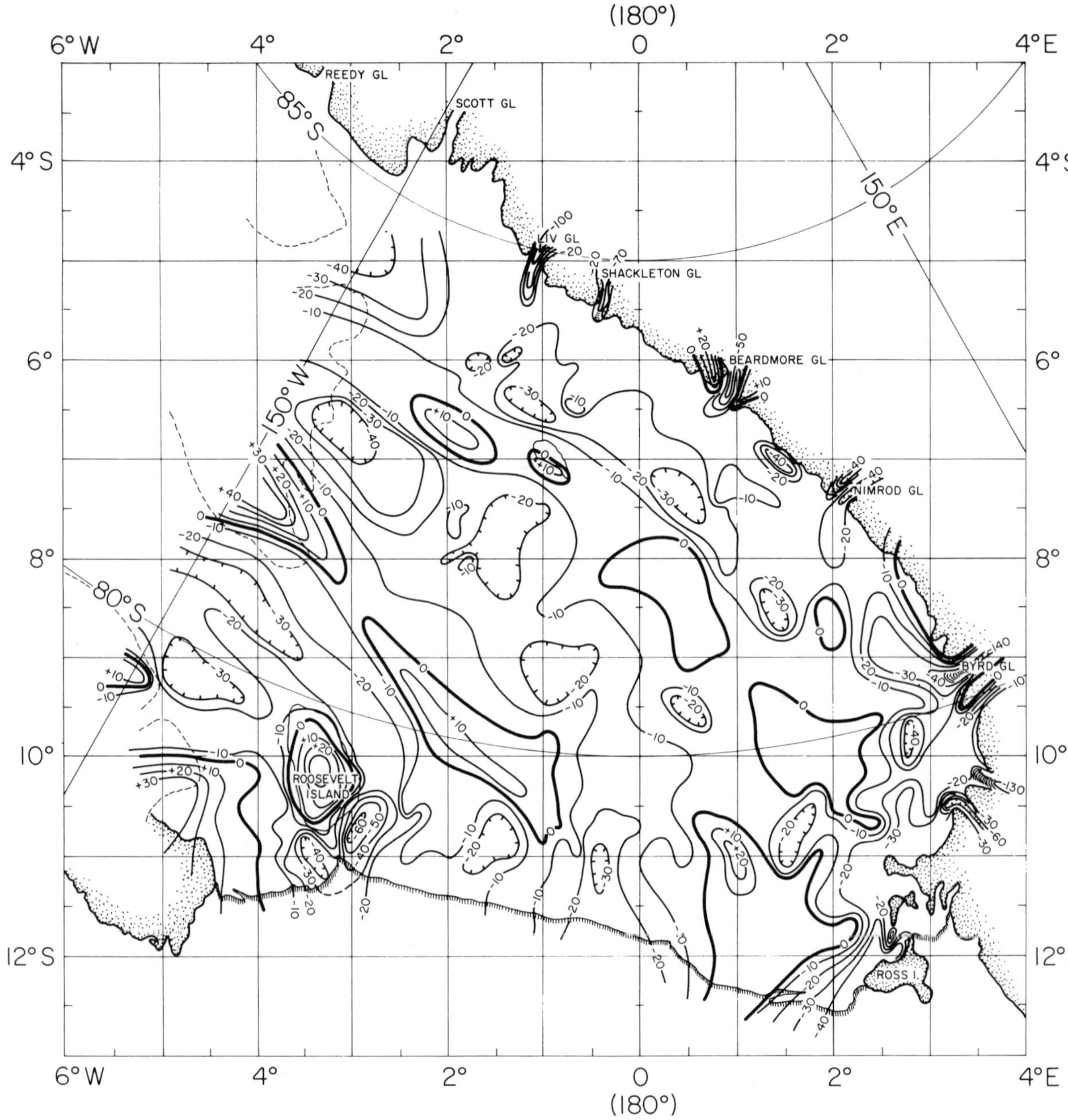

Figure 140.3. Free-air gravity anomalies on the Ross Ice Shelf, in mgal.

the side of the trough and probably have a tectonic significance that lacks surface expression.

Bouguer gravity anomalies

Free-air gravity anomalies have been converted to Bouguer anomalies using a crustal density of 2.67 g/cm³. The resulting Bouguer anomaly map (Figure 140.4) is strongly positive, as is normal for oceanic areas. As one would expect, there is some correlation between the Bouguer anomalies and the sea-bottom topography. A linear positive anomaly overlies the trough next to the Transantarctic Mountains, and the most positive region of Bouguer gravity on the ice shelf coincides with the deep water in and to the southwest of Discovery Deep. The submarine bank south of Roosevelt Island is overlain by a strongly negative, oval anomaly, whereas the trough to the west of the island is relatively positive. The correlation breaks down, however, over the troughs between Roosevelt Island and the Crary Ice Rise. The trough immediately north of the island is mildly negative rather than positive, and a strong, relatively negative anomaly coincides with the depression extending from the middle of the Siple Coast. Even these anomalies, though, are linear and parallel to topography, suggesting that there is a close relationship between sea-bottom topography and subbottom density variations.

Seismic velocities in the sediments

Subbottom reflections of acceptable quality for accurate interval velocity calculations were recorded at three stations (7° S, 0.5° W; 9° S, 2° W; and 8° S, 0.5° W) on the western half of the ice shelf during seismic reflection surveying. The interval velocities at the three locations are 2.5 ± 0.2, 2.6 ± 0.4, and 3.2 ± 0.1 km/sec, respectively, in

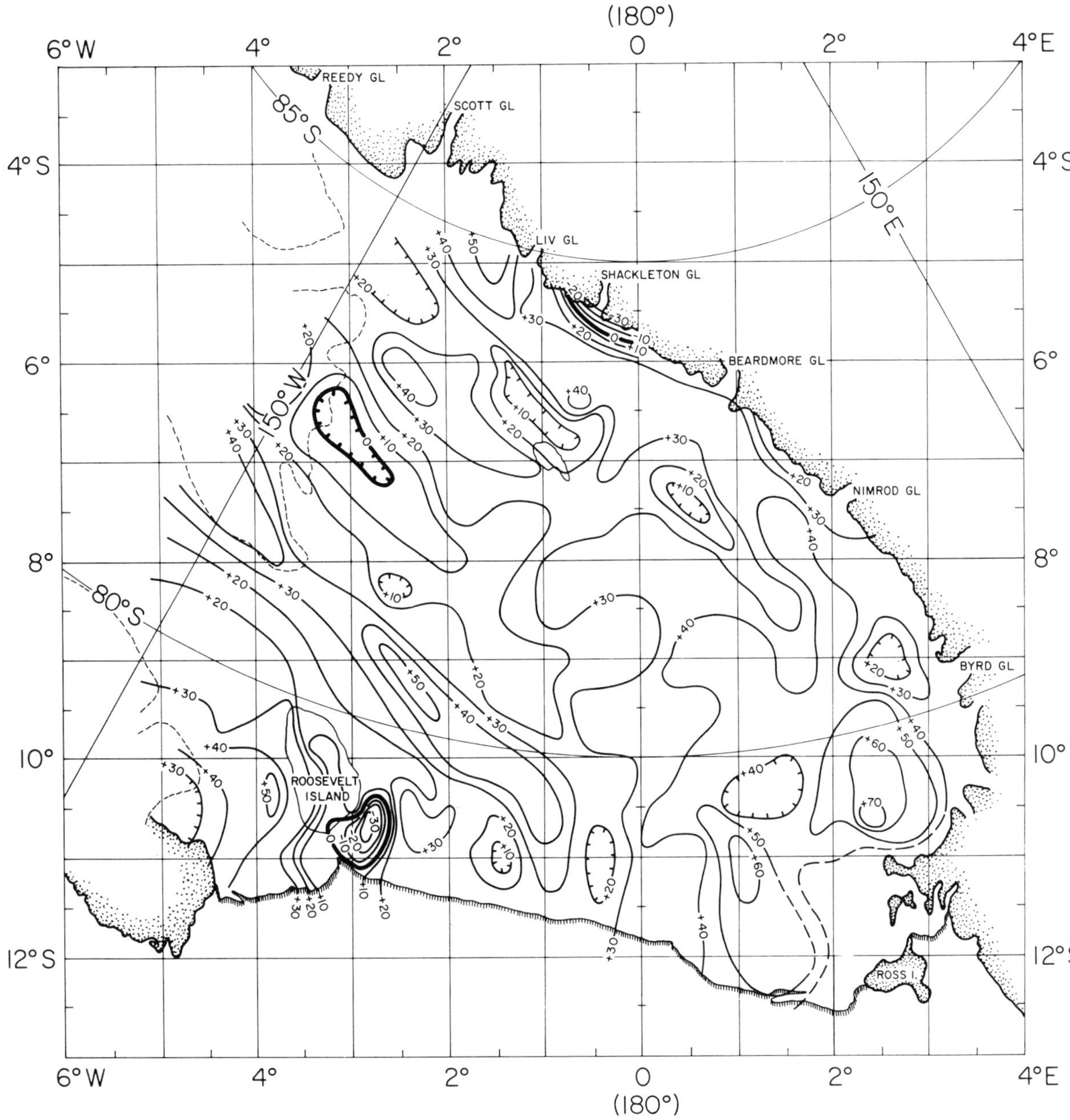

Figure 140.4. Bouguer gravity anomalies on the Ross Ice Shelf, in mgal.

sediment layers immediately below the sea floor with thicknesses of 45, 54, and 156 m, respectively. The first two velocities are consistent with other measurements (Crary, 1961; Houtz and Davey, 1973). The third is substantially higher, indicative of either a different type of sediment or a dipping subbottom reflector.

Many sampling studies have shown that the sediments on the floor of the Ross Sea consist of poorly sorted clastic deposits of glacial origin (Philippi, 1912; Stetson and Upson, 1937; Hough, 1950; Thomas, 1959, 1960). Based on these analyses and, in particular, on a sediment core near Little America V (Thomas, 1960), Crary has correlated the 2.5 km/sec seismic velocity with a mixture of coarse- and fine-grained glacial till. It is likely that the upper submarine layers at RIGGS stations also are glacial till, a conclusion supported by acoustic impedance measurements at nine sites (Robertson, 1975).

The sediment deposits sensed by the subbottom reflections are only a small part of the total sedimentary column below the ice shelf. Long refraction profiling (see below) generally produces estimates in the range of 500–1500 m for the total thickness of the sedimentary section, and seismic profiler sections of the Ross Sea shelf likewise show that the sedimentary column generally is many hundreds of meters thick (Houtz and Meijer, 1970; Houtz and Davey, 1973). The reflector noted here at 50 or 150 m depth within the sedimentary section may correlate with a change in lithology or with an erosional interface. A reflector within a few tens of meters of the sea floor is widespread in the Ross Sea (Houtz and Davey, 1973). The interface was sampled at depths of 20, 20, and 30 m, respectively, in three of the four holes drilled during Leg 28 of the Deep Sea Drilling Project and was determined to be a glacial erosion surface formed between 4 and 5 m.y. ago when the ice shelf was grounded (Scientific Staff, Leg 28, Deep Sea Drilling Project, 1973). The seismic reflector at the RIGGS stations may be a continuation of this late Pliocene erosional surface beneath the ice shelf. If so, the subsequent deposition has been substantially greater than at the Ross Sea stations.

Seismic refraction profiles

Bedrock arrivals were recorded on seismic refraction profiles at Crary Ice Rise, J9, RI, C-16, and Q13 (Figure 140.1). At station BC, a shot 20 km from the recording spread failed to produce a bedrock arrival. The results from Crary Ice Rise, J9, and RI are summarized below; for further details see Robertson (1975). Those at C-16 and Q13, obtained during the 1976–77 field season, will be presented at a future time.

Crary Ice Rise

Crary Ice Rise is particularly well suited to long refraction profiling owing to the absence of a water layer. The apparent velocity in the "basement" from three shots at approximately 2 km intervals recorded on a fixed geophone spread is 5.72 ± 0.03 km/sec (corrected for slope), in good agreement with the average of the individual cross-spread velocities of 5.8 ± 0.1 km/sec. The depth to basement calculated from the time-intercept is 650, 750, or 900 m beneath the bottom of the ice, varying as the velocity in the intervening layer is assumed to be 2.0, 2.4, or 3.0 km/sec, respectively.

Station J9

At J9 (the Ross Ice Shelf Project drill site), the apparent velocity in the basement between two geophone spreads 3 km apart is 6.8 km/sec. This velocity is substantially higher than the cross-spread velocities of 5.1 ± 0.2 km/sec and 5.8 ± 0.2 km/sec that characterize the individual seismograms. The differences apparently are caused by topography along the refractor, rather than the existence of more than one layer. Since the configuration of the topography is not known, the true velocity in the basement cannot be determined. However, the mean of the three apparent velocities is close to that at Crary Ice Rise only 75 km away, so it seems likely that the same type of basement underlies both stations. Using the velocity 5.7 km/sec from Crary Ice Rise and the same three assumed velocities in the intervening layer as before, the calculated depths to basement are 1300, 1500, and 1750 m, respectively, beneath the sea bottom. A topographic configuration that includes slopes of about 4° reconciles the various velocities.

Station RI

Either of two different interpretations of the long refraction profile at station RI is allowable because of poor quality of the seismograms. The major features of the first interpretation are: (1) the refractor has a minimum dip of 2°, corresponding to a minimum true basement velocity of 4.8 km/sec; (2) a layer of low-velocity sediment probably underlies the sea floor so that the dip and velocity are greater than 2° and 4.8 km/sec; and (3) the greater the true basement velocity, the steeper the slope of the basement-sediment interface. If a 5.5 km/sec velocity from nearby results on Roosevelt Island (M. P. Hochstein, personal communication, 1966) is used, this leads to a 6° slope in bedrock and a sediment thickness averaging 850 m. The major features of the second interpretation are: (1) basement velocity is 4.9 km/sec; (2) no dip in the refractor is needed to fit the data; and (3) the thickness of the sediment layer is about 100 m. The first interpretation is preferred because of better agreement with a gravity profile along the seismic line (Whiting, 1975).

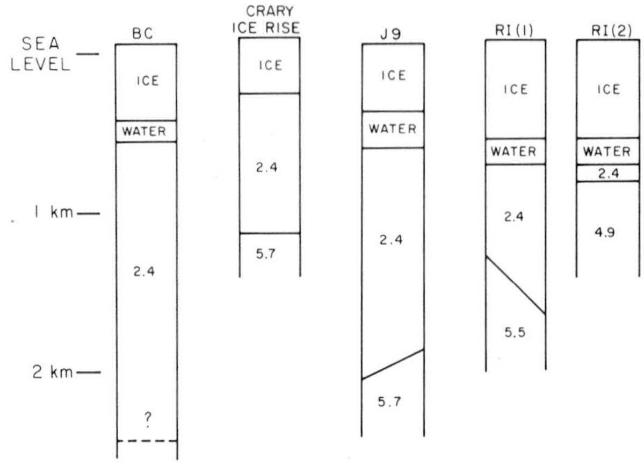

Figure 140.5. Seismic sections at station BC, Crary Ice Rise, station J9, and station RI; RI(1) is the preferred interpretation and RI(2) is an alternate interpretation. Numbers are seismic velocities in km/sec.

There is additional evidence for substantial topographic variation in the basement surface in the Ross Sea region. Houtz and Davey (1973) have presented profiler data showing the existence of numerous topographic features on the upper surface of the basement layer under the Ross Sea. Slopes of as much as several degrees are commonplace on their profiles, in good agreement with the magnitudes of the slopes on the RIGGS seismic sections.

Geologic interpretation

The best estimates of the RIGGS seismic velocity columns are shown in Figure 140.5, assuming a sediment velocity of 2.4 km/sec. The column at Crary Ice Rise is the most definite. Those at stations J9 and RI are suspect in that the estimated bedrock velocities may be in error, as discussed above. Both interpretations of the RI data are included in the figure. For station BC, the indicated sediment thickness is a minimum, based on the assumption that the 20 km shot-spread separation was too short to record a bedrock arrival.

For comparison with the RIGGS values, there is some information on the seismic velocities in bedrock in surrounding areas. Houtz and Davey (1973) have reported that basement velocities range from 5.2 to 6.3 km/sec under the continental shelf of the Ross Sea. All of their data came from basement highs along about long. 180°. A seismic velocity column at Byrd Station (lat. 80° S, long. 120° W) (Bentley and Clough, 1972) consists of a layer with a velocity of 4.3 km/sec (calculated from reflection amplitudes) directly underlying the ice and overlying a 5.9 km/sec layer. In the RIGGS area itself, six seismic refraction profiles have previously been shot, one each at Little America V and C-49 (Crary, 1961) and four on or close to Roosevelt Island (M. P. Hochstein, personal communica-

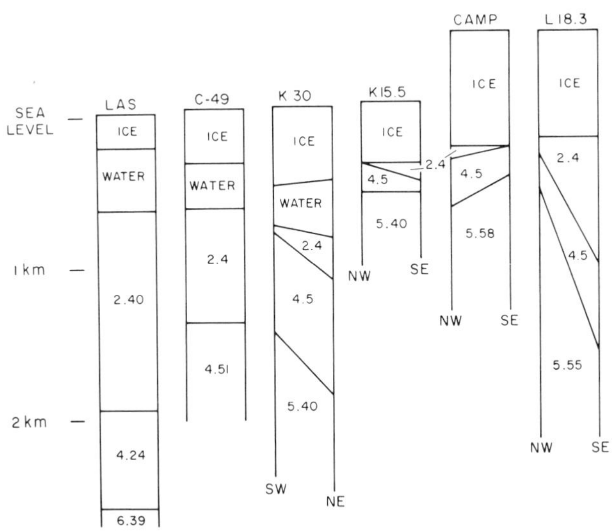

Figure 140.6. Seismic sections on and near Roosevelt Island. Sections for LAS and C-49 are from Crary (1961); others are from Hochstein (personal communication, 1966). Directions on Hochstein's sections are geographic, not grid, directions. See Figure 140.1 for locations of sections. Numbers are seismic velocities in km/sec.

tion, 1966) (Figure 140.6). In his interpretation, Hochstein used constant intermediate-layer velocities of 2.4 and 4.5 km/sec, which are averages computed from seismic reflection shooting.

Layers with a 4.2–4.5 km/sec velocity are not present in the RIGGS columns, although the existence of such layers is reasonably well established by their independent appearances in the work of Crary (1961), Hochstein (personal communication, 1966), and Bentley and Clough (1972). The 4.2–4.5 km/sec velocity appears in all subglacial columns grid-south of a line passing through the Byrd Subglacial Basin and Roosevelt Island, but in none between that line and the Horlick and Transantarctic Mountains. However, the velocity in question is so close to the velocity in the ice (3.8 km/sec) that the corresponding arrival would be masked unless the layer was several km thick (Crary's measurements were made partly on sea ice), so the absence of seismic evidence for the layer does not necessarily mean that the layer is absent. If rocks with such velocities are actually present, depths to "basement" are correspondingly greater than those calculated above. The 4.2–4.5 km/sec velocity is absent from the measurements in the open Ross Sea (Houtz and Davey, 1973).

In general, seismic velocities of 4.2–4.5 km/sec are associated with consolidated sedimentary strata. Crary has tentatively identified the layer at Little America V and C-49 with sandstone from the Beacon Supergroup. Bentley and Clough have stated that the layer at Byrd Station could be upper Paleozoic–Mesozoic sedimentary rock or, alternatively, an interbedded combination of Cenozoic volcanic flows and pyroclastic rocks. Thus, the layer could be a part of the West Antarctic Cenozoic volcanic suite. The velocity of 4.2–4.5 km/sec is also appropriate for low-grade metasedimentary rocks like those of the Rockefeller Mountains or Ford Ranges (Wade, 1969). If the layer is actually present but masked under the ice shelf, association with Beacon-type rocks seems probable. If, however, the layer is truly absent under the shelf except in the grid-southwest corner, one of the other identifications seems more likely.

Seismic velocities of 5.7–5.9 km/sec usually are associated with metamorphic rocks, hard limestones, or felsic intrusive rocks. One such refracting layer along long. 180° in the Ross Sea was found to consist of a gray, foliated marble and calc-silicate gneiss that may be a part of the same lithologic unit as the Koettlitz Marble in the Royal Society Range and the Cambrian limestone in the central Transantarctic Mountains, i.e., a part of the Ross orogenic zone (Scientific Staff, Leg 28, Deep Sea Drilling Project, 1973). The 5.7 km/sec basement layers in the Ross Ice Shelf seismic columns, however, could consist of crystalline rock from any of the geologic provinces around the ice shelf since all contain abundant metasedimentary and felsic intrusive formations. It may seem more likely that the basement at Crary Ice Rise and station J9 is associated with the Ross orogenic province than with the Ellsworth or Andean, simply because of proximity. On the other hand, if West Antarctica has a continental drift history different from that of East Antarctica, some other association is required.

Velocities in the lowest layer under and around Roosevelt Island range from about 5.4 to 5.6 km/sec. Hochstein (personal communication, 1966) has tentatively associated these refractors in the Roosevelt Island seismic sections with the metasedimentary rocks and felsic intrusive rocks of the Rockefeller Mountains and Ford Ranges in the Andean orogenic zone, an association which we find reasonable, particularly in light of the possible correlation between the overlying 4.2–4.5 km/sec layer and the low-grade metasedimentary rocks of the Rockefeller Mountains and Ford Ranges, as discussed above.

The anomalies on the Bouguer gravity anomaly map (Figure 140.4) are caused by lateral density changes below the ocean floor. These changes are primarily a combination of variations in the thickness of the sediment layer between the sea bottom and the basement rock and lateral compositional heterogeneity in the crust related to emplacement of intrusive rocks, block faulting, or other tectonic activity. The best seismic estimates of sediment thickness at Crary Ice Rice, J9, RI, Little America V, and C-49 are, respectively, 750, 1500, 850, 1300, and 750 m, and the corresponding Bouguer anomalies are +29.3, +16.5, +19.1, +34.0, and +22.4 mgal. The poor correlation between gravity anomalies and sediment thickness, which is not removed by isostatic corrections (Bentley and others, Chapter 121, this volume), indicates deep structural control of linear gravity anomaly patterns and therefore deep structural control of linear sea-bottom topographic features, since the two sets of trends are nearly parallel. The structures obviously predate the current episode of glaciation in West Antarctica, suggesting that long-standing geologic conditions have determined such glaciologic phenomena as the present courses of ice streams. The topography of the ocean bottom below the ice shelf is not fundamentally determined by the ice streams themselves.

Acknowledgments. This work was supported by Grants GV-36963 and DPP72-05802 from the National Science Foundation.

References cited

Bennett, H. F. 1964. A gravity and magnetic survey of the Ross Ice Shelf area, Antarctica. *Res. Rep. Ser., Geophys. Polar Res. Center, Univ. Wisconsin,* No. 64-3, 97 pp.

Bentley, C. R. 1976. Analysis of Ross Ice Shelf geophysics. *Antarct. Jnl U.S., 11,* 276–77.

Bentley, C. R., and Clough, J. W. 1972. Antarctic subglacial structure from seismic refraction measurements. (*In* Adie, R. J., *ed. Antarctic Geology and Geophysics.* Oslo, Universitetsforlaget, 683–91.)

Bentley, C. R., and Jezek, K. C. 1977. Ross Ice Shelf Geophysical Survey 1976–77. *Antarct. Jnl U.S., 12,* 142–44.

Bentley, C. R., Clough, J. W., and Robertson, J. D. 1974. RISP Geophysical Work. *Antarct. Jnl U.S., 9,* 157–58.

Clough, J. W., and Robertson, J. D. 1975. RISP geophysical survey. *Antarct. Jnl U.S., 10,* 153.

Crary, A. P. 1961. Marine sediment thickness in the eastern Ross Sea area, Antarctica. *Geol. Soc. Am. Bull., 72,* 787–90.

Crary, A. P., Robinson, E. S., Bennett, H. F., and Boyd, W. W., Jr. 1962. Glaciological studies of the Ross Ice Shelf, Antarctica, 1957–60. *I.G.Y. Glaciol. Rep., 6,* 193 pp.

Hough, J. L. 1950. Pleistocene lithology of Antarctic ocean bottom sediments. *J. Geol., 58,* 254–60.

Houtz, R., and Davey, F. J. 1973. Seismic profiler and sonobuoy measurements in Ross Sea, Antarctica. *J. geophys. Res., 78,* 3448–68.

Houtz, R., and Meijer, R. 1970. Structure of the Ross Sea shelf from profiler data. *J. geophys. Res., 75,* 6592–97.

Philippi, E. 1912. Die grundproben der Deutschen Südpolar-Expedition. (*In* von Drygalski, E., *ed. Deutsche Südpolar-expedition 1901–1903. 2.* Berlin, G. Reimer, 411–616.)

Robertson, J. D. 1975. *Geophysical Studies on the Ross Ice Shelf, Antarctica.* Ph.D. dissertation, University of Wisconsin–Madison, 214 pp.

Scientific Staff, Leg 28, Deep Sea Drilling Project. 1973. Leg 28 deep-sea drilling in the southern oceans. *Geotimes, 18,* 19–24.

Stetson, H. C., and Upson, J. E. 1937. Bottom deposits of the Ross Sea. *J. sedim. Petrol., 7,* 55–66.

Thomas, C. W. 1959. Lithology and zoology of an Antarctic Ocean bottom core. *Deep Sea Res., 6,* 5–15.

Thomas, C. W. 1960. Late Pleistocene and Recent limits of the Ross Ice Shelf. *J. geophys. Res., 65,* 1789–92.

Wade, F. A. 1969. Geology of Marie Byrd Land. (*In* Bushnell, V. C., and Craddock, C., *eds.* Geologic Maps of Antarctica. *Antarct. Map Folio Ser.,* Folio 12, Pl. XVIII.)

Whiting, L. R. 1975. *A Gravity Survey of the Eastern Portion of the Ross Ice Shelf, Antarctica.* Masters Thesis, University of Wisconsin—Madison, 97 pp.

141

Physiographic Analysis and Interpretation of the Ferrar Glacier–Victoria Valley Area, Antarctica

JOHN H. WRENN* and PETER N. WEBB**
Department of Geology, Northern Illinois University, DeKalb, Illinois 60115, U.S.A.

Abstract. Analysis of topographic and subsurface profiles across the area west of McMurdo Sound provides insights into the Cenozoic geologic history of this part of the Transantarctic Mountains. North-south topographic profiles extending across Victoria, Wright, and Taylor Valleys and Ferrar fjord show a progressive lowering of valley-floor elevations toward the south. Thus, the floor of Victoria Valley stands about 600 m higher than that of glaciated Ferrar fjord some 50 km to the south. Paleontologic evidence indicates that during late Miocene and Pliocene time eastern Taylor Valley was a deep ice-covered fjord. During Pliocene time Wright Valley was also an ice-covered fjord of considerable depth. Both paleofjords probably bore a close resemblance to the present-day Ferrar fjord. Stratigraphic and geomorphic evidence points to a longer marine history for eastern Taylor Valley than for Wright Valley. It is unclear whether Victoria Valley was ever a fjord. Emergence (marine regression) occurred earlier in the northern part of the area. Wright Valley emerged above sea level in late Pliocene time and eastern Taylor Valley in late Pleistocene time; Ferrar fjord is currently occupied by marine waters covered by glacial and seasonal ice. East-west topographic profiles through the Taylor Valley, Asgard Range, and Kukri Hills area disclose a sharp eastward decrease in mountain-peak elevations near long. 163° E. Inland of this meridian, Taylor Valley cross sections are V-shaped, while toward the coast they are broad and U-shaped. Numerous small outcrops of Plio-Pleistocene volcanic rocks occur just west of this meridian. The western border of the Miocene–Holocene eastern Taylor Valley sedimentary basin is also located at about 163° E. All these data suggest the presence of a major and long-active north-south fault zone crossing Taylor Valley in the Lake Leon–Mount Coleman area. This is a probable extension of Gunn and Warren's Lister Fault, which extends along the eastern side of the Royal Society Range.

The Ferrar fjord (occupied by Ferrar Glacier) and Victoria Valley area of Antarctica (Figure 141.1) constitutes the major part of the southern Victoria Land ice-free valley network adjacent to McMurdo Sound. The present study is the first attempt to analyze the physiography of the region. More specifically, this study attempts to use physiographic evidence to test the hypothesis of late Cenozoic regional uplift suggested by micropaleontologic evidence gathered from drill cores of the Dry Valley Drilling Project (DVDP) taken in eastern Taylor Valley (Webb and Wrenn 1976*a*, *b*).

The study involved the interpretation of data from eight north-south (NS) (Figures 141.2, 141.3, and 141.4) and three east-west (EW) (Figure 141.5) physiographic profiles. These profiles were prepared from U.S. Geological Survey 1:250,000 scale topographic maps, subglacial radio-echo soundings (Calkin, 1974), subsurface drill-hole data from eastern Taylor Valley (McGinnis, 1974, 1975), and bathymetric (McGinnis, 1973) and seismic data from McMurdo Sound (Northey and others, 1975).

* Present address: Department of Geology, Louisiana State University, Baton Rouge, Louisiana 70803, U.S.A.

** Present address: Department of Geology, The Ohio State University, Columbus, Ohio 43210, U.S.A.

Physiographic profiles

North-south profiles

Profiles NS-1 and NS-2 (Figures 141.1 and 141.2) follow paths parallel to the coast and lie 12 km and 10 km, respectively, offshore. They cross the Wright, Taylor, and Ferrar submarine valleys approximately normal to the valley axes.

Profile NS-3 (Figure 141.1) starts from Hanson Ridge (611 m) and continues southward over Wilson Piedmont Glacier, crossing the subglacial Wright Valley (Figure 141.2), to Cape Bernacchi, New Harbor, and terminates on Bowers Piedmont Glacier. The configuration of the subglacial Wright Valley was revealed by airborne radio-echo soundings (Calkin, 1974).

Westward extrapolation of Calkin's (1974) data for the region crossed by profile NS-3 suggests that the depth of the Ferrar submarine valley may be greater than previously believed. This extrapolated depth is plotted as a dashed line in Figure 141.2, lying beneath the profile surface derived from the bathymetric map of McMurdo Sound (McGinnis, 1973).

Profile NS-4 begins on Staeffler Ridge just north of Greenwood Valley and passes southward over the Wilson Piedmont Glacier, subglacial Wright Valley, and Taylor

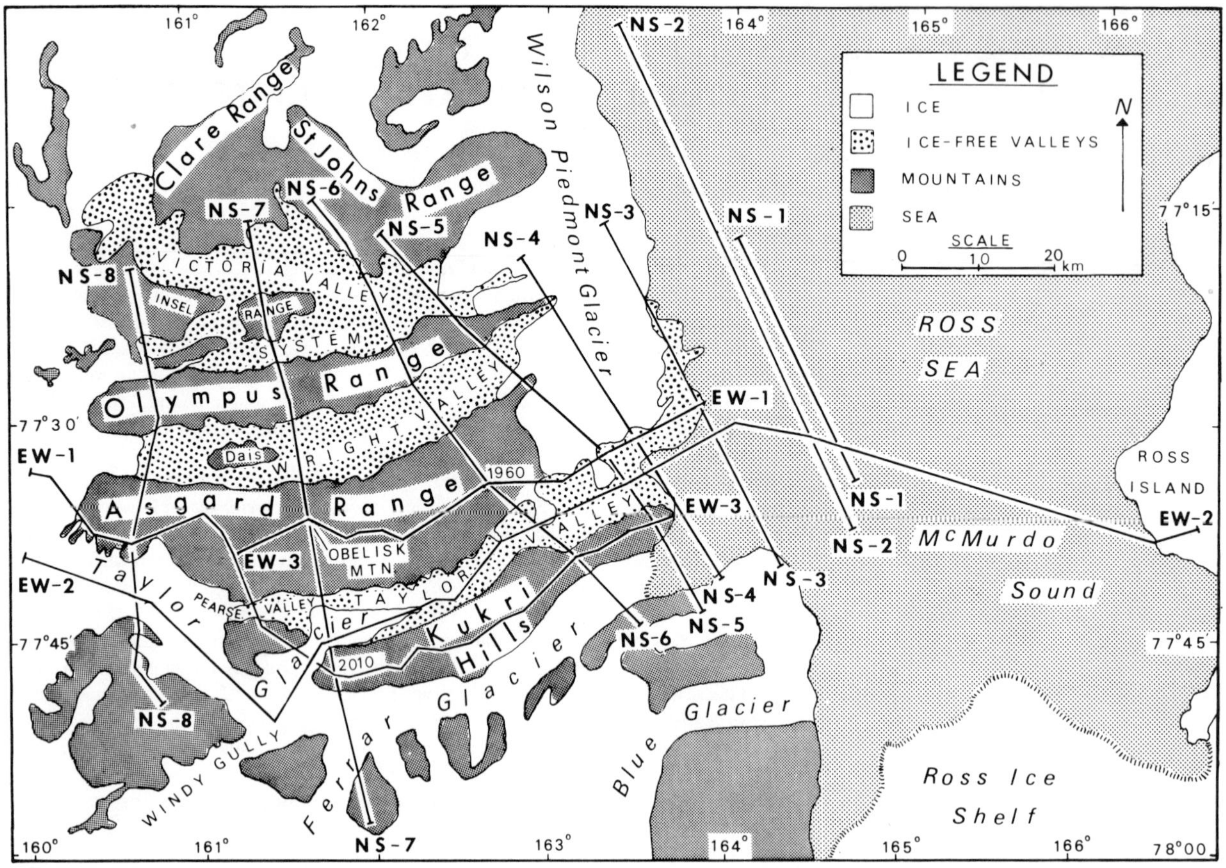

Figure 141.1. Map of the ice-free valleys west of McMurdo Sound, showing locations of the topographic, bathymetric, and subice profiles. Elevations in meters.

and Ferrar submarine valleys before terminating on Bowers Piedmont Glacier (Figures 141.1 and 141.2). It should be noted that the subglacial profile on Wright Valley along NS-4 is broader and more U-shaped than along NS-3. The suggested greater depth of the Ferrar submarine valley is also indicated on NS-4.

Profiles NS-5, NS-6, and NS-7 (Figures 141.1, 141.3, and 141.4) begin on the north side of the Victoria Valley system, run south across the ice-free valleys, and terminate on the south side of the Ferrar fjord. Profile NS-5 crosses Taylor Valley between the sites of DVDP holes 8–10 and DVDP holes 11 (Figures 141.1 and 141.3). Profile NS-6 crosses Taylor Valley through the site of DVDP hole 12.

Profile NS-8 crosses the extreme western end of the ice-free valleys from Gibson Spur in the north to the south side of the upper Taylor Glacier and terminates on West Beacon Heights.

Interpretation of North-South Physiographic Profiles

Valley-floor elevations

There is a marked decrease in valley-floor elevations from north to south across the area (see NS-3 through NS-6). The extreme depth of Wright Valley partially obscures the relationship along NS-7, although the overall southerly decrease is evident. The valley-floor elevations decrease along NS-8 only as far south as Taylor Valley, south of which they are much higher. These high-standing tributary valleys may be structurally controlled.

Post-Jurassic faulting in the southwestern part of the ice-free area (Hamilton, 1965) resulted in the Terra Cotta Hills (to the southeast of Windy Gully) being displaced downward 460 m relative to Beacon Heights to the northwest. A vertical displacement of similar magnitude between the opposite walls of Pearse Valley (referred to as Solitary Dry Valley by Hamilton, 1965) is probably the result of movement along a west-northwest-trending fault under the valley floor. This Pearse Valley fault continues east-southeast under lower Taylor Glacier, terminating west of the glacier's snout. West-northwest of Pearse Valley the fault is hidden beneath upper Taylor Glacier.

The upward displacement of strata on the south side of the Pearse Valley fault and to the north of Terra Cotta Hills suggests the uplift of a single fault block. To the east, this fault block may be bordered by a large fault beneath the Taylor Glacier at the point where it turns north-northeastward across the western end of the Kukri Hills. Hamilton (1965) has noted the presence of small faults cutting across the western end of the Kukri Hills.

The occurrence of an upthrown fault block would explain the anomalously high valley-floor elevations seen at the south end of profile NS-8.

Tectonic uplift and the marine record

Fjords have been features of the southern Victoria Land coastline for much of middle and late Cenozoic time (Crary, 1966; Webb, 1972; Webb and Wrenn, 1976*a,b*). Dated fossiliferous successions and the elevations of major unconformities and marine deposits within each paleofjord provide some indication of the tectonic history of

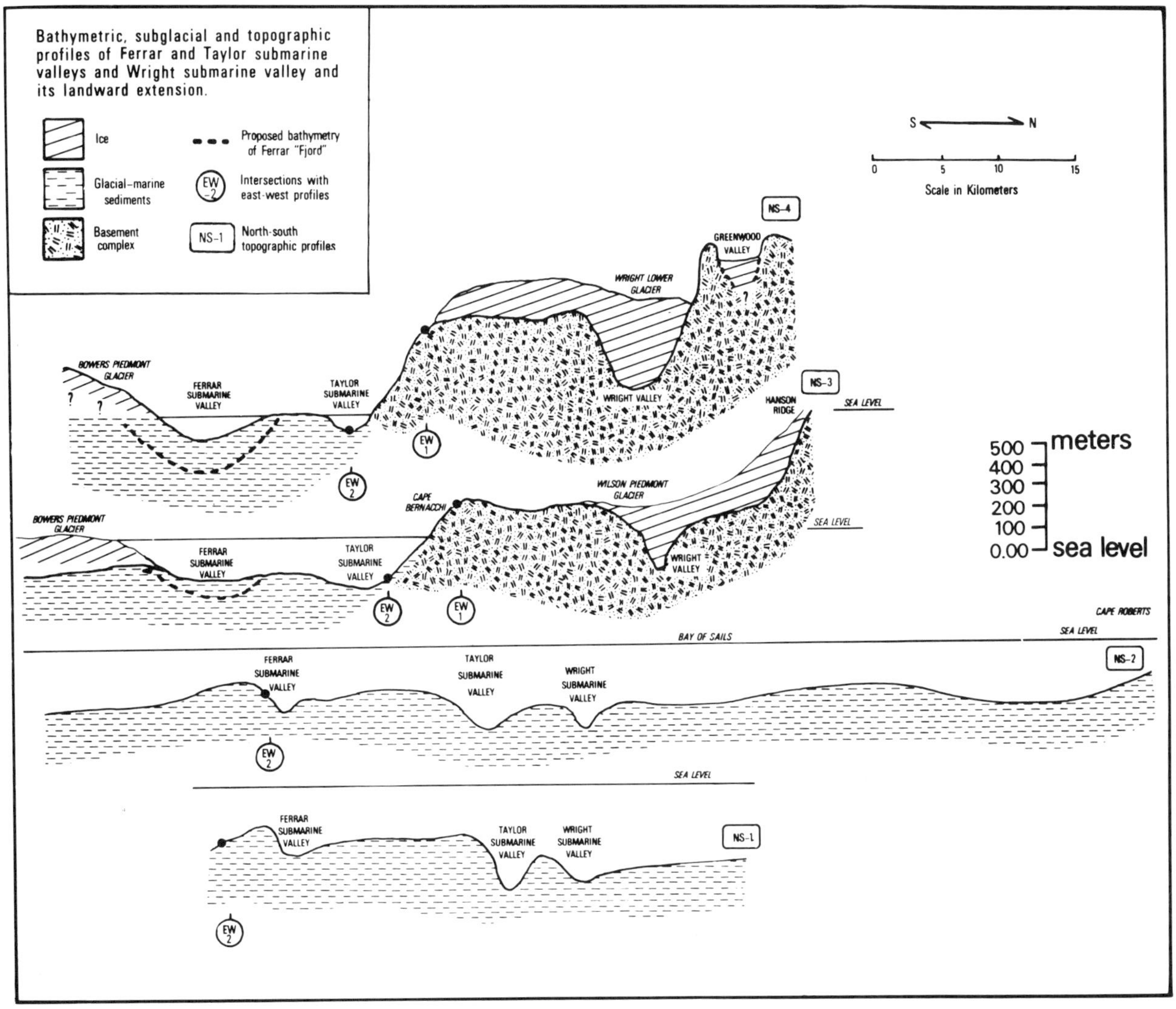

Figure 141.2. North-south profiles NS-1, NS-2, NS-3, and NS-4.

these valley basins and the faulted blocks on which they formed. The age and tectonic history of the Ferrar fjord (Figures 141.3 and 141.4) and its eastern extension, Ferrar submarine valley (Figure 141.2), are not known. This deep fjord is currently occupied by Ferrar Glacier (Figures 141.1, 141.3, and 141.4). The glacier is grounded throughout the fjord except in the east, where it is afloat.

Foraminiferal studies in eastern Taylor Valley (Webb and Wrenn, 1976*a,b*; Wrenn, 1977) indicate the occurrence of deep-water (~600 m) upper Miocene–lower Pliocene faunas in drill holes at depths of only 155 m below present sea level. The sedimentary successions from which these faunas were recovered are unconformably overlain by sediments containing much shallower water (<100 m) middle Pliocene (correlated with the Wright Valley *Pecten* gravels) and Pleistocene faunas. Sometime in early Pliocene time, the coastal basement block (eastern Taylor Valley block) on which these sediments rest was elevated more than 400 m. The block to the immediate west (middle Taylor Valley block) may have been elevated an even greater amount.

To the north, in central Wright Valley, the *in situ* middle Pliocene *Pecten* gravels stand some 300 m higher than their subsurface correlatives in Taylor Valley. These are the youngest marine sediments presently known from Wright Valley. It is deduced that the major part of this differential uplift occurred in late Pliocene or early Pleistocene time and resulted in the regression of marine waters from Wright Valley (Webb, 1972). The uplifted rock threshold at the mouth of Wright Valley is portrayed in profiles NS-3 through NS-5 (Figures 141.2 and 141.3), where Wright Valley's floor elevation is shown to rise inland at a quicker rate than does that of either the Taylor or Ferrar Valleys. There is no geologic evidence to indicate that the Victoria Valley system was ever the location of a fjord or a group of fjords, or even that it was transgressed by marine waters.

The chronology of the emergence (total marine regression) of the fjord floors starts with Wright Valley (late Pliocene time) and ends with the still submerged Ferrar fjord. This history conforms with the prevailing southerly decrease in valley-floor elevations, but it does not necessarily mean that a north-south progression of tectonic events has occurred. As noted above, the early to middle Pliocene uplift in eastern Taylor Valley occurred before, and was probably greater than, the late Pliocene–early Pleistocene uplift of Wright Valley.

The evidence suggests that the Transantarctic Mountains and the McMurdo Sound region are a mosaic of discrete

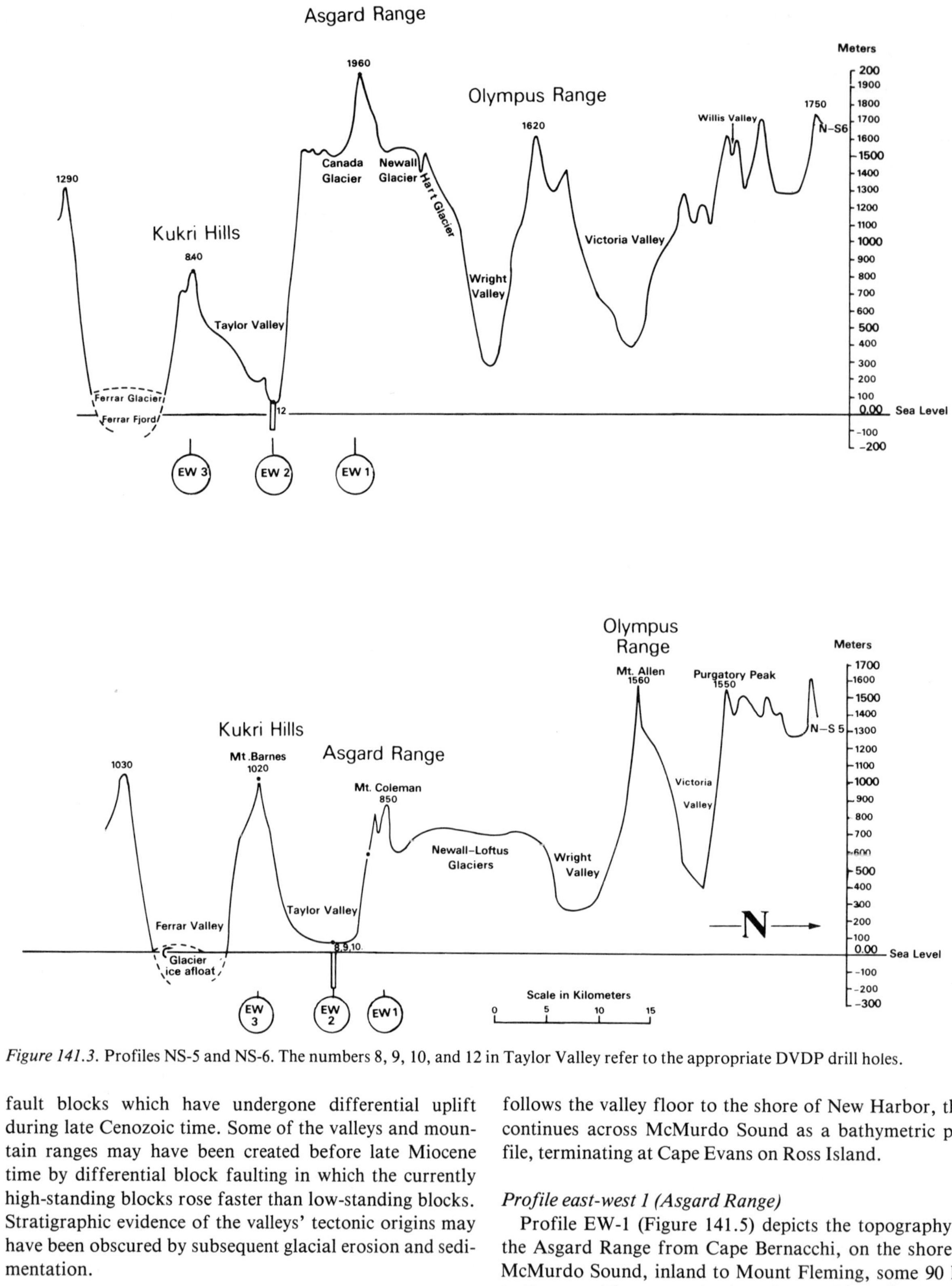

Figure 141.3. Profiles NS-5 and NS-6. The numbers 8, 9, 10, and 12 in Taylor Valley refer to the appropriate DVDP drill holes.

fault blocks which have undergone differential uplift during late Cenozoic time. Some of the valleys and mountain ranges may have been created before late Miocene time by differential block faulting in which the currently high-standing blocks rose faster than low-standing blocks. Stratigraphic evidence of the valleys' tectonic origins may have been obscured by subsequent glacial erosion and sedimentation.

East-west profiles

Two east-west profiles (Figures 141.1 and 141.5) follow the general crest of the Asgard Range (EW-1) and the Kukri Hills (EW-3). A third profile (EW-2) follows the course of Taylor Glacier from its source in the East Antarctic Polar Ice Sheet eastward to the glacier's terminus in Taylor Valley. East of the Taylor Glacier, profile EW-2 follows the valley floor to the shore of New Harbor, then continues across McMurdo Sound as a bathymetric profile, terminating at Cape Evans on Ross Island.

Profile east-west 1 (Asgard Range)

Profile EW-1 (Figure 141.5) depicts the topography of the Asgard Range from Cape Bernacchi, on the shore of McMurdo Sound, inland to Mount Fleming, some 90 km to the west. The mountains have been subjected to intense dissection by glacial erosion for at least 4.5 m.y. (Armstrong, 1978) and perhaps since Eocene time 38–40 m.y. ago (Wilson, 1973).

One of the most striking features of profile EW-1 is the sharp increase in altitude from the relatively low (1200 m), isolated coastal peaks, such as Hjorth Hill and Mount Coleman, to the high-standing (1800 to 2400 m) peaks of

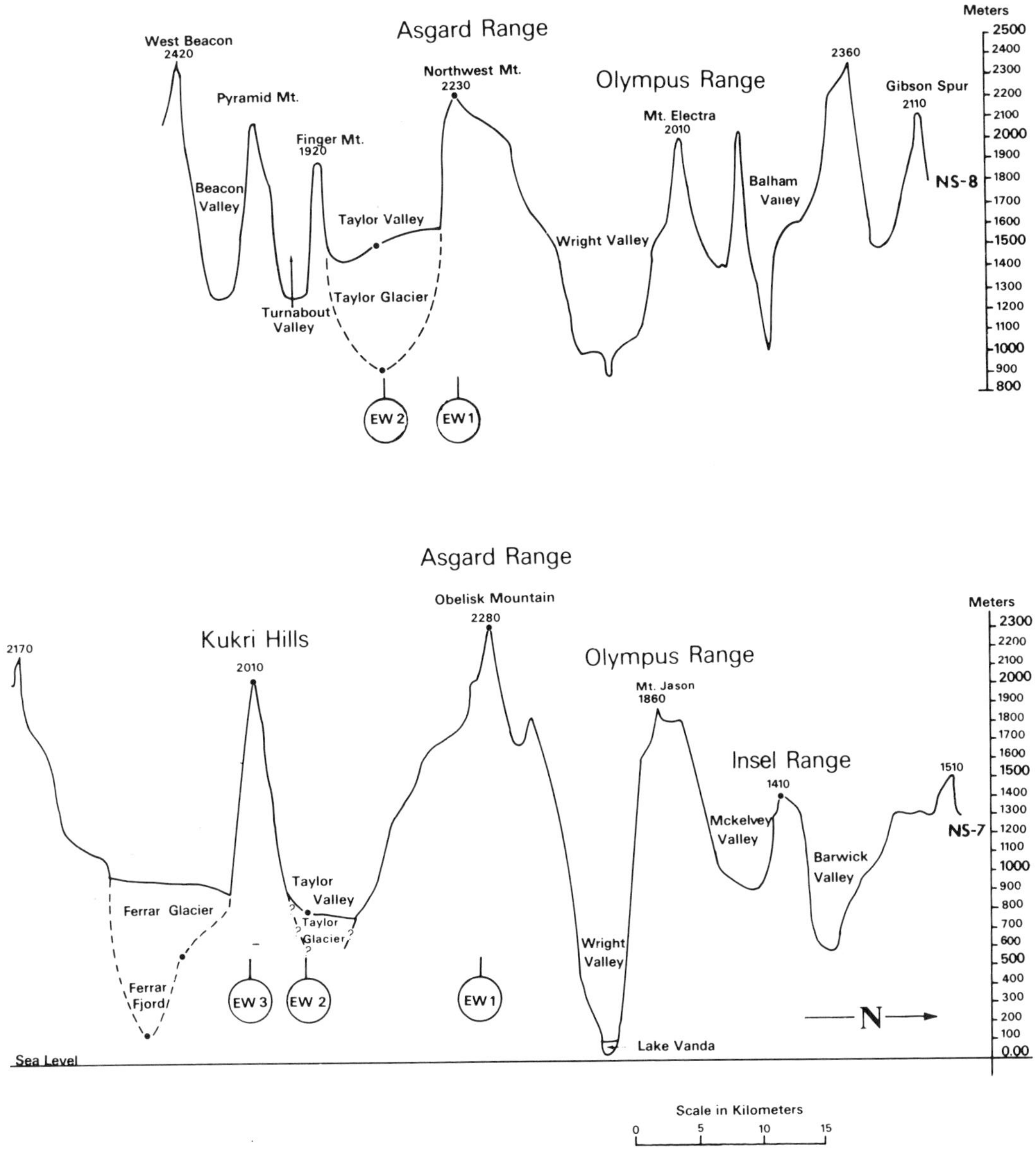

Figure 141.4. Profiles NS-7 and NS-8.

the inland mountain masses, such as Obelisk Mountain and St. Paul's Mountain. This sharp increase in altitude between the coastal and inland mountains is not unique to the Asgard Range but is typical of the McMurdo Sound coastal topography. The major change in elevation occurs between Mount Falconer (810 m) and the unnamed peak designated as having an elevation of 1760 m, i.e., within a distance of approximately 3 km (Figure 141.5). The highest elevation along EW-1 is at Round Mountain (2410 m), approximately 50 km west of the peak at 1760 m. Relief is greatest in the coastal area and least in the central part of the Asgard Range.

Two sets of accordant mountain peaks and valley bottoms are present within the main part of the Asgard Range. Some of these valleys are occupied by alpine glaciers which obscure valley-bottom elevations. The eastern accordant peaks (2150 ± 30 m) are accompanied by accordant valley bottoms at about 1800 m, whereas the higher, more westerly accordant peaks (2380 m) have intervening valley bottoms at about 1690 m.

Profile east-west 3 (Kukri Hills)

The EW-3 (Kukri Hills) topographic profile (Figures 141.1 and 141.5) is very similar to EW-1. Low coastal peaks are present, though they are not isolated as are the comparable peaks along the EW-1 (Asgard Range) profile. The sharp increase in elevation over a short horizontal distance is also evident, though somewhat less marked than that noted along the Asgard Range profile.

The major elevation increase in the Kukri Hills occurs on the west side of Kitticarrara Glacier. Peak elevations east of this point are typically 1000 m or less (Figure 141.5), but elevations to the west are generally greater than 1800 m.

Accordant peaks (2030 m) and valley floors (1500 m) are present in the extreme western Kukri Hills. These features occur at lower elevations than those in the Asgard Range.

Profile east-west 2 (Taylor Valley)

The surface elevation of Taylor Glacier (Figure 141.5) along EW-2 decreases from approximately 2200 m at its

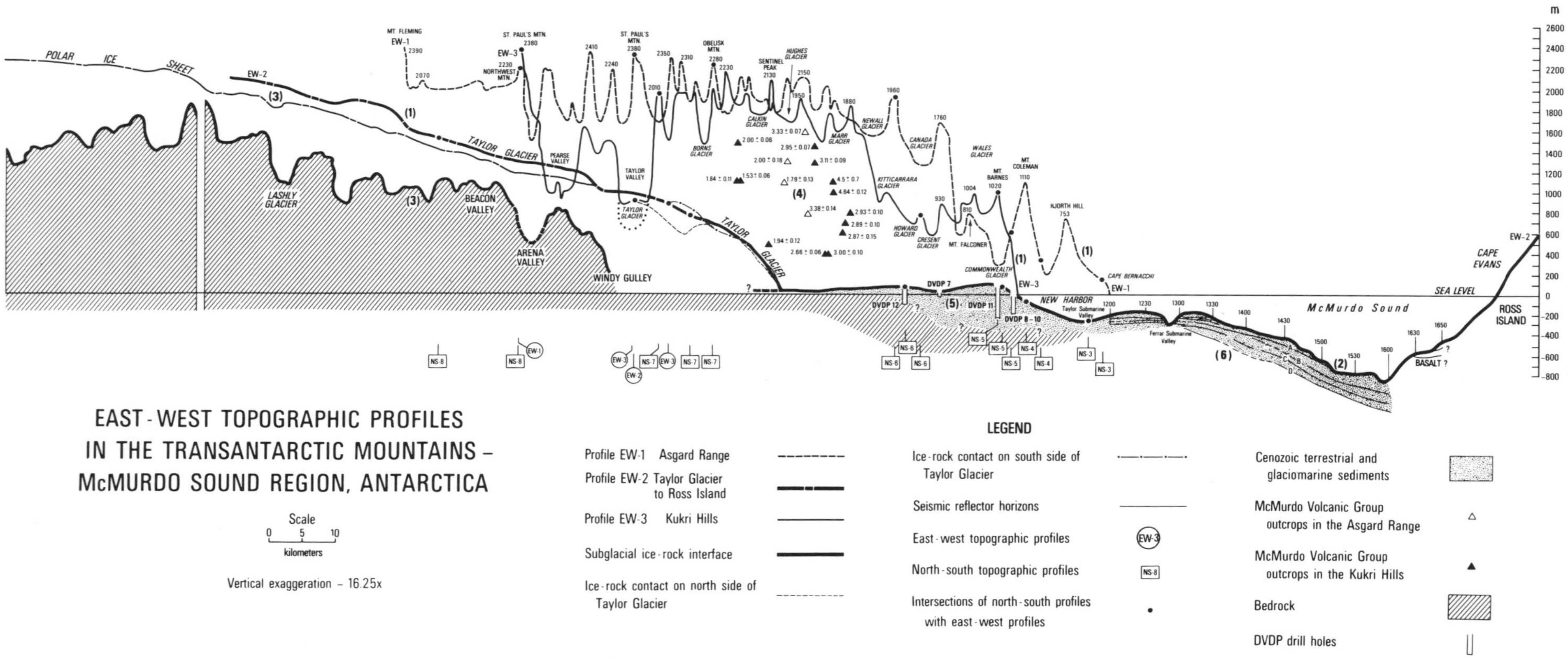

Figure 141.5. East-west physiographic profiles from the Polar Ice Sheet, through the Taylor Valley area, and across McMurdo Sound to Ross Island. The numbers in parentheses labeling various parts of the profiles refer to the following list of data sources: (1) Topographic data: U.S. Geological Survey maps Ross Island ST 57-60/6 and Taylor Glacier ST 57-60/5; (2) Bathymetric data: McGinnis (1973) and Calkin (1974); (3) Surface of Taylor Glacier and subglacial data: Calkin (1974); the gap in the lower left of the diagram indicates missing radio-echo data (Calkin, 1974); (4) McMurdo Volcanic Group data: Denton and Armstrong (1968) and Armstrong (1978); (5) DVDP data: McGinnis (1974, 1975); and (6) Seismic data: Northey and others (1975).

source, the East Antarctic Polar Ice Sheet, to approximately 76 m at its terminus in middle Taylor Valley (Hamilton, 1960). Calkin's (1974) interpretation of airborne radio-echo soundings, collected by Robin and others (1970), shows that the glacier surface elevations follow the eastward decrease in subglacial elevations of the ice-rock interface (Figure 141.5). West of the head of Taylor Glacier, beyond the limits of Figure 141.5, the subice rock surface also decreases in elevation. The ice surface elevation, however, decreases slightly in the same direction. Calkin (1974) has noted that the subglacial west flank of the Transantarctic Mountains consists of many steep-sided plateaus, whose origin he attributes to complex block faulting.

The subglacial surface profile on the east side of the Transantarctic Mountains (Figure 141.5) exhibits a number of round-bottomed notches, or valleys, which can be extrapolated laterally to the sides of Taylor Valley. These are subice extensions of tributary valleys of the upper Taylor Valley and may be floored by Tertiary sedimentary beds not presently known to crop out anywhere in the ice-free valleys. Alternatively, they may contain fossil moraines laid down in the lateral (approximately north-south) valleys prior to the excavation of the upper part of Taylor Valley. Examples of tributary valleys include Arena and Beacon Valleys.

The subglacial rock surface configuration is unknown for Taylor Glacier in the area where it turns northward across the western end of the Kukri Hills (Figure 141.1). However, the traces of the ice-rock contact on the north and south sides of Taylor Glacier are plotted on profile EW-2.

Profile EW-2 continues eastward from the Taylor Glacier terminus (Figure 141.5) to the shore of McMurdo Sound at New Harbor. East of New Harbor, the bathymetric profile follows, for the most part, the continuous seismic profiling track XXII of Northey and others (1975). The geophysical data collected along track XXII are plotted beneath the bathymetric profile. Interpretation of the data indicates that sedimentary layers are approximately 500 m thick in Western McMurdo Sound and more than 1 km thick off Ross Island (Northey and others, 1975). Profile EW-2 continues beyond the eastern end of track XXII to the shore of Ross Island at Cape Evans, where it terminates at a point 600 m above sea level.

Interpretation of east-west profiles

Sedimentary wedge in the eastern Taylor Valley Basin

Recent drilling in eastern Taylor Valley by the Dry Valley Drilling Project (DVDP) shows the present valley floor to be underlain by a seaward-thickening wedge of Miocene to Holocene marine and terrestrial sediments (Figure 141.5). This wedge increases in thickness from 165 m at Lake Leon (DVDP 12) to more than 328 m at the Commonwealth Glacier drill site (DVDP 11). The total thickness of the sedimentary wedge in the eastern Taylor Valley is unknown, although Chapman-Smith and Luckman (1974) have suggested a total thickness of 500 m at the New Harbor drill site (DVDP 8–10). The composition of the bedrock underlying the eastern Taylor Valley Basin is unknown except on its extreme western border, where DVDP 12 penetrated Precambrian or lower Paleozoic crystalline basement rocks (Figure 141.5). These rocks are similar to those that crop out along the valley walls.

McMurdo Volcanic Group in Taylor Valley

Plio-Pleistocene volcanic rocks crop out in central Taylor Valley (Armstrong, 1978). Most of these McMurdo Volcanic Group outcrops are located on the southern side of the valley, in the Kukri Hills. There is a noteworthy coincidence of the volcanic rocks, the region of major increase in peak elevations noted along profiles EW-1 and EW-3, and the western or inland margin of the eastern Taylor Valley Basin. Most of the volcanic rocks in the Kukri Hills are within 5 or 6 km of the peak shown in Figure 141.5 as having an elevation of 1800 m. The volcanic rocks in the Asgard Range are approximately 12 km west of the peak shown in Figure 141.5 as having an elevation of 1960 m.

Furthermore, the Taylor Valley volcanic outcrops fall on a line between the small volcanic outcrops in Wright Valley (Fleck and others, 1972) to the north and the more widespread outcrops in the Mount Dromedary–Howchin Glacier area to the south (McIvers and Gevers, 1970). These volcanic outcrops may lie along a roughly north-northwest crustal fracture.

Fault block origin of the eastern Taylor Valley Basin

The juxtaposition of (1) Plio-Pleistocene volcanic outcrops, (2) the western limit of the thick upper Miocene–Holocene sedimentary wedge in Taylor Valley, and (3) the abrupt change in peak elevations along the Kukri Hills and Asgard Range profiles strongly favors the presence of at least two fault blocks separated by a steep north-south fault zone. The fault zone is believed to cross Taylor Valley in the same area through which profile NS-5 passes (Figure 141.1). This area is bounded on the west by a line running from the peak at 1880 m in the Kukri Hills to the peak at 1760 m in the Asgard Range. The fault zone's eastern boundary is along a line connecting Howard Glacier in the Kukri Hills and Mount Falconer in the Asgard Range (Figure 141.6).

The Taylor Valley fault zone may be an extension of the Lister Fault which trends roughly north-south along the east side of the Royal Society Range (Gunn and Warren, 1962) in an area south of the major ice-free valleys. A northward extension of the proposed Taylor Valley fault zone may be related to the boundaries of the rock threshold across the entrance of Wright Valley.

The development of the eastern Taylor Valley Basin was strongly influenced by differential fault block movement prior to late Miocene time and probably much earlier. Subsequent to its formation, the basin became a locus for glaciofluvial, lacustrine, and glaciomarine sedimentation.

Conclusions

From the evidence presented above the following conclusions are drawn.

1. Valley-floor elevations in the ice-free area decrease from north to south except where fault block movement or extreme erosion have created anomalies in the regional trends.

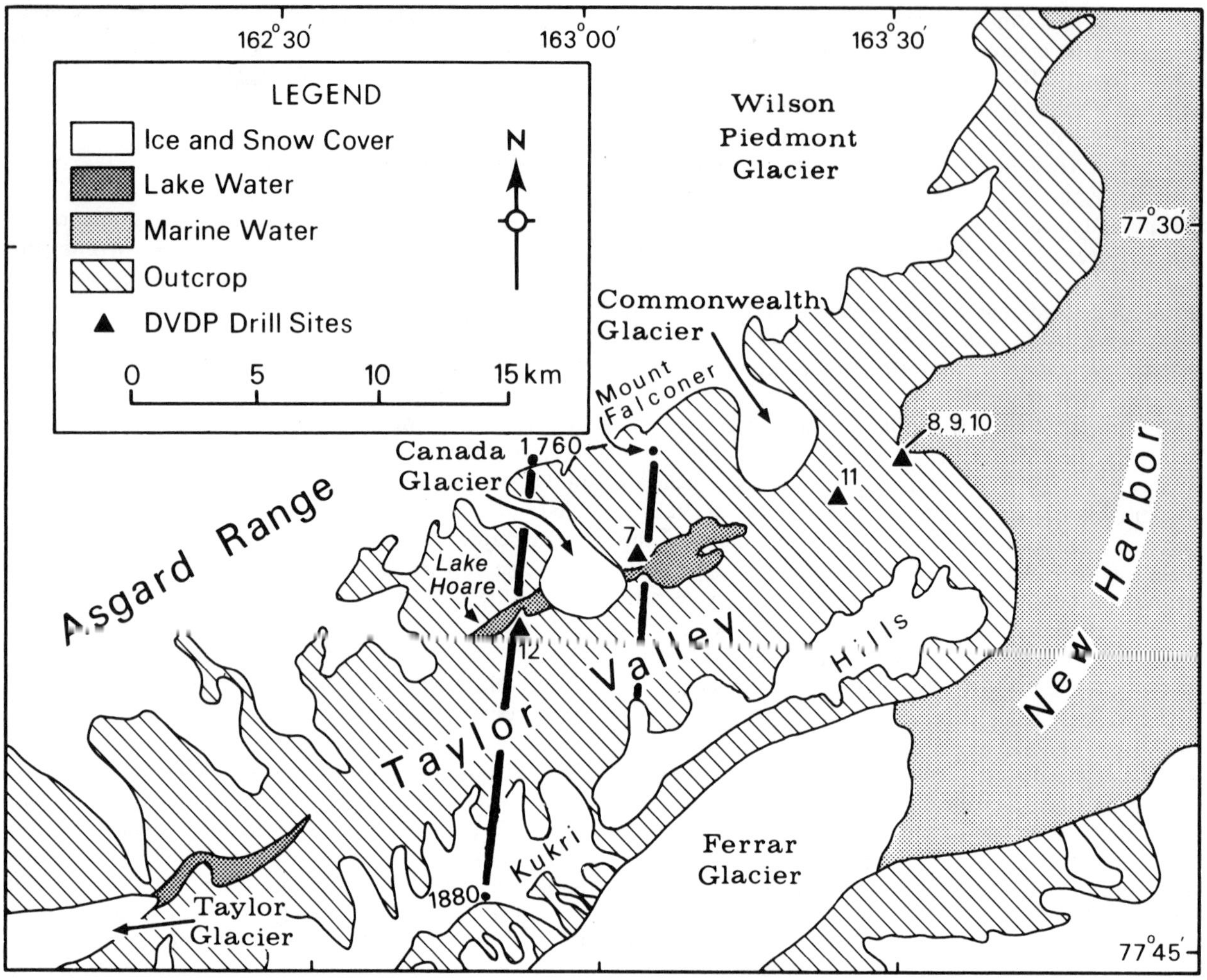

Figure 141.6. Generalized location map of eastern Taylor Valley showing the approximate boundaries of the proposed fault zone (heavy lines) trending roughly north-south across the valley.

2. This north-to-south decrease in valley-floor elevations suggests significant tilting of the Victoria-Wright-Taylor Valleys block(s).
3. Fjord emergence (marine regression) progressed from north to south across the ice-free valleys area.
4. Micropaleontologic evidence indicates that rates and periods of uplift are more complicated than the available fjord emergence sequence and physiographic data suggest. The area is probably composed of a mosaic of small fault blocks with slightly different kinematic histories.
5. The close spatial relationship between Plio-Pleistocene volcanic rocks, the western limit of the upper Cenozoic (upper Miocene–Holocene) sedimentary wedge in eastern Taylor Valley, and the abrupt change in peak elevations noted along the Kukri Hills and Asgard Range profiles is evidence for the presence of a major, steeply dipping, north-south fault zone through eastern Taylor Valley. This fault zone appears to have been active during one or more periods of time between the end of the Jurassic Period and the end of the Pleistocene Epoch.

Acknowledgments. The authors appreciate the assistance of the Cartographic Section of the School of Geosciences, Louisiana State University, and the Cartographic Laboratory at Northern Illinois University. The authors express their thanks to George F. Hart for his critical reading of the manuscript. This work was supported by National Science Foundation Grant OPP 74-22894.

References cited

Armstrong, R. L. 1978. K-Ar dating: McMurdo Volcanics and dry valley glacial history, Victoria Land, Antarctica. *N.Z. Jl Geol. Geophys., 21,* No. 6, 685–98.

Calkin, P. E. 1974. Subglacial geomorphology surrounding the ice-free valleys of southern Victoria Land, Antarctica. *J. Glaciol., 13,* No. 69, 415–29.

Chapman-Smith, M., and Luckman, P. 1974. Late Cenozoic glacial sequence cored at New Harbour, Victoria Land, Antarctica (DVDP 8 and 9). *Bull. Dry Valley Drilling Project,* No. 3, 137–48.

Crary, A. P. 1966. Mechanism for fjord formation indicated by studies of an ice-covered inlet. *Geol. Soc. Am. Bull., 77,* 911–30.

Denton, G., and Armstrong, R. L. 1968. Glacial geology and chronology of the McMurdo Sound region. *Antarct. Jnl U.S., 3,* No. 4, 99–101.

Fleck, R. J., Jones, L. M., and Behling, R. E. 1972. K-Ar dates of the McMurdo Volcanics and their relation to the glacial history of Wright Valley. *Antarct. Jnl U.S., 7,* No. 6, 245–48.

Gunn, B. M., and Warren, G. 1962. Geology of Victoria Land between the Mawson and Mulock Glaciers, Antarctica. *Bull. N.Z. Geol. Surv., 71,* 157 pp.

Hamilton, W. 1960. New interpretation of Antarctic tectonics. Short Papers in the Geological Sciences. *Prof. Pap. U.S. geol. Surv.,* No. 400-B, B379–380.

Hamilton, W. 1965. Diabase sheets of the Taylor Glacier region, Victoria Land, Antarctica. *Prof. Pap. U.S. geol. Surv.,* No. 465-B, 71 pp.

McGinnis, L. D. 1973. McMurdo Sound—a key to the Cenozoic of Antarctica. *Antarct. Jnl U.S., 7,* No. 4, 166–69.

McGinnis, L. D. 1974. Dry Valley Drilling Project Summaries. An overview of DVDP, 1973–74. *Bull. Dry Valley Drilling Project,* No. 3, 1–6.

McGinnis, L. D., 1975. Dry Valley Drilling Project Summaries. An overview of DVDP, 1974–75. *Bull. Dry Valley Drilling Project,* No. 5, 1–4.

McIvers, J. R., and Gevers, T. W. 1970. Volcanic vents below the Royal Society Range, central Victoria Land, Antarctica. *Trans. Geol. Soc. S. Afr., 73,* 65–88.

Northey, D. J., Brown, C., Christoffel, D. A., Wong, H. K., and Barrett, P. J. 1975. A continuous seismic profiling survey in McMurdo Sound, Antarctica—1975. *Bull. Dry Valley Drilling Project,* No. 5, 167–78.

Robin, G. de Q., Swithinbank, C. W. M., and Smith, B. M. E. 1970. Radio echo exploration of the Antarctic ice sheet. (*In* Gow, A. J., Keeler, C., Langway, C. C., and Weeks, W. F., *eds. International Symposium on Antarctic Glaciological Exploration (ISAGE), Gentbrugge, 1968,* 97–115.) [International Association of Scientific Hydrology Publication, No. 86.]

Webb, P. N. 1972. Wright Fjord, Pliocene marine invasion of an Antarctic dry valley. *Antarct. Jnl U.S., 7,* No. 6, 227–34.

Webb, P. N., and Wrenn, J. H. 1976*a*. Interpretation of foraminiferal assemblages from lower Taylor Valley (DVDP 8–12) (abstract). *Bull. Dry Valley Drilling Project,* No. 7, 105–7.

Webb, P. N., and Wrenn, J. H. 1976*b*. Foraminifera from DVDP holes 8–12, Taylor Valley. *Antarct. Jnl U.S., 11,* No. 2, 85–86.

Wilson, A. 1973. The great antiquity of some Antarctic landforms—evidence for an Eocene temperate glaciation in the McMurdo region. (*In* Van Zinderen Bakker, E. M., *ed. Palaeoecology of Africa and of the Surrounding Islands and Antarctica. 8.* Cape Town, South Africa, A. A. Balkema, 23–35.)

Wrenn, J. H. 1977. *Cenozoic Subsurface Micropaleontology and Geology of Eastern Taylor Valley, Antarctica.* Masters thesis, Northern Illinois University, 255 pp.

142

Weathering and Mineral Synthesis in Antarctic Soils

F. C. UGOLINI
College of Forest Resources, University of Washington, Seattle, Washingon 98195, U.S.A.

M. L. JACKSON
Department of Soil Science, University of Wisconsin, Madison, Wisconsin 53706, U.S.A.

Abstract. Studies of the mineralogy of three soils of different ages in the ice-free areas in the Wright Valley of southern Victoria Land, Antarctica, have shown that alteration of minerals is presently occurring, as reported earlier for the Mirny oasis, rather than only the physical weathering often emphasized in studies of the polar regions. Soils developed on glacial till show increases in clay plus silt with age. The oxalate/dithionite extractable iron ratio and the silica/sesquioxide molar ratio indicate more chemical weathering with time. A decrease in feldspar with decreasing particle size and the appearance of montmorillonite and mica-vermiculite intergrades in finer particle size fractions occur in the youngest soil. In the soil of intermediate age, weathering of feldspar through mica and mica-vermiculite intergrade to vermiculite-montmorillonite intergrade and chloritized montmorillonite to chloritized montmorillonite has occurred. In the oldest soil, weathering of feldspar through mica and montmorillonite to chloritized montmorillonite has occurred. The best documentation for neoformation of minerals in Antarctica was obtained from mica in the oldest soil, where an average fission-track age for 11 authigenic micas (sericite-like weathered replacement of plagioclase) was 4.1 ± 0.2 m.y., in contrast to an age of 151 m.y. for the parent-rock mica (biotite). Alteration of the soil mica since then has mainly been limited to some chloritization in the surface horizon. Weathering of the feldspar to mica during a favorable climatic episode 4 m.y. ago is thus indicated. The presence of unfrozen water at mineral surfaces inevitably allows hydrolytic weathering even under the present xeric and frigid conditions. More intense weathering, however, is connected with amelioration of the climate and abundance of melt water during major interglacial episodes.

Weathering in continental Antarctica is limited by low temperatures and the scarcity of liquid water. Under these desert conditions, biological activity is decreased so that weathering occurs with little organic participation. Traditionally in studies of the polar regions, emphasis has been placed on the prevalence of physical rather than chemical weathering, and in Antarctica some results indicate that chemical weathering is of little significance (Kelly and Zumberge, 1961). Other studies in the Mirny oasis and in the ice-free areas of southern Victoria Land provide evidence of mineral alteration in the past and present (Glazovskaia, 1958; Ugolini, 1964; Claridge, 1965; Everett and Behling, 1970; Linkletter, 1971; Behling, 1971). Additional support for chemical weathering has been provided by the similarity in $^{87}Sr/^{86}Sr$ ratio between the water soluble salts in soils and the parent material (Jones and Faure, 1969), by an increase in silt and clay percentages from younger to older soils (Ugolini and Bull, 1965; Linkletter, 1971; Behling, 1971; Ugolini and others, 1973), and by direct evidence of ionic migration and a continuous unfrozen film of water which could support chemical weathering at the surface of the soil particles in frozen Antarctic soils (Ugolini and Grier, 1969; Ugolini and Anderson, 1973). The results in this report indicate that chemical weathering is occurring and that there is a weathering-time relationship. To assess the two possible mechanisms for the formation of clays—(1) synthesis of new minerals and (2) comminution of the existing minerals in the soil—the present study was initiated on a chronologic sequence of a young, an intermediate, and an old soil from Wright Valley, southern Victoria Land.

Materials

Sampling was done in the lower Wright Valley on the moraine complex facing the Wright Lower Glacier and on the north walls of the Wright Valley (Figure 142.1).

Profile 5

The youngest profile (5) is located about 1 km from Wright Lower Glacier, a distributary lobe of the Wilson

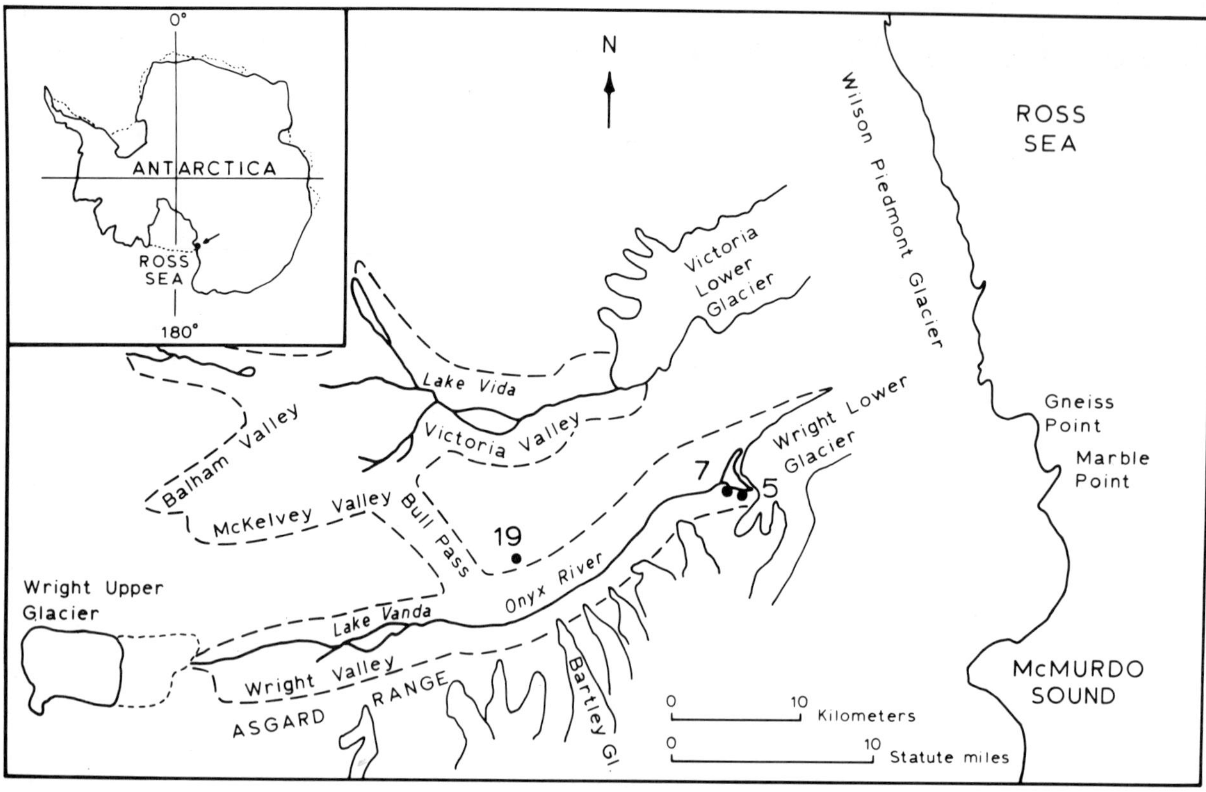

Figure 142.1. Wright Valley region, southern Victoria Land, Antarctica. Numbers indicate locations of the soil profiles. Base map after Jackson and others (1977).

Piedmont Glacier, and is described (with Munsel Soil Color Notation) as follows:

Surface: unvegetated, with angular boulders and desert stone pavement (measured from zero reference at the soil surface). Well-developed, unsorted polygons. Samples were obtained from the center of a polygon. The soil surface temperature fluctuated between −6°C and −16°C, respectively, measured at 0700 and 1300, 21 October 1971.

1–0 cm: desert stone pavement consisting of angular fragments with some free salts on the lower surfaces.

0–33 cm (measured downward from zero at soil surface level): light brown-gray (2.5Y 6/2, dry), gravelly, coarse- and very coarse-grained sand; structureless, single grain, loose, except in areas at irregular depths where the soil is partially cemented, weakly indurated, and dry. Lower boundary is abrupt and smooth.

33+ cm: ice-cemented frost table on 28 October 1971. The temperature of the surface of the ice-cemented permafrost was −20°C. (The depth to the ice-cemented permafrost was 39 cm on 2 November 1969.)

Profile 5 is located on ice-cored moraines that perhaps represents the most recent advance and recession of the Wright Lower Glacier. The Wright Lower Glaciation has been placed within a minimum age of 6000 to 9500 years and a maximum age of 34,000 years or older (Calkin and others, 1970).

Profile 7

This profile (7) is located about 2 km from the Wright Lower Glacier on the western end of the ice-cored moraine. Although no absolute age for the glacial deposits is available, a greater age than profile 5 is suggested by cavernously weathered boulders and numerous ventifacts in the desert pavement. Profile 7 is described as follows:

Surface: unvegetated, covered by cavernous weathered boulders and ventifacts; a few boulders weathered and eroded down to ground level. The soil surface temperatures oscillated between −5°C and −14°C, respectively, measured at 0800 and 1400, 29 October 1971.

1–0 cm: desert pavement containing ventifacts and salt accumulation under the rock fragments.

0–6 cm (measured downward from zero at soil surface level): gray brown (2.5Y 5/2 dry); fine sandy loam; weak coarse crumb, loose to soft; abrupt and wavy lower boundary.

6–18 cm: variegated; sandy loam, structureless, single grain; loose except for areas where firm because frozen; abrupt and wavy lower boundary.

18–44 cm: gray-brown (2.5Y 5/2 dry); gravelly loamy sand to sand, structureless, single grain to massive; loose to soft; abrupt and wavy boundary.

44–49 cm: variegated, gravelly sand; structureless, single grain; loose except where ice-cemented; abrupt and smooth lower boundary.

49+ cm: ice-cemented frost table on 29 October 1971; its temperature was constant at −20°C. The depth to the ice-cemented permafrost varied locally to 54 cm on 1 November 1969.

Profile 19

This soil is apparently derived from the weathering of a porphyroblastic biotite-hornblende granosyenite (Lopatin, 1975), the biotite of which was dated as 151 m.y. old (Jackson and others, 1977). It is situated on a glacial bench carved by the Vanda glacial episode (Nichols, 1971) on the north side wall of the Wright Valley at an elevation of about 965 m. Its description is as follows:

Surface: unvegetated, covered by desert pavement. Deeply cavernous weathered bedrock projections surround the site.

3–0 cm: desert pavement consisting of angular fragments with a thick coating of salts on the undersurfaces.

0–10 cm: strong brown (7.5YR 5/6 dry), very gravelly sand; structureless, single grain; loose; clear and smooth lower boundary.

10–30 cm: red yellow (7.5YR 6/6 dry); very gravelly loamy sand; structureless, single grain; loose to weakly coherent; clear and smooth lower boundary. On the larger fragments, a film of claylike material is present.

30–50 cm: red yellow (7.5YR 6/6 dry) very gravelly loamy sand, structureless, single grain; loose to soft; clear and smooth lower boundary.

50+ cm: Shattered bedrock consisting of angular fragments.

Fleck and others (1972), using K-Ar dates, have suggested that the major valley glaciation, the Vanda stage, occurred as much as 4.2 ± 0.2 m.y. ago. Other K-Ar dates obtained from unglaciated basaltic volcanic cones near Bartley Glacier (Figure 142.1) give a minimum age of 3.7 ± 0.2 m.y. for the Vanda stage (Denton and others, 1970; Nichols, 1971; R. L. Armstrong, personal communication, 1975). Fission tracks in the soil mica particles from three soil horizons of profile 19 give an age of 4.1 ± 0.2 m.y. (Jackson and others, 1977). Since no younger glaciations have reoccupied the glacial bench, weathering and soil formation have been in progress at the site for 4.2 m.y.

Methods

Unless otherwise stated, all the analyses were performed on the less-than-2-mm air-dry soil fraction; all the data are reported on an oven-dry weight basis. Particle-size separation was performed as described by Jackson (1975) after removal of the salts with NaOAc and iron with citrate-bicarbonate-dithionite (Jackson, 1975). Soluble salts were obtained from the saturation extract, and elemental analysis was by Na_2CO_3 fusion (Jackson, 1971). Silica was determined gravimetrically, aluminum by atomic absorption spectrophotometry, and iron and titanium colorimetrically. More active iron oxides were extracted with citrate-bicarbonate-dithionite (Jackson, 1975) and with acid ammonium oxalate (McKeague and Day, 1966). The amount of extracted iron was determined by atomic absorption spectrophotometry.

Very fine grained sand (ground to less than 300 mesh), coarse- and fine-grained silt, and coarse- and fine-grained clay were saturated with Mg^{2+}, treated with ethylene glycol, saturated with K^+, and heated at 350°C and 500°C. The samples were X-rayed on a Picker–Nuclear diffraction unit.

Results and discussion

Particle size

The particle-size analysis of the three profiles (Table 142.1) emphasizes the effect of the parent material on the mechanical constituents of the soil. Because of gains and losses through transfers, deviations of ± 1% to 3% from 100% total of the size fractions was observed. The old soil (19) derived from granitic bedrock has somewhat less clay and silt than the intermediate-age soil (7) formed on moraines where the glacier had provided the initial grinding of the minerals. The young soil (5), however, despite derivation from glacial deposits, has the lowest clay and silt content. In terms of clay distribution, the intermediate and the young soils show the largest accumulation in the upper 20 cm; the old soil, between 10 and 50 cm. For the young and intermediate soils, these results are attributed to increased weathering at the surface, where fluctuations of temperature and moisture are greater (Ugolini, 1963; Behling, 1971). Similar textural trends have been reported by Behling (1971) in the soils of the Meserve Glacier, Wright Valley. For the old soil, the formation of clay may have taken place at the surface, but its transport to greater depths may have occurred during periods of increased water movement. Near Molodezhnaya station on the west coast of Antarctica, where conditions are milder and wetter, MacNamara (1969) has reported such clay migration, which led there to the development of a recognizable B horizon. The other possibility is that the weathering front in the old soil has progressively moved downward with time. This process has been observed by Behling (1971) for soils of different ages in front of the Meserve Glacier.

Soluble salts

Electrical conductivity measurements in the laboratory show a higher soluble salt content within the older soil (Table 142.2). If it is assumed that most of the soluble salts (chlorides and sulfates) have been derived from precipitation and that leaching has been virtually unimportant, then the salt content may be a time-dependent function (Ugolini, 1963; Ugolini and Bull, 1965; Behling, 1971).

Extractable iron

The iron liberated through processes of mineral weathering and soil formation tends to be in amorphous oxide form during the early stages of soil development; with time it becomes incorporated within crystalline materials. Forms of extractable iron have been used for differentiating soil horizons (McKeague and Day, 1966), for assessing degree of soil development (Blume and Schwertmann, 1969; Soil Survey Staff, 1975), and for correlating Quarternary deposits (Alexander, 1974). Citrate-bicarbonate-dithionite extraction usually removes all forms of secondary iron: organically complexed, amorphous inorganic, and crystalline. The acid-ammonium oxalate extracts the amorphous inorganic iron and the iron organically complexed (McKeague and others, 1971). The ratio of oxalate to dithionite extracted iron, Fe_o/Fe_d, should decrease with time. The ratio should be less than 1 unless some ferruginous silicates are decomposed by the acid oxalate, which does occur in several horizons (Table 142.2). The data in Table 142.2 portray a progressive deepening of the weathering release of iron from the young (5) to the old (19) profiles. In the young profile (5) both Fe_d and Fe_o are mostly concentrated in the first 20 cm; in the intermediate-age profile (7) they are distributed through the entire depth of 50 cm; and in the old profile (19) the weathering front not only has progressed to the bedrock, but the amount of released iron (Fe_d) exceeds that in all of the other profiles.

Table 142.1. Particle size distribution data (in %) for profiles 5, 7, and 19, Wright Valley, Antarctica

Depth (cm)	Sand: Very coarse 2–1 (mm)	Sand: Coarse 1–0.5 (mm)	Sand: Medium 0.5–0.25 (mm)	Sand: Fine 0.25–0.10 (mm)	Sand: Very fine 0.10–0.05 (mm)	Sand: Total	Silt: Coarse 50–20 (μm)	Silt: Medium 20–5 (μm)	Silt: Fine 5–2 (μm)	Silt: Total	Clay: Coarse 2–0.2 (μm)	Clay: Fine <0.2 (μm)	Clay: Total
Profile 5													
0–10	19.2	33.5	5.7	30.8	2.9	92.1	2.7	1.9	0.7	5.3	1.3	0.9	2.2
10–20	16.9	32.2	8.0	30.4	3.4	90.9	3.3	1.9	1.4	6.6	1.9	0.8	2.7
20–30	7.2	59.3	3.5	24.6	0.8	95.4	0.8	0.3	0.2	1.3	0.2	0.2	0.4
30–35	19.3	51.1	10.8	11.7	0.8	93.7	0.5	0.1	0.1	0.7	0.1	0.1	0.2
Profile 7													
0–10	18.6	20.3	12.0	17.7	4.6	73.2	4.8	10.8	4.8	20.4	5.0	1.5	6.5
10–20	16.5	19.9	11.5	20.1	5.4	73.4	6.7	6.3	3.7	16.7	6.1	1.4	7.5
20–30	21.4	21.5	11.9	21.2	8.0	84.0	5.0	3.8	9.4	18.2	0.8	0.6	1.4
30–40	26.3	28.7	14.1	18.3	4.6	92.0	2.6	1.2	0.6	4.4	1.4	0.2	1.6
40–50	18.6	25.1	13.6	21.6	7.5	86.37	6.1	2.6	1.3	10.0	2.5	0.3	2.8
Profile 19													
0–10	20.3	21.8	22.1	27.7	2.8	94.7	0.8	0.2	2.7	3.7	0.5	1.0	1.5
10–30	26.2	19.8	11.1	19.0	7.0	82.1	3.6	0.7	9.0	13.3	2.4	1.3	3.7
30–50	37.3	19.3	6.8	11.4	4.2	79.0	7.2	0.5	9.7	17.4	2.0	1.6	3.6

Table 142.2. Selected chemical parameters for profiles 5, 7, and 19, Wright Valley, Antarctica (oxides in %)

Depth (cm)	Electrical conductivity (mmhos/cm)	Total oxides: SiO_2	Total oxides: Al_2O_3	Total oxides: Fe_2O_3	Total oxides: TiO_2	Total oxides: $\frac{SiO_2}{R_2O_3}$	Free oxides: Fe_2O_{3d} (dithion)	Free oxides: Fe_2O_{3o} (oxalate)	Free oxides: $\frac{Fe_2O_{3o}}{Fe_2O_{3d}}$
Profile 5									
0–10	0.33	65.1	14.2	3.7	0.4	7.0	0.3	0.44	1.4
10–20	1.50	66.3	13.1	4.3	0.3	6.8	0.37	0.50	1.3
20–30	0.02	67.6	13.9	3.1	0.2	7.6	0.14	0.06	0.4
30–35	0.01	69.3	13.5	2.8	0.2	7.3	0.12	0.05	0.4
Profile 7									
0–10	3.85	60.8	19.1	4.2	0.8	4.9	0.40	0.30	0.75
10–20	1.70	62.6	14.7	4.3	0.6	5.1	0.40	0.15	0.37
20–30	1.10	63.7	14.8	4.0	0.5	6.2	0.36	0.35	0.97
30–40	0.70	63.6	18.3	4.6	0.5	6.1	0.24	0.40	1.6
40–50	0.70	62.3	19.0	4.9	0.5	4.8	0.30	0.50	1.6
Profile 19									
0–10	22.2	39.5	10.5	6.0	—	5.0	0.70	0.20	0.28
10–30	12.2	39.0	11.0	6.2	—	5.0	2.0	0.20	0.10
30–50	7.5	35.0	11.0	6.1	—	4.5	2.0	0.30	0.15
Bedrock	—	65.0	15.1	6.7	—	6.8	—	—	—

Furthermore, in this profile the Fe_o/Fe_d ratios acquire the lowest values, indicating considerable recrystallization of the liberated iron and thus indirectly indicating the great antiquity of the soil. Loss of silica and accumulation of iron and aluminum traditionally have been evaluated by the $SiO_2/(Al_2O_3 + Fe_2O_3)$ or SiO_2/R_2O_3 molar ratios. These ratios calculated for the three soils (Table 142.2) show that the young soil (5) has, for any depth, a larger ratio than the intermediate (7) and the old (19) soils. Ratios in the old soil show an appreciable decrease compared to the bedrock; loss of silica and retention of the Fe and Al must have occurred as the granite weathered into soil. In terms of absolute molar ratios of SiO_2/R_2O_3, there is a definite trend from young to old soil, revealing that this ratio is time-dependent.

Soil mineralogy

The mineralogy of the young soil profile (5) displays a decrease in feldspar content with decreasing particle size (Table 142.3). Feldspar is the major mineral component of the very fine-grained sand, coarse-grained silt, and partially of the fine-grained silt. Montmorillonite appears as a major mineral in the coarse-grained clay at the top 20 cm only, but it becomes a major mineral all through the profile in the fine-grained clay. Traces of chloritized montmorillonite are present in the coarse-grained clay. The presence of vermiculite in the very fine-grained sand suggests that it has weathered from micas. Mica becomes prominent in the coarse- and fine-grained clay where some of it has expanded to form mica-vermiculite intergrades. Chlorite is a trace component in the coarse-grained clay. The presence of montmorillonite, vermiculite, chloritized-montmorillonite, mica-vermiculite intergrades, and vermiculite-chlorite intergrades in the finer fractions is considered to reflect chemical weathering. Because of the paucity of the sediments in the 5–20 μm size fraction was not sufficient for X-ray analysis, data for it could not be included in Table 142.1.

In the intermediate-age soil profile, feldspar is a major mineralogic constituent in the very fine-grained sand and coarse- and fine-grained silt; it persists as a minor component of the coarse- and fine-grained clay (Table 142.4). Vermiculite and mica are prominent in all fractions; mica-vermiculite intergrades appear, however, in the coarse- and fine-grained clay and fine-grained silt. Montmorillonite-vermiculite intergrades become major components in the coarse-grained clay; montmorillonite is a trace in the fine-grained silt. The fine-grained clay is dominated by mica, mica-vermiculite intergrades, and vermiculite. Minerals in this profile considered authigenic are montmorillonite, vermiculite-chlorite intergrades, mica-vermiculite intergrades, montmorillonite-vermiculite intergrades, and montmorillonite-chlorite intergrades.

Although not fully documented, the weathering sequence in the young and intermediate-age profiles seems to indicate that the feldspar is weathering to mica, the mica to vermiculite, and the vermiculite to montmorillonite. Chloritization of montmorillonite and vermiculite occurs in the coarse- and fine-grained clay and some in the fine-grained silt of the intermediate profile. In the young profile, this alteration is only recorded at the surface in the coarse-grained clay.

The best documentation of *in situ* clay synthesis was found in the old soil (19) (Jackson and others, 1977). For that soil, thin-section study of samples shows that plagioclase feldspar in the granitic rock has weathered into a mica. X-ray diffraction for the very coarse-grained sand, medium- and coarse-grained silt, coarse- and fine-grained clay show a progressive decline in intensity of the feldspar peak with decreasing particle size. Concomitant with the decrease of feldspar, the mica peak becomes increasingly prominent. Clays at all depths have virtually no feldspar and are dominated by mica.

Discussion

Claridge (1965) has presented evidence for clay formation in Antarctica at the present time and has suggested that the process is comparable to that in other arid environments of the world, i.e., the hydration of primary mica into the expandable mineral vermiculite. Montmorillonite was considered to be an authigenic mineral, derived from the weathering of mafic rocks in the presence of high pH and in the absence of leaching. Behling (1971) also has concluded that authigenic clay minerals are forming today in the soils of the Meserve Glacial till. There mica has weathered to vermiculite, biotite and hornblende to chlorite, and feldspar to amorphous aluminosilicate. Vermiculite, chlorite, and pedogenic 2:1–2:2 intergrades are the principal authigenic minerals formed. Linkletter (1971) has reported more rapid weathering of the mafic minerals than of the felsic components and formation of illite and montmorillonite through chemical weathering.

The findings reported here are in general agreement with those of the cited investigators, but we note in addition the weathering of plagioclase into authigenic soil mica, particularly in the old profile (19). Weathering in crystalline rock occurred following a major glacial episode 4 m.y. ago (as documented above) when conditions may have been wetter and warmer. Since soluble potassium is needed for the formation of mica from plagioclase, a relatively closed system must have existed wherein K was preserved from biotite and K-feldspar. Evidence for a warmer episode 4 m.y. ago is also found in the paleontologic record of a deep-sea core obtained off the coast of Antarctica (Kennett, personal communication, 1976).

Unfortunately, a less firm chronology is available for the intermediate-age and young soils developed in till. Lack of fission-track ages for the mica prevents establishment of mica age and determination of whether the mica is detrital or authigenic. However, the mica seems to be weathering into mica-vermiculite intergrade minerals, and vermiculite seems to be further expanding into montmorillonite as a result of chemical weathering. Some montmorillonite could also have originated through weathering of mafic minerals. If the age of the ice-cored moraines in the lower Wright Valley is on the order of 10,000 years, a maximum period for such weathering is indicated. Chloritization of montmorillonite seems to be more prevalent in the intermediate-age and old soils, implying that a longer time is needed; however, formation of chlorite-montmorillonite intergrades has been observed in other parts of southern Victoria Land, in soils estimated not to be older than 8000 or 10,000 years (Ugolini, 1977). Local factors may, therefore, overshadow a simple time-weathering relationship.

Table 142.3. Mineralogy of different size-fractions by X-ray diffraction analysis, profile 5, Wright Valley, Antarctica

Depth (cm)	Fine clay (<0.2μm)			Coarse clay (2–0.2 μm)			Fine silt (5–2μm)			Coarse silt (20–50μm)			Very fine sand (50–100μm)		
	Major	Minor	Trace	Major	Minor	Trace	Major	Minor	Trace	Major	Minor	Trace	Major	Minor	Trace
0–10	Mont Verm Mica	M-V-I Amph Qtz	Amph Feld Qtz	Mont Verm Mica	M-V-I Amph Feld Qtz	Mont-C-I V-C-I Chlor	Verm Feld Qtz	Mica M-V-I Chlor Amph		Feld Qtz	Mica Verm		Feld Qtz		Chlor Verm Mica Amph
10–20	Mont Verm Mica M-V-I	Amph Qtz	Feld	Mont Verm Mica Feld Qtz	M-V-I	Mont-C-I? Chlor V-C-I	Verm	Chlor Mica M-V-I Mont-C-I Feld Qtz		Verm Feld Qtz	Mica Amph		Verm Feld Qtz	Mica Amph	Chlor
20–30	Mont Verm Mica M-V-I	Amph Feld Qtz		Mica Feld Qtz	Mont Verm M-V-I Chlor M-C-I	V-C-I	Verm Feld Qtz	Mica Chlor Amph		Verm Amph Feld Qtz		Mica	Feld Qtz	Verm Mica Amph	Chlor
30–35	Mont Verm	Mica M-V-I Amph Feld Qtz		Feld Qtz	Mont Verm M-V-I Mica Chlor Amph		Feld Qtz	Verm Amph	Chlor	Verm Feld Qtz	Mica Amph		Feld Qtz	Verm Mica Amph	Chlor

Mont: montmorillonite.
Verm: vermiculite.
M-V-I: mica-vermiculite-intergrades.
Mont-C-I: montmorillonite-chlorite-intergrades.
M-C-I: mica-chlorite-intergrades.
Chlor: chlorite.
Amph: amphibole.
Feld: feldspar.
Qtz: quartz.

Table 142.4. Mineralogy of different size-fractions by X-ray diffraction analysis, profile 7, Wright Valley, Antarctica

Depth (cm)	Fine clay (<0.2μm)			Coarse clay (2–0.2 μm)			Fine silt (5–2μm)			Coarse silt (20–50μm)			Very fine sand (50–100μm)		
	Major	Minor	Trace	Major	Minor	Trace	Major	Minor	Trace	Major	Minor	Trace	Major	Minor	Trace
0–10	Mica Verm	M-V-I Feld Qtz	Mont Chlor V-C-I	Mica Qtz	Verm Mont-V-I M-V-I Feld	Mont-C-I M-C-I Amph	Verm Mica Feld Qtz	Chlor Amph	Mont Mont-C-I	Mont Chlor Amph Feld Qtz	Mica		Verm Mica Amph Feld Qtz		Chlor
10–20	Mica Verm M-V-I	Mont Feld Qtz	Chlor V-C-I Amph	Mont-V-I Verm	Mica M-V-I Feld Qtz	Mont-C-I Chlor	Verm Mica Feld Qtz	Chlor M-V-I Amph	Mont Mont-C-I	Amph Feld Qtz	Mica Chlor		Verm Feld Qtz Amph		Chlor
20–30	Mica Verm M-V-I	Mont Feld Qtz	Chlor V-C-I Amph	Mont-V-I Verm Mica Qtz	Mont M-V-I Feld Amph	Mont-C-I Chlor	Verm Mica Feld Qtz	Chlor M-V-I Amph	Mont	Chlor Feld Qtz	Mica Amph		Verm Feld Qtz Amph		Chlor
30–40	Verm	Mica M-V-I Feld Qtz	Chlor Mont V-C-I Amph	Mont-V-I Verm Feld Qtz	Mica Amph	V-C-I Chlor	Verm Mica M-V-I Feld Qtz	Chlor Amph	Mont	Feld Qtz	Chlor Amph		Verm Feld Qtz Amph		Chlor
40–50	Verm M-V-I	Mont Mica Feld Qtz Amph	Chlor Mont-C-I	Mont Verm	Mica M-V-I Mont Feld Qtz Amph	M-C-I	Verm Feld Qtz	Chlor Mica M-V-I Amph	Mont	Chlor Feld Qtz	Mica Amph		Verm Feld Qtz		Chlor

Mont: montmorillonite.
Verm: vermiculite.
Feld: feldspar.
Qtz: quartz.
Chlor: chlorite.
Amph: amphibole.
M-V-I: mica-vermiculite-intergrades.
M-C-I: mica-chlorite-intergrades.
Mont-C-I: montmorillonite-chlorite-intergrades.
Mont-V-I: montmorillonite-vermiculite-intergrades.
V-C-I: vermiculite-chlorite-intergrades.

Conclusions

Because of the presence of unfrozen water on the soil mineral surfaces, the hydrolytic release of ions and chemical weathering attack on minerals is inevitable, even under the present xeric and frigid Antarctic conditions.

Extensive weathering occurs in the ice-free areas of southern Victoria Land during major interglacial episodes, especially those connected with amelioration of the climate and greater abundance of melt water from the ice masses. Hendy and others (1979) have suggested that glaciations in Antarctica are triggered by worldwide warm periods when more water would be available in high latitudes.

Weathering in Antarctica includes the mineralogic progression of feldspar → mica → expansible phyllosilicates → chloritic intergrades, the latter caused by release and then hydrolysis of trivalent ions.

Acknowledgments. We acknowledge the help of K. Trafton for the soil analyses and the U.S. Navy Task Force 43 for logistic support. This research was supported by National Science Foundation Grants G23787, OPP74-20701, and OPP76-23041 to Ugolini and EAR-76-19783 to Jackson. The second author was Distinguished Visiting Professor, second quarter, 1973, at the College of Forest Resources and the Quaternary Research Center, University of Washington, Seattle. This research was also supported in part by the College of Agricultural and Life Sciences, University of Wisconsin, Madison.

References cited

Alexander, E. B. 1974. Extractable iron in relation to soil age on terraces along the Truckee River, Nevada. *Proc. Soil Sci. Soc. Am., 38,* 121–24.

Behling, R. E. 1971. *Pedologic Development on Moraines of Meserve Glacier, Antarctica.* Ph.D. dissertation, Ohio State University, 216 pp.

Blume, H. P., and Schwertmann, U. 1969. Genetic evaluation of profile distribution of aluminum, iron, and manganese oxides. *Proc. Soil Sci. Soc. Am., 33,* 438–44.

Calkin, P. E., Behling, R. E., and Bull, C. 1970. Glacial history of Wright Valley, southern Victoria Land, Antarctica. *Antarct. Jnl U.S., 5,* 22–27.

Claridge, G. G. C. 1965. The clay mineralogy and chemistry of some soils from the Ross Dependency, Antarctica. *N.Z. Jl Geol. Geophys., 8,* 186–220.

Denton, G. H., Armstrong, R. L., and Stuiver, M. 1970. Late Cenozoic glaciation in Antarctica: the record in the McMurdo Sound region. *Antarct. Jnl U.S., 5,* No. 1, 15–21.

Everett, K. R., and Behling, R. E. 1970. Chemical and physical characteristics of Meserve Glacier morainal soils, Wright Valley, Antarctica: an index of relative age? (Abs.) (*In* Gow, A. J., Keeler, C., Langway, C. C., and Weeks, W. F., *eds. International Symposium on Antarctic Glaciological Exploration,* Hanover, N.H., 1968, 459–60.) [International Association of Scientific Hydrology, Publ. No. 86.]

Fleck, R. J., Jones, L. M., and Behling, R. E. 1972. K-Ar dates of the McMurdo volcanics and their relation to the glacial history of Wright Valley. *Antarct. Jnl U.S., 7,* 245–46.

Glazovskaia, M. A. 1958. Vyvetrivanie i pervichnoe pochvoobrazovanie v Antarktike [Weathering and primary soil formation in Antarctica]. *Nauk Dok. Vyss. Sk., Geol. Geogr. Nauk* (Moskva), *1,* 63–76.

Hendy, C. H., Healy, T. R., Rayner, E. M., Shaw, J., and Wilson, A. T. 1979. Late Pleistocene glacial chronology of the Taylor Valley, Antarctica, and the global climate. *Quat. Res., 11,* 172–84.

Jackson, M. L. 1975. *Soil Chemical Analysis—Advanced Course.* 2nd ed. Madison, WI., published by the author, 895 pp.

Jackson, M. L. 1971. *Soil Chemical Analysis.* Madison, WI., published by the author, 494 pp.

Jackson, M. L., Lee, S. Y., Ugolini, F. C., and Helmke, P. A. 1977. Age and uranium content of soil micas from Antarctica by the fission particle track replica method. *Soil. Sci., 123,* 241–48.

Jones, L. M., and Faure, G. 1969. The isotope composition of strontium and cation concentration of Lake Vanda and Lake Bonney in southern Victoria Land, Antarctica. *Rep. Lab. Isotope Geol. Geochem., Inst. polar Stud. Ohio State Univ.,* No. 4, 32 pp.

Kelly, W. C., and Zumberge, J. H. 1961. Weathering of a quartz diorite at Marble Point, McMurdo Sound, Antarctica. *J. Geol., 69,* 433–46.

Linkletter, G. O. 1971. *Weathering and Soil Formation in Antarctic Dry Valleys.* Ph.D. dissertation, University of Washington, 122 pp.

Lopatin, B. C., 1975. The geology of the folded basement in the McMurdo oasis (Victoria Land). (*In* Bugaev, V. A., *ed.* The Antarctic Committee Reports 1969. (Antarktika, Doklady Komissi, 1969.) "Nauka" Publishers, 1971. Translated from Russian. Published for the Office of Polar Programs, NSF, Washington, D.C. by Amerind Publishing Co. Pvt. Ltd., New Delhi, 1975, pp. 75–93. (TT 7252012.)

MacNamara, E. E. 1969. Biological research opportunities at the Soviet Antarctic station Molodezhnaya. *Antarct. Jnl U.S., 4,* 8–12.

McKeague, J. A., and Day, J. H. 1966. Dithionite- and oxalate-extractable Fe and Al as aids in differentiating various classes of soils. *Can. Jnl Soil Scis, 46,* 13–22.

McKeague, J. A., Brydon, J. E., and Miles, N. M. 1971. Differentiation of forms of extractable iron and aluminum in soils. *Proc. Soil Sci. Soc. Am., 35,* 33–38.

Nichols, R. L. 1971. Glacial geology of Wright Valley, McMurdo Sound. (*In* Quam, L. O., *ed.* Research in the Antarctic. *Am. Ass. Adv. Sci.,* Publ. No. 93, 293–340.)

Soil Survey Staff. 1975. *Soil Taxonomy. A Basic System of Soil Classification for Making and Interpreting Soil Surveys.* Washington, Government Printing Office, 754 pp. [Agriculture Handbook 436.]

Ugolini, F. C. 1963. Soil investigation in the lower Wright Valley, Antarctica. (*In Proceedings of an International Conference on Permafrost, 1963.* Washington, National Academy of Sciences/ National Research Council, 55–61.) [Publication 1287.]

Ugolini, F. C. 1964. *A Study of Pedologic Processes in Antarctica.* Final report to National Science Foundation. New Brunswick, N.J., Rutgers University, 82 pp.

Ugolini, F. C. 1977. The Protoranker soils and the evolution of an ecosystem at Kar Plateau, Antarctica. (*In* Llano, G. A., *ed. Adaptation within Antarctic Ecosystems: Proc. 3rd S.C.A.R. Symposium on Antarctic Biology,* Part 7, 1091–110.)

Ugolini, F. C., and Anderson, D. M. 1973. Ionic migration and weathering in frozen Antarctic soils. *Soil Sci., 115,* 461–70.

Ugolini, F. C., and Bull, C. 1965. Soil development and glacial events in Antarctica. *Quaternaria, 7,* 251–69.

Ugolini, F. C., and Grier, C. C. 1969. Biological weathering in Antarctica. *Antarct. Jnl U.S., 4,* 110–11.

Ugolini, F. C., Bockheim, J. G., and Anderson, D. M. 1973. Soil development and patterned ground evolution in Beacon Valley, Antarctica. *Permafrost, North American Contribution, Second International Conference.* Yakutsk, U.S.S.R., July 1973, Washington, National Academy of Sciences, 246–54.

143

Late Cenozoic Marine and Terrestrial Glacial Sedimentation in Eastern Taylor Valley, Southern Victoria Land

B. C. McKELVEY
Department of Geology, University of New England, Armidale, N.S.W. 2351, Australia

Abstract. Upper Cenozoic cores recovered from Dry Valley Drilling Project (DVDP) sites 10 and 11 in Taylor Valley show the valley to have been a fjord, into which floating ice extended from McMurdo Sound westward upvalley for 15 km. The cores consist of more than 100 m of latest Miocene and Pliocene marine tillite overlain by more than 200 m of Pleistocene marine tillite and younger strandline sediments and fluvioglacial deposits. The Pliocene–Pleistocene contact is disconformable because of erosion by grounded ice. Another major erosion surface occurs near the top of the sequence, separating Holocene(?) Taylor Formation sedimentary rocks from younger Quaternary outwash deposits. The marine tillites show considerable textural diversity, reflecting mainly variation in the amount of glacial erosion under differing conditions of dry-based and wet-based glaciation. Some of these tillites show evidence of derivation from older tillites by resedimentation, and abundant intraformational detritus throughout the sequence indicates widespread erosion contemporaneous with sedimentation. The petrography of the upper Cenozoic strata indicates derivation from an eastern source composed largely of the lower Paleozoic basement complex of southern Victoria Land and, to a much lesser extent, the Jurassic Ferrar Dolerite. Basaltic volcanic detritus is present throughout but is abundant only in the youngest Quaternary horizons.

The McMurdo oasis or dry valleys region of southern Victoria Land includes approximately 4000 km^2 of deglaciated mountainous desert, bounded on the east by the coast of southern Victoria Land and on the west by the Antarctic polar plateau. The Taylor and Wright Valleys and the Victoria Valley system make up much of the oasis. The floors of these largely or completely ice-free valleys are mantled with variable hummocky moraines and are further diversified by frozen lakes, saline ponds, areas of sand dunes, and in mid-summer, vigorous meltwater streams and rivers. This paper briefly summarizes some of the results of the Dry Valley Drilling Project's investigations within the Taylor Valley between 1973 and 1975.

It has long been recognized that the oasis has undergone multiple glaciations (Péwé, 1960; Bull and others, 1962; Nichols, 1971). Ice has repeatedly advanced into the oasis, moving both westward from the coast and eastward from the polar plateau. Furthermore, alpine glaciers in the ranges bordering the valleys have fluctuated considerably in extent. These individual advances from different directions have not always been in phase with each other (Denton and others, 1970; Bull and Webb, 1973; Calkin and Bull, 1974).

The Taylor Valley extends from the polar plateau eastward down through the southern Victoria Land coastal mountain ranges for a distance of 100 km to McMurdo Sound. Sites 10, 11, and 12 of the Dry Valley Drilling Project (DVDP) are all located within the deglaciated eastern end of Taylor Valley (Figure 143.1). Site 10 is at the valley mouth on the coast, at an altitude of 1.9 m and approximately 50 m from the edge of the sea ice at New Harbour. DVDP drilling sites 8 and 9 (Chapman-Smith and Luckman, 1974) occupied this same location. Site 11 is approximately 3 km inland, situated on undulating valley-floor moraines at an altitude of 80.2 m, near the terminus of the Commonwealth Glacier. Site 12 (Chapman-Smith, 1975) is located 75 m above sea level, 15 km farther inland from DVDP 11. This paper is confined to a discussion of the results obtained from DVDP 10 and 11, and where appropriate, comparison is made with data obtained from DVDP 12.

The rocks of both the Kukri Hills and the Asgard Range, bordering Taylor Valley, consist of upper Precambrian or lower Paleozoic Skelton Group metamorphic rocks, and plutons and dike swarms of the lower Paleozoic Admiralty Intrusives. Further inland this crystalline basement complex (Lopatin, 1972) is overlain nonconformably by cratonic sedimentary rocks of the Devonian (or older) to Jurassic Beacon Supergroup, which is intruded by Jurassic Ferrar Dolerite sheets and sills (Haskell and others, 1965). The floor of Taylor Valley is mantled by a hummocky surface of Cenozoic moraine diversified by several frozen glacial lakes and scattered basaltic cinder cones (Péwé, 1960; McCraw, 1962, 1967).

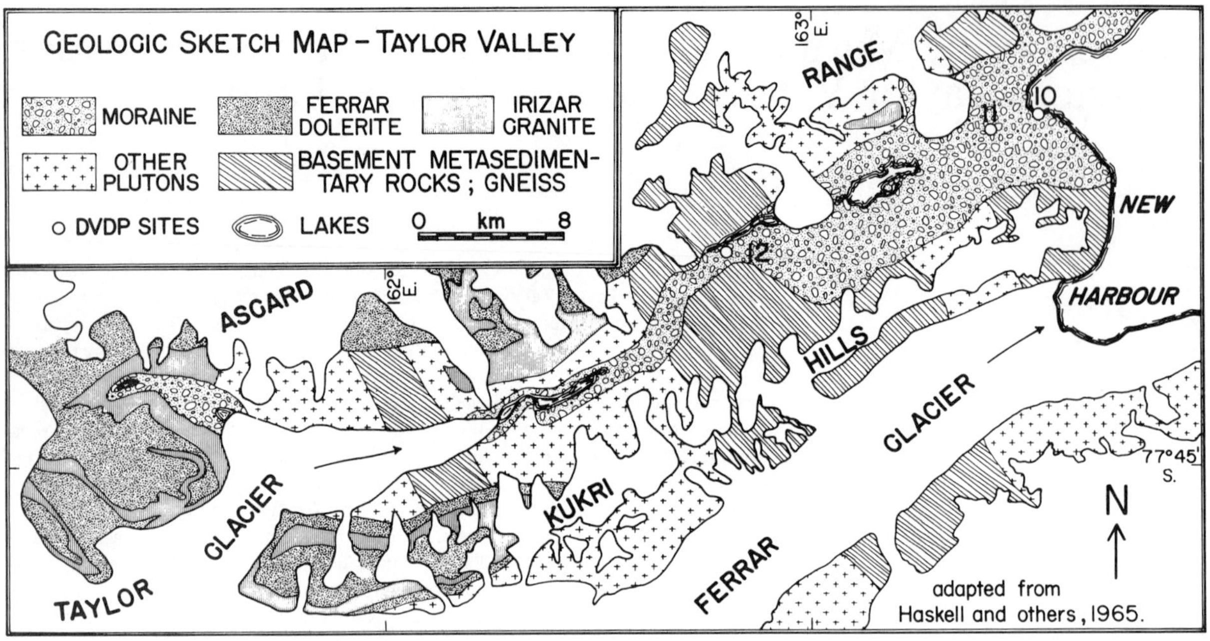

Figure 143.1. Geologic sketch map of Taylor Valley, southern Victoria Land, showing the positions of Dry Valley Drilling Project sites 10, 11, and 12. The Ferrar Dolerite is Jurassic and the Irizar Granite and other plutons are lower Paleozoic. The basement metasedimentary rocks and gneisses are considered to be upper Precambrian or lower Paleozoic. (North is up.)

Dry Valley Drilling Project Site 10

The 185.47 m of core recovered from DVDP 10 consists of five major lithostratigraphic units (Figure 143.2). Sandstone, pebbly sandstone, and conglomerate make up the two youngest units and overlie, at 39 m a highly varied 104 m sequence (Units 3 and 4) of diamictite and interbedded sandstone and pebbly sandstone, associated with minor pebble conglomerate, breccia, and laminated sandy mudstone. The basal Unit 5 is at least 35.8 m thick and consists largely of diamictite with minor amounts of sandstone and mudstone. The diamictite of Unit 5 contrasts with that of younger units in that it is thin-bedded and commonly graded, and the pebbles and dispersed granule and sand-sized detritus is overall very sparse. The diamictite recovered at DVDP 10 is all marine tillite, reflecting profuse rafting, whereas most of the sandstone and other coarser-grained sediments are marginal marine (i.e., strandline) and fluvioglacial outwash deposits.

The petrography of the DVDP 10 core shows derivation of the sequence mainly from the basement complex and to a much lesser extent from the Ferrar Dolerite. As at DVDP 11, basaltic detritus in very small amounts is present throughout the sequence. Especially in the younger horizons, sandstone contains a small but persistent component of angular to subangular intraformational mudstone and fine-grained sandstone clasts. Toward the bottom of the cores, particularly in Unit 5, the mudstone and diamictite show appreciable lithification.

Summary geologic log of DVDP 10, New Harbour

For a more detailed summary log, and the complete log of DVDP 10, see McKelvey (1975, pp. 19–24, 33–42).

Unit 1. (21.21 m thick). Pebbly, very coarse-grained sandstone with minor granule conglomerate passing irregularly downward to pebbly, medium-grained sandstone. Pebble conglomerate (average clast, 4 mm) at 15.5–16.5 m. Intraformational mudstone debris throughout unit. Stratification indistinct above 10 m, sharper in older horizons. Initial dip averages 15°–20° (maximum 30°) above 12.5 m; below 16.0 m, 0°–6°. Pelecypod debris down to 24.1 m.

Unit 2. (14.20 m thick). Granule and pebble conglomerate. Average clast size 1–10 cm; largest clast noted exceeds 22 cm. Interbeds of minor medium- to coarse-grained sandstone above 31 m. Intraformational mudstone debris throughout (up to 5%). Overall, unit becomes finer upward; individual beds also become finer upward.

Unit 3. (73.00 m thick). Diamictite beds, up to 13 m thick, interbedded with minor pebble conglomerate and coarse- to fine-grained sandstone. Rare laminated sandy mudstone (less than 12 cm) occurs below 50.1 m. Most diamictite matrix is indistinctly laminated, poorly sorted, medium- to fine-grained sand. Clast component ranges up to 17%; average is less than 10%. Average clast size is 2–10 cm; largest clast noted exceeds 50 cm. Lithologic contacts are gradational to sharp, and overall bedding attitude is horizontal. Local deformation indicated by strata inclined as much as 20°.

Unit 4. (31.33 m thick). Lithologically diverse unit. Diamictite (clasts 3–30%) interbedded with minor pebble conglomerate, breccia (at 130 m), sandstone, and sandy laminated mudstone. Mudstone in sequences as much as 0.8 m (average 2–10 cm) contains scattered (<1%) pebbles and granules. All gradations between pebbly sandstone and diamictite. Average clast size 1–2 cm; largest clast exceeds 32 cm. Stratification overall horizontal. Locally deformed with dips as high as 30°.

Unit 5. (at least 35.81 m thick). Olive-gray, massive, and laminated sandy mudstone (diamictite) containing widely dispersed pebbles (frameworks average 0.5%), and fine- to medium-grained, poorly sorted sandstone. Strata commonly graded. Minor thin pebble conglomerate. Breccia at

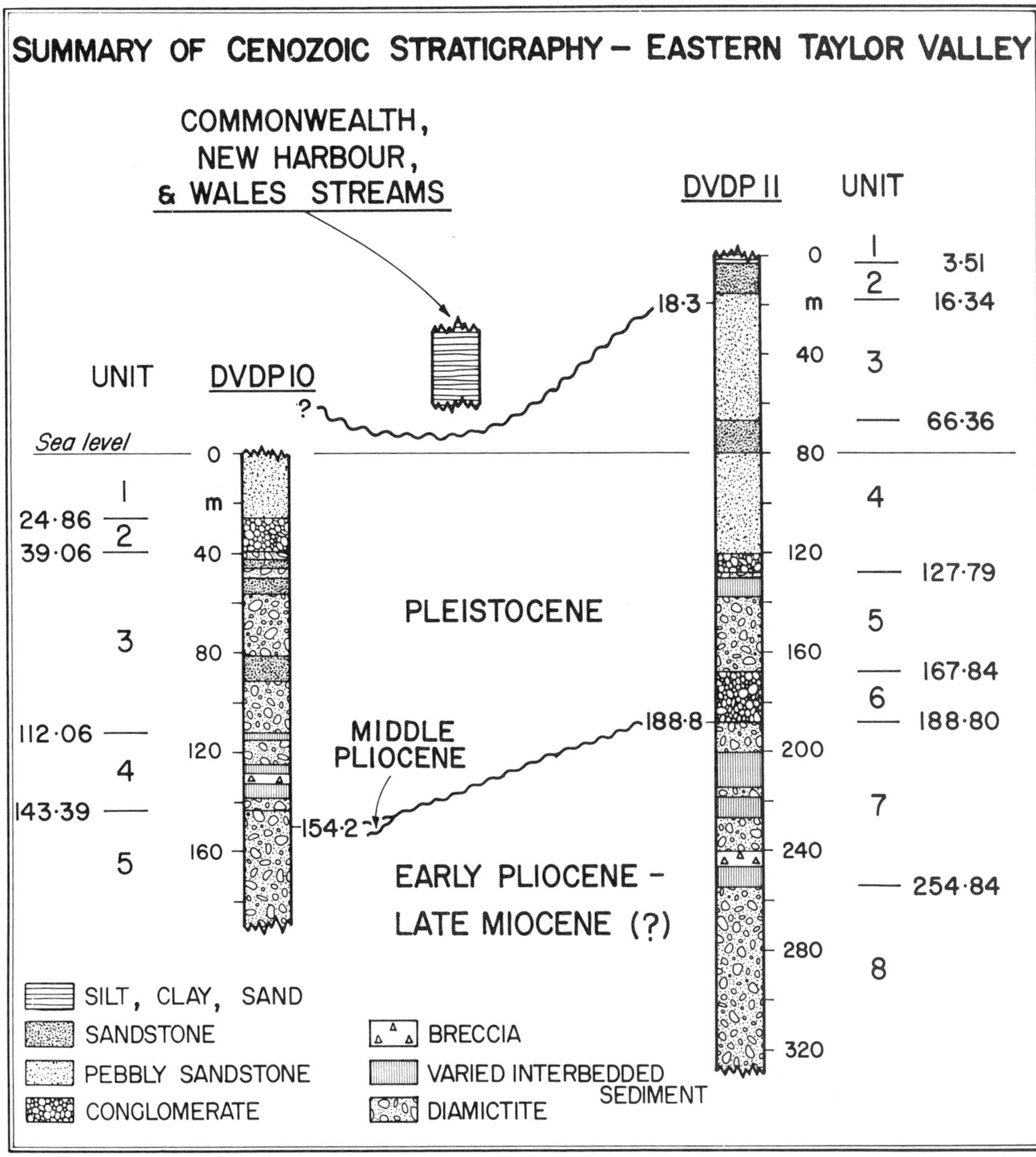

Figure 143.2. Summary of the known Cenozoic stratigraphy in eastern Taylor Valley. The position of the Holocene–Pleistocene boundary is not known. Depth in m.

170.7 m. Overall sequence thinly bedded (less than 0.5 m). Proportion of diamictite increases in older horizons. Scattered bioturbation.

Dry Valley Drilling Project Site 11

The 327.96 m core recovered from DVDP 11 is overall very similar to that from DVDP 10 and shows an irregular transition from pebbly diamictite (Figure 143.3) at the base, to sandstone and pebbly sandstone with subordinate conglomerate (Figure 143.4) at the top. This diamictite is also marine tillite and reflects both suspension current sedimentation and profuse glacial rafting. The sandstone, pebbly sandstone, and conglomerate are the products of traction current sedimentation in marginal marine and fluvioglacial environments.

Eight major lithostratigraphic units are recognized (Figure 143.2). The youngest four, which total over 127 m in thickness, consist of sandstone, pebbly sandstone, and conglomerate. Units 7 and 8, which total 139.2 m in thickness, are predominantly diamictite and minor sandy mudstone, with some conglomerate present near the top of Unit 7. The intervening Units 5 and 6 total 61 m in thickness and together constitute a transitional sequence with interstratification of all the above rock types.

One major petrographic trend is apparent when the overall provenance of this core is considered. Units 1, 2, and 3, along with the younger horizons of Unit 4, seem to

Figure 143.3. Diamictite (marine tillite), Unit 7.17 (a subunit of Unit 7), DVDP 11 (McKelvey, 1975, p. 55). Angular to subrounded clasts of varied basement complex rock types dispersed in an abundant matrix of poorly sorted, fine-grained sand. Field width 60 cm.

Figure 143.4. Conglomerate, Unit 6.2 (a subunit of Unit 6), DVDP 11 (McKelvey, 1975, p. 53). Coarse, poorly sorted pebble conglomerate. Note diversity of clast types and considerable range in clast sizes. Field width 60 cm.

be derived largely from medium-grained hornfelsic source rocks of the coastal metamorphic Skelton Group of the basement complex. In older units, coarse-grained intrusive rocks, particularly hornblende-biotite diorites and varied granodiorites, are more prominent along with similarly coarse-grained, dark biotite gneisses and schists. This change in provenance is gradual.

Considerable lithification of the core is apparent, particularly in Units 5, 7, and 8, where progressively older mudstone beds show increasingly better developed conchoidal fracture. In addition, below 240 m in Unit 8, common, although slight, changes in the hue of the mudstone matrix suggest diagenetic modification of clay-size detritus. Carbonate cementation of sandstone is not uncommon, and near the top of Unit 4, below about 85 m, there is a marked increase in the development of carbonate and other secondary minerals.

SUMMARY GEOLOGIC LOG OF DVDP 11, COMMONWEALTH GLACIER

For a more detailed summary log, and the complete log of DVDP 11, see McKelvey (1975, pp. 24–30, 44–60).

Unit 1. (1.94 m thick). Laminated silty sandstone, mudstone, and granule conglomerate.

Unit 2. (12.83 m thick). Lithologically uniform. Massive medium-grained sandstone with intraformational mudstone debris. Stratification indistinct. Attitude 0°–7°.

Unit 3. (50.02 m thick). Lithologically uniform. Predominantly pebbly, coarse-grained sandstone; pebbles average 7% of total volume. Minor sandstone and granule conglomerate. Sorting moderate to poor. Intraformational mudstone debris. Stratification indistinct. Attitude 0°–35°. Basaltic pebble-granule conglomerate at 26.35–26.80 m.

Unit 4. (61.43 m thick). Lithologically variable. Medium-grained sandstone passing irregularly (at approximately 81 m) through pebbly sandstone to (at approximately 113 m) pebble and cobble conglomerate. Pebbly sandstone predominates; pebbles average 5–7% of total volume. A few thin mudstone beds, with bioturbation and soft sediment deformation, below 85.75 m. Overall stratification distinct. Attitude 0°–20°, maximum 45°, due to deformation. Some cross-stratification.

Unit 5. (40.05 m thick). Lithologically variable. Diamictite (marine tillite) interbedded with sandy mudstone, pebble-granule conglomerate, and minor sandstone. Diamictite highly variable. Clasts average 7–10% of total volume (maximum 25%). Mean clast size 2 cm; largest clast exceeds 1.04 m. Matrix very poorly sorted, average fine- to medium-grained sand. Sandy mudstone shows intraformational brecciation and soft-sediment deformation. No definite bioturbation. Bedding contacts sharp to gradational. Attitude mainly horizontal, but some beds dip 5°–45°, the latter due to deformation.

Unit 6. (20.96 m thick). Lithologically uniform. Poorly sorted granule-pebble conglomerate. Unit coarsens irregularly downward. Mean clast size 1–5 cm; largest clast exceeds 60 cm. At depth of 176 m, clasts greater than 20 cm exceed 40% of total volume. Stratification indistinct, but attitude horizontal; some dips of 10°–15°. Scattered bivalve(?) debris throughout.

Unit 7. (66.04 m thick). Lithologically variable. Diamictite (marine tillite) interbedded with laminated, sandy to silty mudstone. Minor rock types include thin-bedded (<1 m), granule–fine pebble conglomerate, sandstone, and pebbly sandstone. Breccias at base (approximately 240 m). Diamictite clasts average 7–20% of total volume. Mean clast size 2.5–5 cm; largest clast exceeds 85 cm. Overall bedding horizontal, little apparent deformation. Conglomerate beds dip 20° and 15° at 246 and 251 m. Basaltic lapilli tuff and finer ash scattered between 204.26 and 206.5 m. Gastropods (<2 mm) at 205.99 m. Bioturbation widespread.

Unit 8. (at least 73.12 m thick). Lithologically uniform. Diamictite (marine tillite). Clasts average 1–12% of total volume. Mean clast size 1.5 cm. Diamictite matrix indistinctly laminated, ranges from very poorly sorted, fine-grained sandstone to sandy mudstone. Minor rock types include sandy mudstone and thin-bedded (<0.1 m) coarse-grained sandstone and conglomerate. Attitude horizontal, but some soft-sediment deformation. No apparent bioturbation.

Biostratigraphy of DVDP 10 and 11

Webb and Wrenn (Chapter 144, this volume) have recovered well-preserved calcareous foraminiferal assemblages from DVDP sites 8–11, and Brady (Chapter 145, this volume) similarly records both freshwater and marine diatom faunas in DVDP 10 and 11. Many horizons, however, are barren or contain only reworked fossils. Webb and Wrenn recognize in DVDP 8–11 three faunal zones of late Miocene–early Pliocene (I), mid-Pliocene (II), and Pleistocene (III) ages, respectively. The boundaries of these zones are considered disconformable, but their precise locations are not known with certainty, due in part to barren samples and to reworking. The Zone I and II faunas occur in DVDP 10 only below 154.2 m. However, Webb and Wrenn note that the lowest convincing Zone III (i.e., Pleistocene) fauna is at approximately 125 m. Similarly in DVDP 11 where the Pleistocene Zone III fauna disconformably overlies the Miocene Zone I fauna, the oldest Zone III fauna occurs at 170.43 m, but the youngest *in situ* Zone I fauna occurs at 205.96 m. Webb and Wrenn place the Zone I–Zone II boundary in DVDP 11 provisionally at 202.45 m, as a major paleomagnetic reversal is suspected at that depth. In DVDP 10 they place the Zone III–Zone II boundary at 154.2 m.

From a sedimentologic viewpoint, it is noteworthy that in both DVDP 10 and 11 the oldest convincing Zone III faunas (at 125 and 170.43 m) occur just above or within units that herald major changes in sedimentation. In DVDP 10, the diamictite above 124 m (i.e., the base of Unit 4.1, a subunit of Unit 4) contrasts markedly in texture with all older diamictite. Similarly in DVDP 11, the Unit 5 diamictite contrasts with that of Units 7 and 8 and is separated from the latter by the very coarse lag conglomerate (see below) of Unit 6 (167.8–188.8 m). It would seem logical, therefore, to place the base of the Pleistocene sequence in DVDP 10 at 124.5 m and in DVDP 11 at 188.8 m. However, in DVDP 10 Brady records Pleistocene diatoms down to 154.2 m, and so the Pleistocene–Pliocene disconformity is here placed at that level.

Beneath their Pleistocene Zone III sequence, Webb and

Wrenn (Chapter 144, this volume) recognize a pelagic and bethonic late Miocene–early Pliocene fauna and a benthic mid-Pliocene fauna. Brady, however, on diatom evidence, considers that the interval containing Webb and Wrenn's late Miocene–early Pliocene fauna to be wholly earliest Pliocene. The mid-Pliocene benthic fauna of Webb and Wrenn was recovered from only one sample at 154.8 m in DVDP 10. Brady recovered no diatoms from this horizon. In DVDP 8, Zone II faunas occur at 153.88 and 154.51 m.

The apparent absence of mid-Pliocene strata in DVDP 11 suggests convergence between the two sites of the two disconformities present in DVDP 10. Thus at DVDP 11, Pleistocene erosion has completely removed the mid-Pliocene sequence, and only a very thin interval survives at DVDP 10.

Environment of deposition

The distribution of *in situ* foram and diatom faunas in both DVDP 10 and 11 shows the older units in these cores are wholly marine. Clearly the fossiliferous diamictite of Units 4 and 5 of DVDP 10 and of Units 7 and 8 of DVDP 11 is marine tillite. *In situ* foraminiferal faunas have been recovered from all these diamictites (Webb and Wrenn, Chapter 144, this volume), and in addition Brady records *in situ* marine diatoms from sandstone and mudstone interbedded with the diamictite. While no *in situ* forams or diatoms are found in the Unit 5 diamictite of DVDP 11, foram faunas considered by Webb and Wrenn to be probably in place occur in the underlying Unit 6 (with bivalve detritus) and in the base of the overlying Unit 4.

The coarse conglomerate of Unit 6 in DVDP 11 (Figure 143.4), containing broken, thick-walled bivalve fragments, is most simply regarded as reworked (winnowed) diamictite or lag deposits from which interstitial fines have been removed by competent currents associated with a strandline environment. The sturdy nature of the bivale fragments suggests a high-energy environment. Similarly, the basal conglomerate of Unit 4 (Unit 4.13 of McKelvey, 1975) is presumably of the same origin, overlying a diamictite (Unit 5.1 of McKelvey, 1975) which shows abundant evidence of rafting.

Above the youngest diamictite of DVDP 10 and 11, the sequences are almost entirely composed of traction-current sediments. Paleontologic data from these strata are few. Brady (Chapter 145, this volume) reports only reworked marine diatoms, and he finds that some large intervals are barren. In DVDP 11 he records one abundant nonmarine fauna between 89.9 and 115.3 m, and another between 0 and 31.8 m. Webb and Wrenn, however, record *in situ* benthic marine forams in the upper 24 m of DVDP 10, and the long-known (David and Priestley, 1914, p. 272) occurrence of delicately preserved bivalves through approximately the same stratigraphic interval, in Unit 1 sandstone and pebbly sandstone of DVDP 10 (equivalent to the Quaternary Taylor Formation of Speden, 1962), further confirms a marine environment.

The common occurrence of inclined stratification, such as in Unit 1 of DVDP 10 and Units 2, 3, and 4 of DVDP 11, suggests deposition with initial or primary dip, probably a result of outwash streams debouching into a littoral delta-front environment, similar to that where both the Wales and Commonwealth Streams enter New Harbour today. In such a situation, interbedding and mixing of marine, freshwater, and reworked marine fossils is to be expected. Although few if any strata in DVDP 10 seem to be wholly terrestrial, it is likely, particularly in the light of Brady's findings, that Units 1, 2, and 3 of DVDP 11 are largely strandline or terrestrial deposits. The homogeneous sandstone of Unit 2 is very similar to the outwash sand being deposited at present adjacent to the inland margin of the lower Wright Glacier.

Feature of sedimentation

Four major features of the DVDP 10 and 11 cores that merit comment are (1) the marked lateral facies variations, (2) the textural variations in the diamictite, (3) the common occurrence of graded beds in Unit 5 of DVDP 10, and (4) the small but ubiquitous intraformational component present in all varieties of strata.

1. Both cores clearly show the same irregular upward transition from marine tillite at the base to traction current sandstone and pebbly sandstone at the top. However, examination of the detailed logs (McKelvey, 1975) suggests that few, if any, direct lithostratigraphic correlations can be made between the two sequences. Facies variation between the two sites, only 3 km apart, is clearly pronounced.

2. Within DVDP 10, the Pleistocene diamictite above 124.5 m is appreciably coarser than all older Pleistocene, Pliocene, and Miocene diamictite. Not only is the diamictite matrix coarser, but the percentage framework (5–30%), average clast size range (1–4 cm), and maximum clast sizes (42–32 cm) are all appreciably greater than those of older horizons. Below 124.5 m the diamictite frameworks average less than 1% of the bulk volume, and average clast sizes are 0.5–1 cm. The largest clast noted was only 11 cm.

Within DVDP 11, textural contrasts within the diamictite are not so apparent. However, the matrix of the Pleistocene Unit 5 (127.8–167.8 m) is certainly coarser than that of the early Pliocene and Miocene(?) strata in Units 7 and 8. In addition, within Units 7 and 8 an overall downward decrease in textural grade is apparent. Unit 7.17 (a subunit of Unit 7), extending from approximately 226 to 240 m (McKelvey 1975, p. 55), has a dispersed clast component that varies between 10 and 15% of bulk volume. (These clasts are also the largest encountered in DVDP 11, many exceeding 50 cm.) In Unit 8.3 (a subunit of Unit 8, at 261.6–302.8 m), the average clast component generally ranges between 5 and 10%. Still lower in the sequence, in Units 8.5 at 307–328 m (McKelvey, 1975, p. 59), the clast component generally varies between only 5 and 7% and in many intervals is less than 1%.

3. Evidence strongly suggesting resedimentation is apparent within both Pleistocene and Pliocene strata of Unit 5 in DVDP 10. Beneath 150.8 m, no less than six intervals consist of sandy mudstone or sandstone grading down through approximately 1 m to granule or pebble conglomerate and breccia (McKelvey, 1975, pp. 40–41). The basal contacts of these units are sharp. A similar graded unit occurs at the base of Unit 3.4 (a subunit of Unit 3) within the Pleistocene sequence of DVDP 11.

4. Perhaps one of the most striking features of the cores is the nearly ubiquitous occurrence of intraformational sandstone and mudstone clasts. This component is most

easily observed within the traction current sandstone and pebbly sandstone where in some instances it comprises as much as 5% of the total volume. Fragments of older diamicite beds have been observed in Unit 7 of DVDP 11 (at 236.2 m) and in Unit 2 of DVDP 10 (at 37.5 m). These occurrences and the relative abundance of the other intraformational detritus suggest the presence of many disconformities within the sequence. However, some of the intraformational fragments, particularly those in the diamictite, may be rafted pellets originally formed between clear ice crystals in foliation bands within glacier ice (Ovenshine, 1970).

Quaternary strata exposed in Commonwealth, New Harbour, and Wales Streams

An unfolded Quaternary sequence, perhaps as much as 50 m thick, of sandstone, siltstone, and minor diamictite is exposed in the Commonwealth, New Harbour, and Wales Streams near the mouth of Taylor Valley. Webb and Neall (1972) identified reworked Cretaceous foraminifera in these deposits. Referred to previously as the New Harbour Polymicton by Murrell (1973), this sequence is here considered to be disconformably draped over an irregular erosion surface cut on the older Quaternary (Holocene?) Taylor Formation of Speden (1962, p. 758). The collars of both DVDP 10 and 11 were in the Taylor Formation, beneath the disconformity surface.

The younger Quaternary sequence consists predominantly of interbedded black and dull gray (basalt-derived) sandstone and yellowish green and cream colored, poorly sorted sandstone (see Webb and Neall, 1972, Figures 3, 4, and 5). Other rock types include yellow-brown siltstone, claystone, and diamictite. Intraformational siltstone and fine-grained sandstone clasts are abundant within the lighter colored sandstone and siltstone. In some localities the intraformational clasts are so abundant that they form discrete layers of intraformational breccia. Scattered throughout all these younger Quaternary strata are widely dispersed dropstones, some up to 2 m in diameter, of crystalline basement rocks. The diamictite has only been observed in the vicinity of Wales Stream on the south side of the Taylor Valley mouth where it occurs as lensoidal deposits. Its framework of basement-derived cobbles and pebbles comprises a maximum of about 5% of the total volume and is dispersed in a light gray, silty sand matrix.

The dark basalt-derived sandstone is characteristic of the upper horizons of this younger Quaternary sequence. Local minor disconformities are common. Sedimentary structures include current ripples and scour-and-fill bedding contacts. Some stratigraphic intervals as much as 2 m thick show intraformational slumping, with almost complete destruction of the original stratification.

The identification of a major disconformity separating this younger Quaternary sequence from the older Quaternary strata penetrated at the top of DVDP 10 and 11 (Figure 143.2) is based on lithologic criteria. From the altitudes of the drill collar of DVDP 11 and those of the exposures in the three streams, it would appear that the younger Quaternary sequence should be laterally equivalent to Units 1, 2, and 3 of DVDP 11. However, Units 2 and 3 show few lithologic similarities to the younger Quaternary sequence. In particular, Unit 3 contrasts with it in containing no beds of intraformational breccia, no diamictite, and no appreciable basaltic component. Only Unit 2 and the very limited amount of core recovered from Unit 1 of DVDP 11 display the basaltic detritus, and for this reason the disconformity is here tentatively placed near the top of Unit 3 at 18.26 m.

Conclusions

The fossilferous marine sequence recovered from DVDP 10 and 11 shows Taylor Valley to have been a fjord for at least a large part of Neogene and Pleistocene time. In this way, the geologic history of Taylor Valley is similar to that of Wright Valley laying immediately to the north (Webb, 1972; McSaveney and McSaveney, 1972).

Webb and Wrenn (Chapter 144, this volume), on the basis of both benthoic and pelagic forams making up their oldest fauna (i.e., Zone I), conclude that incomplete or only intermittently complete ice cover existed in the fjord. Brady's (chapter 145, this volume) finding, from the same stratigraphic interval, that marine diatom floras are absent from the diamictite but present in the interstratified rock types, further confirms the incomplete or intermittent nature of the ice cover at that time.

The regressive transition upward from glacial diamictite to shallow marine and terrestrial outwash sediment in both DVDP 10 and 11 reflects a post-middle or -late Miocene tectonic uplift, perhaps combined with variations in sea level. Webb and Wrenn have stated that the benthonic component of their Zone I fauna suggests a water depth of between 600 and 900 m. In contrast, their mid-Pliocene and Pleistocene faunas suggest depths of less than 100 m and less than 50 m, respectively. This indicates appreciable uplift subsequent to deposition fo the late Miocene–early Pliocene Zone I fauna.

A significant major paleoclimatic inference can be drawn from (1) the coarser nature of the younger Pleistocene diamictite and (2) the downward decrease in the abundance of the coarse rafted component of the Pliocene and Miocene diamictite in DVDP 10 and 11. These changes suggest an amelioration of climate, with gradual transition from cold, dry-based glaciation in Miocene, Pliocene, and perhaps early Pleistocene time, to wet-based, later Pleistocene glaciation characterized by increased glacial erosion and consequent coarser sedimentation. The controversial relationship between wet-based and dry-based glaciation and the amount of glacial erosion has been discussed at length by Warnke (1970) and Barrett (1975).

The two disconformities beneath both the mid-Pliocene and Pleistocene sequences result from the grounding of a coastal ice shelf and the ensuing erosion. However, the youngest disconformity, that between the younger Quaternary sequence of the Commonwealth, New Harbour, and Wales Streams and the older Quaternary strata penetrated at the top of DVDP 10 and 11 (i.e., Taylor Formation), is presumably a subaerial erosion surface resulting from continued tectonic elevation of the Pleistocene and Tertiary sequences above sea level. The high basaltic component in the upper Quaternary sequence suggests a source up the valley to the west, where basaltic cinder cones are not uncommon (McCraw, 1962).

Although the petrography of the cores shows that their derivation was overwhelmingly from the basement complex of southern Victoria Land and to a lesser extent from the Ferrar Dolerite, one basement complex rock type is nearly absent in the diamictite clasts. No clasts of the petrographically striking Irizar (or Vida) Granite were identified in DVDP 10, and only three clasts (at 187, 287.3, and 295 m) were observed during a reexamination of DVDP 11. Yet this biotite-hornblende granite is widely exposed farther west in Taylor Valley. It is not known to crop out east of DVDP 11 (Figure 143.1). This strongly suggests that the sequences recovered from DVDP 10 and 11 have been largely deposited by ice moving westward into the valley from McMurdo Sound. The Pleistocene, Pliocene, and Miocene sequences therefore represent successive Ross glaciations, in the sense of Denton and others (1970).

From DVDP 12, which penetrated basement (Chapman-Smith, 1975), no marine faunas have been recovered, apart from some reworked forams (Webb and Wrenn, Chapter 144, this volume). Chapman-Smith (1975) considers the DVDP 12 sequence to be terrestrial, and cites the presence of interbedded evaporites, glaciolacustrine tillite, and fluviatile sediment. It then follows that if the DVDP 11 and 12 sequences are broadly equivalent, the strandline of the marine incursions into Taylor Valley in late Miocene, Pliocene, and Pleistocene time was situated somewhere between these two drilling sites.

As neither DVDP 10 nor 11 penetrated basement, it is possible that still older Cenozoic marine tillite, as yet unsampled, exists at depth in Taylor Valley.

Acknowledgments. The author acknowledges the considerable assistance given both in the field and in the laboratory by other members of the Dry Valley Drilling Project. He also wishes to express his appreciation of the many helpful discussions with P. N. Webb, H. Brady, and J. H. Wrenn.

References cited

Barrett, P. J. 1975. Textural characteristics of Cenozoic preglacial and glacial sediments at Site 270, Ross Sea, Antarctica. (*In* Hayes, D. E., Frakes, L. A., and others. *Initial Reports of the Deep Sea Drilling Project,* Vol. 28. Washington [U.S. Government Printing Office], 757–67.)

Bull, C., and Webb, P. 1973. Some recent developments in the investigation of the glacial history and glaciology of Antarctica. (*In* van Zinderen, E. M., *ed. Palaeoecology of Africa and Antarctica. 8.* Capetown, A. A. Balkema, 55–84.)

Bull, C., McKelvey, B. C., and Webb, P. N. 1962. Quaternary glaciations in southern Victoria Land, Antarctica. *J. Glaciol., 4,* No. 31, 63–78.

Calkin, P. E., and Bull, C. 1974. The glacial history of the ice-free area, southern Victoria land, Antarctica. *Polar Rec., 17,* No. 107, 129–37.

Chapman-Smith, M. 1975. Geology of DVDP holes 12 and 14. *Antarct. Jnl U.S., 10,* No. 4, 170–72.

Chapman-Smith, M., and Luckman, P. G. 1974. Late Cenozoic glacial sequence cored at New Harbour, Victoria Land, Antarctica (DVDP 8 and 9). *Bull. Dry Valley Drilling Project,* No. 3, 120–48.

David, T. W. E., and Priestley, R. E. 1914. Glaciology, physiography, stratigraphy, and tectonic geology of south Victoria Land. *Rep. Scient. Invest. Br. Antarct. Exped. 1907–09, Geology, 1,* 319 pp.

Denton, G. H., Armstrong, R. L., and Stuiver, M. 1970. Late Cenozoic glaciation in Antarctica: the record in the McMurdo Sound region. *Antarct. Jnl U.S., 5,* No. 1, 15–21.

Haskell, T. R., Kennett, J. P., Prebble, W. M., Smith, G., and Wilis, I. A. G. 1965. The geology of the middle and lower Taylor Valley of south Victoria Land, Antarctica. *Trans. R. Soc. N.Z., 2,* No. 12, 169–86.

Lopatin, B. G. 1972. Basement complex of the McMurdo 'Oasis,' south Victoria Land. (*In* Adie, R. J., *ed. Antarctic Geology and Geophysics.* Oslo, Universitetsforlaget, 287–92.)

McCraw, J. D. 1962. Volcanic detritus in Taylor Valley, Victoria Land, Antarctica. *N.Z. Jl Geol. Geophys., 5,* No. 5, 740–45.

McCraw, J. D. 1967. Some surface features of McMurdo Sound Region, Victoria Land, Antarctica. *N.Z. Jl Geol. Geophys., 10,* No. 2, 394–417.

McKelvey, B. C. 1975. Preliminary site reports, DVDP Sites 10 and 11, Taylor Valley. *Bull. Dry Valley Drilling Project,* No. 5, 16–60.

McSaveney, M. J., and McSaveney, E. R. 1972. A reappraisal of the Pecten glacial episode, Wright Valley, Antarctica. *Antarct. Jnl U.S., 7,* No. 6, 235–40.

Murrell, B. 1973. Cenozoic stratigraphy in lower Taylor Valley, Antarctica. *N.Z. Jl Geol. Geophys., 16,* No. 2, 225–42.

Nichols, R. L. 1971. Glacial geology of Wright Valley. (*In* Quam, L. O., *ed.* Research in the Antarctic. *Am. Ass. Adv. Sci.,* Publn No. 93, 293–340.)

Ovenshine, A. T. 1970. Observations of iceberg rafting in Glacier Bay, Alaska, and the identification of ancient ice-rafted deposits. *Geol. Soc. Am. Bull., 81,* 891–94.

Péwé, T. L. 1960. Multiple glaciation in the McMurdo Sound region, Antarctica. A progress report. *J. Geol., 68,* No. 5, 498–514.

Speden, I. G. 1962. Fossiliferous Quaternary marine deposits in the McMurdo Sound Region, Antarctica. *N.Z. Jl Geol. Geophys., 5,* No. 5, 746–77.

Warnke, D. A. 1970. Glacial erosion, ice rafting, and glacial-marine sediments: Antarctica and the Southern Ocean. *Am. J. Sci., 269,* 276–94.

Webb, P. N. 1972. Wright Fjord, Pliocene marine invasion of an Antarctic dry valley. *Antarct. Jnl U.S., 7,* No. 6, 226–34.

Webb, P. N., and Neall, V. E. 1972. Cretaceous foraminifera in Quaternary deposits from Taylor Valley. (*In* Adie, R. J., *ed., Antarctic Geology and Geophysics.* Oslo, Universitetsforlaget, 653–57.)

Webb, P. N., and Wrenn, J. H. 1975. Interpretation of foraminiferal assemblages from lower Taylor Valley (DVDP 8–12). *Bull. Dry Valley Drilling Project,* No. 7, 105–7.

144

Upper Cenozoic Micropaleontology and Biostratigraphy of Eastern Taylor Valley, Antarctica

PETER N. WEBB* and JOHN H. WRENN**
Department of Geology, Northern Illinois University, DeKalb, Illinois 60115, U.S.A.

Abstract. Study of drill-core samples from Dry Valley Drilling Project (DVDP) holes 8 through 12 and 15 reveals a wide range of macro- and microfossil material. Sponge spicules, fish scales, scolecodonts, echinoderms, mollusca, plant debris, fecal pellets, foraminifera, ostracods, diatoms, radiolaria, silicoflagellates, and undifferentiated microscopic debris have been recovered during this investigation. Foraminifera and diatoms are the dominant and biostratigraphically most significant groups in DVDP holes 8 through 11. DVDP 12 contains only sporadic occurrencs of microfossils; these are probably not *in situ,* and no biostratigraphic applications have been attempted. Three foraminiferal zones, separated by two major disconformities, are delineated. These are summarized as follows.

Foraminiferal Zone I (upper Miocene–lower Pliocene): This zone includes 155.26–179.97 m in hole 10 and 202.45–327.96 m in hole 11. It contains exceptionally rich foraminiferal assemblages. Sixty-seven species have been recovered, thirty-seven of which are not present in the younger Zones II and III. Rich assemblages composed of a few planktonic taxa were encountered in some beds, but most taxa are minute members of the benthic Miliolidae, Lagenidae, and Glandulindae. Drilling terminated before basement was reached in holes 8 to 11; older Cenozoic microfaunas are undoubtedly present at greater depth in the eastern Taylor Valley sedimentary succession.

Foraminiferal Zone II (middle Pliocene): This zone includes 154.20–155.26 m in hole 10 and is absent in hole 11. Twenty-two species of foraminifera were recovered, most of which also occur in Zones I and III. This zone is separated from the underlying Zone I by a marked disconformity. Similar faunas occur in the *Pecten* gravels of Wright Valley and the Scallop Hill Formation of Black and White Islands.

Foraminiferal Zone III (Pleistocene): This zone includes 0–154.20 m in hole 10 and 0–202.45 m in hole 11. Faunas from the short succession in hole 15 are also placed in this zone. Forty-four species of foraminifera have been recovered; fifteen of these species were not encountered in Zones I or II. Distribution of foraminifera is restricted to a few thin intervals, with the most diverse faunas occurring in the upper 30 m of hole 10. Both *in situ* and reworked assemblages are present. These faunas correlate with Taylor Formation faunas of the McMurdo Sound area. Zone III (Pleistocene) faunas are separated by marked disconformities from underlying Zone II (middle Pliocene) and Zone I (upper Miocene–lower Pliocene) foraminiferal faunas in holes 10 and 11, respectively.

Drill cores from six Dry Valley Drilling Project (DVDP) sites (8 through 12 and 15) (Figures 144.1, 144.2, and 144.3) were examined for foraminifera. Generalized logs of the penetrated sediments and rocks for each site are shown in Figure 144.4. Sample meterage cited here refers to depth below the top of the drilled succession. More detailed lithologic and stratigraphic information on the intervals discussed here is provided in the geologic logs presented by Chapman-Smith and Luckman (1974), McKelvey (1975), and Barrett and others (1976).

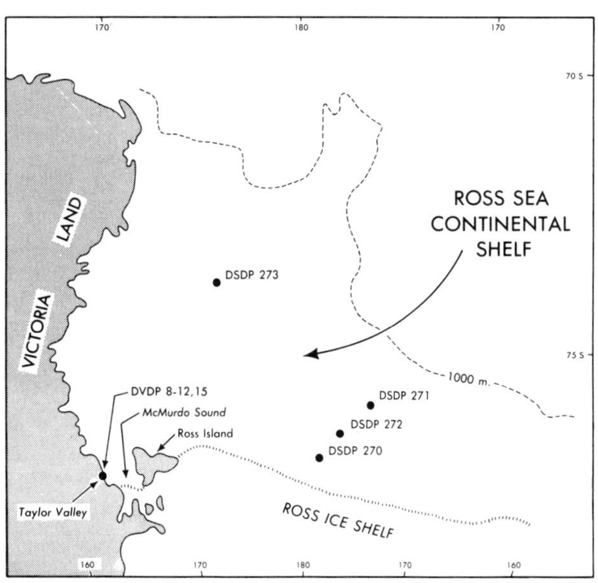

Figure 144.1. Location of DVDP and DSDP drill holes in the Ross Sea area.

Biostratigraphy

Provisional foraminiferal zonation

Antarctic Miocene–Pleistocene benthic foraminifera are commonly long-ranging and not helpful for the develop-

* Present address: Department of Geology and Mineralogy, The Ohio State University, Columbus, Ohio 43210, U.S.A.

** Present address: Department of Geology, Louisiana State University, Baton Rouge, Louisiana 70803, U.S.A.

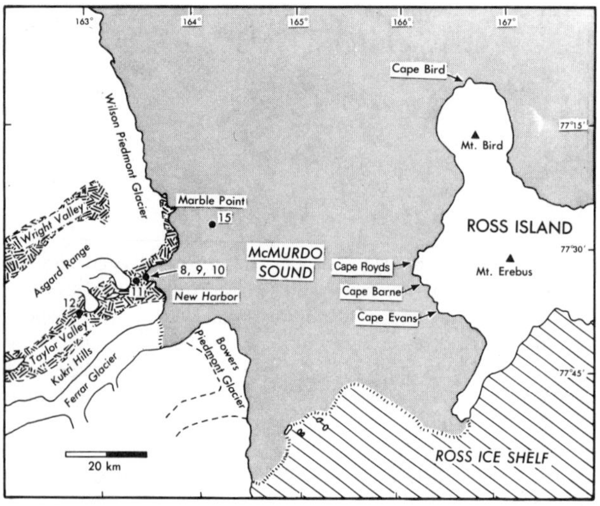

Figure 144.2. Location of drill holes in eastern Taylor Valley and McMurdo Sound area.

ment of biostratigraphically useful bioseries. Diversity in even the richest microfaunas rarely exceeds thirty to forty species, and most taxa are members of long-ranging Lagenidae and Glandulinidae. Planktonic forms are rare, and diversity rarely exceeds two or three species. Useful species of keeled late Cenozoic *Globorotalia* failed to penetrate these cold-water, high-latitude locales. Despite these discouraging factors, calcareous benthic and—to a lesser extent—agglutinated benthic foraminifera, are useful in infraregional biozonation and correlation. Benthic taxa record the ecologic imprint of climatic, tectonic, and bathymetric oscillations through the last 5 to 7 m.y. Long-ranging benthic taxa exhibit subtle changes of morphology with time. In general, tests of older (late Miocene–early Pliocene) microfaunas tend to be very small, while their more recent counterparts have tests four or five times larger. The foraminifera from eastern Taylor Valley have been used to establish three provisional biostratigraphic zones.

ZONE I

Occurrence

Zone I is found in the lower parts of DVDP 10 and 11 (Figure 144.4).

Vertical distribution

In DVDP 10, Zone I foraminifera extend from 155.26 m to 179.97 m. The interval from 179.97 m to the bottom of the hole, at 185.47 m, is barren. A markedly different fauna (Zone II) occurs at 154.80–154.90 m, less than 2 m higher in the succession. Samples from 155.40 m and 155.88 m are barren. The Zone I–Zone II boundary is provisionally placed at the sharp base of unit 5.4, at 155.26 m (McKelvey, 1975).

In DVDP 11, Zone I foraminifera extend from 199.30 m (possibly reworked) down to a sample at the base of the drill hole (327.96 m). The placement of the Zone I–Zone III boundary is complex; Zone II has not been recognized in this succession and has probably been removed by erosion. The lowest Zone III fauna occurs at 170.43 m. On first sight it seems reasonable to place the Zone I–Zone III boundary somewhere between 170.43 m and 199.30 m. Brady (1977) has reported his highest Biozone A diatom flora at 199.30 m and elected to place his Biozone A (lower Pliocene)–Biozone D (Pleistocene) boundary in the base of a pebble conglomerate at 195.22 m. The small foraminif-

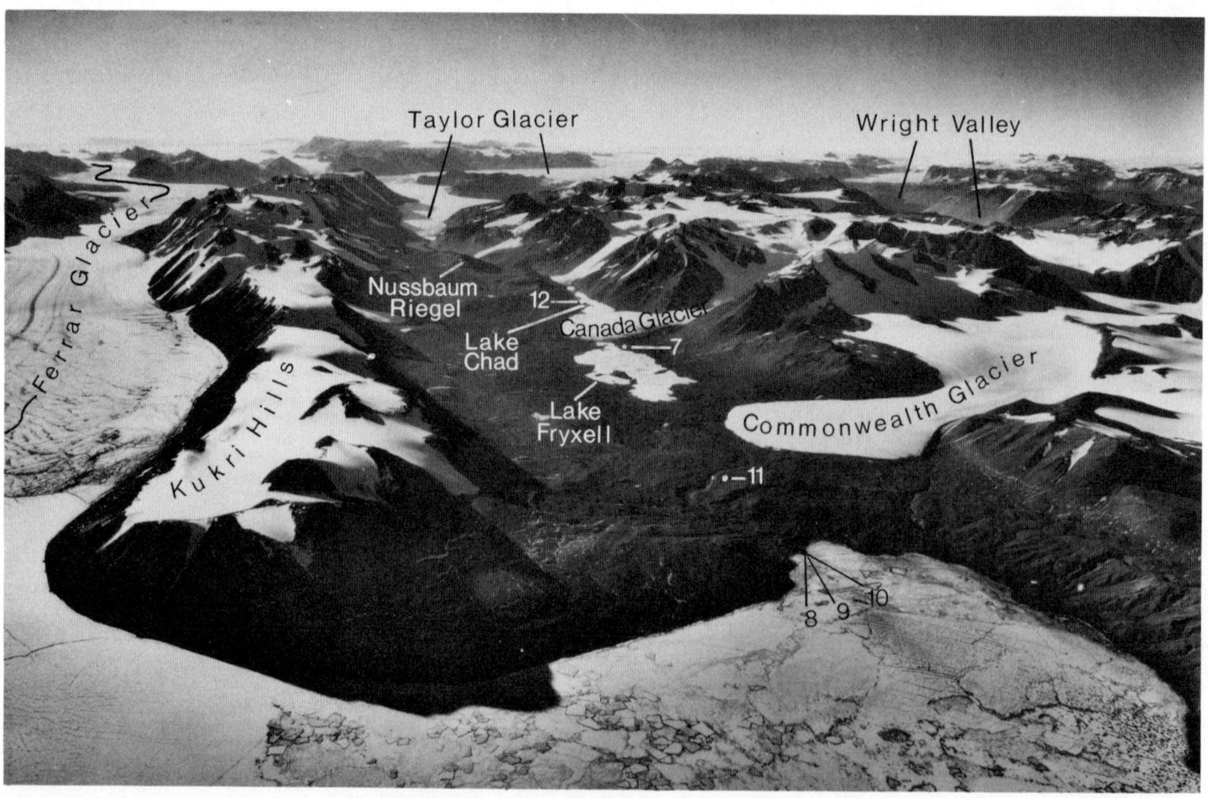

Figure 144.3. Taylor Valley viewed from the east, showing location of DVDP drill holes.

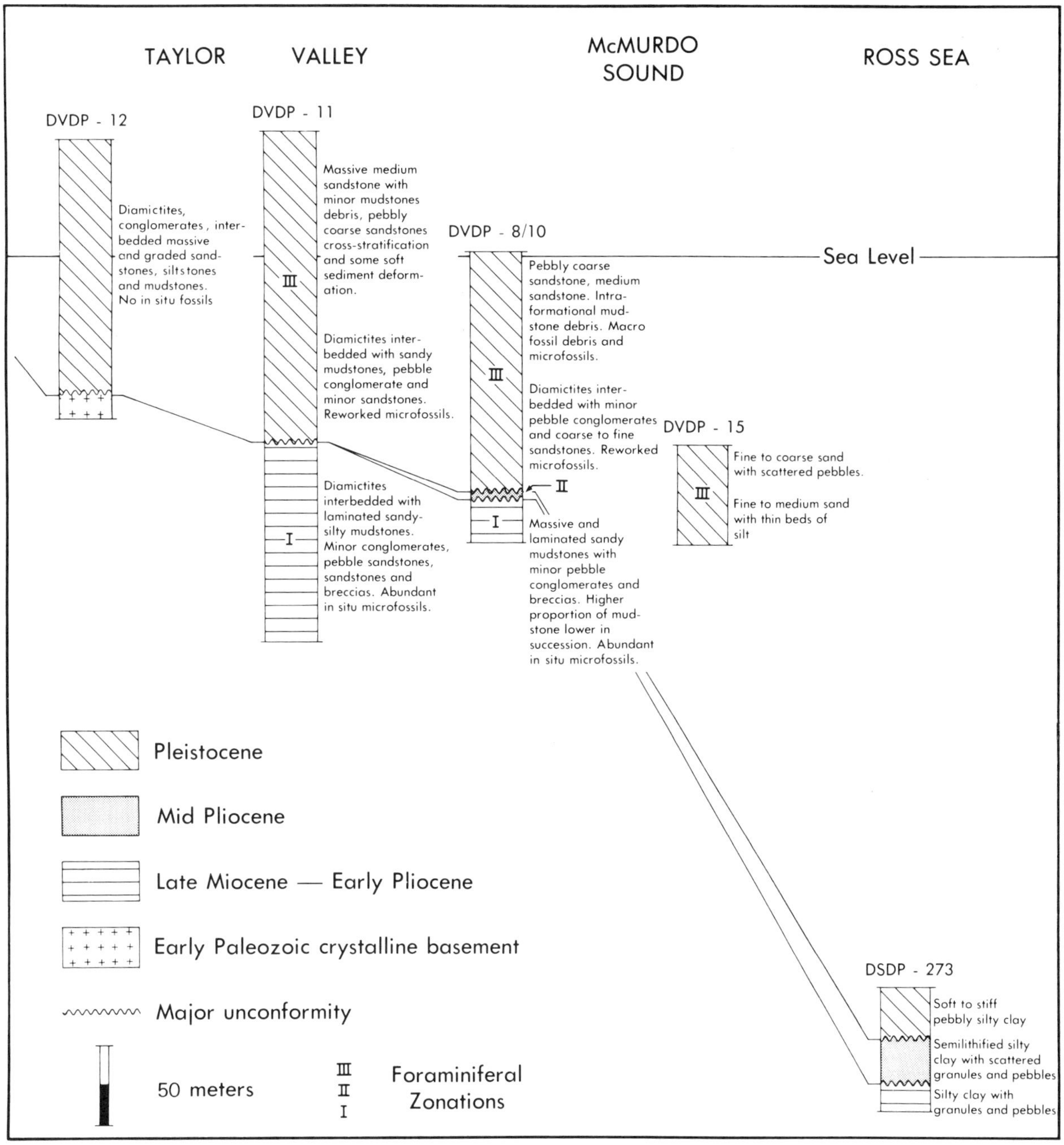

Figure 144.4. DVDP drill holes showing stratigraphic distributoin of foraminiferal zonations.

eral fauna at 199.30 m is thought to be recycled, whereas that at 205.96 m is certainly in place. A major disconformity is, therefore, thought to lie somewhere between 170.43 m and 205.96 m. A major paleomagnetic reversal has been detected between 202.50 m and 203.50 m (Elston and Spall, personal communication) and this is probably related to a major disconformity. The Zone I–Zone III boundary in DVDP 11 is provisionally placed at the base of a diamictite unit at 202.45 m.

Thickness

Zone I has a minimum thickness of 125 m.

Lithology

In Zone I (DVDP 10), the dominant sediment types are mudstone (66%) and sandstone (26%), whereas breccia, conglomerate, and diamictite together constitute only about 8% of the sequence. Proportions of these sediment groups are markedly different in DVDP 11, only 2.5 km upvalley. There, diamictite and mudstone comprise 60% and 26%, respectively, with conglomerate, sandstone, and breccia making up about 14%.

Microfaunal detail

At least sixty-seven species of foraminifera are present. Thirty-seven species (55% of the fauna) are restricted to Zone I. Four species range up into and terminate within Zone II. Twenty-six species range through all three zones. Tests are generally minute, thin walled, and exceedingly well preserved. Most faunas are regarded as being in place. The zone is dominated by members of the Miliolidae, Lagenidae, and Glandulinidae. Rare occurrences of the planktonic taxa *Neogloboquadrina pachyderma* (Ehrenberg), *Globigerinita* c.f. *glutinata* (Egger), and *Globorotalia* c.f. *acostaensis* Blow–c.f. *siakensis* Le Roy are present. The last two taxa do not range into higher zones. Tiny fragile tests of *Trochoelphidiella onyxi* Webb occur throughout Zone I and are precursors to the large thick-walled forms in Zone II. It may be possible to further subdivide Zone I.

Age

Zone I foraminifera closely resemble those in Deep Sea Drilling Project (DSDP) 273 (cores 8–9) and 273A (cores 1A–10A). The latter intervals are regarded as Miocene–Pliocene in age by Webb (*in* Hayes and others, 1975). McCollum (1975) and Ciesielski (1975) have suggested a Miocene age for the same intervals, based on their study of diatoms and silicoflagellates. Based on his study of diatoms from DSDP 266 and Elatanin 34-19, Brady (1977) regards sediments embraced by foraminiferal Zone I in DVDP 10 and 11 as early Gilbert in age (below the "b" event), i.e., 3.97 to 5 m.y. old, or early Pliocene. *Neogloboquadrina pachyderma* (Ehrenberg) is present in some Zone I faunas. This taxon first appears during late Miocene time (about 10.5 m.y. ago) and provides a maximum age for this zone.

Zone II

Occurrence

Zone II is present in DVDP 8 and 10 (Figure 144.4).

Vertical distribution

Zone II faunas occur at 153.88 m and 154.41 m in hole 8. A sample at 155.18 m is barren, and it is not clear whether sediments at this level should be placed in Zone I or II. No Zone III faunas are known immediately above the 153.88 m level; the lowest occur at 125.17–125.28 m. The top of Zone II in DVDP 8 is provisionally set at the base of a fine- to medium-grained gravel (153.61 m).

In DVDP 10 the Zone II fauna occurs at 154.80 m. The lower boundary, with Zone I, is placed at 155.26 m, while the upper boundary, with Zone III, is placed at 154.20 m. Disconformities mark the top and base of Zone II in DVDP 10. A granule conglomerate marks the base of Zone II in DVDP 10.

Thickness

Zone II is less than 2 m thick in both DVDP 8 and 10. The presence of a disconformity at its upper boundary suggests that it is probably an erosional remnant.

Lithology

This very thin succession is composed of sandstone (85%) and conglomerate (15%).

Faunal detail

Twenty-two species of foraminifera occur in Zone II. Only two of these are confined to the zone, namely, *Trochoelphidiella onyxi* Webb (large thick-walled form) and *Fissurina radiata* Seguenza. The lowest *Globocassidulina crassa rossensis* Kennett occurs in Zone II and ranges through to the top of Zone III. No planktonic taxa have been encountered. Most of this fauna also occurs in Zones I and III.

Age

This microfauna is correlated with the middle Pliocene *Pecten* gravels microfauna from Wright Valley (Webb, 1972, 1974) for which Webb proposed an age of between 2.5 and 3.8 m.y. It was noted above that Brady (1977) proposed an early Pliocene age (3.97–5.0 m.y.) for the diatom floras which occur along with the foraminiferal faunas in the underlying Zone I. Zone II foraminifera can also be correlated with those present in DSDP 273 core 5-1 (71–75 cm) to 273 (core 7cc). Diatoms from this same interval are regarded as middle Pliocene in age (Brady, 1977).

Zone III

Occurrence

Zone III is present in five drill holes, DVDP 8 to 11 and 15 (Figures 144.4 and 144.5).

Vertical distribution

Zone III occurs above Zone II in holes 8 and 10 and above Zone I in hole 11 (Figures 144.4 and 144.5). In each case it extends to the top of the hole. In DVDP 8, Zone III extends from 0–153.61 m; in 10, from 0–154.20 m; and in 11, from 0–202.45 m. It must be admitted that at this stage faunal definition of the Zone III lower boundary is poor. While Zone I and Zone II faunas exit abruptly at or very near the provisional Zone III contact, the lowest convincing Zone III faunas appear some 20–30 m above the contact (125 m in 8 and 10 and 170 m in 11). Rather than propose a barren zone, the Zone III lower boundary is brought down to the level of probable major disconformities.

Thickness

The thickness of this zone is 154 m in DVDP 10 and 203 m in DVDP 11.

Lithology

In DVDP 10, diamictite (48%), sandstone (26%), and conglomerate (18%) are the dominant sediment types. In DVDP II, sandstone (54%), conglomerate (22%), and diamictite (18%) predominate. Breccia and mudstone are relatively minor in both successions.

Faunal detail

Zone III microfaunas total forty-four species, fifteen of which are confined to this zone. Three species range up from Zone II and twenty-six from Zone I. The distribution of foraminifera through Zone III is sporadic. While many Zone III faunas have been encountered in DVDP 8, 9, 10, 11, and 15, it is commonly difficult to separate *in situ* and reworked assemblages. Faunas with diversities of as much as twenty-four species occur in the upper 24 m of DVDP 8, 9, and 10. These are probably *in situ* marine faunas. These faunas were not encountered at the same level in nearby DVDP 11. Faunas in the lower and middle parts of Zone III in 8, 10, and 11 are modified to varying degrees and commonly occur with reworked marine diatom floras (Brady, 1977). Faunas at 90.39 m, 125.15 m, and 125.25 m in DVDP 8, and at 125.86 m, 132.84 m, and 170.35 m in DVDP 11 are possibly in place and therefore indicate marine environments. Zone III foraminifera cited from the uppermost part of DVDP 15 (Brady and Webb, 1976) are identical to those in the upper 24 m of DVDP 8, 9, and 10.

Age

Zone III foraminifera are clearly Pleistocene in age. It is not clear, however, whether Zone III embraces a continuous sedimentary succession or only a partial Pleistocene record interrupted by major disconformities and minor diastems. The second interpretation is more likely the case. *Adamussium colbecki* shells from the 23–24 m level in DVDP 8 and 9 have been dated (carbon 14) at 6670 years old (Stuiver and others, 1976). Carbon-14 dating of macrofossil debris from DVDP 11 at 170.35 m provided an age of >26,800 years (M. Stuiver, personal communication). The basal part of Zone III could well prove to be early or middle Pleistocene in age.

FORAMINIFERAL BIOSTRATIGRAPHY

The biostratigraphic relationships for several taxa discussed here differ significantly from results reported from Ross Sea *Eltanin* cores by Fillon (1974). He has maintained that *Globocassidulina biora* is endemic to Gauss (Pliocene) sediments and evolved into *G. crassa rossensis* during Matuyama and early Brunhes time (late Pliocene–middle Pleistocene). Occurrences of *G. biora* in Ross Sea Bruhnes sediments have been explained as a result of reworking. The present study suggests that *G. crassa rossensis* precedes *G. biora,* with the latter occurring only in upper Pleistocene sediments. Fillon has questioned Crespin's age assignment of Pleistocene–Holocene for *G. biora* in fjord terrace sediments in the Vestfold Hills area of Antarctica. The eastern Taylor Valley results, however, support Crespin's age determination.

HIATUSES

The biostratigraphic results presented above portray three zonal units separated by two major temporal breaks. In reconstructing a geologic history in an area subjected to deteriorating and ameliorating climate, hiatuses can be extremely important. Hiatuses or sedimentary disconformities may result from erosive action of terrestrial glaciers or their sublittoral extensions, erosive action by grounded marine ice shelves, or erosive scouring action by currents beneath thick and nearly grounded ice shelves. Such erosion may be coincident over a wide area. This is probably the case with the Ross Sea continental shelf and adjacent land areas. The recognition and accurate calibration of these events is important, particularly in attempts at relating the marine and terrestrial record.

Fillon's (1972, 1974) paleomagnetic and radiolarian results from the Ross Sea suggest the presence of a major disconformity which separates upper Brunhes (upper Pleistocene) sediments from lower Matuyama and Gauss sediments (older than 1.87 m.y., or Pliocene). This disconformity is common across a wide area of the Ross Sea floor in present-day water depths of 285 m to more than 2000 m. Fillon's Brunhes sediments range from 20 cm to 6 m in thickness. It is tempting to relate this hiatus to the disconformity between Zones II (middle Pliocene) and III (Pleistocene) in eastern Taylor Valley.

At DSDP 273 one obvious hiatus has been determined geophysically and micropaleontologically, and it occurs between cores 4 and 5 (Hayes and others, 1975, p. 344). This matches the Zone II–Zone III hiatus in eastern Taylor Valley, which separates middle Pliocene from Pleistocene deposits. A second, lower hiatus has now been determined between the bottom of core 7 and core 8-2 (64–68). It separates upper Miocene–lower Pliocene foraminifera (containing tiny fragile *T. onyxi*) from middle Pliocene microfaunas with large robust specimens of the same species. Core 8 records (Hayes and others, 1975, p. 347) indicate "rare granules along irregular erosion surface at 8-1 (103 cm)." Maximum pebble size is given as 7 mm. This boundary is equated with the Zone I–Zone II hiatus in eastern Taylor Valley.

In conclusion, major hiatuses detected both in the Ross Sea and in Taylor Valley appear to span the same periods and probably derive from some coincident glacial phenomena.

Paleoecology

It was pointed out in the preceding section that the Zone I fauna was both very abundant and very diverse. One of the richest faunas (at 205.96 m in DVDP 11) occurs in a pebble conglomerate and breccia; other very rich assemblages (e.g., at 257.63 m and 275.95 m) occur in a thick diamictite succession. A few faunas (e.g., 249.18 m in DVDP 11) were obtained from thin silty mudstone interbedded with the diamictite. Slighly smaller and less diverse Zone I faunas are found in the sandy mudstone of DVDP 10 (e.g., 160.62 m and 179.87 m).

These faunas are similar to upper Miocene–lower Pliocene faunas found at 600 m below present sea level beneath the Ross Sea continental shelf. According to the bathymetric ranges given for comparable Holocene faunas (Fillon, 1972; Anderson, 1975), the Zone I benthic microfaunas could represent depths of more than 700 m. Zone I faunas show affinities with a fauna reported from a depth of 890 m in Hardangerfjord, Norway (Holtedahl, 1965). The Zone I fauna closely approximates Anderson's (1975) *Fresh Water Shelf Facies,* a facies stated to range between 235 and 713 m.

The composition of the enveloping marine diamictons, the presence of tiny, exceedingly well preserved and diverse calcareous taxa, the occurrence of rich planktonic diatom floras, and the occurrence of some planktonic foraminifera permit the following paleoecologic conclusions. A floating arm of the Ross Ice Shelf extended into eastern Taylor Valley. Ice cover was thin and disintegrated readily. Open water probably existed periodically. Coarse, poorly sorted glaciomarine diamictons settled to the fjord floor from melting shelf ice. Periods of open water allowed the development of biologically fertile photic zones of normal salinity, from which organic material descended to the fjord floor and helped support a diverse benthic fauna. The water depth was great, perhaps as much as 600–900 m. Shelf or glacier ice did not extend to the bottom of this deep fjord, and the bottom sediments were little agitated by currents. Consequently, a minimal amount of faunal restructuring took place.

The occurrence in DVDP 11 of deep-water Zone I microfaunas at depths of as little as 120–130 m below present sea level is anomalous. Their present position can only be explained through the tectonic uplift of these upper Miocene–lower Pliocene beds by 600 m or more (Wrenn and Webb, Chapter 141, this volume).

Zone II microfaunas (middle Pliocene) contrast markedly with those just discussed. Webb (1972, 1974) deduced a bathymetric range of 10 to 100 m for a similar fauna in Wright Valley (the *Pecten* gravels). The sediments and microfauna of Zone II in DVDP 10 (154 m below present sea level) clearly represent a sublittoral fjord environment. Zone II sediments and fauna developed on an eroded surface of the uplifted Zone I succession. Insufficient sedimentary information is available from this very thin succession to aid in deciphering this marginal marine facies. Thin sea or shelf ice probably floated above Zone II sites. Planktonic groups were not present in the water column. According to Webb (1972, 1974), this fauna developed under interglacial conditions.

Zone II sediments were eroded and almost destroyed by Pleistocene glacial activity, which extended to 153 m below present sea level. All Zone III microfaunas discussed above are thought to indicate a water depth of less than 50 m. Intense sampling of the upper parts of DVDP 10 and 11 reveal poorly preserved foraminifera faunas with diversity seldom exceeding 10 species. Although undisturbed faunas may occur at 125.15 m (silt-fine sand) and 90.39 m (diamictite) in DVDP 10, and at 132.84 m (conglomerate) in DVDP 11, most of these Pleistocene sandstone and diamicton units are unfossiliferous, or nearly so. Pleistocene sea level appears to have stood at or near −153 m. The deposition of freshwater and marginal marine sediments kept pace with the rise of sea level during late Pleistocene time so that water depths in eastern Taylor Valley were always shallow. During several phases of expansion and thickening of the Ross Ice Shelf, grounding and erosion occurred (starting with erosion of middle Pliocene sediments at −153 m). Marine microfacies were periodically displaced eastward to be replaced by brackish and freshwater conditions. Brackish and freshwater diatom floras flourished during these periods (Brady, 1977). Thin microfossil-bearing Pleistocene marine units were destroyed and reworked during grounding periods. Some of the fragmentary Zone I, II, and III faunas encountered in Zone III were probably transported by shelf ice within and into Taylor Valley. During warm phases, the shelf thinned, became locally buoyant, and marine waters penetrated a short distance into eastern Taylor Valley.

Very diverse (24 species) foraminiferal faunas in the sandy upper 24 m of DVDP 8 and 10 lived in sublittoral (<50 m) waters probably blanketed by semipermanent sea ice. Freshwater, brackish, and marine diatoms flourished (Brady, 1977), but planktonic foraminifera seem not to have been present. It is important to note that this uppermost fauna (~6000 years old) is not found at the corresponding interval (80–120 m) in DVDP 11, only 2.5 km inland. Thus, the marine penetration of Taylor Valley probably was not extensive during the last 6000 years. A probably *in situ* fauna at 132.84 m in DVDP 11 (equivalent to 52 m in DVDP 10) suggests littoral marine penetration as far west as Commonwealth Glacier just prior to 6000 years ago. Possibly *in situ* microfaunas do occur in DVDP 10 (90.39 m and 95.28 m) and in DVDP 11 at the corresponding interval (170.35 m). These faunas indicate that a short period of floating shelf ice existed in "Taylor Fjord." Absence of microfaunas from any of the DVDP 12 material argues against marine penetration west to Canada Glacier at any time during the Pleistocene Epoch.

Acknowledgments. This work was supported by National Science Foundation Grant DPP74-22894. Drafting for this paper was undertaken in the School of Geosciences, Louisiana State University. Meticulous sampling of the several hundred samples upon which this study is based was executed by Dennis Cassidy at Florida State University, Tallahassee.

References cited

Anderson, J. B. 1975. Ecology and distribution of foraminifera in the Weddell Sea of Antarctica. *Micropaleontology, 21,* No. 1, 69–96.

Barrett, P. J., Treves, S., Barnes, C., Brady, H., McCormack, S., Nakai, N., Oliver, J., and Sillars, K. 1976. Initial Report of DVDP 15, western McMurdo Sound, Antarctica. *Bull. Dry Valley Drilling Project,* No. 7, 1–100.

Brady, H. 1977. *Late Neogene Diatom Biostratigraphy and Paleoecology of the Dry Valleys and McMurdo Sound, Antarctica.* Masters thesis, Northern Illinois University, 172 pp.

Brady, H., and Webb, P. N. 1976. Microfossils from DVDP Hole 15 western McMurdo Sound. *Antarct. Jnl U.S., 9,* No. 2, 81–82.

Chapman-Smith, M., and Luckman, P. 1974. Late Cenozoic glacial sequence cored at New Harbor, Victoria Land, Antarctica (DVDP 8 and 9). *Bull. Dry Valley Drilling Project,* No. 3, 137–48.

Ciesielski, P. F. 1975. Biostratigraphy and paleoecology of Neogene and Oligocene silicoflagellates from cores recovered during Antarctic Leg 28, Deep Sea Drilling Project. (*In* Hayes, D. E., Frakes, L. A., and others. *Initial Reports of the Deep Sea Drilling Project,* Vol. 28. Washington [U.S. Government Printing Office], 625–91.)

Fillon, R. H. 1972. Evidence from the Ross Sea for widespread erosion. *Nature Phys. Sci., 238,* No. 81, 40–42.

Fillon, R. H. 1974. Late Cenozoic foraminiferal paleoecology of the Ross Sea, Antarctica. *Micropaleontology, 20,* No. 2, 129–51.

Hayes, D. E., Frakes, L. A., and others. 1975. *Initial Reports of the Deep Sea Drilling Project,* Vol. 28. Washington (U.S. Government Printing Office), 1017 pp.

Holtedahl, H. 1965. Recent turbidites in the Hardangerfjord, Norway. *Proc. Colston Res. Soc., 17,* 107–40.

Holtedahl, H. 1967. Notes on the formation of fjords and fjord-valleys. *Geogr. Annlr.,* Series A, *49,* No. 2–4, 188–203.

McCollum, D. W. 1975. Diatom stratigraphy of the Southern Ocean. (*In* Hayes, D. E., Frakes, L. A., and others. *Initial Reports of the Deep Sea Drilling Project,* Vol. 28. Washington [U.S. Government Printing Office], 515–38.)

McKelvey, B. C. 1975. Preliminary site reports: DVDP Sites 10 and 11, Taylor Valley. *Bull. Dry Valley Drilling Project,* No. 5, 16–60.

Stuiver, M., Denton, G. H., and Borns, H. W., Jr. 1976. C-14 Dates of *Adammussium colbecki:* (Mollusca) in marine deposits at New Harbor, Taylor Valley. *Antarct. Jnl U.S., 11,* No. 2, 86–88.

Webb, P. N. 1972. Wright Fjord, Pliocene marine invasion of an Antarctic dry valley. *Antarct. Jnl U.S., 7,* No. 6, 226–34.

Webb, P. N. 1974. Micropaleontology, paleoecology and correlations of the Pecten gravels, Wright Valley, Antarctica, and description of *Trochoelphidiella onyxi,* n.gen. n.sp. *J. Foramin. Res., 4,* No. 4, 184–99.

Webb, P. N., and Wrenn, J. H. 1976. Foraminifera from DVDP Holes 8–12, Taylor Valley. *Antarct. Jnl U.S., 11,* No. 2, 85–86.

Wrenn, J. H. 1977. *Cenozoic Subsurface Micropaleontology and Geology of Eastern Taylor Valley, Antarctica.* Masters thesis, Northern Illinois University, 255 pp.

145

Late Cenozoic History of Taylor and Wright Valleys and McMurdo Sound Inferred from Diatoms in Dry Valley Drilling Project Cores

HOWARD T. BRADY*
Department of Geology, Northern Illinois University, DeKalb, Illinois 60115, U.S.A.

Abstract. Sedimentary successions have been recovered from drill holes in Wright Valley, Taylor Valley, and McMurdo Sound. Fossil diatoms, both nonmarine and marine, record the Miocene–early Pliocene fjord history in Wright and Taylor Valley and subsequent periods of lacustrine conditions. At least one of these lakes was associated with the wasting of the Ross Ice Sheet in western McMurdo Sound.

Marine and nonmarine diatoms occur in the sedimentary successions recovered by the Dry Valley Drilling Project from Taylor Valley (DVDP 10 and 11), Wright Valley (DVDP 4A), and McMurdo Sound (DVDP 15) (Figure 145.1). In many intervals, diatoms are the only abundant microfossils present.

Marine diatom zones can be correlated with partial range zones proposed for the Southern Ocean by Donahue (1970), McCollum (1975), and Schrader (1976). Nonmarine diatoms indicate two distinct paleoenvironments: (1) meltpools on the surface of an ablating ice shelf and (2) terrestrial lakes. Diatoms are also useful in the recognition of lacustrine, as distinct from marine, environments.

Diatom zonations of DVDP successions

The use of event names

Multiple glaciations in the McMurdo region were first recognized and named by Péwé (1960). These have been dated as Pleistocene events and named (from oldest to youngest): the McMurdo Glaciation, the Taylor Glaciation, the Fryxell Glaciation, and the Koettlitz Glaciation. Other workers revised, extended, and correlated with Péwé's initial classification (Bull and others, 1962; Nichols, 1964, 1971; Calkin, 1971; Denton and Armstrong, 1968; Denton and others, 1970; Denton and Borns, 1974).

During these various studies of Victoria, Wright, and Taylor Valleys, a separate terminology evolved for the glaciations within each valley system. Authors agree on three sources for these glaciations:

1. glaciations from the polar plateau (e.g., Taylor Glaciations in Taylor Valley);
2. glaciations from the névé fields in the ranges between valleys, e.g., the Kukri Hills, Asgaard Range, Olympus Range, Insel Range, and St. John's Range (Alpine Glaciations); and
3. glaciations that advance inland from the coast, originating either from a piedmont glacier or from expansion of a grounded Ross Ice Shelf (Ross Glaciations).

A summary of current terminology is given in Figure 145.2 (after Calkin and Nichols, 1972).

It is difficult to correlate subsurface DVDP drill-hole successions with defined surface events. For example, it could be misleading to correlate two glaciofluvial events in the uppermost 112 m of DVDP 11 (0–31.83 m and 89.91–111.90 m) with lacustrine phases of Ross I and Ross II Glaciations in Taylor Valley when two lacustrine phases within the one event (e.g., Ross I) may be involved. Because of this difficulty, independent subsurface zonations based on diatom biostratigraphy are proposed for DVDP 4A, 10, 11, and 15.

The nature of diatom zones in DVDP successions

Marine diatom zonations proposed for DSDP successions are based on fossil assemblages which are characterized by the first or last appearance of certain diatom species within the assemblage. The DVDP record (which represents diverse environments) is so fragmentary that zonal definitions need to be based on additional factors. The diatom zones proposed here thus also depend upon:

1. marine or nonmarine diatom species present *in situ,*
2. the presence of reworked diatoms,
3. lithostratigraphic unconformities, and
4. foraminiferal zones (Webb and Wrenn, 1976; Wrenn, 1977).

Because facies changes are so abrupt in the environments represented by DVDP successions, diatom zonation boundaries cannot in all cases be traced over a wide area. DVDP diatom zonations simply reflect the paleoenvironments represented in each local succession.

The term *biozone* is used here for a succession with a particular assemblage of microfossils which are either *in situ* or reworked. The term *interzone* is used here for a succession which is barren of fossil material.

The importance of nonmarine diatoms

It can be misleading to employ assemblages of foraminifera, bryozoa, sponge spicules, and shell material in making paleoecologic interpretations of sedimentary deposits in the McMurdo Sound area. Ice sheets rework these fossils, and floating ice shelves bring them to the ice surface by the processes of bottom growth (incorporating new material from the sea floor) and surface ablation

* Present address: School of Biological Sciences, Macquarie University, North Ryde, New South Wales, Australia.

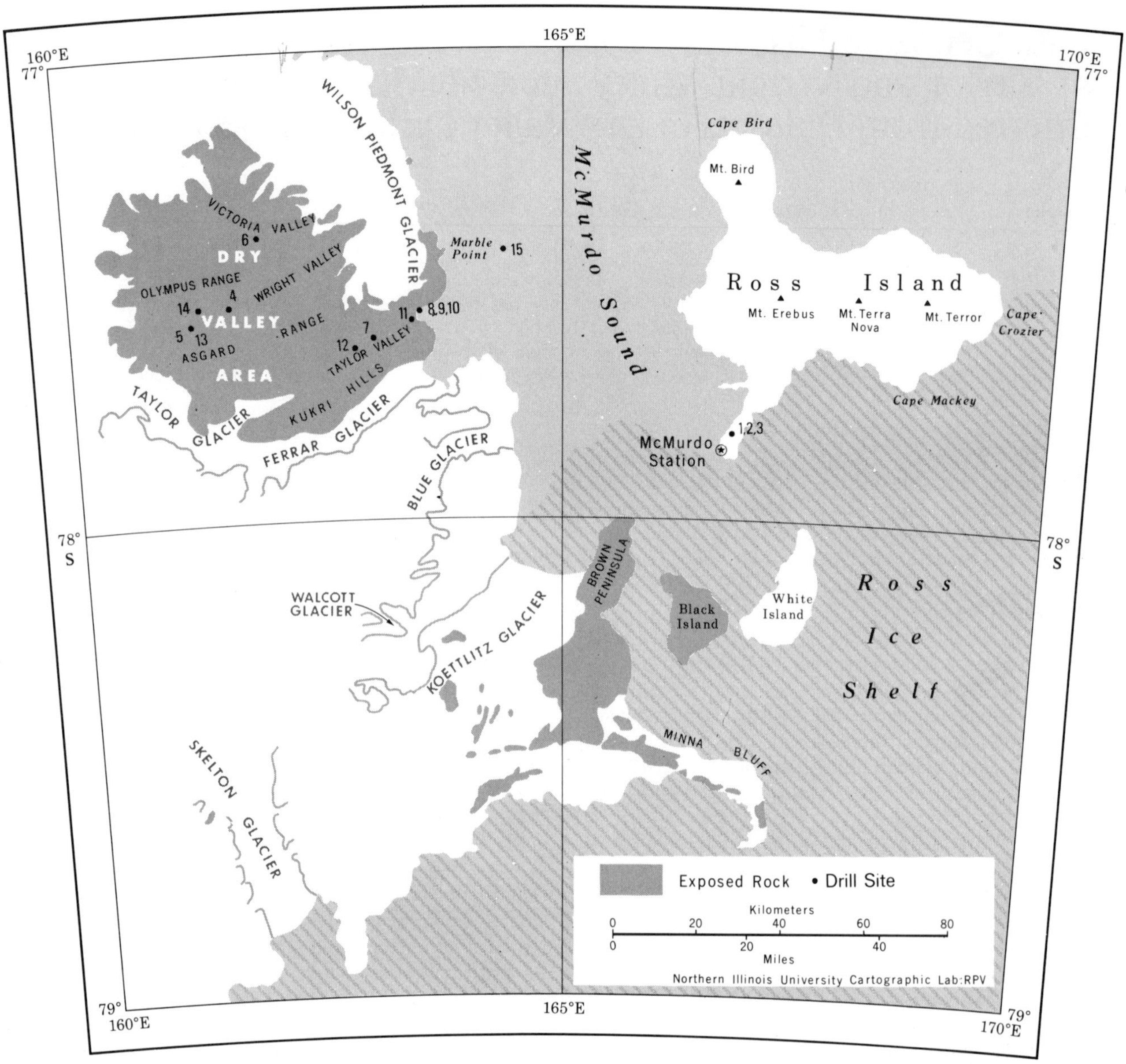

Figure 145.1. Dry Valley Drilling Project drillsites.

(Debenham, 1921). When these marine assemblages are brought to the surface of an ice shelf, they become mixed with the nonmarine diatoms that live in summer meltpools on the surface of the ice shelf. Ice-cored material of the Taylor Formation from an exposure 31 m above sea level at Cape Barne, Ross Island, contains nonmarine diatoms as well as foraminifera, bryozoa, sponge spicules, and shells. This indicates that the deposit is not *in situ* marine but that it has been reworked and transported by an ice shelf. The interpretation by Hendy and others (1969) that this deposit is *in situ* is thought to be incorrect; these authors did not record the presence of nonmarine diatoms.

The recent interpretation of sediments at Salmon Glacier near Cape Chocolate as deltaic by Miagkov and others (1976) also seems to be oversimplified. They record the presence of both marine and nonmarine diatoms in deposits near Salmon Creek, but they interpret the deposit as evidence of former high sea levels. However, there is no evidence that the nonmarine diatoms of the McMurdo Sound region coexist with marine diatoms in waters of reduced salinities. Instead of interpreting this deposit as *in situ* marine in an area of freshwater input, it seems more logical to interpret the deposit as reworked and transported marine material diluted with the nonmarine diatoms that flourish on the surface of an ice shelf. As an ice shelf deposits marine material, freshwater diatoms derived from surface meltwater streams would be mixed with this material. There is also the possibility that freshwater moats might exist between an ice shelf and the land. The Salmon Creek deposits do not provide definite evidence of a higher stand of sea level. The diatoms recorded there come from not one, but two distinct environments.

The Cenozoic history of DVDP successions

Wright Valley

Using the lake ice as a drilling platform, the DVDP rig was able to core the sediments of Lake Vanda. The Lake Vanda sedimentary succession in DVDP 4A (Figure 145.3)

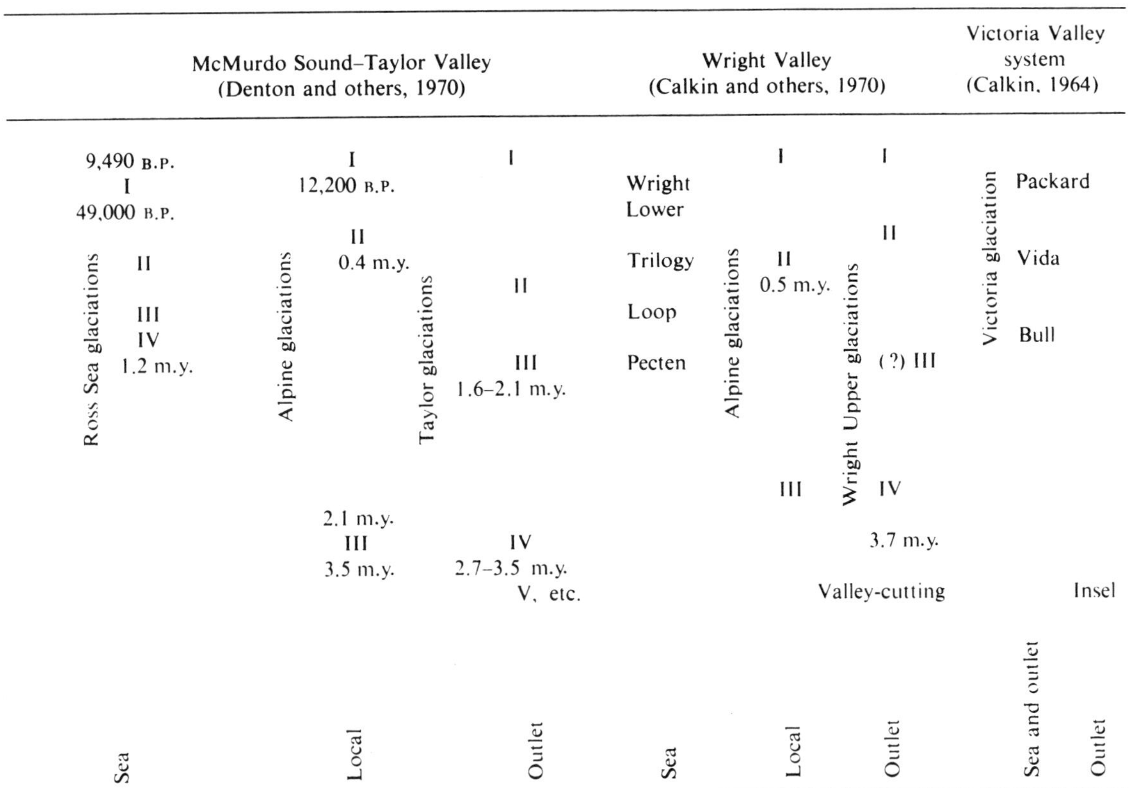

Figure 145.2. Glaciations proposed in the McMurdo Sound region (after Calkin and Nichols, 1972).

consists of 11.2 m of siltstone with minor amounts of sandstone and conglomerate (Figure 145.3). There is no diamictite in the sequence, suggesting that no ice tongues have floated in a lacustrine or marine environment over the Lake Vanda area during the deposition of the Vanda sediments. A major glaciation could, of course, have removed such deposits.

Freshwater nonmarine diatoms are present only in the uppermost 3.84 m of DVDP 4A (Biozone Z) (Brady, 1974, 1975). Gypsum and evaporite layers in this interval indicate several desiccation phases in the lake's history.

In the lowermost sedimentary interval of DVDP 4A (Figure 145.3) only Miocene marine diatoms are present. These include *Denticula dimorpha* Schrader, *Denticula hustedtii* Simonsen and Kanaya, *Denticula lauta* Bailey, and *Trinacria excavata* Heiberg. Other fragments have been tentatively identified as *Nitzschia maleinterpretaria* Schrader and *Nitzschia grossepunctata* Schrader. These marine diatoms are present throughout Biozone Y, but there is a marked drop in their abundance in its uppermost 1.5 m (Figure 145.3).

It is difficult to explain this flora as having been reworked into Wright Valley during early Ross glaciations. The earliest Ross glaciation of Taylor Valley is younger than 1.2 m.y. (Denton and Armstrong, 1968; Denton and others, 1970). Any Ross glaciations since that time should have introduced early Pliocene diatom floras into Wright Valley. Rich floras of this age occur *in situ* at the mouth of Taylor Valley (DVDP 10 and DVDP 11) and are presumed to have been flourishing in McMurdo Sound at the same time. The Miocene diatom floras in the DVDP 4A succession could not have been reworked and redeposited in Wright Valley from McMurdo Sound without being accompanied by these early Pliocene floras. The logical conclusion is that the Miocene diatoms in the Lake Vanda succession are either *in situ* or have been introduced into the Vanda Basin from elsewhere within Wright Valley.

This evidence suggests that Wright Valley was a marine fjord prior to late Miocene time and that the valley glaciations that formed Wright Valley and the adjacent dry valleys are middle Miocene or earlier events.

Taylor Valley

Lower Pliocene sediments are represented in Biozone A (DVDP 10 and 11) and Biozone B (DVDP 10) (Figure 145.4). These two biozones are defined as follows.

Biozone A (DVDP 11, 195.22–327.96 m, and DVDP 10, 155.86–183 m). Early Pliocene marine diatom floras are *in situ*. Floral elements include: *Actinocyclus ingens* Rattray, *Cocconeis imperatrix* Schmidt, *Cocconeis placentula* Ehrenberg, *Denticula hustedtii* Simonsen and Kanaya, *Diploneis crabo* Ehrenberg, *Diploneis smithii* Cleve, *Eucampia balaustium* Castracane, *Grammatophora* spp., *Melosira sulcata* (Ehrenberg) Kutzing, *Nitzschia praeinterfrigidaria* McCollum, *Rhizosolenia barboi* Brusn, *Rhizosolenia hebetata f. hiemalis* (Bailey) Grun, *Rhizosolenia styloformis* Brightwell, *Rouxia antarctica* Heiden and Kolbe, *Rouxia naviculoides* Schrader, *Stephanopyxis turris* (Greville and Arnold) Ralfs, *Thalassionema nitzschoides* Grunow, *Thalassiosira oestrupi* (Ostenfeld) Proshkina-Lavrenko, and *Thalassiosira torokina* Brady n. sp. The foraminiferal fauna is described as late Miocene and is discussed by Webb and Wrenn (1976) and Wrenn (1977).

Biozone B (DVDP 10, 154.80–155.40 m). The middle Pliocene foraminiferal fauna from this interval is discussed by Webb and Wrenn (1976) and Wrenn (1977). No diatoms have been recovered.

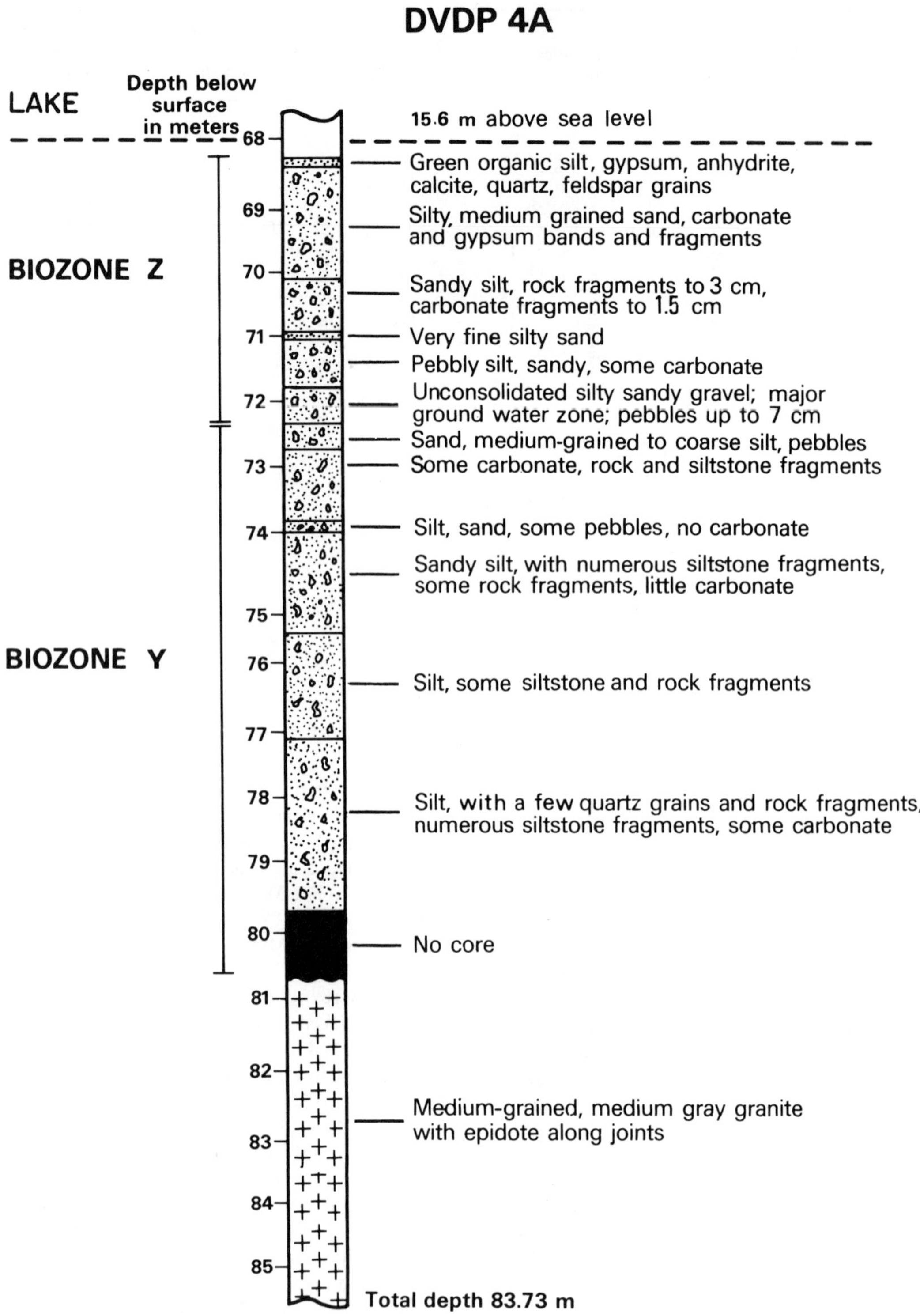

Figure 145.3. Diatom zonations of DVDP 4A, Lake Vanda, Wright Valley (after Brady, 1977).

The lowermost sediments of the DVDP 10 and 11 successions (Biozone A) are dated by diatom biostratigraphy as early Pliocene in age. Worm burrows and abundant well-preserved diatom floras and foraminiferal faunas indicate that the sediments are *in situ*. The presence of *Thallassiosira oestrupi* (Ostenfeld) Proshkina-Lavrenko indicates a Pliocene (DSDP) age, as this taxon does not range into the Miocene in DSDP cores from the Southern Ocean (McCollum, 1975; Schrader, 1976), from the equatorial Pacific Ocean (Burckle, 1972), or from the North Pacific Ocean (Koizumi, 1973; Schrader, 1973). The most abundant taxon is a new early Pliocene diatom species *Thalassiosira torokina* (Brady, 1978*b*), which occurs in the early Gilbert interval of *Eltanin* core 34/19 and in the interval from the "b" event of Gilbert age to the unconformity near the "c" event of Gilbert age in DSDP 266. This author examined the Miocene intervals in DSDP 279 cores, and since this taxon is absent in these successions, it is proposed that *T. torokina* ranges from the "b" event of Gilbert age to the Miocene–Pliocene boundary (Brady, 1978*b*); this evidence,

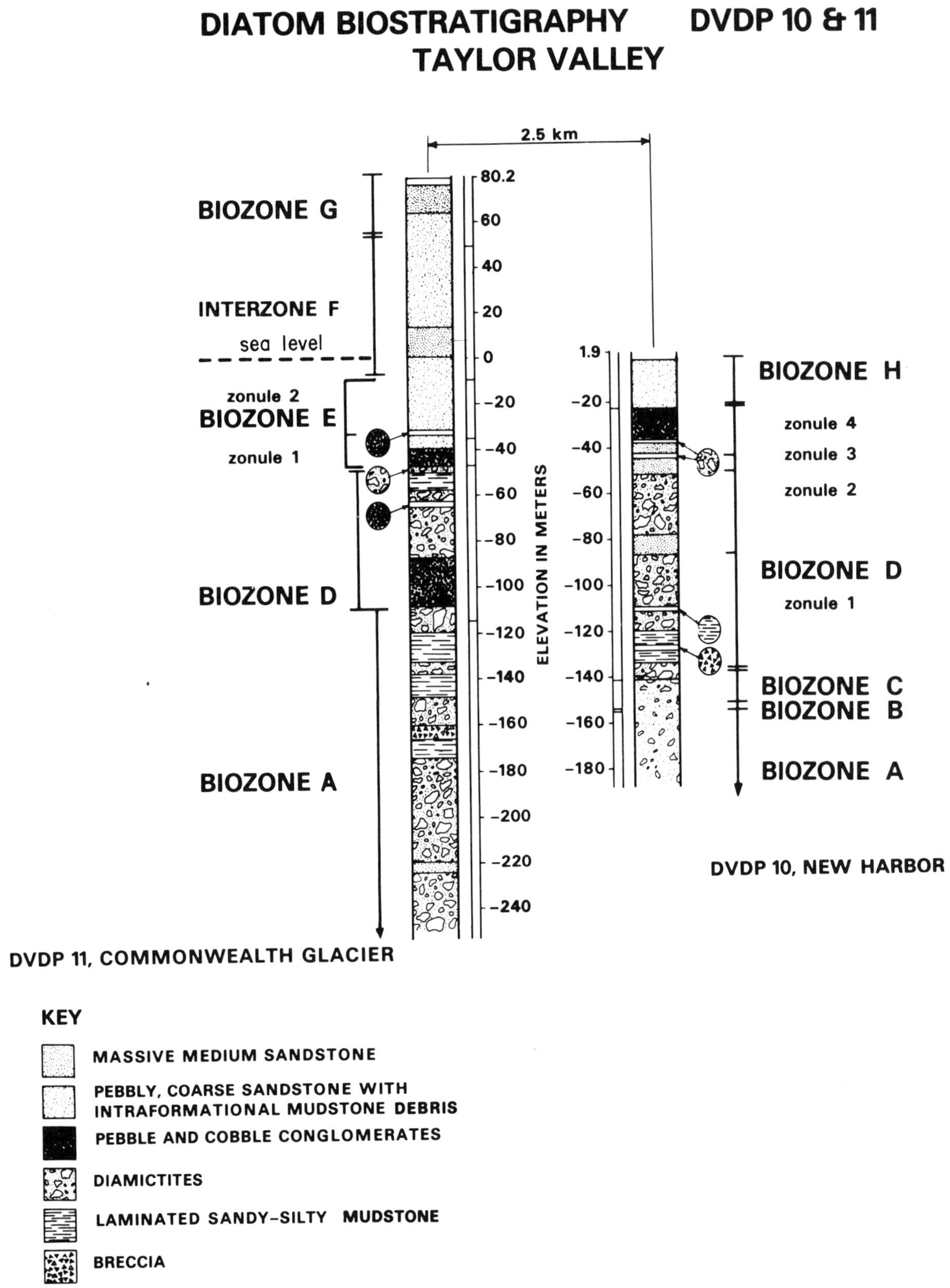

Figure 145.4. Diatom zonations of DVDP 10 and 11, Taylor Valley (after Brady, 1977).

however, is not conclusive. It is possible that lower intervals in DVDP 11 that do not contain *Nitzschia praeinterfrigidaria* or *Thalassiosia oestrupi* (208–324 m) may represent upper Miocene.

Diatom floras indicate that Taylor Valley was a marine fjord in early Pliocene time. The presence of diamictite in Biozone A of DVDP 11 and its absence in the same biozone of DVDP 10, Biozone A, suggests (but does not prove) that the Taylor Glacier was floating at this time near the DVDP 11 site (4 km upvalley) but that it was not present at the mouth of the valley (DVDP 10).

The absence of diatoms in diamictite (thick ice cover)

and their presence in siltstone and mudstone (open water) suggests that the thanatocoenonsis mirrors the original environment.

The only other Pliocene sediments in DVDP 10 and 11 occur in a thin (2 m) interval in DVDP 10 (Biozone B). No diatoms have been recovered from this interval, but Webb and Wrenn (1976) and Wrenn (1977) have described an abundant foraminiferal fauna. The same fauna has been found on Black Island (Scallop Hill Formation) (Speden, 1962) and in Wright Valley (Nichols, 1971; Webb, 1972, 1974). In DVDP 10, the fauna is stratigraphically in order, and so it is possible to interpret it as *in situ*. The oxidized state of the fauna (which is otherwise well preserved) can be explained by subaerial exposure during low stands of late Pliocene–Pleistocene sea level.

The remaining diatom zonations proposed for DVDP 10 and 11 are defined as follows.

Biozone C (DVDP 10, 143.01–154.20 m). An early Pliocene *in situ* marien diatom flora occurs throughout the interval. Floral elements include: *Cocconeis placentula* Ehrenberg, *Coscinodiscus lentiginosus* (Janisch) Hasle, *Coscinodiscus stellaris* var. *symbolophorus* (Grunow) Jorgensen, *Diploneis crabo* Ehrenberg, *Eucampia balaustium* Castracane, *Nitzschia curta* (Van Huerck) Hasle, *Nitzschia obliquecostata* (Van Huerck) Hasle, *Nitzschia praeinterfrigidaria* McCollum, *Nitzschia aff. pseudonama* Hasle, *Nitzschia separanda* (Hustedt) Hasle, *Rhizosolenia styloformis* Brightwell, *Rouxia antarctica* Heiden and Kolbe, *Rouxia californica* Peragallo, *Schimperiella antarctica* Karsten, *Stephanopyxis turris* (Greville and Arnold) Ralfs, *Thalassionema nitzschoides* Grunow, *Thalassiosira gracilis* (Karsten) Hustedt, *Thalassiosira oestrupi* (Ostenfeld) Proshkina-Lavrenko, and *Thalassiosira torokisa* Brady. Near the base of the interval, at 152.10 m, a nonmarine diatom flora is present with the marine flora. Floral elements include: *Navicula murrayi* W. and G. S. West, *Navicula muticopsis f. muticopsis* Van Huerck, *Navicula schackletoni* W. and G. S. West, and *Navicula seminulum* Grun. The basal unit of Biozone C is a sandy breccia bed that overlies Biozone B, and the top of the biozone falls within the lowest diamictite in DVDP 10.

Biozone D (DVDP 11, 127.79–195.22 m, and DVDP 10, 24.86–143.01 m). This zone contains reworked marine diatom floras, reworked Pleistocene foraminiferal faunas (Webb and Wrenn, 1976; Wrenn, 1977) and reworked sponge spicules. Sediment types include diamictite interbedded with conglomerate, sandstone, and mudstone. In DVDP 11 the base of Biozone D directly overlies Biozone A (early Pliocene); a marked unconformity separates the two zones. The top of Biozone D is at the level of the highest diamictite. In DVDP 10 the base of Biozone D marks the lowermost diamictite, and the top marks the uppermost diamictite. In DVDP 10, Biozone D has been subdivided into zonules on the basis of erosional contacts, thick carbonate beds, and thin layers with salt rind on pebbles (McKelvey, 1975). These features give some indication of the fluctuating environment of deposition.

Biozone E (DVDP 11, 89.91–127.79 m). Sediment types include interbedded conglomerate and silty sandstone. Rare reworked marine diatoms and sponge spicules are present. A rich and abundant nonmarine diatom flora occurs in zonule 2 (89.91–115.35 m). Elements of this flora include: *Hantzschia amphioxys* Grun, *Melosira* spp., *Navicula gibbula* Cleve var. *peraustralis* W. and G. S. West, *Navicula murrayi* W. and G. S. West, *Navicula mutica* Kutz *f. cohnii* (Hilse) Grun, *Navicula muticopsis f. muticopsis* Van Huerck, *Nitzschia antarctica* W. and G. S. West, *Navicula seminulum* Grun, and *Pinnularia cymatopleura* W. and G. S. West. The base of the biozone overlies the uppermost diamictite in DVDP 11, and the top underlies a thin bed rich in carbonate (McKelvey, 1975).

Biozone G (DVDP 11, 0–17.15 m). This zone contains reworked marine diatoms and foraminifera and well-preserved, but uncommon, nonmarine diatoms. The flora includes: *Hantzschia amphioxys* Grun, *Navicula muticopsis f. muticopsis* Van Huerck, *Nitzschia antarctica* W. and G. S. West, and *Navicula seminulum* Grun.

Biozone H (DVDP 10, 0–24.86 m). This zone contains mixed assemblages of marine and nonmarine diatoms. Nonmarine diatoms include *Hantzschia amphioxys* Grun, *Navicula murrayi* W. and G. S. West, and *Navicula muticopsis f. muticopsis* Van Huerck. Pleistocene marine diatoms include: *Coscinodiscus lentiginosus* (Janisch) Hasle, *Nitzschia curta* (Van Huerck) Hasle, and *Nitzschia angulata* (O'Meara) Hasle. The Pleistocene foraminifera are discussed by Webb and Wrenn (1976). A few marine shells are present.

The oldest Pleistocene sediments in DVDP cores may occur in DVDP 10, Biozone C (Figure 145.4). Sandstone and mudstone contain foraminiferal faunas and diatom floras. Diatom species such as *Coscinociscus lentiginosus* (Janisch) Hasle and *Nitzschia curta* (Van Huerck) Hasle may indicate a Pleistocene age, but their ranges do extend into the Pliocene. Biozone C represents a marine transgression into Taylor Valley. During the later stages of this transgression (Biozone D), ice tongues from the Ross Ice Shelf brought McMurdo Volcanic Group material into Taylor Valley along with marine fossils. The presence of reworked marine foraminifera and diatoms in Biozones D, E, G, and H indicate a relationship to Ross Glaciations from McMurdo Sound.

The evidence from Biozone D suggests that ice tongues invaded Taylor Valley from McMurdo Sound; at certain times the ice was grounded (depositing subglacial conglomerate), and at other times it floated (depositing diamictite). The fluctuating salinities of Biozone D in DVDP 10 and the low salinities in the corresponding diamictites in DVDP 11 (4 km upvalley) suggest the frequent wedging of marine and freshwater near the mouth of Taylor Valley. (L. McGinnis, personal communication). There is no evidence of *in situ* marine or freshwater diatoms in Biozone D, suggesting that ice tongues were thick and that no open water conditions occurred in the lower Taylor Valley during this time. It is possible that freshwater lakes existed farther upvalley when the Ross Ice Shelf dammed the lower section of the valley.

The remaining history of DVDP 10 and 11 (Biozone E, Interzone F, and Biozones G and H) is explained by the occurrence of grounded glacial ice blocking the mouth of Taylor Valley. Diamictite beds are absent in this upper succession because the moraine threshold at the mouth of Taylor Valley had been built up by the earlier ice sheets that deposited the sediments of Biozone D. It is also possi-

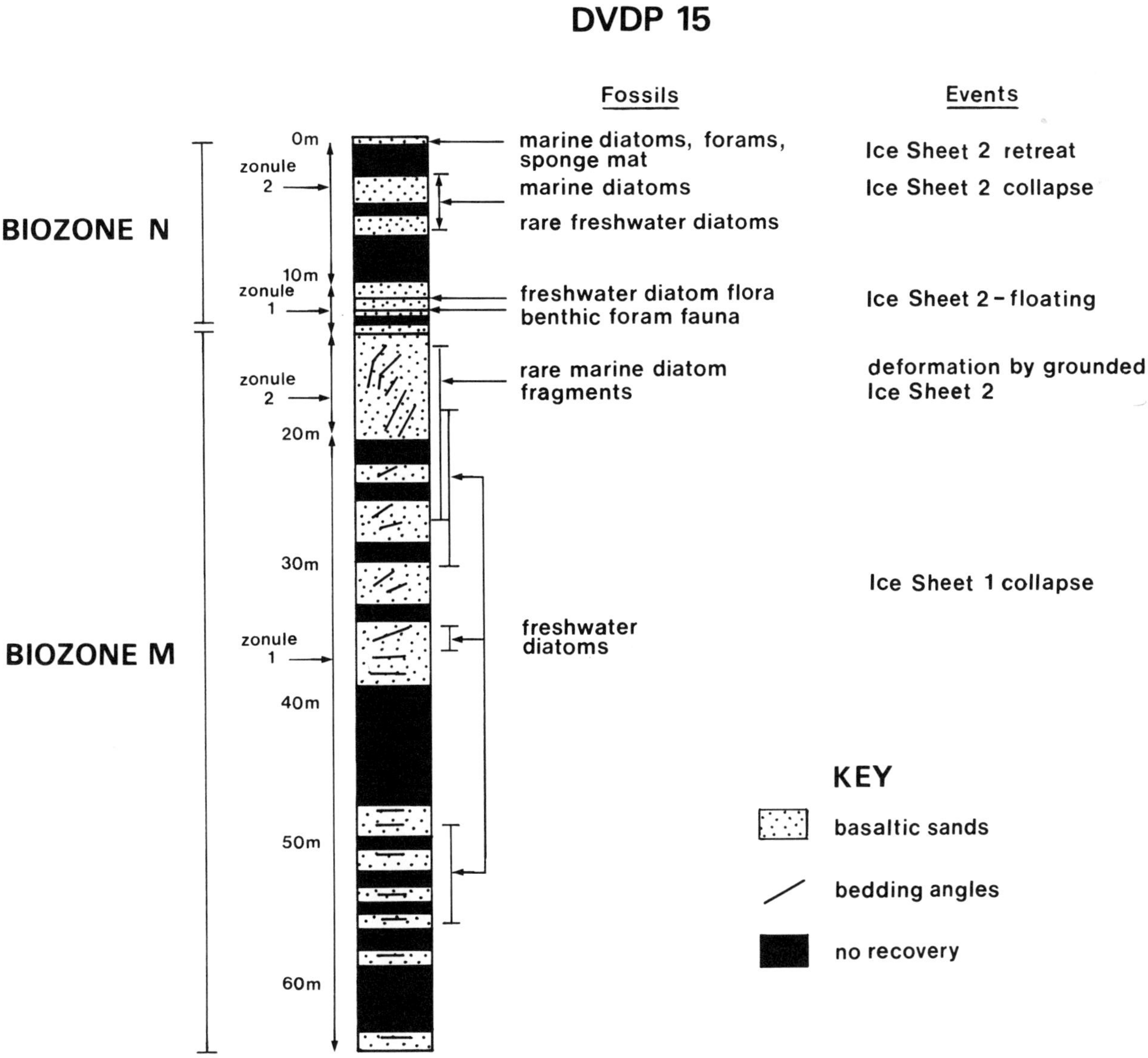

Figure 145.5. Diatom zonations of DVDP 15, McMurdo Sound off Marble Point (after Brady, 1977).

ble (but not proven) that tectonic uplift may have been partly responsible for raising the rock threshold.

Biozones E and G represent two fluvial (possibly lacustrine) intervals. The rich nonmarine diatom flora of Biozone E in beds with inclinations of 20°–30° indicates that the diatom flora was deposited *in situ* in the marginal delta of a freshwater lake. Although nonmarine diatoms are present in Biozone G, no abundant thanatocoenoses occur. The interval may represent either a lacustrine succession or deposits from meltpools on the surface of a disintegrating ice sheet.

The margins of freshwater lakes present during the time periods represented by Biozones E and G are not necessarily near the DVDP 10 or DVDP 11 sites. Margins may have been farther upvalley so that the Ross Ice Shelf that invaded the valley from McMurdo Sound could have deposited subglacial esker sediments at either site (e.g., Interzone F, DVDP 11). These esker systems are being investigated by G. Denton and others (personal communication), and they have been observed by this author.

During the last Ross Ice Shelf invasions of Taylor Valley, the ice margin was slightly upvalley; hence, more deposition occurred at the DVDP 11 site (4 km upvalley) than at the DVDP 10 site (near the mouth of the valley). The uppermost interval of DVDP 10 represents a mixed zone developed after the last retreat of the Ross Ice Shelf from Taylor Valley. The nonmarine diatoms have either developed in a postglacial lake that formed between the terminal moraines and the retreating ice sheet, or they have been deposited subsequently by Wales Stream, which cuts through the terminal moraines at the mouth of Taylor Valley and enters the sea at the DVDP 10 site.

McMurdo Sound

Only one drill hole has penetrated the sediments of McMurdo Sound (DVDP 15) (Figures 145.1 and 145.5). The drill site was located 20 km east of Marble Point, western McMurdo Sound, in a water depth of 122 m.

The age of the uppermost biozone (N) is Quaternary. The foraminiferal faunas (0 m and 12.5 m) can be correlated with Webb and Wrenn's Zone III fauna from DVDP 10 and 11 in Taylor Valley (Webb and Wrenn, 1976) and with the faunas in the elevated deposits of the Taylor Formation (Chapman, 1916; Speden, 1962). The age of the lowermost biozone (M) is uncertain because the range of the freshwater flora is unknown at this time. Diatom zonations proposed for DVDP 15 (Figure 145.5) are summarized below.

Biozone M (13.65–64.6 m). *Zonule 1* (30.25–64.6 m): marine microfossils are absent. Well-preserved nonmarine diatoms include *Melosira* spp., *Navicula seminulum* Grun, *Nitzschia antarctica* W. and G. S. West, and *Navicula capitata* spp. *Zonule 2* (13.65–30.23 m): includes well-preserved nonmarine diatoms with rare marine diatom fragments.

Biozone N (0–13.65 m). *Zonule 1* (11.38–13.65 m): a zonule of mixed marine and nonmarine diatoms. The foraminiferal fauna is dominated by *Ehrenbergina glabra* Heron and Earland (Brady and Webb, 1976). The nonmarine diatom flora elements are the same as in Biozone M. Marine diatom floral elements include *Coscinodiscus lentiginosus* (Janisch) Hasle, *Nitzschia perigalli* Hasle, *Nitzschia curta* (Van Huerck) Hasle, and *Eucampia balaustium* Castracane. *Zonule 2* (0.30–11.38 m): contains rare marine diatoms and foraminifera and extremely rare nonmarine diatoms. *Zonule 3* (0–0.30 m): dominated by the sponge mat of the present-day sea floor. It also includes a rich benthic foraminiferal fauna, rich marine diatom floras, and rare but well-preserved nonmarine diatoms.

The DVDP 15 succession provides important insights into the possible environmental history of McMurdo Sound. Evidence from the lowest beds (Biozone M) indicates that freshwater conditions existed in McMurdo Sound. Only nonmarine diatoms are present and pore water salinities are one-fifth that of seawater (L. McGinnis, personal communication). These deposits formed either during low stands of sea level, in freshwater lakes, or along an upraised continental shelf.

Since freshwater deposits are present 180 m below present-day sea level, and since no recent tectonic movements of that magnitude have been suggested for late Cenozoic time in southern Victoria Land, a tectonic explanation for these sediments seems unlikely. Even if crustal forebulge effects for a glacial maximum are calculated for a distance of 100 miles from the edge of an ice sheet, the uplift is less than 60 m (using equations from McGinnis, 1968).

Biozone M sediments and diatom floras can be explained by suggesting freshwater environments along the continental shelf during low stands of sea level. If marine ice sheets disintegrate prior to deglaciation of the continental ice sheets, then large areas of the continental shelf around Antarctica would be exposed to freshwater or low-salinity lacustrine conditions prior to the elevation of sea level caused by the collapse of the continental ice sheets. This would suggest that marine ice sheets collapse prior to deglaciation and during low sea levels, and not prior to the onset of glaciation and during normal interglacial sea levels (Wilson, 1964, 1966).

This author's preferred explanation for Biozone M is that conditions in McMurdo Sound might be independent of marine sea levels. If McMurdo Sound was blocked by a grounded Ross Ice Sheet to the north of the part of it in McMurdo Sound collapsed, then McMurdo Sound would become a freshwater lake dammed to the north and the southeast by the Ross Ice Sheet, dammed to the east by Ross Island, and dammed to the west by the Antarctic continent (Brady, 1978*a*).

The upper 6 m of Biozone M contains evidence of a marine transgression and subsequent deformation by ice loading; the sigmoidal foreset beds within the plane-parallel beds have dips well above the normal angle of repose (Barrett and others, 1976). This indicates that although the basaltic sediments of the upper sequence (Biozone N) may be derived from the present-day Ross Ice Shelf, the lowermost sediments of DVDP 15 have been derived from an earlier ice sheet which filled McMurdo Sound in pre-Wisconsin time. This earlier ice sheet is designated *Ice Sheet 1* here. Another later ice sheet (perhaps removing some of the marine record in the uppermost interval of Biozone M) deformed the uppermost 6 m of the underlying lacustrine succession. This ice sheet (designated *Ice Sheet 2* here) then thinned and floated, and a benthic foraminiferal and bryozoan community developed in stable but not eutrophic conditions. This fauna is abundant between depths of 12 and 12.5 m. The sediments above this interval are explained by the disintegration of Ice Sheet 2. It is not clear whether this break-up occurred in fully marine conditions. Marine diatoms are rare, and foraminifera are absent in this interval (0.30–12 m). Since nonmarine diatoms are present, together with marine diatoms, it is possible that the rare nonmarine diatoms have been deposited by streams carrying this freshwater flora from meltpools on the surface of a wasting ice shelf. This interval is between 122.30 and 134 m below present-day sea level, a position consistent with low stands of sea level during the Wisconsin glacial maximum.

Conclusions

The inferred geologic history of DVDP cores, as developed from diatom paleoecology, supports the general sequence of events developed in the course of surface geologic surveys of Taylor Valley by Denton and Armstrong (1968), Denton and others (1970), and Denton and Borns (1974); of Wright Valley by Bull and others (1962), Nichols (1971), and Calkin and others (1970); and of Victoria Valley by Calkin (1971). The occurrence of Ross Glaciations and the development of effective ice dams in Taylor Valley are confirmed by the presence of lacustrine nonmarine diatom zones in the Pleistocene successions of DVDP 10 and 11.

Diatom paleoecology and biostratigraphy indicate that no wet-based glaciations from the polar plateau have reached McMurdo Sound through Taylor Valley during the last 4 m.y. The presence of Miocene diatoms in the lowermost sedimentary interval of DVDP 4A (Wright Valley) points to the antiquity of the whole dry valley system and suggests that the main valley glaciations that formed them were middle Miocene or earlier events. This interpretation is consistent with other evidence for Tertiary ice-sheet glaciation in Antarctica (Rutford and others, 1972).

The nonmarine diatom succession in DVDP 15, McMurdo Sound, 180 m below present-day sea level, indicates that western McMurdo Sound was a freshwater oasis at least once, probably during Pleistocene time. Interpretation of the DVDP 15 succession seems to require either ice-sheet disintegration during low stands of sea level or ice-sheet disintegration within McMurdo Sound while thicker sections of the Ross Ice Sheet dammed McMurdo Sound on the north and the southeast.

Acknowledgments. This work was supported by National Science Foundation Grant DPP75-22894, directed by Peter N. Webb.

References cited

Barrett, P., Treves, S., Barnes, C., Brady, H., McCormick, S., Nakai, N., Oliver, J., and Sillars, K. 1976. Initial report of DVDP 15, western McMurdo Sound, Antarctica. *Bull. Dry Valley Drilling Project,* No. 7, 1-100.

Brady, H. 1974. Diatoms from the Lake Vanda cores. *Bull. Dry Valley Drilling Project,* No. 3, 181-84.

Brady, H. 1975. Siliceous sponge material in DVDP 4 (Lake Vanda). *Bull. Dry Valley Drilling Project,* No. 6, 4.

Brady, H. 1977. *Late Neogene Diatom Biostratigraphy and Paleoecology of the Dry Valleys and McMurdo Sound, Antarctica.* Master's thesis, Northern Illinois University, 172 pp.

Brady, H. 1978*a*. Freshwater lakes in Pleistocene McMurdo Sound. *Antarct. Jnl U.S., 12,* No. 4, 117-18.

Brady, H. 1978*b*. *Thalassiosira torokina* n. sp. (diatom) and its significance in late Cenozoic biostratigraphy. *Antarct. Jnl U.S., 12,* No. 4, 122-23.

Brady, H., and Webb, P. N. 1976. Microfossils from DVDP 15, western McMurdo Sound. *Antarct. Jnl U.S., 11,* No. 2, 81-82.

Bull, C., McKelvey, B. C., and Webb, P. N. 1962. Quaternary glaciations in southern Victoria Land, Antarctica. *J. Glaciol., 4,* No. 31, 63-78.

Burckle, L. 1972. Late Cenozoic planktonic diatom zones from the eastern equatorial Pacific. *Nova Hedw. Beih., 39,* 217-46.

Calkin, P. 1964. Glacial geology of the Mt. Gran area, southern Victoria Land, Antarctica. *Geol. Soc. Am. Bull., 75,* 1031-36.

Calkin, P. 1971. Glacial geology of the Victoria Valley system, southern Victoria Land, Antarctica. (*In* Crary, A. P., *ed. Antarctic Snow and Ice Studies 2.* American Geophysical Union, Washington, 363-412.) [Antarctic Research Series, Vol. 16.]

Calkin, P., and Nichols, R. 1972. Quaternary studies in Antarctica. (*In* Adie, R. J., *ed. Antarctic Geology and Geophysics.* Oslo, Universitetsforlaget, 625-43.)

Calkin, P., Behling, R., and Bull, C. 1970. Glacial history of Wright Valley, southern Victoria Land, Antarctica. *Antarct. Jnl U.S., 5,* No. 1, 22-27.

Chapman, F. 1916. Report on the Forminifera and Ostracoda from elevated deposits on the shores of the Ross Sea. *Rep. Scient. Invest. Br. antarc. Exped. 1907-09, Geol., 2,* 27-52.

Debenham, F. 1921. Recent and local deposits of the McMurdo region. *Nat. Hist. Rep. Br. antarc. (Terra Nova) Exped. 1910, Geol., 1,* No. 3, 63-100.

Denton, G. H., and Armstrong, R. L. 1968. Glacial geology and chronology of the McMurdo Sound region. *Antarct. Jnl U.S., 3,* No. 4, 99-101.

Denton, G. H., and Borns, H. 1974. Former grounded ice sheets in the Ross Sea. *Antarct. Jnl U.S., 9,* No. 4, 167.

Denton, G. H., Armstrong, R. L., and Stuiver, M., 1970. Late Cenozoic glaciation in Antarctica: the record in the McMurdo Sound region. *Antarct. Jnl U.S., 5,* No. 1, 15-21.

Donahue, J. 1970. *Diatoms as Quaternary Biostratigraphic and Paleoclimatic Indicators in High Latitudes of the Pacific Ocean.* Ph.D. dissertation, Columbia University, 230 pp.

Hendy, C., Neall, V., and Wilson, A. 1969. Recent marine deposits from Cape Barne, McMurdo Sound, Antarctica. *N.Z. Jl Geol. Geophys., 12,* No. 4, 707-12.

Koizumi, I. 1973. The late Cenozoic diatoms of Sites 183-193. (*In* Creager, J. S., Scholl, D. W., and others. *Initial Reports of the Deep Sea Drilling Project,* Vol. 19. Washington [U.S. Government Printing Office], 805-56.)

McCollum, D. W. 1975. Diatom Stratigraphy of the Southern Ocean. (*In* Hayes, D. E., Frakes, L. A., and others. *Initial Reports of the Deep Sea Drilling Project,* Vol. 28. Washington [U.S. Government Printing Office], 515-71.)

McGinnis, L. 1968. Glacial crustal bending. *Geol. Soc. Am. Bull., 79,* 769-76.

McKelvey, B. C. 1975. DVDP Sites 10 and 11, Taylor Valley. *Bull. Dry Valley Drilling Project,* No. 5, 16-61.

Miagkov, S. M., Nedeshave, G. N., and Riabova, E. I. 1976. McMurdo Sound sea-level changes in the last 50,000 years. *Antarct. Jnl U.S., 11,* No. 4, 233-35.

Nichols, R. 1964. Multiple glaciation in the Wright Valley, McMurdo Sound, Antarctica. *Antarct. Jnl U.S., 7,* No. 4, 97-99.

Nichols, R. 1971. Glacial Geology of Wright Valley. (*In* Quam, L. O., *ed.* Research in the Antarctic. *Am. Ass. Adv. Sci.,* Publ. No. 93, 293-340.)

Péwé, T. L. 1960. Multiple glaciation in the McMurdo Sound region, Antarctica. A progress report. *J. Geol., 68,* 498-514.

Rutford, R. H., Craddock, C., White, C. M., and Armstrong, R. L. 1972. Tertiary glaciation in the Jones Mountains. (*In* Adie, R. J., *ed. Antarctic Geology and Geophysics.* Oslo, Universitetsforlaget, 239-43.)

Schrader, H. 1973. Cenozoic Diatoms from the North Pacific, Leg 18. (*In* Kulm, L. D., von Huene, R., and others. *Initial Reports of the Deep Sea Drilling Project,* Vol. 18. Washington [U.S. Government Printing Office], 673-797.)

Schrader, H. 1976. Cenozoic planktonic diatom stratigraphy of the southern Pacific Ocean. (*In* Hollister, C. D., Craddock, C., and others. *Initial Reports of the Deep Sea Drilling Project,* Vol. 35. Washington [U.S. Government Printing Office], 605-72.)

Speden, I. G. 1962. Fossiliferous Quaternary marine deposits in McMurdo Sound region, Antarctica. *N.Z. Jl Geol. Geophys., 5,* No. 5, 746-77.

Webb, P. N. 1972. Wright Fjord, Pliocene marine invasion of an Antarctic dry valley. *Antarct. Jnl U.S., 7,* No. 6, 227-34.

Webb, P. N. 1974. Micropaleontology, paleoecology and correlations of the Pecten Gravels, Wright Valley, Antarctica, and description of *Trochoelphidiella onyxi,* n. gen., n. sp. *J. Foramin. Res., 4,* No. 4, 184-99.

Webb, P. N., and Wrenn, H. H. 1976. Foraminifera from DVDP holes 8-12, Taylor Valley. *Antarct. Jnl U.S., 11,* No. 2, 85-86.

Wilson, A. 1964. Origin of ice ages: an ice shelf theory for Pleistocene glaciations. *Nature, Lond., 201,* No. 4915, 147-49.

Wilson, A. 1966. Variation in solar insolation to the south polar region as a trigger which induces instability in the Antarctic ice sheet. *Nature, Lond., 210,* No. 5035, 477-78.

Wrenn, J. H. 1977. *Cenozoic Subsurface Micropaleontology and Geology of Eastern Taylor Valley, Antarctica.* Master's thesis, Northern Illinois University, 255 pp.

146

Paleohydrology Inferred from Salinity Measurements on Dry Valley Drilling Project Cores from Taylor Valley, Antarctica

L. D. McGINNIS and D. R. OSBY
Department of Geology, Northern Illinois University, DeKalb, Illinois 60115, U.S.A.

F. A. KOHOUT
U.S. Geological Survey, Woods Hole, Massachusetts 02543, U.S.A.

Abstract. Pore-ice salinity curves, derived from permafrost samples collected in three boreholes drilled as part of the Dry Valley Drilling Project in Taylor Valley, Antarctica, indicate three distinct paleohydrologic environments. The variable salinity curves are explained by a grounded ice shelf advancing westward into Taylor Valley, with the ice margin resting for a long time between boreholes 10 and 11. Sediment beneath the ice remained frozen, but a deep, ice-marginal lake in the valley was heated by solar radiation and produced a permeable window through the permafrost (talik) at borehole 11. Water from the lake percolated downward through the unfrozen sediments, producing a relatively homogeneous groundwater system. Farther to the west, shallower impermeable basement rock prevented deep-water circulation, and alpine glaciers isolated the deep freshwater lake marginal to the Ross Ice Sheet from the inland lakes where more saline terrestrial conditions prevailed.

During the 1974–75 field season, frozen sediments in Taylor Valley, Antarctica, were cored as part of the Dry Valley Drilling Project (DVDP), using diesel oil as drilling fluid. Boreholes were drilled vertically into the subsurface. Between the time of drilling and the time of analyses in early 1976, the cores were preserved in their frozen state. All cores exceeded 2 cm in diameter; the exterior was scraped to prepare a 2 cm^3 sample for study. In 1976, sections of core at 5 m intervals in DVDP boreholes 10, 11, and 12 (Figure 146.1) were allowed to thaw, and the interstitial water was extracted with a hydraulic squeezer apparatus under pressures less than 320 kg/cm^2. Salinity of the water was measured with optical refractometers having a range from 0 to 160 parts per thousand (ppt) (see Manheim and Sayles, 1974, for experimental-analytical techniques).

For convenience in describing the various salinity ranges of the ice, we define some terms arbitrarily as follows:

1. *Fresh ice* or *relatively fresh ice:* interstitial or pore ice, which when melted and squeezed from core materials, has a salinity range of 0–30 ppt;
2. *Sea ice:* similarly defined, but which has a salinity range of 30–40 ppt, thus bracketing the average salinity of seawater (about 35 ppt); and
3. *Brine ice:* similarly defined, but which has a salinity greater than 40 ppt.

Because of the natural exclusion of salt during the freezing process, the remaining brine fraction becomes progressively enriched in salt. After migration, this residual brine also may freeze. Thus, both fresh and brine ice might come from a parent fluid identical in composition to sea water. Fresh ice may also be produced by the freezing of meteoric or precipitated water. We thus use the terms *meteoric* or *nonmarine ice* and *marine ice* to specify the origin of the water, in addition to the salinity terms defined above. Differing salinity patterns may be created depending on the history of fractionation of pore fluids in different environments.

Borehole descriptions

New Harbor

The New Harbor DVDP borehole 10, located near the shoreline and about 1 m above sea level, shows the greatest range in salt content (Figure 146.2). Fresh, sea, and brine ice succeed one another at about 20 m intervals to the bottom of the hole; a thin, relatively fresh-ice layer is present 170 m below mean sea level (msl). Maximum salinity, about 100 ppt, was measured immediately below this layer, at 175 m below msl.

Commonwealth Glacier

Borehole 11 (Commonwealth Glacier site) contains relatively fresh pore ice to 110 m below msl. Successive zones of fresh, sea, and brine pore ice prevail below this depth, with salinities reaching 87 ppt at 175 m below msl. Except for three salinity peaks at +50, +17, and −70 m, the average salinity above the −115 m level is about 3 ppt. Within the three saline zones, the maximum salinity of 17 ppt at +17 m is still well below seawater salinity (35 ppt). Drilling at this site was terminated due to excess loss of drilling fluid at −230 m. The salinity of 70 ppt at the bottom of the hole is the same as that at the bottom of borehole 10, which was terminated when brine of about 70 ppt rose up in the hole and froze.

Lake Leon

The Lake Leon borehole 12 was started at 75 m above msl and several meters above a shallow lake (Lake Leon) impounded west of the Canada Glacier in Taylor Valley.

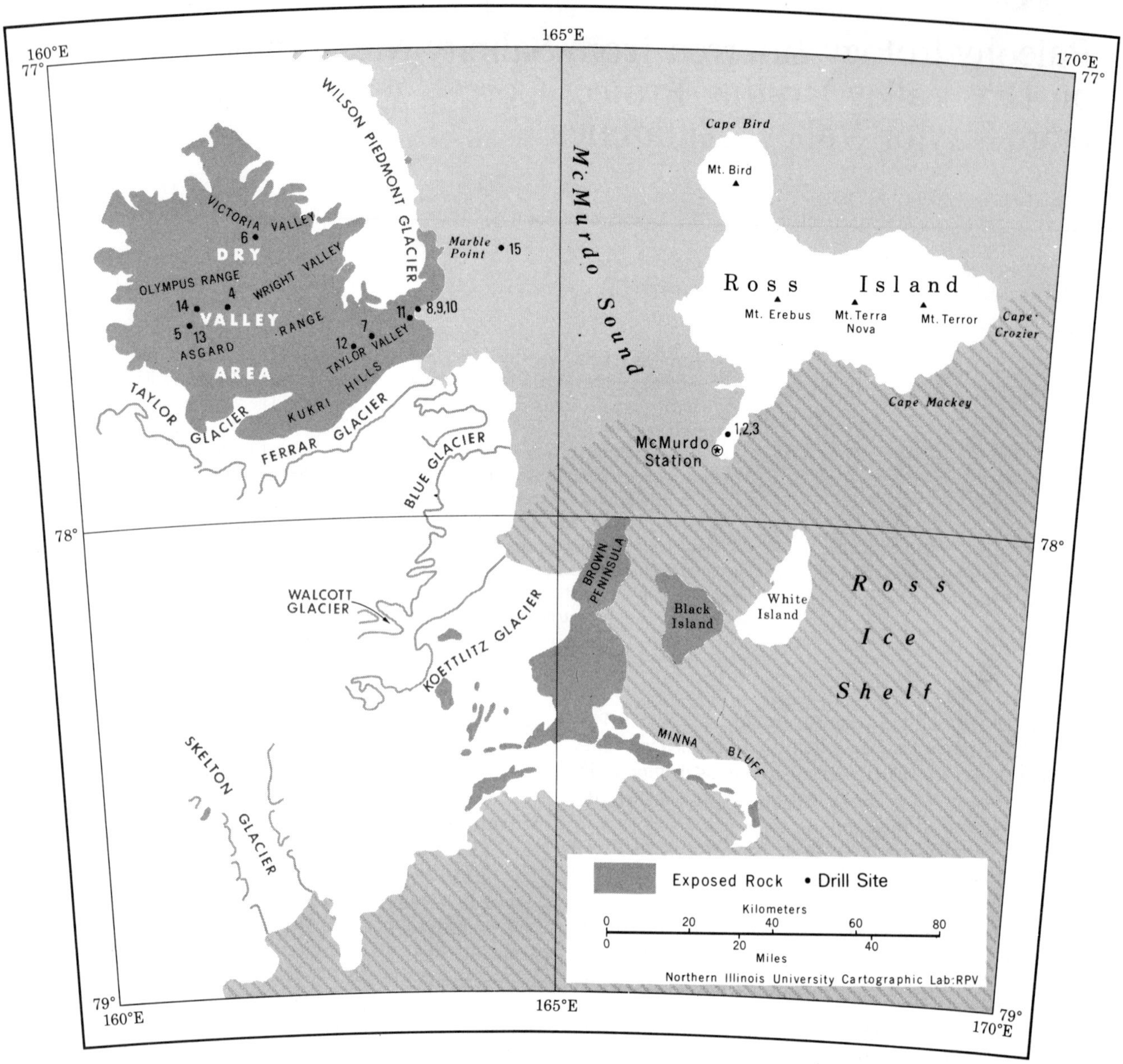

Figure 146.1. The dry valley–McMurdo Sound–Ross Island region showing the location of the 15 boreholes drilled during the Dry Valley Drilling Project. Holes discussed in this paper include DVDP 10, 11, and 12 in Taylor Valley.

The salinity-depth curve derived from the Lake Leon core is considerably different from those for boreholes 10 and 11. Minimum salinity is recorded at the top of the hole, with erratically increasing values to sea level and rather steadily increasing values of salinity down to basement crystalline rock at −100 m. The gradually increasing salinity in DVDP 12 suggests a mode of origin for these pore ice salts different from that at sites 10 and 11.

Discussion

Successive freshwater and saltwater lenses beneath terrain glaciated during Pleistocene time have been recently discovered by Kohout and others (1977) in a well drilled on Nantucket Island, Massachusetts. They have postulated freshwater recharge and a zone of diffusion that extended to abnormally great depths, according to the Ghyben-Herzberg model, during a Pleistocene sea-level lowering. Similarly, successive fresh- and salt-ice layering in DVDP hole 11 in lower Taylor Valley may represent freshwater invasion to great depths during low stands of sea level after deposition of upper Cenozoic fjord sediments. Alternatively, the freshwater-saltwater layering of pore ice may be caused by a rise in level of a freshwater lake impounded behind a more extensive Ross Ice Shelf or by freshwater entrapment caused by crustal uplift near the margin of an ice sheet (the "forebulge effect"; see McGinnis, 1968). The fact that borehole 10 does not display a thick, fresh ice layer in its upper part suggests to the authors that the alternative hypothesis is more likely.

During impoundment of waters by ice or morainal damming, natural lakes similar to those in Taylor Valley at the present time have undoubtedly existed from time to time throughout the millions of years since the valley formed. Pore water contained in lake sediments may have remained frozen since drainage of the lakes. Successive fresh and saline lake conditions could produce the layering observed in the upper valley. Oscillations of salinity values in bore-

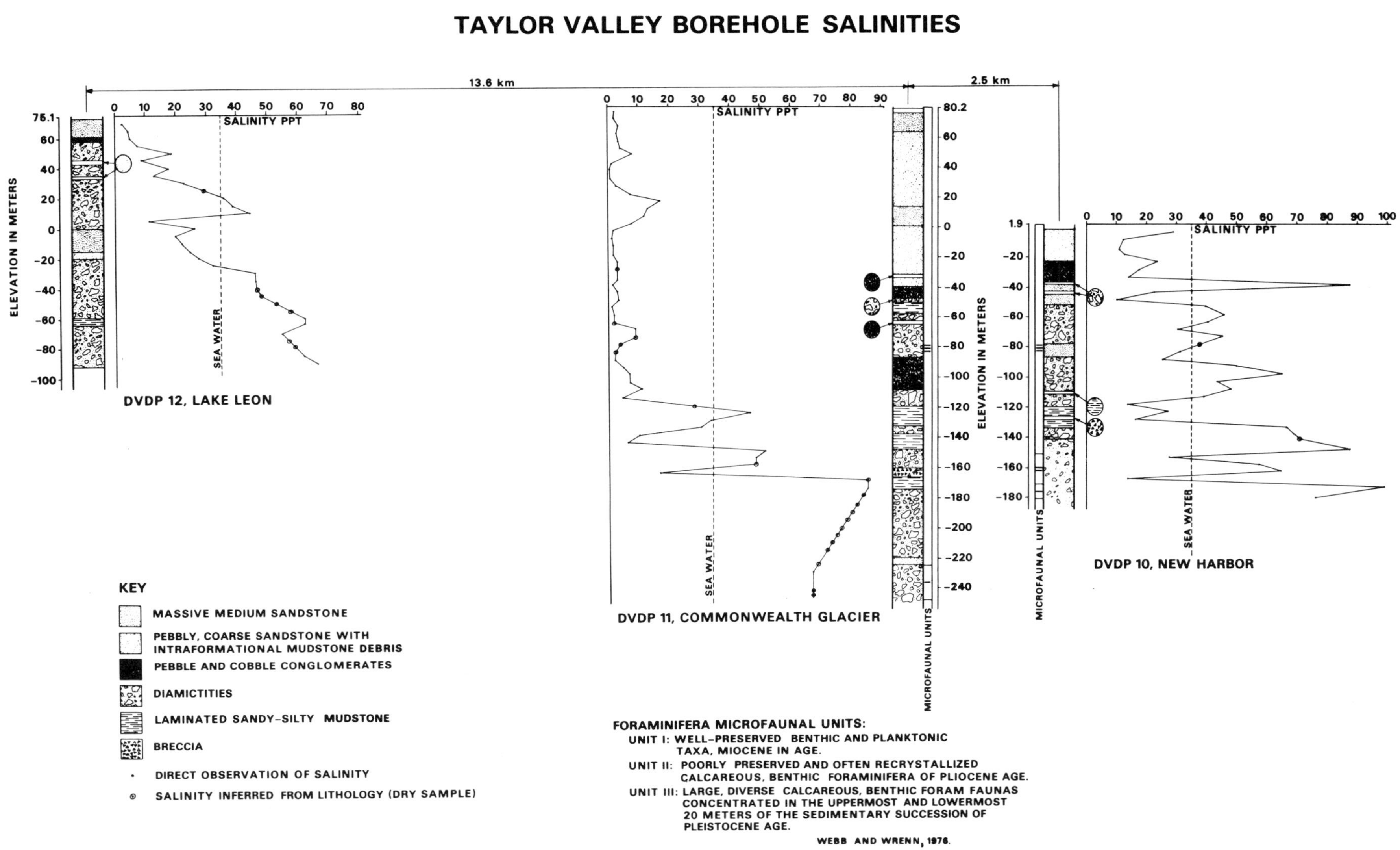

Figure 146.2. Salinity logs in DVDP boreholes 10, 11, and 12 in Taylor Valley.

hole 12 near Lake Leon may represent an effect of this type, whereas the thick fresh zone in borehole 11 appears to correspond more with the Ghyben-Herzberg phenomenon, as suggested by McGinnis (1976) for Ross Island. Similar salinities, twice that of seawater, at the bottom of boreholes 10 and 11, which both penetrated the permafrost layer, suggest a common mechanism of salt enrichment.

During a period of permafrost deepening, salts may migrate downward and away from the advancing permafrost front, thus enriching the unfrozen ground water below. Gradual downward enrichment in pore-ice salinities would be observed where hydraulic conditions were relatively static (no groundwater movement or flushing). Borehole 12 is a case in point, where ice salinities continuously increase with depth, beginning at sea-level elevation. If this enrichment process can be used to explain the curve in borehole 12, which was drilled through the sedimentary beds into crystalline basement, it would indicate that the freezing wave advanced slowly from about sea level downward with little or no escape of salts. This would suggest that the pore water of the 90 m stratigraphic sequence from sea level to basement remained unfrozen and mobile during sediment deposition, as expected at the bottom of an inland lake. A second period of sedimentation, with later freezing, is suggested by salinities in the upper 75 m of the hole.

Conclusions

Correlation of salinity layers between boreholes is suggested only for the deeper layers in boreholes 10 and 11. A cyclic salinity curve, alternating above and below the seawater salinity level, persists to 180 m below sea level in borehole 10 at New Harbor. Since unfrozen water of 72 ppt surged up into the hole at 185 m during drilling, it is presumed that freshwater is not present below 170 m. Brady (1977) has proposed two models by which freshwater conditions could be emplaced below sea level in McMurdo Sound and in the dry valleys. One model invokes lowering of sea level, the other, glacial damming north of Ross Island, with the creation of an inland freshwater lake. The freshwater lens at −170 m is contained in fossiliferous Pliocene (Webb and Wrenn, 1976) diamictite. Presuming that the pore ice is connate, we conclude from the record of seven freshwater lenses that the salinity record at New Harbor (DVDP 10) requires at least one, but more likely a combination of several, of the following events:

1. seven advances of the Ross Ice Shelf into New Harbor so that a lake is impounded, with the first recorded in the lower part of foraminiferal unit II (Pliocene) of Webb and Wrenn (1976) or near Biozone B of Brady (1977 and Chapter 145, this volume);
2. seven lowerings of sea level since Miocene time, with one lowering at least 170 m below present sea level; or
3. seven periods of crustal uplift, with one at least 170 m above the present elevation.

From the data of Webb and Wrenn (1976) and of Brady (1977), it is suggested that the deepest freshwater lens is of early Pliocene age. One of the freshwater lenses (third from the bottom of borehole 10 at about 130 m depth) is associated with the lower part of unit 4 of McKelvey (1975 and Chapter 143, this volume). This unit contains a breccia at 130 m with an initial dip of 20°. Because of its association with freshwater, this breccia may represent a glacial scouring episode from ice moving into Taylor Valley from the east.

The salinity curve of borehole 11 suggests a hydrologic history quite different from that of 10. Fresh ice to 200 m depth (120 m below sea level) occurs in lithologic units as diverse in nature as those found in borehole 10. This relatively simple salinity curve suggests homogeneity of groundwater quality caused by an episode of flushing that would occur according to the Ghyben-Herzberg model (see McGinnis, 1976). Two brine-ice lenses lying above the zone of brecciation in borehole 11 occur in the only fine-grained strata described in the geologic log of McKelvey (1975). These strata would be less subject to flushing and would therefore be most apt to retain the original water of deposition. Webb and Wrenn (1976) and Brady (1977) indicate that the zone of brecciation is early Pleistocene in age.

A model whereby hydraulic conditions could differ by such great extremes as present in boreholes 10 and 11, only 2.5 km apart, must call upon equally diverse environmental conditions. It is here proposed that the hydrologic regimen that produced the salinity curves is one in which a relatively recent glacier tongue of the Ross Ice Shelf advanced into Taylor Valley and rested upon the present site of borehole 10, but with the glacial margin lying between boreholes 10 and 11. A large lake formed west of the glacier tongue (ancient Lake Fryxell), thus covering the site of borehole 11. Bottom-water temperatures may have risen, due to insolation, to values similar to those presently found at the bottom of Lake Vanda (26°C). A downward-advancing, high-temperature front ultimately produced a window (talik) through the permafrost (McGinnis and Jensen, 1971), permitting impounded lake water to move downward into the subsurface and homogenizing the quality of groundwater in the reservoir underlying the lake. If stratified saltwaters were present at the lake bottom as they are in Lake Vanda, they would have been the first to be flushed through the talik and may now be represented by the high-salinity groundwater observed in borehole 10. The position of the ice tongue over borehole 10 would prevent thawing, thus preserving the original salinity signature in borehole 10 strata.

The time of occurrence of this sequence of events must be of sufficient antiquity to provide time for the temperature gradients to decline back to their presently observed levels (Bucher and Decker, 1976). They must also, however, be sufficiently recent geologically to explain the apparent flushing from the surface to about 240 m depth.

The salinity-depth curve of borehole 12 at Lake Leon represents a situation in which the pore ice may again be considered connate. Coarse-grained sediment is associated with fresh ice. Salinities increase to near 70 ppt at the bottom of the hole. As none of the strata at hole 12 appear to be fossiliferous, the high salinity must be caused by ancient polar desert conditions, very much like the present environment.

On the whole, each of the salinity-depth logs of boreholes 10, 11, and 12 in Taylor Valley displays unique and independent characteristics. Because of their independent nature, it is inferred that pore ice in the valley sediments is not derived from a regionally extensive aquifer system but instead from local conditions that may or may not be associated with the environment of deposition of the sediments. It is thus implied that sediment and interstitial water are, in some cases, contemporaneous and that the ice may then be defined as connate. If they are contemporaneous, the biological record contained in the core and the pore salinity, taken together, provide powerful indicators of the environment as far back as late Miocene time.

Acknowledgments. This research was supported by the National Science Foundation, Division of Polar Programs Contract C-642, and by the Marine Geology Branch, U.S. Geological Survey, Woods Hole. Core squeezing for fluid extraction was done at Northern Illinois University with equipment provided by the U.S. Geological Survey.

References cited

Brady, H. 1977. *Late Neogene Diatom Biostratigraphy and Paleoecology of the Dry Valleys and McMurdo Sound, Antarctica.* Master's thesis, Northern Illinois University, 172 pp.

Bucher, G., J., and Decker, E. R. 1976. Geothermal studies in the McMurdo Sound region. *Antarct. Jnl U.S., 11,* No. 2, 88–89.

Kohout, F. A., Hathaway, J. C., Folger, D. W., Bothner, M. H., Walker, E. H., Delaney, D. F., Frimpter, M. H., and others. 1977. Fresh groundwater stored in aquifers under the Continental Shelf: implications from a deep test, Nantucket Island, Massachusetts. *Bull. Am. Water Resour. Ass., 13,* No. 2, 373–86.

Manheim, F. T., and Sayles, F. L. 1974. Composition and origin of interstitial waters of marine sediments, based on deep sea drill cores. (*In* Maxwell, A. E., *ed. The Sea. Marine Chemistry. 5.* John Wiley, 527–68.)

McGinnis, L. D. 1968. Glacial crustal bending. *Geol. Soc. Am. Bull., 79,* 769–76.

McGinnis, L. D. 1976. Application of the Ghyben-Herzberg principle to an understanding of the physical chemistry of Ross Island and New Harbor permafrost core. Northern Illinois University, *Bull. Dry Valley Drilling Project,* No. 7, 113–14.

McGinnis, L. D., and Jensen, T. E. 1971. Permafrost—hydrogeologic regimen in two ice-free valleys, Antarctica, from electrical depth sounding. *Quat. Res., 1,* No. 3, 389–409.

McKelvey, B. C. 1975. DVDP Sites 10 and 11, Taylor Valley. *Bull. Dry Valley Drilling Project,* No. 5, 16–60.

Webb, P. N., and Wrenn, J. H. 1976. Foraminifera from DVDP holes 8–12, Taylor Valley. *Antarct. Jnl U.S., 11,* No. 2, 85–86.

147

Pressure Fluctuations in an Antarctic Aquifer: The Freight-Train Response to a Moving Rock Glacier

HENRY J. H. HARRIS and KEROS CARTWRIGHT
Illinois State Geological Survey, Urbana, Illinois 61801, U.S.A.

Abstract. Don Juan Pond, an unfrozen pond in southern Victoria Land, receives very cold, highly saline groundwater from a confined aquifer. Rapid, transient fluctuations of fluid pressure in the aquifer were recorded at a borehole drilled in the Don Juan basin. The fluctuations were natural hydrogeologic phenomena of a kind which, to our knowledge, has not been reported. They are explained in terms of existing theories for the behavior of fluids in stressed porous media and are considered to be analogous to the effects commonly produced by trains passing over confined aquifers. The fluctuations are interpreted as evidence of movements in a rock glacier lying near the pond. The characteristics of the fluctuations suggest that the causative movements were small and abrupt and involved a large part of the rock glacier.

Don Juan Pond is a small, highly saline, perennially unfrozen water body situated in the south fork of Wright Valley (lat. 77° 33′ 20″ S, long. 161° 09′ 53″ E). The pond lies in a basin 30 km east of the East Antarctic Ice Sheet and 60 km west of the Ross Sea at an elevation 117 m above mean sea level. Wright Valley is an ice-free or "dry" valley, part of an ice-free region with an area of approximately 4000 km². The geomorphic history (Calkin and Bull, 1974) and hydrogeology (Cartwright and Harris, in press) of the region are discussed in detail elsewhere. Figure 147.1 presents a general picture of the geology and topography of the south fork of Wright Valley.

Since its discovery in 1961 (Meyer and others, 1962), Don Juan Pond has attracted attention by virtue of its extraordinary chemical composition. Salinities of almost 400,000 mg/l have been reported; about 90% of the salt is calcium chloride (Mudrey and others, 1973; Torii and others, 1976; Harris and others, 1979). A new mineral, antarcticite ($CaCl_2 \cdot 6H_2O$), was found in the Don Juan basin in 1965 (Torii and Ossaka, 1965). Geophysical studies initiated in 1969 have indicated that sediments underlying the basin contain highly saline and perennially unfrozen water, suggesting the presence of a conduit for liquid groundwater (McGinnis and Jensen, 1971; McGinnis and others, 1973). These phenomena prompted the siting of two Dry Valley Drilling Project (DVDP) boreholes (McGinnis and others, 1972) in the basin, one (DVDP 5) in 1973 and another (DVDP 13) in 1975. Our observations at the holes established that the basin is a discharge zone for very cold (−16°C), saline (200,000 mg/l) groundwater (Cartwright and others, 1974; Cartwright and Harris, 1976*a*; Harris and others, 1979).

Geologic conditions in the basin are depicted in Figure 147.2. DVDP 13 is almost 75 m deep and is cased to about 13 m, 1 m below the contact between the sedimentary sequence and the Ferrar Dolerite. Groundwater having a potential above that of water in Don Juan Pond was present throughout the borehole during drilling and during all subsequent sampling and logging runs in two austral summers (1974–75 and 1975–76). Although zones of groundwater discharge probably occur in other places on the continent (Weand and others, 1975; Cartwright and Harris, in

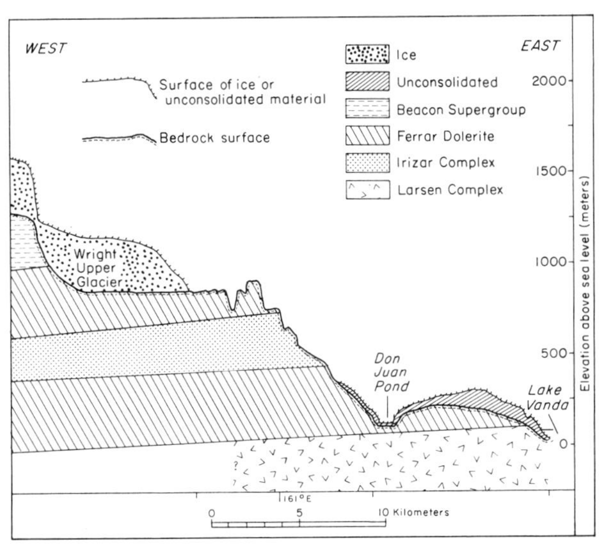

Figure 147.1. Geologic cross section of western Wright Valley (south fork) from Lake Vanda to Wright Upper Glacier icefalls; vertical exaggeration is 10×. Compiled from Barrett and others (1972), McKelvey and Webb (1962), U.S. Geological Survey (unpublished), and our own observations.

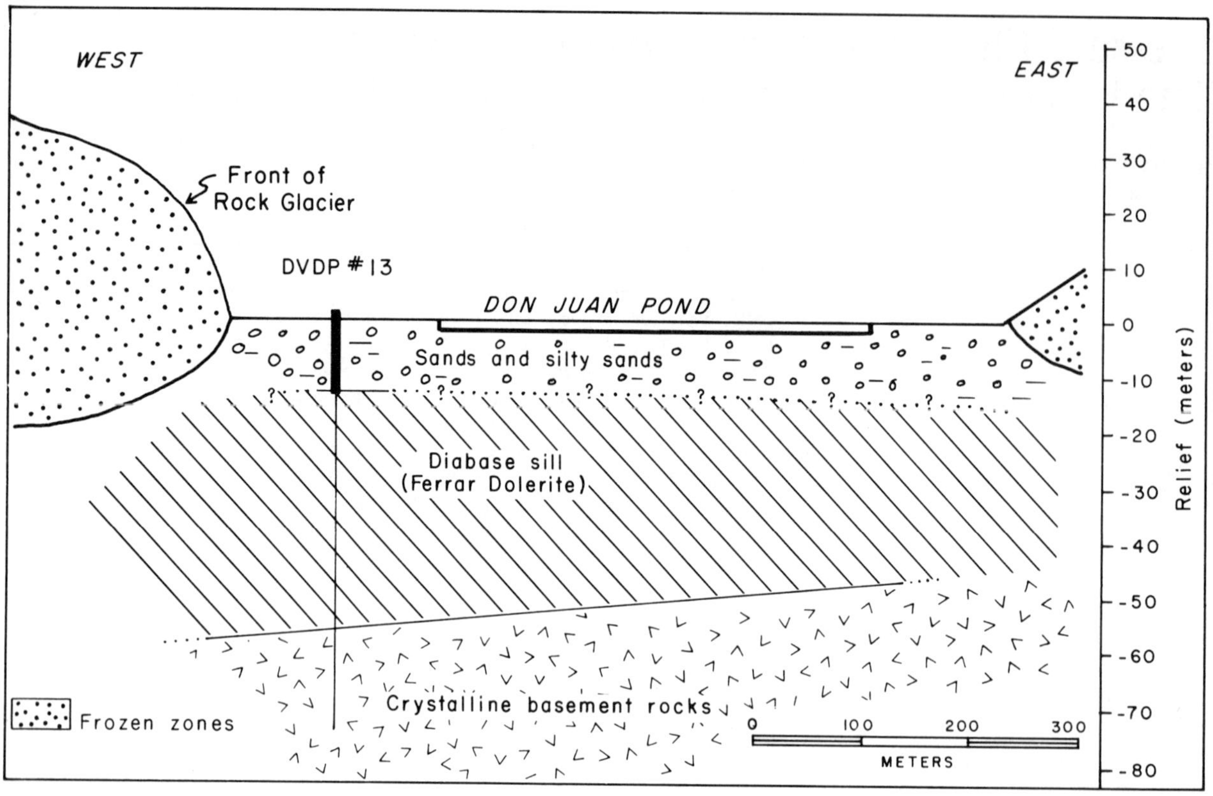

Figure 147.2. Geologic cross section of the Don Juan basin. Compiled from Bucher and Decker (1975), McGinnis and Jensen (1971), McGinnis and others (1973), Mudrey and others (1975), and our own observations. Topography is schematic.

press), Don Juan Pond is at the moment unique, the only substantiated Antarctic example of discharge from a confined aquifer.

Fluctuations in groundwater pressure at DVDP 13

On 5 December 1975 we installed a Stevens water-level recorder (type F, model 68) with a 1:1 pulley, a 20 cm float, and an 8 day clock at DVDP 13. The instrument produced a 700-hour-long hydrograph that ended on 11 January 1976 (Figure 147.3). As expected, the aquifer showed the barometric response characteristic of confined conditions. Completely unexpected, however, were extremely rapid fluctuations in fluid pressure that were recorded as sharp peaks. At least 96 separate events of this kind occurred during the period of record; some occurred in such rapid succession that they could not be distinguished from one another. On at least 11 different occasions, 3 of which transpired when the recorder was not in operation, pressure in the aquifer became great enough to cause water to flow from the top of the well, which was 62 cm above ground surface.

The resolution of the hydrograph is, at best, about 10 minutes, which is insufficient to show much detail in any of the fluctuations. However, several general characteristics of the fluctuations are apparent (Table 147.1 and Figure 147.3): (1) each consists of a rise above and an immediately subsequent and uninterrupted return to a relatively stable or "base" water level; (2) there is no evidence that the rise time and amplitude of each fluctuation are related; rise times are short and are invariably less than or equal to the resolution of the hydrograph; (3) those fluctuations with relatively large amplitudes invariably have relatively long return times; (4) in those with return times longer than the resolution of the hydrograph, the first half of the decline appears to be more rapid than the second half; (5) those with small amplitudes are more numerous than those with large; and (6) they appear to occur in clusters. Fortuitous direct observations of two rapid fluctuations also revealed some of the same characteristics, namely, a rapid increase above a base level, followed immediately by a relatively and increasingly slow decline to the same level. During both fluctuations (which occurred when the recorder was not in operation), fluid level in the well increased to a maximum (about 15 cm above a base level) in less than 30 seconds, whereas the immediately following return to the original level occurred in about 30 minutes. The initial 25% to 50% of the return occurred in less than 3 minutes.

Published descriptions of fluctuations in groundwater pressure are numerous, particularly for regions where water supplies are obtained from confined aquifers. Of the natural agents that produce such fluctuations, only two—wind effects and earthquakes—have to our knowledge

Table 147.1. Amplitudes, frequencies, return times, and overflow frequencies of rapid fluctuations at DVDP borehole 13, from 5 December 1975 through 11 January 1976.

Amplitude of fluctuation (cm)	Number of recorded events	Time to 50% return (minutes)	Time to 90% return (minutes)	Number of overflows
1 to 5	57	≤10	<15 to 30	0
5 to 10	25	≤10	<15 to 30	2
10 to 15	8	≤10 to 30	<15 to 45	3
15 to 20	3	30 to 60	60 to 90	1
20 to 25	0	—	—	—
25 to 30	3	30 to 120	60 to >120	2

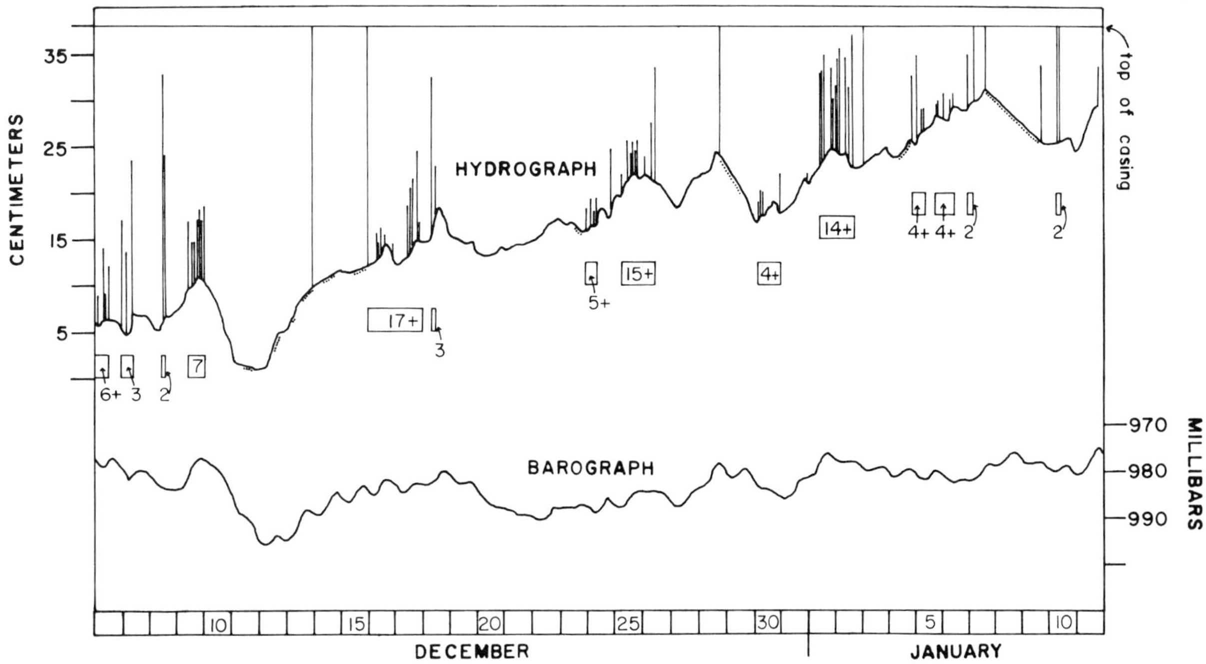

Figure 147.3. Hydrograph (upper curve) of DVDP borehole 13 and barograph (lower curve) from Don Juan basin, 6 December 1975 through 11 January 1976. The numerous vertical lines in the hydrograph are rapid fluctuations; not all of those tabulated (Table 147.1) are shown. Rectangles below the hydrograph indicate periods during which fluctuations occurred at intervals more frequent than 2 hours. Numbers indicate number of fluctuations during period; pluses after numbers indicate that some fluctuations were too close together to resolve. Dotted portions of hydrograph are gaps in the record. Datum for hydrograph (left scale) is arbitrary. Datum for barograph (right scale) is barometric pressure (976MB) at Vanda Station, midnight, 4 December 1975.

been reported to cause changes as rapid as those in DVDP 13 (Parker and Stringfield, 1950). Both can be eliminated from consideration as possible causes of the majority of the events described here. Wind effects are eliminated simply: many fluctuations occurred during periods of atmospheric calm, whereas there were none during some periods of violent wind. In an aquifer affected by an earthquake, compressional seismic waves cause alternating compression and expansion of the fluid and porous medium; fluid pressure shows a corresponding oscillation about a base pressure. The events we recorded do not have

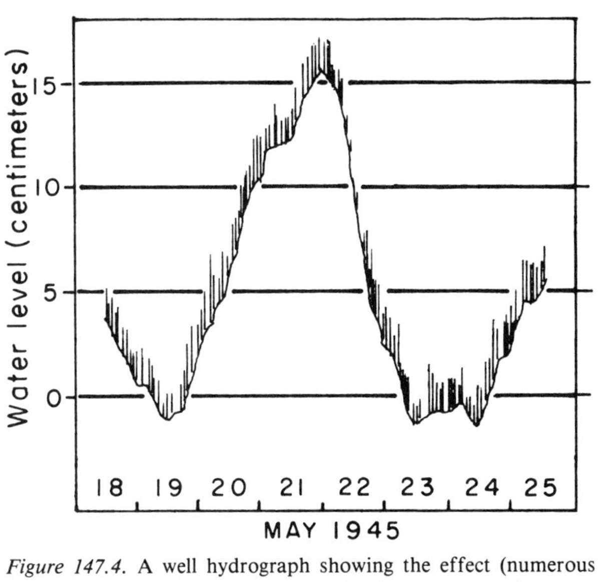

Figure 147.4. A well hydrograph showing the effect (numerous short vertical lines) of passing railroad trains on pressure in a confined aquifer in central Illinois. Modified from Roberts and Romine (1947).

the characteristics of a seismic oscillation. They do not succeed each other periodically, and each occurs only as a fluctuation above the undisturbed or base pressures that precede and follow it, without any complementary downward fluctuation. The apparently complete absence of disturbances caused by earthquakes is consistent with the observation by other workers that earthquakes originating in Antarctica are both small and infrequent (Adams, 1969; Bell, 1966; Evison, 1967).

Laboratory experiments and studies of groundwater in cold regions have shown that elevated fluid pressure can develop when the pore water in a saturated medium freezes and expands (Anderson and Morgenstern, 1973; Ginsburg and Neizvestnov, 1978; Miller, 1973). The extremely cold temperature of water in the Don Juan aquifer makes this at least a possible explanation of the fluctuations, but we know of no studies showing that freezing can cause such rapid fluctuations. Human agency is also an unlikely cause, as there is virtually no human activity within 140 km of the site. However, certain human activities on other continents produce effects remarkably similar to those observed in DVDP 13. For instance, the passage of a train over a confined aquifer may cause rapid water-level fluctuations in any nearby well tapping the aquifer, as shown in Figure 147.4 (Jacob, 1939; Roberts and Romine, 1947).

Theoretical considerations

Discussion of all such transient behavior is best phrased in terms of a general theory for the mechanical behavior of a stressed, saturated porous medium. In the following paragraphs, we present those elements of the theory which describe the "freight-train response" and the apparently similar phenomena observed at DVDP 13. Detailed expositions of the equations we employ are given in standard hydrogeologic texts (e.g., Bear, 1972; Domenico, 1972; Freeze and Cherry, 1979); a lucid exposition is also given in a recent paper by Domenico and Palciauskas (1979).

The stress balance equation

At any point in a deformable, saturated porous medium, the rate of change of pressure (P) in the pore fluid is specified by the stress balance equation

$$\frac{\partial P}{\partial t} = A\nabla^2 P + B\frac{\partial \sigma}{\partial t},$$

where t is time, A is hydraulic diffusivity, B is a pore-pressure coefficient, and σ is total vertical stress. A and B are measurable parameters incorporating properties of both fluid and medium and are here taken to be spatially and temporally invariant. If there is no change in fluid pressure with time ($\partial P/\partial t = 0$) at any point in the medium, the system is in either a "hydrostatic" or a "steady-state flow" condition. Pressures higher *or* lower than those obtaining under hydrostatic or steady-state conditions are called "excess" or "nonequilibrium" pressures.

The stress balance equation in this form states that all temporal changes in pore pressure are caused by the combined effects of two processes, changes in total vertical stress ($B\partial\sigma/\partial t$) and diffusion of pore pressure ($A\nabla^2 P$). Both for pedagogic purposes and in many practical cases, however, the two processes may be considered separately.

Stress changes

If the rate of stress change is high relative to the rate of diffusion, the stress balance equation takes the practical form

$$\frac{\partial P}{\partial t} = B\frac{\partial \sigma}{\partial t}$$

for any period of time that is short relative to the period over which diffusive effects become measurable. In this case, the stress change is called "instantaneous." The pore-pressure coefficient, B, is defined by the relationship

$$B \equiv \frac{1}{1 + n(C_w/C_m)},$$

where n and C_m are, respectively, the porosity and compressibility of the medium, and C_w is the compressibility of the fluid. The compressibility of sediment or fractured rock during an increase in stress ($\partial\sigma/\partial t$ positive) is so much greater than that of water that B is approximately equal to 1. During a decrease in stress, however, the compressibility (or "expansivity") of the medium can be smaller than that of water, so B can have values considerably less than one. Hysteresis of this kind is common in sand and fractured rock, less common in silt, and uncommon in clay (Lambe and Whitman, 1969). An instantaneous change in total stress, then, causes a simultaneous change in pore pressure. If the change in stress is positive, the magnitude of the pressure increase nearly equals that of the stress change; if the change is negative, the magnitude of the pressure decrease can be nearly equal to or less than that of the stress change.

Natural changes in total stress within saturated porous rocks are obviously common and are produced in many ways. For example, loads applied to or removed from the surfaces of rock units are causes of stress changes within the units. Common natural events that apply or remove loads are glaciation, deglaciation, sedimentation, and erosion. Much more rapid stress changes are caused by changes in barometric pressure and by earthquakes. Changes of total stress in any direction (not just the vertical) are capable of causing excess pore pressures (as, for example, in earthquakes); in this discussion we need consider only vertical stresses in media that deform only in a vertical direction.

Of the mechanisms capable of causing stress changes, loading of part of the boundary of a medium is, for present purposes, the most significant. The stress balance equation describes conditions only at a point in the medium. A description of the effect on pore pressure of a boundary load therefore requires some relationship between that load and the stresses it induces throughout the medium. In general, the stresses induced by a boundary load are largest close to the load and diminish greatly with distance from the load; the zone within which stress changes are of practical interest (e.g., measurable) may be called the "stress bulb." Stress gradients within and around the stress bulb are, in general, not uniform spatially ($\nabla^2\sigma \neq 0$).

Following an approach suggested by Jacob (1939), we later employ equations derived by Boussinesq (1885) to calculate stresses and their distribution. Although this approach permits specification of the magnitudes of the stress at all points in the medium for any equilibrium stress condition, it does not allow specification of the rate or mechanism of equilibration. Jacob (1939) has asserted that equilibration may be taken to occur immediately, meaning that for purposes of measurement, no time elapses between the application of a given load and the resulting maximum change in total stress and fluid pressure at distant points. The rate of equilibration is probably on the order of the velocity of a seismic "*P*" wave in the medium. By this token, then, any time dependence in pore-pressure changes resulting directly from stress changes caused by loading must be attributed to time dependence in the loading. We define an "instantaneous" loading as one that produces an instantaneous stress change.

Diffusion

If there are no temporal changes in stress at any point in the medium ($\partial\sigma/\partial t = 0$), the stress balance equation reduces to a diffusion equation,

$$\frac{\partial P}{\partial t} = A\nabla^2 P.$$

Diffusion occurs in any saturated porous medium in which there is a pore-pressure gradient ($\nabla P \neq 0$) that is not uniform throughout the medium ($\nabla^2 P \neq 0$). Diffusion acts to return pore pressures to equilibrium after any disturbance that creates excess pressures; in effect, diffusion transfers pressure from regions of relatively high to regions of relatively low pressure. In general, $\nabla^2 P$ decreases as the diffusion of a particular disturbance proceeds, so that the rate of change of fluid pressure at any point also decreases.

Nonuniform gradients in pore pressure ($\nabla^2 P \neq 0$) are common in saturated rocks and are caused in many ways. Stimuli acting at the boundaries of rock units are one cause; such stimuli are modeled mathematically as boundary conditions for the stress balance equation. If, for example, a precipitation event were to increase the pore pressure in the recharge zone (i.e., at one boundary) of an aquifer, diffusion would distribute that increase throughout the aquifer, effecting a dissipation of excess pressure in the recharge zone and a corresponding rise in pressure in other parts of the aquifer. The rate of these changes would decrease with time after the precipitation event. Nonuniform gradients in pore pressure can also be generated by stress changes. If, for example, a load were applied to part of the upper surface of a horizontal aquifer, the stress graidents within and around the resulting stress bulb would, in general, not be uniform. Nonuniform gradients in pore pressure would therefore occur, causing diffusion.

Additional processes

Processes in addition to diffusion and changes in total vertical stress are capable of causing temporal changes in pore pressure. Mathematically, these processes are analogous to stress changes; they may be modeled by terms of the form $C\partial f/\partial t$ added to the right-hand side of the stress balance equation, where C is a pore-pressure parameter incorporating properties of fluid and medium, and f specifies the variable driving the process. For example, Domenico and Palciauskas (1979) have derived a term of this form to describe the effects of expansion caused by temperature changes ($D\partial T/\partial t$); in this case, D incorporates the thermal expansivity of the fluid.

The freight-train response

The freight-train response, which was first studied in detail by Jacob (1939), is a fluctuation in pore pressure initiated by a load rapidly applied to or removed from part of the upper surface of an aquifer. The response displays the combined effects of stress changes and diffusion, as shown in Figure 147.5. In the archetypal example, a train rolls rapidly onto the upper surface of an aquifer and halts; the weight of the train is, in effect, a vertical load that is instantaneously applied to a part of the surface. The resulting fluctuation is seen in a well tapping the aquifer and situated within the stress bulb produced by the train. During the instant of application, there is an abrupt rise in water pressure at the well and at all other points within the stress bulb. After that instant, when the train is stationary, the load and the stresses produced by it do not change ($\partial\sigma/\partial t = 0$); the high pore pressures generated by the load slowly diffuse away from the stress bulb and into the surrounding aquifer; and pressure at the well slowly returns to base level. The fluctuation caused by the departure of the train is different only in that the causative stress change is negative. Pressure at the well first drops abruptly, then slowly returns to base level as pressure diffuses into the stress bulb.

It is instructive to compare the two responses shown in Figure 147.5 with the numerous responses shown in Figure 147.4. The latter were caused by numerous trains that passed over an aquifer without stopping, apparently giving diffusion no time to occur. Both the rises and the returns were apparently caused by stress changes, the rises when the trains passed onto the aquifer and the returns when the trains departed.

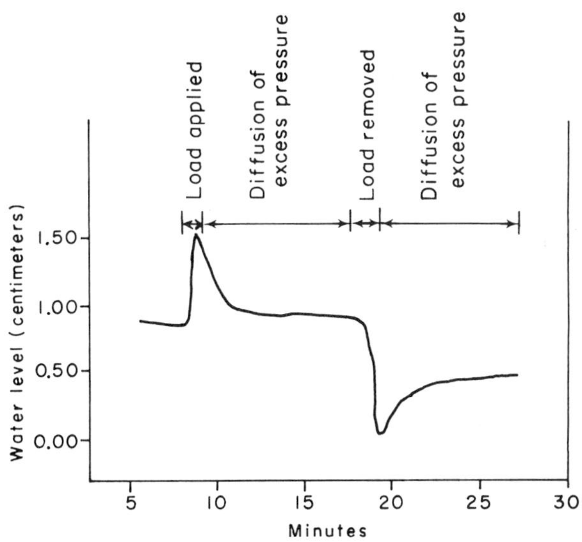

Figure 147.5. A well hydrograph showing the freight-train response in detail. Modified from Jacob (1939).

Summary

Temporal changes in pore pressure are always initiated by a "stimulus" (e.g., a load). There are, in general, only two kinds of process ("direct" and "indirect") whereby a stimulus can effect a pressure change at points distant from the stimulus. Direct processes are those modeled by terms of the form $C\partial f/\partial t$ (e.g., stress change). If a point lies within the zone of influence (e.g., the stress bulb) of the stimulus, pressure at that point is directly affected. The only indirect process is diffusion. Diffusion occurs in response to nonuniform pressure gradients generated by direct processes. Points within the zone of influence of a stimulus are affected by diffusion; diffusion may also cause pressure changes beyond the bounds of the zone of influence.

Loading as the probable cause of the fluctuations

The similarity between the freight-train response and the rapid fluctuations seen at DVDP 13 is now apparent; it is this similarity which encourages the hypothesis that the fluctuations were caused by loads applied at the ground surface in or near the Don Juan basin. However, the information contained in our hydrograph is insufficient for an analytical demonstration that surface loading caused the fluctuations. In principle, many other causes, including those we have not imagined, are possible. In the following paragraphs we discuss some of these causes and argue that, although they are possible, they are also improbable.

Combined stimuli

To this point, we have tacitly assumed that each fluctuation is a single event; that is, that each abrupt rise and subsequent decline together represent the response of the aquifer to a single stimulus. It is possible, however, that each fluctuation is produced by a combination of stimuli. We consider this explanation unlikely because every fluctuation consists of a rise followed immediately by an unin-

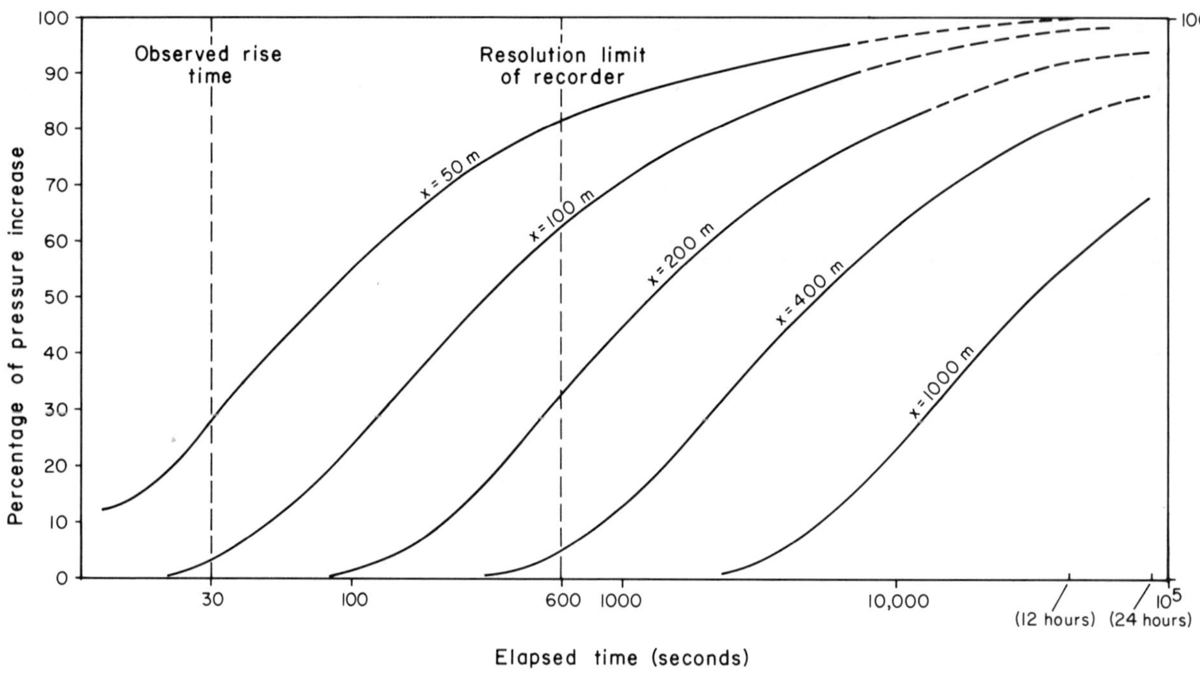

Figure 147.6. Graph showing the rate at which a pressure increase in a vertical plane ($x = 0$) will diffuse over a given horizontal distance (x) in an aquifer having a diffusivity (A) of 3.5×10^2 cm^2/sec.

terrupted decline of the same magnitude. If several stimuli had been at work, one would expect to find in the record declines followed by rises, as well as juxtaposed rises and declines with differing magnitudes. For example, it is unreasonable to postulate an increase in fluid pressure caused by a load applied in one place, immediately followed by removal of a different load in just such a manner as to return the pressure to its initial condition.

FLUCTUATIONS CAUSED SOLELY BY DIFFUSION

It is possible that the fluctuations were caused not directly, but indirectly, entirely by diffusion. A stimulus acting in some distant part of the aquifer might have produced changes in pore pressure that were transmitted to the well by diffusion alone. However, examination of a simple solution for the diffusion equation suggests that this explanation is highly unlikely. Given a horizontal aquifer in which $\nabla P = 0$, if pressure is instantaneously increased in any vertical plane ($x = 0$) of the aquifer by an amount P_0, and if diffusion occurs only in the dimension ($+x$) perpendicular to the plane, the solution of the diffusion equation is given by

$$\frac{P}{P_0} = 1 - \operatorname{erf}\left[\frac{x}{2(At)^{1/2}}\right],$$

where P is the pressure increase resulting at distance x from the plane of disturbance after time t, and *erf* is the error function (when $t = 0$, $P = 0$; when $t > 0$, P_0 at $x = 0$) (Carslaw and Jaeger, 1959). All values of P/P_0 for times up to one day are plotted in Figure 147.6 for five values of x and for $A = 3.5 \times 10^2$ cm^2/sec; we chose the very high diffusivity so as not to underestimate the rapidity of diffusion.

As Figure 147.6 shows, diffusion effects a temporal spreading of the disturbance at points distant from the stimulus; the spreading is more pronounced at greater distances. This strongly suggests that the rapid increases in pressure observed at DVDP 13 were too abrupt to have been the result of diffusion. For example, consider a recharge event as the possible stimulus. Ground surface elevations sufficient to sustain recharge to the aquifer do not occur within 100 m of the well; the observed increases were far too rapid to have been the result of diffusion from a stimulus 100 m distant. Disturbances propagated from a source as distant as the East Antarctic Ice Sheet, which may supply water to the aquifer (McGinnis and Jensen, 1971; Cartwright and others, 1975; Cartwright and Harris, 1976*b*), would be complete only over a period of days.

FLUCTUATIONS CAUSED BY DIRECT PROCESSES

If diffusion alone cannot explain the rapid increases in pressure, some direct process ($C\partial f/\partial t$) must have been the cause. This means that the ''point'' of pressure measurement (i.e., the bottom of the well casing), and probably also other surrounding points, fell within the zone of influence of the stimulus and suggests that the rate of increase in water level during any fluctuation is directly proportional to the rate of the process. Other than stress changes, only two direct processes—thermal expansion ($D\partial T/\partial t$) and freezing—appear to merit attention.

If the value of the coefficient D is taken to be 10 g/cm^{2}°C, then a temperature change at the bottom of the well casing on the order of 1°C/10 min would be required to produce a fluctuation as large and rapid as those recorded. If the two fluctuations observed directly are representative of those recorded, then a temperature change on the order of 1°C/30 sec would be needed. Furthermore, more than 100 of these rapid changes would be required, as there were more than 100 fluctuations. These seem to be extraordinarily rapid, numerous, and intermittent changes in temperature; we cannot imagine stimuli (i.e., heat sources or sinks) that can plausibly account for them.

Although rigorous mathematical models of the freezing

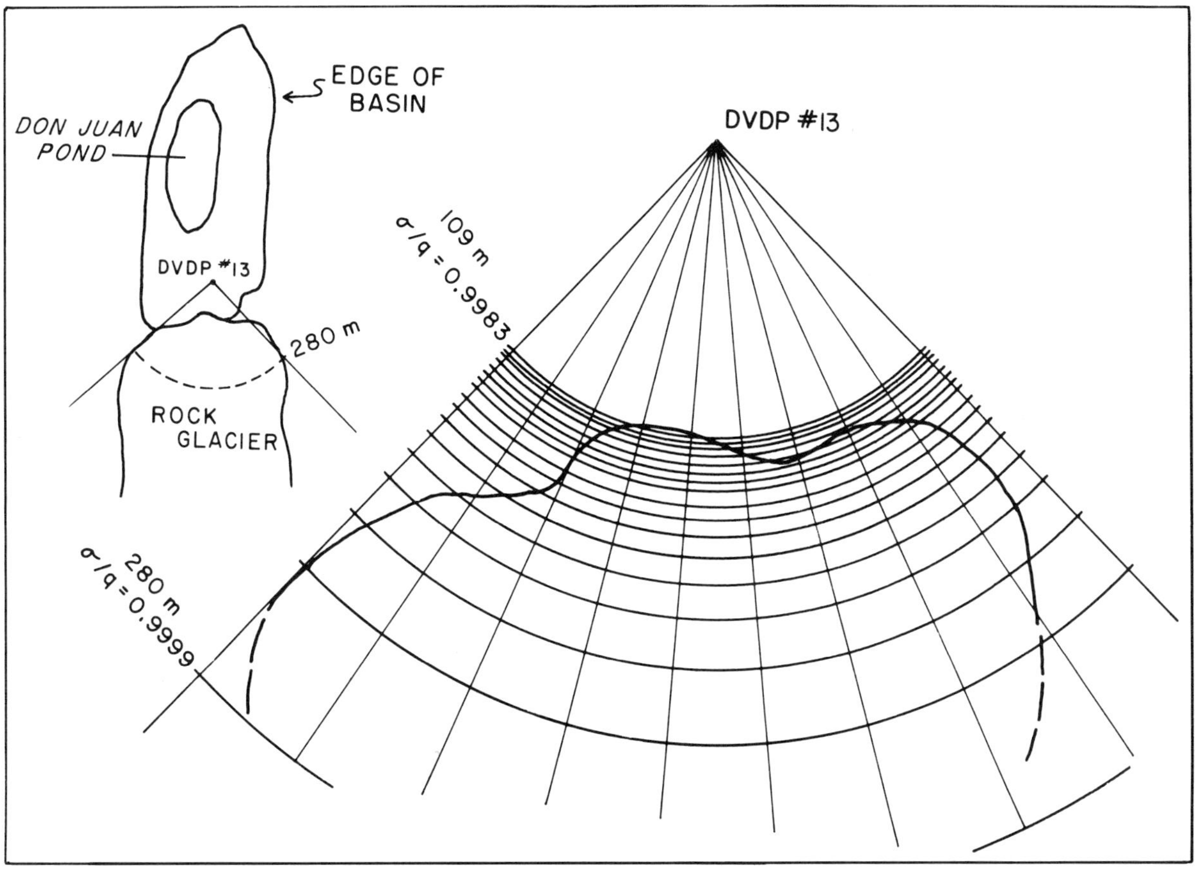

Figure 147.7. Part of a Newmark chart (z = 13 m) centered at DVDP borehole 13 and covering the toe of the rock glacier. Influence value of each segment is about 3 × 10^{-6} (dimensionless, g/cm² at z ÷ g/cm² on segment).

process that are appropriate to this discussion have not been formulated, it appears that extraordinarily rapid, numerous, and intermittent freezing events would be needed to cause fluctuations like those observed at DVDP 13. The stimuli (i.e., heat sinks) that might account for these events are, again, difficult to imagine. A further difficulty is that, without intervening thaws, successive freezing events would plug the bottom of the casing with ice. Although ice plugged the top of the casing in October and November, and small, floating crystals of ice were seen in the well in December, at no time during the period of record was the well obstructed by ice.

Lacking other plausible hypotheses, we suggest that stress changes produced the fluctuations and that the stress changes were effected by loading.

The loading mechanism

If the analogy with the freight-train response is apt, we should seek natural mechanisms which might abruptly apply large loads to the upper surface of the aquifer. There seen to be only two candidates with the mass and potential mobility necessary for this role. The first of these, the glaciers lying south and west of Don Juan Pond, appear to be too distant to have the required effect. The second is a large body of basaltic rock debris—variously termed a "rock glacier" (Cartwright and others, 1974), an inactive "solifluction lobe" (Ugolini and Bull, 1965), the "snout of an almost stagnant glacier, covered by . . . ablation moraine" (Gunn, 1963), a "tongue of . . . drift" (Tedrow and others, 1963), and a "moraine" (Bull and others, 1962, 1964)—lying at the western margin of the Don Juan basin (Figures 147.2 and 147.7). As the range of names indicates, the origin of this feature is a matter of debate. Potter (1972) has defined a rock glacier as a "tonguelike or lobate body usually of angular boulders that resembles a small glacier . . . and has a steep front at the angle of repose." Although we cannot say if the steep front of the feature is at the angle of repose, the rest of the definition applies, and we choose the name "rock glacier." The name is descriptive, but implies the capacity for motion.

Pressures caused by loading

Newmark (1942) devised a simple graphic method of evaluating the stress produced at any point in an elastic solid by a load applied to a part of its surface. The method is based on the Boussinesq relationship

$$\frac{\sigma}{q} = 1 - \left[1 + (r/z)^2\right]^{-1.5},$$

where q is the load applied vertically to a circular region of radius r centered over a point located at depth z beneath the surface of the solid, and σ is the total vertical stress resulting at that point (Coates, 1967). A Newmark chart consists of several concentric circles centered over the point; the annuli formed by every pair of adjacent circles are segmented by lines radial from the center. The chart is constructed on the basis of the Boussinesq relationship in such a way that any segment loaded with a given load per unit area has the same stress effect ("influence value") at the point as any other segment loaded with the same load per unit area.

Figure 147.7 includes a Newmark chart centered on DVDP 13 and constructed for z = 13 m, the depth to which the well is cased; the influence value for each segment of the chart is about 3×10^{-6} (dimensionless, g/cm^2 at z ÷ g/cm^2 on segment). Between the two circles of radii 109 and 280 m, the toe of the rock glacier covers 90 segments of the chart, a region with an influence value of approximately 3×10^{-4}. A mapping error in the location of the front of the rock glacier would affect this value: if the front were 10 m closer to the well, the value would be roughly doubled; if it were 10 m farther away, the value would be halved. The value of the entire region beyond 280 m and between the outermost radii shown in the figure is about 2.5×10^{-5}, so the influence of any area more than 280 m from the borehole may be safely ignored.

To acquire some feeling for the relationship between putative loads applied by the rock glacier and fluid level changes in the well, consider the following calculation. If the rock glacier has an average density (weight per unit volume) of 2.5 g/cm^3 (20% porosity, 0.2×0.9 g/cm^3 for ice + 0.8×2.9 g/cm^3 for basalt) and an average thickness of 50 m, then the average stress it exerts on the surface beneath it is 1.25×10^4 g/cm^2. If this load were applied instantaneously, the resulting increase in vertical stress at the well would be almost 4 g/cm^2 (1.25×10^4 g/cm^2 times influence value 3×10^{-4} = 3.75 g/cm^2). In a fluid having the specific gravity (approximately 1.16) of that found in the Don Juan aquifer, this stress change corresponds to a rise of about 3.5 cm. About 80% of the fluctuations actually recorded had an amplitude between 1 and 10 cm; none exceeded 30 cm (Table 147.1).

By comparison, the Wright Upper Glacier (300 m thick), lying 11 km west of and 800 m above DVDP 13, occupies a region with a calculated influence value of less than 2×10^{-5}. The value is calculated on the assumption that the space between borehole and glacier is filled entirely by an elastic solid. As this is clearly not the case (Figure 147.1), the actual influence value of the region must be far smaller. A calculation using this unrealistically high value shows that loading by the glacier would produce a fluid-level rise of about 5 mm.

These calculations demonstrate that if loading at the land surface had been responsible for the fluctuations, a very large mass must have been involved. Both because of this restriction and because loading must have occurred repeatedly, the mechanism implied in the calculations could not have caused the fluctuations. The calculations imply that the toe of the rock glacier moves horizontally and instantaneously onto the influence region, where a static vertical load is produced by gravity. Such a mechanism, although appropriate for Jacob's freight train, is obviously inappropriate here. The abrupt occupation by the rock glacier of additional area close to the borehole would apply a new vertical load in that area alone; areas already beneath the rock glacier would not be affected anew by lateral motion. The area covered during any single fluid-level change could only be small, both in size and in influence value. Studies on other continents have shown

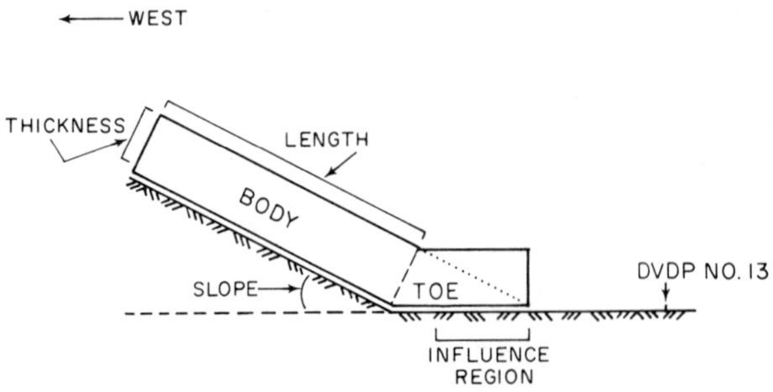

Figure 147.8. Diagrammatic cross section showing the geometry of the rock glacier.

that annual lateral bulk motion of a rock glacier rarely exceeds a few tens of centimeters (Potter, 1972; Warhaftig and Cox, 1959). There is no evidence to suggest that the Don Juan rock glacier might have behaved differently; displacement of the front is not discernible when field measurements made in 1975–76 are compared with large-scale aerial photographs taken in 1964 (U.S. Navy, 1964). In any case, a single abrupt advance of 10 m would, by the implied mechanism, produce a fluctuation of only 3 cm. One such enormous advance would be immediately obvious to anyone in the Don Juan basin; several would noticeably alter the geography of the basin.

A possible mechanism

A second mechanism is suggested by the geometry of the entire rock glacier. Westward from the basin, the body of the rock glacier extends without lateral turns for approximately 2 km, continuing westward thereafter through two broad turns. The surface slope of the body is between 10° and 15° from horizontal; the slope at the base is unknown. Under these conditions, a very large mass is in a position to act upon the influence region (Figure 147.8). If part of the weight of the body of the rock glacier were supported by the toe, a static stress having a vertical component would be applied to the influence region; the magnitude of the stress would be a function of the slope and length of the body. Consider, for example, a body 1000 m long (weight per unit volume = 2.5 g/cm^3) lying on a 25° slope. The component of gravitational acceleration acting parallel to the slope is 414 cm/sec^2, so that the force applied by the body to every square centimeter of the interface between body and toe is 1.06×10^5 g. The vertical component of this force is 4.5×10^4 g, acting upon an area of 2.37 cm^2, which yields a vertical stress of 1.9×10^4 g/cm^2. Applied uniformly to the influence region, this load would produce a rise in fluid level of about 5 cm at DVDP 13.

Fluctuation amplitudes were calculated as in the example above for basal slopes ranging from 10° to 30° and for body lengths ranging from 500 to 2500 m (Figure 147.9); thickness was considered to be uniform (50 m). For each pair of body length and basal slope values, the calculated amplitude is the maximum possible for the mechanism; we have assumed that stress is communicated through the entire body undiminished by internal and basal resistance to shear. Internal resistance along planes parallel to the base would cause a spatial variation in the magnitude of the load applied to the influence region and reduce the amplitude of the resulting fluctuation. Neither internal nor basal resistance to shear would change the basic nature of the

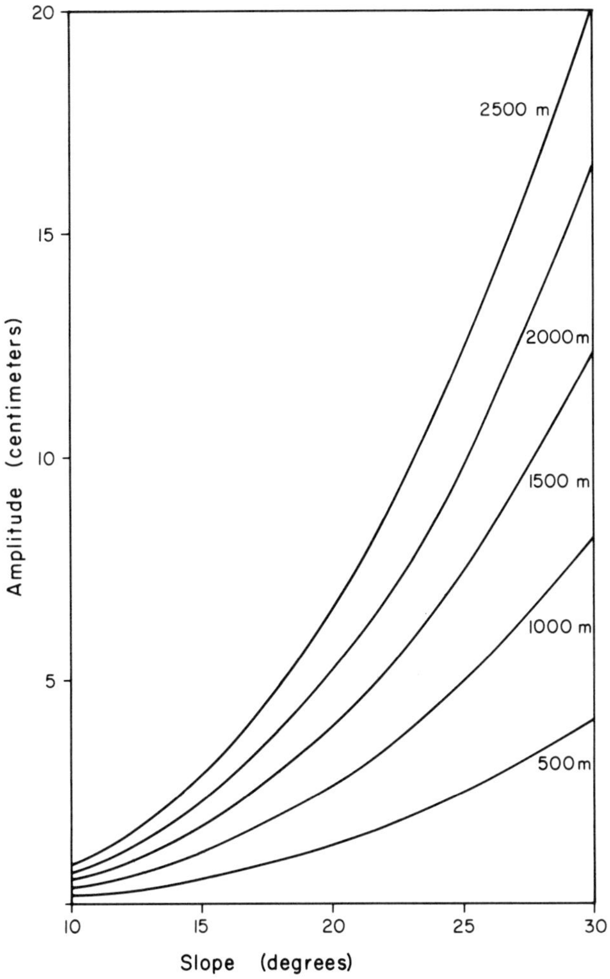

Figure 147.9. Graph showing calculated amplitudes of pressure fluctuations in DVDP borehole 13. Amplitudes are a function of the length and basal slope of the rock glacier.

mechanism. Figure 147.9 demonstrates that this mechanism could apply a load sufficent in magnitude to explain any water-level fluctuation observed at DVDP 13. Furthermore, the application of a sufficiently large load would require only a very small displacement of the rock glacier.

Two characteristics of the water-level fluctuations are especially pertinent to further discussion of the loading mechanism. First, the abundance of fluctuations indicates that the mechanism must have functioned repeatedly, substantially unaltered by earlier episodes of loading. Second, if the fluctuations are indeed analogous to the freight-train response, then the rises were produced by stress changes and the subsequent declines by diffusion alone. This means that, during any given fluctuation, the magnitude of the causative load did not change once the maximum fluid pressure for that fluctuation was attained. If the magnitude had changed once diffusion had begun, pressure would not have returned directly to the level obtaining just prior to loading. Removal of the load upon completion of diffusion might, in contrast, have had little measurable effect on fluid pressure, particularly if the rate of removal had been low. Thus, if loading by the rock glacier is to explain the fluctuations, part of the weight of the body of the rock glacier must have been borne by the toe throughout each fluctuation. In addition, in order for the mechanism to have functioned repeatedly, that weight must have been removed from the toe subsequent to each fluctuation.

It appears, then, that this mechanism involves a complex sequence of movements in both the body and the toe of the rock glacier. The exact nature and causes of the movements are obscure. Perhaps the toe anchors the body: the body "slips," abruptly loading the toe, which then moves horizontally and relatively slowly in response to the load; movement of the toe effects a readjustment of the stress field in body and toe, which entails removal of the load from the influence region and a return to conditions favoring continued slippage of the body. Perhaps the clusters of fluctuations occur when different segments of the body slip in succession. Whatever the exact nature of these putative movements, they are not without existing analogues. Studies on other continents have shown that the motion of a glacier can be "jerky," involving repeated, small ("a few millimeters"), and sudden displacements (Goldthwait, 1973; Van Wormer and Berg, 1973).

Summary and conclusions

Some of the natural fluctuations in fluid pressure observed at DVDP borehole 13 are extraordinarily rapid and are not caused in any usual manner. They are similar to the effects commonly produced by trains passing over confined aquifers. They are too rapid to be the result of diffusion alone and instead are caused by a direct process acting at the bottom of the well casing. Of three direct processes considered, two—freezing and thermal expansion—are implausible, primarily because both appear to require intermittent, very high heat flux. Stress changes are probably the cause of the fluctuations.

By analogy with the freight-train response, we postulate that the stress changes are produced by large loads applied rapidly to the upper surface of the aquifer tapped by DVDP 13. The loads are produced by a rock glacier situated near the borehole. The loads are applied through the toe of the rock glacier and are generated by the weight of the body of the rock glacier. The loading mechanism involves a complex sequence of small movements in both the body and the toe. The body, or segments of the body, move abruptly and intermittently. The toe may anchor the body and move relatively slowly in response to the loads abruptly applied to it by the body.

To our knowledge, active Antarctic rock glaciers have not been reported. We suggest that, even under prevailing climatic conditions, rock glaciers can form and move in the dry valley region. The location of the front of the Don Juan rock glacier should therefore not be taken as an index of the extent of past glacial activity in Wright Valley. We further suggest that the situation of this rock glacier in the discharge zone of an aquifer may not be coincidental. Groundwater under pressure could facilitate lateral motion of the rock glacier by reducing the effective stress at its base.

In principle, other causes we have not imagined could have produced the rapid fluctuations in water level seen at DVDP 13. Nevertheless, motion in the rock glacier is a reasonable explanation. We believe that it is also the most likely one and note that it is, with proper instruments, subject to test.

Acknowledgments. The field work upon which this research is based would have been impossible without the untiring, highly competent, and diverse efforts of Michael Chapman-Smith of the University of Auckland, New Zealand. David Gross of the Illinois State Geological Survey provided invaluable assistance during the early part of field season 1975–76. Noel Potter, Jr. of Dickinson College proved an inexhaustible source of knowledge about rock glaciers. The men and women of U.S. Navy Squadron VXE-6, of New Zealand's Vanda Station, of the U.S. National Science Foundation, and of Holmes and Narver, Inc. all rendered assistance and comfort in Antarctica. Critical reviews by Patrick Domenico of the University of Illinois and by Paul Heigold and Robert Bergstrom of the Illinois State Geological Survey resulted in many improvements in the manuscript; any remaining errors are of course the sole responsibility of the authors. This research was supported in part by National Science Foundation Grant OPP73-05917.

References cited

Adams, R. D. 1969. Small earthquakes in Victoria Land, Antarctica. *Nature, Lond., 224,* 255–56.

Anderson, D. M., and Morgenstern, N. R. 1973. Physics, chemistry, and mechanics of frozen ground: a review. (*In Permafrost, North American Contribution, Second International Conference. Yakutsk, U.S.S.R., July 1973.* Washington, National Academy of Sciences, 257–88.)

Barrett, P. J., Grindley, G. W., and Webb, P. N. 1972. The Beacon Supergroup of East Antarctica. (*In* Adie, R. J., *ed. Antarctic Geology and Geophysics.* Oslo, Universitetsforlaget, 319–32.)

Bear, J. 1972. *Dynamics of Fluids in Porous Media.* New York, Elsevier, 764 pp.

Bell, R. A. I. 1966. A seismic reconnaissance in the McMurdo Sound region, Antarctica. *J. Glaciol., 6,* No. 44, 209–21.

Boussinesq, J. 1885. *Application des Potentials à l'Etude de l'Equilibre et du Movement des Solides Elastiques.* Paris, Gauthier-Villars.

Bucher, G., and Decker, E. 1975. Unpublished geothermal log of Dry Valley Drilling Project Borehole No. 13. Laramie, University of Wyoming, 1 p.

Bull, C., McKelvey, B. C., and Webb, P. N. 1962. Quaternary glaciations in southern Victoria Land, Antarctica. *J. Glaciol., 4,* No. 31, 63–78.

Bull, C., McKelvey, B. C., and Webb, P. N. 1964. Glacial benches in southern Victoria Land. *J. Glaciol., 5,* No. 37, 131–34.

Calkin, P. E., and Bull, C. 1974. The glacial history of the ice-free area, southern Victoria Land, Antarctica. *Polar Rec., 17,* No. 107, 129–37.

Carslaw, H. S., and Jaeger, J. C. 1959. *Conduction of Heat in Solids.* 2nd ed. Oxford, Oxford University Press, 510 pp.

Cartwright, K., and Harris, H. J. H. 1976*a*. Hydrogeology in the dry valleys. *Antarct. Jnl U.S.,* 11, No. 2, 89–90.

Cartwright, K., and Harris, H. J. H. 1976*b*. Ground water at Don Juan Pond, Wright Valley, southern Victoria Land, Antarctica: a probable origin at the base of the East Antarctic Icecap. *Abstr. with Programs, Geol. Soc. Am.,* 804.

Cartwright, K., and Harris, H. J. H. In press. Hydrogeology of the Dry Valley region, Antarctica. (*In* McGinnis, L. D., *ed. Dry Valley Drilling Project.* Washington, American Geophysical Union [Antarctic Research Series].)

Cartwright, K., Harris, H., and Follmer, L. 1975. DVDP hydrogeological studies. *Bull. Dry Valley Drilling Project,* No. 5, 134–38.

Cartwright, K., Treves, S. B., and Torii, T. 1974. Geology of DVDP 5, Don Juan Pond, Wright Valley, Antarctica. *Bull. Dry Valley Drilling Project,* No. 3, 75–91.

Coates, D. F. 1967. *Rock Mechanics Principles.* Ottawa, Department of Energy, Mines, and Resources. [Mines Monograph No. 847.]

Domenico, P. A. 1972. *Concepts and Models in Groundwater Hydrology.* New York, McGraw-Hill, 405 pp.

Domenico, P. A., and Palciauskas, V. V. 1979. Thermal expansion of fluids and fracture initiation in compacting sediments. *Geol. Soc. Am. Bull., 90,* 953–79.

Evison, F. F. 1967. Note on the aseismicity of Antarctica. *N.Z. Jl Geol. Geophys., 10,* No. 2, 479–83.

Freeze, R. A., and Cherry, J. A. 1979. *Groundwater.* Englewood Cliffs, N.J., Prentice-Hall, 604 pp.

Ginsburg, G. D., and Neizvestnov, Y. Y. 1978. Hydrodynamic and hydrochemical processes in the area of the cooling of earth's crust. (*In Permafrost, U.S.S.R. Contribution, Second International Conference. Yakutsk, U.S.S.R., July 1973.* Washington, National Academy of Sciences, 377–81.)

Goldthwait, R. P. 1973. Jerky glacier motion and meltwater. (*In Symposium on the Hydrology of Glaciers. Cambridge, September 1969,* 183–88. [International Association of Scientific Hydrologists, Publ. No. 95.])

Gunn, B. M. 1963. Quaternary glaciations in Victoria Land. *J. Glaciol., 4,* No. 36, 829–30.

Harris, H. J. H., Cartwright, K., and Torii, T. 1979. Dynamic chemical equilibrium in a polar desert pond: a sensitive index of meteorological cycles. *Science, N.Y., 204,* 301–3. (See also erratum, *Science, N.Y., 204,* 909.)

Jacob, C. E. 1939. Fluctuations in artesian pressure produced by passing railroad-trains as shown in a well on Long Island, New York. *Trans. Am. Geophys. Union, 20,* 666–74.

Lambe, T. W., and Whitman, R. W. 1969. *Soil Mechanics.* New York, John Wiley, 553 pp.

McGinnis, L. D., and Jensen, T. E. 1971. Permafrost—hydrogeologic regimen in two ice-free valleys, Antarctica, from electrical depth sounding. *Quat. Res., 1,* No. 3, 389–409.

McGinnis, L. D., Nakao, K., and Clark, C. C. 1973. Geophysical identification of frozen and unfrozen ground, Antarctica. (*In Permafrost, North American Contribution, Second International Conference. Yakutsk, U.S.S.R., July 1973.* Washington, National Academy of Sciences, 136–46.)

McGinnis, L. D., Torii, T., and Webb, P. N. 1972. The Dry Valley Drilling Project. *Antarct. Jnl U.S., 7,* No. 3, 53–56.

McKelvey, B. C., and Webb, P. N. 1962. Geological investigations in southern Victoria Land, Antarctica. Part 3. Geology of Wright Valley. *N.Z. Jl Geol. Geophys., 5,* No. 1, 143–62.

Meyer, G. H., Morrow, M. B., Wyss, O., Berg, T. E., and Littlepage, J. L. 1962. Antarctica: the microbiology of an unfrozen saline pond. *Science, N.Y., 138,* No. 3545, 1103–4.

Miller, R. D. 1973. Soil freezing in relation to pore water pressure and temperature (1). (*In Permafrost, North American Contribution, Second International Conference. Yakutsk, U.S.S.R., July 1973.* Washington, National Academy of Sciences, 344–52.)

Mudrey, M. G., Shimp, N. F., Keighin, C. W., Oberts, G. L., and McGinnis, L. D. 1973. Chemical evolution of water in Don Juan Pond, Antarctica. *Antarct. Jnl U.S., 8,* No. 4, 164–66.

Mudrey, M. G., Torii, T., and Harris, H. 1975. Geology of DVDP No. 13, Don Juan Pond, Wright Valley, Antarctica. *Bull. Dry Valley Drilling Project,* No. 5, 78–99.

Newmark, N. 1942. Influence charts for the computation of stresses in elastic foundations. *Bull. Univ. Illinois Exper. Stat.,* No. 228, 28 pp.

Parker, G. G., and Stringfield, V. T. 1950. Effects of earthquakes, trains, tides, winds and atmospheric pressure changes on water in the geologic formations of southern Florida. *Econ. Geol., 45,* 441–60.

Potter, N., Jr. 1972. Ice-cored rock glacier, Galena Creek, Northern Absaroka Mountains, Wyoming. *Geol. Soc. Am. Bull., 83,* 3025–58.

Roberts, W. J., and Romine, H. E. 1947. Effect of train loading on the water level in a deep glacial-drift well in central Illinois. *Trans. Am. Geophys. Union, 28,* 912–17.

Skempton, A. W. 1954. The pore-pressure coefficients *A* and *B*. *Geotechnique, 4,* 143–47.

Tedrow, J. C. F., Ugolini, F. C., and Janetscheck, H. 1963. An Antarctic saline lake. *N.Z. Jl Sci., 6,* 150–56.

Torii, T., and Ossaka, J. 1965. Antarcticite: a new mineral, calcium chloride hexahydrate, discovered in Antarctica. *Science, N.Y., 149,* No. 3687, 975–77.

Torii, T., Yamagata, N., and Ossaka, J. 1976. Salt balance in the Don Juan basin. *Bull. Dry Valley Drilling Project,* No. 6, 33–40.

Ugolini, F. C., and Bull, C. 1965. Soil development and glacial events in Antarctica. *Quaternaria, 7,* 251–69.

United States Geological Survey. Unpublished working copies of photogrammetrically produced topographic maps of the Dry Valley region, 1:50,000. Reston, Va., U.S.G.S. Topographic Division.

United States Navy. 1964. Unpublished aerial photography, western Wright Valley, TMA #1310, approx. 1:15,000. Reston, Va., U.S. Geological Survey.

Van Wormer, D., and Berg, E. 1973. Seismic evidence for glacier motion. *J. Glaciol., 12,* No. 65, 259–65.

Warhaftig, C., and Cox, A. 1959. Rock glaciers in the Alaska Range. *Bull. geol. Soc. Am., 70,* 383–436.

Weand, B. L., Fortner, R. D., Hoehn, R. C., and Parker, B. C. 1975. Subterranean flow into Lake Bonney. *Antarct. Jnl U.S., 10,* No. 1, 15–19.

148

Petrography and Mineralogy of Dry Valley Drilling Project Hole 1, Hut Point Peninsula, Ross Island

SAMUEL B. TREVES and M. ALI
Department of Geology, University of Nebraska, Lincoln, Nebraska 68588, U.S.A.

Abstract. Dry Valley Drilling Project hole 1 was drilled on the south flank of Twin Craters, an extinct volcano located near McMurdo Station on Ross Island, Antarctica. A total of 196.54 m of core was recovered; core recovery was 99%. Detailed logging indicates that the penetrated sequence consists of 32 flows, 1 paleosol, 7 pyroclastic units, and 2 dikes. The flows are thin and most are oxidized. They contain phenocrystic kaersutite, augite, and olivine, and texturally they range from basalt to blocky breccia; two units show conglomeratic phases. The dike is andesitic. Chemical analyses of the flow units show that most are intermediate in composition, in contrast to the surface units, which are primarily basalt (basanite) and trachyte (phonolite). Chemical and petrographic data obtained from the study of these rocks support previous conclusions concerning the petrogenesis and geologic history and indicate that the rocks are differentiates of an alkaline magma that was mantle-derived.

Dry Valley Drilling Project (DVDP) hole 1 was drilled immediately northeast of McMurdo Station on the Hut Point Peninsula. The exact location of the drill site is lat. 77° 50′ 46″ S, long. 166° 40′ 11″ E. Drilling started on 21 January 1973 and was terminated by a drilling accident on 29 January 1973. During that time, 196.54 m of core were recovered and logged.

On-site logging based on the megascopic features of the core (Treves and Kyle, 1973) yielded 51 units: 41 flows, 1 paleosol, and 9 pyroclastic units (Table 148.1). Subsequent reexamination and relogging of the core after it was returned to the United States yielded 32 flows, 1 paleosol, 7 pyroclastic units, and 2 dikes (Table 148.2).

The flows are generally thin, ranging from 0.50 to 7.24 m thick. Tops and bases are characterized by rocky, scoriaceous, ice-cemented zones as much as 9 cm thick. Cavity walls are lined with radiating, delicate, white crystals of calcite(?) and mirabilite(?) that project into the ice. The scoriaceous zones pass into vesicular zones that, in thick flow units, give way to massive volcanic rock which is usually well jointed. Almost all of the units show some evidence of oxidation; most are shades of red or reddish black. Joint surfaces and vesicle walls commonly show secondary minerals.

Pyroclastic units are thicker than the flows, ranging from 1.49 to 14.69 m thick (Table 148.2). The full thickness of the basal pyroclastic unit (unit 40) is not known because drilling was terminated after penetrating 52.52 m. The rocks of unit 40 resemble the basal hyaloclastite of DVDP holes 2 and 3, which were located 600 m south of DVDP 1 (lat. 77° 53′ 3″ S, long. 116° 40′ 8″ E) (Kyle and Treves, 1974). The full thickness of the hyaloclastite (unit 15) of DVDP hole 3 is also not known, as drilling was terminated at a total depth of 318 m after penetrating 214.17 m of this basal unit.

The pyroclastic units in DVDP 1 consist primarily of red and black tuff, lapilli tuff, and breccia, and they range from well to poorly sorted. In general, sequences of several beds become finer upward. Thicker units are characterized by several sequences of a basal breccia, lapilli tuff, and tuff.

In detail, the stratigraphy of the basal pyroclastic unit is as follows. The uppermost 10 cm consist of lapilli that passes transitionally downward into 5 m of volcanic breccia. The breccia passes gradationally into 13.5 m of black and dark-gray lapilli tuff that, in turn grades into 14.5 m of green palagonitic tuff. The next 8 m consist of interbedded layers of lithic tuff, lapilli tuff, and breccia, ranging from 32 to 190 cm in thickness. Some units are palagonitic. This mixed interval passes into black lapilli breccia which contains some green to brown palagonitic ash. The lithic fragments of the breccias throughout unit 40 are olivine-augite basalt, both megascopically and in thin sections.

The paleosol (unit 5 at 13.52 m) is 0.39 m thick and consists of ash(?) and rounded, black and red fragments of vesicular, aphyric basalt that range up to 6 cm in diameter. These fragments resemble the surface rocks that occur in ice-wedge areas of polygonal or patterned ground.

The dikes (unit 23 at 81.17 and 90.14 m) are 0.78 and 6.61 m thick. Both show sharp, steeply inclined intrusive contacts with the flow rocks. The dike rock is gray and massive and shows streaky flow banding. Phenocrysts of black amphibole are common and are aligned parallel to flow bands. Joints are numerous and are coated with secondary minerals.

Petrography

Thin-section analysis (Table 148.3) of the flow units indicates that they consist of kaersutite basalt, kaersutite andesite, pyroxene-olivine basalt, and trachytic kaersutite andesite or basalt.

Kaersutite basalts (Table 148.3) are the most common flow rocks. They range from holocrystalline rocks to units that contain as much as 40% glass. They are typically fine grained; many are aphyric and some are porphyritic. In

Table 148.1. On-site, megascopic description of geologic units penetrated at DVDP hole 1

Unit number[a]	Description	Depth interval (m)	Thickness (m)
1	Aphyric basalt flow	0.00- 3.03	3.03
2	Aphyric basalt flow	3.03- 5.30	2.27
3	Aphyric basalt flow	5.30- 6.47	1.17
4	Aphyric basalt flow	6.47- 6.96	0.49
5	Augite-olivine basalt flow	6.96- 7.83	0.87
6	Augite basalt flow	7.83- 8.94	1.11
7	Aphyric basalt flow	8.94- 13.52	4.58
8	Paleosol(?)	13.52- 13.91	0.39
9	Olivine-augite basalt flow	13.91- 16.22	2.31
10	Aphyric basalt flow	16.22- 17.97	1.75
11	Aphyric basalt flow	17.97- 19.74	1.77
12	Aphyric basalt flow	19.74- 21.15	1.41
13	Aphyric basalt flow	21.15- 23.51	2.36
14	Aphyric basalt flow	23.51- 26.69	3.18
15	Olivine basalt flow	26.69- 29.79	3.10
16	Hornblende basalt flow	29.79- 32.76	2.97
17	Hornblende basalt flow	32.76- 36.59	3.83
18	Hornblende basalt flow	36.59- 43.83	7.24
19	Hornblende basalt flow	43.83- 45.55	1.72
20	Augite basalt flow	45.55- 49.98	4.43
21	Aphyric basalt flow	49.98- 51.89	1.91
22	Aphyric basalt flow	51.89- 56.43	4.54
23	Hornblende basalt flow	56.43- 58.95	2.52
24	Hornblende basalt flow	58.95- 62.73	3.78
25	Aphyric basalt flow	62.73- 63.61	0.88
26	Pyroclastics	63.61- 78.30	14.69
27	Hornblende basalt flow	78.30- 81.70	3.40
28	Hornblende trachyte flow	81.70- 88.00	6.30
29	Hornblende-augite basalt flow	88.00- 93.44	5.44
30	Olivine-augite basalt flow	93.44- 94.29	0.85
31	Olivine-augite basalt flow	94.29- 96.58	2.29
32	Pyroclastics	96.58-103.64	7.06
33	Augite basalt flow	103.64-104.98	1.34
34	Augite olivine basalt flow	104.98-111.85	6.87
35	Augite-olivine basalt flow	111.85-112.56	0.71
36	Pyroclastics	112.56-116.74	4.18
37	Olivine-augite basalt flow	116.74-123.79	7.05
38	Pyroclastics	123.79-126.63	2.84
39	Pyroclastics	126.63-127.59	0.96
40	Augite-hornblende basalt flow	127.59-131.45	3.86
41	Augite basalt flow	131.45-131.95	0.50
42	Pyroclastics	131.95-133.47	1.52
43	Aphyric basalt flow	133.47-133.83	0.36
44	Pyroclastics	133.83-134.13	0.30
45	Aphyric basalt flow	134.13-134.68	0.55
46	Olivine-augite basalt flow	134.68-137.90	3.22
47	Aphyric basalt flow	137.90-141.71	3.81
48	Aphyric basalt flow	141.71-144.02	2.31
49	Pyroclastics	144.02-145.50	1.48
50	Olivine-augite basalt flow	145.50-146.20	0.70
51	Pyroclastics	146.20-196.54	50.34

[a] Unit numbers in this table do not correspond to unit numbers in Tables 148.2 and 148.3.

Table 148.2. Summary log of DVDP hole 1, Ross Island

Unit number	Depth interval (m)	Thickness (m)	Description
1	0.00- 3.03	3.03	Flow
2	3.03- 6.96	3.93	Flow
3	6.96- 8.94	1.98	Flow
4	8.94- 13.52	4.58	Flow
5	13.52- 13.91	0.39	Paleosol(?)
6	13.91- 16.22	2.31	Flow
7	16.22- 17.97	1.75	Flow
8	17.97- 19.74	1.77	Flow
9	19.74- 21.15	1.41	Flow
10	21.15- 23.51	2.36	Flow
11	23.51- 29.79	6.28	Flow
12	29.79- 32.76	2.97	Flow
13	32.76- 36.59	3.83	Flow
14	36.59- 43.83	7.24	Flow
15	43.83- 49.98	6.15	Flow
16	49.98- 51.89	1.91	Flow
17	51.89- 56.43	4.54	Flow
18	56.43- 58.95	2.52	Flow
19	58.95- 62.73	3.78	Flow
20	62.73- 63.61	0.88	Flow
21	63.61- 78.30	14.69	Pyroclastics
22	78.30- 81.17	2.87	Flow
23	81.17- 87.78	6.61	Dike
22	87.78- 88.10	0.32	Flow
24	88.10- 90.14	2.04	Flow
23	90.14- 90.92	0.78	Dike
24	90.92- 93.44	2.52	Flow
25	93.44- 94.29	0.85	Flow
26	94.29- 96.58	2.29	Flow
27	96.58-101.15	4.57	Pyroclastics
28	101.15-102.50	1.35	Flow
29	102.50-103.99	1.49	Flow
30	103.99-104.98	0.99	Flow
31	104.98-111.85	6.87	Flow
32	111.85-112.55	0.70	Flow
33	112.56-116.74	4.18	Pyroclastics
34	116.74-123.79	7.05	Flow
35	123.79-128.93	5.14	Pyroclastics
36	128.93-131.45	2.52	Flow
37	131.45-131.95	0.50	Flow
38	131.95-141.71	9.56	Pyroclastics
39	141.71-144.02	2.32	Flow
40	144.02-196.54	52.52	Pyroclastics

thin section, the textures range from intergranular to intersertal; a few exhibit hyalo-ophitic texture. These flows consist of abundant plagioclase, clinopyroxene, minor kaersutite, even less olivine, and considerable glass. The mineral phases are discussed in more detail below.

Modal analyses (Table 148.3) indicate six of the flows are kaersutite andesite. Five are holocrystalline, but unit 18 contains 25% glass. Unlike the kaersutite basalts, these rocks usually lack olivine and contain more plagioclase. The plagioclase is andesine, whereas the plagioclase of the basalts is bytownite or labradorite. The rocks are fine grained; some are porphyritic. Five of the flows have intergranular texture; the texture of unit 18 is intersertal. In thin section and in hand specimen, these andesites resemble the kaersutite basalts.

Seven flows (Table 148.3) from the lower part of DVDP 1 are pyroxene-olivine basalts. These rocks are typically fine grained and porphyritic. They lack kaersutite, but contain abundant olivine and glass and more clinopyroxene and less plagioclase than other flow units. The plagioclase of these rocks is bytownite or labradorite. The texture, in thin section, is intersertal. Most of the clasts of the pyroclastic units are also pyroxene-olivine basalt or pyroxene basalt that resembles the flows.

Trachytic kaersutite andesite (Table 148.3) constitutes unit 23, a dike, and units 22 and 24. The dike rocks are fine grained and trachytic, show a few phenocrysts of kaersutite, contain abundant oligoclase and some clinopyroxene, and lack olivine. The dike rocks contain no glass. The flow unit contains 37% glass.

Mineralogy

The flow units are composed of plagioclase, clinopyroxene, kaersutite, olivine, glass, and accessory and secondary minerals (Table 148.3). Plagioclase occurs as micropheno-

Table 148.3. Modal analyses (in wt. %) of flow units of DVDP hole 1, Hut Point Peninsula, Ross Island

Flow unit	Thickness (m)	Thin section depth (m)	Rock type	Plagioclase	Pyroxene	Amphibole	Olivine	Glass
1	3.03	1.00	Kaersutite basalt	28	22	Trace	00	40
2	3.93	4.82	Kaersutite basalt	30	25	02	01	28
2	3.93	5.58	Kaersutite basalt	30	22	08	01	24
2	3.93	6.78	Kaersutite basalt	24	26	09	04	29
3	1.98	7.67	Kaersutite basalt	26	21	04	02	15
3	1.98	8.17	Kaersutite basalt	40	25	02	01	06
4	4.58	12.21	Kaersutite basalt	37	24	02	01	11
4	4.58	13.18	Kaersutite basalt	38	24	02	01	11
6	2.31	15.08	Kaersutite basalt	38	25	02	01	12
7	1.75	17.10	Kaersutite basalt	37	17	05	01	15
8	1.77	18.43	Kaersutite basalt	35	17	05	02	17
9	1.41	20.04	Kaersutite basalt	35	18	05	02	18
10	2.36	22.09	Kaersutite basalt	30	18	01	00	22
11	6.28	23.91	Kaersutite basalt	32	17	01	00	22
11	6.28	27.26	Kaersutite basalt	30	20	06	00	20
12	2.97	32.33	Kaersutite andesite	38	19	01	00	17
13	3.83	32.95	Kaersutite andesite	58	24	02	00	00
14	7.24	39.04	Kaersutite andesite	50	26	02	00	00
14	7.24	41.01	Kaersutite andesite	47	26	03	00	00
15	6.15	44.14	Kaersutite basalt	42	11	04	00	21
15	6.15	48.14	Kaersutite basalt	42	20	03	00	11
16	1.91	50.95	Kaersutite basalt	43	19	03	00	10
17	4.54	54.51	Kaersutite basalt	33	19	04	00	19
18	2.52	57.94	Kaersutite andesite	35	18	03	00	25
19	3.78	59.03	Kaersutite basalt	36	26	03	00	06
19	3.78	62.00	Kaersutite basalt	56	12	05	00	11
20	0.88	63.00	Trachytic kaersutite andesite	34	13	08	00	37
23	6.61	81.53	Kaersutite andesite	43	20	10	02	00
23	6.61	81.62	Trachytic kaersutite andesite	66	16	02	00	00
23	6.61	85.35	Trachytic kaersutite andesite	60	16	06	00	00
23	6.61	87.18	Trachytic kaersutite andesite	64	15	04	00	00
24	4.56	88.55	Pyroxene basalt	60	18	00	01	00
24	4.56	91.43	Kaersutite basalt	53	19	02	01	00
24	4.56	91.45	Kaersutite basalt	53	25	02	01	00
25	0.85	93.93	Pyroxene-olivine basalt	25	29	00	18	18
26	2.29	94.51	Pyroxene-olivine basalt	25	30	00	17	18
30	0.99	104.80	Pyroxene-olivine basalt	24	33	00	16	07
31	6.87	111.37	Pyroxene-olivine basalt	22	29	00	23	06
32	0.70	112.27	Pyroxene-olivine basalt	21	34	00	20	05
34	7.05	121.88	Pyroxene-olivine basalt	22	30	00	17	13
36	2.52	131.36	Kaersutite basalt	61	18	01	00	00
37	0.50	131.84	Trachytic kaersutite basalt	53	10	08	00	20
38	0.36	133.76	Kaersutite basalt	60	19	01	01	00
38	0.55	134.41	Kaersutite basalt	56	22	01	01	00
38	3.22	137.11	Trachytic kaersutite basalt	52	19	02	01	00
38	3.81	138.12	Pyroxene basalt	50	22	00	01	00
39	2.32	142.51	Trachytic kaersutite basalt	57	18	01	01	00
40	52.52	146.68	Pyroxene-olivine basalt	23	30	00	18	17

crysts and microlites of bytownite or labradorite in the basalts, andesine in the andesites, and oligoclase in the trachytic units. Oligoclase microphenocrysts have albitic rims.

The clinopyroxene of the basalts and andesites is pink to brown titanaugite. The clinopyroxene of the trachytic units is aegirine augite. Xenocrystic, green or clear clinopyroxene with rims of titanaugite occurs in some of the basalts and andesites.

Kaersutite is present in small quantities, usually less than 10%, in all of the flow units except the olivine-pyroxene and pyroxene basalts (Table 148.3). It occurs as phenocrysts and microphenocrysts and as reaction rims on olivine. Locally, it is partially or completely replaced by magnetite. In thin section, the kaersutite is pleochroic and red brown to brown, and the angle between *c* and *z* is small.

Olivine occurs in small quantities (1–2%) in many of the flow units, except for the pyroxene-olivine basalts, in which it constitutes from 17 to 23% of the rock. The olivine occurs as phenocrysts, microphenocrysts, and in the groundmass. The 2*V* angle indicates a composition of about Fo_{80}. It shows some slight alteration to iddingsite and is commonly mantled with kaersutite. Resorption features are common.

Glass is an important constituent of almost all flow units. It is brown, tan, or clear and may be charged with an opaque mineral which may be magnetite, ilmenite, pyrite, or pyrrhotite.

Accessory opaque minerals and apatite occur in all of the rocks. Rhönite (Kyle and Price, 1975) occurs in unit 33 and perhaps in other units. Nepheline occurs sparingly in the pyroxene-olivine basalts as clear, interstitial patches. It is difficult to detect in unstained thin sections.

Secondary minerals, particularly calcite and zeolite, are common in veinlets, on joints, and on vesicle walls. Browne (1973) has studied and described a large suite of secondary minerals. He has noted that some are of hydrothermal origin and that they are most probably related to late stage or subsequent igneous activity, but that some crystallized from ice, snow, and/or frozen groundwater.

Conclusions

It is obvious that the rocks of DVDP hole 1 are similar to local surface rocks of the McMurdo volcanic province, as described by Smith (1954), Treves (1967), Goldich and others (1975), and Kyle (1976) and that they are thus part of an alkaline suite. Chemical studies now in progress should indicate the alkaline nature of these rocks and allow their exact chemical classification. This work will probably indicate that some of the basaltic units are basanitoids and that the rocks are basanites, hawaiites, mugearites, and perhaps even benmoreites. The chemical data should also confirm the apparent abundance of intermediate rock types in the core. These data, and some radiometric dates, should allow us to trace the history and evolution of the magma that fed this vent.

Data in hand (Treves and Kyle, 1973; Kyle and Treves, 1973, 1974) indicate that the initial volcanic activity of this area was submarine and that it produced a thick hyaloclastite pile upon which the flow units of DVDP hole 1 accumulated. Unit 40, the thick basal pyroclastic unit, may represent the top of the hyaloclastite pile.

The chemical work and the radiometric dates will be reported in a later paper, which will also treat the petrogenesis and geologic history of these rocks.

References cited

Browne, P. R. L. 1973. Secondary minerals in cores from DVDP 1 and 2. *Bull. Dry Valley Drilling Project,* No. 2, 83–89.

Goldich, S. S., Treves, S. B., Suhr, N. H., and Stuckless, J. S. 1975. Geochemistry of the Cenozoic volcanic rocks of Ross Island and vicinity, Antarctica. *J. Geol., 83,* 415–35.

Kyle, P. R. 1976. *Geology, Mineralogy and Geochemistry of the Late Cenozoic McMurdo Volcanic Group, Victoria Land, Antarctica.* Ph.D. dissertation, Victoria University (Wellington, New Zealand), 444 pp.

Kyle, P. R., and Price, R. C. 1975. Occurrences of Rhönite in alkalic lavas of the McMurdo volcanic group, Antarctica, and Dunedin Volcano, New Zealand. *Am. Miner., 60,* 722–25.

Kyle, P. R., and Treves, S. B. 1973. Review of the geology of Hut Point Peninsula, Ross Island, Antarctica. *Bull. Dry Valley Drilling Project,* No. 2, 1–10.

Kyle, P. R., and Treves, S. B. 1974. Geology of DVDP 3, Hut Point Peninsula, Ross Island, Antarctica. *Bull. Dry Valley Drilling Project,* No. 3, 13–48.

Smith, W. C. 1954. The volcanic rocks of the Ross Archipelago. *Rep. Scient. Invest. Br. antarc. Exped., 1910, Geol.,* 1–107.

Treves, S. B. 1967. Volcanic rocks from Ross Island, Marguerite Bay and Mt. Weaver areas, Antarctica. (*In* Nagata, T., *ed.* Proceedings of the Symposium of Pacific-Antarctic Sciences. *JARE Sci. Rep., Special Issue,* No. 1, 136–49.)

Treves, S. B., and Kyle, P. R. 1973. Geology of DVDP 1 and 2, Hut Point Peninsula, Ross Island, Antarctica. *Bull. Dry Valley Drilling Project,* No. 2, 11–82.

149

The Initiation of Antarctic Glaciation: A Critical Review

Abstract

FRED M. WEAVER
Exxon Production Research Company, Houston, Texas 77001, U.S.A.

JOHN B. ANDERSON
Department of Geology, Rice University, Houston, Texas 77001, U.S.A.

This paper surveys decades of research in the marine and continental areas of Antarctica relating to the initiation and subsequent development of glaciation. Methodologies utilized in the delineation of the beginnings of Antarctic glaciation are compared and reviewed for their relative reliability, and basic assumptions unique to each method are discussed.

Direct and indirect evidence recorded in marine sediments and terrestrial rocks relevant to the initiation of Antarctic glaciation is synthesized and critically evaluated for three Tertiary time intervals: the Eocene, Oligocene, and Miocene. These data are discussed in light of new evidence recently provided by deep-sea drilling in the Southern Ocean region.

150

Migration of Cenozoic Volcanic Activity in Marie Byrd Land

Abstract

W. E. LeMASURIER
Department of Natural and Physical Sciences—Geology, University of Colorado, Denver, Colorado 80202, U.S.A.

D. C. REX
Department of Earth Sciences, University of Leeds, Leeds LS2 9JT, U.K.

The Cenozoic volcanoes in Marie Byrd Land are composed of a highly alkaline, bimodal, basalt-trachyte suite of rocks, of the type that characterizes rift valley or extensional tectonic environments. The volcanoes occur in linear and rectilinear chains that appear to have been controlled by north-south and east-west fracture systems. Contemporaneous volcanism and block faulting are suggested by differential vertical displacements of an early Tertiary peneplain, which have produced deep ice-filled grabens roughly parallel to the volcanic ranges.

Migrations of volcanic activity can be recognized in four different linear ranges on the basis of geomorphic characteristics and K-Ar dating. Migrations of roughly 100 km, over periods averaging about 10 m.y., have taken place, and in one range the migrations can be shown to have been quite systematic.

Efforts to explain these phenomena by drift of a lithospheric plate over a stationary mantle plume encounter serious difficulties. Different ranges show opposing directions of migration, and it seems improbable that the large volumes of petrologically diverse acid rocks in these volcanoes were all erupted from reservoirs below the lithosphere. The migrations could be interpreted as a process related to propagation of fracture systems, that may in turn be related to fragmentation of the Antarctic plate.

151

Magnetic Stratigraphy in DVDP Hole 11, Taylor Valley, Antarctica

Abstract

HENRY SPALL
U.S. Geological Survey, Reston, Virginia 22092, U.S.A.

DONALD P. ELSTON
U.S. Geological Survey, Flagstaff, Arizona 86001, U.S.A.

A total of 204 cores, each 2.5 cm in diameter and length, were collected between the depths of 1.98 and 313.5 m in unfolded Cenozoic glacial and fluvioglacial sediments from Dry Valley Drilling Project hole 11 (DVDP 11), Taylor Valley, Antarctica. Stratigraphic spacings were 1 m, wherever possible. Ice is the principal bonding agent in the upper 200 m of the section. The samples were drilled and measured still in the frozen state.

Natural remanent magnetization (NRM) was measured in a superconducting cryogenic magnetometer that has a noise level of 2×10^{-7} emu or less. Very large soft components of magnetization were observed in samples from the upper part of the hole, marked by intensities of 10^{-3} emu/gm. Storage in a double-walled shield for 24 hours allowed decay of the soft components accompanied by decreases in intensity of one to two orders of magnitude. Repeat measurements after storage gave the same directions and intensities.

NRM intensities are fairly uniform to a depth of 86 m and are about 3×10^{5} emu/gm. Below this, intensities vary considerably, ranging from 10^{-6} to 10^{-4} emu/gm. A marked two-orders-of-magnitude decrease in intensity takes place at depths between 202.5 and 208.9 m. A major discontinuity may exist at this interval.

At least 13 intervals of normal and reversed polarity are seen in the NRM data, indicating a substantial age for the strata. If all the reversals are valid, lithologies in the lower part of the hole would be of Miocene age. Alternating field demagnetization, and susceptibility and anhysteretic remanence (ARM) studies with respect to bedding, must be carried out to determine the validity of this preliminary magnetic stratigraphy.

Index

Author and Subject Index

The names of all authors of papers included in this symposium appear in the index. Names of authors cited in the reference list at the end of each chapter have been omitted. Only main references to subject matter appear, and in order to maintain brevity and accessibility of this index, only relevant place-names have been included.

JACKET DESIGNED BY MIKE JAYNES
COMPOSED BY FOX VALLEY TYPESETTING, MENASHA, WISCONSIN
MANUFACTURED BY KINGSPORT PRESS, KINGSPORT, TENNESSEE
TEXT AND DISPLAY LINES ARE SET IN TIMES ROMAN

Library of Congress Cataloging in Publication Data
Symposium on Antarctic Geology and Geophysics
(1977: Madison, Wis.)
Antarctic geoscience.
(Series B/International Union of Geological
Sciences; no. 4)
Includes bibliographies and index.
1. Earth sciences—Antarctic regions—
Congresses. I. Craddock, Campbell, 1930–.
II. International Council of Scientific Unions.
Scientific Committee on Antarctic Research.
III. International Union of Geological Sciences.
IV. Inter-Union Commission on Geodynamics.
V. Title. VI. Series: Series B (International
Union of Geological Sciences); no. 4.
QE350.S95 1977 559.8′9 80–5113
ISBN 0–299–08410–8 AACR2

IUGS PUBLICATIONS
Series B

Number 1
Adie, R. J. (ed.) *Antarctic Geology and Geophysics.* Universitetsforlaget, Oslo (1972), 876 p. Proceedings of the 2nd Symposium on Antarctic Geology and Solid Earth Geophysics, held in Oslo, August 6–15, 1970.

Number 2
Collins, B. W., and Fraser, R. (eds.) *Recent Crustal Movements and Associated Seismicity.* Royal Society of New Zealand, Bulletin 9 (1971), 247 p. Selected papers of the International Symposium on Recent Crustal Movements and Associated Seismicity, held in Wellington in February 1970.

Number 3
Burger, A., and Dubertret, L. (eds.) *Hydrogeology of Karstic Terrains/Hydrogéologie des terrains karstiques.* Published by the International Association of Hydrogeologists, Paris (1975), 190 p.